P9-DCP-318

Element	Symbol	Atomic Number	Average Atomic Mass[a]
Actinium	Ac	89	[227]
Aluminum	Al	13	26.9815386
Americium	Am	95	[243]
Antimony	Sb	51	121.760
Argon	Ar	18	39.948
Arsenic	As	33	74.92160
Astatine	At	85	[210]
Barium	Ba	56	137.327
Berkelium	Bk	97	[247]
Beryllium	Be	4	9.012182
Bismuth	Bi	83	208.98040
Bohrium	Bh	107	[264]
Boron	B	5	10.811
Bromine	Br	35	79.904
Cadmium	Cd	48	112.411
Calcium	Ca	20	40.078
Californium	Cf	98	[251]
Carbon	C	6	12.0107
Cerium	Ce	58	140.116
Cesium	Cs	55	132.9054519
Chlorine	Cl	17	35.453
Chromium	Cr	24	51.9961
Cobalt	Co	27	58.933195
Copper	Cu	29	63.546
Curium	Cm	96	[247]
Darmstadtium	Ds	110	[271]
Dubnium	Db	105	[262]
Dysprosium	Dy	66	162.500
Einsteinium	Es	99	[252]
Erbium	Er	68	167.259
Europium	Eu	63	151.964
Fermium	Fm	100	[257]
Fluorine	F	9	18.9984032
Francium	Fr	87	[223]
Gadolinium	Gd	64	157.25
Gallium	Ga	31	69.723
Germanium	Ge	32	72.64
Gold	Au	79	196.966569
Hafnium	Hf	72	178.49
Hassium	Hs	108	[277]
Helium	He	2	4.002602
Holmium	Ho	67	164.93032
Hydrogen	H	1	1.00794
Indium	In	49	114.818
Iodine	I	53	126.90447
Iridium	Ir	77	192.217
Iron	Fe	26	55.845
Krypton	Kr	36	83.798
Lanthanum	La	57	138.90547
Lawrencium	Lr	103	[262]
Lead	Pb	82	207.2
Lithium	Li	3	6.941
Lutetium	Lu	71	174.967
Magnesium	Mg	12	24.3050
Manganese	Mn	25	54.938045
Meitnerium	Mt	109	[268]
Mendelevium	Md	101	[258]
Mercury	Hg	80	200.59
Molybdenum	Mo	42	95.94
Neodymium	Nd	60	144.242
Neon	Ne	10	20.1797
Neptunium	Np	93	[237]
Nickel	Ni	28	58.6934
Niobium	Nb	41	92.90638
Nitrogen	N	7	14.0067
Nobelium	No	102	[259]
Osmium	Os	76	190.23
Oxygen	O	8	15.9994
Palladium	Pd	46	106.42
Phosphorus	P	15	30.973762
Platinum	Pt	78	195.084
Plutonium	Pu	94	[244]
Polonium	Po	84	[209]
Potassium	K	19	39.0983
Praseodymium	Pr	59	140.90765
Promethium	Pm	61	[145]
Protactinium	Pa	91	[231.03588]
Radium	Ra	88	[226]
Radon	Rn	86	[222]
Rhenium	Re	75	186.207
Rhodium	Rh	45	102.90550
Roentgenium	Rg	111	[272]
Rubidium	Rb	37	85.4678
Ruthenium	Ru	44	101.07
Rutherfordium	Rf	104	[261]
Samarium	Sm	62	150.36
Scandium	Sc	21	44.955912
Seaborgium	Sg	106	[266]
Selenium	Se	34	78.96
Silicon	Si	14	28.0855
Silver	Ag	47	107.8682
Sodium	Na	11	22.98976928
Strontium	Sr	38	87.62
Sulfur	S	16	32.065
Tantalum	Ta	73	180.94788
Technetium	Tc	43	[98]
Tellurium	Te	52	127.60
Terbium	Tb	65	158.92535
Thallium	Tl	81	204.3833
Thorium	Th	90	232.03806
Thulium	Tm	69	168.93421
Tin	Sn	50	118.710
Titanium	Ti	22	47.867
Tungsten	W	74	183.84
Uranium	U	92	238.02891
Vanadium	V	23	50.9415
Xenon	Xe	54	131.293
Ytterbium	Yb	70	173.04
Yttrium	Y	39	88.90585
Zinc	Zn	30	65.409
Zirconium	Zr	40	91.224

Source: IUPAC, International Union of Pure and Applied Chemistry, http://www.iupac.org/reports/periodic_table/.
[a]Values in brackets give the mass number of the radioactive isotope with the largest half-life.

Chemistry

The Science in Context

Thomas R. Gilbert
NORTHEASTERN UNIVERSITY

Rein V. Kirss
NORTHEASTERN UNIVERSITY

Natalie Foster
LEHIGH UNIVERSITY

Geoffrey Davies
NORTHEASTERN UNIVERSITY

W. W. NORTON & COMPANY
NEW YORK · LONDON

TO OUR FAMILIES AND FRIENDS

W. W. Norton & Company has been independent since its founding in 1923, when William Warder Norton and Mary D. Herter Norton first published lectures delivered at the People's Institute, the adult education division of New York City's Cooper Union. The Nortons soon expanded their program beyond the Institute, publishing books by celebrated academics from America and abroad. By mid-century, the two major pillars of Norton's publishing program—trade books and college texts—were firmly established. In the 1950s, the Norton family transferred control of the company to its employees, and today—with a staff of four hundred and a comparable number of trade, college, and professional titles published each year—W. W. Norton & Company stands as the largest and oldest publishing house owned wholly by its employees.

Copyright © 2009 by W. W. Norton & Company, Inc.

Editor: Erik Fahlgren
Project editor: Carla L. Talmadge
Marketing manager: Rob Bellinger
Editorial assistants: Matthew A. Freeman, Jeff Larson, Erin O'Brien
Director of manufacturing, College: Roy Tedoff
Managing editor, College: Marian Johnson
Book designer: Rubina Yeh
Layout artist: Paul Lacy
Photo researcher: Neil Ryder Hoos
Editorial production assistance: Justine Cullinan
Media editor: April E. Lange
Ancillaries editor: Lisa Rand
Developmental editor: Richard Morel
Composition: TSI Graphics, Precision Graphics
Illustration studio: Precision Graphics
Manufacturing: Courier, Kendallville
Cover design: Scott Idleman/Blink
Cover images: Galaxy: Science Faction/Getty Images; Earth from Space: PLI/Getty Images; Bees: Getty Images; DNA: 3DClinic/Getty Images

Library of Congress Cataloging-in-Publication Data

Chemistry : the science in context. — 2nd ed. / Thomas R. Gilbert . . . [et al.].
   p. cm.
Includes index.
ISBN 978-0-393-92649-1 (hardcover)
1. Chemistry. I. Gilbert, Thomas R. Chemistry.
QD33.2.G55 2009
540—dc22

                                   2007050388

W. W. Norton & Company, Inc., 500 Fifth Avenue, New York, NY 10110
www.wwnorton.com

W. W. Norton & Company Ltd., Castle House, 75/76 Wells Street, London W1T 3QT

1 2 3 4 5 6 7 8 9 0

Brief Contents

Contents

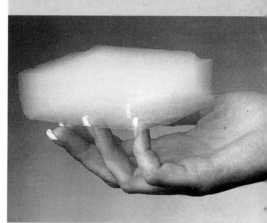

v

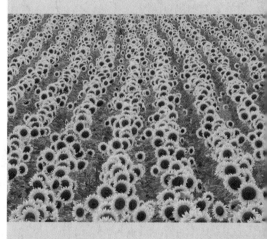

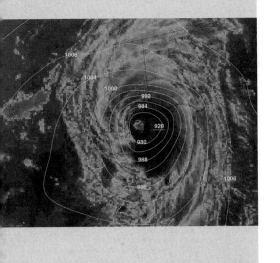

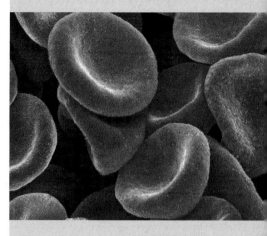

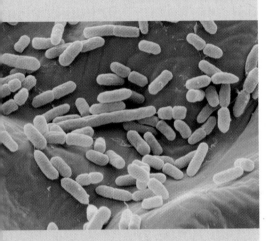

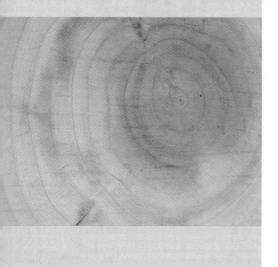

Descriptive Chemistry Boxes

ChemTours

Thomas R. Gilbert has a BS in chemistry from Clarkson and a PhD in analytical chemistry from MIT. After 10 years with the Research Department of the New England Aquarium in Boston, he joined the faculty of Northeastern University, where he is currently an associate professor of chemistry and chemical biology and Associate Dean of the Graduate School of Education. His research interests are in chemical and science education. He teaches general chemistry and science education courses and conducts professional development workshops for K–12 teachers. He has won Northeastern's Excellence in Teaching Award and Outstanding Teacher of First-Year Engineering Students Award.

Rein V. Kirss received both a BS in chemistry and a BA in history as well as an MA in chemistry from SUNY Buffalo. He received his PhD in inorganic chemistry from the University of Wisconsin, Madison, where the seeds for this textbook were undoubtedly planted. After two years of postdoctoral study at the University of Rochester, he spent a year at Advanced Technology Materials, Inc., before returning to academics at Northeastern University in 1989. He is an associate professor of chemistry with an active research interest in organometallic chemistry.

Natalie Foster is an associate professor of chemistry at Lehigh University in Bethlehem, PA. She received a BS in chemistry from Muhlenberg College and MS, DA, and PhD degrees from Lehigh University. Her research interests include studying poly(vinyl alcohol) gels by NMR as part of a larger interest in porphyrins and phthalocyanines as candidate contrast enhancement agents for MRI. She teaches the introductory chemistry class every fall to engineering, biology, and other non-chemistry majors. Natalie also regularly teaches a spectral analysis course at the graduate level.

Geoffrey Davies holds BSc, PhD, and DSc degrees in chemistry from Birmingham University, England. He joined the faculty at Northeastern University in 1971 after postdoctoral research on the kinetics of very rapid reactions at Brandeis University, Brookhaven National Laboratory, and the University of Kent at Canterbury. He is now a Matthews Distinguished University Professor at Northeastern University. His research group has explored experimental and theoretical redox chemistry, alternative fuels, transmetalation reactions, tunable metal–zeolite catalysts and, most recently, the chemistry of humic substances, the essential brown animal and plant metabolites in sediments, soils, and water. He edits a column on experiential and study-abroad education in the *Journal of Chemical Education* and a book series on humic substances. He is a Fellow of the Royal Society of Chemistry and was awarded Northeastern's Excellence in Teaching Award in 1981, 1993, and 1999, and its first Lifetime Achievement in Teaching Award in 2004.

Preface

The challenge facing instructors of general chemistry is to make connections between chemical principles and students' lives. We continue to learn how the world works at the molecular level, and the chemist's point of view—the molecular point of view—is valuable as we seek to understand and change our world. Whatever disciplines our students ultimately pursue in their professional lives, the substance of that discipline will be expressed in the molecular language of chemistry. Health and disease are being understood at the molecular level, as are corrosion, adhesion, conductivity, memory, and thought.

We explain how drugs work in terms of their molecular structure; we make better contact lenses because we understand how the molecules in the natural lenses in our eyes function; we produce tougher, stronger, and lighter materials to make cars more fuel-efficient, to protect athletes from injury, and to enable the construction of environmentally friendly and beautiful buildings because we understand how molecular structure determines physical properties. This text introduces chemical principles in the context of other disciplines, but its heartland is the molecular language of chemistry. Chemists see the world differently from the way others do, and that different point of view—the molecular point of view—is the keystone of this text.

We gauge students' understanding of the molecular point of view largely by their ability to solve problems. They must recall the chemical principles they have learned and apply them to situations which may not seem familiar. Sometimes the hardest part of solving problems is recognizing the necessary information provided, identifying what is needed, and understanding where to begin to work out the answer. We ask our students to do this in homework assignments, on exams, and in the laboratory.

Chemistry: The Science in Context, Second Edition, introduces chemical principles within the context of issues and topics from fields as diverse as biology, environmental science, materials science, astronomy, medicine, and geology. Our experience is that the context makes the chemistry more interesting, more relevant, and more memorable to students. By placing the concepts and facts of chemistry in a meaningful framework, we make them easier for students to recall and draw upon when solving problems. To complement this framework, we provide an easy-to-remember and consistent problem-solving approach for students to use as they develop their problem-solving skills.

In designing *Chemistry: The Science in Context, Second Edition* to be an effective learning tool for all students, we began by recognizing that students use textbooks in different ways. Some students begin with page one of a chapter, and for them we have started each chapter with **A Look Ahead.** This feature is written to capture the students' attention by providing a glimpse of how the upcoming material connects to the world around them. We have used topics familiar to most students, but we place them in chemical contexts that may surprise them.

As students read, they find **key terms** in boldface and defined in the text as well as in the margin. We have deliberately duplicated this material so that students can continue reading without interruption, while those reviewing a chapter can quickly find the important terms and definitions. All key terms are also defined in the Glossary in the back of the book.

We have worked with artists to create molecular art that is both appealing and informative. In most chapters we have chosen specific concepts and figures to illustrate "**What a Chemist Sees**" (see, for example, Figure 5.13 on page 196). Chemists look at the world at the molecular level, and students need to begin to incorporate this perspective as they learn.

In class, we often tie a concept we're teaching to a concept taught earlier in the course. We make these same important connections between topics in the book with **Connection** icons.

Students looking for additional help may take advantage of the approximately 100 **ChemTours** available on StudySpace (see p. xxiii) for the second edition. Denoted by the ChemTour icon, ChemTours demonstrate dynamic processes and help students visualize events at the molecular level. In many cases, students can manipulate variables and observe the resulting changes to a system. Questions at the end of tutorials guide students through step-by-step problem-solving exercises and offer useful feedback.

Concept Tests are short, conceptual questions that are integrated throughout each chapter. They serve as a self-check by asking students to stop and answer a question about something they have just read. Concept Tests are designed so students can see for themselves whether they have grasped a key concept. The answers to all Concept Tests can be found in the back of the book.

The parts of the chapter used most often by students are the **Sample Exercises.** For students reading the entire chapter, the Sample Exercise takes the concept being discussed and shows them how to apply it to a problem. For students who begin a chapter by solving the end-of-chapter Problems, the Sample Exercises are the models they use when they get stuck. Because problem solving is so important, Sample Exercises were a focus of this revision.

As beginning problem-solvers, students need a consistent approach they can use. In Section 1.5 we introduce the acronym **COAST** (**C**ollect and **O**rganize, **A**nalyze, **S**olve, and **T**hink about it) to represent a four-step method for answering questions and solving problems. We then use these steps in every Sample Exercise throughout the book:

COLLECT AND ORGANIZE helps students understand where to begin. In this step we restate the problem, identify the information given, and articulate what the student is asked to find.

ANALYZE shows students how to map out a strategy for solving the problem.

SOLVE takes the information and the strategy and shows students each step of working through the problem. We show and explain each step so students can follow the logic used and each mathematical step required by that logic.

THINK ABOUT IT reminds students to check their answer and think about the result. Does it seem realistic? Are the units correct? Is the number of significant figures appropriate?

We believe that providing an easy-to-remember and consistent model will help students develop good habits as they use the book and that COAST will become the process they use as they solve problems on their own.

 **CONNECTION**

▶❙❙ **CHEMTOUR**

CONCEPT TEST

Which contains more atoms—1 oz of gold (Au) or 1 oz of silver (Ag)?

(Answers to Concept Tests are in the back of the book.)

SAMPLE EXERCISE 4.2 **Calculating Molarity from Mass and Volume**

PVC is widely used in constructing homes and office buildings, but because vinyl chloride may leach from PVC pipe, it is used for drain pipes, not for drinking water supplies. The maximum concentration of vinyl chloride (C_2H_3Cl) allowed in drinking water in the United States is 0.002 mg/L. What is this concentration in terms of molarity?

COLLECT AND ORGANIZE We are asked to convert a concentration from milligrams of solute per liter of solution to moles of solute per liter of solution. This conversion involves changing mass in milligrams into mass in grams and then into an equivalent number of moles.

ANALYZE The conversion factors we need are the molar mass of C_2H_3Cl and the conversion factor that 1000 mg = 1 g:

$$\mathcal{M}_{C_2H_3Cl} = 2 \text{ mol C} \times \frac{12.01 \text{ g C}}{1 \text{ mol C}} + 3 \text{ mol H} \times \frac{1.008 \text{ g H}}{1 \text{ mol H}}$$
$$+ 1 \text{ mol Cl} \times \frac{35.45 \text{ g Cl}}{1 \text{ mol Cl}} = 62.49 \text{ g/mol}$$

SOLVE Setting up the unit conversion steps so that the initial set of units cancels out, we have:

$$0.002 \, \frac{mg}{L} \times \frac{1 \text{ g}}{1000 \text{ mg}} \times \frac{1 \text{ mol}}{62.49 \text{ g}} = 3.2 \times 10^{-8} \, \frac{mol}{L} = 3 \times 10^{-8} \, M \text{ or } 0.03 \, \mu M$$

THINK ABOUT IT This number may at first seem very small, but remember that the mass of PVC (0.002 mg, or 2×10^{-6} g) in 1 L of water is also very small. Sometimes it is helpful to do an "order of magnitude" estimation in a problem like this: estimate the concentration as 1×10^{-6} g/L, the molar mass as 100 g/mol (1×10^2 g/mol), and the approximate concentration is then

$$\frac{1 \times 10^{-6} \text{ g/L}}{1 \times 10^2 \text{ g/mol}} = 1 \times 10^{-8} \text{ mol/L}$$

which is very close to the number we calculated. Our number seems reasonable based on an estimation you can do quite quickly in your head. The extremely low limit of only 0.002 mg/L or 3×10^{-8} M reflects recognition of the danger this substance poses to human health.

Practice Exercise When a 1.00 L sample of water from the surface of the Dead Sea (which is more than 400 meters below sea level and much saltier than ordinary seawater) is evaporated, 179 grams of $MgCl_2$ are recovered. What is the molar concentration of $MgCl_2$ in the original sample?

(Answers to Practice Exercises are in the back of the book.)

Near the end of many chapters is a **Descriptive Chemistry** box. While Chapter 21 covers descriptive chemistry with a biological emphasis, we have integrated these boxes with the context of the chapters to show students relationships between the properties of elements highlighted in the text and their real-life manifestations. We discuss where the substances come from and how they are used in ways that touch our lives and shape our world. Because properties depend upon characteristic periodic features like atomic size, electronegativity, and ionization energy, examples of how and why the different elements within a family are used in nature and how and why people have used them illustrate their properties in a very direct way.

At the end of each chapter we have provided students with two Summaries. The first **Summary** is a brief synopsis of the chapter, organized by Section. The second, the **Problem-Solving Summary**, is unique to this general chemistry book—it succinctly outlines the different types of problems a student should know how to work out after reading a chapter. We believe all students will find this a valuable study tool.

The end-of-chapter **Questions and Problems** have been significantly revised in this second edition. In addition to providing an increased number of Questions and Problems, we have also increased the types of Problems available. The first group consists of **Visual Problems** (see p. 121). Visual Problems ask students to use a figure to solve the Problem. By learning to recognize trends or visual cues, students become better problem-solvers overall.

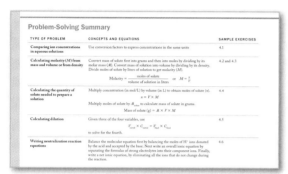

Next, we organize **Concept Review Questions and Problems** by topic. Concept Reviews are qualitative in nature and ask students to explain why or how something happens. Problems are paired and can be quantitative, conceptual, or a combination of both. **Contextual Problems** have a title that describes the context in which the Problem is placed. **Additional Problems** can come from any section or sections in the chapter or use concepts from previous chapters.

Finally, some Problems are marked with an asterisk (*). This denotes a Problem that requires multiple steps to solve. Answers to all odd-numbered Problems are provided in the back of the book.

We used a very rigorous triple-check accuracy program for the second edition. Each end-of-chapter Question and Problem was solved independently by the solutions manual author, Karen Brewer, as well as by two PhD chemists. Karen compared her solution to the other two, and then resolved any discrepancies. This process ensures clearly written Problems as well as an accurate solutions manual.

To support the users of the second edition, we have created a complete learning and teaching program that includes:

2.104. **Stainless Steel** The gleaming metallic appearance of the Gateway Arch (Figure P2.104) in St. Louis, Missouri comes from the stainless steel used in its construction. This steel is made mostly of iron but it also contains 19% by mass chromium and 9% by mass nickel.
a. Stainless steel maintains its metallic sheen because the chromium and nickel in it combine with oxygen from the atmosphere, forming a layer of Cr_2O_3 and NiO that is too thin to detract from the luster of the steel but that protects the metal beneath from further corrosion. What are the names of these two ionic compounds?
b. What are the charges of the cations in Cr_2O_3 and NiO?

FIGURE P2.104

For Students

SmartWork: An Online Tutorial and Homework Program for General Chemistry

wwnorton.com/smartwork/chemistry

Created by chemistry educators, SmartWork is the most intuitive online tutorial and homework-management system available for general chemistry. Powerful engines support an unparalleled range of question types, which include graded molecule drawing, math and chemical equations, and graphs. Answer-specific feedback, hints, and stepwise tutorials coach students through solving problems. Integration of ebook and multimedia content completes this chemistry learning system. Assigning, editing, and administering homework within SmartWork is easy. WYSIWYG (What You See Is What You Get) authoring tools allow instructors to modify existing problems or develop new content.

End-of-chapter Problems from the second edition are available in two formats in SmartWork:

- Simple Feedback Problems anticipate common wrong answers and offer prompts at the right "teachable moment" to help students discover the correct solution. Answer-specific hints are provided to guide students as they work through these problems.

- Guided Problems address more challenging topics. If a student answers a problem incorrectly, SmartWork coaches the student through a series of discrete tutorial steps that map a path to a general solution. Each step is a simple question that the student answers, with hints if necessary. After completing the tutorial, the student returns to the original problem ready to apply this newly obtained knowledge. Note that better-prepared students who are able to answer challenging questions without tutorial help can proceed through their homework quickly and efficiently, because instruction is only presented to students who have asked for it. Students have the option to complete only part of a tutorial sequence and return to the original problem when they are ready.

In both types of problems, SmartWork makes prudent use of algorithmic variables so that students see slightly different versions of the same problem. This makes it more difficult for students to share answers. The guiding principle at the heart of this system is that given enough time and effort, every student should be able to earn an "A" on every assignment.

Assignments are scored automatically. SmartWork includes equally sophisticated and flexible tools for managing class data and determining how assignments are scored.

StudySpace

wwnorton.com/studyspace

This open website is available to all students but is aimed at students looking to further their understanding or check what they have learned. StudySpace offers open access to the approximately 100 ChemTours, a Study Plan for each chapter, diagnostic Concept Test quizzes, flashcards, and links to premium content: SmartWork and the Norton ebook.

STUDENT SOLUTIONS MANUAL BY KAREN BREWER, HAMILTON COLLEGE

This guide provides students with fully worked solutions, using the COAST format, to odd-numbered end-of-chapter Problems.

STUDENT STUDY GUIDE BY STEPHANIE MYERS, AUGUSTA STATE UNIVERSITY, AND RICHARD LAVRICH, COLLEGE OF CHARLESTON

This guide begins with an Overview for each chapter. Next, it provides a Working with Key Equations and Concepts section, which includes additional Sample Exercises to emphasize good problem-solving strategies and additional Self-Test Questions and Problems for practice.

For Instructors

INSTRUCTOR SOLUTIONS MANUAL BY KAREN BREWER, HAMILTON COLLEGE

Following the COAST format, every end-of-chapter Problem is completely worked out and explained. Complete answers to all Concept Review Questions, many using the COAST format, are also included.

INSTRUCTOR RESOURCE MANUAL BY REIN KIRSS, NORTHEASTERN UNIVERSITY

The most complete resource manual for instructors is written by one of the textbook authors. Each chapter begins with a brief overview of the chapter followed by suggestions for integrating the contexts featured in the book into a lecture, suggested sample lecture outlines, and alternate contexts to use with each chapter. Summaries of the animations and tutorials available on the Norton Media Library round out each chapter.

NORTON MEDIA LIBRARY INSTRUCTOR CD-ROM

This CD-ROM includes all line art and tables as JPEG images and PowerPoint slides, lecture outlines, "clicker" questions in PowerPoint, and PowerPoint–ready offline versions of the approximately 100 ChemTours from StudySpace.

NORTON RESOURCE LIBRARY wwnorton.com/instructors

Downloadable resources will include the test bank in rich-text and ExamView formats, all line art and tables as JPEG images and PowerPoint slides, "clicker" questions in PowerPoint, and BlackBoard and WebCT materials.

BLACKBOARD AND WEBCT COURSE CARTRIDGES

Course cartridges for BlackBoard and WebCT include access to the ChemTours, a Study Plan for each chapter, multiple-choice tests, and links to premium content in the ebook and SmartWork.

TEST ITEM FILE BY JOHN GOODWIN, COASTAL CAROLINA UNIVERSITY, AND DUANE SWANK, PACIFIC LUTHERAN UNIVERSITY

More than 2000 Questions and Problems from which to choose. Each Question and Problem is ranked by difficulty and type.

Acknowledgments

The decision to revise a textbook unleashes a storm of activity. Our first order of thanks must go to W. W. Norton for having enough confidence in the idea behind the first edition to commit to the massive labor of the second. The people at W. W. Norton with whom we worked most closely deserve much more than the feeble thanks and first billing we give them here. First, we appreciate the work of our former editor, Vanessa Drake-Johnson, on the first edition and for getting the ball rolling on the second edition. Our editor and director of the storm, Erik Fahlgren, has been an indefatigable source of guidance, help, inspiration, and the occasional bit of prodding with carrots or sticks that was crucial to keep four academicians moving toward a common goal in a timely fashion. Erik's involvement in the project is the single greatest reason for its completion, and our greatest thanks is too small an offering for his unwavering focus. Our developmental editor, Richard Morel, contributed abundant and unfailingly perceptive suggestions on issues of content, tone, level, and clarity of text and illustrations. Our project editor and doyenne of all things artistic in the book, Carla Talmadge, wrangled illustrations, figures, and tables with skill, finesse, and unfailing good humor. Editorial assistants Erin O'Brien, Jeff Larson, and Matthew Freeman in turn kept the paper flowing and all of us on the same page. Thanks as well to Rubina Yeh for a design that is appealing and easy to navigate; Debra Morton Hoyt for a spectacular cover; Neil Hoos for finding great photos; Roy Tedoff for his tireless work behind the scenes; Lisa Rand for managing the print ancillaries; April Lange for her diligence on the media; and Rob Bellinger for his marketing prowess and sales expertise. We thank our copy editors Susan Middleton, Philippa Solomon, Mary Eberle, and Mark Neitlich, and our proofreaders Elizabeth Rosato and Justine Cullinan for their dedication to clearing up our fuzzy thoughts and faulty modifiers, correctly placing commas, and undangling our participles. Their attention to detail has enhanced the readability of this text. The entire Norton team was so highly competent and professional that they actually made the daunting task of bookmaking fun.

This book has benefited greatly from the care and thought that many reviewers, listed here, gave to their readings of earlier drafts. We owe an extra special thanks to Karen Brewer for her dedicated and precise work on the solutions manual. She, along with Duane Swank, Casey Raymond, and Richard Lavrich are the triple-check accuracy team who solved each Problem and during that process helped us refine the Problem sets. We are deeply grateful to Margaret Asirvatham, Kevin Cantrell, Margaret Haak, Craig McLauchlan, Jason Overby, and Michael van Stipdonk for their detailed comments on nearly every chapter of the book. Their insights into teaching, learning, and chemistry have had a profound and positive impact on the content. Finally, we greatly appreciate Patrick Caruana, Kelley J. Donaghy, Dan Durfey, Jordan Fantini, Nancy Faulk, Rebecca Miller, Tim Minger, Jodi O'Donnell, MaryKay Orgill, and Edmund Tisko for checking the accuracy of the myriad facts that form the framework of the science.

Thomas R. Gilbert
Rein V. Kirss
Natalie Foster
Geoffrey Davies

Second Edition Reviewers

William Acree, Jr., University of North Texas
Marsi Archer, Missouri Southern State University
Margaret Asirvatham, University of Colorado, Boulder
Robert Balahura, University of Guelph
Anil Banerjee, Texas A&M University, Commerce
Sandra Banks, Mills College
Mufeed Basti, North Carolina Agricultural & Technical State University
Kevin Bennett, Hood College
H. Laine Berghout, Weber State University
Karen Brewer, Hamilton College
Timothy Brewer, Eastern Michigan University
Andrew Burns, Kent State University
Sharmaine Cady, East Stroudsburg University
Chris Cahill, George Washington University
Kevin Cantrell, University of Portland
Patrick Caruana, State University of New York, Cortland
David Cedeno, Illinois State University
William Cleaver, University of Vermont
Jeffery Coffer, Texas Christian University

Robert Cozzens, George Mason University
Margaret Czrew, Raritan Valley Community College
Laura Deakin, University of Alberta
Anthony Diaz, Central Washington University
Klaus Dichmann, Vanier College
Mauro Di Renzo, Vanier College
Kelley J. Donaghy, State University of New York, College of Environmental Science and Forestry
Michelle Driessen, University of Minnesota
Dan Durfey, Naval Academy Preparatory School
Stefka Eddins, Gardner-Webb University
Nancy Faulk, Blinn College, Bryan
Barbara Gage, Prince George's Community College
Arthur Glasfeld, Reed College
Steve Gravelle, Saint Vincent College
Margaret Haak, Oregon State University
C. Alton Hassell, Baylor University
Brad Herrick, Colorado School of Mines
Angela Hoffman, University of Portland
Shahid Jalil, John Abbot College

David Katz, Pima Community College
Phillip Keller, University of Arizona
Angela King, Wake Forest University
Richard Langley, Stephen F. Austin State University
Richard Lavrich, College of Charleston
Boon Loo, Towson University
Roderick M. Macrae, Marian College
John Maguire, Southern Methodist University
Susan Marine, Miami University, Ohio
Diana Mason, University of North Texas
Garrett McGowan, Alfred University
Craig McLauchlan, Illinois State University
Stephen Mezyk, California State University, Long Beach
Rebecca Miller, Lehigh University
John Milligan, Los Angeles Valley College
Timothy Minger, Mesa Community College
Ellen Mitchell, Bridgewater College
Melanie Nilsson, McDaniel College
Gerard Nyssen, University of Tennessee, Knoxville
Jodi O'Donnell, Siena College
MaryKay Orgill, University of Nevada, Las Vegas
Jason Overby, College of Charleston
Greg Owens, University of Utah
Giuseppe Petrucci, University of Vermont
Julie Peyton, Portland State University
Gretchen Potts, University of Tennessee, Chattanooga
Robert Quandt, Illinois State University
Casey Raymond, State University of New York, Oswego
Beatriz Ruiz Silva, University of California, Los Angeles
Pam Runnels, Germanna Community College
Jerry Sarquis, Miami University, Ohio
Shawn Sendlinger, North Carolina Central University
Susan Shadle, Boise State University
Peter Sheridan, Colgate University
Ernest Siew, Hudson Valley Community College
Roberta Silerova, John Abbot College
Sally Solomon, Drexel University
Duane Swank, Pacific Lutheran University
Keith Symcox, University of Tulsa
Agnes Tenney, University of Portland
Edmund Tisko, University of Nebraska, Omaha
Mike van Stipdonk, Wichita State University
Andrew Vruegdenhill, Trent University
Ed Walton, California State Polytechnic University, Pomona
Stephen Wood, Brigham Young University
Noel Zaugg, Brigham Young University, Idaho
James Zimmerman, Missouri State University
Martin Zysmilich, George Washington University

First Edition Reviewers

R. Allendoefer, State University of New York, Buffalo
Sharon Anthony, The Evergreen State College
Jeffrey Appling, Clemson University
Robert Bateman, University of Southern Mississippi
Eric Bittner, University of Houston
David Blauch, Davidson College
Robert Boggess, Radford University

Simon Bott, University of Houston
Michael Bradley, Valparaiso University
Julia Burdge, Florida Atlantic University
Robert Burk, Carleton University
Nancy Carpenter, University of Minnesota, Morris
Tim Champion, Johnson C. Smith University
Penelope Codding, University of Victoria
Renee Cole, Central Missouri State University
Brian Coppola, University of Michigan
Richard Cordell, Heidelberg College
Dwaine Eubanks, Clemson University
Lucy Eubanks, Clemson University
Tricia Ferrett, Carleton College
Matt Fisher, St. Vincent College
Richard Foust, Northern Arizona University
David Frank, California State University, Fresno
Cynthia Friend, Harvard University
Brian Gilbert, Linfield College
Jack Gill, Texas Woman's University
Frank Gomez, California State University, Los Angeles
John Goodwin, Coastal Carolina University
Tom Greenbowe, Iowa State University
Stan Grenda, University of Nevada
Todd Hamilton, Adrian College
Robert Hanson, St. Olaf College
David Harris, University of California, Santa Barbara
Holly Ann Harris, Creighton University
Donald Harriss, University of Minnesota, Duluth
Dale Hawley, Kansas State University
Vicki Hess, Indiana Wesleyan University
Donna Hobbs, Augusta State University
Tamera Jahnke, Southern Missouri State University
Kevin Johnson, Pacific University
Martha Joseph, Westminster College
John Krenos, Rutgers University
C. Krishnan, State University of New York, Stony Brook
Sandra Laursen, University of Colorado, Boulder
David Laws, The Lawrenceville School
George Lisensky, Beloit College
Jerry Lokensgard, Lawrence University
Heather Mernitz, Tufts University
Stephanie Myers, Augusta State College
Sue Nurrenbern, Purdue University
Jung Oh, Kansas State University, Salinas
Robert Orwoll, College of William and Mary
Alexander Pines, University of California, Berkeley
Robert Pribush, Butler University
Gordon Purser, University of Tulsa
Barbara Sawrey, University of California, Santa Barbara
Truman Schwartz, Macalester College
Estel Sprague, University of Cincinnati
Steven Strauss, Colorado State University
Mark Sulkes, Tulane University
Brian Tissue, Virginia Technological University
William Vining, University of Massachusetts
Charles Wilkie, Marquette University
Ed Witten, Northeastern University

Chemistry

The Science in Context

Matter, Energy, and the Origins of the Universe

1

WHIRLPOOL GALAXY M51, NGC 5194 *A spiral galaxy 31 million light-years from Earth.*

A LOOK AHEAD: Blinded by the Sun

Today we know things about the universe and its origins that our grandparents could not even imagine; someday the same will be true of our grandchildren and their grandchildren. We know about the processes that take place in stars and how those processes release energy and produce elements, including those elements that make up everything on our planet Earth. However, all of this newly acquired knowledge has not diminished the number of questions we still ask about the universe and our place in it. Our tendency to ask how and why is part of our nature as human beings, and one of the truths about scientific explorations is that every answer spawns a host of new questions.

Even though people throughout history have asked how and why natural events occur, only in the last thousand years have they moved away from mythological speculation and toward a rational approach based on developing testable explanations called hypotheses. The process of testing hypotheses by running experiments is fundamental to scientific inquiry. An early example of this process ended a debate in ancient Greece about how vision works. Mathematicians, including Euclid, used geometry to reason that light traveled from the eye to the object; Aristotle and other philosophers proposed that light traveled from the object to the eye. Both theories were logical, but which one was right? Around the year 1000, a Persian philosopher named Ibn-al-Haytham suggested an experiment to resolve the dilemma: stare at the sun. Everyone knew you could not do that; if you stare at the sun, your eyes will be burned; you might even be blinded. This can only happen if light from the sun enters the eye and damages it. This simple experiment resolved a centuries-old debate. Since the time of Ibn-al-Haytham, testing hypotheses has led to profound theories about how the natural world works. No process has been more influential in shaping our modern world.[1]

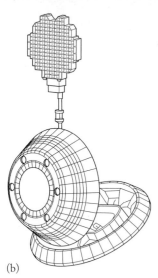

(a)

(b)

(c)

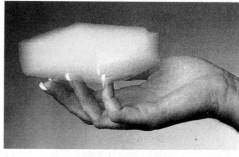

(d)

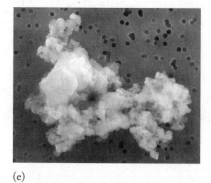

(e)

One way scientists learn about the early universe is to collect particles of cosmic dust. NASA's *Stardust* probe, launched in 1999, traveled to a region beyond the orbit of Earth's moon, where it trapped dust particles streaming into our solar system. *Stardust*'s trap was a filter made of aerogel, a material somewhat like dried Jell-O with very tiny pores. Cosmic dust is thought to be left over from the presolar time, and may be the oldest material ever available for study. This sequence shows (a) an artist's rendition of the probe approaching the tail of a comet; (b) a drawing of the *Stardust* probe with its aerogel dust collector extending like a tennis racket above the capsule; (c) a photo of Jet Propulsion Laboratory scientist Dr. Peter Tsou, holding a *Stardust* sample tray, after the probe's safe return in January 2006; (d) a sample of aerogel; and (e) a photograph of a cosmic dust fragment.

---

[1] Richard Powers, "Best Idea: Eyes Wide Open," *New York Times*, April 18, 1999.

We all use the scientific method. We wonder how things work; we develop experiments to test our ideas; we make more observations; we repeat the cycle. We do this all the time. Someone suggests that you would get better mileage in your car if you were to use premium gas as opposed to regular fuel and that the difference in mileage would more than make up for the difference in price. You fill your tank with premium, determine your mileage, and accept or reject the hypothesis based on the results. You cannot determine whether the hypothesis is correct just by thinking about it; you have to run the experiment.

Experiments require measurements, and it was with careful measurements that modern science began. The evidence on which the synthesis of elements in stars is based comes from studies of known nuclear reactions, of Einstein's theory of the equivalency of mass and energy, and from the measurement and analysis of light that reaches us from the stars. Although we have learned a great deal about where the chemical elements came from and how they behave, many processes are still not understood, so our current explanations will doubtless be revised and refined as more observations lead to more hypotheses, which will be tested by more experiments. Science is a dynamic endeavor, and the science of chemistry provides a useful and valuable way to view our material world, to interpret it, and also to change it.

1.1 Chemistry and the Classes of Matter

Consider the world around us. All things in it that are physically real—from the air we breathe to the ground we walk on to the sun's rays that warm our planet—are forms of matter or energy. For centuries, scientists considered matter and energy completely separate domains in nature. Then, at the beginning of the 20th century, Albert Einstein (1879–1955) proposed that matter (m) and energy (E) were equivalent and mathematically related by the speed of light (c):

$$E = mc^2$$

Einstein's insights into the interconvertibility of matter and energy paved the way for astounding scientific advances during the 20th century, including the development of atomic weapons and harnessing the energy of the atom for peaceful purposes. The link between energy and matter also formed the basis for a theory of how the matter of the universe came into being. Before we address that theory and some of the research on which it is based, let's review some fundamental concepts and definitions about matter and its properties.

Scientists define **matter** as everything in the universe that has **mass** and occupies space. **Chemistry** is the science of matter. Chemists study the composition, structure, and properties of matter. They observe the changes that matter undergoes and measure the energy that is produced or consumed during those changes. Chemistry provides an understanding of many natural events and has led to the synthesis of new forms of matter that have greatly affected the way we live and the planet on which we live. Modern life is difficult to imagine without plastics, computers, photocopy machines, aspirin, and countless other synthetic materials and technological innovations that have been made possible by understanding matter at atomic and molecular levels through the study of chemistry.

All the different forms of matter in our world fall into the classes shown in Figures 1.1 and 1.2. The principal categories are (1) pure substances and (2) mixtures. A **pure substance** is a form of matter that has the same physical and chemical properties, no matter what its source. It can be separated into simpler substances

Matter is the material of which the universe is made; all matter has mass and occupies space.

Mass is a property that defines the quantity of matter in an object. Mass is measured on balances.

Chemistry is the science of matter and the study of its composition, structure, and properties.

A **pure substance** has the same physical and chemical properties independent of its source.

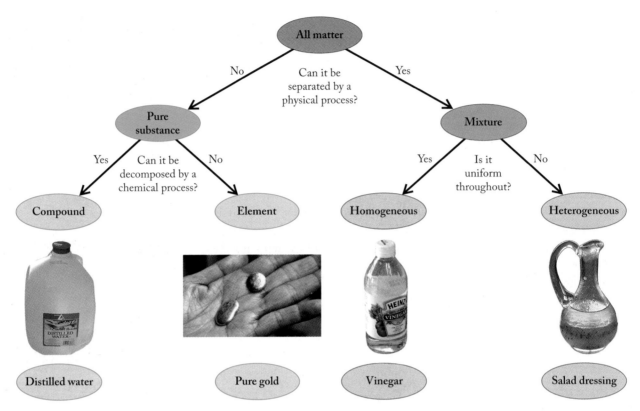

FIGURE 1.1 Matter is classified as shown in this diagram. The major classes are pure substances and mixtures. A pure substance may be an element (such as gold) or a compound, such as water. When the constituent substances in a mixture are distributed uniformly, as they are in vinegar (a solution of acetic acid in water), the mixture is homogeneous. If the constituents are not distributed uniformly, as when solids are suspended in a liquid and then settle to the bottom of the container as they do in Italian salad dressing, the mixture is heterogeneous.

A **mixture** is a combination of pure substances in variable proportions in which the individual substances retain their chemical identities.

A **homogeneous mixture** consists of components that are distributed uniformly throughout and have no visible boundaries or regions.

A **heterogeneous mixture** has distinct regions of different composition, like particles of silt suspended in water.

An **element** is a pure substance that cannot be separated into simpler substances by chemical means.

A **compound** is a substance composed of two or more elements linked together in fixed proportions.

only by chemical reactions (if at all). **Mixtures** are composed of two or more substances in variable proportions in which the pure substances retain their chemical identities. In a **homogeneous mixture**, the constituents are distributed uniformly and the composition and appearance of the mixture are uniform throughout. In contrast, in a **heterogeneous mixture** the individual components can be seen as separate substances. Unlike pure substances, mixtures can be separated into their constituent substances by physical means; chemical reactions are not required.

Pure substances are subdivided into two groups: elements and compounds. An **element** is the simplest kind of material with unique physical and chemical properties. The periodic table inside the front cover shows all the known elements. Only a few of them (such as gold, silver, and sulfur) are found in nature as pure substances. Most elements occur in chemical combination with other elements in the form of compounds. A **compound** is a pure substance that consists of two or

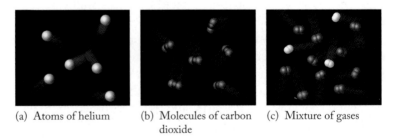

(a) Atoms of helium   (b) Molecules of carbon dioxide   (c) Mixture of gases

FIGURE 1.2 All matter is made up of either pure substances (of which there are few in nature) or mixtures of substances. The examples shown here for each class are (a) an element, helium (He), the second most abundant atom in the universe; (b) a compound, carbon dioxide (CO_2), the gas used in many fire extinguishers; and (c) a homogeneous mixture containing nitrogen (N_2), hydrogen (H_2), and oxygen (O_2).

more elements linked together in characteristic and definite proportions. Compounds typically have characteristics very different from the elements of which they are composed.

Water, for example, is a compound that is composed of two elements: hydrogen and oxygen. When these elements react with each other and form water, a considerable quantity of energy is released. This reaction provides the energy that supplies the thrust of the main engines of the U.S. space shuttles (Figure 1.3). We can force this reaction to run in reverse and convert water back into hydrogen and oxygen, but only if we add back all the energy that was released in the forward reaction. One way to do this is by applying electrical energy using a 6-volt battery as shown in Figure 1.4. The apparatus in the figure includes two stiff wires held in place so that one end of each is directly beneath an inverted test tube filled with water. The other end is connected to one of the two poles of the battery. When the battery is connected, bubbles of oxygen gas form at the end of the wire under the test tube (on the right in Figure 1.4) connected to the positive pole of the battery, and twice as much hydrogen gas by **volume (V)** forms at the end of the wire connected to the negative pole.

This 2:1 volume ratio of hydrogen to oxygen is produced every time a sample of water is decomposed to its elements. This consistency illustrates a principle known as the **law of constant composition**, which states that every sample of a particular compound always contains the same elements combined in the same proportions. The law applies if we express the quantities of the elements based on their masses or, in the case of gases, if we base quantities on their volumes.

FIGURE 1.3 Hydrogen reacts with oxygen, producing water and releasing energy. Energy from this reaction propels the U.S. space shuttles. During liftoff, each of the five main engines of the shuttle consumes about 40 tons per minute of hydrogen and oxygen, which are stored in the large brown tank under the shuttle. Once in space, the electric power needs of the space shuttles are supplied by fuel cells energized by the same chemical reaction between hydrogen and oxygen. The water generated as a product of this reaction is used aboard the shuttle.

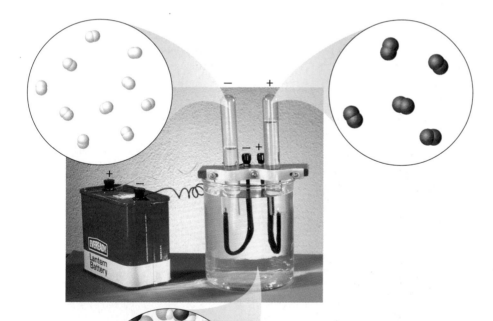

FIGURE 1.4 When an electric current is passed through water, the molecules of water decompose into oxygen and hydrogen. Oxygen gas forms at the end of the wire connected to the positive pole of the battery, and twice as much hydrogen by volume forms at the end of the wire connected to the negative pole. Quantitative observations like this illustrate the law of constant composition.

Volume (V) is the space occupied by matter.

The **law of constant composition** states that all samples of a particular compound always contain the same elements in the same proportions.

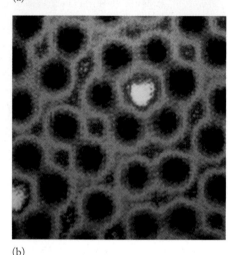

(a)

(b)

FIGURE 1.5 (a) Silicon wafers like these are widely used to make computer chips and photovoltaic (solar) cells. Since the 1980s, scientists have been able to image individual atoms using an instrument called a scanning tunneling microscope (STM). (b) In this STM image of silicon, the fuzzy spheres are individual atoms of silicon that have radii of only 117 picometers (pm), or 117 trillionths of a meter. Individual atoms are the tiniest pieces of a silicon chip that still retain the characteristics of silicon.

1.2 Matter: An Atomic View

Chemists view matter and its properties on an atomic level. They have even developed instruments to produce images of matter at that level. For example, if the surface of a silicon wafer, like those used to make computer chips or photovoltaic (solar) cells (see Figure 1.5a), is magnified over 100 million times using a **scanning tunneling microscope (STM)**, the result is an image of individual silicon atoms as shown in Figure 1.5(b). An **atom** is the smallest representative particle of an element. If, for example, you were to grind up a silicon chip into the finest dust imaginable, there would be a limit to how tiny a particle could be and still be silicon. That limit is an atom of silicon.

Most elements occur in nature combined with other elements in the form of compounds. Just as elements are made of particles called atoms, compounds such as water are made of particles called **molecules**, which are a collection of atoms chemically bonded together in a characteristic pattern and proportion. Let's take a molecular view of the space shuttle launch in Figure 1.3. In Figure 1.6, we represent the reactions using models depicting atoms of hydrogen (the white spheres) and oxygen (the red spheres). Hydrogen and oxygen, like many elements that are gases at room temperature, exist as *diatomic* (two-atom) molecules and therefore are represented by the chemical formulas H_2 and O_2. A **chemical formula** consists of the symbols of the elements that make up a compound and subscripts that indicate the proportions of the elements in the compound. Two molecules of H_2 combine with one molecule of O_2 to form two molecules of water (H_2O). This relation is described in Figure 1.6 by both the molecular models and the chemical equation beneath them. In a **chemical equation**, formulas are used to express the identities and quantities of the substances involved in a **chemical reaction**. We address how to write chemical equations in Chapter 3 in more detail, and the energetics of this chemical reaction are discussed in Chapter 5.

Chemical formulas provide information about the proportions of the elements in a compound, but they do not tell us how the atoms of those elements are connected within a molecule of the compound nor do they specify the shapes of the

(a)

(b)

Electrical power for this garden light is generated by silicon photocells connected to rechargeable batteries. (a) The photocells use the energy from sunlight to recharge the light's batteries during the daytime; (b) after dark, electricity from the batteries automatically powers two highly efficient light sources called light-emitting diodes.

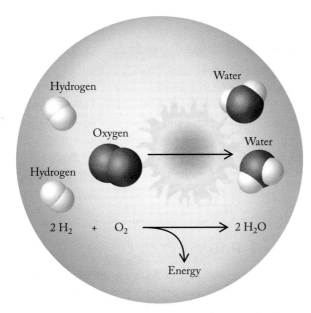

FIGURE 1.6 The reaction between hydrogen and oxygen is depicted here with space-filling models and in the form of a chemical equation.

A **scanning tunneling microscope (STM)** is an instrument that generates images of surfaces at the atomic scale.

An **atom** is the smallest particle of an element that retains the chemical characteristics of that element.

A **molecule** is a collection of atoms chemically bonded together. A molecule is the smallest entity that contains the constituent atoms of a compound in their characteristic proportions.

A **chemical formula** consists of the symbols of the constituent elements in a compound with subscripts to identify the number of atoms of each element present in one molecule.

A **chemical equation** uses chemical formulas to express the identities and quantities of substances involved in a chemical reaction.

A **chemical reaction** is the transformation of one or more substances into different substances.

A **chemical bond** is the force that holds two atoms in a molecule together.

molecules themselves. We use structural formulas and molecular models, such as those in Figure 1.7, to present perspectives of atom-to-atom connections and molecular shapes. Atoms are linked together by **chemical bonds** in a fixed arrangement. A *structural formula* shows the atoms and the bonds between them in a molecule; it shows how the atoms are connected, but it does not necessarily indicate the correct angles between the bonds and the three-dimensional shape of the molecule. When chemists write reactions, they frequently use a short-hand form of structural formula called a *written formula* that depicts the subgroups within a molecule and conveys important information on composition and the connections between subunits in the structure.

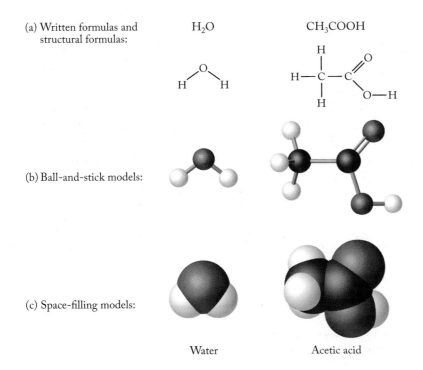

(a) Written formulas and structural formulas:

H_2O CH_3COOH

(b) Ball-and-stick models:

(c) Space-filling models:

Water   Acetic acid

FIGURE 1.7 Alternate ways to represent the arrangement of atoms in the molecules are illustrated for water and acetic acid (a principal ingredient in vinegar): (a) written formulas and structural formulas; (b) ball-and-stick models, where white spheres represent hydrogen; black ones, carbon; and red ones, oxygen atoms; and (c) space-filling models.

Filtration is a process for separating particles suspended in a liquid or a gas from that liquid or gas by passing the mixture through a medium that retains the particles.

Solution is another name for a homogeneous mixture. Solutions are often liquids, but they may also be solids or gases.

Distillation is a separation technique in which the more *volatile* (more easily vaporized) components of a mixture are vaporized and then condensed, thereby separating them from the less volatile components.

Ball-and-stick models use spheres to represent atoms and sticks to represent bonds. The advantage of ball and stick models is that they clearly show the correct angles between the bonds. The relative sizes of the spheres match the relative sizes of the atoms in some ball-and-stick models but do not in others. Moreover, in real molecules the atoms touch each other; they are not separated as they are in ball-and-stick models. *Space-filling models* are based on spheres that are drawn to scale and that abut each other as atoms do in real molecules. Unfortunately, the fidelity to atomic size and proximity in these models sometimes obscures the bond angles between the atoms.

1.3 Mixtures and How to Separate Them

The world around us, and we ourselves, are complex mixtures. Recall that mixtures are combinations of pure substances in variable proportions in which individual substances retain their chemical identities and that, unlike the decomposition of a compound, no chemical reactions are needed to isolate the constituents in a mixture.

Consider, for example, the water we drink. It may be safe to drink, but it is not "pure" water. For many of us, drinking water comes from a river, a lake, or a reservoir. Water drawn from any of these sources contains dissolved substances along with fine particles of soil, sediment, and other matter that does not dissolve. This water is an example of a heterogeneous mixture, because the suspended particles are not likely to be distributed uniformly, and we can see physically distinct regions in the material. The suspended particles may be separated from the water by physical means—allowing them to settle (Figure 1.8) and/or filtering the mixture. After the suspended solids have been removed by **filtration**, the filtered water contains only dissolved substances that are distributed uniformly, making it a homogeneous mixture, which is also called a **solution**. During filtration (Figure 1.9), the water may be either pumped or pulled through a porous membrane or other material that allows the water and substances dissolved in it to pass but that retains particles, including some too small to see.

Filtration is based on the principle that an object cannot pass through a pore that is smaller than it is. Air is often filtered to remove particles suspended in it. Air filters range in size from screens used in doors to prevent insects from entering buildings, to HEPA (high-efficiency particulate air) filters used in clean rooms to prevent particles or biological materials as small as 0.3 μm in diameter from contacting samples such as high-purity silicon chips or cells in culture that must be protected from contamination.

(a)

(b)

FIGURE 1.8 Water samples from a water treatment plant (a) after a coagulant has been added to help remove suspended particles, and (b) after the resulting suspension has been allowed to settle.

FIGURE 1.9 The technique of filtration is applied in daily life to separate spaghetti from the water in which it was cooked. The colander (the filter) traps the spaghetti and allows the water in which the pasta was cooked to pass into the sink.

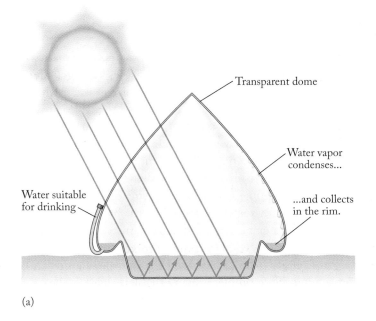

(a)

(b)

The identities and quantities of the dissolved substances in the water that we drink vary, depending on the water's source. This variability in composition is something that drinking water shares with all mixtures. Sometimes the concentrations of dissolved substances are so high, as in seawater, that the water is unfit to drink. One way to render seawater drinkable is through another physical separation method—**distillation**—whereby the salty seawater is heated to a temperature at which water is vaporized. The water vapor flows through a cooling condenser where it is converted back into liquid water and collected as purified *distillate*. Any dissolved salts and suspended particles remain behind in the boiling mixture (Figure 1.10) because they are much less volatile than water.

(a) In a solar still used in survival gear to provide fresh water from seawater, sunlight passes through the transparent dome and heats a pool of seawater. Water vapor rises, contacts the inside of the transparent dome (which is relatively cool), and condenses. The distilled water collects in the depression around the rim and then passes into the attached tube, from which one may drink it. (b) A solar still in use.

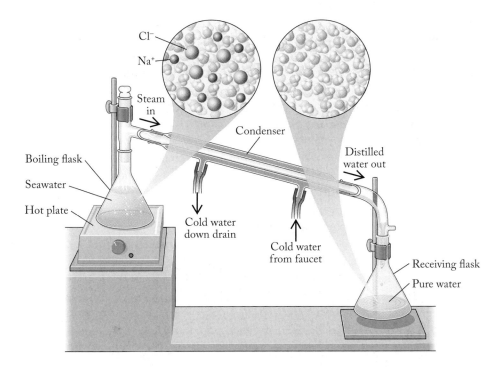

FIGURE 1.10 We can obtain pure water from seawater by heating seawater to the temperature at which it boils (its boiling point) in the flask on the left. Water vapor released from the boiling seawater passes into the condenser, where it condenses into pure liquid water and is collected. Sea salts (mostly sodium chloride) are not so easily vaporized and remain in the flask on the left. Distillation is used to obtain drinking water from seawater in those regions where fresh water is scarce but sources of energy are abundant, such as the area around the Arabian (Persian) Gulf.

1.4 Properties of Matter

We have already used descriptions to differentiate among samples of matter mentioned in this chapter: water is a liquid at room temperature, hydrogen and oxygen are gases, and silicon is a solid. Some substances dissolve in water while others do not. We use information from our senses to describe the world around us, and in doing so we categorize matter in a variety of ways. We breathe the gaseous mixture of matter called air to bring oxygen into our bodies to metabolize food. Oxygen is odorless, as is nitrogen, the main component of air on Earth, so when we smell something, we know we are inhaling some other gas along with the oxygen and nitrogen in normal air. Some matter burns, some puts out fires, some is shiny, some is dull, some is brittle, some is malleable—the list from our personal experiences with the world goes on and on. Because chemistry is the exploration of matter and its properties, it is crucial in the study of chemistry that you develop a clear understanding and ability to describe matter the way scientists do and that you develop an approach to solving problems using that language.

Pure substances have distinctive properties. Some of these properties do not vary from one sample to the next. Consider gold, for example: its color (very distinctive), its hardness (soft for a metal), its malleability (the fact that it can be hammered into very thin sheets called gold leaf), and its melting temperature (1064°C) all apply to any sample of pure gold. These are examples of **intensive properties**, properties that characterize a substance independent of the quantity of the material present. On the other hand, an object made of gold, such as an ingot, has a particular length, width, mass, and volume. These properties of a particular sample of a substance, which depend on how much of the substance is present, are examples of **extensive properties**.

The properties of substances fall into two other general categories: physical properties and chemical properties. **Physical properties** can be observed without changing the substance into another substance. Pure gold, for example, has a distinctive yellow color and metallic luster that we can observe without even touching it. Gold is relatively soft compared with other metals and has a relatively high **density** (d); the ratio of the mass (m) of a gold nugget to its volume (V)

$$d = \frac{m}{V} \qquad (1.1)$$

is high (19.3 g/mL) compared to the densities of many substances found in rocks and minerals.

As noted in Section 1.1, gold is one of the few elements found in nature as a free element. *Free elements* may exist as single atoms like He or as molecules containing atoms of only the element, like O_2 and S_8. They are not bonded in compounds with other elements. Most elements are not free but are found combined with other elements in compounds, like hydrogen in H_2O or sodium in NaCl. As we have already noted, hydrogen combines explosively with oxygen to form water as the product of a chemical reaction. High flammability is a **chemical property** of hydrogen. Like any chemical property, the only way to observe it is to actually react that substance with another substance and thereby produce a different material. A substance's reactivity with other substances, including the rates of the reactions, the identities of the other reacting substances, and the identities of the products that are formed, all define the chemical properties of a substance.

The physical and chemical properties of a compound, such as water, are very different from those of the elements that combine to form it. Water is a liquid at room temperature, whereas hydrogen and oxygen are gases. Water expands when

An **intensive property** of a substance is a characteristic that is independent of the amount of substance present.

An **extensive property** is a characteristic that varies with the quantity of the substance present.

A **physical property** is a characteristic of a substance that can be observed without changing it into another substance.

Density (d) is the ratio of the mass (m) of an object to its volume (V). The density of a pure substance is an intensive physical property of that substance.

A **chemical property** is a characteristic of a substance that can be observed only by reacting it to form another substance.

it freezes at 0°C; hydrogen (freezing point, −259°C) and oxygen (−219°C) do not. Oxygen supports combustion reactions, while hydrogen is a highly flammable fuel. Water neither supports combustion nor is flammable. Indeed, it is widely used to put out fires.

1.5 A Method for Answering Questions and Solving Problems

As we continue with the discussion of matter and its behavior, we will frequently pause in the story to pose questions or problems for you to solve in order to test your comprehension of the material you have just read. Your success in this course depends, in part, on your ability to solve these problems, so it is appropriate for the text to provide some guidance about how to develop answers to questions about chemistry.

Problem solving is like playing an instrument: the more you practice, the better you become. In this section we present a framework to help you approach problem solving systematically. We follow this model of problem solving in the Sample Exercises that appear throughout this book. Each Sample Exercise is followed by a Practice Exercise involving the same type of problem, which can be solved using a similar approach. We strongly encourage you to hone your problem-solving skills by working all the exercises as you read the chapters. You should also find the approach described here useful in solving the Concept Tests in this book as well as problems you encounter in other courses and other contexts.

We use the acronym COAST (**C**ollect and **O**rganize, **A**nalyze, **S**olve, and **T**hink about the answer) to represent the four steps in this approach. As you read about it here and use it, keep in mind that COAST is a framework for solving problems, not a recipe. It should serve as a guide as you develop your own approach to solve each problem. Let's consider what is involved in each step.

COLLECT AND ORGANIZE The first step in solving a problem is to decide how to begin. In COAST you start by collecting information as well as your ideas. This step is based on your understanding of the problem, including the fundamental chemical principles on which it is based. In the process of collecting and organizing the information, you complete these tasks:

- Identify the key concept at the heart of the problem. (You may find it useful to restate the problem in your own words.)

- Identify and define the key terms used to express that concept.

- Sort through the information in the problem, separating what is pertinent to solving the problem from what is not.

- Assemble any supplemental information that may be needed, including equations, definitions, and constants.

ANALYZE The next step is to analyze the information you were given and have collected to determine how to connect it to the answer you seek. Sometimes it is easier to work backward to create these links: consider the nature of the answer first and think about how you might get to it from the information provided in the problem and other sources. If the problem is quantitative and requires a numerical answer, frequently the units of initial values and the final answer help you identify how they are connected and what equation(s) may be useful. This step may include rearranging the terms in equations to solve for an unknown, or setting up conversion factors. (Section 1.9 addresses conversion factors.) For some problems, drawing a sketch based on molecular models or an experimental setup may help you visualize how the starting points and final answer are connected.

SOLVE At this point, for qualitative problems you need to decide how to express your answer. Most likely the correct expressions are contained in the definitions within the text sections you have just read. For quantitative problems you need to insert the starting values and appropriate constants into equations, conversion factors, or other mathematical expressions you have set up, and calculate the answer. In this step, make sure that the units are consistent and cancel out as needed, and that the certainty of the quantitative information you are using is reflected in the number of significant figures in your final answer. (Section 1.7 addresses significant figures.)

THINK ABOUT IT Finally, you need to think about your result and answer such questions as, Does this answer make sense based on my own experience and on what I have just learned? Is the value for a quantitative answer reasonable? Are the units correct? Is the number of significant figures appropriate? Then ask yourself how confident you are that you could solve another problem, perhaps drawn from another context but based on the same chemical concept. Sometimes you will also think about how this problem relates to other observations you may have made about matter in your daily life.

The COAST approach should help you to solve problems in a logical away and to avoid certain pitfalls, such as grabbing an equation that seems to have the right variables in it and plugging in numbers, or resorting to random trial and error. Examine the steps in each Sample Exercise carefully. As you do, try to answer these questions about each step:

- **What** was done in that step?

- **How** was it done?

- **Why** was it done?

Once you have answered these questions, you will be ready to solve the matched Practice Exercises and their related end-of-chapter Questions and Problems in a systematic way. The COAST method has been applied in all the Sample Exercises that follow in this book.

SAMPLE EXERCISE 1.1 **Distinguishing Physical and Chemical Properties**

Which of the following properties of gold are chemical and which are physical?

a. Gold metal can be dissolved by reacting it with a mixture of nitric and hydrochloric acids known as aqua regia;
b. Gold melts at 1064°C.
c. Gold can be hammered into gold leaf sheets that are so thin that light passes through them.
d. Gold can be recovered from gold ore by treating the ore with a solution containing cyanide, which reacts with and dissolves tiny particles of gold in the ore.

COLLECT AND ORGANIZE We are asked to determine whether properties are chemical properties, which describe how a substance reacts with other substances, or physical properties, which can be observed or measured without changing a substance into another one.

ANALYZE Properties (a) and (d) involve chemical reactions because elemental gold, which is insoluble, reacts with substances that convert it into soluble compounds. Properties (b) and (c) describe processes in which gold remains gold. Liquid gold, when it melts, is converted into solid gold again when it cools, and gold leaf is still solid gold.

SOLVE Properties (a) and (d) are chemical properties, and (b) and (c) are physical properties.

THINK ABOUT IT When possible, fall back on your experiences and observations about the world. Ice is just water in the solid phase; water vapor or steam is water in the gas phase. The identity of the component atoms or molecules does not change as a substance melts, freezes, or vaporizes.

Practice Exercise Which of the following properties of water are chemical and which are physical? (a) It freezes at 0.0°C. (b) It is useful for putting out most fires. (c) A cork floats in it but a piece of copper sinks. (d) During digestion, starch reacts with water to form sugar.

(Answers to Practice Exercises are in the back of the book.)

1.6 States of Matter

All the matter in our world exists in one of three physical states: solid, liquid, or gas. You are probably familiar with characteristic properties of these states:

- A **solid** has a definite volume and shape.

- A **liquid** occupies a definite volume, but it does not have a definite shape.

- A **gas** (or *vapor*) has neither a definite volume nor a definite shape. Rather, it expands to occupy the entire volume and shape of its container.

A gas, unlike a solid or a liquid, is compressible, which means that it can be squeezed into a smaller volume if its container is not rigid and additional external pressure is applied to it.

The differences between solids, liquids, and gases can be understood if we view these three states on a molecular level. Consider, for example, the differences in the arrangements and freedom of motion of water molecules in ice, liquid water, and water vapor, as shown in Figure 1.11. The water molecules in ice are mostly surrounded and locked in place in an extended three-dimensional array. Molecules in this structure may vibrate, but they are not free to move past the molecules that surround them. The molecules in liquid water are free to flow past each other, but they are still in close proximity to each other and do not expand to fill a larger volume. In water vapor the molecules are widely separated, and the volume occupied by the molecules is negligible compared to the volume occupied by the gas. This separation accounts for the compressibility of gases. The molecules are relatively independent and move throughout the space the vapor occupies.

A **solid** consists of atoms or molecules in close contact with each other and often in an organized arrangement; solids have a definite shape and volume.

A **liquid** consists of atoms or molecules in close proximity to each other (usually not as close as in a solid); liquids occupy a definite volume, but flow to assume the shape of their containers.

A **gas** (also called a *vapor*) consists of very widely separated atoms or molecules; gases have neither definite volume nor shape, and they expand to fill their containers.

Sublimation is the transformation of a solid directly into a vapor (gas).

Deposition is the transformation of a vapor (gas) directly into a solid.

The air that we breathe is a mixture of gaseous elements, principally nitrogen (N_2), oxygen (O_2), and argon (Ar) together with carbon dioxide (CO_2) and water vapor. Do you think these gases in a sample of air are mixed uniformly, making air a solution of gases?

(Answers to Concept Tests are in the back of the book.)

We can transform water between one physical state and another if we raise or lower its temperature. The ice on the pond in (Figure 1.11a) forms when temperatures drop in early winter and the water freezes. This process runs in reverse when warmer temperatures return in the spring and the ice melts (Figure 1.11b). Similarly, the heat of the sun may vaporize liquid water during the daytime, but at night colder temperatures can result in the condensation of water vapor, producing clouds in the upper atmosphere and fog above the surface of water (Figure 1.11c).

Some solids change directly into gases with no intervening liquid phase. For example, snow may be converted to water vapor on a very cold sunny winter day, even though the air temperature remains well below freezing. This transformation of solid directly to vapor is called **sublimation**. The reverse process—in which, for example, water vapor forms a layer of frost on a cold night—is an example of a gas being transformed directly into a solid without ever being a liquid, a process called **deposition**. The names of the physical changes that matter undergoes when it is transformed from one physical state to another are illustrated in Figure 1.12. The transitions represented by upward-pointing arrows require the addition of heat energy; those represented by downward-pointing arrows release heat.

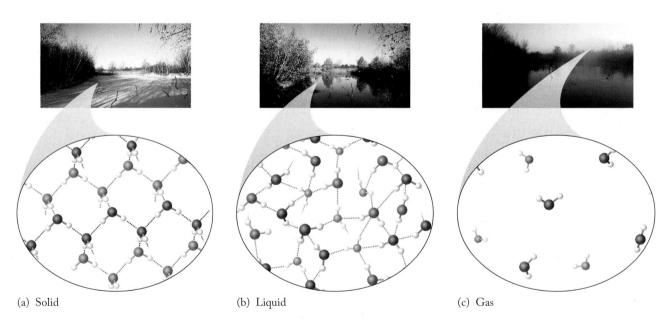

(a) Solid   (b) Liquid   (c) Gas

FIGURE 1.11 Water can exist in three states: (a) solid (ice), (b) liquid (water), or (c) gas (water vapor). The water molecules in ice are each mostly surrounded and locked in place by strong interactions (represented by the dotted lines) with other water molecules in rigid, three-dimensional arrays. Molecules of liquid water are free to tumble over each other, and they are largely independent in water vapor. Water vapor is invisible; however, we can see clouds and fog that form when atmospheric water vapor condenses into tiny drops of liquid H_2O.

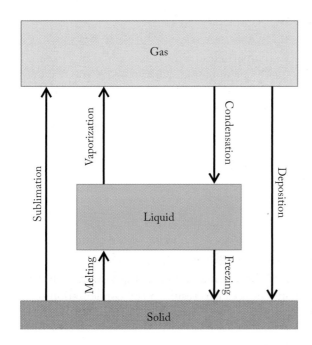

FIGURE 1.12 Transformations of matter between the three physical states of matter are caused by either the addition or removal of heat. The arrows pointing upward represent transitions that require heat; the arrows pointing downward represent transitions that release heat.

SAMPLE EXERCISE 1.2   **Recognizing the Physical States of Matter**

For each pair of boxes below: What is the physical state represented by the particles, and what change of state is associated with the arrow. (The particles could be atoms or molecules.) What change of state would be indicated if the arrow pointed in the opposite direction?

COLLECT AND ORGANIZE The particles in each box represent an arrangement of atoms or molecules in a solid, liquid, or gas. Once we have defined the initial and final states represented by each pair of boxes, we can name the transition using the information in Figure 1.12.

ANALYZE We should look for the following patterns among the particles:

- An ordered arrangement of particles that does not fill or match the shape of the box represents a solid.

- A less-ordered arrangement of particles that partially fills the box and conforms to its shape represents a liquid.

- A dispersed array of particles distributed throughout the box represents a gas.

SOLVE a. The particles in the left box partially fill the box and adopt the shape of the container; therefore they represent a liquid. The particles in the box on the right are ordered and form a shape different from that of the box, so they must represent a solid. The arrow represents the process in which a liquid turns into a solid, and thus the phase change is called freezing. In the reverse process a solid is liquefied—the phase change is called melting.

b. The particles in the box on the left represent a solid because they are ordered and do not adopt the shape of the container. Those in the right box are dispersed throughout the container and so must represent a gas. The process by which a solid goes directly to the vapor state is sublimation. The reverse process is deposition.

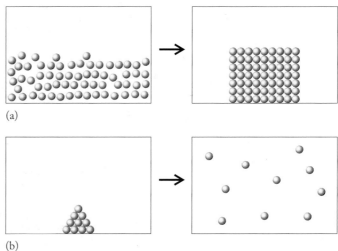

(a)

(b)

THINK ABOUT IT The particles in the boxes could represent either atoms of an element or molecules of a compound. Most pure substances, under the right conditions of temperature and pressure, undergo the transformations illustrated in this exercise. You may have noticed that on a sunny day after a snowfall the snow disappears more quickly from areas in direct sunlight than from shady places, even if temperature does not rise above freezing. The snow sublimes. Even on overcast days, snow in areas not directly exposed to light lasts longer because it does not sublime as rapidly.

Practice Exercise (a) What physical state is represented by the particles in each of these boxes, and what is the name of the change of state associated with the arrow? (b) What change of state would be illustrated if the arrow were pointed in the opposite direction?

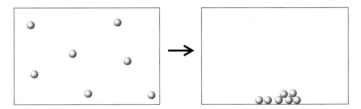

(Answers to Practice Exercises are in the back of the book.)

1.7 The Scientific Method: Starting Off with a Bang

FIGURE 1.13 For centuries, cultures have passed on stories about the creation of the universe. Native American cultures left pictographs (shown here) on canyon walls in the southwestern desert of the United States.

The **scientific method** is an approach to acquiring knowledge that is based on careful observation of phenomena; development of a simple, testable explanation; and conducting additional experiments that test the validity of the hypothesis.

In the previous section we examined the properties and classifications of matter. In this section and those that follow we address where matter came from by examining one explanation of how the universe came into being. For thousands of years people have pondered the question of the origin of the universe. Until very recently they looked for divine guidance to find answers. The ancient Greeks believed that at the beginning of time there was no matter, only a vast emptiness they called Chaos, and that from that emptiness emerged the first supreme being, Gaia (also known as Mother Earth), who gave birth to Uranus (Father Sky). Other cultures and religions have described creation in similar terms: of supernatural beings creating ordered worlds out of vast emptiness. The opening verses of the book of Genesis, for example, describe a darkness that "was without form and void" and from which God created light (energy) and then the heavens and Earth (matter). Similar stories are part of Asian, African, and Native American cultures (Figure 1.13).

Today we have the ability and technological capacity to take a different approach to explaining natural phenomena and answering fundamental questions about how our world was formed and what physical forces control it. Our approach is based on the **scientific method** of inquiry (Figure 1.14).

This approach evolved in the late Renaissance during a time when economic and social stability gave people an opportunity to look more closely at the workings of nature and to question old beliefs. By the early 17th century the English philosopher Francis Bacon (1561–1626) had published his *Novum Organum* (which translates as *New Organ* or *New Instrument*), in which he described the creation of knowledge and understanding through observation, experimentation, and careful reflection. In the process he described, observations lead to tentative explanations, or **hypotheses**, the validity of which can be tested through additional experimentation. Further testing and observation might support a hypothesis or disprove it, or perhaps require that it be modified so that it adequately explains all of the experimental results. A hypothesis that withstands the tests of many experiments over time and that adequately explains the results of extensive observation and experimentation may be elevated to the rank of a **scientific theory**, or **model**.

Let's examine the application of the scientific method by answering a profound question: how did the universe form? In the early 20th century, careful observations with new, increasingly powerful telescopes allowed astronomers to discover that the universe contains billions of galaxies, each one containing billions of stars. They also discovered that (1) the other galaxies in the universe are moving away from our own Milky Way and from each other, and (2) the speeds with which these galaxies are receding are proportional to the distance between them and the Milky Way—the farthest galaxies are moving away the fastest.

The significance of these discoveries can be appreciated if we think what might happen if the motion of the galaxies had somehow been recorded on videotape since the beginning of time and we were able to play the tape in reverse. The images of the other galaxies would then be moving toward ours (and toward one another), with the ones farthest away closing in the fastest, catching up with those galaxies that were closer to start with but moving more slowly. Eventually, near the beginning of the tape, all the galaxies and all the matter in the universe would approach the same point at the same time. As all the matter was compressed into a smaller and smaller volume, its density would become enormous and it would get very hot (just as, for example, compressing air with a bicycle pump makes it hot). At the very beginning the universe would be squeezed into an infinitesimally small space of unimaginably high temperature. Allowing time to run forward from this point would produce an enormous amount of energy, an event that has come to be known as the Big Bang.

The Big Bang hypothesis of how the universe may have begun was proposed in 1927 by Belgian astronomer Georges-Henri Lemaître (1894–1966), and has been the subject of extensive testing—and controversy—ever since. Indeed, the term Big Bang was first used by British astronomer Sir Fred Hoyle (1915–2001) to poke fun at the idea. However, the results of an enormous number of experiments and observations conducted since the 1920s support what is now commonly described as the Big Bang theory; in the next section we examine a few of these.

CONCEPT TEST

If the volume of the universe is expanding and if the mass of the universe is not changing, is the density of universe (a) increasing, (b) decreasing, or (c) constant?

(Answers to Concept Tests are in the back of the book.)

A **hypothesis** is a tentative and testable explanation for an observation or a series of observations.

A **scientific theory** (also called a **model**) is a general explanation of widely observed phenomena that have been extensively tested.

▶❙❙ **CHEMTOUR** Big Bang

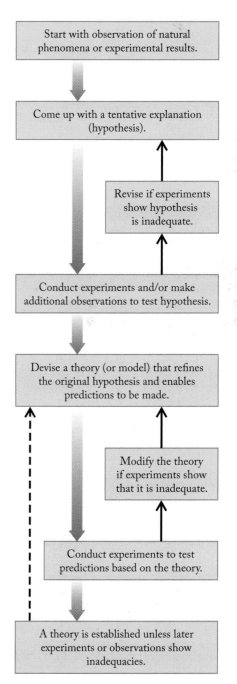

FIGURE 1.14 In the scientific method, observations lead to a tentative explanation, or hypothesis, which in turn leads to more observations and testing, which may lead to the formulation of a succinct, comprehensive explanation called a scientific theory. The overall process, which repeats in a self-correcting fashion, represents a highly reasoned way to understand the dynamics of nature.

1.8 Making Measurements and Expressing the Results

Before we examine the results of experiments designed to test the Big Bang theory, we should note the critical importance of accurate measurements in science, whether we are exploring the nature of the cosmos or determining whether a medication contains the right quantity of a particular drug. Accurate measurements are essential to our ability to characterize the physical and chemical properties of matter. The rise of scientific inquiry in the 17th and 18th centuries brought about a heightened awareness of the need for accurate measurements and for expressing the results of observations and measurements in ways that were understandable to others. Standardization of the units of measurement was essential.

SI Units

In 1791 French scientists proposed a standard unit of length equal to 1/10,000,000 of Earth's quadrant—the distance along an imaginary line running from the North Pole to the equator. Hard work by teams of surveyors led in 1794 to the adoption of a length corresponding to 39.37 inches as the standard meter (after the Greek *metron*, which means "measure").

These French scientists also settled on a decimal-based system for designating lengths that were multiples or fractions of a meter (m). They chose Greek prefixes for the names of units of distance much greater than a meter, such as *deka-* and *kilo-* for lengths of 10 meters and 1000 meters, and they chose Latin prefixes for much smaller lengths, such as *centi-* and *milli-* for lengths that were 1/100 and 1/1000 of a meter (see Table 1.1). Eventually these decimal prefixes carried over

TABLE 1.1 Commonly Used Prefixes for SI Units

PREFIX		VALUE	
Name	Symbol	Numerical	Exponential
zetta	Z	1,000,000,000,000,000,000,000	10^{21}
exa	E	1,000,000,000,000,000,000	10^{18}
peta	P	1,000,000,000,000,000	10^{15}
tera	T	1,000,000,000,000	10^{12}
giga	G	1,000,000,000	10^{9}
mega	M	1,000,000	10^{6}
kilo	k	1,000	10^{3}
hecto	h	100	10^{2}
deka	da	10	10^{1}
deci	d	0.1	10^{-1}
centi	c	0.01	10^{-2}
milli	m	0.001	10^{-3}
micro	μ	0.000001	10^{-6}
nano	n	0.000000001	10^{-9}
pico	p	0.000000000001	10^{-12}
femto	f	0.000000000000001	10^{-15}
atto	a	0.000000000000000001	10^{-18}
zepto	z	0.000000000000000000001	10^{-21}

to the names of standard units for other dimensions. These dimensions included the mass of a cubic centimeter of water at the temperature of its maximum density, which was the basis for the gram (g). The volume corresponding to 1 cubic decimeter (a cube 1/10 of a meter on a side) came to be known as the liter (L). In July 1799 a platinum (Pt) rod 1 meter long and a platinum block weighing 1 kilogram were placed in the French National Archives to serve as the legal standards for length and mass. These objects served as references for the metric system for expressing measured quantities.

Since 1960 scientists have by international agreement used a modern version of the French metric system: *Système international d'unités*, commonly abbreviated SI. The base SI units are listed in Table 1.2. Other units are derived from these base units. For example, the standard unit for volume is the cubic meter (m^3), and the standard unit for velocity is meters per second (m/s). Table 1.3 contains some of these derived units and their equivalents in the English system of units. The SI unit of length is the same as in the original metric system: the meter. For those more familiar with the English system of measures (which is a deceptive name since the system is widely used only in the United States, not in England), the meter is about 10% longer than a yard.

TABLE 1.2 SI Base Units

Quantity or Dimension	Unit Name	Unit Abbreviation
Mass	kilogram	kg
Length	meter	m
Temperature	kelvin	K
Time	second	s
Energy	joule	J
Electric current	ampere	A
Amount of a substance	mole	mol
Luminosity	candela	cd

TABLE 1.3 Conversion Factors for SI and Other Commonly Used Units

Quantity or Dimension	Equivalent Units
Mass	1 kg = 2.205 pounds (lb); 1 lb = 0.4536 kg = 453.6 g 1 g = 0.03527 ounce (oz); 1 oz = 28.35 g
Length (distance)	1 m = 1.094 yards (yd); 1 yd = 0.9144 m 1 m = 39.37 inches (in); 1 foot (ft) = 0.3048 m 1 in = 2.54 cm (exactly) 1 km = 0.6214 miles (mi); 1 mi = 1.609 km
Volume	$1 m^3 = 35.31 ft^3$; $1 ft^3 = 0.02832 m^3$ $1 m^3 = 1000$ liters (L) (exactly) 1 L = 0.2642 gallon (gal); 1 gal = 3.785 L 1 L = 1.057 quarts (qt); 1 qt = 0.9464 L

The needs of modern science require that the length of the meter, as well as the dimensions of other SI units, be known or defined by quantities that are much more constant than the length of a rod of platinum. Two such quantities are the speed of light (*c*) and time. In recent years the length corresponding to 1 meter has been redefined as the distance traveled by the light from a helium–neon laser in 1/299,792,458 of a second.

Significant Figures

All scientific measurements have one thing in common: there is a limit to how well we can know their results. Nobody is perfect, and no analytical method is perfect either. All methods have an inherent limit on their ability to produce accurate results, and we need ways to express experimental results that reflect those limits. We do so by expressing numerical data from experiments to the appropriate number of **significant figures**.

The number of significant figures in a measured value, or in the result of a calculation based on one or more measured values, indicates how certain we are of the value. For example, suppose we determine the mass of a penny on the two balances shown in Figure 1.15. We assume both balances are working properly. The balance on the right can be used to weigh objects to the nearest 0.0001 g; the balance on the left can be used to weigh heavier objects but only to the nearest 0.01 g. According to the balance on the left, our penny weighs 2.53 g; according to the balance on the right, it weighs 2.5271 g. The mass obtained with the left-hand balance has three significant figures: the digits 2, 5, and 3 are considered *significant*, which means that we are confident in their values. The mass obtained using the right-hand balance has five significant figures (the digits 2, 5, 2, 7 and 1). We may conclude that the mass of the penny can be determined with greater certainty when we use the balance on the right.

Now suppose that a puff of air blows across the tops of both balances as someone walks by them. The value displayed by the one on the left will probably not be affected, but the balance on the right is so sensitive that there will likely be a change in the last digit of the value it displays. This occurrence illustrates an important point: for many measured values there is some uncertainty in the rightmost digit. However, the last measurable digit is still considered significant even though we are less certain of its value than we are of the value of the others. The significant figures in a number includes all the digits whose values we know and the first digit to the right of the known digits that is uncertain.

Now consider this experimental result: a small aspirin tablet is placed on the balance on the right, and the display reads 0.0810 g. How many significant figures

> The **significant figures** are the number of digits in a measured value; they include all the digits with known values plus the first uncertain digit to the right of the known digits. The greater the number of digits, the greater the certainty with which the value is known.

FIGURE 1.15 The mass of a penny can be measured to the nearest 0.01 gram with the balance on the left, and to the nearest 0.0001 gram with the balance on the right.

are in this value? You might be tempted to say "five" because that is the number of digits displayed. However, the first two zeros in the value are not considered significant, because they only serve to fix the decimal point. They tell us that this value, expressed in scientific notation, is 8.10×10^{-2} and not, for example, 8.10×10^{-3} g or 8.10×10^{-4} g. In *scientific notation*, only the digits before the exponent indicate how precisely we know the value; the digit(s) in the exponent do not. All three of these values have three significant figures: the digits 8, 1, and 0. (Scientific notation is explained in detail in Appendix 1.)

Why is the rightmost zero in 0.0810 g significant? The answer is related to the capability of the balance to measure masses to the nearest 0.0001 g. If the balance is operating correctly, we may assume that the mass of the tablet is 0.0810 g, not 0.0811 g or 0.0809 g. We may assume that the last digit really is zero, and we need a way to express that. If we dropped the zero and recorded a value of only 0.081 g, we would be implying that we knew the value to only the nearest 0.001 g, which is not the case. The following guidelines will help you handle zeros (highlighted in green) in deciding the number of significant figures in a value:

1. Zeros at the beginning of a value, as in **0.0**592, are never significant. In this example, they just set the decimal place.
2. Zeros at the end of a value and after a decimal point, as in $3.\mathbf{00} \times 10^8$, are always significant.
3. Zeros at the end of a value that contains no decimal point, as in 96,**500**, may or may not be significant. They may be there only to set the decimal place. We should use scientific notation to avoid this ambiguity; 9.65×10^4 (3 significant figures) and 9.6500×10^4 (5 significant figures) are two of the possible interpretations of 96,500.
4. Zeros in between nonzero digits, as in 101.3, are always significant.

CHEMTOUR Significant Figures

CHEMTOUR Scientific Notation

CONCEPT TEST

How many significant figures are there in each value in guidelines 1, 2, and 4 above?

(Answers to Concept Tests are in the back of the book.)

Significant Figures in Calculations

Now let's consider how significant figures are used to express the results of calculations involving measured quantities. An important rule to remember is that significant figures should be enforced at *the end of a calculation*, never on intermediate results. A reasonable guideline to follow is that one digit to the right of the last significant digit should be carried forward in intermediate steps, and the rules about significant figures should be applied in expressing the final answer.

Suppose we believe that a small nugget of yellow metal in a mineral collection is pure gold. We could test our belief by determining the mass and volume of the nugget and then calculating its density. If its density matches that of gold (19.3 g/mL), chances are good the nugget is gold, because few minerals are that dense. We run the experiment and find that the mass of the nugget is 4.72 g and its volume is 0.25 mL. What is the density of the nugget, expressed in the appropriate number of significant figures?

Using Equation 1.1 to calculate the nugget's density (d) from its mass (m) and volume (V), we have

$$d = \frac{m}{V} = \frac{4.72 \text{ g}}{0.25 \text{ mL}} = 18.88 \text{ g/mL}$$

This result appears to be slightly less than that of pure gold. However, we need to answer the question, "How well do we know the result?" The value of the mass is known to three significant figures, but the value of volume is known only to two. At this point we need to invoke the weak-link principle, which is based on the idea that a chain is only as strong as its weakest link. In calculations involving measured values, this principle means that we can know the answer of a calculation involving multiplication or division only as well as we know the least well-known value used in the calculation. In this example the weak link is the value of the volume because it has the fewest—in this case two—significant figures. Since our final answer cannot have more than two significant figures, we must convert this value to only two significant figures. We do this by a process called rounding off.

Rounding off a value means dropping the insignificant digits (all digits to the right of the first uncertain digit) and either rounding up or rounding down the last significant digit. If the next digit to the right is greater than 5, we round up; if it is less than 5, we round down. The question remains of how to handle cases in which the next digit is *exactly* 5. A good rule to follow in those cases is to round to the nearest even number when the first insignificant digit is a 5, and either there are no additional insignificant digits or else all subsequent insignificant digits are 0. For example, rounding off 45.45 or 45.450 to 3 significant figures makes it 45.4; however, we would round off 45.55 or 45.550 to 45.6. Rounding 45.450001 to 3 significant figures makes it 45.5.

In the case of our gold nugget, the density must be rounded to the nearest whole unit (shown in blue below). Because the digit in the first decimal place (shown in red) is greater than 5, we round up the blue eight to nine:

$$18.88 \text{ g/mL} = 19 \text{ g/mL}$$

The density of pure gold, expressed to two significant figures, is also 19 g/mL, so based on how well we know the measured values, we conclude that our hypothesis may be true and the nugget could indeed be made of gold.

CONCEPT TEST

Does the result of the preceding density determination prove that the nugget is pure gold, or does it fail to disprove that the nugget is not pure gold? Explain the difference between these two conclusions, and explain why you prefer one over the other.

(Answers to Concept Tests are in the back of the book.)

The weak-link principle for significant figures also applies to calculations requiring addition and subtraction, but we handle the numbers in a different way than in multiplication and division. To illustrate, let's revisit the determination of the density of a nugget of gold. How might we determine the volume of such an irregular object? One approach, if the object is more dense than water, is to measure the volume of water that it displaces. Suppose we add exactly 50.0 mL of water to a 100 mL graduated cylinder as shown in Figure 1.16. Note that the scale on the cylinder is graduated in milliliters. We can read the value of the volume to the ones place (50 mL is different than either 49 mL or 51 mL), and we can estimate the tenths place. The volume in the cylinder on the left is correctly recorded as 50.0 mL. The number has three significant figures: two that we read directly on the scale and the third one we estimate.

Then we place the chunk of the shiny yellow mineral into the graduated cylinder, being careful not to splash any water out. We read the level of the water and observe that it has increased to 58.0 mL. Again the volume is reported to three significant figures: the two we read directly (5 and 8) and the one we estimate (0 in the tenths place). The volume of the nugget is the same as the volume of the water that it displaced, which is the difference between 50.0 and 58.0 mL:

$$
\begin{array}{r}
58.0 \text{ mL} \\
- \ 50.0 \\
\hline
8.0 \text{ mL}
\end{array}
$$

The result has only two significant figures because the initial and final volumes are known to the nearest tenth of a milliliter, so we can know the difference between them to only the nearest tenth of a milliliter. The location of the decimal point in values that are added or subtracted is important in setting the number of significant digits in the answer. When measured numbers are added or subtracted, the result has the same number of digits to the right of the decimal as the number with the fewest digits to the right of the decimal. Correspondingly, if whole numbers are added or subtracted, the result has the same number of significant digits as in the least precisely known number.

Consider one more example. Suppose the U.S. Mint defines the average mass of a penny to be 2.53 g and stipulates that a roll of pennies must contain 50 pennies. We weigh a roll of pennies, again on the balance on the left side of Figure 1.15. Using the tare feature, which enables us to correct for the mass of the wrapper, we find that the mass of just the pennies in the roll is 124.01 g. Dividing this value by the average penny's mass, we get

$$
\frac{124.01 \ \cancel{g}}{2.53 \ \cancel{g}/\text{penny}} = 49.02 = 49.0 \text{ pennies}
$$

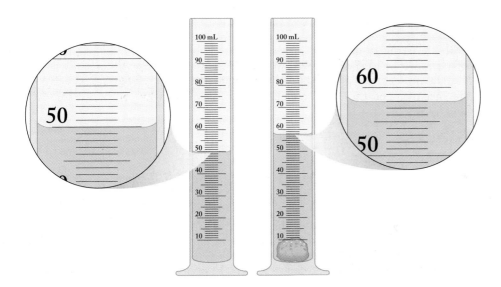

FIGURE 1.16 The volume of a small, irregularly shaped object can be determined by measuring the volume of water that it displaces. A small nugget believed to be pure gold is placed in a graduated cylinder containing 50.0 mL of water. The volume rises to 58.0 mL, which means that the volume of the nugget is 8.0 mL.

If we were measuring uncountable substances, then rounding off to 49.0 would be appropriate (because under these circumstances, the initial value of 2.53 g/penny would be the weak link in this calculation). However, because we only use whole numbers to count items, we round to 49 and conclude that the roll has exactly 49 pennies. That means we have to add one more penny to the roll to make it complete.

Suppose the penny we add is one we know weighs 2.5271 g, because we weighed it on the balance shown on the right in Figure 1.15, which determines masses to ± 0.0001 g (where the symbol ± means "plus or minus" the value that follows it). What is the total mass of all 50 pennies to the appropriate number of significant figures? Adding the mass of the 50th penny to the other 49, we have

$$
\begin{array}{r}
124.01 \text{ g} \\
+ \quad 2.5271 \\
\hline
126.5371 \text{ g}
\end{array}
$$

The total mass of the first 49 pennies has two decimal places, and the mass of the 50th has four decimal places. Therefore, by the weak-link principle we can know the sum of the two numbers to only two decimal places. This makes the last two digits in the sum, shown in red, not significant, so we must round off the total mass to 126.54 g.

SAMPLE EXERCISE 1.3 **Using Significant Figures in Calculations**

A nugget of shiny yellow mineral has a mass of 30.01 g. Its volume is determined by placing it in a 100 mL graduated cylinder partly filled with water. The volume of the water before adding the mineral is 56.3 mL. The volume after the mineral was added is 62.6 mL. Is the nugget made of gold?

COLLECT AND ORGANIZE In response to the question, we make the hypothesis that the object is gold. We are given the mass of the object. The volume of the object is the volume of water that it displaces, which is the difference in the volume of the contents of the graduated cylinder before and after the object is added. Because the initial and final volumes are known to only the nearest tenth of a milliliter, we can know the difference between them to only the nearest tenth of a milliliter.

Density is calculated by dividing the mass of an object by its volume. To either support or reject our hypothesis, we need to know the density of pure gold to compare it with whatever value we determine experimentally. According to the data in Appendix 3, the density of pure gold is 19.3 g/mL.

ANALYZE Using Equation 1.1, the mass of the object, and volume of the object as determined by water displacement, we calculate the density of the object. Then we compare our result to the known density of gold and decide to accept or reject our hypothesis.

SOLVE

$$
d = \frac{m}{V}
$$

$$
= \frac{30.01 \text{ g}}{(62.6 - 56.3)\text{mL}} = \frac{30.01 \text{ g}}{6.3 \text{ mL}} = \frac{4.7635 \text{ g}}{\text{mL}}
$$

Because we know the volume of the object (6.3 mL) to two significant figures, we can know its density to only two significant figures, so we must round off the result to 4.8 g/mL:

$$d = \frac{4.7635 \text{ g}}{\text{mL}} = \frac{4.8 \text{ g}}{\text{mL}}$$

We test the hypothesis by comparing the density of the object with that of gold. The densities do not match, so we reject our hypothesis.

THINK ABOUT IT Measuring density is an easy way to test whether two samples might be made of the same material. Of course, knowing density to only two significant figures may mean that several materials have to be considered. In this case, two significant figures is enough to establish that the sample in not pure gold, thereby refuting the original hypothesis. Another question arises: what else could the object be? The density of the gold-colored mineral iron pyrite, called fool's gold, is 4.7 g/mL, so a new hypothesis could be formulated that our object might be fool's gold. Other physical or chemical properties could be studied to reject or support that hypothesis.

Practice Exercise Express the results of the following calculation to the appropriate number of significant figures:

$$\frac{(0.391)(0.0821)(273 + 25)}{8.401}$$

(Answers to Practice Exercises are in the back of the book.)

CONCEPT TEST

Suggest a way to test the hypothesis that the object in the Sample Exercise 1.3 is fool's gold.

(Answers to Concept Tests are in the back of the book.)

The results of measurements always have some degree of uncertainty in their value, which limits the number of significant figures that can be used to express the value. On the other hand, some values are known exactly, such as 12 eggs in a dozen, 3 feet in a yard, and 60 seconds in a minute. There is no uncertainty in these values. Put another way, we know their value with absolute precision. These values are treated as definitions and are not considered when determining the appropriate number of significant figures in the answer to a mathematical calculation in which they appear.

SAMPLE EXERCISE 1.4 **Distinguishing Exact from Uncertain Values**

Which of the following numerical values associated with the Washington Monument, in Washington, DC, are exact numbers, and which are based on measurements and so are known to only limited precision? (a) The monument is made of 36,941 white marble blocks. (b) The monument is 169 m tall. (c) There are 893 steps to the top. (d) The mass of the aluminum capstone is 2.8 kg. (e) The area of the foundation is 1487 m^2.

COLLECT AND ORGANIZE Exact numbers are those based on quantities of individual objects that are either definitions, such as eggs in a dozen, or are directly counted, like steps in a staircase. Quantities that are obtained from measurements—including length, mass, and area—have an inherent uncertainty.

ANALYZE Our task is to divide the five values in this Sample Exercise into the two groups—either exactly counted or defined numbers, or calculated values—based on their origin.

SOLVE (a) The number of marble blocks and (c) the number of stairs are quantities we can count, so they are exact numbers. The other three quantities are based on measurements of (b) length, (d) mass, and (e) area and therefore are not exact.

THINK ABOUT IT Every time you use a number in a calculation, you need to think about what kind of number it is in terms of significant figures. Exact numbers are definitions or are the result of counting, so they do not contribute to uncertainty. Any measured number is uncertain, and that uncertainty must be taken into account when answers to calculations using that number are reported.

Practice Exercise Which of the following statistics associated with the Golden Gate Bridge in San Francisco, California are exact numbers, and which of them have some inherent uncertainty? (a) The roadway is six lanes wide. (b) The width of the bridge is 27.4 m. (c) The bridge weighs 3.808×10^8 kg. (d) The total length of the bridge is 2740 m. (e) The toll for a car traveling south has been $5.00 since 2002.

(Answers to Practice Exercises are in the back of the book.)

Precision and Accuracy

Two terms—precision and accuracy—are used to describe how well a measured quantity or a value calculated from a measured quantity is known. **Precision** indicates how repeatable a measurement is. Suppose that we used the balance on the left in Figure 1.15 to weigh a penny over and over again, and suppose that, every time we did so, the result was the same: 2.53 g. These results tell us that our determinations of the mass of the penny were precisely 2.53 g. Said another way, the results were precise to the nearest 0.01 g.

Now suppose that we weigh and then reweigh four more times the same penny using the balance on the right in Figure 1.15 and obtain the following results for the five measurements:

Measurement	Mass (g)
1	2.5270
2	2.5271
3	2.5272
4	2.5271
5	2.5271

Note that there is a small variability in the last decimal place. Such variability is not unusual when using a balance that can weigh to the nearest 0.0001 g, as discussed on page 20. These results are quite consistent, and we can say that the balance is precise. In addition to the cause mentioned there—air moving over the pan—particles of dust landing on it, vibration of the laboratory bench under the

Precision is the repeatability of a measurement and the extent to which repeated measurements agree among themselves.

balance, or the transfer of moisture from our fingers to the penny as we pick it up and place it on the pan could all produce a change in mass of 0.0001 g or more. (Using tweezers or rubber gloves to handle objects to be weighed to the nearest 0.0001 g is a good idea.)

One way to express the precision of these results is to cite the range between the highest and lowest values, in this example, 2.5270 to 2.5272. Range can also be expressed using the average value (2.5271 g) and the range above (0.0001) and below (0.0001) the average that includes all the observed results. A convenient way to express the observed range in this case is 2.5271 ± 0.0001.

While precision relates to the agreement among repeated measurements, **accuracy** reflects how close the measured value is to the true value. Using our example of the penny on page 20, let's suppose the true mass of the penny is 2.5267 g. The average result obtained with the right-hand balance in Figure 1.15—2.5271 g—is 0.0004 g too heavy. Thus the measurements with the balance on the right may have been precise to within 0.0001 g, but they were not accurate to within 0.0001 g of the true value. A way to visualize the difference between accuracy and precision is presented in Figure 1.17.

How can we be sure that the results of a measurement are accurate? The accuracy of a balance can be checked by weighing objects of known masses. A thermometer can be calibrated by measuring the temperatures of known phase transitions of pure substances or of systems at known temperatures. A mixture of ice and pure water, for example, should have a temperature of 0.0°C, and the temperature of pure boiling water is 100.0°C at sea level. A measurement that is validated by calibration with an accepted standard material is considered accurate.

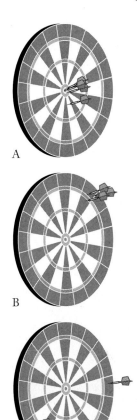

FIGURE 1.17 Accuracy refers to the proximity of a result to the true value. On a target the closer to the bull's-eye the dart lands, the more accurate the throw. Precision refers to the proximity of several results to one another. On the target, proximity is represented by three darts very close to each other. Throws at target A are both accurate and precise; at target B, precise but not accurate; at target C, neither precise nor accurate.

1.9 Unit Conversions and Dimensional Analysis

Throughout this book we frequently encounter the need to convert quantities from one set of units to another. Sometimes it is a simple matter of using the appropriate prefix to express a value, such as the distance between Toronto and Montreal in kilometers instead of meters. Other times it is a matter of converting between English and SI or metric units, such as converting a gasoline price per liter to an equivalent price per gallon.

To do these calculations and many others, we make use of an approach sometimes called the unit factor method but more often known as *dimensional analysis*. The approach makes use of **conversion factors**, which are fractions in which the numerators and denominators have different units but represent equivalent quantities. This equivalency means that multiplying another number by a conversion factor is a bit like multiplying by 1: the intrinsic value that the quantity represents does not change; it is simply expressed in different units.

A key to using conversion factors correctly is to set them up with the correct orientation, with the appropriate units in the numerator and denominator. If, for example, a student who is 66.5 inches tall wishes to express her height in centimeters, she needs to translate the following equivalency from Table 1.3:

$$1 \text{ in (inch)} = 2.54 \text{ cm (centimeters)}$$

into this conversion factor:

$$\frac{2.54 \text{ cm}}{1 \text{ in}}$$

Accuracy is the agreement between an experimental value and the true value.

A **conversion factor** is a fraction in which the numerator is equivalent to the denominator but is expressed in different units.

When she multiplies her height in inches by this conversion factor, the inches in the initial value and in the denominator of the conversion factor cancel each other out, and the answer has the desired units of centimeters:

$$66.5 \; \cancel{\text{in}} \times \frac{2.54 \text{ cm}}{1 \; \cancel{\text{in}}} = 168.9 \text{ cm} = 169 \text{ cm}$$

For the general case in which we convert any initial value expressed in one unit to a final value with the desired units, the generic conversion has this form:

$$\cancel{\text{initial units}} \times \frac{\text{desired units}}{\cancel{\text{initial units}}} = \text{desired units}$$

▶❙❙ **CHEMTOUR** Dimensional Analysis

The use of conversion factors will become clearer as you work through the following Sample and Practice Exercises.

With a mass of 530.20 carats, the Star of Africa is one of the largest cut diamonds in the world. It is mounted in the handle of the Royal Sceptre in the Crown Jewels displayed in the Tower of London.

SAMPLE EXERCISE 1.5 **Converting Unit Prefixes**

The Star of Africa, one of the world's largest diamonds, weighs 106.04 grams. What is its mass in milligrams and in kilograms?

COLLECT AND ORGANIZE We are converting a mass expressed in grams, first to milligrams and second, to kilograms. The prefixes *milli-* and *kilo-* represent units that are 1/1000 and 1000 times the value of the base unit of grams.

ANALYZE Let's express the above relation in conversion factors in which the numerators contains the desired units and the denominators contain the equivalent values in the initial units (grams):

$$\frac{1000 \text{ mg}}{1 \text{ g}} \qquad \frac{1 \text{ kg}}{1000 \text{ g}}$$

SOLVE We multiply the given mass by each of the fractions to obtain the results:

$$106.04 \; \cancel{\text{g}} \times \frac{1000 \text{ mg}}{1 \; \cancel{\text{g}}} = 106{,}040 \text{ mg} \qquad 106.04 \; \cancel{\text{g}} \times \frac{1 \text{ kg}}{1000 \; \cancel{\text{g}}} = 0.10604 \text{ kg}$$

THINK ABOUT IT The results make sense because there should be many more of the much smaller units in a given mass and correspondingly fewer of the larger units. The mass of gems is typically expressed in carats. There are exactly 5 carats in a gram. At 530.20 carats, the Star of Africa is a very large diamond.

Practice Exercise The Eiffel Tower in Paris is 324 m tall, including a 24 m television antenna that was not there when the tower was built in 1889. What is the height of the Eiffel Tower in kilometers and in centimeters?

(Answers to Practice Exercises are in the back of the book.)

SAMPLE EXERCISE 1.6 **Converting English Units to SI Units**

The summit of Mount Washington, in New Hampshire, is famous for its weather: on average it experiences hurricane-force winds 110 days per year. Remarkably, the summit is only 6288 feet above sea level. What is this altitude in meters?

COLLECT AND ORGANIZE Table 1.3 contains the following information about converting units for distance: 1 ft = 0.3048 m. This equivalency contains both the initial and the desired units.

ANALYZE The needed conversion factor with the initial unit in the denominator is

$$\frac{0.3048 \text{ m}}{1 \text{ ft}}$$

A warning to Mount Washington hikers from the U.S. Forest Service.

SOLVE Multiplying the initial value given in the statement of the problem by this conversion factor, we get

$$6288 \text{ ft} \times \frac{0.3048 \text{ m}}{1 \text{ ft}} = 1916.5824 \text{ m}$$

Rounding off the answer to four significant figures, we have 1917 m.

THINK ABOUT IT The decision to round off to four significant figures is made because both values used in the calculation have four significant figures. A foot is a little less than one-third of a meter, so the result makes sense because the answer is about one third of the initial value.

Practice Exercise Perhaps the most famous horse race ever run in America was a 1938 match race between the heavily favored War Admiral and a much smaller horse named Seabiscuit, who stood only 15 hands high. Seabiscuit won the race. The hand used to measure horses is exactly 4 inches. How tall was Seabiscuit in centimeters?

(Answers to Practice Exercises are in the back of the book.)

SAMPLE EXERCISE 1.7 | **English-to-SI Unit Conversions with Multiple Steps**

On April 12, 1934, a wind gust of 231 miles per hour was recorded at the summit of Mount Washington, NH. What is this world-record wind speed in SI units (meters per second)?

COLLECT AND ORGANIZE Speeds and velocities are expressed in units of distance per unit of time. In this conversion, the units on both distance and time must change. Table 1.3 contains the following equivalence for converting distances: 1 km = 0.621 mi (miles). The desired unit of distance is meters, so we also need to convert kilometers to meters (1 km = 1000 m). Finally, we need to convert hours to seconds. However, hour is the unit in the denominator of the initial value. This means that hour cancels out only if it appears in the numerator of the first time-conversion factor.

ANALYZE Here are the needed conversion factors based on the above analysis and a basic knowledge of the units of time:

$$\frac{1 \text{ km}}{0.621 \text{ mi}} \qquad \frac{1000 \text{ m}}{1 \text{ km}} \qquad \frac{1 \text{ hr}}{60 \text{ min}} \qquad \frac{1 \text{ min}}{60 \text{ s}}$$

Note that the first conversion factor for distance is written with the given unit in the denominator; in the second conversion factor, the unit in the denominator is the result of the first conversion, and the desired unit is in the numerator. This same relation between units is also used in the conversion factors for time.

SOLVE Multiplying the initial wind speed by the above conversion factors, we get

$$231 \, \frac{mi}{hr} \times \frac{1 \, km}{0.621 \, mi} \times \frac{1000 \, m}{1 \, km} \times \frac{1 \, hr}{60 \, min} \times \frac{1 \, min}{60 \, s} = 103 \, \frac{m}{s}$$

THINK ABOUT IT The result makes sense because if we multiply only the conversion factors together, we would have a fraction with a numerator of 1000 and a denominator that is about $\frac{2}{3}$ of 60 × 60, or 2400. That fraction, 1000/2400, is a little less than one-half, and the result of our calculation is a little less than half the initial value. The result was rounded off to 3 significant figures because that is how many significant figures were present in the initial measured value of the speed of the wind: 231 mi/hr.

Practice Exercise If light travels exactly 1 meter in 1/299,772,458 of a second, how many kilometers does it travel in one year?

(Answers to Practice Exercises are in the back of the book.)

1.10 Testing a Hypothesis: The Big Bang Revisited

Let's return to our discussion of experiments that have tested the validity of the Big Bang hypothesis. If the matter in the universe was formed during a huge release of energy, and if the universe has been expanding ever since, then theoretically the universe has been cooling throughout time. Just as compressing a gas heats it, allowing a gas to expand cools it. This is the principle on which refrigerators and air-conditioners operate.

▶❚❚ **CHEMTOUR** Temperature Conversion

Temperature Scales

If the universe is still expanding and still cooling, then some residual warmth must be left over from the Big Bang. By the 1950s some scientists not only thought so but also predicted how much leftover warmth there should be in interstellar space: enough to give it a temperature of 2.73 K, where K represents a temperature value on the Kelvin scale.

Several temperature scales are in common general use, and the Kelvin scale is frequently used when calculations using temperature are required, because the Kelvin scale has no negative temperatures. In the United States the Fahrenheit scale is the most popular. In the rest of the world, temperatures are given in degrees Celsius. The Fahrenheit and Celsius scales differ in two ways, as shown in Figure 1.18: First, their zero points are different. Zero degrees Celsius (0°C) is the temperature at which water freezes under normal conditions, but that temperature is 32 degrees on the Fahrenheit scale (32°F). The other difference is in the size of the temperature change corresponding to a degree. The difference between the freezing and boiling points of water is 212 − 32 = 180 degrees on the Fahrenheit scale but only 100 degrees on the Celsius scale. This difference means that a Fahrenheit degree is 100/180, or $\frac{5}{9}$, as large as a Celsius degree. To convert temperatures from Fahrenheit into Celsius, we need to account for the differences in zero point and in degree size. The following equation does both:

$$°C = \frac{5}{9}(°F - 32) \tag{1.2}$$

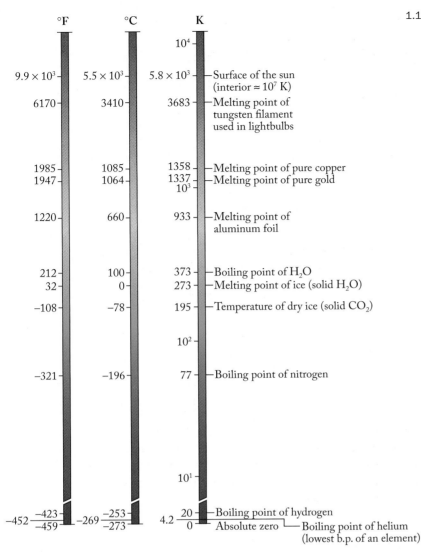

FIGURE 1.18 Three temperature scales are commonly used today, although the Fahrenheit scale is rarely used in chemistry. Fortunately we can readily convert temperature values between the scales.

The SI unit of temperature (see Table 1.2) is the **kelvin (K)**. The zero point on the Kelvin scale is not related to the freezing of a particular substance such as water; rather, it is the coldest temperature—called **absolute zero (0 K)**—that can theoretically exist. It is equivalent to −273.15°C. No one has ever been able to chill matter to absolute zero, nor is it theoretically possible to do so, but scientists have come very close, cooling samples to less than 0.000000001 K. Even interstellar space doesn't reach absolute zero—it has a temperature of 2.73 K.

The zero point of the Kelvin scale differs from that of the Celsius scale, but their increments (degrees) are the same size. The conversion from a Celsius temperature into a Kelvin temperature is simply a matter of adding 273.15 to account for the difference in zero points:

$$K = °C + 273.15 \tag{1.3}$$

After this introduction to the Kelvin scale, we can see that the temperature of outer space, 2.73 K, is extremely cold. We will see just how cold by converting that temperature to equivalent Celsius and Fahrenheit temperatures in the following Sample Exercise.

The **kelvin (K)** is the SI unit of temperature. The Kelvin scale is an absolute temperature scale, with no negative temperatures.

Absolute zero (0 K) is theoretically the lowest temperature possible.

SAMPLE EXERCISE 1.8   **Temperature Conversions**

The temperature of interstellar space is 2.73 K. What is this temperature on the Celsius and Fahrenheit scales?

COLLECT AND ORGANIZE We are asked to convert a temperature in one scale to that in other scales. The Kelvin, Celsius, and Fahrenheit temperature scales are related by Equations 1.2,

$$^\circ C = \frac{5}{9}(^\circ F - 32)$$

and 1.3,

$$K = {}^\circ C + 273.15$$

ANALYZE Equation 1.2 relates Celsius and Fahrenheit temperatures; Equation 1.3 relates Kelvin and Celsius temperatures. Use Equation 1.3 to convert 2.73 K into an equivalent Celsius temperature, and then use Equation 1.2 to calculate an equivalent Fahrenheit temperature.

SOLVE We want to convert a temperature in kelvins to degrees Celsius. We can use Equation 1.3 to do this by solving for degrees Celsius and then inserting the Kelvin temperature of outer space:

$$K = {}^\circ C + 273.15 \qquad \text{or} \qquad {}^\circ C = K - 273.15$$

$$= 2.73 - 273.15 = -270.42^\circ C$$

Now we want to convert degrees Celsius to degrees Fahrenheit. We rearrange Equation 1.2 to solve for degrees Fahrenheit:

$$^\circ C = \frac{5}{9}(^\circ F - 32)$$

Multiplying both sides by 9, and dividing both by 5 gives us

$$\frac{9}{5}{}^\circ C = {}^\circ F - 32 \qquad \text{or} \qquad {}^\circ F = \frac{9}{5}{}^\circ C + 32$$

Inserting the results of the first calculation into this equation, we have

$$^\circ F = \frac{9}{5}{}^\circ C + 32 = \frac{9}{5} \times (-270.42) + 32 = -454.76^\circ F$$

The value 32°F is considered a definition, so it is not used in determining the number of significant figures in the answer. The number that determines the accuracy to which we can know this value is −270.42°C.

THINK ABOUT IT Interstellar space is indeed extremely cold. The much larger negative value of the Fahrenheit temperature than the Celsius temperature makes sense: the size of each Fahrenheit degree is only a little more than half the size of a Celsius degree, so there are many more Fahrenheit degrees between a temperature near absolute zero and zero on the Fahrenheit scale.

Practice Exercise The temperature of the moon's surface varies from −233°C at night to 123°C during the day. What are these temperature ranges on the Kelvin and Fahrenheit scales?

(Answers to Practice Exercises are in the back of the book.)

An Echo of the Big Bang

In the early 1960s, Princeton University physicist Robert Dicke (1916–1997), who had predicted the presence of a residual energy left over from the Big Bang, was anxious to test this hypothesis. He set out to do so by building an antenna that could detect microwaves reaching Earth from outer space. Why did Dicke choose to detect microwaves? Even matter as cold as 2.73 K still emits a "glow" (an energy signature)—but not a glow you can see, like the visible light emitted by the sun, or feel, like the infrared rays emitted by any warm object. The glow from an object that is only 2.73 K is in an energy region called the microwave region, which is the same form of energy we use to cook food and to transmit and receive cell phone signals (see Figure 1.19). The microwave region is a small part of the *electromagnetic spectrum,* a continuous band of energy ranging from low-energy radio waves to high-energy gamma rays.

Dicke's plans to build a cosmic microwave detector never came to fruition because of events that unfolded just a short distance from Princeton. By the early 1960s the United States launched the first communication satellites, called *Echo* and *Telstar.* These early satellites were basically reflective spheres designed to bounce microwave signals from transmitters to receivers back on Earth. An antenna designed to receive such signals had been built at a Bell Laboratories research facility in Holmdel, NJ. Two Bell Labs scientists, Robert W. Wilson and Arno A. Penzias, were working to improve the antenna's reception when they encountered a problem. They observed that no matter where they directed their antenna it picked up a background signal much like the hissing sound radios sometimes make. They hypothesized that the signal was an artifact due to some flaw in their antenna or in the instruments connected to it. At one point they hypothesized that its source was a pair of pigeons roosting on the antenna and coating parts of it with their droppings, but when the droppings were cleaned up, the problem persisted. Additional testing of the equipment led them to ultimately discard this original hypothesis, but it left unanswered the question of where the signal was coming from.

The nuisance signal encountered by Wilson and Penzias matched the microwave echo of the Big Bang Professor Dicke had predicted; this observation supported Dicke's original hypothesis. When the scientists at Bell Labs learned about Dicke's prediction, they realized the significance of the signal they had discovered; others did too. Wilson and Penzias shared the Nobel Prize in Physics in 1978 for discovering the cosmic microwave background radiation of the universe.

Robert Dicke (1916–1997) predicted the existence of cosmic microwave background radiation. His prediction was confirmed by the serendipitous discovery of this radiation by Robert Wilson and Arno Penzias.

FIGURE 1.19 The electromagnetic spectrum includes (but is not limited to) visible sunlight, infrared radiation emitted by warm objects (and thus used to evaluate how much heat is lost from buildings), and the microwaves used for cell phone communications. We examine the properties of these and the other forms of radiation in more detail in Chapter 7.

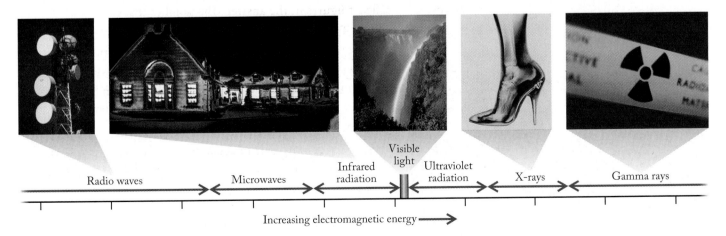

Radio waves   Microwaves   Infrared radiation   Visible light   Ultraviolet radiation   X-rays   Gamma rays

Increasing electromagnetic energy ⟶

In 1965 Robert Wilson (left) and Arno Penzias (right) discovered the microwave echo of the Big Bang while tuning this highly sensitive "horn" antenna in Holmdel, NJ.

FIGURE 1.20 The top image is a map of the cosmic microwave background released in 1992. It is a 360-degree image of the sky made by collecting microwave signals for a year from the microwave telescopes of the *COBE* satellite. The higher-resolution bottom image is based on measurements made in 2002 by NASA's *Wilkinson Microwave Anisotropy Probe (WMAP)* satellite. The colors represent temperature variations, with red indicating regions that are up to 200 mK warmer than the average temperature of 2.73 K and blue indicating regions that are up to 200 mK colder.

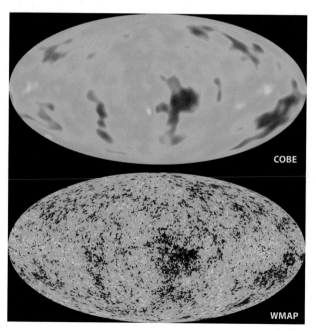

COBE

WMAP

As happens repeatedly in science when one discovery leads to still more questions, the discovery of cosmic microwaves simultaneously reinforced the Big Bang hypothesis and opened up more questions about another mystery. The Bell Labs antenna picked up the same microwave signal no matter where in the sky it was pointed. In other words, the cosmic microwaves appeared to be uniformly distributed throughout the universe. If the afterglow from the Big Bang really was uniform, then how could the galaxies have formed in the first place? Some nonuniformity had to arise in the expanding universe—some clustering of the matter in it—to allow galaxies to form. Scientists doing the same work as Dicke formulated a new hypothesis suggesting that, if that clustering had occurred, a record of it should exist as subtle heterogeneities in the cosmic background radiation.

Unfortunately, a microwave antenna in New Jersey, or indeed in most populated locations, cannot detect such subtle differences in signals. Too many other sources of microwaves interfere with the measurements. One way for scientists to find these differences, if they did exist, would be from a radio antenna in space, where there are fewer interferences. Thus a new experiment was proposed that required an antenna in space, far away from interferences on Earth.

In late 1989 the United States launched such an antenna in the form of the *Cosmic Background Explorer (COBE)* satellite. Its mission was to obtain high-resolution maps of interstellar cosmic microwaves. The observation of these microwaves would test the hypothesis that heterogeneities in the waves did exist. The instruments onboard the satellite could detect differences in microwave signals of as little as 1 part in 10,000, or 0.01%. After many months of collecting data and many more months of analyzing it, the results (the top image in Figure 1.20) were released in 1992. The image appeared on the front pages of newspapers and on magazine covers around the world. This was a major news story, because the predicted heterogeneity had been found. The heterogeneity observed provided direct evidence in support of the Big Bang theory of the origin of the universe. At a news conference in Washington, DC, the lead scientist on the *COBE* project called the map a "fossil of creation."

COBE's measurements and those obtained a decade later by a satellite with even higher resolving power (the bottom image in Figure 1.20) support the hypothesis that the universe did not expand and cool uniformly. The blobs and ripples in the images in Figure 1.20 indicate that galaxy "seed clusters" had already formed early in the history of the universe, when it was only about 300,000 years old. Cosmologists believe that these ripples represent a record of the next step after the Big Bang in the creation of matter. In this phase the first galaxies and stars formed, and within them the remaining natural elements were created. Later in this book we will examine a theory of how the elements may have been formed and continue to be formed as we examine the nuclear reactions that fuel our sun and all the other stars in the universe.

Summary

SECTION 1.1 The physical universe is made of **matter**, which has **mass** and occupies space, and energy. Energy and matter can be inter-converted; quantities involved are related by Einstein's equation:

$$E = mc^2$$

Chemistry is the science of matter. The classes of matter include (1) **pure substances**, which may be **elements** or **compounds** (elements chemically combined together according to the **law of constant composition**), and (2) **mixtures**, which may be **homogeneous** or **heterogeneous**.

SECTION 1.2 An **atom** is the smallest representative particle of an element; a **molecule** is a collection of atoms held together by **chemical bonds** in a characteristic pattern and proportion. A **chemical formula** expresses the elemental composition of a compound. A **chemical equation** is used to describe the proportions of the substances involved in a **chemical reaction**. Space-filling and ball-and-stick models are used to show molecular structure, that is, the three-dimensional arrangement of atoms in a molecule.

SECTION 1.3 The components of mixtures are not present in fixed proportions. They can be separated by physical means, such as **filtration** and **distillation**.

SECTION 1.4 The properties of a substance can be subdivided into **intensive properties**, which are independent of quantity, and **extensive properties**, which are related to the quantity of the substance. Properties are also subdivided into **physical properties** (such as **density**, *d*, the ratio of mass to volume), which can be observed without changing the substance into another one, and **chemical properties** (such as flammability), which can only be observed through chemical reactions involving the substance.

SECTION 1.5 The COAST framework used in this book to solve problems has four components: **C**ollect and **O**rganize information and ideas, **A**nalyze the information to determine how it can be used to obtain the answer, **S**olve the problem (often the math-intensive step), and **T**hink about the answer.

SECTION 1.6 The states of matter include **solid**, in which the particles are locked in place; **liquid**, in which the particles are free to move past each other; and **gas** (or vapor), in which the particles have the most freedom and completely fill their container. The transformation of a solid directly into a gas is called **sublimation**; the reverse process is called **deposition**.

SECTION 1.7 The **scientific method** is an approach to acquiring understanding based on observation, developing a tentative explanation, or **hypothesis**, testing the hypothesis through further experimentation, and formulating **scientific theories (models)**.

SECTION 1.8 The International (SI) System of units, in which the kilogram is the standard unit of mass and the meter is the standard unit of length, evolved from the metric system and is widely used in science to express the results of measurements. Prefixes naming powers of 10 are used with SI base units to express quantities that are much larger or much smaller than the base units. The appropriate number of **significant figures** is used to express the certainty in the result of a measurement or calculation. **Precise** measurements are those that repeatedly yield close to the same value for a given sample, while **accurate** measurements yield close to true values.

SECTION 1.9 Dimensional analysis, also called the unit factor method, uses **conversion factors** (fractions in which the numerators and denominators have different units but represent the same quantity) to convert a value from one set of units into another.

SECTION 1.10 The Fahrenheit temperature scale is widely used in the United States, but the Celsius scale is used in most other countries and by scientists everywhere. The two scales are related by the equation

$$°C = \frac{5}{9}(°F - 32)$$

The Kelvin scale is related to the Celsius scale by the equation

$$K = °C + 273.15$$

where **K** represents units on the Kelvin scale called **kelvins**. Zero on the Kelvin scale is **absolute zero**, the coldest possible temperature. Objects emit characteristic electromagnetic radiation, depending on their temperatures. Cosmic microwave radiation, corresponding to a temperature of 2.73 K, is believed to be a remnant of the energy released by the Big Bang.

Problem-Solving Summary

TYPE OF PROBLEM	CONCEPTS AND EQUATIONS	SAMPLE EXERCISES
Distinguishing physical and chemical properties	The chemical properties of a substance can be determined only by reacting it with another substance; physical properties can be determined without altering the substance's composition.	1.1
Using particles to represent states of matter	Particles in a solid are ordered; particles in a liquid are randomly arranged but close together; particles in a gas are separated by space and entirely fill the volume the gas occupies.	1.2

TYPE OF PROBLEM	CONCEPTS AND EQUATIONS	SAMPLE EXERCISES
Using significant figures in calculations	Apply the weak-link rule: the certainty in a calculated quantity can be no greater than the certainty in the least-certain value used to calculate it.	1.3
Calculating density from mass and volume	The mass of an object can be calculated by multiplying its volume by its density; the volume of an object can be calculated by dividing its mass by its density: $$d = \frac{m}{V}$$	1.3
Distinguishing exact from uncertain values	Measured quantities or conversion factors that are not exact values are inherently uncertain. Quantities that can be counted are exact.	1.4
Doing dimensional analysis and converting units	Converting values from one set of units to another involves multiplication by conversion factors set up so that the original units cancel out.	1.5–1.8

Visual Problems

1.1. For each of the images in Figure P1.1, identify what is being depicted (a pure element or compound) and also identify the physical states.

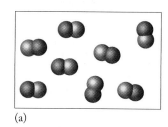

(a)

(b)

FIGURE P1.1

1.2. For each of the images in Figure P1.2, identify what is being depicted (a pure element, a compound, a mixture of elements, or a mixture of compounds) and also identify the physical states.

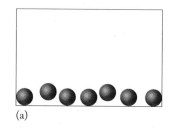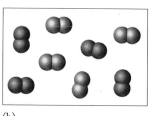

(a)

(b)

FIGURE P1.2

1.3. How would you describe the change depicted in Figure P1.3?
   a. A mixture of two gaseous elements undergoes a chemical reaction, forming a gaseous compound.
   b. A mixture of two gaseous elements undergoes a chemical reaction, forming a solid compound.
   c. A mixture of two gaseous elements undergoes deposition.
   d. A mixture of two gaseous elements condenses.

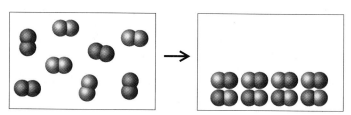

FIGURE P1.3

1.4. How would you describe the change depicted in Figure P1.4?
   a. A mixture of two gaseous elements is cooled to a temperature at which one of them condenses.
   b. A mixture of two gaseous compounds is heated to a temperature at which one of them decomposes.
   c. A mixture of two gaseous elements undergoes deposition.
   d. A mixture of two gaseous elements reacts together to form two compounds, one of which is a liquid.

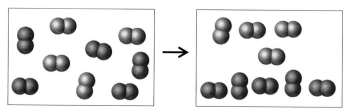

FIGURE P1.4

Questions and Problems

Matter

CONCEPT REVIEW

1.5. Two students get into an argument over whether the sun is an example of matter or energy. Which point of view is correct? Why?

1.6. List three differences and three similarities between a compound and an element.

1.7 List one chemical and four physical properties of gold.

1.8. Describe three physical properties that gold and silver have in common, and three physical properties that distinguish them.

1.9. How might you use filtration to separate a mixture of salt and sand?

1.10. How can distillation be used to desalinate seawater?

1.11. Which of the following processes entails a chemical change? (a) distillation; (b) combustion; (c) filtration; (d) condensation

1.12. Gasohol is a fuel that contains ethanol dissolved in gasoline. Is gasohol a heterogeneous mixture or a homogeneous one?

1.13. Which of the following foods is a heterogeneous mixture? (a) solid butter; (b) a Snickers bar; (c) a hot dog; (d) an uncooked hamburger

1.14. Which of the following foods is a homogeneous mixture? (a) freshly brewed coffee; (b) vinegar; (c) a slice of white bread; (d) a slice of ham

1.15. Which of the following foods is a heterogeneous mixture? (a) apple juice; (b) cooking oil; (c) solid butter; (d) orange juice; (e) tomato juice

1.16. Indicate which of the following is a homogeneous mixture: (a) a wedding ring; (b) sweat; (c) Nile River water; (d) human blood; (e) compressed air in a scuba tank

1.17. Give three properties that enable a person to distinguish between table sugar, water, and oxygen.

1.18. Give three properties that enable a person to distinguish between table salt, sand, and copper.

1.19. Indicate whether each of the following properties is a physical or a chemical property of sodium (Na):
  a. Its density is greater than that of kerosene and less than that of water.
  b. It has a lower melting point than most metals.
  c. It is an excellent conductor of heat and electricity.
  d. It is soft and can be easily cut with a knife.
  e. Freshly cut sodium is shiny, but it rapidly tarnishes in contact with air.
  f. It reacts very vigorously with water to form hydrogen gas (H_2) and sodium hydroxide (NaOH).

1.20. Indicate whether each of the following is a physical or chemical property of hydrogen gas (H_2):
  a. At room temperature, its density is less than that of any other gas.
  b. It reacts vigorously with oxygen (O_2) to form water.
  c. Liquefied H_2 boils at a very low temperature ($-253°C$).
  d. H_2 gas does not conduct electricity.

*1.21. Enzymes are proteins. Proteins are constituents of egg whites. Assume we have a sample of an enzyme dissolved in water. Would filtration or distillation be a suitable way of separating the enzyme from the water?

1.22. Suggest a way of separating ice and water.

1.23. What is the state of each of these elements at ordinary temperature and pressure? Fe, O_2, Hg

1.24. Which of these mixtures is a solid at room temperature and pressure? sea salt, ketchup, ready-to-eat Jell-O

1.25. Can an extensive property be used to identify a substance? Explain why or why not.

1.26. Which of these are intensive properties of a sample of a substance? freezing point, heat content, temperature

The Scientific Method: Starting Off with a Bang

CONCEPT REVIEW

1.27. What kinds of information are needed to formulate a hypothesis?

1.28. How does a hypothesis become a theory?

1.29. Is it possible to disprove a scientific hypothesis?

1.30. Why is the theory that matter consists of atoms universally accepted?

1.31. How do people use the word *theory* in normal conversation?

1.32. Can a theory be proven?

Making Measurements and Expressing the Results; Unit Conversions and Dimensional Analysis

CONCEPT REVIEW

1.33. Describe in general terms how the SI and English systems of units differ.

1.34. Suggest two reasons why English units are so widely used in the United States.

PROBLEMS

NOTE: the physical properties of the elements are in Appendix 3.

1.35. The speed of light in a vacuum is 2.9979×10^8 m/s. Calculate the speed of light in miles per hour.

1.36. **Boston Marathon** To qualify to run in the 2005 Boston Marathon, a distance of 26.2 miles, an 18-year-old woman had to have completed another marathon in 3 hours and 40 minutes or less. To qualify, what must a woman's average speed have been (a) in miles per hour and (b) in meters per second?

1.37. **Olympic Mile** An Olympic "mile" is actually 1500 m. What percentage is an Olympic mile of a real mile (5280 feet)?

*1.38. The price of a popular soft drink is $1.00 for 24 fluid ounces (fl oz) or $0.75 for 0.50 L. Which is a better buy? 1 qt = 32 fl oz.

1.39. **Nearest Star** At a distance of 4.3 light-years, Proxima Centauri is the nearest star to our solar system. What is the distance to Proxima Centauri in kilometers?

*1.40. The level of water in a rectangular swimming pool needs to be lowered 6.0 inches. If the pool is 40 feet long and 16 feet wide, and the water is siphoned out at a rate of 5.2 gal/min (gallons per minute), how long will the siphoning take? $1 ft^3 = 7.48$ gal.

*1.41. If a wheelchair-marathon racer moving at 13.1 miles per hour expends energy at a rate of 665 Calories per hour, how much energy in Calories would be required to complete a marathon race (26.2 miles) at this pace?

1.42. A sports-utility vehicle has an average mileage rating of 18 miles per gallon. How many gallons of gasoline are needed for a 389-mile trip?

1.43. A single strand of natural silk may be as long as 4.0×10^3 m. Convert this length into miles.

*1.44. **Automotive Engineering** The original (1955) Ford Thunderbird (Figure P1.44, left) was powered by a 292-cubic-inch (displacement) engine. The 2005 Thunderbird (Figure P1.44, right) was powered by a 3.9-liter engine. Which engine was bigger?

FIGURE P1.44

1.45. Suppose a runner completes a 10K (10.0 km) road race in 41 minutes and 23 seconds. What is the runner's average speed in miles per hour?

1.46. **Kentucky Derby** The fastest time for the Kentucky Derby is 1 minute and 59 seconds, set in 1973 by a horse named Secretariat. What was Secretariat's average speed (in miles per hour) over the 1.25-mile race?

1.47. What is the mass of a magnesium block that measures 2.5 cm × 3.5 cm × 1.5 cm?

1.48. What is the mass of an osmium block that measures 6.5 cm × 9.0 cm × 3.25 cm? Do you think you could lift it with one hand?

1.49. A chemist needs 35.0 g of concentrated sulfuric acid for an experiment. The density of concentrated sulfuric acid at room temperature is 1.84 g/mL. What volume of the acid is required?

1.50. What is the mass of 65.0 mL of ethanol? (Its density at room temperature = 0.789 g/mL.)

1.51. A brand new silver dollar weighs 0.934 ounces. Express this mass in grams and kilograms. 1 oz = 28.35 g.

1.52. A dime weighs 2.5 g. What is the dollar value of 1 kg of dimes?

1.53. What volume of gold would be equal in mass to a piece of copper with a volume of 125 mL?

*1.54. A small hot-air balloon is filled with 1.00×10^6 L of air ($d = 1.20$ g/L). As the air in the balloon is heated, it expands to 1.09×10^6 L. What is the density of the heated air in the balloon?

1.55. What is the volume of 1.00 lb (pound) of mercury?

1.56. A student wonders whether a piece of jewelry is made of pure silver. She determines that its mass is 3.17 g. Then she drops it into a 10 mL graduated cylinder partially filled with water, and determines that its volume is 0.3 mL. Could the jewelry be made of pure silver?

*1.57. The average density of Earth is 5.5 g/cm³. The mass of Venus is 81.5% of Earth's mass, and the volume of Venus is 88% of Earth's volume. What is the density of Venus?

1.58. Earth has a mass of 6.0×10^{27} g and an average density of 5.5 g/cm³.
   a. What is the volume of Earth in cubic kilometers?
   *b. Geologists sometimes express the "natural" density of Earth after doing a calculation that corrects for gravitational squeezing (compression of the core because of high pressure). Should the natural density be more or less than the observed value calculated in part a?

*1.59. A small plastic cube is 1.2×10^{-5} km on a side and has a mass of 1.10×10^{-3} kg. Water has a density of 1.00 g/cm³. Will the cube float on water?

1.60. **The Sun** The sun is a sphere with an estimated mass of 2×10^{30} kg. If the radius of the sun is 7.0×10^5 km, what is the average density of the sun in units of grams per cubic centimeter? The volume of a sphere is $\frac{4}{3} \pi r^3$.

1.61. Diamonds are measured in carats, where 1 carat = 0.200 g. The density of diamond is 3.51 g/cm³. What is the volume of a 5.0-carat diamond?

*1.62. If the concentration of mercury in the water of a polluted lake is 0.33 μg (micrograms) per liter of water, what is the total mass of mercury in the lake, in kilograms, if the lake has a surface area of 10.0 square miles and an average depth of 45 feet?

1.63. The cartoon in Figure P1.63 applies accuracy and precision to the measurement of body mass.
   a. Give definitions of accuracy and precision.
   b. Is the lawyer using the terms correctly?
   c. Is it possible to be "precisely accurate"?
   d. What does the sign "Precise Weight" say about the uncertainty in the measurements?

FIGURE P1.63

1.64. Healthy Snack? Three different analytical techniques were used to determine the quantity of sodium in a Mars Milky Way Dark candy bar. Each technique was used to analyze five portions of the same candy bar, with the following results (expressed in milligrams of sodium per candy bar):

Technique 1	Technique 2	Technique 3
109	110	114
111	115	115
110	120	116
109	116	115
110	113	115

The actual quantity of sodium in the candy bar was 115 mg. Which techniques would you describe as precise, which as accurate, and which as both? What is the range of the values for each technique?

*1.65. The widths of copper lines in printed circuit boards must be close to a specified value. Three manufacturers were asked to prepare circuit boards with copper lines that are 0.500 μm (micrometers) wide. (1 μm = 1 × 10⁻⁶ m.) Each manufacturer's quality control department reported the following line widths on five sample circuit boards (given in micrometers):

Manufacturer #1	Manufacturer #2	Manufacturer #3
0.512	0.514	0.500
0.508	0.513	0.501
0.516	0.514	0.502
0.504	0.514	0.502
0.513	0.512	0.501

a. What is the range of the data provided by each manufacturer?
b. Can any of the manufacturers justifiably advertise that they produce circuit boards with "high precision"?
c. Is there an instance where this claim is misleading?

*1.66. Temperature measurements are important in the lab, and a typical first assignment in a chemistry lab program is to select a thermometer for subsequent use. Standard temperatures are an ice–water mixture (at 0.0°C) and boiling water (100.0°C at exactly 1 atmosphere of pressure; it varies with atmospheric pressure). You are to select the "best" one of three thermometers A, B, and C for the semester's lab work. Here's how they perform in ice–water and boiling water.

Thermometer	Measured Temperature, °C	True Temperature, °C	Measured Temperature, °C	True Temperature, °C
A	−0.8	0.0	99.4	100.0
B	0.2	0.0	99.8	100.2
C	0.4	0.0	101.0	100.0

Explain your choice of the "best" thermometer for your lab work.

1.67. Which of the following quantities has four significant figures? (a) 0.0592; (b) 0.08206; (c) 8.314; (d) 5420; (e) 5.4 × 10³; (f) 3.752 × 10⁻⁵

1.68. Which of the following numbers have just three significant figures? (a) 7.02; (b) 6.452; (c) 302; (d) 6.02 × 10²³ (e) 12.77; (f) 3.43

1.69. Perform each of the following calculations and express the answer with the correct number of significant figures:
a. 0.6274 × 1.00 × 10³/[2.205 × (2.54)³] =
b. 6 × 10⁻¹⁸ × (1.00 × 10³) × 17.4 =
c. (4.00 × 58.69)/(6.02 × 10²³ × 6.84) =
d. [(26.0 × 60.0)/43.53] / (1.000 × 10⁴) =

1.70. Perform each of the following calculations, and express the answer with the correct number of significant figures:
a. [(12 × 60.0) + 55.3]/(5.000 × 10³) =
b. (2.00 × 183.9)/[6.02 × 10²³ × (1.61 × 10⁻⁸)³] =
c. 0.8161/[2.205 × (2.54)³] =
d. (9.00 × 60.0) + (50.0 × 60.0) + (3.00 × 10¹) =

Testing a Hypothesis: The Big Bang Revisited

CONCEPT REVIEW

1.71. Can a temperature in °C ever have the same value in °F?
1.72. What is meant by an *absolute* temperature scale?

PROBLEMS

1.73. Liquid helium boils at 4.2 K. What is the boiling point of helium in degrees Celsius?
1.74. Liquid hydrogen boils at −253°C. What is the boiling point of H₂ on the Kelvin scale?

1.75. Ethanol (also known as grain alcohol) boils at 78°C. What are the boiling points of ethanol on the Fahrenheit and Kelvin scales?
1.76. The temperature of the dry ice (solid carbon dioxide) in ice cream vending carts is −78°C. What is this temperature on the Fahrenheit and on the Kelvin scales?

1.77. A person has a fever of 102.5°F. What is this temperature in degrees Celsius?
1.78. Physiological temperature, or body temperature, is considered to be 37.0°C. What is this temperature in °F?

1.79. **Record Low** The lowest temperature measured on Earth is −128.6°F, recorded at Vostok, Antarctica, in July 1983. What is this temperature on the Celsius and Kelvin scales?
1.80. **Record High** The highest temperature ever recorded in the United States is 134°F at Greenland Ranch, Death Valley, CA, on July 13, 1913. What is this temperature on the Celsius and Kelvin scales?

1.81. Liquid helium boils at 4.2 K at 1 atmosphere of pressure. What is this temperature on the Fahrenheit scale?
1.82. Silver and gold melt at 962°C and 1064°C, respectively. Convert these two temperatures to the Kelvin scale.

1.83. **Critical Temperature** The discovery of new "high temperature" superconducting materials in the mid-1980s spurred a race to prepare the material with the highest superconducting temperature. The *critical temperature* (T꜀)—the temperatures at which the material becomes superconducting—of $YBa_2Cu_3O_7$, Nb_3Ge, and $HgBa_2CaCu_2O_6$ are 93.0 K, −250.0°C, and −231.1°F, respectively. Convert these temperatures into a single temperature scale, and determine which superconductor has the highest T_c value.

1.84. The boiling point of O_2 is −183°C; the boiling point of N_2 is 77 K. As air is cooled, which gas condenses first?

Additional Problems

*1.85. A farmer applies 1500 kg of a fertilizer that contains 10% nitrogen to his fields each year. Fifteen percent of the fertilizer washes into a river that runs through the farm. If the river flows at an average rate of 0.22 cubic feet per second, what is the additional concentration of nitrogen (expressed in milligrams of nitrogen per liter) in the river water due to the farmer's yearly application of fertilizer?

1.86. Your laboratory instructor has given you two metal cylinders. Both samples have a similar silvery sheen. Your assignment is to determine which one is made of aluminum ($d = 2.699$ g/mL) and which one is made of titanium ($d = 4.54$ g/mL). The mass of each sample was determined on a balance to five significant figures. The volume was determined by immersing the samples in a graduated cylinder as shown in Figure P1.86. The initial volume of water was 25.0 mL. The following data were collected:

	Mass (g)	Height (cm)	Diameter (cm)
Cylinder A	15.560	5.1	1.2
Cylinder B	35.536	5.9	1.3

a. Calculate the volume of each cylinder using the dimensions of the cylinder only.
b. Calculate the volume from the water displacement method.
c. Which volume measurement allows for the greater number of significant figures in the calculated densities?
d. Express the density of each cylinder to the appropriate number of significant figures using the results from parts a and b.

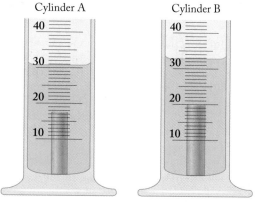

FIGURE P1.86

*1.87. Sodium chloride is used for melting ice in the winter months. This compound contains 1.54 g of chlorine for every 1.00 g of sodium. Which of the following mixtures

would react to leave only sodium chlorine with no sodium and chlorine left over?
a. 11.0 g of sodium and 17.0 g of chlorine
b. 6.5 g of sodium and 10.0 g of chlorine
c. 6.5 g of sodium and 12.0 g of chlorine
d. 6.5 g of sodium and 8.0 g of chlorine

*1.88. Stannous fluoride (SnF_2) is added to toothpastes that help prevent tooth decay. This compound has 3.12 g of tin (Sn) for every 1.00 g of fluorine. Which of the following mixtures would react to leave fluorine mixed with the desired SnF_2 product?
a. 53.0 g of tin and 17.0 g of fluorine
b. 55.0 g of tin and 17.0 g of fluorine
c. 51.0 g of tin and 19.0 g of fluorine
d. 6.0 g of tin and 1.5 g of fluorine

1.89. Slicing an average loaf of bread gives 13 medium-thickness slices, each of which is to be decorated with 0.05 oz of mayonnaise. Each sandwich of two bread slices contains a slice of ham and a slice of cheese. Half the sandwiches also have a sliced olive. How many loaves of bread and how many ounces of mayonnaise are needed to make 243 ham and cheese sandwiches with an olive?

*1.90. Large doses of vitamin C are supposed to cure the common cold. One commercial over-the-counter product consists of 500.0 mg tablets that are 20% by mass vitamin C. How many tablets should one take for a 1.00 g dose of vitamin C?

1.91. We are building bicycles from separate parts. Each bicycle needs a frame, a front wheel, a rear wheel, two pedals, a set of handlebars, a bike chain, and a set each of front and rear brakes. How many complete bicycles can we make from 111 frames, 81 front wheels, 95 rear wheels, 112 pedals, 47 sets of handlebars, 38 bike chains, 17 front brakes, and 35 rear brakes?

1.92. Each Thursday the 11 kindergarten students in Miss Goodson's class are each allowed one slice of pie, one cup of orange juice, and two "doughnut holes." The leftovers will be given to the custodian on the night shift. This Thursday the caterer has left two pies that each can be cut into 8 slices, 18 cups of orange juice, and 24 doughnut holes. How many slices of pie, cups of orange juice, and doughnut holes are left for the custodian?

1.93. Manufacturers of trail mix have to control the distribution of items in their products. Deviations of more than 2% outside specifications cause supply problems and downtime in the factory. A favorite trail mix is designed to contain 67% peanuts and 33% raisins. Bags of trail mix were sampled from the assembly line on different days. The bags were opened and the contents counted, with the following results:

Day	Peanuts	Raisins	Day	Peanuts	Raisins
1	50	32	21	48	34
11	56	26	31	52	30

On which day(s) did the product meet the specification of 65% to 69% peanuts in the bag?

*1.94. Heptane and water do not mix, and heptane has a lower density (0.684 g/mL) than water (1.00 g/mL). A 100 mL graduate cylinder with an inside diameter of 3.2 cm contains 34.0 g of heptane and 34.0 g of water. What is the combined height of the two liquid layers in the cylinder? The volume of a cylinder is $\pi r^2 h$, where r is the radius and h is the height.

Atoms, Ions, and Compounds

2

V838 MONOCEROTIS *Hubble view of the distant star, which is approximately 20,000 light-years away from Earth.*

A LOOK AHEAD: When Artillery Shells Bounced off Tissue Paper

Scientists who study the dynamics of the cosmos estimate the age of the universe at about 13.7 billion years and the age of our solar system at about 4.6 billion years. These estimates are based on careful analyses of the matter that makes up the universe and the way that matter interacts with energy. Within the last 250 years our curiosity about the nature of matter has brought us to an understanding of the properties of the elements that make up our material world. In the 1860s a Russian scientist named Mendeleev proposed a periodic table as an organizational scheme for known elements based on their physical and chemical properties. Questions about what these elements were made of and theories about their origin have occupied scientists ever since.

For centuries since the Greek philosopher Democritus (ca. 400 BC) proposed that everything was composed of indestructible particles, we assumed all matter consisted of atoms that were indivisible. In the early 20th century we found we were wrong. By the 1930s we had learned that the atom itself is divisible, and that all the diversity we saw in both living and inanimate matter was generated by different combinations of three tiny particles called protons, neutrons, and electrons. This stunning revelation was only the first of many to follow.

After accepting that the atom is composed of smaller, more fundamental units (some of which are charged), scientists began to speculate about how these subatomic particles might be arranged within atoms. In 1907, when students of Ernest Rutherford fired a beam of positively charged α particles at a thin gold foil, they were shocked to observe that a few of the particles bounced back at the source of the beam. Rutherford described his amazement at the result: "It is about as incredible as if you had fired a 15-inch shell at a piece of tissue paper and it came back and hit you." In light of this experiment, Rutherford proposed the model of the nuclear atom we still use today.

Hypotheses about atomic structure led to the exploration of the energy needed to hold the fundamental particles together in atoms, and theories were developed about how atoms formed after the Big Bang. People looked to the stars not only for inspiration but also for answers to questions about the origin and distribution of matter and the role of energy in creating and sustaining the material world. The energy we have derived from chemical processes such as the combustion of fossil fuels has served us well since prehistoric times. Our modern inquiries into the origins of matter have helped us understand and use the much larger stores of energy contained within atoms themselves.

2.1 Elements of the Solar System

Chapter 1 ended with the seeds of galactic clusters beginning to coalesce about 300,000 years after the Big Bang. Millions of years later these clusters had turned into the first generation of galaxies. Nuclear processes in the stars in these galaxies created elements that would make up later generations of stars and planets, including our own. We explore these processes in Section 2.9. For now, let's focus on a theory of how our solar system may have formed about 4.6 billion years ago out of elements synthesized in earlier generations of stars.

The theory starts with a gigantic swirling mass of gas called a solar nebula. In cooler regions of the solar nebula, millions of miles from the sun, some of the swirling matter condensed into structures called planetesimals. As these structures collided and fused with each other, the planets of the solar system formed. The distribution of the elements originally in the nebula was not uniform among the planets. The lightest and most volatile elements were swept away from the planets closest to the sun by a combination of solar wind (high-velocity charged particles emitted by the sun) and solar heat. This separation of substances based on their relative volatilities left the planets closest to the sun—Mercury, Venus, Earth, and Mars—rich in those elements that are not volatile or that form non-volatile compounds (among them Fe, O, Si, Mg, Al, and Ni). More easily vaporized elements, including hydrogen and helium, were enriched in the outermost planets. This separation is evident in the elemental compositions of Earth (a terrestrial planet) and Jupiter (one of the so-called gas-giant planets) as shown in Figure 2.1. It turns out that the composition of Jupiter is nearly the same as that of the sun, with hydrogen and helium making up over 98% of its mass. Earth, on the other hand, is made up mostly of four other elements—iron, oxygen, silicon, and magnesium—which account for 92% of its mass.

The early Earth was a hot, molten mass. In this molten state its component elements *fractionated*—that is, separated into distinct regions within the mass by virtue of their different densities and melting points. Dense elements, notably iron and nickel, sank to the center of the planet. A less-dense mantle rich in aluminum, magnesium, silicon, and oxygen formed around the core. As time passed

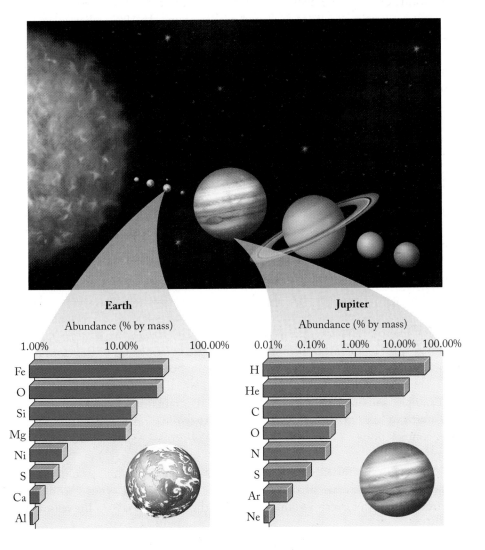

FIGURE 2.1 Earth and the other three terrestrial planets are composed mostly of Fe, O, Si, and Mg (plotted on a log scale), whereas Jupiter and those outer planets of the solar system called gas giants are over 98% by mass hydrogen and helium.

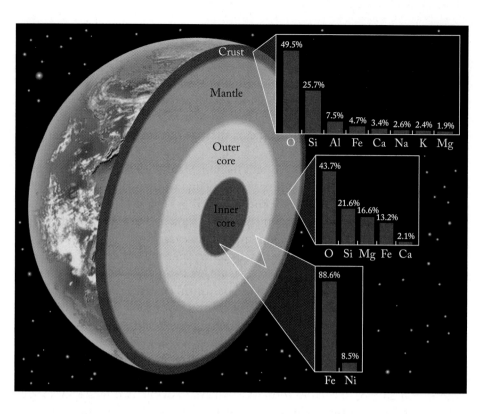

FIGURE 2.2 Earth is composed of a solid inner core consisting mostly of nickel and iron surrounded by a molten outer core of similar composition. A liquid mantle composed mostly of oxygen, silicon, magnesium, and iron lies between the outer core and a relatively thin solid crust.

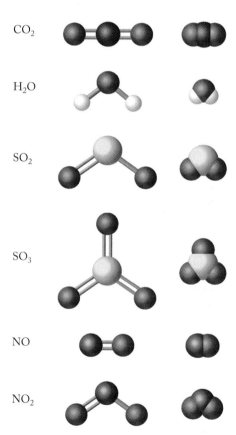

FIGURE 2.3 Chemical formulas, ball-and-stick models, and space-filling models of molecules present in the early atmosphere of Earth.

and Earth cooled, the *mantle* fractionated further, allowing a solid *crust* to form from the components of the mantle that were the least dense and had the highest melting points, and the core to separate into a solid *inner core* and a molten *outer core*. The composition of the resultant regions of inner core, outer core, mantle, and crust are shown in Figure 2.2.

2.2 Compounds and Earth's Early Atmosphere

Earth's early crust was torn by the impact of asteroids and widespread volcanic activity. Gases were released when craters were formed during impact and from volcanic eruptions. This generated a primitive atmosphere with a chemical composition very different from the air that we breathe today. Earth's early atmosphere was probably rich in oxygen-containing compounds but not in oxygen itself. The early atmosphere was mostly carbon dioxide (CO_2), water vapor (H_2O), and other volatile oxides including sulfur dioxide (SO_2), sulfur trioxide (SO_3), nitrogen monoxide (NO), and nitrogen dioxide (NO_2) (Figure 2.3). As noted in Chapter 1, these chemical formulas describe the composition of the compounds by indicating the specific elements and the number of atoms of each element present in one molecule of the substance. The chemical formula of carbon dioxide, for example, tells us that a molecule of CO_2 contains one atom of carbon and two atoms of oxygen. These formulas are further illustrations of the law of constant composition (see Section 1.1); the *atomic ratio* of oxygen to carbon in CO_2 is always 2:1.

We can also examine compounds in terms of the relative masses of their constituent atoms. An analysis of any sample of CO_2 discloses that it contains 8 grams of oxygen for every 3 grams of carbon: the *mass ratio* is 8:3. The reason why the mass ratio differs from the atomic ratio is that an atom of carbon and an atom of oxygen have different masses. In Chapter 3 we will use data from experiments to define the mass ratio of elements in a compound, and from that information and the average atomic masses of the elements we can determine its formula.

The compounds in Earth's early atmosphere illustrate a general property of most compounds, first described by British scientist and mathematician John Dalton (1766–1844). At the beginning of the 19th century, Dalton laid the foundation for modern chemistry when he developed his atomic theory of matter.

According to this theory, compounds are formed from atoms of two or more different elements. In a given compound the ratio of the numbers of atoms of constituent elements can be expressed in small whole numbers. For example, the ratio of oxygen atoms to carbon atoms in carbon dioxide is 2:1. Dalton's atomic view of compounds was based on the experimental results that had led to the law of constant composition. His hypotheses also explained another property of some compounds: if the same two elements (call them X and Y) can form more than one compound, then the two different masses of Y that react with a given mass of X to form two of these compounds can be expressed as the ratio of two small whole numbers. This experimentally observed principle is known as Dalton's **law of multiple proportions**. Consider, for example, SO_2 and SO_3. Suppose we measure the mass of oxygen that reacts with 10 g of sulfur to form SO_2 and find that exactly 10 g of oxygen react. Under different conditions, 10 g of sulfur react completely with 15 g of oxygen to form SO_3. The ratio of the two masses of oxygen is 10:15, or 2:3 expressed in lowest whole numbers, which is indeed the ratio of two small numbers. Similarly, we can confirm experimentally that the mass of oxygen that reacts with a given mass of nitrogen to form NO_2 (10 g of nitrogen react completely with 22.8 g of oxygen) is twice as much as the mass of oxygen that reacts with the same mass of nitrogen to form NO (10 g of nitrogen react completely with 11.4 g of oxygen). These two oxygen ratios (2:3 in the case of the sulfur compounds and 2:1 in the case of nitrogen compounds) are examples of Dalton's law of multiple proportions.

> The **law of multiple proportions** states that the ratio of the masses of one element, Y, that react with a given mass of another element, X, to form any two compounds is the ratio of two small whole numbers.

SAMPLE EXERCISE 2.1 **Using Chemical Formulas and the Law of Multiple Proportions**

Carbon combines with oxygen to form either CO_2 or CO, depending on the conditions of the reaction. If 13 g of oxygen react completely with 5.0 g of carbon to make CO_2, how many grams of oxygen are required to react completely with 5.0 g of carbon under the conditions required to make CO?

COLLECT AND ORGANIZE The two compounds contain the same two elements but in different proportions, so Dalton's law of multiple proportions applies. We have formulas for both compounds, and both reactions involve 5.0 g of carbon.

ANALYZE The masses of oxygen required should be in the ratio of the two whole numbers defined by the chemical formulas: 1 C : 2 O in CO_2 and 1 C : 1 O in CO.

SOLVE The ratios indicate that 2 oxygen atoms are required in CO_2 for every atom of oxygen in CO, which is a ratio of 2:1. Therefore, if 13 g of oxygen are required to react with carbon to make CO_2, half as many grams of oxygen will be required to make CO:

$$\frac{2 \text{ oxygen atoms}}{1 \text{ oxygen atom}} = \frac{13 \text{ g oxygen}}{x} \qquad \text{so } x = 6.5 \text{ g oxygen}$$

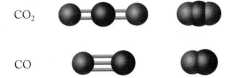

Oxides of carbon in Sample Exercise 2.1.

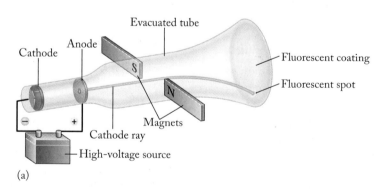

N_2O_5

N_2O_2

Oxides of nitrogen in Practice Exercise 2.1.

THINK ABOUT IT Notice how Dalton's law of multiple proportions is explained by his atomic theory of matter. We used an atomic view of these compounds, expressed by their chemical formulas, to calculate the different masses of oxygen required to react completely with the same mass of carbon to form these two compounds. In practice, analysis of the composition of a compound, usually expressed as mass percentages of each of its constituent elements, is used to determine its chemical formula.

Practice Exercise Predict the mass of oxygen required to react completely with 14 g of nitrogen to make N_2O_5 if 16 g of oxygen are required to react completely with 14 g of nitrogen in order to make N_2O_2.

(Answers to Practice Exercises are in the back of the book.)

2.3 The Rutherford Model of Atomic Structure

By the end of the 19th century, scientists had concluded that atoms were not the smallest particles of matter but were themselves made up of even smaller units, called **subatomic particles**. This realization came in part from studies with a device called a cathode-ray tube by British scientist Joseph John (J. J.) Thomson (1856–1940).

Electrons

When electricity is passed through a glass tube from which most of the air has been removed, a beam of **cathode rays** is generated. The rays are invisible, but if the end of the tube is coated with a fluorescent material, a glowing spot appears where the ray contacts it. Thomson showed that cathode rays were deflected in one direction by a magnetic field (Figure 2.4a) and in the opposite direction by an electric field (Figure 2.4b). The direction of these deflections established that the charge on the particles was negative. By adjusting the strengths of the electric and

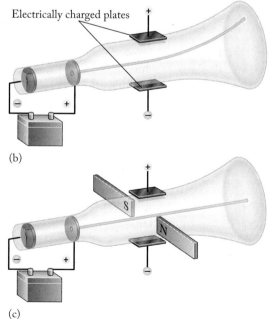

FIGURE 2.4 A cathode ray is generated when a high voltage is applied to the gas in a tube with a partial vacuum. The ray passes through a hole in the anode and makes a bright spot where it contacts a fluorescent material coated on the end of the tube. (a) Cathode ray deflected by a magnetic field; (b) cathode ray deflected by an electric field; (c) electric and magnetic fields tuned to balance out the deflections.

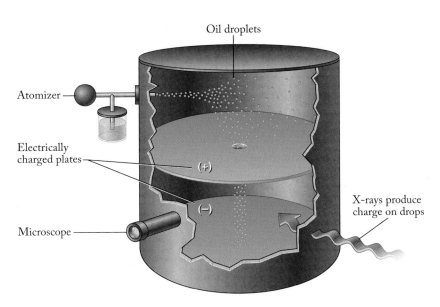

Oil droplets

Atomizer

Electrically charged plates

(+)

(−)

Microscope

X-rays produce charge on drops

FIGURE 2.5 Millikan's oil-droplet experiment.

Atoms are composed of **subatomic particles**, which include neutrons, protons, and electrons.

Cathode rays are streams of electrons emitted by the cathode (negative electrode) in a partially evacuated tube.

An **electron** is a negatively charged subatomic particle.

magnetic fields, Thomson was able to balance out the deflections (Figure 2.4c). From the strengths of the two opposing fields he was able to calculate *the mass-to-charge ratio (m/e)* of the particles in the ray. Thomson and others also observed that particles composing the rays were always the same in terms of their behavior and their mass-to-charge ratio, no matter what their source. This observation established that the particles in the cathode rays, which ultimately became known as **electrons**, are fundamental particles in all matter.

In 1909 the American physicist Robert Millikan advanced Thomson's work in an elegant experiment to determine the charge on the electron. Thomson had defined the mass-to-charge ratio of the electron, and if the charge could be determined, the mass could be calculated as well. Figure 2.5 illustrates the components of Millikan's experiment: highly energetic X-rays remove electrons from molecules of air in the lower chamber and, as an oil droplet falls from the upper chamber into the lower one, it picks up those electrons and develops a charge. Millikan measured the mass of the droplets in the absence of an electric field, when the rate of their fall was governed by gravity. He then turned on and adjusted the electric field to make the droplets fall at different rates. He could even suspend the droplets in midair. From the strength of the field and the rate of descent, he calculated the total charge on the droplets. After measuring hundreds of droplets, Millikan determined that their charge was a whole-number multiple of a minimum charge, which he defined as the charge on one electron. Millikan's value was within 1% of the modern value of -1.602×10^{-19} C. (The coulomb, abbreviated "C," is the SI-derived unit for electric charge; see the inside back cover.) Using Thomson's value of m/e for the electron, he was then able to calculate the mass of the electron as 9.109×10^{-28} g.

The discovery of the electron established that the atom, which was previously thought to be the smallest particle of matter, was itself made of particles. Consequently, further questions arose about the nature of those particles. Everyone knew that matter was neutral, but no one knew how the positive and negative regions of matter were arranged at the atomic level. Thomson formulated the "plum pudding" model, based on the hypothesis that an atom was a diffuse sphere of positive charge with negatively charged electrons embedded in the sphere, like raisins in a plum pudding (Figure 2.6).

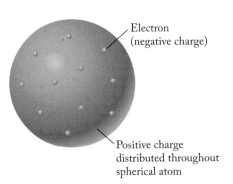

Electron (negative charge)

Positive charge distributed throughout spherical atom

FIGURE 2.6 The plum-pudding model was an early model of the interior structure of atoms proposed by British scientist J. J. Thomson in 1904. In this model a neutral atom of an element consisted of electrons distributed throughout a much more massive, positively charged, but very diffuse sphere. The plum-pudding model lasted less than 10 years before being replaced by a new model of atomic structure as a result of experiments carried out under the direction of Thomson's former student Ernest Rutherford.

J. J. Thomson (1856–1940) discovered electrons in 1897 using a cathode-ray tube, but he was not sure where electrons fit into the structure of atoms.

Radioactivity and the Nuclear Atom

Thomson's plum-pudding model did not last long. Its demise was linked to another scientific discovery in the 1890s: radioactivity. In 1896 French physicist Henri Becquerel (1852–1908) discovered that a sample of *pitchblende*, a mineral containing uranium (mostly UO_2 and UO_3) and other radioactive elements, produced invisible radiation that could be detected on photographic plates. Becquerel and his contemporaries initially thought this nuclear radiation consisted of rays much like the X-rays that had just been discovered by German scientist Wilhelm Conrad Röntgen (1845–1923).[1] Additional experiments—by Becquerel, by the French husband-and-wife team of Marie (1867–1934) and Pierre (1859–1906) Curie, and by British scientist Ernest Rutherford (1871–1937)—led to the conclusion that several types of rays existed, at least two of which were actually composed of particles. One type consisted of **beta (β) particles**, which the Curies found penetrated materials better than a second type of ray composed of **alpha (α) particles**. The observation that β particles could be deflected by a magnetic field proved they were particles and not simply rays of energy. The degree of deflection in a magnetic field allowed the mass-to-charge ratio of β particles to be calculated, and the results exactly matched the mass-to-charge ratio of the electron determined by J. J. Thomson. These data established that a β particle is simply a high-energy electron.

In 1903 Rutherford discovered that the path of an α particle could be deflected by a combination of electric and magnetic fields. Because it was deflected in the opposite direction from that of a β particle, the charge on an α particle had to be positive. The β particle (an electron) was assigned a charge of 1−; based on its observed behavior in electric and magnetic fields, the α particle was assigned a charge of 2+. Alpha particles were also found to have nearly the same mass as an atom of helium, making them over 10^3 times more massive than β particles.

The plum-pudding model had to be discarded because of the results of an experiment directed by Rutherford and carried out by two of his students at Cambridge University: Hans Geiger (for whom the Geiger counter was named) and Ernest Marsden. Geiger and Marsden bombarded a thin foil of gold with α particles emitted from a radioactive source. Rutherford asked the students to measure how many α particles were deflected from their original path and to what extent they were deflected. Rutherford's hypothesis was that, if Thomson's model were correct, most of the α particles would pass straight through the diffuse positive spheres of gold atoms but a few of the particles would interact with the tiny electrons embedded in these spheres and be deflected slightly (Figure 2.7a). Geiger and Marsden observed something completely unexpected. For the most part, the α particles did indeed pass directly through the gold. However, about 1 in every 8000 particles was deflected from the foil through an average angle of 90 degrees (Figure 2.7b), and a very few (perhaps 1 out of 100,000) bounced right back in the direction from which the particles came. Years later Rutherford recounted his astonishment at these results:

> It is about as incredible as if you had fired a 15-inch shell at a piece of tissue paper and it came back and hit you.

A β **particle** is a type of radioactive emission that consists of a high-energy electron.

An α **particle** is a radioactive emission (composed of subatomic particles) with a charge of 2+ and a mass equivalent to that of a helium nucleus.

---

[1] Röntgen discovered X-rays in experiments with a cathode-ray tube much like the apparatus used by J. J. Thomson. After completely encasing the tube in a black carton, Röntgen discovered that invisible rays escaped the carton and were detected by a photographic plate. Because he knew so little about these rays, he called them X-rays.

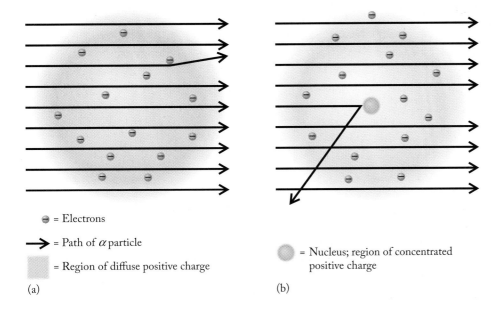

= Electrons

= Path of α particle

= Region of diffuse positive charge

(a)

= Nucleus; region of concentrated positive charge

(b)

FIGURE 2.7 (a) If Thomson's model had been correct, most of the α particles would have passed through the foil and a few would have been deflected slightly. (b) However, in Rutherford's famous 1907 experiment, most α particles passed straight through the foil, but a few were scattered widely. This unexpected result led to the theory that an atom has a small, positively charged nucleus that contains most of the mass of the atom.

The results of the gold-foil experiments required that the plum-pudding model be rejected because it could not account for the large-angle deflections. Rutherford concluded that the few large deflections of positively charged α particles occurred because the particles occasionally encountered small regions characterized by high concentrations of positive charge and high mass (the nuclei of the gold atoms). Rutherford determined that the region of positive charge was only about 1/10,000 of the size of the gold atoms. His model of the atom, shown in Figure 2.8, became the basis for our current understanding of atomic structure. It assumes that atoms consist of massive—but tiny—positively charged nuclei that are far apart from each other and are surrounded by negatively charged electrons. This gives rise to the description of atoms as being mostly empty space occupied by electrons plus a tiny center called the **nucleus**, which contains all of the atom's positive charge and nearly all its mass.

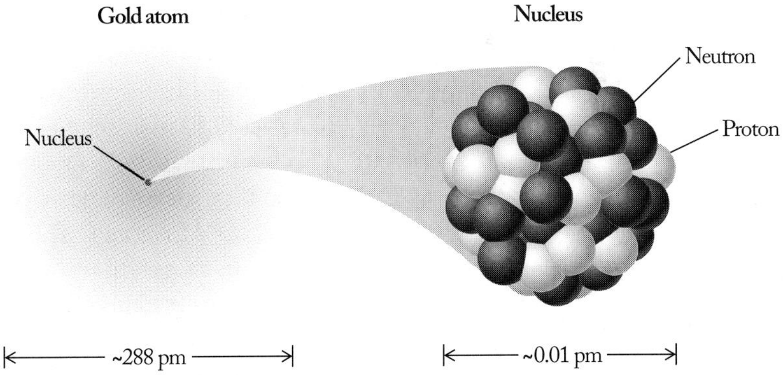

► II **CHEMTOUR** Rutherford Experiment

Gold atom

Nucleus

Nucleus

Neutron

Proton

|← ~288 pm →|

|← ~0.01 pm →|

FIGURE 2.8 Rutherford's model of the atom includes a nucleus that is about 1/10,000 the size of the atom. Note the scales are given in picometers (pm): the nucleus would be too small to see if drawn to scale.

Protons and Neutrons

In the decade following the gold-foil experiments, Rutherford and others observed that bombarding various materials, including gases, with α particles could change the elements in those materials into other elements. This process is called *transmutation*, a process we will explore in more detail in Chapter 20. Here we note that the key feature of transmutation that contributed to understanding atomic structure was the discovery that hydrogen nuclei were also frequently produced during transmutation reactions. By 1920 a consensus was growing that hydrogen nuclei, which Rutherford called **protons** (from the Greek *protos*, meaning "first"), were fundamental building blocks of all nuclei. For example, to account for the mass and charge of an α particle, Rutherford assumed that it was made of four protons, two of which had combined with two electrons to form two neu-

The **nucleus** of an atom contains all the positive charge and nearly all the mass in an atom.

A **proton** is a positively charged subatomic particle present in the nucleus of an atom.

TABLE 2.1 Properties of Subatomic Particles

Particle	Symbol	MASS		CHARGE	
		In Atomic Mass Units (amu)	In Grams (g)	Relative Value	Charge (C[a])
Neutron	$_0^1 n$	$1.00867 \approx 1$	1.67494×10^{-24}	0	0
Proton	$_1^1 p$	$1.00728 \approx 1$	1.67263×10^{-24}	1+	$+1.602 \times 10^{-19}$
Electron	$_{-1}^0 e$	$5.485799 \times 10^{-4} \approx 0$	9.10939×10^{-28}	1−	-1.602×10^{-19}

[a] The *coulomb* (C) is the SI-derived unit of electric charge. When a current of 1 ampere (see Table 1.2) passes through a conductor for 1 second, the quantity of electric charge that moves past any point in the conductor is 1 C.

tral particles, which he called **neutrons**. Repeated attempts to produce neutrons by neutralizing protons with electrons were unsuccessful. However in 1932 one of Rutherford's students, James Chadwick (1891–1974) successfully isolated and characterized free neutrons for the first time.

Table 2.1 provides a summary of the properties of neutrons, protons, and electrons. For convenience, the masses of these tiny particles are expressed in **atomic mass units (amu)**, which make up a relative scale used to express the masses of individual atoms as well as of subatomic particles. The amu scale is defined such that the mass of one atom of carbon—the nucleus of which consists of 6 protons and 6 neutrons—is exactly 12 amu. If you compare the amu values in the table to the masses of the particles in grams, you see that these particles are tiny indeed: 1 amu is only 1.66054×10^{-24} g.

2.4 Isotopes

Even as Thomson investigated the properties of cathode rays, other scientists raised a troublesome issue about them: if matter is neutral, and cathode rays are beams composed of negatively charged particles, shouldn't there also be a corresponding beam of positively charged particles? One of Thomson's former students, Francis W. Aston (1877–1945), built cathode-ray tubes with even less gas in them than in those Thomson had used, and Aston did indeed discover a second beam of positively charged particles. Another pivotal fact emerged from his discovery: whereas the electrons were all the same regardless of what substance was used for the cathode or what kind of gas filled the tube, the properties of positive rays varied with the source material. The particles in positive rays were not individual protons common to all atoms but rather combinations of subatomic particles that were unique to the gas in the tube. They were, in fact, whole atoms that had lost electrons and formed positively charged ions.

Aston used his device, dubbed a positive-ray analyzer (Figure 2.9), to study positively charged beams of a gas by passing the beams through a magnetic field. The particles in the beam were deflected along paths determined by their masses: the greater the mass, the larger the radius of deflection. Aston studied the purest sample of neon gas available and determined that most particles had a mass of 20 amu, but about 1 in 10 had a mass of 22 amu. To explain his data, Aston proposed that neon consists of two different kinds of atoms, or **isotopes**. Both have the same number of protons (10) per nucleus, but one isotope has 10 neutrons in each nucleus, giving it a mass of 20 amu, while the other has 12 neutrons, giving it a mass of 22 amu.

A **neutron** is an electrically neutral or uncharged subatomic particle found in the nucleus of an atom.

Atomic mass units (amu) comprise a relative scale used to express the masses of atoms and subatomic particles. The scale is based on the definition that the mass of 1 atom of carbon with 6 protons and 6 neutrons in its nucleus is exactly 12 amu.

Isotopes are atoms of an element whose nuclei have the same number of protons but different numbers of neutrons.

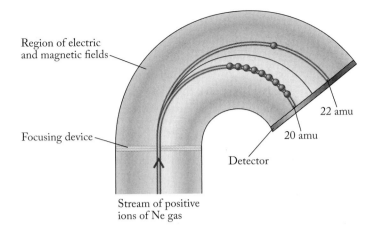

FIGURE 2.9 Aston's positive-ray analyzer: a beam of positively charged ions of neon gas is passed through a focusing slit into a region of an electric and magnetic field. The ions are separated according to their mass: 90% of the ions with a mass of 20 amu hit the detector at a different spot than the 10% with a mass of 22 amu. When the sensitivity of the detector was improved, Ne^+ ions with a mass of 21 amu were also observed. Aston's positive-ray analyzer was the forerunner of the modern mass spectometer.

Since the time of John Dalton, scientists had defined an element as matter composed of identical atoms, all of which had the same mass. Aston's work required a modification of this atomic view of matter: henceforth each element was said to consist of atoms with the same number of *protons* in its nuclei. This number is called the **atomic number (Z)** of the element. The total number of **nucleons** (neutrons and protons) in the nucleus of an atom defines its **mass number (A)**. The isotopes of a given element thus all have the same atomic number but different mass numbers. The general format for identifying a particular isotope is

$$^{A}_{Z}X$$

where A is the mass number, Z is the atomic number, and X is the one- or two-letter symbol for the element. Because both Z and X provide the same information—each by itself identifies the element—the subscript Z is frequently omitted, so that the isotope is denoted as simply ^{A}X. This same information—mass number and element—is frequently spelled out without superscripts: for example, two of the isotopes of neon are neon-20 and neon-22. The modern **periodic table** (see the inside front cover) identifies the elements in order of their atomic numbers. The following Sample Exercise provides practice in linking the numbers of protons and neutrons in the nucleus of an atom to its identity and isotopic symbol.

| SAMPLE EXERCISE 2.2 | **Writing the Symbol of an Isotope** |

Using the periodic table, write the symbols of the isotopes that have the following combinations of protons and neutrons: (a) 6 protons and 6 neutrons; (b) 11 protons and 12 neutrons; (c) 92 protons and 143 neutrons

COLLECT AND ORGANIZE The number of protons in the nucleus of an atom defines its atomic number (Z) and identifies which element it is. The sum of the nucleons (protons plus neutrons) yields the mass number of the atom and defines which isotope of the element it is. The elements in the periodic table are arranged (left to right, top to bottom) in order of increasing atomic number, so by using the table, we can find the symbol of the element that has a given number of protons.

ANALYZE The atomic number and mass number of an isotope together allow us to symbolize the isotope using the form $^{A}_{Z}X$, where A is the mass number, Z is the atomic number, and X is the symbol of the element as it appears in the periodic table.

The **atomic number (Z)** is the number of protons in the nucleus of an atom of an element.

The protons and neutrons in atomic nuclei are called **nucleons**.

The **mass number (A)** is the total number of nucleons (sum of the numbers of protons and neutrons) in one atom of an element.

The **periodic table** identifies the elements in order of their atomic numbers.

SOLVE

a. This isotope has 6 protons; therefore $Z = 6$, so this must be an isotope of carbon. The combination of 6 protons and 6 neutrons gives the isotope a mass number of $6 + 6 = 12$. Therefore the isotope is carbon-12, or $^{12}_{6}\text{C}$.

b. This isotope has 11 protons; therefore $Z = 11$, so this must be an isotope of sodium. The combination of 11 protons and 12 neutrons gives the isotope a mass number of $11 + 12 = 23$. Therefore the isotope is sodium-23, or $^{23}_{11}\text{Na}$.

c. This isotope has 92 protons; therefore $Z = 92$, so this must be an isotope of uranium. The combination of 92 protons and 143 neutrons gives the isotope a mass number of $92 + 143 = 235$. Therefore the isotope is uranium-235, or $^{235}_{92}\text{U}$.

THINK ABOUT IT Among this group of isotopes, notice how the atomic numbers and mass numbers are related to each other: as atomic numbers increase, mass numbers increase too, but by even greater proportions. This pattern holds for most of the isotopes found in nature. The nuclei of isotopes (except hydrogen) have at least as many neutrons as they do protons, and the neutron to proton ratio increases with increasing atomic number.

Practice Exercise Using the periodic table, write the symbols of the isotopes with the following numbers of protons and neutrons: (a) 26 protons and 30 neutrons; (b) 7 protons and 8 neutrons; (c) 17 protons and 20 neutrons; (d) 19 protons and 20 neutrons.

(Answers to Practice Exercises are in the back of the book.)

2.5 Average Atomic Masses

Now let's turn again to the periodic table on the inside front cover of this book. Note that each cell includes the letter symbol of an element at its center and the atomic number of that element at the top. The value at the bottom of each cell is the weighted **average atomic mass** (commonly called the atomic mass or the atomic weight) of the atoms of that element. For neon, for example, this number is 20.1797 amu. To understand the meaning of the weighted average, consider the information about neon isotopes given in Figure 2.9. According to Aston's original data, 90% of neon atoms have a mass of 20 amu, while 10% have a mass of 22 amu. More-accurate measurements with modern **mass spectrometers** have revealed a third isotope of neon with a mass of 21 amu. The following data set shows the accepted mass and **natural abundance** (expressed as a percent) for each of the neon isotopes.

The **average atomic mass** of an element is calculated by multiplying the natural abundance of each isotope by its exact mass in atomic mass units and then summing these products.

A **mass spectrometer** is an instrument that measures precise masses and relative amounts of ions of atoms and molecules.

The **natural abundance** of an isotope is its relative proportion, usually expressed as a percentage, among all the isotopes of that element as found in a natural sample. The total abundances for the isotopes should sum to 100% (or very close to it, due to measurement error).

Isotope	Exact Mass of Isotope (amu)	Natural Abundance (%)
Neon-20	19.9924	90.4838
Neon-21	20.99395	0.2696
Neon-22	21.9914	9.2465

When calculations are carried out using the values of natural abundance, the number is expressed as a decimal. We can use the exact mass and the natural abundance of each of these isotopes to calculate the average atomic mass of neon:

$$\begin{aligned}\text{Average atomic mass of neon} = &(19.9924 \text{ amu} \times 0.904838) \\ &+ (20.99395 \text{ amu} \times 0.002696) \\ &+ (21.9914 \text{ amu} \times 0.092465) \\ = &\ 20.1797 \text{ amu}\end{aligned}$$

This average is the number appearing below the symbol in the periodic table. It is important to note that no neon isotope has this mass: this is the weighted average of all the isotopes.

This method of calculating the average atomic mass of neon works for all elements. The following equation describes in a general way how the exact masses and natural abundances of the isotopes of an element can be used to calculate the element's average atomic mass:

$$m_X = a_1 m_1 + a_2 m_2 + a_3 m_3 + \cdots \qquad (2.1)$$

where m_X is the average atomic mass of an element X, which has isotopes with masses $m_1, m_2, m_3, \ldots$ that have natural abundances (expressed as decimals) $a_1, a_2, a_3, \ldots$. We use Equation 2.1 to calculate average atomic masses in the following Sample Exercise.

SAMPLE EXERCISE 2.3 **Calculating an Average Atomic Mass from Natural Abundances**

The precious metal platinum has six isotopes with the following natural abundances:

Symbol	Mass (amu)	Natural Abundance
^{190}Pt	189.96	0.00014
^{192}Pt	191.96	0.00783
^{194}Pt	193.96	0.32967
^{195}Pt	194.96	0.33832
^{196}Pt	195.96	0.25242
^{198}Pt	197.97	0.07163

Use these data to calculate the average atomic mass of platinum.

COLLECT AND ORGANIZE The average atomic mass of an element is the weighted average of the masses of its isotopes. To calculate a weighted average, we need to find the product of the mass of each isotope times its natural abundance and then sum the products (Equation 2.1).

ANALYZE Notice that this table is the same as the previous data table except that we have added a column to contain the product of the mass of each isotope times its natural abundance. These are the $a_i m_i$ terms to be added together in Equation 2.1. At the same time let's add a row at the bottom of the table for the sums of the natural abundance values as a check that our data set is complete.

Symbol	Mass (amu)	Natural Abundance	Mass × Natural Abundance (amu)
^{190}Pt	189.96	0.00014	?
^{192}Pt	191.96	0.00783	?
^{194}Pt	193.96	0.32967	?
^{195}Pt	194.96	0.33832	?
^{196}Pt	195.96	0.25242	?
^{198}Pt	197.97	0.07163	?
Sum		?	?

SOLVE Now let's complete the table by calculating the product of the mass of each isotope times its natural abundance and the sum of these products (in blue):

Symbol	Mass (amu)	Natural Abundance	Mass × Natural Abundance (amu)
^{190}Pt	189.96	0.00014	0.027
^{192}Pt	191.96	0.00783	1.50
^{194}Pt	193.96	0.32967	63.943
^{195}Pt	194.96	0.33832	65.959
^{196}Pt	195.96	0.25242	49.464
^{198}Pt	197.97	0.07163	14.18
Sum		**0.9997 ≈ 1.000**	**195.07**

The value in the lower right-hand corner of the table, 195.07 amu, is the one we seek: the average atomic mass of all platinum atoms.

THINK ABOUT IT Notice that the natural abundances add up to a value very close to 1, as they should if we have included all the platinum isotopes in our table (we have). The slight difference in the total arises from measurement errors and rounding. The result of the calculation—195.07—is the weighted average atomic mass of platinum. The value given in the inside front cover—195.084 amu—is the most precise value currently available. Remember, however, that no individual Pt atom has that mass; it is the weighted average of all Pt atoms.

Practice Exercise Silver (Ag) has only two stable isotopes: ^{107}Ag, with a mass of 106.90 amu, and ^{109}Ag, with a mass of 108.90 amu. If the atomic mass of silver is 107.87 amu, what is the natural abundance of each of the isotopes? *Hint:* Let x be the natural abundance of one of the isotopes. Then $1 - x$ is the natural abundance of the other.

(Answers to Practice Exercises are in the back of the book.)

2.6 The Periodic Table of the Elements

The Rutherford model of the internal structure of the atom was one of several major advances in the early 20th century that enabled scientists to understand better the nature of matter on an atomic scale and to explain why the elements exhibit their particular physical and chemical properties. Actually, chemists had for many years looked for patterns in the properties of elements, even before they were aware of the existence of subatomic particles and the concept of atomic numbers. Among the pioneers in this effort was Russian chemist Dmitrii Ivanovich Mendeleev (1834–1907).

Mendeleev's Periodic Table

Mendeleev published his version of the periodic table of the elements (Figure 2.10) in 1869. It was a remarkable accomplishment because many elements were still unknown in the mid-19th century. Being a circumspect scientist, Mendeleev knew that there was still much he did not know, so he left open spaces in his periodic table to account for elements not yet discovered. He was able to predict the chemical properties of these missing elements based on the locations of the blank spaces with respect to known elements. These insights greatly facilitated the subsequent discov-

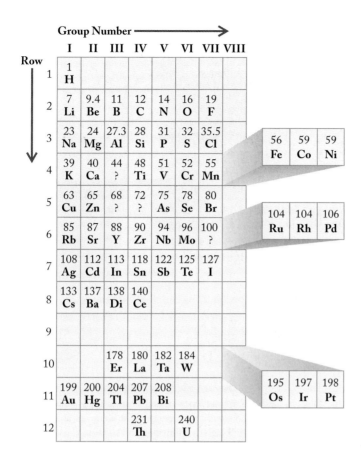

Group Number ⟶

FIGURE 2.10 Mendeleev organized elements in his periodic table on the basis of their chemical and physical properties and atomic masses. He assigned three elements with similar properties to group VIII in rows 4, 6, and 10. Because he did this, the elements in the rows that followed lined up in appropriate groups. In this way, rows 4 and 5 combined, for example, in Mendeleev's table contains spaces for 18 elements, corresponding to the 18 groups in the modern periodic table.

ery of the missing elements by other scientists. Mendeleev arranged the elements in his periodic table in order of increasing atomic mass, which could be estimated in that era, and not atomic number, which was not identified until over 50 years later.

CONCEPT TEST

Find two pairs of elements in the modern periodic table that would have been out of sequence in Mendeleev's version, which was based on atomic mass.

(Answers to Concept Tests are in the back of the book.)

Navigating the Modern Periodic Table

The modern periodic table (Figure 2.11) is also based on a classification of elements in terms of their physical and chemical properties. As we have already indicated, elements appear in the table in order of increasing atomic number, or number of protons in the nucleus of the atom. Further, the table is organized as a series of horizontal rows (called **periods**) and columns (called **groups** or **families**). Elements in the same family have similar chemical properties.

The periods are numbered at the far left of each row. The first period consists of only two elements—hydrogen and helium. The next two periods have eight elements each: lithium through neon, and sodium through argon. The fourth and fifth periods have 18 elements each; elements 21 through 30 and 39 through 48 are the first two rows of the block known collectively as the **transition metals**. The sixth period is actually 32 elements long. In part to accommodate the periodic

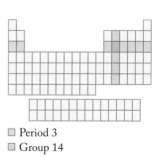

☐ Period 3
☐ Group 14

The horizontal rows in the periodic table are called **periods**.

Columns in the periodic table define elements in the same **family**, or **group**. Elements in the same family have similar chemical properties.

Transition metals are the elements in groups 3 through 12 in the periodic table.

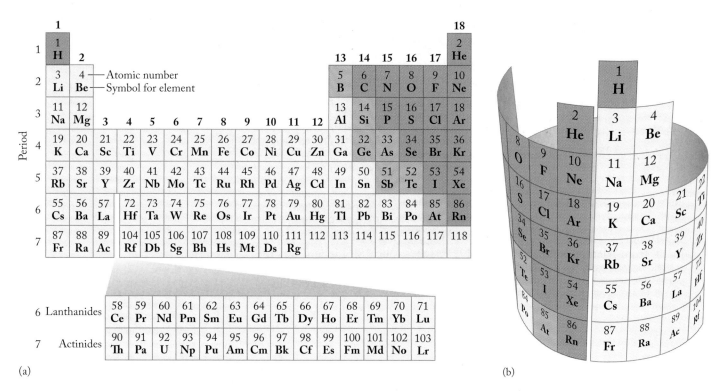

FIGURE 2.11 (a) In the periodic table, the elements are sequenced based on increasing atomic number (Z) and in a pattern related to their physical and chemical properties. The elements on the left (in tan) are classified as metals, those on the right (in blue) as nonmetals, and those in the middle (in green) as metalloids (or semimetals). (b) In this view of the periodic table, the sides have been joined together to illustrate the sequence of the elements from one row to the next.

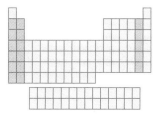

☐ Group 1: Alkali metals
☐ Group 2: Alkaline earth metals
☐ Group 17: Halogens

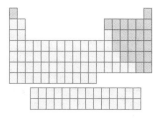

☐ Metals
☐ Nonmetals
☐ Metalloids

table on standard sized paper, 14 of those elements (elements 58 through 71) are placed at the bottom of the table and form a series called the *lanthanides*, from the name of element 57 (lanthanum). Correspondingly, another 14-element series (elements 90 to 103) is "excised" from the seventh period and placed at the bottom of the table. This series is known as the *actinides*, from element 89 (actinium).

The columns are numbered from left to right from 1 to 18. Some of the groups have specific names, typically based on a particular distinctive property of the family members. Group 17, for example, is the **halogen** family; the word *halogen* is derived from the Greek for "salt former." In terms of reactivity, chlorine exemplifies the halogens. Chlorine forms a 1:1 compound with every element in group 1, the **alkali metal** family; an example of this is NaCl, common table salt. In contrast, chlorine forms 2:1 compounds with every element in group 2, the **alkaline earth metals**; the compound $CaCl_2$ is an example of this pattern. The other halogens also form 1:1 compounds with group 1 metals and 2:1 compounds with group 2 metals. This example illustrates the logic Mendeleev used to build his original table and also what is meant by "similar properties."

The table is also divided into three broad categories based on general properties. Elements in tan boxes in Figure 2.11 are **metals**. They tend to conduct heat and electricity well; they tend to be malleable (capable of being shaped by hammering) or ductile (capable of being drawn out in a wire), and all but mercury (Hg) are shiny solids at room temperature. Elements shown in blue boxes are **nonmetals**. They are poor conductors of heat and electricity: most are gases at room temperature, the solids among them tend to be brittle, and bromine (Br) is a low-boiling liquid. The elements in green boxes are **metalloids**, or **semimetals**, so named because they have some metallic and some nonmetallic properties.

In the modern table, groups 1, 2, and 13 through 18 are referred to as the **main group elements**, or **representative elements**. They have certain characteristic properties that differentiate them from the elements in the block formed by groups 3 through 12, known as the **transition metals**. As their name implies, the latter group of elements are true metals, but we will see in subsequent chapters how their chemical behavior and physical properties result in their meriting a subcategory that differentiates them from metals in the main group.

Take a moment to compare Mendeleev's 19th-century version of the periodic table in Figure 2.10 with the modern version in Figure 2.11. For the first 20 elements the arrangements are nearly the same. All the elements in groups 1, 2, and 13 through 17 of the modern table appear in columns headed by a Roman numeral in Mendeleev's table. (Some modern periodic tables acknowledge this older naming system calling these groups IA through VIIA.) However, a group is missing from Mendeleev's table: the **noble gases** (group 18). Helium was the first noble gas to be discovered, and that was not until 1895. The noble gases were elusive substances for early chemists to isolate and identify because they are chemically so unreactive. Because Mendeleev arranged his table largely on the basis of reactivity, he had no reason to predict the existence of a family of elements that had little or no reactivity.

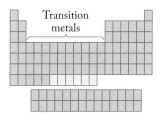

Transition metals

☐ Main group elements
   (representative elements)
☐ Transition elements

| SAMPLE EXERCISE 2.4 | **Navigating the Periodic Table** |

Use the periodic tables below and in Figure 2.11 to identify the following elements; for each element give both its symbol and its name:

a. The alkali metal in the fourth period
b. The halogen with fewer than 16 protons in its nucleus
c. The metalloid in group 14 in the third period
d. The metal in the third period that forms a neutral compound with 2 bromine atoms

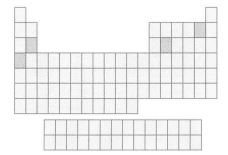

COLLECT AND ORGANIZE The periodic table is divided into vertical columns called groups (or families), some of which have specific names. The rows are called periods. The number of protons in the nucleus gives the atomic number of an element, which appears above the symbol in its cell in the table.

ANALYZE (a) The alkali metals are group 1, (b) the halogens are group 17, (c) this element is in the third row in group 14, and (d) this element must be in group 2 in the third row.

SOLVE (a) K, potassium; (b) F, fluorine; (c) Si, silicon; (d) Mg, magnesium.

THINK ABOUT IT Each element has a unique location in the periodic table determined by its reactivity with other elements and compounds and by its physical properties.

Group 17 is the **halogen** family; group 1 is the **alkali metals**; and group 2 is the **alkaline earth metals**.

Metals are elements on the left-hand side of the periodic table. They are typically shiny solids that conduct heat and electricity well and are malleable and ductile.

Nonmetals have properties opposite to those of metals; they are poor conductors of heat and electricity, and they range in character from brittle solids to gases.

Metalloids (also called **semimetals**) define the border between metals and nonmetals in a row on the periodic table; they have some metallic and some nonmetallic properties.

Main group elements, or **representative elements**, are the elements in groups 1, 2, and 13 through 18.

Transition metals are the elements in groups 3 through 12.

The **noble gases** are the elements in group 18.

Practice Exercise Write the symbol and name the element (or elements) described as follows: (a) the metalloid in family 15 closest in mass to the noble gas krypton; (b) a representative element in the fourth period that is an alkaline earth; (c) a transition metal in the sixth row that you would expect to show some behavior similar to that of zinc ($Z = 30$); (d) a nonmetal in the third row that you would expect to behave similarly to oxygen.

(Answers to Practice Exercises are in the back of the book.)

2.7 Trends in Compound Formation

There are two kinds of chemical compounds: **molecular compounds** and **ionic compounds**. The molecules that make up molecular compounds contain atoms held together by **covalent bonds**. Ionic compounds contain atoms that have lost electrons to form positively charged ions called **cations**, and other atoms that have gained electrons to form negatively charged ions called **anions**. These ions are held together by electrostatic interactions (forces of attraction between oppositely charged particles). The periodic table helps us predict what kinds of compounds pairs of elements form. If the compound is ionic, the table also enables us to predict the charges on the ions.

Binary compounds consist of atoms of two different elements. Let's start by considering binary compounds of the main group elements. First, we use the periodic table to determine whether each element is a metal or a nonmetal, and then to determine what the charges are on any ions that are formed. As we discuss larger molecules and develop increasingly sophisticated ways to view matter, we apply the basic principles we develop here.

In Section 2.2 we described some molecules in the early atmosphere of the Earth: H_2O, CO_2, SO_2, SO_3, NO, and NO_2. Since they are all binary compounds of oxygen with other nonmetals, as a group these compounds are called *nonmetal oxides*. In addition to being nonmetals, the two elements involved are close to each other on the periodic table. Bonds between atoms near each other on the periodic table tend to be covalent. Hence the compounds that form tend to be molecular compounds. The smallest unit of a molecular compound is the molecule itself, and the formula specifying the number of atoms of each element in the molecule is called a **molecular formula**. The models shown in Figure 2.3 all depict covalently bound molecules, as do those in Sample Exercise 2.1 and the accompanying Practice Exercise.

In contrast to compounds of two nonmetals, the chemical combination of a metal and a nonmetal frequently results in the formation of an ionic compound. This is especially true when the metal is in group 1 or 2 and the nonmetal is in group 16 or 17. Metals lose electrons to form cations: for example, metallic elements in group 1 form 1+ cations, while those in group 2 form 2+ cations. Notice that for cations of elements in groups 1 and 2, the charge is the same as the group number. This correspondence changes, however, with higher group numbers: Aluminum in group 13 forms a 3+ cation; the other metals in group 13 form 3+ ions and ions with other charges. The metals in groups 14 and 15 can also form more than one ion. We will deal with that behavior later.

Nonmetallic elements involved in ionic compounds gain electrons to form anions. The number of electrons gained by a nonmetal atom is equal to 18 minus its group number. Thus chlorine (group 17) forms a 1− anion by picking up one electron ($18 − 17 = 1$), as do all the halogens.

Molecular compounds are composed of atoms held together in molecules by covalent bonds.

Ionic compounds are composed of positively and negatively charged ions that are held together by electrostatic attraction (forces of attraction between oppositely charged particles).

Covalent bonds are shared pairs of electrons that chemically bond atoms together.

Ions with negative charges are called **anions**; those with positive charges are called **cations**.

A **molecular formula** describes the exact number and type of atoms present in one molecule of a compound.

Hydrogen is a special case. It is the first element in group 1, but it is a non-metal. When hydrogen is bound to a nonmetal, as it is in HCl or NH_3, the compound is molecular, as we would expect of compounds between two nonmetals (hydrogen and chlorine in the first case, hydrogen and nitrogen in the second). Hydrogen as a nonmetal also forms compounds with metals; LiH is an example. We would predict LiH to behave as an ionic compound, because it is a compound between a metal (lithium) and a nonmetal (hydrogen). What matters in compounds with hydrogen is its character as a nonmetal.

The most common charges on ions derived from some of the elements are shown in the periodic table in Figure 2.12.

As an example of typical behavior, when sodium (a group 1 metal) and chlorine (a group 17 nonmetal) combine to form an ionic compound, each sodium atom gives up an electron to form a sodium cation:

$$Na \rightarrow Na^+ + e^-$$

and each chlorine atom gains an electron to form a chloride anion:

$$Cl + e^- \rightarrow Cl^-$$

All compounds are electrically neutral, so sodium chloride must contain one cationic sodium ion for every one anionic chloride ion. Figure 2.13 shows a crystal of sodium chloride (halite, or table salt). The ratio of sodium ions to chloride ions in the crystal is 1:1, as electrical neutrality requires, but each sodium ion is actually surrounded by six chloride ions, and each chloride ion is in turn surrounded by six sodium ions. *Salts*, ionic compounds like sodium chloride, are not molecular, so it is incorrect to call their formulas (in this case NaCl) molecular formulas. Instead we call them **empirical formulas**. Such a formula describes a **formula unit**, the smallest electrically neutral unit within the larger crystal. Put another way, an empirical formula gives the lowest whole-number ratio of cations to anions in a salt's crystal structure.

The use of empirical formulas is not restricted to salts but can also be used to define the kinds of atoms and their relative numbers present in molecules. For example, the molecule with the molecular formula N_2O_2 has the empirical formula NO.

> An **empirical formula** gives the simplest whole-number ratio of elements in a compound.
>
> The **formula unit** of an ionic compound is the smallest electrically neutral unit within a crystal of the compound. It contains the number and type of ions expressed in the formula of the compound.

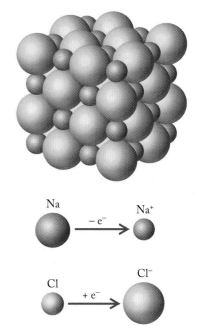

FIGURE 2.13 This is a crystal of halite (also called sodium chloride), an important source of table salt. The cubic shape of the crystal is directly related to the cubic array of Na^+ and Cl^- ions that make up its structure. Its empirical formula, NaCl, describes the smallest whole-number ratio of cations to anions in the structure, which is electrically neutral.

1																	18
H^+	2											13	14	15	16	17	
Li^+														N^{3-}	O^{2-}	F^-	
Na^+	Mg^{2+}	3	4	5	6	7	8	9	10	11	12	Al^{3+}		P^{3-}	S^{2-}	Cl^-	
K^+	Ca^{2+}	Sc^{3+}	Ti^{3+} Ti^{4+}	V^{3+} V^{5+}	Cr^{3+}	Mn^{2+} Mn^{4+}	Fe^{2+} Fe^{3+}	Co^{3+} Co^{2+}	Ni^{2+}	Cu^+ Cu^{2+}	Zn^{2+}	Ga^{3+}			Se^{2-}	Br^-	
Rb^+	Sr^{2+}									Ag^+	Cd^{2+}	In^{3+}	Sn^{2+} Sn^{4+}		Te^{2-}	I^-	
Cs^+	Ba^{2+}									Hg_2^{2+} Hg^{2+}	Tl^+ Tl^{3+}	Pb^{2+} Pb^{4+}					

FIGURE 2.12 This version of the periodic table shows the most common charges on the ions of some common elements. Note that representative elements within a group typically have the ions with the same charge. All the ions have only one atom except for Hg_2^{2+}, the mercury (I) ion.

| SAMPLE EXERCISE 2.5 | Identifying Molecular and Ionic Compounds |

Identify the following binary compounds as ionic compounds or molecular compounds: (a) sodium bromide (NaBr); (b) carbon dioxide (CO_2); (c) lithium iodide (LiI); (d) magnesium fluoride (MgF_2); (e) calcium chloride ($CaCl_2$).

COLLECT AND ORGANIZE Compounds formed by reacting metals with nonmetals tend to be ionic; those that contain only nonmetallic elements are molecular. We can use the periodic table to determine which of the elements in the compounds are metallic and which are nonmetallic.

ANALYZE (a) sodium bromide, (c) lithium iodide, (d) magnesium fluoride, and (e) calcium chloride are all compounds formed between a group 1 or group 2 metal and a group 17 nonmetal. Only carbon dioxide is composed of two nonmetals.

SOLVE (a), (c), (d), and (e) are ionic; (b) is molecular.

THINK ABOUT IT In later chapters we will discover that the world of chemical compounds is not as black and white as painted in this exercise. We will find, for example, that some covalent bonds have a degree of "ionic character," and we will explore a way based on the elements' positions in the periodic table to determine how much ionic character covalent bonds have.

Practice Exercise Which of the following binary compounds are molecular compounds and which are ionic? (a) carbon disulfide (CS_2); (b) carbon monoxide (CO); (c) ammonia (NH_3); (d) water (H_2O); (e) sodium iodide (NaI).

(Answers to Practice Exercises are in the back of the book.)

2.8 Naming Compounds and Writing Their Formulas

At this point we need to establish some rules for naming compounds and writing their chemical formulas. These names and formulas are a foundation of the language of chemistry. The periodic table is an extremely valuable resource when it comes to naming simple chemical compounds, translating the names of compounds into chemical formulas, and translating chemical formulas into names.

Binary Molecular Compounds

Nonmetals may combine to form binary (two-element) molecular compounds. Translating the molecular formula of one of these molecular compounds into its name is straightforward if you follow these guidelines:

1. The first element in the formula has the same name as the element.
2. The second element has the same name as the element except that the final syllable is replaced with the suffix *-ide*.
3. Prefixes (see Table 2.2) are appended to the first and second names to indicate the number of atoms of each type in the molecule. (However, we do not bother with the prefix *mono-* for the first element in a name.)

For example, NO is nitrogen monoxide, whereas NO_2 is nitrogen dioxide. Similarly, SO_2 is sulfur dioxide and SO_3 is sulfur trioxide. Note that when prefixes ending in *o-* or *a-* (like *mono-* and *tetra-*) precede a name that begins with a vowel

TABLE 2.2	Naming Prefixes for Molecular Compounds
one	*mono-*
two	*di-*
three	*tri-*
four	*tetra-*
five	*penta-*
six	*hexa-*
seven	*hepta-*
eight	*octa-*
nine	*nona-*
ten	*deca-*

(such as *oxide*), the *o* or *a* at the end of the prefix is deleted to make the combination of prefix and name easier to pronounce.

This pattern holds for all binary molecular compounds. The order in which the elements are named and given in formulas corresponds to the relative positions of the elements in the periodic table: the element with the lower group number (that is, to the left of the other) appears first in the name and formula. When both elements are in the same column—for example, sulfur and oxygen—then the name of the element with the higher atomic number (that is, lower in the column) appears first.

SAMPLE EXERCISE 2.6   **Naming Binary Molecular Compounds**

Name the following binary compounds of nitrogen and oxygen: (a) N_2O; (b) N_2O_4; (c) N_2O_5

COLLECT AND ORGANIZE These are binary compounds of nonmetals; therefore we can use the appropriate prefixes from Table 2.2 to indicate the number of atoms of each element per molecule.

ANALYZE The first element in all three compounds is nitrogen, so the first word in the compounds is *nitrogen* with the appropriate prefix. The second element in all three compounds is oxygen, so the second word in all their names, again with the appropriate prefix, is *oxide*.

SOLVE

   a. There are two nitrogen atoms per molecule of N_2O and one atom of oxygen, so its name is *dinitrogen monoxide*.
   b. There are two nitrogen atoms and four oxygen atoms in each molecule of N_2O_4, so its name is *dinitrogen tetroxide*.
   c. There are two nitrogen atoms and five oxygen atoms in each molecule of N_2O_5, so its name is *dinitrogen pentoxide*.

THINK ABOUT IT Some compounds have common names that do not follow the naming rules, because they were named before systematic names became standardized. For example, N_2O is also called *nitrous oxide*, or laughing gas.

Practice Exercise Name the following compounds: (a) P_4O_{10}; (b) CO; (c) NCl_3.

(Answers to Practice Exercises are in the back of the book.)

Binary Ionic Compounds

A binary ionic compound consists of a positively charged cation formed by a metallic element and a negatively charged anion formed by a nonmetal. The names of binary ionic compounds begin with the name of the cation, which is simply the name of the parent element. This is followed by the name of the anion, which is the name of the element, except that the last syllable of that element's name is replaced with *-ide*. Consider two examples: the chemical name of table salt (NaCl) is sodium chloride, and the material mixed with it in preparations of iodized salt is potassium iodide (KI).

Prefixes are not used in the names of salts of representative elements because the metal ions in the main groups (group 1, 2, and aluminum in group 13) typically make only one cation. For example, magnesium (group 2) forms only a 2+ cation. Correspondingly the nonmetallic halogens (group 17) as anions have a charge of 1−; nonmetallic anions from group 16 are 2−, and those from group 15 are 3−. Ionic compounds are electrically neutral, so the negative and positive charges of the ions in an ionic compound must balance. Therefore the name "magnesium fluoride" is absolutely unambiguous: it can mean only MgF_2.

SAMPLE EXERCISE 2.7 | **Writing Formulas of Binary Ionic Compounds**

Write the chemical formulas of the following compounds: (a) potassium bromide; (b) calcium oxide; (c) sodium sulfide; (d) magnesium chloride; (e) aluminum oxide

COLLECT AND ORGANIZE Each of the substances contains one metal and one nonmetal from the main group elements combined in an ionic compound. The names of binary compounds like these all end in -*ide*. To write correct formulas of ionic compounds, we assign the charges on the ions based on their group numbers in the periodic table.

ANALYZE In the periodic table (Figure 2.11), locate each of the elements involved in the five compounds listed above, and predict the charge of its most common ion based on its location and its group number. If you have difficulty predicting a charge, refer to Figure 2.12. You should arrive at the following ionic charges: K^+, Br^-, Ca^{2+}, O^{2-}, Na^+, S^{2-}, Mg^{2+}, Cl^-, and Al^{3+}.

SOLVE Balance the positive and negative charges in each compound:

   a. In potassium bromide, the ionic charges are 1+ and 1−, so a 1:1 ratio of the ions is required for electrical neutrality, making the formula KBr.
   b. In calcium oxide the ionic charges are 2+ and 2−; a 1:1 ratio of the ions balances their charges, making the formula CaO.
   c. In sodium sulfide the ionic charges are 1+ and 2−; a 2:1 ratio of Na^+ ions to S^{2-} ions is needed, making the formula Na_2S.
   d. In magnesium chloride the ionic charges are 2+ and 1−; a 1:2 ratio of Mg^{2+} ions to Cl^- ions is needed, making the formula $MgCl_2$.
   e. In aluminum oxide the ionic charges are 3+ and 2−. If we use two Al^{3+} ions for every three O^{2-} ions, the charges will balance. The resulting formula is Al_2O_3.

THINK ABOUT IT Different approaches may be used to work out the formulas of ionic compounds. The basic principle is that the sum of the total positive and negative charges must balance to give a net charge of zero. One approach to writing formulas of compounds with multiple charged ions that have different charges (such as aluminum oxide) is to use the coefficient of the charge of the cation as the subscript for the anion, and vice versa. In the aluminum oxide example, the coefficients of the charges on Al^{3+} and O^{2-} ions are 3 and 2, respectively, so to make the compound electrically neutral, the subscripts after Al and O in the formula must be 2 and 3:

$$Al^{3+}O^{2-} \rightarrow Al_2O_3$$

Practice Exercise Write the chemical formulas of the following compounds: (a) strontium chloride; (b) magnesium oxide; (c) sodium fluoride; (d) calcium bromide.

(Answers to Practice Exercises are in the back of the book.)

Binary Compounds of Transition Metals

Some metallic elements, including many of the transition metals in the middle of the periodic table, form cations with several different charges. For example, most of the copper found in nature is present as Cu^{2+}; however, some copper compounds contain Cu^+. Thus the name *copper chloride* could apply to $CuCl_2$ or $CuCl$. Systematic names are needed to distinguish between the two compounds. One system uses a Roman numeral that defines the charge on the cation after the word *copper* in the name of the compound. Thus copper(II) chloride represents the chloride of Cu^{2+}, which is $CuCl_2$. The formula of copper(I) chloride is $CuCl$. Chemists have for many years also used different names to identify cations of the same element with different charges. Cu^+ compounds are called *cuprous*, and Cu^{2+} compounds are called *cupric*. Similarly, the typical ions of iron—Fe^{2+} and Fe^{3+}—are called *ferrous* and *ferric*, respectively. Note that, in both these pairs of ions, the name of the ion with the lower charge ends in *-ous* and the name of the one with the higher charge ends in *-ic*.

SAMPLE EXERCISE 2.8	**Writing Formulas of Transition Metal Compounds**

(a) Write the chemical formulas of iron(II) sulfide and iron(III) oxide. (b) Write the alternate names for these compounds that do not use Roman numerals to indicate the charge on iron.

COLLECT AND ORGANIZE In each of these ionic compounds of the transition metal iron, the metal cation has a different charge. To write the correct formulas, we need to determine the charges on the anions. Alternate names for these two compounds use the names *ferrous* and *ferric*.

ANALYZE The Roman numerals indicate the value of the positive charges on the iron cations: 2+ for iron(II) and 3+ for iron(III). Oxygen and sulfur are both in group 16 of the periodic table. Therefore the charge on both the sulfide ion and oxide ion is 2−. In the alternate naming system, the 2+ is *ferrous* and the 3+ ion of iron is *ferric*.

SOLVE

a. A charge balance in iron(II) sulfide is achieved with equal numbers of Fe^{2+} and S^{2-} ions, so the chemical formula is FeS. To balance the different charges on the Fe^{3+} and O^{2-} ions in iron(III) oxide, we need three O^{2-} ions for every two Fe^{3+} ions. Thus the formula of iron(III) oxide is Fe_2O_3.
b. The alternate names for FeS and Fe_2O_3 are ferrous sulfide and ferric oxide, respectively.

THINK ABOUT IT You may encounter the system of nomenclature using *-ous* and *-ic* as suffixes frequently in the names of other ions and molecules. Suffix nomenclature has been largely replaced by the convention using Roman numerals because suffixes are inadequate to name compounds in a series like MnO, MnO_2, Mn_2O_3, and Mn_2O_7. However, suffix nomenclature is still frequently used in the names of common salts of transition metal ions in the fourth period.

Practice Exercise Write the formulas of manganese(II) chloride and manganese(IV) oxide.

(Answers to Practice Exercises are in the back of the book.)

Polyatomic ions are charged groups consisting of two or more atoms joined together by covalent bonds.

Oxoanions are polyatomic ions that contain oxygen in combination with one or more other elements.

Polyatomic Ions

Table 2.3 lists the names and chemical formulas of some commonly encountered ions. Several of them are **polyatomic ions**; that is, they consist of more than one kind of atom joined by covalent bonds. The ammonium ion (NH_4^+) is the only common cation among the polyatomic ions; all the others are anions.

Polyatomic ions containing oxygen and one or more other elements are called **oxoanions**. Each oxoanion has a name based on the name of the element that appears first in the formula, but the ending is changed to either *-ite* or *-ate*, depending on the number of oxygen atoms in the formula. Thus, SO_4^{2-} is sulfate whereas SO_3^{2-} is sulfite. The sulfate ion has one more oxygen atom than the sulfite ion. The general rule—that ions with names ending in *-ate* have one more oxygen than those whose names end in *-ite*—applies to naming the oxoanions of nitrogen, nitrate (NO_3^-) and nitrite (NO_2^-), and to naming other sets of oxoanions.

If an element forms more than two kinds of oxoanions, as chlorine and the other group 17 elements do, prefixes are used to distinguish them. The oxoanion with the largest number of oxygen atoms gets the prefix *per-*, and that with the smallest number of oxygen atoms may have the prefix *hypo-* in its name. Table 2.4 applies these rules to the oxoanions of chlorine. Note that these rules do not

TABLE 2.3 Names and Charges of Some Common Ions

Name	Chemical Formula	Name	Chemical Formula
Acetate	CH_3COO^-	Hydride	H^-
Ammonium	NH_4^+	Hydrogen phosphate	HPO_4^{2-}
Azide	N_3^-	Hydroxide	OH^-
Bromide	Br^-	Nitrate	NO_3^-
Carbonate	CO_3^{2-}	Nitride	N^{3-}
Chlorate	ClO_3^-	Nitrite	NO_2^-
Chloride	Cl^-	Oxide	O^{2-}
Chromate	CrO_4^{2-}	Perchlorate	ClO_4^-
Cyanide	CN^-	Permanganate	MnO_4^-
Bicarbonate or hydrogen carbonate	HCO_3^-	Peroxide	O_2^{2-}
Bisulfite or hydrogen sulfite	HSO_3^-	Phosphate	PO_4^{3-}
Dichromate	$Cr_2O_7^{2-}$	Sulfate	SO_4^{2-}
Dihydrogen phosphate	$H_2PO_4^-$	Sulfide	S^{2-}
Disulfide	S_2^{2-}	Sulfite	SO_3^{2-}
Fluoride	F^-	Thiocyanate	SCN^-

TABLE 2.4 Oxoanions of Chlorine and Their Corresponding Acids

Ions		Acids	
ClO^-	hypochlorite	$HClO$	hypochlorous acid
ClO_2^-	chlorite	$HClO_2$	chlorous acid
ClO_3^-	chlorate	$HClO_3$	chloric acid
ClO_4^-	perchlorate	$HClO_4$	perchloric acid

enable you to predict the chemical formula from the name or the charge on the anion. You need to memorize the formulas, charges, and names of the common oxoanions such as those of chlorine, sulfur, and phosphorus. Many of the ions in Table 2.3 are oxoanions.

SAMPLE EXERCISE 2.9   **Writing Formulas of Compounds Containing Oxoanions**

Write the chemical formulas of (a) sodium sulfate and (b) magnesium phosphate.

COLLECT AND ORGANIZE These two substances each contain a metallic element and an oxoanion, so the compounds must be ionic. To write correct formulas of ionic compounds, we need to know the charges on the ions involved. We can determine the charges of sodium and magnesium ions based on the location of these elements in the periodic table, and the charges on the oxoanions from memory.

ANALYZE Consulting the periodic table in Figure 2.11, we find that sodium is in group 1 and magnesium is in group 2. The charges on their ions are 1+ and 2+, respectively. The sulfate ion is SO_4^{2-} and phosphate is PO_4^{3-}.

SOLVE

a. To balance the charges between Na^+ ions and SO_4^{2-} ions, we need twice as many Na^+ ions as sulfate ions. Therefore the formula of sodium sulfate is Na_2SO_4.
b. To balance the charges on the Mg^{2+} and PO_4^{3-} ions, we need three Mg^{2+} ions and two PO_4^{3-}, which gives us $Mg_3(PO_4)_2$.

THINK ABOUT IT We used parentheses around the phosphate ion in writing the formula of magnesium phosphate to make it clear that the subscript 2 applies to the entire phosphate ion. Proficiency in predicting correct formulas from names—and names from chemical formulas—comes from knowing the name, the formula, and the charges of the cation and anion. Charges higher than 3+ for cations and 3− for anions are rare.

Practice Exercise Write the chemical formulas of (a) strontium nitrate and (b) potassium sulfite.

(Answers to Practice Exercises are in the back of the book.)

SAMPLE EXERCISE 2.10   **Naming Compounds Based on Their Chemical Formulas**

What are the names of the following compounds? (a) $CaCO_3$; (b) $LiNO_3$; (c) $MgSO_3$; (d) Rb_2SO_4; (e) $KClO_3$; (f) $NaHCO_3$.

COLLECT AND ORGANIZE Every one of these compounds consists of a metallic element combined with an oxoanion. The common oxoanions are listed in Table 2.3.

ANALYZE You need to memorize the names of the common oxoanions in Table 2.3, or be able to consult the table to solve this Sample Exercise.
    The cations in these compounds are (a) Ca^{2+}, (b) Li^+, (c) Mg^{2+}, (d) Rb^+, (e) K^+, and (f) Na^+. Since these are ionic compounds, the cations are named as the elements.
    The oxoanions in these compounds are (a) CO_3^{2-}, (b) NO_3^-, (c) SO_3^{2-}, (d) SO_4^{2-}, (e) ClO_3^-, and (f) HCO_3^-; they are named as listed in Table 2.3.

Organic compounds include most compounds containing carbon, and commonly include certain other elements such as hydrogen, oxygen, and nitrogen.

A **carboxylic acid** is an organic compound containing the –COOH functional group.

Functional groups are structural subunits in organic molecules that impart characteristic chemical and physical properties.

SOLVE Combining the names of these cations and oxoanions, we get (a) calcium carbonate, (b) lithium nitrate, (c) magnesium sulfite, (d) rubidium sulfate, (e) potassium chlorate, and (f) sodium hydrogen carbonate.

THINK ABOUT IT Sodium hydrogen carbonate is commonly called sodium bicarbonate. The prefix *bi–* is sometimes used to indicate that there is a hydrogen ion (H^+) attached to an oxoanion.

Practice Exercise Name the following compounds: (a) $Ca_3(PO_4)_2$; (b) $Mg(ClO_4)_2$; (c) $LiNO_2$; (d) $NaClO$; (e) $KMnO_4$.

(Answers to Practice Exercises are in the back of the book.)

Acids

Certain compounds have special names that are used to highlight particular properties. Among these are acids. We will discuss acids in greater detail in later chapters, but for now it is sufficient to say that acids release hydrogen ions (H^+)—sometimes referred to as protons—when they dissolve in water. The binary compound HCl is hydrogen chloride. When dissolved in water, this compound produces an acidic solution, which we call hydrochloric acid. In naming this and other binary acids, the following guidelines apply:

1. Affix the prefix *hydro–* to the name of the element other than hydrogen in the molecule.
2. Replace the last syllable in the name with the suffix *-ic* and add *acid*.

The most common binary acids are compounds of hydrogen and the halogens. Their aqueous solutions are called hydrofluoric, hydrochloric, hydrobromic, and hydroiodic acid.

The scheme for naming the acids of oxoanions is illustrated with the names of the oxoanions of chlorine in Table 2.4. If the name of the oxoanion ends in *-ate*, the name of the corresponding acid ends in *-ic*; when the name of the oxoanion ends in *-ite*, the name of the corresponding acid ends in *-ous*. Thus Na_2SO_4 is the formula of sodium sulfate, and H_2SO_4 is the formula of sulfuric acid. Similarly, $NaNO_2$ is the formula of sodium nitrite, and HNO_2 is the formula of nitrous acid.

Other acids exist besides those with oxoanions in their structures. Many of them are produced by processes in living organisms. For example, fermentation reactions produce acetic acid. Vinegar, a common ingredient in salad dressings, is an aqueous solution of acetic acid. Acetic acid is an example of an organic acid, so-called because of its origins in biochemical processes. **Organic compounds** were originally considered the products of living systems only. The definition has broadened considerably and now includes most compounds of carbon. The formula for acetic acid is often given in a manner that describes its structure. The formula of acetic acid is most frequently written as CH_3COOH (Figure 2.14), and –COOH is called the **carboxylic acid** group. Many organic acids have the –COOH group in their structures. It is one of many **functional groups** that are responsible for the chemical and physical properties of organic compounds. We will describe several others later in the book. The hydrogen ion (proton) that acetic acid releases in aqueous solution is the hydrogen atom (proton) in the –COOH group.

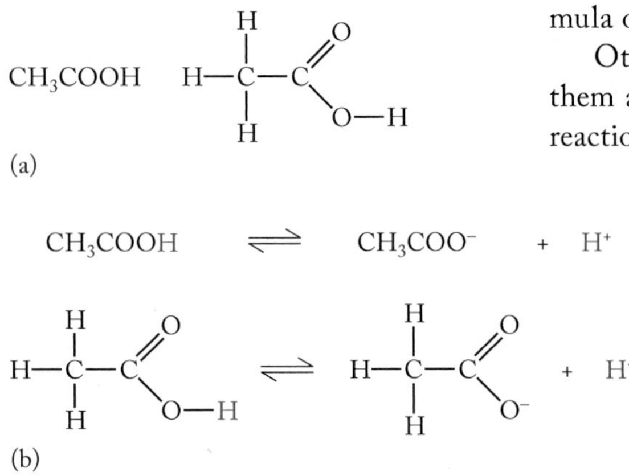

FIGURE 2.14 (a) The written and structural formulas for molecular acetic acid. (b) Only the hydrogen atom on the carboxylic acid group (–COOH) dissociates when molecular acetic acid ionizes in water to form the acetate ion and the hydrogen ion.

| SAMPLE EXERCISE 2.11 | **Naming the Acids of Oxoanions** |

What are the names of the acids formed by the following oxoanions? (a) SO_3^{2-}; (b) ClO_4^-; (c) NO_2^-

COLLECT AND ORGANIZE The question involves linking the name of an acid to the name of the anion formed when an oxoacid has lost all its ionizable protons. When the anion's name ends in *-ite*, the acid's name ends in *-ous*. When the anion's name ends in *-ate*, the acid's name ends in *-ic*.

ANALYZE According to the information in Table 2.3, the names of the three oxoanions in question are (a) sulfite, (b) perchlorate, and (c) nitrite.

SOLVE By changing the endings of the oxoanion base names from *-ite* to *-ous* or from *-ate* to *-ic*, as appropriate, and adding the word *acid*, we get (a) sulfurous acid, (b) perchloric acid, and (c) nitrous acid.

THINK ABOUT IT To name the acids of oxoanions, you must learn the formulas, charges, and names of the oxoanions in Table 2.3.

Practice Exercise What are the names of the following acids? (a) HClO; (b) $HClO_2$; (c) H_2CO_3.

(Answers to Practice Exercises are in the back of the book.)

2.9 Nucleosynthesis

Now, as promised in Section 2.1, we return to the topic of how elements may have formed from the energy released by the Big Bang and in the cores of stars. Theoretical physicists believe that within a few microseconds after the Big Bang its energy had transformed into matter in accordance with Einstein's equation:

$$E = mc^2$$

The first matter to form consisted of the smallest fundamental particles: electrons and **quarks**, as illustrated in Figure 2.15. Less than a millisecond later, the universe had expanded and "cooled" to about 10^{12} K. Even at this extremely high temperature, quarks slowed down enough to combine with each other, forming neutrons and protons. In less than a second the matter in the universe consisted of the three subatomic particles that would eventually make up atoms.

Early Nucleosynthesis

By about 4 minutes after the Big Bang the universe had expanded and cooled to about 10^9 K. In this hot, dense subatomic "soup," protons outnumbered neutrons by about 7:1, and neutrons and protons that collided with each other began to fuse together. The result was the beginning of **nucleosynthesis**, which is the fusing of fundamental and subatomic particles to create atomic nuclei. The early stages of nucleosynthesis are sometimes called *primordial nucleosynthesis*. In the first step, pairs of neutrons and protons fused together to form particles called *deuterons*. These particles are the nuclei of deuterium (^2H), which is an isotope of hydrogen. Note that hydrogen is the only element whose isotopes are given different names: hydrogen (^1H), *deuterium* (^2H, also known as "heavy hydrogen"), and *tritium* (^3H,

Quarks are elementary particles that combine to form neutrons and protons.

Nucleosynthesis is the fusing of fundamental and subatomic particles to create atomic nuclei.

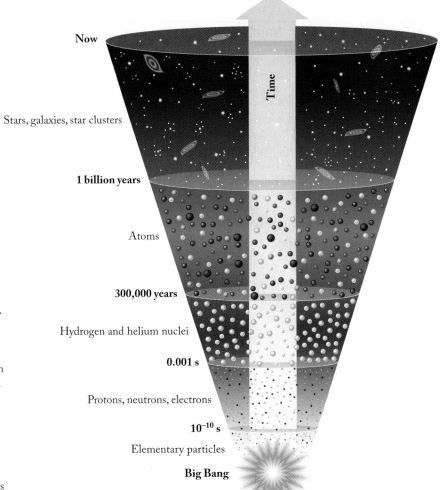

FIGURE 2.15 This timeline depicts the transformations of energy and matter that are believed to have occurred since the universe began. In this model, subatomic particles formed in the first millisecond after the Big Bang, followed by hydrogen and helium nuclei. Atoms of H and He did not form until after 300,000 years of expansion and cooling, and other elements did not form until the first galaxies appeared, around 1 billion years after the Big Bang. According to this model, our solar system, our planet, and all matter including living systems are composed of elements that were synthesized in stars that were born, burned brightly, and then disappeared millions to billions of years after the Big Bang (and millions to billions of years ago).

known as "very heavy hydrogen"). The first two are stable **nuclides**, nuclei of specific isotopes of an element, while the third nuclide is radioactive.

We can write an equation describing this fusion process as follows:

$$\,^1_1\text{H} + \,^1_0\text{n} \rightarrow \,^2_1\text{H} \tag{2.2}$$

In writing this equation, we follow the rules described in Section 2.4 for writing the symbols of isotopes: each superscript is the mass number (the sum of the neutrons and protons) and each subscript is the atomic number of the nucleus. We use a similar convention for writing the symbols of subatomic particles, except that each subscript represents the charge on the particle. For example, the symbol of the neutron is ^1_0n because it has a mass number of 1 but a charge of 0. We use the symbol ^1_1H to represent a proton because the nucleus of ^1H *is* a proton.

Deuteron formation proceeded rapidly, consuming most of the neutrons in the universe in a matter of seconds. No sooner did deuterons form than they too were rapidly consumed: by colliding with each other and fusing to form nuclei of ^4_2He, which are the same as α particles:

$$2\,^2_1\text{H} \rightarrow \,^4_2\text{He} \tag{2.3}$$

As a result of the fusion reactions shown in Equations 2.2 and 2.3, the matter of the universe about 5 minutes after the Big Bang was 75% (by mass) protons and 25% α particles. Less than 0.01% consisted of other products of neutron–proton fusion, including deuterons, *tritons* (the nuclei of ^3H, or tritium), and nuclei of another isotope of helium, ^3_2He.

A **nuclide** is the nucleus of a specific isotope of an element, also often used as a synonym for isotope.

Let's take a second look at Equations 2.2 and 2.3. These nuclear equations are related to the chemical equations we will begin writing in Chapter 3 in that they are *balanced*. This means that in each of these nuclear equations the sum of the masses (superscript numbers) of the particles to the left of the arrow is equal to the sum of the masses of all the particles to the right. Similarly the sum of the charges (subscript numbers) of all the particles on the left side equals the sum of the charges of all the particles on the right. Check Equations 2.2 and 2.3 to confirm that they are indeed balanced in terms of mass and charge.

Perhaps you are wondering why early nucleosynthesis stopped after the synthesis of helium. Why, for example, did α particles and protons not collide and fuse to make ^{5}Li?

$$^4_2\text{He} + ^1_1\text{H} \xrightarrow{?} ^5_3\text{Li}$$

And why did two α particles not fuse to make ^{8}Be?

$$^4_2\text{He} + ^4_2\text{He} \xrightarrow{?} ^8_4\text{Be}$$

Neither of these processes took place because neither ^{5}Li nor ^{8}Be is stable under the conditions present in the early universe. In fact, no stable nuclei with mass numbers of 5 or 8 exist. As the universe continued to expand and cool, its temperature dropped below 10^8, which is too low to sustain nuclear fusion. As a result the elemental composition of the universe was "locked in" at 75% H and 25% He for the better part of a billion years.

Stellar Nucleosynthesis

That matter in the universe today still consists mostly of H and He is strong evidence supporting the theory of the Big Bang. But how were the other elements in the periodic table, including those that make up most of our planet, formed? **Cosmologists**, scientists who study the physical nature and form of the universe, theorize that synthesis of the heavier elements had to wait until nuclear fusion resumed, in the first generation of galaxies. Inside coalescing stars in these galaxies, hydrogen and helium underwent compression heating; pressures increased as temperatures inside the stars reached 10^8 K; then the nuclear reactions in these stars ignited as hydrogen began to fuse, making more helium.[2]

Some of these stars, known as *red giants*, had cores so extraordinarily hot and dense that groups of three α particles could simultaneously collide and fuse together. This *triple-alpha process* produces a nucleus with 6 protons and 6 neutrons, which is the stable isotope carbon-12:

$$3\,^4_2\text{He} \rightarrow ^{12}_6\text{C}$$

With the formation of ^{12}C, the barrier to the nucleosynthesis of larger nuclei had been overcome. At the even higher temperatures occurring in the cores of stars called *supergiants*, ^{12}C nuclei fuse with α particles to form ^{16}O (Figure 2.16). Then ^{16}O fuses with another α particle to form ^{20}Ne, and so on. Additional fusion reactions involving even larger nuclei with increasingly greater positive charges (see Figures 2.16 and 2.17) are possible in these intensely hot (10^9 K) stars because nuclei move fast enough to overcome the electrostatic repulsion experienced by particles with large positive charges.

▶❚❚ CHEMTOUR Synthesis of Elements

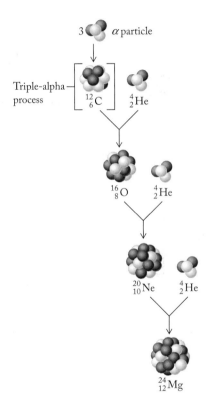

FIGURE 2.16 Shown schematically is the fusion of α particles, which forms carbon-12 in the triple-alpha process, and then the fusion of successively heavier nuclei to form oxygen-16, neon-20, magnesium-24, and so on. These fusion processes release the energy that fuels the nuclear furnaces of giant stars.

Cosmology is the study of the physical nature and the form of the universe as a whole.

---

[2] The hydrogen–helium fusion process in stars such as our sun is not the same as in primordial nucleosynthesis because there are few free neutrons in these stars. Details of the stellar hydrogen fusion process will be described in Chapter 20.

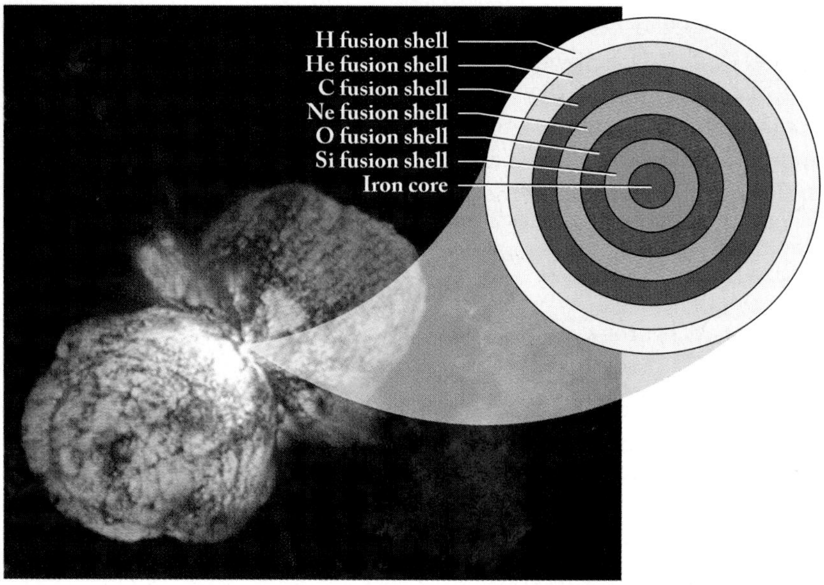

FIGURE 2.17 The star η-Carinae is believed to be evolving toward a supernova explosion. The outer regions of the star are still fueled by the energy released by the fusion of hydrogen isotopes; the star is increasingly hotter and denser closer to the center; this allows the fusion of larger nuclei, and results in the production of ^{56}Fe in the core.

For the past 13 billion years, fusion reactions of the sort described above have simultaneously fueled the nuclear furnaces of giant stars while producing isotopes as heavy as ^{56}Fe. However, once the core of a giant star turns into iron, the star is in trouble because additional fusion reactions involving iron nuclei do not release energy; instead they *consume* it. For example, fusing an α particle with ^{56}Fe to make ^{60}Ni:

$$^{56}_{26}\text{Fe} + {}^{4}_{2}\text{He} \rightarrow {}^{60}_{28}\text{Ni}$$

requires the *addition* of energy. A giant star with an iron core has essentially run out of fuel. Its central nuclear furnace goes out, and the star begins to cool and collapse because of the gravity produced by its enormous mass.

As the star collapses, compression reheats its core to temperatures above 10^9 K. At such temperatures, nuclei begin to disintegrate into free protons and neutrons. Free neutrons create additional pathways for making elements. Neutrons readily fuse with atomic nuclei in a process called **neutron capture**. When a stable nucleus such as ^{56}Fe captures enough neutrons, it becomes unstable, emitting a β particle in a process called **β decay**. Beta (β) decay is the process by which a neutron in a neutron-rich nucleus decays into a proton and a β particle. Let's consider what happens as a nucleus of ^{56}Fe captures first one, then another, and finally a third neutron. The three neutron-capture steps—

$$^{56}_{26}\text{Fe} + {}^{1}_{0}\text{n} \rightarrow {}^{57}_{26}\text{Fe}$$

$$^{57}_{26}\text{Fe} + {}^{1}_{0}\text{n} \rightarrow {}^{58}_{26}\text{Fe}$$

$$^{58}_{26}\text{Fe} + {}^{1}_{0}\text{n} \rightarrow {}^{59}_{26}\text{Fe}$$

—produce two stable iron isotopes: ^{57}Fe and ^{58}Fe. The final isotope produced, ^{59}Fe, is unstable. Iron-59 has 33 neutrons and 26 protons, and its instability arises from its high neutron-to-proton ratio. The ^{59}Fe isotope undergoes β decay to reduce its neutron-to-proton ratio. Let's write a nuclear equation for this reaction starting with what we know,

$$^{59}_{26}\text{Fe} \rightarrow ? + {}^{0}_{1-}\beta \tag{2.4}$$

and then identify the other product. We use the symbol ${}^{0}_{1-}\beta$ to represent a β particle, because it is a high-speed electron and has a charge of 1−. We assign it a

Neutron capture is the absorption of a neutron by a nucleus.

Beta (β) decay is the process by which a neutron in a neutron-rich nucleus decays into a proton and a β particle.

mass number of 0 because an electron has less than 1/1000 the mass of a neutron or proton (see Table 2.1). We can identify the unknown product by balancing the values of the superscripts and subscripts in Equation 2.4. This leaves us with $^{59}_{27}X$ for the unknown isotope product. Checking the periodic table reveals that the element for which $Z = 27$ is cobalt. Therefore the isotope is ^{59}Co, which is stable. The overall equation describing the two processes of neutron capture and β decay is written in equation form this way:

$$^{56}_{26}Fe + 3\,^{1}_{0}n \rightarrow \,^{59}_{27}Co + \,^{0}_{1-}\beta$$

Note that the product of this process, ^{59}Co, has 27 protons, but the starting material was ^{56}Fe with 26 protons. The additional proton was produced during β decay as a neutron disintegrated into a proton, which stayed in the nucleus, and a β particle, which was emitted from the nucleus:

$$^{1}_{0}n \rightarrow \,^{1}_{1}p + \,^{0}_{1-}\beta$$

SAMPLE EXERCISE 2.12   **Predicting the Product of Neutron Capture Followed by β Decay**

A neutron-rich nuclide of krypton, Kr ($Z = 36$), undergoes β decay. Which element is produced?

COLLECT AND ORGANIZE When a neutron-rich nucleus undergoes β decay, it ejects a β particle (high-speed electron), and a neutron becomes a proton. The mass number of the atom does not change, but the atomic number (Z) increases by 1.

ANALYZE Krypton has 36 protons, so it will become a nucleus with 37 protons and an atomic number of 37.

SOLVE The element for which $Z = 37$ is rubidium.

THINK ABOUT IT Heavier elements are formed when dense stars called neutron stars explode in a supernova. The explosion expels matter throughout space. The presence of heavy elements identifies a second-generation star—those that coalesce after such an explosion. Our sun is at least a second-generation star. This means that some atoms in our solar system existed before our solar system formed and are therefore older than the solar system itself.

Practice Exercise A proton-rich nuclide of arsenic ($Z = 33$) undergoes neutron capture. Which element is produced?

(Answers to Practice Exercises are in the back of the book.)

Repeated neutron capture and β decay events in the cores of collapsing stars produce nuclei of the heaviest elements in the periodic table. However, one more chapter in the element-manufacturing story needs to be told: distribution of the products. The enormous heating that occurs when a giant star collapses produces a gigantic explosion. Cosmologists call such an event a *supernova*. In addition to finishing the job of synthesizing the elemental building blocks found in the universe, a supernova serves as its own element-distribution system, blasting its inventory of elements throughout its galaxy (Figure 2.18). The legacies of supernovas are found in the elemental composition of later-generation stars like our sun and in the planets like Earth that orbit these stars. Indeed, everything that exists in our solar system, including all life, is made from the residues of exploding stars.

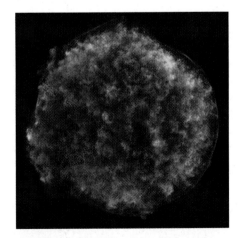

FIGURE 2.18 This colorized picture from the Chandra X-ray Observatory shows the remains of the supernova of the star Tycho in the constellation Cassiopeia. The expanding bubble of debris colored red and green is a cloud of hot ionized gas inside a more rapidly moving shell of extremely high-energy electrons in blue.

Summary

SECTION 2.1 Earth and the other three planets closest to the sun are rich in Fe, O, Si, Mg, Al, Ni and other elements that are either nonvolatile or form nonvolatile compounds. Jupiter and the other gas giants in the outer solar system are made up mostly of hydrogen and helium, and so have an elemental composition much like that of the sun and most of the rest of the universe.

SECTION 2.2 In the early 1800s John Dalton proposed his atomic theory of matter to explain the law of constant composition (discussed in Section 1.1) and his own **law of multiple proportions**. The latter states that two different masses of element Y, reacting with a given mass of element X to form two compounds, can be expressed as the ratio of two small whole numbers.

SECTION 2.3 Atoms are composed of negatively charged **electrons** surrounding a **nucleus**, which contains positively charged **protons** and electrically neutral **neutrons**. The values of charge and mass of electrons were determined by J. J. Thomson's studies using cathode-ray tubes and by Robert Millikan's oil-droplet experiments. The research of Ernest Rutherford's group, who bombarded thin gold foil with α **particles** and determined how the particles were deflected, showed that all the positive charge and nearly all the mass of an atom are contained in its nucleus.

SECTION 2.4 The number of protons in the nucleus of an element defines its **atomic number (Z)**; the number of **nucleons** (protons and neutrons) in the nucleus defines the element's **mass number (A)**. The different **isotopes** of an element consist of atoms with the same number of protons per nucleus but different numbers of neutrons.

SECTION 2.5 **Mass spectrometers** are used to determine precisely the masses of atoms and molecules. The **average atomic mass** of an element is found by multiplying the mass of each of its isotopes times the **natural abundance** of that isotope, and then summing these products.

SECTION 2.6 Elements are arranged in the periodic table in order of increasing atomic number and in a pattern based on their physical and chemical properties. Elements in the same vertical column are said to be in the same **group**, or **family**. Among the **main group** (or **representative) elements** are the **alkali metals** (group 1), the **alkaline earth metals** (group 2), and the **halogens** (group 17). Most of the elements are **metals**, which means that they are malleable, ductile solids (except mercury) and are good conductors of heat and electricity. **Nonmetals** include elements in all three physical states that are poor conductors of heat and electricity. **Metalloids**, or **semimetals**, have many of the physical properties of metals but the chemical prop-

erties of nonmetals; in the periodic table they are located between the metals to the left and the nonmetals to the right.

SECTION 2.7 When metals react with nonmetals, the metals form positively charged ions, called **cations**, and the nonmetals form negatively charged ions, called **anions**. These ions are held together by electrostatic attraction in solid **ionic compounds**. The charge of the ions formed by a main group element can be predicted on the basis of which group the element is in. When nonmetals react with other nonmetals or with metalloids, they form **molecular compounds**, in which atoms are held together by **covalent bonds**. A **molecular formula** describes the exact number and type of atoms present in one molecule of a compound. The **empirical formula** of a molecular or ionic compound gives the simplest whole-number ratio of the atoms (or ions) in it.

SECTION 2.8 For binary compounds, the name or symbol of the element that is to the left or below the other in the periodic table is named first. In the names of molecular compounds, prefixes indicate the number of atoms of the second element in each molecule and the number of atoms of the first one when there is more than one atom per molecule. For binary ionic compounds, the name or symbol of the metallic element is followed by the name or symbol of the nonmetal, but the ending of the name of the nonmetal is changed to *-ide*. Roman numerals in parentheses indicate the charges of cations formed by transition metals. The names of **oxoanions** (**polyatomic ions** containing oxygen atoms) end in *-ate* or *-ite* and may have a prefix of *per-* or *hypo-* or no prefix, depending on the number of oxygen atoms per ion. In most acids hydrogen is written first. The names of binary acids begin with the prefix *hydro-* and end with *-ic acid*. The names of acids based on *-ate* and *-ite* oxoanions end in *-ic* and *-ous*, respectively.

SECTION 2.9 Neutrons and protons consist of fundamental particles called **quarks**. According to current theories of **cosmology**, neutrons, protons, and electrons formed within seconds of the Big Bang. Then **nucleosynthesis** (the creation of atomic nuclei more complex than hydrogen) began: protons and neutrons fused to produce isotopes of hydrogen and helium (early nucleosynthesis). After galaxies formed, the nuclei of atoms with $Z \leq 26$ formed when the nuclei of lighter elements fused together in the cores of giant stars (stellar nucleosynthesis). The nuclei of elements with $Z > 26$ then formed by a combination of **neutron capture**, β **decay**, and other nuclear reactions that occurred during supernovas (explosions of giant stars). As a result of these explosions, the synthesized elements were distributed throughout galaxies for possible inclusion in later-generation stars and in planets such as our own. Stellar nucleosynthesis and supernovas continue today.

Problem-Solving Summary

TYPE OF PROBLEM	CONCEPTS AND EQUATIONS	SAMPLE EXERCISES
Calculating the quantity of an element in one compound based on the quantity of the same element in another compound	Use the chemical formulas of the two compounds and the law of multiple proportions.	2.1

TYPE OF PROBLEM	CONCEPTS AND EQUATIONS	SAMPLE EXERCISES
Writing symbols of isotopes	To the left of the element symbol, place a superscript for the mass number and a subscript for the atomic number.	2.2
Calculating the average atomic mass of an element	Multiply the mass (m) of each stable isotope of the element times the natural abundance (a) of that isotope; then sum these products: $$m_X = a_1m_1 + a_2m_2 + a_3m_3 + \cdots$$	2.3
Identifying ionic and molecular compounds	Ionic compounds form in reactions between metals and nonmetals; molecular compounds form in reactions between nonmetals and metalloids, or between different nonmetals.	2.5
Naming binary inorganic compounds and writing their formulas	Apply the naming rules given in Section 2.8.	2.6 and 2.7
Naming transition metal compounds and writing their formulas	Use a Roman numeral to indicate the charge on the transition metal cation.	2.8
Naming compounds containing oxoanions and writing their formulas	Apply the naming rules in Section 2.8; the names of oxoanions end in -ate or -ite. The names of the corresponding acids end in -ic acid or -ous acid.	2.9–2.11
Predicting the product of neutron capture followed by β decay	Neutron capture adds to the mass number of a nucleus, emission of a β particle increases its atomic number by one.	2.12

Visual Problems

2.1. In Figure P2.1 the blue spheres represent nitrogen atoms and the red spheres represent oxygen atoms. The figure as a whole represents which of the following gases? (a) N_2O_3; (b) N_7O_{11}; (c) a mixture of NO_2 and NO; (d) a mixture of N_2 and O_3

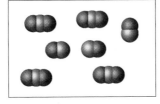

FIGURE P2.1

2.2. In Figure P2.2 the black spheres represent carbon atoms and the red spheres represent oxygen atoms. Which of the following statements about the two equal-volume compartments is or are true?
  a. The compartment on the left contains CO_2; the one on the right contains CO.
  b. The compartments contain the same mass of carbon.
  c. The ratio of the oxygen to carbon in the gas in the left compartment is twice that of the gas in the right compartment.
  d. The pressures inside the two compartments are equal. (Assume that the pressure of a gas is proportional to the number of particles in a given volume.)

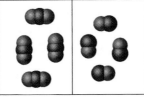

FIGURE P2.2

2.3. Which of the highlighted elements in Figure P2.3 formed, according to the theory of the Big Bang, before the first generation of galaxies formed?

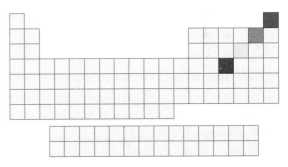

FIGURE P2.3

2.4. Which of the highlighted elements in Figure P2.4 *is not* formed by fusion of lighter elements in the cores of giant stars?

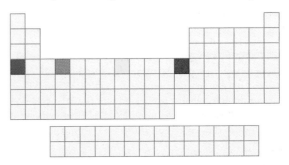

FIGURE P2.4

2.5. Which of the highlighted elements in Figure P2.5 is the following? (a) a reactive nonmetal; (b) a chemically inert gas; (c) a reactive metal

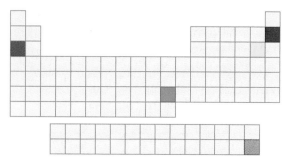

FIGURE P2.5

2.6. Which of the highlighted elements in Figure P2.6 forms monatomic ions with the following charge? (a) 1+; (b) 2+; (c) 3+; (d) 1–; (e) 2–

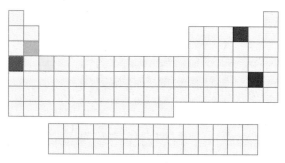

FIGURE P2.6

2.7. Which of the highlighted elements X in Figure P2.7 form or forms an oxide with the following generic formulas? (a) XO; (b) X_2O; (c) XO_2; (d) X_2O_3

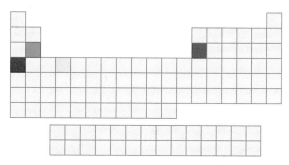

FIGURE P2.7

2.8. Which of the highlighted elements X in Figure P2.8 form or forms oxoanions with the following generic formulas? (a) XO_4^-; (b) XO_4^{2-}; (c) XO_4^{3-}; (d) XO_3^-

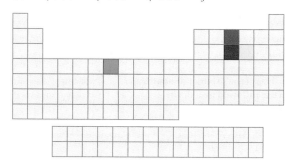

FIGURE P2.8

Questions and Problems

Dalton's View of Matter

CONCEPT REVIEW

2.9. Their names suggest that the law of constant composition and the law of multiple proportions are incompatible. Explain why they are not.

2.10. Explain why the law of constant composition is classified a scientific *law*, whereas Dalton's view of the atomic structure of matter is classified a scientific *theory*.

2.11. How does Dalton's atomic theory of matter explain the fact that when water is decomposed into hydrogen and oxygen gas, the volume of hydrogen is always twice that of oxygen?

2.12. **Pollutants in Automobile Exhaust** In the internal combustion engines that power most automobiles, nitrogen and oxygen may combine to form NO. When NO in automobile exhaust is released into the atmosphere, it reacts with more oxygen, forming NO_2, a key ingredient in smog. How do these reactions illustrate Dalton's law of multiple proportions?

PROBLEMS

2.13. Cobalt forms two sulfides, CoS and Co_2S_3. Predict the ratio of the two masses of sulfur that combine with a fixed mass of cobalt to form CoS and Co_2S_3.

2.14. Lead forms two oxides: PbO and PbO_2. Predict the ratio of the two masses of oxygen that combine with a fixed mass of lead to form PbO and PbO_2.

2.15. Nitrogen monoxide (NO) is 46.7% nitrogen by mass. What is the mass percentage of nitrogen in nitrogen dioxide (NO_2)?

2.16. When 5.0 grams of sulfur are combined with 5.0 grams of oxygen, 10.0 grams of sulfur dioxide are formed. What mass of oxygen would be required to convert 5.0 grams of sulfur into sulfur trioxide?

The Rutherford Model of Atomic Structure

CONCEPT REVIEW

2.17. Explain how the results of the gold-foil experiment led Rutherford to dismiss the plum-pudding model of the atom and create his own model based on a nucleus surrounded by electrons.

2.18. Had the plum-pudding model been valid, how would the results of the gold-foil experiment have differed from what Geiger and Marsden actually observed?

2.19. What properties of cathode rays led Thomson to conclude that they were not pure energy, but rather particles with an electric charge?

*2.20.** Alpha particles, with a charge of 2+ and a mass of 4 amu, are actually helium nuclei. The element helium was first discovered in a sample of pitchblende, an ore of radioactive uranium oxide. How did helium get in the ore?

Isotopes; Average Atomic Masses

CONCEPT REVIEW

2.21. What is meant by a *weighted average*?

2.22. Explain how percent natural abundances are related to average atomic masses.

PROBLEMS

2.23. If the mass number of an isotope is more than twice the atomic number, is the neutron-to-proton ratio less than, greater than, or equal to 1?

2.24. In each of the following pairs of isotopes, which isotope has more protons and which has more neutrons? (a) ^{127}I or ^{131}I; (b) ^{188}Re or ^{188}W; (c) ^{14}N or ^{14}C

2.25. Boron, lithium, nitrogen, and neon each have two stable isotopes. In which of the following pairs of isotopes is the heavier isotope more abundant?
  a. ^{10}B or ^{11}B (average atomic mass, 10.81 amu)
  b. ^{6}Li or ^{7}Li (average atomic mass, 6.941 amu)
  c. ^{14}N or ^{15}N (average atomic mass, 14.01 amu)
  d. ^{20}Ne or ^{22}Ne (average atomic mass, 20.18 amu)

2.26. Naturally occurring copper contains a mixture of 69.17% copper-63 (62.9296 amu) and 30.83% copper-65 (64.9278 amu). What is the average atomic mass of copper?

2.27. Naturally occurring chlorine consists of two isotopes: 75.78% ^{35}Cl, and 24.22% ^{37}Cl. Calculate the average atomic mass of chlorine.

2.28. Naturally occurring sulfur consists of four isotopes: 95.04% ^{32}S (31.97207 amu); ^{33}S (32.97146 amu, 0.75%); ^{34}S (33.96787 amu, 4.20%); and ^{36}S (35.96708 amu, 0.01%). Calculate the average atomic mass of sulfur in atomic mass units.

2.29. Chemistry of Mars The 1997 mission to Mars included a small robot, the Sojourner, which analyzed the composition of Martian rocks. Magnesium oxide from a boulder dubbed "Barnacle Bill" was analyzed and found to have the following isotopic composition:

Exact Mass of MgO (amu)	Natural Abundance (%)
39.9872	78.70
40.9886	10.13
41.9846	11.17

If essentially all of the oxygen in the Martian MgO sample is oxygen-16 (which has an exact mass of 15.9948 amu), is the average atomic mass of magnesium on Mars the same as on Earth (24.31 amu)?

2.30. Using the following table of abundances and masses of the three naturally occurring argon isotopes, calculate the exact mass of ^{40}Ar.

Symbol	Exact Mass (amu)	Natural Abundance (%)
^{36}Ar	35.96755	0.337
^{38}Ar	37.96272	0.063
^{40}Ar	?	99.60
Average	39.948	

2.31. From the following table of abundances and masses of five naturally occurring titanium isotopes, calculate the exact mass of ^{48}Ti.

Symbol	Exact Mass (amu)	Natural Abundance (%)
^{46}Ti	45.95263	8.25
^{47}Ti	46.9518	7.44
^{48}Ti	?	73.72
^{49}Ti	48.94787	5.41
^{50}Ti	49.9448	5.18
Average	47.87	

2.32. Strontium has four isotopes: ^{84}Sr, ^{86}Sr, ^{87}Sr, and ^{88}Sr.
  a. How many neutrons are there in each isotope?
  b. The natural abundances of the four isotopes are 0.56% ^{84}Sr (83.9134 amu), 9.86% ^{86}Sr (85.9094 amu), 7.00% ^{87}Sr (86.9089 amu), and 82.58% ^{88}Sr (87.9056 amu). Calculate the average atomic mass of a Sr atom and compare it to the value in the periodic table on the inside front cover.

The Periodic Table of the Elements; Trends in Compound Formation

CONCEPT REVIEW

2.33. Mendeleev ordered the elements in his version of the periodic table on the basis of their atomic masses instead of their atomic numbers. Why?

2.34. Why did Mendeleev not include the noble gases in his version of the periodic table?

PROBLEMS

2.35. How many protons, neutrons, and electrons are there in the following atoms? (a) ^{14}C; (b) ^{59}Fe; (c) ^{90}Sr; (d) ^{210}Pb

2.36. How many protons, neutrons, and electrons are there in the following atoms? (a) ^{11}B; (b) ^{19}F; (c) ^{131}I; (d) ^{222}Rn

2.37. Fill in the missing information in the following table of four neutral atoms:

Symbol	^{23}Na	?	?	?
Number of protons	?	39	?	79
Number of neutrons	?	50	?	?
Number of electrons	?	?	50	?
Mass number	?	?	118	197

2.38. Fill in the missing information in the following table of four neutral atoms:

Symbol	^{27}Al	?	?	?
Number of protons	?	42	?	92
Number of neutrons	?	56	?	?
Number of electrons	?	?	60	?
Mass number	?	?	143	238

2.39. Fill in the missing information in the following table of ions:

Symbol	$^{37}Cl^-$	?	?	?
Number of protons	?	11	?	88
Number of neutrons	?	12	46	?
Number of electrons	?	10	36	86
Mass number	?	?	81	226

2.40. Fill in the missing information in the following table of ions:

Symbol	$^{137}Ba^{2+}$	?	?	?
Number of protons	?	30	?	40
Number of neutrons	?	34	16	?
Number of electrons	?	28	18	36
Mass number	?	?	32	90

2.41. Seawater The most abundant anion in seawater is the chloride ion. Write the formulas for the chlorides and sulfates of the most abundant cations in seawater: sodium, magnesium, calcium, potassium, and strontium.

2.42 The most abundant cation in seawater is the sodium ion. The evaporation of seawater gives a mixture of ionic compounds containing sodium combined with chloride, sulfate, carbonate, bicarbonate, bromide, fluoride, and tetrahydroxy borate, $B(OH)_4^-$. Write the chemical formulas of all these compounds.

2.43. Which of the following compounds consist of molecules and which consist of ions? (a) CH_3COOH; (b) $SrCl_2$; (c) $MgCO_3$; (d) H_2SO_4

2.44. Which of these compounds consist of molecules, and which consist of ions? (a) $LiOH$; (b) $Ba(NO_3)_2$; (c) HNO_3; (d) $CH_3(CH_2)_3OH$

2.45. Which element is most likely to form a cation with a 2+ charge? (a) S; (b) P; (c) Be; (d) Al

2.46. Which element is most likely to form an anion with a 2− charge? (a) S; (b) P; (c) Be; (d) Al

2.47. Which species contains the greatest number of electrons? (a) F; (b) O^{2-}; (c) S^{2-}; (d) Cl

2.48. Which species contains the lowest number of electrons? (a) F; (b) O^{2-}; (c) S^{2-}; (d) Cl

2.49. Which ion has the same number of electrons as an atom of argon? (a) S^{2-}; (b) P^{3-}; (c) Be^{2+}; (d) Ca^{2+}

2.50. Which ion has the same number of electrons as an atom of krypton? (a) Se^{2-}; (b) As^{3-}; (c) Ca^{2+}; (d) K^+

2.51. Which element is a nonmetal? (a) Si; (b) Br; (c) Ca; (d) Ru

2.52. Which element is a metalloid? (a) Si; (b) Br; (c) Ca; (d) Ru

Naming Compounds and Writing Their Formulas

CONCEPT REVIEW

2.53. Consider a mythical element X, which forms only two oxoanions: XO_2^{2-} and XO_3^{2-}. Which of the two has a name that ends in *-ite*?

2.54. Concerning the oxoanions in Problem 2.53, would the name of either of them require a prefix such as *hypo-* or *per-*? Explain why or why not.

2.55. What is the role of Roman numerals in the names of the compounds formed by transition metals?

2.56. Why do the names of the ionic compounds formed by the alkali metals and by the alkaline earth metals not include Roman numerals?

PROBLEMS

2.57. The following list contains the chemical formulas of eight binary compounds formed by nitrogen and oxygen. Name each compound: (a) NO_3; (b) N_2O_5; (c) N_2O_4; (d) NO_2; (e) N_2O_3; (f) NO; (g) N_2O; (h) N_4O

2.58. More than a dozen binary compounds containing sulfur and oxygen have been identified. Give the chemical formulas for the following six: (a) sulfur monoxide; (b) sulfur dioxide; (c) sulfur trioxide; (d) disulfur monoxide; (e) hexasulfur monoxide; (f) heptasulfur dioxide

2.59. Predict the formula and give the name of the binary ionic compound containing the following: (a) sodium and sulfur; (b) strontium and chlorine; (c) aluminum and oxygen; (d) lithium and hydrogen

2.60. Predict the formula and give the name of the binary ionic compound containing the following: (a) potassium and bromine; (b) calcium and hydrogen; (c) lithium and nitrogen; (d) aluminum and chlorine

2.61. Give the chemical names of the cobalt oxides that have the following formulas: (a) CoO; (b) Co_2O_3; (c) CoO_2

2.62. Give the formula of each of the following copper minerals:
a. cuprite, copper(I) oxide
b. chalcocite, copper(I) sulfide
c. covellite, copper(II) sulfide

2.63. Give the formula and charge of the oxoanion in each of the following compounds: (a) sodium hypobromite; (b) potassium sulfate; (c) lithium iodate; (d) magnesium nitrite

*__2.64.__ Give the formula and charge of the oxoanion in each of the following compounds: (a) potassium tellurite; (b) sodium arsenate; (c) calcium selenite; (d) potassium chlorate

2.65. Give chemical names of the following ionic compounds: (a) $NiCO_3$; (b) $NaCN$; (c) $LiHCO_3$; (d) $Ca(ClO)_2$

2.66. Give names for the following ionic compounds: (a) $Mg(ClO_4)_2$; (b) NH_4NO_3; (c) $Cu(CH_3COO)_2$; (d) K_2SO_3

2.67. Give the name or chemical formula of each of the following acids: (a) HF; (b) $HBrO_3$; (c) phosphoric acid; (d) nitrous acid

2.68. Give the name or chemical formula of each of the following acids: (a) HBr; (b) HIO_4; (c) selenous acid; (d) hydrocyanic acid

2.69. Name these compounds: (a) Na_2O; (b) Na_2S; (c) Na_2SO_4; (d) $NaNO_3$; (e) $NaNO_2$

2.70. Name these compounds: (a) K_3PO_4; (b) K_2O; (c) K_2SO_3; (d) KNO_3; (e) KNO_2

2.71. Write the chemical formulas of these compounds: (a) potassium sulfide; (b) potassium selenide; (c) rubidium sulfate; (d) rubidium nitrite; (e) magnesium sulfate

2.72. Write the chemical formulas of these compounds: (a) rubidium nitride; (b) potassium selenite; (c) rubidium sulfite; (d) rubidium nitrate; (e) magnesium sulfite

2.73. Name these compounds: (a) MnS; (b) V_3N_2; (c) $Cr_2(SO_4)_3$; (d) $Co(NO_3)_2$; (e) Fe_2O_3

2.74. Name these compounds: (a) RuS; (b) $PdCl_2$; (c) Ag_2O; (d) WO_3; (e) PtO_2

2.75. Which compound is sodium sulfite? (a) Na_2S; (b) Na_2SO_3; (c) Na_2SO_4; (d) $NaHS$

2.76. Which compound is calcium nitrate? (a) Ca_3N_2; (b) Ca_2NO_3; (c) $Ca_2(NO_3)_2$; (d) $Ca(NO_3)_2$

2.77. Which element is a halogen? (a) N_2; (b) Cl_2; (c) Xe; (d) H_2

2.78. Which element is a noble gas? (a) N_2; (b) Cl_2; (c) Xe; (d) I_2

2.79. Which element is an alkali metal? (a) Na; (b) Cl_2; (c) Xe; (d) Br_2

2.80. Which element is an alkaline earth metal? (a) Na; (b) Ca; (c) Xe; (d) H_2

Nucleosynthesis

CONCEPT REVIEW

2.81. Write brief (one-sentence) definitions of *chemistry* and *cosmology*, and then give as many examples as you can of how the two sciences are related.

2.82. In the history of the universe, which of these particles formed first and which formed last?
a. deuteron   b. neutron   c. proton   d. quark

2.83. Chemists don't include quarks in the category of subatomic particles—can you think of a reason why?

2.84. Why did primordial nucleosynthesis last such a short time?

2.85. In the current cosmological model, the volume of the universe is increasing with time. How might this expansion affect the density of the universe?

2.86. **Components of Solar Wind** Most of the ions that flow out from the sun in the solar wind (see Section 2.1) are hydrogen ions. The ions of which element should be next most abundant?

*__2.87.__ **Nucleosynthesis in Giant Stars** A star needs a core temperature of about 10^7 K for hydrogen fusion to occur. Core temperatures above 10^8 K are needed for helium fusion. Why does helium fusion require much higher temperatures?

2.88. Why was the triple-alpha process unlikely to happen in a rapidly cooling universe soon after the Big Bang?

*__2.89.__ It takes nearly twice the energy to remove an electron from a helium atom as it does to remove an electron from a hydrogen atom. Propose an explanation for this.

2.90. **Origins of the Elements** Our sun contains carbon even though its core is not hot or dense enough to sustain carbon synthesis through the triple-alpha process. Where could the carbon have come from?

2.91. Early nucleosynthesis produced a universe that was more than 99% hydrogen and helium with less than 1% lithium. Why were the other elements not formed?

2.92. What is the effect of β decay on the ratio of neutrons to protons in a nucleus?

PROBLEMS

2.93. Radioactive ^{137}I decays to ^{137}Xe, which is also radioactive and decays to ^{137}Cs. Do either, or both, of these decay processes involve emission of a β particle?

2.94. **Isotopes in Geochemistry** The relative abundances of the stable isotopes of the elements are not entirely constant. For example, in some geologic samples (soils and rocks) the ratio of ^{87}Sr to ^{86}Sr is affected by the presence of a radioactive isotope of another element, which slowly undergoes β decay to produce more ^{87}Sr. What is this other isotope?

2.95. What nuclide is produced in the core of a giant star by each of the following fusion reactions?
a. $^{12}C + {}^4He \rightarrow$ b. $^{20}Ne + {}^4He \rightarrow$ c. $^{32}S + {}^4He \rightarrow$

2.96. What nuclide is produced in the core of a giant star by each of the following fusion reactions?
a. $^{28}Si + {}^4He \rightarrow$ b. $^{40}Ca + {}^4He \rightarrow$ c. $^{24}Mg + {}^4He \rightarrow$

2.97. What nuclide is produced in the core of a collapsing giant star by each of the following reactions?
   a. $^{56}Fe + 3\,^1n \rightarrow$ ___ $+\,^0_{-1}\beta$
   b. $^{118}Sn + 3\,^1n \rightarrow$ ___ $+\,^0_{-1}\beta$
   c. $^{109}Ag + \,^1n \rightarrow$ ___ $+\,^0_{-1}\beta$

2.98. What nuclide is produced in the core of a collapsing giant star by each of the following reactions?
   a. $^{65}Cu + 3\,^1n \rightarrow$ ___ $+\,^0_{-1}\beta$
   b. $^{68}Zn + 2\,^1n \rightarrow$ ___ $+\,^0_{-1}\beta$
   c. $^{88}Sr + \,^1n \rightarrow$ ___ $+\,^0_{-1}\beta$

Additional Problems

2.99. In April 1897, J. J. Thomson presented the results of his experiment with cathode-ray tubes (Figure P2.99) in which he proposed that the rays were actually beams of negatively charged particles, which he called "corpuscles."
   a. What is the name we use for these particles today?
   b. Why did the beam deflect when passed between electrically charged plates, as shown in Figure P2.99?
   c. If the polarity of the plates were switched, how would the position of the light spot on the phosphorescent screen change?
   d. If the voltage on the plates were reduced by half, how would the position of the light spot change?

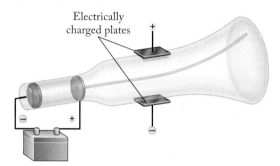

Electrically charged plates

FIGURE P2.99

*2.100. Suppose the electrically charged discs at the end of the cathode-ray tube are replaced with a radioactive source, as shown in Figure P2.100. Also suppose the radioactive material inside the source emits α and β particles, plus rays of energy with no charge. The only way for any of the three kinds of particles or rays to escape the source is through a narrow channel drilled through a block of lead.
   a. How many light spots do you expect to see on the phosphorescent screen?
   b. What are their positions relative to the electrical plates and which particle produces which spot?

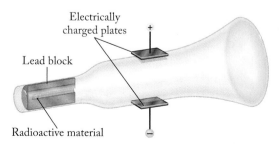

Electrically charged plates

Lead block

Radioactive material

FIGURE P2.100

*2.101. Suppose the radioactive material inside the source in the apparatus shown in Figure P2.100 emits protons and α particles, and suppose both kinds of particles have the same velocities.
   a. How many light spots do you expect to see on the phosphorescent screen?
   b. What are their positions on the screen (above, at, or below the center)? Which particle produces which spot?

2.102. Cosmologists estimate that the matter in the early universe was 75% by mass hydrogen-1 and 25% helium-4 when atoms first formed.
   a. Assuming these proportions are correct, what was the ratio of hydrogen to helium *atoms* in the early universe?
   b. The ratio of hydrogen to helium atoms in our solar system is slightly less than 10:1. Compare this value with the value you calculated in part a.
   c. Propose a hypothesis that accounts for the difference in composition between the solar system and the early universe.
   d. Describe an experiment that would test your hypothesis.

2.103. Potassium can form three compounds with oxygen: K_2O (potassium oxide), K_2O_2 (potassium peroxide), and KO_2 (potassium superoxide). Potassium superoxide is used in self-contained breathing apparatus as a source of oxygen, while potassium peroxide can be used to scrub carbon dioxide from the atmosphere in submarines. Predict the ratio of the masses of oxygen that combine with a fixed mass of potassium in K_2O, K_2O_2, and KO_2.

2.104. **Stainless Steel** The gleaming metallic appearance of the Gateway Arch (Figure P2.104) in St. Louis, Missouri comes from the stainless steel used in its construction. This steel is made mostly of iron but it also contains 19% by mass chromium and 9% by mass nickel.
   a. Stainless steel maintains its metallic sheen because the chromium and nickel in it combine with oxygen from the atmosphere, forming a layer of Cr_2O_3 and NiO that is too thin to detract from the luster of the steel but that protects the metal beneath from further corrosion. What are the names of these two ionic compounds?
   b. What are the charges of the cations in Cr_2O_3 and NiO?

FIGURE P2.104

2.105. Bronze Age Historians and archaeologists often apply the term "Bronze Age" to the period in Mediterranean and Middle Eastern history that began about 4000 BC and ended about 1200 BC when bronze was the preferred material for making weapons, tools, and other metal objects. Ancient bronze was an alloy prepared by blending molten copper (90%) and tin (10%) by mass. What is the ratio of copper to tin atoms in a piece of bronze with this composition?

*2.106. In his version of the periodic table, Mendeleev arranged elements based on the formulas of the compounds they formed with hydrogen and oxygen. The elements in one of his eight groups formed compounds with these generic formulas: MH_3 and M_2O_5, where M is the symbol of an element in the group. Which Roman numeral did Mendeleev assign to this group?

2.107. In the Mendeleev table in Figure 2.10, there are no symbols for elements with predicted atomic masses of 44, 68, and 72.
   a. Which elements are these?
   b. Mendeleev anticipated the later discovery of these three elements and gave them tentative names: ekaaluminum, ekaboron, and ekasilicon, reflecting the probability that their properties would resemble those of aluminum, boron, and silicon respectively. What are the modern names of ekaaluminum, ekaboron, and ekasilicon?
   c. When were these elements finally discovered? To answer this question you may wish to consult a reference such as webelements.com.

2.108. Many common chemical compounds have old fashioned, non-systematic names that are still widely used. Search for the systematic names and chemical formulas of compounds with these non-systematic names: (a) saltpeter; (b) caustic soda; (c) soda ash; (d) baking soda; (e) magnesia; (f) corundum; (g) Epsom salt; (h) lime

*2.109. The ruby shown in Figure P2.109 has a mass of 12.04 carats (1 carat = 200.0 mg). Rubies are made of a crystalline form of Al_2O_3.
   a. What percentage of the mass of the ruby is aluminum?
   b. The density of rubies is 4.02 g/cm³. What is the volume of the ruby?

FIGURE P2.109

2.110. In chemical nomenclature, the prefix *thio-* is used to indicate that a sulfur atom has replaced an oxygen atom in the structure of a molecule of a polyatomic ion.
   a. With this rule in mind, write the formula for the thiosulfate ion.
   b. What is the formula of sodium thiosulfate?

2.111. There are two stable isotopes of gallium. Their masses are 68.92558 and 70.9247050 amu. If the average atomic mass of gallium is 69.7231 amu, what is the natural abundance of the lighter isotope?

2.112. There are two stable isotopes of bromine. Their masses are 78.9183 and 80.9163 amu. If the average atomic mass of bromine is 79.9091 amu, what is the natural abundance of the heavier isotope?

*2.113. Start with the information in the previous question, and then do the following:
   a. Predict the possible masses of individual molecules of Br_2.
   b. Calculate the natural abundance of molecules with each of the masses predicted in part (a) in a sample of Br_2.

*2.114. There are three stable isotopes of magnesium. Their masses are 23.9850, 24.9858, and 25.9826 amu. If the average atomic mass of magnesium is 24.3050 amu and the natural abundance of the lightest isotope is 78.99%, what are the natural abundances of the other two isotopes?

3

Chemical Reactions and Earth's Composition

FIRE, WATER, AND SALT *Nate Smith demonsrates a fire vortex on the Great Salt Lake in Utah. Fire, water, and salts are fundamental to the chemistry of Earth from its origins to the present day.*

A LOOK AHEAD: Purple Salt and Bottled Water

Once the matter in the solar system had cooled sufficiently and planetary masses formed, chemical reactions began to take place among the elements. Scientists believe that an early atmosphere containing carbon dioxide (CO_2), methane (CH_4), and nitrogen (N_2) formed from gases escaping from Earth's interior while the planetary mass was still molten. According to computer simulations based on this theory, this early material was carried away from Earth by solar winds that were much stronger then than they are now.

As the crust formed on Earth because of further cooling, volcanic activity produced another new atmosphere vastly different from the one we know today. Volatile oxides of nonmetallic elements—CO_2, SO_2, SO_3, NO_2—erupted from the interior of the planet, where they had been formed by reactions of carbon, sulfur, and nitrogen with oxygen. The oxide of any hydrogen still present in the planetary mass probably formed as well, and the vapor of H_2O was probably present in this second-generation atmosphere. As the planet continued to cool, the water vapor condensed and the first rain fell on the young Earth. The other nonmetal oxides dissolved in the water and reacted with it to form acidic solutions.

Two questions that have not yet been answered are (1) where all the water on Earth came from and (2) why there is so much more water on Earth than on the other planets in our solar system. Water vapor degassing from the interior of Earth is certainly one source; however, most of the hydrogen in the solar system was swept away from the inner planets by the heat of the primordial sun, so the amount of water that could be formed was limited by the small amount of hydrogen present. One additional source of the water on Earth, beyond that due to degassing, is thought to be water from outer reaches of our own solar system. Water-rich asteroids and comets colliding with Earth could have supplied significant quantities of water to the planet.

Analysis of a meteorite that smashed to the ground in Texas in 1998 provided additional data to support the theory of interstellar water coming to Earth. Within the meteorite, researchers from NASA's Johnson Space Center found minute bubbles of water sealed within purple crystals of sodium chloride: small drops of water in tiny bottles made of purple salt. The crystals of sodium chloride—ordinary table salt—were bright purple as a result of their exposure to the intense radiation in space.

How and where interstellar water is formed is not completely understood, but many comets are known to be balls of ice, and meteorite researchers have for years seen indirect evidence of chemical reactions involving water on asteroids. Large numbers of comets, asteroids, and other objects from the outer reaches of the solar system, like the meteorite that fell in Texas, may have brought to Earth much of the water that now fills our oceans. The chemical reactions of the elements on the early Earth that formed the atmosphere and the reactions of these gases with water are the topics of this chapter.

The *NEAR* (*Near Earth Asteroid Rendezvous*) *Shoemaker* spacecraft landed on this 33-km-long, peanut-shaped asteroid, known as 433 Eros, on February 12, 2001. Instruments onboard the spacecraft determined that the elemental composition of the asteroid was much the same as that of Earth, indicating that the asteroid formed from the same materials and at the same time as our planet—about 4.6 billion years ago.

3.1 Composition of Earth

In Chapter 2 we described briefly how Earth is believed to have formed from the swirling gases comprising the solar nebula some 4.6 billion years ago. We noted that the chemical composition of our young planet was, and still is, much different from most of the matter in the universe, because most of the substances that make up Earth and the other terrestrial planets (Mercury, Venus, and Mars) were left behind after the heat of the primordial sun had blown away more volatile

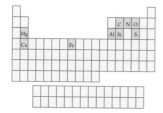

Volcanic eruptions, such as this one at Mount St. Helens, WA, on May 18, 1980, release sulfur and nitrogen oxides including SO_2, SO_3, NO, and NO_2. Volcanologists monitor atmospheric concentrations of SO_2 near volcanic craters to determine when an eruption is likely to occur. The 1980 eruption of Mount St. Helens also released a cubic kilometer of ash. There were reports of falling ash over most of the Rocky Mountain states and western provinces of Canada.

A **chemical equation** describes the identities and quantities of **reactants** (substances that are consumed during a chemical reaction) and **products** (substances that are formed).

A **mole (mol)** of particles (atoms, ions, or molecules) contains **Avogadro's number** ($N_A = 6.022 \times 10^{23}$) of the particles.

substances. These volatile substances included much of the two most abundant elements in the universe: hydrogen and helium. This fractionation process left behind another gaseous element, oxygen, as the second most abundant element (by mass) on Earth after iron. Why was oxygen not also swept into the outer reaches of the solar system? The answer is linked to the chemical reactivity of oxygen, which forms nonvolatile oxides with the elements that would become the principal components of Earth's mantle and crust (see Figure 2.2)—iron, silicon, magnesium, aluminum, and calcium—and volatile oxides with sulfur, nitrogen, and carbon that formed the components of the atmosphere.

The crust of the early Earth was torn by incessant volcanic activity and bombarded by asteroids and other debris from the solar nebula. In time, volcanic gas emissions produced an atmosphere rich in carbon dioxide, water vapor, and several oxides of sulfur and nitrogen. Life as we know it could not have survived in such an environment. Besides lacking free oxygen, the atmosphere was very corrosive: it contained a number of strong acids that form when sulfur and nitrogen oxides combine with water. For example, when sulfur trioxide combines with water vapor, sulfuric acid [$H_2SO_4(\ell)$] is produced, as shown in the molecular models in Figure 3.1. This reaction can also be expressed by the following chemical equation:

$$SO_3(g) + H_2O(g) \rightarrow H_2SO_4(\ell) \qquad (3.1)$$

Reactants   Product

FIGURE 3.1 A molecule of SO_3 and a molecule of H_2O react to produce a molecule of H_2SO_4.

In a **chemical equation** the formulas of the **reactants** (the reacting substances) appear first, followed by a reaction arrow that points toward the formulas of the **products** (the substances produced). Symbols in parentheses after the chemical formulas indicate the physical states of reactants and products: (*g*) for gas, (*ℓ*) for liquid, (*s*) for solid, and (*aq*) for substances dissolved in water.

3.2 The Mole

Equation 3.1 describes a reaction between molecules of sulfur trioxide and water vapor. In our macroscopic (visible) world, chemists rarely work with individual atoms or molecules but deal routinely with larger quantities of reactants and products that they can see, weigh, and manipulate. These measurable quantities contain enormous numbers of particles: atoms, molecules or ions, and so chemists need a unit that is based on an enormous number to relate measurable quantities of substances to the number of particles they contain. That unit is the SI base unit for expressing quantities of substances (see Table 1.1) and is called the **mole (mol)**.

A mole of a substance is defined by international agreement as the quantity of the substance that contains the same number of particles as the number of carbon atoms in exactly 12 grams of the carbon-12 isotope. The number is very large: 6.022×10^{23} atoms, expressed to four significant figures. This value is called **Avogadro's number** (N_A) after the Italian scientist Amedeo Avogadro (1776–1856), whose research enabled other scientists to accurately determine the atomic masses of the elements. Dividing a particular number of particles by Avogadro's number

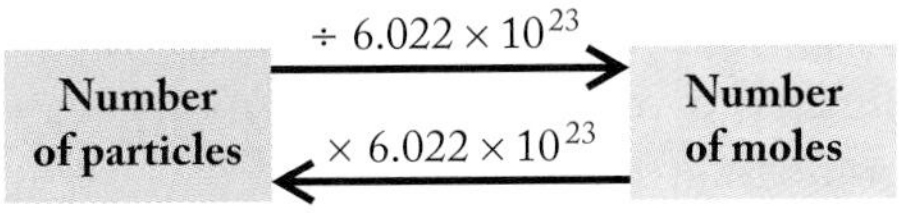

FIGURE 3.2 Converting between a number of particles and an equivalent number of moles (or vice versa) is a matter of dividing (or multiplying) by Avogadro's number.

$$\div\ 6.022 \times 10^{23}$$

Number of particles		Number of moles

$$\times\ 6.022 \times 10^{23}$$

yields the number of moles of those particles; conversely, multiplying a number of moles by Avogadro's number gives us the number of particles in that many moles. These conversions are illustrated in Figure 3.2 and demonstrated in the following Sample Exercise. The quantities of some common elements equivalent to 1 mole are illustrated in Figure 3.3.

SAMPLE EXERCISE 3.1   **Converting Number of Moles into Number of Particles**

It's not unusual for the air above major U.S. cities to contain as much as 5×10^{-10} moles of SO_2 per liter of air. How many molecules of SO_2 are in a liter of such air?

COLLECT AND ORGANIZE The problem states the number of moles of SO_2 in 1 liter of air. Avogadro's number defines the number of particles in 1 mole.

ANALYZE Using Avogadro's number, we can convert the number of moles into the number of molecules.

SOLVE We convert the number of moles to the number of molecules by multiplying the number of moles by Avogadro's number:

$$5 \times 10^{-10}\ \frac{\cancel{\text{mol}}\ SO_2}{\text{liter}} \times 6.022 \times 10^{23}\ \frac{\text{molecules}}{\cancel{\text{mol}}} = 3 \times 10^{14}\ \frac{\text{molecules}\ SO_2}{\text{liter}}$$

THINK ABOUT IT The result tells us that a tiny fraction of a mole of a molecular substance is equivalent to a very large number of molecules, which makes sense given the enormity of Avogadro's number. While it is sometimes useful to know the number of molecules of a substance in a sample, we'll mostly be concerned about knowing how many *moles* are in a particular quantity of a substance.

Practice Exercise One milliliter of seawater contains about 2.5×10^{-14} moles of gold. How many atoms of gold are in 1 milliliter of seawater?

(Answers to Practice Exercises are in the back of the book.)

FIGURE 3.3 The quantities shown are equivalent to 1 mole of each material: 4.003 g of helium, 32.07 g of sulfur, 63.55 g of copper, and 200.59 g of mercury.

▶❚❚ **CHEMTOUR** Avogadro's Number

Molar Mass

The mole provides an important link from the atomic mass values in the periodic table to masses of macroscopic quantities of elements and compounds. The mass numbers on the table are each an average atomic mass of the atoms of an element expressed in atomic mass units (amu). They are also the mass of 1 mole of atoms of that element in grams. Thus the average mass of an atom of helium is 4.003 amu, and the mass of a mole of helium (6.022×10^{23} atoms of He) is 4.003 g. In other words, the **molar mass** ($\mathcal{M}$) of helium is 4.003 g/mol.

The molar mass of any substance can be used to convert between the mass of a sample of the substance and the number of moles in that sample. The mole provides the link between equations that describe how particles react at the atomic and molecular level, and equations relating bulk quantities of materials (those we actually observe and measure.) In Sample Exercises involving molar masses,

The **molar mass** ($\mathcal{M}$) of a substance is the mass of 1 mole of the particles that comprise that substance. The molar mass of an element, in grams per mole, is numerically the same as that element's average atomic mass, in atomic mass units.

the values given in the periodic table are typically rounded to four significant figures: hydrogen is 1.008 g/mol, carbon 12.01 g/mol, mercury 200.6 g/mol, and so forth.

Let's look at the concept of the mole another way: The mole enables us to count by weighing, because it represents both a fixed number of particles and a specific mass of a given substance. We know, for example, that if we were to weigh 137.3 grams (1 mole) of barium atoms (Ba) and 32.07 grams (1 mole) of sulfur atoms (S) into a reaction vessel, we have the same number (Avogadro's number) of reactant atoms of both elements. The atoms react on heating to produce 169.40 grams (1 mole) of barium sulfide, BaS(s).

The following Sample Exercise illustrates how the mass of a sample of a pure substance can be used to find the equivalent number of moles of that substance.

SAMPLE EXERCISE 3.2   **Converting Mass in Grams into Number of Moles**

If the helium balloons sold at an amusement park each contain 1.85 g of He, how many moles of He are in each balloon?

COLLECT AND ORGANIZE This is an exercise in converting grams of a substance (helium) into moles of that substance. According to the periodic table, the average atomic mass of an atom of helium is 4.003 amu, which means that the molar mass of He is 4.003 g/mol.

ANALYZE We want to convert grams of He into moles of He. The conversion factor we should use to convert grams of He into an equivalent number of moles is the reciprocal of the molar mass, or

$$\frac{1 \text{ mol He}}{4.003 \text{ g He}}$$

SOLVE

$$1.85 \text{ g He} \times \frac{1 \text{ mol He}}{4.003 \text{ g He}} = 0.462 \text{ mol He}$$

THINK ABOUT IT This mathematical operation is equivalent to *dividing* the mass of He by its molar mass. The opposite conversion (moles to grams) involves *multiplying* the number of moles by the molar mass.

Practice Exercise A length of copper wire weighs 4.86 g. How many moles of copper are in the wire?

(Answers to Practice Exercises are in the back of the book.)

SAMPLE EXERCISE 3.3   **Converting Number of Moles into Mass in Grams**

Each Tums Ultra® antacid tablet contains 1.00×10^{-2} moles of calcium (as Ca^{2+} ions). How many milligrams of calcium are in each tablet?

COLLECT AND ORGANIZE Answering this question requires us to convert moles to grams. The link is the molar mass of calcium, which according to the periodic table is 40.08 g/mol. Remember that the mass of an electron is insignificant compared with that of protons and neutrons. Therefore the mass of a calcium ion is close enough to the mass of a calcium atom, even though the ion has two fewer electrons than the neutral atom.

ANALYZE We can use the molar mass of calcium to convert moles of calcium ion to grams of calcium ion. The problem requires that the answer be in units of milligrams (mg); converting from grams to milligrams will be an additional mathematical step (1 g = 1000 mg).

SOLVE

$$1.00 \times 10^{-2} \; \text{mol Ca}^{2+} \times \frac{40.08 \; \text{g Ca}^{2+}}{1 \; \text{mol Ca}^{2+}} \times \frac{1000 \; \text{mg}}{\text{g}} = 401 \; \text{mg Ca}^{2+}$$

THINK ABOUT IT Human beings seem most comfortable dealing conversationally with numbers between 1 and 1000. Consequently, quantities are frequently expressed in units that put numbers in this range. Units of milligrams were chosen to express the amount of calcium in the tablet because they enable us to express the quantity in question using numbers in the range of 1 to 1000. The answer here seems appropriate based on the small number of moles of calcium in the tablet.

Practice Exercise How many grams of gold are there in 0.250 moles of gold?

(Answers to Practice Exercises are in the back of the book.)

CONCEPT TEST

Which contains more atoms—1 oz of gold (Au) or 1 oz of silver (Ag)?

(Answers to Concept Tests are in the back of the book.)

Molecular Masses and Formula Masses

The **molecular mass** of a molecular compound is the sum of the masses of the atoms in one molecule of the compound. The molar mass ($\mathcal{M}$) of a molecular compound in grams per mole (g/mol) is numerically the same as its molecular mass in atomic mass units (amu). Thus, the molecular mass of sulfur trioxide, SO_3, is the sum of the masses of 1 sulfur atom (32.07 amu) and 3 oxygen atoms (3 × 16.00 amu):

$$1 \; \text{atom S} \times \frac{32.07 \; \text{amu}}{1 \; \text{atom S}} + 3 \; \text{atoms O} \times \frac{16.00 \; \text{amu}}{1 \; \text{atom O}} = 80.07 \; \text{amu/molecule}$$

The molecular mass of SO_3 is 80.07 amu. As with atoms and atomic mass, we can scale up molecular mass to a quantity we can measure and manipulate using the concept of a mole. The result is that atomic mass units become grams, molecules become moles, and we end up with the molar mass of the compound:

$$\mathcal{M}_{SO_3} = 1 \; \text{mol S} \times \frac{32.07 \; \text{g}}{1 \; \text{mol S}} + 3 \; \text{mol O} \times \frac{16.00 \; \text{g}}{1 \; \text{mol O}} = 80.07 \; \text{g/mol}$$

The concept of molar mass applies to ionic as well as molecular compounds. For example, 1 mole of BaS contains 1 mole of Ba^{2+} ions and 1 mole of S^{2-} ions.

The mass of one molecule of a molecular compound is called its **molecular mass**.

The **formula mass** of a compound is the mass of one formula unit.

Molar mass values (g/mol) are numerically the same as:

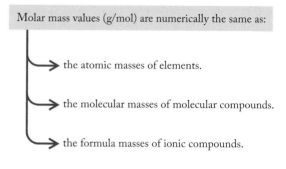

the atomic masses of elements.

the molecular masses of molecular compounds.

the formula masses of ionic compounds.

Therefore, the molar mass of BaS equals the sum of the molar masses of barium and sulfur:

$$1 \; \cancel{\text{mol Ba}^{2+}} \times \frac{137.33 \text{ g}}{1 \; \cancel{\text{mol Ba}^{2+}}} + 1 \; \cancel{\text{mol S}^{2-}} \times \frac{32.07 \text{ g}}{1 \; \cancel{\text{mol S}^{2-}}} = 169.40 \text{ g/mol}$$

Note that there are 2 moles of ions in 1 mole of BaS: 1 mole of Ba^{2+} ions and 1 mole of S^{2-} ions. Also keep in mind that there are no discrete molecules of BaS. As we saw in Chapter 2, crystals of salts such as BaS actually consist of ordered arrays of equal numbers of Ba^{2+} ions and S^{2-} ions. The formula unit (see Figure 3.4) defines the simplest ratio of the positive and negative ions, 1:1 in this case, that describes the composition of such an ionic compound. The mass of one formula unit of an ionic compound is called its **formula mass**. Figure 3.5 summarizes how to use Avogadro's number, chemical formulas, and molar masses to convert between masses, between numbers of moles, and between numbers of particles in elements and compounds.

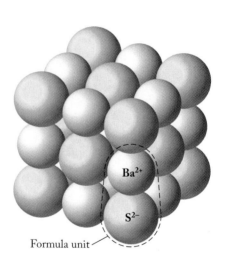

FIGURE 3.4 Crystals of barium sulfide consist of a three-dimensional array of Ba^{2+} ions and S^{2-} ions. The 1:1 ratio of Ba^{2+} ions to S^{2-} ions is represented in the formula unit (circled here) and in the formula of the compound (BaS).

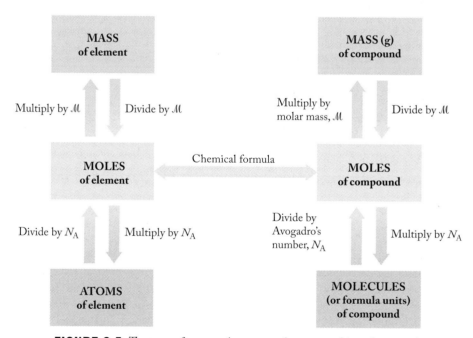

FIGURE 3.5 The mass of a pure substance can be converted into the equivalent number of moles or number of particles (atoms, ions, or molecules) and vice versa.

SAMPLE EXERCISE 3.4 **Calculating the Molar Mass of a Compound**

Calculate the molar masses of (a) H_2O and (b) H_2SO_4.

COLLECT AND ORGANIZE The molar mass of a molecular compound is the sum of the molar masses of the elements in its molecular formula, each multiplied by the number of atoms of that element in one molecule of the compound. The molar masses of elements are numerically the same as the atomic masses given in the periodic table but are expressed in grams/mol.

ANALYZE
a. One mole of H_2O contains 2 moles of H atoms and 1 mole of O atoms.
b. One mole of H_2SO_4 contains 2 moles of H atoms, 1 mole of S atoms, and 4 moles of O atoms.

SOLVE From the atomic masses given in the periodic table, we calculate the following:

a. the mass of 1 mole of H_2O is the sum of the masses of 2 moles of hydrogen atoms and 1 mole of oxygen atoms:
$2(1.008\text{ g}) + 1(16.00\text{ g}) = 18.02\text{ g}$
b. the mass of 1 mole of H_2SO_4 is the sum of the masses of 2 moles of H atoms, 1 mole of S atoms, and 4 moles of O atoms:
$2(1.008\text{ g}) + 1(32.07\text{ g}) + 4(16.00\text{ g}) = 98.09\text{ g}$

THINK ABOUT IT The molar mass of a compound is a fundamental and identifying characteristic of the molecules of that compound. Large molecules have large molar masses.

Practice Exercise Green plants take in water (H_2O) and carbon dioxide (CO_2) and produce glucose ($C_6H_{12}O_6$) and oxygen gas (O_2). Calculate the molar masses of carbon dioxide, oxygen gas, and glucose.

(Answers to Practice Exercises are in the back of the book.)

Moles and Chemical Equations

The concepts of mole and molar mass provide three more interpretations of the reaction between SO_3 and H_2O given in Equation 3.1:

1. One *mole* of SO_3 reacts with 1 *mole* of H_2O, producing 1 *mole* of H_2SO_4.
2. 80.07 *grams* of SO_3 reacts with 18.02 *grams* of H_2O, producing 98.09 *grams* of H_2SO_4.
3. 6.022×10^{23} *molecules* of SO_3 react with 6.022×10^{23} *molecules* of H_2O, forming 6.022×10^{23} *molecules* of H_2SO_4; this means that, in terms of lowest whole-number ratios, 1 *molecule* of SO_3 reacts with 1 *molecule* of H_2O, forming 1 *molecule* of H_2SO_4.

The several interpretations of Equation 3.1 are not limited to considering 1 mole of SO_3 reacting with 1 mole of H_2O. In general, Equation 3.1 tells us that *any* number of moles (*x* moles) of SO_3 will react with an equal number of moles (*x* moles) of water to produce *x* moles of sulfuric acid. This interpretation is valid because the mole ratio of SO_3 to H_2O to H_2SO_4 in the balanced equation is 1:1:1. The mole ratio is specific for a given equation and will be different when different substances combine. For the situation described in Equation 3.1, we can determine either how much of one reactant is needed to react with any quantity of another, or how much product could be made from any quantity of reactants by multiplying *x* by the appropriate molar mass:

$$SO_3(g) \quad + \quad H_2O(g) \quad \rightarrow \quad H_2SO_4(\ell)$$

	1 molecule	+	1 molecule	→	1 molecule
Mole ratios:	1 mol	+	1 mol	→	1 mol
Mass ratios:	80.07 g	+	18.02 g	→	98.09 g
General case (moles):	x mol	+	x mol	→	x mol
General case (masses):	x(80.07 g)	+	x(18.02 g)	→	x(98.09 g)

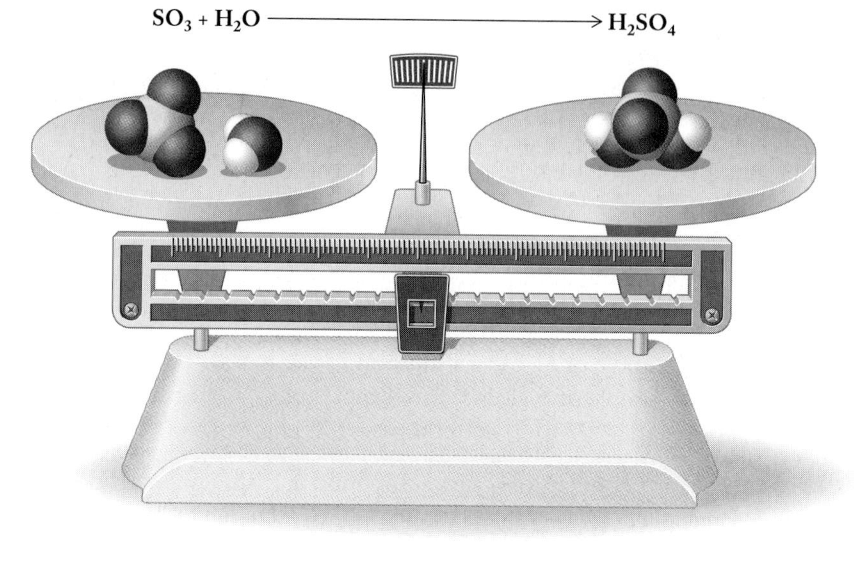

$$SO_3 + H_2O \longrightarrow H_2SO_4$$

FIGURE 3.6 The law of conservation of mass states that the total mass of reactants equals the total mass of products in a chemical reaction. The combined mass of SO_3 and H_2O equals the mass of the H_2SO_4 that forms, as represented by the molecular models on the balance. The numbers of each kind of atom are the same on the left and right sides of the balance.

The quantitative relation between reactants and products involved in a chemical reaction is called its **stoichiometry**. In a reaction where all reactants are completely converted into products, the sum of the masses of the reactants equals the sum of the masses of the products (Figure 3.6). This fact illustrates a fundamental relation known as the **law of conservation of mass**, which applies to all chemical reactions. This law "works" for two reasons: (1) the total number of atoms (and hence moles) of each element on the left-hand side of the reaction arrow must match the total number of atoms (or moles) of that element in the products on the right-hand side, and (2) the identity of atoms does not change in a chemical reaction. Conservation of mass is a key concept applied in writing balanced chemical equations.

3.3 Writing Balanced Chemical Equations

A *balanced* chemical equation has the same number of atoms of each type on both sides of the equation. In this section we look at chemical reactions in the Earth's early atmosphere to learn how to balance chemical equations. Earth's early atmosphere contained nonmetal oxides in addition to SO_3 that form strong acids when they combine with water vapor. Among these binary oxides was dinitrogen pentoxide (N_2O_5). When N_2O_5 combines with water vapor, it produces nitric acid (HNO_3). Putting this information into equation form we have

$$N_2O_5(g) + H_2O(g) \rightarrow HNO_3(\ell) \tag{3.2}$$

Counting up the number of atoms of each type on both sides of the equation reveals that this equation is not balanced:

[2 N atoms + (5 + 1 = 6) O atoms + 2 H atoms ≠ 1 H atom + 1 N atom + 3 O atoms]

Stoichiometry refers to the quantitative relation between the quantities of reactants and products in a chemical reaction.

The **law of conservation of mass** states that the sum of the masses of the reactants in a chemical reaction is equal to the sum of the masses of the products.

There are two nitrogen atoms on the left side but only one on the right, six oxygen atoms on the left side but only three on the right, and two hydrogen atoms on the left side but only one on the right. We have the correct formulas for the substances involved in the reaction, and the formulas never change, so we have to balance the equation by changing the number of molecules of each type that appear in the reaction. We do this by adjusting the *coefficients* in front of the formulas. In the unbalanced equation (3.2), the coefficients are understood to be 1, as the formula itself stands for one molecule (or 1 mole) of the substance.

There are many ways to balance chemical equations. Here is a five-step approach that you may find useful.

1. *Write a preliminary equation using the correct formulas for reactants and products*: Be sure you understand what reaction is occurring. Here you are given the reaction of dinitrogen pentoxide (N_2O_5) with water vapor (H_2O) to make nitric acid (HNO_3). Always check that you have the correct formulas for reactants and products. Then write an equation, including the physical state of the substances involved:

$$N_2O_5(g) + H_2O(g) \rightarrow HNO_3(\ell)$$

Check whether the preliminary equation is balanced by adding up the different types of atoms on each side of the reaction arrow.

2 N atoms + 6 O atoms + 2 H atoms $\neq$ 1 H atom + 1 N atom + 3 O atoms

2. *Assign one of the compounds a coefficient of 1*: It usually works best to pick the compound with the largest number of different elements in it, or simply the one with the most atoms. In this case, let's keep the coefficient of HNO_3 equal to 1.

$$\_\_ N_2O_5(g) + \_\_ H_2O(g) \rightarrow 1\,HNO_3(\ell)$$

3. *Choose coefficients for the other substances so that the numbers of atoms of each element on each side of the equation are equal*: Never change the subscripts in the formulas; work only with the coefficients. (If you change the subscripts, you change the identity of the substances. Once the correct formulas for reactants and products are identified in step 1, you must not change them.)

▶❚❚ **CHEMTOUR** Balancing Equations

$$\tfrac{1}{2} N_2O_5(g) + \tfrac{1}{2} H_2O(g) \rightarrow 1\ HNO_3(\ell)$$

Notice that these coefficients result in [$\tfrac{1}{2}(2) =$] 1 N atom, [$\tfrac{1}{2}(2) =$] 1 H atom, and [$\tfrac{1}{2}(5) + \tfrac{1}{2}(1) =$] 3 O atoms on the left, and 1 N atom, 1 H atom and 3 O atoms on the right. Because the idea of "$\tfrac{1}{2}$ a molecule of water" does not make physical sense, the next step is necessary.

4. *Eliminate any fractional coefficients* by multiplying the entire equation by the smallest integer that converts each fraction into a whole number. Eliminate the 1's that arise, because they are not necessary (the formula by itself always has an implied coefficient of 1). For this reaction, multiply all the coefficients by a factor of 2 (first line below), and simplify the coefficients (second line below):

$$2 \times \tfrac{1}{2} N_2O_5(g) + 2 \times \tfrac{1}{2} [H_2O(g)] \rightarrow 2 \times 1\ HNO_3(\ell)$$

$$N_2O_5(g) + H_2O(g) \rightarrow 2\ HNO_3(\ell)$$

5. *Check that the result is balanced*: Add up the numbers of atoms on both sides:

$$N_2O_5(g)\ +\ H_2O(g)\ \rightarrow\ 2\ HNO_3(\ell)$$

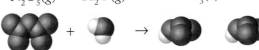

Reactants   Products

2 N atoms + (5 + 1 =) 6 O atoms + 2 H atoms = 2 H atoms + 2 N atoms + 6 O atoms

The equation is balanced.

Hydrolysis is the reaction of water with another material. The hydrolysis of nonmetal oxides produces acids.

Nonmetal oxides are **acid anhydrides**. They react with water to produce acids.

The reaction of N_2O_5 with water is an example of a **hydrolysis** reaction. In all hydrolysis reactions, water reacts with another substance. An important class of hydrolysis reactions is the hydrolysis of nonmetal oxides (such as N_2O_5), which results in acidic solutions. Hydrolysis reactions like this one were the source of the acidic environment on early Earth described in Section 3.1. Because of their behavior, nonmetal oxides are called **acid anhydrides**—literally, acids without water (*a-*, *an-*, "not"; *hydr-*, "water"). To generate an acid, add the corresponding nonmetal oxide to water.

Let's practice with a related reaction. Dinitrogen pentoxide is formed when nitrogen gas reacts with oxygen gas. To write a balanced equation for this reaction, we follow the same steps used earlier to balance the hydrolysis reaction:

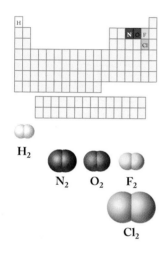

H_2

N_2 O_2 F_2

Cl_2

1. *Write the correct formulas for reactants and products*: For this reaction we need to know that nitrogen and oxygen gases exist in nature as diatomic molecules, so their correct formulas are N_2 and O_2. Indeed, all of the gaseous nonmetallic elements except for the noble gases in group 18 exist in nature as diatomic molecules: H_2, N_2, O_2, F_2, and Cl_2. When these substances are involved in chemical reactions, they must be written in the form of diatomic molecules.

$$N_2(g) + O_2(g) \rightarrow N_2O_5(g)$$

$$2\,N \text{ atoms} + 2\,O \text{ atoms} \neq 2\,N \text{ atoms} + 5\,O \text{ atoms}$$

The equation is not yet balanced.

2. *Assign the molecule with the most elements a coefficient of 1*:

$$\underline{\quad}\,N_2(g) + \underline{\quad}\,O_2(g) \rightarrow 1\,N_2O_5(g)$$

3. *Choose coefficients for the other substances to equalize the numbers of atoms on both sides*:

$$1\,N_2(g) + \tfrac{5}{2}\,O_2(g) \rightarrow N_2O_5(g)$$

4. *Eliminate fractional coefficients*, in this case by multiplying through by 2:

$$2(1)\,N_2(g) + 2(\tfrac{5}{2})\,O_2(g) \rightarrow 2\,N_2O_5(g)$$

5. Count the atoms on both sides to *check that the equation is now balanced*:

$$2\,N_2(g) + 5\,O_2(g) \rightarrow 2\,N_2O_5(g)$$

$$4\,N + 10\,O \rightarrow 4\,N + 10\,O$$

Two molecules of dinitrogen pentoxide (N_2O_5) are formed in a reaction between 2 molecules of nitrogen (N_2) and 5 molecules of oxygen (O_2).

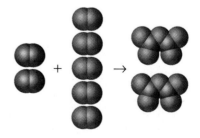

The equation is now balanced.

| | | |

SAMPLE EXERCISE 3.5   **Balancing a Chemical Equation**

Write a balanced chemical equation for the reaction between SO_2 and O_2 that forms SO_3. This reaction, which describes one of the origins of acid rain, occurs when fuels containing sulfur are burned.

COLLECT AND ORGANIZE We are given the correct chemical formulas for the reactants and products.

ANALYZE To write a balanced chemical equation we must add coefficients as needed to make the numbers of atoms of each element on the left and on the right sides of the reaction arrows equal. We determine these coefficients using the method developed for balancing equations.

SOLVE

(1) $SO_2(g) + O_2(g) \rightarrow SO_3(g)$
    1 S atom + 4 O atoms $\neq$ 1 S atom + 3 O atoms

(2) $\_\_ SO_2(g) + \_\_ O_2(g) \rightarrow 1\, SO_3(g)$

(3) $1\, SO_2(g) + \frac{1}{2} O_2(g) \rightarrow 1\, SO_3(g)$

(4) $2(1)\, SO_2(g) + 2\left(\frac{1}{2}\right) O_2(g) \rightarrow 2(1)\, SO_3(g)$

(5) $2\, SO_2(g) + O_2(g) \rightarrow 2\, SO_3(g)$
    2 S atoms + 6 O atoms = 2 S atoms + 6 O atoms

The equation is balanced.

THINK ABOUT IT Oxygen is a very reactive element. It forms two different binary oxides with sulfur and several different binary oxides with nitrogen.

Practice Exercise Balance the chemical equations for two reactions:

a. the reaction of elemental phosphorus [$P_4(s)$] with oxygen to make $P_4O_{10}(s)$
b. the hydrolysis of $P_4O_{10}(s)$ to produce phosphoric acid, $H_3PO_4(\ell)$.

(Answers to Practice Exercises are in the back of the book.)

Nitric, sulfuric, and other acids in Earth's early atmosphere eventually fell to the surface of the planet. This process was accelerated as the atmosphere cooled, and water vapor condensed as liquid water and fell to Earth as rain. Sulfuric acid (H_2SO_4) and nitric acid (HNO_3) readily dissolve in water, forming aqueous solutions that we designate $H_2SO_4(aq)$ and $HNO_3(aq)$ (acid rain). Reactions of these and other acidic compounds with the minerals of Earth's crust dramatically altered the chemical composition of the crust, as we shall see in Chapter 4.

3.4 Combustion Reactions

In addition to carbon dioxide, the Earth's early atmosphere contained significant concentrations of two other carbon-containing compounds: carbon monoxide (CO) and methane (CH_4). As we have noted, molecular oxygen (O_2) was not present. Had it been, it would have been consumed in chemical reactions with CO and CH_4, because the most stable form of carbon in the presence of O_2 is CO_2, not CO or CH_4. Let's complete and balance chemical equations describing the reactions between O_2 and CO and between O_2 and CH_4.

Starting with CO, step 1 (in the method given in Section 3.3) is to write an unbalanced equation with the two reactants and the product:

(1) $$CO(g) + O_2(g) \rightarrow CO_2(g)$$

Combining the next three steps we get,

(2)–(4) $$2 \times [CO(g) + \tfrac{1}{2} O_2(g) \rightarrow 1\, CO_2(g)]$$

Multiplying by 2:

(5) $$2\, CO(g) + O_2(g) \rightarrow 2\, CO_2(g) \hspace{3em} (3.3)$$

Now the equation is balanced.

Methane also reacts with elemental oxygen. We can anticipate that the carbon in methane becomes carbon dioxide during combustion when a sufficient supply of oxygen is available. An inadequate supply of oxygen during the combustion of carbon-containing fuels can result in the production of carbon monoxide, or even particles of unburned carbon, the principal ingredient in soot. In most combustion flames at least some toxic carbon monoxide is produced; inadequate venting of exhaust gases from heaters, furnaces, and engines has led to many tragic deaths.

The hydrogen in methane combines with oxygen to form water. To write a chemical equation for this reaction, we start with what we know about the identities of the reactants and products:

(1) $$CH_4(g) + O_2(g) \rightarrow H_2O(\ell) + CO_2(g)$$

Quick inspection of the equation shows that the number of moles of carbon in this equation is balanced, but those of hydrogen and oxygen are not.

Let's keep the coefficient of the compound with the most atoms, CH_4, 1:

(2) $$1\, CH_4(g) + \underline{}\, O_2(g) \rightarrow \underline{}\, H_2O(\ell) + 1\, CO_2(g)$$

To balance the number of carbon (1) and hydrogen (4) atoms, we need a coefficient of 1 in front of CO_2 and 2 in front of H_2O:

(3a) $$1\, CH_4(g) + O_2(g) \rightarrow 2\, H_2O(\ell) + 1\, CO_2(g)$$

Finally we balance the oxygen atoms:

(3b) $$1\, CH_4(g) + 2\, O_2(g) \rightarrow 2\, H_2O(\ell) + 1\, CO_2(g)$$

Since there are no fractional coefficients to get rid of, we can skip step 4. The balanced equation is thus

(5) $$CH_4(g) + 2\, O_2(g) \rightarrow 2\, H_2O(\ell) + CO_2(g) \hspace{3em} (3.4)$$

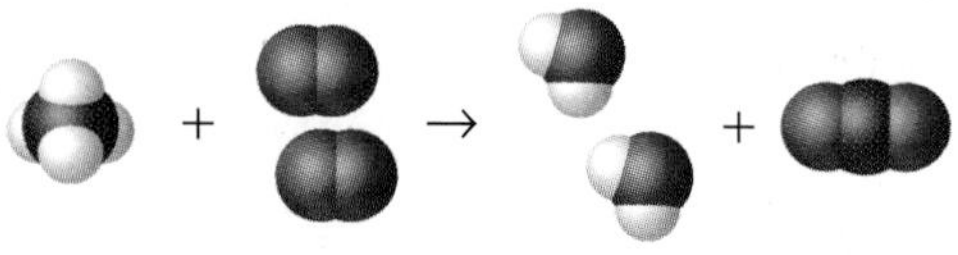

The approach just taken—balancing the moles of C, then H, then O—is useful for many other reactions in which oxygen reacts with compounds containing only carbon and hydrogen. Compounds composed of only carbon and hydrogen are called **hydrocarbons**. In our industrialized world, these reactions are associated with burning natural gas and petroleum-based fuels to produce energy. Equation 3.4 illustrates the principal chemical reaction that takes place when natural gas is burned to heat buildings and to warm water. It is an example of an important class of chemical reactions known as **combustion reactions**. When the combustion of a hydrocarbon is complete, the only products are carbon dioxide and water vapor. The following Sample Exercise provides additional examples of completing and balancing equations describing the combustion of hydrocarbons.

Hydrocarbons are molecular compounds composed of only hydrogen and carbon. Hydrocarbons are a class of organic compounds.

Combustion reactions occur between oxygen and another element or compound. When the other compound is a hydrocarbon, the products of complete combustion are carbon dioxide and water vapor.

SAMPLE EXERCISE 3.6 **Completing and Balancing a Combustion Reaction**

Methane (CH_4) is the principal ingredient in natural gas, but there are also significant concentrations of ethane (C_2H_6) and propane (C_3H_8) in most natural gas samples. Write and balance the equation describing the complete combustion of C_2H_6.

COLLECT AND ORGANIZE Ethane is a hydrocarbon, and *combustion* means that oxygen is the other reactant. The products of complete combustion of a hydrocarbon with oxygen are carbon dioxide and water.

ANALYZE Let's begin by writing the reactants and products of the reaction:

(1) $$C_2H_6(g) + O_2(g) \rightarrow CO_2(g) + H_2O(\ell)$$

SOLVE To write a balanced chemical reaction describing this reaction, balance the numbers of atoms of the three elements that make up the reactant and products: C, H and O, in that order. Start by giving ethane (the compound containing the most atoms) a coefficient of 1:

(2) $$1\, C_2H_6(g) + \text{\_\_}\, O_2(g) \rightarrow \text{\_\_}\, CO_2(g) + \text{\_\_}\, H_2O(\ell)$$

Balance the carbon atoms by giving CO_2 in the product a coefficient of 2:

(3a) $$1\, C_2H_6(g) + \text{\_\_}\, O_2(g) \rightarrow 2\, CO_2(g) + \text{\_\_}\, H_2O(\ell)$$

Balance the hydrogen atoms by giving H_2O a coefficient of 3:

(3b) $$1\, C_2H_6(g) + \text{\_\_}\, O_2(g) \rightarrow 2\, CO_2(g) + 3\, H_2O(\ell)$$

Balance the oxygen atoms by giving O_2 a coefficient $\frac{7}{2}$:

(3c) $$1\, C_2H_6(g) + \frac{7}{2}\, O_2(g) \rightarrow 2\, CO_2(g) + 3\, H_2O(\ell)$$

Eliminate fractional coefficients by multiplying through by 2:

(4) $$2\, C_2H_6(g) + 7\, O_2(g) \rightarrow 4\, CO_2(g) + 6\, H_2O(\ell)$$

Check by counting up all the atoms on both sides of the equation:

(5) 4 C atoms + 12 H atoms + 14 O atoms
$$= 4 \text{ C atoms} + 12 \text{ H atoms} + 14 \text{ O atoms}$$

This equation is balanced.

THINK ABOUT IT Combustion reactions of hydrocarbons are a major source of energy in modern society. The volatile hydrocarbons such as methane (CH_4), ethane (C_2H_6), and propane (C_3H_8) are the constituents of natural gas; butane (C_4H_{10}) is used in disposable lighters; less volatile hydrocarbons make up liquid fuels including gasoline, kerosene, and diesel fuel used to power all manner of vehicles.

Practice Exercise Complete and balance a chemical equation describing the complete combustion of propane (C_3H_8).

(Answers to Practice Exercises are in the back of the book.)

3.5 Stoichiometric Calculations and the Carbon Cycle

The composition of Earth's atmosphere underwent a major change beginning about 600 million years ago with the evolution of green plants and the onset of *photosynthesis*. Equation 3.5 describes this overall chemical reaction, which takes place in several steps and is driven by the energy contained in visible light. Its products are molecular oxygen and the sugar glucose:

$$6\,CO_2(g) + 6\,H_2O(\ell) \rightarrow C_6H_{12}O_6(aq) + 6\,O_2(g) \tag{3.5}$$

The reverse reaction is

$$C_6H_{12}O_6(aq) + 6\,O_2(g) \rightarrow 6\,CO_2(g) + 6\,H_2O(\ell) \tag{3.6}$$

In biological systems the reaction described by Equation 3.6, called *respiration*, is the major source of energy for all living things on Earth.

Photosynthesis and respiration are nearly in balance in Earth's biosphere. They are key chemical reactions in the *carbon cycle*, illustrated in Figure 3.7. If photosynthesis and respiration were exactly in balance, no net change would have taken place in the concentrations of atmospheric carbon dioxide or oxygen in the past 600 million years. However, about 0.01% of the decaying mass of plants and animals (called detritus) is incorporated into sediments and soil when they die and is thereby shielded from exposure to oxygen and is not converted back into CO_2. This proportion may not seem like much, but over hundreds of millions of years it has added up to the burial of about 10^{20} kg of carbon and the removal of an equivalent amount of carbon dioxide from the atmosphere. About 10^{15} kg of this buried carbon is in the form of fossil fuels: coal, petroleum, and natural gas.

As a result of human activity and the combustion of fossil fuels, the natural balance that helped limit the concentration of CO_2 in the atmosphere is being altered. Annually about 6.8 trillion (6.8×10^{12}) kilograms of carbon is reintroduced to the atmosphere as CO_2 as a result of fossil-fuel combustion. Deforestation adds another 2×10^{12} kg/yr. The effects of these additions on global climate have been the subject of considerable debate, and we will examine them again in Chapter 9.

If the combustion of fossil fuel adds 6.8×10^{12} kg of carbon into the atmosphere each year as CO_2, what is the mass of the carbon dioxide that is added? The mass must be more than 6.8×10^{12} kg, because this amount does not take into account the mass of oxygen in CO_2. The following equation describes the combustion of carbon to carbon dioxide in the presence of excess O_2:

$$C(s) + O_2(g) \rightarrow CO_2(g)$$

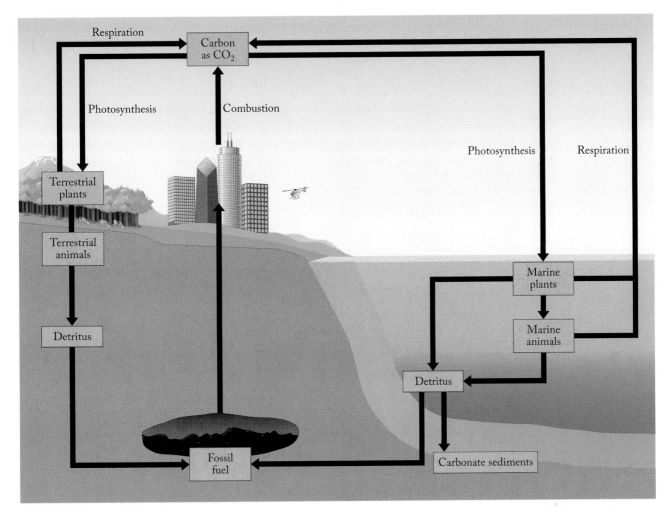

 placeholder labels: Respiration, Carbon as CO$_2$, Photosynthesis, Combustion, Photosynthesis, Respiration, Terrestrial plants, Terrestrial animals, Detritus, Marine plants, Marine animals, Detritus, Fossil fuel, Carbonate sediments

FIGURE 3.7 Carbon is cycled through the environment by many processes. Green plants incorporate CO$_2$ into their biomass. Some of the plant biomass becomes the biomass of animals. As plants and animals respire, they release CO$_2$ back into the environment. When they die, the decay of their tissues releases most of their carbon content as CO$_2$, but about 0.01% is incorporated into carbonate minerals and deposits of coal, petroleum, and natural gas (fossil fuels). Mining and the combustion of fossil fuels for human use is shifting the natural equilibrium that has controlled the concentration of CO$_2$ in the atmosphere.

▶❙❙ **CHEMTOUR** Carbon Cycle

This balanced equation tells us that 1 mole of carbon combines with 1 mole of O$_2$, forming 1 mole of CO$_2$. To make use of this balanced equation to determine the mass of CO$_2$ released, we first convert the mass of carbon into an equivalent number of moles of carbon:

$$6.8 \times 10^{12} \text{ kg C} \times \frac{1000 \text{ g}}{1 \text{ kg}} \times \frac{1 \text{ mol C}}{12.01 \text{ g C}} = 5.7 \times 10^{14} \text{ mol C}$$

Because the mole ratio of CO$_2$ to C is 1:1, the amount of CO$_2$ produced is also 5.7×10^{14} moles:

$$5.7 \times 10^{14} \text{ mol C} \times \frac{1 \text{ mol CO}_2}{1 \text{ mol C}} = 5.7 \times 10^{14} \text{ mol CO}_2$$

To convert from moles of CO$_2$ to the mass of CO$_2$ we need to calculate the molar mass of CO$_2$:

$$\mathscr{M}_{\text{CO}_2} = 1 \text{ mol C} \times \frac{12.01 \text{ g C}}{1 \text{ mol C}} + 2 \text{ mol O} \times \frac{16.00 \text{ g O}}{1 \text{ mol O}} = 44.01 \text{ g/mol}$$

And then we multiply the number of moles of CO$_2$ by its molar mass:

$$5.7 \times 10^{14} \text{ mol CO}_2 \times \frac{44.01 \text{ g}}{\text{mol}} = 2.5 \times 10^{16} \text{ g CO}_2$$

In order to compare the above result directly with the mass of carbon given (6.8×10^{12} kg C), we now convert it into kilograms:

$$2.5 \times 10^{16} \text{ g CO}_2 \times \frac{1 \text{ kg}}{1000 \text{ g}} = 2.5 \times 10^{13} \text{ kg CO}_2$$

The procedure we just followed can be applied to determining the mass of any substance (reactant or product) involved in any chemical reaction if we know (1) the mass of another substance in the reaction, and (2) their *stoichiometric relation*, that is, the mole ratio between the two substances in the balanced chemical equation.

SAMPLE EXERCISE 3.7 **Calculating the Mass of a Product from the Mass of a Reactant**

In 2005, electric power plants in the United States consumed about 1.0×10^{11} kg of natural gas. Natural gas is mostly methane, so we can closely approximate the chemistry of the chemical reaction generating the energy using the following chemical equation:

$$CH_4(g) + 2 O_2(g) \rightarrow 2 H_2O(g) + CO_2(g)$$

How many kilograms of CO_2 were released into the atmosphere from these power plants in 2005?

COLLECT AND ORGANIZE This is a stoichiometry problem. In the balanced equation, 1 mole of carbon dioxide is produced for every mole of methane consumed. We need to convert the mass of CH_4 into the number of moles of CH_4, which in this reaction is equal to the number of moles of CO_2 produced, and then convert the number of moles of CO_2 into kilograms of CO_2.

ANALYZE We check the chemical equation to confirm that it is balanced as written—and it is. To convert between grams and moles of CH_4 and CO_2, we need to calculate the molar masses of these compounds using the atomic masses of carbon, hydrogen, and oxygen found in the periodic table:

$$\mathcal{M}_{CH_4} = 1 \text{ mol C} \times \frac{12.01 \text{ g}}{\text{mol}} + 4 \text{ mol H} \times \frac{1.008 \text{ g}}{\text{mol}} = 16.04 \text{ g/mol}$$

$$\mathcal{M}_{CO_2} = 1 \text{ mol C} \times \frac{12.01 \text{ g}}{\text{mol}} + 2 \text{ mol O} \times \frac{16.00 \text{ g}}{\text{mol}} = 44.01 \text{ g/mol}$$

SOLVE

1. Convert the mass of CH_4 into an equivalent number of moles:

$$1.0 \times 10^{11} \text{ kg CH}_4 \times \frac{1000 \text{ g}}{1 \text{ kg}} \times \frac{1 \text{ mol CH}_4}{16.04 \text{ g CH}_4} = 6.2 \times 10^{12} \text{ mol CH}_4$$

2. Convert the moles of CH_4 into moles of CO_2 using the coefficients from the balanced chemical equation:

$$6.2 \times 10^{12} \text{ mol CH}_4 \times \frac{1 \text{ mol CO}_2}{1 \text{ mol CH}_4} = 6.2 \times 10^{12} \text{ mol CO}_2$$

3. Convert the number of moles of CO_2 into mass of CO_2:

$$6.2 \times 10^{12} \text{ mol CO}_2 \times \frac{44.01 \text{ g CO}_2}{1 \text{ mol CO}_2} \times \frac{1 \text{ kg}}{1000 \text{ g}} = 2.7 \times 10^{11} \text{ kg CO}_2$$

THINK ABOUT IT Once you are confident in solving stoichiometry problems using this method, you can arrange the entire problem in one sequence. To illustrate this idea, notice that the answer to the question posed in each step in this Sample Exercise is the starting point for the next step. Therefore we can combine the calculations of all three steps into a single calculation:

$$1.0 \times 10^{11} \text{ kg } CH_4 \times \frac{1000 \text{ g}}{1 \text{ kg}} \times \frac{1 \text{ mol } CH_4}{16.04 \text{ g } CH_4} \times \frac{1 \text{ mol } CO_2}{1 \text{ mol } CH_4}$$

$$\times \frac{44.01 \text{ g } CO_2}{1 \text{ mol } CO_2} \times \frac{1 \text{ kg}}{1000 \text{ g}} = 2.7 \times 10^{11} \text{ kg } CO_2$$

This quantity of CO_2 represents about 1% of the global CO_2 emissions from all sources. The combustion of all fossil fuels in the United States accounted for 27% of global CO_2 emissions in 2005.

Practice Exercise Disposable lighters burn butane (C_4H_{10}) and produce CO_2 and H_2O. Balance the chemical equation for combustion of butane, and determine how many grams of CO_2 are produced by burning 1.00 g of C_4H_{10}.

(Answers to Practice Exercises are in the back of the book.)

3.6 Determining Empirical Formulas from Percent Composition

Some regions of Earth's crust are rich in minerals that have had a significant effect on human development and economic growth. Since the beginning of the Industrial Revolution, access to geological deposits rich in iron oxides has been key to the development of national and ultimately global economies. The more accessible a particular iron-containing mineral is and the higher its iron content, the more valuable it is. Higher iron content in this case means a greater mass of iron in a given mass of ore. We typically express this content in terms of **percent composition**, that is, the percentage of the total mass of a compound that each of its constituent elements contributes. Let's consider two iron-containing minerals. One is called wustite and has the chemical formula FeO. The other is hematite and has the formula Fe_2O_3. Which of the two has the higher iron content and what is that content, expressed on a mass basis?

A simple way to answer the first question is to examine the Fe-to-O mole ratios in the two minerals. In FeO, that ratio is simply 1:1. In Fe_2O_3, the ratio is 2:3, or 1:1.5. There is a higher proportion of O and a lower proportion of Fe in Fe_2O_3; so FeO must have the higher iron content. Now let's calculate the iron content of FeO to determine just how rich the ore is in iron. Suppose we have 1 mole of FeO. The molar mass of FeO is the sum of the molar masses of Fe and O:

▶‖ **CHEMTOUR** Percent Composition

$$\mathcal{M}_{FeO} = 1 \text{ mol Fe} \times \frac{55.85 \text{ g Fe}}{\text{mol Fe}} + 1 \text{ mol O} \times \frac{16.00 \text{ g O}}{\text{mol O}} = 71.85 \text{ g/mol}$$

Of this 71.85 g/mol, Fe accounts for 55.85 g. Therefore, the iron content of FeO is

$$\frac{\text{mass of Fe}}{\text{total mass}} = \frac{55.85 \text{ g Fe}}{71.85 \text{ g FeO}} = 0.7773 \text{ or } 77.73\%$$

It follows that the O content is

$$100 - 77.73 = 22.27\%$$

Percent composition is the percentage by mass of each element in a compound; it can also refer to the percentage by mass of each component in a mixture.

These values give the percentage by mass of FeO that is Fe and that which is O. Together, they describe the percent composition of FeO. Note that they sum to give 100%. The following Sample Exercise illustrates how to calculate the percent composition of a three-element compound.

SAMPLE EXERCISE 3.8 **Calculating Percent Composition from a Chemical Formula**

What is the percent composition of the mineral forsterite, Mg_2SiO_4?

COLLECT AND ORGANIZE We are given the formula of forsterite and asked to calculate its percent composition, that is, the percentage of its mass that each of its component elements contributes. We will need the molar masses of each of these elements to calculate the molar mass of Mg_2SiO_4 and each of their contributions to it. Molar masses of elements are numerically the same as their average atomic masses in the periodic table.

ANALYZE A mole of forsterite contains 2 moles of Mg, 1 mole of Si, and 4 moles of O. Therefore, the molar mass of the compound is the sum of twice the molar mass of Mg, the molar mass of Si, and four times the molar mass of O. The percentage of the molar mass that each element contributes can be calculated by dividing the mass of each element in 1 mole of the compound by the molar mass.

SOLVE The molar mass of Mg_2SiO_4 is

$$2 \text{ mol Mg} \times \frac{24.31 \text{ g Mg}}{\text{mol Mg}} + 1 \text{ mol Si} \times \frac{28.09 \text{ g Si}}{\text{mol Si}}$$

$$+ 4 \text{ mol O} \times \frac{16.00 \text{ g O}}{\text{mol O}} = 140.71 \text{ g/mol}$$

The percent composition values of the three elements are:

$$\% \text{ Mg} = \frac{48.62 \text{ g Mg}}{140.71 \text{ g}} = 34.55\% \text{ Mg}$$

$$\% \text{ Si} = \frac{28.09 \text{ g Si}}{140.71 \text{ g}} = 19.96\% \text{ Si}$$

$$\% \text{ O} = \frac{64.00 \text{ g O}}{140.71 \text{ g}} = 45.48\% \text{ O}$$

THINK ABOUT IT The percentage of the mass of forsterite that is Mg is nearly twice the percentage that is Si, which makes sense because the molar masses of Mg and Si are not that different and there are 2 moles of Mg for every 1 mole of Si. It also makes sense that oxygen accounts for nearly half the mass of forsterite because, even though O has the smallest mass of the three elements, there are 4 moles of O for every 3 of the other elements. Also note that the three percentages sum to 99.99%, so we have accounted for the total mass of the compound.

Practice Exercise Earth's crust contains a mineral called enstatite, which has the chemical formula $MgSiO_3$. Determine the percent composition of $MgSiO_3$.

(Answers to Practice Exercises are in the back of the book.)

We typically know the percent composition of a substance from the results of an elemental analysis. To determine a formula from such data, we start by converting percent composition values to actual masses by assuming that we have exactly 100.00 grams each of wustite (77.73% Fe) and hematite (69.94% Fe). Because this is not a measured number, it is considered a definition and hence exact. The percent composition values are multiplied by 100.00 grams to obtain:

Wustite: 77.73% Fe in 100.00 g sample = 77.73 g Fe

Hematite: 69.94% Fe in 100.00 g sample = 69.94 g Fe

The only other element in these minerals is oxygen, and we can use the principle of conservation of mass to calculate the oxygen content by determining the difference between the total mass and the mass of Fe present in each sample:

Wustite: 100.00 g − 77.73 g Fe = 22.27 g O

Hematite: 100.00 g − 69.94 g Fe = 30.06 g O

Next we can determine the number of moles of Fe and O in these two quantities of Fe and O by dividing the mass in grams of each element in the sample by the molar mass of that element given in the periodic table:

Wustite: $\dfrac{77.73 \text{ g Fe}}{55.85 \text{ g/mol}} = 1.392 \text{ mol Fe}$ $\dfrac{22.27 \text{ g O}}{16.00 \text{ g/mol}} = 1.392 \text{ mol O}$

Hematite: $\dfrac{69.94 \text{ g Fe}}{55.85 \text{ g/mol}} = 1.252 \text{ mol Fe}$ $\dfrac{30.06 \text{ g O}}{16.00 \text{ g/mol}} = 1.879 \text{ mol O}$

Focusing first on the results for wustite, note that the mole ratio of iron to oxygen is 1.392 : 1.392, or 1:1. This means that in 1 mole (or 1 molecule) of the compound, there is 1 mole of iron atoms (or 1 Fe atom) for every 1 mole of oxygen atoms (or every O atom). Thus the chemical formula of wustite is FeO. The iron-to-oxygen mole ratio of hematite is 1.252 : 1.879. To convert this ratio to one of small whole numbers, we first divide the number of moles of each element by the number of moles of the element present in the smallest amount. This guarantees that 1 will be one of the two numbers in the ratio and that the other number will be greater than 1:

Hematite: $\dfrac{1.252 \text{ mol Fe}}{1.252} = 1.000 \text{ mol Fe}$ $\dfrac{1.879 \text{ mol O}}{1.252} = 1.501 \text{ mol O}$

This step gives us a mole ratio of about 1:1.5. To turn 1:1.5 into a ratio in which *both* numbers are integers, multiply the ratio by a factor that results in two whole numbers. Multiplying 1.5 ($= \frac{3}{2}$) by 2 yields the integer 3; the other term in the ratio, 1, must also be multiplied by 2. This produces an Fe-to-O ratio of 2:3. Thus the chemical formula of hematite is Fe_2O_3.

The two iron oxide formulas we just determined are examples of empirical formulas, which we discussed in Chapter 2 as the typical formulas of salts. (*Empirical*, which is related to the word *experiment*, literally means "derived from experimental data.") Both FeO and Fe_2O_3 may be treated in this context as ionic compounds made up of extended three-dimensional arrays of Fe^{2+} and O^{2-} ions (in FeO) and Fe^{3+} and O^{2-} ions (in Fe_2O_3). **Empirical formulas** represent the proportions of the ions in the formula units. In wustite, for example, Fe^{2+} and O^{2-} ions are arranged in what is often called a *halite structure*, because it is the same as

Empirical formulas represent the proportions of the ions in the formula units.

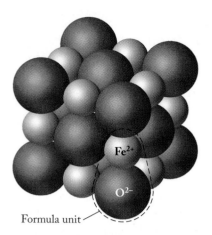

FIGURE 3.8 The Fe^{2+} and O^{2-} ions in FeO adopt the same three-dimensional structure as the Na^+ and Cl^- ions in NaCl (see Figure 2.13); compare this structure also with that in Figure 3.4.

the structure that Na^+ and Cl^- ions adopt in the mineral halite (see Figure 2.13), but one in which Fe^{2+} and O^{2-} ions take the place of Na^+ and Cl^- ions (Figure 3.8). Thus in an ionic compound the chemical formula, the empirical formula, and the formula unit are all the same.

SAMPLE EXERCISE 3.9 **Deriving an Empirical Formula from Percent Composition**

A team of geologists and students from the University of British Columbia encountered this 8-metric-ton (8000 kg) magnetite boulder on Vancouver Island during a 2005 field trip to an abandoned mine site.

Another mineral called magnetite is composed only of iron and oxygen and is 72.36% iron by mass. What is its empirical formula?

COLLECT AND ORGANIZE The statement of the problem gives the percent composition by mass of the compound. Because the only elements in the mineral are iron and oxygen, the percent composition by mass of the oxygen is calculated as 100% − 72.36% = 27.64%. In addition, we need the molar masses of both elements, which we can find in the periodic table.

ANALYZE There are four key steps for deriving an empirical formula from percent composition data: (1) Assume you have 100.00 grams of the substance so that the percent composition values are equivalent to the value of the mass of each element in grams. (2) Convert the mass of each element into an equivalent number of moles. (3) Compute the mole ratio by reducing one of the mole values to 1. (4) Convert the results into ratios of whole numbers if they are not already in that form after step 3.

SOLVE To convert the masses (grams) of Fe and O into moles of Fe and O, we divide by the molar masses of these elements: 55.85 g/mol for iron and 16.00 g/mol for oxygen. Now we are ready for the four-step process of determining the empirical formula of magnetite:

1. Convert percent composition values into masses by assuming a sample size of 100.00 grams:

$$72.36\% \text{ Fe in } 100.00 \text{ g sample} = 72.36 \text{ g Fe}$$
$$27.64\% \text{ O in } 100.00 \text{ g sample} = 27.64 \text{ g O}$$

2. Convert the masses of Fe and O into the equivalent numbers of moles:

$$\frac{72.36 \text{ g Fe}}{55.85 \text{ g/mol}} = 1.296 \text{ mol Fe} \qquad \frac{27.64 \text{ g O}}{16.00 \text{ g/mol}} = 1.728 \text{ mol O}$$

3. Simplify the mole ratio of oxygen to iron by dividing by the smaller value:

$$\frac{1.296 \text{ mol Fe}}{1.296} = 1.000 \text{ mol Fe} \qquad \frac{1.728 \text{ mol O}}{1.296} = 1.333 \text{ mol O}$$

So the mole ratio of Fe to O is $1:1.333$, or $1:\frac{4}{3}$.

4. Convert this fractional mole ratio to a ratio of whole numbers by multiplying through by 3; that is, $3 \times (1:1.333)$, which produces a mole ratio of 3 moles of Fe for every 4 moles of O. Thus the empirical formula of this substance is Fe_3O_4.

THINK ABOUT IT This formula may cause you to wonder about the charge on the Fe ions in magnetite. Assuming the charge on each O is $2-$, then the total negative charge in one formula unit is $4 \times (2-) = 8-$. To achieve electrical neutrality, the average positive charge on each Fe ion must be $\frac{8}{3}+ = 2\frac{2}{3}+$. Can an ion have a fractional charge? Recall that positive ions form when atoms lose electrons, and each electron has a charge of $1-$. Since electrons are not divisible by chemical means, we must work with charges that are integers. An iron atom may lose 2 or 3 electrons, but not $2\frac{2}{3}$. Magnetite has more than one kind of iron ion in its structure. In a crystal of magnetite, $\frac{2}{3}$ of the iron ions are Fe^{3+} ions and $\frac{1}{3}$ are Fe^{2+} ions, which results in a weighted average charge of $2\frac{2}{3}+$. Minerals are frequently mixtures of the same transition metal with different ion charges.

Practice Exercise For thousands of years the mineral chalcocite (pronounced KAL-kuh-site) has been a highly prized source of copper because it is so rich in the metal. Its chemical composition is 79.85% Cu and 20.15% S. What is its empirical formula?

(Answers to Practice Exercises are in the back of the book.)

A photograph of chalcocite, a copper-containing ore.

Many minerals, and indeed many types of substances, contain more than two elements. In such cases the process of finding empirical formulas from the percent composition of these substances proceeds in exactly the same way, as the next Sample Exercise demonstrates.

SAMPLE EXERCISE 3.10   **Deriving an Empirical Formula of a Compound of More than Two Elements**

A sample of the carbonate mineral dolomite (named after a mountain range in northern Italy called the Dolomites) has the following percent composition: 21.73% Ca, 13.18% Mg, 13.03% C, and 52.06% O. What is its empirical formula?

COLLECT AND ORGANIZE We follow the same steps in deriving this empirical formula as we did in Sample Exercise 3.9, except that here we must calculate more than one mole ratio to determine the empirical formula.

ANALYZE As an initial check, we add the mass percentages given in the problem to make sure they total 100%. Because these data are all derived from experiment and some numbers have been rounded, the value of this sum may sometimes be slightly less or slightly more than exactly 100%, but the total should be very close to 100%; in this case, 21.73 + 13.18 + 13.03 + 52.06 = 100.00%. For this exercise we also need the molar masses of the four component elements, which we get from the average atomic masses in the periodic table: rounded to two decimal places they are C, 12.01 g/mol; O, 16.00 g/mol; Mg, 24.31 g/mol; and Ca, 40.08 g/mol.

SOLVE Assuming we have a sample of dolomite that is exactly 100.00 g, we have the following masses of the four elements: 21.73 g Ca, 13.18 g Mg, 13.03 g C, and 52.06 g O. Converting these masses into numbers of moles, we have:

$$\frac{21.73 \text{ g Ca}}{40.08 \text{ g/mol}} = 0.5422 \text{ mol Ca} \qquad \frac{13.18 \text{ g Mg}}{24.31 \text{ g/mol}} = 0.5422 \text{ mol Mg}$$

$$\frac{13.03 \text{ g C}}{12.01 \text{ g/mol}} = 1.085 \text{ mol C} \qquad \frac{52.06 \text{ g O}}{16.00 \text{ g/mol}} = 3.254 \text{ mol O}$$

Next we divide each mole value by the smallest value (0.5422) to obtain a simple ratio of the four elements where at least one of the numbers is 1:

$$\frac{0.5422 \text{ mol Ca}}{0.5422} = 1.000 \text{ mol Ca} \qquad \frac{0.5422 \text{ mol Mg}}{0.5422} = 1.000 \text{ mol Mg}$$

$$\frac{1.085 \text{ mol C}}{0.5422} = 2.001 \text{ mol C} \qquad \frac{3.254 \text{ mol O}}{0.5422} = 6.001 \text{ mol O}$$

All these results are either whole numbers or extremely close to whole numbers, so we can omit step 4 and use these numbers to get a single mole ratio of 1:1:2:6 for Ca to Mg to C to O, which leads directly to the empirical formula $CaMgC_2O_6$.

THINK ABOUT IT Dolomite is a carbonate-containing mineral, which means that the carbon and oxygen atoms in its structure are bonded together in CO_3^{2-} ions. Therefore, a formula more illustrative of the composition of dolomite would be $CaMg(CO_3)_2$. This result illustrates how elemental analysis alone does not provide information about how atoms are bonded together in polyatomic ions (or molecules).

Practice Exercise Determine the empirical formula of a mineral with the following composition: 28.59% O, 24.95% Fe, and 46.46% Cr.

(Answers to Practice Exercises are in the back of the book.)

3.7 Empirical and Molecular Formulas Compared

We noted that early Earth was bombarded by asteroids and other debris that formed from the solar nebula. Even today extraterrestrial particles of various sizes rain down on our planet. Most are so small that they vaporize as they enter the atmosphere, but bigger ones, called meteorites, make it all the way to the surface. About one in 20 of the meteorites that fall to Earth is classified a carbonaceous chondrite. These meteorites are unusual in that they contain volatile substances locked in their structures, including water and a variety of organic compounds.

Some scientists view the presence of these compounds, which we associate with biological processes, as evidence that the molecular building blocks of life on Earth may have come from space, and that perhaps the first life-forms had extraterrestrial origins. Though we will not explore that theory here, we will focus on some of the organic molecules that have been detected by instrumental methods in interstellar gases and that have been found inside meteorites that fell to Earth's surface.

A structural model of one of these compounds, glycolaldehyde, is shown in Figure 3.9. Its presence in space is significant, in that glycolaldehyde readily combines with additional organic molecules that have also been detected in interstellar gases to form sugars and other molecules essential for life. Elemental analysis of glycolaldehyde gives the following percent composition data: 40.00% C, 6.71% H, and 53.28% O. We can use this information to calculate the empirical formula of the compound and then compare our result with the molecular structure in Figure 3.9. Using the four-step method developed in Section 3.6, we obtain these results:

FIGURE 3.9 Glycolaldehyde is one of over 100 organic compounds detected in interstellar gases.

$$\frac{40.00 \text{ g C}}{12.01 \text{ g/mol}} = 3.331 \text{ mol C} \qquad \frac{6.71 \text{ g H}}{1.008 \text{ g/mol}} = 6.66 \text{ mol H}$$

$$\frac{53.28 \text{ g O}}{16.00 \text{ g/mol}} = 3.330 \text{ mol O}$$

The mole ratio of carbon to hydrogen to oxygen is thus 3.33 : 6.66 : 3.33, or simply 1:2:1, which defines an empirical formula of CH_2O.

A comparison of the empirical formula CH_2O with the molecular model shows that the two are related but not equivalent: Counting up the atoms in the model in Figure 3.9, we arrive at a molecular formula of $C_2H_4O_2$, which represents the actual number of atoms in one molecule. In this case, if we multiply each subscript in the empirical formula by 2, we will obtain the molecular formula.

As this example demonstrates, the results of elemental analysis do not necessarily reveal the molecular formula of a compound: they simply provide the simplest mole ratios of the elements in it and hence the empirical formula. Occasionally, the empirical and molecular formulas are identical. For example, formaldehyde—which also has the percent composition 40.00% C, 6.71% H, and 53.28% O—has the molecular formula CH_2O. Formaldehyde is used to preserve biological tissues and has also been detected in interstellar gases. Many other compounds also have the empirical formula CH_2O, as illustrated in Sample Exercise 3.11.

The empirical formula of the molecular compound formaldehyde is identical to its molecular formula.

SAMPLE EXERCISE 3.11

SAMPLE EXERCISE 3.11 | **Relating Empirical and Molecular Formulas**

Shown here are the molecular structures of four common sugars. Which one(s) have the empirical formula CH_2O?

Glucose

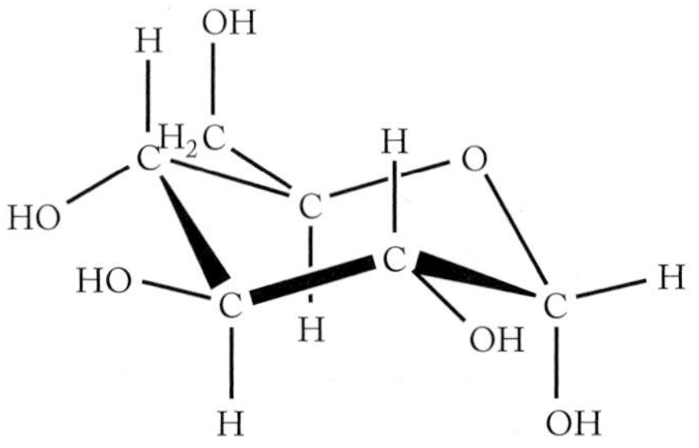

Fructose

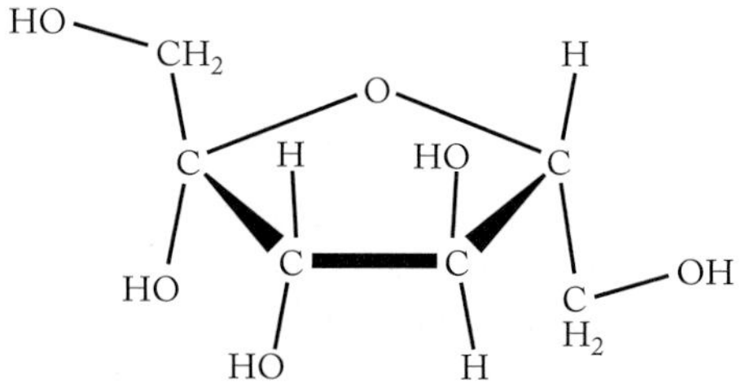

Sucrose

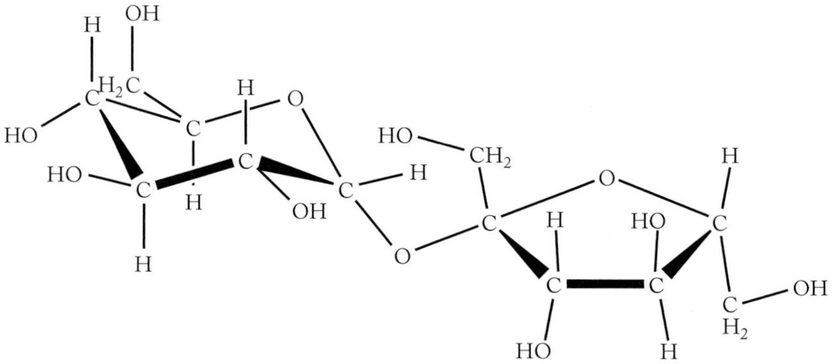

Lactose

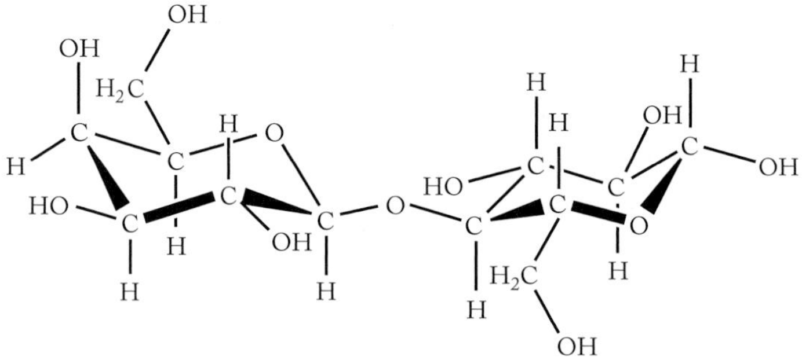

The molecular structures of glucose, fructose, sucrose, and lactose.

COLLECT AND ORGANIZE An empirical formula represents the simplest mole ratio of the elements in a compound, whereas a molecular formula expresses the actual number of atoms of each component element in a molecule of a compound. The problem illustrates the molecular structures of the sugars.

ANALYZE We may determine the molecular formulas of the sugars from the molecular structures and then evaluate whether the ratios of carbon to hydrogen to oxygen in the molecules simplify to 1:2:1, as seen in the empirical formula of formaldehyde (CH_2O).

SOLVE Counting up the atoms in the structures reveals the following molecular formulas:

Name	Molecular Formula
Glucose	$C_6H_{12}O_6$
Fructose	$C_6H_{12}O_6$
Sucrose	$C_{12}H_{22}O_{11}$
Lactose	$C_{12}H_{22}O_{11}$

The carbon-hydrogen-oxygen ratios in glucose and fructose are 6:12:6, which simplify to 1:2:1, resulting in an empirical formula of CH_2O for these two sugars. The ratios for lactose and sucrose, both 12:22:11, clearly do not simplify to 1:2:1, so only fructose and glucose have the empirical formula CH_2O.

THINK ABOUT IT Sugars are a major class of biological compounds called *carbohydrates*. All carbohydrates have the generic molecular formula $C_x(H_2O)_y$. Glucose and fructose are two examples of simple sugars in which the subscripts x and y in their molecular formulas are both 6. Lactose and sucrose are composed of two simple sugar molecules that have combined to produce one molecule of lactose or sucrose and one molecule of water. The loss of this water means that in their generic molecular formula, $y = 11$, which is one less than $x = 12$. This results in a molecular formula for both compounds that is the same as the empirical formula. Thus, $C_{12}H_{22}O_{11}$ describes the composition of a molecule of both compounds and also is the lowest whole-number ratio of atoms in the molecules.

Practice Exercise The molecular formulas of acetylene (C_2H_2) and benzene (C_6H_6) are different, but these compounds have the same percent composition and therefore the same empirical formula. What is their empirical formula?

(Answers to Practice Exercises are in the back of the book.)

Mass Spectrometry and Molecular Mass

In the preceding Sample and Practice Exercises we were given the molecular formulas of compounds and asked to compare them with the empirical formulas of the compounds. In the real world, chemists usually do not know the molecular formulas of compounds they have isolated from a reaction mixture or a biological system. To determine the molecular formulas of such compounds, chemists need more information than just elemental composition. This information typically comes from a technique called mass spectrometry, an early version of which we discussed in Chapter 2. Provided with the percent composition and molecular mass of a substance obtained by mass spectrometry, we can determine the molecular formula of a substance.

All mass spectrometers separate atoms and molecules by first converting them into ions and then separating those ions based on the ratio of their masses (m) to their electric charges (Z). There are several ways to produce ions in mass spectrometers. One of the most widely used methods is illustrated in Figure 3.10. This method involves vaporizing a sample and then bombarding the vapor with a beam of high-energy electrons to knock electrons off the atoms and molecules, converting them into positively charged ions.

FIGURE 3.10 In some mass spectrometers, neutral atoms or molecules of a substance are bombarded with a beam of high-energy electrons to make atomic or molecular ions.

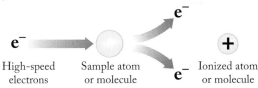

High-speed electrons — Sample atom or molecule — e^- e^- + Ionized atom or molecule

A **molecular ion (M^+)** is formed in a mass spectrometer when a neutral molecule loses an electron after being bombarded with a high-energy beam. The molecular ion has a charge of 1+ and has essentially the same molecular mass as the neutral molecule from which it came.

A **mass spectrum** is a graph of the data from a mass spectrometer, where m/Z ratios of the deflected particles are plotted against the number of particles with a particular mass. Because the charge on the ions typically is 1+, $m/Z = m/1 = m$, and the mass of the particle may be read directly from the m/Z axis.

Mass spectrometry in its early form, as Aston's positive-ray analyzer (Figure 2.9), was used to establish the presence and abundance of isotopes in gaseous samples of the elements. Modern mass spectrometers are used to determine the molecular masses of compounds and also their molecular structures. One of the most useful kinds of ions for determining molecular mass is one that forms when a molecule loses a single electron (of insignificant mass) and forms a **molecular ion (M^+)** as shown in Figure 3.10 and described in the following reaction equation, where M^+ contains an odd number of electrons:

$$M + \text{high-energy beam} \rightarrow M^{+\cdot} + e^-$$

Lasers, electric fields, or atomic or molecular collisions are also used to ionize samples, producing molecular ions.

Once ions have been formed, they are separated based on their mass-to-charge (m/Z) ratios. In magnetic-sector mass spectrometers (Figure 3.11), which are the direct descendents of Aston's positive-ray analyzer, ions are accelerated by an electric field into an evacuated chamber that is bracketed by the poles of a large electromagnet. The magnetic field bends the trajectories of the ions into circular paths with radii that depend on the mass-to-charge ratios of the ions. Ions with larger m/Z values have more momentum because of their greater mass, so they will not bend from their initial path as much as ions with smaller masses and lower m/Z values. (Think about how much easier it would be to deflect a Ping-Pong ball from its flight path than a golf ball moving at the same velocity.) By varying electrical power to the magnet, the strength of the magnetic field is varied. As the field strength changes, ions with different m/Z ratios pass through the detector slit and the number of them striking the detector is recorded.

Data produced by a mass spectrometer for a particular sample are displayed in a graphical record called a **mass spectrum** (plural *spectra*): the mass-to-charge ratios (m/Z) of particles detected are plotted on the *x*-axis against the intensity

FIGURE 3.11 In a magnetic-sector mass spectrometer, ions are pushed by an electric field into an evacuated chamber where a variable-strength magnetic field bends their paths depending on their mass-to-charge (m/Z) ratios. At a given moment, the strength of the field defines the m/Z ratio of only those ions passing through the detector slit. All others collide with the walls of the flight tube and are lost. The mass spectrum consists of peaks whose heights depend on the numbers of ions with different m/Z ratios that struck the detector. Compare this apparatus with Aston's positive-ray analyzer in Figure 2.9.

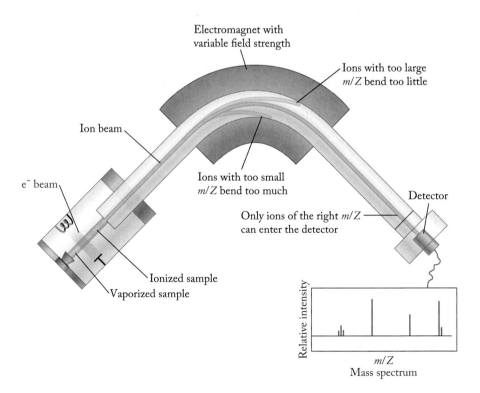

(number of particles) on the *y*-axis. Because the charge on the ions produced is 1+, the *m*/*Z* ratio is *m*/1, so we can read the mass of the particle directly from the position of the peak on the *x*-axis. Figure 3.12 shows examples of spectra for two different compounds, acetylene and benzene. Mass spectra such as these allow scientists to know with high precision and accuracy the molecular mass of compounds. Such spectra also enable scientists to detect subtle variations in isotopic abundances among particular samples, which can provide clues about where the samples came from or how old they are. For now, we will concentrate on *the molecular-ion peak*. But it is important to note that the other peaks in a mass spectrum are also useful. Because they represent masses smaller than the mass of the molecular ion, they can be used to determine the structure of the molecule. Called fragment ions, they arise when the molecule breaks apart into fragments.

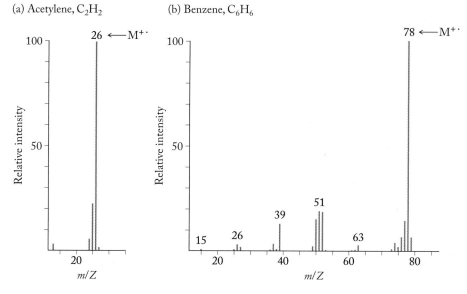

FIGURE 3.12 Mass spectra of (a) acetylene and (b) benzene. The molecular-ion peak in both spectra is labeled M$^{+\cdot}$. The peaks at much smaller *m*/*Z* values are produced when the ionizing beam also breaks apart the molecules into fragments.

Although much higher levels of accuracy are routinely possible, the mass spectra used in this text to determine molecular mass are recorded at *unit-mass resolution*; that is, the data are reported to the nearest whole-mass unit. The molecular-ion peak is not necessarily the tallest peak in the spectrum, but it always is the peak of highest mass (not counting the minor ^{13}C isotope peaks at *m*/*Z* = 27 and 79 in Figure 3.12a and b).

Using Percent Composition and Mass Spectra to Determine Molecular Formulas

Mass spectrometry provides a means of determining the molecular mass of a compound, and percent composition data allows us to determine its empirical formula. With these two pieces of information, we can determine the molecular formula of a compound.

Each subscript in the molecular formula of a compound is a multiple of the corresponding subscript in the empirical formula. For example, a multiplier (*n*) of 6 converts the empirical formula of CH_2O, for glucose and fructose, to their common molecular formula $C_6H_{12}O_6$:

$$(CH_2O)_n = (CH_2O)_6 = C_6H_{12}O_6$$

The key to translating empirical formulas into molecular formulas is to determine the value of *n*, and the key to determining the value of *n* is knowing the molecular mass.

For example, suppose elemental analysis tells us that the empirical formula of a liquid hydrocarbon is C_4H_5. Using the stoichiometric relation given by this formula and the molar masses of the two elements involved, we can then calculate the mass of the formula unit represented by the empirical formula:

$$4 \text{ atoms C} \times \frac{12.01 \text{ amu}}{1 \text{ atom C}} + 5 \text{ atoms H} \times \frac{1.008 \text{ g}}{1 \text{ atom H}} = 53.08 \text{ amu}$$

$$= 53 \text{ amu to unit-mass resolution}$$

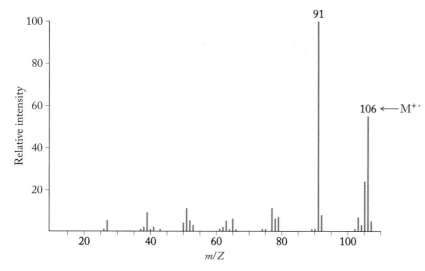

FIGURE 3.13 The mass spectrum of a liquid hydrocarbon.

The formula of the molecule is made up of some number of subunits, all of which have 4 carbon atoms for every 5 hydrogen atoms. A mass spectrum of the compound reveals its molecular mass to be 106 amu (Figure 3.13). How many C_4H_5 subunits are needed to make a molecule with a mass of 106 g?

$$\frac{\text{Molar mass}}{\text{Mass of formula unit}} = \frac{106 \ \cancel{\text{g/mol}}}{53 \ \cancel{\text{g/mol}}} = 2$$

This result represents the multiplier (n) that we must apply to the subscripts of the empirical formula to obtain the molecular formula of the compound:

$$(C_4H_5)_n = (C_4H_5)_2 = C_8H_{10}$$

SAMPLE EXERCISE 3.12   **Determining A Molecular Formula**

Pheromones are substances given off by living things that stimulate responses in members of the same species. Researchers often make use of pheromones to control the numbers or behaviors of certain animal populations. For example, certain pheromones are excreted by one sex in order to attract the other sex for mating. The mating pheromone of the Japanese beetle has been isolated, analyzed, and synthesized in the laboratory for use in traps in an attempt to reduce this insect's population.

Eicosene is a pheromone related to the Japanese beetle pheromone. It has a composition of 85.63% C and 14.37% H. The mass spectrum of eicosene shows a peak for the molecular ion at 280 amu. Determine the molecular formula of eicosene.

COLLECT AND ORGANIZE We can follow the procedure used in Sample Exercise 3.9 to determine the empirical formula of eicosene, and then use the molar masses of the elements in the molecule to determine the value of the multiplier needed to convert the empirical formula into a molecular formula.

ANALYZE In addition to the information provided we will need the molar masses of carbon and hydrogen, which are 12.01 g/mol and 1.008 g/mol.

SOLVE Assuming a 100 g sample, we have 85.63 g of carbon and 14.37 g of hydrogen. Converting these masses into numbers of moles gives us the following:

$$85.63 \ \cancel{\text{g C}} \times \frac{1 \text{ mol C}}{12.01 \ \cancel{\text{g C}}} = 7.130 \text{ mol C}$$

$$14.37 \ \cancel{\text{g H}} \times \frac{1 \text{ mol H}}{1.008 \ \cancel{\text{g H}}} = 14.26 \text{ mol H}$$

The mole ratio of C to H is 7.13 : 14.26, or 1:2, which means the empirical formula is CH_2. To determine the molecular mass, we first determine the mass of the empirical formula unit:

$$1 \ \cancel{\text{atom C}} \times \frac{12.01 \text{ amu}}{1 \ \cancel{\text{atom C}}} + 2 \ \cancel{\text{atoms H}} \times \frac{1.008 \text{ amu}}{1 \ \cancel{\text{atom H}}}$$

$$= 14 \text{ amu to unit-mass resolution}$$

Next we divide the molecular mass by the mass of the empirical formula unit to determine the multiplier (n):

$$n = \frac{\text{molar mass}}{\text{mass of formula unit}} = \frac{280 \text{ g/mol}}{14 \text{ g/mol}} = 20$$

Using 20 as a multiplier, we can now determine the molecular formula of eicosene:

$$(\text{CH}_2)_n = (\text{CH}_2)_{20} = \text{C}_{20}\text{H}_{40}$$

THINK ABOUT IT The molecular formula of a compound is a fundamental identifying characteristic. The ability to combine percent composition data (to determine the ratio of atoms in a molecule) with mass spectrometry (to determine its molar mass) is very important in the process of evaluating unknown compounds from all types of sources—from astronomical samples to geological samples, to samples from living systems.

Practice Exercise Determine the empirical and molecular formulas of a compound that contains 56.36% O and 43.64% P and has a molar mass of 284 g/mol.

(Answers to Practice Exercises are in the back of the book.)

> In **combustion analysis**, a substance is burned completely in oxygen to produce known compounds whose masses are used to determine the composition of the original material.

3.8 Combustion Analysis

As noted in Section 3.4, combustion involves burning substances in oxygen; in **combustion analysis**, complete combustion of organic compounds yields products that enable chemists to determine the chemical composition of the original unknown organic substance. To insure that combustion is complete, the process is carried out in excess oxygen, which means more oxygen is present than the stoichiometric amount needed. In these determinations, chemists rely on the fact that the complete combustion of organic compounds converts all of the carbon in them to CO_2 and all of the hydrogen to H_2O. The chemical equation for the complete combustion of a hydrocarbon in excess oxygen is

$$\text{C}_a\text{H}_b + \text{excess O}_2(g) \rightarrow a\,\text{CO}_2(g) + \tfrac{b}{2}\,\text{H}_2\text{O}(g)$$

Consider the following scenario: A hydrocarbon of unknown composition is burned completely in a chamber through which a stream of pure oxygen flows (Figure 3.14). The gases produced flow through two tubes in sequence. One tube is packed with solid $Mg(ClO_4)_2$ that selectively absorbs all the water vapor produced during combustion; the second contains $NaOH(s)$, which absorbs all of the carbon dioxide. The masses of these tubes are measured before and after the combustion reaction. Suppose the mass of the tube that traps CO_2 increases by

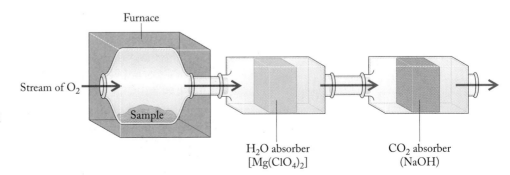

FIGURE 3.14 A carbon/hydrogen elemental analyzer relies on the complete combustion of organic compounds in excess oxygen. The products are CO_2 and H_2O vapor. Water vapor is absorbed by a $Mg(ClO_4)_2$ filter, and carbon dioxide is absorbed by a NaOH filter. The empirical formula of the compound is calculated from the masses of H_2O and CO_2 absorbed.

1.320 g, and the mass of the tube that traps $H_2O(g)$ increases by 0.540 g. How can we use these results to determine the composition and empirical formula of the unknown compound?

First, let's establish what we know about the unknown compound:

1. It is a hydrocarbon, so it contains only carbon and hydrogen.
2. Complete conversion of its carbon content into CO_2 produces 1.320 g of CO_2.
3. Complete conversion of its hydrogen content into H_2O produces 0.540 g of H_2O.

To derive an empirical formula for the unknown compound, we must determine the number of moles of carbon in 1.320 g of carbon dioxide and the number of moles of hydrogen in 0.540 g of water vapor. These quantities are directly related to the number of moles of carbon and hydrogen in the initial sample. Because we are looking for numbers that relate to each other in terms of small whole numbers and because all these data are derived from experiments that are not perfect, we can round the answers we get in this step if they are very close to numbers that will yield simple whole-number ratios:

$$1.320 \text{ g CO}_2 \times \frac{1 \text{ mol CO}_2}{44.01 \text{ g CO}_2} \times \frac{1 \text{ mol C}}{1 \text{ mol CO}_2} = 0.02999 \text{ mol C} \approx 0.0300 \text{ mol C}$$

$$0.540 \text{ g H}_2O \times \frac{1 \text{ mol H}_2O}{18.02 \text{ g H}_2O} \times \frac{2 \text{ mol H}}{1 \text{ mol H}_2O} = 0.05993 \text{ mol H} \approx 0.0600 \text{ mol H}$$

Our next step is to derive an empirical formula based on these proportions of C and H. The best way to do so is to divide both molar amounts by the smaller of the two:

$$\frac{0.0300 \text{ mol C}}{0.0300 \text{ mol C}} = 1 \qquad \frac{0.0600 \text{ mol H}}{0.0300 \text{ mol C}} = 2$$

This ratio corresponds to the empirical formula CH_2.

If we wanted to extend this analysis of this hydrocarbon to include the determination of its molecular formula, we would need additional information including its molecular mass. Suppose we have a mass spectrum of the unknown compound that shows a molecular ion with a mass of 84 g. The mass of the empirical formula unit is 14 g. By dividing the molecular mass by the mass of the empirical formula unit, we get the multiplier:

$$n = \frac{84 \text{ g}}{14 \text{ g}} = 6$$

This in turn gives us a molecular formula of

$$(CH_2)_6 = C_6H_{12}$$

Note that in this problem we did not need to know the initial mass of the sample to determine its empirical formula; we only needed to know that the sample was a hydrocarbon and that during combustion it was completely converted into the amounts of CO_2 and H_2O that were collected and weighed. What if the composition of the compound were totally unknown? What if it were a compound isolated from a tropical plant in the hope that it has beneficial pharmacological properties? Many such compounds are made of carbon, hydrogen, oxygen, and sometimes nitrogen. To calculate the empirical formula of a compound made of C, H, and O, for example, we need to know the proportion of oxygen in it, but there is no simple way of measuring that directly when the compound is burned in an atmosphere of pure oxygen. The data from combustion analysis give the percentages by mass of all atoms in the molecule except oxygen. If you are given data from a combustion analysis, always check first to see whether the percentages add up to 100%. If they do, then all the elements in the unknown compound are accounted for in the results. If they do not, then the missing mass is due to oxygen.

The following Sample Exercise shows how to directly determine the carbon and hydrogen content of a known mass of such a sample and then indirectly determine its oxygen content and its empirical formula.

SAMPLE EXERCISE 3.13   **Determining an Empirical Formula from Combustion Analysis**

Combustion of a 1.000 g sample of an organic compound known to contain carbon, hydrogen, and oxygen produces 2.360 g of CO_2 and 0.640 g of H_2O. What is the empirical formula of the compound?

COLLECT AND ORGANIZE We are given the initial mass of the sample and the masses of CO_2 and H_2O produced during its combustion. This information enables us to determine the masses of C and H in the original sample. The sum of these two masses subtracted from the mass of the original sample should equal the mass of oxygen in the sample. Converting that mass into moles of oxygen atoms will provide the information we need to derive an empirical formula.

ANALYZE First we determine the numbers of moles and the masses of C atoms and H atoms in the CO_2 and H_2O produced during combustion. The only source of these elements was the sample that was burned, so these numbers are equivalent to the numbers of moles and the masses of C and H in the original sample. Once we know the masses of C and H in the original sample, we can apply the law of conservation of mass to determine the mass of O in the original sample. From the mass of oxygen, we can calculate the number of moles of O in the original sample, determine the mole ratio of C to H to O, and finish by converting that ratio into a ratio of small whole numbers.

SOLVE The moles of C and H in the masses of CO_2 and H_2O trapped during combustion are as follows:

$$2.360 \text{ g CO}_2 \times \frac{1 \text{ mol CO}_2}{44.01 \text{ g CO}_2} \times \frac{1 \text{ mol C}}{1 \text{ mol CO}_2} = 0.05362 \text{ mol C}$$

$$0.640 \text{ g H}_2\text{O} \times \frac{1 \text{ mol H}_2\text{O}}{18.02 \text{ g H}_2\text{O}} \times \frac{2 \text{ mol H}}{1 \text{ mol H}_2\text{O}} = 0.07103 \text{ mol H}$$

Using these results, we calculate the corresponding masses of C and H:

$$0.05362 \text{ mol C} \times \frac{12.01 \text{ g C}}{1 \text{ mol C}} = 0.6440 \text{ g C}$$

$$0.07103 \text{ mol H} \times \frac{1.008 \text{ g H}}{1 \text{ mol H}} = 0.07160 \text{ g H}$$

The sum of these two masses (0.6440 g + 0.07160 g = 0.7156 g) is less than the mass of the starting sample (1.000 g). The difference must be the mass of oxygen in the sample:

$$\text{Mass of oxygen} = 1.000\text{g} - 0.7156 \text{ g} = 0.2844 \text{ g O}$$

The number of moles of O atoms in the sample is thus:

$$0.2844 \ \cancel{g \ O} \times \frac{1 \ mol \ O}{16.00 \ \cancel{g \ O}} = 0.01778 \ mol \ O$$

Combining the mole quantities calculated above, we have a mole ratio of the three elements in the sample:

$$0.05362 \ mol \ C : 0.07103 \ mol \ H : 0.01778 \ mol \ O$$

To simplify the ratio, we divide through by the smallest value (0.0178 mol) to get a carbon-hydrogen-oxygen mole ratio of 3:4:1. Therefore, the empirical formula of the sample is C_3H_4O.

THINK ABOUT IT The subscripts in our final answer are relatively small whole numbers, which should give us confidence that our answer is a plausible one.

Practice Exercise Vanillin is the compound containing carbon, hydrogen, and oxygen that gives vanilla beans their distinctive flavor. The combustion of 30.4 mg of vanillin produces 70.4 mg of CO_2 and 14.4 mg of H_2O. Use this information and the mass spectrum below to determine the molecular formula of vanillin.

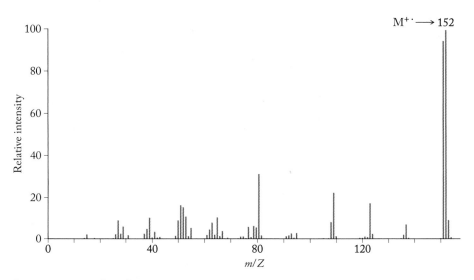

A mass spectrum of vanillin.

(Answers to Practice Exercises are in the back of the book.)

3.9 Limiting Reactants and Percent Yield

Let's return to photosynthesis, the process responsible for the O_2 in our present-day atmosphere and for the food chain that sustains life on Earth's surface. The stoichiometry of the reaction calls for equal parts (on a mole basis) of CO_2 and H_2O.

$$6 \ CO_2(g) + 6 \ H_2O(\ell) \rightarrow C_6H_{12}O_6(aq) + 6 \ O_2(g) \qquad (3.7)$$

In nature there is little likelihood of having exactly equal proportions of reactants at a reaction site. Consider what would happen if more than six molecules of water were available for every six molecules of CO_2 (a common occurrence in biological systems). The photosynthetic production of glucose could continue only until all the CO_2 was consumed, leaving the extra molecules of water unreacted (Figure 3.15). In this example, carbon dioxide is the **limiting reactant**, meaning that the extent to which the reaction proceeds is determined by the quantity of CO_2 available and not by the quantity of H_2O, which is in excess.

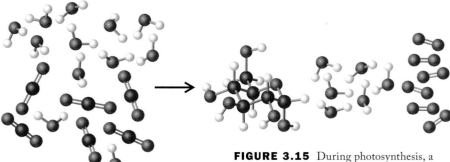

FIGURE 3.15 During photosynthesis, a reaction mixture of six molecules of CO_2 and 12 molecules of H_2O would produce only one molecule of glucose ($C_6H_{12}O_6$). The water is in excess, since six molecules of H_2O remain unreacted. In this case, CO_2 is the limiting reactant.

Calculations Involving Limiting Reactants

In the preceding example it was easy to identify CO_2 as the limiting reactant. However, selecting the limiting reactant in reactant mixtures when masses of substances are given is a real-life situation. Let's return to a reaction, discussed earlier in this chapter, that contributed to the acidity of Earth's early atmosphere and that is a major contributor to acid rain today—the combination of sulfur trioxide and water vapor to form sulfuric acid:

$$SO_3(g) + H_2O(g) \rightarrow H_2SO_4(\ell) \tag{3.8}$$

Suppose a mixture of gases contains 20.00 g of sulfur trioxide and 10.00 g of water vapor. What is the limiting reactant, and how many grams of H_2SO_4 could this mixture produce?

Anytime you are given the quantities of two reactants in a mixture, chances are that one of them is the limiting reactant that defines how much product can be formed. Which is the limiting reactant in the example? It is tempting to select the reactant with the lower mass in the mixture. Avoid this temptation and take a more systematic approach based on the stoichiometry of the reaction. Several approaches are possible; we consider two of them here—one that uses ratios, and another that uses stoichiometric calculations.

The first method consists of these steps, where the generic symbols A and B refer to the two reactants:

- Convert the given masses of reactants A and B into moles.

- Calculate the mole ratio of A to B.

- Compare this mole ratio with the stoichiometric mole ratio from the balanced chemical equation. If

$$\left(\frac{\text{moles of A}}{\text{moles of B}}\right)_{\text{given}} > \left(\frac{\text{moles of A}}{\text{moles of B}}\right)_{\text{stoichiometric}} \tag{3.9}$$

▶❚❚ **CHEMTOUR** Limiting Reactant

then B is the limiting reactant. If the above inequality is reversed (so that the ratio on the left side is less than the ratio on the right), then A is the limiting reactant. If the two ratios are equal, then the given masses are the exact stoichiometric amounts and both reactants will be consumed completely.

Let's try this approach with the masses of SO_3 and H_2O given (just after Equation 3.9). Because we need to compare numbers of moles, first we must determine the molar masses of the two reactants so we can calculate the number of moles from the masses given:

$$\mathscr{M}_{SO_3} = 1 \text{ mol S} \times \frac{32.07 \text{ g}}{\text{mol S}} + 3 \text{ mol O} \times \frac{16.00 \text{ g}}{\text{mol O}} = 80.07 \text{ g/mol}$$

$$\mathscr{M}_{H_2O} = 2 \text{ mol H} \times \frac{1.008 \text{ g}}{\text{mol H}} + 1 \text{ mol O} \times \frac{16.00 \text{ g}}{\text{mol O}} = 18.02 \text{ g/mol}$$

The **limiting reactant** is consumed completely in a chemical reaction. The amount of product formed depends on the amount of the limiting reactant available.

The second step is to calculate the mole ratio of SO_3 to H_2O in the reaction mixture:

$$\left(\frac{\text{mol } SO_3}{\text{mol } H_2O}\right)_{given} = \frac{20.00 \text{ g } SO_3 \times \dfrac{1 \text{ mol } SO_3}{80.07 \text{ g } SO_3}}{10.00 \text{ g } H_2O \times \dfrac{1 \text{ mol } H_2O}{18.02 \text{ g } H_2O}} = \frac{0.2498 \text{ mol } SO_3}{0.5549 \text{ mol } H_2O} = 0.4502$$

In the third step we observe that this ratio is less than the 1:1 stoichiometric ratio of SO_3 to H_2O. Expressing this outcome in equation form, we get

$$\left(\frac{\text{mol } SO_3}{\text{mol } H_2O}\right)_{given} < \left(\frac{\text{mol } SO_3}{\text{mol } H_2O}\right)_{stoichiometric}$$

Since this inequality is the reverse of Equation 3.9, SO_3 (reactant A in Equation 3.9) must be the limiting reactant.

A different approach that does not involve ratios uses two calculations of the amount of product formed in two different reactions: one that assumes that reactant A is the limiting reactant and a second calculation that assumes reactant B is limiting. Using the same example as before, we first calculate how much sulfuric acid would be produced if 20.00 g of SO_3 were to react completely:

$$20.00 \text{ g } SO_3 \times \frac{1 \text{ mol } SO_3}{80.07 \text{ g } SO_3} \times \frac{1 \text{ mol } H_2SO_4}{1 \text{ mol } SO_3} \times \frac{98.09 \text{ g } H_2SO_4}{1 \text{ mol } H_2SO_4}$$
$$= 24.50 \text{ g } H_2SO_4$$

Next we carry out the same calculation to determine how much sulfuric acid would be produced if 10.00 g of water were to react completely:

$$10.00 \text{ g } H_2O \times \frac{1 \text{ mol } H_2O}{18.02 \text{ g } H_2O} \times \frac{1 \text{ mol } H_2SO_4}{1 \text{ mol } H_2O} \times \frac{98.09 \text{ g } H_2SO_4}{1 \text{ mol } H_2SO_4}$$
$$= 54.43 \text{ g } H_2SO_4$$

The SO_3 is the limiting reactant because when it is completely consumed it results in less product being formed than when the amount of water available is completely consumed. Thus when 20.00 g of SO_3 and 10.00 g of H_2O are combined, the maximum amount of product that can form is 24.50 g of H_2SO_4. Making that amount of sulfuric acid completely consumes the available SO_3 but not the H_2O.

The two approaches we have just employed to determine the limiting reactant probably seem similar in terms of the number of steps involved. This is generally the case when the masses of the reactants are given in grams, because converting those masses into equivalent numbers of moles is one step in the process. The advantage of the second method is that it not only determines the identity of the limiting reagent; it also determines the **theoretical yield**, the amount of product that can be produced by the given mixture.

SAMPLE EXERCISE 3.14   **Identifying the Limiting Reactant**

During the launch of a U.S. space shuttle, high-pressure pumps deliver 4400 kg of liquid H_2 fuel and 31,000 kg of liquid O_2 to each of its main engines each minute. Is one of these reactants a limiting reactant, or is the mixture stoichiometric?

COLLECT AND ORGANIZE This is a limiting-reactant problem because we are given quantities of the two reactants (H_2 and O_2) and asked to determine whether

The **theoretical yield** (sometimes called the stoichiometric yield) is the amount of product expected for a chemical reaction given a specific quantity of reactant.

one of them is limiting and if so, which one. One way we can make that determination is to calculate the ratio of H_2 to O_2 in the balanced chemical equation for the reaction.

ANALYZE An unbalanced version of this combustion reaction is

$$H_2(g) + O_2(g) \rightarrow H_2O(g)$$

After balancing this equation, we will compare the ratio of the reactants described in the statement of the problem with the stoichiometric ratio in the balanced equation to determine whether one of the reactants is limiting or if the ratio is stoichiometric.

SOLVE First we need to balance the combustion equation:

$$\_\_\, H_2(g) + \_\_O_2(g) \rightarrow 1\, H_2O(g)$$

$$2 \times [1\, H_2(g) + \tfrac{1}{2} O_2(g) \rightarrow 1\, H_2O(g)]$$

$$2\, H_2(g) + O_2(g) \rightarrow 2\, H_2O(g)$$

Now that the equation is balanced, it tells us that the mole ratio of hydrogen to oxygen in the reaction is 2:1; this is the ratio we'll compare with the ratio that defines the amounts of reactants given in the problem.

We calculate the mole ratio of H_2 to O_2 in the reaction of interest:

$$31,000 \text{ kg } O_2 \times \frac{1000 \text{ g } O_2}{1 \text{ kg } O_2} \times \frac{1 \text{ mol } O_2}{32.00 \text{ g } O_2} = 9.7 \times 10^5 \text{ mol } O_2$$

$$4,400 \text{ kg } H_2 \times \frac{1000 \text{ g } H_2}{1 \text{ kg } H_2} \times \frac{1 \text{ mol } H_2}{2.016 \text{ g } H_2} = 2.2 \times 10^6 \text{ mol } H_2$$

The stoichiometric ratio of H_2 to O_2 is 2:1; the ratio calculated for the quantities specified in the shuttle is $2.2 \times 10^6 : 9.7 \times 10^5$, which is about 2.3:1, so hydrogen is in slight excess and oxygen is the limiting reactant.

THINK ABOUT IT This result implies that some of the hydrogen fuel is wasted because there is not enough O_2 to burn it completely.

Practice Exercise A mixture of fuel and oxygen that contains more oxygen than is needed to burn the fuel completely is called a lean mixture, while a mixture with more fuel and too little oxygen is called a rich mixture. A high-performance heater that burns propane $[C_3H_8(g)]$ is adjusted so that 100.0 g of oxygen gas (O_2) enters the system for every 100.0 g of propane. Is this mixture of fuel and oxygen rich or lean?

(Answers to Practice Exercises are in the back of the book.)

The power that thrusts a space shuttle into orbit comes from two solid fuel booster rockets and three main engines that burn hydrogen; both hydrogen and oxygen are stored in liquid form in the large brown tank under the shuttle.

Actual Yields versus Theoretical Yields

When we calculated the mass of sulfuric acid that was formed by 10.00 g of water and 20.00 g of sulfur trioxide, we defined the value as a theoretical yield: it is the maximum amount of product that could be made from the given quantities of reactants. In nature, industry, or the laboratory, the **actual yield** is often less than the theoretical yield for several reasons. The same reactants may undergo different chemical reactions, yielding different sets of products. Sometimes the rate of a reaction is so low that some reactants remain unreacted even after an

The amount of product obtained as a result of a chemical reaction conducted in the lab is the **actual yield**, which is often less than the theoretical yield.

Hydrogen and Helium: The Bulk of the Universe

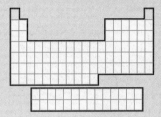

Hydrogen and helium are the only two elements in the first row of the periodic table. They are the least dense of all elements and the most abundant in the universe, accounting for over 99% of all atoms. They are much less abundant on Earth as free elements: Hydrogen gas (H_2) constitutes only about 0.00005% of the atmosphere, by volume. The atmospheric concentration of helium is a factor of 10 higher—0.0005% by volume—which still makes He a very minor component.

These two elements have similar physical properties, but their chemical properties are remarkably different. Helium is chemically inert, while hydrogen is extremely reactive and occurs in more compounds than any other element. Compounds containing carbon and hydrogen are the major components of all living matter, and water (H_2O) is ubiquitous and essential for life on Earth.

Hydrogen is currently being hailed as the fuel of the future, with the potential for large-scale use in internal combustion engines and fuel cells. It is already produced and stored on a large scale for the U.S. space program. The principal advantage of hydrogen over fossil fuels is its high fuel value—the amount of energy it releases per unit of mass. Hydrogen is also a clean fuel, producing only water and energy and not oxides of carbon, nitrogen, or sulfur, which contribute to atmospheric pollution or global warming.

A hydrogen fueling station for vehicles.

Because hydrogen does not occur as the free element in nature in significant amounts, it must be extracted from its compounds. Most of the H_2 produced in the United States and about half that produced worldwide comes from methane (CH_4), the principal component in natural gas. Hydrogen is generated from methane by steam–methane reforming. In this process, CH_4 reacts with excess steam:

$$CH_4(g) + H_2O(g) \rightarrow CO(g) + 3\ H_2(g)$$

Additional H_2 is generated when the CO produced by reforming is reacted with more steam in the water–gas shift reaction:

$$CO(g) + H_2O(g) \rightarrow CO_2(g) + H_2(g)$$

Another way to generate hydrogen is to split H_2O molecules into H_2 and O_2 by *electrolysis*:

$$2\ H_2O(\ell) \xrightarrow{\text{electric current}} 2\ H_2(g) + O_2(g)$$

The electricity used in electrolysis is often generated by power plants that burn fossil fuels. Currently, developing hydrogen as a widely used fuel is hampered by the fact that it cannot be generated in bulk quantities by any efficient means that do not also consume fossil fuels.

Probably the most important use of hydrogen is in producing ammonia in the *Haber–Bosch process*:

$$N_2(g) + 3\ H_2(g) \rightarrow 2\ NH_3(g)$$

The hydrogen used to make ammonia comes directly from steam–methane reforming, so the production of hydrogen and ammonia are tightly linked commercially. Ammonia is used directly as a fertilizer; it is also a source of nitrogen for manufacturing other nitrogen-containing compounds, including explosives.

Another major use of hydrogen is as a reactant with liquid vegetable oils to make solid, edible fats such as margarine. This reaction is called hydrogenation, and the resultant products are called hydrogenated fats. Sometimes this reaction is run in the reverse direction: hydrogen-rich solid fats are dehydrogenated to produce hydrogen gas and liquid oils. The oils from this reverse reaction are trans-fats, which differ in structure from natural vegetable oils and are the object of much scrutiny due to their negative impact on our diets.

The isotopes of hydrogen are also useful. Deuterium (^2H or D) is a naturally occurring isotope of hydrogen that accounts for about 0.015% of hydrogen. Heavy water, with the molecular formula D_2O, is used in nuclear power–generating systems. D_2O is an essential component in nuclear reactors where plutonium is generated from uranium; it is also crucial to the production of nuclear weapons. Heavy water is so important to this technology that its production is monitored closely and the material is kept under strict export controls. Heavy water has a slightly higher boiling point (101.4°C) than normal water (100°C) and in principle can be separated from the latter by distillation. In practice, D_2 is separated from liquefied H_2 and is then used to make compounds that can be purified by subsequent processing. For example, D_2 is reacted to form D_2S, which is treated with H_2O and converted to D_2O. The production of 1 liter of D_2O requires about 340,000 liters of H_2O.

Tritium (^3H) differs from the other two isotopes of hydrogen in that it is radioactive. It can be detected easily, which gives rise to its utility. Tritium is produced naturally in the upper atmosphere when cosmic rays from the sun strike nitrogen molecules. It is also produced in nuclear reactors. Tritium is used in devices where dim light is needed but no source of electricity is available, such as emergency exit signs (in buildings that must function in times of power failure), dials in aircraft, rifle sights, and even in the luminous dials of some watches.

On early Earth, hydrogen was retained because it formed compounds with heavier elements, but helium forms no compounds and was too light to be retained by Earth's gravitational field. Any helium now on Earth is the product of radioactive decay. Most helium in the world is obtained from natural gas deposits in Texas, Oklahoma, Kansas, the Middle East, and Russia. These locations have significant deposits of uranium ore, and the decay of uranium produces helium, which mixes with the underground deposits of natural gas. Helium is separated from the natural gas, in which its concentration may be as high as 0.3% by mass.

Helium has a number of uses beyond keeping the familiar Goodyear blimps, weather balloons, and decorative balloons aloft. It is mixed with oxygen in the air tanks of deep-sea divers to create nitrogen-free gas mixtures. Because nitrogen is soluble in blood, divers breathing nitrogen-laden air run the risk of developing nitrogen narcosis, a disorienting condition in which the nervous system becomes saturated with nitrogen. Helium mixtures also help divers avoid decompression sickness (the bends), a painful and life-threatening condition caused by nitrogen bubbles forming in the blood and joints as divers ascend to the surface. Helium protects divers from these dangers because it is less soluble in blood than nitrogen.

Helium is increasingly being used as a cryogenic liquid. A *cryogen* is a gas that has been liquefied by cooling to a very low temperature. Helium has the lowest boiling point of any element (−268.9°C), and it is used as a cryogen to create and maintain superconducting magnets in instruments ranging from equipment in chemistry and physics laboratories to magnetic-resonance-imaging units (MRIs) in hospitals and clinics. In its liquid phase, helium also has some extraordinary physical properties. It exhibits superfluidity, which means it flows so freely that it actually flows up the walls of containers. Liquid helium is also a nearly perfect conductor of heat and electricity.

(a)

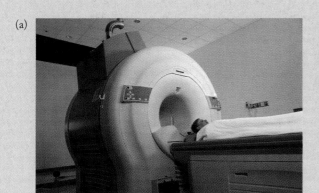

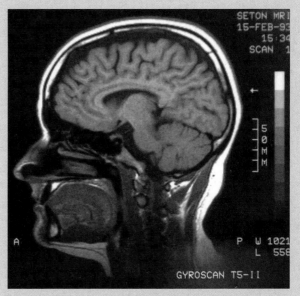

(a) The magnet in a magnetic-resonance-imaging (MRI) unit maintains its superconductivity by being submerged in liquid helium inside the large white cylinder shown here. (b) An MRI image.

(b)

The **percent yield** is the ratio, expressed as a percentage, of the actual yield of a chemical reaction to the theoretical yield.

extended time. Other reactions do not go to completion no matter how long they are allowed to run, yielding a mixture of reactants and products, the composition of which does not change with time. For these and other reasons, it is useful to distinguish between the theoretical and the actual yields of a chemical reaction and to calculate the **percent yield**:

$$\text{Percent yield} = \frac{\text{actual yield}}{\text{theoretical yield}} \times 100\% \qquad (3.10)$$

SAMPLE EXERCISE 3.15 **Calculating Percent Yield**

The industrial process for making the ammonia used in fertilizer, explosives, and many other products is based on the reaction between nitrogen and hydrogen at high temperature and pressure:

$$N_2(g) + 3\,H_2(g) \rightarrow 2\,NH_3(g)$$

If 18.20 kg of NH_3 are produced by a reaction mixture that initially contains 6.00 kg of H_2 and an excess of N_2, what is the percent yield of the reaction?

COLLECT AND ORGANIZE We know how much NH_3 was produced as a result of the reaction: 18.20 kg is the actual yield. We also know that H_2 must be the limiting reactant, because the problem specifies an excess of N_2. We know that the reaction mixture initially contained 6.00 kg of H_2. We need to use this quantity of H_2 to calculate how much NH_3 could have been produced and then divide this value into the quantity of NH_3 that was actually produced to calculate the percent yield.

ANALYZE The first stage in solving this problem is to calculate the theoretical yield of NH_3—the mass of NH_3 that would have been produced had the yield been 100%. This can be done following the steps in Figure 3.5 or Sample Exercise 3.7: (1) convert the mass of H_2 to moles of H_2; (2) convert the moles of H_2 to moles of NH_3; and (3) then convert the moles of NH_3 to the mass of NH_3. We need to work with the molar masses of H_2 and NH_3 and the stoichiometry of the reaction, which tells us that 2 moles of NH_3 are produced for every 3 moles of H_2 that are consumed. Finally we use Equation 3.10 to calculate the percent yield.

SOLVE

1. Calculate the molar masses of H_2 and NH_3:

$$\mathcal{M}_{H_2} = 2\ \text{mol H} \times \frac{1.008\ \text{g}}{\text{mol H}} = 2.016\ \text{g/mol}$$

$$\mathcal{M}_{NH_3} = 1\ \text{mol N} \times \frac{14.01\ \text{g}}{\text{mol N}} + 3\ \text{mol H} \times \frac{1.008\ \text{g}}{\text{mol H}} = 17.03\ \text{g/mol}$$

2. Calculate the theoretical yield of NH_3:

$$6.00\ \text{kg H}_2 \times \frac{1000\ \text{g H}_2}{1\ \text{kg H}_2} \times \frac{1\ \text{mol H}_2}{2.016\ \text{g H}_2} \times \frac{2\ \text{mol NH}_3}{3\ \text{mol H}_2}$$

$$\times \frac{17.03\ \text{g NH}_3}{1\ \text{mol NH}_3} \times \frac{1\ \text{kg NH}_3}{1000\ \text{g NH}_3} = 33.8\ \text{kg NH}_3$$

3. Divide the actual yield by the theoretical yield to calculate the percent yield:

$$\frac{18.20 \text{ kg NH}_3}{33.8 \text{ kg NH}_3} \times 100\% = 53.8\%$$

THINK ABOUT IT A yield of about 54% may seem low, but it is typical for the industrial production of ammonia. Additional factors must be considered in terms of running reactions at the industrial level, such as optimizing the rate at which a reaction proceeds and methods for separating the product from the reactants.

Practice Exercise The combustion of 58 g of butane (C_4H_{10}) produces 158 g of CO_2. What is the percent yield of the reaction?

(Answers to Practice Exercises are in the back of the book.)

Summary

SECTION 3.1 In **chemical equations** the chemical formulas of **reactants** appear first, followed by a reaction arrow and the chemical formulas of the **products**. The proportions of the reactants and products are expressed by coefficients preceding their formulas, and symbols in parentheses are used to indicate their physical states: (g) for gas, (ℓ) for liquid, and (s) for solid.

SECTION 3.2 The **mole (mol)** is the SI base unit for expressing quantities of substances. One mole of particles (atoms, ions, or molecules) is equal to **Avogadro's number** ($N_A = 6.022 \times 10^{23}$) of the particles. The mass of 1 mole of a substance is called its **molar mass** ($\mathcal{M}$); the molar mass of a compound is the sum of the molar masses of the elements that make up the compound, each multiplied by the number of moles of that element in a mole of the compound. The mole ratios of reactants and products in a chemical equation are called the **stoichiometry** of the reaction. The **law of conservation of mass** states that the sum of the masses of the reactants of a chemical reaction is equal to the sum of the masses of the products.

SECTION 3.3 In a balanced chemical equation the number of moles of each element, either as the pure element or in compounds, is the same on both sides of the reaction arrow. An important type of reaction is **hydrolysis**, in which water reacts with another material. The hydrolysis of nonmetal oxides produces acidic solutions.

SECTION 3.4 In the complete *combustion* of compounds made of carbon and hydrogen, called **hydrocarbons**, the hydrogen combines with oxygen to form H_2O, and the carbon combines with oxygen to form CO_2. **Combustion reactions** occur between oxygen and another element or compound.

SECTION 3.5 In the carbon cycle, CO_2 from the atmosphere is converted to glucose during photosynthesis in green plants. Most but not all of the carbon in green plants returns to the atmosphere as CO_2 during respiration.

SECTION 3.6 The **percent composition** is the percentage by mass of each element in a compound. The percentages are determined using a method called elemental analysis. The **empirical formula** of a compound can be determined from percent composition data.

SECTION 3.7 The empirical formula of a molecule, derived from the results of an elemental analysis, is the simplest whole-number mole ratio of the elements in the compound. The empirical formula of a compound may or may not be the same as its molecular formula, which indicates the number of each type of atom in one molecule. To convert an empirical formula into a molecular formula, we need the molar mass of the compound, which can be determined by identifying the **molecular ion (M^+)** in its **mass spectrum**.

SECTION 3.8 During the **combustion analysis** of a weighed sample of an organic compound, the carbon content of the sample is converted into CO_2 and the hydrogen content is converted into H_2O. The resulting masses of CO_2 and H_2O are measured and used to determine the masses of C and H, and then the empirical formula of the original compound.

SECTION 3.9 The **limiting reactant** in a reaction mixture is the reactant in shortest supply that limits how much product can be made. Sometimes the quantity of product is less than that which should have been made given the quantities of reactants available. The ratio of the **actual yield** to the **theoretical yield** is usually expressed as a percentage, called the **percent yield.**

Problem-Solving Summary

TYPE OF PROBLEM	CONCEPTS AND EQUATIONS	SAMPLE EXERCISES
Converting number of particles into number of moles (or vice versa)	Divide (or multiply) by Avogadro's number ($N_A = 6.022 \times 10^{23}$ particles/mol).	3.1
Converting mass of a substance into number of moles (or vice versa)	Divide (or multiply) by molar mass ($\mathcal{M}$) of the substance.	3.2 and 3.3
Calculating molar mass of a compound	Multiply the molar mass of each element in the compound by its subscript in the compound's formula; then add the products.	3.4
Balancing a chemical reaction	Add coefficients to the equation so that the number of atoms of each element are the same on each side of the reaction arrow.	3.5
Writing balanced chemical equations for combustion reactions	C and H in organic compounds react with O_2 to form CO_2 and H_2O, for example: $$CH_4(g) + 2\,O_2(g) \rightarrow 2\,H_2O(\ell) + CO_2(g)$$ Balance the moles of C first, then H, then O.	3.6
Calculating mass of a product from mass of a reactant	Use $\mathcal{M}_{reactant}$ to convert mass of the reactant into moles of the reactant; use the stoichiometry of the reaction to calculate moles of the product, and then $\mathcal{M}_{product}$ to calculate mass of the product.	3.7
Calculating percent composition from a chemical formula	Calculate the molar mass. Note the mass of each element in the formula. Divide each element mass by the formula mass. Express each result as a percent.	3.8
Determining an empirical formula from percent composition	Assume the sample has a mass of 100 g. Divide the mass of each element by its atomic mass to calculate moles of each element. Simplify mole ratios to lowest whole numbers, and use the numbers as subscripts for the appropriate element symbols.	3.9 and 3.10
Relating empirical and molecular formulas	Calculate a conversion factor by dividing the molar mass by the mass of the formula unit. Multiply subscripts in the empirical formula by this conversion factor.	3.11 and 3.12
Deriving an empirical formula from a combustion analysis	For hydrocarbons, convert given masses of CO_2 and H_2O into moles of CO_2 and H_2O, and then to moles of C and H. For compounds containing O, convert moles of C and H into masses of C and H and subtract the sum of these values from the sample mass to calculate mass of O. Convert mass of O to moles of O. Simplify the mole ratios of C to H to O.	3.13
Identifying the limiting reactant	Methods: (1) Convert masses of reactants into moles. Compare mole ratios of reactants to the corresponding mole ratios in balanced chemical equation (Equation 3.9): $$\left(\frac{\text{moles of A}}{\text{moles of B}}\right)_{given} > \left(\frac{\text{moles of A}}{\text{moles of B}}\right)_{stoichiometric}$$ or (2) Calculate how much product each reactant could make; the reactant making the least product is the limiting reactant.	3.14
Calculating percent yield	Calculate theoretical yield of product using mass of limiting reactant. Divide actual yield (given) by theoretical yield (Equation 3.10): $$\text{Percent yield} = \frac{\text{actual yield}}{\text{theoretical yield}} \times 100\%$$	3.15

Visual Problems

3.1. Each of the pairs of containers pictured in Figure P3.1 contains substances composed of elements X (red balls) and Y (blue balls). For each pair, write a balanced chemical equation describing the reaction that takes place. Be sure to indicate the physical states of the reactants and products using the appropriate symbols in parentheses.

3.2. Identify the limiting reactant in each of the pairs of containers pictured in Figure P3.2. The red balls represent atoms of element X, the blue balls are atoms of element Y. Each question mark means that there is unreacted reactant left over.

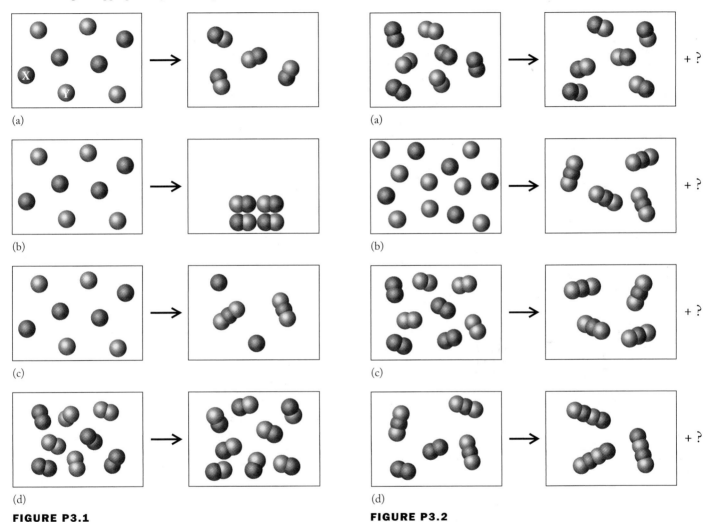

(a)

(b)

(c)

(d)

FIGURE P3.1

(a)

+ ?

(b)

+ ?

(c)

+ ?

(d)

+ ?

FIGURE P3.2

Questions and Problems

Composition of Earth

CONCEPT REVIEW

3.3. On the basis of the distribution of the elements in Earth's layers (see Figure 2.2), which of the following substances should be the most dense? $SiO_2(s)$; $Al_2O_3(s)$; $Fe(\ell)$

3.4. On the basis of the compositions and physical states of Earth's various layers (Figure 2.2), which of the following substances has the highest melting point? Al_2O_3; Fe; Ni; S

3.5. Most of our planet is covered with water. However, any water present as Earth formed should have evaporated and been blown into the outer reaches of the solar system. Where did all the water on Earth today come from?

*3.6. The proportions of the elements that make up the asteroid 433 Eros are similar to those that make up Earth. Scientists believe that this similarity means that the asteroid and Earth formed around the same time. If the asteroid formed after Earth formed a solid crust, and if the asteroid was the product of a collision between Earth and an even larger asteroid, how would the composition of 433 Eros be different from what it actually is?

The Mole

CONCEPT REVIEW

3.7. In principle we could use the more familiar unit **dozen** in place of mole when expressing the quantities of particles (atoms, ions, or molecules) in chemical reactions. What would be the disadvantage in doing so?

3.8. In what way is the molar mass of an ionic compound and its formula mass the same, and in what ways are they different?

3.9. Do molecular compounds containing three atoms per molecule always have a molar mass greater than that of molecular compounds containing two atoms per molecule? Explain.

3.10. Without calculating their molar masses (though you may consult the periodic table), predict which of the following oxides of nitrogen has the larger molar mass: NO_2 or N_2O.

PROBLEMS

3.11. Earth's atmosphere contains many volatile substances that are present in trace amounts. The following quantities of these trace gases were found in a 1.0 mL sample of air. Calculate the number of moles of each gas in the sample:
a. 4.4×10^{14} atoms of $Ne(g)$
b. 4.2×10^{13} molecules of $CH_4(g)$
c. 2.5×10^{12} molecules of $O_3(g)$
d. 4.9×10^9 molecules of $NO_2(g)$

3.12. The following quantities of trace gases were found in a 1.0 mL sample of air. Calculate the number of moles of each compound in the sample:
a. 1.4×10^{13} molecules of $H_2(g)$
b. 1.5×10^{14} atoms of $He(g)$
c. 7.7×10^{12} molecules of $N_2O(g)$
d. 3.0×10^{12} molecules of $CO(g)$

3.13. Moles of Memory The capacities of memory storage devices in computers are represented by the number of bytes of information that can be stored. How many moles of bytes are there in each of the following?
a. a 1.5-terabyte hard drive (where 1 terabyte = 10^{12} bytes)?
b. a 2.0-gigabyte flash drive (where 1 gigabyte = 10^9 bytes)?

3.14. Express the following population estimates for the year 2010 in nanomoles of people: (a) The United States and Canada, 348 million people; (b) Europe, 719 million people; (c) Asia, 4.15 billion people; (d) the world, 6.83 billion people

3.15. How many atoms of titanium are there in 0.125 mole of each of the following? (a) ilmenite, $FeTiO_3$; (b) titanium(IV) chloride; (c) Ti_2O_3; (d) Ti_3O_5

3.16. How many atoms of iron are there in 2.5 moles of each of the following? (a) wolframite, $FeWO_4$; (b) pyrite, FeS_2; (c) magnetite, Fe_3O_4; (d) hematite, Fe_2O_3

3.17. Which substance in each of the following pairs of quantities contains more moles of oxygen? (a) 1 mole of Al_2O_3 or 1 mole of Fe_2O_3; (b) 1 mole of SiO_2 or 1 mole of N_2O_4; (c) 3 moles of CO or 2 moles of CO_2

3.18. Which substance in each of the following pairs of quantities contains more moles of oxygen? (a) 2 moles of N_2O or

1 mole of N_2O_5; (b) 1 mole of NO or 1 mole of calcium nitrate; (c) 2 moles of NO_2 or 1 mole of sodium nitrite

3.19. Elemental Composition of Minerals Aluminum, silicon, and oxygen form minerals known as aluminosilicates. How many moles of aluminum are in 1.50 moles of
a. pyrophyllite, $Al_2Si_4O_{10}(OH)_2$?
b. mica, $KAl_3Si_3O_{10}(OH)_2$?
c. albite, $NaAlSi_3O_8$?

3.20. The uranium used for nuclear fuel exists in nature in several minerals. Calculate how many moles of uranium are in 1 mole of the following:
a. carnotite, $K_2(UO_2)_2(VO_4)_2$
b. uranophane, $CaU_2Si_2O_{11}$
c. autunite, $Ca(UO_2)_2(PO_4)_2$

3.21. How many moles of carbon are there in 500.0 grams of carbon?

3.22. How many moles of gold are there in 2.00 ounces of gold?

3.23. How many moles of Ca^{2+} ions are in 0.25 moles of calcium titanate, $CaTiO_3$? What is the mass in grams of these Ca^{2+} ions?

3.24. How many moles of O^{2-} ions are in 0.55 moles of aluminum oxide, Al_2O_3? What is the mass in grams of these O^{2-} ions?

3.25. How many moles of iron are there are in 1 mole of the following compounds? (a) FeO; (b) Fe_2O_3; (c) $Fe(OH)_3$; (d) Fe_3O_4

3.26. How many moles of sodium are there in 1 mole of the following compounds? (a) $NaCl$; (b) Na_2SO_4; (c) Na_3PO_4; (d) $NaNO_3$

3.27. Calculate the molar masses of the following atmospheric molecules: (a) SO_2; (b) O_3; (c) CO_2; (d) N_2O_5

3.28. Determine the molar masses of the following minerals:
a. rhodonite, $MnSiO_3$ c. ilmenite, $FeTiO_3$
b. scheelite, $CaWO_4$ d. magnesite, $MgCO_3$

3.29. Flavoring Additives Calculate the molar masses of the following common flavors in food:
a. vanillin, $C_8H_8O_3$ c. anise oil, $C_{10}H_{12}O$
b. oil of cloves, $C_{10}H_{12}O_2$ d. oil of cinnamon, C_9H_8O

3.30. Sweeteners Calculate the molar masses of the following common sweeteners:
a. sucrose, $C_{12}H_{22}O_{11}$ c. aspartame, $C_{14}H_{18}N_2O_5$
b. saccharin, $C_7H_5O_3NS$ d. fructose, $C_6H_{12}O_6$

3.31. Suppose pairs of balloons are filled with 10.0 grams of the following pairs of gases. Which balloon in each pair has the greater number of particles? (a) CO_2 or NO; (b) CO_2 or SO_2; (c) O_2 or Ar

3.32. If you had equal masses of the substances in the following pairs of compounds, which of the two would contain the greater number of ions? (a) $NaBr$ or KCl; (b) $NaCl$ or $MgCl_2$; (c) $BaCl_2$ or Li_2CO_3

3.33. How many moles of SiO_2 are there in a quartz crystal (SiO_2) that weighs 45.2 g?

3.34. How many moles of $NaCl$ are there in a crystal of halite that weighs 6.82 g?

3.35. What is the mass of 0.122 mol of $MgCO_3$?

3.36. What is the volume of 1.00 mol of benzene (C_6H_6) at 20°C? The density of benzene is 0.879 g/mL.

---

*3.37. The density of uranium (U; 19.05 g/cm³) is more than five times as great as that of diamond (C; 3.514 g/cm³). If you have a cube (1 cm on a side) of each element, which cube contains more atoms?

*3.38. Aluminum ($d = 2.70$ g/mL) and strontium ($d = 2.64$ g/mL) have nearly the same density. If we manufacture two cubes, each containing 1 mole of one element or the other, which cube will be smaller? What are the dimensions of this cube?

Writing Balanced Chemical Equations

CONCEPT REVIEW

3.39. In a balanced chemical equation, does the number of moles of reactants always equal the number of moles of products?

3.40. In a balanced chemical equation, does the sum of the coefficients for the reactants always equal the sum of the coefficients of the products?

3.41. In a balanced chemical equation, must the sum of the masses of all the gaseous reactants always equal the sum of the masses of the gaseous products?

3.42. In a balanced chemical equation, must the sum of the volumes occupied by the reactants always equal the sum of the volumes occupied by the products?

PROBLEMS

3.43. Balance the following reactions, which are believed to take place in the environment of a protostar, like that of our early sun:
a. $CH_4(g) + H_2O(g) \rightarrow CO(g) + H_2(g)$
b. $NH_3(g) \rightarrow N_2(g) + H_2(g)$
c. $CO(g) + H_2O(g) \rightarrow CO_2(g) + H_2(g)$

3.44. Chemistry of Volcanic Gases Balance the following reactions that occur among volcanic gases:
a. $SO_2(g) + O_2(g) \rightarrow SO_3(g)$
b. $H_2S(g) + O_2(g) \rightarrow SO_2(g) + H_2O(g)$
*c. $H_2S(g) + SO_2(g) \rightarrow S_8(s) + H_2O(g)$

---

*3.45. Chemical Weathering of Rocks and Minerals Balance the following chemical reactions, which contribute to weathering of the iron-silicate minerals ferrosillite ($FeSiO_3$), fayallite (Fe_2SiO_4), and greenalite [$Fe_3Si_2O_5(OH)_4$]:
a. $FeSiO_3(s) + H_2O(\ell) \rightarrow Fe_3Si_2O_5(OH)_4(s) + H_4SiO_4(aq)$
b. $Fe_2SiO_4(s) + CO_2(g) + H_2O(\ell) \rightarrow$
$$FeCO_3(s) + H_4SiO_4(aq)$$
c. $Fe_3Si_2O_5(OH)_4(s) + CO_2(g) + H_2O(\ell) \rightarrow$
$$FeCO_3(s) + H_4SiO_4(aq)$$

*3.46. Copper was one of the first metals used by humans because it can be prepared from a wide variety of copper minerals, such as cuprite (Cu_2O), chalcocite (Cu_2S), and malachite [$Cu_2CO_3(OH)_2$]. Balance the following reactions for converting these minerals into copper metal:
a. $Cu_2O(s) + C(s) \rightarrow Cu(s) + CO_2(g)$
b. $Cu_2O(s) + Cu_2S(s) \rightarrow Cu(s) + SO_2(g)$
c. $Cu_2CO_3(OH)_2(s) + C(s) \rightarrow Cu(s) + CO_2(g) + H_2O(g)$

3.47. Balance the following reactions, which contribute to atmospheric pollution by nitrogen oxides:
a. $N_2(g) + O_2(g) \rightarrow NO(g)$
b. $NO(g) + O_2(g) \rightarrow NO_2(g)$
c. $NO(g) + NO_3(g) \rightarrow NO_2(g)$
d. $N_2(g) + O_2(g) \rightarrow N_2O(g)$

3.48. Chemistry of Geothermal Vents Some scientists believe that life on Earth may have originated near deep-ocean vents. Balance the following reactions, which are among those taking place near such vents:
a. $CH_3SH(\ell) + CO(g) \rightarrow CH_3COSCH_3(\ell) + H_2S(g)$
b. $H_2S(g) + CO(g) \rightarrow CH_3CO_2H(g) + S_8(s)$

---

*3.49. Write a balanced chemical equation for each of the following reactions:
a. Dinitrogen pentoxide reacts with sodium metal to produce sodium nitrate and nitrogen dioxide.
b. A mixture of nitric acid and nitrous acid is formed when water reacts with dinitrogen tetroxide.
c. At high pressure, nitrogen monoxide decomposes to dinitrogen monoxide and nitrogen dioxide.

3.50. Write a balanced chemical equation for each of the following reactions:
a. Carbon dioxide reacts with carbon to form carbon monoxide.
b. Potassium reacts with water to give potassium hydroxide and the element hydrogen.
c. Phosphorus (P_4) burns in air to give diphosphorus pentoxide.

3.51. Write a balanced chemical equation for the combustion of acetylene (C_2H_2).

3.52. Write a balanced chemical equation for the combustion of octane (C_8H_{18}).

Combustion Reactions; Stoichiometric Calculations and the Carbon Cycle

CONCEPT REVIEW

3.53. If the sum of the masses of the reactants in a chemical equation equals the sum of the masses of the products, must the equation be balanced?

*3.54. There are two ways to write the equation for the combustion of ethane:
$$C_2H_6(g) + \tfrac{7}{2}O_2(g) \rightarrow 3\,H_2O(g) + 2\,CO_2(g)$$
$$2\,C_2H_6(g) + 7\,O_2(g) \rightarrow 6\,H_2O(g) + 4\,CO_2(g)$$
Do these two different ways of writing the equation affect the calculation of how much CO_2 is produced from a known quantity of C_2H_6?

PROBLEMS

3.55. Land Management The United Nations Intergovernmental Panel on Climate Change reported in June 2000 that better management of cropland, grazing land, and forests would reduce the amount of carbon dioxide in the atmosphere by 5.4×10^9 kg of carbon per year: (a) How many moles of carbon are present in 5.4×10^9 kg of carbon? (b) How many kilograms of carbon dioxide does this quantity of carbon represent?

3.56. Energy generation results in the addition of an estimated 27 billion metric tons of CO_2 to the atmosphere each year. (a) How many moles of CO_2 does 27 billion tons represent? (b) How many grams of carbon are in 27 billion tons of CO_2?

3.57. When $NaHCO_3$ is heated above 270°C, it decomposes to Na_2CO_3, H_2O, and CO_2. (a) Write a balanced chemical equation for the decomposition reaction. (b) Calculate the mass of CO_2 produced from the decomposition of 25.0 g of $NaHCO_3$.

3.58. **Egyptian Cosmetics** Pb(OH)Cl, one of the lead compounds used in ancient Egyptian cosmetics (see Problem 3.74), was prepared from PbO according to the following ancient recipe:
$$PbO(s) + NaCl(aq) + H_2O(\ell) \rightarrow Pb(OH)Cl(s) + NaOH(aq)$$
How many grams of PbO and how many grams of NaCl would be required to produce 10.0 g of Pb(OH)Cl?

3.59. The manufacture of aluminum includes the production of cryolite (Na_3AlF_6) from the following reaction:
$$6\ HF(g) + 3\ NaAlO_2(s) \rightarrow$$
$$Na_3AlF_6(s) + 3\ H_2O(\ell) + Al_2O_3(s)$$
How much $NaAlO_2$ (sodium aluminate) is required to produce 1.00 kg of Na_3AlF_6?

3.60. Chromium metal can be produced from the high-temperature reaction of Cr_2O_3 [chromium(III) oxide] with silicon or aluminum by each of the following reactions:
$$Cr_2O_3(s) + 2\ Al(\ell) \rightarrow 2\ Cr(\ell) + Al_2O_3(s)$$
$$2\ Cr_2O_3(s) + 3\ Si(\ell) \rightarrow 4\ Cr(\ell) + 3\ SiO_2(s)$$
a. Calculate the number of grams of aluminum required to prepare 400.0 g of chromium metal by the first reaction.
b. Calculate the number of grams of silicon required to prepare 400.0 g of chromium metal by the second reaction.

*3.61. Suppose 25 metric tons of coal that is 3.0% sulfur by mass are burned at an electric power plant. During combustion, the sulfur is converted into sulfur dioxide. How many tons of sulfur dioxide are produced?

3.62. Charcoal (C) and propane (C_3H_8) are used as fuel in backyard grills. (a) Write balanced chemical equations for the complete combustion reactions of C and C_3H_8. (b) How many grams of carbon dioxide are produced from burning 500.0 g of each of the two fuels?

*3.63. The uranium minerals found in nature must be refined and enriched in ^{235}U before the uranium can be used as a fuel in nuclear reactors. One procedure for enriching uranium relies on the reaction of UO_2 with HF to form UF_4, which is then converted into UF_6 by reaction with fluorine:
$$(1)\quad UO_2(g) + 4\ HF(aq) \rightarrow UF_4(g) + 2\ H_2O(\ell)$$
$$(2)\quad UF_4(g) + F_2(g) \rightarrow UF_6(g)$$
a. How many kilograms of HF are needed to completely react with 5.00 kg of UO_2?
b. How much UF_6 can be produced from 850.0 g of UO_2?

3.64. The mineral bauxite, which is mostly Al_2O_3, is the principal industrial source of aluminum metal. How much aluminum can be produced from 1.00 metric ton of Al_2O_3?

*3.65. Chalcopyrite ($CuFeS_2$) is an abundant copper mineral that can be converted into elemental copper. How much Cu could be produced from 1.00 kg of $CuFeS_2$?

*3.66. **Mining for Gold** Unlike most metals, gold is found in nature as the pure element. Miners in California in 1849 searched for gold nuggets and gold dust in streambeds, where the denser gold could be easily separated from sand and gravel. However, larger deposits of gold are found in veins of rock and can be separated chemically in a two-step process:
$$(1)\quad 4\ Au(s) + 8\ NaCN(aq) + O_2(g) + 2\ H_2O(\ell)$$
$$\rightarrow 4\ NaAu(CN)_2(aq) + 4\ NaOH(aq)$$
$$(2)\quad 2\ NaAu(CN)_2(aq) + Zn(s)$$
$$\rightarrow 2\ Au(s) + Na_2[Zn(CN)_4](aq)$$
If a 1.0×10^3 kilogram sample of rock is 0.019% gold by mass, how much Zn is needed to react with the gold extracted from the rock? Assume that reactions (1) and (2) are 100% efficient.

Determining Empirical Formulas from Percent Composition; Empirical and Molecular Formulas Compared

CONCEPT REVIEW

3.67. What is the difference between an empirical formula and a molecular formula?

3.68. Do the empirical and molecular formulas of a compound have the same percent composition values?

3.69. Is the element with the largest atomic mass always the element present in the highest percentage by mass in a compound?

3.70. Sometimes the composition of a compound is expressed as a mole percent or atom percent. Are the values of these parameters likely to be the same for a given compound, or different?

PROBLEMS

3.71. Calculate the percent composition of (a) Na_2O, (b) NaOH, (c) $NaHCO_3$, and (d) Na_2CO_3.

3.72. Calculate the percent composition of (a) sodium sulfate, (b) dinitrogen tetroxide, (c) strontium nitrate, and (d) aluminum sulfide.

3.73. **Organic Compounds in Space** The following compounds have been detected in space. Which of them contains the greatest percentage of carbon by mass? (a) naphthalene, $C_{10}H_8$; (b) chrysene, $C_{18}H_{12}$; (c) pentacene, $C_{22}H_{14}$; (d) pyrene, $C_{16}H_{10}$

3.74. Ancient Egyptians used a variety of lead compounds as white pigments in their cosmetics (see Problem 3.58), including PbS, $PbCO_3$, PbCl(OH), and $Pb_2Cl_2CO_3$. Which of these compounds contains the highest percentage of lead?

3.75. Of the nitrogen oxides—N_2O, NO, N_2O_3, and NO_2—which is more than 50% oxygen by mass?

3.76. Of the sulfur oxides—S_2O, SO, SO_2, and SO_3—which is more than 50% oxygen by mass?

3.77. Do any two of the following compounds, which have been detected in outer space, have the same empirical formula?
   a. naphthalene, $C_{10}H_8$ d. pyrene, $C_{16}H_{10}$
   b. chrysene, $C_{18}H_{12}$ e. benzoperylene, $C_{22}H_{12}$
   c. anthracene, $C_{14}H_{10}$ f. coronene, $C_{24}H_{12}$

3.78. Which, if any, of the following nitrogen oxides have the same empirical formula? (a) N_2O; (b) NO; (c) NO_2; (d) N_2O_2; (e) N_2O_4

3.79. Zircon is a common substitute for diamond in inexpensive jewelry. The percent composition of zircon is 49.76% Zr, 15.32% Si, and the remainder oxygen. Determine the empirical formula of zircon.

3.80. A sample of an iron-containing compound is 22.0% iron, 50.2% oxygen, and 27.8% chlorine by mass. What is the empirical formula of this compound?

3.81. In an experiment, 2.43 g of magnesium reacts with 1.60 g of oxygen, forming 4.03 g of magnesium oxide. (a) Use these data to calculate the empirical formula of magnesium oxide. (b) Write a balanced chemical equation for this reaction.

3.82. Ferrophosphorus (Fe_2P) reacts with pyrite (FeS_2), producing iron(II) sulfide and a compound that is 27.87% P and 72.13% S by mass and has a molar mass of 444.56 g/mol. (a) Determine the empirical and molecular formulas of this compound. (b) Write a balanced chemical equation for this reaction.

3.83. Asbestos was used for years as an insulating material in buildings until exposure to asbestos was found to cause lung cancer. Asbestos is a mineral containing magnesium, silicon, oxygen, and hydrogen. One form of asbestos, chrysolite, has the composition 28.03% magnesium, 21.60% silicon, and 1.16% hydrogen. Determine the empirical formula of chrysolite.

3.84. **Chemistry of Soot** A candle flame produces easily seen specks of soot near the edges of the flame, especially when the candle is moved. A piece of glass held over a candle flame will become coated with soot, which is the result of the incomplete combustion of candle wax. Elemental analysis of a compound extracted from a sample of this soot gave these results: 7.74% H and 92.26% C by mass. Calculate the empirical formula of the compound.

3.85. What is the empirical formula of the compound that is 24.2% Cu, 27.0% Cl, and 48.8% O by mass?

3.86. The compound made of chlorine and oxygen that has been used to kill anthrax spores in contaminated buildings is 52.6% Cl. What is its empirical formula?

Combustion Analysis

CONCEPT REVIEW

3.87. Explain why it is important for combustion analysis to be carried out in an excess of oxygen.

3.88. Why is the quantity of CO_2 obtained in a combustion analysis not a direct measure of the oxygen content of the starting compound?

3.89. Can the results of a combustion analysis ever give the true molecular formula of a compound?

3.90. What additional information is needed to determine a molecular formula from the results of an elemental analysis of an organic compound?

PROBLEMS

3.91. The combustion of 135.0 mg of a hydrocarbon produces 440.0 mg of CO_2 and 135.0 mg H_2O. The molar mass of the hydrocarbon is 270 g/mol. Determine the empirical and molecular formulas of this compound.

3.92. A 0.100 g sample of a compound containing C, H, and O is burned in oxygen, producing 0.1783 g of CO_2 and 0.0734 g of H_2O. Determine the empirical formula of the compound.

3.93. Methylheptenone contains the elements C, H, and O and has a lemon-like odor. The complete combustion of 192 mg of methylheptenone produces 528 mg of CO_2 and 216 mg of H_2O. What is the empirical formula of methylheptenone?

*__3.94.__ The combustion of 40.5 mg of a compound containing C, H, and O, and extracted from the bark of the sassafras tree, produces 110.0 mg of CO_2 and 22.5 mg of H_2O. The molar mass of the compound is 162 g/mol. Determine its empirical and molecular formulas.

Limiting Reactants and Percent Yield

CONCEPT REVIEW

3.95. If a reaction vessel contains equal masses of Fe and S, a mass of FeS corresponding to which of the following could theoretically be produced? (a) the sum of the masses of Fe and S; (b) more than the sum of the masses of Fe and S; (c) less than the sum of the masses of Fe and S

3.96. A reaction vessel contains equal masses of magnesium metal and oxygen gas. The mixture is ignited, forming MgO. After the reaction has gone to completion, the mass of the MgO is less than the mass of the reactants. Is this result a violation of the law of conservation of mass? Explain your answer.

3.97. Explain how the parameters of theoretical yield and percent yield differ.

3.98. Can the percent yield of a chemical reaction ever exceed 100%?

3.99. Give two reasons why the actual yield from a chemical reaction is usually less than the theoretical yield.

3.100. A chemical reaction produces less than the expected amount of product. Is this result a violation of the law of conservation of mass?

PROBLEMS

3.101. A recipe for 1 cup of hollandaise sauce calls for $\frac{1}{2}$ cup of butter, $\frac{1}{4}$ cup of hot water, 4 egg yolks, and the juice of a medium-sized lemon. How many cups of this sauce can be made from a pound (2 cups) of butter, a dozen eggs, 4 medium-size lemons, and an unlimited supply of hot water?

3.102. A factory making toy wagons has 13,466 wheels, 3360 handles, and 2400 wagon beds in stock. What maximum number of wagons can the factory make?

3.103. Potassium superoxide, KO_2, reacts with carbon dioxide to form potassium carbonate and oxygen:
$$4\,KO_2(s) + 2\,CO_2(g) \rightarrow 2\,K_2CO_3(s) + 3\,O_2(g)$$
This reaction makes potassium superoxide useful in a self-contained breathing apparatus. How much O_2 could be produced from 2.50 g of KO_2 and 4.50 g of CO_2?

3.104. A reaction vessel contains 10.0 g of CO and 10.0 g of O_2. How many grams of CO_2 could be produced according to the following reaction?
$$2\,CO(g) + O_2(g) \rightarrow 2\,CO_2(g)$$

3.105. Ammonia rapidly reacts with hydrogen chloride, making ammonium chloride. Write a balanced chemical equation for the reaction, and calculate the number of grams of excess reactant when 3.0 g of NH_3 reacts with 5.0 g of HCl.

3.106. Sulfur trioxide dissolves in water, producing H_2SO_4. How much sulfuric acid can be produced from 10.0 mL of water ($d = 1.00$ g/mL) and 25.6 g of SO_3?

3.107. The reaction of 3.0 g of carbon with excess O_2 yields 6.5 g of CO_2. What is the percent yield of this reaction?

3.108. Baking soda ($NaHCO_3$) can be made in large quantities by the following reaction:
$$NaCl(aq) + NH_3(aq) + CO_2(aq) + H_2O(\ell) \rightarrow$$
$$NaHCO_3(s) + NH_4Cl(aq)$$
If 10.0 g of NaCl reacts with excesses of the other reactants and 4.2 g of $NaHCO_3$ is isolated, what is the percent yield of the reaction?

3.109. Chemistry of Fermentation Yeast converts glucose ($C_6H_{12}O_6$) into ethanol ($d = 0.789$ g/mL) in a process called fermentation. An unbalanced equation for the reaction can be written as follows:
$$C_6H_{12}O_6(aq) \rightarrow C_2H_5OH(\ell) + CO_2(g)$$
a. Write a balanced chemical equation for this fermentation reaction.
b. If 100.0 g of glucose yields 50.0 mL of ethanol, what is the percent yield for the reaction?

*3.110. Composition of Seawater A 1-liter sample of seawater contains 19.4 g of Cl^-, 10.8 g of Na^+, and 1.29 g of Mg^{2+}. (a) How many moles of each ion are present? (b) If we evaporated the seawater, would there be enough Cl^- present to form the chloride salts of all the sodium and magnesium present?

Additional Problems

*3.111. The material often used to make artificial bones is the same material that gives natural bones their structure. Its common name is hydroxyapatite, and its formula is $Ca_5(PO_4)_3OH$.
a. Propose a systematic name for this compound.
b. What is the mass percentage of calcium in it?
c. When treated with hydrogen fluoride, hydroxyapatite becomes fluorapatite [$Ca_5(PO_4)_3F$], a stronger structure. The strengthening is due to the altered mass percentage of calcium in the substance. Does the % mass of Ca increase or decrease as a result of this substitution?

*3.112. As a solution of copper sulfate slowly evaporates, beautiful blue crystals made of Cu(II) and sulfate ions form such that water molecules are trapped inside the crystals. The overall formula of the compound is $CuSO_4 \cdot 5\,H_2O$. (a) What is the percent water in this compound? (b) At high temperatures the water in the compound is driven off as steam. What mass percentage of the original sample of the blue solid is lost as a result?

3.113. Aluminum is mined as the mineral bauxite, which consists primarily of Al_2O_3 (alumina).
a. How much aluminum is produced from 1 metric ton of Al_2O_3?
$$2\,Al_2O_3(s) \rightarrow 4\,Al(s) + 3\,O_2(g)$$
b. The oxygen produced in part a is allowed to react with carbon to produce carbon monoxide. Write a balanced equation describing the reaction of alumina with carbon. How much CO is produced in part a?

*3.114. Chemistry of Copper Production "Native," or elemental, copper can be found in nature, but most copper is mined as oxide or sulfide minerals. Chalcopyrite ($CuFeS_2$) is one copper mineral that can be converted to elemental copper in a series of chemical steps. Reacting chalcopyrite with oxygen at high temperature produces a mixture of copper sulfide and iron oxide. The iron oxide is separated from CuS by reaction with sand. CuS is converted to Cu_2S in the process and the Cu_2S is burned in air to produce Cu and SO_2:
(1) $2\,CuFeS_2(s) + 3\,O_2(g)$
$$\rightarrow 2\,CuS(s) + 2\,FeO(s) + 2\,SO_2(g)$$
(2) $FeO(s) + SiO_2(s) \rightarrow FeSiO_3(s)$
(3) $2\,CuS(s) \rightarrow Cu_2S(s) + S(s)$
(4) $Cu_2S(s) + O_2(g) \rightarrow 2\,Cu(s) + SO_2(g)$
An average copper penny minted in the 1960s weighed about 3.0 g.
a. How much chalcopyrite had to be mined to produce one dollar's worth of pennies?
b. How much chalcopyrite had be mined to produce one dollar's worth of pennies if reaction 1 above had a percent yield of 85% and reactions 2, 3, and 4 had percent yields of essentially 100%?
c. How much chalcopyrite had be mined to produce one dollar's worth of pennies if each reaction involving copper proceeded with an 85% yield?

*3.115. Mining for Gold Gold can be extracted from the surrounding rock using a solution of sodium cyanide. While effective for isolating gold, toxic cyanide finds its way into watersheds, causing environmental damage and harm to human health.
$$4\,Au(s) + 8\,NaCN(aq) + O_2(g) + 2\,H_2O(\ell) \rightarrow$$
$$4\,NaAu(CN)_2(aq) + 4\,NaOH(aq)$$
$$2\,NaAu(CN)_2(aq) + Zn(s) \rightarrow 2\,Au(s) + Na_2[Zn(CN)_4](aq)$$
a. If a sample of rock contains 0.009% gold by mass, how much NaCN is needed to extract the gold from 1 metric ton of rock as $NaAu(CN)_2$?
b. How much zinc is needed to convert $NaAu(CN)_2$ from part a to metallic gold?
c. The gold recovered in part b is manufactured into a gold ingot in the shape of a cube. The density of gold is 19.3 g/cm³. How big is the block of gold?

*3.116. Phosgenite, a lead compound with the formula $Pb_2Cl_2CO_3$, is found in Egyptian cosmetics. Phosgenite was prepared by the reaction of PbO, NaCl, and CO_2. An unbalanced equation of the reactant mixture is

$$PbO(s) + NaCl(aq) + H_2O(\ell) + CO_2(g) \rightarrow$$
$$Pb_2Cl_2CO_3(s) + NaOH(aq)$$

   a. Balance this equation.

   b. How many grams of phosgenite can be obtained from 10.0 g of PbO and 10.0 g NaCl in the presence of excess water and CO_2?

   c. Phosgenite can be considered a mixture of two lead compounds. Which compounds appear to be combined to make phosgenite?

*3.117. Uranium oxides used in the preparation of fuel for nuclear reactors are separated from other metals in minerals by converting the uranium to $UO_x(NO_3)_y(H_2O)_z$, where uranium has a positive charge ranging from 3+ to 6+.

   a. Roasting $UO_x(NO_3)_y(H_2O)_z$ at 400°C leads to loss of water and decomposition of the nitrate ion to nitrogen oxides, leaving behind a product with the formula U_aO_b that is 83.22% by mass. What are the values of a and b? What is the charge on U in U_aO_b?

   b. Higher temperatures produce a different uranium oxide, U_cO_d, with a higher uranium content, 84.8% U. What are the values of c and d? What is the charge on U in U_aO_b?

   c. The values of x, y, and z in $UO_x(NO_3)_y(H_2O)_z$ are found by gently heating the compound to remove all of the water. In a laboratory experiment, 1.328 g of $UO_x(NO_3)_y(H_2O)_z$ produced 1.042 g of $UO_x(NO_3)_y$. Continued heating generated 0.742 g of U_nO_m. Using the information in parts a and b, calculate x, y, and z.

*3.118. Large quantities of fertilizer are washed into the Mississippi River from agricultural land in the Midwest. The excess nutrients collect in the Gulf of Mexico, promoting the growth of algae and endangering other aquatic life.

   a. One commonly used fertilizer is ammonium nitrate. What is the chemical formula of nitrate?

   b. Corn farmers typically use 5.0×10^3 kg of ammonium nitrate per square kilometer of cornfield per year. Ammonium nitrate can be prepared by the following reaction:

$$NH_3(aq) + HNO_3(aq) \rightarrow NH_4NO_3(aq)$$

   How much nitric acid would be needed to make the fertilizer needed for 1 km² of cornfield per year?

   c. The ammonium ions can be converted into NO_3^- by bacterial action.

$$NH_4^+(aq) + 2O_2(g) \rightarrow NO_3^-(aq) + H_2O(\ell) + 2H^+(aq)$$

   If 10% of the ammonium component of 5.0×10^2 kg of fertilizer ends up as nitrate, how much oxygen would be consumed?

3.119. Calculate the number of molecules of compound in each of the following common, over-the-counter medications: (a) ibuprofen, a pain reliever and fever reducer that contains 200.0 mg of the active ingredient, $C_{13}H_{18}O_2$; (b) an antacid containing 500.0 mg of calcium carbonate; (c) an allergy tablet containing 4 mg Chlor-Trimeton ($C_{16}H_{19}N_2Cl$)

3.120. **Chemistry of Pain Relievers** The common pain relievers aspirin ($C_9H_8O_4$), acetaminophen ($C_8H_9NO_2$), and naproxen sodium ($C_{14}H_{13}O_3Na$) are all available in tablets containing 200.0 mg of the active ingredient. Which compound contains the greatest number of molecules per tablet? How many molecules of the active ingredient are present in each tablet?

3.121. **Fiber in the Diet** Dietary fiber is a mixture of many compounds including xylose ($C_5H_{10}O_5$) and methyl galacturonate ($C_7H_{12}O_7$). (a) Do these compounds have the same empirical formula? (b) Write a balanced chemical equation for the combustion of xylose and methyl galacturonate.

3.122. Some catalytic converters in automobiles contain the manganese oxides Mn_2O_3 and MnO_2. (a) Give the names of Mn_2O_3 and MnO_2. (b) Calculate the percent manganese by mass in Mn_2O_3 and MnO_2. (c) Explain how Mn_2O_3 and MnO_2 are consistent with the law of multiple proportions.

*3.123. A number of chemical reactions have been proposed for the formation of organic compounds from inorganic precursors. Here is one of them:

$$H_2S(g) + FeS(s) + CO_2(g) \rightarrow FeS_2(s) + HCO_2H(\ell)$$

   a. Identify the ions in FeS and FeS_2. Give correct names for each compound.

   *b. How much HCO_2H is obtained by reacting 1.00 g of FeS, 0.50 g of H_2S, and 0.50 g of CO_2 if the reaction results in a 50.0% yield?

*3.124. The formation of organic compounds by the reaction of iron(II) sulfide with carbonic acid is described by the following chemical equation:

$$2\,FeS + H_2CO_3 \rightarrow 2\,FeO + \frac{1}{n}(CH_2O)_n + 2\,S$$

   a. How much FeO is produced starting with 1.50 g FeS, and 0.525 moles of H_2CO_3 if the reaction results in a 78.5% yield?

   b. If the carbon-containing product has a molar mass of 3.00×10^2 g/mol, what is the chemical formula of the product?

*3.125. **Marine Chemistry of Iron** On the seafloor, iron(II) oxide reacts with water to form Fe_3O_4 and hydrogen in a process called serpentization.

   a. Balance the following equation for serpentization:

$$FeO(s) + H_2O(\ell) \rightarrow Fe_3O_4(s) + H_2(g)$$

   b. When CO_2 is present, the product is methane, not hydrogen. Balance the following chemical equation:

$$FeO(s) + H_2O(\ell) + CO_2(g) \rightarrow Fe_3O_4(s) + CH_4(g)$$

3.126. Titanium dioxide and zinc oxide are two of the 16 active ingredients approved by the FDA for use in sunscreens. (a) What are the chemical formulas of these compounds? (b) How would you modify the names for the two compounds based on the rules for naming given in Chapter 2? (c) Which compound contains the higher percentage of oxygen by mass?

3.127. The solar wind is made up of ions, mostly protons, flowing out from the sun at about 400 km/s. Near Earth, each cubic kilometer of interplanetary space contains on average 6×10^{15} solar-wind ions. How many moles of ions are in a cubic kilometer of near-Earth space?

3.128. The famous Hope Diamond at the Smithsonian National Museum of Natural History weighs 45.52 carats (see Figure P3.128). Diamond is a crystalline form of carbon. (a) How many moles of carbon are in the Hope Diamond (1 carat = 200 mg)? (b) How many carbon atoms are in the diamond?

FIGURE P3.128

*3.129. E-85 is an alternative fuel for automobiles and light trucks that consists of 85% (by volume) ethanol, C_2H_5OH, and 15% gasoline. The density of ethanol is 0.79 g/mL. How many moles of ethanol are in a gallon of E-85?

3.130. With reference to the previous question, how many moles of carbon dioxide are produced by the complete combustion of the quantity of ethanol in a gallon of E-85 fuel?

*3.131. A 100.00 g sample of white powder (substance A) is heated to 550°C. At that temperature the powder decomposes, giving off colorless gas B, which is denser than air and which is neither flammable nor does it support combustion. The products also include 56 g of a second white powder C. When gas B is bubbled through a solution of calcium hydroxide, substance A reforms. What are the identities of substances A, B, and C?

3.132. A sealed chamber contains 1.604 g of CH_4 and 6.800 g of O_2. The mixture is ignited. How many grams of CO_2 are produced?

*3.133. You are given a 0.6240 g sample of a substance with the generic formula $MCl_2(H_2O)_2$. After completely drying the sample (which means removing the 2 moles of H_2O per mole of MCl_2), the sample weighs 0.5471 g. What is the identity of element M?

3.134. A compound found in crude oil consists of 93.71% C and 6.29% H by mass. The molar mass of the compound is 128 g/mol. What is its molecular formula?

3.135. A reaction vessel for synthesizing ammonia by reacting nitrogen and hydrogen is charged with 6.04 kg of H_2 and excess N_2. A total of 28.0 kg of NH_3 are produced. What is the percent yield of the reaction?

3.136. If a cube of table sugar, which is made of sucrose, $C_{12}H_{22}O_{11}$, is added to concentrated sulfuric acid, the acid "dehydrates" the sugar: removing the hydrogen and oxygen from it and leaving behind a lump of carbon. What percent of the initial mass of sugar that remains is carbon?

*3.137. A power plant burns 1.0×10^2 metric tons of coal that contains 3.0% (by mass) sulfur. The sulfur is converted to SO_2 during combustion.
   a. How many metric tons of SO_2 are produced?
   b. When SO_2 escapes into the atmosphere it may combine with O_2 and H_2O, forming sulfuric acid, H_2SO_4. Write a balanced chemical equation describing this reaction.
   c. How many metric tons of sulfuric acid, a component of acid rain, could be produced from the quantity of SO_2 calculated in part a?

3.138. **Reducing SO₂ Emissions** With respect to the previous question, one way to reduce the formation of acid rain involves trapping the SO_2 by passing smokestack gases through a spray of calcium oxide and O_2. The product of this reaction is calcium sulfate.
   a. Write a balanced chemical equation describing this reaction.
   b. How many metric tons of calcium sulfate would be produced from each ton of SO_2 that is trapped?

3.139. In the early 20th century, Londoners suffered from severe air pollution caused by burning high-sulfur coal. The sulfur dioxide that was emitted into the air mixed with London fog, forming sulfuric acid. For every gram of sulfur that was burned, how many grams of sulfuric acid could have formed?

*3.140. **Gas Grill Reaction** The burner in a gas grill mixes 24 volumes of air for every one volume of propane (C_3H_8) fuel. Like all gases, the volume that propane occupies is directly proportional to the number of moles of it at a given temperature and pressure. Air is 21% (by volume) O_2. Is the flame produced by the burner fuel rich (excess propane in the reaction mixture), fuel lean (not enough propane), or stoichiometric (just right)?

3.141. A common mineral in Earth's crust has the chemical composition 34.55% Mg, 19.96% Si, and 45.49% O. What is its empirical formula?

*3.142. **Ozone Generators** Some indoor air-purification systems work by converting a little of the oxygen in the air to ozone, which oxidizes mold and mildew spores and other biological air pollutants. The chemical equation for the ozone generation reaction is
$$3\,O_2(g) \rightarrow 2\,O_3(g)$$
It is claimed that one such system generates 4.0 g of O_3 per hour from dry air passing through the purifier at a flow of 5.0 L/min. If 1 liter of indoor air contains 0.28 g of O_2, (a) what fraction of the molecules of O_2 is converted to O_3 by the air purifier? (b) What is the percent yield of the ozone generation reaction?

Solution Chemistry and the Hydrosphere

4

ANGEL FISH BY WILLIAM NUTT *This sculpture was carved from Vermont marble, which was formed from the calcium carbonate remains of life in Earth's early oceans.*

A LOOK AHEAD: The Constant Ocean

Millions of years after Earth formed, the temperature of its atmosphere decreased and water vapor condensed, producing torrential rains. As this rain fell, oxides of sulfur, nitrogen, and other abundant nonmetals in the atmosphere reacted with water, making acidic rain. When this rain contacted rocks made of metal oxides and other basic minerals capable of reacting with acids, chemical reactions transformed the rocks by dissolving some and forming new minerals from others.

The oceans of Earth have existed in essentially the same volume since their original formation, and the chemical content of seawater has changed very little over the eons. How can this be, considering the continuous addition of soluble and suspended material from rivers? A physical and chemical system operates on a grand scale to maintain the composition of the oceans. Soluble nutrients in the form of ions constantly dissolve from the land and enter the sea. The ocean is salty because of these materials, but it lacks the soluble nutrients that plants need. Where do the nutrients go? They end up trapped in the mud on the ocean floor. This mud eventually returns to Earth's surface as oceans recede, new coasts appear, and mountain ranges emerge made from the sedimentary rocks formed when the mud of the ocean floor is compressed. This monumental cycle repeats: rain dissolves ions when it falls on land, rivers carry dissolved ions and suspended matter to the sea, and they all ultimately return to the mud on the ocean floor, only to be recycled again as nutrient-rich land with the next shift in Earth's crust.

Our body fluids are saline solutions, though not quite as salty as seawater. The concentration of dissolved matter in the tiny oceans within our cells also stays remarkably constant. Cells contain sodium, potassium, chloride, and hydrogen carbonate ions; when blood tests are done to assess the health of our systems, the concentrations of these ions are a crucial part of the picture. A shift in the usual balance indicates serious problems ranging from kidney disease to drug abuse.

Many other solutions are part of daily life. Vinegar in salad dressing is a solution of acetic acid in water. Those of us who wear contact lenses probably wash them in a solution containing sodium chloride and cleaning agents. The fluid in the battery of our car is a solution of sulfuric acid in water. Most beverages are solutions—even bottled water contains dissolved ions that give it a refreshing flavor.

Identifying the right solution for a particular task requires that the concentration of dissolved material be known. Environmentalists and toxicologists talk of concentrations in parts per million and parts per billion, physicians report normal ions in blood in milligrams per liter (mg/L), nutrients and additives in foods and beverages may be described in percentages or in grams per liter (g/L), oceanographers use milligrams per kilogram (mg/kg) for ions in seawater, and chemists use a special unit called molarity (moles/L). Everyone has the concentration units they need to define the solutions that concern them. In this chapter, we present several ways to express the concentrations of solutions and discuss the major reactions taking place when solutions are combined to prepare new materials or used to analyze the world around us.

4.1 Earth: The Water Planet

In Chapter 3 we discussed a theory of how Earth's early atmosphere may have been formed from the gases released by volcanic activity. A major component of that atmosphere was water vapor. As Earth cooled, this vapor condensed and torrential rains fell. Eventually the rare combination of Earth's atmospheric composition and proximity to the sun allowed Earth to become the "water planet": depressions in its crust gradually filled with about 1.5×10^{21} liters of nature's solvent, liquid H_2O.

Life exists on our planet because liquid water is abundant here. The current debate over the existence of life elsewhere in our solar system and in other planetary systems hinges on the prospect of liquid water existing on those planets. Even the current theories about life on Mars still rely on the presence of water. The belief that Martian meteorites collected in Antarctica may contain fossilized forms of life is linked to evidence that there may have been water on the surface of Mars in the distant past. The Grand Canyon in the southwestern United States was formed by the action of water on rocks, and similar features have been observed on Mars (Figure 4.1). In December 2006, scientists from the Malin Space Science Center in San Diego, CA, showed images from the *Mars Orbital Camera* indicating that water may have flowed down two gullies on the surface of Mars during the past several years (Figure 4.2). This observation lends support to the theory that liquid water may now exist just under the surface of Mars, and where there is water, there could be life. The biochemical reactions of all living cells, from single-celled organisms to human beings, require the presence of liquid water, and understanding the chemistry of life as we know it requires understanding the principles of chemical reactions between substances dissolved in water.

(a)

(b)

FIGURE 4.1 (a) Arizona's Grand Canyon is a dramatic example of how Earth's surface is continually modified by flowing water. (b) Similar topographic features observed on Mars suggest that water once flowed on the Martian surface, and may even flow there now (see Figure 4.2).

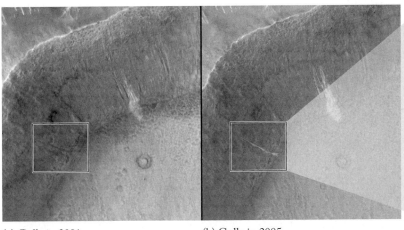

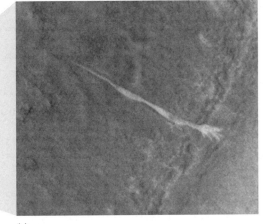

(a) Gully in 2001   (b) Gully in 2005   (c)

FIGURE 4.2 Photos from the *Mars Orbital Camera*: (a) Gully site as it appeared in 2001. (b) Mosaic of two images of the same gully taken in 2005 shows light-colored material that appears to have flowed in what was an otherwise nondescript gully. (c) An enlargement of the 2005 photo showing the new, light-colored deposit in the gully.

(a)

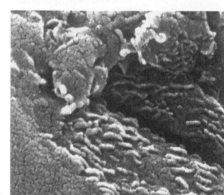

(b)

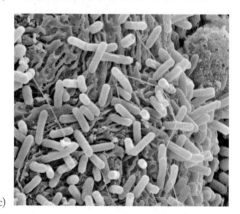

(c)

Antarctica is an ideal place for finding meteorites because of the lack of vegetation. (a) This meteorite, known as ALH84001, is believed to have originated on Mars when the impact of an asteroid millions of years ago led to the ejection of material from that planet. Some scientists believe that microscopic features in the sample are fossilized bacteria. (b) Possible fossilized bacteria in the meteorite ALH84001; (c) a colony of *E. coli*.

FIGURE 4.3 What a Chemist Sees

Adding sugar (the solute) to water (the solvent) produces a homogeneous solution of sugar molecules (represented by red spheres) evenly distributed among water molecules.

All natural waters, whether salt water or fresh, contain ionic and molecular compounds dissolved in H_2O. When an element or compound dissolves in another, a solution forms. **Solutions** are homogeneous mixtures of two or more substances (Figure 4.3). The substance present in a solution in the greatest proportion is called the **solvent**; the other ingredients are called **solutes**. When the solvent is water, the solution is an *aqueous* solution. The solvent may be a liquid, but it does not have to be. In Chapter 3 we considered the complex composition of Earth's crust. Many minerals in the crust are examples of *solid* solutions: uniform mixtures of substances in the solid state that do not have a fixed composition.

Seawater is sometimes called salt water, but that name may be misleading. Seawater is not just an aqueous solution of common table salt (NaCl). In fact, the water in the ocean is not a simple solution at all. Although seawater does contain an array of dissolved ions (Table 4.1) and molecular compounds, it also contains undissolved matter, including fine-grained sediments suspended by wave action near coastlines and soil eroded by rivers and streams. Regions of water containing such suspensions are visually distinct from solutions because they scatter light in ways that true solutions do not. In the open ocean, most of the suspended matter in surface seawater is biological, including microscopic plants known as phytoplankton, which can impart distinctive colors to the sea when their concentrations are unusually high (Figure 4.4). Marine scientists determine these concentrations

Sugar

Water

Homogeneous solution

TABLE 4.1 **Average Concentrations of the 11 Major Constituents of Seawater**

Ions	g/kg	mmol/kg	mmol/L[a]
Na^+	10.781	468.96	480.57
K^+	0.399	10.21	10.46
Mg^{2+}	1.284	52.83	54.14
Ca^{2+}	0.4119	10.28	10.53
Sr^{2+}	0.00794	0.0906	0.0928
Cl^-	19.353	545.88	559.40
SO_4^{2-}	2.712	28.23	28.93
HCO_3^-	0.126	2.06	2.11
Br^-	0.0673	0.844	0.865
$B(OH)_3$	0.0257	0.416	0.426
F^-	0.00130	0.068	0.070
Total	35.169	1119.87	1147.59

[a]Molar solutions are discussed in Section 4.2.

A **solution** is a uniform mixture of two or more substances.

The **solvent** of a solution is the component present in the largest amount. In an aqueous solution, water is the solvent.

The **solutes** of a solution are the components present in smaller amounts than the solvent. A solution may contain one or more solutes.

(a)   (b)

FIGURE 4.4 (a) Very high concentrations (called "blooms") of a phytoplankton called *coccolithophores* in the Bering Sea were captured by a NASA satellite in April 1998. During intense blooms, coccolithophores turn the color of the ocean a milky aquamarine. The color comes from chlorophyll in the phytoplankton, the milkiness from sunlight scattering off the organisms. (b) The many intricately designed plates (or *coccoliths*) made of $CaCO_3$ surround each cell of the organism. These coccoliths are so tiny that 500 of them placed end-to-end would fit in a length of 1 mm.

FIGURE 4.5 Particles in suspension, such as phytoplankton in seawater, can be separated by filtration. Scientists may extract the material trapped on the filter with an organic solvent such as acetone that dissolves the phytoplankton's chlorophyll and other pigments. The intensity of the color of these acetone extracts provides a measure of the concentration of phytoplankton in the original sample.

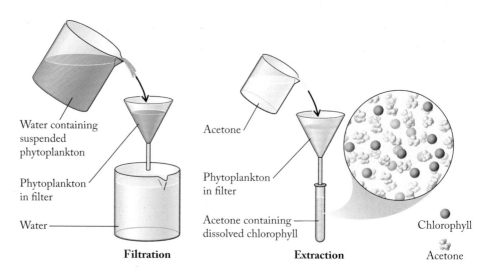

Water containing suspended phytoplankton

Phytoplankton in filter

Water

Filtration

Acetone

Phytoplankton in filter

Acetone containing dissolved chlorophyll

Extraction

Chlorophyll

Acetone

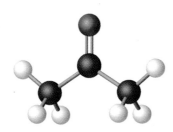

A ball-and-stick model of acetone.

by filtering samples of seawater to collect the suspended matter. During filtration, particulate matter including phytoplankton is trapped on the filter, while salts and molecular solutes pass through it (Figure 4.5).

CONCEPT TEST

Which, if any, of the following forms of matter are solutions—that is, homogeneous mixtures? (a) muddy river water; (b) helium gas; (c) clear cough syrup; (d) filtered air.

(Answers to Concept Tests are in the back of the book.)

4.2 Solution Concentration and Molarity

Table 4.1 lists the major ions in seawater and their concentrations expressed using three sets of units. The left column of data is expressed in grams of solute ion per kilogram (g/kg) of seawater. Many scientists, including those involved in environmental studies, use mass-to-mass ratios such as milligrams per kilogram (mg/kg), or mass-to-volume ratios such as milligrams per liter (mg/L), to express concentrations. When environmental regulatory agencies establish limits[1] on the concentrations of contaminants that are permitted in drinking water, those limits are usually expressed using the mass-to-volume ratio of milligrams of contaminant (solute) per liter of drinking water (solution). These agencies often express the maximum recommended human dosage of these contaminants in units of milligrams of contaminant per kilogram of body mass.

Another convenient set of units for expressing very small concentrations is based on parts per million (ppm) and parts per billion (ppb). A solution with a concentration of 1 ppm contains 1 part solute for every million parts of solution, for example, 1 gram of solute for every million (10^6) grams of solution. Put an-

---

[1] The U.S. Environmental Protection Agency (EPA) sets contaminant concentration limits called *maximum contaminant levels* (MCLs) for air and water, including drinking water. The MCLs of all contaminants in drinking water are given in mg/L. Similar limits have been established by Environment Canada, the World Health Organization, and other environmental agencies.

other way, each gram of a 1 ppm solution contains one-millionth of a gram (or 10^{-6} g = 1 μg) of solute, as shown by the following conversions:

$$1 \text{ ppm} = \frac{1\text{g}}{10^6 \text{ g}} \times \frac{10^{-6}}{10^{-6}} = \frac{10^{-6}\text{ g}}{\text{g}} = \frac{1\text{ μg}}{\text{g}}$$

A concentration of 1 ppm (1 μg/g) is also the same as 1 mg/kg:

$$1 \frac{\text{μg}}{\text{g}} \times \frac{1\text{ mg}}{10^3 \text{μg}} \times \frac{10^3 \text{g}}{\text{kg}} = \frac{1\text{ mg}}{\text{kg}}$$

A concentration of 1 ppb is 1/1000 as concentrated as a 1 ppm solution. Since a 1 ppm solution is the same as 1 mg/kg, a 1 ppb solution is equivalent to 1 μg/kg:

$$1 \text{ ppb} = \frac{1\text{ ppm}}{1000} = \frac{10^{-6}\text{ g}}{1000\text{ g}} = \frac{1\text{ μg}}{\text{kg}}$$

Let's express the smallest concentration value in Table 4.1 (0.00130 g F^-/kg seawater) in ppm. To do so, we need to convert grams of F^- ion into milligrams of F^- ion because 1 mg/kg is the same as 1 ppm:

$$0.00130 \frac{\text{g}}{\text{kg}} F^- \times \frac{10^3\text{ mg}}{\text{g}} = 1.30 \frac{\text{mg}}{\text{kg}} F^- = 1.30 \text{ ppm } F^-$$

As you can see, parts per million is a particularly convenient unit for expressing the concentration of F^- ion in seawater because it avoids the use of exponents or lots of zeroes to set the decimal place.

SAMPLE EXERCISE 4.1   **Comparing Ion Concentrations in Aqueous Solutions**

The average concentration of chloride ion in seawater is 19.353 g/kg. The maximum allowable concentration of chloride ion in drinking water is 250 ppm. How many times as much chloride ion is there in seawater than in the maximum amount allowed in safe drinking water?

COLLECT AND ORGANIZE The concentrations of Cl^- ion in the two samples are given in different units. To compare them, we must express them in the same units.

ANALYZE A link between ppm and mg/kg is the fact that mg/kg has the same denominator as the units in which the seawater concentration is expressed. We can convert the seawater concentration units of g/kg to mg/kg using the conversion factor 1000 mg/g.

SOLVE

1. We convert the concentration of Cl^- ion in seawater to ppm:

$$19.353 \frac{\text{g Cl}^-}{\text{kg}} \times \frac{10^3\text{ mg}}{\text{g}} = 19,353 \frac{\text{mg Cl}^-}{\text{kg}} = 19,353 \text{ ppm Cl}^-$$

2. Next we take the ratio of the two concentrations: (19,353 ppm Cl^-)/(250 ppm Cl^-) = 77.4 times as much chloride ion in seawater as in acceptable drinking water.

THINK ABOUT IT This question could also be answered by converting the suggested standard for chloride ion in drinking water from parts per million to grams per kilogram and comparing that number to 19.353 g/kg. It doesn't matter what

Molarity (*M*) is a concentration unit; it is the amount of solute (in moles) divided by the volume of solution (in liters): $M = n/V$. A 1.0 *M* solution contains 1.0 mole of solute per liter of solution.

▶ǁ CHEMTOUR Molarity

units you choose in order to make the comparison, as long as the units of both numbers are the same.

Practice Exercise The drinking water standard of the World Health Organization (WHO) for arsenic is 10.0 μg/L. Some water from tube wells in Bangladesh was found to contain as much as 1.2 mg/L of arsenic per liter. How many times above the WHO standard is this level?

(Answers to Practice Exercises are in the back of the book.)

Most scientists interested in studying chemical processes in natural waters or in solutions in the laboratory prefer to work with concentrations based on quantities of solute expressed in moles, not mass. When quantities are expressed in moles, comparisons of effects and behavior are based on the numbers of particles of substances. For these scientists, the preferred concentration unit is moles of solute per liter of solution volume. This ratio is called **molarity (*M*)**. A 1.0 molar (1.0 *M*) solution contains 1.0 mol of solute for every liter of solution. We can express this definition in equation form:

$$\text{Molarity} = \frac{\text{moles of solute}}{\text{volume of solution in liters}} \quad \text{or} \quad M = \frac{n}{V} \qquad (4.1)$$

For a 1.0 *M* solution,

$$1.0\ M = \frac{1.0\ \text{mol solute}}{\text{L}}$$

If we know the volume (*V*) of a solution in liters and its molar concentration (*M*), we can readily calculate the number of moles (*n*) of the solute in the solution. First we rearrange the terms in Equation 4.1 as follows:

$$n = V \times M \qquad (4.2)$$

Next, we convert the number of moles of solute into an equivalent mass in grams by multiplying *n* by the molar mass ($\mathcal{M}$):

$$\text{Mass of solute (g)} = n \times \mathcal{M} \qquad (4.3)$$

Combining Equations 4.2 and 4.3 yields this equation:

$$\text{Mass of solute (g)} = \mathcal{M} \times V \times M \qquad (4.4)$$

Equation 4.4 is useful when we need to calculate the mass of a solute needed to prepare a solution of a desired volume and concentration, or when we have a solution of known concentration and want to know the mass of solute present in a given volume of the solution.

Before we get to several Sample Exercises involving calculations of molarity, we should note that in many environmental and biological systems, a 1.0 mol/L solution represents a very high concentration of solute. Consider, for example, the major salts in seawater. The concentrations of these salts make seawater unfit for us to drink, yet the right-most column in Table 4.1 expresses these concentrations not in moles per liter but in millimoles per liter (mmol/L), which for simplicity are denoted m*M* (1 m*M* = 10^{-3} *M*). The concentrations of the minor and trace elements in surface seawater are often expressed in μ*M* ("micromolar," which is equivalent to 10^{-6} mol/L), nM ("nanomolar," or 10^{-9} mol/L), or even pM ("picomolar," or 10^{-12} mol/L). The concentration ranges of many biologically active substances in blood, urine, and other biological liquids are also so small that they are often expressed in units such as these.

The middle column in Table 4.1 contains values based on the number of millimoles of each ion per *kilogram* of seawater. Oceanographers prefer concentration units based on mass of seawater rather than volume because the volume of a given mass of water decreases with increasing depth on account of increasing pressure.

SAMPLE EXERCISE 4.2   **Calculating Molarity from Mass and Volume**

PVC is widely used in constructing homes and office buildings, but because vinyl chloride may leach from PVC pipe, it is used for drain pipes, not for drinking water supplies. The maximum concentration of vinyl chloride (C_2H_3Cl) allowed in drinking water in the United States is 0.002 mg/L. What is this concentration in terms of molarity?

COLLECT AND ORGANIZE We are asked to convert a concentration from milligrams of solute per liter of solution to moles of solute per liter of solution. This conversion involves changing mass in milligrams into mass in grams and then into an equivalent number of moles.

ANALYZE The conversion factors we need are the molar mass of C_2H_3Cl and the conversion factor that 1000 mg = 1 g:

$$\mathcal{M}_{C_2H_3Cl} = 2\text{ mol C} \times \frac{12.01\text{ g C}}{1\text{ mol C}} + 3\text{ mol H} \times \frac{1.008\text{ g H}}{1\text{ mol H}}$$

$$+ 1\text{ mol Cl} \times \frac{35.45\text{ g Cl}}{1\text{ mol Cl}} = 62.49\text{ g/mol}$$

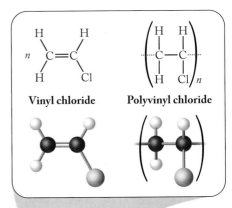

Vinyl chloride, a suspected carcinogen, enters drinking water by leaching from pipes made of polyvinyl chloride (PVC). Vinyl chloride is the common name of a small molecule, called a monomer (Greek for "one unit"), from which the very large molecule called a polymer ("many units") is formed. PVC usually consists of hundreds of molecules of vinyl chloride bonded together.

SOLVE Setting up the unit conversion steps so that the initial set of units cancels out, we have:

$$0.002\ \frac{\text{mg}}{\text{L}} \times \frac{1\text{ g}}{1000\text{ mg}} \times \frac{1\text{ mol}}{62.49\text{ g}} = 3.2 \times 10^{-8}\ \frac{\text{mol}}{\text{L}} = 3 \times 10^{-8}\ M \text{ or } 0.03\ \mu M$$

THINK ABOUT IT This number may at first seem very small, but remember that the mass of PVC (0.002 mg, or 2×10^{-6} g) in 1 L of water is also very small. Sometimes it is helpful to do an "order of magnitude" estimation in a problem like this: estimate the concentration as 1×10^{-6} g/L, the molar mass as 100 g/mol (1×10^2 g/mol), and the approximate concentration is then

$$\frac{1 \times 10^{-6}\text{ g/L}}{1 \times 10^2\text{ g/mol}} = 1 \times 10^{-8}\text{ mol/L}$$

which is very close to the number we calculated. Our number seems reasonable based on an estimation you can do quite quickly in your head. The extremely low limit of only 0.002 mg/L or 3×10^{-8} M reflects recognition of the danger this substance poses to human health.

Practice Exercise When a 1.00 L sample of water from the surface of the Dead Sea (which is more than 400 meters below sea level and much saltier than ordinary seawater) is evaporated, 179 grams of $MgCl_2$ are recovered. What is the molar concentration of $MgCl_2$ in the original sample?

(Answers to Practice Exercises are in the back of the book.)

SAMPLE EXERCISE 4.3   **Calculating Molarity from Density**

A water sample from the Great Salt Lake in Utah contains 83.6 milligrams of Na^+ per 1.000 gram of lake water. What is the molarity of Na^+ if the water has a density of 1.160 g/mL?

COLLECT AND ORGANIZE We are given a concentration in terms of a mass of solute per gram of solution. To convert this concentration into molarity (*moles* of solute per *liter* of solution), we need to convert the mass of solute into an equivalent number of moles, and the mass of solution into an equivalent volume of solution in liters.

ANALYZE Starting with the solute, we need to convert milligrams of Na^+ to grams of Na^+ and then into moles of Na^+. For this calculation, we need the molar mass of sodium ion: 22.99 g/mol. As for the solution, we need to convert its mass in grams into a volume in milliliters by dividing by the density given in the problem, and then convert milliliters into liters.

SOLVE

1. We find the number of moles of solute (sodium ion):

$$83.6 \text{ mg } Na^+ \times \frac{1 \text{ g}}{1000 \text{ mg}} \times \frac{1 \text{ mol } Na^+}{22.99 \text{ g } Na^+} = 3.64 \times 10^{-3} \text{ mol } Na^+$$

2. Then we calculate the volume of solution in liters, given that the sample of solution has a mass of 1.000 g and a density of 1.160 g/L:

$$1.000 \text{ g} \times \frac{1 \text{ mL}}{1.160 \text{ g}} \times \frac{1 \text{ L}}{1000 \text{ mL}} = 8.621 \times 10^{-4} \text{ L}$$

The molarity of sodium ion in the water from the Great Salt Lake is

$$\frac{\text{moles}}{\text{L}} = \frac{3.64 \times 10^{-3} \text{ mol } Na^+}{8.621 \times 10^{-4} \text{ L}} = 4.22 \ M$$

THINK ABOUT IT The data in Table 4.1 indicate that there are about 10 g of Na^+ for every 1 kg of seawater; this is the equivalent to 10 mg of Na^+ for every 1 g of water, which is about 10 mg/mL. The concentration given for sodium in the Great Salt Lake is over 8 times that number, so the concentration should be more than 8 times the concentration given in the table. Our answer is reasonable in light of that comparison. Only the Dead Sea (on the boundary between Israel and Jordan) is a saltier body of water, with Na^+ ion concentrations almost 11 times the average concentration of seawater.

Practice Exercise The density of ocean water depends on depth because pressure compresses the water. If the density of ocean water at a depth of 10,000 m is 1.071 g/mL and if 25.0 g of water at that depth contains 190 mg of potassium chloride, what is the molar concentration of potassium chloride in the sample?

(Answers to Practice Exercises are in the back of the book.)

SAMPLE EXERCISE 4.4 **Calculating the Quantity of Solute Needed to Prepare a Solution**

In setting up a saltwater aquarium, you need a stock solution of calcium hydroxide called Kalkwasser. How many grams of $Ca(OH)_2$ do you need in order to make 500.0 mL of a 0.0225 M solution of calcium hydroxide?

COLLECT AND ORGANIZE We know the volume (V) and molar concentration (M) of the solution. We also know the identity of the solute, which enables us to calculate its molar mass ($\mathcal{M}$). We can then use Equation 4.4

$$\text{Mass (g) of solute} = \mathcal{M} \times V \times M$$

to calculate the mass of solute $Ca(OH)_2$ needed. We also need to convert the volume in milliliters to liters to obtain the mass in grams of solute required.

ANALYZE We calculate the molar mass of $Ca(OH)_2$ as follows:

$$1 \; \text{mol Ca} \times \frac{40.08 \; \text{g}}{\text{mol Ca}} + 2 \; \text{mol O} \times \frac{16.00 \; \text{g}}{\text{mol O}} + 2 \; \text{mol H} \times \frac{1.008 \; \text{g}}{\text{mol H}} = 74.10 \; \text{g/mol}$$

and 500.0 mL converts to 0.5000 L.

SOLVE We combine the molar mass and volume values with the desired concentrations using Equation 4.4 to calculate the mass of $Ca(OH)_2$ needed:

$$\text{Mass (g) of } Ca(OH)_2 = \mathcal{M} \times V \times M$$

$$= 74.10 \; \frac{\text{g}}{\text{mol}} \times 0.0225 \; \frac{\text{mol}}{\text{L}} \times 0.5000 \; \text{L} = 0.834 \; \text{g}$$

THINK ABOUT IT To solve a problem like this, all you really need to remember is the definition of molarity: moles of solute per liter of solution. The solution you want has a concentration of 0.0225 M, or 0.0225 mole/L. You do not need an entire liter of solution; you only need 0.5000 L ($\frac{1}{2}$ L). This means that you do not need 0.0225 mole of $Ca(OH)_2$; you only need $\frac{1}{2}$ that amount (0.0113 mole). Finally, just calculate how many grams of calcium hydroxide are in 0.0113 mole of

The steps for preparing 500.0 mL of 0.0225 M $Ca(OH)_2$ include (a) weighing out and transferring the desired mass of $Ca(OH)_2$ to a 500 mL volumetric flask; (b) adding 200 to 300 mL of water to the flask; (c) swirling the flask to make sure all the solute is dissolved; and then (d) diluting the solution to exactly 500 mL. Stopper the flask and invert it several times to thoroughly mix the solution.

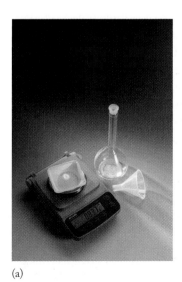

(a)

(b)

(c)

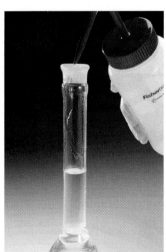

(d)

calcium hydroxide (0.834 g). Calcium hydroxide not only supplies calcium ions, which are essential to the health of creatures in the aquarium, but it also helps to maintain the correct levels of alkalinity (total quantity of basic substances) in the water.

Practice Exercise A solution known as Ringer's lactate is administered intravenously to trauma victims suffering from blood loss or severe burns. The solution contains the chloride salts of sodium, potassium, and calcium and is also 4.0 mM in sodium lactate ($NaC_3H_5O_3$). How many grams of sodium lactate are needed to prepare 10.0 liters of Ringer's lactate?

(Answers to Practice Exercises are in the back of the book.)

CONCEPT TEST

According to the data in Table 4.1, the concentration of Cl^- ion in seawater is 545.88 mmol/kg and 559.40 mmol/L. Does a 1.000 liter sample of seawater weigh more or less than 1.000 kilogram?

(Answers to Concept Tests are in the back of the book.)

4.3 Dilutions

In laboratories that test the quality of drinking water, standard stock solutions of regulated substances, such as pesticides and toxic metals, are available commercially in concentrations that are too high (typically 1000 mg/L) to be used directly in analytical procedures. Instead, these **stock solutions** are diluted to prepare *working standards*, which have concentrations of the substances of interest close to those found in actual samples. **Dilutions** lower concentration by adding more solvent so that the solute in a sample of the stock solution is distributed throughout a larger volume of solvent than before. We can use Equation 4.2 and the concentration of a stock solution to determine how to make a working standard with a desired concentration by dilution.

Suppose, for example, we need to prepare 250.0 mL of a working standard with a concentration of 3.18×10^{-3} M in Cu^{2+} by starting with a standard solution of Cu^{2+} that has a concentration of 0.1000 M. What volume of the standard solution do we need? Using Equation 4.2 we can calculate the number of moles of Cu^{2+} needed in the final solution:

$$n_{final} = V_{final} \times M_{final} = (0.2500 \text{ L}) \times (3.18 \times 10^{-3} \text{ mol/L}) = 7.95 \times 10^{-4} \text{ mol}$$

The subscript "final" in the above equation refers to the *final solution*. Using the same equation, we can calculate the *initial* quantities, that is, what volume of the *initial stock solution* we need to deliver that number of moles:

$$n_{initial} = V_{initial} \times M_{initial} \quad \text{or} \quad V_{initial} = \frac{n_{initial}}{M_{initial}}$$

$$V_{initial} = \frac{7.95 \times 10^{-4} \text{ mol}}{0.1000 \text{ mol/L}} = 7.95 \times 10^{-3} \text{ L} = 7.95 \text{ mL}$$

To make the desired working standard, add enough solvent (water) to 7.95 mL of the stock solution to make a solution with a final volume of 250.0 mL. The images in Figure 4.6 illustrate how this dilution is carried out.

A **stock solution** is a concentrated solution of a substance used to prepare solutions of lower concentration.

Dilution is the process of lowering the concentration of a solution by adding more solvent.

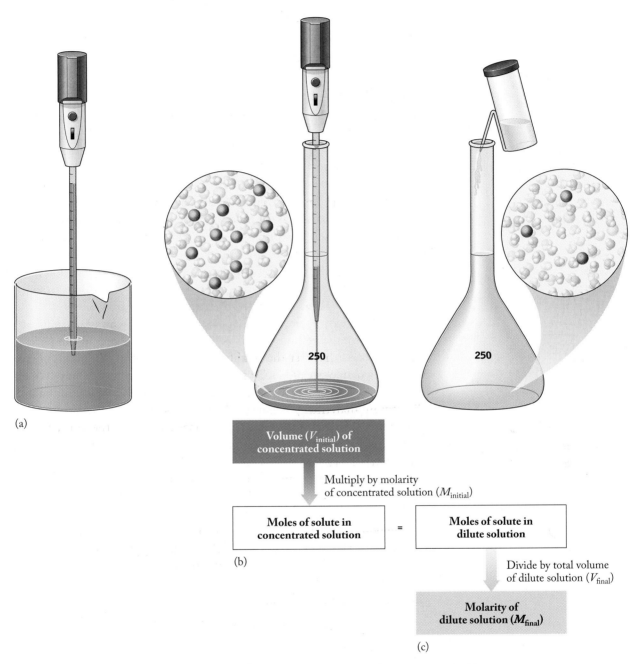

(a)

(b)

Volume ($V_{initial}$) of
concentrated solution

Multiply by molarity
of concentrated solution ($M_{initial}$)

Moles of solute in concentrated solution	=	Moles of solute in dilute solution

Divide by total volume
of dilute solution (V_{final})

Molarity of
dilute solution (M_{final})

(c)

FIGURE 4.6 To prepare 250.0 mL of a solution 3.18×10^{-3} M in Cu^{2+}, (a) a
pipet is used to withdraw 7.95 mL of a 0.1000 M standard solution. (b) The volume is
transferred into a 250.0 mL volumetric flask. (c) Water is then added to bring the total
volume of the new solution to 250.0 mL.

Because the number of moles of solute is the same in both calculations ($n_{initial} = n_{final}$), we can combine these two steps into one calculation in the form of an equation that applies to all situations in which we know any three of the four following variables: an initial volume ($V_{initial}$); an initial solute concentration ($M_{initial} = C_{initial}$); a final volume ($V_{final}$); and a final solute concentration ($M_{final} = C_{final}$). These variables are related by the following equation:

$$V_{initial} \times M_{initial} = n_{initial} = n_{final} = V_{final} \times M_{final}$$

or for the most general case,

$$V_{initial} \times C_{initial} = (\text{amount of solute})_{initial} = (\text{amount of solute})_{final} = V_{final} \times C_{final}$$

Simplifying this produces

$$V_{initial} \times C_{initial} = V_{final} \times C_{final} \tag{4.5}$$

Equation 4.5 works because each side of the equation represents a specific quantity of solute, and that quantity does not change because of dilution. Furthermore, this equation can be used for *any* units of volume and concentration, as long as the units used to express the initial and final volumes are the same, and the units for the initial and final concentrations are the same. The usefulness of Equation 4.5 becomes clear when we apply it to the following calculation for a system on a much larger scale: determining the concentration of Na^+ ions in a coastal bay when the salinity is reduced by fresh water flowing in from rivers.

Where large rivers flow into the sea, the salinity of seawater is reduced because it is diluted by the volume of fresh river water that enters it. Suppose a coastal bay contains 3.8×10^{12} L of ocean water mixed with 1.2×10^{12} L of river water for a total volume of 5.0×10^{12} L. What is the concentration of Na^+ ions in the bay water, assuming the open ocean water is $0.48\ M\ Na^+$? Assume that the fresh river water has an insignificant amount of sodium ion in it and as such does not contribute to the Na^+ ion content of the bay water.

Solving Equation 4.5 for C_{final} and inserting the values just given, we have:

$$C_{final} = \frac{V_{initial} \times C_{initial}}{V_{final}} = \frac{(3.8 \times 10^{12}\ \text{L}) \times 0.48\ M}{5.0 \times 10^{12}\ \text{L}} = 0.36\ M$$

The concentration of sodium ion in the bay is $0.36\ M$. It is lower than the concentration of sodium ion in ocean water ($0.48\ M$) because bay water is diluted by the addition of fresh water from the rivers.

SAMPLE EXERCISE 4.5   **Calculating Dilutions Using** $V_{initial} \times C_{initial} = V_{final} \times C_{final}$

The solution used in hospitals for intravenous infusion is called *physiological saline* ("saline solution"). It is $0.155\ M$ in NaCl. It may be purchased as a more concentrated stock solution: typically 5.28 L of stock are diluted with water to make 60.0 L of $0.155\ M$ NaCl. What is the initial concentration of NaCl in the stock solution?

COLLECT AND ORGANIZE We know the volume ($V_{final} = 60.0$ L) and concentration ($C_{final} = 0.155\ M$) of the diluted solution of NaCl, and the volume of the initial solution ($V_{initial} = 5.28$ L) before dilution. We are asked to calculate the initial concentration of the stock solution before dilution. These four variables are related by Equation 4.5:

$$V_{initial} \times C_{initial} = V_{final} \times C_{final}$$

ANALYZE The two volumes are expressed in the same units so no unit conversions are needed. We do need to rearrange the terms in Equation 4.5 to solve for $C_{initial}$.

SOLVE Solving Equation 4.5 for $C_{initial}$ and inserting the values of $V_{initial}$, V_{final}, and C_{final} gives us the following:

$$C_{initial} = \frac{V_{final} \times C_{final}}{V_{initial}} = \frac{60.0 \text{ L} \times 0.155 \, M_{final}}{5.28 \text{ L}} = 1.76 \, M_{initial}$$

THINK ABOUT IT The answer makes sense because the volume of the stock solution is about one-tenth the volume of the diluted solution, so its concentration should be about 10 times greater than 0.155 M.

Practice Exercise The concentration of Pb^{2+} in a commercially available standard solution is 1.000 mg/mL. What volume of this solution should be diluted to 500.0 mL to produce a solution in which the concentration of Pb^{2+} is 0.0575 mg/L?

(Answers to Practice Exercises are in the back of the book.)

CONCEPT TEST

Bags of physiological saline (see Sample Exercise 4.5) are usually labeled "0.90% w/V sodium chloride," where w/V means that the percentage defines the mass of NaCl per unit volume of solution (g/100 mL). Show that this concentration is approximately 0.155 M.

(Answers to Concept Tests are in the back of the book.)

4.4 Electrolytes and Nonelectrolytes

The high concentrations of NaCl and other salts in seawater (Table 4.1) make it a good conductor of electricity, as demonstrated by the following experiment. Suppose we immerse two *electrodes* (metal wires) in a sample of distilled water (Figure 4.7a) and connect these wires to a battery and a lightbulb. For electricity to flow and the bulb to illuminate, the circuit must be completed by mobile charge carriers (ions) in the solution. The lightbulb does not light up when the electrodes are placed in distilled water, because there are very few ions in distilled water. If the electrodes are immersed in 0.50 M NaCl (used in this demonstration to approximate seawater), the bulb lights up as shown in Figure 4.7(b) because the abundant ions in the solution carry the electric current between the two electrodes.

The saline solution conducts electricity because Na^+ ions are attracted to and migrate toward the electrode connected to the negative terminal of the battery, while Cl^- ions migrate toward the positive electrode. These migrating ions take their electric charges with them, and the migrating charges carry electric current through the water. The electric charge passes through the connecting wires and lightbulb in the form of a flow of electrons.

Any solute that imparts electrical conductivity to an aqueous solution is called an **electrolyte**. Sodium chloride is considered a **strong electrolyte** because it completely dissociates in water into its component ions when it dissolves, creating an abundance of charge carriers. However, not all solutes dissociate completely

An **electrolyte** is a substance that dissociates into ions when it dissolves, enhancing the conductivity of the solvent.

A **strong electrolyte** is a substance that dissociates completely into ions when it dissolves in water.

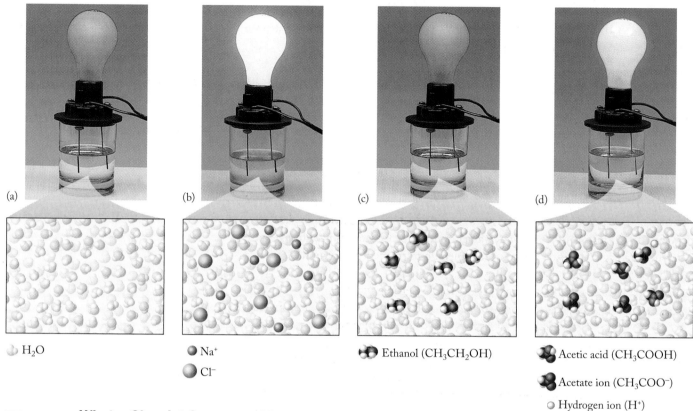

(a)   (b)   (c)   (d)

$\circ$ H$_2$O

$\bullet$ Na$^+$

$\circ$ Cl$^-$

Ethanol (CH$_3$CH$_2$OH)

Acetic acid (CH$_3$COOH)

Acetate ion (CH$_3$COO$^-$)

$\circ$ Hydrogen ion (H$^+$)

FIGURE 4.7 What a Chemist Sees

The solutions in this demonstration all contain the same molar concentration of solute. (a) Pure water (no solute) conducts very poorly because water is a molecular material and contains very few ions. (b) A 0.50 M solution of sodium chloride (NaCl), approximating seawater, conducts electricity very well. (c) A 0.50 M solution of ethanol (CH$_3$CH$_2$OH) does not improve the conductivity of water. (d) A 0.50 M solution of acetic acid (CH$_3$COOH) conducts electricity better than pure water or the ethanol solution but to a much smaller extent than the NaCl solution. A small fraction of the acetic acid molecules dissociate into acetate and H$^+$ ions; this partial ionization is characteristic of weak electrolytes like acetic acid.

into ions upon dissolution. Solutions of substances like sugar or an alcohol such as ethanol (CH$_3$CH$_2$OH) in water (Figure 4.7c) conduct electricity no better than pure water. These solutes do not form ions in aqueous solutions; they are **nonelectrolytes** in water. If we pour vinegar (an aqueous solution of acetic acid) in the apparatus in Figure 4.7(d) to prepare a concentration of acetic acid in the solution equivalent to that of the other solutes shown in the figure, the lightbulb lights up but not very brightly. The acetic acid molecules ionize, forming acetate ions and hydrogen ions, but the process does not go to completion:

$$CH_3COOH(aq) \rightleftharpoons H^+(aq) + CH_3OO^-(aq) \tag{4.6}$$

The small proportion of acetic acid molecules that ionize makes acetic acid a **weak electrolyte.** The arrow with the top half pointing right and the bottom half pointing left in Equation 4.6 indicates that all species are present in solution: the hydrogen ion, the acetate ion, and un-ionized, molecular acetic acid. This type of arrow is always used to symbolize the incomplete ionization of a weak electrolyte.

A **nonelectrolyte** is a substance that does not form ions and does not enhance the conductivity of water when it dissolves.

A **weak electrolyte** only partly ionizes when it dissolves in water.

CONCEPT TEST

Why does the demonstration in Figure 4.7 specify equivalent molar concentrations of solutes in the solutions whose conductivity is being compared?

(Answers to Concept Tests are in the back of the book.)

4.5 Acid–Base Reactions

Volcanic activity introduces a variety of gases into the atmosphere, including volatile oxides of sulfur, nitrogen, and other nonmetals. When these oxides dissolve in rainwater, they produce hydrogen ions in solution. We define acids as materials that increase the concentration of hydrogen ions in aqueous solution. To understand the chemistry of acids more deeply, we focus in this section on the processes that occur in solution when hydrogen ions are produced.

Let's look more closely at the behavior of a binary acid like hydrochloric acid (HCl). Pure hydrogen chloride is a molecular gas. However, when it dissolves in water, it behaves like a strong electrolyte and ionizes completely to give a solution containing $H^+(aq)$ and $Cl^-(aq)$. Generally we use $H^+(aq)$ to describe the cation produced in an aqueous medium by an acid. However, the hydrogen ion (H^+, also called the proton) does not have an independent existence in water because it associates very strongly with a water molecule to form H_3O^+. The **hydronium ion (H_3O^+)** also associates with other water molecules, so it is written $H_3O^+(aq)$. We write the hydrogen ion in water this way when we want to emphasize the role of water, and we describe the ionization of HCl by saying that it donates a proton to a water molecule. **Acids** are *proton donors*. Because the water molecule accepts the proton, water in this case is a *proton acceptor* and so is a **base**. These definitions were originally proposed by two chemists, Johannes Brønsted and Thomas Lowry, and the acids and bases so identified are called *Brønsted–Lowry acids and bases*. We will develop this concept much more completely later, but for now it provides a convenient way to define acids and bases in aqueous solutions of ions. For the reaction between HCl and water,

$$HCl(g) \quad + \quad H_2O(\ell) \quad \rightarrow \quad Cl^-(aq) + H_3O^+(aq) \quad \quad (4.7)$$

$$\text{proton donor} \quad \text{proton acceptor}$$
$$\text{(acid)} \quad \quad \text{(base)}$$

The simplest definition of a base is a substance that produces hydroxide ions in an aqueous solution. With the new definition of *base* as a proton acceptor, the hydroxide ion is still classified as a base, but other substances can now be categorized as bases as well. By focusing on the behavior of the hydrogen ion, we have expanded our definition of acids and bases.

If we look at the reaction that occurs when an aqueous solution of HCl reacts with an aqueous solution of NaOH [which is completely dissociated to give $Na^+(aq)$ and $OH^-(aq)$ ions]

$$HCl(aq) + NaOH(aq) \rightarrow NaCl(aq) + H_2O(\ell) \quad \quad (4.8)$$

we see that the HCl functions as an acid by donating a proton to the hydroxide ion, which functions as a base and accepts the proton; the two ions combine chemically to form a molecule of water. The remaining two ions—the anion characteristic of the acid (Cl^-) and the cation characteristic of the base (Na^+)—form a salt, sodium chloride, which remains dissolved in the aqueous solution. This reaction is called a **neutralization reaction** because both the acid and the base have been eliminated as a result of the reaction. The products of the neutralization of a strong acid like HCl with a strong base like NaOH are always water and a salt. This provides us with a new definition of a **salt** as a substance formed along with water as a product of a neutralization reaction.

⊂⊃ **CONNECTION** We defined hydrolysis reactions in Chapter 3. The nonmetal oxides not only dissolve in water; they also react with water (hydrolyze) to produce ions. Water is both a solvent and a co-reactant.

The **hydronium ion (H_3O^+)** is the form in which the hydrogen ion is found in an aqueous solution. *Hydronium ion, hydrogen ion,* and *proton* are synonymous; all are used to describe the hydrogen ion produced by acids in aqueous solution.

A Brønsted–Lowry **acid** is a proton donor.

A Brønsted–Lowry **base** is a proton acceptor.

A **neutralization reaction** takes place when an acid reacts with a base and produces a solution of a salt in water.

A **salt** is the product of a neutralization reaction; a salt is made up of the cation characteristic of the base and the anion characteristic of the acid in that reaction.

A **molecular equation** describes a reaction in solution in which the reactants are written as undissociated molecules.

An **overall ionic equation** shows all the species, both ionic and molecular, present in a reaction occurring in aqueous solution.

A **net ionic equation** describes the actual reaction taking place in aqueous solution, and is found by eliminating the spectator ions from the overall ionic equation.

Spectator ions are present in, but remain unchanged by, the reaction taking place.

Equation 4.8 shows a neutralization reaction written in the form of a **molecular equation**. Each of the participants in the reaction is written as a neutral molecule. The subscript *aq* after HCl, NaOH, and NaCl indicates that these substances are dissolved in water. Any undissociated molecules in aqueous solution are also typically written with an *aq* subscript to indicate they are *hydrated* (surrounded by water molecules). Molecular equations are sometimes the easiest equations to balance, so reactions involving substances that dissociate to form ions in solution are frequently written first as molecular equations to indicate clearly the formulas of the compounds involved and to simplify balancing the equation.

Another way to write Equation 4.8 indicates the species actually present in solution. Called an **overall ionic equation**, Equation 4.9 shows the ions formed by substances in the reaction that dissociate in water:

$$HCl(aq) + NaOH(aq) \rightarrow NaCl(aq) + H_2O(\ell)$$
$$H^+(aq) + Cl^-(aq) + Na^+(aq) + OH^-(aq) \rightarrow Na^+(aq) + Cl^-(aq) + H_2O(\ell) \quad (4.9)$$

Overall ionic equations clearly distinguish ionic substances from molecular substances in a chemical reaction taking place in solution.

Now think of Equation 4.9 as you would an algebraic equation. Notice that chloride ions and sodium ions appear on both sides as hydrated ions. If this were an algebraic equation, we would cross out the terms that are the same on both sides of the equation. When we do the same thing in chemistry, the result is called a **net ionic equation**. The ions that are removed (in this case, sodium ion and chloride ion) are **spectator ions**, and the resulting equation (4.10) focuses only on the actual chemical change that occurs:

Overall ionic equation:

$$H^+(aq) + \cancel{Cl^-(aq)} + \cancel{Na^+(aq)} + OH^-(aq) \rightarrow \cancel{Na^+(aq)} + \cancel{Cl^-(aq)} + H_2O(\ell)$$

Net ionic equation:

$$H^+(aq) + OH^-(aq) \rightarrow H_2O(\ell) \quad (4.10)$$

Sodium ions and chloride ions are unchanged by the neutralization reaction and remain in the solution. The chemical change that occurs is the reaction between a hydrogen ion and a hydroxide ion to form a molecule of water.

SAMPLE EXERCISE 4.6 | **Writing Neutralization Reaction Equations**

Write the balanced (a) molecular, (b) overall ionic, and (c) net ionic equations that describe the reaction taking place when an aqueous solution of sulfuric acid is neutralized by an aqueous solution of potassium hydroxide.

COLLECT AND ORGANIZE We need the formulas for the two reactants: sulfuric acid is H_2SO_4 and potassium hydroxide is KOH. Both substances are dissolved in water.

ANALYZE The products of the molecular equation are water and a salt. The salt in this case is potassium sulfate, made from the cation characteristic of the base and the anion characteristic of the acid. Write each substance in the balanced molecular equation in the form of ions in solution to generate the overall ionic equation. Finally, we need to pick out the spectator ions and eliminate them to generate the net ionic equation.

SOLVE

a. $H_2SO_4(aq) + KOH(aq) \rightarrow H_2O(\ell) + K_2SO_4(aq)$ (unbalanced)
$H_2SO_4(aq) + 2\,KOH(aq) \rightarrow 2\,H_2O(\ell) + K_2SO_4(aq)$ (balanced)

b. $2\,H^+(aq) + SO_4^{2-}(aq) + 2\,K^+(aq) + 2\,OH^-(aq) \rightarrow$
$$2\,H_2O(\ell) + 2\,K^+(aq) + SO_4^{2-}(aq)$$

c. The spectator ions are potassium ions and sulfate ions. They may be eliminated from both sides of the equation to show reactants and products:

$$2\,H^+(aq) + 2\,OH^-(aq) \rightarrow 2\,H_2O(\ell)$$

THINK ABOUT IT Remember we can treat chemical equations just like algebraic equations and cancel out spectator ions that are unchanged in the reaction and appear on both sides of the reaction arrow in the overall ionic equation. After canceling the spectator ions, we are left with the net ionic equation, which focuses on those species that were changed as a result of the reaction. It is also important to remember that sulfate is a polyatomic ion (see Table 2.3) that retains its identity in solution and does not separate into atoms and ions. Note also that the net ionic equation in this example is equivalent to Equation 4.10.

Practice Exercise Write balanced (a) molecular, (b) overall ionic, and (c) net ionic equations that describe the reaction taking place when an aqueous solution of phosphoric acid, $H_3PO_4(aq)$, is neutralized by an aqueous solution of sodium hydroxide.

(Answers to Practice Exercises are in the back of the book.)

In Carlsbad Caverns, NM, stalagmites of limestone grow up from the cavern floor.

For billions of years, acid-base reactions have played key roles in the chemical transformations of rocks and minerals that geologists call chemical weathering. One type of chemical weathering occurs when carbon dioxide (a nonmetal oxide and an acid anhydride) dissolves in rainwater to create a weakly acidic solution of carbonic acid, $H_2CO_3(aq)$. This solution, an example of mildly acidic rain, preferentially dissolves calcium carbonate, which occurs as chalk, limestone, and marble and is very insoluble in plain water. The reaction of carbonic acid and calcium carbonate is responsible for the formation of caves with stunning structures made by stalactites and stalagmites. This reaction and others caused by even more acidic rain are also responsible for the degradation of statues and the exterior of buildings made of marble. Table 4.2 lists some common nonmetal oxides that are considered pollutants in the atmosphere because they produce acid rain. Table 4.3 lists common basic minerals that are more soluble in aqueous acids than in neutral water.

Carbonic acid is a weak acid and a weak electrolyte. Other nonmetal oxides, dinitrogen pentoxide and sulfur trioxide, for example, form solutions that are much more highly acidic.

Formation of nitric acid:

$N_2O_5(g) + H_2O(\ell) \rightarrow 2\,HNO_3(aq)$ $HNO_3(aq) \rightarrow H^+(aq) + NO_3^-(aq)$

Formation of sulfuric acid:

$SO_3(g) + H_2O(\ell) \rightarrow H_2SO_4(aq)$ $H_2SO_4(aq) \rightarrow H^+(aq) + HSO_4^-(aq)$

TABLE 4.2 **Volatile Nonmetal Oxides and Their Acids**

Oxides	Acids
SO_2, SO_3	H_2SO_3, H_2SO_4
NO_2, N_2O_5	HNO_2, HNO_3
CO_2	H_2CO_3

TABLE 4.3 **Names and Formulas of Some Basic Minerals**

Name[a]	Formula
Calcite	$CaCO_3$
Gibbsite	$Al(OH)_3$
Dolomite	$MgCa(CO_3)_2$

[a]All these minerals are considered weak bases. Strong bases include all the hydroxides of the group 1 and group 2 elements except Mg.

TABLE 4.4 Strong Acids

Acid	Molecular Formula
Hydrochloric acid	HCl
Hydrobromic acid	HBr
Hydroiodic acid	HI
Nitric acid	HNO_3
Sulfuric acid	H_2SO_4
Perchloric acid	$HClO_4$

Atmospheric sulfuric acid, made when $SO_3(g)$ dissolves in rainwater, attacks marble statues and converts the calcium carbonate to calcium sulfate, which is more soluble and is slowly washed away by rain and snow.

A **strong acid** is completely ionized in aqueous solution.

A **weak acid** is a weak electrolyte and only partially ionized in aqueous solution; it has a limited capacity to donate protons to the medium.

A **strong base** is completely ionized in aqueous solution.

A **weak base** is a weak electrolyte and only partially ionized in aqueous solution; it has a limited capacity to accept protons in the medium.

An **amphiprotic** substance can behave as either a proton acceptor or a proton donor.

Both nitric acid and sulfuric acid are strong acids in water. **Strong acids** are strong electrolytes and are completely ionized in water. The common strong acids are listed in Table 4.4. Nitric acid dissociates completely in solution to produce hydrogen ions and the nitrate ion. Because one molecule of nitric acid donates one proton, nitric acid is called a *monoprotic acid*. Hydrochloric acid is another strong monoprotic acid. Sulfuric acid, on the other hand, has two hydrogen atoms that dissociate. Because it can potentially donate two protons, sulfuric acid is a *diprotic acid*. However, only one proton dissociates completely from sulfuric acid in water. This is described by saying that sulfuric acid is a strong acid in terms of the donation of the first proton, which is 100% ionized; this ionization results in the formation of the $HSO_4^-(aq)$ ion (hydrogen sulfate). For any diprotic acid, the donation of the second proton is a weaker process than the donation of the first proton. Because HSO_4^- dissociates only to a small extent, it is therefore a **weak acid** and a weak electrolyte. The equations for the donation of protons by sulfuric acid are written to reflect this behavior:

Strong acid produced:

$$H_2SO_4(aq) \rightarrow H^+(aq) + HSO_4^-(aq) \qquad (\rightarrow \text{means complete ionization})$$

Weak acid produced:

$$HSO_4^-(aq) \rightleftharpoons H^+(aq) + SO_4^{2-}(aq) \qquad (\rightleftharpoons \text{means incomplete ionization})$$

A balance, called a *dynamic equilibrium*, is achieved in solutions of weak electrolytes, at which point the concentrations of the reactants and the products no longer change. Solutions of weak electrolytes, such as weak acids and weak bases, are characterized by reactants and products existing together in dynamic equilibrium.

In a similar fashion to acids, bases are classified as **strong bases** or **weak bases** depending on the extent to which they ionize in aqueous solution. Strong bases include the hydroxides of groups 1 and 2 elements, which are completely ionized when dissolved in water. An example of a weak base is ammonia (NH_3). Ammonia in its molecular form is a gas. When it dissolves in water, it produces a solution that conducts electricity weakly, which means that ammonia is a weak electrolyte. This reaction takes place between ammonia and water:

$$NH_3(aq) + H_2O(\ell) \rightleftharpoons NH_4^+(aq) + OH^-(aq) \qquad (4.11)$$
$$\text{base} \qquad \text{acid}$$

In this reaction, ammonia functions as a proton acceptor, receiving a hydrogen ion donated to it by water, and water acts as an acid, a proton donor.

If you compare Equations 4.7 and 4.11, you will note that water behaves as a base in Equation 4.7 and an acid in Equation 4.11. Is water an acid or a base? The answer depends on what is dissolved in it. Water is called an **amphiprotic** substance, because it can function as either a proton acceptor (base) or a proton donor (acid), depending on what else is present in the solution.

4.6 Precipitation Reactions

Some of the most abundant elements in Earth's crust, including silicon (Si), aluminum (Al), and iron (Fe), are not abundant in seawater. Rocks and minerals must be made of compounds with limited water solubility or they would never survive on the surface as solid substances in the presence of all the water on our planet. The reasons why some compounds, such as SiO_2 (quartz), are insoluble in water while others, such as NaCl, are readily soluble involve many factors that we will discuss in subsequent chapters. For now, Table 4.5 summarizes solubility rules for common ionic compounds. Chemists make use of the relative solubilities of ionic compounds to make materials and to analyze solutions.

A **precipitate** is a solid product formed from a reaction in solution.

TABLE 4.5 Solubility Guidelines for Common Ionic Compounds in Water

All compounds containing the following ions are soluble:
- Cations: Group 1 ions (alkali metals) and NH_4^+
- Anions: NO_3^- and CH_3COO^- (acetate)

Compounds containing the following anions are soluble except as noted:
- Group 17 ions (halides), except the halides of Ag^+, Cu^+, Hg_2^{2+}, and Pb^{2+}
- SO_4^{2-}, except the sulfates of Ba^{2+}, Ca^{2+}, Hg_2^{2+}, Pb^{2+}, and Sr^{2+}

All other compounds are insoluble except the following group 2 (alkaline earth) hydroxides:
- $Ba(OH)_2$, $Ca(OH)_2$, and $Sr(OH)_2$

Making Insoluble Salts

When two solutions containing ions are mixed, a solid product called a **precipitate** may form, and the reaction is called a *precipitation reaction*. We can use Table 4.5 to predict when a precipitate will form. For example, does a precipitate form when a solution of potassium nitrate is mixed with a solution of sodium iodide? To answer this question we need to do the following:

1. *Recognize that both salts are in solution, which indicates that they are both soluble.* Moreover, Table 4.5 specifies that all compounds containing cations from group 1 on the periodic table are soluble. Potassium and sodium are in group 1, so salts containing them are soluble:

    Solution 1: $KNO_3(aq) \rightarrow K^+(aq) + NO_3^-(aq)$

    Solution 2: $NaI(aq) \rightarrow Na^+(aq) + I^-(aq)$

    From this information we can set up the reactant side of the equation for any chemical change that might take place:

    $$K^+(aq) + NO_3^-(aq) + Na^+(aq) + I^-(aq) \rightarrow ?$$

2. *Determine whether any of the possible combinations of ions produces an insoluble product.* KNO_3, KI, NaI, and $NaNO_3$ are all soluble according to Table 4.5, so no precipitate forms when the two solutions are mixed:

    $$K^+(aq) + NO_3^-(aq) + Na^+(aq) + I^-(aq) \rightarrow \text{no reaction}$$

What happens if a solution of the salt lead(II) nitrate $[Pb(NO_3)_2]$ is substituted for the solution of potassium nitrate? All nitrate salts are soluble, so $Pb(NO_3)_2$ dissolves in water and forms ions.

$$Pb^{2+}(aq) + 2 NO_3^-(aq) + Na^+(aq) + I^-(aq) \rightarrow ?$$

The solubility rules indicate that salts of the group 17 ions (the halide ions) form insoluble compounds with $Pb^{2+}(aq)$, so we predict that $PbI_2(s)$ will precipitate

(a)   (b)

FIGURE 4.8 (a) The beakers contain 0.1 M solutions of $Pb(NO_3)_2$ and NaI. Both solutions are colorless. (b) As the NaI solution is poured into the solution of $Pb(NO_3)_2$, a dense yellow precipitate of PbI_2 forms.

(Figure 4.8). Because a reaction takes place, we need to write a balanced equation that describes the process, so let's look first at the molecular equation:

Unbalanced: $Pb(NO_3)_2(aq) + NaI(aq) \rightarrow PbI_2(s) + NaNO_3(aq)$

Balanced: $Pb(NO_3)_2(aq) + 2\,NaI(aq) \rightarrow PbI_2(s) + 2\,NaNO_3(aq)$

From the balanced molecular equation we get the overall ionic equation for the reaction:

$$Pb^{2+}(aq) + 2\,NO_3^-(aq) + 2\,Na^+(aq) + 2\,I^-(aq) \rightarrow$$
$$PbI_2(s) + 2\,Na^+(aq) + 2\,NO_3^-(aq)$$

and the net ionic equation that describes only the chemical change taking place:

$$Pb^{2+}(aq) + 2\,I^-(aq) \rightarrow PbI_2(s)$$

Soluble and *insoluble* are qualitative terms. In principle, all ionic compounds dissolve in water to some extent. In practice, we consider a compound to be insoluble in water if the maximum amount that dissolves gives rise to a concentration of less than 0.01 M. At this level, a solid appears to be insoluble to the naked eye; the tiny amount of solid that dissolves is not noticeable with respect to the amount of solid in contact with the solvent.

SAMPLE EXERCISE 4.7   **Predicting Precipitation Reactions**

A precipitate forms if an aqueous solution of ammonium sulfate is mixed with an aqueous solution of barium chloride. Write the net ionic equation that describes the reaction.

COLLECT AND ORGANIZE The two salts are in solution, so they must be soluble. We need to evaluate the ions in solution to determine which combinations of them produce insoluble solids.

ANALYZE First we need to write correct formulas of the salts and indicate the ions they produce when they dissociate in an aqueous solution. Then we need to check Table 4.5 to determine which of the possible combinations are insoluble in water.

SOLVE

Solution 1: $(NH_4)_2SO_4(aq) \rightarrow 2\,NH_4^+(aq) + SO_4^{2-}(aq)$

Solution 2: $BaCl_2(aq) \rightarrow Ba^{2+}(aq) + 2\,Cl^-(aq)$

The new combinations are $BaSO_4$ and NH_4Cl. Table 4.5 indicates that all ammonium salts are soluble so the ammonium and chloride ions remain in solution, but barium sulfate is insoluble so that salt precipitates. The net ionic equation describes the formation of the precipitate:

$$Ba^{2+}(aq) + SO_4^{2-}(aq) \rightarrow BaSO_4(s)$$

THINK ABOUT IT To identify the precipitation reaction that occurs when two solutions containing ions are mixed, we check to see whether an insoluble material is formed when the ions change partners. In this exercise, the sulfate ion, originally in the salt ammonium sulfate, replaces the chloride ion, originally partners with the barium, forming insoluble barium sulfate. The other new combination, ammonium chloride, remains in solution.

Practice Exercise In each of the following pairs of solutions, does a precipitate form when they are mixed? If a reaction does take place, write the net ionic equation that describes the chemical change: (a) a solution of sodium acetate and a solution of ammonium sulfate; (b) a solution of calcium chloride and a solution of mercury(I) nitrate.

(Answers to Practice Exercises are in the back of the book.)

Precipitation reactions can be used to synthesize insoluble salts. For example, barium sulfate (the precipitate produced in Sample Exercise 4.7) is used clinically in medical procedures to image the gastrointestinal tract. Barium sulfate can be made by the reaction of any soluble barium salt (for example, barium nitrate, barium chloride, or barium acetate) with a solution of a soluble salt containing the sulfate ion (such as sodium sulfate, potassium sulfate, or ammonium sulfate). The precipitate from such a reaction can be collected by filtration, dried, and used for whatever purpose it may have. The soluble compound remaining in solution as hydrated ions can also be collected. The ammonium chloride left in solution in Sample Exercise 4.7 can be isolated by boiling off the water, at which point any ions left in solution can be collected as a solid residue.

CONCEPT TEST

Lead(II) dichromate ($PbCr_2O_7$; the dichromate ion is $Cr_2O_7^{2-}$) is a water-insoluble pigment called schoolbus yellow that is used to paint lines on highways. Design a synthesis of schoolbus yellow that uses a precipitation reaction.

(Answers to Concept Tests are in the back of the book.)

Using Precipitation in Analysis

Chemists make use of the insolubility of ionic compounds in analyses to determine the concentrations of ions in solution. For example, in some climates the use of NaCl to melt ice and snow on roads and sidewalks during the winter causes water supplies to become contaminated and to exceed the limits of sodium and chloride ion concentrations considered safe. Analytical chemists can determine the concentration of chloride ion in a sample of drinking water by reacting it with a solution of silver nitrate, $AgNO_3$. Any chloride ion from salt in the sample combines with Ag^+ to form a precipitate of solid AgCl:

$$NaCl(aq) + AgNO_3(aq) \rightarrow AgCl(s) + NaNO_3(aq) \qquad (4.12)$$

This precipitate can be filtered, dried, and weighed. From its mass, the concentration of chloride in the water sample can be calculated.

Equation 4.12 is the molecular equation describing the formation of the AgCl precipitate. Let's write the net ionic equation for this precipitation reaction. All the reactants are soluble ionic compounds and are completely dissociated into their component ions. Therefore the overall ionic equation is as follows:

$$Na^+(aq) + Cl^-(aq) + Ag^+(aq) + NO_3^-(aq) \rightarrow$$
$$AgCl(s) + Na^+(aq) + NO_3^-(aq) \qquad (4.13)$$

The $Na^+(aq)$ and $NO_3^-(aq)$ ions appearing on both sides of Equation 4.13 are spectators in this reaction. Eliminating them yields the following net ionic equation:

$$Ag^+(aq) + Cl^-(aq) \rightarrow AgCl(s) \qquad (4.14)$$

SAMPLE EXERCISE 4.8   **Calculating a Solute Concentration from the Mass of a Precipitate**

To determine the concentration of chloride ion in a sample of groundwater, a chemist adds 1.0 mL of 1.00 M AgNO$_3$ solution to 100.0 mL of the sample. The mass of the resulting AgCl precipitate is 71.7 mg. What is the concentration of chloride in the original sample (expressed in milligrams of Cl$^-$ per liter)?

COLLECT AND ORGANIZE From the data given, we know the volume of the sample and the mass of precipitate formed by the chloride ion in the sample when $Ag^+(aq)$ ions were added. We may assume that excess silver nitrate was added and that all the chloride in the sample was precipitated as AgCl(s). Our task is to calculate how much chloride is contained within the 71.7 mg of AgCl precipitate, and then convert that quantity to a concentration in the sample. We need a balanced chemical equation that describes the reaction taking place. In principle, we can use the molecular, the overall ionic, or the net ionic equation. Because the net ionic equation focuses on the chemical reaction that takes place, it is an appropriate equation to use, since this problem requires us to analyze the process stoichiometrically.

ANALYZE The net ionic equation for the reaction taking place is Equation 4.14:

$$Ag^+(aq) + Cl^-(aq) \rightarrow AgCl(s)$$

The equation shows that 1 mole of Cl$^-$(aq) reacts with 1 mole of $Ag^+(aq)$ to produce 1 mole of AgCl(s). The mass of 1 mole of Cl$^-$(aq) is the same as the molar mass of chlorine: 35.45 g/mol. The molar mass of AgCl is the sum of the molar masses of Ag and Cl, or 143.32 g. We can use the methods developed in Chapter 3 to determine the amount of chloride ion in the original sample, and from which we can calculate the concentration.

SOLVE

$$Ag^+(aq) + Cl^-(aq) \rightarrow AgCl(s)$$
$$x \text{ mg} \qquad 71.7 \text{ mg}$$

$$0.0717 \text{ g AgCl(s)} \times \frac{1 \text{ mol AgCl(s)}}{143.32 \text{ g AgCl(s)}} \times \frac{1 \text{ mol Cl}^-(aq)}{1 \text{ mol AgCl(s)}} \times \frac{35.45 \text{ g Cl}^-(aq)}{1 \text{ mol Cl}^-(aq)}$$

$$= 0.0177 \text{ g Cl}^-(aq) = 17.7 \text{ mg Cl}^-(aq)$$

The 100.0 mL sample therefore contains 17.7 mg Cl⁻(aq). The Cl⁻(aq) concentration in milligrams per liter is:

$$\frac{17.7 \text{ mg Cl}^-(aq)}{100.0 \text{ mL}} \times \frac{1000 \text{ mL}}{L} = 177 \text{ mg Cl}^-(aq)/L$$

> A **saturated solution** contains the maximum concentration of a solute possible at a given temperature.

THINK ABOUT IT An alternative way to solve this problem is to define the chloride ion content in grams in 1 g of AgCl(s): 1 mol of AgCl(s) contains 1 mol of Cl⁻, so

$$\frac{1 \text{ mol Cl}^-}{1 \text{ mol AgCl}} \times \frac{35.45 \text{ g/mol Cl}^-}{143.32 \text{ g/mol AgCl}} = 0.2473 \text{ g Cl}^-/\text{g AgCl}$$

Applying this factor (0.2473 g of Cl⁻ for every 1 g of AgCl, or 247.3 mg of Cl⁻ for every 1000 mg of AgCl) to the mass of AgCl(s) that precipitated from solution, we get

$$71.7 \text{ mg AgCl} \times \frac{247.3 \text{ mg Cl}^-}{1000 \text{ mg AgCl}} = 17.7 \text{ mg Cl}^- \text{ precipitated as AgCl}(s)$$

From here, the problem proceeds as before.

Practice Exercise The silver compounds of other halides are also insoluble and can be used in precipitation reactions. Addition of excess Ag⁺(aq) to 50.0 mL of a solution containing Br⁻(aq) results in the precipitation of 22.8 mg of AgBr(s). What was the original concentration of Br⁻(aq) in the sample in milligrams per liter?

(Answers to Practice Exercises are in the back of the book.)

CONCEPT TEST

Confirm that the amount of silver nitrate added to the solution in Sample Exercise 4.8 was indeed in excess and that all the chloride ion was precipitated as AgCl(s).

(Answers to Concept Tests are in the back of the book.)

Saturated Solutions and Supersaturation

When a solution contains the maximum concentration of solute that can dissolve in it, the result is a **saturated solution** (Figure 4.9). The amount of solute that can dissolve in a given quantity of solvent depends on the temperature; often, the higher the temperature the greater the solubility. More table sugar, for example, dissolves in hot water than in cold, a fact applied in the making of rock candy: After a large amount of sugar is dissolved in hot water, the solution is slowly cooled.

▶II **CHEMTOUR** Saturated Solutions

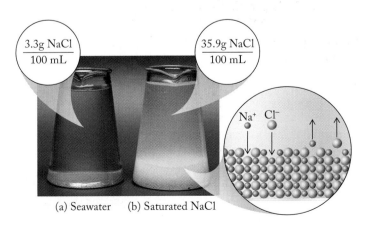
(a) Seawater (b) Saturated NaCl

FIGURE 4.9 (a) The liquid in the beaker, although a concentrated aqueous solution of NaCl, is far from saturated. (b) The second solution is in equilibrium with solid NaCl at 20°C (the white mass at the bottom of the beaker), so the liquid is a saturated solution of NaCl: the solid is constantly dissolving, while ions in the solution are constantly precipitating out of solution. A balance—a dynamic equilibrium—has been achieved whereby the concentration of solute in solvent remains constant.

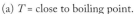

(a) *T* = close to boiling point.

(b) Put in a wooden stirrer.

(c) Cool to room temperature.

FIGURE 4.10 Making rock candy: (a) Sugar is very soluble in hot water; (b) as a hot, saturated solution of sugar in water cools, (c) the solute gradually precipitates.

An object with a slightly rough surface suspended in the solution, like a string or a wooden stirrer, serves as a site for crystallization; as the solution cools below the point when the solubility of the sugar in the water is exceeded, crystals of sugar begin to grow on the rough surface (Figure 4.10). The formation of crystals from the cooling of a hot, saturated solution is another example of precipitation.

Sometimes, more solute temporarily dissolves than is predicted by its solubility, creating a **supersaturated solution**. This can happen when the temperature of a saturated solution drops slowly or when the volume of an unsaturated solution is reduced through slow evaporation. Slow evaporation of the solvent is in part responsible for the formation of stalactites and stalagmites in caves as well as of the salt flats in the American Southwest. Sooner or later, usually in response to a disruption like a change in temperature, mechanical shock, or the addition of a *seed crystal* (a small crystal of the solute that provides a site for further crystallization to take place), the solute in a supersaturated solution rapidly comes out of solution to form a precipitate, as shown in Figure 4.11.

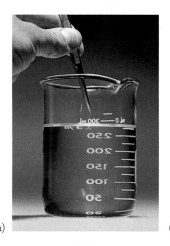
(a)

(b)

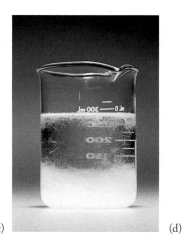
(c)

(d)

FIGURE 4.11 (a) Sodium acetate precipitates from a supersaturated solution when a seed crystal is added. (b) The crystal becomes a site for rapid crystal growth as sodium acetate begins to crystallize; (c, d) crystal growth continues until the solution is no longer supersaturated but merely saturated with sodium acetate.

A **supersaturated solution** contains more than the quantity of a solute that is predicted to be soluble in a given volume of solution at a given temperature.

Ion exchange is a process by which one ion is exchanged for another. As it is usually carried out, 2+ ions in water that contribute to its hardness are removed from the water in exchange for sodium ions.

4.7 Ion Exchange

In precipitation reactions in an aqueous solution, we saw how ions exchanged partners, forming an insoluble precipitate. The basic idea of one ion exchanging for another is behind a process of considerable importance in water purification. Water containing certain metal ions—principally Ca^{2+} and Mg^{2+}—is called "hard" water, and it causes practical problems for industrial or domestic use. Hard water combines with soap to form an ugly gray scum; clothes washed in it appear gray and dull; it forms a hard scale in boilers, pipes, and kettles, diminishing their ability to conduct heat and carry water; and it sometimes has an unpleasant taste. One common method of *water softening* (the process of removing these ions) involves a technique known as **ion exchange**: hard water is passed through a treatment system that replaces problematic ions with innocuous ones.

An ion-exchange system contains cartridges packed with beads of a porous plastic resin (R) to which ionic groups capable of binding ions have been chemically bonded. To bind calcium, magnesium, and iron cations, these sites consist of anionic groups. The carboxylate group is often used; its formula when bonded to a resin is R—COO⁻. Water in the ion-exchange cartridge contains Na^+ to balance the negative charges on the anionic groups in fresh resin. When writing formulas of salts containing organic ions, it is conventional to write the organic ion first. We can therefore think of the resin as beads containing (R—COO⁻)Na^+ units.

As hard water flows through the pores of the resin, Ca^{2+}, Mg^{2+}, and other 2+ ions exchange with, or displace, the sodium ions on the resin. Here is the ion-exchange reaction for ridding water of calcium ion:

$$2\,(R{-}COO^-)Na^+(s) + Ca^{2+}(aq) \rightarrow (R{-}COO^-)_2\,Ca^{2+}(s) + 2\,Na^+(aq)$$

Hard water that has been softened in this way (Figure 4.12) contains increased concentrations of sodium ion. Although this may not be a problem for healthy children and adults, people suffering from high blood pressure often must limit their intake of Na^+, and so should not drink water softened by this kind of ion-exchange reaction.

Many chemical reactions require the use of highly pure, *deionized* water. High-purity water is prepared in laboratories by distillation followed by ion exchange. The ion-exchange systems used in the laboratory setting include cartridges packed with porous resins that have sites for both cation exchange ([RC]⁻ sites in the H^+, or acid, form) and for anion exchange ([RA]⁺ sites in the OH^-, or base, form). As water containing very small concentrations of ions [such as $Na^+(aq)$ and $Cl^-(aq)$] that were not removed by distillation flows past these reaction sites, the following reactions take place:

$$H^+[RC]^-(s) + Na^+(aq) \rightarrow Na^+[RC]^-(s) + H^+(aq)$$

$$[RA]^+OH^-(s) + Cl^-(aq) \rightarrow [RA]^+Cl^-(s) + OH^-(aq)$$

The products of these ion-exchange reactions, $H^+(aq)$ and $OH^-(aq)$, neutralize each other, forming water:

$$H^+(aq) + OH^-(aq) \rightarrow H_2O(\ell)$$

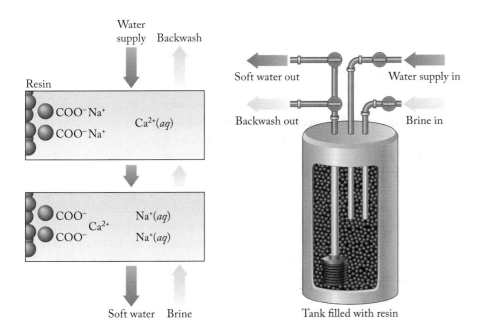

FIGURE 4.12 What a Chemist Sees Residential water softeners use ion exchange to remove 2+ ions (such as Ca^{2+}) that make water hard. The ion-exchange resin contains cation exchange sites that are initially occupied by Na^+ ions. These ions are replaced by 2+ "hardness" ions as water flows through the resin. Eventually most of the ion-exchange sites are occupied by 2+ ions and the system loses its water-softening ability. The resin is then backwashed with a saturated solution of NaCl (*brine*), displacing the hardness ions (which wash down the drain), thus restoring the resin to its Na^+ form.

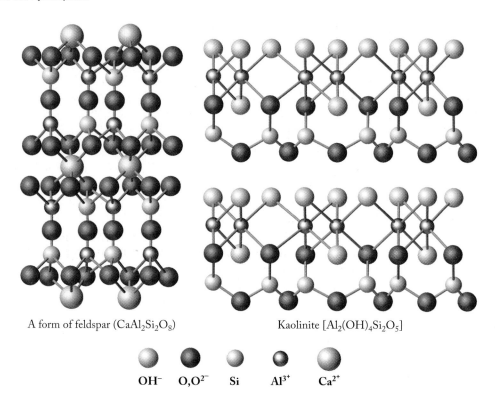

FIGURE 4.13 The structures of kaolinite (a clay mineral used to make fine china) and feldspar (the mineral from which kaolinite forms) feature layers of aluminum and silicon bonded to oxygen.

A form of feldspar ($CaAl_2Si_2O_8$)

Kaolinite [$Al_2(OH)_4Si_2O_5$]

OH⁻   O,O²⁻   Si   Al³⁺   Ca²⁺

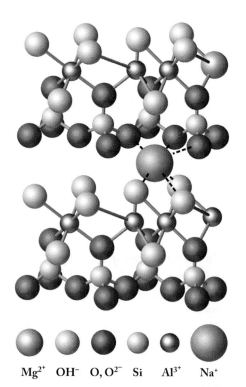

Mg²⁺   OH⁻   O, O²⁻   Si   Al³⁺   Na⁺

FIGURE 4.14 If an ion with a 2+ charge, such as Mg^{2+}, is incorporated into the kaolinite structure in place of an Al^{3+} ion, then the difference in their charges gives the kaolinite a net negative charge. This negative charge is balanced by the inclusion of a cation such as Na^+ in the spaces between the layers of hydroxide ions.

Thus the overall result of the ion-exchange process is the replacement of $Na^+(aq)$ and $Cl^-(aq)$ in the original water supply by $H_2O(\ell)$. Similar reactions can be written for other ionic species. The liquid product that flows from the ion-exchange cartridge is essentially pure, deionized water. The ion-exchange reaction proceeds as long as both $H^+[RC]^-$ and $[RA]^+OH^-$ reaction sites are available. The cartridges are replaced when most of the sites have been consumed by the ion-exchange process (often indicated by a change in the color of the resin).

Some naturally occurring minerals known as aluminosilicates function as ion exchangers, too. **Zeolites** are porous materials in this family that consist of three-dimensional networks of channels made of aluminum, silicon, and oxygen. The channels also contain sodium ions or other 1+ cations. The 1+ cations are not part of the rigid structure and are exchangeable for other ions. When hard water passes through a zeolite bed, one 2+ ion can exchange for two sodium ions and thereby soften the water. Synthetic zeolites are being added to some detergents now for exactly this purpose.

The structure of many other aluminosilicates found in Earth's crust feature layers of two-dimensional sheets made up of silicon and oxygen, or aluminum and oxygen, bonded together (Figure 4.13). The multilayered structure of these minerals gives them distinctive physical and chemical properties. Other properties may be related to imperfections in their molecular structures due to the presence of cations of lower charge in sites normally occupied by Al^{3+} ions.

In many aluminosilicates, Fe^{2+} or Mg^{2+} ions substitute for Al^{3+} in the Al–O layer (Figure 4.14). These substitutions mean that the structure is deficient in positive charge and thus has a negative charge overall. In nature this charge is balanced by additional positively charged ions, such as Na^+ or K^+, that reside in the gaps between aluminosilicate layers, similar to the way 1+ ions reside in the three-dimensional channels of zeolites. Because these ions are not held tightly in place, they can be replaced by other positive ions. For example, H^+ ions in acidic groundwater can replace Na^+ ions in an ion-exchange reaction that removes hydrogen ions from solution, which also makes it a neutralization reaction.

4.8 Oxidation–Reduction Reactions

One of the earliest descriptions of the properties of oxygen came from Leonardo da Vinci in the 15th century: "Where flame cannot live, no animal that draws breath can live." Although oxygen was not isolated and recognized as an element until over 250 years later, da Vinci captured an important property of the substance in his observation. Oxygen makes up about 50% by mass of Earth's crust, it is almost 89% by mass of the water that covers 70% of Earth's surface, and it is about 20% of the volume of Earth's atmosphere. It is essential to the metabolism of all creatures that "draw breath" just as it is essential for combustion reactions. Its name is applied to perhaps the most important type of chemical processes implied in da Vinci's statement: oxidation–reduction reactions.

On early Earth the reactions of oxygen with sulfur and nitrogen to produce acid anhydrides in the atmosphere and with metallic elements to produce the minerals of the crust are illustrations of this type of reaction. Oxidation was first defined as a reaction that increased the oxygen content of a substance. Hence two reactions involving oxygen—with methane [$CH_4(g)$] in the atmosphere of early Earth to produce $CO_2(g)$ and $H_2O(\ell)$, and with elemental iron as Earth's crust formed to produce $Fe_2O_3(s)$—are both oxidations; the products contain more oxygen than CH_4 or Fe:

$$CH_4(g) + 2\,O_2(g) \rightarrow CO_2(g) + 2\,H_2O(\ell)$$

$$4\,Fe(s) + 3\,O_2(g) \rightarrow 2\,Fe_2O_3(s)$$

Reduction reactions were originally defined in a similar fashion—as reactions in which the oxygen content of a substance is reduced. A classic example of a reduction reaction is the processing of iron ore to produce pure metallic iron:

$$Fe_2O_3(s) + 3\,CO(g) \rightarrow 2\,Fe(s) + 3\,CO_2(g)$$

The oxygen content of the ore [$Fe_2O_3(s)$] is reduced as a result of its reaction with $CO(g)$. Note that in this process the carbon atom in $CO(g)$ by this definition is oxidized—the oxygen content of the carbon species has been increased.

Ultimately, these first definitions of oxidation as the addition of oxygen to another substance and reduction as the removal of oxygen proved too restrictive. **Oxidation** now refers to any chemical reaction in which a substance *loses electrons.* Similarly, **reduction** is a *gain of electrons.*[2] Oxidation and reduction always occur together: If one substance loses electrons, then another substance must gain them; if one species is oxidized, then another must be reduced. The crucial event in oxidation-reduction is this movement of electrons from one atom or ion to another. The recognition that oxidation and reduction must occur together gives rise to the common way of referring to them as *redox reactions.*

Oxidation Numbers

Some redox reactions are easy to identify: all combustion reactions, all reactions that generate electricity in batteries or that take place when an electric current is passed through a solution, and all reactions of elements that combine to make compounds are redox reactions. Beyond that, an unambiguous way to determine whether a reaction involves redox is by assigning an oxidation number (O.N.) to each element in the

Zeolites are natural crystalline minerals or synthetic materials consisting of three-dimensional networks of channels that contain sodium or other 1+ cations. They may function as ion exchangers.

Oxidation is a chemical change in which a species loses electrons. The oxidation number of the species increases.

Reduction is a chemical change in which a species gains electrons. The oxidation number of the species decreases.

All fires are redox reactions.

---

[2] The mnemonic OIL RIG may be helpful for remembering these expanded definitions: "Oxidation Is Loss; Reduction Is Gain."

The **oxidation number (O.N.)** of an element in a molecule or ion, also called its **oxidation state**, is a positive or negative number based on the number of electrons that each atom of the element gains or loses when it forms an ion, or that it shares when it forms a covalent bond with another element. Pure elements have an oxidation number of zero.

⊙⊙ **CONNECTION** In Chapter 2 we discussed the naming of compounds containing ions of transition metals that could have more than one charge. The roman numerals in symbols such as Fe(II) and Fe(III) identify the different oxidation numbers of the iron ion.

reactants and products. If the oxidation number of an element changes as a result of a chemical reaction, then the reaction is definitely redox. The **oxidation number (O.N.)**, or **oxidation state**, of an element in a molecule or ion is a positive or negative number whose value depends on the number of electrons that each atom of the element gains or loses when it forms an ion, or the number of electrons that it shares when it forms a covalent bond with another element. The O.N. value represents the charge an atom has, or appears to have, as determined by the following set of guidelines:

1. The oxidation numbers of the elements in a neutral molecule sum to zero; those on the elements in an ion sum to the charge on the ion.
2. Each atom in a pure element has an oxidation number of zero. For example:

$$F_2: \text{O.N. of F} = 0 \qquad Fe: \text{O.N. of Fe} = 0$$

$$O_2: \text{O.N. of O} = 0 \qquad Na: \text{O.N. of Na} = 0$$

3. For monatomic (one-element) ions, the oxidation number is equal to the charge on the ion. Examples:

$$F^-: \text{O.N.} = -1 \qquad Fe^{3+}: \text{O.N.} = +3$$

$$O^{2-}: \text{O.N.} = -2 \qquad Na^+: \text{O.N.} = +1$$

Note that the number precedes the sign of the charge in, for example, Fe^{3+}, but the sign comes first in writing an oxidation number, as in "the O.N. of Fe in Fe_2O_3 is +3."

4. In compounds with other elements, fluorine is *always* −1. Examples:

$$KF: \text{F is} -1; \text{K is} +1$$

$$OF_2: \text{each F atom is} -1; \text{O is} +2$$

$$CF_4: \text{each F atom is} -1; \text{C is} +4$$

5. The oxidation number of hydrogen is +1 and oxygen is −2 in most compounds. Common exceptions are the hydrogen in metal hydrides (for example: LiH) where it is −1; the oxygen in the peroxide ion (O_2^{2-} as in hydrogen peroxide: H_2O_2) where it is −1.
6. Unless they are combined with oxygen or fluorine, the halogens chlorine, fluorine, and bromine are −1. Examples:

$$CaCl_2: \text{O.N. of Cl} = -1; \text{O.N. of Ca} = +2$$

$$ClO_4^-: \text{O.N. of O} = -2; \text{O.N. of Cl} = +7$$

$$IF_5: \text{O.N. of F} = -1; \text{O.N. of I} = +5$$

SAMPLE EXERCISE 4.9   **Determining Oxidation Numbers**

What is the oxidation number of sulfur in each of these compounds? (a) SO_2; (b) Na_2S; (c) $CaSO_4$

COLLECT AND ORGANIZE We are asked to assign O.N. values to sulfur in three of its compounds. To do so we apply the guidelines on this page for determining the oxidation states of the elements in these compounds.

ANALYZE One compound, SO_2, is molecular because its components are both nonmetals. Na_2S is a simple binary ionic compound, and the oxidation number of an element in an ionic compound equals the charge of its ion. Sulfur in $CaSO_4$ is part of the sulfate ion, so the O.N. values of the elements in the ion must add up to the charge of the ion.

SOLVE

a. For the oxygen in SO_2, we automatically assign an O.N. value of -2. For sulfur, let O.N. $= x$. The sum of the O.N. values of the sulfur atom and the two oxygen atoms must add up to zero:

$$x + 2(-2) = 0 \quad \text{or} \quad x = +4$$

b. Sodium is present in all its compounds as Na^+. This means that the O.N. of Na is $+1$. To balance the O.N. values in Na_2S, we can use the following equality, and solve for y, the O.N. of sulfur:

$$2(+1) + y = 0 \quad \text{or} \quad y = -2$$

c. The charge on the calcium ion is always 2+. This means that the charge on the sulfate ion must be 2−. Assigning O.N. values of -2 for oxygen and z for sulfur, we have the following equality for the sulfate ion:

$$z + 4(-2) = -2 \quad \text{or} \quad z = +6$$

THINK ABOUT IT Sulfur has oxidation numbers that range from -2 to $+6$ in various molecules and ions. Unlike H, Na, and Ca, there is no oxidation number rule for S (or for most other elements). We must derive their O.N. values by applying the guidelines and considering the oxidation numbers of the other elements in the molecule or ion.

Practice Exercise Determine the oxidation number of nitrogen in the following compounds: (a) NO_2; (b) N_2O; (c) HNO_3

(Answers to Practice Exercises are in the back of the book.)

Examples of Redox Reactions

The combustion of hydrocarbon fuels is a classic redox reaction because it involves oxygen as a reactant. In the complete combustion of a fuel like methane (CH_4), the carbon is completely converted into $CO_2(g)$ and the hydrogen to $H_2O(\ell)$. Combustion reactions of hydrocarbons can often be balanced by first balancing the moles of carbon on both sides of the reaction arrow, then the moles of hydrogen, and finally the moles of oxygen. However, like all redox reactions, hydrocarbon combustion reactions can also be balanced by balancing the changes in oxidation numbers of the elements in the reactants and products. Let's use this second approach with the combustion of methane. First we assign O.N. values to the atoms in the reactants and products:

$$CH_4(g) + 2\,O_2(g) \rightarrow CO_2(g) + 2\,H_2O(\ell) \tag{4.15}$$

CONNECTION In Chapter 3 we defined complete combustion of a hydrocarbon as a reaction with oxygen that converts a compound of carbon and hydrogen into carbon dioxide and water.

These values tell us that carbon is oxidized because each carbon atom loses 8 electrons in going from a -4 oxidation state in methane to a $+4$ oxidation state

The **oxidizing agent** in a redox reaction accepts electrons from another species, thereby oxidizing that species (increasing its oxidation number). The oxidizing agent itself is reduced (decreasing its oxidation number).

The **reducing agent** in a redox reaction gives up electrons to another species, thereby reducing that species (reducing its oxidation number). The reducing agent itself is oxidized (increasing its oxidation number).

in carbon dioxide. Remember that electrons have a charge of $1-$; so, going from -4 to $+4$ involves a loss of 8 electrons.

Oxygen, on the other hand, is reduced: The O.N. value for oxygen changes from 0 to -2 since each of the 4 oxygen atoms gains 2 electrons. To balance the gains and losses of electrons, we need 4 atoms of O for every atom of C. The coefficient of 2 in front of O_2 in Equation 4.15 gives us the 4:1 ratio we need. We also need a coefficient of 2 in front of H_2O to balance the moles of oxygen and hydrogen. Note that hydrogen is neither oxidized nor reduced in this reaction.

Because oxygen in Equation 4.15 is the substance that takes the electrons from the carbon in methane, oxygen is called the **oxidizing agent** in the combustion reaction. We say that "oxygen oxidizes methane." In the process of functioning as an oxidizing agent, the oxygen is reduced. This is always the case: the agent that takes electrons away from some material is reduced in the process. Any species that is reduced experiences a *reduction in oxidation number*; oxygen goes from O.N. = 0 to O.N. = -2. By the same token, the carbon in methane is the species being oxidized; as such, it gives electrons to another substance and reduces it. Methane in this reaction is called the **reducing agent** because it gives electrons to another species (oxygen in this case), thus causing that species (oxygen) to be reduced. The reducing agent is always oxidized. The material that is oxidized experiences an *increase in oxidation number*; in this reaction, carbon goes from O.N. = -4 to O.N. = $+4$. Every redox reaction has an oxidizing agent and a reducing agent.

Another classic redox reaction that can be balanced by inspection is the reduction of iron ore [as $Fe_2O_3(s)$] to metallic iron. Iron ore is heated with carbon monoxide, and the products of the reaction are elemental iron and carbon dioxide:

$$\overset{+3 \qquad\quad reduction \qquad\quad 0}{Fe_2O_3(s) + 3\,CO(g) \rightarrow 2\,Fe(s) + 3\,CO_2(g)}$$
$$\underset{+2 \qquad oxidation \qquad +4}{}$$

The iron in the iron ore is reduced from $+3$ to 0; the carbon in carbon monoxide is oxidized from $+2$ to $+4$. The iron ore is the oxidizing agent; the carbon monoxide is the reducing agent. Two iron atoms with O.N. = $+3$ become 2 iron atoms with O.N. = 0, for a total gain of 6 electrons; 3 carbon atoms at $+2$ become 3 carbon atoms at $+4$, for a total loss of 6 electrons. The same number of electrons is gained as lost. Note that the oxygen atoms are not involved in this redox reaction; carbon is oxidized, iron is reduced.

SAMPLE EXERCISE 4.10   **Identifying Oxidizing Agents and Reducing Agents**

One of the reactions of oxygen in the atmosphere of early Earth was with $SO_2(g)$ to produce $SO_3(g)$. Identify the species oxidized and the species reduced in this reaction. Also identify the oxidizing agent and the reducing agent.

COLLECT AND ORGANIZE The reactants are oxygen and sulfur dioxide; the product is sulfur trioxide.

ANALYZE The equation of the reaction is:

Unbalanced: $\quad SO_2(g) + O_2(g) \rightarrow SO_3(g)$

Balanced: $\quad 2\,SO_2(g) + O_2(g) \rightarrow 2\,SO_3(g)$

SOLVE By inspection we can see that oxygen is the oxidizing agent because SO_2 is oxidized. That means that oxygen is reduced and SO_2 is the reducing agent. We verify this by assigning oxidation numbers:

$$\overset{+4 \quad \text{oxidation} \quad +6}{2\,SO_2(g) + O_2(g) \rightarrow 2\,SO_3(g)} \qquad (4.16)$$

$$\underset{0 \;\; \text{reduction} \; -2}{}$$

THINK ABOUT IT Note that not all of the atoms in a reaction participate in the transfer of electrons. We tend to identify the specific *atoms* that are oxidized or re- duced, while whole *molecules or ions* are identified as the oxidizing agents or reducing agents. In Equation 4.16, sulfur is oxidized but sulfur dioxide is the reducing agent.

Practice Exercise The following reaction is used to remove dissolved oxygen gas from aqueous solutions:

$$O_2(aq) + N_2H_4(aq) \rightarrow 2\,H_2O(\ell) + N_2(g)$$

Identify the species oxidized and the species reduced in this reaction. Also iden- tify the oxidizing agent and the reducing agent.

(Answers to Practice Exercises are in the back of the book.)

Balancing Redox Reactions Using Half-Reactions

Consider the following reaction in aqueous medium. When a piece of copper wire [elemental copper: $Cu(s)$] is placed in a colorless solution of silver nitrate [$Ag^+(aq)$ + $NO_3^-(aq)$], the solution gradually turns blue, the color of a solution of $Cu^{2+}(aq)$, and beautiful branchlike structures of pure silver [$Ag(s)$] form in the medium (Figure 4.15). The process can be summarized as follows:

$$\overset{+1 \quad \text{reduction} \quad 0}{Ag^+(aq) + Cu(s) \rightarrow Ag(s) + Cu^{2+}(aq)} \qquad (4.17)$$

$$\underset{0 \quad \text{oxidation} \quad +2}{}$$

At first glance, this reaction may seem balanced—and in a material sense it is: the number of silver and copper atoms or ions is the same on each side. However, the charges are not because one electron is involved in the reduction reaction while two are involved in the oxidation reaction. Equal numbers of electrons must be lost and gained, so we must develop a method to balance redox reactions like this one that ac- counts for electron transfer. The method we apply is called the *half-reaction method*.

Think of the reaction in Equation 4.17 as consisting of two reactions: one oxidation and one reduction. Each of these reactions represents *half* of the overall reaction; hence, each is called a **half-reaction**.

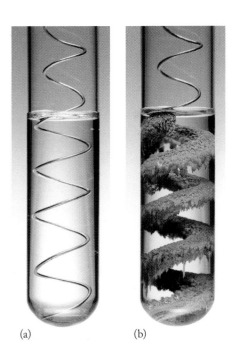

(a)   (b)

FIGURE 4.15 (a) When a helix of Cu wire is immersed in a solution of $AgNO_3$, Cu metal starts to oxidize to Cu^{2+} ions as Ag^+ ions are reduced to Ag metal. (b) A day later the solution has the blue color of a solution of $Cu(NO_3)_2$ and the wire is coated with Ag metal.

Oxidation–reduction reactions consist of two **half-reactions**, one for the oxidation component and one for the reduction component.

1. *Write the separated reactants and products:*

$$\text{Oxidation:} \quad Cu(s) \rightarrow Cu^{2+}(aq)$$

$$\text{Reduction:} \quad Ag^+(aq) \rightarrow Ag(s)$$

2. *Balance the number of particles on both sides of both equations:* This example requires no changes at this point, because both equations are balanced in terms of mass. Each species has an understood coefficient of 1:

$$\text{Oxidation:} \quad Cu(s) \rightarrow Cu^{2+}(aq)$$

$$\text{Reduction:} \quad Ag^+(aq) \rightarrow Ag(s)$$

3. *Balance each half-reaction in terms of charge* by adding electrons to the appropriate side to make the charges on both sides equal. Always add electrons; never subtract them. Remember that electrons have a negative charge:

$$\text{Oxidation:} \quad Cu(s) \rightarrow Cu^{2+}(aq) + 2\,e^-$$

$$\text{total charge} = 0 \quad \text{total charge} = (2+) + (2-) = 0$$

$$\text{Reduction:} \quad 1\,e^- + Ag^+(aq) \rightarrow Ag(s)$$

$$\text{total charge} = (1-) + (1+) = 0 \quad \text{total charge} = 0$$

Note that electrons are added on the reactant side of the reduction half-reaction and on the product side of the oxidation reaction. This will always be the case in this step.

4. The number of electrons lost in oxidation must equal the number gained in reduction. *Multiply each half-reaction by the appropriate whole number* to make the number of electrons in the oxidation half-reaction equal the number in the reduction half-reaction:

$$\text{Oxidation:} \quad 1 \times [Cu(s) \rightarrow Cu^{2+}(aq) + 2\,e^-]$$

$$\text{Reduction:} \quad 2 \times [1\,e^- + Ag^+(aq) \rightarrow Ag(s)]$$

As a result we have the following:

$$\text{Oxidation:} \quad Cu(s) \rightarrow Cu^{2+}(aq) + 2\,e^-$$

$$\text{Reduction:} \quad 2\,e^- + 2\,Ag^+(aq) \rightarrow 2\,Ag(s)$$

5. *Add the two half-reactions together to generate the overall reaction.* The electrons must cancel:

$$\text{Oxidation:} \quad Cu(s) \rightarrow Cu^{2+}(aq) + \cancel{2\,e^-}$$

$$\text{Reduction:} \quad \cancel{2\,e^-} + 2\,Ag^+(aq) \rightarrow 2\,Ag(s)$$

$$\text{Overall reaction:} \quad 2\,Ag^+(aq) + Cu(s) \rightarrow 2\,Ag(s) + Cu^{2+}(aq)$$

The reaction is now balanced in terms of mass and electron transfer. The final equation is a net ionic equation. Only those species that are involved in the redox reaction—those losing or gaining electrons—are included. The nitrate ion mentioned at the beginning of this subsection is a spectator ion in this process because it does not participate in the chemical reaction; hence it does not show up in Equation 4.17 or any of the subsequent half-reactions we used to balance Equation 4.17.

SAMPLE EXERCISE 4.11 **Balancing Redox Reactions by the Half-Reaction Method**

Iodine is slightly soluble in water and makes a pale yellow-brown solution of $I_2(aq)$. When $Sn^{2+}(aq)$ is dissolved in an iodine-containing solution, the solution turns colorless as $I^-(aq)$ forms and $Sn^{2+}(aq)$ is converted into the $Sn^{4+}(aq)$ ion. (a) Is this a redox reaction? (b) Balance the equation that describes the process taking place.

COLLECT AND ORGANIZE The reactants in this process are $I_2(aq)$ and $Sn^{2+}(aq)$; the products are $I^-(aq)$ and $Sn^{4+}(aq)$.

ANALYZE The oxidation numbers change from $Sn^{2+}(aq)$ (O.N. = +2) to $Sn^{4+}(aq)$ (O.N. = +4), and from $I_2(aq)$ (O.N. = 0) to $I^-(aq)$ (O.N. = −1).

SOLVE

a. Because the oxidation numbers change, this is definitely a redox reaction.
b. The unbalanced equation showing reactants and products is

$$Sn^{2+}(aq) + I_2(aq) \rightarrow I^-(aq) + Sn^{4+}(aq)$$

We will balance the equation following the steps for the half-reaction method:

(1) Oxidation: $Sn^{2+}(aq) \rightarrow Sn^{4+}(aq)$

Reduction: $I_2(aq) \rightarrow I^-(aq)$

(2) Oxidation: $Sn^{2+}(aq) \rightarrow Sn^{4+}(aq)$

Reduction: $I_2(aq) \rightarrow 2\,I^-(aq)$

(3) Oxidation: $Sn^{2+}(aq) \rightarrow Sn^{4+}(aq) + 2\,e^-$

Reduction: $2\,e^- + I_2(aq) \rightarrow 2\,I^-(aq)$

(4) (No multiplication needed: numbers of electrons are the same in both reactions.)

(5) Oxidation: $Sn^{2+}(aq) \rightarrow Sn^{4+}(aq) + \cancel{2\,e^-}$

Reduction: $\cancel{2\,e^-} + I_2(aq) \rightarrow 2\,I^-(aq)$

Overall reaction: $Sn^{2+}(aq) + I_2(aq) \rightarrow 2\,I^-(aq) + Sn^{4+}(aq)$

THINK ABOUT IT Notice in Step 4 that the total charge in the oxidation half-reaction does not have to be zero, but it must be the *same* on both sides. Always check the final overall reaction for mass balance (here, 1 Sn + 2I = 2I + 1 Sn) and the charge balance [here the reactant side = 2+, and the product side = 2(−1) + (+4) = +2].

Practice Exercise If a nail made of elemental iron [Fe(s)] is placed in an aqueous solution of a soluble palladium(II) salt [$Pd^{2+}(aq)$], the nail will gradually disappear as the iron enters the solution as $Fe^{3+}(aq)$ and palladium metal [Pd(s)] forms. (a) Is this a redox reaction? (b) Balance the equation that describes this reaction.

(Answers to Practice Exercises are in the back of the book.)

FIGURE 4.16 The orange-red strata in Bryce Canyon consist of oxides of the iron(III) ion that result from weathering of the rock.

(a)

(b)

FIGURE 4.17 Wetland soils: (a) the soil in this wetland has a blue-gray color because of the presence of iron(II) compounds. (b) Orange mottling indicates the presence of Fe(III) oxides formed as a result of O_2 permeation through channels made by plant roots.

Redox in Nature

Just as acid–base reactions and precipitation reactions are involved in the chemical weathering of rocks, redox processes also play a major role in determining the character of rocks and soils. For example, oxygen reacts with iron to form iron oxide minerals such as hematite. The red-colored rocks like those seen in Bryce Canyon National Park in Utah are examples of iron(III) oxide–containing minerals (Figure 4.16). Iron(III) oxide is also the form of iron known as rust, the crumbly red-brown solid that forms on objects made of iron that are exposed to water and oxygen. Rocks containing hematite are relatively weak and easily broken, a fact which gives rise to the fascinating shapes of such deposits.

The color of the soil in an area is also influenced by its mineral content. Wetlands, areas that are saturated or flooded with water during much of the year, are protected environments in several areas of the United States. Defining a given area as a wetland relies in part on the characteristics of the soils in the area, and the color of the soil is an important diagnostic feature in identifying an area as a true wetland (Figure 4.17). The color of soil is strongly influenced by the frequency and duration of periods of water saturation, primarily because soil saturated with water is a reducing environment in which free oxygen is prevented from contacting the minerals. Iron-containing components of the soil are reduced in such an environment: the soil is gray with bluish-green tints (Figure 4.17a), in contrast to the red color associated with Fe(III) compounds. When the roots of plants penetrate this soil, they create pores in the soil. This allows iron in its reduced form to come into contact with oxygen diffusing through the pores; as a result the iron becomes oxidized as the familiar orange- or red-brown iron(III) compounds (Figure 4.17b).

The redox reaction of iron(II) and iron(III) compounds in the soils of wetlands can be modeled in the laboratory. Figure 4.18(a) shows a solution of ferrous ammonium sulfate $[(NH_4)_2Fe(SO_4)_2(aq)]$. The greenish color of the solution is characteristic of the iron(II) ion in its compounds and in an aqueous solution. When a solution of sodium hydroxide is added, $Fe(OH)_2(s)$ precipitates out of solution (Figure 4.18b) as a grayish olive-green solid, very similar in color to the wetland soil shown in Figure 4.17(a). The solid remains that color as long as it is submerged in water. When the solution is poured onto a paper towel (Figure 4.18c), however, the iron(II) compound is exposed to free oxygen in the air. Within about 20 minutes (Figure 4.18d) the green precipitate turns the orange-brown color of iron(III) hydroxide. The iron(II) has been oxidized—exactly the way the iron(II) compounds in a wetland soil are oxidized when exposed to free oxygen.

The color of iron-containing soil is a reliable indicator of the extent to which soil in an area is exposed to free oxygen. Because oxygen gas is not very soluble in water, the soil color in an area is therefore a good index of the saturation level of water in the ground and is useful data for assisting in the classification of true wetlands.

Balancing the reaction for the conversion of iron(II) to iron(III) in an aqueous base requires us to revisit the half-reaction method and add some further steps. Many redox reactions take place in acids and bases, and the acid [as $H^+(aq)$], the base [as $OH^-(aq)$], or even the water in the medium may play a role in the reaction. Let's look at a reaction used in analyzing solutions containing iron before we tackle the one demonstrated in Figure 4.18.

A method for determining the concentration of $Fe^{2+}(aq)$ in an acidic solution involves its oxidation by adding the intensely purple-colored permanganate ion

$[MnO_4^-(aq)]$ to $Fe^{3+}(aq)$. In the course of the reaction, the permanganate ion is reduced to $Mn^{2+}(aq)$, which is very pale pink. Here is the unbalanced reaction:

$$Fe^{2+}(aq) + MnO_4^-(aq) \rightarrow Fe^{3+}(aq) + Mn^{2+}(aq)$$

To balance the equation for this reaction, we follow the steps outlined in the half-reaction method. The first step is the familiar step 1 from before:

(1) Oxidation: $\qquad\qquad\qquad\qquad Fe^{2+}(aq) \rightarrow Fe^{3+}(aq)$

     Reduction: $\qquad\qquad\qquad\qquad MnO_4^-(aq) \rightarrow Mn^{2+}(aq)$

Step 2 needs to be broken down into substeps to account for any role played by the aqueous acid: First we balance all the elements *except hydrogen and oxygen*:

(2a) Oxidation: $\qquad\qquad\qquad\qquad Fe^{2+}(aq) \rightarrow Fe^{3+}(aq)$

     Reduction: $\qquad\qquad\qquad\qquad MnO_4^-(aq) \rightarrow Mn^{2+}(aq)$

In this case the iron and the manganese are already balanced, so we move to the second substep: balancing any oxygen present by adding water to the reaction. This substep is appropriate because it accounts for any role played by water in the reaction:

(2b) Oxidation: $\qquad\qquad\qquad\qquad Fe^{2+}(aq) \rightarrow Fe^{3+}(aq)$

(No oxygen, so this half-reaction stays the same.)

     Reduction: $\qquad\qquad\qquad\qquad MnO_4^-(aq) \rightarrow Mn^{2+}(aq)$

(Oxygen is present . . .)

$$MnO_4^-(aq) \rightarrow Mn^{2+}(aq) + 4\,H_2O(\ell)$$

(. . . so we need to add water.)

In the final substep we balance any hydrogen by adding $H^+(aq)$; this is appropriate because the reaction is run in acidic medium:

(2c) Oxidation: $\qquad\qquad\qquad\qquad Fe^{2+}(aq) \rightarrow Fe^{3+}(aq)$

(No hydrogen, so this half-reaction stays the same.)

     Reduction: $\qquad\qquad MnO_4^-(aq) \rightarrow Mn^{2+}(aq) + 4\,H_2O(\ell)$

(Hydrogen is present . . .)

$$8\,H^+(aq) + MnO_4^-(aq) \rightarrow Mn^{2+}(aq) + 4\,H_2O(\ell)$$

(. . . so we need to add hydrogen ions.)

Steps 3, 4, and 5 are the same as before:

(3) Oxidation: $\qquad\qquad\qquad\qquad Fe^{2+}(aq) \rightarrow Fe^{3+}(aq) + 1\,e^-$

     Reduction: $\quad 5\,e^- + 8\,H^+(aq) + MnO_4^-(aq) \rightarrow Mn^{2+}(aq) + 4\,H_2O(\ell)$

(4) Oxidation: $\qquad\qquad 5 \times [Fe^{2+}(aq) \rightarrow Fe^{3+}(aq) + 1\,e^-]$

     Reduction: $1 \times [5\,e^- + 8\,H^+(aq) + MnO_4^-(aq) \rightarrow Mn^{2+}(aq) + 4\,H_2O(\ell)]$

(5) Oxidation: $\qquad\qquad\qquad 5\,Fe^{2+}(aq) \rightarrow 5\,Fe^{3+}(aq) + \cancel{5\,e^-}$

     Reduction: $\quad \cancel{5\,e^-} + 8\,H^+(aq) + MnO_4^-(aq) \rightarrow Mn^{2+}(aq) + 4\,H_2O(\ell)$

Overall reaction:

$$8\,H^+(aq) + MnO_4^-(aq) + 5\,Fe^{2+}(aq) \rightarrow 5\,Fe^{3+}(aq) + Mn^{2+}(aq) + 4\,H_2O(\ell)$$

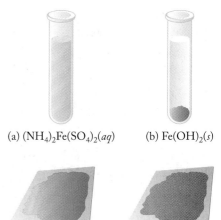

(a) $(NH_4)_2Fe(SO_4)_2(aq)$ (b) $Fe(OH)_2(s)$

(c) $Fe(OH)_2(s)$ (d) $Fe(OH)_3(s)$

FIGURE 4.18 (a) A solution of ferrous ammonium sulfate. (b) The same solution after the addition of $NaOH(aq)$. (c) The resulting suspension immediately after being poured onto a paper towel. (d) The same precipitate after about 20 minutes of exposure to air.

Acidic mine drainage flowing into this stream in western Pennsylvania contains high concentrations of iron that precipitate as orange $Fe(OH)_3$ when the drainage mixes with less-acidic stream water.

Whenever you balance an equation, always check the mass balance and a charge balance as a final check. In this equation, the reactant side contains 8 H, 1 Mn, 4 O, and 5 Fe atoms and the product side shows 8 H, 1 Mn, 4 O, and 5 Fe, so the mass balance is correct. Now let's consider charge: in a net ionic equation the charges on both sides must be equal. On the reactant side the total charge is $(8+) + (1-) + (10+) = 17+$, and on the product side $(15+) + (2+) = 17+$, so the reaction is correctly balanced.

This method works for balancing any redox reaction taking place in an aqueous acid. The demonstration of the oxidation of $Fe(OH)_2(s)$ to $Fe(OH)_3(s)$ by oxygen in the air (Figure 4.18), however, was carried out in a basic medium. How do we account for that? Balancing a redox reaction in a basic medium is simple: we carry out the steps outlined above as though the reaction were run in an acidic medium. Then we add a final step in which sufficient hydroxide ion is added to both sides of the equation to convert any $H^+(aq)$ ions to $H_2O(\ell)$. This step converts the medium to an aqueous basic solution, as specified in the reaction. Let's see how this works.

As always, we start with an unbalanced equation to clearly define the reactants and products:

$$Fe(OH)_2(s) + O_2(g) \rightarrow Fe(OH)_3(s)$$

This reaction takes place in the solid phase in the presence of hydroxide ions in water. The oxidation half-reaction is clear: iron(II) hydroxide [$Fe(OH)_2(s)$] is oxidized to iron(III) hydroxide [$Fe(OH)_3(s)$]. What is the reduction half-reaction? The oxygen must be converted into the additional OH^- ion in the iron(III) hydroxide:

(1) $$Fe(OH)_2(s) \rightarrow Fe(OH)_3(s)$$
$$O_2(g) \rightarrow OH^-(aq)$$

(2a) $$H_2O(\ell) + Fe(OH)_2(s) \rightarrow Fe(OH)_3(s)$$
$$O_2(g) \rightarrow OH^-(aq) + H_2O(\ell)$$

(2b) $$H_2O(\ell) + Fe(OH)_2(s) \rightarrow Fe(OH)_3(s) + H^+(aq)$$
$$3\,H^+(aq) + O_2(g) \rightarrow OH^-(aq) + H_2O(\ell)$$

(3) $$H_2O(\ell) + Fe(OH)_2(s) \rightarrow Fe(OH)_3(s) + H^+(aq) + 1\,e^-$$
$$4\,e^- + 3\,H^+(aq) + O_2(g) \rightarrow OH^-(aq) + H_2O(\ell)$$

(4) $$4 \times [H_2O(\ell) + Fe(OH)_2(s) \rightarrow Fe(OH)_3(s) + H^+(aq) + 1\,e^-]$$
$$1 \times [4\,e^- + 3\,H^+(aq) + O_2(g) \rightarrow OH^-(aq) + H_2O(\ell)]$$

(5) $$4\,H_2O(\ell) + 4\,Fe(OH)_2(s) \rightarrow 4\,Fe(OH)_3(s) + 4\,H^+(aq) + \cancel{4\,e^-}$$
$$\cancel{4\,e^-} + 3\,H^+(aq) + O_2(g) \rightarrow OH^-(aq) + H_2O(\ell)$$

Overall reaction:

$$\overset{3}{\cancel{4}}H_2O(\ell) + 4\,Fe(OH)_2(s) + \cancel{3\,H^+(aq)} + O_2(g) \rightarrow$$
$$4\,Fe(OH)_3(s) + \overset{1}{\cancel{4}}H^+(aq) + OH^-(aq) + \cancel{H_2O(\ell)}$$

We can simplify the equation the same way we would an algebraic one by cancelling out the same species on both sides:

$$\overset{2}{\cancel{3}}H_2O(\ell) + 4\,Fe(OH)_2(s) + O_2(g) \rightarrow 4\,Fe(OH)_3(s) + \underbrace{\cancel{H^+(aq)} + \cancel{OH^-(aq)}}_{\cancel{H_2O(\ell)}}$$

This gives us the final balanced equation:

$$2\,H_2O(\ell) + 4\,Fe(OH)_2(s) + O_2(g) \rightarrow 4\,Fe(OH)_3(s)$$

Checking this equation for mass, we find it is correct:

$$12\,H + 12\,O + 4\,Fe = 12\,H + 12\,O + 4\,Fe$$

as is the charge balance: $0 = 0$.

The reaction in wetland soil is essentially the same as the one we solved here. The absence of oxygen in soil heavily saturated or flooded with water establishes a reducing environment that favors iron(II) compounds. When free oxygen becomes available, either because of dry conditions or because of porosity caused by the action of plant roots, the iron(II) species oxidizes quickly to red- or orange-brown iron(III) compounds.

4.9 Titrations

In Section 4.6 we discussed a method for determining the dissolved concentration of chloride ions in a sample of drinking water by precipitating $Cl^-(aq)$ ions as insoluble $AgCl(s)$ and then weighing the precipitate. Another analytical approach, called a **titration**, is based on the same precipitation reaction, but titrations are based on measured volumes instead of masses. In a precipitation titration to analyze the concentration of chloride ion, we precisely measure the volume of a **standard solution** of $AgNO_3(aq)$ of known concentration, called the **titrant**, needed to completely react with the chloride in a known volume of sample. This information is then used to calculate the concentration of chloride in the sample. The calculations involved are described in Sample Exercise 4.12.

SAMPLE EXERCISE 4.12 **Calculating a Concentration from the Results of a Titration**

In climates with subfreezing winters, elevated concentrations of NaCl in groundwater are often the result of contamination from runoff containing salt used on roads. To determine the concentration of $Cl^-(aq)$ in a sample of drinking water from a well, a sample is titrated with a standard solution of $Ag^+(aq)$. Determining the chloride concentration of a sample in this way is a widely used application of a precipitation titration. The titration reaction is given in Equation 4.12:

$$NaCl(aq) + AgNO_3(aq) \rightarrow AgCl(s) + NaNO_3(aq)$$

What is the concentration of $Cl^-(aq)$ in a 100.0 mL sample of the well water if 24.1 mL of 0.100 M $AgNO_3(aq)$ are needed to react with all the $Cl^-(aq)$ in the sample? Express your answer in molarity (as a chemist would) and in mg $Cl^-(aq)$/L (as environmental scientists tend to do).

COLLECT AND ORGANIZE We know the volume (24.1 mL) and concentration (0.100 M) of the $AgNO_3$ solution and the volume of the sample (100.0 mL). We need to calculate the concentration of $Cl^-(aq)$ ion in the sample. The link between the quantity of $AgNO_3(aq)$ used and the quantity of $Cl^-(aq)$ ions in the sample is the stoichiometry of the titration reaction, in which 1 mol $Ag^+(aq)$ reacts with 1 mol $Cl^-(aq)$ as they precipitate as $AgCl(s)$.

ANALYZE Let's first calculate the quantity of $Ag^+(aq)$ consumed from the volume and concentration of the $AgNO_3$ solution, then convert that to an equivalent

A **titration** is an analytical method for precisely determining the concentration of a solute in a sample by reacting it with a standard solution of known concentration.

A **standard solution** is a solution of known concentration.

The **titrant** is the standard solution added to the sample in a titration.

quantity of Cl⁻(aq), and finally divide that quantity of Cl⁻(aq) by the volume of the sample to obtain the values of Cl⁻(aq) concentration in molarity and then in milligrams per liter.

SOLVE The quantity of $Ag^+(aq)$ consumed is

$$24.1 \text{ mL} \times \frac{0.100 \text{ mol Ag}^+(aq)}{L} = 2.41 \text{ mmol Ag}^+(aq)$$

The quantity of Cl⁻(aq) in the sample must have been

$$2.41 \text{ mmol Ag}^+(aq) \times \frac{1 \text{ mol Cl}^-(aq)}{1 \text{ mol Ag}^+} = 2.41 \text{ mmol Cl}^-$$

and the concentration of Cl⁻(aq) in the sample was

$$\frac{2.41 \text{ mmol Cl}^-(aq)}{100 \text{ mL}} = 2.41 \times 10^{-2}\frac{\text{mol Cl}^-(aq)}{L} = 2.41 \times 10^{-2} \, M$$

Expressing this concentration of Cl⁻(aq) in milligrams per liter, we get

$$2.41 \times 10^{-2} \frac{\text{mol Cl}^-}{L} \times \frac{35.45 \text{ g Cl}^-(aq)}{\text{mol Cl}^-} \times \frac{1000 \text{ mg}}{g} = 854 \text{ mg Cl}^-(aq)/L$$

THINK ABOUT IT To estimate the answer, consider that the balanced equation says that 1 mole of Cl⁻(aq) requires 1 mole of $Ag^+(aq)$ to form AgCl(s). The volume of $Ag^+(aq)$ solution (about 25 mL) used to titrate the sample was about one-quarter the size of the sample (100 mL), which means that the concentration of Cl⁻(aq) in the sample was about one-quarter the concentration of the $Ag^+(aq)$ solution used in the titration. Our answer ($2.41 \times 10^{-2} \, M = 0.0241 \, M$) is about one-quarter the value of the $Ag^+(aq)$ concentration, so it is a very reasonable answer; it is also well above the maximum concentration allowed by drinking-water standards.

Practice Exercise The concentration of $SO_4^{2-}(aq)$ in a sample of river water can be determined using a precipitation titration in which a salt of $Ba^{2+}(aq)$ is the standard solution and $BaSO_4(s)$ is the precipitate formed. What is the concentration of $SO_4^{2-}(aq)$ in a 50.0 mL sample if 6.55 mL of a 0.00100 M $Ba^{2+}(aq)$ solution is needed to precipitate all the $SO_4^{2-}(aq)$ in the sample?

(Answers to Practice Exercises are in the back of the book.)

One of the most common types of titration is acid–base neutralization (see Section 4.5). A typical analysis to which an acid–base titration is often applied consists of quantifying the environmental problem of sulfuric acid contamination of the waters draining from abandoned coal mines. Suppose we have a 100.0 mL sample of drainage water containing an unknown concentration of sulfuric acid. We decide to use an acid–base titration in which the sulfuric acid in the sample is neutralized with a standard solution of 0.00100 M NaOH. The following neutralization reaction takes place:

$$H_2SO_4(aq) + 2 \, NaOH(aq) \rightarrow Na_2SO_4(aq) + 2 \, H_2O(\ell)$$

Suppose that exactly 22.40 mL of the NaOH solution is required to completely react with the H_2SO_4 in the 100.0 mL sample. Because we know the volume (V) and molar concentration (M) of the NaOH solution, we can calculate the number of moles (n) of NaOH consumed in the neutralization reaction using Equation 4.2:

$$n = V \times M$$

$$= 0.02240 \text{ L} \times 0.00100 \frac{\text{mol NaOH}}{\text{L}} = 0.00002240 \text{ mol NaOH}$$

We know from the stoichiometry of the neutralization reaction that 2 moles of NaOH are required to neutralize 1 mole of H_2SO_4, so the number of moles of H_2SO_4 in the sample must be as follows:

$$0.00002240 \text{ mol NaOH} \times \frac{1 \text{ mol H}_2\text{SO}_4}{2 \text{ mol NaOH}} = 0.00001120 \text{ mol H}_2\text{SO}_4$$

If the 100.0 mL sample of mine effluent contains 0.00001120 mol H_2SO_4, then the concentration of H_2SO_4 is

$$\frac{0.00001120 \text{ mol H}_2\text{SO}_4}{0.1000 \text{ L}} = 1.120 \times 10^{-4} \frac{\text{mol H}_2\text{SO}_4}{\text{L}} = 1.12 \times 10^{-4} \, M \, \text{H}_2\text{SO}_4$$

How did we know that exactly 22.40 mL of the NaOH solution were needed to react with the sulfuric acid in the sample? A setup for doing an acid–base titration is illustrated in Figure 4.19. The standard solution of NaOH is poured into a buret, a narrow glass cylinder with volume markings. The solution is gradually added to the sample until a visual indicator that changes color in response to the concentration of $H^+(aq)$ ion in the solution signals that the reaction is complete. The point in the titration when just enough standard solution has been added to completely react with all the solute of interest is called the **equivalence point**. If the correct indicator has been chosen for the reaction, the equivalence point is very close to the **end point**, the point at which the indicator changes color.

To detect the end point in our sulfuric acid–sodium hydroxide titration, we might use a color indicator called phenolphthalein, which is colorless in acidic solutions but changes to pink when a slight excess of base is present. The part of the titration that requires the most skill is adding just enough NaOH solution to reach the end point, which is indicated by a persistent pink color in the solution being titrated. To catch this end point, the standard solution must be added no faster than one drop at a time with thorough mixing in between drops.

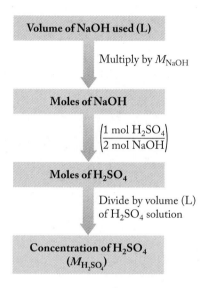

FIGURE 4.19 The concentration of a sulfuric acid solution can be determined through titration with a standard solution of sodium hydroxide. (a) A known volume of the H_2SO_4 solution is placed into the flask. The buret is filled with an aqueous solution of sodium hydroxide of known concentration. A few drops of phenolphthalein indicator solution are added to the flask. (b) Sodium hydroxide is carefully added until the indicator changes from colorless to pink. The concentration of H_2SO_4 is calculated by using the steps in the flowchart above.

(a)　　　　(b)

Calcium: In the Limelight

Calcium is the fifth most abundant element in Earth's crust. Vast deposits of $CaCO_3$ occur over large areas of the planet as limestone, marble, and chalk. These minerals are the fossilized remains of earlier life in the oceans. Calcium carbonate is called a biomineral because it is an inorganic substance synthesized as an integral part of a living organism. Coccolithophores, which produce the coccoliths pictured in Figure 4.4, are among the largest producers of calcium carbonate on Earth today, and probably have been since the Jurassic period (150 million years ago). Corals and sea shells are mainly calcium carbonate, and islands like the Florida Keys and the Bahamas are made of large beds of $CaCO_3$. Calcium sulfate, normally called gypsum, is another important calcium-containing mineral. If you are reading this in a room with plaster walls, you are surrounded by gypsum, as it is the primary component in sheet rock and plaster.

Although mostly known for their use as bulk chemicals in heavy industry and construction, special forms of these two chemicals are valued for their beauty. Pearls (Figure 4.20a) are gemstones consisting mainly of calcium carbonate, in particular a heterogeneous mixture of $CaCO_3(s)$ and a small amount of water suspended within the solid. Light travels differently through the solid material and the dispersed water in a fine pearl. The same phenomenon occurs when a thin layer of oil or gas floats on the surface of water: we see an array of colors. Diffraction of the light gives a high-quality pearl its special iridescence. The substance known as alabaster, prized for its translucence and subtle patterns, is a fine-grained form of calcium sulfate (Figure 4.20b).

Compounds of calcium have myriad other small uses. The sulfate and carbonate of calcium are two common materials used in desiccants, those small packets of granules included in boxes of electronic devices, shoes, pharmaceuticals, and foods. Desiccants (from the Latin *siccus*, "dry") remove water from the air, thereby keeping dry the items with which they are packed. About 0.5% by mass of calcium silicate ($CaSiO_3$) is added to solid table salt to keep it freely flowing in damp weather. It is estimated that calcium silicate can absorb almost 600 times its mass in water and still remain a free-flowing powder. Calcium hypochlorite $[Ca(OCl)_2]$ is one of the sources of chlorine used to disinfect the water in swimming pools.

In terms of bulk consumption, calcium carbonate is an industrial chemical of great utility. High-quality paper contains $CaCO_3(s)$, which improves brightness as well as the paper's ability to bind ink. The pharmaceutical industry sells it as an antacid and in dietary supplements, but $CaCO_3$ has an even greater industrial use as the precursor of two other important materials: quicklime (CaO) and slaked lime $[Ca(OH)_2]$. To produce lime, calcium carbonate (derived usually from mining it or from oyster shells) is roasted.

$$CaCO_3(s) \xrightarrow{\text{heat}} CaO(s) + CO_2(g)$$

Slaked lime is produced from the hydrolysis of calcium oxide:

$$CaO(s) + H_2O(\ell) \rightarrow Ca(OH)_2(s)$$

(a)

(b)

FIGURE 4.20 Prized examples of calcium minerals are (a) pearls (heterogeneous mixtures of calcium carbonate and water) and (b) alabaster (fine-grained calcium sulfate).

To be "in the limelight," is now often used to describe anyone who is the center of attention. The expression derives from the use of lighting systems based on calcium oxide in 19th-century theaters. A British engineer discovered that calcium oxide emits a brilliant but soft white light when heated in a flame of hydrogen mixed with oxygen. By the mid-1860s, limelights fitted with reflectors to direct and spread illumination were in wide use in theaters. In modern times, lime has moved out of the entertainment business and into heavy industry. About 75 kg of lime is used to produce 1 ton of steel. Lime is added to molten iron, from which it removes phosphorus, sulfur, and silicon to purify the metal. It is used in large amounts in the treatment of water supplies, in which it serves to coagulate suspended solids.

Lime is also used in large quantities in smokestack scrubbers to remove oxides of sulfur from the gases emitted by power plants that burn coal. During combustion, sulfur in coal forms $SO_2(g)$ and $SO_3(g)$. When allowed to escape into the atmosphere, these nonmetal oxides mix with atmospheric water to make acid rain. To "scrub" these environmentally harmful gases from stack emissions, calcium oxide is suspended in water through which the plant's exhaust gases pass. Here is the set of scrubber reactions for trapping SO_3:

$$CaO(s) + H_2O(\ell) \rightarrow Ca(OH)_2(aq)$$
(hydrolysis of a metal oxide to make a base)

$$SO_3(g) + H_2O(\ell) \rightarrow H_2SO_4(aq)$$
(hydrolysis of a nonmetal oxide to make an acid)

$$H_2SO_4(aq) + Ca(OH)_2(aq) \rightarrow CaSO_4(s) + 2H_2O(\ell)$$
(neutralization makes water and a salt)

$$CaO(s) + SO_3(g) \rightarrow CaSO_4(s)$$
(net overall reaction)

The SO_2 gas is trapped as solid calcium sulfate (gypsum), which may be disposed of in a landfill (Figure 4.21).

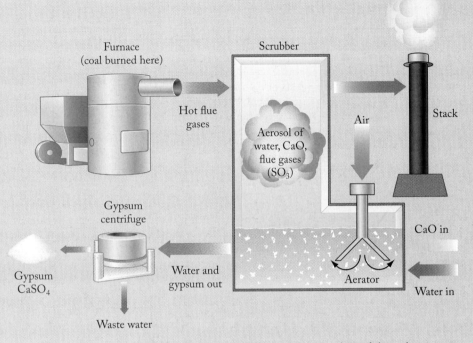

FIGURE 4.21 A scrubber capitalizes on the acid–base chemistry of aqueous solutions of calcium oxide and the oxides of sulfur in order to trap exhaust gases that would otherwise produce strong acid rain if released into the atmosphere.

Calcium is an essential element for life; most animals—including humans—must absorb calcium every day from their diet. Over 95% of the calcium in the human system is found in bones and teeth, and the average concentration of calcium in the blood is 100 mg/L. The body needs calcium to clot blood, transmit nerve impulses, and regulate the heartbeat. If there is insufficient calcium in your blood for these uses, your body will break down bone tissue to extract it. Hence, the amount of calcium ion in the blood does not provide an index of the status of calcium in your system; measurements of bone density are required to determine the overall balance. In addition, bones are not static: even in the absence of demand from other parts of your body, your bones are constantly being remodeled, in which a small amount of old bone is removed each day and new bone formed in its place. After about the age of 35, however, bone is removed faster than it is deposited, resulting in net bone loss as we grow older.

Summary

SECTION 4.1 **Solutions** are homogeneous mixtures of two or more substances. The substance in greatest molar proportion is the **solvent**; all other ingredients are called **solutes**. Insoluble substances may be suspended in a liquid. Such suspensions scatter light in ways that true solutions do not. Filtration can be used to remove suspended solid particles from a liquid, but not solutes from a solution.

SECTION 4.2 The concentration of solute in a solution can be expressed in many ways, including mass of solute per mass of solution, such as g/kg, mg/kg (ppm), and μg/kg (ppb). Other ways to express concentration include mass of solute to volume of solvent (for example, g/L) and moles of solute per liter of solution (mol/L); the latter ratio is called **molarity (M)**.

SECTION 4.3 During **dilution** the quantity of solute in solution does not change, but as the volume of the solution increases, the concentration of solute decreases.

SECTION 4.4 A solute that dissociates into ions when it dissolves in aqueous solution is called an **electrolyte**. Migration of these ions makes the solution a better electrical conductor than pure water. **Strong electrolytes** dissociate completely into ions when they dissolve in water, **weak electrolytes** partially ionize, and **nonelectrolytes** do not ionize at all.

SECTION 4.5 In aqueous solutions a Brønsted–Lowry **acid** is a proton (H^+) donor, and a Brønsted–Lowry **base** is a proton H^+ acceptor. In acid–base **neutralization reactions** in solution, H^+ ions from the acid combine with an equal number of OH^- ions from the base, forming H_2O. The **net ionic equation** of a reaction in solution includes only the species that change during the reaction and omits the **spectator ions**.

SECTION 4.6 In a precipitation reaction, soluble reactants in solution form an insoluble **precipitate**. A **saturated solution** contains the highest-possible concentration of a solute. A **supersaturated solution** contains more than the maximum concentration of a solute.

SECTION 4.7 In **ion-exchange** reactions, ions in solution displace ions held by electrostatic attraction to ion-exchange sites on the surface of a solid. In commercial water softeners, Ca^{2+}, Mg^{2+}, and other cations in water displace Na^+ ions from the surface of an ion-exchange resin. The process is reversed when the resin is recharged and converted back to its Na^+ form.

SECTION 4.8 In a redox reaction, substances either gain electrons (thereby undergoing **reduction**) or they lose electrons (undergoing **oxidation**). These two processes must be balanced so that the number of electrons gained by reduction equals the number lost by oxidation. A reaction is a redox reaction if the **oxidation numbers (O.N.)**, or **oxidation states**, of the elements in the reactants change during the reaction. Electrons are gained in a reduction **half-reaction**; electrons are lost in an oxidation half-reaction.

SECTION 4.9 **Titrations** are precise methods for determining the concentration of a sample by adding just enough volume of a **standard solution**, delivered using a buret, that contains a known concentration of substance that will react with the solute in sample. Titrations are often based on redox, acid–base neutralization, or precipitation reactions.

Problem-Solving Summary

TYPE OF PROBLEM	CONCEPTS AND EQUATIONS	SAMPLE EXERCISES
Comparing ion concentrations in aqueous solutions	Use conversion factors to express concentrations in the same units	4.1
Calculating molarity (M) from mass and volume or from density	Convert mass of solute first into grams and then into moles by dividing by its molar mass ($\mathcal{M}$). Convert mass of solution into volume by dividing by its density. Divide moles of solute by liters of solution to get molarity (M). $$\text{Molarity} = \frac{\text{moles of solute}}{\text{volume of solution in liters}} \quad \text{or} \quad M = \frac{n}{V}$$	4.2 and 4.3
Calculating the quantity of solute needed to prepare a solution	Multiply concentration (in mol/L) by volume (in L) to obtain moles of solute (n). $$n = V \times M$$ Multiply moles of solute by $\mathcal{M}_{solute}$ to calculate mass of solute in grams. $$\text{Mass of solute (g)} = \mathcal{M} \times V \times M$$	4.4
Calculating dilution	Given three of the four variables, use $$V_{initial} \times C_{initial} = V_{final} \times C_{final}$$ to solve for the fourth.	4.5
Writing neutralization reaction equations	Balance the molecular equation first by balancing the moles of H^+ ions donated by the acid and accepted by the base. Next write an overall ionic equation by separating the formulas of strong electrolytes into their component ions. Finally, write a net ionic equation, by eliminating all the ions that do not change during the reaction.	4.6

TYPE OF PROBLEM	CONCEPTS AND EQUATIONS	SAMPLE EXERCISES
Predicting precipitation reactions	Write out all the cations and anions present in the solutions that have been mixed. If any cation/anion pair is the same as present in an insoluble compound, that compound will precipitate from the solution.	4.7
Calculating a solute concentration from the mass of a precipitate	Convert the mass of the precipitate into moles by dividing by its molar mass. Convert moles of precipitate into moles of solute. Calculate the concentration (M) of the solute in the sample by dividing moles of solute by volume (in liters) of sample.	4.8
Determining oxidation numbers (O.N.) of elements in a compound	O.N. is the charge of a monatomic ion. In molecules and molecular ions, assign O.N. values of 0 to elements, +1 to H, and −2 to O; then calculate O.N. of the element that gives the appropriate overall charge (0 for molecules).	4.9
Identifying the oxidizing and reducing agents in a redox reaction	The oxidizing agent has an element whose O.N. decreases during the reaction; the reducing agent has an element whose O.N. increases.	4.10
Balancing redox reactions with half-reactions	Multiply one or both half-reactions by the appropriate coefficient(s) to balance the loss and gain of electrons. Combine the two half-reactions and simplify.	4.11
Calculating a concentration from the results of a titration	Use the balanced chemical equation to set the mole ratio of reactants; use the number of moles of solute and volume of the sample to calculate the concentration.	4.12

Visual Problems

4.1. In Figure P4.1, which shows a solution containing three binary acids, one of the three is a weak acid and the other two are strong acids. Which color sphere represents the ionized weak acid?

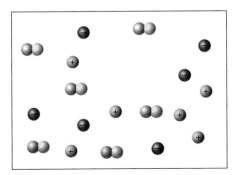

FIGURE P4.1

4.2. Solutions of sodium chloride and silver nitrate are mixed together and vigorously shaken. Which colored spheres in Figure P4.2 represent the following species? NOTE: the silver spheres represent Ag^+ ions. (a) Na^+; (b) Cl^-; (c) NO_3^- ions

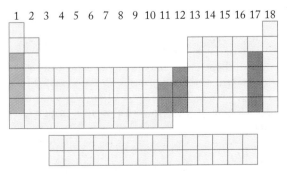

FIGURE P4.2

4.3. Which of the highlighted elements in Figure P4.3 forms an acid with the following generic formula? (a) HX; (b) H_2XO_4; (c) HXO_3; (d) H_3XO_4

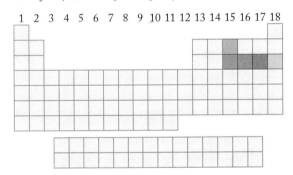

FIGURE P4.3

4.4. In which of the highlighted groups of elements in Figure P4.4 will you find an element that forms the following? (a) insoluble halides; (b) insoluble hydroxides; (c) hydroxides that are soluble; (d) binary compounds with hydrogen that are strong acids

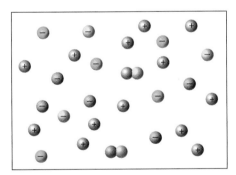

FIGURE P4.4

Questions and Problems

Solution Concentration and Molarity

CONCEPT REVIEW

4.5. How do we decide which component in a solution is the solvent?

4.6. Can a solid ever be a solvent?

4.7. What is the molarity of a solution that contains 1.00 millimole of solute per milliliter of solution?

*4.8. A beaker contains 100 grams of 1.00 M NaCl. If you transfer 50 grams of the solution to another beaker, what is the molarity of the solution remaining in the first beaker?

PROBLEMS

4.9. Calculate the molarity of each of the following solutions:
   a. 0.56 mole of $BaCl_2$ in 100.0 mL of solution
   b. 0.200 mole of Na_2CO_3 in 200.0 mL of solution
   c. 0.325 mole of $C_6H_{12}O_6$ in 250.0 mL of solution
   d. 1.48 mole of KNO_3 in 250.0 mL of solution

4.10. Calculate the molarity of each of the following solutions:
   a. 0.150 mole of urea (CH_4N_2O) in 250.0 mL of solution
   b. 1.46 mole of $NH_4C_2H_3O_2$ in 1.000 L of solution
   c. 1.94 mole of methanol (CH_3OH) in 5.000 L of solution
   d. 0.045 mole of sucrose ($C_{12}H_{22}O_{11}$) in 50.0 mL of solution

4.11. Calculate the molarity of each of the following ions:
   a. 0.33 g Na^+ in 100.0 mL of solution
   b. 0.38 g Cl^- in 100.0 mL of solution
   c. 0.46 g SO_4^{2-} in 50.0 mL of solution
   d. 0.40 g Ca^{2+} in 50.0 mL of solution

4.12. Calculate the molarity of each of the following:
   a. 64.7 g LiCl in 250.0 mL of solution
   b. 29.3 g $NiSO_4$ in 200.0 mL of solution
   c. 50.0 g KCN in 500.0 mL of solution
   d. 0.155 g $AgNO_3$ in 100.0 mL of solution

4.13. How many grams of solute are needed to prepare each of the following solutions?
   a. 1.000 L of 0.200 M NaCl
   b. 250.0 mL of 0.125 M $CuSO_4$
   c. 500.0 mL of 0.400 M CH_3OH

4.14 How many grams of solute are needed to prepare each of the following solutions?
   a. 500.0 mL of 0.250 M KBr
   b. 25.0 mL of 0.200 M $NaNO_3$
   c. 100.0 mL of 0.375 M CH_3OH

4.15. River Water The Mackenzie River in northern Canada contains, on average, 0.820 mM Ca^{2+}, 0.430 mM Mg^{2+}, 0.300 mM Na^+, 0.0200 M K^+, 0.250 mM Cl^-, 0.380 mM SO_4^{2-}, and 1.82 mM HCO_3^-. What, on average, is the total mass of these ions in 2.75 L of Mackenzie River water?

4.16. Zinc, copper, lead, and mercury ions are toxic to Atlantic salmon at concentrations of 6.42×10^{-2} mM, 7.16×10^{-3} mM, 0.965 mM, and 5.00×10^{-2} mM, respectively. What are the corresponding concentrations in milligrams per liter?

4.17. Calculate the number of moles of solute contained in the following volumes of aqueous solutions of four pesticides:
   a. 0.400 L of 0.024 M Lindane
   b. 1.65 L of 0.473 mM Dieldrin
   c. 25.8 L of 3.4 mM DDT
   d. 154 L of 27.4 mM Aldrin

4.18. A sample of crude oil contains 3.13 mM naphthalene, 12.0 mM methylnaphthalene, 23.8 mM dimethylnaphthalene, and 14.1 mM trimethylnaphthalene. What is the total number of moles of all naphthalene compounds combined in 100.0 mL of the oil?

4.19. The toxicity of DDT ($C_{14}H_9Cl_5$) led to a ban on its use in the United States in 1972. Determinations of DDT concentrations in groundwater samples from Pennsylvania between 1969 and 1971 yielded the following results:

Location	Sample Size	Mass of DDT
Orchard	250.0 mL	0.030 mg
Residential	1.750 L	0.035 mg
Residential after a storm	50.0 mL	0.57 mg

Express these concentrations in millimoles per liter.

4.20. Pesticide concentrations in the Rhine River between Germany and France between 1969 and 1975 averaged 0.55 mg/L of hexachlorobenzene (C_6Cl_6), 0.06 mg/L of Dieldrin ($C_{12}H_8Cl_6O$), and 1.02 mg/L of hexachlorocyclohexane ($C_6H_6Cl_6$). Express these concentrations in millimoles per liter.

4.21. Effluent from municipal sewers often contains high concentrations of zinc. A sewer pipe discharges effluent that contains 10 mg Zn^{2+}/L. What is the molar concentration of Zn^{2+} in the effluent?

*4.22. The concentration of copper(II) sulfate in one brand of soluble plant fertilizer is 0.07% by mass. If a 20 g sample of this fertilizer is dissolved in 2.0 L of solution, what is the molar concentration of Cu^{2+}?

*4.23. For which of the following compounds is it possible to make a 1.0 M solution at 0°C?
   a. $CuSO_4 \cdot 5H_2O$, solubility = 23.1 g/100 mL
   b. $AgNO_3$, solubility = 122 g/100 mL
   c. $Fe(NO_3)_2 \cdot 6H_2O$, solubility = 113 g/100 mL
   d. $Ca(OH)_2$, solubility = 0.189 g/mL

4.24. **Gold in the Ocean** About 6×10^9 g of gold is thought to be dissolved in the oceans of the world. If the total volume of the oceans is 1.5×10^{21} L, what is the average molar concentration of gold in seawater?

4.25. The concentration of Mg^{2+} in a sample of coastal seawater is 1.09 g/kg. What is the molar concentration of Mg^{2+} in this seawater with a density of 1.02 g/mL?

4.26. Hemoglobin in Blood A typical adult body contains 6.0 liters of blood. The hemoglobin content of blood is about 15.5 g/100.0 mL of blood. The approximate molar mass of hemoglobin is 64,500 g/mol. How many moles of hemoglobin are present in a typical adult?

4.27. Which of these following solutions has the greatest number of particles (atoms or ions) of solute per liter? (a) 1 M NaCl; (b) 1 M CaCl$_2$; (c) 1 M ethanol; (d) 1 M acetic acid

4.28. Which of the following solutions contains the most particles per liter? (a) 1 M KBr; (b) 1 M Mg(NO$_3$)$_2$; (c) 4 M ethanol; (d) 4 M acetic acid

Dilutions

PROBLEMS

4.29. Calculate the final concentrations of the following aqueous solutions after each has been diluted to a final volume of 25.0 mL:
a. 1.00 mL of 0.452 M Na$^+$
b. 2.00 mL of 3.4 mM LiCl
c. 5.00 mL of 6.42 × 10^{-2} mM Zn^{2+}

4.30. Chemists who analyze samples for trace metals may buy standard solutions that contain 1000.0 mg/L concentrations of the metals. If a chemist wishes to dilute a 1000.0 mg/L stock solution to prepare 500.0 mL of a working standard that has a concentration of 5.00 mg/L, what volume of the 1000.0 mg/L standard solution is needed?

*__4.31.__ A puddle of coastal seawater, caught in a depression formed by some coastal rocks at high tide, begins to evaporate on a hot summer day as the tide goes out. If the volume of the puddle decreases to 23% of its initial volume, what is the concentration of Na$^+$ after evaporation if initially it was 0.449 M?

4.32. What volume of 2.5 M SrCl$_2$ is needed to prepare 500.0 mL of 5.0 mM solution?

*__4.33. Dilutions Due to Rainfall__ In some parts of the United States a "100 year" storm is one that produces 8.5 inches of rain in 24 hours. If this much rain fell on a small pond with an average depth of 66 inches before the storm, by what percentage would the concentration of chloride in the pond decrease? Assume that the rainwater contained an insignificant concentration of chloride compared with the pond water.

*__4.34. Mixing Fertilizer__ The label on a bottle of "organic" liquid fertilizer concentrate states that it contains 8 grams of phosphate per 100.0 mL and that 16 fluid ounces should be diluted with water to make 32 gallons of fertilizer to be applied to growing plants. What is the phosphate concentration in grams per liter in the diluted fertilizer? (1 gallon = 128 fluid ounces.)

Electrolytes and Nonelectrolytes

CONCEPT REVIEW

4.35. A solution of table salt is a good conductor of electricity, but a solution containing an equal molar concentration of table sugar is not. Why?

4.36. Metallic fixtures on the bottom of a ship corrode more quickly in seawater than in fresh water. Why?

4.37. Explain why liquid methanol, CH$_3$OH, cannot conduct electricity whereas molten NaOH can.

4.38. Fuel Cells The electrolyte in an electricity-generating device called a *fuel cell* consists of a mixture of Li$_2$CO$_3$ and K$_2$CO$_3$ heated to 650°C. At this temperature these ionic solids melt. Explain how this mixture of molten carbonates can conduct electricity.

4.39. Rank the following solutions on the basis of their ability to conduct electricity, starting with the most conductive: (a) 1.0 M NaCl; (b) 1.2 M KCl; (c) 1.0 M Na$_2$SO$_4$; (d) 0.75 M LiCl

4.40. Rank the conductivities of 1 M solutions of each of the following solutes, starting with the most conductive: (a) acetic acid; (b) methanol; (c) dextrose (table sugar); (d) hydrochloric acid

PROBLEMS

4.41. What is the molar concentration of Na$^+$ ions in 0.025 M solutions of the following sodium salts in water? (a) NaBr; (b) Na$_2$SO$_4$; (c) Na$_3$PO$_4$

4.42. What is the total molar concentration of all the ions in 0.025 M solutions of the following salts in water? (a) KCl; (b) CuSO$_4$; (c) CaCl$_2$

Acid–Base Reactions

CONCEPT REVIEW

4.43. What name is given to a proton donor?

4.44. What is the difference between a strong acid and a weak acid?

4.45. Give the formulas of two strong acids and two weak acids.

4.46. Why is HSO$_4$$^-$(aq) a weaker acid than H$_2$SO$_4$?

4.47. What name is given to a proton acceptor?

4.48. What is the difference between a strong base and a weak base?

4.49. Give the formulas of two strong bases and two weak bases.

4.50. Write the net ionic reaction for the neutralization of a strong acid by a strong base.

PROBLEMS

4.51. For each of the following acid–base reactions, identify the acid and the base and then write the net ionic equation:
a. H$_2$SO$_4$(aq) + Ca(OH)$_2$(aq) → CaSO$_4$(s) + 2 H$_2$O (ℓ)
b. PbCO$_3$(s) + H$_2$SO$_4$(aq) → PbSO$_4$(s) + CO$_2$(g) + H$_2$O(ℓ)
c. Ca(OH)$_2$(s) + 2 CH$_3$COOH(aq) → Ca(CH$_3$COO)$_2$(aq) + 2 H$_2$O(aq)

4.52. Complete and balance each of the following neutralization reactions, name the products, and write the net ionic equations.
a. HBr(aq) + KOH(aq) →
b. H$_3$PO$_4$(aq) + Ba(OH)$_2$(aq) →
c. Al(OH)$_3$(s) + HCl(aq) →
d. CH$_3$COOH(aq) + Sr(OH)$_2$(aq) →

4.53. Write a balanced molecular equation and a net ionic equation for the following reactions:
   a. Solid magnesium hydroxide reacts with a solution of sulfuric acid.
   b. Solid magnesium carbonate reacts with a solution of hydrochloric acid.
   c. Ammonia gas reacts with hydrogen chloride gas.

4.54. Write a balanced molecular equation and a net ionic equation for the following reactions:
   a. Solid aluminum hydroxide reacts with a solution of hydrobromic acid.
   b. A solution of sulfuric acid reacts with solid sodium carbonate.
   c. A solution of calcium hydroxide reacts with a solution of nitric acid.

4.55. The use of lead(II) carbonate and lead(II) hydroxide as pigments in white paint has been discontinued because they dissolve in acidic solutions and lead is toxic to humans. Using the appropriate net ionic equations, show why lead carbonate and lead hydroxide dissolve in acid.

4.56. Many homeowners treat their lawns with $CaCO_3(s)$ to reduce the acidity of the soil. Write a net ionic equation for the reaction of $CaCO_3(s)$ with a strong acid.

Precipitation Reactions

CONCEPT REVIEW

4.57. What is the difference between a saturated solution and a supersaturated solution?

4.58. What are common solubility units?

4.59. What is a precipitation reaction?

4.60. A precipitate may appear when two completely clear aqueous solutions are mixed. What circumstances are responsible for this event?

4.61. Is a saturated solution always a concentrated solution? Explain.

4.62. Honey is a concentrated solution of sugar molecules in water. Clear, viscous honey becomes cloudy after being stored for long periods. Explain how this transition illustrates supersaturation.

PROBLEMS

4.63. According to the solubility guidelines in Table 4.5, which of the following compounds have limited solubility in water? (a) barium sulfate; (b) barium hydroxide; (c) lanthanum nitrate; (d) sodium acetate; (e) lead hydroxide; (f) calcium phosphate

4.64. The black "smoke" that flows out of deep ocean hydrothermal vents (Figure P4.64) is made of insoluble metal sulfides suspended in seawater. Of the following cations that are present in the water flowing up through these vents, which ones could contribute to the formation of the black smoke? Na^+, Li^+, Mn^{2+}, Fe^{2+}, Ca^{2+}, Mg^{2+}, Zn^{2+}, Pb^{2+}, Cu^{2+}

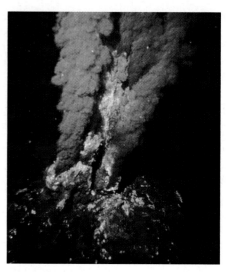

FIGURE P4.64

4.65. Complete and balance the chemical equations for the precipitation reactions, if any, between the following pairs of reactants, and write the net ionic equations:
   a. $Pb(NO_3)_2(aq) + Na_2SO_4(aq) \rightarrow$
   b. $NiCl_2(aq) + NH_4NO_3(aq) \rightarrow$
   c. $FeCl_2(aq) + Na_2S(aq) \rightarrow$
   d. $MgSO_4(aq) + BaCl_2(aq) \rightarrow$

*__4.66.__ Show with appropriate net ionic reactions how Cr^{3+} and Cd^{2+} can be removed from wastewater by treatment with solutions of sodium hydroxide.

*__4.67.__ An aqueous solution containing Ca^{2+}, Cl^-, CO_3^{2-}, and NO_3^- is allowed to evaporate. Which compound will precipitate first?

4.68. Ten milliliters of a 5×10^{-3} M solution of Cl^- ions are reacted with a 0.500 M solution of $AgNO_3$. What is the maximum mass of AgCl that precipitates?

4.69. Calculate the mass of $MgCO_3$ precipitated by mixing 10.0 mL of a 0.200 M Na_2CO_3 solution with 5.00 mL of 0.0500 M $Mg(NO_3)_2$ solution.

4.70. Toxic chromate can be precipitated from an aqueous solution by bubbling SO_2 through the solution. How many grams of SO_2 are required to treat 3.0×10^8 L of 0.050 mM Cr(VI)?

$$2\,CrO_4^{2-}(aq) + 3\,SO_2(g) + 4\,H^+(aq) \rightarrow Cr_2(SO_4)_3(s) + 2\,H_2O(\ell)$$

4.71. Fe(II) can be precipitated from a slightly basic aqueous solution by bubbling oxygen through the solution, which converts Fe(II) to insoluble Fe(III):

$$4\,Fe(OH)^+(aq) + 4\,OH^-(aq) + O_2(g) + 2\,H_2O(\ell) \rightarrow 4\,Fe(OH)_3(s)$$

How many grams of O_2 are consumed to precipitate all of the iron in 75 mL of 0.090 M Fe(II)?

4.109. How many milliliters of 0.100 M HNO_3 are needed to neutralize the following solutions? (a) 45.0 mL of 0.667 M KOH; (b) 58.5 mL of 0.0100 M $Al(OH)_3$; (c) 34.7 mL of 0.775 M NaOH

*4.110. The solubility of slaked lime, $Ca(OH)_2$, in water at 20°C is 0.185 g/100.0 mL. What volume of 0.00100 M HCl is needed to neutralize 10.0 mL of a saturated $Ca(OH)_2$ solution?

4.111. The solubility of magnesium hydroxide, $Mg(OH)_2$, in water is 9.0×10^{-4} g/100.0 mL. What volume of 0.00100 M HNO_3 is required to neutralize 1.00 L of saturated $Mg(OH)_2$ solution?

4.112. A 10.0 mL dose of the antacid in Figure P4.112 contains 830 mg of magnesium hydroxide. What volume of 0.10 M stomach acid (HCl) could one dose neutralize?

FIGURE P4.112

*4.113. **Exercise Physiology** The ache, or "burn," you feel in your muscles during strenuous exercise is caused by the accumulation of lactic acid, which has the structure shown in Figure P4.113. Only the hydrogen atom in the —COOH group is acidic, that is, can release a H^+ ion in aqueous solutions. To determine the concentration of a solution of lactic acid, a chemist titrates a 20.00 mL sample of it with 0.1010 M NaOH, and finds that 12.77 mL of titrant are required to reach the equivalence point. What is the concentration of the lactic acid solution in moles per liter?

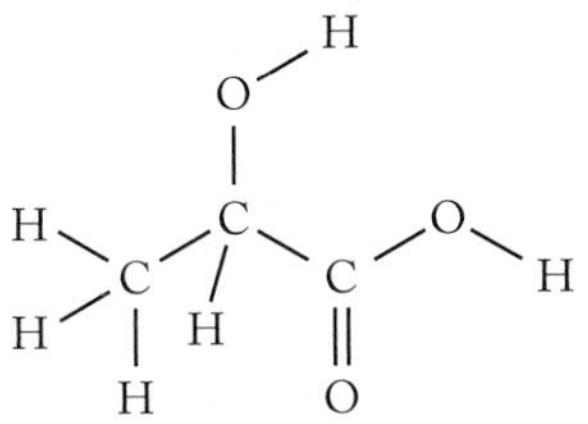

Lactic acid

FIGURE P4.113

Additional Problems

4.114. To determine the concentration of SO_4^{2-} ion in a sample of groundwater, 100.0 mL of the sample is titrated with 0.0250 M $Ba(NO_3)_2$, forming insoluble $BaSO_4$. If 3.19 mL of the $Ba(NO_3)_2$ solution are required to reach the end point of the titration, what is the molar concentration of SO_4^{2-} in the sample?

4.115. **Antifreeze** Ethylene glycol is the common name for the liquid used to keep the coolant in automobile cooling systems from freezing. It is 38.7% carbon, 9.7% hydrogen, and 51.6% oxygen by mass. Its molar mass is 62.07 g/mol, and its density is 1.106 g/mL at 20°C.
a. What is the empirical formula of ethylene glycol?
b. What is the molecular formula of ethylene glycol?
c. In a solution prepared by mixing equal volumes of water and ethylene glycol, which ingredient is the solute and which is the solvent?

*4.116. According to the label on a bottle of concentrated hydrochloric acid, the contents are 36.0% HCl by mass and have a density of 1.18 g/mL.
a. What is the molarity of concentrated HCl?
b. What volume of it would you need to prepare 0.250 L of 2.00 M HCl?
c. What mass of sodium hydrogen bicarbonate would be needed to neutralize the spill if a bottle containing 1.75 L of concentrated HCl dropped on a lab floor and broke open?

4.117. Chlorine was first prepared in 1774 by heating a mixture of NaCl and MnO_2 in sulfuric acid:

$$NaCl(aq) + H_2SO_4(aq) + MnO_2(s) \rightarrow \\ Na_2SO_4(aq) + MnCl_2(aq) + H_2O(\ell) + Cl_2(g)$$

a. Assign oxidation numbers to the elements in each compound, and balance the redox reactions in acid solution.
b. Write a net ionic equation describing the reaction for formation of chlorine.

*4.118. When a solution of dithionate ions ($S_2O_4^{2-}$) is added to a solution of chromate ions (CrO_4^{2-}), the products of the ensuing chemical reaction that occurs under basic conditions include soluble sulfite ions and solid chromium(III) hydroxide. This reaction is used to remove Cr(VI) from wastewater generated by factories that make chrome-plated metals.
a. Write the net ionic equation for this redox reaction.
b. Which element is oxidized and which is reduced?
c. Identify the oxidizing and reducing agents in this reaction.
d. How many grams of sodium dithionate would be needed to remove the Cr(VI) in 100.0 L of wastewater that contains 0.00148 M chromate ions?

4.119. A prototype battery based on iron compounds with large, positive oxidation numbers was developed in 1999. In the following reactions, assign oxidation numbers to the elements in each compound and balance the redox reactions in basic solution:
a. $FeO_4^{2-}(aq) + H_2O(\ell) \rightarrow FeOOH(s) + O_2(g) + OH^-(aq)$
b. $FeO_4^{2-}(aq) + H_2O(\ell) \rightarrow Fe_2O_3(s) + O_2(g) + OH^-(aq)$

4.120. Polishing Silver Silver tarnish is the result of silver metal reacting with sulfur compounds, such as H_2S, in the air. The tarnish on silverware (Ag_2S) can be removed by soaking in a solution of $NaHCO_3$ (baking soda) in a basin lined with aluminum foil.
  a. Write a balanced equation for the tarnishing of Ag to Ag_2S, and assign oxidation numbers to the reactants and products. How many electrons are transferred per mole of silver?
  b. Write a balanced reaction for the reaction of Ag_2S with Al metal and water to produce $Al(OH)_3$, H_2S, H_2, and Ag metal.

4.121. Give the formula and name of the acids formed in the following chemical reactions of chlorine oxides.
  a. $ClO + H_2O \rightarrow$?
  b. $Cl_2O + H_2O \rightarrow HCl +$ ___
  c. $Cl_2O_6 + H_2O \rightarrow$ ___ $+$ ___

4.122. Nonmetal oxides containing second-row elements form acidic solutions. Give the formula and name for the acids produced from the following reactions:
  a. $P_4O_{10} + 6 H_2O \rightarrow$?
  b. $SeO_2 + H_2O \rightarrow$?
  c. $B_2O_3 + 3 H_2O \rightarrow$?

4.123. Write overall and net ionic equations for the reactions that occur when
  a. a sample of acetic acid is titrated with a solution of KOH.
  b. a solution of sodium carbonate is mixed with a solution of calcium chloride.
  c. calcium oxide dissolves in water.

4.124. One way to determine the concentration of hypochlorite ions (ClO^-) in solution is by first reacting them with I^- ions. Under acidic conditions the products of this reaction are I_2 and Cl^- ions. Then the I_2 produced in the first reaction is titrated with a solution of thiosulfate ions ($S_2O_3^{2-}$). The products of the titration reaction are $S_4O_6^{2-}$ and I^- ions. Write net ionic equations for these two reactions.

*4.125. Fluoride Ion in Drinking Water Sodium fluoride is added to drinking water in many municipalities to protect teeth against cavities. The target of the fluoridation is hydroxyapatite, $Ca_{10}(PO_4)_6(OH)_2$, a compound in tooth enamel. There is concern, however, that fluoride ions in water may contribute to skeletal fluorosis, an arthritis-like disease.
  a. Write a net ionic equation for the reaction between hydroxyapatite and sodium fluoride that produces fluorapatite, $Ca_{10}(PO_4)_6F_2$.
  b. The EPA currently restricts the concentration of F^- in drinking water to 4 mg/L. Express this concentration of F^- in molarity.
  c. One study of skeletal fluorosis suggests that drinking water with a fluoride concentration of 4 mg/L for 20 years raises the fluoride content in bone to 6 mg/g, a level at which a patient may experience stiff joints and other symptoms. How much fluoride (in milligrams) is present in a 100 mg sample of bone with this fluoride concentration?

*4.126. Rocket Fuel in Drinking Water Near Las Vegas, NV, improper disposal of perchlorates used to manufacture rocket fuel has contaminated a stream that flows into Lake Mead, the largest artificial lake in the United States and a major supply of drinking and irrigation water for the American Southwest. The EPA has proposed an advisory range for perchlorate concentrations in drinking water of 4 to 18 μg/L. The perchlorate concentration in the stream averages 700.0 μg/L, and the stream flows at an average rate of 161 million gallons per day (1 gal = 3.785 L).
  a. What are the formulas of sodium perchlorate and ammonium perchlorate?
  b. How many kilograms of perchlorate flow from the Las Vegas Stream into Lake Mead each day?
  c. What volume of perchlorate-free lake water would have to mix with the stream water each day to dilute the stream's perchlorate concentration from 700.0 to 4 μg/L?
  d. Since 2003, Maryland, Massachusetts, and New Mexico have limited perchlorate concentrations in drinking water to 0.1 μg/L. Five replicate samples were analyzed for perchlorates by laboratories in each state, and the following data (μg/L) were collected:

MD	MA	NM
1.1	0.90	1.2
1.1	0.95	1.2
1.4	0.92	1.3
1.3	0.90	1.4
0.9	0.93	1.1

Which of the labs produced the most precise analytical results?

*4.127. Water from Mines Water draining from abandoned mines on Iron Mountain in California is extremely acidic, and leaches iron, zinc, and other metals from the underlying rock. One liter of drainage contains as much as 80.0 g of dissolved iron and 6 g of zinc.

FIGURE P4.127 Until 1963, iron, copper, gold, silver, and zinc were mined here at Richmond Mine in Iron Mountain, CA. After it closed, acidic mine drainage washed as much as a ton of these and other metals from the mine into the Sacramento River each day.

a. Calculate the molar concentrations of iron and zinc in the drainage.
b. One source of the dissolved iron is the reaction between water containing H_2SO_4 and solid $Fe(OH)_3$. Complete the following chemical equation, and write a net ionic equation for the process.

$$2\,Fe(OH)_3(s) + 3\,H_2SO_4(aq) \rightarrow ?$$

c. Sources of zinc include the mineral smithsonite, $ZnCO_3$. Write a balanced net ionic equation for the reaction between smithsonite and H_2SO_4 that produces $Zn^{2+}(aq)$.
d. The mineral franklinite is a mixture of Zn(II), Fe(II) and Fe(III) oxides. The generic formula for the mineral is $Zn_xFe_{1-x}O \cdot Fe_2O_3$. If acidic mine waste flowing through a deposit of franklinite contains 80 g of Fe and 6 g of Zn as a result of dissolution of the mineral, what was the value of x in the formula of the franklinite in the deposit?

*4.128. **Making Apple Cider Vinegar** Some people who prefer natural foods make their own apple cider vinegar. They start with freshly squeezed apple juice that contains about 6% natural sugars. These sugars, which all have nearly the same empirical formula, CH_2O, are fermented with yeast in a chemical reaction that produces equal numbers of moles of ethanol (Figure P4.128a) and carbon dioxide. The product of this fermentation, called hard cider, undergoes an acid fermentation step in which ethanol and dissolved oxygen gas react together to form acetic acid (Figure P4.128b) and water. This acetic acid is the principal solute in vinegar.
a. Write a balanced chemical equation describing the fermentation of natural sugars to ethanol and carbon dioxide. You may use in the equation the empirical formula given in the above paragraph.
b. Write a balanced chemical equation describing the acid fermentation of ethanol to acetic acid.
c. What are the oxidation states of carbon in the reactants and products of the two fermentation reactions?
d. If a sample of apple juice contains 1.00×10^2 g of natural sugar, what is the maximum quantity of acetic acid that could be produced by the two fermentation reactions?

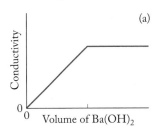

Ethanol
$$CH_3 - CH_2 - OH$$
(a)

Acetic acid
$$CH_3 - COOH$$
(b)

FIGURE P4.128

*4.129. A food chemist determines the concentration of acetic acid in a sample of apple vinegar (see Problem 4.128) by acid–base titration. The density of the sample is 1.01 g/mL. The titrant is 1.002 M NaOH. The average volume of titrant required to titrate 25.00 mL subsamples of the vinegar is 20.78 mL. What is the concentration of acetic acid in the vinegar? Express your answer the way a food chemist probably would: as percent by mass.

*4.130. One way to follow the progress of a titration and detect its equivalence point is by monitoring the conductivity of the titration reaction mixture. For example, consider the way the conductivity of a sample of sulfuric acid changes as it is titrated with a standard solution of barium hydroxide before and then after the equivalence point.
a. Write the overall ionic equation for the titration reaction.
b. Which of the four graphs in Figure P4.130 comes closest to representing the changes in conductivity during the titration? (The zero point on the y-axis of these graphs represents the conductivity of pure water; the break points on the x-axis represents the equivalence point.)

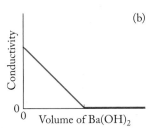

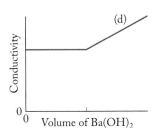

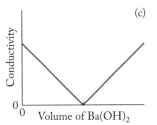

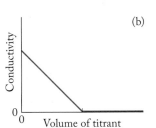

FIGURE P4.130

*4.131. Which of the graphs in Figure P4.131 best represents the changes in conductivity that occur before and after the equivalence point in each of these following titrations:
a. Sample, $AgNO_3(aq)$; titrant, $KCl(aq)$
b. Sample, $HCl(aq)$; titrant, $LiOH(aq)$
c. Sample, $CH_3COOH(aq)$; titrant, $NaOH(aq)$

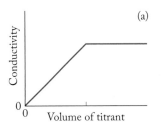

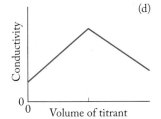

FIGURE P4.131

*4.132. Regular-strength chlorine bleach is a 5.25% solution of NaClO by mass. How many milliliters of 0.02005 M sodium thiosulfate will be required to reach the equivalence point in the titration of a 1.000 mL sample of bleach? *Hint*: in the overall chemistry of the analysis, 2 moles of thiosulfate ions are consumed for each mole of hypochlorite ions in the original sample.

Calcium: In the Limelight

4.133. Rocks in Caves The stalactites and stalagmites in most caves are made of limestone (calcium carbonate) (see page 147). However, in the Lower Kane Cave in Wyoming they are made of gypsum (calcium sulfate). The presence of $CaSO_4$ is explained by the following sequence of reactions:

$$H_2S(aq) + 2\,O_2(g) \rightarrow H_2SO_4(aq)$$

$$H_2SO_4(aq) + CaCO_3(s) \rightarrow CaSO_4(s) + H_2O(\ell) + CO_2(g)$$

a. Which (if either) of these reactions is a redox reaction? How many electrons are transferred?
b. Write a net ionic equation for the reaction of H_2SO_4 with $CaCO_3$.
c. How would the net ionic equation be different if the reaction were written as follows?

$$H_2SO_4(aq) + CaCO_3(s) \rightarrow CaSO_4(s) + H_2CO_3(aq)$$

4.134. The alkaline earth elements react with nitrogen to form nitrides (with the general formula M_3N_2) for M = Be, Mg, Ca, Sr, and Ba. Like the alkaline earth oxides (general formula MO), the nitrides react with water to form alkaline earth hydroxides, $M(OH)_2$. Predict the other product for the reaction, and balance the equation:

$$M_3N_2(s) + H_2O(\ell) \rightarrow M(OH)_2(s) + \underline{\quad}$$

4.135. Which of the following reactions of calcium compounds is or are redox reactions?
 a. $CaCO_3(s) \rightarrow CaO(s) + CO_2(g)$
 b. $CaO(s) + SO_2(g) \rightarrow CaSO_3(s)$
 c. $CaCl_2(s) \rightarrow Ca(s) + Cl_2(g)$
 d. $3\,Ca(s) + N_2(g) \rightarrow Ca_3N_2(s)$

*4.136. HF is prepared by reacting CaF_2 with H_2SO_4:

$$CaF_2(s) + H_2SO_4(\ell) \rightarrow 2\,HF(g) + CaSO_4(s)$$

HF can in turn be electrolyzed when dissolved in molten KF to produce fluorine gas:

$$2\,HF(\ell) \rightarrow F_2(g) + H_2(g)$$

Fluorine is extremely reactive, so it is typically sold as a 5% mixture by volume in an inert gas such as helium. How much CaF_2 is required to produce 500.0 L of 5% F_2 in helium? Assume the density of F_2 gas is 1.70 g/L.

Thermochemistry

5

SUNRISE FROM THE SPACE SHUTTLE *The sun is the ultimate source of energy for most life on Earth.*

A LOOK AHEAD: The Sunlight Unwinding

Energy—to power an automobile, to heat a home, or to support life—is an abstract idea: energy is invisible, it is not a substance, it has no mass, and it occupies no volume. We have witnessed energy changing matter—sunlight melts snow, for example—and energy being transformed from one form to another, the way chemical energy in gasoline is changed into mechanical energy to move a vehicle. A portion of the energy in the gasoline, however, does not contribute to moving the vehicle and is dispersed to the surroundings as heat. If we add up the energy used to cause motion and the energy distributed to the surroundings, we find that the total amount is equal to the energy contained in the gasoline that was burned. In other words, energy is neither created nor destroyed during chemical reactions.

Heat is a familiar form of energy. Adding heat makes water boil. Heat is given off when wood burns in a fireplace or when gasoline burns in an engine. We also know that heat always flows from a hot object into a cold one. If you put a cold hamburger in a hot pan, heat flows from the pan to the hamburger, and we can measure the change in the temperature of the hamburger with a thermometer. We never directly measure the amount of heat in an object, but by measuring changes in temperature and observing the results of heat flowing from object to object, we can calculate the amount of heat involved.

We can cook a hamburger over a campfire using the heat energy from burning logs. But where does the heat energy come from when wood burns? R. Buckminster Fuller, a 20th century architect, inventor, and futurist, described a burning log like this: Trees gather the energy in sunlight and combine it with water and carbon dioxide to make the molecules that compose the wood. When wood is burned, the chemical products are carbon dioxide and water, and the fire is, as Fuller said, "all that sunlight unwinding." The sunlight unwinding is the chemical energy stored in molecules that make up the wood. Sunlight, through the transforming power of green plants, is the source of the chemical energy contained within all the molecules we consume as food and fuel.

Virtually all chemical reactions involve heat that can be thought of as either a product or a reactant, and all physical changes that matter undergoes involve changes in energy. By following the flow of heat, we can explain many events, such as ponds freezing in the winter and thawing in the spring, and determine the answer to many questions, for example, whether methane is a better fuel than propane. Studying the transfer of heat provides us with important insights into the way the natural world works, and helps us address important questions about the impact of human activities on our world.

5.1 Energy: Basic Concepts and Definitions

Like many important chemical reactions, the one in which hydrogen combines with oxygen to form water releases energy in the form of heat: this energy may be converted into motion, as occurs in the main engines of the U.S. space shuttles during liftoff, or into electrical energy in the fuel cells that provide electric power to space shuttles, some buildings, and even some automobiles. Heat energy is a product of this reaction, and the written form of any reaction including heat as a reactant or product, such as the following reaction, is called a **thermochemical equation**:

$$H_2(g) + \tfrac{1}{2}O_2(g) \rightarrow H_2O(\ell) + \text{heat}$$

A **thermochemical equation** is the chemical equation of a reaction that includes heat as a reactant or a product.

The traditional definition of **energy** is the capacity to transfer **heat** or to do **work**. What does this mean? In the most general sense, it means that when transferred from one object to another, energy produces work or heat or a combination of both. Forms of energy that do work include electrical energy, mechanical energy, light energy, sound energy, and many others. Whatever the form, energy used to do work causes an object with mass to move: we observe a change in the location of an object because of the action of energy as work. In contrast, heat energy causes changes in the temperature of an object. We observe changes due to the transfer of heat by noting whether objects become warmer or colder. Heat energy changes can also cause phase changes (such as melting or freezing). The study of energy and its transformation from one form to another is called **thermodynamics**. That part of thermodynamics in which heat energy changes and transformations are associated with chemical reactions is known as **thermochemistry.**

Heat Transfer and Temperature

When you put an ice cube (at $-10°C$, the typical temperature of a freezer) into warm water at room temperature ($25°C$) the ice cube melts and the water cools. Energy as heat moves from the warm water into the cold ice cube. The process of heat moving from one object into another is called **heat transfer**. The difference in temperature defines the direction in which the heat flows when two objects come into contact; heat always flows from a hotter object at a higher temperature into a colder object at a lower temperature (Figure 5.1). Heat transfer changes the temperature of matter; it can also cause changes in *phase* (changes in *state*). The ice cube in this example undergoes a change of state from solid to liquid as it takes up heat from the liquid water. The fluid in the example remains in the liquid phase but drops in temperature as it transfers heat to the ice cube. Ultimately these two portions of matter achieve the same temperature, higher than the initial temperature of the ice cube but lower than the initial temperature of the liquid water. At this point, **thermal equilibrium** has been reached; temperature is constant throughout the material and no heat flow occurs from point to point. On the macroscopic level, no further change in temperature occurs and the temperature throughout the entire sample of water is the same.

Work, Potential Energy, and Kinetic Energy

By definition in the physical sciences, work (w) is done when a force (F) moves an object through a distance (d). The amount of work done is the force exerted on the object times the distance through which the force is exerted:

$$w = F \times d \tag{5.1}$$

Consider Equation 5.1 as it relates to skiers ascending a mountain (Figure 5.2). The work (w) done by, for example, a gondola lift on each skier riding it equals the length of the ride on the lift (d) times the force (F) needed to overcome gravity and transport the skier up the mountain. Some of the work done is stored in the mass and position of the skier as **potential energy (PE)**, which is the energy in an object because of its position. The farther up the mountain the object is, the greater its potential energy. The simple mathematical expression of the skier's potential energy is

$$PE = m \times g \times h \tag{5.2}$$

where m is the mass of the skier, g is the force of gravity, and h is the vertical distance between the location of the skier on the mountain and the skier's starting point. How the skier gets to the position on the mountainside is not important.

Energy is the capacity to transfer heat or to do work.

Heat is energy transferred between objects because of a difference in their temperatures.

Moving an object through a distance is **work**.

Thermodynamics is the study of energy and its transformations.

Thermochemistry is the study of the relation between chemical reactions and changes in heat energy.

Heat transfer is the process of heat energy flowing from one object into another.

Thermal equilibrium is a condition in which temperature is constant throughout a material and no heat flow occurs from point to point.

The **potential energy (PE)** of an object is the energy stored in that object because of its position.

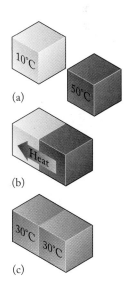

FIGURE 5.1 (a) Two identical blocks at different temperatures (b) are brought into contact. Heat is transferred from the one at higher temperature (dark red) to the one at lower temperature (pale red) until (c) thermal equilibrium (same temperature) is reached.

FIGURE 5.2 Work is done as skiers ascend to the top of a mountain. The amount of work differs, depending on whether a skier (1) rides a gondola on a direct route to the top or (2) hikes to the top along a winding path, because the paths taken are of different lengths and work depends on the distance through which a force acts in moving a mass.

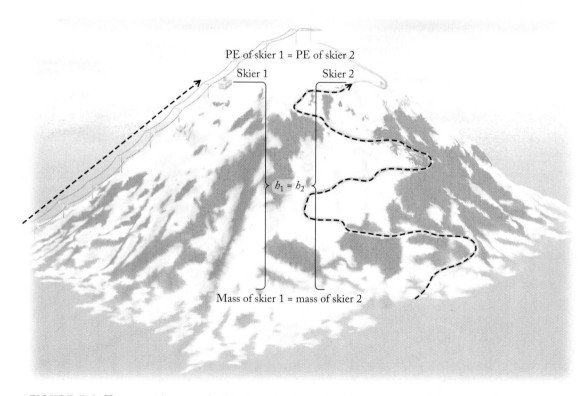

FIGURE 5.3 The potential energy of a skier depends only on the skier's mass and height above the base. If two skiers are at the same height ($h_1 = h_2$) and both skiers have the same mass, then the potential energy of each skier is the same, no matter how each skier got there.

Because the potential energy does not depend on how she has gotten to a particular point means that potential energy is a **state function**, which means it is independent of the path followed to achieve the position. Only the position of the skier is important (Figure 5.3). (Note that *state function* refers to a property of a system that is determined by the position or condition of the system; do not confuse it with changes of state.)

Now consider the potential energy of a ski jumper about to leave the starting gate (Figure 5.4a). The potential energy is converted into **kinetic energy (KE)**, the energy of an object in motion as the skier is pulled down the hill by the force of gravity and gains speed. At any moment (Figure 5.4b) between the start of the run and coming to a stop at the bottom of the hill, the ski jumper's kinetic energy is proportional to the product of her mass (*m*) times the square of her speed (*u*):

$$KE = \tfrac{1}{2}mu^2 \qquad (5.3)$$

A heavier skier moving at the same speed as a lighter skier has more kinetic energy at any given position on the path. Our intuition and experience tell us this as well. If you were standing at the bottom of the ski jump, how would your fate differ if a 300 lb skier going 25 mi/hr ran into you compared to a 100 lb skier going 25 mi/hr? The difference lies in their relative kinetic energies.

According to the **law of conservation of energy**, energy cannot be created or destroyed. However, it can be converted from one form to another, as this example illustrates. Potential energy at the top of the run becomes kinetic energy

A **state function** is a property of an entity based solely on its chemical or physical state or both, but not on how it achieved that state.

The **kinetic energy (KE)** of an object in motion is the energy that it has because of its mass *(m)* and its speed *(u)*: $KE = \tfrac{1}{2}mu^2$.

The **law of conservation of energy** states that energy cannot be created or destroyed.

▶❚❚ **CHEMTOUR** State Functions and Path Functions

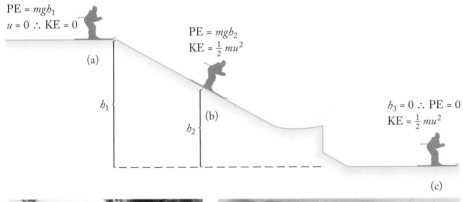

PE = mgh_1
$u = 0$ ∴ KE = 0

(a)

PE = mgh_2
KE = $\tfrac{1}{2}mu^2$

(b)

h_1

h_2

$h_3 = 0$ ∴ PE = 0
KE = $\tfrac{1}{2}mu^2$

(c)

FIGURE 5.4 (a) A skier leaving the starting gate of a ski jump has potential energy (PE) due to her position at the gate (*h*), her mass (*m*), and the force of gravity (*g*): PE = *mgh*. (b) During her run, the jumper's potential energy is converted into kinetic energy (KE), which is the energy of motion related to her mass (*m*) and speed (*u*) by Equation 5.3, namely KE = $\tfrac{1}{2}mu^2$. The skier has both KE and PE. (c) At the end of the run, the jumper's PE is 0 and her KE decreases from its maximum value to 0 as she slows to a stop.

$m_1 > m_2$

Starting gate

$m_1 > m_2$

during the run. The total energy at any position on the hill is the sum of the skier's potential and kinetic energies.

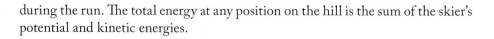

CONCEPT TEST

Two skiers with masses m_1 and m_2, where $m_1 > m_2$, are poised at the starting gate of a downhill course. Is the potential energy (PE) of skier 1 different from or the same as the PE of skier 2? If you think they are different, which skier has more potential energy?

(Answers to Concept Tests are in the back of the book.)

CONCEPT TEST

Two skiers with masses m_1 and m_2, where $m_1 > m_2$, go past the same point on parallel racecourses at the same time. (a) At that moment, is the potential energy of skier m_2 more than, less than, or equal to the PE of skier m_1? (b) If the two are moving at the same speed, which has the greater kinetic energy (KE), or do they have the same KE?

(Answers to Concept Tests are in the back of the book.)

Kinetic Energy and Potential Energy at the Molecular Level

The relation just described between kinetic and potential energy holds for atoms and molecules as well. The analogy with the kinetic and potential energies of the skier breaks down, however, because the gravitational forces that control the skier play virtually no role in the interactions of very small objects. Temperature and charge dominate the relation between kinetic and potential energy at the microscopic level. Temperature governs motion on the atomic-molecular scale. Chemical bonds and differential electric charges cause interactions between particles that give rise to the potential energy stored in the arrangements of atoms and ions in matter.

The kinetic energy of a particle at the microscopic level still depends on its mass and its velocity, but because its velocity depends on temperature, its kinetic energy does, too. As the temperature of a population of molecules increases, the average kinetic energy of the molecules also increases. Consequently, if we have two portions of water vapor ($\mathcal{M}$ = 18.02 g/mol) at room temperature, the two populations of molecules have the same average kinetic energy. The average velocities of the molecules are the same because their masses are identical. If the temperature of one portion of water vapor is increased, that population of molecules will have a higher kinetic energy because the average velocity of the molecules will be higher. An equivalent population of vapor-phase ethanol molecules ($\mathcal{M}$ = 46.07 g/mol) at room temperature has the same average kinetic energy as the water molecules at room temperature, but the average velocity of the ethanol molecules is lower because their mass is higher (Figure 5.5).

This kinetic energy associated with the random motion of molecules is called **thermal energy**, and the thermal energy of a given material is proportional to temperature. However, thermal energy in a sample also depends on the number of particles in that sample. The water in a swimming pool and in a cup of water taken from the pool have the same temperature, so their molecules have the same

Thermal energy is the kinetic energy of atoms and molecules.

$\mathcal{M}$ = 18.02 g/mol $\mathcal{M}$ = 46.07 g/mol

$$T_{\text{water}} = T_{\text{ethanol}}$$

$$\text{KE}_{\text{water}} = \text{KE}_{\text{ethanol}}$$

$$\tfrac{1}{2}\,m(u_{\text{avg.water}})^2 = \tfrac{1}{2}\,m(u_{\text{avg.ethanol}})^2$$

Because $m_{\text{H}_2\text{O}} < m_{\text{ethanol}}$, $u_{\text{avg.H}_2\text{O}} > u_{\text{avg.ethanol}}$

FIGURE 5.5 Two populations of molecules have the same temperature and the same average kinetic energy. Because ethanol molecules have a greater mass than water molecules ($m_{\text{ethanol}} > m_{\text{water}}$), the average velocity of the water molecules in the vapor phase above the liquid is greater than the average velocity of the gaseous ethanol molecules ($u_{\text{water}} > u_{\text{ethanol}}$).

average kinetic energy. The water in the pool has much more thermal energy than the water in the cup, simply because there is a larger number of molecules in the pool. A large number of particles at a given temperature has a higher total thermal energy than a small number of particles at the same temperature.

CONCEPT TEST

If we heat a cup of water from the swimming pool to a temperature close to the boiling point (100°C), do you think its thermal energy will be more than, less than, or the same as the water in the pool?

(Answers to Concept Tests are in the back of the book.)

One of the most important forms of potential energy at the atomic-molecular level arises from electrostatic interactions due to charges on particles. The magnitude of this **electrostatic potential energy (E_{el})**, also known as *coulombic interaction*, is directly proportional to ($\propto$) the product of the charges on the particles (Q_1 and Q_2) and inversely proportional to the distance (d) between them:

$$E_{\text{el}} \propto \frac{Q_1 \times Q_2}{d} \tag{5.4}$$

If both charges in Equation 5.4 are the same (either both positive or both negative), the product of the charges is positive, the particles repel each other, and E_{el}

Electrostatic potential energy (E_{el}) is the energy of a particle because of its position with respect to another particle. It is directly proportional to the product of the charges of the particles and inversely proportional to the distance between them.

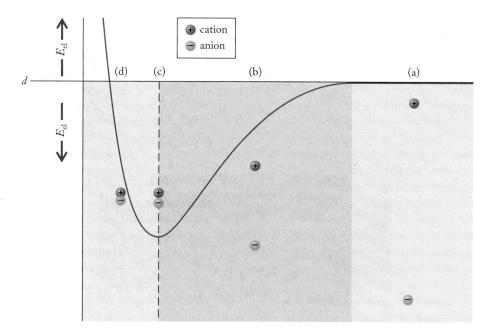

FIGURE 5.6 Electrostatic potential energy (E_{el}): (a) A positive ion and a negative ion are so far apart (d is large) that they do not interact at all. (b) As the particles move closer together (d decreases), the electrostatic potential energy between the two ions becomes more negative. (c) At some point the attraction between oppositely charged particles produces an arrangement that is the most favorable energetically because the particles in that combination have the lowest potential energy. The minimum in the curve corresponds to that most favorable arrangement. (d) If the ions are forced even closer together, they repel each other as their nuclei begin to interact.

Combustion of hydrogen can be used to launch rockets.

is positive. If one particle is positive and the other negative, however, the product of the charges is negative, particles attract each other, and E_{el} in this case is negative (Figure 5.6a–c). Lower energy (a more-negative value of E_{el}) corresponds to greater stability, so particles that attract each other because of their charge form an arrangement characterized by a lower electrostatic potential energy than particles that repel each other. There is a limit, however, to how close two particles can be to each other. Remember that ions, whatever their charge, have positively charged nuclei that begin to repel each other if they are pushed together too closely (Figure 5.6d). At the microscopic level, potential energy is stored in the position of particles with respect to each other, just as the potential energy of skiers is determined by their positions on a slope. Neutral molecules such as water attract each other as well because of distortions in the distribution of electrons about the nuclei of their atoms, and the same analysis applies in terms of their relative potential energy.

Coulombic interactions, not gravitational forces, determine the potential energy of matter at the atomic level. The total energy of matter at the microscopic level is the sum of the kinetic energy due to the random motion of particles and the potential energy in their arrangement.

The energy given off or absorbed during a chemical reaction is due to the difference in potential energy stored in the arrangement of atoms in molecules of reactants and products. For example, when hydrogen molecules are burned in oxygen, the products of the combustion reaction are water and a considerable amount of released energy. The energy given off by this reaction has been used frequently to power rockets to launch satellites and is now being applied to run vehicles such as buses or automobiles. Because significant energy is given off in this process, the molecules of the product must be at a lower potential energy than the molecules of the reactants, as shown graphically in Figure 5.7. The difference between the energy of the two arrangements of atoms is the energy released. In the case of hydrogen combustion this energy is sufficient to run a variety of vehicles now powered by the combustion of fossil fuels.

(a)

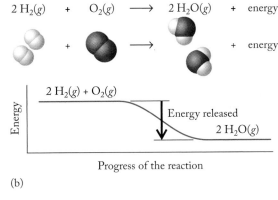

(b)

FIGURE 5.7 (a) Fueling a hydrogen-powered vehicle. (b) Hydrogen reacts with oxygen to produce water. The molecules in the product are at a lower energy than those in the reactants; hence the reaction releases energy.

5.2 Systems, Surroundings, and the Flow of Energy

The study of thermodynamics requires us to define precisely what part of the universe we are evaluating. In thermochemistry and thermodynamics, the specific and limited part of the universe we are studying is called the **system**. Everything else that is not part of the system is called the **surroundings**. Typically, we limit our concern about surroundings to that part of the universe that can exchange energy (or energy and matter) with the system we have defined. Even for a system as small as a red blood cell, the surroundings are the entire universe, but we conventionally limit our consideration to the immediate region around the cell. In evaluating the energy gained or lost in chemical reactions, the system may include the reaction vessel and its contents, or it may be just the atoms, molecules, or ions that are involved in the reaction.

Isolated, Closed, and Open Systems

Although systems can be as large as a galaxy or as small as a living cell, most of the systems we examine quantitatively in this chapter are small and are of a size that fits on a laboratory bench. We limit our concern about surroundings to that part of the universe that can exchange energy with the system we have defined. In discussions involving the transfer of energy as heat, work, or both, three types of systems are common: isolated systems, closed systems, and open systems (Figure 5.8). These designations are important because they define the system we are dealing with and the part of the universe with which the system interacts.

From the point of view of systems and surroundings, hot soup in a thermos bottle is an example of an **isolated system**, which exchanges neither energy nor matter with its surroundings. The insulated bottle does not allow heat or matter to be transferred away from the soup (the system). From the soup's point of view, the soup itself *is* the entire universe; in an isolated system, the system has no surroundings. Of course, even the best thermos cannot maintain the soup over

A **system** is the part of the universe that is the focus of a thermodynamic study.

The **surroundings** include everything that is not part of the system.

An **isolated system** exchanges neither energy nor matter with the surroundings.

FIGURE 5.8 The transfer of energy and matter in isolated, closed, and open systems: (a) Hot soup in a thermos bottle approximates an isolated system: no vapor escapes, no matter is added or removed, and no heat escapes to the surroundings (as long as the thermos behaves as a perfect insulator). An ideal thermos bottle completely isolates the soup from the rest of the universe. (b) Hot soup in a cup with a lid is a closed system; the soup transfers heat to the surroundings as it cools; however, no matter escapes and none is added. (c) Hot soup in a cup is an open system; it transfers both matter (water vapor) and energy (heat) to the surroundings as it cools. Matter in the form of pepper, grated cheese, crackers, or other matter from the surroundings may also be added to the soup (the system).

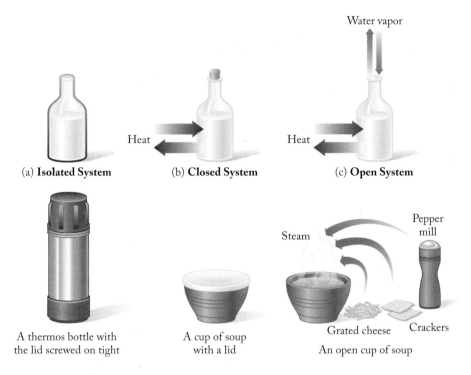

(a) **Isolated System**   (b) **Closed System**   (c) **Open System**

A thermos bottle with the lid screwed on tight   A cup of soup with a lid   An open cup of soup

time as a perfectly isolated system, because heat will leak out and the contents will cool, but for short time periods the soup in a good thermos approximates an isolated system.

Soup in a cup with a lid is an example of a closed system. A **closed system** (the soup) exchanges energy but not matter with the surroundings (the cup, the air, the tabletop). Because the cup has a lid, no vapor escapes from the soup and nothing else can be added. Only heat is exchanged with the surroundings, not matter. The heat from the soup is transferred to the walls of the cup, then into the air and the tabletop, and the soup gradually cools.

Soup in an open cup is an example of an open system. An **open system** exchanges energy and matter with the surroundings. The heat from the soup is transferred to the walls of the cup, to the air, and to the tabletop, and matter from the soup in the form of water vapor leaves the system and enters the air. We may also add matter from the surroundings to the system by sprinkling on a little grated cheese, some ground pepper, or adding a few crackers.

Most of the systems we deal with in introductory thermochemistry are either isolated systems or closed systems, because they are the easiest to study quantitatively. However, many interesting, important systems are open—including cells, living organisms, engines, and Earth itself (Figure 5.9).

Heat Flow

A chemical reaction or a physical change of state that produces a net flow of heat energy from a system to its surroundings is **exothermic** from the system's point of view. This flow can be detected by an increase in the temperature of the surroundings. Combustion reactions that release heat and light are examples of exothermic reactions. In contrast, chemical reactions and physical changes that absorb heat from their surroundings are **endothermic** from the system's point of view. For example, ice cubes (the system) in a glass of warm water (the surroundings) absorb heat from the water, which causes them to melt. From the point of view of the system (ice cubes), the process is endothermic.

A **closed system** exchanges energy but not matter with the surroundings.

An **open system** exchanges both energy and matter with the surroundings.

In an **exothermic** process, heat flows from a system into its surroundings.

In an **endothermic** process, heat flows from the surroundings into the system.

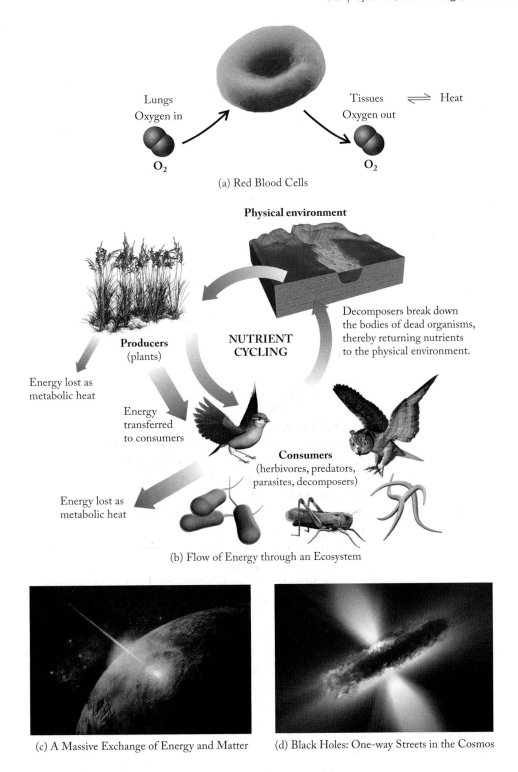

(a) Red Blood Cells

Physical environment

NUTRIENT CYCLING

Producers
(plants)

Energy lost as metabolic heat

Energy transferred to consumers

Energy lost as metabolic heat

Decomposers break down the bodies of dead organisms, thereby returning nutrients to the physical environment.

Consumers
(herbivores, predators, parasites, decomposers)

(b) Flow of Energy through an Ecosystem

(c) A Massive Exchange of Energy and Matter

(d) Black Holes: One-way Streets in the Cosmos

FIGURE 5.9 Living thermodynamic systems and beyond: (a) Red blood cells are open systems. Matter (as O_2 and water) enters and leaves red blood cells. Red blood cells also exchange heat with their surroundings. (b) Energy flows through an ecosystem as heat and light from the sun and CO_2 from the atmosphere are absorbed by green plants, which use these to produce glucose and other plant materials. The energy stored in glucose is transferred to living systems, which lose heat as they oxidize plant materials back to CO_2. When living organisms die, decomposition returns the chemical components of their bodies to the physical environment. Decomposition in the absence of oxygen produces the highly reduced forms of carbon in fossil fuels. (c) Massive exchanges of energy and matter occur when our biosphere, Earth, collides with interstellar matter such as asteroids. Sixty-five million years ago, an asteroid colliding with Earth is believed to have caused massive destruction, climate change, and the extinction of the dinosaurs. (d) Black holes have gravitational fields so strong that matter and light are drawn in and not released. They are one-way systems, where the flow of matter and energy is inexorably in, never out. Black holes stretch our convenient definitions of earthly thermodynamic systems.

In another phase change, water vapor (the system) from the air on a humid day condensing into droplets of liquid water on the outside surface of a glass containing an ice-cold drink (the surroundings) requires that heat flow from the system (the water vapor) to the surroundings (the glass). From the point of view of the water vapor, this process is exothermic. If we suddenly put hot water in the glass, the water droplets on the outside surface would absorb heat and vaporize; from the point of view of the droplets, that process is endothermic. This illustrates an important concept: an exothermic process in one direction (vapor → liquid: *condensation*) is endothermic in the reverse direction (liquid → vapor: *vaporization*) (Figure 5.10).

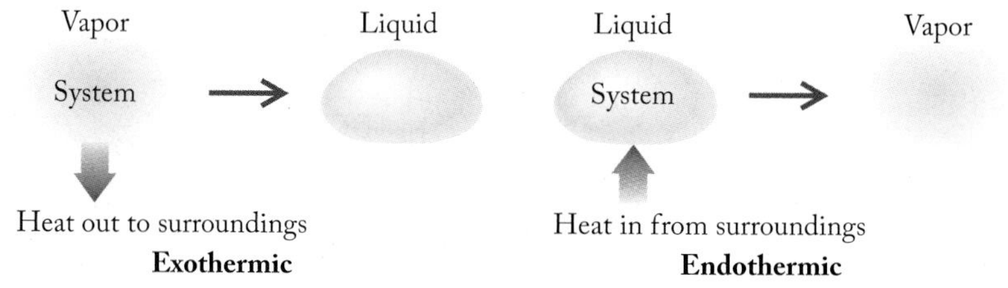

FIGURE 5.10 Reversing a process changes the sign but not the magnitude of the heat change.

Vapor — System → Liquid
Heat out to surroundings
Exothermic

Liquid — System → Vapor
Heat in from surroundings
Endothermic

The flow of heat during physical changes is summarized in Figure 5.11. We use the symbol q to represent heat produced or consumed by a chemical reaction or a physical change. If the reaction or process is exothermic, then q is negative, meaning that the system *loses* heat to its surroundings. If the reaction or process is endothermic, then q is positive, indicating that heat is *gained* as it flows from its surroundings into the system. In Figure 5.11, endothermic changes of state are represented by arrows pointing upward: solid → liquid, liquid → gas, and solid → gas. The opposite changes represented by arrows pointing downward are exothermic. To summarize:

$$\text{Exothermic:} \quad q < 0 \qquad \text{Endothermic:} \quad q > 0$$

FIGURE 5.11 Matter can be transformed from one of the three physical states into another by adding or removing heat. Upward-pointing arrows represent endothermic processes (heat enters the system; $q > 0$). Downward-pointing arrows represent exothermic processes (heat leaves the system and enters the surroundings; $q < 0$).

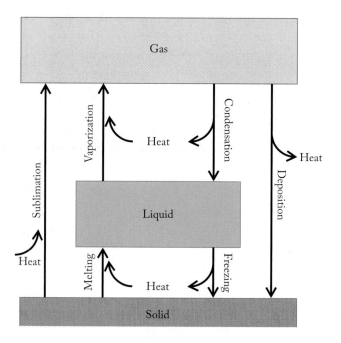

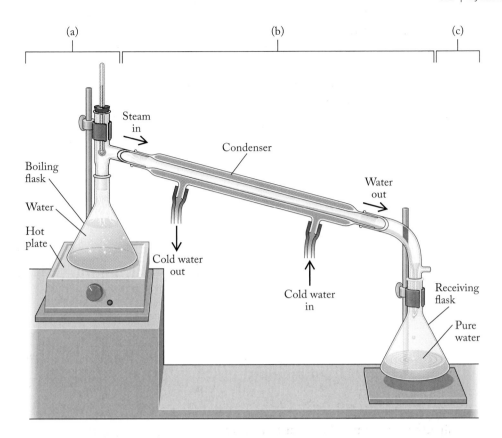

FIGURE 5.12 A laboratory setup for the distillation of water: (a) Water is heated to the boiling point in the distillation flask. (b) Vapors rise and enter the condenser, where they are liquefied. (c) The liquid is collected in the receiving flask.

SAMPLE EXERCISE 5.1   **Identifying Exothermic and Endothermic Processes**

Describe the flow of heat during the purification of water by distillation (Figure 5.12). Consider the water to be the system.

COLLECT AND ORGANIZE Since the water is the system, we must evaluate how the water gains or loses heat during distillation.

ANALYZE In distillation, heat flows in two distinct steps: (a) Liquid water is heated to the boiling point. (b) The resulting vapors are cooled as they pass into the condenser and are then collected.

SOLVE (a) Heat flows into the water to make it boil. The liquid water is converted into water vapor. Heat flows from the surroundings (the heat source) into the system (the water); the process is endothermic.

(b) The vapors from the boiling water enter the condenser. Heat flows from the system (the water vapor) into the surroundings (the walls of the condenser). The process is exothermic.

THINK ABOUT IT Being able to define the system and the surroundings is very important in dealing with heat flow.

Practice Exercise What is the sign of q as (a) a match burns, (b) drops of molten candle wax solidify, and (c) perspiration (water) evaporates? In each case, define the system clearly and indicate which of these processes are endothermic and which are exothermic.

(Answers to Practice Exercises are in the back of the book.)

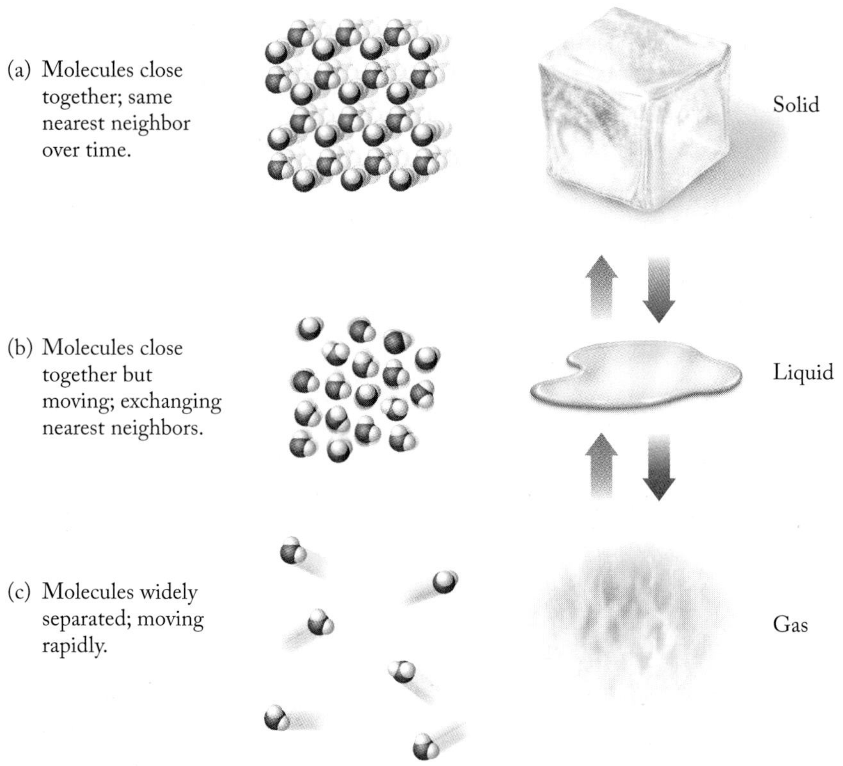

(a) Molecules close together; same nearest neighbor over time.

Solid

FIGURE 5.13 What a Chemist Sees Phase changes: (a) Solid ice absorbs heat and is converted to liquid water. The molecules in the solid are held together in a rigid three-dimensional arrangement; they have the same nearest neighbors over time. (b) When the solid melts, the molecules in the liquid phase exchange nearest neighbors and occupy many more positions with respect to each other than were possible in the solid. (c) The liquid absorbs heat and is converted to a vapor. The molecules in a gas are widely separated and move very rapidly. The reverse of these processes occurs when water in the vapor phase loses heat and condenses to a liquid. The liquid also loses heat when it is converted to a solid.

(b) Molecules close together but moving; exchanging nearest neighbors.

Liquid

(c) Molecules widely separated; moving rapidly.

Gas

▶❚❚ CHEMTOUR Internal Energy

Consider the flow of energy when a tray of ice cubes is inadvertently left on a kitchen counter (Figure 5.13). Let's focus on the ice cubes (the system). As the ice cubes warm and start to melt, the attractive forces that hold molecules of H_2O in place in the solid ice are broken. These molecules now have more positions they can occupy, so they have more potential energy. After all the ice at 0°C has melted into liquid water at 0°C, the temperature of the water (the system) slowly rises to room temperature. Higher temperature is also the result of increasing the kinetic energy of the molecules of water in the tray. This kinetic energy is part of the overall **internal energy (E)** of the system. The absolute internal energy of the system is the sum of the kinetic and potential energies of all the components of the system. The values of these energies are difficult to determine, but *changes* in internal energy (ΔE) are fairly easy to measure, because a change in a system's physical state or temperature provides a measure of the change in its internal energy. (The capital "delta", Δ, is the standard way scientists symbolize change in a quantity.) The change in internal energy (ΔE) is simply the difference between the final internal energy (E_{final}) of the system and the initial internal energy ($E_{initial}$) of the system:

$$\Delta E = E_{final} - E_{initial} \tag{5.5}$$

Internal energy is another example of a state function, because the value of the change in internal energy depends only on the initial state and the final state. How the change occurs in the system does not matter.

Some of the types of molecular motion that contribute to the overall internal energy of a system: (a) translational motion, motion from place to place along a path; (b) rotational motion, motion about a fixed axis; and (c) vibrational motion, movement back and forth from some central position.

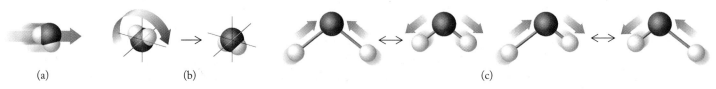

(a)   (b)   (c)

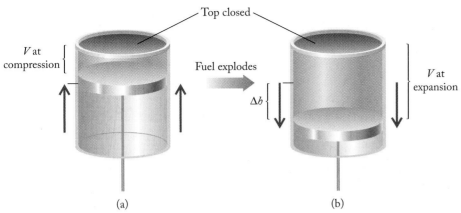

FIGURE 5.14 Work performed by changing the volume of a gas. (a) Highest position of the piston after it compresses the gas in the cylinder, causing the fuel to ignite in a diesel engine. (b) The exploding fuel gives off heat, causing the gas in the cylinder to expand and push the piston down.

> The **internal energy (E)** of a system is the sum of all the kinetic and potential energies of all of the components of the system.
>
> The **first law of thermodynamics** states that the energy gained or lost by a system must equal the energy lost or gained by the surroundings.
>
> The **calorie (cal)** is the amount of heat necessary to raise the temperature of 1 g of water 1°C.
>
> The **joule (J)** is the SI unit of energy; $4.184\,J = 1$ cal.

The law of conservation of energy applies to processes involving the transfer of energy in materials. The total energy change experienced by a given system must be balanced by the total energy change experienced by its surroundings. This is the **first law of thermodynamics**, and it is another statement of the law of conservation of energy. The energy changes of the system and the surroundings are equal in magnitude but opposite in sign, so their sum is zero. Energy is neither created nor destroyed; it is conserved.

Units of Energy

We need to express quantities of energy transferred in terms of some unit. Energy changes that accompany chemical reactions and changes in physical state have been traditionally measured in calories. A **calorie (cal)** is the quantity of heat required to raise the temperature of 1 g of water from 14.5 to 15.5°C. In the Système International the unit of energy is the **joule (J)**; 1 cal = 4.184 J. What is called a "Calorie" (Cal; note capital C) in nutrition is actually the kilocalorie (kcal): 1 Cal = 1 kcal = 1000 cals. The energy value in food is still frequently expressed in Calories in the United States. We routinely use joules throughout this text.

Doing work on a system is a way to add to its internal energy. For example (Figure 5.14), compressing a quantity of gas into a smaller volume, like the action of a piston to compress a mixture of fuel vapors and air in a diesel engine, is work done on the gases that causes their temperature to rise. The total increase in the energy of a system is the sum of the work done on it (w) and the heat (q) that flows into it:

$$\Delta E = q + w \qquad (5.6)$$

When diesel fuel in the cylinder ignites and produces hot gases, the gases (the system) expand and do work by pushing on the piston. As another example of a system doing work on its surroundings, consider the enormous balloons used by adventurers to fly around the world. These balloons typically have a chamber of hot air heated by burners fueled by ethane and propane (Figure 5.15). The burners allow a balloonist to adjust the temperature of the air to control the buoyancy of the balloon and its altitude. Burning propane or ethane causes the air in the balloon above the burner to expand. As the temperature of the gas filling the balloon goes up, the gas expands, lowering its density. The increase in volume of

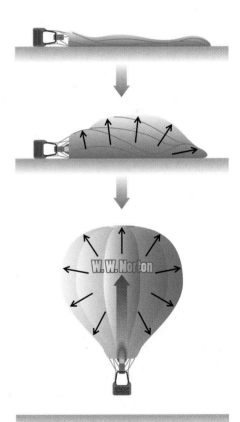

FIGURE 5.15 In a deflated balloon, $V = 0$. Inflating the balloon does work ($P\Delta V$) as the incoming gas causes the balloon to push against the surrounding atmosphere.

Pressure–volume ($P\Delta V$) work is work associated with the expansion or compression of gases.

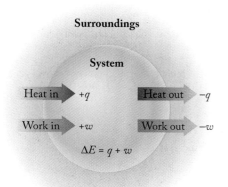

FIGURE 5.16 Heat absorbed by a system from the surroundings and work done on a system by the surroundings (arrows entering the system) are both positive quantities because both increase the internal energy of a system. Heat released by a system to the surroundings and work done by a system on the surroundings (arrows leaving the system) are both negative quantities because both decrease the internal energy of a system.

▶‖ **CHEMTOUR** Pressure–Volume Work

the balloon against the pressure of the atmosphere outside the balloon is a type of work known as **pressure–volume ($P\Delta V$) work**.

The pressure of the atmosphere on the balloon and indeed on all objects is a result of the Earth's gravity pulling the atmosphere toward the surface of Earth, where it exerts a force on all things because of its mass. The pressure at sea level on a nice dry, sunny day is approximately 1 atmosphere (1 atm) of pressure, which equals a force of 14.7 lb/in² pressing on each square inch of surface area. Atmospheric pressure varies slightly with the weather and decreases with elevation above sea level.

Atmospheric pressure (P) remains constant throughout the process of inflating the balloon. If we define the system as the air in the balloon, then the work is done to expand the air trapped in the balloon above the burner. This is $P\Delta V$ work done by the system on its surroundings; the surroundings in this case are the atmosphere. The energy of the system decreases as it performs this $P\Delta V$ work. We can relate the change in internal energy (ΔE) of the system to the heat lost by the system (q) and the work done by the system ($P\Delta V$) with this equation:

$$\Delta E = q + (-P\Delta V) = q - P\Delta V \qquad (5.7)$$

The negative sign in front of $P\Delta V$ is appropriate because when the system expands (positive change in volume ΔV), it loses energy (negative change in internal energy ΔE) as it does work on its surroundings. When the surroundings do work on the system (for example, when a gas is compressed), then $P\Delta V$ has a positive sign. By the same token, the sign of the heat (q) may be either positive or negative. If heat flows from the surroundings into the system, heat is positive ($+q$). Heat has a positive sign in the case of the balloon because the heat flows from the surroundings (the burner) into the air in the balloon (the system). When heat flows from the system into the surroundings, its sign is negative ($-q$) (Figure 5.16). Considering Equation 5.7, the change in internal energy of a system is positive when more energy enters the system than leaves, and negative when more energy leaves the system than enters.

SAMPLE EXERCISE 5.2   **Calculating Changes in Internal Energy**

Think of the figure below as a simplified version of a piston and cylinder in the engine of an automobile. When the fuel in the cylinder burns, it produces heat. Suppose the gases in the cylinder (the system) absorb 155 J of heat from the combustion reaction of the fuel and expand, thereby pushing the piston down. In doing so, the gases do 93 J of work. What is the change in internal energy (ΔE) of the gases?

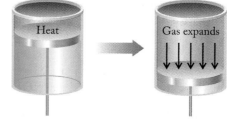

COLLECT AND ORGANIZE Change in internal energy is related to the work done by or on a system and the heat gained or lost by the system (Equation 5.6). First we have to decide whether heat and work are positive or negative from the point of view of the system (the gases in the cylinder).

ANALYZE In this example, the system (the gases in the cylinder) absorbs heat, so $q > 0$ (positive). The system also does work on the surroundings (the piston), so $w < 0$ (negative).

SOLVE The change in internal energy is the sum of the heat and the work, so

$$\Delta E = q + w = (+155\,J) + (-93\,J) = +62\,J$$

THINK ABOUT IT In this example, more energy enters the system than leaves it, so the change in internal energy (ΔE) is positive.

Practice Exercise In a different experiment the piston in Sample Exercise 5.2 compresses the gases in the cylinder and does 64 J of work on the system. As a result, the system gives off 132 J of heat to the surroundings. What is the change in internal energy of the system?

(Answers to Practice Exercises are in the back of the book.)

SAMPLE EXERCISE 5.3 **Calculating $P\Delta V$ Work**

A tank of compressed helium is used to inflate balloons for sale at a carnival on a nice day when the atmospheric pressure is 1.01 atm. If 100 balloons are each inflated from an initial volume of 0.0 L to a final volume of 4.8 L, how much $P\Delta V$ work is done to inflate the balloons? Assume that the pressure remains constant during the filling process.

COLLECT AND ORGANIZE We know that one balloon goes from being completely flat ($V = 0.0\,L$) to a volume of 4.8 L. We are given a value for the pressure, which remains constant, so work is $P\Delta V$.

ANALYZE We are asked about the $P\Delta V$ work done by 100 such balloons, so we have to multiply the volume change of one balloon by 100 to get the total change in volume (ΔV).

SOLVE The total volume change (ΔV) as all the balloons are inflated is

$$100 \text{ balloons} \times 4.8 \text{ L/balloon} = 480 \text{ L}$$

The work (w) to inflate this volume at a pressure of 1.01 atm is

$$w = P\Delta V = 1.01 \text{ atm} \times 480 \text{ L}$$
$$= 480 \text{ L} \cdot \text{atm}$$

Because the system did work on the surroundings by pushing back the atmosphere as the balloons inflated, the work has a negative sign: $w = -480\,L \cdot atm$.

THINK ABOUT IT "L · atm" (liters times atmospheres) may seem strange units for expressing work. The SI unit for energy (the ability to do work) is the joule (J). If we had a conversion factor that related joules to liters times atmospheres, we could convert the answer into the units we are used to seeing for work. From calculations involving gases, this conversion factor has been determined as 101.32 J/(L · atm). Using this factor for the work done in filling the balloons, we get

$$w = -480 \text{ L} \cdot \text{atm} \,[101.32\,J/(L \cdot atm)] = -4.9 \times 10^4\,J = -49\,kJ$$

Practice Exercise The *Spirit of Freedom*, the balloon flown around the world by American Steve Fossett in June and July of 2002, contained 550,000 cubic feet of helium. How much $P\Delta V$ work was done to inflate the balloon, assuming 1.00 atm of pressure? Express your answer in liters times atmospheres (L · atm) and then convert it to joules, given that: 1 m³ = 1000 L = 35.3 ft³, and 1.00 L · atm = 101.32 J.

(Answers to Practice Exercises are in the back of the book.)

Spirit of Freedom

Enthalpy (H) is the sum of the internal energy and the pressure–volume product of a system.

The **change in enthalpy (ΔH)** for a reaction carried out at constant pressure is the heat gained or lost by the system.

5.3 Enthalpy and Enthalpy Changes

Many of the physical and chemical changes discussed so far, as well as many others, take place at constant atmospheric pressure (P). An important thermodynamic variable that relates heat flow into or out of a system during chemical reactions or physical changes at constant pressure is called enthalpy. **Enthalpy (H)** is the sum of a system's internal energy and the product of its pressure and its volume:

$$H = E + PV \qquad (5.8)$$

Suppose a chemical change takes place at constant pressure. This change is accompanied by a change in internal energy (ΔE) and by work done if the system changes volume (ΔV) at constant pressure (P). These changes produce a **change in enthalpy (ΔH)**, described mathematically by inserting Δ in front of the appropriate variables in Equation 5.8:

$$\Delta H = \Delta E + P\Delta V \qquad (5.9)$$

Let's use Equation 5.9 to define enthalpy in terms of work and heat by substituting the mathematical definition of change in internal energy from Equation 5.7 ($q - P\,\Delta V$) for ΔE:

$$\Delta H = (q - P\Delta V) + P\Delta V$$

Simplifying this expression, we get

$$\Delta H = q_\mathrm{P} \qquad (5.10)$$

where q_P stands for heat flow at constant pressure (the usual condition in a laboratory). Equation 5.10 tells us that the enthalpy change of a reaction at constant pressure is equivalent to the heat gained or lost by the system during the reaction.

When heat flows from the system to the surroundings, the change is exothermic and the enthalpy change is negative: $\Delta H < 0$. When heat flows from the surroundings into the system, the change is endothermic and $\Delta H > 0$. For example, at a constant atmospheric pressure, heat flows into a melting ice cube from its surroundings; this process is endothermic ($\Delta H > 0$). However, to make ice cubes in a freezer, heat must be removed from the water in the ice-cube tray ($\Delta H < 0$). The changes in enthalpy for the two processes—melting and freezing—have different signs but, for a given amount of water, they have the same absolute value. For example, $\Delta H_{\mathrm{fus}} = +6.01$ kJ if a mole of ice melts, but $\Delta H_{\mathrm{solid}} = -6.01$ kJ if a mole of water freezes. We add the subscript *fus* or *solid* to ΔH to identify which process is occurring: melting (*fusion*) or freezing (*solidification*). The part of the universe of interest (system or surroundings) may also be indicated as ΔH_{sys} or ΔH_{surr}.

SAMPLE EXERCISE 5.4　　　**Determining the Value and Sign of ΔH**

In between periods of a hockey game a Zamboni machine spreads 855 liters of water across the surface of a hockey rink.

   a. If the system is the water, what is the sign of ΔH_{sys} as it freezes?
   b. To completely freeze this volume of water at 0°C, what quantity of heat is lost? Assume that 6.01 kJ of heat must be removed to freeze 1 mol of water, and the density of water is 1.00 g/mL.

COLLECT AND ORGANIZE The water from the refinishing machine is identified as the system, so we determine the sign of ΔH with respect to heat transfer into or out of the water. Because ΔH values are reported in units of joules per mole (J/mol or kJ/mol) or joules per gram (J/g), we are also given a specific volume of water and its density, so we can calculate the mass of water involved in the phase change.

ANALYZE (a) We may assume that the freezing process takes place at constant pressure, so the sign of the enthalpy change is the same as the sign of the quantity of heat transferred.

(b) To calculate the amount of heat lost, we must convert 855 L into an equivalent number of moles, because the conversion factor between the quantity of heat removed and the quantity of water that freezes is 6.01 kJ/mol.

SOLVE

a. The system (the water) changes from a liquid to a solid—it freezes. For that to happen, heat must be removed from the system. Heat lost is $-q$; therefore the sign of ΔH_{sys} is also negative.

b. We convert the volume of water into number of moles:

$$855 \text{ L} \times \frac{1000 \text{ mL}}{1 \text{ L}} \times \frac{1.00 \text{ g}}{\text{mL}} \times \frac{1 \text{ mol}}{18.02 \text{ g}} = 4.74 \times 10^4 \text{ mol}$$

Calculating q

$$4.74 \times 10^4 \text{ mol} \times \frac{6.01 \text{ kJ}}{1 \text{ mol}} = 2.85 \times 10^5 \text{ kJ of heat lost by the water} = -q$$

$\Delta H_{sys} = -q$ because pressure is constant.

Therefore, $\Delta H_{sys} = -2.85 \times 10^5 \text{ kJ}$

THINK ABOUT IT The value given in the problem for the molar heat of fusion of water applies specifically to conditions of 1 atm of pressure. The value of ΔH changes with pressure, but the magnitude of the change is insignificant for pressures around normal atmospheric pressure.

Practice Exercise The flame in a torch used to cut metal is produced by acetylene burning in an atmosphere of pure oxygen. Assuming the combustion of a mole of acetylene (C_2H_2) releases 1251 kJ of heat, what mass of acetylene is needed to cut through a piece of steel if the process requires 5.42×10^4 kJ of heat?

(Answers to Practice Exercises are in the back of the book.)

5.4 Heating Curves and Heat Capacity

Winter hikers and high-altitude mountain climbers use portable stoves fueled by propane or butane to prepare hot meals where the only source of water may be ice or snow. In this section, we use this scenario to examine the flow of heat into water that begins as snow and ends up as vapor.

Hot Soup on a Cold Day

Let's consider the changes in temperature and changes of state that water undergoes as some hikers set out to prepare hot soup using dry soup mix and melted snow. Suppose they start with a saucepan filled with snow at 0°F (−18°C). They place it above the flame of a portable stove, and heat begins to flow into the snow. The temperature of the snow immediately begins to rise. If the flame of the stove is steady (so that the flow of heat is constant), then the temperature of the snow should follow the series of connected lines segments shown in Figure 5.17. First, heat from the stove warms the snow to its melting point, 0°C. This temperature rise and the accompanying heat flow into the snow are represented by line segment $\overline{AB}$ in Figure 5.17. When the temperature reaches 0°C, it remains steady

FIGURE 5.17 The heat required to melt snow and boil the resultant water is illustrated by the four line segments on the heating curve of water: heating snow to its melting point ($\overline{AB}$); melting the snow to form liquid water ($\overline{BC}$); heating the water to its boiling point ($\overline{CD}$); and boiling the water to convert it to vapor ($\overline{DE}$).

▶‖ CHEMTOUR Heating Curves

Molar heat capacity (c) is the heat required to raise the temperature of 1 mole of a substance by 1°C. When the system is at constant pressure, the symbol used is c_p.

Specific heat (c_s) is the heat required to raise the temperature of 1 gram of a substance by 1°C at constant pressure.

Heat capacity (C_p) is the quantity of heat needed to raise the temperature of some specific object by 1°C at constant pressure.

while the snow continues to absorb heat and the snow melts. This process is a phase change that yields liquid water from snow. Melting, also called fusion, is represented by the constant temperature (flat) line segment $\overline{BC}$.

When all the snow has melted, the temperature of the water again rises (line segment $\overline{CD}$ in the figure) until it reaches 100°C. At 100°C another phase change takes place: liquid water absorbs a quantity of energy (denoted by the length of line segment $\overline{DE}$) and is vaporized. If all of the liquid water in the pot were converted into vapor, the temperature of whatever vapor remained in the pot would begin to rise again as more heat was added.

The differences in the *x*-axis coordinates of the points along the curve in Figure 5.17 indicate how much heat is required in each step in this heating process. In the first step (along line segment $\overline{AB}$), the heat required to raise the temperature of snow from −18°C to 0°C can be calculated if we know how many moles of snow we have and what quantity of heat is required to change the temperature of a mole of snow. This latter quantity, called molar heat capacity, has been determined for many materials; Table 5.1 lists a few of these. **Molar heat capacity (c)** is the quantity of heat required to raise the temperature of 1 mole of a substance by 1°C. Water in the solid phase has a molar heat capacity of 37.1 J/(mol · °C). From the units of molar heat capacity, J/(mol · °C), we see that if we know the number of moles of snow and the temperature change it experiences, we can calculate q for the process:

$$q = nc_p \Delta T \tag{5.11}$$

where q represents heat, n is the number of moles, ΔT is the temperature change in degrees Celsius, and c_p is the molar heat capacity of snow or ice at constant pressure in J/(mol · °C).

In addition to the molar heat capacity, there are other measures of the quantity of heat required to increase the temperature of a substance, each referring to a specific object or a specific amount of the substance. For example, some tables of thermodynamic data list a parameter related to the molar heat capacity, called **specific heat (c_s)**, which is the heat required to raise the temperature of *1 gram of a substance* by 1 degree Celsius and has units of J/(g · °C). Later in this chapter, we make use of the parameter **heat capacity (C_p)**, which is the quantity of energy required to increase the heat of *a particular object* by 1 degree Celsius.

TABLE 5.1 Molar Heat Capacities of Selected Substances

Substance	c_p [J/(mol · °C)]
Water(s)	37.1
Water(ℓ)	75.3
Ethanol(ℓ): CH_3CH_2OH	113.1
Carbon(s): graphite	8.54
Al(s)	24.4
Cu(s)	24.5
Fe(s)	25.1

Let's assume that the hikers decide to cook their meal with 270.0 g of snow. Dividing the mass of snow by the molar mass of water (18.02 g/mol), we find that they are using 14.98 mol of snow. To calculate how much heat is needed to raise the temperature of 14.98 mol of $H_2O(s)$ from −18°C to 0°C, we use Equation 5.11 and insert the quantity of snow (14.98 mol), the temperature change (+18°C), and the molar heat capacity of snow, [+37.1 J/(mol · °C)]:

$$q = nc_p \Delta T$$

$$= 14.98 \text{ mol} \times \frac{37.1 \text{ J}}{\text{mol} \cdot °C} \times (+18°C) = 1.0 \times 10^4 \text{ J} = +10 \text{ kJ}$$

Notice that this value is positive, which means that the system (the snow) gains energy as it warms up.

During the next phase of the heating process, the snow melts, or *fuses*, into liquid water. The heat absorbed during this step, represented by line segment $\overline{BC}$ in Figure 5.17, can be calculated using the amount of heat required to melt 1 mole of snow, called the **molar heat of fusion (ΔH_{fus})** of water (6.01 kJ/mol), in the following equation:

$$q = n\Delta H_{fus} \qquad (5.12)$$

Substituting the given information into the equation, we get

$$q = 14.98 \text{ mol} \times \frac{6.01 \text{ kJ}}{\text{mol}} = +90.2 \text{ kJ}$$

This value is positive, because the snow requires energy to break the attractive forces between the water molecules in the solid phase and become a freely flowing liquid. No factor for temperature appears in this calculation; phase changes of pure substances take place ideally at constant temperature. Snow at 0°C becomes liquid water at 0°C.

When all the snow has melted, the temperature of the water rises from its melting point of 0°C to its boiling point of 100°C as more heat is added. During this stage, the relation between temperature and heat absorbed is defined by Equation 5.11 for liquid water, which has a molar heat capacity of 75.3 J/(mol · °C):

$$q = nc_p \Delta T$$

$$= 14.98 \text{ mol} \times \frac{75.3 \text{ J}}{\text{mol} \cdot °C} \times 100°C = +1.13 \times 10^5 \text{ J} = +1.13 \times 10^2 \text{ kJ}$$

This value is positive because the system takes in energy as its temperature rises.

The **molar heat of fusion (ΔH_{fus})** of a substance is the heat required to convert 1 mole of a solid substance at its melting point to 1 mole of liquid.

The **molar heat of vaporization** (ΔH_{vap}) is the heat required to convert 1 mole of a substance at its boiling point to 1 mole of vapor.

At this point in our story, let's assume that our hiker-chefs accidentally leave the boiling water unattended and it is completely vaporized. The conversion of liquid water into vapor is depicted by line segment $\overline{DE}$ in Figure 5.17. The temperature of the water remains at 100°C until enough energy (represented by the length of horizontal line segment $\overline{DE}$) has been absorbed to vaporize all of it. The enthalpy change associated with changing 1 mole of a liquid to a gas is the **molar heat of vaporization (ΔH_{vap})**. This quantity of energy absorbed in this part of the process is the product of the number of moles of water times the value of ΔH_{vap} for water, which is 40.67 kJ/mol:

$$q = n\Delta H_{vap} \tag{5.13}$$
$$= (14.98 \text{ mol}) (40.67 \text{ kJ/mol}) = 610 \text{ kJ}$$

This value is positive, because the system takes up heat in converting a liquid into a vapor. Again no factor for temperature appears because the phase change takes place ideally at constant temperature.

Only after all the water has completely vaporized does its temperature increase above 100°C along the line above point E. The molar heat capacity of steam [43.1 J/(mol · °C)] is used to calculate the energy required by the system to heat the vapor to a temperature above 100°C.

Note that the c_p values of solid and liquid water and water vapor are all different. This trend is common: nearly all substances have different molar heat capacities in their different physical states.

In Figure 5.17 notice that the horizontal line segment $\overline{DE}$ representing the boiling of water is much longer than the line segment representing the melting of ice or snow. The relative lengths of these lines indicate that the molar heat of vaporization of water (40.67 kJ/mol) is much larger than the molar heat of fusion of ice (6.01 kJ/mol). Why does it take more energy to boil a mole of water than to melt a mole of ice? The answer is related to the extent to which attractive forces between molecules must be overcome in each process and how the potential energy (PE) of the system changes as heat is added (Figure 5.18).

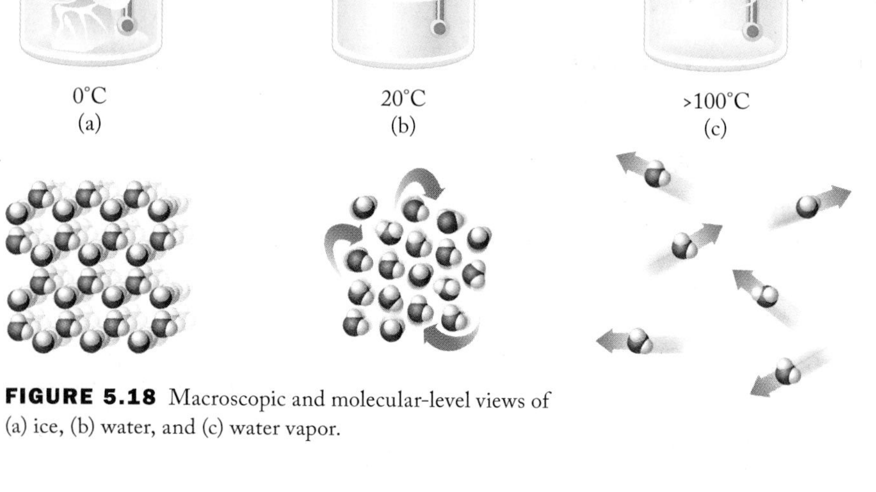

Attractive forces determine the organization of molecules and their relation to their nearest neighbors. In snow or ice, the values of KE (kinetic energy) and PE are low and attractive forces are strong enough to hold molecules in place relative to each other. Melting ice—changing the solid into a liquid—requires that some of these attractive forces between water molecules be overcome. The heat energy provided to the system at the melting point of the solid is sufficient to overcome the attractive forces and change the arrangement (and hence PE) of the molecules.

The attractive forces still exist between molecules of H_2O in liquid water, but the energy added at the melting point goes into increasing the potential energy of the molecules and enables them to occupy positions

0°C
(a)

20°C
(b)

>100°C
(c)

FIGURE 5.18 Macroscopic and molecular-level views of (a) ice, (b) water, and (c) water vapor.

other than those within the rigid framework of the solid with respect to their nearest neighbors. The molecules are still as close together in the liquid as in the solid, but their positions relative to each other change; they no longer have the same nearest neighbor over time. The molecules have higher potential energy in the liquid than in the solid.

Once the solid has completely melted, additional heat causes the temperature to rise, increasing values of both KE and PE. When water vaporizes at the boiling point, essentially all the attractive forces between water molecules must be overcome to separate the molecules by increasing the distance between them as they enter the gas phase. Separating the molecules widely in space is a much larger change in position—in potential energy—for the molecules, and hence more energy is required.

SAMPLE EXERCISE 5.5 **Calculating the Heat Needed to Raise the Temperature**

Calculate the amount of heat required to raise the temperature of 237 g of water (about 1 cup) from 15.0°C to 100.0°C.

COLLECT AND ORGANIZE The amount of heat required can be determined using the molar heat capacity of water (given in Table 5.1) and the number of moles of water (which can be calculated from the mass of water specified). We have the initial and final temperatures of the water, so we can calculate the temperature change (ΔT).

ANALYZE: We can calculate the number of moles in 237 grams of water using the molar mass of water ($\mathcal{M} = 18.02$ g/mol). The molar heat capacity of liquid water is 75.3 J/(mol · °C). The water increases in temperature from 15.0°C to 100.0°C.

SOLVE Calculate the number of moles.

$$n = 237 \; \cancel{g \, H_2O} \times \frac{1 \; mol \; H_2O}{18.02 \; \cancel{g \, H_2O}} = 13.2 \; mol \; H_2O$$

Using the number of moles of H_2O, its molar heat capacity, and the temperature change, we solve for the heat required using Equation 5.11:

$$q = n c_p \, \Delta T$$

$$= (13.2 \; \cancel{mol}) \, \frac{75.3 \; J}{\cancel{mol} \cdot \cancel{°C}} \, (100.0 - 15.0) \cancel{°C}$$

$$= 8.45 \times 10^4 \, J = 84.5 \; kJ$$

THINK ABOUT IT Using the definition of molar heat capacity, we can solve this exercise even without Equation 5.11. Molar heat capacity (c_p) defines the amount of heat needed to raise the temperature of 1 mole of a substance by 1°C. Multiplying that value by the number of moles of substance at issue (in this case, 13.2 moles) gives the amount of heat needed to raise the temperature of 13.2 moles of water by 1°C. We want to raise the temperature by 85°C, so multiplying by 85°C gives us the amount of heat needed to raise 13.2 moles of water from 15°C to 100°C.

Practice Exercise Calculate the heat released when 125 g of water vapor at 100.0°C condenses to liquid water and then cools to 25.0°C.

(Answers to Practice Exercises are in the back of the book.)

Water is an extraordinary substance for many reasons, but its high molar heat capacity is one of the more important. The ability of water to absorb large quantities of heat is one reason it is used as a *heat sink*, both in the radiators of cars and in our own bodies. Weather and climate are driven in large part by cycles involving the retention of heat by large bodies of water.

Cold Drinks on a Hot Day

Let's consider another case in heat transfer. Suppose we throw a party in the summer and plan to chill three cases (72 aluminum cans, each containing 355 mL) of our favorite beverages by placing them in the bottom of an insulated cooler and covering them with ice cubes. If the temperature of the ice (sold in 10-pound bags) is −8.0°C and the temperature of the beverages is initially 25.0°C, how many bags of ice do we need to chill the cans and their contents to 0.0°C (as in "ice cold")?

Perhaps you have confronted this situation in the past and already have an idea that more than 1 bag, but probably fewer than 10, will be needed. With a little mathematical analysis, we can predict more exactly how much ice is required. In doing so, we assume that whatever heat is absorbed by the ice is lost by the beverage cans. As ice absorbs heat from the cans, its temperature will increase from −8.0°C to 0.0°C, and the resulting liquid water should remain at 0.0°C until all the ice has melted (see the heating curve in Figure 5.17). We need enough ice so that the last portion melts just as the temperature of the beverage in the cans reaches that of the melting ice, 0.0°C. The cooling process is a little simpler because there is no phase change, only the cooling of liquid beverage and the solid aluminum cans.

Let's consider the heat lost in the cooling process first. Two materials are to be chilled: 72 aluminum cans and 72 × 355 mL = 25,560 mL of beverage. The beverages are mostly water. The other ingredients are present in such small concentrations that they will not significantly affect our calculation, so we assume that we need to reduce the temperature of what is essentially 25,560 mL of water by 25.0°C. We can calculate the amount of heat lost with Equation 5.11 if we first calculate the number of moles of water in 25,560 mL of water:

$$25,560 \text{ mL } H_2O \times 1.00 \text{ g/mL} \times \frac{1 \text{ mol } H_2O}{18.02 \text{ g } H_2O} = 1418 \text{ mol } H_2O$$

We can use the molar heat capacity of water and Equation 5.11 to calculate the heat lost by 1418 moles of water as its temperature decreases from 25.0°C to 0.0°C:

$$q = nc_p\Delta T$$
$$= 1418 \text{ mol} \times \frac{75.3 \text{ J}}{\text{mol} \cdot °C} \times (-25.0°C)$$
$$= -2.66 \times 10^6 \text{ J}$$

We must also consider the energy needed to lower the temperature of 72 aluminum cans by 25.0°C. Suppose the mass of each can is 12.5 g. The molar heat capacity of solid aluminum (Table 5.1) is 24.4 (J/mol · °C) and its molar mass is 26.98 g/mol. Using these values in Equation 5.11:

$$q = nc_p \Delta T$$

$$= 72 \text{ cans} \times \frac{12.5 \text{ g Al}}{\text{can}} \times \frac{1 \text{ mol}}{26.98 \text{ g}} \times \frac{24.4 \text{ J}}{\text{mol} \cdot {}^\circ\text{C}} \times (-25.0^\circ\text{C})$$

$$= -2.03 \times 10^4 \text{ J}$$

The total quantity of energy that must be removed from the cans of beverage is the sum of the heat removed from the beverage and the heat removed from the cans:

$$q_{\text{total lost}} = q_{\text{beverage}} + q_{\text{cans}} = [(-2.66 \times 10^6) + (-2.03 \times 10^4)] \text{ J}$$

$$= (-2.66 - 0.0203) \times 10^6 \text{ J} = -2.68 \times 10^6 \text{ J}$$

$$= -2.68 \times 10^3 \text{ kJ}$$

This quantity of heat must be absorbed by the ice as it warms to its melting point and then melts. The calculation of the amount of ice that is needed is an algebra problem. Let x be the number of moles of ice needed. Then the heat absorbed is the sum of the heat needed (1) to raise the temperature of x moles of ice from -8.0° to 0.0°C and (2) to melt x moles of ice. The quantities of heat required can be calculated with Equation 5.11 for step 1 and Equation 5.12 for step 2. Note that c_p is in units of joules, while ΔH_{fus} is in kilojoules. We need to make sure both terms are expressed in the same units:

$$q_{\text{total gained}} = q_1 + q_2$$

$$= nc_p \Delta T + n \, \Delta H_{\text{fus}}$$

$$= x\left(\frac{37.1 \text{ J}}{\text{mol} \cdot {}^\circ\text{C}}\right)(8.0^\circ\text{C}) + x \, (6.01 \text{ kJ/mol})$$

$$= x \, (6.31 \text{ kJ/mol})$$

The heat lost by the cans of beverage balances the heat gained by the ice:

$$-q_{\text{total lost}} = +q_{\text{total gained}}$$

$$2.68 \times 10^3 \text{ kJ} = x \, (6.31 \text{ kJ/mol})$$

$$x = 4.25 \times 10^2 \text{ mol ice}$$

Converting 4.25×10^2 moles of ice into pounds gives us

$$(4.25 \times 10^2 \text{ mol}) \times \frac{18.02 \text{ g}}{\text{mol}} \times \frac{1 \text{ lb}}{453.6 \text{ g}} = 17 \text{ lb of ice}$$

Thus we will need at least two 10-pound bags of ice to chill three cases of our favorite beverages.

CONCEPT TEST

The heat lost by the beverage inside the 72 cans in the preceding illustration was more than 100 times the heat lost by the cans themselves. What factors contributed to this large difference between the heat lost by the cans and the heat lost by their contents?

(Answers to Concept Tests are in the back of the book.)

| SAMPLE EXERCISE 5.6 | **Calculating Temperature at Thermal Equilibrium** |

If you add 250.0 g of ice that is initially at −18.0°C to 237 g (1 cup) of freshly brewed tea initially at 100.0°C, and the ice melts, what would be the final temperature of the tea? Assume that the mixture is an isolated system (in an insulated container) and that 237 g of tea has the same molar specific heat as 237 g of water.

COLLECT AND ORGANIZE Before solving this problem, we need to think about the changes that take place during the process in which the hot tea and the ice cubes come into contact. Assuming the ice cubes melt completely, three heat-flows (q_1, q_2, and q_3) consume the heat lost by the hot tea:

1. q_1: raising the temperature of the ice cubes to their melting point (0°C);
2. q_2: melting the ice cubes;
3. q_3: bringing the mixture to thermal equilibrium, in which the temperature of the melted ice rises and the temperature of the tea ($T_{initial} = 100.0$°C) falls to the final temperature of the mixture ($T_{final} = ?$).

ANALYZE Let's begin by noting that the heat gained by the ice cubes equals the heat lost by the tea:

$$q_{ice\ cubes} = -q_{tea}$$

Based on our organization of the information, we know that

$$q_{ice} = q_1 + q_2 + q_3$$

SOLVE The heat lost by the tea as it cools from 100.0°C to its final temperature (T_{final}) is

$$q_{tea} = nc_p \Delta T_{hot\ water}$$

$$= 237\ g \times \frac{1\ mol}{18.02\ g} \times \frac{75.3\ J}{mol \cdot °C} \times (T_{final} - 100.0°C)$$

$$= (990\ J/°C)(T_{final} - 100.0°C)$$

The transfer of this heat is responsible for the changes in the ice. We can treat the process of heat transfer from the hot tea to the ice cubes in terms of the processes we identified. In step 1, the ice is warmed from −18.0°C ($T_{initial}$) to its melting point (0°C; T_{final} in this step):

$$q_1 = nc_p \Delta T$$

$$= 250.0\ g \times \frac{1\ mol}{18.02\ g} \times \frac{37.1\ J}{mol \cdot °C} \times [0°C - (-18.0°C)]$$

$$= 9.26 \times 10^3\ J$$

In step 2, the ice melts, requiring the absorption of the following quantity of heat based on ΔH_{fus}:

$$q_2 = n\Delta H_{fus}$$

$$= 250.0\ g \times \frac{1\ mol}{18.02\ g} \times \frac{6.01\ kJ}{mol} = 83.4\ kJ$$

In step 3, the melted ice water warms to the final temperature (where $\Delta T = T_{final} - T_{initial} = T_{final} - 0°C$):

$$q_3 = n_{H_2O}\, c_{H_2O(\ell)}\, \Delta T_{H_2O}$$

$$= 250.0 \text{ g} \times \frac{1 \text{ mol}}{18.02 \text{ g}} \times \frac{75.3 \text{ J}}{\text{mol} \cdot °C} \times (T_{final} - 0°C)$$

$$= 1045 \text{ J/}°C \times T_{final}$$

As we indicated at the outset, q_{ice}, the sum of the amount of heat absorbed by the ice during steps 1 through 3, must balance the heat lost by the hot tea:

$$q_{ice} = q_1 + q_2 + q_3 = -q_{tea}$$

$$9260 \text{ J} + 83.4 \text{ kJ} + (1045 \text{ J/}°C)(T_{final}) = -[(990 \text{ J/}°C)(T_{final} - 100°C)]$$

Expressing all values in the same units (in this case, in kilojoules), we get:

$$9.26 \text{ kJ} + 83.4 \text{ kJ} + (1.045 \text{ kJ/}°C)(T_{final}) = -[(0.990 \text{ kJ/}°C)(T_{final} - 100°C)]$$

Rearranging the terms to solve for T_{final}, we have:

$$(2.04 \text{ kJ/}°C)T_{final} = -9.26 \text{ kJ} - 83.4 \text{ kJ} + 99.0 \text{ kJ} = 6.34 \text{ kJ}$$

$$T_{final} = 3°C$$

THINK ABOUT IT In the statement of the problem, the setup was described as an isolated system. If you were to carry out this experiment in a real ceramic mug instead of an ideal insulated container, do you think the final temperature of the solution at thermal equilibrium would be 3°C, as your calculation predicted?

Practice Exercise Calculate the final temperature of a mixture of 350.0 g of ice cubes at −18°C and 237 g of water at 100.0°C.

(Answers to Practice Exercises are in the back of the book.)

5.5 Calorimetry: Measuring Heat Capacity and Calorimeter Constants

Up to this point, we have discussed heats of reaction and heats associated with phase changes, but we have not broached the issue of how we know these values. The experimental approach for accurately measuring the quantities of heat associated with chemical reactions and physical changes is called **calorimetry**. The device used to measure the heat released or absorbed during a process is a **calorimeter**.

Determining Heat Capacity and Specific Heat

When we determined the amount of ice needed to cool 72 cans of a beverage, we used the molar heat capacity of aluminum to determine the quantity of heat taken up by the cans. How are heat capacities determined? We can apply the first law of thermodynamics and design an experiment to determine the specific heat of aluminum, from which we can calculate the molar heat capacity. The units of specific heat tell us exactly what we have to do. "Determine the specific heat of aluminum" means determine the amount of heat required to change the temperature of 1 g of aluminum by 1°C. If we determine how much heat is required to change the temperature of any known mass of aluminum by any measured number of

Calorimetry is the measurement of the change in heat that occurs during a physical change or chemical process.

A **calorimeter** is the device used to measure the absorption or release of heat by a physical or chemical process.

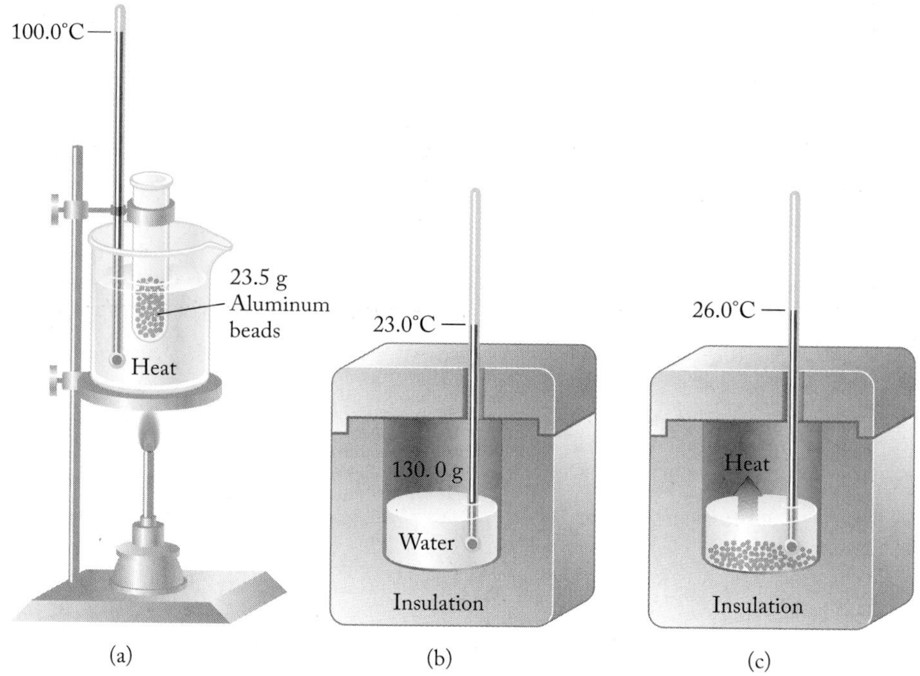

FIGURE 5.19 Experimental setup to determine heat capacity of a metal: (a) a pile of pure aluminum beads with a mass of 23.5 g is heated to 100.0°C in boiling water; (b) 130.0 g of water at a temperature of 23.0°C is in an insulated beaker. (c) The hot Al beads are dropped into the water, and the temperature when thermal equilibrium is reached is 26.0°C.

degrees, we can calculate how much heat is required to change the temperature of 1 g of aluminum by 1°C.

Suppose we have beads of pure aluminum with a total mass of 23.5 g and want to transfer a known amount of heat to or from the aluminum. We can rely on one fact and two definitions to help design the experiment: (1) the specific heat of water [4.184 J/(g · °C)], (2) the definition of thermal equilibrium, and (3) the first law of thermodynamics. Refer to Figure 5.19 for our discussion.

First, we bring a quantity of water to its boiling point (100.0°C) in a beaker containing a test tube holding the aluminum beads (the system). We wait a few minutes while the beads and the boiling water come to the same temperature: 100.0°C (Figure 5.19a).

While the metal is heating, we place a measured amount of water (in this case, 130.0 g) in a Styrofoam block with a cavity and with a Styrofoam lid (Figure 5.19b). The Styrofoam insulation effectively isolates our experiment from the rest of the universe and enables us to define the water in the device as the surroundings of the metal (the system). We place a thermometer in the water, read the temperature (room temperature = 23.0°C), and leave the thermometer in it. When we have waited long enough for the Al beads (the system) to heat up to 100.0°C, we re-move the test tube from the boiling water, drop the aluminum beads into the water (the surroundings) in the insulated container, and close the lid (Figure 5.19c). The temperature of the water rises because it is now in contact with the hot aluminum. Because of the insulation, we assume that all the heat coming from the aluminum beads goes into the water. Suppose the temperature of the aluminum–water mix-ture rises to 26.0°C, at which point it does not change anymore, so the system is at thermal equilibrium.

Heat is supplied from the hot beads of aluminum to the water. From the first law of thermodynamics, we can state that

$$-q_{aluminum} = q_{water}$$

Now we have everything we need to calculate the specific heat of aluminum. Note that the water in Figure 5.19(b) is insulated from the surroundings. The boiling water in Figure 5.19(a) was just the vehicle we used to bring the aluminum to some known temperature. First, we determine the amount of heat gained by the water:

$$q_{water} = nc_p\Delta T_{water}$$

where $\Delta T_{water} = T_{final} - T_{initial} = (26.0 - 23.0)°C = +3.0°C$

$$= (130.0 \text{ g H}_2\text{O})\left(\frac{4.184 \text{ J}}{\text{g} \cdot °\text{C}}\right)(3.0°C)$$

$$= +1600 \text{ J}$$

The heat gained by the water has a positive sign. The water (the surroundings) absorbed heat from the system.

The heat that changed the temperature of the water came from the beads. Therefore,

$$-q_{aluminum} = nc_s\Delta T_{aluminum}$$

where $\Delta T_{aluminum} = T_{final} - T_{initial} = (26.0 - 100.0)°C = -74.0°C$

$$-q_{aluminum} = (23.5 \text{ g})(c_s)(-74.0°C)$$

where c_s, the specific heat of aluminum, is the unknown. The heat lost by the system (the aluminum) has a negative sign. In terms of the heat-transfer part of the experiment, the aluminum starts out at 100.0°C and drops in temperature when it transfers heat to the water. Recalling the first law of thermodynamics, we can write

$$(23.5 \text{ g})(c_s)(-74.0°C) = -(+1600 \text{ J})$$

and solve for c_s:

$$c_s = \frac{0.920 \text{ J}}{\text{g} \cdot °\text{C}}$$

The molar heat capacity of aluminum is the specific heat multiplied by the mass of 1 mole of aluminum:

$$c_s \times \mathcal{M} = \left(\frac{0.920 \text{ J}}{\text{g} \cdot °\text{C}}\right)(26.98 \text{ g/mol}) = \frac{24.8 \text{ J}}{\text{mol} \cdot °\text{C}}$$

CONCEPT TEST

In experiments to calculate very precise values of specific heats, thermometers capable of measuring to the nearest 0.001°C are used. What impact would this change in equipment have in the numerical values reported for specific heats?

(Answers to Concept Tests are in the back of the book.)

Calorimeter Constants

The heat change accompanying a chemical reaction, an important characteristic, is defined by a quantity known as the **enthalpy of reaction (ΔH_{rxn})**, also called the **heat of reaction**. The subscript "rxn" stands for "reaction." The subscript may be changed to reflect the specific type of reaction studied; for example, ΔH_{comb} stands for the enthalpy change of a combustion reaction. Heats of reaction are measured with a device called a **bomb calorimeter** (Figure 5.20). The sample is placed in a sealed vessel (called a *bomb*) capable of withstanding high pressures. Oxygen is introduced into the vessel, and the mixture is ignited with an electric spark. As combustion occurs, heat generated by the reaction flows into the walls of the bomb and the water surrounding the bomb, all of which is inside a heavily insulated container. A good bomb calorimeter contains the system (the chemical reaction) within the bomb and ensures that all heat generated by the reaction is taken up by the calorimeter. From the point of view of the reactants (the system),

The **enthalpy of reaction (ΔH_{rxn})** is the heat absorbed or given off by a chemical reaction. It is also called the **heat of reaction**.

A **bomb calorimeter** is a constant-volume device used to measure the heat of a combustion reaction.

▶II CHEMTOUR Calorimetry

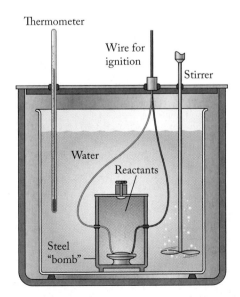

FIGURE 5.20 A bomb calorimeter.

the surroundings (the rest of the universe that needs to be considered) consists of the bomb, the water, the vessel in which they are contained, and anything else in that vessel. Insulation ensures that no heat escapes from the calorimeter; thus from the system's point of view, the rest of the universe (the surroundings) consists of only the calorimeter and its contents.

The calorimeter is defined as everything outside the system. The heat gained by the calorimeter is determined by measuring the temperature of the water before and after the reaction takes place. The water is in thermal equilibrium with the parts of the calorimeter it contacts—the walls of the bomb, the thermometer, and the stirrer—so the temperature change of the water takes the entire calorimeter into account.

Measuring the change in temperature, however, is not the whole story. We also need to know the heat capacity of the surroundings (the calorimeter). Because this is a value unique to every calorimeter, it is frequently referred to as the **calorimeter constant ($C_{calorimeter}$)**. If we know the value of $C_{calorimeter}$ and if we can measure the change in temperature (ΔT), we can calculate the quantity of heat (q) that flows from the reaction into the calorimeter:

$$q = C_{calorimeter} \Delta T \qquad (5.14)$$

Rearranging the terms in Equation 5.14, we have the following expression for the heat capacity of the surroundings:

$$C_{calorimeter} = \frac{q}{\Delta T} \qquad (5.15)$$

Equation 5.15 indicates that heat capacity is expressed in units of heat/temperature, or kilojoules per degree Celsius (kJ/°C). Equation 5.15 can be used to determine the heat capacity of a bomb calorimeter. To determine $C_{calorimeter}$, we must completely burn a quantity of material in the calorimeter that produces a known quantity of heat when it burns. Benzoic acid ($C_7H_6O_2$) is often used for this purpose, because it can be obtained in a form that is both very pure and very dry. The combustion of 1 gram of benzoic acid is complete and produces 26.38 kJ of heat. Once the heat capacity of a calorimeter has been determined, the calorimeter can be used to determine ΔH_{comb} for other substances, and the observed increases in temperature as a result of combustion reactions can be used to calculate the quantities of heat that they produce on a per-gram or per-mole basis. Because there is no change in the volume of the reaction mixture, this technique is referred to as *constant-volume calorimetry*. No $P\Delta V$ work is done, so according to Equation 5.15 the heat lost by the system equals the internal energy lost by the system during combustion:

$$q_{sys} = \Delta E_{comb} \qquad (5.16)$$

Other types of calorimeters allow the volume to change while the pressure remains constant, so q_{sys} in those cases equals the enthalpy of reaction (ΔH_{rxn}). The pressure inside a bomb calorimeter may change as a result of a combustion reaction, and q_{sys} (ΔE_{comb}) in that case is not *exactly* the same as ΔH_{comb}. However, the pressure effects are usually so small that ΔH_{comb} is *nearly* the same as ΔE_{comb}, and we do not need to be concerned about the very small differences. Hence we shall discuss enthalpies of reactions (ΔH_{rxn}) throughout and apply the approximate relation

$$q_{calorimeter} = \Delta H_{comb}$$

or even more generally

$$q_{rxn} = -q_{calorimeter} \qquad (5.17)$$

SAMPLE EXERCISE 5.7   **Determining a Calorimeter Constant**

Before we can determine the heat of reaction, we must determine the heat capacity of the calorimeter being used. What is the calorimeter constant, $C_{calorimeter}$, of a bomb calorimeter assuming that combustion of 1.000 g of benzoic acid causes the temperature of the calorimeter to rise by 7.248°C?

COLLECT AND ORGANIZE We are asked to find the heat capacity of a calorimeter. That means we need to determine the amount of heat required to raise the temperature of the calorimeter by 1°C. We are given data describing how much the temperature of the calorimeter rises when benzoic acid is burned in it. To proceed, we need the heat of reaction of benzoic acid; it was given earlier in this section as 26.38 kJ/g.

ANALYZE The heat capacity of a calorimeter can be calculated using Equation 5.15 and the knowledge that the combustion of 1.000 g of benzoic acid produces 26.38 kJ of heat.

SOLVE
$$C_{calorimeter} = \frac{q}{\Delta T}$$
$$= \frac{26.38 \text{ kJ}}{7.248°C}$$
$$= 3.640 \text{ kJ/°C}$$

Now this calorimeter can be used to determine the heat of combustion of any material.

THINK ABOUT IT The calorimeter constant is determined for a specific calorimeter. If anything changes—if the thermometer breaks and has to be replaced, or if the calorimeter loses any of the water it contains—a new constant must be determined.

Practice Exercise Assume that when 0.500 g of a mixture of volatile hydrocarbons is burned in the bomb calorimeter described in Sample Exercise 5.7, the temperature rises by 6.76°C. How much energy (in kilojoules) is released during combustion? How much energy would be released by the combustion of 1.000 g of the same mixture?

(Answers to Practice Exercises are in the back of the book.)

5.6 Enthalpies of Formation and Enthalpies of Reaction

As noted in Section 5.2, it is impossible to measure the *absolute* value of the internal energy of a substance. The same is true for the enthalpy of a substance. However, we can establish *relative* enthalpy values that are referenced to a convenient standard. This approach is similar to using the freezing point of water as the zero point on the Celsius temperature scale or sea level as the zero point for expressing altitude. The zero point of enthalpy values is the **standard enthalpy of formation (ΔH_f°)**, defined as the enthalpy change that takes place when 1 mole of a substance in its standard state is formed from its constituent elements in their standard states. A reaction that fits this description is known as a **formation reaction**. The **standard state** of an element is its most stable physical form under standard conditions of pressure (1 bar) and some specified temperature. This value of pressure is very close to 1 atmosphere; for the level of precision

> The **standard enthalpy of formation (ΔH_f°)**, also called the *standard heat of formation* (or simply the *enthalpy of formation* or *heat of formation*), is the enthalpy change of a formation reaction.

> A **formation reaction** is the process of forming 1 mole of a substance in its standard state from its component elements in their standard states.

> The **standard state** of a substance is its most stable form under 1 bar pressure and some specified temperature (assumed to be 25.0°C unless otherwise stated).

required in the data used in this book, a standard pressure of 10^5 pascals will be considered equivalent to 1 atm.

Enthalpy change is equivalent to the flow of heat at constant pressure, so ΔH_f° is frequently called the *standard heat of formation* or simply the *heat of formation*. By definition, ΔH_f° of a pure element in its most stable form under standard conditions is zero. In their standard states, oxygen is a gas, water is a liquid, hydrogen is a gas, carbon dioxide is a gas, and carbon is the solid graphite.

Because the definition of a formation reaction specifies 1 mole of product, writing balanced equations for formation reactions may require the use of something we avoided in Chapter 3: fractional coefficients. For example, the reaction for the production of ammonia from nitrogen and hydrogen is usually written

$$N_2(g) + 3\,H_2(g) \rightarrow 2\,NH_3(g) \qquad \Delta H_{rxn}^\circ = -92.2 \text{ kJ}$$

Although all reactants and products in this equation are in their standard states, it is not a formation reaction because 2 moles of product are formed, which is why we denote the heat of this reaction by ΔH_{rxn}° rather than ΔH_f°. To write the formation reaction of ammonia, all coefficients in the equation are divided by 2 (to make the coefficient of $NH_3(g)$ equal to 1). Because heat is stoichiometric, the heat of reaction is divided by 2 as well to determine ΔH_f°:

$$\tfrac{1}{2}\,N_2(g) + \tfrac{3}{2}\,H_2(g) \rightarrow NH_3(g) \qquad \Delta H_f^\circ = -46.1 \text{ kJ}$$

SAMPLE EXERCISE 5.8   **Recognizing Formation Reactions**

Which of the following reactions are formation reactions at 25°C? For those that are not formation reactions, explain why they are not.

a. $H_2(g) + \tfrac{1}{2}\,O_2(g) \rightarrow H_2O(g)$
b. $C_{graphite}(s) + 2\,H_2(g) + \tfrac{1}{2}\,O_2(g) \rightarrow CH_3OH(\ell)$
c. $CH_4(g) + \tfrac{3}{2}\,O_2(g) \rightarrow CO_2(g) + 2\,H_2O(\ell)$
d. $P_4(s) + 2\,O_2(g) + 6\,Cl_2(g) \rightarrow 4\,POCl_3(\ell)$
    ($POCl_3$ is a liquid at room temperature.)

COLLECT AND ORGANIZE For a reaction to be a formation reaction, it must be a reaction that produces 1 mole of a substance in its standard state from its component elements in their standard states. The standard state is the most stable form of a substance at 1 atm pressure and some specified temperature.

ANALYZE Each reaction must be evaluated for the quantity of product and for the state of the substances involved.

SOLVE (a) The reaction shows 1 mole of water being formed from its constituent elements, but water appears as a gas. The standard state of water at 25.0°C and 1 atm pressure is liquid, so this is not a formation reaction; it is not at standard conditions, and the heat of this reaction would be symbolized ΔH_{rxn}.

(b) Methanol (CH_3OH) is a liquid under standard conditions; 1 mole of it is formed from its constituent elements in their standard states. This reaction therefore is a formation reaction, and its heat change would be symbolized ΔH_f°.

(c) This is not a formation reaction because the reactants are not elements in their standard states, and more than one product is being formed. The change in heat of this reaction would be symbolized ΔH_{comb}.

(d) This is not a formation reaction because the product is 4 moles of $POCl_3$, whereas in a formation reaction, by definition, only 1 mole of product is formed. Note, however, that all of the constituent elements and the product are in their standard states and that only one compound is formed. We could therefore call this reaction, as written, a standard heat of reaction (ΔH°_{rxn}). We could easily convert this into a formation reaction by dividing all the coefficients in the balanced equation by 4; the *standard heat of reaction* (ΔH°_{rxn}), also divided by 4, would then be equal to the standard heat of formation (ΔH°_f).

THINK ABOUT IT Just because we can write formation reactions for substances like methanol does not mean that anyone would ever use that reaction to make methanol. Remember that formation reactions are defined to provide a standard against which other reactions can be compared when their thermochemistry is evaluated.

Practice Exercise Write formation reactions for the following substances: (a) $CaCO_3(s)$; (b) $CH_3COOH(\ell)$ (acetic acid); (c) $KMnO_4(s)$.

(Answers to Practice Exercises are in the back of the book.)

Regarding the symbol ΔH°_f, the subscript f stands for "formation," but the meaning of the superscript $^\circ$ is not as obvious. Throughout thermochemistry this superscript refers to the value of a parameter, such as ΔH_{rxn}, under standard conditions. The symbol of the enthalpy change associated with a reaction that takes place under standard conditions is thus ΔH°_{rxn}, and the value is called a **standard enthalpy of reaction**. Implied in our notion of standard states and standard conditions is the assumption that parameters such as ΔH change with temperature and pressure. That assumption is correct, although the changes are so small that we ignore them in many of the calculations in this chapter and those that follow.

Table 5.2 lists thermochemical values for some compounds, primarily hydrocarbons. A more complete list of these values for a wide variety of compounds can be found in Appendix 4. Note that the standard enthalpies of formation of the acetylene and ethylene in Table 5.2 are actually positive, which means that the reactions in which these compounds are formed from their component elements are endothermic. The other hydrocarbon fuels in Table 5.2 have negative enthalpies of formation, but the values for all the fuels are less negative than those for the formation of water and carbon dioxide. Recall from Chapter 3 that the products of the complete combustion of hydrocarbons are carbon dioxide and water (CO_2 and H_2O). The more negative the heat of formation, the more stable the substance. The implication of this fact is something stated earlier: the reaction of fuels with oxygen to produce CO_2 and H_2O is exothermic. The reaction produces heat, which is why hydrocarbons are useful as fuels. Note also that the values for the C_2 to C_4 hydrocarbons become more negative with increasing molar mass.

Knowing the standard enthalpies of formation of substances is useful for predicting heats of reaction. We can calculate enthalpy changes by determining the difference in ΔH°_f between products and reactants in a balanced chemical equation. To see how this approach works, let's calculate ΔH°_{comb} for the combustion of methane

$$CH_4(g) + 2\,O_2(g) \rightarrow CO_2(g) + 2\,H_2O(g)$$

The **standard enthalpy of reaction** (ΔH°_{rxn}) is the heat associated with a reaction that takes place under standard conditions. Also called *standard heat of reaction*.

TABLE 5.2 Thermodynamic Properties at 25°C for Selected Compounds

A. STANDARD ENTHALPIES OF FORMATION

Substance	ΔH°_f (kJ/mol)
Water(g)	−241.8
Water(ℓ)	−285.8
$CH_4(g)$: methane	−74.8
$C_2H_2(g)$: acetylene	226.7
$C_2H_4(g)$: ethylene	52.26
$C_2H_6(g)$: ethane	−84.68
$C_3H_8(g)$: propane	−103.8
$C_4H_{10}(g)$: butane	−125.6
$CO_2(g)$	−393.5
$CO(g)$	−110.5
$NH_3(g)$: ammonia	−46.1
$N_2H_4(g)$: hydrazine	95.4
$N_2H_4(\ell)$: hydrazine	50.63
$CH_3OH(\ell)$: methanol	−238.7
$CH_3CH_2OH(\ell)$: ethanol	−277.7
$CH_3COOH(\ell)$: acetic acid	−484.5

B. STANDARD ENTHALPIES OF COMBUSTION

Substance	ΔH°_{comb} (kJ/mol)
$CO(g)$	−283.0
$C_5H_{12}(\ell)$: pentane	−3535
$C_9H_{20}(\ell)$ (avg. gasoline molecule)	−6160
$C_{14}H_{30}(\ell)$ (avg. diesel molecule)	−7940

CH₄ C : H   1 : 4

Methane (CH₄) is a renewable source of energy because it can be produced by degradation of organic matter by methanogenic bacteria. Methane is produced in swamps, hence commonly called swamp gas. Experiments in the bulk production of methane from natural sources, like the one shown here of a plastic tarp covering a lagoon of animal waste, are being carried out worldwide to augment the fuel supply.

using the values from Table 5.2. In this calculation, we make use of the following equation:

$$\Delta H^{\circ}_{rxn} = \sum n\, \Delta H^{\circ}_{f,products} - \sum m\, \Delta H^{\circ}_{f,reactants} \qquad (5.18)$$

Equation 5.18 states that ΔH°_{rxn} equals the sum (Σ) of the ΔH°_f values for the number of moles of each product minus the sum of the ΔH°_f values for the number of moles of each reactant in the balanced chemical equation. Thus we multiply the values of ΔH°_f for $O_2(g)$ by 2 before summing with ΔH°_f for $CH_4(g)$ because the coefficient for O_2 is 2 in the balanced equation. We do the same with ΔH°_f for water in the product.

Inserting values for the products (CO_2 and H_2O) and the reactants (CH_4 and O_2) and their coefficients from the balanced chemical equation into Equation 5.18, yields the following:

$$\Delta H^{\circ}_{rxn} = [(1\text{ mol } CO_2)(-393.5 \text{ kJ/mol}) + (2\text{ mol } H_2O)(-241.8 \text{ kJ/mol})] -$$
$$[(1\text{ mol } CH_4)(-74.8 \text{ kJ/mol}) + (2\text{ mol } O_2)(0.0 \text{ kJ/mol})]$$

$$= [(-393.5 \text{ kJ}) + (-483.6 \text{ kJ})] - [(-74.8 \text{ kJ}) + (-0.0 \text{ kJ})] = -803.3 \text{ kJ}$$

Calculations of ΔH°_{rxn} from ΔH°_f values can be carried out for all kinds of chemical reactions. Consider the following overall reaction between methane and steam, which yields a mixture of hydrogen and carbon monoxide known as *water gas*:

$$CH_4(g) + H_2O(g) \rightarrow CO(g) + 3\, H_2(g)$$

This reaction is used to synthesize the hydrogen used in the steel industry to remove impurities from molten iron and in the chemical industry to make literally hundreds of compounds including, for example, ammonia, NH_3, for use in fertilizers and industrial refrigerants. It is also important in the manufacture of hydrogen fuel for fuel cells, which are used to generate electricity. Inserting the appropriate values for the compounds from Table 5.2 into Equation 5.18 along with the coefficient 3 for H_2, we have the following:

$$\Delta H^{\circ}_{rxn} = [(1\text{ mol } CO)(-110.5 \text{ kJ/mol}) + (3\text{ mol } H_2)(0.0 \text{ kJ/mol}) -$$
$$[(1\text{ mol } CH_4)(-74.8 \text{ kJ/mol}) + (1\text{ mol } H_2O)(-241.8 \text{ kJ/mol})]$$

$$= +206.1 \text{ kJ}$$

The positive enthalpy change tells us that this reaction is endothermic. Therefore, heat must be added to make the reaction take place. Remember that values for changes in enthalpy are independent of pathway, as Figure 5.21 illustrates: It

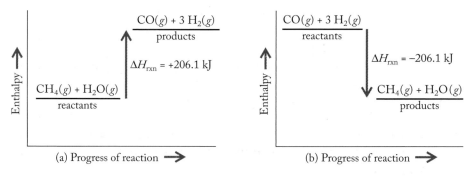

FIGURE 5.21 (a) The reaction of methane with water to produce carbon monoxide and hydrogen is endothermic. (b) The reverse of the reaction, hydrogen plus carbon monoxide to produce methane and water, is exothermic. The value of the enthalpy change of the two reactions has the same magnitude but is opposite in sign.

does not matter what path we take to get from the reactants to the products; the enthalpy difference between them will always be the same. And as we saw in Section 5.2 with phase changes, once we know the value of the heat associated with running a reaction in one direction, we also know the value of the heat associated with running the reaction in the reverse direction because it has the same magnitude but the opposite sign:

$$CO(g) + 3\ H_2(g) \rightarrow CH_4(g) + H_2O(g) \qquad (\Delta H°_{rxn} = -206.1\ kJ)$$

The *steam-reforming* (or *steam–methane reforming*) *process*, as it is called in industry, is typically conducted at temperatures near 1000°C. Although hydrogen is attractive as a fuel because it burns vigorously and produces only water as a product, its production by steam reforming does not reduce the use of fossil fuels but merely shifts them to an early step in the production process. Fossil fuels other than methane (natural gas) are also used as the starting materials in steam reforming; the reaction is still endothermic and requires the input of energy, which means more fossil fuels are consumed; and CO_2 is released into the environment as a result of the production of hydrogen in this way.

SAMPLE EXERCISE 5.9   **Calculating Standard Heats of Reaction**

Calculate $\Delta H°_{rxn}$ for the combustion of propane (C_3H_8) in air using the appropriate values from Table 5.2.

COLLECT AND ORGANIZE A typical combustion reaction, by definition, involves oxygen as a reactant. Propane is a hydrocarbon, so the products are water vapor (steam) and carbon dioxide. The values we need from Table 5.2 are the standard enthalpies of formation of the reactants and products in the equation for the combustion of propane. We also need Equation 5.18, which gives the relation between enthalpies of formation of reactants and products and the standard enthalpy of reaction.

ANALYZE The balanced equation for combustion of propane is

$$C_3H_8(g) + 5\ O_2(g) \rightarrow 3\ CO_2(g) + 4\ H_2O(g)$$

SOLVE Inserting values for the products [$CO_2(g)$ and $H_2O(g)$] and reactants [$C_3H_8(g)$ and $O_2(g)$] from Table 5.2 and the coefficients in the balanced chemical equation into Equation 5.18:

$$\Delta H°_{rxn} = [(3\ mol\ CO_2)(-393.5\ kJ/mol) + (4\ mol\ H_2O)(-241.8\ kJ/mol)] +$$
$$[(1\ mol\ C_3H_8)(-103.8\ kJ/mol) + (5\ mol\ O_2)(0.0\ kJ/mol)]$$

$$= -2043.9\ kJ$$

THINK ABOUT IT Hydrocarbons such as methane, ethane, and propane are components of natural gas and are excellent fuels. They are, however, nonrenewable fuels, and their combustion produces carbon dioxide.

Practice Exercise Calculate $\Delta H°_{rxn}$ for the *water-gas shift reaction*, in which the carbon monoxide formed by the steam-reforming process is reacted with more steam, producing CO_2 and H_2:

$$CO(g) + H_2O(g) \rightarrow CO_2(g) + H_2(g)$$

(Answers to Practice Exercises are in the back of the book.)

What is the standard heat of reaction for the production of C_3H_8 and oxygen from $CO_2(g)$ and $H_2O(\ell)$? (See information in Sample Exercise 5.9.)

(Answers to Concept Tests are in the back of the book.)

5.7 Fuel Values and Food Values

The results of calculations in Section 5.6 show that the enthalpy change is much greater and much more heat is released in the combustion of a mole of propane (-2043.9 kJ) than in the combustion of a mole of methane (-803.3 kJ). Does this make propane an inherently better (that is, higher-energy) fuel? Not necessarily. Expressing ΔH values on a per-mole basis is the only way to insure that we are talking about the same number of molecules. That approach can be very important in certain analyses. However, we do not purchase fuels, or anything for that matter, in units of moles. Depending on the fuel, we buy it in units of mass or volume.

Fuel Value

To calculate the enthalpy change that takes place when 1 gram of methane or 1 gram of propane burns in air, we divide the absolute value of ΔH°_{rxn} (in units of kilojoules per mole) for each of the combustion reactions by the number of grams in 1 mole of the compound to determine the number of kilojoules of heat released per gram of substance:

$$CH_4: \quad \frac{803.3 \text{ kJ}}{\text{mol}} \times \frac{1 \text{ mol}}{16.04 \text{ g}} = 50.08 \text{ kJ/g}$$

$$C_3H_8: \quad \frac{2043.9 \text{ kJ}}{\text{mol}} \times \frac{1 \text{ mol}}{44.10 \text{ g}} = 46.35 \text{ kJ/g}$$

These quantities of energy per gram of fuel are called **fuel values**. If we carry out similar calculations for the other hydrocarbons in natural gas, we can determine the fuel values for the compounds listed in Table 5.2.

Fuel values of hydrocarbons decrease with increasing molar mass. Why should this be the case? The answer lies in the hydrogen-to-carbon ratio in these compounds. As the number of carbon atoms per molecule increases, the hydrogen-to-carbon ratio decreases. Consider that there are 12 times as many H atoms in a gram of hydrogen as there are C atoms in a gram of carbon; this means that, *gram for gram*, 1 g of hydrogen can form six times as many moles of H_2O as 1 g of carbon can form moles of CO_2. More energy is released in a combustion reaction by the formation of a mole of CO_2 (393.5 kJ) than by a mole of $H_2O(g)$ (241.8 kJ), but even if we take this difference into account, *gram for gram* hydrogen has many times the fuel value of carbon.

Without doing any calculations, predict which of the following quantities of fuel releases more heat during combustion in air: (a) 1 mole of CH_4, or 1 mole of H_2; (b) 1 gram of CH_4, or 1 gram of H_2.

(Answers to Concept Tests are in the back of the book.)

Fuel value is the energy released during complete combustion of 1 g of a substance.

| **SAMPLE EXERCISE 5.10** | **Comparing Fuel Values** |

The great majority of automobiles run on either gasoline or diesel fuel. Although both fuels are mixtures, the energy content in regular gasoline can be approximated by considering it to be composed of hydrocarbons with the formula C_9H_{20} ($d = 0.718$ g/mL), while diesel fuel may be considered to be $C_{14}H_{30}$ ($d = 0.763$ g/mL). Using these two formulas as approximations of the two fuels, compare the fuel values (a) per gram of, and the fuel density (b) per liter of, gasoline and diesel fuels.

COLLECT AND ORGANIZE Fuel values can be used to calculate the amount of heat energy available from a given quantity of fuel. The heats of combustion in Table 5.2 are given in units of kilojoules per mole, so to answer this question in terms of grams and liters, some conversions are necessary. For (a), we need molar masses to convert moles to grams and for (b) we need density ($d = m/V$), typically in grams per milliliter, to convert grams to liters.

ANALYZE Taking C_9H_{20} and $C_{14}H_{30}$ as the average molecules in regular gasoline and diesel fuels, respectively, we can determine their molar masses. Then the fuel value per gram can be calculated from the fuel values given on a per-mole basis. Using the densities given in the problem, we can calculate fuel values on a per-liter basis.

SOLVE

a. Gasoline as C_9H_{20}: $6160 \dfrac{kJ}{mol} \times \dfrac{1\ mol}{128.25\ g} = 48.0$ kJ/g

Diesel as $C_{14}H_{30}$: $7940 \dfrac{kJ}{mol} \times \dfrac{1\ mol}{198.38\ g} = 40.0$ kJ/g

b. Gasoline as C_9H_{20}: $\dfrac{48.0\ kJ}{g} \times 0.718 \dfrac{g}{mL} \times \dfrac{1000\ mL}{1\ L} = 34{,}500$ kJ/L

Diesel as $C_{14}H_{30}$: $40.0 \dfrac{kJ}{g} \times 0.763 \dfrac{g}{mL} \times \dfrac{1000\ mL}{1\ L} = 30{,}500$ kJ/L

THINK ABOUT IT In the United States, fuel for cars and trucks is bought by the gallon and its consumption is rated in terms of miles per gallon, whereas in most countries it is bought by the liter (1 gal = 3.7854 L). Because of the way we buy fuels, it makes sense to compare fuel values in terms of volume, which is also called fuel density.

Practice Exercise Another hydrocarbon fuel derived from crude oil is kerosene, used as jet fuel to power high-performance aircraft and also in space heaters to heat homes. Kerosene is intermediate in composition between gasoline and diesel fuel, and may be approximated as $C_{12}H_{26}$ ($d = 0.750$ g/mL; $\Delta H_{comb} = -7050$ kJ/mol). Estimate the fuel value of kerosene on both a per-gram and a per-liter basis.

(Answers to Practice Exercises are in the back of the book.)

Food Value

Food serves in part the same purpose in living systems as fuel does in mechanical systems. The chemical reactions that convert food into heat and energy resemble combustion but consist of many more steps that are much more highly controlled. Carbon dioxide and water are the end products, so in a fundamental way, metabolism of food by a living system and combustion of fuel in an engine are the same process. The **food value** of the material we eat—the amount of heat produced when food is burned completely—can be determined using the same equipment

Food value is the amount of heat produced when a material consumed by an organism for sustenance is burned completely. It is typically reported in Calories (kilocalories).

and applying the same concepts of thermochemistry that we have developed to discuss fuels for vehicles. We can analyze the relative food value of the items we consume the same way we analyzed fuel value: by burning material in a bomb calorimeter and quantifying the heat released.

As an illustration, let's consider the food value of a jelly doughnut. To determine its caloric content, we simply have to burn a jelly doughnut in a calorimeter. Fresh jelly doughnuts may be hard to incinerate, however, because of their high water content, so we prepare the doughnut first by drying it. To get an accurate value of the heat of combustion, we need to make sure that the great majority of the mass we attribute to the sample is due to its carbon-containing components that will burn completely and produce CO_2 and H_2O.

After drying completely, suppose our jelly doughnut weighs 55 g. We then put it in a large calorimeter, whose calorimeter constant is known ($C_{calorimeter}$ = 41.8 kJ/°C) and burn it completely in excess oxygen. We observe that the temperature of the calorimeter (ΔT) rises by 25.0°C in this experiment. What is the food value of the jelly doughnut?

Using Equations 5.14 and 5.17 yields the answer in SI-derived units. First we insert the given values into Equation 5.14 to get the heat value of the calorimeter:

$$q_{calorimeter} = C_{calorimeter} \, \Delta T = (41.8 \text{ kJ/°C})(25.0°C) = 1040 \text{ kJ}$$

Now we insert this result into Equation 5.17 to find the food value of the jelly doughnut:

$$-q_{comb, doughnut} = q_{calorimeter}$$
$$= 1040 \text{ kJ}$$

Notice that the heat flow from the jelly doughnut has the same value—but opposite sign—as the heat absorbed by the calorimeter.

Most of us still think in terms of nutritional Calories (kilocalories), and we can use the definition of a Calorie (abbreviated Cal) from Section 5.1 to convert the heat in kilojoules into that more familiar unit used by nutritionists:

$$\frac{1040 \text{ kJ}}{\text{jelly doughnut}} \times \frac{1 \text{ Cal}}{4.184 \text{ kJ}} = \frac{249 \text{ Cal}}{\text{jelly doughnut}}$$

Between 8 and 10 jelly doughnuts would provide all the calories needed by an adult human being with normal activity in one day. In contrast, a bicyclist riding the Tour de France would need the caloric equivalent of about 32 jelly doughnuts a day.

SAMPLE EXERCISE 5.11   **Calculating Food Value**

Glucose ($C_6H_{12}O_6$) is a simple sugar formed in plants during photosynthesis. The complete combustion of 0.5763 g of glucose in a calorimeter ($C_{calorimeter}$ = 6.20 kJ/°C) raises the temperature of the calorimeter by 1.45°C. What is the food value of glucose in Calories per gram?

COLLECT AND ORGANIZE We are asked to determine the food value of glucose, which means the heat given off when 1 g is burned completely. We have the mass of glucose burned and the calorimeter constant. Equation 5.17 relates the heat given off by the system (the reaction) to heat gained by the surroundings (the calorimeter). We also have Equation 5.14, which relates the heat given off by the reaction to the temperature change and the calorimeter constant.

ANALYZE We use the data from the calorimetry experiment to determine how much heat in kilojoules is given off when the stated amount of glucose is burned. We can convert that quantity into Calories by using the equality 1 Cal = 4.184 J.

SOLVE

$$q_{calorimeter} = C_{calorimeter} \, \Delta T = (6.20 \text{ kJ/}^\circ\text{C})(1.45^\circ\text{C}) = 8.99 \text{ kJ}$$

$$q_{glucose} = -q_{calorimeter} = -8.99 \text{ kJ (for combustion of 0.5763 g of glucose)}$$

To convert this energy to a fuel value, we divide by the sample mass:

$$\frac{-8.99 \text{ kJ}}{0.5763 \text{ g}} = -15.6 \text{ kJ/g}$$

and finally into calories:

$$(15.6 \text{ kJ/g}) \left(\frac{1 \text{ Cal}}{4.184 \text{ kJ}} \right) = 3.73 \text{ Cal/g}$$

THINK ABOUT IT The value of the heat calculated with the data from the calorimetry experiment has a negative sign because combustion of a food material, just like combustion of a fuel, is an exothermic process.

Practice Exercise Sucrose (table sugar) has the formula $C_{12}H_{22}O_{11}$ ($\mathcal{M} = 342.30 \text{ g}$) and a food value of 6.49 kJ/g. Determine the calorimeter constant of the calorimeter in which the combustion of 1.337 g of sucrose raises the temperature by 1.96°C.

(Answers to Practice Exercises are in the back of the book.)

5.8 Hess's Law

In Section 5.6 we described a process for synthesizing hydrogen gas from methane and steam at high temperatures. The overall reaction is carried out on the industrial scale in two steps. The first step is an endothermic reaction between methane and a limited supply of high-temperature steam, producing carbon monoxide and hydrogen gas:

▶‖ **CHEMTOUR** Hess's Law

(1) $\quad CH_4(g) + H_2O(g) \rightarrow CO(g) + 3 H_2(g) \qquad \Delta H_1^\circ = +206 \text{ kJ}$

In the second step the carbon monoxide from the first reaction is allowed to react with more steam, producing carbon dioxide and more hydrogen gas:

(2) $\quad CO(g) + H_2O(g) \rightarrow CO_2(g) + H_2(g) \qquad \Delta H_2^\circ = -41 \text{ kJ}$

We can write an overall reaction equation that combines steps 1 and 2:

$$CH_4(g) + H_2O(g) + CO(g) + H_2O(g) \rightarrow CO_2(g) + H_2(g) + CO(g) + 3 H_2(g)$$

which simplifies to the following overall reaction:

$$CH_4(g) + 2 H_2O(g) \rightarrow CO_2(g) + 4 H_2(g) \qquad \Delta H_3^\circ = ?$$

Just as we obtained the overall chemical equation by adding the equations for steps 1 and 2, we obtain the enthalpy change for the overall reaction by adding the values of ΔH° for steps 1 and 2. The thermochemical equation for the overall reaction is

$$CH_4(g) + 2 H_2O(g) \rightarrow CO_2(g) + 4 H_2(g) \qquad \Delta H_3^\circ = +165 \text{ kJ}$$

In summary:

(1) $\qquad$ $CH_4(g) + H_2O(g) \rightarrow \cancel{CO(g)} + 3\,H_2(g)$ $\qquad$ $\Delta H_1^\circ = +206$ kJ

(2) $\qquad$ $+\;\cancel{CO(g)} + H_2O(g) \rightarrow CO_2(g) + H_2(g)$ $\qquad$ $\Delta H_2^\circ = -41$ kJ

Overall reaction: $\quad CH_4(g) + 2\,H_2O(g) \rightarrow CO_2(g) + 4\,H_2(g)$ $\qquad$ $\Delta H_3^\circ = +165$ kJ

This calculation of ΔH° for a given reaction by summing the enthalpy changes of the steps within the overall process is an application of **Hess's law**. Also known by the longer name *Hess's law of constant heat of summation*, this law states that the enthalpy change of a reaction that is the sum of two or more other reactions is equal to the sum of the enthalpy changes of the constituent reactions. This relation is illustrated in Figure 5.22.

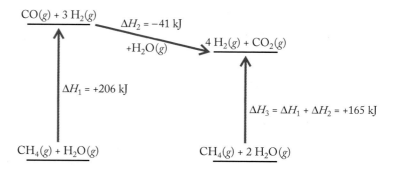

FIGURE 5.22 Hess's law predicts that the enthalpy change for the production of $H_2(g)$ and $CO_2(g)$ from $CH_4(g)$ and $H_2O(g)$ is the sum of the enthalpies of two reactions: the formation of carbon monoxide and hydrogen from methane and water (ΔH_1), and the reaction of carbon monoxide with water to produce carbon dioxide and hydrogen (ΔH_2).

Let's consider another case. Hess's law is perhaps most useful for calculating energy changes that are difficult to measure directly. For example, CO_2 is the principal product of the combustion of carbon (as in charcoal):

$$C(s) + O_2(g) \rightarrow CO_2(g)$$

However, CO, a highly toxic gas, is also produced when the oxygen supply is limited:

$$C(s) + \tfrac{1}{2}O_2(g) \rightarrow CO(g)$$

It is difficult to directly measure the enthalpy of combustion of carbon to carbon monoxide, because as long as any oxygen is present, some of the $CO(g)$ formed reacts with it to form $CO_2(g)$. A mixture of CO and CO_2 results as the product.

Because we can run the reaction of carbon with excess oxygen and thereby force it to go to completion, we can measure the enthalpy of combustion of carbon to form carbon dioxide:

Reaction A: $\qquad C(s) + O_2(g) \rightarrow CO_2(g)$ $\qquad \Delta H_{comb}^\circ = -393.5$ kJ

Because pure samples of $CO(g)$ are available, it is possible to measure the enthalpy of combustion of $CO(g)$ in oxygen to form carbon dioxide:

Reaction B: $\quad CO(g) + \tfrac{1}{2}O_2(g) \rightarrow CO_2(g)$ $\qquad \Delta H_{comb}^\circ = -283.0$ kJ

Hess's law gives us a way to calculate the heat of reaction for the process we cannot measure:

Reaction C: $\qquad C(s) + \tfrac{1}{2}O_2(g) \rightarrow CO(g)$ $\qquad \Delta H_{comb}^\circ = ?$

To calculate the unknown heat of reaction, we must find a way to combine these three equations so that any two of them sum to equal the third. Once we have that combination, because we know two of the enthalpies, we can calculate the one we do not know.

One way to approach this task is to focus on the reactants and products in the reaction whose enthalpy is unknown. Reaction (C) has carbon and oxygen as reactants and carbon monoxide as a product. Reaction B has carbon monoxide as a reactant. If we add reaction B to reaction C, the carbon monoxide cancels out, and we end up with reaction A as the sum of B and C:

Reaction C: $\quad C(s) + \frac{1}{2}O_2(g) \rightarrow CO(g) \qquad \Delta H^\circ_{comb} = ?$

$+$ Reaction B: $\quad CO(g) + \frac{1}{2}O_2(g) \rightarrow CO_2(g) \qquad \Delta H^\circ_{comb} = -283.0 \text{ kJ}$

Reaction A: $\quad C(s) + O_2(g) \rightarrow CO_2(g) \qquad \Delta H^\circ_{comb} = -393.5 \text{ kJ}$

As written,

$$\Delta H^\circ(C) + \Delta H^\circ(B) = \Delta H^\circ(A)$$

Therefore,

$$\Delta H^\circ(C) + (-283.0 \text{ kJ}) = -393.5 \text{ kJ}$$

$$\Delta H^\circ(C) = -110.5 \text{ kJ}$$

Always remember that ΔH is a state function. This means we can manipulate equations in two important ways, should the need arise when applying Hess's law. (1) We may multiply the coefficients in a balanced equation and the heat of reaction by a factor to change the quantity of material we are dealing with. (2) We may reverse an equation (make the reactants the products and the products the reactants) if we also change the sign on ΔH_{rxn}. The use of both techniques is illustrated in Sample Exercise 5.12.

SAMPLE EXERCISE 5.12 **Calculating Enthalpies of Reaction Using Hess's Law**

When hydrocarbons burn in a limited supply of air, they may not burn completely, and $CO(g)$ may be generated. One of the reasons why furnaces and hot-water heaters fueled by natural gas need to be properly vented is that incomplete combustion can produce toxic carbon monoxide:

Reaction A: $\quad 2 CH_4(g) + 3 O_2(g) \rightarrow 2 CO(g) + 4 H_2O(g) \qquad \Delta H^\circ = ?$

Using the following two thermochemical reactions

Reaction B: $\quad CH_4(g) + 2 O_2(g) \rightarrow CO_2(g) + 2 H_2O(g) \qquad \Delta H^\circ = -802 \text{ kJ}$

Reaction C: $\quad 2 CO(g) + O_2(g) \rightarrow 2 CO_2(g) \qquad \Delta H^\circ = -566 \text{ kJ}$

calculate the enthalpy of the reaction for the incomplete combustion of methane to produce carbon monoxide and water.

COLLECT AND ORGANIZE We are given two equations (B and C) with thermochemical data and a third (A) for which we are asked to find ΔH. We may manipulate the two thermochemical equations so that they sum to give the equation for which ΔH is unknown. Then we can calculate ΔH by applying Hess's law.

ANALYZE The reaction of interest (A) has methane on the reactant side. Reaction B shows methane as a reactant. We can use B as written. Reaction A has CO as a product. Reaction C involves CO, but as a reactant, so we have to reverse C in order to get CO on the product side and change the sign of $\Delta H°$ when we do. If the coefficients as given will not enable us to sum the two reactions to yield reaction A, we may have to multiply one or both reactions by other factors.

SOLVE Start with reaction B as written, and add the reverse of reaction C, remembering to change the sign on $\Delta H°_{rxn}(C)$:

B: $CH_4(g) + 2\,O_2(g) \rightarrow CO_2(g) + 2\,H_2O(g)$ $\Delta H°_{rxn}(B) = -802$ kJ

C: $2\,CO_2(g) \rightarrow 2\,CO(g) + O_2(g)$ $\Delta H°_{rxn}(C) = +566$ kJ

Look at

A: $2\,CH_4(g) + 3\,O_2(g) \rightarrow 2\,CO(g) + 4\,H_2O(g)$

and note that methane has a coefficient of 2. That means we should multiply all the terms in reaction B, including $\Delta H°_{rxn}(B)$, by 2:

2B: $2 \times [CH_4(g) + 2\,O_2(g) \rightarrow CO_2(g) + 2\,H_2O(g)]$ $2[\Delta H°_{rxn}(B) = -802$ kJ]

The carbon monoxide in reaction A has a coefficient of 2 as written, so we do not need to multiply reaction C by any factor.

Here is the result of adding 2 × B to C and canceling out common terms:

2B: $2\,CH_4(g) + \overset{3}{\cancel{4}}\,O_2(g) \rightarrow \cancel{2\,CO_2(g)} + 4\,H_2O(g)$ $\Delta H°_{rxn}(2B) = -1604$ kJ

C: $\cancel{2\,CO_2(g)} \rightarrow 2\,CO(g) + \cancel{O_2(g)}$ $\Delta H°_{rxn}(C) = +566$ kJ

A: $2\,CH_4(g) + 3\,O_2(g) \rightarrow 2\,CO(g) + 4\,H_2O(g)$ $\Delta H°_{rxn}(A) = -1038$ kJ

The value of $\Delta H°_{rxn}(A)$ for the incomplete combustion of methane to CO is thus -1038 kJ.

THINK ABOUT IT It does not matter how you assemble the equations. In this Sample Exercise, we could have assembled reactions A and C so that the sum would give reaction B. The calculation would still have resulted in the same value for the unknown $\Delta H°_{rxn}(A)$.

Practice Exercise At high temperatures, such as those in the combustion chambers of automobile engines, molecular nitrogen and oxygen combine in an endothermic reaction to form nitrogen monoxide:

$$N_2(g) + O_2(g) \rightarrow 2\,NO(g) \qquad \Delta H° = +180 \text{ kJ}$$

When NO is released into the environment, it may undergo further oxidation to NO_2 in the following exothermic reaction:

$$2\,NO(g) + O_2(g) \rightarrow 2\,NO_2(g) \qquad \Delta H° = -112 \text{ kJ}$$

Is the overall reaction between N_2 and O_2 forming NO_2

$$N_2(g) + 2\,O_2(g) \rightarrow 2\,NO_2(g) \qquad \Delta H° = ?$$

exothermic or endothermic? What is $\Delta H°$ for this reaction?

(Answers to Practice Exercises are in the back of the book.)

When we used standard enthalpies of formation (ΔH_f°) to determine standard heats of reaction (ΔH_{rxn}°) in Section 5.6, we were actually applying Hess's law. To see that this is the case, consider the reaction of ammonia [$NH_3(g)$] with oxygen to make $NO(g)$ and water in the first step of the industrial process for the synthesis of nitric acid:

$$4\,NH_3(g) + 5\,O_2(g) \rightarrow 4\,NO(g) + 6\,H_2O(\ell)$$

First we write the formation reactions for each reactant and product that is not an element in its standard state (remember $\Delta H_f^\circ = 0$ for elements in their standard states):

$$\tfrac{1}{2}N_2(g) + \tfrac{3}{2}H_2(g) \rightarrow NH_3(g) \qquad\qquad \Delta H_f^\circ = -46.1 \text{ kJ/mol}$$

$$\tfrac{1}{2}N_2(g) + \tfrac{1}{2}O_2(g) \rightarrow NO(g) \qquad\qquad \Delta H_f^\circ = +90.3 \text{ kJ/mol}$$

$$H_2(g) + \tfrac{1}{2}O_2(g) \rightarrow H_2O(\ell) \qquad\qquad \Delta H_f^\circ = -285.8 \text{ kJ/mol}$$

When we manipulate the equations

$$4\,[NH_3(g) \rightarrow \tfrac{1}{2}N_2(g) + \tfrac{3}{2}H_2(g)] \qquad 4(\Delta H_f^\circ = +46.1 \text{ kJ/mol})$$

$$4\,[\tfrac{1}{2}N_2(g) + \tfrac{1}{2}O_2(g) \rightarrow NO(g)] \qquad 4(\Delta H_f^\circ = +90.3 \text{ kJ/mol})$$

$$6\,[H_2(g) + \tfrac{1}{2}O_2(g) \rightarrow H_2O(\ell)] \qquad 6(\Delta H_f^\circ = -285.8 \text{ kJ/mol})$$

their sum yields the reaction of interest:

$$4\,NH_3(g) \rightarrow \cancel{2\,N_2(g)} + \cancel{6\,H_2(g)} \qquad \Delta H_{rxn}^\circ = +184.4 \text{ kJ}$$

$$\cancel{2\,N_2(g)} + 2\,O_2(g) \rightarrow 4\,NO(g) \qquad \Delta H_{rxn}^\circ = +361.2 \text{ kJ}$$

$$+ \cancel{6\,H_2(g)} + 3\,O_2(g) \rightarrow 6\,H_2O(\ell) \qquad \Delta H_{rxn}^\circ = -1714.8 \text{ kJ}$$

$$\overline{4\,NH_3(g) + 5\,O_2(g) \rightarrow 4\,NO(g) + 6\,H_2O(\ell) \qquad \Delta H_{rxn}^\circ = -1169.2 \text{ kJ}}$$

This is exactly the same mathematical operation that results when we use Equation 5.18:

$$\Delta H_{rxn}^\circ = \sum n\,\Delta H_{f,\text{products}}^\circ - \sum m\,\Delta H_{f,\text{reactants}}^\circ$$

$$= [4\text{ mol}(\Delta H_f^\circ)NO(g) + 6\text{ mol}(\Delta H_f^\circ)H_2O(\ell)] -$$
$$[4\text{ mol}(\Delta H_f^\circ)NH_3(g) + 5\text{ mol}(\Delta H_f^\circ)O_2(g)]$$

$$= [4\text{ mol}(+90.3 \text{ kJ/mol}) + 6\text{ mol}(-285.8 \text{ kJ/mol})] -$$
$$[4\text{ mol}(-46.1 \text{ kJ/mol}) + 5\text{ mol}(0 \text{ kJ/mol})] = -1169.2 \text{ kJ/mol}$$

Hess's law is simply a consequence of the fact that enthalpy is a state function. This statement means that for a particular set of reactants and products, the enthalpy change of the reaction is the same whether the reaction takes place in one step or a series of steps. The concept of enthalpy and its expression in Hess's law are very useful because the heats associated with a large number of reactions can be calculated from a few that are actually measured.

Carbon: Diamonds, Graphite, and the Molecules of Life

Carbon is an element central to our existence both because we are carbon-based life-forms and because carbon is the central element in hydrocarbons—currently our most significant sources of energy and environmental changes. We also value carbon in its crystalline forms. One of these is *diamond*; we prize diamonds for their beauty and take advantage of its hardness by using it in drill bits and as abrasives. Another crystalline form, *graphite*, is black and opaque, and prized for its strength and flexibility (fibers are used in golf clubs and tennis rackets), and for its softness and smoothness (flakes are used as lubricants, and mixed and baked with clay to make the "lead" in pencils). A large variety of highly specialized noncrystalline materials known as *activated carbons* are made industrially for wastewater treatment, gas purification, and sugar refining (where activated carbon decolorizes solutions of raw sugar by selectively binding colored impurities to produce white crystals of sugar).

In 1996 the Nobel Prize in chemistry was awarded to Robert Curl, Harold Kroto, and Richard Smalley for their 1985 discovery of a third crystalline form of carbon, buckminsterfullerene. Each molecule consists of a cluster of carbon atoms with the formula C_{60}; they are also known as "buckyballs" because they look like soccer balls and are reminiscent of the architectural designs of Buckminster Fuller. Buckminsterfullerene is the most abundant form of a class of carbon-clustering molecules called *fullerenes*. Their discovery has spawned research into applications ranging from drugs to treat cancer and AIDS to rocket fuels.

Carbon is found in minor amounts in Earth's crust as the free element. Diamonds are found in ancient volcanic pipes (openings in the crust through which lava flows); the exact way they were formed is still an item of debate. Graphite deposits occur worldwide. Over half of the carbon on Earth is held in the form of carbonate minerals (limestone, dolomite, and chalk), in which carbon is in its highest oxidation state as the CO_3^{2-} ion (O.N. +4), and in the fossil fuels (petroleum, coal, natural gas), in which carbon is in its fully reduced state (O.N. −4). These two reservoirs of carbon are interconnected by a dynamic system we introduced you to in Chapter 3, namely the *carbon cycle*, which links free $CO_2(g)$ in the atmosphere and the carbon-based life-forms on Earth's surface with the carbonate minerals and fossil fuels.

Atmospheric CO_2 is the source of all carbon-containing compounds produced in the leaves of green plants by photosynthesis:

$$\text{Sunlight} + 6\,CO_2(g) + 6\,H_2O(\ell) \rightarrow C_6H_{12}O_6(aq) + 6\,O_2(g)$$

The most fundamental carbon-containing product of photosynthesis is the sugar glucose ($C_6H_{12}O_6$), which is the main food for biological systems. Glucose is also the primary structural subunit of cellulose, which is the chief component of plant fibers such as wood and cotton. A single molecule of cellulose may contain over 1500 glucose units bonded together; plants produce over 100 billion tons of cellulose

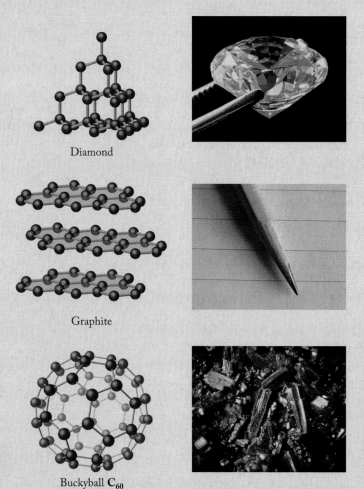

Diamond

Graphite

Buckyball C_{60}

Three crystalline forms of carbon: diamond, graphite and buckminsterfullerene ("buckyball").

Cotton is a major source of cellulose.

each year. The oxidation number of carbon in glucose is 0, so green plants may be thought of as carbon-reducing machines. Other life-forms eat plants and derive large amounts of energy from metabolism through a series of oxidation reactions that are the reverse of photosynthesis, which takes carbon at O.N. 0 (glucose) back to O.N. +4 (carbon dioxide). Life-forms are carbon-oxidizing machines.

When plants and animals die on the surface of Earth, their carbon-containing molecules are oxidized back to CO_2 and water if decomposition is complete. However, plant material deprived of oxygen may be further reduced by specialized bacteria that produce the hydrocarbon fossil fuels petroleum and coal. The hydrocarbons that make up these fuels consist of carbon in its most reduced form. When a hydrocarbon fuel reacts with oxygen in a combustion reaction, energy is released because carbon at O.N. −4 becomes carbon at O.N. +4 when the hydrocarbons are converted into CO_2.

Some CO_2 is returned to the atmosphere as a result of respiration by animals and plants, but our burning of fos-

hensible variety of structures that in turn can link together to form gigantic molecules containing hundreds and even thousands of carbon atoms. Over 16 million compounds of carbon are known, and over 90% of the thousands of new materials synthesized each year contain carbon. The number of compounds that contain carbon is far greater than the number of compounds that do not. An entire branch of chemistry, called *organic chemistry*, is devoted to the study of their structures and properties.

The chemistry of carbon is the chemistry of pharmaceuticals, soft contact lenses, paints, synthetic fibers, plastics, and gasoline. It is the chemistry of the food we eat and the fuel that powers our cars and airplanes. Fossil fuels such as petroleum are the primary source of molecules that industry modifies to make the items we regard as part of modern life, so petroleum can be regarded as fuel for the manufacturing industry as well as for the transportation industry. Combined in precise ways with a small number of other elements—such as sulfur, phosphorous, oxygen, nitrogen, and hydrogen—carbon serves as the foundation of life.

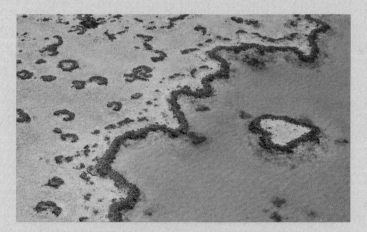

Atmospheric CO_2 affects coral reefs.

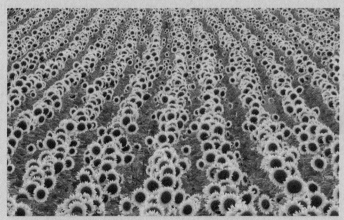

Sunflowers are the source of many useful compounds.

sil fuels inserts carbon that was formerly sequestered underground into the atmosphere as CO_2, which adds to the carbon dioxide produced by respiration. Atmospheric CO_2 dissolves in seawater to form carbonic acid, which increases the acidity of the ocean. Changing the acidity affects the solubility of $CaCO_3$ in seawater, which in turn has potentially disastrous consequences for marine life, including the many oceanic-life forms that build shells, reefs, and other structures from $CaCO_3$ and other carbonates.

In addition to its critical role in the biosphere, carbon forms a vast number of compounds. Carbon atoms attach themselves to other carbon atoms to an extent not possible for atoms of any other element. The result is the formation of chains and rings that combine to create an almost incompre-

Corn provides alternative fuels.

Summary

SECTION 5.1 **Energy** is the capacity to transfer **heat** or to do **work** (w). A **thermochemical equation** includes the heat energy absorbed or released in a chemical reaction or physical change. **Thermodynamics** is the study of energy and its transformations. Differences of temperature result in **heat transfer** from one object to another. **Thermal equilibrium** is established when objects have the same temperature and net heat transfer ceases. **Potential energy (PE)** is the energy of position and is a **state function**. **Kinetic energy (KE)** is the energy of motion. The **law of conservation of energy** states that energy cannot be created or destroyed. Heating a sample of molecules increases their average kinetic and **thermal energies**. **Electrostatic potential energy (E_{el})** results from interaction of charged particles. Energy is stored in compounds, and heat is absorbed or released when they are transformed into different compounds.

SECTION 5.2 A **system** is the part of the universe under study. Everything not part of the system is called the **surroundings**. **Isolated systems** exchange neither heat nor matter with their surroundings, **closed systems** exchange only energy, and **open systems** exchange energy and matter. In an **exothermic** process the system loses heat ($-q$) to its surroundings; in an **endothermic** process the system absorbs heat ($+q$) from its surroundings. The sum of the kinetic and potential energies of a system is called its **internal energy (E)**. The **first law of thermodynamics** states that the energy gained or lost by a system equals the energy lost or gained by the surroundings. Common energy units are **calories (cal)** and **joules (J)**. The internal energy of a system is increased ($\Delta E = E_{final} - E_{initial}$ is positive) when it is heated ($+q$) and work is done on it ($+w$).

SECTION 5.3 The **enthalpy (H)** of a system is given by $H = E + PV$. The **enthalpy change (ΔH)** of a system is equal to the heat (q_p) added to or removed from the system at constant pressure. From the system's point of view, $\Delta H > 0$ for endothermic reactions and $\Delta H < 0$ for exothermic reactions.

SECTION 5.4 Several concepts are useful in discussing the amounts of heat involved in heating and cooling objects at constant pressure. The **heat capacity (C_p)** of an object is the amount of heat required to increase the temperature of the object by 1.00°C. The **molar heat capacity (c)** of a substance is the amount of heat required to increase the temperature of 1 mole by 1.00°C. The **specific heat (c_s)** of an object is the amount of heat required to increase the temperature of 1 gram of the object by 1.00°C. The **molar heat of fusion (ΔH_{fus})** of a substance is the amount of heat required to convert 1 mole of the solid substance to a liquid at the melting point of the substance. The **molar heat of vaporization (ΔH_{vap})** of a substance is the amount of heat required to convert 1 mole of the liquid substance to a vapor at the boiling point of the substance.

SECTION 5.5 A **calorimeter** is a device used to measure the amounts of heat involved in physical and chemical processes. A **bomb calorimeter** is a device used to measure the heat of a combustion reaction (ΔH_{comb}). The quantity of heat associated with a process is called the enthalpy change or **enthalpy of reaction (ΔH_{rxn})**.

SECTION 5.6 Standard conditions are 1 atm of pressure and some specified temperature. The **standard enthalpy of formation** of a substance, ΔH_f°, is the amount of heat involved in a **formation reaction**, in which 1 mole of the substance is made from its constituent elements in their **standard states** (under standard conditions). The values of the standard enthalpies of formation of all elements are defined as zero. Enthalpy changes in processes such as combustion can be calculated from the enthalpies of formation of the reactants and products.

SECTION 5.7 The **fuel value** is the amount of heat released on complete combustion of 1 gram of the fuel. Methane, CH_4 (fuel value 50 kJ/g), has the highest chemical fuel value of the hydrocarbons. **Food value** is the amount of heat released when a material consumed by an organism for sustenance is burned completely; nutritionists often express food values in Calories (kilocalories) rather than the SI-derived unit kilojoules.

SECTION 5.8 **Hess's law** states that the enthalpy change of a reaction that is the sum of two or more reactions is equal to the sum of the enthalpy changes of the constituent reactions. It can be used to calculate the heat changes of reactions that are hard or impossible to measure.

Problem-Solving Summary

TYPE OF PROBLEM	CONCEPTS AND EQUATIONS	SAMPLE EXERCISES
Identifying endothermic and exothermic processes, and calculating internal energy change (ΔE) and $P\Delta V$ work	For the system: $\Delta E = q + w$ (where $w = -P\Delta V$).	5.1–5.3
Determining the value and sign of ΔH associated with changing the temperature of a substance or phase change	Freezing a liquid at its freezing point: $\Delta H = -n\,\Delta H_{fus}$ (where ΔH_{fus} = heat of fusion, and n = number of moles). Heating a liquid: $q = nc_p\Delta T$ (where c_p = heat capacity of the liquid, and ΔT = temperature change).	5.4–5.6

TYPE OF PROBLEM	CONCEPTS AND EQUATIONS	SAMPLE EXERCISES
Measuring the heat capacity (calorimeter constant) of a calorimeter	$C_{calorimeter} = q/\Delta T$ (where $C_{calorimeter}$ = heat capacity of calorimeter, q = amount of heat released by a standard combustion reaction, and ΔT = temperature change of calorimeter).	5.7
Recognizing formation reactions	In a formation reaction, the reactants are elements in their standard states and the product is one mole of a single compound in its standard state.	5.8
Calculating standard heats of reaction from heats of formation	$\Delta H^\circ_{rxn} = \Sigma\, n\, \Delta H^\circ_{f,\,products} - \Sigma\, m\, \Delta H^\circ_{f,\,reactants}$	5.9
Calculating the fuel values of fuels in mass and volume units; calculating food value of foods	The total fuel value or food value of a sample is its mass or volume times the appropriate term defining heat content, such as kJ/unit mass or Cal/unit volume.	5.10–5.11
Calculating enthalpies of reaction using Hess's law	Reorganize the information so that the reactions add together as desired. Reversing a reaction changes the sign of the heat change. Multiplying the coefficients in a reaction by a factor has the same effect on the standard heat change involved.	5.12

Visual Problems

5.1. A brick lies perilously half off the edge of the flat roof of a building (Figure P5.1). The roof edge is 50 feet above street level, and the brick has 500 J of potential energy with respect to street level. Someone (we don't know who) edges the brick off the roof, and it begins to fall. What is the brick's kinetic energy when it is 35 feet above street level? What is its kinetic energy when it hits the street surface?

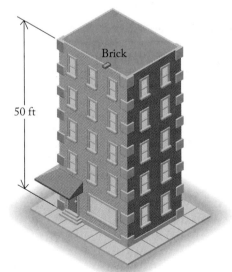

FIGURE P5.1

5.2. Figure P5.2 diagrams pairs of cations and anions in contact. The energy of each interaction is proportional to Q_1Q_2/d, where Q_1 and Q_2 are the charges on the cation and anion, and d is the distance between their nuclei. Which pair has the greatest interaction energy and which has the smallest?

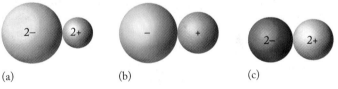

(a)   (b)   (c)

FIGURE P5.2

5.3. Figure P5.3 shows a sectional view of an assembly consisting of a gas trapped inside a stainless steel cylinder with a stainless steel piston and a block of iron on top of the piston to keep it in place. (a) Sketch the situation after the assembly has been heated for a few seconds with a blowtorch. (b) Is the piston higher or lower in the cylinder? (c) Has heat (q) been added to the system? (d) Has the system done work (w) on its surroundings, or have the surroundings done work on the system?

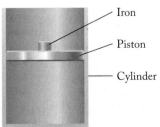

FIGURE P5.3

5.4. Figure P5.4 represents the energy change in a chemical reaction. The reaction results in a very small decrease in volume. (a) Has the internal energy of the system decreased or increased as a result of the reaction? (b) What sign is given to ΔE for the reaction system? (c) Is the reaction endothermic or exothermic?

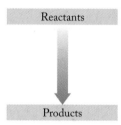

FIGURE P5.4

5.5. The closed rigid metal box lying on the wooden kitchen table in Figure P5.5 is about to be involved in an accident. (a) Are the contents of the box an isolated system, a closed system, or an open system? (b) What will happen to the internal energy of the system if the table catches fire and burns? (c) Will the system do any work on the surroundings while the fire is burning or after the table has burned away and collapsed?

FIGURE P5.5

5.6. The diagram in Figure P5.6 shows how a chemical reaction in a cylinder with a piston affects the volume of the system. (a) In this reaction, does the system do work on the surroundings? (b) If the reaction is endothermic, does the internal energy of the system increase or decrease when the reaction is proceeding?

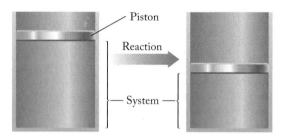

FIGURE P5.6

5.7. The enthalpy diagram in Figure P5.7 indicates the enthalpies of formation of four compounds made from the elements listed on the "zero" line of the vertical axis. (a) Why are the elements all put on the same horizontal line? (b) Why is $C_2H_2(g)$ sometimes called an "endothermic" compound? (c) How could the data be used to calculate the heat of the reaction that converts a stoichiometric mixture of $C_2H_2(g)$ and $O_2(g)$ to $CO(g)$ and $H_2O(g)$?

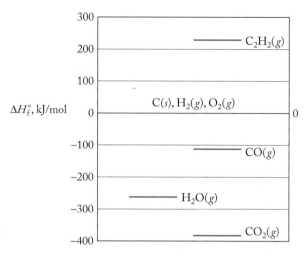

FIGURE P5.7

5.8. The process illustrated in Figure P5.8 takes place at constant pressure. (The figure is not to scale; the molecules are actually much, much smaller than shown here.) (a) Write a balanced equation for the process. (b) Is w positive, negative, or zero for this reaction? (c) Using data from Appendix 4, calculate ΔH°_{rxn} for the formation of 1 mole of the product.

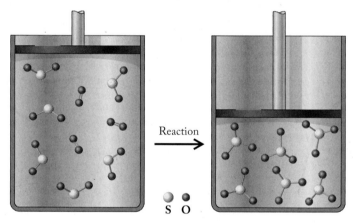

FIGURE P5.8

5.9. Explain how the information in Figure P5.9 can be used to calculate the heat change labeled ΔH_C. Write an equation that corresponds to your answer.

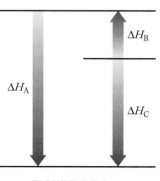

FIGURE P5.9

Questions and Problems

Energy: Basic Concepts and Definitions

CONCEPT REVIEW

5.10. How are energy and work related?

5.11. Explain the difference between potential energy and kinetic energy.

5.12. Explain what is meant by a state function.

5.13. Are kinetic and potential energies both state functions?

5.14. Explain the nature of the potential energy in the following:
a. the new battery for your portable CD player.
b. a gallon of gasoline.
c. the crest of a wave before it crashes onto shore.

5.15. Explain the kinetic energy in a stationary ice cube.

Systems, Surroundings, and the Flow of Energy

CONCEPT REVIEW

5.16. What is meant by the terms *system* and *surroundings*?

5.17. What is the difference between an exothermic process and an endothermic one?

5.18. Give two ways of increasing the internal energy of a gas sample.

5.19. In each of the following processes describe the system and give the sign of q_{system}: (a) combustion of methane; (b) freezing water to make ice; (c) touching a hot stove.

5.20. In each of the following processes describe the system and give the sign of q_{system}: (a) driving an automobile; (b) applying ice to a sprained ankle; (c) cooking a hot dog.

PROBLEMS

5.21. Which of the following processes are exothermic, and which are endothermic? (a) a match burns; (b) a molten metal solidifies; (c) rubbing alcohol feels cold on the skin

5.22. Which of the following processes are exothermic, and which are endothermic? (a) ice cubes solidify in the freezer; (b) water evaporates from a glass left on a windowsill; (c) dew forms on grass overnight

5.23. What happens to the internal energy of a liquid at its boiling point when it vaporizes?

5.24. What happens to the internal energy of a gas when it expands (with no heat flow)?

5.25. How much $P\Delta V$ work does a gas system do on its surroundings at a constant pressure of 1.00 atm if the volume of gas triples from 250.0 mL to 750.0 mL? Express your answer in L · atm and joules (J).

5.26. An expanding gas does 150.0 J of work on its surroundings at a constant pressure of 1.01 atm. If the gas initially occupied 68 mL, what is the final volume of the gas?

5.27. Calculate ΔE for the following situations:
a. $q = 100.0$ J; $w = -50.0$ J
b. $q = 6.2$ kJ; $w = 0.70$ J
c. $q = -615$ J; $w = -325$ J

5.28. Calculate ΔE for a system that absorbs 726 kJ of heat from its surroundings and does 526 kJ of work on its surroundings.

5.29. Calculate ΔE for the combustion of a gas that releases 210.0 kJ of heat to its surroundings and does 65.5 kJ of work on its surroundings.

5.30. Calculate ΔE for a chemical reaction that produces 90.7 kJ of heat but does no work on its surroundings.

*5.31. The following reactions take place in a cylinder equipped with a movable piston at atmospheric pressure (Figure P5.31). Which reactions will result in work being done on the surroundings? Assume the system returns to an initial temperature of 110°C. *Hint:* The volume of a gas is proportional to number of moles (*n*) at constant temperature and pressure.
a. $CH_4(g) + 2 O_2(g) \rightarrow CO_2(g) + 2 H_2O(g)$
b. $C_3H_8(g) + 5 O_2(g) \rightarrow 3 CO_2(g) + 4 H_2O(g)$
c. $N_2(g) + 2 O_2(g) \rightarrow 2 NO_2(g)$

FIGURE P5.31

*5.32. In which direction will the piston described in Figure 5.31 move when the following reactions are carried out at atmospheric pressure inside the cylinder and after the system has returned to its initial temperature of 110°C? *Hint:* The volume of a gas is proportional to number of moles (*n*) at constant temperature and pressure.
a. $N_2(g) + 3 H_2(g) \rightarrow 2 NH_3(g)$
b. $C(s) + O_2(g) \rightarrow CO_2(g)$
c. $C_2H_5OH(g) + 3 O_2(g) \rightarrow 2 CO_2(g) + 3 H_2O(g)$

Enthalpy and Enthalpy Changes

CONCEPT REVIEW

5.33. What is meant by an *enthalpy change*?

5.34. Describe the difference between an internal energy change (ΔE) and an enthalpy change (ΔH).

5.35. Why is the sign of ΔH negative for an exothermic process?

5.36. What happens to the magnitude and sign of the enthalpy change when a process is reversed?

PROBLEMS

5.37. A Clogged Sink Adding Drano to a clogged sink causes the drainpipe to get warm. What is the sign of ΔH for this process?

5.38. A Cold Pack Breaking a small pouch of water inside a larger bag containing ammonium nitrate activates chemical cold packs. What is the sign of ΔH for the process taking place in the cold pack?

5.39. Break a Bond The stable form of oxygen at room temperature and pressure is the diatomic molecule O_2. What is the sign of ΔH for the following process?

$$O_2(g) \rightarrow 2\,O(g)$$

5.40. Plaster of Paris Gypsum is the common name of calcium sulfate dihydrate ($CaSO_4 \cdot 2\,H_2O$). When gypsum is heated to 150°C, it loses most of the water in its formula and forms plaster of Paris ($CaSO_4 \cdot 0.5\,H_2O$):

$$2\,CaSO_4 \cdot 2\,H_2O(s) \rightarrow 2\,CaSO_4 \cdot 0.5\,H_2O(s) + 3\,H_2O(g)$$

What is the sign of ΔH for making plaster of Paris from gypsum?

5.41. Metallic Hydrogen A solid with metallic properties is formed when hydrogen gas is compressed under extremely high pressures. Predict the sign of the enthalpy change for the following reaction:

$$H_2(g) \rightarrow H_2(s)$$

5.42. Kitchen Chemistry A simple "kitchen chemistry" experiment requires placing some vinegar in a soda bottle. A deflated balloon containing baking soda is stretched over the mouth of the bottle. Adding the baking soda to the vinegar starts the following reaction and inflates the balloon:

$$NaHCO_3(aq) + CH_3CO_2H(aq) \rightarrow$$
$$CH_3CO_2Na(aq) + CO_2(g) + H_2O(\ell)$$

If the contents of the bottle are considered the system, is work being done on the surroundings or on the system?

Heating Curves and Heat Capacity

CONCEPT REVIEW

5.43. What is the difference between *specific heat* and *heat capacity*?

5.44. What happens to the heat capacity of a material if its mass is doubled? Is the same true for the specific heat?

5.45. Are the heats of fusion and vaporization of a given substance usually the same?

5.46. An equal amount of heat is added to pieces of metal A and metal B having the same mass. Does the metal with the larger heat capacity reach the higher temperature?

*__5.47.__ Cooling an Automobile Engine** Most automobile engines are cooled by water circulating through them and a radiator. However, the original Volkswagen Beetle had an air-cooled engine. Why might car designers choose water cooling over air cooling?

*__5.48.__ Nuclear Reactor Coolants** The reactor-core cooling systems in some nuclear power plants use liquid sodium as the coolant. Sodium has a thermal conductivity of 1.42 J/(cm · s · K), which is quite high compared with that of water [6.1×10^{-3} J/(cm · s · K)]. The respective molar heat capacities are 28.28 J/(mol · K) and 75.31 J/(mol · K). What is the advantage of using liquid sodium over water in this application?

PROBLEMS

5.49. How much heat is needed to raise the temperature of 100.0 g of water from 30.0°C to 100.0°C?

5.50. 100.0 grams of water at 30°C absorbs 290.0 kJ of heat from a mountain climber's stove at an elevation where the boiling

point of water is 93°C. Is this amount of energy sufficient to heat the water to its boiling point?

5.51. Use the following data to sketch a heating curve for 1 mole of methanol. Start the curve at −100°C and end it at 100°C.

Boiling point	65°C
Melting point	−94°C
ΔH_{vap}	37 kJ/mol
ΔH_{fus}	3.18 kJ/mol
Molar heat capacity (ℓ)	81.1 J/(mol · °C)
(g)	43.9 J/(mol · °C)
(s)	48.7 J/(mol · °C)

5.52. Use the following data to sketch a heating curve for 1 mole of octane. Start the curve at −57°C and end it at 150°C.

Boiling point	125.7°C
Melting point	−56.8°C
ΔH_{vap}	41.5 kJ/mol
ΔH_{fus}	20.7 kJ/mol
Molar heat capacity (ℓ)	254.6 J/(mol · °C)
(g)	316.9 J/(mol · °C)

5.53. Keeping an Athlete Cool During a strenuous workout, an athlete generates 2000.0 kJ of heat energy. What mass of water would have to evaporate from the athlete's skin to dissipate this much heat?

5.54. The same quantity of energy is added to 10.00 g pieces of gold, magnesium, and platinum, all initially at 25°C. The molar heat capacities of these three metals are 25.41 J/(mol · °C), 24.79 J/(mol · °C), and 25.95 J/(mol · °C), respectively. Which piece of metal has the highest final temperature?

*__5.55.__ Exactly 10.0 mL of water at 25.0°C are added to a hot iron skillet. All of the water is converted into steam at 100.0°C. The mass of the pan is 1.20 kg and the molar heat capacity of iron is 25.19 J/(mol · °C). What is the temperature change of the skillet?

*__5.56.__ A 20.0 g piece of iron and a 20.0 g piece of gold at 100.0°C were dropped into 1.00 L of water at 20.0°C. The molar heat capacities of iron and gold are 25.19 J/(mol · °C) and 25.41 J/(mol · °C), respectively. What is the final temperature of the water and pieces of metal?

Calorimetry: Measuring Heat Capacity and Calorimeter Constants

CONCEPT REVIEW

5.57. Why is it necessary to know the heat capacity of a calorimeter?

5.58. Could an endothermic reaction be used to measure the heat capacity of a calorimeter?

5.59. If we replace the water in a bomb calorimeter with another liquid, do we need to redetermine the heat capacity of the calorimeter?

5.60. When measuring the heat of combustion of a very small amount of material, would you prefer to use a calorimeter having a heat capacity that is small or large?

PROBLEMS

5.61. Calculate the heat capacity of a calorimeter if the combustion of 5.000 g of benzoic acid led to a temperature increase of 16.397°C.

5.62. Calculate the heat capacity of a calorimeter if the combustion of 4.663 g of benzoic acid led to an increase in temperature of 7.149°C.

5.63. The complete combustion of 1.200 g of cinnamaldeyde (C_9H_8O, one of the compounds in cinnamon) in a bomb calorimeter ($C_{calorimeter}$ = 3.640 kJ/°C) produced an increase in temperature of 12.79°C. Calculate the molar enthalpy of combustion of cinnamaldehyde (ΔH_{comb}) in kilojoules per mole of cinnamaldehyde.

5.64. Aromatic Spice The aromatic hydrocarbon cymene ($C_{10}H_{14}$) is found in nearly 100 spices and fragrances including coriander, anise, and thyme. The complete combustion of 1.608 g of cymene in a bomb calorimeter ($C_{calorimeter}$ = 3.640 kJ/°C) produced an increase in temperature of 19.35°C. Calculate the molar enthalpy of combustion of cymene (ΔH_{comb}) in kilojoules per mole of cymene.

5.65. Phthalate Plasticizers Phthalates used as plasticizers in rubber and plastic products are believed to act as hormone mimics in humans. The value of $\Delta H°_{comb}$ for dimethylphthalate ($C_{10}H_{10}O_4$) is 4685 kJ/mol. Assume that 1.00 g of dimethylphthalate is combusted in a calorimeter whose heat capacity ($C_{calorimeter}$) is 7.854 kJ/°C at 20.215°C. What is the final temperature of the calorimeter?

5.66. Flavorings The flavor of anise is due to anethole, a compound with the molecular formula $C_{10}H_{12}O$. The ΔH_{comb} value for anethole is 5541 kJ/mol. Assume 0.950 g of anethole is combusted in a calorimeter whose heat capacity ($C_{calorimeter}$) is 7.854 kJ/°C at 20.611°C. What is the final temperature of the calorimeter?

Enthalpies of Formation and Enthalpies of Reaction

CONCEPT REVIEW

5.67. Oxygen and ozone are both forms of elemental oxygen. Are the standard enthalpies of formation of oxygen and ozone the same?

5.68. Why are the standard enthalpies of formation of elements in their standard states assigned a value of zero?

PROBLEMS

5.69. For which of the following reactions does $\Delta H°_{rxn}$ represent an enthalpy of formation?
a. $C(s) + O_2(g) \rightarrow CO_2(g)$
b. $CO_2(g) + C(s) \rightarrow 2\, CO(g)$
c. $CO_2(g) + H_2(g) \rightarrow H_2O(g) + CO(g)$
d. $2\, H_2(g) + C(s) \rightarrow CH_4(g)$

5.70. For which of the following reactions does $\Delta H°_{rxn}$ also represent an enthalpy of formation?
a. $2\, N_2(g) + 3\, O_2(g) \rightarrow 2\, NO_2(g) + 2\, NO(g)$
b. $N_2(g) + O_2(g) \rightarrow 2\, NO(g)$
c. $2\, NO_2(g) \rightarrow N_2O_4(g)$
d. $N_2(g) + 2\, O_2(g) \rightarrow 2\, NO_2(g)$

5.71. Methanogenesis Use enthalpies of formation to calculate the standard enthalpy of reaction for the following methane-generating reaction of methanogenic bacteria:

$$4\, H_2(g) + CO_2(g) \rightarrow CH_4(g) + 2\, H_2O(g)$$

5.72. Methanogenesis Use enthalpies of formation to calculate the standard enthalpy of reaction for the following methane-generating reaction of methanogenic bacteria, given ΔH_f of $CH_3NH_2(g)$ = −22.97 J/mol:

$$4\, CH_3NH_2(g) + 2\, H_2O(\ell) \rightarrow$$
$$3\, CH_4(g) + CO_2(g) + 4\, NH_3(g)$$

5.73. Ammonium nitrate decomposes to N_2O and water vapor at temperatures between 250 and 300°C. Write a balanced chemical reaction describing the decomposition of ammonium nitrate, and calculate the standard heat of reaction using the appropriate enthalpies of formation.

5.74. Above 300°C, ammonium nitrate decomposes to N_2, O_2, and H_2O. Write a balanced chemical reaction describing the decomposition of ammonium nitrate, and determine the standard heat of reaction by using the appropriate enthalpies of formation.

5.75. Explosives Mixtures of fertilizer (ammonium nitrate) and fuel oil (a mixture of long-chain hydrocarbons similar to decane, $C_{10}H_{22}$) are the basis for a powerful explosion. Determine the standard enthalpy change of the following explosive reaction by using the appropriate enthalpies of formation ($\Delta H°_{f,C_{10}H_{22}}$ = 249.7 kJ/mol):

$$3\, NH_4NO_3(s) + C_{10}H_{22}(\ell) + 14\, O_2(g) \rightarrow$$
$$3\, N_2(g) + 17\, H_2O(g) + 10\, CO_2(g)$$

*5.76. **A Little TNT** Trinitrotoluene (TNT) is a highly explosive compound. The thermal decomposition of TNT is described by the following chemical equation:

$$2\, C_7H_5N_3O_6(s) \rightarrow 12\, CO(g) + 5\, H_2(g) + 3\, N_2(g) + 2\, C(s)$$

If ΔH_{rxn} for this reaction is −10,153 kJ/mol, how much TNT is needed to equal the explosive power of 1 mol of ammonium nitrate in Problem 5.75?

Fuel Values and Food Values

CONCEPT REVIEW

5.77. What is meant by *fuel value*?

5.78. What are the units of fuel values?

5.79. How are fuel values calculated from molar heats of combustion?

5.80. Is the fuel value of liquid propane the same as that of propane gas?

PROBLEMS

5.81. If all the energy obtained from burning 1.00 pound of propane is used to heat water, how many kilograms of water can be heated from 20.0°C to 45.0°C?

5.82. A 1995 article in *Discover* magazine on world-class sprinters contained the following statement: "In one race, a field of eight runners releases enough energy to boil a gallon jug of ice at 0.0°C in ten seconds!" How much "energy" do the runners release in 10 seconds? Assume that the ice weighs 128 ounces.

5.83. The Joys of Camping Lightweight camping stoves typically use *white gas*, a mixture of C_5 and C_6 hydrocarbons.
  a. Calculate the fuel value of C_5H_{12}, given that $\Delta H^\circ_{comb} = -3535$ kJ/mol.
  b. How much heat is released during the combustion of 1.00 kg of C_5H_{12}?
  c. How many grams of C_5H_{12} must be burned to heat 1.00 kg of water from 20.0°C to 90.0°C? Assume that all the heat released during combustion is used to heat the water.

5.84. The heavier hydrocarbons in white gas are hexanes (C_6H_{14}).
  a. Calculate the fuel value of C_6H_{14}, given that $\Delta H^\circ_{comb} = -4163$ kJ/mol.
  b. How much heat is released during the combustion of 1.00 kg of C_6H_{14}?
  c. How many grams of C_6H_{14} are needed to heat 1.00 kg of water from 25.0°C to 85.0°C? Assume that all of the heat released during combustion is used to heat the water.
  d. Assume white gas is 25% C_5 hydrocarbons and 75% C_6 hydrocarbons; how many grams of white gas are needed to heat 1.00 kg of water from 25.0°C to 85.0°C?

Hess's Law

CONCEPT REVIEW

5.85. How is Hess's law consistent with the law of conservation of energy?

5.86. Why is the enthalpy of formation of $CO(g)$ difficult to measure experimentally?

5.87. Explain how the use of ΔH°_f to calculate ΔH°_{rxn} is an example of Hess's law.

5.88. Why is it important for Hess's law that enthalpy is a state function?

PROBLEMS

5.89. How can the first two of the following reactions be combined to obtain the third reaction?
  a. $CO(g) + \frac{1}{2}O_2(g) \rightarrow CO_2(g)$
  b. $C(s) + O_2(g) \rightarrow CO_2(g)$
  c. $C(s) + \frac{1}{2}O_2(g) \rightarrow CO(g)$

5.90. How can the enthalpy of formation of $CO(g)$ be calculated from the enthalpy of formation of $CO_2(g)$ and the heat of combustion of $CO(g)$?

5.91. Calculate the enthalpy of formation of $SO_2(g)$ from the standard enthalpy changes of the following reactions:

$2 SO_2(g) + O_2(g) \rightarrow 2 SO_3(g)$ $\Delta H^\circ_{rxn} = -196$ kJ

$\frac{1}{4} S_8(s) + 3 O_2(g) \rightarrow 2 SO_3(g)$ $\Delta H^\circ_{rxn} = -790$ kJ

$\frac{1}{4} S_8(s) + O_2(g) \rightarrow SO_2(g)$ $\Delta H^\circ_f = ?$

5.92. Ozone Layer The destruction of the ozone layer by chlorofluorocarbons (CFCs) can be described by the following reactions:

$ClO(g) + O_3(g) \rightarrow Cl(g) + 2 O_2(g)$ $\Delta H^\circ_{rxn} = -29.90$ kJ

$2 O_3(g) \rightarrow 3 O_2(g)$ $\Delta H^\circ_{rxn} = 24.18$ kJ

Determine the value of heat of reaction for the following:

$Cl(g) + O_3(g) \rightarrow ClO(g) + O_2(g)$ $\Delta H^\circ = ?$

5.93. The mineral spodumene ($LiAlSi_2O_6$) exists in two crystalline forms called α and β. Use Hess's law and the following information to calculate ΔH°_{rxn} for the conversion of α-spodumene into β-spodumene:

$Li_2O(s) + 2 Al(s) + 4 SiO_2(s) + \frac{3}{2}O_2(g) \rightarrow$
$\qquad\qquad 2 \alpha\text{-}LiAlSi_2O_6(s)$ $\Delta H^\circ = -1870.6$ kJ

$Li_2O(s) + 2 Al(s) + 4 SiO_2(s) + \frac{3}{2}O_2(g) \rightarrow$
$\qquad\qquad 2 \beta\text{-}LiAlSi_2O_6(s)$ $\Delta H^\circ = -1814.6$ kJ

5.94. Use the following data to determine whether the conversion of diamond into graphite is exothermic or endothermic:

$C_{diamond}(s) + O_2(g) \rightarrow CO_2(g)$ $\Delta H^\circ = -395.4$ kJ

$2 CO_2(g) \rightarrow 2 CO(g) + O_2(g)$
$\qquad\qquad\qquad \Delta H^\circ = 566.0$ kJ

$2 CO(g) \rightarrow C_{graphite}(s) + CO_2(g)$
$\qquad\qquad\qquad \Delta H^\circ = -172.5$ kJ

$C_{diamond}(s) \rightarrow C_{graphite}(s)$ $\Delta H^\circ = ?$

5.95. You are given the following data:

$\frac{1}{2}N_2(g) + \frac{1}{2}O_2(g) \rightarrow NO(g)$ $\Delta H^\circ_{rxn} = -90.3$ kJ

$NO(g) + \frac{1}{2}Cl_2(g) \rightarrow NOCl(g)$ $\Delta H^\circ_{rxn} = -38.6$ kJ

$2 NOCl(g) \rightarrow N_2(g) + O_2(g) + Cl_2(g)$
$\qquad\qquad\qquad \Delta H^\circ_{rxn} = ?$

  a. Which of the ΔH_{rxn} values represent enthalpies of formation?
  b. Determine ΔH_{rxn} for the decomposition of NOCl.

5.96. The enthalpy of decomposition of NO_2Cl is −114 kJ. Use the following data to calculate the heat of formation of NO_2Cl from N_2, O_2, and Cl_2:

$NO_2Cl(g) \rightarrow NO_2(g) + \frac{1}{2}Cl_2(g)$
$\qquad\qquad\qquad \Delta H^\circ_{rxn} = -114$ kJ

$\frac{1}{2}N_2(g) + O_2(g) \rightarrow NO_2(g)$ $\Delta H^\circ_f = +33.2$ kJ

$\frac{1}{2}N_2(g) + O_2(g) + \frac{1}{2}Cl_2(g) \rightarrow NO_2Cl(g)$ $\Delta H^\circ_f = ?$

Additional Problems

*5.97. Laundry forgotten and left outside to dry on a clothesline in the winter slowly dries by "ice vaporization" (sublimation). The increase in internal energy of water vapor produced by sublimation is less that the amount of heat absorbed. Explain.

5.98. Chlorofluorocarbons (CFCs) such as CF_2Cl_2 are refrigerants whose use has been phased out because of their destructive effect on Earth's ozone layer. The standard enthalpy of evaporation of CF_2Cl_2 is 17.4 kJ/mol, compared with $\Delta H^\circ_{vap} = 41$ kJ/mol for liquid water. How many grams of liquid CF_2Cl_2 are needed to cool 200.0 g of water from 50.0 to 40.0°C? The specific heat of water is 4.184 J/(g · °C).

5.99. A 100.0 mL sample of 1.0 *M* NaOH is mixed with 50.0 mL of 1.0 *M* H_2SO_4 in a large Styrofoam coffee cup; the cup is fitted with a lid through which passes a calibrated thermometer. The temperature of each solution before mixing is 22.3°C. After adding the NaOH solution to the coffee cup and stirring the mixed solutions with the thermometer, the maximum temperature measured is

31.4°C. Assume that the density of the mixed solutions is 1.00 g/mL, that the specific heat of the mixed solutions is 4.18 J/(g · °C), and that no heat is lost to the surroundings. (a) Write a balanced chemical equation for the reaction that takes place in the Styrofoam cup. (b) Is any NaOH or H_2SO_4 left in the Styrofoam cup when the reaction is over? (c) Calculate the enthalpy change per mole of H_2SO_4 in the reaction.

5.100. Varying the scenario in Problem 5.99 assumes this time that 65.0 mL of 1.0 M H_2SO_4 are mixed with 100.0 mL of 1.0 M NaOH and that both solutions are initially at 25.0°C. Assume that the mixed solutions in the Styrofoam cup have the same density and specific heat as in Problem 5.99 and that no heat is lost to the surroundings. What is the maximum measured temperature in the Styrofoam cup?

*5.101. An insulated container is used to hold 50.0 g of water at 25.0°C. A sample of copper weighing 7.25 g is placed in a dry test tube and heated for 30 minutes in a boiling water bath at 100.1°C. The heated test tube is carefully removed from the water bath with laboratory tongs and inclined so that the copper slides into the water in the insulated container. Given that the specific heat of solid copper is 0.385 J/(g · °C), calculate the maximum temperature of the water in the insulated container after the copper metal is added.

5.102. The mineral magnetite (Fe_3O_4) is magnetic, whereas iron(II) oxide is not. (a) Write and balance the chemical equation for the formation of magnetite from iron(II) oxide and oxygen. (b) Given that 318 kJ of heat are released for each mole of Fe_3O_4 formed, what is the enthalpy change of the balanced reaction of formation of Fe_3O_4 from iron(II) oxide and oxygen?

5.103. Which of the following substances has a standard heat of formation ΔH_f° of zero? (a) Pb at 1000°C; (b) C_3H_8 (g) at 25.0°C and 1 atm pressure; (c) solid glucose at room temperature; (d) $N_2(g)$ at 25.0°C and 1 atm pressure

*5.104. The standard molar heat of formation of water is −285.8 kJ/mol. (a) What is the significance of the negative sign associated with this value? (b) Why is the magnitude of this value so much larger than the heat of vaporization of water ($\Delta H_{vap}^\circ = 41$ kJ/mol)? (c) Calculate the amount of heat produced in making 50.0 mL of water from its elements under standard conditions.

5.105. Endothermic compounds are unusual: they have positive heats of formation. An example is acetylene, C_2H_2 ($\Delta H_f^\circ = 226.7$ kJ/mol). The combustion of acetylene is used to melt and weld steel. Use Appendix 4 to answer the following questions. (a) Give the chemical formulas and names of three other endothermic compounds. (b) Calculate the standard molar heat of combustion of acetylene.

*5.106. Balance the following chemical equation, name the reactants and products, and calculate the standard enthalpy change using the data in Appendix 4.

$$FeO(s) + O_2(g) \rightarrow Fe_2O_3(s)$$

*5.107. Add reactions 1 and 2, and label the resulting reaction 3. Consult Appendix 4 to find the standard enthalpy change for balanced reaction 3.

(1) $Zn(s) + \frac{1}{8} S_8(s) \rightarrow ZnS(s)$

(2) $ZnS(s) + 2 O_2(g) \rightarrow ZnSO_4(s) + SO_2(g)$

5.108. Conversion of 0.90 g of liquid water to steam at 100.0°C requires 2.0 kJ of heat. Calculate the molar enthalpy of evaporation of water at 100.0°C.

5.109. The specific heat of solid copper is 0.385 J/(g · °C). What thermal energy change occurs when the temperature of a 35.3 g sample of copper is cooled from 35.0°C to 15.0°C? Be sure to give your answer the proper sign. This amount of heat is used to melt solid ice at 0.0°C. The molar heat of fusion of ice is 6.01 kJ/mol. How many moles of ice are melted?

*5.110. Methanol decomposes according to the following reaction:

$$90.2 \text{ kJ} + CH_3OH(g) \rightarrow CO(g) + 2 H_2(g)$$

a. Is this reaction endothermic or exothermic?
b. What is the value of ΔH_{rxn} for this reaction?
c. How much heat would be absorbed or released if 60.0 g of methanol were decomposed in this reaction?
d. Would you expect ΔH_{rxn} for the decomposition of $CH_3OH(\ell)$ to give $CO(g)$ and $H_2(g)$ to be larger or smaller than 128 kJ?

5.111. Using the information given below, write an equation for the enthalpy change in reaction 3:

(1) $B(g) + A(s) \rightarrow C(g)$ ΔH_1

(2) $C(g) \rightarrow C(s)$ ΔH_2

(3) $C(s) \rightarrow A(s) + B(g)$ ΔH_3

5.112. From the information given below, write an equation for the enthalpy change in reaction 3:

(1) $F(g) \rightarrow D(s) + E(g)$ ΔH_1

(2) $F(s) \rightarrow F(g)$ ΔH_2

(3) $D(s) + E(g) \rightarrow F(s)$ ΔH_3

5.113. The reaction of $CH_3OH(g)$ with $N_2(g)$ to give $HCN(g)$ and $NH_3(g)$ requires 164 kJ/mole of heat.
a. Write a balanced chemical equation for this reaction.
b. Should the thermal energy involved be written as a reactant or as a product?
c. How much heat is involved in the reaction of 60.0 g of $CH_3OH(g)$ with excess $N_2(g)$ to give $HCN(g)$ and $NH_3(g)$ in this reaction?

5.114. Calculate ΔH_{rxn}° for the reaction

$$2 \text{ Ni}(s) + \frac{1}{4} S_8(s) + 3 O_2(g) \rightarrow 2 \text{ NiSO}_3(s)$$

from the following information:

(1) $NiSO_3(s) \rightarrow NiO(s) + SO_2(g)$ $\Delta H_{rxn}^\circ = 156$ kJ

(2) $\frac{1}{8} S_8(s) + O_2(g) \rightarrow SO_2(g)$ $\Delta H_{rxn}^\circ = -297$ kJ

(3) $Ni(s) + \frac{1}{2} O_2(g) \rightarrow NiO(s)$ $\Delta H_{rxn}^\circ = -241$ kJ

5.115. Use the following information to calculate the amount of heat involved in the complete reaction of 3.0 g of carbon to form $PbCO_3$ (s) in reaction 4. Be sure to give the proper sign (positive or negative) with your answer.

(1) $Pb(s) + \frac{1}{2} O_2(g) \rightarrow PbO(s)$ $\Delta H_{rxn}^\circ = -219$ kJ

(2) $C(s) + O_2(g) \rightarrow CO_2(g)$ $\Delta H_{rxn}^\circ = -394$ kJ

(3) $PbCO_3(s) \rightarrow PbO(s) + CO_2(g)$ $\Delta H_{rxn}^\circ = 86$ kJ

(4) $Pb(s) + C(s) + 1.5 O_2(g) \rightarrow PbCO_3(s)$

*5.116. **Ethanol as Automobile Fuel** Brazilians are quite familiar with fueling their automobiles with ethanol, a fermentation product from sugarcane. Calculate the standard molar enthalpy for the complete combustion of liquid ethanol using the standard enthalpies of formation of the reactants and products as given in Appendix 4.

5.117. **Formation of CO$_2$** Baking soda decomposes on heating as follows, creating the holes in baked bread:

$$2 \, NaHCO_3(s) \rightarrow Na_2CO_3(s) + CO_2(g) + H_2O(\ell)$$

Calculate the standard molar enthalpy of formation of NaHCO$_3(s)$ from the following information:

$$\Delta H^\circ_{rxn} = -129.3 \text{ kJ} \quad \Delta H^\circ_f [Na_2CO_3(s)] = -1131 \text{ kJ/mol}$$

$$\Delta H^\circ_f [CO_2(g)] = -394 \text{ kJ/mol} \quad \Delta H^\circ_f [H_2O(\ell)] = -286 \text{ kJ/mol}$$

*5.118. **Specific Heats of Metals** In 1819, Pierre Dulong and Alexis Petit reported that the product of the atomic mass of a metal times its specific heat is approximately constant, an observation called the *law of Dulong and Petit*. Use the following data to answer the following questions.

Element	$\mathcal{M}$ (g/mol)	c_s [J/(g · °C)]	$\mathcal{M} \times c_s$
Bismuth		0.120	
Lead	207.2	0.123	25.5
Gold	197.0	0.125	
Platinum	195.1		
Tin	118.7	0.215	
		0.233	
Zinc	65.4	0.388	
Copper	63.5	0.397	
		0.433	
Iron	55.8	0.460	
Sulfur	32.1		
		Average value:	

a. Complete each row in the table by multiplying each given molar mass and specific heat pair (one result has been entered in the table). What are the units of the resulting values in column 4?
b. Next, calculate the average of the values in column 4.
c. Use the mean value from part b to calculate the missing atomic masses in the table. Do you feel confident in identifying the element from the calculated atomic mass?
d. Use the average value from part (b) to predict the missing specific heat values in the table.

*5.119. **Unpleasant Smells** Urine odor gets worse with time because urine contains the metabolic product urea [CO(NH$_2$)$_2$], a compound that is slowly converted to ammonia and carbon dioxide:

$$CO(NH_2)_2(aq) + H_2O(\ell) \rightarrow CO_2(aq) + 2 \, NH_3(aq)$$

This reaction is much too slow for the enthalpy change to be measured directly using a temperature change. Instead, the enthalpy change for the reaction may be calculated from the following data:

Compound	ΔH°_f(kJ/mol)
Urea(aq)	−319.2
CO$_2$(aq)	−412.9
H$_2$O(ℓ)	−285.8
NH$_3$(aq)	−80.3

Calculate the standard molar enthalpy for the reaction.

5.120. **Experiment with a Metal** Explain how the specific heat of a metal sample could be measured in the lab. *Hint*: you'll need a test tube, a boiling water bath, a Styrofoam cup calorimeter containing a weighed amount of water, a calibrated thermometer, and a weighed sample of the metal.

*5.121. **Rocket Fuels** The payload of a rocket includes a fuel and oxygen for combustion of the fuel. Reactions 1 and 2 describe the combustion of dimethylhydrazine and hydrogen, respectively. Pound for pound, which is the better rocket fuel, dimethylhydrazine or hydrogen?

(1) $(CH_3)_2NNH_2(\ell) + 4 \, O_2(g) \rightarrow$
$N_2(g) + 4 \, H_2O(g) + 2 \, CO_2(g) \quad \Delta H^\circ_{rxn} = -1694 \text{ kJ}$

(2) $H_2(g) + \frac{1}{2} O_2(g) \rightarrow H_2O(g) \qquad \Delta H^\circ_{rxn} = -286 \text{ kJ}$

Carbon: Diamonds, Graphite, and the Molecules of Life

*5.122. **Industrial Use of Cellulose** Research is being carried out on cellulose as a source of chemicals for the production of fibers, coatings, and plastics. Cellulose consists of long chains of glucose molecules (C$_6$H$_{12}$O$_6$), so for the purposes of modeling the reaction we can consider the conversion of glucose to formaldehyde.
a. Is the reaction to convert glucose into formaldehyde an oxidation or a reduction?
b. Calculate the heat of reaction for the conversion of 1 mole of glucose into formaldehyde, given the following thermochemical data:

$$\Delta H^\circ_{comb} \text{ of formaldehyde gas} \qquad -572.9 \text{ kJ/mol}$$

$$\Delta H^\circ_f \text{ of solid glucose} \qquad -1274.4 \text{ kJ/mol}$$

$$C_6H_{12}O_6(s) \rightarrow 6 \, CH_2O(g)$$

$$\text{Glucose} \qquad \text{Formaldehyde}$$

5.123. **Converting Diamond to Graphite** The standard state of carbon is graphite. $\Delta H^\circ_{f, diamond}$ is +1.896 kJ/mol. Diamonds are normally weighed in units of carats, where 1 carat = 0.20 g. Determine the standard enthalpy of the reaction for the conversion of a 4-carat diamond into an equivalent mass of graphite. Is this reaction endothermic or exothermic?

Properties of Gases: The Air We Breathe

SCUBA DIVER AND SEA GOLDIES *The gas exhaled by the diver forms bubbles that increase in size as they rise to the surface.*

A LOOK AHEAD: An Invisible Necessity

Gases are easy to overlook. We do not think much about breathing, but the exchange of gases between our lungs and the surrounding atmosphere is crucial to our lives. One reason we may not notice gases like oxygen—a colorless, odorless, tasteless, invisible gas—is that we do not have to pay for the oxygen needed to stay alive. We pay for the fuel to drive cars, but the air required to burn the fuel is free, so we rarely think about it. For every gallon of gasoline an engine burns, it consumes the oxygen in an average-sized room. The same quantity of air would supply the oxygen needed for an entire day by three people breathing normally. In a hospital operating room, managing the delicate balance of gases entering and leaving a patient can mean the difference between transitioning to a normal conscious life or irreversible coma. Having the right mixture of gases in our bodies is so important that during surgery, anesthesiologists constantly and carefully monitor blood levels of oxygen and carbon dioxide to ensure that safe levels of these gases are maintained.

Life as we know it in our biopshere requires oxygen. Fish, insects, birds, mammals, and even plants must bring $O_2(g)$ into their systems and expel $CO_2(g)$. All the food we eat depends ultimately on plants' use of sunlight; during photosynthesis they combine carbon dioxide and water vapor to form sugars that contain the chemical energy required to support most of the life on our planet.

The thin shell of gases that surrounds Earth and forms its atmosphere is a unique blend not found on other planets in our solar system. Mercury, the planet closest to the sun, has no significant atmosphere. The atmosphere of Venus consists mostly of carbon dioxide (96%) and nitrogen (3%) and features thick clouds of sulfuric acid. The atmospheric pressure on Venus is 90 times that on Earth. The Earth's early atmosphere was modified to its current composition (21% oxygen and 78% nitrogen) when plant life appeared and began to give off oxygen as a product of photosynthesis. Mars' atmosphere, with a pressure only about 1/100 of Earth's, is composed of carbon dioxide (95%), nitrogen (3%), and less than 0.2% oxygen. Beyond Mars, the gas giants—Jupiter, Saturn, Uranus, and Neptune—are essentially huge balls of gas with frozen cores. The atmosphere of each gas giant is mostly hydrogen, helium, and small amounts of methane and other hydrocarbons, with ammonia and hydrogen sulfide particles suspended by perpetual winds. Only Earth has significant free oxygen in its atmosphere.

Our dependence on our atmosphere is so great that if we venture very far away from Earth's surface, we have to take the right blend of gases with us to sustain life. Hence, a commercial airplane has a pressurized cabin, an undersea explorer carries a mixture of compressed gases in a scuba tank, and a mountain climber takes canisters of oxygen to breathe when ascending far above sea level.

Because of their physical properties as well as their chemical reactivity, gases are used in many ways that touch our lives. Our ability to understand how gases fit into our lives—whether in living rooms, scuba tanks, tires, airplanes, or balloons—is based on our knowledge of how gases behave in response to changing volume, temperature, and pressure—the subject of this chapter.

6.1 The Gas Phase

Earth is surrounded by a layer of gases 50 km thick. By volume it is composed primarily of nitrogen (78%), oxygen (21%), and lesser amounts of other gases (Table 6.1). To put the thickness of the atmosphere in perspective, if Earth were

TABLE 6.1 Composition of Earth's Atmosphere

Compound	% (by volume)	Mole Fraction[a]
Nitrogen	78.08	0.7808
Oxygen	20.95	0.2095
Argon	0.934	0.00934
Carbon dioxide	0.033	0.00033
Methane	2×10^{-4}	2×10^{-6}
Hydrogen	5×10^{-5}	5×10^{-7}

[a]Mole fraction is defined in Section 6.6.

the size of an apple, the atmosphere would be about as thick as the apple's skin. This layer of gases, which is held around our planet by gravity, exerts pressure on Earth's surface and on all sides of objects on the surface. This pressure decreases with elevation above sea level.

As discussed in Chapter 1, gases have neither a definite volume nor a definite shape and expand to occupy the entire volume and shape of their container. This property is only one of the differences between gases and other states of matter. Let's review some of the other properties of gases that distinguish them from liquids and solids under conditions we might encounter in our everyday lives:

CONNECTION The physical properties of gases were introduced in Chapter 1.

1. Unlike liquids and solids, the volume occupied by a gas changes significantly with pressure. If we were to carry a balloon from sea level (0 m) to the top of a 1600 m (5249 ft) mountain, the volume of the balloon would increase by about 20%. The volume of a liquid or solid would be unchanged under these conditions.

2. The volume of a gas also changes with temperature. For example, a balloon filled with room-temperature air decreases in size when it is taken outside on a cold winter's day. A temperature decrease of 20°C leads to a volume decrease of about 7%, whereas the volume of a liquid or solid remains practically unchanged by such modest temperature changes.

3. Gases are *miscible* with each other, which means that they can be mixed together in any proportion (unless they chemically react with each other). A hospital patient experiencing respiratory difficulties may be given a mixture of nitrogen and oxygen in which the relative amount of oxygen is much higher than its proportion in the air we ordinarily breathe. A scuba diver may leave the ocean's surface with a tank of air containing 17% oxygen, 34% nitrogen, and 49% helium, which form a homogeneous mixture. In contrast, many pairs of liquids are immiscible, such as the oil and water in salad dressing.

4. Gases are typically much less dense than liquids or solids. The density of dry air at 20°C at typical atmospheric pressures is 1.20 g/L. Notice that the density of gases is expressed in units of grams per *liter*. The density of liquids, such as water, is expressed in units of grams per *milliliter*. Thus, the density of liquid water, 0.9986 g/mL (998.6 g/L) at 20°C, is over 800 times greater than the density of dry air.

These four observations about gases are consistent with the idea that the molecules or atoms are further apart in gases than in solids and liquids. The larger spaces between molecules in air, for example, account for the ready compressibility of air

Pressure (*P*) is defined as the ratio of force to surface area.

Atmospheric pressure (*P*atm) is the force exerted by the gases surrounding Earth on Earth's surface and on all surfaces of all objects.

A **barometer** is an instrument that measures atmospheric pressure.

The **standard atmosphere (1 atm)** is the pressure capable of supporting a column of mercury 760 mm high in a barometer. It is also called simply 1 atmosphere.

Pressure is measured in **millimeters of mercury (mmHg)**, also known as **torr**, where 1 atm = 760 mmHg = 760 torr.

into cylinders used by scuba divers and into tank trucks that transport gases for industrial use. Greater distances between molecules account for the lower densities and observed miscibility of gases. We will explore each of these general properties of gases in more detail, including mathematical relations that allow us to determine exactly how pressure and temperature affect volume and density.

6.2 Atmospheric Pressure

The gases in Earth's atmosphere are pulled toward the surface of the planet by gravity. The resulting weight of these gases exerts a force that is spread across the surface (Figure 6.1). The ratio of the force (*F*) to surface area (*A*) is called **pressure (*P*)**, and the force exerted by the gases on Earth's surface is called **atmospheric pressure (*P*atm)**:

$$P = \frac{F}{A} \tag{6.1}$$

Atmospheric pressure is routinely measured with an instrument called a **barometer**. A simple but effective barometer design is shown in Figure 6.2. It consists of a tube nearly a meter long and closed at one end. The tube is filled with mercury, and its open end is inverted in a pool of mercury open to the atmosphere. Gravity pulls the mercury in the tube downward, creating a vacuum at the top of the tube, but atmospheric pressure pushes the mercury in the pool up into the tube. The net effect of these opposing forces is indicated by the height of the mercury in the tube, which provides a measure of atmospheric pressure.

Atmospheric pressure varies from place to place and with changing weather conditions. For reference purposes, the **standard atmosphere (1 atm)** of pressure is the pressure capable of supporting a column of mercury 760 mm high in a barometer. The pressure unit **millimeters of mercury (mmHg)** is also called the **torr** in honor of Evangelista Torricelli (1608–1647), the Italian mathematician and physicist who invented the barometer. Thus we have the following relation:

$$1 \text{ atm} = 760 \text{ mmHg} = 760 \text{ torr}$$

The SI unit of pressure is the *pascal (Pa)*, named in honor of French mathematician and physicist Blaise Pascal (1623–1662). He was the first to propose that atmospheric pressure decreases with increasing altitude. We can explain this phenomenon by noting that pressure is related to the total mass of molecules of gas *above* a given altitude (Figure 6.3). As altitude increases, the mass of the gases remaining above it must decrease, and less mass means less pressure.

The pascal is an example of a derived SI unit; it is defined using the following SI base units (see Table 1.1):

$$1 \text{ Pa} = \frac{1 \text{ kg}}{\text{m} \cdot \text{s}^2}$$

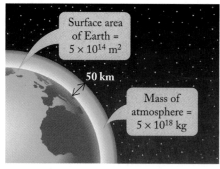

(a)

FIGURE 6.1 (a) Atmospheric pressure results from the force exerted by the atmosphere acting on the surface of Earth. (b) If you stretch out your hand, palm facing upward, the mass of the column of air above your palm is about 100 kg (220 lb). This textbook weighs about 5 lb, so the mass of the atmosphere on your palm is equivalent to about 40 textbooks.

(b)

To understand the logic of this combination of units, consider the law in physics that states that pushing an object of mass (m) with a force (F) causes the object to accelerate at a rate a:

$$F = ma \qquad (6.2)$$

Combining Equations 6.1 and 6.2 we have

$$\frac{F}{A} = P = \frac{ma}{A} \qquad (6.3)$$

If the units of the terms in the right-hand side of Equation 6.3 are kilograms (kg) for mass (m), square meters (m^2) for area (A), and meters per square second (m/s^2) for acceleration (a), then pressure (P) is given in units of

$$\frac{kg \cdot m}{m^2 \cdot s^2} \quad \text{which simplifies to} \quad \frac{kg}{m \cdot s^2} \quad \text{or Pa}$$

The relation between atmospheres and pascals is

$$1 \text{ atm} = 101{,}325 \text{ Pa}$$

Clearly, 1 pascal is a tiny quantity of pressure. In many applications it is more convenient to express pressure in kilopascals (1 kPa = 1000 Pa); the above relation now becomes

$$1 \text{ atm} = 101.325 \text{ kPa}$$

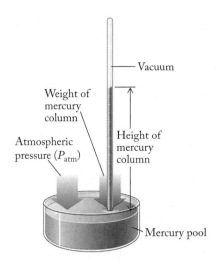

FIGURE 6.2 The height of the mercury column in this simple barometer designed by Evangelista Torricelli is proportional to atmospheric pressure.

FIGURE 6.3 Atmospheric pressure decreases with increasing altitude as the mass of the column of air above a given area decreases with increasing altitude. One atmosphere of pressure at sea level corresponds to 0.83 atm in Denver and 0.35 atm at the summit of Mt. Everest.

FIGURE 6.4 Small differences in atmospheric pressure are associated with major changes in weather. Adjacent isobars on this map of the eye of Hurricane Katrina differ by 4 mbar.

For many years, meteorologists have expressed atmospheric pressure in *millibars (mbar)*. The weather maps prepared by the U.S. National Weather Service show changes in atmospheric pressure by using contour lines of constant pressure, called *isobars*, that are 4 mbar apart (Figure 6.4). There are exactly 10 mbar in 1 kPa; thus at standard atmospheric pressure,

$$1 \text{ atm} = (101.325 \text{ } \cancel{kPa})(10 \text{ mbar/}\cancel{kPa})$$

$$= 1013.25 \text{ mbar}$$

The relations between different units for pressure are summarized in Table 6.2.

SAMPLE EXERCISE 6.1   **Calculating Atmospheric Pressure**

The mass of Earth's atmosphere is estimated to be 5.3×10^{18} kg. The surface area of Earth is 5.1×10^{14} m². The acceleration due to gravity is 9.8 m/s². From these values calculate an average atmospheric pressure in kilopascals (kPa).

COLLECT AND ORGANIZE We are given the total mass of the atmosphere in kilograms, the surface area of Earth in square meters, and the acceleration due to gravity in meters per squared seconds (m/s²). These units can be combined to calculate atmospheric pressure in pascals.

ANALYZE The force of gravity is exerted on the mass of the atmosphere (5.3×10^{18} kg). The area is the surface area of Earth, 5.1×10^{14} m². We can use Equation 6.3 and the known acceleration due to gravity to calculate atmospheric pressure.

SOLVE Substituting the values given in Equation 6.3 gives us the following:

$$F = \frac{ma}{A}$$

$$= \frac{(5.3 \times 10^{18} \text{ kg})(9.8 \text{ } \cancel{m}\text{/s}^2)}{5.1 \times 10^{14} \text{ } \cancel{m^2}}$$

$$= \frac{1.0 \times 10^5 \text{ kg}}{\text{m} \cdot \text{s}^2} = 1.0 \times 10^5 \text{ Pa}$$

$$= 1.0 \times 10^2 \text{ kPa}$$

THINK ABOUT IT The results of this calculation are consistent with the average value of atmospheric pressure at sea level: 101.325 kPa.

Practice Exercise Calculate the pressure exerted on a tabletop by a cube of iron that is 1.00 cm on each side and weighs 7.87 g.

(Answers to Practice Exercises are in the back of the book.)

Other units of pressure are derived from masses and areas in the English system. The pressure of 1 atm is equivalent to 14.7 pounds per square inch (lb/in²; also abbreviated psi). Tire pressure in the United States is most often expressed in psi units. Weather reports in the United States also tend to give readings of atmospheric pressure in inches of Hg, rather than mmHg. A pressure of 1 atm is equivalent to 29.92 inches of Hg. The relation between different units for pressure is illustrated in Sample Exercise 6.2 and in Table 6.2. The subsequent Sample Exercise illustrates how a barometer measures atmospheric pressure.

TABLE 6.2 **Units for Expressing Pressure**

Unit	Value
Atmosphere (atm)	1 atm
Pascal (Pa)	1 atm = 1.01325×10^5 Pa
Kilopascal (kPa)	1 atm = 101.325 kPa
Millimeter of mercury (mmHg)	1 atm = 760 mmHg
Torr	1 atm = 760 mmHg = 760 torr ✓
Bar	1 atm = 1.01325 bar
Millibar (mbar)	1 atm = 1013.25 mbar
Pounds per square inch (psi)	1 atm = 14.7 psi

SAMPLE EXERCISE 6.2   **Expressing Pressure Using Different Units**

The lowest atmospheric pressure ever recorded at sea level was 25.69 inches of Hg. This pressure was measured on October 12, 1979, in the eye of typhoon Tip, northwest of Guam in the Pacific Ocean. What was the pressure in typhoon Tip in units of (a) mmHg, (b) atm, (c) kPa, and (d) torr? There are 2.54 cm in 1 inch.

COLLECT AND ORGANIZE We are given the atmospheric pressure in inches of mercury and asked to express the pressure in a variety of other units. We need selected conversion factors from Table 6.2.

ANALYZE This is an exercise in converting units of pressure. (a) We can convert the pressure given in inches of mercury to millimeters of mercury with the conversion factor given in the statement of the problem. (b) From there we can calculate pressure in atmospheres. (c) Pressure in atmospheres is easily converted to pressure in torr, (d) from which we can change the pressure units to pascals and kilopascals (kPa).

SOLVE The following conversions provide the desired new units of pressure:

a. Converting from inches of Hg to mmHg requires us to convert from inches to millimeters. We can do this in two steps, using the conversion factor relating inches and centimeters (1 in = 2.54 cm), followed by the conversion factor for centimeters to millimeters (10 mm/cm):

$$(25.69 \text{ inHg}) \left(\frac{2.54 \text{ cm}}{1 \text{ in}} \right) \left(\frac{10 \text{ mm}}{1 \text{ cm}} \right) = 652.5 \text{ mmHg}$$

b. Using the conversion factor 1 atm = 760 mmHg, we can convert the units of a pressure measurement from millimeters of mercury to atmospheres:

$$(652.5 \text{ mmHg}) \left(\frac{1 \text{ atm}}{760 \text{ mmHg}} \right) = 0.8586 \text{ atm}$$

c. Earlier, 1 atm was defined as 101.325 kPa, so multiplying the pressure in atmospheres by 101.325 kPa/atm yields the pressure in kilopascals:

$$(0.8586 \text{ atm}) \left(\frac{101.325 \text{ kPa}}{\text{atm}} \right) = 87.00 \text{ kPa}$$

d. Finally, we recognize that pressure units of mmHg are the same as those in torr, so the conversion factor between these units is 1 mmHg = 1 torr:

$$(652.5 \text{ mmHg}) \left(\frac{1 \text{ torr}}{1 \text{ mmHg}} \right) = 652.5 \text{ torr}$$

A **manometer** is an instrument for measuring the pressure exerted by a gas.

A satellite infrared image of Hurricane Katrina.

THINK ABOUT IT Let's think about the unit conversions we just completed. Typhoons and hurricanes are associated with extremely low pressures, so we would expect each answer to be *less* than the value for the atmospheric pressure we calculated in the text just prior to this Sample Exercise: 1 atm = 760 mmHg = 760 torr = 101.325 kPa. Looking at the results of our calculations, we find that the pressure in each case is indeed lower—652.5 mmHg < 760 mmHg, 0.8586 atm < 1 atm, 87.00 kPa < 101.325 kPa, and 652.5 torr < 760 torr—so our answers all make sense.

Practice Exercise Atmospheric pressure in the eye of Hurricane Katrina, which devastated New Orleans in August 2005, was 90.2 kPa. What is this pressure in (a) atm, (b) mmHg, and (c) torr?

Scientists conducting experiments with gases must often measure the pressures exerted by gases. A **manometer** is an instrument used to measure gas pressures in closed systems. Two types of manometers are illustrated in Figure 6.5. In each case, a U-shaped tube filled with mercury (or another dense liquid) is attached to a flask containing a gas. The difference between the two manometers is whether the other end of the tube is sealed or open to the atmosphere. In a closed-end manometer, the difference in the height of the columns on either side of the U-tube is a direct measure of the gas pressure. In a manometer in which the U-tube is open, the difference in height represents the difference between the pressure in the flask and atmospheric pressure. When flask pressure is greater than atmospheric pressure, the level in the arm attached to the flask is lower than

FIGURE 6.5 A closed-end manometer (a,b): (a) The air has been removed from the flask and the mercury levels in the arms of the U-shaped tube are equal, because both ends of the U-tube are exposed to a vacuum. (b) Gas is allowed to enter the flask; it exerts pressure on the mercury, causing the mercury level to drop in the left-hand arm and rise in the right-hand arm. The difference in the height of the mercury columns (Δh) is the pressure exerted by the gas. An open-end manometer (c–e): (c) The right-hand arm is open to the atmosphere, while the left is connected to the flask. The gas in the flask is at atmospheric pressure, so the mercury level is the same in both arms. (d) The pressure of gas in the flask is less than atmospheric pressure, so the mercury level in the left-hand arm is higher than the right: $P_{gas} = P_{atm} - \Delta h$. (e) Here the pressure of gas in the flask is *greater* than atmospheric pressure, so the mercury level is *lower* in the left-hand arm than in the right: $P_{gas} = P_{atm} + \Delta h$.

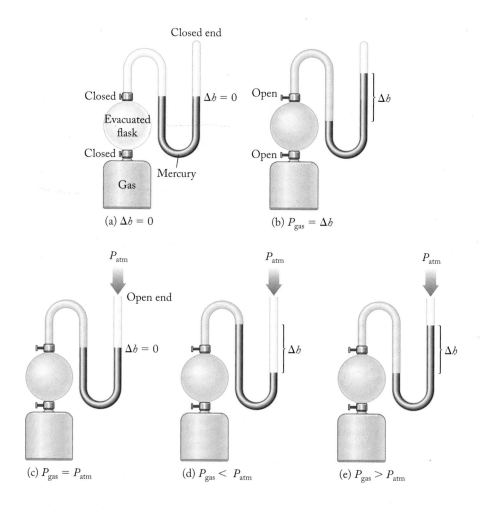

(a) $\Delta h = 0$

(b) $P_{gas} = \Delta h$

(c) $P_{gas} = P_{atm}$

(d) $P_{gas} < P_{atm}$

(e) $P_{gas} > P_{atm}$

the level in the arm exposed to atmospheric pressure; when the pressure in the flask is lower, the level in the arm attached to the flask is higher.

SAMPLE EXERCISE 6.3 **Measuring Gas Pressure with a Manometer**

Oyster shells are composed of calcium carbonate ($CaCO_3$). When they are roasted (heated) on an industrial scale to produce calcium oxide [CaO(s), quicklime], $CO_2(g)$ is also a product of the reaction. A chemist roasts an oyster shell in a flask (from which most of the air has been removed) that is attached to a closed-end manometer (see part a of the accompanying figure). When the roasting is complete and the system has cooled to room temperature, the difference in levels of mercury (Δh) in the arms of the manometer is 143.7 mm (part b of the figure). Calculate the pressure of $CO_2(g)$ in (a) torr, (b) atmospheres, and (c) kilopascals.

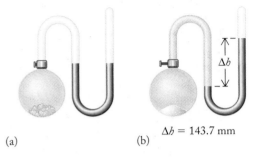

(a)   (b) Δh = 143.7 mm

Roasting $CaCO_3$ in a closed manometer: (a) An evacuated flask containing oyster shells. (b) The same setup after roasting the shells, which releases $CO_2(g)$ in the flask.

COLLECT AND ORGANIZE The pressure of the gas in the flask equals the difference in the height of the mercury in the two columns in a closed manometer. We are given that in the problem.

ANALYZE The difference is measured in millimeters of mercury, so we must convert that to the requested units of torr and atmospheres. The appropriate conversion factors are 1 mmHg = 1 torr and 760 torr = 1 atm.

SOLVE

a. First we convert millimeters of Hg to torr:

$$143.7 \text{ mmHg} \times \frac{1 \text{ torr}}{1 \text{ mmHg}} = 143.7 \text{ torr}$$

b. Then we convert torr to atmospheres:

$$143.7 \text{ torr} \times \frac{1 \text{ atm}}{760 \text{ torr}} = 0.1891 \text{ atm}$$

c. Finally we convert atmospheres to kilopascals:

$$0.1891 \text{ atm} \times \frac{101.325 \text{ kPa}}{1 \text{ atm}} = 19.16 \text{ kPa}$$

THINK ABOUT IT Because millimeters of Hg and torr are equivalent, the numerical value of a measurement expressed in one unit is the same as in the other. The other numbers make sense: a pressure less than 760 torr should also be less than 1 atm; by the same token, a pressure less than 1 atm should also be less than 101.325 kPa. The answers fit these criteria.

Practice Exercise If the same quantity of $CO_2(g)$ is produced in an open-end manometer (see accompanying figure) when the atmospheric pressure is 760 torr, indicate where the mercury levels should be at the conclusion of the experiment. What is the numerical value of Δh?

Before roasting

CONCEPT TEST

If you are running an experiment using an open-ended manometer, how do you determine the value of atmospheric pressure?

(Answers to Concept Tests are in the back of the book.)

Boyle's law states that the volume of a given amount of gas at constant temperature is inversely proportional to its pressure.

FIGURE 6.6 Weather balloons are used to carry instruments aloft in the atmosphere.

FIGURE 6.7 Because gases are important commercial items, large quantities of gas are compressed into small volumes for more efficient transport in trucks and railcars.

6.3 The Gas Laws

In Section 6.1, we summarized some of the properties of gases and described the effect of pressure (P) and temperature (T) on volume (V) mostly in qualitative terms. Our knowledge of the quantitative relationships between P, V, and T goes back more than three centuries, to a time before the field of chemistry as we now recognize it even existed. Many years of experiments led to an understanding of the behavior of gases in terms of their response to changes in temperature, pressure, and amount of gas present. Some of these experiments were driven by human interest in flight, specifically using hot-air balloons to rise above Earth's surface. Conclusions from these experiments have stood the test of time and scientific scrutiny. Centuries later, we now understand the relationships between P, V, and T on a molecular level. Balloons, which were so much a part of early experiments in responses of gases to changing conditions, are still used extensively in the study of weather and atmospheric phenomena (Figure 6.6), and their successful and safe use requires applying an understanding of gas properties that was first gained in the 17th and 18th centuries.

Boyle's Law: Relating Pressure and Volume

Air, like any gas or mixture of gases, is compressible. The compressibility of air allows us to pump up bicycle or automobile tires and store large amounts of air in relatively small metal cylinders in compressed form. The ability to compress air is the cornerstone of the compressed-gas industry because it makes the packaging and transport of large quantities of gas in a small space economical (Figure 6.7).

The relation between the pressure (P) and volume (V) of a fixed quantity of gas (constant value of n, where n is the number of moles) at constant temperature (T) was investigated by British chemist Robert Boyle (1627–1691) and is known as **Boyle's law**. Boyle observed that the volume of a given amount of a gas is inversely proportional to its pressure when kept at a constant temperature, as shown in Figure 6.8 and symbolized below:

$$P \propto \frac{1}{V} \qquad (T \text{ and } n \text{ fixed}) \qquad (6.4)$$

In other words, as the pressure on a constant amount of gas increases, the volume of the gas decreases, and conversely, as the pressure on the gas decreases, the volume of a given amount of gas increases.

(a)   (b)

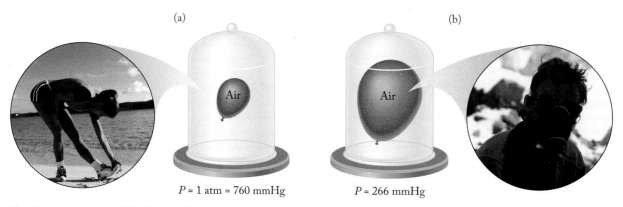

$P = 1 \text{ atm} = 760 \text{ mmHg}$ $P = 266 \text{ mmHg}$

The change in the size of the balloon demonstrates the relation between pressure and volume. (a) The balloon inside the bell jar is at a pressure of 1 atm (760 mmHg), comparable to atmospheric pressure on a nice day at sea level. (b) A slight vacuum applied to the bell jar drops the interior pressure to 266 mmHg, comparable to the pressure at the top of Mt. Everest (over 29,000 ft above sea level), and the volume of the balloon increases.

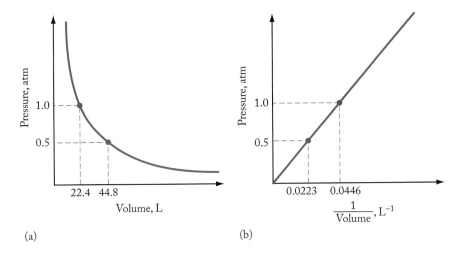

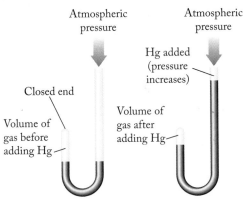

FIGURE 6.8 The graphs show the relation between pressure and volume. (a) At a constant temperature, the pressure of a given quantity of gas is inversely proportional to the volume that it occupies. This relation means that (b) a plot of P versus $1/V$ is a straight line.

Boyle's law describes why a balloon expands (increases its volume) when carried up a mountain. How can we use Boyle's law to predict exactly how much a balloon will expand when the pressure decreases? In his experiments Boyle used a U-shaped tube like the one shown in the manometer in Figure 6.5. One end of the tube was sealed, and mercury was poured into the open arm of the tube. The pressure exerted on the gas trapped in the tube depends on the mass of the mercury added. The amount of gas in his experiment remained constant, and the temperature of the apparatus was also held constant (Figure 6.9). The pressure on the gas in the tube was changed by varying the amount of mercury poured into the tube. Remember that the more mercury used, the greater the force exerted by the mercury ($F = ma$), and the greater the force per unit area ($P = F/A$). As Boyle added mercury to the U-tube, the volume occupied by the gas decreased in a way described by Equation 6.4. Mathematically we can replace the proportionality symbol with an equals sign and a constant,

$$PV = \text{constant} \tag{6.5}$$

where the value of the constant depends on the mass of gas in the sample and the temperature.

Because the value of the product PV in Equation 6.5 does not change for a given mass of gas at constant temperature, any two combinations of pressure and volume are related as follows:

$$P_1V_1 = P_2V_2 \tag{6.6}$$

This relationship is illustrated by the dotted lines in Figure 6.8. When 44.8 L (V_1) of gas is held at a pressure of 0.500 atm (P_1), then

$$P_1V_1 = (0.500 \text{ atm})(44.8 \text{ L}) = 22.4 \text{ L} \cdot \text{atm}$$

Equation 6.6 enables us to calculate the pressure required to compress this quantity of gas to a volume, for example, of 22.4 L:

$$P_1V_1 = P_2V_2$$

$$\frac{P_1V_1}{V_2} = P_2 = \frac{(0.500 \text{ atm})(44.8 \text{ L})}{22.4 \text{ L}}$$

$$= 1.00 \text{ atm}$$

The products of P and V are the same for each point on the graph. Equation 6.6 can also be used to calculate the change in volume that takes place at constant

FIGURE 6.9 Boyle used a U-shaped tube similar to the one shown here for his experiments on the relation between P and V. The volume of gas in the closed end of the U-tube changes in proportion to the difference in height of the mercury columns as mercury is added to the open arm of the tube.

temperature when the pressure of a quantity of gas changes. The result in the following Sample Exercise confirms the statement in Section 6.1 that taking a balloon from sea level to 1600 m leads to an increase in volume of some 20%.

SAMPLE EXERCISE 6.4   **Using Boyle's Law**

A balloon is partly inflated with 5.00 liters of helium at sea level, where the atmospheric pressure is 1.00 atm. The balloon ascends to an altitude of 1600 meters, where the pressure is 0.83 atm. (a) Describe what happens to the balloon as it rises. (b) What is the volume of the helium in the balloon at the higher altitude? (c) What is the percent increase in the volume of the balloon? Assume that the temperature of the gas in the balloon does not change in the ascent.

COLLECT AND ORGANIZE We are given the volume ($V_1 = 5.00$ L) of a gas at a given pressure ($P_1 = 1.00$ atm), and we are asked to find the volume that same quantity of gas would occupy ($V_2 = $ unknown) when the pressure changes ($P_2 = 0.83$ atm). The problem further states that the temperature of the gas does not change and that the amount of gas in the balloon remains constant.

ANALYZE The balloon contains a fixed amount of gas and its temperature is constant, so we have a pressure–volume relation, or an application of Boyle's law. Boyle's law tells us that pressure and volume are inversely proportional, and that PV for a quantity of gas is constant as long as n and T are constant.

SOLVE

a. Since the pressure decreases as the balloon rises, its volume should increase.

b. Equation 6.6 ($P_1V_1 = P_2V_2$) relates the pressure and volume of a given quantity of gas under two sets of conditions. We let the pressure at sea level be P_1 and the volume inside the balloon at sea level V_1, so the pressure at 1600 m is P_2 and the volume of the balloon at that altitude V_2. We know the values for P_1, V_1, and P_2, and we need to calculate V_2. Rearranging Equation 6.6 to solve for V_2 and inserting the given values for P_1, P_2, and V_1 gives us the following:

$$V_2 = \frac{P_1 V_1}{P_2}$$

$$V_2 = \frac{(1.00 \text{ atm})(5.00 \text{ L})}{0.83 \text{ atm}} = 6.0 \text{ L}$$

c. To calculate the percent increase in the volume we must determine the difference between V_1 and V_2 and compare the result with V_1:

$$\frac{V_2 - V_1}{V_1} \times 100 = \frac{6.0 \text{ L} - 5.0 \text{ L}}{5.0 \text{ L}} \times 100 = 20\%$$

THINK ABOUT IT The prediction we made in part a is confirmed by the calculation in part b.

Practice Exercise A scuba diver exhales 3.5 L of air while swimming at a depth of 20 m where the sum of atmospheric pressure and water pressure is 3.0 atm. By the time the bubbles of air rise to the surface, where the pressure is 1.0 atm, what is their total volume?

At the time Boyle discovered the relationship between P and V in the 17th century, scientists lacked any understanding of atoms or molecules. Given what we now know about matter, how can Boyle's law be explained on a molecular basis? Consider the effects of compressing a collection of molecules of gas into a smaller space. The molecules in the gas phase are constantly in random motion, which means that they collide with one another and with the interior surface of their container. The collisions with the walls of its container are responsible for the pressure exerted by a volume of gas. If molecules are squeezed into a smaller space—if the volume of the container is decreased—more collisions occur per unit time (Figure 6.10). The more frequent the collisions, the greater the force exerted by the molecules against the walls of their container, and thus the greater the pressure. Dividing the force (F) of the collisions by the area (A) over which they occur yields the pressure (P) (Equation 6.1). The area (A) of the walls of the container decreases when the size of the container is decreased, which in turn causes F/A to increase. Therefore the value of P increases as the size of the container decreases; as volume goes down, pressure goes up, because the frequency of collisions between gas molecules and container walls increases. The converse is also true: as the volume of a container is increased, A increases, the frequency of collisions between the molecules and the walls decreases, and pressure drops; as volume goes up, pressure goes down.

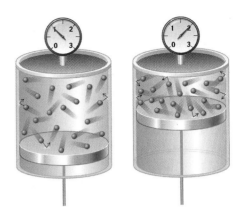

FIGURE 6.10 Gas molecules are in constant random motion, exerting pressure through collisions with their surroundings. When a quantity of gas is squeezed into half its original volume, the frequency of collisions per unit surface area on the inside of the container increases by a factor of 2, and so does the pressure.

Charles's Law: Relating Volume and Temperature

Nearly a century after Boyle's discovery of the inverse relation between pressure and volume, French scientist Jacques Charles (1746–1823) first documented the linear relation between the volume and temperature of a fixed quantity of gas at constant pressure. Now known as **Charles's law**, the relation states that the volume (V) of a fixed quantity of gas is directly proportional to the absolute temperature (T, in kelvins) of the gas at a constant pressure:

$$V \propto T \qquad (P \text{ and } n \text{ fixed}) \qquad (6.7)$$

The effects of Charles's law can be seen in Figure 6.11, where a balloon has been attached to a flask, trapping a fixed amount of gas in the apparatus. Heating the flask causes the gas to expand, inflating the balloon. The higher the temperature, the greater the volume occupied by the gas, and the bigger the balloon.

As in the case of Boyle's law, we can replace the proportionality symbol in Equation 6.7 by an equals sign if we include a proportionality constant:

$$V = \text{constant} \times T$$

This time, let's solve for the constant (by dividing each side of the equation by T):

$$\frac{V}{T} = \text{constant} \qquad (6.8)$$

The value of the constant in Equation 6.8 depends on the pressure (P) of the gas and the number of moles of gas (n) present in the sample. The ratio V/T does not change for a given quantity of gas at constant pressure, and any two combinations of volume and temperature are related as follows:

$$\frac{V_1}{T_1} = \frac{V_2}{T_2} \qquad (6.9)$$

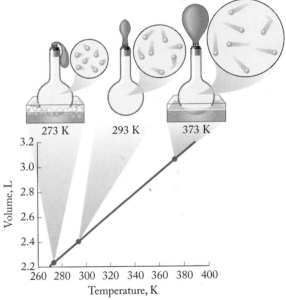

FIGURE 6.11 What a Chemist Sees A deflated balloon attached to a flask inflates as the temperature of the gas inside the flask increases from 273 K to 373 K. This behavior is described by Charles's law.

Charles's law states that the volume of a fixed quantity of gas at constant pressure is directly proportional to its absolute temperature.

Applying Charles's law, as described by Equation 6.9, allows us to make the statement in Section 6.1 that a decrease in temperature of 20°C leads to a decrease in volume of about 7% for a balloon. Let's see how we arrive at these numbers. Consider what happens when a balloon containing 2.00 L (V_1) of air at 20°C (T_1) is taken outside on a winter's day when the temperature is 0°C (T_2). The amount of gas in the balloon is fixed, and the atmospheric pressure is constant. Our own experience tells us that the volume of the balloon will decrease; Charles's law enables us to quantify that change. Substituting the values we are given into Equation 6.9, we can calculate the final volume (V_2), provided that we express the temperatures T_1 and T_2 in kelvins, not degrees Celsius. We must use absolute temperatures in gas equations. Because the lowest temperature on the absolute scale (Kelvin scale) is absolute zero (0 K), temperatures expressed in kelvins are always positive numbers:

$$\frac{V_1}{T_1} = \frac{2.00 \text{ L}}{(20 + 273) \text{ K}} = \frac{V_2}{0 + 273 \text{ K}}$$

Solving for V_2:

$$V_2 = \frac{2.00 \text{ L} \times 273 \text{ K}}{293 \text{ K}} = 1.86 \text{ L}$$

As predicted, the volume of the balloon decreased when the temperature decreased. The percent change in the volume of the balloon is calculated as:

$$\% \text{ decrease} = \frac{V_1 - V_2}{V_1} \times 100 = \frac{2.00 \text{ L} - 1.86 \text{ L}}{2.00 \text{ L}} \times 100 = 7\%$$

Sample Exercise 6.5 provides another example of calculations using Charles's law.

SAMPLE EXERCISE 6.5   **Using Charles's Law**

Charles and his compatriot Joseph Gay-Lussac (1778–1850) were drawn to the study of gases because of the interest in hot-air balloons in the late 18th century. We can imagine an early experiment these two scientists might have performed: what temperature is required to increase the volume of a sealed balloon from 2.00 L to 3.00 L if the initial temperature is 15°C and the atmospheric pressure is constant?

COLLECT AND ORGANIZE We are given the volume of a gas at an initial temperature and asked to calculate the temperature needed to make the gas reach a different volume. The container (a balloon) is sealed, so n is constant, as is atmospheric pressure.

ANALYZE We are given a volume ($V_1 = 2.00$ L) at a specific temperature ($T_1 = 15°C$), and are asked to calculate the temperature needed ($T_2 = $ unknown) to increase the volume of the gas to a new value ($V_2 = 3.00$ L). The relation between volume and temperature is summarized by Charles's law. One form of Charles's law is given by Equation 6.9 when temperatures are expressed in kelvins:

$$\frac{V_1}{T_1} = \frac{V_2}{T_2}$$

SOLVE First rearrange Equation 6.9 to solve for T_2:

$$T_2 = \frac{V_2 T_1}{V_1}$$

Substitute for V_1, V_2, and T_1, and remember to convert temperature to kelvins:

$$T_2 = \frac{V_2 T_1}{V_1} = \frac{(3.00\ \text{L})(288\ \text{K})}{2.00\ \text{L}} = 432\ \text{K}$$

To convert the final temperature to degrees Celsius, we must subtract 273 from the temperature in kelvins:

$$T_2 = 432\ \text{K} - 273 = 159°\text{C}$$

THINK ABOUT IT Notice that to increase the volume by 1.5 times, the temperature in kelvins must also increase 1.5 times [432 K = (1.5)(288K)], as expected from the statement that volume is directly proportional to temperature. The volume of the gas is not directly proportional to temperature in degrees Celsius.

Practice Exercise Hot expanding gases can be used to move a piston and to perform useful work in a cylinder with a movable piston. If the temperature of the gas is raised from 245°C to 560°C, what is the ratio of the initial volume to the final volume if pressure remains constant?

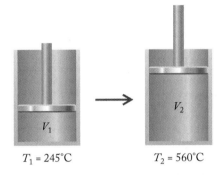

$T_1 = 245°C$ $T_2 = 560°C$

Charles's law helps explain why a balloon inflated with hot air rises. A given mass of hot air occupies more volume than does the same mass of cooler ambient air. Therefore a balloon filled with hot air displaces a mass of ambient air larger than its own mass, giving the balloon buoyancy. Put another way, the density of a given quantity of gas decreases with increasing temperature.

In the following way, Charles's law also allows us to determine that the value of absolute zero (0 K) is equal to −273.15°C. Consider what happens if we plot the volume of a fixed quantity of gas at constant pressure as a function of temperature. Some typical results for 28 cm³ (28 mL) of gas initially at 293 K (20°C) are shown in Figure 6.12. The decrease in volume with decreasing temperature is linear. In the laboratory we are limited by what low temperatures we can reach before the gas liquefies; at this point the gas laws no longer describe its behavior because it is no longer a gas. However, we can extrapolate from our data to the point where the volume of a gas would reach zero if liquefaction did not occur. As we saw in Chapter 1, that temperature is defined as absolute zero.

FIGURE 6.12 The volume of a gas (*n* and *P* constant) is plotted against temperature in both the Celsius and Kelvin scales. As predicted by Charles's law, volume decreases as the temperature decreases; the relationship is linear on both scales. The dashed line shows the extrapolation from the data to a point corresponding to a zero volume. This temperature is known as absolute zero: −273°C on the Celsius scale and 0 K on the Kelvin scale. Numerically, volume is directly proportional to temperature only on the Kelvin scale.

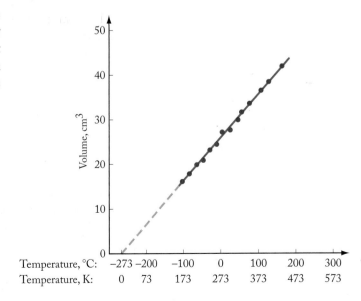

Avogadro's law states that the volume of a gas at a given temperature and pressure is proportional to the quantity of the gas.

Amontons's law states that, as the absolute temperature of a fixed amount of gas increases, the pressure increases as long as the volume and quantity of gas remain constant.

Avogadro's Law: Relating Volume and Moles of Gas

Both Boyle's and Charles's experiments involved a constant quantity of gas in a confined sample. Now consider what happens when a balloon is inflated. The more gas is blown into the balloon, the larger the balloon gets (until it bursts). Clearly, the volume (V) depends on the amount of gas (n). If a hole is poked in the balloon, the gas escapes, decreasing both the amount of gas and the volume. The relation between pressure (P) and the amount of gas (n) can be illustrated in several ways.

Consider a bicycle tire that is inflated (holds its shape) but is nevertheless too soft to ride on. We can increase the pressure inside the tire by pumping more air into it. In the process the tire does not expand much because the heavy rubber is designed to be much more rigid than the material in a balloon. Another example of the relation between P and n is the carnival vendor who sells helium-filled balloons. The helium tank has a pressure gauge attached to the top. When sales are brisk, it is possible to observe the pressure decreasing (from falling numbers on the pressure gauge) as the vendor fills more and more balloons. Filling balloons means that the pressure in the cylinder decreases as the amount of gas in the cylinder decreases.

A few decades after Charles discovered the relation between V and T, Amedeo Avogadro (1776–1856), who we know from Avogadro's number (N_A), recognized the relation between V and n. This relation is described by **Avogadro's law** (Figure 6.13), which states that the volume (V) of a gas at a given temperature (T) and pressure (P) is proportional to the quantity (n, in moles) of the gas:

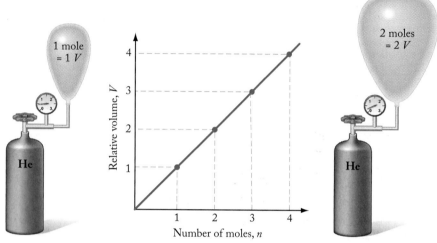

$$V \propto n \quad \text{or} \quad \frac{V}{n} = \text{constant} \qquad (P \text{ and } T \text{ fixed}) \qquad (6.10)$$

FIGURE 6.13 The weather balloon on the right has twice the volume of the balloon on the left because it is filled with twice the quantity of helium. This direct relation between the volume (V) and quantity (n) of a gas at constant temperature (T) and pressure (P) is known as Avogadro's law.

Amontons's Law: Relating Pressure and Temperature

We have followed the discovery of the relationships between pressure and volume (Boyle's law) as well as volume and temperature (Charles's law). Both of these laws describe effects on volume, but is there a relation between pressure and temperature? It turns out that pressure is directly proportional to temperature when n and V are constant:

$$P \propto T \quad \text{or} \quad \frac{P}{T} = \text{constant} \qquad (V \text{ and } n \text{ fixed}) \qquad (6.11)$$

The relation stated in Equation 6.11 means that as the absolute temperature (T in kelvins) of a fixed amount of gas increases, the pressure (P) of the gas increases as long as the volume and quantity of the gas are kept constant. It is sometimes referred to as **Amontons's law** in honor of the French physicist Guillaume Amontons (1663–1705), a contemporary of Robert Boyle who constructed a thermometer based on the observation that the pressure of a gas is directly proportional to its temperature.

Where do we see evidence of Amontons's law? Consider a bicycle tire inflated to the prescribed pressure, say 100 psi (6.8 atm), on a rare hot autumn afternoon when the temperature has soared to 25°C. The next morning, following an early freeze in which the temperature falls from 25° to 0°C, the pressure in the tire drops to 6.2 atm (~91 psi). Did the tire leak? Perhaps, but the decrease in pressure can also be explained by Amontons's law.

We can confirm that the decrease in tire pressure is consistent with the decrease in temperature by applying the logic we used to develop Equations 6.6 and 6.9, namely, relating the initial pressure (P_1) and initial temperature (T_1) to the final pressure (P_2) and temperature (T_2):

$$\frac{P_1}{T_1} = \frac{P_2}{T_2} \qquad (6.12)$$

Substitution into Equation 6.12 yields the equation

$$\frac{6.8 \text{ atm}}{298 \text{ K}} = \frac{P_2}{273 \text{ K}}$$

Solving for P_2, we get

$$P_2 = \frac{(6.8 \text{ atm})(273 \text{ K})}{298 \text{ K}} = 6.2 \text{ atm}$$

Why is pressure directly proportional to temperature? Like pressure, temperature is directly related to molecular motion. The average speed at which a population of molecules moves increases with increasing temperature. For a given number of molecules, increasing temperature increases motion and therefore increases the frequency and force with which gas molecules collide with the walls of their container. Because pressure is related to the frequency and force of these collisions, then it follows that higher temperatures produce higher pressures, as long as other factors such as volume and quantity of gas are constant (Figure 6.14). Sample Exercise 6.6 examines how pressure–temperature relations come into play when we handle pressurized products.

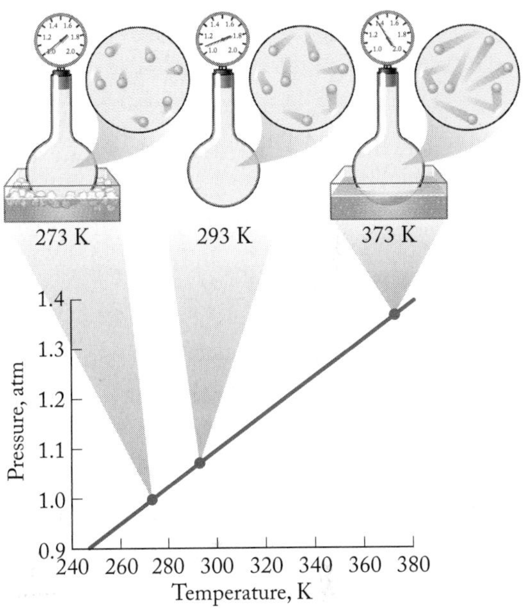

FIGURE 6.14 The pressure (P) of a given quantity of gas (n) is directly proportional to its absolute temperature (T) at constant volume (V). Each flask has the same volume and contains the same number of molecules. The relative velocities of the molecules (represented by the length of their "tails") increase with increasing temperature, causing more frequent and more forceful collisions with the walls of the containers and hence higher pressures.

SAMPLE EXERCISE 6.6   **Pressure–Temperature Relations for a Gas**

Labels on aerosol cans caution against incineration because the cans explode when pressures inside them exceed 3.00 atm. (a) Why does the pressure inside the can increase as its temperature increases? (b) At what temperature (in degrees Celsius) will an aerosol can burst if initially the pressure inside the can is 2.20 atm at 25°C?

COLLECT AND ORGANIZE (a) We are first asked to explain why the pressure increases in the can as the temperature of the can increases. (b) Then we are given the temperature (25°C) and pressure (2.20 atm) of a gas and asked to determine the temperature at which the pressure will exceed 3 atm.

ANALYZE In this problem, the quantity of gas and its volume are constant. (a) Amontons's law tells us that, in this situation, the pressure of a gas increases as temperature increases. Once the pressure exceeds the strength of the can, the can will rupture (explode).

We can express Amontons's law using Equation 6.12:

$$\frac{P_1}{T_1} = \frac{P_2}{T_2}$$

CONNECTION In Chapter 5 we discussed kinetic and potential energies.

We are given the initial pressure (P_1) and temperature (T_1) for the can before it is heated. We are also told that the can will rupture if the final pressure (P_2) exceeds 3 atm. Using these three values, we can calculate the temperature at which the can explodes. Before proceeding, we should recognize that the initial and final temperatures are in degrees Celsius, but Equation 6.12 for gases works only with temperatures expressed in kelvins. Therefore we need to convert from the given temperature values from degrees Celsius to kelvins to solve the problem, and then back again into degrees Celsius to express the answer in the units requested.

SOLVE

a. The pressure of a confined gas increases as temperature increases because the average kinetic energy of the molecules increases. Volume is constant in this scenario, so the area (A) of the walls inside the container does not change. Increasing molecular motion produces both more frequent and more forceful collisions between the gas molecules and their container. More frequent and forceful collisions on the same area result in an increase in pressure.

b. To solve for the temperature at which the can will explode, we rearrange Equation 6.12 to solve for T_2:

$$T_2 = \frac{T_1 P_2}{P_1}$$

Inserting the values given in the statement of the problem, and converting T_1 from degrees Celsius to kelvins, we have

$$T_2 = \frac{(25 + 273)\text{K} \times 3.00 \text{ atm}}{2.20 \text{ atm}} = 406 \text{ K}$$

Converting T_2 from kelvins back into degrees Celsius gives us

$$T_2 = 406 \text{ K} - 273 = 133°\text{C}$$

THINK ABOUT IT This temperature is certainly higher than the original temperature, so the answer makes sense. Notice that the predicted temperature is easily reached in a closed car on a hot summer day, especially in a place like Arizona or Texas.

Practice Exercise The air pressure in the tires of an automobile is adjusted to 28 psi at a gas station in San Diego, CA, where the air temperature is 68°F (20°C). The air in the tires is at the same temperature as the atmosphere. The automobile is then driven east along a hot desert highway. Along the way, the temperature of the tires reaches 140°F (60°C). What is the pressure in the tires?

Gases that behave in accordance with the linear relations discovered by Boyle, Charles, Avogadro, and Amontons are called **ideal gases**. Most gases exhibit ideal behavior at the pressures and temperatures typically encountered in the atmosphere. Under these conditions, the volumes occupied by the molecules or atoms of gas themselves are insignificant compared with the volume the gas as a whole occupies. All of the open space between gas molecules makes gases compressible. In a sample of an ideal gas, atoms or molecules are also assumed not to interact with one another; rather, they move independently with velocities that are related to the temperature of the gas and to their masses.

An **ideal gas** is a gas whose behavior is predicted by the linear relations defined by Boyle's, Charles's, Avogadro's, and Amontons's laws.

6.4 The Ideal Gas Law

Think back to the scenario described in Section 6.1 in which we considered the effect on its volume of carrying a balloon from sea level to 1600 m. What if we launched a weather balloon from the surface of Earth and allowed it to drift to an elevation of 10,000 m? The volume of the balloon would expand as the atmospheric pressure decreased, but the air temperature would decrease as the balloon ascended. How could we determine the final volume of the balloon when we are changing three variables (P, V, and T) while keeping just one variable (n, the amount of gas) fixed? Taken individually, none of the four gas laws and their accompanying mathematical relations fit this situation: Boyle's law relates P and V, Charles's law relates V and T, Avogadro's law relates V and n, while Amontons's law relates P and T. Nevertheless, we can derive a relation, known as the **ideal gas equation**, that combines all four variables (P, V, n, and T):

$$PV = nRT \qquad (6.13)$$

where R is the **universal gas constant**. Its values and units are listed in Table 6.3. Equation 6.13 is also called the **ideal gas law**.

R has different values depending on which units are used for pressure, volume, and temperature. For many calculations in chemistry, it is convenient to use the first value of R listed in Table 6.3: 0.08206 has units of $\text{L} \cdot \text{atm}/(\text{mol} \cdot \text{K})$. We can use that value in Equation 6.13 only when the quantity of gas is expressed in moles (n), the volume in liters, the pressure in atmospheres, and the temperature in kelvins. The units of the quantities P, V, and T must be the same as the units of R.

Where does the ideal gas equation come from? Boyle's law expresses that volume and pressure are inversely proportional. Charles's law tells us that volume is directly proportional to temperature. Putting Boyle's and Charles's laws together, (Equations 6.5 and 6.8) allows us to write a new equation relating P, V, and T and combining the constants:

$$PV = \text{constant} \qquad \text{(for constant } n \text{ and } T\text{)}$$

$$\frac{V}{T} = \text{constant} \qquad \text{(for constant } n \text{ and } P\text{)}$$

$$\frac{PV}{T} = \text{combined constant} \qquad (6.14)$$

Avogadro's law (Equation 6.10: $V \propto n$, or $V/n = \text{constant}$) tells us that volume is also directly proportional to the amount of gas present (in moles) when T and P are constant, so we can include n in Equation 6.14 and combine its constant with the other constants:

$$\frac{PV}{nT} = \text{combined constant} \quad \text{or} \quad PV = \text{combined constant} \times nT \qquad (6.15)$$

We turn Equation 6.15 into a meaningful equality by inserting the appropriate constant of proportionality. By convention, this constant of proportionality is the universal gas constant R, giving us the ideal gas law (Equation 6.13):

$$PV = nRT$$

The ideal gas law expresses all possible relationships between the four variables P, V, T, and n.

The **ideal gas equation**, also called the **ideal gas law**, relates the pressure (P), volume (V), number of moles (n), and temperature of an ideal gas, and is expressed as $PV = nRT$, where R is a constant.

In the ideal gas equation, the numerical constant R is called the **universal gas constant**. Its value and units depend on the units used for the variables in the ideal gas equation.

TABLE 6.3 Values for the Universal Gas Constant (R)

Value of R	Units
0.08206	$\text{L} \cdot \text{atm}/(\text{mol} \cdot \text{K})$
8.314	$\text{kg} \cdot \text{m}^2/(\text{s}^2 \cdot \text{mol} \cdot \text{K})$
8.314	$\text{J}/(\text{mol} \cdot \text{K})$
8.314	$\text{m}^3 \cdot \text{Pa}/(\text{mol} \cdot \text{K})$
62.37	$\text{L} \cdot \text{torr}/(\text{mol} \cdot \text{K})$

▶❙❙ **CHEMTOUR** The Ideal Gas Law

The **general gas equation**, also called the **combined gas law**, is based on the ideal gas law; it is used when one or more of the four gas variables are held constant while the remaining variables change.

We can derive another relation involving these four variables that is useful to consider when a system starts in an initial state (P_1, V_1, T_1, n_1) and moves to a final state (P_2, V_2, T_2, n_2) as a result of changing conditions:

$$P_1V_1 = n_1RT_1 \quad \text{so} \quad \frac{P_1V_1}{n_1T_1} = R \quad \text{and} \quad P_2V_2 = n_2RT_2 \quad \text{so} \quad \frac{P_2V_2}{n_2T_2} = R \qquad (6.16)$$

Because both the initial-state and final-state expressions are equal to R, they are also equal to each other, allowing us to simplify Equation 6.16 to the following:

$$\frac{P_1V_1}{n_1T_1} = \frac{P_2V_2}{n_2T_2} \qquad (6.17)$$

which is known as the **general gas equation**, or **combined gas law**. As an example of how to use it, let's apply it in calculating the volume of the weather balloon described at the beginning of this section. Because no gas is added or released from the balloon, n is constant, so we can rewrite the general gas equation as follows:

$$\frac{P_1V_1}{T_1} = \frac{P_2V_2}{T_2} \qquad (n \text{ fixed})$$

Sample Exercise 6.7 illustrates the effect of P and T on the volume of a weather balloon. Different reduced versions of Equation 6.17 result when other variables besides n are fixed; for another example, see Figure 6.15.

FIGURE 6.15 Breathing illustrates the relation between P, V, and n. When you inhale, your rib cage expands and your diaphragm moves down, increasing the volume (V) of your lungs. Increased volume decreases the pressure (P) inside your lungs, in accord with Boyle's law. Decreased pressure allows more air to enter until the pressure inside your lungs matches atmospheric pressure. At high altitudes, atmospheric pressure is low and so is the pressure in your lungs. Therefore the quantity (n) of air (and hence of oxygen) that enters your lungs is smaller than normal.

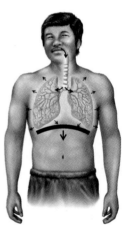

SAMPLE EXERCISE 6.7 **Calculations Involving Changes in *P*, *V*, and *T***

A weather balloon filled with 100.0 liters of He is launched from sea level ($T = 20°C$, $P = 755$ torr). No gas is added or removed from the balloon during its flight. (a) Predict the effect on the volume of the balloon when it rises to an altitude of 10 km, where the temperature of the atmosphere and the gas in the balloon are both $-52°C$ and atmospheric pressure is 195 torr. (b) Calculate the volume of the balloon at an altitude of 10 km.

COLLECT AND ORGANIZE We are given values for the initial temperature (T_1), pressure (P_1), and volume (V_1) of a gas, and are asked to determine the final volume (V_2) after the pressure and temperature have both changed.

ANALYZE The quantity of gas does not change during the flight; however, the pressure (P_2), volume (V_2), and temperature (T_2) differ from the initial conditions. We can use this version of Equation 6.17:

$$\frac{P_1 V_1}{T_1} = \frac{P_2 V_2}{T_2}$$

We are also asked to predict the effect of decreasing the temperature and pressure on the gas in the weather balloon.

(a) The decrease in temperature will lead to a decrease in volume, but the decrease in pressure will lead to an increase in volume. We need to express the temperatures in kelvins and estimate which variable will dominate the change.

(b) Because we are asked to find the final volume, we need to solve for V_2 in the general gas equation.

$$\frac{P_1 V_1}{T_1} = \frac{P_2 V_2}{T_2} \quad \text{or} \quad V_2 = \frac{P_1 V_1 T_2}{T_1 P_2}$$

$$T_1 = (20°C + 273) = 293 \text{ K} \quad \text{and} \quad T_2 = -52°C + 273 = 221 \text{ K}$$

SOLVE

a. The decrease in pressure to 195 torr (about 25% of the starting value of 755 torr) is considerably larger than the decrease in temperature from 20°C (293 K) to −52°C (221 K) (about 75% of the starting value). Therefore we may conclude that pressure is the dominant of the two variables and that the volume of the balloon increases as it attains its final altitude.

b. Inserting the given values into the expression we derived above for V_2, we have:

$$V_2 = \frac{(755 \text{ torr})(100.0 \text{ L})(221 \text{ K})}{(293 \text{ K})(195 \text{ torr})} = 292 \text{ L}$$

THINK ABOUT IT The volume of the balloon increases nearly threefold as it ascends to 10 km. This result makes sense because atmospheric pressure decreases to about one-quarter of its sea-level value during the ascent. The balloon's volume would have increased more than threefold if not for the countervailing effect of the lower temperature at 10 km.

Practice Exercise The balloon in this Sample Exercise is designed to continue its ascent to an altitude of 30 km, where it bursts, releasing a package of meteorological instruments that parachute back to Earth. If the atmospheric pressure at 30 km is 28.0 torr and the temperature is −45°C, what is the volume of the balloon when it bursts?

The ideal gas law describes the relation between pressure, volume, and temperature for any gas, provided it behaves as an ideal gas. At typical atmospheric pressures, many gases obey the ideal gas law. This allows us to apply the ideal gas equation to many situations outside the laboratory. For example, in Sample Exercise 6.8, we calculate the mass of oxygen in an alpine climber's compressed-oxygen cylinder. The answer might prove critical to the success of an ascent: does the climber have sufficient oxygen?

SAMPLE EXERCISE 6.8   **Applying the Ideal Gas Law**

Bottles of compressed O_2 carried by alpine climbers have an internal volume of 5.90 L. Assume that such a bottle has been filled with O_2 to a pressure of 2025 psi at 25.0°C. (a) How many moles of O_2 are in the bottle? (b) What is the mass (in grams) of O_2 in the bottle? Assume that O_2 behaves as an ideal gas and that 1 atm = 14.7 psi.

COLLECT AND ORGANIZE We are given the pressure, volume, and temperature of a gas and asked to determine the corresponding amount of oxygen in grams.

ANALYZE Before we can calculate the mass, we need to find the number of moles of oxygen in the tank. The ideal gas equation enables us to use P, V, and T to calculate n, the number of moles of O_2. Then we can use its molar mass (32.00 g/mol) to calculate the mass of O_2 in the bottle.

Let's start with the ideal gas equation and rearrange the terms to solve for n:

$$PV = nRT \quad \text{so} \quad n = \frac{PV}{RT}$$

SOLVE Before using this expression for n, we need to convert pressure into atmospheres and temperature into kelvins:

$$P = (2025 \ \cancel{psi})\left(\frac{1 \ atm}{14.7 \ \cancel{psi}}\right) = 138 \ atm \qquad T = 25.0°C + 273 = 298 \ K$$

We insert these values and the volume given into the above expression for n:

$$n = \frac{(138 \ \cancel{atm})(5.90 \ \cancel{L})}{\left(0.08206 \ \dfrac{\cancel{L} \cdot \cancel{atm}}{mol \cdot \cancel{K}}\right)(298 \ \cancel{K})} = 33.3 \ mol$$

Converting moles into grams is a matter of multiplying by the molar mass:

$$(33.3 \ \cancel{mol})\left(\frac{32.00 \ g}{\cancel{mol}}\right) = 1066 \ g = 1.07 \times 10^3 \ g$$

THINK ABOUT IT This is our first use of the ideal gas equation. If we use the universal gas constant $R = 0.08206 \ L \cdot atm/(mol \cdot K)$ in our calculations, then pressure must be in atmospheres, volume in liters, n in moles, and the temperature of the gas sample in kelvins.

Practice Exercise Starting with the moles of O_2 calculated in this Sample Exercise, calculate the volume of O_2 that the bottle could deliver to a climber near the summit of Mt. Everest. Assume that the temperature at that altitude is −38°C and the atmospheric pressure is 0.35 atm.

SAMPLE EXERCISE 6.9   **Combining Stoichiometry and the Ideal Gas Law**

Oxygen generators in some airplanes are based on a chemical reaction between sodium chlorate and iron:

$$NaClO_3(s) + Fe(s) \rightarrow O_2(g) + NaCl(s) + FeO(s)$$

How many grams of $NaClO_3$ are needed to produce 125 liters of O_2 at 1.00 atm and 20.0°C?

COLLECT AND ORGANIZE We are given the volume of gas we need to prepare at a particular pressure and temperature. This information can be used to determine the mass of $NaClO_3$ needed based on the stoichiometric relations in the balanced chemical equation.

ANALYZE The solution requires two separate calculations. We are asked to determine the quantity of sodium chlorate (in grams) that will produce 125 L of $O_2(g)$ at 1.00 atm of pressure and 20.0°C. We start by recognizing that the balanced chemical equation tells us that 1 mole of $NaClO_3$ is needed to produce 1 mole of oxygen. If we can determine how many moles of O_2 occupy a volume of 125 L at 1.00 atm pressure and 20.0°C, then we can determine the number of moles of $NaClO_3$ we need. Remember from Chapter 3 that we can easily convert moles of $NaClO_3$ to grams of $NaClO_3$. To determine the moles of oxygen (n_{O_2}), we can use the ideal gas law (Equation 6.13), which includes terms for P, V, T, and n:

$$P_{O_2}V_{O_2} = n_{O_2}RT_{O_2}$$

First we calculate the number of moles of oxygen in a volume of 125 L at 1.00 atm and 20.0°C, using the ideal gas law. Then we use the value for number of moles of oxygen required along with the balanced chemical equation to determine the number of moles and number of grams of $NaClO_3(s)$ required. The stoichiometry of the balanced chemical equation indicates that 1 mole of $NaClO_3(s)$ is required for every mole of $O_2(g)$ produced.

To find the number of moles (n), we rearrange the ideal gas equation ($PV = nRT$) to get

$$n = \frac{PV}{RT}$$

We also need to verify that the statement of the problem gives us the balanced chemical equation, and not just the reactants and products.

SOLVE We use the rearranged ideal gas equation to solve for the moles of O_2:

$$n_{O_2} = \frac{P_{O_2}V_{O_2}}{RT_{O_2}}$$

Inserting the oxygen volume, temperature in kelvins, and pressure in atmospheres yields

$$n = \frac{(1.00 \text{ atm})(125 \text{ L})}{\left(0.08206 \dfrac{\text{L} \cdot \text{atm}}{\text{mol} \cdot \text{K}}\right)(273 + 20.0) \text{ K}} = 5.20 \text{ mol } O_2$$

To convert moles of O_2 into an equivalent mass of $NaClO_3$, we first check that the equation given is balanced; it is. Then we use the stoichiometry of the reaction to calculate the equivalent number of moles of $NaClO_3$, and use the conversion factor of molar mass to determine the number of grams of $NaClO_3$ required:

$$5.20 \text{ mols } O_2 \times \frac{1 \text{ mol } NaClO_3}{1 \text{ mol } O_2} \times \frac{106.44 \text{ g } NaClO_3}{\text{mol } NaClO_3} = 5.53 \times 10^2 \text{ g } NaClO_3$$

THINK ABOUT IT When dealing quantitatively with any chemical reaction that involves a gas as either reactant or product and the volume of the gas is at issue, we need to know the conditions under which the reaction occurs. If V, T, and P

An automobile air bag inflates when solid NaN_3 rapidly decomposes, producing N_2 gas.

are known, we can use the ideal gas equation to determine the number of moles of gas in the system. Once we know that, we can use stoichiometric calculations as described in Chapter 3.

Practice Exercise Automobile air bags are caused to inflate during a crash or sudden stop by the rapid generation of nitrogen gas from sodium azide, according to the following reaction:

$$2\ NaN_3(s) \rightarrow 2\ Na(s) + 3\ N_2(g)$$

How many grams of sodium azide are needed to fill a $45 \times 45 \times 25$ cm bag to a pressure of 1.20 atm at a constant temperature of 15°C?

A useful reference point in studying the properties of gases is **standard temperature and pressure (STP)**, defined as 0°C and 1 atm.[1] Another reference point is the **molar volume** of an ideal gas, which is the volume that a mole of a gas occupies at STP. We can calculate the molar volume from the ideal gas equation,

$$PV = nRT \qquad \text{or} \qquad V = \frac{nRT}{P}$$

by inserting the values of n, P, and T at standard conditions:

$$V = \frac{(1\ \text{mol})\left(0.08206\ \dfrac{\text{L} \cdot \text{atm}}{\text{mol} \cdot \text{K}}\right)(273\ \text{K})}{1\ \text{atm}} = 22.4\ \text{L}$$

Many chemical and biochemical processes take place at pressures near 1 atm and at temperatures between 0°C and 40°C. Within this range of conditions the volume that a mole of gaseous reactant or product occupies will be no more than about 15% greater than the molar volume, so volumes can be estimated easily if molar amounts are known. An important feature of molar volume is that it applies to any ideal gas, independent of its chemical composition. In other words, 1 mole of helium at STP occupies the same volume as 1 mole of methane (CH_4), of carbon dioxide (CO_2), or even of a compound like uranium hexafluoride (UF_6), which has a molar mass almost 90 times that of helium.

Figure 6.16 is a summary of the four individual laws that are combined in the ideal gas law. The ideal gas law can be used to determine many relations involving the physical properties of gases.

6.5 Gas Density

Carbon dioxide (CO_2) is a relatively minor component of our atmosphere. It is produced by the combustion of fuels. The increased levels of CO_2 in the atmosphere are of great concern because of their impact on climate change. Carbon dioxide is colorless, odorless, and denser than air, and sudden releases of large quantities of $CO_2(g)$ from areas of volcanic activity are life-threatening events. On the Dieng Plateau in Indonesia in 1979, 142 people died from asphyxiation in a valley after a massive, sudden release of carbon dioxide from one of many volcanic centers occurred in a large, low-lying area about 9 miles (14 km) long by 4 miles (6 km) wide. The dense carbon dioxide settled in the valley, displacing the less-dense air containing the oxygen necessary for life.

In the United States, **standard temperature and pressure (STP)** is a temperature of 0°C and a pressure of 1 atm. Elsewhere, STP is defined as 0°C and a pressure of 1 bar.

The **molar volume** of an ideal gas is the volume occupied by 1 mole of a gas at STP and is equal to 22.4 L.

[1] The definition of STP given by the International Union of Pure and Applied Chemistry (IUPAC) is a pressure of 1 bar (100 kPa) and a temperature of 0°C.

(a) Boyle's law: volume inversely proportional to pressure; n and T constant

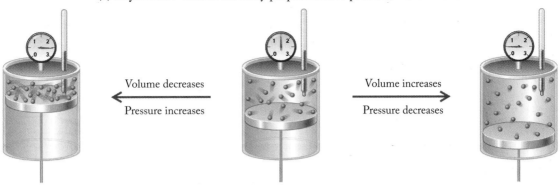

(b) Charles's law: volume directly proportional to temperature; n and P constant

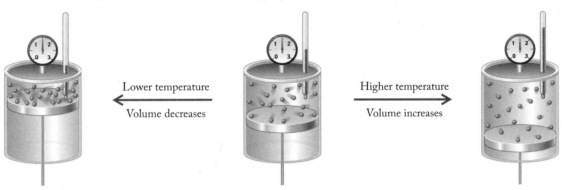

(c) Amontons's law: pressure directly proportional to temperature; n and V constant

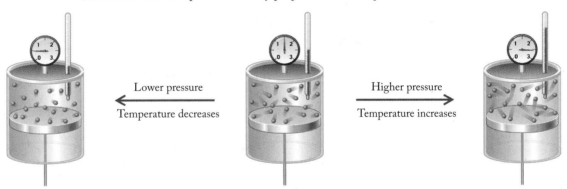

(d) Avogadro's law: volume directly proportional to number of moles; T and P constant

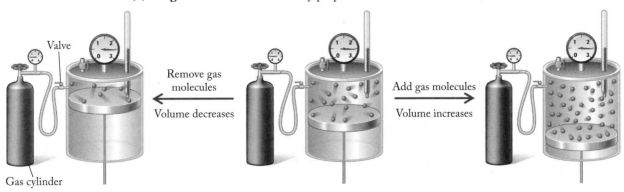

FIGURE 6.16 Relations among the four gas variables. (a) Increasing or decreasing V at constant n and T: Boyle's law. (b) Increasing or decreasing T at constant n and P: Charles's law. (c) Increasing or decreasing P at constant n and V: Amontons's law. (d) Increasing or decreasing n at constant T and P: Avogadro's law.

The release of CO_2 from volcanic centers on the Dieng Plateau killed so many because the dense gas flowed over the valley floor, effectively displacing the air (and oxygen) that the inhabitants and their livestock needed to survive.

CONNECTION We defined density in Chapter 1 and have used it in subsequent chapters to characterize matter. Just as liquids and solids vary in density, so do gases, causing some to float on others the way oil floats on water. Density differences can cause gases to layer if they are not well mixed.

The density of a gas at STP can be calculated from its molar mass and molar volume. Carbon dioxide, for example, has a molar mass of 44.0 g/mol. Therefore the density of CO_2 at STP is

$$\frac{44.0 \text{ g/mol}}{22.4 \text{ L/mol}} = 1.96 \text{ g/L}$$

The density of air is about 1.2 g/L, so it is not surprising that the CO_2 in the Dieng Plateau disaster was concentrated close to the surface of the land. Remember that under normal conditions densities of gases are very much less than densities of liquids: $d_{gas} << d_{liquid}$.

CONCEPT TEST

Which of the following gases has the highest density at STP? CH_4, Cl_2, Kr, or C_3H_8?

The density of an ideal gas can be calculated for any combination of temperature and pressure using the ideal gas equation. Because density is the mass of the sample divided by its volume, we need to identify the terms in the ideal gas equation that represent the density. The mass of the sample can be determined from the number of moles (n) of gas, while the V term in Equation 6.13 represents the volume. Let's rearrange Equation 6.13 so that the ratio of amount (n) to volume (V) is on one side:

$$\frac{P}{RT} = \frac{n}{V}$$

For a given sample, the term n can be defined as the mass of a substance (in grams) divided by its molar mass ($\mathcal{M}$): $m/(\mathcal{M})$. We can substitute ($m/\mathcal{M}$) into the equation for n and solve the resulting expression for m/V:

$$PV = \frac{m \times RT}{\mathcal{M}} \qquad \text{so} \qquad \frac{\mathcal{M}P}{RT} = \frac{m}{V}$$

Look at the units associated with the terms on the left side of the right-hand equation above: $\mathcal{M}$ is in grams per mole, P is in atmospheres, R is in $L \cdot atm/(K \cdot mol)$, and T is in kelvins:

$$\frac{\left(\dfrac{\text{g}}{\text{mol}}\right)(\text{atm})}{\left(\dfrac{\text{L} \cdot \text{atm}}{\text{K} \cdot \text{mol}}\right)(\text{K})} = \text{g/L} = \text{density}$$

We can write an expression for the density of a gas based on the pressure (P), temperature (T), and molar mass ($\mathcal{M}$) of any gas by simply defining n and solving for grams per liter (density):

$$d = \frac{\mathcal{M}P}{RT} \qquad (6.18)$$

Figure 6.17 illustrates the relationship between density and molar mass for three pure gases and air. Note how the balloon containing $CO_2(g)$ sinks to the tabletop because of its greater density relative to air. In the absence of mixing, carbon dioxide moves to the lowest level accessible to it, which is why it filled the valley in the Dieng Plateau disaster and cut the inhabitants off from their supply of oxygen.

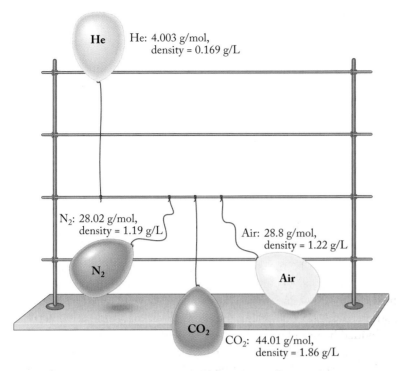

He: 4.003 g/mol,
density = 0.169 g/L

N₂: 28.02 g/mol,
density = 1.19 g/L

Air: 28.8 g/mol,
density = 1.22 g/L

CO₂: 44.01 g/mol,
density = 1.86 g/L

FIGURE 6.17 The densities of the gases in the balloons relative to air (at 15°C and 1 atm) determine whether they float (helium), hover just slightly above the bench top (nitrogen and air, which is four-fifths nitrogen), or sink (carbon dioxide).

SAMPLE EXERCISE 6.10 **Calculating the Density of a Gas**

Calculate the density of air at 1.00 atm and 302 K. Assume that air has an average molar mass of 28.8 g/mol. (The average molar mass of air is the weighted average of the molar masses of the gases present in air.)

COLLECT AND ORGANIZE We are provided with values for average molar mass, temperature, and atmospheric pressure of air. These three values will be used to calculate the density of air.

ANALYZE We are given the molar mass of air, the atmospheric pressure, and a temperature to use in the calculation of density. These quantities are related by Equation 6.18:

$$d = \frac{\mathcal{M}P}{RT}$$

SOLVE Inserting the values of P, T, and $\mathcal{M}$ into Equation 6.18, we have:

$$d_{air} = \frac{\mathcal{M}_{air}P_{air}}{RT_{air}} = \frac{\left(28.8\,\frac{g}{mol}\right)(1.00\,\text{atm})}{(0.08206\,\text{L} \cdot \text{atm/mol} \cdot \text{K})(302\,\text{K})} = 1.16\ g/L$$

THINK ABOUT IT The solution confirms that the density of air cited in Section 6.1 is indeed ~1.2 g/L. Note that the density calculated at 302 K is slightly less than the value at 288 K in Figure 6.17.

Practice Exercise Air is a mixture of mostly nitrogen and oxygen. A balloon is filled with oxygen and released in a room full of air. Will it sink to the floor or float to the ceiling?

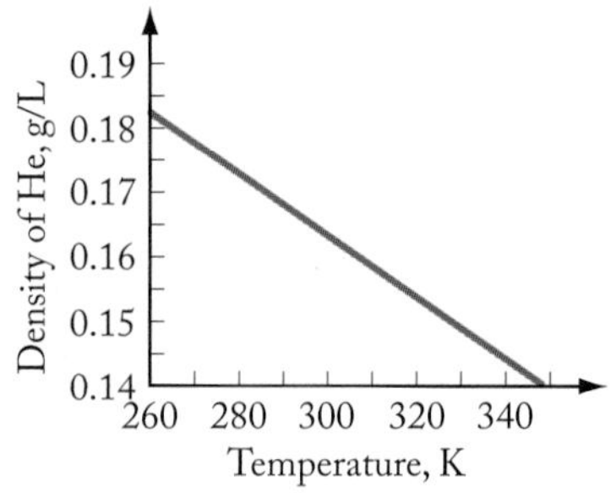

FIGURE 6.18 At 1 atm, the density of helium decreases with increasing temperature. The density of any ideal gas at constant pressure is inversely proportional to its temperature in kelvins.

CONNECTION In Chapter 3 we learned that molar mass is an important, fundamental property of a compound, and we saw how mass spectrometry is used to establish the molecular mass of a substance. Measuring mass, V, T, and P of a quantity of gas and applying the ideal gas law is a way to measure the molar mass of a gaseous substance.

CONCEPT TEST

Which of the following changes will result in the greater decrease in the density of a gas? (a) decreasing its pressure from 2.00 to 1.00 atm; (b) increasing its temperature from 20°C to 40°C

The density of a gas also depends on temperature (Figure 6.18). Equation 6.18 tells us that density is inversely proportional to temperature, which means that density decreases as temperature increases. Thus we can use this equation to calculate the molar mass of a gas from its density at any temperature and pressure. To do so, we rearrange the terms in Equation 6.18 to solve for molar mass:

$$\mathcal{M} = \frac{dRT}{P} \tag{6.19}$$

Gas density can be measured with a specially designed glass bulb of known volume attached to a vacuum pump. The mass of the bulb is determined when it has essentially no gas in it and again when it is filled with the test gas. The difference in the masses divided by the internal volume of the bulb is the density of the gas.

SAMPLE EXERCISE 6.11 **Calculating Molar Mass from Density**

Vent pipes at solid-waste landfills often emit foul-smelling gases; these may be relatively pure substances or mixtures of several gases. A sample of such an emission is found to have a density of 0.65 g/L at 25°C and 757 mmHg. What is the molar mass of the emission? (Note that if the gaseous sample is a mixture, the answer will be the weighted average of molar masses of the individual gases in the mixture.)

COLLECT AND ORGANIZE We are given the density, temperature, and pressure of a gas and asked to calculate the molar mass of the gaseous sample.

ANALYZE Equation 6.19 relates molar mass ($\mathcal{M}$) to the density, pressure, and temperature of a gas, so we can use it to calculate the molar mass of a pure gas or the weighted average mass of a mixture of gases. We need to convert temperature from °C to kelvins:

$$T = 25°C + 273 = 298 \text{ K}$$

We need pressure units expressed in atmospheres, so we must convert pressure from millimeters of Hg to atmospheres:

$$P = (757 \text{ mmHg})\left(\frac{1 \text{ atm}}{760 \text{ mmHg}}\right)$$

$$= 0.996 \text{ atm}$$

SOLVE Inserting these values into Equation 6.19 yields the following:

$$\mathcal{M} = \frac{dRT}{P} = \left(\frac{0.65 \text{ g}}{\text{L}}\right)\left(\frac{0.08206 \text{ L} \cdot \text{atm}}{\text{mol} \cdot \text{K}}\right)\left(\frac{298 \text{ K}}{0.996 \text{ atm}}\right)$$

$$= 16.0 \text{ g/mol}$$

THINK ABOUT IT A principal ingredient in the gases emitted by decomposing solid waste is methane (CH_4), which has a molar mass of 16.0 g/mol. The gases coming from the vent could also be a mixture of gases (possibly including methane), in which case the density would have been the weighted average of the mixture's components.

Practice Exercise When solutions of hydrochloric acid, HCl(*aq*), and sodium bicarbonate, NaHCO$_3$(*aq*), are mixed together, a chemical reaction takes place in which a gas is one of the products. A sample of the gas has a density of 1.81 g/L at 1.00 atm and 23°C. What is the molar mass of the gas? Can you identify the gas?

6.6 Dalton's Law and Mixtures of Gases

We noted early in this chapter that air is a mixture of mostly N$_2$ and O$_2$ and smaller proportions of other gases (see Table 6.1). Each gas in the mixture exerts its own pressure, called a **partial pressure**. Atmospheric pressure is the sum of the partial pressures of all of the gases in the air:

$$P_{atm} = P_{N_2} + P_{O_2} + P_{Ar} + P_{CO_2} + \ldots$$

A similar expression can be written for any mixture of gases:

$$P_{total} = P_1 + P_2 + P_3 + P_4 + \ldots \tag{6.20}$$

Dalton's law of partial pressures states that the total pressure of any mixture of gases equals the sum of the partial pressures of each component gas in the mixture. Summing partial pressures to calculate the total pressure of a mixture of gases as in Equation 6.20 illustrates Dalton's law.

CHEMTOUR Dalton's Law

You might guess that the most abundant gases in a mixture have the greatest partial pressures and contribute the most to the total pressure of the mixture. The mathematical term used to express the abundance of each component is its **mole fraction (X$_x$)**, which is the ratio of the moles of the component gas to the total number of moles in the mixture:

$$X_x = \frac{n_x}{n_{total}} \tag{6.21}$$

The concentration scale using mole fractions has three characteristics worth noting: (a) unlike molarity, mole fractions have no units; (b) unlike molarity, mole fractions are mass based and can be used for any kind of mixture or solution (solid, liquid, or gas); and (c) the mole fractions of all the components of a mixture add up to exactly 1.

To see how mole fractions work, consider a portion of the atmosphere (air) that contains 100.0 moles of atmospheric gases. Contained in these 100.0 moles are 20.9 moles of O$_2$ and 78.1 moles of N$_2$. The mole fraction of O$_2$ (X_{O_2}) in air is the ratio of the number of moles of O$_2$ to the total number of moles in the mixture:

$$X_{O_2} = \frac{n_{O_2}}{n_{total}} = \frac{20.9}{100.0}$$

$$= 0.209$$

Similarly, the mole fraction of N$_2$ is

$$X_{N_2} = \frac{n_{N_2}}{n_{total}} = \frac{78.1}{100.0}$$

$$= 0.781$$

A **partial pressure** is the contribution to the total pressure made by a component gas in a gas mixture.

Dalton's law of partial pressures states that the total pressure of any mixture of gases equals the sum of the partial pressures of each gas in the mixture.

The **mole fraction (X$_x$)** of a substance is the ratio of the number of moles of a component in a mixture to the total number of moles in a mixture.

CONCEPT TEST

Can the partial pressure of one component of a gas be measured directly?

The partial pressure of O_2 (P_{O_2}) in air is the product of the mole fraction of O_2 times total atmospheric pressure. If the total pressure is 1.00 atm, then the partial pressures of O_2 and N_2 are as follows:

$$P_{O_2} = \chi_{O_2} P_{total} \qquad\qquad P_{N_2} = \chi_{N_2} P_{total}$$
$$= (0.209)(1.00 \text{ atm}) \qquad\qquad = (0.781)(1.00 \text{ atm})$$
$$= 0.209 \text{ atm} \qquad\qquad\quad = 0.781 \text{ atm}$$

These two equations are examples of the general equation for the partial pressure of a gas in a mixture of gases:

$$P_x = \chi_x P_{total} \qquad\qquad (6.22)$$

The partial pressure of a gas is proportional to the quantity of the gas in a given volume and does not depend on the identity of the gas (Figure 6.19).

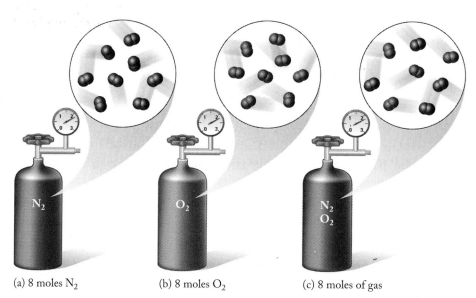

(a) 8 moles N_2 (b) 8 moles O_2 (c) 8 moles of gas

FIGURE 6.19 Pressure is directly proportional to number of moles of an ideal gas independent of the identity of the gas. All three containers have the same volume and are at the same temperature. (a) Eight moles of nitrogen exert the same pressure as (b) 8 moles of oxygen and as (c) an 8 mole mixture of gases (4 moles of nitrogen and 4 moles of oxygen).

SAMPLE EXERCISE 6.12   **Calculating Mole Fraction**

Scuba divers who descend more than 130 feet below the surface may breathe a gas mixture called Trimix, which is 11.7% He, 56.2% N_2, and 32.1% O_2 by mass. Calculate the mole fraction of each gas in this mixture.

COLLECT AND ORGANIZE We are given the composition of a gas mixture in terms of mass percentages to be converted to mole fractions of each component. Mole fraction is defined in Equation 6.21, and to calculate it we need the total number of moles of gas in the mixture and the number of moles of each component.

ANALYZE If we consider a 100.0 gram sample of Trimix, we can calculate the mass of each gas in the sample from the mass percentages given in the problem. Then we can convert each mass to the number of moles, from which we can determine the total number of moles in the sample and the mole fraction of each gas in Trimix.

SOLVE Trimix is a solution of three gases. We can pick any quantity of sample we would like to do this calculation; because we are given mass percentages, let's consider 100.0 g of Trimix. One hundred grams of Trimix contains 11.7 g of He, 56.2 g of N_2, and 32.1 g of O_2. Using the molar mass of each gas to convert these quantities into moles yields the following:

$$(11.7 \text{ g He})\left(\frac{1 \text{ mol He}}{4.003 \text{ g He}}\right) = 2.92 \text{ mol He}$$

$$(56.2 \text{ g } N_2)\left(\frac{1 \text{ mol } N_2}{28.01 \text{ g } N_2}\right) = 2.01 \text{ mol } N_2$$

$$(32.1 \text{ g } O_2)\left(\frac{1 \text{ mol } O_2}{32.00 \text{ g } O_2}\right) = 1.00 \text{ mol } O_2$$

Mole fraction values are calculated with Equation 6.21: $X_x = n_x/n_{total}$. The total number of moles in the 100.0 gram sample of Trimix is the sum of the number of moles of the constituent gases: He, N_2, and O_2.

$$n_{total} = 2.92 + 2.01 + 1.00 = 5.93 \text{ mol}$$

Thus, X_{He}, X_{N_2}, and X_{O_2} are:

$$X_{He} = \frac{2.92 \text{ mol He}}{5.93 \text{ mol}} = 0.492$$

$$X_{N_2} = \frac{2.01 \text{ mol } N_2}{5.93 \text{ mol}} = 0.339$$

$$X_{O_2} = \frac{1.00 \text{ mol } O_2}{5.93 \text{ mol}} = 0.169$$

Check the answer by summing the mole fractions; they should add up to 1.

$$X_{He} + X_{N_2} + X_{O_2} = 1$$

$$0.492 + 0.339 + 0.169 = 1.00$$

THINK ABOUT IT Alternatively, we could calculate the mole fraction of oxygen by subtracting the sum of the mole fractions of He and N_2 from 1.00:

$$X_{O_2} = 1 - (X_{He} + X_{N_2}) = 1 - (0.492 + 0.339) = 0.169$$

Note that mole fraction is a concentration scale based on number of moles of material present. While nitrogen is present in the greatest mass in this Sample Exercise, the concentrations of the three gases indicate there are more helium atoms in the mixture than molecules of either nitrogen or oxygen.

Practice Exercise A different gas mixture, 6.11% O_2 and 93.89% He by mass, is known as Heliox and is used in scuba tanks for descents more than 190 feet below the surface. Calculate the mole fraction of He and O_2 in this mixture.

Explain why the sum of the mole fractions in a mixture must add up to 1.

Let's consider the implications of Equation 6.22 in the context of the "thin" air at high altitudes. The mole fraction of oxygen in air is 0.209, which does not change significantly with increasing altitude. However, the total pressure of the atmosphere, and therefore the partial pressure of oxygen, decreases with increasing altitude, as illustrated in the following Sample Exercise.

SAMPLE EXERCISE 6.13   **Calculating Partial Pressure**

Calculate the partial pressure (in atmospheres) of O_2 in the air outside an airplane cruising at an altitude of 33,000 feet (about 10 km), where the atmospheric pressure is 190.0 mmHg. Assume that the mole fraction of O_2 in air is 0.209.

COLLECT AND ORGANIZE We are given the mole fraction of oxygen at a given total atmospheric pressure and asked to calculate the partial pressure of oxygen. Equation 6.22 relates those two values.

ANALYZE Using Equation 6.22, the partial pressure of a gas in a gas mixture can be determined from its mole fraction in the mixture and the total gas pressure. In the calculation we do not need to use the elevation, which is provided only as "background"—an indicator of where we might encounter the conditions described in the problem.

SOLVE We insert the given values into Equation 6.22 and solve:

$$P_{O_2} = \chi_{O_2} P_{total} = (0.209)(190.0 \text{ mmHg})\left(\frac{1 \text{ atm}}{760 \text{ mmHg}}\right)$$

$$= 0.0523 \text{ atm}$$

THINK ABOUT IT The partial pressure of a gas is the contribution that gas makes to the total pressure of a gas mixture. As we move upward from the surface of Earth, the atmosphere becomes thinner (less dense). Our answer of 0.0523 atm makes sense because, at an altitude of 10 km, the air is much less dense than at sea level, where oxygen has a much higher partial pressure of 0.209 atm.

Practice Exercise Assume a scuba diver is working at a depth where the total pressure is 5.0 atm (about 50 m below the surface). What is the mole fraction of oxygen necessary so that the partial pressure of oxygen in an artificial gas mixture that the diver breathes will be 0.21 atm?

Dalton's law of partial pressures is useful in the laboratory when we need to know the pressure exerted by a gaseous product of a chemical reaction. One way to collect a gaseous product is in an inverted vessel filled with water (Figure 6.20). For example, heating potassium chlorate ($KClO_3$) causes it to decompose into $KCl(s)$ and $O_2(g)$. The oxygen gas produced by this reaction can be collected by bubbling the gas into an inverted bottle that is initially filled with water. As the reaction proceeds, O_2 displaces the water in the bottle. When the reaction is complete, the volume of water displaced provides a measure of the volume of O_2 produced. If the temperature of the water and the barometric pressure are known, then the number of moles of O_2 produced can be calculated using the ideal gas law.

In this section we examine the **kinetic molecular theory** of gases, a unifying theory developed in the 19th century that explains the relations described by the ideal gas law and its predecessors, the laws of Boyle, Charles, Dalton, Amontons, and Avogadro.

Here are the main assumptions of the kinetic molecular theory:

1. Gas molecules have tiny volumes compared with the collective volume they occupy. Their individual volumes are so small as to be considered negligible, allowing particles in a gas to be treated as so-called point masses—masses with essentially no volume. Gas molecules are separated by large distances; hence a gas is mostly empty space.

2. Gas molecules move constantly and randomly throughout the volume they collectively occupy.

3. The motion of these molecules is associated with an average kinetic energy that is proportional to the absolute temperature of the gas. All populations of gas molecules at the same temperature have the same average kinetic energy.

4. Gas molecules continuously collide with one another and with their container walls. These collisions are *elastic*; that is, they result in no net transfer of energy to the walls. Therefore, the average kinetic energy of gas molecules is not affected by these collisions and remains constant as long as there is no change in temperature.

5. Each molecule acts independently of all the others in the sample. We assume there are no forces of attraction or repulsion between the molecules.

To illustrate and test the kinetic molecular theory of gases, let's consider the examples provided by air (a mixture primarily of nitrogen and oxygen), and of the mixture of nitrogen, oxygen, and helium in the tank carried by a scuba diver on a dive.

> The **kinetic molecular theory** of gases is a model that describes the behavior of gases. It is based on a set of assumptions, and all the equations that define the relations between pressure, volume, temperature, and number of moles of gases (P, V, T, and n) can be derived from the theory.

> **CONNECTION** We talk about the kinetic energy of molecules in the same way we described the kinetic energy of skiers in Chapter 5. As with large objects, the kinetic energy of small particles such as molecules is energy of motion.

Explaining Boyle's, Dalton's, and Avogadro's Laws

We already have a picture in our minds that describes the origin of gas pressure. Collisions between gas molecules and the walls of their container generate a force. The more frequent the collisions, the greater the force and thus the greater the pressure. Compressing more molecules into a smaller space by reducing the volume of the container means more collisions take place per unit time, so the pressure increases (Boyle's law: $P \propto 1/V$). We can also increase the number of collisions by increasing the number of gas molecules (number of moles, n) in a container of fixed volume (because $P \propto n$). We can increase n by adding more of the same or a different gas to a container of fixed volume. If we add two different gases, such as N_2 and O_2, to a container in a $4:1$ mole ratio, then 4 times as many collisions will occur involving nitrogen molecules than oxygen molecules. Therefore, the pressure due to nitrogen molecules (P_{N_2}) will be 4 times that from oxygen molecules (P_{O_2}), and the total pressure will be $P_{total} = P_{O_2} + P_{N_2}$. This statement is precisely what Dalton proposed in his law of partial pressures.

Avogadro's law is also explained by the kinetic molecular theory. Avogadro observed that the volume of a gas at constant pressure is directly proportional to the number of moles of the gas. Because additional gas molecules in a given volume lead to increased numbers of collisions and an increase in pressure, the only way to reduce the pressure is to allow the gas to expand (Figure 6.21).

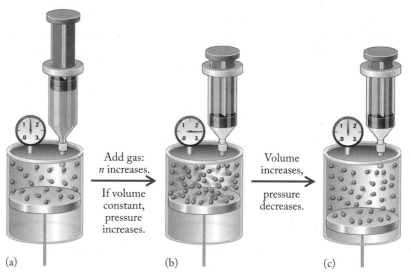

FIGURE 6.21 (a) Gas contained in a cylinder with a movable piston exerts a pressure equal to atmospheric pressure ($P_{gas} = P_{atm}$). (b) More gas is added to the cylinder (n increases), and P_{gas} increases because the number of collisions on the walls of the container increases. (c) For the pressure to remain the same, the volume (V_{gas}) must increase; hence, the gas expands.

Explaining Charles's and Amontons's Laws

The laws of both Amontons and Charles depend on temperature. Kinetic molecular theory predicts that, as temperature increases,

- gas molecules move faster and collide more frequently with the walls of their container; and

- gas molecules collide with greater average kinetic energy, resulting in a greater pressure in the container.

In other words, as the temperature of a gas increases, the pressure the gas exerts also increases, precisely as stated by Amontons's law. Figure 6.22(a, b) illustrates Amontons's law on a molecular level.

How do increased collisions and a higher average kinetic energy explain Charles's law? Remember that Charles's law relates volume and temperature *at constant pressure*. As the pressure increases with increasing temperature, the system must expand—increase its volume—in order to maintain a constant pressure (Figure 6.22b, c).

FIGURE 6.22 (a) As the temperature (T_{gas}) of a given amount of gas (n) in a cylinder increases and the volume remains constant, the number and forcefulness of molecular collisions with the walls of the container increase, causing (b) the pressure (P_{gas}) to increase, as stated by Amontons's law. As the temperature of the gas increases, if the pressure is to remain the same, (c) the volume must increase until $P_{gas} = P_{atm}$. Hence, as temperature increases at constant pressure, volume increases, as stated by Charles's law.

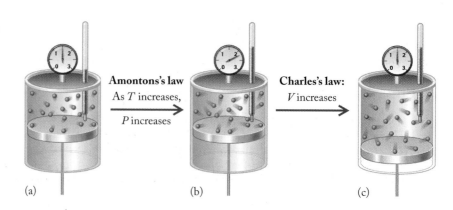

Molecular Speeds and Kinetic Energy

Kinetic molecular theory tells us that all populations of gas molecules at a given temperature have the same *average* kinetic energy. The kinetic energy of a single molecule or atom of a gas can be calculated by using the following equation:

$$KE = \tfrac{1}{2}mu^2$$

where m is the mass of a molecule of the gas and u is its velocity. Not all molecules within a population, however, are traveling at exactly the same speed at the same time. Even elastic collisions between two molecules may result in one molecule moving with a greater speed than the other after the collision. One molecule might even stop completely. Thus collisions between molecules result in gas molecules in any sample having a range of velocities.

Figure 6.23(a) shows a typical distribution of velocities in a population of gas molecules. The peak in the curve represents the *most probable speed* (u_m) of molecules within a population. It is the speed that characterizes the largest number of molecules in the gas. The distribution of all the speeds is not symmetrical; as a result the *average speed* (u_{avg}), which is simply the arithmetic average of all the different speeds of all the molecules in the population, is a little higher than the most probable speed. A very important quantity is the **root-mean-square speed** (**u_{rms}**); this is the speed of a molecule possessing the average kinetic energy. At a given temperature the entire population of molecules in a gas has the same *average* kinetic energy (KE) as every other population of gas molecules at that same temperature. The value of the average kinetic energy is defined as

$$KE_{avg} = \tfrac{1}{2}m(u_{rms})^2 \qquad (6.23)$$

The root-mean-square speed (u_{rms}) of a gas with molar mass $\mathcal{M}$ at temperature T is defined as

$$u_{rms} = \sqrt{\frac{3RT}{\mathcal{M}}} \qquad (6.24)$$

The **root-mean-square speed** (u_{rms}) is the square root of the average of the squared speeds of all the molecules in a population of gas molecules. A molecule possessing the average kinetic energy moves at this speed.

▶❚❚ **CHEMTOUR** Molecular Speed

◉◉ **CONNECTION** The equation for kinetic energy was introduced in Chapter 7, where m was the mass of an object. Here m is always the molar mass of the gas being described.

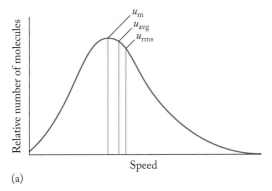

(a)

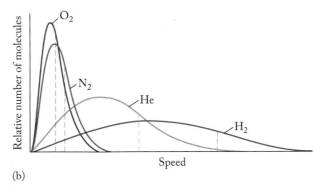

(b)

FIGURE 6.23 The speeds of gas atoms or molecules at a constant temperature cover a range of values. (a) The most probable speed (u_m), associated with the highest point in the plot, characterizes the largest fraction of molecules in the population; more molecules have the most probable speed than have any other speed. The average speed (u_{avg}), a little higher than the most probable speed, is simply the arithmetic average of all the speeds. The root-mean-square speed (u_{rms}), a little higher still, is directly proportional to the square root of the temperature of the gas (in kelvins) and inversely proportional to the square root of its molar mass. (b) The curves describing the relative numbers of molecules in a population plotted against speed at constant temperature show that the lower the molar mass, the higher the root-mean-square speed (u_{rms}) and the broader the distribution of speeds. (Approximate u_{rms} values are indicated by dashed lines.)

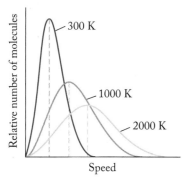

FIGURE 6.24 The most probable speed (u_m, dashed lines) increases with increasing temperature. As the temperature increases, notice that the distributions broaden, a smaller portion of atoms or molecules moves at any given speed, and more speeds are represented by a significant portion of the population.

The value of u_{rms} is typically expressed in meters per second and the temperature is expressed in kelvins, but care must be taken in choosing the units for R in Equation 6.24. What value of the gas constant R should we use? One value of R (see Table 6.3) that includes meters among its units is

$$R = 8.314 \text{ kg} \cdot \text{m}^2/(\text{s}^2 \cdot \text{mol} \cdot \text{K})$$

Using this value for R, however, requires that we express the molar mass of the gas in units of kilograms per mole rather than in the more common units of grams per mole.

Equation 6.24 indicates that, because different gases have different molar masses, heavier molecules, such as N_2 (28.01 g/mol = 2.801×10^{-2} kg/mol) move more slowly at a given temperature than lighter molecules, such as helium (4.003 g/mol = 4.003×10^{-3} kg/mol). Figure 6.23(b) shows the different distributions of speeds for several gases at a constant temperature.

Figure 6.24 shows how the most probable speed and the kinetic energy of gases increase with temperature. How fast do gas molecules move? Gas molecules move very rapidly at or slightly above room temperature—up to 1000 mi/hr and even faster (depending on temperature).

SAMPLE EXERCISE 6.15   **Calculating Root-Mean-Square Speeds**

Calculate the root-mean-square speed of nitrogen at 300.0 K. Express the answer in (a) meters per second and (b) miles per hour.

COLLECT AND ORGANIZE Given only the temperature and identity of a gas (N_2 in this case), we are asked to calculate the root-mean-square speed of nitrogen. According to Equation 6.24, we need the molar mass of the gas in question in addition to the temperature.

ANALYZE In this case, the gas is nitrogen (N_2), which has a molar mass of 28.01 g/mol. We also need the correct value of R. The second value listed in Table 6.3, $R = 8.314 \text{ kg} \cdot \text{m}^2/(\text{s}^2 \cdot \text{mol K})$, is the correct one for this calculation because it will give us a speed in meters per second. Having chosen this value of R, we then need to express the molar mass in kilograms; for N_2 this value is 0.02801 kg/mol.

SOLVE

a. Inserting the appropriate values into Equation 6.24 yields the root-mean-square speed in meters per second:

$$u_{rms,N_2} = \sqrt{\frac{3RT}{\mathcal{M}}} = \sqrt{\frac{3\left[\dfrac{8.314 \; \cancel{\text{kg}} \cdot \text{m}^2}{\text{s}^2/(\cancel{\text{mol}} \cdot \cancel{\text{K}})}\right](300.0 \; \cancel{\text{K}})}{0.02801 \; \dfrac{\cancel{\text{kg}}}{1 \; \cancel{\text{mol}}}}}$$

$$= 516.9 \text{ m/s}$$

b. To convert speed from meters per second into miles per hour (mi/hr) requires the use of several conversion factors:

$$1 \text{ mi} = 1.6093 \text{ km} = 1.6093 \times 10^3 \text{ m} \quad \text{and} \quad 1 \text{ hr} = 3600 \text{ s}$$

Using these factors gives us the requested result in miles per hour:

$$u_{N_2} = \left(516.9 \; \frac{\cancel{\text{m}}}{\cancel{\text{s}}}\right)\left(\frac{1 \text{ mi}}{1.6093 \times 10^3 \; \cancel{\text{m}}}\right)\left(3600 \; \frac{\cancel{\text{s}}}{\text{hr}}\right)$$

$$= 1156 \text{ mi/hr} = 1.156 \times 10^3 \text{ mi/hr}$$

THINK ABOUT IT It's quite amazing to find that the average N_2 molecule at a temperature close to room temperature moves at more than 1000 mi/hr.

Practice Exercise Calculate the root-mean-square speed of helium at 300 K. Express the answer in meters per second, and compare your result with the root-mean-square speed of nitrogen.

CONCEPT TEST

Put the following gas molecules in order of increasing root-mean-square speed at a temperature of 20°C: H_2, CO_2, Ar, SF_6, UF_6, Kr.

It is not immediately obvious why the pressure exerted by two different gases, such as N_2 and He, is the same at a constant temperature when their root-mean-square speeds are quite different. The answer lies in the assumption in the kinetic molecular theory that all populations of gas molecules have the same average kinetic energy at a given temperature, independent of their molar mass. Can we show that two gases, such as nitrogen and helium, have the same average kinetic energy at 300 K? Let's calculate the average kinetic energy of 1 mole of N_2 and 1 mole of He at 300 K. The solution to Sample Exercise 6.15 reveals that the u_{rms} of N_2 molecules at 300 K is 517 m/s. The corresponding root-mean-square speed for helium atoms is 1370 m/s (the result you should have gotten for the last Practice Exercise). For 1 mole of each gas we can substitute into Equation 6.23 as follows:

$$KE_{N_2} = \frac{1}{2} m_{N_2} (u_{rms,N_2})^2 = \frac{1}{2}(2.801 \times 10^{-2} \text{ kg/mol})(517 \text{ m/s})^2$$

$$= 3.75 \times 10^3 \text{ kg} \cdot \text{m}^2/(\text{s}^2 \cdot \text{mol})$$

$$KE_{He} = \frac{1}{2} m_{He} (u_{rms,He})^2 = \frac{1}{2}(4.003 \times 10^{-3} \text{ kg/mol})(1370 \text{ m/s})^2$$

$$= 3.75 \times 10^3 \text{ kg} \cdot \text{m}^2/(\text{s}^2 \cdot \text{mol})$$

This calculation demonstrates that the same quantities (same values of n) of the two *different* gases have the identical average kinetic energy. Therefore at the same temperature (T) they exert the same pressure (P). The slower nitrogen molecules collide with the container walls less often but exert a greater force than the faster helium atoms because nitrogen molecules are heavier.

Equation 6.24 enables us to compare the relative root-mean-square speeds of two gases at the same temperature. Consider an equimolar mixture (one with the same number of moles) of nitrogen and helium. We have just demonstrated that at a constant temperature (300 K) their average kinetic energies are the same:

$$KE_{N_2} = \frac{1}{2} m_{N_2} (u_{rms,N_2})^2 = \frac{1}{2} m_{He} (u_{rms,He})^2 = KE_{He}$$

or

$$m_{N_2} (u_{rms,N_2})^2 = m_{He} (u_{rms,He})^2$$

Rearranging the terms to express the ratio of the root-mean-square speeds of the two gases in terms of the ratio of their molar masses, and then taking the square root of each side, gives

$$\frac{(u_{rms,He})^2}{(u_{rms,N_2})^2} = \frac{m_{N_2}}{m_{He}}$$

$$\frac{u_{rms,He}}{u_{rms,N_2}} = \sqrt{\frac{m_{N_2}}{m_{He}}} = \sqrt{\frac{28.01 \text{ g}}{4.003 \text{ g}}} = 2.65$$

The root-mean-square speed of helium atoms at 300 K is 2.65 times higher than that of nitrogen molecules. We can test this using the values for u_{rms} from the calculation of average kinetic energies:

$$\frac{u_{rms,He}}{u_{rms,N_2}} = \frac{1370 \text{ m/s}}{517 \text{ m/s}} = 2.65$$

We can write a generic form of the preceding equation that relates to any pair of gases (x and y) to their masses:

$$\frac{u_{rms,x}}{u_{rms,y}} = \left(\frac{m_y}{m_x}\right)^{1/2} \tag{6.25}$$

Equation 6.25 holds only when we have the same number of particles of each gas, say 1 mol of gas x and 1 mol of gas y. Equation 6.25 leads to the conclusion that the root-mean-square speed of more massive particles is lower than u_{rms} of lighter particles.

Graham's Law: Effusion and Diffusion

Let's observe what happens to two balloons, one filled with nitrogen and another with helium (Figure 6.25). The volume, temperature, and pressure of the gas in each balloon are identical. The pressure inside each balloon, however, is greater than the external atmospheric pressure. Over time we observe that the volume of the helium balloon decreases significantly while the nitrogen-filled balloon has nearly the same volume as it had at the start of our experiment. The skin of any balloon is slightly permeable; that is, it has microscopic holes that allow gas to escape, reducing the pressure inside the balloon. Why does the helium balloon leak faster than the balloon filled with nitrogen? We now know the relation between the root-mean-square speeds and molar masses of two different gases, and we have calculated that helium atoms move about 2.65 times faster than nitrogen atoms. It appears that the gas with the greater root-mean-square speed (He) leaks out of the balloon at a higher rate. This process of moving out of a space at higher pressure through a small opening into a region of lower pressure is called **effusion**. Helium effuses through the imperfections in a rubber balloon into a region of lower pressure more rapidly than nitrogen.

In the 19th century, Scottish chemist Thomas Graham (1805–1869) recognized that the rate of effusion of a gas is related to its molar mass. Today this relation is known as **Graham's law of effusion**, which states that the rate of effusion of gases is inversely proportional to the square root of their molar masses. We can derive a mathematical representation of Graham's law by substituting rate (r) for root-mean-square speed in Equation 6.25:

$$\frac{r_x}{r_y} = \frac{u_{rms,x}}{u_{rms,y}} = \sqrt{\frac{m_y}{m_x}} \tag{6.26}$$

From Equation 6.24 we know that:

$$u_{rms} = \sqrt{\frac{3RT}{\mathcal{M}}}$$

Effusion is the process by which a gas escapes from its container through a tiny hole into a region of lower pressure.

Graham's law of effusion states that the rate of effusion of a gas is inversely proportional to the square root of its molecular mass.

By substituting r for u_{rms} in the above expression we obtain

$$\frac{r_x}{r_y} = \frac{\sqrt{\dfrac{3RT}{\mathcal{M}_x}}}{\sqrt{\dfrac{3RT}{\mathcal{M}_y}}}$$

which simplifies to the usual statement of Graham's law:

$$\frac{r_x}{r_y} = \sqrt{\frac{\mathcal{M}_y}{\mathcal{M}_x}} \tag{6.27}$$

Using Equation 6.27 in the example of the helium ($\mathcal{M}_{He} = 4.003$ g/mol) and nitrogen ($\mathcal{M}_{N_2} = 28.01$ g/mol) balloons, we can show that the rate of effusion of helium is 2.64 times that of nitrogen:

$$\frac{r_{He}}{r_{N_2}} = \sqrt{\frac{\mathcal{M}_{N_2}}{\mathcal{M}_{He}}} = \sqrt{\frac{28.01 \text{ g/mol}}{4.003 \text{ g/mol}}} = 2.64$$

Notice that this expression has the same form—and hence the ratio has the same value—as the ratio of the root-mean-square speeds of the two gas molecules.

How does the kinetic molecular theory explain Graham's law of effusion? The escape of a gas molecule from a balloon (see Figure 6.25) requires that the molecule collide with the pinhole. The faster a gas molecule moves, the more likely it is to find the hole.

The effusion of gas through a tiny hole in a container is related to the process of **diffusion**, the spread of one substance through another. A couple of examples are the diffusion of perfume molecules from a person throughout a room, and the smell of bread baking from the kitchen throughout a house. The rates of diffusion depend on the average speeds of the molecules of gas and on their molar masses. Graham's law applies equally well to the diffusion of gases.

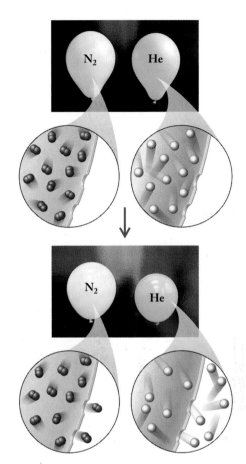

FIGURE 6.25 Two balloons at the same temperature and pressure and filled with equal volumes of nitrogen (N_2) and helium (He) both decrease in volume over time as gas effuses through small pores in the membranes of the balloons. The helium atoms are lighter than the nitrogen molecules and therefore have higher root-mean-square speeds. As a consequence, the helium atoms effuse faster, causing the helium-containing balloon to shrink more quickly than the nitrogen balloon.

SAMPLE EXERCISE 6.16 **Applying Graham's Law to the Diffusion of Gases**

An odorous gas emitted by a hot spring was found to diffuse 2.92 times faster than helium. What is the molar mass of this gas?

COLLECT AND ORGANIZE We are asked to determine the molar mass of an unknown gas based on its rate of diffusion relative to helium.

ANALYZE Graham's law (Equation 6.27) relates the relative rate of diffusion of two gases to their molar masses. We know that helium ($\mathcal{M}_{He} = 4.003$ g/mol) diffuses 2.92 times faster than the unidentified gas (y).

SOLVE We insert the atomic mass of He and the ratio of their diffusion rates into Equation 6.27 and solve for molar mass of the unknown gas:

$$\frac{r_{He}}{r_y} = \sqrt{\frac{\mathcal{M}_y}{\mathcal{M}_{He}}}$$

$$2.92 = \sqrt{\frac{\mathcal{M}_y}{4.003 \text{ g/mol}}}$$

$$\mathcal{M}_y = (4.003 \text{ g/mol})(2.92)^2 = 34.1 \text{ g/mol}$$

Diffusion is the spread of one substance (usually a gas or liquid) through another.

THINK ABOUT IT Two things about Equation 6.27 for Graham's law are worth noting: (1) like the root-mean-square speed in Equation 6.25, Graham's law involves a square root. (2) With labels x and y for two gases in a mixture, the left side of Equation 6.27 has the x term in the numerator and the y term in the denominator, while the right side of the equation has the opposite: y term in the numerator and x term in the denominator. Keeping the labels straight is the key to solving problems involving Graham's law. The molar mass of the unidentified gas is 34.1 g/mol. Coupled with other data, such as the distinctive odor of the compound, its density, and its source, a scientist might suspect it to be hydrogen sulfide (H_2S).

Practice Exercise Helium effuses 3.16 times as fast as which other noble gas?

CONCEPT TEST

Put the following gas molecules in order of increasing rate of effusion from a container at a temperature of 20°C: H_2, CO_2, Ar, SF_6, UF_6, Kr.

6.8 Real Gases

Up to now we have treated all gases as ideal gases. Under typical atmospheric pressures and temperatures, most gases *do* behave as ideal gases. We have also assumed, according to the kinetic molecular theory, that the volume occupied by individual gas molecules is negligible compared with the total volume occupied by the gas. In addition, we have assumed that no interactions occur between gas molecules other than random elastic collisions. This situation changes, however, when we begin to compress gases into increasingly smaller volumes or cool them to lower temperatures. Gas properties then begin to deviate from ideal behavior.

Deviations from Ideality

Let's start by considering the behavior of 1.0 mole of a gas as we substantially increase the pressure on the gas. From the ideal gas law, we know that $PV/RT = n$, so for 1 mole of gas, PV/RT should remain equal to 1.0 regardless of how we change the pressure. This relation between PV/RT and P for an ideal gas is shown by the purple line in Figure 6.26. However, plots of PV/RT versus P for CH_4, H_2, and CO_2 at very high pressure do not have flat, straight lines. Not

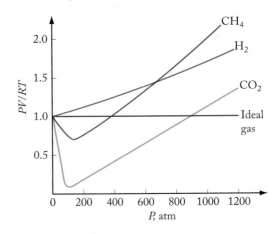

FIGURE 6.26 The effect of high pressure on the behavior of real and ideal gases: The curves diverge from ideal behavior in a manner unique to each gas. When gases deviate from ideal behavior, their identity matters.

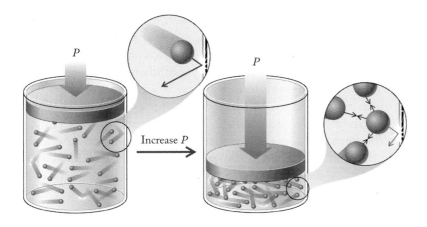

FIGURE 6.27 At high pressures, more of the volume of a gas is occupied by the atoms or molecules of the gas. The greater density of particles also results in more interactions between them (black arrows in expanded view). These interactions reduce the frequency and force of collisions with the walls of the container (red arrow), thereby reducing the pressure.

only do the curves diverge from the ideal value of PV/RT but also the curve is different for each gas, indicating that in these circumstances the identity of the gas does matter.

Why don't real gases behave like ideal gases at high pressure? One reason is that the ideal gas law considers gas molecules to have so little volume compared with the volume of their container that they are assumed to have no volume at all. However, at high pressures, more molecules are squeezed into a given volume of space (Figure 6.27). Under these conditions the volume of the individual molecules can become significant. What is the impact of this occurrence on the relation $PV/RT = 1$?

In the expression PV/RT, the volume (V) actually refers to the *free volume* ($V_{\text{free volume}}$), the empty space not occupied by gas molecules. Because $V_{\text{free volume}}$ is difficult to measure, we instead measure V_{total}, the total volume of the container holding the gas:

$$V_{\text{total}} = V_{\text{free volume}} + V_{\text{molecules}}$$

In an ideal gas the molecules have "no" volume ($V_{\text{molecules}} = 0$), so $V_{\text{total}} = V_{\text{free volume}}$. In a real gas, when the external pressure is low and the volume occupied by the gas is large, the volume contributed by all the molecules collectively is so minuscule that $V_{\text{molecules}}$ is essentially 0. Under these conditions, then, the ideal gas equation is a reasonable predictor of the behavior of a real gas.

Now consider what happens to a real gas in a closed but flexible container (like a balloon) as the conditions are changed. Increasing the external pressure on the container causes the container—and hence the volumes occupied by the gas—to shrink. As the external pressure increases, the assumption that $V_{\text{molecules}} = 0$ becomes less and less valid because the proportion of the container's volume taken up by the molecules themselves becomes increasingly significant ($V_{\text{molecules}} > 0$). The relation $V_{\text{total}} = V_{\text{free volume}} + V_{\text{molecules}}$ still applies, but $V_{\text{free volume}}$ can no longer be approximated by V_{total}. Since $V_{\text{total}} > V_{\text{free volume}}$, the ratio PV/RT is also larger for a real gas than for an ideal gas:

$$\frac{PV_{\text{total}}}{RT} > \frac{PV_{\text{free volume}}}{RT}$$

As a consequence, the curve for a real gas in Figure 6.26 diverges upward from the ideal gas line.

A second situation also causes the PV/RT ratio to diverge from 1. Kinetic molecular theory assumes that molecules do not interact, but molecules do attract

The **van der Waals equation** attempts to account for the behavior of real gases by including experimentally determined factors that quantify the contributions of molecular volume and intermolecular interactions to the properties of gases.

each other. The attractive forces are much weaker than the bonds that hold atoms together in molecules, but they arise in all matter because of imbalances in the distribution of electrons. These attractive forces function over short distances, so the assumption that molecules behave independently is a good one as long as the molecules are far apart. As pressure increases on a population of gas molecules, pushing them closer and closer together, intermolecular attractive forces can become significant. This causes the molecules to associate with each other, which decreases the force of their collisions with the walls of their container, thereby decreasing the pressure exerted by the gas. If the value of P in PV/RT is smaller than the ideal gas value, then the value of the ratio decreases and the curve for a real gas diverges from the ideal gas line, moving below it on the graph.

The van der Waals Equation for Real Gases

Because the ideal gas equation does not hold at high pressures, it is useful to find a relation that can be used under nonideal conditions. In determining a new equation for real gases, we must consider that

1. the true volume will be less than the volume of an ideal gas since the molecules themselves occupy significant space; and
2. the observed pressure will be less than the pressure of an ideal gas because of molecular attractions.

The **van der Waals equation**

$$\left(P + \frac{n^2 a}{V^2}\right)(V - nb) = nRT \qquad (6.28)$$

includes terms to correct for the pressure ($n^2 a/V^2$) and for the volume (nb). The symbols a and b are called *van der Waals constants*; their values have been determined experimentally for each gas. Table 6.5 lists the van der Waals constants for common atmospheric gases. Both a and b increase with increasing molar mass and the number of atoms in the molecule.

TABLE 6.5 Van der Waals Constants of Selected Gases

Substance	a (L^2 · atm/mol^2)	b (L/mol)
He	0.0341	0.02370
Ar	1.34	0.0322
H_2	0.244	0.0266
N_2	1.39	0.0391
O_2	1.36	0.0318
CH_4	2.25	0.0428
CO_2	3.59	0.0427
CO	1.45	0.0395
H_2O	5.46	0.0305
NO	1.34	0.02789
NO_2	5.28	0.04424
HCl	3.67	0.04081
SO_2	6.71	0.05636

SAMPLE EXERCISE 6.17   **Calculating Pressure with the van der Waals Equation**

Calculate the pressure of 1.00 mol of N_2 in a 1.00 L container at 300.0 K using the van der Waals equation. Compare the result of this calculation with the pressure of N_2 calculated using the ideal gas equation.

COLLECT AND ORGANIZE We are given the amount of nitrogen, its volume, and its temperature. We are asked to calculate its pressure using the ideal gas equation and the van der Waals equation. Calculations with the van der Waals equation will require the use of the constants listed in Table 6.5.

ANALYZE Using the van der Waals equation (6.28) requires knowing the identity of the gas (N_2 in this case) and having values for the van der Waals constants a and b for N_2. Table 6.5 tells us these values are 1.39 $L^2 \cdot atm/mol^2$ and 0.0391 L/mol, respectively. Recall that a and b represent experimentally determined corrections for the interactions between molecules and the volume of the gas molecules themselves.

We rearrange the terms in Equation 6.28

$$\left(P + \frac{n^2 a}{V^2}\right)(V - nb) = nRT$$

to solve for pressure:

$$P = \frac{nRT}{V - nb} - \frac{n^2 a}{V^2}$$

SOLVE We insert the appropriate values of n, V, T, a, and b given in the problem and in Table 6.5 into the above equation:

$$P_{N_2} = \frac{(1.00 \text{ mol})\left(0.08206 \frac{L \cdot atm}{mol \cdot K}\right)(300.0 \text{ K})}{1.00 \text{ L} - (1.00 \text{ mol})0.0391 \frac{L}{mol}} - \frac{(1.00 \text{ mol})^2\left(1.39 \frac{L^2 \cdot atm}{mol^2}\right)}{(1.00 \text{ L})^2}$$

$$= 24.2 \text{ atm}$$

If nitrogen behaved as an ideal gas, then we would have

$$P = \frac{nRT}{V}$$

$$P_{N_2} = \frac{(1.00 \text{ mol})\left(0.08206 \frac{L \cdot atm}{mol \cdot K}\right)(300.0 \text{ K})}{1.00 \text{ L}}$$

$$= 24.6 \text{ atm}$$

THINK ABOUT IT The pressure in this example is many times greater than normal atmospheric pressure, so we would expect some difference in the calculated pressures. Furthermore, because of the likely deviation from ideal behavior, we would expect the pressure calculated using the ideal gas equation to be too high. On the basis of the ideal gas equation we predict the N_2 gas pressure to be 24.6 atm, while the actual value is 24.2 atm, or 0.4/24.6 = 2% lower than predicted.

Practice Exercise Assuming the same set of conditions stated in this Sample Exercise, calculate the ideal pressure occupied by 1.00 mol He in a 1.00 L container and the pressure predicted by the van der Waals equation. Based on your calculations, which gas behaves more ideally at 300 K, He or N_2?

Nitrogen is the most abundant element in the atmosphere, making up 78% of the atmosphere by volume. It is the most abundant free element available on Earth and the sixth most abundant element in the universe. However, it is a relatively minor constituent of Earth's crust (only about 0.003% N by mass). Nitrogen exists in the crust mostly in deposits of potassium nitrate (KNO_3) or sodium nitrate ($NaNO_3$). Common names for KNO_3 include saltpeter and niter. The latter name is the source of the name of the element itself: in Greek *nitro-* and *-gen* mean "niter forming."

Nitrogen in combined forms is essential to all life: it is the nutrient needed in greatest quantity by food crops. The continuous exchange of nitrogen between the atmosphere and the biosphere (Figure 6.28) is mediated by species of bacteria called diazotrophs ("nitrogen eaters") that live in the roots of leguminous plants such as peas, beans, and peanuts (Figure 6.29). These bacteria have the ability to convert atmospheric nitrogen (N_2) into ammonia (NH_3) under conditions of normal temperature and pressure in a process called *nitrogen fixation*. Other microorganisms then oxidize NH_3 to produce the oxides of nitrogen. Lightening, fires, and other events also produce relatively small quantities of nitrogen oxides. Nitrogen fixation is necessary to sustain life on Earth because ammonia is required for the formation of biologically essential nitrogen-containing molecules, such as proteins and DNA. The process of nitrogen fixation has enormous agronomic and economic significance.

A simple laboratory preparation of nitrogen is the thermal decomposition of ammonium nitrite:

$$NH_4NO_2(s) \rightarrow N_2(g) + 2\,H_2O(g)$$

Industrially, the only large-scale production of nitrogen is carried out by distilling liquid air. Nitrogen boils at $-196°C$, so it distills before oxygen, which boils at $-183°C$. Liquid nitrogen is a cryogen, which (as we learned in Chapter 3) means a very cold fluid. Because of this property, it is used widely in commerce where refrigeration is required. Nitrogen's relative lack of reactivity in the absence of high temperature and pressure coupled with its availability make it the agent of choice as a protective gas in the semiconductor industry to keep oxygen away from sensitive materials, and for several industrial processes involving welding or soldering. Oil companies pump nitrogen under high pressure into oil deposits to force crude oil to the surface (Figure 6.30).

In the early 20th century, mined nitrates were widely used in the manufacture of gunpowder and other explosives. For example, a mixture of potassium nitrate, sulfur, and carbon known as *black powder* is still the best fuse material ever discovered. It reacts exothermically and explosively, producing nitrogen and carbon dioxide. Rapid expansion of these hot gases adds to the explosive character of the reaction:

$$2\,KNO_3(s) + \tfrac{1}{8}S_8(s) + 3\,C(s) \rightarrow K_2S(s) + N_2(g) + 3\,CO_2(g)$$

When naval blockades cut off Germany's supplies of KNO_3 during World War I, German chemist Fritz Haber devel-

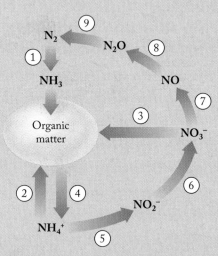

FIGURE 6.28 The nitrogen cycle describes the relationship between atmospheric nitrogen and the compounds of nitrogen necessary to support life on Earth. At any time, a large portion of nitrogen is contained in biomass: living organisms and their dead remains. When dead biomass decomposes, ammonium ion and other simple nitrogen compounds are released. (1) Nitrogen fixation provides ammonia to organisms as a major source of nitrogen. (2, 3) Microorganisms take up NH_4^+ and NO_3^-. (4) NH_4^+ is released during decomposition of organism remains. (5, 6) Microorganisms oxidize NH_4^+ to various oxides of nitrogen. (7–9) Microorganisms in anaerobic conditions (no oxygen present) use NO_3^- as an oxidizing agent and convert nitrate ion back into N_2.

FIGURE 6.29 The root system of a soybean plant, like peas and other legumes, contains nodules caused by the presence of nitrogen-fixing bacteria that live in a symbiotic relation with the plants.

oped a process for making nitrates, starting with the synthesis of ammonia from hydrogen gas and nitrogen from the atmosphere. The *Haber–Bosch process* relies on high temperature and pressure to promote the reaction:

$$N_2(g) + 3 H_2(g) \rightarrow 2 NH_3(g)$$

In this reaction the very strong nitrogen–nitrogen bond in N_2 must be broken, which requires an energy investment. Today, the *Haber–Bosch process* is the principal source of ammonia and the nitrogen compounds derived from ammonia that are widely used in industry and agriculture.

The first step in converting ammonia into other important nitrogen compounds entails its oxidation to NO_2 and H_2O:

$$4 NH_3(g) + 5 O_2(g) \rightarrow 4 NO(g) + 6 H_2O(g)$$

$$2 NO(g) + O_2(g) \rightarrow 2 NO_2(g)$$

Nitrogen dioxide dissolves in water, producing a mixture of nitrous and nitric acids:

$$2 NO_2(g) + H_2O(\ell) \rightarrow HNO_2(aq) + HNO_3(aq)$$

Heating the mixture converts nitrous acid into nitric acid, as steam and NO are given off:

$$3 HNO_2(aq) \rightarrow HNO_3(aq) + H_2O(g) + 2 NO(g)$$

About 75% of nitric acid produced in this way is combined with more ammonia to produce ammonium nitrate:

$$NH_3(g) + HNO_3(\ell) \rightarrow NH_4NO_3(s)$$

Ammonium nitrate is an important industrial chemical and the source of water-soluble nitrogen in many formulations of fertilizers.

All of this chemistry described for industrial use—some of which takes place under conditions of high temperature and pressure and requires energy derived from fossil fuels—also takes place in bacteria but under conditions of normal temperature and pressure. Much research is being carried out on the mechanism of these reactions in bacteria in an effort to find alternative methods of producing the large quantities of nitrogen-containing substances needed by commerce and agriculture in ways that minimize the high costs associated with industrial production.

Nitrogen has a wide and varied chemistry beyond the synthesis and uses of ammonia. Hydrazine (N_2H_4) is used as a rocket fuel. The reaction of metals with nitrogen produces metal nitrides containing the N^{3-} anion. Iron nitride (FeN) is used to harden the surface of steel. Sodium azide is a bactericide used in many biology laboratories to control bacterial growth. Metal salts containing the azide ion (N_3^-) are often explosive and are used as a ready source of nitrogen

gas in some automobile air bags. The overall reaction in an air bag is:

$$20 NaN_3(s) + 6 SiO_2(s) + 4 KNO_3(s) \rightarrow$$
$$32 N_2(g) + 5 Na_4SiO_4(s) + K_4SiO_4(s)$$

Azides are contact explosives: mechanical contact causes them to detonate. The mechanism in a steering column that causes the azide to explode and produce nitrogen to inflate the bag involves the transmission of the force of an accident to the azide contained in the air bags.

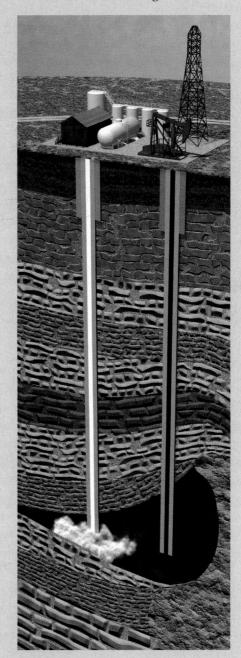

FIGURE 6.30 Nitrogen under high pressure is pumped into oil deposits to bring crude oil to the surface.

Summary

SECTION 6.1 Gases occupy the entire volume of their container. The volume occupied by a gas changes significantly with pressure and temperature. Gases are miscible, mixing in any proportion. They are much less dense than liquids or solids.

SECTION 6.2 The mass of the gases in Earth's atmosphere combined with the force of gravity results in an **atmospheric pressure (P_{atm})** on the surface of the planet. **Pressure (P)** is defined as the ratio of force to surface area and is measured with **barometers** and **manometers**. Pressure can be expressed in interconvertible units of **atmosphere (atm)**, **torr**, **mmHg**, millibars (mbar), and pascals (Pa).

SECTION 6.3 The volume of a gas sample increases with decreasing pressure at constant temperature (**Boyle's law**), but it increases with increasing temperature at constant pressure (**Charles's law**). Increasing the number of moles of gas at a given pressure produces a proportionate increase in volume at constant temperature (**Avogadro's law**). Pressure also increases with temperature at constant volume (**Amontons's law**).

SECTION 6.4 Boyle's, Charles's, Avogadro's, and Amontons's laws can be combined into the **ideal gas law**, also called the **ideal gas equation**: $PV = nRT$ where R is the **universal gas constant**. The ideal gas equation accurately describes the behavior of most gases under normal conditions. The **molar volume** of any gas is 22.4 liters at **standard temperature and pressure (STP)**, which in the United States is defined as 0°C and 1 atm.

SECTION 6.5 Gas densities increase with increasing molar mass and pressure, and they decrease with increasing temperature. The ideal gas law allows the straightforward calculation of gas density from molar mass and measurements of P and T. Calculating the molar mass of a gas is possible from the mass, volume, pressure, and temperature of a sample of the gas.

SECTION 6.6 In a gas mixture the contribution each component gas makes to the total gas pressure is called its **partial pressure**. **Dalton's law of partial pressures** states that the total pressure of a gas mixture is the sum of the partial pressures of its constituents. Dalton's law allows us to calculate the partial pressure (P_x) of any constituent gas x in a gas mixture if we know its **mole fraction (X_x)** and the total pressure. The total pressure of a gas sample collected over a liquid is the sum of the vapor pressure of the liquid and the pressure of the gas itself.

SECTION 6.7 The behavior of gases is explained by the **kinetic molecular theory**. Gases are composed of particles in constant random motion, moving at **root-mean-square speeds (u_{rms})** that are inversely proportional to the square root of their molar masses and directly proportional to the square root of their temperature. The root-mean-square speed of the molecules increases with increasing temperature and decreasing molar mass. Pressure arises from elastic collisions between gases and the walls of their container. The kinetic molecular theory of gases explains **Graham's law of effusion**, which states that the rate of **effusion** (escape through a pinhole) or **diffusion** (spreading) of a gas at a fixed temperature is inversely proportional to its molar mass.

SECTION 6.8 Ideal gas behavior is observed at moderate temperatures and low pressures where $PV = nRT$ (the ideal gas equation). Ideal gas behavior is characterized by elastic collisions and an absence of attractive forces between gas molecules. At low temperatures and high pressures, the behavior of real gases deviates from the predictions of the ideal gas law. The **van der Waals equation**, a modified form of the ideal gas equation, accounts for real gas properties.

Problem-Solving Summary

TYPE OF PROBLEM	CONCEPTS AND EQUATIONS	SAMPLE EXERCISES
Calculating pressure and atmospheric pressure	Divide force by area using the equation $$P = \frac{F}{A}$$	6.1, 6.3
Expressing pressure using different units	Multiply or divide by the appropriate conversion factor: 1 atm = 760 torr = 760 mmHg = 101.325 kPa	6.2
Calculating changes in P, V, and/or T in response to changing conditions	Substitute given values of P, T, and V into the equation $$\frac{P_1 V_1}{T_1} = \frac{P_2 V_2}{T_2}$$ and solve for missing value.	6.4–6.7
Determining n from P, V, and T	Substitute the given values of P, T, and V into the ideal gas equation $$PV = nRT$$ and solve for n.	6.8, 6.9
Calculating the density of a gas	Substitute values for pressure, temperature, and molar mass into the equation $$d = \frac{\mathcal{M}P}{RT}$$ and solve for density.	6.10

TYPE OF PROBLEM	CONCEPTS AND EQUATIONS	SAMPLE EXERCISES
Calculating molar mass ($\mathcal{M}$) from density (d)	Substitute values for pressure, temperature, and density into the equation $$\mathcal{M} = \frac{dRT}{P}$$ and solve for molar mass.	6.11
Calculating mole fraction	Divide the number of moles of the component gas by the total number of moles in the mixture $$\chi_x = \frac{n_x}{n_{total}}$$	6.12
Calculating partial pressure	Substitute the mol fraction of the component and the total pressure in the equation $P_1 = \chi_1 P_{total}$. Add the partial pressures of the component gases in the mixture $$P_{total} = P_1 + P_2 + P_3 + P_4 + \ldots$$	6.13, 6.14
Calculating root-mean-square speeds	Substitute temperature and molar mass in the equation $$u_{rms} = \sqrt{\frac{3RT}{\mathcal{M}}}$$ and solve for root-mean-square speed.	6.15
Calculating relative rate of effusion or diffusion from molar masses	Substitute values for molar masses into the equation $$\frac{r_x}{r_y} = \sqrt{\frac{\mathcal{M}_y}{\mathcal{M}_x}}$$ and solve for the relative rate.	6.16
Calculating P and V for real gases with the van der Waals equation	Substitute the number of moles of gas and the values of the van der Waals constants into the equations $$P_{real} = \left(P + \frac{n^2 a}{V^2}\right)$$ and $$V_{real} = (V - nb)$$ and solve for the new pressure and volume.	6.17

Visual Problems

6.1. Shown in Figure P6.1 are three barometers. The one in the center is located at sea level. Which barometer is most likely to reflect the atmospheric pressure in Denver, CO, where the elevation is approximately 1500 m? Explain your answer.

6.2. A rubber balloon is filled with helium gas. Which of the drawings in Figure P6.2 most accurately reflects the gas in the balloon on a molecular level? The blue spheres represent helium atoms. Explain your answer.

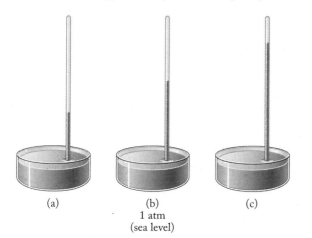

(a)   (b) 1 atm (sea level)   (c)

FIGURE P6.1

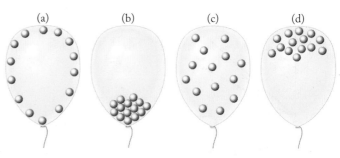

(a)   (b)   (c)   (d)

FIGURE P6.2

6.3. Which of the three changes shown in Figure P6.3 best illustrates what happens when the atmospheric pressure on a helium-filled rubber balloon is increased at constant temperature?

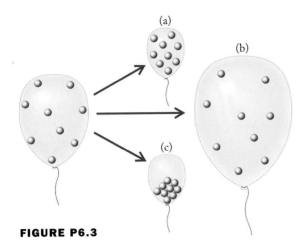

FIGURE P6.3

6.4. Which of the drawings in Figure P6.3 best illustrates what happens when the temperature of a helium-filled rubber balloon is increased at constant pressure?

6.5. Which of the three changes shown in Figure P6.5 best illustrates what happens when the amount of gas in a helium-filled rubber balloon is increased at constant temperature and pressure?

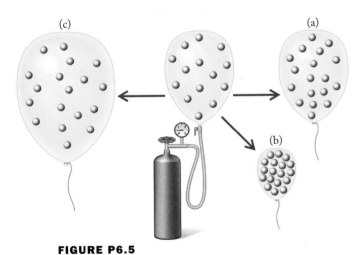

FIGURE P6.5

6.6. Which line plotting volume (V) versus reciprocal pressure ($1/P$) in Figure P6.6 corresponds to the higher temperature?

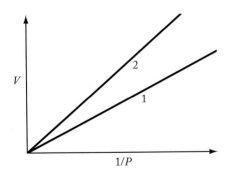

FIGURE P6.6

6.7. In Figure P6.7, which line of volume (V) versus temperature (T) represents a gas at higher pressure? Is the x-axis an absolute temperature scale?

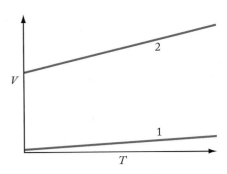

FIGURE P6.7

6.8. In Figure P6.8, which of the two plots of volume versus pressure at constant temperature is not consistent with the ideal gas law?

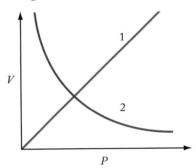

FIGURE P6.8

6.9. In Figure P6.9, which of the two plots of volume versus temperature at constant pressure is not consistent with the ideal gas law?

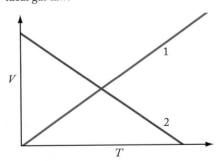

FIGURE P6.9

6.10. In Figure P6.10, which line in the following graph of density (d) versus pressure (P) at constant temperature for methane (CH_4) and nitrogen (N_2) should be labeled *methane*?

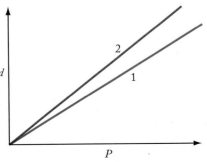

FIGURE P6.10

6.11. Add lines showing the densities of He and NO as a function of pressure to a copy of the graph in Figure P6.10.

6.12. Which of the drawings in Figure P6.12 best depicts the arrangement of molecules in a mixture of gases? The red and blue spheres represent atoms of two different gases such as helium and neon.

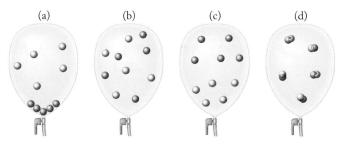

(a) (b) (c) (d)

FIGURE P6.12

6.13. The drawings in Figure P6.13 illustrate four mixtures of two diatomic gases (such as N_2 and O_2) in flasks of identical volume and at the same temperature. Is the total pressure in each flask the same? Which flask has the highest partial pressure of nitrogen, depicted by the blue molecules?

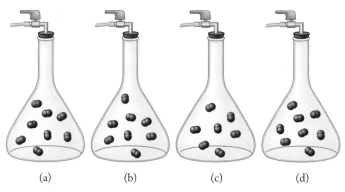

(a) (b) (c) (d)

FIGURE P6.13

6.14. Figure P6.14 shows the distribution of molecular speeds of CO_2 and SO_2 molecules at 25°C. Which curve is the profile for SO_2? Which of these profiles should match that of propane (C_3H_8), a common fuel in portable grills?

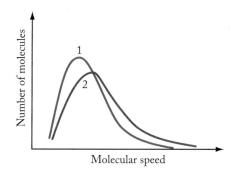

FIGURE P6.14

6.15. How would a graph showing the distribution of molecular speeds of CO_2 at −100°C differ from the curve for CO_2 shown in Figure P6.14?

6.16. A container with a pinhole leak contains a mixture of the elements highlighted in Figure P6.16. Which element leaks the slowest from the container?

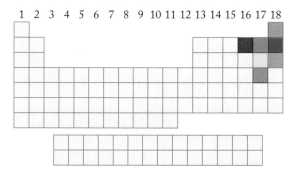

FIGURE P6.16

6.17. A container with a pinhole leak contains a mixture of the highlighted elements in Figure P6.16. Which element has the smallest root-mean-square speed?

6.18. Which of the highlighted elements in Figure P6.16 has the smallest van der Waals *b* constant?

6.19. Which of the two outcomes diagrammed in Figure P6.19 more accurately illustrates the effusion of helium from a balloon at constant atmospheric pressure?

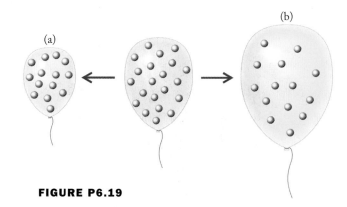

FIGURE P6.19

6.20. Which of the two outcomes shown in Figure P6.20 more accurately illustrates the effusion of gases from a balloon at constant atmospheric pressure if the red spheres have a greater root-mean-square speed than the blue spheres?

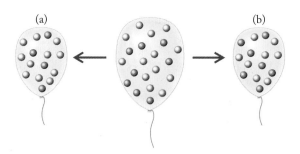

FIGURE P6.20

Questions and Problems

The Gas Phase; Atmospheric Pressure

CONCEPT REVIEW

6.21. Describe the difference between force and pressure.

6.22. How does Torricelli's barometer measure atmospheric pressure?

6.23. What is the relation between *torr* and *atmospheres* of pressure?

6.24. What is the relation between *millibars* and *pascals* of pressure?

6.25. Three barometers based on Torricelli's design are constructed using water (density d = 1.00 g/mL), ethanol (d = 0.789 g/mL), and mercury (d = 13.546 g/mL). Which barometer contains the tallest column of liquid?

6.26. In constructing a barometer, what advantage is there in choosing a dense liquid?

6.27. Why does an ice skater exert more pressure on ice when wearing newly sharpened skates than when wearing skates with dull blades?

6.28. Why is it easier to travel over deep snow when wearing boots and snowshoes (Figure P6.28) rather than just boots?

FIGURE P6.28

6.29. Why does atmospheric pressure decrease with increasing elevation?

*6.30. Pieces of different metals have exactly the same mass but different densities. Could these objects ever exert the same pressure?

PROBLEMS

6.31. Calculate the downward pressure due to gravity exerted by the bottom face of a 1.00 kg cube of iron that is 5.00 cm on a side.

6.32. The gold block represented in Figure P6.32 has a mass of 38.6 g. Calculate the pressure exerted by the block when it is on (a) a square face, and (b) a rectangular face.

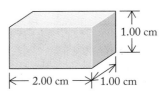

FIGURE P6.32

6.33. Convert the following pressures into atmospheres: (a) 2.0 kPa; (b) 562 mmHg

6.34. Convert the following pressures into mm Hg (torr): (a) 0.541 atm; (b) 2.8 kPa

6.35. Record High Atmospheric Pressure The highest atmospheric pressure recorded on Earth was measured at Tosotsengel, Mongolia, on December 19, 2001, when the barometer read 108.6 kPa. Express this pressure in (a) millimeters of Hg, (b) atmospheres, and (c) millibars.

6.36. **Record Low Atmospheric Pressure** Despite the destruction from Hurricane Katrina in September 2005, the lowest pressure for a hurricane in the Atlantic Ocean was measured several weeks after Katrina. Hurricane Wilma registered an atmospheric pressure of 88.2 kPa on October 19, 2005, about 2 kPa lower than Hurricane Katrina. What was the *difference* in pressure between the two hurricanes in (a) millimeters of Hg, (b) atmospheres, and (c) millibars?

The Gas Laws

CONCEPT REVIEW

6.37. From the molecular perspective, why is pressure directly proportional to temperature at fixed volume (Amontons's law)?

6.38. How do we explain Boyle's law on a molecular basis?

6.39. A balloonist is rising too fast for her taste. Should she increase the temperature of the gas in the balloon, or decrease it?

6.40. Could the pilot of the balloon in Problem 6.39 reduce her rate of ascent by allowing some gas to leak out of the balloon? Explain your answer.

6.41. A quantity of gas is compressed into half its initial volume as it cools from 20°C to 10°C. Does the pressure of the gas increase, decrease, or remain the same?

6.42. If the volume of gasoline vapor and air in an automobile engine cylinder is reduced to 1/10 of its original volume before ignition, by what factor does the pressure in the cylinder increase? (Assume there is no change in temperature.)

PROBLEMS

6.43. The volume of 1.00 mol of ammonia gas at 1.00 atm of pressure is gradually decreased from 78.0 mL to 39.0 mL. What is the final pressure of ammonia if there is no change in temperature?

6.44. The pressure on a sample of an ideal gas is increased from 715 mmHg to 3.55 atm at constant temperature. If the initial volume of the gas is 485 mL, what is the final volume of the gas?

6.45. A scuba diver releases a balloon containing 153 L of helium attached to a tray of artifacts at an underwater archaeological site (Figure P6.45). When the balloon reaches the surface, it has expanded to a volume of 352 L. The pressure at the surface is 1.00 atm; what is the pressure at the underwater site? Pressure increases by 1.0 atm for every 10 m of depth; at what depth was the diver working? Assume the temperature remains constant.

FIGURE P6.45

6.46. Breath-Hold Diving The world record for diving without supplemental air tanks ("breath-hold diving") is about 125 meters, a depth at which the pressure is about 12.5 atm. If a diver's lungs have a volume of 6 L at the surface of the water, what is their volume at a depth of 125 m?

6.47. Use the following data to draw a graph of the volume (V) of 1 mole of H_2 as a function of the reciprocal of pressure ($1/P$) at 298 K:

P (mmHg)	V (L)
100	186
120	155
240	77.5
380	48.9
500	37.2

Would the graph be the same for the same number of moles of argon?

6.48. The following data for P and V were collected for 1 mole of argon at 300 K. Draw a graph of the volume (V) of 1 mole of Ar as a function of the reciprocal of pressure ($1/P$). Does this graph look like the graph you drew for Problem 6.47?

P (atm)	V (L)
0.10	246.3
0.25	98.5
0.50	49.3
0.75	32.8
1.0	24.6

6.49. Use the following data to draw a graph of the volume (V) of He as a function of temperature for 1.0 mol of He gas at a constant pressure of 1.00 atm:

V (L)	T (K)
7.88	96
3.94	48
1.97	24
0.79	9.6
0.39	4.8

How would the graph change if the amount of gas were halved?

6.50. Use the following data to draw a graph of the volume (V) of He as a function of temperature (T) for 0.50 mol of He gas at a constant pressure of 1.00 atm:

V (L)	T (K)
3.94	96
1.97	48
0.79	24
0.39	9.6
0.20	4.8

Does this graph match your prediction from Problem 6.49?

6.51. A cylinder with a piston (Figure P6.51) contains a sample of gas at 25°C. The piston moves in response to changing pressure inside the cylinder. At what gas temperature would the piston move so that the volume inside the cylinder doubled?

FIGURE P6.51

6.52. The temperature of the gas in Problem 6.51 is reduced to a temperature at which the volume inside the cylinder has decreased by 25% from its initial volume at 25.0°C. What is the new temperature?

6.53. A 2.68 L sample of gas is warmed from 250 K to a final temperature of 398 K. Assuming no change in pressure, what is the final volume of the gas?

6.54. A 5.6 L sample of gas is cooled from 78°C to a temperature at which its volume is 4.3 L. What is this new temperature? Assume no change in pressure of the gas.

6.55. Balloons for a New Year's Eve party in Fargo, ND, are filled to a volume of 2.0 L at a temperature of 22°C and then hung outside where the temperature is −22°C. What is the volume of the balloons after they have cooled to the outside temperature? Assume that atmospheric pressure inside and outside the house is the same.

6.56. The air inside a hot-air balloon is heated to 45°C ($T_{initial}$) and then cools to 25°C (T_{final}). By what percentage does the volume of the balloon change?

6.57. Which of the following actions would produce the greatest increase in the volume of a gas sample? (a) lowering the pressure from 760 mmHg to 720 mmHg at constant temperature, or (b) raising the temperature from 10°C to 40°C at constant pressure?

6.58. Which of the following actions would produce the greatest increase in the volume of a gas sample: (a) doubling the amount of gas in the sample at constant temperature and pressure, or (b) raising the temperature from 244°C to 1100°C?

*6.59. What happens to the volume of gas in a cylinder with a movable piston under the following conditions?
   a. Both the absolute temperature and the external pressure on the piston double.
   b. The absolute temperature is halved, and the external pressure on the piston doubles.
   c. The absolute temperature increases by 75%, and the external pressure on the piston increases by 50%.

*6.60. What happens to the pressure of a gas under the following conditions?
   a. The absolute temperature is halved and the volume doubles.
   b. Both the absolute temperature and the volume double.
   c. The absolute temperature increases by 75%, and the volume decreases by 50%.

6.61. A 150.0 L weather balloon filled with 6.1 moles of helium has a small leak. If the helium leaks at a rate of 10 mmol/hr, what is the volume of the balloon after 24 hours?

6.62. Which has the greater effect on the volume (1.0 L) of a gas at constant temperature: doubling the number of moles, or doubling the pressure?

6.63. Temperature Effects on Bicycle Tires A bicycle racer inflates his tires to 7 atm on a warm autumn afternoon when temperatures reached 27°C. By morning the temperature has dropped to 5°C. What is the pressure in the tires if

we assume that the volume of the tire does not change significantly?

*6.64. A balloon vendor at a street fair is using a tank of helium to fill her balloons. The tank has a volume of 150.0 L and a pressure of 120.0 atm at 25°C. After awhile she notices that the valve has not been closed properly. The pressure had dropped to 110.0 atm. How many moles of gas have been lost?

The Ideal Gas Law

CONCEPT REVIEW

6.65. What is meant by standard temperature and pressure (STP)? What is the volume of 1 mole of an ideal gas at STP?

6.66. Which of the following are not characteristics of an ideal gas? (a) The molecules of gas have little volume compared with the volume that they occupy. (b) Their volume is independent of temperature. (c) The density of all ideal gases is the same. (d) Gas atoms or molecules do not interact with one another.

*6.67. What does the slope represent in a graph of pressure as a function of $1/V$ at constant temperature for an ideal gas?

*6.68. How would the graph in Problem 6.7 change if we increased the temperature to a larger but constant value?

PROBLEMS

6.69. How many moles of air must there be in a bicycle tire with a volume of 2.36 L if it has an internal pressure of 6.8 atm at 17.0°C?

6.70. At what temperature will 1.00 mole of an ideal gas in a 1.00 L container exert a pressure of 1.00 atm?

6.71. What is the pressure inside a 500 mL cylinder containing 1.00 g of deuterium gas (D_2) at 298 K?

6.72. What is the volume of 100 g of H_2O vapor at 120°C and 1.00 atm?

6.73. A weather balloon with a volume of 200.0 L is launched at 20°C at sea level, where the atmospheric pressure is 1.00 atm. The balloon rises to an altitude of 20,000 m, where atmospheric pressure is 63 mmHg and the temperature is 210 K. What is the volume of the balloon at 20,000 m?

6.74. For some reason, a skier decides to ski from the summit of a mountain near Park City, UT (elevation = 9970 ft, $T = -10°C$, and $P_{atm} = 623$ mmHg) to the base of the mountain (elevation = 6920 ft, $T = -5°C$, and $P_{atm} = 688$ mmHg) with a balloon tied to each of her ski poles. If each balloon is filled to a volume of 2.00 L at the summit, what is the volume of each balloon at the base?

6.75. Hydrogen holds promise as an "environment friendly" fuel. How many grams of H_2 gas are present in a 50.0 L fuel tank at a pressure of 2850 lb/in² (psi) at 20°C? Assume that 1 atm = 14.7 psi.

6.76. Liquid Nitrogen-Powered Car Students at the University of North Texas and the University of Washington built a car propelled by compressed nitrogen gas. The gas was obtained by boiling liquid nitrogen stored in a 182 L tank. What volume of N_2 is released at 0.927 atm of pressure and 25°C from a tank full of liquid N_2 ($d = 0.808$ g/mL)?

6.77. Miners' Lamps Before the development of reliable batteries, miners' lamps burned acetylene produced by the reaction of calcium carbide with water:

$$CaC_2(s) + H_2O(\ell) \rightarrow C_2H_2(g) + CaO(s)$$

A lamp uses 1.00 L of acetylene per hour at 1.00 atm pressure and 18°C.
a. How many moles of C_2H_2 are used per hour?
b. How many grams of calcium carbide must be in the lamp for a 4 hr shift?

6.78. Acid precipitation dripping on limestone produces carbon dioxide by the following reaction:

$$CaCO_3(s) + 2\,H^+(aq) \rightarrow Ca^{2+}(aq) + CO_2(g) + H_2O(\ell)$$

If 15.0 mL of CO_2 were produced at 25°C and 760 mmHg, then
a. how many moles of CO_2 were produced?
b. how many milligrams of $CaCO_3$ were consumed?

6.79. Oxygen is generated by the thermal decomposition of potassium chlorate:

$$2\,KClO_3(s) \rightarrow 2\,KCl(s) + 3\,O_2(g)$$

How much $KClO_3$ is needed to generate 200.0 L of oxygen at 0.85 atm and 273 K?

6.80. Calculate the volume of carbon dioxide at 20°C and 1.00 atm produced from the complete combustion of 1.00 kg of methane. Compare your result with the volume of CO_2 produced from the complete combustion of 1.00 kg of propane (C_3H_8).

6.81. Reducing CO₂ in Submarines The CO_2 that builds up in the air of a submerged submarine can be removed by reacting it with sodium peroxide:

$$2\,Na_2O_2(s) + 2\,CO_2(g) \rightarrow 2\,Na_2CO_3(s) + O_2(g)$$

If a sailor exhales 150.0 mL of CO_2 per minute at 20°C and 1.02 atm, how much sodium peroxide is needed per sailor in a 24 hr period?

6.82. Rescue Breathing Devices Self-contained self-rescue breathing devices, like the one shown in Figure P6.82, convert CO_2 into O_2 according to the following reaction:

$$2\,KO_2(s) + CO_2(g) \rightarrow K_2CO_3(s) + O_2(g)$$

How many grams of KO_2 are needed to produce 100.0 L of O_2 at 20°C and 1.00 atm?

FIGURE P6.82 People exploring underwater caverns often choose a rebreathing apparatus over air tanks, since the apparatus is smaller and produces more oxygen.

Gas Density

CONCEPT REVIEW

6.83. Do all gases at the same pressure and temperature have the same density? Explain your answer.

6.84. Birds and sailplanes take advantage of thermals (rising columns of warm air) to gain altitude with less effort than usual. Why does warm air rise?

6.85. How does the density of a gas sample change when (a) its pressure is increased, and (b) its temperature is decreased?

6.86. How would you measure the density of a gas sample of known molar mass?

PROBLEMS

6.87. Radon Hazards in Homes Radon is hazardous because it is easily inhaled and it emits α particles when it undergoes radioactive decay.
a. Calculate the density of radon at 298 K and 1 atm of pressure.
b. Are radon concentrations likely to be greater in the basement or on the top floor of a building?

6.88. Four balloons, each with a mass of 10.0 g, are inflated to a volume of 20.0 L, each with a different gas: helium, neon, carbon monoxide, or nitrogen monoxide. If the density of air at 25°C and 1.00 atm is 0.00117 g/mL, will any of the balloons float in this air?

6.89. A 150.0 mL flask contains 0.391 g of a volatile oxide of sulfur. The pressure in the flask is 750 mmHg, and the temperature is 22°C. Is the gas SO_2 or SO_3?

6.90. A 100.0 mL flask contains 0.193 g of a volatile oxide of nitrogen. The pressure in the flask is 760 mmHg at 17°C. Is the gas NO, NO_2, or N_2O_5?

6.91. The density of an unknown gas is 1.107 g/L at 300 K and 740 mmHg. Could this gas be CO or CO_2?

6.92. The density of a gas containing chlorine and oxygen has a density of 2.875 g/L at 756 mmHg and 11°C. What is the most likely molecular formula of the gas?

Dalton's Law and Mixtures of Gases

CONCEPT REVIEW

6.93. What is meant by the *partial* pressure of a gas?

6.94. Can a barometer be used to measure just the partial pressure of oxygen in the atmosphere? Why or why not?

6.95. Which gas sample has the largest volume at 25°C and 1 atm pressure? (a) 0.500 mol of dry H_2; (b) 0.500 mol of dry N_2; (c) 0.500 mol of wet H_2 (H_2 collected over water)

6.96. Two identical balloons are filled to the same volume at the same pressure and temperature. One balloon is filled with air and the other with helium. Which balloon contains more particles (atoms and molecules)?

PROBLEMS

6.97. A gas mixture contains 0.70 mol of N_2, 0.20 mol of H_2, and 0.10 mol of CH_4. What is the mole fraction of H_2 in the mixture?

6.98. A gas mixture contains 7.0 g of N_2, 2.0 g of H_2, and 16.0 g of CH_4. What is the mole fraction of H_2 in the mixture?

6.99. Calculate the pressure of the gas mixture and the partial pressure of each constituent gas in Problem 6.97 if the mixture is in a 10.0 L vessel at 27°C.

6.100. Calculate the pressure of the gas mixture and the partial pressure of each constituent gas in Problem 6.98 if the mixture is in a 1.00 L vessel at 0°C.

6.101. A sample of oxygen was collected over water at 25°C and 1.00 atm. If the total sample volume was 0.480 L, how many moles of O_2 were collected?

6.102. Water was removed from the O_2 sample in Problem 6.101. What is the volume of the dry O_2 gas sample at 25°C and 1.00 atm?

6.103. The following reactions were carried out in sealed containers. Will the total pressure after each reaction is complete be greater than, less than, or equal to the total pressure before the reaction? Assume all reactants and products are gases at the same temperature.
a. $N_2O_5(g) + NO_2(g) \rightarrow 3\,NO(g) + 2\,O_2(g)$
b. $2\,SO_2(g) + O_2(g) \rightarrow 2\,SO_3(g)$
c. $C_3H_8(g) + 5\,O_2(g) \rightarrow 3\,CO_2(g) + 4\,H_2O(g)$

6.104. In each of the following gas-phase reactions, determine whether the total pressure at the end of the reaction (carried out in a sealed, rigid vessel) will be greater than, less than, or equal to the total pressure at the beginning. Assume all reactants and products are gases at the same temperature.
a. $H_2(g) + Cl_2(g) \rightarrow 2\,HCl(g)$
b. $4\,NH_3(g) + 5\,O_2(g) \rightarrow 4\,NO(g) + 6\,H_2O(g)$
c. $2\,NO(g) + O_2(g) \rightarrow 2\,NO_2(g)$

6.105. High-Altitude Mountaineering Alpine climbers use pure oxygen near the summits of 8000 m peaks, where $P_{atm} = 0.35$ atm (Figure P6.105). How much more O_2 is there in a lung full of pure O_2 at this elevation than in a lung full of air at sea level?

FIGURE P6.105

6.106. Scuba Diving A scuba diver is at a depth of 50 m, where the pressure is 5.0 atm. What should be the mole fraction of O_2 in the gas mixture the diver breathes to replicate the same gas mixture at sea level?

6.107. Carbon monoxide at a pressure of 680 mmHg reacts completely with O_2 at a pressure of 340 mmHg in a sealed vessel to produce CO_2. What is the final pressure in the flask?

6.108. Ozone reacts completely with NO, producing NO_2 and O_2. A 10.0 L vessel is filled with 0.280 mol of NO and 0.280 mol of O_3 at 350 K. Find the partial pressure of each product and the total pressure in the flask at the end of the reaction.

*6.109. Ammonia is produced industrially from the reaction of hydrogen with nitrogen under pressure in a sealed reactor. What is the percent decrease in pressure of a sealed reaction vessel during the reaction between 3.60×10^3 mol of H_2 and 1.20×10^3 mol of N_2 if half of the N_2 is consumed?

*6.110. A mixture of 0.156 mol of C is reacted with 0.117 mol of O_2 in a sealed, 10.0 L vessel at 500 K, producing a mixture of CO and CO_2. The total pressure is 0.640 atm. What is the partial pressure of CO?

The Kinetic Molecular Theory of Gases and Graham's Law

CONCEPT REVIEW

6.111. What is meant by the *root-mean-square speed* of gas molecules?

6.112. Why don't all molecules in a sample of air move at exactly the same speed?

6.113. How does the root-mean-square speed of the molecules in a gas vary with (a) molar mass and (b) temperature?

6.114. Does pressure affect the root-mean-square speed of the molecules in a gas? Explain your answer.

6.115. How can Graham's law of effusion be used to determine the molar mass of an unknown gas?

6.116. Is the ratio of the rates of effusion of two gases the same as the ratio of their root-mean-square speeds?

6.117. What is the difference between *diffusion* and *effusion*?

6.118. If gas X diffuses faster in air than gas Y, is gas X also likely to effuse faster than gas Y?

PROBLEMS

6.119. Rank the gases SO_2, CO_2, and NO_2 in order of increasing root-mean-square speed at 0°C.

6.120. In a mixture of CH_4, NH_3, and N_2, which gas molecules are, on average, moving fastest?

6.121. At 286 K, three gases, A, B, and C, have root-mean-square speeds of 360 m/s, 441 m/s, and 472 m/s, respectively. Which gas is O_2?

6.122. Air is approximately 21% O_2 and 78% N_2 by mass. Calculate the root-mean-square speed of each gas at 273 K.

6.123. Calculate the root-mean-square speed of Ne atoms at the temperature at which their average kinetic energy is 5.18 kJ/mol.

6.124. Determine the root-mean-square speed of CO_2 molecules that have an average kinetic energy of 4.2×10^{-21} J per molecule.

6.125. What is the ratio of the root-mean-square speed of D_2 to that of H_2 at constant temperature?

6.126. **Enriching Uranium** The two isotopes of uranium, ^{238}U and ^{235}U, can be separated by diffusion of the corresponding UF_6 gases. What is the ratio of the root-mean-square speed of $^{238}UF_6$ to that of $^{235}UF_6$ at constant temperature?

6.127. Molecular hydrogen effuses 4 times as fast as gas X at the same temperature. What is the molar mass of gas X?

6.128. Gas Y effuses half as fast as O_2 at the same temperature. What is the molar mass of gas Y?

6.129. If an unknown gas has one-third the root-mean-square speed of H_2 at 300 K, what is its molar mass?

6.130. A flask of ammonia is connected to a flask of an unknown acid HX by a 1.00 m glass tube. As the two gases diffuse down the tube, a white ring of NH_4X forms 68.5 cm from the ammonia flask. Identify element X.

6.131. Studies of photosynthesis in green plants often make use of $^{13}CO_2$. (a) Calculate the relative rates of diffusion of $^{13}CO_2$ and $^{12}CO_2$. (b) Specify which gas diffuses faster.

6.132. At fixed temperature, how much faster does NO effuse than NO_2?

6.133. Two balloons were filled with H_2 and He, respectively. The person responsible for filling them neglected to label them. After 24 hours the volumes of both balloons had decreased but by different amounts. Which balloon contained hydrogen?

6.134. Compounds sensitive to oxygen are often manipulated in *glove boxes* that contain a pure nitrogen or pure argon atmosphere. A balloon filled with carbon monoxide was placed in a glove box. After 24 hours, the volume of the balloon was unchanged. Did the glove box contain N_2 or Ar?

Real Gases

CONCEPT REVIEW

6.135. Explain why real gases behave nonideally at low temperatures and high pressures.

6.136. Under what conditions is the pressure exerted by a real gas *less* than that predicted for an ideal gas?

6.137. Explain why the values of the van der Waals constant b of the noble gas elements increase with atomic number.

6.138. Explain why the constant a in the van der Waals equation generally increases with the molar mass of the gas.

PROBLEMS

6.139. The graphs of PV/RT versus P (see Figure 6.26) for a mole of CH_4 and H_2 differ in how they deviate from ideal behavior. For which gas is the effect of the volume occupied by the gas molecules more important than the attractive forces between molecules?

6.140. Which noble gas is expected to deviate the most from ideal behavior in a graph of PV/RT versus P?

6.141. At high pressures, real gases do not behave ideally. (a) Use the van der Waals equation and data in the text to calculate the pressure exerted by 40.0 g H_2 at 20°C in a 1.00 L container. (b) Repeat the calculation assuming that the gas behaves like an ideal gas.

6.142. (a) Calculate the pressure exerted by 1.00 mol of CO_2 in a 1.00 L vessel at 300 K, assuming that the gas behaves ideally. (b) Repeat the calculation by using the van der Waals equation.

Additional Problems

6.143. A student places a piece of solid CO_2 (dry ice) in an open, preweighed flask with an internal volume of 153.6 mL. The dry ice is allowed to sublimate. Any excess gas exits the flask through a pinhole in a piece of aluminum foil over the mouth of the flask. Just when the last of the dry ice has sublimated, the flask is weighed again. At this point the student calculates the mass of the CO_2 inside the flask to be 0.285 g. The student also records the temperature (18.5°C) and atmospheric pressure (771 mmHg) in the laboratory.
  a. How many moles of CO_2 are in the flask?
  b. Use the given data to calculate the molar mass of CO_2.
  c. Describe how the student might have determined the internal volume of the flask to the nearest 0.1 mL.

6.144. A flask with an internal volume of exactly 267.5 mL contains a few milliliters of a volatile (boiling point < 100°C) hydrocarbon. The flask is covered with a piece of foil. A tiny hole is pricked in the foil, and the flask is placed in boiling water (100°C). After all the liquid had vaporized, the flask is weighed and found to contain 0.728 g of hydrocarbon vapor at 755 mmHg.
  a. Calculate the molar mass of the compound.
  b. Combustion analysis of the compound reveals that the compound is 85.7% C and 14.3% H. What is the molecular formula of the compound?

6.145. The cylinders of a particular sports-car engine contain 828 mL of air and gasoline vapor before the motion of a piston increases the pressure inside the cylinder 9.6 times. What is the volume of the air and gasoline vapor mixture after this compression, assuming the temperature of the mixture does not change significantly?

6.146. A balloon is partly inflated with 5.00 liters of helium at sea level where the atmospheric pressure is 1008 mbar. The balloon ascends to an altitude of 3000 meters, where the pressure is 855 mbar. What is the volume of the helium in the balloon at the higher altitude? Assume that the temperature of the gas in the balloon does not change in the ascent.

6.147. The volume of a sample of a gas is compressed at constant temperature. How does this change in volume affect the following?
  a. average kinetic energy of the molecules
  b. average speed of the molecules
  c. number of collisions of the gas molecules with the container wall per unit of time

6.148. Tropical Storms The severity of a tropical storm is related to the depressed atmospheric pressure at its center. Figure P6.148 is a photograph of Typhoon Odessa taken from the space shuttle *Discovery* in August 1985, when the maximum winds of the storm were about 90 mi/hr and the pressure was 40 mbar lower at the center than normal atmospheric pressure. In contrast, the central pressure of Hurricane Andrew was 90 mbar lower than its surroundings when it hit southern Florida with winds as high as 165 mi/hr. If a small weather balloon with a volume of 50.0 L at a pressure of 1.0 atmosphere was deployed at the center of Andrew, what was the volume of the balloon when it reached the center?

FIGURE P6.148

6.149. Planetary Atmospheres Saturn's largest moon (Titan) has a surface atmospheric pressure of 1220 torr. The atmosphere consists of 82% N_2, 12% Ar, and 6% CH_4 by volume. Calculate the partial pressure of each gas in Titan's atmosphere. Chilly temperatures aside, could life as we know it exist on Titan?

6.150. Designing Tennis Balls Tennis players are familiar with the problem that tennis balls, once removed from a pressurized can, begin to lose their bounce as they lose their internal pressure. If you could choose from among the following gases to fill tennis balls with to keep them from losing their bounce, which would you choose? Explain your selection. (a) He; (b) H_2; (c) CO_2; (d) Ne; (e) N_2

6.151. Scientists have used laser light to slow atoms to speeds corresponding to temperatures below 0.00010 K. At this temperature, what is the root-mean-square speed of argon atoms?

6.152. Blood Pressure A typical blood pressure in a resting adult is "120 over 80," meaning 120 mmHg with each beat of the heart and 80 mmHg of pressure between heartbeats. Express these pressures in the following units: (a) torr; (b) atm; (c) bar; (d) kPa.

*6.153. A popular scuba tank is the "aluminum 80," so named because it can deliver 80 cubic feet of air at "normal" temperature (72°F) and pressure (1.00 atm, or 14.7 psi) when filled with compressed air at a pressure of 3000 psi. A particular aluminum 80 tank weighs 33 pounds empty. How many pounds does it weigh when filled with air at 3000 psi?

6.154. The flame produced by the burner of a gas (propane) grill is a pale blue color when enough air mixes with the propane (C_3H_8) to burn it completely. For every gram of propane that flows through the burner, what volume of air is needed to burn it completely? Assume that the temperature of the burner is 200°C, the pressure is 1.00 atm, and the mole fraction of O_2 in air is 0.21.

6.155. Which noble gas effuses at about half the effusion rate of O_2?

6.156. Anesthesia A common anesthesia gas is halothane, with the formula $C_2HBrClF_3$ and the structure shown in Figure P6.156. Liquid halothane boils at 50.2°C and 1.00 atm. If halothane behaved as an ideal gas, what volume would 10.0 mL of liquid halothane (d = 1.87 g/mL) occupy at 60°C and 1.00 atm of pressure? What is the density of halothane vapor at 55°C and 1.00 atm of pressure?

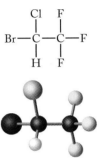

FIGURE P6.156

6.157. A cotton ball soaked in ammonia and another soaked in hydrochloric acid were placed at opposite ends of a 1.00 m glass tube (Figure P6.157). The vapors diffused toward the middle of the tube and formed a white ring of ammonium chloride where they met.
   a. Write the chemical equation for this reaction.
   b. Should the ammonium chloride ring be closer to the end of the tube with ammonia or the end with hydrochloric acid? Explain your answer.
   c. Calculate the distance from the ammonia end to the position of the ammonium chloride ring.

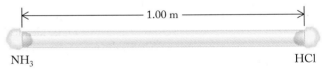

NH_3 HCl

FIGURE P6.157

*6.158. The same apparatus described in Problem 6.157 was used in another series of experiments. A cotton ball soaked in either hydrochloric acid (HCl) or acetic acid (CH_3COOH) was placed at one end. Another cotton ball soaked in one of three amines (a class of organic compounds)—[CH_3NH_2, $(CH_3)_2NH$, or $(CH_3)_3N$]—was placed in the other end (Figure P6.158).
   a. In one combination of acid and amine, a white ring was observed almost exactly halfway between the two ends. Which acid and which amine were used?
   b. Which combination of acid and amine would produce a ring closest to the amine end of the tube?

c. Do any two of the six combinations result in the formation of product at the same position in the ring? Assume measurements can be made to the nearest centimeter.

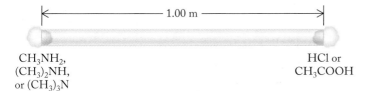

FIGURE P6.158

CH$_3$NH$_2$,
(CH$_3$)$_2$NH,
or (CH$_3$)$_3$N

HCl or
CH$_3$COOH

6.159. A flashbulb of volume 2.6 mL contains $O_2(g)$ at a pressure of 2.3 atm and a temperature of 26°C. How many grams of $O_2(g)$ does the flashbulb contain?

6.160. The pressure in an aerosol can is 1.5 atm at 27°C. The can will withstand a pressure of about 2 atm. Will it burst if heated in a campfire to 450°C?

6.161. An expandable container at 50.0°C and 1.0 atm pressure contains 30.0 g of $CO(g)$. How many times larger is the volume of the gas when 51.0 g of the gas is in the container at the same temperature and pressure?

6.162. A sample of 11.4 liters of an ideal gas at 25.0°C and 735 torr is compressed and heated so that the volume is 7.9 liters and the temperature is 72.0°C. What is the pressure in the container?

6.163. A sample of argon gas at STP occupies 15.0 L. What mass of argon is present in the container?

6.164. A sample of a gas weighs 2.889 g and has a volume of 940 mL at 735 torr and 31°C. What is its molar mass?

6.165. What pressure is exerted by a mixture of 2.00 g of H_2 and 7.00 g of N_2 at 273°C in a 10.0 L container?

6.166. Uranus has a total atmospheric pressure of 130 kPa and consists of the following gases: 83% H_2, 15% He, and 2% CH_4 by volume. Calculate the partial pressure of each gas in Uranus's atmosphere.

6.167. A sample of $N_2(g)$ requires 240 seconds to diffuse through a porous plug. It takes 530 seconds for an equal number of moles of an unknown gas X to diffuse through the plug under the same conditions of temperature and pressure. What is the molar mass of gas X?

6.168. The rate of effusion of an unknown gas is 2.0 ft/min and the rate of effusion of $SO_3(g)$ is 1.1 ft/min under identical experimental conditions. What is the molar mass of the unknown gas?

6.169. Why do gas bubbles exhaled by a scuba diver get larger as they rise to the surface?

6.170. Derive an equation that expresses the ratio of the densities (d_1 and d_2) of a gas under two different combinations of temperature and pressure: (T_1, P_1) and (T_2, P_2).

6.171. Denitrification in the Environment In some aquatic ecosystems, nitrate (NO_3^-) is converted to nitrite (NO_2^-), which then decomposes to nitrogen and water. As an example of this second reaction, consider the decomposition of ammonium nitrite:

$$NH_4NO_2(aq) \rightarrow N_2(g) + 2\,H_2O(\ell)$$

What would be the change in pressure in a sealed 10.0 L vessel due to the formation of N_2 gas when the ammonium nitrite in 1.00 L of 1.0 M NH_4NO_2 decomposes at 25°C?

*6.172. When sulfur dioxide bubbles through a solution containing nitrite, chemical reactions that produce gaseous N_2O and NO may occur.
   a. How much faster on average would NO molecules be moving than N_2O molecules in such a reaction mixture?
   b. If these two nitrogen oxides were to be separated based on differences in their rates of effusion, would unreacted SO_2 interfere with the separation? Explain your answer.

*6.173. **Using Wetlands to Treat Agricultural Waste** Wetlands can play a significant role in removing fertilizer residues from rain runoff and groundwater; one way they do this is through denitrification, which converts nitrate ions to nitrogen gas:

$$2\,NO_3^-(aq) + 5\,CO(g) + 2\,H^+(aq) \rightarrow$$
$$N_2(g) + H_2O(\ell) + 5\,CO_2(g)$$

Suppose 200.0 g of 2 NO_3^- flow into a swamp each day. What volume of N_2 would be produced at 17°C and 1.00 atm if the denitrification process were complete? What volume of CO_2 would be produced? Suppose the gas mixture produced by the decomposition reaction is trapped in a container at 17°C; what is the density of the mixture assuming P_{total} = 1.00 atm?

6.174. Ammonium nitrate decomposes upon heating. The products depend on the reaction temperature:

$$NH_4NO_3(s) \xrightarrow{\ >300°C\ } N_2(g) + \tfrac{1}{2}\,O_2(g) + 2\,H_2O(g)$$
$$\xrightarrow{\ 200–260°C\ } N_2O(g) + 2\,H_2O(g)$$

A sample of NH_4NO_3 decomposes at an unspecified temperature, and the resulting gases are collected over water at 20°C.
   a. Without completing a calculation, predict whether the volume of gases collected can be used to distinguish between the two reaction pathways. Explain your answer.
   *b. The gas produced during the thermal decomposition of 0.256 g of NH_4NO_3 has displaced 79 mL of water at 20°C and 760 mmHg of atmospheric pressure. Is the gas N_2O or a mixture of N_2 and O_2?

6.175. Calculate the pressure (in pascals) exerted by the atmosphere on a 1.0 m^2 area in La Paz, Bolivia, given a mass of gases of 6.6×10^3 kg.

Nitrogen: Feeding Plants and Inflating Air Bags

6.176. Write balanced chemical equations for the following reactions: (a) nitrogen reacts with hydrogen; (b) nitrogen dioxide dissolves in water.

6.177. Write balanced chemical equations for reactions of ammonia with (a) oxygen and (b) nitric acid.

6.178. This is the chemical reaction that occurs when gunpowder is ignited:

$$2\,KNO_3(s) + \tfrac{1}{8}\,S(s) + 3\,C(s) \rightarrow K_2S(s) + N_2(g) + 3\,CO_2(g)$$

(a) Identify the oxidizing and reducing agents.
(b) How many electrons are transferred per mole of KNO_3?

6.179. A simple laboratory-scale preparation of nitrogen involves the thermal decomposition of ammonium nitrite (NH_4NO_2).
   a. Write a balanced chemical equation for the decomposition of NH_4NO_2 to give nitrogen and water.
   b. Is the thermal decomposition of NH_4NO_2 a redox reaction?

6.180. Oklahoma City Bombing The 1995 explosion that destroyed the Murrah Federal Office Building in Oklahoma City was believed to be caused by the reaction of ammonium nitrate with fuel oil:

$$3\,NH_4NO_3(s) + C_{10}H_{22}(\ell) + 14\,O_2(g) \rightarrow$$
$$3\,N_2(g) + 17\,H_2O(g) + 10\,CO_2(g)$$

   a. What common use for ammonium nitrate made it easy for the conspirators to acquire this compound in large quantities without suspicion?
   b. Which elements are oxidized and which are reduced in the reaction?
   *c. Write a chemical equation for the thermal decomposition of ammonium nitrate in the absence of fuel and oxygen.

*6.181. Automobile Air Bags Sodium azide is used in automobile air bags as a source of nitrogen gas in the reaction

$$2\,NaN_3(s) \rightarrow 2\,Na(\ell) + 3\,N_2(g)$$

The reaction is rapid at 350° but produces elemental sodium. Potassium nitrate and silica (SiO_2) are added to remove the liquid sodium metal formed in the reaction. Why is it necessary from a safety perspective to avoid molten alkali metals in an air bag?

6.182. Manufacturing a Proper Air Bag Here is the overall reaction in an automobile air bag:

$$20\,NaN_3(s) + 6\,SiO_2(s) + 4\,KNO_3(s) \rightarrow$$
$$32\,N_2(g) + 5\,Na_4SiO_4(s) + K_4SiO_4(s)$$

Calculate how many grams of sodium azide (NaN_3) are needed to inflate a $40 \times 40 \times 20$ cm bag to a pressure of 1.25 atm at a temperature of 20°C. How much more sodium azide is needed if the air bag must produce the same pressure at 10°C?

6.183. The first rocket-propelled aircraft used a mixture of hydrogen peroxide and hydrazine as the propellant. Using the thermochemical data in the appendices and ΔH_f° of $N_2H_4(\ell) = 50.63$ kJ/mol, calculate the enthalpy change for the reaction:

$$2\,H_2O_2(\ell) + N_2H_4(\ell) \rightarrow N_2(g) + 4\,H_2O(g)$$

*6.184. In Problem 6.76, nitrogen gas obtained by boiling liquid nitrogen ($d = 0.808$ g/mL) from a 182 L tank was used to power a car. How much hydrazine (in grams) is needed to produce an equivalent amount of N_2 gas by reaction with hydrogen peroxide in Problem 6.183?

Electrons in Atoms and Periodic Properties

7

POURING MOLTEN METAL AT A STEEL PLANT *Hot molten steel emits a broad range of electromagnetic radiation including visible light.*

A LOOK AHEAD: Can Nature Be as Absurd as It Seems?

In terms of chemistry, the 18th century was the century of mass. Most of the ideas of stoichiometry that are fundamental to chemistry as a quantitative science were developed in experiments by accounting for the mass of substances. Chemistry in the 19th century was the century of energy and how matter and energy interact. It included observations of the interaction of matter with electric and magnetic fields that led to the discovery of electrons and protons as the fundamental particles from which all matter is made. The 20th century, especially the first 30 years, was the century of the electron.

At the end of the 19th century, many people thought that everything about the natural world had been discovered and all that was left to do was to solve a few lingering problems. One of them, called the photoelectric effect, involved the flow of electrons caused by light shining on a metal surface. Another involved the range of energies given off by hot objects. A third dealt with the light emitted from gases such as neon when an electric current was passed through them. In this chapter we will see that 20th-century solutions to these and related problems resulted in a new vision of matter at the atomic level and a new theory called quantum mechanics, which explains the structure of atoms and interactions between matter and energy. In addition to giving us powerful insights into the way the world works on the atomic level, these developments forever altered our sense of what we can really know about the world around us.

The world at the microscopic level is governed by an entirely different set of rules than those describing the behavior of the visible, macroscopic world that we inhabit. The theory that has been developed to explain the properties and behavior of electrons and light at the microscopic level is called quantum theory, and it is a towering intellectual achievement. We now know how electrons and light behave, but the behavior of things on a very small scale is like nothing any of us have seen. To express the uniqueness of the world at the atomic scale, some

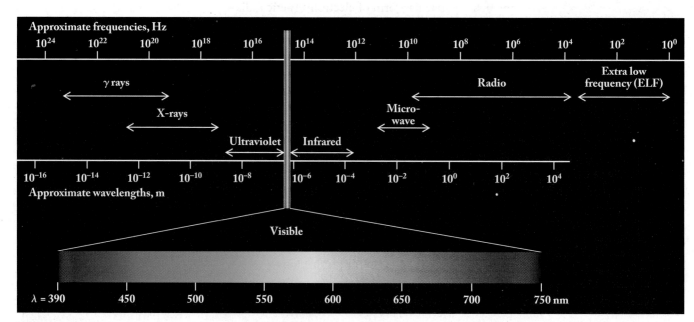

FIGURE 7.1 The visible light region between about 390 and 750 nm is a tiny fraction of the electromagnetic spectrum that ranges from very short, high frequency γ (gamma) rays to very long, low-frequency radio waves. Note the inverse relation between frequency and wavelength: frequencies increase from right to left, while wavelengths increase from left to right.

scientists say that if you can visualize behavior on the quantum mechanical level, you are wrong.

Two of the giants of quantum theory were Niels Bohr and Werner Heisenberg. These scientists worked together at the University of Copenhagen for a brief period, and after a particularly heated discussion about a feature of quantum theory, Heisenberg posed the question to himself, "Can nature possibly be as absurd as it seems?" The answer is probably yes. In this chapter we begin our exploration of light, electrons, and the quantum theory that has shaped the modern world.

7.1 Waves of Light

In Chapter 1 we discussed the Big Bang theory and the discovery by American astronomer Edwin Hubble that we live in a universe that has been expanding for billions of years. We did not describe how Hubble determined the universe to be expanding. To understand his conclusion, we need to examine the nature of the light emitted by our sun and all the stars in the universe.

Properties of Waves

Visible light is a small part of the **electromagnetic spectrum** shown in Figure 7.1. This spectrum consists of a broad, continuous range of radiant energy that extends from low-energy radio waves through visible light to X-rays and ultrahigh-energy γ (gamma) rays. All of these forms of radiant energy are examples of **electromagnetic radiation**.

The term *electromagnetic* comes from the theory proposed by Scottish scientist James Clerk Maxwell (1831–1879) that radiant energy consists of *waves* which have two components: an oscillating electric field and an oscillating magnetic field. These two fields are perpendicular to each other and travel together through space, as shown in Figure 7.2. Maxwell derived a set of equations based on his oscillating-wave model that accurately described nearly all the observed properties of light.

Maxwell's wave model means that a particular form of electromagnetic radiation has a characteristic **wavelength** (the distance from crest to crest or trough to trough, λ) and **frequency** (the number of times one wave passes a point of reference in 1 second, ν), as shown in Figure 7.3. Wavelengths have units of distance; for visible light the most convenient unit is the nanometer (nm): 1 nm = 10^{-9} m.

The **electromagnetic spectrum** is a continuous range of radiant energy that includes radio waves, infrared radiation, visible light, ultraviolet radiation, X-rays, and gamma rays.

Electromagnetic radiation is any form of radiant energy in the electromagnetic spectrum.

Wavelength (λ) is the distance from crest to crest or trough to trough on a wave.

The **frequency (ν)** of a wave expresses the number of times a wave passes a given point in some unit of time (typically 1 second).

CONNECTION In Chapter 1 we discussed the background microwave radiation of the universe, discovered by Penzias and Wilson, that provided evidence for the Big Bang.

CHEMTOUR Electromagnetic Radiation

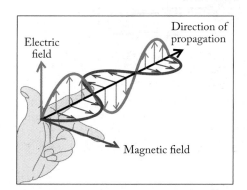

FIGURE 7.2 Visible light and other forms of electromagnetic radiation travel through space as oscillating electric and magnetic fields. The fields oscillate in planes at right angles to each other.

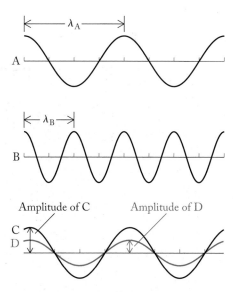

FIGURE 7.3 Waves of electromagnetic radiation have a characteristic wavelength (λ), frequency (ν), and amplitude (intensity). Wave A has a longer wavelength and thus a lower frequency than wave B; wave C has a greater intensity than wave D. Note that waves C and D have the same wavelength and frequency; only their amplitudes differ.

The SI unit of frequency is the **hertz (Hz)**, which is expressed in the unit s^{-1}: $1\text{ Hz} = 1\text{ s}^{-1} = 1$ cycle per second (cps).

The **amplitude** of a wave is the height of the crest or the depth of the trough with respect to the center line of the wave. The intensity of a wave is related to its amplitude.

Frequencies have units of **hertz (Hz)**, also called *cycles per second* (cps): $1\text{ Hz} = 1$ *cps* $= 1\text{ s}^{-1}$, the passage of one wave by a point in 1 second. The product of wavelength (λ) and frequency (ν) of any electromagnetic radiation is the universal constant c, the *speed of light* in a vacuum (2.99792458×10^8 m/s, or 3.00×10^8 m/s to three significant figures) in Equation 7.1. Thus the wavelength and the frequency of electromagnetic radiation have a reciprocal relationship: as the wavelength decreases, the frequency increases:

$$\lambda\nu = c \qquad (7.1)$$

Another characteristic of a wave is its **amplitude**, the height of the crest or the depth of the trough with respect to the center line of the wave (see Figure 7.3). The intensity of the energy of a wave is related to its amplitude. In the case of visible light, a wave of bright light has greater amplitude than a wave of dim light.

SAMPLE EXERCISE 7.1   **Calculating Frequency from Wavelength**

What is the frequency of the yellow-orange light ($\lambda = 589$ nm) produced by sodium-vapor streetlights?

COLLECT AND ORGANIZE We are given light of a specific wavelength in nm and asked to find the frequency. Frequency and wavelength are related by Equation 7.1 ($\lambda\nu = c$).

ANALYZE The speed of light (c) has units of meters per second (m/s) and wavelength has units of nanometers (nm). We need to convert wavelength from nanometers to meters so that the distance units of wavelength and the speed of light match: $1\text{ nm} = 10^{-9}$ m.

SOLVE Let's rearrange Equation 7.1 to solve for frequency

$$\nu = \frac{c}{\lambda}$$

and then convert the wavelength units from nanometers to meters:

$$589 \text{ nm} \times \frac{10^{-9}\text{ m}}{1 \text{ nm}} = 589 \times 10^{-9}\text{ m} = 5.89 \times 10^{-7}\text{ m}$$

By inserting the value of the speed of light and the wavelength into the equation, we can calculate the frequency:

$$\nu = \frac{3.00 \times 10^8 \text{ m/s}}{5.89 \times 10^{-7} \text{ m}}$$

$$= 5.09 \times 10^{14} \text{ s}^{-1}$$

THINK ABOUT IT The large value of the result makes sense: the product of the wavelength and frequency equals the speed of light. The magnitude of the wavelength is much smaller than the value of c, so the magnitude of the frequency has to be much larger than the value of c. Remember that λ and ν are reciprocal: as one increases, the other decreases.

Practice Exercise If a radio station operates at a frequency of 90.9 MHz (where $1\text{ MHz} = 10^6$ Hz), what is the wavelength of its transmission?

(Answers to Practice Exercises are in the back of the book.)

CONCEPT TEST

The ultraviolet (UV) region of the electromagnetic spectrum contains waves with wavelengths from about 10^{-7} m to 10^{-9} m in length; the infrared (IR) region contains waves with wavelengths from about 10^{-4} m to 10^{-6} m in length. Are waves in the UV region higher in frequency or lower in frequency than waves in the IR region?

(Answers to Concept Tests are in the back of the book.)

The Behavior of Waves

In the 17th century, Sir Isaac Newton (1642–1727) used glass prisms to separate sunlight into its component colors (Figure 7.4a). This separation process is known as **refraction** because the path of a beam of light entering a prism is bent, or *refracted,* as it moves from air into the prism. Different colors of light are bent through different angles, as shown in Figure 7.4(b): violet light is bent the most, and red light is bent the least. When Newton placed two prisms back to back, the spectrum of colors produced by the first prism was recombined back into *white light* (light that contains all the colors in the visible spectrum) by the second prism. From the results of such experiments Newton concluded that refraction separates but does not change colors, and that white light is a mixture of all colors of light. Refraction is not unique to visible light; any electromagnetic radiation is refracted when it passes from a substance with one density into a substance with a different density.

▶❙❙ **CHEMTOUR** Light Diffraction

An additional characteristic of electromagnetic radiation is that it undergoes the related processes of **diffraction** and **interference**. Both of these processes are illustrated by what happens when a beam of light passes through two narrow slits as shown in Figure 7.5(a). As the waves emerge from the slits they spread out in circles. This pattern means that the waves passing through the slits bend around, or are *diffracted* by, the slits. As the two sets of circular waves spread out, they run into and *interfere* with one another, much like the spreading waves produced when two pebbles are simultaneously dropped in a pool (Figure 7.5b). In both cases, the overlapping waves produce interference patterns: the light and dark areas on the screen in Figure 7.5(a) and the ripples in (b).

◉◉ **CONNECTION** Diffraction of light caused by water suspended within layers of calcium carbonate was originally mentioned in Chapter 4 as the phenomenon responsible for the iridescence of pearls.

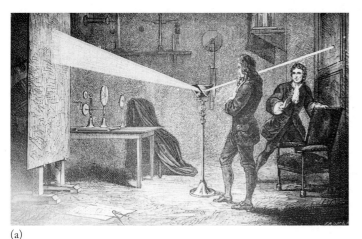

(a)

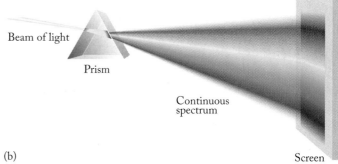

Beam of light

Prism

Continuous spectrum

(b)

Screen

FIGURE 7.4 The components of light: (a) Sir Isaac Newton used a prism to separate visible light into a continuous spectrum containing all the colors of the rainbow. (b) Light of different wavelengths is refracted as it moves from air into the medium of the prism. Red light is refracted the least, violet light the most.

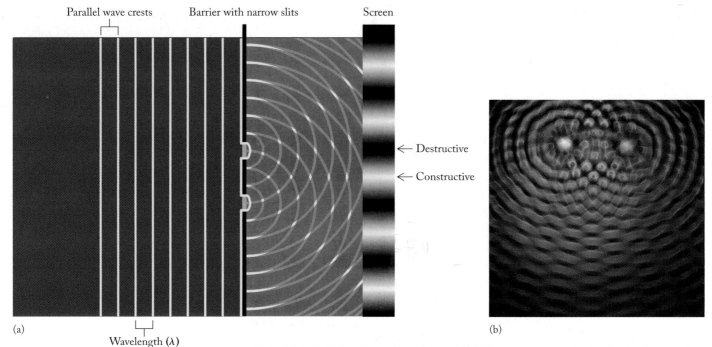

Parallel wave crests   Barrier with narrow slits   Screen

← Destructive

← Constructive

(a)   (b)

Wavelength (λ)

FIGURE 7.5 Diffraction and interference: (a) When waves approach a barrier that has gaps in it about the size of the wavelength, the waves are bent around the edges of the gaps and radiate out in circular wave patterns. When the two crests overlap, they enhance each other; when a crest overlaps with a trough, they cancel each other out as shown in more detail in Figure 7.6. (b) When two pebbles are dropped in a pool they create concentric waves that intersect, forming regions of constructive and destructive interference and this interference pattern.

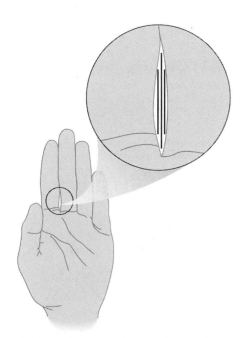

Hold your hand up in front of a light with your palm facing you. Slowly bring fingers together until they form a narrow slit. You should see one or more dark lines in the gap between your fingers, representing an interference pattern generated by diffraction of the light through the slit formed by your fingers.

The effect of interference can be either *constructive* or *destructive* (Figure 7.6). Consider two waves with the same wavelength and frequency. When the crests of the two identical waves overlap, the waves are said to be *in phase*: they combine constructively so that the amplitude of the resulting wave is twice the amplitude of the two original waves. On the other hand, when the crest of one wave overlaps with the trough of another, the waves are said to be *out of phase*: the two waves combine destructively, the amplitude of the resultant wave is zero, and the wave disappears. The pattern that results when the combined waves collide with some large feature in their path is called a *diffraction pattern*.

CONCEPT TEST

When two identical waves of red light (λ = 660 nm) interfere constructively, what happens to the wavelength and the frequency of the light?

Colors Missing from Sunlight

In 1800, English scientist William Hyde Wollaston (1766–1828) made a startling discovery about sunlight. Using carefully ground prisms to resolve the different colors in the spectrum of the sun, Wollaston discovered that the spectrum was not continuous: some of the spectrum was missing. Narrow dark lines were evident as gaps in the band of color. Using even better quality prisms, German

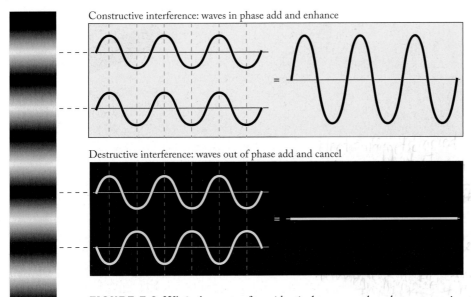

FIGURE 7.6 When the crests of two identical waves overlap, the waves are in phase and experience constructive interference; they enhance each other. When a crest from one pattern overlaps a trough of the other, the waves are out of phase and experience destructive interference; they cancel each other. Constructive interference results in bright bands on the screen, while destructive interference results in dark bands. The bright and dark bands form a diffraction pattern.

physicist and optical craftsman Joseph von Fraunhofer (1787–1826) resolved and mapped the wavelengths of over 500 missing lines, now called **Fraunhofer lines**, in the solar spectrum (Figure 7.7).

The Expanding Universe

Fraunhofer lines were crucial to Hubble's determination that we live in an expanding universe, because the same pattern of lines is missing from the light produced by distant galaxies. Hubble discovered, however, that the wavelengths of the lines in galactic spectra are not exactly the same as those in sunlight. The pattern of lines is the same, but the lines themselves are all shifted to longer wavelengths. Because the color of the longest wavelength of visible light is red, scientists call this frequency shift a *red shift*.

To understand why red shifts occur, let's consider the different sounds of a moving train engine's whistle. As a train approaches, the pitch of its engine whistle seems higher; after it passes, the pitch seems lower. Such motion-induced shifts in frequency are called the *Doppler effect*. Hubble explained the red-shifted light from distant galaxies using the Doppler effect and suggested that these galaxies are moving away from Earth at velocities that are a significant fraction of the speed of light. Indeed, the most-distant galaxies are moving away from us at nearly the speed of light.

CHEMTOUR Doppler Effect

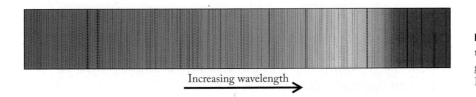

Increasing wavelength

FIGURE 7.7 The spectrum of light from the sun is not continuous but actually contains gaps, which appear as narrow dark lines called Fraunhofer lines.

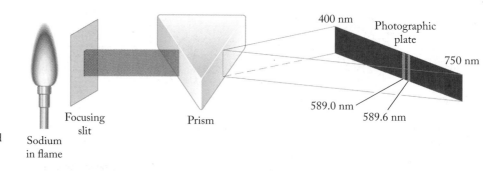

FIGURE 7.8 Sodium atoms heated in a flame emit a bright yellow-orange light. When this light is passed through a prism, the result is a bright line spectrum consisting of lines that exactly match the wavelength and frequency of some of the Fraunhofer lines in the solar spectrum.

▶❙❙ **CHEMTOUR** Light Emission and Absorption

7.2 Atomic Spectra

Fraunhofer and his contemporaries knew that there were narrow lines of light missing from the sun's spectrum, but they did not know why the lines were missing. Nearly a half century later, German chemist Robert Wilhelm Bunsen (1811–1899) and physicist Gustav Robert Kirchhoff (1824–1887) collaborated on extensive studies of the light emitted (given off) by elements when vaporized in the transparent flame of a burner designed by Bunsen. Unlike the spectrum of the sun, which displays dark lines or gaps in the nearly continuous spectrum of all colors, the spectra produced by the elements in these flames consist of only a few bright lines on a dark background. Bunsen and Kirchoff discovered that these **atomic emission spectra** (or **bright-line spectra**) consist of lines in *exactly the same place* as the wavelengths of some of the dark Fraunhofer lines in the sun's spectrum. For example, Fraunhofer's D line (actually a pair of closely spaced lines at 589.0 and 589.6 nm) exactly matched the wavelengths of yellow-orange light produced by hot sodium vapor (Figure 7.8).

These experiments and others employing light sources called gas discharge tubes (Figure 7.9) showed that each element *emits* a unique electromagnetic spectrum when its atoms are heated to sufficiently high temperature. The same element

∞ **CONNECTION** In Chapter 2 we discussed studies by J. J. Thomson that determined the mass-to-charge ratio of the electron, and by F. Aston that led to the discovery of stable isotopes and the technique of mass spectrometry. These studies were carried out using gas discharge tubes like those shown in Figure 7.9.

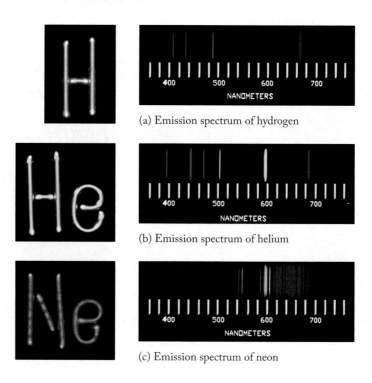

(a) Emission spectrum of hydrogen

(b) Emission spectrum of helium

(c) Emission spectrum of neon

FIGURE 7.9 The colored light emitted from gas discharge tubes, here formed in shapes of the chemical symbols of the gases they contain, produces bright-line spectra that are characteristic of the element. The gas discharge tubes shown are filled with (a) hydrogen, (b) helium, and (c) neon. Beside each tube is the emission spectrum of the gas.

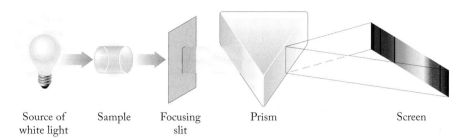

Source of white light   Sample   Focusing slit   Prism   Screen

Absorption spectrum of hydrogen:

Absorption spectrum of helium:

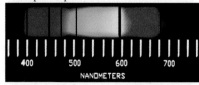

Absorption spectrum of neon:

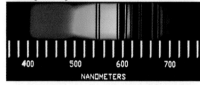

FIGURE 7.10 When gaseous atoms of the same elements whose bright-line spectra shown in Figure 7.9 are illuminated by an external source of white light with a continuous spectrum, the resultant atomic absorption spectrum contains dark lines that are characteristic of the element. The dark lines in the atomic absorption spectra of the elements have the same wavelength and frequency as the bright lines in their atomic emission spectra.

also *absorbs* electromagnetic radiation of the same wavelengths when its atoms are illuminated by an external source of radiation, such as sunlight (Figure 7.10). Dark lines appear in the solar spectrum because gaseous atoms around the sun absorb radiation as sunlight passes through them on its way to Earth. The absorption of energy from some external source by atoms produces **atomic absorption spectra** (or **dark-line spectra**), which consist of dark gaps in the otherwise continuous spectrum. Thus the two processes we know as *emission* and *absorption* involve electromagnetic radiation of the same wavelength and frequency.

CONCEPT TEST

A set of five visible spectra are given in the following set. The first one is the emission spectrum of mercury vapor. Select the absorption spectrum of mercury vapor from the other four spectra.

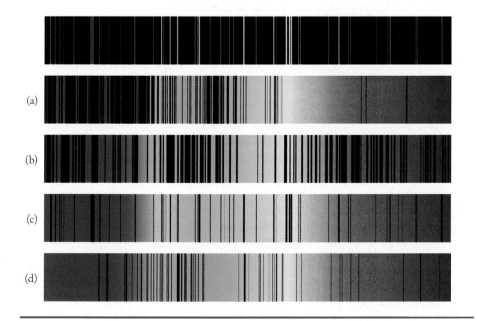

(a)

(b)

(c)

(d)

Atomic emission spectra (also called **bright-line spectra**) consist of bright lines on a dark background; the lines appear at specific wavelengths.

Atomic absorption spectra (or **dark-line spectra**) consist of characteristic series of dark lines produced when free, gaseous atoms are illuminated by external sources of radiation.

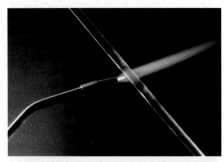

FIGURE 7.11 When a metal filament is heated, the object first glows red, then orange, and finally becomes white-hot as light of all colors is emitted.

A **quantum** is the smallest discrete quantity of a particular form of energy.

Quantum theory is based on the idea that energy is absorbed and emitted in discrete quanta.

Something that is **quantized** has values that are restricted to whole-number multiples of a specific base value. The base unit of energy is the quantum.

Planck's constant (h) is the proportionality constant between the energy and frequency of electromagnetic radiation, as expressed in the relation $E = h\nu$. Its value is $6.6260755 \times 10^{-34}$ J·s, which we will typically round to 6.626×10^{-34} J·s.

7.3 Particles of Light and Quantum Theory

As studies of electromagnetic radiation progressed in the 19th century, scientists discovered that there were limits to the wave model. One limit was encountered when they tried to account for radiation given off by objects heated to very high temperatures. Consider, for example, what happens when a metal filament is heated (Figure 7.11). At low temperature the filament gives off only heat (infrared radiation). With a little more heating the metal begins to glow a dull red. With more heating the intensity of the red glow increases; the color begins to shift to red-orange, orange, yellow, and eventually, with maximum heat, intensely white hot as the metal emits all visible colors. Emissions diminish in the ultraviolet range and do not occur beyond it.

This phenomenon was well known at the end of the 19th century; however, none of the equations based on the wave properties of radiation that Maxwell developed could account for the emission spectrum produced by such a filament in the visible range with diminishing emissions at higher frequencies. Another explanation—indeed, another model of radiation behavior—was required.

Quantum Theory

In 1900, German scientist Max Planck (1858–1947) proposed such a model. It was based on the view that light could have both *wavelike* and *particle-like* properties. Planck called particles of radiant energy **quanta**. In his model, which we call **quantum theory**, quanta are the smallest amounts of radiant energy in nature, the sort that a single atom or molecule might absorb or emit. An object made of a large, discrete number of atoms or molecules could therefore emit only a discrete number of particles of light. In other words, light from such a source must be **quantized**.

Planck proposed that the energy of a quantum was related to the frequency of the corresponding wave of radiation by a constant (the value of which is 6.626×10^{-34} J·s), which is now called **Planck's constant (h)**:

$$E = h\nu \tag{7.2}$$

When we combine Equations 7.1 and 7.2, we find that the energy of a quantum of energy is inversely proportional to its wavelength:

$$\nu\lambda = c \qquad \text{so} \qquad \nu = \frac{c}{\lambda} \qquad \text{and}$$

$$E = \frac{hc}{\lambda} \tag{7.3}$$

SAMPLE EXERCISE 7.2   **Calculating the Energy of a Quantum of Light**

What is the energy of a quantum of red light that has a wavelength of 656 nm?

COLLECT AND ORGANIZE Equation 7.3 ($E = hc/\lambda$) relates the energy of a quantum to its wavelength. The speed of light has units of meters per second, and its wavelength is given in nanometers. We need to convert nanometers to meters so the distance units of wavelength and the speed of light match.

ANALYZE The wavelength is 656 nm, and 1 nm = 10^{-9} m. The value of Planck's constant (h) is 6.626×10^{-34} J·s, and the speed of light (c) is 3.00×10^{8} m/s. We can use Equation 7.3 and solve directly for the energy.

SOLVE

$$E = \frac{hc}{\lambda} = \frac{(6.626 \times 10^{-34}\,\text{J}\cdot\text{s})\left(3.00 \times 10^8\,\frac{\text{m}}{\text{s}}\right)}{656\,\text{nm} \times \dfrac{10^{-9}\,\text{m}}{1\,\text{nm}}}$$

$$= 3.03 \times 10^{-19}\,\text{J}$$

THINK ABOUT IT This quantity of energy is extremely small, as it should be, because a quantum is a particle of radiant energy at the atomic level, where particles such as protons and neutrons have extremely small masses and charges.

Practice Exercise Some instruments discriminate between individual quanta of electromagnetic energy based on their energies. Assume such an instrument has been adjusted to detect quanta with 1.00×10^{-16} J of energy. What is the wavelength of the detected radiation?

To visualize the meaning of Planck's quanta, consider the two ways you might get from the sidewalk to the entrance of a building (Figure 7.12). If you walked up the steps (Figure 7.12a), there would be discrete heights—each the rise of a single step—between the sidewalk and the entrance. You could not stand at an elevation between two adjacent steps; there would be nothing to stand on at that height. However, if you walked up the ramp (Figure 7.12b), you could stop at any height between the sidewalk and the entrance. The discrete height changes represented by the steps in a flight of stairs are a model of Planck's hypothesis that energy is released (analogous to walking down the steps) or absorbed (walking up the steps) in discrete packets, or quanta, of energy.

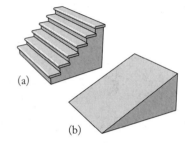

FIGURE 7.12 Quantized and unquantized heights: (a) A flight of stairs exemplifies quantization: each step rises by a discrete height to the next step. (b) In contrast, the heights on a ramp are not quantized.

CONCEPT TEST

Which of the following quantities vary continuously (are not quantized) and which vary by discrete values (are quantized)?

a. The volume of water that evaporates from a lake each day during a summer heat wave

b. The quantity of eggs remaining in a carton in a refrigerator

c. The time it takes you to get ready for class in the morning

d. The number of red lights encountered when driving the length of Fifth Avenue in New York City

Although the equations Planck developed based on his quantum theory fit the emission spectra of hot objects, Planck had no experimental evidence to support the existence of quanta. That evidence, supplied by Albert Einstein, came just 5 years after Planck published his quantum theory.

The Photoelectric Effect

If light shines on a metallic surface, the surface may give off an electric current (a flow of electrons). This phenomenon is the basis for photoelectric cells used in some older automatic door openers. As you approach the door, you block the beam of light that shines on a metal surface thereby causing the flow of electricity to cease, which in turn triggers the door to open. Because light produces the current, this phenomenon is called the **photoelectric effect**, and the electrons emitted are called *photoelectrons*.

Current only flows, however, if the frequency of the light is above some minimum **threshold frequency (ν_0)** (Figure 7.13). A source of radiation with frequencies less than the threshold value produces no photoelectrons, no matter how intense the source. On the other hand, even a dim source of radiation with frequencies equal to or greater than the threshold frequency produces at least a few photoelectrons. Einstein used Planck's quantum theory to explain this unexpected behavior. He proposed that the threshold frequency was the frequency of the minimum quantum of energy:

$$E = h\nu_0 \tag{7.4}$$

needed to remove a single electron from the surface of a metal. When a metal surface is illuminated by a beam of these quanta (Einstein called these quanta of light **photons**), individual quanta are absorbed by individual electrons, giving the electrons enough kinetic energy to be dislodged from the surface.

The quantity $h\nu_0$ of energy needed to dislodge the electron from the surface of a metal is known as the **work function (Φ)** of the metal. This energy is related to the strength of the attractive force between metal nuclei on the surface and the surrounding electrons. If the surface is irradiated with light of a frequency above the threshold frequency ($\nu > \nu_0$), the quantity $h\nu_0$ is used to dislodge the electron, and any energy in excess of that value is imparted to the electron as kinetic energy:

$$KE_{electron} = h\nu - h\nu_0 = h\nu - \Phi \tag{7.5}$$

Considering light as photons provides a consistent explanation of the photoelectric effect. The higher the frequency of the photon above the threshold frequency, the faster the ejected electron moves. If the frequency of the photon is below the

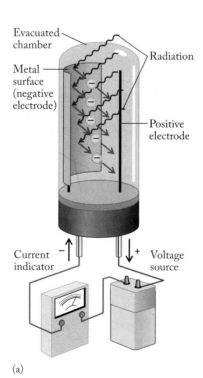
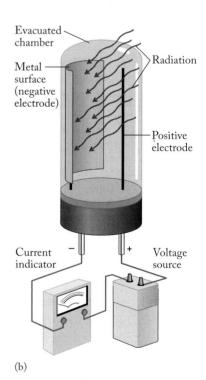

FIGURE 7.13 A device called a phototube includes a positive electrode and a metallic negative electrode. (a) If radiation of high enough frequency and energy (shown in violet) illuminates the negative electrode, electrons are dislodged from the surface and flow toward the positive electrode. This flow of electrons completes the circuit and produces an electric current. The size of the current is proportional to the intensity of the radiation—that is, to the number of photons per unit time striking the negative electrode. (b) Photons of lower energy and frequency (shown in red) do not have sufficient energy to dislodge electrons and so do not produce the photoelectric effect no matter how many of them bombard the surface of the metal. The circuit is not complete and no current flows.

threshold frequency, no electron is ejected, no matter how many photons (i.e. no matter how intense the light) hit the metal surface.

SAMPLE EXERCISE 7.3   **Using the Work Function**

The work function (Φ) for mercury is 7.22×10^{-19} J. (a) What is the minimum frequency of light that can be used to cause photoelectrons to be ejected from the surface of mercury? (b) Could we use light in the visible range of the electromagnetic spectrum to produce the photoelectric effect in this element?

COLLECT AND ORGANIZE The value of the work function is defined by Equation 7.4 ($E = h\nu_0$), where ν_0 is the minimum frequency of light needed to produce the photoelectric effect, and h is Planck's constant. We can use Equation 7.4 to calculate the frequency and use Figure 7.1 to determine the region of the electromagnetic spectrum in which that energy is found.

ANALYZE Because the work function defines the minimum frequency of light required, we have to compare the frequency of light in the visible region of the spectrum to the value we calculate to see if the calculated value is higher or lower than the frequency range of visible light.

SOLVE (a) Using Equation 7.4, we calculate the threshold frequency for Hg:

$$E = h\nu_0 = 7.22 \times 10^{-19} \text{ J} = (6.626 \times 10^{-34} \text{ J}\cdot\text{s})\, \nu_0$$

$$\nu_0 = \frac{7.22 \times 10^{-19} \text{ J}}{6.626 \times 10^{-34} \text{ J}\cdot\text{s}} = 1.09 \times 10^{15}\text{s}^{-1}$$

(b) According to Figure 7.1, the highest frequency of light in the visible region of the electromagnetic spectrum is about 8×10^{14} s^{-1} (Hz), which is a lower frequency than required by the work function. Therefore no light in the visible region can produce the photoelectric effect with mercury. A threshold frequency of 1.09×10^{15} s^{-1} falls in the UV region of the electromagnetic spectrum.

THINK ABOUT IT The value for the work function is very small ($\sim 10^{-19}$ J$\cdot$s), but remember that it defines the energy required to remove only one electron from the metal in question. The frequency of the light needed is a very large number ($\sim 10^{15}$ s^{-1}), but it is near the middle of the whole electromagnetic spectrum in Figure 7.1. The numbers in this exercise are typical values associated with the photoelectric effect, and it is important to become accustomed to dealing with such very large and very small values.

Practice Exercise The work function (Φ) of silver is 7.59×10^{-19} J. What is the longest wavelength of electromagnetic radiation that can eject an electron from the surface of a piece of silver?

CONCEPT TEST

Consult Figure 7.1 to answer the following without making a calculation: if a photon of red light has sufficient energy to eject an electron from the surface of a metal, will a photon of violet light have enough energy to overcome the work function of that metal?

Based on the behavior of electromagnetic radiation in the photoelectric effect, the conclusion that it can exhibit both wavelike and particle-like properties is inescapable. The dual nature of light played a key role in the early 20th century as scientists developed our modern view of atomic structure and the behavior of electrons inside atoms.

7.4 The Hydrogen Spectrum and the Bohr Model

In formulating his quantum theory, Max Planck was influenced by the results of investigations of the line spectra produced by free atoms: results that led him to question whether any spectrum, even that of an incandescent lightbulb, was truly continuous. Among those earlier results was a discovery made in 1885 by a Swiss school teacher named Johann Balmer (1825–1898).

Lines in the Hydrogen Spectrum

Balmer determined that the frequencies of the four lines in the visible spectrum of hydrogen (Figure 7.9a) fit the simple equation

$$\nu = (3.2881 \times 10^{15} \text{ s}^{-1})\left(\frac{1}{2^2} - \frac{1}{n^2}\right) \tag{7.6}$$

where n is a whole number greater than 2; in particular, 3 for the red line, 4 for the green, 5 for the blue, and 6 for the violet line in the spectrum. Without having seen any other lines in the spectrum of hydrogen, Balmer predicted there should be one more line (for $n = 7$) at the edge of the violet region, and indeed such a line was later discovered.

Balmer also predicted that other sets of hydrogen lines should exist in other regions of the electromagnetic spectrum, corresponding to frequency values calculated by replacing $1/2^2$ in his formula with $1/1^2$, $1/3^2$, $1/4^2$, and so forth. He was right. In 1908 the German physicist Friedrich Paschen (1865–1947) discovered infrared emission lines in the hydrogen spectrum that corresponded to $1/3^2$ in Balmer's formula. A few years later, Theodore Lyman at Harvard University discovered lines for hydrogen in the UV region corresponding to $1/1^2$. By the 1920s the $1/4^2$ and $1/5^2$ lines had been discovered. Like the $1/3^2$ lines, they were in the infrared region.

Later, Swedish physicist Johannes Robert Rydberg (1854–1919) revised Balmer's equation by changing frequencies to *wave numbers* ($1/\lambda$), that is, the number of wavelengths per unit of distance:

$$\frac{1}{\lambda} = [1.097 \times 10^{-2} \text{ (nm)}^{-1}]\left(\frac{1}{n_1^{\,2}} - \frac{1}{n_2^{\,2}}\right) \tag{7.7}$$

where n_1 is a whole number that remains fixed for a series of calculations in which n_2 is also a whole number with values of $n_1 + 1$, $n_1 + 2$, . . . for successive lines in the spectrum.

When Balmer and Rydberg derived their equations, no one knew what process gave rise to these lines, but scientists were beginning to think about what the relation between emission spectra and absorption spectra was saying about the structure of matter at the atomic level. The spectra of discrete frequencies (energies) indicated that only certain amounts of energy were available to atoms and

German scientist Max Karl Ernst Ludwig Planck is considered the father of quantum physics. He won the 1918 Nobel Prize in Physics for his pioneering work on the quantized nature of electromagnetic radiation. Planck was revered by his colleagues for his personal qualities as well as his scientific accomplishments.

molecules, and the fact that the equations describing the lines in the hydrogen spectrum involved ratios of small whole numbers was intriguing. The explanation of what caused spectral lines would coincide with further exploration of the concepts being developed within quantum theory.

The Bohr Model of Hydrogen

Scientists in the early 20th century faced yet another dilemma. Ernest Rutherford's work had established a picture of a nuclear atom that was mostly empty space occupied by negatively charged electrons with a tiny center called the nucleus containing virtually all the mass and all of the positive charge. What kept the electrons from falling into the nucleus? Rutherford himself had suggested that the electrons had to be in motion and theorized that they might orbit the nucleus the way planets orbit the sun. However, classical physics predicted that negative electrons orbiting a positive nucleus would emit energy in the form of electromagnetic radiation and eventually spiral into the nucleus. This does not happen.

The problem of explaining the stability of negative electrons with respect to the positive nucleus was addressed by the Danish theoretician Niels Bohr (1885–1962), who was well acquainted with the issue because he had studied with Rutherford. Bohr designed a theoretical model of the hydrogen atom from the point of view of quantum theory. He dealt with the kinetic energy of the electron in a fashion that allowed it to change in discrete amounts (quanta) rather than continuously as classical physics required. From these considerations, Bohr derived the following formula for the possible energy differences (ΔE) between any pair of orbits with values n_1 and n_2:

$$\Delta E = \frac{2\pi^2 m e^4}{h^2}\left(\frac{1}{n_1^2} - \frac{1}{n_2^2}\right) \tag{7.8}$$

CONNECTION We discussed Rutherford's gold-foil experiment and the development of the idea of the nuclear atom in Chapter 2.

▶❚❚ **CHEMTOUR** Bohr Model of the Atom

where m and e are the mass and charge of the electron, respectively.

Compare this expression to Equations 7.6 and 7.7. They are much alike. The coefficients are different because the units used to express frequency, reciprocal wavelength, and energy are different. The key point is that the equations developed to fit the appearance of lines in the absorption and emission spectra of hydrogen are the same as the equation developed by Bohr using a combination of classical theory and quantum theory to explain why the electron does not fall into the nucleus. Emission and absorption spectra must be reporting information about the energy of electrons in atoms.

In 1913, Bohr published a paper describing a model of the hydrogen atom based on the notion that its electron must reside in one of an array of discrete energy states called *allowed energy levels*. Here are the key features of Bohr's hydrogen model:

- The electron in a hydrogen atom occupies a discrete energy level and may exist only in the available energy levels.
- The electron may move from one energy level to another by either acquiring or releasing an amount of energy equal to the difference in the energies of the two levels.
- Each energy level is designated by a specific value for n, called the *principal quantum number*. The level closest to the nucleus has the lowest value for n; that is, $n = 1$. The next closest level has a value of $n = 2$, and so on.

Wolfgang Pauli (left) and Neils Bohr are apparently amused by the behavior of a toy called a "tippe top," which, when spun on its base, will tip itself over automatically and spin on its stem. The flip is caused by a combination of friction and the tippe top's angular momentum. In Bohr's model of the hydrogen atom, the angular momentum of its electron is quantized. No such quantization is needed to explain the motion of a tippe top.

An **energy level** is an allowed state that an electron can occupy in an atom.

Movements of electrons between energy levels are called **electron transitions**.

The link between line spectra and electrons is that the lines in the spectrum of hydrogen correspond to energy emitted or absorbed when electrons move from one **energy level** to another. This type of electron movement is called an **electron transition**. Figure 7.14 is an *energy-level diagram* for the electron in a hydrogen atom. An energy level diagram shows the allowed energy states for an electron in an atom.

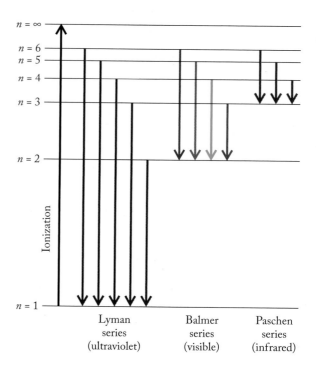

FIGURE 7.14 An energy-level diagram showing the electron transitions for the electron in the hydrogen atom.

SAMPLE EXERCISE 7.4   **Calculating a Wavelength of a Line in the Hydrogen Spectrum**

What is the wavelength of the line in the visible hydrogen spectrum corresponding to $n_1 = 2$ and $n_2 = 3$?

COLLECT AND ORGANIZE Lines in the visible spectrum of hydrogen correspond to the energy required for electron transitions. We will use the general form of the equation that relates wavelength to energy level in the hydrogen atom, Equation 7.7 to calculate λ:

$$\frac{1}{\lambda} = [1.097 \times 10^{-2}\,(\text{nm})^{-1}]\left(\frac{1}{n_1{}^2} - \frac{1}{n_2{}^2}\right)$$

ANALYZE We are given n_1 and n_2 in the statement of the problem.

SOLVE Inserting the values of n_1 and n_2 into Equation 7.7 and solving for λ gives us the wavelength we are looking for:

$$\frac{1}{\lambda} = [1.097 \times 10^{-2}\,(\text{nm})^{-1}]\left(\frac{1}{2^2} - \frac{1}{3^2}\right) = [1.097 \times 10^{-2}\,(\text{nm})^{-1}]\left(\frac{1}{4} - \frac{1}{9}\right)$$

$$= [1.097 \times 10^{-2}\,(\text{nm})^{-1}](0.1389) = 1.524 \times 10^{-3}\,(\text{nm})^{-1}$$

$$\lambda = 656.4\ \text{nm}$$

THINK ABOUT IT If we check our result against the continuous spectrum in Figure 7.1, we find that 656 nm is in the visible region. Thus the answer is reasonable.

Practice Exercise What are the wavelengths of the lines of the hydrogen spectrum corresponding to $n_1 = 2$ and $n_2 = 4$, 5, and 6? Before calculating them, predict which value of n corresponds to the longest wavelength of the three.

The lowest energy level available to an electron in an atom is the **ground state**.

An **excited state** of an electron in an atom is any energy state above the ground state.

Energies of Photons and Electron Transitions

The energies of the photons absorbed or emitted by a hydrogen atom must match the differences in allowed energies of the electron within the atom as defined by the energy levels. Let's derive an equation for calculating these energies, starting with the energies of the photons themselves. We can do this using Equations 7.3 and 7.7 and inserting the nanometers-to-meters conversion step:

$$E = \frac{hc}{\lambda} = hc\frac{1}{\lambda}$$

$$= (6.626 \times 10^{-34}\,\text{J}\cdot\text{s})\left(\frac{3.00 \times 10^8\,\text{m}}{\text{s}}\right)\left(\frac{10^9\,\text{nm}}{\text{m}}\right)\left(\frac{1.097 \times 10^{-2}}{\text{nm}}\right)\left(\frac{1}{n_1^{\,2}} - \frac{1}{n_2^{\,2}}\right)$$

$$= (2.18 \times 10^{-18}\,\text{J})\left(\frac{1}{n_1^{\,2}} - \frac{1}{n_2^{\,2}}\right) \tag{7.9}$$

The Bohr model says that energy values calculated using Equation 7.9 must equal the changes in energy (ΔE) that the electron in a hydrogen atom undergoes as it experiences electron transitions from an initial energy level n_i to a final energy level n_f within the atom. This equality is achieved when the energy of each level (E_n) is as follows:

$$E_n = (-2.18 \times 10^{-18}\,\text{J})\left(\frac{1}{n^2}\right) \tag{7.10}$$

According to Equation 7.10, the energy of the electron in a hydrogen atom is always negative. A hydrogen atom has its lowest (most negative) internal energy when its electron is in the $n = 1$ energy level closest to the nucleus:

$$E_n = (-2.18 \times 10^{-18}\,\text{J})\left(\frac{1}{1^2}\right) = -2.18 \times 10^{-18}\,\text{J}$$

This level of energy represents the **ground state** of the atom. The value of E_n increases (becomes less negative) as n increases because the value of the fraction $1/n^2$ decreases. As the value of n approaches infinity ($n \rightarrow \infty$), E_n approaches zero ($E_n \rightarrow 0$). At $n = \infty$, the electron is no longer in the hydrogen atom. In other words, the hydrogen atom has ionized, producing a free electron and leaving behind a proton.

All the energy states above the $n = 1$ ground state are called **excited states**. Bohr proposed that the electron in a hydrogen atom can move from the ground state to one of the excited states (for example, $n_1 \rightarrow n_3$), from one excited state to another (such as $n_5 \rightarrow n_4$), or from an excited state back to the ground state ($n_2 \rightarrow n_1$). The H electron can make one of these transitions only if it absorbs or emits a quantum of energy that exactly matches the energy difference (ΔE) between the two states. We can express this energy difference as

$$\Delta E = E_f - E_i \tag{7.11}$$

where E_i is the energy of the initial state (with principal quantum number n_i) and E_f is the energy of the final state (with principal quantum number n_f). Combining Equations 7.10 and 7.11 yields

$$\Delta E = \left[(-2.18 \times 10^{-18}\,\text{J})\left(\frac{1}{n_f^2}\right)\right] - \left[(-2.18 \times 10^{-18}\,\text{J})\left(\frac{1}{n_i^2}\right)\right]$$

which can be simplified by combining the two terms in brackets on the right side of the equation:

$$\Delta E = (-2.18 \times 10^{-18}\,\text{J})\left(\frac{1}{n_f^2} - \frac{1}{n_i^2}\right) \qquad (7.12)$$

SAMPLE EXERCISE 7.5   **Calculating the Energy of an Electron Transition in a Hydrogen Atom**

How much energy is required to ionize a ground-state hydrogen atom?

COLLECT AND ORGANIZE We are asked to determine the energy required to remove an electron from a hydrogen atom starting from its lowest possible energy state ($n = 1$). Equation 7.12 enables us to calculate the energy change associated with an electron transition.

ANALYZE To use Equation 7.12, we need to identify the initial and final states of the electron. The electron starts at the ground state ($n = 1$). The electron is removed and the atom is ionized when $n = \infty$.

SOLVE We can use these two values (1 and ∞) for n_i and n_f, respectively, in Equation 7.12 to calculate the value of the required energy:

$$\Delta E = (-2.18 \times 10^{-18}\,\text{J})\left(\frac{1}{n_f^2} - \frac{1}{n_i^2}\right)$$

$$= (-2.18 \times 10^{-18}\,\text{J})\left(\frac{1}{\infty^2} - \frac{1}{1^2}\right)$$

Dividing by ∞^2 yields zero, so the bracket value simplifies to -1, which gives us the following:

$$\Delta E = 2.18 \times 10^{-18}\,\text{J}$$

THINK ABOUT IT The answer is a small amount of energy, but if you compare it to results we calculated for phenomena such as the photoelectric effect, where we were removing an electron from the surface of a metal, this answer is in the same range, so it is reasonable.

Practice Exercise Calculate the energy required to ionize a hydrogen atom from the $n = 3$ excited state. Before doing the calculation, predict whether this energy is greater than or less than the energy required to ionize a ground-state hydrogen atom.

One of the strengths of the Bohr model of the hydrogen atom is that it accurately predicts the energy needed to remove the electron. This energy is called the *ionization energy* of the hydrogen atom. We will examine the ionization energies of other elements later in this chapter. Most elements require considerably less energy to ionize than does hydrogen, for reasons having to do with the way their electrons are distributed in their atoms.

Equation 7.12 looks a lot like Equation 7.9. A major difference is the negative sign on the right-hand side of Equation 7.12. This difference makes sense if we consider the kind of energy these equations allow us to calculate. Equation 7.12 is used to calculate changes in the internal energies of atoms as electrons move between energy levels. If the electron moves to a higher energy level ($n_f > n_i$) the term ($1/n_f^2 - 1/n_i^2$) is negative and ΔE is positive (the energy of the electron has increased). However, when an electron moves from an excited state to a lower energy level, then n_f is less than n_i, the atom loses energy, so ΔE is negative.

Equation 7.9 is used to calculate the energy of the photons absorbed or emitted by hydrogen atoms as their internal energies change due to electron transitions. A photon cannot have a negative energy, so Equation 7.9 is written to insure that all calculated values are positive. This is done by requiring that n_2 always be greater than n_1. That way the value of ($1/n_1^2 - 1/n_2^2$) is always positive, as is the calculated energy of the photon. We need to keep in mind that a photon of radiation described by Equation 7.12 can be absorbed (producing a positive ΔE in the H atom that absorbed it, as given by Equation 7.12) or emitted (producing a negative ΔE in Equation 7.12). These options for either absorption or emission are illustrated in the energy-level diagram (Figure 7.14). The black arrow pointing upward represents absorption of sufficient energy to completely remove an electron from a hydrogen atom. The downward-pointing colored arrows represent decreases in the internal energy of the hydrogen atom that occur when photons are emitted. (If the colored arrows pointed up, they could represent *absorption* of photons leading to *increases* in the internal energy of the atom.) In every case the energy of the photon matches the absolute value of ΔE.

The Bohr model provided a major step forward in our understanding of matter at the atomic level. Bohr's application of quantum theory to defining the energy levels of electrons in atoms and his linking of data from absorption and emission spectra to energy levels within atoms are two concepts we still use. But Bohr's treatment also has a severe limitation: it applies only to hydrogen atoms and to ions such as He^+ with a single electron. The model does not account for the observed spectra of multielectron elements and ions. In addition, the movement of electrons in atoms is much less clearly defined than Bohr allowed. The picture of the atom provided by Bohr's model is limited, but it started scientists down the road toward using ideas of quantum theory to explain the behavior of matter at the atomic level.

7.5 Electrons as Waves

A decade after Bohr published his model of the hydrogen atom, a French graduate student named Louis de Broglie (1892–1977) provided a theoretical basis for the stability of electron orbits and greatly affected our view of the structure of matter. His approach incorporated yet another significant change in the way that early 20th-century scientists viewed the behavior of atoms and subatomic particles—namely, to think of them not only as particles of matter but also as waves.

De Broglie Wavelengths

After Planck and Einstein established the dual nature of light as both wave and particle, de Broglie proposed that, if light, which we normally think of as a wave, has particle-like properties, perhaps the electron, which we normally think of as a particle, can have wavelike properties. He predicted that the wavelength of a moving particle like the electron could be calculated using an equation that he derived from Einstein's statement of the equivalence of mass and energy ($E = mc^2$) and Planck's equation for the energy of a photon [$E = h\nu = h(c/\lambda)$]:

$$\lambda = \frac{h}{mu} \qquad (7.13)$$

▶❙❙ CHEMTOUR De Broglie Wavelength

where m is the mass of the electron in kg, u is its velocity in m/s, and h is Planck's constant. The wavelength of a particle calculated in this way is often called the *de Broglie wavelength*. De Broglie's equation tells us that any moving particle has wavelike properties; it is a **matter wave**. De Broglie predicted that moving particles much bigger than electrons, such as atomic nuclei, molecules, and even tennis balls, airplanes in flight, and Earth revolving around the sun, have characteristic wavelengths that can be calculated using Equation 7.13. The wavelengths of such large objects must be extremely small, given the tiny size of Planck's constant in the numerator of Equation 7.13. Because the wavelengths are so small, we never notice the wavelike nature of large objects in motion.

SAMPLE EXERCISE 7.6 **Calculating the Wavelength of a Particle in Motion**

Compare the wavelength of a baseball weighing 142 g thrown by a major league pitcher at 98 mi/hr (98 mi/hr = 44 m/s) to the wavelength of an electron ($m_e = 9.109 \times 10^{-31}$ kg) in a hydrogen atom moving at the speed of light. For the purposes of comparison, consider a baseball to have a radius of about 3.5 cm (7×10^{-2} m) and a hydrogen atom to have a radius of 5.3×10^{-11} m.

COLLECT AND ORGANIZE We have the mass and velocity of both moving objects. De Broglie's equation (Equation 7.13) relates those terms to the wavelength.

ANALYZE Because h has units of joule-seconds (J · s) and wavelength has units of length only, we need a conversion factor in our calculation that allows us to cancel out the joules: 1 J = 1 (kg · m²)/s². We also must express the mass of the baseball in kilograms (to cancel those units). We know about how big a baseball is; we are given an approximate radius of a hydrogen atom so that we can compare the size of the wavelength to the size of the space in which an electron may be found. We do not need the dimensions of the objects for the calculation.

SOLVE

A **matter wave** is the wave associated with any particle.

A **standing wave** is confined to a given space and has a wavelength that is related to the length (L) of the space by the equation $L = n(\lambda/2)$, where n is a whole number.

A **node** is a location in a standing wave that experiences no displacement.

■ For the baseball:

$$\lambda = \frac{h}{mu} = \frac{6.626 \times 10^{-34} \text{ J} \cdot \text{s}}{(142 \text{ g})(1 \text{ kg}/1000 \text{ g})(44 \text{ m/s})} \times \frac{(\text{kg} \cdot \text{m} \cdot \text{m})/\text{s}^2}{\text{J}}$$

$$= 1.06 \times 10^{-34} \text{ m}$$

■ For the electron:

$$\lambda = \frac{h}{mu} = \frac{6.626 \times 10^{-34} \text{ J} \cdot \text{s}}{(9.109 \times 10^{-31} \text{ kg})(3.00 \cdot 10^8 \text{ m/s})} \times \frac{(\text{kg} \cdot \text{m} \cdot \text{m})/\text{s}^2}{\text{J}}$$

$$= 2.42 \times 10^{-12} \text{ m kg}$$

THINK ABOUT IT The wavelength calculated for the baseball is extremely small compared with the size of the object. The wave is much too small to be observed, so its character contributes nothing to the behavior of the baseball. We expected that. Our experience with objects in the world is that they behave "like matter," not waves. For the electron, however, the wavelength is within a factor of 25 of the size of the hydrogen atom, so wavelike properties should be significant contributors to the behavior of the electron.

Practice Exercise The speed of the electron in the ground state of the hydrogen atom is 2.2×10^6 m/s. What is the wavelength of such an electron?

De Broglie then applied the following reasoning to explain the stability of the electron levels in Bohr's model of the hydrogen atom. He proposed that the electron in a hydrogen atom behaves as a wave, and in particular as a circular wave oscillating around the nucleus. To understand the implications of this statement, we need to examine what it takes to make a stable, circular wave. Consider the motion of a vibrating violin string of length L (Figure 7.15a). Because it is fixed at both ends, its vibration has no motion at the ends and is a maximum in the middle. The resulting motion is an example of a **standing wave**, that is, a wave that oscillates back and forth within a fixed space. The ends of the string are points of zero wave displacement; that is, they do not move during vibration. Such points are called **nodes**. The sound wave produced in this way is called the *fundamental* of the string. The wavelength (λ) of the fundamental is $2L$. It has the lowest frequency and longest wavelength possible for that string. The length of the string is one-half the wavelength of the fundamental ($L = \lambda/2$).

If the string is held down in the middle (creating a third node) and plucked halfway between the middle and one end, a new, higher-frequency wave called the *first harmonic* is produced. The wavelength of this harmonic is equal to L, and $L = 2(\lambda/2)$. We can continue generating higher frequencies with wavelengths that are related to the length of the string by the following equation:

$$L = \frac{n\lambda}{2}$$

where n is a whole number: $3(\lambda/2)$, $4(\lambda/2)$, $5(\lambda/2)$, and so on.

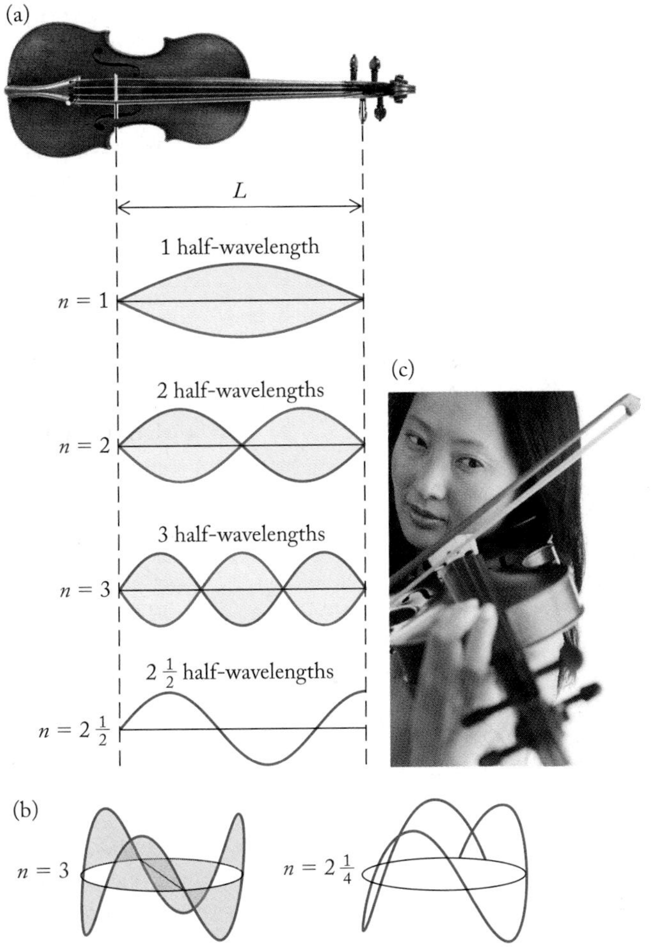

FIGURE 7.15 Linear and circular standing waves: (a) The wavelength (λ) of a standing wave in a violin string is related to the distance between the ends of the string (L) by the equation $L = n(\lambda/2)$. In the standing waves shown, $n = 1$, 2, and 3. An n value of $2\frac{1}{2}$ does not produce a standing wave because there can be no string motion at either end. (b) The circular standing waves proposed by de Broglie account for the stability of the energy levels in Bohr's model of the hydrogen atom. Each stable wave must have a circumference equal to $n\lambda$. In the first image $n = 3$. If the circumference is not an exact multiple of λ, as shown in the second image, there is no continuous wave form, and no standing wave. (c) A vibrating violin string is an example of standing waves.

CONCEPT TEST

Explain why the waveforms possible in a violin string are frequently cited as an example of quantization.

The standing-wave pattern for a circular wave differs slightly from the model of an oscillating violin string in that there are no defined stationary ends. Instead, the electron oscillates in an endless series of waves but only if, as shown in Figure 7.15(b), the circumference (cf) of the circle equals a whole-number multiple of the wavelength of the electron:

$$\text{cf} = n\lambda \tag{7.14}$$

Equation 7.14 gives a new meaning to Bohr's quantum number n: it represents the number of characteristic wavelengths at a given energy level.

De Broglie's research created a quandary for the graduate faculty at the University of Paris, where he studied. Bohr's model of electrons moving between allowed energy levels had been widely criticized as an arbitrary suspension of well-tested physical laws. De Broglie's rationalization of Bohr's model seemed even more outrageous to many scientists. Before the faculty would accept his thesis, they wanted another opinion, so they sent it to Albert Einstein for review. Einstein wrote back that he found the young man's work "quite interesting." That was a good-enough endorsement for the faculty: de Broglie's thesis was accepted in 1924 and immediately submitted for publication.

The Heisenberg Uncertainty Principle

Bohr treated electrons as particles in his theory of atomic structure. Once de Broglie and others established that electrons exhibited both particle-like and wavelike behavior, certain issues arose about the impact of wavelike behavior on our ability to locate the electron. A wave by its very nature is spread out in space. The question "Where is the electron?" has a different answer if we treat the electron as a wave rather than a particle. This issue was addressed by the German physicist Werner Heisenberg (1901–1976), who proposed the following thought experiment: watch an electron with a hypothetical γ-ray microscope to "see" what the path of an electron around an atom would be. The microscope (if it existed) would need to use γ rays for illumination because they are the only part of the electromagnetic spectrum with wavelengths short enough to match the diminutive size of electrons. However, the short wavelengths and high frequencies of γ rays also mean that they have enormous energies—so large that, if a γ ray were to strike an electron, it would knock the electron off course. The only way not to affect the electron's motion would be to use a much lower-energy, longer-wavelength source of radiation to illuminate it, but then we would not be able to see the tiny electron clearly.

This situation presents a quantum mechanical dilemma. The only means for clearly observing an electron also make it impossible to know its motion, or, more precisely, its momentum. Therefore we can never know exactly both the position and the momentum of the electron simultaneously. This conclusion is known as the **Heisenberg uncertainty principle**, which is mathematically expressed by

$$\Delta x \cdot m\,\Delta u \geq \frac{h}{4\pi} \tag{7.15}$$

where Δx is the uncertainty in the position, m is the mass of the particle, Δu is the uncertainty in the speed, and h is Planck's constant. To Heisenberg, this uncertainty is the essence of quantum mechanics. Its message for us is that there are limits to what we can observe, measure, and therefore know.

> The **Heisenberg uncertainty principle** says that you cannot determine both the position and the momentum of an electron in an atom at the same time.

SAMPLE EXERCISE 7.7   **Calculating Uncertainty**

Return to Sample Exercise 7.6 and compare the uncertainty in the speed of the baseball with the uncertainty in the speed of the electron. Assume that the position of the baseball is known to within one wavelength of red light ($\Delta x_{\text{baseball}} = 680$ nm) and the position of the electron is known to within the radius of the hydrogen atom ($\Delta x_{e} = 5.3 \times 10^{-11}$ m).

COLLECT AND ORGANIZE To solve for the uncertainty in speed (Δu) using Equation 7.15, we need to know the mass of the particles in question. (From Sample Exercise 7.6 we know that the baseball is 0.142 kg and the electron is 9.109×10^{-31} kg).

ANALYZE We have the terms needed to use Equation 7.15. We need to solve the equation for the uncertainty in speed (Δu):

$$\Delta u \geq \frac{h}{4\pi \Delta x m}$$

SOLVE

- For the baseball:

$$\Delta u \geq \frac{6.626 \times 10^{-34} \, \cancel{J} \cdot \cancel{s}}{4\pi (6.80 \times 10^{-7} \, \cancel{m})(0.142 \, \cancel{kg})} \times \frac{(\cancel{kg} \cdot m \cdot \cancel{m})/(s \cdot \cancel{s})}{\cancel{J}}$$

$$\geq 5.46 \times 10^{-28} \text{ m/s}$$

- For the electron:

$$\Delta u \geq \frac{6.626 \times 10^{-34} \, \cancel{J} \cdot \cancel{s}}{4\pi (5.3 \times 10^{-11} \, \cancel{m})(9.109 \times 10^{-31} \, \cancel{kg})} \times \frac{(\cancel{kg} \cdot m \cdot \cancel{m})/(s \cdot \cancel{s})}{\cancel{J}}$$

$$\geq 1.09 \times 10^6 \text{ m/s}$$

THINK ABOUT IT The numbers make clear that the uncertainty in the measurement of the speed of the baseball is so minuscule as to be insignificant. This result is expected for objects in the macroscopic world. The uncertainty in the measurement of the speed of the electron, however, is huge—over 1 million meters per second—which is what one must expect at the atomic level, where "particles" like the electron are also waves.

Practice Exercise What is the uncertainty in the position of an electron moving near a nucleus at a speed of 8×10^7 m/s. Assume the uncertainty in the speed of the electron is 1% of its value—that is, $\Delta u = (0.01)(8 \times 10^7$ m/s).

When Heisenberg proposed his uncertainty principle, he was working with Niels Bohr at the University of Copenhagen. The two scientists had widely different views about the significance of the uncertainty principle. To Heisenberg it was a fundamental characteristic of nature. To Bohr, the uncertainty principle was merely a mathematical consequence of the wave–particle duality of electrons; there was no physical meaning to an electron's position and path. The debate between these two gifted scientists was heated at times. Heisenberg later wrote about one particularly emotional debate:

> [A]t the end of the discussion I went alone for a walk in the neighboring park [and] repeated to myself again and again the question: "Can nature possibly be as absurd as it seems?" [1]

---

[1] Werner Heisenberg, *Physics and Philosophy: The Revolution in Modern Science* (Harper & Row, 1958), p. 42.

The description of the wavelike behavior of particles on the atomic level is called **wave mechanics** or **quantum mechanics**.

The **Schrödinger wave equation** describes the electron in hydrogen as a matter wave and indicates how it varies with location and time around the nucleus. Solutions to the wave equation are the energy levels of the hydrogen atom.

A **wave function (ψ)** is a solution to the Schrödinger equation.

Orbitals, defined by the square of the wave function (ψ^2), are regions around the nucleus of an atom where the probability of finding an electron is high. Each orbital is identified by a unique combination of three integers called **quantum numbers**.

The **principal quantum number (n)** is a positive integer that describes the relative size and energy of an atomic orbital or group of orbitals in an atom.

The **angular momentum quantum number (ℓ)** is an integer that may have any value from 0 to $n - 1$. It defines the shape of an orbital.

7.6 Quantum Numbers and Electron Spin

Many of the leading scientists of the 1920s were unwilling to accept the dual wave–particle nature of electrons proposed by de Broglie until it could be used accurately to predict the features of the hydrogen spectrum. Such application required the development of equations describing the behavior of electron waves. Over his Christmas vacation in 1925, Austrian physicist Erwin Schrödinger (1887–1961) did just that, developing in a few weeks the mathematical foundation for what came to be called **wave mechanics**, or **quantum mechanics**.

Schrödinger's mathematical description of electron waves is called the **Schrödinger wave equation**. The equation will not be discussed in detail in this book. However, you should know that solutions to the Schrödinger wave equation are called **wave functions**: mathematical expressions represented by the Greek letter psi (**ψ**) that describe how the matter wave of the electron varies with location and time around the nucleus. The solutions to the Schrödinger equation define the energy levels in the hydrogen atom. Wave functions can be simple trigonometric functions, such as sine or cosine waves, or they can be very complex.

What is the physical significance of a wave function? Actually there is none. However, the *square of a wave function (ψ^2)*, does have physical meaning. Initially Schrödinger believed that a wave function depicted the smearing of an electron through three-dimensional space. This notion of subdividing a discrete particle was later rejected in favor of the model developed by German physicist Max Born (1882–1970), who proposed that ψ^2 defines an **orbital**, the space around the nucleus in an atom where the probability of finding an electron is high. Born later showed that his interpretation could be used to calculate the probability of a transition between two orbitals, as happens when an atom absorbs or emits a photon.

To help you visualize the probabilistic meaning of ψ^2, consider what happens when you spray ink from a perfume sprayer onto a flat surface (Figure 7.16). If you then draw a circle encompassing most of the ink spots, you are identifying the region of maximum probability for finding the ink spots.

It is important to understand that quantum mechanical orbitals are not concentric circles; they are three-dimensional regions of space with distinctive shapes, orientations, and average distances from the nucleus. Each orbital is a solution to Schrödinger's wave equation and is identified by a unique combination of three integers called **quantum numbers**, whose values flow directly from the mathematical solutions to the wave equations. The quantum numbers are:

- The **principal quantum number (n)** is like Bohr's quantum number n for the hydrogen atom in that it is a positive integer that indicates the relative size and energy level of an orbital or a group of orbitals in an atom. Orbitals with the same value of n are in the same *shell*. Orbitals with larger values of n are farther from the nucleus and, in the hydrogen atom, represent higher energy levels (consistent with Bohr's model of the hydrogen atom). Energy levels of orbitals are more complex in multielectron atoms, but generally they increase with increasing values of n.

- The **angular momentum quantum number (ℓ)** is an integer with a value ranging from zero to $n - 1$, and defines the shape of an orbital. Orbitals with the same values of n and ℓ are said to be in the same *subshell* and

FIGURE 7.16 The probability of finding an ink spot in the pattern produced by a source of ink spray decreases with increasing distance from the center of the pattern—in much the way that electron density in the 1s orbital decreases with increasing distance from the nucleus.

represent equivalent energy levels. Orbitals with a given value of ℓ are identified with a letter according to the following scheme:

Value of ℓ	0	1	2	3	4
Letter Identifier	s	p	d	f	g

- The **magnetic quantum number (m_ℓ)** is an integer with a value from $-\ell$ to $+\ell$. It defines the orientation of an orbital in the space around the nucleus of an atom.

Each subshell within an atom has a two-part designation containing the appropriate numerical value of n and a letter designation for ℓ. For example, orbitals with $n = 3$ and $\ell = 1$ are in the $3p$ subshell and so are called $3p$ orbitals. Electrons in $3p$ orbitals are called $3p$ electrons. How many $3p$ orbitals are there? We can answer this question by finding all possible values of m_ℓ. Because p orbitals are those for which $\ell = 1$, they have m_ℓ values of -1, 0, and $+1$. These three values mean that there are three orbitals in a p subshell, each with a unique combination of n, ℓ, and m_ℓ values. A more complete view of all the possible combinations of these three quantum numbers for the orbitals of the first four shells is given in Table 7.1.

The **magnetic quantum number (m_ℓ)** is an integer that may have any value from $-\ell$ to $+\ell$. It defines the orientation of an orbital in space.

▶❚❚ **CHEMTOUR** Quantum Numbers

TABLE 7.1 Quantum Numbers of the Orbitals in the First Four Shells

Values of n	Allowed Values of ℓ	Subshell Labels	Allowed Values of m_ℓ	Number of Orbitals in: Subshell	Number of Orbitals in: Shell
1	0	s	0	1	1
2	0	s	0	1	
	1	p	$-1, 0, +1$	3	4
3	0	s	0	1	
	1	p	$-1, 0, +1$	3	
	2	d	$-2, -1, 0, +1, +2$	5	9
4	0	s	0	1	
	1	p	$-1, 0, +1$	3	
	2	d	$-2, -1, 0, +1, +2$	5	
	3	f	$-3, -2, -1, 0, +1, +2, +3$	7	16

SAMPLE EXERCISE 7.8 | **Identifying the Subshells and Orbitals in an Energy Level**

(a) What are the designations of all the subshells in the $n = 4$ shell? (b) How many orbitals are in these subshells?

COLLECT AND ORGANIZE The designations of subshells are based on the possible values of quantum numbers n and ℓ. We are asked about the shell with principal quantum number $n = 4$.

ANALYZE The allowed values of ℓ depend on the value of n, in that ℓ is an integer between 0 and $n - 1$. The number of orbitals in a subshell depends on the number of possible values of the m_ℓ quantum number. These include all integral values between and including $-\ell$ to $+\ell$.

SOLVE (a) The allowed values of ℓ for $n = 4$ range from 0 to $n - 1$ and are 0, 1, 2, and 3. These values of ℓ correspond to the subshell designations *s, p, d,* and *f.* The appropriate subshell names are thus *4s, 4p, 4d,* and *4f.*

(b) The possible values of m_ℓ from $-\ell$ to $+\ell$ are as follows:

- $\ell = 0; m_\ell = 0$: This combination of ℓ and m_ℓ values represents a single *4s* orbital.

- $\ell = 1; m_\ell = -1, 0,$ or $+1$: These three combinations of ℓ and m_ℓ values represent the three *4p* orbitals.

- $\ell = 2; m_\ell = -2, -1, 0, +1,$ or $+2$: These five combinations of ℓ and m_ℓ values represent the five *4d* orbitals.

- $\ell = 3; m_\ell = -3, -2, -1, 0, +1, +2,$ or $+3$: These seven combinations of ℓ and m_ℓ values represent the seven *4f* orbitals.

Thus there is a grand total of $1 + 3 + 5 + 7 = 16$ orbitals in the $n = 4$ shell.

THINK ABOUT IT Remember that the values of the quantum numbers are related: n determines the possible values of ℓ; ℓ determines the possible values of m_ℓ.

Practice Exercise How many orbitals are there in the second shell ($n = 2$)?

Several relationships are worth noting in the quantum numbering system (Figure 7.17):

- There are n subshells in the nth shell: one subshell (*1s*) in the first shell, two subshells (*2s* and *2p*) in the second shell, three subshells (*3s, 3p,* and *3d*) in the third shell, and so on.

- There are n^2 orbitals in the nth shell: one in the first shell, four in the second, nine in the third, and so on.

- There are $(2\ell + 1)$ orbitals in each subshell: one *s* orbital ($2 \times 0 + 1 = 1$), three *p* orbitals ($2 \times 1 + 1 = 3$), five *d* orbitals ($2 \times 2 + 1 = 5$), and so on.

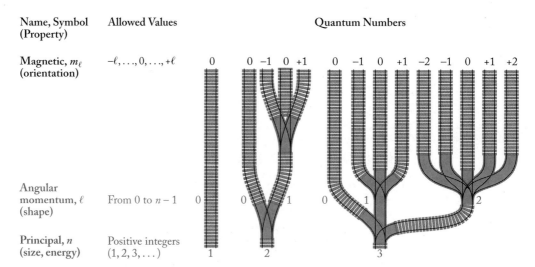

Name, Symbol (Property)	Allowed Values	Quantum Numbers

FIGURE 7.17 Quantum numbers are related by a set of rules. To illustrate, consider the options presented by three levels (n) of train tracks whose points of origin are labeled 1, 2, and 3 in red. The higher the value of n, the more unique addresses (end points along the top) that the trains can serve. Track $n = 1$, the lowest level, serves only one destination; its track does not split ($\ell = 0$). Track $n = 2$ branches into two additional tracks: $\ell = 0$ continues to one address while $\ell = 1$ serves three new addresses ($-1, 0, +1$). Track $n = 3$ splits into three additional tracks: $\ell = 0$ (serving one address); $\ell = 1$ (three addresses), and $\ell = 2$ (five addresses). Values of n up to 7 (not shown) split into ever more tracks to serve ever more addresses.

As we have seen, the mathematical solution to the Schrödinger equation yields the three quantum numbers that describe the orbitals in which an electron may reside. As we shall see shortly, although the Schrödinger wave equations and the quantum numbers they define were developed for a one-electron system, they may also be applied to multielectron systems. The Schrödinger equation accounts for the spectral features of many elements, but it does not explain a specific feature of the emission spectrum of several substances. The spectrum of hydrogen, for example, actually has a pair of red lines at 656 nm where Balmer thought there was only one. These lines appear at 656.272 nm and 656.285 nm as shown in Figure 7.18. Other pairs of lines were discovered in the spectra of multielectron atoms with a single electron in the outermost shell.

In 1925, two students at the University of Leiden in the Netherlands, Samuel Goudsmit (1902–1978) and George Uhlenbeck (1900–1988), proposed that the pairs of lines, called *doublets*, could result if the electrons were assumed to spin in one of two directions, thereby producing two energy states not accounted for by Schrödinger's equation. According to quantum mechanics, the resulting magnetic fields produced by spinning electrons can have two orientations: up and down. Goudsmit and Uhlenbeck proposed a fourth quantum number, called the **spin magnetic quantum number** (m_s), to account for these two spin orientations. The values of m_s are $+\frac{1}{2}$ for spin up and $-\frac{1}{2}$ for spin down.

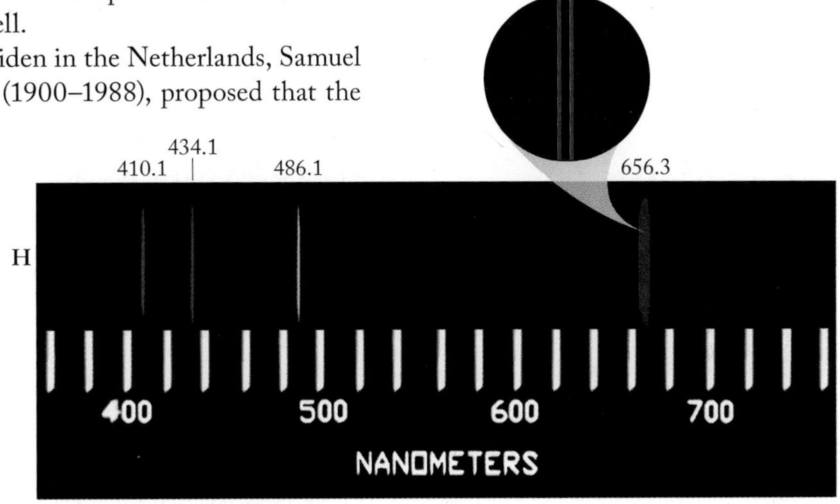

The direction of spin of a hydrogen atom's electron influences how the atom behaves in a magnetic field. Even before Goudsmit and Uhlenbeck proposed the electron-spin hypothesis, two other scientists, Otto Stern (1888–1969) and Walther Gerlach (1889–1979), observed the effect of electron spin in a multielectron system when they shot a beam of silver ($Z = 47$) atoms through a nonuniform magnetic field (Figure 7.19). The field split the beam in two. Those atoms in which the net electron spin was up were deflected one way by the magnetic field; those in which the net electron spin was down were deflected in the opposite direction.

> The **spin magnetic quantum number** (m_s) of an electron in an atom is either $+\frac{1}{2}$ or $-\frac{1}{2}$, indicating that the electron-spin orientation is either up or down.

434.1
410.1 | 486.1
656.3

H

400   500   600   700

NANOMETERS

FIGURE 7.18 The Schrödinger equation does not account for the appearance of closely spaced pairs of lines in atomic spectra, such as the red lines at 656.272 and 656.285 nm in the spectrum of hydrogen.

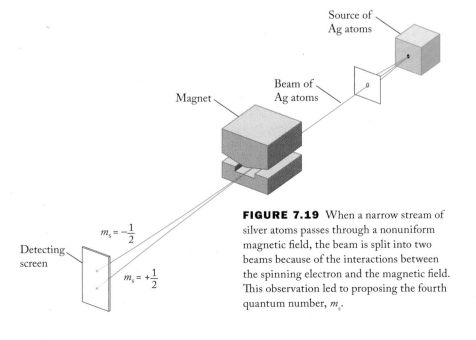

Source of
Ag atoms

Magnet

Beam of
Ag atoms

$m_s = -\frac{1}{2}$

Detecting
screen

$m_s = +\frac{1}{2}$

FIGURE 7.19 When a narrow stream of silver atoms passes through a nonuniform magnetic field, the beam is split into two beams because of the interactions between the spinning electron and the magnetic field. This observation led to proposing the fourth quantum number, m_s.

The **Pauli exclusion principle** states that no two electrons in an atom can have the same set of four quantum numbers.

In 1925, Austrian physicist Wolfgang Pauli (1900–1958) proposed that no two electrons in a multielectron atom have the same set of four quantum numbers. This idea is known as the **Pauli exclusion principle**. The three quantum numbers derived from solutions to Schrödinger's wave equation define the orbitals where an atom's electrons are likely to be. The two allowed values of the spin magnetic quantum number indicate that each orbital can hold two electrons, one with $m_s = +\frac{1}{2}$ and the other with $m_s = -\frac{1}{2}$. Thus, each electron in an atom has a unique "quantum address" defined by a particular combination of n, ℓ, m_ℓ, and m_s values.

SAMPLE EXERCISE 7.9   **Identifying Possible Sets of Quantum Numbers**

Choose the set(s) of quantum numbers from the following list that are possible because they correctly describe an electron in an atom. For the sets that are incorrect, indicate the value or values that make the set invalid.

	n	ℓ	m_ℓ	m_s
(a)	1	0	−1	$+\frac{1}{2}$
(b)	3	2	−2	$+\frac{1}{2}$
(c)	2	2	0	0
(d)	2	0	0	$-\frac{1}{2}$
(e)	−3	−2	−1	$-\frac{1}{2}$

COLLECT AND ORGANIZE Possible values for the first three quantum numbers for atoms in the ground state are related; the spin quantum number has a value of either $+\frac{1}{2}$ or $-\frac{1}{2}$.

ANALYZE We need to look at each set and determine if the value for ℓ is possible considering the value for n; if the value for m_ℓ is possible given the value for ℓ; and if m_s is either $+\frac{1}{2}$ or $-\frac{1}{2}$.

SOLVE

     a. If n is 1, ℓ can only have a value of 0 [$\ell = 0$ to $(n − 1)$], which means m_ℓ must be zero ($m_\ell = +\ell$ to $-\ell$). The value of n is 1, so ℓ can be 0, but m_ℓ cannot be −1; set a is invalid. The spin quantum number is valid. To make this set possible, m_ℓ has to be 0.

     b. The value of n is 3, so ℓ can be 2, m_ℓ can be −2, and m_s can be $+\frac{1}{2}$; set b is valid.

     c. The value of n is 2; ℓ cannot be 2, so this set is invalid. To make the set valid, ℓ could be either 0 or 1. If ℓ were either 0 or 1, then $m_\ell = 0$ would be possible; however, m_s is never 0, so that number is also incorrect. The value of m_s must be either $+\frac{1}{2}$ or $-\frac{1}{2}$.

     d. The value of n is 2, so ℓ can be 0; in turn, m_ℓ can be 0, and m_s can be $-\frac{1}{2}$. This set is possible.

     e. The value of n is −3, which is impossible; n must be a positive integer. The set is invalid. If n were +3, the remaining quantum numbers would be a valid set.

THINK ABOUT IT Key facts to remember are that the values of n, ℓ, and m_ℓ are related and that the value of m_s can only be either $+\frac{1}{2}$ or $-\frac{1}{2}$. Only certain combinations of quantum numbers are possible, and each valid set describes the region of space in which a given electron may be found. The quantum numbers are analogous to the address of an electron.

Practice Exercise Write all the possible sets of quantum numbers for an electron in the third shell ($n = 3$) that have an angular momentum quantum number $\ell = 1$ and a spin quantum number $m_s = +\frac{1}{2}$.

7.7 The Sizes and Shapes of Atomic Orbitals

Atomic orbitals have three-dimensional shapes that are graphical representations of ψ^2. Using the solutions to Schrödinger's wave equation, we can build up the orbitals in a one-electron atom such as hydrogen.

s Orbitals

Figure 7.20 provides several representations of the $1s$ orbital of the hydrogen atom. In Figure 7.20(a), ψ^2 is plotted against distance from the nucleus. Note that the scale of the x-axis is in picometers (10^{-12} m). The curve seems to suggest that the most likely place to find the electron is in the nucleus, yet we know that cannot be true. Is there another interpretation of ψ^2? It turns out that Figure 7.20(a) is really a plot of *probable electron density*. To obtain a more useful view of where the electron is likely to be, we need to adjust the curve by including a term for volume, which must be in the denominator of any expression of density.

Figure 7.20(b) provides a more useful profile of electron location. To understand why, think of a hydrogen atom as being like an onion, made of many layers all of the same thickness, as shown in Figure 7.20(c). What is the probability of finding an electron in one of these layers? A layer very close to the nucleus has a very small radius (r), so it accounts for only a small fraction of the total volume of the atom. A layer with a larger radius makes up a much larger fraction of the volume of the atom, because the volumes of these layers increase as a function of r^2. Even though electron densities are high close to the nucleus, the volumes of the layers there are so small that the chances of the electron being near the center of an atom are extremely low. Farther from the nucleus, electron densities are lower but the volumes of the layers are much larger, so the probability of the electron being in one of these layers actually increases. At even greater distances, layer volumes are very much larger, but ψ^2 drops to nearly zero (see Figure 7.20a); the chances of finding the electron there decrease with distance from the nucleus.

CONNECTION In Chapter 1 we defined density as the ratio of mass to volume (m/V). The same concept applies here: electron density is the ratio of the probability of finding the electron in a region of space to the volume of the space (ψ^2/V).

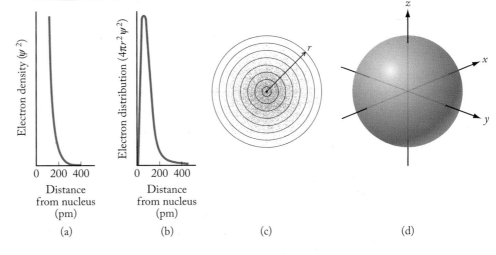

FIGURE 7.20 Probable electron density in the $1s$ orbital (ground state) of the hydrogen atom is represented here by (a) a plot of electron density (ψ^2) versus distance from the nucleus; (b) a radial distribution profile, which indicates the probability of an electron being in (c) a thin spherical shell with radius r; and (d) a boundary–surface representation, a sphere within which the probability of finding a $1s$ electron is 90%.

Thus the curve in Figure 7.20(b) represents the net result of combining two competing factors—increasing layer volume but decreasing electron probability—to produce a *radial distribution profile*. It is not a plot of ψ^2 versus distance (r), but rather of $4\pi r^2 \psi^2$ versus r. In geometry, $4\pi r^2$ is the formula for the area of a sphere, but here it represents the volume of one of the thin spherical layers in Figure 7.20(c).

A significant feature of the curve in Figure 7.20(b) is that it has a maximum value corresponding to the distance between the nucleus and the radial distance where the electron is most likely to be. The value of r corresponding to this maximum for the 1s (ground-state) orbital of hydrogen is 53 pm.

Figure 7.20(d) provides a view of the spherical shape of this (or any) s orbital. The surface of the sphere encloses the volume within which the probability of finding a hydrogen 1s electron is 90%. This type of depiction, called a *boundary–surface representation*, is one of the most useful ways to view the relative size, shape, and orientation of orbitals. All s orbitals are spheres, which have only one orientation and in which electron density depends only on distance from the nucleus. As spheres, s orbitals have no angular dependence of electron density ($\ell = 0$). For s orbitals, boundary surfaces are simply a useful way to depict relative size.

The sizes of 1s, 2s, and 3s orbitals are shown in Figure 7.21. Note that their sizes increase with increasing values of the principal quantum number (n). Also note that in each profile a local maximum in electron density is located close to the nucleus. The heights of these local maxima decrease with increasing n, and they are separated from other local maxima farther from the nucleus by nodes. Nodes have the same meaning here as they did for one-dimensional standing waves: places where the wave has a value of zero. In the context of electrons as three-dimensional matter waves, nodes are locations at which electron density

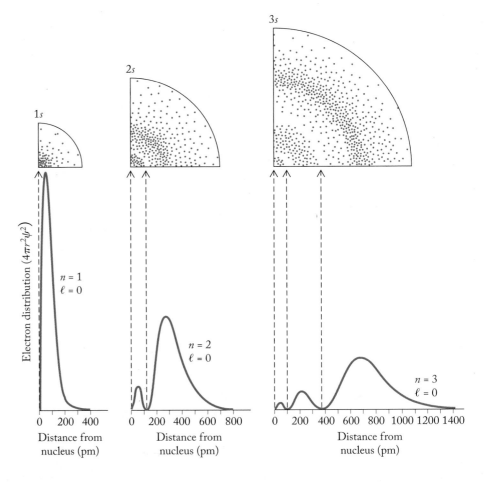

FIGURE 7.21 These radial distribution profiles of 1s, 2s, and 3s orbitals have 0, 1, and 2 nodes, respectively, identifying locations of zero electron density. Electrons in all three of these s orbitals have the probability of being close to the nucleus, but 3s electrons are more likely to be farther away from the nucleus than 2s electrons, which in turn are more likely to be farther away than 1s electrons.

goes to zero. The radial distribution profiles in Figure 7.21 show how electrons in *s* orbitals, even those with high values of *n*, have some probability of being close to the nucleus.

p and *d* Orbitals

We have noted that *s* orbitals are spherical, so in these orbitals electron density depends only on distance from the nucleus, not on angle of orientation. Let's now consider the shapes of *p* and *d* orbitals, shown as boundary surfaces in Figures 7.22 and 7.23, respectively. All shells with $n \geq 2$ have a subshell containing three *p* orbitals ($\ell = 1$; $m_\ell = -1, 0, +1$). Each of these orbitals has two balloon-shaped lobes on either side of the nucleus along one of the three axes. These orbitals are designated p_x, p_y, or p_z, depending on the axis along which the lobes are aligned. The two lobes of a *p* orbital are sometimes labeled with plus and minus signs, indicating that the sign of the wave function defining them has negative and positive values. Remember that one *p* orbital has two lobes, and an electron in a *p* orbital occupies both lobes. Because a node of zero probability separates the two lobes, you may wonder how an electron gets from one lobe to the other. One way to think about how this happens is to remember that an electron behaves as a three-dimensional standing wave, and waves have no difficulty passing through nodes. After all, there is a node right in the middle of a one-dimensional standing wave in a violin string separating the positive and negative displacement regions in the vibrating string (see Figure 7.15).

The five *d* orbitals ($m_\ell = -2, -1, 0, +1, +2$) found in shells of $n \geq 3$ all have different orientations. Four of them are shaped like a four-leaf clover. Three of those four have an array of four lobes that lie between, not on, the *x*-, *y*-, and *z*-axes. These orbitals are designated d_{xy}, d_{xz}, and d_{yz}. The fourth orbital in this set, designated $d_{x^2-y^2}$, has four lobes that lie along the *x*- and *y*-axes. The fifth *d* orbital, designated d_{z^2}, is mathematically equivalent to the other four but has a much different shape, with two major lobes oriented along the *z*-axis and a doughnut shape called a *torus* in the *x*-*y* plane that surrounds the middle of the two lobes. We will not address the shapes of other types of orbitals because they are much less important than *s*, *p*, and *d* to the discussions of chemical bonding in the chapters to come.

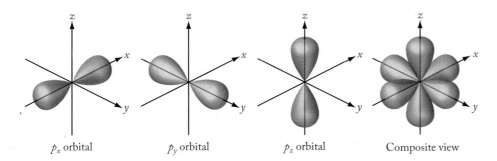

p_x orbital   *p_y* orbital   *p_z* orbital   Composite view

FIGURE 7.22 Stylized drawings of boundary–surface views of a set of three *p* orbitals showing their orientation along the *x*-, *y*-, and *z*-axes. The nucleus is located at the origin.

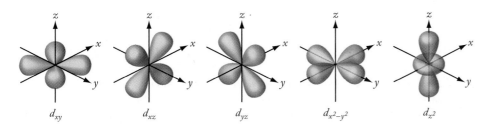

d_{xy} d_{xz} d_{yz} $d_{x^2-y^2}$ d_{z^2}

FIGURE 7.23 Boundary–surface views of five *d* orbitals are projected onto *x*-, *y*-, and *z*-axes.

7.8 The Periodic Table and Filling the Orbitals of Multielectron Atoms

Now that we have developed a model for distributing electrons among the regions in atoms called orbitals, we can use that model and fill the orbitals, using the periodic table as a guide to building the electron configuration of atoms containing more than one electron. An **electron configuration** is a set of symbols describing the distribution of electrons in an atom or ion typically in its ground state. To determine electron configurations of atoms, we start at the beginning of the periodic table with hydrogen and consider how each successive electron is added to the lowest-energy orbitals available as the atomic numbers increase in order. One proton and one or more neutrons are added to the nucleus as we move from element to element across the rows. This method is called the **aufbau principle** (from the German *aufbauen*, "to build up"). We build the electron configuration of atoms one electron at a time.

To decide which orbitals contain electrons and to ensure that the electron configurations we build describe atoms in their ground states, we start with the following simple rules:

1. Electrons always go into the lowest energy orbital available.
2. Each orbital has a maximum occupancy of two electrons.

Using these rules, let's build the electron configurations of specific elements, starting with hydrogen ($Z = 1$). We know that the single electron in a ground-state atom of hydrogen is in the $1s$ orbital. We represent this arrangement as $1s^1$, where the superscript indicates that there is *one* electron in the $1s$ orbital. Because there are no more electrons to add, $1s^1$ is the complete electron configuration of hydrogen.

The next element in the periodic table is helium (He). Its atomic number of 2 tells us there are two protons in the nucleus and therefore two electrons in the neutral atom. Using the aufbau principle, we simply add an electron to the electron configuration of hydrogen. Which orbital should this second electron go in? The electron must occupy the lowest-energy orbitals available (rule 1), which in this case is the $1s$ orbital. This orbital has only one other electron, so space for a second electron is available there because each orbital can accommodate up to two electrons (rule 2). Therefore helium has the electron configuration $1s^2$. With two electrons, the $1s$ orbital is filled to capacity. In addition, because the first ($n = 1$) shell contains only this single $1s$ orbital, the first shell is also completely filled. The idea of a *filled shell* is an important concept in understanding the chemical properties of the elements: those elements composed of atoms with filled s and p subshells in their outermost shells are chemically stable and generally unreactive. Helium is such an element, as are all the other elements in group 18 of the periodic table.

Lithium (Li; $Z = 3$) is the next element after helium in the periodic table and the first element in the second row (also called the second period) of the periodic table. This location is a signal that an atom of lithium has an electron in its second ($n = 2$) shell. The row numbers in the periodic table correspond to the n values of the outermost shells of the atoms in the rows.

The second shell has four orbitals (one $2s$ orbital and three $2p$ orbitals) and can hold eight electrons. Notice that there are eight elements in the second row. This observation illustrates the orbital basis of the periodic table's organization and

An **electron configuration** describes the distribution of electrons among the orbitals of an atom or ion.

The **aufbau principle** is the method of building electron configurations one electron at a time. One electron is added to the lowest-energy orbitals of a ground-state atom (and one proton and one or more neutrons are added to the atom's nucleus). The electron configurations of atoms are built in sequence as atomic number increases in order across the rows of the periodic table.

reveals an important concept. Remember that the periodic table was originally arranged based on considerations of physical and chemical properties. Because the electron configurations also vary periodically, the location of electrons in an atom must have a great deal to do with how that atom behaves.

Which of the four orbitals in the second shell contains the third electron of an atom of lithium? The energy levels of the subshells in a given shell are related to their shapes and the degree to which electrons in them feel the pull of the positive charge of the nucleus. These factors mean that the electron in a $2s$ orbital has a lower average energy than an electron in a $2p$ orbital. Therefore, the third electron in a lithium atom is in the $2s$ orbital making the electron configuration $1s^2 2s^1$.

The electrons in the outermost shell of an atom—the shell with the highest n value—are those that most influence its chemical behavior, so they are given a special name: **valence electrons**. An abbreviated form of the electron configuration, known as a *condensed electron configuration*, is helpful for highlighting the arrangement of the valence electrons. This is done by replacing the symbols for the configuration of the **core electrons**, the full inner shells of an atom, with the symbol in brackets of the group 18 (noble gas) element that has that configuration. For lithium, the group 18 element that precedes it is He, so the condensed electron configuration for Li is $[\text{He}]2s^1$. Notice that Li has a single electron in its valence shell s orbital as does hydrogen, the element directly above it on the periodic table: both may be described as having the valence-shell configuration ns^1, where n represents both the row number in which the atom is located on the table and the principal quantum number of the valence shell in the atom:

Element	Electron Configuration	Condensed Electron Configuration
Lithium	$1s^2 2s^1$	$[\text{He}]2s^1$

Beryllium (Be) is the fourth element in the periodic table and the first in group 2. Its electron configuration is $1s^2 2s^2$ (full) or $[\text{He}]2s^2$ (condensed). The other elements in group 2 also have two spin-paired electrons in the s orbital of their outermost shell. The second shell is not full at this point because it also has three p orbitals, which are all empty and will be filled next.

Boron (B; $Z = 5$) is the first element in group 13. Its fifth electron is in one of its three $2p$ orbitals, resulting in the condensed electron configuration $[\text{He}]2s^2 2p^1$. Which of the three $2p$ orbitals contains the fifth electron is not important because all three orbitals have the same energy; they are **degenerate**.

The next element after boron is carbon (C). It has four electrons in its valence shell, so its abbreviated electron configuration is $[\text{He}]2s^2 2p^2$. Note that there are two electrons in $2p$ orbitals. Are they both in the same orbital? Remember that all electrons have a negative charge and repel each other. Thus they tend to occupy orbitals that allow them to be as far away from each other as possible. The two $2p$ electrons in carbon occupy separate $2p$ orbitals. This distribution pattern is an example of the application of **Hund's rule**, which states that for degenerate orbitals the lowest-energy electron configuration is the one with the maximum number of unpaired valence electrons, all of which have the same spin. Hund's rule is the third rule we apply when determining electron configurations. By convention, the first electron placed in an orbital has a positive spin. When electrons fill a degenerate set of orbitals, Hund's rule requires that an electron with a positive spin occupies each valence orbital before any electron with a negative spin enters any orbital. Electrons do not pair in degenerate orbitals until they have to.

Valence electrons are the electrons in the outermost shell of an atom and have the most influence on the atom's chemical behavior.

Core electrons are electrons in the filled, inner-shells in an atom and are not involved in chemical reactions.

Orbitals that have the exact same energy level are **degenerate**.

Hund's rule states that the lowest-energy electron configuration of an atom is the configuration with the maximum number of unpaired electrons in degenerate orbitals, all having the same spin.

Orbital diagrams are one way of showing the arrangement of electrons in an atom or ion using boxes to represent orbitals.

Another way to express electron configurations enables us to illustrate the occupancy of degenerate orbitals. **Orbital diagrams** show how electrons, represented by single-headed arrows, are distributed among orbitals, which are represented by boxes. An arrow pointing upward represents an electron with an up spin ($m_s = +\frac{1}{2}$), while a downward-pointing arrow represents an electron with the opposite spin orientation, spin down ($m_s = -\frac{1}{2}$). To obey Hund's rule the orbital diagram for carbon must be

$$\boxed{\uparrow\downarrow} \quad \boxed{\uparrow\downarrow} \quad \boxed{\uparrow\,|\,\uparrow\,|\,}$$
$$1s \qquad 2s \qquad 2p$$

because the two $2p$ electrons are unpaired and spinning in the same (spin-up) direction.

The next element is nitrogen (N; $Z = 7$). According to Hund's rule, the third $2p$ electron resides alone in the third $2p$ orbital. Nitrogen has full and condensed electron configurations of $1s^2 2s^2 2p^3$ and $[He]2s^2 2p^3$, and the electrons are distributed among its orbitals as follows:

$$\boxed{\uparrow\downarrow} \quad \boxed{\uparrow\downarrow} \quad \boxed{\uparrow\,|\,\uparrow\,|\,\uparrow}$$
$$1s \qquad 2s \qquad 2p$$

As we proceed across the second row to neon (Ne; $Z = 10$), we continue to fill the $2p$ orbitals as shown in Figure 7.24. The last three electrons added (in oxygen, fluorine, and neon) pair up with the first three, so in an atom of neon, all $2p$ orbitals are filled to capacity. At this point the second shell is filled as well. Note that neon is directly below helium in group 18. Helium, neon, and all the noble gases in group 18 have filled s and p orbitals in their valence shells. Elements in the same column of the periodic table have the same valence-shell configuration.

Helium is chemically unreactive and has a filled valence (first) shell. Neon is inert and also has a completely filled valence (second) shell. The electron configurations of argon (Ar), krypton (Kr), xenon (Xe), and radon (Rn) also have completely filled s and p orbitals in their outermost shells, and they also are chemically inert gases at room temperature. Thus chemical inertness accompanies completely filled valence shell s and p orbitals.

	Orbital Diagram			Electron Configuration	Condensed Configuration
	$1s$	$2s$	$2p$		
H	$\boxed{\uparrow}$	$\boxed{}$	$\boxed{\,|\,\,|\,}$	$1s^1$	
He	$\boxed{\uparrow\downarrow}$	$\boxed{}$	$\boxed{\,|\,\,|\,}$	$1s^2$	
Li	$\boxed{\uparrow\downarrow}$	$\boxed{\uparrow}$	$\boxed{\,|\,\,|\,}$	$1s^2 2s^1$	$[He]2s^1$
Be	$\boxed{\uparrow\downarrow}$	$\boxed{\uparrow\downarrow}$	$\boxed{\,|\,\,|\,}$	$1s^2 2s^2$	$[He]2s^2$
B	$\boxed{\uparrow\downarrow}$	$\boxed{\uparrow\downarrow}$	$\boxed{\uparrow\,|\,\,|\,}$	$1s^2 2s^2 2p^1$	$[He]2s^2 2p^1$
C	$\boxed{\uparrow\downarrow}$	$\boxed{\uparrow\downarrow}$	$\boxed{\uparrow\,|\,\uparrow\,|\,}$	$1s^2 2s^2 2p^2$	$[He]2s^2 2p^2$
N	$\boxed{\uparrow\downarrow}$	$\boxed{\uparrow\downarrow}$	$\boxed{\uparrow\,|\,\uparrow\,|\,\uparrow}$	$1s^2 2s^2 2p^3$	$[He]2s^2 2p^3$
O	$\boxed{\uparrow\downarrow}$	$\boxed{\uparrow\downarrow}$	$\boxed{\uparrow\downarrow\,|\,\uparrow\,|\,\uparrow}$	$1s^2 2s^2 2p^4$	$[He]2s^2 2p^4$
F	$\boxed{\uparrow\downarrow}$	$\boxed{\uparrow\downarrow}$	$\boxed{\uparrow\downarrow\,|\,\uparrow\downarrow\,|\,\uparrow}$	$1s^2 2s^2 2p^5$	$[He]2s^2 2p^5$
Ne	$\boxed{\uparrow\downarrow}$	$\boxed{\uparrow\downarrow}$	$\boxed{\uparrow\downarrow\,|\,\uparrow\downarrow\,|\,\uparrow\downarrow}$	$1s^2 2s^2 2p^6$	$[He]2s^2 2p^6 = [Ne]$

FIGURE 7.24 Orbital diagrams and electron configurations of the first 10 elements show that each orbital (indicated by a square) holds a maximum of two electrons of opposite spin. The orbitals are filled in order of increasing quantum numbers n and ℓ. A complete list of electron configurations for all elements is given in Appendix 3.

Sodium (Na; $Z = 11$) follows neon in the periodic table. It is the third element in group 1 and the first element in the third period. The 11th electron in its atoms is in the next lowest-energy orbital above $2p$, which is $3s$, so the condensed electron configuration of Na is $[Ne]3s^1$. Just as we have been writing condensed electron configurations that provide detailed information on only the outermost shell, we can diagram an abbreviated orbital diagram for Na this way:

$$[Ne] \; \boxed{\uparrow}$$
$$3s$$

This diagram reinforces the message that the electron configuration of a sodium atom consists of the neon core plus a single electron in the $3s$ orbital in the outermost shell. The sodium atom has the same valence-shell configuration as lithium and hydrogen, namely ns^1, where n is the row number. This pattern for elements in the same group continues, as the electron configuration of magnesium is $[Ne]3s^2$, and the electron configuration of every element in group 2 consists of the immediately preceding noble gas core followed by ns^2.

Configurations of the next six elements in the periodic table—from aluminum (Al: $[Ne]3s^23p^1$) to argon (Ar: $[Ne]3s^23p^6$)—show a pattern of increasing numbers of $3p$ electrons, which continues until all these orbitals are filled. Thus, argon atoms have completely filled valence-shell p orbitals; argon is also chemically inert, as predicted by its position in group 18.

After argon comes potassium (K; $Z = 19$) in group 1 of row 4 ($[Ar]4s^1$), followed by calcium (Ca; $Z = 20$) in group 2 ($[Ar]4s^2$). At this point, the $4s$ orbital is filled but the $3d$ orbitals are still empty. Why were the $3d$ orbitals not filled before $4s$?

Our energy-based orbital-filling process would be very straightforward were it not for the fact that the differences in energy between shells get smaller as n gets larger (see Figure 7.14). These smaller differences result in orbitals with large ℓ values in one shell having energies similar to orbitals with small ℓ values in the next higher shell. This is clearer in Figure 7.25, which provides more detail than is shown in Figure 7.14; for example, the energy of the $4s$ orbital is slightly lower than that of the $3d$ orbitals, so the $4s$ orbital is filled first, followed by the five $3d$ orbitals.

The next element ($Z = 21$) in the periodic table is scandium (Sc). It is the first element in the center region of the table populated by transition metals. The properties of scandium and its place after calcium in the periodic table indicate that scandium has the abbreviated electron configuration $[Ar]3d^14s^2$. Note that the $3d$ orbitals are filled in the transition metals from scandium to zinc in the fourth row. This pattern of filling the d orbitals of the shell whose principal quantum number is 1 less than the row number, $(n-1)d$, is followed throughout the periodic table: the $4d$ orbitals are filled in the transition metals of the fifth row, and so on (Figure 7.26).

The next element after Sc is titanium (Ti; $Z = 22$), which has one more d electron than Sc, and its condensed electron configuration is $[Ar]3d^24s^2$. At this point, you may feel that you can accurately predict the electron configurations of the remaining transition metals in the fourth row. However, because the energies of the $3d$ and $4s$ orbitals are similar, the sequence of d-orbital filling deviates in two spots from the pattern you might expect. Vanadium (V; $Z = 23$) has the expected configuration $[Ar]3d^34s^2$, but the next element, chromium (Cr; $Z = 24$), has the configuration $[Ar]3d^54s^1$. The reason for this difference is that a half-filled set of d orbitals is an energetically favored configuration. Apparently, the energy needed to raise a $4s$ electron to a $3d$ orbital is compensated by the stability of

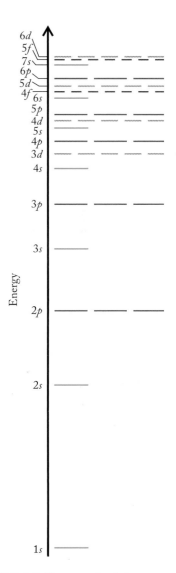

FIGURE 7.25 The energy levels in multielectron atoms increase with increasing values of the principal quantum number (n) and with increasing values of ℓ within a shell. The differences in energy between shells decrease with increasing values of n, which means that the energies of their subshells may overlap. For example, electrons in $3d$ orbitals have slightly more energy than those in the $4s$ orbital resulting in the order of subshell filling $4s^2 \rightarrow 3d^{10} \rightarrow 4p^6$ as predicted by following the periodic table. The colors of the orbitals' energy levels correspond to the colored blocks in the periodic table of Figure 7.26.

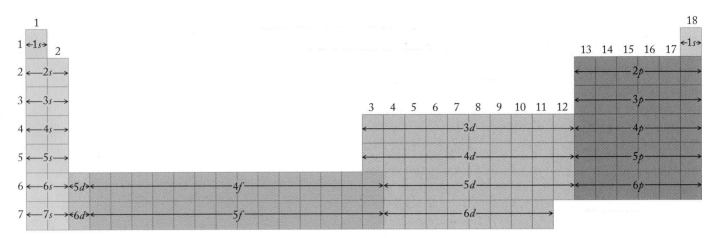

FIGURE 7.26 The organization of the periodic table based on the chemical properties of the elements is related to their electron configurations. The color coded blocks are made up of elements in which the highest-energy electrons are in either *s* (green), *p* (red), *d* (orange), or *f* (purple) orbitals. The aufbau process of filling orbitals by moving left to right across a given period (row) makes clear the order in which orbitals are filled. Note, for example, how the 4*s* orbital fills before the 3*d*, and how the 6*s* orbitals fill before the 4*f*.

 CONNECTION We described in Section 2.6 how Russian chemist Dmitrii Mendeleev developed the first useful periodic table of the elements, not only decades before de Broglie gave us electron waves and Schrödinger pioneered quantum mechanics, but even before atoms were known to consist of electrons, protons, and neutrons.

having five half-filled 3*d* orbitals. As a result, [Ar]$3d^54s^1$ is a more stable electron configuration than [Ar]$3d^44s^2$. Another variation in the expected filling pattern is observed near the end of a row of transition metals. Copper (Cu; $Z = 29$) has the electron configuration [Ar]$3d^{10}4s^1$ instead of the expected [Ar]$3d^94s^2$ because a completely filled set of *d* orbitals also represents a stable electron configuration.

We can use the periodic table to help us write the electron configuration of any element. Take barium (Ba; $Z = 56$) as an example. Barium is in group 2 in the "*s* block." That means its valence-shell configuration must be ns^2; because Ba is in the sixth row, $n = 6$. The noble gas immediately before barium at the end of row 5 on the table is xenon (Xe), so the condensed configuration for barium is [Xe]$6s^2$.

Now consider germanium (Ge; $Z = 32$). Germanium is in group 14, which means its outermost shell is ns^2np^2; because Ge is in row 4, $n = 4$. The noble gas immediately before Ge is argon (Ar; $Z = 18$); however, note as we progress across row 4 from Ar to Ge that we pass through the first row of transition metals and through the $(n - 1)d$ block. Those electrons are not in the argon core and are not in the valence shell, but their presence in Ge atoms must still be indicated. Hence the condensed electron configuration of germanium is [Ar]$3d^{10}4s^24p^2$. In this text we use the convention for writing electron configurations of ordering the shells by increasing values of n; thus, even though the actual filling order is $4s^2 \rightarrow 3d^{10} \rightarrow 4p^2$, we express the configuration in the numerical order of the principal quantum numbers.

After the 4*s* and all the 3*d* orbitals are filled, the 4*p* orbitals are filled for elements with atomic numbers 31 to 36, leaving us with the chemically stable configuration of a full outer shell for krypton. The pattern emerging from our orbital-filling exercise can be summarized as follows:

- *s* orbitals of the shell corresponding to the row number (n) are filled in groups 1 and 2.

- *d* orbitals of a shell one number less than the row number ($n - 1$) are filled in groups 3 through 12.

- *p* orbitals of the shell corresponding to the row number (n) are filled in groups 13 through 18.

SAMPLE EXERCISE 7.10   **Writing Electron Configurations of Atoms**

Write the condensed electron configuration of an atom of silver ($Z = 47$).

COLLECT AND ORGANIZE A condensed electron configuration provides a picture of the distribution of electrons in the orbitals of the outermost shell and represents the inner core of filled shells and subshells with the symbol of the preceding noble gas. Silver (Ag) is a transition metal in the fifth row of the periodic table; krypton (Kr) is the immediately preceding noble gas at the end of the fourth row.

ANALYZE Ag ($Z = 47$) has 11 more electrons than Kr ($Z = 36$). Therefore we need to account for the orbital locations of these 11 electrons.

SOLVE Ordinarily the first two of the additional 11 electrons would be in the $5s$ orbital and the next nine would be in $4d$ orbitals, resulting in a condensed electron configuration of $[Kr]4d^9 5s^2$. However, a completely filled set of d orbitals is more stable than a partially filled set, so Ag, like Cu, has 10 electrons in its outermost set of d orbitals, and only one outer-shell s electron. This arrangement gives Ag the condensed electron configuration of $[Kr]4d^{10} 5s^1$.

THINK ABOUT IT We can generate a tentative electron configuration by simply using the aufbau principle and moving across a row until we come to the atom of interest. However, in the transition metals, we have to remember the special stabilities of half-filled and filled d orbitals and make the appropriate adjustments in our initial configuration.

Practice Exercise Write the condensed orbital diagram of an atom of cobalt (Co; $Z = 27$).

7.9 Electron Configurations of Ions

We can write electron configurations of ions just as we do of neutral atoms; in doing so we gain additional insight into the quantum mechanical basis for the periodic arrangement of the elements.

Ions of Main Group Elements

We begin our examination of ion configurations by looking at cations of the main group elements. Consider the group 1 alkali metals. They all form 1+ cations, which means they all lose one electron. What is the electron configuration of their ions?

To write the electron configuration of an ion, we always begin with the electron configuration of the neutral atom. All the group 1 metals have $[NG]ns^1$ as their ground-state electron configuration, where NG stands for the noble gas immediately preceding the metal on the periodic table. When a group 1 atom forms a 1+ ion by losing a valence electron, the electron configuration of the resulting cation is that of the noble gas core:

$$[NG]ns^1 \rightarrow [NG]^+ + 1e^-$$

When group 2 atoms form cations, they lose 2 valence electrons:

$$[NG]ns^2 \rightarrow [NG]^{2+} + 2e^-$$

Similarly, group 13 atoms lose 3 valence electrons:

$$[NG]ns^2 np^1 \rightarrow [NG]^{3+} + 3e^-$$

CONNECTION In the previous chapters, we learned the charges on the ions of common elements. Electron configurations now help us understand why these ions have the charges they do.

CONNECTION In Chapter 2 when we discussed navigating the periodic table, we defined main group elements as those elements in groups 1, 2, and 13 through 18 in the periodic table. We can now think of them as the elements having electrons in ns and np orbitals.

> Atoms and ions that are **isoelectronic** with each other have identical numbers and configurations of electrons.

All of the cations of the group 1 and 2 elements and many of the group 13 elements are **isoelectronic** with the noble gases; that is, they have the same electron configuration as the noble gas that precedes them in the periodic table.

What happens to the electron configurations of the main group elements that form anions? They tend to be isoelectronic with the noble gas that immediately follows them on the periodic table. Take, for example, elements in group 17. Their neutral atoms each have seven valence electrons and the configuration $[NG]ns^2np^5$. These atoms all tend to gain one electron to form 1− anions:

$$[NG]ns^2np^5 + 1e^- \rightarrow ([NG]\ ns^2np^6)$$

which is isoelectronic with the noble gas that immediately follows the element. In the case of fluorine, for example,

$$[He]ns^2np^5 + 1e^- \rightarrow ([He]\ ns^2np^6)$$

Because of these trends involving the formation of cations and anions of main group elements, a special stability appears to be associated with achieving an electron configuration isoelectronic with a noble gas core. We will also see this trend in the next chapter when we talk about molecules that bond by sharing electrons.

SAMPLE EXERCISE 7.11 | **Determining Isoelectronic Species among Ions of Main Group Elements**

(a) Determine the electron configuration of each ion in the following compounds, and (b) indicate which ions are isoelectronic with atoms of neon: NaF, $MgCl_2$, CaO, and KBr.

COLLECT AND ORGANIZE *Isoelectronic* means having the same number of electrons. Therefore the question is asking us to (a) determine the number of electrons and the electron configuration of each ion in these binary ionic compounds, and then (b) compare that number and configuration to those terms for a neon atom.

ANALYZE The elements in the given compounds include

- two from group 1: Na and K, which form cations with a charge of 1+;
- two from group 2: Mg and Ca, which form cations with a charge of 2+;
- one from group 16: O, which forms an anion with a charge of 2−;
- three from group 17: F, Cl, and Br, which form anions with a charge of 1−.

Let's organize these elements in a table that includes the charge on each ion, the number of electrons per ion, and the ion's electron configuration. Remember that an atom loses valence electrons in becoming a cation and gains valence electrons in becoming an anion:

Element	Electron Configuration of Atom	Atomic Number (Z)	Charge on Ion	Electrons/Ion	Electron Configuration of Ion
Na	$[Ne]3s^1$	11	1+	10	$[Ne]$
K	$[Ar]4s^1$	19	1+	18	$[Ar]$
Mg	$[Ne]3s^2$	12	2+	10	$[Ne]$
Ca	$[Ar]4s^2$	20	2+	18	$[Ar]$
O	$[He]2s^22p^4$	8	2−	10	$[He]2s^22p^6$
F	$[He]2s^22p^5$	7	1−	10	$[He]2s^22p^6$
Cl	$[Ne]3s^23p^5$	17	1−	18	$[Ne]3s^23p^6$
Br	$[Ar]3d^{10}4s^24p^5$	35	1−	36	$[Ar]3d^{10}4s^24p^6$

SOLVE

a. The electron configurations for the eight ions in question—Na^+, K^+, Mg^{2+}, Ca^{2+}, O^{2-}, F^-, Cl^-, and Br^-—are given in the right column.

b. Of the ions formed by these eight elements, four have the same number of electrons as an atom of neon (10), and so are isoelectronic with Ne (and with each other). Those four are Na^+, Mg^{2+}, O^{2-}, F^-.

THINK ABOUT IT Atoms of the elements in groups 1, 2, 16, and 17 can achieve the electron configurations of noble gases by forming ions with charges of 1+, 2+, 2−, and 1−, respectively. The monatomic ions of elements with atomic numbers that are within ±3 of the atomic number of a particular noble gas are isoelectronic with the neutral atoms of that noble gas.

Practice Exercise What are the electron configurations of the ions in the following compounds, and which ions are isoelectronic with Ar? KI, BaO, Rb_2O, and $AlCl_3$

Transition Metal Ions

Having examined the electron configurations of main group ions, we turn to the configurations of transition metal cations. As with main group ions, we always begin with the neutral ion. Let's consider a common ion of manganese: Mn^{2+}. The electronic configuration of neutral Mn atoms is $[Ar]3d^54s^2$. We might expect Mn atoms to lose their slightly higher-energy 3d electrons before losing their 4s electrons, under the reasonable expectation that the last orbitals to be filled should be the first to be emptied when an atom forms a positive ion. If that happened, the resulting electron configuration would be $[Ar]3d^34s^2$. However, the results of analyzing spectra and magnetic measurements (which we will examine in detail in Chapter 17) indicate that Mn^{2+} ions have five unpaired electrons. To achieve that arrangement, each Mn atom has to lose its two 4s electrons, leaving Mn^{2+} ions with the electronic configuration $[Ar]3d^5$. Transition metal ions have the maximum number of d electrons consistent with their charge.

This preferential retention of d electrons in the outermost shell is a common trait among the transition metal ions and explains why the most frequently encountered charge on their ions is 2+. In the ions the 3d orbitals appear to be slightly lower in energy than the 4s orbitals, whereas the reverse is true for the two subshells in the corresponding atoms. Many transition metal atoms may also lose one or more d electrons as they form ions with charges ≥ 2+. Atoms of scandium, for example, lose both their 4s electrons and their lone 3d electron as they form Sc^{3+} ions. The chemistry of titanium is dominated by its tendency to lose both its 4s and 3d electrons as it forms Ti^{4+} ions.

When determining electron configurations for transition metal ions, note that we remove electrons first from the outermost s orbital, then from the outermost d subshell, until the charge on the ion in question is achieved.

SAMPLE EXERCISE 7.12 **Writing Electron Configurations of Transition Metal Atoms and Ions**

What are the electron configurations of the iron atom (Fe^0) and the ions Fe^{2+} and Fe^{3+}?

COLLECT AND ORGANIZE Iron is a transition metal in row 4 of the $(n-1)$ block on the periodic table.

ANALYZE The location of iron on the periodic table tells us that the atom has $4s$ and $3d$ electrons built on an argon core. The ions form by retaining the maximum number of $3d$ electrons.

SOLVE The electron configuration of the neutral iron atom is

$$Fe = [Ar]3d^6 4s^2$$

The electron configurations of the ions are

$$Fe^{2+} = [Ar]3d^6 \quad \text{and} \quad Fe^{3+} = [Ar]3d^5$$

THINK ABOUT IT For transition metals, the order of *filling* shells is different than the order in which electrons are *lost* from atoms to form ions. Just follow the periodic table to get the order of filling, and remember that the correct electron configuration of an ion is generated by retaining the maximum number of d electrons.

Practice Exercise Write the electron configurations for the neutral manganese atom and the ions Mn^{3+} and Mn^{4+}.

We have yet to consider the two rows at the bottom of the periodic table. As mentioned in Chapter 2 elements 58 through 71 are the lanthanide (also called *rare earth*) elements, and those of $Z = 90$ through 103 are the actinide (or *transuranic*) elements. These two rows of elements have partly filled $4f$ (lanthanide) and $5f$ (actinide) orbitals. There are 14 elements in each row, reflecting the capacity of the seven orbitals in each f subshell ($\ell = 3$), with m_ℓ values of -3, -2, -1, 0, $+1$, $+2$, and $+3$. As you can see in Figure 7.26, the $4f$ orbitals are not filled until after the $6s$ orbital has been filled and one electron has entered the $5d$ orbitals. This order of filling is due to the similar energies of $6s$ and $4f$ orbitals. Similarly, the $5f$ orbitals are filled after the $7s$ orbital is filled and one electron has entered the $6d$ orbital. The block containing these elements is referred to as the $(n - 2)f$ block. The actinides are elements with no stable isotopes, and most of them are not found in nature.

The periodic table is a useful reference for predicting the physical and chemical properties of elements. Our quantum mechanical perspectives on atomic structure provide us with a theoretical basis for explaining why particular groups (families) of elements behave similarly. The original table was based on periodic trends in observable chemical properties. Now we know that the chemical properties of

FIGURE 7.27 A comparison of covalent, metallic, and ionic radii. (a) A covalent radius is half the distance between the nuclei in a diatomic molecule such as Cl_2. (b) A metallic radius is based on the distance of closest approach of adjacent atoms in a solid metal. (c) An ionic radius is determined by a series of different comparisons among ionic compounds containing the ions of interest.

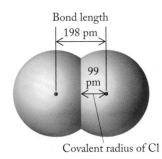

Bond length
198 pm
99 pm

(a) Covalent radius of Cl

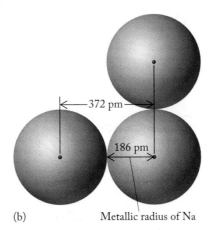

372 pm
186 pm

(b) Metallic radius of Na

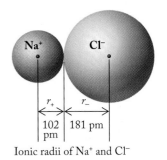

Na⁺ Cl⁻
r_+ r_-
102 pm 181 pm

(c) Ionic radii of Na⁺ and Cl⁻

an element are closely linked to the electron configurations of their atoms. One illustration of periodic behavior in physical properties is the variation in the size of atoms with respect to their location in their family.

7.10 The Sizes of Atoms and Ions

The size of an atom or ion influences both physical and chemical properties of substances. Because of the lack of certainty with which we know the location of electrons, the outer boundary of an atom or ion cannot really be defined. Therefore we define the size of an atom by considering the distance between two nuclei that are bonded together in a chemical compound. The radius of each atom involved in the bond is then defined in terms of the internuclear distance (Figure 7.27). For nonmetals that exist as diatomic molecules the atomic (covalent) radius is defined as half the distance between the nuclear centers in the molecule (Figure 7.27a). For metals the radius is defined as half the distance between the nuclear centers in a crystal of the element (Figure 7.27b). Based on these definitions, the relative sizes of atoms with respect to their locations in the periodic table are shown in Figure 7.28. When the sizes of ions are considered, ionic radii are defined by considering the distance between nuclear centers in ionic crystals (Figure 7.27c).

FIGURE 7.28 Radial sizes (in picometers) of neutral atoms for selected elements in the periodic table.

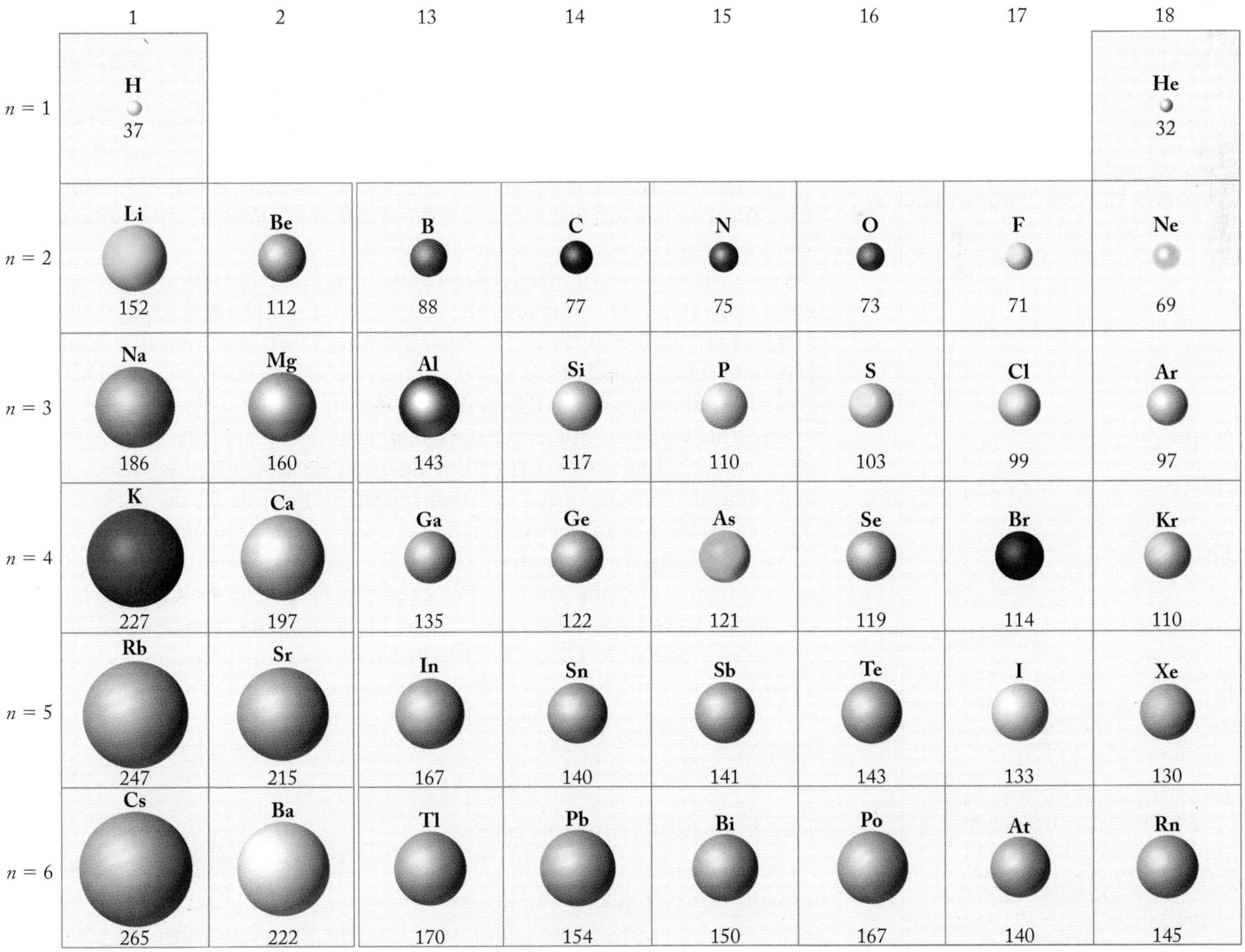

	1	2	13	14	15	16	17	18
$n = 1$	H 37							He 32
$n = 2$	Li 152	Be 112	B 88	C 77	N 75	O 73	F 71	Ne 69
$n = 3$	Na 186	Mg 160	Al 143	Si 117	P 110	S 103	Cl 99	Ar 97
$n = 4$	K 227	Ca 197	Ga 135	Ge 122	As 121	Se 119	Br 114	Kr 110
$n = 5$	Rb 247	Sr 215	In 167	Sn 140	Sb 141	Te 143	I 133	Xe 130
$n = 6$	Cs 265	Ba 222	Tl 170	Pb 154	Bi 150	Po 167	At 140	Rn 145

Atomic size varies from top to bottom in a given family (group) and from left to right across a given row (period). Two opposing factors contribute to these variations. The first is the principal quantum number n, which determines the most probable distance of the electrons from the nucleus. As n increases, the probability increases that the electrons are farther from the nucleus. The second factor is the charge on the nucleus, which determines the attractive force holding the electrons in the atom. As nuclear charge increases, the positive charge felt by the electrons increases, and the electrons are pulled closer to the nucleus. However, focusing on simple nuclear charge alone is insufficient in multielectron atoms because electrons are also repelled by other electrons. We consider the combined interactions of an electron with the positively charged nucleus and other negatively charged electrons and assess the resulting force felt by the electrons.

Orbital Penetration and Effective Nuclear Charge

Let's consider, for example, an electron in a $2s$ orbital. Figure 7.29 shows that the $2s$ orbital has some electron density close to the nucleus. This is an example of **orbital penetration**, which occurs when an electron in an outer orbital has the probability of being close to the nucleus. We can think of it as though an electron in the $2s$ orbital enters the space occupied by the electrons in the $1s$ orbital. Orbital penetration is important when outer-shell electrons are separated from the nucleus by one or more filled inner shells. Electrons in orbitals that penetrate closer to the nucleus experience more of the positive charge of the nucleus. Thus they are attracted more strongly to the nucleus than are electrons in orbitals that do not penetrate inner orbitals.

The profiles in Figure 7.29 show that electrons in a $2p$ orbital penetrate less effectively than those in a $2s$ orbital. Penetration by electrons in shells farther from the nucleus decreases among the different orbitals within the same shell according to the following order: $s > p > d > f$. Since greater penetration means lower energy, the energies of the orbitals in a given shell are $s < p < d < f$. Because the electrons in the ground state of a multielectron atom occupy the lowest-energy orbitals available, the order of filling orbitals in a given shell is s first, then (if available) p, then d, then f. In other words, electrons in orbitals that penetrate inner orbitals have a lower (more negative) energy because they have greater exposure to the charge on the nucleus.

Another situation influences the charge perceived by an electron in a multielectron atom. Electrons lying between a given electron and the nucleus **shield**, or **screen**, that electron from experiencing the full charge on the nucleus. From the point of view of an electron, nuclear charge is reduced from its actual value to that of an **effective nuclear charge (Z_{eff})**. The Z_{eff} experienced by outer-shell electrons is increased if the orbitals they occupy penetrate lower-lying orbitals. For example, a $3s$ electron that penetrates close to the nucleus experiences a higher Z_{eff} than does a $3d$ electron that has little penetration of inner-shell orbitals. Trends in properties that depend upon the strength of attraction between electrons and nuclei are determined by how effectively electrons are screened from nuclear charge.

Trends in Atom and Ion Sizes

As one moves down the column of a family in the periodic table, two trends with opposite effects influence atomic size. In the first, the principal quantum number n increases from row to row. As n increases, the probability that the outermost electrons are farther from the nucleus increases. However, Z_{eff} also increases from

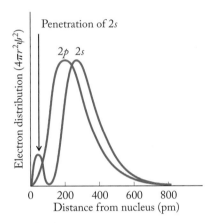

FIGURE 7.29 The $2s$ orbital appears to be farther from the nucleus than the $2p$ orbital. However, the $2s$ orbital is of lower energy because electrons in it penetrate more closely to the nucleus. The result is that $2s$ electrons experience a greater effective nuclear charge and so have lower energy than $2p$ electrons.

Orbital penetration occurs when an electron in an outer orbital has some probability of being as close to the nucleus as an electron in an inner shell.

Screening (or **shielding**) is the effect when inner-shell electrons protect outer-shell electrons from experiencing the total nuclear charge.

Effective nuclear charge (Z_{eff}) is the attractive force toward the nucleus experienced by an electron in an atom. Its value is the positive charge on the nucleus reduced by the extent to which other electrons in the atom shield it from the nucleus.

row to row, causing electrons to be pulled closer to the nucleus because of increasing nuclear charge. Despite this, the net effect of progression down a family is that n dominates and atomic radii increase.

As one moves from left to right across a row in the main group elements, n stays the same. Z_{eff} increases, because nuclear charge increases. Z_{eff} also increases because shielding by the core electrons does not change, and valence electrons are less effective at shielding other valence electrons from the nuclear charge. Across a row, Z_{eff} dominates and atoms generally decrease in size. The major variation in this trend is seen within the transition metals, where size initially decreases for the first two elements but then decreases so gradually that it is virtually constant until the last element in the series. The d electrons that fill in the transition series enter an inner shell $(n - 1)$ and hence do not cause the size to change significantly.

An additional effect of the d electrons causes a deviation from periodic behavior when going down a column in the periodic table. In group 13, for example, we would predict gallium atoms to be larger than aluminum atoms; however, Ga atoms are actually smaller. This is rationalized by noting that the d electrons are not very good at shielding nuclear charge, so the valence electrons in gallium experience an increase in Z_{eff} because 10 additional protons have been added to the nucleus of the atoms in addition to 10 d electrons.

In considering trends in the sizes of ions, the most important observations come from comparing trends in neutral-atom sizes with sizes of their respective ions. Figure 7.30 compares ionic radii with the radii of neutral atoms for several elements. When a cation forms, electrons are removed from the valence shell of a neutral atom, leaving behind a noble gas core. The outermost electrons in that core are in a shell of lower n, so they are on average closer to the nucleus than the valence electrons. In addition, taking sodium as an example, the neutral atom contains 11 total electrons and 11 protons; the sodium cation contains 10 electrons that are held by the same 11 protons. The attractive force per electron increases, while electron–electron repulsion decreases, so the remaining electrons are held more tightly. All of these factors combine to make cations smaller than the neutral atoms from which they arise.

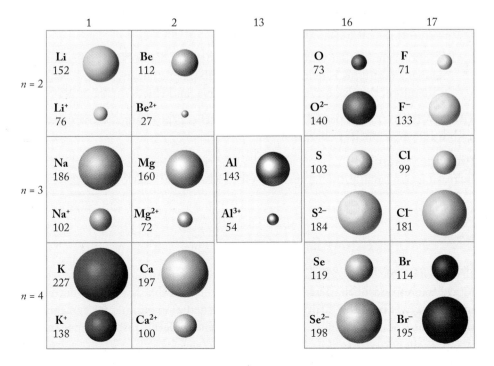

FIGURE 7.30 The radii of a few neutral atoms are compared with the radii of their common ions (all values given in picometers, where 1 pm = 10^{-12} m).

In the formation of anions, the opposite effect applies. Atoms gain electrons, which enter the valence shell of the neutral atom, while the nuclear charge stays the same. The number of electrons in the anion is larger than the number of protons; hence, the attractive force per electron has decreased while the electron–electron repulsion has increased. Anions are thus always larger than their corresponding neutral atoms.

As one moves down a family, the size of ions increases for the same reasons the size of neutral atoms increases. When moving across a row, cations and anions must be considered separately. Cations decrease as the charge on the ion increases; for example, in terms of size, $Na^+ > Mg^{2+} > Al^{3+}$ (see Figure 7.30) because the higher the charge on the cation, the greater the reduction in electron repulsion and the higher the Z_{eff} value. The next atom in the row that tends to form an ion is phosphorus, which forms a 3– anion. The P^{3-} anion is much larger than any of the cations in the row because of the tremendous increase in electron–electron repulsion and the decrease in Z_{eff}. The S^{2-} anion is smaller than P^{3-}, and Cl^- is smaller than S^{2-}, because the electron–electron repulsion is lower and the Z_{eff} value is higher, as fewer electrons are added to make the anions. The largest anion in a row is the anion of highest negative charge, and the largest cation in a row is the cation of lowest positive charge.

SAMPLE EXERCISE 7.13   **Ordering Atoms and Ions by Size**

Arrange the following sets of species by size, largest to smallest: (a) O, P, S; (b) Na^+, Na, K.

COLLECT AND ORGANIZE We need the periodic table to determine the relative locations of the species given.

ANALYZE Size decreases across a row (from left to right) and increases down a column. In addition, a cation is always smaller than the neutral atom from which it is made.

SOLVE

a. S is below O in group 16 on the periodic table, so in terms of atomic size, S > O. S is to the right of P, so P > S. For the set O, P, and S, the size order is thus P > S > O.

b. Cations are smaller than their neutral atoms, so Na > Na^+. Size increases down a family; K is below Na in the alkali metals family, so K > Na. Therefore, the size order of the set Na^+, Na, and K is K > Na > Na^+.

THINK ABOUT IT The periodic trends in size reflect the change in the value of n and the effective nuclear charge (Z_{eff}).

Practice Exercise Arrange the following sets of species in order of their increasing size (smallest to largest): (a) Cl^-, F, Li^+; (b) P^{3-}, Al^{3+}, Mg^{2+}.

7.11 Ionization Energies

In developing electron configurations in the previous section, we followed a theoretical framework for the arrangement of electrons in orbitals that was developed in the early years of the 20th century. Is there actual experimental evidence for the existence of orbitals representing different energy levels inside atoms? Indeed there is. It includes the measurements of the energies needed to remove electrons from atoms and their ions.

Ionization energy (IE) is defined as the energy needed to remove 1 mole of electrons from a mole of free (gas-phase) atoms or ions in their ground states. Removing these electrons always costs energy because a negatively charged electron is attracted to the positively charged nucleus, and energy is required to overcome that attractive force. The amount of energy needed to remove one mole of electrons per mole of atoms to make 1 mole of cations with a 1+ charge is called the *first ionization energy* (IE_1); the energy needed to remove a second mole of electrons to make 1 mole of 2+ cations is the *second ionization energy* (IE_2), and so forth. For example:

$$Mg(g) \rightarrow Mg^+(g) + 1e^- \qquad (IE_1 = 738 \text{ kJ/mol})$$

$$Mg^+(g) \rightarrow Mg^{2+}(g) + 1e^- \qquad (IE_2 = 1451 \text{ kJ/mol})$$

The total energy required to make one mole of Mg^{2+} cations from one mole of magnesium atoms in the gas phase is the sum of these two ionization energies:

$$Mg(g) \rightarrow Mg^{2+}(g) + 2e^-$$

$$\text{Total energy required} = (738 + 1451) \text{ kJ/mol}$$

$$= 2189 \text{ kJ/mol}$$

Figure 7.31 shows how the first ionization energies vary among the representative elements (groups 1, 2, and 13 through 18). We will consider the IE_1 values of the first 10 elements in detail and then note how the pattern in these values is repeated among the heavier elements.

Hydrogen's IE_1 value is 1312 kJ/mol. Helium's is nearly twice as big: 2372 kJ/mol. This difference makes sense because He atoms have two protons per nucleus whereas H atoms have only one. Twice the nuclear charge leads to twice the nuclear attraction, and in turn twice the energy to pull an electron away from the nucleus. In general, first ionization energies increase from left to right across a row on the periodic table. Thus the easiest element to ionize in a row is the first one (leftmost), and the hardest is the last one (rightmost). This pattern makes sense because the charge of the nucleus increases across a row and so does the attraction between it and the surrounding electrons.

Ionization energy (IE) is the amount of energy needed to remove 1 mole of electrons from a mole of ground-state atoms or ions in the gas phase.

▶❚❚ **CHEMTOUR** The Periodic Table

FIGURE 7.31 The first ionization energies of the representative (main group) elements generally increase from left to right in a period and decrease from top to bottom in a group of the periodic table, as illustrated here in (a) a two-dimensional plot and (b) a three-dimensional histogram.

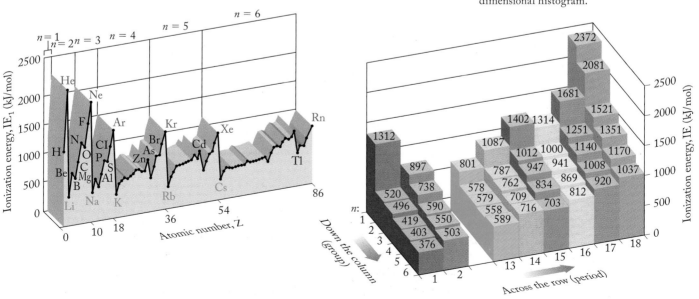

(a)　　　　　　　　　　　　　　　　　　(b)

Note, however, that there are two anomalies in this trend starting in the second row: IE_1 values actually decrease between elements in groups 2 and 13 and between elements in groups 15 and 16 in the same rows. The decreases between groups 2 and 13 happen because p electrons in a given shell experience less effective nuclear charge than s electrons and so are more easily ionized. To understand the slight decreases between groups 15 and 16, think back to the stability associated with half-filled d orbitals in the transition metals. As with d orbitals, some enhanced stability comes with half-filled sets of p orbitals, too. For example, when oxygen loses a $2p$ electron, it achieves the slightly more stable configuration of a half-filled set of $2p$ orbitals:

$$\boxed{\uparrow\downarrow}\quad \boxed{\uparrow\downarrow}\quad \boxed{\uparrow\uparrow\uparrow\uparrow} \;\rightarrow\; \boxed{\uparrow\downarrow}\quad \boxed{\uparrow\downarrow}\quad \boxed{\uparrow\uparrow\uparrow} \;+\; 1\,e^-$$
$$\quad 1s \qquad 2s \qquad 2p \qquad\qquad 1s \qquad 2s \qquad 2p$$

The relative stability of the product of ionization means that the ionization process requires less energy.

Now let's compare the IE_1 values of Li and H. A lithium atom has three times the nuclear charge as an atom of hydrogen, so we might expect its ionization energy to be about three times larger. However, its IE_1 value is only 520 kJ/mol, or less than half that of hydrogen. Why is it less? Our quantum mechanical model of atomic structure explains the much smaller IE_1 for Li this way: the most easily ionized (highest-energy) electron in a Li atom is in a $2s$ orbital, which means that electron is farther from the nucleus than the $1s$ electron is in a H atom; the $2s$ electron is also shielded from the nucleus by the two electrons in the filled $1s$ orbital. Thus the effective nuclear charge felt by a Li valence electron is much less than that felt by the electron in a H atom. This combination of larger atomic size and more shielding by inner-shell electrons leads to decreasing IE_1 values from top to bottom in every group of the periodic table.

Another perspective on the different energy levels of orbitals is provided by looking at successive ionization energies (IE_1, IE_2, IE_3, and so on) for each element. Consider the trend in successive ionization energies for the first 10 elements in Table 7.2. For multielectron atoms, the energy needed to remove a second electron is always greater than that needed to remove the first, because a second, negatively charged electron is being removed from an ion that already has a posi-

TABLE 7.2 Successive Ionization Energies of the First 10 Elements in kJ/mol

Element	Z	IE_1	IE_2	IE_3	IE_4	IE_5	IE_6	IE_7	IE_8	IE_9	IE_{10}
H	1	1312									
He	2	2372	5249								
Li	3	520	7296	12040							
Be	4	897	1758	15050	21070						
B	5	801	2426	3660	24682	32508					
C	6	1087	2348	4617	6201	37926	46956				
N	7	1402	2860	4581	7465	9391	52976	64414			
O	8	1314	3383	5298	7465	10956	13304	71036	84280		
F	9	1681	3371	6020	8428	11017	15170	17879	92106	106554	
Ne	10	2081	3949	6140	9391	12160	15231	19986	23057	115584	131236

tive charge. The energy needed to remove a third electron is greater still because it is being removed from an ion with a 2+ charge. Superimposed on this trend is a much more dramatic increase in ionization energy (defined by the red line in Table 7.2) when all the valence electrons have been removed and so the next electron must come from an inner shell. The core electrons are clearly held much more tightly by the nucleus and are not likely to take part in chemical reactions.

SAMPLE EXERCISE 7.14   **Identifying Trends in Ionization Energies**

Arrange the following elements in order of increasing first ionization energies: Ar, Mg, P.

COLLECT AND ORGANIZE Ionization energy is the energy required to remove one mole of electrons from a mole of atoms or ions. We can use the periodic table to identify the trend.

ANALYZE First ionization energy increases from left to right across the periodic table, and decreases down a family. It is the energy required to remove one mole of electrons from a mole of neutral atoms in the gas phase.

SOLVE All three elements are in the same row. Their order across the row is $Mg \rightarrow P \rightarrow Ar$, hence the predicted order of increasing first ionization energies is $Mg < P < Ar$.

THINK ABOUT IT Magnesium (Mg) tends to form a positive ion, so one would expect its ionization energy to be smaller than the other two elements. Argon (Ar) is a noble gas with a stable valence-shell electron configuration, so its first ionization energy would logically be the highest in the set.

Practice Exercise Arrange this set of elements in order of decreasing first ionization energies: Cs, Ca, Ne.

CONCEPT TEST

Why do ionization energies decrease with increasing atomic number for the three noble gases He, Ne, and Ar?

To end this chapter, let's return to the momentous advances in chemistry and physics of the first three decades of the 20th century. They are due to the brilliant discoveries of Einstein, Planck, Rutherford, de Broglie, Schrödinger, Bohr, and others, which have forever changed scientists' view of the fundamental structure of matter and the universe. Figure 7.32 summarizes and connects some of these advances.

Consensus in the scientific community did not come easily. We have seen how hot sodium atoms in the excited state emit photons of yellow-orange light as they fall to the ground state. This phenomenon, called *spontaneous emission*, was a concept that Einstein puzzled over for several years before deciding that the exact moment when spontaneous emission occurs and the direction of the photon emitted could not be predicted exactly. He concluded that quantum theory only allows us to calculate the probability of a spontaneous electron transition; the details of the event are left to chance. No force of nature actually causes a hot sodium atom to fall to a lower energy level at a particular instant.

A Noble Family: Special Status for Special Behavior

In the descriptive chemistry boxes in previous chapters, we have presented information about one or two elements whose properties and behavior were discussed in the chapter. Now that we have developed electron configurations and highlighted their role in determining the periodic behavior of elements, we can use these boxes to discuss those features of the chemical and physical behavior of elements that vary periodically for entire families. The six elements in the rightmost column of the periodic table (group 18) are a family known as the noble gases. We have encountered several of them in this and earlier chapters. The members of this family were originally called inert gases, but in the 1960s chemists discovered how to make compounds of several members of the family with very reactive substances like fluorine. After this discovery, the family name was changed to the noble gases. As in the past, noble families tended not to interact with common people; the noble gases do not react with common elements.

Helium is the second element in the periodic table and the second most abundant element in the universe (hydrogen is the most abundant). Its presence in the absorption spectrum of gases around the sun was detected in 1868 as a set of lines that did not correspond with those of any of the known elements on Earth. These lines were assigned to helium, whose name derives from *helios*, the Greek word for "sun." The element was isolated on Earth by Sir William Ramsay (1852–1916) in 1895.

Argon was isolated from the atmosphere the year before Ramsay discovered helium. The name of this element is from the Greek *argos* meaning "lazy," in reference to the chemical inertness of the element and the others in group 18. Argon is the most abundant of the group 18 elements, making up about 0.94% (by volume) of the atmosphere, and is the 12th most abundant element in the universe. Most of the argon in the atmosphere is the product of radioactive decay: It is produced in rocks and minerals when nuclei of radioactive ^{40}K capture one of their own electrons and a proton is transformed into a neutron. The low rate of the decay process allows scientists to date the ages of rocks on the basis of their ^{40}Ar-to-^{40}K ratio. Eventually the argon leaks into the atmosphere, from which it is isolated by first liquefying the air and then separating Ar from N_2 and O_2 by distillation. Argon is used to provide an inert atmosphere in arc-welding certain metals such as stainless steel. Argon is also used in lightbulbs to improve the lifetime of the filaments and in photoelectric devices.

Neon was discovered by Ramsay and his coworkers as an impurity in argon in 1898. Neon takes its name from the Greek *neos*, which means "new." Although it is found in some minerals as a result of radioactive decay, its principal industrial source is the atmosphere. However, its tiny concentration in the atmosphere (0.0018% by volume) means that nearly 100 kg of air must be liquefied to obtain 1 g of neon. When an electric current is passed through a tube containing neon at low pressure, the gas emits the charac-

(a)

(b)

Both hydrogen and helium gases have been used to fill lighter-than-air craft, but (a) the hydrogen-filled Hindenburg ended its last trans-Atlantic flight in flames in Lakehurst, New Jersey, in 1937. (b) Modern blimps such as this one get their buoyancy from helium, which is chemically inert and hence not a hazard as far as fire and explosion are concerned.

teristic red orange light that we associate with neon signs. Electrical discharge tubes filled with the other noble gases emit their own unique combinations of emission lines and characteristic colors. Neon was the gas analyzed by Aston in his positive-ray analyzer that led to the discovery of stable isotopes and to the development of mass spectrometry (see Chapter 2).

Gas discharge tubes filled with noble gases.

Krypton (from the Greek *kryptos*, "hidden") and xenon (from the Greek *xenos*, "foreign") are even rarer components of the atmosphere. The concentration of krypton is only 1.1 parts per million (ppm) by volume, which means that there are, on average, 1.1 atoms of krypton for every million atoms and molecules of the other ingredients in the atmosphere. The atmospheric concentration of xenon is even smaller: only 86 parts per billion (ppb) by volume. Because these gases have higher boiling points than do the principal components of the atmosphere, they can be separated by fractional distillation of liquefied air. These two elements are recovered in the least-volatile fraction.

All of the noble gases have completely filled outer shells. The lack of reactivity of these elements is a direct result of their stable electron configurations. Helium has a filled 1*s* orbital, which corresponds to a full first shell, and only two valence electrons; the other noble gases have completely filled *s* and *p* orbitals and eight valence electrons. In terms of the periodic properties that characterize the noble gas family, each noble gas has the highest ionization energy of all the elements in its row (period) of the periodic table. For example, it takes about 20% more energy to remove

an electron from neon than from the element immediately preceding it (fluorine) and 400% more energy than lithium. The ionization energies of the six noble gases decrease with increasing atomic number because the increasing numbers of core electrons shield the outermost electron from the positive charge on the nucleus. The sizes of noble atoms increase from top to bottom in column 18 of the periodic table: as one moves down the family from element to element, the valence electrons that determine the size of the atom are in shells of increasing average distance from the nucleus. Because covalent radii are determined experimentally with respect to internuclear distances in diatomic molecules and compounds, the values of the radii of helium, neon, and argon, for which no stable compounds are known, are estimations.

The noble gases do not easily share electrons to form compounds with other atoms. However, xenon and krypton do react with fluorine to form, for example, XeF_2, XeF_4, XeF_6, and KrF_2. The three xenon compounds react with water, yielding compounds containing xenon, fluorine, and oxygen.

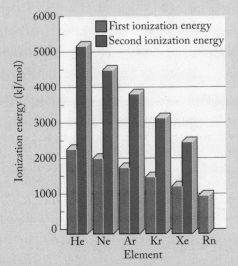

The first ionization energies of the group 18 elements decrease down the group because the inner electrons shield the outer electrons from the nuclear charge. The larger, second ionization energies are due to the greater amount of energy required to remove an electron from an ion with a positive charge.

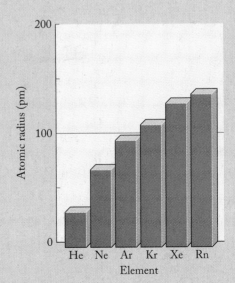

The atomic radii of the group 18 elements increase down the group, corresponding to the increase in atomic number, which in turn corresponds to electrons occupying orbitals found, on average, farther from the nucleus.

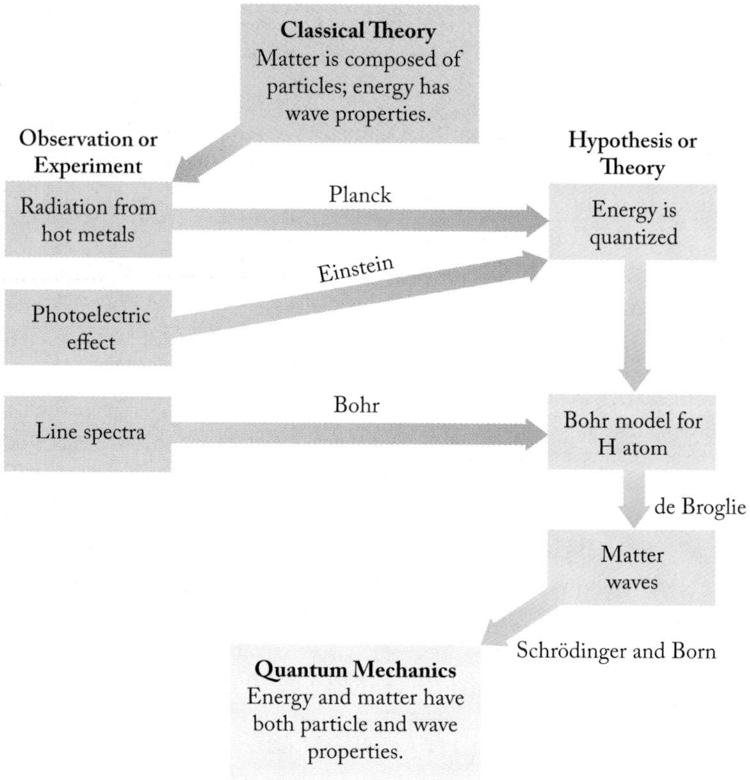

FIGURE 7.32 During the first decades of the 20th century, quantum theory evolved from classical (19th-century) theories of the nature of matter and energy. These boxes and arrows show the interplay between observations and formulating and testing hypotheses. Follow the arrows to trace the development of modern quantum theory, which assumes that energy and mass are equivalent, showing properties of both particles and waves.

Classical Theory
Matter is composed of particles; energy has wave properties.

Observation or Experiment

Hypothesis or Theory

Radiation from hot metals

Planck

Energy is quantized

Photoelectric effect

Einstein

Line spectra

Bohr

Bohr model for H atom

de Broglie

Matter waves

Schrödinger and Born

Quantum Mechanics
Energy and matter have both particle and wave properties.

How different is this behavior from the laws governing the behavior of large objects? A pebble picked up and dropped immediately falls to the ground. However, electrons remain in excited states for indeterminate (though usually short) times before falling to their ground states. This lack of determinacy bothered Einstein and many of his colleagues. Had they discovered an underlying theme of nature—that some processes could not be described or known with certainty? Are there fundamental limits to how well we can know and understand our world and the events that change it?

Many scientists in the early decades of the 20th century did not care to make such admissions. They preferred the Newtonian view of the world as a place where events occurred for a reason and where there were causes and effects. They believed that the more they studied nature with ever more sophisticated tools, the more they would understand why things happened as they did. Soon after Max Born published a probabilistic interpretation of Schrödinger's wave functions in 1926, Einstein wrote Born a letter:

> Quantum Mechanics is very impressive. But an inner voice tells me that it is not yet the real thing. The theory produces a great deal but hardly brings us closer to the secret of the Old One. I am at all events convinced that He does not play dice.[2]

There do seem to be limits to what we can know about the world.

---

[2] Letter to Max Born, 12 December 1926; quoted in R. W. Clark, *Einstein: The Life and Times* (New York: HarperCollins, 1984), p. 880.

Summary

SECTION 7.1 Light is one form of **electromagnetic radiation**. As such, light has wavelike properties described by characteristic visible **wavelengths** and **frequencies**. In the **electromagnetic spectrum**, visible light is surrounded by forms of radiation with longer wavelengths, such as infrared (IR), microwave, and radio-wave radiation, and forms with shorter wavelengths, such as ultraviolet (UV) radiation, X-rays, and γ rays. A beam of sunlight passing through a prism is dispersed into all its colors because of **refraction**. Waves also experience **diffraction** and **interference**. The sun's spectrum contains narrow gaps, called **Fraunhofer lines**, that are also present in the light from distant galaxies. In galactic spectra these lines are shifted to longer wavelengths due to the Doppler effect, which indicates that the galaxies are rapidly moving away from Earth.

SECTION 7.2 Free atoms in flames and in gas discharge tubes produce **atomic emission spectra** or **bright-line spectra**. When a continuous source of radiation passes through a gas composed of free atoms, absorption of a fraction of the radiation produces a spectrum of dark lines, called an **atomic absorption spectrum**. The lines of an element's atomic absorption and emission spectra are at exactly the same wavelengths.

SECTION 7.3 According to the **quantum theory** developed by Max Planck, light is composed of atomic-scale energy particles called **quanta**. Albert Einstein called these particles of light **photons** in his explanation of the **photoelectric effect**: a process in which a metallic surface emits electrons when illuminated by electromagnetic radiation that is at or above the required **threshold frequency** (ν_0). The radiant energy required to dislodge an electron from the surface of a metal is called the **work function (Φ)** of the metal.

SECTION 7.4 Johann Balmer derived an equation that accounted for the frequencies of all the visible spectral lines emitted by a hydrogen discharge tube and that accurately predicted the existence of hydrogen lines in the UV and IR regions of the electromagnetic spectrum. Niels Bohr proposed that the frequencies predicted by Balmer's equation are related to allowed electron **energy levels** inside the hydrogen atom. A **ground state** atom or ion has all of its electrons in the lowest possible energy level. Higher energy levels are called **excited states**. **Electron transitions** from higher to lower states cause the emission of particular frequencies of electromagnetic radiation; absorption of the same frequencies accompanies transitions from the same lower to higher electron energy levels.

SECTION 7.5 Louis de Broglie proposed that electrons in atoms, and indeed all moving particles, have wavelike properties. He explained the stability of the electron orbits in the Bohr model of the hydrogen atom in terms of **standing waves**: the circumferences of the allowed orbits had to be whole-number multiples of the hydrogen electron's characteristic wavelength. The **Heisenberg uncertainty principle** states that the position and momentum of an electron cannot both be precisely known at the same time.

SECTION 7.6 Solutions to **Schrödinger's wave equation** are mathematical expressions called **wave functions (ψ)**, which define allowed electron energy levels within atoms. Though wave functions have no physical meaning, ψ^2 defines **orbitals**, the three-dimensional regions inside an atom that describe the probability of finding an electron at a given distance from the nucleus. Each orbital has a unique set of three **quantum numbers**: the **principal quantum number (n)**, which largely defines the orbital's size and energy level (or shell); the

angular momentum quantum number (ℓ), which defines the orbital's shape (and subshell); and the **magnetic quantum number (m_ℓ)**, which defines the orbital's orientation in space. A fourth quantum number, the **spin magnetic quantum number (m_s)**, is necessary to explain certain characteristics of emission spectra. The two electrons that occupy one orbital have opposite spins ($m_s = +\frac{1}{2}$ and $-\frac{1}{2}$). The **Pauli exclusion principle** states that no two electrons in an atom can have the same four values of the quantum numbers n, ℓ, m_ℓ, and m_s.

SECTION 7.7 Orbitals have characteristic three-dimensional sizes, shapes, and orientations that are depicted by boundary–surface representations. All s orbitals are spheres that increase in size with increasing values of n. The three p orbitals in any shell of $n \geq 2$ have two balloon-shaped lobes on one of the x-, y- or z-axes. The five d orbitals found in shells of $n \geq 3$ come in two different forms: four are shaped like a four-leaf clover, and the fifth has two lobes oriented along the z-axis and a torus surrounding the middle of the two lobes.

SECTION 7.8 According to the **aufbau principle**, electrons fill the lowest-energy atomic orbitals of a ground-state atom first. An **electron configuration** is a set of symbols used to express the number of electrons in each occupied orbital in an atom. The periodic table is very useful as a guide to develop electron configurations. The electrons in the outermost shell of an atom are called **valence electrons**. They are the electrons lost, gained, or shared in chemical reactions. The orbitals of a given p, d, or f subshell are **degenerate**; that is, they have exactly the same energy as other orbitals in that subshell. **Hund's rule** states that an s, p, d, or f subshell must be completely occupied by unpaired electrons before any of these orbitals can have two electrons.

SECTION 7.9 Atoms of the elements in groups 1 and 2 tend to lose electrons and form ions with charges of 1+ and 2+, respectively. In so doing they achieve the same electron configurations as (and thus are **isoelectronic** with) the noble gas elements in the preceding row of the periodic table. Atoms in groups 16 and 17 tend to gain electrons to form ions with charges of 2– and 1–, respectively, thereby achieving the electron configurations of the noble gases in their rows of the periodic table. When transition metals form ions, the ions retain the maximum number of d electrons consistent with their charge.

SECTION 7.10 The **effective nuclear charge (Z_{eff})** is the net nuclear charge felt by an electron because other electrons closer to the nucleus **screen** (or **shield**) the electron from experiencing the full nuclear charge. Trends in Z_{eff} account in part for trends in atomic sizes; **orbital penetration** also plays a role. Atomic sizes of the elements decrease with increasing Z_{eff} across the rows of the periodic table as increasing nuclear charge draws the the outermost electrons inward. Atomic sizes increase with increasing Z_{eff} down a column as electrons fill shells that are farther from the nucleus. Anions are larger than their parent atoms due to additional electron–electron repulsion, but cations are smaller than their parent atoms—sometimes much smaller when all the electrons in the valence shells are lost.

SECTION 7.11 **Ionization energies (IE)** generally increase with increasing Z_{eff} across the rows of the periodic table. Exceptions can be explained by the relatively higher energy of electrons in p orbitals than in the s orbital in a given shell, and by the stability of half-filled sets of p and d orbitals. The energy differences between atoms in different shells and subshells are also apparent in the values of successive ionization energies (IE_1, IE_2, IE_3, . . .) for a given element.

Problem-Solving Summary

TYPE OF PROBLEM	CONCEPTS AND EQUATIONS	SAMPLE EXERCISES
Calculating frequency from wavelength	$\lambda \nu = c$	7.1
Calculating the energy of a quantum of light	$E = \dfrac{hc}{\lambda}$	7.2
Using the work function	$\Phi = h\nu_0$	7.3
Calculating the wavelength of a line in the hydrogen spectrum	$\dfrac{1}{\lambda} = \left[1.097 \times 10^{-2}(\text{nm})^{-1}\right]\left(\dfrac{1}{n_1^{\,2}} - \dfrac{1}{n_2^{\,2}}\right)$	7.4
Calculating the energy of an electron transition in a hydrogen atom	$\Delta E = (-2.18 \times 10^{-18}\ \text{J})\left(\dfrac{1}{n_f^{\,2}} - \dfrac{1}{n_i^{\,2}}\right)$	7.5
Calculating the wavelength of a particle in motion	$\lambda = \dfrac{h}{mu}$	7.6
Calculating uncertainty	$\Delta x \cdot m\, \Delta u \geq \dfrac{h}{4\pi}$	7.7
Identifying the subshells and orbitals in an energy level	n is the shell number, ℓ defines the subshell, or type of orbital; for $n = 5$, ℓ can have values of 0, 1, 2, 3, and 4, with corresponding subshell labels s, p, d, f, and g.	7.8
Identifying possible sets of quantum numbers	An orbital has a unique combination of allowed n, ℓ, and m_ℓ values. ℓ is any integer from 0 to $n - 1$; m_ℓ is any integer from $-\ell$ to $+\ell$, including zero.	7.8, 7.9
Writing electron configurations of atoms	Follow the sequence below (superscripts represent maximum numbers of electrons) until all the electrons are accounted for. $1s^2 2s^2\ 2p^6 3s^2 3p^6 4s^2\ 3d^{10} 4p^6\ 5s^2 4d^{10} 5p^6\ 6s^2\ 4f^{14} 5d^{10} 6p^6 \ldots$	7.10
Determining isoelectronic species among ions of main group elements	Atoms and ions with the same number of electrons are isoelectronic.	7.11
Writing electron configurations of transition metal atoms and ions	In forming ions, electrons are removed to maximize the number of d electrons; there is enhanced stability in half-filled d subshells.	7.12
Ordering atoms and ions by size	Size decreases left to right across a row and increases down a column; cations are smaller and anions are larger than their parent atoms.	7.13
Identifying trends in ionization energies	The first ionization energy increases across a row (period) and decreases down a column (family).	7.14

Visual Problems

7.1. Which of the elements highlighted in Figure P7.1 consists of atoms with
   a. a single *s* electron in their outermost shells? (More than one answer is possible.)
   b. filled sets of *s* and *p* orbitals in their outermost shells?
   c. filled sets of *d* orbitals?
   d. half-filled sets of *d* orbitals?
   e. two *s* electrons in their outermost shells?

7.2. Which of the highlighted elements in Figure P7.1 has the greatest number of unpaired electrons per atom?

7.3. Which of the highlighted elements in Figure P7.1 form(s) common monatomic ions that are larger than their parent atoms?

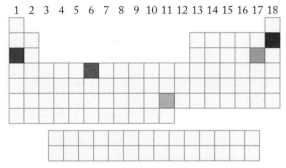

FIGURE P7.1

7.4. Which of the elements highlighted in Figure P7.4 forms monatomic ions by
   a. losing an *s* electron?
   b. losing two *s* electrons?
   c. losing two *s* electrons and a *d* electron?
   d. adding an electron to a *p* orbital?
   e. adding electrons to two *p* orbitals?

7.5. Which of the highlighted elements in Figure P7.4 form(s) common monatomic ions that are smaller than their parent atoms?

7.6. Rank the elements highlighted in Figure P7.4 based on increasing size of their atoms.

7.7. Rank the highlighted elements in Figure P7.4 based on increasing size of their most common monatomic ions.

7.8. Which of the elements highlighted in Figure P7.4 has the largest first ionization energy, IE_1?

7.9. Which of the elements highlighted in Figure P7.4 has the largest second ionization energy, IE_2?

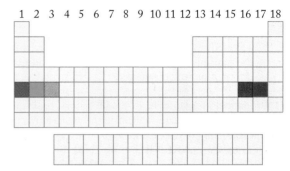

FIGURE P7.4

7.10. Consider the results of an experiment shown in Figure P7.10. The screen containing two narrow slits is illuminated by a distant source of light off to the left. Instead of just the images of the two slits appearing on the screen on the right, a whole series of slit images (more than the five shown here) appear. Explain how there could be many more than two slit images on the screen on the right.

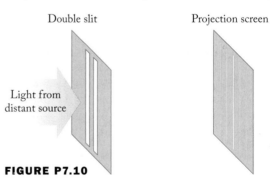

FIGURE P7.10

Questions and Problems

Waves of Light

CONCEPT REVIEW

7.11. Why are the various forms of radiant energy called *electromagnetic* radiation?

7.12. Explain with a sketch why the frequencies of long-wavelength waves of electromagnetic radiation are lower than those of short-wavelength waves.

7.13. Dental X-rays When X-ray images are taken of your teeth and gums in the dentist's office, your body is covered with a lead shield. Explain the need for this precaution.

7.14. Ultraviolet radiation causes skin damage that may lead to cancer, but exposure to infrared radiation does not seem to cause skin cancer. Why do you think this is so?

7.15. Refraction in Raindrops Sketch a figure showing how sunlight could be refracted by a drop of rain falling to Earth.

7.16. Explain how Hubble's observations led to the belief that our universe is expanding.

*7.17. The Doppler effect is described by the equation
$$\frac{(\nu - \nu^1)}{\nu} = \frac{u}{c}$$
where ν is the unshifted frequency, ν^1 the perceived frequency, c is the speed of light, and u is the speed at which the object is moving. If hydrogen in a galaxy that is receding from Earth at half the speed of light emits radiation with a wavelength of 656 nm, will the radiation still be in the visible part of the electromagnetic spectrum when it reaches Earth?

*7.18. If light consists of waves, why don't things look "wavy" to us?

PROBLEMS

7.19. A neon light emits radiation of $\lambda = 616$ nm. What is the frequency of this radiation?

7.20. **Submarine Communications** In the 1990s the Russian and American navies developed extremely low frequency communications networks to send messages to submerged submarines. The frequency of the carrier wave of the Russian network was 82 Hz, while the Americans used 76 Hz.
   a. What was the ratio of the wavelengths of the Russian network to the American network?
   *b. To calculate the actual underwater wavelength of the transmissions in either network, what additional information would you need?

7.21. **Broadcast Frequencies** FM radio stations broadcast at different frequencies. Calculate the wavelengths corresponding to the broadcast frequencies of the following radio stations: (a) KKNB (Lincoln, NE), 104.1 MHz; (b) WFNX (Boston, MA), 101.7 MHz; (c) KRTX (Houston, TX), 100.7 MHz

7.22. Which radiation has the longer wavelength? (a) radio waves from an AM radio station broadcasting at 680 kHz, or (b) infrared radiation emitted by the surface of Earth ($\lambda = 15$ mm)

7.23. Which radiation has the lower frequency? (a) radio waves from an AM radio station broadcasting at 1090 kHz or (b) the green light ($\lambda = 550$ nm) from an LED (light-emitting diode) on a stereo system

7.24. Which radiation has the higher frequency? (a) the red light on a bar-code reader at a grocery store, or (b) the green light on the battery charger for a lap top computer

7.25. **Speed of Light** How long does it take light to reach Earth from the sun? (The distance from the sun to Earth is 93 million miles.)

7.26. **Exploration of the Solar System** How long would it take an instruction to a Martian rover to travel from a NASA site on Earth to Mars? Assume the signal is sent when Earth and Mars are 75 million kilometers apart.

Atomic Spectra

CONCEPT REVIEW

7.27. Describe the similarities and differences in the atomic emission and absorption spectra of hydrogen.

7.28. Are the Fraunhofer lines in the spectra of stars the result of atomic emission or atomic absorption?

7.29. How did the study of the atomic emission spectra of elements lead to the identification of the Fraunhofer lines in sunlight?

*7.30. What would happen in terms of the appearance of the Fraunhofer lines in the solar spectrum if sunlight were passed through a flame containing high-temperature calcium atoms and then analyzed?

Particles of Light and Quantum Theory

CONCEPT REVIEW

7.31. What is a quantum?

7.32. What is a photon?

7.33. A variable power supply is connected to an incandescent lightbulb. At the lowest power setting, the bulb feels warm to the touch but produces no light. At medium power, the lightbulb filament emits a red glow. At the highest power, the lightbulb emits white light. Explain this emission pattern.

*7.34. Has a photon of radiation that is red-shifted 10% actually lost 10% of its energy, where $E = h\nu$?

PROBLEMS

7.35. Which of the following have quantized values? Explain your selections.
   a. The number of eggs remaining in a open carton of eggs
   b. The elevation up a handicapped access ramp
   c. The elevation up a flight of stairs
   d. The speed of an automobile

7.36. Which of the following have quantized values? Explain your selections.
   a. The pitch of a note played on a slide trombone
   b. The pitch of a note played on a flute
   c. The wavelengths of light produced by the heating elements in a toaster
   d. The wind speed at the top of Mt. Everest

7.37. When a piece of metal is irradiated with UV radiation ($\lambda = 162$ nm), electrons are ejected with a kinetic energy of 5.34×10^{-19} J. What is the work function of the metal?

7.38. The first ionization energy of a gas-phase atom of a particular element is 6.24×10^{-19} J. What is the maximum wavelength of electromagnetic radiation that could ionize this atom?

*7.39. Thin layers of potassium ($\Phi = 3.68 \times 10^{-19}$ J and sodium ($\Phi = 4.41 \times 10^{-19}$ J) are exposed to radiation of wavelength 300 nm. Which metal emits electrons with the greater velocity? What is the velocity of these electrons?

7.40. Photovoltaic cells convert solar energy into electricity. Could tantalum ($\Phi = 6.81 \times 10^{-19}$ J) be used to convert visible light to electricity? Assume that most of the electromagnetic energy from the sun is in the visible region near 500 nm.

7.41. With reference to Problem 7.40, could tungsten ($\Phi = 7.20 \times 10^{-19}$ J) be used to construct solar cells?

7.42. Titanium ($\Phi = 6.94 \times 10^{-19}$ J) and silicon ($\Phi = 7.24 \times 10^{-19}$ J) surfaces are irradiated with UV radiation with a wavelength of 250 nm. Which surface emits electrons with the longer wavelength? What is the wavelength of the electrons emitted by the titanium surface?

7.43. The power of a red laser ($\lambda = 630$ nm) is 1.00 watt (abbreviated W, where 1 W = 1 J/s). How many photons per second does the laser emit?

*7.44. The energy density of starlight in interstellar space is 10^{-15} J/m³. If the average wavelength of starlight is 500 nm, what is the corresponding density of photons per cubic meter of space?

The Hydrogen Spectrum and the Bohr Model

CONCEPT REVIEW

7.45. Why should hydrogen have the simplest atomic spectrum of all the elements?

7.46. For an electron in a hydrogen atom how is the value of n of its orbit related to its energy?

7.47. Does the energy of electromagnetic energy emitted by an excited state H atom depend on the individual values of n_1 and n_2, or only on the difference between them $(n_1 - n_2)$?

7.48. Explain the difference between a ground-state H atom and an excited-state H atom.

7.49. Without calculating any wavelength values, predict which of the following electron transitions in the hydrogen atom is associated with radiation having the shortest wavelength.
  a. From $n = 1$ to $n = 2$ c. From $n = 3$ to $n = 4$
  b. From $n = 2$ to $n = 3$ d. From $n = 4$ to $n = 5$

7.50. Without calculating any frequency values, rank the following transitions in the hydrogen atom in order of increasing frequency of the electromagnetic radiation that could produce them.
  a. From $n = 4$ to $n = 6$ c. From $n = 9$ to $n = 11$
  b. From $n = 6$ to $n = 8$ d. From $n = 11$ to $n = 13$

7.51. Electron transitions from $n = 2$ to $n = 3$, 4, 5, or 6 in hydrogen atoms are responsible for some of the Fraunhofer lines in the sun's spectrum. Are there any Fraunhofer lines due to transitions that start from $n = 3$ in hydrogen atoms?

7.52. Are there any visible lines in the atomic emission spectrum of hydrogen due to electron transitions to the $n = 1$ state?

7.53. Balmer observed a hydrogen line for the transition from $n = 6$ to $n = 2$, but not for the transition from $n = 7$ to $n = 2$. Why?

*7.54. In what ways should the emission spectra of H and He$^+$ be alike, and in what ways should they be different?

PROBLEMS

7.55. What is the wavelength of the photons emitted by hydrogen atoms when they undergo transitions from $n = 4$ to $n = 3$? In which region of the electromagnetic spectrum does this radiation occur?

7.56. What is the frequency of the photons emitted by hydrogen atoms when they undergo transitions from $n = 5$ to $n = 3$? In which region of the electromagnetic spectrum does this radiation occur?

*7.57. The energies of the photons emitted by one-electron atoms and ions fit the equation

$$E = (2.18 \times 10^{-18}\ \text{J})Z^2 \left(\frac{1}{n_1^2} - \frac{1}{n_2^2} \right)$$

where Z is the atomic number, n_2 and n_1 are positive integers, and $n_2 > n_1$.
  a. As the value of Z increases, does the wavelength of the photon associated with the transition from $n = 2$ to $n = 1$ increase or decrease?
  b. Can the wavelength associated with the transition from $n = 2$ to $n = 1$ ever be observed in the visible region of the spectrum?

*7.58. Can transitions from higher energy states to the $n = 2$ state in He$^+$ ever produce visible light? If so, for what values of n_2? (*Hint:* The equation in Problem 7.57 may be useful.)

7.59. The transition from $n = 3$ to $n = 2$ in a hydrogen atom produces a photon with a wavelength of 656 nm. What is the wavelength of the transition from $n = 3$ to $n = 2$ in a Li^{2+} ion? (Use the equation in Problem 7.57.)

7.60. The hydrogen spectrum includes an atomic emission line with a wavelength of 92.3 nm.
  a. Is this line associated with a transition between different excited states, or between an excited state and the ground state?
  b. What is the value of n_i in this transition?
  c. What is the energy of the longest-wavelength photon that a ground-state hydrogen atom can absorb?

Electrons as Waves

CONCEPT REVIEW

7.61. Identify the symbols in the de Broglie relation $\lambda = h/mu$, and explain how the relation links the properties of a particle to those of a wave.

7.62. Explain how the observation of electron diffraction supports the description of electrons as waves.

7.63. Would the density or shape of an object have an effect on its de Broglie wavelength?

7.64. How does de Broglie's hypothesis that electrons behave like waves explain the stability of the electron orbits in a hydrogen atom?

PROBLEMS

7.65. Calculate the wavelengths of the following objects:
  a. A muon (a subatomic particle with a mass of 1.884×10^{-25} g) traveling at 325 m/s
  b. Electrons ($m_e = 9.10939 \times 10^{-28}$ g) moving at 4.05×10^6 m/s in an electron microscope
  c. An 80 kg athlete running a 4-minute mile
  d. Earth (mass $= 6.0 \times 10^{27}$ g) moving through space at 3.0×10^4 m/s

7.66. Two objects are moving at the same speed. Which (if any) of the following statements about them are true?
  a. The de Broglie wavelength of the heavier object is longer than that of the lighter one.
  b. If one object has twice as much mass as the other, its wavelength is one-half the wavelength of the other.
  c. Doubling the speed of one of the objects will have the same effect on its wavelength as doubling its mass.

7.67. Which (if any) of the following statements about the frequency of a particle is true?
  a. Heavy, fast-moving objects have lower frequencies than those of lighter, faster-moving objects.
  b. Only very light particles can have high frequencies.
  c. Doubling the mass of an object and halving its speed result in no change in its frequency.

7.68. How rapidly would each of the following particles be moving if they all had the same wavelength as a photon of red light ($\lambda = 750$ nm)?
  a. An electron of mass 9.10939×10^{-28} g
  b. A proton of mass 1.67262×10^{-24} g
  c. A neutron of mass 1.67493×10^{-24} g
  d. An α particle of mass 6.64×10^{-24} g

7.69. Particles in a Cyclotron The first cyclotron was built in 1930 at the University of California, Berkeley, and was used to accelerate molecular ions of hydrogen, H_2^+, to a velocity of 4×10^6 m/s. (Modern cyclotrons can accelerate particles to nearly the speed of light.) If the uncertainty in the velocity of the H_2^+ ion was 3%, what was the uncertainty of its position?

7.70. Radiation Therapy An effective treatment for some cancerous tumors involves irradiation with "fast" neutrons. The neutrons from one treatment source have an average velocity of 3.1×10^7 m/s. If the velocities of individual neutrons are known to within 2% of this value, what is the uncertainty in the position of one of them?

Quantum Numbers and Electron Spin; The Sizes and Shapes of Atomic Orbitals

CONCEPT REVIEW

7.71. How does the concept of an orbit in the Bohr model of the hydrogen atom differ from the concept of an orbital in quantum theory?

7.72. What properties of an orbital are defined by each of the three quantum numbers n, ℓ, and m_ℓ?

7.73. How many quantum numbers are needed to identify an orbital?

7.74. How many quantum numbers are needed to identify an electron in an atom?

PROBLEMS

7.75. How many orbitals are there in an atom with each of the following principal quantum numbers? (a) 1; (b) 2; (c) 3; (d) 4; (e) 5

7.76. How many orbitals are there in an atom with the following combinations of quantum numbers?
a. $n = 3, \ell = 2$ b. $n = 3, \ell = 1$ c. $n = 4, \ell = 2, m_\ell = 2$

7.77. What are the possible values of quantum number ℓ when $n = 4$?

7.78. Which are the possible values of m_ℓ when $\ell = 2$?

7.79. What set of orbitals corresponds to each of the following sets of quantum numbers?
a. $n = 2, \ell = 0$ c. $n = 4, \ell = 2$
b. $n = 3, \ell = 1$ d. $n = 1, \ell = 0$

7.80. What set of orbitals corresponds to each of the following sets of quantum numbers?
a. $n = 2, \ell = 1$ c. $n = 3, \ell = 2$
b. $n = 5, \ell = 3$ d. $n = 4, \ell = 3$

7.81. How many electrons could occupy orbitals with the following quantum numbers?
a. $n = 2, \ell = 0$ c. $n = 4, \ell = 2$
b. $n = 3, \ell = 1, m_\ell = 0$ d. $n = 1, \ell = 0, m_\ell = 0$

7.82. How many electrons could occupy an orbital with the following quantum numbers?
a. $n = 3, \ell = 2$ c. $n = 3, \ell = 0$
b. $n = 5, \ell = 4$ d. $n = 4, \ell = 1, m_\ell = 1$

7.83. Which of the following combinations of quantum numbers are allowed?
a. $n = 1, \ell = 1, m_\ell = 0, m_s = +\frac{1}{2}$
b. $n = 3, \ell = 0, m_\ell = 0, m_s = -\frac{1}{2}$
c. $n = 1, \ell = 0, m_\ell = 1, m_s = -\frac{1}{2}$
d. $n = 2, \ell = 1, m_\ell = 2, m_s = +\frac{1}{2}$

7.84. Which of the following combinations of quantum numbers are allowed?
a. $n = 3, \ell = 2, m_\ell = 0, m_s = -\frac{1}{2}$
b. $n = 5, \ell = 4, m_\ell = 4, m_s = +\frac{1}{2}$
c. $n = 3, \ell = 0, m_\ell = 1, m_s = +\frac{1}{2}$
d. $n = 4, \ell = 4, m_\ell = 1, m_s = -\frac{1}{2}$

The Periodic Table and Filling the Orbitals of Multielectron Atoms; Electron Configurations of Ions

CONCEPT REVIEW

7.85. What is meant when two or more orbitals are said to be degenerate?

7.86. Explain how the electron configurations of the group 2 elements are linked to their location in the periodic table developed by Mendeleev (see Figure 2.10).

7.87. How do we know from examining the periodic table's structure that the $4s$ orbital is filled before the $3d$ orbital?

7.88. Explain why so many transition metals form ions with a 2+ charge.

PROBLEMS

7.89. List the following orbitals in order of increasing energy in a multielectron atom:
a. $n = 3, \ell = 2$ c. $n = 3, \ell = 0$
b. $n = 5, \ell = 4$ d. $n = 4, \ell = 1, m_\ell = -1$

7.90. Place the following orbitals in order of increasing energy in a multielectron atom:
a. $n = 2, \ell = 1$ c. $n = 3, \ell = 2$
b. $n = 5, \ell = 3$ d. $n = 4, \ell = 3$

7.91. What are the electron configurations of Li, Li^+, Ca, F^-, Na^+, Mg^{2+}, and Al^{3+}?

7.92. Which species listed in Problem 7.91 are isoelectronic with Ne?

7.93. What are the condensed electron configurations of K, K^+, S^{2-}, N, Ba, Ti^{4+}, and Al?

7.94. In what way are the electron configurations of H, Li, Na, K, Rb, and Cs similar?

7.95. Write the electron configurations of the following species: Na, Cl, Mn, and Mn^{2+}.

7.96. Write the electron configurations of the following species: C, S, Ti, and Ti^{4+}.

7.97. How many unpaired electrons are there in the following ground-state atoms and ions? (a) N; (b) O; (c) P^{3-}; (d) Na^+

7.98. How many unpaired electrons are there in the following ground-state atoms and ions? (a) Sc; (b) Ag^+; (c) Cd^{2+}; (d) Zr^{4+}

7.99. Identify the atom whose electron configuration is $[Ar]3d^24s^2$. How many unpaired electrons are there in the ground state of this atom?

7.100. Identify the atom whose electron configuration is $[Ne]3s^23p^3$. How many unpaired electrons are there in the ground state of this atom?

7.101. Which monatomic ion has a charge of $1-$ and the electron configuration $[Ne]3s^23p^6$? How many unpaired electrons are there in the ground state of this ion?

7.102. Which monatomic ion has a charge of $1+$ and the electron configuration $[Kr]4d^{10}5s^2$? How many unpaired electrons are there in the ground state of this ion?

7.103. Predict the charge of the monatomic ions formed by Al, N, Mg, and Cs.

7.104. Predict the charge of the monatomic ions formed by S, P, Zn, and I.

7.105. Which of the following electron configurations represent an excited state?
a. $[He]2s^12p^5$ c. $[Ar]3d^{10}4s^24p^5$
b. $[Kr]4d^{10}5s^25p^1$ d. $[Ne]3s^23p^24s^1$

7.106. Which of the following electron configurations represent an excited state?
a. $[Ne]3s^23p^1$ c. $[Kr]4d^{10}5s^15p^1$
b. $[Ar]3d^{10}4s^14p^2$ d. $[Ne]3s^23p^64s^1$

7.107. In which subshell are the highest-energy electrons in a ground-state atom of the isotope ^{131}I? Are the electron configurations of ^{131}I and ^{127}I the same?

7.108. Although no currently known elements contain electrons in g orbitals, such elements may be synthesized some day. What is the minimum atomic number of an element whose ground-state atoms have an electron in a g orbital?

The Sizes of Atoms and Ions

CONCEPT REVIEW

7.109. Sodium atoms are much larger than chlorine atoms, but in NaCl sodium ions are much smaller than chloride ions. Why?

7.110. Why does atomic size tend to decrease with increasing atomic number across a row of the periodic table?

7.111. Which of the following group 1 elements has the largest atoms? Li, Na, K, Rb. Explain your selection.

7.112. Which of the following group 17 elements has the largest monatomic ions? F, Cl, Br, I. Explain your selection.

Ionization Energies

CONCEPT REVIEW

7.113. How do ionization energies change with increasing atomic number (a) down a group of elements in the periodic table, and (b) from left to right across a period of elements?

7.114. The ionization energies of the main group elements are given in Figure 7.1. Explain the differences in ionization energy between the following pairs of elements: (a) He and Li; (b) Li and Be; (c) Be and B; (d) N and O.

7.115. Explain why it is more difficult to ionize a fluorine atom than a boron atom.

7.116. Do you expect the ionization energies of anions of group 17 elements to be lower or higher than for neutral atoms of the same group?

7.117. Which of the following elements should have the smallest *second* ionization energy? Br, Kr, Rb, Sr, Y

7.118. Why is the first ionization energy (IE_1) of Al ($Z = 13$) less than the IE_1 of Mg ($Z = 12$) *and* less than the IE_1 of Si ($Z = 14$)?

Additional Problems

7.119. Interstellar Hydrogen Astronomers have detected hydrogen atoms in interstellar space in the $n = 732$ excited state. Suppose an atom in this excited state undergoes a transition from $n = 732$ to $n = 731$.
a. How much energy does the atom lose as a result of this transition?
b. What is the wavelength of radiation corresponding to this transition?
c. What kind of telescope would astronomers need in order to detect radiation of this wavelength? (*Hint:* it would not be one designed to capture visible light.)

*7.120. When an atom absorbs an X-ray of sufficient energy, one of its 2s electrons may be emitted, creating a hole that can be spontaneously filled when an electron in a higher-energy orbital—a 2p, for example—falls into it. A photon of electromagnetic radiation with an energy that matches the energy lost in the $2p \rightarrow 2s$ transition is emitted. Predict how the wavelengths of $2p \rightarrow 2s$ photons would differ between (a) different elements in the fourth row of the periodic table, and (b) different elements in the same column (for example, between the noble gases from Ne to Rn).

*7.121. Two helium ions (He^+) in the $n = 3$ excited state emit photons of radiation as they return to the ground state. One ion does so in a single transition from $n = 3$ to $n = 1$. The other does so in two steps: $n = 3$ to $n = 2$ and then $n = 2$ to $n = 1$. Which of the following statements about these two pathways is true?
a. The sum of the energies lost in the two-step process is same as the energy lost in the single transition from $n = 3$ to $n = 1$.
b. The sum of the wavelengths of the two photons emitted in the two-step process is equal to the wavelength of the single photon emitted in the transition from $n = 3$ to $n = 1$.
c. The sum of the frequencies of the two photons emitted in the two-step process is equal to the frequency of the single photon emitted in the transition from $n = 3$ to $n = 1$.
d. The wavelength of the photon emitted by the He^+ ion in the $n = 3$ to $n = 1$ transition is shorter than the wavelength of a photon emitted by an H atom in an $n = 3$ to $n = 1$ transition.

*7.122. Use your knowledge of electron configurations to explain the following observations:

a. Silver tends to form ions with a charge of 1+, but the elements to the left and right of silver in the periodic table tend to form ions with 2+ charges.

b. The heavier group 13 elements (Ga, In, Tl) tend to form ions with charges of 1+ or 3+ but not 2+.

c. The heavier elements of group 14 (Sn, Pb) and group 4 (Ti, Zr, Hf) tend to form ions with charges of 2+ or 4+.

7.123. Trends in ionization energies of the elements as a function of the position of the elements in the periodic table are a useful test of our understanding of electronic structure.

a. Should the same trend in the first ionization energies for elements with atomic numbers $Z = 31$ through $Z = 36$ be observed for the second ionization energies of the same elements? Explain why or why not.

b. Which element should have the greater second ionization energy: Rb ($Z = 37$) or Kr ($Z = 36$)? Why?

7.124. **Chemistry of "Photo-Gray" Glasses** "Photo-gray" lenses for eyeglasses darken in bright sunshine because the lenses contain tiny, transparent AgCl crystals. Exposure to light removes electrons from Cl^- ions forming a chlorine atom in an excited state (indicated below by the asterisk):

$$Cl^- + h\nu \rightarrow Cl^* + e^-$$

The electrons are transferred to Ag^+ ions, forming silver metal:

$$Ag^+ + e^- \rightarrow Ag$$

Silver metal is reflective, giving rise to the "photo-gray" color.

a. Write abbreviated electron configurations of Cl^-, Cl, Ag, and Ag^+.

b. What do we mean by the term *excited state*?

c. Would more energy be needed to remove an electron from a Br^- ion or from a Cl^- ion? Explain your answer.

*d. How might substitution of AgBr for AgCl affect the light sensitivity of photo-gray lenses?

7.125. Tin (in group 14) forms both Sn^{2+} and Sn^{4+} ions, but magnesium (in group 2) forms only Mg^{2+} ions.

a. Write abbreviated ground-state electron configurations for the ions Sn^{2+}, Sn^{4+}, and Mg^{2+}.

b. Which neutral atoms have ground-state electron configurations identical to Sn^{2+} and Mg^{2+}?

c. Which 2+ ion is isoelectronic with Sn^{4+}?

7.126. **Oxygen Ions in Space** The *Far Ultraviolet Spectroscopic Explorer (FUSE)* satellite has been analyzing the spectra of emission sources within the Milky Way. Among the satellite's findings are interplanetary clouds containing oxygen atoms that have lost five electrons.

a. Write an electron configuration for these highly ionized oxygen atoms.

b. Which electrons have been removed from the neutral atoms?

c. The ionization energies corresponding to removal of the third, fourth, and fifth electrons are 4581 kJ/mol, 7465 kJ/mol, and 9391 kJ/mol, respectively. Explain why removal of each additional electron requires more energy than removal of the previous one.

d. What is the maximum wavelength of radiation that will remove the fifth electron from O^{4+}?

*7.127. Effective nuclear charge (Z_{eff}) is related to atomic number (Z) by a parameter called the shielding parameter (σ) according to the equation $Z_{eff} = Z - \sigma$.

a. Calculate Z_{eff} for the outermost s electrons of Ne and Ar given $\sigma = 4.24$ (for Ne) and 11.24 (for Ar).

b. Explain why the shielding parameter is much greater for Ar than for Ne.

7.128. **Fog Lamp Technology** Sodium fog lamps and street lamps contain gas-phase sodium atoms and sodium ions. Sodium atoms emit yellow-orange light at 589 nm. Do sodium ions emit the same yellow-orange light? Explain why or why not.

7.129. How can an electron get from the (+) lobe of a p orbital to the (−) lobe without going through the node between the lobes?

7.130. Einstein did not fully accept the uncertainty principle, remarking that "He [God] does not play dice." What do you think Einstein meant? Niels Bohr allegedly responded by saying, "Albert, stop telling God what to do." What do you think Bohr meant?

A Noble Family: Special Status for Special Behavior

*7.131. Compounds have been made from xenon and krypton with fluorine and oxygen, but not from the other noble gases. Suggest a reason, based on periodic properties, why xenon and krypton can be forced to form compounds but helium, neon, and argon cannot.

7.132. The successive ionization energies of the noble gases increase uniformly, in comparison to the successive ionization energies of the other elements, which at some place in the sequence show a rather large jump between values. Why is this the case?

*7.133. Helium is the only element that was discovered extraterrestrially before it was found on Earth. Suggest a reason for this.

7.134. The first ionization energy of the noble gas neon is 2081 kJ/mol. The sodium ion (Na^+) is isoelectronic with neon, but the energy required to take one electron from the sodium ion is 4560 kJ/mol. Why is it so much harder to remove an electron from a sodium ion than from a neon atom if the two are isoelectronic?

Chemical Bonding and Atmospheric Molecules

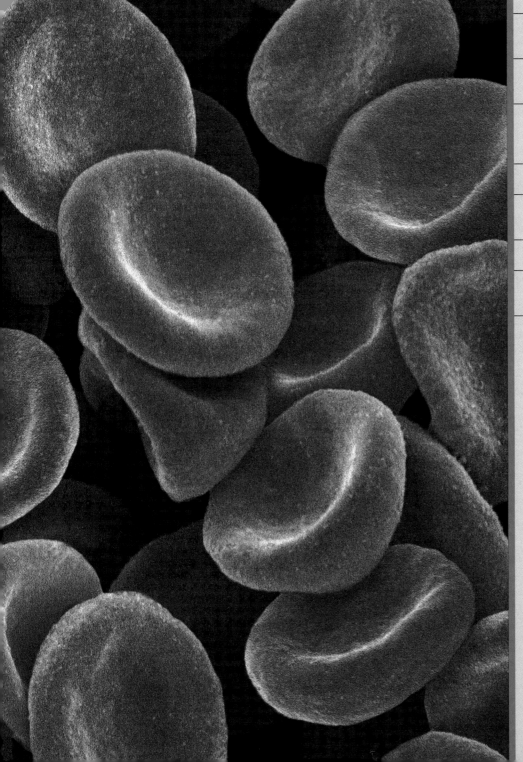

HEALTHY RED BLOOD CELLS
Red blood cells contain hemoglobin, which binds oxygen and transports it from the lungs to tissue.

A LOOK AHEAD: Take a Deep Breath

The air we breathe is a mixture of many different substances, but nitrogen and oxygen are dominant. Every breath we inhale consists of about 78% nitrogen (N_2) and 21% oxygen (O_2). Both are diatomic molecules, similar in size but very different in behavior in the living system. Hemoglobin, a large iron-containing molecule in our blood, captures oxygen from the air we inhale and transports it to cells, where it reacts in the highly controlled oxidative processes of metabolism. The iron ion in hemoglobin binds oxygen and ignores nitrogen. Nitrogen is not reactive in our bodies. Nitrogen is the bulk of what we inhale, and it is exhaled unchanged.

If carbon monoxide (CO), which is very similar in size to N_2 and O_2, is present in the air we breathe at a level as low as 1.2%, it can cause immediate unconsciousness and death in 1 to 3 minutes. Carbon monoxide can be produced whenever incomplete combustion of fossil fuels occurs. It is the most common toxic gas generated indoors worldwide, emitted from heaters, fireplaces, and cars idling in garages. CO kills by asphyxiation: It diminishes the capacity of blood to carry oxygen by attaching itself to the iron in hemoglobin; it does this over 200 times more readily than oxygen does, thereby excluding oxygen from the transport system. Blood containing CO-bound hemoglobin cannot deliver oxygen to cells, which then suffocate and die. In addition, CO causes damage through its effect on other small molecules besides O_2, notably nitric oxide (NO).

Nitric oxide (NO) is another molecule similar in size to N_2, O_2, and CO. It has both beneficial and lethal effects on the body. Small quantities of nitric oxide that are essential for life are produced in the body in a highly controlled series of reactions. This NO is bound at one specific site on hemoglobin for transport to blood vessels when they must dilate (widen) to enhance blood flow to regions of the body. If NO is not highly controlled and we are exposed to free nitric oxide, for example by inhaling it, it is extremely dangerous to life. Like carbon monoxide and oxygen, NO attaches to the iron in hemoglobin. Like carbon monoxide, it stops oxygen transport in the body, so it can also kill by asphyxiation. But unlike CO, whose toxicity is associated with binding to hemoglobin, NO not bound to hemoglobin or other molecules that control its behavior is also extremely damaging to brain cells and other tissues. This leads to another avenue of toxicity by CO. Carbon monoxide can push essential NO off its normal binding site on hemoglobin and set it free in the body. Therefore the presence of CO causes the concentration of free NO in tissues to rise and produces lethal damage.

These four gases—N_2, O_2, CO, and NO—all contain two small atoms. Why does their behavior in our bodies differ so widely? Why is nitrogen inert whereas oxygen is required in relatively large amounts to maintain life? Why is nitric oxide essential in very small amounts but otherwise deadly, while carbon monoxide is toxic because it interferes with the normal management of O_2 and NO? The answers to all of these questions lie in the nature of the distribution of electrons in molecules, which depends in large part on the connections between the atoms. Shape and electron distribution at the molecular level are crucial in determining properties at the macroscopic level. In this chapter we begin the discussion of where electrons are found in molecules and ions, and we take the first steps toward answering questions concerning the origin of the different chemical behavior of these small gaseous molecules.

TABLE 8.1	Average Concentrations of Major and Minor Atmospheric Constituents		
Major Constituents			**Volume (%)**
Nitrogen			78.08
Oxygen			20.95
Argon			0.934
Water			Variable
Minor Constituents			**ppm**
Carbon dioxide			375
Methane			2
Nitrous oxide			0.3
Ozone			0.03
Nitrogen dioxide			0.025
Nitric oxide			0.025
Hydrogen			0.5
Sulfur dioxide			0.001

8.1 Ionic, Covalent, and Metallic Bonds

Earth's atmosphere contains a variety of gases whose molecules have two or more atoms. Some of them are listed in Table 8.1. Most are present in trace concentrations, and some are compounds considered to be pollutants. Traveling upward in the atmosphere toward outer space, we find that the composition changes in favor of helium, nitrogen, and oxygen atoms and ions, not molecules (Figure 8.1). The difference between atoms of oxygen and molecules of oxygen (O_2) is the presence of a **chemical bond** in molecular oxygen. As we learned in Chapter 3, in the most general terms a chemical bond results from two or more atoms sharing electrons or two ions of opposite charge attracting each other (Figure 8.2).

Bonds that result from the sharing of a pair of electrons between atoms are called **covalent bonds**. Covalent bonds form because sharing negative electrons by more than one positively charged nucleus is energetically favorable. Not just the gases listed on Table 8.1 but *all* molecules and polyatomic ions formed by sharing one or more pairs of electrons are held together by covalent bonds. Covalent bonds typically form between atoms of nonmetallic elements.

When a metal binds with a nonmetal, electrons are transferred from the metal atom, making it a cation, to the nonmetal atom, which now is an anion. Metals have low ionization energies and tend to lose electrons easily; they do not gain electrons easily. The electrostatic attraction of the cation for the anion results in the second general type of bond: an **ionic bond**. The resulting compound is called an ionic compound.

The third type of bond results when a metal binds to another metal in the solid state. A **metallic bond** results when metal atoms in a solid sample pool their electrons, producing a "sea" of electrons. A given electron is not associated with one nucleus but is shared by all nuclei in the sample. The shared electrons are mobile, which gives metals their characteristic properties of electrical and thermal conductivity.

The attractive forces between atoms in molecules result in the formation of **chemical bonds**.

A **covalent bond** is formed when two atoms share a pair of electrons.

An **ionic bond** results from the electrostatic attraction of a cation for an anion.

A **metallic bond** consists of the nuclei of metal atoms surrounded by a "sea" of shared electrons.

CONNECTION In Chapter 2 we described covalent bonds between atoms giving rise to molecular compounds, which are described by molecular formulas. Remember that molecular formulas reflect the actual numbers of atoms in a molecule.

CONNECTION In Chapter 2 we described ionic compounds in terms of their empirical formulas, which give the lowest whole-number ratio of cations to anions in the substance.

▶❙❙ **CHEMTOUR** Bonding

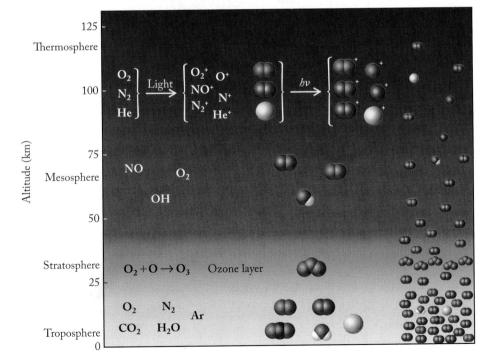

FIGURE 8.1 The species present in the atmosphere and the reactions they undergo vary with altitude. Radiation from the sun dissociates atmospheric molecules and ionizes both atoms and molecules. At ground level the major species are molecular. Ozone is present in the stratosphere, where it absorbs UV radiation from the sun that would otherwise be harmful to life on Earth. Above the stratosphere, radiation from the sun provides molecules with the energy to react to form ions, and ion–molecule reactions take place.

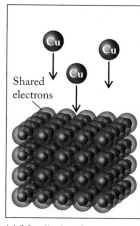

(a) Covalent bonding

(b) Ionic bonding

(c) Metallic bonding

FIGURE 8.2 Three general kinds of bonding take place between atoms: (a) covalent bonding between atoms of nonmetals (bromine); (b) ionic bonding between a nonmetal (chlorine) and a metal (sodium); and (c) metallic bonding between metal atoms (copper) in the solid phase. Each type is distinguished by the behavior of the valence electrons in the atoms in forming the bond.

In this chapter we concentrate on covalent and ionic bonding. Metallic bonding will be discussed in greater detail when we explore modern materials in Chapter 11.

8.2 An Introduction to Lewis Theory

In 1916, American chemist Gilbert N. Lewis (1875–1946) proposed that atoms form bonds by sharing their electrons. Through this sharing process, each atom acquires enough electrons to mimic the electron configuration of a noble gas. For the atoms in many compounds, Lewis's idea translates to atoms sharing, gaining, or losing electrons such that each atom achieves a set of 8 valence electrons called an **octet**. Because covalent bonds involve pairs of electrons, each atom with an octet has four electron pairs in its valence shell. Today we call this guideline the **octet rule** and associate it with filled *s* and *p* orbitals of an atom's outermost shell; such an electron configuration is isoelectronic with a noble gas. Lewis's theory still dominates our view of the formation of covalent bonds among atoms of the main group elements. Hydrogen is an exception to the octet rule. Its outer shell is the 1*s* orbital, which has a maximum occupancy of 2 electrons. Consequently, hydrogen atoms achieve a **duet** of electrons when their outermost shell is full.

Lewis Structures for Molecules with Single Bonds

The structures for molecules in large part determine their physical and chemical properties. **Lewis structures** are flat, two-dimensional representations of molecules that are useful for providing an initial picture of the most fundamental aspect of structure—the *connections* between atoms. Because covalent bonds are pairs of valence electrons shared between two atoms, Lewis structures focus on

⊙⊙ CONNECTION In Chapter 2 we defined groups 1 and 2 and 13 through 18 on the periodic table as the main group elements. In Chapter 7 we observed that an atom in the main group with filled *s* and *p* orbitals in its outermost shell is isoelectronic with a noble gas.

An **octet** is a set of 8 electrons in the outermost (valence) shell of an atom.

The **octet rule** states that atoms of main group elements make bonds by gaining, losing, or sharing electrons to achieve an outer shell containing 8 electrons, or four electron pairs.

The outermost shell of a hydrogen atom ($n = 1$) can contain a maximum of 2 electrons called a **duet**.

A **Lewis structure** is a two-dimensional representation of a molecule that provides a view of the connections between its atoms; valence electrons are depicted as dots around the atomic symbol.

1								18
·H								:He
	2		13	14	15	16	17	
·Li	·Be·		·B·	·Ċ·	·N̈·	·Ö·	:Ḟ·	:N̈e:
·Na	·Mg·		·Äl·	·S̈i·	·P̈·	·S̈·	:C̈l·	:Är:
·K	·Ca·		·Ga·	·Ge·	·As·	·Se·	:Br·	:Kr:
·Rb	·Sr·		·In·	·Sn·	·Sb·	·Te·	:I·	:Xe:
·Cs	·Ba·		·Tl·	·Pb·	·Bi·	·Po·	:At·	:Rn:
·Fr	·Ra·							

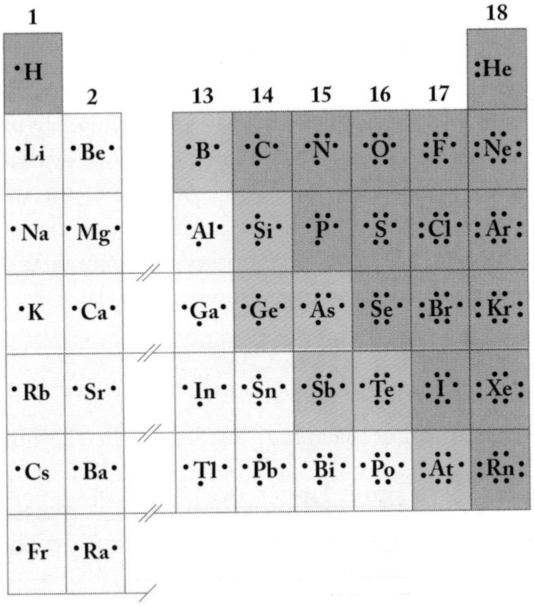

FIGURE 8.3 Lewis symbols for the main group elements of the periodic table. Because elements in a family have the same outer-shell electron configuration, they have the same Lewis symbol. The dots represent the valence electrons.

A **Lewis symbol** is the chemical symbol for an atom surrounded by one or more dots representing the valence electrons.

A **single bond** results when two atoms share one pair of electrons.

electron pairs, which are represented by pairs of dots around an atomic symbol. A Lewis structure helps us consider where the electrons are in a molecule by showing how the valence electrons are distributed among its atoms. Lewis structures are sometimes sufficient for understanding the structure of a molecule; frequently, they are just the starting point for more extensive analyses of structure.

The starting point for drawing the Lewis structure for a molecule is to draw Lewis symbols for the atoms involved, in which electrons are portrayed using dots. The **Lewis symbol** for an atom is the atomic symbol surrounded by the number of dots equal to the number of valence electrons in the neutral atom. The dots are placed around the symbol one at a time on the four sides of the symbol (top, bottom, right, and left). The order in which they are placed does not matter as long as one dot is placed on each side before any dots are paired. The number of unpaired dots indicates the capacity of the atom to form bonds. Figure 8.3 shows the Lewis symbols for the main group elements. Because the valence-electron configuration of each element in a group is the same, all atoms in a group have the same Lewis symbol.

When Lewis structures are drawn for molecules, an electron pair is symbolized like this:

$$:$$

so the Lewis structure for a hydrogen molecule containing two hydrogen atoms looks like this:

H:H

Both hydrogen atoms in the H_2 molecule have achieved a duet by sharing:

(H:H)

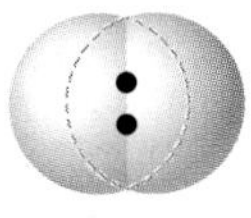

The circles enclose the electrons associated with each atom. They overlap to indicate that electrons in bonds are shared by two atoms.

The second convention in standard notation is to use a line (—) for a **single bond** involving the sharing of two electrons, so the hydrogen molecule is also symbolized in a line-bond formula as

H—H

Hydrogen

 CHEMTOUR Lewis Dot Structures

A **lone pair,** or unshared pair, of electrons is a pair of electrons that is not shared.

A **bonding pair** of electrons is a pair of electrons shared between two atoms.

Starting with these two conventions for constructing Lewis structures, let's develop a stepwise approach to drawing Lewis structures by looking at three covalent molecules: $HCl(g)$, aqueous solutions of which are called hydrochloric acid; $CH_4(g)$, a major hydrocarbon fuel; and $NH_3(g)$, widely used as a fertilizer. Essentially, we consider the number of valence electrons available and distribute them according to a set of guidelines. Notice our use of the word *guideline*. The statements are suggestions, not rules. We will modify and extend them as we deal with larger molecules with more diverse options for arrangements. Here are the first two guidelines arising directly from Lewis's model:

1. Hydrogen tends to form one single bond containing 2 electrons.
2. Other elements tend to form bonds to achieve an octet of electrons.

Using a system of electron bookkeeping, we determine the number of electrons each element brings to the molecule and compare that to the number of electrons each atom needs to complete its outermost shell. The difference between what an atom has and what it needs is related to the number of covalent bonds it must form to achieve a complete outer shell.

EXAMPLE A: HCl (HYDROGEN CHLORIDE)

a. *Initial electron bookkeeping:* Look at the number of valence electrons each atom has and compare that with the number each atom needs to have either a duet (hydrogen) or octet (all others) of electrons. For HCl, we use this step to define the number of covalent bonds in the molecule:

$$\text{Need:} \quad (1\,H \times 2\,e^-) + (1\,Cl \times 8\,e^-) = 10\,e^-$$

$$\text{Have:} \quad (1\,H \times 1\,e^-) + (1\,Cl \times 7\,e^-) = 8\,e^-$$

The difference between what the atoms have and what Lewis theory suggests they need is achieved by sharing 2 electrons, which corresponds to one covalent bond:

$$\text{Difference:} \quad (10\,e^- - 8\,e^-) = 2\,e^-; \quad \frac{2\,e^-}{2\,e^-/\text{bond}} = 1 \text{ bond}$$

Putting those two electrons between the hydrogen and the chlorine makes 1 covalent bond:

$$\text{H:Cl} \quad \text{or} \quad \text{H—Cl}$$

b. We have 6 electrons left over that are not involved in bonding. These leftover electrons are distributed around the atoms *in pairs* to satisfy the octet (or duet) rule. Hydrogen already has its duet through sharing. The leftover electrons are given to chlorine to complete its octet:

$$\text{H:}\ddot{\text{C}}\text{l:} \quad \text{or} \quad \text{H—}\ddot{\text{C}}\text{l:}$$

c. *Final electron bookkeeping:* We have used all 8 electrons. The chlorine atom has achieved an octet and the hydrogen atom a duet by sharing two electrons in a covalent bond:

Hydrogen chloride

The electron pairs on chlorine that are not involved in bonding are called **lone pairs** (or *unshared pairs*) to distinguish them from **bonding pairs** that are shared by two atoms.

EXAMPLE B: CH₄ (METHANE)

Methane has more atoms than HCl, but we carry out exactly the same steps. In a polyatomic molecule or ion, we need to identify the "central" atom—the atom to which the others are bound. Another guideline is:

3. The first atom in a molecular formula is often the central atom.

a. *Initial electron bookkeeping:* To develop the Lewis structure for methane, we calculate the number of covalent bonds required. To accomplish this, we first total the number of valence electrons in all the atoms in the molecule:

$$(1\,C \times 4\,e^-) + (4\,H \times 1\,e^-) = 8\,e^-$$

Next we apply guidelines 1 and 2:

The carbon atom needs an octet: $1\,C = 1 \times 8\,e^- = 8\,e^-$
Each hydrogen atom needs a duet: $4\,H = 4 \times 2\,e^- = 8\,e^-$
_____
Total electrons wanted: $16\,e^-$

The atoms in the structure need a total of $16\,e^-$, but only $8\,e^-$ are available. The difference is: $16\,e^- - 8\,e^- = 8\,e^-$. That means the Lewis structure needs 4 covalent bonds:

$$\frac{8\,e^-}{2\,e^-/\text{bond}} = 4 \text{ bonds}$$

b. We use a pair of electrons to form a bond between each pair of bound atoms. How can we tell which atoms are bound? C comes first in the formula, so we assign it the position of central atom. Also remember the first two guidelines: H tends to form 1 bond to achieve a duet; other elements tend to share as many electrons as they need to achieve an octet. The carbon atom has 4 electrons; it tends to pick up 4 more electrons by sharing, so it tends to make 4 bonds:

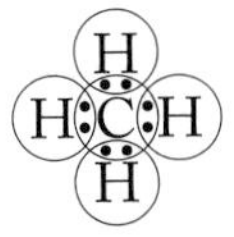

Hydrogens achieve duets;   Lewis structures 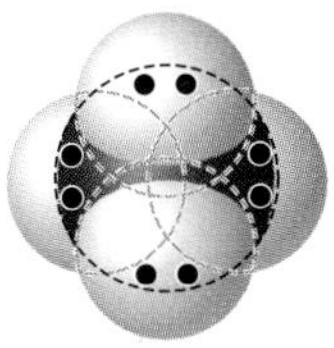   Methane
carbon achieves octet.

c. *Final electron bookkeeping:* The atoms contributed a total of 8 electrons; the structure contains 8 electrons (2 in each single bond). The carbon has achieved an octet; each hydrogen has a duet; therefore all the electrons have been used.

EXAMPLE C: NH₃ (AMMONIA)

a. *Initial electron bookkeeping:*

Need: $(1\,N \times 8\,e^-) + (3\,H \times 2\,e^-) = 14\,e^-$

Have: $(1\,N \times 5\,e^-) + (3\,H \times 1\,e^-) = 8\,e^-$
_____
Difference: $6\,e^- = 3 \text{ bonds}$

b. Three covalent bonds require $6\,e^-$. We have $2\,e^-$ left over. Each hydrogen atom has a duet, but the nitrogen atom only has $6\,e^-$ at this point:

$$\text{H:N:H}$$
$$\overset{..}{\text{H}}$$

Therefore the remaining $2\,e^-$ must be a lone pair on the nitrogen:

$$\text{H:}\overset{..}{\text{N}}\text{:H} \quad \text{or} \quad (\text{H:}\overset{..}{\text{N}}\text{:H}) \quad \text{or} \quad \text{H}-\overset{..}{\text{N}}-\text{H}$$
$$\overset{}{\text{H}} \qquad\qquad \overset{}{\text{H}} \qquad\qquad\quad \overset{|}{\text{H}}$$

Ammonia

c. *Final electron bookkeeping:* We started with 8 electrons and the final structure has 8 electrons. Nitrogen has an octet (three bonding pairs of electrons plus one lone pair); each hydrogen atom has a duet.

SAMPLE EXERCISE 8.1 **Drawing Lewis Symbols for Atoms**

Draw the Lewis symbol for argon, the third most abundant gas in the atmosphere.

COLLECT AND ORGANIZE We are asked to draw the Lewis symbol for argon, an element in group 18 of the periodic table.

ANALYZE We need to determine the number of valence electrons in an atom of argon and then arrange them as Lewis dots around the argon symbol (Ar).

SOLVE The condensed electron configuration of argon is $[\text{Ne}]3s^23p^6$, telling us that argon has 8 valence electrons. Arranging these 8 electrons around the symbol for an argon atom gives the result:

$$:\!\overset{..}{\underset{..}{\text{Ar}}}\!:$$

THINK ABOUT IT Argon is in group 18, the noble gases. It has a full outer shell and a complete octet of electrons.

Practice Exercise The interaction of chlorine-containing compounds with light in the atmosphere can lead to the formation of chlorine atoms. Draw the Lewis symbol for a chlorine atom.

(Answers to Practice Exercises are in the back of the book.)

SAMPLE EXERCISE 8.2 **Drawing Lewis Structures for Molecules**

Chloroform ($CHCl_3$) is a low-boiling liquid that was used as an early anesthetic in surgery. It does not interact with hemoglobin, but breathing its vapors induces unconsciousness. What is the Lewis structure for chloroform?

COLLECT AND ORGANIZE We need to locate each element in the chloroform molecule on the periodic table and determine the number of valence electrons it has. We can then use the step-by-step process outlined earlier in this section to determine the Lewis structure.

ANALYZE Carbon is written first in the formula, so we assume it is the central atom. This makes sense because there are four other atoms in chloroform (1 H atom and 3 Cl atoms) and carbon tends to make 4 bonds.

SOLVE We apply the conventions for Lewis structures and take the steps developed in Examples A through C to get the following:

a. *Initial electron bookkeeping:*

Need: $(1\,C \times 8\,e^-) + (1\,H \times 2\,e^-) + (3\,Cl \times 8\,e^-) = 34\,e^-$

Have: $(1\,C \times 4\,e^-) + (1\,H \times 1\,e^-) + (3\,Cl \times 7\,e^-) = 26\,e^-$

Difference: $\qquad\qquad\qquad\qquad\qquad\qquad\qquad 8\,e^- = 4\text{ bonds}$

b. If carbon is the central atom, then four pairs of electrons (4 covalent bonds) can be located about it. Each bond extends from the carbon to one of the other atoms:

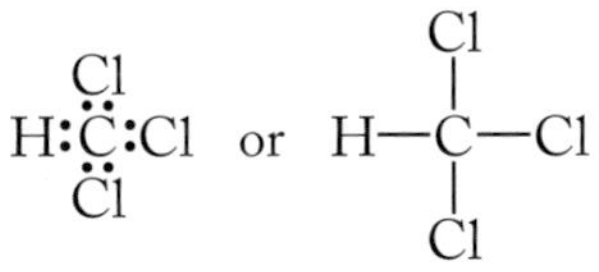

We have now used 8 electrons, leaving us $26\,e^- - 8\,e^- = 18\,e^-$ to accommodate. The hydrogen atom already has its duet of electrons, and the carbon has its octet, so we put the remaining electrons in pairs around the chlorine atoms to achieve octets for the three chlorines:

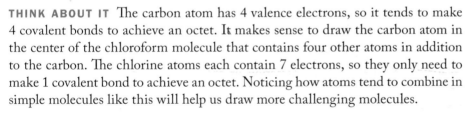

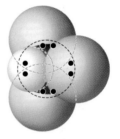

Chloroform

c. *Final electron bookkeeping:* The atoms in $CHCl_3$ contributed a total of 26 electrons to the structure for the molecule. The Lewis structure as drawn contains 26 electrons. The hydrogen atom has a duet of electrons; each carbon and chlorine atom has achieved an octet of electrons.

THINK ABOUT IT The carbon atom has 4 valence electrons, so it tends to make 4 covalent bonds to achieve an octet. It makes sense to draw the carbon atom in the center of the chloroform molecule that contains four other atoms in addition to the carbon. The chlorine atoms each contain 7 electrons, so they only need to make 1 covalent bond to achieve an octet. Noticing how atoms tend to combine in simple molecules like this will help us draw more challenging molecules.

Practice Exercise Hydrogen peroxide, H_2O_2, is used to disinfect cuts and to bleach hair. The H_2O_2 molecule contains an $O—O$ bond. Draw the Lewis structure for H_2O_2.

Lewis Structures for Molecules with Multiple Bonds

Since Chapter 2 we have shown structures for molecules such as acetic acid (CH_3COOH) and formaldehyde (CH_2O) that have a pair of lines connecting some of the atoms:

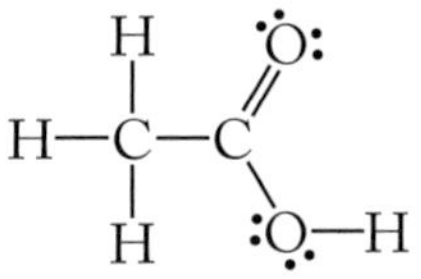

Acetic acid
(CH_3COOH)

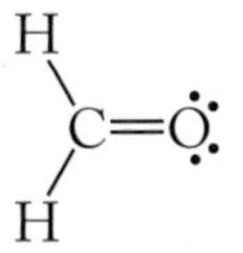

Formaldehyde
(CH_2O)

Lewis structures also describe molecules such as these with *multiple* bonds between pairs of atoms. When two atoms share two pairs of electrons (4 electrons, 2 covalent

A **double bond** results when two atoms share two pairs of electrons.

A **triple bond** results when two atoms share three pairs of electrons.

Bond length is the distance between the nuclear centers of two atoms joined together in a bond.

bonds), the bond is called a **double bond** ($=$); when two atoms share three pairs of electrons, the bond is called a **triple bond** ($\equiv$). In Section 8.7 we discuss the characteristics of these bonds in detail, but for now we just need to understand the concept of bond length. **Bond length** is the distance between the nuclear centers of two atoms that are joined together in a bond. If we compare single, double, and triple bonds between the same two atoms, the bond lengths decrease in the following order:

Single bond > double bond > triple bond

EXAMPLE D: O₂ (OXYGEN)

a. *Initial electron bookkeeping:*

$$\text{Need: } 2O \times 8e^- = 16e^-$$
$$\text{Have: } 2O \times 6e^- = 12e^-$$
$$\overline{\text{Difference:} \qquad\qquad 4e^- = 2 \text{ bonds}}$$

b. Two covalent bonds require that 4 electrons be shared between the two oxygen atoms in a double bond:

O::O   or   O=O

That leaves $12e^- - 4e^- = 8$ electrons (four electron pairs) to be arranged about the oxygen atoms to achieve an octet:

        :Ö=Ö:

c. *Final electron bookkeeping:* All 12 electrons have been used, and each oxygen atom has achieved an octet using two lone pairs of electrons and two bonding pairs.

Oxygen

Notice that we have drawn the lone pairs on the oxygen atoms at an angle of about 120° from each other and the bonding pairs. Electrons are negatively charged and repel each other, so we draw them as far apart from each other as possible. This logic will be an important part of drawing three-dimensional structures of molecules in Chapter 9.

EXAMPLE E: C₂H₄ (ETHYLENE)

a. *Initial electron bookkeeping:*

$$\text{Need: } (2C \times 8e^-) + (4H \times 2e^-) = 24e^-$$
$$\text{Have: } (2C \times 4e^-) + (4H \times 1e^-) = 12e^-$$
$$\overline{\text{Difference:} \qquad\qquad 12e^- = 6 \text{ bonds}}$$

b. The idea of a central atom breaks down a bit here. If we think about it, however, clearly we need a carbon–carbon bond; remember, hydrogen only makes 1 bond, so we cannot have a hydrogen atom in the middle of a molecule connecting two atoms. If we start with a carbon–carbon bond, we can then attach two hydrogen atoms to each carbon. This yields a structure with 5 bonds:

●●● **CONNECTION** In Chapter 5 we discussed many fossil fuels that consist of chains of carbon atoms.

H H   H   H
:C:C: C—C
H H   H   H

However, our formula says we need 6 bonds. As we saw with O$_2$ in Example D, multiple bonds are possible. Atoms joined by a single bond share 2 electrons; atoms joined by a double bond ($=$) share 4 electrons. The hydrogen atoms in the structure all have duets; the carbon atoms only have 6 electrons, so each could share the remaining 2 electrons as another bonding pair. That makes ethylene:

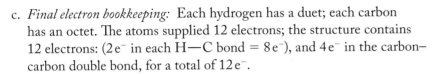

Ethylene

c. *Final electron bookkeeping:* Each hydrogen has a duet; each carbon has an octet. The atoms supplied 12 electrons; the structure contains 12 electrons: (2 e$^-$ in each H—C bond = 8 e$^-$), and 4 e$^-$ in the carbon–carbon double bond, for a total of 12 e$^-$.

This example illustrates the fourth guideline we use in drawing Lewis structures: carbon can bond with itself to form chains and rings.

SAMPLE EXERCISE 8.3   **Drawing Lewis Structures with Multiple Bonds**

Draw the Lewis structure for nitrogen gas (N$_2$).

COLLECT AND ORGANIZE Nitrogen is in group 15 in the periodic table. Each nitrogen atom has 5 valence electrons.

ANALYZE Because nitrogen has 5 valence electrons, we may expect it to make 3 covalent bonds to achieve an octet.

SOLVE

a. *Initial electron bookkeeping:*

Need: 2 N $\times$ 8 e$^-$ = 16 e$^-$

Have: 2 N $\times$ 5 e$^-$ = 10 e$^-$

Difference: 6 e$^-$ = 3 bonds

b. Three covalent bonds use 6 of the available electrons:

N:::N   N$\equiv$N

—and leave 4 electrons (two pairs) to be accommodated. Locating one lone pair on each nitrogen atom gives each nitrogen an octet:

:N:::N: or   :N:::N: or   :N$\equiv$N:

Nitrogen

c. *Final electron bookkeeping:* The structure contains 10 electrons; each nitrogen atom has an octet consisting of three bonding pairs and one lone pair of electrons.

THINK ABOUT IT As we predicted, nitrogen makes 3 covalent bonds. Nitrogen tends to make 3 bonds in molecules in which it binds to other atoms as well.

Practice Exercise Draw the Lewis structure for acetylene (C$_2$H$_2$), the hydrocarbon fuel used in oxyacetylene torches for welding.

Examples D and E and Sample Exercise 8.3 illustrate the fifth guideline we use in drawing Lewis structures: atoms may make multiple bonds by sharing more than one electron pair.

CONCEPT TEST

Look at the Lewis structures for oxygen (Example D) and nitrogen (Sample Exercise 8.3), and identify a feature that might contribute to the differences in their behavior with hemoglobin.

(Answers to Concept Tests are in the back of the book.)

In summary, the initial set of guidelines we use to draw Lewis structures for covalent molecules and polyatomic ions is:

1. Hydrogen atoms share electrons to achieve a duet.
2. Atoms of all other main group elements share electrons to achieve an octet.
3. The central atom is often the element that comes first in a molecular formula.
4. Carbon can bond with itself to make chains and rings.
5. Atoms may make multiple bonds by sharing more than one electron pair.

We will add to or modify these guidelines as necessary when we encounter larger molecules and experimental evidence for certain structural features in more challenging molecules later in the chapter.

Lewis Structures for Ionic Compounds

We can also use the octet rule and Lewis symbols to rationalize the composition of many common ionic compounds such as sodium chloride (NaCl). By giving away a valence electron, a sodium atom becomes the positively charged ion Na^+ and achieves the same electron configuration as Ne: $1s^2 2s^2 2p^6$. In other words, Na^+ has a filled outer shell and a complete octet. If a chlorine atom with the electron configuration $[Ne]3s^2 3p^5$ gains an electron to form Cl^-, it achieves a filled outer shell and is isoelectronic with the noble gas argon. We can illustrate this behavior using Lewis symbols as follows:

$$\text{Na}\cdot \ + \ \cdot \overset{\cdot\cdot}{\underset{\cdot\cdot}{\text{Cl}}}\!: \ \rightarrow \ \text{Na}^+ \left[:\overset{\cdot\cdot}{\underset{\cdot\cdot}{\text{Cl}}}\!: \right]^-$$

The square brackets around the chloride ion are used to emphasize that all 8 valence electrons are associated with the chloride ion. The compound is held together by the attractive force between two oppositely charged ions.

All alkali and alkaline earth elements tend to lose, rather than share, their valence electrons, in part because they have low ionization energies. However, the cations that they form still obey the octet rule because they have the electron configurations of the noble gases that precede the parent elements in the periodic table.

CONNECTION Remember the periodic trend in ionization energy from Chapter 7: ionization energies increase from left to right across a row in the periodic table.

SAMPLE EXERCISE 8.4 **Drawing Lewis Symbols for Monatomic Ions**

Draw Lewis symbols for the most common ions formed by (a) oxygen and (b) sulfur.

COLLECT AND ORGANIZE Oxygen and sulfur are in group 16 in the periodic table. They both have 6 valence electrons and have the same outer shell electron configuration: ns^2np^4.

ANALYZE Both oxygen and sulfur are nonmetals, so they tend to form anions. Because they have the same outer-shell configuration, they both acquire 2 electrons to fill their outer shells and form ions with a 2− charge.

SOLVE Adding 2 electrons to the Lewis symbol for both atoms yields the following:

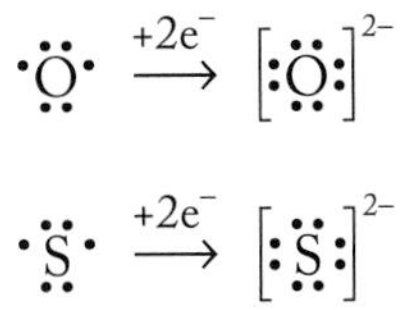

$$\ddot{\underset{\cdot\cdot}{\overset{\cdot\cdot}{O}}}\cdot \overset{+2e^-}{\longrightarrow} \left[:\ddot{\underset{\cdot\cdot}{O}}:\right]^{2-}$$

$$\cdot\ddot{\underset{\cdot\cdot}{S}}\cdot \overset{+2e^-}{\longrightarrow} \left[:\ddot{\underset{\cdot\cdot}{S}}:\right]^{2-}$$

THINK ABOUT IT The members of a group in the periodic table all have the same outer–shell configuration, so they all have the same number of electrons in their Lewis symbols for neutral atoms and for ions having the same charge.

Practice Exercise Draw the Lewis symbols for the monatomic anions formed by (a) bromine and (b) iodine.

CONCEPT TEST

Using X as the symbol for all nonmetallic elements in group 15 of the periodic table, write the Lewis symbol for the anion you predict is formed by all these elements. Using Y as the symbol for metallic elements in group 2 of the periodic table, write the Lewis symbol for the cation you predict is formed by all these elements.

8.3 Unequal Sharing, Electronegativity, and Other Periodic Properties

Ionic bonds result when an atom of a metal donates one or more electrons from its outer shell to an atom of a nonmetal. The electron is not shared; it is transferred from one atom to the other. In contrast, when two identical atoms, such as two hydrogen atoms, combine to form a molecule of H_2, the electrons in the covalent bond are shared equally by the two hydrogen atoms. The two hydrogen nuclei exert the same attractive force on both electrons, and the electrons are distributed evenly between the two atoms.

Electronegativity

Gilbert Lewis realized that electron sharing in many covalent bonds does not necessarily mean *equal* sharing. For example, he knew from the chemical properties of HCl that it is a **polar** molecule, which means that its hydrogen and chloride "ends" are electrically polarized in much the way that the ends of a bar magnet have north and south magnetic poles. He explained this *polarity*, or the separation of charge, by assuming that the shared electrons and their negative

A **polar** molecule contains bonds that have an uneven distribution of charge because electrons in the bonds are not shared equally by the two atoms.

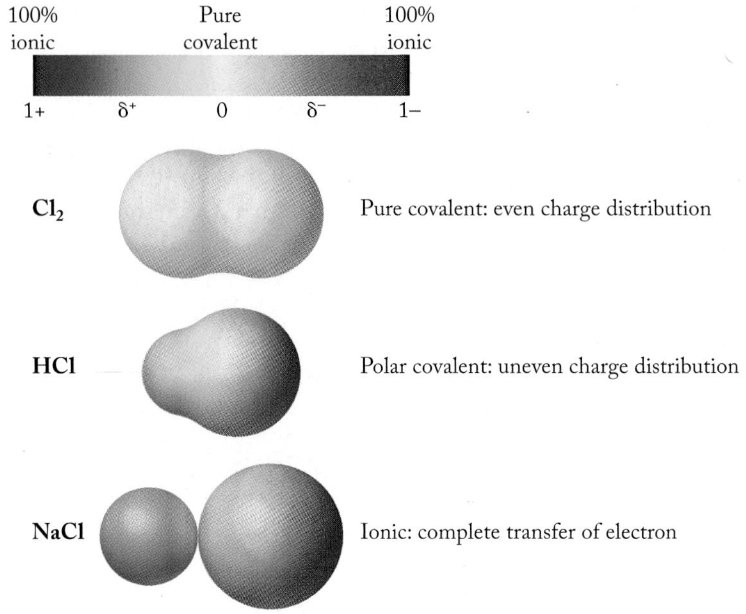

FIGURE 8.4 The electron density in molecules can be calculated. In the pure covalent bond in Cl_2, the electron density around both atoms is yellow, indicating that they are electrically neutral. Unequal sharing of electrons occurs in HCl. The more electronegative chlorine atom attracts more of the electrons, as shown by the larger electron cloud around Cl and the partial negative charge (δ^- yellow-orange). The H atom has a partial positive charge (δ^+ yellow-green). In the ionic compound NaCl, the blue sodium ion has a full 1+ charge and the dark red chloride ion a 1− charge.

charges are closer to the chlorine end of the molecule than the hydrogen end. This unequal distribution of electrons in the covalent bond makes the HCl molecule a *dipole* (from *di-*, "two": one pole is slightly negative, the other slightly positive). Figure 8.4 illustrates the range of sharing of electrons by comparing the electron density in the **nonpolar** molecule chlorine (Cl_2) with the polar molecule HCl and with the ionic compound NaCl. As a result of unequal sharing, the electrons in HCl are drawn more closely to the chlorine end of the molecule, which means that the H—Cl bond is a **polar covalent bond**. The complete transfer of an electron from a sodium atom to a chlorine atom results in nearly complete charge separation and an ionic bond.

Another American chemist, Linus Pauling (1901–1994), developed the concept of electronegativity to enable the prediction of the polarity of bonds. Pauling proposed an electronegativity scale (Figure 8.5) based on the idea that bonds between atoms of different elements are neither 100% covalent nor 100% ionic, but somewhere on a continuum between these two extremes. The location of a bond along this continuum depends on differences in the ability of the two atoms in a bond to attract electrons. Pauling's **electronegativity** scale provides a relative measure of the ability of an atom in a bond to attract shared electrons.

Electronegativity is a periodic property, which means that electronegativity is related to an element's position in the periodic table. Electronegativity values generally increase from left to right and decrease from top to bottom (Figure 8.5). Note that the most electronegative elements, including fluorine, oxygen, and nitrogen, are found in the upper right region of the table, and the least electronegative elements are the metals in the lower left.

Electronegativity values are dimensionless (have no units) because they are expressed on a relative scale. However, electronegativity is related to parameters that do have absolute values, such as the size of atoms (Figure 8.6) and the ionization energies of the elements (Figure 8.7). For the main group elements, there is an inverse relation between atomic size and electronegativity: the smaller the atom, the greater its electronegativity. Within a period (row), elements on the right-hand side have more protons in their nuclei than do those on the left, but their valence electrons are in the same shell as those on the left. Therefore the attractive force experienced by the valence electrons is greater in atoms of ele-

A **nonpolar** molecule contains bonds that have an even distribution of charge; electrons in the bonds are shared equally by the two atoms. Pure covalent bonds give rise to nonpolar diatomic molecules.

A **polar covalent bond** results from unequal sharing of bonding pairs of electrons between atoms.

Electronegativity is a relative measure of the ability of an atom in a bond to attract electrons to itself.

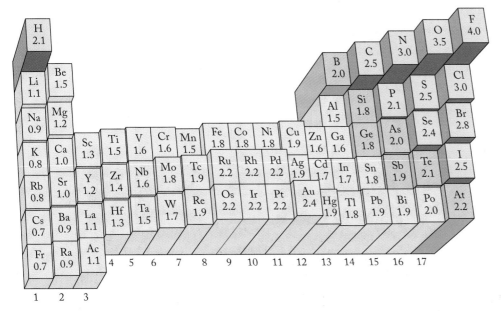

FIGURE 8.5 Electronegativity increases from left to right within rows of the periodic table and decreases from top to bottom within families of elements. The greater the electronegativity value, the greater that atom's ability to attract electrons in a bond. The Pauling electronegativity values are unitless numbers that define the relative ability of an atom in a bond to attract electrons toward it.

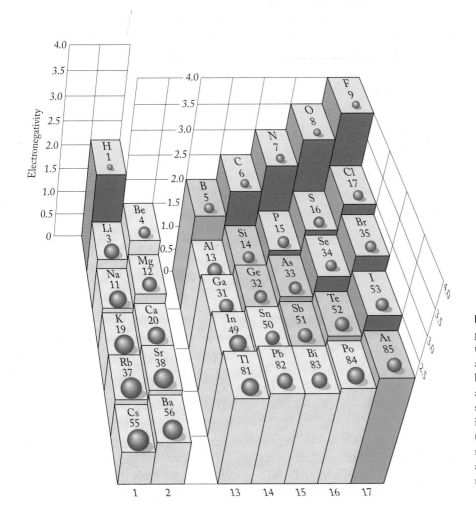

FIGURE 8.6 Sizes of atoms of the main group elements (shown as the size of spheres under the symbols) increase with increasing atomic number from top to bottom in a family but decrease with increasing atomic number across rows. Electronegativities (box heights) show the opposite trend, decreasing with increasing atomic number down a column (family) but increasing with increasing atomic number across a row. These opposing trends are consistent with the trends in effective nuclear charge.

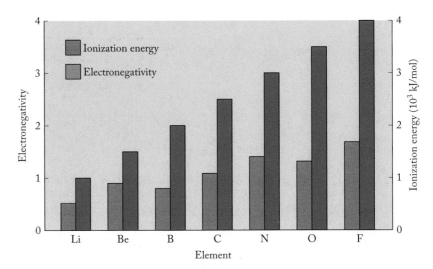

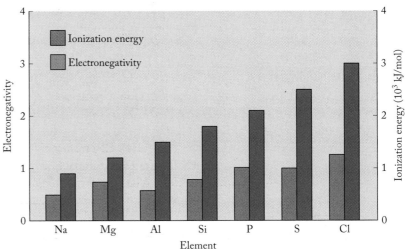

FIGURE 8.7 The trend in ionization energies among the main group elements follows the trend in their electronegativities. The columns on the left for each element represent the first ionization energies. The columns on the right represent electronegativity values. Elements in the second and third row are shown.

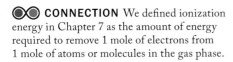

CONNECTION We defined ionization energy in Chapter 7 as the amount of energy required to remove 1 mole of electrons from 1 mole of atoms or molecules in the gas phase.

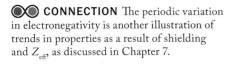

CONNECTION The periodic variation in electronegativity is another illustration of trends in properties as a result of shielding and Z_{eff}, as discussed in Chapter 7.

ments on the right-hand side than in atoms on the left-hand side of the periodic table. Greater attraction to the nucleus draws the valence electrons of an atom inward, making the size of the atoms decrease from left to right across a row. Greater attractive force is also experienced by bonding electrons, so electronegativity increases from left to right across a row.

CONCEPT TEST

Why does Figure 8.5 not include electronegativity values of the noble gases?

Within a group of elements, differences in electronegativity are related to increasing atomic size with increasing atomic number. From the top down within a group, the size of the outermost orbitals increases as the value of the principal quantum number (n) increases. The valence electrons of the elements toward the bottom of a group in the periodic table are shielded from their nuclei by greater distances and by more layers of inner-shell electrons than are the elements toward the top of the group. More shielding and greater atomic size lead to less pull on bonding pairs of electrons; consequently electronegativity decreases with increasing atomic number.

The correlation between electronegativity and ionization energy is easy to rationalize. The same factors that make it more difficult to remove an outer-shell electron from an atom also make that atom more effective at attracting shared valence electrons. Electronegativity values follow the same pattern as ionization energies.

Bond Polarity

Electronegativity allows us to determine which end of a covalent bond is electron-rich and which is electron-poor because we can quantify the difference in electronegativity between the two atoms involved in a bond. It also allows us to estimate the covalent character of a bond by determining a relative **bond polarity** as a measure of the degree of unequal sharing of electrons in a covalent bond. The greater the difference in electronegativity, the more uneven the distribution of electrons between the two atoms, and the more polar the bond (Figure 8.8). There is no simple rule for all compounds, but in general we assume that ionic compounds are formed between two elements when the absolute value of the electronegativity difference between the two atoms is 2.0 or greater. Such differences frequently exist in compounds formed between metallic and nonmetallic elements. For example, calcium oxide is considered an ionic compound because the difference in electronegativity between Ca (1.0) and O (3.5) is 2.5. On the other hand, the bonds formed between different nonmetals are covalent bonds. These bonds vary in polarity as a function of the difference in the electronegativities of the elements. For example, a H—C bond is not as polar as a H—O bond because the difference in electronegativity between C and H (2.5 − 2.1 = 0.4) is less than the difference in electronegativity between O and H (3.5 − 2.1 = 1.4). Electronegativity differences make covalent bonds polar. Figure 8.8 illustrates the impact of decreasing electronegativity on polarity in the series of hydrogen halides. From top to bottom, electronegativity decreases from F to Cl to Br to I, and the polarity of the covalent bond also decreases from H—F to H—Cl to H—Br to H—I.

Figure 8.9 lists values for the polarity of bonds and introduces the use of some new symbolism: The arrow with the plus sign embedded in its tail (⊢→) is placed over a bond to indicate the *direction of polarity*: it points toward the more electronegative, electron-rich atom in the bond, and the position of the plus sign indicates the more positive, electron-poor atom. The polarity of the bond may also be symbolized by placing δ^- (lowercase delta minus) next to the electron-rich atom and δ^+ (lowercase delta plus) next to the electron-poor atom in the structure. The presence of a δ means that the charges are less than a full 1+ or 1−, indicating some polarity in the bond (uneven distribution) but not complete transfer of an electron.

> **Bond polarity** is a measure of the extent to which bonding electrons are shared between two atoms in a covalent bond. The less equally they are shared, the more uneven the distribution, the more polar the bond. Differences in electronegativity determine bond polarity.

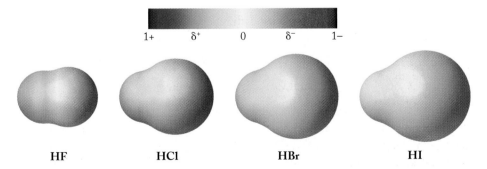

| 1+ | δ^+ | 0 | δ^- | 1− |

| HF | HCl | HBr | HI |

FIGURE 8.8 The main group elements whose atoms have the highest electronegativity values tend to be at the top of a group. Electronegativity decreases down the group. In this case, F is the most electronegative element in group 17, with I the least electronegative.

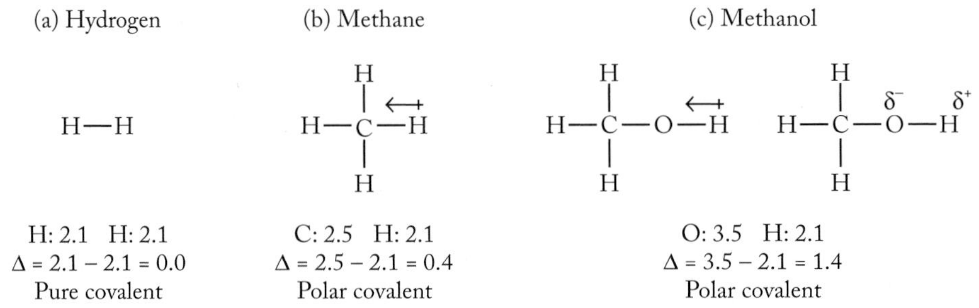

(a) Hydrogen   (b) Methane   (c) Methanol

$$H—H$$

$$
\begin{array}{c}
H \\
| \\
H—C—H \\
| \\
H
\end{array}
\leftrightarrow
$$

$$
\begin{array}{c}
H \\
| \\
H—C—O—H \\
| \\
H
\end{array}
\leftrightarrow
\qquad
\begin{array}{c}
H \\
| \quad \delta^- \quad \delta^+ \\
H—C—O—H \\
| \\
H
\end{array}
$$

H: 2.1 H: 2.1   C: 2.5 H: 2.1   O: 3.5 H: 2.1
$\Delta = 2.1 - 2.1 = 0.0$ $\Delta = 2.5 - 2.1 = 0.4$ $\Delta = 3.5 - 2.1 = 1.4$
Pure covalent   Polar covalent   Polar covalent

FIGURE 8.9 Three ways to indicate the polarity of a bond: (a) Using Pauling electronegativity values, we can calculate the difference in electronegativity between atoms in a bond. (b) An arrow with a plus sign on its tail, placed next to the bond, illustrates the direction of polarity between the two atoms. (c) Using δ^+ and δ^-, we can indicate the electron-poor and electron-rich atoms in the bond.

CONCEPT TEST

Describe the polarity of the CO bond relative to the O_2 bond, and suggest a possible reason for the different way the two substances interact with hemoglobin.

We can use the relative electronegativity of the atoms in a molecule to decide which atom is the central atom when drawing Lewis structures. In Section 8.2 we indicated that the central atom is often placed first in a molecular formula. For example, C is the central atom in CO_2. Carbon is also the least electronegative element in the molecule. The least electronegative element is usually the central atom in a Lewis structure.

SAMPLE EXERCISE 8.5   **Comparing the Polarity of Bonds**

Rank the bonds formed between the following pairs of elements in order of increasing polarity: O and C; Cl and Ca; N and S; O and Si. Do any of these pairs form an ionic bond?

COLLECT AND ORGANIZE We are given four pairs of atoms and asked to determine whether any pair forms an ionic bond. We need to refer to the Pauling electronegativities (Figure 8.5) to judge the relative polarity.

ANALYZE Polarity is related to the difference in electronegativity of the atoms in a chemical bond. If the electronegativity difference is 2.0 or greater, then the bond is considered ionic.

SOLVE Calculating electronegativity differences for each pair of atoms, we get these values:

$$O \text{ and } C: \quad 3.5 - 2.5 = 1.0$$

$$Cl \text{ and } Ca: \quad 3.0 - 1.0 = 2.0$$

$$N \text{ and } S: \quad 3.0 - 2.5 = 0.5$$

$$O \text{ and } Si: \quad 3.5 - 1.8 = 1.7$$

These differences are proportional to the polarity of the bonds formed between the pairs of atoms. Therefore, ranking them in order of increasing polarity we have

$$N \text{ and } S < O \text{ and } C < O \text{ and } Si < Cl \text{ and } Ca$$

The bond between Cl and Ca is ionic because the electronegativity difference between the elements is 2.0.

THINK ABOUT IT Ca and Cl are a metal and a nonmetal, so the result indicating that the bond between them is ionic is reasonable. All the other pairs are two nonmetals, and we expect those bonds to be covalent. Furthermore, O and Si are further apart on the periodic table than are O and C or N and S, so it is reasonable that that bond is the most polar of the three pairs of nonmetals.

Practice Exercise Which of following pairs of elements forms the most polar bond? O and S; Be and Cl; N and H; C and Br. Is the bond between that pair considered ionic?

| SAMPLE EXERCISE 8.6 | **Drawing Structures Showing Bond Polarity** |

Draw Lewis structures for ammonia (NH_3) and arsine (AsH_3), and use the convention of the arrow with the plus sign (shown in Figure 8.9) to indicate the direction of polarity of the bonds. How does the polarity of the N—H bond differ from that of the As—H bond?

COLLECT AND ORGANIZE We can use the Lewis structure for NH_3 that we determined in Section 8.2. We need the Pauling electronegativities in Figure 8.5 for As, N, and H to determine the polarity of the bonds in each molecule.

ANALYZE Consulting the periodic table, we see that arsenic and nitrogen are in the same group and each has 5 valence electrons; hydrogen has 1 valence electron. Because As and N have the same number of valence electrons, our electron bookkeeping will be exactly the same for AsH_3 as for NH_3, so we already have the Lewis structures for both compounds. The electronegativities are As, 2.0; N, 3.0; and H, 2.1.

SOLVE The Lewis structures for NH_3 and AsH_3 and the bond polarities are as follows:

$$\text{Ammonia} \qquad\qquad \text{Arsine}$$

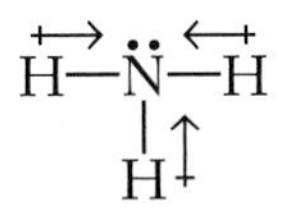

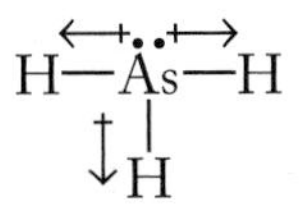

$$\overset{\longrightarrow}{\text{H}}\!-\!\overset{\displaystyle\cdot\cdot}{\text{N}}\!-\!\overset{\longleftarrow}{\text{H}} \qquad\qquad \overset{\longleftarrow}{\text{H}}\!-\!\overset{\displaystyle\cdot\cdot}{\text{As}}\!-\!\overset{\longrightarrow}{\text{H}}$$
$$\underset{\text{H}}{\big\uparrow} \qquad\qquad\qquad \underset{\text{H}}{\big\updownarrow}$$

$$\text{N: 3.0 \quad H: 2.1} \qquad \text{As: 2.0 \quad H: 2.1}$$
$$\Delta = 3.0 - 2.1 = 0.9 \qquad \Delta = 2.1 - 2.0 = 0.1$$

The N—H bond is more polar than the As—H bond. Another difference is also apparent: In the N—H bond the N is more electronegative (δ^-), whereas in the As—H bond the H is more electronegative (δ^-). The direction of polarity differs in the two bonds.

THINK ABOUT IT Notice that nitrogen is written first in the formula for NH_3. We expect it to be the central atom because the only other atom in the molecule, hydrogen, makes only 1 bond. This is a case where the central atom appears first in the formula but is not the least electronegative atom in the molecule.

Practice Exercise Draw the Lewis structures for HBr and ICl, and determine which molecule is more polar. Using arrows above the bond, indicate the direction of polarity.

8.4 Resonance

⊙⊙ **CONNECTION** In Chapter 7 we
discussed the regions of the electromagnetic
spectrum. Ultraviolet (UV) light is radiation
with shorter wavelengths, higher frequencies,
and therefore higher energies than visible
light.

FIGURE 8.10 Lightning strikes contain
sufficient energy to break oxygen–oxygen
double bonds. The O atoms formed in this
fashion collide with O_2 molecules, forming
ozone (O_3), an allotrope of oxygen.

Allotropes are different forms of the same
element, such as oxygen (O_2) and ozone (O_3).

Resonance occurs when two or more
equivalent Lewis structures can be drawn
for one compound.

A **resonance structure** is one of two
or more Lewis structures with the same
arrangement of atoms but different
arrangements of electrons for a molecule.

Ozone (O_3) is a compound present in trace amounts in the lower atmosphere,
where it is occasionally called "bad ozone" because high levels damage crops, harm
trees, and lead to human health problems. Ozone is present in larger concentrations in the upper atmosphere, where it is referred to as "good ozone" because it
shields life on Earth from potentially harmful UV radiation from the sun.

Ozone is a *triatomic* (three-atom) molecule produced naturally by lightning
(Figure 8.10) and accounts for the pungent odor sometimes detectable after a
severe thunderstorm. Ozone (O_3) and diatomic oxygen (O_2) have the same empirical formula: O. Different molecular forms of the same element, such as O_2 and
O_3, are called **allotropes** and have different chemical and physical properties.
Ozone, for example, is an acrid, pale-blue gas that is toxic even at low concentrations, whereas O_2 is a colorless, odorless gas essential for life.

Their different molecular formulas tell us that they must also have different molecular structures. The Lewis structure for O_2 was shown in Section 8.2;
let's now develop the structure for ozone. First we need to do our electron
bookkeeping:

a. *Initial electron bookkeeping:*

$$\text{Need: } 3\,O \times 8\,e^- = 24\,e^-$$

$$\text{Have: } 3\,O \times 6\,e^- = 18\,e^-$$

$$\text{Difference: } \qquad 6\,e^- = 3 \text{ bonds}$$

b. At this point, we are faced with a choice: there are two structures we
can draw that have three bonds between the oxygen atoms and have the
remaining $12\,e^-$ distributed around the atoms as electron pairs:

(We are drawing ozone as a bent molecule here because that is what its
structure is known to be. We will develop the means to predict this shape
in Chapter 9.)

c. *Final electron bookkeeping:* both structures contain the $18\,e^-$ to which we
had access, and both have octets around all three atoms.

Which structure is correct? Neither one alone, but they both are when taken
together. This is due to an important concept in Lewis theory: **resonance** occurs
when more than one valid Lewis structure can be written for a molecule. Resonance
between equivalent structures is indicated with a double-headed arrow:

To generate alternative **resonance structures** (also called *resonance forms*), start
with one of the resonance structures and evaluate whether there are alternative
arrangements for the electrons. Do not move the atoms; just redistribute the electrons. Note that the two structures of ozone are equivalent, that each oxygen
atom has an octet, and that the resonance structures are generated by moving
pairs of electrons:

The red arrows indicate the movement of electrons: one pair of electrons in the double bond becomes a new lone pair on the oxygen atom on the left, and one lone pair of electrons on the oxygen atom on the right becomes a bonding pair.

It is necessary to invoke resonance because experimental evidence indicates that two oxygen–oxygen bonds in ozone are equivalent and are neither single nor double but somewhere between the two in length (Figure 8.11). The molecule does not spend half its time in one form and half in the other; rather, the single structure (which for most molecules with resonance cannot easily be drawn) is in between the extremes defined by its resonance structures.

Leaf damaged by exposure to ozone.

| **SAMPLE EXERCISE 8.7** | **Drawing Resonance Structures for Molecules** |

Sulfur dioxide (SO_2) is produced as a byproduct of the combustion of coal and other fossil fuels. Draw Lewis structures for all the resonance forms of SO_2.

COLLECT AND ORGANIZE We are asked to draw all Lewis structures for SO_2, so we need to consult a periodic table to find the number of valence electrons in each atom.

ANALYZE Both S and O are in group 16, so both have 6 valence electrons. We can use the method developed in Section 8.2 to draw the Lewis structures.

SOLVE We carry out the usual electron bookkeeping calculations:

Need: $(1\,S \times 8\,e^-) + (2\,O \times 8\,e^-) = 24\,e^-$

Have: $(1\,S \times 6\,e^-) + (2\,O \times 6\,e^-) = 18\,e^-$

Difference: $\qquad\qquad\qquad\qquad 6\,e^- = 3$ bonds

Sulfur is listed first in the formula, indicating that it is the central atom. S is below O in group 16, so S is less electronegative than O. These factors suggest we should build the structure with the two oxygen atoms connected to the central sulfur atom. The three bonding pairs and requisite lone pairs can be arranged in two different ways:

$$:\!\overset{..}{O}\!=\!\overset{..}{\underset{..}{S}}\!-\!\overset{..}{\underset{..}{O}}\!: \quad \longleftrightarrow \quad :\!\overset{..}{\underset{..}{O}}\!-\!\overset{..}{\underset{..}{S}}\!=\!\overset{..}{O}\!:$$

Therefore, two resonance structures are possible for SO_2.

THINK ABOUT IT The resonance forms of ozone and sulfur dioxide have the same electron structure. Sulfur and oxygen are in the same group in the periodic table and have the same number of valence electrons. We can expect similarities in the bonding in O_3 and SO_2, just as we saw in Sample Exercise 8.6 for NH_3 and AsH_3.

Practice Exercise Draw Lewis structures for all resonance forms of sulfur trioxide (SO_3).

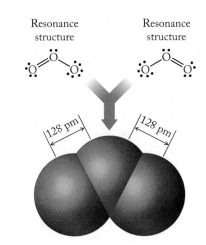

FIGURE 8.11 The molecular structure of ozone, shown here with a space-filling model, is an average of the two resonance structures at the top. Both bonds in ozone are 128 picometers (10^{-12} m) long, a value between the average length of an O—O single bond (148 pm) and an O=O double bond (121 pm), which indicates that these bonds are not single bonds or double bonds but something in between.

▶❙❙ **CHEMTOUR** Resonance

Up to this point we have drawn Lewis structures primarily for neutral molecules. Just as we drew structures for monatomic ions, however, we can also draw them for polyatomic ions. The only change to make in our electron bookkeeping is to account for the charge on the ion; we do this by adding electrons to the total in the case of polyatomic anions, and subtracting electrons from the total in the case of polyatomic cations.

◐◑ **CONNECTION** In Chapter 2 we introduced polyatomic ions as groups of atoms held together by covalent bonds and having a charge. While many polyatomic anions have been mentioned, the only polyatomic cations we have seen are ammonium (NH_4^+), and hydronium (H_3O^+).

SAMPLE EXERCISE 8.8   **Drawing Resonance Structures for Ions**

Draw Lewis structures for all resonance forms of the nitrate ion (NO_3^-).

COLLECT AND ORGANIZE We need to consult the periodic table to determine the number of valence electrons in each atom in the ion.

ANALYZE A nitrate ion contains one nitrogen atom (group 15) with 5 valence electrons; three oxygen atoms (group 16), each with 6 valence electrons; and has an overall negative charge of 1−, which means we have to add 1 electron to the total in our electron bookkeeping to account for all the electrons in the structure.

SOLVE Using the standard method for generating Lewis structures and adding 1 electron to account for the charge of 1− on the ion, we get the following:

Need: $(1\,N \times 8\,e^-) + (3\,O \times 8\,e^-)$ $= 32\,e^-$

Have: $(1\,N \times 5\,e^-) + (3\,O \times 6\,e^-) + (1\,e^-) = 24\,e^-$

Difference: $8\,e^- = 4$ bonds

Drawing a structure that contains 4 covalent bonds and accommodates the remaining $16\,e^-$ as lone pairs yields:

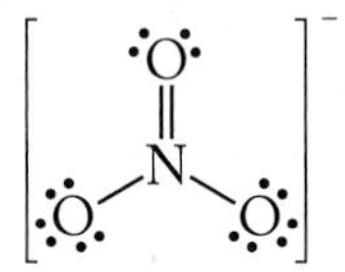

The double bond can be between the N and any of the three oxygen atoms. This means there are three equivalent ways of drawing the Lewis structure for nitrate, and thus three resonance forms:

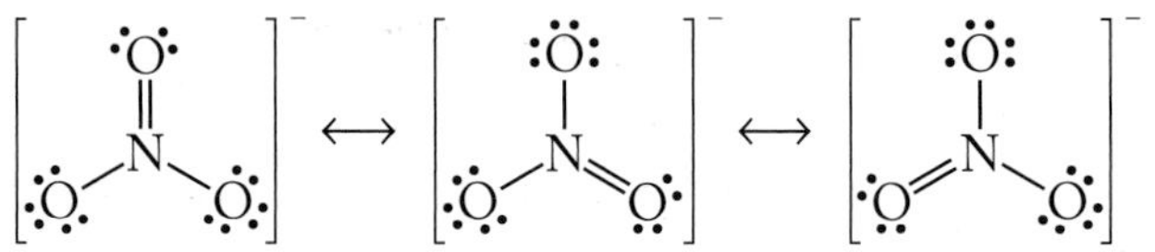

THINK ABOUT IT Remember that the nitrate ion does not contain one double bond and two single bonds. All the N—O bonds in NO_3^- are the same length and somewhere between the lengths of N—O and N=O bonds in other molecules

Practice Exercise Draw all the resonance forms of the azide ion (N_3^-) and the nitronium ion (NO_2^+).

Resonance is a common feature among certain organic molecules that have alternating single and double bonds in their structures. One such molecule is the compound benzene (C_6H_6), a six-membered ring of carbon atoms with alternating single and double bonds:

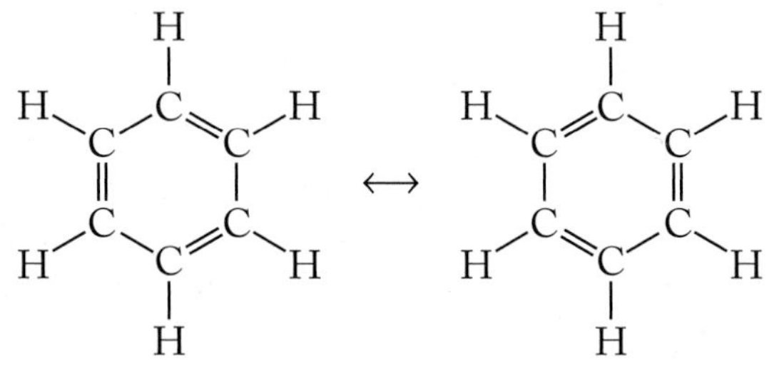

If we fix the atoms in place in benzene, we see that there are two ways the electrons can be arranged. Resonance in organic compounds such as benzene is responsible for its *aromatic* properties, which include reactivities quite different from the reactivities of organic compounds that do not have resonance. To depict the presence of resonance in the benzene molecule, chemists frequently draw it as a six-membered ring with a circle in the center. This symbol emphasizes that the six bonds are all identical and intermediate in character between single and double bonds.

Resonance in a benzene ring is sometimes illustrated with a circle inside a hexagon.

8.5 Formal Charges: Choosing among Lewis Structures

The two resonance forms we drew for ozone in the previous section are equivalent; each contains one O=O double bond and one O—O single bond. The two resonance forms for SO_2 in Sample Exercise 8.7 and the three for NO_3^- in Sample Exercise 8.8 also differ only in the location of a double bond. The nitrate ion, however, has a negative charge, so the question arises, Which atom bears the negative charge in the Lewis structures for NO_3^-?

The apparent charges on each atom in NO_3^- (and other molecules and ions) can be determined by assigning a formal charge to each atom in the structure. A **formal charge** is not a real charge; it is a charge calculated using Equation 8.1 and based on the number of electrons formally assigned to an atom in a Lewis structure according to the following procedure:

1. Determine the number of valence electrons in the atom.
2. Count the number of electrons in lone pairs about an atom.
3. Add to it *half* the number of electrons in bonding pairs.
4. Subtract that total from the number of valence electrons in a single atom.

The result is the formal charge (FC) on the atom in the Lewis structure. Expressed as an equation:

Formal change =

$$\left(\begin{matrix} \text{number of} \\ \text{valence electrons} \end{matrix}\right) - \left(\begin{matrix} \text{number of electrons} \\ \text{in lone pairs} \end{matrix} + \begin{matrix} \frac{1}{2} \text{ number of electrons} \\ \text{in bonding pairs} \end{matrix}\right) \quad (8.1)$$

The calculation of formal charge assumes that each atom has exclusive title to the electrons in its lone pairs and shares its bonding electrons equally; half the bonding electrons are formally assigned to each atom in the bond.

To illustrate the method, let's calculate the formal charges on the atoms in carbon dioxide:

$$\ddot{O}::C::\ddot{O}: \qquad (\ddot{O}:C:\ddot{O}:)$$

Formal charge = 0 | Formal charge = 0
Formal charge = 0

The formal charge on each oxygen is equal to $6 - [4 + \frac{1}{2}(4)] = 0$; the formal charge on carbon is equal to $4 - [0 + \frac{1}{2}(8)] = 0$. The sum of the formal charges on the atoms in a neutral compound must equal zero, and that is the case here. When the species in question is an ion, the sum of the formal charges must equal the charge on the ion.

> The **formal charge** on an atom in a molecule equals the number of valence electrons on the free atom in the Lewis structure minus the sum of the number of electrons in its lone pairs plus half the number of electrons in its bonding pairs.

To calculate the formal charges on the atoms in the nitrate ion, we can use any of the three resonance structures for NO_3^- because they are all equivalent. One of the resonance structures is shown.

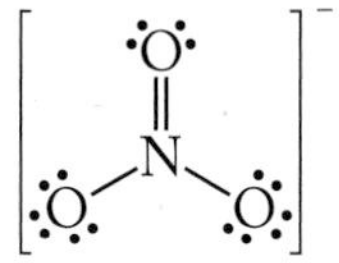

Using Equation 8.1 and the above Lewis structure, we calculate the formal charge for nitrogen:

$$\text{Formal charge of N} = 5 - [0 + \tfrac{1}{2}(8)] = +1$$

Regarding oxygen, two different situations are present in the ion. Two oxygen atoms are identical and have single bonds to the nitrogen; the third oxygen atom has a double bond. Treating the oxygen atoms with the single bond first using Equation 8.1:

$$\text{Formal charge of O}_{\text{single bond}} = 6 - [6 + \tfrac{1}{2}(2)] = -1$$

The oxygen atom with the double bond to nitrogen:

$$\text{Formal charge of O}_{\text{double bond}} = 6 - [4 + \tfrac{1}{2}(4)] = 0$$

The sum of the formal charges for all four atoms in NO_3^- (or any molecule) must equal the charge on the molecule or ion. The nitrogen atom has a formal charge of +1, the two oxygen atoms connected to the nitrogen atom by a single bond are each −1, and the oxygen atom connected by the double bond has a formal charge of 0, so

$$+1 + 2(-1) + 0 = -1$$

which equals the charge on the ion, 1−. The formal charges also indicate that the negative charge on the ion is located on two oxygen atoms while the nitrogen atom has a formal charge of +1. Since oxygen is the more electronegative element, it is reasonable that any negative charge in the molecule be assigned to oxygen.

Let's turn our attention to a different atmospheric molecule, dinitrogen monoxide or N_2O, also known as nitrous oxide. The consideration of formal charges in this compound helps us choose between alternative Lewis structures that are possible for the molecule.

Average nitrous oxide concentrations in the lower atmosphere are relatively low, approximately 300 to 500 ppb, or approximately 10 times greater than the average concentration of ozone in air. Nitrous oxide is produced naturally by bacterial action on soils and through human activities such as agriculture and sewage treatment. It is known as laughing gas because people who inhale it usually laugh spontaneously. Nitrous oxide was an early anesthetic used in dentistry, because breathing it has a narcotic effect. It is occasionally in the news because its concentration in the troposphere (the atmosphere at ground level) has been increasing. Along with other gases, it may be altering global weather patterns.

To draw a Lewis structure for nitrous oxide, we start by putting one nitrogen atom at the center of the molecule (because nitrogen is less electronegative than oxygen) and then applying our process of electron bookkeeping:

Need: $(2N \times 8e^-) + (1O \times 8e^-) = 24e^-$

Have: $(2N \times 5e^-) + (1O \times 6e^-) = 16e^-$

Difference: $8e^- = 4$ bonds

After making 4 covalent bonds, we have $8e^-$ left over to distribute among the atoms in order to provide each atom with an octet:

$$:\!\ddot{N}\!=\!N\!=\!\ddot{O}\!: \;\longleftrightarrow\; :N\!\equiv\!N\!-\!\ddot{\underset{..}{O}}\!: \;\longleftrightarrow\; :\!\ddot{\underset{..}{N}}\!-\!N\!\equiv\!O:$$

Each structure has a total of 16 valence electrons, and each atom has a complete octet. Unlike the two resonance forms for O_3 or the three resonance forms for SO_2 and NO_3^-, the resonance structures for N_2O are not equivalent. The structure on the left contains a $N\!=\!N$ double bond and a $N\!=\!O$ double bond, while the middle structure has a $N\!\equiv\!N$ triple bond and a $N\!-\!O$ single bond. The structure on the right hand side has a $N\!-\!N$ single bond and a $N\!\equiv\!O$ triple bond.

We can use formal charges to determine whether one of the three possible structures is more reasonable than the others. Our criteria for the evaluation are that the "best" resonance structure has

1. formal charges minimized (zero or close to zero), and

2. negative charges on the most electronegative atom(s).

We start by assigning formal charges to each atom in each resonance structure. Using the method described for CO_2, we determine the following formal charges for the atoms in the three resonance structures for N_2O:

	(a)			(b)			(c)		
	$:\!\ddot{N}\!=\!N\!=\!\ddot{O}\!:$		$\longleftrightarrow$	$:N\!\equiv\!N\!-\!\ddot{O}\!:$		$\longleftrightarrow$	$:\!\ddot{N}\!-\!N\!\equiv\!O:$		
(Number of valence electrons)	5	5	6	5	5	6	5	5	6
− (Number of valence e⁻ assigned)	6	4	6	5	4	7	7	4	5
Formal charge	−1	+1	0	0	+1	−1	−2	+1	+1

The formal charge on the central N atom in all three resonance structures is the same (+1), but the formal charges on the noncentral atoms differ. In structure (a), the oxygen atom has a formal charge of 0. In structure (b), the formal charge on O is −1, and in structure (c) it is +1. Considering the two criteria for judging the best resonance structure, we find that in none of the three structures do all three atoms have zero formal charge. The formal charges in structures (a) and (b), however, are closer to zero than those in structure (c). Furthermore, only in structure (b) does the more electronegative element, oxygen, have a negative formal charge. We conclude that structure (b) probably contributes the most to the bonding in N_2O and is the best of the three resonance structures drawn. We predict that the bond between the two nitrogen atoms is likely an average between a double and a triple bond while the nitrogen–oxygen bond is the average between a single and a double bond.

Without actually calculating a formal charge value using Equation 8.1, predict the formal charge on a nitrogen atom that has three lone pairs of electrons and one bonding pair.

SAMPLE EXERCISE 8.9   **Selecting Lewis Structures Based on Formal Charges**

A Lewis structure for carbon dioxide was determined earlier in this section. There are actually three resonance forms in all for CO_2:

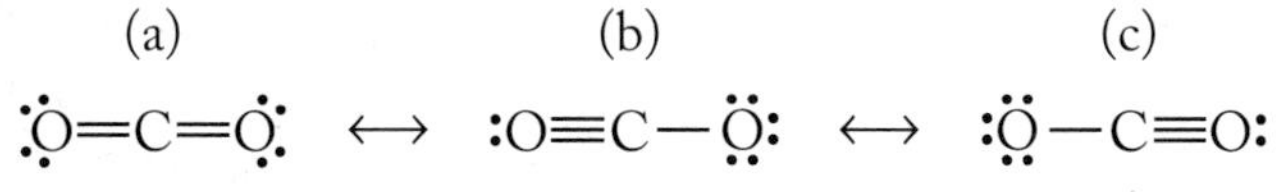

    (a)   (b)   (c)

$$\ddot{O}=C=\ddot{O} \;\; \longleftrightarrow \;\; :O\equiv C-\ddot{\underset{..}{O}}: \;\; \longleftrightarrow \;\; :\ddot{\underset{..}{O}}-C\equiv O:$$

Assign formal charges, and explain why the structure with two C=O double bonds is preferred.

COLLECT AND ORGANIZE We are given three resonance forms for carbon dioxide and asked to assign formal charges to each atom in each structure and to use the results to support our choice for the preferred structure.

ANALYZE The preferred structure is one in which the formal charges are closest to zero and any negative formal charges are on the more electronegative atom. In this case, O is the more electronegative atom.

SOLVE To find the formal charge on each atom we can use Equation 8.1. For each atom:

$$\text{Formal charge} = \begin{pmatrix}\text{number of} \\ \text{valence } e^-\end{pmatrix} - \begin{pmatrix}\text{number of } e^- & \frac{1}{2}\text{ the number } e^- \\ \text{in lone pairs} & \text{in bonding pairs}\end{pmatrix}$$

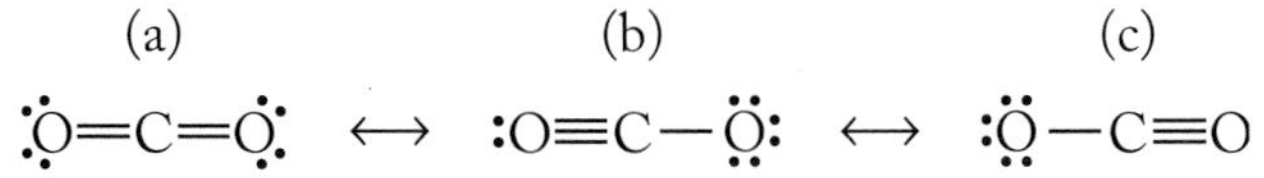

	(a)			(b)			(c)		
	$\ddot{O}=C=\ddot{O}$			$:O\equiv C-\ddot{O}:$			$:\ddot{O}-C\equiv O$		
(Number of valence electrons)	6	4	6	6	4	6	6	4	6
− valence e^- assigned) (Number of	6	4	6	5	4	7	7	4	5
Formal charge	0	0	0	+1	0	−1	−1	0	+1

Structure (a), with zero formal charge on all its atoms, is the preferred form. In both structures (b) and (c), one oxygen atom has a negative formal charge but the other has a positive formal charge.

THINK ABOUT IT Notice that the sum of the formal charges is zero in all three resonance forms. A net value of zero is expected because the sum of the formal charges should equal the actual charge, which is zero for a neutral molecule.

Practice Exercise Which resonance forms of the azide ion (N_3^-) and the nitronium ion (NO_2^+) contribute the most to bonding in their respective structures?

Taking a final look at all the structures for which we have determined formal charges, you may have noticed a connection between the formal charge on an atom and its **bonding capacity** (or *valency*). Bonding capacity reflects the number of covalent bonds an element forms when it has a formal charge of zero (0). For example, in the Lewis structures for NO_3^-, one oxygen atom forms a double bond, which counts as 2 bonds because 2 electron pairs are shared. The oxygen atom is *divalent*. (A single bond is 1 bond; a triple bond, 3 bonds.) This number matches oxygen's bonding capacity, and the formal charge on this oxygen atom is zero. This outcome is in accord with an important principle: when the number of bonding pairs of electrons to an atom matches its bonding capacity, the formal charge on the atom is zero. If the number of bonding pairs is 1 more than the atom's bonding capacity (such as O atoms with 3 bonds), the formal charge is +1, and if the number of bonding pairs is 1 less than the bonding capacity (such as O atoms with 1 bond), the formal charge is −1.

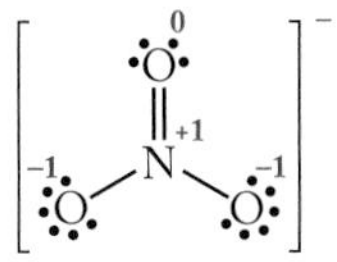

Notice how this rule applies to the three resonance structures for N_2O:

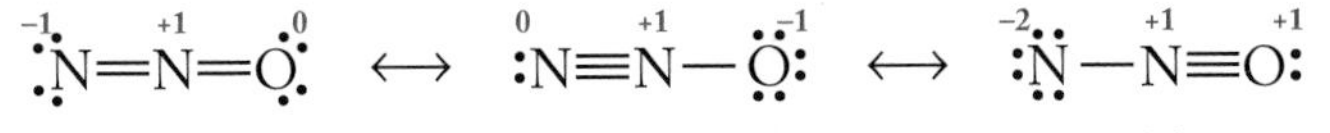

$$\overset{-1}{:N}=\overset{+1}{N}=\overset{0}{\ddot{O}:} \quad \longleftrightarrow \quad :\overset{0}{N}\equiv\overset{+1}{N}-\overset{-1}{\ddot{O}:} \quad \longleftrightarrow \quad :\overset{-2}{\ddot{N}}-\overset{+1}{N}\equiv\overset{+1}{O}:$$

$$\text{(a)} \qquad\qquad \text{(b)} \qquad\qquad \text{(c)}$$

The bonding capacity for N is 3, but the central nitrogen in all the structures forms 4 bonds, 1 more than the bonding capacity, so it is assigned a formal charge of +1. However, the terminal nitrogens vary. In structure (a), the terminal nitrogen forms only 2 bonds, 1 less than its bonding capacity, so the formal charge on this atom is −1. In structure (b), the terminal nitrogen forms 3 bonds, equaling its bond capacity, so it has a formal charge of zero. Finally, in structure (c), the terminal nitrogen forms only 1 bond, 2 less than its bonding capacity, so its formal charge is −2. Especially for atoms in row 2 of the periodic table, bonding capacity is a useful concept to consider when drawing Lewis structures.

8.6 Exceptions to the Octet Rule

Earth's atmosphere contains several compounds that illustrate the limitations of the octet rule. For example, sulfur hexafluoride (SF_6) contains six sulfur–fluorine covalent bonds, and nitrogen dioxide (NO_2) and nitric oxide (NO) produced in automobile engines have an odd number of valence electrons. These molecules represent exceptions to the octet rule.

Molecules with Less than an Octet

In some cases, atoms of main group elements other than hydrogen may have fewer than 8 valence electrons in a Lewis structure. Boron atoms often have 6 valence electrons. These compounds with central atoms that have less than an octet are rarely found in the atmosphere, but they are common in chemistry laboratories. One example is boron trifluoride (BF_3), a gas at room temperature and a compound used on the kiloton scale in the chemical industry worldwide, mostly to make pure boron. Boron trifluoride has a total of 24 valence electrons; 3 from boron and 21 from the three fluorines. A Lewis structure for BF_3 developed using the method in Section 8.2 has complete octets for each fluorine atom and 6 valence electrons on boron. There are a total of 24 valence electrons in the molecule, and each atom has a formal charge of zero (shown in red):

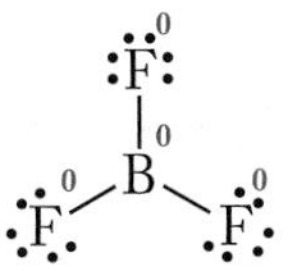

Bonding capacity reflects the number of covalent bonds an element forms when it has a formal charge of 0. The bonding capacity of carbon is 4 (it is tetravalent); nitrogen is 3 (trivalent); oxygen is 2 (divalent); fluorine is 1 (monovalent).

Electron-deficient compounds are substances whose central atoms (other than H) in Lewis structures have fewer than four electron pairs (that is, less than an octet of electrons).

We can also draw three resonance structures for BF_3 in which the molecule has one B=F double bond. While these structures allow boron to have a complete octet, the F that is bonded to B with a double bond now has a formal charge of +1 and the boron has a formal charge of −1.

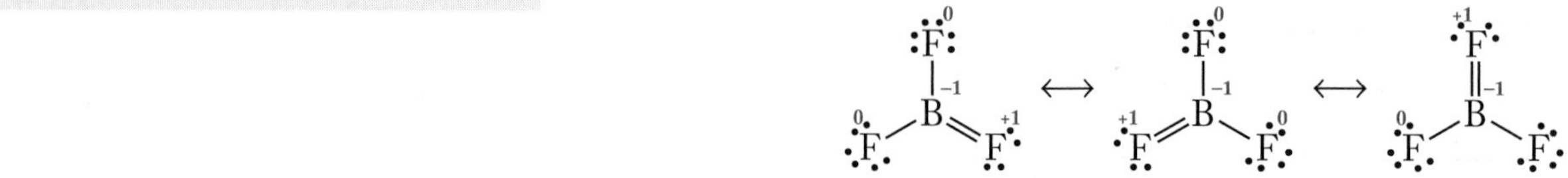

Which structure for BF_3 is most likely? As before, electronegativity helps us evaluate these structures. Fluorine is the more electronegative element (see Figure 8.5), so the three structures with a positive fluorine atom are less acceptable than the structure with only 6 electrons around the central boron atom.

The presence of only 6 valence electrons around boron makes BF_3 very reactive with molecules containing a lone pair of electrons, such as ammonia (NH_3). Reaction between BF_3 and NH_3 generates the compound BF_3NH_3, which contains a covalent boron–nitrogen bond (shown in red) and octets of electrons around B, F, and N:

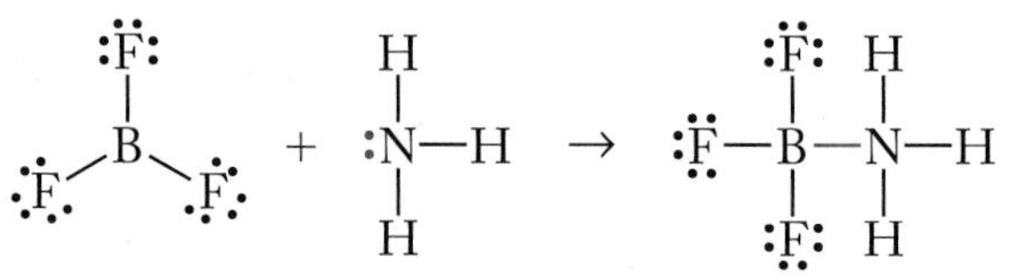

Compounds like BF_3 are called **electron-deficient compounds** because they contain atoms other than hydrogen that do not have an octet of electrons. Typically this behavior arises with small atoms such as boron and beryllium in row 2 of the periodic table. Aluminum also forms some electron-deficient compounds. Considerations of formal charge again make it possible for us to predict such compounds.

Molecules with More than an Octet

Sulfur hexafluoride (SF_6) is used as an insulator in electrical transformers because it is quite unreactive, surviving temperatures up to 500°C and not reacting with water. Sulfur hexafluoride, however, is on the list of chemical compounds that may be affecting global temperatures because it can escape from transformers into the atmosphere.

To draw the Lewis structure for SF_6, we apply our electron bookkeeping method:

Need: $(1\,S \times 8\,e^-) + (6\,F \times 8\,e^-) = 56\,e^-$

Have: $(1\,S \times 6\,e^-) + (6\,F \times 7\,e^-) = 48\,e^-$

Difference: $8\,e^- = 4$ bonds

This is clearly problematic: seven atoms must be bound together in SF_6, yet our formula predicts only 4 covalent bonds. Sulfur is the less electronegative atom and the one written first in the formula, so we pick sulfur for our central atom. To accommodate the six fluorine atoms about a central sulfur atom, the sulfur must *expand its octet* to accommodate more than 8 electrons:

Electrical transformers use sulfur hexafluoride as an insulator. Leaks of SF_6 have led to concern that SF_6 is contributing to global warming.

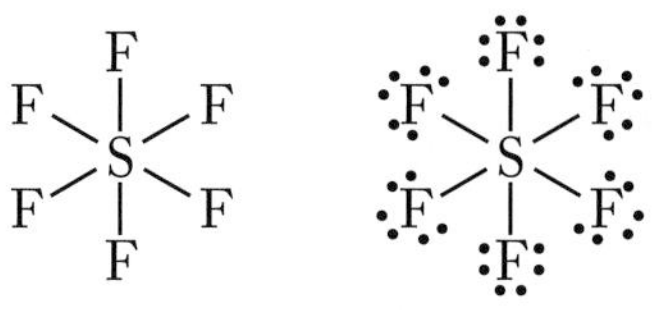

The presence of six S—F bonds and 12 valence electrons around S leaves the SF_6 molecule with zero formal charges on each atom:

$$\text{Formal charge of S} = 6 - [0 + \tfrac{1}{2}(12)] = 0$$
$$\text{Formal charge of F} = 7 - [6 + \tfrac{1}{2}(2)] = 0$$

Based on our criteria for judging Lewis structures, this one is very good. But how can a sulfur atom accommodate more than 8 valence electrons?

Larger atoms of the elements in the third row and below in the periodic table have the ability to expand their valence shells by using their empty d orbitals. The ground-state electron configuration of S is $[Ne]3s^23p^4$. From the discussion of quantum mechanics (Section 7.6) we know that, for principal quantum number $n = 3$ and beyond, the values of the quantum number ℓ include $\ell = 2$, or d orbitals. Elements such as N and O do not have d orbitals in their outermost $n = 2$ shell, so they have no space to accommodate more than 8 electrons. The increasing size of atoms as atomic number increases means that more electron pairs can fit around a larger atom such as sulfur. Atoms in row 2 do not expand their octet, but atoms in rows 3 and higher can accommodate more than four electron pairs and hence can form more than 4 covalent bonds.

Having the ability to expand their octets does not mean that these atoms *always* expand their octets. They tend to do so when:

1. they form compounds with strongly electronegative elements, particularly F, O, and Cl; or
2. smaller formal charges are the result.

Electron bookkeeping alerts us to situations in which we must consider expanding an atom's octet, in particular, when the number of bonds calculated is smaller than the number of bonds needed to build the molecule.

To illustrate a situation in which expanding the octet of an atom results in smaller formal charges, let's consider the structure of the sulfate ion (SO_4^{2-}). This ion is the anion of sulfuric acid, which is generated in the atmosphere when high-sulfur coal is burned, releasing SO_2 into the atmosphere. The reaction of sulfur dioxide with atmospheric oxygen yields sulfur trioxide (SO_3):

$$2\,SO_2(g) + O_2(g) \rightarrow 2\,SO_3(g)$$

Sulfur trioxide reacts with water vapor:

$$SO_3(g) + H_2O(g) \rightarrow H_2SO_4(\ell)$$

Sulfuric acid (H_2SO_4) is a principal component of acidic precipitation in the eastern United States and Europe. Sulfuric acid has the potential to dissociate into the sulfate ion and two hydrogen ions in aqueous media:

$$H_2SO_4(\ell) \rightarrow 2\,H^+(aq) + SO_4^{2-}(aq)$$

The Lewis structure for the sulfate ion (SO_4^{2-}) has S as the central atom because it is less electronegative than oxygen. Applying our electron bookkeeping method we can draw the following structure, which obeys the octet rule:

CONNECTION In Chapter 4 we discussed the reaction of SO_3 with water vapor in the atmosphere; we also learned about scrubbers that use chemical reactions to remove SO_2 and SO_3 gases from exhausts so they do not enter the atmosphere and produce acid rain.

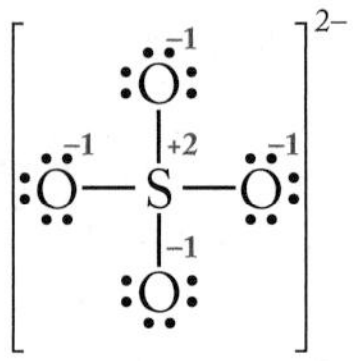

$$\left[\begin{array}{c} \overset{-1}{:\ddot{O}:} \\ | \\ :\ddot{O} - \overset{+2}{S} - \overset{-1}{\ddot{O}:} \\ | \\ \overset{-1}{:\ddot{O}:} \end{array} \right]^{2-}$$

Calculating the formal charges, we find each oxygen atom has a formal charge of −1 and the sulfur atom has a formal charge of +2. The total charge on the sulfate ion is 1(+2) + 4(−1) = 2−. While this calculation does yield the correct ionic charge, remember that the goal is to minimize formal charges in Lewis structures, which we can now do by expanding the sulfur atom's octet:

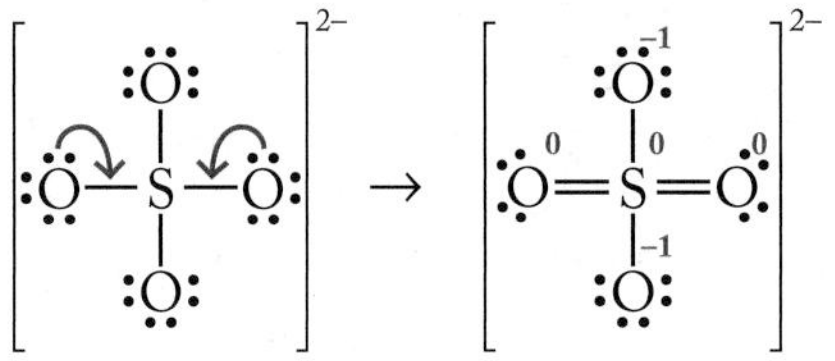

$$\left[\begin{array}{c} :\ddot{O}: \\ | \\ :\ddot{O} - S - \ddot{O}: \\ | \\ :\ddot{O}: \end{array} \right]^{2-} \rightarrow \left[\begin{array}{c} \overset{-1}{:\ddot{O}:} \\ | \\ :\ddot{O} \overset{0}{=} \overset{0}{S} \overset{0}{=} \ddot{O}: \\ | \\ \overset{-1}{:\ddot{O}:} \end{array} \right]^{2-}$$

Note that in this Lewis structure for SO_4^{2-}, the oxygen atoms still have complete octets but now the sulfur has expanded its octet and formal charges are minimized. Because we could draw the two double bonds at any location around the sulfur atom in the Lewis structure for the sulfate ion, this structure has several resonance forms (not shown).

If we now wanted to draw the Lewis structure for sulfuric acid, we could bond two hydrogen ions (H^+) to the two negative oxygen atoms in the structure:

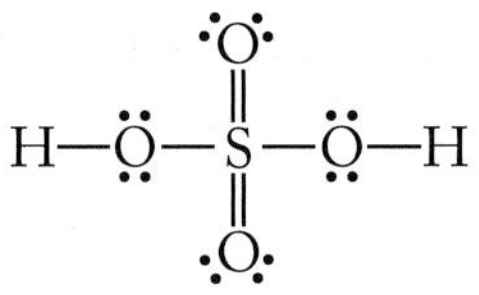

$$H - \ddot{O} - \underset{\underset{\cdot\ddot{O}\cdot}{\|}}{\overset{\overset{\cdot\ddot{O}\cdot}{\|}}{S}} - \ddot{O} - H$$

Each hydrogen atom has achieved a duet of electrons, each oxygen atom has an octet, and the central sulfur has expanded its octet to minimize its formal charge.

SAMPLE EXERCISE 8.10 **Drawing Lewis Structures for Ions with Expanded Octets**

Draw the Lewis structure for the phosphate ion (PO_4^{3-}) that minimizes the formal charges on its atoms.

COLLECT AND ORGANIZE The phosphate ion contains one phosphorus atom (group 15), four oxygen atoms (group 16), and has a charge of 3−. Drawing a Lewis structure that obeys the octet rule first enables us to assign formal charges to all the atoms in the phosphate ion. We can then determine whether expanding the octet of the central atom would lower the formal charge on any atom in the ion.

ANALYZE Phosphorus is the less electronegative atom, so we select it as the central atom to which the oxygen atoms are attached.

SOLVE We apply our electron bookkeeping method:

Need: $(1\,P \times 8\,e^-) + (4\,O \times 8\,e^-)$ $= 40\,e^-$

Have: $(1\,P \times 5\,e^-) + (4\,O \times 6\,e^-) + 3\,e^- = 32\,e^-$

Difference: $8\,e^- = 4$ bonds

After making the 4 bonds, $24\,e^-$ are left over to complete the octets in the atoms:

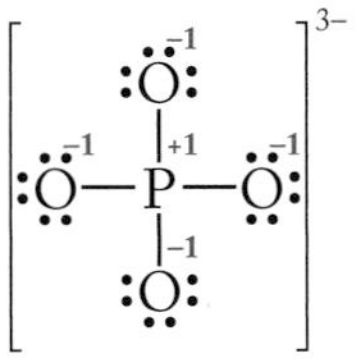

Calculating formal charges, we find that each oxygen atom has a formal charge of -1 and the formal charge on the phosphorus atom is $+1$. Phosphorus is in row 3 in the periodic table, so it can expand its octet. We can create more zero formal charges if we use one of the lone pairs on an oxygen atom to form another bond to P:

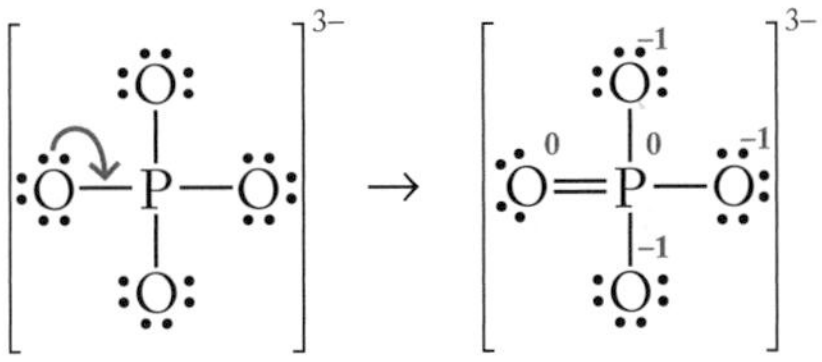

Now there are three single-bonded oxygen atoms, each with a formal charge of -1, which corresponds to an overall charge on the ion of $3-$. The formal charges on the phosphorus atom and the double-bonded oxygen atom are both zero. The structure with one $P{=}O$ double bond results in lower formal charges than the structure containing all $P{-}O$ single bonds.

THINK ABOUT IT The final structure puts 10 valence electrons around the phosphorus atom. This exception to the octet rule is allowed because phosphorus is in the third period and has *d* orbitals available with which to expand its octet.

Practice Exercise Sulfur makes several compounds with fluorine, including SF_6 and SF_4. Draw the Lewis structure for SF_4 and assign formal charges to the atoms.

CONCEPT TEST

The noble gases Xe and Kr make some compounds with fluorine, whereas no compounds of He and Ne have yet been made. Suggest a reason for this difference in behavior.

Odd-Electron Molecules

Nitric oxide is produced in the upper atmosphere (Figure 8.1), but it is also produced at ground level when high temperatures in vehicle engines lead to a chemical reaction between nitrogen and oxygen:

$$N_2(g) + O_2(g) \rightarrow 2\,NO(g) \qquad (8.2)$$

Free radicals are molecules having an odd number of valence electrons and hence unpaired electrons in their Lewis structures.

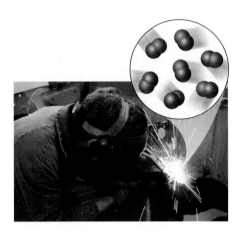

Arc welding produces significant amounts of nitric oxide (NO). The Environmental Protection Agency limits NO exposure to an average of 25 ppm, or 0.0025% of air by volume.

Nitric oxide reacts rapidly with oxygen to produce nitrogen dioxide (NO_2):

$$2\ NO(g) + O_2(g) \rightarrow 2\ NO_2(g) \qquad (8.3)$$

Nitric oxide in the air is toxic to humans at very low levels, and the Lewis structure for NO helps us to understand its reactivity with oxygen as well as its damaging effect on tissue.

When we make our electron bookkeeping calculation for drawing a Lewis structure for nitric oxide, we immediately encounter a problem. The molecule has an odd number of valence electrons, resulting in a structure requiring a fractional number of covalent bonds:

$$\text{Need: } (1\,N \times 8\,e^-) + (1\,O \times 8\,e^-) = 16\,e^-$$

$$\text{Have: } (1\,N \times 5\,e^-) + (1\,O \times 6\,e^-) = 11\,e^-$$

$$\overline{\text{Difference: } \qquad\qquad\qquad 5\,e^- = 2\tfrac{1}{2}\ \text{bonds}}$$

If we draw the possible Lewis structures:

$$:\!N\!\!=\!\!\ddot{O}: \qquad :\!\ddot{N}\!\!=\!\!\ddot{O}: \qquad :\!N\!\!\equiv\!\!O:$$

we find that each structure has a total of 11 valence electrons, but none of the structures obey the octet rule. A triple bond between nitrogen and oxygen allows us to complete the octets of both atoms but leaves 1 electron left over. If we draw a double bond between nitrogen and oxygen, there are not enough valence electrons to satisfy the octet rule for each atom. Applying the criteria involving formal charge, we can judge which structure is the best.

If we are to draw the Lewis structure for an odd-electron molecule, we need to decide which atom gets the odd number—in this case, 7 valence electrons instead of 8. The approach that makes the most sense is to put an octet around the more electronegative element and to shortchange the less electronegative element. Therefore the nitrogen atom in the Lewis structure for NO gets only 7 electrons and has a $N\!=\!O$ double bond. Assigning formal charges (shown in red) to the atoms in the three Lewis structures supports our choice for best structure, which has formal charges of zero (shown in red) for both atoms:

$$\overset{0}{:\!N}\!\!=\!\!\overset{0}{\ddot{O}:} \qquad \overset{-1}{:\!\ddot{N}}\!\!=\!\!\overset{+1}{\ddot{O}:} \qquad \overset{-1}{:\!N}\!\!\equiv\!\!\overset{+1}{O:}$$

Lewis structures are unfortunately often limited in their ability to handle odd-electron molecules; this is especially true when the criteria of formal charge and electronegativity are used to pick the best among several possible structures.

Compounds that have odd numbers of valence electrons are called **free radicals**. They are typically very reactive species: it is energetically favorable for them to pick up an electron from another molecule or ion. This characteristic makes them excellent oxidizing agents and is responsible for the damage they cause to the human body and other materials with which they come in contact.

CONCEPT TEST

Nitric oxide has an important role in the human body as a transmitter of signals from cell to cell, but it is very tightly controlled by other molecules such as hemoglobin that bind it and transport it. How do Lewis structures help explain the behavior of NO in terms of its reactions with hemoglobin compared with that of O_2, N_2, and CO?

SAMPLE EXERCISE 8.11   **Drawing Lewis Structures for Odd-Electron Molecules**

Draw Lewis structures for nitrogen dioxide (NO_2), assign formal charges, and choose the best structure.

COLLECT AND ORGANIZE Nitrogen dioxide contains one atom of nitrogen (group 15) and two atoms of oxygen (group 16).

ANALYZE NO_2 has a total of 17 valence electrons, 5 from the N atom and 12 from the two oxygen atoms. It is an odd-electron molecule, so we must expect to analyze possible structures by assigning formal charges to select the best one. Since nitrogen is less electronegative than oxygen and listed first in the formula, we pick it as the central atom, connected to the two oxygen atoms.

SOLVE Our electron bookkeeping method yields the following:

Need: $(1\,N \times 8\,e^-) + (2\,O \times 8\,e^-) = 24\,e^-$

Have: $(1\,N \times 5\,e^-) + (2\,O \times 6\,e^-) = 17\,e^-$

Difference: $7\,e^- = 3\frac{1}{2}$ bonds

The need for a fractional number of covalent bonds confirms that NO_2 is a free radical. The possible Lewis structures have one atom with 7 valence electrons. We can draw at least three Lewis structures, each of which has a resonance form (not shown):

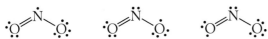

The leftmost structure, with an odd number of electrons on N, is preferred because oxygen is the more electronegative element. The nitrogen atom has a +1 formal charge and one oxygen atom has a formal charge of −1. Because the formal charge of −1 occurs on the more electronegative atom, the leftmost resonance Lewis structure above seems to be the best choice. It is shown below as two resonance forms:

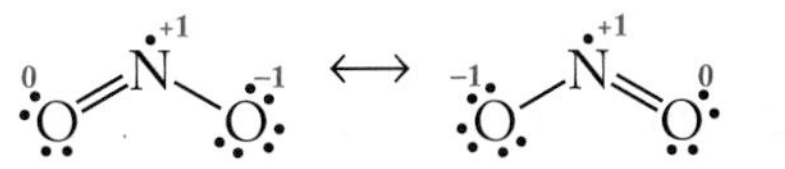

Determining formal charges for the atoms in the other two Lewis structures:

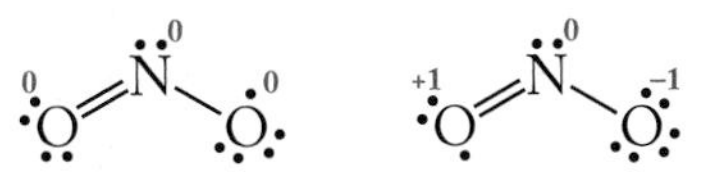

we find that the structure on the right, with an odd number of valence electrons on the oxygen atom that forms a $N\!=\!O$ double bond, has a formal charge of +1 on the oxygen; this structure is not favorable because the more electronegative element has a positive formal charge. However, the structure on the left, with an odd number of electrons on the oxygen that forms a N—O single bond, has formal charges equal to zero on each atom.

Using the criteria for selecting among possible structures, we have to select two: the one with the unpaired electron on nitrogen that has the lowest formal charge, and the one that has a formal charge of zero on each atom but the unpaired electron on oxygen.

THINK ABOUT IT This example illustrates some of the limitations of the octet rule and of formal charges. An argument can be made for two structures based on formal charges and the relative electronegativity of the elements. The two structures, however, are quite different with respect to which atom has an odd number of valence electrons.

Practice Exercise Chlorine dioxide (ClO_2) is a compound used in water treatment and to bleach paper. Draw a Lewis structure for ClO_2.

8.7 The Length and Strength of Bonds

CONNECTION In this section we will apply some of the reactions and techniques we discussed in Chapter 5 on thermochemistry to determine the energies associated with forming covalent bonds.

Two features of a molecule that can be measured experimentally are useful in determining the real suitability of a Lewis structure: bond length and bond strength. Bond length is measured by a technique called X-ray diffraction, and bond strength is measured in thermochemical experiments.

Bond Length

Bond length depends on the identity of the atoms as well as on the number of bonds between them. For example, the H—Cl bond length in HCl (127 picometers) differs from the H—C bond length in CH_4 (110 pm). The carbon–oxygen bond length in carbon monoxide (113 pm), which has a triple bond between the two atoms, is shorter than the carbon–oxygen bond length in carbon dioxide (123 pm), which has a double bond between the two atoms.

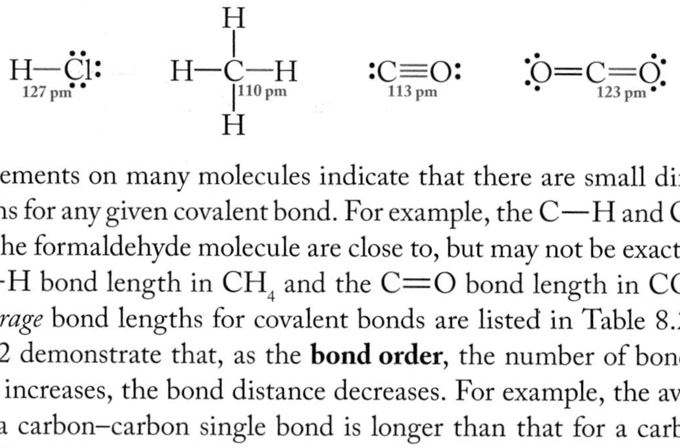

Measurements on many molecules indicate that there are small differences in bond lengths for any given covalent bond. For example, the C—H and C=O bond lengths in the formaldehyde molecule are close to, but may not be exactly the same as, the C—H bond length in CH_4 and the C=O bond length in CO_2. For this reason, *average* bond lengths for covalent bonds are listed in Table 8.2. The data in Table 8.2 demonstrate that, as the **bond order**, the number of bonds between two atoms, increases, the bond distance decreases. For example, the average bond length for a carbon–carbon single bond is longer than that for a carbon–carbon double bond, which in turn is longer than a carbon–carbon triple bond.

The values of average bond lengths help us to understand resonance structures in molecules such as ozone. The bond lengths between the oxygen atoms in O_3 are equal: 128 pm. We can compare this information with the bond lengths between oxygen atoms in O_2 (121 pm), which has an O=O double bond, and with the O—O single bond in hydrogen peroxide (H_2O_2; 148 pm):

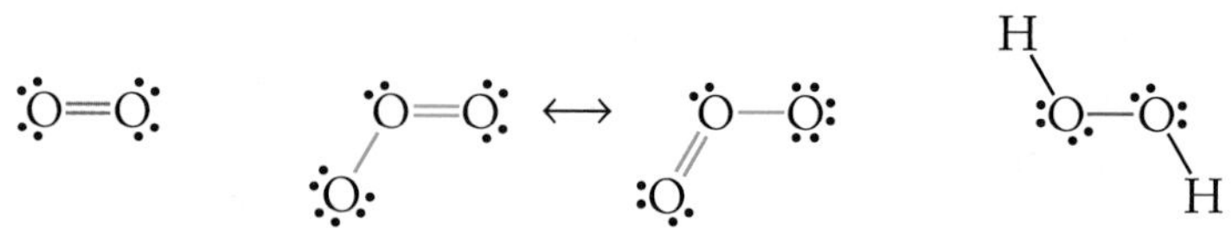

Formaldehyde

The **bond order** is the number of bonds between atoms. The bond orders are 1 for a single bond, 2 for a double bond, and 3 for a triple bond.

The energy needed to break 1 mole of a covalent bond in the gas phase is the **bond energy** of that bond.

| 121 pm | 128 pm | 148 pm |
| Oxygen | Ozone | Hydrogen peroxide |

Note that the oxygen–oxygen bond length in ozone falls between the values for a single and double bond. The *average bond order* for the O—O bond in ozone is $1\frac{1}{2}$.

Bond Energies

The energy changes associated with chemical reactions described in Chapter 5 involve the making and breaking of bonds. For example, if we assume that all the bonds in the reactant molecules are broken and all the bonds in the product molecules are reformed, then the reaction between methane and oxygen to form carbon dioxide and water requires that four C—H bonds are broken in CH_4 and two O=O bonds are broken in two molecules of O_2. In return, two C=O bonds are formed in CO_2 and four O—H bonds are formed in two molecules of H_2O:

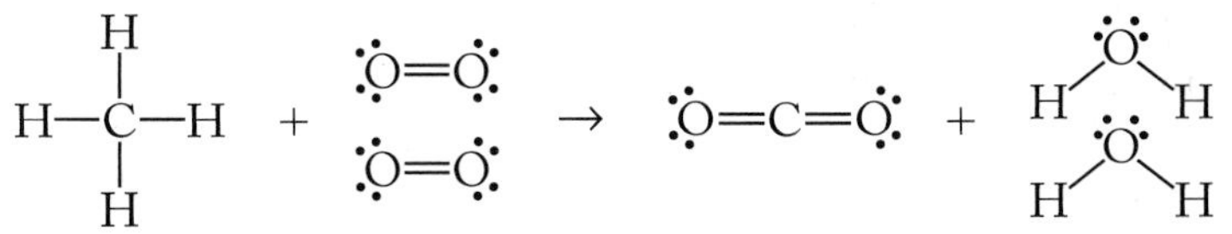

Breaking bonds consumes energy, and forming them releases energy. An important equality to consider in this analysis is that the quantity of energy needed to break a chemical bond equals the quantity of energy released when the same bond forms. If a chemical reaction is exothermic, as with burning methane, then more energy is released in forming the bonds of the products than is required to break the bonds in the reactants.

Bond energy, or *bond strength*, is usually expressed in terms of the enthalpy change (ΔH) required to break 1 mole of bonds in the gas phase. Bond energies for some common covalent bonds are given in Table 8.2. Like the list of bond lengths, the bond energies in Table 8.2 are average values. Bond energies are always positive quantities because breaking bonds requires energy. Bond energies can vary for a given bond, depending on molecular structure. For example, the bond energy of a C=O bond in carbon dioxide is 799 kJ/mol but is closer to 743 kJ/mol in formaldehyde. Another view of the variability in bond energy values for the same type of bond comes from the step-by-step decomposition of CH_4:

Decomposition Step	Energy Needed (kJ/mol)
$CH_4 \rightarrow CH_3 + H$	435
$CH_3 \rightarrow CH_2 + H$	453
$CH_2 \rightarrow CH + H$	425
$CH \rightarrow C + H$	339
Total:	1652
Average:	413

These results tell us that the chemical environment of a bond affects the energy required to break it: breaking the first C—H bond in methane is easier than breaking the second but more difficult than breaking the third or fourth. The total energy needed to break all four C—H bonds is 1652 kJ/mol, for an average of 1652/4, or 413 kJ/mol per bond.

Also note the differences in bond energies for O—O and O=O bonds given in Table 8.2. At 495 kJ/mol, the bond energy of the oxygen–oxygen double bond is more than three times the bond energy of the oxygen–oxygen single bond. This correlation between bond order and bond energy is true for other pairs of atoms:

TABLE 8.2 Selected Average Covalent Bond Lengths and Their Average Energies

Bond	Bond Length (pm)	Bond Energy (kJ/mol)
C—C	154	348
C=C	134	614
C≡C	120	839
C—N	143	293
C=N	138	615
C≡N	116	891
C—O	143	358
C=O	123	743[a]
C≡O	113	1072
C—H	110	413
C—F	133	485
C—Cl	177	328
N—H	104	388
N—N	147	163
N=N	124	418
N≡N	110	941
N—O	136	201
N=O	122	607
N≡O	106	678
O—O	148	146
O=O	121	495
O—H	96	463
S—O	151	265
S=O	143	523
S—S	204	266
S—H	134	347
H—H	75	436
H—F	92	567
H—Cl	127	431
H—Br	141	366
H—I	161	299
F—F	143	155
Cl—Cl	200	243
Br—Br	228	193
I—I	266	151

[a]The bond energy of the C=O bond in CO_2 is 799 kJ/mol.

greater bond order means greater bond energy. The bond energy of the nitrogen–nitrogen triple bond (941 kJ/mol) is one of the largest in Table 8.2. It is more than twice the bond energy of the nitrogen–nitrogen double bond and more than four times that of the nitrogen–nitrogen single bond. The large amount of energy that must be consumed in breaking a nitrogen–nitrogen triple bond is one of the reasons why N_2 participates in so few chemical reactions.

An examination of the reaction between CH_4 and oxygen (Figure 8.12) reveals that, during the combustion of 1 mole of CH_4, 4 moles of C—H bonds and 2 moles of O=O bonds must be broken. The formation of 1 mole of CO_2 and 2 moles of H_2O requires the formation of 2 moles of C=O bonds and 4 moles of O—H bonds. The net change in energy resulting from these bonds breaking and forming can be calculated from the following data:

Type of Bond	Number of Bonds (mol)	Bond Energy (kJ/mol)	Sign of ΔH in Calculation	ΔH (kJ)
C—H	4	413	+	$+(4 \times 413) = +1652$
O=O	2	495	+	$+(2 \times 495) = +990$
O—H	4	463	–	$-(4 \times 463) = -1852$
C=O	2	799	–	$-(2 \times 799) = -1598$
				$\Delta H_{rxn} = -808$

▶❙❙ **CHEMTOUR** Estimating Enthalpy Changes

Here is the general equation for calculating enthalpies of reaction from bond energies:

$$\Delta H_{rxn} = \sum \Delta H_{\text{bond breaking}} + \sum \Delta H_{\text{bond forming}}$$

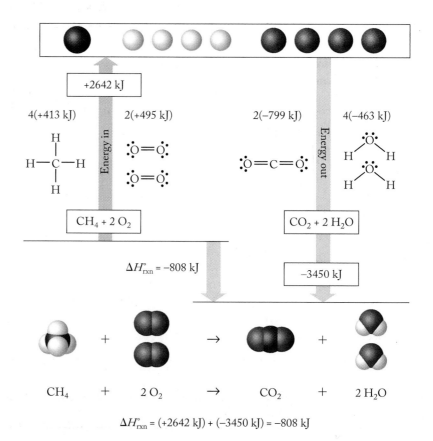

FIGURE 8.12 The combustion of 1 mole of methane requires that 4 moles of C—H bonds and 2 moles of O=O bonds be broken. These processes require an enthalpy change of +2642 kJ/mol. Formation of 4 moles of O—H bonds and two moles of C=O bonds results in the release of −3450 kJ/mol. The overall reaction is exothermic by −808 kJ/mol.

SAMPLE EXERCISE 8.12   **Calculating Heats of Reaction from Average Bond Energies**

Use the average bond energies in Table 8.2 to estimate ΔH_{rxn} for forming HCl(g) from $H_2(g)$ and $Cl_2(g)$:

$$H\!-\!H \quad + \quad \ddot{\underset{..}{Cl}}\!-\!\ddot{\underset{..}{Cl}}\!: \quad \rightarrow \quad 2\,H\!-\!\ddot{\underset{..}{Cl}}\!:$$

COLLECT AND ORGANIZE From the appropriate bond energies for hydrogen, chlorine, and hydrogen chloride in Table 8.2, we can determine the enthalpy of formation of hydrogen chloride.

ANALYZE The reaction between H_2 and Cl_2 requires that 1 mole of H—H bonds and 1 mole of Cl—Cl bonds must be broken. The formation of 2 moles of HCl requires the formation of 2 moles of H—Cl bonds.

SOLVE The corresponding bond energies for H—H, Cl—Cl, and H—Cl bonds from Table 8.2 are 436 kJ/mol, 243 kJ/mol, and 431 kJ/mol:

$$\underset{+436\ kJ/mol}{H\!-\!H} \quad + \quad \underset{+243\ kJ/mol}{\ddot{\underset{..}{Cl}}\!-\!\ddot{\underset{..}{Cl}}\!:} \quad \rightarrow \quad \underset{2(-431\ kJ/mol)}{2\,H\!-\!\ddot{\underset{..}{Cl}}\!:}$$

In calculating ΔH_{rxn} we must remember that breaking bonds requires energy (a positive value) but making a bond releases energy (a negative value). A table of the bond energies involved helps us calculate the enthalpy of the reaction:

Type of Bond	Number of Bonds (mol)	Bond Energy (kJ/mol)	Sign of ΔH in Calculation	ΔH (kJ)
H—H	1	436	+	$+(1 \times 436) = +436$
Cl—Cl	1	243	+	$+(1 \times 243) = +243$
H—Cl	2	431	–	$-(2 \times 427) = -862$
				$\Delta H_{rxn} = -183$

THINK ABOUT IT The formation of 2 moles of HCl from H_2 and Cl_2 produces much less energy than the combustion of methane. On a per mole basis, the enthalpy of formation of HCl is

$$\Delta H_f = \frac{-183\ kJ}{2\ mol} = -91.5\ kJ/mol$$

This value is close to the experimentally determined value for the heat of formation of HCl found in Appendix 4, −92.3 kJ/mol.

Practice Exercise Use average bond energies to calculate ΔH for the reaction of H_2 and N_2 to form ammonia (NH_3):

$$:N\!\equiv\!N\!: \quad + \quad 3(H\!-\!H) \quad \rightarrow \quad 2\,H\!-\!\overset{..}{N}\!-\!H \atop \quad\quad\quad\quad\quad\quad | \atop \quad\quad\quad\quad\quad\quad H$$

CONCEPT TEST

Considering bond energies, suggest a reason why oxygen can be used in reactions in the human body while nitrogen is inert.

Fluorine and Oxygen: Location, Location, Location

Fluorine is the most reactive substance known. It forms compounds with every element in the periodic table except He, Ne, and Ar, and it does not occur as the free element in nature. It attacks anything with which it comes into contact: if fluorine gas is allowed to flow over the surface of water, the water actually burns:

$$5\ F_2(g) + 5\ H_2O(\ell) \rightarrow$$
$$8\ HF(g) + O_2(g) + H_2O_2(\ell) + OF_2(g)$$

Wood, plastic, and even some metals burst into white-hot, intense flame in an atmosphere of fluorine gas. Even "fireproof" asbestos burns in fluorine. Fluorine reacts with hydrocarbons to yield products called fluorocarbons that are completely fluorinated; all the hydrogen atoms in the hydrocarbon are replaced by fluorine atoms.

Although fluorine is extremely reactive itself, fluorocarbons are among the most stable compounds known because of the stability of the carbon–fluorine bond. Fluorocarbons are very rare in nature; most are manufactured. They are generally unreactive toward most chemical reagents, inert to solvents, and nonflammable. The stability of the carbon–fluorine bond has been used to advantage in certain commercial products. The addition of a single carbon–bound fluorine to certain pharmaceutical compounds improves their potency, because it makes the molecules more difficult for the body to degrade. The anesthetic halothane is a fluorine-containing molecule that was introduced in 1957 as the first modern inhaled anesthetic. The low toxicity, low flammability, and high stability of halothane caused it to replace the much more hazardous anesthetics such as ether ($CH_3CH_2OCH_2CH_3$) and chloroform ($CHCl_3$). A synthetic polymer (a very large molecule) made when molecules of tetrafluoroethylene bind together:

$$n \quad \underset{F}{\overset{F}{C}} = \underset{F}{\overset{F}{C}} \quad \rightarrow \quad \left[\begin{matrix} F & F \\ | & | \\ C & C \\ | & | \\ F & F \end{matrix} \right]_n \quad \text{where } n \approx 1000$$

Tetrafluoroethylene   Teflon

is commonly known as Teflon, a substance so stable to moderate temperatures and so nonreactive compared with other nonmetallic materials that it is used on no-stick cookware and electrical insulation. Fluorine itself is so reactive, you might wonder what kind of containers hold it. Fluorine is frequently shipped in metal cylinders lined with Teflon.

In contrast to the diatomic element F_2, the fluoride ion (F^-) is common and relatively nontoxic. For decades, small quantities of fluoride salts have been added to municipal water supplies in the United States to prevent cavities in adults and children. Sodium fluoride is also added in small amounts (about 0.15% by weight) to toothpaste for the same reason.

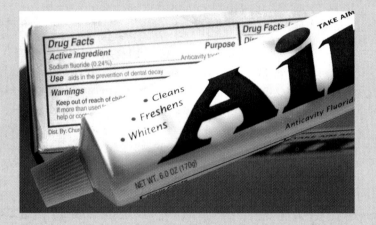

Oxygen is the second most reactive element, but O_2 is so much less reactive under normal conditions than F_2 that it exists in large quantities in the atmosphere. It is essential to life on Earth. Physical activity for human beings becomes difficult if the percentage of oxygen in the air drops from the normal value of 21% to around 12%, and breathing air that contains only 6% oxygen can result in death in 6 to 8 minutes.

In addition to being required for life, O_2 is an important commercial bulk chemical, heavily used in the steel industry and in medicine. It is frequently transported and stored as a cryogenic (very cold) liquid referred to as LOX (*l*iquid *ox*ygen). Although very cold (around −183°C), LOX is extremely reactive and must be handled with great care. Metallic iron burns so vigorously in liquid oxygen that the iron melts from the heat of combustion even though it is surrounded by a liquid that is colder than −180°C. Many materials normally considered noncombustible burn at explosive rates in liquid oxygen; and oils, greases, paint, and paper burn immediately on contact. LOX penetrates into porous materials such as wood and asphalt; and even when combustion does not occur immediately, the material remains potentially hazardous. Asphalt has been known to burst into flames days after LOX has been spilled on it. Large tanks containing cryogenic oxygen at hospitals or manufacturing sites are always mounted on concrete slabs, not asphalt, as a safety precaution.

The stories of oxygen and fluorine intersect in the upper atmosphere of Earth. The molecules at the intersection are ozone (O_3) and chlorofluorocarbons (CFCs) from fire extinguishers (halons) and refrigerants (Freons). CFCs such as CF_2Cl_2 have been banned because of their deleterious effect on the ozone layer in the stratosphere. Halons were used for years as "clean" fire-extinguishing agents in rooms containing sensitive electronic equipment (aircraft interiors, computers) that needs to be protected in the advent of fire. Freons were used for decades as refrigerants and aerosol propellants. Very useful at ground level, these CFCs become extremely deleterious in the upper atmosphere.

How ozone is regarded also depends on location. At ground level, ozone is very dangerous, especially if inhaled in concentrations as low as 0.1 ppm. Its reactivity makes it damaging to cells, which in turn makes it a very useful antiseptic and antibacterial agent and even bleach, but ozone is an inherently unstable gas. It cannot be shipped from place to place; when needed, it typically is generated on site from oxygen by a device called an ozonizer. In contrast, ozone can exist for long periods of time in the upper atmosphere, where the decreased density of the air diminishes the likelihood that ozone will encounter something with which it can react. The amount of ozone in the upper atmosphere is not large—if it were compressed by atmospheric pressure at Earth's surface, it would be squashed to a band only a few millimeters thick—but in the upper atmosphere it exists in a layer more than 20 km thick that absorbs UV radiation from the sun and protects life on Earth from damage. Ozone is produced in the upper atmosphere when oxygen molecules react in the presence of UV light to yield O_3 and an oxygen atom:

$$2\,O_2(g) \xrightarrow{h\nu} O_3(g) + O(g)$$

The conversion of ozone back to O_2 occurs but is slow under conditions in the upper atmosphere, so a protective ozone layer remains in place unless disrupted by other agents.

When compounds like the halons and Freons are released in the environment at ground level, they persist because of the very nonreactivity that makes them so useful. Wind and air currents carry them to the upper reaches of the atmosphere, where UV radiation provides sufficient energy to break very stable bonds and form free radicals, very reactive odd-electron species:

$$CF_2Cl_2(g) \xrightarrow{h\nu} CF_2Cl(g) + Cl(g)$$

The chlorine free radicals (chlorine atoms) react with ozone to make chlorine monoxide (another free radical) and oxygen:

$$Cl(g) + O_3(g) \rightarrow ClO(g) + O_2(g)$$

Chlorine monoxide reacts with more ozone, generating more oxygen and another chlorine free radical:

$$ClO(g) + O_3(g) \rightarrow Cl(g) + 2\,O_2(g)$$

If we add the previous two equations together and cancel out species that appear on both sides, the result is the conversion of ozone to oxygen:

$$\cancel{Cl(g)} + O_3(g) \rightarrow \cancel{ClO(g)} + O_2(g)$$
$$\cancel{ClO(g)} + O_3(g) \rightarrow \cancel{Cl(g)} + 2\,O_2(g)$$
$$\overline{2\,O_3(g) \rightarrow 3\,O_2(g)}$$

Not only does this process take place much faster than the spontaneous conversion of ozone to oxygen in the absence of other reactive free radicals, but chlorine free radicals are not consumed in the overall reaction. One chlorine free radical can destroy thousands of ozone molecules before it is eliminated by a reaction with another free radical in the stratosphere.

The Montreal Protocol, an international agreement now signed by over 180 nations, mandated the phasing out of synthetic substances responsible for depletion of Earth's ozone layer. Research continues to search for new substances that will be as useful for fire extinguishers and refrigerants as CFCs were but without damaging the ozone layer in the stratosphere.

Summary

SECTION 8.1 A **chemical bond** results from two atoms sharing electrons (a **covalent bond**) or two ions being attracted to each other (an **ionic bond**). Metal atoms in solid pieces of metal pool their electrons in a **metallic bond**.

SECTION 8.2 **Lewis symbols** are useful in accounting for the **valence** (outermost) **electrons** in an atom or monatomic ion. Chemical stability is achieved in many of the atoms in molecules and ions that have 8 valence electrons, following the **octet rule**. Hydrogen deviates from the octet rule, as it can accommodate only 2 valence electrons, or a **duet**. The **Lewis structures** for molecules show how pairs of valence electrons (**bonding pairs**) may be shared in **single**, **double**, and **triple** covalent bonds. The **bond length**—the distance between the nuclear centers of bonded atoms—decreases in following order: single bond > double bond > triple bond. The **lone pairs** of electrons on atoms in a molecule do not contribute to bonding.

SECTION 8.3 Unequal electron sharing between atoms of different elements results in **polar covalent bonds** and **polar** molecules. Unequal sharing of the electrons in covalent bonds results from different **electronegativities** of the bonded atoms. The electronegativities of elements decrease down a column (group) in the periodic table and increase from left to right along a row. They generally increase with increasing ionization energies and with decreasing atomic size. Thus, fluorine is the most electronegative element. **Bond polarity** is related to the difference in electronegativity between two bonded atoms.

SECTION 8.4 **Resonance** occurs when two or more equivalent Lewis structures can be drawn for one molecule or ion. Lewis structures with the same arrangement of atoms but different arrangements of electrons are **resonance structures**. The presence of resonance structures indicates that the actual electron arrangement in a molecule is an average of two or more arrangements.

SECTION 8.5 The most favored structure of a molecule is one in which the **formal charges** on its atoms are closest to zero, and any negative formal charges are on the more-electronegative atoms. Formal charge of an atom in a Lewis structure is the difference between the number of valence electrons in the free atom and the number of valence electrons assigned to the atom in the structure. The **bonding capacity** of an element—the number of bonds it makes when its formal charge is zero—is useful to consider when drawing Lewis structures.

SECTION 8.6 **Electron-deficient compounds** can have less than 8 valence electrons around one of their atoms (as around boron in certain boron compounds), while compounds containing elements with $Z \geq 12$ (for example, sulfur) such as SF_6 can expand their octets by using empty valence-shell d orbitals. Odd-electron molecules (**free radicals**), such as NO, have an odd number of valence electrons on the less electronegative atom.

SECTION 8.7 The bond lengths between atoms in molecules and polyatomic ions can be determined from experimental measurements. The **bond order** of a covalent bond reflects the number of bonding pairs involved in the bond. **Bond energies** measure the energy required to break 1 mole of a covalent bond in the gas phase. As the bond order between two atoms increases, the bond length decreases and the bond energy increases.

Problem-Solving Summary

TYPE OF PROBLEM	CONCEPTS AND EQUATIONS	SAMPLE EXERCISES
Drawing Lewis symbols for atoms	Arrange the valence electrons around the atom without exceeding 8 valence electrons.	8.1
Drawing Lewis structures for molecules and monatomic ions	Connect the atoms with covalent bonds, distributing the valence electrons to give each atom (except H) 8 valence electrons (2 for H); introduce multiple bonds where necessary.	8.2–8.4
Drawing structures and comparing the polarity of bonds	Assess the relative electronegativities of the two atoms in a covalent bond.	8.5–8.6
Drawing resonance structures of molecules and ions	Include all possible arrangements of covalent bonds in the molecule if more than one equivalent structure can be drawn.	8.7–8.8
Selecting Lewis structure based on formal charges	Formal charge = $\left(\begin{array}{c}\text{number of valence}\\ \text{e}^- \text{ for the atom}\end{array}\right) - \left(\begin{array}{c}\text{number of electrons}\\ \text{in lone pairs}\end{array} + \frac{1}{2}\begin{array}{c}\text{number of electrons}\\ \text{in bonding pairs}\end{array}\right)$ Select structures with formal charges closest to zero and with negative formal charges on more-electronegative atoms.	8.9

TYPE OF PROBLEM	CONCEPTS AND EQUATIONS	SAMPLE EXERCISES
Drawing Lewis structures for molecules containing atoms with expanded octets	Distribute the valence electrons in the Lewis structure allowing elements in rows 3 and below to have more than 8 valence electrons if more than 4 bonds are needed or if the structure with the expanded octet results in lower formal charges.	8.10
Drawing Lewis structures for odd-electron molecules	Distribute the valence electrons in the Lewis structure to leave the most electronegative atom(s) with 8 valence electrons.	8.11
Calculating heats of reaction from average bond energies	$\Delta H = \Sigma$ (bond energies for bonds broken) + Σ (bond energies for bonds formed).	8.12

Visual Problems

8.1. Which group highlighted in Figure P8.1 contains atoms that have the following? (a) 1 valence electron; (b) 4 valence electrons; (c) 6 valence electrons

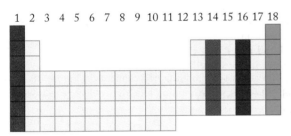

FIGURE P8.1

8.2. Which of the groups highlighted in Figure P8.2 contains atoms with the following? (a) 2 valence electrons; (b) 5 valence electrons; (c) 3 valence electrons

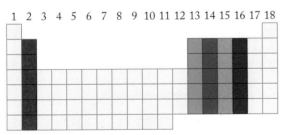

FIGURE P8.2

8.3. Which of the following Lewis symbols correctly portrays the most stable ion of magnesium?

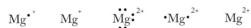

8.4. Which of the following Lewis symbols are correct?

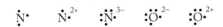

8.5. Which of the highlighted elements in Figure P8.5 has the greatest bonding capacity?

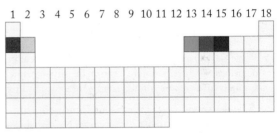

FIGURE P8.5

8.6. Which two of the highlighted elements in Figure P8.6 is the pair that forms the bond with the most ionic character?

8.7. Referring to the periodic table in Figure P8.6, which of the highlighted elements has the greatest electronegativity?

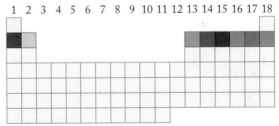

FIGURE P8.6

NOTE: The color scale used in Problems 8.8 and 8.9 is the same as in Figure 8.4, where dark blue is a charge of 1+, red is a charge of 1−, and yellow is 0. The larger the size, the greater the electron density.

8.8. Which of the drawings in Figure P8.8 is the best description of the electron density in CO?

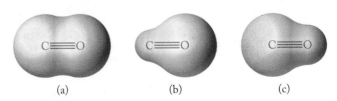

(a)   (b)   (c)

FIGURE P8.8

*8.9. Which of the drawings in Figure P8.9 best describes the electron density in LiF?

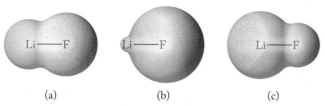

(a)   (b)   (c)

FIGURE P8.9

8.10. Are the following three structures resonance forms of the thiocyanate ion (SCN^-)?

$$\left[:\ddot{S}-N\equiv C:\right]^- \quad \left[:N\equiv C-\ddot{S}:\right]^- \quad \left[:\ddot{N}-S\equiv C:\right]^-$$

8.11. Why are the structures below not all resonance forms of the molecule S_2O?

$$:\ddot{S}=\ddot{S}-\ddot{O}: \qquad :\ddot{S}-\ddot{O}=\ddot{S}:$$

$$:\ddot{S}=\overset{\ddot{S}}{\underset{}{}}-\ddot{O}: \qquad :\ddot{S}-\overset{\ddot{O}}{\underset{}{}}=\ddot{S}:$$

*8.12. Which of the drawings in Figure P8.12 most accurately describes the distribution of electron density in ozone? Explain your answer. The color scale is defined in Problem 8.8.

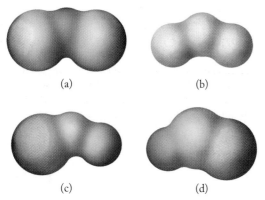

(a)   (b)

(c)   (d)

FIGURE P8.12

*8.13. Which of the drawings in Figure P8.13 most accurately describes the distribution of electron density in SO_2? Explain your answer. The color scale is defined in Problem 8.8.

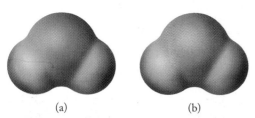

(a)   (b)

FIGURE P8.13

8.14. Which groups among main group elements in Figure P8.14 have an odd number of valence electrons?

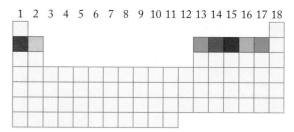

FIGURE P8.14

8.15. Krypton and xenon form compounds with only the most reactive of other elements. Which of the highlighted elements in Figure P8.15 is one of these highly reactive elements?

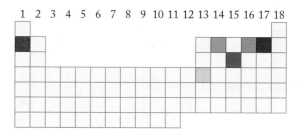

FIGURE P8.15

8.16. Which of the highlighted elements in Figure P8.16 expands its octet when bonding to a highly electronegative element?

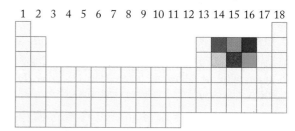

FIGURE P8.16

8.17. Which of the highlighted groups in Figure P8.17 are most likely to have negative formal charges in diatomic compounds with hydrogen, HX?

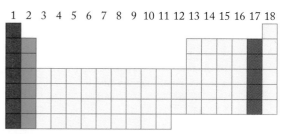

FIGURE P8.17

Questions and Problems

Ionic, Covalent, and Metallic Bonds; An Introduction to Lewis Theory

CONCEPT REVIEW

8.18. Which electrons in an atom are considered the *valence* electrons?

8.19. Does the number of valence electrons in a neutral atom ever equal the atomic number?

8.20. Does the number of valence electrons in a neutral atom ever equal the group number?

8.21. Do all the elements in a group in the periodic table have the same number of valence electrons?

8.22. Describe the differences in bonding in *covalent* and *ionic* compounds.

8.23. Some of his critics described Gilbert N. Lewis's approach to explaining covalent bonding as an exercise in double counting and therefore invalid. Explain the basis for this criticism.

8.24. Does the octet rule mean that a diatomic molecule has 16 valence electrons?

8.25. Does the octet rule mean that a diatomic molecule has 8 valence electrons?

8.26. Why is the bonding pattern in water H—O—H and not H—H—O?

*8.27. Does each atom in a pair that is covalently bonded always contribute the same number of valence electrons to form the bonds between them?

PROBLEMS

8.28. Draw Lewis symbols for atoms of lithium, magnesium, and aluminum.

8.29. Draw Lewis symbols for atoms of nitrogen, oxygen, fluorine, and chlorine.

8.30. Draw Lewis symbols for Na^+, In^+, Ca^{2+}, and S^{2-}.

8.31. Draw Lewis symbols for the most stable ion formed by lithium, magnesium, aluminum, and fluorine.

8.32. Which of the following ions have a complete valence-shell octet? B^{3+}, I^-, Ca^{2+}, or Pb^{2+}

8.33. Draw Lewis symbols for Xe, Sr^{2+}, Cl, and Cl^-. How many valence electrons are in each atom or ion?

8.34. Draw the Lewis symbol for an ion that has the following:
 a. 1+ charge and 1 valence electron
 b. 3+ charge and 0 valence electrons

8.35. Draw the Lewis symbol for an ion that has the following:
 a. 1− charge and 8 valence electrons
 b. 1+ charge and 5 valence electrons

8.36. How many valence electrons does each of the following species contain? (a) BN; (b) HF; (c) OH^-; (d) CN^-

8.37. How many valence electrons does each of the following species contain? (a) N_2^+; (b) CS^+; (c) CN; (d) CO

8.38. Draw Lewis structures for the following diatomic molecules and ions: (a) CO; (b) O_2^-; (c) ClO^-; (d) CN^-

8.39. Draw Lewis structures for the following diatomic molecules and ions: (a) F_2; (b) NO^+; (c) SO; (d) HI

8.40. How many electron pairs are shared in each of the molecules and ions in Problem 8.38?

8.41. How many covalent bonds are there in each of the molecules and ions in Problem 8.39?

8.42. Chlorofluorocarbons (CFCs) are linked to the depletion of stratospheric ozone. They are also greenhouse gases. Draw Lewis structures for the following CFCs:
 a. CF_2Cl_2 (Freon 12)
 b. Cl_2FCCF_2Cl (Freon 113, containing a C—C bond)
 c. C_2Cl_3F (Freon 1113, containing a C—C bond)

8.43. The replacement of one halogen in a CFC by hydrogen makes the compound more environmentally "friendly." Draw Lewis structures for the following compounds:
 a. CHF_2Cl (Freon 22)
 b. $CHBr_3$ (bromoform, used as a solvent in geology)
 c. CH_2Cl_2 (methylene chloride, a common laboratory solvent)

8.44. **Skunks and Rotten Eggs** Many sulfur-containing organic compounds have characteristically foul odors: butane thiol ($CH_3CH_2CH_2CH_2SH$) is responsible for the odor of skunks, and rotten eggs smell the way they do because they produce tiny amounts of pungent hydrogen sulfide, H_2S. Draw the Lewis structures for $CH_3CH_2CH_2CH_2SH$ and H_2S.

8.45. Formic acid, HC(O)OH, is the smallest organic acid. Draw its Lewis structure given the connectivity of the atoms as shown in Figure P8.45.

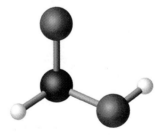

FIGURE P8.45

8.46. Chlorine combines with oxygen in several proportions. Dichlorine monoxide (Cl_2O) is used in the manufacture of bleaching agents. Potassium chlorate ($KClO_3$) is used in oxygen generators aboard aircraft. Draw the Lewis structures for Cl_2O and ClO_3^-. Cl is the central atom in each case.

8.47. Dangers of Mixing Cleansers Labels on household cleansers caution against mixing bleach with ammonia (Figure P8.47) because the reaction produces monochloramine (NH_2Cl), hydrazine (N_2H_4), and the hypochlorite ion (OCl^-):

$$NH_3(aq) + OCl^-(aq) \rightarrow NH_2Cl(aq) + OH^-(aq)$$

$$NH_2Cl(aq) + NH_3(aq) + OH^-(aq) \rightarrow N_2H_4(aq) + H_2O(\ell)$$

Draw the Lewis structures for monochloramine, hydrazine, and the hypochlorite ion.

FIGURE P8.47

Unequal Sharing, Electronegativity, and Other Periodic Properties

CONCEPT REVIEW

8.48. Which of the following periodic properties can be determined experimentally: electronegativity or ionization energy?

8.49. How can we use electronegativity to predict whether a bond between two atoms is likely to be covalent or ionic?

8.50. How do the electronegativities of the elements change across a period and down a group?

8.51. Explain on the basis of atomic structure why trends in electronegativity are related to trends in atomic size.

8.52. Is the element with the most valence electrons in a period also the most electronegative?

8.53. What is meant by the term *polar covalent bond*?

8.54. Why are the electrons in bonds between different elements not shared equally?

PROBLEMS

8.55. Which of the following bonds are polar? C—Se, C—O, Cl—Cl, O=O, N—H, C—H. In the bond or bonds that you selected, which atom has the greater electronegativity?

8.56. Which is the least polar of the following bonds? C—Se, C=O, Cl—Br, O=O, N—H, C—H

8.57. Which of the binary compounds formed by the following pairs of elements contain polar covalent bonds, and which are considered ionic compounds?
a. C and S
c. Al and Cl
b. C and O
d. Ca and O

8.58. Which of the beryllium halides, if any, are considered ionic compounds?

Resonance

CONCEPT REVIEW

8.59. Explain the concept of resonance.

8.60. Does resonance help to stabilize a molecule or an ion?

8.61. Explain why SO_2 exhibits resonance, whereas H_2S does not.

8.62. What features do all the resonance forms of a molecule or ion have in common?

*8.63. What role does resonance play in our understanding of bond lengths?

8.64. How can you recognize when resonance forms are needed to account for the bonding in a molecule or ion?

PROBLEMS

8.65. Draw two Lewis structures showing the resonance that occurs in cyclobutadiene (C_4H_4), a cyclic molecule with a structure that includes a ring of four carbon atoms.

*8.66. Pyridine (C_5H_5N) and pyrazine ($C_4H_4N_2$) have structures similar to benzene's. Both compounds have structures with six atoms in a ring. Draw Lewis structures for pyridine and pyrazine showing all resonances forms. The N atoms in pyrazine are across the ring from each other as shown in Figure P8.66.

FIGURE P8.66

*8.67. Oxygen and nitrogen combine to form a variety of nitrogen oxides, including the following two unstable compounds each with two nitrogen atoms per molecule: N_2O_2 and N_2O_3. Draw Lewis structures for these molecules showing all resonance forms.

*8.68. Oxygen and sulfur combine to form a variety of different sulfur oxides. Some are stable molecules and some, including S_2O_2 and S_2O_3, decompose when they are heated. Draw Lewis structures for these two compounds showing all resonance forms.

8.69. Draw Lewis symbols for hydrocyanic acid (HCNO) showing all resonance forms.

8.70. Draw Lewis symbols for hydroazoic acid (HN_3) showing all resonance forms.

8.71. Draw three Lewis structures showing the resonance that occurs in the carbonate ion (CO_3^{2-}).

8.72. Nitrogen-fixing bacteria convert urea [$H_2NC(O)NH_2$] into nitrite ions (NO_2^-). Draw Lewis structures for these two species. Include all resonance forms. (*Hint*: There is a C=O bond in urea.)

Formal Charges: Choosing among Lewis Structures

CONCEPT REVIEW

8.73. Describe how formal charges are used to choose between possible molecular structures.

8.74. How do the electronegativities of elements influence the selection of which a Lewis structure is favored?

8.75. In a molecule containing S and O atoms, is a structure with a negative formal charge on sulfur more likely to contribute to bonding than an alternative structure with a negative formal charge on oxygen?

8.76. In a cation containing N and O, why do Lewis structures with a positive formal charge on nitrogen represent electron arrangements that contribute more to the picture of actual bonding in the molecule than do those structures with a positive formal charge on oxygen?

*8.77. Does the more electronegative atom in a diatomic molecule made up of two different atoms always have a negative formal charge?

PROBLEMS

8.78. Hydrogen isocyanide (HNC) has the same elemental composition as hydrogen cyanide (HCN) but the H in HNC is bonded to the nitrogen. Draw a Lewis structure for HNC, and assign formal charges to each atom. How do the formal charges on the atoms differ in the Lewis structures for HCN and HNC?

8.79. Molecules in Interstellar Space Hydrogen cyanide (HCN) and cyanoacetylene (HC_3N) have been detected in the interstellar regions of space. Draw Lewis structures for these molecules, and assign formal charges to each atom. The hydrogen atom is bonded to carbon in both cases.

8.80. Origins of Life The discovery of polyatomic organic molecules such as cyanamide (H_2NCN) in interstellar space has led some scientists to believe that the molecules from which life began on Earth may have come from space. Draw Lewis structures for cyanamide, and select the preferred structure on the basis of formal charges.

8.81. Complete the Lewis structures, and assign formal charges to the atoms in five of the resonance forms of thionitrosyl azide (SN_4). Indicate which of your structures should be most stable. The molecule is linear with S at one end.

*8.82. Nitrogen is the central atom in molecules of nitrous oxide (N_2O). Draw Lewis structures for another possible arrangement: N—O—N. Assign formal charges, and suggest a reason why this structure is not likely to be stable.

8.83. More Molecules in Space Formamide ($HCONH_2$) and methyl formate (HCO_2CH_3) also have been detected in space. Draw Lewis structures for these compounds, based on the following arrangements of atoms:

*8.84. Nitromethane (CH_3NO_2) reacts with hydrogen cyanide to produce $CNNO_2$ and CH_4:

$$HCN(g) + CH_3NO_2(g) \rightarrow CNNO_2(g) + CH_4(g)$$

a. Draw Lewis structures for CH_3NO_2, showing all resonance forms.

b. Draw Lewis structures for $CNNO_2$, showing all resonance forms, based on the following two possible skeletal structures for it:

Assign formal charges, and predict which structure is more likely to exist.

c. Are the two structures of $CNNO_2$ resonance forms of each other?

8.85. Use formal charges to determine which resonance form of each of the following ions is preferred: CNO^-, NCO^-, and CON^-.

Exceptions to the Octet Rule

CONCEPT REVIEW

8.86. Why do C, N, O, and F atoms in covalently bonded molecules and ions have no more than 8 valence electrons?

8.87. Are all odd-electron molecules exceptions to the octet rule?

8.88. Do atoms in rows 3 and below always expand their octets? Explain your answer.

PROBLEMS

8.89. In which of the following molecules does the sulfur atom have an expanded octet? (a) SF_6; (b) SF_5; (c) SF_4; (d) SF_2

8.90. In which of the following molecules and ions does the phosphorus atom have an expanded octet? (a) $POCl_3$; (b) PF_5; (c) PF_3; (d) P_2F_4 (which has a P—P bond).

8.91. How many electrons are there in the covalent bonds surrounding the sulfur atom in the following species? (a) SF_4O; (b) SOF_2; (c) SO_3; (d) SF_5^-

8.92. How many electrons are there in the covalent bonds surrounding the phosphorus atom in the following species? (a) $POCl_3$; (b) H_3PO_4; (c) H_3PO_3; (d) PF_6^-

*8.93. Draw Lewis structures for NOF_3 and POF_3 in which the group 15 element is the central atom and the other atoms are bonded to it. What differences are there in the types of bonding in these molecules?

*8.94. The phosphate anion (PO_4^{3-}) is common in minerals. The corresponding nitrogen-containing anion NO_4^{3-}, is unstable but can be prepared by reacting sodium nitrate with sodium oxide at 300°C. Draw Lewis structures for each anion. What are the differences in bonding between these ions?

8.95. Dissolving NaF in selenium tetrafluoride (SeF_4) produces $NaSeF_5$. Draw Lewis structures for SeF_4 and SeF_5^-. In which structures does Se have more than 8 valence electrons?

8.96. Reaction between NF_3, F_2, and SbF_3 at 200°C and 100 atm pressure gives the ionic compound NF_4SbF_6:

$$NF_3(g) + 2 F_2(g) + SbF_3(g) \rightarrow NF_4SbF_6(s)$$

Draw Lewis structures for the ions in this product.

8.97. Ozone Depletion The compound Cl_2O_2 may play a role in ozone depletion in the stratosphere. In the laboratory, reaction of $FClO_2$ with aluminum chloride produces Cl_2O_2 and $AlCl_2F$:

$$FClO_2(g) + AlCl_3(s) \rightarrow Cl_2O_2(g) + AlFCl_2(s)$$

Draw a Lewis structure for Cl_2O_2 based on the arrangement of atoms in Figure P8.97:

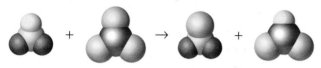

FIGURE P8.97

Does either of the chlorine atoms in the structure have an expanded octet?

8.98. Cl_2O_2 decomposes to chlorine and chlorine dioxide as shown in Figure P8.98:

$$2\,Cl_2O_2(g) \rightarrow Cl_2(g) + 2\,ClO_2(g)$$

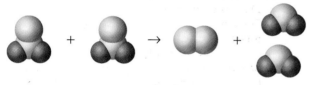

FIGURE P8.98

Draw the Lewis structure for ClO_2. Which atom in the structure does not have 8 valence electrons?

8.99. Which of the following chlorine oxides are odd-electron molecules? (a) Cl_2O_7; (b) Cl_2O_6; (c) ClO_4; (d) ClO_3; (e) ClO_2

8.100. Which of the following nitrogen oxides are odd-electron molecules? (a) NO; (b) NO_2; (c) NO_3; (d) N_2O_4; (e) N_2O_5

8.101. In the following species, which atom is most likely to have an unpaired electron? (a) SO^+; (b) NO; (c) CN; (d) OH

8.102. In the following molecules, which atom is most likely to have an unpaired electron? (a) NO_2; (b) CNO; (c) ClO_2; (d) HO_2

8.103. Which of the following Lewis structures contributes most to the bonding in CNO?

a. $\cdot\ddot{C}-N\equiv O\colon$ c. $\colon C\equiv N-\ddot{O}\cdot$

b. $\colon\ddot{C}=N=\ddot{O}\colon$ d. $\cdot C\equiv N-\ddot{O}\colon$

8.104. Why is the following Lewis structure unlikely to contribute much to the bonding in NCO?

$$\colon\ddot{N}-C\equiv O\cdot$$

The Length and Strength of Bonds

CONCEPT REVIEW

8.105. Do you expect the nitrogen–oxygen bond length in the nitrate ion to be the same as in the nitrite ion?

8.106. Why is the oxygen–oxygen bond length in O_3 not the same as in O_2?

8.107. Explain why the nitrogen–oxygen bond lengths in N_2O_4 (which has a nitrogen–nitrogen bond) and N_2O are nearly identical (118 and 119 pm, respectively).

8.108. Do you expect the sulfur–oxygen bond lengths in sulfite (SO_3^{2-}) and sulfate (SO_4^{2-}) ions to be about the same? Why?

8.109. Rank the following ions in order of increasing nitrogen–oxygen bond lengths: NO_2^-, NO^+, and NO_3^-.

8.110. Rank the following ions in order of increasing carbon–oxygen bond lengths: CO, CO_2, and CO_3^{2-}.

8.111. Rank the following ions in order of increasing nitrogen–oxygen bond energy: NO_2^-, NO^+, and NO_3^-.

8.112. Rank the following ions in order of increasing carbon–oxygen bond energy: CO, CO_2, and CO_3^{2-}.

8.113. Why must the stoichiometry of a reaction be known in order to estimate the enthalpy change from bond energies?

8.114. Why must the structures of the reactants and products be known in order to estimate the enthalpy change of a reaction from bond energies?

*8.115. When calculating the enthalpy change for a chemical reaction using bond energies, why is it important to know the phase (solid, liquid, or gaseous) for every compound in the reaction?

*8.116. If the energy needed to break 2 moles of $C=O$ bonds is greater than the sum of the energies needed to break the $O=O$ bonds in a mole of O_2 and vaporize 1 mole of carbon, why does the combustion of pure carbon release heat?

PROBLEMS

8.117. Use the average bond energies in Table 8.2 to estimate the enthalpy changes of the following reactions:
a. $2\,N_2(g) + 3\,H_2(g) \rightarrow 2\,NH_3(g)$
b. $N_2(g) + 2\,H_2(g) \rightarrow H_2NNH_2(g)$
c. $2\,N_2(g) + O_2(g) \rightarrow 2\,N_2O(g)$

8.118. Use the average bond energies in Table 8.2 to estimate the enthalpy changes of the following reactions:
a. $CO_2(g) + H_2(g) \rightarrow H_2O(g) + CO(g)$
b. $N_2(g) + O_2(g) \rightarrow 2\,NO(g)$
*c. $C(s) + CO_2(g) \rightarrow 2\,CO(g)$
NOTE: The heat of sublimation of graphite, $C(s)$, is 719 kJ/mol.

8.119. The combustion of carbon monoxide releases 283 kJ/mol of CO. If the $O=O$ and $C=O$ bond energies are 495 and 799 kJ/mol, respectively, what is the predicted bond energy of carbon monoxide?

$$2\,CO(g) + O_2(g) \rightarrow 2\,CO_2(g) \qquad \Delta H_{comb} = -566\;\text{kJ}$$

8.120. The reaction of H_2 with F_2 produces HF with $\Delta H = -269$ kJ/mol of HF. If the H—H and H—F bond energies are 436 and 567 kJ/mol, respectively, what is the F—F bond energy?

$$H_2(g) + F_2(g) \rightarrow 2\,HF(g)$$

8.121. The reaction between one molecule of CS_2 and three molecules of O_2 is shown in Figure P8.121. If the enthalpy change for the reaction of one *mole* of CS_2 with three *moles* of O_2 is $\Delta H = -1102$ kJ/mol, calculate the average bond energy for the carbon–sulfur bond in a single molecule of CS_2 using the data in Table 8.2 for C=O and O=O bond energies. The yellow, black, and red spheres represent S, C, and O atoms, respectively

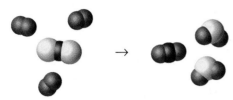

FIGURE P8.121

*8.122. The enthalpy change for the reaction between two molecules of carbon oxysulfide (COS) to form one molecule of CO_2 and one molecule of CS_2, as shown in Figure P8.122, is -3.2×10^{-24} kJ per molecule. Are the apparent bond energies of the carbon–sulfur and carbon–oxygen bonds in COS stronger or weaker than in CS_2 and CO_2, respectively?

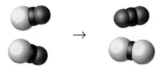

FIGURE P8.122

8.123. Use the average bond energies in Table 8.2 to estimate how much less energy is released during the incomplete combustion of 1 mole of methane to carbon monoxide and water vapor than in the complete combustion to carbon dioxide and water vapor.

8.124. Use the average bond energies in Table 8.2 to estimate how much more energy is released by the reaction

$$C(s) + O_2(g) \rightarrow CO_2(g)$$

than by the reaction

$$C(s) + O_2(g) \rightarrow CO(g)$$

*8.125. Carbon and oxygen form three oxides: CO, CO_2, and carbon suboxide (C_3O_2). Draw a Lewis structure for C_3O_2 in which the three carbon atoms are bonded to each other, and predict whether the carbon–oxygen bond lengths in the carbon suboxide molecule are equal.

*8.126. Draw a Lewis structure for the molecule N_4O. Spectroscopic analysis of the molecule reveals that the nitrogen–oxygen bond length is 120 pm and that there are three N—N bond lengths: 148, 127, and 115 pm. Draw the Lewis structure for N_2O consistent with these observations.

*8.127. Starting with the bond energies in Table 8.2, estimate $\Delta H°$ for the following reaction:

$$4\,NH_3(g) + 7\,O_2(g) \rightarrow 4\,NO_2(g) + 6\,H_2O(g)$$

*8.128. The value of $\Delta H°_{rxn}$ for the reaction

$$2\,H_2S(g) + 3\,O_2(g) \rightarrow 2\,SO_2(g) + 2\,H_2O(g)$$

is -1036 kJ. Estimate the energy of the bonds in SO_2 from the appropriate average bond energies in Table 8.2.

Additional Problems

8.129. The unpaired dots in Lewis symbols for the elements represent valence electrons available for covalent bond formation. Which of the following options for placing dots around the symbol for each element is preferred?

    a. Be⁚ or ·Be·   b. ⁚Al· or ·Ȧl·

    c. ·Ċ· or ⁚C·   d. ⁚Ḧe⁚ or ·He·

8.130. Based on the following Lewis symbols, predict to which group in the periodic table element X belongs:

    a. ·Ẋ·   b. ·Ẋ⁚   c. ⁚Ẋ⁚   d. ·Ẍ⁚

8.131. Use formal charges to predict whether the atoms in carbon disulfide are arranged CSS or SCS.

8.132. Use formal charges to predict whether the atoms in hypochlorous acid are arranged HOCl or HClO.

*8.133. **Chemical Weapons** Phosgene is a poisonous gas first used in chemical warfare during World War I. It has the formula $COCl_2$ (C is the central atom).
    a. Draw its Lewis structure.
    b. Phosgene kills because it reacts with water in nasal passages, in the lungs, and on the skin to produce carbon dioxide and hydrogen chloride. Write a balanced chemical equation for this process showing the Lewis structures for reactants and products.

8.134. The dinitramide anion $[N(NO_2)_2^-]$ was first isolated in 1996. The arrangement of atoms in $N(NO_2)_2^-$ is shown below:

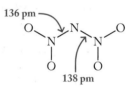

    a. Complete the Lewis structure for $N(NO_2)_2^-$ including any resonance forms, and assign formal charges.
    b. Explain why the nitrogen–oxygen bond lengths in $N(NO_2)_2^-$ and N_2O should (or should not) be similar.
    c. $N(NO_2)_2^-$ was isolated as $[NH_4^+][N(NO_2)_2^-]$. Draw the Lewis structure for NH_4^+.

8.135. Silver cyanate (AgOCN) is a source of the cyanate ion (OCN⁻), which reacts with a number of small molecules. Under certain conditions the species OCN is an anion with a charge of 1−; under others it is a neutral, odd-electron molecule, OCN. The products of these reactions pose interesting questions about covalent bonding.
    a. Two molecules of OCN combine to form OCNNCO. Draw the Lewis structures for this molecule including all resonance forms.
    b. The OCN⁻ ion reacts with BrNO, forming the unstable molecule OCNNO. Draw the Lewis structures for BrNO and OCNNO including all resonance forms.
    c. The OCN⁻ ion reacts with Br_2 and NO_2 to produce N_2O, CO_2, BrNCO, and OCN(CO)NCO. Draw three of the resonance forms of OCN(CO)NCO, which has the following arrangement of atoms:

$$\begin{array}{c} \qquad\quad O \\ \qquad\quad \| \\ O-C-N-C-N-C-O \end{array}$$

*8.136. During the reaction of the cyanate ion (OCN^-) with Br_2 and NO_2, a very unstable substance called an *intermediate* forms and then quickly falls apart. Its formula is O_2NNCO.

a. Draw three of the resonance forms for O_2NNCO, assign formal charges, and predict which of the three contributes the most to the bonding in O_2NNCO. The connectivity of the atoms is shown below:

$$
\begin{array}{c}
O \\
\diagdown \\
N-N-C-O \\
\diagup \\
O
\end{array}
$$

b. Which bond in O_2NNCO must break in the reaction with Br_2 to form BrNCO? What other product forms?

c. Draw Lewis structures for a different arrangement of the N, C, and O atoms in O_2NNCO:

$$
\begin{array}{c}
\qquad\qquad\qquad O \\
\qquad\qquad\qquad \diagup \\
O-N-N-C \\
\qquad\qquad\qquad \diagdown \\
\qquad\qquad\qquad O
\end{array}
$$

8.137. A compound with the formula Cl_2O_6 decomposes to a mixture of ClO_2 and ClO_4. Draw two Lewis structures for Cl_2O_6: one with a chlorine–chlorine bond and one with a Cl–O–Cl arrangement of atoms. Draw a Lewis structure for ClO_2.

$$Cl_2O_6 \rightarrow ClO_2 + ClO_4$$

*8.138. A compound consisting of chlorine and oxygen, Cl_2O_7, decomposes by the following reaction:

$$Cl_2O_7 \rightarrow ClO_4 + ClO_3$$

a. Draw two Lewis structures for Cl_2O_7: one with a chlorine–chlorine bond and one with a Cl–O–Cl arrangement of atoms.

b. Draw a Lewis structure for ClO_3.

*8.139. The odd-electron molecule CN dimerizes to give cyanogen (C_2N_2).

a. Draw a Lewis structure for CN, and predict which arrangement for cyanogen is more likely: NCCN or CNNC.

b. Cyanogen reacts slowly with water to produce oxalic acid ($H_2C_2O_4$) and ammonia; the Lewis structure for oxalic acid is:

Compare this structure to your answer in (a). When the actual structures of molecules have been defined experimentally, the structures have been used to refine Lewis structures. Does this structure increase your confidence that the structure you selected in (a) may be the better one?

8.140. The odd-electron molecule SN forms S_2N_2, which has a cyclic structure (the atoms form a ring).

a. Draw a Lewis structure for SN and complete the following possible Lewis structures for S_2N_2:

$$
\begin{array}{cc}
S-N \qquad\qquad & S-N \\
|\ \ \ | \qquad\qquad & |\ \ \ | \\
S-N \qquad\qquad & N-S
\end{array}
$$

b. Which of the two is the preferred structure for S_2N_2?

*8.141. In electron diffraction, an interference pattern is produced based on the location of atoms when a beam of electrons is fired at a sample. The electrons are diffracted by the arrangement of the atoms in a crystal just the way water waves are diffracted by passing through gaps in a barrier, as we saw in Chapter 7. The position of the atoms can be determined from the interference pattern. The molecular structure of sulfur cyanide trifluoride (SF_3CN) has been shown by electron diffraction studies to have the following arrangement of atoms with the indicated bond lengths:

$$
\begin{array}{c}
\qquad\qquad\qquad N \\
116\ pm \longrightarrow |\!\!| \\
\qquad\qquad\qquad C \\
\qquad\qquad \longleftarrow 174\ pm \\
F-S-F \\
\qquad | \quad \longrightarrow 160\ pm \\
\qquad F
\end{array}
$$

Using the observed bond lengths as a guide, complete the Lewis structure for SF_3CN and assign formal charges.

8.142. **Strike-Anywhere Matches** Heating phosphorus with sulfur gives P_4S_3, a solid used in the heads of strike-anywhere matches (Figure P8.142). P_4S_3 has the following Lewis structure framework:

FIGURE P8.142

Complete this Lewis structure so that each atom has the optimum formal charge.

*8.143. The heavier group 16 elements can expand their octets. The $TeOF_6^{2-}$ anion was first prepared in 1993. Draw the Lewis structure for $TeOF_6^{2-}$.

*8.144. **Sulfur in the Environment** Sulfur is cycled in the environment through compounds such as dimethylsulfide (CH_3SCH_3), hydrogen sulfide (H_2S), and sulfite and sulfate ions. Draw Lewis structures for these four species. Are expanded octets needed to optimize the formal charges for any of these species?

*8.145. **Compounds in Comets** Comets, such as Hyakutake (shown in Figure P8.145 in the Arizona sky in March 1996), contain hydrogen isocyanide (HNC). Draw a Lewis structure for HNC, and propose an explanation for why hydrogen cyanide (HCN) might be more stable. Suggest a reason why these less stable compounds can exist in space.

FIGURE P8.145

8.146. Consider a hypothetical structure of ozone that is cyclic (the atoms form a ring) such that its three O atoms are at the corners of a triangle. Draw the Lewis structure for this molecule.

*8.147. How might electron diffraction (see Problem 8.141) help characterize the bonding in a compound with the formula A_2X in terms of the following:
  a. Distinguishing between these two bonding patterns: X–A–A and A–X–A,:
  b. Distinguish between these resonance forms: $A—X≡A$, $A=X=A$, and $A≡X—A$.

8.148. The highly explosive N_5^+ cation was first isolated in 1999 by reaction of N_2F^+ with HN_3:

$$N_2F^+ + HN_3 \rightarrow [N_5^+] + HF$$

Draw the Lewis structures for the reactants and products, including all resonance forms.

*8.149. **Jupiter's Atmosphere** The ionic compound NH_4SH was detected in the atmosphere of Jupiter (Figure P8.149) by the Galileo space probe in 1995. Draw the Lewis structure for NH_4SH. Why couldn't there be a covalent bond between the nitrogen and sulfur atoms, making NH_4SH a molecular compound?

FIGURE P8.149

8.150. **Tums** Antacid tablets commonly contain calcium carbonate and/or magnesium hydroxide. Draw the Lewis structures for calcium carbonate and magnesium hydroxide.

*8.151. An allotrope of nitrogen, N_4, was reported in 2002. The compound has a lifetime of $1\,\mu s$ at $298\,K$ and was prepared by adding an electron to N_4^+. Since the compound cannot be isolated, its structure is unconfirmed experimentally.
  a. Draw the Lewis structures for all the resonance forms of linear N_4. (Linear means that all four nitrogens are in a straight line.)
  b. Assign formal charges, and determine which structure is the best description of N_4.
  c. Draw a Lewis structure for a ring (cyclic) form of N_4, and assign formal charges.

*8.152. Scientists have predicted the existence of O_4 even though this compound has never been observed. However, O_4^{2-} has been detected. Draw the Lewis structures for O_4 and O_4^{2-}.

8.153. Draw a Lewis structure for $AlFCl_2$, the second product in the synthesis of Cl_2O_2 in the reaction

$$FClO_2(g) + AlCl_3(s) \rightarrow Cl_2O_2(g) + AlFCl_2(s)$$

*8.154. Draw Lewis structures for BF_3 and $(CH_3)_2BF$. The B–F distance in both molecules is the same (130 pm). Does this observation support the argument that all the boron–fluorine bonds in BF_3 are single bonds?

8.155. Which of the following molecules and ions contains an atom with an expanded octet? (a) Cl_2; (b) ClF_3; (c) ClI_3; (d) ClO^-

8.156. Which of the following molecules contains an atom with an expanded octet? (a) XeF_2; (b) $GaCl_3$; (c) ONF_3; (d) SeO_2F_2

*8.157. A linear nitrogen anion, N_5^-, was isolated for the first time in 1999.
  a. Draw the Lewis structures for four resonance forms of linear N_5^-.
  b. Assign formal charges to the atoms in the structures in part a, and identify the structures that contribute the most to the bonding in N_5^-.
  c. Compare the Lewis structures for N_5^- and N_3^-. In which ion do the N–N bonds have the higher average bond order?

*8.158. Carbon tetraoxide (CO_4) was discovered in 2003.
  a. The four atoms in CO_4 are predicted to be arranged as shown below:

$$\begin{array}{c} O \\ | \\ C \\ O{\diagdown}{\diagup}O \\ O \end{array}$$

  Complete the Lewis structure for CO_4.
  b. Are there any resonance forms for the structure you drew that have zero formal charges on all atoms?
  c. Can you draw a structure in which all four oxygen atoms in CO_4 are bonded to carbon?

8.159. Plot the electronegativities of elements with $Z = 3$ to 9 (y-axis) versus their first ionization energy (x-axis). Is the plot linear? Use your graph to predict the electronegativity of neon, whose first ionization energy is 2080 kJ/mol.

8.160. Plot electronegativity as a function of ionization energy for the main group elements of the fifth row of the periodic table. Estimate the electronegativity of xenon, whose first ionization energy is 1170 kJ/mol. Use the data in Figure 8.5 for electronegativity values and those in Figure 7.31 for ionization energies.

8.161. The cation N_2F^+ is isoelectronic with N_2O.
  a. What does it mean to be isoelectronic?
  b. Draw the Lewis structure for N_2F^+. (*Hint:* The molecule contains a nitrogen–nitrogen bond.)
  c. Which atom has the +1 formal charge in the structure you drew in part b?
  d. Does N_2F^+ have resonance forms?
  e. Could the middle atom in the N_2F^+ ion be a fluorine atom? Explain your answer.

8.162. **Ozone Depletion** Methyl bromide (CH_3Br) is produced naturally by fungi. Methyl bromide has also been used in agriculture as a fumigant, but this use is being phased out because the compound has been linked to ozone depletion in the atmosphere.
  a. Draw the Lewis structure for CH_3Br.
  b. Which bond in CH_3Br is more polar, carbon–hydrogen or carbon–bromine?

*8.163. Why is SF_4 a stable molecule, while the molecule OF_4 does not exist?

*8.164. Why is PF_5 a stable molecule, while the molecule NF_5 does not exist?

8.165. How many pairs of electrons does xenon share in the following molecules and ions? (a) XeF_2; (b) $XeOF_2$; (c) XeF^+; (d) XeF_5^+; (e) XeO_4

8.166. How many pairs of valence electrons do bromine atoms have in the following molecules and ions? (a) BrF; (b) BrF_2^+; (c) BrF_6^+; (d) BrF_2^-; (e) BrF_5

Fluorine and Oxygen: Location, Location, Location

8.167. Write a balanced equation for the reaction of fluorine gas with water.
  a. Identify the oxidizing and reducing agents.
  b. Draw Lewis structures for all the compounds involved in the reaction, and place arrows on each polar bond to indicate the direction of the polarity.

8.168. Why are He, Ne, and Ar nonreactive with fluorine, while Kr and Xe do form compounds with fluorine?

*8.169. All the carbon–fluorine bonds in tetrafluoroethylene:

Tetrafluoroethylene

are polar, but experimental results show that the molecule as a whole is nonpolar. Draw arrows on each bond to show the direction of polarity, and suggest a reason why the molecule is nonpolar.

*8.170. Free radicals increase in stability, and hence decrease in reactivity, if they have more than one atom in their structure that can carry the unpaired electron.
  a. Draw Lewis structures for the three free radicals formed from CF_2Cl_2 in the stratosphere—ClO, Cl, and CF_2Cl—and use this principle to rank them in order of reactivity.
  b. Free radicals are "neutralized" when they react with each other, forming an electron pair (a single covalent bond) between two atoms. This is called a *termination reaction*, because it shuts down any reaction that was powered by the free radical. Predict the formula of the molecules that are produced when the chlorine free radical reacts with each of the free radicals in part a.

Molecular Geometry and Bonding Theories

9

THE PASTERZE GLACIER
Comparison of this Austrian glacier between 1900 (top) and 2000 (bottom) shows the effects of global warming.

A LOOK AHEAD: Aromas, Pharmaceuticals, and the Greenhouse Effect

Hold your hands out in front of you, palms up, fingers extended. Your hands are virtually identical in most major respects. Rotate your wrists inward so that your thumbs point toward the ceiling and your palms are perpendicular to the floor. Your right hand looks just like the image your left hand makes in a mirror. Despite the fact that your two hands are very similar—both have four fingers and a thumb and are the same size—they are also profoundly different. For example, you cannot put your right hand in a glove made for your left, a fact that illustrates the importance of three-dimensional shape in the macroscopic world. Shape is also important at the molecular level, and the consequences of this fact are remarkable.

The refreshing aroma of spearmint is due to an organic molecule containing 10 carbon atoms, 14 hydrogen atoms, and 1 oxygen atom. The molecule responsible for the musty aroma of caraway seeds has the same formula, $C_{10}H_{14}O$. Not only are the molecular formulas the same, their Lewis structures are absolutely identical, the same atoms are connected in both molecules, and both have the same number of covalent bonds. We have to envision their shape in three-dimensions to understand how they differ, and the only difference between the two molecules is the orientation in three-dimensional space of two groups that are attached to one of the 10 carbon atoms in the structure. We perceive a difference in their aromas because each molecule has a unique site where it attaches to our nasal membranes. Just as a left hand fits only in a left glove, the spearmint molecule fits only in the spearmint-shaped site, and the caraway molecule fits only in the caraway-shaped site.

Living systems discriminate among many biologically active substances based on their three-dimensional molecular shapes. The activity of the tuberculosis drug diambutol depends on the shape of the molecule. If two of the atoms attached to one of the carbons in the structure change places, a totally inactive material results. The two different forms of diambutol have the same Lewis structures; the same atoms are bonded together; the three-dimensional shape determines if the molecule cures tuberculosis or is ineffective.

Shape matters just as much in the simpler molecules that compose our atmosphere. An important property of atmospheric gases is their ability to contribute to the greenhouse effect, which is a natural process by which the atmosphere warms the Earth's surface. Although the greenhouse effect makes the temperature on Earth hospitable to life as we know it, we are currently concerned that we may have too much CO_2 in the atmosphere. Interestingly, the major constituents of the atmosphere, nitrogen (N_2) and oxygen (O_2), are not greenhouse gases; but other atmospheric molecules—CO_2, H_2O, CH_4 (methane), the chlorofluorocarbons (CFCs), and SF_6 (sulfur hexafluoride)—are greenhouse gases. Why is this? Two features of molecular structure determine whether or not a molecule is a greenhouse gas: one is the identity of the atoms bonded together, and the other is the three-dimensional shape of the molecule. In this chapter, we begin the study of the three-dimensional shapes of molecules and explore the impact of molecular shape on the properties of substances, including greenhouse gases.

9.1 Molecular Shape

Molecular shape affects many properties of a compound: its physical state at any temperature, its solubility in fluids, its chemical reactivity, its biological activity, and its distribution in the environment. The characteristics of molecules that lead to taste and smell in food, the ability of many pharmaceutical agents to influence physiological processes, and the capacity of some molecules in the atmosphere to behave as greenhouse gases all arise from features of their molecular structures.

We have written the Lewis structures for a number of molecules, but Lewis structures are two-dimensional representations of the arrangement of electrons in molecules. Lewis structures enable us to determine connectivity (what atom is connected to what other atom), an absolutely fundamental characteristic, but they do not convey three-dimensional shape.

Consider how we have drawn the Lewis structures for two greenhouse gases, carbon dioxide and water:

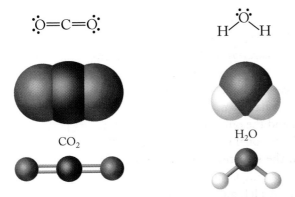

The Lewis structure we have drawn for carbon dioxide shows a linear arrangement of three atoms with the two oxygen atoms attached to a central carbon and an O—C—O bond angle of 180°. The Lewis structure for water also contains three atoms, but the water molecule is bent. All three atoms in both molecules lie in the same plane, but we have drawn the carbon dioxide as linear and the water molecule as a bent molecule because analytical techniques provided data from which we can determine their structures. On the basis of Lewis structures alone, we could just as reasonably have drawn water as a linear molecule or carbon dioxide as bent. In this chapter, we learn why water is a bent molecule as we develop a method for predicting the three-dimensional shapes of molecules.

To accurately predict the shape of a small molecule we need to know the angles between the bonds in the compound. The **bond angle** is the angle (in degrees) defined by lines joining the centers of two atoms to a third atom to which they are covalently bonded, as shown here for CO_2 and H_2O.

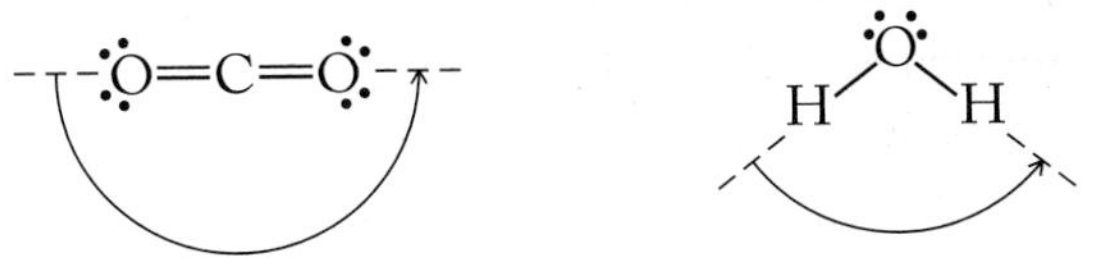

O—C—O bond angle = 180°
Molecular shape = linear

H—O—H bond angle = 104.5°
Molecular shape = bent

Why are the bond angles in CO_2 and H_2O different? To begin to address this question we need to consider why bonds form where they do about an atom. Molecular shape, or **molecular geometry**, is governed by energy. The geometry of a molecule is determined by the arrangement of its atoms that results in the lowest potential energy. The lowest energy state is the one in which the repulsions between bonding and lone pairs of electrons are minimized, as we shall discuss in the next section.

A **bond angle** is the angle (in degrees) defined by lines joining the centers of two atoms to a third atom to which they are covalently bonded.

The **molecular geometry** or shape of a molecule is defined by the lowest energy arrangement of its atoms in three-dimensional space.

⦿⦿ **CONNECTION** In Chapter 8 we defined bond lengths of covalent bonds and related them to the sizes of atoms.

9.2 Valence-Shell Electron-Pair Repulsion Theory (VSEPR)

All electrons are negatively charged, so all electrons repel each other. **Valence-shell electron-pair repulsion (VSEPR)** theory is a model that enables us to rationalize molecular geometry on the basis of minimizing electron repulsions between valence electrons about a central atom. VSEPR theory applies to molecules of main group elements that are held together by covalent bonds. To predict molecular geometry, we predict the location of covalent bonds (pairs of electrons that hold atoms to the central atom) and any lone pairs of electrons about a central atom. The molecular geometry is the result of the arrangement of atoms that minimizes the repulsive forces between the electrons.

Central Atoms with No Lone Pairs

To determine the shape of a molecule, we start by writing a correct Lewis structure. From the Lewis structure, we determine the **steric number** (SN) of the central atom, which is the sum of the number of atoms bonded to the central atom and the number of lone pairs of electrons on the central atom:

Steric number (SN) =

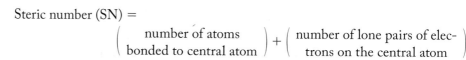

Let's initially focus on molecules in which the central atom has no lone pairs, so the *steric number equals the number of atoms bonded to the central atom.* In evaluating these structures, we will generate a set of five common shapes that are the basis for describing the molecular geometries of many simple covalent compounds.

To generate arrangements in which repulsive forces between the electron pairs about the central atom are minimized, we can do the following thought experiment. Think of an atom (the central atom in a molecule) in the center of a sphere with other atoms floating on the surface of the sphere (Figure 9.1) linked to the central atom by shared pairs of electrons, i.e., covalent bonds.

If the central atom has only two pairs of electrons oriented towards the atoms (SN = 2), how do the electron pairs arrange themselves to minimize their mutual repulsion? The answer is: as far apart as possible—on opposite sides of the sphere. This gives a **linear** arrangement, corresponding to a bond angle of 180°. For SN = 3, the bonding electron pairs all lie in a plane 120° apart in a **trigonal planar** arrangement, and for SN = 4, they are oriented 109.5° apart in a **tetrahedral** geometry. For SN = 5, two pairs reside on the north and south poles of the sphere 180° apart, and three reside on the equator 120° apart. The resultant arrangement is called **trigonal bipyramidal.** Finally, for SN = 6, all the positions are equivalent. We can think of the electron pairs in three sets, each set with a bond angle of 180°, and oriented at 90° to the other two pairs, just like the axes of the x-, y-, z-coordinate system. On the sphere, four atoms lie in the plane of the equator 90° apart, and two lie at the north and south poles, 180° apart from each other and 90° from each pair on the equator. This arrangement defines an **octahedral** molecular geometry. Figure 9.2 shows the repulsion of balloons to illustrate these structures and summarizes the five basic geometric shapes. The shapes are named for the solid objects outlined when the positions of the electron pairs are connected in three-dimensional space.

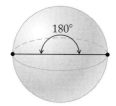

Linear SN = 2

Trigonal planar SN = 3

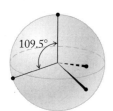

Tetrahedral SN = 4

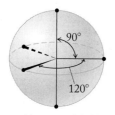

Trigonal bipyramidal SN = 5

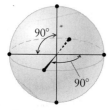

Octahedral SN = 6

FIGURE 9.1 Different numbers of atoms floating on the surface of a sphere and bonded to a central atom at the center of the sphere adopt these molecular geometries with the bond angles shown.

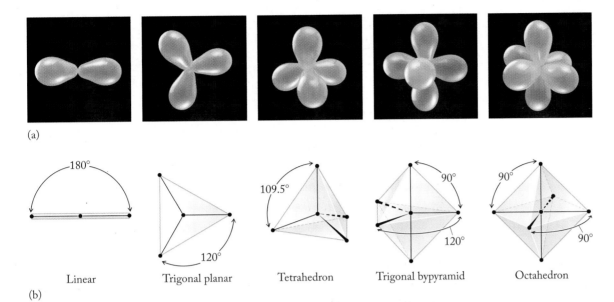

(a)

180°   109.5°   90°   90°   120°   90°

Linear   Trigonal planar   Tetrahedron   Trigonal bypyramid   Octahedron

(b)

FIGURE 9.2 Geometric forms. (a) Balloons charged with static electricity illustrate the locations of electron pairs about a central point. (b) Connecting the positions of minimum energy defines the solid forms that give the molecular shapes their names.

Now let's look at five simple molecules that have only atoms and no lone pairs about the central atom and apply VSEPR theory to predict their shapes. To do so we

1. Draw the Lewis structure for the molecule.
2. Determine the steric number of the central atom.
3. Use the SN to locate the atoms about the central atom.
4. Define the resultant molecular geometry by considering the location of the atoms.

EXAMPLE I: CARBON DIOXIDE, CO_2

1. The Lewis structure for CO_2 is $\ddot{O}=C=\ddot{O}$
2. The central carbon atom has SN = 2 (2 atoms bonded to it and no lone pairs of electrons; SN = 2 + 0 = 2).
3. SN = 2 tells us the O—C—O bond angle is 180°.
4. When the bond angle between three atoms is 180° we say the molecule is linear.

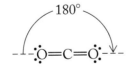

180°

Bond angle = 180°
Molecular geometry = linear

Carbon dioxide, CO_2

The **valence-shell electron-pair repulsion** or **VSEPR** model predicts the arrangement of valence electron pairs around a central atom that will minimize their mutual repulsion to produce the lowest energy orientations. VSEPR theory may be applied to covalently bound molecules and polyatomic ions of main group elements.

The **steric number** (SN) of a central atom is the number of atoms bonded to the central atom plus the number of lone pairs of electrons on the central atom.

A **linear** molecular geometry results when the bond angle between three atoms is 180°.

A steric number of 3 for the central atom of a molecule with no lone pairs results in a **trigonal planar** structure with bond angles of 120°.

A steric number of 4 for the central atom of a molecule with no lone pairs corresponds to a **tetrahedral** structure of atoms about a central atom with bond angles of 109.5°.

A steric number of 5 for the central atom of a molecule with no lone pairs corresponds to a **trigonal bipyramid**, in which three atoms occupy equatorial sites in the plane around the central atom with ideal bond angles of 120° and two other atoms occupy axial sites above and below the central atom with a bond angle of 180°.

A steric number of 6 for the central atom of a molecule with no lone pairs gives rise to an **octahedral** arrangement in which all six sites are equivalent. The ideal bond angle for atoms in this arrangement is 90°.

∞ **CONNECTION** We learned in Chapter 8 that some compounds have central atoms that do not have an octet of electrons. The boron atom has only three pairs of electrons in BF_3, but because this distribution results in formal charges of 0 on all of the atoms, it is the preferred structure.

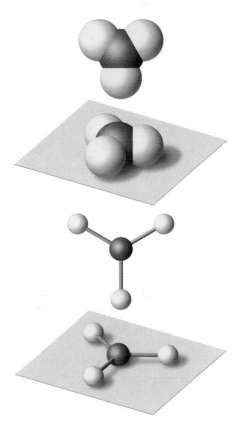

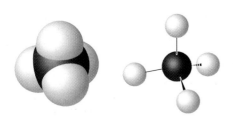

Boron trifluoride, BF_3

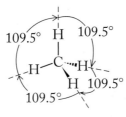

Methane, CH_4

EXAMPLE II: BORON TRIFLUORIDE, BF₃

1. The Lewis structure for boron trifluoride is

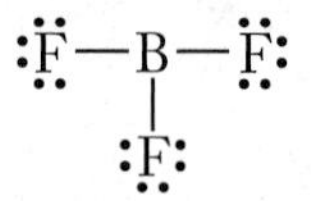

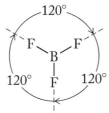

2. The central boron atom has SN = 3 (3 atoms bonded to it and no lone pairs; SN = 3 + 0 = 3).

3. SN = 3 means the three atoms bonded to the central boron atom and the central atom itself lie in a plane. The F—B—F bond angles are all 120°.

4. The molecule has a trigonal planar structure.

Bond angles = 120°   Molecular geometry = trigonal planar

EXAMPLE III: METHANE, CH₄

1. The Lewis structure for methane is

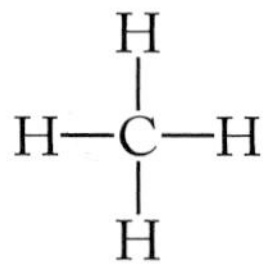

$$
\begin{array}{c}
\text{H} \\
| \\
\text{H—C—H} \\
| \\
\text{H}
\end{array}
$$

2. SN = 4

3. The hydrogen atoms are located at the vertices of a tetrahedron and all four H—C—H bond angles are 109.5°.

4. The molecule has a tetrahedral structure.

Bond angles = 109.5°   Molecular geometry = tetrahedral   Tetrahedron

Before we develop the two remaining basic structures, some comments about the conventions used in drawing three-dimensional structures on two-dimensional paper are in order. To convey the structure of a molecule in three dimensions, we use slender single lines (—) to indicate a bond connecting two atoms that lie in the plane of the paper. A solid wedge (—) indicates a bond that comes out of the paper and toward the viewer. The solid wedge in the structure of methane means that the hydrogen in that position points toward the viewer at a downward angle. The dashed wedge (⋯) indicates a bond that goes away from the viewer behind the paper. The dashed wedge in the structure of methane connects the central carbon to a hydrogen atom behind the plane of the paper, pointing away from the viewer at a downward angle. Note in the structure of methane that two carbon–hydrogen bonds are drawn in the plane of the paper and two are not.

EXAMPLE IV: PHOSPHORUS PENTAFLUORIDE, PF₅

1. The Lewis structure for phosphorus pentafluoride is

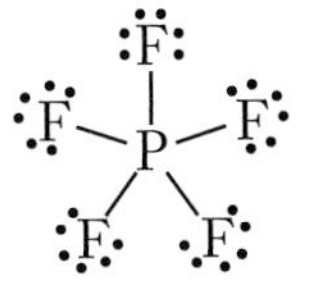

2. SN = 5

▶❙❙ CHEMTOUR VSEPR Model

3. The fluorine atoms define the vertices of a trigonal bipyramid. Three fluorine atoms lie in *equatorial* (*eq*) positions, because if the molecule were enclosed in a sphere, they would lie along the equator. Two others are in *axial* (*ax*) positions; they lie along a vertical axis of rotation of the sphere. The bond angles are: eq–eq (120°), ax–ax (180°), and ax–eq (90°).

4. The molecule forms a trigonal bipyramid.

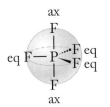

All eq–eq bond angles = 120°
All ax–eq bond angles = 90°
ax–ax bond angle = 180°

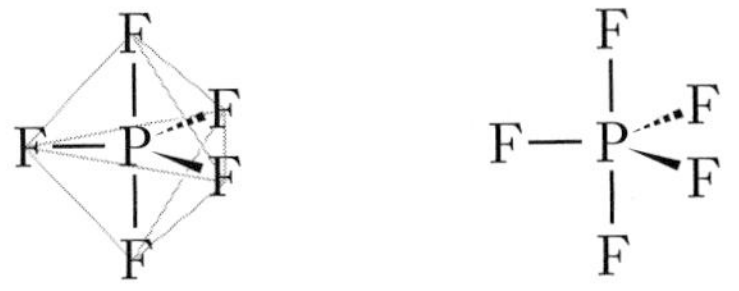

Molecular geometry = trigonal bipyramidal

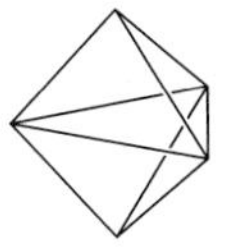

Trigonal bipyramid

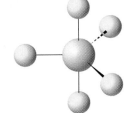

Phosphorus pentafluoride, PF₅

EXAMPLE V: SULFUR HEXAFLUORIDE, SF₆

1. The Lewis structure for sulfur hexafluoride is

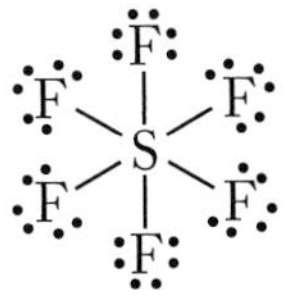

2. SN = 6

3. The fluorine atoms are located at the vertices of an octahedron. All six positions are equivalent; each fluorine atom is 90° from its four nearest neighbors. Although we can define axial and equatorial positions for any given orientation, these positions are interchangeable by simply rotating the molecule.

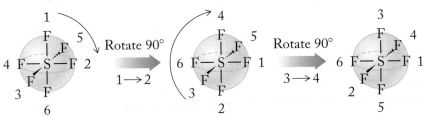

Positions 1 and 6 = axial
Positions 2, 3, 4, 5 = equatorial

Positions 4 and 2 = axial
Positions 1, 3, 5, 6 = equatorial

Positions 3 and 5 = axial
Positions 1, 2, 4, 6 = equatorial

4. The molecule has an **octahedral** shape.

All eq–eq bond angles = 90°
All ax–eq bond angles = 90°
ax–ax bond angle = 180°

Molecular geometry = octahedral

Octahedron

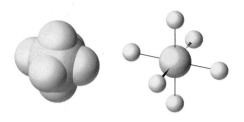

Sulfur hexafluoride, SF₆

| SAMPLE EXERCISE 9.1 | **Predicting Molecular Geometry** |

Formaldehyde, CH_2O, is a gas (bp −21°C) that is used in aqueous solutions to preserve biological samples. It is also a product of incomplete combustion of hydrocarbons. Predict the structure of formaldehyde using VSEPR theory.

COLLECT AND ORGANIZE Formaldehyde consists of atoms of main group elements, so we may apply VSEPR theory to the problem.

ANALYZE We have the formula of formaldehyde and can draw the Lewis structure. Once we determine the SN of the central atom, we can define the molecular geometry.

SOLVE

Lewis structure

SN = (3 atoms bound to a central carbon atom) + (0 lone pairs)
   = 3

Molecular geometry = trigonal planar

Because the SN = 3 and the central atom has no lone pairs, we predict that formaldehyde is trigonal planar.

THINK ABOUT IT Predicting the correct three-dimensional structure for a molecule begins with drawing the correct Lewis structure. Once we have that, the steric number defines the arrangement of atoms about the central atom, and from that we can assign the overall shape of the molecule.

Practice Exercise Determine the shape of the chloroform molecule ($CHCl_3$) using the conventions discussed in the text and name its molecular shape.

(Answers to Practice Exercises are in the back of the book.)

H—C—H bond angle < 120°

FIGURE 9.3 The electrons in the C=O double bond in formaldehyde occupy more space than those in the C—H single bonds. The larger electron cloud repels the clouds of the single bonds and forces them slightly closer together; the H—C—H bond angle is slightly less than the ideal angle of 120°.

Applying VSEPR to formaldehyde (CH_2O; see Sample Exercise 9.1), we predict the molecule to have a trigonal planar geometry.

Measurements of the actual structure of formaldehyde (see Figure 9.3) show that the H—C—H bond angle is actually slightly smaller than the ideal angle of 120° in a trigonal planar arrangement. Remember that the C=O double bond consists of two electron pairs, and as such it occupies more space than one electron pair in a single covalent bond. Because of its larger demand for space, the C=O double bond repels the neighboring C—H bonds and diminishes the size of the H—C—H bond angle around the central atom. VSEPR does not enable us to determine the actual value of the angles in structures containing double bonds, but in structures like formaldehyde we can predict the effect of a double bond on bond angles.

Central Atoms with Lone Pairs and Bonding Pairs of Electrons

Let's revisit the structures defined for each steric number and explore what happens when some of the bonding pairs of electrons are replaced by lone pairs. For SN = 2, the only option of interest is the one with two atoms bound to a central atom. If we replaced one of the two bonding pairs with a lone pair, we would have a molecule with only two atoms, no "central" atom, and no bond angle. Such diatomic molecules must have a linear geometry. Therefore, we begin the discussion of structures containing lone pairs of electrons with SN = 3.

Consider the molecular geometry of SO_2, one of the gases produced when high sulfur coal is burned. The Lewis structure for SO_2 is

$$:\ddot{O}-\ddot{S}=\ddot{O}: \longleftrightarrow :\ddot{O}=\ddot{S}-\ddot{O}:$$

The sulfur atom has a steric number of 3 because 2 atoms are bonded to it and it has 1 lone pair of electrons on the central atom. When SN = 3, the arrangement of atoms and lone pairs about the sulfur atom is trigonal planar. To determine the molecular structure, we draw the central atom and put the atoms and lone pairs around it in the correct arrangement:

Electron-pair geometry = trigonal planar   Molecular geometry = bent

We call the arrangement of atoms and lone pairs the **electron-pair geometry**, because it describes where the electrons in bonds and the electrons in lone pairs are located about the central atom. In molecules with no lone pairs of electrons, the electron-pair geometry is the same as the molecular geometry. In molecules such as SO_2 in which the central atom has both lone pairs and atoms around it, the molecular geometry is different from the electron-pair geometry, because molecular geometry is defined by where the *atoms* are in a molecule. Hence, the electron-pair geometry of SO_2 is trigonal planar, but the molecular geometry is *bent*.

Each bond angle in a trigonal planar structure is theoretically 120°. Experimental measurements establish that SO_2 is indeed a bent molecule as VSEPR theory predicts, but the O—S—O bond angle in SO_2 is a little smaller than 120°. We explain the smaller angle using VSEPR theory by considering the different amount of space occupied by electrons involved in a bond compared to the space occupied by electrons in a lone pair (Figure 9.4). The electrons in the bond are attracted to two nuclei. They have a high probability of being located between the two atomic centers that share them. In contrast, a lone pair of electrons is not shared with a second atom and is more diffuse. VSEPR theory is based on consideration of electron–electron repulsions, and the electrons in a lone pair repel electrons in neighboring bonds more than electrons in neighboring bonds repel each other. This increased repulsion results in the lone pair pushing the bonding pairs closer together, thereby reducing the bond angle. Again, VSEPR theory does not enable us to predict the actual numerical value of the bond angle in a structure containing electrons in bonds as well as lone pairs, but we can predict that in most cases the bond angles in a molecule with lone pairs on the central atom are slightly smaller than the theoretical angles. When predicting actual bond angles, keep in mind that repulsion between two lone pairs (lp–lp) is greater than repulsion between a lone pair and a bonding pair (lp–bp); repulsion between

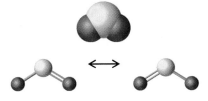

Sulfur dioxide, SO_2

⊙⊙ **CONNECTION** In Chapter 4 we discussed how SO_2 and SO_3 can be removed from flue gases by a process called scrubbing.

O—S—O bond angle < 120°

FIGURE 9.4 The lone pair of electrons on the central sulfur atom in SO_2 occupies more space than the electron pairs in the S–O bonds. The increased repulsion forces the atoms closer together; the O—S—O bond angle is slightly less than the ideal angle of 120°.

bonding pairs (bp–bp) is the smallest: lp–lp > lp–bp > bp–bp. Repulsions caused by a lone pair are also greater than the repulsions associated with a double bond, so the interactions of a lone pair with the rest of the electrons dominate the predictions of deviations from ideal structures.

Sulfur dioxide has two resonance forms, but both resonance forms have the same steric number and hence the same molecular geometry.

SAMPLE EXERCISE 9.2 **Predicting Relative Sizes of Bond Angles**

The central atoms in SO_3 (sulfur trioxide) and O_3 (ozone) both have SN = 3. Define the geometry of each molecule and compare the size of the O—S—O bond angle to that of the O—O—O bond angle.

COLLECT AND ORGANIZE We are given the steric number of both central atoms, so we can draw structures with the correct electron-pair geometry and, from those, determine the molecular geometry.

ANALYZE In SO_3, three oxygen atoms are bonded to the central sulfur atom, so the sulfur has no lone pairs of electrons. Because only two oxygen atoms are bound to the central oxygen in O_3, there must be one lone pair on the central oxygen.

SOLVE Sulfur trioxide has three resonance structures, and ozone has two. However, for both substances the resonance structures are equivalent in terms of steric number of the central atom, so we need only deal with one resonance structure, as shown in the accompanying figure:

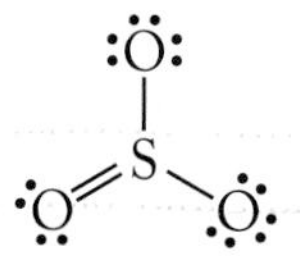

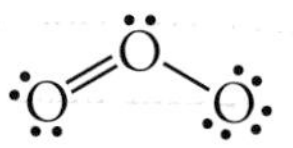

SN = 3
Three atoms bonded to the central atom

Electron-pair geometry = trigonal planar

Molecular geometry = trigonal planar

SN = 3
Two atoms and one lone pair bonded to the central atom

Electron-pair geometry = trigonal planar

Molecular geometry = bent

The theoretical bond angle in trigonal planar arrangements is 120°. Since sulfur trioxide has no lone pairs, we predict the O—S—O bond angle in sulfur trioxide to be 120°. We expect the bond angle in O_3 to be slightly smaller than the ideal angle because a lone pair occupies more space than a bonding pair and forces the bonding pairs closer together; therefore, we predict that the O—O—O bond angle in ozone is slightly smaller than the ideal value of 120°.

THINK ABOUT IT In many cases the presence of lone pairs on a central atom pushes atoms bonded to the central atom closer together and decreases the bond angles.

Practice Exercise The central sulfur atom in sulfur dioxide (SO_2) has SN = 3. Predict how the O—S—O bond angle in SO_2 compares in size to the bond angles in SO_3 and O_3.

CONNECTION In Chapter 8 we discussed the value of the ozone layer that protects life on Earth from potentially damaging UV radiation.

CONCEPT TEST

Why is CO_2 a linear molecule, but SO_2 is bent?

(Answers to Concept Tests are in the back of the book.)

For SN = 4, there are three possible combinations of atoms and lone pairs about the central atom: four atoms and no lone pairs (CH_4), three atoms and one lone pair, and two atoms and two lone pairs.

SN = 4, ELECTRON-PAIR GEOMETRY = TETRAHEDRAL		
No. of Bonded Atoms	No. of Lone Pairs	Molecular Geometry
4	0	Tetrahedral
3	1	Trigonal pyramidal
2	2	Bent

To illustrate the second situation, consider ammonia (NH_3). The Lewis structure for ammonia shows three atoms and one lone pair of electrons about the central nitrogen atom, so SN = 4. That means the electron-pair geometry of ammonia is tetrahedral.

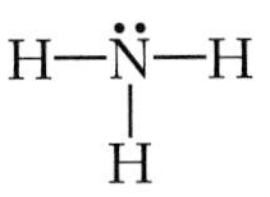

Lewis structure · SN = 4   Electron-pair geometry = tetrahedral 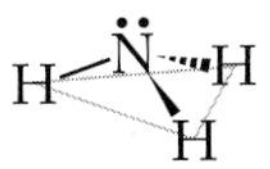   Molecular geometry = trigonal pyramidal 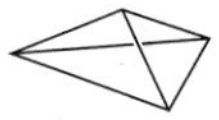   Trigonal pyramid

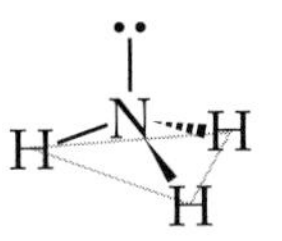

Predicted bond angle < 109.5°

To determine the molecular geometry, we consider only the location of the atomic centers. If you connect the one nitrogen atom and three hydrogen atoms, the hydrogens form a triangular base from which three triangular faces extend to the nitrogen apex; this corresponds to a geometric shape called a *trigonal pyramid*; the molecular geometry of ammonia is trigonal pyramidal. Because of the lone pair on the central nitrogen, the actual H—N—H bond angle in ammonia is slightly smaller than the theoretically predicted 109.5°. Experimental measurements establish that the actual bond angle in ammonia is 107°.

The central oxygen atom in the water molecule also has SN = 4, and the Lewis structure for water reveals the central atom is surrounded by two atoms and two lone pairs of electrons. The electron-pair geometry is tetrahedral; the molecular geometry is bent. The theoretical bond angle in a tetrahedral arrangement is 109.5°, but we predict that the actual H—O—H bond angle in water is smaller than that due to repulsions between the two lone pairs plus the lone-pair/bonding-pair repulsion. The actual measured bond angle in water is 104.5°.

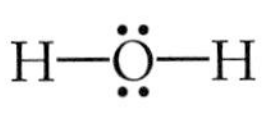

Lewis structure · SN = 4

Two bonded atoms and two lone pairs

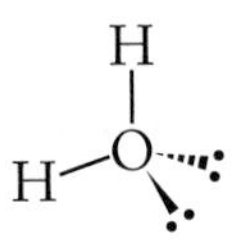

Electron-pair geometry = tetrahedral

Ideal bond angle = 109.5°

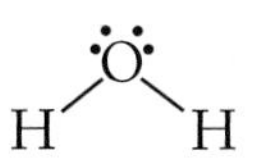

Molecular geometry = bent

Predicted bond angle < 109.5°

CONCEPT TEST

The bond angles in silane (SiH_4), phosphine (PH_3), and hydrogen sulfide (H_2S) are 109.5°, 93.6°, and 92.1°, respectively (shown on the next page). Explain the trend in bond angles using the VSEPR model.

H—C(—H)(H)(H) 109.5°   H—N(—H)(H) 107°   H—O(—H) 104.5°

Molecules having a central atom with SN = 5 can have the following geometries:

SN = 5, ELECTRON-PAIR GEOMETRY = TRIGONAL BIPYRAMIDAL		
No. of Bonded Atoms	**No. of Lone Pairs**	**Molecular Geometry**
5	0	Trigonal bipyramidal
4	1	Seesaw
3	2	T-shaped
2	3	Linear

The electron-pair geometry in all cases is trigonal bipyramidal, but the molecular geometry varies with the number of lone pairs of electrons.

Atoms and lone pairs may occupy two geometrically distinct locations in a trigonal bipyramid, either axial or equatorial. VSEPR theory enables us to predict the locations of atoms by identifying geometries that minimize repulsions felt by the lone pairs. Repulsive forces decrease in the following order: 90° > 120° > 180°. To determine the most likely geometry, we minimize 90° repulsions between lone pairs because those repulsive forces are by far the largest. This prediction is confirmed by comparing known structures of molecules.

One lone pair of electrons in a trigonal bipyramidal arrangement experiences minimum repulsion when it is located in an equatorial position:

Equatorial lone pair
Two 90° lp–bp repulsions

Axial lone pair
Three 90° lp–bp repulsions

The lone pair in the equatorial position has two 90° repulsions, compared to three if the lone pair is axial, so equatorial lone pairs are preferred. The molecular geometry that results in this case is called *seesaw*, because the shape resembles a seesaw on a playground. (The formal name for this shape is *disphenoidal*.)

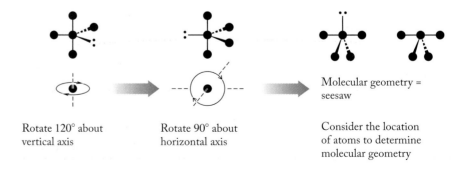

Rotate 120° about vertical axis

Rotate 90° about horizontal axis

Molecular geometry = seesaw

Consider the location of atoms to determine molecular geometry

Furthermore, considering the influence of lone pairs on bond angles, we predict that all the bond angles in the seesaw would be slightly smaller than their counterparts in the ideal trigonal bipyramid: ax–ax < 180°; eq–eq < 120°; and eq–ax < 90°.

Two lone pairs in a trigonal bipyramid occupy two equatorial sites, again based on minimizing 90° repulsions.

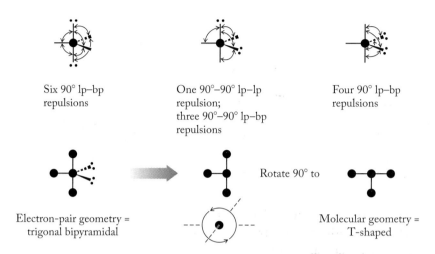

Six 90° lp–bp
repulsions

One 90°–90° lp–lp
repulsion;
three 90°–90° lp–bp
repulsions

Four 90° lp–bp
repulsions

Electron-pair geometry =
trigonal bipyramidal

Rotate 90° to

Molecular geometry =
T-shaped

A 90° repulsion between two lone pairs is very strong, which eliminates the middle figure from consideration. The two lone pairs in the equatorial position have fewer 90° lp–bp interactions than the lone pairs in the axial position. The molecular geometry that results from this arrangement is called *T-shaped*. The bond angles in the T-shaped molecule are also slightly less than the ideal angles in the trigonal bipyramid: ax–ax < 180°, and eq–ax < 90°.

Finally for SN = 5, when there are three lone pairs and two bonded atoms, the lone pairs occupy the three equatorial sites, and the resultant molecular geometry is linear.

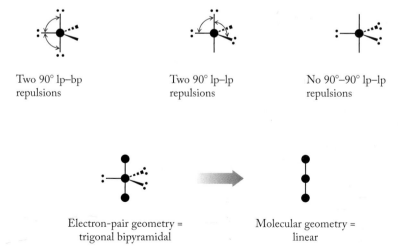

Two 90° lp–bp
repulsions

Two 90° lp–lp
repulsions

No 90°–90° lp–lp
repulsions

Electron-pair geometry =
trigonal bipyramidal

Molecular geometry =
linear

CONCEPT TEST

The bond angle in the linear molecule derived from trigonal bipyramidal geometry is 180°. Why is this angle not decreased by the presence of the lone pairs?

Molecules having a central atom with a SN = 6 can have the following geometries:

SN = 6, ELECTRON-PAIR GEOMETRY = OCTAHEDRAL		
No. of Bonded Atoms	No. of Lone Pairs	Molecular Geometry
6	0	Octahedral
5	1	Square pyramidal
4	2	Square planar
3	3	Although these arrangements are possible, we will not encounter any molecules with these arrangements.
2	4	

When one lone pair of electrons is present in a molecule with SN = 6, only one molecular geometry results, because all the sites in an octahedron are equivalent. The molecular geometry that results is called *square pyramidal*. The geometric solid object corresponding to the molecular shape has a square base and four triangular sides.

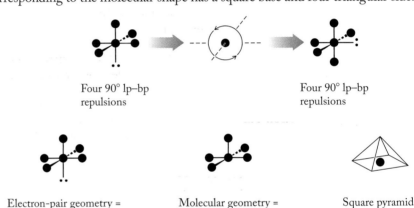

Four 90° lp–bp repulsions

Four 90° lp–bp repulsions

Electron-pair geometry = octahedral

Molecular geometry = square pyramidal

Square pyramid

Because the lone pair of electrons occupies more space than the bonding pairs, we predict the bond angles in the square pyramid would be slightly less than the ideal angle of 90°. The molecule BrF_5, for example, has a square pyramidal shape, and its F_{eq}—Br—F_{ax} bond angles are 85°.

When two lone pairs are present at the central atom in a molecule with octahedral electron-pair geometry, two possibilities arise:

One 90° lp–lp repulsion

No 90° lp–lp repulsions

The structure with no 90° lp–lp repulsions is clearly preferable, and when two lone pairs are present about a central atom with SN = 6, they always occupy vertices opposite each other on the octahedron. The resultant molecular geometry is called *square planar* because the molecule is shaped like a square and all five atoms reside in the same plane.

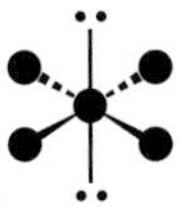

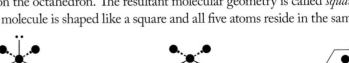

Electron-pair geometry = octahedral

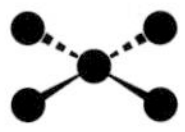

Molecular geometry = square planar

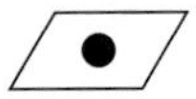

Square plane

We further predict that the presence of the lone pairs would not cause any distortion in the bond angles of a square planar molecule.

| SAMPLE EXERCISE 9.3 | **Using VSEPR Theory to Predict Molecular Geometry** |

The Lewis structures for nitrous oxide (N_2O) and sulfur tetrafluoride (SF_4) are shown below.

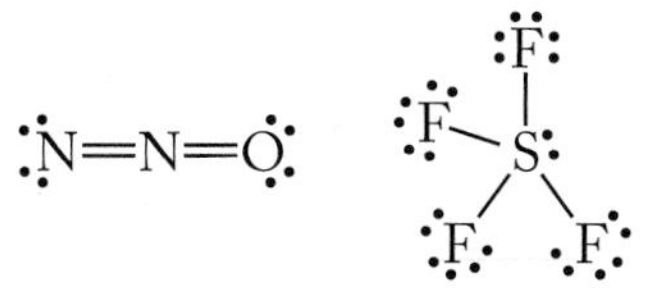

Use the VSEPR model to predict the molecular geometries of N_2O and SF_4 and the values of the bond angles in both molecules.

COLLECT AND ORGANIZE We are given the Lewis structures for N_2O and SF_4, so we can immediately determine the steric number of the central atoms and hence the electron-pair geometry of both molecules.

ANALYZE The central nitrogen in N_2O has SN = 2, so its electron-pair geometry is linear. The central sulfur in SF_4 has SN = 5, so its electron-pair geometry is trigonal bipyramidal.

SOLVE For N_2O, the electron-pair geometry and molecular geometry are both linear because the central nitrogen has no lone pairs of electrons. Its molecular structure is

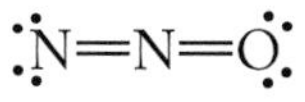

and the N—N—O bond angle is 180°.

A molecule of SF_4 has a trigonal bipyramidal electron-pair geometry that includes one lone pair of electrons. The lone pair is placed in an equatorial position, where it experiences fewer 90° lp–bp interactions than if it were placed axially.

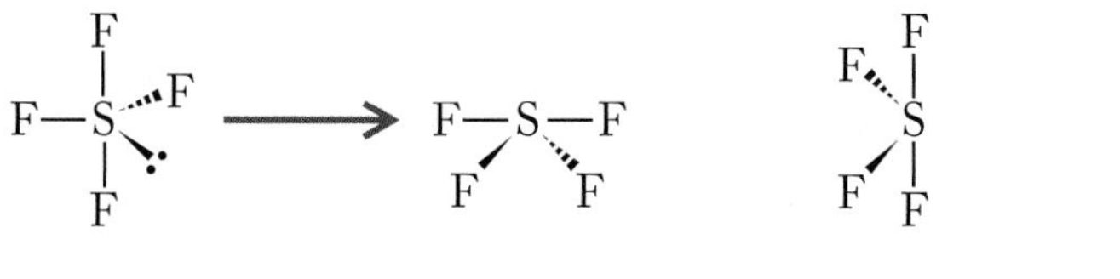

| Electron-pair geometry = trigonal bipyramidal | Molecular geometry = seesaw | Frequently drawn from this perspective as well |

The ax–ax bond angle is < 180°; the eq–eq bond angles are < 120°; and the ax–eq bond angles are < 90°. The molecule has a seesaw shape.

THINK ABOUT IT N_2O is isoelectronic with CO_2 and therefore has the same Lewis structure. Consequently, the two molecules have the same molecular geometry. Using the same reasoning, we see that any molecule with SN = 5 and one lone pair should have the same geometry as SF_4.

Practice Exercise Determine the molecular geometries and describe the bond angles of (a) NOF (which has a central N atom) and (b) SO_2Cl_2 (in which S is the central atom).

CONNECTION We used the term isoelectronic for the first time in Chapter 7 when we compared isoelectronic atoms, that is, atoms that have the same number of electrons.

Two covalently bonded atoms with different electronegativities have partial electrical charges of opposite sign creating a **bond dipole.**

○●○ **CONNECTION** In Chapter 8 we introduced electronegativity and the unequal distribution of electrons that gives rise to polar covalent bonds.

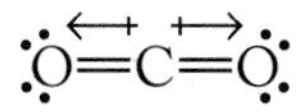

▶‖ **CHEMTOUR** Partial Charges and Bond Dipoles

(a)

Nonpolar

(b)

Dipole of each bond

Dipole of molecule

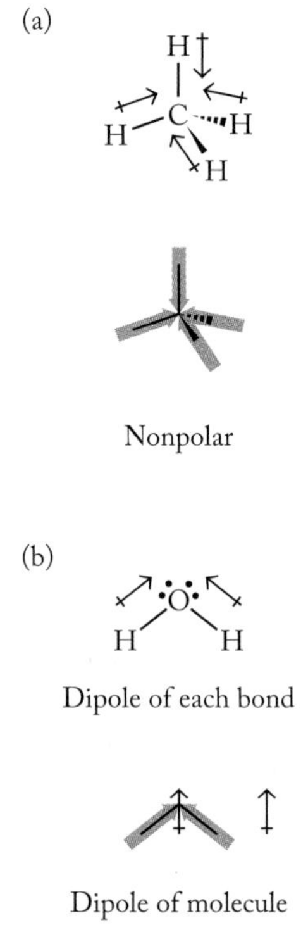

FIGURE 9.6 The influence of shape on polarity. (a) The four identical C—H bond dipoles offset each other in a tetrahedral molecule, causing CH_4 to be nonpolar. (b) The two bond dipoles in the water molecule have the same magnitude but do not offset each other because of their direction. The water molecule is polar.

9.3 Polar Bonds and Polar Molecules

Of all the minor gases in the atmosphere, H_2O is the only one that exists in nature as a liquid at standard temperature and pressure. Why is H_2O a liquid at room temperature, whereas other compounds of comparable molar mass, such as N_2, O_2, CO_2, and CH_4, are gases? The answer in part is that water is a polar molecule, but the others are not. The polarity of a molecule depends on the polarities of its bonds and its geometry. If a molecule contains only nonpolar covalent bonds, the molecule is nonpolar. This is the case with N_2 and O_2. If a molecule contains polar bonds, it may be polar or nonpolar depending on the molecular geometry.

The individual C=O double bonds in CO_2 are polar because carbon and oxygen have different electronegativities. The carbon in the molecule has a partial positive charge and the oxygen atoms have a partial negative charge. When two bonded atoms have an unequal distribution of charge, a **bond dipole** is created. As we saw in Chapter 8, bond dipoles can be represented by arrows pointing in the direction of higher density of the bonding electrons. The points of the arrows represent partial negative charges (δ^-) and the tails represent partial positive charges (δ^+), as shown here for carbon dioxide:

$$\overset{\longleftarrow \; \longrightarrow}{\ddot{O}=C=\ddot{O}}$$

The overall polarity of a molecule can be determined by summing individual bond dipoles, which are vectors having both direction and magnitude. This summing process takes into consideration the identities of the atoms to determine the magnitude of the polarity of each bond, and the geometry of the molecule to determine whether the polarities of the bonds offset each other. In CO_2 the magnitude of the bond dipole of one C=O bond is identical to the other C=O bond, and because the molecule is linear, one dipole is offset exactly by the other so that overall CO_2 is nonpolar (Figure 9.5).

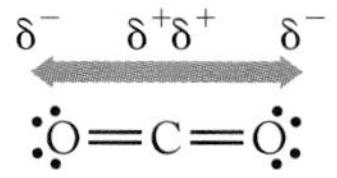

$$\overset{\delta^- \qquad \delta^+\delta^+ \qquad \delta^-}{\xleftarrow{\hspace{3cm}}\xrightarrow{\hspace{3cm}}}$$
$$\ddot{\ddot{O}}=C=\ddot{\ddot{O}}$$

FIGURE 9.5 Both C=O bonds are polar. Because the polarities are identical in magnitude and opposite in direction in a linear molecule, they offset each other and the molecule is nonpolar.

The C—H bonds in CH_4 are polar bonds because carbon and hydrogen have different electronegativities. (Consult Figure 8.5 to check relative electronegativity values to assign the direction of dipoles.) The carbon atom is more electronegative than hydrogen, so it has a small partial negative charge and the hydrogen atoms each have a small partial positive charge. The magnitudes of the C—H bond dipoles are all the same because the same atoms are involved. The tetrahedral geometry of CH_4 means that the bond dipoles of the C—H bonds offset each other, so that overall the CH_4 molecule is nonpolar (Figure 9.6a).

On the other hand, even though the two O—H bond dipoles have the same magnitude, the bent geometry of a water molecule means that the bond dipoles of the two O—H bonds do not offset each other and do not cancel. Instead, an overall dipole is directed toward the oxygen in the molecule, as shown in Figure 9.6(b). The presence of this overall dipole means that water is a polar molecule. This polarity leads to interactions between molecules based on attractions between the positive poles of some molecules and the negative poles of others. We discuss these interactions in detail in Chapter 10.

| SAMPLE EXERCISE 9.4 | **Determining Polarity of Bonds and Molecules** |

List A: N_2 O_2 List B: CO_2 H_2O CH_4 SF_6

Gases in list A are not greenhouse gases; gases in list B are greenhouse gases. Describe the polarity of the individual bonds in the molecules of each gas and the overall polarity of each molecule. Suggest a feature that may be responsible for a gas participating in the greenhouse effect based on your description.

COLLECT AND ORGANIZE We have determined the molecular geometry of all these compounds elsewhere in the chapter.

ANALYZE We can estimate bond polarity using Figure 8.5 to determine relative electronegativities of atoms bonded together. To determine the overall polarity of molecules, we sum the individual bond polarities as vectors.

SOLVE List A: N_2 and O_2 are diatomic molecules made up of the same atoms. They are both nonpolar, because electrons are shared equally in pure covalent bonds.

List B: In CO_2 the individual C=O bonds are polar, but the vectors sum to zero because the molecule is linear and the bond dipoles have the same magnitude.

In H_2O the individual H—O bonds are polar and have the same magnitude. Because the molecule is bent, the dipole vectors do not offset each other and the molecule is polar.

In CH_4 the individual H—C bonds are polar and have the same magnitude. In tetrahedral geometry, the vectors offset each other so that their sum is zero. The overall molecule is nonpolar.

In SF_6 the individual S—F bonds are polar and have the same magnitude. SF_6 is octahedral:

Each bond is polar and has the same magnitude.

Vectors offset each other; molecule is nonpolar.

The vectors offset each other and the overall molecule is nonpolar.

The molecules in list A have nonpolar covalent bonds and are nonpolar molecules. List B contains both polar and nonpolar molecules, but all the molecules have polar bonds. Based on this, we suggest that the presence of polar bonds in molecules makes it possible for them to function as greenhouse gases, even if the overall molecule is nonpolar.

THINK ABOUT IT If all the bonds in a molecule are nonpolar, as they are in N_2 and O_2, the molecule is nonpolar. The presence of polarity in bonds may give rise to polar or nonpolar molecules, depending on the overall three-dimensional shapes of the molecules.

Practice Exercise Predict which of these gaseous substances are potential greenhouse gases: H_2, Ne, NH_3, Cl_2O, CF_4.

⊙⊙ **CONNECTION** In Chapter 8 we discussed the impact of bond polarity on the physical properties of compounds like the hydrogen halides.

The permanent **dipole moment (μ)** is a measured value that defines the extent of separation of positive and negative charge centers in a covalently bonded molecule. It is a quantitative expression of the polarity of a molecule.

CONCEPT TEST

If the atmosphere of a planet contained gaseous phosphorus as P_4 tetrahedra, would those molecules function as greenhouse gases?

In the examples we have seen thus far, all dipoles within each molecule have had the same magnitude because all have involved the same atoms. What happens when a central atom has different atoms attached to it? The magnitude of any bond dipole depends on the electronegativities of the bonded pairs of atoms, and the arrangement of the bonded atoms determines the overall polarity of the molecule. Actually, the polarity of a molecule can be determined experimentally by measuring its permanent **dipole moment (μ)**. The permanent dipole moment is a measure of the extent of separation of positive and negative charge in covalently bonded molecules. This separation is characteristic of the atoms bonded together and the three-dimensional shape of the molecule and so is a characteristic physical property of a molecular substance.

When water or any polar compound is placed in an electrical field created by two charged plates (Figure 9.7), its molecules align with the field so that their negative poles are oriented toward the positive plate and their positive poles are oriented toward the negative plate. The more polar the molecules are, the more strongly they align with the field. Dipole moments of molecules are usually expressed in units of *debyes* (D), where $1 \, D = 3.34 \times 10^{-30} \, C \cdot m$. The dipole moment of water is 1.85 D compared to a dipole moment of 1.47 D for ammonia. The larger value for the dipole moment of water means that water is more polar than ammonia.

Methane (CH_4) is tetrahedral, has four identical polar bonds, and is nonpolar: $\mu_{methane} = 0 \, D$. Chloroform ($CHCl_3$) is also tetrahedral (Figure 9.8a), but its dipole moment is 1.04 D. Chloroform has a dipole moment because its polar

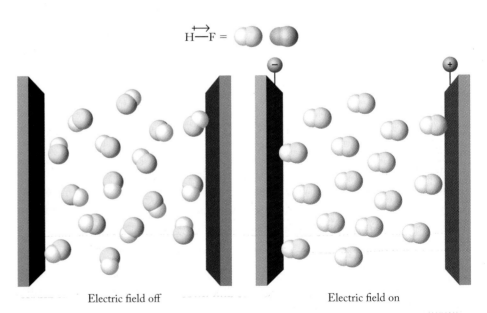

FIGURE 9.7 Polar molecules such as hydrogen fluoride (HF) adopt a random arrangement of their dipoles in the absence of an electric field but align themselves when an electric field is applied to two metal plates. The negative end of each dipole (in this case, the fluorine atom) is directed toward the positively charged plate; the positive end of each dipole (the hydrogen atom), toward the negative plate.

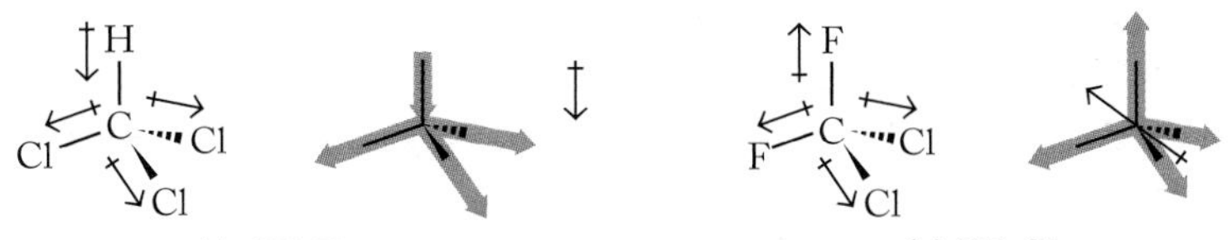

FIGURE 9.8 (a) The four bonds in chloroform ($CHCl_3$) are not equivalent in terms of direction or magnitude of polarity. The molecule has a dipole moment. (b) The two CF bond dipoles do not offset the two CCl bond dipoles. CF_2Cl_2 also has a dipole moment.

(a) $CHCl_3$

(b) CF_2Cl_2

bonds differ in *degree* of polarity and in their *direction* of polarity with respect to the central atom. The chlorine atoms are more electronegative than the central carbon atom, while the hydrogen atom is less electronegative. Consequently the electron distribution is uneven and an overall dipole moment results.

As another illustration of this concept (Figure 9.8b), consider the chlorofluoro-carbon CF_2Cl_2. All the bonds in this molecule are polarized in the same direction with respect to the central carbon, because fluorine and chlorine are both more electronegative than carbon. However, the C—F bonds are more polar than the C—Cl bonds because fluorine is more electronegative than chlorine. As a result, CF_2Cl_2 has a dipole moment. The net direction of the dipole is shown by the dipole arrow in the structure on the right.

SAMPLE EXERCISE 9.5   **Predicting the Presence of a Permanent Dipole**

Phosgene ($COCl_2$) is a highly toxic gas used in the preparation of foam rubber. Does it have a permanent dipole moment?

COLLECT AND ORGANIZE We need to generate a Lewis structure, predict the three-dimensional shape of the molecule using VSEPR theory, and then determine the polarity of each bond and whether the molecule has a permanent dipole moment.

ANALYZE Carbon is the least electronegative compound in the formula, so we assume it is the central atom. Oxygen and chlorine are both more electronegative than carbon, so the individual bonds in the molecule must be polar. Oxygen is more electronegative than chlorine. The three-dimensional shape will determine if the molecule has a permanent dipole.

SOLVE For $COCl_2$:

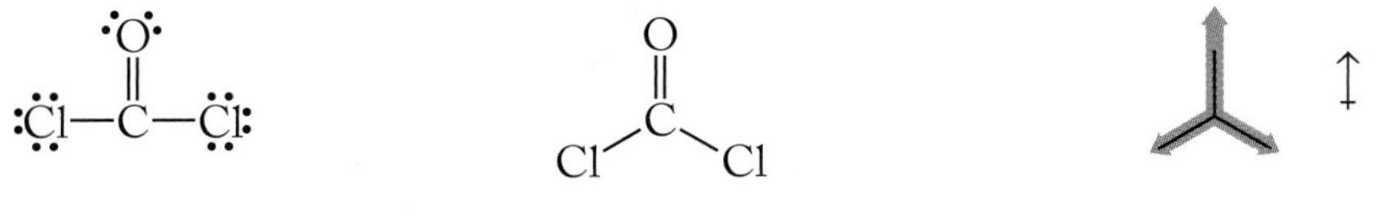

Lewis structure
SN = 3

Molecular geometry =
trigonal planar

Magnitude and direction
of dipole moment vectors

The individual bonds in the molecule are polar, and the dipoles of the bonds do not offset each other, so the molecule has a permanent dipole moment.

THINK ABOUT IT Individual bonds in a molecule may be polar, but the three-dimensional structure determines the size and direction of the permanent dipole moment.

Practice Exercise Carbon disulfide (CS_2) is present in small amounts in crude petroleum. Does CS_2 have a dipole moment? If it is released into the atmosphere, does it behave as a greenhouse gas?

CONNECTION We first heard about vibrational motion of molecules in Chapter 6 when we discussed what happens to molecules when their temperature is raised. In Chapter 7 we learned that IR radiation is the region of the continuous spectrum with longer wavelengths and lower energies than visible light.

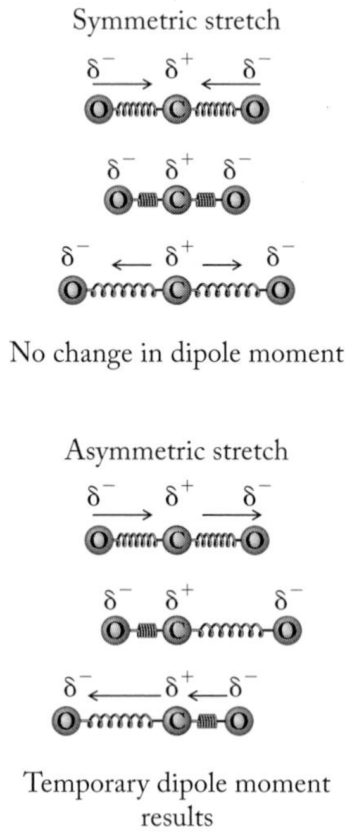

Symmetric stretch

No change in dipole moment

Asymmetric stretch

Temporary dipole moment results

FIGURE 9.9 When the two C=O bonds in CO_2 stretch in the opposite direction (symmetrical stretch), the dipole moment vectors cancel each other, and no net change in dipole moment of the molecule occurs. When the bonds vibrate in the same direction (asymmetrical stretch), a temporary dipole moment results and the molecule can absorb or emit IR radiation.

CONCEPT TEST

Water (H_2O) and hydrogen sulfide (H_2S) both have a bent geometry with dipole moments of 1.85 D and 0.98 D, respectively. Why is the permanent dipole moment in H_2S less than the permanent dipole moment of water?

In working Sample Exercise 9.4, we found out that for molecules to act as greenhouse gases, they need to have polar bonds, but do not need to have a permanent dipole moment. What happens at the molecular level that makes them behave as greenhouse gases?

Greenhouse gases absorb and emit infrared (heat) radiation. These processes are quantized. It so happens that the energy in the infrared (IR) region of the spectrum is in the same range as the energy required to make covalent bonds vibrate. For absorption or emission of IR radiation to take place, a molecule has to experience a change in its dipole moment as it vibrates. Oxygen and nitrogen molecules have nonpolar covalent bonds. They neither absorb nor emit IR radiation because they do not have a dipole moment nor undergo a change in dipole moment as their bonds vibrate. In contrast, molecules with polar bonds—even if the molecules do not have permanent dipoles—absorb and emit in the IR region because their dipole moments change as a result of bond vibrations. This is the molecular mechanism behind the greenhouse effect.

Let's take a look at CO_2 and see how its dipole moment changes as its covalent bonds vibrate. Several kinds of vibrational motions are possible, but for the purposes of this discussion, we consider the vibration of a bond to involve a displacement in space of one atom with respect to another. Think of a bond as a spring holding two atoms together and bond vibration as the motion of the two atoms back and forth on the spring (Figure 9.9).

When the two C=O bonds in CO_2 stretch in opposite directions, the dipole moment vectors cancel out and no change in dipole moment occurs. When the bonds stretch in the same direction, the vectors do not offset each other and a momentary dipole moment exists. This dipole moment is not a permanent feature of the molecule and is called a *temporary dipole*. The presence of a temporary dipole constitutes a change in the dipole moment of a molecule and is sufficient to enable the absorption and emission of IR energy.

CONCEPT TEST

Bonds in oxygen and nitrogen vibrate. Why do they not develop a temporary dipole because of these vibrations?

9.4 Valence Bond Theory

Drawing Lewis structures and applying the VSEPR model enable us to predict structures of many molecules reliably. Up to now, however, we have avoided making a connection between molecular geometry and the picture of electrons in atoms that quantum mechanics provides; namely, that electrons in atoms exist in atomic orbitals. The absence of a connection between the electronic structures of atoms and that of molecules led to the development of bonding theories using atomic orbitals that account for the observed molecular geometries predicted by VSEPR theory. One bonding theory that reliably explains observed geometries of many molecules is valence bond theory.

Orbital Overlap and Hybridization

Valence bond theory arose in the late 1920s, largely as a result of the genius and efforts of Linus Pauling. Pauling studied quantum mechanics with Erwin Schrödinger and Niels Bohr. After this experience, Pauling developed a theory of molecular bonding based on quantum mechanics. Valence bond theory assumes that chemical bonding results from the **overlap** of atomic orbitals on different atoms. The theory also assumes that the greater the overlap, the stronger and more stable the bond. Shared electrons in a chemical bond are attracted to the nuclei of two atoms, not just one atom, and this greater attraction leads to lower energy and greater stability. This view of chemical bonding is especially useful for considering chemical reactions among covalent substances as well as their physical properties, because it provides a picture of where the electrons are in an entire molecule. Just as the locations of electrons in atoms define atomic properties, so too do the locations of electrons in molecules enable us to understand properties of molecules.

Let's begin by applying Pauling's valence bond theory to some simple diatomic molecules like H_2, HCl, and Cl. The Lewis structures for these molecules are

$$\text{H—H} \qquad \text{H—}\ddot{\underset{..}{\text{C}}}\text{l:} \qquad :\ddot{\underset{..}{\text{C}}}\text{l—}\ddot{\underset{..}{\text{C}}}\text{l:}$$

In H_2, each hydrogen atom has a single electron in a $1s$ atomic orbital. Valence bond theory explains that the overlap between two half-filled $1s$ orbitals produces the single bond that joins the two hydrogen atoms together in H_2 (Figure 9.10). Valence bond theory explains that the bond between hydrogen and chlorine in HCl results when the s orbital of the H atom overlaps with the one half-filled p orbital of the Cl atom, as shown in Figure 9.10. (Recall that the electron configuration of Cl is $[Ne]3s^23p^5$.)

Overlapping the $1s$ orbital on H and the half-filled $3p$ orbital on Cl increases the electron density along the axis connecting the nuclei of the two atoms. Whenever the region of highest electron density lies along the bond axis between two atoms in a molecule, the resulting covalent bond is called a **sigma (σ) bond.**

Valence bond theory assumes that covalent bonds form when orbitals on different atoms **overlap** or occupy the same region in space.

A **sigma (σ) bond** is a covalent bond in which the highest electron density lies between the two atoms along the bond axis connecting them.

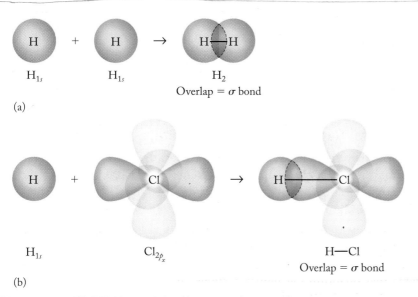

FIGURE 9.10 Valence bond theory explains bonding in terms of orbital overlap. (a) The overlap of $1s$ orbitals on two hydrogen atoms produces a single sigma (σ) bond that holds the two hydrogen atoms together. (b) The $3p$ orbital on chlorine containing a single electron overlaps with the half-filled $1s$ orbital on hydrogen to produce a single σ bond that holds the two atoms together in the HCl molecule.

In valence bond theory, **hybridization** is the mixing of atomic orbitals to generate new sets of orbitals that then are available to overlap and form covalent bonds with other atoms.

A **hybrid atomic orbital** is one of a set of equivalent orbitals about an atom created when specific atomic orbitals are mixed.

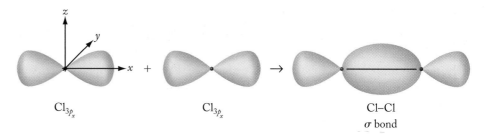

Cl_{3p_x} + Cl_{3p_x} → Cl–Cl
σ bond

FIGURE 9.11 In terms of valence bond theory, the bond in Cl_2 consists of one σ bond formed when two $2p_x$ orbitals on two chlorine atoms overlap and produce increased electron density on the axis between the two atoms.

One possible description of the electron distribution in the p orbitals of chlorine is $2p_x^1 2p_y^2, 2p_z^2$. In this case, overlap between two $2p_x$ orbitals on two chlorine atoms forms a σ bond, as shown in Figure 9.11 for atoms approaching each other along the x-axis.

Unfortunately, when we consider other molecules, an inconsistency arises between the molecular shapes that we predict and the shapes and orientations of atomic orbitals described in Chapter 7. Methane (CH_4), for example, has a carbon atom at its center. A carbon atom has the electron configuration $[He]2s^22p^2$ and forms four covalent bonds to complete its octet. The atomic orbitals of the valence electrons on carbon consist of a full $2s$ orbital and two $2p$ orbitals, each containing one electron. How can four equivalent bonds oriented at 109.5° to one another in a tetrahedral arrangement form from a filled $2s$ orbital and two partly filled $2p$ orbitals? Furthermore, how can two double bonds at an angle of 180° from each other form around the central carbon atom in CO_2, or how can one double bond and two single bonds at an angle of 120° form around the central carbon in formaldehyde (CH_2O)? Remember that the $2s$ orbital is spherical and the three $2p$ orbitals, p_x, p_y, and p_z, are at an angle of 90° to one another (Figure 9.12).

To account for the geometries of methane, carbon dioxide, formaldehyde, and many other molecules, valence bond theory calls for selected atomic orbitals of different shapes and energies to be mixed together, or **hybridized**, to form **hybrid atomic orbitals**. Covalent bonds then result from the overlap of a hybrid orbital on one atom with an unhybridized orbital on another, or from the overlap of two hybrid orbitals on two atoms. Let's begin by looking at the types of hybrid orbitals we can form on carbon and how these orbitals account for observed molecular geometries.

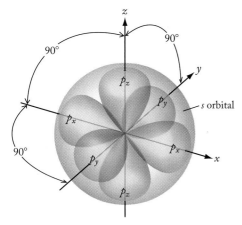

FIGURE 9.12 Relative orientation in space of $2s$ orbital (spherical) and $2p$ orbitals (three orbitals at 90° angles to each other) along axes of coordinate system.

Tetrahedral Geometry: sp^3 Hybrid Orbitals

We know that a methane molecule contains a carbon atom bonded to four hydrogen atoms and that the hydrogen atoms are located 109.5° apart at the vertices of a tetrahedron. In terms of atomic orbitals, carbon's electron configuration of $[He]2s^22p^2$ indicates that the s orbital is filled and two of the $2p$ orbitals have one electron each. Note in the orbital box diagram (Figure 9.13a) that the p orbitals are slightly higher in energy than the s orbital. This arrangement does not look like one that can make four covalent bonds, but hybridization provides a way around this problem.

First we promote one of the electrons from the $2s$ orbital to the empty $2p$ orbital. We now have four orbitals, each of which could overlap with the $1s$ orbital on a hydrogen atom to form a sigma bond, but the geometry would be wrong for methane. The p orbitals are at right angles to each other, so three of the bond

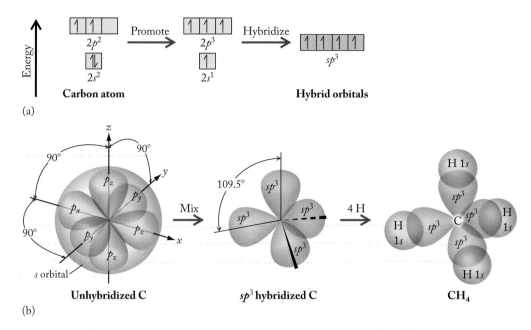

(a)

(b)

FIGURE 9.13 (a) When s and p atomic orbitals are hybridized in carbon, one electron is promoted from the full $2s$ orbital to an unoccupied $2p$ orbital. The four orbitals are then mixed to create four new sp^3 hybrid orbitals. (b) sp^3 hybridized carbon atom in methane (CH_4).

angles would be 90°, and the position of the sigma bond formed by overlap with the $2s$ orbital is unclear. This is where hybridization comes in.

The four atomic orbitals are mixed and averaged (hybridized) to make a set of four new hybrid orbitals, each of which contains one electron. The steric number of an atom from the VSEPR model always indicates the number of orbitals that must be mixed to generate the hybrid set, and the number of hybrid orbitals in a set is always equal to the number of atomic orbitals mixed. Methane has SN = 4, so four atomic orbitals are mixed to form four new hybrid orbitals.

The new orbitals are called ***sp³* hybrid orbitals**, because they result from the combination of one s and three p orbitals. They are equivalent in energy and are a blend of 25% s character and 75% p character. Their energy level is the weighted average of the energies of the s and p orbitals that were used to form them. In terms of their arrangement in three-dimensional space, the angles between the orbitals are 109.5°, and the orbitals point toward the vertices of a tetrahedron (Figure 9.13b). Hybridizing the s and all the p orbitals in the valence shell around an atom gives rise to new orbitals having a tetrahedral arrangement. Four hydrogen atoms can form σ bonds with an sp^3 hybridized carbon. The s orbitals of the hydrogen atoms overlap with the sp^3 hybrid orbitals of the carbon and result in a molecule with tetrahedral geometry.

According to valence bond theory, any atom with a tetrahedral orientation of its valence electrons has a set of four equivalent sp^3 hybrid orbitals. This includes atoms in which one or more hybrid orbitals are filled before any bonding takes place. For example, sp^3 hybridization of the valence electrons on the nitrogen atom in ammonia (NH_3) produces three orbitals that are half-filled and one orbital that is completely filled (Figure 9.14a). The three sp^3 orbitals containing one electron each form the three σ bonds in ammonia by overlapping with the $1s$ orbitals of three hydrogen atoms. The orbital that was already filled contains the lone pair on nitrogen. Similarly, the oxygen atom in water has four sp^3 hybrid

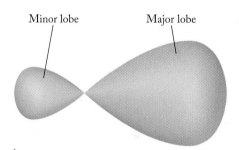

Minor lobe   Major lobe

*A single sp^3 hybrid orbital. Hybridization always produces orbitals that each have major and minor lobes. Because the minor lobes are not involved in orbital overlap leading to bond formation, the minor lobes have been omitted in the orbital models in this section.

A tetrahedral orientation of valence electrons is achieved by forming four **sp³ hybrid orbitals** from one s and three p atomic orbitals.

= # of atoms + # of lone e⁻ pairs surround central atom

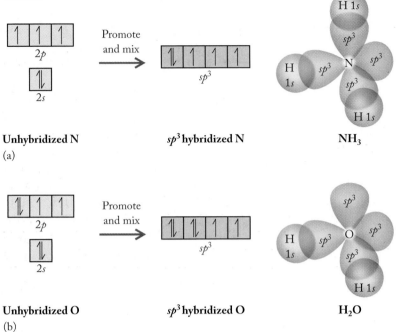

FIGURE 9.14 The four sp^3 hybrid orbitals are equivalent and point toward the vertices of a tetrahedron. (a) sp^3 hybridized nitrogen atom in ammonia (NH_3); (b) sp^3 hybridized oxygen atom in water (H_2O).

orbitals in its valence shell, two of which are filled before any bonding takes place, and two of which are half-filled and available for bond formation. The latter two are the ones that overlap with $1s$ orbitals from hydrogen atoms and form the two O—H σ bonds in H_2O (Figure 9.14b). A carbon atom making four σ bonds (SN = 4), a nitrogen making three σ bonds and having one lone pair of electrons (SN = 4), and an oxygen making two σ bonds and having two lone pairs of electrons (SN = 4) are all sp^3 hybridized atoms.

Trigonal Geometry: sp^2 Hybrid Orbitals

We have also seen carbon form trigonal planar molecules such as formaldehyde:

σ bonds
π bond

A different hybridization scheme must be used to generate an orbital array that gives rise to this known geometry. The VSEPR model defines SN = 3 for the carbon in formaldehyde, so three atomic orbitals must be mixed to make three new hybrid orbitals. To produce a trigonal planar array, the $2s$ orbital on carbon is mixed with two $2p$ orbitals, and one $2p$ orbital is left unhybridized (Figure 9.15).

Mixing and averaging one s and two p orbitals generates three hybrid orbitals called **sp^2 hybrid orbitals**. The resulting hybrid orbitals are 33% s in character and 67% p in character and their relative energy level is slightly lower than that of the sp^3 orbitals. The sp^2 hybrid orbitals lie in the same plane and are separated by angles of 120°. The two lobes of the unhybridized p orbital lie above and below the plane of the triangle defined by the sp^2 hybrid orbitals (Figure 9.15). An sp^2

hybridized carbon with this array of orbitals can form three σ bonds with its three hybridized orbitals.

Valence electrons in s and p orbitals in other atoms can also be sp^2 hybridized. To complete the valence bond picture of formaldehyde, the oxygen atom must be sp^2 hybridized as well. This points out another difference between VSEPR theory, which considers only the central atom in a molecule like formaldehyde, and valence bond theory, which applies the idea of hybridization to all the atoms in the molecule.

The valence bond view of the bonding in formaldehyde (Figure 9.15) shows the $1s$ orbitals of two hydrogen atoms overlapping with two of the sp^2 hybrid orbitals on the carbon atom, and the remaining sp^2 hybrid orbital on the carbon atom overlapping with one sp^2 hybrid orbital on the oxygen atom. These overlapping orbitals constitute the σ bonding framework of the molecule. The unhybridized p orbital on carbon is parallel to the unhybridized p orbital on the oxygen atom and the overlap of these two orbitals above and below the plane of the σ bonding network forms a π **bond**. Pi bonds have the greatest electron density above and below the internuclear axis (or in front of and in back of the internuclear axis). Pi bonds can be formed only by the overlap of partially filled p orbitals that are parallel to each other and perpendicular to the σ bond joining the atoms. The lone pairs of electrons on the oxygen atom are located in the two filled sp^2 hybrid orbitals, at an angle of about $120°$ in the plane of the molecule.

The trigonal planar orientation of valence electrons is achieved by mixing one s and two p orbitals resulting in three **sp^2 hybrid orbitals**.

In a covalent **pi (π) bond**, electron density is greatest above and below or in front of and behind the bonding axis.

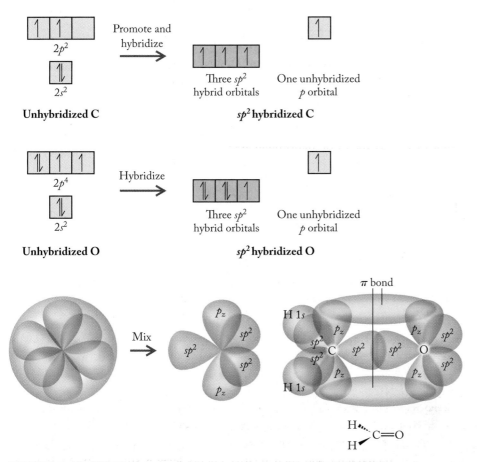

FIGURE 9.15 When one s orbital on carbon and two p orbitals are hybridized, an sp^2 hybridized carbon atom results. An sp^2 hybridized carbon forms three σ bonds and one π bond, as illustrated here where carbon bonds to an sp^2 hybridized oxygen and two hydrogen atoms to form the trigonal planar molecule formaldehyde (CH_2O).

Mixing one *s* and one *p* orbital forms two **sp hybrid orbitals** at an angle of 180° on the hybridized atom.

Nitrogen can also form sp^2 hybrid orbitals, in which case the atom forms two σ bonds and one π bond. A carbon atom forming three σ bonds and one π bond (SN = 3), a nitrogen atom forming two σ bonds and one π bond and having one lone pair of electrons (SN = 3), and an oxygen atom forming one σ bond and one π bond and having two lone pairs of electrons (SN = 3) are all sp^2 hybridized atoms. This suggests that in O=O, each O is sp^2 hybridized.

Linear Geometry: *sp* Hybrid Orbitals

We have seen hybrid orbitals generated by mixing one *s* orbital with three *p* orbitals and one *s* orbital with two *p* orbitals. The remaining mixture of one *s* orbital with one *p* orbital forms the final set of important hybrid orbitals made from *s* and *p* atomic orbitals. Again using a compound of carbon as an example, we examine the linear molecule acetylene, C_2H_2. In acetylene both carbon atoms have SN = 2 according to the VSEPR model. That means we must mix two atomic orbitals to make two hybrid orbitals on each carbon atom.

$$H—C\equiv C—H$$

The results of mixing one *s* and one *p* orbital on carbon and leaving two *p* orbitals unhybridized are shown in Figure 9.16. The two **sp hybrid orbitals** formed are 50% *s* in character and 50% *p*, which makes two hybrid orbitals at an angle of 180° on the carbon atom. One set of the unhybridized *p* orbitals form lobes above and below the axis of the *sp* hybrid orbitals; the second set is in front and in back of the plane of the hybrid orbitals (Figure 9.16).

The valence bond description of acetylene shows a σ bonding framework consisting of a 1*s* orbital on a hydrogen atom overlapping with one *sp* hybrid orbital on a carbon atom, and the other *sp* hybrid orbital on that carbon atom overlapping with its carbon counterpart that is also bonded to a hydrogen atom (Figure 9.16). This arrangement brings the two sets of unhybridized *p* orbitals on both carbons into parallel alignment with each other. They then overlap to form the two π bonds between the two carbon atoms. A carbon atom that is *sp* hybridized forms two σ bonds and two π bonds.

FIGURE 9.16 When one *s* orbital and one *p* orbital on carbon are hybridized, an *sp* hybridized carbon atom results. An *sp* hybridized carbon forms two σ and two π bonds, as illustrated here where two *sp* hybridized carbon atoms bond to each other and two hydrogen atoms to form the linear molecule acetylene (C_2H_2).

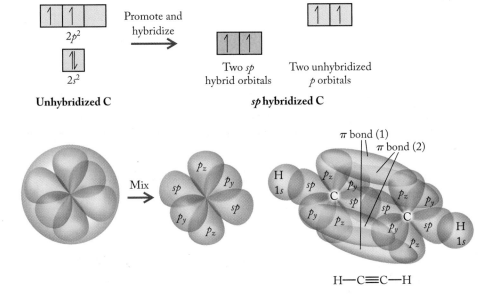

SAMPLE EXERCISE 9.6   **Describing Bonding in a Molecule**

We know that carbon dioxide is a linear molecule with two equivalent C=O double bonds. Using valence bond theory, account for the linear shape of CO_2, determine the hybridization state of carbon and oxygen, and describe the orbitals that overlap to form the bonds between carbon and oxygen. Draw the molecule, clearly showing the orbitals that overlap to form the bonds.

COLLECT AND ORGANIZE We know that CO_2 is linear and the structure of the molecule is

$$\ddot{\text{O}}\!=\!\text{C}\!=\!\ddot{\text{O}}$$

ANALYZE From the structure we can see that the carbon atom forms two σ bonds and two π bonds. Each oxygen atom is making one σ bond and one π bond.

SOLVE The hybridization state of the carbon must be sp, because sp hybrid orbitals are at an angle of 180° and give rise to linear molecules. The oxygen atoms are sp^2 hybrids because sp^2 hybridized oxygen atoms make one σ and one π bond. The σ bond framework is formed when each sp orbital on the carbon atom overlaps with one sp^2 orbital on an oxygen atom. The π bonds are formed when an unhybridized p orbital on an oxygen atom overlaps with a parallel unhybridized p orbital on the adjacent carbon atom.

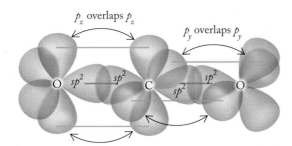

Notice that unhybridized p orbitals of the carbon atom are at an angle of 90° to one another, and the plane containing the three sp^2 orbitals of one oxygen atom is rotated 90° with respect to the plane containing the three sp^2 orbitals of the other oxygen atom. This orientation is necessary for the formation of the two π bonds, because the two orbitals that overlap to form a π bond must be parallel. In addition, for each oxygen atom, the unshared pairs of electrons and the σ bond to carbon lie in a trigonal plane 120° apart. The sp^2 hybridization of the oxygen accounts for all these features.

THINK ABOUT IT We use valence bond theory to rationalize the shape of a molecule and describe how bonds are formed. Because CO_2 is linear, the carbon atom in it must be sp hybridized. Any oxygen atom that forms a double bond must be sp^2 hybridized.

Practice Exercise In which of the following molecules does the central atom have sp^3 hybrid orbitals: (a) CCl_4; (b) HCN; (c) SO_2; (d) PH_3.

CONCEPT TEST

Why can a carbon atom with sp^3 hybrid orbitals not form π bonds?

Mixing one *s* orbital, three *p* orbitals, and two *d* orbitals from the same shell gives six equivalent **sp³d² hybrid orbitals** that point toward the vertices of an octahedron.

Mixing one *s* orbital, three *p* orbitals, and one *d* orbital from the same shell yields five equivalent **sp³d hybrid orbitals** with lobes that point toward the vertices of a trigonal bipyramid.

▶❚❚ **CHEMTOUR** Expanded Valence Shells

Octahedral and Trigonal Bipyramidal Geometry: *sp³d²* and *sp³d* Hybrid Orbitals

Hybridization can also be used to describe the shapes and bonding in compounds where the central atom has more than eight valence electrons. For example, the molecule SF_6 is octahedral, and the central sulfur atom expands its octet to bind six fluorine atoms. As we discussed in Chapter 8, only atoms with *d* orbitals in their valence shells can expand their octets. Valence bond theory provides a way to describe the expansion of an octet by including these *d* orbitals in hybridization schemes. Six atomic orbitals on sulfur—one 3*s* orbital, three 3*p* orbitals, and two 3*d* orbitals—can be hybridized to produce six equivalent *sp³d²* **hybrid orbitals** that have the appropriate orientation (Figure 9.17a).

Other molecules, such as phosphorus pentachloride (PCl_5), have a trigonal bipyramidal geometry, and the central phosphorus atom has 10 valence electrons. Five atomic orbitals—one 3*s* orbital, three 3*p* orbitals, and one 3*d* orbital—can be hybridized to produce five equivalent *sp³d* **hybrid orbitals** with lobes that point toward the vertices of a trigonal bipyramid (Figure 9.17b). A summary of the orientation of different hybrid orbitals is given in Figure 9.18.

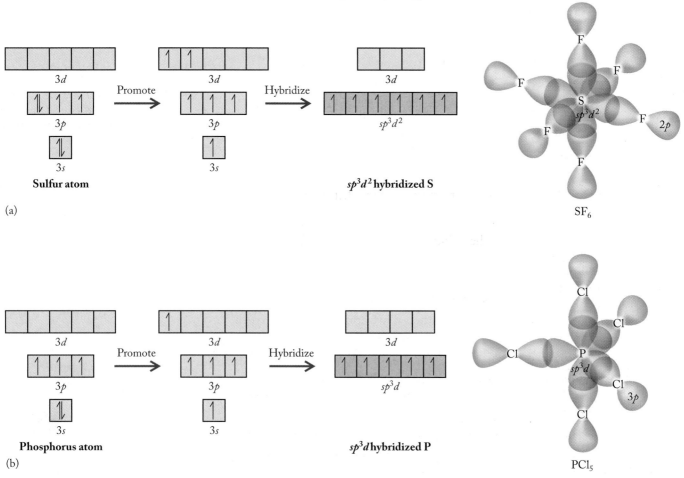

FIGURE 9.17 Hybrid orbitals can be generated by combining *d* orbitals with *s* and *p* orbitals in atoms that expand their octet. (a) One *s*, three *p*, and two *d* orbitals mix on the central sulfur atom in SF_6 to form six *sp³d²* hybrid orbitals. (b) One *s*, three *p*, and one *d* orbitals mix on the central phosphorus atom in PCl_5 to form five *sp³d* hybrid orbitals. Note in both structures that to simplify the figure we show only the *p* orbital on the chlorine or fluorine atom that overlaps with the hybrid orbital on the central atom.

Hybridization	Orientation of Hybrid Orbitals	Number of σ Bonds	Molecular Geometries	Angles between Hybrid Orbitals
sp		2	Linear	180°
sp^2		3 2	Trigonal planar Bent	120°
sp^3		4 3 2	Tetrahedral Trigonal pyramidal Bent	109.5°
sp^3d		5 4 3 2	Trigonal bipyramidal Seesaw T shape Linear	ax–ax 180° eq–eq 120°
sp^3d^2		6 5 4	Octahedral Square pyramidal Square planar	90°

FIGURE 9.18 The hybridization schemes and orientation of orbitals derived from them account for the bonding and the geometry in small molecules.

SAMPLE EXERCISE 9.7 **Recognizing Hybridized Atoms in Molecules**

Examine the Lewis structure for the thiocyanate ion (SCN⁻). (a) Identify the hybridization of each atom in the structure and (b) discuss the orbitals that overlap to form the bonds shown.

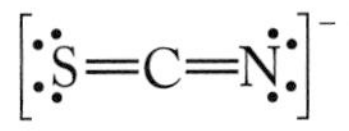

$$\left[\, \ddot{\underset{\cdot\cdot}{S}}=C=\ddot{\underset{\cdot}{N}}\, \right]^-$$

COLLECT AND ORGANIZE We are provided with the Lewis structure for the thiocyanate (SCN⁻) ion. Although other Lewis structures may be drawn, this one is favored because it locates the negative charge on the most electronegative atom (the nitrogen), and has a formal charge of 0 on sulfur and carbon. It is therefore the one we use in this analysis.

ANALYZE We can use VSEPR theory to establish the molecular geometry of SCN⁻. That information plus the Lewis structure is sufficient to determine the hybridization for each atom.

SOLVE The central carbon atom has SN = 2, so VSEPR theory predicts a structure that is the same as the Lewis structure drawn. The ion is linear, and the carbon atom in the structure has two double bonds, which means it makes two σ bonds and two π bonds. Both of these facts mean the carbon atom is sp hybridized;

it therefore has two *sp* hybrid orbitals at an angle of 180° and two unhybridized *p* orbitals, one with lobes above and below the σ bond joining C and S, and one with lobes in front and in back of it.

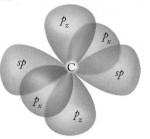

sp **hybridized C**

The sulfur atom in the thiocyanate ion makes one double bond to the carbon atom (which means one σ bond and one π bond) and has two lone pairs of electrons. This means it is *sp²* hybridized; it has three *sp²* hybrid orbitals at an angle of 120° in a plane, and one unhybridized p_z orbital with lobes above and below the plane. The way the orbitals around the sulfur atom are drawn in the accompanying figure, the unhybridized p_z orbital can overlap with the unhybridized p_z orbital on the *sp* hybridized carbon atom to form the π bond.

▶❚❚ **CHEMTOUR** Hybridization

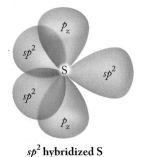

sp² **hybridized S**

The nitrogen atom in the ion makes one double bond to the carbon atom (that is, one σ bond and one π bond) and has two lone pairs of electrons. It is also *sp²* hybridized and has three *sp²* hybrid orbitals and one unhybridized *p* orbital. The nitrogen atom has the same arrangement of orbitals as the sulfur atom. However, in order for the nitrogen atom to be aligned correctly to form a π bond with the central carbon atom, it must have its unhybridized *p* orbital oriented at 90° to the π bond between carbon and sulfur atoms. The unhybridized nitrogen *p* orbital is perpendicular to the plane of the paper and parallel to the remaining unhybridized p_x orbital on the carbon atom.

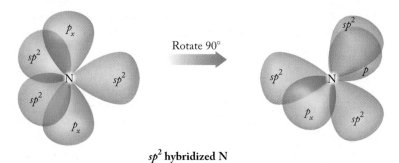

Rotate 90°

sp² **hybridized N**

The bonding in this structure is explained by allowing one *sp* hybrid orbital on the carbon atom to overlap with an *sp²* hybrid orbital on the sulfur atom and the other sp orbital on the carbon atom to overlap with an sp^2 orbital on the nitrogen atom. This accounts for the σ bond framework in the drawing. The unhybridized p_z orbital on the sulfur atom overlaps with the unhybridized p_z orbital parallel to it on the carbon atom to form a π bond; the unhybridized p_x orbital on the nitrogen atom is parallel to the remaining p_x orbital on the carbon atom. These two orbitals also overlap and form another π bond.

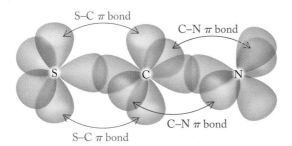

THINK ABOUT IT The sp, sp^2, and sp^3 designations for a particular atom in a molecule define its geometry, and the combination of the atomic shapes defines the molecular geometry.

Practice Exercise Determine the hybridization of the carbon atoms in acetic acid and in carbon disulfide, CS_2.

9.5 Shapes and Bonding in Larger Molecules

Up to now, most of the molecules and ions we have considered have small molar masses and typically have a single central atom bonded to two or more atoms. It is important to learn to apply the skills we have developed in looking at small molecules to larger molecules, because molecular shape is an important factor in determining physical and chemical properties of all substances. For example, molecular shape is involved in our perception of pleasant or unpleasant odors due to low concentrations of volatile molecules only slightly larger than those we have already considered. We sense these odorous molecules because they activate our sense of smell when they associate with sites in our nasal membranes called *receptors*. Our sense of smell tells us when bread is baking in the kitchen, when someone has eaten garlic or onions, and when the soup we stored in the refrigerator was there too long and has spoiled. The particular odor and taste of food, beverages, and other things are due to molecules with specific three-dimensional shapes.

Many living things respond to molecules in the air that interact with receptors in their systems. For instance, a banana in a paper or plastic bag ripens faster than on a kitchen counter. Ethylene, the molecular trigger given off by many plants and fruits as they ripen, is trapped in the bag as it is emitted by the banana and speeds the ripening process. Ethylene has the molecular formula C_2H_4 and the structure shown in Figure 9.19, which contains a $C=C$ double bond and four $C-H$ single bonds. Both carbon atoms in C_2H_4 have SN = 3. This means that the geometry around each carbon atom is trigonal planar and that the carbons are both sp^2 hybridized. Taken together, the two trigonal planar carbon atoms produce an overall planar geometry for ethylene; all six atoms lie in the same plane.

To describe the bonding in ethylene and to account for the overall shape of the molecule, remember that each carbon atom has three sp^2 hybrid orbitals that lie in the same plane as the carbon and are 120° apart. Two of these orbitals on each carbon atom are used to form σ bonds to hydrogen by overlapping with hydro-

Acetic acid: CH_3COOH

SN = 3 — Trigonal planar   SN = 3 — Trigonal planar

(a)   Two triangular planes

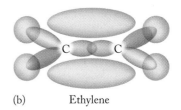

(b)   Ethylene

FIGURE 9.19 Structure and bonding in ethylene. (a) VSEPR theory predicts a flat molecule with a double bond between the two carbon atoms. Both the carbon atoms are at the center of two overlapping triangular planes. (b) Valence bond theory defines exactly the same geometry for the molecule but gives a slightly different picture about where the electrons are in the molecule with respect to the bonded atoms.

gen *s* orbitals. The third *sp²* hybrid orbital on each carbon atom is used to form the C–C σ bond. The C=C double bond is formed by overlap of the remaining half-filled *p* orbital on each carbon atom. The structure of ethylene described by valence bond theory is exactly the same as the structure we would predict based on VSEPR theory, but note the difference in the picture of the molecule provided by the two different approaches. VSEPR predicts the location of *atoms* in three dimensions about a central atom; valence bond theory rationalizes the geometry through a picture of how the *valence electrons* are distributed in hybridized orbitals. Both views are very important when we think about the three-dimensional space occupied by molecules.

SAMPLE EXERCISE 9.8 **Comparing Structures Using Valence Bond Theory**

The molecule ethane, CH_3—CH_3, is not a molecular trigger for the ripening process because it does not interact with the ethylene receptor in plant tissue. Use valence bond theory to describe the structure and bonding in ethane and compare them to the structure and bonding in ethylene.

COLLECT AND ORGANIZE We are given the molecular formula of ethane and asked to describe its structure and bonding and then compare its structure and bonding to that of ethylene.

ANALYZE We can use Lewis and VSEPR theories to determine ethane's molecular structure and then establish the hybridization of its carbon atoms. Then we can compare the location of the atoms and the electron distribution in the two molecules.

SOLVE There are four single bonds and no lone pairs around each carbon atom in ethane, which means that both carbon atoms have SN = 4, a tetrahedral geometry, and *sp³* hybridization. Compare the structures of ethylene and ethane by looking at the locations of the electrons in the σ and π bonds.

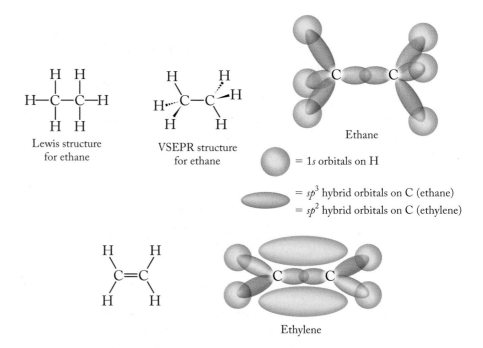

Lewis structure for ethane

VSEPR structure for ethane

Ethane

= 1*s* orbitals on H

= *sp³* hybrid orbitals on C (ethane)
= *sp²* hybrid orbitals on C (ethylene)

Ethylene

The atoms in ethylene all lie in the same plane with the π bond above and below the plane. This is a very different structure from ethane, in which the atoms do not lie in a plane and there are only σ bonds.

THINK ABOUT IT Remember the analogy at the beginning of the chapter about how molecules fit receptors like hands fit gloves. It's clear from the structures that ethylene and ethane would fit into very different gloves.

Practice Exercise Diazene (N_2H_2) and hydrazine (NH_2NH_2) are highly reactive nitrogen compounds. Compare the bonding in both using valence bond theory, and describe the differences between them.

The electrons in the π bonds in a system with alternating single and double bonds can be **delocalized** over several atoms or even an entire molecule.

To humans, ethylene is odorless, but a related compound acrolein is one of the components of barbeque smoke that contributes to the distinctive odor of a cookout. At high concentrations, it is believed to be a cancer-causing compound. Acrolein (Figure 9.20) contains features of formaldehyde (C=O double bond) and ethylene (C=C double bond). The molecule contains three trigonal planar, sp^2 hybridized carbon atoms. The pattern of alternating single and double bonds gives rise to a property sometimes called *conjugation*, which means that the electrons in the π bonds are **delocalized** over the three carbon atoms and the oxygen atom. Delocalization can occur when the atoms involved are all carbon atoms, or when double bonds in the system involve two different atoms, as in acrolein.

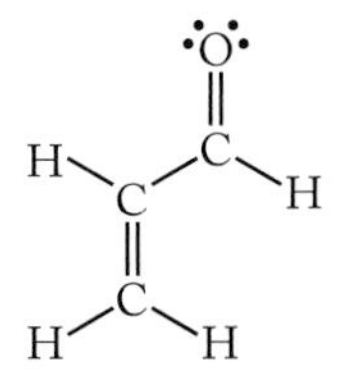

Benzene, C_6H_6, is another molecule in the complex mixture of chemical compounds found in barbeque smoke. While the use of benzene in laboratories is highly regulated because it is a carcinogen, benzene is ubiquitous in the environment as a product of combustion. Benzene is composed of a hexagon of six carbon atoms each bonded to a single hydrogen atom. There are also three C=C bonds in benzene. The resonance Lewis structures for benzene are shown in Figure 9.21 along with depictions of the π bonding system described by valence bond theory. Each carbon in benzene has a trigonal planar geometry and forms sp^2 hybrid orbitals. In benzene, as in ethylene, the carbon framework of the molecules is made of σ bonds formed by overlap of sp^2 hybrid orbitals on adjacent carbon atoms, and C—H bonds are formed by overlap of carbon sp^2 hybrid orbitals with hydrogen s orbitals.

The two resonance forms of benzene represented by Lewis structures correspond to shifts in the locations of the π bonds described by valence bond theory. The C=C π bonds form by overlap of carbon p orbitals. However, because all the carbon p orbitals are identical, they are all equally likely to overlap with their neighbors, and the π bonds in benzene are delocalized over all six carbon atoms rather than being in an alternating arrangement of the single and double carbon–carbon bonds. The distinction between localized and delocalized π bonds is shown in Figure 9.21. The presence of delocalized π bonds in benzene is often represented by a circle in the middle of the hexagon of carbon atoms.

FIGURE 9.20 The compound acrolein contains three sp^2 hybridized carbon atoms and one sp^2 hybridized oxygen atom.

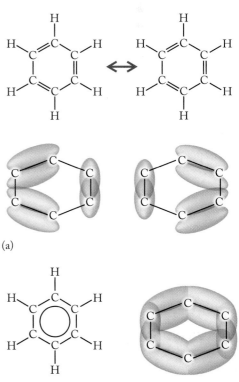

FIGURE 9.21 Two resonance structures are required in Lewis theory to describe the electron distribution in benzene. (a) The electron distributions in the two structures are equivalent. Both resonance hybrids can be drawn using the valence bond model showing localized π bonds between adjacent carbon atoms. (b) Valence bond theory depicts resonance in the π system in benzene by showing the electrons delocalized over all carbon atoms in the structure. Note that in all the valence bond structures hydrogen atoms have been excluded to simplify the graphic.

Aromatic compounds are molecules consisting of flat rings with π electron clouds delocalized above and below the plane of the molecules.

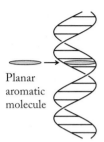

Planar
aromatic
molecule

DNA double helix
Intercalation of PAH in DNA

Benzene

PAH = polycyclic aromatic
hydrocarbons

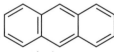

Naphthalene

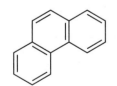

Anthracene

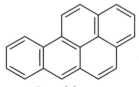

Phenanthrene

Benzo[a]pyrene

CONNECTION We learned about the emission of light energy from atoms in the form of bright-line spectra in Chapter 7.

CONCEPT TEST

In which of the molecules and polyatomic ions in Figure 9.22 are the π bonds delocalized?

(a)   (b)   (c)

FIGURE 9.22 Organic compounds with double bonds: (a) butadiene, used to make polymers; (b) 2,4-pentanedione, a reactive organic compound used in synthesis; and (c) the oxalate ion, present in spinach.

In acrolein, the delocalization of electrons in π orbitals is restricted to the part of the molecule containing the alternating single and double bonds. The π bonds in benzene delocalize over the entire structure because all the carbon atoms are part of the alternating single and double bond pattern, and the ring is flat. Compounds with this kind of structure—a flat ring with delocalized π electrons above and below the plane of the molecule—are called **aromatic compounds**. Compounds consisting of several benzene rings joined together form a family known as *polycyclic aromatic hydrocarbons* (PAHs), all of which are flat. PAHs are formed any time coal, oil, gas, and most hydrocarbon fuels are burned, and they are found in cigarette smoke and vehicle exhaust. In 2004 they were even discovered in interstellar space. Their shape gives rise to a particular health hazard. After we inhale or ingest them, some PAHs can associate with DNA in a process called *intercalation*. Because they are flat, they can intercalate DNA—they slide into the two strands of DNA. Once there, they sometimes form covalent bonds, altering or completely preventing replication and thereby damaging or killing cells. Intercalation is one step in the process by which PAHs induce cancer.

9.6 Molecular Orbital (MO) Theory

Sometimes the skies at high northern and southern latitudes brighten with a shimmering glow known as the *aurora borealis* or northern lights in the north and the *aurora australis* in the southern hemisphere (Figure 9.23). Auroras form in the upper atmosphere in a layer called the thermosphere that extends from 65 to 500 km above Earth. The colors of the aurora can be traced to oxygen atoms, nitrogen molecules (N_2), and molecular ions of nitrogen (N_2^+). Auroras occur when these atoms, molecules, and molecular ions absorb UV radiation from the sun or when they collide with the high-velocity electrons and positive particles in the solar wind. These absorptions and collisions produce excited-state molecules and ions that emit the characteristic colors of aurora displays as they return to their ground states. To understand how N_2 and N_2^+ absorb and emit electromagnetic radiation to produce the dramatic light show of an aurora, we need to take a closer look at how we describe the bonding in these species.

Lewis structures and valence bond theory help us understand the bonding capacities of elements and bonding patterns in molecules and molecular ions, while VSEPR and valence bond theories account for the observed molecular geometries of a large number of molecules and molecular ions. However, none of these models enables us to explain why O_2 is attracted to a magnetic field while N_2 is repelled slightly, to determine the location of unpaired electrons in odd electron molecules like NO, or to account for the emission of light by molecules in an aurora.

The need to explain the magnetic behavior and spectra of molecular compounds led to the development of another bonding theory known as **molecular orbital (MO) theory**. The historical roots of MO theory can be traced to the same decades as Pauling's pioneering work in valence bond theory, but the theory did not appeal to a wide number of chemists until the late 1950s and early 1960s. German physicist Friedrich Hund (1896–1997), who developed Hund's rule for filling atomic orbitals (Section 7.10), and American chemist Robert S. Millikan (1896–1986) were instrumental in developing MO theory.

Like valence bond theory, molecular orbital theory invokes the mixing of atomic orbitals. Valence bond theory uses hybrid atomic orbitals, while molecular orbital theory is based on **molecular orbitals**. A key difference between the two models is that hybrid atomic orbitals are associated only with a particular atom in the molecule while molecular orbitals are spread out, or *delocalized*, over all the atoms in a molecule. By analogy, if atoms have atomic orbitals, then molecules have molecular orbitals. Molecular orbitals represent discrete energy states in molecules just as atomic orbitals represent allowed energy states in free atoms. As with atomic orbitals, electrons enter and fill the lowest energy MO first, and higher energy levels are filled as more electrons are added. Electrons in molecules can be raised to higher energy MOs by absorbing quanta of electromagnetic radiation. When they return to lower energy MOs, they emit distinctive wavelengths of UV and visible radiation, including some of the shimmering colors in the aurora.

According to MO theory, when two atomic orbitals combine, they form two molecular orbitals. One of the two molecular orbitals is a **bonding orbital**. When electrons occupy a bonding orbital, they serve to hold the molecule together by increasing the electron density between the atoms. The second type of molecular orbital formed is an **antibonding orbital**. The antibonding orbital is higher in energy than its corresponding bonding orbital, and when electrons reside in an antibonding orbital, they destabilize the molecule and do not assist in holding the atoms together. To understand the distinction between bonding and antibonding molecular orbitals, we begin by applying MO theory to the simplest molecular compound, hydrogen gas.

Molecular Orbitals for H_2

According to MO theory, molecules of hydrogen are formed when the $1s$ orbitals on two hydrogen atoms combine to form two molecular orbitals. Molecular orbital theory stipulates that mixing two atomic orbitals creates two molecular orbitals, and that these two orbitals represent two different energy states (Figure 9.24). As a general rule, in MO theory *the total number of molecular orbitals formed in a molecule equals the number of atomic orbitals combined in the mixing process.*

When two $1s$ orbitals are combined, the lower energy, bonding molecular orbital formed is oval in shape and spans the two atomic centers. Its shape corresponds to enhanced electron density between atoms. This enhanced electron density is a covalent bond. When two electrons occupy a bonding MO, a single bond is formed. When the region of highest density lies along the bond axis, as it does in the bonding MO in H_2, the MO is designated a **sigma (σ) molecular orbital** and the resulting covalent bond is a sigma (σ) bond. The σ bonding molecular orbital in H_2 is labeled σ_{1s} in Figure 9.24(a) because it is formed by mixing two $1s$ atomic orbitals.

FIGURE 9.23 Auroras are spectacular displays of color produced when the solar wind collides with Earth's upper atmosphere.

Molecular orbital theory or **MO theory** is a bonding theory based on the mixing of atomic orbitals of similar shapes and energies to form molecular orbitals that belong to the molecule as a whole.

A **molecular orbital** is a region of characteristic shape and energy where electrons in a molecule are located.

Electrons in **bonding orbitals** serve to hold atoms together in molecules by increasing the electron density between nuclear centers.

Electrons in **antibonding orbitals** in a molecule destabilize the molecule because they do not increase the electron density between nuclear centers and therefore do not participate in holding the molecule together.

When the region of highest density lies along the bond axis between two nuclear centers, the molecular orbital is designated a **sigma (σ) molecular orbital**. Electrons in σ molecular orbitals form sigma (σ) bonds.

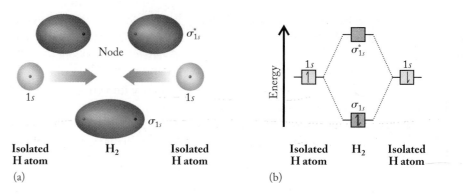

FIGURE 9.24 Mixing the $1s$ orbitals of two hydrogen atoms creates two molecular orbitals: a filled bonding σ_{1s} and an empty antibonding σ_{1s}^*. (a) The shapes of the molecular orbitals. (b) A molecular orbital diagram showing the relative energies of the atomic and molecular orbitals.

The second hydrogen molecular orbital that forms is a higher energy (less stable) antibonding molecular orbital designated σ_{1s}^*. Scientists call such an orbital "sigma star" to verbally note the presence of the asterisk and to indicate that it is an antibonding orbital. This antibonding orbital has two separate lobes of electron density and a region of zero electron density (a node) between the two hydrogen atoms, as shown in Figure 9.24(a).

Figure 9.24(b) is an example of a **molecular orbital diagram** analogous to the energy level diagrams for atomic orbitals (Section 7.5). It shows the relative energies of the molecular orbitals. The σ_{1s} bonding MO is lower in energy and therefore more stable than the $1s$ atomic orbitals by nearly the same amount that the σ_{1s}^* antibonding MO is raised in energy relative to the $1s$ atomic orbitals. Therefore, in the absence of electrons the formation of the two MOs does not significantly change the total energy of the system. Hydrogen (H_2) has a total of two valence electrons. Both valence electrons in H_2 reside in the lower energy σ_{1s} bonding orbital, because that is the lowest energy orbital available, and the electrons have opposite spins. The energy of the σ_{1s} orbital (and hence the electrons in that MO) is lower than that of two separate H atomic orbitals ($1s^1$), giving the molecule a lower energy electron configuration, which is written $(\sigma_{1s})^2$. The electron configuration describing the arrangement of electrons in a molecule consists of the type of bonds present, written in order of increasing energy, with a subscript denoting the atomic orbitals that were combined to form the MO with the number of electrons in that MO as a superscript. The lower energy of the $(\sigma_{1s})^2$ electron configuration compared to the energy of the electrons in two nonbonded hydrogen atoms explains why hydrogen gas exists as H_2 molecules rather than as H atoms.

Hydrogen is a diatomic gas, but helium exists as free atoms and not as molecular He_2. MO theory enables us to explain why. One helium atom has two valence electrons in a $1s$ atomic orbital. Mixing two He $1s$ orbitals yields the same molecular orbital diagram we generated for H_2 as shown in Figure 9.25. Unlike H_2, the two helium atoms have a total of four valence electrons. Each molecular orbital in Figure 9.25 has a maximum capacity of two electrons. Placing four valence electrons into the molecular orbital diagram in Figure 9.25 requires us to fill both the bonding and the antibonding molecular orbitals. The presence of two electrons in the higher energy σ_{1s}^* antibonding orbital cancels the stability gained from having two electrons in the lower energy σ_{1s} bonding orbital. Because there is no net gain in stability when He_2 forms from two separate He atoms, the molecule He_2 does not exist. Another way of comparing the bonding in H_2 and He_2 is to look at the bond order in the two molecules. In MO theory, the bond order in a diatomic molecule can be calculated using Equation 9.1:

$$\text{Bond order} = \tfrac{1}{2}(\text{number of bonding electrons} - \text{number of antibonding electrons}) \quad (9.1)$$

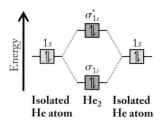

FIGURE 9.25 This molecular orbital diagram for He_2 indicates that the same number of electrons occupy antibonding orbitals and bonding orbitals. Therefore, the bond order is 0 and the molecule is not stable and does not exist.

A **molecular orbital diagram** or energy level diagram shows the relative energies and electron occupancy of the molecular orbitals for a molecule.

For H_2, this calculation establishes a bond order of 1, as there are two electrons in bonding MOs and none in antibonding MOs. A bond order of 1 corresponds to a single covalent bond. For He_2, the bond order is 0 because an equal number of electrons (2) reside in bonding and antibonding orbitals:

$$\text{Bond order} = \tfrac{1}{2}(2-2) = 0$$

A bond order of 0 means that no covalent bond exists between the two atoms. In general, the greater the bond order, the stronger the bond and the more stable the molecule.

CONNECTION We first discussed bond order in Chapter 8 when we related length and strength of bonds to the number of electron pairs shared by two atoms.

SAMPLE EXERCISE 9.9 **Using MO Diagrams to Determine Bond Order**

Develop an MO diagram for the molecular H_2^- ion. Determine the bond order of H_2^-. Does the diagram predict a stable molecular structure?

COLLECT AND ORGANIZE We must apply MO theory to determine the bond order in H_2^- and to assess the stability of the molecular ion. In the MO diagram (Figure 9.24) for the neutral molecule H_2 there are two electrons in the σ_{1s} orbital and the MO description is $(\sigma_{1s})^2$.

ANALYZE We should be able to base the MO diagram for H_2^- on the MO diagram for H_2 because H_2^- ion has only one more electron than H_2 and the empty σ_{1s}^* orbital in H_2 can accommodate up to two more.

SOLVE The σ_{1s} orbital is filled in H_2 and the next available orbital is the antibonding molecular orbital σ_{1s}^*. The third valence electron in H_2^- is in the σ_{1s}^* orbital and the ion's MO description is $(\sigma_{1s})^2(\sigma_{1s}^*)^1$. The bond order in H_2^- is

$$\text{Bond order} = \tfrac{1}{2}(2-1) = \tfrac{1}{2}$$

The molecule is not as stable as H_2, but it is stable.

THINK ABOUT IT H_2^- has a bond order of $\tfrac{1}{2}$, which means that the bond between the two atoms in H_2^- is weaker than that in H_2.

Practice Exercise Use MO theory to predict whether the H_2^+ ion can exist.

Molecular Orbitals for Homonuclear Diatomic Molecules

Molecular orbital diagrams for other homonuclear (same atom) diatomic molecules like N_2 and O_2 are more complex than that of H_2 because of the greater number and variety of atomic orbitals available in N_2 and O_2. Not all combinations of atomic orbitals result in effective bonding, but we can establish general guidelines for determining a molecular orbital diagram for any molecule. These guidelines include:

1. The total number of molecular orbitals formed in a molecule equals the number of atomic orbitals used in the mixing process.

2. Orbitals with similar energy and shape mix more effectively than do those that have different energies and shapes. For example, an s orbital mixes more effectively with another s orbital than with a p orbital.

3. Orbitals of different principal quantum numbers (for example, $1s$ and $2s$) have different sizes and energies, resulting in less effective mixing than

two 1*s* or two 2*s* orbitals. Better mixing leads to a larger energy difference between bonding and antibonding orbitals and greater stabilization of the bonding MOs.

4. A molecular orbital can accommodate a maximum of two electrons; two electrons in the same MO have opposite spins.

5. Electrons are placed in molecular orbitals according to Hund's rule.

Taking N_2 and O_2 as examples, we first consider the mixing of their 2*s* orbitals. We only consider electrons in the valence shells of atoms, because the core electrons do not participate in bonding. As shown in Figure 9.24, mixing *s* orbitals results in σ and σ^* molecular orbitals. With N_2 and O_2, these MOs are labeled σ_{2s} and σ_{2s}^*, to denote that they were formed by mixing two 2*s* atomic orbitals.

Next we move on to the *p* orbitals. The different orientations of the $2p_x$, $2p_y$, and $2p_z$ atomic orbitals in space result in different kinds of MOs. To orient ourselves in three-dimensional space, we define the *x*, *y*, and *z*-axes as indicated in Figure 9.26. This figure shows how the *p* orbitals from two atoms are mixed to form MOs. The $2p_x$ orbitals of the bonded atoms point toward each other. When

FIGURE 9.26 Two atoms are brought together and their *p* orbitals are allowed to mix to form molecular orbitals. (a) The $2p_x$ orbitals combine along the axis between the two atomic centers. This gives rise to a σ bonding orbital and a σ^* antibonding orbital. (b) The $2p_y$ orbital on one atom is parallel to the $2p_y$ on the other, as are the $2p_z$ orbitals on the two atoms, so they mix to form two π_{2p} bonding orbitals and two π_{2p}^* antibonding orbitals to complete the set of molecular orbitals.

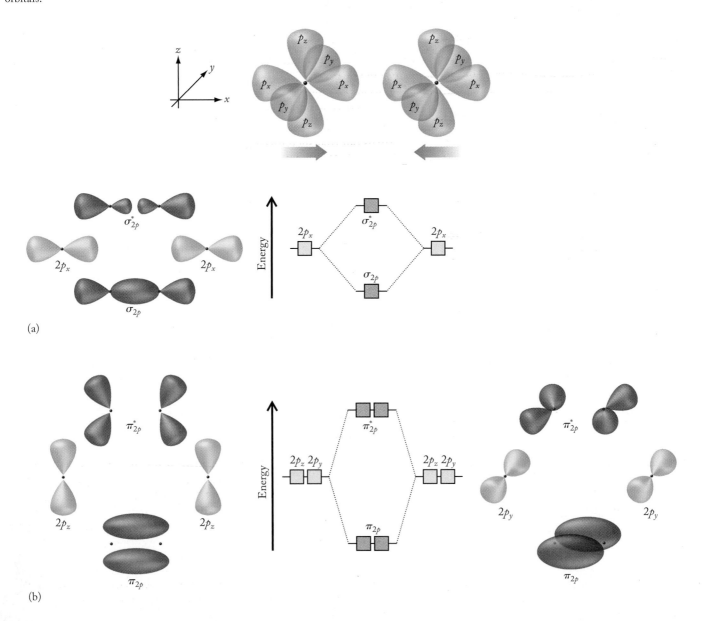

the two atomic orbitals mix, two molecular orbitals form, a bonding σ_{2p} orbital and a σ_{2p}^* antibonding orbital, in accordance with the first guideline on the list. The lobes of the $2p_y$ and $2p_z$ atomic orbitals are oriented at 90° to the bonding axis. When the $2p_z$ orbital on one atom mixes with the $2p_z$ orbital on the other, they form a set of **pi (π and π^*) molecular orbitals**. When electrons occupy a π orbital, they form a π bond. Recall that the electron density in a π bond is greatest above and below the bonding axis (Figure 9.26). Similar mixing can occur between two $2p_y$ orbitals, forming a second set of π and π^* orbitals, which have lobes in front of and behind the bonding axis. If a second pair of electrons occupies this π orbital, then a second π bond can form. Note that mixing two $2p_y$ and two $2p_z$ atomic orbitals creates a total of $2 + 2 = 4$ molecular orbitals. The two bonding MOs are labeled π_{2p} and the two antibonding MOs are labeled π_{2p}^*, as shown in Figure 9.26.

How do we populate the molecular orbitals with the available electrons? Just as in atoms, electrons in molecules in the ground state enter the lowest energy orbital available. In addition, our guidelines indicate that we also follow Hund's rule, just as we did when filling in orbitals around atoms, so electrons do not pair in orbitals if there are unoccupied orbitals of the same energy available.

The relative energies of σ and π molecular orbitals are indicated in the MO diagrams for N_2 and O_2 shown in Figure 9.27. In both molecules, the MOs derived from mixing $2s$ orbitals (σ_{2s} and σ_{2s}^*) are lower in energy than the energy of σ_{2p} for the same reason that a $2s$ atomic orbital is lower in energy than a $2p$ atomic orbital. For all diatomic species in the periodic table, up to and including N_2, the π_{2p} orbitals are lower in energy than the σ_{2p} orbital. To rationalize why this is the case, we may consider the specific case of mixing the $2s$ orbital of one nitrogen atom with the $2p_x$ orbital of the other nitrogen atom. It turns out that such mixing occurs in N_2 because of the stability (and lower energy) of the half-filled $2p$ orbitals in N atoms, which brings their energy closer to that of the $2s$ orbital. Mixing $2s$ and $2p_x$ orbitals tends to lower the energy of the σ_{2s} orbital and raise the energy

Pi (π and π^*) molecular orbitals form by the mixing of atomic orbitals that are oriented above and below, or in front of or behind, the bonding axis in a molecule. Electrons occupying π orbitals form π bonds.

CONNECTION In Chapter 7 we defined Hund's rule, which states that the lowest energy electron configuration of an atom is the one with the maximum number of unpaired electrons, all having the same spin.

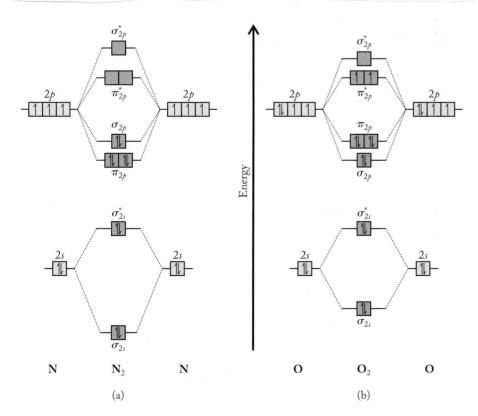

N N_2 N   O O_2 O

(a)   (b)

FIGURE 9.27 The relative energies of s and p molecular orbitals are given for (a) N_2 and (b) O_2 molecules. The σ and π orbitals of all diatomic forms of elements up to and including N_2 ($Z \leq 7$) have the order shown for nitrogen. The molecular orbitals for diatomic molecules of oxygen and all elements beyond it ($Z \geq 8$) have the order shown for oxygen. The difference between the two is the relative energy of the σ_{2p} and π_{2p} orbitals.

of the σ_{2p} orbital. The relative energies of the σ_{2s} and σ_{2p} orbitals of N_2 are shown in Figure 9.27(a). The MOs are filled by the 10 valence electrons in N_2 from the lowest energy to the highest in the following order:

$$\sigma_{2s} < \sigma_{2s}^* < \pi_{2p} < \sigma_{2p}$$

The resulting valence-shell electron configuration is

$$(\sigma_{2s})^2(\sigma_{2s}^*)^2(\pi_{2p})^4(\sigma_{2p})^2$$

These 2s–2p interactions are strong enough to cause this ordering of molecular orbitals in all homonuclear diatomic species of atoms with seven or fewer total electrons. Hence B_2 and C_2 have the same ordering of molecular orbitals as N_2.

The array of molecular orbitals for O_2 and for any homonuclear diatomic molecules of elements beyond it in the periodic table (Figure 9.28) differs from that for N_2. In molecules of the heavier elements, interaction between the 2s and 2p orbitals of the two atoms is not as strong as in N_2 because the energy difference between the 2s and 2p orbitals is greater. As a result, the σ_{2p} orbital is lower in energy than the π_{2p} orbital.

Lewis theory and valence bond theory describe the bonding between atoms, and MO theory does, too. If we look at the electron configurations of O_2 and N_2 (Figure 9.27), molecules of O_2 have two more valence electrons than do molecules of N_2. These two electrons must occupy π_{2p}^* antibonding orbitals because all bonding orbitals lower in energy are filled. The two π_{2p}^* antibonding orbitals are equivalent in energy, so each of them is half-filled, with one electron in each according to Hund's rule. The resulting electron configuration of O_2 is

$$(\sigma_{2s})^2(\sigma_{2s}^*)^2(\sigma_{2p})^2(\pi_{2p})^4(\pi_{2p}^*)^2$$

We can calculate the bond order for both N_2 and O_2 using Equation 9.1. In N_2, a total of eight electrons occupy bonding orbitals and two electrons occupy antibonding orbitals. Substituting these numbers into Equation 9.1, we get

$$\text{Bond order} = \tfrac{1}{2}(8 - 2) = 3$$

In O_2, eight electrons occupy bonding orbitals and four electrons occupy antibonding orbitals, so the calculation of bond order yields

$$\text{Bond order} = \tfrac{1}{2}(8 - 4) = 2$$

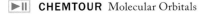 ▶❙❙ **CHEMTOUR** Molecular Orbitals

FIGURE 9.28 Molecular orbital diagrams of the valence electrons and magnetic properties of the homonuclear diatomic molecules of elements in the second row of the periodic table.

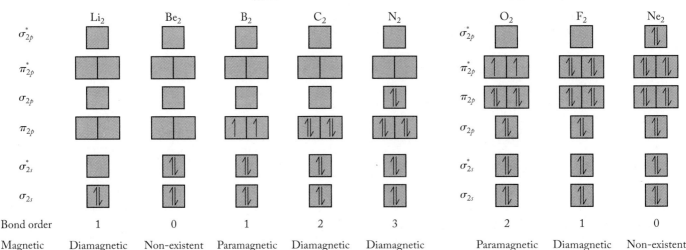

	Li_2	Be_2	B_2	C_2	N_2		O_2	F_2	Ne_2
Bond order	1	0	1	2	3		2	1	0
Magnetic properties	Diamagnetic	Non-existent	Paramagnetic	Diamagnetic	Diamagnetic		Paramagnetic	Diamagnetic	Non-existent

On the basis of their Lewis structures, we predicted a triple bond in N_2 and a double bond in O_2, and molecular orbital theory leads us to the same conclusions.

In analyzing the orbital makeup of the triple bond in N_2, we see that the four electrons from the two $2s$ atomic orbitals fill the σ_{2s} bonding orbital and the σ^*_{2s} antibonding orbital, and no net bond results from $2s$ overlap. However, the six $2p$ electrons fill three bonding MOs: the two π_{2p} orbitals and the σ_{2p} orbital. These three filled bonding MOs between two atomic centers constitute a triple bond.

To carry out the same analysis for the double bond in O_2, the same three bonding MOs are occupied, but there is one electron in each of the two π^*_{2p} antibonding orbitals. The presence of these two electrons in antibonding MOs partially offsets the stabilizing effect of six electrons in lower energy bonding MOs, and a net of $6 - 2 = 4$ electrons occupy bonding orbitals. Dividing four bonding electrons by two electrons per bond, we calculate that a double bond connects the two atoms in O_2.

SAMPLE EXERCISE 9.10   **MO Diagrams and Bond Order**

In which of the homonuclear diatomic molecules in Figure 9.28 is there an increase in bond order when one electron is removed from the molecule?

COLLECT AND ORGANIZE We are asked to determine the bond order for the series of diatomic molecular cations formed when one electron is removed from the molecules in Figure 9.28. We are to compare the bond order in these molecular ions with those for the starting diatomic molecules to determine which cases have an increase in bond order.

ANALYZE Removing an electron from B_2, C_2, N_2, O_2, and F_2 will result in the molecular ions B_2^+, C_2^+, N_2^+, O_2^+, and F_2^+. The MO diagrams for these molecular ions have the same order of orbitals as shown in Figure 9.28. The difference is only in the number of electrons that occupy the MOs. Furthermore, because the molecular ions B_2^+, C_2^+, N_2^+, O_2^+, and F_2^+ each have one electron less than the molecules in Figure 9.28, we can create the MO diagrams for the ions from the molecular diagrams for the neutral molecules by removing one electron from the highest energy molecular orbital. We then calculate bond order using Equation 9.1.

SOLVE Removing one electron from each of the MO diagrams in Figure 9.28 leaves us with the following electron configurations for the molecular cations:

Bond order $= \frac{1}{2}$(number of bonding electrons $-$ number of antibonding electrons)

B_2^+: $(\sigma_{2s})^2(\sigma^*_{2s})^2(\pi_{2p})^1$ bond order $= \frac{1}{2}(3 - 2) = 0.5$

C_2^+: $(\sigma_{2s})^2(\sigma^*_{2s})^2(\pi_{2p})^3$ bond order $= \frac{1}{2}(5 - 2) = 1.5$

N_2^+: $(\sigma_{2s})^2(\sigma^*_{2s})^2(\pi_{2p})^4(\sigma_{2p})^1$ bond order $= \frac{1}{2}(7 - 2) = 2.5$

O_2^+: $(\sigma_{2s})^2(\sigma^*_{2s})^2(\sigma_{2p})^2(\pi_{2p})^4(\pi^*_{2p})^1$ bond order $= \frac{1}{2}(8 - 3) = 2.5$

F_2^+: $(\sigma_{2s})^2(\sigma^*_{2s})^2(\sigma_{2p})^2(\pi_{2p})^4(\pi^*_{2p})^3$ bond order $= \frac{1}{2}(8 - 5) = 1.5$

The bond orders for B_2, C_2, N_2, O_2, and F_2 are listed in Figure 9.28 as 1, 2, 3, 2, and 1, respectively. Thus, the bond order increases for O_2^+ and F_2^+ relative to O_2 and F_2, but decreases for B_2^+, C_2^+, and N_2^+ relative to B_2, C_2, and N_2.

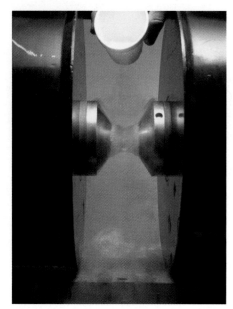

FIGURE 9.29 Molecular orbital theory explains the behavior of liquid oxygen in a magnetic field. The unpaired electrons in the π^*_{2p} orbitals make O_2 paramagnetic, which means it is attracted to a magnetic field, as shown in this picture of liquid oxygen being poured between the poles of a large magnet. Substances composed of diamagnetic molecules in which all electrons are paired do not exhibit this behavior.

⬤⬤ **CONNECTION** In Chapter 7 we discussed the spin magnetic quantum number (m_s), which has a value of either $+\frac{1}{2}$ or $-\frac{1}{2}$ and defines the orientation of an electron in a magnetic field.

Diamagnetic atoms, ions, and molecules have no unpaired electrons and are weakly repelled by a magnetic field.

Paramagnetic atoms, ions, and molecules contain at least one unpaired electron and are attracted by an external magnetic field. The strength of the attraction increases as the number of unpaired electrons increases.

THINK ABOUT IT Removing an electron from O_2 and F_2 reduces the number of electrons in antibonding molecular orbitals while leaving the number of electrons in bonding molecular orbitals unchanged. The result is an increase in bond order. In the other three homonuclear diatomic molecules, removing an electron reduces the number of electrons in bonding molecular orbitals while leaving the number of electrons in antibonding orbitals unchanged. The result is a reduction in bond order for B_2^+, C_2^+, and N_2^+.

Practice Exercise Which of the homonuclear diatomic molecules in Figure 9.28 shows an increase in bond order when one electron is added to the molecule?

One of the strengths of MO theory is that it also enables us to explain properties of matter that are not disclosed by Lewis structures or by VSEPR or valence bond theory. The magnetic behavior of diatomic molecules of the elements is a case in point. All matter that contains electrons is weakly repelled by a magnetic field because of the interaction of the field with the negatively charged electrons. Remember that the need for the fourth quantum number, m_s or the spin magnetic quantum number, arises from an experiment indicating that electrons have two possible orientations in a magnetic field. If all the electrons in a molecule or ion are paired, the species is **diamagnetic**; the effect of the electron spins cancels out, and the molecules or ions are repelled slightly by a magnetic field. In contrast, if a molecule or ion has unpaired electrons, it is **paramagnetic**, and the molecules or ions are attracted by a magnetic field. The more unpaired electrons, the greater the paramagnetism. Most molecules that we have seen thus far are diamagnetic; oxygen is paramagnetic (Figure 9.29). Only MO theory accounts for the magnetic behavior of oxygen.

In Lewis theory, VSEPR theory, and valence bond theory, all the electrons in the structures we have drawn for O_2 are paired. The MO diagram for O_2 indicates that the molecule has two unpaired electrons. In this case, MO theory correctly explains the magnetic properties of O_2 while the other theories do not. MO theory accounts for the magnetic properties of all diatomic compounds of elements in the second row of the periodic table and for many other substances as well.

CONCEPT TEST

Can liquid N_2 and O_2 be separated with a magnet?

Figure 9.28 contains MO-based electron configurations for the homonuclear diatomic molecules formed by the elements in the second row of the periodic table. These configurations enable us to make several predictions about the behavior of these molecules. First, we would predict that Be_2 and Ne_2 do not exist for the same reason that He_2 does not exist: They have as many antibonding electrons as they have bonding electrons and therefore have a net bond order of 0. Second, Li_2, B_2, and F_2 have a bond order of 1, whereas C_2 has a bond order of 2. Like O_2, B_2 is paramagnetic, whereas Li_2, C_2, N_2, and F_2 are diamagnetic.

Molecular Orbitals for NO and Heteronuclear Diatomic Species

Molecular orbital theory also enables us to account for the bonding in diatomic species formed between two different atoms, so-called *heteronuclear* diatomic molecules. Some of these have proven difficult to account for in terms of the other bonding theories we have explored. For example, it is often difficult to draw a single Lewis structure for an odd-electron molecule like NO. In the previous chapter, we considered a number of different possibilities for arrangements of electrons in NO, such as

$$:N\!\!=\!\!\overset{\cdot\cdot}{O}\!\!\cdot \qquad \cdot\overset{\cdot\cdot}{N}\!\!=\!\!\overset{\cdot\cdot}{O}\!\!\cdot$$

Our earlier comparison of nitrogen and oxygen led us to predict that, as the more electronegative element, oxygen was more likely to have a complete octet of valence electrons. In addition, experimental evidence allowed us to rule out structures with unpaired electrons on the oxygen atom. The preferred structure is the one shown in red; however, the bond distance in NO (115 pm) is considerably shorter than the value in Table 8.2 for an average $N\!\!=\!\!O$ double bond (122 pm).

Molecular orbital theory turns out to be useful for explaining the bonding in odd-electron molecules such as NO. The MO diagram of nitric oxide (NO) is different from the diagrams of homonuclear diatomic gases. Nitrogen and oxygen atoms have different numbers of protons and electrons, and the difference in effective nuclear charge in N and O atoms means that their atomic orbitals have different energies. In fact, the $2s$ and $2p$ orbitals on nitrogen have higher energies than do the same orbitals on oxygen (Figure 9.30). In constructing an MO diagram for NO, the guidelines described previously still apply. For example, the number of MOs formed must equal the number of atomic orbitals combined, and the energy and orientation of the atomic orbitals being mixed must be considered. One additional factor arises for heteronuclear diatomic molecules: when we mix atomic orbitals of the same type (for example, two p_x orbitals) from different atoms, each of the resulting MOs will have a greater contribution from one of the atoms than the other. The atom whose atomic orbitals are closer in energy to the resulting MO makes the greatest contribution to that MO. Bonding MOs in heteronuclear diatomic molecules tend to be closer in energy to the atomic orbitals of the more electronegative element from which they are derived, and the antibonding MOs are closer in energy to the atomic orbitals of the less electronegative element.

What does this mean for the molecular orbital diagram for NO shown in Figure 9.30? The π_{2p} bonding MOs in NO are closer in energy to the $2p$ orbitals of oxygen, and the antibonding π_{2p}^* MOs are closer in energy to the $2p$ orbitals of nitrogen. The proximity of the nitrogen atomic orbitals to the π_{2p}^* MOs gives these MOs more nitrogen character. Greater nitrogen character suggests that the single electron in the π_{2p}^* MOs is more likely to be on nitrogen than on oxygen. When the 11 valence-shell electrons in NO are placed in the MOs according to Hund's rule, the valence electron configuration is $(\sigma_{2s})^2(\sigma_{2s}^*)^2(\sigma_{2p})^2(\pi_{2p})^4(\pi_{2p}^*)^1$. The unpaired electron is in a π_{2p}^* antibonding orbital. This observation is consistent with our earlier prediction that the odd electron in NO is on the nitrogen atom.

Molecular orbital theory also enables us to rationalize the bond distance in NO. If we take half the difference in the number of valence-shell electrons in bonding and antibonding MOs, we calculate a bond order of $\frac{1}{2}(8 - 3) = 2.5$. This value is in between the value for a $N\!\!=\!\!O$ double bond and an $N\!\!\equiv\!\!O$ triple bond and is consistent with a bond distance (115 pm) that is between the bond distances for a nitrogen–oxygen double bond (122 pm) and triple bond (106 pm).

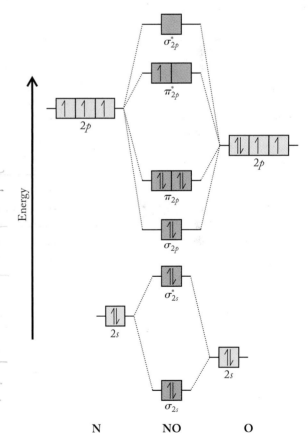

FIGURE 9.30 The molecular orbital diagram for the odd-electron molecule NO (nitric oxide) shows that the odd electron occupies a π^* antibonding orbital that is closer in energy to the atomic orbitals of nitrogen than to those of oxygen. Consequently, the electron has more nitrogen character and is more likely to be located on the nitrogen atom than on the oxygen atom.

| SAMPLE EXERCISE 9.11 | **Using MO Diagrams for Heteronuclear Diatomic Molecules** |

Nitric oxide (NO) reacts with many transition metals, including the iron in our blood. In these compounds, the NO is sometimes considered to be NO^+, while in other situations it is considered to be NO^-. Use the MO diagram in Figure 9.30 to predict the bond order of NO^+ and NO^-.

COLLECT AND ORGANIZE We can use the molecular orbital diagram in Figure 9.30 to help us determine the bond order in the two molecular ions, NO^+ and NO^-.

ANALYZE Using the MO diagram for NO, we can remove one electron from the diagram to arrive at the MO diagram for NO^+, and add an electron to the diagram to have the correct MO diagram for NO^-. Once again, bond order is calculated using Equation 9.1.

SOLVE Removing one electron from NO generates NO^+, which has 10 valence electrons. Removing an electron from the highest occupied MO (the π_{2p}^* molecular orbital in Figure 9.30) leaves NO^+ with an electron configuration:

$$NO^+: (\sigma_{2s})^2(\sigma_{2s}^*)^2(\sigma_{2p})^2(\pi_{2p})^4$$

Adding an electron to NO generates NO^-, which has 12 valence electrons. Adding an electron to the π_{2p}^* orbital in Figure 9.30 results in the electron configuration for NO^-:

$$NO^-: (\sigma_{2s})^2(\sigma_{2s}^*)^2(\sigma_{2p})^2(\pi_{2p})^4(\pi_{2p}^*)^2$$

The bond order of each molecular ion is calculated from Equation 9.1:

Bond order = $\frac{1}{2}$(number of bonding electrons − number of antibonding electrons)

$$NO^+: \text{bond order} = \frac{1}{2}(8 − 2) = 3$$

$$NO^-: \text{bond order} = \frac{1}{2}(8 − 4) = 2$$

These calculations indicate a bond order of 3 for NO^+ and a bond order of 2 for NO^-.

THINK ABOUT IT NO^+ is isoelectronic with N_2, so its bond order should be the same as in N_2. NO^- is isoelectronic with O_2. While it is hard to draw Lewis structures for NO, we can easily draw Lewis structures for NO^+ and NO^-:

$$[:N\equiv O:]^+ \qquad [:N=\ddot{O}:]^-$$

Practice Exercise Using the MO diagram for NO in Figure 9.30 as a guide, draw an MO diagram for carbon monoxide (CO), and determine the bond order for the carbon–oxygen bond.

Molecular Orbitals for N_2^+ and Spectra of Auroras

In addition to predicting the magnetic properties of molecules, MO theory is particularly useful for predicting their spectroscopic properties. In our discussion of the hydrogen atom in Section 7.4, we noted that the light emitted by excited state hydrogen atoms is quantized and can be related to the movement of electrons between atomic orbitals. Broadly speaking, the same is true in molecules; electrons can move from one molecular orbital to another by absorbing or emitting light. The principal species that contribute to the colors of the aurora are listed in Table 9.1. Excited states of two nitrogen-containing species, N_2^{+*} and N_2^*, produce the blue-violet (391–470 nm) and deep crimson red (650–680 nm) waves in auroras, respectively. The MO diagrams for these excited states are shown in Figure 9.31. Compare the MO diagram of N_2^* with that of N_2 in its ground state (see Figure 9.31a *right*). Note that an electron has been raised from the σ_{2p} MO in N_2 to a π_{2p}^* orbital in N_2^*, leaving an unpaired σ_{2p} electron behind. The corresponding excited state N_2^{+*} of the molecular ion N_2^+ (Figure 9.31b, right) also has one electron in a π_{2p}^* orbital, but the σ_{2p} orbital is empty because the other σ_{2p} electron was lost when the molecule was ionized. As the π_{2p}^* electrons return from their antibonding orbital in the excited state to the bonding (ground state) σ_{2p} orbital, the distinctive emissions of N_2^{+*} and N_2^* appear.

TABLE 9.1 **Origins of Colors in the Aurora**

Wavelength (nm)	Color	Chemical Species
650–680	Deep red	N_2^*
630	Red	O^*
558	Green	O^*
391–470	Blue violet	N_2^{+*}

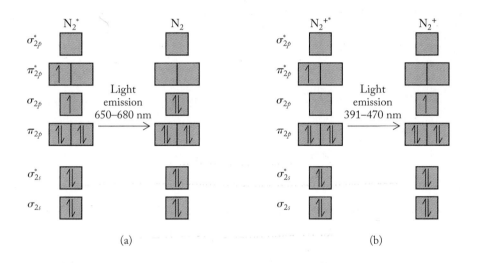

(a)   (b)

FIGURE 9.31 Molecular orbital diagrams for (a) N_2 and (b) N_2^+ show the electronic transitions that result in the emission of visible light. Collisions with ions in the solar wind result in the promotion of an electron from the σ_{2p} orbital in N_2 to the π_{2p}^* orbital, thereby creating an N_2^* molecule in the excited state. When the electron returns to the ground state, the difference in energy between the states is emitted as light in the 650 to 680 nm range. Higher energy collisions create N_2^{+*} molecular ions, which also have electrons in π_{2p}^* orbitals. When these electrons return to the ground state, light is emitted in the wavelength range of 391 to 470 nm.

CONCEPT TEST

Are the bond orders in excited state N_2^{+*} and N_2^* the same as in ground state N_2^+ and N_2?

9.7 A Bonding Retrospective

In Chapters 8 and 9 we have presented four theories of chemical bonding. Each theory has its strengths and weaknesses. The best one to apply in a given situation depends upon the question being asked and the level of sophistication required in the answer. Molecular orbital theory may provide the most complete picture of covalent bonding, but it is also the most difficult to apply to situations dealing with molecules larger than the simple diatomic ones discussed here, and it does not account for molecular shape.

Although we have focused heavily on the small gas phase molecules found in the lower atmosphere—nitrogen, oxygen, water, carbon dioxide, methane, ozone, and others—it is important to realize that the principles described in Chapter 9 apply to many larger and more complex molecules, and that the shapes of molecules are important in determining their properties at microscopic and macroscopic levels.

The chalcogens, oxygen, sulfur, and the other group 16 elements of the periodic table, have valence-shell electronic configurations ns^2np^4. Consequently, the formation of structures in which there are two bonds on each atom determines the common chemical properties of these elements, since this enables the atoms of group 16 to fill their valence octets.

We have already discussed oxygen in comparison to its neighbor fluorine to the right in the periodic table. In this section, we compare oxygen, the first of the chalcogens, to sulfur, the next member of the same family, located immediately below oxygen in the periodic table. A significant difference between the two elements was illustrated in many of the compounds discussed in this chapter. In compounds, oxygen tends to make two covalent bonds and achieve an octet of electrons in its valence shell. Sulfur does that too, but it also can expand its octet to accommodate five or even six electron pairs in its valence shell. Hence the types of compounds formed by sulfur are much more diverse than those of oxygen, and the shapes of its compounds show greater variability. This in turn affects the properties of the compounds.

lutely essential for life. The hydrides of sulfur (S), selenium (Se), and tellurium (Te) are all gases under ordinary conditions, have positive heats of formation, have bond angles close to 90°, and in addition are foul-smelling and poisonous. Hydrogen sulfide is responsible for the smell of rotten eggs. It is especially dangerous because it tends to very quickly dull and fatigue the sensory organs responsible for detecting it. This means that the intensity of the odor is a very poor guide to the concentration of H_2S in the air. Headache and nausea set in at air concentrations of H_2S as low as 5 ppm, and at 100 ppm paralysis and death result.

Organic compounds of sulfur similarly have properties that differ from those of their oxygen-containing counterparts, and many of them have characteristic odors. Methanol is an organic alcohol with the formula CH_3—OH. It is a liquid at room temperature and has an odor usually described as slightly alcoholic. Methanethiol, CH_3—SH, is a gas at room temperature and has the pungent odor of rotten cabbage. It is produced in the intestinal tract of animals by the action of bacteria on proteins and is one of the sulfur compounds responsible for the characteristic aroma of a feedlot or a barnyard.

If we go up one more carbon unit in size, the oxygen-containing compound is a liquid at room temperature called ethanol (beverage grade alcohol), CH_3—CH_2—OH. The corresponding sulfur compound is a very low boiling liquid at room temperature called ethanethiol, which has a penetrating and unpleasant odor like very powerful green onions. The human nose can detect the presence of ethanethiol at levels as low as 1 ppb (part per billion) in the air. This gives rise to its use as an odorant in natural gas. Natural gas has no odor, and leaks of natural gas are such enormous

Compounds of hydrogen with oxygen, sulfur, and the other group 16 elements provide interesting contrasts with respect to molecular shape and properties.

The data on this table show that water is clearly different from the other compounds in many respects. It is a liquid at ordinary temperatures and pressures, it has a negative heat of formation, and it has a much larger bond angle than the other three hydrides. We also know that water is odorless and is abso-

Hydride[a]	Melting Point (°C)	Boiling Point (°C)	Bond Length (pm)	Bond Angle (degrees)	Heat of Formation (kJ/mol)
H_2O	0	100	96	104.5	−285.9
H_2S	−86	−60	134	92	20.1
H_2Se	−66	−41	146	91	73.0
H_2Te	−51	−4	169	90	99.6

[a]H_2Po is excluded; too little is known of its chemistry. Polonium has no stable isotopes and is present on Earth only in very small quantities.

451

fire hazards that ethanethiol is added to natural gas streams to make leaks immediately detectable.

If we rearrange the atoms in ethanol and ethanethiol, we produce two new compounds. In the case of ethanol we get $CH_3—O—CH_3$, dimethyl ether, a colorless gas used in refrigeration systems. Its counterpart, dimethyl sulfide, $CH_3—S—CH_3$, is one of the compounds responsible for the odor of the ocean.

Three of the sulfur compounds described—hydrogen sulfide, methanethiol, and dimethyl sulfide—are referred to as volatile sulfur compounds (VSCs) by dentists. They are produced by the bacteria in the mouth and are the principal compounds responsible for bad breath. One of the reasons the odors of these compounds differ from those of their oxygen counterparts is that their shape is slightly different in terms of bond angle and bond length. Also their polarities differ because of the electronegativity difference between oxygen and sulfur. In the Look Ahead that opened this chapter, we discussed the importance of molecular shape in determining the extent of interaction of a compound with a receptor in nasal membranes. In part, the vast differences in odor and sensory detectability of these compounds are due to their shapes and electron distributions.

Not all sulfur compounds have an odor, but many odiferous compounds do contain sulfur. The characteristic and unpleasant smell of urine produced by some people after eating asparagus results from the inability of their bodies to convert odiferous sulfur compounds into odor-free sulfate ions. Not all people are able to convert the sulfur compounds in asparagus to sulfate, and not all people are able to smell the odiferous sulfur compounds. Apparently, genetic differences exist in the population in terms of metabolism as well as in the ability to detect odors; it is estimated that about 40% of the population produces the odor, while only about 10% of the population can detect it.

The odor of skunk is due to butanethiol, $CH_3—CH_2—CH_2—CH_2—SH$, and the odor of well-used athletic shoes is primarily due to the presence of sulfur compounds produced by bacteria. Not all sulfur compounds have aromas as unpleasant as these, however. A compound with the formula $C_{10}H_{18}S$ is responsible for the aroma of grapefruit. If the orientation of two atoms on one of the carbon atoms in this molecule is switched, the resulting molecule has the same Lewis structure but now has no aroma at all, despite the presence of sulfur.

Summary

SECTION 9.1 The shape of a molecule, or its **molecular geometry**, reflects the arrangement of the atoms in three-dimensional space and is determined largely by characteristic **bond angles**. The valence electrons shared by two atoms in covalently bonded molecules are bonding pairs, while the valence electrons that remain unshared and associated with a single atom are lone (or nonbonding) pairs.

SECTION 9.2 Minimizing repulsions between pairs of valence electrons (the **VSEPR** model) around an atom results in the lowest energy orientations of bonding and nonbonding electron pairs and accounts for the observed molecular geometry of the molecule. The shape of a molecule can be determined by its **steric number** (the sum of the number of bonded atoms and lone pairs around a central atom) and the **electron-pair geometry**, or arrangement of atoms and lone pairs. Molecules with a steric number equal to two and no lone pairs on the central atom have a **linear** geometry (bond angle 180°), while the shapes of molecules with steric numbers 3 to 6 are **trigonal planar** (bond angle 120°), **tetrahedral** (bond angle 109.5°), **trigonal bipyramidal** (bond angles 90°, 120°, and 180°), and **octahedral** (bond angle 90° and 180°), respectively. Bent (or angular), trigonal pyramidal, seesaw, T-shaped, square pyramidal, and square planar geometries result from different combinations of bonded atoms and lone pairs about a central atom. The observed bond angles in molecules deviate from the ideal values as a result of unequal repulsions between lone pairs and bonding pairs of electrons.

SECTION 9.3 Two covalently bonded atoms with different electronegativities have partial electrical charges of opposite sign creating a **bond dipole**. If the individual bond dipoles in a molecule do not offset each other, then the molecule is a polar molecule. The geometry of the molecule is important in determining whether the bond dipoles offset each other. If the individual bond dipoles in a molecule offset each other, then the molecule is a nonpolar molecule. A polar molecule has a permanent **dipole moment (μ)**, which is a quantitative measure of the polarity of the molecule.

Covalent bonds behave more like flexible springs than rigid rods. They can undergo a variety of bond vibrations upon absorbing certain wavelengths (energies) of radiation. Compounds whose bonds absorb energy in the infrared (IR) region act as greenhouse gases, reflecting heat (IR radiation) back into the planet's atmosphere, warming the atmosphere and the planet below. Greenhouse gases contain polar bonds but may be nonpolar molecules.

SECTION 9.4 Valence bond theory is one model of covalent bonding in molecules. According to valence bond theory, the **overlap** of orbitals results in a covalent bond between two atoms in a molecule. Valence bond theory accounts for the observed shape (geometry) of a molecule by mixing, or **hybridizing**, atomic orbitals to create **hybrid atomic orbitals**. Mixing one s and one p orbital forms two sp **hybrid orbitals**. Overlap between two sp hybrid orbitals results in a **sigma (σ) bond** where the highest electron density lies along the bond axis with a linear orientation of valence electrons. Covalent bonds in which the electron density is greatest above and below, or in front of and behind, the bonding axis are called **pi (π) bonds**. Mixing one s and two p orbitals forms three sp^2 **hybrid orbitals**. Overlap between sp^2 orbitals and other atomic or hybrid orbitals results in up to three σ bonds and a trigonal planar orientation of valence electrons. Mixing one s and three p orbital forms four sp^3 **hybrid orbitals**. Overlap between sp^3 orbitals

and other atomic or hybrid orbitals results in up to four σ bonds and a tetrahedral orientation of valence electrons. Mixing an s orbital, three p orbitals, and two d orbitals gives six equivalent sp^3d^2 **hybrid orbitals** that point toward the vertices of an octahedron. Overlap between sp^3d^2 orbitals and other atomic or hybrid orbitals results in up to six σ bonds. Mixing a 3s orbital, three 3p orbitals, and one 3d orbital yields five equivalent sp^3d **hybrid orbitals** with lobes that point toward the vertices of a trigonal bipyramid. Overlap between sp^3d orbitals and other atomic or hybrid orbitals results in up to five σ bonds.

SECTION 9.5 The shape of a molecule with more than one central atom reflects the geometries around each atom and can be difficult to describe. Molecules with exclusively sp^2 hybridized central atoms have a linear or planar geometry. Electrons in π bonds can be spread, or **delocalized**, over several atoms or even an entire molecule.

SECTION 9.6 Molecular orbital (MO) theory is a second bonding theory based on the formation of **molecular orbitals** that belong to the molecule as a whole. Molecular orbital theory can provide a better description than valence bond theory of the location of unpaired electrons in molecules. MO theory explains the magnetic and spectroscopic properties of molecules but does not help explain their shapes. Atomic orbitals with similar energies and shapes mix to give molecular orbitals delocalized over the entire molecule. Mixing two atomic orbitals creates a **bonding orbital** and an **antibonding orbital**. A bonding molecular orbital can contain a maximum of two electrons. A single covalent bond is formed when two electrons occupy a bonding molecular orbital. Whenever two atomic orbitals combine to form a bonding molecular orbital, a second higher energy antibonding orbital must also be formed. An antibonding molecular orbital can also contain a maximum of two electrons. Electrons in antibonding orbitals weaken the covalent bonds in a molecule. When the region of highest density in a bonding molecular orbital lies along the bond axis, the molecular orbital is designated as a **sigma (σ) molecular orbital**. Electrons in σ molecular orbitals form sigma (σ) bonds. Antibonding molecular orbitals that accompany sigma (σ) molecular orbitals are called σ^* ("sigma star") molecular orbitals. When the region of highest density in a bonding molecular orbital is not oriented along the bonding axis in a molecule, the molecular orbital is designated as a **pi (π) molecular orbital**. Electrons occupying π orbitals form π bonds. Antibonding molecular orbitals that accompany pi (π) molecular orbitals are called π^* ("pi star") molecular orbitals. The bond order in a molecular orbital description of a molecule or ion equals one-half the difference between the number of electrons in bonding MOs and the number of electrons in antibonding MOs. A **molecular orbital diagram** shows the relative energies of the molecular orbitals of a molecule including all σ, σ^*, π, and π^* **molecular orbitals**. Each molecular orbital can be occupied by a maximum of two electrons. We can write electron configurations of molecules using the designations σ, σ^*, π, and π^* to describe the type of molecular orbitals occupied by electrons, subscripts to identify the atomic orbitals that led to the MOs, and superscripts to indicate the number of electrons in each molecular orbital; for example, $(\sigma_{1s})^2$ $(\sigma^*_{1s})^2(\sigma_{2s})^2(\sigma^*_{2s})^2(\pi_{2p})^4(\sigma_{2p})^2$ for N_2. Atoms, ions, and molecules with no unpaired electrons are called **diamagnetic** and are slightly repelled by an applied magnetic field. Atoms, ions, and molecules containing at least one unpaired electron are called **paramagnetic** and are attracted by an external magnetic field.

Problem-Solving Summary

TYPE OF PROBLEM	CONCEPTS AND EQUATIONS	SAMPLE EXERCISES
Predicting molecular geometry	Draw a Lewis structure for the molecule. Determine the steric number of the molecule by summing the number of bonded atoms and lone pairs on the central atom. Choose a geometry that minimizes the repulsions between electron pairs.	9.1
Predicting the relative sizes of bond angles	Draw a Lewis structure for the molecule and determine the molecular geometry using the VSEPR model. Consider the repulsions between bonded atoms and lone pairs; lone pairs of electrons on a central atom always push atoms bonded to the central atom closer together and decrease the bond angles.	9.2
Using VSEPR theory to predict molecular geometry	Draw a Lewis structure for the molecule. Determine the steric number of the molecule by summing the number of bonded atoms and lone pairs. Choose a geometry that minimizes the repulsions between electron pairs.	9.3
Determining the polarity of bonds and molecules	Assign dipoles to each bond and use molecular geometry to determine whether the dipoles offset each other.	9.4, 9.5
Describing the bonding in molecules using hybrid orbitals and recognizing hybrid orbitals	Identify the hybrid orbitals in molecules that result from mixing different numbers of s, p, and d orbitals that result in the observed molecular geometry: $s + p$ = two sp hybrid orbitals $s + $ two p = three sp^2 $s + $ three p = four sp^3 $s + $ three $p + d$ = five sp^3d $s + $ three $p + $ two d = six sp^3d^2 hybrid orbitals	9.6–9.8
Determining the bond order of a homonuclear or heteronuclear diatomic molecule or molecular ion	Bond order = $\frac{1}{2}$(number of electrons in bonding MOs − number of electrons in antibonding MOs)	9.9–9.11

Visual Problems

9.1. Two compounds with the same formula, S_2F_2, have been isolated. The structures in Figure P9.1 show the arrangements of the atoms in these different compounds. Can these two compounds be distinguished by their dipole moments?

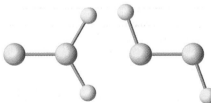

FIGURE P9.1

9.2. Could you distinguish between the two structures of N_2H_2 shown in Figure P9.2 by the magnitude of their dipole moments?

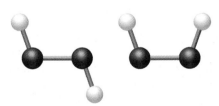

FIGURE P9.2

9.3. Which of the molecules shown in Figure P9.3 are planar, that is, have all atoms in a single plane? Are there delocalized π electrons in any of these molecules?

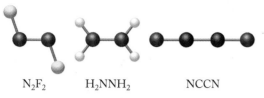

N₂F₂   H₂NNH₂   NCCN

FIGURE P9.3

9.4. Which of the molecules shown in Figure P9.4 is *not* planar? Are there delocalized π electrons in any of these molecules?

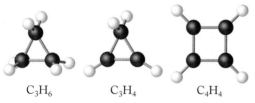

C₃H₆   C₃H₄   C₄H₄

FIGURE P9.4

9.5. Use the MO diagram in Figure P9.5 to predict whether O_2^+ has more or fewer electrons in antibonding molecular orbitals than O_2^{2+}.

σ_{2p}^*

π_{2p}^*

π_{2p}

σ_{2p}

σ_{2s}^*

σ_{2s}

FIGURE P9.5

9.6. Under appropriate conditions, I_2 can be oxidized to I_2^+, which is bright blue. The corresponding anion, I_2^-, is not known. Use the molecular orbital diagram in Figure P9.6 to explain why I_2^+ is more stable than I_2^-.

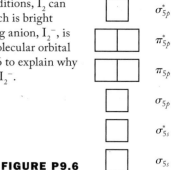

σ_{5p}^*

π_{5p}^*

π_{5p}

σ_{5p}

σ_{5s}^*

σ_{5s}

FIGURE P9.6

9.7. The molecular geometry of ReF_7 is an uncommon structure called a pentagonal bipyramid, which is shown in Figure P9.7. What are the bond angles in a pentagonal bipyramid?

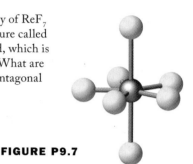

FIGURE P9.7

*9.8. The molecular geometry of the transition metal–containing anions MF_8^{2-} (M = Mo and W) is not known. What would be the F—M—F bond angles in MF_8^{2-} if the anion has a cubic geometry as shown in Figure P9.8?

FIGURE P9.8

Questions and Problems

Molecular Shape; Valence-Shell Electron-Pair Repulsion Theory (VSEPR)

CONCEPT REVIEW

9.9. Why is the shape of a molecule determined by repulsions between electron pairs and not by repulsions between nuclei?

9.10. Do all resonance forms of a molecule have the same molecular geometry? Explain your answer.

9.11. How can SO_3 and BF_3 have different numbers of bonds but the same trigonal planar geometry?

9.12. Account for the range of bond angles from about 104° to 180° in triatomic molecules.

9.13. In a molecule of ammonia, why is the repulsion between the lone pair and a bonding pair of electrons on nitrogen greater than the repulsion between two N—H bonding pairs?

9.14. Why is it important to draw a correct Lewis structure for a molecule before predicting its geometry?

9.15. Why does the seesaw structure have lower energy than a trigonal pyrimidal structure derived by removing an apical atom from a trigonal bipyramidal AB_5 molecule?

*9.16. Which geometry do you predict will have lower energy: a square pyramid or a trigonal bipyramid?

PROBLEMS

9.17. Rank the smallest bond angle for the following molecular geometries in order of increasing bond angle: (a) trigonal planar; (b) octahedral; (c) tetrahedral.

9.18. Rank the smallest bond angle for the following molecular geometries in order of increasing bond angle: (a) seesaw; (b) tetrahedral; (c) square pyramidal.

9.19. Which of the molecular geometries discussed in this chapter have more than one characteristic bond angle?

*9.20. Which molecular geometries for molecules of the general formula AB_x (x = 2 − 6) discussed in this chapter have the same bond angles when lone pairs replace one or more atoms?

9.21. Which of the following molecular geometries does not lead to linear triatomic molecules after removing one or more atoms? (a) tetrahedral; (b) octahedral; (c) T-shaped

9.22. Which of the following molecular geometries does not lead to linear triatomic molecules after removing one or more atoms? (a) trigonal bipyramidal; (b) seesaw; (c) trigonal planar

***9.23.** Describe the molecular geometries that result from removing one atom from an AB_7 molecule with a pentagonal bipyramidal geometry. (See Figure P9.7 for the shape of a pentagonal bipyramid.)

***9.24.** Which atoms would you have to remove from the cubic AB_8 molecule to create a geometry that approximates an octahedron? (See Figure P9.8 for the shape of a cubic molecule.)

9.25. Determine the molecular geometries of the following molecules: (a) GeH_4; (b) PH_3; (c) H_2S; (d) $CHCl_3$.

9.26. Determine the molecular geometries of the following molecules and ions: (a) NO_3^-; (b) NO_4^{3-}; (c) S_2O; (d) NF_3.

9.27. Determine the molecular geometries of the following ions: (a) NH_4^+; (b) CO_3^{2-}; (c) NO_2^-; (d) XeF_5^+.

9.28. Determine the geometries of the following ions: (a) SCN^-; (b) $CH_3PCl_3^+$ (P is the central atom and this cation contains a C—P bond); (c) ICl_2^-; (d) PO_3^{3-}.

9.29. Determine the geometries of the following ions and molecules: (a) $S_2O_3^{2-}$; (b) PO_4^{3-}; (c) NO_3^-; (d) NCO.

9.30. Determine the geometries of the following molecules: (a) ClO_2; (b) ClO_3; (c) IF_3; (d) SF_4.

9.31. Which of the following triatomic molecules, O_3, SO_2, and CO_2, have the same molecular geometry?

9.32. Which of the following species, N_3^-, O_3, and CO_2, have the same molecular geometry?

9.33. Which of the following ions, SCN^-, CNO^-, and NO_2^-, have the same geometry?

9.34. Which of the following molecules, N_2O, S_2O, and CO_2, have the same molecular geometry?

9.35. The Venusian Atmosphere A number of sulfur oxides not found in Earth's atmosphere have been detected in the atmosphere of Venus (Figure P9.35), including S_2O and S_2O_2. Draw Lewis structures for S_2O and S_2O_2, and determine their molecular geometries.

FIGURE P9.35

9.36. The structures of NOCl, NO_2Cl, and NO_3Cl were determined in 1995. They have the skeletal structures shown in Figure P9.36. Draw Lewis structures for these three compounds and predict the bonding geometry at each nitrogen atom.

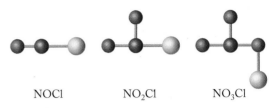

NOCl NO_2Cl NO_3Cl

FIGURE P9.36

***9.37.** For many years, it was believed that the noble gases could not form covalently bonded compounds. However, xenon reacts with fluorine and oxygen. Reaction between xenon tetrafluoride and fluoride ions produces the pentafluoroxenate anion:

$$XeF_4 + F^- \rightarrow XeF_5^-$$

Draw Lewis structures for XeF_4 and XeF_5^-, and predict the geometry around xenon in XeF_4. The crystal structure of XeF_5^- compounds indicates a pentagonal bipyramidal orientation of valence pairs around Xe. Sketch the structure for XeF_5^-.

***9.38.** The first compound containing a xenon–sulfur bond was isolated in 1998. Draw a Lewis structure for HXeSH and determine its molecular geometry.

***9.39. Chemical Terrorism** In 1995 a Japanese cult attacked the Tokyo subway system with the nerve gas Sarin and focused world attention on the dangers of chemical warfare agents. The structure in Figure P9.39 shows the connectivity of the atoms in the Sarin molecule. Complete the Lewis structure by adding bonds and lone pairs as necessary. Assign formal charges to the P and O atoms, and determine the molecular geometry around P.

$$
\begin{array}{c}
\text{H} \\
| \\
\text{H} \quad \text{O} \quad \text{H—C—H} \\
| \quad\quad | \quad\quad\quad | \\
\text{H—C—P———C—H} \\
| \quad\quad | \quad\quad\quad | \\
\text{H} \quad \text{F} \quad \text{H—C—H} \\
| \\
\text{H}
\end{array}
$$

Sarin

FIGURE P9.39

9.40. Determine the bonding geometry around the nitrogen atom in the following unstable nitrogen oxides: (a) N_2O_2; (b) N_2O_5; (c) N_2O_3. (N_2O_2 and N_2O_3 have N–N bonds; N_2O_5 does not.)

Polar Bonds and Polar Molecules

CONCEPT REVIEW

9.41. Explain the difference between a polar bond and a polar molecule.

9.42. Must a polar molecule contain polar covalent bonds? Why?

9.43. Can a nonpolar molecule contain polar covalent bonds?

9.44. What does a dipole moment measure?

9.45. Why does infrared radiation cause bonds to vibrate but not break (as UV radiation can)?

9.46. Argon is the third most abundant species in the atmosphere. Why isn't it a greenhouse gas?

*__9.47.__ Would the energy required to cause the bond in CO to vibrate be more or less than that required by the carbon–oxygen bond in CO_2?

*__9.48.__ Which compound absorbs IR radiation of a longer wavelength—NO or NO_2?

PROBLEMS

9.49. The following molecules contain polar covalent bonds. Which of them are polar molecules and which are nonpolar? (a) CCl_4; (b) $CHCl_3$; (c) CO_2; (d) H_2S; (e) SO_2

9.50. Photolysis of Cl_2O_2 is thought to produce compounds with the skeletal structures shown in Figure P9.50. Do the two compounds have the same dipole moment?

FIGURE P9.50

9.51. Freon Bar Compounds containing carbon, chlorine, and fluorine are known as Freons or chlorofluorocarbons (CFCs). As we saw in Chapter 8, the widespread use of these substances was banned because of their effect on the ozone layer in the upper atmosphere. Which of the following chlorofluorocarbons (CFCs) are polar and which are nonpolar? (a) Freon 11 ($CFCl_3$); (b) Freon 12 (CF_2Cl_2); (c) Freon 113 (Cl_2FCCF_2Cl)

9.52. Which of the following chlorofluorocarbons (CFCs) are polar and which are nonpolar? (a) Freon C318 (C_4F_8, cyclic structure); (b) Freon 1113 (C_2ClF_3); (c) $Cl_2HCCClF_2$

9.53. Predict which molecule in each of the following pairs is the more polar: (a) Freon 13 ($CClF_3$) or Freon 13B1 ($CBrF_3$); (b) Freon 12 (CF_2Cl_2) or Freon 22 (CHF_2Cl); (c) Freon 113 (Cl_2FCCF_2Cl) or Freon 114 (ClF_2CCF_2Cl).

9.54. Which molecule in each of the following pairs is more polar? (a) NH_3 or PH_3; (b) CCl_2F_2 or CBr_2F_2

9.55. A series of carbonyl dihalide compounds of formula COX_2 (where X = I, Cl, or Br) has been prepared. Place these compounds—COI_2, $COCl_2$, and $COBr_2$—in order of increasing polarity. Explain your reasoning.

9.56. Simple diatomic molecules detected in interstellar space include CO, CS, SiO, SiS, SO, and NO. Arrange these molecules in order of increasing dipole moment based on the location of the constituent elements in the periodic table, and then calculate the electronegativity differences from the data in Figure 8.5.

Valence Bond Theory

CONCEPT REVIEW

9.57. Are hybrid orbitals ever constructed from atomic orbitals with different principal quantum numbers?

9.58. Why aren't the orbitals on free atoms hybridized?

9.59. Do all resonance forms of N_2O have the same hybridization at the central N atom?

PROBLEMS

9.60. What is the hybridization of nitrogen in each of the following ions and molecules? (a) NO_2^+; (b) NO_2^-; (c) N_2O; (d) N_2O_5; (e) N_2O_3

9.61. What is the hybridization of sulfur in each of the following molecules? (a) SO; (b) SO_2; (c) S_2O; (d) SO_3

9.62. Airbags Azides such as sodium azide, NaN_3, are used in automobile airbags as a source of nitrogen gas. Another compound with three nitrogen atoms bonded together is N_3F. What differences are there in the arrangement of the electrons around the nitrogen atoms in the azide ion (N_3^{2-}) and N_3F? Is there a difference in the hybridization of the central nitrogen atom?

9.63. N_3F decomposes to nitrogen and N_2F_2 by the following reaction:

$$2 N_3F \rightarrow 2 N_2 + N_2F_2$$

N_2F_2 has two possible structures as shown in Figure P9.63. Are the differences between these structures related to differences in the hybridization of nitrogen in N_2F_2? Identify the hybrid orbitals that account for the bonding in N_2F_2. Are they the same as those in acetylene, C_2H_2?

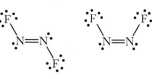

FIGURE P9.63

9.64. How does the hybridization of the sulfur atom change in the series SF_2, SF_4, SF_6?

9.65. How does the hybridization of the central atom change in the series CO_2, NO_2, O_3, and ClO_2?

*__9.66.__ Draw the Lewis structure of the chlorite ion, ClO_2^-, which is used as a bleaching agent. Include all resonance structures in which formal charges are closest to zero. What is the shape of the ion? Suggest a hybridization scheme for the central chlorine atom that accounts for the structures you have drawn.

*__9.67.__ Draw the Lewis structure of the perchlorate ion, ClO_4^-, which is used as an oxidizer in solid rocket fuel. Include all resonance structures in which formal charges are closest to zero. What is the shape of the ion? Suggest a hybridization scheme for the central chlorine atom that accounts for this shape.

9.68. Draw a Lewis structure for Cl_3^+. Determine its molecular geometry and the hybridization of the central Cl atom.

9.69. Synthesis of the first compound of argon was reported in 2000. HArF was made by reacting Ar with HF. Draw a Lewis structure for HArF, and determine the hybridization of Ar in this molecule.

9.70. The Lewis structure for N_4O, with the skeletal structure O–N–N–N–N contains one N–N single bond, one N=N double bond, and a N≡N triple bond. Is the hybridization of all the nitrogen atoms the same?

9.71. The trifluorosulfate anion (Figure P9.71) was isolated in 1999 as the tetramethylammonium salt $(CH_3)_4NSOF_3$. Determine the geometry around sulfur in the anion and describe the bonding according to valence bond theory.

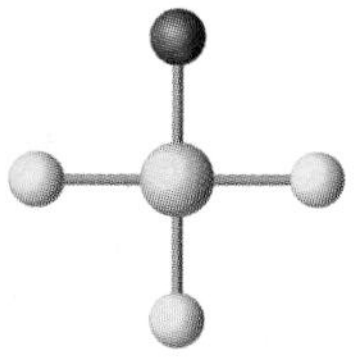

FIGURE P9.71

Shapes and Bonding in Larger Molecules

CONCEPT REVIEW

9.72. Why is it difficult to assign a single geometry to a molecule with more than one central atom?

9.73. Can molecules with more than one central atom have resonance forms?

*9.74. Can hybrid orbitals be associated with more than one atom?

*9.75. Are resonance structures examples of electron delocalization? Explain your answer.

PROBLEMS

9.76. Cyclic structures exist for many compounds of carbon and hydrogen. Describe the molecular geometry and hybridization around each carbon atom in benzene (C_6H_6), cyclobutane (C_4H_8), and cyclobutene (C_4H_6) (Figure P9.76).

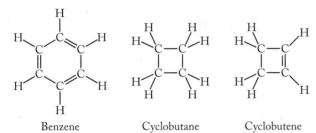

Benzene   Cyclobutane   Cyclobutene

FIGURE P9.76

9.77. The two nitrogen atoms in the nitramide anion are connected with two oxygen atoms on one terminal nitrogen and two hydrogens connected to the other nitrogen atom (Figure P9.77). What is the molecular geometry of each nitrogen atom in the nitramide anion? Is the hybridization of both of the nitrogen atoms the same?

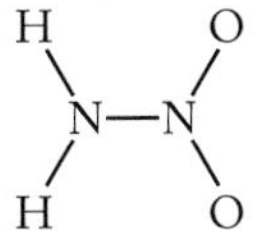

FIGURE P9.77

9.78. What is the geometry around each sulfur atom in the disulfate anion shown in Figure P9.78? What is the hybridization of the central oxygen atom?

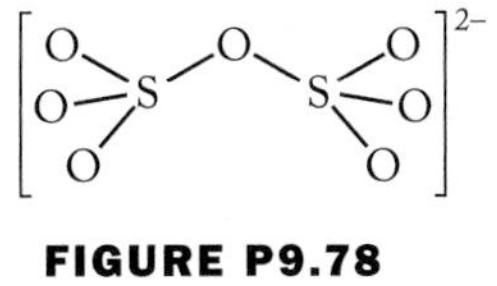

FIGURE P9.78

9.79. What is the molecular geometry around sulfur and nitrogen in the sulfamate anion shown in Figure P9.79? Which atomic or hybrid orbitals overlap to form the S–O and S–N bonds in the sulfamate anion?

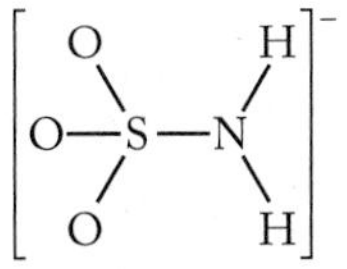

FIGURE P9.79

Molecular Orbital (MO) Theory; A Bonding Retrospective

CONCEPT REVIEW

9.80. Which atomic orbitals are more likely to mix to form a molecular orbital—a $2s$ and a $3p$ orbital or a $4s$ and a $5p$ orbital?

9.81. Which better explains molecular geometry: valence bond theory or molecular orbital theory?

9.82. Which better explains the magnetic properties of a diatomic molecule: valence bond theory or molecular orbital theory?

9.83. Do all σ molecular orbitals result from the overlap of s atomic orbitals?

9.84. Do all π molecular orbitals result from the overlap of p atomic orbitals?

9.85. Are s atomic orbitals with different principal quantum numbers (n) as likely to overlap and form MOs as s atomic orbitals with the same value of n?

9.86. Are atomic orbitals with the same principal quantum number (n) but different angular momentum quantum numbers (ℓ) as likely to overlap and form MOs as orbitals with the same values of n and ℓ?

PROBLEMS

9.87. Make a sketch showing how two $1s$ orbitals overlap to form a σ_{1s} bonding molecular orbital and a σ_{1s}^* antibonding molecular orbital.

9.88. Make a sketch showing how two $2p_y$ orbitals overlap "sideways" to form a π_{2p} bonding molecular orbital and a π_{2p}^* antibonding molecular orbital.

9.89. Use MO theory to predict the bond orders of the following molecular ions: N_2^+, O_2^+, C_2^+, and Br_2^{2-}. Do you expect any of these species to exist?

9.90. Diatomic noble gas molecules, such as He_2 and Ne_2, do not exist. Would removing an electron create molecular ions, such as He_2^+ and Ne_2^+, that are more stable than He_2 and Ne_2?

9.91. Which of the following molecular ions is expected to have one or more unpaired electrons? (a) N_2^+; (b) O_2^+; (c) C_2^{2+}; (d) Br_2^{2-}

9.92. Which of the following molecular ions is expected to have one or more unpaired electrons? (a) O_2^-; (b) O_2^{2-}; (c) N_2^{2-}; (d) F_2^+

9.93. Which of the following anions have electrons in π antibonding orbitals? (a) C_2^{2-}; (b) N_2^{2-}; (c) O_2^{2-}; (d) Br_2^{2-}

9.94. Which of the following molecular cations have electrons in π antibonding orbitals? (a) N_2^+; (b) O_2^+; (c) C_2^{2+}; (d) Br_2^{2+}

9.95. For which of the following diatomic molecules does the bond order increase with the gain of two electrons, forming the corresponding anion with a 2− charge?
a. $B_2 + 2\,e^- \rightarrow B_2^{2-}$ c. $N_2 + 2\,e^- \rightarrow N_2^{2-}$
b. $C_2 + 2\,e^- \rightarrow C_2^{2-}$ d. $O_2 + 2\,e^- \rightarrow O_2^{2-}$

9.96. For which of the following diatomic molecules does the bond order increase with the loss of two electrons, forming the corresponding cation with a 2+ charge?
a. $B_2 \rightarrow B_2^{2+} + 2\,e^-$ c. $N_2 \rightarrow N_2^{2+} + 2\,e^-$
b. $C_2 \rightarrow C_2^{2+} + 2\,e^-$ d. $O_2 \rightarrow O_2^{2+} + 2\,e^-$

9.97. Do the 1+ cations of homonuclear diatomic molecules of the second row elements always have shorter bond lengths than the corresponding neutral molecules?

9.98. Do any of the anions of the homonuclear diatomic molecules formed by B, C, N, O, and F have shorter bond lengths than those of the corresponding neutral molecules? Consider only the anions with 1− and 2− charge.

Additional Problems

9.99. Draw the Lewis structure for the two ions in ammonium perchlorate (NH_4ClO_4), which is used as a propellant in solid fuel rockets, and determine the molecular geometries of the two polyatomic ions.

9.100. **Pressure-Treated Lumber** By December 31, 2003, concerns over arsenic contamination had prompted the manufacturers of pressure-treated lumber (Figure P9.100) to voluntarily cease producing lumber treated with CCA (chromated copper arsenate) for residential use. CCA-treated lumber has a light greenish color and was widely used to build decks, sand boxes, and playground structures. Draw the Lewis structure for the arsenate ion (AsO_4^{3-}) that yields the most favorable formal charges. Predict the angles between the arsenic–oxygen bonds in the arsenate anion.

FIGURE P9.100

9.101. Consider the molecular structure of the amino acid glycine in Figure P9.101. What is the angle between the N—C—C bonds in this structure? What are the O—C—O and C—O—H bond angles?

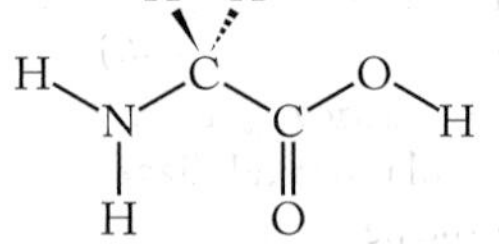

FIGURE P9.101

9.102. **Global Warming** Global warming occurs when the quantity of greenhouse gases in the atmosphere increases. This has happened in large part because of human activities related to energy production from fossil fuels, the combustion of which releases large quantities of CO_2 into the air. Eight states and New York City filed a global-warming lawsuit in July 2004, accusing five major power plant operators of emitting gases that cause global warming and threaten our planet's water, air, and living creatures. What specific scientific evidence based on the concepts in this chapter will the lawyers for the plaintiffs need to present to make their case?

9.103. Methane is produced by a variety of natural processes, but, like carbon dioxide, its atmospheric concentrations have been rising in recent years as a result of human activity. Do a literature and/or internet search to find out what kinds of activities cause increasing methane concentrations and rank their relative importance in contributing to global warming.

9.104. Cl_2O_2 may play a role in ozone depletion in the stratosphere. In the laboratory, a reaction between ClO_2F and $AlCl_3$ produces Cl_2O_2 and $AlCl_2F$. Draw the Lewis structure for Cl_2O_2 based on the skeletal structure in Figure P9.104. What is the geometry about the central chlorine atom?

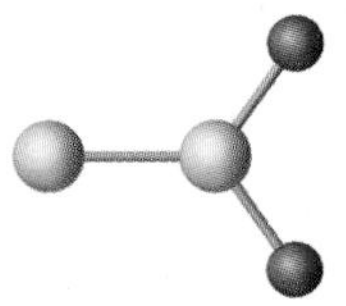

FIGURE P9.104

9.105. Photolysis of Cl_2O_2, which was also the subject of Problem 104, is thought to produce the two compounds with the skeletal structures shown in Figure P9.105. Draw the Lewis structures for these compounds. Do both of these molecules have linear geometry?

FIGURE P9.105

*9.106. Complete the Lewis structure for the cyclic structure of Cl_2O_2 shown in Figure P9.106. Is the cyclic Cl_2O_2 molecule planar?

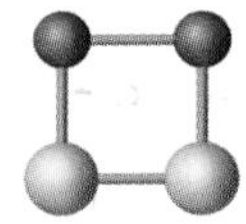

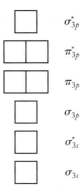

FIGURE P9.106

9.107. In 1999, the ClO^+ ion, a potential contributor to stratospheric ozone depletion, was isolated in the laboratory. (a) Draw the Lewis structure for ClO^+. (b) Using the molecular orbital diagram for ClO^+ in Figure P9.107, determine the order of the Cl–O bond in ClO^+.

$$\sigma_{3p}^*$$
$$\pi_{3p}^*$$
$$\pi_{3p}$$
$$\sigma_{3p}$$
$$\sigma_{3s}^*$$
$$\sigma_{3s}$$

FIGURE P9.107

9.108. The first compound containing a xenon–sulfur bond, HXeSH, was isolated in 1998. Draw the Lewis structure of HXeSH and predict the hybridization of the Xe atom.

9.109. Life on Earth The photochemical reaction of sodium hydrogen phosphite with formaldehyde is illustrated in Figure P9.109. This reaction may have played a role in the formation of nucleic acids before life existed on Earth. Complete the Lewis structure for the hydrogen phosphite ion. Complete the Lewis structure for the product, hydroxymethylphosphonic acid.

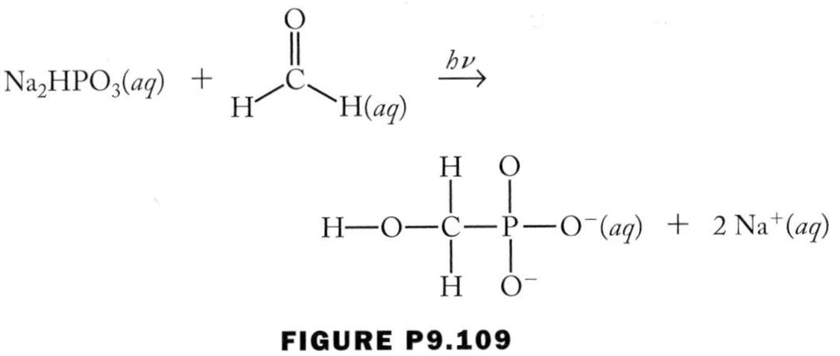

FIGURE P9.109

9.110. Cola Beverages Phosphoric acid imparts a tart flavor to cola beverages. The skeletal structure of phosphoric acid is shown in Figure P9.110. Complete the Lewis structure for phosphoric acid in which formal charges are closest to zero. What is the molecular geometry around the phosphorus atom in your structure?

FIGURE P9.110

9.111. Fluoroaluminate anions AlF_4^- and AlF_6^{3-} have been known for over a century but the structure of the pentafluoroaluminate ion, AlF_5^{2-}, was not determined until 2003. Draw the Lewis structures for AlF_3, AlF_4^-, AlF_5^{2-}, and AlF_6^{3-}. Determine the molecular geometry of each molecule or ion. Describe the bonding in AlF_3, AlF_4^-, AlF_5^{2-}, and AlF_6^{3-} using valence bond theory.

*****9.112.** Thermally unstable compounds can sometimes be synthesized using matrix isolation methods in which the compounds are isolated in a nonreactive medium such as frozen argon. The reaction of boron with carbon monoxide produces compounds with these skeletal structures: B–B–C–O and O–C–B–B–C–O. For each of these compounds, draw the Lewis structure that minimizes formal charges. Do any of your structures contain atoms with incomplete octets? Predict the molecular geometries of BBCO and OCBBCO.

*****9.113.** The products of the reaction between boron (B) and NO can be trapped in solid Ar matrices. Among the products is BNO. Draw the Lewis structure for BNO including any resonance forms. Assign formal charges and predict which structure provides the best description of the bonding in this molecule. Do any of your structures contain atoms without complete octets? Predict the molecular geometry of BNO.

9.114. Horseradish Mustard and horseradish are known for their strong taste and smell. The chemical compounds that impart these properties are called isothiocyanates. The simplest isothiocyanate is methyl isothiocyanate CH_3NCS. Draw the Lewis structure for CH_3NCS, including all resonance forms. Assign formal charges and determine which structure is likely to contribute the most to bonding. Predict the molecular geometry of the molecule at both carbon atoms.

9.115. Methyl thiocyanate (CH_3SCN) is used as an agricultural pesticide and fumigant. Draw the Lewis structure for methyl thiocyanate. Assign formal charges and predict which structure would be the most stable. Predict the molecular geometry of the molecule at both carbon atoms.

9.116. Borazine, $B_3N_3H_6$ (a cyclic compound with alternating B and N atoms in the ring), is isoelectronic with benzene (C_6H_6). Are there delocalized π electrons in borazine?

9.117. Draw a molecular orbital diagram for F_2. How many electrons are found in antibonding molecular orbitals in F_2?

*****9.118.** Some chemists think HArF consists of H^+ ions and ArF^- ions. Using an appropriate MO diagram, determine bond order of the Ar–F bond ArF^-.

*****9.119.** Assuming HArF is a molecular compound:
 a. Draw its Lewis structure.
 b. What are the formal charges on Ar and F in the structure you drew?
 c. What is the shape of the molecule?
 d. Is HArF polar?

9.120. Which of the following unstable nitrogen oxides, N_2O_2, N_2O_5, and N_2O_3, are polar molecules? (N_2O_2 and N_2O_3 have N–N bonds; N_2O_5 does not.)

9.121. Explain why O_2 is paramagnetic.

9.122. Using an appropriate molecular orbital diagram, show that the bond order in the disulfide anion S_2^{2-} is equal to 1. Is S_2^{2-} diamagnetic or paramagnetic?

9.123. Use molecular orbital diagrams to determine the bond order of the peroxide (O_2^{2-}) and superoxide ions (O_2^-). Are these bond order values consistent with those predicted from Lewis structures?

9.124. Elemental sulfur has several allotropic forms including cyclic S_8 molecules. What is the orbital hybridization of sulfur atoms in this allotrope? The bond angles are about 108°.

From Alcohol to Asparagus: The Nose Knows

9.125. All of the group 16 elements form compounds with the generic formula H_2E (E = O, S, Se, or Te). Which compound is the most polar? Which compound is the least polar?

*9.126. Ozone (O_3) has a permanent dipole moment (0.54 D). How can a molecule with only one kind of atom have a permanent dipole moment?

*9.127. The bond angle in H_2O is 104.5°; the bond angle in H_2S, H_2Se, and H_2Te is very close to 90°. Which theory would you apply to describe the geometry in H_2S, H_2Se, and H_2Te: VSEPR? Valence bond without invoking hybrid orbitals? Valence bond theory using hybrid orbitals? Why?

9.128. **Garlic** Garlic contains the molecule alliin (Figure P9.128). When garlic is crushed or chopped, a reaction occurs that converts alliin into the molecule allicin, which is primarily responsible for the aroma we associate with garlic. (a) Describe the molecular geometry about the sulfur atoms in both compounds. (b) Do any of the sulfur atoms in allicin have the same geometry as the sulfur atoms in the volatile sulfur compounds that cause bad breath (H_2S, CH_3—SH, and CH_3—S—CH_3)?

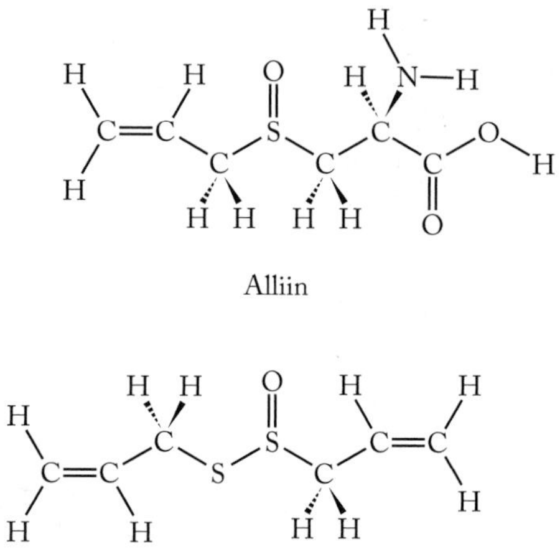

Alliin

Allicin

FIGURE P9.128

Forces between Ions and Molecules and Colligative Properties

10

SEA SPRAY *Ocean waves crashing on a shoreline produce a spray that carries the ions dissolved in the ocean into the air.*

A LOOK AHEAD: A Breath of Fresh Air

With every breath we take, oxygen gas enters tiny, moist sacs in our lungs where it dissolves and is transported into the blood. Oxygen moves from the gas phase into the liquid phase, and this transition requires that it be soluble in water. We have talked about the shape of oxygen molecules and the bonding between atoms. The ways in which oxygen molecules interact with other molecules like water affects most of life on earth.

About 21% of air is oxygen, which is used by living systems and is essential for metabolism of food and production of energy. Although oxygen is essential to the life of aerobic (oxygen-using) organisms, too much oxygen can be hazardous. In humans high concentrations of O_2 are toxic, and exposure to pure oxygen at pressures between 1 and 2 atmospheres can cause brain damage and death.

Too little oxygen is also dangerous. Humans die if deprived of oxygen for several minutes, and mountain climbers can become nauseous, disoriented, and unconscious when the partial pressure of oxygen decreases at high altitudes. The role of pressure is a major factor in determining solubility of a gas in a liquid.

Fish and other aquatic creatures also need oxygen to survive. They extract oxygen from water, where it is present at much lower concentrations than in the air. Oxygen is soluble in water at about 10 mg/L at normal atmospheric pressure. This concentration is about one-twentieth of the concentration of oxygen in air, but access to that low level is sufficient to sustain aquatic life.

In water, changes in temperature can disastrously affect the lives of fish. You may have seen photographs of dead fish floating in water that has become too warm. It is not heat that kills the fish but rather that the solubility of oxygen in water decreases as the temperature of the water increases. The temperature does not have to increase much before the level of oxygen, which is already very low in water, drops below the concentration necessary for survival.

The availability of oxygen for many forms of life, including humans, hinges on the solubility of oxygen in water: in the water that accounts for the bulk of fluid in blood, and the water in which fish live. Once in our systems, oxygen binds to the iron in hemoglobin for transport throughout our bodies. When humans go to extreme environments—such as high altitudes on mountains or deep under water—they must attend to the impact of the surroundings on the availability of oxygen and the solubility of gases due to pressure changes. As altitude increases, pressure decreases, and although the air is still 21% oxygen, less air is available, so additional oxygen must be provided to supplement the amount in the prevailing atmosphere. Under water, the pressure experienced by scuba divers goes up by about 1 atm for every 10 m increase in depth. Increased pressure increases the solubility of all gases in blood. If divers return to the surface too quickly, gases bubble out of their blood just as they do from a soda can when the top is popped and the pressure is suddenly released. The result is a serious medical condition (the bends) that is both painful and life-threatening. To diminish the likelihood of the bends, divers breathe a mixture of helium (He) and oxygen, rather than the nitrogen–oxygen mixture we inhale normally, because He is less soluble than N_2 in water (and therefore blood) at any given pressure. Because it does not dissolve in blood to the same degree as N_2, it poses less of a problem in bubbling out as a diver returns to an environment at lower pressure.

Our lungs are not the only organs that take in oxygen. The corneas in our eyes lack blood vessels but, as living tissue, they rely on oxygen to function. Oxygen is normally directly dissolved from the atmosphere into the cornea. Contact lenses rest over the corneas of eyes, and if they are not specially crafted, they can block the dissolution of oxygen into the tissue. Extended-wear contact lenses must be designed to allow oxygen at the molecular level to move from the air into the corneas on which they rest. Hard contact lenses were originally made of a polymer that repelled water and did not dissolve oxygen, but modern soft contact lenses are made of a material called a hydrogel that absorbs water, which makes them both more flexible and also better able to transmit the oxygen needed by the cornea. Understanding the factors that influence solubility was essential in developing lenses that allow oxygen to enter the corneas.

Life depends on the presence of oxygen in a watery environment. Life also relies on the solubility of other compounds and nutrients and their ability to be transported as solutes through blood. We have previously discussed the aqueous solubility of many materials—gases, salts, molecular solids, and liquids such as sugar and ethanol—but we have not yet addressed the features of structure at the atomic-molecular level that influence solubility; nor have we discussed the impact of the presence of solutes in solutions on the properties of those solutions. In this chapter we discuss the influence of structure on solubility and the properties of solutions.

10.1 Sea Spray and Salts

Ocean waves crashing on a rocky shoreline create plumes of sea spray carrying small drops of seawater into the atmosphere, where they evaporate. As these small drops evaporate and shrink, their concentrations of sea salts, Cl^-, Na^+, Mg^{2+}, Ca^{2+} and other ions, increase. Eventually, the decreased volume of the drops produces supersaturated solutions of the sea salts, which begin to precipitate. Among the first of the solids to form is $CaSO_4$. Among the last is NaCl, which does not precipitate until 90% of the seawater has evaporated. This sequence takes place even though the concentrations of Ca^{2+} and SO_4^{2-} ions are much lower than the concentrations of Na^+ and Cl^- ions. Why does $CaSO_4$ precipitate before NaCl? Put another way: Why is NaCl more soluble in water than $CaSO_4$? And why are NaCl and $CaSO_4$ both solids under ordinary conditions of temperature and pressure, while water is a liquid? To answer these questions, we need to examine the attractive forces between the particles that make up these substances.

Think about the three common states of matter (Figure 10.1) and how they differ based on the forces between their atomic-scale building blocks. In solids, there are strong attractions between adjacent particles that have little freedom to move. In liquids, particles have weaker interactions with their neighbors and more freedom of motion. In the gas phase, particles have the greatest freedom of motion and the weakest forces between one another.

A substance made of particles that interact relatively strongly has high melting and boiling points. Substances composed of particles with strong attractive forces require more kinetic energy to overcome those forces and melt or vaporize. A substance with strong interactive forces is likely to be a solid at room temperature and normal pressure. Under the same conditions, substances with somewhat weaker interactions probably are liquids, and those with the least interaction are likely to be gases. In general, the state of a substance is a reflection of the strength of the attractions between particles in the substance and the kinetic energies of the particles.

Solid

Liquid

Gas

FIGURE 10.1 In ice, molecules of water lack sufficient motion to overcome attractive forces. In liquid water, increased molecular motion overcomes some of the attractive forces. In water vapor, the molecules are widely separated and interact less frequently.

 CONNECTION The discussion of solutions in Chapter 4 began with a look at the concentration of ions in sea water.

CONNECTION In Chapter 5, we looked at the flow of energy accompanying phase changes of water (solid ice ⇌ liquid water ⇌ water vapor) in terms of the changes in kinetic and potential energies of the molecules.

Ion–Ion Interactions and Lattice Energy

Ionic compounds are among the substances most likely to be solids at room temperature. They are solids because the attraction between ions of opposite charge is the strongest kind of interactive force between particles at the level of atoms and molecules. The strength of ion–ion interactions is defined by **Coulomb's law**, which we used in Chapter 5 when talking about the potential energy of particles:

$$E \propto \frac{(Q_1 Q_2)}{d} \qquad (10.1)$$

The value of E in Equation 10.1 is negative when Q_1 and Q_2 have opposite signs. Negative values of E make sense for salts because the cations and anions composing a salt have opposite charges and attract each other to form an arrangement characterized by a lower potential energy. The combination of cations and anions to make a salt releases energy and is an exothermic process. Correspondingly, the separation of a cation from an anion requires energy and is endothermic. The amount of energy involved in both processes is the same; the sign of the energy changes to indicate whether energy is released as the ions combine, or absorbed when the ions are pulled apart.

CONNECTION In Chapter 5 we learned that if we know the amount of energy associated with a process, the reverse of that process involves the same amount of energy; only the sign changes to indicate whether the system gives off energy (negative sign) or absorbs energy (positive sign).

Coulomb's law specifically states that the strength of the interaction between two ions is directly proportional to the product of the charges of the two ions (Q_1 and Q_2) and is inversely proportional to the distance (d) between their nuclei. The distance (d) between ions is the bond length. We can use these relations to make a qualitative comparison of the relative strengths of interactive forces between ions in salts if we know the charges on the ions and their radii. On the basis of Coulomb's law, we can state that the attractive force between two ions increases as ionic charge increases and as ionic size decreases.

The bond distance d for ionic compounds is the sum of their ionic radii. Recall that size of atoms and ions is a periodic property. In Chapter 7 we learned that cations are always smaller than the neutral atoms from which they are formed, and anions are always larger. All cations (of the same charge) and anions increase in size down a group of the periodic table. The radii for monatomic cations and anions are given in Figure 10.2. Sometimes we will use actual numerical values of ionic radii to evaluate the energy of ion-ion interactions, while in other cases we will use the general trends of radii defined with the aid of the periodic table.

CONNECTION In Chapter 7, we discussed the calculation of the distance between the nuclei of two ions in a salt from ionic radii.

To begin addressing the question of why $CaSO_4$ precipitates from sea spray before NaCl, we can use Coulomb's law to make a qualitative comparison of the relative energies of the interactions between the two pairs of ions in the two salts. The ions in NaCl are Na^+ and Cl^-, so Q_1 and Q_2 are 1+ and 1−, respectively. The numerator in the expression is therefore $(1+)(1-) = 1-$. The ions in $CaSO_4$ are 2+ and 2−, so the numerator in the expression has a value of 4−. Based on the charge on the ions, $CaSO_4$ has a stronger energy of interaction by a factor of 4, but we must also compare bond distances and include their impact on the attractive forces in our analysis.

The radii are 102 pm for Na^+ and 181 pm for Cl^-, so d for NaCl is $102 + 181 = 283$ pm. The calcium cation (100 pm) is almost the same size as the sodium ion. The sulfate ion is a polyatomic ion, and its radius is not given in Figure 10.2. Table 10.1 gives estimated values for the

TABLE 10.1 Estimated Radii of Polyatomic Ions

Polyatomic Ion	X—O Bond Length = Radius
CO_3^{2-}	164 pm
NO_3^{-}	165 pm
SO_4^{2-}	244 pm
PO_4^{3-}	238 pm

Values from James E. Huheey, Ellen A. Keiter, and Richard L. Keiter, *Inorganic Chemistry: Principles of Structure and Reactivity* (New York: HarperCollins, 1993), p. 118.

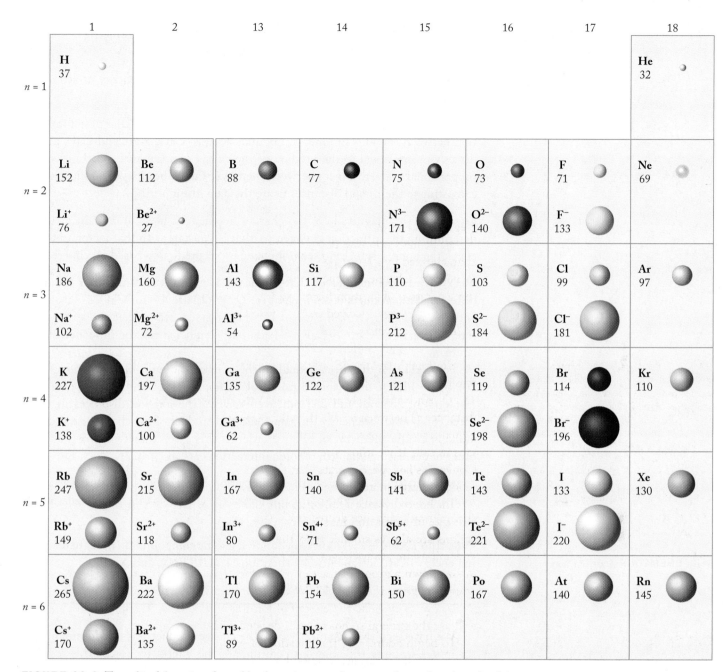

FIGURE 10.2 The radii of the anions formed by the main group elements are larger than the radii of their parent atoms. The radii of cations are smaller than those of their parent atoms. All values are in picometers (1 pm = 10^{-12} m). Values from N. N. Greenwood and A. Earnshaw, *Chemistry of the Elements*, 2nd ed. (Boston: Butterworth-Heinemann, 1997); Catherine E. Housecroft and Alan G. Sharpe, *Inorganic Chemistry*, 3rd ed. (Upper Saddle River, NJ: Pearson Prentice Hall, 2008).

common polyatomic ions discussed in this chapter. The estimated radius of the sulfate ion is 244 pm, which is just slightly larger than the chloride ion, making *d* for $CaSO_4$ 344 pm. The denominator of the expression for calcium sulfate is only about 1.2 times the size of that for sodium chloride, so the factor of 4 difference in the numerators dominates this comparison. As a result, we predict that the calcium ions and sulfate ions experience a stronger attraction than the sodium ions and chloride ions. The larger attraction between its ions is not the only reason why $CaSO_4$ precipitates first from sea spray, but it is a major factor contributing to that behavior.

SAMPLE EXERCISE 10.1 **Predicting the Relative Strengths of Attractive Forces**

List the following ionic compounds in order of decreasing energy of attraction between their ions: CaO, NaF, and CaF_2.

COLLECT AND ORGANIZE According to Coulomb's law, the energies of interaction between the ions in ionic compounds such as CaO, NaF, and CaF_2 are (a) directly proportional to the product of the charges on the ions and (b) inversely proportional to the distances between them. We know the charges on the ions; we can estimate their bond distances using the data in Figure 10.2.

ANALYZE The magnitude of the charges on the ions is the dominant factor in determining the attraction between particles. We start with that factor and then check the bond distances to see if it changes our prediction based on charge.

SOLVE The combinations of charges on the ions in these compounds are quite different. The products of their charges ($Q_1 \times Q_2$) are 4− for CaO, 1− for NaF, and 2− for CaF_2. On the basis of this factor alone, we predict the strength of interactions to be CaO > CaF_2 > NaF, but we must check the influence of d to confirm.

For CaO, the distance is equal to the sum of the radii of Ca^{2+} and O^{2-} = 100 pm + 140 pm = 240 pm. Similar calculations for NaF and CaF_2 give ion–ion distances of 235 pm and 233 pm, respectively. The ion–ion distances in these three compounds are nearly the same, and the product of the charges on the ions alone is a valid predictor of relative order in this case. Therefore, the predicted order of decreasing strength of ionic interactions is CaO > CaF_2 > NaF.

THINK ABOUT IT The strength of interaction between ions increases (becomes more negative) as the value of the product of the charges on the ions increases and the interionic distance decreases.

Practice Exercise Arrange the following ionic compounds in order of decreasing energy of attraction between their ions: $CaCl_2$, BaO, and NaCl.

(Answers to Practice Exercises are in the back of the book.)

▶‖ **CHEMTOUR** Lattice Energy

The **lattice energy** (*U*) of an ionic compound is the energy released when one mole of the ionic compound forms from its free ions in the gas phase.

We have seen two instances where the charge product was the dominant factor in determining the relative magnitude of the interactive force. This result is general. The ionic radii of cations in Figure 10.2 range from 31 pm to 169 pm, and those of monatomic anions and polyatomic anions (Table 10.1) range from 136 pm to 238 pm. If we put together the smallest anion and cation and the largest cation and anion, the range of distances is 167 to 407 ppm, which is less than a factor of 2.5. Moreover, the radii of common ions such as those in CaO, NaF, and CaF_2 cover a much smaller range of 100 to 140 pm, which is barely a factor of 1.5. We have seen the charge product $Q_1 \times Q_2$ range from 1− to 4− (a factor of 4) in Sample Exercise 10.1, and for common ionic compounds it is indeed the charge product $Q_1 \times Q_2$ that is nearly always responsible for large differences of the strengths of ionic interactions.

The strength of ion–ion interactions is also the major factor affecting the **lattice energy** (*U*) of an ionic compound. Lattice energy is the energy released (−*E*, exothermic process) when free gaseous ions combine to form one mole of a solid ionic compound. The lattice energies of some common binary ionic compounds are given in Table 10.2. The formula for lattice energy

$$U = \frac{k(Q_1 Q_2)}{d} \qquad (10.2)$$

resembles the expression for Coulomb's law, except that it includes a proportionality constant (k), the value of which depends on the structure of the ionic solid.

We examine the structures of solids in detail in Chapter 11, at which point we will learn about the different arrangements possible for ions in a crystal. Different arrangements give rise to different values of k. For now, we can consider the charges on ions and the distances between them and use Coulomb's law to predict relative values of lattice energies, just as we used it for relative attractive forces.

Differences in lattice energies affect the properties of ionic solids. For one thing, they affect the temperatures at which the solids melt. An ionic structure that is more tightly held together should require more energy and a higher temperature to melt than a less tightly held structure—a trend we observe experimentally. Consider two ionic compounds: LiF ($U = -1047$ kJ/mol) and MgO ($U = -3791$ kJ/mol). The greater lattice energy of MgO, nearly four times that of LiF, is reflected by its much higher melting point, 2825°C versus 848°C for LiF, and its much higher boiling point, 3600°C versus 1673°C for LiF.

The solubilities of ionic compounds in water are also affected by the strengths of ion–ion interactions. The positive and negative ends of the water dipole associate with anions and cations in a crystalline lattice and pull those ions into solution. The greater the interactive force between ions in the crystal, the greater the force required for water molecules to pull apart the ions in a crystal of solute, and the less soluble the solute is likely to be. We might predict that LiF is more soluble than MgO on the basis of the greater lattice energy of MgO, and that prediction is correct: at 20°C, the solubility of LiF in water is 2.7 g/L and the solubility of MgO is only 0.006 g/L.

TABLE 10.2	Lattice Energies (U) of Common Binary Ionic Compounds
Compound	**U (kJ/mol)**
LiF	−1047
LiCl	−864
NaCl	−790
KCl	−720
KBr	−691
$MgCl_2$	−2540
MgO	−3791

CONNECTION The solubility trends for ionic compounds given in Section 4.6 are related to the strengths of intermolecular forces.

SAMPLE EXERCISE 10.2 **Ranking Lattice Energies and Melting Points**

Rank the following three ionic compounds in order of increasing lattice energy and increasing melting point: NaF, KF, and RbF. Assume that these compounds have the same solid structure, which means they all have the same value of k in Equation 10.2.

COLLECT AND ORGANIZE The relative values of lattice energy and melting point are both affected by the strength of ion–ion interactive forces, which are in turn directly proportional to the product of the charges on the ions and inversely proportional to bond lengths. The three compounds we are given have a common anion; they are all fluoride salts of alkali metals.

ANALYZE The compound with the greatest attractive force should have the greatest lattice energy and the highest melting point.

SOLVE All of the cations are alkali metal cations and have a charge of 1+; all of the anions are the same (the fluoride ion) and have a charge of 1−. Therefore, any differences in lattice energy must be related to differences in the distance between ions because the product of the charges is 1− for all three compounds. Because the fluoride ion is common to all the salts, the variation in distance between the two ions depends on the size of the cations.

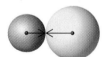
$Na^+ = 102$ pm
$F^- = 133$ pm

$K^+ = 138$ pm
$F^- = 133$ pm

$Rb^+ = 149$ pm
$F^- = 133$ pm

Size is a periodic property that increases as one moves down a family in the periodic table, and arranging the three cations in order of increasing size gives $Na^+ < K^+ < Rb^+$. This means the compounds in order of increasing bond length (*d*) are NaF < KF < RbF. As bond length increases, the attractive force decreases, so the RbF must have the lowest attractive force between its ions, followed by KF, followed by NaF with the highest. The lattice energies and melting points follow the same trend as the attractive forces, so from lowest lattice energy and melting point to highest lattice energy and melting point, the prediction is: RbF < KF < NaF.

This predicted order is confirmed by the actual melting points:

Compound	Melting Point (°C)
NaF	988
KF	846
RbF	775

THINK ABOUT IT The relationship defined by Coulomb's law enables us to predict differences in properties such as melting points and boiling points of simple ionic compounds. Generally, large negative lattice energies lead to high melting points because they reflect the forces of attraction that have to be overcome to convert a solid substance into a liquid.

Practice Exercise Predict which of the following compounds should have the highest melting point: $CaCl_2$, $PbBr_2$, or TiO_2. All three compounds have nearly the same structure and therefore the same value of *k* in Equation 10.2. The radius of Ti^{4+} is 60.5 pm.

Calculating Lattice Energies Using the Born–Haber Cycle

Lattice energies are difficult to measure directly, but they can be calculated for a binary compound by considering its formation from the reaction between its constituent elements. Hess's law is used to treat the energies associated with a series of steps that take the elements from their standard states to ions in the gas phase and then to the ionic crystal lattice.

CONNECTION We used Hess's law in Chapter 5 to determine heats of reactions that are difficult to measure experimentally.

Consider the reaction between sodium metal and chlorine gas to form NaCl:

$$Na(s) + \tfrac{1}{2} Cl_2(g) \rightarrow NaCl(s)$$

The reaction is exothermic and produces heat, which can be measured with a calorimeter. The heat released corresponds to an enthalpy of reaction (ΔH_{rxn}) of −411 kJ/mol NaCl produced in the process. This enthalpy change can be viewed as the net result of all the energy changes associated with a series of five steps which together form a **Born–Haber cycle** (Figure 10.3):

A **Born–Haber cycle** is a series of steps with corresponding enthalpy changes that describes the formation of an ionic solid from its constituent elements.

1. Vaporization of a mole of sodium metal into gaseous sodium atoms; energy = ΔH_{vap} (the enthalpy of vaporization of sodium)

2. Breaking the covalent bonds in half a mole of Cl_2 molecules to make one mole of Cl atoms in the gas phase; energy = $\tfrac{1}{2} \Delta H_{BE}$ [the bond energy of $Cl_2(g)$]

3. Ionization of a mole of Na atoms, forming a mole of Na^+ ions in the gas phase and a mole of electrons; energy = ΔH_{IE_1} (the first ionization energy of sodium)

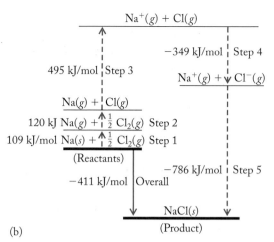

(a)   (b)   (Product)   (c)

FIGURE 10.3 (a) The violent reaction between sodium metal and chlorine gas releases more than 400 kJ per mole of NaCl product. (b) The Born–Haber cycle for the formation of NaCl from its elements shows that the principal reason for this violently exothermic reaction is the energy lost when free sodium ions and chloride ions combine to form solid NaCl. The energy lost is the lattice energy U of (c) the solid NaCl produced.

4. Combining one mole of chlorine atoms with one mole of electrons to form one mole of chloride ions; energy = ΔH_{EA_1} (the first electron affinity of chlorine)

5. Forming one mole of solid sodium chloride from one mole of gas-phase Na^+ ions and one mole of gas-phase Cl^- ions; energy = $\Delta H_{lattice}$ (the lattice energy of NaCl)

One term in this series of steps, **electron affinity**, defines the energy of a process we have not yet discussed: Electron affinity is the energy change that occurs when one mole of electrons is added to one mole of atoms or ions in the gas phase. For example, the energy associated with adding electrons to a mole of chlorine atoms in the gas phase is:

$$Cl(g) + e^- \rightarrow Cl^-(g) \qquad \Delta H_{EA_1} = -349 \text{ kJ/mol}$$

The electron affinities of many atoms are exothermic (see Table 10.3) because the association of a negative electron with a positively charged nucleus is energetically favorable. An examination of the values of electron affinity in Table 10.3 reveals that the trends in this property are not as regular as the trends in size and ionization energy that we saw in Chapter 7. Electron affinity increases down a column only for the group 1 metals; other groups do not display a definite trend. In general, electron affinity becomes more negative as you move across a row to the right on the periodic table, but that is only a generality and several inconsistencies are evident in the data.

Successive electron affinities for ions may be defined for each electron added to a species. For example, to make the oxide ion with a charge of 2−:

$$O(g) + e^- \rightarrow O^-(g) \qquad \Delta H_{IE_1} = \text{first electron affinity} = -141 \text{ kJ/mol}$$
$$O^-(g) + e^- \rightarrow O^{2-}(g) \qquad \Delta H_{IE_2} = \text{second electron affinity} = +744 \text{ kJ/mol}$$

The overall energy required to produce the O^{2-} anion in the gas phase is

$$\Delta H_{EA_1} + \Delta H_{EA_2} = (-141 \text{ kJ/mol}) + (+744 \text{ kJ/mol})$$
$$= +603 \text{ kJ/mol}$$

The formation of the 2− anion of oxygen in the gas phase is a highly endothermic process, but as we shall see in our discussion of the Born–Haber cycle, the energy required by highly endothermic processes like this is offset by other highly exothermic steps in the process.

Electron affinity is the energy change occurring when one mole of electrons combines with one mole of atoms or ions in the gas phase.

TABLE 10.3 **Values of First Electron Affinity (kJ/mol)**

Group 1	Group 2	Group 13	Group 14	Group 15	Group 16	Group 17
Li −59.6	Be >0	B −26.7	C −122	N +7	O −141	F −328
Na −52.9	Mg >0	Al −42.5	Si −134	P −72	S −200	Cl −349
K −48.4	Ca −2.4	Ga −28.9	Ge −11	As −78.2	Se −195	Br −325
Rb −46.9	Sr −5.0	In −28.9	Sn −107	Sb −103	Te −190	I −295

To calculate the lattice energy of $NaCl(s)$ using the Born–Haber cycle, we recognize that the sum of the individual energy changes in the five steps equals the overall energy change measured with a calorimeter. In this data set, the only unknown value is the lattice energy of sodium chloride.

Step	Process	Energy Change (kJ)
1	$Na(s) \rightarrow Na(g)$	$\Delta H_{vap} = +109$
2	$\frac{1}{2} Cl_2(g) \rightarrow Cl(g)$	$\frac{1}{2}\Delta H_{BE} = \frac{1}{2}(240) = +120$
3	$Na(g) \rightarrow Na^+(g) + e^-$	$\Delta H_{IE_1} = +495$
4	$Cl(g) + 1e^- \rightarrow Cl^-(g)$	$\Delta H_{EA_1} = -349$
5	$Na^+(g) + Cl^-(g) \rightarrow NaCl(s)$	$\Delta H_{rxn} = U$

Overall

$$Na(s) + \frac{1}{2} Cl_2(g) \rightarrow NaCl(s) \qquad \Delta H_{rxn} = -411 \text{ kJ}$$

The overall energy change is

$$-417 \text{ kJ} = (+109 \text{ kJ}) + \frac{1}{2}(240 \text{ kJ}) + (+495 \text{ kJ}) + (-349 \text{ kJ}) + U$$

Calculating the unknown lattice energy yields

$$U = -786 \text{ kJ}$$

The Born–Haber cycle is frequently used to calculate lattice energies as we have shown here. In fact, it may also be used to calculate any one of the factors among the steps that is unknown. For example, electron affinities can be very difficult to measure, and if the thermochemical values are known for all other steps in the cycle, the Born–Haber cycle can be applied to the calculation of electron affinities.

SAMPLE EXERCISE 10.3   **Calculating Lattice Energy**

The enthalpy change in the reaction between a mole of lithium metal and half a mole of fluorine gas is −617 kJ. Calculate the lattice energy of LiF, given:

$$\Delta H_{vap} \text{ Li}(s) = +161 \text{ kJ/mol}$$

$$\Delta H_{BE} \text{ F}_2(g) = +154 \text{ kJ/mol} \times \frac{1}{2}$$

$$\Delta H_{IE_1} \text{ Li}(g) = +520 \text{ kJ/mol}$$

$$\Delta H_{EA_1} \text{ F}(g) = -328 \text{ kJ/mol}$$

COLLECT AND ORGANIZE We are asked to calculate the lattice energy using enthalpy values for processes that can be summed to describe an overall process that produces a salt from its constituent elements. We can use the Born–Haber cycle in Figure 10.3 as a guide to enable us to calculate the unknown lattice energy. We also need a balanced chemical equation to ensure we are including the correct stoichiometric quantities of heat for each step in the overall process.

ANALYZE The overall balanced thermochemical equation for the reaction is:

$$Li(s) + \tfrac{1}{2}F_2(g) \rightarrow LiF(s) \qquad \Delta H_{rxn} = -617 \text{ kJ/mol}$$

This value is the enthalpy change for the entire process. The other thermochemical values add up to this value. Note that the fluorine molecule has a coefficient of $\tfrac{1}{2}$; this means the enthalpy for breaking the F—F bonds to make atomic fluorine will have to be multiplied by a factor of $\tfrac{1}{2}$.

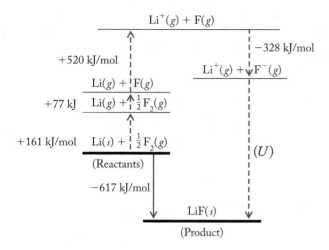

SOLVE The overall energy change in the reaction is equal to:

$$\Delta H_{rxn} = \Delta H_{vap}Li(s) + \Delta H_{IE_1}Li(g) + \tfrac{1}{2}\Delta H_{BE}F_2(g) + \Delta H_{EA_1}F(g) + U(LiF)(s)$$

Substituting the given values of these energies gives:

$$-617 \text{ kJ} = (+161 \text{ kJ}) + (+520 \text{ kJ}) + \tfrac{1}{2}(154 \text{ kJ}) + (-328 \text{ kJ}) + U$$

Solving for U, the lattice energy of $LiF(s)$:

$$U = (-617 \text{ kJ}) - (+161 \text{ kJ} + 520 \text{ kJ} + 77 \text{ kJ} - 328 \text{ kJ})$$

$$= -1047 \text{ kJ/mol } LiF(s) \text{ formed}$$

THINK ABOUT IT Considering Coulomb's law, we would predict that the lattice energy of LiF is larger than that of NaCl, so the calculated value is reasonable. The large lattice energy more than compensates for the energy that must be expended to vaporize, ionize, and break covalent bonds in the constituent elements.

Practice Exercise Burning magnesium metal in air produces MgO and a very bright white light, making the reaction popular in fireworks and signaling devices:

$$Mg(s) + \tfrac{1}{2}O_2(g) \rightarrow MgO(s) + \text{light}$$

The energy change that accompanies this reaction is -602 kJ/mol MgO.

Calculate the lattice energy of MgO from the following energy changes:

Process	Enthalpy Change (kJ/mol)	Process	Enthalpy Change (kJ/mol)
$Mg(s) \rightarrow Mg(g)$	150	$Mg(g) \rightarrow Mg^{2+}(g) + 2e^-$	2188
$O_2(g) \rightarrow 2O(g)$	499	$O(g) + 2e^- \rightarrow O^{2-}(g)$	603

An **ion–dipole** interaction occurs between an ion and the partial charge of a molecule with a permanent dipole.

The cluster of water molecules that surround an ion in aqueous medium is a **sphere of hydration**. The general term applied to such a cluster forming in any solvent is a sphere of solvation.

Lattice energies relate directly to the physical properties of salts, because the magnitude of the forces between ions directly affects many aspects of behavior. As we have seen, among salts with similar arrangements of ions, melting points increase as lattice energy increases, because more energy is required to overcome attractive forces and separate the ions. In terms of solubility in water, solubility tends to increase as lattice energy decreases, because the interaction of water molecules with the ions in the crystal lattice provides the energy to separate the ions, and the smaller the ion–ion interactions, the less energy required. Another property related to lattice energy is hardness. Hardness is resistance to scratching, and the higher the lattice energy, the more mechanical force is needed to remove ions from the crystal and scar the surface.

CONCEPT TEST

Aluminum oxide is an extremely hard material with a high melting point used in industry as an abrasive and polishing agent. Would you predict those properties based on its chemical formula? Why?

(Answers to Concept Tests are in the back of the book.)

10.2 Interactions Involving Polar Molecules

Ionic and covalent bonds involve very strong interactions that hold the ions together in a crystal lattice and the atoms together in a molecule. Typical bond energies for covalent and ionic compounds range from several hundred to several thousand kilojoules per mole. In contrast, interactions between molecules and ions, such as those responsible for solubility of salts in water, are 10- to 100-fold weaker than bonding interactions. Even though these interactions are much weaker than bonds, they are of great consequence in determining properties and behavior of both pure substances and mixtures. For example, covalent substances like water interact with ionic substances if their molecules have a dipole (a uneven charge distribution) that makes association with charged ions energetically favorable. These interactions between ions and polar molecules are responsible for the solubility of salts in water, as we saw in the previous section. Covalently bonded molecules with dipoles also attract each other, and these dipole–dipole interactions are responsible for the range of physical properties observed among molecular materials.

Ion–Dipole Interactions

As we have already discussed, the solubilities of ionic compounds in water are affected by lattice energies and other factors. These other factors include the attraction between an ion and the region of the polar water molecule with the opposite partial charge. This interaction is an **ion–dipole** interaction, and it occurs between ions and polar molecules in all kinds of mixtures and solutions.

When ionic solids dissolve in water, ion–dipole interactions between the ions and water molecules pull solute ions from the solid substance into solution (Figure 10.4). As an ion is pulled away from its solid-state neighbors, it is surrounded by water molecules that form a **sphere of hydration**. If the solvent is something other than water, the cluster is called a *sphere of solvation*. These dissolved ions are said to be *hydrated* or, for other solvents, *solvated*. When the

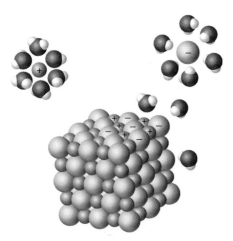

FIGURE 10.4 The positive ends of the water molecules' dipoles are attracted to the negative chloride ions, while the negative ends of the water molecules' dipoles are attracted to the sodium ions. The attraction overcomes the lattice forces at the surface of the crystal and causes the solid NaCl to dissolve.

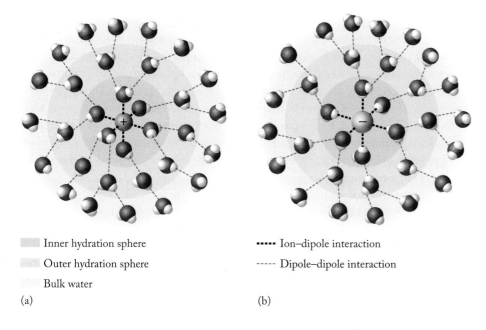

Inner hydration sphere
Outer hydration sphere
Bulk water

(a)

▪▪▪▪ Ion–dipole interaction
----- Dipole–dipole interaction

(b)

FIGURE 10.5 Each hydrated ion is surrounded by six water molecules in an inner hydration sphere characterized by ion–dipole interactions. (a) Positive sodium ions attract the negative (oxygen) end of the water dipole. (b) Negative chloride ions attract the positive (hydrogen) end of the water dipole. Water molecules surround the inner hydration sphere and are oriented with respect to it by dipole–dipole interactions. Dipole–dipole interactions also occur between the outer sphere and molecules in bulk water.

strengths of many ion–dipole interactions are sufficient to overcome the ion–ion interactions of an ionic compound, these spheres of hydration (solvation) form and the compound dissolves.

Within a sphere of hydration, the water molecules closest to an ion are oriented so that their oxygen atoms (negative poles) are directed toward a cation (such as Na^+ in Figure 10.5) and their hydrogen atoms (positive poles) are directed toward an anion (such as Cl^- in Figure 10.5). The number of molecules oriented in this way depends on the size of the ion. Typically six water molecules hydrate an ion, but the number can range from four to nine. Six water molecules surround the Na^+ and Cl^- ions in an aqueous solution of NaCl.

CONNECTION In previous chapters, we have indicated the presence of a hydration sphere around an ion in aqueous solution by placing the symbol (*aq*) after the symbol for the ion.

Dipole–Dipole Interactions

Surrounding the water molecules closest to the ions in Figure 10.5 are other water molecules. The latter are more loosely bound than the layer next to the ion, but are not randomly oriented. The ordering of these outer-sphere water molecules is caused by another intermolecular force: **dipole–dipole** interactions. Dipole–dipole interactions operate between molecules that have permanent dipole moments. In water molecules, the partial electrical charges on oxygen and hydrogen atoms result in attractions between a hydrogen atom of one molecule and the oxygen atom of another. Dipole–dipole interactions are not as strong as ion–dipole interactions, because dipole–dipole interactions involve only partial charges caused by unequal sharing of electrons, whereas one of the particles in an ion–dipole interaction has completely lost or gained electrons and has a full positive or negative charge. The magnitude of the dipole moment of a compound reflects the strength of the dipole–dipole interactions between its molecules.

In a solution of a salt such as NaCl, water molecules are oriented about the ions in the inner and outer spheres of hydration. However, all the water molecules interact with each other because they are polar. Polar molecules containing O—H and N—H bonds and the molecule HF have large dipole moments and stronger than average dipole–dipole interactions. The hydrogen atoms in these molecules are bonded to small, highly electronegative atoms and have a large partial positive charge. Therefore, the unequal distribution of the electrons in

 CHEMTOUR Intermolecular Forces

CONNECTION In Chapter 9 we learned that dipole moments are experimentally measured values expressed in units of debyes (D) that define the polarities of molecules.

Dipole–dipole interactions are attractive forces between polar molecules.

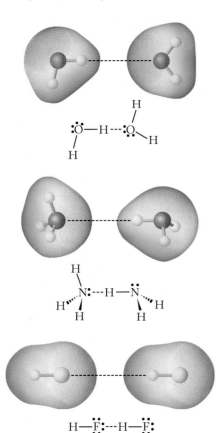

FIGURE 10.6 Hydrogen bonds (dashed lines) occur between hydrogen atoms bonded to O, N, or F in one molecule and an O, N, or F in an adjacent molecule.

these molecules gives rise to large dipole moments. Because of its strength, the interaction between the partially positive H atom on one molecule and a partially negative O, N, or F atom with unshared pairs of electrons on an adjacent molecule merits special distinction: it is called a **hydrogen bond** (Figure 10.6). Hydrogen bonds are the strongest dipole–dipole interactions. About one-tenth the strength of covalent bonds (hydrogen bonds range in strength from about 5 to about 25 kJ/mol), hydrogen bonds in water play a key role in defining the remarkable behavior of H_2O (described in detail in Section 10.8). One physical property strongly influenced by dipole–dipole interactions is boiling point. Water, HF, and ammonia all have abnormally high boiling points compared to other compounds of similar mass (Figure 10.7) or with elements in the same periodic group, primarily because of the very strong dipole–dipole interactions due to hydrogen bonds.

The **boiling point** is the temperature where vaporization occurs throughout a liquid. Transforming a liquid into a gas requires that the closely associated molecules in the liquid be separated from each other to form the much less dense gas phase. Boiling points depend on the molecular mass of particles and also on intermolecular attractions that tend to hold the molecules together and make them more difficult to separate.

The possibility of hydrogen bonds occurs whenever –OH or –NH groups are present in a molecule. They influence the physical properties of a large number of substances because of the extensive networks that are established between molecules that form hydrogen bonds (Figure 10.8).

Hydrogen bonds also have a defining impact on the overall three-dimensional structures of many polymers and large biological molecules such as proteins and DNA. They are responsible in large part for stabilizing the shapes of these molecules, and as we have seen with small molecules, shape often determines behav-

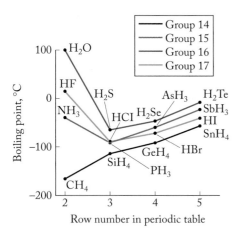

FIGURE 10.7 The boiling points of binary hydrides are plotted against row number on the periodic table. The boiling points of H_2O (18.02 g/mol), NH_3 (17.03 g/mol), and HF (20.01 g/mol) are not only significantly higher than that of CH_4 (16.04 g/mol), but they are also much higher than one would predict based on periodic properties.

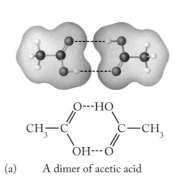

(a)   A dimer of acetic acid

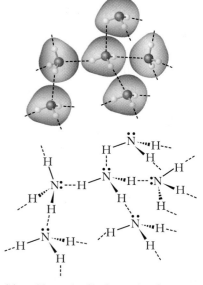

(b)   Network of hydrogen bonds in ammonia

FIGURE 10.8 Hydrogen bonds occur whenever –NH and –OH groups are present in molecules. (a) Hydrogen bonding interactions are so strong in a carboxylic acid like acetic acid that two molecules stay together as a unit called a dimer in the liquid phase. (b) Hydrogen bonds between nitrogen and hydrogen in ammonia form extensive three-dimensional networks.

ior. The presence of hydrogen bonds also gives rise to certain aspects of function. The ability of the two strands of DNA to stay together in a double helix is due to many hydrogen bonds, which hold the molecule together in aqueous solution. The bonds are weaker than covalent bonds, however, which makes it possible for the DNA strands to be pulled apart relatively easily for replication.

A **hydrogen bond** is the strongest kind of dipole–dipole interaction. It occurs between a hydrogen atom bonded to a small, highly electronegative element (O, N) and an atom of oxygen or nitrogen in another molecule. The molecule HF also experiences hydrogen bonds.

The **boiling point** of a substance is the temperature at which vaporization occurs throughout a liquid.

| SAMPLE EXERCISE 10.4 | **Explaining Relative Boiling Points** |

The gas dimethyl ether (C_2H_6O) has a molar mass of 46.07 g/mol and a boiling point of −24.9°C. The liquid ethanol (C_2H_6O) has the same formula and therefore the same molar mass as diethyl ether, but ethanol has a much higher boiling point of 78.5°C. Explain the large difference in the relative boiling points of the two liquids. Their structures are

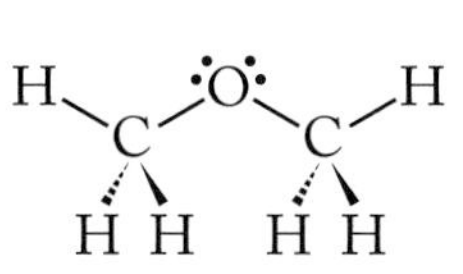

Dimethyl ether
CH_3—O—CH_3

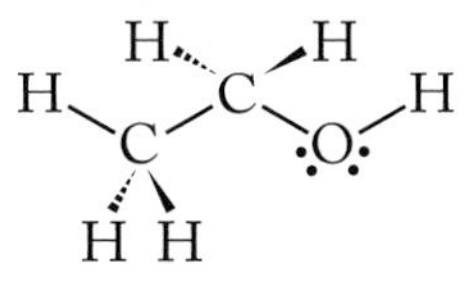

Ethanol
CH_3—CH_2—OH

COLLECT AND ORGANIZE We are asked to explain the large difference in boiling point of two compounds that have the same molar mass. The molar masses of the two molecules are the same, and therefore intermolecular interactions must account for the difference of the boiling points.

ANALYZE Both molecules contain a very electronegative oxygen atom and are polar, but ethanol has one –OH group.

SOLVE The crucial difference in the structures of the two molecules is that ethanol has an –OH group while dimethyl ether does not. Molecules of ethanol experience hydrogen bonding, which is the strongest dipole–dipole interaction, while molecules of dimethyl ether experience weaker dipole–dipole interactions.

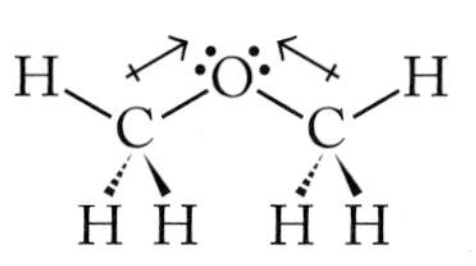

–OH group

Dimethyl ether
Polar

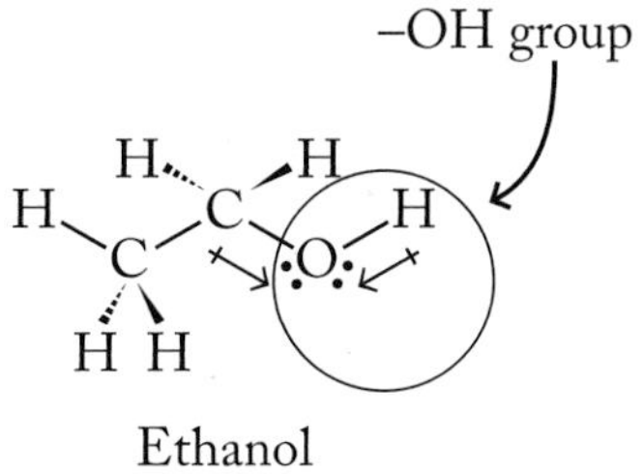

Ethanol
Polar and capable
of hydrogen bonding

The ethanol molecules, as shown at the top of the next page, are held together much more tightly because of the hydrogen bonds between the molecules and therefore require the input of more energy to make them vaporize, as is reflected in the higher boiling point of ethanol.

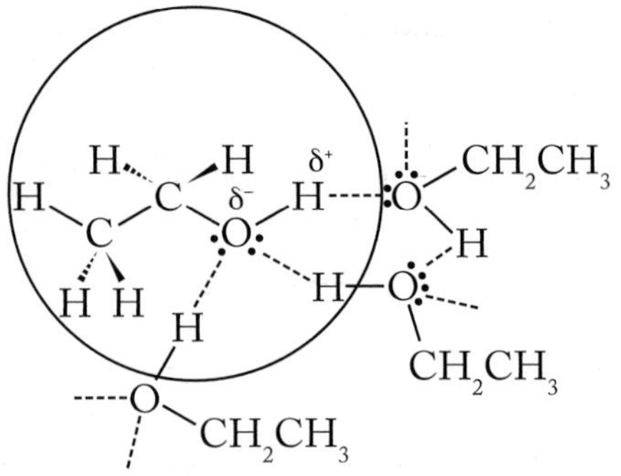

A portion of the hydrogen-bonding
network in ethanol

THINK ABOUT IT Hydrogen bonding is the strongest intermolecular attractive force between covalently bonded molecules.

Practice Exercise Suggest a reason why ethylene glycol (molar mass 62.07 g), used in radiators as antifreeze, boils at 196°C, while isopropanol (molar mass 60.01 g), used as a cooling rub for sore muscles, boils at 82°C, which is more than 100° lower.

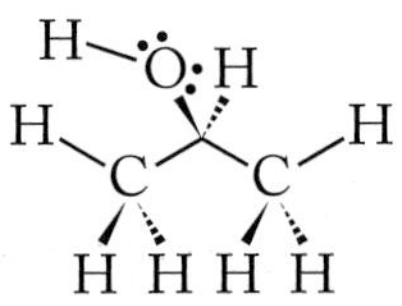

Isopropanol 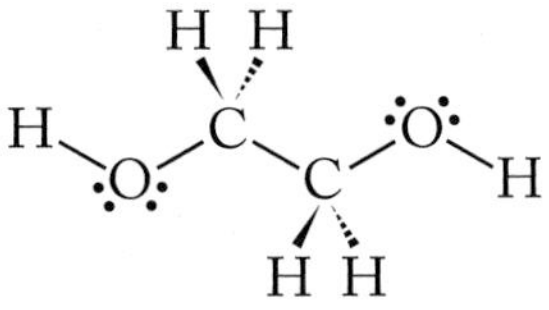   Ethylene glycol

---

CONCEPT TEST

Dimethyl ether, CH_3OCH_3, and acetone, $CH_3C(O)CH_3$,

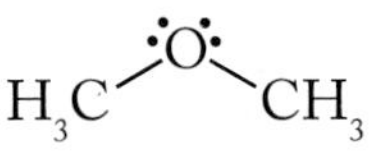

Dimethyl ether 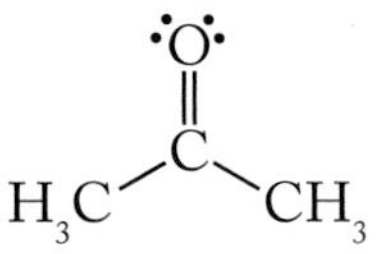   Acetone

have similar formulas and molar masses (46 g/mol and 58 g/mol, respectively). However, the dipole moments of dimethyl ether and acetone are quite different: 1.30 and 2.88 D, respectively. Predict which compound has the higher boiling point.

10.3 **Interactions Involving Nonpolar Molecules**

Attractive forces also exist between nonpolar molecules. They are called **dispersion forces** or **London forces** in honor of German-American physicist Fritz London (1900–1954). His work explained attractions between atoms of the noble gases and between nonpolar molecules caused by **temporary dipoles** (or **induced dipoles**), produced by momentary changes in electron distributions within atoms or molecules. These forces are weaker than ion–ion and ion–dipole forces, but they

Dispersion forces (**London forces**) are intermolecular forces caused by the presence of temporary dipoles in molecules.

A **temporary dipole** (or **induced dipole**) is a separation of charge produced in an atom or molecule by a momentary uneven distribution of electrons.

operate between all molecules, and in molecules with large hydrophobic regions they can be stronger than dipole–dipole forces.

Temporary dipoles occur, for example, when two noble-gas atoms approach one another and mutual repulsion between their electron clouds perturbs the distributions of the electrons in these clouds, inducing nonuniformity and creating temporary dipoles (Figure 10.9). The temporary dipoles have slightly positive and negative regions. The resultant interaction with other particles having transient dipoles causes the atoms to associate more tightly than they would in the absence of the temporary charge. The magnitude of an induced dipole depends on the ease with which the electron cloud surrounding a molecule, ion, or atom can be perturbed. The relative ability of an electron cloud to be distorted by another particle is its **polarizability**.

Temporary dipoles can also arise in mixtures of polar molecules and nonpolar molecules with polarizable electron clouds. Polar molecules induce dipoles in nonpolar molecules by causing a distortion in electron distribution (Figure 10.10). The interactions between the dipole and the induced dipole that results are weaker than the dipole–dipole forces between molecules of a polar substance, and they are on the same order of magnitude as dispersion (London) forces between nonpolar molecules.

What factors determine how easily the cloud of electrons around an atom or molecule can be polarized? To answer this question, let's examine how dispersion forces influence the boiling points of nonpolar compounds.

A comparison of the boiling points of the monatomic noble gases and the nonpolar diatomic halogens (Table 10.4) discloses the same trend for both groups: boiling points increase with increasing atomic size and molar mass. The molecular interpretation of this trend is twofold. First, heavier molecules require more energy and hence higher temperatures to move from the liquid phase into the gas phase. Second, particles having larger atomic or molar masses have larger electron clouds, and larger electron clouds are more easily polarized than smaller ones, because they are farther from the nucleus and held less tightly by it. Greater polarizability leads to stronger temporary dipoles and stronger intermolecular interactions, so dispersion forces become stronger with increasing molar mass. Therefore as molar mass increases, boiling points increase.

The overall shape of a molecule also plays a role in determining the magnitude of dispersion forces. This point is illustrated by three hydrocarbons that have exactly the same formula and therefore the same molar mass but differ-

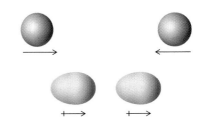

FIGURE 10.9 Two atoms of a noble gas approach one another and create temporary dipoles when their electron clouds repel each other.

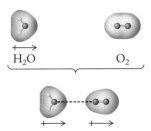

FIGURE 10.10 The approach of a water molecule induces a temporary dipole in a nonpolar oxygen molecule by distorting its electron cloud.

TABLE 10.4 **Molar Masses and Boiling Points of the Halogens and Noble Gases**

Halogen	$\mathcal{M}$ (g/mol)	Boiling Point (K)	Noble Gas	$\mathcal{M}$ (g/mol)	Boiling Point (K)
			He	4	4
F_2	38	85	Ne	20	27
Cl_2	71	239	Ar	40	87
Br_2	160	332	Kr	84	120
I_2	254	457	Xe	131	165
			Rn	222	211

Polarizability is a term that describes the relative ease with which an electron cloud is distorted by an external charge.

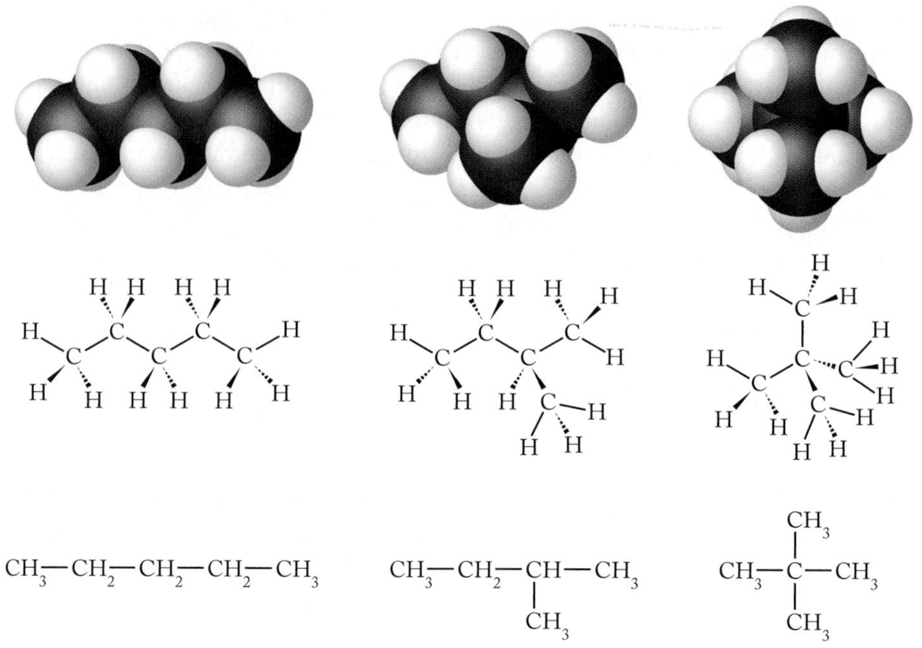

FIGURE 10.11 Three different molecular structures are possible for C_5H_{12}.

$CH_3-CH_2-CH_2-CH_2-CH_3$

n-Pentane:
boiling point 36°C

$CH_3-CH_2-CH-CH_3$
               |
              CH_3

2-Methylbutane:
boiling point 28°C

          CH_3
           |
CH_3-C-CH_3
           |
          CH_3

2,2-Dimethylpropane:
boiling point 9°C

ent shapes (Figure 10.11). Three different arrangements of atoms correspond to the formula C_5H_{12}. Although these three molecules have exactly the same molar mass (72.15 g/mol), their boiling points differ over a range of 27°C. All these molecules are nonpolar, so the only interactions involved must be based on dispersion forces. We can think of molecules of *n*-pentane as being relatively long and straight; think of them as pencils. They can interact with each other over a much larger surface area and therefore have more possibilities for dispersion forces to hold them together. In contrast, molecules of 2,2-dimethylpropane are almost spherical; think of them as Ping-Pong balls. They have less surface area to interact with adjacent molecules; less contact means fewer interactions, which in turn means weaker attractive forces, resulting in a lower boiling point than that of either *n*-pentane or 2-methylbutane. 2-methylbutane boils at a temperature lower than that of *n*-pentane, but higher than that of 2,2-dimethylpropane; in terms of structure, it is neither as straight as *n*-pentane nor as spherical as 2,2-dimethylpropane, but is in between the two. Therefore, it is logical that its boiling point is between the two.

Dispersion forces are weak, but they add up, and for larger molecules they can provide considerable force to hold molecules together. A large number of weak interactions can sometimes dominate a much smaller number of strong interactions in a system, as we will see in the discussion of polarity and solubility in Section 10.5.

SAMPLE EXERCISE 10.5   **Explaining Trends in Boiling Points**

Figure 10.12 gives the formulas, molar masses, and boiling points of hydrocarbons and alcohols with comparable molar masses. Structures of the compounds in row 4 are given after the table. Explain the trends in values of boiling points (1) in both columns going down the table, and (2) as one moves across a row from a hydrocarbon to an alcohol of comparable molar mass.

HYDROCARBON			ALCOHOL		
Molecular Formula	Molar Mass (g)	Boiling Point (°C)	Molecular Formula	Molar Mass (g)	Boiling Point (°C)
1. CH_4	16.04	−161.5			
2. CH_3CH_3	30.07	−88	CH_3OH	32.04	64.5
3. $CH_3CH_2CH_3$	44.09	−42	CH_3CH_2OH	46.07	78.5
4. $CH_3CH(CH_3)CH_3$	58.12	−11.7	$CH_3CH(OH)CH_3$	60.09	82
5. $CH_3CH_2CH_2CH_3$	58.12	−0.5	$CH_3CH_2CH_2OH$	60.09	97

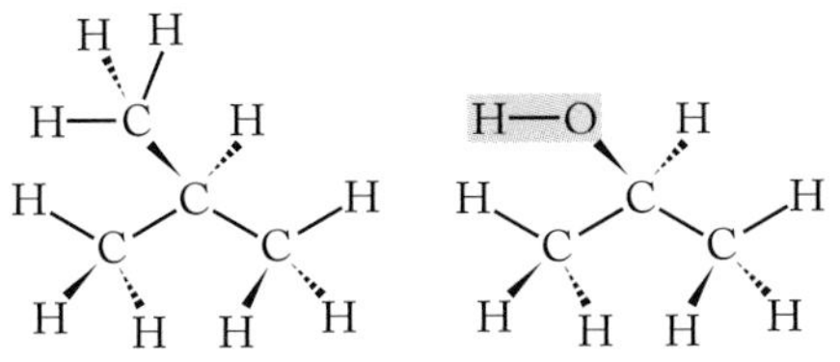

$CH_3CH(CH_3)CH_3$
Boiling point −11.7°C

$CH_3CH(OH)CH_3$
Boiling point 82°C

FIGURE 10.12 Data table for Sample Exercise 10.5 and molecular structures of two compounds.

COLLECT AND ORGANIZE We are asked to explain trends in the boiling points of two different kinds of compounds as their molar masses increase. Then we are asked to compare the boiling points of different compounds that have similar molar masses.

ANALYZE Two changes affect the boiling points of substances: as molar mass and size of electron clouds increase, boiling points increase; as intermolecular forces increase in strength, boiling points increase. We must determine which effect is operating in the data sets defined.

SOLVE 1. As we go down both columns in the table, the boiling point increases. The mass increases for four of the five hydrocarbons in the left column as well as for three of the four alcohols in the right column. However, since two compounds in each column have the same molar mass (compounds in rows 4 and 5), something else must be contributing to changes in the boiling point. In both cases, the molar masses are the same but the shapes differ. Based on the structures shown, the molecules in row 4 are more spherical and have smaller surface areas than the compounds with longer, straighter shapes shown in row 5. The compounds in row 4 have fewer points of contact and therefore dispersion forces are weaker. Both compounds in row 4 boil at lower temperatures than their counterparts in row 5 with the same molar mass because dispersion forces are weaker.

2. As we go across a row in Figure 10.12, the molar masses of the compounds change very little; the molar masses of ethane and methanol are very similar, as are the molar masses of each of the other pair of compounds. The alcohols all have much higher boiling points than their hydrocarbon counterparts, and differences in the magnitudes of intermolecular forces must be the cause. All the hydrocarbons are nonpolar and interact via weak dispersion forces. All the alcohols are not only polar with permanent dipole moments, but they are also capable of hydrogen bonding. Hydrogen bonds are the strongest dipole–dipole forces, and their presence in the alcohol molecules accounts for the much higher boiling points.

THINK ABOUT IT Identifying the type of intermolecular forces between molecules enables us to rank the relative strength of the forces and therefore their relative contribution to determining the boiling point. Shape matters, too, especially in terms of the relative contribution of dispersion forces to molecular attraction.

Practice Exercise For the substances listed, predict which one has (1) the largest dipole–dipole interactive forces, (2) the largest dispersion forces, (3) the lowest boiling point: (a) H_2NNH_2; (b) H_2C=CH_2; (c) Ne; (d) $CH_3CH_2CH_2CH_2CH_3$.

Explain why CF_4 is a gas at room temperature but CCl_4 is a liquid.

10.4 Real Gases Revisited

Chapter 6 ended with a discussion of how the ideal gas equation ($PV = nRT$) is modified to account for the behavior of real gases such as H_2 and CO_2. Real gases have pressures and volumes that do not fit the ideal gas equation, especially under conditions of high pressure and/or low temperature. High pressures push molecules closer together and low temperatures reduce their average speeds, meaning that they take longer to pass by one another and hence have more chance to interact. Both factors favor intermolecular interactions that are not accounted for in the ideal gas theory that leads to the equation $PV = nRT$. The van der Waals equation [$(P + n^2a/V^2)(V - nb) = nRT$; see Section 6.8] relates the pressure, volume, and temperature of n moles of a real gas. It includes two terms (a and b) that account for the behavior of real gas molecules.

The first correction term, n^2a/V^2, accounts for interactions between gas molecules. Because attractive forces between molecules are considered in the van der Waals equation for real gases, the term **van der Waals forces** is frequently used collectively to indicate all types of attractive forces: dipole–dipole interactions, interactions between dipoles and induced-dipoles, and dispersion forces. Notice that the term containing a increases in magnitude with the number of moles of gas in a sample and when the volume of that sample is decreased. The second correction term, $-nb$, accounts for the fact that real molecules have a volume of their own. Bigger molecules have larger molecular volumes. Table 6.5 contains values of a and b for some real gases.

CONNECTION The role of the $-nb$ term in influencing the free volume of a real gas was discussed in detail in Chapter 6.

SAMPLE EXERCISE 10.6 **Assessing Empirical Parameters of Real Gases**

The van der Waals constant a of SO_2, 6.71 $L^2 \cdot$ atm/mol^2, is nearly twice the value of a of CO_2, which is 3.59 $L^2 \cdot$ atm/mol^2. Suggest a reason for this difference.

COLLECT AND ORGANIZE We are asked to suggest a reason why the value of a in the van der Waals equation for real gases of SO_2 is almost twice the value of a for CO_2. Because a reflects the attractive force between molecules in a real gas, we need to consider the strengths of the intermolecular forces in the two gases.

ANALYZE For molecules of similar mass and size, in order of relative strength, intermolecular forces are hydrogen bonding, dipole–dipole, or dispersion forces. Neither molecule has hydrogen attached to an electronegative atom, so hydrogen bonding is not a factor. We need to look at structures of the molecules to assess whether either has a permanent dipole.

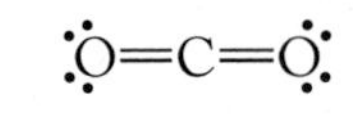

The term **van der Waals forces** is frequently used to refer collectively to all types of attractive forces: dipole–dipole interactions and dispersion forces.

No net dipole; nonpolar molecule, $\mathcal{M} = 44$ g/mol

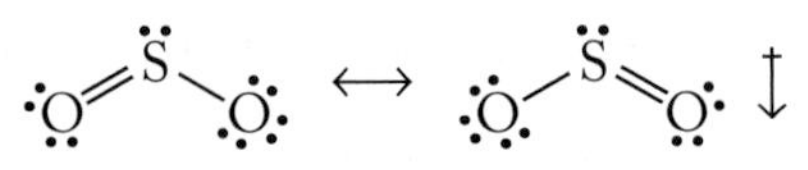

Net dipole; polar molecule, $\mathcal{M} = 64$ g/mol

SOLVE Carbon dioxide molecules are linear and nonpolar, so they interact only through dispersion (London) forces. Sulfur dioxide molecules are bent and have a permanent dipole moment.

The intermolecular forces in SO_2 are dipole–dipole interactions, which should be stronger than the dispersion forces between CO_2 molecules. Therefore, SO_2 molecules experience a greater attractive force, which is reflected in the larger value of a. SO_2 also has a larger molar mass than CO_2, increasing dispersion forces.

THINK ABOUT IT Large values of parameter a in the van der Waals equation are associated with relatively strong intermolecular forces.

Practice Exercise Without looking up the values in Chapter 6, place the following substances in order of increasing value of the van der Waals constant a: H_2O, O_2, and CO.

10.5 Polarity and Solubility

The solubilities of ionic salts in water are explained in Section 10.2 in terms of ion–dipole interactions. Solubilities of liquids in other liquids and of gases in liquids can also be explained by considering intermolecular forces. In Section 10.2 we considered the case where the solute was an ionic solid and the solvent was water. Now we consider the situation where the solvent is water or some other liquid and the solute is a covalent molecule. To understand and predict solubilities, we look at the balance between solute–solute interactions (interactions of molecules of solute with themselves), solvent–solvent interactions (interactions of molecules of the solvent with themselves), and solvent–solute interactions (interactions of solvent molecules with solute).

Just as ionic compounds dissolve in polar solvents because of strong ion–dipole interactions, polar solutes tend to dissolve in polar solvents because of dipole–dipole interactions between solute and solvent molecules. On the other hand, nonpolar solutes tend not to dissolve in polar solvents because the solvent–solute interactions that promote dissolution are weaker than those that keep solute molecules together in one phase and solvent molecules together in another. This observation is the source of a common phrase used to describe solubility: like dissolves like. Factors other than polarity influence solubility: temperature and pressure are two very important ones. Nevertheless, we can use the type of van der Waals forces present in any mixture of molecules as a guide to predict the relative solubilities of different solutes in a given solvent.

SAMPLE EXERCISE 10.7   **Predicting Water Solubility**

Which of these compounds should be highly water soluble and which should have limited solubility? (a) CCl_4; (b) NH_3; (c) HF; (d) O_2

COLLECT AND ORGANIZE We are asked to predict the solubilities of several compounds in water. We can make our predictions based on polarity. Because water is polar, compounds soluble in it will likely also be polar.

ANALYZE To determine whether each of these compounds is polar, we must draw their structures, evaluate each bond for its polarity, and see whether the molecule has a permanent dipole moment.

If two or more liquids are **miscible**, they form a homogeneous solution when mixed in any proportion.

SOLVE The structures and net dipole moments for the given molecules are indicated in the accompanying figure.

Nonpolar   Polar   Polar   Nonpolar

On the premise that "like dissolves like," we predict that the polar molecules (NH_3 and HF) would be very soluble in water while the nonpolar ones (CCl_4 and O_2) would be much less soluble.

THINK ABOUT IT The ability of ammonia and hydrogen fluoride to participate in hydrogen bonding makes them even more likely to be water soluble. That makes them much like water and hence very soluble.

Practice Exercise In Chapter 6, we learned that deep-sea divers breathe a mixture of gases rich in helium to diminish the possibility of experiencing the bends, a painful condition caused by nitrogen gas dissolving in blood. The statement was made that helium is less soluble in blood than is nitrogen. Assuming that blood behaves like water in terms of dissolving substances, why should helium be less soluble in water than nitrogen is?

Water contains sufficient dissolved oxygen to sustain a variety of aquatic life.

In Sample Exercise 10.7 we predicted that ammonia (NH_3) and hydrogen fluoride (HF) are water soluble, while carbon tetrachloride (CCl_4) and oxygen (O_2) are much less soluble. These predictions are essentially accurate. At a given temperature and pressure, ammonia is very soluble in water, but NH_3 also reacts with water, which complicates the determination of solubility. Hydrogen fluoride is **miscible** in water, meaning it is soluble in all proportions. Hydrogen fluoride also reacts with water to a small degree. A substance that is miscible in another is sometimes referred to as being *infinitely soluble*. Ammonia and hydrogen fluoride both make hydrogen bonds, just like water, and this enhances their interactions with water. In contrast, the nonpolar molecules carbon tetrachloride and oxygen are very limited in their water solubility. Their limited solubility results from interactions between dipoles and induced dipoles that are of the same order of magnitude as the dispersion (London) forces between molecules of a nonpolar solute. Nonpolar molecules such as O_2 are very sparingly soluble in polar solvents such as H_2O. Even so, for aquatic organisms, the limited solubility of oxygen in water, about 10 mg/L at normal atmospheric pressure, is sufficient to sustain life.

Calculating the Solubility of Gases in Water: Henry's Law

The O_2 level in natural waters is normally sufficient to support life, but you may have noticed fish in rivers gasping for air in very warm weather. The actual solubility of a gas in a liquid depends on both temperature and pressure. In terms of temperature, the solubility of O_2 and most gases in water decreases with increasing temperature.

Solubility of a gas also depends on its partial pressure, and the low partial pressure of oxygen at high altitudes can mean lower than normal concentrations of oxygen in liquids, including the blood and tissues of humans and animals in those regions. For example, climbers in the Himalayas may become weak and unable to think clearly because of a condition known as *anoxia*. Anoxia occurs because the solubility of O_2 in tissues is proportional to its partial pressure. This relationship applies to all sparingly soluble gases that do not react chemically with the solvent and is known as **Henry's law**. In equation form it is written as follows

$$C_{gas} = k_H P_{gas} \qquad (10.3)$$

where C_{gas} represents the concentration of dissolved gas and k_H is the Henry's law constant for the gas in that particular solvent. When dissolved concentrations are expressed in molarity, the units of the Henry's law constant are mol/(L · atm). Equivalent values based on mol/kg · mmHg of liquid are listed in Table 10.5.

Henry's law predicts that the concentration of dissolved oxygen in blood is proportional to the partial pressure of oxygen, and thus proportional to atmospheric pressure. This is an accurate statement of Henry's law, but residents of Denver, Colorado or Kimberly, Canada (average atmospheric pressure 0.85 atm) do not live with less blood oxygen than residents of New York City or Rome, Italy (average atmospheric pressure 1.00 atm) because of the way the oxygen transport system in our bodies responds to local conditions.

The amount of oxygen in the blood is related to the concentration of hemoglobin and to the fraction of the hemoglobin sites that contain oxygen as the blood leaves the lungs. This saturation of binding sites depends on the partial pressure of oxygen as well as proper lung function. For most people breathing air with $P_{O_2} >$ 85 mmHg (85 torr) results in nearly 100% saturation of hemoglobin bonding sites. If P_{O_2} decreases to about 50 mmHg (as it does on high mountains), the percent saturation decreases to about 80%. Over several weeks the body responds to lower oxygen partial pressures by producing more red blood cells and more hemoglobin. This increase in O_2 carriers compensates for the lower partial pressure of O_2.

> **Henry's law** states that the solubility of a sparingly soluble, chemically unreactive gas in a liquid is proportional to the partial pressure of the gas.

TABLE 10.5 Henry's Law Constants for Gas Solubility in Water at 20°C

Gas	k_H [mol/(L · atm)]	k_H [mol/(kg · mmHg)]
He	3.5×10^{-4}	5.1×10^{-7}
O_2	1.3×10^{-3}	1.9×10^{-6}
N_2	6.7×10^{-4}	9.7×10^{-7}
CO_2	3.5×10^{-2}	5.1×10^{-5}

SAMPLE EXERCISE 10.8 **Calculating Gas Solubility Using Henry's Law**

Calculate the solubility of oxygen in water in moles per liter at 1.00 atm pressure and 20°C. The mole fraction of O_2 in air is 0.209.

COLLECT AND ORGANIZE We are asked to determine the solubility of oxygen gas in water. We are given temperature, pressure, and the mole fraction of oxygen in air. We also have Equation 10.3 for Henry's law, which relates these factors, and the Henry's law constant for oxygen in Table 10.5.

ANALYZE Henry's law (Equation 10.3) describes the relation between the solubility of oxygen and its partial pressure.

$$C_{gas} = k_H P_{gas}$$

SOLVE We need the partial pressure of oxygen for Henry's law. Partial pressure is related to the mole fraction (X) of a gas and the total pressure:

$$P_{O_2} = X_{O_2} P_{total} = (0.209)(1.00 \text{ atm}) = 0.209 \text{ atm}$$

Substituting this value for P_{O_2} and k_H for O_2 in water in Equation 10.3 gives

$$C_{O_2} = k_H P_{O_2} = \left(\frac{1.3 \times 10^{-3} \text{ mol}}{\text{L} \cdot \text{atm}}\right)(0.209 \text{ atm}) = 2.7 \times 10^{-4} \text{ mol/L}$$

THINK ABOUT IT The answer seems reasonable in terms of predictions we can make based on relative solubilities. We predict that oxygen is not very soluble in water because the O_2 molecule is nonpolar. Our numerical answer agrees with that prediction.

Practice Exercise Calculate the solubility of oxygen in water at an atmospheric pressure of 0.35 atm (a typical atmospheric pressure at the top of Mt. Everest). The mole fraction of O_2 in air is 0.209.

Combinations of Intermolecular Forces

More than one type of intermolecular force may need to be considered when larger molecules dissolve. Consider the solubilities in water of three organic alcohols with the structures shown below:

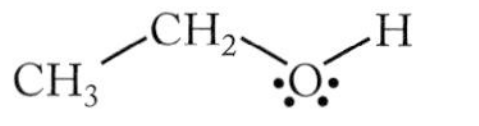

Ethanol

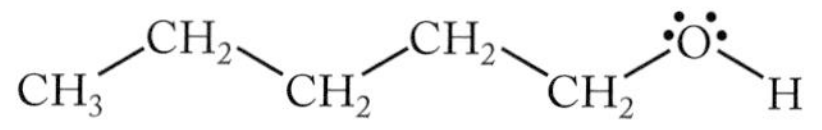

n-Pentanol

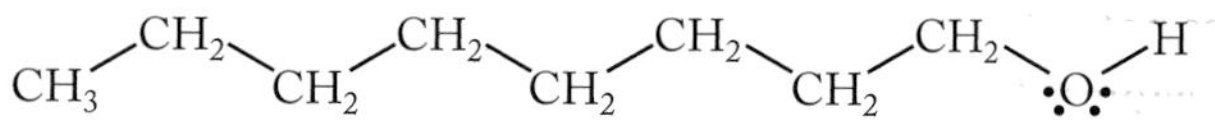

n-Octanol

Alcohol	Water Solubility at 20°C
CH_3CH_2OH Ethanol	Miscible in all proportions
$CH_3CH_2CH_2CH_2CH_2OH$ n-Pentanol	2.6 g/100 mL water
$CH_3CH_2CH_2CH_2CH_2CH_2CH_2CH_2OH$ n-Octanol	5.8×10^{-7} g/100 mL water (sparingly soluble)

All three compounds contain an –OH group, and in that sense they are "water-like" and make hydrogen bonds. They are all polar, and the dipole of water is attracted to the dipole of the –OH group in each of them. However, their solubilities differ greatly.

Ethanol is miscible with water; it is soluble in all proportions. Mixing any amount of ethanol and water produces a homogeneous solution. Pentanol has a defined solubility, which means that once a solution of pentanol in water is saturated, adding more pentanol results in a heterogeneous mixture, in which two layers of fluid would be visible. Octanol is sparingly soluble in water, and almost any measurable amount of octanol added to water results in a heterogeneous mixture.

All three molecules have a water-like dipole, the –OH group, but the feature that differentiates them is the nature of the rest of the molecule. The long series of –CH_2– groups in octanol makes that part of the molecule very nonpolar and very

CONNECTION We discussed homogeneous and heterogeneous mixtures in Chapter 1.

much unlike water. The dispersion forces that arise between the long chains of adjacent octanol molecules are very strong (Figure 10.13). The sum of the attractive forces keeps the octanol molecules together, even though the polar ends of the molecules are attracted to molecules of water. The miniscule solubility of octanol in water illustrates a situation where dispersion forces contribute more to intermolecular interactions than do the dipole–dipole forces that are localized in one small region of a molecule. In ethanol, the polarity and hydrogen bonding ability of the –OH group dominates the interaction of ethanol with water because the hydrocarbon portion of the molecule is too short (only 2 carbon atoms long) to have significant dispersion forces. The solubility of pentanol (a chain 5 carbon atoms long) is less than that of ethanol but much greater than that of octanol (a chain 8 carbon atoms long). The competing dispersion forces that limit solubility and the hydrogen-bonding interactions that promote it combine to produce moderate solubility.

Nonpolar interactions like the dispersion forces between hydrocarbon chains are called **hydrophobic** (literally, "water-fearing") interactions, whereas interactions that promote solubility in water are called **hydrophilic** ("water-loving") interactions. For molecules like these alcohols having polar and nonpolar groups, the ultimate solubility in water is due to the balance between hydrophilic and hydrophobic interactions. As the hydrophobic portion of the molecule increases in size, the entire molecule becomes more hydrophobic.

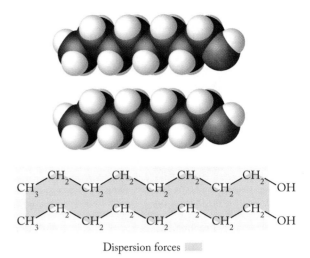

Dispersion forces

FIGURE 10.13 Dispersion forces between long hydrocarbon chains provide sufficient attractive force to hold octanol molecules together.

CONCEPT TEST

Do you think an attractive force exists between ions and nonpolar molecules? How would you describe such an interaction?

10.6 Vapor Pressure

Water in a glass left on a countertop slowly disappears. Of course it does not actually disappear, it just enters the gas phase and we can no longer see it. Molecules on the surface of the liquid **vaporize** or **evaporate** over time; they escape from the surface of the liquid and enter the gas phase. The rate at which the molecules make this transformation depends on the temperature, the surface area of the liquid, and the strength of the intermolecular forces that hold the molecules together in the liquid phase:

1. The higher the temperature, the greater the numbers of molecules in a population with sufficient kinetic energy to break the attractive forces that hold them together in the liquid phase and enable them to enter the gas phase.
2. The greater the surface area of the liquid, the larger the number of molecules on the surface in a position to enter the gas phase.
3. The stronger the intermolecular forces, the greater the kinetic energy needed for a molecule to escape the surface, and the smaller the number of molecules in the population that actually have this energy.

A **hydrophobic** ("water-fearing") interaction repels water and diminishes water solubility.

A **hydrophilic** ("water-loving") interaction attracts water and promotes water solubility.

Vaporization or **evaporation** is the transformation of molecules in the liquid phase into the gas phase.

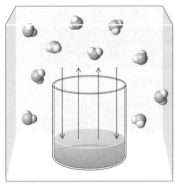

↓ Condensation

↑ Evaporation

🌕 Water vapor

FIGURE 10.14 A glass of water sealed in a closed container achieves a dynamic equilibrium in which the amount of water lost from the glass to evaporation equals the amount returned by condensation.

⦿⦿ **CONNECTION** The terms *condensation, volatile,* and *nonvolatile* were defined in Chapter 1.

Vapor pressure is the force exerted at a given temperature by a vapor in equilibrium with its liquid phase.

If that same glass of water is sealed in a closed container (Figure 10.14), a different situation arises. The molecules still vaporize, but now they are confined within the container. Some of them *condense* at the surface of the liquid and return to the liquid phase. In a short time, the two processes of evaporation and condensation equalize, and the same number of molecules leave the surface as reenter it. At this point, called a *dynamic equilibrium,* no further change takes place in the level of the liquid in the glass, although molecules continue to evaporate and condense constantly. In such a situation at constant temperature, the pressure exerted by the vapor in equilibrium with its liquid phase is called its **vapor pressure**.

Seawater versus Pure Water

An open beaker of seawater on a laboratory bench also evaporates over time. The water is volatile. A residue of nonvolatile sea salts remains in the beaker that contained the seawater. However, if a beaker of seawater and a beaker of distilled water are placed in a sealed chamber, as shown in Figure 10.15, a curious phenomenon takes place: The volume in the beaker of the seawater actually increases over time, and the volume of distilled water decreases at the same rate. Ultimately all the water in both beakers ends up in the beaker that originally contained the seawater. These changes are all due to the presence of dissolved ions in the seawater.

As water in the beakers in Figure 10.15 evaporates, the concentration of water molecules in the air space in the container increases. As the concentration of water vapor increases, the pressure of the water vapor increases. At constant temperature, this pressure eventually stabilizes at a value equal to the vapor pressure of water at that temperature. At this point, the total rate of evaporation from the beakers is equal to the total rate of water vapor condensation. If the rates of evaporation and condensation are the same in both beakers, the liquid levels in the beakers will not change over time.

The liquid levels in the two beakers in Figure 10.15 do change until there is no distilled water left. Because the molecules of water vapor in the chamber are free to condense into either beaker, the rate of condensation of water vapor into both beakers is the same. For the transfer of water from one beaker to the other to occur, the distilled water must have a higher rate of evaporation than seawater. This conclusion leads to another: if the seawater evaporates at a lower rate, it must have a lower vapor pressure than the distilled water at the same temperature. This is indeed the case, and this is a clear illustration of a more general observation: at a given temperature, the vapor pressure of the solvent in a solution containing nonvolatile solutes is less than the vapor pressure of the pure solvent.

Because the vapor pressure of the distilled water is greater than the vapor pressure of the seawater, more water vapor enters the air in the chamber from the distilled water beaker than from the seawater. The condensation rates (which determine how much water is returned to both beakers in a given time) are the same, so the distilled water beaker over time loses more water than is restored to it by condensation, while the seawater beaker gains more water than it loses by evaporation. The seawater gains the same amount by condensation as the distilled water, but because it loses less to evaporation, it receives more in return than it gives off. This imbalance results in all the liquid water ultimately ending up in the beaker of seawater.

The vapor pressure of water or any liquid increases with increasing temperature, as shown in the graph in Figure 10.16. As we learned in Chapters 5 and 6, molecular motion increases with increasing temperature. At higher tempera-

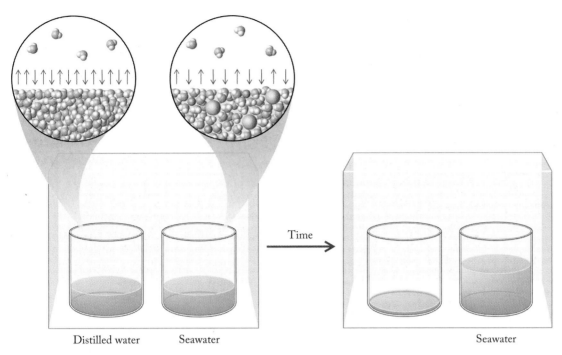

FIGURE 10.15 When a beaker containing seawater is placed in a sealed container with a beaker of distilled water, the slightly higher vapor pressure of distilled water leads to a net transfer of water from the beaker of distilled water to the beaker containing seawater.

tures, a greater fraction of the water molecules in the liquid phase have enough kinetic energy to break their attraction to other molecules at the surface and enter the vapor phase. At 100°C, the vapor pressure of pure water is 1 atm (760 torr), and water boils. At this temperature, essentially all the water molecules have enough energy to enter the vapor phase. In the general case, the *normal* **boiling point** of any liquid is the temperature at which its vapor pressure equals 1 atm (Figure 10.16). The seawater has a lower vapor pressure than pure water at all temperatures, so it follows than the solution has a higher boiling point than the distilled water.

▶‖ **CHEMTOUR** Molecular Motion

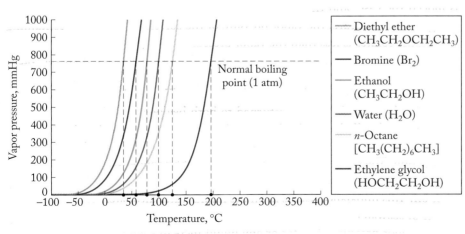

FIGURE 10.16 A graph of vapor pressure versus temperature for six liquids shows that vapor pressure increases with increasing temperature. The temperature at which the vapor pressure equals 760 torr (1 atm) is the normal boiling point of the liquid.

The *normal* **boiling point** of a liquid is the temperature at which its vapor pressure equals 1 atmosphere (760 torr).

Vapor Pressures of Solutions: Raoult's Law

The connection between the vapor pressure of solutions and their concentrations of nonvolatile solutes was studied extensively by French chemist François Marie Raoult (1830–1901). He discovered the following relation between the vapor pressure of a solution, $P_{solution}$, and that of the pure solvent, $P_{solvent}$, a relation called **Raoult's law**:

$$P_{solution} = \chi_{solvent} P_{solvent} \qquad (10.4)$$

Here, $\chi_{solvent}$ is the mole fraction of solvent: the moles of solvent divided by the total number of moles of solvent and solutes.

Let's take another look at the two beakers in Figure 10.15. At 20°C (293 K), the vapor pressure of water produced by evaporation of the pure water is 0.0231 atm. The vapor pressure of a pure substance is an intensive property that depends on the substance and its temperature, but not on the amount of substance. What is the vapor pressure produced by evaporation of water from seawater if the mole fraction of water is 0.980? We can use the mole fraction to calculate the vapor pressure of the seawater solution:

$$P_{solution} = (0.980)(0.0231 \text{ atm}) = 0.0226 \text{ atm}$$

Because seawater has a slightly lower vapor pressure than does distilled water, fewer molecules leave the salt water for the vapor phase compared to the number that leave the distilled water; the evaporation rates differ.

The lowering of the vapor pressure of a solution compared to the vapor pressure of pure solvent depends only on the concentration of solute particles in the water, not on their identity. Properties of solutions that depend only on the concentration of particles and not on their identity are called **colligative properties**. We will examine them in detail in Section 10.9.

What we have just described in terms of the direct relation between concentration and vapor pressure is the behavior of an **ideal solution**, where an ideal solution is one that obeys Raoult's law. Deviations from this ideal behavior occur because of interactions between solvent and solute molecules at the surface of a liquid. If interactions between solvent molecules and solute particles (solvent–solute interactions) are stronger than interactions between solvent molecules (solvent–solvent interactions), then vaporization of the solvent is decreased and vapor pressure is less than predicted by Raoult's law. Because the solvent molecules are held at the surface by favorable interactions with solute, more energy is required to separate them from the surface and fewer vaporize at a given temperature.

In contrast, if solvent–solute interactions are weaker than solvent–solvent interactions, less energy is required to separate solvent molecules from the surface, and more solvent molecules vaporize. In this case the actual vapor pressure at a given temperature is greater that that predicted by Raoult's law.

Departures from ideal behavior in Raoult's law and other expressions of colligative properties occur. The definition of colligative properties states that the identity of the solute particle does not matter, but that is not always the case if intermolecular forces intervene. We will treat solutions for the most part as ideal systems, but you should be aware that deviations from ideal behavior exist in the liquid phase just as they do in the gas phase.

CONNECTION We first defined mole fraction during our discussion of partial pressure of gases in Chapter 6. Remember that the sum of all the mole fractions of substances in a mixture equals 1.

▶❚❚ **CHEMTOUR** Raoult's Law

Raoult's law states that the vapor pressure of a solution containing nonvolatile solutes is proportional to the mole fraction of the solvent.

Colligative properties of solutions depend on the concentration and not the identity of particles dissolved in the solvent.

An **ideal solution** is one that obeys Raoult's law.

SAMPLE EXERCISE 10.9   **Calculating the Vapor Pressure of a Solution**

The fluid used in automobile cooling systems is prepared by dissolving ethylene glycol ($HOCH_2CH_2OH$, molar mass 62.07 g/mol) in water. What is the vapor pressure of a solution prepared by mixing 1.00 L of ethylene glycol (density = 1.114 g/mL) with 1.00 L of water (density = 1.000 g/mL) at the normal boiling point of the solvent? Assume that the mixture obeys Raoult's law and that ethylene glycol is nonvolatile.

COLLECT AND ORGANIZE We are asked to calculate the vapor pressure of a solution of ethylene glycol in water. We are told to assume that ethylene glycol is a nonvolatile solute and that the resultant solution obeys Raoult's law, which means it behaves ideally.

ANALYZE Raoult's law defines the vapor pressures of solutions by considering the mole fraction of solute particles present. We need to calculate the mole fraction of water (X_{water}) in the solution. This requires the conversion of the volumes given into equivalent numbers of moles. Then we can calculate the mole fraction of water and the vapor pressure of the solution with Raoult's law: $P_{solution} = X_{solvent}P_{solvent}$.

SOLVE The number of moles of solute and solvent in 1.00 L of each can be calculated from their densities and molar masses.

For ethylene glycol: $\dfrac{1.114 \text{ g}}{1 \text{ mL}} \times 1000 \text{ mL} \times \dfrac{1 \text{ mol}}{62.07 \text{ g}} = 17.95 \text{ mol}$

For water: $\dfrac{1.000 \text{ g}}{1 \text{ mL}} \times 1000 \text{ mL} \times \dfrac{1 \text{ mol}}{18.02 \text{ g}} = 55.49 \text{ mol}$

Calculating the mole fraction of water yields

$$X_{water} = \frac{(\text{moles of water})}{(\text{moles of water} + \text{moles of ethylene glycol})} = \frac{55.49 \text{ mol}}{(55.49 \text{ mol} + 17.95 \text{ mol})}$$

$$= 0.756$$

Because the solution is at its normal boiling point and ethylene glycol is considered nonvolatile for the purposes of our calculation, the vapor pressure of the solution is due only to the solvent, and $P_{solvent} = P_{H_2O} = 1.00$ atm. Using this value and the calculated mole fraction of water in the mixture yields

$$P_{solution} = X_{H_2O} \times P_{H_2O} = (0.756)(1.00 \text{ atm}) = 0.756 \text{ atm}$$

THINK ABOUT IT The presence of a nonvolatile solute causes the vapor pressure of the solution to be less than 1 atm in this example. That means the solution does not boil at 100°C, but boils at a higher temperature. That is what we should expect for a solution used in a car radiator.

Practice Exercise Glycerol ($HOCH_2CHOH)CH_2O_2$ is a nonvolatile water-soluble material. Its density is 1.25 g/mL. Predict the vapor pressure of a solution of 275 mL of glycerol in 375 mL of water at the normal boiling point of water.

A **phase diagram** is a graphic representation of the dependence of the stabilities of the physical states of a substance on temperature and pressure.

The **melting point** is the temperature at which a solid transforms into a liquid at the same rate at which the liquid transforms into the solid. The melting point of a substance is identical to its **freezing point**.

Sublimation is the direct conversion of a solid to a gas without an intermediate liquid phase.

Deposition is the direct conversion of a gas into a solid without an intermediate liquid phase.

In Sample Exercise 10.9 we calculated that a solution of ethylene glycol and water does not have a vapor pressure of 1 atm at 100°C, the normal boiling point of pure water. The solution's vapor pressure is lower than 1 atm at that temperature, so for the solution to boil, it must be taken to a temperature higher than the solvent's boiling point and where the solution's vapor pressure is 1 atm. A temperature higher than the solvent's boiling point is necessary for the vapor pressure of this antifreeze solution to be greater than 0.756 atm. This is exactly how a coolant works in an automobile engine. It raises the temperature at which water boils, making it possible for the water to remove more heat from the engine.

10.7 Phase Diagrams: Intermolecular Forces at Work

The strengths of forces between particles and the mass of particles control whether a substance exists as a solid, liquid, or gas at room temperature. With increasing temperature, a substance is more likely to be a liquid or gas and less likely to be a solid. However, temperature is not the only factor that influences the state of a substance. Pressure also plays a role. Scientists use **phase diagrams** such as the one for water in Figure 10.17 to show which physical states are stable at various combinations of temperature and pressure.

Physical States and Phase Transformations

The phase diagram of water, like those of many pure substances, has three lines separating regions corresponding to the three states of matter. The line separating the solid and liquid states is a series of **freezing points** or **melting points**; the points on these lines are combinations of temperature and pressure at which the solid and liquid states are in equilibrium with each other, meaning melting and freezing proceed at the same rate and no net change occurs. These lines are sometimes called *equilibrium lines*. The equilibrium line separating liquid and gaseous states is a series of boiling points or liquefaction points at different pressures, and the equilibrium line separating the solid and gaseous states represents a series of sublimation points or deposition points. **Sublimation** is the direct conversion of a solid into a gas without an intermediate liquid phase. The reverse process is **deposition**, the direct conversion of a gas into a solid without an intermediate liquid phase.

Notice that the red line representing boiling points in Figure 10.17 curves from the lower left to the upper right. A liquid boils when its vapor pressure equals atmospheric pressure. If atmospheric pressure increases, then the temperature required to overcome it must increase. The shape of the solid–gas curve also makes sense: higher atmospheric pressures make it more difficult for molecules of ice to leave the solid phase and become water vapor.

The shapes of the boiling point (red line) and sublimation curves (green line) in Figure 10.17 can also be explained by the much lower density of water vapor (and gases, in general) compared to liquid water or solid ice. Liquids and solids are hundreds to thousands of times more dense than the corresponding gases. Applying pressure to a gas forces its molecules closer together. At some point the pressure may change the gas into a liquid or solid that takes up much less volume.

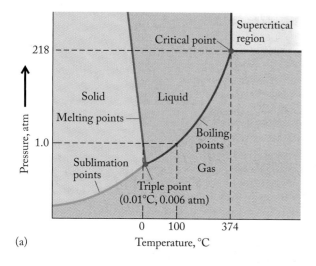

(a)

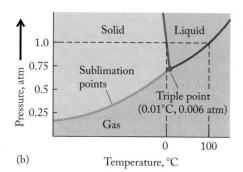

(b)

FIGURE 10.17 (a) The phase diagram for water indicates which phase is stable at various combinations of pressure and temperature. (b) An expanded view of the phase diagram at low pressure and temperature.

Why do directions for cooking pasta call for longer cooking times at high altitudes?

The blue line representing melting points in Figure 10.17 slopes in the opposite direction to the line representing boiling points. It extends from lower right to upper left, which means that the temperature at which ice melts decreases as pressure increases. Another way of looking at this is that, at a given temperature, you can melt ice by increasing the pressure. This trend is opposite to what is observed for almost all other substances (see, for example, Figure 10.18).

For water, the unusual dependence of melting point on pressure arises because water expands when it freezes. Most other substances contract when they freeze because most substances are more dense in their solid phase than in their liquid phase. Water expands because hydrogen bonds between molecules of water in the solid phase create a more open structure than that found in liquid water, making ice less dense. Applying enough pressure to ice forces it into a physical state (liquid water) in which it takes up less volume. When you pack snow to make snowballs, this can happen. A small amount of snow melts under pressure, and then refreezes to make a solid snowball.

A point of interest on a phase diagram is the point where all three lines describing the phase transitions meet. Known as the **triple point**, it identifies the temperature and pressure at which all three states (liquid, solid, and vapor) coexist. For water, the triple point is just above the normal melting point at 0.010°C, but at a very low pressure: 0.0060 atm (4.6 torr).

Another point of interest is the place where the curve separating the liquid and gaseous states ends at specific values of temperature and pressure. At the **critical point**, the liquid and gaseous states are indistinguishable. This point is reached because thermal expansion at these high temperatures causes the liquid form to become less dense, while higher pressures mean that the gas phase is compressed into a small volume, making it more dense. At the critical point the densities of the liquid and gaseous states are equal so one cannot be distinguished from the other and the system consists of one phase.

Above its critical temperature and critical pressure, a substance exists as a **supercritical fluid**. Supercritical fluids have properties different from those of either gases or liquids under standard conditions. They can penetrate like a gas while at the same time they can dissolve material like a liquid. Supercritical carbon dioxide is used in the food-processing industry to decaffeinate coffee and remove fat from potato chips. Its dissolving power is similar to that of nonpolar hydrocarbons like hexane (C_6H_{14}). Supercritical carbon dioxide and water are sometimes mixed to generate a range of fluids able to discriminate among solutes and selectively dissolve materials of interest from a sample while leaving other components untouched.

Reading a Phase Diagram

The phase diagram of CO_2 is shown in Figure 10.18. Note that the triple point, where all three phases coexist, is at −57°C and 5.1 atm. The dashed line at P = 1.0 atm defines the phases that exist at 1 atm pressure, and its location indicates

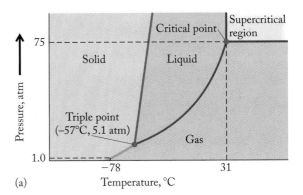

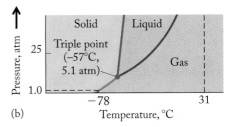

FIGURE 10.18 (a) The phase diagram for carbon dioxide. (b) Expanded view of phase diagram at low pressure.

▶❚❚ **CHEMTOUR** Phase Diagrams

The **triple point** defines the temperature and pressure where all three phases of a substance coexist. Freezing and melting, boiling and liquefaction, and sublimation and deposition all proceed at the same rate, so no net change takes place in the system.

The **critical point** is that specific temperature and pressure at which the liquid and gas phases of a substance have the same density and are indistinguishable from each other.

A **supercritical fluid** is a substance at conditions above its critical temperature and pressure, where the liquid and vapor phases are indistinguishable. It has some characteristics of both a liquid and a gas.

that solid dry ice does not melt and form a liquid phase at normal temperatures. Rather, it sublimes directly to CO_2 gas. This behavior gives rise to its common name, *dry ice*, because it produces no liquid phase as it warms under ordinary pressures. Liquid CO_2 does not exist except at very high pressures. The critical point of CO_2 is at 31.1°C and 73.0 atm, a pressure easily achieved with compressors in laboratories, factories, and food-processing plants, which means it is readily available for use as a supercritical fluid.

SAMPLE EXERCISE 10.10   **Reading a Phase Diagram**

Describe the phase changes that take place as the pressure on a quantity of water at 0°C is increased from 0.001 atm to 100 atm.

COLLECT AND ORGANIZE We are asked about the phase changes water undergoes at one temperature as the pressure is increased. We need to consult the phase diagram for water (Figure 10.17) to address this request.

ANALYZE The changes of interest are the changes in phase as one moves along a line corresponding to a temperature of 0°C and running parallel to the pressure axis.

SOLVE Water at 0°C and 0.001 atm is a vapor, as indicated by the phase diagram. This makes sense from a molecular point of view because low pressures favor the least dense phase. In the phase diagram, the vertical line corresponding to pressure at 0°C indicates that the vapor solidifies to ice at about 0.005 atm. The transformation of a vapor directly into a solid with no intervening liquid phase is deposition. Increasing the pressure further to 1 atm causes the ice to melt and produce liquid water. Water remains a liquid at pressures up to and beyond 100 atm at 0°C.

THINK ABOUT IT To read a phase diagram, we need to remember that every point is characterized by a temperature and a pressure, and that every time we cross an equilibrium line, the phase changes.

Practice Exercise Describe the phase change that occurs when the temperature of CO_2 is increased from −100°C to 200°C at a pressure of 25 atm.

CONCEPT TEST

Look at the phase diagram in Figure 10.18. Does solid CO_2 float on liquid CO_2?

▶❚❚ **CHEMTOUR** Hydrogen Bonding in Water

10.8 The Remarkable Behavior of Water and Properties of Liquids

Water has many remarkable properties. Its melting and boiling points are much higher than those of all other molecular substances with similar molar masses (Table 10.6). Ice is less dense than liquid water so ice floats on water. And water has a high surface tension, so some insects that are more dense than water, such as water striders, can walk on it. Water can rise 100 meters and more through the trunks of tall trees. All these phenomena are related to the strength of the hydrogen bonds that attract water molecules to one another.

TABLE 10.6		Melting Points and Boiling Points of Common Substances		
	Substance	Molar Mass	Melting Point	Boiling Point
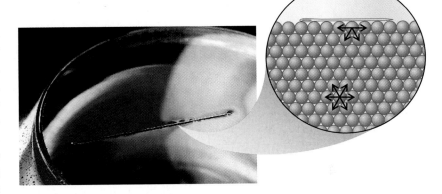	H_2O	18.02 g/mol	0°C	100°C
	HF	20.01 g/mol	−83°C	19.5°C
	NH_3	17.03 g/mol	−78°C	−33°C
	CH_4	16.04 g/mol	−182°C	−164°C

The **surface tension** of a liquid is the energy needed to separate the molecules in a unit area at the surface of a liquid.

An **interface** is a boundary between two phases.

The **meniscus** is the curved surface of a liquid.

Surface Tension and Viscosity

The strength of the hydrogen bonds between molecules at the surface of liquid water accounts for its high **surface tension**, which is the resistance of a liquid to an increase in its surface area. Surface tension represents the energy required to move molecules apart so that an object can break through the surface. Hydrogen bonding creates such a high surface tension for water, 7.29×10^{22} J/m² at 25°C, that a carefully placed steel needle can float on the surface (Figure 10.19). The same needle would sink in oil or gasoline.

FIGURE 10.19 Intermolecular forces, such as hydrogen bonding, are the same in all directions in the interior of a liquid. At the surface, however, they are not equal in all directions. If the magnitude of the surface tension exceeds the downward force due to gravity exerted by a small object such as a needle, the object floats.

CONCEPT TEST

A cold needle floats on water, but a hot needle sinks. Can you explain why?

Another illustration of the forces in play at the surface of a liquid is provided by the behavior of liquids at an **interface**, a boundary between two phases (Figure 10.20). Water in a graduated cylinder or pipette forms a concave surface called a **meniscus** at the interface between the liquid and the glass. Liquid mercury, in contrast, forms a convex surface at the interface. The curved surface is the result of competing *cohesive* forces: hydrogen bonds between water molecules and dispersion forces between mercury atoms; and *adhesive* forces, dipole–dipole interactions between water molecules and polar Si—O—H groups on the glass surface, and between induced dipoles in the mercury atoms and the polar Si—O—H groups on the glass surface. The adhesive forces between water and the glass are strong enough to cause the water to climb upward. The forces between mercury atoms and the glass are much weaker, so mercury atoms are more attracted to each other. Water molecules on the surface and next to the container wall have

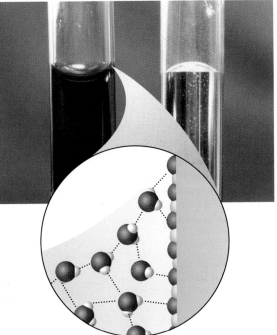

FIGURE 10.20 What a Chemist Sees Hydrogen bonds (dotted lines) between water molecules and oxygen atoms on the surface of silicon dioxide are a cohesive force that causes the molecules to adhere to the glass at the interface and form a concave meniscus. Atoms of mercury have no such attraction for the silicon dioxide surface and experience a more favorable attractive force for each other. Mercury forms a convex meniscus.

FIGURE 10.21 Water rises in a thin glass capillary tube and also in a stick of celery because of capillary action.

▶️II **CHEMTOUR** Capillary Action

FIGURE 10.22 High viscosity results from strong intermolecular forces between large polar molecules. The red fluid is more viscous than the green, which pours freely.

Capillary action is the rise of a liquid up a narrow tube as a result of adhesive forces between the liquid and the tube and cohesive forces within the liquid.

Viscosity is a measure of the resistance to flow of a fluid.

less contact with other water molecules and are drawn upward by adhesive forces. Interactions between mercury atoms are stronger than those between mercury atoms and the wall, so mercury atoms on the surface are more attracted to each other and less attracted to the container wall. As a result they do not adhere to the wall and form a convex surface.

This behavior in water has consequences in other situations. In very narrow tubes such as capillaries, adhesion to the capillary walls draws the outer layer of water molecules upward as shown in Figure 10.21. Water molecules adjacent to those attracted to the capillary walls are also drawn upward by **capillary action** as a result of cohesive interactions with the outer layer of molecules. The maximum height of the column is reached when the force of gravity balances the adhesive and cohesive forces. When a doctor takes a sample of your blood after a pin prick in your finger, the blood (essentially an aqueous solution) spontaneously moves into a capillary tube because of capillary action.

CONCEPT TEST

Do you expect that mercury will be spontaneously drawn into a glass capillary?

The **viscosity**, or resistance to flow, of liquids is another property related to the strength of intermolecular forces (Figure 10.22). Among nonpolar compounds, for example, petroleum products like the fuels pentane and hexane, viscosity increases with increasing molar mass. Lubricating oil (the average molecule of which contains 18 carbon atoms) has a higher viscosity than gasoline (the average molecule contains 9 carbon atoms) because of stronger dispersion forces among the larger molecules in lubricating oil. Stronger interactions mean that the molecules do not slide past one another as easily as the shorter chain hydrocarbons in gasoline, and so bulk quantities of the larger molecules do not flow as easily when poured. On the other hand, water is more viscous than gasoline, even though water molecules are much smaller than the nonpolar compounds in gasoline. The remarkable viscosity of water is another property directly related to the hydrogen bonds between water molecules.

Is the viscosity of seawater greater than that of distilled water? Why or why not?

The density of liquid water increases as it is cooled to 4°C, as shown in Figure 10.23. This pattern is observed for most liquids and solids and for all gases. However, as water is cooled from 4°C to 0°C, it expands and its density decreases. As water freezes, its density drops even more, to about 0.92 g/mL for ice, causing ice to float on liquid water (Figure 10.24). This unusual behavior is caused by the formation of a network of hydrogen bonds in ice. With each oxygen atom covalently bonded to two hydrogen atoms and hydrogen-bonded to two others, molecules of H_2O form an extensive and open hexagonal network in ice. Because of the space between the molecules in the solid network, the same number of molecules occupies more volume in ice than in liquid water. When ice melts, some of the hydrogen bonds in the rigid array break, allowing the molecules in the liquid to be arranged more compactly.

Water and Aquatic Life

The expanded structure of ice plays a crucial ecological role in temperate and polar climates. The lower density of ice means that lakes, rivers, and polar oceans freeze from the top down, allowing fish and other aquatic life to survive in the liquid waters below. Each fall, surface waters cool first, and their density increases. The denser, colder water sinks to the bottom, bringing warmer water from the bottom to the surface. This autumnal turnover of the water stirs up dissolved nutrients, including nitrates and phosphates in bottom waters, making them available for life in the sunlit surface during the next growing season. When all of the water has reached 4°C, the surface water begins to cool further, and ice eventually forms. The layer of ice effectively insulates the 4°C water beneath it in lakes and deep ponds from colder winter temperatures, allowing aquatic life to go on.

In spring, the ice melts and the surface water warms to 4°C. At this temperature, the entire column of water has nearly the same temperature and density; dissolved nutrients for plant growth are evenly distributed, and the stage is set for a burst of photosynthesis and biological activity called the spring bloom.

Further warming of the surface waters during the summer months creates a warm upper layer separated from colder, denser deep water by a *thermocline*, a sharp change in temperature between the two layers. Biological activity depletes the pool of nutrients above the thermocline as decaying biomass settles into the bottom layer. Consequently, photosynthetic activity drops from its spring maximum during the summer, even though there is much more energy available from the sun. The thermocline persists until the autumn turnover mixes the water column and the cycle begins anew.

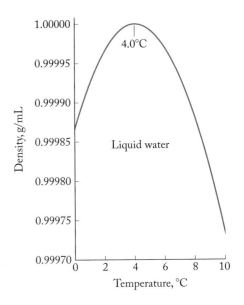

FIGURE 10.23 As water is cooled, its density increases to reach a maximum value at 4°C; as water further cools to its freezing point at 0°C, the density decreases.

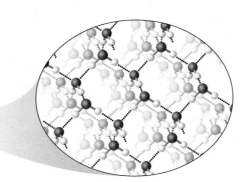

FIGURE 10.24 Because of the changes in water density near the freezing point and because ice has a lower density (0.92 g/mL) than water, ice and ice-cold water float on top of warmer water (4°C) in lakes and rivers. In ice, each oxygen atom is in a rigid tetrahedral environment formed by two O—H covalent bonds and two O···H hydrogen bonds.

10.9 Colligative Properties of Solutions

We have learned in previous sections that adding a nonvolatile solute such as NaCl to a pure solvent, such as water, changes the physical properties of the solvent. Solutions generally have greater densities than the solvent alone. In addition, other, less obvious changes also take place. As we noted for solutions of ethylene glycol in water (Sample Exercise 10.9), dissolving a nonvolatile solute in a solvent decreases the vapor pressure of the solvent. Water containing ethylene glycol boils at a temperature above 100°C. The solution also has a lower freezing point than pure water. The phase diagrams of water and an aqueous solution of a nonvolatile solute are shown in Figure 10.25. They indicate a lower vapor pressure, lower freezing point, and higher boiling point for the solution than for the solvent. Boiling point elevation and melting point depression are both colligative properties. For colligative properties, only the number of particles in solution determines the impact of the solute on the properties of the solvent.

Boiling Point Elevation and Freezing Point Depression

The sea is the source of drinking water in several desert countries bordering the Persian Gulf. Seawater must be desalinated before it is suitable to drink, and one way to desalinate saltwater is to distill it. It takes a great deal of energy to heat seawater to its boiling point and convert it to steam, and seawater actually boils at a slightly higher temperature than pure water does. The boiling point of seawater is higher than that of pure water for the same reason (discussed in Section 10.6) that the vapor pressure of a solution containing a nonvolatile solute at a given temperature is lower than that of the pure solvent, so the solvent boils above its normal boiling point (Figures 10.15 and 10.25). The greater the concentration of solute, the higher the boiling point of the solution. The boiling point of the solution depends on the concentration of dissolved particles, be they molecules or ions, and not the identity of the particles.

This dependence on concentration alone makes *boiling point elevation* a colligative property of the solvent. It is described in equation form as

$$\Delta T_b = K_b m \tag{10.5}$$

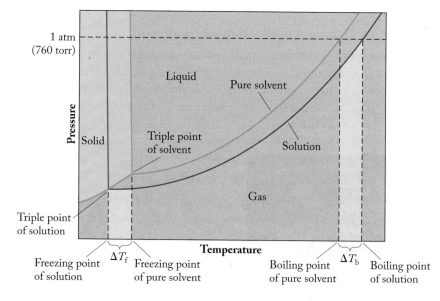

FIGURE 10.25 This graph shows the phase diagram for pure water (light-colored lines) and the phase diagram for a solution of a nonvolatile solvent in water (dark-colored lines). Notice the higher boiling point and lower freezing point of the solution compared to the pure solvent.

where ΔT_b is the increase in boiling point, K_b is the boiling-point-elevation constant of the solvent (0.52°C/m for water), and m is a new concentration unit called **molality**, which is the number of moles of solute per kilogram of solvent. In equation form,

$$m = \frac{n_{solute}}{\text{kg solvent}}$$ (10.6)

where n is the number of moles of solute and the mass of the solvent is 1 kg. Molality (m) must not be confused with molarity (M). The difference in the two terms is that molarity is the number of moles of solute *per liter of solution*, whereas molality is the number of moles of solute *per kilogram of solvent*. Figure 10.26 summarizes the calculation of molality. Molality is used in this case because mass does not change with temperature but volume does. Therefore a concentration expressed in molality does not change with temperature, but one expressed in molarity might if the solution swells or contracts as temperature rises or falls. Molality is the concentration unit of choice when experiments involve solutions and temperature changes.

To examine the difference between the molarity and molality of solutions, let's calculate the molality of an aqueous solution of sodium chloride that is 0.558 M in NaCl. This approximates the concentration of sodium chloride in seawater. To calculate molality, we need the density of the solution so that we can convert liters of solution into kilograms of solvent. The density of the solution is 1.022 g/mL at 25°C, so 1 L of solution has a mass of

$$\frac{1.022}{\text{mL}} \text{ g} \times \frac{1000 \text{ mL}}{1 \text{ L}} = 1022 \text{ g/L}$$

Of this 1022 g, the mass of dissolved NaCl is

$$\frac{0.558 \text{ mol NaCl}}{1 \text{ L solution}} \times \frac{58.44 \text{ g NaCl}}{1 \text{ mol NaCl}} = 32.6 \text{ g NaCl in 1 L of solution}$$

If 32.6 g of the solution is due to the dissolved NaCl, then the amount of water in the solution is (1022 g − 32.6 g) = 989 g = 0.989 kg. From this value, we can now calculate the molality (the number of moles of solute in 1 kg of solution):

$$\frac{0.558 \text{ mol NaCl}}{0.989 \text{ kg solvent}} = 0.564 \text{ } m$$

The two concentration terms are close, but the molality is slightly higher than the molarity. Remember the reason we use molality: volumes change in response to temperature, whereas masses do not. Therefore concentrations expressed in molarity are only accurate at a given temperature. When we need to know the number of particles in a solution in situations when temperatures change, as they do when freezing points or boiling points are determined, we have to use a concentration term that does not change over the course of the experiment. Molality is the concentration unit of choice in such situations.

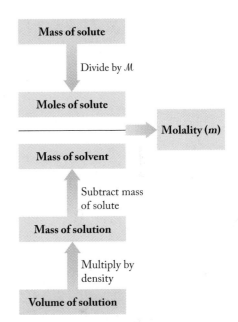

FIGURE 10.26 This flow diagram illustrates the calculation of molality.

CONCEPT TEST

The differences between the molar and molal concentration values of dilute aqueous solutions are small. Why?

SAMPLE EXERCISE 10.11   **Calculating Molal Concentrations**

How many grams of Na_2SO_4 should be added to 275 mL of water to prepare a 0.750 *m* solution of Na_2SO_4? Assume the density of water is 1.000 g/mL.

COLLECT AND ORGANIZE We are asked to determine the mass of a solute, sodium sulfate, needed to prepare a solution with a concentration of 0.750 *m*. Molality is defined in Equation 10.6 as moles of solute per kilogram of solvent. The volume of water, the solvent, is 275 mL.

ANALYZE We can use the density of water to convert 275 mL of water to kilograms. We can calculate the number of moles of Na_2SO_4 needed and convert that number of moles into grams using the molar mass of Na_2SO_4.

SOLVE We can convert 275 mL of water to kilograms of water:

$$275 \text{ mL water} \times \frac{1 \text{ kg water}}{1000 \text{ mL water}} = 0.275 \text{ kg water}$$

We can rearrange Equation 10.6 to determine the number of moles of solute needed:

$$molality = \frac{\text{mol solute}}{\text{kg solvent}}$$

$$\text{mol solute} = molality \times \text{kg of solvent}$$

Inserting the values in the exercise into this equation gives

$$\text{mol solute} = \frac{0.750 \text{ mol } Na_2SO_4}{\text{kg water}} \times 0.275 \text{ kg water} = 0.206 \text{ mol } Na_2SO_4$$

The molar mass of Na_2SO_4 is 142.04 g per mole, and the number of grams of Na_2SO_4 needed is

$$0.206 \text{ mol } Na_2SO_4 \times \frac{142.04 \text{ g}}{1 \text{ mol}} = 29.3 \text{ g } Na_2SO_4$$

THINK ABOUT IT Calculations of molality always involve considerations of the mass of solvent. Remember that molality is the concentration unit of choice for systems when the temperature changes, because mass does not vary with temperature.

Practice Exercise What is the molality of a solution prepared by dissolving 78.2 g of ethylene glycol, $HOCH_2CH_2OH$, in 1.50 L of water? Assume the density of water is 1.00 g/mL.

The temperature at which seawater boils at 1 atm pressure is higher than 100°C because of the presence of nonvolatile salts in the water. The *boiling-point-elevation* constant K_b for water in Equation 10.5 specifies that, for every mole of particles in 1 kg of seawater, the boiling point of the solution rises 0.52°C.

SAMPLE EXERCISE 10.12   **Predicting the Boiling Point of a Solution**

What is the increase in the boiling point of seawater compared to pure water if the total concentration of ions in seawater is 1.14 *m*? The boiling point elevation constant of water is $K_b = 0.52°C/m$.

COLLECT AND ORGANIZE We are asked to calculate the increase in the boiling point of an aqueous solution that has a total concentration of ions of 1.14 *m*. We are given the boiling-point-elevation constant, and we have Equation 10.5.

ANALYZE Equation 10.5 relates boiling point elevation to concentration:

$$\Delta T_b = K_b m$$

SOLVE Inserting the given values for K_b and m, we have

$$\Delta T_b = \frac{0.52°C}{m} \times 1.14\ m$$

$$= 0.59°C$$

We predict the seawater would boil 0.59°C higher than normal.

THINK ABOUT IT The boiling point of a solution is always higher than the boiling point of the pure solvent. At 1 atmosphere pressure, the boiling point of the seawater is predicted to be (100.00 + 0.59)°C = 100.59°C. Remember that the value of K_b depends only on the identity of the solvent.

Practice Exercise When crude oil is pumped out of the ground it may be accompanied by "formation water" that contains high concentrations of NaCl. If the boiling point of a sample of formation water is 2.3°C above the boiling point for pure water, what is the molality of particles in the sample?

Not only does adding a solute to water raise its boiling point, but it also lowers its freezing point. The *freezing point depression* is a colligative property that is put to good use in car radiators. Antifreeze is added to the water in a car's radiator to depress its freezing point and ensure that it does not freeze in winter time in cold climates. The magnitude of freezing point depression is directly proportional to the molal concentration of dissolved solute. The equation describing this dependence

$$\Delta T_f = K_f m \tag{10.7}$$

is very similar to Equation 10.5. The term ΔT_f is the change in the freezing temperature of the solvent; K_f is the freezing-point-depression constant for the solvent; and *m* is the concentration of solute expressed in molality. The chart in Figure 10.27 summarizes calculations of freezing point depression and boiling point elevation.

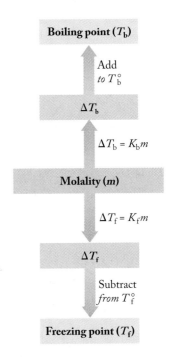

FIGURE 10.27 This flow chart summarizes how to calculate freezing points and boiling points for solutions of known molality ($T_b°$ = normal boiling point; $T_f°$ = normal freezing point).

SAMPLE EXERCISE 10.13 **Calculating the Freezing Point of a Solution**

Ethylene glycol ($HOCH_2CH_2OH$) is a common additive to the water in a car's radiator. In Sample Exercise 10.9 we have already predicted ethylene glycol's influence on the vapor pressure and hence on the boiling point of water in a radiator. What is the freezing point of radiator fluid prepared by mixing 1.00 L of ethylene glycol with 1.00 L of water? The density of ethylene glycol is 1.114 g/mL; the density of water is 1.000 g/mL; and the freezing-point-depression constant of water, K_f, is 1.86°C per molal unit.

COLLECT AND ORGANIZE We are asked to determine the freezing point of a solution of ethylene glycol in water. We know the volume of the components of the solution; we are given the densities of ethylene glycol and water; we know the

$HOCH_2CH_2OH$

Ethylene glycol

freezing point depression constant for the solvent, and we have Equation 10.7. From the formula of ethylene glycol we can calculate its molar mass.

ANALYZE To determine the freezing point, we need to calculate ΔT_f for the solution. This requires converting the quantity of solute given into moles and the mass of the solvent into kilograms so we can calculate molality.

SOLVE For the solvent, we have 1 L water, which is 1000 mL; the density is 1.000 g/mL, so the mass of solvent is 1 kg. For the solute

$$1 \text{ L solute} \times \frac{1000 \text{ mL}}{1 \text{ L}} \times \frac{1.114 \text{ g}}{1 \text{ mL}} \times \frac{1 \text{ mol}}{62.07 \text{ g}} = 17.95 \text{ mol solute}$$

Therefore the molal concentration

$$m = \frac{17.95 \text{ mol solute}}{1 \text{ kg solvent}} = 17.95 \; m$$

Using Equation 10.8

$$\Delta T_f = K_f m$$

$$= \frac{1.86°C}{m} \times 17.95 \; m = 33.4°C$$

Applying this change to the normal freezing point of water, 0.0°C, we have

$$\text{Freezing point of radiator fluid} = 0.0°C - 33.4°C = -33.4°C$$

THINK ABOUT IT The answer makes sense because the reason we add antifreeze to radiators is to depress the freezing point of water, and the solution we evaluated here certainly has a freezing point lower than that of pure water. Remember that the value for K_f depends only on the solvent, not the solute.

Practice Exercise What is the boiling point of the automobile radiator fluid in the preceding Sample Exercise? The K_b of water is 0.52°C/m.

The van't Hoff Factor

Experiments show that the changes in melting point and boiling point depend directly on the number of moles of particles dissolved in a given quantity of solvent and not on the identity of the solute. Think about what this means in terms of the types of substances that are water soluble. The dissolution of 1 mole of a salt such as NaCl in a given quantity of solvent produces the same changes in freezing point and boiling point as 1 mole of KNO_3, even though the latter has a much higher formula mass. Both of these salts add 2 moles of particles (1 mole of cations and 1 mole of anions) to the solvent for every 1 mole of salt that dissolves. If a nonvolatile molecular substance like the ethylene glycol used in several of the Sample Exercises is dissolved in water, 1 mole of a molecular solute produces 1 mole of particles (1 mole of molecules) in the solution. To produce the same changes in boiling point and freezing point in 1 kg of solvent that are produced by 1 mole of NaCl, one would have to dissolve 2 moles of ethylene glycol. Ethylene glycol does not dissociate but remains intact in solution, so 2 moles of ethylene glycol are needed to add 2 moles of particles to the solvent.

The Dutch chemist Jacobus van't Hoff (1852–1911) studied colligative properties and defined a term i, now called the **van't Hoff factor** (or just the i **factor**), that is a ratio of the measured value of a colligative property to the value expected for that property if the solute were molecular. For example, suppose we mixed a solution that was 0.100 m in NaCl by dissolving 5.844 g of NaCl in 1 kg of water.

▶❙❙ **CHEMTOUR** Boiling and Freezing Points

The **van't Hoff factor** (or i **factor**) is a ratio of the measured value of a colligative property to the value expected for that property if the solute were molecular.

According to Equation 10.7, the freezing point of that solution should be lower than that of pure water by

$$\Delta T_f = K_f m = 0.186°C$$

However, the actual freezing point of the solution measured in the laboratory is $-0.372°C$, or $0.372°C$ lower than the normal freezing point. The ratio of the actual value to the calculated value

$$\frac{\Delta T_{f,\text{measured}}}{K_f m} = \frac{0.372°C}{0.186°C} = 2 = i$$

is the van't Hoff factor for sodium chloride. The significance of the i factor is based on the definition of colligative properties: changes due to the number of particles present in solution. When 1 mole of solid sodium chloride dissolves, it produces 2 moles of ions in solution:

$$NaCl(s) \rightarrow Na^+(aq) + Cl^-(aq)$$

and therefore 2 moles of particles. Thus, we should expect that the freezing point depression (or boiling point elevation) that results from the dissolution of a given number of moles of NaCl should be twice as great as that produced when the same number of moles of a molecular substance dissolves.

Equations 10.5 and 10.7 should be modified to include the i factor:

$$\Delta T_b = iK_b m \tag{10.8}$$

$$\Delta T_f = iK_f m \tag{10.9}$$

If the solute dissolved in the solvent is molecular like ethylene glycol, then $i = 1$. If it is ionic like NaCl, i = the number of ions in one formula unit; for NaCl, $i = 2$. For Na_2SO_4, $i = 3$.

SAMPLE EXERCISE 10.14 **Using the van't Hoff Factor**

The salt lithium perchlorate ($LiClO_4$) is one of the most water-soluble salts known. Predict the temperature at which a 12.0 m solution of $LiClO_4$ would freeze. The K_f of water is 1.86°C/m.

COLLECT AND ORGANIZE We are asked to predict the freezing point of a solution. We know the formula of the solute, its molal concentration, and the K_f of the solvent. We know that the freezing point of the pure solvent is 0.00°C. Equations 10.7 and 10.9 relate freezing point depression to solute concentration.

ANALYZE Lithium perchlorate dissolves in aqueous solution to form 2 moles of ions for every mole of solute that dissolves: 1 mole of Li^+ cations and 1 mole of ClO_4^- anions. Therefore, $i = 2$. We need Equation 10.9 to solve for the freezing point depression.

SOLVE Using Equation 10.9, we calculate:

$$\Delta T_f = iK_f m = (2)(1.86°C/m)(12.0\,m) = 44.6°C$$

The freezing point of the solution is $(0.00 - 44.6)°C = -44.6°C$.

THINK ABOUT IT To determine the value of i for a salt, it is necessary that we recognize the polyatomic ions, so we can accurately predict the number of ions formed in solution when a salt dissolves.

Practice Exercise Predict the boiling point of a 1.75 m solution of barium nitrate, $Ba(NO_3)_2$, in water. The K_b of water is 0.52°C/m.

An **ion pair** is a cluster formed when a cation and an anion associate with each other in solution.

CONCEPT TEST

Which aqueous solution has the lowest freezing point: (a) 3 m glucose ($C_6H_{12}O_6$), (b) 2 m potassium iodide (KI), or (c) 1 m sodium sulfate (Na_2SO_4)?

Using the i factor to calculate changes in freezing point depressions or boiling point elevations for concentrated electrolyte solutions often gives values smaller than the experimentally measured values. The positive and negative ions produced when strong electrolytes dissolve may not be totally independent of each other if the solution has a concentration much above 1 m. At high concentrations, positive and negative ions may interact with one another to form ionic clusters. The simplest cluster is called an **ion pair**, which consists of a cation and an anion that associate in solution. An ion pair acts as a single particle. Thus, the overall concentration of particles is reduced when ion pairs form. As a consequence, if ion pairs form in the solution, the freezing point depression and boiling point elevations are smaller than ideally predicted (Figure 10.28).

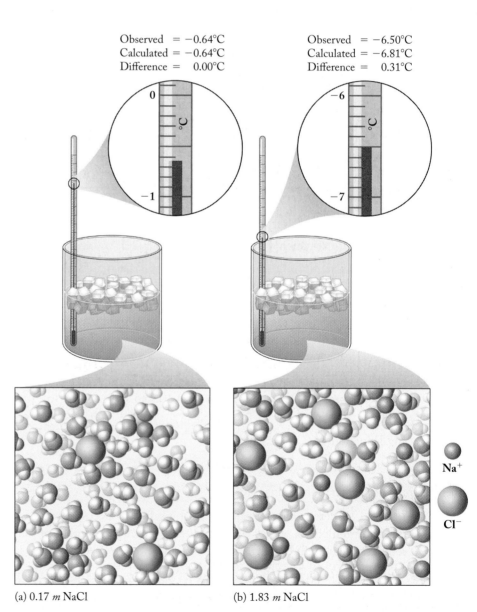

Observed   = −0.64°C
Calculated = −0.64°C
Difference =   0.00°C

Observed   = −6.50°C
Calculated = −6.81°C
Difference =   0.31°C

Na$^+$

Cl$^-$

(a) 0.17 m NaCl

(b) 1.83 m NaCl

FIGURE 10.28 (a) The observed freezing point of a 0.17 m solution of NaCl is very close to the calculated value. (b) In contrast, that for a 1.83 m solution is about 0.3°C higher than calculated because some of the Na$^+$ and Cl$^-$ ions form ion pairs. The formation of ion pairs causes the total concentration of particles in the solution to be less than the ideal number, and the van't Hoff factor for the solution is less than 2.

The extent to which free ions form when an electrolyte dissolves is expressed by the van't Hoff factor. The van't Hoff factor for NaCl in water is 2 if the solution behaves ideally, because ideally there are 2 moles of ions produced in solution for each mole of NaCl dissolved. However, the value of the ratio is actually 1.97 for 0.0100 m NaCl and only 1.87 for 0.100 m NaCl. Figure 10.29 summarizes some calculated and observed values of the van't Hoff factor. When the value of i for a particular solution is calculated, a non-integer value for i indicates that particles are associating in solution and the behavior is nonideal.

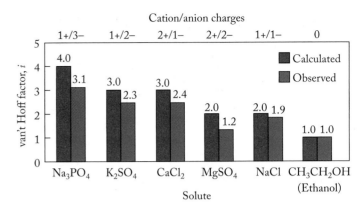

FIGURE 10.29 The vertical bars represent theoretical and experimentally derived values for the van't Hoff factors for 0.1 m solutions of several electrolytes and the nonelectrolyte ethanol. The differences between theoretical and experimental values increase with increasing charge on the ions.

SAMPLE EXERCISE 10.15 **Assessing Particle Interactions in Solution**

The observed freezing point of a 1.83 m aqueous solution of NaCl is −6.50°C. If we assume that the freezing point of pure water is 0.00°C, what is the value of the van't Hoff factor for this solution? Is the solution behaving ideally, or is there evidence that solute particles are interacting? The freezing point depression constant of water is $K_f = 1.86°C/m$.

COLLECT AND ORGANIZE We are asked to calculate the i factor for a solution of known molality. We are given the freezing point of the solution and the K_f value for the solvent. We are also asked whether the solution is behaving ideally. We are given $\Delta T_f = 6.50°C$, $K_f = 1.86°C/m$, and $m = 1.83$.

ANALYZE We can determine the value of i for a solution from the ratio of the observed value of freezing point depression to the ideal value. The ideal value is given by the product of m and K_f of the solvent. If the value is not an integer and the solute is a salt, the solution is not behaving ideally because ion pairs are forming.

SOLVE Solving for i, using Equation 10.9, we take the ratio

$$i = \frac{\Delta T_f}{K_f m}$$

Substitution of the given values results in

$$i = \frac{6.50°C}{\left(\frac{1.86°C}{m}\right) 1.83\ m} = 1.91$$

The value of i is 1.91. We would have predicted $i = 2$ for NaCl, so the solution is not behaving ideally and ion pairs must be forming.

THINK ABOUT IT The ideal value of i is 2 in this case, and, if ion pairing exists, that value would decrease. A slightly smaller value than 2 makes sense as an answer.

Practice Exercise The van't Hoff factor for a 0.050 m solution of magnesium sulfate is 1.3. What is the freezing point of the solution?

The extent of ion pairing in an electrolyte solution is generally predicted to increase with the solute concentration as the solution "runs out of water" to hydrate the ions (see Figure 10.28). The ideal value of i for a salt is an upper limit of possible values. If ion pairing occurs and results in nonideal behavior, then the measured value for i must be smaller than the ideal value.

In **osmosis**, solvent passes through a semipermeable membrane to balance the concentration of solutes in solutions on both sides of the membrane. Solvent flow proceeds from the more dilute solution into the more concentrated one.

Osmosis and Osmotic Pressure

More than 97% of the water on Earth is seawater, and none of it is fit for us to drink. We cannot drink seawater because if we do, the cells in our bodies actually expel water, and dehydration occurs on a massive scale. Drinking seawater does not supply our cells with water; it results in water leaving our cells. The presence of dissolved solutes in seawater is responsible for this behavior. The total concentration of solutes in seawater is at least three times the concentration in blood plasma and in most of the cells in our bodies. This difference in concentration of solutes within our cells compared to seawater results in a process known as osmosis that causes dehydration if we drink seawater.

To get an idea of what osmosis is and how it works, we can study what happens if a human blood sample is mixed with seawater. Upon exposure to seawater, cells suspended in the blood, such as the red blood cells that carry oxygen, collapse (Figure 10.30). They collapse because the water inside them flows from the cells through the cell membrane into the surrounding seawater. This process is **osmosis**. During osmosis, solvent passes through a membrane from a region of low solute concentration into a region of higher solute concentration to balance the concentrations of solutes on both sides. As water flows out of the cells and their volumes decrease, they shrivel and finally cease to function. Why does osmosis occur? One reason is that cell membranes are *semipermeable*, which means that water molecules can pass through them but most solutes cannot. Osmotic flow of solvent proceeds from the more dilute solution into the more concentrated one.

When a cell is surrounded by an aqueous solution with a solute concentration different from that in the cell, water molecules naturally tend to flow through the membrane in an effort to bring the total solute concentrations inside and outside the cell into balance. For a red blood cell bathed in seawater to achieve this balance, nearly three-fourths of the water in it passes through its membrane into the surrounding seawater, causing the cell to shrink. The concentrations of solutes inside it correspondingly increase until they match the concentrations of the salts in the surrounding seawater.

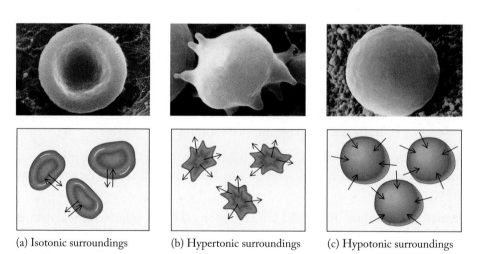

(a) Isotonic surroundings   (b) Hypertonic surroundings   (c) Hypotonic surroundings

FIGURE 10.30 The membranes of a red blood cell are semipermeable. Water easily flows in and out of the cells to equalize the concentration of solutes on both sides. (a) The flow is in balance when the cells are bathed in a solution of NaCl with a concentration equal to the concentration inside the cell (isotonic conditions). (b) When the concentration of solutes is higher (hypertonic conditions), water flows from the cells to the surrounding fluid. (c) When the cells are immersed in pure water (hypotonic conditions), additional water enters and the cells expand.

CONCEPT TEST

Why do cucumbers shrivel when pickled in brine (salt water), and why do wilted flowers revive when placed in fresh water?

Water molecules migrate through the membrane of a blood cell or through any semipermeable membrane because a force makes it happen—a force derived from the different concentrations of solutes on the two sides of the membrane. When we divide the magnitude of this force (F) by the surface area (A) of the membrane, we get pressure (P): $P = F/A$. To stop molecules of water from moving across the membrane, we could apply an opposing pressure equal to F/A. This pressure is called **osmotic pressure** (Figure 10.31). The Greek letter pi (π) is used as the symbol for osmotic pressure to distinguish it from other kinds of pressure exerted by gases that we encountered in Chapter 6.

The magnitude of the osmotic pressure required to stop the net flow of solvent across a membrane separating the pure solvent from a solution is proportional to the concentration of the solution expressed in molarity (M) of solute. Because osmotic pressure is a colligative property, the molarity term is multiplied by the van't Hoff factor for the solute to enable consideration of both molecular and ionic solutes. It is also proportional to absolute temperature (T). We can write an equation connecting these variables by incorporating a constant, R, into the relations. The value of R is 0.0821 L·atm/(mol·K). This is the same constant R that we use in the ideal gas equation. The resulting equation relating osmotic pressure to concentration and temperature is

$$\pi = iMRT \qquad (10.10)$$

The units of R are L·atm / mol·K. This includes the inverse of molarity (L/mol) and the inverse of temperature (1/K). Therefore, multiplying M by $R \times T$ leaves the unit "atm" (pressure in atmospheres) as the unit of π.

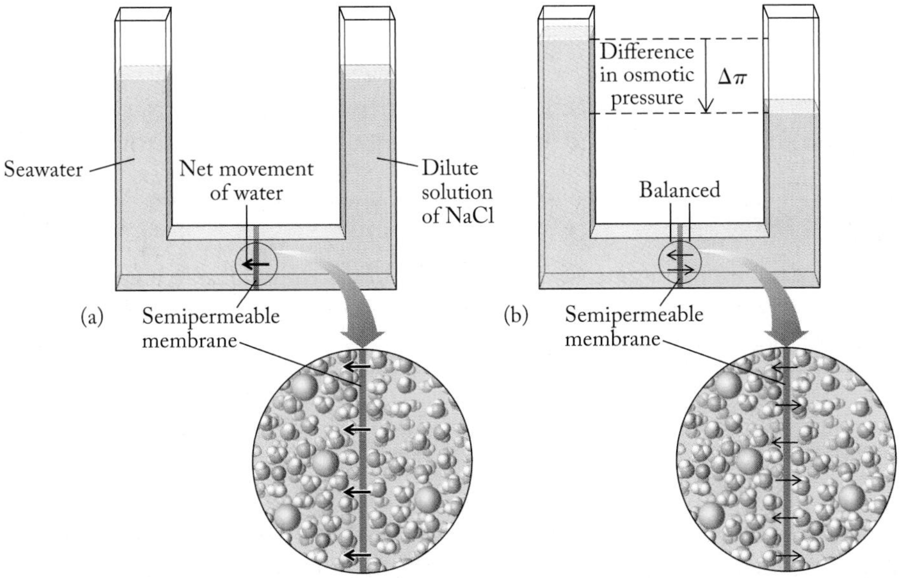

Osmotic pressure (π) is the pressure that has to be applied across a semipermeable membrane to stop the flow of solvent from the compartment containing pure solvent or a less concentrated solution toward a more concentrated solution. The osmotic pressure of a solution increases with the solute concentration M and the solution temperature T.

▶❚❚ **CHEMTOUR** Osmotic Pressure

FIGURE 10.31 If equal volumes of seawater and a dilute solution of NaCl are separated by a semipermeable membrane: (a) a net flow of water from the dilute solution to the seawater is observed; (b) the volume on the side of the dilute solution decreases until it is balanced by the pressure produced by the difference in the heights of the fluid levels. This pressure is the difference in osmotic pressures ($\Delta\pi$) between the two solutions.

SAMPLE EXERCISE 10.16   **Calculating Osmotic Pressure**

Calculate the osmotic pressure across a semipermeable membrane separating pure water from seawater at 25°C. The sum of the concentrations of all the ions in seawater is 1.14 M.

COLLECT AND ORGANIZE We are asked to calculate an osmotic pressure across a membrane that separates pure water from seawater. We are given the temperature and the total concentration of the ions in seawater. We can use Equation 10.10 and the known value of the constant R to solve this problem. The value of iM in the equation is 1.14 M.

ANALYZE Osmosis is the result of the natural tendency of a solution to mix with a more dilute solution when the two solutions are in contact through a semipermeable membrane. Osmotic pressure is related to the total concentration of all particles in solution and the temperature in kelvins.

SOLVE First we need to convert the temperature in Celsius to kelvins: $T(°C) + 273 = T(K)$. Inserting the values of iM, R, and T in Equation 10.10 we have:

$$\pi = iMRT = \frac{1.14 \text{ mol}}{L} \times \frac{0.0821 \text{ L} \cdot \text{atm}}{\text{mol} \cdot \text{K}} \times (25 + 273) \text{ K} = 27.9 \text{ atm}$$

THINK ABOUT IT The pressure across a semipermeable membrane separating seawater from pure water is 27.9 atm, or 28 times normal atmospheric pressure. Similar large pressures can exist across cell membranes, causing the cells to burst or shrink depending on the surrounding conditions (see Figure 10.30).

Practice Exercise Calculate the osmotic pressure across a semipermeable membrane separating seawater (1.14 M total particles) from a solution of normal saline (0.31 M total particles) at 25°C.

Osmotic pressures across membranes can be very large. The pressure of almost 28 atm calculated in Sample Exercise 10.16 is more than ten times the air pressure in a typical automobile tire and over four times the water pressure experienced by a diver at a depth of 82 meters (268 feet). A pressure differential of 28 atm across the walls of a steel-reinforced concrete building is more than sufficient to collapse the building.

CONCEPT TEST

If a solution 1.0 M in glucose is on one side of a semipermeable membrane and a solution that is 1.0 M in total ion concentration is on the other side, in which direction does the water flow?

Osmotic pressure is another colligative property of a solution. Thus a 1.0 M solution of NaCl produces the same osmotic pressure as 1.0 M KCl or 1.0 M $NaNO_3$ because all three solutions are 2.0 M in total ions; the ideal i factor for these solutes is 2. These solutions have twice the osmotic pressure of a 1.0 M solution of glucose because glucose is a molecular substance ($i = 1$) and produces a solution that is only 1.0 M in dissolved particles (glucose molecules).

During a medical emergency, fluids may need to be administered to a patient intravenously (Figure 10.32), and it is crucial that the osmotic pressure of intravenous solutions be identical with that of blood plasma. Such solutions are said to be *isotonic* because they exert the same osmotic pressure. Solutions with higher (*hypertonic*) or lower (*hypotonic*) concentrations cause dehydration or swelling (edema), respectively (see Figure 10.30). Two different solutions may be used for intravenous administration, depending on the clinical situation, but whatever solution is used, it should be isotonic with blood plasma, so the two solutions should both have the same concentration of particles.

Physiological saline is a salt solution containing 0.92% NaCl by weight: 0.92 g NaCl for every 100 g of solution. The density of dilute aqueous solutions is close to 1.00 g/mL, so 100 g of this solution has a volume of 100 mL, and a concentration of 0.92 g/100 g is nearly the same as 0.92 g/100 mL. Scaling up the mass of solute and volume of solution by 10 so that we can compare it to the value of 35 g/L for seawater, we get 9.2 g/L. This value is less than a third the concentration of sea salts in seawater.

The second solution emergency medical technicians may use with trauma patients is a solution called D5W. The acronym stands for 5.5% solution by weight of dextrose (another name for glucose and blood sugar; molar mass = 180.06 g/mol) in water. The use of this solution suggests that it contains the same concentration of solute particles as 9.2 g/L NaCl. Using the same approach we applied to the NaCl solution, we calculate that a 5.5% by weight solution of sugar is equivalent to 55 g dextrose /L.

To compare the solute levels of these two solutions, we can convert each of their concentrations in grams per liter into moles per liter:

$$\text{NaCl solution:} \quad \frac{9.2 \text{ g}}{1 \text{ L}} \times \frac{1 \text{ mol NaCl}}{58.44 \text{ g NaCl}} = 0.16 \text{ } M$$

$$\text{D5W:} \quad \frac{55 \text{ g}}{1 \text{ L}} \times \frac{1 \text{ mol D}}{180.06 \text{ g D}} = 0.31 \text{ } M$$

The molar concentration of dextrose is about twice that of NaCl. How can both match the solute levels in blood? The answer comes again from the definition of a colligative property and the role of the i factor. We must look at the number of particles in solution in both cases. Sodium chloride is a salt with $i = 2$. Dextrose is a sugar, a molecular material, and its molar concentration directly reflects the number of sugar molecules in solution; $i = 1$. The total particle concentration in the NaCl solution is two times the molarity, or about 0.32 M, which is very close to the concentration of particles (molecules) in the dextrose solution of 0.31 M. Therefore, the total solute particle concentration and osmotic pressure in 0.92% NaCl is the same as in 5.5% D5W.

Reverse Osmosis

If an opposing pressure equal to the osmotic pressure can stop osmosis, an even greater opposing pressure can reverse osmotic flow, forcing solvent to flow from a more concentrated solution to a more dilute solution. This concept is the basis of a technique called **reverse osmosis**. Reverse osmosis is a way of purifying salt water. Seawater is retained in a chamber by a semipermeable membrane. Because water molecules can pass through the membrane and most solute particles cannot, if a pressure greater than osmotic pressure is placed on the seawater side, water flows from the high-pressure side of the membrane to the low-pressure side. During this process, the seawater becomes more concentrated, while the water that flows through the membrane has a lower solute concentration (Figure 10.33). Some municipal water-supply systems use reverse osmosis to make saline (brackish) water fit to drink. Some industries use this method to purify conventional tap water, and it is a common way for ships at sea to desalinate ocean water.

FIGURE 10.32 A solution of physiological saline has a concentration of 0.92 g NaCl/L. The concentration of ions in this solution is equal to the concentration of ions inside the cells of the body.

Reverse osmosis is a water purification process in which water is forced through semipermeable membranes, leaving dissolved impurities behind.

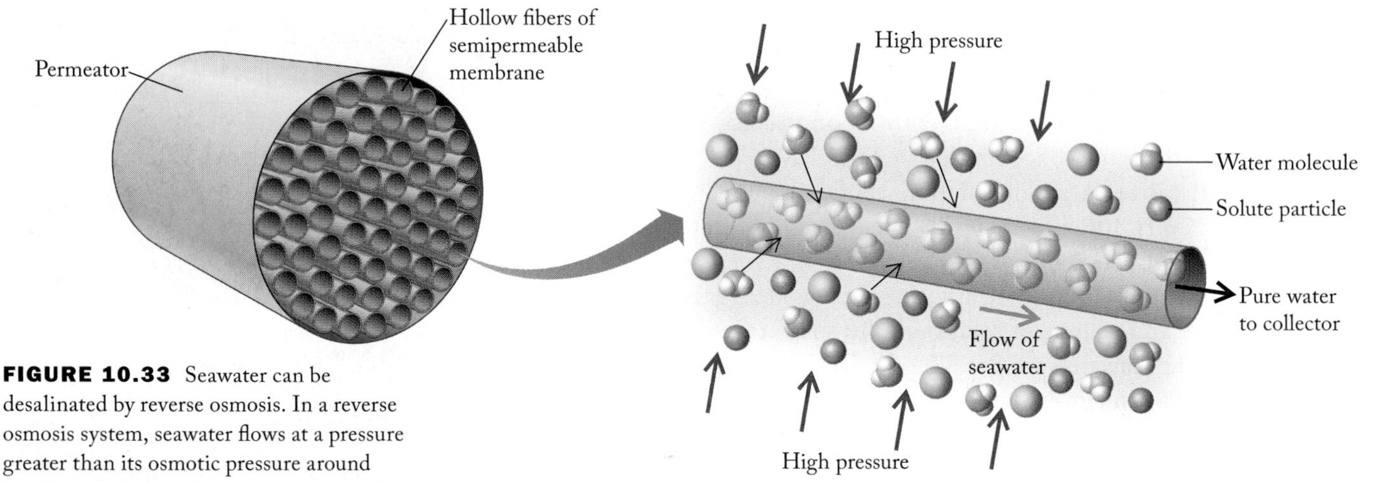

FIGURE 10.33 Seawater can be desalinated by reverse osmosis. In a reverse osmosis system, seawater flows at a pressure greater than its osmotic pressure around bundles of tubes with semipermeable walls. Water molecules pass from the seawater into the tubes, leaving dissolved salts behind. Pure water flows from the insides of the tubes and is collected.

SAMPLE EXERCISE 10.17   **Calculating Pressure for Reverse Osmosis**

What is the reverse osmotic pressure required at 20°C to purify brackish well water containing 0.355 M dissolved particles if the product water is to contain no more than 87 mg of dissolved solids (as NaCl) per liter?

COLLECT AND ORGANIZE We are asked to calculate the pressure needed to purify water to a stated standard by reverse osmosis. We are given the temperature, the concentration of particles in the water to be purified, and the amount of solid (expressed as NaCl) that is tolerable in the final product. We have Equation 10.10 to calculate osmotic pressure.

ANALYZE The molarity of the more concentrated solution is 0.355 M. The molarity of the less concentrated product water must be calculated from the information given. We can calculate the osmotic pressure once we know the difference of molarities of the solutions on either side of the membrane.

SOLVE If the product water is to contain no more than 87 mg NaCl/L, the maximum acceptable molar concentration of NaCl can be calculated from the molar mass of NaCl:

$$\frac{87 \text{ mg NaCl}}{L} \times \frac{1.000 \text{ g}}{1000 \text{ mg}} \times \frac{1 \text{ mol NaCl}}{58.44 \text{ g NaCl}} = \frac{1.5 \times 10^{-3} \text{ mol NaCl}}{L}$$

$$= 1.5 \times 10^{-3} \, M \, \text{NaCl}$$

The total ion concentration for a $1.5 \times 10^{-3} \, M$ NaCl solution is $2(1.5 \times 10^{-3} \, M) =$ 0.00300 M in total ions. The difference in the concentration of total ions in the two solutions is

$$0.355 \, M \text{ (brackish water)}$$

$$\underline{-0.003 \, M \text{ (product water)}}$$

$$0.352 \, M \text{ (across membrane)}$$

We can calculate the osmotic pressure required to generate this concentration difference:

$$\pi = MRT = \frac{0.352 \text{ mol}}{L} \times \frac{0.0821 \text{ L} \cdot \text{atm}}{\text{mol K}} \times (273 + 20) \text{ K} = 8.47 \text{ atm}$$

Therefore, a pressure greater than 8.47 atm will force water from the brackish water sample through a reverse osmotic membrane to give water containing less than 87 mg NaCl per liter. This assumes that, in practice, some Na^+ and Cl^- ions pass through the membrane.

THINK ABOUT IT The pressure to reverse the natural osmotic flow is well above atmospheric pressure, which is reasonable because we are reversing a process that runs spontaneously at atmospheric pressure.

Practice Exercise Calculate the pressure that must be applied to the brackish water of the preceding Sample Exercise if the maximum dissolved concentration allowed in the product water is twice that of the purified water in the Sample Exercise.

Reverse osmosis is a technology suitable for the production of drinking water from seawater. However, tough semipermeable membranes are a necessity since reverse osmosis systems operate at high pressure. These membranes are difficult to make and are expensive.

10.10 Measuring the Molar Mass of a Solute Using Colligative Properties

In principle, the molar mass of a solute can be determined by measuring the effect of the dissolved solute on the physical properties of a solvent. Applying colligative properties to this task works only for nonelectrolytes, which have a van't Hoff factor of 1. Freezing-point-depression measurements, for example, can be used to find the molar mass of a molecular compound if it is sufficiently soluble in a solvent whose K_f value is known, as illustrated in Sample Exercise 10.18.

CONNECTION In Chapter 3 we used molar masses determined by mass spectrometry to convert empirical formulas derived from elemental analyses to molecular formulas.

SAMPLE EXERCISE 10.18 **Determining the Molar Mass of a Solute**

The freezing point of a solution of 100 mg eicosene (a molecular compound and a nonelectrolyte) in 1.00 g benzene was lower by 1.87°C than the freezing point of pure benzene. Determine the molar mass of eicosene. (K_f for benzene is 4.90°C/m.)

COLLECT AND ORGANIZE We are asked to determine the molar mass of a compound. We are given the mass of the substance that lowers the freezing point of a solvent by a known amount, and we have the K_f of the solvent. Because the solute is molecular, $i = 1$. Equation 10.9 describes the relationship among the factors governing the magnitude of freezing point depression.

ANALYZE From the information given we can calculate the molality of the solution. Since we are told the mass of the solvent (1.00 g of benzene), we can calculate the moles of eicosene in the solution. Since the sample weighs 100 mg, we can calculate the molar mass of eicosene.

SOLVE We first calculate the molality of the solution from ΔT_f and Equation 10.9. Eicosene is molecular and a nonelectrolyte so $i = 1$.

$$\Delta T_f = iK_f m = K_f m$$

Solve for m:

$$m = \Delta T_f / K_f = \frac{1.87\,°\!C}{\underset{m}{\underline{4.90\,°\!C}}} = 0.382\ m \text{ eicosene}$$

Molality is moles of solute per kilogram of solvent. We know the mass of solvent used to prepare the solution, and we can calculate the moles of eicosene (the solute) in the sample:

$$m = \frac{\text{moles solute}}{\text{mass solvent}}$$

$$0.382\ m = \frac{\text{moles eicosene}}{1.00 \times 10^{-3}\,\text{kg benzene}}$$

$$\text{Moles of eicosene} = \frac{0.382\ \text{mol eicosene}}{\text{kg benzene}} \times 1.00 \times 10^{-3}\ \text{kg benzene}$$

$$= 3.82 \times 10^{-4}\ \text{mol eicosene}$$

Finally, recall that molar mass is mass of eicosene per mole of eicosene. Since 100 mg of eicosene was used to prepare the solution,

$$\text{Molar mass} = \frac{\text{mass eicosene}}{\text{moles of eicosene}} = \frac{0.100\ \text{g eicosene}}{3.82 \times 10^{-4}\ \text{mol}} = 262\ \text{g/mol}$$

THINK ABOUT IT The calculated mass is within the range we have previously seen for molecular compounds, so it seems reasonable in that regard. The solvent used in this exercise, benzene, has a larger K_f than that of water, so a smaller quantity of solute can be used to detect a measurable change in freezing point.

Practice Exercise A solution prepared from 360 mg of a sugar (a molecular material and a nonelectrolyte) in 1.00 g of water froze at $-3.72°C$. What is the molar mass of this sugar? The value of K_f is $1.86°C/m$.

Osmotic pressure is the best of the three methods for determining the molar mass of water-soluble substances for a number of reasons: First, the K_f and K_b parameters of water are much smaller than those of other solvents (Table 10.7). Thus, more concentrated molal solutions are needed to observe the same ΔT_f or ΔT_b value with water than with other solvents. Second, biomaterials such as proteins and carbohydrates are nearly always available only in small quantities, and they often are not very soluble in nonaqueous solvents with larger K_f and K_b values. These biomaterials often have high molar masses, which means that large quantities are needed to give high enough molal concentrations for reliable ΔT_f or ΔT_b measurements. Third, a solute might need to be recovered unchanged for

TABLE 10.7 **Molal Freezing-Point-Depression and Boiling-Point-Elevation Constants for Selected Solvents**

Solvent	Freezing Point (°C)	K_f (°C/m)	Boiling Point (°C)	K_b (°C/m)
Water (H_2O)	0.0	1.86	100.0	0.52
Benzene (C_6H_6)	5.5	4.90	80.1	2.53
Ethanol (CH_3CH_2OH)	−114.6	1.99	78.4	1.22
Carbon tetrachloride (CCl_4)	−22.3	29.8	76.8	5.02

other uses, and so boiling-point-elevation measurements are ruled out for heat-sensitive solutes. In contrast, very small osmotic pressures can be measured precisely, the equipment can be miniaturized, and the measurements can be made at room temperature.

SAMPLE EXERCISE 10.19 **Using Osmotic Pressure to Determine Molar Mass**

A molecular nonelectrolyte was isolated from a South African tree. A 47-mg sample was dissolved in water to make 2.50 mL of solution at 25°C. The measured osmotic pressure of the solution was 0.489 atm. Calculate the molar mass of the compound.

COLLECT AND ORGANIZE We are given the mass of a substance, the volume of its aqueous solution, the temperature, and its osmotic pressure. We can relate these parameters using Equation 10.10.

ANALYZE We can calculate the molar concentration of the solution from the information given and Equation 10.10. Since we know the solution's volume, we can calculate the number of moles of the solute from the molarity. Since we know the mass of this number of moles, we can calculate the molar mass of the solute.

SOLVE The osmotic pressure (π) of this nonelectrolyte solution at 25°C is 0.489 atm, and $\pi = iMRT$. Rearrangement and substitution of $\pi = 0.489$ atm, $i = 1$, $R = 0.0821$ L $\cdot$ atm (mol $\cdot$ K)$^{-1}$, and $T = 273 + 25 = 298$ K gives

$$M = \pi/RT = \frac{0.489 \text{ atm}}{[0.0821 \text{ L} \cdot \text{atm (mol} \cdot \text{K)}^{-1} \times 298 \text{ K]}}$$

$$= \frac{2.00 \times 10^{-2} \text{ mol}}{\text{L}} = 2.00 \times 10^{-2} M$$

The molar concentration of the solution is given by

$$M = n/V$$

where n is the number of moles of solute and V is the volume of the solution (in liters). Solving for n, with $M = 2.00 \times 10^{-2}$ mol/ L, and $V = 2.50$ mL $= 2.50 \times 10^{-3}$ L, we have

$$n = MV = \frac{2.00 \times 10^{-2} \text{ mol}}{\text{L}} \times 2.50 \times 10^{-3} \text{ L} = 5.00 \times 10^{-5} \text{ mol}$$

According to the information, this number of moles of solute has a mass of 47 mg. If we convert this mass to grams, we can calculate the unknown molar mass

$$\text{Molar mass} = \frac{\text{g solute}}{\text{moles of solute}}$$

$$= \frac{47 \times 10^{-3} \text{ g}}{5.08 \times 10^{-5} \text{ mol}} = 9.4 \times 10^{2} \text{ g/mol}$$

THINK ABOUT IT A small amount of material was dissolved in the solvent, but even for a rather large molecule, the osmotic pressure was sufficiently large to enable us to calculate the molar mass. The compound isolated from the tree has relatively large molecules with a molar mass over 900 g/mol.

Practice Exercise A solution was made by dissolving 5.00 mg of hemoglobin in water to give a final volume of 1.00 mL. The osmotic pressure of this solution was 1.91×10^{-3} atm at 25°C. Calculate the molar mass of hemoglobin.

The halogen family consists of four common elements: fluorine, chlorine, bromine, and iodine. A fifth, astatine, is rarely encountered: all of its 24 isotopes are radioactive and have brief existences. Nothing is known of the bulk physical properties of astatine, and it may actually be the rarest naturally occurring terrestrial element. Estimates suggest that the outermost kilometer of Earth's crust contains less than 45 mg of astatine. We discussed fluorine in the descriptive chemistry section in Chapter 8, and we address the remaining halogens here. The word *halogen* means salt-former, and although it is the chemistry of the halogens as components of salts that we frequently think of first, they have a rich molecular chemistry as well.

Physical properties of the common halogens follow the expected trends, as illustrated in the accompanying table.

	Fluorine (F$_2$)	Chlorine (Cl$_2$)	Bromine (Br$_2$)	Iodine (I$_2$)
Color	Pale yellow	Yellow-green	Red-brown	Violet-black
State under standard conditions	Gas	Gas	Liquid	Solid
Melting point (°C)	−219	−101	−7	114
Boiling point (°C)	−188	−34	59	185
Atomic radius (pm)	71	99	114	133
Ionic radius (pm)	133	181	196	220
Energy of HX bond (kJ/mol)	565	431	366	299

Fluorine and chlorine are both gases under standard conditions, and bromine and iodine have high vapor pressures. Dark-red vapor always accompanies liquid bromine at room temperature, and violet vapor is frequently visible above solid iodine. Bromine is the only liquid nonmetallic element.

All the halogens have strong, penetrating aromas and are quite hazardous to human health. A concentration of chlorine in the air greater than 1 ppm is damaging to health, and a few breaths of air containing 1,000 ppm of chlorine are fatal. Bromine is extremely corrosive to human tissue, and exposure of the skin to bromine causes burns that are very painful and slow to heal. Iodine is an essential trace element; however, ingestion of large amounts of iodine is dangerous and consumption of 2 to 3 g is fatal to humans.

The halogens may be the most important group in the periodic table in terms of general industrial use. By far the most important use of chlorine is in the manufacture of chemicals and materials that contain one or more atoms of chlorine in their structure: agents to sterilize water for drinking and swimming pools; bleach for the paper industry; explosives; dyes; insecticides; cleaning agents, and plastics. The synthetic sweetener Splenda is made from sucrose by replacing three −OH groups with chlorine atoms. The molecule retains its sweetness but is not metabolized in our bodies and hence is considered calorie-free. Chloride ions are ubiquitous in living systems and play a vital role in biology. They balance the charge on the sodium and potassium ions found in body fluids and maintain electrical neutrality.

Bromine is used industrially much like chlorine: in making fumigants and insecticides, dyes, compounds used to purify water, and flame-proofing agents. Iodine has important medical uses. Tincture of iodine, typically part of emergency survival and first aid kits, is a solution of up to 10% iodine in ethanol used to disinfect wounds and purify water. A radioactive isotope of iodine, ^{131}I, is used to treat thyroid cancer and diseases of the thyroid. Two other radioactive isotopes, ^{123}I and ^{125}I, are used as imaging agents to evaluate thyroid function. Salt for human consumption is often enriched with iodide ion (iodized salt) to insure proper production of thyroid hormones that regulate metabolism. Iodine is necessary for human health at the dietary level of 150 μg per day. Iodine deficiency in infants is a leading cause of preventable mental retardation and is a serious health problem in developing nations.

None of the halogens exist in nature as free diatomic elements. One source of chloride, bromide, and iodide ions is the ocean, from which they are extracted and then oxidized to produce the pure elements. It is estimated that 10^{16}

Liquid bromine and solid iodine both have high vapor pressures at typical room temperature.

tons of chloride ion are available in the Earth's oceans. If this amount of solid NaCl were formed into square columns 1 km on a side, a total of 47 columns could be made from the available salt, each of which would be tall enough to extend from the Earth to the moon. However, only about one-third of the NaCl used commercially is claimed from the currently existing oceans; the bulk of it is mined from rock salt deposits, which are the residues from the evaporation of ancient seas.

Chlorine is produced industrially by the electrolysis of molten NaCl or of aqueous solutions of NaCl. In the first case, sodium metal is produced; in the second, hydrogen gas and the base sodium hydroxide are produced, giving rise to the name chlor-alkali process for the reaction.

$$2 \, NaCl(\ell) \rightarrow 2 \, Na(s) + Cl_2(g)$$

$$2 \, NaCl \, (aq) + 2 \, H_2O \, (\ell) \rightarrow H_2(g) + Cl_2(g) + 2 \, NaOH(aq)$$

Bromine is obtained from normal ocean water and also from highly concentrated brine sources like the Dead Sea. The bromide ion is oxidized by treatment with Cl_2 to produce Br_2:

$$2 \, Br^-(aq) + Cl_2(g) \rightarrow Br_2(\ell) + 2 \, Cl^-(aq)$$

Elemental iodine is obtained from natural brines as well, and its collection involves the oxidation of I^- to I_2. Another industrial source is Chilean saltpeter, in which iodine is found in the form of iodate salts (IO_3^-). To produce iodine from the iodate ion, the iodate is first reduced to I^-; then the iodide ion is oxidized to I_2 by treatment with more IO_3^-:

$$2 \, IO_3^-(aq) + 6 \, HSO_3^-(aq) \rightarrow$$
$$2 \, I^-(aq) + 6 \, SO_4^{2-}(aq) + 6 \, H^+(aq)$$

$$5 \, I^-(aq) + IO_3^-(aq) + 6 \, H^+(aq) \rightarrow 3 \, I_2(s) + 3 \, H_2O(\ell)$$

Elemental chlorine reacts with water to form a mixture of hydrochloric and hypochlorous acids:

$$Cl_2(g) + H_2O(\ell) \rightarrow HCl(aq) + HOCl(aq)$$

All halogens react with water in this fashion to make the corresponding acids. The addition of NaOH to solutions of HOCl (a neutralization reaction) produces aqueous NaOCl, which is the active ingredient in household bleach:

$$HOCl(aq) + NaOH(aq) \rightarrow NaOCl(aq) + H_2O(\ell)$$

Chlorine gas is a powerful bleach on its own, but because it is a highly toxic gas and hence difficult to handle, most bleaches contain the hypochlorite ion (OCl^-), either in solution or in solid form as a salt. Hypochlorites are used as laundry bleach, in the paper industry to bleach wood pulp to produce white paper, and to sterilize water in swimming pools. The chlorine odor of pools is due to compounds called chloramines, produced when hypochlorous acid reacts with ammonia and nitrogen-containing compounds in bacteria:

$$HOCl(aq) + NH_3(aq) \rightarrow NH_2Cl(aq) + H_2O(\ell)$$

Chloramine itself is also a disinfectant and is used in municipal water treatment plants as an alternative to direct chlorination. Salts containing chlorate (ClO_3^-) and chlorite (ClO_2^-) ions are also used as bleaching agents. The perchlorate ion (ClO_4^-) reacts with organic matter rapidly, and some perchlorate salts are extremely reactive as contact explosives or as oxidizing agents when mixed with other substances that are easily oxidized. The oxidizers in many fireworks are perchlorate salts. When potassium perchlorate is mixed with charcoal the material ignites spontaneously:

$$KClO_4(s) + 2 \, C(s) \rightarrow KCl(s) + 2 \, CO_2(g)$$

Mixtures of potassium perchlorate, sulfur, and aluminum provide the white flash and noise in fireworks. The flash powder used in stage shows is a mixture of magnesium metal and $KClO_4$. Ammonium perchlorate decomposes explosively with heat and is also shock sensitive:

$$2 \, NH_4ClO_4(s) \rightarrow N_2(g) + Cl_2(g) + 4 \, H_2O(g) + 2 \, O_2(g)$$

This behavior gives rise to its use as a solid propellant in the booster rockets of the U.S. space shuttles. Oxoanions of the other halogens, including bromate, iodate, perbromate, and periodate, have chemistries that are similar to those of the corresponding chlorine oxoanions.

Salt formations at the shore of the Dead Sea indicate the high salt content of the water.

Summary

SECTION 10.1 The strengths of intermolecular attractive forces determine whether a compound will be a gas, a liquid, or a solid under normal conditions. Strong attractive ion–ion interactions hold ionic solids together. The strengths of these ion–ion interactions and the magnitudes of the resulting **lattice energies** depend on the charges of the cation and anion and the distance between them. High lattice energies correlate with high melting and boiling points and low aqueous solubilities of ionic compounds. The energy change that occurs when one mole of an atom or ion combines with one mole of electrons in the gas phase is called its **electron affinity**. Lattice energies can be calculated with a **Born–Haber cycle**, an application of Hess's law of constant heat summation (Chapter 5).

SECTION 10.2 Ions interact with water through **ion–dipole** forces. **Dipole–dipole** forces exist between water and between other polar molecules, but water's properties indicate the existence of stronger dipole–dipole interactions called **hydrogen bonds**. These bonds form between other polar molecules containing O—H, N—H or F—H covalent bonds.

SECTION 10.3 **Dispersion (London) forces** are due to the **polarization** of atoms and molecules and the existence of **temporary** (or **induced) dipoles**. These interactions are weak compared with ion–ion and ion–dipole interactions. The strongest dispersion forces exist between the largest atoms, ions, and molecules.

SECTION 10.4 The van der Waals equation (Chapter 6) accounts for the behavior of gaseous substances at high pressures and low temperatures. The magnitudes of its parameters a and b depend on the substance and reflect the strength of intermolecular forces and molecular size, respectively. These parameters are largest for relatively large polar molecules.

SECTION 10.5 Polar solutes dissolve in polar solvents when the dipole–dipole interactions between solute and solvent molecules offset the interactions that keep solute or solvent molecules together. The limited solubility of nonpolar solutes in polar solvents is a result of interactions between dipoles and induced dipoles. **Hydrophilic** substances are more soluble in water than are **hydrophobic** substances which prefer nonpolar solvents. **Henry's law** describes the solubility of sparingly soluble, chemically unreactive gases in liquids. Often more than one type of intermolecular force is responsible for the physical properties of a molecular substance.

SECTION 10.6 During **evaporation**, molecules at the surface of a liquid break intermolecular interactions with neighboring molecules and enter the vapor phase. These vapor-phase molecules are the reason for the liquid's **vapor pressure**, which is an intensive property of a substance. A greater proportion of liquid molecules enter the vapor phase as the temperature increases, leading to higher vapor pressures at higher temperatures. The presence of particles of a nonvolatile solute in a solution decreases the vapor pressure of the solvent. This phenomenon is called a **colligative property** of the solvent. **Raoult's law** relates the vapor pressure of a solution to its composition and the vapor pressure of the solvent.

SECTION 10.7 The **phase diagram** of a substance indicates whether it exists as a solid, liquid, gas, or **supercritical fluid** at a particular pressure and temperature. All three states (solid, liquid, and gas) exist in equilibrium at the **triple point**. Above their critical temperatures and critical pressures, substances exist as **supercritical fluids**.

SECTION 10.8 The remarkable behavior of water, including its high melting and boiling points, **surface tension**, **viscosity**, and its **capillary action**, result from the strength of intermolecular hydrogen bonds.

SECTION 10.9 Solutes in solution increase the solvent's **boiling point** and decrease its **freezing point**. **Osmosis** is a process in which solvent flows through a semipermeable membrane from a solution of lower solute concentration into a solution of higher solute concentration. Like freezing point and boiling point, **osmotic pressure** is a colligative property of a solvent. **Reverse osmosis** is used to purify water. The concentration scales used for colligative property measurements include molarity and **molality**, which is the number of moles of solute per kilogram of solvent. The more concentrated a solution is, the higher is its boiling point, the lower its freezing point, and the higher its osmotic pressure. The **van't Hoff factor** accounts for the colligative properties of electrolytes and the formation of solute **ion pairs** in concentrated solutions.

SECTION 10.10 The molar mass of a compound can be determined by measuring the freezing point depression, boiling point elevation, or osmotic pressure of a known mass of compound in a solution of known concentration.

Problem-Solving Summary

TYPE OF PROBLEM	CONCEPTS AND EQUATIONS	SAMPLE EXERCISES
Predicting the relative magnitudes of attractive forces, lattice energies, and melting points of ionic compounds	Use Coulomb's law: $E \propto Q_1Q_2/d$, where Q_1 and Q_2 are the charges on the ions and d is the distance between the nuclei (calculated by adding the radii of the cation and anion). More negative values of E correspond to stronger ion–ion attraction. Use Coulomb's law to predict lattice energies and relative melting points. Compounds with the same solid structure and the most negative value of Q_1Q_2/d have the highest melting points.	10.1, 10.2

TYPE OF PROBLEM	CONCEPTS AND EQUATIONS	SAMPLE EXERCISES
Calculating lattice energy with a Born–Haber cycle	The heat of formation of an ionic compound such as LiF(s) is the sum of (a) the heat needed to vaporize one mole of solid Li, plus (b) the heat needed to atomize half a mole of $F_2(g)$, plus (c) the ionization energy of Li(g), plus (d) the electron affinity of F(g) atoms, plus (e) the lattice energy of LiF(s).	10.3
Explaining the relative boiling points of liquids and trends in boiling points of pure substances	Large molecules usually have higher boiling points than smaller molecules, with notable exceptions. The presence of polar –OH and –NH groups in molecules of a liquid leads to intermolecular hydrogen bonding that markedly increases the liquid's boiling point.	10.4, 10.5
Assessing empirical parameters of real gases	Molecular size, polarity, and intermolecular hydrogen bonding all tend to increase the parameters a and/or b in the van der Waals equation for real gases.	10.6
Predicting the water solubility of substances	Molecular polarity and intermolecular hydrogen bonding enhance the solubility of substances in water. Like dissolves like.	10.7
Calculating the solubility of an unreactive gas using Henry's law	Use Henry's law: $C_{gas} = k_H P_{gas}$, where C_{gas} is the solubility of the gas, k_H depends on the gas, the solvent, and the temperature, and P_{gas} is the pressure of the gas (or partial pressure if the gas is part of a gas mixture). The units of C_{gas} depend on the units of k_H.	10.8
Calculating the vapor pressure of a solution	Use Raoult's law: $P_{solution} = X_{solvent} P_{solvent}$, where $P_{solution}$ is the vapor pressure of the solution, $X_{solvent}$ is the mole fraction of the solvent in the solution, and $P_{solvent}$ is the vapor pressure of the pure solvent at the same temperature.	10.9
Reading a phase diagram	Note carefully whether the temperature, pressure, or both are specified or varied. Temperature changes are on the horizontal axis, pressure changes on the vertical. Phase changes occur when a line in the phase diagram is crossed.	10.10
Calculating molal concentrations	Molality: $m = n_{solute}/\text{kg solvent}$, where n is the number of moles of solute.	10.11
Predicting the boiling point or freezing point of a solution	For the boiling point of nonelectrolytes, use $\Delta T_b = K_b m$; for the freezing point, use $\Delta T_f = K_f m$. The term ΔT_b is the increase in the boiling point of the solvent, K_b is a constant that depends only on the solvent, and m is the molality of the solution. ΔT_f is the decrease in the freezing point of the solvent, K_f is a constant that depends only on the solvent, and m is the molality of the solution.	10.12, 10.13
Assessing interactions among particles in solution by comparing the ideal value of the van't Hoff factor i with the value calculated from real data	For electrolytes, use $\Delta T_b = i K_b m$ and $\Delta T_f = i K_f m$, where i accounts for the dissociation of ionic solutes in solution. Examples of ideal i values for electrolytes: $i = 2$ for NaCl, 3 for $CaCl_2$, and 4 for Na_3PO_4. For real solutions of electrolytes, rearrange $\Delta T_b = i K_b m$ and $\Delta T_f = i K_f m$ to calculate the value of i and compare the result with the theoretical value.	10.14, 10.15
Calculating osmotic pressure	Use $\pi = iMRT$, where π is the osmotic pressure (atm), M is the molar concentration of the solution, R is the ideal gas constant, T is the absolute temperature of the solution, and i is the van't Hoff factor.	10.16, 10.17
Determining the molar mass of a nonelectrolyte solute from boiling point elevation, freezing point depression, or osmotic pressure data	Use the appropriate equation ($\Delta T_b = K_b m$, $\Delta T_f = K_f m$, or $\pi = MRT$) to calculate the molality or molarity (for osmotic pressure) of the solution. Use the calculated concentration to determine the moles of solute and then the molar mass of the solute.	10.18, 10.19

Visual Problems

10.1. Look at the pairs of ions in the structures of KF and KI represented in Figure P10.1. Which substance has the stronger cation–anion attractive forces and the higher melting point?

FIGURE P10.1

10.2. In Figure P10.2, identify the physical state (solid, liquid or gas) of xenon and classify the attractive forces between the xenon atoms.

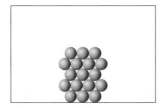

FIGURE P10.2

10.3. Figure P10.3 depicts molecules of XH_3 and YH_3 (not shown to scale) and the boiling points at 1 atm pressure of XH_3 and YH_3, respectively. One substance is phosphine (PH_3) and the other substance is ammonia (NH_3). Which molecule is phosphine? Explain your answer.

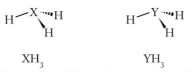

FIGURE P10.3 XH_3 YH_3
Boiling point −88°C   Boiling point −33°C

10.4. Figure P10.4 shows representations of the molecules *n*-pentane, C_5H_{12}, and *n*-decane, $C_{10}H_{22}$. Which substance has the lower freezing point? Explain your answer and use the Web to verify your prediction.

FIGURE P10.4   *n*-Pentane   *n*-Decane

10.5. Use the graph in Figure P10.5 to estimate the normal boiling points of substances X and Y. Which substance has the stronger intermolecular forces?

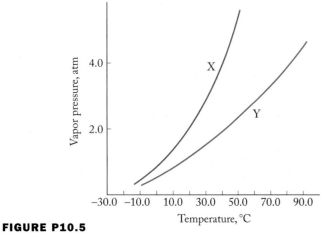

FIGURE P10.5

10.6. Examine the phase diagram of substance Z in Figure P10.6. Does the freezing point of the substance increase or decrease with increasing pressure?

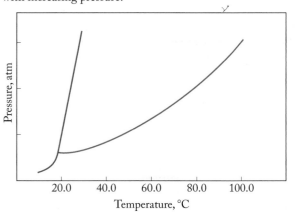

FIGURE P10.6

10.7. The graph in Figure P10.7 shows the decrease in the freezing point of water ΔT_f for solutions of two different substances A (triangles) and B (circles) in water. Explain how you can reasonably conclude that (a) A and B are nonelectrolytes and (b) the freezing point depression constant K_f of water is independent of the solute's identity.

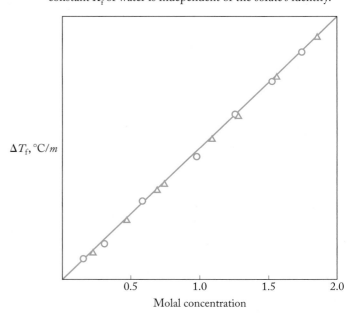

FIGURE P10.7

10.8. The arrow in Figure P10.8 indicates the direction of solvent flow through a semipermeable membrane in equipment designed to measure osmotic pressure. Which solution, A or B, is the more concentrated? Explain your answer.

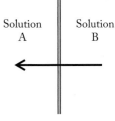

FIGURE P10.8

Questions and Problems

Sea Spray and Salts

CONCEPT REVIEW

10.9. Indicate the substance that contains the largest anion.
(a) $BaCl_2$; (b) AlF_3; (c) KI; (d) $SrBr_2$

10.10. Indicate the substance that contains the most negatively charged anion.
(a) $BaCl_2$; (b) Al_2O_3; (c) Mg_3N_2; (d) SrS

10.11. Which substance has the most negative lattice energy?
(a) MgO; (b) CaO; (c) SrO; (d) BaO

10.12. Why is $CaSO_4$ less soluble in water than NaCl?

10.13. Explain how the Born–Haber cycle relates to Hess's law of constant heat summation and the law of conservation of energy.

10.14. List all the experimental enthalpy data that are needed to calculate the lattice energy of strontium oxide, SrO.

PROBLEMS

10.15. Rank the following ionic compounds in order of increasing attraction between their ions: KBr, $SrBr_2$, and CsBr.

10.16. Rank the following ionic compounds in order of increasing attraction between their ions: BaO, $BaCl_2$, and CaO.

10.17. How do the melting points of the series of sodium halides NaX (X = F, Cl, Br, I) relate to the atomic number of X?

10.18. Rank the following ionic compounds in order of increasing melting point: BaF_2, $CaCl_2$, $MgBr_2$, and SrI_2.

10.19. Use a Born–Haber cycle to calculate the lattice energy of potassium chloride (KCl) from the following data:
Ionization energy of K(g) = 425 kJ/mol
Electron affinity of Cl(g) = −349 kJ/mol
Energy to vaporize K(s) = 89 kJ/mol
$Cl_2(g)$ bond energy = 240 kJ/mol
Energy change for the reaction =
$K(s) + \frac{1}{2}Cl_2(g) \rightarrow KCl(s)$ is −438 kJ/mol KCl

10.20. Calculate the lattice energy of sodium oxide (Na_2O) from the following data:
Ionization energy of Na(g) = 495 kJ/mol
Electron affinity of O(g) for 2 e⁻ = 603 kJ/mol
Energy to vaporize Na(s) = 109 kJ/mol
$O_2(g)$ bond energy = 499 kJ/mol
Energy change for the reaction =
$2 Na(s) + \frac{1}{2}O_2(g) \rightarrow Na_2O(s)$ is −416 kJ/mol Na_2O

Interactions Involving Polar Molecules

CONCEPT REVIEW

10.21. How are the water molecules preferentially oriented around the anion in an aqueous solution of sodium chloride?

10.22. How are the water molecules preferentially oriented around the cation in an aqueous solution of potassium bromide?

10.23. Why are dipole–dipole interactions generally weaker than ion–dipole interactions?

10.24. Two liquids—one polar, one nonpolar—have the same molar mass. Which one is likely to have the higher boiling point?

10.25. Why are hydrogen bonds considered a special class of dipole–dipole interactions?

10.26. Can all polar hydrogen-containing molecules form hydrogen bonds?

PROBLEMS

10.27. In an aqueous solution containing chloride and iodide salts, which anion would you expect to be the more strongly hydrated?

10.28. In an aqueous solution containing Fe(II) and Fe(III) salts, which cation would you expect to be the more strongly hydrated?

10.29. Explain why the melting point of methyl fluoride, CH_3F (−142°C) is higher than the melting point of methane, CH_4 (−182°C)?

10.30. Explain why the boiling point of Br_2 (59°C) is lower than that of iodine monochloride, ICl (97°C), even though they have nearly the same molar mass.

10.31. Why doesn't methane (CH_4) exhibit hydrogen bonding while methanol, CH_3OH, does?

10.32. The boiling point of phosphine (PH_3) (−88°C) is lower than that of ammonia (NH_3) (−33°C) even though PH_3 has twice the molar mass of NH_3. Why?

10.33. In which of the following compounds do the molecules experience the strongest dipole–dipole attractions? (a) CF_4; (b) CF_2Cl_2; (c) CCl_4

10.34. Which of the following compounds, CO_2, NO_2, SO_2, or H_2S, is expected to have the weakest interactions between its molecules?

Interactions Involving Nonpolar Molecules

CONCEPT REVIEW

10.35. Which type of intermolecular force exists in all substances?

10.36. Why do the strengths of London (dispersion) forces generally increase with increasing molecular size?

PROBLEMS

10.37. The dipole moment of CH_2F_2 (1.93 debyes) is larger than that of CH_2Cl_2 (1.60 debyes), yet the boiling point of CH_2Cl_2 (40°C) is much higher than that of CH_2F_2 (−52°C). Why?

10.38. How is it that the dipole moment of HCl (1.08 D) is larger than the dipole moment of HBr (0.82 D), yet HBr boils at a higher temperature?

10.39. In each of the following pairs of molecules, which compound experiences the stronger London (dispersion) forces? (a) CCl_4 or CF_4? (b) CH_4 or C_3H_8?

10.40. What kinds of intermolecular forces must be overcome as (a) solid CO_2 sublimes; (b) $CHCl_3$ boils; (c) ice melts?

Real Gases Revisited

CONCEPT REVIEW

10.41. Why do gases behave nonideally at high pressures and low temperatures?

10.42. What properties of real gas molecules are associated with parameters a and b of the van der Waals equation?

PROBLEMS

10.43. Explain why the van der Waals constant a for Ar is greater than it is for He.

10.44. The van der Waals constant a for CO_2 is $3.59 \; L^2 \cdot atm/mol^2$. Would you expect the value of a for CS_2 to be larger or smaller than $3.59 \; L^2 \cdot atm/mol^2$?

Polarity and Solubility

CONCEPT REVIEW

10.45. What is the difference between the terms *miscible* and *insoluble*?

10.46. Which substances are essentially insoluble in water? (a) benzene, $C_6H_6(\ell)$; (b) KBr(s); (c) $Br_2(\ell)$

10.47. Why does the solubility of sparingly soluble gases in most liquids increase with increasing gas pressure?

10.48. Why does the solubility of most gases in most liquids increase with decreasing temperature?

10.49. Which term, k_H or P, in Henry's law is affected by temperature?

10.50. Air is primarily a mixture of nitrogen and oxygen. Is the Henry's law constant for the solubility of air in water the sum of k_H for N_2 and k_H for O_2? Explain why or why not.

10.51. Why is the Henry's law constant for CO_2 so much larger than for N_2 and O_2 at the same temperature? *Hint:* Elsewhere in the text, check the reactivity of CO_2 with water.

10.52. A student observes bubbles while heating a sample of water in a beaker at 60°C. What are the gases in the bubbles and where did they come from?

10.53. In what context do the terms *hydrophobic* and *hydrophilic* relate to the solubilities of substances in water?

10.54. How does the presence of increasingly longer hydrocarbon chains in the structure affect the solubility of a series of structurally related molecules in water?

PROBLEMS

10.55. In each of the following pairs of compounds, which compound is likely to be more soluble in water?
 a. CCl_4 or $CHCl_3$
 b. CH_3OH or $C_6H_{11}OH$
 c. NaF or MgO
 d. CaF_2 or BaF_2

10.56. In each of the following pairs of compounds, which compound is likely to be more soluble in CCl_4?
 a. Br_2 or NaBr
 b. CH_3CH_2OH or CH_3OCH_3
 c. CS_2 or KOH
 d. I_2 or CaF_2

*10.57.** **Arterial Blood** Arterial blood contains about 0.25 g of oxygen per liter at 37°C and standard atmospheric pressure. What is the Henry's law constant (mol/L · atm) for O_2 dissolution in blood? The mole fraction of O_2 in air is 0.209.

10.58. The solubility of O_2 in water is 6.5 mg/L at an atmospheric pressure of 1 atm and temperature of 40°C. Calculate the Henry's law constant of O_2 at 40°C.

*10.59.** Use the Henry's law constant for O_2 dissolved in arterial blood from Problem 10.57 to calculate the solubility of O_2 in the blood of (a) a climber on Mt. Everest (P_{atm} = 0.35 atm) and (b) a scuba diver at 100 feet ($P \approx 3$ atm).

*10.60.** The solubility of air in water is approximately 2.1×10^{-3} M at 20°C and 1.0 atm. Calculate the Henry's law constant for air. Is the value of k_H for air equal to the sum of k_H for O_2 and for N_2?

*10.61.** Use the graph of solubility of O_2 versus temperature in Figure P10.61 to calculate the value of the Henry's law constant k_H for O_2 at 10°C, 20°C, and 30°C.

10.62. Based on the data in Figure P10.61, which has a greater effect on the solubility of oxygen in water: (a) decreasing the

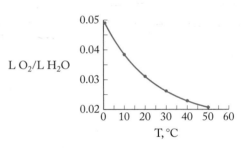

FIGURE P10.61

temperature from 20°C to 10°C or (b) raising the pressure from 1.00 atm to 1.25 atm?

10.63. Which of the following compounds is likely to be the most soluble in water? (a) NaCl; (b) KI; (c) $Ca(OH)_2$; (d) CaO

10.64. Which sulfur oxide would you predict to be more soluble in nonpolar solvents: SO_2 or SO_3? *Hint:* Draw the Lewis structures of SO_2 and SO_3.

10.65. Which of these substances is the least soluble in water?
 (a) $CH_3(CH_2)_2CH_2OH$; (b) $CH_3(CH_2)_4CH_2OH$;
 (c) $CH_3(CH_2)_6CH_2OH$; (d) $CH_3(CH_2)_8CH_2OH$

10.66. Which of these substances is the most soluble in water?
 (a) $CH_3(CH_2)_2CH_2NH_2$; (b) $CH_3(CH_2)_4CH_2Cl$;
 (c) $CH_3(CH_2)_6CH_2Br$; (d) $CH_3(CH_2)_8CH_2I$

Vapor Pressure

CONCEPT REVIEW

10.67. Explain the term *nonvolatile solute*.

10.68. Which has the higher vapor pressure at constant temperature, pure water or seawater? Explain your answer.

10.69. Why does the vapor pressure of a liquid increase with increasing temperature?

10.70. What happens when the vapor pressure of a liquid is equal to or greater than atmospheric pressure?

10.71. Generally speaking, how is the vapor pressure of a liquid affected by the strength of intermolecular forces?

10.72. Is vapor pressure an intensive or extensive property of a substance?

PROBLEMS

10.73. An experiment like that shown in Figure 10.15 is set up with the beaker containing pure water full to the brim and the beaker containing seawater half-full. Explain why the beaker that contained the pure water will be empty after a long experimental time period.

10.74. Explain to a nonscientist how the water gets from one beaker to the other in the experiment depicted in Figure 10.15.

10.75. Rank the following compounds in order of increasing vapor pressure at 298 K: (a)CH_3CH_2OH, (b)CH_3OCH_3, and (c)$CH_3CH_2CH_3$.

10.76. Rank the compounds in Figure P10.76 in order of increasing vapor pressure at 298 K.

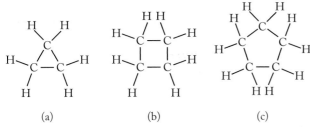

(a)   (b)   (c)

FIGURE P10.76

10.77. A solution contains 3.5 mol of water and 1.5 mol of nonvolatile glucose ($C_6H_{12}O_6$). What is the mole fraction of water in this solution? What is the vapor pressure of the solution at 25°C, given that the vapor pressure of pure water at 25°C is 23.8 torr?

10.78. A solution contains 4.5 mol of water, 0.3 mol of sucrose ($C_{12}H_{22}O_{11}$), and 0.2 mol of glucose. Sucrose and glucose are nonvolatile. What is the mole fraction of water in this solution? What is the vapor pressure of the solution at 35°C, given that the vapor pressure of pure water at 35°C is 42.2 torr?

Phase Diagrams: Intermolecular Forces at Work

CONCEPT REVIEW

10.79. Explain the difference between sublimation and evaporation.

10.80. Can ice be melted merely by applying pressure? Explain your answer.

10.81. Explain what is meant by the term *equilibrium line*.

10.82. Explain how the solid–liquid line in the phase diagram of water differs in character from the solid–liquid line in the phase diagrams of most other substances, such as CO_2.

10.83. Which phase of a substance (gas, liquid, or solid) is more likely to be the stable phase: (a) at low temperatures and high pressures; (b) at high temperatures and low pressures?

10.84. At what temperatures and pressures does a substance behave as a supercritical fluid?

10.85. Preserving Food Freeze-drying is used to preserve food at low temperature with minimal loss of flavor. Freeze-drying works by freezing the food and then lowering the pressure with a vacuum pump to sublime the ice. Must the pressure be lower than the pressure at the triple point of H_2O?

10.86. Solid helium cannot be converted directly into the vapor phase. Does the phase diagram of He have a triple point?

PROBLEMS

For help in answering Problems 87 through 90, consult Figures 10.17 and 10.18.

10.87. List the steps you would take to convert a 10.0 g sample of water at 25°C and 1 atm pressure to water at its triple point.

10.88. List the steps you would take to convert a 10.0 g sample of water at 25°C and 2 atm pressure to ice at 1 atm pressure. At what temperature would the water freeze?

10.89. What phase changes, if any, does liquid water at 100°C undergo if the initial pressure of 5.0 atm is reduced to 0.5 atm at constant temperature?

10.90. What phase changes, if any, occur if CO_2 initially at −80°C and 8.0 atm is allowed to warm to −25°C at 5.0 atm?

For help in answering Problems 91 and 92, consult Figure 10.18.

10.91. Below what temperature can solid CO_2 (dry ice) be converted into CO_2 gas simply by lowering the pressure?

10.92. What is the maximum pressure at which solid CO_2 (dry ice) can be converted into CO_2 gas without melting?

For help in answering Problems 93 and 94, consult Figure 10.17.

10.93. Predict the phase of water that exists under the following conditions:
  a. 2 atm of pressure and 110°C
  b. 0.5 atm of pressure and 80°C
  c. 6.0×10^{-3} atm of pressure and 0°C

10.94. Which phase or phases of water exist under the following conditions?
  a. 0.32 atm and 70°C
  b. 300 atm and 400°C
  c. 1 atm and 0°C

The Remarkable Behavior of Water and Properties of Liquids

CONCEPT REVIEW

10.95. Explain why a needle floats on the surface of water but sinks in a container of methanol (CH_3OH).

10.96. Explain why different liquids do not reach the same height in capillary tubes of the same diameter.

10.97. Explain why pipes filled with water are in danger of bursting when the temperature drops below 0°C.

10.98. A hot needle sinks when put on the surface of cold water. Will a cold needle float in hot water?

10.99. The meniscus of mercury in a thermometer (Figure P10.99) is convex, rather than concave. Explain why.

FIGURE P10.99

*10.100. The mercury level in a capillary tube placed in a dish of mercury is actually below the surface of the mercury in the dish. Explain why.

10.101. Describe the origin of surface tension at the molecular level.

10.102. What is the origin of the high viscosity of molasses?

10.103. Describe how the surface tension and viscosity of a liquid are affected by increasing temperature.

10.104. Explain how strong intermolecular forces are expected to result in relatively high surface tension and viscosity of a liquid.

PROBLEMS

10.105. One of two glass capillaries of the same diameter is placed in a dish of water and the other in a dish of ethanol (CH_3CH_2OH). Which liquid will rise higher in its capillary?

10.106. Would you expect water to rise to the same height in a tube made of a polyethylene plastic as it does in a glass capillary of the same diameter? The molecular structure of polyethylene is shown in Figure P10.106.

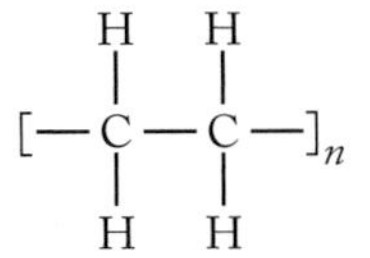

$$[-\overset{\overset{\displaystyle H}{|}}{\underset{\underset{\displaystyle H}{|}}{C}}-\overset{\overset{\displaystyle H}{|}}{\underset{\underset{\displaystyle H}{|}}{C}}-]_n$$

FIGURE P10.106

10.107. The normal boiling points of liquids A and B are 75.0°C and 151°C, respectively. Which of these liquids would you expect to have the higher surface tension and viscosity at 25°C? Explain your answer.

10.108. One beaker contains pure water and the other beaker contains pure methanol at the same temperature. Which liquid has the higher surface tension and viscosity? Explain your answer.

Colligative Properties of Solutions

CONCEPT REVIEW

10.109. What is the difference between molarity and molality?

10.110. As a solution of NaCl becomes more concentrated, does the difference between its molarity and its molality increase or decrease?

10.111. Explain why seawater has a lower freezing point than freshwater.

10.112. The thermostat in a refrigerator filled with cans of soft drinks malfunctions and the temperature of the refrigerator drops below 0°C. The contents of the cans of diet soft drinks freeze, rupturing many of the cans and causing an awful mess. However, none of the cans containing regular, nondiet soft drinks rupture. Why?

10.113. Why is it important to know if a substance is a strong electrolyte before predicting its effect on the boiling and freezing points of a solvent?

10.114. What role does the van't Hoff factor play in describing colligative properties of solutions?

10.115. Explain how the theoretical value of the van't Hoff factor i for substances such as CH_3OH, NaBr, and K_2SO_4 can be predicted from their formulas.

10.116. Explain why it is possible for an experimentally measured value of a van't Hoff factor to be less than the theoretical value.

10.117. What is a semipermeable membrane?

10.118. A pure solvent is separated from a solution containing the same solvent by a semipermeable membrane. In which direction does the solvent flow across the membrane, and why?

10.119. A dilute solution is separated from a more concentrated solution containing the same solvent by a semipermeable membrane. In which direction does the solvent tend to flow across the membrane, and why?

10.120. How is the osmotic pressure of a solution related to its molar concentration and its temperature?

10.121. What is reverse osmosis? List the basic components of equipment used to purify seawater by reverse osmosis.

10.122. Explain how the minimum pressure for purification of seawater by reverse osmosis can be estimated from its composition.

PROBLEMS

10.123. Calculate the molality of each of the following solutions:
a. 0.875 mol of glucose ($C_6H_{12}O_6$) in 1.5 kg of water
b. 11.5 mmol of acetic acid (CH_3COOH) in 65 g of water
c. 0.325 mol of baking soda ($NaHCO_3$) in 290.0 g of water

*10.124. Table 4.1 lists molar concentrations of major ions in seawater. Using a density of 1.022 g/mL for seawater, convert the concentrations into molalities.

10.125. What mass of the following solutions contains 0.100 mol of solute? (a) 0.334 m NH_4NO_3; (b) 1.24 m ethylene glycol, $HOCH_2CH_2OH$; (c) 5.65 m $CaCl_2$

10.126. How many moles of solute are there in the following solutions?
a. 0.150 m glucose solution made by dissolving the glucose in 100.0 kg of water
b. 0.028 m Na_2CrO_4 solution made by dissolving the Na_2CrO_4 in 1000.0 g of water
c. 0.100 m urea solution made by dissolving the urea in 500.0 g of water

10.127. Fish Kills High concentrations of ammonia (NH_3), nitrite ion, and nitrate ion in water can kill fish. Lethal concentrations of these species for rainbow trout are 1.1 mg/L, 0.40 mg/L, and 1361 mg/L, respectively. Express these concentrations in molality units, assuming a solution density of 1.00 g/mL.

10.128. The concentrations of six important elements in a sample of river water are: 0.050 mg/kg of Al^{3+}, 0.040 mg/kg of Fe^{3+}, 13.4 mg/kg of Ca^{2+}, 5.2 mg/kg of Na^+, 1.3 mg/kg of K^+, and 3.4 mg/kg of Mg^{2+}. Express each of these concentrations in molality units.

10.129. Cinnamon Cinnamon owes its flavor and odor to cinnamaldehyde (C_9H_8O). Determine the boiling point elevation of a solution of 100 mg of cinnamaldehyde dissolved in 1.00 g of carbon tetrachloride ($K_b = 5.03$°C/m).

10.130. Spearmint Determine the boiling point elevation of a solution of 125 mg of carvone ($C_{10}H_{14}O$, oil of spearmint) dissolved in 1.50 g of carbon disulfide ($K_b = 2.34°C/m$).

10.131. What molality of a nonvolatile, nonelectrolyte solute is needed to lower the melting point of camphor by 1.000°C ($K_f = 39.7°C$)?

10.132. What molality of a nonvolatile, nonelectrolyte solute is needed to raise the boiling point of water by 7.60°C ($K_b = 0.52°C/m$)?

10.133. Saccharin Determine the melting point of an aqueous solution made by adding 186 mg of saccharin ($C_7H_5O_3NS$) in 1.00 mL of water (density = 1.00 g/mL, $K_f = 1.86°C/m$).

10.134. Determine the boiling point of an aqueous solution that is 2.50 m ethylene glycol ($HOCH_2CH_2OH$); K_b for water is 0.52°C/m. Assume that the boiling point of pure water is 100.00°C.

10.135. Which aqueous solution has the lowest freezing point: 0.5 m glucose, 0.5 m NaCl, or 0.5 $mCaCl_2$?

10.136. Which aqueous solution has the highest boiling point: 0.5 m glucose, 0.5 m NaCl, or 0.5 m $CaCl_2$?

10.137. Which one of the following aqueous solutions should have the highest boiling point: 0.0200 m CH_3OH, 0.0125 m KCl, or 0.0100 m $Ca(NO_3)_2$?

10.138. Which one of the following aqueous solutions should have the lowest freezing point: 0.0500 m $C_6H_{12}O_6$, 0.0300 m KBr, or 0.0150 m Na_2SO_4?

10.139. Arrange the following aqueous solutions in order of increasing boiling point:
a. 0.06 m $FeCl_3$ ($i = 3.4$)
b. 0.10 m $MgCl_2$ ($i = 2.7$)
c. 0.20 m KCl ($i = 1.9$)

10.140. Arrange the following solutions in order of increasing freezing point depression:
a. 0.10 m $MgCl_2$ in water, $i = 2.7$, $K_f = 1.86°C/m$
b. 0.20 m toluene in diethyl ether, $i = 1.00$, $K_f = 1.79°C/m$
c. 0.20 m ethylene glycol in ethanol, $i = 1.00$, $K_f = 1.99°C/m$

10.141. The following pairs of aqueous solutions are separated by a semipermeable membrane. In which direction will the solvent flow?
a. A = 1.25 M NaCl; B = 1.50 M KCl
b. A = 3.45 M $CaCl_2$; B = 3.45 M NaBr
c. A = 4.68 M glucose; B = 3.00 M NaCl

10.142. The following pairs of aqueous solutions are separated by a semipermeable membrane. In which direction will the solvent flow?
a. A = 0.48 M NaCl; B = 55.85 g of NaCl dissolved in 1.00 L of solution
b. A = 100 mL of 0.982 M $CaCl_2$; B = 16 g of NaCl in 100 mL of solution
c. A = 100 mL of 6.56 mM $MgSO_4$; B = 5.24 g of $MgCl_2$ in 250 mL of solution

10.143. Calculate the osmotic pressure of each of the following aqueous solutions at 20°C:
a. 2.39 M methanol (CH_3OH)
b. 9.45 mM $MgCl_2$
c. 40.0 mL of glycerol ($C_3H_8O_3$) in 250.0 mL of aqueous solution (density of glycerol = 1.265 g/mL)
d. 25 g of $CaCl_2$ in 350 mL of solution

10.144. Calculate the osmotic pressure of each of the following aqueous solutions at 27°C:
a. 10.0 g of NaCl in 1.50 L of solution
b. 10.0 mg/L of $LiNO_3$
c. 0.222 M glucose
d. 0.00764 M K_2SO_4

10.145. Determine the molarity of each of the following solutions from its osmotic pressure at 25°C. Include the van't Hoff factor for the solution when the factor is given.
a. $\pi = 0.674$ atm for a solution of ethanol (C_2H_5OH)
b. $\pi = 0.0271$ atm for a solution of aspirin ($C_9H_8O_4$)
c. $\pi = 0.605$ atm for a solution of $CaCl_2$, $i = 2.47$

10.146. Determine the molarity of each of the following solutions from its osmotic pressure at 25°C. Include the van't Hoff factor for the solution when the factor is given.
a. $\pi = 0.0259$ atm for a solution of urea (CH_4N_2O)
b. $\pi = 1.56$ atm for a solution of sucrose ($C_{12}H_{22}O_{11}$)
c. $\pi = 0.697$ atm for a solution of KI, $i = 1.90$

10.147. Is the following statement true or false? For solutions of the same reverse osmotic pressure at the same temperature, the molarity of a solution of NaCl will always be less than the molarity of a solution of $CaCl_2$. Explain your answer.

10.148. Suppose you have 1.00 M aqueous solutions of each of the following solutes: glucose ($C_6H_{12}O_6$), NaCl, and acetic acid (CH_3COOH). Which solution has the highest pressure requirement for reverse osmosis?

Measuring the Molar Mass of a Solute Using Colligative Properties

CONCEPT REVIEW

10.149. What effect does dissolving a solute have on the following properties of a solvent? (a) its osmotic pressure; (b) its freezing point; (c) its boiling point

10.150. How can measurements of osmotic pressure, freezing point depression, and boiling point elevation be used to find the molar mass of a solute? Why is it important to know whether the solute is an electrolyte or a nonelectrolyte?

PROBLEMS

10.151. Throat Lozenges A 188 mg sample of a nonelectrolyte isolated from throat lozenges was dissolved in enough water to make 10.0 mL of solution at 25°C. The osmotic pressure of the resulting solution was 4.89 atm. Calculate the molar mass of the compound.

*10.152. An unknown compound (152 mg) was dissolved in water to make 75.0 mL of solution. The solution did not conduct electricity and had an osmotic pressure of 0.328 atm at 27°C. Elemental analysis revealed the substance to be 78.90% C, 10.59% H, and 10.51% O. Determine the molecular formula of this compound.

*10.153. **Cloves** Eugenol is one of the compounds responsible for the flavor of cloves. A 111-mg sample of eugenol was dissolved in 1.00 g of chloroform ($K_b = 3.63°C/m$), increasing the boiling point of chloroform by 2.45°C. Calculate eugenol's molar mass. Eugenol is 73.17% C, 7.32% H, and 19.51% O by mass. What is the molecular formula of eugenol?

*10.154. **Caffeine** The freezing point of a solution prepared by dissolving 150 mg of caffeine in 10.0 g of camphor is lower by 3.07°C than that of pure camphor ($K_f = 39.7°C/m$). What is the molar mass of caffeine? Elemental analysis of caffeine yields the following results: 49.49% C, 5.15% H, 28.87% N, and the remainder is O. What is the molecular formula of caffeine?

Additional Problems

10.155. Which substance contains the smallest cation?
(a) CsF; (b) RbF; (c) KCl; (d) NaBr

10.156. Which substance contains the most positively charged cation?
(a) $MgCl_2$; (b) AlF_3; (c) KI; (d) SrO

10.157. Which substance has the least negative lattice energy?
(a) MgI_2; (b) $MgBr_2$; (c) $MgCl_2$; (d) MgF_2

10.158. Explain why the boiling point of pure sodium chloride is much higher than the boiling point of an aqueous solution of sodium chloride.

10.159. Why does methanol (CH_3OH) boil at a lower temperature than water, even though CH_3OH has the greater molar mass?

10.160. Why is methanol (CH_3OH) miscible with water, while CH_4 is practically insoluble in water?

10.161. Does the sublimation point of ice increase or decrease with increasing pressure?

10.162. Sketch a phase diagram for element X, which has a triple point (152 K, 0.371 atm), a boiling point of 166 K at a pressure of 1.00 atm, and a normal melting point of 161 K.

*10.163. The melting point of hydrogen is 14.96 K at 1.00 atm pressure. The temperature of its triple point is 13.81 K. Does H_2 expand or contract when it freezes?

10.164. Explain why water climbs higher in a narrow glass capillary than in a test tube.

10.165. Explain why ice floats on water.

10.166. Predict how changing temperature might affect the capillary action of water. Explain your answer.

10.167. Do nonpolar liquids have a surface tension?

*10.168. **Melting Ice** $CaCl_2$ is often used to melt ice on sidewalks. Could $CaCl_2$ melt ice at −20°C? Assume that the solubility of $CaCl_2$ at this temperature is 70.1 g $CaCl_2$/100.0 g of H_2O and that the van't Hoff factor for a saturated solution of $CaCl_2$ is 2.5.

*10.169. **Making Ice Cream** A mixture of table salt and ice is used to chill the contents of hand-operated ice-cream makers. What is the melting point of a mixture of 2.00 lb of NaCl and 12.00 lb of ice if exactly half of the ice melts. Assume that all the NaCl dissolves in the melted ice and that the van't Hoff factor for the resulting solution is 1.44.

10.170. The freezing points of 0.0935 m ammonium chloride and 0.0378 m ammonium sulfate in water were found to be −0.322°C and −0.173°C, respectively. What are the values of the van't Hoff factors for these salts?

10.171. The following data were collected for three compounds in aqueous solution. Determine the value of the van't Hoff factor for each salt (K_f for water = 1.86°C/m).

Compound	Concentration	Observed ΔT_f (°C)
LiCl	5.0 g/kg	0.410
HCl	5.0 g/kg	0.486
NaCl	5.0 g/kg	0.299

10.172. **Physiological Saline** 100.0 mL of a solution of physiological saline (0.92% NaCl by weight) is diluted by the addition of 250.0 mL of water. What is the osmotic pressure of the final solution at 37°C? Assume that NaCl dissociates completely into $Na^+(aq)$ and $Cl^-(aq)$.

10.173. 100.0 mL of 2.50 mM NaCl is mixed with 80.0 mL of 3.60 mM $MgCl_2$ at 20°C. Calculate the osmotic pressure of each starting solution and that of the mixture, assuming that the volumes are additive and that both salts dissociate completely into their component ions.

10.174. A solution of 7.50 mg of a small protein in 5.00 mL aqueous solution has an osmotic pressure of 6.50 torr at 23.1°C. What is the molar mass of the protein?

The Halogens: The Salt of the Earth

10.175. What are the major natural sources of chlorine and bromine?

10.176. Under normal conditions, bromine is a volatile red liquid and iodine is a violet solid. Why are Br_2 and I_2 so volatile?

10.177. How do the boiling points of the halogens change in the progression F_2, Cl_2, Br_2, I_2?

10.178. Which is by far the least abundant of the halogens?

10.179. Give two industrial uses each of (a) chlorine and (b) bromine.

10.180. Give the chemical formula of chloramine. What shape is the chloramine molecule? Would you expect chloramine to be soluble in water? Why?

10.181. Which of each halide pair is the most easily oxidized?
(a) fluoride, iodide; (b) bromide, chloride; (c) chloride, fluoride; (d) iodide, bromide

10.182. Give the formula of each of the following halogen species: (a) hypochlorite; (b) chlorite; (c) chlorate; (d) perchlorate.

10.183. Give the species with the higher oxidation number of the halogen in each of the following pairs: (a) hypochlorous acid, Cl^-; (b) chlorous acid, $HClO_3$; (c) chloric acid, BrO_2^-; (d) perchloric acid, IO^-.

10.184. Balance the equation for the reduction of iodate by hydrogen sulfite ions to give iodide and sulfite in basic aqueous solution.

The Chemistry of Solids

ALLOYS FOR STRENGTH *When the Leonard P. Zakim Bunker Hill Bridge across the Charles River in Boston opened for traffic in 2003, it was the widest (56 m and 10 lanes of traffic) cable-stayed bridge in the world. Each metal cable can support over 5 million kilograms.*

A LOOK AHEAD: Stronger, Tougher, Harder

Did you know that gold rings are not made of pure gold? Because gold is one of the softer metals, it is very easy to bend a ring made of pure gold (also called "24-karat gold") out of shape. In contrast, rings made of gold blended with other metals are more resistant to physical damage and can last a lifetime. The 22-karat gold used to make most wedding rings is a homogeneous mixture, called an alloy, that is 92% by mass gold (22/24 = 0.92, or 92%) and 8% other metals. Different colors of gold result from different blends: White gold typically includes from 25% to 60% palladium and silver, and rose gold contains from 20% to 55% copper. An especially beautiful alloy of 96% copper and 4% gold called shakudo is prized in Japan for its dark purplish-blue sheen.

There are over 500,000 different alloys that have been made and characterized. Some are solid solutions that are homogeneous at the atomic level just as liquid solutions are; others are heterogeneous. The purpose of making alloys, and the reason why so many are possible, is that their properties can be controlled by varying the proportions of components in the mixture. Alloys may have properties similar to or very different from the pure metals from which they are made.

The many types of steel are alloys of iron, the nonmetal carbon, and other elements. Steel that is 0.4% by weight carbon is much harder and more durable than pure iron. Other elements may be added to give steel certain properties. For example, stainless steel that is 18% chromium by weight resists rusting and corrosion. Stainless steel is used in the knives, forks, and spoons of standard tableware and in other products where appearance matters and strength is required.

The construction industry uses alloys in many ways, from steel I-beams for basic structures, to other metals for architectural features and interior fixtures. Brass is an alloy of copper and zinc that is both harder and stronger than either element alone. It is commonly used in doorknobs and hinges, but over time they tarnish, looking dull, even dirty. Because stainless steel doorknobs look clean and sleek, they are frequently used in public places such as hospitals and schools, where the appearance of cleanliness is important. Recent studies have found, however, that brass may be the better choice because it actually kills bacteria that come in contact with it. This bactericidal effect is due to the presence of copper, which has long been known to exhibit antimicrobial activity. This property of pure copper metal is retained in the alloy.

Alloys are of great importance in the transportation industry. Aluminum alloys are ideal for making aircraft because they are strong and lightweight. The wingtip of a new 747-400 airplane is about 1.8 m longer than the usual 747's wing, but it is actually lighter because it is made of a new alloy of 95% aluminum blended with copper, manganese, and iron, and the longer wing weighs over 2 metric tons less.

Even sports equipment benefits from the development of new alloys. A patented five-metal alloy of nickel, zirconium, titanium, copper, and beryllium is advertised as possibly changing the game of golf forever because the alloy produces a stronger, lighter, and more resilient club that enables a golfer to transfer more energy from the swing into the ball.

In this chapter we examine why metal alloys are so much stronger, tougher, and harder than the pure metals from which they are made. We will explore the links between the physical properties of solids at the macroscopic level and the structures of these solids at the atomic level. We will start with metals and their alloys and conclude with an equally important class of nonmetallic compounds called ceramics. 陶瓷(的)

11.1 Metallic Bonds and Conduction Bands

Most of the elements in the periodic table are metals, which means they typically are hard, shiny, malleable (easily shaped), ductile (easily drawn out), and conduct heat and electricity. In this section we explore *why* they conduct heat and electricity so well and display some of their other properties. All of these properties can be explained by examining the structures of metals at the atomic level and by exploring models that account for the nature of the bonds that hold metal atoms together.

According to valence bond theory, covalent bonds form between pairs of atoms when partially filled orbitals—one from each atom—overlap. Our focus in Chapter 9 was on bonding in molecules, principally in the gas phase. One focus of this chapter is the bonding between atoms that are densely packed together in metallic solids. Dense packing means that the valence orbitals of one atom overlap with orbitals of as many as 12 nearest neighbors, as shown in Figure 11.1. Thus the bonds that link a metal atom to its nearest neighbors extend in virtually all directions. This large number of interactions keeps metal atoms together and makes metals strong. At the same time the sharing of a limited number of valence electrons with many bonding partners means that the bond linking a particular pair of metal atoms is relatively weak.

To understand this point, consider the nature of the bond that would form if 2 copper atoms came together in the gas phase and formed a molecule of Cu_2. Copper atoms have the electron configuration $[Ar]3d^{10}4s^1$. When the partially filled $4s$ orbitals of 2 copper atoms overlap, they form a diatomic molecule held together by a single covalent bond (Figure 11.1a). Now consider an atom of copper inside a piece of copper wire (Figure 11.1b). It is surrounded by and bonded to 12 other Cu atoms, which means it must share its $4s$ electron with 12 other atoms, not just 1. Inevitably, this dispersion of bonding electrons weakens the Cu–Cu bond between each pair of Cu atoms.

Adding to the diffuse nature of this bonding, metallic elements have lower electronegativities than nonmetals. Recall from Chapter 8 that differences in electronegativity are important for defining how bonding pairs of electrons are distributed between pairs of atoms. In a pure metal there are no such differences in electronegativity. However, the low electronegativities of metals mean that the valence electrons are not held tightly to the nuclei of individual atoms. On an energy scale where free electrons have zero energy and electrons tightly bound to a nucleus have a much lower (more negative) energy, the energies of the electrons in metallic bonds are *less* negative than those in covalent bonds. Higher energy coupled with the diffuse nature of metallic bonding means that individual atoms in a metal have a very tenuous hold on their valence-shell electrons.

In Chapter 8 we introduced the concept of metallic bonding in terms of metal atoms "floating" in seas of mobile bonding electrons. The diffuse nature of

⊙⊙ **CONNECTION** In Chapter 9 we discussed the valence bond theory of chemical bond formation.

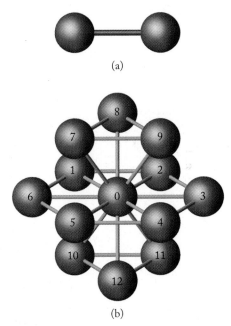

FIGURE 11.1 Covalent bonds differ from metallic bonds: (a) This ball-and-stick model of a molecule of Cu_2 is based on the assumption that each Cu atom shares its $4s$ electron in forming a single covalent bond. (b) In solid copper, the atom labeled 0 shares its $4s$ electron with 12 other atoms. As a result, the bonds in copper and other metals are much more diffuse than the covalent bonds in small molecules.

Band theory is an extension of molecular orbital theory that describes bonding in solids.

Bands of orbitals that are filled or partially filled by valence electrons are called **valence bands**.

Higher-energy unoccupied bands in which electrons are free to migrate are called **conduction bands**.

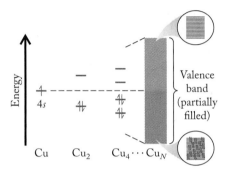

FIGURE 11.2 What a Chemist Sees As the half-filled 4s orbitals of an increasing number of Cu atoms overlap, their energies are split into a half-filled valence band. Electrons can move from the filled half (purple) of the valence band to the slightly higher-energy upper half (red), where they are free to migrate from one empty orbital to another across the entire solid.

metallic bonding described in the preceding paragraphs certainly fits the sea-of-electrons model in Chapter 8, but there is a more sophisticated approach, called **band theory**, to explain the bonding in metals and other solids. To develop an understanding of band theory, which is an extension of molecular orbital theory, let's start with copper atoms in the gas phase. The energy of an electron in a 4s orbital of a free Cu atom is equal to the negative of the first ionization energy of copper: −745 kJ/mol. When the 4s orbitals on 2 Cu atoms overlap to form Cu_2, the atomic orbitals split into two molecular orbitals with different energies equally spaced above and below the initial value, as shown in Figure 11.2. This splitting process is analogous to the formation of low-energy bonding and high-energy antibonding molecular orbitals (see Section 9.6). If another 2 Cu atoms join the first 2 to form a molecule of Cu_4, the two molecular orbitals split into four. If we add another 4 atoms to make Cu_8, the four molecular orbitals split into eight. In all these molecules the lower-energy orbitals are filled with the available 4s electrons and the upper orbitals are empty. If we apply this model to the enormous number of atoms in a piece of copper wire, all this energy splitting produces a continuous *band* of available energies. Because the band comes from overlapping valence-shell orbitals, we call it a **valence band**.

Band theory explains the conductivity of copper and many other metals by assuming that there is essentially no gap between the energy of the occupied lower portion of the valence band and the empty upper portion. Therefore, valence electrons can move easily from the lower portion to the upper portion, where they are free to move from one empty orbital to the next and flow throughout the solid. for metals

This partially filled valence-shell model explains the conductivity of many metals, but not all. Consider one of copper's next-door neighbors in the periodic table, zinc. Its electron configuration ($[Ar]3d^{10}4s^2$) tells us that all of its valence-shell electrons reside in filled orbitals. This means that the valence band in solid zinc is completely filled, too. With no empty orbitals to move into, the valence-shell electrons in Zn might appear to be immobile. However, they actually *do* migrate; zinc is a good conductor of heat and electricity. Band theory explains why. It assumes *all* atomic orbitals of comparable shape and energy, including zinc's empty 4p orbitals, can mix together to form energy bands. The energy band produced by splitting empty 4p orbitals is broad enough to overlap the valence band produced by splitting the filled 4s orbitals, as shown in Figure 11.3. This means that it is easy for electrons from the filled valence band to move up to the empty orbitals of what is called a **conduction band**, where they are free to migrate and conduct electricity.

Overlapping occupied (valence) and unoccupied (conduction) bands make metals good conductors of electricity and also heat. For example, when atoms on the outside of a copper pot are heated, they vibrate faster. Electrons absorb this thermal energy and are raised from the filled portion of the valence band to the unfilled portion, where they are free to migrate. The higher the temperature, the more electrons move into the unfilled portion. Thermal energy in the form of a large population of mobile electrons moves rapidly through a metal pot from the source of heat to the contents of the pot.

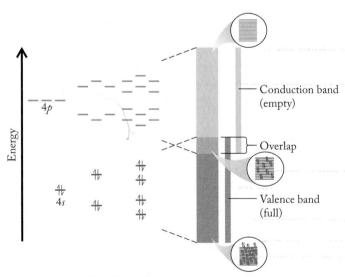

FIGURE 11.3 What a Chemist Sees As the filled 4s orbitals of an increasing number of Zn atoms overlap, they form a filled valence band (shown in purple). An empty conduction band (gray) is produced by the overlap of empty 4p orbitals on a large number (N) of Zn atoms. Once N is large enough so that these bands overlap each other, electrons can move easily from the filled valence band to the empty conduction band.

Is the electrical conductivity of magnesium metal best explained in terms of overlapping conduction and valence bands, or in terms of a partially filled valence band? Explain your answer.

(Answers to Concept Tests are in the back of the book.)

11.2 Semimetals and Semiconductors

To the right of the metals in the periodic table there is a "staircase" of elements that tend to have the physical properties of metals and chemical properties of nonmetals. These semimetals (or metalloids) are not as good at conducting electricity as metals, but they are much better at it than the nonmetals. We can explain this "in between" behavior using band theory. Let's begin with the most abundant of the semimetals, silicon. When we apply band theory to the bonding in pure silicon we find that there is an energy gap of about 100 kJ/mol between the occupied valence band and the empty conduction band (Figure 11.4a). The difference is called, logically, a **band gap** (E_g).

Generally, only a few of the electrons in the valence band of solid Si have sufficient energy to move to the conduction band, which limits silicon's ability to conduct heat and electricity. However, as we are about to see, the conductivity of silicon and the other semimetals can be enhanced by many orders of magnitude with relatively small changes in chemical composition. This varying ability to conduct electricity, in between that of metals and nonmetals, makes the semimetals **semiconductors**.

We can change the conductivity of solid Si (or any elemental semiconductor) by replacing some of the Si atoms in its solid structure with atoms of another element of similar atomic radius but with a different number of valence electrons. The replacement process is called *doping*, and the added element is called a *dopant*. Suppose the dopant is a group 15 element such as phosphorus: each phosphorus atom adds one extra valence electron to the structure. These electrons have a different energy from the silicon electrons and populate a band within the silicon band gap (Figure 11.4b). This arrangement effectively diminishes the size of the band gap, making it easier for valence electrons to reach the conduction band. This **n-type** semiconductor is so called because the dopant contributes extra negative charges (electrons) to the structure of the host element.

The conductivity of solid silicon can also be enhanced by substituting Si atoms with atoms of a group 13 element, such as gallium. These substitutions mean that there are fewer valence electrons in the solid. The result is an empty band in the band gap (Figure 11.4c). Less energy is needed to promote an electron from the filled valence band of silicon to the new empty (conduction) band created by the group 13 element, increasing electrical conductivity. This type of semiconductor, in which there is a reduction in the number of valence electrons (or negative charges) is called a **p-type** semiconductor because a reduction in the number of negatively charged electrons is equivalent to the presence of positively charged holes. The transistors used in solid-state electronics are combinations of n- and p-type semiconductors.

Semiconductors can also be prepared from mixtures of group 13 and group 15 elements. For example, gallium arsenide (GaAs) is a semiconductor that emits red ($\lambda = 874$ nm) light when connected to an electrical circuit. This emission is

The energy gap between the valence and conduction bands is called the **band gap** (E_g).

A **semiconductor** is a substance whose conductivity can be made to vary over several orders of magnitude by altering its chemical composition.

An **n-type** semiconductor contains excess electrons contributed by electron-rich dopant atoms.

A **p-type** semiconductor contains electron-poor dopant atoms that cause a reduction in the number of electrons, which is equivalent to the presence of positively charged holes.

CONNECTION We introduced the semimetals (metalloids) in Section 2.6 when we described the structure of the periodic table.

both of them increase the conductivity.

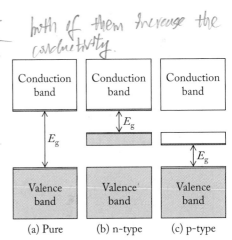

FIGURE 11.4 Semiconductors such as silicon have variable conductivities that can be greatly enhanced by adding other elements. (a) Pure Si has a band gap (E_g) of about 100 kJ/mol. (b) Adding an element that has more valence electrons per atom than Si creates a filled band in the band gap of an n-type semiconductor. (c) Adding an element with fewer valence electrons than the host material creates an empty conduction band in the band gap of a p-type semiconductor.

used in devices such as laser pointers, bar-code readers in stores, and CD players. The energy of each photon of red light corresponds to the energy gap between the valence and conduction bands in GaAs. When electrical energy is applied to the material, electrons are raised to the conduction band. When they fall back to the valence band, they emit light. If aluminum is substituted for gallium in GaAs, the band gap increases, and predictably the wavelength of emitted light decreases. For example, a material with the empirical formula $AlGaAs_2$ emits yellow-orange light ($\lambda = 620$ nm). Many of the multicolored indicator lights in electronic devices such as sound system components rely on light-emitting diodes (LEDs) based on aluminum gallium arsenide semiconductors.

SAMPLE EXERCISE 11.1   **Distinguishing Between p- and n-Type Semiconductors**

What kind of semiconductor—n-type or p-type—does doping germanium with arsenic create?

COLLECT AND ORGANIZE We are asked to determine the type of semiconductor formed when germanium (Ge, group 14) is doped with arsenic (As, group 15).

ANALYZE The fact that As is in group 15 means that its atoms have one more valence electron than the atoms of Ge.

SOLVE When the dopant has more valence electrons than the host semimetal, the two form an n-type semiconductor.

THINK ABOUT IT Arsenic is a good candidate for making an n-type semiconductor because its atoms are nearly the same size as those of germanium and so fit easily into the structure of solid Ge.

Practice Exercise Gallium arsenide (GaAs) is a semiconductor used in optical scanners in retail stores. GaAs can be made an n-type or a p-type semiconductor by replacing some of the As with another element. Which element—Se or Sn—would form an n-type semiconductor with GaAs, and which would form a p-type semiconductor?

(Answers to Practice Exercises are in the back of the book.)

11.3 Metallic Crystals

In Section 11.1 we observed that each Cu atom in a piece of copper metal touches 12 other copper atoms. Let's explore how the atoms in different metals adopt characteristic stacking patterns.

Stacking Patterns

When copper is heated above its melting point and then slowly allowed to cool, its atoms arrange themselves in ordered three-dimensional arrays, forming a **crystalline solid**. Think of these arrays as stacked layers (two-dimensional planes) of Cu atoms packed together as tightly as possible. Each atom in a layer touches 6 others in that layer. The atoms in the layer directly above nestle into the spaces created by the atoms below, as shown in Figure 11.5. Similarly, the atoms in a third layer nestle among those in the second. However, two different alignments are possible

A **crystalline solid** is made of an ordered array of atoms, ions, or molecules.

(a)   (b)

Displays of (a) oranges in a grocery store or (b) cannonballs at an 18th-century fort illustrate closest-packed arrays of spherical objects.

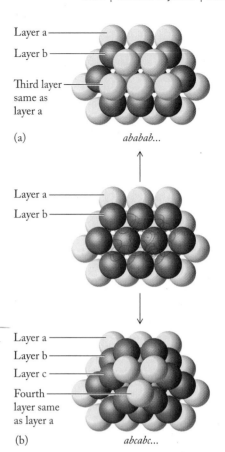

FIGURE 11.5 Two efficient ways to stack layers of solid spheres; the difference between them stems from the arrangement of the third layer: (a) The spheres in the third layer are directly above those in the first, giving an *ababab* . . . stacking pattern. (b) The alignment of spheres in the third layer is different from either of the two layers just below. A fourth layer does align with the first, generating an *abcabc* . . . stacking pattern.

for the atoms in the third layer. They could be located directly above the atoms in the first layer (Figure 11.5a). However, they could fit just as snugly above the atoms of the second layer in such a way that the third layer of atoms is not directly above the first layer (Figure 11.5b). When a fourth layer is nestled into the spaces of the third layer in Figure 11.5(b), the atoms of the fourth layer do lie directly above those in the first. The three layers with their distinctive alignments are labeled *a*, *b*, and *c*. Together they form a repeating layering pattern *abcabcabc* . . . throughout the crystal. On the other hand, the layers in Figure 11.5(a), where the third layer is aligned with the first (layer *a*), have a simpler *ababab* . . . repeating pattern.

Packing Structures and Unit Cells

Let's take a closer look at the *ababab* . . . stacking pattern, and focus on a cluster of atoms in this pattern (Figure 11.6). The cluster forms a *hexagonal* (six-sided) prism of closely packed spheres. In fact, they are as tightly packed as they can be, so the structure is called **hexagonal closest-packed (hcp)**. The atoms in 16 metallic elements (see Figure 11.7) have hcp structures. Within these structures the 15-atom cluster in Figure 11.6(c) serves as an atomic-scale building block—a pattern of atoms repeated over and over again in all three dimensions inside the solid. We

▶❚❚ CHEMTOUR Crystal Packing

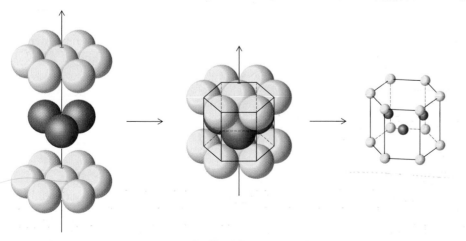

(a) *ababab*... layering   (b) Hexagonal closest-packed structure   (c) Hexagonal unit cell

FIGURE 11.6 (a,b) Hexagonal closest-packed (hcp) structures and (c) their corresponding hexagonal unit cells represent a highly efficient way to pack atoms in crystalline solids. Sixteen metals, including all those in groups 3 and 4, have hcp structures.

Hexagonal closest-packed (hcp) describes a crystal structure in which the layers of atoms or ions in hexagonal unit cells have an *ababab* . . . stacking pattern.

The **unit cell** is the basic repeating unit of the arrangement of atoms, ions, or molecules in a crystalline solid.

A **crystal lattice** is the three-dimensional array of particles (atoms, ions, or molecules) in a crystalline solid.

Cubic closest-packed (ccp) describes a crystal structure in which the layers of atoms, ions, or molecules in face-centered cubic unit cells have an *abcabcabc . . .* stacking pattern.

A **face-centered cubic (fcc) unit cell** has the same kind of particle (atom, ion, or molecule) at the 8 corners of a cube and at the center of each face.

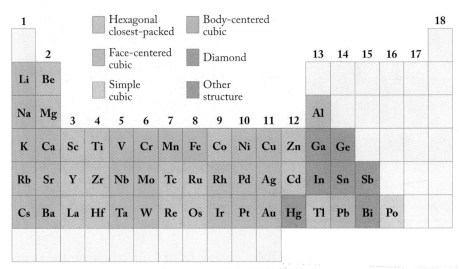

FIGURE 11.7 Summary of the crystal structures of metals in the periodic table. Five metals ("other structure" in the color key) have structures more complicated than can easily be described in this book: mercury (Hg), gallium (Ga), indium (In), antimony (Sb), and bismuth (Bi).

▶❚❚ CHEMTOUR Unit Cell

call these crystalline building blocks **unit cells**. A unit cell represents the minimum repeating pattern that describes the three-dimensional array of atoms in a crystalline structure, or **crystal lattice**, of a solid. Think of unit cells as microscopic analogues of the repeating patterns you may have seen in fabrics, wrapping paper, or even a checkerboard, as illustrated in Figure 11.8. Look carefully at the image in Figure 11.8a to confirm that the outlined portion represents the minimum repeating pattern. A unit cell serves a similar function: it is the three-dimensional pattern of atoms, ions, or molecules that is repeated throughout a crystalline solid. The six-sided cell in Figure 11.6 is called a *hexagonal unit cell*.

Atoms in solid copper and in a dozen other metals adopt the *abcabc . . .* stacking pattern when they solidify. To understand the sort of crystals they form, consider what happens when we take a cluster of 14 spheres stacked in an *abcabc . . .* pattern and turn the cluster about 45° as shown in Figure 11.9. The atoms form a cube: 8 of them at the 8 corners of the cube and 6 at the centers of each of the 6 faces. Because the atoms are stacked together as closely as possible, this structure is called **cubic closest-packed (ccp)**, and the corresponding unit cell is called a **face-centered cubic (fcc) unit cell**. In the fcc unit cell, the atoms at the corners do not touch each other, but they do touch along the face diagonal, a line that connects the centers of the atoms at opposite corners of any face of the cube. The edges of the fcc unit cell are all of equal length, and the angle between any two edges is 90°.

FIGURE 11.8 (a) The highlighted "unit cell" of a checkerboard is the minimum set of squares that defines the pattern repeated over the entire board. (b) This wrapping paper has a more complex pattern. One unit cell is highlighted. Can you locate others?

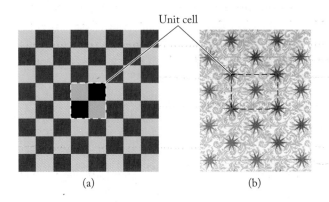

Unit cell

(a)   (b)

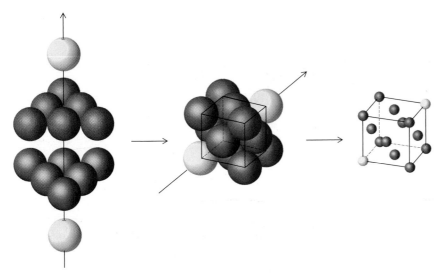

(a) *abcabc...* layering   (b) Cubic closest-packed structure   (c) Face-centered cubic unit cell

FIGURE 11.9 Solids with cubic closest-packed (ccp) structures have face-centered cubic (fcc) unit cells, which are more easily seen if the *abcabc*... layers are tipped about 45°. The packing efficiency in a ccp structure is the same as in a hexagonal closest-packed (hcp) structure (see Figure 11.6).

So far we have introduced two *closest-packed* lattice structures—hexagonal and ccp—along with their associated unit cells (hexagonal and fcc). There are also structures in which the atoms or ions are arranged close together but not as efficiently as hcp and ccp lattices. Figure 11.10 shows two other types of cubic unit cells. The **body-centered cubic (bcc) unit cell** has an atom at each of the 8 corners and 1 in the middle of the cell (Figure 11.10b). All the group 1 metals and many transition metals have bcc unit cells. The **simple cubic unit cell** consists of just 8 corner atoms (Figure 11.10c). It is the least efficiently packed of the three structures in the cubic crystal system and quite rare among metals: only radioactive polonium (Po) forms a simple cubic unit cell.

You may be wondering why some metals adopt one kind of unit cell and other metals another. We will address that question later in the chapter. For now, keep in mind that crystalline solids with cubic unit cells form crystals that are cubic in appearance on a macroscopic scale, whereas crystalline solids with hexagonal unit cells tend to form hexagonal crystals.

Unit-Cell Dimensions

The cutaway views in Figure 11.10 show the geometric and mathematical relationships between the edge length (ℓ), face-diagonal, and body-diagonal dimensions in fcc and bcc unit cells. These views also provide us with a way to determine how many atoms are actually inside each type of cubic unit cell.

Let's start with the simplest unit cell, the cutaway view of the simple cubic unit cell (Figure 11.10c). Note how only a fraction of each corner atom is actually inside the unit-cell boundaries. Each corner atom is shared by the 8 unit cells meeting at that corner, and so only one-eighth of each corner atom is inside each unit cell. There are 8 corners in a cube, so there is a total of

$$\tfrac{1}{8} \times 8 = 1 \text{ corner atom per simple cubic unit cell}$$

(Actually, the calculation $\tfrac{1}{8} \times 8 = 1$ applies to the corner atoms in any cubic unit cell.) Note that the atoms in Figure 11.10(c) touch along each edge. Therefore the edge length (ℓ) in the simple cubic unit cell of an element is equal to twice the atomic radius of the element. *that's why.*

A **body-centered cubic (bcc) unit cell** has atoms at the 8 corners of a cube and at the center of the cell.

A **simple cubic unit cell** has atoms only at the 8 corners of a cube.

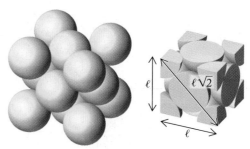

(a) Face-centered cubic

(b) Body-centered cubic

FIGURE 11.10 Cubic crystals may form from a variety of cubic unit cells. Shown here are whole-atom and cutaway views of (a) face-centered cubic (fcc), (b) body-centered cubic (bcc), and (c) simple cubic unit cells. Note how the atoms in an fcc unit cell touch along the face diagonal, which forms a right triangle with two adjoining edges of length ℓ. This relationship means that the length of the face diagonal is $\sqrt{\ell^2 + \ell^2} = \ell\sqrt{2}$. In a bcc cell the atoms touch along a body diagonal. Another right triangle forms from an edge, a face diagonal, and a body diagonal, making the body diagonal $\sqrt{3}$ times the unit-cell edge length.

(c) Simple cubic

In an fcc unit cell (Figure 11.10a) there are 8 corner atoms and 1 atom in the center of each of the 6 faces. Each face atom is shared by the 2 unit cells that abut each other at that face. Therefore each cell gets credit for half of each face atom, making a total of

$$\frac{1}{2} \times 6 = 3 \text{ face atoms}$$

and a grand total of

$$\left(\frac{1}{8} \times 8 = 1\right) \text{corner atom} + 3 \text{ face atoms} = 4 \text{ atoms per fcc unit cell}$$

To relate the size of these atoms to the dimensions of the cell, note that the atoms touch along the face diagonal in the cutaway view of Figure 11.10(a). Look carefully and see how the diagonal spans a radius of 2 corner atoms and the diameter (2 radii) of the atom in the center of the face. Therefore the face diagonal is equivalent to $1 + 2 + 1 = 4$ atomic radii (r). The face diagonal connects the ends of two edges and forms a right triangle. Therefore, according to the Pythagorean theorem, the face diagonal is the square root of the sum of the squares of the two edge-length (ℓ) distances:

$$\text{Face diagonal} = 4r = \sqrt{\ell^2 + \ell^2} = \ell\sqrt{2}$$

$$\text{so} \quad r = \frac{\ell\sqrt{2}}{4} = 0.3536\,\ell \tag{11.1}$$

Now let's focus on the cutaway view of the bcc unit cell in Figure 11.10(b). In addition to the corner atoms there is also 1 atom in the center of the cell that is entirely within the cell. This means that there are a total of:

$$\left(\frac{1}{8} \times 8 = 1\right) \text{corner atom} + 1 \text{ center atom} = 2 \text{ atoms per bcc unit cell}$$

Relating the unit-cell edge length to the atomic radius of the atoms in a bcc cell is complicated by the fact that the atoms do not touch along the edges, nor do they touch along face diagonals. However, each corner atom does touch the atom in the center of the cell. This means that the *body diagonal* spans (1) the radius of a corner atom, (2) the diameter (2 radii) of the atom in the center, and (3) the radius of the atom in the opposite corner. This makes the body diagonal equivalent to $1 + 2 + 1 = 4$ atomic radii (r).

We can again use the Pythagorean theorem to define the edge length (ℓ) with respect to the body diagonal. A right triangle is formed by an edge, a face diagonal, and a body diagonal serving as the hypotenuse of the triangle. In such a triangle the body diagonal relates to the edge length as follows:

$$\text{Body diagonal} = 4r = \sqrt{(\text{edge length})^2 + (\text{face diagonal})^2} = \sqrt{\ell^2 + (\ell\sqrt{2})^2} = \ell\sqrt{3}$$

so
$$r = \frac{\ell\sqrt{3}}{4} = 0.4330\,\ell \qquad (11.2)$$

In the following Sample Exercises we use Equations 11.1 and 11.2 to calculate the radii of the atoms in a unit cell and to calculate the densities of elements that crystallize in the unit cells depicted in Figure 11.10. For both calculations we need the edge lengths of the unit cell, which are obtained using an analytical technique called X-ray diffraction (described in detail later in Section 11.9). As you do these calculations, you may wish to consult the summary in Table 11.1 of how atoms in different locations in fcc, bcc, and simple cubic unit cells contribute to the contents of those cells.

SAMPLE EXERCISE 11.2 **Calculating Atomic Radius from Crystal Structure**

At room temperature the most stable crystal structure of iron has a body-centered cubic unit cell with an edge length of 287 pm (1 pm = 10^{-12} m). What is the radius in picometers of the iron atoms?

COLLECT AND ORGANIZE The crystal structure is based on a bcc unit cell (Figure 11.10b). This means that the iron atoms do not touch along cell edges or face diagonals but do touch along the body diagonals. We know the edge length of the cell: 287 pm.

ANALYZE Atoms touch along the body diagonals of a bcc unit cell, so the radius (r) of the atoms is related to the cell's edge length (ℓ) by Equation 11.2:

$$r = 0.4330\,\ell$$

SOLVE We insert the edge-length value into this equation to solve for r:

$$r = 0.4330 \times 287\text{ pm} = 124\text{ pm}$$

THINK ABOUT IT The average atomic radius of iron atoms is 126 pm, so the result of this calculation is reasonable. It should be because the values for atomic radii are derived from analyses of crystal structures in the first place.

Practice Exercise At 1070°C the most stable structure of iron has an fcc unit cell with an edge length of 361 pm. What is the atomic radius of iron at this temperature?

TABLE 11.1 Contributions of Atoms to Cubic Unit Cells

Position of Atom in Unit Cell	Contribution to Unit Cell	Unit-Cell Type
Center	1	bcc
Face	$\frac{1}{2}$	fcc
Corner	$\frac{1}{8}$	fcc, bcc, simple cubic

SAMPLE EXERCISE 11.3 **Using Crystal Structure to Calculate Density**

Calculate the density of iron in grams per cubic centimeter (g/cm³) at 25°C, given that its crystal structure is based on a bcc unit cell, and that the edge length of the cell is 287 pm.

COLLECT AND ORGANIZE We are asked to calculate a density of iron knowing its unit-cell geometry and edge length. We assume that the density of the unit cell is the same as the density of solid Fe. The fact that Fe has a bcc unit cell means that there are 2 atoms of Fe per cell. Recall that density is the ratio of mass to volume. The molar mass of Fe is 55.84 g/mol. The volume of a cubic cell is the cube of its edge length: $V = \ell^3$.

ANALYZE The density of an Fe bcc unit cell is the mass of 2 Fe atoms divided by the volume of the cell (287 pm)³. We need to calculate the mass of 2 Fe atoms starting with the molar mass (55.84 g/mol). The conversion includes dividing by Avogadro's number to calculate the mass of each Fe atom in grams. To obtain a cell volume in cubic centimeters (cm³), we make use of the definitions for *pico-* and *centi-*:

$$1 \text{ pm} = 10^{-12} \text{ m} \qquad 1 \text{ cm} = 10^{-2} \text{ m}$$

These may be combined to produce the conversion factor

$$1 \text{ pm} = 10^{-10} \text{ cm}$$

SOLVE
We calculate the mass (*m*) of 2 Fe atoms first—

$$m = \frac{55.84 \text{ g Fe}}{\text{mol Fe}} \times \frac{1 \text{ mol}}{6.022 \times 10^{23} \text{ atoms}} \times 2 \text{ atoms} = 1.855 \times 10^{-22} \text{ g}$$

and then the volume of the cell in cubic centimeters:

$$V = \ell^3 = (287 \text{ pm})^3 \times \frac{(10^{-10} \text{ cm})^3}{\text{pm}^3} = 2.364 \times 10^{-23} \text{ cm}^3$$

Taking the ratio of mass to volume, we have:

$$d = \frac{m}{V} = \frac{1.855 \times 10^{-22} \text{ g}}{2.364 \times 10^{-23} \text{ cm}^3} = 7.85 \text{ g/cm}^3$$

THINK ABOUT IT According to the data in Appendix 3, the density of iron is 7.874 g/mL, so the result of the calculation is reasonable.

Practice Exercise Silver and gold both crystallize in a face-centered cubic unit cell with very similar edge lengths: 407.7 and 407.0 pm. Calculate the density of each metal and compare your answers with the densities listed in Appendix 3.

11.4 Alloys

People have been using metals for tools, weapons, currency, and decorations (particularly jewelry) for tens of thousands of years. For most of that time the available metals were only three group 11 elements—Cu, Ag and Au. These were the only ones found as free metals in Earth's crust. Even so, most silver and copper occurs not as Ag and Cu but in ores such as argentite (Ag_2S), chalcopyrite ($CuFeS_2$), and chalcocite (Cu_2S). **Ores** are naturally occurring compounds or mixtures of compounds from which elements can be extracted.

Around 6000 years ago metal technology took a giant leap forward when artisans in the Tigris-Euphrates river valley (present-day Iraq) and the Indus River valley (now Pakistan) discovered how to convert copper ore, principally $CuFeS_2$, to copper metal. The process involved pulverizing the ore and then baking it in an oven. Baking initiated a chemical reaction with O_2 (from air), which converted the Cu in $CuFeS_2$ to CuO. The second step in the process allows CuO to react with carbon monoxide, which was produced by burning wood or charcoal (mostly carbon) in a furnace with an insufficient supply of air:

$$CuO(s) + CO(g) \rightarrow Cu(s) + CO_2(g) \tag{11.3}$$

One disadvantage of making tools and weapons from copper is that the metal is very malleable, which means that copper tools and weapons are easily bent and damaged. We can explain the malleability of Cu (and other metals) in terms of the diffuse metallic bonds between its atoms (Section 11.1) and their cubic closest-packed (ccp) structure (Section 11.3). A solid piece of copper has an enormous number of layers of tightly packed, two-dimensional planes of copper atoms that are only weakly bonded to each of their nearest neighbors. This arrangement gives the atoms in one plane the ability, under stress, to slip past atoms in adjacent planes (Figure 11.11). When the stress is relieved and the atoms stop slipping, many have different atoms as nearest neighbors but the overall crystal structure is still cubic closest-packed. The ease with which copper atoms slip past each other made it easy for prehistoric metalworkers to hammer copper metal into spear points and shields, but it also meant that those objects could easily be damaged in the heat of battle.

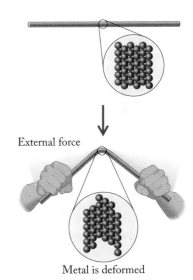

External force

Metal is deformed

FIGURE 11.11 Metals such as copper are malleable because their atoms are stacked in two-dimensional planes that can slip past each other under stress. Slippage is possible because of the diffuse nature of metallic bonds and the relatively weak interactions between pairs of atoms in adjoining planes.

Substitutional Alloys

About 5500 years ago, people living around the Aegean Sea discovered that mixing molten tin and copper together produced bronze, a new material that was much stronger than either tin or copper alone. Its discovery ushered in the Bronze Age. Bronze is an **alloy**, a metallic material made when a host metal is blended with one or more other elements, which may or may not be metals, changing the properties of the host metal. Alloys can be classified into categories according to their composition. Bronze is an example of a homogeneous alloy, a solid solution (homogeneous mixture) in which the atoms of the added element(s) (in this case tin) are randomly but uniformly distributed among the atoms of the host (cop-

Ores are naturally occurring compounds or mixtures of compounds from which elements can be extracted.

An **alloy** is a blend of a host metal and one or more other elements (which may or may not be metals) that are added to change the properties of the host metal.

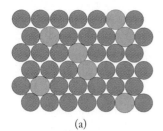

(a)

(b)

FIGURE 11.12 Two atomic-scale views of one type of bronze, a substitutional alloy: (a) A plane (layer) of close-packed Cu atoms with a random substitution of Sn atoms. (b) An fcc unit cell of Cu atoms in which Sn atoms have replaced two of the Cu atoms.

FIGURE 11.13 The larger Sn atoms in bronze perturb the crystal structure of Cu atoms, producing atomic-scale bumps in the slip plane (wavy line) between layers of Cu atoms. These bumps make it more difficult for Cu atoms to slide by each other when an external force is applied, and that makes bronze tools and weapons harder and stronger than those made of pure copper.

per, for example). Some alloys are *heterogeneous alloys* because a sample has small "islands" within the larger matrix where different metals are in contact. Still others are *intermetallic compounds* because they have a reproducible stoichiometry and constant composition.

At the concentrations of tin used in Bronze Age tools and weapons (up to 30% by mass), the alloy formed by tin and copper is also known as a *substitutional alloy* because the atoms of one metal replace atoms in the crystal lattice of the other. Substitutional alloys may form between metals that have the same crystal structure and atomic radii that are within about 15% of each other.

CONCEPT TEST

Is it accurate to call bronze a *solution* of tin *dissolved* in copper? Explain why or why not.

Figure 11.12 illustrates the structure of bronze in which tin (Sn) atoms have replaced some of the copper atoms in what would have been a face-centered cubic unit cell of Cu. The radii of copper and tin atoms are similar—128 pm and 140 pm, respectively. Inserting the slightly larger Sn atoms in the cubic closest-packed structure of copper atoms disturbs the Cu lattice structure a little, making the planes of copper atoms "bumpy" instead of uniform (Figure 11.13). This atomic-scale roughness makes it more difficult for the copper atoms to slip past one another. Less slippage makes bronze less malleable than copper, but being less malleable also means that bronze is harder and stronger.

CONCEPT TEST

Does the face-centered cube of Cu and Sn atoms in Figure 11.12(b) represent the unit cell of the alloy?

There are many other substitutional alloys besides bronze. Some are also copper based, including brass (zinc alloyed with copper) and pewter (tin alloyed with copper and antimony). However, most substitutional alloys in the modern world are *ferrous alloys*, which means that the host metal is iron. An important class of these alloys are rust-resistant *stainless steels*, which contain about 10% nickel and up to 20% chromium. When atoms of Cr on the surface of a piece of stainless steel combine with oxygen, they form a layer of Cr_2O_3 that tightly bonds to the surface and protects the metallic material beneath from further oxidation. This resistance to surface discoloration due to corrosion means that the surfaces of these alloys "stain less" than those of pure iron.

Interstitial Alloys

The Bronze Age began to fade about 3000 years ago with the discovery that iron oxides could be reduced to iron metal in wood or charcoal fires by limiting the air supplied to the fire. As in copper smelting, the reducing agent for iron oxide is carbon monoxide:

$$Fe_2O_3(s) + 3\,CO(g) \rightarrow 2\,Fe(s) + 3\,CO_2(g) \qquad (11.4)$$

Iron quickly replaced bronze as the metallic material of choice for fabricating tools and weapons both because iron ore is much more abundant in Earth's crust than the ores of copper and tin, and because tools and weapons made of iron and iron-containing alloys are much stronger than those made of bronze.

Today, the reduction of iron ore is done in blast furnaces, enormous reaction vessels (some more than 50 m tall) that operate continuously at about 1600°C (Figure 11.14). Iron ore, hot carbon (coke), and limestone are added to the top of the vessel and molten iron and solid by-products (called *slag*) are harvested from the bottom (Figure 11.14a). Blast furnaces get their name from blasts of air preheated to about 1000°C that are injected through nozzles near the bottom and that suspend the reactants until the iron reduction process is complete. It may take as long as eight hours for the reactants to fall to the bottom of a blast furnace. On their way down, O_2 in the hot air partially oxidizes coke (carbon) to carbon monoxide, and the CO reduces the iron in iron ore as described in Equation 11.4. Limestone ($CaCO_3$) decomposes to calcium oxide (CaO) or lime:

$$CaCO_3(s) \rightarrow CaO(s) + CO_2(g) \tag{11.5}$$

CaO in turn reacts with silica impurities in the ore to form calcium silicate:

$$CaO(s) + SiO_2(s) \rightarrow CaSiO_3(\ell) \tag{11.6}$$

Calcium silicate becomes part of the slag that floats on the denser molten iron at the bottom of the furnace. The largest blast furnaces, operating continuously, produce 10,000 to 15,000 metric tons of iron a week in response to increasing

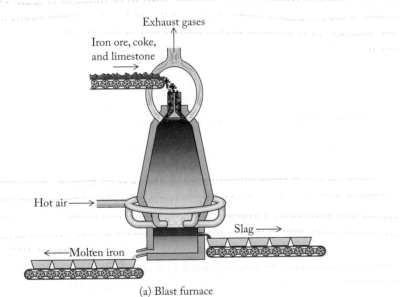

(a) Blast furnace

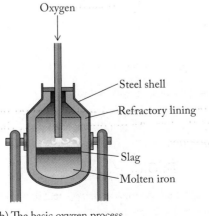

(b) The basic oxygen process

FIGURE 11.14 (a) Blast furnaces operate continuously at temperatures near 1600°C to convert iron ore into iron (see Equation 11.4) at rates as high as 2000 tons of iron per day. Injected O_2 converts C to CO_2, and lime (CaO) is added to react with Si and P impurities. The products of these reactions become part of the slag layer. (b) In the basic oxygen process the carbon content of steel is reduced by injecting oxygen. The reaction vessel pivots on its side arms, allowing the purified iron to be tapped off below the slag layer.

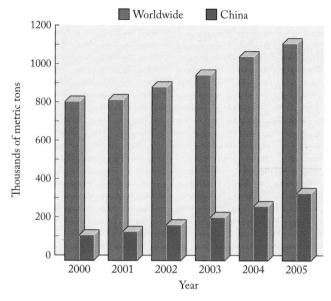

FIGURE 11.15 Worldwide production of steel has risen steadily in recent years due mainly to the rapidly growing steel industry in China.

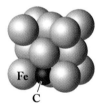

FIGURE 11.16 Carbon steel is an interstitial alloy of carbon in iron. The fcc form of iron (austentite) that forms at high temperatures can accommodate carbon atoms in its octahedral holes.

FIGURE 11.17 Close-packed atoms in adjacent layers of a crystal lattice produce tetrahedral and octahedral holes between clusters of 4 and 6 adjacent spheres, respectively. Octahedral holes are larger than tetrahedral holes and so can accommodate larger atoms in interstitial alloys.

world demand for products made of steel. Figure 11.15 shows annual worldwide steel production since the beginning of the 21st century.

Molten iron produced in a blast furnace may contain up to 5% carbon. To reduce the carbon content, the molten iron is transferred to a second furnace into which hot O_2 and additional CaO are injected. In this *basic oxygen process* (Figure 11.14b), some of the carbon is oxidized to CO_2 and any remaining silicon impurities form a slag of $CaSiO_3$. Sulfur and phosphorus impurities in the fuels used in the blast furnace also end up in the slag. Worldwide about 60% of the iron used to make steel is refined using the basic oxygen process.

When molten iron cools to its melting point of 1538°C, it crystallizes in a face-centered cubic structure called *austentite*. With further cooling the structure changes into a body-centered cubic structure called *ferrite*. The spaces, or *holes*, between iron atoms in austentite can accommodate carbon atoms forming an *interstitial alloy* because the carbon atoms occupy spaces, or *interstices*, between the iron atoms (Figure 11.16).

There are two kinds of holes between the atoms in closest-packed crystal structures (Figure 11.17). The smaller of the two are located between clusters of 4 host atoms and are called *tetrahedral holes*. The larger holes are surrounded by clusters of 6 host atoms in the shape of an octahedron and are called *octahedral holes*. The data in Table 11.2 provide guidelines on which holes are more likely to be occupied based on the relative sizes of the smaller atoms in the holes and larger host atoms. According to the data in Appendix 3, the atomic radii of C and Fe are 77 and 124 pm, respectively. The ratio of the two is 77/124 = 0.62. According to the guidelines in Table 11.2, this ratio means that C atoms should fit in the octahedral holes of austentite, as shown in Figure 11.16, but not in the tetrahedral holes.

As austentite iron cools to room temperature, it turns into body-centered ferrite iron. The octahedral holes in ferrite are smaller than those in austentite: too small to accommodate carbon atoms. As a result, carbon precipitates as clusters of carbon atoms, or it may react with iron forming iron carbide, Fe_3C. The clusters of carbon and Fe_3C disrupt ferrite's crystal structure and inhibit the host iron atoms from slipping past each other when a stress is applied. This resistance to slippage, which is much like that experienced by the copper atoms in bronze (Figure 11.12), makes iron–carbon alloys, known as carbon steel, much harder and stronger than pure iron. In general, the higher the carbon concentration, the stronger the steel, as indicated by the information in Table 11.3. Note from the values in the table that increased strength and hardness comes at the cost of increased brittleness.

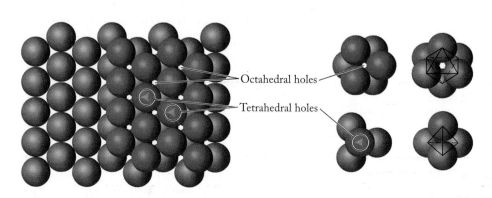

Octahedral holes

Tetrahedral holes

TABLE 11.2 Atomic Size Ratios and the Location of Atoms in Unit Cells

Packing	Type of Hole for Smaller Atom	Radius Ratio[a]
hcp or ccp	Tetrahedral	0.22–0.41
hcp or ccp	Octahedral	0.41–0.73
Simple cubic	Cubic	0.73–1.00

[a]Using radius ratios as predictors of solid-state structures is limited because atoms are not truly solid spheres with a constant radius; the radius of an atom may differ in different compounds, so these ranges are approximate.

TABLE 11.3 Effect of Carbon Content on the Properties of Steel

Carbon Content (%)	Designation	Properties	Used to Make
0.05–0.19	Low carbon	Malleable, ductile	Nails, cables
0.20–0.49	Medium carbon	High strength	Construction girders
0.5–3.0	High carbon	Hard but brittle	Cutting tools

SAMPLE EXERCISE 11.4 **Predicting the Structure of a Two-Element Alloy**

As noted at the beginning of this chapter, pure gold is too soft to use in jewelry. So too is pure silver. However, sterling silver, which is 93% Ag and 7% Cu by mass, is widely used in jewelry. The presence of Cu inhibits tarnishing and strengthens the alloy. Is this copper–silver alloy a substitutional or an interstitial alloy? Silver has a cubic closest-packed (ccp) structure with face-centered cubic (fcc) unit cells.

COLLECT AND ORGANIZE We are asked to decide whether Cu alloys with Ag by forming a substitutional alloy or an interstitial alloy. Elements form substitutional alloys when the atoms of the elements have similar sizes. According to Table 11.2, interstitial alloys may form only when atoms of an alloying element are less than about 73% the size of the atoms in a closest-packed lattice of the host element. We are given that the atoms in solid Ag form fcc unit cells and are closest-packed. The atomic radii of Ag and Cu are 144 and 128 pm, respectively.

ANALYZE The ratio of the atomic radii of Cu to Ag is 128 pm / 144 pm = 0.89, or 89%.

SOLVE Atoms of Cu can substitute for atoms of Ag in a ccp structure of the latter with some room to spare. Both elements form fcc unit cells, so little disruption to the Ag lattice should result from incorporating Cu atoms, which are nearly the same size. Copper atoms are too big, however, to fit into the octahedral holes in the lattice of silver atoms. Thus, the Cu–Ag alloy, sterling silver, is a substitutional alloy.

THINK ABOUT IT The guideline for two metals forming a substitutional alloy is that their atomic radii are within 15% of each other. The radii of Ag and Cu differ by only 11%, so we would expect copper and silver to form a substitutional alloy. Moreover, copper and silver are in the same group in the periodic table and have the same crystal structures (Figure 11.7).

Practice Exercise Would you expect gold (atomic radius = 144 pm) to form substitutional alloys with silver (atomic radius = 144 pm) or copper (atomic radius = 128 pm)?

11.5 Aluminum and Its Alloys: Light Weight and High Performance

Over the last century, aluminum and its alloys have replaced steel in applications where high strength-to-weight ratios and corrosion resistance are paramount, as in airplanes, motor vehicles, and the facades of buildings. The most important aluminum ore is called bauxite. It is a mixture of aluminum minerals containing aluminum(III) oxide. To convert bauxite to aluminum metal, the ore is crushed and extracted with a concentrated solution of NaOH in a reaction called the Bayer process. The product of this reaction is a soluble complex ion that has an Al^{3+} ion at its center surrounded by 4 OH^- ions:

$$Al_2O_3(s) + 2\ OH^-(aq) + 3\ H_2O(\ell) \rightarrow 2\ Al(OH)_4^-(aq) \tag{11.7}$$

Impurities in bauxite, such as $CaCO_3$, Fe_2O_3, and TiO_2, are insoluble in concentrated NaOH and can be removed by filtration. Acid is then added to neutralize the NaOH in the filtered solution, resulting in the precipitation of high-purity $Al(OH)_3(s)$:

$$Al(OH)_4^-(aq) + H^+(aq) \rightarrow Al(OH)_3(s) + H_2O(\ell) \tag{11.8}$$

The precipitate is then converted to pure Al_2O_3 by heating, which drives off steam:

$$2\ Al(OH)_3(s) \xrightarrow{\Delta} Al_2O_3(s) + 3\ H_2O(g) \tag{11.9}$$

In 1886, two 23-year-old chemists, Charles Hall and Paul-Louis-Toussaint Héroult (both 1863–1914), working independently in the United States (Hall) and France (Héroult), developed the same industrial process for converting Al_2O_3 (called alumina) to aluminum. The process is based on passing an electric current between carbon electrodes immersed in a solution of aluminum oxide dissolved in molten cryolite (Na_3AlF_6), as shown in Figure 11.18. As electricity passes through the solution, aluminum ions are reduced to aluminum metal at the negative electrode while the positively charged carbon electrode is oxidized to carbon dioxide. The process is described by the following reaction:

$$2\ Al_2O_3\ \text{(in molten Na}_3\text{AlF}_6\text{)} + 3\ C(s) \rightarrow 4\ Al(\ell) + 3\ CO_2(g) \tag{11.10}$$

The high cost of the electricity needed to produce aluminum in this way makes recycling aluminum metal a good idea both economically and environmentally. Let's compare the energy requirements of producing Al from Al_2O_3 with recycling products made of Al metal. The principal energy cost of the Hall–Héroult process is the electricity needed to reduce Al_2O_3. The major cost in recycling is the energy required to melt aluminum metal. We can estimate the energy required for the Hall–Héroult process by calculating the standard enthalpy change for the reaction in Equation 11.10. Starting with standard enthalpies of formation, we have:

$$\Delta H^\circ_{rxn} = [3(\Delta H^\circ_{f,CO_2}) + 4(\Delta H^\circ_{f,Al})] - [2(\Delta H^\circ_{f,Al_2O_3}) + 3(\Delta H^\circ_{f,C})]$$

$$= \left[(3\ \text{mol CO}_2)\left(\frac{-393.5\ \text{kJ}}{\text{mol CO}_2}\right) + (4\ \text{mol Al})\left(\frac{0.0\ \text{kJ}}{\text{mol Al}}\right) \right]$$

$$- \left[(2\ \text{mol Al}_2\text{O}_3)\left(\frac{-1675.7\ \text{kJ}}{\text{mol Al}_2\text{O}_3}\right) + (3\ \text{mol C})\left(\frac{0.0\ \text{kJ}}{\text{mol C}}\right) \right]$$

$$= 2171\ \text{kJ}$$

Dividing this value by the 4 moles of Al produced by the reaction, we get 542.7 kJ per mole of Al.

Charles M. Hall and Paul-Louis-Toussaint Héroult independently developed the same electrolytic process for producing aluminum metal from alumina. Hall's sister, Julia, also a chemistry major at Oberlin College, assisted her brother in the lab, and her business skills helped make their aluminum production company a financial success. The company eventually became the Aluminum Company of America, shortened to Alcoa, Inc.

Now let's calculate the energy required to recycle 1 mole of Al by heating it from 25°C to its melting point (660°C) until all the Al melts. The energy needed to heat 1 mole of aluminum from 25°C to 660°C can be calculated from its molar heat capacity [24.31 J/(mol · °C)] and Equation 5.11:

$$q = nc_p\Delta T$$

$$= 1.00 \text{ mol} \times \frac{24.20 \text{ J}}{\text{mol} \cdot °C} \times (660 - 25)°C$$

$$= 15.37 \text{ kJ}$$

Once the Al is at its melting point, the heat required to melt 1 mole of Al is its heat of fusion (ΔH_{fus} = 10.79 kJ/mol):

$$\frac{10.79 \text{ kJ}}{\text{mol}} \times 1.00 \text{ mol} = 10.79 \text{ kJ}$$

Thus the total energy needed to heat and melt 1.00 mole of Al is

$$15.37 \text{ kJ} + 10.79 \text{ kJ} = 26.16 \text{ kJ}$$

This value represents

$$\frac{26.16 \text{ kJ}}{542.7 \text{ kJ}} \times 100\% = 4.82\%$$

of the energy needed to electrochemically reduce 1 mole of Al^{3+} to Al^0.

There are other energy costs associated with aluminum production and recycling, but overall, recycling saves aluminum manufacturers about 95% of the energy required to produce the metal from aluminum ore. This energy savings has inspired the rapid growth of a global aluminum recycling industry. (Aluminum recycling in the United States is a $1 billion per year business.) Currently, junkyards in the United States recycle 85% of the aluminum in cars (5×10^8 kg/yr). Over 50% of the aluminum in food and beverage containers is recycled. The value of scrap aluminum is about 10 times that of scrap steel because of these energy considerations and because it takes less energy to melt aluminum than it does to melt the same mass of scrap steel.

Aluminum forms alloys with many elements. Alloys with Mg, Si, Cu, and Zn are the most common and are widely used in aircraft construction. The tabs on beverage cans are manufactured from an aluminum alloy containing Mg and Mn.

FIGURE 11.18 This sketch of Hall's first aluminum production facility shows large trays made of cast iron that held the molten reaction mixture. Positively charged carbon electrodes were attached by copper rods to copper plates suspended over the trays. Molten aluminum formed at negatively charged carbon electrodes at the bottom of the trays and was poured into the ingot molds shown on the floor.

A **covalent network solid** consists of atoms held together by extended arrays of covalent bonds.

Aluminum alloys containing Li are attractive for applications where light weight is essential, because Li has the smallest molar mass (6.941 g/mol) and lowest density (0.534 g/mL) of all the metallic elements.

Aluminum and aluminum alloys are widely used as building materials and in fuselages of airplanes because they are corrosion resistant. This resistance is noteworthy because aluminum is actually a very reactive metal. However, when the surface of a piece of aluminum starts to oxidize in air, a thin layer of Al_2O_3 is produced that adheres strongly to the aluminum metal below it and acts as a protective shield against further oxidation. This corrosion protection is similar to the way a thin coating of Cr_2O_3 protects stainless steel from rusting.

11.6 The Structures of Some Nonmetals

Figure 11.7 describes the crystal structures of metallic elements. Notice (from the color key) that it includes a structure type labeled "diamond." Diamonds are not metals. They are a rare and thus expensive crystalline form of the nonmetal carbon, although they owe their distinctive colors and some of their value to the presence of metal impurities in their crystal structures.

CONNECTION Allotropes are structurally different forms of the same physical state of an element, as explained in Chapter 8.

Diamond is one of three allotropes of carbon (Figures 11.19 and 11.20), the other two being graphite and fullerenes. Diamond is a crystalline **covalent network solid**, because it consists of atoms held together in an extended three-dimensional network of covalent bonds (Figure 11.19a). Each carbon atom in diamond is bonded by overlapping sp^3 hybrid orbitals with 4 neighboring carbon atoms, creating a network of carbon tetrahedra. The carbon atoms in these tetrahedra are connected by localized σ bonds, making diamonds poor electrical conductors. The three-dimensional network of these σ bonds is extremely rigid, making diamond the hardest natural material known. The atoms of other group 14 elements, particularly silicon, germanium, and tin, also form covalent network solids based on the diamond structure.

CHEMTOUR Allotropes of Carbon

Natural diamond forms from graphite under intense heat (> 1700 K) and pressure ($> 50,000$ atm) deep in Earth. Industrial diamonds are also synthesized at high temperatures and pressures from graphite or any source rich in carbon. Synthetic diamonds are used as abrasives and for coating the tips and edges of cutting tools. Diamond has the highest thermal conductivity of any natural substance (five times higher than copper and silver, the most thermally conductive metals), so tools made from diamond do not become overheated. Most industrial diamonds lack the size and optical clarity of gemstones.

FIGURE 11.19 Carbon crystallizes as (a) diamond, a three-dimensional covalent network solid made of carbon atoms, each connected by σ bonds to 4 adjacent carbon atoms; and (b) graphite, in which carbon atoms are connected by σ bonds and delocalized π bonds. The structure of a third allotrope of carbon is shown in Figure 11.20.

By far the most abundant allotrope of carbon is graphite, another covalent network solid, frequently the principal ingredient in soot and smoke and used to make pencils, lubricants, and gunpowder. Graphite contains sheets of carbon atoms, in which each atom is connected by overlapping sp^2 hybrid orbitals to 3 neighboring carbon atoms in a two-dimensional covalent network of 6-membered rings (Figure 11.19b). Each carbon–carbon bond is 142 pm, shorter than the C–C bond in diamond. Overlapping unhybridized p orbitals on the carbon atoms form a network of π bonds that are delocalized

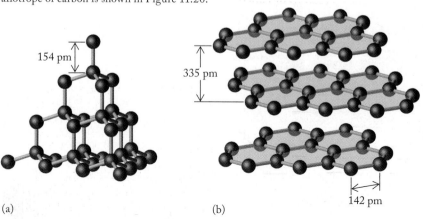

154 pm

335 pm

142 pm

(a)   (b)

across the plane defined by the rings. The mobility of these delocalized electrons makes graphite a conductor of electricity. *that's why.*

As shown in Figure 11.19(b), the two-dimensional sheets in graphite are 335 pm apart. This distance is much too long to be a covalent bonding distance. Instead, the sheets are held together only by London (dispersion) forces. These intermolecular interactions are much weaker than covalent bonds. This weakness allows adjacent sheets to slide past each other, making graphite soft, flexible, and a good lubricant.

> Crystals of **molecular solids** are formed by neutral covalently bonded molecules held together by intermolecular attractive forces.

CONNECTION We introduced London forces between molecules in Chapter 10.

CONCEPT TEST

This phase diagram of carbon shows the temperatures and pressures at which carbon vapor, liquid carbon, and two solid forms (graphite and diamond) are stable. Which region (a, b, c, or d) represents diamond?

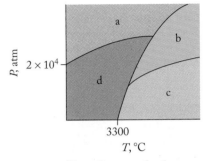

Phase diagram of carbon.

A third allotrope of carbon was discovered in the 1980s. Networks of 5- and 6-atom carbon rings form molecules of 60, 70, or more carbon atoms that look like miniature soccer balls (Figure 11.20a). They are called *fullerenes* because their shape resembles the geodesic domes designed by American architect R. Buckminster Fuller (1895–1983). Many chemists call them "bucky balls" for the same reason. In terms of their size and their properties, bucky balls are too small to be covalent network solids but too large to be molecular solids (discussed in the following paragraphs). They fall in an ambiguous zone between small molecules and large networks and are classified as *clusters.*

When fullerenes were first discovered they were believed to be a form of carbon rarely found in nature. In recent years, analyses of soot and emission spectra from giant stars have disclosed that fullerenes are present in trace amounts throughout the universe. Other nonmetals have cagelike crystal structures too, and may be considered clusters, including a crystalline form of boron that is made of closest-packed arrays of 12-sided molecules called *icosahedra* (singular *icosahedron*; Figure 11.20b).

Other nonmetals form crystalline **molecular solids**. Molecular solids consist of neutral, covalently bonded molecules held together by intermolecular forces. Ice, CO_2, glucose, and all organic molecules that crystallize are molecular solids. The structures of the two most common allotropes of phosphorus, white and red phosphorus (Figure 11.21), are both based on tetrahedral P_4 molecules that crystallize in ordered arrays. White phosphorus is a molecular solid, consisting of discrete P_4 tetrahedra arranged in a cubic array. It is a waxy material that oxidizes rapidly under standard conditions and gives off a yellow-green light called *phosphorescence.* Red phosphorus is a covalent network solid made of chains of P_4 tetrahedra connected by covalent phosphorus–phosphorus bonds. Both red and white phosphorus melt to give the same liquid consisting of symmetrical P_4 tetrahedral molecules.

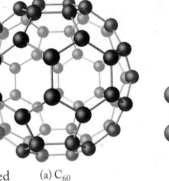

(a) C_{60} (b) B_{12}

FIGURE 11.20 Some solids are described as clusters: (a) Buckminsterfullerene (C_{60}) is a molecular solid containing 60 sp^2-hybridized carbon atoms. Both 5- and 6-membered rings are required to construct the nearly spherical molecule. (b) One crystalline form of boron consists of a close-packed array of B_{12} molecules.

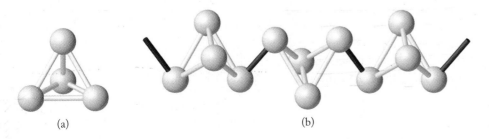

(a)   (b)

FIGURE 11.21 The two most common allotropes of phosphorus are molecular solids based on P_4 tetrahedra: (a) White phosphorus consists of discrete P_4 molecules; it is waxy and soft enough to be cut with a knife. (b) Red phosphorus is a polymer formed from white phosphorus when one bond in each P_4 tetrahedron is broken and a covalent P–P bond is formed between adjacent tetrahedra.

Ionic solids consist of monatomic or polyatomic ions held together by ionic bonds.

FIGURE 11.22 One form of sulfur is a molecular solid based on puckered 8-membered S_8 rings. Different allotropes of sulfur have different stacking patterns of these rings.

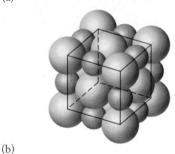

(a)

(b)

(c)

FIGURE 11.23 (a) Sodium chloride forms cubic crystals that sometimes reach impressive size. (b) The NaCl crystal structure is based on an fcc unit cell that the Cl^- ions adopt, with Na^+ ions occupying the octahedral holes. (c) Ions at the corners and faces and along the edges are shared with other unit cells. As shown in this cutaway view, only fractional portions of ions at these positions are assigned to a single unit cell.

Sulfur has more allotropic forms than any other element. Most are molecular solids. This variety of forms arises because sulfur atoms form cyclic (ring) compounds of different sizes, which in turn gives rise to the possibility of different crystalline arrangements of the molecules. The most common allotropes of sulfur consist of puckered rings (Figure 11.22) containing 8 sulfur atoms covalently bonded together. Dispersion forces hold one ring to another in solid sulfur. The weakness of these interactions is the reason why elemental sulfur is soft and melts at only 119.6°C. The packing of S_8 rings together in the crystalline molecular solid results in a very complex structure that cannot be described in terms of any of the lattice arrangements we have discussed.

11.7 Salt Crystals: Ionic Solids

Most of Earth's crust is composed of **ionic solids** consisting of monatomic or polyatomic ions held together by ionic bonds. Most of these solids are crystalline. The simplest structures are those of binary salts, such as NaCl (Figure 11.23). The cubic shape of salt crystals (Figure 11.23a) is reflected in the cubic shape of the NaCl unit cell, which is based on a face-centered cubic array of the larger Cl^- ions (Figure 11.23b). Smaller Na^+ ions occupy the 12 octahedral holes along the edges of the unit cell and the single octahedral hole in the middle of the cell.

Let's take an inventory of the ions in a unit cell of NaCl. Like the metal atoms in the fcc structure in Figure 11.10(a), there are the equivalent of 4 Cl^- ions in the unit cell in Figure 11.23(c). As for Na^+ ions, there is 1 in the central octahedral hole, and 1 in each of the 12 octahedral holes along the edges of the cell. Each Na^+ ion along an edge is shared by 4 unit cells so only one-fourth of it is in each cell. Only the Na^+ in the center belongs completely to 1 unit cell. Therefore the total number of Na^+ ions in the unit cell is

$$\left(12 \times \tfrac{1}{4}\right) + 1 = 4 \ Na^+ \text{ ions}$$

The ratio of Na^+ to Cl^- ions in the unit cell is 4:4 (or 1:1), which is consistent with the chemical formula of NaCl. Because the 4 Na^+ ions occupy all the octahedral holes in the unit cell, the result of this calculation also means that each fcc unit cell contains the equivalent of 4 octahedral holes.

Note in Figure 11.23(c) that the Cl^- ions do not touch along the face diagonal the way the metal atoms do in Figure 11.10(a). This is because the Cl^- ions have to spread out a little to accommodate the Na^+ ions in the octahedral holes. Sodium and chloride ions touch along each edge of the unit cell in NaCl. This means that each Na^+ ion in NaCl touches 6 Cl^- ions and each Cl^- ion touches 6 Na^+ ions. This arrangement of positive and negative ions is common enough among binary ionic compounds to be assigned its own name: the *rock salt* structure.

In other binary ionic solids, the smaller ion is small enough to fit into the tetrahedral holes formed by the larger ions. For example, in the unit cell of the mineral sphalerite (zinc sulfide) the larger S^{2-} anions (ionic radius, 184 pm) are arranged in an fcc structure (Figure 11.24), and half of the eight tetrahedral holes inside the cell are occupied by smaller Zn^{2+} cations (74 pm). Therefore the unit cell contains four Zn^{2+} ions which balance the charges on the four S^{2-} ions in the unit cell. This pattern of half-filled tetrahedral holes in an fcc structure is sometimes called the *sphalerite structure*. Many other compounds, particularly those

(a)

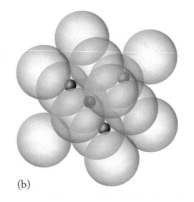

(b)

FIGURE 11.24 Sphalerite: (a) Many crystals of the mineral sphalerite, like the largest one in this photograph, have a tetrahedral shape. (b) The crystal structure of sphalerite is based on an fcc unit cell of S^{2-} ions with Zn^{2+} ions in 4 of the 8 tetrahedral holes.

formed between transition metal cations with a 2+ charge and anions of the group 6 elements with a 2− charge, have sphalerite structures.

The structure of the mineral fluorite (CaF_2) is based on an fcc unit cell of larger Ca^{2+} ions in which all 8 tetrahedral holes are filled by smaller F^- ions. Because there is a total of 4 Ca^{2+} ions in each unit cell, this arrangement satisfies the 1:2 mole ratio of Ca^{2+} ions to F^- ions. This structure is so common that it too has its own name: the *fluorite structure* (Figure 11.25). Other compounds having this structure include SrF_2, $BaCl_2$, CaF_2, and PbF_2. Another group of compounds with a cation-to-anion mole ratio of 2:1 has an *antifluorite structure*. In the crystal lattices of these compounds, which include Li_2O and K_2S, the smaller cations occupy the tetrahedral holes formed by cubic closest packing of the larger anions.

SAMPLE EXERCISE 11.5	Calculating an Ionic Radius from a Unit-Cell Dimension

Lithium chloride (LiCl) crystallizes with the same structure as NaCl except the Li^+ ions are small enough to allow the Cl^- ions to touch along the face diagonals in LiCl. If the edge length (ℓ) of the fcc cell is 513 pm, what is the radius of the Cl^- ion in LiCl?

COLLECT AND ORGANIZE We know that the Cl^- ions are in a face-centered cubic unit cell such as that shown in Figure 11.10(a). We also know the edge length of the unit cell: 513 pm.

ANALYZE In a unit cell where the particles (Cl^- ions in this case) touch along the face diagonal, the relationship between edge length (513 pm) and the radius (r) of the ions is given by Equation 11.1:

$$r = 0.3536\, \ell$$

SOLVE

$$r = 0.3536 \times 513 \text{ pm} = 181 \text{ pm}$$

THINK ABOUT IT The average ionic radii values in Figure 10.2 were determined from crystal structures. The value in Figure 10.2 for Cl^- ions is 181 pm, so our calculation is correct.

Practice Exercise Assuming the radius of the Cl^- ion in NaCl is also 181 pm, what is the radius of the Na^+ ion in NaCl? Assume that the edge length of the unit cell of NaCl shown in Figure 11.23(c) is 564 pm.

(a)

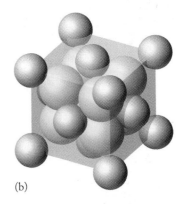

(b)

FIGURE 11.25 (a) The cubic structure of the ions in the mineral fluorite (CaF_2) can produce beautiful, nearly transparent cubic crystals. (b) The crystal structure of CaF_2 is based on an fcc array of Ca^{2+} ions, with F^- ions occupying all 8 tetrahedral holes. However, F^- ions are bigger than Ca^{2+} ions, so they do not fit in the normal tetrahedral holes of a cubic closest-packed array of Ca^{2+} ions. Instead, the Ca^{2+} ions spread out to accommodate the larger F^- ions while maintaining an fcc structure.

SAMPLE EXERCISE 11.6 **Calculating the Density of a Salt from Its Unit-Cell Dimensions**

What is the density of LiCl at 25°C if the edge length of its fcc unit cell is 513 pm?

COLLECT AND ORGANIZE We are given the edge length of an fcc unit cell of a crystalline salt and are asked to calculate its density. As in Sample Exercise 11.3, we have to assume that the density of the unit cell is the same as the density of the overall solid. The fact that LiCl has an fcc unit cell means that there are 4 Cl⁻ ions and 4 Li⁺ ions in the cell. The molar masses of these ions are 35.45 and 6.941 g/mol, respectively. Density is the ratio of mass to volume, and the volume of a cubic cell is the cube of its edge length: $V = \ell^3$.

ANALYZE The density of an fcc unit cell of LiCl is the sum of the masses of 4 Cl⁻ ions and 4 Li⁺ ions divided by the volume of the cell (513 pm)³. As in Sample Exercise 11.3, we need to convert molar masses into the masses of individual ions by dividing by Avogadro's number. To obtain a cell volume in cubic centimeters, we use the conversion factor:

$$1 \text{ pm} = 10^{-10} \text{ cm}$$

SOLVE Calculating the masses of 4 Cl⁻ ions, we get:

$$m = \frac{35.45 \text{ g } \cancel{Cl^-}}{\cancel{mol\ Cl^-}} \times \frac{1 \cancel{mol}}{6.022 \times 10^{23} \cancel{ions}} \times 4 \cancel{ions} = 2.355 \times 10^{-22} \text{ g}$$

and the masses of 4 Li⁺ ions:

$$m = \frac{6.941 \text{ g } \cancel{Li^+}}{\cancel{mol\ Li^+}} \times \frac{1 \cancel{mol}}{6.022 \times 10^{23} \cancel{ions}} \times 4 \cancel{ions} = 0.461 \times 10^{-22} \text{ g}$$

Next we combine the masses of the two kinds of ions in the unit cell:

$$2.355 \times 10^{-22} \text{ g}$$
$$+ 0.461 \times 10^{-22} \text{ g}$$
$$\overline{2.816 \times 10^{-22} \text{ g}}$$

The volume of the cell in cubic centimeters is

$$V = \ell^3 = (513 \cancel{pm})^3 \times \frac{(10^{-10} \text{ cm})^3}{\cancel{pm^3}} = 1.350 \times 10^{-22} \text{ cm}^3$$

Taking the ratio of mass to volume, we have

$$d = \frac{m}{V} = \frac{2.816 \times 10^{-22} \text{ g}}{1.350 \times 10^{-22} \text{ cm}^3} = 2.08 \text{ g/cm}^3$$

THINK ABOUT IT The result is reasonable in that most minerals are more dense than water but less dense than common metals.

Practice Exercise What is the density of NaCl at 25°C if the edge length of its fcc unit cell is 564 pm?

11.8 Ceramics: Insulators to Superconductors

A **ceramic** is a solid inorganic compound or mixture of compounds that has been heated to transform it to a harder and more heat-resistant material. The use of ceramic materials preceded metal technology by many thousands of years. The first ceramics were probably made of clay: the very fine-grained products of the physical and chemical weathering of igneous rocks (rocks of volcanic origin). Moist clay is easily molded into the desired shape and then hardened over fires or in wood-burning kilns. This ancient process is still in use today. In this section we examine some of the physical properties of both primitive earthenware (ceramics fired at low temperatures) and modern ceramic materials (typically fired at high temperatures), and relate those properties to the chemical structures of these materials and to the chemical changes that occur when they are heated.

Polymorphs of Silica

One of the most abundant families of minerals found in igneous rocks has the chemical composition SiO_2 and is called silica or, to be chemically correct, silicon dioxide. Silica is a covalent network solid in which each silicon atom is bonded to 4 oxygen atoms, forming oxygen tetrahedra with silicon atoms at their centers and O—Si—O bond angles of 109° (Figure 11.26). Each oxygen atom is bonded to 2 silicon atoms, thereby linking the tetrahedra into extended three-dimensional networks. Because each corner oxygen atom is bonded to 2 silicon atoms, each silicon atom gets only half "ownership" of the 4 oxygen atoms to which it is bonded—hence the formula SiO_2.

At least eight different minerals have the chemical composition SiO_2. The members of a family of substances with the same composition but different structures and properties are called *polymorphs*. The most abundant silica polymorph is quartz, a type of SiO_2 that can form impressively large, nearly transparent crystals (Figure 11.26a). Note how the hexagonal ordering of the SiO_2 tetrahedra (Figure 11.26b) translates into hexagonal crystals.

Most, but not all, silica is crystalline. When lava containing molten SiO_2 flows from a volcano into the sea or a lake, it cools quickly. During rapid cooling, Si and O atoms may not have enough time to achieve an ordered structure as the viscous (slow-moving) lava solidifies. A product of rapid solidification is an *amorphous* (disordered, noncrystalline) form of silica known as volcanic glass, or obsidian (Figure 11.27). *Glass* is a term scientists and engineers use to describe any solid that has no crystalline structure or only very tiny crystals surrounded by disor-

A **ceramic** is a solid inorganic compound or mixture that has been transformed into a harder, more heat-resistant material by heating.

(a)

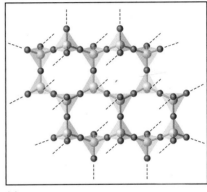
(b)

FIGURE 11.26 Quartz: (a) The hexagonal shape of quartz crystals where (b) the hexagonal arrays of silicon–oxygen tetrahedra share oxygen atoms in their crystalline structures. (Here and in Figures 11.27–11.30, the Si [silver] and O [red] atoms are drawn undersized to make the arrangement of tetrahedra easier to see.) These "hexagons of tetrahedra" are covalently bonded to others above and below them in an extended three-dimensional network (not shown).

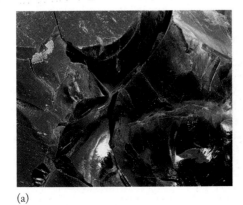
(a)

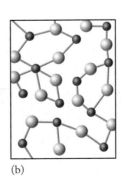
(b)

FIGURE 11.27 (a) Obsidian (volcanic glass) is an unusual form of silica in that it is not crystalline. (b) It contains mostly amorphous silica with random arrangements of silicon and oxygen atoms.

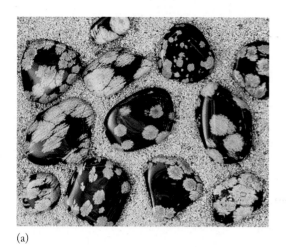

(a)

(b)

FIGURE 11.28 Cristobalite: (a) Deposits of α-cristobalite (white) formed here on pieces of volcanic glass. (b) In the crystal structure of α-cristobalite, four silica tetrahedra reside within an expanded fcc unit cell of silicon atoms.

dered arrays of atoms. This definition applies to laboratory glassware and the drinking glasses we use at home, as well as to solids of the type discussed here that have no crystalline structure.

Other members of the silica group have slightly different crystal structures. For example, in the unit cell of the mineral α-cristobalite (Figure 11.28), a common ingredient of volcanic rocks, four silicon–oxygen tetrahedra reside within a face-centered cubic lattice of silicon atoms. There are also β-cristobalite and β-quartz structures that differ from the α-cristobalite and quartz structures described here.

As you may have noticed, there are many structural forms of silica in nature. As solid silica minerals form from cooling molten SiO_2, they take on different structures depending on the pressures and temperatures at which they crystallize. Denser forms, such as β-cristobalite and β-quartz, are more likely to form at higher pressures; less dense forms occur at lower pressures. The presence of these different structures provides geologists with clues about the conditions under which igneous rocks in a region formed.

Ionic Silicates

In addition to covalent silica, igneous rocks also contain ionic minerals made of silicon and oxygen. These minerals have some of the tetrahedral structure of silica, but not all the oxygen corner atoms are bonded to 2 Si atoms. Instead some of these O atoms have an extra electron. The result is a *silicate* anion. One of the most common kinds of silicates consists of sheets of linked silicon–oxygen tetrahedra that form hexagonal clusters of six tetrahedra each (Figure 11.29). Each tetrahedron has 3 O atoms that it shares with other tetrahedra and 1 O atom—the one with the extra electron—that it does not share. Thus the basic tetrahedral repeating unit consists of 1 Si atom and $\left[1 + 3\left(\frac{1}{2}\right)\right] = 2.5$ O atoms as well as a negative charge. This gives the sheet the empirical formula $SiO_{2.5}^{-}$. We generally use whole-number subscripts in chemical formulas when possible. In this case, multiplying $SiO_{2.5}^{-}$ by 2 gives us the empirical formula $Si_2O_5^{2-}$ for these silicate sheets. The subscript n in the formula in Figure 11.29 indicates that there are many empirical formula units in a single crystal.

FIGURE 11.29 (a) Chrysotile is one of the two principal minerals in asbestos. The ease with which thin fibers of chrysotile can flake off is related to (b) its layered crystalline structure and the relatively weak intermolecular interactions between layers. Chrysotile fibers are curled, slightly soluble, and apparently noncarcinogenic. Crodidolite, the second and less commonly used form of asbestos, makes straighter, insoluble fibers (not shown); these do pose a health risk when breathed in regularly, leading to asbestosis and lung cancer.

(a)

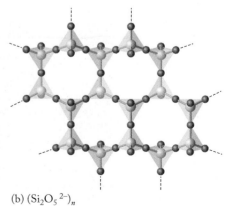

(b) $(Si_2O_5^{2-})_n$

Of course, silicate minerals are neutral materials, so they must contain cations to neutralize the negative charges on the $Si_2O_5^{2-}$ layers. When this cation is Al^{3+}, the minerals are called *aluminosilicates*. One of the most common examples of an aluminosilicate is a clay mineral called kaolinite (Figure 11.30). At least a little kaolinite is found in practically every soil, but rich deposits of nearly pure, brilliantly white kaolinite are found in highly weathered soils. For centuries these deposits have been mined to make fine china and white porcelain. Today the greatest demand for kaolinite is in the production of glossy white paper, which is used in most magazines and books (including this one).

Common metal ions found in igneous rocks—including Na^+, K^+, Ca^{2+}, Mg^{2+}, and Fe^{3+}—are largely absent in kaolinite. Their absence indicates that kaolinite deposits form under acidic weathering conditions. Under these conditions H^+ ions displace other cations from ion-exchange sites. For example, $-O^-Na^+$ sites exchange H^+ for Na^+, leaving behind $-OH$ groups such as those shown in Figure 11.30.

Kaolinite has the formula $Al_2(Si_2O_5)(OH)_4$. The foundation of its crystal structure is a silicate layer that fits the $Si_2O_5^{2-}$ formula and structure shown in Figure 11.29(b). Aluminum(III) ions form a second layer in which each ion resides at the center of an octahedron of 6 oxygen atoms: 3 from the Si–O tetrahedra in the layer below, and 3 from hydroxide ions (OH^-) that form a third layer above. The OH^- ions are in turn hydrogen-bonded to O atoms of a second layer of silicate tetrahedra. This entire layering pattern repeats itself, forming an extended three-dimensional structure. The strengths of the ionic interactions and hydrogen bonds between layers are so strong that the layers in kaolinite are not easily separated; unlike with most clays, water molecules cannot penetrate between them. This property has made kaolinite pots handy vessels for carrying water for thousands of years. Because water cannot penetrate between its layers, kaolinite does not expand when water is added. For the same reason, it does not shrink as much as most clays when dehydrated at high temperatures. This is another property that has made kaolinite a desirable starting material for ceramics. High-purity kaolinite is also bright white, making it the material of choice for the manufacture of

CONNECTION Ion-exchange reactions were discussed in Chapter 4.

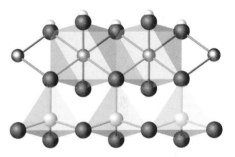

FIGURE 11.30 This edge-on view of the layered structure of kaolinite shows a bottom layer of silicate ($Si_2O_5^{2-}$) represented by silver and red spheres, much like the bottom layer in the top-down view in Figure 11.29(b). Next comes a layer of Al^{3+} ions (gray spheres) that interact with lone pairs of valence electrons on the O atoms of the silicate layer below and with a layer of hydroxide ions in the top layer. Each Al^{3+} ion interacts with 3 O atoms from each layer, and so resides in a kind of octahedral hole surrounded by 6 O atoms. The empirical formula of this structure is $Al_2(Si_2O_5)(OH)_4$.

This enormous kaolinite mine in Bulgaria contributes to a worldwide production of about 44 trillion (10^{12}) metric tons of the mineral per year. The mineral is used in many products, from ceramics to glossy white paper.

the porcelain used in bathroom fixtures, fine china, and other ceramics. Finally, moist kaolinite is *plastic*, meaning that it can be molded into a shape, and it keeps that shape during heating and cooling.

From Clay to Ceramic

Creating ceramic objects from kaolinite and other clays takes several steps. First, moist clay is formed into pots, bricks, and other objects on a potter's wheel or in molds or presses. Drying at just above 100°C removes much of the water that made the clay plastic. Further heating to about 450°C removes water that was adsorbed onto the surfaces of the clay particles or between the layers of non-kaolinite clays. Between 450°C and 650°C the hydroxide ions of the octahedra surrounding the Al^{3+} ions react with each other to form water and a solid that is richer in Al and Si:

$$Al_2Si_2O_5(OH)_4(s) \xrightarrow{450-650°C} 2\,H_2O(g) + Al_2Si_2O_7(s) \tag{11.11}$$

The next structural change occurs just below 1000°C when $Al_2Si_2O_7$ is converted into another aluminosilicate, $Al_4Si_3O_{12}$:

$$2\,Al_2Si_2O_7(s) \xrightarrow{\sim950°C} Al_4Si_3O_{12}(s) + SiO_2(s) \tag{11.12}$$

In this transformation, SiO_2 is excluded from the aluminosilicate structure, forming a second solid phase. At even higher temperatures $Al_4Si_3O_{12}$ continues to lose SiO_2:

$$Al_4Si_3O_{12}(s) \xrightarrow{950°C} 2\,Al_2SiO_5(s) + SiO_2(s) \tag{11.13}$$

$$3\,Al_2SiO_5(s) \xrightarrow{1350°C} Al_6Si_2O_{13}(s) + SiO_2(s) \tag{11.14}$$

The last product, $Al_6Si_2O_{13}$, is called mullite. The formula of mullite is sometimes written as $3\,Al_2O_3 \cdot 2\,SiO_2$ indicating that it is a blend of Al_2O_3 (alumina) and SiO_2 (silica) structures. Mullite is a widely used ceramic material in the manufacture of products that must tolerate temperatures as high as 1700°C: furnaces, boilers, ladles, and kilns. These products are used as containers of molten metals and in the glass, chemical, and cement industries. Mullite is also very hard and is widely used as an abrasive.

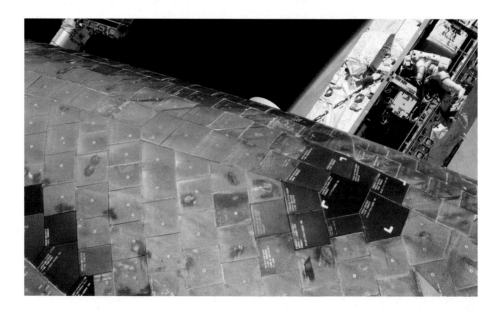

FIGURE 11.31 About 25,000 heat-resistant tiles cover much of the surface of each U.S. space shuttle, protecting it from re-entry temperatures of over 1100°C. Most tiles are made of high-purity, amorphous SiO_2 fibers. Ninety percent of each tile is open space between the fibers, which means that the tiles are both lightweight and highly resistant to high temperatures.

Other ceramics are used in high-temperature applications ranging from cookware to fireplace bricks to the tiles on the U.S. space shuttles (Figure 11.31). Ceramics are well suited to these uses because of their high melting points and because they are good thermal and electrical insulators. For example, the thermal conductivity of aluminum metal at 100°C is over eight times that of alumina (Al_2O_3).

Structure and Bonding: Superconductors

Band theory can be used to describe the properties of, and bonding in, all solids, including the compounds in ceramic materials. Earlier in the chapter we noted that metals are good conductors because of the ease with which their valence electrons can move to conduction bands, whereas semimetals are semiconductors because they have significant energy gaps between their valence and conduction bands (though these gaps can be reduced by the process of doping). Similarly, the insulating properties of ceramics can be explained by the large energy gaps between their valence and conduction bands (see Figure 11.4). As a result, their valence electrons have very limited mobility.

In conventional metallic conductors the vibration of metal atoms in their crystalline lattices can interfere with the flow of free electrons: the higher the temperature, the greater the atomic vibration and the greater the resistance to electron flow. However, at temperatures approaching absolute zero these vibrations become very small. In the early 20th century, scientists discovered that the lattice vibrations in mercury and some metal alloys are so small at temperatures below about 20 K, called **critical temperatures (T_c)**, that electrons can pass freely through these materials, and they become **superconductors**. Today superconducting alloys such as Nb_3Sn are widely used in devices that require very high electrical currents, such as the electromagnets of magnetic resonance imaging (MRI) instruments in hospitals.

Unfortunately it is difficult and expensive to chill materials to 20 K. In 1986 it was discovered that $YBa_2Cu_3O_7$ and several other ceramic materials become superconductors when cooled to temperatures just above the boiling point of liquid nitrogen (77 K). This represented a tremendous increase in the critical temperature for the onset of superconductivity. Since then, scientists from around the world have produced a variety of superconducting ceramic materials with critical temperatures as high as 133 K (−140°C). It is much easier and cheaper to cool a material with liquid nitrogen than with liquid helium (boiling point = 4 K), which is used to achieve superconductivity in Hg and metal alloys.

Let's take a closer look at the structure of one of the so-called "High-T_c" superconductors, $YBa_2Cu_3O_7$. We will start with the unit cell shown in Figure 11.32(a). It is called a *perovskite* unit cell, and is adopted by many compounds that have the generic formula ABO_3. An ion of metal "A," which is represented by the Ba^{2+} ion in Figure 11.32(a), is in the center of the cell; eight ions of a metal "B" (Cu^{2+} in this case) occupy the corners, and 12 oxide ions are located along the edges. Figure 11.32(b) contains a stack of three such perovskite structures, two containing a Ba^{2+} ion and one containing a Y^{4+} ion at its center.

Applying our usual practice of assigning fractions of atoms and ions to unit cells, we find that the array of 3 unit cells in Figure 11.32(b) has 9 oxide ions ($\frac{1}{4}$ of the 20 that occupy edges of the 3-cell array + $\frac{1}{2}$ of the 8 ions that occupy faces—9 in all) and 3 copper ions ($\frac{1}{8}$ of the 8 ions on the top and bottom surfaces + $\frac{1}{4}$ of the 8 ions around the middle—3 in all). Thus, the material with the structure shown

FIGURE 11.32 The crystal structure of $YBa_2Cu_3O_7$, a high-temperature superconductor: (a) A perovskite unit cell. (b) The process of building a yttrium–barium–copper superconductor begins by stacking 3 unit cells with a central barium ion in the top and bottom cells, and a central yttrium ion in the middle cell. The formula corresponding to this structure is $YBa_2Cu_3O_9$. (c) This modified perovskite structure has 8 fewer oxide ions in edge positions, which reduces by 2 the number assignable to the structure, and produces the formula $YBa_2Cu_3O_7$.

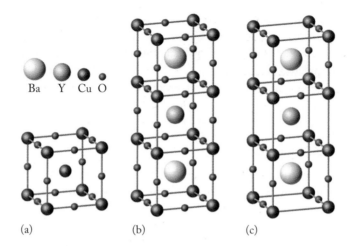

Ba Y Cu O

(a)   (b)   (c)

▶❙❙ **CHEMTOUR** Superconductors

in Figure 11.32(b) has the chemical formula $YBa_2Cu_3O_9$, but it is not superconducting. However, it becomes superconducting when it loses 8 oxygen atoms from edge sites to form $YBa_2Cu_3O_7$ (Figure 11.32c).

Yttrium–barium–copper oxides and related materials behave as superconductors because of the formation of electron pairs called *Cooper pairs* in their crystals (Figure 11.33). When an electron is introduced into a superconductor, electrostatic forces produce a slight displacement in the positively charged cations in the crystal. The attraction between electrons and these positively charged cations can bring two electrons into close proximity, resulting in formation of a Cooper pair. The movement of Cooper pairs rather than individual electrons through the material accounts for its superconductivity. To understand the superconductivity of a Cooper pair, consider a pair of horses harnessed together to pull a wagon

FIGURE 11.33 (a) An electron moving through a lattice of positively charged cations causes (b) a distortion in the lattice as the cations are attracted to the electron's negative electrical charge. (c) This localized increase in positive charge brings another electron toward the first, creating a Cooper pair.

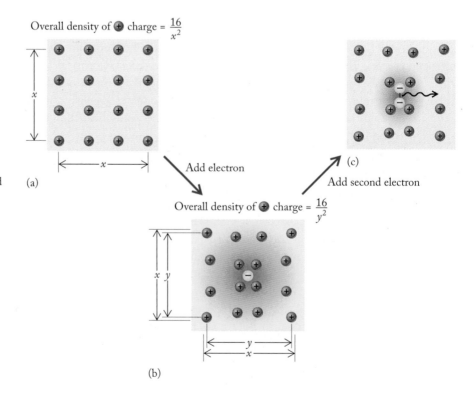

(or a three-legged race at a child's birthday party). The team of horses will travel forward, avoiding obstacles in their path but never separating from each other. Single animals, however, will take divergent paths, scattering over a wide area. Similarly, single electrons are easily deflected by the atoms or ions in a solid, but electrons scatter much less when harnessed in a Cooper pair. Less scattering lowers electrical resistance.

Superconductors are of technological interest because, in principle, high currents can flow through them with no resistance. This property is attractive for the design of extremely fast computers. Superconductors also have the ability to exclude magnetic fields—a property known as the *Meissner effect* (Figure 11.34). A small cylinder of superconductive material cooled below its critical temperature floats in a magnetic field because magnetic lines of force are excluded from it. Because of this exclusion, the magnet below the superconductor repels it, suspending it in air. This magnetic levitation could in principle be used to both float a train above electromagnetic tracks and propel it along the tracks at relatively low cost and at much higher speeds than a conventional train. Levitating trains, however, remain a goal for the future.

FIGURE 11.34 The Meissner effect: The magnetic field produced by the magnet (background) cannot penetrate the small cylinder of superconductive material that has been chilled below its critical temperature. As a result, the magnet repels the superconductor, causing it to float above the magnet.

11.9 X-ray Diffraction: How We Know Crystal Structures

Earlier in the chapter we pointed out that the dimensions of unit cells are determined using **X-ray diffraction (XRD)**. In an XRD instrument a narrow beam of X-rays is directed at the surface of a crystalline sample, as shown in Figure 11.35. The angle between the surface and the beam is called the angle of incidence and given the symbol θ. Some of the X-rays hit atoms on the surface of the crystal and bounce off them. These reflected X-rays leave the surface with an angle of reflection that is also θ. The total change in direction of the reflected beam is the sum of the angles of incidence and reflections, or 2θ. This value is called the *angle of diffraction.* 2θ.

Now suppose some of the incident X-rays pass through the surface and bounce off atoms (or ions) in a second layer a distance (d) deeper into the crystal. They will also undergo an angle of diffraction 2θ. These X-rays and those reflected from the first row of atoms are much like the rays of light passing through two narrow slits (see Figure 7.5). This means they can undergo constructive or destructive interference and so produce an interference pattern. To interfere constructively, the two beams of reflected X-rays must be in phase, as shown in Figure 11.35(b). To be in phase, the greater distance traveled by the beam reflecting off the second row must be some multiple of the wavelength. The greater distance is the sum of the lengths of line segments $\overline{XY}$ and $\overline{YZ}$. The two right triangles incorporating these line segments share a hypotenuse equal to d. Geometry tells us that the angles opposite $\overline{XY}$ and $\overline{YZ}$ are not only the same but also equal to θ. According to trigonometry, the ratio of either $\overline{XY}$ or $\overline{YZ}$ to d is the sine of θ:

$$\frac{\overline{XY}}{d} = \frac{\overline{YZ}}{d} = \sin\theta$$

Rearranging these terms, we find that the difference in distance traveled ($\overline{XY} + \overline{YZ}$) is

$$d\sin\theta + d\sin\theta = 2d\sin\theta$$

▶❚❚ **CHEMTOUR** X-ray Diffraction

X-ray diffraction (XRD) is a technique for determining the arrangement of atoms or ions in a crystal by analyzing the pattern that results when X-rays are scattered after bombarding the crystal.

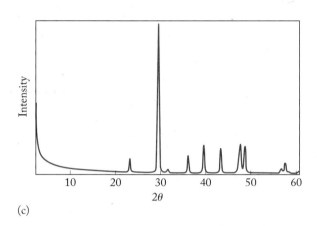

(a)

(b)

(c)

FIGURE 11.35 X-ray diffraction: (a) An X-ray diffractometer is used to determine the crystal structures of solids. A source of X-rays and a detector are mounted so that they can rotate around the sample. (b) A beam of X-rays of wavelength λ is directed at the surface of a sample. The beam is diffracted by the sample, reaching the detector when the angle of diffraction (2θ) satisfies the Bragg equation. (c) Moving the source and detector around the sample as X-ray signals are recorded produces a scan such as this one for quartz.

When this difference equals some whole-number multiple (n) of the wavelength (λ) of the X-ray beam, the following relation, called the **Bragg equation**, holds true:

$$n\lambda = 2d\sin\theta \qquad (11.15)$$

Under these conditions, the crests and valleys of the two X-rays are in phase with each other as they emerge from the crystal (Figure 11.35b). When waves of electromagnetic radiation are aligned in this way, they undergo constructive interference, which means that a detector aligned θ degrees to the surface of the sample detects X-rays diffracted by the sample. The greater the constructive interference, the greater the intensity at a given value of 2θ on the scan. An XRD scan such as the one for quartz in Figure 11.35(c) allows scientists to determine the spacing between layers of atoms or ions in crystals. They convert the 2θ values that correspond to the peaks in the scan to θ values and then use these values in Equation 11.15 to calculate values of d.

With this information, scientists may be able to identify which crystalline solid is present in a sample and determine its crystal structure. Sample Exercise 11.7 illustrates how we use X-ray diffraction to find the distance between the layers in a crystal of copper metal.

The **Bragg equation** relates the angle of diffraction (2θ) of X-rays to the spacing (d) between the layers of ions or atoms in a crystal: $n\lambda = 2d\sin\theta$.

SAMPLE EXERCISE 11.7   **Determining Interlayer Distances by X-ray Diffraction**

An XRD analysis (λ = 154 pm) of a sample of copper has peaks at 2θ = 24.64°, 50.54°, and 79.62°. What is the distance (d) between layers of Cu atoms that could produce this diffraction pattern?

COLLECT AND ORGANIZE We are given a series of diffraction angles (2θ) and asked to find the distance (d) between the layers of copper atoms in a crystalline sample. Equation 11.15 ($n\lambda = 2d\sin\theta$) allows us to calculate the distance d in a crystal structure on the basis of those angles of diffraction associated with constructive interference in a beam of scattered X-rays.

ANALYZE To use Equation 11.15, we must first convert the 2θ values into θ values by dividing by 2, which produces θ values of 12.32°, 25.27°, and 39.81°. We can also rearrange the terms in Equation 11.15 to solve for the parameter we are after: d.

$$d = \frac{n\lambda}{2\sin\theta}$$

The additional unknown in Equation 11.15 is the wavelength multiplier (n). Its value must be evaluated before we can solve for d. The key to determining the value of n is to look for a pattern in the values of θ. Notice that the higher values of θ are approximately two and three times the lowest value, 12.32°. This pattern suggests that the values of n for this set of data are 1, 2, and 3 for θ = 12.32°, 25.27°, and 39.81°.

SOLVE Let's use these combinations of n and θ to see whether they all give the same value of d:

$$d = \frac{(1)(154 \text{ pm})}{2 \sin 12.32°} = \frac{154 \text{ pm}}{(2)(0.2134)} = 361 \text{ pm}$$

$$d = \frac{(2)(154 \text{ pm})}{2 \sin 25.27°} = \frac{308 \text{ pm}}{(2)(0.4269)} = 361 \text{ pm}$$

$$d = \frac{(3)(154 \text{ pm})}{2 \sin 39.81°} = \frac{462 \text{ pm}}{(2)(0.6402)} = 361 \text{ pm}$$

We do indeed get the same value of d, so d = 361 pm is the distance between Cu atoms that produced the three peaks.

THINK ABOUT IT These consistent results mean that our assumption about the values of n for the three values of θ was correct. Also, the value of d is in the range of the edge lengths of unit cells we used in several Sample Exercises earlier in the chapter, and so is reasonable.

Practice Exercise An X-ray diffraction analysis (λ = 71.2 pm) of CsCl has a prominent peak at 2θ = 19.9°. If this peak corresponds to n = 2, what is the spacing (d) between the layers of ions in a crystal of CsCl? What are the corresponding 2θ values for n = 3 and n = 4?

As we have seen throughout this text, the arrangement of atoms, ions, and molecules in a sample of matter is responsible for many of the properties of materials. XRD is a powerful tool for analyzing a wide range of materials, including metals, minerals, polymers, plastics, ceramics, pharmaceuticals, and semiconductors. The technique is indispensable for scientific research and industrial production.

Silicon, Silica, Silicates, Silicone: What's in a Name?

Group 14 of the periodic table is headed by carbon, an element so important to life, energy, and commerce that it merited its own descriptive chemistry section in Chapter 5. The element below carbon in group 14 is silicon, an element that is central to both ancient and modern technology and to the entire solid-state revolution; it too deserves to be highlighted.

Silicon is the second most abundant element in Earth's crust; oxygen is the first. Elemental silicon never occurs free in nature. Mostly it is present as silica (SiO_2, a covalent network solid) and silicate minerals, or silicates (solids containing silicon–oxygen groups and metals). Silicates almost invariably consist of silicon–oxygen tetrahedra units bonded together in chains, rings, sheets, and three-dimensional arrays similar to those found in carbon structures (Section 11.6). Silica constitutes about 60% by weight of Earth's crust and is of great geological and commercial importance. It is most familiar to us as quartz crystals and beach sand.

Humans have used silica and silicates since prehistoric times in simple tools, arrowheads, and flint knives. The glass most familiar to us as window panes and bottles, called soda-lime glass, is principally a mixture of SiO_2 (73%), CaO (11%), and Na_2O (13%) with the balance composed of other metal oxides. Leaded crystal stemware is made by adding PbO (24%) and K_2O (15%) to silica (60%). Laboratory glassware is composed primarily of silica (81%), B_2O_3 (11%), and Na_2O (5%). In addition to the glass industry, natural silicates are the backbone of the ceramic and concrete industries. Silica is also used in detergents; in the pulp and paper industry; as anticaking agents in powdered foods such as cocoa, sugar, and spices; and as an ingredient in paints to provide a matte finish. Small packs of material labeled "desiccant" in packaged electronics, pharmaceuticals, and some foods may contain porous silica, because it can absorb about 40% of its weight in water and thereby keep materials sensitive to moisture dry. Silica is a major component in toothpaste, functioning as both a mild abrasive to clean teeth and a thickening agent to hold the components in pastes and gels together.

Most silica is now used in the production of highly purified elemental silicon, the most basic material of the microelectronics industry. The name "Silicon Valley" to describe the semiconductor industry around Palo Alto, CA, is an acknowledgment of the importance of the element to the industry. Computers, calculators, liquid-crystal displays, cell phones, video games, solar cells, and space vehicles are all products of the solid-state revolution that began in 1947 with the discovery of the transistor effect at Bell Laboratories in Murray Hill, NJ. The theoretical work and practical development of transistors and semiconductors by Walter H. Brattain, John Bardeen, and William B. Shockley resulted in them sharing the Nobel Prize in Physics in 1956.

Silicon is the element most widely used as a starting material in the production of semiconductors. (Germanium, the element below silicon in the periodic table, is second). To function correctly in electronic devices, silicon must contain less than 1 part per billion (ppb) of impurities. Such ultrapure silicon is produced in a series of reactions starting with the reaction of silicon dioxide with carbon:

$$SiO_2(s) + C(s) \rightarrow Si(s) + CO_2(g)$$

Silicon prepared in this way is about 98% pure. Before it can be used to make microchips, it must be refined further, through reaction with chlorine to form $SiCl_4$:

$$Si(s) + 2\,Cl_2(g) \rightarrow SiCl_4(g)$$

Impurities that do not form volatile chlorides are removed, and the $SiCl_4$ is heated above 700°C, where it decomposes to silicon and chlorine gas:

$$SiCl_4(g) \rightarrow Si(s) + 2\,Cl_2(g)$$

The final step in processing silicon results in both an ultrapure material (one that contains only silicon atoms) and a solid that is free of crystalline imperfections (every position in the lattice is occupied by a silicon atom, and no gaps or vacancies exist). A semiconductor is usually made from a single crystal of silicon grown by a process called "pulling." A small seed crystal is allowed to contact the surface of a pool of melted silicon. The seed crystal is withdrawn from the melt at a rate that allows material from the melt to solidify on the seed crystal. Single crystals of silicon 25 cm long or longer are made in this fashion, and semiconductor devices are made from pieces of these single crystals. Crystalline silicon has the same structure as diamond.

Silicones are another commercially important family of silicon compounds; they are made of carbon, hydrogen, oxygen, and silicon and are hence called organosilicon compounds. No natural organosilicon compounds exist, but they are the basis of a major industry. A silicone polymer is the mate-

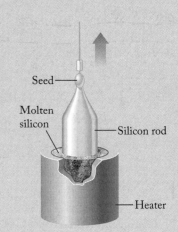

"Pulling" a silicon crystal from a melt yields ultrapure and perfectly crystalline silicon.

Seed

Molten silicon

Silicon rod

Heater

rial applied to the windshields of automobiles and aircraft to prevent bugs and grime from clouding vision; to reduce glare; and to disperse rain, sleet, and snow. The most common silicone used in this application (polydimethyl siloxane, or PDMS) has this as the repeating unit:

$$\left[\begin{array}{c} CH_3 \\ | \\ -O-Si-O- \\ | \\ CH_3 \end{array}\right]$$

Glass contains silicates, which interact with water, as we discussed in Chapter 10 in terms of the meniscus formation. PDMS molecules adhere readily to glass because their structure is similar to that of the silicates, but the $-CH_3$ groups make PDMS much more hydrophobic. When water contacts the PDMS coating, it does not adhere to the surface; raindrops are repelled.

Silicones are available as liquids, gels, and flexible solids. They are stable over a range of temperatures from below $-40°C$ to greater than $150°C$. Thousands of applications for silicones beyond coatings for glass surfaces include their use as moisture-proof sealants for ignition cables, spark plugs, medical appliances and catheters, and as protective surfaces for everything from integrated circuits to textiles and skyscrapers. PDMS is also a key ingredient in Silly Putty.

The chemistry of the other elements in group 14, germanium, tin, and lead, resembles that of silicon, and their periodic properties vary uniformly down the family. Germanium (like silicon) is a metalloid, whereas tin and lead are two of the oldest metals known in human experience. All three form oxides with the formula MO_2 (where M is the group 14 metal or metalloid) and react with halogens to form tetrahalides (MX_4, where X = F, Cl, Br, or I). The dichlorides of tin and lead, $SnCl_2$ and $PbCl_2$, also are stable. Both germanium and one of the allotropes of tin, α-tin, have the diamond structure, but β-tin has a more complicated structure. Elemental lead has a face-centered cubic structure. Germanium and α-tin are semiconductors, but β-tin and lead are conductors. Tin is used extensively in metallurgy in combination with other metals. Tin oxides are used to make ceramics. Lead and lead(IV) oxide are the electrode materials in most automobile batteries.

Rain beads on metal treated with a silicone polymer.

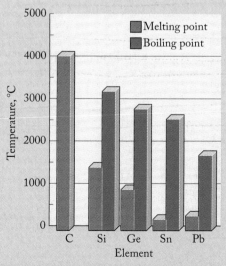

Melting points and boiling points of the group 14 elements; the boiling point of carbon has not yet been determined.

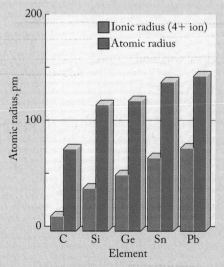

Atomic and ionic radii of the group 14 elements.

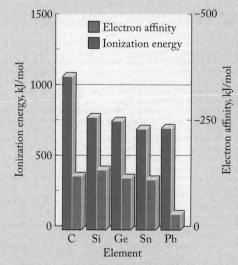

The electron affinities and ionization energies of the group 14 elements.

Summary

SECTION 11.1 Most metals are malleable and ductile. The conductivity of metals can be explained by **band theory** as the easy movement of electrons from the **valence band** to the **conduction band**.

SECTION 11.2 Semimetals are **semiconductors**, intermediate in conducting ability between metals and nonmetals. The energies of the conduction band and the valence band are separated by a **band gap**. Substituting electron-rich atoms into a semiconductor (such as group 15 elements into silicon) results in **n-type** semiconductors. Substituting electron-poor atoms (such as group 13 elements into silicon) results in **p-type** semiconductors. Both types of substitution increase the conductivity of the semiconductor by decreasing its band gap.

SECTION 11.3 Many metallic crystals are based on **cubic closest-packed (ccp)** and **hexagonal closest-packed (hcp)** atoms, which are the two most efficient ways of packing atoms in a solid. **Crystalline solids** contain repeating units called **unit cells**. They include **simple cubic**, **body-centered cubic (bcc)**, and **face-centered cubic (fcc) unit cells**. The crystal structure of a metal can be used to determine the radius of its atoms and predict its density.

SECTION 11.4 **Alloys** are blends of a host metal and one or more other elements (which may or may not be metals) that are added to enhance the properties of the host including strength, hardness, and corrosion resistance. In substitutional alloys atoms of the added elements replace atoms of the host metal in its crystal structure. In interstitial alloys atoms of added elements are located in the tetrahedral and/or octahedral holes between atoms of the host metal.

SECTION 11.5 Aluminum and aluminum alloys are highly desirable for applications requiring corrosion resistance and light weight. Aluminum is produced by the reduction of aluminum oxide using an electric current. The energy cost of aluminum manufacture makes recycling aluminum profitable.

SECTION 11.6 The allotropes of carbon include graphite and diamond, which form **covalent network solids**. Many nonmetals form **molecular solids**, including sulfur, which forms puckered rings of 8 sulfur atoms, and phosphorus, which forms P_4 tetrahedra.

SECTION 11.7 Many **ionic solids** consist of crystals with 1 ion forming one of the close-packed or closest-packed unit cells (fcc, bcc, simple cubic, or hexagonal) with the other ion occupying octahedral and tetrahedral holes in the crystal lattice. Unit-cell structures include rock salt, sphalerite, fluorite, and antifluorite. The unit-cell edge lengths and crystal structures of ionic solids can be used to calculate ionic radii and predict densities.

SECTION 11.8 Heating selected solid inorganic compounds (such as clays, which are aluminosilicate minerals) to high temperature alters their chemical composition and makes the materials harder, denser, and stronger. The resulting heat- and chemical-resistant materials (**ceramics**) are electrical insulators due to the large energy gap (E_g) between their filled valence and empty conduction bands. The polymorphs of silica consist of SiO_4 tetrahedra, each of which can share some or all its oxygen atoms with other tetrahedra, forming Si—O—Si bridges and two- and three-dimensional covalent networks. In **superconducting** ceramics, the electrical resistance of the material drops to zero below its **critical temperature (T_c)**.

SECTION 11.9 **X-ray diffraction** is an analytical method that records constructive and destructive interference of X-rays reflecting off different layers of atoms or ions in crystals. The distances between layers can be calculated using the **Bragg equation**. X-ray diffraction makes it possible to determine the solid-state structure of crystalline solids.

Problem-Solving Summary

TYPE OF PROBLEM	CONCEPTS AND EQUATIONS	SAMPLE EXERCISES
Distinguishing between p- and n-type semiconductors	Determine whether the dopant has more (n-type) or fewer (p-type) valence electrons than the host semiconductor	11.1
Calculating atomic radii from crystal structures	Determine the length of a unit-cell edge, face diagonal, or body diagonal along which the atoms touch. Calculate the number of radii that correspond to this distance.	11.2
Using crystal structure to calculate the density of a solid	$d = m/V$ Determine the mass of the atoms in the unit cell from their molar mass and the volume of the unit cell from the unit-cell edge length.	11.3
Predicting the type of alloy two elements form	Compare the radii of the alloying elements: similarly sized radii (within 15%) predict a substitutional alloy. When the radius of the smaller atom is < 73% of the radius of the larger atom, an interstitial alloy forms.	11.4
Calculating ionic radii from unit-cell dimensions	Determine the dimension (edge or diagonal) along which the ions touch. Use the geometric relationship between the edge or diagonal and the ionic radius (r) to calculate the value of r.	11.5

TYPE OF PROBLEM	CONCEPTS AND EQUATIONS	SAMPLE EXERCISES
Calculating the density of a salt from its unit-cell dimensions	Determine the mass of the ions in the unit cell and the volume of the unit cell from the unit-cell edge length. Then apply $d = m/V$.	11.6
Determining interlayer distances by X-ray diffraction	Apply the Bragg equation: $n\lambda = 2d\sin\theta$.	11.7

Visual Problems

11.1. In Figure P11.1, which drawing is analogous to a crystalline solid, and which is analogous to an amorphous solid?

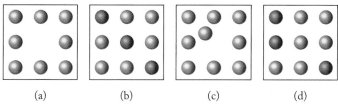

(a) (b) (c) (d)

FIGURE P11.1

11.2. The unit cells in Figure P11.2 continue infinitely in two dimensions. Draw a box around the unit cell in each pattern. How many light squares and how many dark squares are in each unit cell?

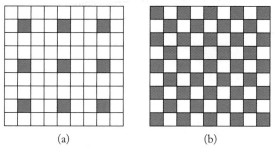

(a) (b)

FIGURE P11.2

11.3. The pattern in Figure P11.3 continues indefinitely in three dimensions. Draw a box around the unit cell. If the red circles represent element A and the blue circles element B, what is the chemical formula of the compound?

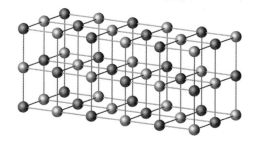

FIGURE P11.3

11.4. How many total cations (A and B) and anions (X) are there in the unit cell in Figure P11.4?

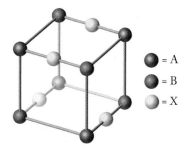

= A
= B
= X

FIGURE P11.4

11.5. How many equivalent atoms of elements A and B are there in the portion of a unit cell in Figure P11.5?

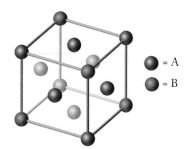

= A
= B

FIGURE P11.5

11.6. What is the chemical formula of the compound a portion of whose unit cell is shown in Figure P11.6?

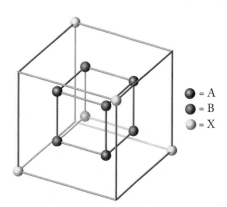

= A
= B
= X

FIGURE P11.6

11.7. What is the chemical formula of the ionic compound a portion of whose unit cell is shown in Figure P11.7? (A and B are cations, X is an anion.)

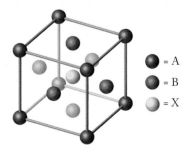

= A
= B
= X

FIGURE P11.7

11.8. When amorphous red phosphorus is heated at high pressure, it is transformed into the allotrope black phosphorus, which can exist in one of several forms. One form consists of 6-membered rings of phosphorus atoms (Figure P11.8a). Why are the 6-atom rings in black phosphorus puckered, whereas the 6-atom rings in graphite are planar (Figure P11.8b)?

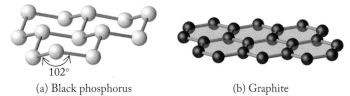

102°

(a) Black phosphorus   (b) Graphite

FIGURE P11.8

11.9. The distance between atoms in a cubic form of phosphorus is 238 pm (Figure P11.9). Calculate the density of this form of phosphorus.

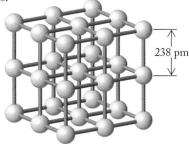

238 pm

FIGURE P11.9

11.10. How many of the large spheres and how many of the small ones are assignable to the unit cell in Figure P11.10?

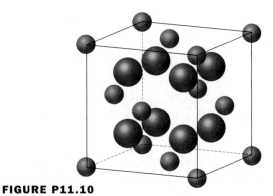

FIGURE P11.10

11.11. What is the formula of the compound that crystallizes with aluminum ions occupying half of the octahedral holes and magnesium ions occupying one-eighth of the tetrahedral holes in a cubic closest-packed arrangement of oxide ions? See Figure P11.11.

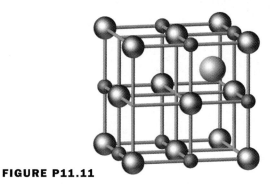

FIGURE P11.11

11.12. What is the formula of the compound that crystallizes with zinc atoms occupying half of the tetrahedral holes in a cubic closest-packed arrangement of sulfur atoms? See Figure P11.12.

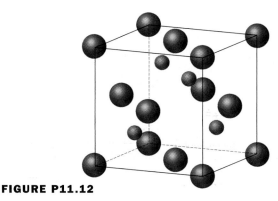

FIGURE P11.12

11.13. What is the formula of the compound that crystallizes with lithium ions occupying all of the tetrahedral holes in a cubic closest-packed arrangement of sulfide ions? See Figure P11.13.

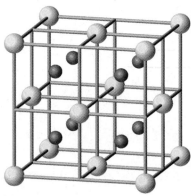

FIGURE P11.13

11.14. Figure P11.14 shows the unit cell of CsCl. From the information given and the radius of the chloride (corner ions) of 181 pm, calculate the radius of Cs^+ ions.

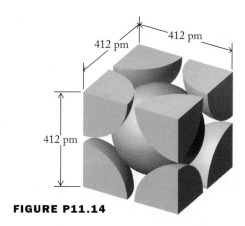

FIGURE P11.14

11.15. A number of metal chlorides adopt the rock salt structure in which the metal ions occupy all the octahedral holes in a face-centered cubic array of chloride ions. For at least one (maybe more) of the metallic elements highlighted in Figure P11.15, this crystal structure is not possible. Which one(s) are they?

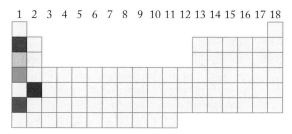

FIGURE P11.15

11.16. A number of metal fluorides adopt the fluorite structure in which the fluoride ions occupy all the tetrahedral holes in a face-centered cubic array of metal ions. For at least one (maybe more) of the highlighted metallic elements in Figure P11.16, this crystal structure is not possible. Which one(s)?

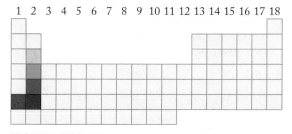

FIGURE P11.16

*11.17. Superconducting Materials In 2000, magnesium boride was observed to behave as a superconductor. Its unit cell is shown in Figure P11.17. What is the formula of magnesium boride? A boron atom is in the center of the unit cell (on the left), which is part of the hexagonal structure (right).

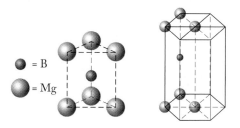

FIGURE P11.17

*11.18. Superconducting Materials The 1987 Nobel Prize in physics was awarded to G. Bednorz and K. A. Müller for their discovery of superconducting ceramic materials such as $YBa_2Cu_3O_7$. Figure P11.18 shows the unit cell of another yttrium–barium–copper oxide.
a. What is the chemical formula of this compound?
b. Eight oxygen atoms must be removed from the unit cell shown here to produce the unit cell of $YBa_2Cu_3O_7$. Does it make a difference which oxygen atoms are removed?

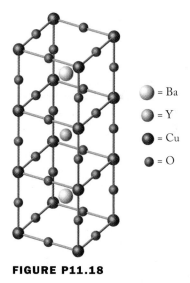

FIGURE P11.18

Questions and Problems

Metallic Bonds and Conduction Bands

CONCEPT REVIEW

11.19. How does the sea-of-electrons model (Chapter 8) explain the high electrical conductivity of gold?
*11.20. How does band theory explain the high electrical conductivity of mercury?

11.21. The melting and boiling points of sodium metal are much lower than those of sodium chloride. What does this difference reveal about the relative strengths of metallic bonds and ionic bonds?
*11.22. Which metal do you expect to have the higher melting point—Al or Na? Explain your answer.

11.23. Some scientists believe that the solid hydrogen that forms at very low temperatures and high pressures may conduct electricity. Is this hypothesis supported by band theory?

11.24. Would you expect solid helium to conduct electricity?

Semimetals and Semiconductors

CONCEPT REVIEW

11.25. Which groups in the periodic table contain metals with filled valence bands?

11.26. Insulators are materials that do not conduct electricity; conductors are substances that allow electricity to flow through them easily. Rank the following in order of increasing band gap: semiconductor, insulator, conductor.

11.27. Why is it important to keep phosphorus out of silicon chips during their manufacture?

11.28. How might doping of silicon with germanium affect the conductivity of silicon?

*11.29. Antimony (Sb) combines with sulfur to form the semiconductor compound Sb_2S_3. In which group of the periodic table might you find elements for doping Sb_2S_3 to form a p-type semiconductor?

*11.30. In which group of the periodic table might you find elements for doping Sb_2S_3 to form an n-type semiconductor?

PROBLEMS

11.31. Thin films of doped diamond hold promise as semiconductor materials. Trace amounts of nitrogen impart a yellow color to otherwise colorless pure diamonds.
   a. Are nitrogen-doped diamonds examples of semiconductors that are p-type or n-type?
   b. Draw a picture of the band structure of diamond to indicate the difference between pure diamond and N-doped (nitrogen-doped) diamond.
   *c. N-doped diamonds absorb violet light at about 425 nm. What is the magnitude of E_g that corresponds to this wavelength?

11.32. Trace amounts of boron give diamonds (including the Smithsonian's Hope Diamond) a blue color (Figure P11.32).
   a. Are boron-doped diamonds examples of semiconductors that are p-type or n-type?
   b. Draw a picture of the band structure of diamond to indicate the difference between pure diamond and B-doped diamond.
   *c. What is the band gap in energy if blue diamonds absorb orange light with a wavelength of 675 nm?

FIGURE P11.32

*11.33. The nitride ceramics AlN, GaN, and InN are all semiconductors used in the microelectronics industry. Their band gaps are 580.6, 322.1, and 192.9 kJ/mol, respectively. Which if any of these energies correspond to radiation in the visible region of the spectrum?

*11.34. Calculate the wavelengths of light emitted by the semiconducting phosphides AlP, GaP, and InP, which have band gaps of 241.1, 216.0, and 122.5 kJ/mol used in LEDs and are like those in Figure P11.34.

FIGURE P11.34

Metallic Crystals

CONCEPT REVIEW

11.35. Explain the difference between cubic closest-packed and hexagonal closest-packed arrangements of identical spheres.

*11.36. Is it possible to have a closest-packed structure with four different repeating layers, *abcdabcd . . .*?

11.37. Which unit cell has the greater packing efficiency, simple cubic or body-centered cubic?

11.38. Consult Figure 11.10 to predict which unit cell has the greater packing efficiency: body-centered cubic, or face-centered cubic.

11.39. Iron metal adopts either an fcc or a bcc structure, depending on its temperature (see Sample Exercise 11.2). Do these two crystal structures represent allotropes of iron? Explain your answer.

*11.40. Calcium metal crystallizes in two different unit cells, an fcc and a bcc. Why might calcium adopt a closest-packed arrangement at low temperature (fcc unit cell) but the nonclosest-packed structure at higher temperature (bcc)?

PROBLEMS

11.41. Derive the edge length in bcc and fcc unit cells in terms of the radius (r) of the atoms in the cells in Figure 11.10.

*11.42. Derive the length of the body diagonal in simple cubic and fcc unit cells in terms of the radius (r) of the atoms in the unit cells in Figure 11.10.

11.43. Europium is one of the lanthanide elements used in television screens. Europium crystallizes in a bcc structure with a unit-cell edge of 240.6 pm. Calculate the radius of a europium atom.

11.44. Nickel has an fcc unit cell with an edge length of 350.7 pm. Calculate the radius of a nickel atom.

11.45. What is the length of an edge of the unit cell when barium (atomic radius = 222 pm) crystallizes in a bcc structure?

11.46. What is the length of an edge of the unit cell when aluminum (atomic radius = 143 pm) crystallizes in an fcc structure?

11.47. A crystalline form of copper has a density of 8.95 g/cm^3. If the radius of copper atoms is 127.8 pm, what is the copper unit cell? (a) simple cubic; (b) body-centered cubic; or (c) face-centered cubic

11.48. A crystalline form of molybdenum has a density 10.28 g/cm^3 at a temperature at which the radius of a molybdenum atom is 139 pm. Which unit cell is consistent with these data? (a) simple cubic; (b) body-centered cubic; or (c) face-centered cubic

Alloys

CONCEPT REVIEW

11.49. Is there a difference between a solid solution and a homogeneous alloy?

11.50. White gold was originally developed to give the appearance of platinum. One formulation of white gold contains 25% nickel and 75% gold. Which is more malleable, white gold or pure gold?

*11.51. Explain why an alloy that is 28% Cu and 72% Ag melts at a lower temperature than the melting points of either Cu or Ag.

11.52. Is it possible for an alloy to be both substitutional and interstitial?

11.53. The interstitial alloy tungsten carbide (WC) is one of the hardest materials known. It is used on the tips of cutting tools. Without consulting a table of radii, decide which element you think is the host and which occupies the holes.

11.54. Why are the alloys that second-row nonmetals—such as B, C, and N—form with transition metals more likely to be interstitial than substitutional?

PROBLEMS

11.55. The unit cell of a substitutional alloy consists of a face-centered cube that has an atom of element X at each corner and an atom of element Y at the center of each face.
a. What is the formula of the alloy?
b. What would the formula of the alloy be if the positions of the two elements were reversed in the unit cell?

11.56. The bcc unit cell of a substitutional alloy has atoms of element A at the corners of the unit cell and an atom of element B at the center of the unit cell.
a. What is the formula of the alloy?
b. What would the formula of the alloy be if the positions of the two elements were reversed in the unit cell?

11.57. Vanadium reacts with carbon to form vanadium carbide, an interstitial alloy. Given the atomic radii of V (135 pm) and C (77 pm), which holes in a cubic closest-packed array of vanadium atoms do you think the carbon atoms are more likely to occupy—octahedral or tetrahedral?

11.58. What is the minimum atomic radius required for a cubic closest-packed metal to accommodate boron atoms (radius 88 pm) in its octahedral holes?

11.59. Dental Fillings Dental fillings are mixtures of several alloys including one with the formula Ag$_3$Sn. Silver (radius 144 pm) and tin (140 pm) both crystallize in an fcc unit cell. Is this alloy likely to be a substitutional alloy or an interstitial alloy?

11.60. An alloy used in dental fillings has the formula Sn$_8$Hg. The radii of tin and mercury atoms are 140 pm and 151 pm, respectively. Which alloy has a smaller mismatch (percent difference in atomic radii), Sn$_3$Hg or bronze (Cu/Sn alloys)?

*11.61. Hardening Metal Surfaces Plasma nitriding is a process for embedding nitrogen atoms in the surfaces of metals that hardens the surfaces and makes them more corrosion resistant. Do the nitrogen atoms in the nitrided surface of a sample of cubic closest-packed iron fit in the octahedral holes of the crystal lattice? (Assume that the atomic radii of N and Fe are 75 and 126 pm, respectively.)

*11.62. Hydrogen Storage A number of transition metals (including titanium, zirconium, and hafnium) can store hydrogen as metal hydrides for use as fuel in a hydrogen-powered vehicle. Which metal or metals are most likely to accommodate H atoms (radius 37 pm) with the least distortion, given that their atomic radii are 147, 160, and 159 pm, respectively?

11.63. An interstitial alloy is prepared from metals A and B where B has the smaller atomic radius. Metal A crystallizes in an fcc structure. What is the formula of the alloy if B occupies the following? (a) all of the octahedral holes; (b) half of the octahedral holes; (c) half of the tetrahedral holes

11.64. An interstitial alloy was prepared from two metals. Metal A with the larger atomic radius has a hexagonal closest-packed structure. What is the formula of the alloy if atoms of metal B occupy the following? (a) all of the tetrahedral holes; (b) half of the tetrahedral holes; (c) half of the octahedral holes

11.65. An interstitial alloy contains 1 atom of B for every 5 atoms of host element A, which has an fcc structure. What fraction of the octahedral holes is occupied in this alloy?

11.66. If the B atoms in the alloy described in Problem 11.65 occupy tetrahedral holes in A, what percentage of the holes would they occupy?

The Structures of Some Nonmetals

CONCEPT REVIEW

11.67. S$_8$ is not a flat octagon—why?

*11.68. Selenium exists as Se$_8$ rings or in a crystalline structure with helical chains of Se atoms. Are these two structures of selenium allotropes? Explain your answer.

*11.69. If the carbon atoms in graphite are replaced by alternating B and N atoms, would the resulting structure contain puckered rings (see Figure P11.8) like black phosphorus or flat ones like graphite?

*11.70. Cyclic allotropes of sulfur containing up to 20 sulfur atoms have been isolated and characterized. Propose a reason why the bond angles in S$_n$ (where n = 10, 12, 18, and 20) are all close to 106°?

PROBLEMS

11.71. Ice is a network solid. However, theory predicts that, under high pressure, ice (solid H$_2$O) becomes an ionic compound composed of H$^+$ and O^{2-} ions. The proposed structure of ice under these conditions is a body-centered cubic arrangement of oxygen ions with hydrogen ions in holes.
a. How many H$^+$ and O^{2-} ions are in each unit cell?
b. Draw a Lewis structure for "ionic" ice.

11.72. **Ice Under Pressure** Kurt Vonnegut's novel *Cat's Cradle* describes an imaginary, high-pressure form of ice called "ice nine." With the assumption that ice nine has a cubic closest-packed arrangement of oxygen atoms with hydrogen atoms in the appropriate holes, what type of hole will accommodate the H atoms?

11.73. A chemical reaction between H_2S_4 and S_2Cl_2 produces cyclic S_6. What are the bond angles in S_6?

11.74. Reaction between S_8 and six equivalents of AsF_5 yields $[S_4^{2+}][AsF_6^-]_2$ by the reaction

$$S_8 + 6\,AsF_5 \rightarrow 2\,[S_4^{2+}][AsF_6^-]_2 + AsF_3$$

(The brackets identify complex ions in the structure, in this case indicating that the S_4 unit has a 2+ charge and the AsF_6 unit has a 1− charge.) The S_4^{2+} ion has a cyclic structure. Are all 4 sulfur atoms in one plane?

Salt Crystals: Ionic Solids

CONCEPT REVIEW

11.75. Crystals of both LiCl and KCl have the rock salt crystal structure. In the unit cell of LiCl the Cl^- ions touch each other. In KCl they don't. Why?

11.76. Can $CaCl_2$ have the rock salt crystal structure?

***11.77.** In some books the unit cell of CsCl is described as being body-centered cubic (Figure P11.77); in others, as simple cubic (see Figure 11.10c). Explain how CsCl might be assigned to both structures.

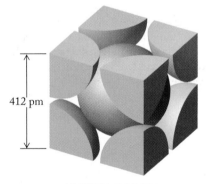

412 pm

FIGURE P11.77

11.78. In the crystals of ionic compounds, how do the relative sizes of the ions influence the location of the smaller ions?

***11.79.** Instead of describing the unit cell of NaCl as an fcc array of Cl^- ions with Na^+ ions in octahedral holes, might we describe it as an fcc array of Na^+ ions with Cl^- ions in octahedral holes? Explain why or why not.

11.80. Why isn't crystalline sodium chloride considered a network solid?

11.81. If the unit cell of a substitutional alloy of copper and tin has the same unit-cell edge as the unit cell of copper, will the alloy have a greater density than copper?

11.82. If the unit cell of an interstitial alloy of vanadium and carbon has the same unit-cell edge as the unit cell of vanadium, will the alloy have a greater density than vanadium?

***11.83.** As the cation–anion radius ratio increases for an ionic compound with the rock salt structure, is the calculated density more likely to be greater than, or less than, the measured value?

11.84. As the cation–anion radius ratio increases for an ionic compound with the rock salt structure, is the length of the unit-cell edge calculated from ionic radii likely to be greater than, or less than, the observed unit-cell edge length?

PROBLEMS

11.85. What is the formula of the oxide that crystallizes with Fe^{3+} ions in one-fourth of the octahedral holes, Fe^{3+} ions in one-eighth of the tetrahedral holes, and Mg^{2+} in one-fourth of the octahedral holes of a cubic closest-packed arrangement of oxide ions (O^{2-})?

11.86. What is the formula of the compound that crystallizes with Ba^{2+} ions occupying half of the cubic holes in a simple cubic arrangement of fluoride ions?

11.87. **The Vinland Map** At Yale University there is a map, believed to date from the 1400s, of a landmass labeled "Vinland" (Figure P11.87). The map is thought to be evidence of early Viking exploration of North America. Debate over the map's authenticity centers on yellow stains on the map paralleling the black ink lines. One analysis suggests the yellow color is from the mineral anatase, a form of TiO_2 that was not used in 15th-century inks.
   a. The structure of anatase is approximated by a ccp arrangement of oxide ions with titanium(IV) ions in holes. Which type of hole are Ti^{4+} ions likely to occupy?
   b. What fraction of these holes are likely to be occupied? (The radius of Ti^{4+} is 60.5 pm.)

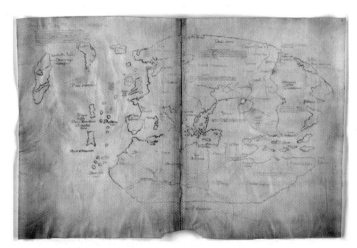

FIGURE P11.87

***11.88.** The structure of olivine—M_2SiO_4 (M = Mg, Fe)—can be viewed as a ccp arrangement of oxide ions atoms with Si(IV) in tetrahedral holes and the metal ions in octahedral holes.
   a. What fraction of each type of hole is occupied?
   b. The unit-cell volumes of Mg_2SiO_4 and Fe_2SiO_4 are 2.91×10^{-26} cm^3 and 3.08×10^{-26} cm^3. Why is the unit-cell volume of Fe_2SiO_4 larger?

***11.89.** In nature, cadmium(II) sulfide (CdS) exists as two minerals. One of them, called *hawleyite*, has a sphalerite crystal structure, and its density at 25°C is 4.83 g/cm^3. A hypothetical form of CdS with the rock salt structure would have a density of 5.72 g/cm^3. Why should the rock salt structure of CdS be denser? The ionic radii of Cd^{2+} and S^{2-} are 95 pm and 184 pm, respectively.

11.90. There are two crystalline forms of manganese(II) sulfide (MnS): the α form has a NaCl structure; the β form has a sphalerite structure.
   a. Describe the differences between the two structures of MnS.
   b. The ionic radii of Mn^{2+} and S^{2-} are 67 and 184 pm. Which type of cubic hole in a ccp lattice of sulfide ions could theoretically accommodate an Mn^{2+} ion?

11.91. The unit cell of rhenium trioxide (ReO_3) consists of a cube with rhenium atoms at the corners and an oxygen atom on each of the 12 edges. The atoms touch along the edge of the unit cell. The radii of Re and O atoms in ReO_3 are 137 and 73 pm. Calculate the density of ReO_3.

11.92. With reference to Figure P11.77, calculate the density of simple cubic CsCl.

11.93. Magnesium oxide crystallizes in the rock salt structure. Its density is 3.60 g/cm^3. What is the edge length of the fcc unit cell of MgO?

11.94. Crystalline potassium bromide (KBr) has a rock salt structure and a density of 2.75 g/cm^3. Calculate its unit-cell edge length.

Ceramics: Insulators to Superconductors

CONCEPT REVIEW

11.95. Which of the following properties are associated with ceramics and which are associated with metals? Ductile; thermal insulator; electrically conductive; malleable

11.96. Many ceramics such as TiO_2 are electrical insulators. What differences are there in the band structure of TiO_2 compared with Ti metal that account for the different electrical properties?

11.97. Replacement of Al^{3+} ions in kaolinite [$Al_2(Si_2O_5)(OH)_4$] with Mg^{2+} ions yields the mineral antigorite. What is its formula?

11.98. What is the formula of the silicate mineral talc, obtained by the replacement of Al^{3+} ions in pyrophyllite [$Al_4Si_8O_{20}(OH)_4$] with Mg^{2+} ions?

PROBLEMS

11.99. Kaolinite [$Al_2(Si_2O_5)(OH)_4$] is formed by weathering of the mineral $KAlSi_3O_8$ in the presence of carbon dioxide and water, as described by the following unbalanced reaction:

$$KAlSi_3O_8(s) + H_2O(\ell) + CO_2(aq) \rightarrow$$
$$Al_2(Si_2O_5)(OH)_4(s) + SiO_2(s) + K_2CO_3(aq)$$

Balance the reaction and determine whether or not this is a redox reaction.

11.100. Albite, a feldspar mineral with an ideal composition of $KAlSi_3O_8$, can be converted to jadeite ($NaAlSi_2O_6$) and quartz. Write a balanced chemical equation describing this transformation.

11.101. Under the high pressures in Earth's crust, the mineral anorthite ($CaAl_2Si_2O_8$) is converted to a mixture of three minerals: grossular [$Ca_3Al_2(SiO_4)_3$], kyanite (Al_2SiO_5), and quartz (SiO_2). (a) Write a balanced chemical equation describing this transformation. (b) Determine the charges and formulas of the silicate anions in anorthite, grossular, and kyanite.

*11.102. The calcium silicate mineral grossular is also formed under pressure in a reaction between anorthite ($Ca_3Al_2Si_2O_8$), gehlenite ($Ca_2Al_2SiO_7$), and wollastonite ($CaSiO_3$):

$$Ca_3Al_2Si_2O_8 + Ca_2Al_2SiO_7 + CaSiO_3 \rightarrow Ca_3Al_2(SiO_4)_3$$

   Anorthite   Gehlenite   Wollastonite   Grossular

   a. Balance this chemical equation.
   b. Express the composition of gehlenite the way mineralogists often do: as the percentage of the metal and semimetal oxides in it, that is, %CaO, %Al_2O_3, and %SiO_2.

11.103. The ceramic material barium titanate ($BaTiO_3$) is used in devices that measure pressure. The radii of Ba^{2+}, Ti^{4+}, and O^{2-} are 135, 60.5, and 140 pm, respectively. If the O^{2-} ions are in a closest-packed structure, which hole(s) can accommodate the metal cations?

11.104. The mixed metal oxide $LiMnTiO_4$ has a structure with cubic closest-packed oxide ions and metal ions in both octahedral and tetrahedral holes. Which metal ion is most likely to be found in the tetrahedral holes? The ionic radii of Li^+, Mn^{2+}, Ti^{4+}, and O^{2-} are 76, 67, 60.5, and 140 pm, respectively.

X-ray Diffraction: How We Know Crystal Structures

CONCEPT REVIEW

11.105. Why does an amorphous solid not produce an XRD scan with sharp peaks?

11.106. X-ray diffraction cannot be used to determine the structures of compounds in solution—why?

11.107. Why are X-rays rather than microwaves chosen for diffraction studies of crystalline solids?

11.108. The radiation sources used in X-ray diffraction can be changed. Figure P11.108 shows a diffraction pattern made by a short-wavelength source. How would changing to a longer-wavelength source affect the pattern?

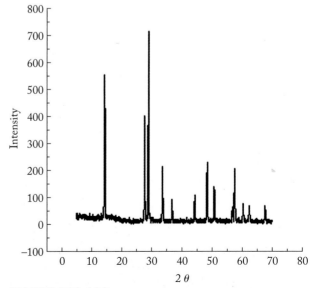

FIGURE P11.108

*11.109. Why might a crystallographer (a scientist who studies crystal structures) use different X-ray wavelengths to determine a crystal structure? (*Hint*: consider what mechanical limits are inherent in the design of the instrument depicted in Figure 11.35, and how those limits impact the 2θ scanning range.)

*11.110. Where in earlier chapters have we seen diffraction used to acquire structural information?

PROBLEMS

11.111. The spacing between the layers of ions in sylvite (the mineral form of KCl) is larger than in halite (NaCl). Which crystal will diffract X-rays of a given wavelength through larger 2θ values?

11.112. Silver halides are used in black-and-white photography. In which compound would you expect to see a larger distance between ion layers, AgCl or AgBr? Which compound would you expect to diffract X-rays through larger values of 2θ if the same wavelength of X-ray were used?

11.113. Galena, Illinois, is named for the rich deposits of lead(II) sulfide (PbS) found nearby. When PbS is exposed to X-rays with $\lambda = 71.2$ pm, strong reflections from a single crystal of PbS are observed at $13.98°$ and $21.25°$. Determine the values of n to which these reflections correspond, and calculate the spacing between the crystal layers.

11.114. Pigments in Ceramics Cobalt(II) oxide is used as a pigment in ceramics. It has the same type of crystal structure as NaCl. When cobalt(II) oxide is exposed to X-rays with $\lambda = 154$ pm, reflections are observed at $42.38°$, $65.68°$, and $92.60°$. Determine the values of n to which these reflections correspond, and calculate the spacing between the crystal layers.

11.115. Pyrophyllite $[Al_2Si_4O_{10}(OH)_2]$ is a silicate mineral with a layered structure. The distances between the layers is 1855 pm. What is the smallest angle of diffraction of X-rays with $\lambda = 154$ pm from this solid?

11.116. Minnesotaite $[Fe_3Si_4O_{10}(OH)_2]$ is a silicate mineral with a layered structure similar to that of kaolinite. The distance between the layers in minnesotaite is 1940 ± 10 pm. What is the smallest angle of diffraction of X-rays with $\lambda = 154$ pm from this solid?

Additional Problems

11.117. A unit cell consists of a cube that has an ion of element X at each corner, an ion of element Y at the center of the cube, and an ion of element Z at the center of each face. What is the formula of the compound?

11.118. The unit cell of an oxide of uranium consists of cubic closest-packed uranium ions with oxide ions in all the tetrahedral holes. What is the formula of the oxide?

*11.119. The density of pure silicon is 2.33 g/mL. What is the packing efficiency (percentage of the space inside the unit cell occupied by atoms) of the Si atoms in pure Si if the radius of one Si atom is 117 pm? Packing efficiency (%) =

$$\frac{\text{volume occupied by Si atoms}}{\text{volume of unit cell}} \times 100$$

*11.120. **The Composition of Light-Emitting Diodes** The colored lights on many electronic devices are light-emitting diodes (LEDs). One of the compounds used to make them is aluminum phosphide (AlP), which crystallizes in a sphalerite structure.
   a. If AlP were an ionic compound, would the ionic radii of Al^{3+} and P^{3-} be consistent with the size requirements of the ions in a sphalerite structure?
   b. If AlP were a covalent compound, would the atomic radii of Al and P be consistent with the size requirements of atoms in a sphalerite structure?

11.121. Under the appropriate reaction conditions, small cubes of molybdenum, 4.8 nm on a side, can be deposited on carbon surfaces. These "nanocubes" are made of bcc arrays of Mo atoms.
   a. If the edge of each nanocube corresponds to 15 unit-cell lengths, what is the effective radius of a molybdenum atom in these structures?
   b. What is the density of each molybdenum nanocube?
   c. How many Mo atoms are in each nanocube?

11.122. In one form of the fullerene C_{60}, molecules of C_{60} form a cubic closest-packed array of spheres with a unit-cell edge length of 1410 pm.
   a. What is the density of crystalline C_{60}?
   b. What is the radius of the C_{60} molecule?
   c. C_{60} reacts with alkali metals to form M_3C_{60} (where M = Na or K). The structure of M_3C_{60} contains cubic closest-packed spheres of C_{60} with metal ions in holes. If the radius of a K^+ ion is 138 pm, which type of hole is a K^+ ion likely to occupy? What fraction of the holes will be occupied?
   d. Under certain conditions, a different substance, M_6C_{60}, can be formed in which the C_{60} molecules have a bcc structure. Calculate the density of a crystal of M_6C_{60}.

11.123. The center of Earth is composed of a solid iron core within a molten iron outer core. When molten iron cools, it crystallizes in different ways depending on pressure. It crystallizes in a bcc unit cell at low pressure and in an hcp array at high pressures like those at Earth's center.
   a. Calculate the density of bcc iron given that the radius of iron is 126 pm.
   b. Calculate the density of hcp iron given a unit-cell volume of 5.414×10^{-23} cm³.
   *c. Seismic studies suggest that the density of Earth's solid core is only about 90% of that of hcp Fe. Laboratory studies have shown that up to 4% by mass of Si can be substituted for Fe without changing the hcp crystal structure. Calculate the density of such a crystal.

11.124. An alloy with a 1:1 ratio of magnesium and strontium crystallizes in the bcc CsCl structure (see Problem 11.77). The unit-cell edge of MgSr is 390 pm. What is the density of MgSr?

11.125. Gold and silver can be separately alloyed with zinc to form AuZn (unit-cell edge = 319 pm) and AgZn (unit-cell edge = 316 pm). The two alloys have the same crystalline structure. Which alloy is more dense?

*11.126. Removing two electrons from cyclo-S_8 yields the dication cyclo-S_8^{2+}. Will all of the sulfur atoms be in one plane in the S_8^{2+} cation?

11.127. The packing efficiency for a unit cell can be calculated by the following equation:

$$\text{Packing efficiency (\%)} = \frac{\text{volume occupied by spheres}}{\text{volume of unit cell}} \times 100$$

What is the packing efficiency in a simple cubic unit cell?

11.128. Using the equation in Problem 11.127, calculate the packing efficiency in body-centered cubic unit cells.

11.129. Manganese steels are a mixture of iron, manganese, and carbon. Is the manganese likely to occupy holes in the austentite structure or are manganese steels substitutional alloys?

11.130. Aluminum forms alloys with lithium (LiAl), gold ($AuAl_2$), and titanium (Al_3Ti). Based on their crystal structures, each of these alloys is considered to be a substitutional alloy.
   a. Do these alloys fit the general size requirements for substitutional alloys? The atomic radii for Li, Al, Au, and Ti are 152, 143, 144, and 147 pm, respectively.
   b. If LiAl adopts the bcc CsCl structure (see Problem 11.77), what is the density of LiAl?

*11.131. The aluminum alloy Cu_3Al crystallizes in a bcc unit cell. Propose a way that the Cu and Al atoms could be allocated between bcc unit cells that is consistent with the formula of the alloy.

Silicon, Silica, Silicates, Silicone: What's in a Name?

11.132. Write a balanced chemical equation for each of the following reactions: (a) silicon dioxide reacts with carbon; (b) silicon reacts with chlorine; (c) germanium reacts with bromine.

11.133. Write balanced chemical equations for each of the following reactions: (a) tin reacts with chlorine to make a tetrahalide; (b) lead reacts with chlorine to make a dihalide; (c) silicon(IV) chloride is heated above 700°C.

11.134. Calcium silicide ($CaSi_2$) has a structure consisting of graphitelike layers of Si atoms with Ca atoms between the layers. In an X-ray diffraction analysis of a sample of $CaSi_2$, X-rays with a wavelength of 154 nm produced signals at $2\theta = 29.86°$, $45.46°$, and $62.00°$ for $n = 2, 3,$ and 4.
   a. What is the distance between the Si layers?
   b. If the Ca^{2+} ion lies exactly halfway between the layers, what is the Ca–Si distance?
   c. $CaSi_2$ is sometimes used as a "deoxidizer" in converting iron ore to steel. Explain how $CaSi_2$ might fill this role. (*Hint*: What is the most common oxidation state of Si?)

11.135. The ionization energies and electron affinities of the group 14 elements have values close to the corresponding average values for all the elements in their rows in the periodic table. How does this "average" behavior explain the tendency for the group 14 elements to form covalent bonds rather than ionic bonds?

12

Organic Chemistry: Fuels and Materials

OLYMPIC POLYMERS *The skis and body armor used by ski racers are made of synthetic organic polymers.*

A LOOK AHEAD: The Stuff of Daily Life

Think for a moment about the stuff of everyday life. We drive cars powered by gasoline or diesel fuel. We cook chicken on a grill that burns propane. We live in homes heated directly by the burning of natural gas or fuel oil, or in electrically heated homes, for which most electricity is produced by the combustion of a fossil fuel. All of these fuels are organic compounds, and they are used as fuels because their combustion reactions with oxygen to produce CO_2 and H_2O are highly exothermic.

The great majority of medicines used to treat everything from headaches to strokes, to control diabetes, and to combat cancer are organic compounds. Whether they are natural products derived from living plants or animals, or synthetic materials prepared in the laboratory, the products of the pharmaceutical industry are compounds of carbon.

Your favorite soft T-shirt is made of cotton; the soles of your running shoes are made of a synthetic rubber. Cotton is an organic natural fiber produced by a plant. Because of the composition, shape, and orientation of the large molecules that form cotton fibers, the material is soft and absorbent. Synthetic rubber is a polymer, manufactured from compounds of carbon and hydrogen derived from crude oil. Because of the size, shape, and orientation of the polymer molecules in synthetic rubber, the material is springy, nonabsorbent, and resistant to abrasive contact with surfaces.

Fans at a hockey game sit on molded plastic seats and watch goalies wearing helmets of impact-resistant Kevlar. We pour sodas out of plastic bottles and drink them from Styrofoam cups while eating sandwiches wrapped in plastic wrap. Plastics, Kevlar, Styrofoam, and cling wrap are made from compounds derived from crude oil. The plastic seat is strong, light, and capable of bearing considerable weight without breaking. The goalie's helmet must be light-weight and able to withstand the impact of a hockey puck moving at speeds close to 100 miles per hour. Soda bottles must resist punctures and prevent CO_2 from escaping, the Styrofoam cup must be a poor conductor of heat so that beverages in it stay hot or cold, and the sandwich wrap must be flexible and prevent oxygen from reaching the sandwich. All these properties can be designed into the materials at the molecular level, because the physical properties of these materials are directly linked to their structure and composition.

Apart from our bones and teeth, and the water and electrolytes that form the basis of bodily fluids, we humans are all composed of carbon compounds. The great bulk of the food we eat is made up of carbon-containing molecules, too. Compounds of carbon are everywhere, and they are so varied in size, shape, and properties that an entire field within chemistry is devoted to their study. This chapter is a brief introduction to that field, which is known as organic chemistry. A knowledge of organic chemistry is fundamental and essential for a scientific understanding of fuels, foods, pharmaceutical agents, plastics, fibers, living creatures, plants, indeed almost everything we are, almost everything we need to survive, and almost everything we have produced that makes our lives easier in the modern world. This chapter introduces a systematic study of the incredible range of organic compounds, which are classified in accordance with their structure, reactivity, and physical properties. We also discuss the relationships between the structures of organic compounds and their function.

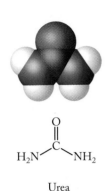

Urea

∞ **CONNECTION** In Chapter 10, we saw that compounds with the same chemical formula but different arrangements of atoms differed in physical properties such as melting point and boiling point.

The term **in vivo** is Latin for "in life." When applied to the origin of a molecule, it means the molecule was produced in a living system. The term **in vitro** literally means "in glass" and refers to materials produced in a test tube, that is, in the laboratory.

Organic chemistry is the study of the compounds of carbon, in which carbon atoms are bound to other carbon atoms, to hydrogen atoms, as well as to atoms of other elements.

A **functional group** is an atom or group of atoms within a molecule that imparts characteristic properties to the organic compounds containing that group.

The symbol **R** in a general formula stands for an organic group that has one available bond. It is used to indicate the variable part of a structure so that the focus is placed on the functional group.

Polymers or *macromolecules* are very large molecules with high molar masses. Polymer literally means "many units."

12.1 Carbon: The Scope of Organic Chemistry

The designation "organic" for carbon-containing compounds was once limited to substances produced only by living organisms, but that definition has been broadened for two fundamental reasons. First, since Friedrich Wöhler (1800–1882) discovered how to prepare urea in the laboratory "without the intervention of a kidney" (1828), scientists have learned to synthesize many materials *in vitro* that were previously thought to be the products only of living systems. Second, chemists have also learned how to synthesize many materials based on carbon that have never been produced *in vivo*. The study of **organic chemistry** today encompasses the chemistry of all carbon compounds, regardless of their origin.

Families Based on Functional Groups

Carbon is an integral part of a vast number of compounds because of its ability to bond to itself as well as to other elements in several different ways. Much of the variety in the chemistry of carbon arises from a carbon atom binding to another carbon atom, which in turn binds to another, to make compounds containing a few to many thousands of carbon atoms. Additional variety is introduced because alternative arrangements are possible for the carbon atoms in these molecules; they do not simply form long, straight chains. Just as we saw in Chapter 9, three-dimensional structure is as important as composition in determining physical and chemical properties. The combination of wide-ranging structural variety for a given atomic composition with almost unending variability of composition creates a scenario in which the number of existing and possible carbon compounds vastly surpasses the possibilities for any other element.

Managing the wealth of information about millions of organic compounds requires some organizing concepts, and scientists have developed two major principles to aid them in this task. For the first, recall that the periodic table groups elements in families on the basis of similar chemical and physical properties. Using the table, we can predict the behavior of elements by using our knowledge of periodic trends. Chemists handle the overwhelming number of organic compounds in a similar fashion by grouping them in families or classes of compounds defined on the basis of **functional groups**—subunits of structure that confer upon molecules particular chemical and physical properties. We have already discussed some classes of compounds and functional groups: alkanes, composed of carbon and hydrogen atoms connected by single bonds; the –OH group in alcohols; the –NH$_2$ group in amines; and the –COOH group in carboxylic acids. In this chapter, we deal with these and the other groups summarized in Table 12.1. When functional groups are discussed, it is conventional to use the symbol **R** to indicate the remaining hydrocarbon part of the molecule so that we can focus on the identity of the functional group. Unless otherwise specified R stands for any other organic group(s). It literally means the rest of the molecule.

Small Molecules, Oligomers, and Polymers

A second organizing principle for organic compounds is based on a physical property as opposed to a chemical one. Scientists divide the universe of carbon compounds into one of two categories based simply on size. Small molecules, with molar masses typically less than 500–1000 g/mol, are one category; **polymers** or *macromolecules* are the second. These size boundaries are somewhat arbi-

TABLE 12.1 Functional Groups of Organic Compounds

Name	Structural Formula of Group	Example and Name	
Alkane	R—H	$CH_3CH_2CH_3$	Propane
Alkene	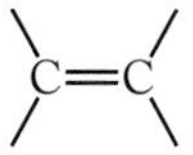 C=C	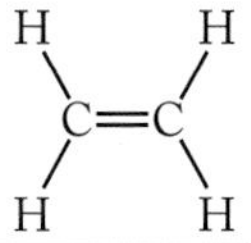 C=C	Ethylene (ethene)
Alkyne	—C≡C—	H—C≡C—H	Acetylene (ethyne)
Aromatic	e.g., 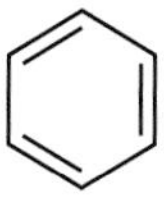 ⬡	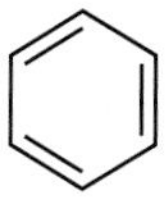 ⬡	Benzene
Alcohol	R—OH	CH_3CH_2OH	Ethanol
Ether	R—O—R	$CH_3CH_2OCH_2CH_3$	Diethyl ether
Aldehyde	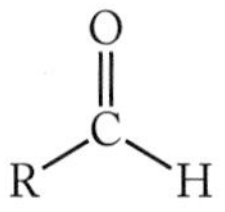 R—C(=O)—H	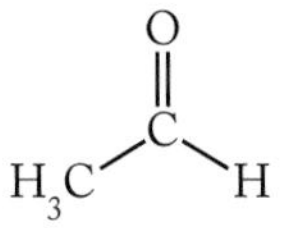 H_3C—C(=O)—H	Acetaldehyde
Ketone	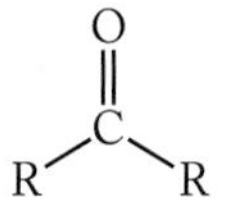 R—C(=O)—R	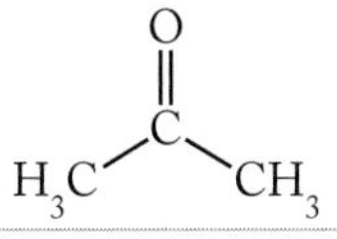 H_3C—C(=O)—CH_3	Acetone
Carboxylic acid	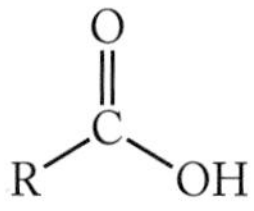 R—C(=O)—OH	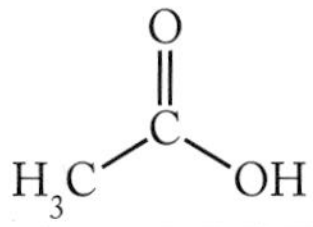 H_3C—C(=O)—OH	Acetic acid
Ester	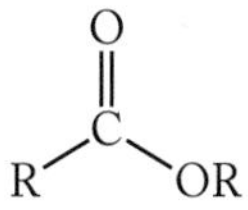 R—C(=O)—OR	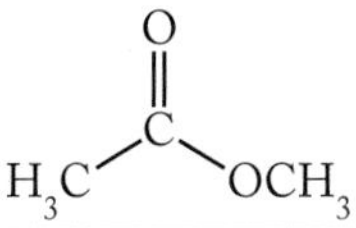 H_3C—C(=O)—OCH_3	Methyl acetate
Amide	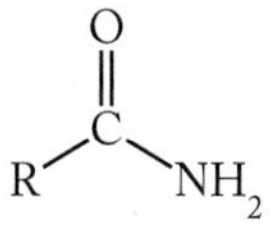 R—C(=O)—NH_2	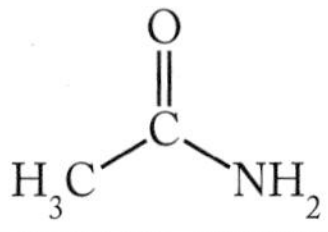 H_3C—C(=O)—NH_2	Acetamide
Amine	R—NH_2 R—NHR R—NR_2	H_3C—NH_2	Methylamine

trary, and in between small molecules and polymers is the realm of **oligomers**, which are materials that are not quite polymers but do contain multiple structural units. We have already seen several examples of polymers in previous chapters: poly(vinylchloride) used to make drain pipes and other building materials; Teflon used in nonstick cookware and lubricants; cellulose from plants used to make paper. Polymers are large molecules with high molar masses ranging from several thousand to over 1,000,000 g/mol. Polymers are composed of small structural units called **monomers**.

Oligomers are molecules that contain a few mers; "oligos" is Greek for "few." The boundary between small molecules and polymers is not distinct, and oligomers occupy the middle ground.

Monomers are small molecules that bond together to form polymers. The root word *meros* is Greek for "part" or "unit," so a monomer is one unit.

Synthetic polymers are macromolecules made in the laboratory and often produced industrially for commercial use.

The properties of polymers depend on the functional groups in the monomers from which they are made, but the relationship of their size and shape to their behavior distinguishes them as a class from small molecules. Polymers may be formed from more than one type of monomer unit, they may have more than one functional group, and they may have shapes other than long, straight chains. However, the single feature that distinguishes them as a class is their large size.

Small Molecules versus Polymers: Physical Properties

Polymers are extraordinarily useful materials. Much of what we say about their behavior in this chapter reflects the influence of composition and structure on their properties. With that concept in mind, let's first discuss the difference between small organic molecules and organic polymers. For now, we restrict our discussion to **synthetic polymers** made in the laboratory.

⊗⊗ **CONNECTION** We have already introduced many organic molecules in Chapters 4 through 10: acetic acid; methane, ethane, and other hydrocarbons as fuels; acetylene used in welding; ethanol and other alcohols.

We have already seen examples of many important compounds made of small organic molecules. All of them have well-defined properties: for example, they have constant composition, and their phase transitions take place at well-defined temperatures (their melting points and boiling points). These features are typical characteristics of pure samples of substances made of small molecules.

On the other hand, many commercially produced polymers are not materials with constant composition and well-defined properties. Most synthetic polymers are really mixtures of very similar, large molecules that differ only in the number of monomers in their chains. Their physical properties depend heavily on the range and distribution of molecular weights of the molecules in the bulk sample. Figure 12.1 shows two examples of common items made from synthetic polymers.

In Chapter 10 we discussed the influence of mass and intermolecular attractive forces on melting and boiling points. Polymers are made of molecules so large that a molecule cannot take on sufficient kinetic energy to enable it to enter the gas phase. Polymers typically do not have well-defined melting points and boiling points. As temperature increases, some polymers gradually soften and become more malleable. Some may eventually melt and become very viscous liquids, while others remain solids up to temperatures at which their covalent bonds break, causing the material to decompose. For example, plastic grocery bags are made of polyethylene, the monomer unit for which is ethylene ($CH_2 = CH_2$). The melting point of ethylene is $-169°C$; its boiling point is $-104°C$. Polyethylene is a tough and flexible solid at room temperature. It softens gradually over a range of temperatures from $85°C$ to $110°C$, and at higher temperatures tends to decompose into molecules of low molar-mass organic gases. For the purposes of our initial discussion, it is sufficient for you to understand that matter composed of polymers has a physical behavior very different from matter composed of small molecules, and that this difference is mostly because the molecules are so much larger.

No element rivals carbon for the sheer number and variety of compounds it forms. From small molecules to macromolecules, carbon is the element beyond all others that builds the fabric of our lives. In this chapter, we use two means of classification to describe the huge array of carbon compounds. We

FIGURE 12.1 Items made of synthetic polymers are ubiquitous in modern society.

start with small molecules and define the physical and chemical properties of most of the important families of compounds based on functional groups. We then describe the polymers that arise when monomers containing these functional groups combine to make materials ranging from plastic wrap and coffee cups to artifical grass and drain pipes.

12.2 Hydrocarbons

The first three functional groups discussed here define the families of compounds called alkanes, alkenes, and alkynes (Table 12.1). We saw examples of these families in Chapter 5 among the compounds in hydrocarbon fuels. Alkanes consist of carbon in its most reduced state. Methane, propane, octane, and other alkanes are the principal ingredients in several common fuels. When they are converted from alkanes to carbon dioxide and water during complete combustion, they give off large amounts of energy in the process.

All alkanes, alkenes, and alkynes have only carbon and hydrogen in their structures. The three classes are distinguished by the types of carbon–carbon bonds in their molecules: **alkanes** have only carbon–carbon single bonds; **alkenes** have one or more carbon–carbon double bonds; **alkynes** have one or more carbon–carbon triple bonds. The double and triple bonds in alkenes and alkynes (between sp^2 and sp hybridized carbons, respectively) are the functional groups that differentiate them from alkanes, which have only sp^3 hybridized carbon atoms. Alkanes are also known as **saturated hydrocarbons** because they have the maximum ratio of hydrogen atoms to carbon atoms in their structure. The generic molecular formula of alkanes is C_nH_{2n+2}, where n is the number of carbon atoms per molecule. The hydrogen-to-carbon ratio in alkanes $[(2n + 2)/n]$ is as large as it can be, given the limit of four bonds per carbon atom. The influence of the double and triple bonds on the hydrogen-to-carbon ratio can be seen by looking at similar molecules in the three families. The two-carbon alkane is ethane, CH_3—CH_3, in which the H-to-C ratio = 3:1; the simplest alkene is ethene (more commonly known as ethylene), CH_2=CH_2, in which the H-to-C ratio = 2:1; the simplest alkyne is ethyne (more commonly known as acetylene), CH≡CH, in which the H-to-C ratio = 1:1.

Alkenes and alkynes are **unsaturated hydrocarbons**, because their double or triple bonds can react with hydrogen to incorporate more hydrogen into their molecular structures. The reaction in which hydrogen is added to a molecule is called **hydrogenation**. No general formula can be written to describe the hydrogen-to-carbon ratios in alkenes and alkynes, because that ratio will vary depending on how many unsaturations are in each molecule. However, all unsaturated hydrocarbons react with hydrogen and can be converted into saturated hydrocarbons (Figure 12.2). Hydrogenation reactions result in the addition of one or more molecules of hydrogen to each molecule of unsaturated hydrocarbon. A molecule with one double bond is described as having one *unsaturation* and combines with one molecule of H_2; a molecule with one triple bond has two *unsaturations* and combines with two molecules of H_2. If a molecule had one double bond and one triple bond, it would have three unsaturations. This feature is referred to as the *degree of unsaturation*, and determining the degree of unsaturation of an unknown compound using a hydrogenation reaction can help identify its structure.

Alkanes are hydrocarbons in which all the bonds are single bonds.

Alkenes are hydrocarbons containing one or more carbon–carbon double bonds.

Alkynes are hydrocarbons containing one or more carbon–carbon triple bonds.

A **saturated hydrocarbon** is one in which each carbon is bonded to four other atoms.

Alkenes and alkynes are **unsaturated hydrocarbons**. They contain one or more carbon–carbon double or carbon–carbon triple bonds and therefore contain less than the maximum amount of hydrogen possible per carbon atom.

The reaction of an unsaturated hydrocarbon with hydrogen is called **hydrogenation**.

CONNECTION In Chapter 5, we defined hydrocarbons as compounds composed of only carbon and hydrogen atoms.

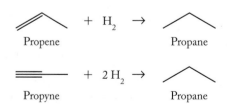

FIGURE 12.2 Carbon-skeleton structures of propene, propyne, and propane each contain 3 carbon atoms. One mole of the alkene reacts completely with 1 mole of hydrogen to produce 1 mole of propane; propyne reacts completely with 2 moles of hydrogen to produce 1 mole of propane.

SAMPLE EXERCISE 12.1	**Distinguishing between Alkanes, Alkenes, and Alkynes**

We are given 1 mole quantities of three unidentified hydrocarbons, each having 4 carbon atoms in their structures. We are told one is an alkane, one an alkene (with one double bond), one an alkyne (with one triple bond). Design an experiment based on their reactivity with hydrogen to determine which is the alkane, which is the alkene, and which is the alkyne.

COLLECT AND ORGANIZE We are asked to classify three hydrocarbons based on their reactivity with hydrogen. We have 1 mole of each compound, and we are directed to use their reactivities with hydrogen gas to classify them.

ANALYZE The alkane is a saturated hydrocarbon; it already contains the maximum number of hydrogen atoms possible in its structure. It does not react with hydrogen. The alkene and alkyne are unsaturated, so they can both react with hydrogen. One mole of the alkene, with one double bond, will react with 1 mole of hydrogen. One mole of the alkyne, with one triple bond, will react with 2 moles of hydrogen.

SOLVE We can classify the compounds by measuring the amount of hydrogen gas they consume in a reaction. If we allow 1 mole of each substance to react with 2 moles of hydrogen and measure how much hydrogen is consumed in the process, we can identify the compounds. We pick 2 moles because that is the maximum amount that any of the unknowns require for complete reaction.

$$1 \text{ mol alkane} + 2 \text{ mol } H_2 \xrightarrow{\text{no reaction}} 2 \text{ mol } H_2 \text{ left}$$

$$1 \text{ mol alkene} + 2 \text{ mol } H_2 \rightarrow 1 \text{ mol alkane} + 1 \text{ mol } H_2 \text{ left}$$

$$1 \text{ mol alkyne} + 2 \text{ mol } H_2 \rightarrow 1 \text{ mol alkane} + 0 \text{ mol } H_2 \text{ left}$$

We can classify the compounds because the alkane is the unknown that does not react at all with hydrogen, while 1 mole of the alkene consumes 1 mole of hydrogen, and 1 mole of the alkyne consumes 2 moles of hydrogen.

THINK ABOUT IT Measuring the amount of hydrogen that reacts with a hydrocarbon is a way to distinguish saturated from unsaturated hydrocarbons. Unsaturated hydrocarbons may be distinguished if their degree of saturation differs.

Practice Exercise The labels have fallen off two containers in the lab. One label has the structure of compound A printed on it; the other, the structure of compound B. Can you distinguish between the two compounds in the unlabeled bottles using hydrogenation reactions? What are the structures of the products of the hydrogenation reactions in both cases?

$$CH_3-CH_2-CH{=}CH-CH{=}CH-CH_3$$

Compound A

$$CH_3-CH{=}CH-CH{=}CH-CH{=}CH_2$$

Compound B

(Answers to Practice Exercises are in the back of the book.)

Could we use information from hydrogenation reactions to distinguish a hydrocarbon with one triple bond from a hydrocarbon with two double bonds?

(Answers to Concept Tests are in the back of the book.)

Alkanes: Physical Properties and Structure

Alkanes are also known as paraffins, a name derived from the Latin meaning "little affinity." This is a perfect description of the alkanes, which tend to be rather unreactive compared to the other hydrocarbon families. Despite their lack of reactivity, alkanes are compounds of great importance because of their use as fuels, oils, and lubricants. "Unreactive" may not immediately seem like the correct term to apply to compounds that are fuels, but alkanes do not react readily, even with oxygen. Most fuels need some source of energy, such as a spark from a spark plug in an engine, to initiate combustion.

In Chapter 10 we discussed how to predict relative boiling points of compounds based on molar mass and the strengths of intermolecular forces. In doing so, we looked at a series of alkanes as a function of their molecular weight and their shape. Table 12.2 contains a data set for a **homologous series**. A homologous series of alkanes consists of compounds in which one member differs from the next by the addition of a $-CH_2-$ unit, called a **methylene group**. The terminal $-CH_3$ group is called a **methyl group**. You can think of a methyl group as simply

TABLE 12.2 Melting Points and Boiling Points for Selected Alkanes

Condensed Structural Formula[a]	Use	Melting Point (°C)	Normal Boiling Point (°C)
$CH_3CH_2CH_3$	Gaseous fuels	−190	−42
$CH_3(CH_2)_2CH_3$		−138	−0.5
$CH_3(CH_2)_3CH_3$		−130	36
$CH_3(CH_2)_4CH_3$	Gasoline	−95	69
$CH_3(CH_2)_5CH_3$		−91	98
$CH_3(CH_2)_6CH_3$		−57	126
$CH_3(CH_2)_7CH_3$		−54	151
$CH_3(CH_2)_{10}CH_3$ through $CH_3(CH_2)_{16}CH_3$	Diesel fuel and heating oil	−10 / 28	216 / 316
$CH_3(CH_2)_{18}CH_3$ through $CH_3(CH_2)_{32}CH_3$	Paraffin candle wax	37 / 72–75	343 / na
$CH_3(CH_2)_{34}CH_3$ and higher homologs	Asphalt	72–76	na

[a]See Figure 12.3.
na = not available; compound decomposes before boiling at 1 atm pressure.

A **homologous series** is a set of related organic compounds that differ from one another by the number of subgroups, such as $-CH_2-$, in their molecular structures.

A **methylene group** $(-CH_2-)$ is a structural unit that can make two bonds.

A **methyl group** $(-CH_3)$ is a structural unit that can make only one bond.

A **straight-chain alkane** is a hydrocarbon in which the carbon atoms are bonded together in one continuous line. Linear hydrocarbon chains have a methyl group at each end with methylene groups connecting them.

⊙⊙ **CONNECTION** In Chapter 10 we identified dispersion forces as the main attractive force between molecules such as the hydrocarbons, and we related these to the differences in boiling points as a function of linear or branched structures.

methane (CH_4) with one hydrogen atom missing. All of the alkanes in Table 12.2 are **straight-chain alkanes** with two terminal methyl groups and a continuous zig-zag chain with an ever-increasing number of methylene groups. Straight-chain alkanes are identified by "*n-*" (for *normal* alkanes) preceding the chemical name. As we will see in the next section on drawing organic molecules, linear chains of sp^3 hybridized carbon atoms are kinked because of the carbon–carbon bond angles. The term *straight-chain alkane* simply means a continuous, linear sequence of carbon atoms.

The data sets given in Table 12.2 show similar trends in melting and boiling points: as alkanes increase in molar mass, their melting and boiling points increase. This trend in physical properties is typical of all homologous series.

Drawing Organic Molecules

It soon gets tedious to write Lewis structures for alkanes or any organic molecule. Figure 12.3 shows examples of several systems chemists use to convey structures in organic chemistry. For example, a condensed structural formula does not show individual C—H or single C—C bonds the way Lewis structures do, as illustrated below for *n*-octane:

Lewis structure:

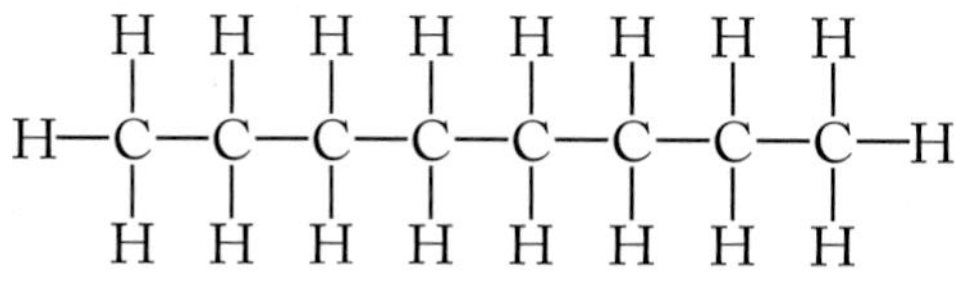

Condensed structural formula: $CH_3CH_2CH_2CH_2CH_2CH_2CH_2CH_3$

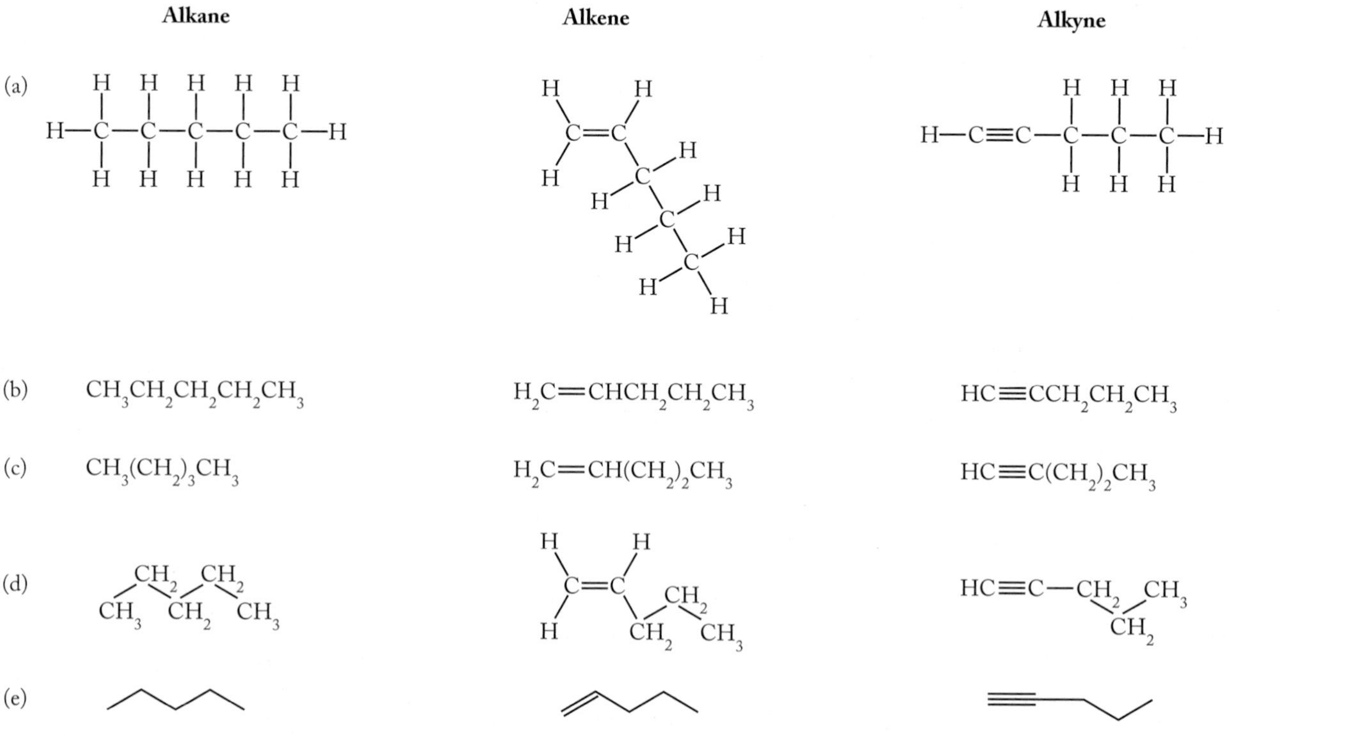

FIGURE 12.3 These structures of alkanes, alkenes, and alkynes illustrate common ways to represent molecular structures: (a) Lewis structures; (b), (c), and (d) are forms of condensed structural formulas; and (e) carbon-skeleton structures.

Condensed structures can be shortened even further by indicating the number of times a particular subgroup is repeated in a structure, such as the $-CH_2-$ groups in octane:.

$$CH_3(CH_2)_6CH_3$$

The numerical subscript after the parenthetical methylene group means that 6 of these groups separate the 2 methyl groups. Eight carbon atoms form the backbone of this molecule.

Another common short-hand system does not even use the symbols of the elements. Short, connected lines symbolize the carbon–carbon bonds of a molecule in a carbon-skeleton structure, as shown here for butane:

Condensed strucutural formula for butane:

$$CH_3CH_2CH_2CH_3 \quad \text{or} \quad CH_3(CH_2)_2CH_3$$

Carbon-skeleton structure for butane:

The group at the end of a single line in a skeletal structure is a $-CH_3$ group and the group at the intersection of two single lines is a $-CH_2-$ group. Using these conventions, octane is:

$$CH_3CH_2CH_2CH_2CH_2CH_2CH_2CH_3 \quad \text{or} \quad CH_3(CH_2)_6CH_3 \quad \text{or}$$

Carbon-skeleton structures are favored for drawing the simplest representation of a molecule and especially to indicate the bond angles between the carbon atoms. VSEPR theory and valence bond theory both predict a tetrahedral bond orientation around each carbon atom. The complete three-dimensional structure of butane showing all carbon and hydrogen atoms is

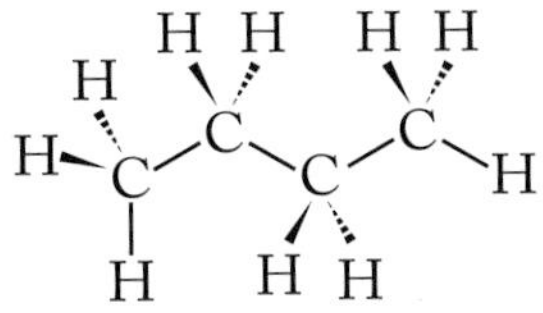

The three-dimensional structure is much more complicated to draw but the same information about the carbon framework is conveyed by the carbon-skeleton structure. Hydrogen atoms are not shown in the skeletal structure because all carbon atoms are known to make four bonds, and any bonds not shown are understood to be C—H bonds. In the carbon-skeleton structures of the alkanes in Figure 12.3, all the carbon atoms are drawn in the plane of the paper, but some hydrogen atoms are understood to be either in front of (solid wedge) or in back of (dashed wedge) that plane, as shown above in the more complete three-dimensional structure of butane.

SAMPLE EXERCISE 12.2   **Drawing Structures**

Write a condensed structural formula and draw the carbon-skeleton structure for the 3-carbon straight-chain hydrocarbon propane and the 12-carbon straight-chain hydrocarbon *n*-dodecane.

COLLECT AND ORGANIZE Condensed structural formulas do not show C—H or single C—C bonds, and the carbon-skeleton structures use short lines to represent the carbon–carbon bonds in molecules.

ANALYZE The compounds are straight-chain hydrocarbons, so all the carbon atoms will be in one continuous line.

SOLVE Propane only has 3 carbon atoms, and there is only one way to connect them. Two short straight lines suffice for the skeleton structure; both ends are methyl groups and there is one methylene group in the middle.

Condensed
structural formula: $CH_3CH_2CH_3$ Carbon-skeleton structure:

n-Dodecane is a continuous chain of 12 carbon atoms. There must be 2 methyl groups at the ends with 10 methylene groups between them.

Condensed
structural formula:

$$CH_3CH_2CH_2CH_2CH_2CH_2CH_2CH_2CH_2CH_2CH_2CH_3$$
$$\text{or } CH_3(CH_2)_{10}CH_3$$

Carbon-skeleton
structure:

THINK ABOUT IT When drawing structures, we pick the style that best illustrates the point we want to make.

Practice Exercise Draw the carbon-skeleton structure of *n*-hexane, $CH_3(CH_2)_4CH_3$, and write two condensed structural formulas of *n*-heptane.

n-Heptane

Structural Isomers

The family of alkanes would be huge if it consisted of only straight-chain molecules. However, another structural possibility arises that makes the family even larger. We first saw this possibility in Chapter 10 when we compared the boiling points of compounds having the same molecular formula but different shapes. Butane, for example, has the formula C_4H_{10}, but the straight-chain arrangement of atoms we discussed (Table 12.3, column 1) is not the only one possible. An alternative structure (Table 12.3, column 2) is not a straight chain but rather is branched. A branch is a side chain attached to the main carbon skeleton. Note that the formula of the side chain (a methyl group in this structure) is written in parentheses in the condensed structural formula.

The formula of both structures is C_4H_{10}, but they represent different compounds with different properties. In recognition of these differences, the two compounds have different formal names (*n*-butane and 2-methylpropane), but

TABLE 12.3 Comparing Structural Isomers

	n-Butane	2-Methylpropane
Condensed structural formula:	$CH_3CH_2CH_2CH_3$	$CH_3CH(CH_3)CH_3$ or $CH_3\underset{\underset{CH_3}{\mid}}{CH}CH_3$
Carbon-skeleton structure:		
Melting point (°C)	−138	−160
Normal boiling point (°C)	0	−12
Density of gas (g/L)	0.5788	0.5934

the issue for you is not to be able to name them but to be able to recognize them as **structural isomers**: compounds with the same formula but with atoms that are connected in different ways. This difference makes them chemically distinct species. 2-Methylpropane is an example of a **branched-chain hydrocarbon**, which means that its molecular structure contains at least one side chain.

To determine whether or not two condensed structural formulas or carbon-skeleton structures represent hydrocarbons that are identical or structural isomers of each other, follow these steps:

- Translate the structural information into molecular formulas and compare them. If they are different, then the compounds are different.

- If their molecular formulas are the same, then compare their structures. You may have to reverse or rotate the structures to determine whether they really are structural isomers.

For example, structures 1 and 2 below may seem to be different:

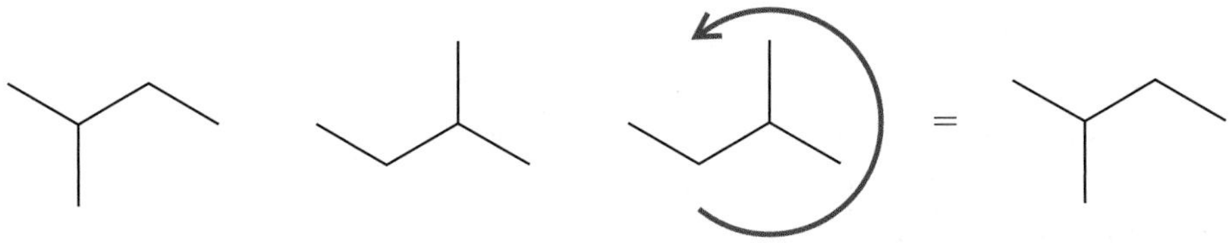

(1)   (2)   180° rotation   After rotation

However, if we rotate structure (2) 180°, we see that it is the same as structure (1). The two represent the same compound: 2-methylbutane. To make it easier to compare hydrocarbon structures, the longest chain is drawn from left to right with any branches drawn above or below it. To determine whether two structures are identical, draw the structures with the longest chain horizontal, and then check to see whether the same side chains are attached at the same positions along the longest chain. For example, structures (1) and (2) both have a four-carbon chain with a methyl group on the second carbon when we number the carbon atoms starting with the end closer to the side chain.

SAMPLE EXERCISE 12.3   **Recognizing Structural Isomers**

Do the two structures in each set describe the same compound, two isomers, or compounds with different molecular formulas?

(a) $(CH_3)_2CHCH_2CH(CH_3)_2$

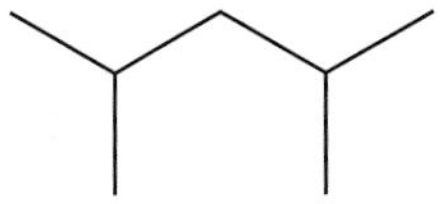

(b)

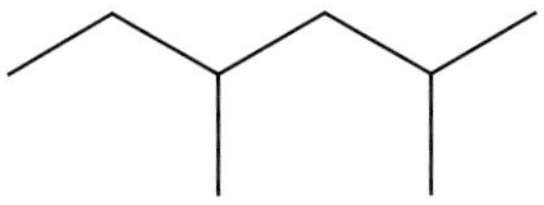

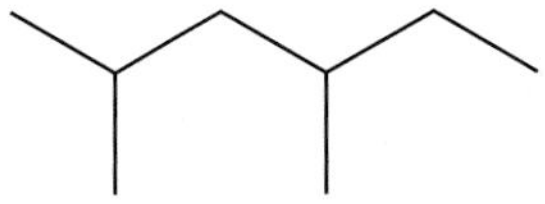

(c)

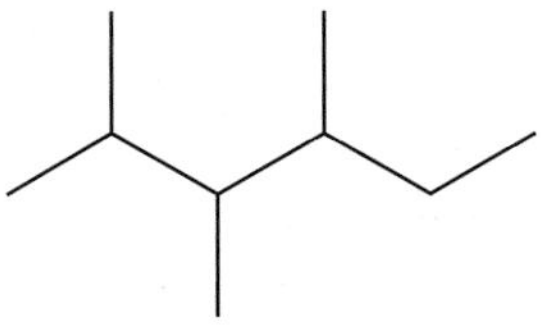

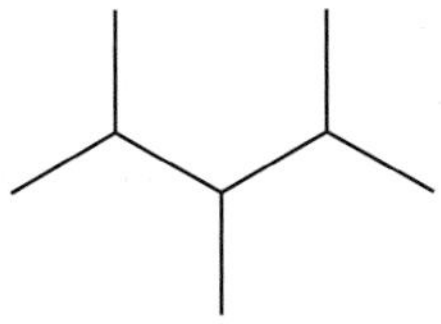

> **Structural isomers** have the same formula but different structures. They are different compounds and have different chemical and physical properties.
>
> A **branched-chain hydrocarbon** is an organic molecule in which the chain of carbon atoms is not linear.

COLLECT AND ORGANIZE We are asked to determine if the structures represent compounds that are identical (same formula, same connectivity of the atoms), are isomers (same formula but different connectivity), or are compounds with different formulas.

ANALYZE In each pair, we check first to see if the formulas are the same. That means we count the number of carbon atoms and hydrogen atoms. If they are different, they represent different compounds. If they are the same, we may have one compound drawn two ways, or two isomers with different structures.

If the carbon skeletons of the molecules are the same, they represent the same hydrocarbon.

SOLVE

a. The formulas are the same: C_7H_{16}. Converting the condensed structural formula of the compound on the left to a carbon-skeleton structure:

$$(CH_3)_2CHCH_2CH(CH_3)_2 \rightarrow$$

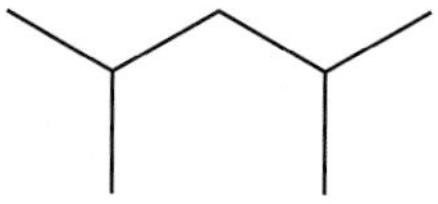

we get a structure that is identical to the one on the right. Therefore the formula and structure represent the same molecule.

b. Both have the same formula: C_8H_{18}. Both have six carbon atoms in the longest continuous chains. Both structures have a methyl group attached to the second carbon from one end of the chain, and another methyl group is attached to the third carbon atom from the other end. Therefore, they must describe the same compound.

c. The first structure contains nine carbon atoms; the second, eight. Therefore, they have different formulas and represent different compounds.

THINK ABOUT IT When we draw structures of organic compounds, such as carbon-skeleton structures, we usually draw them with the longest carbon sequence horizontal to make it easy to compare structures.

Practice Exercise Do the two structures in each set describe the same compound, two isomers, or compounds with completely different formulas?

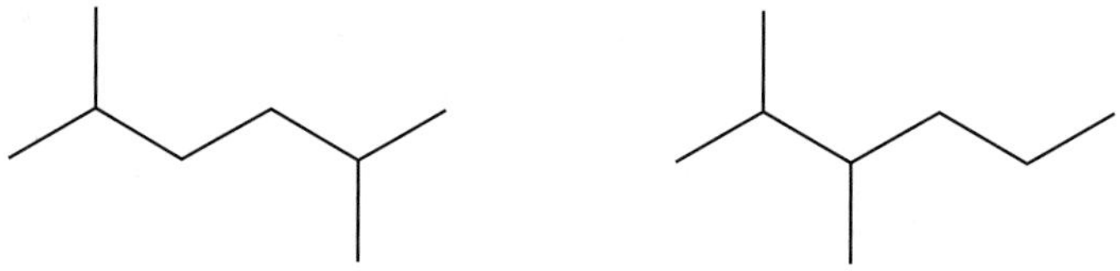

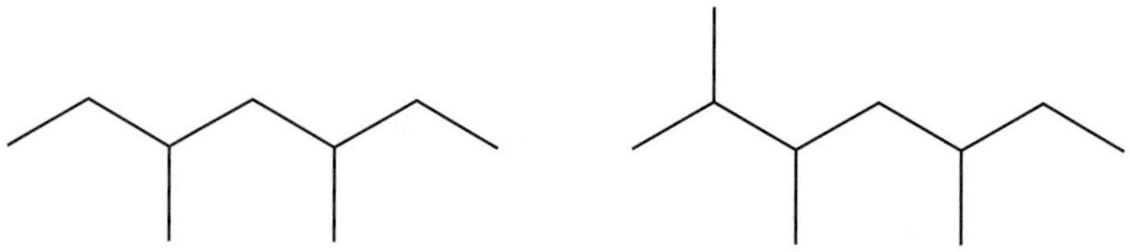

When the heats of combustion of the isomers in Table 12.3 are estimated using bond energies, the two isomers have the same value. The experimentally determined values of heat of combustion differ. Why?

Naming Alkanes

Now that we know how to draw the structures of alkanes, let's look at a few simple principles for naming them systematically. Since the system follows the same pattern for all families of organic compounds, we begin with rules for the alkanes and will add a few additional rules when we discuss alkenes and alkynes. These rules are called IUPAC rules after the International Union of Pure and Applied Chemistry, the organization that defined them. Appendix 7 contains a more extensive treatment of nomenclature for organic compounds that you may consult as required.

The first four members of the alkane family are methane (1 carbon atom: a C_1 alkane), ethane (2 carbon atoms: a C_2 alkane), propane (a C_3 alkane), and butane (a C_4 alkane). The names of alkanes with more than 4 carbon atoms are derived from the Latin or Greek prefix for the number of carbon atoms per molecule followed by -*ane*. Table 12.4 lists the prefixes for C_1 through C_{10} alkanes. Molecules with more than 10 carbon atoms in their parent structure are named with the appropriate Latin prefix for the number: C_{11} is undecane, C_{12} dodecane, and so forth.

Because every alkane larger than propane has structural isomers, we need an unambiguous way to name specific structures. We have already introduced the use of the prefix *n*- before a name to indicate a straight-chain alkane. To name branched-chain alkanes, follow these steps to build up a systematic name:

1. Select the longest continuous chain and use the prefixes in Table 12.4 to name this as the parent structure. For the structure that follows, the parent name is pentane because the longest chain is 5 carbon atoms long.

$$CH_3CH(CH_3)CH_2CH_2CH_3$$

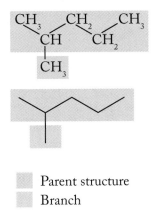

- ▨ Parent structure
- ▨ Branch

TABLE 12.4	Prefixes for Naming *n*-Alkanes	
Prefix	**Formula**	**Name**
Meth-	CH_4	Methane
Eth-	CH_3CH_3	Ethane
Prop-	$CH_3CH_2CH_3$	Propane
But-	$CH_3(CH_2)_2CH_3$	Butane
Pent-	$CH_3(CH_2)_3CH_3$	Pentane
Hex-	$CH_3(CH_2)_4CH_3$	Hexane
Hept-	$CH_3(CH_2)_5CH_3$	Heptane
Oct-	$CH_3(CH_2)_6CH_3$	Octane
Non-	$CH_3(CH_2)_7CH_3$	Nonane
Dec-	$CH_3(CH_2)_8CH_3$	Decane

2. Identify the branches and name them with the prefix from Table 12.4 that defines the number of carbons in the branch; append the suffix -*yl* to the prefix. A $-CH_3$ group is a methyl group, a $-CH_2CH_3$ group is an ethyl group, and so forth. The structure in step 1 has a $-CH_3$ (methyl) group as a branch; the name of the branch comes before the name of the parent structure, and the two are written together as one word: methylpentane.

3. To indicate the point where the branch is attached, we number the carbons in the parent chain so that the branch (or branches, if more than one is present) has the lowest possible number. In the case of the molecule in step 1, we start numbering from the left so that the methyl group is on carbon atom 2 in the parent chain.

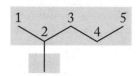

We show the position of the branch by a 2 followed by a hyphen (2-), in front of the name generated in step 2. The unambiguous name for this compound is 2-methylpentane.

4. If the same group is attached more than once to the parent structure, we use the prefix di-, tri-, tetra- and so forth to indicate the number of groups present. The position of each group is indicated by the appropriate number before the group name. These numbers are separated by commas. The structure that follows is the compound 2,4-dimethylpentane.

$$CH_3CH(CH_3)CH_2CH(CH_3)CH_3$$
$$1 \quad 2 \qquad 3 \quad 4 \qquad 5$$

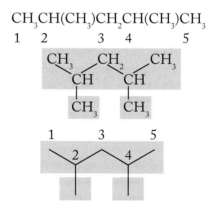

5. If different groups are attached to a parent chain, they are named in alphabetical order. The following compound is 3-ethyl-4-methylheptane

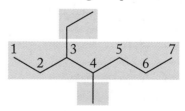

or drawn as condensed structural formulas:

$$CH_3CH_2CH(CH_2CH_3)CH(CH_3)CH_2CH_2CH_3$$
$$1 \quad 2 \quad 3 \qquad\qquad 4 \qquad 5 \quad 6 \quad 7$$

Note that we begin numbering the parent chain in this structure at the carbon on the far left, in accord with the instruction in step 2 so that the branch locations have the lowest possible numbers. In naming the compound, the side groups are written in alphabetical order. Prefixes are disregarded in determining alphabetical order: **e**thyl comes before **m**ethyl, and **e**thyl also comes before **d**imethyl.

> **Cycloalkanes** are ring-containing alkanes with the general formula C_nH_{2n}.

CONCEPT TEST

A useful way of determining whether structures are the same is to name the compound according to the IUPAC rules. Why is this the case?

Cycloalkanes

Alkanes can also form ring structures (Figure 12.4). Compounds with this structure are called **cycloalkanes**. They have the general formula C_nH_{2n}, which is different from the general formula for straight-chain alkanes because cycloalkanes have one more carbon–carbon bond and 2 fewer hydrogen atoms per molecule than the *n*-alkanes with the same number of carbon atoms. Although these compounds with rings are alkanes, and all the carbon atoms in cycloalkane rings are *sp*^{3} hybridized, the compounds are actually unsaturated. In principle, 1 mole of a cycloalkane can react with 1 mole of hydrogen to make 1 mole of a straight-chain compound. In practice, only the three-carbon and four-carbon rings actually react with hydrogen, but all the cycloalkanes are considered unsaturated.

The most abundant cycloalkane in the fraction of crude oil used as gasoline is cyclohexane, C_6H_{12}. Figure 12.4 shows the condensed structural formula of cyclohexane drawn in two dimensions, in which the C—C—C bond angles appear to be 120°. Because the carbon atoms have *sp*^{3} hybrid orbitals, we expect the angles to be 109.5°, and indeed they are in this molecule. A more accurate representation of the structure of cyclohexane is shown in the adjacent structures, in which the ring is puckered instead of flat and the bond angles are 109.5°. There are two possible pucker patterns: one is called the *chair* form, and the other is called the *boat* form. In the boat form, 2 hydrogen atoms across the ring from each other are pointed toward each other. Repulsion between these 2 atoms causes this form to be less stable than the chair form, and so the chair form is the preferred, lower-energy form. Figure 12.4 also shows the carbon-skeleton forms of the three structures.

▶❚❚ **CHEMTOUR** Structure of Cyclohexane

▶❚❚ **CHEMTOUR** Cyclohexane in 3-D

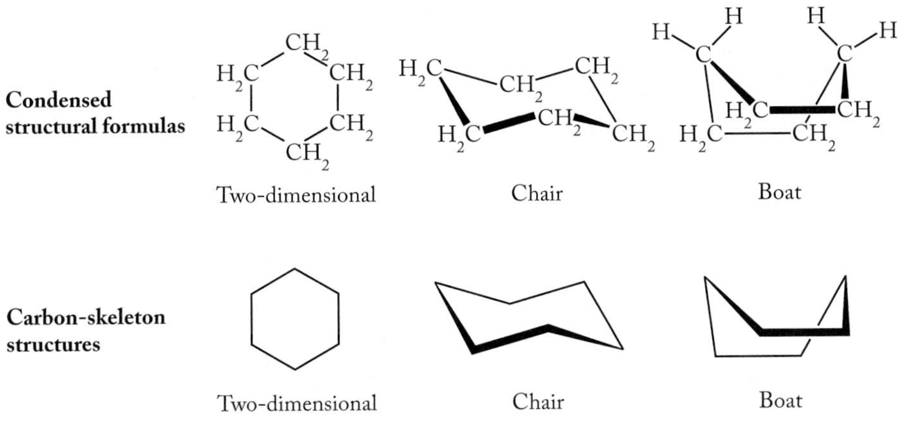

FIGURE 12.4 The six-membered ring of cyclohexane is drawn either flat or in styles that show its three-dimensional puckered rings.

Ring structures are possible for other alkanes, but smaller rings are less stable than the six-membered one. One of the important features that determines the relative stability of rings is the size of the C—C—C bond angle. For example, cyclopropane (a C_3 ring) exists, but it is a very reactive species because the interior

Cyclopropane   Cyclobutane   Cyclopentane

angle in an equilateral triangle is only 60°, which is very far from the ideal bond angle of 109.5° for an sp^3 hybridized carbon atom. No puckering is possible in a three-membered ring to relieve the strain in this system, and cyclopropane tends to react in a fashion that opens up the ring and relieves the strain. Rings of four and five carbons (cyclobutane and cyclopentane) are progressively less strained. Rings of six sp^3 hybridized carbons and beyond are essentially the same as straight-chain alkanes in terms of bond angle and have no ring strain. Six-membered rings are the most favored, because other thermodynamic factors destabilize rings with seven or more carbons.

12.3 Sources of Alkanes

As noted in Chapter 5, carbon compounds have been accumulating in Earth's crust since the onset of biological activity. When decaying aquatic plant and animal matter is buried in rapidly accumulating sediments, its carbon and hydrogen content is shielded from oxidation to CO_2 and H_2O. The matter is slowly converted into natural gas and a complex mixture of water-insoluble organic compounds called **kerogen**. As kerogen is buried by more sediment, increasing pressure and temperature promotes additional chemical transformations. When these transformations take place at depths between about 1 and 4 km below Earth's surface and at temperatures between 75°C and 150°C, the products include **crude oil**, a mixture of hundreds of compounds composed mostly of carbon and hydrogen. Crude oil contains hydrocarbons with 5 or more carbon atoms in their molecular structures. The hydrocarbons with 1 to 4 carbon atoms are usually found in deposits of natural gas, although they are also dissolved in deposits of crude oil.

Kerogen is a mixture of a large number of liquid organic compounds formed during the conversion of plant matter into crude oil in the absence of oxygen.

Crude oil is a combustible liquid mixture of hundreds of hydrocarbons and other organic molecules formed under the Earth's surface by the processing of plant matter under conditions of high temperature and pressure.

The **Clausius–Clapeyron equation** relates the vapor pressure of a substance at different temperatures to its heat of vaporization.

CONCEPT TEST

The nonpolar constituents of natural gas may dissolve in crude oil, which consists of mostly nonpolar liquid components. As crude oil is pumped to the surface from thousands of meters deep in Earth's crust, how is the solubility of natural gas affected?

Volatility and the Clausius–Clapeyron Equation

After crude oil has been extracted from the ground, it is separated at refineries into fractions, or *cuts,* on the basis of the volatilities of the components in the solution. As we learned in Chapter 10, the vapor pressure of a liquid increases as tempera-

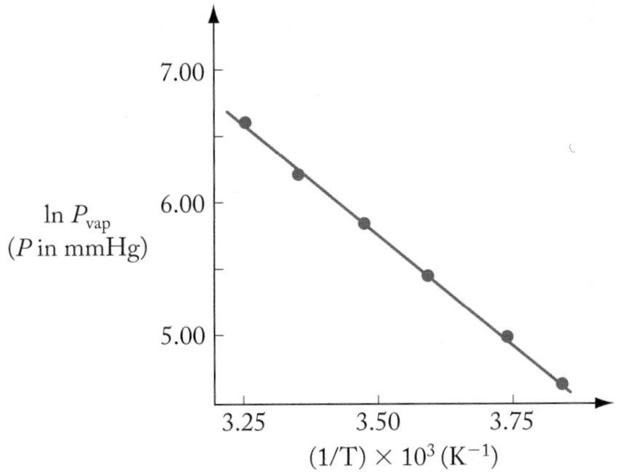

FIGURE 12.5 Plotting the natural log of the vapor pressure (here in torr or mmHg) versus the reciprocal of temperature in K gives a straight line described by the Clausius–Clapeyron equation. The graph shows the plot for *n*-pentane.

ture increases. Before we can understand what happens when a solution of volatile substances is distilled, we need to examine the vapor pressure of a pure substance as a function of temperature.

The relationship between the vapor pressure of a pure substance and absolute temperature is not linear. However, if we graph the natural logarithm of the vapor pressure [$\ln(P_{vap})$] versus $1/T$, we get a straight line (Figure 12.5). This line can be described by the equation

$$\ln(P_{vap}) = -\frac{\Delta H_{vap}}{R}\left(\frac{1}{T}\right) + C$$

where ΔH_{vap} is the enthalpy of vaporization, R is the gas constant, and C is a constant that depends on the identity of the liquid. We can solve this equation for C and write it in terms of two different temperatures:

$$\ln(P_{vap,T_1}) + \left(\frac{\Delta H_{vap}}{RT_1}\right) = C = \ln(P_{vap,T_2}) + \left(\frac{\Delta H_{vap}}{RT_2}\right)$$

which can then be rearranged to

$$\ln\left(\frac{P_{vap,T_1}}{P_{vap,T_2}}\right) = \frac{\Delta H_{vap}}{R}\left(\frac{1}{T_2} - \frac{1}{T_1}\right) \qquad (12.1)$$

Equation 12.1 is called the **Clausius–Clapeyron equation**, and it can be used to calculate ΔH_{vap} if the vapor pressures at two temperatures are known, or to calculate the vapor pressure (P_2) at any given temperature (T_2) or the temperature (T_2) at any given pressure (P_2) if ΔH_{vap} and P_1 and T_1 are known. Because the units of ΔH_{vap} are typically joules (J) or kilojoules (kJ), the value of the gas constant R used in the Clausius–Clapeyron equation is 8.314 J/mol · K.

SAMPLE EXERCISE 12.4 **Calculating Vapor Pressure**

At its normal boiling point of 126°C, octane [$CH_3(CH_2)_6CH_3$] has a vapor pressure of 760 torr. What is its vapor pressure at room temperature (25°C)? The enthalpy of vaporization (ΔH_{vap}) of octane is 39.07 kJ/mol.

COLLECT AND ORGANIZE We are asked to determine the vapor pressure of octane at 25°C. We are given its enthalpy of vaporization and its vapor pressure at its normal boiling point.

CONNECTION In Chapter 10 we defined the normal boiling point as the temperature at which the vapor pressure of a liquid reaches 1 atm (760 torr).

ANALYZE We can use the Clausius–Clapeyron equation (Equation 12.1) to answer this question. The temperatures must be in K, and the form of the gas constant R used must be 8.314 J/mol · K. Because of the units of R, ΔH_{vap} must be expressed in joules.

SOLVE The Clausius–Clapeyron equation relates vapor pressure, temperature, and heat of vaporization:

$$\ln\left(\frac{P_{vap,T_1}}{P_{vap,T_2}}\right) = \frac{\Delta H_{vap}}{R}\left(\frac{1}{T_2} - \frac{1}{T_1}\right)$$

Considering the values we are given, the only unknown in this expression is the vapor pressure at 25°C. Expressed in K, the two temperatures given are

$$T_1 = 126°C + 273 = 399\ K \qquad \text{and} \qquad T_2 = 25°C + 273 = 298\ K$$

and the value of ΔH_{vap} in joules is

$$39.07\ \frac{kJ}{mol} \times \frac{1000\ J}{1\ kJ} = 39{,}070\ \frac{J}{mol}$$

$$\ln\left(\frac{760\ torr}{P_{vap,T_2}}\right) = \frac{39{,}070\ \frac{J}{mol}}{8.314\ \frac{J}{mol \cdot K}}\left(\frac{1}{298\ K} - \frac{1}{399\ K}\right)$$

Solving the equation for P_2 yields 14.1 torr as the vapor pressure of octane at 25°C.

THINK ABOUT IT We expect octane to have a low vapor pressure at a temperature much lower than its boiling point, so this number seems reasonable.

Practice Exercise Pentane [$CH_3(CH_2)_3CH_3$] boils at 36°C. What is its molar heat of vaporization in kilojoules per mole if its vapor pressure at room temperature (25°C) is 505 torr?

CONCEPT TEST

Diesel fuel is made of hydrocarbons with an average of 13 carbon atoms per molecule, and gasoline is made of hydrocarbons with an average of 7 carbon atoms per molecule. Which fuel do you predict has the higher vapor pressure at room temperature?

CONNECTION We saw simple distillation in Chapter 1 as a way to purify sea water.

▶‖ **CHEMTOUR** Fractional Distillation

Fractional distillation is a method of separating a mixture of compounds on the basis of their different boiling points.

Volatility at Work: Fractionating Crude Oil

A process called **fractional distillation** (Figure 12.6) is used to separate the components of crude oil. Fractional distillation can separate a mixture of volatile substances because the vapors that arise from a solution of volatile substances are always enriched in the component with the lowest boiling point. If the vapors from a boiling solution are collected and then redistilled, the products of that distillation will be even more enriched in the component with the lowest boiling point. In a fractional distillation apparatus, repeated distillation steps allow components with only slightly different boiling points to be separated from one another.

To see how fractional distillation works, let's examine a series of graphs describing the behavior of pure substances and of a solution of two volatile liquids

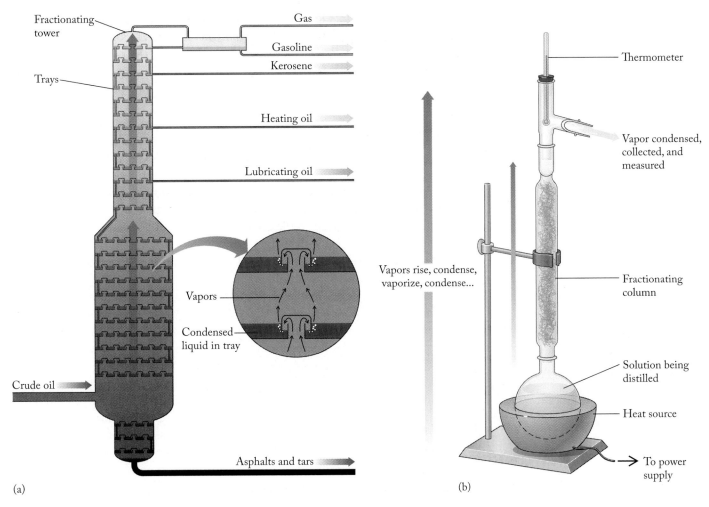

FIGURE 12.6 Fractional distillation is carried out to separate mixtures on (a) an industrial scale and (b) on the laboratory scale. In both cases, vapors rise through a fractionating column where they repeatedly condense and vaporize. On the industrial scale, the more volatile components rise higher in the column; components of the mixture are collected at different levels. On the laboratory scale, the more volatile components distill first, followed by increasingly less volatile, higher-boiling components. The identity of the material distilling is determined by the boiling point observed.

in simple distillation. The graph in Figure 12.7(a) is an idealized description of the behavior of a pure substance when boiled. The temperature remains constant during the phase change, and the composition of the vapor is constant because the substance is pure. If we carry out a simple distillation of a mixture of two volatile liquids, the graph in Figure 12.7(b) shows that the temperature is not constant during the course of the distillation. The two components co-distill over a range of temperatures. Figure 12.7(c) shows the composition of the vapor phase above a series of boiling solutions with a range of concentrations, and herein lies the key to how distillation can separate mixtures. The vapor (red line) in equilibrium with a solution of two volatile substances is enriched in the lower boiling (more volatile) component. The vapor has a different composition than the liquid (blue line).

Now let's carry out a thought experiment to understand how fractional distillation works. Suppose we heat 100 mL of a solution that is 50% by volume *n*-heptane (bp 98°C) and 50% *n*-octane (bp 126°C) in a miniature version of the distillation column in Figure 12.6. The graph in Figure 12.7(c) tells us that the boiling point of the solution will be initially nearly 110°C (point 1 on the graph). The graph also tells us that the vapor that comes from the solution the moment it begins to boil (point 2) is not a 50:50 mixture of *n*-heptane and *n*-octane, but rather it is enriched in the component with the lower boiling point: it is about 67% *n*-heptane and only 33% *n*-octane.

Suppose this vapor rises up in the distillation column, cools, and condenses in the first tray, a process represented by the red arrow from points 2 to 3 in Figure 12.7(c). The liquid in the tray is also 67% *n*-heptane and 33% *n*-octane. Continued heating of the column warms the liquid in the tray and it vaporizes. The vapor that initially forms from the solution in the first tray would be even more enriched in the *n*-heptane: about 80%, according to the coordinates of point 4 in Figure 12.7(c). The vapor rises up to the second tray, where it cools and condenses, and the redistillation process is repeated.

If we continue this process of redistilling mixtures with increasing concentrations of *n*-heptane and then cooling and condensing the vapors they produce in trays higher and higher in the distillation column, we will eventually obtain a condensate that is pure *n*-heptane in the uppermost tray. If we could monitor the temperature at which vapors condense at the very top of our distillation column, we would see a profile of temperature vs. volume of distillate produced like that shown in Figure 12.8. The first liquid to be produced would be pure *n*-heptane, which has a boiling point of 98°C. Over time all 50 mL of *n*-heptane in the original sample would be recovered. As we continued to add heat at the bottom of the column, the temperature at the top would rise to 126°C, signaling that the second component (pure *n*-octane, bp 126°C) is being collected.

Figure 12.8 is highly idealized. Fractional distillation does not work as well as this graph implies, but the graph indicates the best separation one could hope for if the system behaved perfectly.

This thought experiment using two alkanes describes the basis of the fractional distillation of crude oil. The cuts taken when crude oil is fractionated are not pure materials but solutions of several liquids that have desirable properties. The cuts are taken over a range of temperatures where mixtures of similar molecules distill together. The most volatile fraction from crude oil after any natural gas is removed is used as gasoline. The next most volatile cut is used as cleaning

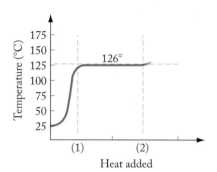

(a) Distillation of *n*-octane
(1) Octane begins to distill
(2) Octane finishes distilling

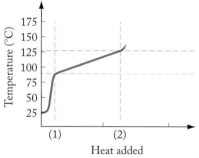

(b) Distillation of a mixture of *n*-heptane and *n*-octane
(1) Solution begins to distill
(2) Solution finishes distilling

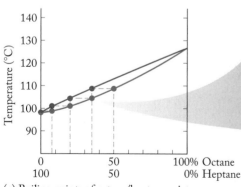

(c) Boiling points of octane/heptane mixtures (blue curve) and the composition of the vapors produced at those boiling points (red curve).

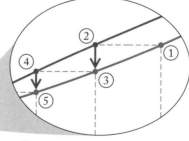

① Solution boils at this temperature
② Vapor has this composition
③ Vapor condenses, then boils
④ Vapor has this composition
⑤ Vapor condenses, then boils

FIGURE 12.7 Simple and fractional distillation graphs (a) and (b) describe the behavior of boiling liquids. (a) Simple distillation of a pure substance takes place at a constant temperature. (b) Simple distillation of a mixture of two volatile substances takes place over a range of temperatures. Graph (c) shows the results of many experiments in which the composition of vapor from a boiling liquid was analyzed. The blue line shows the temperatures where solutions of a given composition boil. The red line shows the composition of the vapor arising from those solutions. The stair-step line illustrates what happens in a fractionating column.

solvents and as paint thinner; the next most volatile after that is kerosene and diesel fuel. Compounds with even higher boiling points are viscous liquids and are used as lubricants.

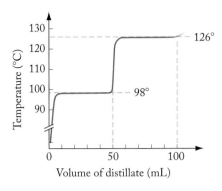

FIGURE 12.8 Fractional distillation of a mixture of 50 mL of *n*-heptane and 50 mL of *n*-octane produces two plateaus at the boiling points of the two components. If fractionation is perfect, the first 50 mL of distillate is pure *n*-heptane and the second 50 is pure *n*-octane.

SAMPLE EXERCISE 12.5	**Interpreting Data from Fractional Distillation**

Suppose that we are studying the composition of gasoline and are given the task of separating mixtures of *n*-octane [CH$_3$(CH$_2$)$_6$CH$_3$; normal boiling point 126°C] and *n*-nonane [CH$_3$(CH$_2$)$_7$CH$_3$; normal boiling point 151°C] by fractional distillation. To test our technique, we are given 100 mL of a solution consisting of 75 mL of *n*-octane and 25 mL of *n*-nonane to fractionate. Draw an idealized graph that describes the temperature measured at the top of the fractional distillation unit as a function of volume of distillate collected.

COLLECT AND ORGANIZE We are asked to draw an idealized graph of temperature versus volume of distillate collected in a fractional distillation. The solution being distilled contains 75 mL of *n*-octane and 25 mL of *n*-nonane. We have the normal boiling point of each component when pure.

ANALYZE Fractional distillation separates solutions of volatile liquids into pure substances on the basis of their different boiling points. The components distill in the order of their boiling points, with the lowest boiling component distilling first.

SOLVE The first component that distills from the solution is the lower boiling *n*-octane. If our system were perfect, 75 mL of *n*-octane would distill at its normal boiling point of 126°C. When the *n*-octane was completely removed from the solution, the only remaining component (25 mL of *n*-nonane) would distill at 151°C. The idealized graph shows the boiling points of the two substances along the *y*-axis and the volume of each distilled along the *x*-axis.

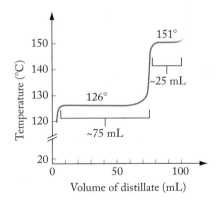

THINK ABOUT IT Our graph shows the best results we could achieve. In a "real world" system, a cut that was a mixture of *n*-octane and *n*-nonane would probably distill at temperatures between the boiling points of the two pure components.

Practice Exercise Draw an idealized graph that describes the behavior of temperature versus volume of distillate collected when a solution consisting of 30 mL of *n*-hexane (boiling point 69°C) and 50 mL of *n*-heptane (boiling point 98°C) is fractionally distilled.

⊙⊙ **CONNECTION** In Chapter 10 we used Raoult's law to describe the increase in boiling point of a solvent when a nonvolatile solute was dissolved in it.

Now that we know how fractional distillation works, the next step is to understand *why* fractional distillation works. We return to Raoult's law, which we used in Chapter 10 to describe the influence of nonvolatile solutes like salts on the boiling point of a pure solvent. Raoult's law also applies to homogeneous mixtures of volatile compounds, such as crude oil. Because the solutes in a solution of crude oil are volatile, they contribute to the vapor pressure. For a solution of volatile substances, the total vapor pressure equals the sum of the vapor pressures of each pure component (P_x°) at the temperature of interest multiplied by the mole fraction of that component in the solution (X_x) as shown in Equation 12.2.

$$P_{total} = X_1 P_1^\circ + X_2 P_2^\circ + X_3 P_3^\circ + \ldots \qquad (12.2)$$

SAMPLE EXERCISE 12.6   **Calculating the Vapor Pressure of a Solution**

Calculate the vapor pressure of a solution prepared by dissolving 13 grams of *n*-heptane (C_7H_{16}) in 87 grams of *n*-octane (C_8H_{18}) at 25°C. By what factor is the more volatile component enriched in the vapor phase compared with the liquid? The vapor pressures of octane and heptane at 25°C are 11 torr and 31 torr, respectively.

COLLECT AND ORGANIZE We can use Equation 12.2 to determine the total vapor pressure of the solution from the vapor pressures of the components after we determine the composition of the solution in terms of mole fractions.

ANALYZE To calculate mole fractions, we need the molar masses of *n*-heptane and *n*-octane. The mole fraction is then equal to the number of moles of each component divided by the total number of moles of material in the solution.

SOLVE The number of moles of each component is

$$87 \text{ g } C_8H_{18} \times \frac{1 \text{ mol } C_8H_{18}}{114.23 \text{ g } C_8H_{18}} = 0.76 \text{ mol } C_8H_{18}$$

$$13 \text{ g } C_7H_{16} \times \frac{1 \text{ mol } C_7H_{16}}{100.20 \text{ g } C_7H_{16}} = 0.13 \text{ mol } C_7H_{16}$$

The mole fraction of each component in the mixture is therefore

$$X_{octane} = \frac{0.76 \text{ mol}}{(0.76 + 0.13) \text{ mol}} = 0.85$$

$$X_{heptane} = 1 - X_{octane} = 0.15$$

Using these mole fraction values and the vapor pressures of the two alkanes in Equation 12.2, we have

$$P_{total} = X_{heptane} P_{heptane}^\circ + X_{octane} P_{octane}^\circ$$

$$= 0.15(31 \text{ torr}) + 0.85(11 \text{ torr})$$

$$= 4.6 \text{ torr} + 9.4 \text{ torr} = 14 \text{ torr}$$

To calculate how enriched the vapor phase is in the more-volatile component (that is, the component with the greater vapor pressure, *n*-heptane), we need to recall Dalton's law of partial pressures (see Section 6.6): The partial pressure of a gas in a mixture of gases is proportional to its mole fraction in the mixture. In equation form, this principle is

$$P_x = X_x P_{total}$$

Therefore, the ratio of the mole fraction of *n*-heptane to that of *n*-octane in the vapor phase is the ratio of their two vapor pressures:

$$\frac{4.6 \text{ torr}}{9.4 \text{ torr}} = 0.49$$

The mole ratio of *n*-heptane to *n*-octane in the liquid mixture is

$$\frac{0.13 \text{ mol}}{0.76 \text{ mol}} = 0.17$$

Therefore, the vapor phase is enriched in *n*-heptane by a factor of

$$\frac{0.49}{0.17} = 2.9$$

THINK ABOUT IT This result illustrates how fractional distillation works. The vapor is enriched in the lower boiling component, and fractional distillation makes it possible to separate a mixture into its pure components.

Practice Exercise Benzene (C_6H_6) is a trace component of gasoline. What is the mole ratio of benzene to *n*-octane in the vapor above a solution of 10% benzene and 90% *n*-octane by mass at 25°C? The vapor pressures of *n*-octane and benzene at 25°C are 11 torr and 95 torr, respectively.

Solutions such as the alkanes in crude oil obey Raoult's law when the strengths of intermolecular interactions between solute and solvent, solvent and solvent, and solute and solute are similar. Under these conditions, a solution behaves as an *ideal solution*. However, if intermolecular interactions between solute and solvent molecules are stronger than solvent–solvent or solute–solute interactions, the solute inhibits the solvent from vaporizing and the solvent inhibits the solute from vaporizing. This situation produces negative deviations from the vapor pressures predicted by Raoult's law, as shown in Figure 12.9(a).

On the other hand, if solute–solvent intermolecular interactions are weaker than solvent–solvent and solute–solute interactions, it is easier for solvent and solute molecules to vaporize from the solution, which leads to greater-than-predicted vapor pressures and positive deviations from Raoult's law, as shown in Figure 12.9(b). A mixture of hydrocarbons such as octane and heptane is expected to behave like an ideal solution because intermolecular interactions between the components are all London forces acting on molecules of similar structure and size.

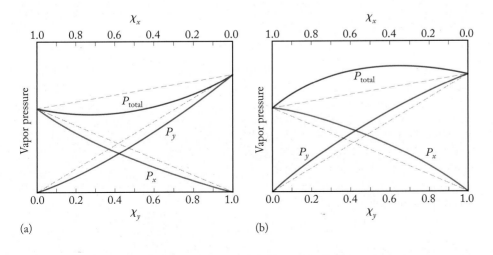

FIGURE 12.9 In a mixture of two volatile substances *x* and *y*, P_x and P_y may deviate from ideal behavior predicted by Raoult's law and described by the dashed line in both graphs. (a) If solute–solvent interactions are stronger than solvent–solvent or solute–solute interactions, negative deviations from Raoult's law arise. (b) If solute–solvent interactions are weak, positive deviations may occur.

Cuts of Crude Oil

If a crude oil sample contains a mixture of one- to four-carbon hydrocarbons (C_1–C_4, the common fuels methane, ethane, propane, and butane), these small molecules distill first because they are by far the most volatile. Methane is produced in this way, but methane is also produced during bacterial decomposition of vegetable matter in the absence of air, a condition that frequently arises in swamps. Hence, methane's common name is swamp gas or marsh gas. Frequently in the reductive environment of a marsh, a second material known as phosphine (PH_3) is also formed; phosphine and methane together spontaneously ignite. This produces a ghostly flame known as "will-o'-the-wisp" that features prominently in some legends and gothic mysteries. Methane can also be produced in coal mines, where it can be especially dangerous because it forms explosive mixtures with humid air, which gives rise to another common name for the gas: fire-damp.

The next fractions from the fractional distillation of crude oil are used in gasoline, which is a mixture of hydrocarbon isomers from C_5 through C_{12}. The next most volatile cuts are kerosene, C_9 to C_{16}, and diesel fuels, C_9 to C_{20} hydrocarbons. Compounds with even higher boiling points are viscous liquids used as lubricating oils. The low-melting solids (beyond C_{20}) that come off next are the paraffins, used in candles and in manufacturing matches. The material left in the distillation vessel consists typically of very heavy hydrocarbon gums and solid residues (C_{36} and up) that are collected and used as asphalt for paving roads.

CONNECTION In Chapter 9 we discussed methane's role as one of the greenhouse gases.

12.4 Alkenes and Alkynes

In the introduction to alkanes in Section 12.2, we described other hydrocarbons whose molecules contain fewer hydrogen atoms per carbon atom than do the alkanes. This subgroup known as unsaturated hydrocarbons is further divided into the family of alkenes, compounds having one or more carbon–carbon double bonds (two or more sp^2 hybridized carbons), and the family of alkynes, compounds having one or more carbon–carbon triple bonds (two or more sp hybridized carbons).

Alkenes and alkynes are minor components of crude oil, and alkenes are actually produced during the high-temperature processes that separate petroleum into commercially useful fractions. Alkynes, in contrast, are manufactured by the controlled oxidation of alkanes. When the simplest alkane, methane, is oxidized completely, the products are CO_2 and H_2O. But if this oxidation is carried out in a highly controlled process, the simplest alkyne, acetylene (the gas used in welding), can be produced.

$$6\,CH_4(g) + O_2(g) \rightarrow 2\,HC\equiv CH(g) + 2\,CO(g) + 10\,H_2(g)$$

This reaction points out an important difference between alkanes and the unsaturated hydrocarbons. Alkanes are the most reduced form of carbon. The sp^2 hybridized carbon atoms in alkenes are in a higher oxidation state than sp^3 hybridized carbons, and the sp hybridized carbons in alkynes are in a higher oxidation state than both alkanes and alkenes. Alkenes and alkynes are more reactive than alkanes, and the carbon–carbon double bond in alkenes is one of the most versatile functional groups in organic chemistry in terms of the number and types of reactions it undergoes.

CONCEPT TEST

We studied the energy changes associated with the combustion reactions of hydro-carbon fuels in Chapter 5. In principle, we could generate tremendous quantities of fuel by combining CO_2 and water vapor and chemically converting them back into CH_4 or other hydrocarbon fuels. Why is this not done in practice?

Alkenes and alkynes share many features of alkanes. The melting and boiling points of homologous series of these compounds (Table 12.5) vary with molar mass and size just as do those of the alkanes. Alkenes and alkynes exist in both linear and branched forms. In addition, because of the double bond, alkenes form geometric isomers (discussed in the following section).

Molecules may have more than one alkene or alkyne group in their structures. Indeed many molecules found in crude oil or produced by living systems have several double bonds. We discuss molecules with many double bonds in the chapter on biochemistry, but for now we concentrate on the properties associated with small molecules containing only one or a small number of double or triple bonds.

Chemical Reactivities of Alkenes and Alkynes

Figure 12.10 shows the electron distributions in the π-bonding orbitals in alkenes and alkynes in three dimensions. The electrons in these orbitals, where electron

TABLE 12.5 Melting Points and Normal Boiling Points of Homologous Series of Alkenes and Alkynes

Formula: Alkene	Melting Point (°C)[a]	Normal Boiling Point (°C)
$H_2C=CHCH_3$	−185	−47
$H_2C=CHCH_2CH_3$	−185	−6
$H_2C=CH(CH_2)_2CH_3$	−138	30
$H_2C=CH(CH_2)_3CH_3$	−140	63
$H_2C=CH(CH_2)_4CH_3$	−119	94
$H_2C=CH(CH_2)_5CH_3$	−104	123
$H_2C=CH(CH_2)_6CH_3$	−81	146
$H_2C=CH(CH_2)_7CH_3$	−87	171
Formula: Alkyne		
$HC\equiv CCH_3$	−102	−23
$HC\equiv CCH_2CH_3$	−126	8
$HC\equiv C(CH_2)_2CH_3$	−90	40
$HC\equiv C(CH_2)_3CH_3$	−132	71
$HC\equiv C(CH_2)_4CH_3$	−81	100

[a]Melting points increase with molar mass, but also depend on how molecules fit into crystal lattices. Melting points of alkenes with even numbers of carbon atoms form one series that follows this trend; alkenes with odd numbers of carbon atoms (gray scale) form another series.

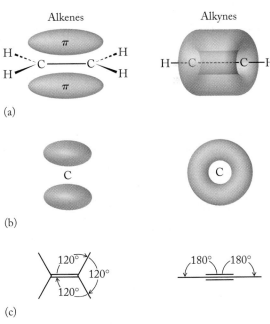

FIGURE 12.10 The characteristic structural feature of alkenes and alkynes is the presence of electrons in π orbitals. (a) The electron distribution in a C–C double bond and a C–C triple bond. (b) View of electron distribution, looking down the C–C bond axis. (c) The idealized H—C—C bond angles in double and triple bonds.

CONNECTION In Chapter 9 we introduced σ and π bonds in terms of valence-bond theory and hybridization of orbitals.

density is above and below the plane of the carbon skeleton, are more accessible to reactants than the electrons in the σ bonds of alkanes, where electron density is greatest between the atoms. Remember that chemical reactions entail the making and breaking of bonds, a process which requires electron density to shift between pairs of atoms. Electrons in π orbitals are more accessible to approaching reactants than electrons in σ bonds, so the reactivity of molecules containing unsaturation is very different from that of alkanes.

As an illustration of this difference in reactivity, consider the behavior of alkenes with compounds such as the hydrogen halides (HCl, HBr, HI). These compounds react with the double bond in alkenes to make alkyl halides (alkanes in which a halogen has been substituted for one of the hydrogen atoms). For example, gaseous HBr combines with ethylene in the following **addition reaction**:

$$HBr(g) + H_2C{=}CH_2(g) \rightarrow CH_3CH_2Br(\ell)$$

This reaction is called an addition reaction because two reactants combine to form one product. As noted above the other hydrogen halides also react with alkenes in this way, as does hydrogen itself (see Sample Exercise 12.1). However, addition reactions do not occur with normal alkanes such as ethane:

$$HBr + CH_3CH_3 \rightarrow no\ reaction$$

Alkynes, like alkenes, react with hydrogen halides. They differ, however, in that two molecules of a hydrogen halide react with one triple bond and yield alkanes that bear 2 halogen atoms as products. Both double and triple bonds react with many other reagents, which is why compounds containing carbon–carbon double and triple bonds in their molecular structure are useful substances in the industrial production of other compounds.

Isomers

Molecules that contain alkene and alkyne functional groups can have straight or branched chains, with the same types of structural isomers we saw with alkanes. One additional facet of structural isomerization involves the location of the double or triple bond within a molecule. For example, consider just the straight-chain isomers of an unsaturated hydrocarbon that contains 5 carbon atoms and one double bond. The possible structures are:

a. $H_2C{=}CH{-}CH_2{-}CH_2{-}CH_3$
b. $CH_3{-}CH{=}CH{-}CH_2{-}CH_3$
c. $CH_3{-}CH_2{-}CH{=}CH{-}CH_3$
d. $CH_3{-}CH_2{-}CH_2{-}CH{=}CH_2$

Applying the same test used in Section 12.2 for alkanes, structures (a) and (d) are the same. This is easier to see if we draw them as carbon-skeleton structures:

(a)   (d)

In both cases the double bond is between carbon atoms 1 and 2.

Structures (b) and (c) are structural isomers of (a) because they have the same chemical formula but have the double bond in a different location. Drawing the carbon-skeleton structures of (b) and (c), however, presents us with a new situation. When drawing structure (b), two options present themselves. If a straight line is drawn through the double bond between the second and third carbon

In an **addition reaction**, two molecules couple together and form one product.

A **cis isomer** (or **Z isomer**) has two like groups (such as two R groups or 2 hydrogen atoms) on the same side of a line drawn through the double bond. *Cis* is Latin for "on this side"; Z comes from the German word *zusammen* (together).

A **trans isomer** (or **E isomer**) in an alkene has two like groups (such as two R groups or 2 hydrogen atoms) on opposite sides of a line drawn through the double bond. *Trans* is Latin for "across"; E comes from the German word *entgegen* (opposite).

Geometric isomers are structures in which molecules with the same connectivity have atoms in different positions because of restricted rotation. In alkenes, there is restricted rotation about a double bond.

atoms in the chain, the longer carbon chain can be placed on the same side as the methyl group (b1) or on the opposite side (b2).

(b1)   (b2)

These two molecules are isomers; they have the same formula, but have different structures and therefore different properties. The isomer on the left is called the **Z isomer** or the *cis* **isomer** (*cis* is Latin for "on this side"), and the one on the right, the **E isomer** or the *trans* isomer (*trans* is Latin for "across"). These isomers are called **geometric isomers**. They arise because the double bond does not rotate freely.

Recall from Chapter 9 that the double bond is formed from the overlap of two unhybridized *p* orbitals on adjacent carbon atoms. As Figure 12.11 shows, for the carbon atoms joined in a double bond to rotate freely, the 2 atoms have to twist about their bond axis, eliminating the orbital overlap and breaking the bond. Breaking a π bond in 1 mole of an alkene costs about 290 kJ of energy, and that much energy is not available to the molecule at room temperature. This situation gives rise to restricted rotation about a carbon–carbon double bond and to the existence of geometrical isomers.

Now let's examine the geometric isomers of structure (c). In the isomer on the left, the 2 hydrogen atoms are on the same side of the double bond; this is the *cis* isomer. In the isomer on the right, the 2 hydrogen atoms are on the opposite sides of the double bond; this is the *trans* isomer. The two molecules are geometric isomers.

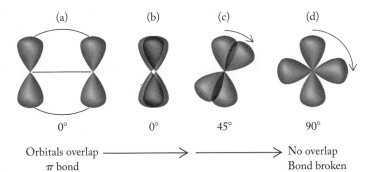

FIGURE 12.11 To form a π bond (a) p_z orbitals overlap to establish a region of shared electron density above and below the plane of the C–C bond (red). (b) If you look down the C–C bond axis, the orbitals line up. (c) If you rotate one carbon while keeping the other fixed, the *p* orbitals are no longer parallel and do not overlap. (d) If you rotate it far enough, the bond breaks.

⬤⬤ **CONNECTION** In Chapter 5 we introduced rotations about bonds as one of the types of motion molecules experience as part of their overall kinetic energy.

(c1)   (c2)

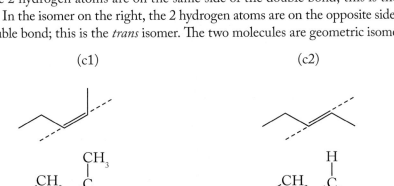

The hydrogens on the double bond are *cis*.   The hydrogens on the double bond are *trans*.
The R groups on the double bond are *cis*.   The R groups on the double bond are *trans*.

A comparison of the geometric isomers for (b) $CH_3—CH=CH—CH_2—CH_3$ and (c) $CH_3—CH_2—CH=CH—CH_3$ shows that the two *cis* structures (b1) and (c1) are identical, as are the two *trans* structures (b2) and (c2). Therefore, the straight-chain alkenes with 5 carbon atoms exist as three isomers. All three isomers are chemically distinct species.

Naming Alkenes and Alkynes

To name simple alkenes and alkynes, the prefixes in Table 12.4 are used to identify the length of the chain in the parent structure. The suffix -*ene* is appended if the compound is an alkene; -*yne* is used if the substance is an alkyne. The carbon atoms in the chain are numbered so that the first carbon atom in the double or triple bond has the lowest number, and that number precedes the name, followed by a dash. Geometric isomers are identified by writing *cis*- or *trans*- before the number. Thus the compounds (b1) and (b2) are *cis*-2-butene and *trans*-2-butene, respectively.

SAMPLE EXERCISE 12.7   **Identifying and Naming Geometric and Structural Isomers**

Draw and name the five possible isomers of a linear chain of six carbon atoms containing one double bond. Draw the isomers as condensed structural formulas and carbon-skeleton structures.

COLLECT AND ORGANIZE We are asked to draw and name the five possible isomers of a molecule consisting of six carbon atoms in a continuous chain with one double bond.

ANALYZE Two kinds of isomers arise: structural isomers that depend on where in the chain the double bond is located, and geometric isomers, that depend on the orientation of groups about the double bond (*cis/trans* isomers).

SOLVE Let's start with the isomers having the double bond at different locations and then draw the *cis/trans* isomers where possible:

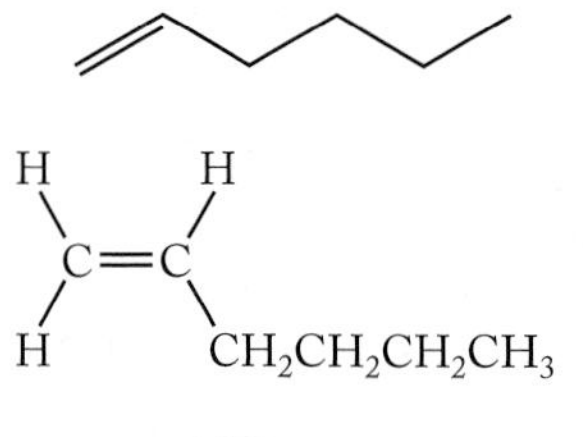

1-Hexene 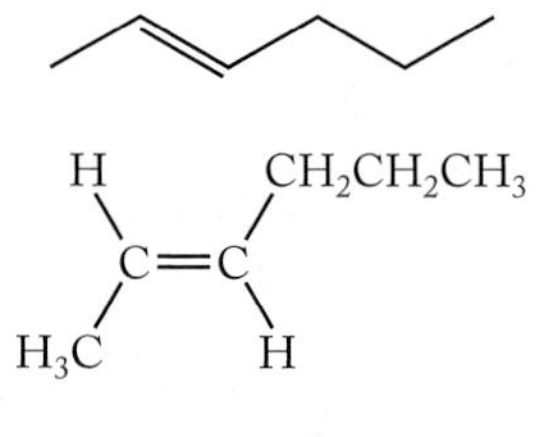   *trans*-2-Hexene 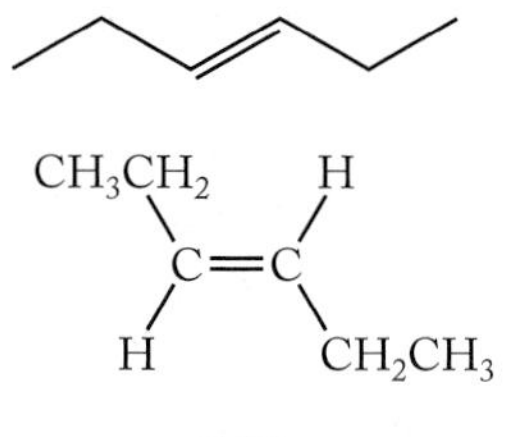   *trans*-3-Hexene

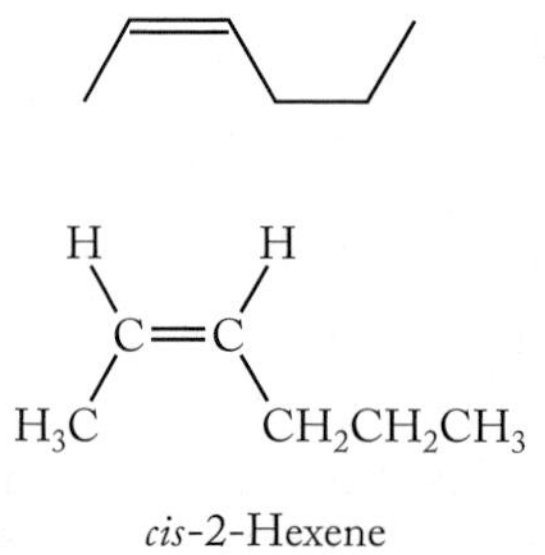

cis-2-Hexene 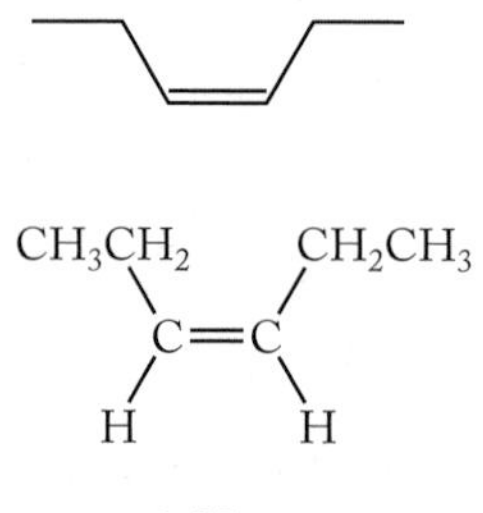   *cis*-3-Hexene

The names of each of the five isomers appear beneath its structure.

THINK ABOUT IT Being aware of the different kinds of isomers that are possible for a compound helps when we are asked to draw all of them.

Practice Exercise Draw the five possible isomers of the molecule with this carbon skeleton and one double bond:

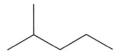

> A **homopolymer** is a polymer composed of only one kind of monomer unit.

CONCEPT TEST

(a) Does a straight-chain hydrocarbon with a terminal double bond, like 1-pentene, $CH_2\!\!=\!\!CHCH_2CH_2CH_3$, have geometric isomers? (b) Do alkynes have geometric isomers?

Polymers of Alkenes

As we indicated, the substances in crude oil are not only useful materials as isolated but also serve as the starting materials for the industrial production of many other commercial compounds. Alkenes are especially useful as monomers in polymerization reactions. Some widely used polymeric alkanes are produced industrially from small molecules derived from petroleum. The simplest among these in terms of structure is linear polyethylene (PE), produced from the ethylene monomer ($CH_2\!\!=\!\!CH_2$) under conditions of high temperature and pressure:

$$n\,CH_2\!\!=\!\!CH_2 \rightarrow \{CH_2CH_2\}_n$$

PE is a **homopolymer**, which means that it is composed of only one type of monomer. The condensed structural formula of PE is $CH_3(CH_2)_nCH_3$, but there are so many methylene groups compared to methyl groups that the structure is frequently written as shown in the equation to highlight the structure and composition of the monomer.

In most products made of polyethylene, n is a very large number ranging from 1000 to almost 1 million. Polyethylene has a wide range of properties that are correlated to the density of the polymer. The properties depend upon the value of n, and hence molar mass, and on whether the polymer chains are linear or branched. Low-density PE (LDPE) is an easily stretchable, soft plastic used in films and wrappers. The chains in LDPE are branched. When the bagger at the grocery store asks "paper or plastic?" the plastic in question is LDPE.

When the molar mass of the PE polymers ranges from 100,000 to 500,000 and the chains are linear, then the polymer is still PE but it has different physical properties from those of grocery bags. The straight-chain polymer is a rigid, translucent solid called high-density polyethylene (HDPE). HDPE is used in milk containers, electrical insulation, and toys. It was the first plastic adapted for use in artificial hip joints. Abrasion between moving parts is a major problem with artificial joints, and although HDPE is very tough, it is not as resistant to abrasion as this application demands. To make the joints last longer, some artificial ball and socket joints are now coated with a relatively new polymer called ultrahigh molecular weight PE (UHMWPE; $n > 100,000$), which is a very tough material and highly resistant to wear. UHMWPE consists of linear molecules with an average molar mass over 3 million.

HDPE   **LDPE**

Products made of polyethylene bear a recycle symbol that identifies them as linear (high-density) or branched (low-density).

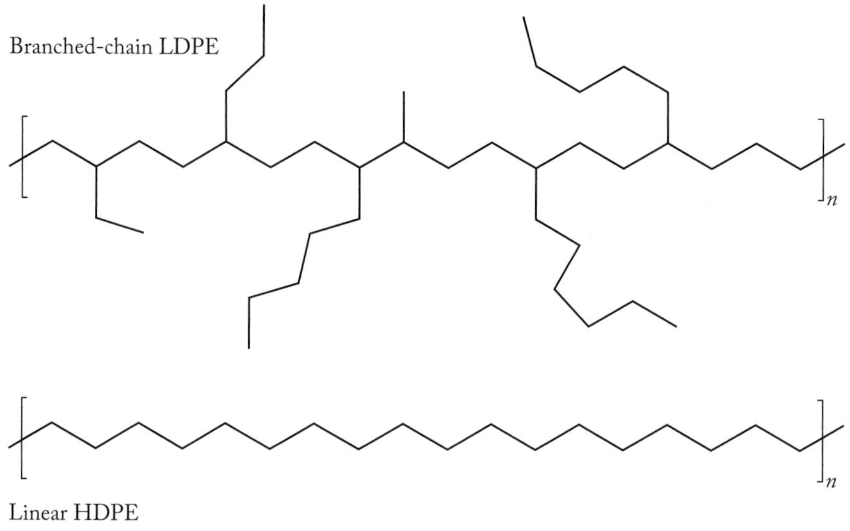

Branched-chain LDPE

FIGURE 12.12 Low-density polyethylene (LDPE) consists of branched chains, while high-density polyethylene (HDPE) consists of linear chains.

Linear HDPE

Why does branching lead to such different properties on the part of a bulk polymer? Figure 12.12 shows a branched-chain molecule of LDPE and a linear molecule of HDPE. Think of the branched polymer as very much like a tree branch that has lots of smaller branches attached to it. The polymer can have branches that come out of the plane of the paper, so it has three dimensions, like a tree branch. In contrast, the straight-chain polymer is like a long straight bamboo pole. Suppose you had a pile of 100 tree branches and a second pile of 100 bamboo poles, and your task was to stack each pile into the smallest possible space to fit into a truck. You can certainly pile the branches on top of each other, but they will not fit together neatly, and probably the best you can do is to make the pile a bit more compact. In contrast, you can line up the bamboo poles and stack them into a very compact pile. You could stack many more bamboo poles in a truck than tree branches. The same situation arises with the branched and linear molecules of polyethylene. Branched PE is LDPE because the molecules do not line up neatly. Their density is low compared to HDPE, primarily because the branched molecules stack less efficiently. This makes the LDPE more deformable and softer; HDPE is more rigid and has regions that are actually crystalline because the packing is so uniform. UHMWPE is an even tougher material, because not only are the molecules linear, they are significantly larger than the molecules in HDPE. The different forms of polyethylene provide a clear illustration of the impact of the size and shape of molecules on the physical properties of a material.

CONCEPT TEST

Articles made from low-density polyethylene are separated from those made from high-density polyethylene for recycling. Why?

Of course, chemical composition also plays a role in determining properties. If the hydrogen atoms in PE are all replaced with fluorine atoms, the resultant polymer is chemically very unreactive, capable of withstanding high temperatures, and has a very low coefficient of friction, which means other things do not stick to it. This polymer is Teflon, $\{CF_2CF_2\}_n$, most familiar for its use as a nonstick surface in cookware. The analogous material cannot be formed with chlorine. However, a polymer does exist in which every other $-CH_2-$ along the backbone is

$$\begin{array}{ccc} \{CH_2-CH_2\}_n & \{CF_2-CF_2\}_n & \{CH_2-CCl_2\}_n \\ \text{Polyethylene} & \text{Teflon} & \text{Saran} \end{array}$$

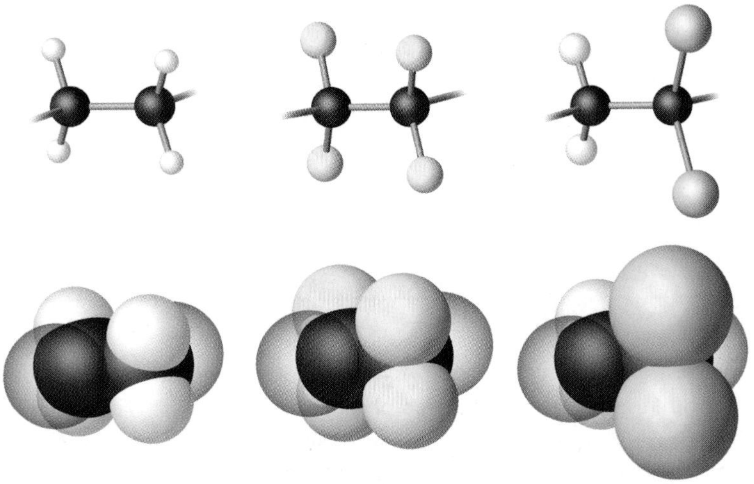

FIGURE 12.13 Repeating units of common PE-type polymers: PE, Teflon, and Saran.

a $-CCl_2-$. Its repeating monomer unit is $\{CH_2CCl_2\}_n$, and the polymer is Saran, the familiar plastic wrap. Usually the composition of a polymer is defined simply by showing the repeating monomer units, as shown in Figure 12.13.

SAMPLE EXERCISE 12.8   **Identifying Monomers**

Teflon $\{CF_2CF_2\}_n$ and Saran $\{CH_2CCl_2\}_n$ are molecular relatives of polyethylene (PE). What are the monomers from which these two polymers are made?

COLLECT AND ORGANIZE We are asked to give the structure of the monomers from which two polymers are made. We are given the repeating units in the polymers.

ANALYZE The polymers are related to polyethylene, which means the monomer units must be substituted ethylene molecules.

SOLVE These two polymers can both be made from substituted ethylenes. To make a polymer like Teflon in which all the hydrogen atoms in PE are replaced by fluorine atoms, the starting material must be ethylene with 4 fluorine atoms instead of 4 hydrogen atoms.

Monomer for Teflon: $CF_2=CF_2$

Saran is $\{CH_2CCl_2\}_n$, so its monomer must be an ethylene with the 2 hydrogen atoms on 1 carbon atom replaced by 2 chlorine atoms.

Monomer for Saran: $CH_2=CCl_2$

THINK ABOUT IT The functional group in the monomer that reacts to form the polymer is the carbon–carbon double bond of the alkene.

Practice Exercise Poly(methyl methacrylate) (PMMA) is the polymer used in making Plexiglas and Lucite as shatterproof replacements for glass. The repeating unit in PMMA is shown on the right; draw the structure of the monomer used to make PMMA.

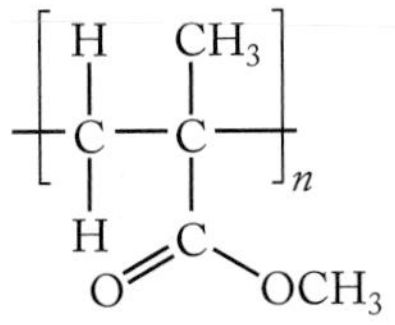

The subgroup CH$_2$=CH– is called the **vinyl group**. The family of polymers formed from monomers containing that group comprise the **vinyl polymers**.

PVC

This is the recycle symbol for PVC.

All polymers based on addition reactions of monosubstituted ethylene are called **vinyl polymers**, because the CH$_2$=CH– subunit is called the **vinyl group**, a name derived from vinum (Latin: wine). The name was given to the group by 18th century chemists who prepared ethylene (CH$_2$=CH$_2$) from ethanol (CH$_3$CH$_2$OH).

Poly(vinyl chloride) is usually referred to as PVC. It is a material widely used in commercial articles ranging from plastic pipes for plumbing to computer cases. The so-called "classic vinyl" phonograph records are made from PVC. The formulas for the monomer and the polymer are

$$n\, CH_2 = CHCl \rightarrow \{CH_2CHCl\}\, n$$

CONCEPT TEST

Suggest a structural reason why Saran, with two chlorine atoms on every other carbon atom, is a soft, flexible polymer, while PVC, with one chlorine on every other carbon atom in its monomer, is more rigid.

Synthetic polymers are materials of major commercial importance in the modern world. Alkenes derived from crude oil are the primary source of monomers for the production of polymers used in construction, in fabrics, as wrapping and packaging material, and in medical devices. Vinyl polymers are the world's second largest selling plastics materials, and polymers in this category are extraordinarily versatile. You probably contact or make use of five to ten vinyl polymers before you leave your room in the morning: vinyl shower curtains, vinyl drain pipes, vinyl flooring, vinyl insulation around electrical conduits. As we explore more organic functional groups, the basic concepts developed for the vinyl polymers will apply to polymers in other categories: the features of functional group, size, and shape determine the chemical and physical properties of these extraordinarily useful materials.

12.5 Aromatic Compounds

Among the components of gasoline that play an important role in increasing octane ratings (a measure of the ignition temperature of the fuel and its ability to resist engine "knock") is a class of compounds called aromatic hydrocarbons. As their class name implies, aromatic hydrocarbons have distinctive odors. However, "aromaticity" from a chemist's perspective is associated with specific features of molecular and electronic structure. The most distinctive features of aromatic structures are planar, hexagonal rings in which 6 sp^2 hybridized carbon atoms are joined by a combination of σ and π bonds as shown in Figure 12.14. Aromatic compounds are relatives of alkenes because they contain carbon–carbon double bonds. However, because their chemical and physical properties are unique and distinct from those of alkenes, they merit designation as a separate family.

The most common of these molecules consists of one six-membered ring and is called benzene, C$_6$H$_6$. The different ways we view the bonding in benzene using

◉◉ **CONNECTION** In Chapters 8 and 9 we introduced benzene as an aromatic compound and described bonding in the benzene molecule using Lewis theory and valence bond theory.

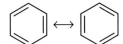
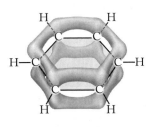

(a) Carbon-skeleton structures showing resonance forms of benzene

Skeletal symbol of benzene ring

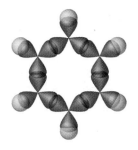
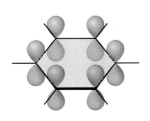

(b) Sigma bonds in benzene

(c) Unhybridized p orbitals of carbon atoms

(d) Delocalized π cloud of electrons above and below plane of ring

FIGURE 12.14 Different views of the bonding in benzene. (a) Skeletal structures showing resonance and double-bond delocalization. (b) Valence bond diagram of overlapping σ bond structure. (c) p_z orbitals on sp^2 hybridized carbons. (d) Delocalized π cloud of electrons above and below the ring.

Lewis theory and valence bond theory are summarized in Figure 12.14. The delocalization of electrons leads to considerable stability on the part of benzene and all aromatic structures.

The stability of aromatic systems because of resonance has an impact on their chemical reactivity. In Section 12.4, we saw that alkenes react rapidly with hydrogen halides to make halogenated alkanes. In contrast, benzene does not react at all if HBr gas is passed through it. Alkenes and aromatic compounds differ completely with respect to this and many other reactions, so the classification of aromatic systems as a unique family is justified.

●● **CONNECTION** In Chapters 8 and 9 we introduced benzene as an aromatic compound and described bonding in the benzene molecule, using Lewis theory and valence bond theory.

●● **CONNECTION** In Chapter 8 we defined resonance in Lewis structures.

▶❙❙ **CHEMTOUR** Structure of Benzene

Structural Isomers of Aromatic Compounds

Many different compounds can be formed by replacing the hydrogen atoms in aromatic rings with other substituents. For example, when one methyl group replaces a hydrogen atom, the resultant compound is methylbenzene, also known by its common name, toluene. All the positions around the benzene ring are equivalent, so it does not matter which carbon atom is bonded to the methyl group.

There are three options for attaching two methyl groups to a benzene ring, so there are three structural isomers of dimethylbenzene, also known as xylene.

Toluene

Again, the existence of isomers, as well as the great variety of substituents that can be attached to an aromatic system, contributes to the structural variety of organic compounds.

Petroleum also contains compounds that have structures in which benzene rings are fused together by sharing one or more of their hexagonal sides. Three such compounds are naphthalene, anthracene, and phenanthrene:

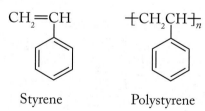

Napthalene   Anthracene   Phenanthrene

As we learned in Section 9.5, compounds with two or more fused aromatic rings are called polycyclic aromatic hydrocarbons (PAHs). Extensive delocalization of the π electrons throughout their rings makes these structures particularly stable. In addition to being in fossil fuels, they may be formed during the incomplete combustion of hydrocarbons and are present in particularly high concentrations in the soot from incinerators and diesel engines. They have also been identified in interstellar dust clouds and in blackened portions of grilled meat. When introduced into the environment, these compounds persist and are among the most long-lived of hydrocarbons.

Polymers Containing Aromatic Rings

Individual aromatic rings as well as fused ring systems are flat molecules, and this physical feature gives rise to several characteristic properties of them as a family. They tend to stack together, as shown in Figure 12.15, and this tendency gives rise to useful properties in materials that incorporate aromatic systems. The polymer polystyrene is a good example. If one of the hydrogen atoms on ethylene is replaced by a benzene ring, the resultant monomer is styrene, and the polymer made from that monomer is polystyrene (PS).

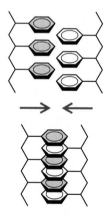

FIGURE 12.15 The aromatic rings on neighboring chains in polystyrene stack together and provide strength to the material.

$$CH_2{=}CH \qquad {+}CH_2CH{+}_n$$

Styrene   Polystyrene

Depending on the molar mass of its molecules, polystyrene displays a range of properties. Solid PS is a colorless, hard, inflexible plastic. In this form it is used for compact disc cases and plastic cutlery. Its most common use is as an expanded solid, made by blowing gas (such as CO_2 or pentane) into molten polystyrene, which then expands and retains voids in its structure when it solidifies. One form of this polymer is Styrofoam, the familiar material of coffee cups, take-out food containers, "peanuts" used as packing materials, and building insulation (Figure 12.16).

The difference in properties between the clear, inflexible polystyrene and the white, pliable, lightweight Styrofoam can be explained by considering the role of the aromatic ring in aligning the polymer chains. Branches in the chains have the same effect that we saw with polyethylene, but the aromatic rings and their tendency to stack provide additional associations that make chain alignment more

PS

This is the recycle symbol for polystyrene.

favorable energetically. The aromatic rings along two neighboring chains can stack, and this interdigitation provides more strength to the aligned chains than in polyethylene (see Figure 12.15). When the chains are blown apart by a gas, the stacking is disrupted and the chains open to make cavities that fill with air, characteristic of the familiar light-weight, opaque material that is a good thermal insulator and packing material.

FIGURE 12.16 Styrofoam can be made into rigid or foamed products, and it is used for plates, cups, compact disc jackets, carry-out food containers, packing materials, and thermal insulation in buildings.

12.6 Alcohols, Ethers, and Reformulated Gasoline

In January 1995, air-quality regulations went into effect in many U.S. cities that mandated reductions in atmospheric concentrations of pollutants associated with emissions from gasoline-fueled engines. The new regulations led to the wide-spread use of "reformulated" gasoline containing additives to promote complete combustion and boost octane ratings. These additives are not hydrocarbons but are organic compounds that contain other atoms in addition to hydrogen and carbon. These atoms are called **heteroatoms**, and they are the defining components of functional groups in other important families of organic molecules. Among the compounds added to gasoline to promote combustion are additives that have oxygen in their molecular structures in the form of either alcohol (R—OH) or ether (R—O—R′) functional groups.

CONCEPT TEST

Explain why many alcohols and ethers with low molar masses are more soluble in polar solvents, such as water, than are hydrocarbons.

Alcohols: Methanol and Ethanol

Alcohols have the general formula R—OH, where R is any alkyl group. Alcohols can be part of molecules with straight chains, branched chains, or rings and have all the structural variability associated with those types of molecules. The special chemical properties of alcohols stem from the –OH group. The physical properties of alcohols can be understood if we recognize that an alcohol looks like a combination of an alkane and water:

R—H	H—OH	R—OH
Alkane	Water	Alcohol

As we saw in Chapter 10, if the carbon subunit in the molecule is small, the alcohol behaves like water; as the carbon subunit gets larger, the alcohol behaves more like a hydrocarbon. As an illustration of this behavior, the data in Table 12.6 show the water solubilities of a homologous series of alcohols. As the hydrocarbon portion of the alcohol gets larger, the solubility of the alcohol in water decreases. The role of the polar –OH group in terms of making the compound "water-like" and hence soluble decreases until, beyond about C_8, the solubility in water is as negligible as that of the corresponding hydrocarbon.

Alcohols react with many other types of molecules and can be used as starting materials for the preparation of a vast array of other substances. The simplest alcohols in terms of structure, methanol and ethanol, are increasingly used as fuel additives. As these two names indicate, the chemical names of alcohols end in

A **heteroatom** is any atom other than carbon or hydrogen in an organic compound.

An **alcohol** contains the –OH functional group and has the general formula R—OH, where R in this case is any alkyl group.

TABLE 12.6 Solubilities of a Homologous Series of Alcohols in Water at 20°C	
Formula	**Water Solubility (g/100 mL)**
CH_3OH	Miscible in all proportions
CH_3CH_2OH	Miscible in all proportions
$CH_3(CH_2)_2OH$	Miscible in all proportions
$CH_3(CH_2)_3OH$	7.9
$CH_3(CH_2)_4OH$	2.3
$CH_3(CH_2)_5OH$	0.6
$CH_3(CH_2)_6OH$	0.2
$CH_3(CH_2)_7OH$	0.05

"-ol"; the suffix identifies the compound being discussed as an alco<u>hol</u>. The following describes the preparation and properties of these two important industrial alcohols.

Methanol (CH_3OH) is known as methyl alcohol or wood alcohol because it was once made by collecting the vapors given off when wood is heated to the point of decomposition in the absence of oxygen. It is used industrially in quantities exceeding those of most other organic chemicals, and it is the starting material in the preparation of several organic compounds used to make polymers. It has been suggested for use as an additive in gasoline. Its industrial synthesis is based on reducing carbon monoxide with hydrogen:

$$CO(g) + 2H_2(g) \rightarrow CH_3OH(\ell)$$

CONNECTION We discussed steam reforming in Chapter 5.

The CO and H_2 used to make methanol come from the reaction of methane and water in the presence of a metal catalyst at high temperatures in a process called steam reforming:

$$CH_4(g) + H_2O(g) \rightarrow CO(g) + 3H_2(g)$$

Methanol burns according to the following thermochemical equation:

$$2CH_3OH(\ell) + 3O_2(g) \rightarrow 2CO_2(g) + 4H_2O(\ell) \qquad \Delta H° = -1454 \text{ kJ}$$

If we divide the absolute value of $\Delta H°$ by twice the molar mass of methanol (because the reaction consumes 2 moles of methanol), we get a fuel value for methanol of

$$\frac{1454 \text{ kJ}}{\left(\dfrac{32.0 \text{ g}}{\text{mol}}\right)(2 \text{ mol})} = 22.7 \text{ kJ/g}$$

To compare the fuel value of methanol with that of octane:

$$2C_8H_{18}(\ell) + 25O_2(g) \rightarrow 16CO_2(g) + 18H_2O(\ell) \qquad \Delta H° = -1.091 \times 10^4 \text{ kJ}$$

$$\frac{1.091 \times 10^4 \text{ kJ}}{\left(\dfrac{114.2 \text{ g}}{\text{mol}}\right)(2 \text{ mol})} = 47.8 \text{ kJ/g}$$

The fuel value of methanol is less than half that of octane (and most of the other hydrocarbons in gasoline). Why is this the case? The answer involves the composition of methanol. The amount of energy released during combustion depends on the number of carbon atoms available for forming $C={=}O$ bonds and the number of hydrogen atoms available for forming $O-H$ bonds as CO_2 and H_2O are produced. The presence of oxygen in CH_3OH adds significantly to its mass (methanol is 50% oxygen by mass) but adds nothing to its fuel value. The oxygen content of a combustible substance essentially dilutes its energy value. The more "oxygenated" a fuel is, the lower its fuel value.

CONNECTION We introduced fuel values in Chapter 5 as the amount of heat given off when one gram of fuel is burned.

CONCEPT TEST

Predict, on the basis of their molecular formulas, which alcohol has the greater fuel value: methanol or ethanol.

Ethanol (CH_3CH_2OH, also known as ethyl alcohol) is the alcohol in alcoholic beverages. It is formed by the fermentation of sugar from an amazing variety of vegetable sources. Indeed, any plant matter containing sufficient sugar may be used to produce ethanol. Grains are commonly used, from which ethanol derives its trivial name of grain alcohol. Ethanol may be the oldest synthetic organic chemical used by human beings, and it is still one of the most important. For industrial purposes, ethanol is prepared by the reaction of water and ethylene.

Ethanol currently makes up about 3% of the gasoline used in the United States, and this number is increasing annually. In some regions, a formulation of gasoline called *gasohol* contains up to 10% ethanol. Most of the ethanol used in gasoline is produced by fermentation of sugar derived from corn. Ethanol burns readily in air:

$$CH_3CH_2OH(\ell) + 3\,O_2(g) \rightarrow 2\,CO_2(g) + 3\,H_2O(\ell) \qquad \Delta H° = -1367 \text{ kJ}$$

SAMPLE EXERCISE 12.9   **Comparing the Energy Content of Fuel Mixtures**

Suppose the hydrocarbons in gasoline can be represented by *n*-nonane, $CH_3(CH_2)_7CH_3$, which has a standard molar heat of combustion ($\Delta H°_{comb}$) of -6160 kJ/mol. The $\Delta H°_{comb}$ of ethanol is -1367 kJ/mol. How much energy in the form of heat is available from a mixture of 10% ethanol / 90% gasoline by weight compared with the amount available from pure gasoline? Carry out the calculation based on 2.70×10^3 g of mixture, which is about the mass of 1 gallon of gasoline.

COLLECT AND ORGANIZE We are given 2.70×10^3 g of a mixture of two fuels and asked to determine how much heat is given off when the mixture is burned. We have the heats of combustion of both components and the composition by mass of the mixture.

ANALYZE The heats of combustion are given in terms of moles of material, and the mass of the mixture is in grams, so to use the thermochemical values we need to calculate the number of moles of each component. To convert grams to moles, we need molar masses, which we can determine from the formulas: ethanol is CH_3CH_2OH and has a molar mass of 46.07 g; *n*-nonane is $CH_3(CH_2)_7CH_3$ and has a molar mass of 128.25 g.

SOLVE Let's calculate how many grams of each component are in the mixture and then convert that to moles:

$$10\% \text{ ethanol: } 0.10(2.70 \times 10^3 \text{ g}) = 2.70 \times 10^2 \text{ g ethanol}$$

$$90\% \text{ } n\text{-nonane: } 0.90(2.70 \times 10^3 \text{ g}) = 2.43 \times 10^3 \text{ g } n\text{-nonane}$$

$$\text{Moles ethanol: } 2.70 \times 10^2 \text{ g} \times \frac{1 \text{ mol}}{46.07 \text{ g}} = 5.861 \text{ mol}$$

$$\text{Moles } n\text{-nonane: } 2.43 \times 10^3 \text{ g} \times \frac{1 \text{ mol}}{128.25 \text{ g}} = 18.95 \text{ mol}$$

The amount of heat given off by the mixture is the sum of the heat given off by both components:

$$5.86 \text{ mol} \left(\frac{-1368 \text{ kJ}}{\text{mol}} \right) + 18.95 \text{ mol} \left(\frac{-6160 \text{ kJ}}{\text{mol}} \right) = -124.750 \text{ kJ}$$

When 2.70×10^3 g of n-nonane are burned, the total amount of heat given off is:

$$2.70 \times 10^3 \text{ g} \times \frac{1 \text{ mol}}{128.25 \text{ g}} \times \frac{-6160 \text{ kJ}}{\text{mol}} = -129{,}700 \text{ kJ}$$

Expressing the difference as a percent:

$$\frac{-(129{,}700 - 124{,}750) \text{ kJ}}{-129{,}700 \text{ kJ}} \times 100\% = 4\%$$

The ethanol/gasoline mixture produces about 4% less energy than gasoline by itself.

THINK ABOUT IT The presence of oxygen in ethanol adds to its mass but does not add to its fuel value. It is logical that a blend of a hydrocarbon and an alcohol has a slightly lower-energy content than the hydrocarbon itself.

Practice Exercise A fuel called E-85 is a mixture of 85% ethanol and 15% gasoline by volume. How much energy in the form of heat is available from this mixture compared with the amount from pure gasoline? Assume the densities of ethanol and gasoline are 0.789 and 0.737 g/mL, respectively, and that n-nonane is an appropriate model hydrocarbon for gasoline.

The first internal combustion engines built in the 1870s burned pure ethanol, and early automobiles, including the Model T Ford, could be modified to run on either gasoline or pure ethanol. The use of alcohol as a gasoline additive resulted in a sharp increase in ethanol production toward the end of the 20th century, with an annual rate of production in the United States estimated as 6 billion gallons in 2007. However, several challenges limit the wide use of ethanol as an automobile fuel. Like methanol, ethanol has less fuel value than that of a comparable mass or volume of gasoline because of the presence of oxygen in its molecular structure. Furthermore, considerable energy and money are consumed in its production: growing and harvesting corn, converting corn starch into sugar, converting the sugar into alcohol, and finally distilling the alcohol from the fermentation mixture. It is estimated that more than two-thirds of the energy released in the combustion of ethanol derived from corn is consumed in its production. Ethanol produced in this way is more expensive than gasoline, even at today's prices. Fuels are extraordinarily complex in terms of their composition, their combustion, the emissions they produce, and the issue of renewability, and this brief discussion addresses a very small part of a challenging problem.

> An **ether** has the general formula R—O—R, where R is any alkyl group or aromatic ring. The two R fragments may be different.

Ethers: Diethyl Ether

Ethers have the general formula R—O—R, where R is any alkyl group or an aromatic ring. Just as with alcohols, you can think of ethers structurally as a combination of two organic groups (R and R', which may be the same or different) and water:

$$R—H \quad H—O—H \quad H—R' \quad R—O—R'$$

Because the C—O—C bond angle is not 180° but is close to the tetrahedral bond angle of 109.5°, the dipole moments of the two C—O bonds do not cancel each other and ethers are polar molecules. These structural features give rise to the properties of typical ethers: their water solubility is comparable to alcohols of similar molar mass, but their boiling points are about the same as alkanes of comparable molar mass (Table 12.7).

The most important ether industrially is diethyl ether, $CH_3CH_2OCH_2CH_3$. You may have heard of this, first, as the material simply called "ether" that has had wide use in medicine as an anesthetic since 1842. Although the actual mechanism of action of anesthetic agents in terms of dulling nerves and putting patients to sleep is still unknown, certain properties of diethyl ether play a role in determining its behavior as a medicinal agent. Because diethyl ether has a low boiling point of 35°C, it vaporizes easily, and a patient can inhale it. Because diethyl ether has a significant solubility in water, it is soluble in bodily fluids like blood, so once inhaled, it can be transported around the body easily. Its low polarity and short saturated hydrocarbon chains combine to make it soluble in cell membranes where it blocks stimuli coming into nerves. Ether has the unfortunate side effect of inducing nausea and headaches, and has been replaced by new anesthetics in modern hospitals, but for many years ether was the anesthetic of choice for surgical procedures.

TABLE 12.7 Functional Groups Affect Physical Properties

	Molar Mass (g/mol)	Normal Boiling Point (°C)	Solubility in Water (g/100 mL at 20°C)
$CH_3CH_2—O—CH_2CH_3$ 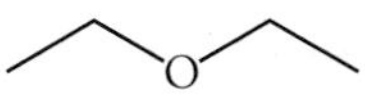 Diethyl ether	74	35	6.9
$CH_3CH_2CH_2CH_2CH_3$  n-Pentane	72	36	0.0038
$CH_3CH_2CH_2CH_2OH$ 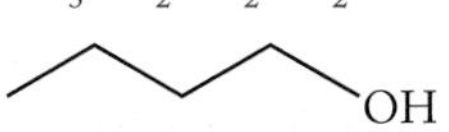 n-Butanol	74	117	7.7

A second common use of ether stems from another property that caused difficulty in the clinical setting. Diethyl ether is extremely flammable and has a low ignition point. This property is used to advantage when ether is sprayed in diesel engines to start them in cold weather when it is too cold for diesel fuel to ignite.

One member of the ether family widely used as gasoline additive in the 1990s was methyl *tert*-butyl ether (MTBE). (The term *"tert-"* in the name is an abbreviation for tertiary, referring to a carbon atom in the structure that is bonded to 3 other carbon atoms.) MTBE was added to gasoline to promote complete combustion.

Unlike the nonpolar hydrocarbons in gasoline, MTBE is soluble in water. Consequently, gasoline spills, leakage from storage tanks, and releases from watercraft can produce extensive MTBE contamination of groundwater and drinking water supplies and have done so since the early 1990s. After toxicity tests showed MTBE to be a possible carcinogen (cancer-causing agent), several states, including California, where more than 25% of the world's production of MTBE was used in gasoline, banned the use of MTBE as a gasoline additive. Most oil companies stopped adding MTBE to their gasolines in 2006. These changes raised the question of which additive(s) would replace MTBE. The leading candidate to date is ethanol.

MTBE

CONCEPT TEST

Predict the relative fuel values of diethyl ether, MTBE, methanol, and ethanol.

Polymers of Alcohols and Ethers

Poly(vinyl alcohol) (PVAL) is another polymer in the vinyl family. Over 400,000 tons of PVAL are produced annually in the United States. It is used by itself or in chemical and physical combinations with other polymers in fibers, adhesives, and in materials known as sizing that change the surface properties of textiles and paper to make them less porous, less able to absorb liquids, and smooth. PVAL is the material of choice for laboratory gloves that are resistant to organic solvents. Because its chains are studded with –OH groups, its surface is very polar and very water-like, and hydrocarbon solvents that are not soluble in water do not penetrate PVAL barriers. PVAL is also impenetrable to carbon dioxide, and this property has led to the use of PVAL in soda bottles. PVAL is blended with another polymer called PETE [poly(ethylene terephthalate)]. The two polymers do not mix and separate into layers (Figure 12.17). In soda containers the PETE makes the bottle strong, so it can bear pressure changes due to temperature change and survive the impact of falling off tables; however, CO_2, the gas dissolved in soda that makes it fizz, passes readily through PETE. The PVAL keeps the CO_2 in the container so the soda does not go flat. Polymers with different properties are frequently combined to create new materials with desired properties.

The monomer from which PVAL is made is not the one you would predict based on the discussion in Section 12.4. From the repeat unit shown, you might expect the molecule "vinyl alcohol" to be the monomer, but that molecule does not exist. Instead, vinyl acetate is polymerized to make poly(vinyl acetate) (PVAC), which is then reacted further to produce PVAL.

The blend of PVAL and PETE in early soda bottles was a physical mixture of the two polymers. New materials can also be made by combining differ-

Repeating unit in PVAL

Repeating unit in PETE

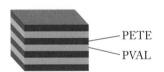

FIGURE 12.17 Layers of the polymers PVAL and PETE are used to make soda bottles.

"Vinyl alcohol"

Vinyl acetate

ent monomer units in one polymer molecule. This type of molecule is called a **heteropolymer**, and it is made from two or more different monomer units. An example of a heteropolymer is a material called EVAL, a **copolymer** of ethylene and vinyl alcohol. The EVAL polymers are used in food wrappings when preservation of aroma and flavor are required. Food usually deteriorates in the presence of oxygen, and packages made of EVAL provide an excellent barrier to the entry of oxygen while retaining the flavors and fragrances of food within the package.

If PVAC as a homopolymer could be symbolized as ⦋A–A–A–A–A⦌ and polyethylene as ⦋B–B–B–B–B–B⦌ then a copolymer of the two could be

$$⦋A–B–A–B–A–B–A–B–A–B⦌$$

in which case the resultant structure is called an *alternating copolymer*. Another possibility is

$$⦋A–A–A–A–B–B–B–B–A–A–A–A–B–B–B–B⦌$$

which is called a *block copolymer*. The block copolymers vary further by having blocks of different lengths. Finally, random copolymers are possible, too:

$$⦋A–A–B–A–B–B–A–B–A–A–A–A–B–B–A–B–B–B⦌$$

EVAL is a random copolymer of PE and PVAC. Other arrangements of the same monomer units provide materials with different properties and uses.

The most commercially important polyethers are poly(ethylene glycol) (PEG) and poly(ethylene oxide) (PEO). Both polymers consist of exactly the same subunit (Figure 12.18). PEG is a low-molar-mass liquid oligomer made from ethylene glycol, and PEO is a higher molar-mass polymer made from ethylene oxide. As a polyether, PEG has properties closely related to those of diethyl ether, in that it is soluble in both polar and nonpolar liquids. It is a common component in toothpaste because it associates both with water and with the water-insoluble materials in the paste and keeps the toothpaste uniform both in the tube and during use. PEGs of many lengths are finding increasing use as attachments to pharmaceutical agents to improve their solubility and biodistribution.

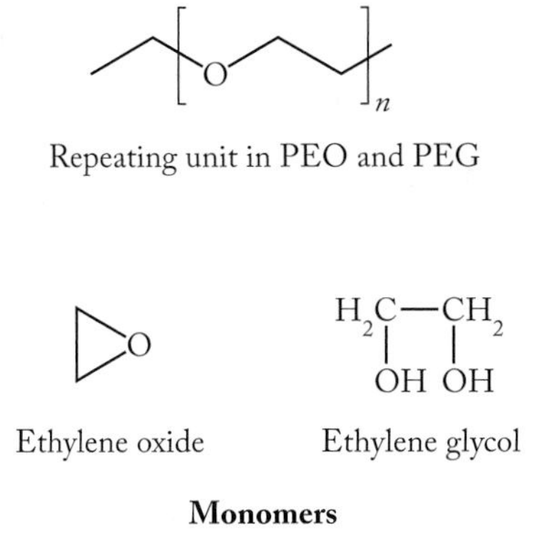

> A **heteropolymer** is made of two or more different monomer units. When the heteropolymer is made of only two different monomers, it is called a **copolymer**.

Repeating unit in PEO and PEG

Ethylene oxide   Ethylene glycol

Monomers

FIGURE 12.18 Poly(ethylene glycol) (PEG) and poly(ethylene oxide) (PEO) have the same repeating unit. The two polymers differ only in terms of molar mass. PEG is typically made from ethylene glycol, while ethylene oxide is the monomer of choice for making PEO.

SAMPLE EXERCISE 12.10 **Assessing Properties of Polymers**

The polymer poly(ethylene glycol) (PEG) is used to blend materials that are not soluble in each other. It is soluble in both water, a polar solvent, and benzene, (C_6H_6), a nonpolar solvent. Describe the structural features of PEG that make it soluble in liquids with very different polarities.

COLLECT AND ORGANIZE We are asked to describe the intermolecular forces that make PEG soluble in polar and nonpolar liquids.

ANALYZE In terms of solubility, we learned in Chapter 10 that like dissolves like. We need to evaluate the structure of PEG and see if one part is water-like and another benzene-like.

SOLVE The structure of PEG consists of $-CH_2CH_2-$ groups connected by oxygen atoms. The oxygen atoms are capable of hydrogen bonding with water molecules, so the attractive force between PEG and water is due to hydrogen bonding. Benzene is nonpolar and is attracted to the $-CH_2CH_2-$ groups in the polymer.

Nonpolar groups interact via London dispersion forces, so those forces must be responsible for the solubility of PEG in nonpolar benzene.

THINK ABOUT IT The ability of PEG to interact with both polar and nonpolar groups gives rise to its use in stabilizing blends of substances that would normally not mix well together.

Practice Exercise PEG is reportedly added to some soft drinks as an anti-foaming agent to keep CO_2, responsible for the fizz in sodas, in solution longer when the soda is poured. What intermolecular attractive forces between PEG and CO_2 might make this use possible?

Different properties arise from different chain lengths, different compositions of chains, and different arrangements of monomer units within chains. Again, the possibilities are virtually endless. From just these few examples, you can perhaps appreciate why synthetic polymers pervade modern life. An understanding of the interplay of composition, structure, and properties enables scientists to design totally new materials with specific properties. The properties may not come out exactly as planned the first time, but chemists can usually pick a good starting point based on knowledge of the behavior of materials that have already been studied. From the starting materials provided by fossil fuels, scientists learn how to make materials to substitute for almost all products that used to be produced from animals and plants and to vastly extend the range of materials available for use in normal daily life as well as in extreme environments.

12.7 More Oxygen-Containing Functional Groups: Aldehydes, Ketones, Carboxylic Acids, Esters, Amides

All five of these functional groups—aldehydes, ketones, carboxylic acids, esters, and amides—contain a subunit called the **carbonyl group**: a carbon atom with a double bond to an oxygen atom.

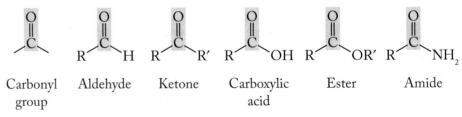

| Carbonyl group | Aldehyde | Ketone | Carboxylic acid | Ester | Amide |

The R groups may be any organic group. Aldehydes and ketones are collectively referred to as *carbonyl compounds*, because the carbonyl group determines their chemistry. Carboxylic acids, as their name implies, are acidic in nature, and their chemistry is determined by the –COOH subunit, referred to as a carboxylic acid group. Esters and amides can be made from carboxylic acids by allowing them to react with alcohols and amines (see Section 12.8).

Aldehydes and Ketones

Aldehydes and ketones closely resemble each other in both chemical and physical properties. An **aldehyde** contains a carbonyl group bound to one R group and one hydrogen atom; its general formula is RCHO or R(C=O)H. A **ketone** contains a carbonyl group bound to two R groups; its general formula is RCOR′ [or

CONNECTION We first saw carboxylic acids in our discussion of acids and bases in Chapter 2.

The **carbonyl group** is a carbon atom with a double bond to an oxygen atom.

An **aldehyde** contains a carbonyl group bonded to one R group and one hydrogen; its general formula is RCHO.

A **ketone** contains a carbonyl group bonded to two R groups; its general formula is RCOR′.

R(C=O)R']. The R groups may be the same or different in a ketone. The carbonyl group contains a double bond, and accounts for the reactivity of aldehydes and ketones. It is different than the double bond in an alkene, however, because it is polar (Figure 12.19). The electronegative oxygen pulls electron density toward itself, and as a consequence the chemistry of aldehydes and ketones is linked to the presence of a δ^+ carbon atom and a δ^- oxygen atom. Other polar species tend to react with carbonyls by having a negative, electron-rich region that approaches the δ^+ carbon of the carbonyl.

In terms of physical properties, aldehydes and ketones are polar, and they tend to parallel the ethers with respect to water solubility. As with ethers, they cannot hydrogen-bond with themselves because they contain only carbon-bonded hydrogen atoms, so they have lower boiling points than alcohols of comparable molar mass.

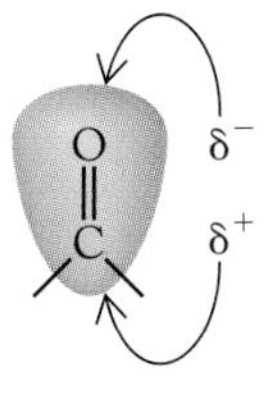

Because of the double bond between the carbon and the oxygen, the aldehydes and ketones are at a higher oxidation state than alcohols, and indeed many of the smaller aldehydes and ketones are made by oxidizing alcohols of the same carbon number. Aldehydes and ketones do not polymerize through their carbonyl groups. Many polymers have carbonyl functional groups as part of their structure, but these groups themselves do not react to form long chains.

In an **ester**, the –OH of a carboxylic acid group is replaced by –OR, where R can be any organic group.

In an **amide**, the –OH of a carboxylic acid group is replaced by an –NH$_2$, –NHR, or –NR$_2$, where R can be any organic group.

FIGURE 12.19 The electron distribution in a carbonyl group is skewed toward the oxygen end of the bond because oxygen is more electronegative than carbon.

Carboxylic Acids, Esters, and Amides

Carboxylic acids are organic compounds that are proton donors. The R group in the general structure may be any organic subunit. The –OH group attached to the δ^+ carbon of the carbonyl is polarized, which explains two characteristics of carboxylic acids. First, the hydrogen atom on the –OH group of one molecule can hydrogen-bond to a neighboring carboxylic acid, either at the oxygen atom in the –OH group or the carbonyl oxygen atom. This association results in high boiling points compared to other organic compounds of comparable molar mass.

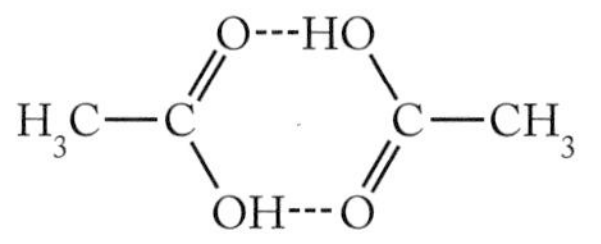

Second, the process of donating a proton results in a negatively charged oxygen whose electron density is delocalized over the carbonyl group. This delocalization contributes to the stability of the carboxylate anion. The common carboxylic acids are weak acids, which means that they are present in aqueous solutions as mostly neutral molecules, a small fraction of which are ionized, donating H$^+$ ions to molecules of water, as shown here for acetic acid.

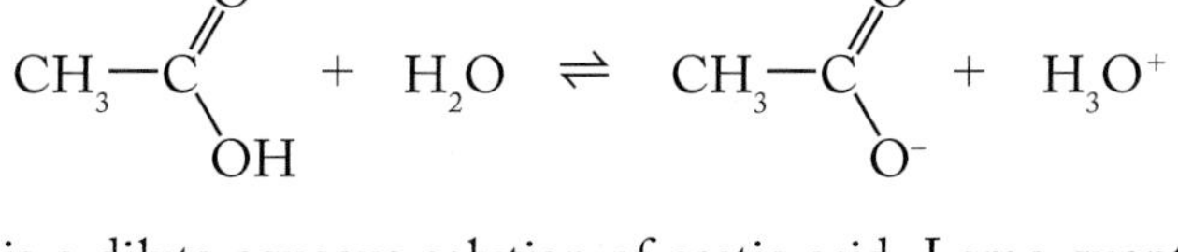

Vinegar is a dilute aqueous solution of acetic acid. Large quantities of vinegar are produced commercially by the air oxidation of ethanol in the presence of enzymes from *Acetobacter* bacteria.

A number of chemical families are closely related to the carboxylic acids. We consider only two of them here: esters and amides. In **esters**, the –OH of the carboxylic acid is replaced by –OR, where R can be any organic group. In **amides**, the –OH is replaced by an –NH$_2$, –NHR, or –NR$_2$. The presence of the carbonyl group makes esters polar, and their boiling points are comparable to those of aldehydes and ketones of similar size. Amides are also polar and capable of intermolecular hydrogen bonding. The hydrogen atoms on the –NH$_2$ group can hydrogen-bond with the oxygen in the carbonyl of an adjacent molecule. This causes their boiling points to be considerably higher than those of esters of comparable size.

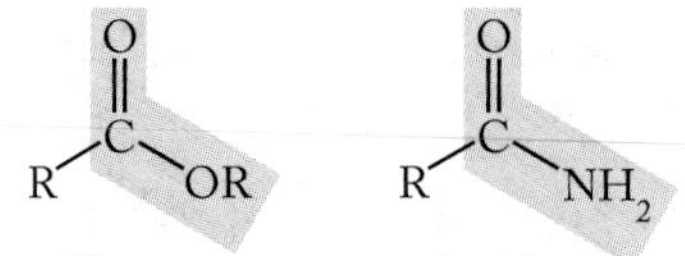

In a **condensation reaction** two molecules combine to form a larger molecule and a small molecule (typically water).

Esters frequently have very pleasant fragrances, totally different from the acids from which they are derived. For example, butyric acid, with a straight chain of 5 carbon atoms, is responsible for the odor of rancid butter. The ethyl ester of butyric acid (ethyl butyrate) is prepared by the *esterification* reaction of the acid with ethanol as shown below. It is the molecule responsible for the aroma of ripe pineapples. Esters are heavily used in the personal products industry to provide pleasant scents for products like shampoos and soaps.

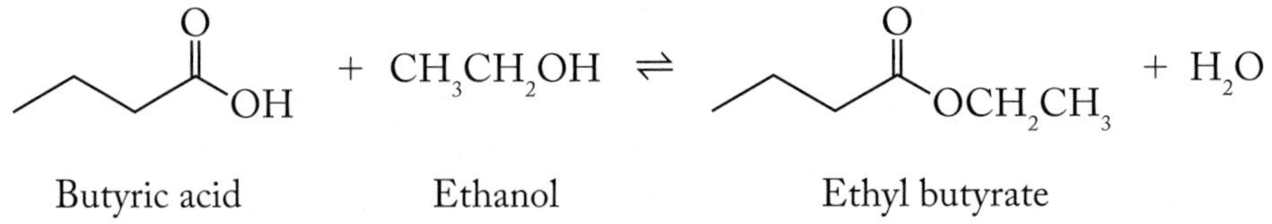

▶❙❙ **CHEMTOUR** Polymers

Butyric acid   Ethanol   Ethyl butyrate

The esterification reaction is an example of a **condensation reaction**. In a condensation reaction, two molecules combine to create a larger molecule while a small molecule (typically water) is also formed in the process.

Amides are made from carboxylic acids by several methods, but the net result is a condensation reaction in which the −OH of the carboxylic acid is replaced by the −NH_2 group of the amide. Water is also formed in the process. From acetic acid, one can prepare acetamide:

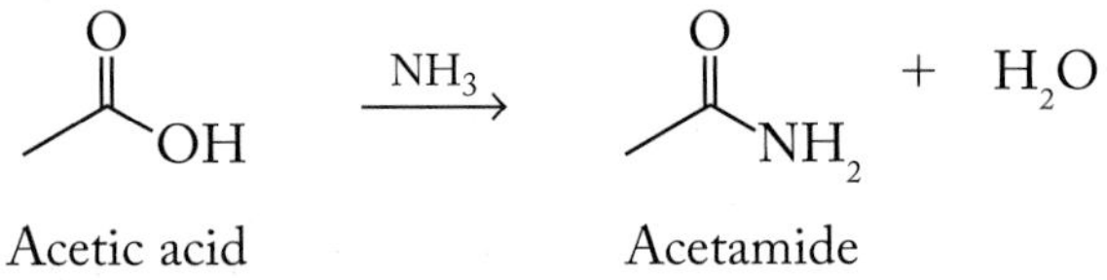

Acetic acid   Acetamide

Polyesters and Polyamides

Prior to this point, all of the compounds we examined have been monofunctional; they all had only one functional group that identifies their family. With the polymers of carboxylic acids and their derivatives, we enter the world of *difunctional* molecules, which are molecules with two functional groups. The key point to remember is that the functional groups for the most part still retain their own distinct chemical reactivity even if they are in a molecule with another functional group. Also remember that the same features we enumerated for all other polymers still apply here: for polymers, function is determined by composition, structure, and size.

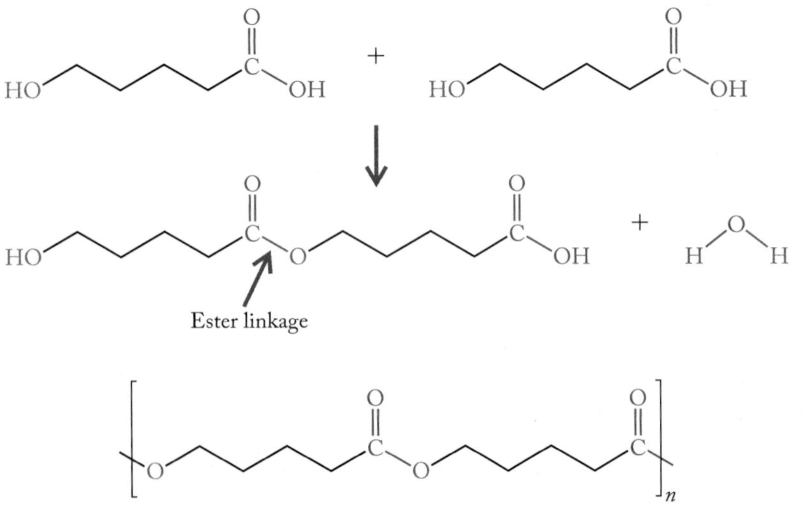

Look back at the esterification reaction in which ethyl butyrate is formed from butyric acid and ethanol. The −COOH group on the acid reacts with the −OH group of the alcohol to form a new carbon–oxygen single bond and release a molecule of water. Think about what could happen at the molecular level if we had a single compound that contained a carboxylic acid functional group at one end and an alcohol functional group at the other (Figure 12.20). The carboxylic acid group of one molecule could react with the alcohol group on another in a condensation reaction to generate a larger molecule that still has a carboxylic acid group at one end,

FIGURE 12.20 Synthesis of a polyester from identical difunctional molecules, each one containing an alcohol and a carboxylic acid functional group. Water is also a product of the reaction.

an alcohol at the other, and an ester linkage in between. If this reaction happens repeatedly, a monomer containing one carboxylic acid and one hydroxy group (a hydroxy acid) polymerizes to form a polyester, as shown in Figure 12.20.

SAMPLE EXERCISE 12.11   **Making a Polymer**

Show how a polyester can be synthesized from this difunctional alcohol:

$$HOCH_2CH_2CH_2OH$$

and this difunctional carboxylic acid:

$$HOOCCH_2CH_2CH_2COOH$$

COLLECT AND ORGANIZE We are asked to show how a polyester, a polymer with a repeating unit that is an ester, can be made. We have an alcohol and a carboxylic acid to react to make the ester monomer.

ANALYZE Because the starting materials we have are difunctional, the alcohol can react with two molecules of carboxylic acid, and the acid can react with two molecules of alcohol.

SOLVE Reacting one molecule of the alcohol with one molecule of the carboxylic acid yields:

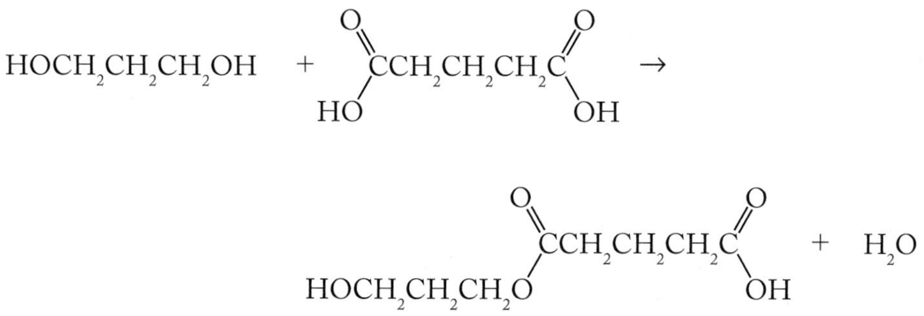

This molecule has an alcohol group on one end that can react with another molecule of carboxylic acid and a carboxylic acid group on the other end that can react with another molecule of alcohol. Continuing these reactions results in the formation of a polymer whose repeating unit is

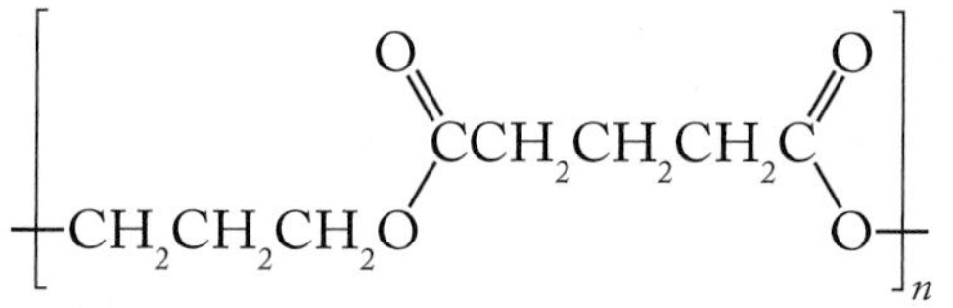

THINK ABOUT IT The repeating unit in the polyester contains one section that came from the alcohol and a second section that came from the carboxylic acid.

Practice Exercise It is possible for one molecule to contain two different functional groups. The molecule shown has both an alcohol functional group and a carboxylic acid functional group. Draw the repeating unit of the polyester made from this molecule:

Many synthetic fabrics are polyesters. Dacron is one example, and its repeat unit is shown in Sample Exercise 12.12. The structure of the repeating unit in Dacron reveals that it must be made from a difunctional alcohol and a difunctional carboxylic acid.

SAMPLE EXERCISE 12.12   **Comparing Properties of Polymers**

Clothes made from Dacron can be more uncomfortable in hot weather than clothes made of cotton because they do not absorb perspiration as effectively as the natural polymeric fiber. The repeating units in cotton and Dacron are shown below:

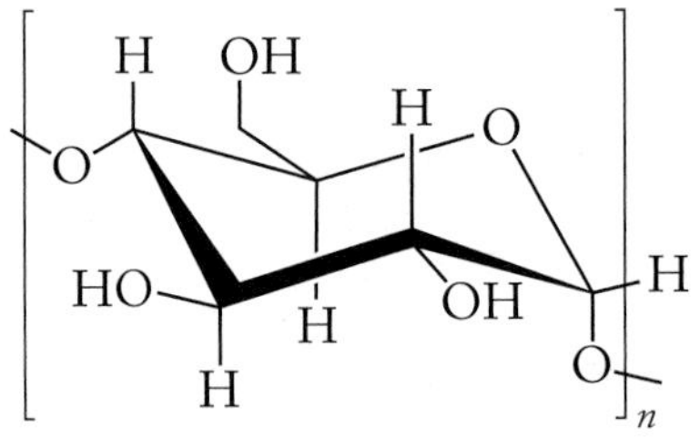

Repeating unit in cotton: $n \sim 10,000$ 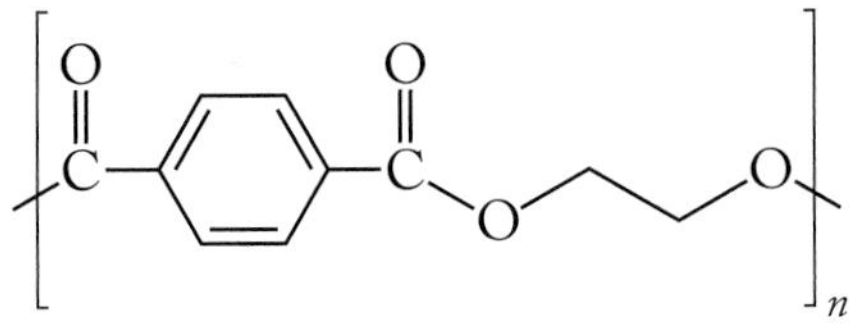   Repeating unit in Dacron

Suggest a structural reason why cotton absorbs perspiration (water) better than Dacron.

COLLECT AND ORGANIZE We are asked to explain why one polymer absorbs water better than another. We are given the repeating units of both polymers.

ANALYZE Water is a polar molecule and is attracted to polar groups. We need to compare the groups in each monomer to see which is more water-like.

SOLVE Each glucose monomer in cotton has three –OH groups attached to it. Each of those groups is polar and very water-like. The monomer in Dacron has oxygen atoms in it, but no –OH groups. Although regions in the Dacron monomer are polar, they are not nearly as polar as the –OH groups in glucose, so the presence of the –OH groups in the cotton monomer that are capable of hydrogen bonding to water molecules makes cotton much more attractive than Dacron to water molecules.

THINK ABOUT IT The principle of like interacting with like works for polymers just as it does for small molecules.

Practice Exercise Gloves made of a woven blend of cotton and polyester fibers protect the hands from exposure to oil and grease but are comfortable to wear because they "breathe"—they allow perspiration to evaporate and pass through them, thereby cooling the skin. Suggest how these gloves work at the molecular level.

The technique of allowing difunctional molecules to react to make polyesters can be used to make another class of very useful materials called polyamides. Polymers can be formed in condensation reactions from identical monomer units, each of which contains one carboxylic acid group and one amine ($-NH_2$) group (Figure 12.21). (The amine family is discussed in Section 12.8.) Water is the other product of the reaction.

Just as we saw with polyesters, it is also possible to form a polyamide by reacting one monomer that contains two carboxylic acid groups (a dicarboxylic acid) with a second monomer containing two amine groups (a diamine). Probably the most famous polyamide is nylon-6,6, a polymer made from the reaction of a straight-chain dicarboxylic acid containing 6 carbon atoms (adipic acid) and a straight-chain diamine molecule containing 6 carbons atoms (hexamethylenediamine) (Figure 12.22). The polymer is called nylon-6,6 to indicate the number of carbon atoms in each monomer.

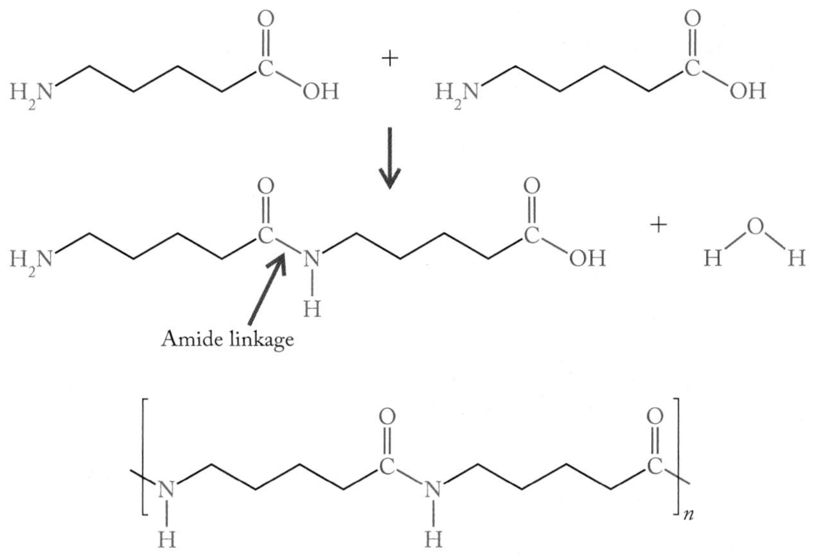

FIGURE 12.21 Synthesis of a polyamide from identical molecules, each containing a carboxylic acid functional group and an amine functional group.

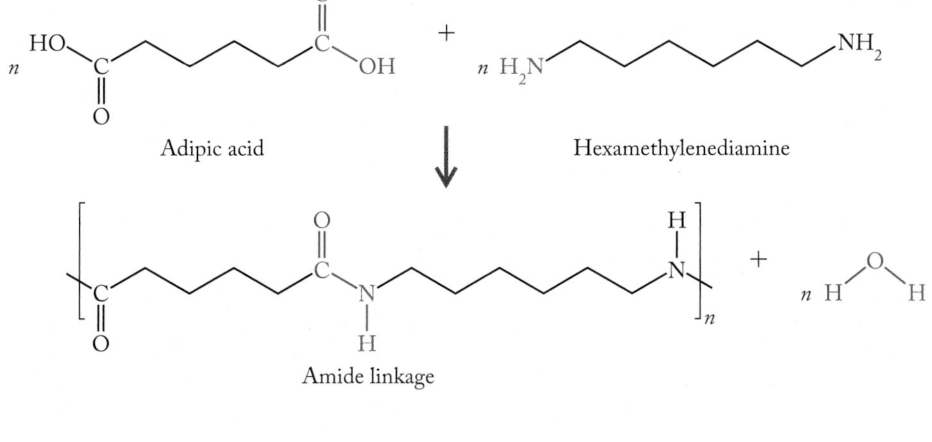

FIGURE 12.22 Synthesis of the polyamide nylon-6,6 from adipic acid (a dicarboxylic acid) and hexamethylenediamine (a diamine). An amide bond links the monomers.

SAMPLE EXERCISE 12.13 **Identifying Monomers**

Another form of nylon is simply named nylon-6, indicating it is made from the reaction of one monomer with itself. By analogy with the polyesters in Sample Exercise 12.11 and the accompanying Practice Exercise, suggest the structure of a six-carbon-atom-long molecule that could polymerize to make nylon-6.

COLLECT AND ORGANIZE We are asked to suggest the structure for a monomer that could react with itself to form a polyamide with a repeating unit six carbon atoms long. The exercise refers us to an example with polyesters to use as an analogy.

ANALYZE By analogy to Sample Exercise 12.11 and its accompanying Practice Exercise, we should suggest a difunctional molecule that has a carboxylic acid (–COOH) on one end and an amine (–NH$_2$) on the other, because they are the two functional groups that react to form the amide linkage in a polyamide.

SOLVE We can build the structure of the molecule required by starting with one of the functional groups. It doesn't matter which one, so let's begin with the amine:

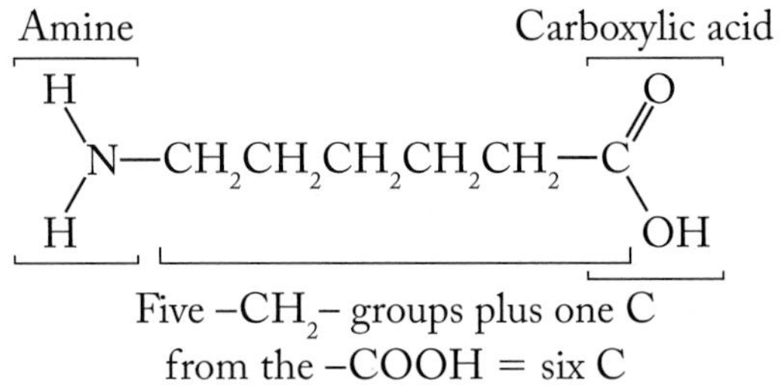

Amine   Carboxylic acid

$$\underset{\text{Five} -CH_2- \text{ groups plus one C}}{\underset{\text{from the} -COOH = \text{six C}}{ \overset{H}{\underset{H}{N}} - CH_2CH_2CH_2CH_2CH_2 - \overset{O}{C}_{OH} }}$$

We then add a chain of five –CH$_2$– units because the name nylon-6 indicates that 6 carbon atoms separate the ends of the repeat unit. Finally, we add the carboxylic acid functional group as the second functional group and the sixth carbon atom in the chain.

THINK ABOUT IT Many nylons with different properties can be made by varying the length of the carbon chain in a molecule like the one in this exercise, or by varying the lengths of chains in both the difunctional acid and difunctional amine in Figure 12.22.

Practice Exercise Draw the structures of two monomers that could react with each other to make nylon-5,4. Draw the structure of the repeating unit in the polymer. (Note: The first number refers to the acid monomer; the second refers to the amine monomer.)

The polymer nylon makes long, straight fibers that are quite strong and excellent for fabrics. A special variety of nylon called Kevlar, invented by Stephanie Kwolek of Dupont in 1965, is so strong that it is used in bullet-proof vests, puncture-resistant tires, and NHL goalies' face masks. The extraordinary combination of strength and flexibility in Kevlar arises directly from its molecular structure.

Nylon is flexible and stretchable because the hydrocarbon chains can bend and curl around much like a telephone cord or a slinky. To produce a stronger nylon, researchers recognized they had to find some way to reduce the ability of the chains to form coils. They discovered this could be done using monomer units containing functional groups that made it difficult for the chains to bend. The result of this work was Kevlar, an *aramide* fiber, which is a special class of polyamide formed from aromatic carboxylic acids and amines. The name of this class of polymers

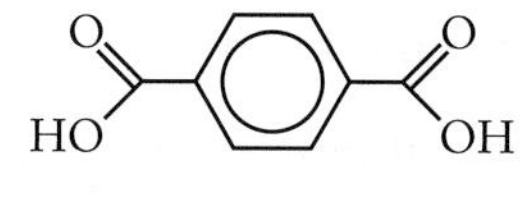

Benzene-1,4-dicarboxylic acid
(terephthalic acid)

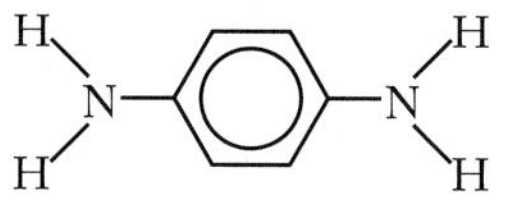

1,4-diaminobenzene

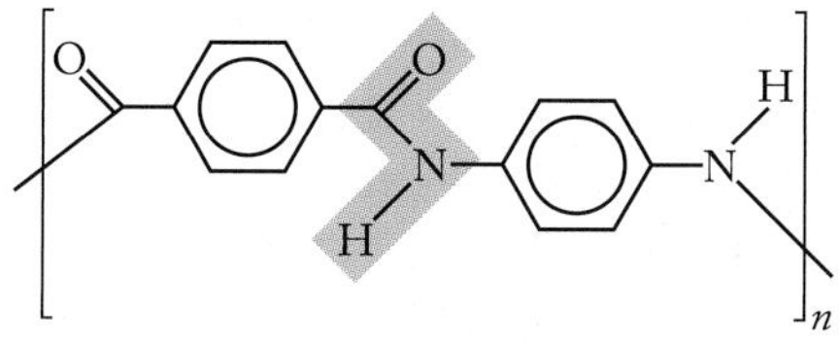

Repeating unit in Kevlar

FIGURE 12.23 The monomers used to make Kevlar are a dicarboxylic acid and a diamine. The amide bond in the repeating unit is highlighted.

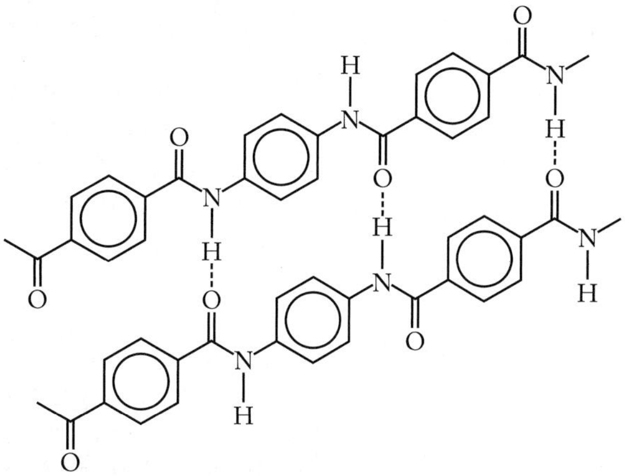

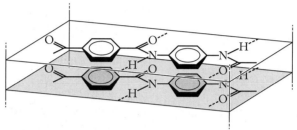

Hydrogen bonding between chains in the same plane

Stacking of benzene rings between chains in layered planes

FIGURE 12.24 Interactions between groups in Kevlar.

indicates that they are made from an *ar*omatic *amide*. The monomers in Kevlar are a dicarboxylic acid of benzene and a diamine of benzene (Figure 12.23). When these two monomers polymerize, the flat and rigid aromatic rings keep the chains straight. As a result, Kevlar forms fibers that are long and straight. In addition, two specific associations are possible that make the chains very regular in their arrangement and hold them together very tightly (Figure 12.24). First, the hydrogen atoms on the −NH of the amide form a hydrogen bond with the oxygen atoms of the carbonyls on an adjacent chain. Second, the rings stack as shown on top of each other (just as they do in polystyrene) and provide additional interactions that hold the chains together in parallel arrays. The result is a fiber that is very strong but still flexible. Fabrics and helmets made of Kevlar resist puncture, even by bullets and hockey pucks fired at them, and are also resistant to flames and reactive chemicals (Figure 12.25).

FIGURE 12.25 Illustration of the strength of Kevlar: a bullet fired point-blank at a sheet of Kevlar does not puncture the fabric.

12.8 Hydrocarbon Fuels from Biomass

Although biomass based on carbohydrates has less fuel value than hydrocarbon fuels, biological processes can convert biomass into the hydrocarbon with the highest fuel value, methane. The digestive systems of many animals, including cows, produce methane through bacterial action. The conversion of biomass into methane is an attractive future source of hydrocarbons, provided the complexities of bacterial action can be adapted for large-scale production. Thus, converting organic matter into hydrocarbon fuel may be possible without waiting millennia for the anaerobic processes deep within the earth to do so.

Methane from Plants

The production of methane from plant residues that are mostly cellulose requires the sequential action of several types of bacteria. In the first stage, hydrolytic and transitional bacteria break up cellulose into mixtures of small molecules including hydrogen, carbon dioxide, and, depending on the bacterial strain, alcohols (methanol, ethanol, and others) or two carboxylic acids: formic and acetic acids.

In the later stages of methane production, the products of the first stage undergo several reactions promoted by the metabolism of **methanogenic bacteria** that rely on hydrogen and simple organic compounds for energy. These bacteria are quite different from most organisms on Earth that rely on oxygen and more complex organic compounds for energy. Acetic acid is the reactant in reaction 2; formic acid is the reactant in reaction 3. Methane is produced in all four of these reactions:

$$(1) \ 4 \, H_2(g) + CO_2(g) \rightarrow CH_4(g) + 2 \, H_2O(\ell)$$

$$(2) \ CH_3COOH(aq) \rightarrow CH_4(g) + CO_2(g)$$

$$(3) \ 4 \, HCOOH(aq) \rightarrow CH_4(g) + 3 \, CO_2(g) + 2 \, H_2O(\ell)$$

$$(4) \ 4 \, CH_3OH(aq) \rightarrow 3 \, CH_4(g) + CO_2(g) + 2 \, H_2O(\ell)$$

Significant quantities of $CO_2(g)$ are also formed in reactions 2, 3, and 4 that use starting materials other than hydrogen gas. Notice that the carbon atoms in acetic acid, formic acid, and methanol are being both oxidized and reduced in these reactions: the products are the most reduced form of carbon (CH_4) and the most oxidized form (CO_2). The hydrogen and carbon in the starting materials are converted into a hydrocarbon as well as into water and carbon dioxide.

Amines

Some strains of methanogenic bacteria produce methane from a class of nitrogen-containing organic compounds called **amines**. We mentioned amines in Section 12.7 when we discussed polyamides such as nylon and Kevlar. Amines as a family are responsible for much of the foul odor of decaying biological tissue. The functional group in amines is a nitrogen atom that forms three bonds. Amines may be viewed as derivatives of ammonia, NH_3. If one hydrogen atom in ammonia is replaced by an R group, the compound is called a *primary amine*. If 2 hydrogen atoms are replaced by R groups, the compound is a *secondary amine*; if all 3 hydrogen atoms are replaced by R groups, a *tertiary amine*. The R groups may be the same or they may be different organic subunits.

Formic acid   Acetic acid

Methanogenic bacteria rely on simple organic compounds and hydrogen for energy. Their respiration produces methane, carbon dioxide, and water, depending on the compounds they consume.

Amines are organic compounds that show appreciable basicity. Their general formula is RNH_2, R_2NH, or R_3N, where R is any organic subgroup.

$$\underset{\text{Primary amine}}{RNH_2} \qquad \underset{\text{Secondary amine}}{R_2NH} \qquad \underset{\text{Tertiary amine}}{R_3N}$$

Amines are organic bases. Their basicity is their defining chemical characteristic. They all react with water to some extent to produce hydroxide ions and protonated cations, just like ammonia:

$$NH_3(aq) + H_2O(\ell) \rightleftharpoons NH_4^+(aq) + OH^-(aq)$$

$$RNH_2(aq) + H_2O(\ell) \rightleftharpoons RNH_3^+(aq) + OH^-(aq)$$

Amines of all three types are polar and form hydrogen bonds with water. Primary and secondary amines can hydrogen-bond with themselves; tertiary amines cannot because their nitrogen atom bears no hydrogen atoms. A particularly important type of amine is one in which the nitrogen atom is part of a ring. We deal with these in greater detail in the biochemistry chapter, but the key point is that, whether in a ring or a straight chain, an amine retains its most significant characteristic—its basicity.

Biochemical processes in the cells of bacteria of the *Methanosarcina* genus convert primary, secondary, and tertiary methylamines to methane, carbon dioxide, and ammonia:

$$4\,CH_3NH_2(aq) + 2\,H_2O(\ell) \rightarrow 3\,CH_4(g) + CO_2(g) + 4\,NH_3(aq)$$

$$2\,(CH_3)_2NH(aq) + 2\,H_2O(\ell) \rightarrow 3\,CH_4(g) + CO_2(g) + 2\,NH_3(aq)$$

$$4\,(CH_3)_3N(aq) + 6\,H_2O(\ell) \rightarrow 9\,CH_4(g) + 3\,CO_2(g) + 4\,NH_3(aq)$$

These reactions describe another pathway by which methane can be produced from the decay of biomass. Amines are much less plentiful than cellulosic material in plants, however, and the industrial development of these processes for fuel production has been slow because fossil fuels are still plentiful enough and not expensive enough to make such development cost-effective. Current research on the biochemistry of methanogenic bacteria, coupled with the power of genetic engineering, may enable the conversion of a broad range of biological materials into CH_4 and other high-energy fuels at reasonable costs. Current approaches to deriving larger amounts of fuel from biological sources involve the production from vegetable oils of fuel known as biodiesel.

A Final Note

In this chapter, we introduced the major functional groups in organic chemistry and examined a few reactions of those groups that give rise to several types of polymeric materials common in the modern world. This treatment can give only a brief taste of the usefulness and importance of organic compounds. Chemical Abstracts Service (CAS), an organization that tracks and collects all chemical information published worldwide, now lists over 28.1 million chemical substances in its registry. Over 26 million of these substances contain carbon. In addition, the simple fact that 0.82 million (over 97%) of the 0.84 million new compounds registered with CAS in 2005 contain carbon gives you some idea of why the study of organic chemistry occupies a special place within the discipline.[1]

---

[1] Personal communication: CAS Customer Care, CIC #96072, May 22, 2005.

Summary

SECTION 12.1 **Organic chemistry** encompasses the study of all carbon compounds, classified on the basis of **functional groups**—subunits of structure that confer upon molecules specific and typical chemical and physical properties. Organic compounds are also differentiated based on size. **Polymers** or macromolecules have molar masses from several thousand to over 1,000,000 g/mol). **Oligomers** are not quite polymers but do have multiple subunits in their molecular structures.

SECTION 12.2 The carbon atoms in **alkanes** (or **saturated hydrocarbons**) bear the maximum number of hydrogen atoms that is possible and contain carbon–carbon single bonds. A **homologous series** of alkanes is generated by sequential addition of $-CH_2-$ units (**methylene groups**) into the chain terminated by a $-CH_3$ (**methyl group**). **Straight-chain alkanes** may have **structural isomers**, hydrocarbons with the same formula but having different arrangements of C–C bonds (**branched-chain hydrocarbons**) and physical properties. **Cycloalkanes** are alkanes containing rings of carbon atoms. **Alkenes** and **alkynes** are **unsaturated hydrocarbons**, because their respective double and triple bonds can be **hydrogenated** to incorporate more hydrogen into their molecular structures.

SECTION 12.3 **Crude oil** is composed primarily of hydrocarbons that can be separated by **fractional distillation** into gasoline and other useful products. The vapor pressure of a pure substance as a function of temperature is determined by the **Clausius–Clapeyron equation**. The vapor pressure of a solution of volatile compounds follows Raoult's law.

SECTION 12.4 Alkenes are found as structural isomers as well as **geometric isomers**: **E** or *trans* isomers and **Z** or *cis* isomers depending on the arrangement of the groups around the double bond. Alkenes can also be polymerized to **homopolymers** used in construction, in fabrics, as wrapping and packaging material, and in medical devices.

SECTION 12.5 Aromatic hydrocarbons are characterized by planar rings in which sp^2-hybridized carbon atoms are joined by a combination of σ and π bonds. The π bond electrons are delocalized over all the carbon atoms in the ring. Compounds with two or more aromatic rings belong to a class of compounds called polycyclic aromatic hydrocarbons. The stability of aromatic systems because of resonance changes their chemical reactivity compared to alkenes.

SECTION 12.6 The **alcohol** (R—OH) and **ether** (R—O—R), functional groups (where R is an alkyl group or an aromatic ring) represent two ways of incorporating the **heteroatom** oxygen into organic compounds. Different monomer units can be chemically combined to make **heteropolymers** or **copolymers** whose properties depend on the arrangement of the monomer units in the polymer.

SECTION 12.7 Organic compounds may also contain carbon–oxygen double bonds in a subunit called the carbonyl group. The carbonyl group is found in **aldehydes** (RCHO), **ketones** (RCOR'), carboxylic acids (RCOOH), **esters** (RCOOR') and **amides** ($RCONH_2$) where R represents an alkyl or aromatic group. The chemical reactivity of aldehydes and ketones centers on the C=O bond. In carboxylic acids, the COOH group imparts acidic properties to the molecules. Carboxylic acids react with alcohols to form esters and with ammonia or amines to form amides in **condensation reactions**. Condensation reactions are used to prepare polymeric esters (polyesters) and amides (polyamides) from difunctional compounds for use in fabrics under the familiar names Dacron and nylon.

SECTION 12.8 Hydrocarbons are produced from biomass through the action of **methanogenic bacteria** on a variety of substrates including CO_2 and H_2, alcohols, carboxylic acids, and **amines**. Amines are organic compounds with appreciable basicity and the general formula RNH_2, R_2NH or R_3N, where R is any organic subgroup.

Problem-Solving Summary

TYPE OF PROBLEM	CONCEPTS AND EQUATIONS	SAMPLE EXERCISES
Distinguishing between alkanes, alkenes and alkynes	Alkanes are saturated hydrocarbons and do not react with H_2. Alkenes contain at least one C=C double bond that can combine with a molecule of H_2. Alkynes contain at least one C≡C triple bond that can combine with two molecules of H_2.	12.1
Drawing structures of organic compounds using carbon-skeleton structures	Use lines to depict single covalent bonds between atoms other than hydrogen. Carbon atoms are omitted but other atoms (e.g., O, N) are shown. A sufficient number of H atoms to complete the valency of the atom is assumed.	12.2
Recognizing structural isomers	Establish that the compounds have the same formula and look for different arrangements of C–C bonds.	12.3

TYPE OF PROBLEM	CONCEPTS AND EQUATIONS	SAMPLE EXERCISES
Calculating vapor pressure at any temperature	Substitute the appropriate values of P_{T_1}, T_1, T_2, ΔH_{vap}, and R into the Clausius–Clapeyron equation: $$\ln\left(\frac{P_{vap,T_1}}{P_{vap,T_2}}\right) = \frac{\Delta H_{vap}}{R}\left(\frac{1}{T_2} - \frac{1}{T_1}\right)$$ and solve for the vapor pressure, P_{T_2}.	12.4
Interpreting fractional distillation data	Draw an idealized graph showing distillation temperature versus volume of distillate.	12.5
Calculating the vapor pressure of a solution of volatile substances	Determine the mole fractions and vapor pressures of each component of the solution. Use Raoult's law to calculate the vapor pressure of the solution: $P_{total} = X_1 P_1^\circ + X_2 P_2^\circ + X_3 P_3^\circ + \cdots$	12.6
Identifying and naming geometric isomers	Molecules with two groups on the same side of a C=C bond are *cis*; those with groups on opposite sides are *trans* isomers. 	12.7
Identifying monomers in polymers	Find the smallest portion of the polymer that is repeated.	12.8, 12.13
Comparing the energy content of fuel mixtures	Determine the number of moles of each component. Multiply the number of moles by the corresponding value of ΔH°_{comb}. The total amount of energy available equals the sum of the heats available from each component.	12.9
Assessing properties of polymers	Identify the functional groups in the polymer. Evaluate the polarity of the functional groups and assess the relative importance of dipole–dipole forces, van der Waals forces, and hydrogen bonds.	12.10, 12.12
Making a polymer	For a polyester, combine monomers with alcohol functional groups (ROH) and carboxylic acid functional groups (R′COOH) to form water and ester groups (R′COOR).	12.11

Visual Problems

12.1. Which fractional distillation cut of crude oil is most likely to contain the hydrocarbons shown in Figure P12.1?

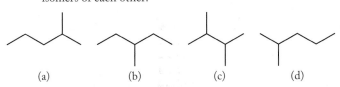

(a)   (b)   (c)   (d)

FIGURE P12.1

12.2. Which of the hydrocarbons in Figure P12.2 are structural isomers of each other?

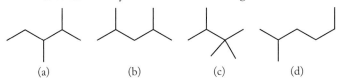

(a)   (b)   (c)   (d)

FIGURE P12.2

12.3. In Figure P12.3 are the line structures of four organic compounds found in nature as fragrant oils. Which are alkenes?

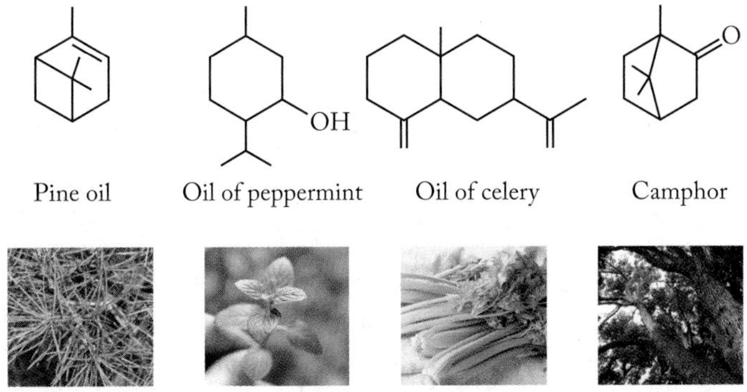

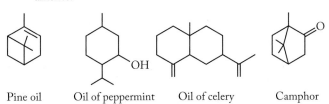

Pine oil   Oil of peppermint   Oil of celery   Camphor

FIGURE P12.3

12.4. Figure P12.4 shows three molecules: acrylonitrile (found in barbeque smoke), capillin (an antifungal drug), and pargyline (an antihypertensive drug). Which of these molecules does not contain the alkyne functional group?

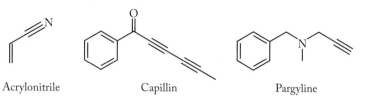

Acrylonitrile   Capillin   Pargyline

FIGURE P12.4

12.5. Which molecules in Figure P12.5 are considered to be aromatic compounds?

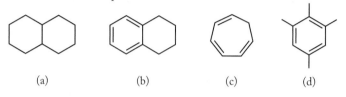

(a)   (b)   (c)   (d)

FIGURE P12.5

12.6. Carvone, benzyl acetate, and cinnamaldehyde are all naturally occurring oils. Their line structures are shown in Figure P12.6. Which ones contain an aromatic ring?

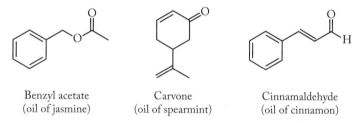

Benzyl acetate (oil of jasmine)   Carvone (oil of spearmint)   Cinnamaldehyde (oil of cinnamon)

FIGURE P12.6

12.7. In addition to the aromatic ring, what other functional groups can you identify in the molecules in Problem 12.6?

12.8. Which of the drawings in Figure P12.8 best describes the distribution of a 1:1 mixture of C_7H_{16} and C_8H_{18} (by volume) in the vapor phase above the solution at the boiling point of the mixture? The blue spheres represent the molecules of C_7H_{16} and the red spheres represent the molecules of C_8H_{18}.

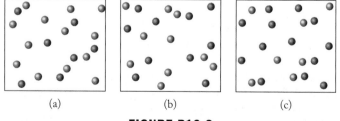

(a)   (b)   (c)

FIGURE P12.8

12.9. The graphs in Figure P12.9 illustrate the vapor pressure of a solution of two liquids, A and B, as a function of mole fraction of A and B. Describe the conditions under which one would expect to see the behavior shown in each graph.

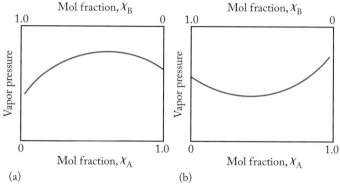

(a)   (b)

FIGURE P12.9

12.10. Hot Mulled Cider Hot apple cider is often flavored with cloves, giving the beverage a distinctive aroma. One of the compounds that contributes to the smell of cloves is a simple ketone called 2-heptanone. Figure P12.10 shows the graph of vapor pressure of 2-heptanone as a function of $1/T$. Use the graph to calculate the heat of vaporization (ΔH_{vap}) of 2-heptanone.

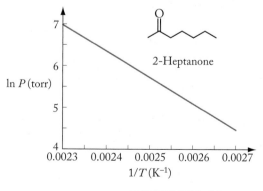

2-Heptanone

FIGURE P12.10

*12.11. The three polymers shown in Figure P12.11 are widely used in the plastics industry. In which of them are the intermolecular forces per mole of monomer the strongest?

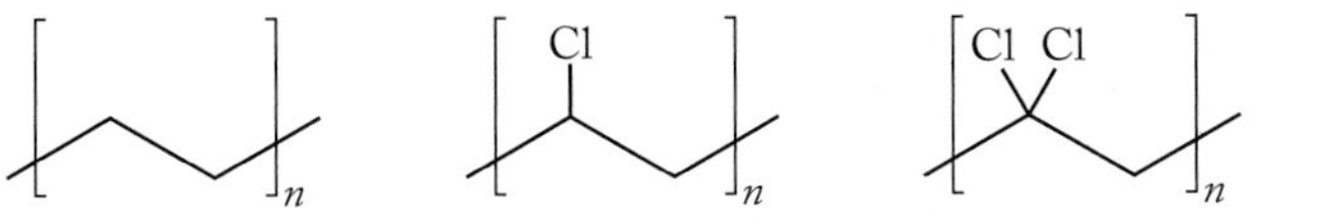

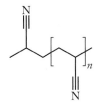

Polyethylene
(a)

Poly(vinyl chloride)
(b)

Poly(vinylidene chloride)
(c)

FIGURE P12.11

Polyacrylonitrile

FIGURE P12.13

*12.12. **Silly Putty** Silly Putty is a condensation polymer of dihydroxydimethylsilane (Figure P12.12). Draw the condensed molecular structure of the repeating monomeric unit in Silly Putty.

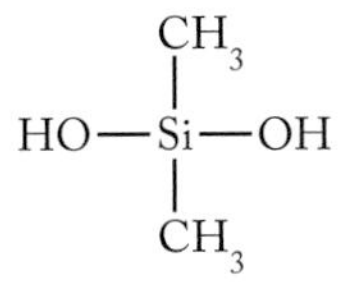

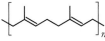

Dihydroxydimethylsilane

FIGURE P12.12

12.13. **Orlon and Acrilon Fibers** Figure P12.13 shows the condensed molecular structure of polyacrylonitrile, which is marketed as Orlon and Acrilon. Identify the monomeric reactant that produces this polymer.

12.14. Rubber is a polymer of isoprene. It is sometimes called polyisoprene. There are two forms of polyisoprene: *cis*-polyisoprene is the soft, flexible material we associate with the term "rubber"; gutta-percha, or *trans*-polyisoprene, is a much harder material. Draw the monomeric units of *cis*- and *trans*-polyisoprene which are shown in Figure P12.14.

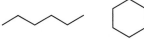

cis-Polyisoprene

trans-Polyisoprene

FIGURE P12.14

Questions and Problems

Carbon: The Scope of Organic Chemistry

CONCEPT REVIEW

12.15. Describe three ways in which carbon atoms can form bonds to other carbon atoms using hybrid orbitals (valence bond theory).

12.16. Can a macromolecule be composed of more than one type of monomer?

*12.17. The structure of the interstitial alloy tungsten carbide (WC) was described in Chapter 11. Do you consider tungsten carbide to be an organic compound?

*12.18. Calcium carbide, CaC_2 was used in miner's lamps. Reaction of CaC_2 with water yields acetylene which, when ignited, gives light. Is calcium carbide considered an organic compound?

PROBLEMS

12.19. What functional groups were introduced in Chapter 8?

12.20. Find an example of a small molecule with more than one functional group in Chapter 9.

12.21. Polyethylene is prepared from the monomer ethylene, C_2H_4. About how many monomers are needed to make a polymer with a molar mass of 100,000 g/mol?

12.22. Synthetic rubber is prepared from butadiene, C_4H_6. About how many monomers are needed to make a polymer with a molar mass of 100,000 g/mol?

Hydrocarbons

CONCEPT REVIEW

12.23. Do linear and branched alkanes with the same number of carbon atoms all have the same empirical formula?

12.24. If an alkane and a cycloalkane have equal numbers of carbon atoms per molecule, do they have the same number of hydrogen atoms?

12.25. What is the hybridization of carbon in alkanes?

12.26. Figure P12.26 shows the structures of *n*-hexane and cyclohexane. Are *n*-hexane and cyclohexane structural isomers?

n-Hexane   Cyclohexane

FIGURE P12.26

12.27. Why isn't cyclohexane a planar molecule?

12.28. Which of the simple cycloalkanes (C_nH_{2n}, $n = 3 - 8$) has a planar geometry?

12.29. Are cycloalkanes saturated hydrocarbons?

12.30. Do structural isomers always have the same molecular formula?

12.31. Do structural isomers always have the same chemical properties?

12.32. Are structural isomers members of a homologous series?

PROBLEMS

12.33. Draw and name all the structural isomers of C_5H_{12}.

12.34. Draw and name all the structural isomers of C_6H_{14}.

12.35. Which of the molecules in Figure P12.35 are structural isomers of *n*-octane (C_8H_{18})? Name these molecules.

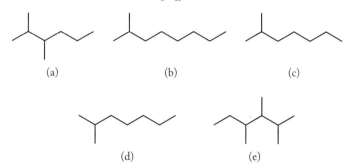

(a)   (b)   (c)

(d)   (e)

FIGURE P12.35

12.36. Which of the molecules in Figure P12.36 are structural isomers of *n*-heptane (C_7H_{16})? Name these molecules.

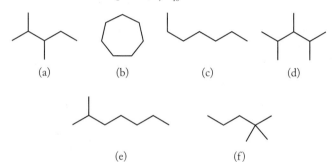

(a)   (b)   (c)   (d)

(e)   (f)

FIGURE P12.36

12.37. Convert the line structures in Problem 12.35 to chemical formulas.

12.38. Convert the line structures in Problem 12.36 to chemical formulas.

12.39. Using the average bond strengths given in Appendix 4, estimate the molar heat of hydrogenation, $\Delta H_{hydrogenation}$, for the conversion of C_2H_4 to C_2H_6.

$$CH_2{=}CH_2(g) + H_2(g) \rightarrow CH_3CH_3(g)$$

12.40. Using the average bond strengths given in Appendix 4, estimate the molar heat of hydrogenation, $\Delta H_{hydrogenation}$, for the conversion of C_2H_2 to C_2H_6.

$$CH{\equiv}CH(g) + 2H_2(g) \rightarrow CH_3CH_3(g)$$

12.41. Place the following molecules in order of increasing boiling point: C_3H_8, $C_{14}H_{30}$, cyclooctane (C_8H_{16}).

*12.42. Rank the molecules in Figure P12.42 in order of decreasing van der Waals forces.

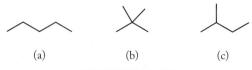

(a)   (b)   (c)

FIGURE P12.42

Sources of Alkanes

CONCEPT REVIEW

12.43. What physical property of the components of crude oil is used to separate them?

12.44. What is the difference between distillation and fractional distillation?

12.45. In an equimolar mixture of C_5H_{12} and C_7H_{12}, which compound is present in higher concentration in the vapor above the solution?

12.46. Why does the boiling point of a mixture of volatile hydrocarbons increase over time during a distillation?

PROBLEMS

12.47. Pine Oil The smell of fresh cut pine is due in part to the cyclic alkene pinene, whose structure is shown in Figure P12.47. Use the data in the table to calculate the heat of vaporization, ΔH_{vap}, of pinene.

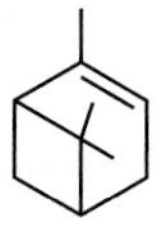

Pinene

FIGURE P12.47

Vapor Pressure (Torr)	Temperature (K)
760	429
515	415
340	401
218	387
135	373

12.48. Almonds and Cherries Almonds and almond extracts are common ingredients in baked goods. Almonds contain the compound benzaldehyde (shown in Figure P12.48), which accounts for the odor of the nut. Benzaldehyde is also responsible for the aroma of cherries. Use the data in the table to calculate the heat of vaporization, ΔH_{vap}, of benzaldehyde.

Benzaldehyde

FIGURE P12.48

Vapor Pressure (Torr)	Temperature (K)
50	373
111	393
230	413
442	433
805	453

12.49. At 20°C, the vapor pressure of ethanol is 45 torr and the vapor pressure of methanol is 92 torr. What is the vapor pressure at 20°C of a solution prepared by mixing 25 g of methanol and 75 g of ethanol?

12.50. A bottle is half-filled with a 50:50 (mole-to-mole) mixture of heptane and octane at 25°C. What is the mole ratio of heptane vapor to octane vapor in the air space above the liquid in the bottle? The vapor pressures of heptane and octane at 25°C are 31 torr and 11 torr, respectively.

12.51. High-Octane Gasoline Gasoline is a complex mixture of hydrocarbons, including different isomers of octane. Gasoline is sold with a variety of octane ratings that are based on the comparison of the gasoline with the combustion properties of isooctane, a structural isomer of C_8H_{18}. The structure of isooctane and another isomer of C_8H_{18} are shown in Figure P12.51, along with their normal boiling points and heats of vaporization. Determine the vapor pressure of each isomer on a very hot summer day when the temperature is 38°C.

Isooctane
bp = 98.2°C
ΔH_{vap} = 35.8 kJ/mol

Tetramethylbutane
bp = 106.5°C
ΔH_{vap} = 43.3 kJ/mol

FIGURE P12.51

12.52. Stove Fuel Portable lanterns and stoves used for camping and backpacking often use a mixture of C_5 and C_6 hydrocarbons known as "white gas." White gas is easy to transport but stoves that burn white gas are harder to light in the cold. Figure P12.52 shows the structure of n-pentane, a typical C_5 alkane, along with its normal boiling point and heat of

vaporization. Determine the vapor pressure of n-pentane on a cold autumn morning when the temperature is 5°C.

n-Pentane
bp = 36.0°C
ΔH_{vap} = 27.6 kJ/mol

FIGURE P12.52

Alkenes and Alkynes

CONCEPT REVIEW

12.53. How do structural isomers differ from geometric isomers?

12.54. Explain why alkanes don't have geometric isomers.

12.55. Can combustion analysis distinguish between an alkene and a cycloalkane containing the same number of carbon atoms?

12.56. Can combustion analysis data be used to distinguish between an alkyne and a cycloalkene containing the same number of carbon atoms?

12.57. Why don't the alkenes in Figure P12.57 have cis and trans isomers?

FIGURE P12.57

12.58. Why don't alkynes have cis and trans isomers?

*12.59. Figure P12.59 shows the structure of carvone, which is found in the oil of spearmint. Why doesn't the molecule carvone have cis and trans isomers?

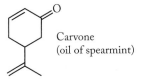

Carvone
(oil of spearmint)

FIGURE P12.59

*12.60. Figure P12.60 shows the structure of the antifungal compound capillin. Are the π electrons in capillin delocalized?

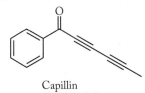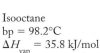

Capillin

FIGURE P12.60

12.61. Ethylene reacts quickly with HBr at room temperature, but polyethylene is chemically unreactive toward HBr. Explain why these related substances have such different properties.

*12.62. Polymerization of butadiene (CH_2=CHCH=CH_2) does not yield the same polymer as polymerization of ethylene (CH_2=CH_2). How could we convert poly(butadiene) into poly(ethylene)?

PROBLEMS

12.63. Label the isomers of cinnamaldehyde (oil of cinnamon) in Figure P12.63 as *cis* or *trans* and E or Z.

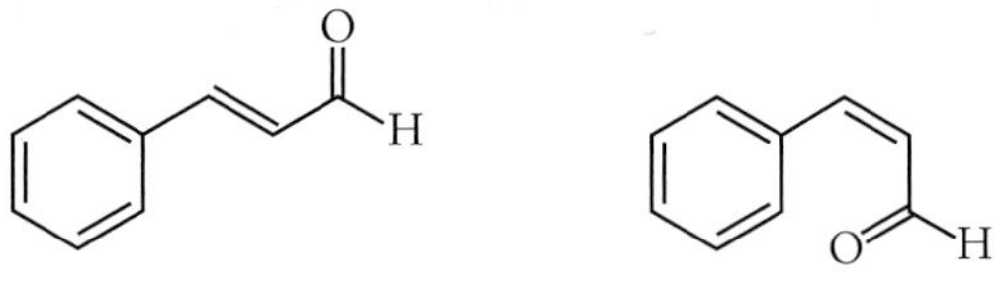

(a)   (b)

FIGURE P12.63

12.64. Prostaglandins are a naturally occurring class of compounds in our bodies that affect blood pressure and cause inflammation and other physiological responses. The synthesis of prostaglandins in our bodies starts with arachidonic acid (Figure P12.64), a long, unsaturated hydrocarbon containing four C=C double bonds and an acid functional group (–COOH). The geometric isomer with all *cis* double bonds is shown. How many geometric isomers of arachidonic acid are possible? Draw the isomer with all *trans* double bonds.

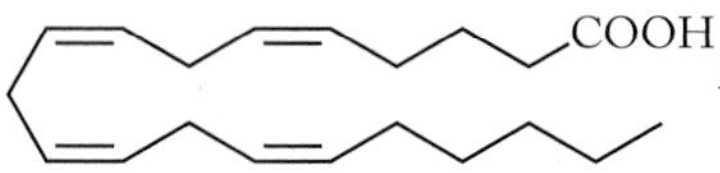

Arachidonic acid

FIGURE P12.64

12.65. Using data in Appendix 4, calculate ΔH_{rxn} for the production of acetylene from controlled combustion of methane:

$$6\,CH_4(g) + O_2(g) \rightarrow 2\,C_2H_2(g) + 2\,CO(g) + 10\,H_2(g)$$

Is this an endothermic or an exothermic reaction?

12.66. Using data in Appendix 4, calculate ΔH_{rxn} for the production of acetylene from reaction between calcium carbide and water given ΔH_f° of CaC_2 is −59.8 kJ/mol

$$CaC_2(s) + 2\,H_2O(\ell) \rightarrow C_2H_2(g) + Ca(OH)_2(s)$$

Is this an endothermic or an exothermic reaction?

12.67. Materials for Computer Disks Computer diskettes are made of poly(vinyl acetate). Draw the condensed molecular structure of this polymer. The structure of its monomer is shown in Figure P12.67.

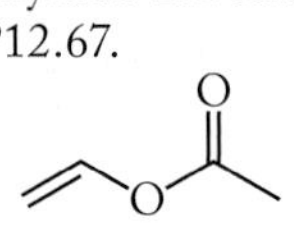

Vinyl acetate

FIGURE P12.67

12.68. The 2000 Nobel Prize in Chemistry was awarded for research on the electrically conductive polymer polyacetylene.
   a. Draw the molecular structure of three monomeric units of the addition polymer that results from polymerization of acetylene, HC≡CH.
   *b. There are two possible geometric isomers of polyacetylene. Describe the two isomeric forms.

Aromatic Compounds

CONCEPT REVIEW

12.69. Why is benzene a planar molecule?

12.70. Why are aromatic molecular structures stable?

12.71. Do tetramethylbenzene and pentamethylbenzene have structural isomers?

12.72. Figure P12.72 shows the structures of butadiene (C_4H_6) and 1,3-cyclohexadiene (C_6H_8). Why aren't butadiene and 1,3-cyclohexadiene considered aromatic molecules?

Butadiene   1,3-Cyclohexadiene

FIGURE P12.72

*12.73. Pyridine (Figure P12.73) has the molecular formula C_6H_5N. Is pyridine an aromatic molecule?

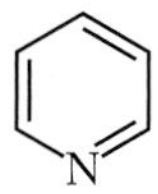

Pyridine

FIGURE P12.73

*12.74. Is graphite (see Chapter 11) an aromatic compound?

PROBLEMS

12.75. Draw all the structural isomers of trimethylbenzene.

12.76. Draw all the structural isomers of dimethylnaphthalene.

12.77. Calculate the fuel values of benzene (C_6H_6) and ethylene (C_2H_4). Does 1 mole of benzene have a higher or lower fuel value than 3 moles of ethylene?

12.78. Does 1 mole of benzene (C_6H_6) have a higher or lower fuel value than 3 moles of acetylene (C_2H_2)?

Alcohols, Ethers, and Reformulated Gasoline

CONCEPT REVIEW

12.79. Why are the fuel values of ethanol and dimethylether (Figure P12.79) lower than that of ethane?

Dimethyl ether   Ethanol

FIGURE P12.79

12.80. Would you expect the fuel value of alcohols to increase or decrease as the number of carbon atoms in the alcohol increases?

12.81. Why do ethers typically boil at lower temperatures than alcohols with the same molecular formula?

12.82. Figure P12.82 shows the structures of MTBE and 2,2-dimethylbutane. Which of these do you expect to be more soluble in water? Explain your answer.

MTBE   2, 2-Dimethylbutane

FIGURE P12.82

*12.83. Disposable wipes used to clean the skin prior to a getting an immunization shot contain ethanol. After wiping your arm, your skin feels cold. Why?

*12.84. In cold climates, condensation of water in the gas tank in the winter months reduces engine performance. An auto mechanic recommends adding "dry gas" to the tank during your next fill-up. "Dry gas" is typically an alcohol that dissolves in gasoline and absorbs water. Based on the structures shown in Figure P12.84, which product would you predict would do a better job—methanol or 2-propanol?

CH_3OH OH

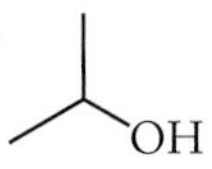

Methanol   2-Propanol

FIGURE P12.84

PROBLEMS

12.85. Which of the compounds in Figure P12.85 are alcohols and which ones are ethers? Place them in the order of increasing boiling point.

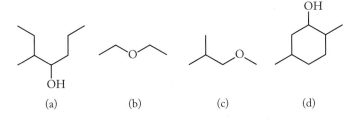

(a)   (b)   (c)   (d)

FIGURE P12.85

12.86. Which of the compounds in Figure P12.86 are alcohols and which ones are ethers? Place them in the order of increasing vapor pressure at 25°C.

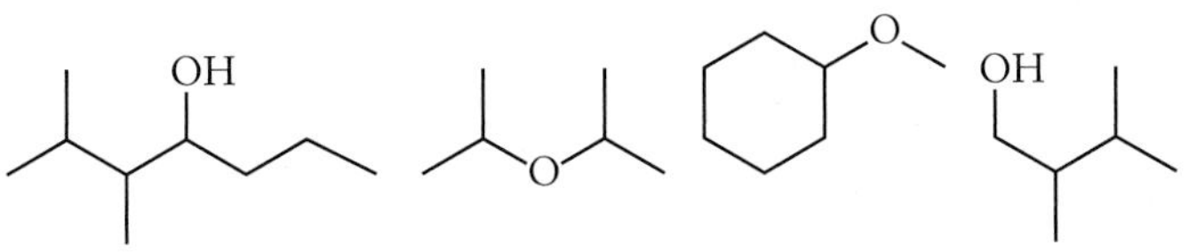

(a)   (b)   (c)   (d)

FIGURE P12.86

Consult tables of the thermochemical data in Appendix 4 for any values you may need to solve Problems 12.87 and 12.88.

12.87. Calculate the fuel value of butanol and diethylether (Figure P12.87). Which has the higher fuel value?

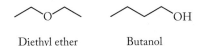

Diethyl ether   Butanol

FIGURE P12.87

12.88. Problem 12.80 asked you to predict whether the fuel value of alcohols increased or decreased with the number of carbon atoms in the alcohol. Calculate the fuel values of methanol and ethanol (Figure P12.88). Does your answer support the prediction you made in Problem 12.80?

CH_3OH OH

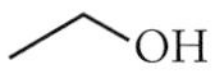

Methanol   Ethanol

FIGURE P12.88

More Oxygen-Containing Functional Groups: Aldehydes, Ketones, Carboxylic Acids, Esters, Amides

CONCEPT REVIEW

12.89. Explain why carboxylic acids tend to be more soluble in water than aldehydes with the same number of carbon atoms.

12.90. The structures of diethyl ether and 2-butanone are shown in Figure P12.90. In reference books, diethyl ether is usually listed as "slightly soluble" in water but 2-butanone is "very soluble." Why do you suppose 2-butanone is more soluble?

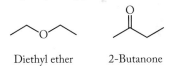

Diethyl ether   2-Butanone

FIGURE P12.90

12.91. Are butanal and 2-butanone (Figure P12.91) structural isomers?

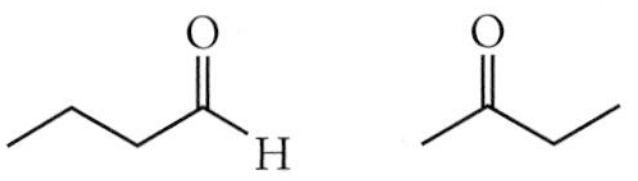

Butanal   2-Butanone

FIGURE P12.91

12.92. **Apples** The two esters shown in Figure P12.92 are both found in apples and contribute to the flavor and aroma of the fruit. Are the two compounds identical, structural isomers, geometric isomers, or do they have different formulas?

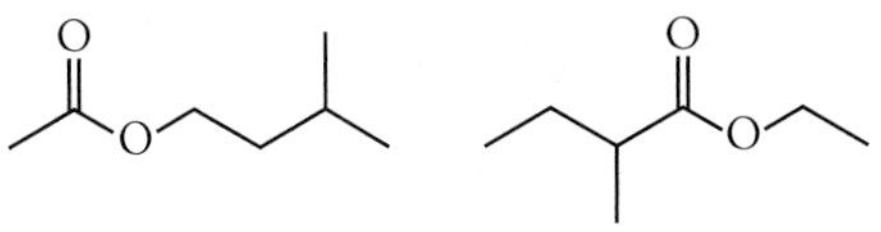

FIGURE P12.92

12.93. Can we distinguish between ketones and aldehydes with the same number of carbon atoms by combustion analysis?

12.94. Can we distinguish between ethers and ketones with the same number of carbon atoms by combustion analysis?

12.95. Resonance forms for acetic acid are shown in Figure P12.95. Which one contributes more to the bonding picture? Explain your choice.

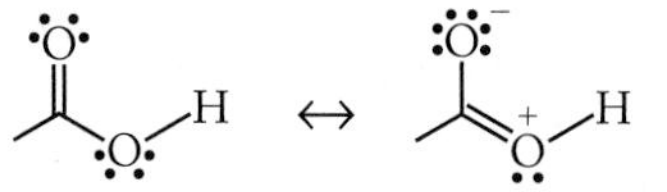

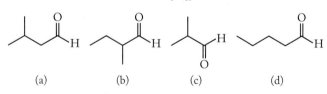

(a)   (b)

FIGURE P12.95

12.96. Figure P12.96 shows resonance forms for acetamide and acetic acid. Does the resonance form (a) containing the C=N double bond contribute more to the bonding picture of acetamide than does the resonance form (b) to the bonding picture in acetic acid? Explain your answer.

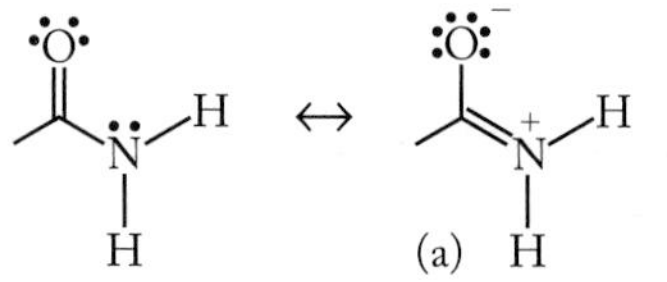

Acetamide

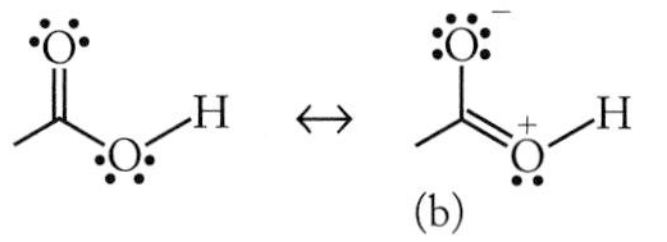

(a)

Acetic acid

(b)

FIGURE P12.96

PROBLEMS

12.97. Which of the compounds in Figure P12.97 are structural isomers of the aldehyde $C_5H_{10}O$?

(a)   (b)   (c)   (d)

FIGURE P12.97

12.98. Each of the natural products in Figure P12.98 contains more than one functional group. Which of the compounds is an aldehyde?

(a)   (b)   (c)

FIGURE P12.98

12.99. Which of the compounds in Figure P12.99 is a ketone?

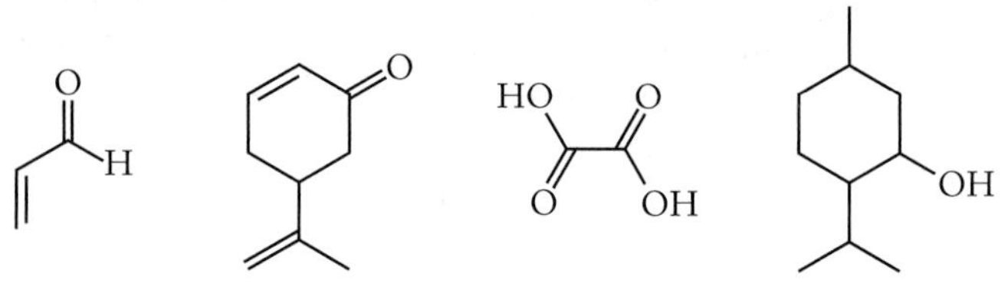

(a)   (b)   (c)   (d)

FIGURE P12.99

12.100. Propanal and acetone (2-propanone) have the same molecular formula, C_3H_6O, but different structures (Figure P12.100). Which compound is a ketone?

Propanal   Acetone

FIGURE P12.100

12.101. Plot the carbon-to-hydrogen ratio in aldehydes with one to six carbons as a function of the number of carbon atoms. Does this graph correlate better with the plot of C:H ratios for alkanes or for alkenes?

12.102. Plot the carbon to hydrogen ratio in carboxylic acids with one to six carbons as a function of the number of carbon atoms. Does this graph correlate better with the plot of C:H ratios for alkanes or for alkenes?

12.103. Esters are responsible for the odors of fruits, including apples, bananas, and pineapples. Figure P12.103 shows three esters from these fruits. Identify the alcohol and carboxylic acid that react to form these compounds.

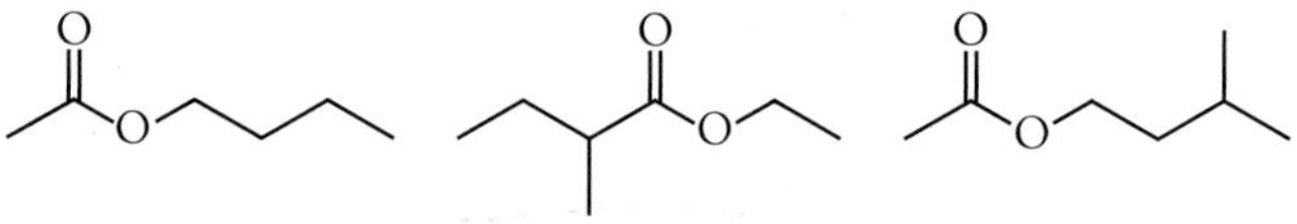

(a) Pineapples   (b) Bananas   (c) Apples

FIGURE P12.103

12.104. Beeswax (Figure P12.104) is an ester composed of an alcohol and a carboxylic acid with a long hydrocarbon chain. Identify the alcohol and acid in beeswax.

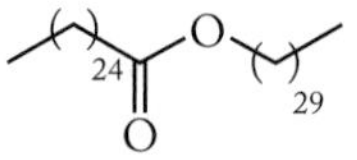

FIGURE P12.104

Consult tables of thermochemical data in Appendix 4 for any values you may need to solve Problems 12.105 and 12.106.

12.105. Calculate the fuel values of formaldehyde and formic acid (Figure P12.105). Which has the higher fuel value?

Formaldehyde   Formic acid

FIGURE P12.105

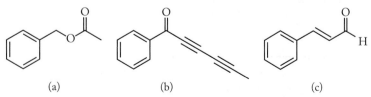

12.106. Calculate the fuel value of formamide and methylformate (Figure P12.106), which have ΔH_f° values of -251 and -391 kJ/mol, respectively. Assume $NO_2(g)$ is a product of formamide combustion.

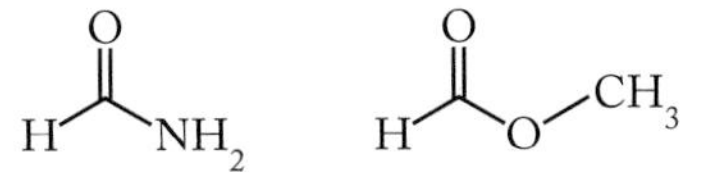

Formamide   Methylformate

FIGURE P12.106

12.107. Reactions between 1,6-diaminohexane, $H_2N(CH_2)_6NH_2$, and different dicarboxylic acids, $HO_2C(CH_2)_nCO_2H$, are used to prepare polymers with structures similar to that of nylon. How many carbon atoms (n) were in the dicarboxylic acids used to prepare the polymers with the repeating units shown in Figure P12.107?

(a)

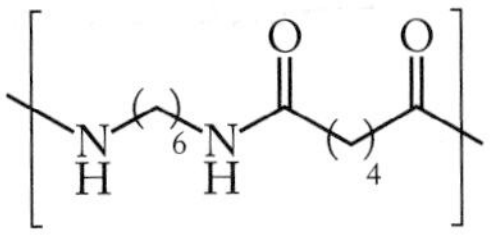

(b)

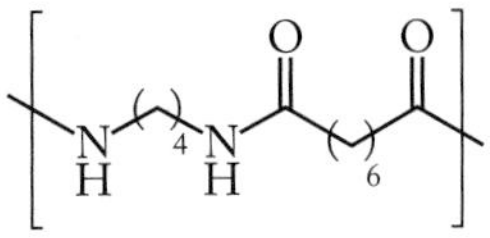

(c)

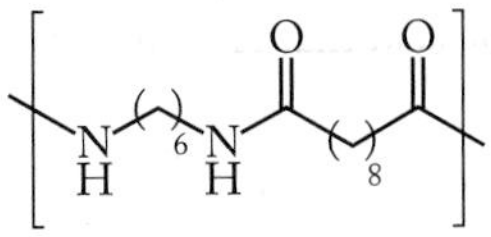

FIGURE P12.107

12.108. The two polymers in Figure P12.108 have the same empirical formula.
a. What pairs of monomers could be used to make each of them?
b. How might the physical properties of these two polymers differ?

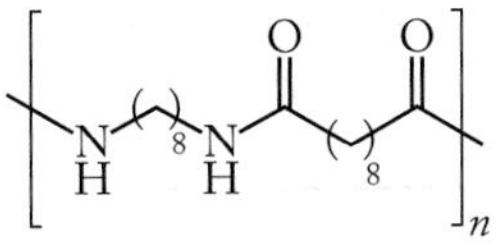

Polymer I

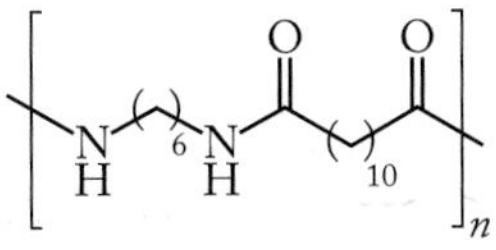

Polymer II

FIGURE P12.108

12.109. The polyester called Kodel is made with polymeric strands prepared by the condensation of dimethyl terephthalate with 1,4-di(hydroxymethyl)cyclohexane (Figure P12.109).

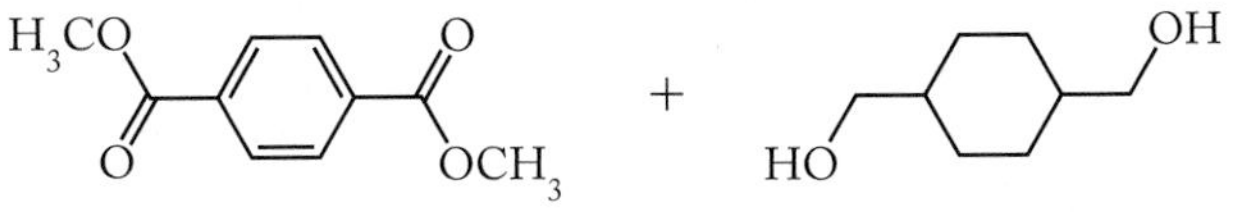

Dimethyl terephthalate   1,4-Di(hydroxymethyl)cyclohexane
(dimethyl benzene-1,4-dicarboxylate)

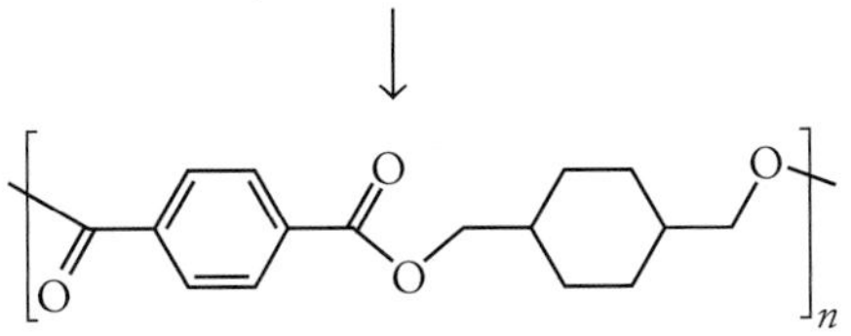

Kodel

FIGURE P12.109

a. Is Kodel a condensation polymer or an addition polymer? What is the other product of the reaction?
*b. Dacron (see Sample Exercise 12.12) is made from dimethyl terephthalate and ethylene glycol. What properties of Kodel fibers might make them better than Dacron as a clothing material?

12.110. Lexan is a polymer belonging to the class of materials called polycarbonates. Figure P12.110 shows the polymerization reaction for Lexan.

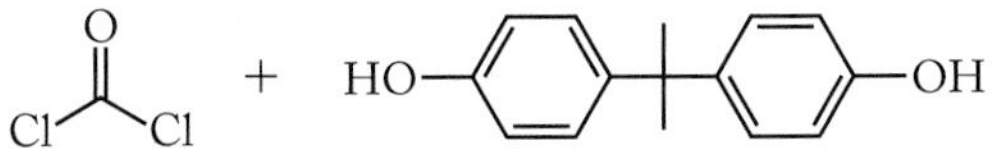

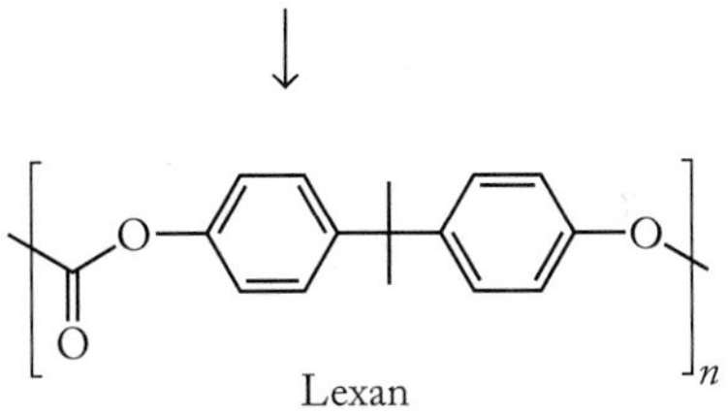

Lexan

FIGURE P12.110

a. What other compound is formed in the polymerization reaction?
*b. Why is Lexan called a "polycarbonate"?

Hydrocarbon Fuels from Biomass

CONCEPT REVIEW

12.111. Explain why methylamine (CH_3NH_2) is more soluble in water than n-butylamine [$CH_3(CH_2)_3NH_2$].

12.112. What distinguishes an amine from an amide?

***12.113.** Why can't we use tertiary amines to prepare amides?

12.114. Combustion of hydrocarbons in air yields carbon dioxide and water. What other product is expected in the combustion of amines?

PROBLEMS

12.115. Nicotine is a stimulant found in tobacco. Valium is a tranquilizer. Both molecules contain two nitrogen atoms in addition to other functional groups. In Figure P12.115 identify the nitrogen atoms shown in blue as belonging to an amine or an amide.

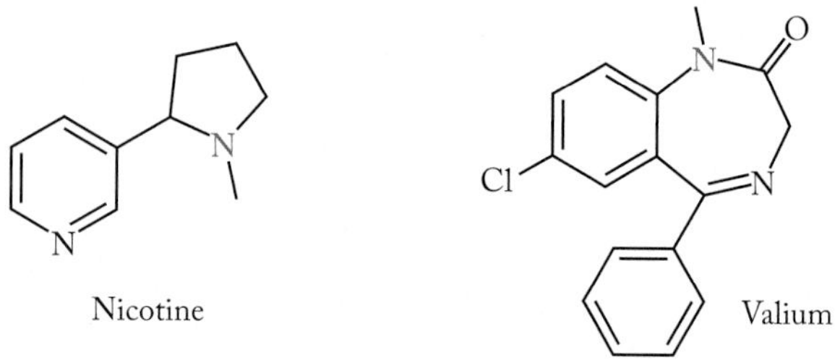

Nicotine   Valium

FIGURE P12.115

12.116. Serotonin and amphetamine both contain the amine functional group (Figure P12.116). Serotonin is responsible, in part, for signaling that we have had enough to eat. Amphetamine, an addictive drug, can be used as an appetite suppressant. Identify the primary and secondary amine functional groups in these molecules.

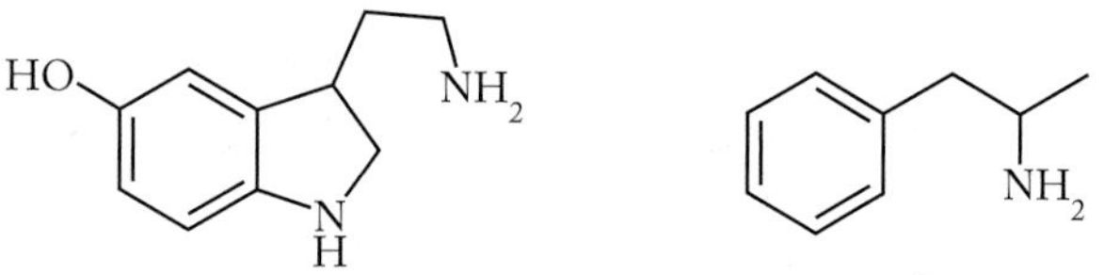

Serotonin   Amphetamine

FIGURE P12.116

12.117. Use the data in Appendix 4 to calculate ΔH°_{rxn} for the following reactions of methanogenic bacteria:

(1) $CH_3COOH(aq) \rightarrow CH_4(g) + CO_2(g)$

(2) $4HCOOH(aq) \rightarrow CH_4(g) + 3CO_2(g) + 2H_2O(\ell)$

12.118. Use the data in Appendix 4 to calculate ΔH°_{rxn} for the following reactions of methanogenic bacteria:

(1) $4H_2(g) + CO_2(g) \rightarrow CH_4(g) + 2H_2O(\ell)$

(2) $4CH_3OH(\ell) \rightarrow 3CH_4(g) + CO_2(g) + 2H_2O(\ell)$

12.119. Bacteria of the genus *Methanosarcina* convert amines to methane. Their action helps make methane a renewable energy source. Determine the standard enthalpy of the following reaction from the appropriate standard enthalpies of formation ($\Delta H^{\circ}_{f,CH_3NH_2} = -23.0$ kJ/mol):

$4CH_3NH_2(g) + 2H_2O(\ell) \rightarrow 3CH_4(g) + CO_2(g) + 4NH_3(g)$

12.120. Determine the ΔH°_{rxn} values of these combustion reactions of methylamine.

$4CH_3NH_2(g) + 13O_2(g) \rightarrow 4CO_2(g) + 4NO_2(g) + 10H_2O(\ell)$

$4CH_3NH_2(g) + 6O_2(g) \rightarrow 4CO_2(g) + 4NH_3(g) + 4H_2O(\ell)$

Additional Problems

12.121. How many grams of methanol must be combusted to raise the temperature of 454 g of water from 20.0°C to 50.0°C? Assume that the transfer of heat to the water is 100% efficient. How many grams of carbon dioxide are produced in this combustion reaction?

12.122. How many grams of methylamine must be combusted to raise the temperature of 454 g of water from 20°C to 50°C? Assume that the transfer of heat to the water is 100% efficient. How many grams of carbon dioxide are produced in this combustion reaction? Also assume $NO_2(g)$ is a product of the combustion of methylamine.

12.123. Two compounds, both with molar masses of 74.0 g/mol, were combusted in a bomb calorimeter with $C_{calorimeter} = 3.640$ kJ/°C. Combustion of 0.9842 g of compound A led to an increase in temperature of 10.33°C, while combustion of 1.110 g of compound B caused the temperature to rise 11.03 °C. Which compound is butanol and which is diethyl ether?

12.124. Why should methane be more soluble in decane ($C_{10}H_{22}$) than in water?

12.125. **Salsa** Salsa has antibacterial properties because it contains dodecenal (Figure P12.125), a compound found in the cilantro used to make salsa.
   a. How many carbon atoms are in dodecenal?
   b. What functional groups are present in dodecenal?
   c. What types of isomerism are possible in dodecenal?

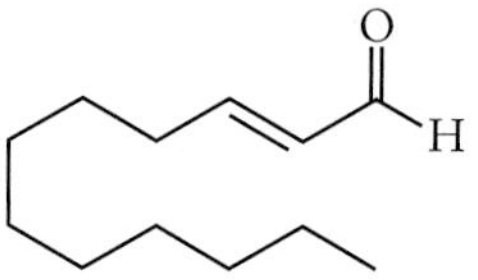

Dodecenal

FIGURE P12.125

12.126. **Turmeric, A Spice** Turmeric is commonly used as a spice in Indian and Southeast Asian dishes. Turmeric contains a high concentration of curcumin (Figure P12.126), a potential anticancer drug and a possible treatment for cystic fibrosis.
   a. Are the substituents on the C=C double bonds in *cis* or *trans* configurations?
   b. Draw two other geometric isomers of this compound.
   c. List all the types of valence shell hybridization of the carbon atoms in curcumin.

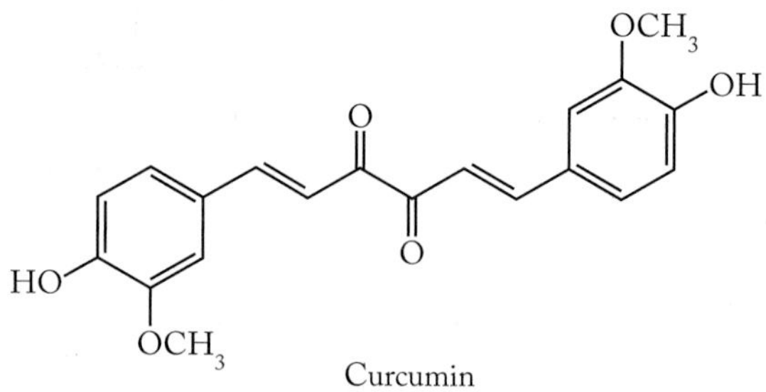

Curcumin

FIGURE P12.126

12.127. Polycyclic aromatic hydrocarbons are potent carcinogens. They are produced during combustion of fossil fuels and have also been found in meteorites. Can we use combustion analysis to distinguish between naphthalene and anthracene (Figure P12.127)?

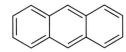

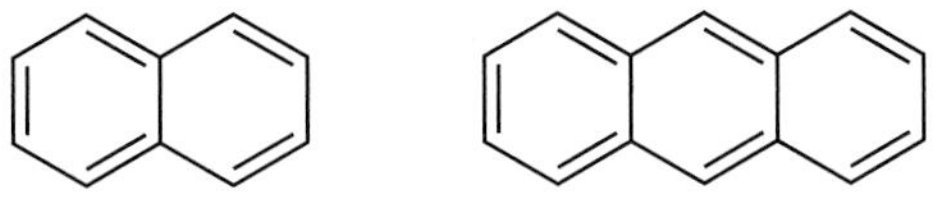

Naphthalene   Anthracene

FIGURE P12.127

12.128. Identify the reactants in the polymerization reactions that produce the polymers shown in Figure P12.128.

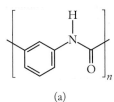
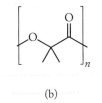

(a)   (b)

FIGURE P12.128

*****12.129.** "Waterproof" nylon garments have a coating to prevent water from penetrating the hydrophilic fibers. Which functional groups in the structure of nylon make it hydrophilic?

12.130. Use a line molecular structure to describe the condensation polymer of $H_2N(CH_2)_6COOH$. How does this polymer compare with nylon-6?

12.131. Putrescine, $H_2N(CH_2)_4NH_2$, is one of the compounds that form in rotting meat.
  a. Draw the structures of all the trimers (a molecule formed from three monomers) that can be formed from putrescine, adipic acid, and terephthalic acid (Figure P12.131). The three monomers forming the trimer do not have to be different from one another.
  b. A chemist wishes to make a putrescine polymer containing a 1:1 ratio of adipic acid to terephthalic acid. What should be the mole ratio of the three reactants?

HOOCCH₂CH₂CH₂CH₂COOH

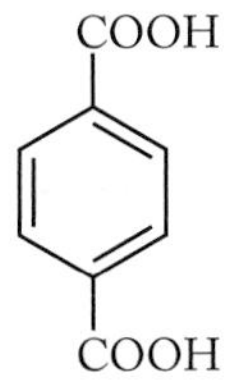

Adipic acid   Terephthalic acid
                (benzene-1,4-dicarboxylic acid)

FIGURE P12.131

*****12.132.** Polymer chemists can modify the physical properties of polystyrene by copolymerizing divinylbenzene with styrene (Figure P12.132). The resulting polymer has strands of polystyrene cross-linked with divinylbenzene. Predict how the physical properties of the copolymer might differ from those of 100% polystyrene.

Divinylbenzene (DVB)   Styrene (S)

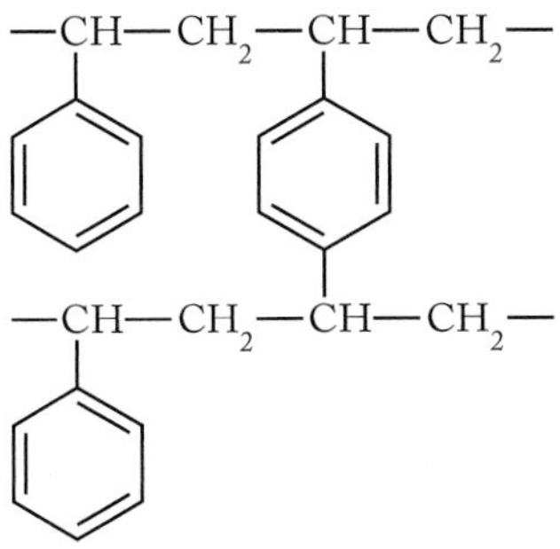

—CH—CH₂—CH—CH₂—

—CH—CH₂—CH—CH₂—

S cross-linked with DVB

FIGURE P12.132

12.133. Styrene and maleic anhydride (Figure P12.133) form a polymer with alternating units of each monomer.

Maleic anhydride   Styrene

FIGURE P12.133

  a. Draw two structural repeating units of the polymer.
  b. Based on the structure of the copolymer, predict how its physical properties might differ from those of polystyrene.

12.134. **Superglue** The active ingredient in "Superglue" is methyl 2-cyanoacrylate (Figure P12.134). The liquid glue rapidly hardens when methyl 2-cyanoacrylate polymerizes. This happens when it contacts a surface containing traces of water or other compounds containing –OH or –NH– groups. Draw the structure of two repeating units of poly(methyl 2-cyanoacrylate).

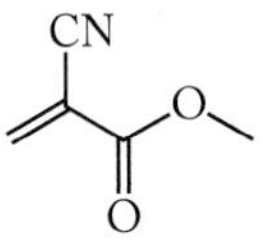

CN

Methyl 2-cyanoacrylate

FIGURE P12.134

*****12.135.** Silicones are polymeric materials with the formula $[R_2SiO]_n$ (Figure P12.135). They are prepared by reaction of R_2SiCl_2 with water yielding the polymer and aqueous HCl. Consider this reaction as taking place in two steps: (1) water reacts with 1 mole of R_2SiCl_2 to produce a new monomer and 1 mole of HCl(*aq*); (2) one new monomer molecule reacts with another new monomer molecule to

eliminate one molecule of HCl and make a dimer with a
Si—O—Si bond.
a. Suggest two balanced equations describing these
reactions that occur over and over again to produce a
silicone polymer.
b. Why are silicones water repellent?

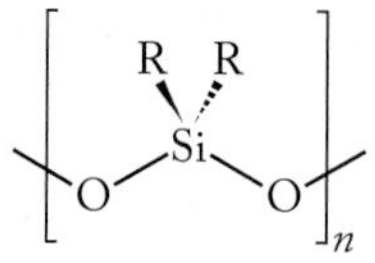

Silicone

FIGURE P12.135

12.136. Piperine and capsaicin are the spicy ingredients of black
and red pepper, respectively (Figure P12.136). Both
compounds contain an amide functional group.

a. Draw the amine and the carboxylic acid that could
react to form these two compounds.
b. Are the double bonds in these molecules *cis* or *trans*?
c. Name the functional groups that contain the oxygen
atoms in these compounds.

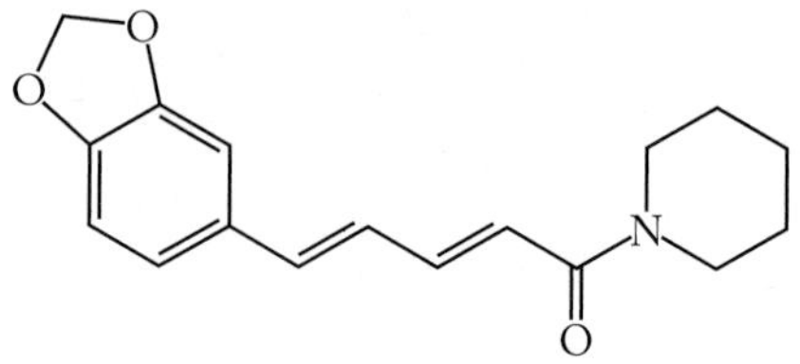

Piperine

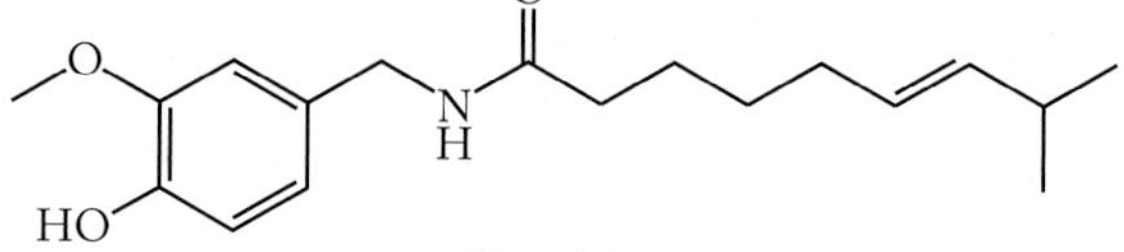

Capsaicin

FIGURE P12.136

Thermodynamics: Spontaneous Processes, Entropy, and Free Energy

13

CHATEAU DE VILLANDRY, FRANCE *Plants do not grow spontaneously in the highly ordered arrangements of a formal garden.*

A LOOK AHEAD: The Game of Energy

If you puncture a tire on a car, the air usually rushes out of the hole and the tire becomes flat. In all of recorded automotive history, no one has ever seen air rush back into the hole and reinflate the tire. If you leave a shiny iron nail outside on the ground, it rusts. No one has ever reported a rusty nail left on the ground turning back into a shiny nail. If you put an ice cube on a countertop at room temperature, the ice cube melts. No one has ever seen an ice cube in this situation become colder and the countertop become warmer. These observations are all very different: a gas moves from one place to another, a chemical reaction converts iron into iron oxide, a solid becomes a liquid. However, these events have an important thing in common: they are all spontaneous. They all happen without ongoing intervention. The reverse of each of these processes is *not* spontaneous—none of them has ever been observed to occur on its own.

These events and many others are so familiar to us that they fall into the realm we enter with the words "As everyone knows . . . ," but a satisfactory explanation for why these events occur spontaneously is crucial for an understanding of how the world works. An explanation of seemingly simple occurrences such as a tire going flat, a nail rusting, and an ice cube melting—as well as more awesome events such as forest fires, hurricanes, and our own sun being reduced someday in the future to a burned out cinder—can be found in the second law of thermodynamics and the concept called entropy.

Think for a moment about a tire going flat. If you saw a movie in which a flat tire reinflated itself, what would you conclude? You would immediately assume that the movie was running backwards. In addition to helping us come to grips with how the world works, the second law of thermodynamics gives rise to our sense of time passing in only one direction: living things age and ultimately die. The passage from young to old in terms of life is part of our sense of time, and conceptualizing this progression is part of the second law.

Mathematical expressions of the second law of thermodynamics and the meaning of entropy are essential, but in this chapter we forgo all but the most fundamental equations. We start with some everyday occurrences that allow us to develop the core ideas behind the flow of energy in all spontaneous processes. Remember that the first law of thermodynamics tells us that energy cannot be created or destroyed. In other words, in the "game" of energy used to do work, the first law says that the best you can do is break even; you cannot create new energy. The second law says that the amount of energy available to us to do useful work is constantly decreasing; you not only cannot win, you cannot break even.

If energy cannot be destroyed, where does the energy go that is unavailable to do useful work? Energy naturally "spreads out," becoming less concentrated over time. Entropy is a measure of the energy that has spread out in such a way that gathering it up again—concentrating it so we could use it to do useful work—requires more energy that we could recover in the process. Spontaneity, the second law of thermodynamics, entropy, and how energy "spreads out" are the subjects of this chapter.

13.1 Spontaneous Processes and Entropy

Many exothermic reactions are **spontaneous**, which means they occur without continuous outside intervention. A spontaneous process may need an initial "push" of energy to get it started. It may also go very slowly. Spontaneous does not mean fast; it means favorable.

The rusting of an iron nail is a spontaneous, exothermic process. The reverse process—converting rust, $Fe_2O_3(s)$, into pure iron, $Fe(s)$—is not impossible; it is just **nonspontaneous**. The nonspontaneous reaction is carried out routinely to produce iron from iron ore, but it requires the continuous input of energy. When the addition of energy stops, the conversion of rust to iron stops, and the spontaneous process of rusting begins again. If a process is spontaneous (such as iron rusting), the reverse process [reducing iron(III) oxide to pure metal] is nonspontaneous.

Just because a process is spontaneous, however, does not mean it necessarily occurs. As we said earlier, sometimes spontaneous reactions need a little boost to get going. For example, a fire may need to be ignited by some source, but then it continues without the input of additional energy as long as fuel and oxidizer are available. Once a spontaneous reaction starts, it keeps going on its own.

Although it is true that most exothermic reactions are spontaneous, some endothermic reactions are spontaneous, too. A good example is the reaction that makes a chemical cold pack cold (see page 643). The reaction is endothermic and so has a positive ΔH. The reaction that spontaneously lowers the temperature of the cold pack when its ingredients mix together illustrates the point that the sign of ΔH—positive if a reaction takes in heat, negative if it gives off heat—cannot be used as the only criterion of spontaneity.

> A **spontaneous** process proceeds in a given direction without outside intervention
>
> A **nonspontaneous** process only occurs for as long as energy is continually added to the system. Nonspontaneous processes are the reverse of spontaneous ones.
>
> **Entropy (S)** is a measure of the distribution of energy in a system at a specific temperature.
>
> The **second law of thermodynamics** states that the total entropy of the universe increases in any spontaneous process.

CONCEPT TEST

Critique these two statements: (1) If a process is spontaneous, it is fast. (2) If a process is nonspontaneous, it does not occur.

(Answers to Concept Tests are in the back of the book.)

▶II **CHEMTOUR** Entropy

Why are some chemical and physical processes spontaneous? The feature that all spontaneous processes have in common is that localized energy is dispersed (spread out) as a result of the process. **Entropy (S)**, a thermodynamic property, is a measure of the distribution of energy at a specific temperature. The **second law of thermodynamics** states that the total entropy *of the universe* increases in any spontaneous process. To develop a clearer picture of entropy and the second law, and what is meant by the energy "dispersing" or "spreading out," let's examine some examples of spontaneous processes at the macroscopic level.

CONNECTION In Chapter 5 we introduced the idea of the universe in thermodynamics. The *universe* means the entire universe: the system we are studying and its surroundings. However, we usually approximate *the surroundings* as local—that part of the larger universe capable of exchanging energy with the system we are studying.

CONNECTION In Chapter 5 we introduced the first law of thermodynamics: energy can neither be created nor destroyed.

Entropy and Microstates: Flat Tires

Let's examine the process of air leaving a punctured tire. In doing so, we make a few assumptions that we know are not entirely true but will help us simplify the initial discussion. We will correct them later in the chapter once the language of thermodynamics is clearer.

Let's start with these assumptions: (1) The air is an ideal gas. (2) The tire, the gas inside, and the outside air are all at the same temperature. (3) The temperature of the gas does not change as it escapes the tire. Because the internal energy of a gas depends only on its temperature, it does not change as the gas escapes, and the enthalpy change (ΔH) is zero. (The temperature does fall somewhat when gas escapes, which is a property that allows air-conditioning systems to work. However, the temperature drop does not significantly affect the overall conclusions about this system, so for now we ignore it.) When the gas escapes from the tire, it expands to fill a larger volume at a lower pressure. The kinetic energy of the molecules, previously contained within the tire, is now spread throughout a greater volume. Because the energy has been dispersed, the entropy of the gas

CONNECTION We introduced enthalpy (H) in Chapter 5 as a thermodynamic quantity that describes heat flow into or out of a system.

(a) Translational motion

(b) Rotational motion

(c) Vibrational motion

FIGURE 13.1 A single diatomic molecule has three different fundamental motions: (a) translational motion, (b) rotational motion, and (c) vibrational motion.

CONNECTION Vibration of the C=O bonds in its molecules makes atmospheric CO_2 a potent greenhouse gas, as we discussed in Section 9.3.

CONNECTION In Chapter 6 we calculated the root-mean-square speed of molecules in the gas phase from their temperature and molar mass.

An **energy state**, also called an **energy level**, is an allowed value of energy.

once contained in the tire has increased. The deflation of the tire was spontaneous, and the spontaneous process led to an increase in entropy.

This description provides a macroscopic view of a spontaneous process taking place with a large quantity of material. A full treatment of entropy requires that we consider what happens at the molecular level as well. To accomplish this, we need to look more closely at how molecules move. Of course, the gas in the tire is a mixture, but to make our story simpler, let's focus on the types of motion available to one oxygen molecule within the population of molecules in the tire.

A single O_2 molecule has three different motions (Figure 13.1): in translational motion the entire molecule changes location (Figure 13.1a); in rotational motion the molecule spins about an axis (Figure 13.1b); and in vibrational motion the individual atoms in the molecule move back and forth like balls connected by a spring (Figure 13.1c). All three motions increase with increases in thermal energy, so the higher the temperature, the greater the motion.

Quantum mechanics teaches that energy is not continuous at the atomic scale; only certain energy levels are possible for electrons within atoms. The motions of atoms and molecules also are quantized, which means different states, separated by characteristic, specific energies, exist for the three types of motion.

Translational energy levels are the closest together; the quanta of energy that separate translational levels are very small. This means that not much energy is required to make an isolated molecule move translationally. The quantized difference in terms of translational energy levels is so small that at temperatures above absolute zero, and especially at room temperature and above, translational energy levels are considered for practical purposes to form a continuum; we do not routinely consider them quantized.

Figure 13.2 describes the **energy states**, or **energy levels**, available to an individual oxygen molecule at room temperature. (A *state* is just a condition; we must derive from the context of the discussion if the condition implies a level of energy, as it does here, or the physical form of a macroscopic sample, as in the solid, liquid, or gas state.) Figure 13.2(a) shows the energies associated with translation, rotation, and vibration, and, for comparison purposes, the energies required to raise a molecule of O_2 to its first electronic excited state, O_2^*, and to dissociate the molecule into two free O atoms. The rotational, vibrational, and electronic states are superimposable, as shown in Figure 13.2(b). The relatively small quanta involved in transitions between rotational and vibrational energy levels, as well as between kinetic energy levels, mean that an O_2 molecule at room temperature is vibrating *and* rotating *and* changing location. All possible energy levels a molecule can occupy at a given temperature are called *accessible states*. We can use Figure 13.2 to envision states accessible to an O_2 molecule as we probe what is meant by entropy and the "distribution of energy" even in a space as small as that occupied by a single O_2 molecule.

In Chapter 6 we noted that the average speed of an oxygen molecule at room temperature is about 1000 mi/hr. Some oxygen molecules in a macroscopic sample move slower; others move faster. They collide with other molecules more than a trillion times a second, and the speed of an individual molecule changes as it bangs into other molecules. The range of velocities for a population of molecules is characterized by a *Boltzmann distribution* (Figure 13.3), developed by Ludwig Boltzmann (1844–1906). An individual oxygen molecule in a sample slows down, speeds up, stops, and speeds up again, depending on its collisions with other molecules, but the overall distribution of velocities for the entire population of molecules stays the same at a given temperature. Because the velocity of a particle is directly related to its kinetic energy ($KE = \frac{1}{2} mu^2$, where u is the velocity), the

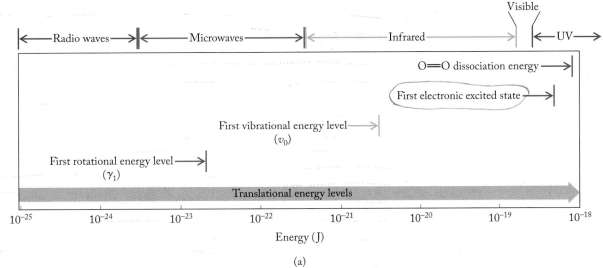

(a)

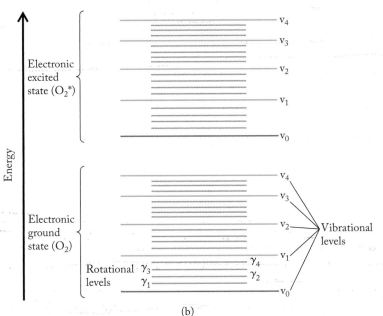

(b)

FIGURE 13.2 Several kinds of energy in an O_2 molecule are related to different types of motion. (a) Kinetic (translational) energy levels are quantized, but the differences between them are so small that kinetic energy seems to vary continuously. Rotational energy levels correspond to electromagnetic energies in the microwave region, while vibrational energy levels are 10^2 times larger and correspond to electromagnetic energies in the infrared region. The energy of a photon of UV radiation is needed to raise an O_2 molecule to its first excited state (O_2^*) and about twice that much energy is needed to break the molecule into two O atoms. (b) Super-imposing rotational, vibrational, and electronic energy levels produces a pattern like this. Note that not all the available rotational and vibrational states are shown and the gaps between them are not to scale.

kinetic energies of the particles vary widely too, since every molecule vibrates, rotates, and translates simultaneously.

With these observations in mind, let's consider a hypothetical situation for one oxygen molecule in a sample of molecules at room temperature. Figure 13.2 shows only a few of the energy levels available to any one molecule in the system, but these are nevertheless sufficient to illustrate the situation. To follow the fate of our oxygen molecule, imagine we are taking photographs of it over time. As it collides and exchanges energy with other molecules in the system, it gains and loses kinetic energy. Should it ever stop completely, it would be in the lowest translational energy state, a rare situation. At room temperature the molecule has enough energy to occupy many states, a common situation. Every so often the molecule picks up a lot of energy, allowing it to occupy a high state, but that situation is also rare. The very highest states may be inaccessible. If all the photographs of all the possible states the individual molecule occupies over a period of time are superimposed, the resulting graph is called a *Boltzmann distribution*. This graph has a profile similar to the ones in Figure 13.3, except that instead of showing the number of molecules moving at different speeds,

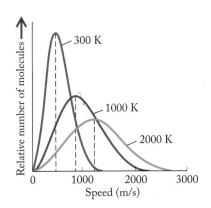

FIGURE 13.3 The range of speeds of O_2 molecules in a sample at three different temperatures is described by a Boltzmann distribution. The dashed lines identify the most probable speeds at each temperature.

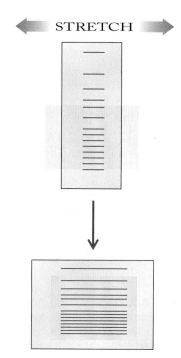

STRETCH

FIGURE 13.4 Accessible states in this system are those visible through the yellow "window." When the size of the system is increased, as illustrated by stretching the elastic sheet horizontally, the energy levels move closer together vertically. The window does not change in size, but the distance between the states becomes smaller as the sheet is stretched, and more states fall within the window and become accessible.

A **microstate** is a unique distribution of particles among energy levels.

it describes the number of times a molecule occupies a particular state among a range of accessible states.

Now let's think about what happens when dealing with a mole of molecules at room temperature. The energy of vibrational, rotational, and translational motion of all the molecules is distributed among the accessible levels in much the way it was distributed for one molecule: a few molecules are in the low levels, more are in the middle levels, and fewer (but not zero) molecules are in the higher levels. As we did with the single molecule, we can imagine photographing the distribution of molecules over time. Each photograph captures the distribution of molecules among the accessible energy levels that result when each molecule in the larger population experiences a collision and changes its level of energy. Every photograph shows a slightly different distribution, although all the photos symbolize the same total energy in the system. Their superimposition is the Boltzmann distribution for the population of molecules at the specific temperature.

The individual photographs—the different distributions of energy—identify microstates. A **microstate** in this example is a unique distribution of particles among energy levels. Because a mole contains so many particles, each having trillions of collisions per second, a mole of gas molecules at room temperature has a huge number of accessible microstates. Microstates are the key to understanding entropy at the molecular level.

This is the point where quantum mechanics comes back into the discussion. Whenever the volume of a system increases, the energy levels of the system move closer together. Imagine printing Figure 13.2(b) on a rubber sheet. If you pull on the sides of the sheet, just like pulling on a rubber band, the width of the sheet increases while the height decreases, which brings the lines closer together (Figure 13.4). That illustrates what happens when you increase the volume of a system: The differences in energy separating the levels decrease.

When the separation between energy states decreases and the temperature remains the same, the number of microstates to which the molecules have access increases. The population of molecules still has the same total energy because the temperature does not change, but when higher levels drop in energy, more molecules have access to them. This means the number of accessible microstates increases—that is what "energy spreading out" means. More microstates means more possibilities for arrangements, because more configurations with the same energy of motion are possible for the system. The energy is less localized because it can be in any one of a greater number of accessible states within a larger space. In terms of the tire, the gas that was previously constrained to a small volume expands to occupy a larger volume, thereby increasing the number of accessible microstates in which the energy can be localized. The increase in the number of microstates is what an increase in entropy means at a molecular level.

If we return to the assumption we made with the deflating tire—that the gas does not drop in temperature as it expands—we can see that even if the temperature drops as the gas escapes from the tire, the drop in kinetic energy of the molecules is more than offset by the increase in the number of microstates available to the molecules. Thus the process *increases* in entropy even though the temperature drops.

Statistical Entropy: A Mathematical View of Microstates

So far in this section we have described the macroscopic view of entropy (energy spreading out) and the microscopic view of entropy (an increase in microstates)—

but only in words. Boltzmann derived the following mathematical equation to relate entropy directly to the number of microstates:

$$S = k \ln W \qquad (13.1)$$

where S is entropy, W is the number of microstates, and k is the *Boltzmann constant*, which has a value of 1.38×10^{-23} J/K. The value of k, called the *gas constant per molecule*, is equal to R/N_A (where R is the universal gas constant and N_A is Avogadro's number). As we discuss later in the chapter, the entropy of some substances has been determined experimentally. If we know the value of entropy, then by using Equation 13.1 we can calculate the number of microstates.

Boltzmann's equation provides a connection between the macroscopic view of entropy, which relies on temperature, and the microscopic picture of entropy, which depends on the position and motion of every particle in a system. The link between these two views is the idea of microstates. Equation 13.1 is sometimes referred to as *statistical entropy*. The equation tells us that entropy increases as the number of microstates increases.

CONCEPT TEST

In his pioneering work in genetics, Gregor Mendel cross-pollinated pea plants. The peas they produced were either yellow or green and either wrinkled or smooth. Based on their appearance, how many "microstates" were accessible to the peas in Mendel's experiments?

13.2 Thermodynamic Entropy

When anything cold comes into contact with warmer surroundings, the cold object spontaneously absorbs heat from the surroundings. As we mentioned earlier, this is an example of an endothermic process that is spontaneous. Let's look at what happens at the microscopic level when a simple spontaneous process—the melting of ice—takes place at room temperature.

Isothermal and Nonisothermal Processes

When ice cubes melt at room temperature, thermal energy flows from the surroundings into the ice cube and thereby becomes more dispersed. Heat enters a region that it did not occupy previously, so it spreads out.

Now consider ice melting at the microscopic level (Figure 13.5). From the point of view of the system (the ice cube), the molecules of water before they melt occupy fixed positions in the ice cube and experience only vibrational motion (Figure 13.5a). As the ice melts, the attractive forces that hold the molecules together in the crystal lattice are overcome, the molecules in the liquid pick up rotational and translational motion as well (Figure 13.5b), and the molecules can occupy more spatial positions. The number of microstates increases. Furthermore, as the liquid warms above the melting point and the temperature of the system increases, even more microstates become available because the molecules have more kinetic energy. When the liquid evaporates and the molecules enter the gas phase, many more microstates become available (Figure 13.5c). More available microstates means higher entropy. In summary, when the ice cube melts, its entropy increases; when the liquid water warms above the melting point, its entropy increases, too; and when liquid water evaporates and becomes a gas, its entropy also increases.

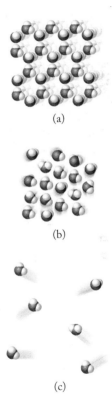

(a)

(b)

(c)

FIGURE 13.5 What a Chemist Sees When ice melts to form liquid water and liquid water vaporizes, the molecules experience different types of motion. (a) Molecules in ice exist in a rigid crystal lattice; they have vibrational motion only. (b) Molecules in liquid water have translational, rotational, and vibrational motion, but their translational motion is restricted because they are in a condensed phase. (c) Molecules in a gas have translational, rotational, and vibrational motion; their translational motion is unrestricted in the vapor phase.

A process that is **isothermal** takes place at constant temperature.

As we saw in Chapter 5, ice cubes melt at 0°C (273 K) and, while going through the change of phase from solid to liquid, they remain at 0°C. Phase changes of pure materials—such as solids melting (fusion) at their melting point, or liquids vaporizing (boiling) at their boiling point—are **isothermal** processes, which means they take place at constant temperature. Melting and vaporizing are also endothermic processes. The *entropy change* (ΔS) for a process such as ice melting is defined as

$$\Delta S = S_{final} - S_{initial}$$

Because entropy (S) is a state function just as enthalpy (H) is, the change in entropy experienced by a system depends only on its initial and final states, not on the path taken during the change.

For the special case of phase changes taking place isothermally, the entropy change (ΔS) associated with the process can be calculated using the equation

$$\Delta S = \frac{q_{rev}}{T} \qquad (13.2)$$

The symbol q_{rev} means the flow of heat as the process is carried out *reversibly*, in very small steps and very slowly.

It takes 6.01 kJ to melt a mole of ice at 0°C. Therefore the entropy change associated with melting 1 mole of ice at 0°C (273 K) is

$$\Delta S = \frac{(1 \text{ mol})(6.01 \times 10^3 \text{ J/mol})}{273 \text{ K}} = +22.0 \text{ J/K}$$

Because heat is added to the ice, q is positive and ΔS is also positive. A positive value for ΔS means that the entropy of the system (ice) increases as a result of melting.

 CONNECTION In Chapter 10 we defined the melting point as the temperature at which a solid transforms into a liquid at the same rate at which the liquid transforms into the solid.

If no heat were added to a mixture of ice and water at 0°C, solid ice would melt at the same rate at which liquid water would freeze. Both states would coexist, and no net change in the quantity of either ice or water would occur over time. What does this mean in terms of the entropy of the system (in this case, ice and water)? The melting of 1 mole of ice at 0°C results in an increase in entropy of +22.0 J/K; the reverse process, the freezing of 1 mole of water at 0°C, results in a decrease in entropy of −22.0 J/K. The net entropy change on the part of the system is 0.0 J/K. Because this process occurs isothermally, the surroundings experience no change in temperature (T remains constant); the surroundings also experience no change in entropy. The total change in entropy of everything, system plus surroundings, is the total entropy change of the universe, which in this case is zero:

$$\Delta S_{univ} = \Delta S_{sys} + \Delta S_{surr} = 0 \qquad (13.3)$$

What happens in terms of entropy in a process that is not isothermal? Let's look at two situations: (1) the melting of 1 mole of ice at 0°C on a countertop at room temperature (22°C; 295 K) to form a pool of water at 0°C, which then warms to room temperature; and (2) the melting of 1 mole of ice at 0°C on the burner of a stove at 90°C (363 K) to form a pool of water at 0°C, which then warms to 90°C. We can consider the changes in each scenario as taking place in two steps: the first step is the isothermal melting of the solid, and the second is the liquid warming to match the temperature of its surroundings.

In the first step of scenario 1, the heat required to melt 1 mole of ice is the same, $q = +6.01 \times 10^3$ J, and the entropy change of the system (the ice) is the same as in the previous example: +22.0 J/K. The heat q_{surr} lost by the countertop in contact with the ice cube (the ice cube's immediate surroundings) must be the same magnitude but opposite in sign: $q_{surr} = -6.01 \times 10^3$ J/mol. The countertop is at a higher temperature, so the entropy change of the surroundings is

$$\Delta S_{surr} = \frac{(1 \text{ mol})(-6.01 \times 10^3 \text{ J/mol})}{295 \text{ K}} = -20.4 \text{ J/K}$$

they are all iso thermal step?

The total entropy change as a result of this spontaneous process is positive:

$$\Delta S_{univ} = \Delta S_{sys} + \Delta S_{surr} = (+22.0 \text{ J/K}) + (-20.4 \text{ J/K})$$
$$= +1.6 \text{ J/K}$$

Notice the relation between the two entropy values (system and surroundings). The cooler object (the system in this case) experiences an increase in entropy that is larger in magnitude than the drop in entropy experienced by the surroundings, which are at a higher temperature.

What is the effect on the magnitude of the entropy change when the temperature difference is larger, as is the case when the ice comes in contact with the burner on the stove at 90°C (scenario 2)? In this case, q_{surr}/T is even smaller in magnitude

$$\Delta S_{surr} = \frac{(1 \text{ mol})(-6.01 \times 10^3 \text{ J/mol})}{363 \text{ K}} = -16.6 \text{ J/K}$$

and the resultant change in total entropy is larger:

$$\Delta S_{univ} = \Delta S_{sys} + \Delta S_{surr} = (+22.0 \text{ J/K}) + (-16.6 \text{ J/K})$$
$$= +5.4 \text{ J/K}$$

Ice melting on a countertop (above) and in a hot pan on the stove (below).

The key points here are that the process of melting for a system that is cooler than its surroundings is spontaneous, and the greater the temperature difference between system and surroundings, the greater the entropy increase in the isothermal step. *

In step 2 of the process, the water formed at 0°C warms up to room temperature or stove temperature and thereby experiences another increase in entropy. Increasing the average kinetic energy of the molecules in a sample increases the number of accessible microstates, so we predict that the water warmed to 90°C would experience a greater increase in entropy in part 2 of the process than the water warmed to 22°C.

In the examples thus far, we have considered spontaneous processes involving heat flowing from a warm object into a cooler object. How does entropy change when the surroundings for a sample of liquid water are suddenly made colder? Suppose a sample of water is collected in a tray and placed in a freezer at −10°C (263 K). We know the water will freeze. The surroundings (the freezer) are lower in temperature than the water. Heat would spontaneously flow from the water at 0°C into the surroundings at −10°C. The net entropy change for this process is:

$$\Delta S_{univ} = \Delta S_{sys} + \Delta S_{surr} = \frac{(1 \text{ mol})(-6.01 \times 10^3 \text{ J/mol})}{273 \text{ K}} + \frac{(+6.01 \times 10^3 \text{ J/mol})}{263 \text{ K}}$$

$$= (-22.0 \text{ J/K}) + (+22.8 \text{ J/K})$$

$$= +0.8 \text{ J/K}$$

Once again there is an increase in the entropy of the universe as heat flows spontaneously from the warmer object (the water) into the colder surroundings (the freezer). The ice that forms at 0°C ultimately cools further to the temperature of the freezer.

You may wonder why we do not bother to consider the change in temperature of the surroundings in either of these examples. The answer lies within your experience: how much does the temperature of the surroundings—in thermodynamic terms, the entire rest of the universe (or for practical purposes, the room, or even just the entire countertop, or the freezer)—change as a result of the melting (or freezing) of one ice cube? It does not change at all, or at least not by a measurable amount. The surroundings (the universe other than the system) may be treated in this example, and indeed in most examples, as a huge constant-temperature heat source or heat sink, depending on the direction in which heat flows in the process being examined. The surroundings are so vast (the rest of the universe) that systems studied in chemical thermodynamics can take small quantities of heat away or dump small quantities of heat into the universe and not change its temperature at all.

A Closer Look at Reversible Processes

In Equation 13.2, we specified that q is the heat from a reversible process. *Reversibility* comes up frequently in thermodynamics, so it is helpful to understand what it means and why it is used.

Whereas the Boltzmann equation (Equation 13.1) relates values of entropy to numbers of microstates, a second approach to calculating entropy deals with quantities of heat and temperature. It derives from 19th-century studies that focused on understanding the work done by steam engines. This approach assesses entropy changes (ΔS) during isothermal processes. The entropy change in this approach is based on the heat exchanged in a process (q_{rev}) carried out at a constant temperature (T, in kelvins). The subscript "rev" signals that q_{rev} is the heat associated with a reversible process: $\Delta S = q_{rev}/T$. A *reversible* process is a process that, after taking place, can be reversed so that no net heat flows to the system or its surroundings if the system is restored to its original state. It is important to understand that reversible processes are idealizations; real processes never quite work this way. But the *idea* of a reversible process provides us with a starting point for calculations involving real processes, especially for determining energy efficiency. By energy efficiency we mean the ratio of the work done by a system and the total free energy available to the system to do work. Ideally, energy efficiency would be 100%; in reality it is always less

Real processes approach reversibility if they are carried out in a series of infinitesimally small steps. In considering entropy, we can approximate a reversible process by breaking the process into small temperature gradients and then seeing what happens when heat is transferred across them. For example, if we transfer heat from surroundings at 273.0001 K to an object at 273.0000 K, that small step approximates a reversible process; the changes in temperature are so small that "no" change really occurs. We could in principle heat an object reversibly by going through a series of small steps like this one. Of course, it would take a very long time, but in thermodynamics time is not important.

Why do we bother to think about this task that is essentially not possible? Considering reversible processes makes the analysis simpler. Just remember, a reversible process is not truly real, but rather ideal, and evaluating the ideal can be a very useful and insightful exercise. (Recall how in our discussion of air escaping from the tire we assumed there was no temperature change. In effect, we carried out a thought experiment in which the process occurred reversibly.)

The changes in entropy calculated for reversible processes are always idealized; they are always minimum changes in entropy for a given process. When the actual process takes place, the changes in entropy are larger. That is why calculated efficiencies of mechanical devices are considered maximum efficiencies. Efficiency for an engine, for example, expresses how much useful work we get out for the amount of energy we put in. If an engine does its work reversibly ("ideally"), we get maximum useful work. Because no engine works ideally, devices are always less efficient than we calculate them to be. Even in the ideal world, we always lose in thermodynamics; in the real world, we lose even more.

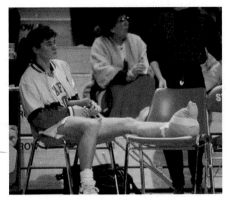

First-aid procedures for many athletic injuries include a bag of ice administered at the point of injury. When a supply of ice is unavailable, chemical cold packs based on an endothermic dissolution process may be used. Many of these packs consist of a bag of water with a pouch containing solid ammonium nitrate inside it. The pack is typically activated by breaking the pouch, causing the NH_4NO_3 to dissolve. Because this is an endothermic process, the solution (the system) is chilled.

Entropy Changes for Other Common Processes

Let's consider the meaning of Equation 13.3 in the context of another spontaneous process, the dissolution of ammonium nitrate (NH_4NO_3) in a cold pack. Among the typical first-aid supplies carried by athletic trainers are cold packs: plastic bags that contain water and a separate compartment filled with solid ammonium nitrate. If an athlete is injured, the interior pouch can be easily broken, mixing solid NH_4NO_3 with the water (Figure 13.6). The dissolution of the salt is spontaneous, but the process is endothermic, so the water cools down considerably, providing the trainer with a cold compress to ease the pain of the injury and reduce swelling.

Why is this endothermic process spontaneous? The dissolution of any solid solute in a solvent increases the freedom of motion of the atoms, ions, or molecules that comprise the solid. The ordered arrangement of dipoles in water is disrupted by the presence of dissolved ions, but the solvent molecules become reordered as water molecules orient themselves around the ions. The net effect is an overall positive change in entropy of the system ($\Delta S_{sys} > 0$).

Similarly, dilution of a concentrated solution by the addition of more solvent is a spontaneous process. If you add water to a concentrated solution of antifreeze before putting it in your car's radiator, the water spontaneously dilutes the antifreeze. There is no significant change in the temperature of water and antifreeze as the two liquids dissolve in each other, so the spontaneity of the process is not

CONNECTION The ion–dipole interactions that promote the solubility of ionic compounds in water are described in Chapter 10.

▶️‖ **CHEMTOUR** Dissolution of Ammonium Nitrate

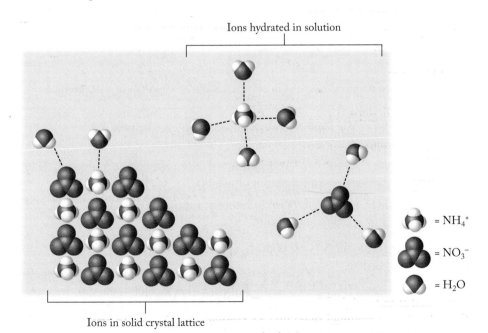

Ions hydrated in solution

= NH_4^+

= NO_3^-

= H_2O

Ions in solid crystal lattice

FIGURE 13.6 What a Chemist Sees Cold packs make use of the endothermic character of the dissolution of ammonium nitrate in water. Dissolving the ionic solute increases the freedom of motion of the ions, which are hydrated by water molecules. Even though the process is endothermic ($\Delta H > 0$), the dissolution process is spontaneous because the result is a positive change in entropy ($\Delta S > 0$).

related to a flow of heat into, or out of, the solution. If there is no significant heat flow, then there is no significant change in the dispersion of energy in the surroundings. This means there is little change in the entropy of the surroundings. Therefore, the spontaneity of the process ($\Delta S_{univ} > 0$) must be linked to the increase in entropy that happens when solute and solvent molecules mix together, dispersing into a larger overall volume ($\Delta S_{sys} > 0$).

Spontaneous processes must result in an increase in the entropy of the universe. No violation of this concept has ever been observed, and it is so widely accepted as true that most people and virtually all scientists have stopped looking for violations. This is why so-called perpetual motion machines (also called free-energy devices), which store or put out more energy than is put into them, are all hoaxes; everything ultimately runs down. The combinations of ΔS_{sys} and ΔS_{surr} that produce a positive ΔS_{univ} are shown on the left side of Figure 13.7 and summarized in mathematical terms as follows:

1. If $\Delta S_{sys} > 0$ and $\Delta S_{surr} > 0$, then $\Delta S_{univ} > 0$ (Figure 13.7a).
2. If $\Delta S_{sys} < 0$, then ΔS_{surr} must be greater than zero, and of such a magnitude that $\Delta S_{sys} + \Delta S_{surr} > 0$ (Figure 13.7b).
3. If $\Delta S_{sys} > 0$, then ΔS_{surr} may be less than zero as long as $\Delta S_{sys} + \Delta S_{surr} > 0$ (Figure 13.7c).

Note that the right side of the figure shows various scenarios for nonspontaneous processes (Figure 13.7d–f).

SAMPLE EXERCISE 13.1   **Predicting the Sign of Entropy Changes**

Predict whether ΔS_{sys} is positive or negative for each of these spontaneous processes taking place at constant temperature:

a. $H_2O(\ell) \rightarrow H_2O(g)$ c. $Pb^{2+}(aq) + 2\,Cl^-(aq) \rightarrow PbCl_2(s)$

b. $NH_3(g) + HCl(g) \rightarrow NH_4Cl(s)$ d. $C_{12}H_{22}O_{11}(s) \xrightarrow{H_2O} C_{12}H_{22}O_{11}(aq)$

COLLECT AND ORGANIZE We are asked to predict whether the entropy change for the system is positive or negative.

ANALYZE

a. A mole of water molecules becomes a mole of vapor molecules, which increases their freedom of motion.

b. Two moles of gas form one mole of solid, which results in a decrease in the freedom of motion.

c. Three moles of ions in solution form one mole of a solid compound, which represents a decrease in freedom of motion.

d. One mole of a solid compound dissolves, forming a mole of molecules dispersed in an aqueous solution.

SOLVE a. $\Delta S_{sys} > 0$ b. $\Delta S_{sys} < 0$ c. $\Delta S_{sys} < 0$ d. $\Delta S_{sys} > 0$

THINK ABOUT IT Entropy increases when solids melt and liquids vaporize because of increased freedom of motion of the particles that make up these substances. Similarly, when solids dissolve, entropy increases, but when gas dissolves in a liquid, it loses freedom of motion and undergoes a decrease in entropy.

Practice Exercise What, if anything, can you conclude about the signs of ΔS_{sys}, ΔS_{surr}, and ΔS_{univ} in each of the four processes in this Sample Exercise?

(Answers to Practice Exercises are in the back of the book.)

Identify a process in which (a) ΔS_{sys} and ΔS_{surr} are positive and another process in which (b) ΔS_{sys} and ΔS_{surr} are negative.

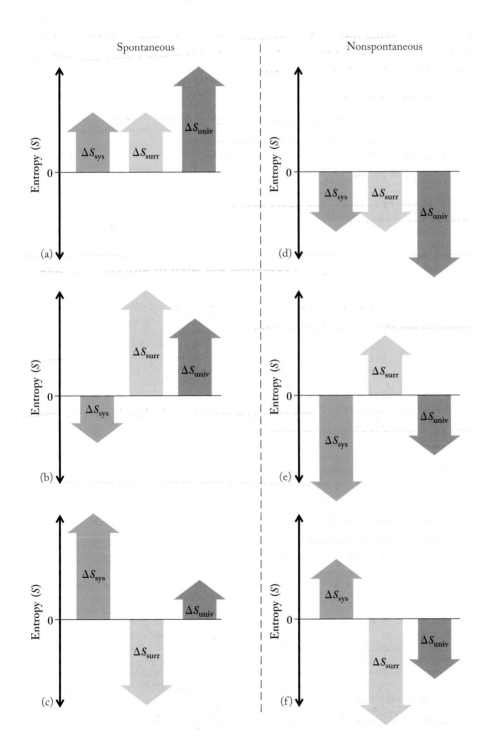

FIGURE 13.7 The relationships between ΔS_{sys}, ΔS_{surr}, and ΔS_{univ} expressed graphically, with spontaneous processes on the left and nonspontaneous processes on the right. A reaction or process is spontaneous (a) if ΔS_{sys} and ΔS_{surr} are both positive; (b) if $\Delta S_{surr} > 0$, $\Delta S_{sys} < 0$, and $|\Delta S_{surr}| > |-\Delta S_{sys}|$; or (c) if $\Delta S_{sys} > 0$, $\Delta S_{surr} < 0$, and $|\Delta S_{sys}| > |\Delta S_{surr}|$. A reaction or process is nonspontaneous if (d) ΔS_{sys} and ΔS_{surr} are both negative; (e) if $\Delta S_{sys} < 0$, $\Delta S_{surr} > 0$, and $|\Delta S_{sys}| > |\Delta S_{surr}|$; or (f) if $\Delta S_{surr} < 0$, $\Delta S_{sys} > 0$, and $|\Delta S_{surr}| > |\Delta S_{sys}|$.

646 | **Chapter Thirteen** | Thermodynamics: Spontaneous Processes, Entropy, and Free Energy

The **third law of thermodynamics** states that the entropy of a perfect crystal is zero at absolute zero.

Absolute entropy is the entropy change of a substance taken from $S = 0$ (at $T = 0$ K) to some other temperature.

Standard molar entropy ($S°$) is the absolute entropy of 1 mole of a substance in its standard state.

CONNECTION In Chapter 5 we defined the *standard state* of a pure substance in its most stable form at 1 bar (100 kPa) of pressure and some specified temperature, often 25.0°C. However, most thermodynamic data in the U.S. (and those given in Appendix 4) are for the slightly higher pressure of 1 atm.

13.3 Absolute Entropy, the Third Law of Thermodynamics, and Structure

The entropy of a substance depends on its temperature. We have seen this relation in terms of the examples given earlier in the chapter; this is also predicted by the kinetic molecular theory: higher temperatures mean higher kinetic energies of particles. Higher kinetic energy means more energy of motion: more vibrational, rotational, and translational motion; more accessible microstates, and therefore more entropy. Conversely, lower temperatures mean less of all these quantities. If we lower the temperature of a substance to absolute zero, in principle all motion ceases. If the particles of a crystalline solid are perfectly aligned and motionless at 0 K, then only one microstate exists, and from Equation 13.1 we know that entropy is zero ($S = k \ln W = k \ln 1 = 0$).

This situation is addressed by the **third law of thermodynamics**, which states that the entropy of a perfect crystal is zero at absolute zero (0 K). By *perfect* we mean that all particles are exactly aligned with one another in a crystal structure (Figure 13.8) and there are no gaps or imperfections in the crystal. Only one arrangement that is perfect can exist. Also, by definition, at absolute zero molecular motion ceases, so no changes in the position of particles occur because there is no movement. By setting a zero point on the entropy scale, scientists can compute **absolute entropy** values of pure substances.

Table 13.1 lists **standard molar entropy ($S°$)** values at 298 K for several solids, liquids, and gases. The values for liquid water [69.9 J/(mol·K)] and water vapor [188.8 J/(mol·K)] illustrate an important difference in the entropies of liquids and gases: The molecules in a gas under standard conditions are much more dispersed. On this basis alone we should expect water vapor to have more entropy at the same temperature than liquid water. All of the gases in Table 13.1 have higher standard molar entropies than their corresponding liquids. The entropies of the different phases of a given substance at a given temperature follow the order $S_{solid} < S_{liquid} < S_{gas}$.

TABLE 13.1 Select Standard Molar Entropy Values (1 atm, 298 K)[a]

Formula	$S°$, J/(mol·K)	Formula	Name	$S°$, J/(mol·K)
$Br_2(g)$	245.5	$CH_4(g)$	Methane	186.2
$Br_2(\ell)$	152.2	$CH_3CH_3(g)$	Ethane	229.5
$C_{diamond}(s)$	2.4	$CH_3OH(g)$	Methanol	239.7
$C_{graphite}(s)$	5.7	$CH_3OH(\ell)$		126.8
$CO(g)$	197.7	$CH_3CH_2OH(g)$	Ethanol	282.6
$CO_2(g)$	213.8	$CH_3CH_2OH(\ell)$		160.7
$H_2(g)$	130.6	$CH_3CH_2CH_3(g)$	Propane	269.9
$N_2(g)$	191.5	$CH_3(CH_2)_2CH_3(g)$	n-Butane	310.0
$O_2(g)$	205.0	$CH_3(CH_2)_2CH_3(\ell)$		231.0
$H_2O(g)$	188.8	$C_6H_6(g)$	Benzene	269.2
$H_2O(\ell)$	69.9	$C_6H_6(\ell)$		172.8
$NH_3(g)$	192.3	$C_{12}H_{22}O_{11}(s)$	Sucrose	360.2

[a] Values for additional substances are given in Appendix 4.

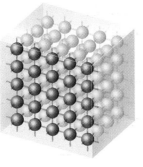
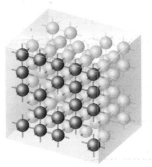
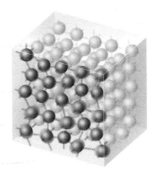

(a) Perfect crystal at 0 K:
$S = 0$

(b) Imperfect crystal at 0 K:
$S > 0$

(c) $T > 0$ K: $S > 0$

FIGURE 13.8 A perfect crystal at 0 K is the basis for defining absolute entropy. (a) In a perfect crystal at absolute zero, all the atoms or molecules are arranged perfectly in their sites within the lattice and $S = 0$. (b) If a crystal has any defects, even at absolute zero, $S > 0$. Here the defects are empty sites. (c) As the temperature increases above absolute zero, the particles in the crystal develop vibrational motion, so entropy increases.

CONCEPT TEST

Think about the ways molecules in solid, liquid, and gas phases are free to move, and explain the ordering of entropies in terms of the number of microstates accessible to molecules in each phase.

We have concluded that increases in temperature lead to changes of state that are linked to changes in entropy. The changes in entropy that occur as ice at a temperature below its melting point is heated are shown in Figure 13.9. Note the jump in entropy as the ice melts and the even bigger jump (almost five times bigger) when liquid water vaporizes. Also note that the lines between the phase changes are curved. The change in entropy with temperature is not linear because heating a substance already at a high temperature produces a smaller increase in entropy than adding the same quantity of heat to the same substance at a lower temperature, as predicted by Equation 13.2.

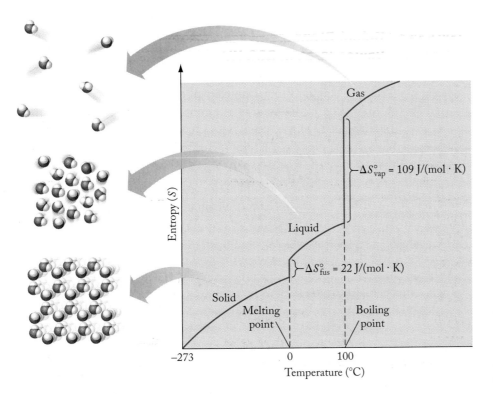

FIGURE 13.9 The entropy of a quantity of water is linked to the motion of its molecules, which increases with increasing temperature. Abrupt increases in entropy accompany changes of state, with the greatest increase during the transition from liquid to gas.

CH₄

CH₃CH₃

CH₃CH₂CH₃

CH₃CH₂CH₂CH₃

The data in Table 13.1 suggest another important factor that influences the absolute entropy of a substance: the complexity of its structure. Among the liquids listed in the table, the entropy of water is less than the entropy of methanol, which in turn is less than the entropy of benzene (water < methanol < benzene). To explore this observation, consider the following standard molar entropies of a closely related series of compounds: the C_1 to C_4 alkanes of natural gas:

Compound:	CH_4	CH_3CH_3	$CH_3CH_2CH_3$	$CH_3(CH_2)_2CH_3$
$S°$ [in J/(mol·K)]:	186	230	270	310

Note that entropy increases with increasing molecular size. We can explain this trend by considering the internal motion of the atoms in these molecules. The more bonds there are in a molecule, the more opportunities there are for internal motion, and so the greater the absolute entropy of the molecule.

CONCEPT TEST

Draw the possible vibrational motions of the diatomic molecule H_2 and the triatomic molecule H_2O. How would the number of possible vibrational motions of a molecule with three bonds (such as NH_3) compare with the number for H_2 and H_2O?

Another structural feature that is important in terms of entropy is rigidity. Diamond and graphite are both polymeric network solids with a large number of carbon–carbon bonds. They have different standard molar entropies, however, because of their very different physical properties. Diamond is the hardest natural material known, whereas graphite is soft and flexible. The more-rigid diamond molecule has less entropy than the less-rigid graphite form of carbon. For this reason, diamond, with its extended tetrahedral network of C—C bonds, has a smaller $S°$ [+2.4 J/(mol·K)] than does graphite [$S° = +5.7$ J/(mol·K)] with its networks of hexagons arranged in sheets that can easily slide by one another (Figure 13.10).

CONCEPT TEST

Consider the values of standard molar entropy of graphite and diamond. Is the reaction for the conversion of diamond to graphite spontaneous? What does this predict about the rate at which the reaction proceeds?

Let's summarize the factors that affect entropy. Entropy increases when (1) temperature increases, (2) volume increases, or (3) the number of independently moving particles increases. Entropy is increased in these events because they each increase the number of microstates. We can often make qualitative predictions about entropy changes based on these three factors, even if we have no experience with the system under examination.

When evaluating a reaction, remember that the reaction is the system. For example, when a hydrocarbon such as propane (C_3H_8) burns in air—

$$C_3H_8(g) + 5\,O_2(g) \rightarrow 3\,CO_2(g) + 4\,H_2O(g)$$

Diamond

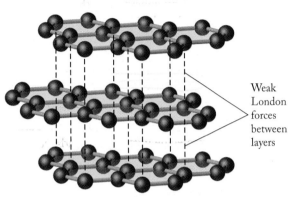

Weak London forces between layers

FIGURE 13.10 Graphite has higher molar entropy than diamond. The four covalent bonds to each sp^3 hybridized carbon atom in diamond make the three-dimensional array of atoms rigid and resistant to deformation. The sheets of sp^2 hybridized carbon atoms in graphite can slide past each other because they are held together by relatively weak dispersion (London) forces, not covalent bonds.

Graphite

—6 moles of gaseous reactants combine to form 7 moles of gaseous products. The number of moles of gas present increases. (Gases have high entropy.) When the reactants and products are at the same temperature and pressure, 7 moles of product gases occupy a greater volume than 6 moles of reactant gases; volume increases. Thus, the reaction produces an increase in the number of particles and the volume they occupy. These facts point to an increase in entropy.

SAMPLE EXERCISE 13.2 **Predicting the Sign of Entropy Changes in the System**

Predict whether these reactions result in an increase or decrease in entropy of the system. Assume the reactants and products are at the same temperature.

a. $CaCO_3(s) + 2\,HCl(aq) \rightarrow CaCl_2(aq) + CO_2(g) + H_2O(\ell)$
b. $NH_3(g) + H_2O(\ell) \rightarrow NH_4^+(aq) + OH^-(aq)$

COLLECT AND ORGANIZE We are given two reactions and asked to predict whether they result in an increase or a decrease in the entropy of the system. The reaction in each case is the system.

ANALYZE Entropy increases if temperature, volume, or the number of independently moving particles increases. The reactants and products are at the same temperature, so temperature is not a factor. We must evaluate each reaction in terms of volume and the number of particles involved.

SOLVE

   a. In this reaction, a total of 3 moles of reactants in the solid or liquid phase (condensed phases) yields 2 moles of products in the liquid phase (a condensed phase) and, more significantly from the point of view of entropy, 1 mole of gaseous product. The high entropy of gases ensures that this reaction has a positive $\Delta S_{sys} (\Delta S_{sys} > 0)$.

   b. This reaction describes the chemical changes that take place when ammonia gas dissolves in water. The key factor here is the loss of 1 mole of gas to the condensed, aqueous phase. Therefore this reaction results in a loss in system entropy ($\Delta S_{sys} < 0$). This example differs from the cold-pack reaction in that it entails the dissolution of a high-entropy gas, not a low-entropy solid.

THINK ABOUT IT To determine whether the entropy of a system (S_{sys}) increases when that system is a chemical reaction, we need to evaluate changes in temperature, changes in volume, and changes in the number of independently moving particles that occur as reactants are converted into products.

Practice Exercise Which process or processes result in an increase in entropy of the system? (a) Amorphous sulfur crystallizes. (b) Solid carbon dioxide sublimes at room temperature. (c) Iron rusts.

13.4 Calculating Entropy Changes in Chemical Reactions

We can calculate the change in entropy under standard conditions for any chemical reaction (ΔS_{rxn}°) from the difference in the standard molar entropies of m moles of reactants and n moles of products:

$$\Delta S_{rxn}^{\circ} = \sum n S_{products}^{\circ} - \sum m S_{reactants}^{\circ} \qquad (13.4)$$

The symbol Σ represents the sum of all of the entropies of the reactants or products under standard conditions. Each standard entropy S° is multiplied by the appropriate coefficient for that substance in the balanced chemical equation. For example, 2 moles of a reactant or product have twice the entropy of 1 mole. In other words, just as we discovered about ΔH° in Chapter 5, entropy is an extensive thermodynamic property that depends on the amount of a substance present or the amount consumed or produced in a reaction. Standard molar entropies for selected substances are tabulated in Appendix 4.

SAMPLE EXERCISE 13.3   **Calculating Changes in Entropy**

Calculate ΔS_{rxn}° for the dissolution of ammonium nitrate, given the following standard entropy values:

$$NH_4NO_3(s) \quad \rightarrow \quad NH_4^+(aq) \quad + \quad NO_3^-(aq)$$

| $S^{\circ}[J/(mol \cdot K)]$ | 151.0 | 112.8 | 146.4 |

COLLECT AND ORGANIZE We are given a balanced chemical equation for the reaction and asked to calculate the entropy change associated with the reaction. We are also given the standard molar entropies for the species involved in the reaction.

ANALYZE We can calculate ΔS_{rxn}° using the given S° values and Equation 13.4.

SOLVE

$$\Delta S^{\circ}_{rxn} = \sum n S^{\circ}_{products} - \sum m S^{\circ}_{reactants}$$

$$= \left[1 \text{ mol} \times \left(\frac{112.8 \text{ J}}{\text{mol} \cdot \text{K}} \right) + 1 \text{ mol} \times \left(\frac{146.4 \text{ J}}{\text{mol} \cdot \text{K}} \right) \right] - 1 \text{ mol} \times \left(\frac{151.0 \text{ J}}{\text{mol} \cdot \text{K}} \right)$$

$$= 108.2 \text{ J/K}$$

The value of ΔS°_{rxn} is positive, so the entropy of the system increases.

THINK ABOUT IT The reaction is the dissolution of a solid salt in water. Because the number of independently moving particles increases, we expect the entropy of the system to increase, so our answer is logical.

Practice Exercise Calculate the standard entropy change (ΔS°_{rxn}) for the combustion of methane gas using the appropriate S° values in Appendix 4. Before carrying out the calculation, predict whether the entropy of the system increases or decreases. Assume that liquid water is one of the products.

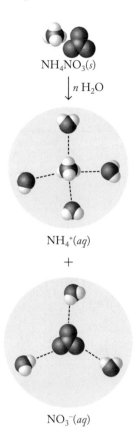

$NH_4NO_3(s)$

$\downarrow n\, H_2O$

$NH_4^+(aq)$

+

$NO_3^-(aq)$

The entropy changes calculated using Equation 13.4 are all entropy changes of the system. As we have repeatedly seen, the surroundings matter as well, even though we tend to treat them as a large heat source or sink, depending on the direction of heat flow during a process. The entropy change experienced by the surroundings of a chemical reaction depends on whether the reaction is exothermic (heat flows into the surroundings) or endothermic (heat flows from the surroundings into the system).

If we assume the reactions we are studying are isothermal processes occurring at constant pressure, then the heat flow is reversible (q_{rev}), allowing us to use Equation 13.2 to determine the change in entropy of the surroundings. Under these conditions, q_{rev} is equal to the enthalpy change (ΔH) of the reaction:

$$\boxed{\Delta H_{rxn} = q_{rev}} \qquad (13.5)$$

CONNECTION In Chapter 5 we defined a change in enthalpy (ΔH) as the heat gained or lost in a reaction carried out at constant pressure.

We can calculate ΔH for the reaction in Sample Exercise 13.3 by using values from Appendix 4 in Equation 5.18 from Chapter 5, which we repeat here:

$$\Delta H^{\circ}_{rxn} = \sum n \,\Delta H^{\circ}_{f,products} - \sum m \,\Delta H^{\circ}_{f,reactants} \qquad (13.6)$$

For the dissolution of 1 mole of $NH_4NO_3(s)$—recall that ammonium nitrate is the solid in a cold pack—Equation 13.6 gives a value of $\Delta H^{\circ} = +26.3$ kJ/mol. The process is endothermic. To calculate the entropy change of the surroundings for the dissolution taking place under standard conditions at 25°C (298 K):

$$\Delta S_{surr} = \frac{q_{surr}}{T} = -\frac{q_{sys}}{T} = -\frac{(1 \text{ mol}) \times \left(\dfrac{26.3 \times 10^3 \text{ J}}{\text{mol}} \right)}{298 \text{ K}}$$

$$= -88.2 \text{ J/K}$$

Note that the sign of the entropy change for the surroundings is negative (-88.2 J/K) while that for the system (calculated in Sample Exercise 13.3) is positive (108.2 J/K). Both terms must be considered to determine how the entropy of the universe changes during this process:

$$\Delta S_{univ} = \Delta S_{sys} + \Delta S_{surr} = (108.2 \text{ J/K}) + (-88.2 \text{ J/K})$$

$$= 20.0 \text{ J/K}$$

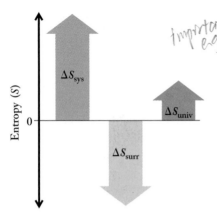

FIGURE 13.11 If the surroundings experience a decrease in entropy, the system must experience an increase in entropy that more than offsets the decrease if the process is spontaneous.

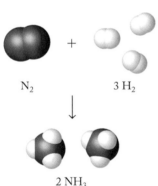

N_2 $3\,H_2$

$\downarrow$

$2\,NH_3$

The entropy change of the universe has a positive sign, so entropy increases. The process is spontaneous even though it is endothermic. For this to be the case, the decrease in entropy of the surroundings must be offset by an increase in entropy of the system (Figure 13.11).

Notice two other features of this discussion. First, because the reaction is endothermic, it takes heat energy from the surroundings. As discussed in Chapter 5, the amount of heat that the reaction absorbs equals the amount of heat taken from the surroundings:

$$\Delta H_{sys} = -\Delta H_{surr}$$

As we learned earlier in the chapter, the entropy change in the system is not the opposite of the entropy change in the surroundings:

$$\Delta S_{sys} \neq \Delta S_{surr}$$

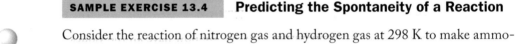

Second, enthalpy changes are typically expressed in kilojoules (kJ) whereas entropy changes in many chemical reactions are routinely expressed in joules per kelvin. For reactions with absolute ΔS values of hundreds of J/K and that occur at temperatures of hundreds of kelvins, $T\Delta S$ values, like ΔH values, are more conveniently expressed in kJ.

SAMPLE EXERCISE 13.4   **Predicting the Spontaneity of a Reaction**

Consider the reaction of nitrogen gas and hydrogen gas at 298 K to make ammonia at the same temperature:

$$N_2(g) + 3\,H_2(g) \rightarrow 2\,NH_3(g)$$

a. Before doing any calculations, predict the sign of the entropy change for the system.

b. Use the data in Table 13.1 and/or Appendix 4 to calculate $\Delta S°$ for the system.

c. Use data in Appendix 4 to calculate ΔS_{surr}.

d. Is the reaction, as written, spontaneous at 298 K and 1 atm pressure?

COLLECT AND ORGANIZE We are asked to predict the sign and calculate the value of the entropy change for a reaction. We are also asked to calculate the entropy change for the surroundings and to determine whether the reaction is spontaneous under the stated conditions.

ANALYZE (a) We have a balanced equation, so we can consider the volume and number of particles to determine the sign of the entropy change for the system. (b) We need to look up $\Delta S°$ for the reactants and products and use Equation 13.4 to calculate $\Delta S°$ of the system. (c) To calculate ΔS_{surr} we need to know the heat the reaction gives off to the surroundings. We can calculate that from the heats of formation ($\Delta H_f°$) of the reactants and products found in Appendix 4. (d) We can use Equation 13.3 to calculate the total entropy change for the universe and predict spontaneity based on the sign of that value.

SOLVE

a. Four volumes of gas (4 moles of particles) on the reactant side of the chemical equation are converted into 2 volumes of gas (2 moles of particles) on the product side. We predict that this decrease in volume and decrease in number of particles will result in a negative value of the entropy change for the reaction.

b. Using data from Table 13.1 we calculate the entropy change for the system (the reaction) as ΔS°_{rxn} using Equation 13.4:

$$\Delta S^\circ_{rxn} = \sum nS^\circ_{products} - \sum mS^\circ_{reactants}$$

$$= \left[2\ \text{mol} \times \left(\frac{192.3\ \text{J}}{\text{mol} \cdot \text{K}}\right)\right] - \left[1\ \text{mol} \times \left(\frac{191.5\ \text{J}}{\text{mol} \cdot \text{K}}\right) + 3\ \text{mol} \times \left(\frac{130.6\ \text{J}}{\text{mol} \cdot \text{K}}\right)\right]$$

$$= -198.7\ \text{J/K}$$

The entropy change for the reaction is negative, as predicted in part a.

c. If we consider the process to be carried out very slowly so that the exchange of heat with the surroundings is reversible, then we can calculate the entropy change of the surroundings from the enthalpy change of the reaction.

$$\Delta H^\circ_{surr} = -\Delta H^\circ_{rxn} = -\left[\sum nH^\circ_{f,products} - \sum mH^\circ_{f,reactants}\right]$$

$$= -\left\{\left[2\ \text{mol} \times \left(-46.1\ \frac{\text{kJ}}{\text{mol}}\right)\right] - \left[1\ \text{mol} \times \left(0.0\ \frac{\text{kJ}}{\text{mol}}\right) + 3\ \text{mol} \times \left(0.0\ \frac{\text{kJ}}{\text{mol}}\right)\right]\right\}$$

$$= 92.2\ \text{J/K heat absorbed by the surroundings}$$

Since we were given the temperature (298 K), we now have all the values we need to calculate the entropy change of the surroundings. Combining Equations 13.2 and 13.6 from the perspective of the surroundings:

$$\Delta S_{surr} = \frac{\Delta H_{surr}}{T} = \frac{92,200\ \text{J}}{298\ \text{K}} = 309\ \text{J/K}$$

The surroundings experience an increase in entropy.

d. We can use Equation 13.3 to calculate the total entropy change for the universe and predict spontaneity based on the sign of that value:

$$\begin{array}{ccccc} \Delta S_{sys} & + & \Delta S_{surr} & = & \Delta S_{univ} \\ -198.7\ \text{J/K} & + & 309\ \text{J/K} & = & 110\ \text{J/K} \end{array}$$

The entropy change of the universe is positive, so the reaction is spontaneous as written.

THINK ABOUT IT As is always the case, just because a reaction is spontaneous does not mean that it actually proceeds rapidly. Even though the reaction is spontaneous, nitrogen and hydrogen at room temperature and 1 atm pressure do not react without a significant "push" in terms of heat and pressure.

Practice Exercise Consider the reaction of hydrogen gas and oxygen gas at 298 K to produce water:

$$2\,H_2(g) + O_2(g) \rightarrow 2\,H_2O(\ell)$$

a. Predict the sign of the entropy change for this process. <0

b. Use the data in Table 13.1 and Appendix 4 to calculate ΔS° for the system.

c. Use the data in Appendix 4 to calculate ΔS_{surr}.

d. Is the reaction as written spontaneous at 298 K and 1 atm pressure?

13.5 Free Energy and Free-Energy Change

So far in this text, we have identified two driving forces that make chemical reactions happen:

1. The formation of low-energy products from high-energy reactants in exothermic reactions ($\Delta H < 0$).

2. The formation of products that have more entropy than the reactants ($\Delta S > 0$).

A reaction that is both exothermic and has a positive entropy ($\Delta S_{sys} > 0$) must be spontaneous. The heat flow produced by the exothermic reaction leads to a positive value for ΔS_{surr}. When ΔS_{sys} and ΔS_{surr} are both positive, then ΔS_{univ} is also positive (see Equation 13.3 and Figure 13.12a); according to the second law of thermodynamics, a reaction under these conditions is spontaneous. On the other hand, a reaction that cools its surroundings ($\Delta S_{surr} < 0$) because it is endothermic ($\Delta H > 0$) and that also results in a loss of entropy on the part of the system ($\Delta S_{sys} < 0$) will never be spontaneous, because ΔS_{univ} is negative in this case (Figure 13.12b). Figure 13.7 summarizes all the possible combinations of positive and negative entropies of systems and surroundings. The key point to remember is that spontaneous reactions increase the entropy of the universe ($\Delta S_{sys} + \Delta S_{surr} > 0$; Figure 13.7a–c), whereas nonspontaneous reactions decrease the entropy of the universe ($\Delta S_{sys} + \Delta S_{surr} < 0$; Figure 13.7d–f).

Determining the Entropy Change in the Universe

Left unaddressed in this analysis is whether an endothermic reaction with a positive entropy ($\Delta S_{sys} > 0$) or an exothermic reaction with a negative entropy ($\Delta S_{sys} < 0$) is spontaneous. Either combination may be spontaneous or not, depending on the relative magnitudes of ΔS_{surr} and ΔS_{sys} (see Figure 13.7).

Because it is not practical to measure changes in the entropy of the universe, we need a parameter based on the system that enables us to predict when a process will produce a positive value for ΔS_{univ} and therefore will be spontaneous. The American mathematical physicist J. Willard Gibbs (1839–1903) is credited with developing the mathematical foundation of modern thermodynamics. Among his accomplishments, he developed a way to use ΔH_{sys} and ΔS_{sys} to predict the spontaneity of a reaction occurring at constant pressure and temperature. In doing so, he defined a new state function, for which in 1877 he coined the term and symbol **free energy (G)**. The *free* in *free energy* does not mean "at no cost"; it means the energy that is available to do useful work—that portion of energy that can be dispersed into the universe.

Free energy provides a way of predicting the spontaneity of a reaction by relating G to the entropy and enthalpy changes of the system. Let's see how by deriving that relation. Based on our discussion of entropy, we know that $\Delta S_{univ} = \Delta S_{sys} + \Delta S_{surr}$. Furthermore, we learned in Chapter 5 that $-q_{sys} = +q_{surr}$ and have learned in this chapter that

$$\Delta S_{surr} = \frac{-q_{rev}}{T} = \frac{-\Delta H_{sys}}{T}$$

Therefore, it is mathematically correct to write

$$\Delta S_{univ} = \Delta S_{sys} + \frac{(-\Delta H_{sys})}{T} \qquad (13.7)$$

CONNECTION Determining the enthalpy change of reactions was discussed in detail in Chapter 5.

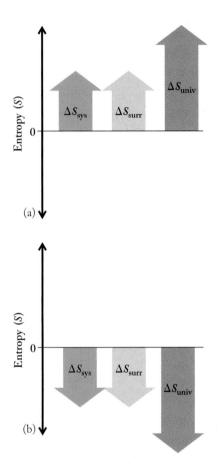

FIGURE 13.12 The magnitudes and signs of ΔS_{sys} and ΔS_{surr} determine the sign and magnitude of ΔS_{univ}. (a) When ΔS_{sys} and ΔS_{surr} are both positive, ΔS_{univ} is always positive and the process is spontaneous. (b) When ΔS_{sys} and ΔS_{surr} are both negative, ΔS_{univ} is always negative and the process is nonspontaneous.

Free energy (G) is an indication of the energy available to do useful work. Free energy is a thermodynamic state function that provides a criterion for spontaneous change.

Equation 13.7 is important because it identifies the change in entropy of the universe exclusively in terms of the system. To reformat Equation 13.7 and eliminate the quotient, we multiply both sides by $-T$:

$$-T\Delta S_{univ} = -T\Delta S_{sys} + \Delta H_{sys}$$

$$= \Delta H_{sys} - T\Delta S_{sys}$$

The **free-energy change (ΔG)** for a process occurring at constant pressure and temperature is defined as the term $-T\Delta S_{univ}$, and

$$\boxed{\Delta G = \Delta H - T\Delta S} \tag{13.8}$$

$G = -T\Delta S_{univ}$

This is the expression for Gibbs free energy. The lack of subscripts means that all the terms refer to the system. If ΔG is negative, then ΔS_{univ} must be positive (remember the definition: $\Delta G = -T\Delta S_{univ}$). Therefore the following conditions hold:

1. If ΔG for a process is negative ($\Delta G < 0$), the process as written is spontaneous.
2. If ΔG for a process is positive ($\Delta G > 0$), the process as written is not spontaneous.
3. If ΔG is zero ($\Delta G = 0$), no net change is occurring.

Equation 13.8 mathematically expresses our understanding of the two thermodynamic forces that drive chemical reactions: enthalpy and entropy. Note that ΔG is negative (and thus a reaction is spontaneous) if ΔH has a large negative value, if $T\Delta S$ has a large positive value, or both. Table 13.2 summarizes the effects of the signs of ΔH and ΔS on ΔG and on reaction spontaneity.

Free-energy change (ΔG) is the change in free energy of a process. For spontaneous processes at a constant temperature and pressure, $\Delta G < 0$.

▶❙❙ **CHEMTOUR** Gibbs Free Energy

TABLE 13.2   Effects of ΔH, ΔS, and T on ΔG and Spontaneity

	SIGN OF		
ΔH	ΔS	ΔG	Spontaneity
−	+	Always −	Always spontaneous
−	−	− at low temperatures	Spontaneous at low temperatures
+	+	− at high temperatures	Spontaneous at high temperatures
+	−	Always +	Never spontaneous

The Meaning of Free Energy

We have already defined free energy as the energy available to do useful work. Let's explore what this means using the combustion of gasoline in an automobile engine. The combustion reaction releases energy (ΔE) in the form of heat (q) and work (w) as defined in Chapter 5 by Equation 5.6:

$$\Delta E = q + w$$

Much of the heat q produced by combustion of fuel is wasted energy that flows from the engine to its surroundings as heat. Automobile engines have cooling systems to manage this heat, but the key issue is that this heat energy does nothing to move the car. The useful energy that propels the car is derived from the rapid expansion of the gaseous products of combustion. The product of the pressure (P) that is exerted by these gases on the engine's pistons times the volume change

$\Delta V = A \Delta h$

FIGURE 13.13 Thermal expansion of the gases in a cylinder in a car engine pushes down on a piston with a pressure (P) represented by the blue arrow. The product of P times the change in volume of the gases (ΔV) is the work done by the expanding gases that propels the car.

(ΔV) produced by expansion of the gases (as the pistons are pushed down) is useful ($P\Delta V$) work (Figure 13.13).

The concept of free energy relates to our attempts to obtain useful work from chemical reactions. The free energy released by a spontaneous chemical reaction at constant T and P is a measure of the maximum amount of energy that is free to do useful work. To see how this amount compares with the total energy released, let's rearrange the terms in Equation 13.8 to show this:

$$\Delta H = \Delta G + T\Delta S \qquad (13.9)$$

Equation 13.9 tells us that the enthalpy change (ΔH) that comes from the making and breaking of bonds during a chemical process may be divided into two parts. One part, ΔG, is the portion of the energy that devices such as internal combustion engines—or steam generators, batteries, and fuel cells—convert into motion, light, or some other manifestation of the interaction of energy with matter. This is what the statement means that ΔG is energy available to do work. The other part of the enthalpy change, $T\Delta S$, is not usable: it is the temperature-dependent change in entropy, the portion of energy that "spreads out," thereby increasing the entropy of the universe.

In the case of an automobile engine (or any engine for that matter), the conversion of chemical energy stored in bonds (ΔH) into useful mechanical energy to move a car (ΔG) is never 100% efficient. Some portion of the energy released by spontaneous processes such as burning fuels is always wasted. Remember the axiom given at the outset of the chapter: in the game of energy, we cannot win (energy cannot be created) nor can we even break even (every time we use energy, some is wasted).

We can think of the efficiency of an engine as the ratio of work done to energy consumed. Efficiency is usually inversely related to the rate at which a spontaneous reaction takes place. The slower the reaction, the greater the amount of free energy likely to be harvested; the faster the reaction, the lesser the amount of free energy likely to be used to do work and the more that will be wasted in heating up the environment. Recall our earlier discussion of reversible processes: the more slowly a process is carried out, the more closely it approaches complete reversibility, at which point it would be at maximum efficiency.

Reporting the efficiencies of energy-using devices in the media is becoming increasingly common. We can now understand what these numbers mean in terms of our discussion here. Mechanical efficiency is frequently expressed in relative percentages of work done per amount of energy consumed: water wheels are 90%; fuel cells, 80%; jet engines, 60%; car engines, 30%; fluorescent lights, 20%; and incandescent lightbulbs, 5%. To reiterate: not only is all of the $T\Delta S$ energy from a reaction wasted, but the amount of ΔG that is actually used must always be less than 100% of the available free energy. Additional heat losses always flow into the surroundings without doing anything useful.

Calculating Free-Energy Changes

The free-energy change in a reaction can be calculated from Equation 13.8. It can also be calculated from the standard free energies of formation of the products and reactants. The **standard free energy of formation (ΔG_f°)** is the change in free energy associated with the formation of 1 mole of a compound in its standard state from its elements in their standard states. See Appendix 4 for a list of ΔG_f° values.

The **standard free energy of formation** (ΔG_f°) is the change in free energy associated with the formation of 1 mole of a compound in its standard state from its elements.

We can use free energies of formation to calculate $\Delta G°_{rxn}$ using the following equation:

$$\Delta G°_{rxn} = \sum n \Delta G°_{f,products} - \sum m \Delta G°_{f,reactants} \qquad (13.10)$$

Note the similarities between Equation 13.10 and other expressions for $\Delta H°_{rxn}$ (Equation 5.18/13.6) and $\Delta S°_{rxn}$ (Equation 13.4):

$$\Delta H°_{rxn} = \sum n \Delta H°_{f,products} - \sum m \Delta H°_{f,reactants}$$

$$\Delta S°_{rxn} = \sum n S°_{products} - \sum m S°_{reactants}$$

According to Equation 13.10, the free-energy change for a chemical reaction forming n moles of products from m moles of reactants is the difference in the sums of the free energies of formation of the products and the reactants. By definition, the free energy of formation of an element in its standard state is zero.

SAMPLE EXERCISE 13.5   **Predicting ΔG° of Reactions**

The physical properties of structural isomers of C_8 alkanes (see Chapter 12) are significantly different. So, too, are their free energies of formation, as shown in the following data:

C_8 Alkane	$\Delta G°_f$ (kJ/mol)	Structure
n-Octane	16.3	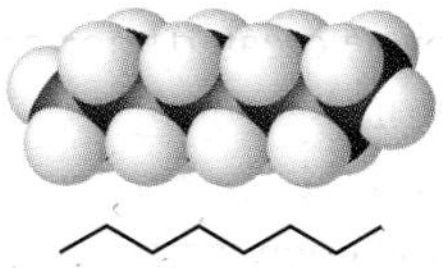
2-Methylheptane	11.7	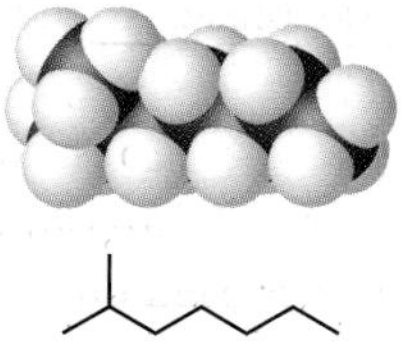
3,3-Dimethylhexane	12.6	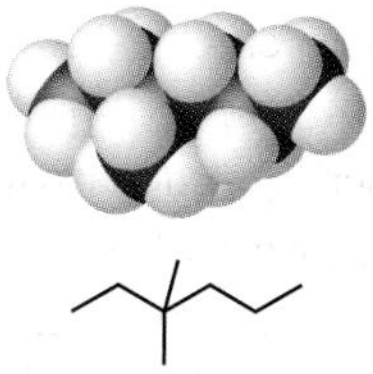

Each compound burns according to the following chemical equation:

$$2C_8H_{18}(\ell) + 25 O_2(g) \rightarrow 16 CO_2(g) + 18 H_2O(g)$$

Predict whether the $\Delta G°_{rxn}$ values for the combustion reactions of these isomers are all the same or different.

COLLECT AND ORGANIZE We are asked to predict whether the values of $\Delta G°$ for the combustion reaction of structural isomers are the same or different.

ANALYZE The free energy of combustion can be calculated according to Equation 13.10; however, there is actually no need to do the calculation. The products of the reaction are the same for each isomer, so $\Delta G°_{f,products}$ will be the same in each combustion reaction.

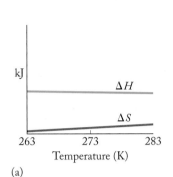

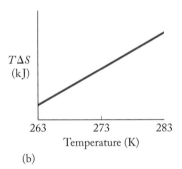

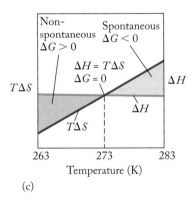

FIGURE 13.14 Changes in entropy (ΔS), enthalpy (ΔH), the quantity $T\Delta S$, and free energy (ΔG) for ice melting (fusion) at temperatures ranging from $-10°C$ (263 K) to $+10°C$ (283 K).

 CONNECTION In Chapter 10 we discussed equilibrium between phases of a substance and used phase diagrams to illustrate which physical states are stable at various combinations of temperature and pressure.

SOLVE Because $\Delta G_f^°$ of O_2 is zero by definition, and $\Delta G_f^°$ is different for each isomer, $\Delta G_{f,reactants}^°$ and $\Delta G_{rxn}^°$ for the combustion reaction are different for each isomer:

$$\Delta G_{rxn}^° = \sum n\,\Delta G_{f,products}^° - \sum m\,\Delta G_{f,reactants}^°$$

$$\Delta G_{rxn}^° = [\text{same products}] - [\Delta G_{f,isomer}^° + (25\ \text{mol} \times 0.0\ \frac{kJ}{\text{mol}})]$$

The notation "[same products]" means that each isomer produces the same products when it is burned; the balanced equations for the combustion of the isomers yield the same quantities of $CO_2(g)$ and $H_2O(g)$.

THINK ABOUT IT The free energy is different for each isomer, so changes in both enthalpy and entropy must arise because the arrangement of atoms within each molecule is unique.

Practice Exercise Using data from the table in Sample Exercise 13.5 and Appendix 4, calculate $\Delta G_{rxn}^°$ for the combustion of 1 mole of *n*-octane.

Temperature, Spontaneity, and Free Energy

Let's reconsider the isothermal process of ice melting, this time in light of the full complement of thermodynamic quantities discussed in this chapter. The influence of temperature on ΔH, ΔS, $T\Delta S$, and ΔG as ice is warmed from $-10°C$ to $+10°C$ (263 to 283 K) is presented graphically in Figure 13.14. The first graph (Figure 13.14a) shows that the enthalpy of fusion (green line) changes little, while the entropy of the system (red line) gradually increases as the temperature rises. In the second graph (Figure 13.14b) $T\Delta S$ increases rapidly as the temperature increases. The third graph (Figure 13.14c) shows the combined influence of ΔH and $T\Delta S$ on ΔG as temperature changes (remember: $\Delta G = \Delta H - T\Delta S$). This graph defines conditions under which melting is either nonspontaneous or spontaneous.

At temperatures above $0°C$, ice melts spontaneously. In the language of thermodynamics, the larger $T\Delta S$ term in Equation 13.8 for the process of melting more than offsets the positive ΔH term. As a result, ΔG is negative, indicating that ice melts spontaneously at temperatures above $0°C$. At temperatures below $0°C$, ice does not melt spontaneously, but the opposite process—liquid water freezing—*is* spontaneous. Again, in thermodynamic terms: when the $T\Delta S$ term in Equation 13.8 is smaller than the positive ΔH term, the free-energy change for melting is positive. Thus, below $0°C$, melting is not spontaneous while freezing is. The important point here is the same as with other thermodynamic quantities: the free-energy change of a process or reaction is equal in magnitude but opposite in sign to the free-energy change of the reverse process or reaction.

Another significant aspect of Figure 13.14(c) is the point on the graph where $\Delta G = 0$. The temperature value at that point defines the melting (or freezing) point, which by definition is the temperature at which the solid melts at the same rate as the liquid freezes. The two phases are at *equilibrium*. At $0°C$, no net change takes place and both phases coexist.

A similar situation exists for water at $100°C$ and 1 atm pressure, the temperature and pressure at which the ΔG values for the vaporization of water and the condensation of water vapor both equal zero. At $100°C$, liquid water vaporizes and water vapor condenses at the same rate; the two phases exist in equilibrium with each other.

The temperature at which the free-energy change for a process equals zero and the process achieves equilibrium can be calculated from Equation 13.8 if we know the values of ΔH and ΔS for the process in question. For example, the changes in standard enthalpy and standard entropy for the fusion of water (melting of ice; forward reaction as written) are

$$H_2O(s) \rightleftharpoons H_2O(\ell) \qquad (\Delta H° = 6.01 \times 10^3 \text{ J}; \Delta S° = 22.1 \text{ J/K})$$

We can use this information and Equation 13.8 to calculate the temperature at which ΔG for ice melting is zero. In doing this calculation, we assume that the values of $\Delta H°$ and $\Delta S°$ do not change significantly with small changes in temperature. For this process and for most physical and chemical changes, such an assumption is acceptable. Therefore we can assume that $\Delta G = \Delta H - T\Delta S \approx \Delta H° - T\Delta S°$. Inserting the values of $\Delta H°$ and $S°$ and using the fact that $\Delta G = 0$ for a process at equilibrium, we get the following:

$$\Delta G = (6.01 \times 10^3 \text{ J}) - T(22.1 \text{ J/K}) = 0$$

Solving for T—

$$T = \frac{6.01 \times 10^3 \text{ J}}{22.1 \text{ J/K}} = 273 \text{ K} = 0°C$$

—we obtain the familiar value for the melting point of ice.

SAMPLE EXERCISE 13.6   **Calculating Free-Energy Changes**

Calculate $\Delta G°$ for the dissolution of 1 mole of ammonium nitrate (NH_4NO_3) in water (total volume = 1 liter) at 298 K, given $\Delta H° = 26.4$ kJ/mol and $\Delta S° = 108.2$ J/(mol·K).

COLLECT AND ORGANIZE We are asked to find the free-energy change for NH_4NO_3 dissolving in water. We are given the change in enthalpy and the change in entropy for the process under standard state conditions.

ANALYZE We can use Equation 13.8 for this calculation.

SOLVE Substituting the values of $\Delta H°$ and $\Delta S°$ for ΔH and ΔS in Equation 13.8 allows us to calculate $\Delta G°$:

$$\Delta G° = \Delta H° - T\Delta S° = 26.4 \text{ kJ} - (298 \text{ K})\left(\frac{0.108 \text{ kJ}}{\text{K}}\right)$$

$$= -5.8 \text{ kJ}$$

THINK ABOUT IT $\Delta G°$ is negative, and the reaction is spontaneous. The decrease in standard free-energy is the result of an unfavorable increase in enthalpy that is more than offset by a favorable increase in entropy: the endothermic enthalpy and the positive entropy change.

Practice Exercise Predict the signs of $\Delta H°$, $\Delta S°$, and $\Delta G°$ for the combustion of methane:

$$CH_4(g) + 2 O_2(g) \rightarrow CO_2(g) + 2 H_2O(\ell)$$

Like physical processes, chemical reactions also can have zero change in free energy. As a spontaneous chemical reaction proceeds ($\Delta G < 0$), reactants become products and ΔG becomes less negative (for reasons that will become clear in a later chapter). In fact, ΔG may reach zero before all the reactants are consumed. In this case, no more products form and no more reactants are consumed. Rather, reactants and products coexist in equilibrium with each other. Reactants still react and products are still formed, but the reverse reaction, in which products re-form reactants, proceeds at the same rate as the forward reaction. This state, called chemical equilibrium, is the subject of Chapters 15 and 16. No net change in the amounts of reactants and products occurs once equilibrium is reached.

SAMPLE EXERCISE 13.7 **Relating Reaction Spontaneity to ΔH and ΔS for a Chemical Reaction**

A chemical reaction is spontaneous at low temperatures but not spontaneous at high temperatures. Use Equation 13.8 ($\Delta G = \Delta H - T\Delta S$) to determine whether ΔH and ΔS of the reactions are greater or less than zero.

COLLECT AND ORGANIZE We are asked to determine the signs of ΔH and ΔS based on the change in spontaneity of a reaction as a function of temperature. This means we need to think about how ΔG varies with temperature depending on the signs of ΔH and ΔS.

ANALYZE Following the relations summarized in Table 13.2, a reaction is spontaneous at all temperatures when $\Delta H < 0$ and $\Delta S > 0$, and a reaction will never be spontaneous when $\Delta H > 0$ and $\Delta S < 0$. There are only two other possibilities for this reaction: (1) both ΔH and ΔS are positive, or (2) both ΔH and ΔS are negative. The importance of ΔS increases with increasing temperature, because the product of the two parameters ($T\Delta S$) appears in Equation 13.8. If the reaction is not spontaneous at high temperatures where the magnitude of $T\Delta S$ is more likely to overwhelm an unfavorable ΔH value, then entropy works against the reaction being spontaneous. On the other hand, the reaction is spontaneous at low temperatures where the impact of an unfavorable (negative) ΔS is more than offset by a favorable decrease in enthalpy ($\Delta H < 0$).

SOLVE The reaction must have negative ΔS and negative ΔH values.

THINK ABOUT IT The summary in Table 13.2 confirms our prediction that a process that is spontaneous only at high temperatures is one in which there is a decrease in both entropy and enthalpy.

Practice Exercise The synthesis of ammonia from N_2 and H_2 is described by the following chemical equation

$$N_2(g) + 3\,H_2(g) \rightarrow 2\,NH_3(g)$$

Use the appropriate thermodynamic data in Appendix 4 to determine whether (a) the reaction is exothermic or endothermic; (b) the reaction is spontaneous at all temperatures.

CONCEPT TEST

How can a reaction be spontaneous when ΔS°_{rxn} is negative?

13.6 Driving the Human Engine: Coupled Reactions

The laws of thermodynamics describe the chemical reactions that power the human engine, and the same rules apply in living systems as apply in the inanimate world (Figure 13.15). Organisms transform the energy contained in the molecules that make up food and use that energy to do work. Just like mechanical engines, humans and other life forms are far from 100% efficient. Life requires the continuous input of energy in terms of the caloric content of the food we eat. Living systems are therefore thermodynamically open systems that continuously take in food (starting materials, or reactants) and give off products of metabolism (CO_2, water, and heat). Because of this constant input of reactants, elimination of products, and exchange of heat, metabolic reactions in living systems do not come to equilibrium. We as humans constantly absorb energy in the form of food and release heat and waste products to our surroundings.

Two concepts are used to describe reactions depending on the sign of ΔG: spontaneous reactions ($\Delta G < 0$) involving a decrease in free energy are called **exergonic** reactions, while nonspontaneous processes ($\Delta G > 0$) requiring the constant input of energy to proceed are called **endergonic** reactions. Exergonic reactions are typical of those dealing with breaking down foods, while endergonic reactions are typical of those involved in building substances needed by the body from the components supplied by food.

Living systems use the energy provided from exergonic reactions to run endergonic reactions; we say that exergonic reactions are *coupled* (linked or paired) to endergonic reactions in living organisms. Part of the study of *biochemistry* involves deciphering the molecular mechanisms that enable the coupling of exergonic reactions—like the breakdown of glucose—to the endergonic reactions our bodies carry out every moment of our lives, from replicating DNA to building proteins in muscle to making memories. This topic will be dealt with in more detail in Chapter 19, but for now it is sufficient to know that elegant molecular processes have evolved to enable living systems to manage the energy of exergonic reactions and direct that energy to drive the endergonic reactions that create and maintain the structures and functions we associate with life.

In Chapter 5 we looked at the energy content of a jelly doughnut, which contains a little protein and a lot of fats and carbohydrates. These three chemical groups—carbohydrates, fats, and proteins—are the primary constituents of the human diet. Young women have an average daily nutritional need of 2100 kilocalories (kcal, or Cal); for young men the figure is 2900 kcal, or a little more than 12,000 kJ each day. This level of caloric intake provides the energy we need in order to function at all levels, from getting out of bed in the morning to thinking, and the wasted energy our bodies give off as heat makes our normal temperature about 37°C.

All of the food we consume to sustain our bodies has sunlight as its ultimate source. Energy flows into the living world in the form of sunlight, which green plants store in their tissues in molecules, such as glucose, that they produce from CO_2 and H_2O during photosynthesis. Animals then consume plant products, use the energy stored in the chemical bonds of glucose and other molecules, and release CO_2 and H_2O back into the environment. They also release heat, which increases the entropy of their surroundings. Because the heat is wasted energy—energy not available to do useful work—all living things participate in the ultimate spreading out of energy. With all its organization, life increases the entropy of the universe by converting chemical and physical energy into heat, and then releasing that heat into the surroundings.

The breakdown of glucose—

$$C_6H_{12}O_6(s) + 6\,O_2(g) \rightarrow 6\,CO_2(g) + 6\,H_2O(\ell)$$

—is highly exergonic; $\Delta G°$ for this reaction is -880 kJ/mol. The production of 1 mole of glucose by a plant is endergonic (nonspontaneous), which is why green

Exergonic reactions are spontaneous reactions.

Endergonic reactions are nonspontaneous reactions.

CONNECTION An open thermodynamic system is one that freely exchanges matter and energy with its surroundings, as we discussed in Chapter 5.

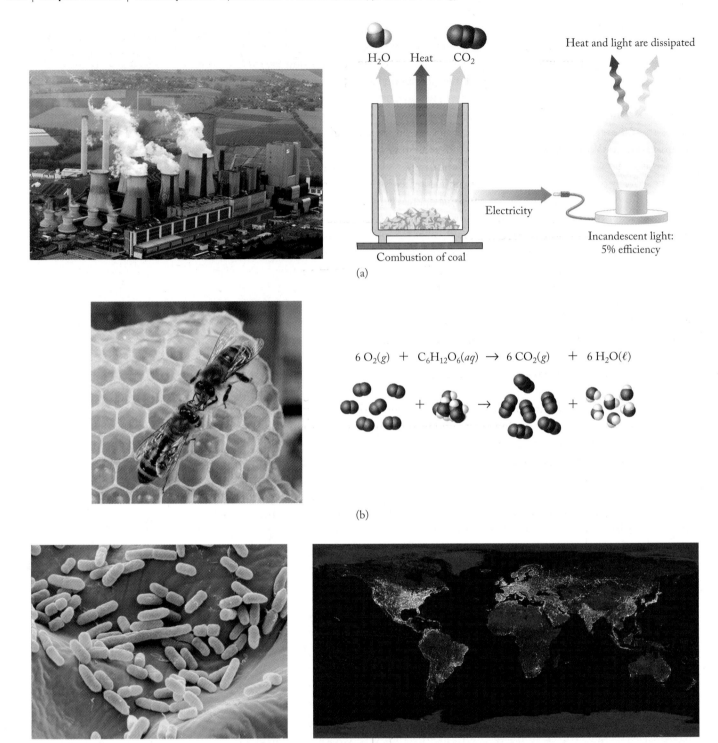

FIGURE 13.15 The rules of thermodynamics that apply in the physical world at all levels also apply to all living systems. (a) A coal-fueled power plant that generates electricity increases entropy as it burns fuel, releasing CO_2, H_2O, and waste heat into the environment. The electricity it generates contributes further to the entropy of the universe, because all machines and appliances powered by electricity lose energy whenever they use energy. (b) Living organisms, such as honey bees, extract energy from nutrients to support life. Bees store energy as honey, a mixture of sugars they produce; honey is then consumed by bees, human beings, and other animals to generate energy. When honey is consumed, heat, water, and carbon dioxide are released, increasing the entropy of the universe. (c) Microscopic organisms—such as the *E. coli* that live in our gastrointestinal tracts—consume nutrients to live and generate heat in the process. Their expenditure of energy also increases the entropy of the universe. (d) Growth of the human population in the last 150 years has been accompanied by dramatic increases in entropy as fossil fuels are increasingly burned to power vehicles and produce electricity, as reflected in this composite night-time photo of Earth.

plants require the energy of sunlight to produce glucose. When we consume glucose and convert it into energy, CO_2, and H_2O, the process is exergonic. This spontaneous process in living systems is highly controlled, so that the energy it produces can be directed into the endergonic molecular processes essential to organisms that don't photosynthesize.

Millions of years of evolution have resulted in a complex series of chemical reactions that mediate the conversion of sugar (glucose) into carbon dioxide and water in our cells. Several sequences of chemical reactions are required to manage this process and to use the considerable energy it releases to drive endergonic reactions. We will look at a portion of one of those sequences to discuss coupled reactions. In the sequence known as **glycolysis**, each mole of glucose ($C_6H_{12}O_6$) is converted into 2 moles of pyruvic acid (Figure 13.16). Pyruvic acid—or more precisely, the pyruvate ion (CH_3COCOO^-)—is a reactant in the *Krebs cycle*, a series of reactions in which carbon in molecules or ions is converted into CO_2. The hydrogen atoms in these molecules are eventually converted into H_2O.

To understand how coupled reactions work, let's first consider an early stage in glycolysis: the conversion of glucose into glucose 6-phosphate. The sequences of chemical reactions that are part of metabolism are not presented here for you to memorize. Rather, the intent is to aid your understanding of how changes in free energy allow spontaneous reactions to drive nonspontaneous reactions, and to illustrate operationally what "coupled reactions" actually means.

The conversion of glucose into glucose 6-phosphate is an example of a **phosphorylation** reaction (Figure 13.17). Glucose reacts with hydrogen phosphate ion (HPO_4^{2-}), producing glucose 6-phosphate and water. This reaction is not spontaneous ($\Delta G° = +13.8$ kJ). The energy needed to make this reaction happen comes from a compound called adenosine triphosphate (ATP). ATP functions in our cells both as a storehouse of energy and as an energy-transfer agent: it hydrolyzes to adenosine diphosphate (ADP) in a reaction (Figure 13.18) that produces a hydrogen phosphate ion and energy: $\Delta G°_{rxn} = -30.5$ kJ. (Note that ATP and ADP are both anions; the phosphate groups are ionized under typical conditions in the living system. ATP has a charge of 4−, ADP a charge of 3−. The presence of the phosphate group in glucose 6-phosphate gives the compound a 2− charge under conditions in the living system.)

Glycolysis is a series of reactions that converts glucose into pyruvate. It is a major anaerobic (no oxygen required) pathway for the metabolism of glucose in the cells of almost all living organisms.

A **phosphorylation** reaction results in the addition of a phosphate group to an organic molecule.

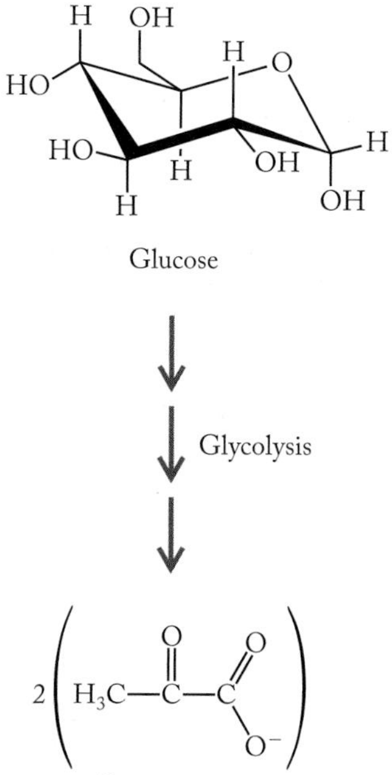

FIGURE 13.16 In glycolysis, one molecule of glucose is converted into two pyruvate ions.

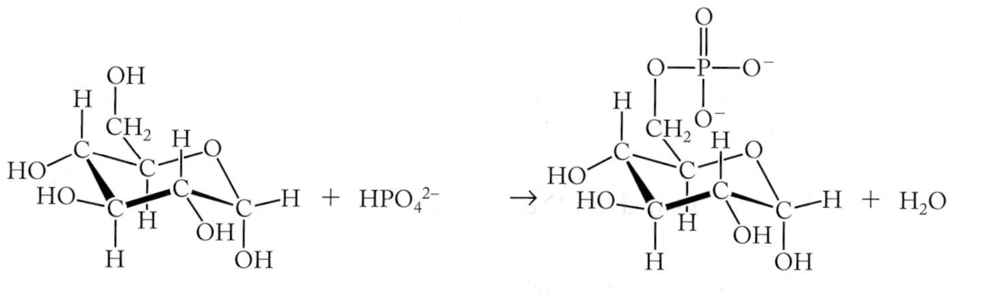

Glucose(aq) + HPO_4^{2-}(aq) → (glucose 6-phosphate)$^{2-}$(aq) + H_2O(ℓ)

Glu(aq) + HPO_4^{2-}(aq) → Glu—O—P(O)(O$^-$)—O$^-$ + H_2O(ℓ)

FIGURE 13.17 The conversion of glucose into glucose 6-phosphate is an early step in glycolysis. This reaction is endergonic (nonspontaneous). $\Delta G°_{rxn} = 17.8$ kJ/mol.

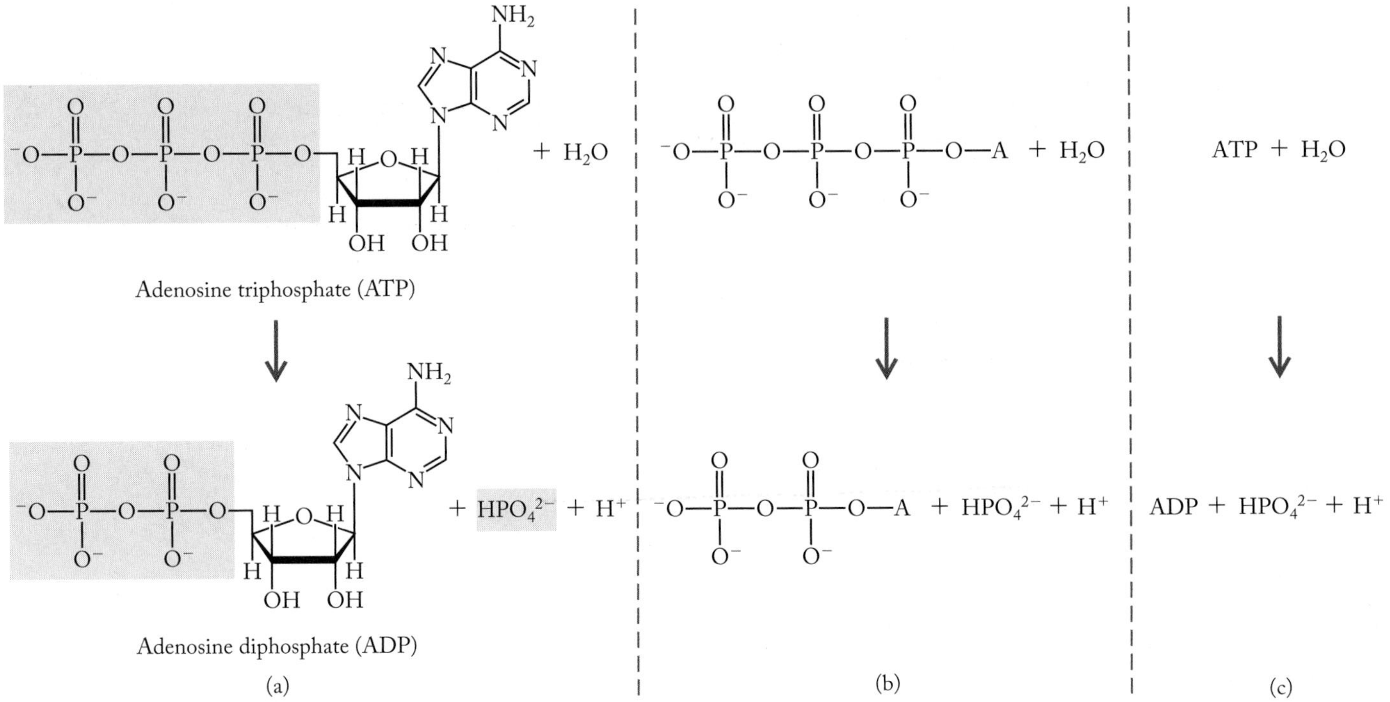

FIGURE 13.18 The hydrolysis of ATP to ADP is an exergonic (spontaneous) reaction; $\Delta G° = -34.5$ kJ/mol of ATP. This reaction is coupled to endergonic reactions to supply energy needed for many processes *in vivo*. The products of the hydrolysis of ATP are ADP, hydrogen phosphate, and a hydrogen ion. Note that ATP, with three phosphate groups, has a charge of 4− under *in vivo* conditions, while ADP has a charge of 3−, because it has only two phosphate groups. The same reaction is shown vertically in three different formats: (a) with complete structures, (b) highlighting the hydrolysis of one phosphate group from adenosine (A), and (c) using the acronyms common in biochemistry and biology for these molecules.

CONNECTION In Chapter 5 we used Hess's law (the enthalpy change of a reaction that is the sum of two or more reactions equals the sum of the enthalpy changes of the constituent reactions) to calculate ΔH. We apply a similar principle here when summing $\Delta G°$ values for coupled reactions.

In the living system, the spontaneous hydrolysis of ATP consumes water and produces HPO_4^{2-} and H^+, whereas the nonspontaneous phosphorylation of glucose consumes HPO_4^{2-} and produces water. The two reactions are coupled together *in vivo*: the exergonic ATP → ADP reaction supplies the energy that drives the endergonic formation of glucose 6-phosphate (Figure 13.19).

This example illustrates another important general point about reactions and $\Delta G°$ values. The $\Delta G°$ values for coupled reactions (or sequential reactions) are additive. This is true for any set of reactions, not just those occurring *in vivo*. For the first two steps in glycolysis:

(1) $ATP^{4-}(aq) + H_2O(\ell) \rightarrow ADP^{3-}(aq) + HPO_4^{2-}(aq) + H^+(aq)$
$$\Delta G° = -30.5 \text{ kJ}$$

(2) $C_6H_{12}O_6(aq) + HPO_4^{2-}(aq) \rightarrow C_6H_{11}O_6PO_3^{2-}(aq) + H_2O(\ell)$
$$\Delta G° = 13.8 \text{ kJ}$$

If we add the reactions in steps 1 and 2 together and add their free energies of reaction, we get

$ATP^{4-}(aq) + \cancel{H_2O(\ell)} + C_6H_{12}O_6(aq) + \cancel{HPO_4^{2-}(aq)} \rightarrow$
$\quad ADP^{3-}(aq) + \cancel{HPO_4^{2-}(aq)} + C_6H_{11}O_6PO_3^{2-}(aq) + \cancel{H_2O(\ell)} + H^+(aq)$

and

$$\Delta G° = (-30.5 + 13.8) \text{ kJ/mol} = -16.7 \text{ kJ/mol}$$

Because equal quantities of H_2O and HPO_4^{2-} appear on both sides of the combined equation, they cancel out, leaving this net overall reaction:

$C_6H_{12}O_6(aq) + ATP^{4-}(aq) \rightarrow ADP^{3-}(aq) + C_6H_{11}O_6PO_3^{2-}(aq) + H^+(aq)$
$$\Delta G° = -16.7 \text{ kJ}$$

Since $\Delta G°$ for the net reaction is negative, the coupled reaction is spontaneous.

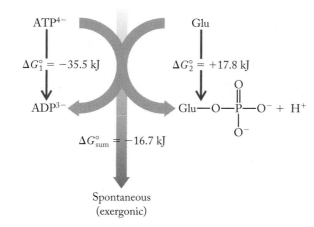

FIGURE 13.19 The hydrolysis of the energy source ATP is coupled to the phosphorylation of glucose, which requires energy. The overall reaction—the sum of the two individual reactions—is exergonic (spontaneous) because $\Delta G^\circ_{sum} < 0$.

SAMPLE EXERCISE 13.8 **Calculating ΔG° of Coupled Reactions**

The body would rapidly run out of ATP if there were not some process by which it could be regenerated from ADP. The hydrolysis of 1,3-diphosphoglycerate^{4-} (1,3-DPG^{4-}) to 3-phosphoglycerate^{3-} (3-PG^{3-}) drives the conversion of ADP^{3-} into ATP^{4-}. The hydrolysis reaction has a large negative ΔG° value. Calculate ΔG° for the reaction

ADP^{3-} + 1,3-diphosphoglycerate^{4-} → 3-phosphoglycerate^{3-} + ATP^{4-}

from the following changes in standard free energy:

(1) 1,3-DPG^{4-}(aq) + H$_2$O(ℓ) →
$\qquad$ 3-PG^{3-}(aq) + HPO$_4{}^{2-}$(aq) + H$^+$(aq) $\quad \Delta G^\circ = -49.0$ kJ

(2) ADP^{3-}(aq) + HPO$_4{}^{2-}$(aq) + H$^+$(aq) →
$\qquad\qquad\qquad$ ATP^{4-}(aq) + H$_2$O(ℓ) $\quad \Delta G^\circ = 30.5$ kJ

COLLECT AND ORGANIZE We are asked to calculate ΔG° for a reaction that is the sum of two reactions. If we add the reactions together, then ΔG° for the overall reaction will be the sum of the ΔG° values for the individual reactions.

ANALYZE The overall reaction between ADP and 1,3-diphosphoglycerate^{4-} is the sum of the reactions describing the hydrolysis of 1,3-diphosphoglycerate^{4-} and the phosphorylation of ADP.

SOLVE We sum the ΔG° values for steps 1 and 2 to determine ΔG° for the overall reaction:

$$\Delta G^\circ_{overall} = \Delta G^\circ_1 + \Delta G^\circ_2 = [(-49.0) + (30.5)] \text{ kJ} = -18.5 \text{ kJ}$$

THINK ABOUT IT The hydrolysis of 1,3-diphosphoglycerate^{4-} provides more than sufficient energy for the conversion of ADP into ATP.

Practice Exercise The conversion of glucose into lactic acid drives the phosphorylation of 2 moles of ADP to ATP:

C$_6$H$_{12}$O$_6$(aq) + 2 HPO$_4{}^{2-}$(aq) + 2 ADP^{3-}(aq) + 2 H$^+$(aq) →
$\qquad\qquad$ 2 CH$_3$CH(OH)COOH(aq) + 2 ATP^{4-}(aq) + 2 H$_2$O(ℓ)
$\qquad\qquad\qquad\qquad\qquad\qquad\qquad\qquad \Delta G^\circ = -135$ kJ/mol

What is ΔG° for the conversion of glucose into lactic acid?

C$_6$H$_{12}$O$_6$(aq) → 2 CH$_3$CH(OH)COOH(aq)

1,3-Diphosphoglycerate^{4-}
(1,3-DPG^{4-})

3-Phosphoglycerate^{3-}
(3-PG^{3-})

Glucose

Lactic acid

The ATP produced as a result of the breaking down (metabolizing) of food can be used to drive endergonic (nonspontaneous) cellular reactions. The metabolism of fats and proteins relies on a series of cycles, all of which involve coupled reactions that handle the chemical processing of the molecules. The ATP–ADP system is a carrier of chemical energy because ADP requires energy to accept a phosphate group and thus is coupled to reactions that yield energy, while ATP donates a phosphate group, releases energy, and is coupled to reactions that require energy. All energy changes in living systems are governed by the first and second laws of thermodynamics, as are all the energy changes in the physical world.

Summary

SECTION 13.1 Entropy (S) is a thermodynamic property that measures the distribution of energy in a system at a specific temperature; changes in entropy (ΔS) are the basis for understanding spontaneity. According to the **second law of thermodynamics**, processes that are **spontaneous** occur without a constant input of energy and result in an increase in entropy of the universe, whereas **nonspontaneous** processes require continuous energy input in order to occur. Forming a gas from a solid or a liquid, converting a solid into a liquid, or dissolving a solid in a liquid are processes accompanied by an increase in system entropy. Most exothermic reactions are spontaneous, but endothermic reactions may also be spontaneous. All particles in a system can occupy one of a large number of **energy states**, or **energy levels**. Each different arrangement of particles is called a **microstate**. The entropy of a system increases as the number of available microstates increases and is generally highest for gases.

SECTION 13.2 Phase changes of pure substances take place over a constant temperature and are **isothermal** processes. For a spontaneous process such as ice melting at room temperature, the entropy change of the universe must be positive: $\Delta S_{univ} > 0$. The entropy change of the universe equals the sum of the entropy change of the system and the entropy change of the surroundings: $\Delta S_{univ} = \Delta S_{sys} + \Delta S_{surr}$. Reversible processes are ideal processes that take place in very small steps and very slowly. No change in entropy takes place in a reversible process. This concept enables us to calculate maximum efficiency, which is the most work available from a given quantity of energy. If $\Delta S_{sys} < 0$ in a spontaneous process, then $\Delta S_{surr} > 0$ and ΔS_{surr} must be large enough to insure that $\Delta S_{sys} + \Delta S_{surr} > 0$. If $\Delta S_{sys} > 0$, then ΔS_{surr} must have a value so that $\Delta S_{sys} + \Delta S_{surr} > 0$.

SECTION 13.3 According to the **third law of thermodynamics**, perfect crystals of a pure substance have zero entropy at absolute zero.

All substances have positive entropies at temperatures above absolute zero. **Standard molar entropies ($S°$)** are entropy values under standard state conditions. The entropy of a system increases with increasing molecular complexity and temperature.

SECTION 13.4 The entropy change in a reaction under standard conditions can be calculated from the standard entropies of the products and reactants and their coefficients in the balanced chemical equation. If we study isothermal processes carried out reversibly, we can calculate the entropy change of the surroundings by dividing the heat exchanged ($\bar{q}_{rev}$) by the temperature at which the process occurs. We can then also calculate the entropy change of the universe and determine whether the process is spontaneous by noting the sign of ΔS_{univ}.

SECTION 13.5 The **free-energy change (ΔG)** in a process is a state function giving the maximum useful work the system can do on its surroundings. Spontaneous processes have negative ΔG values. Reversing a process changes the sign of ΔG. For a system at equilibrium, $\Delta G = 0$. The standard $\Delta G°$ free-energy change in a process can be calculated from the **standard free energies of formation ($\Delta G_f°$)** of the products and reactants or from the enthalpy and entropy changes. Values for $\Delta G_f°$ refer to standard state conditions, and for the formation of elements, $\Delta G_f° = 0$. The temperature range over which a process is spontaneous depends on the values of ΔH and ΔS.

SECTION 13.6 Exergonic reactions are spontaneous and **endergonic** processes are not. Many important biochemical processes, including **glycolysis** and **phosphorylation**, are made possible by coupled exergonic and endergonic reactions. The free energy released in such processes (such as the hydrolysis of ATP to form ADP) is used in our bodies to drive endergonic processes.

Problem-Solving Summary

TYPE OF PROBLEM	CONCEPTS AND EQUATIONS	SAMPLE EXERCISES
Predicting the sign of the entropy change in a physical or chemical process	Look for the net removal of gas molecules or precipitation of a solute from solution ($\Delta S < 0$ for both). The reverse processes have positive entropy changes.	13.1 and 13.2

TYPE OF PROBLEM	CONCEPTS AND EQUATIONS	SAMPLE EXERCISES
Calculating the entropy change of a chemical reaction	Use Equation 13.4 $$\Delta S^{\circ}_{rxn} = \Sigma nS^{\circ}_{products} - \Sigma mS^{\circ}_{reactants}$$ where $S^{\circ}_{products}$ and $S^{\circ}_{reactants}$ are the standard molar entropies of the products and reactants and n and m are the stoichiometric coefficients for the process.	13.3
Predicting the spontaneity of a process	A process is spontaneous if $$\Delta S_{univ} = \Delta S_{sys} + \Delta S_{surr} > 0$$	13.4
Calculating the free-energy change of a physical or chemical process	Use Equation 13.10 $$\Delta G^{\circ}_{rxn} = \Sigma n\Delta G^{\circ}_{f,products} - \Sigma m\Delta G^{\circ}_{f,reactants}$$ where $\Delta G^{\circ}_{f,products}$ and $\Delta G^{\circ}_{f,reactants}$ are the standard molar free energies of formation of the products and reactants, and n and m are the stoichiometric coefficients for the process. Use Equation 13.8, $$\Delta G^{\circ} = \Delta H^{\circ} - T\Delta S^{\circ}$$ where ΔH° and ΔS° are the standard enthalpy and entropy changes for the reaction, respectively.	13.5 and 13.6
Relating reaction spontaneity to ΔH and ΔS	Use Equation 13.8: $$\Delta G = \Delta H - T\Delta S$$	13.7
Calculating the free-energy change of coupled reactions	Like enthalpy and entropy changes, free-energy changes are additive. If adding two reactions gives the desired reaction, add the respective free-energy changes to obtain the free-energy change of the desired reaction.	13.8

Visual Problems

13.1. More air is added to the partially filled party balloon in Figure P13.1. Does the entropy of the balloon and its contents increase or decrease?

but the sample in (a) is cooled so that it solidifies, which sample has the higher entropy?

FIGURE P13.1

13.2. The two cubic containers on the right (Figure P13.2) contain two gas samples at the same temperature and pressure. Which has the higher entropy—the sample in (a) or the sample in (b)? If the sample in (b) is left unchanged

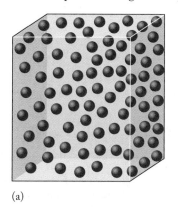
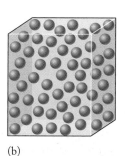

(a) (b)

FIGURE P13.2

13.3. Figure P13.3 shows a glass tube connecting two bulbs containing a mixture of ideal gases A (red spheres) and B (blue spheres). Is the probability high or low that gas A will collect in the left-hand bulb and gas B will collect in the right-hand bulb? Explain your answer in terms of the entropy changes involved.

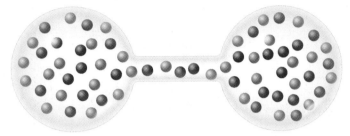

FIGURE P13.3

13.4. The box in Figure P13.4(a) contains a gas. Complete the box in Figure P13.4(b) to illustrate deposition of the gas as a solid. Suggest whether ΔS, ΔH, and/or ΔG are positive or negative for this spontaneous process. Does the entropy of the surroundings increase or decrease?

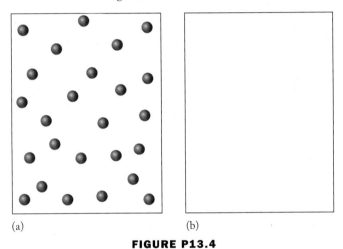

(a)   (b)

FIGURE P13.4

13.5. Figure P13.5 shows the plots of ΔH and $T\Delta S$ for a reaction as a function of temperature. What is the significance of their point of intersection? Over what temperature range is the reaction nonspontaneous?

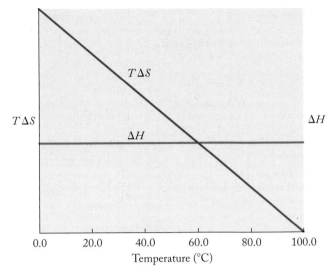

FIGURE P13.5

13.6. Figure P13.6 is based on the ΔG_f° values of elements and compounds selected from Appendix 4. Which of the following conversions are spontaneous? (a) $C_6H_6(\ell)$ to $CO_2(g)$ and $H_2O(\ell)$; (b) $CO_2(g)$ to $C_2H_2(g)$; (c) $H_2(g)$ and $O_2(g)$ to $H_2O(\ell)$. Explain your reasoning.

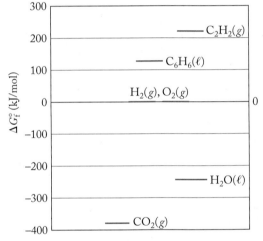

FIGURE P13.6

Questions and Problems

Spontaneous Processes and Entropy; Thermodynamic Entropy

CONCEPT REVIEW

13.7. What happens to the sign of the entropy change when a process is reversed?

13.8. Identify the following processes as spontaneous or nonspontaneous, and explain your choice: (a) the battery in a cell phone discharging; (b) a helium balloon rising above Earth's surface; (c) radioactive decay.

13.9. You have three coins that you flip. If "heads" (H) is assigned the value +1, and "tails" (T) the value −1, then the sequence HHT (+1 +1 −1) is one possible outcome, or microstate. Note that it differs from THH (−1 +1 +1), though both sequences sum to the same value. How many different microstates are possible from flipping the three coins? Which value or values for the sum of the coins in the microstates are the most likely?

13.10. Adding solid Drano to water causes the temperature of the water to increase. If Drano is the system, what are the signs of ΔS_{sys} and ΔS_{surr}?

13.11. Ice cubes melt in a glass of lemonade, cooling the lemonade from 10.0°C to 0.0°C. If the ice cubes are the system, what are the signs of ΔS_{sys} and ΔS_{surr}?

PROBLEMS

13.12. Which of the following combinations of entropy changes for a process are possible?
  a. $\Delta S_{sys} < 0, \Delta S_{surr} > 0, \Delta S_{univ} > 0$
  b. $\Delta S_{sys} < 0, \Delta S_{surr} < 0, \Delta S_{univ} > 0$
  c. $\Delta S_{sys} < 0, \Delta S_{surr} > 0, \Delta S_{univ} < 0$

13.13. Which of the following combinations of entropy changes for a process are possible?
  a. $\Delta S_{sys} > 0, \Delta S_{surr} > 0, \Delta S_{univ} > 0$
  b. $\Delta S_{sys} > 0, \Delta S_{surr} < 0, \Delta S_{univ} > 0$
  c. $\Delta S_{sys} > 0, \Delta S_{surr} > 0, \Delta S_{univ} < 0$

13.14. The spontaneous reaction $A \rightarrow B + C$ increases the system entropy by 132.0 J/K. What is the minimum value of the entropy change of the surroundings?

13.15. The nonspontaneous reaction $D + E \rightarrow F$ decreases the system entropy by 48.0 J/K. What is the maximum value of the entropy change of the surroundings?

13.16. In each of the following pairs, which alternative has the greater entropy?
  a. A pound of ice cubes or a pound of liquid water.
  b. A spoonful of sugar or a spoonful of sugar dissolved in a cup of coffee.
  c. A cup of hot water or a cup of cold water.
  d. A mole of cyclohexane, C_6H_{12}, or a mole of 1-hexene, $CH_2{=}CH(CH_2)_3CH_3$.

13.17. In each of the following pairs, which alternative has the greater entropy?
  a. A fish when it is alive or a fish after it has just died.
  b. Wet paint or dry paint.
  c. One mole of $SO_2(g)$ or 1 mole of $SO_3(g)$.
  d. An aquarium with fish or the same aquarium without fish.

Absolute Entropy, the Third Law of Thermodynamics, and Structure

CONCEPT REVIEW

13.18. In living cells, small molecules react together to make much larger ones. Are these processes accompanied by increases or decreases in entropy of the molecules?

13.19. Which physical state of a substance—solid, liquid, or gas—has the highest standard molar entropy?

13.20. Does the dissolution of a gas in a liquid result in an increase or a decrease in entropy?

*13.21. Diamond and the fullerenes are two allotropes of carbon. On the basis of their different structures and properties, predict which has the higher standard molar entropy.

13.22. The 1996 Nobel Prize in physics was awarded to Douglas Osheroff, Robert Richardson, and David Lee for discovering *superfluidity* (apparently frictionless flow) in ^{3}He. When ^{3}He is cooled to 2.7 mK, the liquid settles into an *ordered* superfluid state. What is the predicted sign of the entropy change for the conversion of liquid ^{3}He into its superfluid state?

PROBLEMS

13.23. Rank the compounds in each of the following groups in order of increasing standard molar entropy ($S°$):
  a. $CH_4(g)$, $CF_4(g)$, and $CCl_4(g)$
  b. $CH_3OH(\ell)$, $CH_3CH_2OH(\ell)$, and $CH_3CH_2CH_2OH(\ell)$
  c. $HF(g)$, $H_2O(g)$, and $NH_3(g)$

13.24. Without referring to either Table 13.1 or Appendix 4, rank the compounds in each of the following groups in order of increasing standard molar entropy ($S°$):
  a. $CH_4(g)$, $CH_3CH_3(g)$, and $CH_3CH_2CH_3(g)$
  b. $CCl_4(\ell)$, $CHCl_3(\ell)$, and $CH_2Cl_2(\ell)$
  c. $CO_2(\ell)$, $CO_2(g)$, and $CS_2(g)$

13.25. Predict the sign of ΔS_{sys} for each of the following processes:
  a. A bricklayer builds a wall out of a random pile of bricks.
  b. You rake a yard full of leaves into a single pile.
  c. $Ag^+(aq) + Cl^-(aq) \rightarrow AgCl(s)$
  d. $Zn(s) + 2HCl(aq) \rightarrow H_2(g) + ZnCl_2(aq)$

13.26. Predict the sign of ΔS_{sys} for each of the following processes:
  a. Sweat evaporates.
  b. Solid silver chloride dissolves in aqueous ammonia.
  c. $CH_3CH_2CH_3(g) + 5O_2(g) \rightarrow 3CO_2(g) + 4H_2O(\ell)$
  d. $N_2O_5(g) \rightarrow NO_2(g) + NO_3(g)$

Calculating Entropy Changes in Chemical Reactions

CONCEPT REVIEW

13.27. The products of a process have lower entropy than the reactants. Is the entropy change of the process positive or negative?

13.28. Why might the entropy change of a process increase with increasing temperature?

13.29. A reaction of $A(s)$ to make $B(\ell)$ is studied at different temperatures. Would the entropy change of the reaction be markedly different at temperatures higher than the melting point of reactant A?

*13.30. What stoichiometric and molecular factors affect the standard entropy change for the conversion of ozone to oxygen?

$$2O_3(g) \rightarrow 3O_2(g)$$

PROBLEMS

13.31. Smog Use the standard molar entropies in Appendix 4 to calculate $\Delta S°$ values for each of the following atmospheric reactions that contribute to the formation of photochemical smog, a haze caused by reactions of NO released from car and truck engines with molecules in the atmosphere:
  a. $N_2(g) + O_2(g) \rightarrow 2NO(g)$
  b. $2NO(g) + O_2(g) \rightarrow 2NO_2(g)$
  c. $NO(g) + \frac{1}{2}O_2(g) \rightarrow NO_2(g)$
  d. $2NO_2(g) \rightarrow N_2O_4(g)$

13.32. Use the standard molar entropies in Appendix 4 to calculate the $\Delta S°$ value for each of the following reactions of sulfur compounds.
  a. $H_2S(g) + \frac{3}{2}O_2(g) \rightarrow H_2O(g) + SO_2(g)$
  b. $2SO_2(g) + O_2(g) \rightarrow 2SO_3(g)$
  c. $SO_3(g) + H_2O(\ell) \rightarrow H_2SO_4(aq)$
  d. $S(g) + O_2(g) \rightarrow SO_2(g)$

13.33. Ozone Layer The following reaction plays a key role in the destruction of ozone in the atmosphere:

$$Cl(g) + O_3(g) \rightarrow ClO(g) + O_2(g)$$

The standard entropy change (ΔS°_{rxn}) is 19.9 J/mol·K. Use the standard molar entropies (S°) in Appendix 4 to calculate the S° value of $ClO(g)$.

13.34. Calculate the ΔS° value for the conversion of ozone to oxygen—

$$2O_3(g) \rightarrow 3O_2(g)$$

—in the absence of Cl atoms, and compare it with the ΔS° value in Problem 13.33.

Free Energy and Free-Energy Change

CONCEPT REVIEW

13.35. The 19th-century scientist Marcillin Berthelot stated that all exothermic reactions are spontaneous. Is this statement correct?

13.36. Under what conditions does an increase in temperature turn a nonspontaneous process into a spontaneous one?

13.37. If a reaction has a negative free-energy change, is it spontaneous?

13.38. If a reaction has a positive free-energy change, does it proceed slowly?

13.39. What can we say if the calculated free energy of a process is positive?

13.40. What can we say if the calculated free energy of a process is zero?

13.41. Are exothermic reactions spontaneous only at low temperature? Explain your answer.

13.42. Are endothermic reactions never spontaneous at low temperature? Explain your answer.

PROBLEMS

13.43. What are the signs of ΔS, ΔH, and ΔG for the sublimation of dry ice (solid CO_2) at 25°C?

13.44. What are the signs of ΔS, ΔH, and ΔG for the formation of dew on a cool night?

13.45. Indicate whether each of the following processes is spontaneous:
a. The fragrance of a perfume spreads through a room.
b. A broken clock is mended.
c. An iron fence rusts.
d. An ice cube melts in a glass of water.

13.46. Indicate whether each of the following processes is spontaneous:
a. Charcoal is converted to carbon dioxide when ignited in air.
b. Steam condenses on a cold window.
c. Sugar dissolves in hot water.
d. $CH_4(g)$ and $O_2(g)$ are formed from $CO_2(g)$ and $H_2O(\ell)$.

13.47. Calculate the free-energy change for the dissolution in water of 1 mole of NaBr and 1 mole of NaI at 298 K, given that the values of ΔH°_{soln} are −1 and −7 kJ/mol for NaBr and NaI. The corresponding values of ΔS°_{soln} are 57 and 74 J/(mol·K).

13.48. The values of ΔH°_{rxn} and ΔS°_{rxn} for the reaction

$$2NO(g) + O_2(g) \rightarrow 2NO_2(g)$$

are −12 kJ and −147 J/K. Calculate ΔG° at 298 K for this reaction. Why do you think the value of ΔS° is negative?

*13.49. A mixture of $CO(g)$ and $H_2(g)$ is produced by passing steam over charcoal:

$$H_2O(g) + C(s) \rightarrow H_2(g) + CO(g)$$

Calculate the ΔG° value for the reaction from the data in Appendix 4, and predict the lowest temperature at which the reaction is spontaneous.

13.50. Consider the following combustion reactions:

(1) $2CH_3OH(g) + 3O_2(g) \rightarrow 2CO_2(g) + 4H_2O(g)$

(2) $CH_4(g) + 2O_2(g) \rightarrow CO_2(g) + 2H_2O(\ell)$

(3) $2H_2(g) + O_2(g) \rightarrow 2H_2O(g)$

a. For each reaction, predict the sign of ΔS° before calculating its value.
b. For each reaction, calculate ΔS° and ΔG° from the data in Appendix 4.

13.51. Use the data in Appendix 4 to calculate ΔH° and ΔS° for the following process:

$$H_2O(\ell) \rightarrow H_2O(g)$$

Assume that the calculated values are independent of temperature, and calculate the boiling point of water at $P = 1.00$ atm.

13.52. Chlorofluorocarbons (CFCs) are no longer used as refrigerants because they help destroy the ozone layer. Trichlorofluoromethane (CCl_3F) boils at 23.8°C and its molar heat of vaporization is 24.8 kJ/mol. Calculate the molar entropy of evaporation of $CCl_3F(\ell)$.

*13.53. Deposits of elemental sulfur are often seen near active volcanoes. Their presence there may be due to the following reaction of SO_2 with H_2S:

$$SO_2(g) + 2H_2S(g) \rightarrow 3S(s) + 2H_2O(g)$$

Calculate ΔH° and ΔS° for this reaction, and predict the temperature range over which the reaction is spontaneous.

13.54. Methanogenic bacteria convert acetic acid (CH_3COOH) into $CO_2(g)$ and $CH_4(g)$. (a) Is this process endothermic or exothermic under standard conditions? (b) Is the reaction spontaneous under standard conditions?

13.55. Use the data in Appendix 4 to calculate ΔG° for each of the following reactions. Is each reaction spontaneous at 355 K?
a. $N_2(g) + O_2(g) \rightarrow 2NO(g)$
b. $2NO(g) + O_2(g) \rightarrow 2NO_2(g)$
c. $NO(g) + \frac{1}{2}O_2(g) \rightarrow NO_2(g)$
d. $2NO_2(g) \rightarrow N_2O_4(g)$

13.56. Use the data in Appendix 4 to calculate ΔG° for each of the following reactions. Is each reaction spontaneous at 211 K?
a. $2H_2S(g) + 3O_2(g) \rightarrow 2H_2O(g) + 2SO_2(g)$
b. $2SO_2(g) + O_2(g) \rightarrow 2SO_3(g)$
c. $SO_3(g) + H_2O(\ell) \rightarrow H_2SO_4(\ell)$
d. $S(g) + O_2(g) \rightarrow SO_2(g)$

13.57. Which of the reactions in Problem 13.55 are spontaneous at (a) high temperature? (b) low temperature? (c) all temperatures?

13.58. Which of the reactions in Problem 13.56 are spontaneous at (a) high temperature? (b) low temperature? (c) all temperatures?

13.59. Use the free energies of formation from Appendix 4 to calculate the standard free-energy change for the decomposition of ammonia in the following reaction:

$$2\,NH_3(g) \rightarrow N_2(g) + 3\,H_2(g)$$

Is the reaction spontaneous under standard conditions?

*13.60. Use heats of formation and entropies of reactants and products from Appendix 4 to calculate the standard free-energy change for the decomposition reaction in Problem 13.59. Compare these results with those you obtained in Problem 13.59. Assume that ΔH_{rxn} and ΔS_{rxn} are independent of temperature, and calculate the temperature at which ΔG_{rxn} is zero. What is the significance of this temperature for this reaction?

*13.61. A reaction

$$A(g) \rightarrow B(g) + C(g)$$

has the following standard thermodynamic parameters:
$\Delta H^{\circ}_{rxn} = 40.0$ kJ/mol and $\Delta S^{\circ}_{rxn} = 80.0$ J/mol·K.
a. Is the reaction exothermic or endothermic?
b. Does the positive entropy change make sense?
c. Is the reaction spontaneous at all temperatures? If not, explain.
d. Calculate the temperature at which the reaction becomes spontaneous.

*13.62. A reaction $C(g) + D(g) \rightarrow E(g)$ has the following standard thermodynamic parameters: $\Delta H^{\circ}_{rxn} = -35.0$ kJ and $\Delta S^{\circ}_{rxn} = -65.0$ J/mol·K.
a. Is the reaction exothermic or endothermic?
b. Does the negative entropy change make sense?
c. Is the reaction spontaneous at all temperatures? If not, explain.
d. Calculate the temperature at which the reaction becomes nonspontaneous.

Driving the Human Engine: Coupled Reactions

CONCEPT REVIEW

13.63. The second step in glycolysis converts glucose 6-phosphate into fructose 6-phosphate (Figure P13.63). Suggest a reason why ΔG° for this reaction is close to zero.

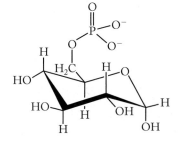

Glucose 6-phosphate

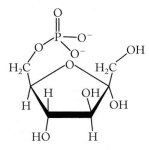

Fructose 6-phosphate

FIGURE P13.63

13.64. Why is it important that at least some of the exergonic steps in glycolysis convert ADP to ATP?

13.65. How do we calculate the overall free-energy change of a process consisting of two steps?

13.66. Make a statement in the style of Hess's law that describes how to calculate ΔG° for two coupled reactions.

PROBLEMS

13.67. The hydrolysis of 1 mole of the sugar maltose ($\Delta G^{\circ}_f = -2246.6$ kJ/mol) produces 2 moles of glucose ($\Delta G^{\circ}_f = -1274.5$ kJ/mol):

$$Maltose + H_2O \rightarrow 2\ glucose$$

If the value of ΔG°_f of water is -237.2 kJ/mol, what is the standard free-energy change of the hydrolysis reaction?

13.68. The standard free-energy change in the reaction of water with ethyl acetate ($CH_3COOC_2H_5$) to give ethanol (CH_3CH_2OH; $\Delta G^{\circ}_f = -487.0$ kJ/mol) and acetic acid (CH_3COOH) is -19.7 kJ. From this value and the values of the standard free energy of formation of liquid ethanol, acetic acid, and water (see Appendix 4), calculate ΔG°_f for ethyl acetate.

Additional Problems

13.69. Explain what part or parts of the following statement are wrong and why: "Almost all substances contract when you cool them down. At some point of cooling they freeze, and continuing to cool the sample increases its entropy." Now revise the statement so that it is incorrect.

13.70. Rewrite the following statement so that it is factually correct: "A dollar bill and a penny in a cash box have higher entropy than 101 penny coins in the same box at the same temperature."

*13.71. At what temperature is the free-energy change for the following reaction equal to zero?

$$NH_4Cl(s) \rightarrow NH_3(g) + HCl(g)$$

13.72. Which of these processes result in an entropy decrease of the system?
a. Diluting hydrochloric acid with water
b. Boiling water
c. $2\,NO(g) + O_2(g) \rightarrow 2\,NO_2(g)$
d. Making ice cubes in the freezer

*13.73. Calculate the standard free-energy change of the following reaction. Is it spontaneous?

$$2\,NO(g) + 2\,H_2(g) \rightarrow N_2(g) + 2\,H_2O(g)$$

13.74. At fixed temperature, which has the higher molar entropy, methane or propane?

13.75. A reaction has a negative enthalpy change and a positive entropy change. Is the reaction spontaneous at any temperature?

*13.76. Estimate the free-energy change of the following reaction at 225°C:

$$C_2H_4(g) + 3\,O_2(g) \rightarrow 2\,CO_2(g) + 2\,H_2O(g)$$

13.77. Explain why the standard molar entropies of elements are positive and not zero.

13.78. Explain why the standard free energies of formation of elements are zero.

*13.79. Show that hydrogen cyanide (HCN) is a gas at 25°C by estimating its normal boiling point from the following data:

	ΔH_f°, kJ/mol	$S°$, J/(mol·K)
HCN(ℓ)	108.9	113
HCN(g)	135.4	202

*13.80. Show that hydrogen peroxide (H_2O_2) is a liquid at 25°C by estimating its normal boiling point from the following data:

	ΔH_f°, kJ/mol	$S°$, J/(mol·K)
$H_2O_2(\ell)$	−187.8	110
$H_2O_2(g)$	−136.3	233

13.81. Keeping Cool Carbon tetrachloride (CCl_4) was a favored refrigerant until it was found to be carcinogenic. The heat of vaporization of $CCl_4(\ell)$ is 32.5 kJ/mol and its normal boiling point is 76.7°C. Estimate the entropy of vaporization of $CCl_4(\ell)$. Should the answer be positive or negative?

13.82. Making Methanol The element hydrogen is not abundant in nature, but it is a useful reagent in, for example, the potential synthesis of the liquid fuel methanol from gaseous carbon monoxide:

$$3 H_2(g) + CO(g) \rightarrow CH_3OH(\ell) + H_2O(\ell)$$

Under what temperature conditions is this reaction spontaneous?

13.83. Lightbulb Filaments Tungsten (W) is the favored metal for lightbulb filaments, in part because of its high melting point of 3410°C. The enthalpy of fusion of tungsten is 35.4 kJ/mol. What is its entropy of fusion?

*13.84. Over what temperature range is the reduction of tungsten(VI) oxide by hydrogen to give metallic tungsten and water spontaneous? The standard heat of formation of $WO_3(s)$ is −843 kJ/mol, and its standard molar entropy is 76 J/(mol · K).

13.85. Two allotropes (A and B) of sulfur interconvert at 369 K and 1 atm pressure:

$$S_8(s, A) \rightleftharpoons S_8(s, B)$$

The enthalpy change in this transition is 297 J/mol. What is the entropy change?

*13.86. Copper forms two oxides, Cu_2O and CuO.
a. Name these oxides.
b. Predict over what temperature range this reaction is spontaneous using the following thermodynamic data:

$$Cu_2O(s) \rightarrow CuO(s) + Cu(s)$$

	ΔH_f°, kJ/mol	$S°$, J/(mol · K)
$Cu_2O(s)$	−170.7	92.4
$CuO(s)$	−156.1	42.6

c. Why is the standard molar entropy of $Cu_2O(s)$ larger than that of CuO(s)?

*13.87. **Lime** Enormous amounts of lime (CaO) are used in steel industry blast furnaces to remove impurities from iron. Lime is made by heating limestone and other solid forms of $CaCO_3(s)$. Why is the standard molar entropy of $CaCO_3(s)$

higher than that of CaO(s)? At what temperature is the pressure of $CO_2(g)$ over $CaCO_3(s)$ equal to 1.0 atm?

	ΔH_f°, kJ/mol	$S°$, J/(mol · K)
$CaCO_3(s)$	−1207	93
CaO(s)	−636	40
$CO_2(g)$	−394	214

*13.88. *Trouton's rule* says that the ratio $\Delta H_{vap}^\circ/T_b$ for a liquid is approximately 80 J/K. Here, ΔH_{vap}° is the molar enthalpy of vaporization of a liquid and T_b is its normal boiling point.
a. What idea suggests that $\Delta H_{vap}^\circ/T_b$ for a range of liquids should be approximately constant?
b. Check Trouton's rule against the data in Figure P13.88. Which liquids deviate from Trouton's rule, and why?

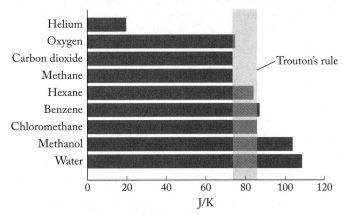

FIGURE P13.88

*13.89. **Melting DNA** When a solution of DNA in water is heated, the DNA double helix separates into two single strands:

$$1 \text{ DNA double helix} \rightleftharpoons 2 \text{ single strands}$$

a. What is the sign on ΔS for the forward process as written?
b. The DNA double helix re-forms as the system cools. What is the sign of ΔS for the process by which two single strands re-form the double helix?
c. The melting point of DNA is defined as the temperature at which $\Delta G = 0$. At that temperature, the forward reaction produces two single strands as fast as two single strands recombine to form the double helix. Write an equation that defines the melting temperature (T) of DNA in terms of ΔH and ΔS.

*13.90. **Melting Organic Compounds** When dicarboxylic acids (compounds with two −COOH groups in their structures) melt, they frequently decompose to produce 2 moles of CO_2 gas for every 1 mole of dicarboxylic acid melted (shown in Figure P13.90).

$$\left(\begin{array}{c} O \\ \| \\ C-[CH_2]_n-C \\ HO \qquad\qquad OH \end{array} \right)(s) \rightarrow H-[CH_2]_n-H(\ell) + 2 CO_2(g)$$

FIGURE P13.90

a. What are the signs of ΔH and ΔS for the process as written?
b. Problem 13.89 describes the DNA double helix re-forming when the system cools after melting. Do you think the dicarboxylic acid will re-form when the melted material cools? Why or why not?

Chemical Kinetics

CLEARING THE AIR *A light snowfall and change in wind direction dramatically affect visibility and air quality over Salt Lake City, Utah.*

A LOOK AHEAD:
How Quickly Can We Clean the Air?

The industrial revolution began in the late 18th century when a combination of new technologies for making steel and the invention of the modern steam engine led to the development of machines for manufacturing goods. The ensuing social and economic transformations that swept Europe and North America had an impact on where people lived and how they made a living. The rise of large cities as manufacturing centers had the unintended consequence of concentrating the production of energy for industrial and residential purposes. When local weather conditions limited the dispersion of air pollutants formed by, for example, the combustion of sulfur-containing coal (the first fossil fuel), the results were sometimes deadly. By the 1900s Londoners had invented a new word to describe the smoky fog that frequently blanketed their city: *smog*. A particularly severe smog in December 1952 caused the deaths of more than 10,000 people in London. Most were young children and adults already suffering from chronic respiratory diseases. This event, sometimes called *The Big Smoke*, triggered efforts to protect air quality in Britain and in other industrialized countries.

Today a combination of switching to less polluting fuels and installation of pollution-abatement equipment has dramatically reduced the concentrations of sulfurous smog over London and most industrialized cities in western Europe and North America. However, there is another kind of smog that persists over many cities, as illustrated by the photo on the previous page. It is called *photochemical smog*. It forms when sunny weather and light winds allow emissions from the internal combustion engines that power cars and trucks to collect over urban centers. Scientists, engineers, and urban planners are still working on how to clean up photochemical smog. Some of their approaches are a bit futuristic, such as electric cars powered by high-performance batteries and/or fuel cells (technologies we discuss in Chapter 18). However, there is an air pollution–control device that has removed many billions of tons of smog-forming pollutants from the air above U.S. cities since it was developed in the 1970s: the catalytic converter.

Development of catalytic converters to remove pollutants from engine exhaust required extensive studies of the chemical reactions taking place in engines burning fossil fuels. The problematic gaseous products of combustion include hydrocarbons, CO, and oxides of nitrogen (NO_x). Pollution due to these emissions can be reduced if the hydrocarbons and CO are both oxidized to CO_2 and H_2O and if all NO_x compounds are reduced to N_2 and O_2. Scientists and engineers have designed devices in which substances in the exhaust stream of an engine are simultaneously and selectively oxidized and reduced. Unfortunately, the necessary redox reactions all take place at different rates. To make their devices succeed, researchers had to find ways to make the reactions take place in the short time between generation in the engine and departure from the vehicle through the exhaust system. Understanding the conditions that influence the rates at which reactions take place and what happens to substances at the molecular level has led to the development of successful catalytic converters.

Problems of air pollution caused by vehicles are ongoing, but so too is our commitment to clean the air. In uncovering details of the chemical reactions taking place in internal combustion engines and in studying the overall system of chemical reactions occurring in the atmosphere, we have learned much about the air quality problems generated from automotive exhaust. In addition, we now understand the molecular basis of other atmospheric problems, first among them the ozone hole in the stratosphere, caused by other materials we have released into the air.

The atmosphere is a dynamic and integrated system, and we now know that what we do in our immediate environment on the Earth's surface has consequences both around the globe and at high altitudes. Studying the reactions taking place in the atmosphere has helped us recognize what we can do to clean the air and restore a healthy environment. A crucial part of these studies has been the examination of rates of reactions—how rapidly they proceed under a variety of conditions. Rates of reactions are the focus of this chapter.

14.1 Cars, Trucks, and Air Quality

The proliferation of automotive vehicles over the past 50 years has created a relatively new atmospheric problem known as **photochemical smog**. This form of pollution in the lower atmosphere is due to the interaction of sunlight with nitrogen oxides (NO_x) produced in the internal combustion engines of cars and trucks, and volatile organic compounds (VOCs) from man-made sources such as gasoline, paints, and solvents and even from biological sources such as vegetation. The interaction of NO_x, VOCs, and oxygen in the atmosphere produces ozone and an important class of pollutants called peroxyacyl nitrates, or PANs. PANs are extremely powerful respiratory and eye irritants that cause symptoms when present in the air at concentrations of only a few parts per billion. PANs are also relatively stable compounds that can be carried by prevailing winds from their urban origins into less densely populated areas, thereby spreading the geographical impact of photochemical smog. The most common of the PAN species is peroxyacetyl nitrate (Figure 14.1), which forms in the atmosphere when nitrogen oxides react with VOCs.

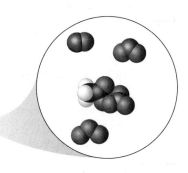
Photochemical smog, like the orange-brown layer in this photograph over Denver, CO, contains a mixture of compounds including NO, NO_2, O_3, and peroxyacetyl nitrate ($CH_3CO_2NO_3$) or PAN.

FIGURE 14.1 (a) Acetaldehyde is a common VOC in automobile exhaust and smoke from wood fires. It is also a product of the respiration of plants and the ripening of fruit. In the atmosphere, acetaldehyde reacts with O_2 to produce (b) acetyl peroxide that in turn reacts with NO_2 to make (c) peroxyacetyl nitrate (PAN).

Let's examine some of the reactions involved in photochemical smog formation. We start with the combustion of gasoline in an automobile engine. The principal products of complete combustion are carbon dioxide and water vapor, but the combustion of gasoline in an engine is usually not complete. Incomplete combustion yields large numbers of compounds including unburned and partially oxidized hydrocarbons and carbon monoxide. Other reactions that take place at the high temperatures inside internal combustion engines include formation of nitrogen monoxide from N_2 and O_2:

$$N_2(g) + O_2(g) \rightarrow 2NO(g) \qquad \Delta H° = 180.6 \text{ kJ} \qquad (14.1)$$

Photochemical smog is a mixture of gases formed in the lower atmosphere when sunlight interacts with compounds produced in internal combustion engines.

The reaction $N_2(g) + O_2(g) \rightarrow 2\,NO(g)$ is spontaneous only at very high temperatures. What is the sign of ΔS°_{rxn}?

(Answers to Concept Tests are in the back of the book.)

Once NO escapes into the atmosphere, it reacts with more oxygen, producing brown nitrogen dioxide gas:

$$2\,NO(g) + O_2(g) \rightarrow 2\,NO_2(g) \qquad \Delta H^\circ = -114.2 \text{ kJ} \qquad (14.2)$$

Radiant energy ($h\nu$) from the sun provides sufficient energy to break the N—O bonds in NO_2, forming NO and very reactive oxygen atoms:

$$NO_2(g) \xrightarrow{h\nu} NO(g) + O(g) \qquad (14.3)$$

This photochemically generated atomic oxygen combines with molecular oxygen, producing ozone:

$$O_2(g) + O(g) \rightarrow O_3(g) \qquad (14.4)$$

Oxygen atoms also react with water vapor in the atmosphere to produce hydroxyl radicals:

$$O(g) + H_2O(g) \rightarrow 2\,OH(g) \qquad (14.5)$$

All these reactions and many others occur simultaneously in the air and combine to produce a mixture whose composition varies throughout the day (Figure 14.2). If we summarize just this simplified group of reactions in terms of where they occur and what the reactants and products are, we see immediately how challenging is the task of understanding the formation of photochemical smog: most materials in photochemical smog are both reactants and products in the process. This means that the relations among the concentrations, even in this simplified view, are more complex than others we have seen in previous chapters.

Ozone is an important ingredient in photochemical smog for two reasons: it poses considerable risks to human health as an irritant, and it participates in chemical reactions that form other noxious and irritating compounds. Volatile hydrocarbons such as acetaldehyde (CH_3CHO) and methane (CH_4) react with ozone, hydroxyl radicals, and oxygen atoms to form a series of compounds with progressively higher oxygen content. These compounds also react with NO_2 to form PANs and similar compounds (Figure 14.3). The pollutants are formed by a series of reactions including those shown in Equations 14.1 to 14.4 and further involving the VOC acetaldehyde (CH_3CHO). Equation 14.6 summarizes the process:

$$N_2(g) + 3\,O_2(g) + OH(g) + CH_3CHO(g) \rightarrow$$
$$CH_3C(O)O_2NO_2(g) + H_2O(g) + NO_2(g) \qquad (14.6)$$

To understand NO_x production in engines and the production of photochemical smog in the environment, we need to understand how reactions producing NO_x are linked. We also need to know how rapidly each reaction takes place, that is, we need to know its **reaction rate**. Reaction rates are the rates at which reactants are consumed or products are formed. Therefore, reaction rates may be expressed in terms of the rate of change in concentration, or partial pressure, of a reactant or product per unit of time. With respect to the reactions involved in smog formation, the products of some reactions are the reactants in others. Therefore, the relative rates of these reactions influence when pollutants appear, how long they persist, and what their concentrations are.

CONNECTION We introduced species like OH and O in Chapter 8 in the discussion of odd-electron molecules and free radicals.

The **reaction rate** describes how rapidly a reaction occurs. It is related to rates of change in the concentrations of reactants and products over time.

Figure 14.2 shows how concentrations of NO, NO$_2$, and O$_3$ rise and fall in the atmosphere during a photochemical smog event. Note that a maximum in the concentration of NO occurs just after the morning rush hour. Later in the morning, the concentration of NO$_2$ reaches a maximum. NO is a precursor to NO$_2$ formation, meaning that NO forms first (Equation 14.1) and is then oxidized to NO$_2$ (Equation 14.2). The rise and fall in the concentrations of NO and NO$_2$ are linked. The highest ozone concentrations are reached in the middle of the afternoon when the photochemical reactions of NO$_2$ (Equations 14.2 and 14.3) are in full swing, producing a supply of oxygen atoms for the formation of ozone (Equation 14.4), hydroxyl radicals (Equation 14.5), and, indirectly, PANs (Equation 14.6). Ozone also reacts with NO to form NO$_2$ and O$_2$:

$$O_3(g) + NO(g) \rightarrow O_2(g) + NO_2(g) \qquad (14.7)$$

The NO$_2$ concentration drops in the afternoon, however, because NO$_2$, ozone, and hydrocarbons react to form an array of compounds including those in the PAN family.

The catalytic converter, now a standard feature on most automobiles, helps to combat the production of photochemical smog by removing NO and unburned or partially oxidized hydrocarbons from the exhaust gases of gasoline-fueled engines. Automakers have also refined engines and fuel systems to further reduce emissions of volatile compounds that contribute to smog formation. Knowledge of **chemical kinetics**, the study of the rates of change of the concentrations of substances involved in chemical reactions, has provided the understanding required to develop technologies like the catalytic converter and to direct other work aimed at cleaning our air and keeping it clean.

14.2 Reaction Rates

The amount of NO that forms inside an automobile engine depends on how rapidly the reaction in Equation 14.1 proceeds. We can express reaction rate as the increase in the concentration of NO ($\Delta[NO]$) that occurs over an interval of time (Δt). The rate of change of [NO] is the difference between the final NO concentration ($[NO]_{final}$) and the initial concentration ($[NO]_{initial}$) divided by the interval of time ($\Delta t = t_{final} - t_{initial}$):

$$\text{Rate of formation of NO} = \frac{\Delta[NO]}{\Delta t} = \frac{[NO]_{final} - [NO]_{initial}}{t_{final} - t_{initial}} \qquad (14.8)$$

Rate Units and Relative Reaction Rates

The rate of the reaction between N$_2$ and O$_2$ can also be expressed as the rate of consumption of either reactant. The rate of consumption of a reactant has a negative sign because its concentration decreases as the reaction proceeds. Reaction rates are positive quantities because they describe the rate at which reactants form products. Because reactant concentrations decrease as the reaction proceeds, a negative sign is placed in front of the change of concentration of a reactant, such as N$_2$, to obtain a positive value for the reaction rate:

$$\text{Rate} = -\frac{\Delta[N_2]}{\Delta t} = -\frac{[N_2]_{final} - [N_2]_{initial}}{t_{final} - t_{initial}} \qquad (14.9)$$

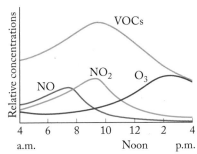

FIGURE 14.2 In photochemical smog, NO from engine exhaust builds up in the morning, then decreases as it reacts with O$_2$ in the atmosphere, forming NO$_2$. Photodecomposition of NO$_2$ leads to the formation of O$_3$. Because smog-forming reactions depend upon temperature and sunlight, smog tends to be severe on hot, sunny days. VOC concentrations also change throughout the day as a function of driving patterns and rising and falling temperatures.

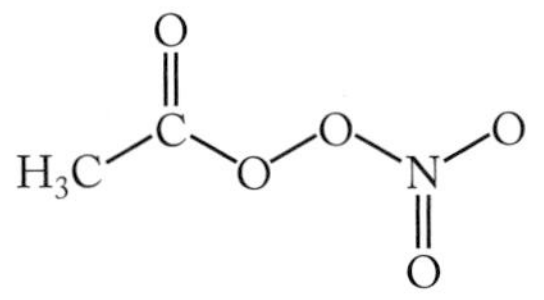

Peroxyacetyl nitrate

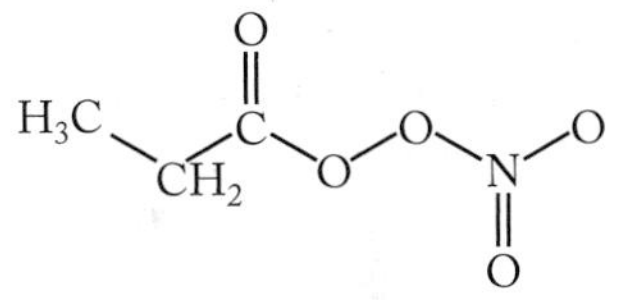

Peroxypropionyl nitrate

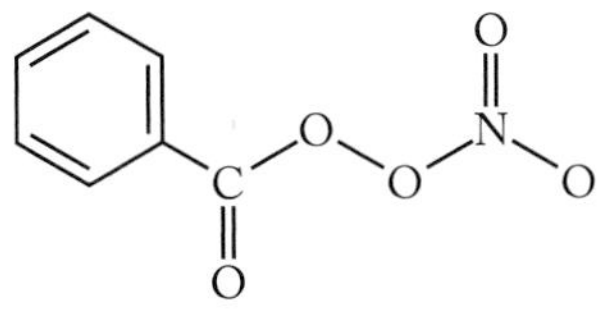

Peroxybenzoyl nitrate

FIGURE 14.3 Peroxyacetyl nitrate (top structure) accounts for about 90% of the PANs in photochemical smog, peroxypropionyl nitrate (middle structure) makes up about 9%, and peroxybenzoyl nitrate (bottom structure) is usually a part of the remaining 1%.

▶︎❙❙ **CHEMTOUR** Reaction Rate

Chemical kinetics is the study of the rates of change of concentrations of substances involved in chemical reactions.

Similarly, the rate of disappearance of oxygen can be used to define the reaction rate, and a negative sign is again placed in front of the change in concentration to produce a positive value for the reaction rate:

$$\text{Rate} = -\frac{\Delta[O_2]}{\Delta t} = -\frac{([O_2]_{final} - [O_2]_{initial})}{(t_{final} - t_{initial})} \qquad (14.10)$$

The rate in both cases is the change in concentration divided by the change in time, so the units on the value are "concentration per unit time," such as molarity per second (M/s).

Information from the balanced equation for the reaction also enters into the expressions of *relative* reaction rates. Two moles of NO are formed from 1 mole of N_2 and 1 mole of O_2:

$$N_2(g) + O_2(g) \rightarrow 2\,NO(g)$$

This fact means that the rate of disappearance of N_2 is the same as the rate of disappearance of O_2, and that the rate of appearance of NO is twice the rate of disappearance of either N_2 or O_2. These relations are expressed in equation form as follows:

$$-2\frac{\Delta[N_2]}{\Delta t} = -2\frac{\Delta[O_2]}{\Delta t} = \frac{\Delta[NO]}{\Delta t} \qquad (14.11)$$

or, if we divide all the terms in Equation 14.11 by 2:

$$\text{Rate} = -\frac{\Delta[N_2]}{\Delta t} = -\frac{\Delta[O_2]}{\Delta t} = \frac{1}{2}\frac{\Delta[NO]}{\Delta t} \qquad (14.12)$$

CONCEPT TEST

Can a reaction have a negative rate?

In Equation 14.12 the value in the denominator in the coefficient of the $\Delta[NO]/\Delta t$ term matches the coefficient of NO in the balanced chemical equation. This pattern applies to all chemical reactions and simplifies writing expressions for relative reaction rates: we use the coefficients from the balanced chemical equation in the denominators of fractions in front of the terms that describe changes of concentration over time. Notice that the **numerators** in Equation 14.13 all have a value of 1 and the numbers in the **denominators** have values equal to the coefficients in the balanced chemical equation:

$$1\,N_2(g) + 1\,O_2(g) \rightarrow 2\,NO(g)$$

$$\text{Rate} = -\frac{1}{1}\frac{\Delta[N_2]}{\Delta t} = -\frac{1}{1}\frac{\Delta[O_2]}{\Delta t} = \frac{1}{2}\frac{\Delta[NO]}{\Delta t} \qquad (14.13)$$

Note that we added "1" coefficients to the chemical equation to make it clear where the numerators in the rate equation came from.

A balanced chemical equation enables us to predict the *relative* rates at which reactants are consumed and products are formed. However, Equations 14.10 to 14.13 provide no information on the actual numerical value of the rate of the reaction. Rate values can only be obtained experimentally, as we see later in this section.

SAMPLE EXERCISE 14.1   **Writing a Relative Reaction Rate Expression**

How is the rate of formation of NH_3 related to the rates of consumption of N_2 and H_2 in the synthesis of ammonia?

COLLECT AND ORGANIZE We need a balanced chemical equation to develop expressions for the relative rates of consumption of reactants and formation of products. The chemical equation for the production of ammonia from nitrogen and hydrogen is

$$N_2(g) + 3H_2(g) \rightarrow 2NH_3(g)$$

ANALYZE Our task is to use the coefficients of this chemical equation to determine the relative rates at which the concentrations of N_2, H_2, and NH_3 change during the reaction. Nitrogen and hydrogen are consumed in the reaction, and ammonia is generated.

SOLVE We can write the expressions for the consumption of reactants and the formation of products with the correct signs and insert the coefficients from the balanced equation in the denominators:

$$\text{Rate} = -\frac{\Delta[N_2]}{\Delta t} = -\frac{1}{3}\frac{\Delta[H_2]}{\Delta t} = \frac{1}{2}\frac{\Delta[NH_3]}{\Delta t}$$

The negative signs in front of the N_2 and H_2 terms indicate that their concentrations decrease as the concentration of NH_3 increases.

THINK ABOUT IT The coefficients make sense because the balanced chemical equation indicates that 2 moles of NH_3 are formed for every 1 mole of N_2 consumed. That means the rate of consumption of N_2 should be half the rate of formation of ammonia. Similarly, the stoichiometry of the reaction tells us that 3 moles of H_2 are consumed for every mole of N_2 consumed. Therefore, the rate at which N_2 is consumed should be one-third the rate at which H_2 is consumed.

Practice Exercise In the oxidation of carbon monoxide to carbon dioxide,

$$2CO(g) + O_2(g) \rightarrow 2CO_2(g)$$

which reactant is consumed at a higher rate: CO or O_2? How is the reaction rate related to the rates of change in the concentrations of CO_2 and O_2?

(Answers to Practice Exercises are in the back of the book.)

Experimentally Determined Rates: Actual Values

Up to this point we have determined relative rates of a reaction: how rapidly a reactant (or product) disappears (or appears) *relative* to another substance involved in the reaction. Actual numerical values for reaction rates are determined experimentally, and once we know the rate of a reaction with respect to one substance and the balanced chemical equation for the process, we can express the rate of a reaction in terms of any component in the chemical equation. The information in Sample Exercise 14.1 can serve as an example. If the rate of the formation of ammonia ($\Delta[NH_3]/\Delta t$) under a given set of conditions is determined to be 0.472 *M*/s, we can calculate the rate of consumption of either reactant by using the mathematical expression derived in the exercise:

$$\text{Rate} = -\frac{\Delta[N_2]}{\Delta t} = -\frac{1}{3}\frac{\Delta[H_2]}{\Delta t} = \frac{1}{2}\frac{\Delta[NH_3]}{\Delta t}$$

The rate of consumption of nitrogen ($-\Delta[N_2]/\Delta t$) is half the rate at which ammonia forms ($\Delta[NH_3]/\Delta t$):

$$-\frac{\Delta[N_2]}{\Delta t} = \frac{1}{2}\frac{\Delta[NH_3]}{\Delta t} =$$

$$\frac{1}{2}(0.472 \; M/s) = 0.236 \; M/s$$

Thus, the rate of the reaction is 0.236 M/s.

SAMPLE EXERCISE 14.2   Converting Reaction Rates

Suppose that during the reaction in the atmosphere between NO and O_2 to form NO_2,

$$2\,NO(g) + O_2(g) \rightarrow 2\,NO_2(g)$$

$\Delta[O_2]/\Delta t = -0.033 \; M/s$. What is the rate of formation of NO_2?

COLLECT AND ORGANIZE We are asked to determine the rate of formation of a product in a reaction. We are given two items that we can use to determine this: the balanced chemical equation and the rate of disappearance of one of the reactants.

ANALYZE The coefficients of NO_2 and O_2 are 2 and 1, respectively. This stoichiometric relation means that the rate of consumption of O_2 is half the rate of production of NO_2.

SOLVE We can write an equation that expresses the relative rates of change in $[NO_2]$ and $[O_2]$ from the coefficients in the balanced chemical equation:

$$\frac{1}{2}\frac{\Delta[NO_2]}{\Delta t} = -\frac{\Delta[O_2]}{\Delta t}$$

The negative sign is needed because the concentration of reactant O_2 decreases with time as the concentration of NO_2 increases. By inserting the value of the rate of change of $[O_2]$ and solving for the rate of change of $[NO_2]$, we get

$$\frac{\Delta[NO_2]}{\Delta t} = -2\,(-0.033 \; M/s) = 0.066 \; M/s$$

THINK ABOUT IT This result makes sense on two counts:

1. It is positive, as it should be because the concentration of a product increases as the reaction proceeds.
2. It is twice the magnitude of $\Delta[O_2]/\Delta t$, which is consistent with the stoichiometry of the reaction: 2 moles of NO_2 are produced for each mole of O_2 consumed.

Practice Exercise The gas NO reacts with H_2, forming N_2

$$2\,NO(g) + 2\,H_2(g) \rightarrow 2\,H_2O(g) + N_2(g)$$

If $\Delta[NO]/\Delta t = -21.5 \; M/s$ under a given set of conditions, what are the rates of change of $[N_2]$ and $[H_2O]$?

Average Reaction Rates and the Formation of NO

Suppose we are studying improvements in internal combustion engines and need to determine the rate of formation of NO in an automobile by running an experiment in the laboratory. We use a reaction vessel as hot as the combustion chambers inside the engine and obtain the data in Table 14.1, which is plotted in Figure 14.4. The data in Table 14.1 can be used to calculate the *average* reaction rate based on the change in the concentration of any of the substances in the reaction, $[N_2]$, $[O_2]$, or $[NO]$, over a particular time interval. The average rate with respect to the formation of NO is the slope of a straight line defined by the difference in concentration values ($\Delta[NO]$) at the beginning and end of the selected time interval (Δt), as shown in Figure 14.4. For example, the average rate of change in $[NO]$ between 5 and 10 μs is

$$\frac{\Delta[NO]}{\Delta t} = \frac{[NO]_{final} - [NO]_{initial}}{t_{final} - t_{initial}} = \frac{(14.8 - 7.8)\ \mu M}{(10 - 5)\ \mu s}$$

$$= 1.40\ M/s$$

During the same interval the average rate of change of $[N_2]$ (or $[O_2]$) is

$$\frac{\Delta[N_2]}{\Delta t} = \frac{[N_2]_{final} - [N_2]_{initial}}{t_{final} - t_{initial}} = \frac{(9.6 - 13.1)\ \mu M}{(10 - 5)\ \mu s}$$

$$= -0.70\ M/s$$

The results of these calculations give us two values for expressing the rate of the reaction. Which one should we use? We use the rate of change of the reactant or product with a coefficient of "1" (one) in the reaction equation as the basis for expressing the rate of the reaction. When we discuss rate laws in Section 14.3, you will appreciate that a reaction can have only one rate under a particular set of conditions no matter how the rate is measured. Therefore only one rate characterizes the reaction under the given conditions. In this example, the rate is for the reactant or product that has the coefficient "1," which refers to $[N_2]$ or $[O_2]$, but not NO_2. The average rate of this reaction is

$$\text{Rate} = -\frac{\Delta[N_2]}{\Delta t} = -(-0.70\ M/s) = 0.70\ M/s$$

TABLE 14.1	Changing Concentrations of Reactants and Products with Time for the Reaction $N_2(g) + O_2(g) \rightarrow 2\,NO(g)$	
Time (μs)	**$[N_2]$, $[O_2]$ (μM)**	**$[NO]$ (μM)**
0	17.0	0.0
5	13.1	7.8
10	9.6	14.8
15	7.6	18.6
20	5.8	22.2
25	4.5	24.8
30	3.6	26.7

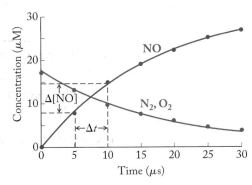

FIGURE 14.4 The graph shows the changes in the concentrations of N_2, O_2, and NO over 30 μs for the reaction $N_2(g) + O_2(g) \rightarrow 2\,NO(g)$.

The **instantaneous rate** of a reaction is the rate of a reaction at a specific time (a specific point on the graph of concentration versus time) in the course of the reaction.

Instantaneous Reaction Rates and the Formation of NO_2

Now let's consider the conversion of the NO in engine exhaust into NO_2 (Equation 14.2), the gas responsible for much of the brown color in the photos of photochemical smog at the beginning of this chapter. The conversion takes place when NO reacts with oxygen in the air:

$$2\,NO(g) + O_2(g) \rightarrow 2\,NO_2(g)$$

The rate of this reaction is described by the data in Table 14.2 and in particular by the rate of consumption of O_2 plotted in Figure 14.5. In this example we express the rate of a reaction in another form called the **instantaneous rate**, which is the reaction rate at a particular point in time or at a particular point along the curve in Figure 14.5. The difference between an average rate and an instantaneous rate of a chemical reaction is analogous to the difference between the average speed and instantaneous speed of a runner. If a competitor in a marathon runs from mile 10 to mile 20 in 1 hour, the runner's average speed over that distance is 10 mi/hr. At a given moment during the run, however, her instantaneous speed could be 12 mi/hr while going downhill or 8 mi/hr going up a hill.

The instantaneous reaction rate at any time in the course of the reaction can be determined from the slope of a line tangent to the curve on the graph of concentration versus time (Figure 14.5). For example, if we wish to determine the instantaneous rate of change of $[O_2]$ at $t = 2000$ s, we first draw a line tangent to the curve for $[O_2]$ at $t = 2000$ s and then pick two convenient points, say at $t = 1000$ and 3000 s, along that tangent line. From the differences in the y and in the x coordinates of these points (the rise of the line over the run), we calculate the slope of the line, which is a measure of $\Delta[O_2]/\Delta t$ and the instantaneous reaction rate at $t = 2000$ s:

$$\text{Rate} = -\frac{\Delta[O_2]}{\Delta t} = -\frac{(0.0072 - 0.0084)\,M}{(3000 - 1000)\,s} = 6 \times 10^{-7}\,M/s$$

Note that we substitute the concentration values corresponding to time points along the tangent line, not along the curve drawn from the data points.

FIGURE 14.5 The instantaneous rate of change in $[O_2]$ in the reaction $2\,NO(g) + O_2(g) \rightarrow 2\,NO_2(g)$ can be determined from a tangent to the curve of $[O_2]$ versus time. For example, the instantaneous rate of change in $[O_2]$ at $t = 2000$ s is equal to the slope of the tangent to the curve at $t = 2000$ s.

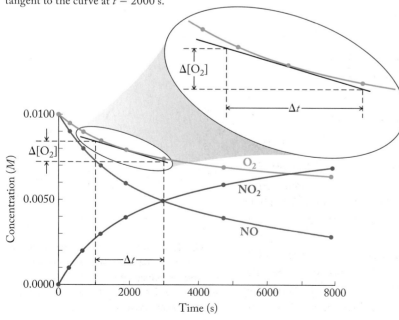

TABLE 14.2 Changing Concentrations of Reactants and Products with Time at 25°C for the Reaction $2\,NO(g) + O_2(g) \rightarrow 2\,NO_2(g)$

Time (s)	[NO] (M)	[O₂] (M)	[NO₂] (M)
0	0.0100	0.0100	0.0000
285	0.0090	0.0095	0.0010
660	0.0080	0.0090	0.0020
1175	0.0070	0.0085	0.0030
1895	0.0060	0.0080	0.0040
2975	0.0050	0.0075	0.0050
4700	0.0040	0.0070	0.0060
7800	0.0030	0.0065	0.0070

SAMPLE EXERCISE 14.3 **Determining Instantaneous Rate**

> The **initial rate** of a reaction is the rate at $t = 0$, immediately after the reactants are mixed.

What is the instantaneous rate of change of [NO] at $t = 2000$ s in the experiment that produced the data in Table 14.2?

COLLECT AND ORGANIZE We are asked to determine the instantaneous rate of a reaction of NO from data describing the concentration of NO versus time.

ANALYZE The instantaneous rate of change in [NO] at the stated time can be determined from a graph of the data.

SOLVE First we plot [NO] versus time and draw a tangent to the curve at the time designated in the statement of the problem (2000 s).

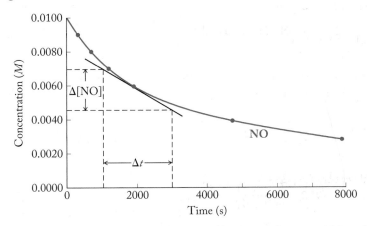

We then choose two points along the tangent, for example, $t = 1000$ and 3000 s, and determine the concentrations corresponding to those times along the y-axis. By using those values, we calculate the slope of the line:

$$\frac{\Delta[\text{NO}]}{\Delta t} = \frac{(0.0046 - 0.0070)\ M}{(3000 - 1000)\ \text{s}} = -1.2 \times 10^{-6}\ M/s$$

THINK ABOUT IT The sign of this $\Delta[\text{NO}]/\Delta t$ is negative because NO is a reactant whose concentration decreases with time. From the reaction stoichiometry, we can determine that the absolute value of the rate of change of [NO] is twice the reaction rate (6×10^{-7} M/s) based on the rate of change of [O_2]. This difference is what we would predict from the stoichiometry of the reaction: 2 moles of NO are consumed for every mole of O_2.

Practice Exercise What is the instantaneous rate of change in [NO_2] at $t = 2000$ s in the experiment that produced the data in Table 14.2?

14.3 Effect of Concentration on Reaction Rate

Figure 14.6 shows a typical result when concentration of a reactant is plotted as a function of time. Tangents have been drawn to the line at three time points: (a) immediately at the beginning of the reaction, (b) when the reaction is about half-way to completion, and (c) when the reaction is nearly over. Point (a) is a special point that defines the **initial rate** of the reaction, which is the rate that occurs as soon as the reactants are mixed at $t = 0$. The key observation in Figure 14.6 is that the slopes of the tangents decrease the longer the reaction runs. When the reaction is over, no more change occurs in the concentration of any remaining reactant, and the slope of a tangent to any time point along the line is zero. This

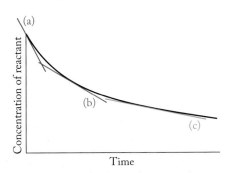

FIGURE 14.6 The graph is a typical plot of the concentration of a reactant as a function of time: (a) tangent at $t = 0$; (b) tangent at the midpoint of the reaction; (c) tangent close to the end of the reaction.

●●○ **CONNECTION** We discussed kinetic molecular theory, a model that describes the behavior of gases, in Chapter 6.

▶❚❚ **CHEMTOUR** Reaction Order

The **order of a reaction** is an experimentally determined number defining the dependence of the reaction rate on the concentration of a reactant.

behavior is typical and describes a general observation: the most rapid changes in the concentrations of reactants and products take place early in the progress of reactions. The graphs in Figures 14.4 through 14.6 illustrate this trend: the rates of these reactions decrease as the reactions proceed and reactants are consumed.

Kinetic molecular theory and our picture of molecules in the gas phase provide us with a way to explain this trend. If we assume that reactions such as the one between NO and O_2 take place as a result of collisions between molecules of reactants, the more molecules there are in a given space, such as a reaction flask, the more collisions take place per unit time, and the more opportunities there are for reactants to turn into products. As the concentration of reactants decreases, fewer reactant molecules occupy the space in the flask, so the frequency of their collisions decreases, and the rate of conversion of reactants to products slows down. The picture based on collisions of molecules is fundamental to the discussion in Section 14.5 on reaction mechanisms and how reactions proceed on a molecular level.

Reaction Order and Rate Constants

Both observations and theoretical considerations tell us that reaction rates depend on the concentrations of reactants. However, they do not tell us *to what extent* rates depend on reactant concentrations. For example, if the concentration of a reactant doubles, should the reaction rate double? The answer to such a question is contained in the **order of a reaction**, which is derived from the results of experiments that tell us the dependence of reaction rate on reactant concentrations. Knowing the order of a reaction provides insights into *how* the reaction takes place: which molecules collide with which others as their bonds break, new bonds form, and the reactants are transformed into products. Let's look at how the order of a reaction is determined by revisiting the reaction (Equation 14.2) between oxygen and nitric oxide

$$2\,NO(g) + O_2(g) \rightarrow 2\,NO_2(g)$$

In a series of experiments to evaluate the kinetics of this reaction, different starting concentrations of NO and O_2 are introduced into a reaction vessel at 25°C, and the values of the initial rates of the reaction are determined at $t = 0$ (Figure 14.7). Table 14.3 contains the results of four such determinations of initial reaction rate.

To interpret the data in Table 14.3, we select pairs of experiments in which the concentrations of one reactant differ, but the concentrations of the other are the same. For example, $[NO]_0$ is the same in experiments 1 and 2, but the value of $[O_2]_0$ in experiment 1 is twice that in experiment 2. (The zero subscripts indicate that the concentrations of NO and O_2 are the values at the start of the experiments when $t = 0$.) The reaction rate in experiment 1 is twice that in experiment 2; doubling the concentration of O_2 while the concentration of NO is held constant causing the reaction rate to double. We conclude that the rate of the reaction is proportional to the concentration of O_2:

$$\text{Rate} \propto [O_2]$$

In experiments 2 and 4 in Table 14.3, $[O_2]_0$ is the same, but $[NO]_0$ in experiment 2 is twice that in experiment 4. Comparing the reaction rates for experiments 2 and 4, we find that the rate in experiment 2 is four times that in experiment 4. Thus, the reaction rate is proportional to [NO] squared:

$$\text{Rate} \propto [NO]^2$$

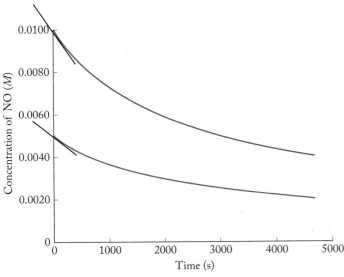

FIGURE 14.7 Initial reaction rates can be determined from the slopes of lines that are tangent to plots of reactant concentration versus time at $t = 0$. In this figure, initial rates for the reaction between NO and O_2 are determined for initial NO concentration ($[NO]_0$) values of 0.0100 (in red) and 0.0050 M (in blue). In both experiments $[O_2]_0 = 0.0100\ M$.

TABLE 14.3 Effect of Reactant Concentrations on Initial Rates at 25°C for the Reaction $2\,NO(g) + O_2(g) \rightarrow 2\,NO_2(g)$

Experiment	$[NO]_0$ (M)	$[O_2]_0$ (M)	Initial Reaction Rate $\left(\dfrac{-\Delta[NO]}{\Delta t},\dfrac{M}{s}\right)$
1	0.0100	0.0100	2.0×10^{-6}
2	0.0100	0.0050	1.0×10^{-6}
3	0.0050	0.0100	5.0×10^{-7}
4	0.0050	0.0050	2.5×10^{-7}

We combine these two rate expressions to get an overall rate expression for the reaction by multiplying the right sides of the rate proportionalities for the two reactants:

$$\text{Rate} \propto [NO]^2[O_2] \qquad (14.14)$$

To understand why we multiply the concentration terms together, look at the images in Figure 14.8. Each cylinder in the figure represents a microscopic volume of NO and O_2 molecules reacting in the gas phase. The double-headed arrows represent all the possible collisions between different pairs of NO and O_2 molecules in the volumes of gas; remember for molecules to react, they must collide. Each NO molecule can collide with each O_2 molecule, so the total number of possible collisions (the total number of arrows) is the product of the number of each type of molecule available. When we run reactions at the macroscopic level, we express relative numbers of molecules as concentrations. Because concentration is directly proportional to the number of molecules in a sample, the product of the concentrations is directly proportional to the number of possible collisions and hence to the rate of the reaction. This is the relation between reaction rate and reactant concentration expressed in Equation 14.14.

FIGURE 14.8 These illustrations show how increasing the concentrations of NO and O_2 increases the number of collisions (double-headed arrows) and the number of potential reaction events. (a) One molecule of each reactant (NO and O_2) may collide only with the other, giving a relative reaction rate of "1" (1×1). (b) Each of the two NO molecules may collide with the single O_2 molecule to produce twice the relative reaction rate ($2 \times 1 = 2$). (c) Two molecules of each of the reactants ($2 \times 2 = 4$) yields four times the relative reaction rate. (d) Two molecules of NO can collide with three molecules of O_2 to yield six times the relative reaction rate (2×3). (e) Three molecules of NO can collide with three molecules of O_2 to achieve nine times the relative reaction rate ($3 \times 3 = 9$). The dependence of relative reaction rate on concentration can be accounted for if the rate is proportional to the product of the numbers of reacting molecules present, which is analogous to the product of the concentrations of the reactants.

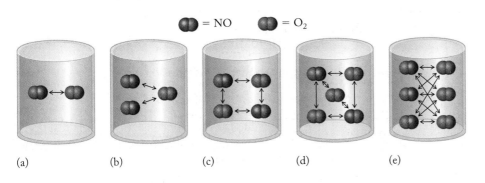

(a)   (b)   (c)   (d)   (e)

The **rate law** for a chemical reaction is an equation that defines the experimentally determined relation between the concentration of reactants and the rate of a reaction.

The **rate constant** is the proportionality constant that relates the rate of a reaction to the concentrations of reactants.

The **overall order** of a reaction is the sum of the exponents of the concentration terms in the rate law.

Similar patterns occur with all chemical reactions whose rate depends on the concentration of more than one reactant. We can write a form of the expression in Equation 14.14 called a **rate law**, which defines the relation between concentrations of reactants and how rapidly the reaction proceeds. To make the expression an equation rather than a proportionality, we include a proportionality constant k called the **rate constant** for the reaction:

$$\text{Rate} = k[\text{NO}]^2[\text{O}_2] \qquad (14.15)$$

The most general form of a rate law for a reaction involving two species such as NO and O_2 is

$$\text{Rate} = k[\text{A}]^m[\text{B}]^n \qquad (14.16)$$

where m and n are the powers to which the concentration of a species is raised in the rate law. In the case of the reaction of NO and O_2, $m = 2$ and $n = 1$. The power to which the concentration term is raised is the order of the reaction in terms of the reactant specified: the reaction of NO and O_2 is *second order* in NO and *first order* in O_2. The **overall order** for a reaction is the sum of the powers in the rate equation, so the reaction of NO and O_2 is *third* order overall.

The exponents of concentration terms in the rate laws for other reactions may be greater than one; they may be fractions; they may even be zero. In rare cases, the exponent may be negative. It is important to remember that these rate laws and orders are different from relative rates and must be determined experimentally. They cannot be predicted from the coefficients in an overall balanced chemical equation for a process. The significance of reaction order in describing how a reaction takes place is addressed in detail in Section 14.5.

It is also important to note that the value of the rate constant k is unique to the particular reaction at a given temperature, and right now we may consider k simply as a proportionality constant. It does not change with concentration; the rate of a reaction depends on the concentration of the reactants, but the rate constant does not. The value of the rate constant only changes with changing temperature or in the presence of a catalyst. Every reaction we run on a given system involving changing concentrations at the same temperature is governed by the same value of k. For a given temperature, we calculate the value of k by using initial reaction rate data such as those in Table 14.3. Let's do so for the reaction of O_2 and NO by selecting the results of one experiment. Which experiment we use doesn't matter, as long as the temperatures are the same. The value of k is constant, so each experiment should give the same value of k. To determine k, we can use Equation 14.15 and insert the data from experiment 1 (rate = 2.0×10^{-6} *M*/s, [NO] = 0.0100 *M*, and [O_2] = 0.0100 *M*) and the rate-law exponents $m = 2$ and $n = 1$ to get

$$2.0 \times 10^{-6} \, M/\text{s} = k(0.0100 \, M)^2(0.0100 \, M)$$

Solving for k gives us

$$k = \frac{2.0 \times 10^{-6} \, M/\text{s}}{(0.0100 \, M)^2(0.0100 \, M)} = 2.0 \, M^{-2}\text{s}^{-1}$$

and substituting this value of k in the rate-law expression yields

$$\text{Rate} = 2.0 \, M^{-2} \, \text{s}^{-1} \, [\text{NO}]^2 \, [\text{O}_2]$$

We need to keep in mind that the value for the rate constant for this reaction is valid only at the temperature at which the experiments in Table 14.3 were carried out.

CONCEPT TEST

Why doesn't it matter which experiment in Table 14.3 is used to calculate the value of k for the reaction $NO(g) + O_2(g) \rightarrow 2\,NO_2(g)$?

The rate of a reaction is always expressed in terms of the rate of appearance of a product or disappearance of a reactant and so should have units of concentration change per unit time interval, such as molarity per second (M/s, or $M\,s^{-1}$). For a first-order reaction in which concentration is expressed in molarity and time in seconds, the units of k must be per second (s^{-1}):

$$\text{Rate} = k[X]^1$$

or

$$k = \frac{\text{rate}}{[X]}$$

$$= \frac{M/s}{M} = s^{-1}$$

The units of the rate constant of a reaction that is second order overall can be derived in a similar way. If a reaction is first order in reactants X and Y,

$$\text{Rate} = k\,[X][Y]$$

so that

$$k = \frac{\text{rate}}{[X][Y]}$$

$$= \frac{M/s}{M^2} = M^{-1}\,s^{-1}$$

The rate constants for second-order reactions have units of $M^{-1}s^{-1}$. Similarly, the units of the rate constant of a reaction that is third order overall are $M^{-2}\,s^{-1}$, as we determined for the reaction of NO and O_2. Chemical reactions each have a characteristic rate that is observed at a given temperature when the concentrations of all reactants are 1 M. Because the number 1 raised to any power is 1, the reaction order is irrelevant in this special case. The rate of the reaction in this situation is called the *specific rate* of a reaction. It is numerically the same as the rate constant, but has the units M/s.

SAMPLE EXERCISE 14.4   **Writing a Rate Law Expression and Calculating an Overall Reaction Order**

If the reaction $2\,NO(g) + O_2(g) \rightarrow 2\,NO_2(g)$ is second order in NO and first order in O_2:

   a. What is the rate law for the reaction?

   b. What is the overall reaction order?

   c. What are the units of the rate constant?

COLLECT AND ORGANIZE We are asked to determine the equation that defines the relation between the concentration of reactants and the rate of the reaction (the rate law), the overall reaction order, and the units in which the rate constant is expressed. We are given the order of the reaction in terms of each individual reactant. That is all we need to answer these questions.

ANALYZE The rate law of a chemical reaction is an equation that relates the rate of the reaction (with units of concentration per unit of time, or M/s) to the product of a rate constant multiplied by one or more reactant concentrations, each raised to an appropriate power that is called the *order*. The overall order of the reaction is the sum of the orders of the individual reactants.

SOLVE

a. A general form of the rate law is written as follows:

$$\text{Rate} = k[\text{NO}]^m[\text{O}_2]^n$$

The reaction is second order in NO and first order in O_2; so $m = 2$ and $n = 1$. Therefore the rate law is

$$\text{Rate} = k[\text{NO}]^2[\text{O}_2]$$

b. The overall reaction order is $m + n = 2 + 1 = 3$, or third order overall.

c. The units of the rate constant must be those that convert the units of the combined concentration terms (M^3) to reaction rate in M/s. Therefore the required units of k are

$$\frac{M/s}{M^3} = M^{-2}\text{s}^{-1}$$

THINK ABOUT IT The units of the rate constant must be consistent with the fact that the rate of the reaction is expressed in molarity per second when concentrations are given in molarity units.

Practice Exercise In an experiment, the following reaction is determined to be second order overall and first order in both A and B.

$$\text{A} + \text{B} + \text{C} \rightarrow \text{D} + 2\,\text{E}$$

a. Determine the reaction order with respect to C.

b. Write an expression for the rate law and determine the units of the rate constant for this reaction.

SAMPLE EXERCISE 14.5   **Deriving a Rate Law from Initial Reaction Rate Data**

The reaction of N_2 and O_2 in an internal combustion engine to produce NO is a key reaction in the formation of photochemical smog. Write the rate law for the reaction of N_2 with O_2 on the basis of the following data, and determine the rate constant for the reaction.

$$\text{N}_2(g) + \text{O}_2(g) \rightarrow 2\,\text{NO}(g)$$

Experiment	$[\text{N}_2]_0$ (*M*)	$[\text{O}_2]_0$ (*M*)	Initial Rate (*M*/s)
1	0.040	0.040	1000
2	0.040	0.010	500
3	0.010	0.010	125

COLLECT AND ORGANIZE We are asked to determine the expression for the rate law and the rate constant for a reaction at a specific temperature. We are given the initial rates of the reaction (note the subscript zero on the concentration terms) for three different initial concentrations.

ANALYZE The rate law for the reaction has the form:

$$\text{Rate} = k[\text{N}_2]^m[\text{O}_2]^n$$

We can use the experimental data given to find the values of k, m, and n for the reaction of N_2 with O_2.

SOLVE To determine the value of m, we can use the data from experiments 2 and 3, in which the concentration of O_2 is the same but the concentration of N_2 is four times as great in experiment 2 as in experiment 3. The rate of the reaction is four times larger in experiment 2, which enables us to conclude that the reaction is first order in N_2 and the value of m is 1.

The data in experiments 1 and 2 enable us to calculate the value of n. In these two experiments, the concentration of N_2 is the same, but the concentration of O_2 is four times larger in experiment 1 than it is in experiment 2. The reaction rate in experiment 3 is only two times as great as that in experiment 2. To solve for the order with respect to O_2, we need to answer the question, "Four raised to what power equals two?" The answer is the $\frac{1}{2}$ power; the reaction is only $\frac{1}{2}$ order in O_2 and $\frac{3}{2}$ order overall:

$$\text{Rate} = k\,[\text{N}_2][\text{O}_2]^{1/2}$$

Substituting the data from experiment 1 gives

$$1000\ M/s = k(0.040\ M)(0.040\ M)^{1/2} = k(0.0080)$$

$$k = 1.25 \times 10^5\ M^{-1/2}\ s^{-1}$$

THINK ABOUT IT A key point to remember is that the important components of a rate law—the rate, the order of the reaction with respect to the reactants, and the rate constant—must be determined experimentally. When deriving a rate law from initial rates, it is necessary to compare experiments in which the concentration of one reactant is constant so that the effect of concentration of a single reactant on the rate will be unambiguous.

Practice Exercise Nitric oxide reacts rapidly with unstable nitrogen trioxide (NO_3) to form NO_2:

$$\text{NO}(g) + \text{NO}_3(g) \rightarrow 2\,\text{NO}_2(g)$$

The following data were collected at 298 K:

Experiment	$[\text{NO}]_0\,(M)$	$[\text{NO}_3]_0\,(M)$	Initial Rate (M/s)
1	1.25×10^{-3}	1.25×10^{-3}	2.45×10^4
2	2.50×10^{-3}	1.25×10^{-3}	4.90×10^4
3	2.50×10^{-3}	2.50×10^{-3}	9.80×10^4

Determine the rate law for the reaction and calculate the value of the rate constant.

In Sample Exercise 14.5, we calculated a reaction order of $\frac{1}{2}$ for O_2. What does a reaction order of $\frac{1}{2}$ mean? The meaning of whole number, zero, and fractional reaction orders is addressed at the molecular level when we examine the mechanisms of reactions in Section 14.5. For now we should simply appreciate that mechanisms describe the stepwise changes that go on at the molecular level when reactions take place. Many chemical reactions take place in two or more steps, only one of which controls the overall rate of the reaction. The concentrations of the reactants in this key step and their relations to the initial reactants determine the rate-law expression for the overall reaction. We need to understand the order of a reaction from experimental data to propose likely steps in the process of breaking bonds and making bonds during chemical reactions, and in this section we concentrate on how to gain that mathematical understanding.

Deriving a Rate Law from a Single Experiment

We can determine the rate law for some reactions from plots of concentration versus time, essentially from the results of just one experiment or one data set. Some reactions have a single reactant, and their reaction rates depend on the concentration of only one substance. An important class of such reactions begins with the absorption of photons of electromagnetic radiation. As noted in Section 14.1, the reactions that lead to photochemical smog formation involve the photochemical decomposition of NO_2. Energy provided by sunlight breaks one of the N–O bonds in NO_2 to form NO and an oxygen atom:

$$NO_2(g) \xrightarrow{h\nu} NO(g) + O(g)$$

Atomic oxygen produced by this reaction combines with molecular oxygen to form ozone:

$$O(g) + O_2(g) \rightarrow O_3(g)$$

Ozone itself undergoes photochemical decomposition. As each molecule of ozone absorbs a photon of UV radiation, one of the oxygen–oxygen bonds breaks. The result is a molecule of oxygen and a free oxygen atom:

$$O_3(g) \xrightarrow{h\nu} O_2(g) + O(g)$$

This decomposition reaction can be studied in the laboratory by using a high-intensity source of ultraviolet light such as a mercury-vapor lamp. One study of the photochemical decomposition of ozone yielded the results listed in Table 14.4. The change in concentration of ozone with time is plotted in Figure 14.9(a). Because ozone is the only chemical reactant, the rate law for the decomposition reaction should depend only on the concentration of ozone. In our interpretation of the data in Table 14.4, we can start with the assumption that the reaction is first order in O_3. If it is, then the rate-law expression would be this:

$$\text{Rate} = -\frac{\Delta[O_3]}{\Delta t} = k[O_3]$$

This rate-law expression can be transformed (with the use of calculus) into an expression that relates the concentration of ozone, $[O_3]$, at any time in the course of the reaction to its concentration at the beginning of the reaction, $[O_3]_0$:

$$\ln \frac{[O_3]}{[O_3]_0} = -kt \tag{14.17}$$

The transformed version of the rate-law equation is called the **integrated rate law** (because integral calculus is used to derive it), and it describes the change in concentration of a reactant in a chemical reaction with time. The general integrated rate law for any reaction that is first order in reactant X is

$$\ln \frac{[X]}{[X]_0} = -kt \tag{14.18}$$

An **integrated rate law** is a mathematical expression that describes the change in concentration of a reactant in a chemical reaction with time.

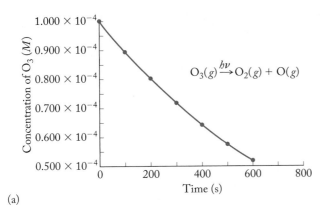

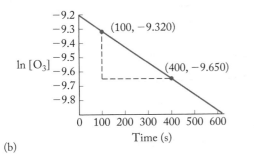

TABLE 14.4	Rate of Photochemical Decomposition of Ozone	
Time (s)	**$[O_3]$ (M)**	**ln $[O_3]$**
0	1.000×10^{-4}	-9.2103
100	0.896×10^{-4}	-9.320
200	0.803×10^{-4}	-9.430
300	0.719×10^{-4}	-9.540
400	0.644×10^{-4}	-9.650
500	0.577×10^{-4}	-9.760
600	0.517×10^{-4}	-9.870

FIGURE 14.9 Data for the decomposition of O_3: (a) plot of $[O_3]$ versus time; (b) plot of ln $[O_3]$ versus time. The line in (b) is straight, indicating that the decomposition reaction is first order in O_3.

In deriving the integrated law we assume that the reaction of ozone with light is first order in ozone. We need to check our assumption to see if it is actually true. To do that, we rearrange the terms in Equation 14.17 by splitting the quotient on the left side of Equation 14.17 into the difference of two logarithmic terms:

$$\ln [O_3] - \ln [O_3]_0 = -kt$$

Rearranging these terms, we get

$$\ln [O_3] = -kt + \ln [O_3]_0 \qquad (14.19)$$

Equation 14.19 is the equation of a straight line because it has the form

$$y = mx + b$$

$$\ln [O_3] = -kt + \ln [O_3]_0$$

where ln $[O_3]$ is the y variable and time (t) is the x variable. The slope of the line (m) is $-k$, and the y intercept (b) is ln $[O_3]_0$.

The plot of the change in the logarithmic concentration of ozone versus time in Figure 14.9(b) is indeed a straight line. This linearity means that the reaction is first order in O_3. If the reaction were not first order, the line would be curved, not straight. Calculating the slope of the line, $\Delta(\ln [O_3])/\Delta t$, from two points on it yields the value of the reaction rate constant (k):

$$k = -\text{slope} = -(-1.1 \times 10^{-3} \text{ s}^{-1}) = 1.1 \times 10^{-3} \text{ s}^{-1}$$

SAMPLE EXERCISE 14.6   **Deriving a Rate Law Using Data from a Single Data Set**

One of the less common nitrogen oxides in the atmosphere is dinitrogen pentoxide (N_2O_5). One reason why concentrations of N_2O_5 are small is that N_2O_5 is unstable and rapidly decomposes to N_2O_4 and O_2:

$$2\,N_2O_5(g) \rightarrow 2\,N_2O_4(g) + O_2(g)$$

A kinetic study of the decomposition of N_2O_5 at a particular fixed temperature yields the following data:

Time (s)	$[N_2O_5]$ (M)
0	0.1000
50	0.0707
100	0.0500
200	0.0250
300	0.0125
400	0.00625

Assume that the decomposition of N_2O_5 is first order in N_2O_5 and determine the value of the rate constant. Test the validity of your assumption.

COLLECT AND ORGANIZE The statement in the exercise instructs us to assume that the reaction studied is first order; we are asked to calculate the rate constant. We are given experimental data showing reactant concentration as a function of time.

ANALYZE If the decomposition of N_2O_5 is first order, then the plot of $\ln [N_2O_5]$ versus time will be a straight line.

SOLVE To find out if the plot gives a straight line, we need to calculate $\ln [N_2O_5]$ values from the data given. We then add these values to the table of data:

Time (s)	$[N_2O_5]$ (M)	$\ln [N_2O_5]$
0	0.1000	−2.303
50	0.0707	−2.649
100	0.0500	−2.996
200	0.0250	−3.689
300	0.0125	−4.382
400	0.00625	−5.075

The plot of $\ln [N_2O_5]$ versus t is shown in the graph at the left. The fact that the plot is a straight line indicates that the reaction is indeed first order in N_2O_5.

By selecting two convenient sets of data points along the line, at $t = 100$ and 300 s, we can calculate the slope of the line:

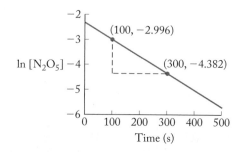

$$\text{Slope} = \frac{\Delta y}{\Delta x} = \frac{-4.382 - (-2.996)}{300 - 100} = \frac{-1.386}{200} = -0.00693$$

The slope of the line equals $-k$. Therefore, the rate constant $k = 0.00693$ s^{-1}. The y intercept is equal to −2.3, which is the natural log of the initial concentration of N_2O_5.

THINK ABOUT IT By plotting ln $[N_2O_5]$ versus time, we tested the assumption that the reaction is first order in N_2O_5. The fact that the plot is a straight line tells us that the assumption was correct. Had the line been curved, the test would have failed, and we would have to conclude that the reaction is not first order in N_2O_5. The linear plot validates the assumption, and a few more experiments with different initial reactant concentrations at the same temperature would also give straight-line plots with the same slope and thus the same rate constant k.

Practice Exercise Hydrogen peroxide (H_2O_2) decomposes into water and oxygen:

$$H_2O_2(\ell) \rightarrow H_2O(\ell) + \tfrac{1}{2}O_2(g)$$

The following data were collected for the decomposition of H_2O_2 at a constant temperature. Determine whether the decomposition of H_2O_2 is first order in H_2O_2, and calculate the value of the rate constant at the temperature of the experiments.

Time (s)	$[H_2O_2]$ (M)
0	0.500
100	0.460
200	0.424
500	0.330
1000	0.218
1500	0.144

Reaction Half-Lives

A parameter frequently cited in kinetic studies is the **half-life ($t_{1/2}$)** of the reaction, which is the interval during which the concentration of a reactant decreases by half. The half-life is related to the rate constant of a chemical reaction, and an inverse relationship exists between the rate of a reaction and its half-life: the higher the rate, the less time it takes for half the reactants to be consumed, and the shorter the half-life.

Let's consider this concept in the context of another nitrogen oxide found in the atmosphere: dinitrogen monoxide (N_2O), also called nitrous oxide or laughing gas, an anesthetic sometimes used by dentists. Atmospheric concentrations of this potent greenhouse gas have been increasing in recent years, but the principal source of atmospheric N_2O is biological: bacterial degradation of nitrogen compounds in soil. Dinitrogen monoxide is not a product of combustion reactions in internal combustion engines because at typical engine temperatures any N_2O formed rapidly decomposes into nitrogen and oxygen:

$$N_2O(g) \rightarrow N_2(g) + \tfrac{1}{2}O_2(g)$$

A rapid reaction rate for this first-order reaction means that it has a short half-life. The passage of several such half-lives is shown in Figure 14.10. We can derive a mathematical relation between the half-life and the rate constant of this or any first-order reaction by starting with the integrated rate law for a reaction that is first order in generic reactant X (Equation 14.18):

> The **half-life ($t_{1/2}$)** of a reaction is the time during which the concentration of a reactant decreases by half.

$$\ln\frac{[X]}{[X]_0} = -kt$$

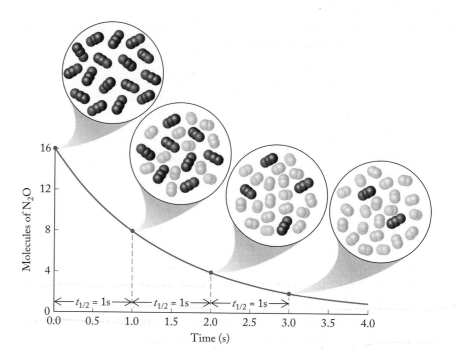

FIGURE 14.10 The decomposition reaction is first order in N_2O. If the half-life ($t_{1/2}$) of the reaction is 1.0 s, then, on average, half of a population of 16 N_2O molecules decomposes in 1.0 s, half of the remaining 8 molecules decomposes in the next 1.0 s, and so on.

After one half-life has passed ($t = t_{1/2}$), the concentration of X is half its original value: $[X] = \frac{1}{2}[X]_0$. That is exactly what the definition of half-life means. Inserting this information into Equation 14.18 yields

$$\ln \frac{\frac{1}{2}[X]_0}{[X]_0} = -kt_{1/2}$$

or

$$\ln\left(\frac{1}{2}\right) = -kt_{1/2}$$

The natural log of $\frac{1}{2}$ is -0.693, so

$$-0.693 = -kt_{1/2}$$

or

$$t_{1/2} = \frac{0.693}{k} \tag{14.20}$$

Thus, the value of the half-life of a first-order reaction is inversely proportional to the rate constant. Whether the concentration of reactant is high or low, half of it is consumed in one half-life.

SAMPLE EXERCISE 14.7 **Calculating the Half-Life of a First-Order Reaction**

The rate constant (k) for the thermal decomposition of N_2O_5 at a particular temperature is 7.8×10^{-3} s^{-1}. What is the half-life of N_2O_5 at that temperature?

COLLECT AND ORGANIZE We are asked to determine the half-life of N_2O_5 in its decomposition reaction.

ANALYZE The half-life of a first-order reaction is independent of the concentration. Equation 14.20 relates the rate constant (k) of a first-order reaction to its half-life ($t_{1/2}$). Note that this relation only applies to first-order reactions. We already showed in Sample Exercise 14.6 that the decomposition reaction of N_2O_5 is first order, so we use that fact here to solve the problem.

SOLVE Inserting the given value of the rate constant k into Equation 14.20 gives

$$t_{1/2} = \frac{0.693}{k} = \frac{0.693}{7.8 \times 10^{-3}\ s^{-1}} = 89\ s$$

THINK ABOUT IT Remember that this relation between half-life and rate constant is only valid for first-order reactions and that the initial concentration of N_2O_5 was not needed here.

Practice Exercise Environmental scientists studying the transport and fate of pollutants often calculate the half-life of these pollutants in the ecosystems into which they are released. They define a *transport rate constant* for the process by which a compound moves out of a system that is equivalent to a reaction rate constant. In a study of the gasoline additive MTBE in Donner Lake, California, scientists from the University of California, Davis, found that the half-life of MTBE in the lake in the summer was 28 days. Assume that the transport process is first order. What was the transport rate constant of MTBE out of Donner Lake during the study? Express your answer in reciprocal days.

Second-Order Reactions

At high temperatures or in response to exposure to sunlight, nitrogen dioxide decomposes into NO and O_2:

$$2\ NO_2(g) \rightarrow 2\ NO(g) + O_2(g)$$

Because only one type of molecule appears on the left-hand side of the equation, as was the case in the reactions for the decomposition of N_2O_5 and H_2O_2, we might expect this reaction to be first order just as the other two reactions were. In that case, the rate-law expression for this reaction would take the form:

$$Rate = k[NO_2]^m \tag{14.21}$$

where $m = 1$. Table 14.5 contains experimental data on the change in concentration of NO_2 over time. When we use these data to evaluate the order (m) of the reaction and the rate constant (k), we generate the plot of $\ln [NO_2]$ versus time for the thermal decomposition of NO_2 shown in Figure 14.11(a). The plot is clearly not linear, which tells us that the thermal decomposition of NO_2 is *not* first order as we originally assumed.

TABLE 14.5 **Rate of Decomposition of NO_2**

Time (s)	$[NO_2]$ (M)	$\ln [NO_2]$	$1/[NO_2]$ (1/M)
0	1.00×10^{-2}	−4.605	100
100	6.48×10^{-3}	−5.039	154
200	4.79×10^{-3}	−5.341	209
300	3.80×10^{-3}	−5.573	263
400	3.15×10^{-3}	−5.760	317
500	2.69×10^{-3}	−5.918	372
600	2.35×10^{-3}	−6.057	426

FIGURE 14.11 Pure NO_2 slowly decomposes at high temperatures into NO and O_2. (a) A plot of ln $[NO_2]$ versus time is not linear, indicating the reaction is not first order. (b) A plot of $1/[NO_2]$ versus time is linear, indicating that the reaction is second order in NO_2. The slope of the line in (b) equals the second-order rate constant (k).

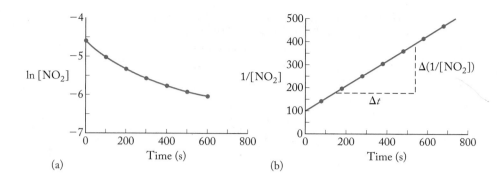

(a)

(b)

What is the order of the reaction? The answer is hidden in how the reaction takes place. If single molecules of NO_2 simply fall apart, the decomposition reaction would be first order, much like the decomposition of N_2O_5. However, if the reaction happens as a result of collisions between pairs of NO_2 molecules, then the reaction would be first order in both reactants and second order overall. The thermal decomposition of NO_2 depends on the collision between pairs of molecules that just happen to be molecules of the same substance, NO_2. The decomposition of NO_2 is second order overall, and, because there is only one reactant, the reaction is second order in that reactant.

How can we determine if the thermal decomposition of NO_2 is really second order? One way is to assume that it is and then test that assumption. The test entails transforming Equation 14.21 into the corresponding integrated rate-law expression for a second-order reaction, again by using calculus. The result of the transformation is:

$$\underbrace{\frac{1}{[NO_2]} = kt + \frac{1}{[NO_2]_0}}_{y = mx + b} \tag{14.22}$$

Equation 14.22 is that of a straight line, with $1/[NO_2]$ the y variable and time the x variable. Such a linear plot is shown Figure 14.11(b). Therefore, the thermal decomposition of NO_2 is second order. The slope of the line provides a direct measure of k, the second-order rate constant, which in this case is 0.544 $M^{-1}\,s^{-1}$. A general form of Equation 14.22 that applies to any reaction that is second order in substance X is:

$$\frac{1}{[X]} = kt + \frac{1}{[X]_0} \tag{14.23}$$

SAMPLE EXERCISE 14.8 | **Distinguishing between First- and Second-Order Reactions**

Chlorine monoxide (ClO) accumulates in the stratosphere above Antarctica in the winter and plays a key role in the formation of the ozone hole above the South Pole in the spring of each year. Eventually, ClO decomposes according to the equation

$$2\,ClO(g) \rightarrow Cl_2(g) + O_2(g)$$

The kinetics of this reaction were studied in a laboratory experiment at 298 K, and the following data were obtained. Determine the order of the decomposition reaction, the rate law, and the value of the rate constant at 298 K.

Time (ms)	[ClO](M)
0	1.50×10^{-8}
10	7.19×10^{-9}
20	4.74×10^{-9}
30	3.52×10^{-9}
40	2.81×10^{-9}
100	1.27×10^{-9}
200	0.66×10^{-9}

COLLECT AND ORGANIZE We are asked to determine the order of the decomposition reaction of ClO. We are given experimental data describing the variation of concentration of ClO with time at 298 K. That data set is all we need to assign the reaction order.

ANALYZE To distinguish between a first- and second-order reaction in which there is a single reactant, we plot ln [ClO] versus time and 1/[ClO] versus time. If the plot of ln [ClO] versus time is linear, then the reaction is first order; if a plot of 1/[ClO] versus time is linear, then the reaction is second order.

SOLVE To evaluate the two possibilities, we need to calculate the natural logarithm and reciprocal values of the tabulated concentrations of ClO:

Time (ms)	[ClO] (M)	ln [ClO]	1/[ClO] (1/M)
0	1.50×10^{-8}	−18.015	6.67×10^7
10	7.19×10^{-9}	−18.751	1.39×10^8
20	4.74×10^{-9}	−19.167	2.11×10^8
30	3.52×10^{-9}	−19.465	2.84×10^8
40	2.81×10^{-9}	−19.690	3.56×10^8
100	1.27×10^{-9}	−20.484	7.89×10^8
200	0.66×10^{-9}	−21.139	1.51×10^9

The next step is to plot ln [ClO] and 1/[ClO] versus time, as shown in the following graphs:

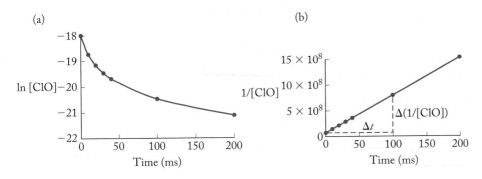

The plot of ln [ClO] versus time (graph a) is not linear, but the plot of 1/[ClO] versus time (graph b) is. Therefore, the reaction is second order in ClO and has the following rate-law expression:

$$\text{Rate} = k[\text{ClO}]^2$$

Applying the generic integrated rate-law expression (Equation 14.23) to this second-order reaction gives us:

$$\frac{1}{[ClO]} = kt + \frac{1}{[ClO]_0}$$

The slope of a plot of $1/[ClO]$ versus t equals the rate constant. By choosing two convenient pairs of data (at 0 and 100 ms), we can calculate the slope (and k) as follows:

$$k = \text{slope} = \frac{\Delta y}{\Delta x} = \frac{\Delta\left(\frac{1}{[ClO]}\right)}{\Delta t}$$

$$= \frac{(7.89 - 0.67) \times 10^8 \ M^{-1}}{(100 - 0) \times 10^{-3} \ s}$$

$$= 7.22 \times 10^9 \ M^{-1} \ s^{-1}$$

THINK ABOUT IT Always remember that the order of a reaction must be determined experimentally. The stoichiometry of the reaction does not identify the order; only the analysis of experimental data enables us to determine order.

Practice Exercise The reaction of NO_2 with CO produces NO and CO_2:

$$NO_2(g) + CO(g) \rightarrow NO(g) + CO_2(g)$$

Experimental evidence shows that the reaction rate is independent of the CO concentration and depends only on the concentration of NO_2. Determine whether the reaction is first or second order in NO_2, and calculate the rate constant from the data in the table to the left, which were obtained at a reaction temperature of 488 K.

Time (hr)	$[NO_2]$ (M)
0.00	0.250
1.39	0.198
3.06	0.159
4.72	0.132
6.39	0.114
8.06	0.099
9.72	0.088
11.39	0.080

CONCEPT TEST

Many reactions are first order, fewer are second order in a single reactant, and third-order reactions in a single reactant are practically nonexistent. Can you suggest why?

The concept of half-life can also be applied to second-order reactions. The relation between the second-order rate constant (k) and the half-life ($t_{1/2}$) for the thermal decomposition of NO_2 can be derived from Equation 14.23 if we first rearrange the terms to solve for kt:

$$kt = \frac{1}{[NO_2]} - \frac{1}{[NO_2]_0}$$

After one half-life has elapsed ($t = t_{1/2}$), $[NO_2]$ has decreased to half the initial concentration: $[NO_2] = \frac{1}{2}[NO_2]_0$. Substituting these terms into the preceding equation, we have

$$kt_{1/2} = \frac{1}{\frac{1}{2}[NO_2]_0} - \frac{1}{[NO_2]_0}$$

$$= \frac{2}{[NO_2]_0} - \frac{1}{[NO_2]_0} = \frac{1}{[NO_2]_0}$$

or

$$t_{1/2} = \frac{1}{k[NO_2]_0} \tag{14.24}$$

An equation of similar form can be written for any reaction that is second order in substance X:

$$t_{1/2} = \frac{1}{k[X]_0} \qquad\qquad (14.25)$$

SAMPLE EXERCISE 14.9 **Calculating the Half-Life of a Second-Order Reaction**

Calculate the half-life of the second-order thermal decomposition of NO_2 if the rate constant is $0.543\ M^{-1}\ s^{-1}$ at a particular temperature and the initial concentration of NO_2 is $0.0100\ M$.

COLLECT AND ORGANIZE We are given the rate constant and initial concentration of NO_2 and asked to calculate the half-life of the decomposition reaction. We also know the reaction is second order.

ANALYZE The half-life of a second-order reaction in which there is only one reactant is given by Equation 14.25.

SOLVE Through substituting the values for k and $[NO_2]_0$ into Equation 14.25, we obtain

$$t_{1/2} = \frac{1}{k[NO_2]_0} = \frac{1}{(0.543\ M^{-1}\ s^{-1})(1.00 \times 10^{-2}\ M)} = 184\ s$$

THINK ABOUT IT If the initial concentration $[NO_2]_0$ is doubled, the half-life is cut in half to 92 s. If the reaction were first order, changing the initial concentration would not have changed the half-life because the concentration of the reactant does not appear in the half-life equation for first-order reactions (Equation 14.20).

Practice Exercise The second-order rate constant for the decomposition of ClO (Sample Exercise 14.8) is $7.22 \times 10^9\ M^{-1}\ s^{-1}$ at a particular temperature. Determine the half-life of ClO when its initial concentration is $1.50 \times 10^{-8}\ M$.

Pseudo-First-Order Reactions

Equation 14.25 applies only to reactions that are second order in a single reactant. It does not apply to reactions that are second order overall, but first order in two reactants, such as the reaction between O_3 and NO:

$$NO(g) + O_3(g) \rightarrow NO_2(g) + O_2(g)$$

or the reaction between O_3 and NO_2:

$$O_3(g) + NO_2(g) \rightarrow NO_3(g) + O_2(g)$$

The integrated rate law for a reaction that is first order in two reactants is quite complicated. To avoid a more complicated analysis, people who study kinetics frequently set conditions in the study so that a simple rate-law expression can be used. A simple rate-law expression for such a reaction can be derived when one of the reactants is present at much higher concentration than the other. This condition is common for many of the components in the urban atmosphere where, for example, the concentrations of ozone are often hundreds to thousands of times greater than the concentrations of NO. With such a large excess of ozone, the

A **pseudo-first-order** reaction is one in which all the reactants but one are present at such high concentrations that they do not decrease significantly during the course of the reaction, so that reaction rate is controlled by the concentration of the limiting reactant.

concentration of ozone remains virtually constant (it does not change significantly) in the course of the reaction:

$$O_3(g) + NO(g) \rightarrow O_2(g) + NO_2(g)$$

In this case, the rate law for the reaction

$$\text{Rate} = k[NO][O_3]$$

may be simplified to

$$\text{Rate} = k'[NO] \tag{14.26}$$

where

$$k' = k[O_3]_0 \tag{14.27}$$

and where $[O_3]_0$ is the initial concentration of ozone, which remains virtually constant throughout the reaction.

Equation 14.26 looks like the rate law for a first-order reaction. It is considered a **pseudo-first-order** rate law because it appears to obey first-order kinetics. A pseudo-first-order reaction has the same integrated rate law as a first-order reaction, but its rate depends on the concentration of more than one reactant. The pseudo-first-order rate constant, k', in Equation 14.26 can be determined from a plot of ln [NO] versus time. The value of k can then be determined by dividing k' by $[O_3]_0$ as shown in the following Sample Exercise.

SAMPLE EXERCISE 14.10 **Deriving a Pseudo-First-Order Rate Law**

The following data were obtained in a study of the oxidation of trace levels of NO by a large excess of ozone at a particular temperature (note that the NO concentrations are expressed in numbers of molecules per cubic centimeter):

Time (μs)	Concentration of NO (molecules/cm³)
0	1.00×10^9
100	8.36×10^8
200	6.98×10^8
300	5.83×10^8
400	4.87×10^8
500	4.07×10^8
1000	1.65×10^8

a. Verify that the reaction is pseudo-first-order and determine the pseudo-first-order rate constant, k', for the reaction.

b. If the initial concentration of ozone is 100 times the initial concentration of NO, or 1.00×10^{11} molecules/cm³, what is the second-order rate constant k for the reaction?

COLLECT AND ORGANIZE We are asked to verify that the reaction is pseudo-first-order and to determine a pseudo-first-order rate constant. We are given experimental data showing the change of concentration of NO with time. We are given the actual concentration of O_3 and are asked to calculate the second-order rate constant for the reaction.

ANALYZE We can anticipate pseudo-first-order kinetics and determine a pseudo-first-order rate constant because ozone is in great excess and its concentration does not change significantly throughout the course of the reaction. We treat the data as we would for a first-order reaction by plotting ln [NO] versus time. The plot should be a straight line.

SOLVE

a. To determine a pseudo-first-order rate constant for the reaction, we need to calculate the natural logarithms of the concentration values.

Time (μs)	[NO] (molecules/cm³)	ln [NO]
0	1.00×10^9	20.723
100	8.36×10^8	20.544
200	6.98×10^8	20.364
300	5.83×10^8	20.184
400	4.87×10^8	20.004
500	4.07×10^8	19.824
1000	1.65×10^8	18.915

The next step is to plot ln [NO] versus time as in the accompanying graph.

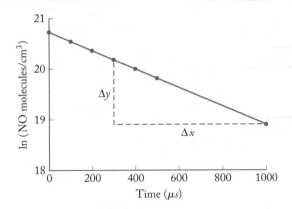

The linear plot shows that the reaction is indeed pseudo-first-order.

The slope of the line ($\Delta y/\Delta x$) is calculated in the usual way from a convenient pair of data points, such as (300, 20.184) and (1000, 18.915):

$$\text{Slope} = \frac{\Delta y}{\Delta x} = \frac{(20.184 - 18.915)}{(300 - 1000)\text{ s}} = -0.00181\text{ s}^{-1}$$

The value of the pseudo-first-order rate constant (k') is 0.00181 s^{-1}.

b. We calculate the second-order rate constant by using Equation 14.25:

$$k' = k[O_3]_0$$

$$k = \frac{k'}{[O_3]} = \frac{0.00181\text{ s}^{-1}}{1.00 \times 10^{11}\text{ molecules/cm}^3} = \frac{1.81 \times 10^{-14}\text{ cm}^3}{\text{molecules} \cdot \text{s}}$$

To obtain a value of k with the conventional units ($M^{-1}\text{ s}^{-1}$), we need to convert the reciprocal concentration units of cm³/molecule into M^{-1}:

$$k = \frac{1.81 \times 10^{-14}\text{ cm}^3\text{ s}^{-1}}{\text{molecules}} \times \frac{6.022 \times 10^{23}\text{ molecules}}{\text{mol}} \times \frac{1\text{ L}}{1000\text{ cm}^3}$$

$$= 1.09 \times 10^7 \frac{\text{L} \cdot \text{s}^{-1}}{\text{mol}} = 1.09 \times 10^7\ M^{-1}\text{ s}^{-1}$$

THINK ABOUT IT Once we determine a pseudo-first-order rate constant for a reaction, we can calculate the second-order rate constant if we have the initial concentration of the substance that is in excess.

Practice Exercise The following reaction is known to be first-order in both reactants.

$$Cl(g) + O_3(g) \rightarrow ClO(g) + O_2(g)$$

Determine the pseudo-first-order and second-order rate constants for the reaction at the temperature at which the results in the table were obtained. The initial concentration of ozone was $8.5 \times 10^{-11}\ M$.

Time (μs)	[Cl] (*M*)
0	5.60×10^{-14}
100	5.27×10^{-14}
600	3.89×10^{-14}
1200	2.69×10^{-14}
1850	1.81×10^{-14}

Zero-Order Reactions

In the Practice Exercise accompanying Sample Exercise 14.8, we were introduced to the reaction of NO_2 with CO:

$$NO_2(g) + CO(g) \rightarrow NO(g) + CO_2(g)$$

We were told that the experimental data show that the rate of the reaction does not depend on the concentration of CO. The observed rate law for this reaction is

$$Rate = k[NO_2]^2$$

It contains no term for [CO], even when reaction conditions are such that the concentrations of CO and NO_2 are comparable. This situation is very different from the situation we just discussed in pseudo-first-order reactions where the rate of the reaction does not depend on the concentration of one of the reactants because it is present in large excess. One interpretation of this situation is that the rate law contains a [CO] term to the power of zero, making the reaction *zero order* in that reactant. Because any value raised to the zero power is one, the rate expression would be

$$Rate = k[NO_2]^2[CO]^0 = k[NO_2]^2(1) = k[NO_2]^2$$

Reactions with a true zero-order rate law are rare, but let's consider a generic reaction involving a single reactant "A" that is zero order in reactant A and zero order overall.

$$A \rightarrow B$$

The rate law for this reaction is

$$Rate = -\Delta[A]/\Delta t = k[A]^0 = k$$

The integrated rate law for this reaction is

$$[A] = -kt + [A]_0$$

The slope of a plot of reactant concentration versus time (Figure 14.12) equals the negative of the zero-order rate constant, k. We can calculate the half-life of a zero-order reaction by substituting $t = t_{1/2}$ and $[A] = [A]_0/2$ into the integrated rate law:

$$[A]_0/2 = -kt_{1/2} + [A]_0$$

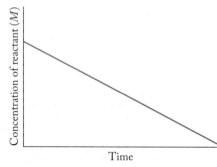

FIGURE 14.12 The change in concentration of a reactant in a zero-order reaction is constant over time.

Rearranging the equation leads to

$$kt_{1/2} = [A]_0 - [A]_0/2 = [A]_0/2$$

or

$$t_{1/2} = [A]_0/2k$$

Right now we leave the discussion of zero-order reactions with this purely mathematical treatment. We return to the discussion of zero-order reactions and examine their meaning at the molecular level in Section 14.5.

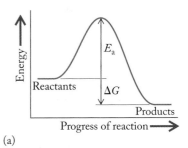
(a)

14.4 Reaction Rates, Temperature, and the Arrhenius Equation

In Section 14.3, we introduced the idea that chemical reactions take place in the gas phase when molecules collide with sufficient energy to break some of the chemical bonds in the reactants while forming bonds in the products. This energy is called the **activation energy (E_a)**, and every chemical reaction has a characteristic activation energy, usually expressed in kJ/mol. Activation energy is an energy barrier that must be overcome if a reaction is to proceed—like the hill that must be climbed by the cyclist in Figure 14.13. Generally speaking, the greater the activation energy, the slower the reaction.

According to kinetic molecular theory, the fraction of molecules with kinetic energies greater than a given activation energy increases with increasing temperature, as shown in Figure 14.14(a). Therefore, the rates of chemical reactions should increase with increasing temperature, and indeed they do (Figure 14.14b).

The mathematical connections between temperature, the value of the rate constant (k) of a reaction, and its activation energy (E_a) are given by the **Arrhenius equation**, where R is the gas constant (in units of J/mol·K) and T is the reaction temperature in kelvins.

$$k = Ae^{-E_a/RT} \tag{14.28}$$

FIGURE 14.13 The energy profile (a) of an exergonic reaction includes an activation energy barrier (E_a) that is analogous to (b) the hill that a bicyclist must climb en route from a starting point on a high plateau to a destination on a lower plain.

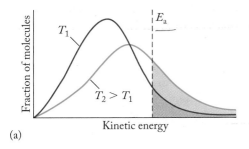
(a)

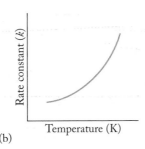
(b)

■ Fraction of molecules in sample with sufficient energy to react at T_1
■ Increase in number of molecules in sample with sufficient energy to react at T_2; $T_2 > T_1$

FIGURE 14.14 According to kinetic molecular theory, some fraction of a population of reactant molecules [the shaded areas in (a)] has kinetic energies equal to or greater than the activation energy (E_a) of the reaction. As temperature increases from T_1 to T_2, the number of molecules with energies exceeding E_a increases, leading to an increase in reaction rate. (b) The rate constant for any reaction increases with increasing temperature.

▶‖ **CHEMTOUR** Arrhenius Equation

Activation energy (E_a) is the minimum energy molecules need to react when they collide.

The **Arrhenius equation** relates the rate constant of a reaction to absolute temperature (T), the activation energy of the reaction (E_a), and the frequency factor (A).

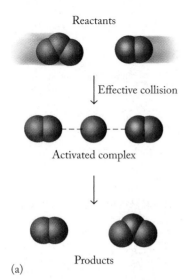

Reactants

Effective collision

Activated complex

Products

(a)

Reactants

Ineffective collision

No products

(b)

FIGURE 14.15 The importance of molecular orientation on reaction rate is illustrated by the reaction between NO and O_3. (a) Collisions in which the nitrogen on NO contacts an oxygen atom of O_3 may lead to an activated complex that yields NO_2 and O_2. (b) Collisions between oxygen atoms on NO and O_3 molecules produce no activated complex and no reaction.

▶❙❙ **CHEMTOUR** Collision Theory

The **frequency factor** (A) is the product of the frequency of molecular collisions and a factor that expresses the probability that the orientation of the molecules is appropriate for a reaction to occur.

An **activated complex** is a species formed in a chemical reaction when molecules have enough potential energy to react with each other.

A **transition state** in a chemical reaction is a high-energy state between reactants and products.

The term A is called the **frequency factor**. The frequency factor is the product of collision frequency and a term that corrects for the fact that not every collision results in a chemical reaction. Some collisions do not lead to products because the colliding molecules are not oriented with respect to each other in the right way. To examine the importance of molecular orientation during collisions, let's revisit a reaction (Equation 14.7) that sustains the level of NO_2 in polluted urban air after the evening rush hour:

$$O_3(g) + NO(g) \rightarrow O_2(g) + NO_2(g)$$

Two of the ways in which ozone and nitric oxide molecules might approach each other are shown in Figure 14.15. Only one of these orientations, the one in which an oxygen atom of an O_3 molecule approaches the nitrogen atom of NO, leads to a chemical reaction between the two molecules.

A collision between O_3 and NO molecules with the correct orientation and enough kinetic energy may result in the formation of an **activated complex** in which some of the bonds that hold these molecules together break and the bonds that hold O_2 and NO_2 together form. Activated complexes are formed by reacting species that have acquired enough potential energy to react with each other. The energies of these complexes are much higher than those of reactants and products, and represent **transition states** between reactants and products. In fact, the energies of these transition states define the heights of the activation energy barriers of chemical reactions.

Consider the energy profile for the reaction between nitric oxide and ozone in Figure 14.16. The x-axis of the right-hand profiles represents the progress of the reaction. Distances along the y-axis represent changes in chemical energy. The size of the activation energy barrier depends on the direction from which it is approached. If a reaction is spontaneous ($\Delta G° < 0$) in the forward direction, such as the one illustrated in Figure 14.16(a), then E_a is smaller in the forward direction than in the reverse direction (Figure 14.16b). A smaller activation energy barrier means that the spontaneous reaction proceeds at a higher rate than the reverse, nonspontaneous reaction.

One of the many uses of the Arrhenius equation is to calculate the value of E_a for a chemical reaction. To see how this procedure works, we must first rewrite Equation 14.28 by taking the natural logarithm of both sides:

$$\ln k = -\frac{E_a}{R}\left(\frac{1}{T}\right) + \ln A \qquad (14.29)$$

$$y = mx + b$$

Equation 14.29 fits the general equation of a straight line ($y = mx + b$) if we make ($\ln k$) the y variable and ($1/T$) the x variable. We can calculate E_a by determining the rate constants for a reaction at several temperatures. Plotting $\ln k$ versus $1/T$ should give a straight line, the slope of which is $-E_a/R$. Table 14.6 and Figure 14.17 show the results of measurements of the rate of the reaction between NO and O_3 at six different temperatures. The straight line that best fits the points in Figure 14.17 has a slope of -1260 K. There are no units for the y values (they are logarithms), and the units of the x values are 1/K. Therefore, the units of the slope $\Delta y/\Delta x$ will be 1/(1/K), or simply K. The slope equals $-E_a/R$, so

$$E_a = -\text{slope} \times R$$

$$= -(-1260 \text{ K}) \times \left(\frac{8.314 \text{ J}}{\text{mol} \cdot \text{K}}\right) = 1.05 \times 10^4 \text{ J/mol}$$

$$= 10.5 \text{ kJ/mol}$$

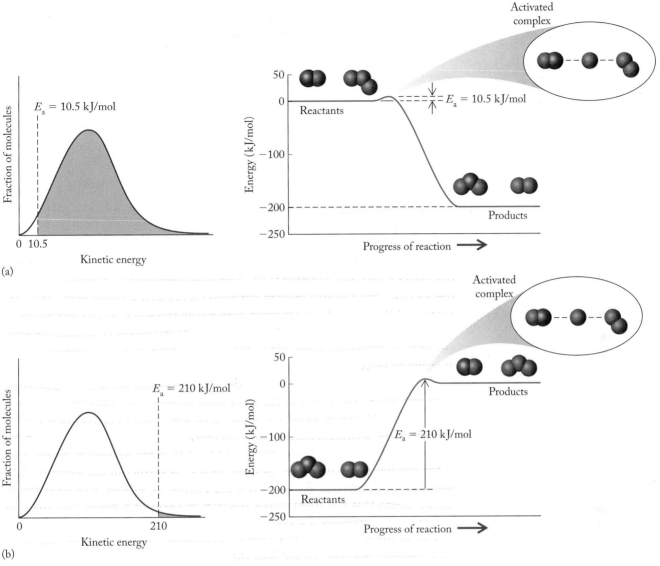

FIGURE 14.16 (a) The energy profile for the spontaneous reaction $O_3(g) + NO(g) \rightarrow O_2(g) + NO_2(g)$ includes an activation energy barrier of 10.5 kJ/mol. (b) The reverse reaction has an activation energy of 210 kJ/mol. In the left-hand plots of kinetic energy versus fraction of molecules, the shaded areas indicate the molecules with sufficient energy (E_a) to react. As shown by the different sizes of the shaded areas under the kinetic energy curves, fewer molecules have the kinetic energy needed to overcome the larger activation energy of the reverse reaction. The reverse reaction proceeds much more slowly than the forward reaction.

TABLE 14.6 Temperature Dependence of the Rate of Reaction for the Reaction $NO(g) + O_3(g) \rightarrow NO_2(g) + O_2(g)$

T (K)	k (M⁻¹ s⁻¹)	ln k	1/T (K⁻¹)
300	1.21×10^{10}	23.216	3.33×10^{-3}
325	1.67×10^{10}	23.539	3.08×10^{-3}
350	2.20×10^{10}	23.814	2.86×10^{-3}
375	2.79×10^{10}	24.052	2.67×10^{-3}
400	3.45×10^{10}	24.264	2.50×10^{-3}
425	4.15×10^{10}	24.449	2.35×10^{-3}

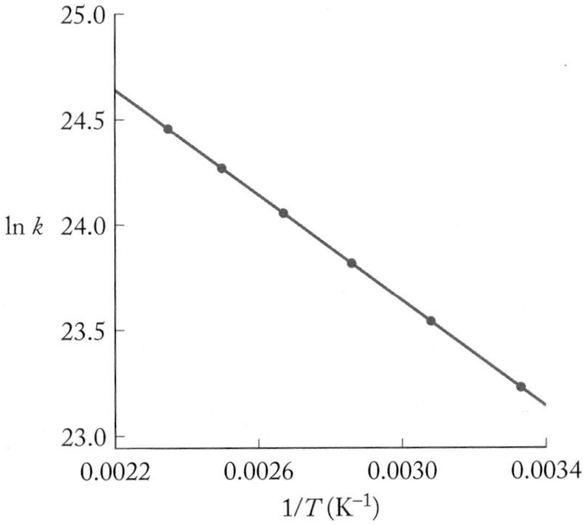

FIGURE 14.17 A graph of the natural logarithm of the rate constant (ln k) versus $1/T$ yields a straight line with a slope equal to $-E_a/R$ and a y intercept equal to the natural logarithm of the frequency factor (ln A).

The y intercept ($1/T = 0$) of the line that best fits the data is 27.4. This value represents $\ln A$. The value of A is

$$A = e^{27.4} = 7.9 \times 10^{11}$$

We can use the values of E_a and A to calculate k at any temperature. For example, at $T = 250$ K

$$k = Ae^{-E_a/RT}$$
$$= 7.9 \times 10^{11} \, e^{\left[\frac{-(1.05 \times 10^4 \text{ J/mol})}{(8.314 \text{ J/mol} \cdot \text{K})(250 \text{ K})}\right]}$$
$$= 5.1 \times 10^9 \, M^{-1} \, \text{s}^{-1}$$

SAMPLE EXERCISE 14.11 **Calculating the Activation Energy from Rate Constants**

The following data were collected in a study of the effect of temperature on the rate of decomposition of ClO into Cl_2 and O_2 in the reaction

$$2\,ClO(g) \rightarrow Cl_2(g) + O_2(g)$$

Determine the activation energy for the reaction.

k (M^{-1} s^{-1})	T (K)
1.9×10^9	238
3.1×10^9	258
4.9×10^9	278
7.2×10^9	298

COLLECT AND ORGANIZE We are asked to determine the activation energy (E_a) for a reaction. We are given values of the rate constant k as a function of absolute temperature.

ANALYZE The activation energy for any reaction can be calculated from rate constants for the reaction at a series of temperatures by using Equation 14.29:

$$\ln k = -\frac{E_a}{R}\left(\frac{1}{T}\right) + \ln A$$

The y variable is ($\ln k$), so we need to convert k values to ($\ln k$) values first. The x and y variables in the Arrhenius equation are ($1/T$) and ($\ln k$). We must convert temperature and rate constant values to these variables before plotting the data.

SOLVE Expanding the data table to include columns of ($1/T$) and ($\ln k$) yields

k (M^{-1} s^{-1})	$\ln k$	T (K)	$1/T$ (K^{-1})
1.9×10^9	21.365	238	4.20×10^{-3}
3.1×10^9	21.855	258	3.90×10^{-3}
4.9×10^9	22.313	278	3.60×10^{-3}
7.2×10^9	22.697	298	3.36×10^{-3}

A plot of (ln k) versus (1/T) yields a straight line:

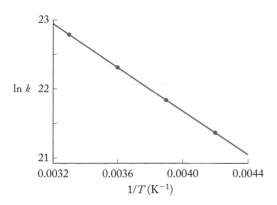

the slope of which is −1564 K. According to the Arrhenius equation, the slope is equal to $-E_a/R$, where E_a is the activation energy and R is the ideal gas constant with units of J/(mol · K).

By using the values of the slope (−1564 K) and R [8.314 J/(mol · K)], we can calculate the value of E_a:

$$E_a = -\text{slope} \times R$$

$$= -(-1564 \text{ K}) \times \left(\frac{8.314 \text{ J}}{\text{mol} \cdot \text{K}}\right) = 1.30 \times 10^4 \text{ J/mol}$$

$$= 13.0 \text{ kJ/mol}$$

THINK ABOUT IT The activation energy is the height of a barrier that has to be overcome before a reaction can proceed. Generally speaking, the higher the barrier, the lower the rate constant and the slower the reaction. The activation energy for ClO decomposition is relatively low, consistent with its very large rate constants. The values of the rate constant are very large, so the reaction is very rapid.

Practice Exercise Atomic bromine reacts with ozone to form BrO and O_2:

$$Br(g) + O_3(g) \rightarrow BrO(g) + O_2(g)$$

The rate constant for the reaction was determined at four temperatures ranging from 238 to 298 K. Calculate the activation energy for this reaction.

T (K)	k [cm³/(molecule · s)]
238	5.9×10^{-13}
258	7.7×10^{-13}
278	9.6×10^{-13}
298	1.2×10^{-12}

Earlier in this section, we noted that activation energy barriers tend to be lower when approached from the reaction direction that has a negative ΔG. This observation raises a question: Do spontaneous reactions always proceed more

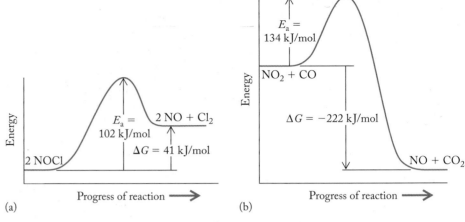

FIGURE 14.18 Spontaneous reactions are not necessarily rapid reactions. Here are reaction-energy profiles for two reactions: (a) the nonspontaneous ($\Delta G > 0$) decomposition of NOCl to NO and Cl_2, which has an activation energy of 102 kJ/mol, and (b) the spontaneous reaction NO_2 + $CO \rightarrow NO + CO_2$, which has an activation energy of 134 kJ/mol. The spontaneous reaction may actually be the slower of the two because its activation energy barrier is higher.

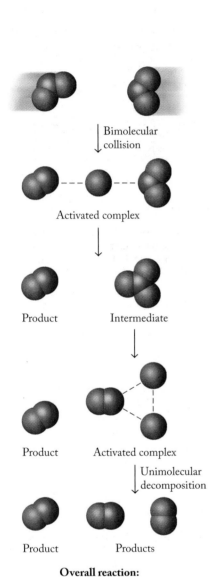

FIGURE 14.19 The decomposition of NO_2 begins when two NO_2 molecules collide, producing NO and NO_3. NO_3 rapidly decomposes into NO and O_2. The second step is faster than the first. The first step is the rate-determining step, and the overall reaction is second order in NO_2.

rapidly than nonspontaneous reactions? The answer is: not necessarily. Consider the reaction-energy profiles in Figure 14.18. The nonspontaneous ($\Delta G > 0$) decomposition of NOCl to NO and Cl_2 will occur only if we add energy. On the other hand, the reaction of NO_2 with CO to form NO and CO_2 is spontaneous under standard conditions and 298 K. However, if both reactions are run at a temperature at which NOCl decomposes, the NOCl reaction may proceed faster than the reaction between NO_2 and CO because the latter reaction has a larger activation energy barrier: 134 kJ/mol vs. 102 kJ/mol. Thus, a reaction with a large decrease in free energy may be "spontaneous" from a thermodynamic point of view, but *spontaneous does not necessarily mean the reaction occurs rapidly*. In fact, the rates of some spontaneous processes, such as the transformation of diamond into graphite, are so slow that they do not proceed at all: they are favored thermodynamically, but not kinetically.

14.5 Reaction Mechanisms

To think about how a reaction proceeds at the molecular level, let's revisit the chemical equation describing the thermal decomposition of NO_2:

$$2\,NO_2(g) \rightarrow 2\,NO(g) + O_2(g)$$

We noted in Section 14.3 that this reaction is second order in NO_2 because it takes place as a result of the collisions of *pairs* of molecules of NO_2. How exactly do we picture the atoms inside two colliding molecules of NO_2 rearranging themselves to form two molecules of NO and a single molecule of O_2? The answer to this question is contained in the mechanism of the reaction. A **reaction mechanism** describes the stepwise manner in which the bonds in molecules of reactants break and the bonds that hold together molecules of the product form.

Elementary Steps

A mechanism proposed for the thermal decomposition of NO_2 is shown in Figure 14.19. In the first step of the mechanism, a collision between a pair of NO_2 molecules produces a very short-lived activated complex in which the two molecules share an oxygen atom. Activated complexes such as this one represent midway points in chemical reactions. They have extremely brief lifetimes and rapidly fall apart, either forming products or re-forming reactants. To form products, the shared oxygen atom is transferred from one NO_2 molecule to the other, forming a molecule of NO and a molecule of NO_3 that rapidly decomposes by forming another activated complex. The bonds in this second complex rearrange so that two oxygen atoms become bonded together, forming a molecule of O_2 and leaving behind a molecule of NO.

NO_3 is considered an **intermediate** in this mechanism because it is produced in one step and then consumed in the next. Intermediates are not considered reactants or products and do not appear in the overall equation describing a reaction.

The reaction mechanism in Figure 14.19 is a combination of two **elementary steps** that provide detailed molecular views of the reaction. Elementary steps may involve one or more molecules. Those that involve a single molecule are called **unimolecular**. Elementary steps based on collisions between two molecules are called **bimolecular**. They are much more common than **termolecular** (three-molecule) reactions because the chances of three molecules colliding simultaneously are much less than those of bimolecular collisions. The terms *uni-*, *bi-*, and *termolecular* are used by chemists to describe the **molecularity** of an elementary step, which refers to the number of atomic-scale particles (free atoms, ions, or molecules) that collide with one another in that step.

A valid mechanism must be consistent with the stoichiometry of the overall reaction. In other words, the sum of the processes shown in Figure 14.19 must be consistent with the observed proportions of reactants and products. In this case, the sum of the two elementary steps matches the overall stoichiometry:

(1) $\quad 2NO_2(g) \rightarrow NO(g) + NO_3(g)$

(2) $\quad NO_3(g) \rightarrow NO(g) + O_2(g)$

Summing steps 1 and 2 yields

$$2NO_2(g) + \cancel{NO_3(g)} \rightarrow 2NO(g) + \cancel{NO_3(g)} + O_2(g)$$

Simplifying, we get the overall reaction

$$2NO_2(g) \rightarrow 2NO(g) + O_2(g)$$

Nitrogen trioxide, NO_3, is an intermediate in this mechanism. It appears in both elementary steps but not in the overall reaction.

Before ending our discussion of the thermal decomposition of NO_2, let's consider how the concept of activation energy applies to a two-step reaction such as this one. The two elementary steps produce an energy profile with two maxima, as shown in Figure 14.20. In elementary step 1, collisions between pairs of NO_2 molecules result in the formation of an activated complex associated with the first transition state in Figure 14.20. As this activated complex transforms into NO and NO_3, the energy of the system drops, reaching the bottom of the valley between the two maxima. The top of the second energy barrier represents the transition state associated with a second activated complex in which the bonds in NO_3 rearrange. Two N–O bonds break and an O–O double bond forms to give the final products, NO and O_2.

For the reaction in the forward direction, the activation energy for elementary step 1 is much greater than that for elementary step 2. This difference is consistent with the relative rates of the two elementary steps: the first elementary step is slower than the second one. When the reaction proceeds in the reverse direction (as NO and O_2 react, forming NO_2) the first energy barrier is the smaller of

A **reaction mechanism** proposes a set of steps that describe how the reaction occurs. The mechanism must be consistent with the rate law for the reaction.

An **intermediate** is a species produced in one step of a reaction and consumed in a following step.

An **elementary step** is a molecular-level view of a single process taking place in a chemical reaction.

A **unimolecular step** in a mechanism or reaction involves only one molecule.

A **bimolecular step** in a mechanism or reaction involves a collision between two molecules.

A **termolecular step** in a mechanism or reaction involves a collision between three molecules.

The **molecularity** of a step in a mechanism refers to the number of particles (ions, atoms, or molecules) that collide in that step.

▶❚❚ **CHEMTOUR** Reaction Mechanisms

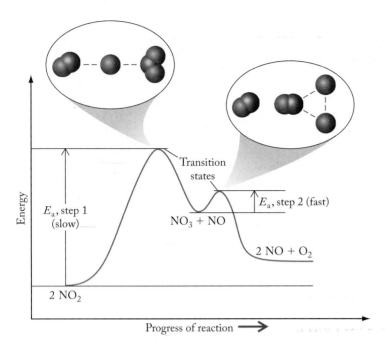

FIGURE 14.20 The reaction-energy profile for the decomposition of NO_2 to NO and O_2 features activation energy barriers for both elementary steps in the overall reaction mechanism. The activation energy of the first step is larger than that of the second step, so the first step is the slower of the two and thus is the rate-determining step.

The **rate-determining step** is the slowest one in a multi-step chemical reaction.

the two, so we expect a faster first elementary step in the reverse process. Experimental evidence supports this expectation.

Rate Laws and Reaction Mechanisms

The observed rate law for the decomposition of NO_2 is second order in NO_2:

$$\text{Rate} = k[NO_2]^2$$

For a mechanism to be accepted as providing a possible view of a chemical reaction, it must be consistent with the rate law derived from experimental data. Is the proposed mechanism in Figure 14.20 consistent with the rate law? For it to be consistent, the molecularity of one of the elementary steps in the mechanism must be the same as the order of the reaction expressed in the rate law. Remember that we cannot write a rate law from a balanced chemical equation. The individual elementary steps that show how a reaction proceeds and account for the formation and reaction of intermediates, however, are different from the overall balanced equation. A rate law for an elementary step can be written directly from its balanced equation. For the decomposition of NO_2, therefore, the rate law for the first elementary step is

$$\text{Rate}_1 = k_1[NO_2]^2$$

and for the second step

$$\text{Rate}_2 = k_2[NO_3]$$

where k_1 and k_2 represent the rate constants for the respective elementary steps.

Comparing the rate laws for the two elementary steps with the observed rate law for the overall reaction shows that the rate law for elementary step 1 is exactly the same as the experimentally determined rate law. This means that step 1 defines how rapidly the entire reaction proceeds. It is the **rate-determining step** in the reaction. The rate-determining step is the slowest elementary step in a chemical reaction. The rate of the slowest elementary step controls the rate of the overall reaction. In the case of the decomposition of NO_2, elementary step 1 is the slower of the two steps.

One way of visualizing the concept of a rate-determining step is to analyze the flow of people through a busy airport. Typically, the number of people in the line outside the security point greatly exceeds the number of available security gates. The time required to make it to your flight depends on the time needed to pass through security. Security screening is an example of the rate-determining step on your way to your flight.

Which step is rate determining in the decomposition of NO_2? The observed rate law is identical to the rate law written for step 1. Therefore, we conclude that the first step is the rate-determining step that governs how rapidly the overall reaction proceeds. The rate constant (k_1) is smaller than the rate constant for step 2 (k_2).

There is actual experimental evidence that $k_1 \ll k_2$. Chemists have studied the kinetics of the formation and decomposition of NO_3 extensively and have determined that the rate constant of step 1 is only about $1 \times 10^{-10} \ M^{-1} \ s^{-1}$ at 300 K. The rate constant of step 2 is enormous by comparison: $6.3 \times 10^4 \ s^{-1}$. Therefore, as soon as any NO_3 is formed in step 1, it rapidly falls apart, forming NO and O_2 in step 2. Similarly, at the airport, as soon as people get through security (the rate-determining step), they move rapidly to the gates for their flights.

Now let's consider the reverse of the thermal decomposition of NO_2—namely, the formation of NO_2 by the reaction of NO and O_2, a key step in the formation of photochemical smog:

$$2 NO(g) + O_2(g) \rightarrow 2 NO_2(g)$$

The observed rate law for this reaction is second order in NO and first order in O_2:

$$\text{Rate} = k[NO]^2[O_2]$$

One proposed mechanism includes two elementary steps, as shown in Figure 14.21.

(1) $\quad NO(g) + O_2(g) \rightarrow NO_3(g) \qquad \text{Rate} = k_1[NO][O_2]$

(2) $\quad NO_3(g) + NO(g) \rightarrow 2 NO_2(g) \qquad \text{Rate} = k_2[NO_3][NO]$

If the first step were the rate-determining step, then the reaction would be first order in NO and O_2, but that is not what the experimentally determined rate law indicates. If the second step were the rate-determining step, then the reaction would be first order in NO and NO_3, but that is not what the rate law stipulates either. Some other option must be suggested that gives the same picture as the rate law.

Consider what happens if the first step is fast *and reversible*. In that case, during the reaction, NO_3 is rapidly produced from NO and O_2 but decomposes rapidly back into NO and O_2. In other words, the rate of step 1 in the forward direction matches the rate of step 1 in reverse, and both are rapid. Expressed mathematically, we have

$$\text{Rate forward reaction} = k_f[NO][O_2] = \text{fast}$$

$$\text{Rate of reverse reaction} = k_r[NO_3] = \text{equally fast}$$

where k_f and k_r are the rate constants for the forward and reverse reactions, respectively. Combining the two rate-law expressions, we have

$$k_f[NO][O_2] = k_r[NO_3]$$

Solving for $[NO_3]$, we have

$$[NO_3] = \frac{k_f}{k_r}[NO][O_2] \qquad (14.30)$$

Remember that the rate law for the rate-determining step in the mechanism must be the same as the experimentally determined rate law. The rate law for the first step does not match the observed rate law. What about the rate law for step 2? The rate law for step 2 is

$$\text{Rate} = k_2[NO_3][NO] \qquad (14.31)$$

Equation 14.31 still doesn't match the observed rate law for the reaction:

$$\text{Rate} = k[NO]^2[O_2]$$

(1) Formation of intermediate

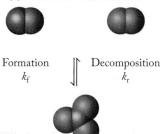

Formation k_f ⇅ Decomposition k_r

(2) Reaction of intermediate

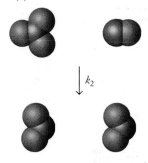

k_2

Overall reaction:

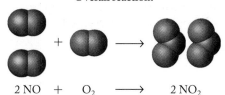

$$2 \, NO \ + \ O_2 \ \longrightarrow \ 2 \, NO_2$$

FIGURE 14.21 A mechanism for the formation of NO_2 from NO and O_2 has two steps: (1) in a fast, reversible bimolecular reaction, NO and O_2 form NO_3, and (2) in a slower rate-limiting bimolecular reaction, NO_3 reacts with a second molecule of NO, forming two molecules of NO_2.

However, if we combine Equations 14.30 and 14.31 by replacing the $[NO_3]$ term in Equation 14.31 with the right-hand side of Equation 14.30, we get

$$\text{Rate} = k_2 \frac{k_f}{k_r}[NO]^2[O_2]$$

The three rate constants can be combined into one overall rate constant

$$k_{\text{overall}} = k_2 \frac{k_f}{k_r}$$

and the rate-law expression for the reaction becomes

$$\text{Rate} = k_{\text{overall}}[NO]^2[O_2]$$

This expression matches the observed order of the reaction, so the proposed mechanism, a fast reversible first step followed by a slow second step, may be valid. The following exercises provide additional practice in relating experimental rate laws to proposed reaction mechanisms.

SAMPLE EXERCISE 14.12   **Linking Reaction Mechanisms to Experimental Rate Laws**

At high temperature, NO reacts with hydrogen, producing nitrogen and water vapor:

$$2\,NO(g) + 2\,H_2(g) \rightarrow N_2(g) + 2\,H_2O(g)$$

The experimentally determined rate law for the reaction is

$$\text{Rate} = k[NO]^2[H_2]$$

It has been proposed that this reaction takes place in two elementary steps:

    (1) $2\,NO(g) + H_2(g) \rightarrow N_2O(g) + H_2O(g)$

    (2) $N_2O(g) + H_2(g) \rightarrow N_2(g) + H_2O(g)$

Is the proposed reaction mechanism consistent with the stoichiometry of the reaction and the rate-law expression? If so, which step in the proposed mechanism is the rate-determining step?

COLLECT AND ORGANIZE We are first asked if the proposed mechanism is consistent with the stoichiometry of the reaction. We are given the two steps in the proposed mechanism, and we can use them to answer that question. We are then asked if the mechanism is consistent with the experimentally determined rate law. We can look at the molecularity of the steps and the rate law and answer that question. Finally, we are asked to identify the rate-determining step in the mechanism.

ANALYZE To answer the questions posed in this exercise we need to do the following:

    1. Determine whether the chemical equations of the elementary steps add up to the overall reaction equation.

    2. Compare the rate-law expressions of the elementary steps to the experimental rate law of the overall reaction to determine which step is rate determining.

SOLVE Let's test whether the elementary steps add up to the overall reaction:

$$(1) \qquad 2\,NO(g) + H_2(g) \rightarrow N_2O(g) + H_2O(g)$$

$$(2) \qquad N_2O(g) + H_2(g) \rightarrow N_2(g) + H_2O(g)$$

Summing these reactions gives

$$2\,NO(g) + 2\,H_2(g) + N_2O(g) \rightarrow N_2O(g) + N_2(g) + 2\,H_2O(g)$$

Simplifying leads to

$$2\,NO(g) + 2\,H_2(g) + \cancel{N_2O(g)} \rightarrow \cancel{N_2O(g)} + N_2(g) + 2\,H_2O(g)$$

or

$$2\,NO(g) + 2\,H_2(g) \rightarrow N_2(g) + 2\,H_2O(g)$$

which is indeed the same equation as that of the overall reaction. The elementary steps are consistent with the stoichiometry of the overall reaction.

Next we need to focus on the reaction mechanism. The first step involves two molecules of NO colliding with one molecule of H_2 in a termolecular reaction. The second reaction is a bimolecular reaction between the N_2O produced in the first step and another molecule of H_2. Now we make use of the fact that the rate laws for elementary steps can be written directly from their stoichiometries. Therefore:

$$(1) \qquad \text{Rate} = k_1[H_2][NO]^2$$

$$(2) \qquad \text{Rate} = k_2[H_2][N_2O]$$

The rate law of step 1 matches the observed rate law of the overall reaction. Therefore, the proposed two-step mechanism is consistent with the experimental rate law, and step 1, by definition, is the rate-determining step.

THINK ABOUT IT Remember that the mechanism in this Sample Exercise is only a proposed mechanism. We do not have direct proof that it is the correct mechanism. Nevertheless, it is consistent with the available data. An acceptable mechanism for a reaction must contain a balanced equation whose stoichiometry is consistent with the rate law for the overall process. This step is the rate-determining step in the reaction. Notice that no intermediates can appear in the overall rate law but may appear in individual steps.

Practice Exercise Here is another proposed mechanism for the reduction of NO by H_2:

$$(1) \qquad H_2(g) + NO(g) \rightarrow N(g) + H_2O(g)$$

$$(2) \qquad N(g) + NO(g) \rightarrow N_2(g) + O(g)$$

$$(3) \qquad H_2(g) + O(g) \rightarrow H_2O(g)$$

Is this a valid mechanism?

As a final point regarding reaction mechanisms, even though the proposed reaction mechanism is consistent with the reaction's overall stoichiometry and with the experimentally derived rate-law expression, these consistencies do not *prove* that the proposed mechanism is correct. We need more experimental evidence, such as detecting the presence of an intermediate (such as N_2O in Sample Exercise 14.12) in the reaction mixture. On the other hand, not finding a reactive (and transient) intermediate would not necessarily disprove a reaction mechanism.

Mechanisms and Zero-Order Reactions

Before we leave this section on reaction mechanisms, let's consider the mechanism of the reaction between NO_2 and CO producing NO and CO_2 that we last saw at the end of Section 14.3:

$$NO_2(g) + CO(g) \rightarrow NO(g) + CO_2(g)$$

$$Rate = k[NO_2]^2$$

The observed rate law for this reaction is zero order in CO, second order in NO_2, and second order overall. What do these experimentally determined facts tell us about the reaction? First, remember that the rate law describes the process taking place in the rate-determining step. The rate law does not include a term for CO; that means that CO is not involved in that step in the mechanism. CO is clearly involved in the reaction—it is converted into CO_2. But whatever step involves CO must be fast compared to the rate-determining step. This leads us to the conclusion that the reaction must involve at least two elementary steps, one rate determining and one not. The elementary step that is rate determining must involve two molecules of NO_2—the rate law tells us that because it is second order in NO_2.

In light of these facts, it has been proposed that this reaction takes place in two elementary steps:

(1) $\qquad 2\,NO_2(g) \rightarrow NO_3(g) + NO(g) \qquad Rate = k_1[NO_2]^2$

(2) $\quad NO_3(g) + CO(g) \rightarrow NO_2(g) + CO_2(g) \qquad Rate = k_2[NO_3][CO]$

The experimental rate law matches the rate law for the first step, which must be the slower, rate-determining step.

14.6 Catalysis

We noted in Section 14.4 that spontaneous reactions may be slow if they have a high activation energy. Suppose we wanted to increase the rate of such a reaction. How could we do it? One way is to increase the temperature of the reaction mixture. However, in some chemical reactions, elevated temperatures can lead to undesired products or lower product yields. Another way is to add another substance, called a **catalyst**, to the reaction mixture. A catalyst is a substance that increases the rate of a reaction but is not consumed by it. In this section, we consider catalysts that exist in the atmosphere and have an impact on the levels of ozone in the stratosphere over Antarctica. We also examine the catalysts that were developed by chemists to reduce air pollution from vehicle exhaust, thereby addressing the problems with photochemical smog that we described in Section 14.1. Later in this book, we return to catalysis in the biological context when we discuss the roles enzymes play in all life forms.

Catalysts and the Ozone Layer

As we discussed in Chapter 8, the way we think about ozone depends on where the ozone is located. Ozone in the stratosphere between 10 and 40 km above Earth's surface is necessary to protect life from harmful UV radiation; ozone at ground level is hazardous to human health and life in general. In this section, we discuss the role of catalysis in the breakdown of ozone in the stratosphere that has led to the formation of the ozone hole.

A **catalyst** is a substance added to a reaction that increases the rate of the reaction but is not consumed in the process.

In 1974, two American scientists, Sherwood Rowland and Mario Molina, predicted significant depletion of stratospheric ozone because chlorofluorocarbons (CFCs) released at ground level ultimately enter the stratosphere. We will examine how CFCs can deplete stratospheric ozone later in this section. For now just note that Rowland and Molina's predictions were later supported by experimental evidence of a thinning of the ozone layer over Antarctica during springtime in the Southern Hemisphere. This region of lowered ozone concentrations is known as the ozone hole. By 2000, stratospheric ozone concentrations over Antarctica were less than half of what they were in the late 1970s (Figure 14.22), and the ozone hole covered nearly all of Antarctica and the tip of South America. Less severe thinning of stratospheric ozone was observed in the Northern Hemisphere.

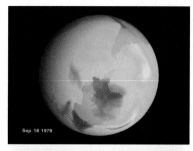

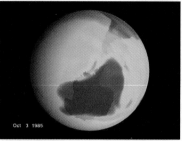

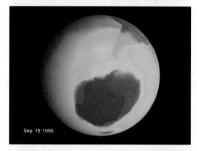

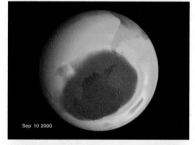

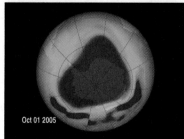

NASA photos showing the change in the ozone hole from September 1979 through October 2005. The colorized scale is in Dobson units (DU), where purple is low ozone (100 DU) and red is high ozone (500 DU). The ozone hole is considered to be wherever the ozone concentration is less than 220 Dobson units.

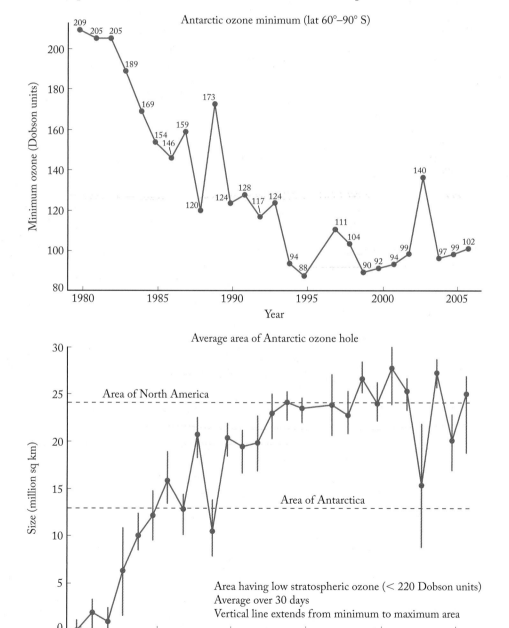

FIGURE 14.22 Graphs describing the Antarctic ozone hole. (a) The change in the minimum O_3 concentration over time from 1979 to early 2005 in Dobson units, a measure of ozone in the atmosphere. (b) The average area experiencing a decreased ozone concentration over the same period. Areas of North America and Antarctica shown by dashed lines for reference.

FIGURE 14.23 Graph showing annual variations in stratospheric ozone concentration during Antarctic winter and spring from 1990 through 2006.

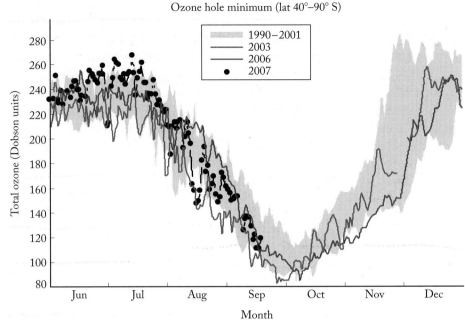

Ozone hole minimum (lat 40°–90° S)

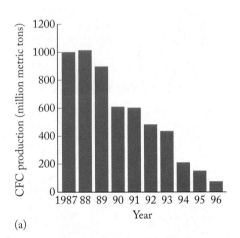

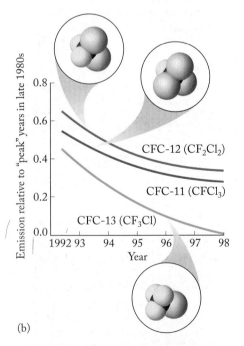

FIGURE 14.24 (a) Worldwide production of CFCs. (b) Release of CFCs into the atmosphere. Both production and release of CFCs decreased since ratification of the Montreal Protocol in 1986. However, the chemical stability of these volatile compounds means that their concentrations in the atmosphere decrease much more slowly.

In 1987, an international agreement known as the Montreal Protocol called for reduced production of ozone-depleting compounds, including CFCs. Revisions in 1992 called for an end to the production of most of these compounds by 1996. The Montreal Protocol has had a dramatic effect on CFC production and emission into the atmosphere. However, these compounds last for many years in the atmosphere, and recovery of the ozone layer may take most of the twenty-first century. Furthermore, despite all these precautionary measures, in September and October of 2006, a colder spring than usual resulted in the largest ozone hole on record (Figure 14.23).

Three of the more widely used CFCs were CF_2Cl_2, $CFCl_3$, and CF_3Cl (Figure 14.24). The chemical inertness of CFCs means that they persist in the environment and so can be swept up to the stratosphere by wind and air currents in the troposphere. In the stratosphere, CFCs encounter UV rays with enough energy to break their C—Cl bonds, releasing chlorine atoms:

$$CFCl_3(g) \rightarrow CFCl_2(g) + Cl(g) \tag{14.32}$$

Free chlorine atoms react with ozone to make chlorine monoxide if both reactants are in the gas phase:

$$Cl(g) + O_3(g) \rightarrow ClO(g) + O_2(g) \tag{14.33}$$

Chlorine monoxide then reacts with more ozone to yield oxygen and regenerate atomic chlorine:

$$ClO(g) + O_3(g) \rightarrow Cl(g) + 2O_2(g) \tag{14.34}$$

If we add Equations 14.33 and 14.34 together, we get

$$Cl(g) + 2O_3(g) + ClO(g) \rightarrow ClO(g) + Cl(g) + 3O_2(g) \tag{14.35}$$

Cancelling species that appear on both sides of the reaction arrow gives

$$2O_3(g) \rightarrow 3O_2(g) \tag{14.36}$$

The rates of the above reactions increase in the presence of fluid droplets such as those present in the clouds that cover much of Antarctica each spring. Actually, these clouds form as crystals of ice and tiny droplets of nitric acid in the winter months, when there is no sunlight and temperatures in the lower stratosphere dip to −80°C. The clouds become collection sites for HCl, ClO, and other compounds containing chlorine. When the sun returns in August (at the end of the Antarctic winter), sunlight melts the ice and ClO is free to migrate to the surface of the droplets of liquid water. These droplets catalyze reactions by *adsorbing* (binding to the surface) the reactants. The droplets are not reactants in themselves, but they provide a surface on which the reactants collect. The resulting proximity of the reactants increases the likelihood of interaction between them and decreases activation energies for the process.

Evidence supporting the occurrence of these reactions in the stratosphere is presented in Figure 14.25. The concentration of ClO decreases sharply in satellite images of the South Pole taken only one month apart, consistent with the reaction between ClO and O_3 shown in Equation 14.34. Although it is difficult to track the concentration of Cl atoms directly, some of them end up as the hydrogen chloride, HCl, present in the icy polar clouds that form during the winter and melt in August. The images in Figure 14.25 confirm that a decrease in ClO concentration between the end of Antarctic winter (late August) and the Antarctic spring (late September) is accompanied by an increase in HCl concentration, which reflects the rise in Cl concentration.

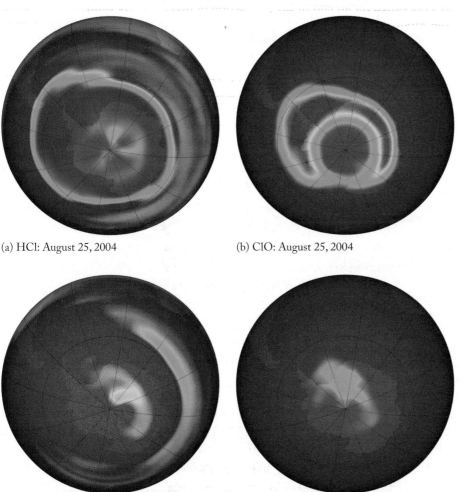

(a) HCl: August 25, 2004   (b) ClO: August 25, 2004

(c) HCl: September 24, 2004   (d) ClO: September 24, 2004

FIGURE 14.25 NASA satellite images of the South Pole showing the stratospheric concentrations of (a) hydrogen chloride (HCl) and (b) chlorine monoxide (ClO) in late August 2004. One month later, as spring arrives in the Southern Hemisphere, the concentrations of (c) HCl and (d) ClO have changed. The concentrations of the two compounds are inversely correlated. The concentration of ClO decreases as it reacts with ozone; the reaction yields chlorine atoms that in turn lead to an increased concentration of HCl. The blue color indicates a low concentration and red indicates a high concentration.

A **homogeneous catalyst** is in the same phase as the reactants.

A **heterogeneous catalyst** is in a different phase than the reactants.

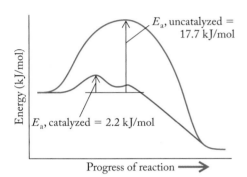

FIGURE 14.26 The decomposition of O_3 in the presence of chlorine atoms has a smaller activation energy (2.2 kJ/mol) than the photodecomposition of O_3 to O_2 (17.7 kJ/mol). The catalytic effect of chlorine is key to the depletion of stratospheric ozone and the formation of an ozone hole over the South Pole.

In terms of reactants and products, Equation 14.36 is the same as the one describing the naturally occurring photodecomposition of ozone in the stratosphere. The natural pathway for ozone destruction involves light:

$$2\,O_3 \xrightarrow{\ h\nu\ } 3\,O_2$$

The natural photochemical conversion of ozone to oxygen begins with the absorption of a photon of UV radiation and the generation of atomic oxygen:

$$(1) \qquad O_3(g) \rightarrow O_2(g) + O(g)$$

Oxygen atoms generated in this fashion can react with additional ozone molecules in a second step to form two more molecules of oxygen:

$$(2) \qquad O_3(g) + O(g) \rightarrow 2\,O_2(g)$$

The rate of the second step is slowed by an activation energy barrier of 17.7 kJ/mol of O_3, as shown in Figure 14.26. Remember that the higher the activation energy, the slower the reaction. The chlorine-catalyzed destruction of ozone occurs much more rapidly because it follows a different reaction pathway, and the activation energy of the pathway that involves Cl is only 2.2 kJ/mol. Moreover, the Cl atoms (from CFCs) are not consumed in the overall process. Rather, they are consumed in an early step but then regenerated in a later step. Thus, atomic chlorine meets our definition of a catalyst in this sytem. Chlorine is a catalyst, not an intermediate, because in a reaction mechanism, a catalyst is consumed in an early step before it is regenerated in a later one. However, an intermediate is produced before it is consumed. A single chlorine atom can catalyze the destruction of hundreds to thousands of stratospheric O_3 molecules before it combines with other atoms and forms a less reactive molecule.

The catalyst (Cl) and reacting species (O_3) are both atmospheric gases and so exist in the same physical phase. When a catalyst and the reacting species are in the same phase, we call the catalyst a **homogeneous catalyst**. When the catalyst is the fluid droplet on which gas molecules adsorb, the catalyst is in a different phase than the reacting species and is called a **heterogeneous catalyst**. Both types of catalyst play a role in the reactions that diminish the amount of ozone in the stratosphere.

SAMPLE EXERCISE 14.13 **Identifying Catalysts in Reaction Mechanisms**

The following reactions have been proposed for the decomposition of ozone in the presence of NO at high temperatures. If the rate of the overall reaction is more rapid than the uncatalyzed decomposition of ozone to oxygen, is NO a catalyst in the reaction?

$$(1) \qquad O_3(g) + NO(g) \rightarrow O_2(g) + NO_2(g)$$
$$(2) \qquad NO_2(g) \rightarrow NO(g) + O(g)$$
$$(3) \qquad O(g) + O_3(g) \rightarrow 2\,O_2(g)$$

COLLECT AND ORGANIZE We are asked to determine if NO is a catalyst in a reaction. The statement of the problem tells us the reaction is more rapid in the presence of NO. We have the steps of the reaction. If NO is not consumed in the overall process, it is a catalyst; if it is consumed, it is not.

ANALYZE We can sum the reactions to determine the overall reaction and establish whether NO is consumed in the overall process.

SOLVE If we sum the reactions in steps 1 through 3, we obtain the overall reaction

$$2O_3(g) \rightarrow 3O_2(g)$$

The reaction does not include NO. The rate of the reaction is higher when NO is present. NO is consumed in the first step and regenerated mole for mole in the second, but does not appear in the overall reaction. It serves as a catalyst because it increases the rate of the reaction and is neither consumed nor produced in the overall reaction.

THINK ABOUT IT The problem indicates that the rate of the reaction is greater than the rate of ozone decomposition in the absence of NO, so one of the conditions of a catalyst is met by this statement. To determine that a substance is a catalyst for a reaction, we need to show that the substance is neither produced nor consumed in the overall reaction. Not every substance will catalyze a reaction even if it does not appear in the overall reaction. In a reaction mechanism, a catalyst is consumed in one step before it is regenerated in a later one, whereas an intermediate is produced before it is consumed.

Practice Exercise The combustion of fossil fuels results in the release of SO_2 into the atmosphere, where it may react with NO_2, forming SO_3 and NO:

$$2NO_2(g) + 2SO_2(g) \rightarrow 2NO(g) + 2SO_3(g)$$

The NO that is formed is then oxidized to NO_2:

$$2NO(g) + O_2(g) \rightarrow 2NO_2(g)$$

Is NO_2 a catalyst in the following reaction?

$$2SO_2(g) + O_2(g) \rightarrow 2SO_3(g)$$

Catalysts and Catalytic Converters

Significant progress has been made in the past 30 years in reducing the frequency and intensity of photochemical smog by reducing the concentrations of NO_x and other pollutants in automobile exhaust. One of the most important advances is the continuing development of *catalytic converters*. The design and development of the catalytic converter required a thorough understanding of the kinetics of the reactions by which NO_x are reduced and CO and hydrocarbons are oxidized. Figure 14.27 provides a view of a catalytic converter and where it is located in

(b)

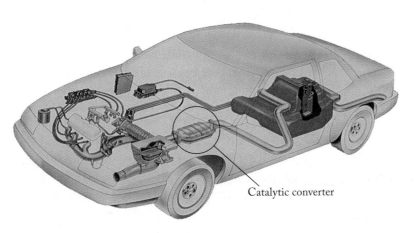

(a)

(c)

FIGURE 14.27 Catalytic converters. (a) In an automobile the converter is located close to the engine because it works best at high temperatures before gases have a chance to cool. (b) The structure of a catalytic converter. (c) The porous ceramic honeycomb support for the metal catalysts.

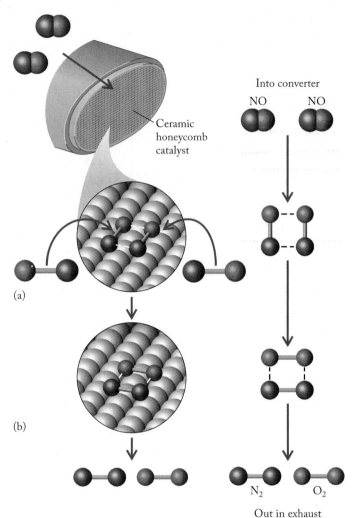

Into converter

NO   NO

N₂   O₂

Out in exhaust

FIGURE 14.28 Catalytic converters in automobiles reduce emissions of NO by lowering the activation energy of its decomposition into N_2 and O_2. (a) NO molecules are adsorbed onto the surface of metal clusters where their NO bonds are broken, and (b) pairs of O atoms and N atoms form O_2 and N_2. The O_2 and N_2 desorb from the surface and are released to the atmosphere.

the exhaust system of a car. Figure 14.28 illustrates how it removes NO from engine exhaust. The catalysis that occurs is heterogeneous, because compounds in exhaust gas flow over the surface of the solid catalyst in the converter.

When hot exhaust gases flow through the converter, they pass through a fine honeycomb mesh of a ceramic that serves as a support for a coating of platinum (Pt) and rhodium (Rh). The ceramic substrate may be alumina (Al_2O_3) or a compound called cordierite, a ceramic made from SiO_2, talc (a silicate), and kaolin clay. The metal catalysts have two roles: they speed up the oxidation of unburned hydrocarbons to CO_2 and water vapor, and they reduce oxides of nitrogen (NO and NO_2) to N_2 and O_2. The catalysts are selective in terms of the molecules they interact with and specific in the reactions they carry out. For example, molecules of NO are adsorbed onto the catalytic surfaces. As a result of the adsorption process, N—O bonds are weakened and break, forming free atoms of N and O. Nitrogen atoms combine with other nitrogen atoms to form N_2 and pairs of oxygen atoms form O_2. The catalytic converters increase the rate of the reaction

$$2\,NO(g) \rightarrow N_2(g) + O_2(g)$$

Rhodium plays a role in reducing the oxides of nitrogen, and Rh and Pt together oxidize hydrocarbons and CO. Palladium-only catalytic converters have also been developed, and indeed Pd is currently the metal used most in catalytic converters. In fabricating the converters, the metals are dissolved as metal salts and dispersed on the surface of the ceramic support. They are then reduced to small metal clusters 2 to 10 nm in diameter. The large surface area of the metal clusters provides sites where the oxidation and reduction of the gases take place.

The catalysts not only speed up the reactions, they also allow the reactions to take place at a lower temperature. For example, CO reacts rapidly with O_2 above 700°C, but the catalysts enable this reaction to take place rapidly at the much lower temperature of automobile exhaust, about 250°C. The catalysts have similar effects on the reduction reactions taking place in the converters.

The catalysts in catalytic converters must also have considerable selectivity for reactions. Over 150 hydrogen- and carbon-containing species are present in the exhaust of a gasoline-powered vehicle, and even if they all do not react, they should at least not hinder the reactions that destroy those compounds targeted for removal from the exhaust stream. Selectivity arises because several reactions are possible in the converter, but one proceeds more rapidly than the others. For example, in the environment of the converter, the preferred reduction reaction is the conversion of the nitrogen in NO to N_2, and not to N_2O or to NH_3. Carbon monoxide and hydrocarbons are potential reductants of NO. The presence of O_2 in the system, however, could lead to oxidation of CO and the hydrocarbons, which would remove both from the system. If that were to happen, the hydrocarbons and CO would be oxidized, but the NO_x would not be reduced and the converter would do only half its job. Although both processes take place in a catalytic converter, the reduction of NO by CO and the hydrocarbons is much faster than the reactions of CO and the hydrocarbons with O_2. As a result, NO reduction is essentially complete before any of the reductants are consumed by reaction with oxygen.

The Platinum Group: Catalysts, Jewelry, and Investment

The platinum group metals (PGMs) are a block of six elements in rows 5 and 6 and columns 8, 9, and 10 of the periodic table: platinum, palladium, rhodium, ruthenium, osmium, and iridium. In addition to having exceptional catalytic properties, all the PGMs have high melting points and are corrosion resistant. Besides their use in catalytic converters, they function as catalysts in fuel cells and in the industrial production of nitric acid and chlorine gas; they are alloyed with other metals in spark plugs and electronic devices; and several of them are key components of cancer chemotherapy drugs.

Platinum and palladium are used to catalyze a variety of chemical reactions. In this chapter, we learned about the use of Pt and Rh in catalytic converters, where a key aspect of their catalytic activity is the ability to bind NO_x, CO, and hydrocarbons to the surfaces of small metallic clusters. Since 1996, Pd has become the most widely used PGM in this application. A fundamental measure of the binding ability of a heterogeneous catalyst is a parameter called the *heat of adsorption*, a measure of the strength with which a substance bonds to the surface of a material. If a substance is adsorbed (bound to the surface) too tightly, it cannot react with other adsorbed species. If it is bonded too weakly, it will desorb (move away from the surface) before it has a chance to react. Platinum and palladium operate between these two extremes, binding a wide range of substances with moderate strength. Both metals bind many of the gases we discussed throughout this book with a moderate heat of adsorption, and this is the key to their catalytic activity.

The technological importance of PGMs extends far beyond their prowess as heterogeneous catalysts of gas-phase reactions. They are very resistant to corrosion by acids, alkalis, and salts; they are stable at high temperatures, and they do not oxidize readily. As a result, they are frequently used as linings or coatings to protect other metals. The rigidity and hardness of individual PGMs can be improved by alloying them with other metals in the PGM family. Platinum is frequently alloyed with Rh, Ir, and Ru to produce hard, highly corrosion resistant solids.

The PGMs are rare, so their heavy use in technologically important applications like automotive catalytic converters has led to recycling efforts to recover the metals. In the United States the most important source of Pt, Pd, and Rh is their collection from scrapped catalytic converters. It can be challenging to recover the PGMs from the ceramic support. One of the techniques used involves the selective dissolution of the metals by treatment with *aqua regia* ("water of kings"), a mixture of hydrochloric and nitric acids so named because it is capable of dissolving gold.

All of the PGMs have some use in medical devices and implants. Because of their lack of reactivity and tolerance for solutions that corrode most metals, they are highly biocompatible. Platinum alloyed with iridium is used in the electrodes of pacemakers and defibrillators and also in the tips of the guidewires surgeons use to direct the placement of catheters. One mode of cancer therapy uses radioactive ^{192}Ir wire wrapped in platinum as an implant to deliver radiation doses *in vivo*.

Platinum is the PGM most commonly used for jewelry. In Japan it is actually preferred over gold in decorative items. As we saw with gold, pure platinum is also too soft for jewelry, so it is usually alloyed with iridium or ruthenium to increase its wear resistance. Palladium is used in small quantities as a whitener in gold jewelry. Pure platinum is also made into coins and ingots and has considerable value as an investment.

This platinum model of the Japanese robot Gundam is 12.5 cm tall and is valued at $250,000.

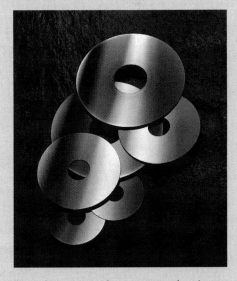

Discs for computer drives are coated with platinum.

Summary

SECTION 14.1 The photochemical reactions that follow the reaction between nitrogen and oxygen in vehicle engines ultimately lead to smog, a major air pollutant in urban areas. Familiarity with **reaction rates** or **chemical kinetics** is important in understanding how natural processes and those caused by human activity occur.

SECTION 14.2 The rate of a reaction is the rate of change in the concentrations of reactants or products. It can be expressed as either an average rate or an **instantaneous rate**. The rates of disappearance of reactants and appearance of products are related by the stoichiometry of the reaction and are typically expressed in molarity per unit time. Reaction rates can be known only from experimental measurements. In most reactions, the reaction rate decreases with increasing reaction time.

SECTION 14.3 The rate of a reaction, such as $a\,A + b\,B \rightarrow c\,C$, depends on the concentrations of A and B as determined experimentally and expressed in the **rate law** for the reaction: rate $= k[A]^m[B]^n$ where the powers m and n are the **order of reaction** with respect to A and B, respectively, and k is the **rate constant**. The units of a rate constant depend on the **overall reaction order**, which is the sum of the reaction orders with respect to individual reactants. The order of a reaction and rate law can be determined from differences in the **initial rates** of reaction observed in reaction mixtures with different concentrations of reactants or from the results of single kinetics experiments where reactant concentrations are plotted versus time (an **integrated rate law**). The **half-life ($t_{1/2}$)** of a reaction is the time required for the concentration of a reactant to decrease to one-half of its starting concentration.

SECTION 14.4 Increasing the temperature increases the rate of a chemical reaction. A reaction's **activation energy (E_a)** is a barrier that separates the sum of the internal energies of the reactants from the energies of the products. The top of the energy barrier is the **transition state** related to the internal energy of a short-lived **activated complex**. Reactions with large activation energies are usually slow. Measuring the rate constant (k) of a reaction at different temperatures allows the calculation of activation energies using the **Arrhenius equation**.

SECTION 14.5 The study of rates gives insight into **reaction mechanisms** and what is happening at a molecular level. The mechanism of a chemical reaction consists of one or more elementary steps that describe on a molecular level how the reaction takes place. The balanced overall reaction is the sum of these **elementary steps**. Elementary steps that involve one, two, or three molecules are said to be **unimolecular, bimolecular,** and **termolecular**, respectively. The rate law for a reaction applies to the slowest elementary step, which is called the **rate-determining step**. The proposed mechanism for any reaction must be consistent with the observed rate law and stoichiometry of the overall reaction.

SECTION 14.6 **Catalysts** increase the rates of reactions by decreasing the activation energy by changing the mechanism of a reaction. **Homogeneous catalysts** are in the same phase as the reactants, whereas **heterogeneous catalysts** (for example, catalytic converters in vehicles) are in a separate phase from that of the reactants.

Problem-Solving Summary

TYPE OF PROBLEM	CONCEPTS AND EQUATIONS	SAMPLE EXERCISES
Writing an expression for relative reaction rate	The relative rates are determined from the stoichiometry of the balanced chemical equation.	14.1
Converting reaction rates	Use the balanced chemical equation and the rate of disappearance of one of the reactants to determine the rate of appearance of the product.	14.2
Determining instantaneous rate	Determine the slope of a line tangent to a point on the plot of concentration versus time.	14.3
Writing a rate-law expression and calculating an overall reaction order	The overall order is the sum of the orders for the individual reactants. The order with respect to a reactant equals its exponent in the rate law. Units of the rate constant depend on the overall order of a reaction.	14.4
Deriving a rate law from initial rate data	Compare the change in rate when the concentration of one reactant is changed (concentrations of other reactants kept constant) to determine the reaction order (usually whole numbers) with respect to that reactant.	14.5
Deriving a rate law using data from a single data set	A linear plot of the natural logarithm of concentration versus time indicates a first-order reaction with a slope of $-k$.	14.6
Calculating the half-life of a first-order reaction	In a first-order reaction, $t_{1/2} = 0.693/k$.	14.7
Distinguishing between the first- and second-order reactions	A linear plot of the natural logarithm of concentration versus time indicates a first-order reaction, whereas a linear plot of $1/C$ (where $C =$ concentration) versus time indicates a second-order reaction.	14.8

TYPE OF PROBLEM	CONCEPTS AND EQUATIONS	SAMPLE EXERCISES
Calculating the half-life of a second-order reaction	In a second-order reaction, $t_{1/2} = 1/k[X]_0$.	14.9
Deriving a pseudo-first-order rate law	A plot of the natural logarithm of concentration of the limiting reagent versus time is linear. The slope of the plot = k[excess reactant] = k'.	14.10
Calculating the activation energy from rate constants	Using the Arrhenius equation, plot $\ln k$ versus $1/T$. The slope is $-E_a/R$.	14.11
Linking reaction mechanisms to experimental rate laws	The order of each reactant in an elementary step equals its coefficient in that step. The rate law for the mechanism must be the same as the observed rate law and does not include intermediates.	14.12
Identifying catalysts in reaction mechanisms	Sum the reactions to determine the overall reaction and establish whether a potential catalyst increases the rate of reaction and is initially consumed and then regenerated in the process.	14.13

Visual Problems

14.1. Nitrous oxide decomposes to nitrogen and oxygen in the following reaction:

$$2\,N_2O(g) \rightarrow 2\,N_2(g) + O_2(g)$$

In Figure P14.1, which curve represents [N_2O] and which curve represents [O_2]?

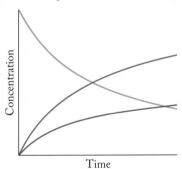

FIGURE P14.1

14.2. Sulfur trioxide is formed in the reaction

$$SO_2(g) + \tfrac{1}{2}O_2(g) \rightarrow SO_3(g)$$

In Figure P14.2, which curve represents [SO_2] and which curve represents [O_2]? All three gases are present initially.

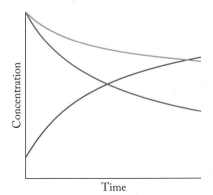

FIGURE P14.2

14.3. The rate law for the reaction $2\,A \rightarrow B$ is second order in A. Figure P14.3 represents samples with different concentrations of A; the red spheres represent molecules of A. In which sample will the reaction $A \rightarrow B$ proceed most rapidly?

(a) (b) (c)

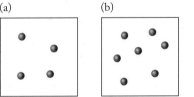

FIGURE P14.3

14.4. The rate law for the reaction $A + B \rightarrow C$ is first order in both A and B. Figure P14.4 represents samples with different concentrations of A (red spheres) and B (blue spheres). In which sample will the reaction $A + B \rightarrow C$ proceed most rapidly?

(a) (b) (c)

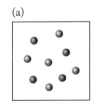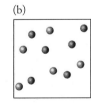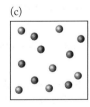

FIGURE P14.4

14.5. Which of the reaction profiles in Figure P14.5 represents the slowest reaction?

(a) (b) (c)

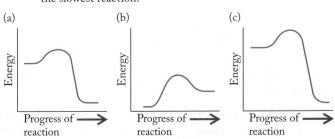

FIGURE P14.5

14.6. Which of the reaction profiles in Figure P14.6 represents the fastest reaction?

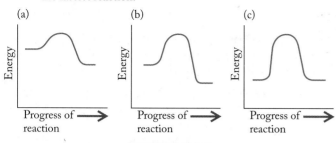

(a) (b) (c)

FIGURE P14.6

14.7. Which of the following mechanisms is consistent with the reaction profile shown in Figure P14.7?

a. $2A \xrightarrow{slow} B$
   $B \xrightarrow{fast} C$

b. $A + B \rightarrow C$

c. $2A \xrightarrow{fast} B$
   $B \xrightarrow{slow} C$

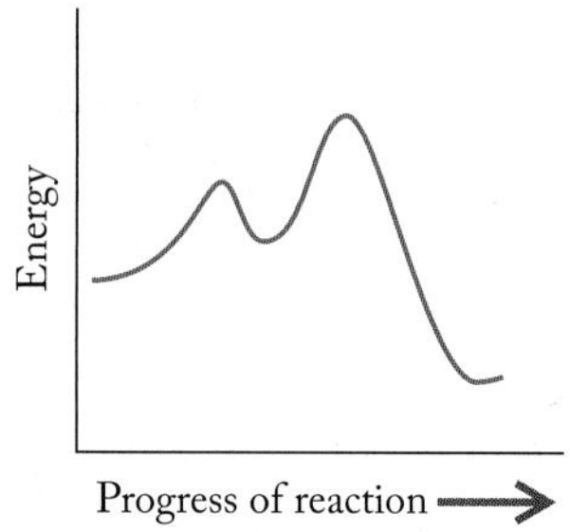

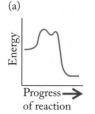

FIGURE P14.7

14.8. Which of the following mechanisms is consistent with the reaction profile shown in Figure P14.8?

a. $A + B \xrightarrow{slow} C$
   $C \xrightarrow{fast} D$

b. $A + B \rightarrow C$

c. $2A \xrightarrow{fast} C$
   $B + C \xrightarrow{slow} D$

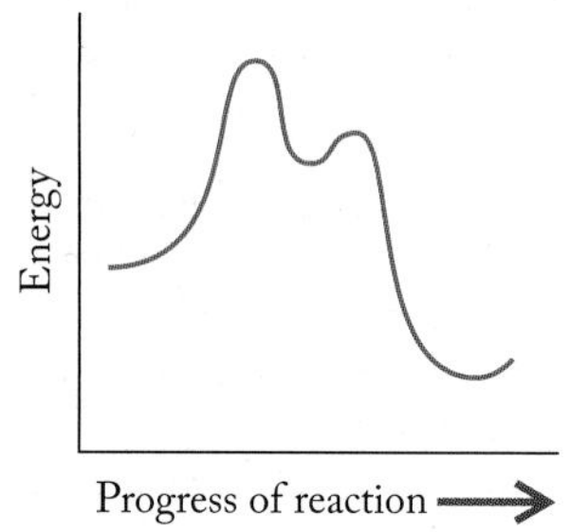

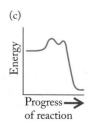

FIGURE P14.8

14.9. Which of the reaction profiles in Figure P14.9 represents the effect of a catalyst on the rate of a reaction?

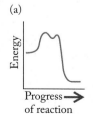

Uncatalyzed reaction

(a) (b) (c)

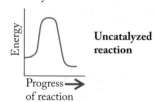

FIGURE P14.9

*14.10. Which of the reaction profiles in Figure P14.10 represents the effect of a catalyst on the rate of a reaction?

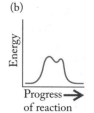

Uncatalyzed reaction

(a) (b) (c)

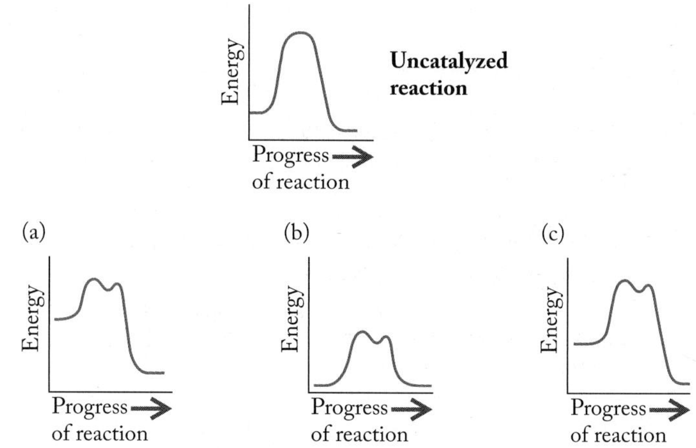

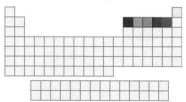

FIGURE P14.10

14.11. Which of the highlighted elements in Figure P14.11 forms volatile oxides associated with photochemical smog formation?

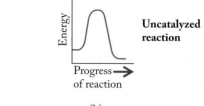

FIGURE P14.11

14.12. Which of the highlighted elements in Figure P14.12 forms noxious oxides that are removed from automobile exhaust as it passes through a catalytic converter?

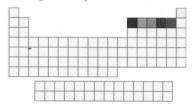

FIGURE P14.12

14.13. Which of the highlighted elements in Figure P14.13 are widely used as heterogeneous catalysts?

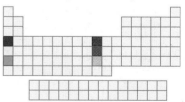

FIGURE P14.13

14.14. Which of the highlighted elements in Figure P14.14 forms volatile, odd-electron oxides that catalyze the destruction of stratospheric ozone?

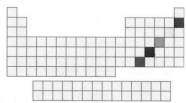

FIGURE P14.14

Questions and Problems

Cars, Trucks, and Air Quality

CONCEPT REVIEW

14.15. Why does the maximum concentration of ozone in Figure 14.2 occur much later in the day than the maximum concentration of NO and NO_2?

14.16. If we plot the concentration of reactant and products as a function of time for any sequence of two spontaneous chemical reactions, such as

$$A \rightarrow B \rightarrow C$$

will the maximum concentration of final product C always appear after the maximum concentration of B?

14.17. Why isn't there an increase in NO concentration after the evening rush hour?

14.18. If ozone can react with NO to form NO_2, why does the ozone concentration reach a maximum in the early afternoon?

PROBLEMS

14.19. By using data in Appendix 4, calculate $\Delta H°$ for the reaction

$$2 NO(g) + O_2(g) \rightarrow 2 NO_2(g)$$

14.20. By using data in Appendix 4, calculate $\Delta H°$ for the reaction

$$O_3(g) + NO(g) \rightarrow O_2(g) + NO_2(g)$$

14.21. Nitrogen and oxygen can combine to form different nitrogen oxides that play a minor role in the chemistry of smog. Write balanced chemical equations for the reaction of N_2 and O_2 that produce (a) N_2O and (b) N_2O_5.

14.22. Nitrogen oxides such as N_2O and N_2O_5 are present in the air in low concentrations, in part because of their reactivity. Write balanced chemical equations for (a) the conversion of N_2O to NO_2 in the presence of oxygen and (b) the decomposition of N_2O_5 to NO_2 and O_2.

Reaction Rates

CONCEPT REVIEW

14.23. Explain the difference between the average rate and the instantaneous rate of a chemical reaction.

14.24. Can the average rate and instantaneous rate of a chemical reaction ever be the same?

14.25. Why do the average rates of most reactions change with time?

14.26. Does the instantaneous rate of a chemical reaction change with time?

PROBLEMS

14.27. Bacterial Degradation of Ammonia *Nitrosomonas* bacteria convert ammonia into nitrite in the presence of oxygen by the following reaction:

$$2 NH_3(aq) + 3 O_2(g) \rightarrow 2 H^+(aq) + 2 NO_2^-(aq) + 2 H_2O(\ell)$$

a. How are the rates of appearance of H^+ and NO_2^- related to the rate of disappearance of NH_3?
b. How is the rate of appearance of NO_2^- related to the rate of disappearance of O_2?
c. How is the rate of disappearance of NH_3 related to the rate of disappearance of O_2?

14.28. Catalytic Converters and Combustion Catalytic converters in automobiles combat air pollution by converting NO and CO into N_2 and CO_2:

$$2 CO(g) + 2 NO(g) \rightarrow N_2(g) + 2 CO_2(g)$$

a. How is the rate of appearance of N_2 related to the rate of disappearance of CO?
b. How is the rate of appearance of CO_2 related to the rate of disappearance of NO?
c. How is the rate of disappearance of CO related to the rate of disappearance of NO?

14.29. Write expressions for the rate of appearance of products and the rate of disappearance of reactants in each of the following reactions:
a. $H_2O_2(g) \rightarrow 2 OH(g)$
b. $ClO(g) + O_2(g) \rightarrow ClO_3(g)$
c. $N_2O_5(g) + H_2O(g) \rightarrow 2 HNO_3(g)$

14.30. Write expressions for the rate of appearance of products and the rate of disappearance of reactants in each of the following reactions:
a. $Cl_2O_2(g) \rightarrow 2 ClO(g)$
b. $N_2O_5(g) \rightarrow NO_2(g) + NO_3(g)$
c. $2 INO(g) \rightarrow I_2(g) + 2 NO(g)$

14.31. Power-Plant Emissions Sulfur dioxide emissions in power-plant stack gases may react with carbon monoxide as follows:

$$SO_2(g) + 3 CO(g) \rightarrow 2 CO_2(g) + COS(g)$$

Write an equation relating the rates in each of the following:
a. The rate of appearance of CO_2 to the rate of disappearance of CO
b. The rate of appearance of COS to the rate of disappearance of SO_2
c. The rate of disappearance of CO to the rate of disappearance of SO_2

14.32. Reducing Nitric Oxide Emissions from Power Plants Nitric oxide (NO) can be removed from gas-fired power-plant emissions by reaction with methane as follows:

$$CH_4(g) + 4 NO(g) \rightarrow 2 N_2(g) + CO_2(g) + 2 H_2O(g)$$

Write an equation relating the rates in each of the following:
a. The rate of appearance of N_2 to the rate of appearance of CO_2
b. The rate of appearance of CO_2 to the rate of disappearance of NO
c. The rate of disappearance of CH_4 to the rate of appearance of H_2O

14.33. Stratospheric Ozone Depletion Chlorine monoxide (ClO) plays a major role in the creation of the ozone holes in the stratosphere over Earth's polar regions.
a. If $\Delta[ClO]/\Delta t$ at 298 K is -2.3×10^7 M/s, what is the rate of change in $[Cl_2]$ and $[O_2]$ in the following reaction?

$$2 ClO(g) \rightarrow Cl_2(g) + O_2(g)$$

b. If $\Delta[ClO]/\Delta t$ is -2.9×10^4 M/s, what is the rate of appearance of oxygen and ClO_2 in the following reaction?

$$ClO(g) + O_3(g) \rightarrow O_2(g) + ClO_2(g)$$

14.34. The chemistry of smog formation includes NO_3 as an intermediate in several reactions.

    a. If $\Delta[NO_3]/\Delta t$ is -2.2×10^5 mM/min in the following reaction, what is the rate of appearance of NO_2?

$$NO_3(g) + NO(g) \rightarrow 2\,NO_2(g)$$

    b. What is the rate of change of $[NO_2]$ in the following reaction if $\Delta[NO_3]/\Delta t$ is -2.3 mM/min?

$$2\,NO_3(g) \rightarrow 2\,NO_2(g) + O_2(g)$$

14.35. Nitrite ion reacts with ozone in aqueous solution, producing nitrate ion and oxygen:

$$NO_2^-(aq) + O_3(g) \rightarrow NO_3^-(aq) + O_2(g)$$

The following data were collected for this reaction at 298 K. Calculate the average reaction rate between 0 and 100 μs (microseconds) and between 200 and 300 μs.

Time (μs)	$[O_3]$ (M)
0	1.13×10^{-2}
100	9.93×10^{-3}
200	8.70×10^{-3}
300	8.15×10^{-3}

14.36. Dinitrogen pentoxide (N_2O_5) decomposes as follows to nitrogen dioxide and nitrogen trioxide:

$$N_2O_5(g) \rightarrow NO_2(g) + NO_3(g)$$

Calculate the average rate of this reaction between consecutive measurement times in the following table.

Time (s)	$[N_2O_5]$ (molecules/cm³)
0	1.500×10^{12}
1.45	1.357×10^{12}
2.90	1.228×10^{12}
4.35	1.111×10^{12}
5.80	1.005×10^{12}

14.37. The following data were collected for the dimerization of ClO to Cl_2O_2 at 298 K.

Time (s)	$[ClO]$ (molecules/cm³)
0	2.60×10^{11}
1	1.08×10^{11}
2	6.83×10^{10}
3	4.99×10^{10}
4	3.93×10^{10}
5	3.24×10^{10}
6	2.76×10^{10}

Plot $[ClO]$ and $[Cl_2O_2]$ as a function of time and determine the instantaneous rates of change in both at 1 s.

14.38. Tropospheric Ozone Tropospheric ozone is rapidly consumed in many reactions, including

$$O_3(g) + NO(g) \rightarrow NO_2(g) + O_2(g)$$

Use the following data to calculate the instantaneous rate of the preceding reaction at $t = 0.000$ s and $t = 0.052$ s.

Time (s)	$[NO]$ (M)
0.000	2.0×10^{-8}
0.011	1.8×10^{-8}
0.027	1.6×10^{-8}
0.052	1.4×10^{-8}
0.102	1.2×10^{-8}

Effect of Concentration on Reaction Rate

CONCEPT REVIEW

14.39. Can two different chemical reactions have the same rate-law expression?

14.40. Why are the units of the rate constants different for reactions of different order?

14.41. Does the half-life of a second-order reaction have the same units as the half-life for a first-order reaction?

14.42. Does the half-life of a first-order reaction depend on the concentration of the reactants?

14.43. What effect does doubling the initial concentration of a reactant have on the half-life of a reaction that is second order in the reactant?

14.44. Two first-order decomposition reactions of the form A → B + C have the same rate constant at a given temperature. Do the reactants in the two reactions have the same half-lives at this temperature?

PROBLEMS

14.45. For each of the following rate laws, determine the order with respect to each reactant and the overall reaction order.

    a. Rate = $k[A][B]$

    b. Rate = $k[A]^2[B]$

    c. Rate = $k[A][B]^3$

14.46. Determine the overall order of the following rate laws and the order with respect to each reactant.

    a. Rate = $k[A]^2[B]$

    b. Rate = $k[A]^2[B][C]$

    c. Rate = $k[A][B]^3[C]^{1/2}$

14.47. Write rate laws and determine the units of the rate constant (by using the units M for concentration and s for time) for the following reactions:

    a. The reaction of oxygen atoms with NO_2 is first order in both reactants.

    b. The reaction between NO and Cl_2 is second order in NO and first order in Cl_2.

    c. The reaction between Cl_2 and chloroform ($CHCl_3$) is first order in $CHCl_3$ and one-half order in Cl_2.

    *d. The decomposition of ozone (O_3) to O_2 is second order in O_3 and an order of -1 in O atoms.

14.48. Compounds A and B react to give a single product, C. Write the rate law for each of the following cases and determine the units of the rate constant by using the units M for concentration and s for time:
  a. The reaction is first order in A and second order in B.
  b. The reaction is first order in A and second order overall.
  c. The reaction is independent of the concentration of A and second order overall.
  d. The reaction is second order in both A and B.

14.49. Predict the rate law for the reaction $2\,BrO(g) \rightarrow Br_2(g) + O_2(g)$ if the following conditions hold true:
  a. The rate doubles when [BrO] doubles
  b. The rate quadruples when [BrO] doubles
  c. The rate is halved when [BrO] is halved
  d. The rate is unchanged when [BrO] is doubled

14.50. Predict the rate law for the reaction $NO(g) + Br_2(g) \rightarrow NOBr_2(g)$ if the following conditions apply:
  a. The rate doubles when [NO] is doubled and [Br$_2$] remains constant
  b. The rate doubles when [Br$_2$] is doubled and [NO] remains constant
  c. The rate increases by 1.56 times when [NO] is increased 1.25 times and [Br$_2$] remains constant
  d. The rate is halved when [NO] is doubled and [Br$_2$] remains constant

14.51. In the reaction of NO with ClO,

$$NO(g) + ClO(g) \rightarrow NO_2(g) + Cl(g)$$

the initial rate of reaction quadruples when the concentrations of both reactants are doubled. What additional information do we need to determine whether the reaction is first order in each reactant?

14.52. The reaction between chlorine monoxide and nitrogen dioxide

$$ClO(g) + NO_2(g) + M(g) \rightarrow ClONO_2(g) + M(g)$$

produces chlorine nitrate (ClONO$_2$). A third molecule (M) takes part in the reaction but is unchanged by it. The reaction is first order in NO$_2$ and in ClO.
  a. Write the rate law for this reaction.
  b. What is the reaction order with respect to M?

14.53. Rate Laws for Destruction of Tropospheric Ozone The reaction of NO$_2$ with ozone produces NO$_3$ in a second-order reaction overall:

$$NO_2(g) + O_3(g) \rightarrow NO_3(g) + O_2(g)$$

  a. Write the rate law for the reaction if the reaction is first order in each reactant.
  b. The rate constant for the reaction is $1.93 \times 10^4\,M^{-1}s^{-1}$ at 298 K. What is the rate of the reaction when $[NO_2] = 1.8 \times 10^{-8}\,M$ and $[O_3] = 1.4 \times 10^{-7}\,M$?
  c. What is the rate of the appearance of NO$_3$ under these conditions?
  d. What happens to the rate of the reaction if the concentration of O$_3(g)$ is doubled?

14.54. Sources of Nitric Acid in the Atmosphere The reaction between N$_2$O$_5$ and water

$$N_2O_5(g) + H_2O(g) \rightarrow 2\,HNO_3(g)$$

is a source of nitric acid in the atmosphere.
  a. The reaction is first order in each reactant. Write the rate law for the reaction.
  b. When [N$_2$O$_5$] is 0.132 mM and [H$_2$O] is 230 mM, the rate of the reaction is 4.55×10^{-4} mM^{-1} min^{-1}. What is the rate constant for the reaction?

14.55. Each of the following reactions is first order in the reactants and second order overall. Which reaction is fastest if the initial concentrations of the reactants are the same? All reactions are at 298 K.
  a. $ClO_2(g) + O_3(g) \rightarrow ClO_3(g) + O_2(g)$
$$k = 3.0 \times 10^{-19}\ cm^3/(molecule \cdot s)$$
  b. $ClO_2(g) + NO(g) \rightarrow NO_2(g) + ClO(g)$
$$k = 3.4 \times 10^{-13}\ cm^3/(molecule \cdot s)$$
  c. $ClO(g) + NO(g) \rightarrow Cl(g) + NO_2(g)$
$$k = 1.7 \times 10^{-11}\ cm^3/(molecule \cdot s)$$
  d. $ClO(g) + O_3(g) \rightarrow ClO_2(g) + O_2(g)$
$$k = 1.5 \times 10^{-17}\ cm^3/(molecule \cdot s)$$

14.56. Two reactions in which there is a single reactant have nearly the same magnitude rate constant. One is first order; the other is second order.
  a. If the initial concentrations of the reactants are both 1.0 mM, which reaction will proceed at the higher rate?
  b. If the initial concentrations of the reactants are both 2.0 M, which reaction will proceed at the higher rate?

14.57. In the presence of water, the species NO and NO$_2$ react to form nitrous acid (HNO$_2$) by the following reaction:

$$NO(g) + NO_2(g) + H_2O(\ell) \rightarrow 2\,HNO_2(aq)$$

When the concentration of NO or NO$_2$ is doubled, the initial rate of reaction doubles. If the rate of the reaction does not depend on [H$_2$O], what is the rate law for this reaction?

14.58. Hydroperoxyl Radicals in the Atmosphere During a smog event, trace amounts of many highly reactive substances are present in the atmosphere. One of these is the hydroperoxyl radical, HO$_2$, which reacts with sulfur trioxide, SO$_3$. The rate constant for the reaction

$$2\,HO_2(g) + SO_3(g) \rightarrow H_2SO_3(g) + 2\,O_2(g)$$

at 298 K is $2.6 \times 10^{11}\,M^{-1}\,s^{-1}$. The initial rate of the reaction doubles when the concentration of SO$_3$ or HO$_2$ is doubled. What is the rate law for the reaction?

14.59. Disinfecting Municipal Water Supplies Chlorine dioxide (ClO_2) is a disinfectant used in municipal water-treatment plants (Figure P14.59). It dissolves in basic solution, producing ClO_3^- and ClO_2^-:

$$2\,ClO_2(g) + 2\,OH^-(aq) \rightarrow ClO_3^-(aq) + ClO_2^-(aq) + H_2O(\ell)$$

The following kinetic data were obtained at 298 K for the reaction:

Experiment	$[ClO_2^-]_0$ (M)	$[OH^-]_0$ (M)	Initial Rate (M/s)
1	0.060	0.030	0.0248
2	0.020	0.030	0.00827
3	0.020	0.090	0.0247

Determine the rate law and the rate constant for this reaction at 298 K.

FIGURE P14.59 Many municipal water-treatment plants use chlorine dioxide as a disinfectant. It dissolves in basic solution to form a mixture of ClO_2^- and ClO_3^-.

14.60. The following kinetic data were collected at 298 K for the reaction of ozone with nitrite ion, producing nitrate and oxygen:

$$NO_2^-(aq) + O_3(g) \rightarrow NO_3^-(aq) + O_2(g)$$

Experiment	$[NO_2^-]_0$ (M)	$[O_3]_0$ (M)	Initial Rate (M/s)
1	0.0100	0.0050	25
2	0.0150	0.0050	37.5
3	0.0200	0.0050	50.0
4	0.0200	0.0200	200.0

Determine the rate law for the reaction and the value of the rate constant.

14.61. Hydrogen gas reduces NO to N_2 in the following reaction:

$$2\,H_2(g) + 2\,NO(g) \rightarrow 2\,H_2O(g) + N_2(g)$$

The initial reaction rates of four mixtures of H_2 and NO were measured at 900°C with the following results:

Experiment	$[H_2]_0$ (M)	$[NO]_0$ (M)	Initial Rate (M/s)
1	0.212	0.136	0.0248
2	0.212	0.272	0.0991
3	0.424	0.544	0.793
4	0.848	0.544	1.59

Determine the rate law and the rate constant for the reaction at 900°C.

14.62. The rate of the reaction

$$NO_2(g) + CO(g) \rightarrow NO(g) + CO_2(g)$$

was determined in three experiments at 225°C. The results are given in the following table:

Experiment	$[NO_2]_0$ (M)	$[CO]_0$ (M)	Initial Rate $-\Delta[NO_2]/\Delta t$ (M/s)
1	0.263	0.826	1.44×10^{-5}
2	0.263	0.413	1.44×10^{-5}
3	0.526	0.413	5.76×10^{-5}

a. Determine the rate law for the reaction.
b. Calculate the value of the rate constant at 225°C.
c. Calculate the rate of appearance of CO_2 when $[NO_2] = [CO] = 0.500\ M$.

14.63. Nitrogen trioxide decomposes to NO_2 and O_2 in the following reaction:

$$2\,NO_3(g) \rightarrow 2\,NO_2(g) + O_2(g)$$

The following data were collected at 298 K:

Time (min)	$[NO_3]$ (μM)
0	1.470×10^{-3}
10	1.463×10^{-3}
100	1.404×10^{-3}
200	1.344×10^{-3}
300	1.288×10^{-3}
400	1.237×10^{-3}
500	1.190×10^{-3}

Calculate the value of the second-order rate constant at 298 K.

14.64. Two structural isomers of ClO_2 are shown in Figure P14.64:

$$O\!-\!Cl\!-\!O \qquad\qquad Cl\!-\!O\!-\!O$$

FIGURE P14.64

The isomer with the Cl–O–O skeletal arrangement is unstable and rapidly decomposes according to the reaction $2\,ClOO(g) \rightarrow Cl_2(g) + 2\,O_2(g)$. The following data were collected for the decomposition of ClOO at 298 K:

Time (μs)	$[ClOO]$ (M)
0.00	1.76×10^{-6}
0.67	2.36×10^{-7}
1.3	3.56×10^{-8}
2.1	3.23×10^{-9}
2.8	3.96×10^{-10}

Determine the rate law for the reaction and the value of the rate constant at 298 K.

14.65. At high temperatures, ammonia spontaneously decomposes into N_2 and H_2. The following data were collected at one such temperature:

Time (s)	[NH$_3$] (*M*)
0	2.56×10^{-2}
12	2.47×10^{-2}
56	2.16×10^{-2}
224	1.31×10^{-2}
532	5.19×10^{-3}
746	2.73×10^{-3}

Determine the rate law for the decomposition of ammonia and the value of the rate constant at the temperature of the experiment.

14.66. Atmospheric Chemistry of Hydroperoxyl Radicals Atmospheric chemistry involves highly reactive, odd-electron molecules such as the hydroperoxyl radical HO_2, which decomposes into H_2O_2 and O_2. Determine the rate law for the reaction and the value of the rate constant at 298 K by using the following data obtained at 298 K.

Time (μs)	[HO$_2$] (μ*M*)
0	8.5
0.6	5.1
1.0	3.6
1.4	2.6
1.8	1.8
2.4	1.1

14.67. Laughing Gas Nitrous oxide (N_2O) is used as an anesthetic (laughing gas) and in aerosol cans to produce whipped cream. It is a potent greenhouse gas and decomposes slowly to N_2 and O_2:

$$2\,N_2O(g) \rightarrow 2\,N_2(g) + O_2(g)$$

a. If the plot of ln [N_2O] as a function of time is linear, what is the rate law for the reaction?
b. How many half-lives will it take for the concentration of the N_2O to reach 6.25% of its original concentration?

14.68. The unsaturated hydrocarbon butadiene (C_4H_6) dimerizes to 4-vinylcyclohexene (C_8H_{12}). When data collected in studies of the kinetics of this reaction were plotted against reaction time, plots of [C_4H_6] or ln [C_4H_6] produced curved lines, but the plot of $1/[C_4H_6]$ was linear.
a. What is the rate law for the reaction?
b. How many half-lives will it take for the [C_4H_6] to decrease to 3.1% of its original concentration?
[*Hint:* The amount of reactant remaining after time t (A_t) is related to the amount initially present (A_0) by the equation $A_t/A_0 = (0.5)^n$, where n is the number of half-lives in time t.]

14.69. Tracing Phosphorus in Organisms Radioactive isotopes such as ^{32}P are used to follow biological processes. The following radioactivity data (in picocuries) were collected for a sample containing ^{32}P:

Time (days)	Radioactivity (pCi)
0	10.0
1	9.53
2	9.08
5	7.85
10	6.16
20	3.79

a. Write the rate law for the decay of ^{32}P.
b. Determine the value of the first-order rate constant.
c. Determine the half-life of ^{32}P.

14.70. Nitrous acid slowly decomposes to NO, NO_2, and water in the following second-order reaction:

$$2\,HNO_2(aq) \rightarrow NO(g) + NO_2(g) + H_2O(\ell)$$

a. Use the data below to determine the rate constant for this reaction at 298 K:

Time (min)	[HNO$_2$] (μ*M*)
0	0.1560
1000	0.1466
1500	0.1424
2000	0.1383
2500	0.1345
3000	0.1309

b. Determine the half-life for the decomposition of HNO_2.

14.71. The dimerization of ClO,

$$2\,ClO(g) \rightarrow Cl_2O_2(g)$$

is second order in ClO. Use the following data to determine the value of k at 298 K:

Time (s)	[ClO] (molecules/cm^3)
0	2.60×10^{11}
1	1.08×10^{11}
2	6.83×10^{10}
3	4.99×10^{10}
4	3.93×10^{10}

Determine the half-life for the dimerization of ClO.

14.72. Kinetic data for the reaction $Cl_2O_2(g) \rightarrow 2\,ClO(g)$ are summarized in the following table. Determine the value of the first-order rate constant.

Time (μs)	[Cl$_2$O$_2$] (*M*)
0	6.60×10^{-8}
172	5.68×10^{-8}
345	4.89×10^{-8}
517	4.21×10^{-8}
690	3.62×10^{-8}
862	3.12×10^{-8}

Determine the half-life for the decomposition of Cl_2O_2.

14.73. Kinetics of Sucrose Hydrolysis The metabolism of table sugar (sucrose, $C_{12}H_{22}O_{11}$) begins with the hydrolysis of the disaccharide to glucose and fructose (both $C_6H_{12}O_6$):

$$C_{12}H_{22}O_{11}(aq) + H_2O(\ell) \rightarrow 2\,C_6H_{12}O_6(aq)$$

The kinetics of the reaction were studied at 24°C in a reaction system with a large excess of water, so the reaction was pseudo-first-order in sucrose. Determine the rate law and the pseudo-first-order rate constant for the reaction from the following data:

Time (s)	$[C_{12}H_{22}O_{11}]$ (M)
0	0.562
612	0.541
1600	0.509
2420	0.484
3160	0.462
4800	0.4417

14.74. Hydroperoxyl radicals react rapidly with ozone to produce oxygen and OH radicals:

$$HO_2(g) + O_3(g) \rightarrow OH(g) + 2\,O_2(g)$$

The rate of this reaction was studied in the presence of a large excess of ozone. Determine the pseudo-first-order rate constant and the second-order rate constant for the reaction from the following data:

Time (ms)	$[HO_2]$ (M)	$[O_3]$ (M)
0	3.2×10^{-6}	1.0×10^{-3}
10	2.9×10^{-6}	1.0×10^{-3}
20	2.6×10^{-6}	1.0×10^{-3}
30	2.4×10^{-6}	1.0×10^{-3}
80	1.4×10^{-6}	1.0×10^{-3}

Reaction Rates, Temperature, and the Arrhenius Equation

CONCEPT REVIEW

14.75. Why are some spontaneous chemical reactions slow?

14.76. Is the activation energy of a spontaneous reaction in the forward direction greater than, less than, or equal to the activation energy for the reverse reaction?

14.77. Which, if any, of the following statements is true?
a. Exothermic reactions are always fast.
b. Reactions with $\Delta G > 0$ are slow.
c. Endothermic reactions are always slow.
d. Reactions accompanied by an increase in entropy are fast.

*14.78. Which, if any, of the following statements is true?
a. Reactions with $\Delta G < 0$ are always fast.
b. Reactions with $\Delta H > 0$ are always fast.
c. Reactions with $\Delta S < 0$ are always slow.
d. Reactions with $\Delta H < 0$ are fast only at low temperature.

*14.79. The order of a reaction is independent of temperature, but the value of the rate constant varies with temperature. Why?

14.80. Why is the value of E_a for a spontaneous reaction less than the E_a value for the same reaction running in reverse?

*14.81. Two first-order reactions have activation energies of 15 and 150 kJ/mol. Which reaction will show the larger increase in rate as temperature is increased?

14.82. According to the Arrhenius equation, does the activation energy of a chemical reaction depend on temperature? Explain your answer.

PROBLEMS

14.83. The rate constant for the reaction of ozone with oxygen atoms was determined at four temperatures. Calculate the activation energy and frequency factor A for the reaction

$$O(g) + O_3(g) \rightarrow 2\,O_2(g)$$

given the following data:

T (K)	k [cm³/(molecule · s)]
250	2.64×10^{-4}
275	5.58×10^{-4}
300	1.04×10^{-3}
325	1.77×10^{-3}

14.84. The rate constant for the reaction

$$NO_2(g) + O_3(g) \rightarrow NO_3(g) + O_2(g)$$

was determined over a temperature range of 40 K, with the following results:

T (K)	k (M⁻¹s⁻¹)
203	4.14×10^5
213	7.30×10^5
223	1.22×10^6
233	1.96×10^6
243	3.02×10^6

a. Determine the activation energy for the reaction.
b. Calculate the rate constant of the reaction at 300 K.

14.85. Activation Energy for Smog-Forming Reactions The initial step in the formation of smog is the reaction between nitrogen and oxygen. The activation energy of the reaction can be determined from the temperature dependence of the rate constants. At the temperatures indicated, values of the rate constant of the reaction

$$N_2(g) + O_2(g) \rightarrow 2\,NO(g)$$

are as follows:

T (K)	k (M⁻¹ᐟ²s⁻¹)
2000	318
2100	782
2200	1770
2300	3733
2400	7396

a. Calculate the activation energy of the reaction.
b. Calculate the frequency factor for the reaction.
c. Calculate the value of the rate constant at ambient temperature, $T = 300$ K.

14.86. Values of the rate constant for the decomposition of N_2O_5 gas at four different temperatures are as follows:

T (K)	k (s⁻¹)
658	2.14×10^5
673	3.23×10^5
688	4.81×10^5
703	7.03×10^5

a. Determine the activation energy of the decomposition reaction.
b. Calculate the value of the rate constant at 300 K.

14.87. Activation Energy of Stratospheric Ozone Destruction Reactions The kinetics of the reaction between chlorine dioxide and ozone is relevant to the study of atmospheric ozone destruction. The activation energy of the reaction can be determined from the temperature dependence of the rate constant. The value of the rate constant for the reaction between chlorine dioxide and ozone was measured at four temperatures between 193 and 208 K. The results are as follows:

T (K)	k (M⁻¹s⁻¹)
193	34.0
198	62.8
203	112.8
208	196.7

Calculate the values of the activation energy and the frequency factor for the reaction.

14.88. Chlorine atoms react with methane, forming HCl and CH_3. The rate constant for the reaction is $6.0 \times 10^7 \, M^{-1}s^{-1}$ at 298 K. When the experiment was repeated at three other temperatures, the following data were collected:

T (K)	k (M⁻¹s⁻¹)
303	6.5×10^7
308	7.0×10^7
313	7.5×10^7

Calculate the values of the activation energy and the frequency factor for the reaction.

Reaction Mechanisms

CONCEPT REVIEW

14.89. The rate law for the reaction between NO and H_2 is second order in NO and third order overall, whereas the reaction of NO with Cl_2 is first order in each reactant and second order overall. Do these reactions proceed by similar mechanisms?

14.90. The rate law for the reaction of NO with Cl_2 (rate = $k[NO][Cl_2]$) is the same as that for the reaction of NO_2 with F_2 (rate = $k[NO_2][F_2]$). Is it possible that these reactions have similar mechanisms?

*14.91. Under what reaction conditions does a bimolecular reaction obey pseudo–first-order reaction kinetics?

*14.92. If a reaction is zero-order in a reactant, does that mean the reactant is never involved in collisions with other reactants? Explain your answer.

PROBLEMS

14.93. The hypothetical reaction A → B has an activation energy of 50.0 kJ/mol. Draw a reaction profile for each of the following mechanisms:
a. A single elementary step.
b. A two-step reaction in which the activation energy of the second step is 15 kJ/mol.
c. A two-step reaction in which the activation energy of the second step is the rate-determining barrier.

14.94. For the spontaneous reaction A + B → C → D + E, draw three reaction profiles, one for each of the following mechanisms:
a. C is an activated complex.
b. The reaction has two elementary steps; the first step is rate determining and C is an intermediate.
c. The reaction has two elementary steps; the second step is rate determining and C is an intermediate.

14.95. Write the rate laws for the following elementary steps and identify them as uni-, bi-, or termolecular steps:
a. $SO_2Cl_2(g) \rightarrow SO_2(g) + Cl_2(g)$
b. $NO_2(g) + CO(g) \rightarrow NO(g) + CO_2(g)$
c. $2NO_2(g) \rightarrow NO_3(g) + NO(g)$

14.96. Write the rate laws for the following elementary steps and identify them as uni-, bi-, or termolecular steps:
a. $Cl(g) + O_3(g) \rightarrow ClO(g) + O_2(g)$
b. $2NO_2(g) \rightarrow N_2O_4(g)$
c. $^{14}_{6}C \rightarrow \, ^{14}_{7}N + \, ^{\,\,0}_{-1}\beta$

14.97. Write the overall reaction that consists of the following elementary steps:

$$N_2O_5(g) \rightarrow NO_3(g) + NO_2(g)$$
$$NO_3(g) \rightarrow NO_2(g) + O(g)$$
$$2O(g) \rightarrow O_2(g)$$

14.98. What overall reaction consists of the following elementary steps?

$$ClO^-(aq) + H_2O(\ell) \rightarrow HClO(aq) + OH^-(aq)$$
$$I^-(aq) + HClO(aq) \rightarrow HIO(aq) + Cl^-(aq)$$
$$OH^-(aq) + HIO(aq) \rightarrow H_2O(\ell) + IO^-(aq)$$

*14.99. In the following mechanism for NO formation, oxygen atoms are produced by breaking O=O bonds at high temperature in a fast reversible reaction. If $\Delta[NO]/\Delta t = k[N_2][O_2]^{1/2}$, which step in the mechanism is the rate-determining step?

(1) $O_2(g) \rightarrow 2O(g)$

(2) $O(g) + N_2(g) \rightarrow NO(g) + N(g)$

(3) $N(g) + O(g) \rightarrow NO(g)$

Overall: $N_2(g) + O_2(g) \rightarrow 2NO(g)$

14.100. A proposed mechanism for the decomposition of hydrogen peroxide consists of three elementary steps:

$$H_2O_2(g) \rightarrow 2\,OH(g)$$

$$H_2O_2(g) + OH(g) \rightarrow H_2O(g) + HO_2(g)$$

$$HO_2(g) + OH(g) \rightarrow H_2O(g) + O_2(g)$$

If the rate law for the reaction is first order in H_2O_2, which step in the mechanism is the rate-determining step?

14.101. At a given temperature, the rate of the reaction between NO and Cl_2 is proportional to the product of the concentrations of the two gases: $[NO][Cl_2]$. The following two-step mechanism was proposed for the reaction:

(1) $\quad NO(g) + Cl_2(g) \rightarrow NOCl_2(g)$

(2) $\quad NOCl_2(g) + NO(g) \rightarrow 2\,NOCl(g)$

Overall: $2\,NO(g) + Cl_2(g) \rightarrow NOCl_2(g)$

Which step must be the rate-determining step if this mechanism is correct?

14.102. **Mechanism of Ozone Destruction** Ozone decomposes thermally to oxygen in the following reaction:

$$2\,O_3(g) \rightarrow 3\,O_2(g)$$

The following mechanism has been proposed:

$$O_3(g) \rightarrow O(g) + O_2(g)$$

$$O(g) + O_3(g) \rightarrow 2\,O_2(g)$$

The reaction is second order in ozone. What properties of the two elementary steps (specifically, relative rate and reversibility) are consistent with this mechanism?

14.103. **Mechanism of NO$_2$ Destruction** The rate laws for the thermal and photochemical decomposition of NO_2 are different. Which of the following mechanisms are possible for the thermal decomposition of NO_2, and which are possible for the photochemical decomposition of NO_2? The rate of thermal decompostion $= k[NO_2]^2$ and the rate of photochemical decomposition $= k[NO_2]$.

a. $\qquad NO_2(g) \xrightarrow{\text{slow}} NO(g) + O(g)$
$\quad O(g) + NO_2(g) \xrightarrow{\text{fast}} NO(g) + O_2(g)$

b. $NO_2(g) + NO_2(g) \xrightarrow{\text{fast}} N_2O_4(g)$
$\quad N_2O_4(g) \xrightarrow{\text{slow}} NO(g) + NO_3(g)$
$\quad NO_3(g) \xrightarrow{\text{fast}} NO(g) + O_2(g)$

c. $NO_2(g) + NO_2(g) \xrightarrow{\text{slow}} NO(g) + NO_3(g)$
$\quad NO_3(g) \xrightarrow{\text{fast}} NO(g) + O_2(g)$

14.104. The rate laws for the thermal and photochemical decomposition of NO_2 are different. Which of the following mechanisms are possible for the thermal decomposition of NO_2, and which are possible for the photochemical decomposition of NO_2? The rate of thermal decomposition $= k[NO_2]^2$ and the rate of photochemical decomposition $= k[NO_2]$.

a. $NO_2(g) + NO_2(g) \xrightarrow{\text{slow}} N_2O_4(g)$
$\quad N_2O_4(g) \xrightarrow{\text{fast}} N_2O_3(g) + O(g)$
$\quad N_2O_3(g) + O(g) \xrightarrow{\text{fast}} N_2O_2(g) + O_2(g)$
$\quad N_2O_2(g) \xrightarrow{\text{fast}} 2\,NO(g)$

b. $NO_2(g) + NO_2(g) \xrightarrow{\text{slow}} NO(g) + NO_3(g)$
$\quad NO_3(g) \xrightarrow{\text{fast}} NO(g) + O_2(g)$

c. $\qquad NO_2(g) \xrightarrow{\text{slow}} N(g) + O_2(g)$
$\quad N(g) + NO_2(g) \xrightarrow{\text{fast}} N_2O_2(g)$
$\quad N_2O_2(g) \xrightarrow{\text{slow}} 2\,NO(g)$

Catalysis

CONCEPT REVIEW

14.105. Does a catalyst affect both the rate and the rate constant of a reaction?

14.106. Is the rate law for a catalyzed reaction the same as that for the uncatalyzed reaction?

14.107. Does a substance that increases the rate of a reaction also increase the rate of the reverse reaction?

14.108. The rate of the reaction between NO_2 and CO is independent of [CO]. Does this mean that CO is a catalyst for the reaction?

14.109. Why doesn't the concentration of a homogeneous catalyst appear in the rate law for the reaction it catalyzes?

*14.110. The rate of a chemical reaction is too slow to measure at room temperature. We could either raise the temperature or add a catalyst. Which would be a better solution for making an accurate determination of the rate constant?

PROBLEMS

14.111. Is NO a catalyst for the decomposition of N_2O in the following two-step reaction mechanism, or is N_2O a catalyst for the conversion of NO to NO_2?

(1) $\quad NO(g) + N_2O(g) \rightarrow N_2(g) + NO_2(g)$

(2) $\qquad 2\,NO_2(g) \rightarrow 2\,NO(g) + O_2(g)$

14.112. **NO as a Catalyst for Ozone Destruction** Explain why NO is a catalyst in the following two-step process that results in the depletion of ozone in the stratosphere:

(1) $\quad NO(g) + O_3(g) \rightarrow NO_2(g) + O_2(g)$

(2) $\quad O(g) + NO_2(g) \rightarrow NO(g) + O_2(g)$

Overall: $O(g) + O_3(g) \rightarrow 2\,O_2(g)$

14.113. On the basis of the frequency factors and activation energy values of the following two reactions, determine which one will have the larger rate constant at room temperature (298 K).

$$O_3(g) + O(g) \rightarrow O_2(g) + O_2(g)$$
$A = 8.0 \times 10^{-12}\ cm^3/(molecule \cdot s) \qquad E_a = 17.1\ kJ/mol$

$$O_3(g) + Cl(g) \rightarrow ClO(g) + O_2(g)$$
$A = 2.9 \times 10^{-11}\ cm^3/(molecule \cdot s) \qquad E_a = 2.16\ kJ/mol$

14.114. On the basis of the frequency factors and activation energy values of the following two reactions, determine which one will have the larger rate constant at room temperature (298 K).

$$O_3(g) + Cl(g) \rightarrow ClO(g) + O_2(g)$$
$A = 2.9 \times 10^{-11}\ cm^3/(molecule \cdot s) \qquad E_a = 2.16\ kJ/mol$

$$O_3(g) + NO(g) \rightarrow NO_2(g) + O_2(g)$$
$A = 2.0 \times 10^{-12}\ cm^3/(molecule \cdot s) \qquad E_a = 11.6\ kJ/mol$

Additional Problems

14.115. A student inserts a glowing wood splint into a test tube filled with O_2. The splint quickly catches on fire (Figure P14.115). Why does the splint burn so much faster in pure O_2 than in air?

FIGURE P14.115

*14.116. A backyard chef turns on the propane gas to a barbecue grill. Even though the reaction between propane and oxygen is spontaneous, the gas does not begin to burn until the chef pushes an igniter button to produce a spark. Why is the spark needed?

14.117. On average, someone who falls through the ice covering a frozen lake is less likely to experience anoxia (lack of oxygen) than someone who falls into a warm pool and is underwater for the same length of time. Why?

*14.118. The rates of chemical reactions increase significantly with increasing temperature, yet temperature has little effect on radioactive decay rates and half-life values. Can you think of a reason why?

14.119. During the decomposition of dinitrogen pentoxide,

$$2\,N_2O_5(g) \rightarrow 4\,NO_2(g) + O_2(g)$$

how is the rate of disappearance of N_2O_5 related to the rate of formation of NO_2 and O_2?

14.120. In the reaction between nitrogen dioxide and ozone

$$2\,NO_2(g) + O_3(g) \rightarrow N_2O_5(g) + O_2(g)$$

how are the rates of change in the concentrations of the reactants and products related?

14.121. Determine the order of the decomposition reaction of N_2O_5, by using the initial rate data from the following table:

Experiment Number	$[N_2O_5]_0$ (M)	Initial Rate (M/s)
1	0.050	1.8×10^{-5}
2	0.100	3.6×10^{-5}

14.122. At the temperature at which the experiments were carried out in the previous problem, what is the rate constant for the decomposition of N_2O_5? Write the complete rate-law expression for the decomposition reaction.

14.123. The table below contains kinetics data for the reaction

$$2\,NO(g) + Cl_2(g) \rightarrow 2\,NOCl(g)$$

Experiment	Initial $[NO]_0$ (M)	Initial $[Cl_2]_0$ (M)	Initial Rate (M/s)
1	0.20	0.10	0.63
2	0.20	0.30	5.70
3	0.80	0.10	2.58
4	0.40	0.20	?

Predict the initial rate of reaction in experiment 4.

14.124. An important reaction in the formation of photochemical smog is the reaction between ozone and NO

$$NO(g) + O_3(g) \rightarrow NO_2(g) + O_2(g)$$

The reaction is first order in NO and O_3. The rate constant of the reaction is 80 $M^{-1}\,s^{-1}$ at 25°C and 3000 $M^{-1}\,s^{-1}$ at 75°C.
  a. If this reaction were to occur in a single step, would the rate law be consistent with the observed order of the reaction for NO and O_3?
  b. What is the value of the activation energy of the reaction?
  c. What is the rate of the reaction at 25°C when $[NO] = 3 \times 10^{-6}M$ and $[O_3] = 5 \times 10^{-9}M$?
  d. Predict the values of the rate constant at 10°C and 35°C.

14.125. Ammonia reacts with nitrous acid to form an intermediate, ammonium nitrite (NH_4NO_2), which decomposes to N_2 and H_2O:

$$NH_3(g) + HNO_2(aq) \rightarrow NH_4NO_2(aq) \rightarrow N_2(g) + 2\,H_2O(\ell)$$

  a. The reaction is first order in ammonia and second order in nitrous acid. What is the rate law for the reaction? What are the units on the rate constant if concentrations are expressed in molarity and time in seconds?
  b. The rate law for the reaction has also been written as

$$\text{Rate} = k\,[NH_4^+][NO_2^-][HNO_2]$$

  Is this expression equivalent to the one you wrote in part a?
  c. With the data in the Appendix 4, calculate the value of ΔH°_{rxn} of the overall reaction (ΔH°_f, $HNO_2 = -43.1$ kJ/mol).
  d. Draw a reaction-energy profile for the process with the assumption that E_a of the first step is lower than E_a of the second step.

*14.126. When ionic compounds such as NaCl dissolve in water, the sodium ions are surrounded by six water molecules. The bound water molecules exchange with those in bulk solution as described by the reaction involving ^{18}O-enriched water:

$$Na(H_2O)_6^+(aq) + H_2^{18}O(\ell) \rightarrow Na(H_2O)_5(H_2^{18}O)(aq) + H_2O(\ell)$$

  a. The following reaction mechanism has been proposed:

(1) $\qquad Na(H_2O)_6^+(aq) \rightarrow Na(H_2O)_5^+(aq) + H_2O(\ell)$

(2) $Na(H_2O)_5^+(aq) + H_2^{18}O(\ell) \rightarrow Na(H_2O)_5(H_2^{18}O)^+(aq)$

  What is the rate law if the first step is the rate-determining step?
  b. If you were to sketch a reaction-energy profile, which would you draw with the higher energy, the reactants or the products?

14.127. Lachrymators in Smog The combination of ozone, volatile hydrocarbons, nitrogen oxide, and sunlight in urban environments produces peroxyacetylnitrate (PAN), a potent lachrymator. PAN decomposes to acetyl radicals and nitrogen dioxide in a process that is second order in PAN, as shown in Figure P14.127:

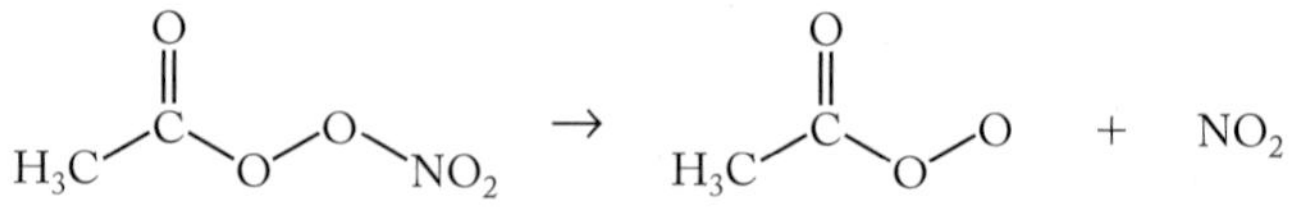

FIGURE P14.127

a. The half-life of the reaction, at 23°C and $P_{CH_3CO_3NO_2} = 10.5$ torr, is 100 hr. Calculate the rate constant for the reaction.
b. Determine the rate of the reaction at 23°C and $P_{CH_3CO_3NO_2} = 10.5$ torr.
c. Draw a graph showing P_{PAN} as a function of time from 0 to 200 hr starting with $P_{CH_3CO_3NO_2} = 10.5$ torr.

14.128. Nitric Oxide in the Human Body Nitric oxide (NO) is a gaseous free radical that plays many biological roles including regulating neurotransmission and the human immune system. One of its many reactions involves the peroxynitrite ion (ONOO$^-$):

$$NO(g) + ONOO^-(aq) \rightarrow NO_2(g) + NO_2^-(aq)$$

a. Use the following data to determine the rate law and rate constant of the reaction at the experimental temperature at which these data were generated.

Experiment	[NO]$_0$ (M)	[ONOO$^-$]$_0$ (M)	Rate (M/s)
1	1.25×10^{-4}	1.25×10^{-4}	2.03×10^{-11}
2	1.25×10^{-4}	0.625×10^{-4}	1.02×10^{-11}
3	0.625×10^{-4}	2.50×10^{-4}	2.03×10^{-11}
4	0.625×10^{-4}	3.75×10^{-4}	3.05×10^{-11}

b. Draw the Lewis structure of peroxynitrite ion (including all resonance forms) and assign formal charges. Note which form is preferred.
c. Use the average bond energies in Table 8.2 to estimate the value of $\Delta H°_{rxn}$ using the preferred structure from part b.

14.129. Kinetics of Protein Chemistry In the presence of O$_2$, NO reacts with sulfur-containing proteins to form S-nitrosothiols, such as C$_6$H$_{13}$SNO. This compound decomposes to form a disulfide and NO:

$$2\,C_6H_{13}SNO(aq) \rightarrow 2\,NO(g) + C_{12}H_{26}S_2(aq)$$

a. The following data were collected for the decomposition reaction at 69°C.

Time (min)	[C$_6$H$_{13}$SNO]
0	1.05×10^{-3}
10	9.84×10^{-4}
20	9.22×10^{-4}
30	8.64×10^{-4}
60	7.11×10^{-4}

Calculate the value of the first-order rate constant for the reaction.
b. Which amino acids might act as sources of S-nitrosothiols?

14.130. Solutions of nitrous acid, HNO$_2$, in ^{18}O-labeled water undergo isotope exchange:

$$HNO_2(aq) + H_2^{18}O(\ell) \rightarrow HN^{18}O_2(aq) + H_2O(\ell)$$

a. Use the following data at 24°C to determine the dependence of the reaction rate on the concentration of HNO$_2$.

Time (min)	[HNO$_2$]
0	5.4×10^{-2}
20	1.5×10^{-3}
40	7.7×10^{-4}
60	5.2×10^{-4}

b. Does the reaction rate depend on the concentration of H$_2^{18}$O?

14.131. Ethylene (C$_2$H$_4$) reacts with ozone to form 2 mol of formaldehyde (a probable human carcinogen) per mole of ethylene as shown in Figure P14.131:

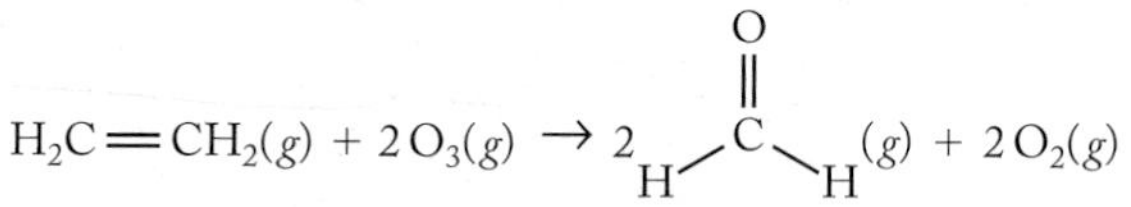

FIGURE P14.131

The following kinetic data were collected at 298 K.

Experiment	[O$_3$]$_0$ (M)	[C$_2$H$_4$]$_0$ (M)	Rate (M/s)
1	0.86×10^{-2}	1.00×10^{-2}	0.0877
2	0.43×10^{-2}	1.00×10^{-2}	0.0439
3	0.22×10^{-2}	0.50×10^{-2}	0.0110

a. Determine the rate law and the value of the rate constant of the reaction at 298 K.
b. The rate constant was determined at several additional temperatures. Calculate the activation energy of the reaction from the following data.

T (K)	k (M^{-1}s^{-1})
263	3.28×10^2
273	4.73×10^2
283	6.65×10^2
293	9.13×10^2

14.132. Reducing NO Emissions Adding NH$_3$ to the stack gases at an electric power generating plant can reduce NO$_x$ emissions. This selective noncatalytic reduction (SNR) process depends on the reaction between NH$_2$ (an odd-electron compound) and NO:

$$NH_2(g) + NO(g) \rightarrow N_2(g) + H_2O(g)$$

The following kinetic data were collected at 1200 K.

Experiment	[NH$_2$]$_0$ (M)	[NO]$_0$ (M)	Rate (M/s)
1	1.00×10^{-5}	1.00×10^{-5}	0.12
2	2.00×10^{-5}	1.00×10^{-5}	0.24
3	2.00×10^{-5}	1.50×10^{-5}	0.36
4	2.50×10^{-5}	1.50×10^{-5}	0.45

a. What is the rate-law expression for the reaction?
b. What is the value of the rate constant at 1200 K?

Chemical Equilibrium

DISRUPTING AN EQUILIBRIUM
Opening a can of soda relieves the pressure that kept CO_2 in solution.

A LOOK AHEAD: The Eyes Have It

People occasionally speak of losing their equilibrium, by which they mean their sense of balance and well-being. Stress is frequently the cause. Whether the stress is mental or physical, people respond to stress by trying to relieve it and thereby return to equilibrium. Economists describe economic equilibrium, when supply and demand are matched and prices are stable. When accidents of life or business intrude on the economic system, equilibrium is disrupted and prices rise or fall in response to the stress. When chemists talk about reactions or processes being at equilibrium, they mean equilibrium at the molecular level: a state of balance in which concentrations of reactants and products remain constant over time. Chemical equilibria are dynamic: reactants are constantly being converted into products, and products are being converted back into reactants. At equilibrium the rates of both of these processes are the same, so no net change takes place.

Our own eyes provide an example of a dynamic equilibrium, the disruption of which can lead to impaired vision and even blindness. An important measurement taken during a routine eye examination is the pressure of the fluid inside the eye. That pressure maintains the shape of the eyeball and also holds the retina in its proper place. The pressure is maintained by a dynamic process in which fluid is constantly produced by clusters of cells just behind the iris. The fluid drains out of the eye into the bloodstream through a network of tiny vessels. In a healthy eye, the amount of fluid produced equals the amount that drains out, so the system maintains a dynamic equilibrium at a pressure where vision functions properly. If the system is disturbed, if too much fluid is produced or not enough drains away, higher pressures develop, and a new equilibrium may be established at a pressure that represents a danger to vision. Dangerously high pressure results in glaucoma, the second leading cause of blindness in the United States. The higher pressure damages the optic nerve that carries visual information to the brain. Treatment consists of reestablishing a safe equilibrium pressure through medication or surgery. The physical equilibrium in our eyes is analogous to chemical equilibrium; both are dynamic processes. Systems at equilibrium remain so unless conditions change. When conditions change, a different equilibrium is established that reflects the altered conditions.

To understand and make best use of a physical or chemical process we need to know about the conditions under which a system reaches equilibrium, and how rapidly the position of equilibrium is established. These two concepts are separate, but knowledge of both is essential. In Chapter 13 we saw that chemical reactions happen for a reason: high-energy reactants form lower-energy products. With spontaneous reactions there is always a decrease in free energy under standard conditions ($\Delta G° < 0$). In Chapter 14 we learned that spontaneous does not necessarily mean rapid; kinetics is as important as thermodynamics in determining which reactions happen and which do not. In this chapter we discuss how the free energy that drives spontaneous reactions is consumed as reactants turn into products. Eventually the free energy is completely used; ΔG goes to 0, and the reaction reaches equilibrium. This chapter together with Chapters 13 (free energy) and 14 (kinetics) gives us the tools to answer three major questions about chemical processes in general: Why do they go at all? Why do they sometimes take a while? And why do they sometimes not go to completion?

15.1 Understanding Equilibrium: A Qualitative View

Being in a state of equilibrium implies a balance of opposing forces. Think about this idea of balance in terms of an unopened can of soda. The soda contains dissolved carbon dioxide, and some CO_2 gas occupies the air space above the liquid in the container. As we have discussed in several earlier chapters, molecules of dissolved gases constantly leave the liquid phase and enter the vapor phase, while other gas molecules leave the vapor phase and enter the solution. At a given temperature and pressure these two opposing processes are in balance and no net change in the dissolved concentration of CO_2 takes place.

When we open the can, suddenly this balance is disrupted and changes occur. The pressure inside the can is relieved, and the soda fizzes and bubbles as the amount of CO_2 leaving the liquid vastly exceeds the amount redissolving. We drink the soda as this process continues and before the soda goes flat and a new balance is established, when much of the CO_2 that was dissolved has escaped into the air outside the can.

The release of CO_2 from a carbonated beverage is based on the solubility of a gas, and so is mostly a physical equilibrium. Chemical reactions also achieve equilibrium. A **chemical equilibrium** is dynamic: reactants are constantly converted into products, and products are converted back into reactants. This reversibility is represented by double-headed reaction arrows:

$$\text{Reactants} \rightleftharpoons \text{products}$$

At equilibrium the rates of both of these processes are the same, so no net change takes place over time:

$$\text{Forward rate (rate}_f\text{)} = \text{reverse rate (rate}_r\text{)}$$

Any chemical reaction carried out in a closed vessel (a closed system) will reach equilibrium, although some may take a long time to do so. For reactions where the position of equilibrium favors products, we say the equilibrium *lies to the right* (the direction of the forward reaction arrow). All or nearly all of the reactants are converted into products. For reactions that occur only to a slight extent, we say the equilibrium *lies to the left* (the direction of the reverse reaction arrow), and the reaction may not produce much product at all. Nothing can be inferred from the position of the equilibrium about the time it takes to achieve equilibrium. Studies of equilibrium reveal the extent to which a reaction proceeds, not how rapidly it proceeds. For that we need kinetics. When we combine our understanding of energy, spontaneity, and ΔG from Chapter 13 with concepts of equilibrium that we develop in this chapter and principles of kinetics from Chapter 14, we then have the answer to the three key questions we posed previously about chemical processes: Does a reaction take place? How quickly does it go? And to what extent does it proceed?

To see how the idea of equilibrium applies to chemical reactions, let's look at a process of considerable commercial importance, the production of hydrogen gas. In Chapter 14 we discussed how catalytic converters reduce photochemical smog by decreasing NO_x emissions from internal combustion engines. The production of nitrogen oxides is highly temperature-dependent, and fuels that burn at a lower temperature than hydrocarbons produce less NO_x. In part because it

▶|| **CHEMTOUR** Equilibrium

A **chemical equilibrium** is a dynamic process in which the concentrations of reactants and products remain constant over time and the rate of a reaction in the forward direction matches its rate in the reverse direction.

○○ **CONNECTION** We introduced the steam-reforming of methane and the water-gas shift reaction in the descriptive chemistry section in Chapter 3.

burns at a lower temperature, hydrogen has been suggested as a possible alternative fuel. Unfortunately hydrogen does not occur in nature as the free element, but rather in compounds such as H_2O and hydrocarbons. If hydrogen is to be used as a fuel it must be recovered from these compounds. A widely used recovery process starts with methane, the most abundant hydrocarbon in natural gas. Hydrogen is obtained from methane in a two-step process. In the first step, which is called the *steam-reforming* reaction, CH_4 reacts with steam on a catalyst made of nickel or iron oxide at temperatures between 700 and 1100°C. The reaction can be written

$$CH_4(g) + H_2O(g) \rightleftharpoons CO(g) + 3\,H_2(g) \qquad (15.1)$$

The carbon monoxide produced in the steam-reforming reaction reacts with more steam in the presence of a Cu/ZnO catalyst at temperatures around 200°C and a pressure of 50 atm to produce carbon dioxide and more hydrogen gas:

$$H_2O(g) + CO(g) \rightleftharpoons H_2(g) + CO_2(g) \qquad (15.2)$$

This reaction is called the *water–gas shift* reaction. The equilibrium arrows in both reactions mean that they are reversible and so may proceed in both the forward and reverse directions.

Let's look at the dynamics of the water–gas shift reaction. If we put an equal number of moles of water vapor and carbon monoxide in a closed chamber and allow them to react, the concentrations of CO and H_2O initially fall as the concentrations of H_2 and CO_2 increase. Because H_2O and CO are present in equimolar quantities and they react in a 1:1 stoichiometric ratio, their concentrations are always equal in the system. Similarly, H_2 and CO_2 are formed in a 1:1 mole ratio, and so they form in equimolar amounts. Figure 15.1 shows the changing concentrations of reactants and products over time. Because the reaction is reversible, products are constantly being formed from reactants, while reactants are constantly being reformed from products. At the point where reactant and product concentrations no longer change, chemical equilibrium has been reached. Note how the concentrations of CO and H_2O do not go to zero. This means that the yield of the reaction is less than 100%.

Now let's think about the *rates* of the forward and reverse reactions as the water–gas shift reaction takes place (Figure 15.2). When the reactants are initially mixed, their concentrations are as high as they will ever be in this system. As reactant molecules collide and form product molecules, reactant concentrations drop. The rate of the forward reaction (rate$_f$) gradually decreases (Figure 15.2) because the likelihood of collisions between reactant molecules decreases as their concentrations decrease.

In contrast, the concentration of products is initially zero and so is the rate of the reverse reaction. As molecules of products form, the likelihood of their colliding with one another to reform reactant molecules increases and so does the rate of the reverse reaction (rate$_r$). Ultimately the reverse rate equals the forward rate and equilibrium is achieved.

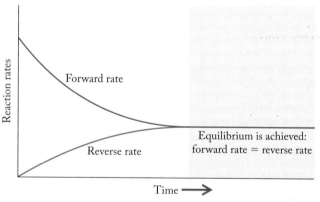

FIGURE 15.1 Concentrations of reactants and products in the water–gas shift reaction change over time until equilibrium is reached.

FIGURE 15.2 The rates of forward and reverse reactions are the same when equilibrium is achieved. At equilibrium in the water–gas shift reaction, H_2 and CO_2 are being formed at the same rate at which they are reacting to reform H_2O and CO.

○○ **CONNECTION** In Chapter 14 we discussed how reactions occur when molecules collide; the higher the concentrations of reactants, the more frequently molecules collide, and the faster a reaction proceeds.

15.2 The Equilibrium Constant Expression and *K*

We have just looked at how a reversible reaction comes to equilibrium as decreasing concentrations of reactants and increasing concentrations of products lead to rates of the forward and reverse reactions that eventually equal each other. Now let's take a more quantitative approach to examining chemical equilibria. We start with data derived from four experiments involving the water–gas shift reaction (see Table 15.1). Each experiment is carried out in a rigid, closed reaction vessel heated to 500 K and is allowed to proceed until there is no further change in the chemical composition of the reaction mixture.

In Experiment 1 the reaction vessel initially contains 0.0200 *M* H_2O and 0.0200 *M* CO. As the water–gas shift reaction proceeds, some, but not all, of these reactants are converted into H_2 and CO_2. In Experiment 2 the reaction vessel initially contains 0.0200 *M* concentrations of H_2 and CO_2. A portion of these gases react with each other, forming H_2O and CO. Look closely at the equilibrium concentrations of reactants and products in Experiments 1 and 2. They are exactly the same. This result indicates that the position of the equilibrium state for a reversible chemical reaction is independent of whether equilibrium is achieved by the reaction proceeding in the forward direction or in reverse. In Experiments 3 and 4, reactants are converted into products even though the initial reaction mixtures contained comparable concentrations of products (Experiment 4) or even higher concentrations of products (Experiment 3) than reactants.

TABLE 15.1 Initial and Equilibrium Concentrations of the Reactants and Products in the Water–Gas Shift Reaction [$H_2O(g)$ + CO(g) $\rightleftharpoons$ $H_2(g)$ + $CO_2(g)$] at 500 K

Experiment	INITIAL CONCENTRATION (*M*)				EQUILIBRIUM CONCENTRATION (*M*)			
	[H_2O]	[CO]	[H_2]	[CO_2]	[H_2O]	[CO]	[H_2]	[CO_2]
1	0.0200	0.0200	0	0	0.0034	0.0034	0.0166	0.0166
2	0	0	0.0200	0.0200	0.0034	0.0034	0.0166	0.0166
3	0.0100	0.0200	0.0300	0.0400	0.0046	0.0146	0.0354	0.0454
4	0.0200	0.0100	0.0200	0.0100	0.0118	0.0018	0.0282	0.0182

The equilibrium compositions of the reaction mixtures in Table 15.1 have three different sets of values; however, all three sets have something in common: if we calculate the product of the equilibrium concentrations of the two products and divide it by the product of the equilibrium concentrations of the two reactants

$$\frac{[H_2][CO_2]}{[H_2O][CO]}$$

we get the same value for every set of experimental data. To prove this point, let's calculate the ratio of the equilibrium concentrations from Experiments 1 and 2 and compare it to the ratio from Experiment 3:

Experiments 1 and 2
$$\frac{[H_2][CO_2]}{[H_2O][CO]} = \frac{(0.0166)(0.0166)}{(0.0034)(0.0034)} = 24$$

Experiment 3
$$\frac{[H_2][CO_2]}{[H_2O][CO]} = \frac{(0.0354)(0.0454)}{(0.0046)(0.0146)} = 24$$

It turns out that we would get the same quotient if we used the data from Experiment 4, or equilibrium concentration data from *any* reaction mixture of these four gases at 500 K.

The constancy of such a ratio of product-to-reactant concentrations at equilibrium is a phenomenon that has been known since the mid-19th century, when Norwegian chemists Cato Guldberg (1836–1902) and Peter Waage (1833–1900)

The **law of mass action** states that an expression relating the concentrations of products to reactants at chemical equilibrium has a characteristic value at a given temperature when each concentration term in the expression is raised to a power equal to the coefficient of that substance in the balanced chemical equation for the reaction.

The **mass action expression** or **equilibrium constant expression** describes the relation between the concentration (or partial pressure) terms of reactants and products when a system is at equilibrium.

The **equilibrium constant (K)** is the value of the ratio of concentration (or partial pressure) terms in the equilibrium constant expression at a specific temperature.

discovered that any reversible reaction eventually reaches a state in which the ratio of the concentrations of products to reactants, each value raised to a power corresponding to the coefficient for that substance in the balanced chemical equation for the reaction, has a characteristic value at a given temperature. They called this phenomenon the **law of mass action**. The ratio of concentration terms for a reaction mixture is called the **mass action expression** of that mixture. It is also called the **equilibrium constant expression**. Its value for a particular reaction at equilibrium at a particular temperature is called the **equilibrium constant (K)** of that reaction at that temperature. For the water–gas shift reaction at 500 K the value of K and the equilibrium constant expression are

$$K = \frac{[CO_2][H_2]}{[CO][H_2O]} = 24$$

Keep in mind that the law of mass action applies to all reaction systems that have reached equilibrium. As we shall see in Section 15.4, mass action expressions are also useful in evaluating the direction of reactions that have yet to achieve equilibrium, but that are on their way to doing so.

The message in Figure 15.2 is that chemical equilibrium is achieved when the rate of a reaction in its forward direction matches the rate of the reaction in the reverse direction. Let's examine that message using the concept of rate laws from Chapter 14. We start with a reaction that has a familiar reactant: NO_2, the gas that gives photochemical smog its distinctive brown color. If we could establish a high concentration of NO_2 in a transparent container, we would see a deep-brown-colored gas. However, over time the color would fade to a less intense brown because of this reversible reaction:

$$2\,NO_2(g) \rightleftharpoons N_2O_4(g)$$
$$\text{(brown)} \quad \text{(colorless)}$$

The color fades because some of the brown NO_2 gas combines to form a colorless dimer, N_2O_4. The fact that some brown color persists tells us that the reaction comes to an equilibrium state at which some NO_2 is still present. Chemists know from experimental data that the rate laws of the forward and reverse reactions are:

$$\text{Rate}_f = k_f[NO_2]^2 \quad \text{and} \quad \text{rate}_r = k_r[N_2O_4]$$

where k_f and k_r are the forward and reverse rate constants. When the reaction achieves chemical equilibrium

$$\text{Rate}_f = \text{rate}_r$$

which means that

$$k_f[NO_2]^2 = k_r[N_2O_4]$$

Let's rearrange this equation so that both rate constants are on the left side

$$\frac{k_f}{k_r} = \frac{[N_2O_4]}{[NO_2]^2}$$

The ratio of two constants is another constant, but in this case not just *any* constant. The ratio on the right is in fact the equilibrium constant expression for the reaction, which means that the ratio of the two *rate* constants at equilibrium is actually the *equilibrium* constant K of the reaction:

$$\frac{k_f}{k_r} = K = \frac{[N_2O_4]}{[NO_2]^2}$$

Remember that the exponents in the equilibrium expression must be the same as the coefficients in the balanced equation. The exponents in rate laws are frequently not the same as the coefficients in the balanced equation, because they describe the stoichiometry of the rate-determining step in the mechanism of the reaction. *So K is not necessarily the ratio of two R's.*

Writing Equilibrium Constant Expressions

The value of the equilibrium constant of a reaction can be calculated from the concentrations of reactants and products in the mass action expression when the reaction is at chemical equilibrium. An equilibrium constant calculated using molar concentrations is identified as K_c. Proportions of reactants and products in a system at equilibrium may also be defined in other ways. For example, equilibrium constants for gas phase reactions may be based on partial pressures of reactants and products and are identified by the symbol K_p. As we will see shortly, values of K_c and K_p for a given reaction may differ, so it is important to use the correct symbol.

The mass action expression in its most general form for the reaction

$$w A + x B \rightleftharpoons y C + z D \qquad (15.3)$$

in terms of molar concentrations is

$$K_c = \frac{[C]^y[D]^z}{[A]^w[B]^x} \qquad (15.4)$$

The square brackets indicate molar concentrations of the species involved in the reaction, and each concentration is raised to a power equal to its coefficient in the balanced chemical equation. Products appear in the numerator and reactants appear in the denominator. If the substances involved in the reaction are all gases (as they are in the steam-reforming and the water–gas shift reactions), then the equilibrium constant can also be expressed in terms of partial pressures:

$$K_p = \frac{(P_C)^y(P_D)^z}{(P_A)^w(P_B)^x} \qquad (15.5)$$

As we have seen in Chapter 6 and Chapter 10, it is common practice to describe amounts of gases in a mixture in terms of their partial pressures. Equation 15.5 is the generalized mass action expression written with respect to partial pressures.

CONNECTION In Chapter 10 we discussed how the partial pressure of O_2 in the atmosphere is related to its concentration in blood, and the impact of this relationship on the oxygenation of tissues at high altitudes.

▶Ⅱ **CHEMTOUR** Equilibrium in the Gas Phase

SAMPLE EXERCISE 15.1   **Writing Equilibrium Constant Expressions**

At high pressures the dark brown gas NO_2 exists in equilibrium with the colorless gas N_2O_4, which is a dimer of NO_2. Write expressions for K_c and K_p for the reaction of NO_2 to produce N_2O_4.

COLLECT AND ORGANIZE We are given the reactant, NO_2, and the product, N_2O_4, and asked to write equilibrium constant expressions for K_c and K_p for this system. These expressions relate the concentrations (K_c) or partial pressures (K_p) of products and reactants.

ANALYZE To write expressions for K, we need a balanced chemical equation. For this reaction, the equation is

$$2 NO_2(g) \rightleftharpoons N_2O_4(g)$$

SOLVE The expression for K_c written in terms of molar concentrations is

$$K_c = \frac{[N_2O_4]}{[NO_2]^2}$$

The expression for K_p written in terms of partial pressures is

$$K_p = \frac{(P_{N_2O_4})}{(P_{NO_2})^2}$$

THINK ABOUT IT The mass action expression can be written for any reaction directly from the balanced chemical equation. K_c is used when concentrations are of interest. K_p is used specifically for gas-phase reactions, when partial pressures of gases are of interest.

Practice Exercise Write the equilibrium constant expressions, K_c and K_p, for the steam-reforming reaction (Equation 15.1):

$$CH_4(g) + H_2O(g) \rightleftharpoons CO(g) + 3\,H_2(g)$$

(Answers to Practice Exercises are in the back of the book.)

CONCEPT TEST

With reference to the water–gas shift reaction on page 739, suppose an equal number of moles of steam, carbon dioxide, carbon monoxide, and hydrogen gas are injected into a rigid, sealed reaction vessel that is heated to 500 K. The mixture of gases is allowed to come to chemical equilibrium. Which of the following expressions about the equilibrium reaction mixture is true?

a. $[CO] = [H_2] = [CO_2] = [H_2O]$
b. $[CO_2] = [H_2] > [CO] = [H_2O]$
c. $[CO_2] = [H_2] < [CO] = [H_2O]$
d. $[H_2O] = [H_2] > [CO_2] = [CO]$
e. $[H_2O] = [H_2] < [CO_2] = [CO]$

(Answers to Concept Tests are in the back of the book.)

Remember that the value of K for a given system at a given temperature is constant. As is almost always the case with actual experimental data, you must be prepared for slight variation in calculated answers because of variations in the precision of the experiment and rounding.

It is important to note at this point that equilibrium constants are reported without units because they involve measurements that account for nonideal behavior of substances participating in the reaction. Concentrations and partial pressures corrected for nonideal behavior are called *activities*, and when equilibrium constants are determined from thermodynamic data, they are defined in terms of activities. For concentration, activity is the ratio of the molar concentration of a substance to a standard concentration of 1 *M*. For pressure, activity is a ratio of the partial pressure of a gas in atmospheres to a standard pressure of 1 atm. Therefore each concentration or pressure in an equilibrium expression is actually a ratio in which the units cancel. We use concentrations and partial pressures, not activities, in this text, which means we assume that the behavior of substances is ideal. Because each term is understood to be a ratio of a concentration or partial pressure to a standard value, the units on each term cancel, which results in the value of K having no units.

CONNECTION In Chapter 5, we defined the standard pressure as 1 bar, but stated that, because that value is so close to 1 atmosphere, any differences in values resulting from the substitution of 1 atm for 1 bar are almost always so small that they may be ignored.

CONNECTION In Chapter 10 we discussed nonideal behavior of salts in solution because of attractive forces between ions, and we calculated the *i* factor to quantify this nonideal behavior in terms of colligative properties.

SAMPLE EXERCISE 15.2   **Calculating K_c**

Table 15.2 contains data from four experiments at 100°C for the dimerization of NO_2 in a rigid, closed container at a high pressure. Calculate the value of K_c for each experiment.

TABLE 15.2 Data at 100°C for the Reaction $2NO_2(g) \rightleftharpoons N_2O_4(g)$

Experiment	INITIAL CONCENTRATION (M)		EQUILIBRIUM CONCENTRATION (M)	
	$[NO_2]$	$[N_2O_4]$	$[NO_2]$	$[N_2O_4]$
1	0.0200	0.0000	0.0172	0.00140
2	0.0300	0.0000	0.0244	0.00280
3	0.0400	0.0000	0.0310	0.00452
4	0.0000	0.0200	0.0310	0.00452

COLLECT AND ORGANIZE We are given a data set that contains initial and equilibrium concentrations of a reactant and product. We are asked to determine the value of the equilibrium constant in each experiment.

ANALYZE We balanced the equation and wrote the equilibrium constant expression for this reaction in Sample Exercise 15.1:

$$2NO_2(g) \rightleftharpoons N_2O_4(g)$$

Because we are given data in terms of concentrations, we use the expression for K_c here.

SOLVE

$$K_c = \frac{[N_2O_4]}{[NO_2]^2}$$

Experiment 1: $K_c = \dfrac{0.00140}{(0.0172)^2} = 4.73$

Experiment 2: $K_c = \dfrac{0.00280}{(0.0244)^2} = 4.70$

Experiment 3: $K_c = \dfrac{0.00452}{(0.0310)^2} = 4.70$

Experiment 4: $K_c = \dfrac{0.00452}{(0.0310)^2} = 4.70$

THINK ABOUT IT The values of K_c are nearly the same. We expect them to be the same because they are all based on experiments run at the same temperature.

Practice Exercise Mixtures of gaseous CO and H_2 called synthesis gas can be used commercially to prepare methanol (CH_3OH), a compound considered an alternative fuel to gasoline. Under equilibrium conditions at 700 K, $[H_2] = 0.074$ mol/L, $[CO] = 0.025$ mol/L, and $[CH_3OH] = 0.040$ mol/L. What is the value of K_c for this reaction at 700 K?

If all reactants and products are gases and we know their partial pressures in the system at equilibrium, we can use those values to calculate the equilibrium constant (K_p) for the reaction.

SAMPLE EXERCISE 15.3   **Calculating K_p from Partial Pressures**

A sealed chamber contains an equilibrium mixture of NO_2 and N_2O_4 at 300°C. The partial pressures of the two gases are P_{NO_2} = 0.101 atm and $P_{N_2O_4}$ = 0.074 atm. What is the value of K_p for the equilibrium between NO_2 and N_2O_4 under these conditions?

COLLECT AND ORGANIZE We are asked to determine the value of the equilibrium constant for the reaction of NO_2 to form N_2O_4. We are given partial pressures at equilibrium of both gases, so the constant is symbolized K_p.

ANALYZE The equation for the reaction is

$$2\,NO_2(g) \rightleftharpoons N_2O_4(g)$$

so the mass action expression for the reaction based on partial pressures is

$$K_p = \frac{(P_{N_2O_4})}{(P_{NO_2})^2}$$

SOLVE Inserting the given partial pressures and solving for K_p we have

$$K_p = \frac{0.074}{(0.101)^2} = 7.3$$

THINK ABOUT IT The mass action expression in terms of K_p has the same form as that for K_c; partial pressures for products are in the numerator and partial pressures for reactants are in the denominator. K_p is calculated using partial pressures instead of concentrations.

Practice Exercise A reaction vessel contains an equilibrium mixture of SO_2, O_2, and SO_3. Given the following partial pressures of these gases: P_{SO_2} = 0.0018 atm, P_{O_2} = 0.0032 atm, and P_{SO_3} = 0.0166 atm, calculate the value of K_p for the reaction:

$$2\,SO_2(g) + O_2(g) \rightleftharpoons 2\,SO_3(g)$$

The size of K gives an indication of how far a reaction proceeds toward making products under a stated set of conditions. Values for equilibrium constants span a huge range. The ones we have seen thus far—K_c for the water–gas shift reaction at 500 K and both K_p and K_c for the dimerization of NO_2 at 100°C—are considered intermediate in size. Their values are close to 1, and comparable concentrations of reactants and products are present at equilibrium. In contrast, the reaction between H_2 and O_2 to form water proceeds explosively, and when the reaction is over, very little starting material is left. As a result, K at 25°C is very large.

$$2\,H_2(g) + O_2(g) \rightleftharpoons 2\,H_2O(g) \qquad K_c = 3 \times 10^{81}$$

On the other side of the range, the decomposition of CO_2 into CO and O_2 at 25°C proceeds hardly at all. K for this equilibrium is very small and virtually no product is formed.

$$2\,CO_2(g) \rightleftharpoons 2\,CO(g) + O_2(g) \qquad K_c = 3 \times 10^{-92}$$

Relationship between K_c and K_p

The values of K_c and K_p for a given reaction may differ numerically, but they are related to each other via the ideal gas equation, $PV = nRT$. If volume (V) is expressed in liters, we can rearrange this equation to express pressure (P) in terms of molar concentration ($n/V = M$):

$$P = \frac{n}{V}RT = MRT$$

For NO_2 and N_2O_4 in the equilibrium in Sample Exercise 15.3

$$P_{NO_2} = \frac{n_{NO_2}}{V}RT = [NO_2]RT$$

and

$$P_{N_2O_2} = \frac{n_{N_2O_4}}{V}RT = [N_2O_4]RT$$

Substituting these values into the expression for K_p we get

$$K_p = \frac{(P_{N_2O_4})}{(P_{NO_2})^2} = \frac{[N_2O_4]RT}{([NO_2]RT)^2} = \frac{K_c}{RT}$$

or

$$K_p(RT) = K_c$$

This equation defines the specific relationship between K_c and K_p for this reaction, but a more general expression can be derived from Equation 15.5

$$K_p = \frac{(P_C)^y(P_D)^z}{(P_A)^w(P_B)^x}$$

In terms of concentrations, the general expression is

$$K_p = \frac{([C]RT)^y([D]RT)^z}{([A]RT)^w([B]RT)^x}$$

$$= \frac{[C]^y[D][C]^y[O]^z}{[A]^w[B]^x} \times \frac{(RT)^{y+z}}{(RT)^{w+x}}$$

$$= K_c(RT)^{(y+z)-(w+x)} \tag{15.6}$$

To understand what this means, look back at the general equation. The term $(y + z)$ is the sum of the coefficients of the gaseous products, and the term $(w + x)$ is the corresponding value for the gaseous reactants. The difference between these two numbers is the power to which RT is raised in Equation 15.6. We can symbolize that term as Δn (change in the number of moles of gas) and write Equation 15.6 as

$$K_p = K_c(RT)^{\Delta n} \tag{15.7}$$

Looking at the steam-reforming and water–gas shift reactions

$$CH_4(g) + H_2O(g) \rightleftharpoons CO(g) + 3\,H_2(g)$$

$$H_2O(g) + CO(g) \rightleftharpoons H_2(g) + CO_2(g)$$

we see that steam-reforming of methane has 4 moles of products $(1 + 3)$ and 2 moles of reactants $(1 + 1)$, while the water–gas shift reaction has the same number of moles of products $(1 + 1 = 2)$ as reactants $(1 + 1 = 2)$. For steam reforming, $\Delta n = 2$, so $K_p = K_c(RT)^2$, and for the shift reaction $\Delta n = 0$ and $K_p = K_c$, because $(RT)^0 = 1$.

SAMPLE EXERCISE 15.4 **Calculating K_c from K_p**

In Sample Exercise 15.3, we calculated K_p for the dimerization of NO_2 to N_2O_4 in a sealed chamber at 300°C. What is the value of K_c for this reaction at 300°C ($K_p = 7.2$)?

COLLECT AND ORGANIZE We are given K_p and asked to calculate K_c for the reaction under the same conditions.

ANALYZE We need a balanced chemical equation to determine the change in the number of moles of gas in going from reactant to product.

$$2NO_2(g) \rightleftharpoons N_2O_4(g)$$

We can determine Δn and use Equation 15.7 to calculate K_c.

SOLVE The balanced equation shows 1 mole of product and 2 moles of reactant, so

$$\Delta n = (\text{moles of product} - \text{moles of reactant}) = 1 - 2 = -1$$

$$K_p = K_c(RT)^{\Delta n}$$

$$T = 300 + 273 = 573\,K$$

$$7.2 = K_c(0.08206 \times 573)^{-1} = \frac{K_c}{(0.08206)(573)}$$

$$K_c = (7.2)(0.08206)(573) = 339$$

THINK ABOUT IT The value of K_c differs from the value of K_p because the total number of moles of gaseous product is not the same as the total number of moles of gaseous reactant.

Practice Exercise For the reaction that produces ammonia from hydrogen and nitrogen at 30°C, the value of K_c is 2.8×10^{-9}. What is K_p for this reaction under the same conditions?

15.3 Manipulating Equilibrium Constant Expressions

CONNECTION We carried out many of these mathematical operations in Chapter 5 in calculating how much heat is generated or absorbed by chemical reactions.

Equilibrium constant expressions may be written in different ways for a given reaction just as chemical equations may be written differently. We can write equilibrium constant expressions for reactions running in reverse, for chemical equations that have been multiplied or divided by a value that gives a key component a coefficient of one, and for overall reactions that are combinations of other reactions.

K for Reverse Reactions

The K values for forward and reverse reactions in an equilibrium are related. For example, an equilibrium system involving A and B could be written

$$A \rightleftharpoons B$$

or

$$B \rightleftharpoons A$$

For example, in Sample Exercise 15.1, we derived the mass action expression for the dimerization reaction

$$2\,NO_2(g) \rightleftharpoons N_2O_4(g)$$

as

$$K_c = \frac{[N_2O_4]}{[NO_2]^2}$$

The equation for the reverse reaction in which NO_2 forms from N_2O_4

$$N_2O_4(g) \rightleftharpoons 2\,NO_2(g)$$

gives rise to the following mass action expression for the decomposition reaction of the dimer:

$$K_c = \frac{[NO_2]^2}{[N_2O_4]}$$

Let's compare the two mass action expressions: the equilibrium constant for the reverse reaction is the reciprocal of the equilibrium constant for the forward reaction. Expressing this relation mathematically gives

$$K_f = \frac{1}{K_r} \qquad (15.8)$$

where the subscripts "f" and "r" indicate the *forward* and *reverse* reaction directions. If a reaction has a large equilibrium constant, then the system at equilibrium contains mostly product and little reactant. In the reverse reaction, there should be little product and mostly reactant at equilibrium and a correspondingly small equilibrium constant. For example, the equilibrium constant (K_c) for the dimerization of NO_2 to N_2O_4 is 4.7 at 100°C. Using Equation 15.8, the value of the equilibrium constant (K_c) for the reverse reaction, the decomposition of N_2O_4 to NO_2, at 100°C is

$$K_r = \frac{1}{K_f} = \frac{1}{4.7} = 0.21$$

The equilibrium constant for the reverse of a reaction is the reciprocal of the equilibrium constant for the reaction in the forward direction. We must always specify the reactants, products, and temperature when reporting K_c or K_p.

SAMPLE EXERCISE 15.5 **Calculating *K* for a Reverse Reaction**

When NO is formed during combustion in automobile engines, it reacts with available O_2 to form NO_2. The reverse reaction is the decomposition of NO_2 to produce NO and O_2. At 184°C, the value of K_c for the forward reaction is 1.48×10^4. Write the mass action expressions for both reactions and determine the value of K_c for the reverse reaction.

COLLECT AND ORGANIZE We are asked to write mass action expressions for a forward and reverse reaction for a process at equilibrium and to calculate the equilibrium constant for the reverse reaction. We are given the reactants and products and the value of K_c for the forward reaction.

ANALYZE We need a balanced chemical equation describing the formation of NO_2 from the reaction of NO and O_2. The reverse reaction is the decomposition of NO_2 to reform the reactants.

Forward reaction: $2\,NO(g) + O_2(g) \rightleftharpoons 2\,NO_2(g)$

Reverse reaction: $2\,NO_2(g) \rightleftharpoons 2\,NO(g) + O_2(g)$

SOLVE The mass action expressions for the forward and reverse reactions are

$$\text{Forward reaction:} \quad K_f = \frac{[NO_2]^2}{[NO]^2[O_2]}$$

$$\text{Reverse reaction:} \quad K_r = \frac{[NO]^2[O_2]}{[NO_2]^2}$$

$$K_r = \frac{1}{K_f} = \frac{1}{1.48 \times 10^4} = 6.8 \times 10^{-5}$$

THINK ABOUT IT The value of K_f is large, so it makes sense that its reciprocal, K_r, is small.

Practice Exercise At 303 K, the equilibrium constant for the gas phase reaction of N_2 and H_2 to form NH_3 is 2.8×10^{-9}. What is the equilibrium constant for the decomposition of ammonia at the same temperature to produce hydrogen and nitrogen?

K for Equations Multiplied by a Number

As we saw in Chapter 5, sometimes reactions are written with fractional coefficients to enable calculation of properties that reflect the behavior of 1 mole of a substance. As an illustration of this, consider the forward reaction in Sample Exercise 15.5:

$$2NO(g) + O_2(g) \rightleftharpoons 2NO_2(g)$$

We could also write this reaction to reflect the preparation of 1 mole of NO_2 by multiplying each coefficient by $\frac{1}{2}$ to give

$$NO(g) + \tfrac{1}{2}O_2(g) \rightleftharpoons NO_2(g) \tag{15.9}$$

Because the mass action expression must be written to match the balanced chemical equation, its form for Equation 15.9 is

$$K_c = \frac{[NO_2]}{[NO][O_2]^{1/2}}$$

If we compare this expression to the mass action expression written for the forward reaction in Sample Exercise 15.5:

$$K_f = \frac{[NO_2]^2}{[NO]^2[O_2]}$$

we see that the equilibrium constant expression for Equation 15.9 is equal to the expression for K_f to the $\frac{1}{2}$ power (the square root of K_f):

$$K_c = \frac{[NO_2]}{[NO][O_2]^{1/2}} = \sqrt{\frac{[NO_2]^2}{[NO]^2[O_2]}} = \sqrt{K_f}$$

In general terms, if the balanced equation for a reaction is multiplied by some factor n, then the value of K for that reaction is raised to the nth power. In the illustration for the reaction of NO_2, the original equation is multiplied by a factor of $\frac{1}{2}$, so the equilibrium constant of the original equation is raised to the $\frac{1}{2}$ power.

| SAMPLE EXERCISE 15.6 | **Calculating *K* for Different Coefficients** |

An important reaction in the formation of atmospheric aerosols of sulfuric acid (and acid rain) is the oxidation of SO_2 to SO_3, which combines with H_2O to form H_2SO_4. One way to write a chemical equation for the oxidation reaction is

Equation 1: $$SO_2(g) + \tfrac{1}{2}O_2(g) \rightleftharpoons SO_3(g)$$

The value of K_c for this reaction at 298 K is 2.8×10^{12}. What is the value of K_c at 298 K for the following reaction?

Equation 2: $$2\,SO_2(g) + O_2(g) \rightleftharpoons 2\,SO_3(g)$$

COLLECT AND ORGANIZE We are given K_c for a reaction that contains fractional coefficients and describes the production of 1 mole of SO_3. The reaction is rewritten so that the coefficients are all whole numbers. We are asked to calculate K_c for the rewritten reaction.

ANALYZE The first equation is multiplied by 2 to generate the second equation. This means that K_c for the first equation must be squared (raised to the second power) to generate K_c for the rewritten equation.

SOLVE

$$K_2 = (K_1)^2$$
$$= (2.8 \times 10^{12})^2 = 7.8 \times 10^{24}$$

THINK ABOUT IT The concentrations (or partial pressures) of reactants and products are raised to a power defined by the coefficients in a balanced chemical equation. Therefore, doubling the coefficients means squaring the value of *K*.

Practice Exercise At 1000 K, the reaction

$$N_2(g) + 3\,H_2(g) \rightleftharpoons 2\,NH_3(g)$$

has a K_c value of 2.4×10^{-3}. What is the value of K_c at 1000 K for the following reaction:

$$\tfrac{1}{3}N_2(g) + H_2(g) \rightleftharpoons \tfrac{2}{3}NH_3(g)$$

Writing these equations may cause you to wonder how the same reaction having the same reactants and products can have two or more equilibrium constant values. Surely the same ingredients should be present in the same proportions at equilibrium no matter how we choose to write a balanced equation for their reaction. They are. The difference in *K* values is not chemical, it is mathematical. It is related to how we use the equilibrium concentrations in calculating their ratios. It is, for example, about choosing to use [NO] and not [NO]² in such a calculation. That choice does not affect the actual concentration of NO at equilibrium. Always remember that the description of an equilibrium must be defined with respect to a specific balanced equation, and that the coefficients in that equation are the powers to which concentrations or partial pressures are raised in the specific expression for *K* in the system described.

Air quality in Los Angeles, California has improved since this photo of "brown LA haze" was taken. The color was caused by high concentrations of NO_2.

Combining *K* Values

In Chapter 5, we applied Hess's law to calculate the enthalpies of combined reactions. We carry out a similar process with equilibrium constants to determine the overall *K* for a reaction that is the sum of two or more other reactions. This technique is especially useful in situations such as the one we faced in the chapter on kinetics involving photochemical smog, where the product of one reaction was frequently the reactant in another reaction occurring at the same time.

Consider two reactions that we discussed in Chapter 14 that are involved in the formation of photochemical smog. The NO produced from N_2 and O_2 by car and truck engines may be further oxidized with additional O_2 to form NO_2. If we sum the two reactions to obtain the overall reaction for the formation of NO_2, we get

$$(1) \quad N_2(g) + O_2(g) \rightleftharpoons 2\,\cancel{NO(g)}$$

$$(2) \quad 2\,\cancel{NO(g)} + O_2(g) \rightleftharpoons 2\,NO_2(g)$$

$$\text{Overall} \quad N_2(g) + 2\,O_2(g) \rightleftharpoons 2\,NO_2(g)$$

The mass action expression for the overall reaction is

$$K_c = \frac{[NO_2]^2}{[N_2][O_2]^2}$$

We can derive this expression from the equilibrium constant expressions for reactions 1 and 2:

$$K_1 = \frac{[NO]^2}{[N_2][O_2]} \quad \text{and} \quad K_2 = \frac{[NO_2]^2}{[NO]^2[O_2]}$$

if we multiply K_1 by K_2

$$K_1 \times K_2 = \frac{[\cancel{NO}]^2}{[N_2][O_2]} \times \frac{[NO_2]^2}{[\cancel{NO}]^2[O_2]} = \frac{[NO_2]^2}{[N_2][O_2]^2} = K_{\text{overall}}$$

This approach works for all series of reactions, and as a general rule

$$K_{\text{overall}} = K_1 \times K_2 \times K_3 \times K_4 \times \ldots \times K_n \qquad (15.10)$$

The overall equilibrium constant for a combination of two or more reactions is the product of the equilibrium constants of the reactions.

The value of K_c for the overall reaction for the formation of NO_2 from N_2 and O_2 at 1000 K is the product of the equilibrium constants for reactions 1 and 2:

$$K_1 = \frac{[NO]^2}{[N_2][O_2]} = 7.2 \times 10^{-9}$$

$$K_2 = \frac{[NO_2]^2}{[NO]^2[O_2]} = 0.020$$

Using Equation 15.10, $K_{\text{overall}} = K_1 \times K_2 = 7.2 \times 10^{-9} \times 0.020 = 1.4 \times 10^{-10}$.

Remember that the equilibrium constant expression for the overall reaction must have the appropriate product and reactant terms for that reaction. Just as with Hess's law in thermodynamics, we may need to reverse an equation or multiply an equation by a factor when we combine it with another to create the equation of interest. If we reverse a reaction, we must take the reciprocal of its *K*; if we multiply a reaction by a constant, we must raise its *K* to that power.

SAMPLE EXERCISE 15.7   **Calculating _K_ by Combining Equilibria**

At 1000 K, the equilibrium constant K_c of the reaction

$$N_2O_4(g) \rightleftharpoons 2NO_2(g)$$

is 1.5×10^6. On the basis of this value and the value of the equilibrium constant for the reaction

$$N_2(g) + 2O_2(g) \rightleftharpoons 2NO_2(g)$$

at 1000 K ($K_c = 1.4 \times 10^{-10}$), calculate the value of the equilibrium constant at 1000 K for the reaction

$$N_2(g) + 2O_2(g) \rightleftharpoons N_2O_4(g)$$

COLLECT AND ORGANIZE We are given two reactions and the values of their equilibrium constants. We need to combine the two reactions in such a way that N_2 and O_2 are on the reactant side of the overall equation and N_2O_4 is on the product side.

ANALYZE The overall reaction is the sum of the reverse of the first reaction and the second reaction as written:

$$\cancel{2NO_2(g)} \rightleftharpoons N_2O_4(g)$$
$$N_2(g) + 2O_2(g) \rightleftharpoons \cancel{2NO_2(g)}$$
$$\overline{\text{Overall } N_2(g) + 2O_2(g) \rightleftharpoons N_2O_4(g)}$$

SOLVE When a reaction is the sum of two other reactions, then $K_{overall} = K_1 \times K_2$. We are given K_2 in the problem for the reaction as used in the combination, but we reversed the first reaction. Therefore we must use the reciprocal of the value of K_1 in our calculation:

$$K_{overall} = \frac{1}{K_1} \times K_2 = \frac{1}{1.5 \times 10^6} \times 1.4 \times 10^{-10} = 9.3 \times 10^{-17}$$

THINK ABOUT IT Remember that values of K always relate to a specific reaction and how it is written. If we alter a reaction by writing its reverse, or multiply it by some factor, we must always make corresponding changes in the value of the equilibrium constant.

Practice Exercise Calculate the value at K_c for the reaction

$$Q(g) + X(g) \rightleftharpoons M(g)$$

from the following information:

$$2M(g) \rightleftharpoons Z(g) \qquad\qquad K_c = 6.2 \times 10^{-4}$$
$$Z(g) \rightleftharpoons 2Q(g) + 2X(g) \qquad K_c = 5.6 \times 10^{-2}$$

To summarize the key points in manipulating equilibrium constants:

- K values are unitless.
- The value of K for a reverse reaction in an equilibrium is the reciprocal of the K of the forward reaction.
- If the original equation describing an equilibrium is multiplied by some factor n, then K for the new expression is calculated by raising the original K to the nth power.

> The **reaction quotient** (Q) is the numerical value of the mass action expression for any values of the concentrations (or partial pressures) of reactants and products. At equilibrium, the value of the reaction quotient (Q) equals that of the equilibrium constant (K).

15.4 Equilibrium Constants and Reaction Quotients

If we were in the business of making hydrogen gas from the water–gas shift reaction,

$$H_2O(g) + CO(g) \rightleftharpoons H_2(g) + CO_2(g)$$

we could start with any combination of the products and reactants and let them come to equilibrium. In point of fact, if we were using this reaction to produce hydrogen, we would start with a reaction mixture that does not contain any H_2. If the concentration of one of the reactants or products in a reversible reaction is zero, the reaction will proceed in the direction that produces some of that substance. If all the reactants and products are present, however, we must apply the law of mass action to determine the direction in which the reaction will proceed. Let's examine the water–gas shift reaction (Figure 15.3a) with this idea in mind. If we sampled the reaction mixture at some time point after CO and H_2O were initially mixed but before equilibrium was achieved, the concentrations of H_2 and CO_2 would be less than their equilibrium concentrations. If we were to solve the mass action expression using the concentrations of substances present at this point, the numerator would be smaller than at equilibrium and the denominator would be larger. Consequently, the value of the expression would be smaller than the equilibrium constant. The value of the mass action expression written for nonequilibrium concentrations is called the **reaction quotient** (Q), and its size relative to K indicates the direction of a reaction on its way to equilibrium. In the example of the water–gas shift reaction in Figure 15.3, there is too much reactant and not enough product in the reaction mixture for the system to be at equilibrium, so the corresponding value of Q is less than K. As the reaction continues, reactants are consumed and products are formed, and the value of Q approaches K. When the reaction reaches equilibrium, $Q = K$ (Figure 15.3b).

The reaction quotient Q may also be larger than K. For the water–gas shift reaction, Q is larger than K if the ratio of the concentration of products to reactants is larger than at equilibrium. This happens, for example, when we start with only products in the reaction vessel. Under those conditions, the numerator in the mass action equation is too large and the reaction runs in the reverse of the written direction (Figure 15.3c), reducing the concentration of H_2 and CO_2 and increasing the concentrations of H_2O and CO until equilibrium is achieved. The system shifts to the left (reactant) side of the equation. The relative values of Q and K and their consequences are summarized in Table 15.3.

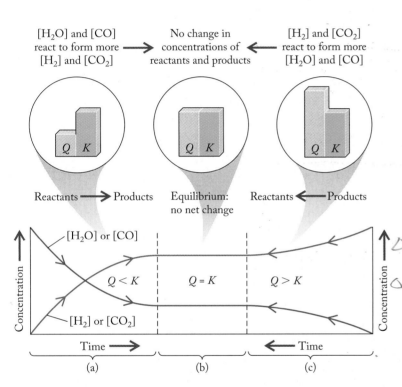

FIGURE 15.3 The value of the reaction quotient Q relative to the equilibrium constant K for the water–gas shift reaction: (a) Reactants are present in higher concentrations than equilibrium concentrations, and products are present in lower concentrations; $Q < K$, and the reactants form more products. (b) Equilibrium concentrations are achieved; $Q = K$, and no net change in concentrations takes place. (c) Products are present in concentrations higher than equilibrium concentrations, and reactants are present in lower concentrations; $Q > K$, and products react to form more reactants.

TABLE 15.3 Comparison of *Q* and *K* Values

Value of Q	What It Means
$Q < K$	Reaction proceeds in the forward direction ($\rightarrow$)
$Q = K$	Reaction is at equilibrium ($\rightleftharpoons$)
$Q > K$	Reaction proceeds in the reverse direction ($\leftarrow$)

Sample Exercise 15.8 shows how we can tell whether a system is at equilibrium or not. If it is not at equilibrium, we can predict the direction in which the system will shift to attain equilibrium.

SAMPLE EXERCISE 15.8   **Using Q and K**

A reaction vessel at 2300 K contains the following molar concentrations of gases:

$[N_2]$	$[O_2]$	$[NO]$
0.50	0.25	0.0042

If $K_c = 1.5 \times 10^{-3}$ for the reaction

$$N_2(g) + O_2(g) \rightleftharpoons 2\,NO(g)$$

at 2300 K, is the reaction mixture at equilibrium? If not, in what direction will the reaction proceed to reach equilibrium?

COLLECT AND ORGANIZE We are asked to evaluate if a reaction is at equilibrium given molar concentrations of the substances involved and the value of K_c under the stated conditions.

ANALYZE The mass action expression for this reaction based on concentrations is

$$K_c = \frac{[NO]^2}{[N_2][O_2]}$$

If we insert the given concentration values into the mass action expression, we can determine the value of Q. Comparing Q to K will enable us to determine if the reaction is at equilibrium or the direction in which the reaction will proceed if it is not.

SOLVE

$$Q_c = \frac{(0.0042)^2}{(0.50)(0.25)} = 1.4 \times 10^{-4}$$

The value of Q_c is less than the value of K_c, so the reaction mixture is not at equilibrium. To achieve equilibrium more of the product must form and the concentration of the reactants must decrease, which happens if the reaction proceeds in the forward direction.

THINK ABOUT IT Determining the reaction quotient Q_c and comparing it to K_c enables us to predict if a reaction proceeds and whether it proceeds in the forward or reverse direction under a given set of concentrations.

Practice Exercise The value of K_c for the reaction

$$2\,NO_2(g) \rightleftharpoons N_2O_4(g)$$

is 4.7 at 373 K. Is a mixture of the two gases in which $[NO_2] = 0.025\ M$ and $[N_2O_4] = 0.0014\ M$ in chemical equilibrium? If not, in which direction does the reaction proceed to achieve equilibrium?

15.5 Equilibrium and Thermodynamics

We have seen that the value of the equilibrium constant for a reaction does not depend on the kinetics of the reaction. Knowing how rapidly a reaction proceeds tells us nothing about the position of equilibrium; correspondingly, knowing the value of the equilibrium constant tells us nothing about how rapidly equilibrium is achieved. However, the equilibrium constant is directly related to the thermodynamics of the reaction and, in particular, to the change in free energy, ΔG.

In Chapter 13 we introduced the change in Gibbs free energy, ΔG, as a thermodynamic function that indicates the tendency of a system to change under conditions of constant temperature and pressure. For example, when two phases are at equilibrium, as are the solid and liquid phases of a pure material at its melting point, the solid melts as rapidly as the liquid freezes, no net change takes place, and the system is at equilibrium. In this situation, $\Delta G = 0$. The same considerations apply to chemical systems, where the rates of forward and reverse reactions at equilibrium are equal, no net change in concentrations of reactants and products takes place, and $\Delta G = 0$.

Because the sign of ΔG indicates whether a process as written is spontaneous, the sign of ΔG for a reaction mixture tells us the direction in which the reaction proceeds to achieve equilibrium. If ΔG is negative, then a reaction is spontaneous as written and proceeds in the forward direction. If ΔG is positive, then the reaction as written is not spontaneous; the reverse of the reaction has a negative ΔG value and is spontaneous, and the reaction of products to form reactants proceeds. As a spontaneous reaction under constant temperature and pressure proceeds, the concentrations of reactants and products change, and the free energy of the system changes as well. At equilibrium, ΔG is zero, no free energy remains to do useful work, and the reaction has achieved equilibrium.

The magnitude of ΔG—how far it is from zero in either a negative or positive direction—indicates how far a system is from its equilibrium position. Whether Q is larger or smaller than K also tells us how far a system is from equilibrium, so it is reasonable to think that Q and ΔG may be related. They are, and their mathematical relationship is one of the most important connections in chemistry because it enables us to relate experimentally measurable properties of substances to the equilibrium composition of a chemical system.

The thermodynamic view of equilibrium and the relationship between ΔG and Q are described by the following equation:

$$\Delta G = \Delta G° + RT \ln Q \qquad (15.11)$$

Let's explore what this means by looking at a reaction we have examined several times in this chapter, the dissociation of N_2O_4 to NO_2:

$$N_2O_4(g) \rightleftharpoons 2\,NO_2(g)$$

We can calculate the change in standard free energy for the reaction ($\Delta G°_{rxn}$) for the conversion of N_2O_4 into NO_2 as we did in Chapter 13 using the formula

$$\Delta G°_{rxn} = \sum \Delta G°_{f,prod} - \sum \Delta G°_{f,react} \qquad (15.12)$$

$$= 2 \text{ mol } (51.3 \text{ kJ/mol}) - 1 \text{ mol } (97.8 \text{ kJ/mol}) = +4.8 \text{ kJ}$$

The positive value of $\Delta G°_{rxn}$ indicates that the reaction, as written, is not spontaneous at 298 K; but understand that this value of $\Delta G°_{rxn}$ applies only under standard conditions when P_{NO_2} and $P_{N_2O_4}$ equal 1 atm. That is not the case when NO_2 and N_2O_4 partial pressures vary.

▶II CHEMTOUR Equilibrium and Thermodynamics

CONNECTION In Chapter 13 we defined Gibbs free energy (G) as the energy available to do useful work. ΔG is a state function that provides a criterion for spontaneous change.

Let's explore what happens in a reaction vessel at 298 K that initially contains 1 mole of N_2O_4 at a partial pressure of 1 atm. While 1 mole of N_2O_4 will never completely turn into 2 moles of NO_2 under these conditions, some fraction (x) of 1 mole of N_2O_4 will decompose into $2x$ moles of NO_2. In other words, the decomposition reaction proceeds a little and then stops as chemical equilibrium is reached. Figure 15.4(a) shows how free energy (G) changes as the amounts of product and reactant change in the reaction mixture during the conversion of N_2O_4 to NO_2 at 25°C and 1 atm pressure. The minimum in the plot of free energy for the reaction (point ③) corresponds to a free energy lower than that of either pure reactant (point ①) or pure product (point ②). In this system under the stated conditions, this minimum occurs for a gaseous mixture of 0.828 moles of N_2O_4 and 0.345 moles of NO_2. To achieve such a mixture, (1.00 − 0.828) or 0.172 moles of N_2O_4 decompose into 0.345 moles of NO_2. In other words, $x = 0.172$ mol.

Now let's continue the story by asking what happens if we begin with 2 moles of pure NO_2. The reaction mixture will approach equilibrium from the other direction, as some amount (2 − y) moles of NO_2 reacts to form $\left(\frac{1}{2}y\right)$ moles of N_2O_4. In this system under the stated conditions, the minimum free energy occurs at exactly the same composition: a gaseous mixture of 0.828 moles of N_2O_4 and 0.345 moles of NO_2. To achieve such a mixture, 1.655 moles of NO_2 combine to form 0.828 moles of N_2O_4, leaving 0.345 moles of NO_2 unreacted.

Both the forward and reverse reactions are spontaneous. When we start with pure N_2O_4, the reaction mixture has too much N_2O_4 and no NO_2. The reaction quotient is smaller than the equilibrium constant ($Q < K$), the system responds to form NO_2 spontaneously, and the reaction proceeds. Correspondingly, when we start with pure NO_2 and no N_2O_4, $Q > K$, the system responds to form N_2O_4 spontaneously, and the reverse reaction proceeds. The reverse reaction goes farther than the forward reaction because it has a more negative $\Delta G°$.

As a reaction takes place spontaneously, free energy always decreases until it reaches its minimum value at the equilibrium concentration of reactants and products. At this point no further change in the composition of the reaction mixture takes place.

To explore how ΔG varies with composition, let's look at the impact of the changing value of Q_c in Equation 15.11 as the decomposition of N_2O_4 proceeds (Figure 15.4b). We can think of ΔG as the "distance" in terms of free energy of a given composition of reactant and product from the equilibrium concentrations. For pure N_2O_4,

$$Q_c = \frac{[NO_2]^2}{[N_2O_4]} = \frac{0}{1} = 0$$

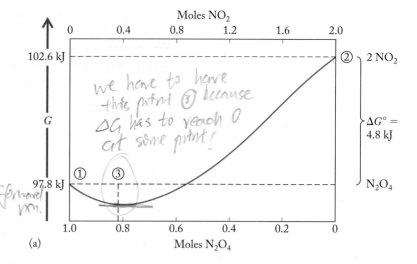

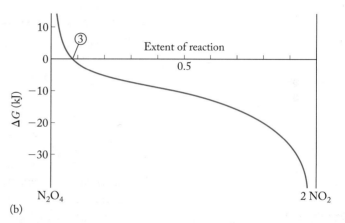

FIGURE 15.4 The value of the change in standard free energy $\Delta G°_{rxn}$ is a constant for a given reaction. (a) The point where free energy is at a minimum defines the composition of a system at equilibrium. At point ①, the system consists of reactants only; at point ②, the system consists of products only. (b) ΔG is the "distance" from equilibrium in terms of free energy. As a spontaneous reaction proceeds, the composition of the system changes and ΔG approaches 0 (point ③), at which point no further change in composition occurs.

CONNECTION The method for calculating free-energy changes from standard free energies of formation is described in Chapter 13. Reversing a reaction changes the sign of its free-energy change.

and, because ln (0) is mathematically undefined, $\Delta G = \Delta G° + RT \ln (0)$ is also undefined. If we think about what happens to the value of the $RT \ln x$ term as x approaches zero, it becomes a larger and larger negative number. Correspondingly for pure NO_2 the value of Q_c is undefined, since

$$Q_c = \frac{[NO_2]^2}{[N_2O_4]} = \frac{2}{0}$$

If we think about what happens to the value of Q_c as the denominator approaches 0, Q becomes a larger and larger positive number, and $\Delta G = \Delta G° + RT \ln Q_c$ increases similarly.

When the system achieves equilibrium

$$\Delta G_{rxn} = \Delta G°_{rxn} + RT \ln \frac{(0.345)^2}{0.828}$$

$$= 4.8 \frac{kJ}{mol} + \left(8.314 \frac{J}{mol \cdot K}\right) \times 298 \, K \times \ln (0.144) \times \left(\frac{1 kJ}{1000 J}\right)$$

$$= (4.8 - 4.8) \frac{kJ}{mol}$$

$$= 0.0 \frac{kJ}{mol}$$

Note that the ΔG value obtained in this calculation is expressed in units of kJ/mol. However, in Chapter 13 and elsewhere in this book we express $\Delta G°$ values as simply kJ. The source of the additional "per mol" unit here, and in calculations of $\Delta G°$, $\Delta H°$, and $\Delta S°$ later in this section, is the units with which we express the constant R. In calculations that relate values of Q or K to changes in enthalpy, entropy, or free energy of a reaction, "per mol" means per mole of the substance(s) in the balanced chemical equation for the reaction with a coefficient of 1. Thus, in the reaction

$$2 NO_2(g) \rightleftharpoons N_2O_4(g)$$

"per mol" means per mole of N_2O_4.

At equilibrium, the value of the reaction quotient Q matches that of the equilibrium constant K and Equation 15.11 becomes

$$\Delta G = \Delta G° + RT \ln K = 0$$

or

$$\boxed{\Delta G° = -RT \ln K} \tag{15.13}$$

Equation 15.13 can be rearranged to allow us to calculate the value of the equilibrium constant for a reaction from its change in standard free energy and absolute temperature. First, we rearrange the terms:

$$\boxed{\ln K = \frac{-\Delta G°}{RT}} \tag{15.14}$$

and then take the antilogs of both sides:

$$\boxed{K = e^{-\Delta G°/RT}} \tag{15.15}$$

Equation 15.15 allows us to calculate the equilibrium constant for a reaction at any temperature from the standard free-energy change. It also provides the following interpretation of reaction spontaneity: if $\Delta G°$ is negative, then $(-\Delta G°/RT)$ is positive. Then $e^{-\Delta G°/RT}$ is greater than one, making K greater than 1. It follows that a reaction with a positive value of $\Delta G°$ has an equilibrium constant K that is less than 1.

To illustrate the connection between $\Delta G°$ and K, let's revisit the formation of NO from molecular nitrogen and oxygen, but this time at room temperature. A balanced equation for the reaction may be written:

$$\tfrac{1}{2}N_2(g) + \tfrac{1}{2}O_2(g) \rightleftharpoons NO(g)$$

The value of $\Delta G°$ for the reaction is equal to the standard free energy of formation of NO, because the reaction describes the formation of one mole of NO from its elements. Thus, the value of $\Delta G°$ at 298 K is the value of $\Delta G°_f$ for NO, which is +86.6 kJ/mol.

We can use Equation 15.13 and this value of $\Delta G°$ to calculate the equilibrium constant for the reaction at 298 K. First, we need to calculate the value of the exponent in Equation 15.15:

$$-\frac{\Delta G°}{RT} = \frac{-\left(\dfrac{86.6\ \cancel{kJ}}{\cancel{mol}}\right)\left(\dfrac{1000\ J}{\cancel{kJ}}\right)}{\left(\dfrac{8.314\ J}{\cancel{mol}\cdot\cancel{K}}\right)(298\ \cancel{K})}$$

$$= -35.0$$

Inserting this value into Equation 15.15 gives

$$K = e^{-\Delta G°/RT} = e^{-35.0} = 6.3 \times 10^{-16}$$

Thus, the positive value of $\Delta G°$ for the reaction (86.6 kJ/mol) can be interpreted two ways:

1. The reaction is not spontaneous.
2. It proceeds in the forward direction to such a small degree (because K is much less than 1) that only a very minute quantity of NO forms.

On the other hand, the reverse reaction in which NO decomposes to give N_2 and O_2

$$NO(g) \rightleftharpoons \tfrac{1}{2}N_2(g) + \tfrac{1}{2}O_2(g)$$

has a negative $\Delta G°_{rxn}$ value at 298 K: −86.6 kJ/mol. This value also can be interpreted in one of two ways:

1. The reaction is spontaneous.
2. It proceeds in the forward direction to such a large degree that only a tiny concentration of NO is present when equilibrium is achieved ($K \gg 1$).

Figure 15.5 shows graphical images of chemical equilibria in three reactions with $\Delta G°$ values that are positive, negative and equal to zero. The relationship between $\Delta G°$ and K is summarized as follows:

$\Delta G°$	K
Zero	1
Negative	> 1
Positive	< 1

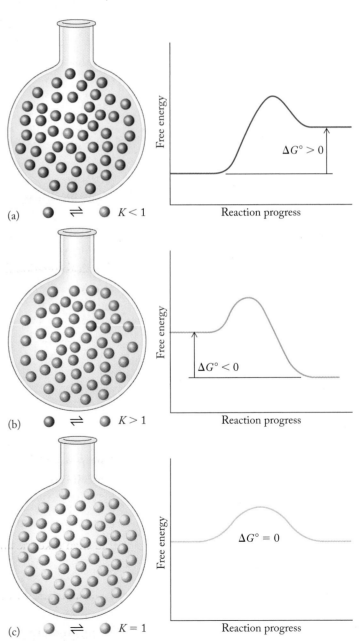

FIGURE 15.5 The value of K for a reaction is related to the change in standard free energy during the course of the reaction. (a) If $K < 1$, then $\Delta G° > 0$. (b) If $K > 1$, then $\Delta G° < 0$. (c) If $K = 1$, $\Delta G° = 0$.

SAMPLE EXERCISE 15.9   **Calculating K from ΔG_f°**

Using the appropriate ΔG_f° values, calculate (a) ΔG_{rxn}° and (b) the equilibrium constant for the formation of NO_2 from NO and O_2 at 298 K.

$$NO(g) + \tfrac{1}{2}O_2(g) \rightleftharpoons NO_2(g)$$

COLLECT AND ORGANIZE We are asked to calculate ΔG_{rxn}° and K for a reaction. We need thermodynamic data from Appendix 4 to calculate ΔG_{rxn}° and Equation 15.14 to calculate K. We can then use the value of ΔG_{rxn}° calculated from Equation 15.14 to calculate the equilibrium constant K.

ANALYZE Calculating the free-energy change of a chemical reaction from free energies of formation (ΔG_f°) involves taking the difference in the sum of the ΔG_f° values of its product(s) minus reactant(s) (Equation 15.12).

$$\Delta G_{rxn}^\circ = \sum n\Delta G_{f,prod}^\circ - \sum m\Delta G_{f,react}^\circ$$

The relation between the standard free-energy change for a reaction and its equilibrium constant is Equation 15.15:

$$K = e^{-\Delta G^\circ/RT}$$

SOLVE Inserting the appropriate values of ΔG_f° from Appendix 4 in Equation 15.12 gives

$$
\begin{aligned}
\Delta G_{rxn}^\circ &= [\Delta G_f^\circ(NO_2)] - [\Delta G_f^\circ(NO) + \tfrac{1}{2}\Delta G_f^\circ(O_2)] \\
&= [1 \text{ mol } (51.3 \text{ kJ/mol})] - [1 \text{ mol } (86.6 \text{ kJ/mol}) + \tfrac{1}{2} \text{ mol } (0.0 \text{ kJ/mol})] \\
&= (51.3 - 86.6) \text{ kJ} \\
&= -35.3 \text{ kJ}
\end{aligned}
$$

or $-35{,}300$ J per mole of NO_2 produced. We can use this value in Equation 15.15 to calculate the equilibrium constant K. First we calculate the value of the exponent at 298 K

$$-\frac{\Delta G^\circ}{RT} = -\frac{\left(\dfrac{-35{,}300 \text{ J}}{\text{mol}}\right)}{\left(\dfrac{8.314 \text{ J}}{\text{mol}\cdot\text{K}}\right)(298 \text{ K})} = 14.2$$

and then we calculate the equilibrium constant

$$K = e^{-\Delta G^\circ/RT} = e^{14.2} = 1.5 \times 10^6$$

THINK ABOUT IT The more negative the free energy change, the larger the equilibrium constant. A moderately large negative free-energy change (like -35.3 kJ here) corresponds to a moderately large equilibrium constant (1.5×10^6).

Practice Exercise The ΔG_f° value for ammonia gas is -16.5 kJ/mol at 298 K. What is the value of the equilibrium constant for the following reaction at 298 K? Note that the equation as written produces 2 moles of NH_3.

$$N_2(g) + 3H_2(g) \rightleftharpoons 2NH_3(g)$$

CONCEPT TEST

What would be the value of K in Sample Exercise 15.9 if ΔG_{rxn}° were equal to $+35.3$ kJ/mol of reactant? Would it be larger or smaller than 1?

We have yet to specify which kind of equilibrium constant, K_c or K_p, is related to $\Delta G°$ by Equation 15.13. The symbol $\Delta G°$ represents a change in free energy under standard conditions. For a gaseous reactant or product, standard conditions mean that its *partial pressure* is 1 atm. Thus, the $\Delta G°$ of a reaction in the gas phase is linked by Equation 15.13 to its K_p value. However, standard conditions for reactions in solution (the focus of Chapter 16) mean that all dissolved reactants and products are present at a *concentration* of 1.00 M. Thus, the $\Delta G°$ of a reaction in solution is related by Equation 15.13 to its K_c value.

> **Homogeneous equilibria** involve reactants and products in the same phase.
>
> **Heterogeneous equilibria** involve reactants and products in more than one phase.

15.6 Heterogeneous Equilibria

Thus far in this chapter we have focused on reactions in the gas phase. However, the principles of chemical equilibrium also apply to reactions in the liquid phase, particularly reactions in solution. Equilibria in which products and reactants are all in the same phase are called **homogeneous equilibria**. Reactions in which reactants and products are in more than one phase at equilibrium are called **heterogeneous equilibria**. In Section 15.3, we considered the equilibrium associated with the oxidation of SO_2 to SO_3, a key step in the formation of H_2SO_4 in the atmosphere and in acid rain. One way to prevent this reaction from happening is to "scrub" SO_2 from the exhaust gases from power plants and factories where sulfur-containing fuels are burned. Solid lime, CaO, is a widely used scrubbing agent. It is sprayed into the exhaust gases and combines in a heterogeneous reaction with SO_2 to form solid calcium sulfite:

> **CONNECTION** Scrubbing exhaust gases with CaO was discussed in Chapter 4.

$$CaO(s) + SO_2(g) \rightarrow CaSO_3(s)$$

The large quantities of lime needed for this reaction and for many other industrial uses come from heating pulverized limestone, $CaCO_3$:

$$CaCO_3(s) \rightleftharpoons CaO(s) + CO_2(g) \qquad \Delta H° = 178.1 \text{ kJ}$$

If we were to write the mass action expression for the decomposition reaction in the usual way we would have

$$K_c = \frac{[CaO][CO_2]}{[CaCO_3]}$$

What are the concentrations of solid CaO and $CaCO_3$? Any pure solid or liquid substance has a constant concentration; the mass per unit volume is always the same. How much CaO or $CaCO_3$ is present while the reaction is coming to equilibrium—as long as some is there—has no influence on the equilibrium (Figure 15.6). Consequently, we remove them from the mass action expression and incorporate them into the value of the equilibrium constant, creating the equilibrium constant we measure:

$$K_c = [CO_2]$$

This expression means that, as long as some CaO and $CaCO_3$ are present, the concentration of CO_2 does not vary at a given temperature. Heating $CaCO_3$ in a closed vessel results in the formation of CaO and a molar concentration of CO_2 specified by the value of K_c at the experimental temperature. If the volume of the vessel is doubled at the same temperature, then twice as much $CaCO_3$ will decompose and $[CO_2]$ will rise until it is equal to K_c. The equilibrium value of $[CO_2]$ is independent of the amounts of $CaCO_3$ and CaO as long as both of these solids are present.

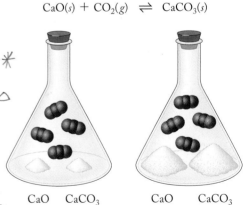

$$CaO(s) + CO_2(g) \rightleftharpoons CaCO_3(s)$$

CaO $CaCO_3$ CaO $CaCO_3$

FIGURE 15.6 The position of a heterogeneous equilibrium between $CaCO_3$, CaO, and CO_2 at constant temperature depends only on the concentration of gas present. As long as some solid is in the system, the equilibrium concentration of CO_2 remains the same, regardless of the quantity of solid present.

The convention of assigning activity values of "1" to pure solids also applies to pure liquids. Thus, mass action expressions of most reactions in aqueous solutions do not include a term for [H_2O] even when water is a reactant or product. The reason is that the concentration of water usually does not change significantly. However, the concentrations of solutes dissolved in water (or other solvents) are included in mass action expressions because the values of these concentrations do change as the reactions proceed. In writing equilibrium constant expressions for heterogeneous equilibria, we follow the same rules as for any other equilibrium, with the additional rule that pure liquids and solids do not appear in the expression.

SAMPLE EXERCISE 15.10   **Writing Mass Action Expressions for Heterogeneous Equilibria**

Write K_c expressions for the following reactions:

a. $CaO(s) + SO_2(g) \rightleftharpoons CaSO_3(s)$
b. $CO_2(g) + H_2O(\ell) \rightleftharpoons H_2CO_3(aq)$

COLLECT AND ORGANIZE We are asked to write mass action expressions for equilibria involving reactants and/or products in more than one phase. We need to identify pure liquids and pure solids that may be involved in the equilibrium. While the quantities of these substances change, their concentrations do not, and concentration terms for them do not appear in the equilibrium constant expression.

ANALYZE The first equilibrium involves two pure solids: CaO and $CaSO_3$. The second involves pure liquid H_2O.

SOLVE

a. If we apply the conventional approach to writing a K_c expression for the reaction, we have

$$K_c = \frac{[CaSO_3]}{[CaO][SO_2]}$$

However, the concentrations of the two solids are constants, and they do not belong in a K_c expression, leaving only the [SO_2] term:

$$K_c = \frac{1}{[SO_2]}$$

b. In the equilibrium constant expression for this reaction, [H_2O] is a constant and is not included, leaving

$$K_c = \frac{[H_2CO_3]}{[CO_2]}$$

THINK ABOUT IT The equilibrium constant expression includes all concentrations that can be changed or can vary. This excludes all pure liquids and pure solids, whose concentrations are constant.

Practice Exercise Write K_p expressions for the following reactions:

a. $C(s) + CO_2(g) \rightleftharpoons 2\,CO(g)$
b. $CO_2(g) + H_2(g) \rightleftharpoons CO(g) + H_2O(\ell)$

Explain why $K_c = [H_2O(g)]$ is the mass action expression for the following equilibrium:

$$H_2O(\ell) \rightleftharpoons H_2O(g)$$

15.7 Le Châtelier's Principle

We can "perturb" chemical reactions at equilibrium by adding more of a reactant or product. For example, increasing the concentration of a product causes the reaction to run in reverse, converting some of the added product to reactants. This response to perturbation, or stress, is described in a principle named after French chemist Henri Louis Le Châtelier (1850–1936). **Le Châtelier's principle** states that if a system at equilibrium is subjected to a stress, the position of the equilibrium shifts in the direction that relieves that stress.

Effects of Adding or Removing Reactants or Products

When a reactant or product is added or removed, a system at chemical equilibrium is perturbed and responds to restore balance under the new conditions. To explore how industrial chemists make use of this phenomenon, let's revisit the water–gas shift reaction to produce hydrogen gas:

▶ll **CHEMTOUR** Le Châtelier's Principle

$$H_2O(g) + CO(g) \rightleftharpoons H_2(g) + CO_2(g)$$

To shift the equilibrium toward the production of H_2, the reaction mixture is passed through a scrubber containing a concentrated aqueous solution of K_2CO_3, which combines with and removes CO_2:

$$CO_2(g) + H_2O(\ell) + K_2CO_3(aq) \rightleftharpoons 2\,KHCO_3(s)$$

Removing CO_2 from the equilibrium of the water–gas shift reaction causes the reaction to shift to the right and form more product. Ultimately the steam and carbon monoxide can be converted completely to carbon dioxide and hydrogen.

If we examine the mass action expression for the water–gas shift reaction

$$K = \frac{[H_2][CO_2]}{[H_2O][CO]}$$

we can see that if CO_2 is removed from a reaction mixture at equilibrium, the numerator will be too small ($Q < K$) and the reaction will respond by shifting to the right, consuming CO and H_2O and producing more CO_2 and H_2 to restore equilibrium ($Q = K$). Continuously removing CO_2 allows the reaction producing H_2 to be driven to completion.

SAMPLE EXERCISE 15.11   **Adding or Removing Reactants or Products to Stress an Equilibrium**

Consider the reaction for the production of ammonia from N_2 and H_2:

$$N_2(g) + 3\,H_2(g) \rightleftharpoons 2\,NH_3(g)$$

Suggest three ways the production of ammonia could be increased by changing the concentration of reactants or products.

Le Châtelier's principle states that a system at equilibrium responds to a stress in such a way that it relieves that stress.

COLLECT AND ORGANIZE We are asked to suggest three ways to increase the production of ammonia by changing concentrations of reactants and products. We can write the mass action expression for the process to help us with our suggestions.

ANALYZE The mass action expression for the reaction is

$$K = \frac{[NH_3]^2}{[N_2][H_2]^3}$$

The concentration of ammonia is in the numerator; those of nitrogen and hydrogen are in the denominator. Any action that decreases the size of the numerator or increases the size of the denominator makes Q for the new conditions smaller than K and stresses the equilibrium in the direction of forming more product.

SOLVE If we (1) increase the concentration of nitrogen or (2) increase the concentration of hydrogen, the equilibrium will move to the right and produce more product. If we (3) remove NH_3 from the reaction as it runs, the equilibrium will also move to the right and more NH_3 will be produced.

THINK ABOUT IT Decreasing the concentration of a substance involved in an equilibrium shifts the equilibrium toward the production of more of that substance. Correspondingly, increasing the concentration of a substance shifts the equilibrium so that more of that substance is consumed in the reaction.

Practice Exercise Describe the changes that occur in the ammonia equilibrium in Sample Exercise 15.11 if the following changes are carried out: (a) The equilibrium mixture is chilled to a temperature at which ammonia liquefies; (b) ammonia is added to the system; (c) the concentrations of both hydrogen and nitrogen are decreased.

Effects of Changes in Pressure and Volume

Being able to predict the behavior of an equilibrium system under stress is an important skill. Let's examine some stresses to chemical systems at equilibrium and the responses that the stresses induce. Figure 15.7 illustrates one system's response. This one is based on the equilibrium between NO_2 and its dimer, N_2O_4, discussed in Sample Exercise 15.1:

$$2 NO_2(g) \rightleftharpoons N_2O_4(g)$$

The images in Figure 15.7 show a reaction mixture of the two gases at equilibrium. The brown color indicates the presence of some NO_2, but it is not the dark brown of pure NO_2. Some colorless N_2O_4 must also be present. When the mixture is compressed, the pressure increases and the volume decreases. The equilibrium is stressed, and the system responds to reduce the pressure and reestablish a new equilibrium position. According to the ideal gas law, the pressure of a gas at constant temperature and volume is proportional to the number of moles of gas in the mixture:

$$PV = nRT \quad \text{or} \quad P = \left(\frac{RT}{V}\right)n$$

FIGURE 15.7 (a) A gas-tight syringe contains an equilibrium reaction mixture of NO_2 gas, which is brown, and N_2O_4 gas, which is colorless. (b) The plunger is pushed in, increasing the pressure. The color of the gas mixture is temporarily darker as the molecules of NO_2 are compressed into a smaller volume. (c) As time passes, the color of the mixture fades as brown NO_2 forms colorless N_2O_4. Two moles of reactant (NO_2) are consumed for every one mole of N_2O_4 formed. The total number of moles of gas in the syringe is reduced, partly relieving the increase in pressure.

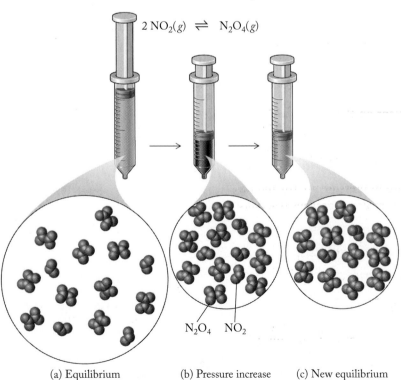

$$2 NO_2(g) \rightleftharpoons N_2O_4(g)$$

N_2O_4 NO_2

(a) Equilibrium   (b) Pressure increase   (c) New equilibrium

Therefore, one way to reduce the total pressure of a mixture of gases is to reduce the total number of moles of gas in the mixture. Because 2 moles of NO_2 combine to form only 1 mole of N_2O_4, making N_2O_4 reduces the pressure. Consequently, the number of moles of gas in a mixture of NO_2 and N_2O_4 can be reduced if the equilibrium shifts in the direction of making more N_2O_4. Compressing the mixture produces an instantaneous darkening of its color (Figure 15.7b) as more molecules of NO_2 are squeezed into a smaller volume, thereby increasing their concentration. Then the color lightens, as shown in Figure 15.7(c), as the equilibrium shifts to form more N_2O_4 and partly relieves the pressure.

Changing pressure shifts the equilibrium of any reaction of gases in which there is a change in the number of moles of gas as the reaction proceeds. *Increased* pressure shifts the equilibrium toward the side of the reaction with *fewer* moles of gas. A *decrease* in pressure shifts the equilibrium toward the side of the reaction with *more* moles of gas.

SAMPLE EXERCISE 15.12 **Assessing Pressure Effects on Gas Phase Equilibria**

only gas

In which of the following equilibria would an increase in ~~pressure~~ promote the formation of more product(s)?

a. $N_2(g) + O_2(g) \rightleftharpoons 2NO(g)$

b. $2NO(g) + O_2(g) \rightleftharpoons 2NO_2(g)$ ✓

c. $N_2O_4(g) \rightleftharpoons 2NO_2(g)$

d. $H_2O(\ell) + CO_2(g) \rightleftharpoons H_2CO_3(aq)$

e. $CaCO_3(s) \rightleftharpoons CaO(s) + CO_2(g)$

COLLECT AND ORGANIZE We are asked to identify the reactions for which an increase in pressure causes an increase in product formation.

ANALYZE Gas phase equilibria are sensitive to pressure changes when there are different numbers of moles of gaseous reactants and products in the balanced chemical equation. We should look for such differences in the reactions in question.

SOLVE Comparing the number of moles of gas present on the reactant side and product side in each reaction, we see that

(a) $2 \rightarrow 2$; (b) $3 \rightarrow 2$; (c) $1 \rightarrow 2$; (d) $1 \rightarrow 0$; and (e) $0 \rightarrow 1$.

The only two forward reactions favored by an increase in pressure are (b) and (d) because they are the only two in which there are more moles of gaseous reactants than products.

THINK ABOUT IT Reactions (c) and (e) will shift from right to left to increase the partial pressures of the reactants if the total pressure on their systems is increased. The only reaction that is not affected by pressure changes is reaction (a), which involves equal numbers of moles of reactant and product.

Practice Exercise How does an increase in pressure affect the equilibrium $N_2(g) + 3H_2(g) \rightleftharpoons 2NH_3(g)$?

CONNECTION In Chapter 5 we defined exothermic reactions as those giving off heat. They have a negative enthalpy change ($\Delta H < 0$), while endothermic reactions absorb heat ($\Delta H > 0$).

Effect of Temperature Changes

If we think of heat, as we did in Chapter 5, as either a reactant or a product in a chemical reaction, then every equilibrium involves one exothermic reaction and one endothermic reaction. For example, in the reaction to produce ammonia, the forward reaction is exothermic,

$$N_2(g) + 3H_2(g) \rightleftharpoons 2NH_3(g) + \text{heat}$$

and the reverse reaction is endothermic. If we think about heat as a product in this reaction, we can use Le Châtelier's principle to predict the result of a temperature change. Increasing the temperature favors the reverse reaction; decreasing the temperature favors the forward reaction. One major difference arises, however; unlike changes in concentration or pressure, temperature changes cause changes in the value of K. An increase in the temperature would cause the concentration of ammonia to drop and the concentrations of nitrogen and hydrogen to rise. A new equilibrium characterized by a smaller K would be established.

In general, the value of K for exothermic reactions decreases with increasing temperature and increases if the temperature decreases. The equilibrium constant for endothermic reactions has the opposite behavior: K increases with increasing temperature and decreases as the temperature falls. At the end of this chapter, we will look in greater detail at the influence of temperature changes on K values, but for now this general analysis will enable us to explain stress on equilibria caused by changing temperature.

SAMPLE EXERCISE 15.13 **Predicting Changes in Equilibria with Temperature**

The color of an aqueous acidic solution of $CoCl_2$ depends on the temperature. In aqueous HCl, the solution is pink when ice-cold, magenta at room temperature, and dark blue at elevated temperature. The equilibrium can be described as

$$\text{Pink} \rightleftharpoons \text{blue}$$

Is the reaction as written exothermic or endothermic?

COLLECT AND ORGANIZE We are asked to determine if an equilibrium as written is exothermic or endothermic (is heat a reactant or product?).

ANALYZE If heat is a product, the reaction is exothermic; if it is a reactant, the reaction is endothermic. If we examine how temperature changes alter the position of equilibrium, we can determine the role of heat in the reaction.

SOLVE When heat is added to the system, the pink color turns blue. That means heat is a reactant, because increasing it pushes the reaction to the right. The reaction as written must be endothermic:

$$\text{Heat} + \text{pink} \rightleftharpoons \text{blue}$$

THINK ABOUT IT To determine the effect of increasing or decreasing temperature on an equilibrium, we can treat heat as a reactant or product if we know whether the reaction as written is endothermic or exothermic.

Practice Exercise Predict how the value of the equilibrium constant of the following reaction changes with increasing temperature:

$$N_2(g) + O_2(g) \rightleftharpoons 2NO(g) \qquad \Delta H° = 181 \text{ kJ}$$

Temperature = 5°C

Temperature = 75°C

$$\underbrace{Co(H_2O)_6{}^{2+}(aq) + 4Cl^-(aq)}_{\text{Pink}} \rightleftharpoons \underbrace{CoCl_4{}^{2-}(aq) + 6H_2O(\ell)}_{\text{Royal blue}}$$

Two different forms of cobalt, one pink and one blue, are in equilibrium in aqueous hydrochloric acid solution. The position of equilibrium shifts as the temperature is changed, causing the color of the solution to change.

Table 15.4 lists examples of stress on chemical equilibria and how a system at equilibrium responds for an exothermic reaction in which 2 moles of gas "A" combine to make 1 mole of gas "B".

TABLE 15.4 Responses of an Exothermic Reaction [2 A(g) ⇌ B(g)] at Equilibrium to Different Kinds of Stress

Kind of Stress	How Stress Is Relieved	Direction of Shift
Add A	Remove A	To the right
Remove A	Add A	To the left
Remove B	Add B	To the right
Add B	Remove B	To the left
Increase temperature by adding heat	Remove some of the heat	To the left
Decrease temperature by removing heat	Add heat	To the right
Increase pressure	Remove A to relieve pressure increase	To the right
Decrease pressure	Add A to maintain equilibrium pressure	To the left

CONCEPT TEST

Describe the effects of removing or adding a reactant to an equilibrium mixture. Indicate how the composition of the mixture changes as the system relieves the stresses caused by these changes.

Let's return to the industrial preparation of ammonia

$$N_2(g) + 3\,H_2(g) \rightleftharpoons 2\,NH_3(g) \qquad \Delta H^\circ_{rxn} = -92.2 \text{ kJ}$$

in which manufacturers take advantage of changes in concentration, pressure, and temperature and apply Le Châtelier's principle to obtain more of the desired product. The process typically uses reaction vessels that operate at 200 atm of pressure. To understand why, note that a total of 4 moles of reactant gases combines to make only 2 moles of product. When the reaction is carried out in a rigid, sealed chamber, the pressure inside the chamber decreases as the reaction proceeds because the total number of moles of gas in the chamber decreases. Running the reaction at high pressure puts a stress on the system that is relieved by making more ammonia.

The reaction is also typically run at high temperatures, which may seem counterintuitive considering the effect of temperature on the equilibrium. The reaction is exothermic, so we might initially assume that cooling the reaction would produce more product. Remember this analysis is based solely on equilibrium. However, kinetics also plays an important role. At room temperature the rate of the reaction is extremely slow. Because the yield of product (equilibrium) and the rate of production of product (kinetics) must both be considered, a compromise must be found between kinetics and equilibrium.

To make a commercially viable amount of ammonia, a hot (400°C) reaction mixture is pumped through a system of chilled condensers, where ammonia, which has a much higher boiling point than either hydrogen or nitrogen, condenses as a liquid and is removed from the gaseous reaction mixture. The removal of ammonia stresses the equilibrium toward the production of more ammonia; the high pressure stresses the equilibrium in the same direction; the high temperature stresses the equilibrium in the reverse direction but means that more molecules collide with enough energy to overcome the activation energy barrier.

Catalysts and Equilibrium

The industrial production of ammonia described in the previous section was developed by the German chemists Fritz Haber (1868–1934) and Carl Bosch (1874–1940) in the early 20th century and is still widely referred to as the Haber–Bosch process. What makes the process commercially feasible is the use of special catalysts that accelerate the reaction. As discussed in Chapter 14, a catalyst increases the rate of a chemical reaction by lowering its activation energy. The question is this: If a catalyst increases the rate of a reaction, does a catalyst affect the equilibrium constant of a reaction?

To answer this question, consider the catalyzed and uncatalyzed energy profiles of the exothermic reaction in Figure 15.8. The catalyst increases the rate of the reaction by decreasing the height of the activation energy barrier. The height of the barrier is reduced by the same amount whether the reaction proceeds in the forward direction or in reverse. As a result, the increases in reaction rates produced by the catalyst are the same in both directions. Therefore a catalyst has no effect on the equilibrium constant of a reaction or the composition of an equilibrium reaction mixture, but it does decrease the time it takes for a reaction to reach equilibrium.

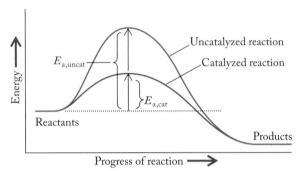

FIGURE 15.8 The graph shows the effect of the presence of a catalyst on a reaction. A catalyst lowers the activation energy barrier by lowering the energy of the transition state; therefore the rate of the reaction increases. Both forward and reverse reactions occur more rapidly, but the position of equilibrium does not change. The system simply comes to equilibrium more rapidly.

CONNECTION In Chapter 14 we discovered that a catalyst simultaneously increases the rate of a reaction in both the forward and the reverse directions.

15.8 Calculations Based on K

Reference books and the tables in Appendix 5 of this book contain lists of equilibrium constants for chemical reactions. These values are used in several kinds of calculations, including those in which

1. We know the equilibrium concentrations of the reactants and products and we wish to calculate the equilibrium constant, as we did for K_c in the formation of N_2O_4 from NO_2 (Sample Exercise 15.2)
2. We want to determine whether a reaction mixture has reached equilibrium, as in Sample Exercise 15.8
3. We know the value of K and the starting concentrations or partial pressures of reactants and products, and we want to calculate the reactant and product concentrations or pressures at equilibrium.

CHEMTOUR Solving Equilibrium Problems

In this section we'll focus on the third kind of calculation. In our first example, we will calculate how much nitrogen monoxide (NO) forms in a sample of air heated to a temperature at which the value of K_p is 1.00×10^{-5} for NO formation from N_2 and O_2:

$$N_2(g) + O_2(g) \rightleftharpoons 2NO(g)$$

The initial partial pressures of N_2 and O_2 are 0.79 and 0.21 atm, respectively, and we assume there is no NO present initially.

The equilibrium constant expression for the formation of NO from N_2 and O_2 is

$$K_p = \frac{(P_{NO})^2}{(P_{N_2})(P_{O_2})}$$

An ICE table is an excellent tool for calculating equilibrium concentrations. The acronym ICE stands for the *i*nitial partial pressures or concentrations of reactants and products, the *c*hanges in their pressures or concentrations as the reaction proceeds to equilibrium, and their pressures or concentrations after *e*quilibrium has been achieved. I, C, and E are the rows of the table, and each reactant and product has a column. Our ICE table for the reaction to produce NO has three rows and three columns.

We start by filling in the initial partial pressures of the reactants and product in the first row from the information given in the example:

	P_{N_2} (atm)	P_{O_2} (atm)	P_{NO} (atm)
Initial	0.79	0.21	0
Change			
Equilibrium			

We know the equilibrium will shift from left to right because there is no NO initially present, so $Q = 0$ and $Q \ll K$.

We need to do some basic algebra to fill in the remaining rows. We don't know how much N_2 or O_2 will be consumed or how much NO will be made. We can define the change in N_2 as $-x$, because N_2 is consumed during the reaction. The mole ratio of N_2 to O_2 in the reaction is 1:1; so the change in O_2 is also $-x$. Two moles of NO are produced from each mole of N_2 and O_2, so the change in P_{NO} is $+2x$. Inserting these values in the second row of the ICE table, we have

	P_{N_2} (atm)	P_{O_2} (atm)	P_{NO} (atm)
Initial	0.79	0.21	0
Change	$-x$	$-x$	$+2x$
Equilibrium			

Combining the *initial* and *change* rows of the ICE table, we obtain the following partial pressures of all three gases at equilibrium:

	P_{N_2} (atm)	P_{O_2} (atm)	P_{NO} (atm)
Initial	0.79	0.21	0
Change	$-x$	$-x$	$+2x$
Equilibrium	$0.79 - x$	$0.21 - x$	$2x$

The last row of partial-pressure terms can be substituted into the K_p equation for the reaction:

$$K_p = \frac{(P_{NO})^2}{(P_{N_2})(P_{O_2})}$$

$$= \frac{(2x)^2}{(0.79 - x)(0.21 - x)}$$

$$= 1.00 \times 10^{-5}$$

Multiplication of the binomial terms in the denominator and setting the resulting fraction equal to the value of K_p gives

$$K_p = 1.00 \times 10^{-5} = \frac{4x^2}{0.1659 - 1.00x + x^2}$$

Cross-multiplying, we get

$$1.659 \times 10^{-6} - (1.00 \times 10^{-5})x + (1.00 \times 10^{-5})x^2 = 4x^2$$

Combining the x^2 terms and rearranging, we have

$$3.99999x^2 + (1.00 \times 10^{-5})x - 1.659 \times 10^{-6} = 0$$

You may recognize this equation as one that fits the general form of a quadratic equation:

$$ax^2 + bx + c = 0$$

which can be solved for x with a scientific calculator or using the quadratic formula

$$x = \frac{-b \pm \sqrt{b^2 - 4ac}}{2a}$$

Two values are possible for x, but only one is positive: 6.428×10^{-4} atm. A gas can't have a negative partial pressure, so we use only the positive value to calculate equilibrium partial pressures:

$$P_{O_2} = 0.21 - x$$

$$= 0.21 - (6.428 \times 10^{-4}) = 0.21 \text{ atm (to two significant figures)}$$

$$P_{N_2} = 0.79 - x$$

$$= 0.79 - (6.428 \times 10^{-4}) = 0.79 \text{ atm}$$

$$P_{NO} = 2x$$

$$= 2(6.428 \times 10^{-4}) = 1.2855 \times 10^{-3} = 0.0013 \text{ atm}$$

The equilibrium partial pressures of the reactants hardly differ from their initial partial pressures. The tiny decreases make sense given the small value (1.0×10^{-5}) of the equilibrium constant. Because the x terms in the denominator of the equilibrium constant equation are small compared with the initial partial pressures, a much simpler approach to calculating P_{NO} is possible: we ignore the x terms in the denominator and use the initial values for P_{N_2} and P_{O_2} instead, to give

$$K_p = 1.0 \times 10^{-5} = \frac{(P_{NO})^2}{(P_{N_2})(P_{O_2})}$$

$$= \frac{4x^2}{(0.79 - x)(0.21 - x)} \cong \frac{4x^2}{(0.79)(0.21)} = \frac{4x^2}{0.1659}$$

$$4x^2 = (0.79)(0.21)(1.0 \times 10^{-5})$$

$$= 1.659 \times 10^{-6}$$

$$x^2 = 4.148 \times 10^{-7}$$

$$x = 6.440 \times 10^{-4} \text{ atm}$$

Note that the difference between this value of x and that obtained from the solution to the quadratic equation (6.428×10^{-4}) is

$$(6.440 \times 10^{-4}) - (6.428 \times 10^{-4}) = 1.2 \times 10^{-6} \text{ atm}$$

Dividing this difference by the original value of x, we have a relative difference of only

$$\frac{1.2 \times 10^{-6} \text{ atm}}{6.428 \times 10^{-4} \text{ atm}} \times 100\% = 0.19\%$$

A small difference like this is not significant when we know the initial partial pressures to only two significant figures. Generally speaking, we can ignore the x component of the concentration or partial-pressure terms for reactants if the value of x is less than 5% of the smallest initial value. Based on the small value of K_p (1.00×10^{-5}) for this reaction, we can say that $0.79 - x \approx 0.79$ and $0.21 - x \approx 0.21$ to avoid more complicated math and get very close to the correct answer.

Another situation where the math can be simplified is when the initial concentrations of the reactants are the same. For example, suppose we have a vessel containing a mixture of 0.100 M concentrations of N_2 and O_2 at a temperature where K_c for the same reaction is 0.100. What is the equilibrium concentration of NO?

$$N_2(g) + O_2(g) \rightleftharpoons 2\,NO(g)$$

Our ICE table looks like this:

	$[N_2]$ (M)	$[O_2]$ (M)	[NO] (M)
Initial	0.100	0.100	0
Change	$-x$	$-x$	$+2x$
Equilibrium	$0.100 - x$	$0.100 - x$	$2x$

Inserting the values from the E line into the expression for K_p gives

$$K_c = \frac{(2x)^2}{(0.100 - x)(0.100 - x)} = 0.100$$

There are two possibilities for simplifying the math and getting to the value of x. One is to "drop the x," which means we would put $0.10 - x \approx 0.10$ in the denominator of the K_c expression. This should not be done, considering that K_c is 1.00×10^{-1}, which means that the equilibrium is further to the right than it is with the $K_c = 1.00 \times 10^{-5}$ we had before. Another alternative is to note that the K_c expression has the form $(X)^2/(Y)^2 = Z$, where $X = 2x$, $Y = 0.10 - x$ and $Z = K_c$. Taking the square root of each side gives

$$\frac{2x}{0.100 - x} = 0.316$$

Collecting the x terms leads to $2.316\,x = 0.0316$, which gives $x = 0.0136$ M. The equilibrium concentrations are thus $[N_2] = [O_2] = 0.086$ M and $[NO] = 2x = 0.0272$ M. It's a good idea to check that these concentrations are consistent with the known value of K_c. Substituting into the K_c expression we get

$$K_c = \frac{(0.0272)^2}{(0.086)^2} = 0.100$$

SAMPLE EXERCISE 15.14 **Using an ICE Table to Calculate Equilibrium Concentrations**

Much of the H_2 used in the Haber–Bosch process is produced by steam-reforming of methane followed by the water–gas shift reaction:

$$CO(g) + H_2O(g) \rightleftharpoons CO_2(g) + H_2(g)$$

If a reaction vessel at 400°C is filled with an equimolar mixture of CO and steam such that $P_{CO} = P_{H_2O} = 2.00$ atm, what will be the partial pressure of H_2 at equilibrium? The equilibrium constant $K_p = 10$ at 400°C.

COLLECT AND ORGANIZE We are asked to find the partial pressure of a gas at equilibrium given the chemical equation and the initial concentrations of reactants. One way to approach this problem involves (1) setting up an ICE table, (2) using the partial pressures from the E line of the ICE table in the equilibrium constant expression, and (3) solving for P_{H_2}.

ANALYZE Let's first look at the reactant and product quantities. This system initially contains no product. This means that the reaction quotient Q is equal to zero, which is less than K. Therefore, the reaction proceeds from left to right, decreasing the reactant concentrations and increasing the product concentrations. We can set up the ICE table for this reaction with the initial partial pressures specified.

SOLVE Let x be the increase in partial pressure of H_2 as a result of the reaction. The change in P_{CO_2} will also be x, and the changes in P_{CO} and P_{H_2} will be $-x$. Inserting these numbers into the ICE table and using them to develop equilibrium terms for the partial pressures of the reactants and products, we have

	P_{CO} (atm)	P_{H_2O} (atm)	P_{CO_2} (atm)	P_{H_2} (atm)
Initial	2.00	2.00	0.00	0.00
Change	$-x$	$-x$	$+x$	$+x$
Equilibrium	$2.00 - x$	$2.00 - x$	x	x

Inserting these equilibrium terms into the equilibrium constant expression for the reaction gives

$$K_p = \frac{(P_{CO_2})(P_{H_2})}{(P_{CO})(P_{H_2O})} = \frac{(x)(x)}{(2.00 - x)(2.00 - x)} = 10$$

The equation has the form $X^2/Y^2 = Z$, and so it can be simplified by taking the square root of both sides:

$$\frac{x}{2.00 - x} = \sqrt{10} = 3.16$$

Solving for x after rearrangement gives $x = 1.52$ atm, which is the equilibrium partial pressure of H_2 and CO_2.

THINK ABOUT IT We should always check the result when the calculation is finished. Substituting $P_{H_2O} = P_{CO} = 2.00 - 1.52 = 0.48$ atm and $P_{H_2} = P_{CO_2} = 1.52$ atm into the equilibrium constant expression gives $K_p = (1.52)^2/(0.48)^2 = 10.03$, which is close enough to show we made no errors in the calculation. It is also a good idea to look for an $X^2/Y^2 = Z$ situation to avoid having to use the quadratic equation to find x.

Practice Exercise The balanced chemical equation for the formation of hydrogen iodide from H_2 and I_2 is

$$H_2(g) + I_2(g) \rightleftharpoons 2\,HI(g)$$

The value of K_p for the reaction is 50 at 450°C. What is the partial pressure of HI in a sealed reaction vessel at 450°C if the initial partial pressures of H_2 and I_2 are both 0.100 atm and there is no HI present?

SAMPLE EXERCISE 15.15 **Calculating Equilibrium Concentrations**

Let's revisit the water–gas shift reaction:

$$CO(g) + H_2O(g) \rightleftharpoons CO_2(g) + H_2(g)$$

Suppose a reaction vessel at 400°C contains a mixture of CO, steam, and H_2 at the following initial partial pressures: $P_{CO} = 2.00$ atm, $P_{H_2O} = 2.00$ atm, and $P_{H_2} = 0.15$ atm. What is the partial pressure of H_2 at equilibrium, given $K_p = 10$ at 400°C?

COLLECT AND ORGANIZE We are asked to calculate the partial pressure of a product at equilibrium. We know the initial partial pressures of CO, steam, and H_2 and the value of K_p. The difference between this and the preceding Sample Exercise is that some H_2 (a product) is already present before the reaction starts. Comparing the reaction quotient Q with the value of K lets us know the direction (left or right) that the reaction moves to attain equilibrium.

ANALYZE Because there is no CO_2 initially present, the reaction quotient Q is zero. Therefore, the reaction proceeds in the forward direction. We can set up the ICE table for this reaction with the initial partial pressures specified.

SOLVE Let x be the increase in partial pressure of H_2 as a result of the reaction. The change in P_{CO_2} will also be x, and the changes in P_{CO} and P_{H_2O} will be $-x$. Inserting these numbers into the ICE table, we get

	P_{CO} (atm)	P_{H_2O} (atm)	P_{CO_2} (atm)	P_{H_2} (atm)
Initial	2.00	2.00	0.00	0.15
Change	$-x$	$-x$	$+x$	$+x$
Equilibrium	$2.00 - x$	$2.00 - x$	x	$0.15 + x$

Inserting these equilibrium terms into the equilibrium constant expression for the reaction gives

$$K_p = \frac{(P_{CO_2})(P_{H_2})}{(P_{CO})(P_{H_2O})}$$

$$= \frac{(x)(0.15 + x)}{(2.00 - x)(2.00 - x)} = 10$$

Solving for x gives two values, $x = 1.50$ and 2.96. Of the two solutions, one of them (2.96) is not physically possible because inserting it into the equilibrium row of the ICE table gives negative partial pressures of CO and H_2O. Therefore, $x = 1.50$ and the partial pressure of H_2 at equilibrium is $1.50 + 0.15 = 1.65$ atm.

THINK ABOUT IT From the ICE table, the equilibrium partial pressures of CO, H_2O, H_2, and CO_2 are 0.50, 0.50, 1.65, and 1.50, respectively. Using these values, the calculated value of K_p = (1.65)(1.50)/(0.5)(0.5) = 9.9, which is equal to 10 when we consider the rounding errors. Notice that 25% of the reactants initially present remain at equilibrium. This much unreacted reactant reflects the relatively small value of K_p.

Practice Exercise The value of K_c for the reaction

$$N_2O_4(g) \rightleftharpoons 2\,NO_2(g)$$

is 0.21 at 373 K. If a reaction vessel at that temperature initially contains 0.030 M concentrations of both NO_2 and N_2O_4, what will be the concentrations of the two gases when equilibrium is achieved?

15.9 Changing *K* with Changing Temperature

Throughout this discussion we have repeatedly stated that values of *K* depend on temperature. Now that we have explored concepts and relationships involving equilibria in some detail, let's look more closely at how temperature influences *K*. In the course of the discussion, we will develop several important equations that express useful ideas.

Temperature, *K*, and Δ*G*°

In Section 15.5 we derived Equation 15.14, which relates the change in standard free energy of a chemical reaction, Δ*G*°, and its equilibrium constant, *K*:

$$\ln K = \frac{-\Delta G°}{RT} \qquad \textit{①}$$

In Chapter 13, we introduced the following equation relating the changes in standard free energy, enthalpy, and entropy of a chemical reaction:

$$\Delta G° = \Delta H° - T\Delta S° \qquad \textit{②}$$

We can combine these two equations to derive one that relates *K* to Δ*H*° and Δ*S*°.

$$\textit{① + ②} : \quad \boxed{\ln K = -\frac{\Delta H°}{RT} + \frac{T\Delta S}{RT} = -\frac{\Delta H°}{RT} + \frac{\Delta S°}{R}} \qquad (15.16)$$

Note how a negative value of Δ*H*° and/or a positive value of Δ*S*° contributes to a large value of *K*. These dependencies make sense because negative values of Δ*H*° and positive values of Δ*S*° are the two factors that combine to make reactions spontaneous.

Because we are discussing the influence of temperature on the value of *K*, let's identify the factors on which *T* exerts its influence. Equation 15.16 indicates that the influence of Δ*S*° on the magnitude of *K* is independent of temperature. (The term in the equation that contains Δ*S*° does not include temperature as a factor.) On the other hand, the influence of the Δ*H*° term does depend on temperature: the higher the temperature, the smaller the $1/T$ term and the smaller the influence of a negative Δ*H*°.

⊙⊙ CONNECTION We discussed the relation between free energy, enthalpy, entropy, and temperature in Chapter 13.

This temperature dependence makes sense if we think of heat as a product in exothermic reactions and a reactant in endothermic reactions. Applying Le Châtelier's principle:

Exothermic reaction:
Reactants $\rightleftharpoons$ products + heat
Increase in *T* causes a decrease in *K*

Endothermic reaction:
Reactants + heat $\rightleftharpoons$ products
Increase in *T* causes an increase in *K*

If the values of $\Delta H°$ and $\Delta S°$ do not vary much with temperature, then Equation 15.16 predicts that ln *K* is a linear function of 1/*T*. Furthermore, we expect the line on the graph of ln *K* versus 1/*T* to have a positive slope for an exothermic process and a negative slope for an endothermic process. We can determine $\Delta H°$ from the slope of the line and $\Delta S°$ from its intercept (the value on the vertical axis of the plot when 1/*T* is zero). This means that we can calculate fundamental thermodynamic values of a reaction at equilibrium by determining its equilibrium constant at different temperatures.

As an example of this method, let's determine thermodynamic values for the system

$$2\,CO_2(g) \rightleftharpoons 2\,CO(g) + O_2(g)$$

The equilibrium constant K_c has the values shown below at three different temperatures:

Temperature (K)	K_c
1500	5.5×10^{-9}
2500	4×10^{-1}
3000	40.3

We can see from the data that K_c increases as temperature increases, so we know heat is a reactant and the reaction is endothermic. We can use these data to establish how endothermic the reaction is by plotting ln K_c versus 1/*T* and calculating $\Delta H°_{rxn}$. We can also establish if the entropy change in the reaction is positive or negative and determine $\Delta S°_{rxn}$. Equation 15.16, rewritten here in the form of the equation for a straight line, links K_c to temperature and to $\Delta H°_{rxn}$:

$$\ln K = -\frac{\Delta H°}{R}\left(\frac{1}{T}\right) + \frac{\Delta S°}{R}$$

$$y = m\,(x) + b$$

We can supplement the table we have been given by calculating ln *K* and 1/*T*.

Temperature (K)	K_c	ln K_c	1/T
1500	5.5×10^{-9}	−19.02	6.67×10^{-4}
2500	4.0×10^{-1}	−0.92	4.00×10^{-4}
3000	40.3	3.70	3.33×10^{-4}

If we plot $\ln K$ versus $1/T$, we see that the plot is a straight line.

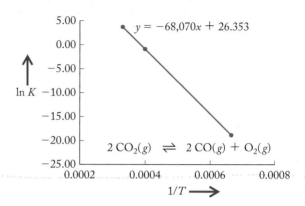

The slope of this line is $-68{,}070$ K $= -\Delta H°_{rxn}/R$, from which we can calculate the value of $\Delta H°_{rxn}$ as follows:

$$-\Delta H°_{rxn} = \frac{68{,}070 \text{ K} \times 8.314 \text{ J}}{\text{mol}\cdot\text{K}}$$

$$= 566{,}000 \text{ J/mol} = 566 \text{ kJ/mol}$$

The y-intercept is

$$26.4 = \Delta S°_{rxn}/R$$

from which we can calculate the value of $\Delta S°_{rxn}$:

$$\Delta S°_{rxn} = \frac{26.4 \times 8.314 \text{ J}}{\text{mol}\cdot\text{K}} = 219 \text{ J/mol}\cdot\text{K}$$

As anticipated, based on the increasing value of K_c as T increased, the reaction is endothermic, as shown by the positive value of $\Delta H°_{rxn} = 566$ kJ/mol. It is accompanied by a positive entropy change ($\Delta S°_{rxn} = 219$ J/mol·K), which is understandable because the forward reaction converts 2 moles of gaseous CO_2 to 3 moles of gaseous products. From the values of $\Delta H°_{rxn}$ and $\Delta S°_{rxn}$ we can predict the value of K_c at room temperature (298 K). The result is $K_c = 2 \times 10^{-88}$, which means that there is virtually no probability that the CO_2 we exhale will be converted to even trace amounts of poisonous carbon monoxide.

In addition to enabling us to calculate useful values like enthalpy and entropy changes for reactions, Equation 15.16 expresses mathematically a number of principles. One principle is that a large gain in entropy of the system (positive $\Delta S°$) contributes to making $\ln K$ positive. This principle means that entropy favors equilibria when there are more moles of products (with a corresponding larger number of microstates) than of reactants.

In addition to using Equation 15.16 to determine thermodynamic values, we can also use it to examine the effect of temperature on a reaction. Let's revisit an important reaction in automobile engines: the decomposition of NO into N_2 and O_2. We can calculate the value of the equilibrium constant of this reaction at the temperature at which an automobile engine operates (500°C) if we know the value of K for a reference temperature. We are looking for K for the decomposition reaction of NO at 500°C, so we start by writing Equation 15.16 in the form

$$\ln K = -\frac{\Delta H°}{R}\left(\frac{1}{T}\right) + \frac{\Delta S°}{R}$$

As we just saw, this equation is that of a straight line ($y = mx + b$) where the y variable is $\ln K$ and the x variable is $1/T$. The slope of the line is $-\Delta H°/R$ and the intercept is $\Delta S°/R$. If we select two points along such a line (temperatures T_1 and T_2) at which the values of the equilibrium constant are K_1 and K_2, respectively, then

$$\ln K_1 = -\frac{\Delta H°}{R}\left(\frac{1}{T_1}\right) + \frac{\Delta S°}{R}$$

$$\ln K_2 = -\frac{\Delta H°}{R}\left(\frac{1}{T_2}\right) + \frac{\Delta S°}{R}$$

The difference between $\ln K_1$ and $\ln K_2$ is related to the difference between T_1 and T_2:

$$\ln K_1 - \ln K_2 = -\frac{\Delta H°}{R}\left(\frac{1}{T_1}\right) + \frac{\Delta H°}{R}\left(\frac{1}{T_2}\right)$$

Rearranging the terms, we have

$$\ln\left(\frac{K_1}{K_2}\right) = -\frac{\Delta H°}{R}\left(\frac{1}{T_1} - \frac{1}{T_2}\right)$$

or

$$\boxed{\ln\left(\frac{K_2}{K_1}\right) = -\frac{\Delta H°}{R}\left(\frac{1}{T_2} - \frac{1}{T_1}\right)} \tag{15.17}$$

Equation 15.17 is called the *Clausius-Clapeyron equation*, which we used in Chapter 12 in our discussion of the fractional distillation of crude oil. We can use it in the context of this chapter to calculate K_p for the decomposition of NO

$$NO(g) \rightleftharpoons \tfrac{1}{2}N_2(g) + \tfrac{1}{2}O_2(g)$$

in the catalytic converter of an automobile at 500°C.

To use Equation 15.17, we need a reference value for K_p. We also need the standard enthalpy change ($\Delta H°$) of the reaction. For the formation of NO

$$\tfrac{1}{2}N_2(g) + \tfrac{1}{2}O_2(g) \rightleftharpoons NO(g)$$

$K_p = 6.6 \times 10^{-16}$ at 298 K. This reaction is the reverse of the decomposition reaction, so $K_p = 6.6 \times 10^{-16}$ is the reciprocal of the value we need. For the reaction

$$NO(g) \rightleftharpoons \tfrac{1}{2}N_2(g) + \tfrac{1}{2}O_2(g)$$

the equilibrium constant $K = 1/K_p = 1.5 \times 10^{15}$. The equilibrium constant for the decomposition reaction at 298 K is a large number, so this reaction is highly favorable at that temperature.

The value of $\Delta H°$ can be found in Appendix 4 for the formation of NO from its elements in their standard states:

$$\tfrac{1}{2}N_2(g) + \tfrac{1}{2}O_2(g) \rightleftharpoons NO(g) \qquad \Delta H_f° = 90.3 \text{ kJ}$$

Thus, $\Delta H°$ for the decomposition of NO is the negative of this value or -90.3 kJ per mole of NO.

CONNECTION In Chapter 14 we discussed the effect of temperature on the rates of reactions occurring in a catalytic converter.

We now have all the terms we need to use Equation 15.17 to calculate K_p for the reaction at 500°C (symbolized as K_2 in the equation). We let $T_1 = 298$ K, $T_2 = 773$ K, $K_1 = 1.5 \times 10^{15}$, and we solve for K_2:

$$\ln\left(\frac{K_2}{K_1}\right) = -\frac{\Delta H°}{R}\left(\frac{1}{T_2} - \frac{1}{T_1}\right)$$

$$\ln\left(\frac{K_2}{1.5 \times 10^{15}}\right) = -\frac{-90.3\ \frac{\cancel{kJ}}{\cancel{mol}} \times \frac{1000\ \cancel{J}}{\cancel{kJ}}}{8.314\ \frac{\cancel{J}}{\cancel{mol} \cdot \cancel{K}}}\left(\frac{1}{773\ \cancel{K}} - \frac{1}{298\ \cancel{K}}\right)$$

$$\ln\left(\frac{K_2}{1.5 \times 10^{15}}\right) = -22.40$$

$$\left(\frac{K_2}{1.5 \times 10^{15}}\right) = e^{-22.40} = 1.87 \times 10^{-10}$$

$$K_2 = 2.8 \times 10^5$$

The equilibrium constant at 773 K is ten orders of magnitude smaller than the value at 298 K because the decomposition reaction is exothermic ($\Delta H° = -90.3$ kJ/mol NO) and heat is a product of the reaction. This result corresponds with the prediction based on Le Châtelier's principle: the equilibrium constant of an exothermic reaction decreases with increasing temperature.

The smaller value of K at the higher temperature in the engine indicates that more NO should remain in engine exhaust at 500°C than if the temperature were lower, assuming chemical equilibrium is achieved. That is a big assumption. The catalytic converter's job is to help achieve chemical equilibrium rapidly by increasing the rate of the slow decomposition of NO.

SAMPLE EXERCISE 15.16 **Calculating Equilibrium Constants at a Specific Temperature**

Ammonia is synthesized from nitrogen gas and hydrogen gas in an exothermic chemical reaction that requires high temperatures to increase its rate. Use the data in Appendix 4 to calculate the equilibrium constant (K_p) for the reaction

$$N_2(g) + 3\,H_2(g) \rightleftharpoons 2\,NH_3(g)$$

at 298 K and at 773 K.

COLLECT AND ORGANIZE We are asked to calculate K_p at two different temperatures for the synthesis of ammonia from nitrogen and hydrogen.

ANALYZE We can calculate the value of K_p for this reaction at 298 K using Equation 15.15 and the value of $\Delta G_f°$ for NH_3 at 298 K from Appendix 4. We need to remember that the $\Delta G_f°$ value (–16.5 kJ/mol) is for the formation of 1 mole of NH_3, but 2 moles of NH_3 are formed in the reaction.

SOLVE The value of $\Delta G_{rxn}°$ for the reaction is twice $\Delta G_f°$, or

$$2 \text{ mol NH}_3 \times -16.5 \text{ kJ/mol NH}_3 = -33.0 \text{ kJ, or } -33,000 \text{ J}$$

To obtain K_p we substitute in Equation 15.15:

$$K_p = e^{-\Delta G°/RT} = \exp\left(\frac{-\Delta G°}{RT}\right)$$

$$= \exp\left(\frac{-\left(-33{,}000\ \frac{\cancel{J}}{\cancel{mol}}\right)}{8.314\ \frac{\cancel{J}}{\cancel{mol}\cdot\cancel{K}}\times 298\ \cancel{K}}\right)$$

$$= 6.1\times 10^{5}$$

Once we have the value of K_p at 298 K, we can calculate K_p for the reaction at 773 K by using Equation 15.17. To use Equation 15.17 we must first calculate the value of ΔH°_{rxn}, which is twice the ΔH°_f of NH_3 (−46.1 kJ/mol):

$$\Delta H^{\circ}_{rxn} = (2\ \cancel{mol\ NH_3})\left(-46.1\ \frac{kJ}{\cancel{mol\ NH_3}}\right)$$

$$= -92.2\ kJ,\ or\ -92{,}200\ J$$

After substituting $K_1 = 6.1\times 10^5$, $T_1 = 298$, and $T_2 = 773$ into the equation, we solve for K_2:

$$\ln\left(\frac{K_2}{6.1\times 10^5}\right) = -\frac{-\left(-92{,}200\ \frac{\cancel{J}}{\cancel{mol}}\right)}{8.314\ \frac{\cancel{J}}{\cancel{mol}\cdot K}}\left(\frac{1}{773\ K} - \frac{1}{298\ K}\right)$$

$$\ln\left(\frac{K_2}{6.1\times 10^5}\right) = -22.86$$

$$\left(\frac{K_2}{6.1\times 10^5}\right) = e^{-22.86} = 1.2\times 10^{-10}$$

$$K_2 = 7.1\times 10^{-5}$$

THINK ABOUT IT The equilibrium constant for this equilibrium decreases markedly with increasing temperature. We expect it to decrease because the reaction is exothermic.

Practice Exercise Use the data in Appendix 4 to calculate the equilibrium constant (K_p) for the reaction

$$2\,N_2(g) + O_2(g) \rightleftharpoons 2\,N_2O(g)$$

at 298 K and 2000 K.

The equilibrium constant (K_p) determined in Sample Exercise 15.16 for the reaction describing the synthesis of ammonia is greater than 10^5 at room temperature, but only about 10^{-5} at 773 K. That's ten orders of magnitude smaller. We now have a more complete picture of the preparation of ammonia that we discussed in Section 15.7. Ammonia is usually synthesized from N_2 and H_2 at temperatures near 400°C. These high temperatures *increase the rate* of this exothermic reaction, even though raising the temperature significantly *decreases* K_p and so decreases how much ammonia can be made when chemical equilibrium is achieved. In this case, the practical benefit of a more favorable reaction rate outweighs the less favorable thermodynamics of running the reaction at high temperature. In practice, the ammonia product is continuously removed, shifting the equilibrium toward the formation of more product as predicted by Le Châtelier's principle.

Using *K* and *T* to Determine Δ*H*° and Δ*S*°

Heats of reaction and entropy changes are important thermodynamic quantities, and Equation 15.16 enables us to calculate them if we have the equilibrium constants for a reversible reaction at three or more temperatures. (Remember we need three data points to define a straight line reliably.) We also need to remember that the mass action expression for a reaction at equilibrium is written and its value is determined with reference to a specific reaction under a given set of conditions and at a specific temperature. If we collect data on a reaction over a range of temperatures, it is important to know that only one process—only one reaction—is actually taking place over that range. To use a symbolic example of what this means, suppose at a low temperature (T_1)

$$A + B \rightleftharpoons C$$

but at a higher temperature (T_2) a second process occurs such that

$$A + 2B \rightleftharpoons D$$

Two different equilibria would exist in the system over the temperature range T_1 to T_2. If we have equilibrium constants for a reversible process determined over a temperature range, we can plot ln *K* versus $1/T$, and if the plot is a straight line, then we can state that the same equilibrium exists over the experimental temperature range. We can then use Equation 15.16 to calculate Δ*H*° and Δ*S*° for the reaction.

As we have seen throughout this chapter, each chemical reaction has a unique equilibrium constant under a given set of conditions and at a specific temperature. If we know the equilibrium constant for a reaction, we can calculate concentrations of reactants and products that result from any combination of concentrations when a reaction is initiated. We can carry out other calculations that enable us to optimize the results of a chemical process and to predict the response of a system to changing conditions.

Summary

SECTION 15.1 Physical or **chemical equilibrium** can be approached from either reaction direction and is achieved when the forward and reverse reaction rates are the same. Chemical equilibria may correspond to comparable amounts of reactants and products or lie to the left (reactant favored) or the right (product favored).

SECTION 15.2 The **mass action expression** or the expression for the **equilibrium constant** K_c for a chemical reaction is the ratio of the equilibrium molar concentrations of the products divided by the equilibrium molar concentrations of the reactants, each raised to the respective stoichiometric coefficient in the balanced equilibrium equation. The **equilibrium constant** K_p for equilibria involving gases uses the equilibrium partial pressures of products and reactants in place of molar concentrations. The equilibrium constant for a specific chemical equilibrium varies only with the temperature.

SECTION 15.3 The reverse of a reaction has an equilibrium constant that is the reciprocal of the *K* of the forward reaction. If reactions are summed to give an overall reaction, their equilibrium constants are multiplied together to obtain an overall equilibrium constant. Doubling the coefficients of a reaction equation means that the value of the equilibrium constant is squared.

SECTION 15.4 The **reaction quotient (*Q*)** is the value of the mass action expression for concentrations (or partial pressures) of reactants and products that may or may not be at equilibrium. If $Q < K$, the reaction proceeds in the forward direction. If $Q > K$, it runs in reverse. At equilibrium, $Q = K$. The equilibrium constant is a thermodynamic property of a chemical reaction: the greater the decrease in standard free energy, the larger the value of the equilibrium constant.

SECTION 15.5 The free-energy change Δ*G* of a reversible reaction and its reaction quotient *Q* are related: $\Delta G = \Delta G° + RT \ln Q$, where Δ*G*° is the standard free-energy change. At equilibrium, Δ*G* is zero and $Q = K$. This leads to $\Delta G° = -RT \ln K$, which can be used with a known value of Δ*G*° to calculate the equilibrium constant *K* at any temperature. A negative value of Δ*G*° corresponds to $K > 1$ and a positive Δ*G*° corresponds to $K < 1$.

SECTION 15.6 **Heterogeneous equilibria** involve more than one phase (for example, gas and solid). The concentrations of pure liquids and solids do not change during a reaction and so are omitted from equilibrium constant expressions.

SECTION 15.7 According to **Le Châtelier's principle**, systems (including chemical reactions) at equilibrium respond to stress by shifting position to relieve the stress. Adding or removing a reactant or product, or applying pressure to a reaction mixture that includes gases creates stress that shifts the position of a chemical equilibrium. A catalyst decreases the time it takes a system to achieve equilibrium, but it does not change the value of the equilibrium constant.

SECTION 15.8 Equilibrium constants K_c or K_p can be calculated from known equilibrium concentrations or equilibrium partial pres-

sures. Equilibrium concentrations or pressures of reactants and products can be calculated from initial concentrations or pressures, the reaction stoichiometry, and the value of the equilibrium constant.

SECTION 15.9 Higher reaction temperatures increase the equilibrium constant of an endothermic reaction but decrease the equilibrium constant of an exothermic reaction. The slope and intercept of a plot of $\ln K$ versus $1/T$ for an equilibrium system are used to determine the standard enthalpy and standard entropy changes for the equilibrium, respectively.

Problem-Solving Summary

TYPE OF PROBLEM	CONCEPTS AND EQUATIONS	SAMPLE EXERCISES
Writing equilibrium constant expressions	For the chemical reaction $$w\text{A} + x\text{B} \rightleftharpoons y\text{C} + z\text{D}$$ $$K_c = \frac{[\text{C}]^y[\text{D}]^z}{[\text{A}]^w[\text{B}]^x} \quad \text{and} \quad K_p = \frac{(P_\text{C})^y(P_\text{D})^z}{(P_\text{A})^w(P_\text{B})^x}$$	15.1
Calculating K_c or K_p	Insert equilibrium molar concentrations (or partial pressures) into the equilibrium constant expression.	15.2 and 15.3
Calculating K_c from K_p	Use the relation $$K_p = K_c(RT)^{\Delta n}$$ rearranged as necessary to give $$K_c = K_p(RT)^{-\Delta n}$$ where $R = 0.08206$ L·atm/mol·K, T is the absolute temperature, and Δn is the number of moles of product gas minus the number of moles of reactant gas in the balanced chemical equation.	15.4
Calculating K for a reverse reaction, or by combining reactions	To calculate the value of K of the reverse of a reaction, take the reciprocal of K of the forward reaction. If reactions are summed to give an overall reaction, their equilibrium constants are multiplied together to obtain an overall K. If all the coefficients in a chemical equation are multiplied by n, the value of K increases by the power of n.	15.5–15.7
Using Q and K	The reaction quotient Q is a ratio of concentrations or partial pressures of products to reactants based on the mass action expression, but it is usually applied to nonequilibrium conditions. If $Q < K$, the reaction proceeds in the forward direction ($\rightarrow$) to make more products; if $Q = K$, the reaction is at equilibrium; if $Q > K$, the reaction proceeds in the reverse direction ($\leftarrow$) to make more reactants.	15.8
Calculating K from $\Delta G°$	Use the relation $\Delta G° = -RT \ln K$. Be sure to convert $\Delta G°$ to J/mol and use $R = 8.314$ J/mol·K.	15.9
Writing mass action expressions for heterogeneous equilibria	Molar concentrations of pure liquids and pure solids are omitted from equilibrium constant expressions because such concentrations are constants.	15.10
Adding or removing reactants or products to stress on equilibrium	Decreasing the concentration of a substance involved in an equilibrium shifts the equilibrium toward the production of more of that substance. Increasing the concentration of a substance shifts the equilibrium so that some of that substance is consumed in the reaction.	15.11
Assessing pressure effects on gas phase equilibria	Equilibria involving different numbers of gaseous reactants and products shift in response to an increase (or decrease) in pressure toward the side with fewer (or more) moles of gases, respectively.	15.12

TYPE OF PROBLEM	CONCEPTS AND EQUATIONS	SAMPLE EXERCISES
Predicting changes in equilibria with temperature	The value of K of an endothermic reaction increases with increasing temperatures; the value of K of an exothermic reaction decreases.	15.13
Using an ICE table to calculate equilibrium concentrations	An ICE table has a column for each reactant and product concentration (or partial pressure) and three rows labeled *Initial*, *Change*, and *Equilibrium*. Calculate Q to see which way an equilibrium shifts under the specified conditions. Call one of the shifts x and write the others as decreases or increases in terms of x, taking into account the reaction stoichiometry. Add the I and C rows to get the expression for each reactant and product concentration in the E row. Substitute these data into the expression for K and solve for x.	15.14 and 15.15
Calculating equilibrium constants at a specific temperature	Use the relation $$\ln\left(\frac{K_2}{K_1}\right) = -\frac{\Delta H°}{R}\left(\frac{1}{T_2} - \frac{1}{T_1}\right)$$ where T_1 and T_2 are Kelvin temperatures and the value of K_1 or K_2 at the respective temperature is known. Be sure to convert $\Delta H°$ to J/mol to match the units of R.	15.16

Visual Problems

15.1. Figure P15.1 shows the energy profiles of reactions $A \rightleftharpoons B$ and $C \rightleftharpoons D$, respectively. Which reaction has the larger forward rate constant? Which reaction has the smaller reverse rate constant? Which reaction has the larger equilibrium constant K_c?

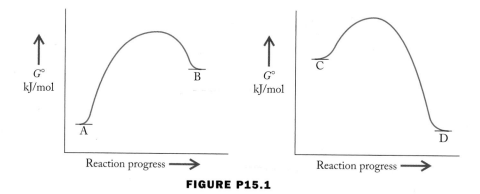

FIGURE P15.1

15.2. The progress with time of a reaction system is depicted in Figure P15.2. Red spheres represent the molar concentration of substance A and blue spheres represent the molar concentration of substance B.
(a) Does the system reach equilibrium? (b) In which direction $(A \rightarrow B$ or $B \rightarrow A)$ is equilibrium attained? (c) What is the value of the equilibrium constant K_c?

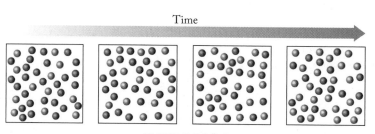

FIGURE P15.2

15.3. In Figure P15.3 the red spheres represent reactant A and the blue spheres represent product B in equilibrium with A.
(a) Write a chemical equation that describes the equilibrium.
(b) What is the value of the equilibrium constant K_c?

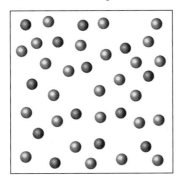

FIGURE P15.3

15.4. The equilibrium constant K_c for the reaction

$$A \text{ (red spheres)} + B \text{ (blue spheres)} \rightleftharpoons AB$$

is 3.0 at 300.0 K. Does the situation depicted in Figure P15.4 correspond to equilibrium? If not, in what direction (to the left or to the right) will the system shift to attain equilibrium?

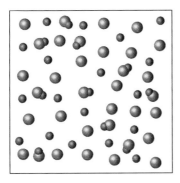

FIGURE P15.4

15.5. The top and bottom diagrams in Figure P15.5 represent equilibrium states of the reaction.

$$A \text{ (red spheres)} + B \text{ (blue spheres)} \rightleftharpoons AB$$

at 300 K and 400 K, respectively. Is this reaction endothermic or exothermic? Explain.

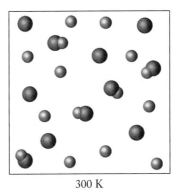

300 K

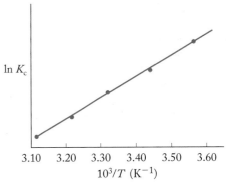

400 K

FIGURE P15.5

15.6. Figure P15.6 shows a plot of $\ln K_c$ versus $1/T$ for the reaction $A + 2B \rightleftharpoons AB_2$. Is the reaction endothermic or exothermic?

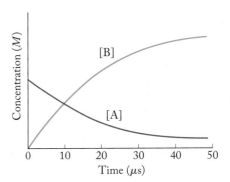

FIGURE P15.6

Questions and Problems

Understanding Equilibrium: A Qualitative View

CONCEPT REVIEW

15.7. How are forward and reverse reaction rates related in a system at chemical equilibrium?

15.8. Describe an example of a dynamic equilibrium that you experienced today.

15.9. Does the reaction $A \rightarrow 2B$ represented in Figure P15.9 reach equilibrium in 20 μs? Explain your answer.

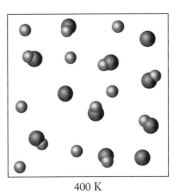

FIGURE P15.9

15.10. At equilibrium, is the sum of the concentrations of all the reactants always equal to the sum of the concentrations of the products? Explain.

15.11. Suppose the forward rate constant of the reaction $A \rightleftharpoons B$ is greater than the rate constant of the reverse reaction at a given temperature. Is the value of the equilibrium constant less than, greater than, or equal to 1?

15.12. Explain how it is possible for a reaction to have a large equilibrium constant but small forward and reverse rate constants.

PROBLEMS

15.13. In a study of the reaction

$$2\,N_2O(g) \rightleftharpoons 2\,N_2(g) + O_2(g)$$

quantities of all three gases were injected into a reaction vessel. The N_2O consisted entirely of isotopically labeled $^{15}N_2O$. Analysis of the reaction mixture after 1 day revealed the presence of compounds with molar masses 28, 29, 30, 32, 44, 45, and 46 g/mol. Identify the compounds and account for their appearance.

15.14. A mixture of ^{13}CO, $^{12}CO_2$, and O_2 in a sealed reaction vessel was used to follow the reaction

$$2\,CO(g) + O_2(g) \rightleftharpoons 2\,CO_2(g)$$

Analysis of the reaction mixture after 1 day revealed the presence of compounds with molar masses 28, 29, 32, 44, and 45 g/mol. Identify the compounds and account for their appearance.

15.15. Suppose the reaction $A \rightleftharpoons B$ in the forward direction is first order in A and the rate constant is $1.50 \times 10^{-2}\ s^{-1}$. The reverse reaction is first order in B and the rate constant is $4.50 \times 10^{-2}\ s^{-1}$ at the same temperature. What is the value of the equilibrium constant for the reaction $A \rightleftharpoons B$ at this temperature?

15.16. At 700 K the equilibrium constant K_c for the gas phase reaction between NO and O_2 forming NO_2 is 8.7×10^6. The rate constant for the reverse reaction at this temperature is $0.54\ M^{-1}\ s^{-1}$. What is the value of the rate constant for the forward reaction at 700 K?

The Equilibrium Constant Expression and *K*

CONCEPT REVIEW

15.17. Under what conditions are the numerical values of K_c and K_p equal?

15.18. At 298 K, is K_p greater than or less than K_c if there is a net increase in the number of moles of gas in a reaction and if $K_c > 1$. Explain your answer.

15.19. Nitrogen oxides play important roles in air pollution. Write expressions for K_c and K_p for the following reactions involving nitrogen oxides.
a. $N_2(g) + 2\,O_2(g) \rightleftharpoons N_2O_4(g)$
b. $3\,NO(g) \rightleftharpoons NO_2(g) + N_2O(g)$
c. $2\,N_2O(g) \rightleftharpoons 2\,N_2(g) + O_2(g)$

15.20. Write expressions for K_c and K_p for the following reactions, which contribute to the destruction of stratospheric ozone.
a. $Cl(g) + O_3(g) \rightleftharpoons ClO(g) + O_2(g)$
b. $2\,ClO(g) \rightleftharpoons Cl_2(g) + O_2(g)$
c. $2\,O_3(g) \rightleftharpoons 3\,O_2(g)$

PROBLEMS

15.21. Use the graph in Figure P15.21 to estimate the value of the equilibrium constant K_c for the reaction

$$N_2O(g) \rightleftharpoons N_2(g) + \tfrac{1}{2}O_2(g).$$

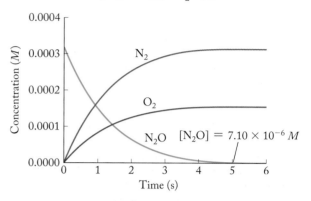

FIGURE P15.21

15.22. Estimate the value of the equilibrium constant K_c for the reaction

$$2\,NO(g) + O_2(g) \rightleftharpoons 2\,NO_2(g)$$

from the data in Figure P15.22.

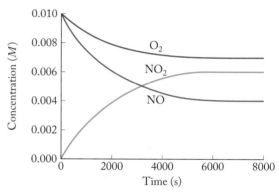

FIGURE P15.22

15.23. At 1200 K the partial pressures of an equilibrium mixture of H_2S, H_2, and S are 0.020, 0.045, and 0.030 atm, respectively. Calculate the value of the equilibrium constant K_p at 1200 K.

$$H_2S(g) \rightleftharpoons H_2(g) + S(g)$$

15.24. At 1045 K the partial pressures of an equilibrium mixture of H_2O, H_2, and O_2 are 0.040, 0.0045, and 0.0030 atm, respectively. Calculate the value of the equilibrium constant K_p at 1045 K.

$$2\,H_2O(g) \rightleftharpoons 2\,H_2(g) + O_2(g)$$

15.25. At equilibrium, the concentrations of gaseous N_2, O_2, and NO in a sealed reaction vessel are $[N_2] = 3.3 \times 10^{-3}\ M$, $[O_2] = 5.8 \times 10^{-3}\ M$, and $[NO] = 3.1 \times 10^{-3}\ M$. What is the value of K_c for the reaction

$$N_2(g) + O_2(g) \rightleftharpoons 2\,NO(g)$$

at the temperature of the reaction mixture?

15.26. Analyses of an equilibrium mixture of gaseous N_2O_4 and NO_2 gave the following results: $[NO_2] = 4.2 \times 10^{-3}\ M$ and $[N_2O_4] = 2.9 \times 10^{-3}\ M$. What is the value of the equilibrium constant K_c for the following reaction at the temperature of the mixture?

$$2\,NO_2(g) \rightleftharpoons N_2O_4(g)$$

15.27. A sealed reaction vessel initially contains 1.50×10^{-2} moles of water vapor and 1.50×10^{-2} moles of CO. After the following reaction

$$H_2O(g) + CO(g) \rightleftharpoons H_2(g) + CO_2(g)$$

has come to equilibrium, the vessel contains 8.3×10^{-3} moles of CO_2. What is the value of the equilibrium constant K_c of the reaction at the temperature of the vessel?

*15.28. A 100 mL reaction vessel initially contains 2.60×10^{-2} moles of NO and 1.30×10^{-2} moles of H_2. At equilibrium, the concentration of NO in the vessel is $0.161\ M$. At equilibrium the vessel also contains N_2, H_2O, and H_2. What is the value of the equilibrium constant K_c for the following reaction?

$$2\,H_2(g) + 2\,NO(g) \rightleftharpoons 2\,H_2O(g) + N_2(g)$$

15.29. The equilibrium constant K_p for the following equilibrium is 32 at 298 K. What is the value of K_c for this same equilibrium at 298 K?

$$A(g) + B(g) \rightleftharpoons AB(g)$$

15.30. The equilibrium constant K_c for the following equilibrium is 6.0×10^4 at 500 K. What is the value of K_p for this same equilibrium at 500 K?

$$CD(g) \rightleftharpoons C(g) + D(g)$$

15.31. At 500°C, the equilibrium constant K_p for the synthesis of ammonia

$$N_2(g) + 3\,H_2(g) \rightleftharpoons 2\,NH_3(g)$$

is 1.45×10^{-5}. What is the value of K_c?

15.32. If the value of the equilibrium constant, K_c, for the following reaction is 5×10^5 at 298 K, what is the value of K_p at 298 K?

$$2\,CO(g) + O_2(g) \rightleftharpoons 2\,CO_2(g)$$

15.33. For which of the following reactions are the values of K_c and K_p equal?
a. $2\,SO_2(g) + O_2(g) \rightleftharpoons 2\,SO_3(g)$
b. $Fe(s) + CO_2(g) \rightleftharpoons FeO(s) + CO(g)$
c. $H_2O(g) + CO(g) \rightleftharpoons H_2(g) + CO_2(g)$

15.34. For which of the following reactions are the values of K_c and K_p different?
a. $2\,NO_2(g) \rightleftharpoons N_2O_4(g)$
b. $2\,NO(g) + O_2(g) \rightleftharpoons 2\,NO_2(g)$
c. $2\,O_3(g) \rightleftharpoons 3\,O_2(g)$

15.35. Bulletproof Glass Phosgene ($COCl_2$) is used in the manufacture of foam rubber and bulletproof glass. It is formed from carbon monoxide and chlorine in the following reaction:

$$Cl_2(g) + CO(g) \rightleftharpoons COCl_2(g)$$

The value of K_c for the reaction is 5.0 at 327°C. What is the value of K_p at 325°C?

15.36. If the value of K_p for the following reaction

$$SO_2(g) + NO_2(g) \rightleftharpoons NO(g) + SO_3(g)$$

is 3.45 at 298 K, what is the value of K_c for the reverse reaction?

Manipulating Equilibrium Constant Expressions

CONCEPT REVIEW

15.37. How is the value of the equilibrium constant affected by scaling up or down the coefficients of the reactants and products in the chemical equation describing the reaction?

15.38. Why must a written form of an equilibrium and the temperatures be given when reporting the value of an equilibrium constant?

PROBLEMS

15.39. The equilibrium constant K_c for the reaction

$$A_2(g) + B_2(g) \rightleftharpoons 2\,AB(g)$$

is 4 at 300 K. What is the value of K_c for the equilibrium

$$\tfrac{1}{2}A_2(g) + \tfrac{1}{2}B_2(g) \rightleftharpoons AB(g)$$

at 300 K?

15.40. The equilibrium constant K_c for the reaction

$$2\,X(g) + 2\,Y(g) \rightleftharpoons 2\,XY(g)$$

is 3.6×10^3 at 405 K. What is the value of K_c for the equilibrium $X(g) + Y(g) \rightleftharpoons XY(g)$ at 405 K?

15.41. The following reaction is one of the elementary steps in the oxidation of NO:

$$NO(g) + NO_3(g) \rightleftharpoons 2\,NO_2(g)$$

Write an expression for equilibrium constant K_c for this reaction and for the reverse reaction:

$$2\,NO_2(g) \rightleftharpoons NO(g) + NO_3(g)$$

How are the two K_c expressions related?

15.42. **Making Ammonia** The value of the equilibrium constant K_p for the formation of ammonia,

$$N_2(g) + 3\,H_2(g) \rightleftharpoons 2\,NH_3(g)$$

is 4.5×10^{-5} at 450°C. What is the value of K_p for the following reaction?

$$2\,NH_3(g) \rightleftharpoons N_2(g) + 3\,H_2(g)$$

15.43. Air Pollutants Sulfur oxides are major air pollutants. The reaction between sulfur dioxide and oxygen can be written in two ways:

$$SO_2(g) + \tfrac{1}{2}O_2(g) \rightleftharpoons SO_3(g)$$

and

$$2\,SO_2(g) + O_2(g) \rightleftharpoons 2\,SO_3(g)$$

Write expressions for the equilibrium constants for both reactions. How are they related?

15.44. At a given temperature, the equilibrium constant K_c for the reaction

$$2\,NO(g) + 2\,H_2(g) \rightleftharpoons N_2(g) + 2\,H_2O(g)$$

is 0.11. What is the equilibrium constant for the following reaction?

$$NO(g) + H_2(g) \rightleftharpoons \tfrac{1}{2}N_2(g) + H_2O(g)$$

15.45. At a given temperature, the equilibrium constant K_c for the reaction

$$2\,SO_2(g) + O_2(g) \rightleftharpoons 2\,SO_3(g)$$

is 2.4×10^{-3}. What is the value of the equilibrium constant for each of the following reactions at that temperature?
a. $SO_2(g) + \tfrac{1}{2}O_2(g) \rightleftharpoons SO_3(g)$
b. $2\,SO_3(g) \rightleftharpoons 2\,SO_2(g) + O_2(g)$
c. $SO_3(g) \rightleftharpoons SO_2(g) + \tfrac{1}{2}O_2(g)$

15.46. If the equilibrium constant K_c for the reaction

$$2\,NO(g) + O_2(g) \rightleftharpoons 2\,NO_2(g)$$

is 5×10^{12}, what is the value of the equilibrium constant of each of the following reactions at the same temperature?
a. $NO(g) + \tfrac{1}{2}O_2(g) \rightleftharpoons NO_2(g)$
b. $2\,NO_2(g) \rightleftharpoons 2\,NO(g) + O_2(g)$
c. $NO_2(g) \rightleftharpoons NO(g) + \tfrac{1}{2}O_2(g)$

15.47. Calculate the value of the equilibrium constant K for the reaction

$$2\,D \rightleftharpoons A + 2\,B$$

from the following information:

$$A + 2\,B \rightleftharpoons C \qquad K_c = 3.3$$
$$C \rightleftharpoons 2\,D \qquad K_c = 0.041$$

15.48. Calculate the value of the equilibrium constant K for the reaction

$$E + F \rightleftharpoons G$$

from the following information:

$$2\,G \rightleftharpoons H \qquad K_c = 3.1 \times 10^{-4}$$
$$H \rightleftharpoons 2\,E + 2\,F \qquad K_c = 2.8 \times 10^{-2}$$

Equilibrium Constants and Reaction Quotients

CONCEPT REVIEW

15.49. What is a reaction quotient?
15.50. For a given equilibrium system how, in most cases, do the equilibrium constant and the reaction quotient differ?
15.51. What does it mean when the reaction quotient Q is numerically equal to the equilibrium constant K?
15.52. Explain how knowing Q and K for an equilibrium system enables you to say whether it is at equilibrium or whether it will shift in one direction or another.

PROBLEMS

15.53. If the equilibrium constant K_c for the reaction $A(aq) \rightleftharpoons B(aq)$ is 22 at a given temperature, and if $[A] = 0.10\ M$ and $[B] = 2.0\ M$ in a reaction mixture at that temperature, is the reaction at chemical equilibrium? If not, in which direction will the reaction proceed to reach equilibrium?

15.54. The equilibrium constant K_c for the reaction

$$2\,C \rightleftharpoons D + E$$

is 3×10^{-3}. At a particular time, the composition of the reaction mixture is $[C] = [D] = [E] = 5 \times 10^{-4}\ M$. In which direction will the reaction proceed to reach equilibrium?

15.55. Suppose the value of the equilibrium constant K_p of the following reaction

$$A(g) + B(g) \rightleftharpoons C(g)$$

is 1.00 at 300 K. Are either of the following reaction mixtures at chemical equilibrium at 300 K?
a. $P_A = P_B = P_C = 1.0$ atm
b. $[A] = [B] = [C] = 1.0\ M$

15.56. In which direction will the following reaction proceed to reach equilibrium under the conditions given?

$$A(g) + B(g) \rightleftharpoons C(g) \qquad K_p = 1.00 \text{ at } 300 \text{ K}$$

a. $P_A = P_C = 1.0$ atm, $P_B = 0.50$ atm
b. $[A] = [B] = [C] = 1.0\ M$

15.57. If the equilibrium constant K_c for the reaction

$$N_2(g) + O_2(g) \rightleftharpoons 2\,NO(g)$$

is 1.5×10^{-3}, in which direction will the reaction proceed if the partial pressures of the three gases are all 1.00×10^{-3} atm?

15.58. At 650 K, the value of the equilibrium constant K_p for the ammonia synthesis reaction

$$N_2(g) + 3\,H_2(g) \rightleftharpoons 2\,NH_3(g)$$

is 4.3×10^{-4}. If a vessel contains a reaction mixture in which $[N_2] = 0.010\ M$, $[H_2] = 0.030\ M$, and $[NH_3] = 0.00020\ M$, will more ammonia form?

15.59. The equilibrium $X + Y \rightleftharpoons Z$ has $K_c = 1.00$ at 350 K. If the initial molar concentrations of X, Y, and Z in a solution are all 0.2 M, in which direction will the reaction shift to reach equilibrium?
a. To the left, making more X and Y
b. To the right, making more Z
c. The system is at equilibrium and the concentrations will not change

15.60. In Problem 15.59, when the equilibrium shifts, does the concentration of X increase or decrease?

Equilibrium and Thermodynamics

CONCEPT REVIEW

15.61. Do all reactions with equilibrium constants < 1 have values of $\Delta G° > 0$?
15.62. The equation $\Delta G° = -RT \ln K$ relates the value of K_p, not K_c, to the change in standard free energy for a reaction in the gas phase. Explain why.
15.63. Starting with pure reactants, in which direction will an equilibrium shift if $\Delta G° < 0$?
15.64. Starting with pure products, in which direction will an equilibrium shift if $\Delta G° < 0$?

PROBLEMS

15.65. Which of the following reactions has the largest equilibrium constant at 25°C?
   a. $Cl_2(g) + F_2(g) \rightleftharpoons 2\,ClF(g)$ $\Delta G° = 115.4$ kJ
   b. $Cl_2(g) + Br_2(g) \rightleftharpoons 2\,ClBr(g)$ $\Delta G° = -2.0$ kJ
   c. $Cl_2(g) + I_2(g) \rightleftharpoons 2\,ICl(g)$ $\Delta G° = -27.9$ kJ

15.66. Glycolysis Problems 15.66–15.69 focus on glycolysis, the multi-step biochemical process by which sugar is metabolized to create energy in the body. Which of the following steps in glycolysis has the largest equilibrium constant?
   a. Fructose 1,6-diphosphate $\rightleftharpoons$
           2 glyceraldehyde–3-phosphate $\Delta G° = 24$ kJ
   b. 3-Phosphoglycerate $\rightleftharpoons$ 2-phosphoglycerate $\Delta G° = 4.4$ kJ
   c. 2-Phosphoglycerate $\rightleftharpoons$ phosphoenolpyruvate $\Delta G° = 1.8$ kJ

15.67. The value of $\Delta G°$ for the phosphorylation of glucose in glycolysis is 113.8 kJ/mol. Using Equation 15.13 what is the value of the equilibrium constant for the reaction at 298 K?

15.68. In glycolysis, the hydrolysis of ATP to ADP drives the phosphorylation of glucose:

Glucose + ATP $\rightleftharpoons$ ADP + glucose 6-phosphate $\Delta G° = -17.7$ kJ

What is the value of K_c for this reaction at 298 K?

15.69. Sucrose enters the series of reactions in glycolysis after its hydrolysis into glucose and fructose:

$$\text{Sucrose} + H_2O \rightleftharpoons \text{glucose} + \text{fructose}$$

$$K_c = 5.3 \times 10^{12} \text{ at 298 K}$$

What is the value of $\Delta G°$ for this process?

15.70. The value of the equilibrium constant K_p for the reaction

$$H_2(g) + CO_2(g) \rightleftharpoons H_2O(g) + CO(g)$$

is 0.534 at 700°C.
   a. Calculate the value of $\Delta G°$ of the reaction using Equation 15.13.
   b. Using values from Appendix 4, calculate the value of $\Delta G°$ of the reaction and compare the result with that obtained in part a.

15.71. Use the following data to calculate the value of K_p at 298 K for the reaction

$$N_2(g) + 2\,O_2(g) \rightleftharpoons 2\,NO_2(g)$$

$N_2(g) + O_2(g) \rightleftharpoons 2\,NO(g)$ $\Delta G° = 173.2$ kJ

$2\,NO(g) + O_2(g) \rightleftharpoons 2\,NO_2(g)$ $\Delta G° = -69.7$ kJ

15.72. Under the appropriate conditions, NO forms N_2O and NO_2:

$$3\,NO(g) \rightleftharpoons N_2O(g) + NO_2(g)$$

Use the values for $\Delta G°$ for the following reactions to calculate the value of K_p for the above reaction at 500°C.

$2\,NO(g) + O_2(g) \rightleftharpoons 2\,NO_2(g)$ $\Delta G° = -69.7$ kJ

$2\,N_2O(g) \rightleftharpoons 2\,NO(g) + N_2(g)$ $\Delta G° = -33.8$ kJ

$N_2(g) + O_2(g) \rightleftharpoons 2\,NO(g)$ $\Delta G° = 173.2$ kJ

Heterogeneous Equilibria

CONCEPT REVIEW

15.73. Write the K_c expression for the following reaction:

$$CuS(s) \rightleftharpoons Cu^{2+}(aq) + S^{2-}(aq)$$

15.74. Write the K_c expression for the following reaction:

$$Al_2O_3(s) + 3\,H_2O(\ell) \rightleftharpoons 2\,Al^{3+}(aq) + 6\,OH^-(aq)$$

15.75. Why does the K_c expression for the reaction

$$CaCO_3(s) \rightleftharpoons CaO(s) + CO_2(g)$$

not contain terms for the concentrations of $CaCO_3$ and CaO?

15.76. Use the equations $\Delta G° = \Delta H° - T\Delta S°$ and $\Delta G° = -RT \ln K$ to explain why the value of K_p for the reaction

$$CaCO_3(s) \rightleftharpoons CaO(s) + CO_2(g)$$

increases with increasing temperature.

Le Châtelier's Principle

CONCEPT REVIEW

15.77. Does adding reactants to a system at equilibrium increase the value of the equilibrium constant?

15.78. Increasing the concentration of a reactant shifts the position of chemical equilibrium towards formation of more products. What effect does adding a reactant have on the rates of the forward and reverse reactions?

15.79. Carbon Monoxide Poisoning Patients suffering from carbon monoxide poisoning are treated with pure oxygen to remove CO from the hemoglobin (Hb) in their blood. The two relevant equilibria are

$$Hb + 4\,CO(g) \rightleftharpoons Hb(CO)_4$$

$$Hb + 4\,O_2(g) \rightleftharpoons Hb(O_2)_4$$

The value of the equilibrium constant for CO binding to Hb is greater than that for O_2. How, then, does this treatment work?

15.80. Is the equilibrium constant K_p for the reaction

$$2\,NO_2(g) \rightleftharpoons N_2O_4(g)$$

in air the same in Los Angeles as in Denver if the atmospheric pressure in Denver is lower but the temperature is the same?

15.81. Henry's law (Chapter 10) predicts that the solubility of a gas in a liquid increases with its partial pressure. Explain Henry's law in relation to Le Châtelier's principle.

*__15.82.__ For the reaction

$$2\,CO(g) + O_2(g) \rightleftharpoons 2\,CO_2(g)$$

why does adding an inert gas such as argon to an equilibrium mixture of CO, O_2, and CO_2 in a sealed vessel increase the total pressure of the system but not affect the position of the equilibrium?

PROBLEMS

15.83. Which of the following equilibria will shift towards formation of more products if an equilibrium mixture is compressed into half its volume?
a. $2 N_2O(g) \rightleftharpoons 2 N_2(g) + O_2(g)$
b. $2 CO(g) + O_2(g) \rightleftharpoons 2 CO_2(g)$
c. $N_2(g) + O_2(g) \rightleftharpoons 2 NO(g)$
d. $2 NO(g) + O_2(g) \rightleftharpoons 2 NO_2(g)$

15.84. Which of the following equilibria will shift towards formation of more products if the volume of a reaction mixture at equilibrium increases by a factor of 2?
a. $2 SO_2(g) + O_2(g) \rightleftharpoons 2 SO_3(g)$
b. $NO(g) + O_3(g) \rightleftharpoons NO_2(g) + O_2(g)$
c. $2 N_2O_5(g) \rightleftharpoons 2 NO_2(g) + O_2(g)$
d. $N_2O_4(g) \rightleftharpoons 2 NO_2(g)$

15.85. What would be the effect of the changes listed on the equilibrium concentrations of reactants and products in the following reaction?

$$2 O_3(g) \rightleftharpoons 3 O_2(g)$$

a. O_3 is added to the system.
b. O_2 is added to the system.
c. The mixture is compressed to one tenth its initial volume.

15.86. How will the changes listed affect the position of the following equilibrium?

$$2 NO_2(g) \rightleftharpoons NO(g) + NO_3(g)$$

a. The concentration of NO is increased.
b. The concentration of NO_2 is increased.
c. The volume of the system is allowed to expand to 5 times its initial value.

15.87. How would reducing the partial pressure of $O_2(g)$ affect the position of the equilibrium in the following reaction?

$$2 SO_2(g) + O_2(g) \rightleftharpoons 2 SO_3(g)$$

*15.88. Ammonia is added to a gaseous reaction mixture containing H_2, Cl_2, and HCl that is at chemical equilibrium. How will the addition of ammonia affect the relative concentrations of H_2, Cl_2, and HCl if the equilibrium constant of reaction 2 is much greater than the equilibrium constant of reaction 1?
(1) $H_2(g) + Cl_2(g) \rightleftharpoons 2 HCl(g)$
(2) $HCl(g) + NH_3(g) \rightleftharpoons NH_4Cl(s)$

15.89. In which of the following equilibria does the product yield increase with increasing temperature?
a. $A + 2B \rightleftharpoons C$ $\Delta H > 0$
b. $A + 2B \rightleftharpoons C$ $\Delta H = 0$
c. $A + 2B \rightleftharpoons C$ $\Delta H < 0$

15.90. In which of the following equilibria does the product yield decrease with increasing temperature?
a. $2X + Y \rightleftharpoons Z$ $\Delta H > 0$
b. $2X + Y \rightleftharpoons Z$ $\Delta H = 0$
c. $2X + Y \rightleftharpoons Z$ $\Delta H < 0$

Calculations Based on *K*

CONCEPT REVIEW

15.91. Explain how checking the value of Q for a specified equilibrium system such as $A \rightleftharpoons B$ enables you to decide whether the reactant or product concentration increases as the system shifts to attain equilibrium.

15.92. If $Q < K$ in a reversible reaction, will the reactant concentration increase or decrease as the system shifts to attain equilibrium?

PROBLEMS

15.93. For the reaction

$$PCl_5(g) \rightleftharpoons PCl_3(g) + Cl_2(g) \qquad K_p = 23.6 \text{ at } 500 \text{ K}$$

a. Calculate the equilibrium partial pressures of the reactants and products if the initial pressures are $P_{PCl_5} = 0.560$ atm and $P_{PCl_3} = 0.500$ atm.
b. If more chlorine is added after equilibrium is reached, how will the concentrations of PCl_5 and PCl_3 change?

15.94. Enough NO_2 gas is injected into a cylindrical vessel to produce a partial pressure, P_{NO_2}, of 0.900 atm at 298 K. Calculate the equilibrium partial pressures of NO_2 and N_2O_4, given

$$2 NO_2(g) \rightleftharpoons N_2O_4(g) \qquad K_p = 4 \text{ at } 298 \text{ K}$$

15.95. The value of K_c for the reaction between water vapor and dichlorine monoxide

$$H_2O(g) + Cl_2O(g) \rightleftharpoons 2 HOCl(g)$$

is 0.0900 at 25°C. Determine the equilibrium concentrations of all three compounds if the starting concentrations of both reactants are 0.00432 *M* and no HOCl is present.

15.96. The value of K_p for the reaction

$$3 H_2(g) + N_2(g) \rightleftharpoons 2 NH_3(g)$$

is 4.3×10^{-4} at 648 K. Determine the equilibrium partial pressure of NH_3 in a reaction vessel that initially contained 0.900 atm N_2 and 0.500 atm H_2 at 648 K.

15.97. The value of K_p for the reaction

$$NO(g) + \tfrac{1}{2} O_2(g) \rightleftharpoons NO_2(g)$$

is 2×10^6 at 25°C. At equilibrium, what is the ratio of P_{NO_2} to P_{NO} in air at 25°C? Assume that $P_{O_2} = 0.21$ atm and does not change.

*15.98. Water Gas The water-gas reaction is a source of hydrogen. Passing steam over hot carbon produces a mixture of carbon monoxide and hydrogen:

$$H_2O(g) + C(s) \rightleftharpoons CO(g) + H_2(g)$$

The value of K_c for the reaction at 1000°C is 3.0×10^{-2}.
a. Calculate the equilibrium partial pressures of the products and reactants if $P_{H_2O} = 0.442$ atm and $P_{CO} = 5.0$ atm at the start of the reaction. Assume that the carbon is in excess.
b. Determine the equilibrium partial pressures of the reactants and products after sufficient CO and H_2 are added to the equilibrium mixture in part a to initially increase the partial pressures of both gases by 0.075 atm.

15.99. The value of K_p for the reaction

$$CO_2(g) + C(s) \rightleftharpoons 2\,CO(g)$$

is 1.5 at 700°C. Calculate the equilibrium partial pressures of CO and CO_2 if initially $P_{CO_2} = 5.0$ atm and $P_{CO} = 0.0$. Pure graphite is present initially and when equilibrium is achieved.

15.100. **Jupiter's Atmosphere** Ammonium hydrogen sulfide (NH_4SH) was detected in the atmosphere of Jupiter subsequent to its collision with the comet Shoemaker–Levy. The equilibrium between ammonia, hydrogen sulfide, and NH_4SH is described by the following equation:

$$NH_4SH(g) \rightleftharpoons NH_3(g) + H_2S(g)$$

The value of K_p for the reaction at 24°C is 0.126. Suppose a sealed flask contains an equilibrium mixture of gaseous NH_4SH, NH_3 and H_2S. At equilibrium, the partial pressure of H_2S is 0.355 atm. What is the partial pressure of NH_3?

*15.101. A flask containing pure NO_2 was heated to 1000 K, a temperature at which the value of K_p for the decomposition of NO_2 is 158.

$$2\,NO_2(g) \rightleftharpoons 2\,NO(g) + O_2(g)$$

The partial pressure of O_2 at equilibrium is 0.136 atm.
a. Calculate the partial pressures of NO and NO_2.
b. Calculate the total pressure in the flask at equilibrium.

15.102. The equilibrium constant K_p of the reaction

$$2\,SO_3(g) \rightleftharpoons 2\,SO_2(g) + O_2(g)$$

is 7.69 at 830°C. If a vessel at this temperature initially contains pure SO_3 and if the partial pressure of SO_3 at equilibrium is 0.100 atm, what is the partial pressure of O_2 in the flask at equilibrium?

*15.103. **NO_x Pollution** In a study of the formation of NO_x air pollution, a chamber heated to 2200°C was filled with air (0.79 atm N_2, 0.21 atm O_2). What are the equilibrium partial pressures of N_2, O_2, and NO if $K_p = 0.050$ for the following reaction at 2200°?

$$N_2(g) + O_2(g) \rightleftharpoons 2\,NO(g)$$

*15.104. The equilibrium constant K_p for the thermal decomposition of NO_2

$$2\,NO_2(g) \rightleftharpoons 2\,NO(g) + O_2(g)$$

is 6.5×10^{-6} at 450°C. If a reaction vessel at this temperature initially contains 0.500 atm NO_2, what will be the partial pressures of NO_2, NO, and O_2 in the vessel when equilibrium has been attained?

15.105. The value of K_c for the thermal decomposition of hydrogen sulfide

$$2\,H_2S(g) \rightleftharpoons 2\,H_2(g) + S_2(g)$$

is 2.2×10^{-4} at 1400 K. A sample of gas in which $[H_2S] = 6.00\ M$ is heated to 1400 K in a sealed high-pressure vessel. After chemical equilibrium has been achieved, what is the value of $[H_2S]$? Assume that no H_2 or S_2 was present in the original sample.

15.106. **Urban Air** On a very smoggy day, the equilibrium concentration of NO_2 in the air over an urban area reaches $2.2 \times 10^{-7}\ M$. If the temperature of the air is 25°C, what is the concentration of the dimer N_2O_4 in the air? Given:

$$N_2O_4(g) \rightleftharpoons 2\,NO_2(g) \qquad K_c = 6.1 \times 10^{-3}$$

*15.107. **Chemical Weapon** Phosgene, $COCl_2$ gained notoriety as a chemical weapon in World War I. Phosgene is produced by the reaction of carbon monoxide with chlorine

$$CO(g) + Cl_2(g) \rightleftharpoons COCl_2(g)$$

The value of K_c for this reaction is 5.0 at 600 K. What are the equilibrium partial pressures of the three gases if a reaction vessel initially contains a mixture of the reactants in which $P_{CO} = P_{Cl_2} = 0.265$ atm and $P_{COCl_2} = 0.000$ atm?

*15.108. At 2000°C, the value of K_c for the reaction

$$2\,CO(g) + O_2(g) \rightleftharpoons 2\,CO_2(g)$$

is 1.0. What is the ratio of [CO] to $[CO_2]$ in an atmosphere in which $[O_2] = 0.0045\ M$?

*15.109. The water–gas shift reaction is an important source of hydrogen. The value of K_c for the reaction

$$CO(g) + H_2O(g) \rightleftharpoons CO_2(g) + H_2(g)$$

at 700 K is 5.1. Calculate the equilibrium concentrations of the four gases if the initial concentration of each of them is 0.050 M.

*15.110. Sulfur dioxide reacts with NO_2, forming SO_3 and NO:

$$SO_2(g) + NO_2(g) \rightleftharpoons SO_3(g) + NO(g)$$

If the value of K_c for the reaction is 2.50, what are the equilibrium concentrations of the products if the reaction mixture was initially 0.50 M SO_2, 0.50 M NO_2, 0.0050 M SO_3, and 0.0050 M NO?

Changing *K* with Changing Temperature

CONCEPT REVIEW

15.111. The value of the equilibrium constant of a reaction decreases with increasing temperature. Is this reaction endothermic or exothermic?

15.112. The reaction

$$2\,CO(g) + O_2(g) \rightleftharpoons 2\,CO_2(g)$$

is exothermic. Does the value of K_p increase or decrease with increasing temperature?

15.113. The value of K_p for the water–gas shift reaction

$$CO(g) + H_2O(g) \rightleftharpoons H_2(g) + CO_2(g)$$

increases as the temperature decreases. Is the reaction exothermic or endothermic?

15.114. Does the value of K_p for the reaction

$$CH_4(g) + H_2O(g) \rightleftharpoons 3\,H_2(g) + CO(g) \qquad \Delta H° = 206\ kJ$$

increase, decrease, or remain unchanged as temperature increases?

PROBLEMS

15.115. **Air Pollution** Automobiles and trucks pollute the air with NO. At 2000°C, K_c for the reaction

$$N_2(g) + O_2(g) \rightleftharpoons 2\,NO(g)$$

is 4.10×10^{-4}, and $\Delta H° = 180.6$ kJ. What is the value of K_c at 25°C?

15.116. At 400 K the value of K_p for the reaction

$$N_2(g) + 3\,H_2(g) \rightleftharpoons 2\,NH_3(g)$$

is 41, and $\Delta H° = -92.2$ kJ. What is the value of K_p at 700 K?

15.117. The equilibrium constant for the reaction

$$NO(g) + O_2(g) \rightleftharpoons 2\,NO_2(g)$$

decreases from 1.5×10^5 at 430°C to 23 at 1000°C. From these data, calculate the value of $\Delta H°$ for the reaction.

15.118. The value of K_c for the reaction A $\rightleftharpoons$ B is 0.455 at 50°C and 0.655 at 100°C. Calculate $\Delta H°$ for the reaction.

Additional Problems

*15.119. **CO as a Fuel** Is carbon dioxide a viable source of the fuel CO? Pure carbon dioxide ($P_{CO_2} = 1$ atm) decomposes at high temperatures. For the system

$$2\,CO_2(g) \rightleftharpoons 2\,CO(g) + O_2(g)$$

the percentage of decomposition of $CO_2(g)$ changes with temperature as follows:

Temperature (K)	Decomposition (%)
1500	0.048
2500	17.6
3000	54.8

Is the reaction endothermic? Calculate the value of K_p at each temperature and discuss the results. Is the decomposition of CO_2 an antidote for global warming?

15.120. Ammonia decomposes at high temperatures. In an experiment to explore this behavior, 2.000 moles of gaseous NH_3 are sealed in a rigid 1 liter vessel. The vessel is heated at 800 K and some of the NH_3 decomposes in the following reaction:

$$2\,NH_3(g) \rightleftharpoons N_2(g) + 3\,H_2(g)$$

The system eventually reaches equilibrium and is found to contain 0.004 mole of NH_3. What are the values of K_p and K_c for the decomposition reaction at 800 K?

*15.121. Elements of group 16 form hydrides with the generic formula H_2X. When gaseous H_2X is bubbled through a solution containing 0.3 M hydrochloric acid, the solution becomes saturated and $[H_2X] = 0.1$ M. The following equilibria exist in this solution:

$$H_2X(aq) + H_2O(\ell) \rightleftharpoons HX^-(aq) + H_3O^+(aq) \qquad K_1 = 8.3 \times 10^{-8}$$

$$HX^-(aq) + H_2O(\ell) \rightleftharpoons X^{2-}(aq) + H_3O^+(aq) \qquad K_2 = 1 \times 10^{-14}$$

Calculate the concentration of X^{2-} in the solution.

*15.122. **Homogeneous Catalysis** Nitrogen dioxide catalyzes the reaction of SO_2 with O_2 to form SO_3:

$$NO_2(g) + SO_2(g) \rightleftharpoons NO(g) + SO_3(g)$$

An equilibrium mixture is analyzed at a certain temperature and found to contain $[NO_2] = 0.100$ M, $[SO_2] = 0.300$ M, $[NO] = 2.00$ M, and $[SO_3] = 0.600$ M. At the same temperature, extra $SO_2(g)$ is added to make $[SO_2(g)] = 0.800$ M. Calculate the composition of the mixture when equilibrium has been reestablished.

*15.123. Carbon disulfide is a foul-smelling solvent that dissolves sulfur and other nonpolar substances. It can be made by heating sulfur in an atmosphere of methane. For the reaction

$$4\,CH_4(g) + S_8(s) \rightleftharpoons 4\,CS_2(g) + 8\,H_2(g)$$

$\Delta H°_{rxn} = 150.3$ kJ per mole of CH_4 and $\Delta S°_{rxn} = 336$ J per mole of CH_4. Calculate the equilibrium constant at 25°C and 500°C.

*15.124. **Making Hydrogen** Debate continues on the practicality of H_2 gas as a fuel. The equilibrium constant K_c for the reaction

$$CO(g) + H_2O(g) \rightleftharpoons CO_2(g) + H_2(g)$$

is 1.0×10^5 at 25°C. Starting with this value, calculate the value of $\Delta G°_{rxn}$ at 25°C, and, without doing any calculations, guess the sign of $\Delta H°_{rxn}$.

*15.125. **Smokestack Cleanup** Calcium oxide is used to remove the pollutant SO_2 from smokestacks and exhaust gases. The $\Delta G°$ of the overall reaction

$$CaO(s) + SO_2(g) + \tfrac{1}{2}O_2(g) \rightleftharpoons CaSO_4(s)$$

is -418.6 kJ. What is P_{SO_2} in equilibrium with air ($P_{O_2} = 0.21$ atm) and solid CaO?

*15.126. **Volcanic Eruptions** During volcanic eruptions, gases as hot as 700°C and rich in SO_2 are released into the atmosphere. As air mixes with these gases, the following reaction converts some of this SO_2 into SO_3:

$$2\,SO_2(g) + O_2(g) \rightleftharpoons 2\,SO_3(g)$$

Calculate the value of K_p for this reaction at 700°C. What is the ratio of P_{SO_2} to P_{SO_3} in equilibrium with $P_{O_2} = 0.21$ atm?

Equilibrium in the Aqueous Phase

16

SHADES OF PURPLE, PINK, AND BLUE FLOWERS *The color of hydrangea flowers depends on the acid content of the soil in which they are grown.*

A LOOK AHEAD: A Balancing Act

Relative acidity or alkalinity is a crucially important characteristic of many systems. Farmers adjust the acid content of soil to enhance the growth of crops, and gardeners recognize the need of plants like blueberries and azaleas for an acidic environment. If the soil in which these plants are grown is insufficiently acidic, the plants cannot make use of nutrients, even if they are abundant in the ground. Most plants grow best in soils that are neutral to very slightly acidic because the microorganisms that play key roles in making nutrients available to plants thrive in such soils.

Hydrangeas vary in color with the acidity of the soil in which they are grown. The acid content of the soil affects the ability of aluminum ions to be taken up by the plant and to interact with pigment molecules called anthocyanins present in the flowers. Aluminum ions are soluble in acidic solutions, and hydrangeas grown in acid soil have blue flowers. In neutral to slightly alkaline soil, however, aluminum precipitates as aluminum hydroxide and is not available to plants. Hydrangeas grown in these soils produce pink blooms.

The relative acidity or alkalinity of soil and ground water depends on mineral content and the characteristics of the rainwater that falls on it. Normal rainwater is slightly acidic because of the presence of dissolved CO_2, which reacts with water and produces hydrogen ions in solution. Acid rain, on the other hand, is formed when oxides of nitrogen and sulfur enter the air, typically as byproducts of the combustion of fossil fuels. Compared to CO_2, NO_x and SO_2 react to a greater extent with water. Air containing NO_x and SO_2 yields rainwater with a hydrogen ion content that is 10 to 1000 times more acidic than that of rainwater containing only dissolved CO_2. The more acidic rainwater alters the chemical balance in soils and interferes with plants' abilities to process nutrients. As a result, plants don't grow well. Sometimes they die.

Carbon dioxide and carbonate compounds play important roles in determining the relative acidity and alkalinity in biological systems. We produce CO_2 during metabolism, and we eliminate it every time we exhale. Our breathing maintains a balance of relative acidity and alkalinity by controlling the amount of dissolved CO_2 in our systems. We actually need more controlled CO_2 levels than O_2 levels. Carbon dioxide in the fluids surrounding cells and in particular in blood maintains the acid–base balance our bodies need. This balance is vital because hydrogen ions interact with amino acids and alter the charge on proteins. Changes in charge disturb protein structure and function. The body can tolerate only a small change in hydrogen ion concentration in the blood, and deviations outside the normal range lead to coma and death.

Ocean water is naturally alkaline and typically has a hydrogen ion concentration nearly 10 times lower than in human plasma. At the oceans' natural level of alkalinity, mollusks such as clams and oysters are able to form solid calcium carbonate shells and corals can form reefs of calcium carbonate, which has limited solubility under these conditions. However, if ocean water becomes more acidic because more CO_2 or other nonmetal oxides dissolve in it, the solubility of $CaCO_3$ increases, which could have disastrous consequences for marine life.

The balance between acidity and alkalinity in rainwater, groundwater, and seawater clearly has a direct or indirect impact on all life forms. In this chapter we examine processes that disturb this balance and others that help restore it.

16.1 Acids and Bases

Acids and bases were originally identified as unique classes of compounds because of their distinctive properties. You must *never* taste or even casually smell unknown materials, but early chemists frequently used taste and aroma as identifying characteristics of substances, and the practice sometimes had fatal consequences. Acids taste sour; bases taste bitter. Aqueous solutions of bases have a slippery, soapy feel. Acids and bases can change the color of certain vegetable dyes. Litmus, a substance extracted from lichen, turns red in the presence of an acid; a base turns red litmus blue. Acids and bases react with each other in a process called neutralization because the reaction causes the acidic and basic properties of the two substances to disappear. Some of the acids and bases that we frequently encounter are listed in Table 16.1.

Acids and bases are compounds of profound importance to industry, every living system, and the environment. Their commercial uses range from the production of steel, fertilizer, fabrics, soaps, pharmaceuticals, and cleaner-burning gasoline to enhancing oil recovery from wells, flavoring foods and beverages, and cleaning metals. In our bodies, hydrolysis of CO_2 produces a weak acid that plays a major role in maintaining the acid–base balance necessary to assure normal functioning.

In Chapter 4 we described the transformation of early Earth by reactions between the acids—formed when volcanic gases such as oxides of nitrogen and sulfur dissolved in rainwater—and basic minerals in the Earth's crust. In Chapters 14 and 15 we learned how internal combustion and diesel engines produce NO and how combustion of fossil fuels containing sulfur produces SO_2. Once in the atmosphere, these nitrogen and sulfur oxides react with additional O_2 and water vapor to produce nitric acid (HNO_3), sulfuric acid (H_2SO_4), and other acids that fall to Earth in the form of acid rain.

The usefulness as well as the potential hazards associated with acids and bases depend in large part on the behavior of these compounds in water. Some acids and bases are completely dissociated in water; the behavior of others is characterized by equilibria that are governed by the same principles we explored for reactions in the gas phase in Chapter 15. In this chapter we apply the lessons we learned in Chapter 15 to an examination of the vitally important chemistry of acids and bases in aqueous solutions.

CONNECTION We introduced acids, bases, and neutralization reactions in Chapter 4, where we defined acids as proton donors and bases as proton acceptors. Neutralization reactions occur between acids and bases to produce water and a salt.

Litmus, which changes color in response to the presence of acid or base, is a natural product extracted from this lichen.

▶❚❚ **CHEMTOUR** Acid Rain

CONNECTION Hydrolysis reactions, which are the reactions of substances with water, were first introduced in Chapters 3 and 4.

CONNECTION In Chapter 3, we identified nonmetal oxides like $CO_2(g)$, $NO_2(g)$, and $SO_3(g)$ as acid anhydrides, because they hydrolyze in water to produce acidic solutions.

TABLE 16.1 Acids and Bases Encountered in Daily Life

ACIDS		BASES	
Chemical Name	**Common Name**	**Chemical Name**	**Common Name**
Acetic acid	Vinegar	Ammonia	Ammonia (household cleaner)
Hydrochloric acid	Gastric fluid and muriatic acid	Sodium hydroxide	Drain cleaner
Citric acid	Lemon juice is 6%–8% citric acid	Sodium bicarbonate	Baking soda
Acetyl salicylic acid	Aspirin	Magnesium hydroxide	Milk of magnesia

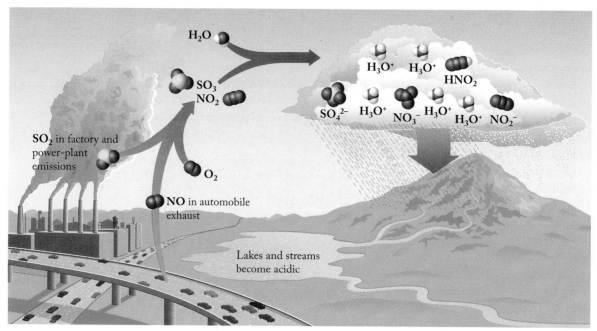

Acid rain forms when volatile nonmetal oxides such as NO and SO_2 are further oxidized in the atmosphere and dissolve in rain to form nitric (HNO_3), nitrous (HNO_2), and sulfuric (H_2SO_4) acids. These acids ionize, forming NO_3^-, NO_2^-, and SO_4^{2-} ions, respectively, and the hydronium (H_3O^+) ions that make acid rain acidic.

16.2 The Brønsted–Lowry Model

CONNECTION In Chapter 4 we introduced three ways to refer to the hydrogen ions produced by acids in aqueous solution: as the proton [$H^+(aq)$], the hydrogen ion [also $H^+(aq)$], or the hydronium ion [$H_3O^+(aq)$]. The hydronium ion most closely represents the species actually present in solution.

In Chapter 4 we defined acids as substances that donate H^+ ions and bases as substances that accept H^+ ions. These descriptions of acids and bases were developed independently by Danish chemist Johannes Brønsted (1879–1947) and English chemist Thomas Lowry (1874–1936) and published in the same year, 1923. Today these descriptions are known as the **Brønsted–Lowry model** of acids and bases.

Strong and Weak Acids

We classify acids as *strong* or *weak* depending on the extent to which they donate H^+ ions. Strong acids are completely ionized in water; weak acids are not. Table 16.2 lists the names and formulas of some common strong acids; the table also presents chemical equations describing how these substances ionize when they dissolve in water. The strong acids are of two types: those formed by binary molecular compounds of the generic formula HX, where X is any group 17 element except fluorine (HF is a weak acid), and those called *oxoacids* with the generic formula H_mXO_n, where X is a nonmetal.

CHEMTOUR Acid–Base Ionization

Ionization of a Brønsted–Lowry acid in water involves the transfer of a hydrogen ion from the acid (the proton donor) to a base (the proton acceptor). Let's look at the ionization of a typical strong acid like nitric acid in water:

$$HNO_3(aq) \rightarrow H^+(aq) + NO_3^-(aq)$$

Let's rewrite this equation to highlight the role of water in this process as a Brønsted–Lowry base:

$$\underset{\substack{\text{acid}\\\text{(proton donor)}}}{HNO_3(aq)} + \underset{\substack{\text{base}\\\text{(proton acceptor)}}}{H_2O(\ell)} \rightarrow H_3O^+(aq) + NO_3^-(aq) \quad (16.1)$$

The **Brønsted–Lowry model** defines acids as H^+ ion donors and bases as H^+ ion acceptors.

TABLE 16.2 **Strong Acids and Bases and Their Ionization Reactions in Water**

Name of Acid	Formula	Reaction in Water
Hydrochloric	HCl	$HCl(aq) + H_2O(\ell) \rightarrow H_3O^+(aq) + Cl^-(aq)$
Hydrobromic	HBr	$HBr(aq) + H_2O(\ell) \rightarrow H_3O^+(aq) + Br^-(aq)$
Hydroiodic	HI	$HI(aq) + H_2O(\ell) \rightarrow H_3O^+(aq) + I^-(aq)$
Nitric	HNO_3	$HNO_3(aq) + H_2O(\ell) \rightarrow H_3O^+(aq) + NO_3^-(aq)$
Sulfuric	H_2SO_4	$H_2SO_4(aq) + H_2O(\ell) \rightarrow H_3O^+(aq) + HSO_4^-(aq)$
		$HSO_4^-(aq) + H_2O(\ell) \rightleftharpoons H_3O^+(aq) + SO_4^{2-}(aq)^a$
Perchloric	$HClO_4$	$HClO_4(aq) + H_2O(\ell) \rightarrow H_3O^+(aq) + ClO_4^-(aq)$

^{a}The first ionization reaction of H_2SO_4 is complete, but the second ionization reaction may be incomplete, depending on the concentration of H_2SO_4.

Nitric acid is the proton donor, and a molecule of water is the proton acceptor and base. Nitric acid is a strong acid, which means that virtually every molecule of HNO_3 in aqueous solution gives up its proton to a water molecule. Consequently, the reaction in Equation 16.1 is not characterized by an equilibrium constant because the reaction goes to completion.

The ionization of a strong acid in water may be viewed on a molecular basis as shown in Figure 16.1 for the ionization of HCl. Hydrogen bonding between the δ^- oxygen atom in a polar H_2O molecule and a δ^+ hydrogen atom in a polar HCl molecule results in the O atom donating a lone pair of electrons to form a bond to the H^+ ion that forms as a molecule of HCl ionizes. A product of this interaction is the hydronium ion, $H_3O^+(aq)$.

We need to think about hydronium ions (H_3O^+) as we describe acid–base reactions in terms of acids donating and bases accepting H^+ ions in water, because free protons (H^+ ions) do not exist in aqueous solutions. Each H^+ ion is bonded to at least one water molecule. Actually, additional ion–dipole interactions between the electronegative oxygen atoms in water molecules and H_3O^+ ions lead to the formation of clusters of molecules of H_2O around H_3O^+ ions as shown in Figure 16.2. As we describe concentrations and transfers of H^+ ions in acid–base reactions, we really mean concentrations and transfers of hydronium ions with water molecules clustered around them.

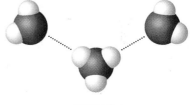

$H(H_2O)_3^+$

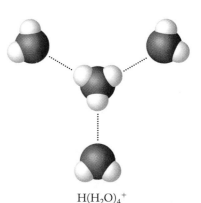

$H(H_2O)_4^+$

FIGURE 16.2 Water molecules cluster around hydronium (H_3O^+) ions, forming species with the general formula $H(H_2O)_n^+$. The clusters shown here have the formulas $H(H_2O)_3^+$ and $H(H_2O)_4^+$.

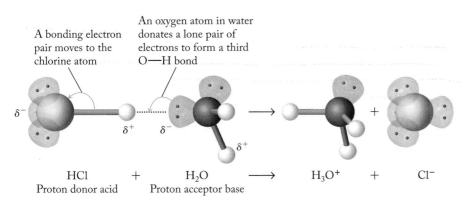

A bonding electron pair moves to the chlorine atom

An oxygen atom in water donates a lone pair of electrons to form a third O—H bond

| HCl | + | H_2O | $\longrightarrow$ | H_3O^+ | + | Cl^- |
| Proton donor acid | | Proton acceptor base | | | | |

FIGURE 16.1 When hydrogen chloride dissolves in water, strong hydrogen bonding between HCl and H_2O molecules results in ionization of HCl and formation of Cl^- and H_3O^+ ions.

TABLE 16.3 Common Weak Acids and Their Ionization Reactions in Water

Name	Formula	Reaction in Water
Nitrous acid	HNO_2	$HNO_2(aq) + H_2O(\ell) \rightleftharpoons H_3O^+(aq) + NO_2^-(aq)$
Hypochlorous acid	$HClO$	$HClO(aq) + H_2O(\ell) \rightleftharpoons H_3O^+(aq) + ClO^-(aq)$
Hydrofluoric acid	HF	$HF(aq) + H_2O(\ell) \rightleftharpoons H_3O^+(aq) + F^-(aq)$
Formic acid	$HCOOH$	$HCOOH(aq) + H_2O(\ell) \rightleftharpoons H_3O^+(aq) + HCOO^-(aq)$
Acetic acid	CH_3COOH	$CH_3COOH(aq) + H_2O(\ell) \rightleftharpoons H_3O^+(aq) + CH_3COO^-(aq)$

In contrast to the behavior of strong acids, the ionization of weak acids in water does not go to completion. In the reaction of hydrofluoric acid with water,

$$HF(aq) + H_2O(\ell) \rightleftharpoons H_3O^+(aq) + F^-(aq) \tag{16.2}$$

the hydrogen fluoride molecule functions as an acid and donates a proton to water, which functions as a base. The reaction in Equation 16.2, however, does not go to completion, and the four species involved reach a state of equilibrium. This means that the products (H_3O^+ and F^-) of this reaction have some tendency to react with each other and reform the reactants (H_2O and HF). The extent to which the forward reaction takes place is characterized by an equilibrium constant K, whose magnitude indicates the relative strength of the acid. The dissociations of all weak acids are equilibrium processes. Table 16.3 shows some common weak acids and their ionization reactions in water.

Conjugate Acid–Base Pairs

Both the forward and the reverse reactions in Equation 16.2 involve proton transfers. In the reverse reaction, $H_3O^+(aq)$ functions as an acid and $F^-(aq)$ functions as a base. In the Brønsted–Lowry model, the acid $HF(aq)$ and the base $F^-(aq)$ are related and differ only by the presence or absence of the proton. Acids and bases related by proton transfer in this manner are called *conjugate pairs*. The fluoride ion is the **conjugate base** of the hydrofluoric acid. Together, HF and F^- form a **conjugate acid–base pair**. The fluoride ion conjugate base is formed when the hydrofluoric acid transfers its proton. In an analogous way, the basic water molecule (H_2O) on the reactant side is related by proton transfer to the acidic hydronium ion (H_3O^+) on the product side. The hydronium ion is the **conjugate acid** of the water molecule. It is formed when the water molecule acts as a base and accepts a proton, and together they form another conjugate acid–base pair. Equation 16.3 summarizes the relations as defined by the Brønsted–Lowry model:

When a Brønsted–Lowry acid donates a H^+ ion, it forms its **conjugate base**.

Conjugate acid–base pairs are a Brønsted–Lowry acid and base differing from each other only by the presence or absence of a proton.

When a Brønsted–Lowry base accepts a proton, it becomes its **conjugate acid**.

gives up a proton

acid $\xrightarrow{\hspace{2cm}}$ conjugate base

$$HF(aq) \quad + \quad H_2O(\ell) \quad \rightleftharpoons \quad H_3O^+(aq) \quad + \quad F^-(aq)$$

base $\xrightarrow{\hspace{1cm}}$ conjugate acid

accepts a proton

$$\text{acid} + \text{base} \rightleftharpoons \text{conjugate acid} + \text{conjugate base} \tag{16.3}$$

SAMPLE EXERCISE 16.1   **Identifying Conjugate Acid–Base Pairs**

Identify the conjugate acid–base pairs in the reactions that result when the following acids dissolve in water: (a) $HClO_4$; (b) HCOOH.

COLLECT AND ORGANIZE We are asked to identify the conjugate acid–base pairs that form in aqueous solutions of two acids. Perchloric acid is a strong acid; formic acid is a weak acid. We need to write equations for the reactions and identify the species that donate protons and those that accept protons.

ANALYZE A conjugate acid–base pair is an acid and a base differing from each other only by the presence or absence of a proton. The reaction for the perchloric acid goes to completion, so we use a single reaction arrow between reactants and products. The reaction for formic acid is an equilibrium, so we use equilibrium arrows in that equation.

SOLVE For perchloric acid:

acid   conjugate base

$$HClO_4(aq) + H_2O(\ell) \;\rightarrow\; H_3O^+(aq) + ClO_4^-(aq)$$

base   conjugate acid

One conjugate pair consists of perchloric acid and its conjugate base, the perchlorate ion; the other conjugate pair consists of water as the base and the hydronium ion as its conjugate acid.

For formic acid:

acid   conjugate base

$$HCOOH(aq) + H_2O(\ell) \;\rightleftharpoons\; H_3O^+(aq) + HCOO^-(aq)$$

base   conjugate acid

One conjugate pair consists of formic acid and its conjugate base, the formate ion; the other conjugate pair consists of water as the base and its conjugate acid, the hydronium ion.

THINK ABOUT IT Identifying conjugate pairs requires us to focus on the transfer of a proton from an acid to a base in a Brønsted–Lowry acid–base reaction.

Practice Exercise Identify the conjugate acid–base pairs in the reaction that takes place when acetic acid (CH_3COOH) dissolves in water.

(Answers to Practice Exercises are in the back of the book.)

Strong and Weak Bases

Common strong bases in water are listed in Table 16.4. They are hydroxides of group 1 and 2 metals except for magnesium. As the reactions in Table 16.4 indicate, when these strong bases dissolve in water, they dissociate completely into their component cations and anions. Therefore, the dissociation process for a strong base is not characterized by an equilibrium constant. In terms of the Brønsted–Lowry

TABLE 16.4	**Strong Bases in Water**	
Name of Base	**Formula**	**Reaction in Water**
Lithium hydroxide	LiOH	$LiOH(aq) \rightarrow Li^+(aq) + OH^-(aq)$
Sodium hydroxide	NaOH	$NaOH(aq) \rightarrow Na^+(aq) + OH^-(aq)$
Potassium hydroxide	KOH	$KOH(aq) \rightarrow K^+(aq) + OH^-(aq)$
Calcium hydroxide	$Ca(OH)_2$	$Ca(OH)_2(aq) \rightarrow Ca^{2+}(aq) + 2OH^-(aq)$
Barium hydroxide	$Ba(OH)_2$	$Ba(OH)_2(aq) \rightarrow Ba^{2+}(aq) + 2OH^-(aq)$
Strontium hydroxide	$Sr(OH)_2$	$Sr(OH)_2(aq) \rightarrow Sr^{2+}(aq) + 2OH^-(aq)$

model, the hydroxide ions produced when substances like NaOH dissolve in water are very effective H^+ acceptors and hence are strong Brønsted–Lowry bases.

When a weak base dissolves in water, the reaction can also be described by applying the Brønsted–Lowry model. Ammonia is a weak base. In aqueous solution, it accepts a hydrogen ion from water to form its conjugate acid, the ammonium ion NH_4^+ (Equation 16.4). In the process, the water molecule that transferred the proton is converted into OH^-. Note that water in solution with ammonia behaves as an acid whose conjugate base is the hydroxide ion:

$$NH_3(aq) + H_2O(\ell) \rightleftharpoons NH_4^+(aq) + OH^-(aq) \qquad (16.4)$$

base — conjugate acid; acid — conjugate base

Relative Strengths of Acids and Bases

We can use the concepts we have developed for describing equilibria to discuss acid–base reactions in aqueous medium and to evaluate the relative strengths of acids and bases. Let's examine the ionization of a strong acid such as HCl in greater detail:

$$HCl(aq) + H_2O(\ell) \rightarrow H_3O^+(aq) + Cl^-(aq)$$

HCl is a strong acid, which means that this reaction goes to completion. It also means that the reverse reaction essentially does not happen at all. This, in turn, means that a Cl^- ion (the conjugate base of HCl) must be a very weak base, because it has no tendency to accept a proton from H_3O^+. The ability to compare the relative strengths of acids and bases in this fashion gives rise to the observation that, for all conjugate pairs, strong acids have extremely weak conjugate bases. In between these two extremes we find many substances that are intermediate in strength.

Nitrous acid, for example, is a weak acid that has a weak conjugate base, NO_2^-.

$$HNO_2(aq) + H_2O(\ell) \rightleftharpoons H_3O^+(aq) + NO_2^-(aq)$$

acid — base — conjugate acid — conjugate base

Thus, solutions of HNO_2 are weakly acidic because HNO_2 transfers its proton to water to a limited extent.

The complementary strengths of a number of the common acid–base conjugate pairs are shown in Figure 16.3. The strongest acids are at the top on the left, and the strongest bases are at the bottom on the right. This figure illustrates an important generalization about Brønsted–Lowry acids and bases: the stronger an acid, the weaker its conjugate base.

All the strong acids in Figure 16.3 ionize completely in water; they all have the same strength, and they all behave as solutions of H_3O^+. Because all strong acids are equal in water, water is said to *level* these acids, and the **leveling effect** means that H_3O^+ is the strongest acid that exists in water. One mole of HNO_3 in water, for example, is completely converted into 1 mole of H_3O^+ ions and NO_3^- ions; molecules of HNO_3 do not exist in water. One mole of HCl is likewise completely ionized; molecules of HCl do not exist in water. The two acids are of equal strength. There is no difference between the two in terms of their ability to donate a hydrogen ion to H_2O; their strengths are *leveled*. Weaker acids, however, may be discriminated by their relative abilities to donate H^+ ions to water. Acids higher on the list in the left-hand column are more able to donate H^+ ions to water than acids lower on the list. Their strengths can be discriminated in water; they are not leveled.

Similarly, the strongest base that exists in water is OH^-. Any bases stronger than OH^- hydrolyze in water to produce OH^-. The oxide ion (O^{2-}), for example, reacts with water and is completely converted into OH^-:

$$O^{2-}(aq) + H_2O(\ell) \rightarrow 2\,OH^-(aq)$$

Strong bases are converted completely to OH^- in aqueous solution; they are leveled in water. Bases weaker than OH^- can be discriminated in water; they are not leveled.

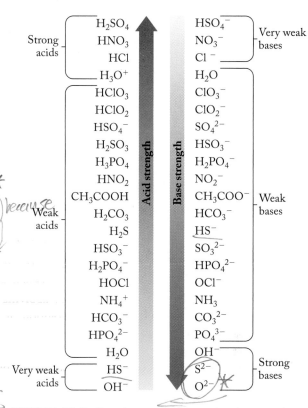

FIGURE 16.3 Opposing trends characterize the strength of acids and their conjugate bases: the stronger the acid, the weaker its conjugate base.

16.3 pH and the Autoionization of Water

The relative acidities of solutions may be expressed in terms of the concentrations of H_3O^+ ions in their solutions. In this section we will examine an alternative scale for expressing acidity or basicity. To understand how the scale works, we need first to understand the *autoionization* of water. As this term implies, molecules of H_2O have the ability to ionize each other. A very small number of hydrogen ions and hydroxide ions are always present in pure water. The ions form by the ionization of water:

$$H_2O(\ell) + H_2O(\ell) \rightleftharpoons H_3O^+(aq) + OH^-(aq) \qquad (16.5)$$

On a molecular scale, the process is driven by hydrogen bonds between hydrogen and oxygen atoms of different water molecules (Figure 16.4). These hydrogen bonds are strong enough to ionize some water molecules.

In the autoionization reaction in Equation 16.5, one water molecule (acting as an acid) donates a hydrogen ion to another water molecule (acting as a base). The first molecule forms its conjugate base (OH^-), and the second forms its conjugate acid (H_3O^+). We have already encountered this dual acid–base nature of water

The **leveling effect** of water refers to the observation that strong acids all have the same strength in water and are completely converted into solutions of H_3O^+. Strong bases are likewise leveled in water and are completely converted into solutions of OH^-.

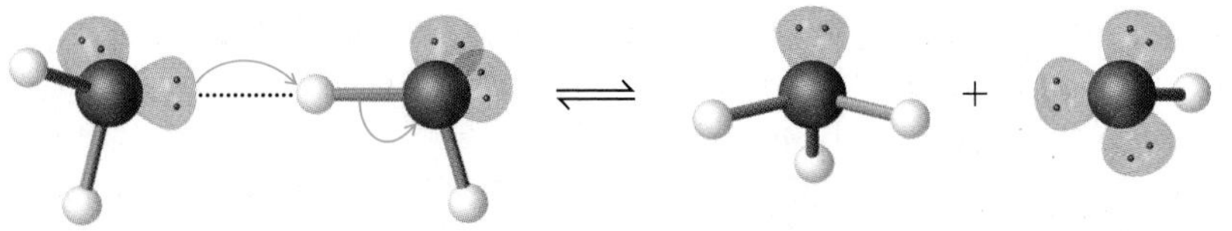

FIGURE 16.4 The autoionization of water takes place when strong hydrogen bonds result in the transfer of a proton from one water molecule to another.

in reactions in which water acts as a proton acceptor (base) in solutions of acids (Equation 16.3) but acts as an acid in a solution of bases (Equation 16.4). When a substance can take on either acidic or basic characteristics—donating hydrogen ions in basic solutions and accepting hydrogen ions in acidic solutions—the substance is said to be **amphoteric**. The autoionization of water is an example of amphoteric behavior.

Because autoionization is a process that establishes an equilibrium, we can discuss it quantitatively by using the concepts we developed in Chapter 15. The equilibrium constant for the autoionization of water can be written as

$$K_c = \frac{[H_3O^+][OH^-]}{[H_2O][H_2O]}$$

CHEMTOUR Self-Ionization of Water

Water is ionized to such a small extent that the concentration of water is essentially a constant and is not shown in the mass action expression. Furthermore, we can substitute $[H^+]$ for $[H_3O^+]$ because these terms represent the same species. Doing so simplifies the equilibrium constant expression, which we identify as K_w:

$$K_w = [H^+][OH^-] \tag{16.6}$$

Measurements of $[H^+]$ and $[OH^-]$ in pure water at 25°C establish that both have the same value: $1.00 \times 10^{-7} \ M$. When we insert these values in Equation 16.6,

$$K_w = [H^+][OH^-]$$

$$= (1.00 \times 10^{-7})(1.00 \times 10^{-7})$$

$$= 1.00 \times 10^{-14}$$

or

$$K_w = [H^+][OH^-] = 1.00 \times 10^{-14} \tag{16.7}$$

Such a tiny value of K_w indicates that autoionization does not happen to a great extent. Rather, the reverse of autoionization—namely, the reaction between $[H^+]$ and $[OH^-]$ to produce H_2O—has an equilibrium constant of $1/K_w$ or 1.00×10^{14} and describes a reaction that essentially goes to completion:

$$H^+(aq) + OH^-(aq) \rightarrow H_2O(\ell) \qquad K = 1/K_w = 1.00 \times 10^{14}$$

The value for K_w in Equation 16.7 applies to all aqueous solutions, not just pure water. Therefore, in any aqueous sample, an inverse relation exists between $[H^+]$ and $[OH^-]$: as the value of one increases, the value of the other must decrease so that the product of the two is always 1.0×10^{-14}. A solution in which $[H^+]$

An **amphoteric** substance is capable of behaving either as an acid or as a base.

is greater than [OH⁻] is considered acidic; a solution in which [H⁺] is less than [OH⁻] is basic; and a solution in which the concentrations of the two are equal ($1.00 \times 10^{-7}\,M$) is neither acidic nor basic; it is neutral.

The tiny value of K_w means that the autoionization of water does not contribute significantly to [H⁺] in solutions of acids or to [OH⁻] in solutions of bases, and so we can ignore the contribution of autoionization in most calculations of acid or base strength. Just keep in mind that autoionization means that the initial values of [H⁺] and [OH⁻] before an acid or base ionizes are never truly zero; they are just too small ($1.00 \times 10^{-7}\,M$) to be significant. However, if acids or bases are extremely weak or if their concentrations are extremely dilute, the autoionization of H_2O may need to be taken into account.

$1 \times 10^{-8}\,M\,HCl.$

> **pH** is the negative logarithm of the hydrogen ion concentration of a solution.

The pH Scale

In the early 1900s, scientists developed an electrochemical device called the hydrogen electrode to directly determine the [H⁺] of solutions. The electrical voltage, or *potential*, produced by the hydrogen electrode, is a linear function of the logarithm of [H⁺]. This relation led Danish biochemist Søren Sørenson (1868–1939) to propose a scale for expressing acidity and basicity based on what he termed "the potential of the hydrogen ion," abbreviated **pH**. Mathematically, we define pH as the negative logarithm to the base 10 of [H⁺], or

$$pH = -\log[H^+] \qquad (16.8)$$

For example, for a solution of a strong acid such as nitric acid that has a concentration of 0.0050 *M*,

$$[H^+] = 0.0050\,M = 5.0 \times 10^{-3}\,M$$

and the pH of that solution is

$$pH = -\log(5.0 \times 10^{-3}) = 2.3$$

Sørenson's pH scale has several attractive features. Because it is a logarithmic scale, there are no exponents, as are commonly found in values of [H⁺]. The logarithmic scale also means that a change of one pH unit corresponds to a 10-fold change in [H⁺]. A solution with a pH of 5.0 has 10 times the [H⁺] and is 10 times as acidic as a solution with a pH of 6.0. Similarly, a solution with a pH of 12.0 has 1/10th the [H⁺], or 10 times the [OH⁻], as a solution with a pH of 11.0. The negative sign in front of the logarithmic term means that most pH values, except for concentrated solutions of strong acids or bases, are between 0 and 14. One thing about the pH scale to keep in mind is that large pH values correspond to small values of [H⁺]. Acidic solutions have pH values less than 7.00 ([H⁺] $> 1.00 \times 10^{-7}\,M$), and basic solutions have pH values greater than 7.00 ([H⁺] $< 1.00 \times 10^{-7}\,M$). The pH values for some common materials are shown in Figure 16.5.

A note about expressing pH values to the appropriate number of significant figures is in order. Remember that pH values are the negative log of the hydrogen ion concentration. The first number in a pH value actually defines the location of the decimal point in the concentration term. As such, it is not considered when determining the number of significant figures in a pH. Consequently, a hydrogen ion concentration of 2.7×10^{-4} (with two significant figures in the coefficient of the power of 10) is between 10^{-3} and 10^{-4}. The corresponding pH is 3.57, where the "3" tells us that the [H⁺] is between 10^{-3} and 10^{-4} and the two significant figures (the "5" and "7") tell us exactly where.

▶❚❚ **CHEMTOUR** pH Scale

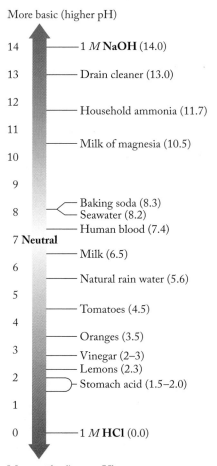

FIGURE 16.5 The pH scale is a convenient way to express the range of acidic or basic properties of some common materials.

SAMPLE EXERCISE 16.2   **Calculating pH and [H⁺]**

Does solution A with a pH of 9.58 have a higher or lower hydrogen ion concentration than solution B whose hydrogen ion concentration is 4.3×10^{-10} M?

COLLECT AND ORGANIZE We are asked if solution A with a given pH has a higher or lower hydrogen ion concentration than solution B with a known hydrogen ion concentration. To answer this, we need to convert the pH value into a concentration so the two values may be compared.

ANALYZE We can use the definition of pH in Equation 16.8 to convert the given pH value to a hydrogen ion concentration.

SOLVE $pH = -\log[H^+]$, so for solution A,

$$9.58 = -\log[H^+]$$

$$[H^+] = 10^{-9.58} = 2.6 \times 10^{-10} \ M$$

$$2.6 \times 10^{-10} \ M < 4.3 \times 10^{-10} \ M$$

so solution A has a lower $[H^+]$ than solution B.

THINK ABOUT IT We could also answer the question by converting the concentration of solution B into pH units and remembering that the solution with the higher pH has the lower hydrogen ion concentration.

Practice Exercise What is the pH of a 6.92×10^{-3} M aqueous solution of hydrochloric acid?

CONCEPT TEST

Is the pH of a 1.00 M solution of a weak acid higher or lower than the pH of a 1.00 M solution of a strong acid?

(Answers to Concept Tests are in the back of the book.)

pOH

Using the prefix "p" to mean "negative logarithm" is not confined to hydrogen ion concentrations. Just as every aqueous solution has a pH value, it also has a **pOH** value. The latter is particularly useful in describing the strength of a basic solution. Equation 16.9 gives the mathematical definition of pOH:

$$pOH = -\log[OH^-] \tag{16.9}$$

A solution's pH and pOH values are related because its $[H^+]$ and $[OH^-]$ are connected by K_w. We have seen the relation between $[H^+]$ and $[OH^-]$ in Equation 16.7, and we can use that equation to derive the relation between pH and pOH. In doing so, we also introduce the term pK_w, the negative logarithm of K_w:

$$K_w = [H^+][OH^-] = 1.00 \times 10^{-14}$$

$$-\log K_w = -\log\{[H^+][OH^-]\} = -\log(1.00 \times 10^{-14})$$

$$pK_w = -\{\log[H^+] + \log[OH^-]\} = -(-14.00)$$

$$pK_w = pH + pOH = 14.00 \tag{16.10}$$

pOH is the negative logarithm of the hydroxide ion concentration in an aqueous solution.

Taking the negative logarithm of K_w converts it into pK_w. Many tables of equilibrium constants list pK values, rather than K values, because doing so does not require the use of exponential notation and is more convenient.

SAMPLE EXERCISE 16.3   **Relating [H⁺], [OH⁻], pH, and pOH**

Because the hydrolysis of dissolved $CO_2(aq)$ forms an acidic solution (called carbonic acid), the pH of normal rainwater is about 5.6, whereas the pH of acid rain may drop to 4.3 or even lower because the dissolved oxides of nitrogen and sulfur also undergo hydrolysis. The pH of rain in Washington, DC, for example, averages between 4.2 and 4.4. Determine [H⁺], [OH⁻], and pOH of pH 4.3 and pH 5.6 rain.

COLLECT AND ORGANIZE We are given pH values and asked to determine the concentrations of hydrogen ion and hydroxide ion and the pOH of two samples of rainwater.

ANALYZE From the definition of pH (Equation 16.8) we can calculate [H⁺], and with Equation 16.10 we can determine pOH. Once we know pOH, we can calculate [OH⁻].

SOLVE From Equation 16.8,

normal rain: pH = 5.6 = $-\log$[H⁺], so [H⁺] = $10^{-5.6}$ = 3×10^{-6} M

acid rain: pH = 4.3 = $-\log$[H⁺], so [H⁺] = $10^{-4.3}$ = 5×10^{-5} M

From Equation 16.10 (pK_w = pH + pOH = 14.00),

normal rain: 5.6 + pOH = 14.00, so pOH = 8.4

acid rain: 4.3 + pOH = 14.00, so pOH = 9.7

To calculate [OH⁻], then,

normal rain: pOH = 8.4 = $-\log$[OH⁻], so [OH⁻] = $10^{-8.4}$ = 4×10^{-9} M

acid rain: pOH = 9.7 = $-\log$[OH⁻], so [OH⁻] = $10^{-9.7}$ = 2×10^{-10} M

THINK ABOUT IT Remember that the pH scale is logarithmic. One unit difference in pH means a 10-fold difference in [H⁺] or [OH⁻]. Because normal rainwater and acid rain differ in pH by a bit more than one pH unit, their hydrogen and hydroxide ion concentrations differ by more than a factor of 10.

Practice Exercise Household ammonia used as a cleaning agent has a pH of about 11.9. What are the concentrations of hydrogen ion and hydroxide ion in household ammonia?

16.4 Calculations with pH and K

If we know the pH and thereby the [H⁺] of a solution of a weakly acidic or basic substance, we can calculate the acid or base equilibrium constant for that substance. First we write a balanced chemical equation for the acid or base reaction, and then we write the corresponding mass action expression. We label such an equilibrium constant K_a or K_b where the subscripts indicate that the constant describes the ionization

K_a and K_b are the values of the equilibrium constants for weak acids and bases, respectively.

of a weak acid or weak base. On the other hand, if we know K_a or K_b of a weak acid or base, we can calculate the pH of an aqueous solution of it. Let's see how.

Weak Acids

The vast majority of the acids we encounter in our environment are classified as weak acids. For example, the nitrous acid formed when NO_2 dissolves in water rapidly reaches an equilibrium, as described by the following chemical equation:

$$HNO_2(aq) + H_2O(\ell) \rightleftharpoons H_3O^+(aq) + NO_2^-(aq) \qquad (16.11)$$

At equilibrium, the rate at which nitrous acid ionizes (the forward reaction) matches the rate at which the ions recombine to form HNO_2.

We may begin the quantitative analysis of the equilibrium by writing the mass action expression for the reaction in Equation 16.11 as

$$K = \frac{[H_3O^+][NO_2^-]}{[HNO_2][H_2O]}$$

However, as was the case for the autoionization of water, the concentration of water does not significantly change during the course of reactions such as this, so we do not include an $[H_2O]$ term. We also usually leave out the water molecule in the hydronium ion and replace $[H_3O^+]$ with $[H^+]$. Therefore the equilibrium constant expression for the ionization of nitrous acid is usually written

$$K_a = \frac{[H^+][NO_2^-]}{[HNO_2]}$$

The subscript "a" indicates that K is an acid-ionization equilibrium constant. A general expression for the ionization of any weak acid, HA, is

$$HA \rightleftharpoons H^+ + A^-$$

and a general expression for the equilibrium constant of that ionization reaction is

$$K_a = \frac{[H^+][A^-]}{[HA]} \qquad (16.12)$$

where the term [HA] is the molar concentration of the acid and $[A^-]$ is the concentration of its conjugate base.

Let's use Equation 16.12 to calculate the K_a of a weak acid, HNO_2, starting with the pH of a solution of the acid. We employ an ICE table for the calculation just as we did in Chapter 15. The table is based on a simplified version of Equation 16.11, describing the ionization of HNO_2 in aqueous solution:

$$HNO_2(aq) \rightleftharpoons H^+(aq) + NO_2^-(aq)$$

In assigning an initial value to $[H^+]$ in the table we make an important assumption: the ionization of HNO_2 is the only significant source of H^+ ions in the solution. Therefore, the initial concentration of H^+ will be tiny (essentially zero) compared with $[H^+]$ at equilibrium. This assumption is valid in most instances. If we let x be $[H^+]$ at equilibrium, then the change in $[H^+]$ is $+x$. On the basis of the stoichiometry of the reaction, the change in $[NO_2^-]$ is also $+x$, and the change in $[HNO_2]$ is $-x$. Suppose we determine that the pH of a 0.100 M solution of HNO_2 is 2.20. The hydrogen ion concentration at equilibrium is

$$-\log[H^+] = 2.20 \text{ or } [H^+] = 6.31 \times 10^{-3} \, M$$

Next we use the equality $[H^+] = x = 6.31 \times 10^{-3}$ to fill in the cells of the table.

	$[HNO_2]$ (M)	$[H^+]$ (M)	$[NO_2^-]$ (M)
Initial	0.100	0	0
Change	$-x = -6.31 \times 10^{-3}$	$+x = 6.31 \times 10^{-3}$	$+x = 6.31 \times 10^{-3}$
Equilibrium	0.094	6.31×10^{-3}	6.31×10^{-3}

When we use the equilibrium values to calculate K_a and make the final adjustment for significant figures, we get

$$K_a = \frac{[H^+][NO_2^-]}{[HNO_2]} = \frac{(6.31 \times 10^{-3})(6.31 \times 10^{-3})}{(0.094)} = 4.25 \times 10^{-4} = 4.2 \times 10^{-4}$$

The relatively small K_a for HNO_2 tells us that it is indeed a weak acid and that, at equilibrium, $[H^+]$ and $[NO_2^-]$ are much less than $[HNO_2]$.

The partial ionization of HNO_2 contrasts with the complete ionization of a strong acid such as HNO_3. Figure 16.6 provides graphic interpretations of the different strengths of these two acids. The complete ionization of HNO_3 (Figure 16.6a) results in the complete conversion of molecular HNO_3 into hydrogen ions and nitrate ions. In contrast, the incomplete ionization of the weak acid HNO_2 leaves most of the molecular substance intact. The ratio of the concentration of H^+ ions (the red bar in the right-hand graph of Figure 16.6b) to the initial concentration of HNO_2 (the green bar in the left-hand graph of Figure 16.6b) represents the **degree of ionization** of HNO_2 in a 0.100 M solution of the acid. The degree of ionization is the ratio of the quantity of a substance that is ionized to its concentration before ionization, sometimes expressed as **percent ionization**. For any weak

Degree of ionization is the ratio of the quantity of a substance that is ionized to the concentration of the substance before ionization. When expressed as a percentage, it is called **percent ionization**.

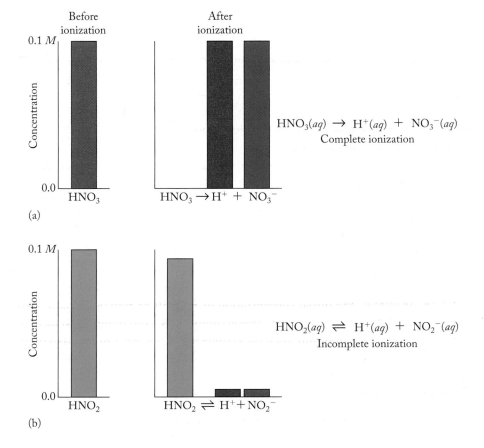

(a)

(b)

FIGURE 16.6 These graphs illustrate the differences in the degrees to which 0.100 M solutions of (a) a strong acid (HNO_3) and (b) a weak acid (HNO_2) are ionized.

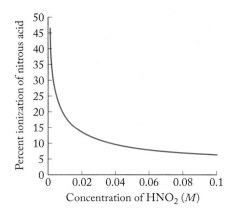

FIGURE 16.7 The degree of ionization of a weak acid increases with decreasing acid concentration. Here the degree of ionization of nitrous acid increases from about 6% in a 0.100 M solution to 18% in a 0.01 M solution to 46% in a 0.001 M solution.

acid (HA) percent ionization is the ratio of $[A^-]$ at equilibrium to $[HA]$ initially present. Assuming $[A^-] = [H^+]$ at equilibrium then:

$$\text{Percent ionization} = \frac{[H^+]_{\text{equilibrium}}}{[HA]_{\text{initial}}} \times 100\% \qquad (16.13)$$

Equation 16.13 is particularly useful if we know the pH of a solution of a weak acid, because from pH we can calculate $[H^+]$ and then percent ionization. Doing so for 0.100 M HNO_2, we have:

$$\text{Percent ionization} = \frac{6.31 \times 10^{-3} \, M}{0.100 \, M} \times 100\% = 6.3\%$$

The degree of ionization of HNO_2, or any weak acid, is related to the value of its K_a: the larger the value of K_a, the larger the degree of ionization. We can describe a weak acid quantitatively by using the value of K_a, and we can use the pH (or $[H^+]$) of a solution of it to calculate its percent ionization.

A plot of the results of several calculations of percent ionization as a function of the initial concentration of nitrous acid is shown in Figure 16.7. The pattern of the results of such calculations is observed for all weak acids: the degree to which they ionize increases as their concentrations decrease. The graph in Figure 16.7 shows the dependence of the percent ionization of HNO_2 on $[HNO_2]$. Weak acids are increasingly ionized as their molar concentrations decrease.

SAMPLE EXERCISE 16.4 **Calculating K_a and Percent Ionization of a Weak Acid**

The pH of a 1.000 M solution of formic acid (HCOOH), an organic acid found in red ants and responsible for the sting of their bite, is 1.88. What is the K_a value of the acid? What is its percent ionization?

COLLECT AND ORGANIZE We are asked to determine the K_a value and the percent ionization of a weak acid in a solution of known concentration of the acid. We are also given the pH. The equilibrium constant for the ionization of a generic weak acid (HA) is described by Equation 16.12. Applying Equation 16.12 to the ionization of formic acid:

$$HCOOH(aq) \rightleftharpoons H^+(aq) + HCOO^-(aq) \text{ and } K_a = \frac{[H^+][HCOO^-]}{[HCOOH]}$$

The percent ionization is the ratio of the concentration of hydrogen ion at equilibrium to the original concentration of the acid.

ANALYZE We can calculate $[H^+]$ from pH, and we can use an ICE table to help us organize the equilibrium concentrations of the species present.

SOLVE The pH of the solution is 1.88, so the $[H^+]$ is

$$1.88 = -\log[H^+]; [H^+] = 1.32 \times 10^{-2} \, M$$

	[HCOOH] (M)	**[H$^+$] (M)**	**[HCOO$^-$] (M)**
Initial	1.000	0	0
Change	-1.32×10^{-2}	1.32×10^{-2}	1.32×10^{-2}
Equilibrium	0.987	1.32×10^{-2}	1.32×10^{-2}

$$K_a = \frac{[H^+][HCOO^-]}{[HCOOH]} = \frac{(1.32 \times 10^{-2})(1.32 \times 10^{-2})}{(0.987)} = 1.8 \times 10^{-4}$$

$$\text{Percent ionization} = \frac{1.32 \times 10^{-2}}{1.000} = 0.0132 = 1.32\%$$

THINK ABOUT IT The K_a for formic acid is smaller than the K_a for nitrous acid, which means that formic acid is the weaker of the two acids.

Practice Exercise The value of $[H^+]$ in a 0.050 M solution of an organic acid is 5.9×10^{-3} M. What is the value of the K_a of the acid?

CONCEPT TEST

Three weak acids have the following K_a values: acid A, 3.6×10^{-5}; acid B, 4.9×10^{-4}; acid C, 9.2×10^{-4}. Order the acids in terms of their relative strength from strongest to weakest.

Just as we used the pH of a solution of known concentration of a weak acid to calculate K_a for that acid, if we know the K_a of an acid and its concentration, we can calculate its pH. As an example, we can calculate the pH of a 0.100 M solution of acetic acid, the weak acid that is the principal ingredient in vinegar. The equation describing the ionization of acetic acid in water is

$$CH_3COOH(aq) \rightleftharpoons H^+(aq) + CH_3COO^-(aq)$$

and the mass action expression and value of K_a are

$$K_a = \frac{[H^+][CH_3COO^-]}{[CH_3COOH]} = 1.76 \times 10^{-5}$$

We can set up an ICE table for the reaction. Initially, $[CH_3COOH] = 0.100$ M, and $[CH_3COO^-] = 0.000$ M. We assume that the initial $[H^+]$ is insignificant (essentially zero) compared with the equilibrium concentration, which we define as x. Given the 1:1 stoichiometry between H^+ and CH_3COO^-, the changes in the concentrations of both H^+ and CH_3COO^- are $+x$, and the change in $[CH_3COOH]$ is $-x$. The resulting ICE table is

	[CH₃COOH] (M)	**[H⁺] (M)**	**[CH₃COO⁻] (M)**
Initial	0.100	0.000	0.000
Change	$-x$	$+x$	$+x$
Equilibrium	$0.100 - x$	x	x

The mass action expression for K_a is

$$K_a = \frac{[H^+][CH_3COO^-]}{[CH_3COOH]} = 1.76 \times 10^{-5}$$

and substituting terms from the ICE table gives

$$\frac{(x)(x)}{(0.100 - x)} = 1.76 \times 10^{-5}$$

In approaching this calculation, we can make an important simplifying assumption. The value of K_a is quite small (1.76×10^{-5}), and the concentration of the acid is relatively large (0.100 M). Under these conditions, the value of x will probably be small compared with 0.100 M. If that is the case, then $(0.100 - x)$ is approximately the same as 0.100, which means that the denominator of the chemical equilibrium expression simplifies to

$$\frac{(x)(x)}{(0.100)} = 1.76 \times 10^{-5}$$

After solving this expression for x^2 and then x, we have

$$x = [H^+] = 1.33 \times 10^{-3} \, M$$

Finally, we take the negative logarithm of $[H^+]$ to obtain the pH:

$$pH = -\log[H^+] = -\log(1.33 \times 10^{-3}) = 2.88$$

Before accepting this answer, we must check the validity of our assumption that x is small compared with 0.100 M. After dividing x by the original concentration of acetic acid, 0.100 M, and expressing the result as a percentage, we have

$$\frac{1.33 \times 10^{-3} \, \cancel{M}}{0.100 \, \cancel{M}} = 0.0133 = 1.33\%$$

The percent ionization of acetic acid in this solution is 1.33%. As a rule, simplifying assumptions that produce an x value of less than 5% of the initial concentration of a reactant are acceptable because they produce negligible errors in the value of x. Our assumption in this calculation is therefore valid, and we may report 2.88 as the pH of the solution. If the simplifying assumption fails—that is, if the percent ionization of a weak acid is greater than 5% of the initial concentration of a reactant—then we must include the $-x$ term in the denominator of the K_a expression. This requirement necessitates the use of a problem-solver program or the quadratic formula to solve for x.

SAMPLE EXERCISE 16.5 **Calculating the pH of a Solution of Weak Acid**

Municipal water supplies and swimming pools are disinfected with solutions containing hypochlorous acid (HClO), the K_a of which is 2.9×10^{-8}. What is the pH of a 0.125 M solution of HClO?

COLLECT AND ORGANIZE We are asked to determine the pH ($-\log [H^+]$) of a solution of a weak acid. We are given the value of K_a for the acid, which is equal to the ratio of concentrations of products to that of reactants in the mass action expression, and we know the concentration of the acid in the solution.

ANALYZE We can use an ICE table to collect the expressions for the concentrations of the reactants and products at equilibrium based on the chemical equation for the ionization of the acid. HClO is a weak acid, so we can use a simplifying assumption to calculate the hydrogen ion concentration without the use of the quadratic equation. We must check the validity of that assumption when we have solved the problem.

SOLVE The equation for the ionization of HClO and the expression for K_a are

$$HClO(aq) \rightleftharpoons H^+(aq) + ClO^-(aq)$$

and

$$K_a = \frac{[H^+][ClO^-]}{[HClO]} = 2.9 \times 10^{-8}$$

The ICE table for this system is

	[HClO] (M)	[H⁺] (M)	[ClO⁻] (M)
Initial	0.125	0	0
Change	$-x$	$+x$	$+x$
Equilibrium	$0.125 - x \approx 0.125$	x	x

$$K_a = \frac{(x)(x)}{0.125} = 2.9 \times 10^{-8}$$

Therefore $x = 6.0 \times 10^{-5}\ M$ and $\mathrm{pH} = -\log(6.0 \times 10^{-5}) = 4.22$.

THINK ABOUT IT To check the simplifying assumption, let's calculate the percent ionization of the acid:

$$\frac{6.0 \times 10^{-5}\ \cancel{M}}{0.125\ \cancel{M}} \times 100\% = 0.048\%$$

This value is much less than 5%, so the assumption is valid, and we may report the pH as 4.22.

Practice Exercise The brown color of most swamp water is due to a family of compounds called tannic acids ($K_a = 5 \times 10^{-6}$). What is the pH of 0.050 M tannic acid?

Weak Bases

Now let's consider what happens when a weakly basic compound dissolves in water. The weak base ammonia accepts hydrogen ions from water as described in the following chemical equation:

$$NH_3(aq) + H_2O(\ell) \rightleftharpoons NH_4^+(aq) + OH^-(aq) \qquad (16.14)$$

This reaction is the result of strong intermolecular forces that lead to covalent bonds breaking and new bonds forming, as shown in Figure 16.8. Hydrogen bonds form between the N atoms of NH_3 molecules and the H atoms of H_2O molecules, leading to the formation of covalent N–H bonds and ionization of O–H bonds, as shown in Figure 16.8. The products of these bonding changes are ammonium ions, NH_4^+, and hydroxide ions. The lone pair of electrons on each N atom is shared with a transferred H⁺ ion to make the fourth N–H bond in each NH_4^+ ion.

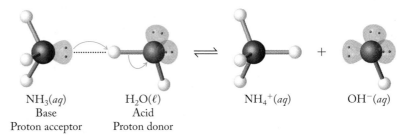

NH₃(aq)
Base
Proton acceptor

H₂O(ℓ)
Acid
Proton donor

NH₄⁺(aq)

OH⁻(aq)

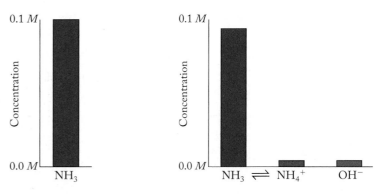

FIGURE 16.8 Ammonia reacts with water, producing ammonium ions and hydroxide ions in solution. These graphs illustrate the small extent to which the reaction proceeds because ammonia is a weak base.

In a solution of ammonia, not all NH_3 molecules accept hydrogen ions. Instead, the reaction described by Equation 16.14 reaches an equilibrium in which most ammonia molecules are present as NH_3 rather than NH_4^+. The limited strength of ammonia as a base is reflected in the small value of the equilibrium constant for Equation 16.14 at 25°C:

$$K_b = \frac{[NH_4^+][OH^-]}{[NH_3]} = 1.76 \times 10^{-5}$$

Here we use the subscript "b" to indicate that the reaction is the ionization of a base that results from the acceptance of hydrogen ions from H_2O. No term for the concentration of H_2O appears in the expression for K_b. As with the ionization of nitrous acid (Equation 16.11), the concentration of water does not change significantly in the course of the reaction. Therefore, just as with a weak acid, we can calculate K_b for a solution of weak base if we know the initial concentration of the base and the pH of the solution. Correspondingly, we can calculate the pH if we know the K_b of the base by using an approach analogous to calculating the pH of solutions of weak acids. One additional step is necessary: the unknown in this equilibrium calculation will be $[OH^-]$ instead of $[H^+]$, but once we know $[OH^-]$ we can take the negative logarithm of it to calculate pOH and then convert that result to pH using Equation 16.10.

SAMPLE EXERCISE 16.6   **Calculating the pH of a Solution of a Weak Base**

The concentration of NH_3 in household ammonia used for cleaning windows ranges between 50 and 100 g/L, or from about 3 M to almost 6 M. What is the pH of a 3.0 M solution of NH_3 ($K_b = 1.76 \times 10^{-5}$)?

COLLECT AND ORGANIZE We are asked to determine the pH of a 3.0 M solution of a weak base. We are given K_b. We can solve for $[OH^-]$ on the basis of the ionization reaction between ammonia and water:

$$NH_3(aq) + H_2O(\ell) \rightleftharpoons NH_4^+(aq) + OH^-(aq)$$

and then convert $[OH^-]$ to pOH and finally to pH. The equilibrium constant for this reaction is

$$K_b = \frac{[NH_4^+][OH^-]}{[NH_3]} = 1.76 \times 10^{-5}$$

ANALYZE Initially, $[NH_3] = 3.0$ M and $[NH_4^+] = 0.0$ M. We may assume that the initial $[OH^-]$ is insignificant (essentially zero) compared with the equilibrium concentration, which we let be x. Given the 1:1:1 stoichiometry of NH_3, NH_4^+, and OH^-, the changes in the concentrations of both NH_4^+ and OH^- are $+x$, and the decrease in $[NH_3]$ is $-x$. The result is shown in the following ICE table:

	$[NH_3]$ (*M*)	$[NH_4^+]$ (*M*)	$[OH^-]$ (*M*)
Initial	3.0	0.0	0.0
Change	$-x$	$+x$	$+x$
Equilibrium	$3.0 - x$	x	x

SOLVE The value of K_b is small (1.76×10^{-5}) and the initial concentration of the base is relatively high (3.0 M). We can make the simplifying assumption that the value of x is small compared with 3.0 M, and so $3.0 - x$ is approximately 3.0. After inserting the appropriate values in the equilibrium constant expression, we have

$$K_b = \frac{[NH_4^+][OH^-]}{[NH_3]} = \frac{(x)(x)}{(3.0)} = 1.76 \times 10^{-5}$$

Solving for x gives us

$$x = [OH^-] = 7.3 \times 10^{-3}\ M$$

Taking the negative logarithm of $[OH^-]$ to calculate pOH we get

$$pOH = -\log[OH^-] = -\log(7.3 \times 10^{-3}\ M) = 2.14$$

Then we subtract this value from 14.00 to obtain the pH:

$$pH + pOH = 14.00$$

$$pH = 14.00 - pOH = 14.00 - 2.14 = 11.86$$

To check our simplifying assumption before we accept this answer, we determine whether the percent ionization is less than 5%:

$$\frac{7.3 \times 10^{-3}}{3.0} = 0.0024 = 0.24\%$$

so our assumption is justified.

THINK ABOUT IT This basic pH value makes sense given the small K_b value but relatively high initial concentration of ammonia. The calculated value of $[OH^-]$ is about 0.2% of the initial concentration of ammonia, making our assumptions and the results of our calculation valid.

Practice Exercise Methylamine is a weak base ($K_b = 4.4 \times 10^{-4}$). Calculate the value of $[OH^-]$ in a 0.200 M CH_3NH_2 solution and determine the pH of the solution.

16.5 Polyprotic Acids

Up to this point, we have dealt with **monoprotic acids** that have only one ionizable hydrogen atom per molecule. Many acids, called **polyprotic acids**, have more than one ionizable hydrogen. Table 16.5 lists several common diprotic acids and the ionization constants that define the equilibria associated with the donation of all the ionizable hydrogen atoms in their structures.

Let's first look at the behavior of sulfuric acid, the most common *diprotic* (two-proton) acid. We have classified sulfuric acid as a strong acid (Table 16.1) but only in terms of donating the first proton. The first ionization step of sulfuric acid is complete (Equation 16.15), so no K is defined for that process:

$$H_2SO_4(aq) \rightarrow HSO_4^-(aq) + H^+(aq) \qquad (16.15)$$

The second ionization step is incomplete, so the donation of the second proton involves an equilibrium and has a characteristic K_a.

$$HSO_4^-(aq) \rightleftharpoons SO_4^{2-}(aq) + H^+(aq) \qquad K_{a_2} = 1.2 \times 10^{-2} \qquad (16.16)$$

The combination of complete and incomplete ionization steps means that most solutions of $H_2SO_4(aq)$ contain more than 1 mol but less than 2 mol of $H^+(aq)$ for

Monoprotic acids have one ionizable hydrogen atom per molecule, whereas **polyprotic acids** have two or more.

TABLE 16.5 Ionization Equilibria for Three Diprotic Acids

Acid	Ionization Equilibria	K_a
Carbonic acid	Step 1: $H_2CO_3(aq) \rightleftharpoons H^+(aq) + HCO_3^-(aq)$	$K_{a_1} = 4.3 \times 10^{-7}$
	Step 2: $HCO_3^-(aq) \rightleftharpoons H^+(aq) + CO_3^{2-}(aq)$	$K_{a_2} = 4.7 \times 10^{-11}$
Sulfurous acid	Step 1: $H_2SO_3(aq) \rightleftharpoons H^+(aq) + HSO_3^-(aq)$	$K_{a_1} = 1.7 \times 10^{-2}$
	Step 2: $HSO_3^-(aq) \rightleftharpoons H^+(aq) + SO_3^{2-}(aq)$	$K_{a_2} = 6.2 \times 10^{-8}$
Sulfuric acid	Step 1: $H_2SO_4(aq) \rightarrow H^+(aq) + HSO_4^-(aq)$	$K_{a_1} \gg 1$
	Step 2: $HSO_4^-(aq) \rightleftharpoons H^+(aq) + SO_4^{2-}(aq)$	$K_{a_2} = 1.2 \times 10^{-2}$

every mole of H_2SO_4. We can be more quantitative about this observation and determine the pH of a 0.100 M solution of H_2SO_4. The starting point in this pH calculation is a solution in which all the H_2SO_4 has ionized to HSO_4^- and H^+, as described in Equation 16.15. Therefore, as the second step begins, $[HSO_4^-] = [H^+] = 0.100 \ M$. Ionization of HSO_4^- (Equation 16.16) produces additional H^+. To analyze the effect of the second ionization, we can set up an ICE table as we have done for other weak acids in which $+x$ is the change in $[H^+]$ produced by the second step. Given the 1:1:1 stoichiometry of the reactant and products in the second step, the increase in $[SO_4^{2-}]$ will also be $+x$, and the change in $[HSO_4^-]$ will be $-x$. When we insert these values in the ICE table and complete the third row, we have

	$[HSO_4^-]$ (M)	$[H^+]$ (M)	$[SO_4^{2-}]$ (M)
Initial	0.100	0.100	0.000
Change	$-x$	$+x$	$+x$
Equilibrium	$0.100 - x$	$0.100 + x$	x

Then when we insert the equilibrium-concentration terms in the equilibrium constant expression (K_{a_2}), we have

$$K_{a_2} = \frac{[H^+][SO_4^{2-}]}{[HSO_4^-]}$$

$$= \frac{(0.100 + x)(x)}{(0.100 - x)} = 1.2 \times 10^{-2}$$

Because K_{a_2} is large for the ionization of a weak acid, it is doubtful that we can use the simplifying assumption we have frequently applied to avoid solving a quadratic equation. Therefore we solve for x in this expression:

$$x^2 + 0.112x - (1.2 \times 10^{-3}) = 0$$

$$x = +0.010 \text{ or } -0.122$$

The negative value for x has no physical meaning, because it corresponds to a negative concentration, so we use only the positive value. At equilibrium,

$$[H^+] = (0.100 + x) \ M = (0.100 + 0.010) \ M = 0.110 \ M$$

The corresponding pH is

$$pH = -\log[H^+] = -\log(0.110 \ M) = 0.96$$

As predicted, the value of [H$^+$], 0.110 M, is between one and two times the initial concentration of H_2SO_4 (0.100 M). The percent ionization of HSO_4^- is greater than 5%:

$$\frac{[SO_4^{2-}]_{equilibrium}}{[HSO_4^-]_{initial}} = \frac{0.010\ \cancel{M}}{0.100\ \cancel{M}} = 0.10 = 10\%$$

so we were correct in our decision not to use the simplifying assumption to avoid solving a quadratic equation.

Table 16.5 lists the K_{a_1} and K_{a_2} values for three diprotic acids found in acid rain. Carbonic acid is present because (1) the atmosphere is about 0.036% (by volume) CO_2, and (2) CO_2 is slightly soluble in water. When CO_2 dissolves in water, it hydrolyzes to form carbonic acid, H_2CO_3:

$$CO_2(g) + H_2O(\ell) \rightleftharpoons H_2CO_3(aq)$$

The H_2CO_3 molecule is not stable in aqueous solutions, but we write it as a convenience to show the result of dissolving CO_2 in water to produce an acidic solution. The net result of CO_2 dissolving in water is actually

$$CO_2(g) + H_2O(\ell) \rightleftharpoons HCO_3^-(aq) + H^+(aq)$$

This ionization process may also be written as it is in Table 16.5.

$$H_2CO_3(aq) \rightleftharpoons HCO_3^-(aq) + H^+(aq) \tag{16.17}$$

In each pair of K_{a_1} and K_{a_2} values in Table 16.5, the value of K_{a_2} is smaller than that of K_{a_1}. Why is the value of K_{a_2} for a diprotic acid much less than that of K_{a_1}? We can rationalize the difference on the basis of electrostatic attractions between oppositely charged ions. The first ionization step produces a negatively charged oxoanion. The second ionization step requires that a positive ion (H$^+$) dissociate from a negative ion to produce an even more negatively charged oxoanion. Separating oppositely charged ions that are naturally attracted to each other is not a process that we would expect to be favored, and the smaller values for K_{a_2} confirm our expectations. For this reason, the K_{a_2} of any diprotic acid is less, and often much less, than K_{a_1}.

The large differences between K_{a_1} and K_{a_2} values for the two weak diprotic acids in Table 16.5 have an important consequence: the second ionization step usually has an insignificant effect on [H$^+$] in solutions of these acids. Essentially all of the limited strength of these acids is due to the first ionization step. Notice in the case of H_2CO_3 (Equation 16.17) that the K_{a_1} and K_{a_2} values differ by a factor of about 10^4. That difference is large enough that we can ignore the contribution of the dissociation of the second proton to the pH of a solution of this weak acid.

SAMPLE EXERCISE 16.7   **Calculating the pH of a Weak Diprotic Acid Solution**

What is the normal pH of rain at 25°C due to the presence of dissolved carbon dioxide and the ionization of carbonic acid? Assume [H_2CO_3] is a constant that is controlled by the solubility of atmospheric CO_2 in water and is $1.2 \times 10^{-5}\ M$.

COLLECT AND ORGANIZE We are asked to determine the pH of a dilute ($1.2 \times 10^{-5}\ M$) solution of carbonic acid. The large difference between K_{a_1} and K_{a_2} allows us to focus on the first ionization step:

$$H_2CO_3(aq) \rightleftharpoons HCO_3^-(aq) + H^+(aq) \qquad K_{a_1} = 4.3 \times 10^{-7}$$

as the one that controls pH.

ANALYZE We can set up an ICE table based on the first ionization step with columns for the concentrations of $H_2CO_3(aq)$, $H^+(aq)$, and $HCO_3^-(aq)$. If we let x be the value of $[H^+]$ at equilibrium, then $[HCO_3^-]$ is also x, and the changes in their concentrations are both $+x$.

Any decrease in $[H_2CO_3]$ due to acid ionization will, according to Le Châtelier's principle, result in more dissolution of atmospheric CO_2 into rain so that $[H_2CO_3]$ is restored to 1.2×10^{-5} M. Therefore, $[H_2CO_3]$ does not change in the course of the reaction, and the ICE table looks like this:

	$[H_2CO_3]$ (M)	$[H^+]$ (M)	$[HCO_3^-]$ (M)
Initial	1.2×10^{-5}	0	0
Change	0	$+x$	$+x$
Equilibrium	1.2×10^{-5}	x	x

SOLVE Using the equilibrium terms from the ICE table in the equilibrium constant expression for K_{a_1}, we have

$$K_{a_1} = \frac{[HCO_3^-][H^+]}{[H_2CO_3]} = \frac{(x)(x)}{1.2 \times 10^{-5}} = 4.3 \times 10^{-7}$$

After solving for x^2 and then x, we have

$$x = [H^+] = 2.3 \times 10^{-6} \ M$$

Taking the negative logarithm of $[H^+]$ to calculate pH gives

$$pH = -\log[H^+] = -\log(2.3 \times 10^{-6} \ M) = 5.64$$

THINK ABOUT IT Carbonic acid is a weak acid, and so obtaining a pH value that is only about 1.4 units below neutral pH is reasonable.

Practice Exercise The relatively high pH calculated in Sample Exercise 16.7 for carbonic acid raises the question of whether the autoionization of water contributes significantly to $[H^+]$ in an acid as weak (and dilute) as this one. One way to check on this is to use an initial value for $[H^+]$ of 1.00×10^{-7} instead of zero and recalculate pH. Recalculate the pH of a 1.2×10^{-5} M solution of H_2CO_3 after accounting for the autoionization of water as a second source of H^+. Assume that $[H_2CO_3]$ does not change. (*Hint*: let $[H^+]_{initial} = 1.00 \times 10^{-7}$ M.)

Several important common acids have *three* ionizable hydrogen atoms per molecule. Commercially and biologically important ones include phosphoric acid, H_3PO_4, and citric acid, the acid responsible for the tart flavor of fruits. Their K_{a_1}, K_{a_2}, and K_{a_3} values are listed in Table 16.6. Note how the progression of the K_a values for both triprotic acids is $K_{a_1} > K_{a_2} > K_{a_3}$. This pattern is much like that for the $K_{a_1} > K_{a_2}$ values of diprotic acids and for the same reason: it is more difficult to remove a second H^+ ion from the negatively charged molecular ion formed after the first H^+ ion is removed, and it is even more difficult to remove a third H^+ ion from an ion with a 2− charge.

TABLE 16.6 Ionization Equilibria for Two Triprotic Acids

Phosphoric Acid

(1) $HO-\overset{\overset{O}{\|}}{\underset{\underset{OH}{|}}{P}}-OH \rightleftharpoons HO-\overset{\overset{O}{\|}}{\underset{\underset{OH}{|}}{P}}-O^- + H^+$ $K_{a_1} = 7.11 \times 10^{-3}$

(2) $HO-\overset{\overset{O}{\|}}{\underset{\underset{OH}{|}}{P}}-O^- \rightleftharpoons {}^-O-\overset{\overset{O}{\|}}{\underset{\underset{OH}{|}}{P}}-O^- + H^+$ $K_{a_2} = 6.32 \times 10^{-8}$

(3) ${}^-O-\overset{\overset{O}{\|}}{\underset{\underset{OH}{|}}{P}}-O^- \rightleftharpoons {}^-O-\overset{\overset{O}{\|}}{\underset{\underset{O^-}{|}}{P}}-O^- + H^+$ $K_{a_3} = 4.5 \times 10^{-13}$

Citric Acid

(1) $HO-\overset{\overset{CH_2COOH}{|}}{\underset{\underset{CH_2COOH}{|}}{C}}-COOH \rightleftharpoons HO-\overset{\overset{CH_2COO^-}{|}}{\underset{\underset{CH_2COOH}{|}}{C}}-COOH + H^+$ $K_{a_1} = 7.44 \times 10^{-4}$

(2) $HO-\overset{\overset{CH_2COO^-}{|}}{\underset{\underset{CH_2COOH}{|}}{C}}-COOH \rightleftharpoons HO-\overset{\overset{CH_2COO^-}{|}}{\underset{\underset{CH_2COOH}{|}}{C}}-COO^- + H^+$ $K_{a_2} = 1.73 \times 10^{-5}$

(3) $HO-\overset{\overset{CH_2COO^-}{|}}{\underset{\underset{CH_2COOH}{|}}{C}}-COO^- \rightleftharpoons HO-\overset{\overset{CH_2COO^-}{|}}{\underset{\underset{CH_2COO^-}{|}}{C}}-COO^- + H^+$ $K_{a_3} = 4.02 \times 10^{-7}$

CONCEPT TEST

Do you expect the second and third acid ionization steps in phosphoric acid and citric acid to influence the pH of 0.100 M solutions of either acid?

16.6 Acid Strength and Molecular Structure

We have noted that nitric acid (HNO_3) is a strong acid, whereas nitrous acid (HNO_2) is weak. Similarly, sulfurous acid (H_2SO_3), which is formed when sulfur dioxide hydrolyzes in water

$$SO_2(g) + H_2O(\ell) \rightleftharpoons H_2SO_3(aq)$$

is a weak acid, whereas sulfuric acid (H_2SO_4) is a strong acid. The reason for this behavior lies in the differences in the molecular structures of the acids. Consider the differences between the Lewis structures of H_2SO_4 and H_2SO_3 shown in Figure 16.9. The ionizable hydrogen atoms in both molecules are bonded to oxygen atoms that are also bonded to the central sulfur atoms. In each case, the central sulfur atom is also bonded to one (in H_2SO_3) or two (in H_2SO_4) other oxygen atoms. Oxygen is the second most electronegative element. Oxygen atoms bonded to the central atom of oxoacids, such as H_2SO_3 and H_2SO_4, attract electron density toward themselves and away from the rest of the molecule. The rest of the molecule in these cases consists of the hydrogen atoms of O–H groups. The

H_2SO_3

H_2SO_4

FIGURE 16.9 Sulfuric acid (H_2SO_4) is a stronger acid than sulfurous acid (H_2SO_3) because of the greater stability that comes with delocalizing the negative charge of a SO_4^{2-} ion over more atoms (shown by the curved arrows).

Acid	Structure	Oxidation Number of Cl	K_a
Hypochlorous	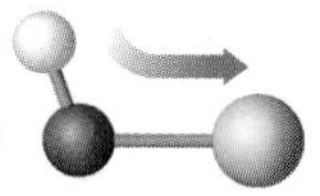	+1	2.9×10^{-8}
Chlorous	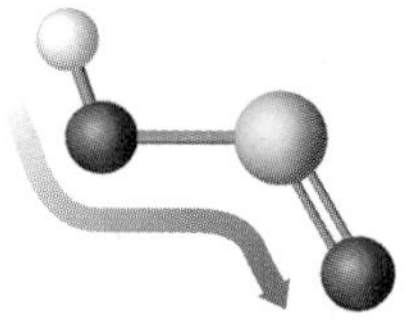	+3	1.1×10^{-2}
Chloric	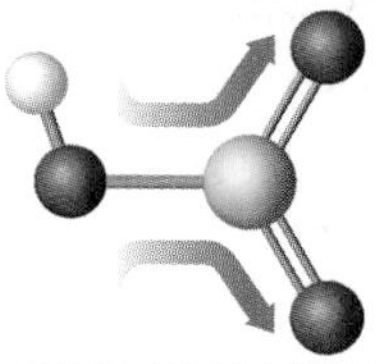	+5	~1
Perchloric	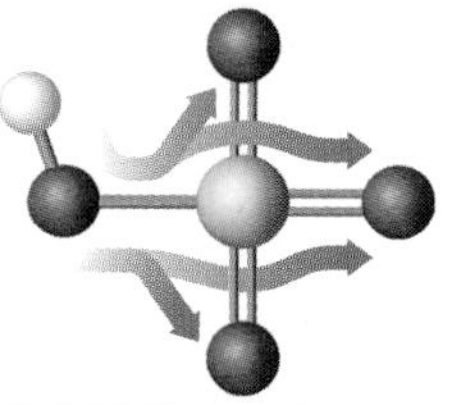	+7	Strong acid

FIGURE 16.10 The strengths of the oxoacids of chlorine increase with increasing oxidation number of chlorine and the corresponding greater delocalization of the negative electrical charge from ionizing the O–H bond.

▶❚❚ **CHEMTOUR** Acid Strength and Molecular Structure

◯◯ **CONNECTION** The periodic trend in electronegativities of the elements is described in Chapter 7. Fluorine is the element with highest electronegativity, followed by oxygen.

Acid	Structure	Electronegativity of Halogen Atom	K_a
Hypochlorous	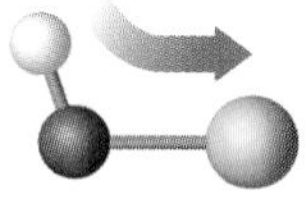	3.0	2.9×10^{-8}
Hypobromous	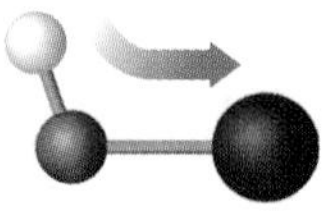	2.8	2.3×10^{-9}
Hypoiodous	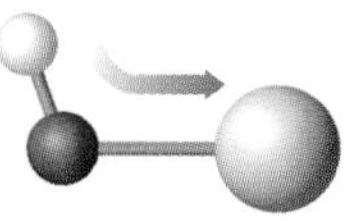	2.5	2.3×10^{-11}

FIGURE 16.11 The strengths of these three hypohalous acids are related to the electronegativities of their halogen atoms. The shifts in electron density (shown by the blue arrows) away from the hydrogen end of the molecule make the OX⁻ conjugate bases more stable (where X is Cl, Br, or I).

more electrical charge that is drawn away from these O–H groups, the easier it is for the anion formed when the H is ionized to have its negative charge spread throughout the remaining molecule. Spreading out charge over more atoms has a stabilizing effect. Thus, sulfuric acid, with three electron-withdrawing oxygen atoms attached to the central sulfur atom, is a stronger acid than sulfurous acid, with only two oxygen atoms attached to the central sulfur atom.

This trend of increasing acid strength with increasing numbers of oxygen atoms bonded to the central atom (i.e., with increasing oxidation number of the central atom) is true for all oxoacids. The trend is exemplified by the strongly acidic properties of HNO_3 and the weak acidity of HNO_2. It is also illustrated in the strengths of the oxoacids of chlorine, as shown in Figure 16.10.

The connection between the electron-withdrawing power of the central atom in an oxoacid and the strength of the acid is also evident when we compare the strengths of acids having different central atoms but otherwise similar structures. Consider, for example, the relative strengths of the three hypohalous acids in Figure 16.11. The differences between them are the identities and electronegativities of the halogen atoms in their structures. The most electronegative of the three (Cl) has the greatest attraction for the pair of electrons that it shares with oxygen. This attraction draws electron density toward chlorine, which in turn pulls electron density away from hydrogen and toward the oxygen end of the already polar O–H bond. These shifts in electron density make the hypochlorite (ClO⁻) ions better able to bear a negative charge, because the charge is delocalized, that is, spread more uniformly over the atoms. Thus, HClO(aq) is the strongest of the three acids, followed by hypobromous acid [HBrO(aq)] and hypoiodous acid [HIO(aq)].

16.7 pH of Salt Solutions

Seawater and the fresh water in many rivers and lakes have pH values that range from neutral to weakly alkaline. How can these waters have pH values a unit or more above (less acidic) the rain that serves, directly or indirectly, as their water supply? The answer is that, when rain falls to Earth and becomes surface or groundwater, its pH changes as it flows over rocks and through soils that contain basic components. Before we look at these more challenging systems that arise in nature when acidic rain interacts with minerals, let's examine the behavior of some simple aqueous solutions of salts.

Consider the following three salts that dissolve in water and dissociate completely: (A) NaCl, (B) NaF, and (C) NH_4Cl. Suppose a 0.01 M solution of each salt is prepared. When the pH of each solution is measured, only solution A is neutral. Solution B is basic, and solution C is acidic. The implication of this observation is that ions formed in two of the solutions react with water to produce either hydrogen ions or hydroxide ions.

Solution A is neutral, which means that neither $Na^+(aq)$ or $Cl^-(aq)$ compete successfully with water molecules for protons. This lack of reactivity is to be expected because the chloride ion is the conjugate base of a strong acid (HCl). Therefore, Cl^- ions must be a very weak base that cannot pull H^+ ions away from water molecules. NaF on the other hand generates sodium ions and fluoride ions in solution. Sodium ions do not hydrolyze, but the fluoride ion is the conjugate base of the weak acid HF and competes with water for protons:

$$F^-(aq) + H_2O(\ell) \rightleftharpoons HF(aq) + OH^-(aq) \qquad (16.18)$$

The reaction in Equation 16.18 accounts for the slightly basic pH of solutions of NaF.

SAMPLE EXERCISE 16.8 — **Distinguishing Acidic, Basic, and Neutral Salts**

Are aqueous solutions of NaClO acidic, basic, or neutral?

COLLECT AND ORGANIZE We are asked to describe the acid or base properties of NaClO. When this salt dissolves in water, it dissociates into Na^+ ions and ClO^- ions.

ANALYZE Many neutral salts consist of cations and anions that do not react with water (hydrolyze) when they dissolve. The cations in acidic salts are the conjugate acids of weak bases. The anions in basic salts are the conjugate bases of weak acids. The Na^+ ions in NaClO do not hydrolyze; however, the ClO^- ion is the conjugate base of HClO, which, according to Table 16.4, is a weak acid.

SOLVE The cyanide ion hydrolyzes in water to produce hydroxide ion in solution:

$$ClO^-(aq) + H_2O(\ell) \rightleftharpoons HClO(aq) + OH^-(aq)$$

The fact that the hydrolysis of $ClO^-(aq)$ produces hydroxide means that solutions of NaClO are basic.

THINK ABOUT IT When a salt containing the conjugate base of a weak acid dissolves in water, the anion reacts with water and produces a basic solution.

Practice Exercise Write a chemical equation for the hydrolysis reaction that explains why an aqueous solution of NH_4Cl is acidic.

Now that we can qualitatively understand the pH of salt solutions by looking at hydrolysis reactions of salts, we can explore quantitative aspects of salt solution pH as well. By using the same techniques we developed for weak acids, we can calculate the pH of a solution of a salt if we know the values of K for reactions of its ions in aqueous media. Equilibria involving the carbonate ion are of profound importance geologically and biologically, so with that in mind let's examine the basic properties of the carbonate ion. For example, if we dissolve sodium carbonate in water, it dissociates completely into sodium ions, which do not hydrolyze, and carbonate ions, which do. The carbonate ion is the conjugate base of $HCO_3^-(aq)$, so it hydrolyzes in water to produce a basic solution:

$$CO_3^{2-}(aq) + H_2O(\ell) \rightleftharpoons HCO_3^-(aq) + OH^-(aq)$$

The equilibrium for the carbonate ion that is behaving as a base is defined by the mass action expression:

$$K_{b_1} = \frac{[HCO_3^-][OH^-]}{[CO_3^{2-}]}$$

If we knew the value of K_{b_1} for the hydrolysis of carbonate ion, we could formulate an ICE table to calculate $[OH^-]$, determine pOH, and from that, pH. We do not have a value for the K_{b_1} of this reaction, but we can calculate it from information we know about the equilibria involving $H_2CO_3(aq)$. In carrying out this analysis, we will develop a general method that we can use for any ion in solution that is the conjugate base of a weak acid of known K_a.

Table 16.5 lists the equations and values of K_{a_1} and K_{a_2} for carbonic acid. The second ionization step is described by the following equation and corresponding equilibrium constant expression:

$$HCO_3^-(aq) \rightleftharpoons CO_3^{2-}(aq) + H^+(aq)$$

$$K_{a_2} = \frac{[CO_3^{2-}][H^+]}{[HCO_3^-]} = 4.7 \times 10^{-11}$$

Now let's compare the K_{b_1} expression of the carbonate ion and the K_{a_2} expression of carbonic acid:

$$K_{b_1} = \frac{[HCO_3^-][OH^-]}{[CO_3^{2-}]} \qquad K_{a_2} = \frac{[CO_3^{2-}][H^+]}{[HCO_3^-]} = 4.7 \times 10^{-11}$$

They have some terms in common. Their similarity becomes more apparent if we write the reciprocal of the K_{a_2} expression:

$$K_{b_1} = \frac{[HCO_3^-][OH^-]}{[CO_3^{2-}]} \qquad \frac{1}{K_{a_2}} = \frac{[HCO_3^-]}{[CO_3^{2-}][H^+]} = \frac{1}{(4.7 \times 10^{-11})}$$

The only difference between K_{b_1} and $1/K_{a_2}$ is the $[OH^-]$ term in the K_{b_1} expression and the $[H^+]$ term in the reciprocal of K_{a_2}. This difference is the key to relating the two equations. As we have seen, $[H^+]$ and $[OH^-]$ are linked by K_w (Equation 16.7). Consider what happens when we multiply K_w by the reciprocal of K_{a_2} for carbonic acid:

$$K_w \times \left(\frac{1}{K_{a_2}}\right) = (\cancel{[H^+]}[OH^-])\left(\frac{[HCO_3^-]}{\cancel{[H^+]}[CO_3^{2-}]}\right) = \frac{[HCO_3^-][OH^-]}{[CO_3^{2-}]}$$

The resulting expression matches the K_{b_1} expression for the carbonate ion. Therefore, from the values of K_w and K_{a_2}, we can calculate the value of K_{b_1} for $CO_3^{2-}(aq)$:

$$K_{b_1} = \frac{K_w}{K_{a_2}} = \frac{1.0 \times 10^{-14}}{4.7 \times 10^{-11}} = 2.1 \times 10^{-4}$$

The inverse relation between the K_a of an acid (HCO_3^-, in this case) and the K_b of its conjugate base (CO_3^{2-}) holds for all conjugate pairs. The relation expressed in equation form is

$$K_b = \frac{K_w}{K_a} \quad \text{or} \quad K_w = K_a \times K_b \qquad (16.19)$$

Equation 16.19 reinforces the complementary nature of an acid and its conjugate base: as the strength (K_a) of the acid increases, the strength (K_b) of its conjugate base decreases, and vice versa. In the case under discussion, the carbonate ion is a stronger base ($K_{b_1} = 2.1 \times 10^{-4}$) than the hydrogen carbonate ion is an acid ($K_{a_2} = 4.7 \times 10^{-11}$).

Now we have a value we can use (namely, K_{b_1}) to calculate the pH of a 0.100 M solution of Na_2CO_3. In setting up an ICE table, we assume that the only important source of OH^- is the carbonate reaction with water and that the autoionization of water does not contribute significantly to $[OH^-]$ at equilibrium. Let x be the equilibrium value of $[OH^-]$. Then $[HCO_3^-]$ also is x. The changes in the two must both be $+x$, and the change in $[CO_3^{2-}]$ must be $-x$. Completing the ICE table, we have

	$[CO_3^{2-}]$ (M)	$[HCO_3^-]$ (M)	$[OH^-]$ (M)
Initial	0.100	0	0
Change	$-x$	$+x$	$+x$
Equilibrium	$0.100 - x$	x	x

By using the equilibrium terms in the K_{b_1} expression for carbonate, we have

$$K_{b_1} = 2.1 \times 10^{-4} = \frac{[HCO_3^-][OH^-]}{[CO_3^{2-}]} = \frac{(x)(x)}{(0.100 - x)}$$

Let's simplify the rest of the calculation by assuming that the value of x will be much less than 0.100 M:

$$\frac{x^2}{0.100} = 2.1 \times 10^{-4}$$

$$x^2 = 2.1 \times 10^{-5}$$

$$x = 4.6 \times 10^{-3} \, M = [OH^-]$$

We must always remember to check the validity of our two assumptions. The calculated $[OH^-]$ is much greater than $[OH^-]$ in pure water ($1.0 \times 10^{-7} \, M$), so the assumption that the autoionization of water could be ignored is valid. As for the assumption that there was no significant decrease in $[CO_3^{2-}]$ as the reaction proceeded to equilibrium, there was in fact a 4.6% decrease:

$$\frac{[CO_3^{2-}]_{change}}{[CO_3^{2-}]_{initial}} = \frac{4.6 \times 10^{-3} \, M}{0.100 \, M} \times 100\% = 4.6\%$$

This value is within our 5% acceptable error guideline, so our simplifying assumption is acceptable.

Just to see the effect of our assumption, however, let's redo the calculation, leaving $-x$ in the denominator:

$$\frac{x^2}{(0.100 - x)} = 2.1 \times 10^{-4}$$

$$x^2 + 2.1 \times 10^{-4}x - 2.1 \times 10^{-5} = 0$$

Solving this quadratic equation yields a positive solution of 4.5×10^{-3}. This value is quite close to the initial result of 4.6×10^{-3}, so we conclude that our simplification is acceptable.

To calculate pH from [OH⁻], we calculate pOH:

$$pOH = -\log[OH^-] = -\log(4.6 \times 10^{-3}) = 2.34$$

and by using Equation 16.10, we have

$$pH = pK_w - pOH = 14.00 - 2.34 = 11.66$$

A pH value of 11.66 is quite basic. If you swam in a pool of water at that pH, you would experience skin irritation and painful burning in your eyes. Sodium carbonate *is* used to adjust the pH of pools, but to make sure the right amount is added to a pool, the pH of the water is tested to ensure that the pool is not too basic.

The final salt solution of the three we considered at the start of this discussion is ammonium chloride. It is the salt that is produced when NH_3 (a weak base) is neutralized with a strong acid (HCl):

$$NH_3(aq) + HCl(aq) \rightarrow NH_4Cl(aq) \qquad (16.20)$$

We know the chloride ion, as the conjugate base of a strong acid, does not hydrolyze at all in water. The acidic pH of the solution must result from the ammonium ion, which is the conjugate acid of the base NH_3 in Equation 16.20. We can take a similar approach to the one we used with Na_2CO_3 to calculate the pH of a solution of NH_4Cl.

SAMPLE EXERCISE 16.9 **Calculating the pH of a Solution of an Acidic or Basic Salt**

What is the pH of a 0.25 M solution of NH_4Cl?

COLLECT AND ORGANIZE The acidic properties of NH_4Cl are derived from the ability of the ammonium ion, NH_4^+, the conjugate acid of NH_3, to act as a H⁺ donor:

$$NH_4^+(aq) \rightleftharpoons NH_3(aq) + H^+(aq)$$

As we have seen, the Cl⁻ ion is an extremely weak H⁺ ion acceptor, so it does not contribute to the acid–base properties of NH_4Cl.

ANALYZE The equilibrium constant expression for the reaction is

$$K_a = \frac{[NH_3][H^+]}{[NH_4^+]}$$

It is related to the equilibrium constant expression for the conjugate base of NH_4^+, which is NH_3, as follows:

$$K_b = \frac{[NH_4^+][OH^-]}{[NH_3]} = 1.76 \times 10^{-5}$$

According to Equation 16.19:

$$K_w = K_a \times K_b$$

After rearranging the terms in Equation 16.19 to solve for K_a and then inserting the numerical value of K_b for NH_3, we have

$$K_a = \frac{K_w}{K_b} = \frac{1.0 \times 10^{-14}}{1.76 \times 10^{-5}} = 5.6 \times 10^{-10} = \frac{[NH_3][H^+]}{[NH_4^+]}$$

We can set up the ICE table for the reaction in which we make the usual assumptions that the reaction is the only significant source of H^+ and that $x = [H^+]$ at equilibrium:

	$[NH_4^+]$ (*M*)	$[NH_3]$ (*M*)	$[H^+]$ (*M*)
Initial	0.25	0	0
Change	$-x$	$+x$	$+x$
Equilibrium	$0.25 - x$	x	x

SOLVE Inserting the equilibrium concentrations into the expression for K_a gives

$$K_a = 5.6 \times 10^{-10} = \frac{[NH_3][H^+]}{[NH_4^+]} = \frac{(x)(x)}{(0.25 - x)}$$

After making the usual simplifying assumption, we get

$$\frac{x^2}{0.25} = 5.6 \times 10^{-10}$$

$$x^2 = 1.4 \times 10^{-10}$$

$$x = 1.2 \times 10^{-5} = [H^+]$$

Taking the negative logarithm of $[H^+]$ to obtain the pH of the solution, we have

$$pH = -\log[H^+] = -\log(1.2 \times 10^{-5}) = 4.92$$

THINK ABOUT IT This result is reasonable: a solution of a very weak acid ($K_a = 5.6 \times 10^{-10}$) should be weakly acidic. The calculated $[H^+]$ is more than 100 times that produced by the autoionization of water (1.0×10^{-7} *M*), so ignoring the contribution of the latter process is permitted, as is assuming that $[H^+]$ is small compared with 0.25 *M*.

Practice Exercise The acetate ion is the conjugate base of acetic acid and reacts with water according to the following reaction equation:

$$CH_3COO^-(aq) + H_2O(\ell) \rightleftharpoons CH_3COOH(aq) + OH^-(aq)$$

Calculate the pH of a 0.25 *M* solution of sodium acetate (CH_3COONa).

The reactions of acidic or basic salts are responsible for neutralization of acid rain in some areas. If acid rain containing dilute sulfuric acid soaks into soil containing $CaCO_3$, the acid is converted into either environmentally more benign carbonic acid

$$CaCO_3(s) + H_2SO_4(aq) \rightleftharpoons CaSO_4(s) + H_2CO_3(aq)$$

or, if enough $CaCO_3$ is available—as limestone, marble, or shellfish shells—into calcium sulfate and soluble calcium hydrogen carbonate:

$$2\,CaCO_3(s) + H_2SO_4(aq) \rightleftharpoons CaSO_4(s) + Ca(HCO_3)_2(aq)$$

As long as carbonates and other basic substances are present in soils and in the sediments of rivers and lakes, nature has the capacity to neutralize some of the acid in acid rain and maintain pH in a range that supports aquatic life.

CONNECTION The reaction of acidic groundwater with calcium carbonate and its connection to the formation of limestone caves are described in Chapter 4.

16.8 The Common-Ion Effect

Aqueous solutions that contain a number of ions from several sources are often encountered in practical situations. When several equilibria are possible under a given set of conditions, the concentration of all ions present must satisfy the equilibrium constant expressions for whatever equilibria are involved. If more of any given ion is added to a solution, then according to Le Châtelier's principle, the equilibria respond by shifting in a fashion that restores equilibrium.

Suppose, for example, that a sample of river water contains carbonic acid at a concentration of 1.2×10^{-5} M as a result of atmospheric carbon dioxide dissolving in the water. Let's also suppose that the sedimentary material carried by the river and/or that forms the river's banks and bed contains calcite or other carbonate minerals. The carbonate ion is an effective base, and this property allows an acid–base reaction to take place in which solid calcium carbonate from the sedimentary material reacts with carbonic acid to form soluble calcium bicarbonate:

$$CaCO_3(s) + H_2CO_3(aq) \rightleftharpoons Ca(HCO_3)_2(aq)$$

Other acidic substances in the river water also would be neutralized by carbonate and would produce more hydrogen carbonate (*bicarbonate*) ions. Suppose that the total concentration of bicarbonate ion produced by all the reactions of acids with carbonate was 1.0×10^{-4} M. What effect does the additional bicarbonate ion have, and what would the pH of the water be?

We know the concentrations of both carbonic acid and bicarbonate ion. These substances are related by the first acid-ionization reaction of carbonic acid

$$H_2CO_3(aq) \rightleftharpoons HCO_3^-(aq) + H^+(aq)$$

and by the following mass action expression:

$$K_{a_1} = \frac{[HCO_3^-][H^+]}{[H_2CO_3]} = 4.3 \times 10^{-7}$$

The difference between the river water in this example and a prepared solution of water and carbonic acid is that the HCO_3^- in a dilute solution is due solely to the ionization of H_2CO_3. The river water, however, has an additional source of HCO_3^-. According to Le Châtelier's principle, adding HCO_3^- should shift the acid-ionization reaction to the left. This shift would lower $[H^+]$ in the river water and thereby raise its pH to values above the 5.6 calculated earlier for natural rainwater.

This phenomenon illustrates a principle known as the **common-ion effect**, when the addition of an ion taking part in a reaction causes a shift in the position of an equilibrium. In any ionic equilibrium, the reaction that produces an ion is suppressed when another source of the same ion is added to the system. In the river-water sample, the ionization of H_2CO_3 is suppressed when HCO_3^- from

CONNECTION We noted in Chapter 2 that *bicarbonate* is a more common name for the HCO_3^- ion than *hydrogen carbonate*.

The **common-ion effect** is the shift in the position of an equilibrium caused by the addition of an ion taking part in the reaction.

carbonate minerals is present. To assess the effect of the added bicarbonate, we insert the given values for $[H_2CO_3]$ and $[HCO_3^-]$ into the K_{a_1} expression (and we let $x = [H^+]$):

$$K_{a_1} = \frac{[HCO_3^-][H^+]}{[H_2CO_3]} = \frac{(1.0 \times 10^{-4})(x)}{(1.2 \times 10^{-5})} = 4.3 \times 10^{-7}$$

By solving for x, we find that $[H^+] = 5.2 \times 10^{-8}\ M$. Taking the negative logarithm of this value indicates that the pH of the river water is 7.28:

$$-\log[H^+] = -\log(5.2 \times 10^{-8}) = 7.29$$

Thus, the presence of bicarbonate results in a solution with a pH nearly two units above that of carbonic acid alone.

Now let's consider the common-ion effect for any equilibrium involving a weak acid and its conjugate base. We begin with a generic equilibrium equation,

$$\text{acid} \rightleftharpoons H^+ + \text{base}$$

which has the equilibrium constant expression:

$$K_a = \frac{[H^+][\text{base}]}{[\text{acid}]}$$

If we take the negative logarithm of both sides of the mass action expression, we transform $[H^+]$ into pH and K_a into pK_a (Section 16.3):

$$pK_a = pH - \log\frac{[\text{base}]}{[\text{acid}]}$$

After rearranging the terms to solve for pH, we have

$$pH = pK_a + \log\frac{[\text{base}]}{[\text{acid}]} \tag{16.21}$$

Equation 16.21 is particularly useful. It is called the **Henderson–Hasselbalch equation**, and it enables us to calculate the pH of a solution in which we know the concentration of an acid and its conjugate base (or a base and its conjugate acid).

Consider what happens to the logarithmic term in the Henderson–Hasselbalch equation when the concentration of the acid is the same as that of its conjugate base. Then the numerator and denominator in the log term are equal, and the value of the fraction is 1. The log of 1 is 0, and under those specific conditions, $pH = pK_a$. This equality serves as a handy reference point in an acid–conjugate base system. If the concentration of the basic component is greater than that of the acid, then the logarithmic term is greater than zero, so $pH > pK_a$. If the concentration of the basic component is less than that of the acid, then the logarithmic term is less than zero, and $pH < pK_a$. Consider the case in which the concentration of the base is 10 times the concentration of the acid, or $[\text{base}] = 10[\text{acid}]$. Substituting this ratio into the Henderson–Hasselbalch equation, we have

$$pH = pK_a + \log\frac{[\text{base}]}{[\text{acid}]}$$

$$= pK_a + \log\frac{10[\text{acid}]}{[\text{acid}]}$$

$$= pK_a + \log 10 = pK_a + 1$$

The **Henderson–Hasselbalch equation** is used to calculate the pH of a solution in which the concentrations of acid and conjugate base are known.

Therefore, a 10-fold higher concentration of base than acid produces a pH one unit above the pK_a value. Similarly, if the concentration of the acid component is 10 times that of the base, pH is one unit below pK_a.

To demonstrate how the Henderson–Hasselbalch equation simplifies pH calculations when we know the concentrations of a weak acid and its conjugate base, let's use that equation to recalculate the pH of our river-water sample. Our starting information includes

$$[H_2CO_3] = 1.2 \times 10^{-5}\ M$$

$$[HCO_3^-] = 1 \times 10^{-4}\ M$$

and the equilibrium and equilibrium constant

$$H_2CO_3(aq) \rightleftharpoons HCO_3^-(aq) + H^+(aq) \qquad K_{a_1} = 4.3 \times 10^{-7}$$

First we take the negative log of K_{a_1}:

$$pK_{a_1} = -\log K_{a_1} = -\log(4.3 \times 10^{-7}) = 6.37$$

and then insert pK_{a_1} and the concentration values of $[H_2CO_3]$ and $[HCO_3^-]$ into the Henderson–Hasselbalch equation. Solving the equation gives

$$pH = pK_a + \log \frac{[\text{base}]}{[\text{acid}]}$$

$$= 6.37 + \log \frac{1.0 \times 10^{-4}}{1.2 \times 10^{-5}}$$

$$= 7.29$$

This result matches the one we calculated.

As we have seen, the Henderson–Hasselbalch equation can be used to calculate the pH of a solution containing a weak acid and its conjugate base. It can also be used to calculate the pH of a solution of a base and its conjugate acid. There is an extra step involved in that calculation because we may know the K_b of the base but not the K_a of its conjugate acid. Fortunately, the conversion from K_b to K_a, or K_a to K_b, for a conjugate pair makes use of the simple relation between K_b and K_a (Equation 16.19):

$$K_a \times K_b = K_w$$

Because we use pK_a values in the Henderson–Hasselbalch equation, we can take the negative logarithms of both sides of Equation 16.19. We first rearrange Equation 16.19 and then insert the value of K_w at 25°C:

$$K_b = \frac{K_w}{K_a} = \frac{1.00 \times 10^{-14}}{K_a}$$

$$-\log K_b = -\log(1.00 \times 10^{-14}) - (-\log K_a)$$

$$pK_b = 14.00 - pK_a$$

or

$$pK_b + pK_a = 14.00 \qquad\qquad (16.22)$$

SAMPLE EXERCISE 16.10 **Calculating the pH of a Solution of a Base and Its Conjugate Acid**

Calculate the pH of a solution that is 0.200 M in NH_3 and 0.300 M in NH_4Cl. The K_b of NH_3 is 1.76×10^{-5}.

COLLECT AND ORGANIZE We are asked to calculate the pH of a solution containing a weak base (NH_3) and a salt of its conjugate acid (NH_4^+). The Henderson–Hasselbalch equation,

$$pH = pK_a + \log \frac{[base]}{[acid]}$$

is used to calculate the pH of such a solution, but first we need to calculate the pK_a of the acid in the NH_4^+ solution.

ANALYZE The pK_a of NH_4^+ is related to the pK_b of NH_3 by the following equation, which applies to all acid–base conjugate pairs in aqueous solutions at 25°C:

$$pK_b + pK_a = 14.00$$

We are given the value of K_b of NH_3: 1.76×10^{-5}. Our approach involves taking the negative logarithm of K_b to obtain pK_b and then converting pK_b to a pK_a.

SOLVE We calculate pK_b first:

$$pK_b = -\log K_b = -\log(1.76 \times 10^{-5}) = 4.75$$

After inserting this value of pK_b into Equation 16.22, we then solve for pK_a:

$$pK_a = 14.00 - pK_b = 14.00 - 4.75 = 9.25$$

By inserting this value and the given concentrations of NH_3 and NH_4^+ into the Henderson–Hasselbalch equation, we have

$$pH = pK_a + \log \frac{[base]}{[acid]}$$

$$= 9.25 + \log \frac{0.200}{0.300} = 9.07$$

THINK ABOUT IT The K_b value of NH_3 is greater than the K_a we calculated for NH_4^+. This statement makes sense because ammonia is a stronger base than the ammonium ion is an acid. It is reasonable that a solution containing similar concentrations of both should be slightly basic.

Practice Exercise Calculate the pH of a solution that is 0.150 M in benzoic acid and 0.100 M in sodium benzoate, its conjugate base.

16.9 pH Buffers

A solution that has the capacity to resist pH changes by neutralizing small additions of acid (or base) is called a **pH buffer**. Buffers are an important component of natural water systems; indeed, they are also essential to life itself. The internal pH of most living cells is regulated by buffer systems. The carbonic acid–bicarbonate system just described is one of the buffer systems that acts in living organisms as well as in rivers. When pH stabilization provided by these buffers is disturbed, proteins do not function optimally, and the health of individual cells and whole organisms is in jeopardy.

A **pH buffer** is a solution of acidic and basic solutes that resists changes in its pH when acids or bases are added to it.

An Environmental Buffer

The presence of bicarbonate ion in the river water in the preceding section gives the water a capacity to resist pH change when, for example, acid rain falls into it. We can use the Henderson–Hasselbalch equation and other acid–base relations to examine the effect of adding a small amount of a strong acid to river water. For purposes of comparison, we can measure the effect on the pH of the same volume of pure (pH = 7.00) water at 25°C. We start with 1.00 L each of river water and pure water and add 10.0 mL of a $1.0 \times 10^{-3} M$ HNO_3 solution to both.

Adding acid to pure water is an exercise in dilution, which is a concept we introduced in Section 4.3. In this case, 10.0 mL of a $1.0 \times 10^{-3} M$ HNO_3 solution is diluted to a final volume of 1.01 L. To calculate the final concentration of HNO_3 (and H^+ ions) we use Equation 4.5:

$$V_{initial} \times C_{initial} = V_{final} \times C_{final}$$

where $C_{initial}$ and C_{final} are the initial and final concentrations of the acid and $V_{initial}$ and V_{final} are its initial and final volumes. Expressing 10.0 mL as 0.0100 L so that the initial and final volume units match and then substituting the given information into Equation 4.5 leads us to

$$0.0100 \text{ L} \times (1.0 \times 10^{-3} M) = 1.01 \text{ L} \times C_{final}$$

$$C_{final} = [H^+] = 9.9 \times 10^{-6} M$$

Nitric acid is completely dissociated, so the final concentration of H^+ must also be $9.9 \times 10^{-6} M$. The corresponding pH value is $-\log(9.9 \times 10^{-6} M) = 5.00$, or 2.00 pH units lower than the initial pH of the pure water sample.

To calculate the change in pH when the same quantity of nitric acid is added to one liter of our pH 7.29 river water, we need to reexamine the carbonic acid–bicarbonate equilibrium:

$$H_2CO_3(aq) \rightleftharpoons HCO_3^-(aq) + H^+(aq)$$

When strong acid (H^+) is added to a mixture of H_2CO_3 and HCO_3^-, it reacts with HCO_3^- to produce H_2CO_3. The ionization reaction runs in the reverse direction:

$$H^+(aq) + HCO_3^-(aq) \rightarrow H_2CO_3(aq)$$

until essentially all of the added H^+ ions have been consumed. We need to determine the number of moles of H^+ that were added so that we can calculate how many moles of HCO_3^- were consumed and, with that number, how many moles are left. There are 1.0×10^{-5} mol of H^+ in 0.010 L of a $1.00 \times 10^{-3} M$ HNO_3 solution. Given the 1:1:1 stoichiometric ratio of the reactants and products, what was added is completely consumed. In the process 1.0×10^{-5} mol of HCO_3^- are consumed. The number of moles of bicarbonate present initially was

$$(1.0 \times 10^{-4} M) \times (1.00 \text{ L}) = 1.0 \times 10^{-4} \text{ mol}$$

After subtracting the number of moles consumed, we have 9.0×10^{-5} mol of HCO_3^- in a final volume of 1.01 L. The resulting pH can be predicted from the K_{a_1} for carbonic acid. In the calculation, we let $x = [H^+]$ and we assume that $[H_2CO_3]$ is controlled by the solubility of CO_2 in water and does not change from its initial value of $1.2 \times 10^{-5} M$. Inserting these values into the mass action expression for K_{a_1} of H_2CO_3 gives

$$K_{a_1} = \frac{[HCO_3^-][H^+]}{[H_2CO_3]} = \frac{\left(\dfrac{9.0 \times 10^{-5} \text{ mol}}{1.01 \text{ L}}\right)x}{1.2 \times 10^{-5} M} = 4.3 \times 10^{-7}$$

Solving for x gives

$$x = [H^+] = 5.7 \times 10^{-8}\ M$$

After taking the negative logarithm of this value we have

$$pH = -\log(5.7 \times 10^{-8}) = 7.24$$

Recall that the original pH was 7.29. Therefore, the addition of nitric acid lowered the pH of the river water by only 0.05 pH units because of the action of the carbonic acid–hydrogen carbonate buffer. We can compare this result to the reduction by 2.00 pH units when the same amount of nitric acid was added to the same volume of pure unbuffered water.

SAMPLE EXERCISE 16.11   **Calculating the Response to Addition of Acid or Base to a Buffer**

Calculate the change in pH upon addition of 1.0 mL of a 1.0 M HCl solution to 100 mL of a solution that is 0.10 M sodium acetate and 0.10 M acetic acid at 25°C.

COLLECT AND ORGANIZE A solution of a weak acid and a salt of its conjugate base functions as a pH buffer. We know the volume and composition of the buffer and of the strong acid added to it. We need to calculate the composition of the reaction mixture after the addition. Equation 16.16 can be used to calculate the pH of such a mixture once we know the new concentration ratio of base (acetate ion) to acid (acetic acid).

ANALYZE The H^+ ions in the added strong acid react with acetate ions, forming unionized acetic acid

$$H^+(aq) + CH_3COO^-(aq) \rightarrow CH_3COOH$$

The product of the volume times the concentration of the strong acid can be used to calculate the number of moles of CH_3COO^- that is converted into CH_3COOH.

SOLVE The initial quantities of CH_3COO^- and CH_3COOH in the buffer are both:

$$100\ mL \times \frac{0.100\ mol}{L} = 10.0\ mmol$$

The quantity of $H^+(aq)$ that was added

$$1.0\ mL \times \frac{1.00\ mol}{L} = 1.00\ mmol$$

represents the quantity of CH_3COO^- converted into CH_3COOH. Therefore the quantities of both after the addition of acid are $(10.0 - 1.0) = 9.0$ mmol CH_3COO^- and $(10.0 + 1.0) = 11.0$ mmol CH_3COOH. Both these quantities are in the same total volume. Therefore we can use the ratio of the two of them in lieu of concentration values in Equation 16.16 to calculate pH:

$$pH = pK_a + \log\frac{[CH_3COO^-]}{[CH_3COOH]} = 4.75 + \log\frac{9.0}{11.0} = 4.66$$

THINK ABOUT IT The result makes sense because adding strong acid should lower the pH from the initial value of the buffer in which $CH_3COO^- = CH_3COOH$. This value is the same as the pK_a of acetic acid (4.75).

Practice Exercise Calculate the change in pH after addition of 10.0 mL of a 0.100 M solution of NaOH to 1.00 L of a solution that is 1.00 M in sodium acetate and 1.00 M in acetic acid.

Suggest a specific situation in which it may be advisable to have more weak acid than conjugate base in a buffer.

A Physiological Buffer

The carbonate buffer system illustrated by the sample of river water is also of great importance in living systems. The rates of most biochemical reactions and their equilibrium constants are sensitive to changes in pH. Many biochemical reactions involved in life-sustaining processes like metabolism, respiration, the transmission of nerve impulses, and muscle contraction and relaxation take place only within a narrow pH range.

The carbonic acid/bicarbonate conjugate acid–base pair form an important buffering system in blood and cellular fluids. The pH of blood can be altered by ingestion of acidic or basic substances, and the carbonate/bicarbonate buffer system compensates for such additions and maintains pH within the required range. This buffering system is intimately tied to our respiration, and an exceptional feature of pH control by this system is the role of breathing to maintain pH balance.

Let's first consider how this buffer system is established in the body. Carbon dioxide is a normal product of metabolism. It is transported to the lungs, where it is eliminated from the body every time we exhale. When CO_2 dissolves in blood plasma (or any aqueous medium), the following equilibria are established:

$$CO_2(aq) + H_2O(\ell) \rightleftharpoons H_2CO_3(aq) \rightleftharpoons H^+(aq) + HCO_3^-(aq)$$

If biochemical processes in the body increase $[H^+]$ in the blood, bicarbonate ions react with the added hydrogen ions and are transformed into carbonic acid, which increases the concentration of dissolved CO_2 in the blood. Respiration increases, and more CO_2 is expelled from the lungs. These shifts are summarized in the following diagram.

$$H^+(aq) + HCO_3^-(aq) \rightleftharpoons H_2CO_3(aq) \rightleftharpoons CO_2(aq) + H_2O(\ell) \rightleftharpoons CO_2(g)$$

increase in $[H^+]$ $\qquad\qquad\qquad\qquad\qquad$ CO_2 exhaled

If biochemical processes consume H^+ ions, then the following equilibrium shifts occur:

$$H^+(aq) + HCO_3^-(aq) \rightleftharpoons H_2CO_3(aq) \rightleftharpoons CO_2(aq) + H_2O(\ell) \rightleftharpoons CO_2(g)$$

decrease in $[H^+]$ $\qquad\qquad\qquad\qquad\qquad$ more CO_2 retained in the blood

The buffer system compensates for this change by producing more H^+ ions as more carbon dioxide dissolves and more carbonic acid ionizes to restore the $[H^+]$ in the blood to the equilibrium concentration. This response requires more CO_2 to be dissolved in the blood, so respiration decreases and more CO_2 is retained in the lungs, which restores the bicarbonate ion concentration in the system.

Because the pH of the blood is in large part under respiratory control, simple alterations in normal breathing can change the pH of the blood. The state of *respiratory acidosis* arises as a result of impaired lung function and hypoventilation. Slow, shallow breathing causes more CO_2 to be retained in the lungs, which in turn causes more CO_2 to be dissolved, more carbonic acid to be formed, and the concentration of bicarbonate and hydronium ion in the blood to rise. Hyperventilation induces the opposite situation—respiratory *alkalosis*.

SAMPLE EXERCISE 16.12   **Describing a Buffer System**

The mice used in biomedical experiments are sometimes euthanized by exposing them to an atmosphere of CO_2. How would exposure to such an atmosphere affect the pH of the blood of the mice?

COLLECT AND ORGANIZE We are asked to predict the effect of inhaling an atmosphere of CO_2 on the blood pH of mice. We need to consider how the position of the equilibrium involving CO_2 and bicarbonate

$$CO_2(g) + H_2O(\ell) \rightleftharpoons H^+(aq) + HCO_3^-(aq)$$

is influenced by an increase in $[CO_2]$.

ANALYZE An increase in $[CO_2]$ shifts the equilibrium

$$CO_2(g) + H_2O(\ell) \rightleftharpoons H^+(aq) + HCO_3^-(aq)$$

to the right.

SOLVE The predicted equilibrium shift produces an increase in $[H^+]$ and a decrease in pH.

THINK ABOUT IT Breathing an atmosphere rich in CO_2 produces a state of respiratory acidosis, which is lethal to mice and other mammals, including humans.

Practice Exercise How would the blood pH of a mouse exposed to CO_2 atmosphere for a short time change if the mouse were suddenly exposed to air containing no $CO_2(g)$? Provide a chemical reason for your answer based on considerations of the appropriate equilibria.

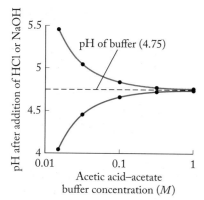

FIGURE 16.12 The changes in pH of buffered solutions caused by adding strong acid (red curve) or strong base (blue curve) increase as the concentrations of the buffers decrease. In this illustration, five 100 mL solutions that are 0.015, 0.03, 0.10, 0.30, and 1.00 M acetic acid and sodium acetate all have an initial pH of 4.75. The graph shows the pH values of these solutions after adding 1.00 mL of 1.00 M HCl or 1.00 mL of 1.00 M NaOH.

Buffer Range and Capacity

As was inferred in Sample Exercise 16.11, to act as a good buffer a solution must maintain a relatively constant pH when either acid or base is added. Two considerations are made when a buffer is selected for a specific task:

1. What pH do we want to maintain? The appropriate buffer is one whose weak acid has a pK_a that is within one pH unit of the desired pH. The Henderson–Hasselbalch equation tells us that over this pH range the concentration ratio of the base to the acid component of the buffer varies from 10:1 to 1:10. Therefore, both components are available to maintain the desired pH.

2. How much acid or base does the system need to consume without a large change in pH? The answer to this question defines the **capacity** of the buffer. A typical buffer is best able to resist changes in pH when the initial concentration of acid and conjugate base are comparable to each other and greater than the concentration of acid or base to be added. The greater the concentration of the components of the conjugate acid–base pair, the greater the ability of the buffer to resist changes in pH (Figure 16.12).

16.10 Acid–Base Titrations and Indicators

Pool owners routinely check the pH of their pools by using colored pH indicators. These indicators are substances that change color with changing pH. One such compound is phenol red. It functions as a *colorimetric* ("based on color") **pH indicator** because it changes color depending on the pH. It is a weak acid (pK_a = 7.6), and its un-ionized form (which, for convenience, we assign the generic formula HIn) is actually yellow, but the ionized (In⁻) form is violet (Figure 16.13). At a pH one unit above the pK_a, that is a pH of 8.6, the ratio of In⁻ to HIn is 10:1 according to the Henderson–Hasselbalch equation, and a phenol-red solution is violet. At pH less than 6.6, phenol red is largely un-ionized, and so a solution of the indicator is yellow. In the pH range from about 6.8 to 8.4,

The **capacity** of a buffer defines the concentration of components needed in solution to consume the acid or base added without experiencing a significant change in pH.

A **pH indicator** is a water-soluble weak organic acid (HIn) that changes color when it ionizes, for example,

$$HIn(aq) \rightleftharpoons H^+(aq) + In^-(aq)$$

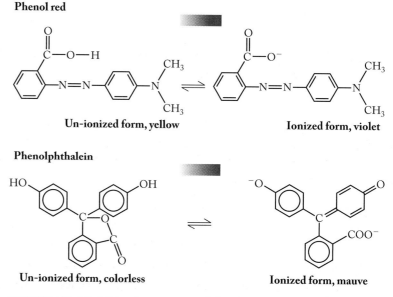

FIGURE 16.13 Molecular structures of phenol red and phenolphthalein.

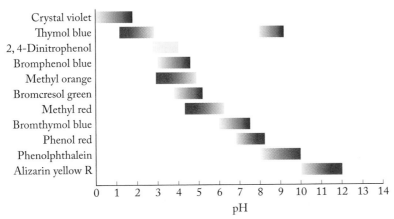

FIGURE 16.14 The color shown by pH indicators depends on their chemistry and the pH of the tested solution. The pH indicators are used to determine the pH of solutions and pH changes in titrations. An indicator is useful within a range of 1 pH unit above and below the pK_a value of the indicator.

(a)

(b)

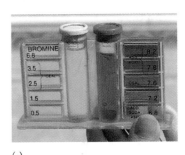

(c)

Many pool test kits include the pH indicator phenol red. A few drops are added to a sample of pool water collected in the tube with the red cap. (a) After a rainstorm, the pH of the pool water is 6.8 (or less), as indicated by the yellow color of the sample. (b) Sodium carbonate is added to the pool to raise the pH. (c) A follow-up test produces a red-orange color, indicating the pH of the pool has been properly adjusted.

the color of a phenol-red solution changes from yellow to orange to red to violet with increasing pH.

These pH-dependent color changes allow the pH of a pool to be adjusted to the optimal pH for a swimmer's comfort (~7.4) to within about 0.2 pH units. The analysis consists of adding a few drops of concentrated indicator to about 5 to 10 mL of sample and comparing the color of the sample with a pH color chart. If the pH of the water is too low (the usual problem, which is generally caused by the addition of rainwater to the pool), Na_2CO_3 is added to the pool and the pH test is repeated.

Phenol red is suitable for testing pool water because its pK_a is close to the pH desired for the pool water, and so the pH range over which it changes color is centered on the desired value. Other color indicators are used to detect changes in pH over other pH ranges, as shown in Figure 16.14. All of them have a useful pH range from 1 pH unit below to 1 pH unit above their individual pK_a values; that is,

$$\text{pH range} = pK_a \pm 1.0$$

Indicators may be used to determine the pH of solutions, and, as we are about to see, they may also be used to detect the large changes in pH that occur in determinations of the concentrations of acids and bases in acid–base titrations.

Acid–Base Titrations

We first discussed titration methods in Chapter 4 (Figure 4.19), and we summarize their mechanics here. Figure 16.15 shows a typical titration apparatus. Titrations require five principal steps:

1. Accurately transfer a known volume of sample to a flask.

2. Fill a buret with a solution, called the *titrant*, of known concentration of a substance that will react with a solute, called the *analyte*, in the sample.

3. Slowly add the standard solution in small amounts to the sample.

4. Monitor the course of the reaction by recording the volumes of titrant added and by using an indicator that changes color when the reaction is

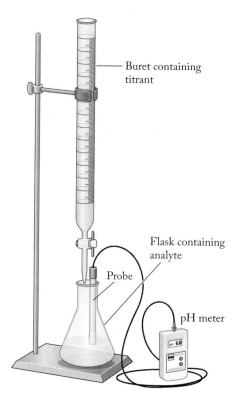

FIGURE 16.15 A typical setup showing a digital pH meter to measure pH during the course of a titration.

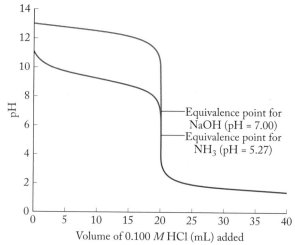

FIGURE 16.16 The titrations, using 0.100 *M* HCl, for 20.0 mL of 0.100 *M* ammonia, a weak base (NH₃, red), and 20.0 mL of 0.100 *M* NaOH, a strong base (blue), yield curves that differ in the region before (to the left of) the equivalence point. The titration reaction involving the weaker base has lower pH values until the equivalence point is reached at 20.0 mL of HCl added, where the pH = 5.27.

complete or a probe such as the electrode described in Section 16.3 that senses changes in [H⁺].

5. Record the minimum volume of titrant needed to completely consume the analyte.

We use the minimum volume of titrant to calculate the concentration of solute in the analytical sample.

The neutralization titrations in Chapter 4 involved titrating strong acids with strong bases and vice versa. An alkalinity titration is more challenging in that it involves the titration of the salt of a weak, diprotic acid. Let's work our way up to dealing with the complexity of an alkalinity titration by starting with titrations of weak and strong monoprotic acids and monobasic bases.

The term *alkalinity* is sometimes used to indicate the buffer capacity of a solution. The term is especially useful to describe the capacity of natural waters to buffer pH against changes due to additions of strong acids, as we discussed in Section 16.9. Although the buffer mixtures found in natural waters have more species involved than we included in our model of river water, the same carbonic acid–hydrogen carbonate buffer system that we have discussed several times is often the most important one.

For our first example, let's consider the titration, using 0.100 *M* HCl, of two 20.0 mL samples: one of 0.100 *M* NH₃, a weak base (the red titration curve in Figure 16.16), and the second of 0.100 *M* NaOH, a strong base (the blue curve). The titration curves in Figure 16.16 are plots of pH vs. volume of HCl titrant. As we carry out the two titrations, we record the volumes of HCl added; we determine these volumes through the use of the graduation marks on the barrel of the buret. We also monitor the pH by taking readings from a pH electrode in the sample.

In both titrations, the pH of the sample does not change much when acid is first added. The initial pH plateau in the NaOH titration curve is higher than that for NH₃ because NaOH is the stronger base and its solutions have higher pH values. However, when just enough titrant has been added to consume all the base present in the samples, the pH values of both samples drop sharply at their equivalence points—a change that can be detected by the signal from a pH electrode or by a change in the color of a pH indicator.

For both samples the equivalence point is reached with the addition of 20.0 mL of titrant, because both samples contained the same quantity of base. The titration curves differ in their starting pH values and the size of the pH change at their equivalence points. The starting pH value is lower and the change in pH near the equivalence point is smaller for NH₃ than for NaOH. Another difference is the pH that corresponds exactly to the equivalence point, which is the inflection point on the sharply falling line. It is 7.00, or neutral, in the NaOH curve, but only 5.27 in the NH₃ curve. To understand why the neutralization of NH₃ with HCl does not produce a neutral solution, think about what ions are in solution at the equivalence point in the ammonia titration. The titration reaction is

$$HCl(aq) + NH_3(aq) \rightarrow NH_4^+(aq) + Cl^-(aq)$$

When a 0.100 *M* solution of HCl neutralizes the ammonia in an equal volume of a 0.100 *M* solution of NH₃, we have a 0.0500 *M* solution of NH₄Cl. (Remember that mixing two solutions dilutes both of them, and as a result of this titration, the volumes of both solutions have been doubled. That means the concentrations of all species in solution have been cut in half.) As we discovered in the results of the Practice Exercise accompanying Sample Exercise 16.8, the pH values of solutions of NH_4Cl are slightly acidic because the ammonium ion is the conjugate acid of the weak base ammonia.

Just as the concentrations of basic compounds can be determined by titrating them with known quantities of strong acids, the concentrations of acidic compounds can be determined by titrating them with standard solutions of strong bases. Figure 16.17 illustrates two such applications: in one of them, 20.0 mL of a 0.100 M solution of HCl is titrated with a 0.100 M solution of NaOH (the red curve); in the other, the sample is 20.0 mL of a 0.100 M solution of acetic acid (the blue curve). As shown in Figure 16.17, the two titration curves differ until their equivalence points are reached (pH values are consistently higher for the weak acid), but they overlap beyond their equivalence points. In the region after the equivalence points, pH is controlled by the increasing concentration of the same NaOH titrant used in both experiments.

Two pH values along the titration curve of acetic acid deserve our attention. The first is at the equivalence point (pH = 8.73). This value is well above pH 7.0, even though it is the point in the titration at which just enough NaOH has been added to exactly neutralize the acetic acid in the sample. The pH of the equivalence point is slightly basic because the product of the titration reaction

$$CH_3COOH(aq) + NaOH(aq) \rightarrow CH_3COONa(aq) + H_2O(\ell)$$

is a 0.0500 M solution of sodium acetate (CH_3COONa). The acetate ion is the conjugate base of a weak acid, and so its aqueous solutions are slightly basic.

The other important milestone in the titration of acetic acid is the point halfway to the equivalence point. At that point, half of the acetic acid initially in the sample has been converted into acetate ion. Therefore the concentrations of the acetic acid still in solution and the acetate that has been produced are the same. This equality means that the logarithmic term in the Henderson–Hasselbalch equation,

$$pH = pK_a + \log \frac{[\text{base}]}{[\text{acid}]}$$

is zero, and so

$$pH = pK_a$$

The pK_a of acetic acid is

$$-\log(1.76 \times 10^{-5}) = 4.75$$

which is indeed the pH halfway to the equivalence point of the acetic acid titration.

To review the calculations we carried out in Chapter 4 for acid–base titrations, consider the following titration of a sample of vinegar to determine its acetic acid content. We'll use a pipet to add 10.00 mL of vinegar to a 125 mL Erlenmeyer flask and then add distilled water to bring the total volume to about 50 mL. We then add a few drops of pH indicator solution and titrate the diluted vinegar with a 0.1050 M solution of NaOH. We repeat this procedure two more times and calculate the average volume of NaOH solution needed to reach the equivalence point. Let's assume in this case the volume required is 16.24 mL. We can use that quantity to find the concentration of acetic acid in the vinegar.

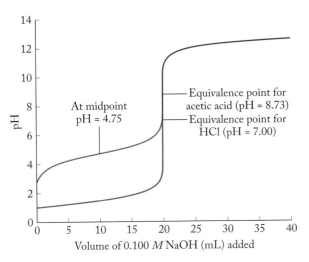

FIGURE 16.17 The titrations, using a 0.100 M solution of NaOH, for 20.0 mL of 0.100 M acetic acid (CH_3COOH, blue) and 20.0 mL of 0.100 M HCl (red) differ in the region before (to the left of) the equivalence point. The titration reaction involving the weaker acid has higher pH values until the equivalence point is reached at 20.0 mL of the NaOH solution added, where the pH = 8.73.

▶❚❚ **CHEMTOUR** Strong Acid and Strong Base Titration

▶❚❚ **CHEMTOUR** Titrations of Weak Acids

A **millimole** is 10^{-3} mol. It is a useful quantity in calculations related to titrations.

The key pieces of information in this titration are the original volume of sample (10.00 mL), the volume of titrant (16.24 mL), and the concentration of titrant (0.1050 M). We also need the balanced equation for the titration reaction, which is

$$CH_3COOH(aq) + NaOH(aq) \rightarrow CH_3COONa(aq) + H_2O(\ell)$$

This stoichiometry tells us that 1 mol of acetic acid is consumed for every mole of NaOH. Because we are working with milliliter volumes, a more useful quantity is the **millimole** (abbreviated mmol), which is 10^{-3} mol. Remember the definition of molarity (M) is moles of solute per liter of solution. We can convert moles to millimoles and liters to milliliters by dividing both by 1000:

$$M = \frac{\text{mol}}{\text{L}} = \frac{\text{mol}/1000}{\text{L}/1000} = \frac{\text{mmol}}{\text{mL}}$$

so moles per liter is equivalent to millimoles per milliliter. Because titrations done in the laboratory usually involve volumes in milliliters, it is often easier to work in terms of mmol. In the case of the reaction of acetic acid and NaOH, we can therefore express the following equality:

$$\text{mmol } CH_3COOH = \text{mmol NaOH}$$

The number of millimoles of a solute in solution is equal to the volume of the solution in milliliters times the molarity of the solute. Thus, the preceding equality can be expressed as

$$(V_{CH_3COOH})(M_{CH_3COOH}) = (V_{NaOH})(M_{NaOH}) \tag{16.23}$$

or

$$M_{CH_3COOH} = \frac{(V_{NaOH})(M_{NaOH})}{V_{CH_3COOH}}$$

After inserting the experimental data, we have

$$M_{CH_3COOH} = \frac{(16.24 \text{ mL})(0.1050 \text{ } M)}{10.00 \text{ mL}}$$

$$= 0.1705 \text{ } M$$

We can write an equation for calculating the concentration of any acidic or basic solute from the results of a titration by using a general version of Equation 16.23:

$$V_A M_A = \frac{n_A}{n_B} V_B M_B \tag{16.24}$$

in which V_A and M_A represent the volume and molarity of the acid, V_B and M_B represent the volume and molarity of the base, and n_A/n_B is the ratio of the moles of acid to moles of base in the balanced chemical equation describing the reaction. For example, in a titration of a solution of sulfuric acid with NaOH,

$$H_2SO_4(aq) + 2\,NaOH(aq) \rightarrow Na_2SO_4(aq) + 2\,H_2O(\ell)$$

the number of moles of NaOH consumed is twice the number of moles of H_2SO_4 consumed. Therefore,

$$\frac{n_A}{n_B} = \frac{1}{2}$$

and Equation 16.24 becomes

$$(V_{H_2SO_4})(M_{H_2SO_4}) = \tfrac{1}{2}(V_{NaOH})(M_{NaOH})$$

Alkalinity Titrations

Let's return to the application with which we began this section on titrations: the determination of the alkalinity of a sample of natural water. We start by precisely dispensing a known volume of the sample into a titration flask and then adding a few drops of a pH color indicator. Then we add a strongly acidic standard solution from a buret. If carbonate is present in the sample, then the first additions of strong acid will convert carbonate into bicarbonate:

$$H^+(aq) + CO_3^{2-}(aq) \rightarrow HCO_3^-(aq)$$
$$\text{(titrant)} \quad \text{(sample)}$$

The conversion of carbonate into bicarbonate is the first stage of the alkalinity titration (Figure 16.18). In the second stage, the bicarbonate formed in the first stage plus any bicarbonate present in the original sample reacts with additional acid titrant:

$$H^+(aq) + HCO_3^-(aq) \rightarrow H_2CO_3(aq)$$
$$\text{(titrant)} \quad \text{(sample)}$$

to form carbonic acid. If more carbonic acid is produced than is soluble [remember carbonic acid is really a solution of $CO_2(aq)$ in water], then the carbonic acid leaves the solution in the form of bubbles of carbon dioxide:

$$H_2CO_3(aq) \rightarrow H_2O(\ell) + CO_2(g)$$

In the second stage of the alkalinity titration, that is, between the first and second equivalence points (Figure 16.18), the titration curve has a region in which added acid has little effect on pH. This is a buffering region where $HCO_3^-(aq)$ and $CO_3^{2-}(aq)$ function effectively as a weak acid–conjugate base pair. Then, as the bicarbonate in the sample is completely consumed, pH again drops sharply, producing a second equivalence point that is detected by the change in color of a second indicator.

Note that the initial pH of the sample in Figure 16.18 is slightly above 10, which is quite alkaline and above the pH range tolerated by many species of aquatic life. Such alkaline water may be found in arid regions, such as the U.S. Southwest, where rocks containing calcite, gypsum, and other basic minerals are in contact with water. The pH of most seawater and fresh water is 8.2 or less, which corresponds to parts of the curve after the first equivalence point in Figure 16.18. Little carbonate is present in these waters, and most of their alkalinity is due to the presence of bicarbonate. The alkalinity titration curves for samples of seawater and fresh water have only one equivalence point at the pH of the second equivalence point in Figure 16.18.

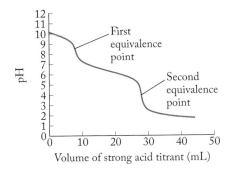

FIGURE 16.18 The results of an alkalinity titration show two equivalence points: The first marks the complete conversion of carbonate into bicarbonate, and the second marks the conversion of bicarbonate into carbonic acid.

Lakes and springs in the desert southwest of the United States may have highly alkaline water because of the presence of basic minerals.

CONCEPT TEST

In the alkalinity titration of a sample that initially contains both CO_3^{2-} and HCO_3^-, the volume of titrant required to reach the first equivalence point is less than that required to go from the first to the second. Why?

How can we determine which color indicators could be used to detect the equivalence points in an alkalinity titration? The phenol-red indicator used to test swimming-pool water would not be a good candidate, because it changes color between pH 6.8 and 8.4. This range is just below the pH of the first equivalence point and well above the second. To detect the first equivalence point, we need a color indicator with a pK_a near the pH of the first equivalence point (8.5). One candidate is phenolphthalein (pK_a = 8.8), which is pink or mauve in its basic form and colorless at low pH (see Figures 16.13 and 16.14). To detect the second equivalence point, we could add bromcresol green (pK_a = 4.6) after the first equivalence point has been reached. We would not add it before the first equivalence point, because its blue–green color in basic solutions would obscure the pink-to-colorless transition of phenolphthalein. On the other hand, phenolphthalein would not obscure the color change obtained by using bromcresol green because phenolphthalein is colorless in acidic solutions.

SAMPLE EXERCISE 16.13   **Interpreting Results of an Acid–Base Titration**

Suppose a 50.00 mL sample of water from an alkaline (pH = 10.0) hot spring is titrated with a 0.02075 M solution of HCl. A few drops of phenolphthalein are added at the beginning of the titration, and the solution turns pink. It takes 11.21 mL of titrant to reach the pink-to-clear equivalence point. Then a few drops of bromcresol green are added, and it takes an additional 32.28 mL of titrant before its blue–green color changes to yellow. What were the initial concentrations of carbonate and bicarbonate in the spring-water sample?

COLLECT AND ORGANIZE We are asked to determine the concentrations of two analytes in one sample from the results of a single titration with a monoprotic strong acid, HCl. These determinations are based on the volumes of titrant needed to reach two equivalence points: an initial one at which any CO_3^{2-} in the sample has been converted to HCO_3^-:

$$HCl(aq) + CO_3^{2-}(aq) \rightarrow HCO_3^-(aq) + Cl^-(aq)$$

and a second one in which HCO_3^- has been converted to H_2CO_3

$$HCl(aq) + HCO_3^-(aq) \rightarrow H_2CO_3(aq) + Cl^-(aq)$$

ANALYZE This HCO_3^- includes that which was there in the original sample plus that produced during the first step in the titration. If there were no HCO_3^- present initially in the sample, another 11.21 mL of titrant would have been needed just to titrate the HCO_3^- produced in the first stage. After subtracting this volume from that required to reach the second equivalence point

$$32.28 \text{ mL} - 11.21 \text{ mL} = 21.07 \text{ mL}$$

we find that 21.07 mL of HCl was needed to titrate the HCO_3^- initially present in the sample.

The stoichiometry of the initial titration reaction indicates that the acid and carbonate react in a 1:1 mole ratio, so, at the first equivalence point,

millimoles of HCl added = millimoles of CO_3^{2-} consumed

In titrations, the moles of reactants are calculated by multiplying the volumes and concentrations of their solutions. Therefore, we can express the 1:1 ratio this way:

$$(V_{HCl})(M_{HCl}) = (V_{CO_3^{2-}})(M_{CO_3^{2-}})$$

After solving for ($M_{CO_3^{2-}}$), we have

$$M_{CO_3^{2-}} = \frac{(V_{HCl})(M_{HCl})}{V_{CO_3^{2-}}}$$

The analogous equation for the second step in the titration is

$$M_{HCO_3^-} = \frac{(V_{HCl})(M_{HCl})}{V_{HCO_3^-}}$$

SOLVE We insert the values and equations given in the problem and developed in the previous step and do the math to get

$$M_{CO_3^{2-}} = \frac{(11.21 \text{ mL})(0.02075 \ M)}{50.00 \text{ mL}}$$

$$= 4.652 \times 10^{-3} \ M$$

and

$$M_{HCO_3^-} = \frac{(V_{HCl})(M_{HCl})}{V_{HCO_3^-}}$$

$$= \frac{(21.07 \text{ mL})(0.02075 \ M)}{50.00 \text{ mL}}$$

$$= 8.744 \times 10^{-3} \ M$$

THINK ABOUT IT The titration results indicate that there was almost twice as much bicarbonate present in the original sample than carbonate. In other words, there was more of the acid than the conjugate base in the second ionization step of carbonic acid:

$$HCO_3^-(aq) \rightleftharpoons H^+(aq) + CO_3^{2-}(aq) \qquad K_{a_2} = 4.7 \times 10^{-11}$$

To check on the validity of these results, let's use the Henderson–Hasselbalch equation to calculate what the pH of the sample should have been given the results of the titration and the value of pK_{a_2} ($-\log(4.7 \times 10^{-11}) = 10.32$):

$$pH = pK_a + \log \frac{[\text{base}]}{[\text{acid}]}$$

$$= 10.32 + \log \frac{4.652 \times 10^{-3} \ M}{8.744 \times 10^{-3} \ M}$$

$$= 10.05$$

This theoretical pH value agrees with the pH (10.0) of the sample given in the problem.

Practice Exercise Your job is to determine the concentration of ammonia in a solution for cleaning windows. In the titration of a 25.00 mL sample of the solution, the equivalence point is reached after 10.49 mL of a 1.155 M solution of HCl has been added.

   a. What is the concentration of ammonia in the solution?
   b. Which of the pH indicators in Figure 16.14 would be suitable for detecting the equivalence point in this titration?

Titrations are used routinely in laboratories and industrial plants to determine the pH of solutions. Many modern chemistry laboratories use automated equipment with sensors that respond to hydrogen ion concentration in carrying out titrations.

Two strong acids—sulfuric acid (H_2SO_4) and nitric acid (HNO_3)—are among the most commonly produced industrial chemicals throughout the world. Sulfuric acid ranks first, with a worldwide production of more than 150 million tons (40 million tons in the United States alone) each year. Nitric acid production is 13th on the industrial chemicals list (U.S. production was 8.0 million tons in 2006). About 70% of the H_2SO_4 and 75% of the HNO_3 produced in the United States is used to make fertilizer. The rest is used in a variety of chemical manufacturing processes, including the preparation of synthetic fibers described in Chapter 12.

Pure sulfuric acid is a dense, colorless, oily liquid. When heated, it fumes as it partly decomposes into H_2O and SO_3. The residual solution is 98.3% H_2SO_4 and 1.7% H_2O. This solution, which is 18 M H_2SO_4, is the liquid sold as "concentrated" sulfuric acid. Concentrated sulfuric acid is very hygroscopic (that is, it absorbs water) and is used as a drying agent and to remove water from many compounds. It can totally dehydrate sugar and thus turn this carbohydrate into carbon. Sulfuric acid dissolves in water in a process so exothermic that the solution may boil, which is why concentrated sulfuric acid must be diluted by slowly adding it to cold water. Never add water to concentrated sulfuric acid.

The synthesis of sulfuric acid starts with the combustion of sulfur or sulfide minerals to sulfur dioxide followed by the oxidation of SO_2 to SO_3:

$$S_8(s) + 8\,O_2(g) \rightarrow 8\,SO_2(g)$$
$$\underline{+\ 8\,SO_2(g) + 4\,O_2(g) \rightleftharpoons 8\,SO_3(g)}$$
$$\text{Overall:} \quad S_8(s) + 12\,O_2(g) \rightleftharpoons 8\,SO_3(g)$$

Both reactions are equilibrium processes, but the equilibrium constant for the formation of SO_2 is large and the reaction essentially goes to completion. The equilibrium between SO_2, O_2, and SO_3 in the second step is strongly affected by Le Châtelier's principle. The reaction is exothermic but slow at ambient temperature. Higher temperatures speed the rate of the reaction but decrease the equilibrium concentration of the product. The yield is improved by (1) increasing the pressure of the reactants, (2) using an excess of O_2, and (3) harvesting SO_3 during the reaction. Vanadium oxide, V_2O_5, is used as a reaction catalyst, allowing the reaction to proceed at an acceptable rate at moderate temperatures. Sulfuric acid itself is produced by reaction between sulfur trioxide and water:

$$SO_3(g) + H_2O(\ell) \rightarrow H_2SO_4(\ell)$$

Note that the steps in the production of sulfuric acid are the same as those that lead to acid rain in the environment.

The production of nitric acid is linked to the production of ammonia because NH_3 is a reactant in the synthesis

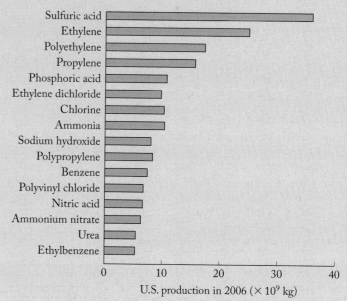

The top 16 industrial chemicals produced in the United States in 2006 included sulfuric acid (1st) and nitric acid (13th). (*Source:* "Facts and Figures of the Chemical Industry," *Chemical and Engineering News*, July 2, 2007, pp. 55–64.)

of HNO_3. The controlled, selective oxidation of ammonia to NO and subsequent conversion into nitric acid is known as the Ostwald process. Developing it earned Wilhelm Ostwald (1853–1932) the Nobel Prize in chemistry in 1909. The three steps in the process are as follows:

(1) $\quad 4\,NH_3(g) + 5\,O_2(g) \rightarrow 4\,NO(g) + 6\,H_2O(g)$

(2) $\quad 2\,NO(g) + O_2(g) \rightarrow 2\,NO_2(g)$

(3) $\quad 3\,NO_2(g) + H_2O(\ell) \rightarrow 2\,HNO_3(\ell) + NO(g)$

Note that the oxidation number of nitrogen increases from -3 (in NH_3) to $+2$ (in NO) to $+4$ (in NO_2) to $+5$ (in HNO_3) during the process. A catalyst composed of platinum or platinum and rhodium and a reaction temperature of 850°C are needed to achieve a rapid-rate conversion of ammonia into nitric acid.

Wilhelm Ostwald was a professor of physical chemistry at Leipzig University in Germany from 1887 until 1906. During that time, he mentored several brilliant students. Most notable are two whose names should be familiar to you: Jacobus Henricus van't Hoff, who won the Nobel Prize in chemistry in 1901, and Svante August Arrhenius, who won the Nobel Prize in chemistry in 1903.

Summary

SECTION 16.1 Acids and bases have distinctive physical properties, such as the slippery feel of bases and the sour taste of acids. Some of the acids present in our environment occur naturally. Others, including H_2SO_4 and HNO_3, are also the result of pollution released by our industrialized society and contribute to the formation of acid rain.

SECTION 16.2 The **Brønsted–Lowry model** of acids and bases defines acids as H^+ ion donors and bases as H^+ ion acceptors. Strong acids include binary (HX where X is a group 17 element other than fluorine) acids and oxoacids formed when nonmetal oxides combine with water. All strong acids are completely ionized in water. The H^+ ions they release combine with water molecules to form hydronium (H_3O^+) ions. Strong bases are those group 1 and 2 hydroxides that are soluble in water and dissociate completely when they dissolve. Most acids, including the organic acids in biological samples, are weak acids that are only partially ionized in water. When weak acid "HA" undergoes ionization, it forms its **conjugate base**, A^-. When base "B" acquires a H^+ ion, it forms its **conjugate acid**, HB^+. The strongest acid in water is the H_3O^+ ion; the strongest base is the OH^- ion.

SECTION 16.3 Water is an **amphoteric** substance in that it is capable of behaving both as an acid and as a base. This behavior is evident in the autoionization of water in which strong hydrogen bonding between water molecules results in formation of a hydronium ion and a hydroxide ion by one pair of water molecules out of every 10 million at 25°C:

$$2\,H_2O(\ell) \rightleftharpoons H_3O^+(aq) + OH^-(aq)$$

In a neutral solution, $[H^+] = [OH^-] = 1.00 \times 10^{-7}$. **pH** is a logarithmic scale for expressing the acidic or basic strength of solutions. Acidic solutions have pH values less than 7.00; basic solutions have pH values greater than 7.00. An increase in one pH unit represents a decrease in $[H^+]$ to 1/10th of its initial value.

SECTION 16.4 Calculations of the pH of weak acids and bases employ ICE tables listing the initial concentrations of reactants and products, the changes in concentration that they undergo, and their concentrations at equilibrium. The **degree of ionization** of a weak acid (HA) is the ratio of $[A^-]$ to the initial (total) acid concentration and is usually expressed as a percentage.

SECTION 16.5 Weak **polyprotic acids** can undergo more than one acid-ionization reaction, but the first one is always the one that controls pH.

SECTION 16.6 The strength of an oxoacid is related to the stability of the anion that is formed as a result of losing a H^+ ion. This stability increases as the electronegativity of the central atom increases and as the number of oxygen atoms double bonded to the central atom increases.

SECTION 16.7 A salt is an acidic salt if its formula contains a cation that is the conjugate acid of a weak base and an anion that is the conjugate base of a strong acid. A salt is a basic salt if its formula includes an anion that is the conjugate base of a weak acid, and a cation that is the ion of a group 1 or 2 element.

SECTION 16.8 Adding a salt of a weak acid (HA) to a solution of the acid provides a second source of the conjugate base (A^-). As predicted by Le Châtelier's principle, part of the added A^- is consumed as the acid-ionization reaction

$$HA + H_2O \rightleftharpoons H_3O^+ + A^-$$

runs in reverse to consume H_3O^+ and raise pH. Similarly, adding a salt of a weak base to a solution of the base provides a second source of its conjugate acid, which lowers pH. These shifts are examples of the **common-ion effect**.

SECTION 16.9 A **pH buffer** is a solution that contains either a weak acid and a salt of its conjugate base or a weak base and a salt of its conjugate acid. Buffer solutions resist pH change when strong acids or bases are added. Additions of acid are neutralized by the basic component of a buffer; additions of base are neutralized by the acidic component. The result is a change in the concentration ratio in the Henderson–Hasselbalch equation, but the changes in the logarithm of the ratio, and the impact on pH, tend to be small as long as neither component is completely consumed.

SECTION 16.10 A **pH indicator** is a weak acid or base that has a color that is different from that of its conjugate base or acid. Indicators can be used to determine the pH (±0.2 pH units) of a solution that is within 1.0 pH unit of the pK_a or pK_b of the indicator. These indicators are also used to detect the equivalence points in pH titrations, which are highly precise methods for determining the concentration of a weak or strong acid or base, called the analyte, in an aqueous sample. A known volume of the sample is titrated with a known concentration of a solution, called the titrant, of strong base or acid. Titrant is added until all the analyte has been consumed—a point called the equivalence point in the titration.

Problem-Solving Summary

TYPE OF PROBLEM	CONCEPTS AND EQUATIONS	SAMPLE EXERCISES
Identifying conjugate acid–base pairs	Determine whether the formula of the base is the formula of the acid less one H^+ ion.	16.1
Calculating pH and $[H^+]$ and relating $[H^+]$, $[OH^-]$, pH, and pOH	Use the following: $$pH = -\log[H^+]$$ $$pOH = -\log[OH^-] \text{ and}$$ $$pH + pOH = 14.00$$	16.2 and 16.3
Calculating the value of K_a of a weak acid (HA) from the values of $[H^+]$ and $[HA]$ in solution; calculating percent ionization of a weak acid (HA)	Use the following: $$K_a = \frac{[H^+][A^-]}{[HA]}$$ $$\text{Percent ionization} = \frac{[A^-]_{equilibrium}}{[HA]_{initial}} \times 100\%$$	16.4
Calculating the pH of a solution of a weak acid (HA) of total concentration [HA]	Set up an ICE table based on the equilibrium $$HA \rightleftharpoons H^+ + A^-$$ Let $x = [H^+] = [A^-]$ at equilibrium. $$K_a = \frac{x^2}{[HA] - x}$$ Then calculate $pH = -\log[H^+]$.	16.5
Calculating the pH of a solution of a weak base (B) of total	Set up an ICE table based on the equilibrium $$B + H_2O \rightleftharpoons HB^+ + OH^-$$ Let $x = [OH^-] = [HB^+]$ at equilibrium. Calculate x by using $$K_b = \frac{x^2}{[B] - x}$$ Then calculate pOH and pH by using $$pOH = -\log[OH^-] \text{ and}$$ $$pH = 14.00 - pOH.$$	16.6
Calculating the pH of a solution of a weak polyprotic acid	Set up an ICE table based on the K_{a_1} equilibrium $$H_2A + H_2O \rightleftharpoons H_3O^+ + HA^-$$ Let $x = [H_3O^+] = [HA^-]$ at equilbrium. Calculate x by using $$K_{a_1} = \frac{x^2}{[H_2A] - x}$$	16.7
Distinguishing acidic, basic, and neutral salts	The cations in acidic salts are the conjugate acids of weak bases. The anions in basic salts are the conjugate bases of weak acids.	16.8
Calculating the pH of a solution of an acidic salt (AHX)	Assume the salt completely dissociates into AH^+ and X^-. Set up an ICE table for the equilibrium $$AH^+ + H_2O \rightleftharpoons A + H_3O^+$$ Let $x = [H_3O^+] = [A]$ at equilibrium. Calculate x by using $$K_a = \frac{K_w}{K_b} = \frac{x^2}{[AH^+] - x}$$ Then calculate pH.	16.9
Calculating the pH of a basic salt (MA)	Assume that salt completely dissociates. Set up an ICE table for the equilibrium $$A^- + H_2O \rightleftharpoons HA + OH^-$$ Let $x = [OH^-]$; calculate its value by using $$K_b = \frac{K_w}{K_a} = \frac{x^2}{[A^-] - x}$$	16.9

TYPE OF PROBLEM	CONCEPTS AND EQUATIONS	SAMPLE EXERCISES
Calculating the pH of a solution of a base and its conjugate acid (or acid and its conjugate base); calculate the response of a buffer; describe behavior of a buffer	Use the relation $$pH = pK_a + \log \frac{[\text{base}]}{[\text{acid}]}$$	16.10, 16.11, 16.12
Interpreting the results of a pH titration	Use the relation $$V_A M_A = \frac{n_A}{n_B} V_B M_B$$ where V_A and M_A are the volume and molarity of the acid, V_B and M_B are the volume and molarity of the base, and n_A and n_B are the coefficients of the acid and base, respectively, in the balanced titration equation.	16.13

Visual Problems

16.1. Which of the lines in Figure P16.1 best represents the dependence of the degree of ionization of acetic acid on its concentration in aqueous solution?

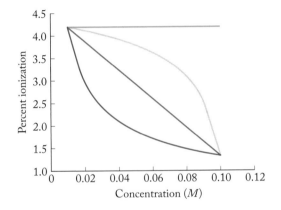

FIGURE P16.1

16.2. The bar graph in Figure P16.2 shows the degree of ionization of 1×10^{-3} M solutions of three hypohalous acids: HOCl, HOBr, and HOI. Which bar is the one for HOI?

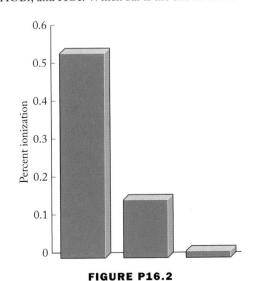

FIGURE P16.2

16.3. The graph in Figure P16.3 shows the titration curves of a 1 M solution of a weak acid with a strong base and a 1 M solution of a strong acid with the same base. Which curve is which?

16.4. Estimate to within one pH unit the pH of a 0.5 M solution of the sodium salt of the weak acid in Problem 16.3.

16.5. Suppose you have four color indicators to choose from for detecting the end point of the titration reaction represented by the red curve (upper curve on the left side of the plot) in Figure P16.3. The pK_a values of the four indicators are 3.3, 5.0, 7.0, and 9.0. Which indicator would be the best one to choose?

16.6. What is the pK_a value of the weak acid in Figure P16.3?

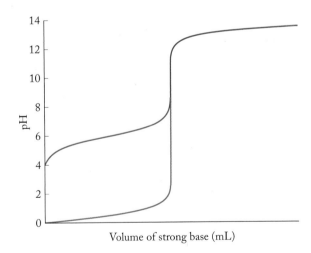

FIGURE P16.3

16.7. One of the titration curves in Figure P16.7 represents the titration of an aqueous sample of Na_2CO_3 with strong acid; the other represents the titration of an aqueous sample of $NaHCO_3$ with the same acid. Which curve is which?

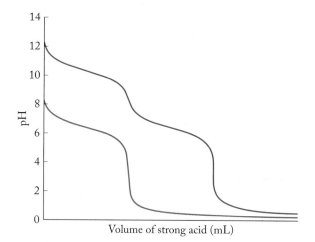

FIGURE P16.7

16.8. Consider the three beakers in Figure P16.8. Each contains a few drops of the color indicator bromthymol blue, which is yellow in acidic solutions and blue in basic solutions. One beaker contains a solution of ammonium chloride, one contains ammonium acetate, and the third contains sodium acetate. Which beaker contains which salt?

FIGURE P16.8

Questions and Problems

Acids and Bases and the Brønsted–Lowry Model

CONCEPT REVIEW

16.9. In an aqueous solution of HBr, which compound acts as a Brønsted–Lowry acid and which is the Brønsted–Lowry base?

16.10. In an aqueous solution of HNO_3, which compound acts as a Brønsted–Lowry acid and which is the Brønsted–Lowry base?

16.11. In an aqueous solution of NaOH, which species acts as a Brønsted–Lowry acid and which is the Brønsted–Lowry base?

16.12. Both NaOH and $Ca(OH)_2$ are strong bases. Does this mean that solutions of the two compounds with the same molarity have the same capacity to neutralize strong acids? Why or why not?

16.13. Identify the acids and bases in the following reactions:
a. $HNO_3(aq) + NaOH(aq) \rightarrow NaNO_3(aq) + H_2O(\ell)$
b. $CaCO_3(s) + 2\,HCl(aq) \rightarrow$
$$CaCl_2(aq) + CO_2(g) + H_2O(\ell)$$
c. $NH_3(aq) + HCN(aq) \rightarrow NH_4CN(aq)$

16.14. Identify the acids and bases in the following reactions:
a. $NH_2^-(aq) + H_2O(\ell) \rightleftharpoons NH_3(aq) + OH^-(aq)$
b. $HClO_4(aq) + H_2O(\ell) \rightleftharpoons ClO_4^-(aq) + H_3O^+(aq)$
c. $HSO_4^-(aq) + CO_3^{2-}(aq) \rightleftharpoons SO_4^{2-}(aq) + HCO_3^-(aq)$

16.15. Identify the conjugate base of each of the following compounds: HNO_2, HOCl, H_3PO_4, and NH_3.

16.16. Identify the conjugate acid of each of the following species: NH_3, ClO_2^-, SO_4^{2-}, and OH^-.

PROBLEMS

16.17. What is the concentration of H^+ ions in a 1.50 M solution of HNO_3?

16.18. What is the concentration of H^+ ions in a solution of hydrochloric acid that was prepared by diluting 20.0 mL of concentrated (11.6 M) HCl to a final volume of 500 mL?

16.19. What is the value of $[OH^-]$ in a 0.0800 M solution of $Sr(OH)_2$?

16.20. A particular drain cleaner contains NaOH. What is the value of $[OH^-]$ in a solution produced when 5.0 g of NaOH dissolves in enough water to make 250 mL of solution?

16.21. Describe how you would prepare 2.50 L of a NaOH solution in which $[OH^-] = 0.70$ M, starting with solid NaOH.

16.22. How many milliliters of a 1.00 M solution of NaOH do you need to prepare 250 mL of a solution in which $[OH^-] = 0.0200$ M?

pH and the Autoionization of Water

CONCEPT REVIEW

16.23. Explain why pH values decrease as acidity increases.

16.24. Solution A is 100 times more acidic than solution B. What is the difference in the pH values of solution A and solution B?

16.25. Under what conditions is the pH of a solution negative?

*16.26. In principle, ethanol (CH_3CH_2OH) can undergo autoionization. Propose an explanation for why the value of the equilibrium constant for the autoionization of ethanol is much less than that of water.

PROBLEMS

16.27. Calculate the pH and pOH of the solutions with the following hydrogen ion or hydroxide ion concentrations. Indicate which solutions are acidic, basic, or neutral.
a. $[H^+] = 3.45 \times 10^{-8}\ M$
b. $[H^+] = 2.0 \times 10^{-5}\ M$
c. $[H^+] = 7.0 \times 10^{-8}\ M$
d. $[OH^-] = 8.56 \times 10^{-4}\ M$

16.28. Calculate the pH and pOH of the solutions with the following hydrogen ion or hydroxide ion concentrations. Indicate which solutions are acidic, basic, or neutral.
a. $[OH^-] = 7.69 \times 10^{-3}\ M$
b. $[OH^-] = 2.18 \times 10^{-9}\ M$
c. $[H^+] = 4.0 \times 10^{-8}\ M$
d. $[H^+] = 3.56 \times 10^{-4}\ M$

16.29. Calculate the pH of stomach acid in which $[HCl] = 0.155\ M$.

16.30. Calculate the pH of a $0.00500\ M$ solution of HNO_3.

16.31. Calculate the pH and pOH of a $0.0450\ M$ solution of NaOH.

16.32. Calculate the pH and pOH of a $0.160\ M$ solution of KOH.

16.33. Calculate the pH of a $1.33\ M$ solution of HNO_3.

*16.34. Calculate the pH of a $6.9 \times 10^{-8}\ M$ solution of HBr.

Calculations with pH and *K*

CONCEPT REVIEW

16.35. One-molar solutions of the following acids are prepared. Rank them in order of decreasing $[H^+]$.

Acid	K_a
CH_3COOH	1.8×10^{-5}
HNO_2	4.5×10^{-4}
$HOCl$	3.5×10^{-8}
$HOCN$	3.5×10^{-4}

16.36. On the basis of the following degree-of-ionization data for $0.100\ M$ solutions, select which acid has the smallest K_a.

Acid	Degree of Ionization (%)
C_6H_5COOH	2.5
HF	8.5
HN_3	1.4
CH_3COOH	1.3

16.37. A $1.0\ M$ aqueous solution of $NaNO_2$ is a much better conductor of electricity than is a $1.0\ M$ solution of HNO_2. Explain why.

16.38. Hydrogen chloride and water are molecular compounds, yet a solution of HCl dissolved in H_2O is an excellent conductor of electricity. Explain why.

16.39. Hydrofluoric acid is a weak acid. Write the mass action expression for its acid-ionization reaction.

16.40. In the formula of formic acid, HCOOH, one H atom is ionizable. Write the mass action expression for the acid-ionization equilibrium of formic acid.

*16.41. The K_a values of weak acids depend on the solvent in which they dissolve. For example, the K_a of alanine in aqueous ethanol is less than its K_a in water.
a. In which solvent does alanine ionize to the largest degree?
b. Which is the stronger Brønsted–Lowry base: water or ethanol?

*16.42. The K_a of proline is 2.5×10^{-11} in water, 2.8×10^{-11} in an aqueous solution that is 28% ethanol, and 1.66×10^{-8} in aqueous formaldehyde at 25°C.
a. In which solvent is proline the strongest acid?
b. Rank these compounds on the basis of their strengths as Brønsted–Lowry bases: water, ethanol, and formaldehyde.

16.43. When methylamine, CH_3NH_2, dissolves in water, the resulting solution is slightly basic. Which compound is the Brønsted–Lowry acid and which is the base?

*16.44. When 1,2-diaminoethane, $H_2NCH_2CH_2NH_2$, dissolves in water, the resulting solution is basic. Write the formula of the ionic compound that is formed when hydrochloric acid is added to a solution of 1,2-diaminoethane.

PROBLEMS

16.45. Sore Muscles The muscle fatigue felt during strenuous exercise is caused by the buildup of lactic acid in muscle tissues. In a $1.00\ M$ aqueous solution, 2.94% of lactic acid is ionized. What is the value of its K_a?

16.46. **Rancid Butter** The odor of spoiled butter is due in part to butanoic acid, which results from the chemical breakdown of butter fat. A $0.100\ M$ solution of butanoic acid is 1.23% ionized. Calculate the value of K_a for butanoic acid.

16.47. At equilibrium, the value of $[H^+]$ in a $0.250\ M$ solution of an unknown acid is $4.07 \times 10^{-3}\ M$. Determine the degree of ionization and the K_a of this acid.

16.48. Nitric acid (HNO_3) is a strong acid that is completely ionized in aqueous solutions of concentrations ranging from 1% to 10% (1.5 M). However, in more concentrated solutions, part of the nitric acid is present as un-ionized molecules of HNO_3. For example, in a 50% solution (7.5 M) at 25°C, only 33% of the molecules of HNO_3 dissociate into H^+ and NO_3^-. What is the value of K_a of HNO_3?

16.49. Ant Bites The venom of biting ants contains formic acid, HCOOH, $K_a = 1.8 \times 10^{-4}$ at 25°C. Calculate the pH of a $0.060\ M$ solution of formic acid.

16.50. **Gout** Uric acid can collect in joints, giving rise to a medical condition known as gout. If the pK_a of uric acid is 3.89, what is the pH of a $0.0150\ M$ solution of uric acid?

16.51. Acid Rain A weather system moving through the American Midwest produced rain with an average pH of 5.02. By the time the system reached New England, the rain it produced had an average pH of 4.66. How much more acidic was the rain falling in New England?

16.52. **Acid Rain** A newspaper reported that the "level of acidity" in a sample taken from an extensively studied watershed in New Hampshire in February 1998 was "an astounding 200 times lower than the worst measurement" taken in the preceding 23 years. What is this difference expressed in units of pH?

16.53. The K_b of dimethylamine $[(CH_3)_2NH]$ is 5.9×10^{-4} at 25°C. Calculate the pH of a $1.20 \times 10^{-3} M$ solution of dimethylamine.

16.54. **Pain Killer** Morphine is an effective pain killer but is also highly addictive. Calculate the pH of a 0.115 M solution of morphine if its $pK_b = 5.79$.

16.55. Pain Killer II Codeine is a popular prescription pain killer because it is much less addictive than morphine. Codeine contains a basic nitrogen atom that can be protonated to give the conjugate acid of codeine. Calculate the pH of a $3.42 \times 10^{-4} M$ solution of codeine if the pK_a of the conjugate acid is 8.21.

16.56. Pyridine (C_5H_5N) is a particularly foul-smelling substance used in manufacturing pesticides and plastic resins. Calculate the pH of a 0.125 M solution of pyridine $(K_b = 1.4 \times 10^{-9})$.

Polyprotic Acids

CONCEPT REVIEW

16.57. Why is the K_{a_2} value of phosphoric acid less than its K_{a_1} value but greater than its K_{a_3} value?

16.58. In calculating the pH of a 1.0 M solution of sulfurous acid, we can ignore the H^+ ions produced by the ionization of the bisulfite ion; however, in calculating the pH of a 1.0 M solution of sulfuric acid, we cannot ignore the H^+ ions produced by the ionization of the bisulfate ion. Why?

PROBLEMS

16.59. What is the pH of a 0.300 M solution of H_2SO_4 $(K_{a_2} = 1.2 \times 10^{-2})$?

16.60. What is the pH of a 0.150 M solution of sulfurous acid? Given: $K_{a_1} = 1.7 \times 10^{-2}$, $K_{a_2} = 6.2 \times 10^{-8}$?

16.61. Ascorbic acid (vitamin C) is a diprotic acid $(K_{a_1} = 8.0 \times 10^{-5}$ and $K_{a_2} = 1.6 \times 10^{-12})$. What is the pH of a 0.250 M solution of ascorbic acid?

16.62. The leaves of the rhubarb plant contain high concentrations of diprotic oxalic acid (HOOCCOOH) and must be removed before the stems are used to make rhubarb pie. If $pK_{a_1} = 1.23$ and $pK_{a_2} = 4.19$, what is the pH of a 0.0288 M solution of oxalic acid?

16.63. Addiction to Tobacco Nicotine is responsible for the addictive properties of tobacco. If $K_{b_1} = 1.05 \times 10^{-6}$ and $K_{b_2} = 1.32 \times 10^{-11}$, determine the pH of a $1.00 \times 10^{-3} M$ solution of nicotine.

16.64. 1,2-Diaminoethane, $H_2NCH_2CH_2NH_2$, is used extensively in the synthesis of compounds containing transition metals in water. If $pK_{b_1} = 3.29$ and $pK_{b_2} = 6.44$, what is the pH of a $2.50 \times 10^{-4} M$ solution of 1,2-diaminoethane?

16.65. **Malaria Treatment** Quinine occurs naturally in the bark of the cinchona tree. For centuries it was the only treatment for malaria. Quinine contains two weakly basic nitrogen atoms, with $K_{b_1} = 3.31 \times 10^{-6}$ and $K_{b_2} = 1.35 \times 10^{-9}$ at 25°C. Calculate the pH of a 0.01050 M solution of quinine in water.

16.66. Dozens of pharmaceuticals ranging from Cyclizine for motion sickness to Viagra for impotence are derived from the organic compound piperazine, whose structure is shown in Figure P16.66:

FIGURE P16.66

a. Solutions of piperazine are basic $(K_{b_1} = 5.38 \times 10^{-5}$; $K_{b_2} = 2.15 \times 10^{-9})$. What is the pH of a 0.0133 M solution of piperazine?
*b. Draw the structure of the ionic form of piperazine that would be present in stomach acid (about 0.15 M HCl).

Acid Strength and Molecular Structure

CONCEPT REVIEW

16.67. Explain why the K_{a_1} of H_2SO_4 is much greater than the K_{a_1} of H_2SeO_4.

16.68. Explain why the K_{a_1} of H_2SO_4 is much greater than the K_{a_1} of H_2SO_3.

16.69. Predict which acid in the following pairs of acids is the stronger acid: (a) H_2SO_3 or H_2SeO_3; (b) H_2SeO_4 or H_2SeO_3.

16.70. Predict which acid in the following pairs of acids is the stronger acid: (a) HOBr or HOBrO; (b) HOCl or HOBr.

pH of Salt Solutions

CONCEPT REVIEW

*16.71. The pK_a values of the conjugate acids of pyridine derivatives shown in Figure P16.71 increase as more methyl groups are added. Do more methyl groups increase or decrease the strength of the parent pyridine bases?

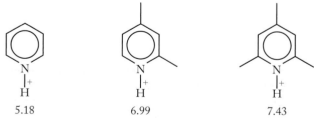

FIGURE P16.71

16.72. Why is it unnecessary to publish tables of K_b values of the conjugate bases of weak acids whose K_a values are known?

16.73. Which of the following salts produces an acidic solution in water: ammonium acetate, ammonium nitrate, or sodium formate?

16.74. Which of the following salts produces a basic solution in water: NaF, KCl, NH_4Cl?

16.75. Neutralizing the Smell of Fish Trimethylamine, $(CH_3)_3N$, $K_b = 6.5 \times 10^{-5}$ at 25°C, is a contributor to the "fishy" odor of not-so-fresh seafood. Some people squeeze fresh lemon juice (which contains a high concentration of citric acid) on cooked fish to reduce the fishy odor. Why is this practice effective?

16.76. Nutritional Value of Beets Beets contain high concentrations of the calcium salt of a dicarboxylic acid with the common name malonic acid and the formula $HOOCCH_2COOH$. Its acid-ionization constants are $K_{a_1} = 1.5 \times 10^{-3}$ and $K_{a_2} = 2.0 \times 10^{-6}$ at 25°C. Could the presence of the calcium salt of malonic acid affect the pH balance of beets? If so, in which direction? Explain why.

PROBLEMS

16.77. If the K_a of the conjugate acid of the artificial sweetener saccharin is 2.1×10^{-11}, what is the pK_b for saccharin?

16.78. If the K_{a_1} value of chromic acid (H_2CrO_4) is 0.16 and its K_{a_2} value is 3.2×10^{-7}, what are the values of K_{b_1} and K_{b_2} of the CrO_4^{2-} anion?

16.79. Dental Health Sodium fluoride is added to many municipal water supplies to reduce tooth decay. Calculate the pH of a 0.00339 M solution of NaF, given that the K_a of HF = 6.8 $\times 10^{-4}$ at 25°C.

16.80. Calculate the pH of a 1.25×10^{-2} M solution of the decongestant ephedrine hydrochloride if the pK_b of ephedrine (its conjugate base) is 3.86.

The Common-Ion Effect and pH Buffers

CONCEPT REVIEW

16.81. Why is a solution of sodium acetate and acetic acid a much better pH buffer than is a solution of sodium chloride and hydrochloric acid?

16.82. Why does a solution of a weak base and its conjugate acid act as a better buffer than does a solution of the weak base alone?

PROBLEMS

16.83. Calculate the pH of a buffer that is 0.244 M acetic acid and 0.122 M sodium acetate. The K_a of acetic acid is 1.76×10^{-5} at 25°C. What is the pH of this mixture at 0°C $(K_a = 1.64 \times 10^{-5})$?

16.84. Calculate the pH of a buffer that is 0.100 M pyridine $(pK_b = 5.25)$ and 0.275 M pyridinium chloride.

16.85. Calculate the pH and pOH of 500.0 mL of a phosphate buffer that is 0.225 M HPO_4^{2-} and 0.225 M PO_4^{3-}; K_a for $HPO_4^{2-} = 4.2 \times 10^{-13}$ at 25°C.

16.86. Determine the pH and pOH of 0.250 L of a buffer that is 0.0200 M boric acid and 0.0250 M sodium borate; pK_{a_1} for $H_3BO_3 = 9.00$ at 25°C.

16.87. What is the ratio of acetate ion to acetic acid $(K_a = 1.76 \times 10^{-5})$ in a buffer containing these compounds at pH = 3.56?

16.88. What is the ratio of lactic acid $(K_a = 1.37 \times 10^{-4})$ to lactate in a solution with pH = 4.00?

16.89. What is the pH of a solution that results from mixing together equal volumes of a 0.05 M solution of ammonia and a 0.025 M solution of hydrochloric acid?

16.90. What is the pH of a solution that results from mixing together equal volumes of a 0.05 M solution of acetic acid and a 0.025 M solution of sodium hydroxide?

---

*16.91. How much 10 M HNO_3 must be added to 1.00 L of a buffer that is 0.010 M acetic acid and 0.10 M sodium acetate to reduce the pH to 5.00?

*16.92. How much 6.0 M NaOH must be added to 0.500 L of a buffer that is 0.0200 M acetic acid and 0.0250 M sodium acetate to raise the pH to 5.75?

---

*16.93. Calculate the pH of 1.00 L of a buffer that is 0.120 M HNO_2 and 0.150 M $NaNO_2$ before and after the addition of 1.00 mL of 12.0 M HCl.

*16.94. Calculate the pH of 100.0 mL of a buffer that is 0.100 M NH_4Cl and 0.100 M NH_3 before and after the addition of 1.0 mL of 6 M HNO_3.

Acid–Base Titrations and Indicators

CONCEPT REVIEW

16.95. What are the differences between the titration curve of a strong acid titrated with a strong base and that of a weak acid titrated with a strong base?

16.96. Do all titrations of a strong base with a strong acid have the same pH at the equivalence point?

16.97. Do all titrations of a weak acid with a strong base have the same pH at the equivalence point?

16.98. What properties must a compound have to serve as an acid–base indicator?

PROBLEMS

16.99. A 25.0 mL sample of a 0.100 M solution of acetic acid is titrated with a 0.125 M solution of NaOH. Calculate the pH of the titration mixture after 10.0, 20.0, and 30.0 mL of base have been added.

16.100. A 25.0 mL sample of a 0.100 M solution of aqueous trimethylamine is titrated with a 0.125 M solution of HCl. Calculate the pH of the solution after 10.0, 20.0, and 30.0 mL of acid have been added; pK_b of $(CH_3)_3N = 4.19$ at 25°C.

16.101. What is the concentration of ammonia in a solution if 22.35 mL of a 0.1145 M solution of HCl are needed to titrate a 100.0 mL sample of the solution?

16.102. In an alkalinity titration of a 100.0 mL sample of water from a hot spring, 2.56 mL of a 0.0355 M solution of HCl is needed to reach the first equivalence point (pH = 8.3) and another 10.42 mL is needed to reach the second equivalence point (pH = 4.0). If the alkalinity of the spring water is due only to the presence of carbonate and bicarbonate, what are the concentrations of each?

16.103. How much 0.0100 M HCl is required to titrate 250 mL of 0.0100 M Na_2CO_3 (to the first endpoint) and 250 mL of 0.0100 M $NaHCO_3$?

16.104. How much 0.0100 M HCl is required to titrate 250 mL of 0.0100 M Na_2CO_3 and 250 mL of 0.0100 M HCO_3^-?

16.105. In the titration of a solution of weak monoprotic acid with a 0.1025 M solution of NaOH, the pH half way to the equivalence point was 4.44. In the titration of a second solution of the same acid, exactly twice as much of a 0.1025 M solution of NaOH was needed to reach the equivalence point. What was the pH half way to the equivalence point in this titration?

16.106. A 125.0 mg sample of an unknown, monoprotic acid was dissolved in 100.0 mL of distilled water and titrated with a 0.050 M solution of NaOH. The pH of the solution was monitored throughout the titration, and the following data were collected. Determine the K_a of the acid.

Volume of OH⁻ Added (mL)	pH
0	3.09
5	3.65
10	4.10
15	4.50
17	4.55
18	4.71
19	4.94
20	5.11
21	5.37
22	5.93
22.2	6.24
22.6	9.91
22.8	10.2
23	10.4
24	10.8
25	11.0
30	11.5
40	11.8

16.107. Sketch a titration curve for the titration of 50.0 mL of 0.250 M HNO_2 with 1.00 M NaOH. What is the pH at the equivalence point?

16.108. Red cabbage juice is a sensitive acid–base indicator; its colors range from red at acidic pH to yellow in alkaline solutions. What color would red cabbage juice have when 25 mL of a 0.10 M solution of acetic acid is titrated with 0.10 M NaOH to its end point?

16.109. Sketch a titration curve for the titration of the malaria drug quinine if 40.0 mL of a 0.100 M solution of quinine is titrated with a 0.100 M solution of HCl, given that K_{b_1} for quinine = 3.31×10^{-6} and $K_{b_2} = 1.35 \times 10^{-9}$.

16.110. Sketch a titration curve for the titration of 100 mL of 1.25×10^{-2} M ascorbic acid with 1.00×10^{-2} M NaOH, given that $K_{a_1} = 8.0 \times 10^{-5}$ and $K_{a_2} = 1.6 \times 10^{-12}$. How many equivalence points should the curve have and what color indicator(s) could be used?

Additional Problems

16.111. Describe the intermolecular forces and changes in bonding that lead to the formation of a basic solution when methylamine (CH_3NH_2) dissolves in water.

16.112. Describe the chemical reactions of sulfur that begin with the burning of high-sulfur fossil fuel and that end with the reaction between acid rain and building exteriors made of marble ($CaCO_3$).

16.113. The value of K_{a_1} of phosphorous acid, H_3PO_3, is nearly the same as the K_{a_1} of phosphoric acid, H_3PO_4.
a. Draw the Lewis structure of phosphorous acid.
b. Identify the ionizable hydrogen atoms in the structure.
c. Explain why the K_{a_1} values of phosphoric and phosphorous acid are similar.

16.114. The value of K_{a_1} of hypophosphorous acid, H_3PO_2, is nearly the same as the K_{a_1} of phosphoric acid, H_3PO_4.
a. Draw a Lewis structure for phosphoric acid that is consistent with this behavior.
b. Identify the ionizable hydrogen atoms in the structure.

*16.115. **pH of Baking Soda** A cook dissolves a teaspoon of baking soda ($NaHCO_3$) in a cup of water; then discovers that the recipe calls for a tablespoon, not a teaspoon. So, the cook adds two more teaspoons of baking soda to make up the difference. Does the additional baking soda change the pH of the solution? Explain why or why not.

*16.116. **Antacid Tablets** contain a variety of bases such as $NaHCO_3$, $MgCO_3$, $CaCO_3$, and $Mg(OH)_2$. Only $NaHCO_3$ has appreciable solubility in water.
a. Write a net ionic equation for the reaction of each antacid with aqueous HCl.
b. Explain how insoluble substances can act as effective antacids.

*16.117. **pH of Natural Waters** In a 1985 study of Little Rock Lake in Wisconsin, 400 gallons of 18 M sulfuric acid were added to the lake over six years. The initial pH of the lake was 6.1 and the final pH was 4.7. If none of the acid was consumed in chemical reactions, estimate the volume of the lake.

16.118. **pH of Natural Waters II** Between 1993 and 1995, sodium phosphate was added to Seathwaite Tarn in the English Lake District to increase its pH. Explain why addition of this compound increased pH.

16.119. **Acid–Base Properties of Pharmaceuticals** Zoloft is a common prescription drug for the treatment of depression. It is sold as a salt of HCl.
a. In the reaction shown in Figure P16.119, which structure is that of the acid salt?
b. When Zoloft dissolves in water, will the resulting solution be acidic or basic?

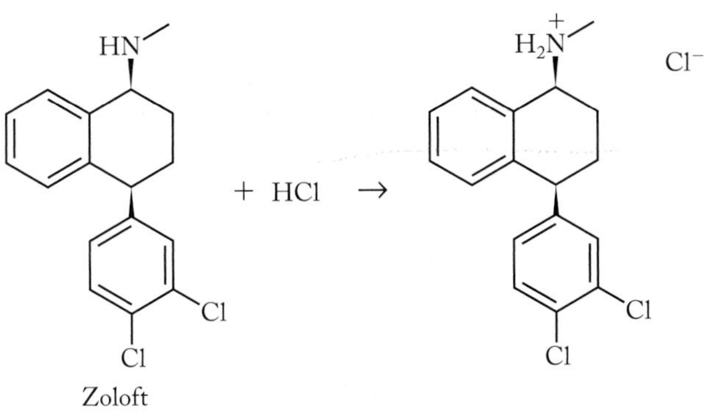

Zoloft

FIGURE P16.119

16.120. Acid–Base Properties of Pharmaceuticals II Prozac is another popular antidepressant drug. Its structure is given in Figure P16.120.
a. Is a solution of Prozac in water likely to be slightly basic or slightly acidic? Explain your answer.
b. Prozac is also sold as a salt of HCl. Which atom, N or O, is most likely to react with HCl?
c. Prozac is sold as a salt of HCl because the solubility of the salt in water is higher than Prozac itself. Why is the salt more soluble?

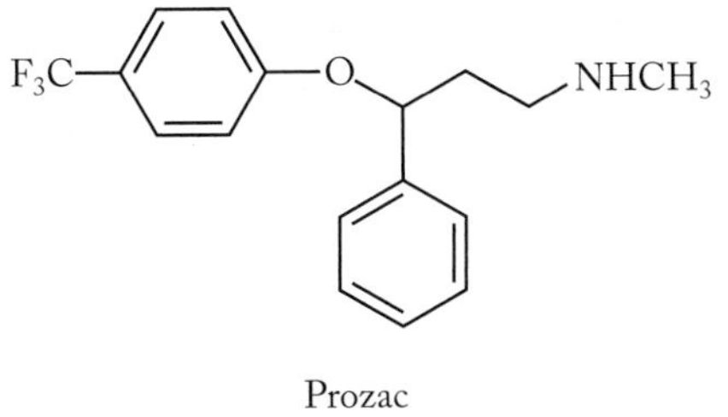

Prozac

FIGURE P16.120

16.121. Hydrogen fluoride (HF) behaves as a weak acid in aqueous solution. Two equilibria influence which fluorine-containing species are present in solution.

$$HF(g) + H_2O\,(\ell) \rightleftharpoons H_3O^+(aq) + F^-(aq) \qquad K_a = 1.1 \times 10^{-3}$$

$$F^-(aq) + HF(g) \rightleftharpoons HF_2^-(aq) \qquad K = 2.6 \times 10^{-1}$$

a. Is fluoride in pH 7.00 drinking water more likely to be present as F^- or HF_2^-?
b. What is the equilibrium constant for this equilibrium?

$$2\,HF(g) + H_2O(\ell) \rightleftharpoons H_3O^+(aq) + HF_2^-(aq)$$

c. What is the pH and equilibrium concentration of HF_2^- in a 0.150 M solution of HF?

*16.122. Pentafluorocyclopentadiene, which has the structure shown in Figure P16.122, is a strong acid.

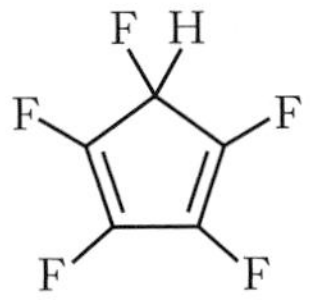

FIGURE P16.122

a. Draw the conjugate base of C_5F_5H.
b. Why is the compound so acidic when most organic acids are weak?

16.123. Naproxen (a.k.a. Aleve) is an anti-inflammatory drug used to reduce pain, fever, inflammation, and stiffness caused by conditions such as osteoarthritis and rheumatoid arthritis. Naproxen is an organic acid; its structure is shown in Figure P16.123. Naproxen has limited solubility in water, so it is sold as its sodium salt.

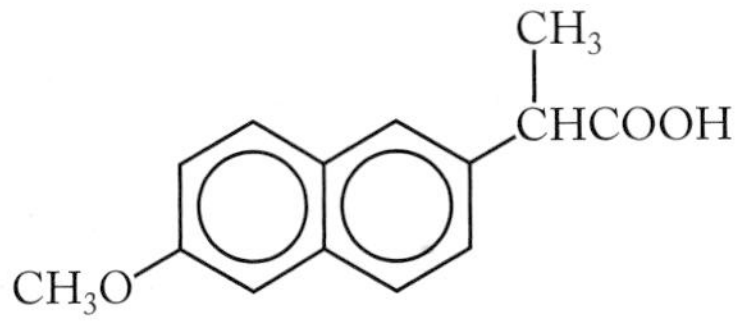

FIGURE P16.123

a. Draw the molecular structure of the sodium salt.
b. Should a solution of the salt be acidic or basic? Explain why.
c. Explain why the salt is more soluble than Naproxen itself.

*16.124. **Greenhouse Gases and Ocean pH** Some climate models predict a decrease in the pH of the oceans of 0.77 pH units because of increases in atmospheric carbon dioxide.
a. Explain, by using the appropriate chemical reactions and equilibria, how an increase in atmospheric CO_2 could produce a decrease in oceanic pH.
b. How much more acidic (in terms of $[H^+]$) would the oceans be if their pH dropped this much?
c. Oceanographers are concerned about the impact of a drop in oceanic pH on the survival of coral reefs. Why?

The Chemistry of Two Strong Acids: Sulfuric and Nitric Acids

16.125. Complete the following chemical equations with the appropriate product(s).
a. $SO_3(g) + H_2O(g) \rightarrow$
b. $3\,NO_2(g) + H_2O(g) \rightarrow$
c. $4\,NH_3(g) + ? \rightarrow ? + 6\,H_2O(g)$

16.126. In the Ostwald process, which steps should have a higher yield at higher total pressure?

16.127. In the Ostwald process, which steps should have higher yields at higher temperature?

16.128. Given ΔH_f° for $SO_2(g)$ (−296.8 kJ/mol) and $SO_3(g)$ (−295.7 kJ/mol), calculate ΔH_{rxn}° for the conversion of SO_2 to SO_3.

16.129. Write balanced chemical equations that correspond to $\Delta H_{f,SO_2}^\circ$ and $\Delta H_{f,SO_3}^\circ$. Show how Hess's law can be used to determine ΔH° for the reaction

$$2\,SO_2(g) + O_2(g) \rightleftharpoons 2\,SO_3(g)$$

*16.130. Reaction of sodium nitrate with sodium oxide at high temperature produces Na_3NO_4, which contains the NO_4^{3-} (orthonitrate) anion. However, the corresponding acid, H_3NO_4, is unknown. Would you expect H_3NO_4 to be a stronger or weaker acid than HNO_3?

*16.131. The Henry's law constant for CO_2 dissolved in water is 3.5×10^{-2} M/atm. Do you expect the corresponding constants for SO_3 and NO_2 to be greater than or less than this value? Explain your answer.

16.132. Adding nitric acid to water is quite exothermic. Calculate the final temperature of 100 mL of water ($d = 1.00$ g/mL) at 20°C [$c_p = -75.3$ J/(mol·°C)] after 10.0 mL of concentrated HNO_3 (16 M, $\Delta H^\circ_{soln} = -33.3$ kJ/mol) is added to it.

16.133. How much ice at 0°C ($c_p = 37.1$ J/(mol·°C), $\Delta H^\circ_{fus} = 6.01$ kJ/mol) is needed to cool the solution in the previous exercise to 5°C, given a molar heat capacity of 80.8 J/(mol·°C) for the solution?

16.134. Write a balanced chemical equation to describe the following reactions of sulfuric acid and nitric acid:
a. Nitric acid reacts with ammonia.
b. Sulfuric acid reacts with ammonia.
c. Sulfuric acid dissolves in water.

*16.135. Thiosulfuric acid, $H_2S_2O_3$, can be prepared by the reaction of H_2S with HSO_3Cl:

$$HSO_3Cl(\ell) + H_2S(g) \rightarrow HCl(g) + H_2S_2O_3(\ell)$$

a. Draw a Lewis symbol for $H_2S_2O_3$, given that it is isostructural with H_2SO_4.
b. Do you expect $H_2S_2O_3$ to be a stronger or weaker acid than H_2SO_4? Explain your answer.

*16.136. Sulfuric acid reacts with nitric acid as shown below:

$$HNO_3(aq) + 2H_2SO_4(aq) \rightarrow$$
$$NO_2(g) + H_3O^+(aq) + 2HSO_4^-(aq)$$

a. Is the reaction a redox process?
b. Identify the acid, base, conjugate acid, and conjugate base in the reaction. (*Hint*: Draw the Lewis structures for each.)

The Colorful Chemistry of Transition Metals

17

AMETHYST *The distinctive violet colors of amethyst are due to trace concentrations of iron in crystalline silicon dioxide.*

A LOOK AHEAD: The Company They Keep

Many of the metallic elements in the periodic table are essential to good health. For example, copper, zinc, and cobalt play key roles in protein function; iron is needed to transport oxygen from our lungs to all the cells of our body, and nearly everyone knows the importance of calcium in building strong teeth and bones.

These and other essential metals should be present in our diets, or at least in the supplements many of us rely on for balanced nutrition. They must also be present in chemical (or biochemical) forms we can digest. Swallowing an 18 mg steel pellet as if it were an aspirin tablet would not be a good way for a young woman to get her recommended daily allowance (RDA) of iron. Chewing on a gram of calcium metal would be even less pleasant. If we are to benefit from consuming essential metals in food and nutritional supplements, the metals need to be in compounds and not free elements, and these compounds must be absorbable by the body during digestion.

All the essential metals occur in nature in ionic compounds, but not all ionic forms are absorbed equally well. For example, most of the iron in fish, poultry, and red meat is readily available because it is present in a form called *heme* iron. However, the iron in plant sources is mostly nonheme and not as readily absorbed. On the other hand, eating a meal that includes both meat and vegetables improves the absorption of the nonheme iron in the vegetables, as does consuming foods high in vitamin C. All these dietary factors work together at the molecular level to provide living systems with the nutrients they need to survive. Metals are essential, but the ions and neutral molecules that accompany them are equally important in making metallic ions chemically reactive and biologically available. Interactions between metal ions and these other ions and molecules influence the metal's solubility, which is a key factor that determines its availability to and activity in plants and animals. These associations also influence other physical properties of the metals, including the wavelengths of visible light that the metal species absorb and therefore the colors of their solutions. In this chapter we explore the consequences of the interactions of metals with substances that change the way the metals behave, how we perceive them, and how living systems use them.

17.1 Lewis Acids and Bases

In this chapter we focus on the interactions between metal ions in solutions and the ions and molecules that surround them in these solutions. If we are to understand the nature of these interactions, we need to reconsider our definitions of acids and bases developed in Chapter 16. Let's begin by revisiting what happens when ammonia gas dissolves in water. The reaction between NH_3 and H_2O is described by this chemical equation:

$$NH_3(g) + H_2O(\ell) \rightleftharpoons NH_4^+(aq) + OH^-(aq)$$

The top of Figure 17.1 provides a Brønsted–Lowry interpretation of this reaction: in donating H^+ ions to ammonia, H_2O acts as a Brønsted–Lowry acid, and, in accepting them, NH_3 acts as a Brønsted–Lowry base.

Another way to view this reaction is illustrated in the bottom portion of Figure 17.1. Rather than focusing on the transfer of hydrogen ions, look at the role of chemical species as *donors or acceptors of electron pairs*. The N atom in a molecule of ammonia donates its lone pair of electrons to the H atom in a water molecule.

In the process the bond between the H atom and the O atom in H_2O is broken in such a way that the bonding pair of electrons remains with the O atom. The donated pair from the N atom forms a fourth N—H covalent bond to the nitrogen in ammonia. The result of this movement of electron pairs is the same as in the Brønsted–Lowry model: a molecule of NH_3 bonds to a hydrogen atom without its electron, a H^+ ion, forming an NH_4^+ ion. A molecule of H_2O loses a H^+ ion, forming an OH^- ion.

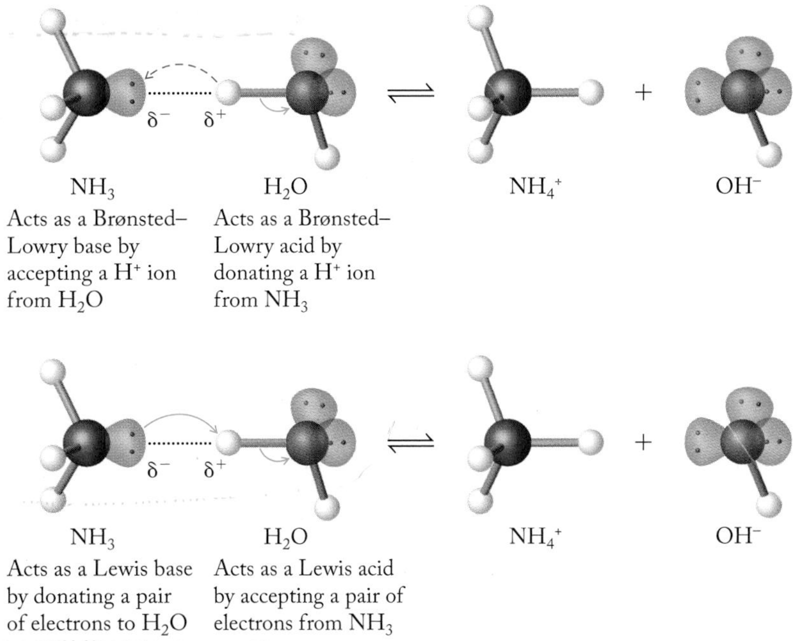

Viewing this process as a donation of an electron pair provides another basis for defining acids and bases:

- A **Lewis base** is a substance that *donates* a pair of electrons in a chemical reaction.
- A **Lewis acid** is a substance that *accepts* a pair of electrons in a chemical reaction.

These definitions are named after their developer, Gilbert N. Lewis, who pioneered research into the nature of chemical bonds (see Section 8.2). The Lewis definition of a base is consistent with the Brønsted–Lowry model (a base is a hydrogen-ion acceptor) because a substance must be able to donate a pair of electrons if it is to form a bond to a hydrogen ion, which has no electrons. However, the Lewis definition of an acid encompasses species that have no hydrogen ions to donate but that can accept electrons. One such compound is boron trifluoride, BF_3.

The Lewis acid BF_3 has only six valence electrons around the boron atom, giving it the capacity to accept another pair to complete its octet. NH_3 is a suitable electron-pair donor, as shown in the reaction and molecular models in Figure 17.2. There is no transfer of H^+ ions in this reaction and so it is not an acid–base reaction according to the Brønsted–Lowry model. However, NH_3 donates a pair of electrons and BF_3 accepts them, so it is an acid–base reaction according to the broader Lewis model.

Many important Lewis bases are anions. Among them are the halide ions, OH^-, and O^{2-}. To see how the O^{2-} ion functions as a Lewis base, let's revisit the reaction we discussed in Section 15.6 in the context of removing SO_2 from the

NH₃
Acts as a Brønsted–Lowry base by accepting a H^+ ion from H_2O

H₂O
Acts as a Brønsted–Lowry acid by donating a H^+ ion from NH_3

NH₄⁺

OH⁻

NH₃
Acts as a Lewis base by donating a pair of electrons to H_2O

H₂O
Acts as a Lewis acid by accepting a pair of electrons from NH_3

NH₄⁺

OH⁻

FIGURE 17.1 The top set of molecular models provides a Brønsted–Lowry view of the reaction between H_2O and NH_3. In the bottom set of models H_2O acts as a Lewis acid and NH_3 acts as a Lewis base.

CONNECTION Lewis's pioneering theories of the nature of covalent bonding are described in Section 8.2.

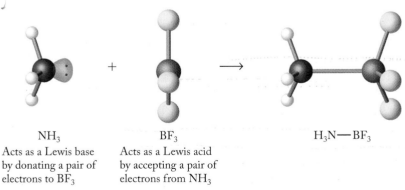

NH₃
Acts as a Lewis base by donating a pair of electrons to BF_3

BF₃
Acts as a Lewis acid by accepting a pair of electrons from NH_3

H₃N—BF₃

FIGURE 17.2 In the reaction between NH_3 and BF_3, NH_3 acts as a Lewis base and BF_3 acts as a Lewis acid.

A **Lewis base** is a substance that *donates* a pair of electrons in a chemical reaction.

A **Lewis acid** is a substance that *accepts* a pair of electrons in a chemical reaction.

stack gases of power stations and smelters. In the removal process a slurry of lime (CaO) powder is mixed with the stack gases, and CaO combines with SO_2 forming calcium sulfite, $CaSO_3$. Writing this reaction using the Lewis structures of the reactants and product gives the following:

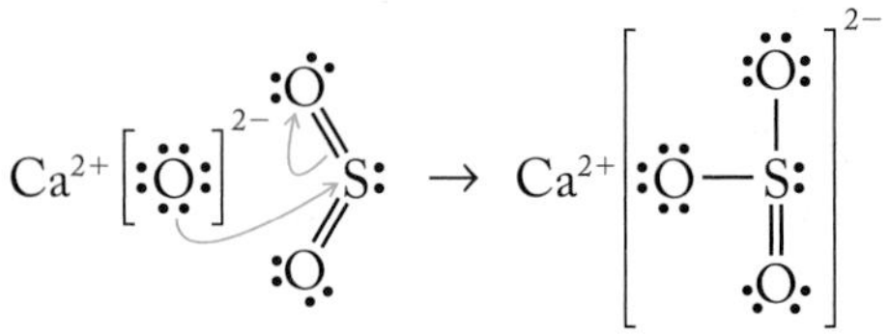

$$ Ca^{2+} \left[:\ddot{O}: \right]^{2-} \ddot{\underset{:\ddot{O}:}{\overset{:\ddot{O}:}{S}}}: \rightarrow Ca^{2+} \left[:\ddot{O}-\underset{:\ddot{O}:}{\overset{:\ddot{O}:}{S}}: \right]^{2-} $$

In this example, the oxide ion in CaO is the electron-pair donor and Lewis base. Sulfur dioxide is the electron-pair acceptor and Lewis acid. Note how one of the S=O double bonds becomes a single bond and a bonding pair becomes a lone pair to accommodate the third O atom. These bonding changes are necessary to produce a structure in which the formal charges on the atoms are as close to zero as possible. In the structure of the SO_3^{2-} ion the formal charges on S and on the double-bonded O atom are both zero; those on the two single-bonded O atoms are 1−. This gives an overall charge of 2−.

Many important reactions in organic chemistry and in aqueous solution chemistry involve the interaction of Lewis acids and bases. In the remainder of this chapter we will explore a class of reactions in which lone pairs of electrons on ammonia, oxide ions, and other Lewis bases are donated to the ions of transition metals.

CONNECTION The concept of formal charge and its calculation are described in Section 8.5.

SAMPLE EXERCISE 17.1 **Identifying Lewis Acids and Bases**

In the following reaction, which species is a Lewis acid and which is a Lewis base?

$$ AlCl_3 + Cl^- \rightarrow AlCl_4^- $$

COLLECT AND ORGANIZE We are to identify which of the two reactants is acting like a Lewis acid, meaning an electron-pair acceptor, and which is acting like a Lewis base, meaning an electron-pair donor.

ANALYZE A Cl^- ion has a complete octet in its valence shell and so is not likely to accept additional pairs of electrons. However, it could *donate* one of its four pairs to form a bond to the Al atom in $AlCl_3$. The Lewis structure of $AlCl_3$ is

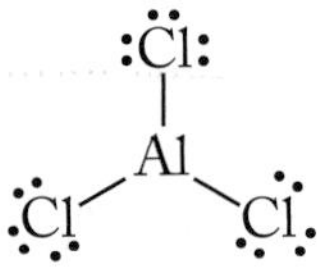

$$ \underset{:\ddot{C}l \quad \ddot{C}l:}{\overset{:\ddot{C}l:}{Al}} $$

There are only three pairs of electrons in the valence shell of the Al atom. Therefore, it could accept one more pair and act like a Lewis acid.

SOLVE In the reaction, $AlCl_3$ is a Lewis acid and Cl^- ions are Lewis bases.

THINK ABOUT IT We can use the Lewis structures of the reactants and product to illustrate the way the reaction proceeds:

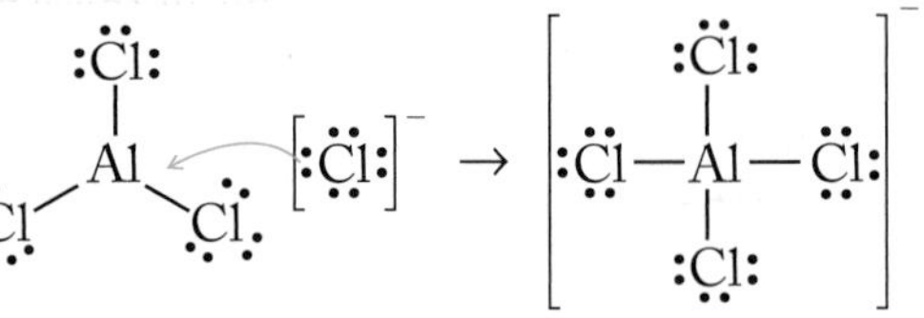

$$ \underset{:\ddot{C}l \quad \ddot{C}l:}{\overset{:\ddot{C}l:}{Al}} \quad \left[:\ddot{C}l: \right]^- \rightarrow \left[:\ddot{C}l-\underset{:\ddot{C}l:}{\overset{:\ddot{C}l:}{Al}}-\ddot{C}l: \right]^- $$

These structures assume that $AlCl_3$ is a molecular compound and not, like most metal chlorides, ionic. This assumption is supported by the physical properties of $AlCl_3$ and particularly by the fact that it sublimes at 178°C and 1 atm pressure. Most other metal halides do not even melt until heated to temperatures many hundreds of degrees higher than that.

Practice Exercise Which reactant is the Lewis acid and which is the Lewis base in the following reaction?

$$CO_2(g) + CaO(s) \rightarrow CaCO_3(s)$$

(Answers to Practice Exercises are in the back of the book.)

17.2 Complex Ions

In Chapter 4 and again in Chapter 10, we described how ions dissolved in water are *hydrated*; that is, surrounded by molecules of water oriented so that their hydrogen atoms and (+) dipoles are directed toward anions and their oxygen atoms and (−) ends of their dipoles are directed toward cations. Ion–dipole interactions in some of these hydrated cations lead to the sharing of lone pairs of electrons on the oxygen atoms of water molecules with empty valence-shell orbitals on the cations. These shared electron pairs meet our definition of covalent bonds (see Section 8.1). These particular bonds are examples of coordinate covalent bonds, or simply **coordinate bonds**, which are bonds that form when a molecule or a negative ion (see Figure 17.3) donates a pair of electrons to an empty orbital in the valence shell of an atom, molecule, or positive ion. Once formed, a coordinate bond is indistinguishable from any other kind of covalent bond.

CONCEPT TEST

Draw the Lewis structures of NH_3 and BF_3, which combine to form H_3N-BF_3. Is the N—B bond an example of a coordinate bond? Explain your answer.

(Answers to Concept Tests are in the back of the book.)

When molecules and/or negative ions form coordinate bonds with a metal ion, the resulting species is called a **complex ion**. The ions or molecules bonded to the central metal ion in a complex ion are called **ligands**. Each ligand in Figure 17.3 donates one pair of electrons to form a coordinate bond with the central atom. Direct bonding to the metal ion means that ligands occupy the **inner coordination sphere** of the metal. Later in this chapter we encounter ligands that have more than one *donor group* (electron pair) per molecule.

To explore the meaning of these and other terms related to complex-ion formation, let's consider what happens when a Zn(II) compound, such as $Zn(NO_3)_2$, dissolves in a solution containing ammonia. As Zn^{2+} ions dissolve they are surrounded by NH_3 molecules and bond to the N atoms of these molecules, as shown in Figure 17.4. A total of four NH_3 molecules occupy the inner coordination sphere of each Zn^{2+} ion, forming a complex ion with the formula $Zn(NH_3)_4{}^{2+}$. We can explain the tetrahedral shape of the complex this way: each Zn^{2+} ion has the electron configuration $[Ar]3d^{10}$. Its d orbitals are full, but its $4s$ and $4p$ orbitals

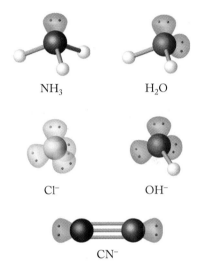

FIGURE 17.3 These models of five molecules and ions show the orientation of covalent bonds and lone pairs of valence-shell electrons based on valence bond theory. Note how each of them has the capacity to donate a pair of electrons to a metal ion. Several have more than one pair of electrons, but only one pair at a time can be directed toward a single metal ion.

A **coordinate bond** forms when one anion or molecule donates a pair of electrons to another ion or molecule to form a covalent bond.

A **complex ion** is an ionic species consisting of a metal ion bonded to one or more Lewis bases.

A **ligand** is a Lewis base bonded to the central metal ion of a complex ion.

The **inner coordination sphere** of a metal consists of the ligands that are bound directly to the metal via coordinate covalent bonds.

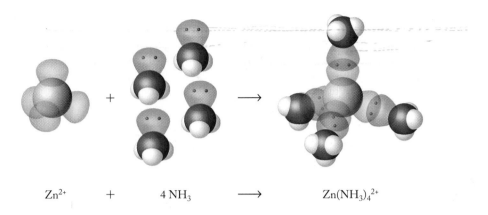

FIGURE 17.4 These molecular models show how the lone pairs of electrons on four ammonia molecules form coordinate bonds with four empty sp^3 orbitals on a Zn^{2+} ion. The resulting complex ion has the formula $Zn(NH_3)_4^{2+}$.

$$Zn^{2+} \quad + \quad 4\ NH_3 \quad \longrightarrow \quad Zn(NH_3)_4^{2+}$$

are empty. According to valence bond theory, these orbitals could mix together forming a set of four empty sp^3 hybrid orbitals:

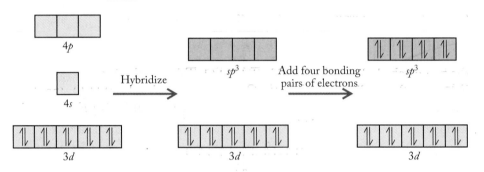

As with all sp^3 systems, the four orbitals are directed to the four corners of a tetrahedron as shown in Figure 17.4. This orientation means they can form a tetrahedral array of four coordinate bonds with electron pairs donated by four ammonia molecules as shown in the orbital diagram above and in the structure on the right in Figure 17.4.

The 2+ charges on the cationic complex ions $Zn(NH_3)_4^{2+}$ in the solution are balanced by NO_3^- ions. The nitrate ions in this solution serve as **counter ions**, a term that means ions of opposite charge that are not part of the inner coordination sphere. These NO_3^- ions are not bonded directly to the Zn^{2+} ions but are electrostatically attracted to them. The overall formula of the solute in this solution is written

$$[Zn(NH_3)_4](NO_3)_2$$

where the brackets separate the formula of the complex ion from the formula of its counter ions. This overall formula is that of a **coordination compound**, which is any compound that contains a complex ion.

The total number of electron pairs shared by ligands with the central metal ion in a complex defines the **coordination number**. Among the more common coordination numbers are 6, 4, and 2. Table 17.1 provides information about the shapes of complex ions with these coordination numbers and some examples of each. The tetrahedral shape of $Zn(NH_3)_4^{2+}$ ions can be explained by sp^3 hybridization of the valence shell orbitals of Zn^{2+} ions. However, we need a different hybridization scheme that has four orbitals oriented toward the corners of a square to account for the square planar geometry of other four-coordinate complexes. As noted in Table 17.1, dsp^2 hybridization creates orbitals with this orientation.

Six-coordinate complexes are octahedral. This is the shape we associate with d^2sp^3 hybridization (see Section 9.4), although there is another hybridization option that would give an octahedral shape. Consider, for example, $Fe(H_2O)_6^{3+}$,

Counter ions are the ions that balance the electrical charges of complex ions in coordination compounds.

Coordination compounds are made up of at least one complex ion.

The **coordination number** of a metal ion identifies the number of electron pairs surrounding it in a complex.

TABLE 17.1 Common Coordination Numbers and Shapes for Complex Ions

Coordination Number	Shape	Hybridization	Structure	Examples
6	Octahedral	sp^3d^2 or d^2sp^3		$Fe(H_2O)_6^{3+}$ $Ni(H_2O)_6^{2+}$ $Co(H_2O)_6^{3+}$
4	Tetrahedral	sp^3		$Zn(H_2O)_4^{2+}$
4	Square planar	dsp^2		$Pt(NH_3)_4^{2+}$
2	Linear	sp		$Ag(NH_3)_2^+$

the first octahedral complex listed in Table 17.1. This complex ion consists of one Fe^{3+} ion with coordinate bonds to the O atoms in six molecules of H_2O. The electron configuration of the Fe^{3+} ion is $[Ar]3d^5$. If Hund's rule is obeyed, then each of the five $3d$ orbitals is half-filled:

$$\boxed{\uparrow}\,\boxed{\uparrow}\,\boxed{\uparrow}\,\boxed{\uparrow}\,\boxed{\uparrow}$$
$$3d$$

and none of them can accept *two* more electrons to form a coordinate bond. Therefore, d^2sp^3 hybridization that involves two of these $3d$ orbitals is not an option. To explain the bonding in $Fe(H_2O)_6^{3+}$ and its octahedral shape we need to include not $3d$ but rather $4d$ orbitals in our hybridization scheme, as shown in this orbital diagram:

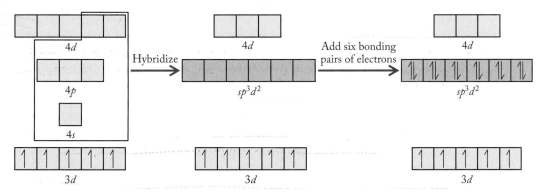

The product of mixing one $4s$, three $4p$, and two $4d$ orbitals is a set of six equivalent sp^3d^2 orbitals. Note that the d term in the hybridization name is last instead of first because $4d$ orbitals were used instead of $3d$ orbitals.

Now let's consider the bonding in another Fe(III) complex ion, $Fe(CN)_6^{3-}$. The $1-$ charge on each of the six CN^- ions gives the complex ion an overall charge of $3-$. Given the electron configuration of Fe^{3+}: $[Ar]3d^5$ and the octahedral shape of the complex ion, you might assume sp^3d^2 hybridization for the reasons discussed in the previous paragraph. However, there is experimental evidence that the hybridization in $Fe(CN)_6^{3-}$ is not sp^3d^2. The evidence comes from measurements of the magnetic properties of the complex, which allow scientists to determine the number of unpaired electrons per ion. Keep in mind that spinning electrons produce tiny magnetic fields.

CONNECTION According to the valence-shell electron-pair repulsion model described in Section 9.2, the shape of a molecule (or a complex ion) depends on the number of lone pairs of electrons and the number of atoms bonded to the central atom or ion.

CONNECTION We introduced the magnetic behavior of matter in Chapter 9 in our discussion of molecular orbital theory.

When two electrons are paired in an orbital, they spin in opposite directions and their magnetic fields cancel out. When an atom or ion has unpaired electrons, it is paramagnetic, and the more unpaired electrons it has, the more paramagnetic it is.

The magnetic properties of $Fe(H_2O)_6^{3+}$ confirm the presence of five unpaired electrons per ion. However, the magnetic properties of $Fe(CN)_6^{3-}$ indicate that there is only one unpaired electron per ion. We can explain this result by assuming that Hund's rule is not obeyed in $Fe(CN)_6^{3-}$. Instead, four of the five $3d$ electrons are paired, producing two empty $3d$ orbitals that are available for hybridization:

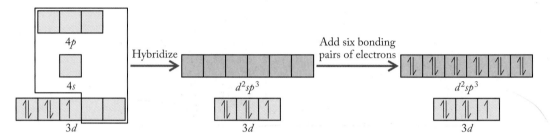

The result is d^2sp^3 hybridization of Fe^{3+} in $Fe(CN)_6^{3-}$ instead of sp^3d^2. You may be wondering why the five $3d$ electrons were unpaired and Hund's rule was obeyed in $Fe(H_2O)_6^{3+}$, but not in $Fe(CN)_6^{3-}$. The answer lies in the different strengths of interaction between the lone pairs of electrons on different ligands and the electrons in the valence shell orbitals of metal ions. We explore these differences and how they have an impact on the distribution of electrons in the d orbitals of complexed metal ions in Section 17.8.

CONCEPT TEST

Is the hybridization of Xe in XeF_4 d^2sp^3 or sp^3d^2? Explain your answer.

17.3 Complex-Ion Equilibria

Let's investigate the formation of complex ions using some of the mathematical tools developed in Chapters 15 and 16. These tools are appropriate because complex formation processes are reversible and usually reach chemical equilibrium rapidly. We start this investigation with another colorful experimental example of complex ion formation. This experiment involves two aqueous solutions, one containing copper(II) sulfate ($CuSO_4$), the other NH_3. The color of the $CuSO_4$ solution (see Figure 17.5) is robin's-egg blue, which is characteristic of $Cu^{2+}(aq)$ ions. Ammonia solutions are colorless. When ammonia solution is added to the copper sulfate solution, the two together turn very dark blue as shown on the right in Figure 17.5. This is the color of tetraamminecopper(II), $Cu(NH_3)_4^{2+}$. The change in color provides visual evidence that the complex ion is formed, and the intensity of the color compared to solutions of known concentrations of $Cu(NH_3)_4^{2+}(aq)$ indicates that the equilibrium position for the reaction

$$Cu^{2+}(aq) + 4\,NH_3(aq) \rightleftharpoons Cu(NH_3)_4^{2+}(aq)$$

is far to the right, favoring formation of the complex ion.

This conclusion is reinforced mathematically by the large value of the equilibrium constant for the complex-ion formation reaction:

$$K_f = \frac{[Cu(NH_3)_4^{2+}]}{[Cu^{2+}][NH_3]^4} = 5.0 \times 10^{13}$$

FIGURE 17.5 What a Chemist Sees The beaker on the left contains a solution of $Cu^{2+}(aq)$, which is a characteristic robin's-egg blue. As a colorless solution of ammonia is added (from the bottle in the middle), the mixture of the two solutions turns a dark navy blue (beaker on the right), which is the color of the $Cu(NH_3)_4^{2+}$ complex ion.

The subscript "f" is used to indicate that this equilibrium constant describes the formation of a complex ion. In fact, K_f values are called simply **formation constants.** For the general case in which a mole of metal ions (M^{m+}) combines with n moles of ligand X^{x-} to form complex ion $MX_n^{(m-nx)+}$, the formation constant expression is

$$K_f = \frac{[MX_n^{(m-nx)+}]}{[M^{m+}][X^{x-}]^n}$$

Formation constants may be used to calculate the concentration of a complex ion that forms from particular initial concentrations of a metal ion and a ligand. They may also be used to calculate the concentration of a free, uncomplexed metal ion, $M^{n+}(aq)$, in equilibrium with a given (often much larger) concentration of ligand, as in the following Sample Exercise.

SAMPLE EXERCISE 17.2 **Calculating a Free Metal Ion Concentration in Equilibrium with a Complex**

Ammonia gas is dissolved in a $1.00 \times 10^{-4}\ M$ solution of $CuSO_4$, so that initially $[NH_3] = 2.00 \times 10^{-3}\ M$. Calculate the concentration of $Cu^{2+}(aq)$ ions in the solution after the reaction mixture has come to chemical equilibrium.

COLLECT AND ORGANIZE The reaction is based on the formation of the tetraamminecopper(II) complex ion:

$$Cu^{2+}(aq) + 4NH_3(aq) \rightleftharpoons Cu(NH_3)_4^{2+}(aq)$$

The values of $[NH_3]$, $[Cu^{2+}]$, and $[Cu(NH_3)_4^{2+}]$ are related by the formation constant expression

$$K_f = \frac{[Cu(NH_3)_4^{2+}]}{[Cu^{2+}][NH_3]^4} = 5.0 \times 10^{13}$$

Initially $[NH_3] = 20\ [Cu^{2+}]$.

A **formation constant** (K_f) is the equilibrium constant describing the mass action expression for forming a metal complex from a free metal ion and its ligands.

ANALYZE There is more than enough NH_3 present to convert all the Cu^{2+} ions into $Cu(NH_3)_4^{2+}$. Because the value of K_f is large, we can assume that nearly all the Cu^{2+} ions are converted and that only a tiny concentration (call it x) of free Cu^{2+} ions remain uncomplexed at equilibrium. Therefore, at equilibrium:

$$[Cu^{2+}] = x$$

and

$$[Cu(NH_3)_4^{2+}] = (1.00 \times 10^{-4}) - x$$

According to the stoichiometry of the reaction, 4 moles of NH_3 are consumed for every mole of $Cu(NH_3)_4^{2+}$ produced. If the change in $[Cu(NH_3)_4^{2+}]$ is $+ (1.00 \times 10^{-4} - x)$, then the change in $[NH_3]$ is (-4) times that.

SOLVE We can use the above values and relationships to create the following ICE table:

	$[Cu^{2+}]$ (M)	$[NH_3]$ (M)	$[Cu(NH_3)_4^{2+}]$ (M)
Initial	1.00×10^{-4}	2.00×10^{-3}	0
Change	$-(1.00 \times 10^{-4} - x)$	$-4(1.00 \times 10^{-4} - x)$	$+ (1.00 \times 10^{-4} - x)$
Equilibrium	x	$1.60 \times 10^{-3} + 4x$	$1.00 \times 10^{-4} - x$

Now we make the simplifying assumption that the value of x is small compared to 1.00×10^{-4} M. If it is, then $4x$ must be small compared to 1.60×10^{-3} M. Therefore, the x terms in the equilibrium values of $[NH_3]$ and $[Cu(NH_3)_4^{2+}]$ can be ignored. Using the simplified concentration terms in the K_f expression,

$$K_f = \frac{[Cu(NH_3)_4^{2+}]}{[Cu^{2+}][NH_3]^4}$$

$$= \frac{1.00 \times 10^{-4}}{x \cdot (1.60 \times 10^{-3})^4} = 5.0 \times 10^{13}$$

Solving for x,

$$x = \frac{1.00 \times 10^{-4}}{(1.60 \times 10^{-3})^4(5.0 \times 10^{13})}$$

$$= 3.0 \times 10^{-7} \, M = [Cu^{2+}]$$

Thus, the concentration of $Cu^{2+}(aq)$ ions is 3.0×10^{-7} M.

THINK ABOUT IT This result confirms our assumption, based on the large value of K_f, that $[Cu^{2+}(aq)]$ is much less than the concentration of the complex ion. In fact, more than 99% of the Cu(II) in the solution is present as $Cu(NH_3)_4^{2+}$.

Practice Exercise Calculate the equilibrium concentration of $Ag^+(aq)$ in a solution that is 0.100 M $AgNO_3$ and 0.800 M NH_3 where

$$Ag^+(aq) + 2 NH_3(aq) \rightleftharpoons Ag(NH_3)_2^+(aq) \qquad K_f = 1.7 \times 10^7$$

17.4 Hydrated Metal Ions as Acids

In Chapter 16 we explored how electronegative atoms and groups of atoms that are bonded to $-OH$ groups stabilize the anions formed when the $-OH$ hydrogen atom ionizes forming $-O^-$ and H^+, thereby promoting the ionization process. Similarly with a hydrated metal ion having the generic formula $M(H_2O)_6^{n+}$, the

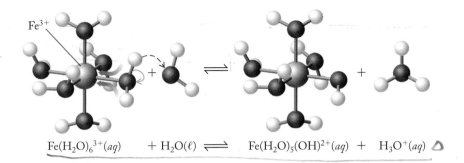

$$Fe(H_2O)_6^{3+}(aq) \quad + H_2O(\ell) \rightleftharpoons \quad Fe(H_2O)_5(OH)^{2+}(aq) \ + \ H_3O^+(aq)$$

FIGURE 17.6 Hydrated 3+ cations such as Fe^{3+} can donate one or more H^+ ions to the surrounding water molecules, leaving behind an OH^- ion in place of an inner sphere H_2O molecule.

bonding electrons of the surrounding water molecules are attracted to the positively charged central metal ion and away from their hydrogen atoms. This delocalization stabilizes the charge of the hydroxide ion that is formed when one of the inner sphere water molecules donates a H^+ ion. Many hydrated metal ions, and particularly those with charges of 2+ and higher, are Brønsted–Lowry acids. The acidic properties of these ions are reflected in the K_a values in Table 17.2.

To illustrate this acidic behavior, let's consider what happens when $FeCl_3$ dissolves in water. The compound dissociates as it dissolves, forming $Fe(H_2O)_6^{3+}$ and Cl^- ions in a 1:3 mole ratio. $Fe(H_2O)_6^{3+}$ ions undergo acid ionization. First, one of the inner sphere water molecules donates a H^+ ion to a water molecule in the bulk solution as shown in Figure 17.6 and described by the chemical equation:

$$Fe(H_2O)_6^{3+}(aq) + H_2O(\ell) \rightleftharpoons Fe(H_2O)_5(OH)^{2+}(aq) + H_3O^+(aq)$$

The products include a complex ion in which one water molecule has been replaced by a hydroxide ion. We can simplify this equation by not including the inner sphere water molecules in the formulas and instead using the (*aq*) symbol to indicate that the ions are in an aqueous solution:

$$Fe^{3+}(aq) + H_2O(\ell) \rightleftharpoons Fe(OH)^{2+}(aq) + H^+(aq)$$

The $Fe(OH)^{2+}$ ion in turn may act as an acid, resulting in a second acid ionization reaction:

$$Fe(OH)^{2+}(aq) + H_2O(\ell) \rightleftharpoons Fe(OH)_2^+(aq) + H^+(aq)$$

A third acid dissociation step is possible, which produces neutral iron(III) hydroxide. Like most transition metal hydroxides, $Fe(OH)_3(s)$ has very limited solubility:

$$Fe(OH)_2^+(aq) + H_2O(\ell) \rightleftharpoons Fe(OH)_3(s) + H^+(aq)$$

These three acid ionization reactions are not unique to Fe(III). Other cations with 3+ charges display similar behavior. These include chromium(III) and aluminum hydroxides; however, these trivalent metals form neutral hydroxide compounds with an unusual solubility pattern. It turns out that $Cr(OH)_3$ and $Al(OH)_3$ are more soluble in strongly basic solutions than in weakly basic solutions because solid $Cr(OH)_3$ and $Al(OH)_3$ accept an additional OH^- ion at high pH, forming soluble anionic complex ions:

$$Cr(OH)_3(s) + OH^-(aq) \rightleftharpoons Cr(OH)_4^-(aq)$$

$$Al(OH)_3(s) + OH^-(aq) \rightleftharpoons Al(OH)_4^-(aq)$$

Formation of these complexes means that hydrated Cr^{3+} and Al^{3+} ions are soluble in *acidic* solutions, where they exist as $Cr^{3+}(aq)$ and $Al^{3+}(aq)$; in *weakly acidic* solutions they exist as positively charged complex ions with generic formulas such as

TABLE 17.2 K_a **Values of Hydrated Metal Ions**

Ion	K_a
$Fe^{3+}(aq)$	3×10^{-3}
$Cr^{3+}(aq)$	1×10^{-4}
$Al^{3+}(aq)$	1×10^{-5}
$Cu^{2+}(aq)$	3×10^{-8}
$Pb^{2+}(aq)$	3×10^{-8}
$Zn^{2+}(aq)$	1×10^{-9}
$Co^{2+}(aq)$	2×10^{-10}
$Ni^{2+}(aq)$	1×10^{-10}

Acid strength

$M(OH)^{2+}(aq)$ and $M(OH)_2^+(aq)$; and in *basic solutions* they exist as $M(OH)_4^-(aq)$. Enhanced solubility at high pH is also observed for $Zn(OH)_2$, but not for most transition metal hydroxides. However, Al and nearly all the transition metals share one common characteristic: they exist as hydrated ions, $M^{n+}(aq)$, only in acidic solutions. They occur as complex ions in aqueous solutions that range from weakly acidic to weakly basic, which includes most environmental waters and biological fluids.

17.5 Solubilities of Ionic Compounds and K_{sp}

In the previous section we noted that transition metal hydroxides have limited solubility in water. In this section we explore ways to express exactly how limited these solubilities are and apply the results of that exploration to describing the solubility of other ionic compounds.

The Solubility Product K_{sp}

Let's begin with a nontransition metal compound found in many medicine cabinets: magnesium hydroxide, the active ingredient in the antacid called milk of magnesia. This liquid appears "milky" because it is an aqueous suspension of $Mg(OH)_2$, a white solid that is only slightly soluble in water. We can express the limited solubility of $Mg(OH)_2$ using an equilibrium constant expression based on the dissolution process:

$$Mg(OH)_2(s) \rightleftharpoons Mg^{2+}(aq) + 2\,OH^-(aq)$$

Because $Mg(OH)_2$ is a solid, its effective concentration does not change during dissolution, at least not until it is all gone. As long as there is still some of it in suspension, its influence on the equilibrium does not change. As described in Chapter 15, we express the constant presence of a pure solid in a chemical equilibrium by assigning it a concentration value of one. Therefore, the equilibrium constant for the dissolution of $Mg(OH)_2$ is

$$K_{sp} = \frac{[Mg^{2+}][OH^-]^2}{1} = [Mg^{2+}][OH^-]^2$$

where K_{sp} represents an equilibrium constant called the **solubility–product constant** or, simply, the **solubility product**.

The K_{sp} values of $Mg(OH)_2$ and several other slightly soluble compounds are listed in Appendix 5. We can use these values to calculate the solubilities of ionic compounds. Solubility (s) values may be expressed in molarity but frequently other concentration units are used, often based on the mass of a compound that dissolves in a particular volume of solution. Let's use the K_{sp} value of $Mg(OH)_2$, 5.6×10^{-12}, to calculate how many moles of $Mg(OH)_2$ dissolve in one liter of aqueous solution and how many grams of it dissolve in 50.0 mL at 25°C. We start with the K_{sp} expression:

$$K_{sp} = [Mg^{2+}][OH^-]^2 = 5.6 \times 10^{-12}$$

If we let s be the number of moles of $Mg(OH)_2$ that dissolves in one liter of solution at 25°C, then s moles of Mg^{2+} ions and $2s$ moles of OH^- ions are produced:

$$Mg(OH)_2(s) \rightleftharpoons Mg^{2+}(aq) + 2\,OH^-(aq)$$
$$\quad\quad -s \quad\quad\quad +s \quad\quad\quad +2s$$

The **solubility–product constant** (or **solubility product**, K_{sp}) is an equilibrium constant that describes the formation of a saturated solution of a slightly soluble salt.

Because these quantities are dissolved in one liter, $[Mg^{2+}] = s$ and $[OH^-] = 2s$. Substituting the algebraic terms in the K_{sp} expression,

$$K_{sp} = [Mg^{2+}][OH^-]^2 = (s)(2s)^2 = s(4s^2) = 4s^3 = 5.6 \times 10^{-12}$$

$$s = 1.1 \times 10^{-4}\ M$$

Note that the entire algebraic expression for $[OH^-]$ is squared in this calculation. Forgetting to square the coefficient is a common mistake. Also note that the molar solubility of $Mg(OH)_2$ is much greater than its solubility product. This difference is true for all sparingly soluble ionic compounds because each K_{sp} is the product of small concentration values multiplied together, producing an even smaller overall K_{sp} value.

If the solubility of $Mg(OH)_2$ is $1.1 \times 10^{-4}\ M$, how many grams of $Mg(OH)_2$ dissolve in 50.0 mL of solution? To answer this question, we convert the moles of solute that dissolve in 1 liter of solution into an equivalent number of grams [the molar mass of $Mg(OH)_2$ is 58.32 g/mol] and then into the mass that dissolves in 50.0 mL of solution:

$$\frac{1.1 \times 10^{-4}\ \cancel{mol}}{\cancel{L}} \times \frac{58.32\ g}{\cancel{mol}} \times \frac{1\ \cancel{L}}{1000\ \cancel{mL}} \times 50.0\ \cancel{mL} = 3.2 \times 10^{-4}\ g$$

We can derive a general equation relating the solubility of any ionic compound M_mX_x to its K_{sp} value. We start with the dissolution process:

$$M_mX_x(s) \rightleftharpoons mM^{n+}(aq) + xX^{y-}(aq)$$

If s moles of M_mX_x dissolve in 1 liter of solution, then $(s \times m)$ moles of M and $(s \times x)$ moles of X are produced. Inserting these values into the corresponding equilibrium constant expression for M_mX_x,

$$K_{sp} = [M]^m[X]^x = (s \times m)^m(s \times x)^x$$

or

$$K_{sp} = (m^m x^x)s^{(m+x)} \tag{17.1}$$

Let's use Equation 17.1 to calculate the solubility of calcium phosphate, $Ca_3(PO_4)_2$, the principal ingredient in the mineral apatite:

$$K_{sp} = (m^m x^x)s^{(m+x)} = (3^3 2^2)s^{(3+2)} = 36s^5$$

Inserting the value of the K_{sp} of $Ca_3(PO_4)_2$ listed in Appendix 5 and solving for s,

$$K_{sp} = 2.1 \times 10^{-33} = 36s^5$$

$$s = \sqrt[5]{\frac{2.1 \times 10^{-33}}{36}} = 1.4 \times 10^{-7}\ M$$

CONNECTION In Chapter 15 we learned that concentrations in mass action expressions are raised to powers equal to the coefficients of species in balanced chemical equations.

SAMPLE EXERCISE 17.3 **Calculating Solubility from K_{sp}**

The mineral barite is mostly barium sulfate ($BaSO_4$) and is widely used in industry and in medical imaging of the digestive system. Calculate the solubility (in molarity) of $BaSO_4$ in pure water and in seawater in which the concentration of sulfate ions is 2.8 g/L of $SO_4{}^{2-}$. The K_{sp} of $BaSO_4$ is 9.1×10^{-11} at 25°C.

COLLECT AND ORGANIZE We are asked to calculate the solubility of $BaSO_4$ in both pure water and in seawater that contains sulfate ions. We are given the K_{sp} value of the compound. The stoichiometry of the dissolution of $BaSO_4$

$$BaSO_4(s) \rightleftharpoons Ba^{2+}(aq) + SO_4{}^{2-}(aq)$$

indicates that 1 mole of barium ions and 1 mole of sulfate ions form from each mole of $BaSO_4$ that dissolves.

ANALYZE If s moles of $BaSO_4$ dissolve per liter of pure water, then $[Ba^{2+}] = s$ and $[SO_4^{2-}] = s$. However, if s moles of $BaSO_4$ dissolve in seawater the value of $[SO_4^{2-}]$ will be greater than s by the amount we can calculate from the given concentration, 2.8 g/L of SO_4^{2-}, and the molar mass of sulfate ions. According to Le Châtelier's principle and the common-ion effect, additional sulfate ion shifts the dissolution equilibrium to the left, and less $BaSO_4$ dissolves in seawater.

SOLVE In the case of pure water, we insert s into the K_{sp} expression for $[Ba^{2+}]$ and $[SO_4^{2-}]$

$$K_{sp} = [Ba^{2+}][SO_4^{2-}] = (s)(s) = 9.1 \times 10^{-11}$$

$$s^2 = 9.1 \times 10^{-11}$$

$$s = 9.5 \times 10^{-6} \, M$$

In the case of seawater, we account for the sulfate already present by calculating the background value of $[SO_4^{2-}]$

$$\text{background } [SO_4^{2-}] = \frac{2.8 \text{ g}}{L} \times \frac{1 \text{ mol}}{96.06 \text{ g}} = \frac{0.029 \text{ mol}}{L}$$

and adding this value to the $[SO_4^{2-}]$ term in the K_{sp} expression:

$$K_{sp} = [Ba^{2+}][SO_4^{2-}] = (s)(0.029 + s) = 9.1 \times 10^{-11}$$

Solving for s is simplified if we ignore its contribution to the total concentration of sulfate. Doing so is reasonable because s is likely to be much less than 0.029 M given the limited solubility of $BaSO_4$ in pure water. Therefore,

$$(s)(0.029) > 9.1 \times 10^{-11}$$

$$s = 3.1 \times 10^{-9} \, M$$

THINK ABOUT IT In seawater, the concentration of sulfate added by the dissolution of $BaSO_4$ is much smaller than the background concentration, and so our simplifying assumption is justified. This calculation is another illustration of the common-ion effect. In this example, the dissolution process is suppressed by the additional sulfate ions present in seawater.

Practice Exercise What is the solubility in pure water at 25°C of $MgCO_3$, the principal component of the mineral dolomite? $K_{sp} = 6.8 \times 10^{-6}$ for $MgCO_3$.

Complexation, pH, and Solubility

We saw in Section 17.4 that the solubilities of $Fe(OH)_3$ and $Al(OH)_3$ in pure water are very limited. This is illustrated by the miniscule sizes of their K_{sp} values: 1.1×10^{-36} and 1.9×10^{-33}, respectively. However, we must keep in mind that these tiny K_{sp} values apply to equilibrium constant expressions that are based on the concentrations of *free metal ions*—for example,

$$K_{sp} = [Al^{3+}][OH^-]^3 = 1.9 \times 10^{-33}$$

If we assume that the solvent is pure water at pH = 7.00, then the concentration of Al^{3+} ions can be no greater than:

$$[Al^{3+}] = \frac{K_{sp}}{[OH^-]^3} = \frac{1.9 \times 10^{-33}}{(1.0 \times 10^{-7})^3} = 1.9 \times 10^{-12} \, M$$

This result suggests that aluminum(III) compounds are essentially insoluble at neutral pH, but what it really means is that $[Al^{3+}]$ is very small at neutral pH. However, in Section 17.4 we saw that most aluminum(III) in solution at pH 7.0 is not $Al^{3+}(aq)$, but rather a mixture of hydroxo complexes. The formation of these and other complexes allows the overall solubility of aluminum(III) in neutral solutions to be much higher than 1.9×10^{-12} M. Moreover, the solubility of aluminum(III) compounds increases at high pH, where charged, soluble $Al(OH)_4^-(aq)$ is the dominant species in solution, and at low pH, where $Al(OH)^{2+}(aq)$ and $Al(OH)_2^+(aq)$ predominate.

The greater solubility of $Al(OH)_3$ and the hydroxides and oxides of some other metals at high and low pH is a result of the capacity of these compounds to behave as both acids and bases. In a strongly basic solution $Al(OH)_3$ is a hydrogen-ion donor, a behavior that is easy to describe if we add the waters of hydration to the formulas of the complex ions:

$$Al(H_2O)_3(OH)_3(s) + OH^-(aq) \rightleftharpoons Al(H_2O)_2(OH)_4^-(aq) + H_2O(\ell)$$

In an acidic solution, $Al(OH)_3$ acts like a basic salt and hydrogen-ion acceptor:

$$Al(H_2O)_3(OH)_3(s) + H_3O^+(aq) \rightleftharpoons Al(H_2O)_4(OH)_2^+(aq) + H_2O(\ell)$$

In the following Sample Exercise we see how low pH affects the solubility of another basic salt, calcium fluoride.

⊙⊙ CONNECTION In Chapter 16 we learned that the solutions of some salts are acidic or basic because of the reaction of their ions with water.

SAMPLE EXERCISE 17.4   **Calculating the Effect of pH on Solubility**

What is the molar solubility of CaF_2 at 25°C in (a) pure water and (b) an acidic buffer in which $[H^+]$ is a constant 0.050 M? The K_{sp} of CaF_2 is 3.9×10^{-11}.

COLLECT AND ORGANIZE We are asked to calculate the solubility of CaF_2 in both pure water and in a strong acid. The dissolution process and solubility are described by the following chemical equation and K_{sp} expression:

(1) $$CaF_2(s) \rightleftharpoons Ca^{2+}(aq) + 2\,F^-(aq)$$

$$K_{sp} = [Ca^{2+}][F^-]^2 = 3.9 \times 10^{-11}$$

Calcium fluoride is a basic salt because the hydroxide of the cation is a strong base, $Ca(OH)_2$, but the fluoride ion is the conjugate base of a weak acid, HF. Therefore, on balance, solutions of CaF_2 are basic.

ANALYZE To assess the impact of pH on the solubility of a fluoride compound, we need to consider the chemical equilibrium in which the fluoride ion acts as a base (hydrogen-ion acceptor):

(2) $$F^-(aq) + H^+(aq) \rightleftharpoons HF(aq)$$

This reaction is the reverse of the acid ionization of HF:

$$HF(aq) \rightleftharpoons F^-(aq) + H^+(aq)$$

The equilibrium constant for the reaction is the reciprocal of the K_a of HF:

$$K = \frac{[HF]}{[H^+][F^-]} = \frac{1}{6.8 \times 10^{-4}} = 1.47 \times 10^3$$

As reaction (2) proceeds, F^- ions are consumed, which shifts the position of the CaF_2 dissolution equilibrium [reaction (1)] to the right and so increases solubility.

SOLVE (a) In pure water: If we let x be the number of moles of CaF_2 that dissolves in 1.000 L of solution, then x moles of Ca^{2+} and $2x$ moles of F^- are produced.

$$CaF_2(s) \rightleftharpoons Ca^{2+}(aq) + 2\,F^-(aq)$$

$$ \quad x \qquad\qquad x \qquad\quad 2x$$

Therefore, $[Ca^{2+}] = x$ and $[F^-] = 2x$. Substituting these terms in the K_{sp} expression,

$$K_{sp} = [Ca^{2+}][F^-]^2 = (x)(2x)^2 = 4x^3$$

$$4x^3 = 3.9 \times 10^{-11}$$

$$x = s = 2.1 \times 10^{-4}\ M$$

(b) In the acidic buffer ($[H^+] = 0.050\ M$) the proportion of fluoride ions that combine with the H^+ ions is calculated as follows:

$$K = \frac{[HF]}{[0.050][F^-]} = \frac{1}{6.8 \times 10^{-4}}$$

and

$$\frac{[HF]}{[F^-]} = 73.5$$

which means

$$[HF] = 73.5\ [F^-]$$

Therefore, most of the fluoride ion produced when calcium fluoride dissolves is converted into HF molecules. The fraction of the total that remains as F^- ions is

$$\frac{[F^-]}{[F^-] + [HF]} = \frac{[F^-]}{[F^-] + 73.5[F^-]} = \frac{[F^-]}{74.5[F^-]} = 0.0134$$

As before, if x moles of CaF_2 dissolve in the acid, then x moles of Ca^{2+} and $2x$ moles of F^- are produced, and $[Ca^{2+}] = x$. However, most of the $2x$ moles of fluoride ions are converted into HF, and the concentration of free fluoride ions, $[F^-]$, is only 2 (0.0134 x) = 0.0268 x. Substituting these terms in the K_{sp} expression,

$$K_{sp} = [Ca^{2+}][F^-]^2 = (x)(0.0268\ x)^2 = 7.18 \times 10^{-4}\ x^3$$

$$7.20 \times 10^{-4}\ x^3 = 3.9 \times 10^{-11}$$

$$x = s = 3.8 \times 10^{-3}\ M$$

THINK ABOUT IT A comparison of the two results shows that CaF_2 is more soluble in the acid than in pure water, as we predicted. In fact, it is more than 10 times more soluble.

Practice Exercise What is the solubility of $ZnCO_3$ in a buffer solution with a pH of 10.33? (*Hint:* The K_{sp} of $ZnCO_3$ and the K_{a2} of H_2CO_3 are given in Appendix 5.)

A **monodentate ligand** is a species that forms only a single coordinate bond to a metal ion in a complex.

A **polydentate ligand** is a species that can form more than one coordinate bond per molecule.

Chelation is the interaction of a metal with a polydentate ligand (chelating agent). One molecule of the ligand occupies two or more coordination sites on the central metal.

CONCEPT TEST

In the second calculation in the preceding Sample Exercise we assumed that $[H^+]$ did not change significantly as a result of the reaction:

$$HF(aq) \rightleftharpoons F^-(aq) + H^+(aq)$$

If, however, the value of H^+ had dropped significantly, how would the solubility of CaF_2 have been affected?

17.6 Polydentate Ligands

All the ligands we have discussed so far, including all those in Figure 17.3 and many in Table 17.3, can donate only one pair of electrons to a metal ion. Even those with more than one lone pair, such the O^{2-} ion, can donate only one pair at a time to a given metal ion because their other lone pairs are oriented too far away from the metal ion (for O the angle is 109.5°). Because these ligands have effectively only one donor group, they are called **monodentate ligands**, which literally means "single-toothed."

Molecules larger than those in Figure 17.3 may have more than one donor group and form more than one coordinate bond to a central metal ion. Ligands in this category are called **polydentate ligands**, or more specifically bidentate, tridentate, and so on, depending on whether they donate two, three, or more electron pairs per molecule to a central metal ion. Among these ligands are the members of the polyamine family of compounds. One member of this family is a compound known best by its common name: ethylenediamine, which has the structure

$$H_2\ddot{N} \qquad \ddot{N}H_2$$
$$H_2C - CH_2$$

Note the presence of the lone pairs of electrons on the two $-NH_2$ groups that are separated from each other by two $-CH_2-$ groups. This combination means that a molecule of ethylenediamine can partially encircle a metal ion so that a lone pair of electrons on each of the nitrogen atoms can form a covalent bond to the metal ion.

The skeletal structure of an ethylenediamine complex ion based on $Ni(H_2O)_6^{2+}$ is shown in Figure 17.7(a). The two bonding orbitals of the metal ion have to be on the same side of the octahedron because the amino groups in ethylenediamine are too close together to straddle opposite corners. However, two more ethylenediamine molecules could bond to other pairs of bonding sites, displacing additional pairs of water molecules and forming a complex ion that is essentially surrounded by ethylenediamine molecules as shown in Figure 17.7(b).

Note that each ethylenediamine molecule forms a five-atom ring with the metal ion. If the ring were a perfect pentagon (with all bond lengths and bond angles exactly the same), all bond angles would be 108°. However, the ring is not a perfect pentagon, but the preferred bond angles of 90° for the N–Ni–N bonds and 107° to 109° for all the others can be accommodated with only a little strain on the natural bond angles.

The structure of a larger polyamine called diethylenetriamine $(H_2NCH_2CH_2NHCH_2CH_2NH_2)$ is shown in Figure 17.8. The lone pairs of electrons on its three nitrogen atoms give diethylenetriamine the capacity to form three coordinate bonds to a metal ion and so act as a tridentate ligand (Figure 17.8b).

As you may imagine, there are even larger molecules with even more atoms per molecule that can donate multiple pairs of electrons to a single metal ion. The interaction of a metal ion with a ligand that has more than one donor atom is called **chelation** (from the Greek *chele*, meaning "claw"), and the polydentate ligands that form them are called *chelating agents*. Many chelating agents have more than one kind of electron-pair-donating group in their structures. One family of such compounds is called *aminocarboxylic acids*. The most important of them is ethylenediaminetetraacetic acid, better known as EDTA. The molecular structure of EDTA is shown in Figure 17.9(a). Note that one molecule of EDTA

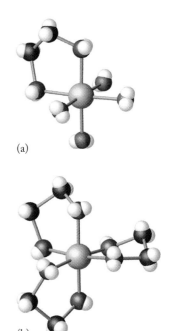

(a)

(b)

FIGURE 17.7 Bidentate chelation. (a) The bidentate ligand ethylenediamine contains two lone-pair donor groups that can donate their lone pairs of electrons to empty *d* orbitals and occupy adjacent octahedral sites on $Ni^{2+}(aq)$ ions in solution. Note how ethylenediamine and the Ni^{2+} ion form a five-atom ring. The other four coordination sites are still occupied by H_2O molecules. (b) Three ethylenediamine molecules can occupy all six octahedral sites.

$$H_2\ddot{N} \diagdown\diagup \overset{\ddot{}}{N}H \diagdown\diagup \ddot{N}H_2$$

(a)

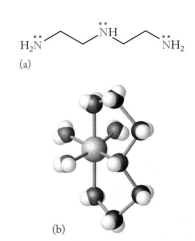

(b)

FIGURE 17.8 Tridentate chelation. (a) The tridentate ligand diethylenetriamine contains three donor groups, which are highlighted in blue. (b) These groups donate their lone pairs of electrons to empty *d* orbitals and occupy three octahedral sites on the Ni^{2+} ion in solution.

TABLE 17.3 Structure of Common Ligands

Ligand	Name within Complex Ion[a]	Structure	Charge	Number of Donor Groups
Iodide	Iodo	I^-	1−	1
Bromide	Bromo	Br^-	1−	1
Chloride	Chloro	Cl^-	1−	1
Nitrate	Nitrato	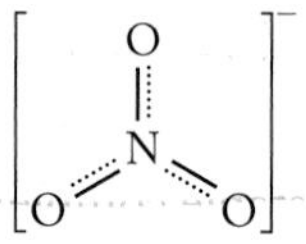	1−	1
Fluoride	Fluoro	F^-	1−	1
Hydroxide	Hydroxo	$[O-H]^-$	1−	1
Water	Aqua	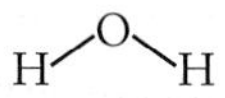	Neutral	1
Pyridine	Pyridine	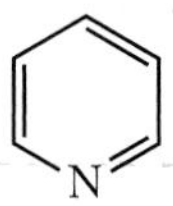	Neutral	1
Ammonia	Ammine	NH_3	Neutral	1
Ethylenediamine (en)	Ethylenediamine	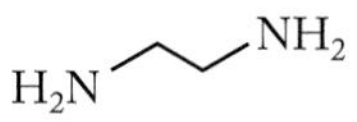	Neutral	2
2,2′-Bipyridine (bipy)	Bipyridyl	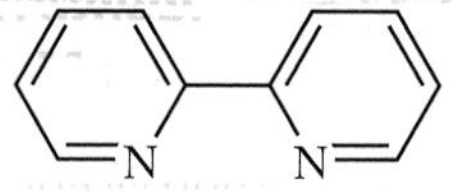	Neutral	2
1,10-Phenanthroline (phen)	1,10-Phenanthroline (phen)	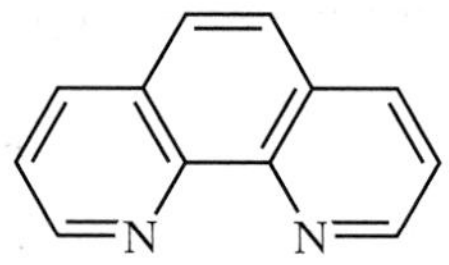	Neutral	2
Cyanide[b]	Cyano	$[C\equiv N]^-$	1−	1
Carbon monoxide[b]	Carbonyl	$C\equiv O$	Neutral	1

[a] The names of neutral ligands in complexes (except for water and ammonia) are the same as the names of the molecules.

[b] Carbon atoms are the lone pair donors in these ligands.

contains two amine groups and four carboxylic acid groups. When the acid groups release their H^+ ions, they form four carboxylate anions, $-COO^-$, that can act as Lewis bases because an oxygen atom on each of them can donate a pair of electrons to a central metal ion. When O atoms on all four groups do so and the two amine groups also form coordinate bonds, six octahedral bonding sites around a metal ion can be occupied as shown in Figure 17.9(b).

(a)

(b)

FIGURE 17.9 (a) Hexadentate ligand EDTA contains two amine groups and four carboxylic acid groups. Ionization of the acid groups produces six coordinate bonding sites. (b) All six Lewis base groups in ionized EDTA can coordinate with a single metal ion, as shown here for Co^{3+}.

EDTA forms very stable complex ions and is used as a metal ion **sequestering agent**, that is, a chelating agent that binds metal ions so tightly that they are prevented from reacting with other substances. For example, EDTA is used as a preservative in many beverages and prepared foods. It performs this function because it sequesters iron, copper, zinc, manganese, and other transition metal ions that are often present in these foods at trace concentrations (less than 1 part per million). Even at such low concentrations, these metals can catalyze the degradation of ingredients in the foods, including those that impart desirable taste and color. Many foods are fortified with ascorbic acid (vitamin C), which is particularly vulnerable to metal-catalyzed degradation because it is also a polydentate ligand and is more likely to be oxidized when chelated to one of the above metal ions. EDTA effectively shields vitamin C from these ions. We explore the preferential binding of metal ions to different ligands in the next section.

⊙⊙ **CONNECTION** The use of units such as parts per million to express trace concentrations of solutes is described in Section 4.2.

SAMPLE EXERCISE 17.5   **Identifying the Donor Groups in a Molecular Structure**

The skeletal structure of nitrilotriacetic acid (NTA) is shown below. How many donor groups (as in "teeth") does this polydentate molecule have?

COLLECT AND ORGANIZE The molecule has a nitrogen atom at its center that is bonded to three carbon atoms, which means that the nitrogen atom is part of a tertiary amine group. The molecule also has three carboxylic acid groups (–COOH).

ANALYZE Because there are three single bonds around the N atom in the center, the N atom's fourth sp^3 orbital contains a lone pair of electrons. When all three carboxylic acid groups are ionized, there are three carboxylate groups in the molecule, with three oxygen atoms that could also donate up to three pairs of electrons to a metal ion. If any one of these O atoms and the center N atom simultaneously formed coordinate bonds with a metal ion, a five-atom ring would form, as in EDTA complexes.

A **sequestering agent** is a polydentate ligand like EDTA that forms very stable complexes with metal ions and thereby separates them from other substances with which they would otherwise react.

SOLVE The central N atom and an O atom from each of the three carboxylate groups form a total of four coordinate bonds. Therefore, NTA is potentially a tetradentate ligand.

THINK ABOUT IT The tetradentate capacity of NTA makes sense because, like EDTA, it is an aminocarboxylic acid. It has one fewer amino and one fewer carboxylic acid group than the hexadentate EDTA.

Practice Exercise The skeletal structure of citric acid, a component of all citrus fruits and a widely used preservative in the food industry, is shown below:

How many chelating "teeth," if any, does it or the citrate ion have?

17.7 Ligand Strength and the Chelation Effect

Let's begin our discussion of ligand strength by exploring the relative affinity of Ni^{2+} for two common monodentate ligands, H_2O and NH_3. We start by dissolving crystals of nickel(II) chloride hexahydrate in water. The formula of this compound is $NiCl_2 \cdot 6\,H_2O$. The dot connecting the two halves of the formula and the prefix *hexa* in the name tell chemists that crystals of nickel(II) chloride contain Ni^{2+} ions, each surrounded by six water molecules, called *waters of hydration*. Brilliant green crystals of $NiCl_2 \cdot 6\,H_2O$ form green solutions when they dissolve in water as shown in Figure 17.10(a) and (b). From this consistency of color we may conclude that the same clustering of water molecules around Ni^{2+} ions occurs in both solid $NiCl_2 \cdot 6\,H_2O$ and in aqueous solutions of Ni^{2+} ions.

(a)

(b)

FIGURE 17.10 (a) When the green solid, nickel(II) chloride hexahydrate, dissolves in water, the resulting solution (b) has the same green color. When ammonia gas is bubbled through the solution, its color changes to blue as NH_3 replaces H_2O in the inner coordination spheres of the Ni^{2+} ions.

Now let's bubble ammonia gas through a solution of $NiCl_2 \cdot 6H_2O$. As shown in Figure 17.10(b) the green solution turns blue. The color change means that different ligands are bonded to the Ni^{2+} ions. We may conclude that molecules of ammonia have displaced at least some molecules of water surrounding the Ni^{2+} ions. If they displaced them all, a coordination complex with the formula $Ni(NH_3)_6^{2+}$ would be formed. We can write a chemical equation describing this change as follows:

$$Ni^{2+}(aq) + 6NH_3(g) \rightleftharpoons Ni(NH_3)_6^{2+}(aq)$$

Keep in mind that the symbol $Ni^{2+}(aq)$ in the above equation represents hydrated Ni^{2+} ions that have six water molecules bonded to each Ni^{2+} ion. A more complete formula of this species is $Ni(H_2O)_6^{2+}$. We could have written the equation:

$$Ni(H_2O)_6^{2+}(aq) + 6NH_3(aq) \rightleftharpoons Ni(NH_3)_6^{2+}(aq) + 6H_2O(\ell)$$

However, we usually don't write "H_2O" terms in chemical equations for reactions in aqueous solution unless molecules of water are consumed or produced in the reaction.

This ligand displacement reaction tells us that Ni^{2+} ions have a greater affinity for ammonia than they do for water molecules. This affinity is reflected in the large value for the formation constant (K_f) for the reaction:

$$Ni^{2+}(aq) + 6NH_3(aq) \rightleftharpoons Ni(NH_3)_6^{2+}(aq) \qquad K_f = 5 \times 10^8$$

Many other transition metal ions also have a greater affinity for ammonia than for water. This reflects the fact that ammonia is inherently a stronger Lewis base and better electron donor than water.

Now suppose we add ethylenediamine to the blue solution of ammonia and Ni^{2+} ions. The color of the solution shifts to purple. This color change indicates that ethylenediamine molecules can displace ammonia molecules from the inner coordination sphere of Ni^{2+} ions. This greater affinity of Ni^{2+} ions for ethylenediamine molecules over ammonia is reflected in the value of the formation constant for $Ni(en)_3^{2+}$ (where "en" represents ethylenediamine)

$$Ni^{2+}(aq) + 3en(aq) \rightleftharpoons Ni(en)_3^{2+}(aq) \qquad K_f = 1.1 \times 10^{18}$$

which is more than 10^9 times greater than that for forming the $Ni(NH_3)_6^{2+}$ complex ion.

Why should the affinity of Ni^{2+} ions for ethylenediamine be so much greater than their affinity for ammonia? After all, the coordinate bonds are formed by lone pairs of electrons on N atoms in both ligands. At one level it is reasonable to argue that two donor groups are better than one. However, thermodynamics can explain why ethylenediamine forms more stable complex ions than ammonia, and in general why polydentate ligands bond so strongly to metal ions. Let's write a chemical equation describing the displacement of ammonia ligands in $Ni(NH_3)_6^{2+}$ ions by ethylenediamine molecules:

$$Ni(NH_3)_6^{2+}(aq) + 3en(aq) \rightarrow Ni(en)_3^{2+}(aq) + 6NH_3(aq)$$

The color change from blue to violet tells us that this reaction is spontaneous. As we discussed in Chapter 13, spontaneous reactions are those in which there is a decrease in free energy ($\Delta G < 0$). Furthermore, ΔG under standard conditions ($\Delta G°$) is related to the changes in enthalpy and entropy that accompany the reaction under standard conditions:

$$\Delta G° = \Delta H° - T\Delta S°$$

A negative value for the change in free energy under standard conditions means that $\Delta H°$ must be negative and/or $T\Delta S°$ must be positive. It turns out that the displacement of NH_3 by ethylenediamine is exothermic, but only slightly ($\Delta H° = -12$ kJ/mol). More importantly, the value of $\Delta S°$ for this reaction is $+185$ J/(K·mol). This means that the $T\Delta S°$ term at 25°C is

$$T\Delta S° = 298\ \text{K} \times \frac{185\ \text{J}}{\text{mol}\cdot\text{K}} \times \frac{1\ \text{kJ}}{1000\ \text{J}} = 55.1\ \text{kJ/mol}$$

To understand why there is such a large increase in entropy, consider that there are a total of 4 moles of dissolved ionic and molecular reactants and 7 moles of dissolved products. Nearly doubling the number of moles of dissolved products over reactants translates into a large gain in entropy and is the primary driving force for this reaction and for many complexation reactions that involve polydentate ligands. The resulting affinity of metal ions for polydentate ligands is called the **chelate effect**.

CONCEPT TEST

The following reaction is spontaneous:

$$Ni(H_2O)_6^{2+}(aq) + 6\,NH_3(aq) \rightleftharpoons Ni(NH_3)_6^{2+}(aq) + 6\,H_2O(\ell)$$

Is this spontaneity due principally to a negative ΔH or a positive ΔS? Explain your answer.

CHEMTOUR Crystal Field Splitting

The **chelate effect** is the greater affinity of metal ions for polydentate ligands compared to the corresponding monodentate ligands.

Crystal field splitting is the separation of a set of d orbitals into subsets with different energies as a result of interactions between electrons in those orbitals and pairs of electrons in ligands surrounding the orbitals.

Crystal field splitting energy (Δ) is the difference in energy between subsets of d orbitals split by interactions in a crystal field.

17.8 Crystal Field Theory

We have seen that the formation of complex ions can dramatically alter the colors of solutions of transition metals. We have yet to address why these color changes happen. The colors of transition metal compounds and transition metal ions in solution are due to transitions of electrons in d orbitals. Let's explore the role of d orbitals using Cr^{3+} as our model transition metal ion.

The Cr^{3+} ion has the electron configuration $[Ar]3d^3$. If a Cr^{3+} ion (or any atom or ion) is in the gas phase, all of the orbitals in each of its subshells have the same energy. In Section 7.8 we introduced the concept of *degeneracy* to describe the energy equality of orbitals in the same subshell. However, when a Cr^{3+} ion is in solution and surrounded by an octahedral array of ligands such as six water molecules in $Cr(H_2O)_6^{3+}$, the energies of its $3d$ orbitals are not the same. As the six oxygen atoms of these water molecules approach the Cr^{3+} ion during coordinate bond formation, repulsions arise between the electrons already in the $3d$ orbitals of Cr^{3+} and the lone pairs of electrons on the oxygen atoms. These repulsions raise the energies of all the d orbitals, and some more than others. This difference happens because repulsions are strongest for electrons in the $3d_{z^2}$ and $3d_{x^2-y^2}$ orbitals whose lobes point directly toward the oxygen atoms (the ligands) at the corners of the octahedron in Figure 17.11. These repulsions raise the energies of the $3d_{z^2}$ and $3d_{x^2-y^2}$ orbitals above those of $3d_{xy}$, $3d_{xz}$, and $3d_{yz}$ because the lobes of the latter three orbitals do not point directly toward the corners of the octahedron. These two different degrees of repulsion split the d orbitals into two subsets with different energies: the $3d_{z^2}$ and $3d_{x^2-y^2}$ are the higher-energy subset, and $3d_{xy}$, $3d_{xz}$, and $3d_{yz}$ are the lower one, as shown in Figure 17.11. This process is known as **crystal field splitting**, and the difference in energy between any subsets thereby created is called **crystal field splitting energy (Δ)**. The name was originally used

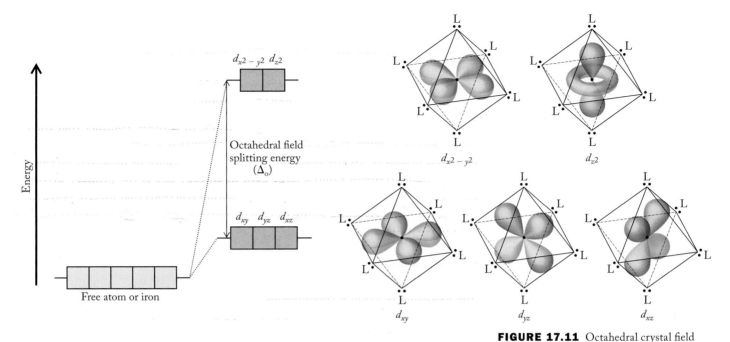

FIGURE 17.11 Octahedral crystal field splitting. In free atoms and ions, all orbitals in a subshell are degenerate. Repulsions between electrons in these d orbitals and those of ligand donor groups raise the energy of these orbitals in an octahedral field. The most repulsion is experienced by electrons in the d_{z^2} and $d_{x^2-y^2}$ orbitals, which are directed at the corners of the octahedron and so are closest to the lone pairs of electrons on the ligands. The lobes of the lower energy d_{xy}, d_{xz}, and d_{yz} orbitals are directed between the corners of the octahedron.

to describe the splitting of d-orbital energies in ionic crystals, but the theory is also routinely applied to species in aqueous solutions and interactions other than strictly electrostatic ones, including the donation and acceptance of lone pairs of electrons that result in coordinate bond formation.

In a Cr^{3+} ion there are three electrons distributed among five $3d$ orbitals. Because three of the orbitals in an octahedral field have lower energy than the other two, it is logical that each of the three electrons should occupy one of the three lower-energy orbitals leaving the two higher-energy orbitals unoccupied. The energy difference between the two subsets of orbitals is symbolized by Δ_o (Figure 17.11). The subscript "o" indicates that the separation of the set of d orbitals (Δ) is caused by an *o*ctahedral field of electron repulsions.

Now let's consider what happens when a photon of electromagnetic radiation whose energy is exactly equal to the energy gap between the two sets of d orbitals strikes a Cr^{3+} ion. The photon might be absorbed as a $3d$ electron moves from a lower-energy orbital to a higher-energy orbital, as shown in Figure 17.12. The wavelength of the absorbed radiation would be related to the energy difference between the two orbitals by Equation 7.3:

$$E = \frac{hc}{\lambda}$$

As we discussed in Chapter 7, the energy and wavelength of a photon of radiation are inversely proportional. Therefore, the larger the ligand field splitting, the shorter the wavelength of radiation absorbed by the complex ion.

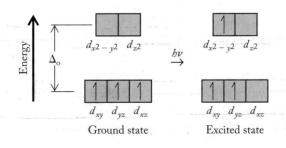

Ground state   Excited state

FIGURE 17.12 When a Cr^{3+} ion absorbs a photon of light with an energy equal to Δ_o, a $3d$ electron in a lower-energy d orbital moves to a higher-energy orbital.

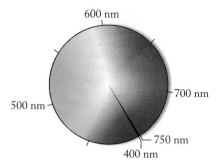

FIGURE 17.13 The color wheel can be used to predict the color of a solution or a solid based on the color of light absorbed. One perceives the color complementary to that absorbed—the color appearing opposite the absorbed color on the color wheel.

⊙⊙ **CONNECTION** Interactions between electromagnetic radiation and matter that lead to the absorption of particular wavelengths (colors) of radiation are described in Chapter 7.

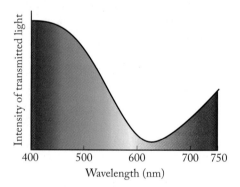

FIGURE 17.14 The visible light transmitted by a solution of $Cu(NH_3)_3^{2+}$ ions is missing much of the yellow, orange, and red portions of the spectrum because of a broad absorption band centered at 620 nm. Our eyes and brain perceive the transmitted colors as navy blue.

⊙⊙ **CONNECTION** Octahedral and tetrahedral holes occur between adjacent atoms and ions tightly packed together in crystalline solids as described in Chapter 11.

The size of the energy gap between split d orbitals frequently corresponds to electromagnetic radiation in the visible region of the spectrum. This means that the color of solutions of metal complexes relates directly to the interaction between the metal and its surrounding ligands. If white light containing all colors of the spectrum passes through a solution containing a metal complex, energy may be absorbed, causing an electron transition. If the absorbed energy corresponds to light of a particular color, the light that passes through the solution will be missing that particular wavelength. We perceive that the solution has a characteristic color due to the wavelengths that were not absorbed; we perceive the color of the transmitted light. The color of a solution (or any object) is not the color that it absorbs, but rather the color(s) that it transmits (if it is transparent, like a solution) or the colors that it reflects (if it is opaque, like many solids). To relate the color of a solution to the wavelengths of light it absorbs, we need to consider complementary colors as defined by an artist's color wheel (Figure 17.13). For example, red and green are complementary colors; therefore a solution that absorbs green light appears red to us because our eyes and brains process the transmitted colors: red, orange, yellow, blue, and violet, as the average of those colors, which is red. For the same reason, a solution of $Cr^{3+}(aq)$ that absorbs yellow-orange light has a distinctive violet color.

Color-averaging also occurs when a substance absorbs a broad band of different colors (as many transition metal solutions do). For example, a solution of $Cu(NH_3)_4^{2+}$ ions has a distinctive deep blue color, as we saw in Figure 17.5. The absorption spectrum (Figure 17.14) of such a solution features an absorption band that spans yellow, orange, and red wavelengths with a maximum absorbance at 620 nm. Our eyes sense the range that is transmitted, which spans violet to blue-green. Then our brain processes this band of transmitted colors and signals to us the average of these colors, a deep navy blue.

Most of the colored solutions we have examined up to this point contained octahedral complexes surrounded by six ligands. We have also seen the intense blue color of the $Cu(NH_3)_4^{2+}$ complex ion (Figure 17.5) whose formula indicates that four donor groups occupy the inner coordination sphere of the Cu^{2+} ion. The presence of four donor groups instead of six means that the deep blue color of this complex ion is caused by a different set of ligand–metal interactions. Four ligands surrounding a central metal can provide either a tetrahedral or square planar field of ligands (Table 17.1). The $Cu(NH_3)_4^{2+}$ complex is square planar, which means that the strongest interactions occur between the $3d$ orbitals on Cu^{2+} and the nitrogen atom lone pairs of electrons at the four corners of the equatorial plane of the octahedron, as shown in Figure 17.15. The $3d$ orbital of the Cu^{2+} ion with the strongest interactions and so the highest energy is the $d_{x^2-y^2}$ orbital because its lobes are oriented directly at the four corners of the plane. The d_{xy} orbital has slightly less energy because its lobes, though in the xy plane, are directed 45° away from the corners. Electrons in d orbitals with lobes out of the xy plane interact even less with the lone pairs of electrons on atoms in the equatorial corners and so have even lower energies. Square planar geometries tend to be limited to the transition metals with nearly filled valence-shell d orbitals, particularly those with d^8 or d^9 electron configurations.

Finally, let's consider the d orbital splitting that occurs in a tetrahedral field. The greatest electron–electron repulsions are experienced by the electrons in the d_{xy}, d_{xz}, and d_{yz} orbitals because the lobes of these orbitals are oriented most directly to the corners of the tetrahedron occupied by donor groups, as shown in Figure 17.16. The other d orbitals are less affected because their lobes do not point toward the tetrahedral corners. The resulting splitting of d orbital energies pro-

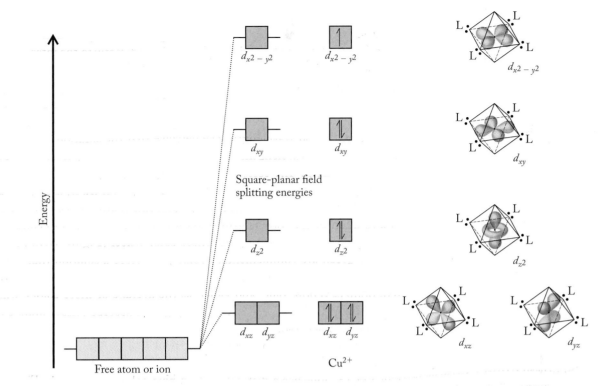

FIGURE 17.15 Square-planar crystal field splitting. The *d* orbitals of a transition metal ion in a square-planar field split into several energy levels, depending on their proximity to the lone pairs of electrons at the corner of the square. The $d_{x^2-y^2}$ orbital has the highest energy because its lobes are directed right at the four corners of the square.

duces a pattern that is an upside-down version of that produced in an octahedral field. The difference in energies between the two subsets of *d* orbitals resulting from the tetrahedral interactions is labeled Δ_t.

Before ending this discussion of the origins of the colors of transition metal ions, let's revisit the color changes that occur when first ammonia and then ethylenediamine are added to a solution of Ni^{2+} ions. Recall that the changes observed were

$$Ni(H_2O)_6^{2+} \rightarrow Ni(NH_3)_6^{2+} \rightarrow Ni(en)_3^{2+}$$
$$\text{Green} \qquad\qquad \text{Blue} \qquad\quad \text{Violet}$$

Let's think of these observed colors in terms of the colors these solutions might *absorb*: a solution of $Ni^{2+}(aq)$ ions is green because it absorbs colors such as red, on the opposite side of the color wheel from green. Similarly, a solution of $Ni(NH_3)_6^{2+}$ is blue because it absorbs colors opposite blue that are centered on orange, and a solution of $Ni(en)_3^{2+}$ is violet because it absorbs colors that are centered on the complement of violet, which is yellow. Note how the colors of light that the three

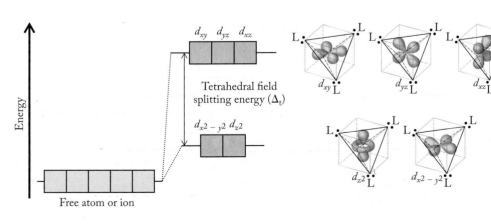

FIGURE 17.16 Tetrahedral crystal field splitting. The lobes of the higher-energy set of *d* orbitals, d_{xy}, d_{xz}, and d_{yz}, are closer to the ligands at the four corners of the tetrahedron (one of the four is hidden in these structures) than those in the lower-energy set.

A **spectrochemical series** is a list of ligands rank-ordered on their abilities to split the energies of the d orbitals of transition metal ions.

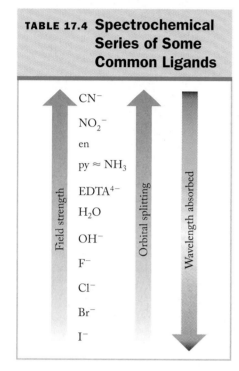

TABLE 17.4 Spectrochemical Series of Some Common Ligands

CN⁻
NO₂⁻
en
py ≈ NH₃
EDTA⁴⁻
H₂O
OH⁻
F⁻
Cl⁻
Br⁻
I⁻

Field strength — Orbital splitting — Wavelength absorbed

FIGURE 17.17 (a) Low-spin and high-spin complexes. The ground state of a free Fe^{3+} ion has a degenerate, half-filled set of 3d orbitals. (b) In a strong octahedral field the energy of the 3d orbitals are split enough (Δ_o > electron pairing energy) to produce the low-spin state. (c) A weak octahedral field (Δ_o < electron pairing energy) produces the high-spin state. The Fe^{3+} ions in crystals of magnetite and (d) aquamarine are high spin.

experimental solutions absorb—red, orange, and yellow—are in a sequence of longest-to-shortest wavelengths. Recall from Chapter 7 that the energy of a photon of electromagnetic radiation is inversely proportional to its wavelength. Therefore, the ability of these three ligands to split the energies of the d orbitals of Ni^{2+} ions is en > NH_3 > H_2O. Chemists use the parameter *field strength* to describe the relative abilities of ligands to split the energies of d orbitals in metal ions with which they form complexes. Chemists rank-order ligands on the basis of their field strengths in what is called a **spectrochemical series**. Table 17.4 contains one such series. The ligands at the top of the chart create the strongest fields, those at the bottom create the weakest. As the field strength of the ligand increases, the splitting energy (Δ) increases. Consequently, the strong-field ligands form complexes that absorb shorter wavelengths of light, whereas complexes of weak-field ligands absorb longer wavelengths.

17.9 Magnetism and Spin States

In addition to determining the color of transition metal ions, crystal field splitting can influence their magnetic properties because these properties depend on the number of unpaired electrons in the valence shell d orbitals; the larger this number, the more paramagnetic are the ions. For example, an Fe^{3+} ion has five 3d electrons. In an octahedral field there are two ways to distribute these electrons among the five 3d orbitals. One way, and the one that conforms to Hund's rule, places a single electron in each orbital, leaving them all unpaired. However, when the ion is in a strong octahedral field where the value of Δ_o is relatively large, then all five electrons may go into the three orbitals of lower energy, leaving the upper two orbitals empty. This pattern occurs when the energy needed to pair the electrons in the lower-energy orbitals is less that the energy needed to promote an electron to one of the two higher-energy orbitals. In this configuration, only one electron is left unpaired. The first option with all five 3d electrons unpaired is called *high spin* because the spin on all five electrons is in the same direction and the resulting magnetic field produced by their spins is maximal. When four of the five electrons are paired and only one is left unpaired, the electron configuration is called *low spin* (Figure 17.17). Both configurations are paramagnetic because both have at least one unpaired electron. However, the high-spin state is much more paramagnetic—a property that can be measured with an instrument called a *magnetometer*.

Other transition metal ions can have high-spin and low-spin states, but not all do. Consider, for example, Cr^{3+} ions in an octahedral field. Each ion has only

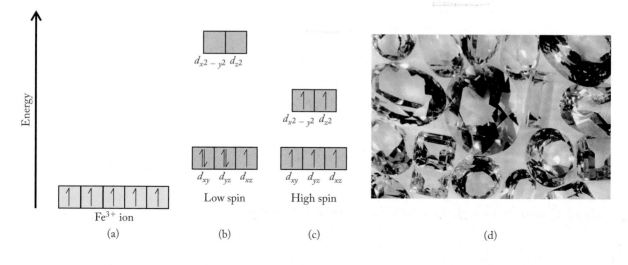

d_{x2-y2} d_{z2}

d_{x2-y2} d_{z2}

d_{xy} d_{yz} d_{xz}
Low spin

d_{xy} d_{yz} d_{xz}
High spin

Energy

Fe^{3+} ion

(a)   (b)   (c)   (d)

three electrons distributed among its 3d orbitals. Each will be unpaired whether the orbitals are split a lot or only a little by the field. Therefore, for Cr^{3+} ions there is only one spin state. Other metal ions with full, or nearly full, sets of d orbitals have only one spin state because there are no, or only one or two, unpaired electrons in these orbitals. The following Sample Exercise illustrates the spin state limitations of having only few or near the maximum number of electrons in a set of d orbitals.

◉◉ CONNECTION Substances made of atoms, ions, or molecules that contain unpaired electrons are paramagnetic (See Section 9.6).

SAMPLE EXERCISE 17.6   **Predicting Spin States**

Not all transition metal ions can have high-spin and low-spin configurations. Determine which of the following ions could have high-spin and low-spin configurations in an octahedral field: (a) Mn^{4+}; (b) Mn^{2+}; (c) Co^{3+}; (d) Cu^{2+}.

COLLECT AND ORGANIZE To determine whether high- and low-spin states are possible, we first need to determine the number of d electrons in each of the ions. Then we need to distribute them among sets of d orbitals split by an octahedral field to see if the ions can have different high- and low-spin states. Mn, Co, and Cu are in groups 7, 9, and 11, respectively, of the periodic table. Also, in an octahedral field, a set of five d orbitals splits into two subsets: a low-energy subset of three orbitals, and a high-energy subset of two orbitals.

ANALYZE The electron configurations of the atoms of the three elements are

Mn: $[Ar]3d^54s^2$ Co: $[Ar]3d^74s^2$ Cu: $[Ar]3d^{10}4s^1$

When they form cations, the atoms of these transition metal ions retain as many of their 3d electrons as possible. Therefore, the numbers of d electrons in the four ions are as follows:

Ion	Mn^{4+}	Mn^{2+}	Co^{3+}	Cu^{2+}
3d electrons	3	5	6	9

SOLVE

a. Mn^{4+}: Putting the electrons into the lowest-energy orbitals available, and keeping them unpaired as much as possible means placing each of the three in its own lower-energy orbital:

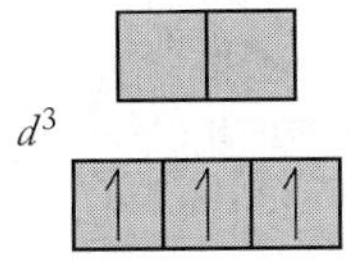

There is no low-spin option for Mn^{4+} if Hund's rule is obeyed.

b. Mn^{2+}: There are two options for distributing five electrons among the five orbitals:

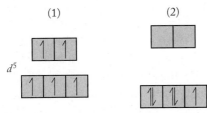

The pattern in (1) represents a high-spin state; the one in (2) is a low-spin state.

c. Co^{3+}: Distributing the six electrons as evenly as possible (1) produces a high-spin state (four unpaired electrons). Placing electrons preferentially in the lower-energy orbitals (2) produces a low-spin state (no unpaired electrons):

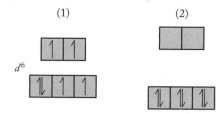

d. Cu^{2+}: The nine $3d$ electrons completely fill the lower-energy orbitals and nearly fill the higher-energy ones. There is no other arrangement possible than the one shown, and so Cu^{2+} has only one spin state:

THINK ABOUT IT In an octahedral field, metal ions with 4, 5, 6, or 7 d electrons can exist in high-spin and low spin states. The magnitude of the crystal field splitting energy, Δ_o, determines which spin state the ion occupies.

Practice Exercise Which, if any, of the spin configurations in Sample Exercise 17.6 are diamagnetic?

Whether a transition metal ion is in a high-spin or a low-spin configuration depends on the relative energy cost of pairing two electrons in a lower-energy d orbital compared with the energy cost of promoting one of them to a higher-energy d orbital (Δ_o). Several factors affect the size of Δ_o. We have already discussed a major one in the context of the spectrochemical series (see Table 17.4): a ligand that is a stronger Lewis base produces a stronger field and larger Δ_o than a ligand that is a weaker Lewis base. For this reason, hydrated metal ions are more likely to be in high-spin states, and metals surrounded by nitrogen-based ligands are more likely to be in low-spin states because the N atoms are stronger Lewis bases; they induce larger d orbital splitting, and so more electrons are paired up in the lower-energy subset of orbitals.

Other factors affecting spin state include the oxidation state of the metal ion. The higher the oxidation number (and ionic charge), the stronger the attraction of the electron pairs on the ligands for the ion. Greater attraction leads to more d orbital–ligand interaction and a larger Δ_o. The size of the ion counts, too. For example, fifth-row transition metal ions are larger than fourth-row ions with the same charge. For the same ligand, the larger the ion, the greater the extension of its d orbitals into the space around its nucleus. This leads to greater interaction with the lone pairs of electrons on a given ligand and to larger Δ_o values.

Our discussion of high- and low-spin states has focused entirely on d orbitals split by octahedral fields. What about spin states in tetrahedral fields? It turns out that most tetrahedral complexes are high spin because tetrahedral fields are created by only four ligands instead of the six in an octahedral field, and so they are weaker. Therefore, d orbital splitting tends to be less in a tetrahedral field; it is usually not large enough to offset the energies associated with pairing electrons in the same orbitals.

CONCEPT TEST

Explain the following: (a) Mn(pyridine)$_6^{2+}$ is a high-spin complex ion, but Mn(CN)$_6^{4-}$ is low spin; (b) Fe(NH$_3$)$_6^{2+}$ is high spin, but Ru(NH$_3$)$_6^{2+}$ is low spin.

17.10 Naming Complex Ions and Coordination Compounds

It is important to be able to name complex ions and coordination compounds so that we can express the identity and oxidation state of the central metal ion, the names and numbers of ligands bound to the inner coordination sphere of the metal ion, the overall charge of the complex ion, and the identities of the counter ions surrounding the complex. To convey all of this information, we need to follow some simple naming rules.

Complexes with a Positive Charge

1. Start with the name(s) of the ligand(s). The names of common ligands appear in Table 17.3. If there is more than one kind of ligand, put their names in alphabetical order.
2. Use the usual prefixes to indicate the number of each type of ligand in the complex ion:

Number of Ligands	Prefix
2	Di-
3	Tri-
4	Tetra-
5	Penta-
6	Hexa-

3. Next write the name of the central metal ion using a Roman numeral to indicate the oxidation state of the transition metal ion.

Following these rules, the names of the following three six-coordinate ions are

Ni(H$_2$O)$_6^{2+}$	Hexaaquanickel(II)
Co(NH$_3$)$_6^{3+}$	Hexaamminecobalt(III)
[Cu(NH$_3$)$_4$(H$_2$O)$_2$]$^{2+}$	Tetraamminediaquacopper(II)

It may seem strange having two *a*'s together in these names, but it is permitted under current naming rules. In the third name, *ammine* comes before the *aqua* because, alphabetically, *am* comes before *aq*. The prefixes are ignored in determining alphabetical order. In these three examples the ligands are all neutral molecules, and so the charge of the complex ion is the same as the charge on the central metal ion, which also matches the oxidation state expressed by the Roman numeral. When the ligands include anions, determining the oxidation state of the central metal ion requires a bit more work, as we shall see in the next examples.

Complexes with a Negative Charge

1. Follow the steps we use to name positively charged complexes.
2. Use an -*ate* ending on the name of the central metal ion to indicate the negative charge of the complex (just as we use -*ate* to end the names of oxoanions). For some metals, the base name changes, too. The two most common examples are iron, which becomes *ferrate*, and copper, which becomes *cuprate*.

Following these rules, the names of the following complex anions are as follows:

$Fe(CN)_6^{3-}$	Hexacyanoferrate(III)
$[Fe(H_2O)(CN)_5]^{3-}$	Aquapentacyanoferrate(II)
$[Al(H_2O)_2(OH)_4]^-$	Diaquatetrahydroxoaluminate

∞∞ **CONNECTION** We do not need a Roman numeral in the name of the aluminum complex because the oxidation number of Al in all its compounds is +3 (see Section 2.8).

In the first two examples we must determine the oxidation state of Fe, that is, the charge on its ion. To do so we start with the overall charge on the complex ion and then take into account the charge on the ligand anions to calculate the charge on the metal ion. For example, the overall charge of the aquapentacyanoferrate(II) ion is 3−. It contains five CN^- ions. To reduce the combined charge of 5− from these cyanide ions to an overall charge of 3−, the charge on Fe must be 2+.

Coordination Compounds

1. If the counter ion of the complex ion is a cation, the cation name goes first, followed by the name of the anionic complex ion.
2. If the counter ion of the complex ion is an anion, the name of cationic complex goes first followed by the name of the anion.

Applying these rules, the names of the following coordination compounds are:

$[Ni(NH_3)_6]Cl_2$	Hexaamminenickel(II) chloride
$K_3Fe(CN)_6$	Potassium hexacyanoferrate(III)
$[Co(NH_3)_5(H_2O)]Br_2$	Pentaammineaquacobalt(II) bromide

A key to naming coordination compounds is to recognize from their formulas that they *are* coordination compounds. For help with this, look for formulas that have:

1. The atomic symbols of one or more metallic elements followed by the formula of one or more of the ligands in Table 17.3.
2. The symbol of a metallic element and one or more of the ligands in Table 17.3, all in brackets, followed by the symbol of an anion.

The following Sample Exercise provides practice in applying these rules.

SAMPLE EXERCISE 17.7 **Naming Coordination Compounds**

Name the following coordination compounds:

(a) $Na_4[Co(CN)_6]$; (b) $[Co(NH_3)_5Cl](NO_2)_2$

COLLECT AND ORGANIZE Our task is to write a name for each compound that unambiguously identifies the composition of its complex ion and its overall composition. The formulas of the complex ions appear in brackets in both compounds.

Cobalt is the central metal ion in both. Cobalt is a transition metal and so we express its oxidation state using Roman numerals. The names of common ligands are given in Table 17.3.

ANALYZE It can be useful to take an inventory of the ligands and counter ions shown in this table.

Compound	Counter Ion	LIGAND			
		Formula	Name	Number	Prefix
$Na_4[Co(CN)_6]$	Na^+	CN^-	Cyano	6	Hexa-
$[Co(NH_3)_5Cl](NO_2)_2$	NO_2^-	NH_3	Ammine	5	Penta-
		Cl^-	Chloro	1	—

The charges of the cobalt ions in the two compounds can be calculated by setting the sum of the charges on all the ions in both compounds equal to zero:

a. Ions: (4 Na^+ ions) + (1 Co ion) + (6 CN^- ions)
Charges: 4+ + x + 6− = 0
$$x = 2+$$

b. Ions: (1 Co ion) + (1 Cl^- ion) + (2 NO_2^- ions)
Charges: x + 1− + 2− = 0
$$x = 3+$$

SOLVE
a. The name of the compound is sodium hexacyanocobaltate(II).
b. The name of the compound is pentaamminechlorocobalt(III) nitrite.

THINK ABOUT IT Naming coordination compounds requires you to (1) distinguish between ligands and counter ions and (2) recall which ligands are neutral and which are anions. Naming a coordination compound correctly shows that you understand these features of their structures.

Practice Exercise Identify the ligands and counter ions in the following compounds and name each compound:

(a) $[Zn(NH_3)_4]Cl_2$; (b) $[Co(NH_3)_4(H_2O)_2](NO_3)_3$

17.11 Isomerism in Coordination Compounds

We introduced the concept of structural and geometric isomerism in Chapter 12. To review:

- Structural isomers are compounds that have the same chemical formula but different arrangements of the bonds in their molecules. For example, these two hydrocarbons are structural isomers because they have the same formula (C_4H_8) but the C=C double bonds are in different locations in their structures:

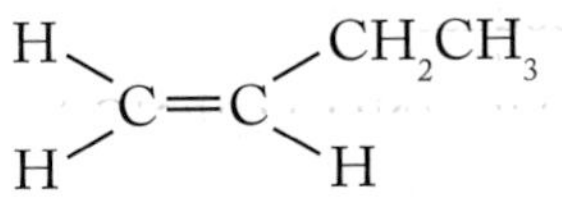

1-Butene 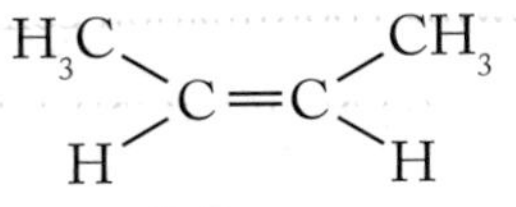   2-Butene

- Geometric isomers, also called *stereoisomers*, are compounds with the same formulas *and* the same bonding pattern, but the orientation of the groups connected by those bonds is different. For example, these two hydrocarbons are geometric isomers:

trans-2-Butene   *cis*-2-Butene

CONNECTION The rigidity of double bonds is described in Section 12.4.

In *trans*-2-butene the −CH$_3$ groups are on opposite sides of the C=C double bonds, but in *cis*-2-butene they are on the same side. Remember that C=C bonds are rigid so that the ends of molecules cannot rotate around the C=C axis as they can around C−C single bonds.

Geometric Isomers

Both structural and geometric isomers occur in complex ions, but geometric isomerism is the focus of our discussion here. Let's begin with the square planar Pt^{2+} coordination compound that has the formula Pt(NH$_3$)$_2$Cl$_2$. Both molecules of ammonia and the two chloride ions are coordinately bonded to the inner coordination sphere of the Pt^{2+} ion, so the name of the compound is the name of the complex: diamminedichloroplatinum(II). There are no counter ions because there is no net charge on the complex.

Two geometric isomers of diamminedichloroplatinum(II) are possible because there are two ways to orient the two pairs of ligands around the square plane of the complex: each pair could be at the same side of the structure, or they could be at opposite corners, as shown in Figure 17.18(a) and (b), respectively. The isomer with each pair on the same side is called *cis*-diamminedichloroplatinum(II), and the isomer with the pairs in opposite corners is *trans*-diamminedichloroplatinum(II). Note that *cis* and *trans* have the same meaning in coordination chemistry as in organic nomenclature.

To illustrate the importance of geometric isomerism, consider this: *cis*-diamminedichloroplatinum(II) is a widely used anticancer drug with the common name *cisplatin*, but the *trans*- isomer is ineffective in fighting cancer. The therapeutic power of cisplatin comes from changes in its structure when it is administered to a cancer patient. The first change is replacement of a chloro ligand (with a 1− charge) with a neutral water molecule, forming the *cis*-diammineaquachloroplatinum(II) ion with a 1+ charge. This compound is very reactive and binds to a nitrogen atom on DNA. Once bound, it readily undergoes another ligand exchange reaction in which the second chloride ion (with a 1− charge) is replaced by another neutral water molecule, forming another ion that is now directly attached to the DNA strand. This reactive ion forms another bond to a nitrogen atom on the DNA strand and causes distortion in the structure of the molecule. This two-point attack on the cancer cell's DNA has a remarkable biological effect: it prevents the DNA from replicating, which causes the cells to die. The *trans* isomer can also undergo ligand replacement of a chloride ion for water molecules, but the resultant ion cannot form links between nitrogen atoms on a DNA strand and is inactive as an anticancer agent.

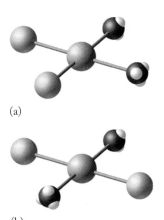

(a)

(b)

FIGURE 17.18 There are two ways to orient the two Cl$^-$ ions and the two NH$_3$ molecules in the square planar coordination compound Pt(NH$_3$)$_2$Cl$_2$: (a) each of the two pairs of ligands could be on the same side of the square, as they are in *cis*-diamminedichloroplatinum(II), or (b) the pairs could be in opposite corners, as in *trans*-diamminedichloroplatinum(II).

CONCEPT TEST

How many isomers does the square-planar coordination compound Pt(NH$_3$)$_3$Cl have?

Geometric isomerism is also possible in six-coordinate octahedral complexes containing more than one type of ligand. For example, there are two possible geometric isomers of the coordination compound $[Co(NH_3)_4Cl_2]Cl$ as shown in the structures in Figure 17.19. The two chloro ligands in the $[Co(NH_3)_4Cl_2]^+$ complex ion are either on the same side of the complex with a 90° Cl—Co—Cl bond angle, or they are across from each other with the metal atom in between. When the two chloro ligands are on the same side, the structure is called *cis*. The isomer with the chloro ligands on opposite sides of the complex is called *trans*. The full names of the two isomeric compounds are *cis*-tetraamminedichlorocobalt(III) chloride, which is violet, and *trans*-tetraamminedichlorocobalt(III) chloride, which is a lovely shade of green.

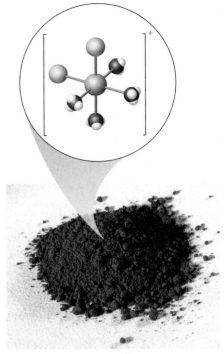

(a) *cis*-tetraamminedichlorocobalt(III) chloride

SAMPLE EXERCISE 17.8   **Identifying Geometric Isomers of Coordination Compounds**

Sketch the structures and name the geometric isomers of $Ni(NH_3)_4Cl_2$.

COLLECT AND ORGANIZE The formula contains no brackets, so there are no counter ions. This means that all four ammonia molecules and two chloride ions must occupy inner coordination sphere bonding sites. Both ligands are monodentate.

ANALYZE The 1− charges on the two chloride ions must be balanced by a charge of 2+ on the nickel ion. A total of six ligands means that the complex must be octahedral.

SOLVE There are two ways to orient the two chloride ions: opposite each other with a Cl—Ni—Cl bond angle of 180° or on the same side of the octahedron with a Cl—Ni—Cl bond angle of 90°:

$$
\begin{array}{cc}
\text{Cl} \quad \text{NH}_3 & \text{H}_3\text{N} \quad \text{Cl}\\
\text{H}_3\text{N}-\text{Ni}-\text{NH}_3 & \text{H}_3\text{N}-\text{Ni}-\text{Cl}\\
\text{H}_3\text{N} \quad \text{Cl} & \text{H}_3\text{N} \quad \text{NH}_3
\end{array}
$$

The first isomer is *trans*-tetraamminedichloronickel(II); the second is *cis*-tetraamminedichloronickel(II).

THINK ABOUT IT Although we could draw other tetraamminedichloronickel(II) complex ions that might not initially look like these two isomers, if we consider their three-dimensional structures, we would see that each of them is the same as one of these two.

Practice Exercise Sketch the geometric isomers of $[CoBr_2(en)(NH_3)_2]^+$ and name them.

(b) *trans*-tetraamminedichlorocobalt(III) chloride

FIGURE 17.19 What a Chemist Sees There are two geometric isomers of the coordination compound with the formula $[Co(NH_3)_4Cl_2]Cl$: (a) *cis*-tetraamminedichlorocobalt(III) chloride and (b) *trans*-tetraamminedichlorocobalt(III) chloride.

Enantiomers

Another kind of geometric isomerism is possible in complex ions and coordination compounds. Consider the octahedral Co(III) complex ion containing two ethylenediamine (en) molecules and two chloride ions. There are two ways to arrange the chloride ions: on adjacent bonding sites in a *cis* isomer or on opposite sides of the octahedron in a *trans* isomer. The *cis* option is shown in the molecular models in Figure 17.20. Note that the name of the isomer includes a prefix, *bis*, that we have not encountered before. In naming complexes with polydentate

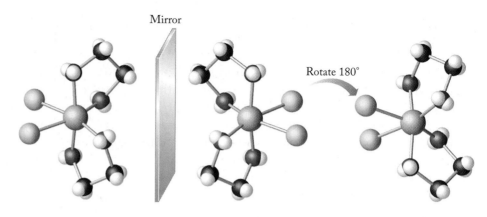

FIGURE 17.20 The complex ion *cis*-dichlorobis(ethylenediamine)cobalt(III) is chiral, which means that the mirror image of one of its chiral forms, or enantiomers, is not superimposable on the original. To illustrate this point we have rotated the mirror image 180° in the structure on the right so that it looks much like the original. However, note that the top ethylenediamine ligand is located behind the plane of this page in the original, but in front of the plane in the rotated mirror image. Thus, the mirror images are not superimposable.

ligands we use *bis-* in place of *di-* to indicate that two molecules of the multitoothed ligand are present in the complex.

Now let's look more closely at the *cis* isomer—"through a looking glass" as it were. Figure 17.20 shows that the *cis*-Co(en)$_2$Cl$_2^+$ ion has a mirror image that is not identical to the original. The difference is demonstrated by the fact that there is no way to rotate the mirror image so that the atoms in its structure align exactly with those in the original. In other words, the two structures are not *superimposable*. You may wish to build models of the complex and its mirror image to satisfy yourself that this is so. Two such nonsuperimposable mirror images are said to be **enantiomers**, and such an ion or molecule is said to be **chiral**, from the Greek word *cheir*, meaning "hand" (the reflection of a left hand in a mirror looks like a right hand—but they are not identical, as anyone knows who tries to put their left hand into a right-hand glove). We will return to this concept of chirality and its profound implications in biochemistry and the life sciences in Chapter 21.

CONCEPT TEST

Would four different ligands arranged in a square planar geometry produce a chiral complex ion? What about four different ligands in a tetrahedral geometry?

17.12 Metal Complexes in Biomolecules

At the beginning of this chapter we noted some of the important roles metals play in human health, and the importance of the presence of these metals in foods in chemical forms that can be readily absorbed by the body during digestion. In this chapter we have discussed how polydentate ligands such as EDTA can surround metal ions and prevent them from reacting with other species in solution. In this section we explore some biological polydentate ligands that involve metal ions in processes essential to nutrition and good health.

Let's begin with the base of our food chain and the process we know as photosynthesis. Green plants can harness solar energy because they contain large biomolecules we call collectively *chlorophyll*. All molecules of chlorophyll have within them ring-shaped polydentate ligands called *chlorins* (Figure 17.21a). The structure of chlorin is very similar to that of another type of polydentate ligand found in biological systems called a **porphyrin** (Figure 17.21b). Note that the structures of the two differ by the presence of a C=C double bond in porphyrin (highlighted in red in Figure 17.21b) that is a single bond in chlorin. Their structures make them members of a class of compounds known generally as **macrocyclic ligands**. (A *macrocycle* is, literally, a *big ring*.)

Enantiomers are geometric isomers that are nonsuperimposable mirror images of each other. Enantiomeric ions and molecules are called **chiral**.

Porphyrin is a type of tetradentate macrocyclic ligand.

A **macrocyclic ligand** is a ring containing multiple electron-pair donors that bind to a metal ion.

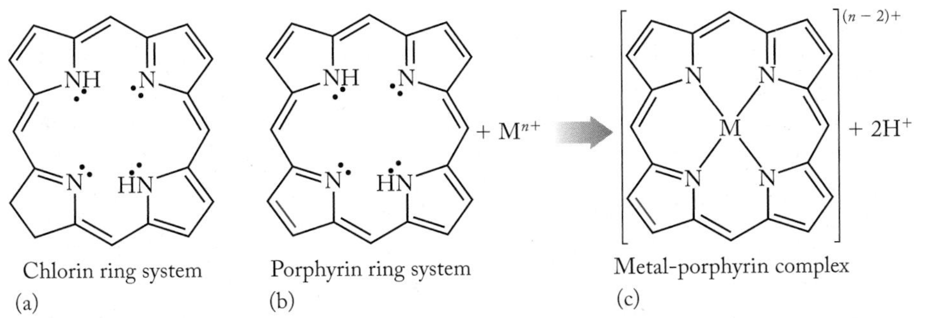

Chlorin ring system
(a)

Porphyrin ring system
(b)

$+ M^{n+} \longrightarrow$

Metal-porphyrin complex
(c)

$\left[\right]^{(n-2)+} + 2H^+$

FIGURE 17.21 Chlorin (a) and porphyrin (b) rings are biologically important tetradentate ligands that have similar core structures: the principal difference is a C=C double bond in the porphyrin structure [highlighted in red in (b)] that is a single bond in chlorin rings. The innermost atoms in the rings are four nitrogen atoms whose lone pairs of electrons are pointed toward the center of the rings. When the two NH groups ionize, all four N atoms are free to form coordinate bonds with a metal ion, as shown with the porphyrin ring in (c). Metal ions at the centers of these rings occupy four sites in the equatorial planes of octahedral fields. The two remaining sites in the axial positions above and below the plane can be occupied by other ligands.

Two of the four nitrogen atoms in the porphyrin and chlorin rings are sp^3 hybridized and are also bound to hydrogen atoms, whereas the other two are sp^2 hybridized with no hydrogen atoms (see Figure 17.21). When a ring coordinates with a metal ion, the two hydrogen atoms ionize, giving the ring a charge of $2-$. In the ionized structure there are four lone pairs of electrons oriented toward the center of the ring. They can occupy the four coordination sites in the equatorial plane of a six-coordinate central metal ion or all four coordination sites in a square planar complex. Octahedral (six-coordinate) metal ions surrounded by the ring still have their two axial sites available for bonding to other ligands. Depending on the charge of the coordinated metal ion, the metal–ring complex may be ionic or neutral.

Porphyrin and chlorin rings are widespread in nature and play many different biochemical roles. Their particular chemical and physical properties depend on

1. The identity of the coordinated metal ion.
2. The groups in the axial coordination sites.
3. The number and identity of organic groups attached to the outside of the rings.

The chlorin ring in chlorophyll has a Mg^{2+} ion at its center (Figure 17.22). Delocalized p electrons in the conjugated double bonds within and around the ring stabilize it and give chlorophyll the ability to absorb wavelengths of red and blue-violet radiation (Figure 17.23). Because chlorophyll absorbs these colors, the color of chlorophyll and that of plants that contain it corresponds to the portion

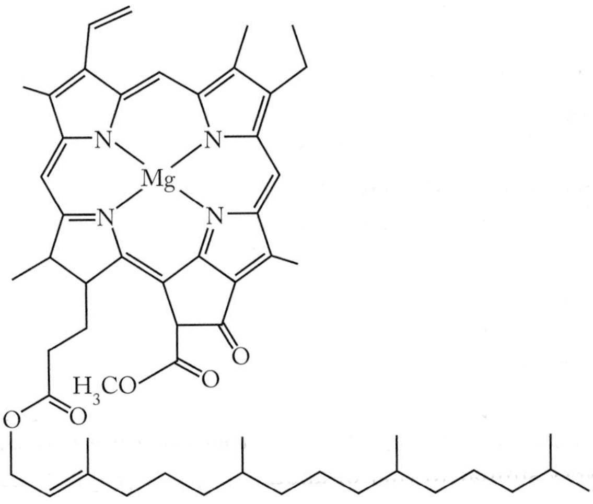

Chlorophyll *a*

FIGURE 17.22 Chlorophyll absorbs sunlight and, through a series of reactions, converts sunlight (electromagnetic energy), carbon dioxide, and water into chemical energy stored in the bonds of carbohydrates. All forms of chlorophyll, including chlorophyll *a* shown here, have a Mg^{2+} ion in a chlorin ring within their molecular structures.

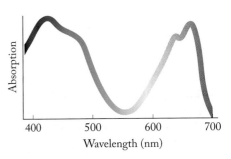

FIGURE 17.23 This is the combined absorption spectrum of chlorophyll and other pigments in a typical green leaf. These pigments absorb most of the visible radiation emitted by the sun except the yellow-green color of the leaves themselves.

FIGURE 17.24 The colors of fall in southern Vermont are characterized by the reds, yellows, and gold colors of leaves that have lost their green pigments and are about to fall to Earth.

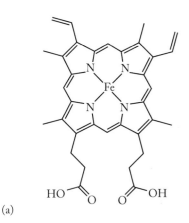

(a)

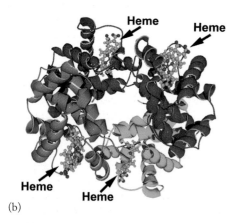

(b)

FIGURE 17.25 Heme. (a) In this heme complex the four nitrogen atoms in a porphyrin ring occupy equatorial positions in an octahedral complex with an Fe^{2+} ion. (b) Below the ring a fifth bond is formed between the Fe^{2+} ion and a lone pair of electrons on another nitrogen atom in the protein hemoglobin. The sixth ligand above the porphyrin ring is a loosely bonded molecule of O_2 as oxygenated hemoglobin leaves the lungs.

of the visible spectrum that is not absorbed, namely, green. In temperate climates chlorophyll is lost from the leaves of trees such as maples and oaks at the end of each growing season, which makes visible the colors of other pigments in the beautiful leaves of autumn (Figure 17.24).

An important porphyrin complex has an Fe^{2+} ion in its center. This type of porphyrin is called a heme group (Figure 17.25). This group of atoms enables the protein called hemoglobin to transport O_2 in the blood and a related protein, myoglobin, to transport O_2 in muscle tissues. The four nitrogen atoms in the porphyrin ring of a heme group occupy equatorial positions in an octahedral complex of an Fe^{2+} ion. Below the ring a fifth bond is formed between the Fe^{2+} ion and a lone pair of electrons on another nitrogen atom in the protein. The sixth ligand, located above the porphyrin ring, is typically a molecule of O_2, as in the oxygenated forms of hemoglobin in blood leaving the lungs.

Each atom in an oxygen molecule has a lone pair of electrons that it can donate to the iron in heme. This coordinate covalent bond is strong enough to enable oxygen to be transported from the lungs to the cells in our bodies, but weak enough that it can be easily broken when the oxygen is delivered to a cell. Other ligands of similar size can bind to the Fe^{2+} ion in heme, and problems arise when some of them do. Carbon monoxide is such a ligand. Similar in size to oxygen, it easily fits into the sixth binding site. Unfortunately, it binds about 200 times more strongly than O_2. If a person breathes air containing carbon monoxide, CO prevents O_2 from being taken up by blood flowing through the lungs.

Another class of proteins, called cytochromes, also contain heme groups (Figure 17.26). Cytochromes mediate oxidation and reduction processes connected with energy production in living cells. The heme group in cytochrome proteins serves as a conveyer of electrons as the half-reaction

$$Fe^{3+} + e^- \rightleftharpoons Fe^{2+}$$

rapidly and reversibly consumes or releases electrons needed in the biochemical reactions that sustain life. There are many kinds of cytochrome proteins with different substituents on the porphyrin rings and different axial ligands. Each of these variables influences the function of the metal–porphyrin complex.

The association of metal ions with ligands results in complexes that vary in properties, appearance, and consequently in their use. The ions and neutral mol-

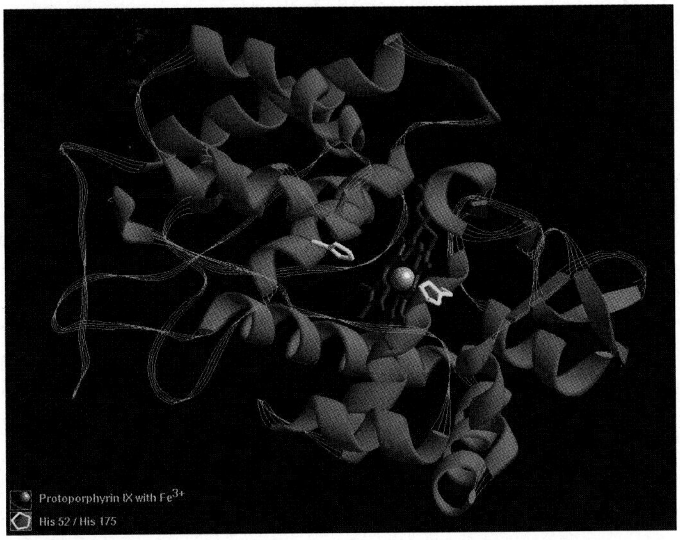

FIGURE 17.26 The structures of cytochrome proteins, such as cytochrome *c* (shown here), include heme complexes that mediate energy production and redox reactions in living cells. Different cytochromes have a variety of different ligands occupying the sixth octahedral coordination site, and different groups are attached to the porphyrin ring.

ecules that interact with metal ions in forming these complexes make the difference between a metal ion fulfilling an essential need in a living system or exerting a toxic effect and killing the organism. The same can be said for metals that act as catalysts in inorganic systems; ligands play an important role in determining the properties of the metal complex.

Summary

SECTION 17.1 A **Lewis base** is a substance that donates pairs of electrons to a **Lewis acid** during a chemical reaction. This donated pair forms a new covalent bond. In some of these reactions, such as that between CaO and SO_2, other bonds must break to accommodate the new one.

SECTION 17.2 Transition metal ions form **complex ions** when **ligands** donate pairs of electrons to empty valence-shell orbitals on the metal ion, thereby forming **coordinate bonds**. The number of coordinate bonds in a complex defines the **coordination number** of the metal ion. The most common coordination numbers are 6, 4, and 2. The metal ion in a complex ion acts as a Lewis acid (electron-pair acceptor), and the ligands act as Lewis bases (electron-pair donors). Compounds that contain complex ions are called **coordination compounds**; the net charges on complex ions are balanced by charges from **counter ions**. Ligands occupy binding sites in the **inner coordination sphere** of a metal ion. Counter ions surround complex ions in their compounds and dissociate from the complex ions when the compounds dissolve.

SECTION 17.3 The stability of complex ions is expressed mathematically by their **formation constants (K_f)**, which can be used to calculate the equilibrium concentration of free metal ions, $M^{n+}(aq)$, in solutions of their complex ions.

SECTION 17.4 Hydrated metal ions with charges of 2+ or greater can act as Brønsted–Lowry acids, which is why solutions of their soluble salts are acidic. Most transition metal ions have limited solubility in strongly basic solutions. An exception is Cr(III), which forms an anionic complex with the formula $Cr(OH)_4^-$.

SECTION 17.5 The solubility of slightly soluble compounds is described by their K_{sp} or **solubility product**, which is the value of the equilibrium constant for their dissolution.

SECTION 17.6 A molecule or ion of a **monodentate ligand** donates only one pair of electrons in a complex ion; **polydentate ligands** donate more than one in a process called **chelation**. EDTA is a particularly effective chelating and **sequestering agent**. It prevents metal ions in solution from reacting with other substances.

SECTION 17.7 Polydentate ligands are particularly effective in forming complex ions. This phenomenon is called the **chelate effect** and can be explained by the increase in entropy that accompanies the chelation process.

SECTION 17.8 The colors of transition metals in crystalline solids and in solutions can be explained by the interactions between electrons in different *d* orbitals and the lone pairs of electrons on surrounding ligands. Different interactions create **crystal field splitting** of the energies (labeled Δ) of the *d* orbitals. A **spectrochemical series** ranks ligands on the basis of their field strengths and the wavelengths of electromagnetic radiation absorbed by their complex ions; the stronger the field the ligand produces, the shorter the wavelength of radiation the complex absorbs. The color of a complex ion in solution or in a crystalline solid is the complement of the color(s) it absorbs.

SECTION 17.9 Strong repulsions and large values of Δ can lead to electron pairing in lower-energy orbitals and an electron configuration called a **low-spin state**. Metals and their ions are less paramagnetic in low-spin states than when their *d* electrons are evenly distributed across all the *d* orbitals in their valence shell—a configuration called a **high-spin state**.

SECTION 17.10 The names of complex ions and coordinate compounds provide information about the identities and numbers of ligands, the identity and oxidation state of the central metal ion, and the identity of any counter ions whose charges balance the net charge of the complex ion.

SECTION 17.11 Complex metal ions containing more than one type of ligand may form geometric isomers. For example, when one type occupies two adjacent corners of a square planar complex, the complex is a *cis* isomer; when the same ligand occupies opposite corners, it is a *trans* isomer. The chemical and biochemical properties of coordination compounds that are geometric isomers can be very dif-

ferent, as illustrated by the difference between the chemotherapeutic power of cisplatin and that of its *trans* isomer.

SECTION 17.12 Metal ion complexes play key roles in many biochemical processes. Among these are photosynthesis, which is mediated by chlorophyll, in which tetradentate chlorin rings coordinately bond to central Mg^{2+} ions; oxygen transport in the body based on the reversible bonding of O_2 molecules to heme groups of Fe^{2+} ions in **porphyrin** rings; and energy production in cells, which is mediated by molecules called cytochromes that contain metals in different oxidation states.

Problem-Solving Summary

TYPE OF PROBLEM	CONCEPTS AND EQUATIONS	SAMPLE EXERCISES
Identifying Lewis acids and bases	Determine which reactant donates a pair of electrons (the base) and which one accepts them (the acid).	17.1
Calculating the free metal ion concentration in a solution of one of its complex ions	Set up an ICE table based on formation of the complex. Let x = the concentration of M^{n+} that *does not* form the complex. If the value of K_f is large (it usually is), you may assume x is much less than the other concentrations.	17.2
Calculating solubility using K_{sp}	Use the mass action expression describing the solubility of a slightly soluble compound to determine concentrations in solution.	17.3 and 17.4
Identifying the donor groups in a molecule or structure	Look for functional groups that have atoms with lone pairs of electrons, such as $-NH_2$, $-NH-$, and $-COO^-$. Multiple coordinate bonds are possible if the groups can form stable ring structures with metal ions.	17.5
Predicting spin states	Sketch a *d*-orbital diagram based on crystal field splitting. Fill the lowest-energy orbitals with valence electrons. If more of them are paired than if they had been distributed evenly over all five *d* orbitals, multiple spin states are possible.	17.6
Naming coordination compounds	Follow the naming rules in Section 17.10.	17.7
Identifying geometric isomers of coordination compounds	Isomers are possible only if there are at least two types of ligands. If ligands of one type are all on the same side of the complex ion, it is a *cis* isomer, if on opposite sides, it is a *trans* isomer.	17.8

Visual Problems

17.1. The chlorides of two of the four highlighted elements in Figure P17.1 are colored. Which ones?

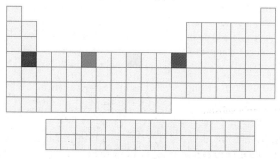

FIGURE P17.1

17.2. Which of the highlighted transition metals in Figure P17.2 form M^{2+} cations that cannot have high-spin and low-spin states?

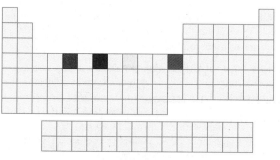

FIGURE P17.2

17.3. Which of the highlighted transition metals in Figure P17.3 have M^{2+} cations that form colorless tetrahedral complex ions?

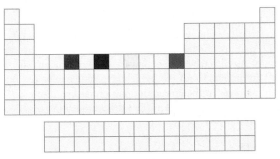

FIGURE P17.3

17.4. Smoky quartz has distinctive lavender and purple colors due to the presence of manganese impurities in crystals of silicon dioxide. Which of the orbital diagrams in Figure P17.4 best describes the Mn^{2+} ion in a tetrahedral field?

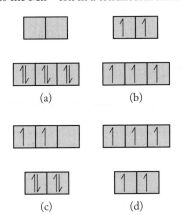

FIGURE P17.4

17.5. Chelation Therapy The compound with the structure shown in Figure P17.5 is widely used in chelation therapy to remove excessive lead or mercury in patients exposed to these metals. (a) How many electron-pair donor groups ("teeth") does the sequestering agent have in this structure, and (b) how many does it have when the carboxylic acid groups are ionized?

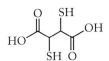

FIGURE P17.5

17.6. Chelation Therapy The compound with the structure shown in Figure P17.6 has been used to treat people exposed to plutonium, americium, and other actinide metal ions. (a) How many donor groups does the sequestering agent have in this structure, and (b) how many does it have when the carboxylic acid groups are ionized?

FIGURE P17.6

17.7. The three beakers in Figure P17.7 contain solutions of polyatomic Co(III) ions, $[CoF_6]^{3-}$, $[Co(NH_3)_6]^{3+}$, and $[Co(CN)_6]^{3-}$. Based on the colors of the three solutions, which compound is present in each of the beakers?

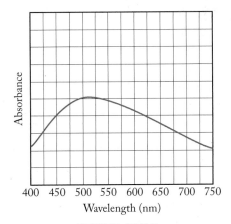

FIGURE P17.7

17.8. Figure P17.8 shows the absorption spectrum of a solution of $Ti(H_2O)_6^{3+}$. What color is the solution?

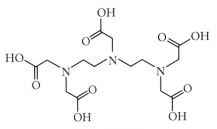

FIGURE P17.8

Questions and Problems

Lewis Acids and Bases

CONCEPT REVIEW

17.9. Can a substance be a Lewis base and not a Brønsted–Lowry base? Explain why or why not.

17.10. Can a substance be a Brønsted–Lowry acid and not be a Lewis acid? Explain why or why not.

17.11. Why is BF_3 a Lewis acid but not a Brønsted–Lowry acid?

17.12. Would you expect NH_3 or H_2O to be a stronger Lewis base? Explain your selection.

PROBLEMS

17.13. Use Lewis structures to show how electron pairs move and bonds form in the following reaction, and identify the Lewis acid and Lewis base.

$$BF_3(g) + F^-(aq) \rightarrow BF_4^-(aq)$$

17.14. Use Lewis structures to show how electron pairs move and bonds form and break in this reaction, and identify the Lewis acid and Lewis base.

$$MgO(s) + CO_2(g) \rightarrow MgCO_3(s)$$

17.15. Use Lewis structures to show how electron pairs move and bonds form and break in this reaction, and identify the Lewis acid and Lewis base.

$$CO_2(g) + H_2O(\ell) \rightarrow H_2CO_3(aq)$$

17.16. Use Lewis structures to show how electron pairs move and bonds form and break in this reaction, and identify the Lewis acid and Lewis base.

$$SO_3(g) + H_2O(\ell) \rightarrow H_2SO_4(aq)$$

17.17. Use Lewis structures to show how electron pairs move and bonds form and break in this reaction, and identify the Lewis acid and Lewis base.

$$B(OH)_3(aq) + H_2O(\ell) \rightarrow B(OH)_4^-(aq) + H^+(aq)$$

*17.18.** Use Lewis structures to show how electron pairs move and bonds form and break in this reaction, and identify the Lewis acid and Lewis base. **NOTE:** $HSbF_6$ is an ionic compound and one of the strongest Brønsted–Lowry acids known.

$$SbF_5(s) + HF(g) \rightarrow HSbF_6(s)$$

Complex Ions

CONCEPT REVIEW

17.19. When NaCl dissolves in water, which molecules or ions occupy the inner coordination sphere around the Na^+ ions?

17.20. When $CrCl_3$ dissolves in water, which of the following species are nearest the Cr^{3+} ions: (a) other Cr^{3+} ions, (b) Cl^- ions, (c) molecules of H_2O with their O atoms closest to the Cr^{3+} ions, (d) molecules of H_2O with their H atoms closest to the Cr^{3+} ions?

17.21. When $Ni(NO_3)_2$ dissolves in water, what molecules or ions occupy the inner coordination sphere around the Ni^{2+} ions?

17.22. When $[Ni(NH_3)_6]Cl_2$ dissolves in water, which molecules or ions occupy the inner coordination sphere around the Ni^{2+} ions?

17.23. Which ion is the counter ion in the coordination compound $Na_2[Zn(CN)_4]$?

17.24. Which ion is the counter ion in the coordination compound $[Co(NH_3)_4Cl_2]NO_3$?

Complex-Ion Equilibria

CONCEPT REVIEW

17.25. Trace Metal Toxicity Dissolved concentrations of Cu^{2+} as low 10^{-6} M are toxic to phytoplankton (microscopic algae). However, a solution that is 10^{-3} M in $Cu(NO_3)_2$ and 10^{-3} M in EDTA is not toxic. Why?

17.26. When a strong base is added to a solution of $CuSO_4$, which is pale blue, a precipitate forms and the solution above the precipitate is colorless. When ammonia is added, the precipitate dissolves and the solution turns a deep navy blue. Use appropriate chemical equations to explain why the observed changes occurred.

17.27. A lab technician cleaning glassware that contains residues of AgCl washes the glassware with an aqueous solution of ammonia. The AgCl, which is insoluble in water, rapidly dissolves in the ammonia solution. Why?

17.28. The procedure used in the previous question dissolves AgCl, but not AgI. Why?

PROBLEMS

NOTE: Appendix 5 contains formation constant (K_f) values that may be useful in solving the following problems.

17.29. One millimole of $Ni(NO_3)_2$ dissolves in 250 mL of a solution that is 0.500 M in ammonia. (a) What is the concentration of $Ni(NO_3)_2$ in the solution? (b) What is the concentration of $Ni^{2+}(aq)$ in the solution?

17.30. A 1.00 L solution contains 3.00×10^{-4} M $Cu(NO_3)_2$ and 1.00×10^{-3} M ethylenediamine. What is the concentration of $Cu^{2+}(aq)$ in the solution?

17.31. Suppose a 500 mL solution contains 1.00 millimoles of $Co(NO_3)_2$, 100 millimoles of NH_3, and 100 millimoles of ethylenediamine. What is the concentration of $Co^{2+}(aq)$ in the solution?

*17.32.** To a 250 mL volumetric flask are added 1.00 mL volumes of three solutions: 0.0100 M $AgNO_3$, 0.100 M NaBr, and 0.100 M NaCN. The mixture is diluted with deionized water to the mark and shaken vigorously. Are the contents of the flask cloudy or clear? Support your answer with the appropriate calculations. (*Hint:* the K_{sp} of AgBr is 5.0×10^{-13}.)

Hydrated Metal Ions as Acids

CONCEPT REVIEW

17.33. Which, if any, aqueous solutions of the following compounds are acidic: (a) $CaCl_2$; (b) $CrCl_3$; (c) NaCl; (d) $FeCl_2$.

17.34. If 0.1 M aqueous solutions of each of these compounds were prepared, which one has the lowest pH? (a) $BaCl_2$; (b) $NiCl_2$; (c) KCl; (d) $TiCl_4$

17.35. When ozone is bubbled through a solution of iron(II) nitrate, dissolved Fe(II) ions are oxidized to Fe(III) ions. How does the oxidation process affect the pH of the solution?

17.36. As an aqueous solution of KOH is slowly added to a stirred solution of $AlCl_3$, the mixture becomes cloudy, but then clears when more KOH is added. (a) Explain the chemical changes responsible for the changes in the appearance of the mixture. (b) Would you expect to observe the same changes if KOH were added to a solution of $FeCl_3$? Explain why or why not.

17.37. Chromium(III) hydroxide is amphoteric. Write chemical equations showing how an aqueous suspension of this compound reacts to the addition of a strong acid and a strong base.

17.38. Zinc hydroxide is amphoteric. Write chemical equations showing how an aqueous suspension of this compound reacts to the addition of a strong acid and a strong base.

17.39. Refining Aluminum To remove impurities such as calcium and magnesium carbonates and Fe(III) oxides from aluminum ore (which is mostly Al_2O_3), the ore is treated with a strongly basic solution. In this treatment Al(III) dissolves but the other metal ions do not. Why?

*17.40. The same quantity of $FeCl_3$ dissolves in 1 M aqueous solutions of each of these acids: HNO_3, HNO_2, H_2SO_3, and CH_3COOH. Is the concentration of $Fe^{3+}(aq)$ the same in all four solutions? Explain why or why not.

PROBLEMS

17.41. What is the pH of 0.50 M $Al(NO_3)_3$?

17.42. What is the pH of 0.25 M $CrCl_3$?

17.43. What is the pH of 0.100 M $Fe(NO_3)_3$?

17.44. What is the pH of 1.00 M $Cu(NO_3)_2$?

17.45. Sketch the titration curve (pH versus volume of 0.50 M NaOH) for a 25 mL sample of 0.5 M $FeCl_3$.

17.46. Sketch the titration curve that results from the addition of 0.50 M NaOH to a sample containing 0.5 M $KFeSO_4$.

Solubilities of Ionic Compounds and K_{sp}

CONCEPT REVIEW

17.47. What is the difference between *solubility* and *solubility product*?

17.48. Describe how the common-ion effect limits the dissolution of a sparingly soluble ionic compound.

17.49. Which of the following cations will precipitate first as a carbonate mineral from an equimolar solution of Mg^{2+}, Ca^{2+}, and Sr^{2+}?

17.50. If the solubility of a compound increases with increasing temperature, does K_{sp} increase or decrease?

17.51. The K_{sp} of strontium sulfate increases from 2.8×10^{-7} at 37°C to 3.8×10^{-7} at 77°C. Is the dissolution of strontium sulfate endothermic or exothermic?

17.52. How will adding concentrated NaOH(*aq*) affect the solubility of an Al(III) salt?

17.53. Chemistry of Tooth Decay Tooth enamel is composed of a mineral known as hydroxyapatite with the formula $Ca_5(PO_4)_3OH$. Explain why tooth enamel can be eroded by acidic substances released by bacteria growing in the mouth.

17.54. **Fluoride and Dental Hygiene** Fluoride ions in drinking water and toothpaste convert hydroxyapatite in tooth enamel into fluorapatite:

$$Ca_5(PO_4)_3OH(s) + F^-(aq) \rightleftharpoons Ca_5(PO_4)_3F(s) + OH^-(aq)$$

Why is fluorapatite less susceptible than hydroxyapatite to erosion by acids?

PROBLEMS

17.55. At a particular temperature the value of $[Ba^{2+}]$ in a saturated solution of barium sulfate is 1.04×10^{-5} M. Starting with this information, calculate the K_{sp} value of barium sulfate at this temperature.

17.56. Suppose a saturated solution of barium fluoride contains 1.5×10^{-2} M F^-. What is the K_{sp} value of BaF_2?

17.57. What are the equilibrium concentrations of Cu^+ and Cl^- in a saturated solution of copper(I) chloride if $K_{sp} = 1.02 \times 10^{-6}$?

17.58. What are the equilibrium concentrations of Pb^{2+} and F^- in a saturated solution of lead fluoride if the K_{sp} value of PbF_2 is 3.2×10^{-8}?

17.59. What is the solubility of calcite ($CaCO_3$) in grams per milliliter at a temperature at which its $K_{sp} = 9.9 \times 10^{-9}$?

17.60. What is the solubility of silver iodide in grams per milliliter at a temperature at which its $K_{sp} = 1.50 \times 10^{-16}$.

17.61. What is the pH at 25°C of a saturated solution of silver hydroxide, given $K_{sp} = 1.52 \times 10^{-8}$?

17.62. **pH of Milk of Magnesia** What is the pH of a saturated solution of magnesium hydroxide (the active ingredient in the antacid milk of magnesia)?

17.63. Suppose you have 100 mL of each of the following solutions. In which will the most $CaCO_3$ dissolve? (a) 0.1 M NaCl; (b) 0.1 M Na_2CO_3; (c) 0.1 M NaOH; (d) 0.1 M HCl

17.64. In which of the following solutions will CaF_2 be most soluble? (a) 0.010 M $Ca(NO_3)_2$; (b) 0.01 M NaF; (c) 0.001 M NaF; (d) 0.10 M $Ca(NO_3)_2$

17.65. Composition of Seawater The average concentration of sulfate in surface seawater is about 0.028 M. The average concentration of Sr^{2+} is 9×10^{-5} M. If the K_{sp} value of strontium sulfate is 3.4×10^{-7}, is the concentration of strontium in the sea probably controlled by the insolubility of its sulfate salt?

17.66. **Fertilizing the Sea to Combat Global Warming** Some scientists have proposed adding Fe(III) compounds to large expanses of the open ocean to promote the growth of phytoplankton that would in turn remove CO_2 from the atmosphere through photosynthesis. The average pH of open ocean water is 8.1.
   a. What is the maximum value of $[Fe^{3+}]$ in pH 8.1 seawater if the K_{sp} value of $Fe(OH)_3$ is 1.1×10^{-36}?
   *b. If you were to use the result from part a. to predict the solubility (*s*) of $Fe(NO_3)_3$ in seawater using the equation $K_{sp} = (m^m x^x)s^{(m+x)}$, would the prediction be accurate? Explain your answer.

Polydentate Ligands; Ligand Strength and the Chelation Effect

CONCEPT REVIEW

17.67. What is meant by the term *sequestering agent*? What properties makes a substance an effective sequestering agent?

17.68. The condensed molecular structures of two compounds that each contain two $-NH_2$ groups are shown in Figure P17.68. The one on the left is ethylenediamine, a bidentate ligand. Does the molecule on the right have the same ability to donate two pairs of electrons to a metal ion? Explain why you think it does or does not.

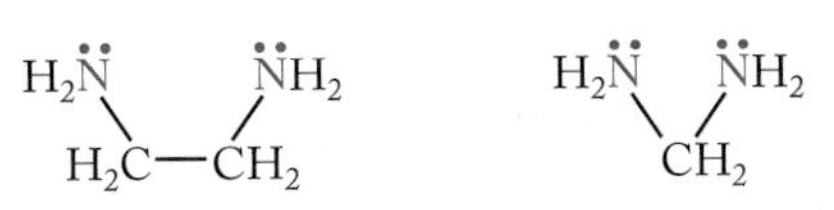

FIGURE P17.68

17.69. How does the chelating ability of an *aminocarboxylate* vary with changing pH?

*17.70. The EDTA that is widely used as a food preservative is added to food, not as the undissociated acid, but rather

as the calcium disodium salt: $Na_2[CaEDTA]$. This salt is actually a coordination compound with a Ca^{2+} ion at the center of a complex ion. Draw a line structure of this compound.

Crystal Field Theory

CONCEPT REVIEW

17.71. Explain why the compounds of most of the first-row transition metals are colored.

17.72. Unlike the compounds of most transition metal ions, those of Ti^{4+} are colorless. Why?

17.73. Why is the d_{xy} orbital higher in energy than the d_{xz} and d_{yz} orbitals in a square-planar crystal field?

17.74. On average, the d orbitals of a transition metal ion in an octahedral field are higher in energy than they are when the ion is in the gas phase. Why?

PROBLEMS

17.75. Aqueous solutions of one the following complex ions of Cr(III) are violet; solutions of the other are yellow. Which is which? (a) $Cr(H_2O)_6^{3+}$; (b) $Cr(NH_3)_6^{3+}$

17.76. Which of the following complex ions should absorb the shortest wavelengths of electromagnetic radiation? (a) $Cu(Cl)_4^{2-}$; (b) $Cu(F)_4^{2-}$; (c) $Cu(I)_4^{2-}$; (d) $Cu(Br)_4^{2-}$

17.77. The octahedral crystal field splitting energy Δ_o of $Co(phen)_3^{3+}$ is 5.21×10^{-19} J/ion. What is the color of a solution of this complex ion?

17.78. The octahedral crystal field splitting energy Δ_o of $Co(CN)_6^{3-}$ is 6.74×10^{-19} J/ion. What is the color of a solution of this complex ion?

17.79. Solutions of $NiCl_4^{2-}$ and $NiBr_4^{2-}$ absorb light at 702 and 756 nm, respectively. In which ion is the split of d-orbital energies greater?

17.80. Chromium(III) chloride forms six-coordinate complexes with bipyridine, including cis-$Cr(bipy)_2Cl_2$, which reacts slowly with water to produce two products, cis-$Cr(bipy)_2(H_2O)Cl^+$ and cis-$Cr(bipy)_2(H_2O)_2^{2+}$. In which of these complexes should Δ_o be the largest?

Magnetism and Spin States

CONCEPT REVIEW

17.81. What determines whether a transition metal ion is in a *high-spin* configuration or a *low-spin* configuration?

*17.82. In Section 7.2 we noted that the bonding in $Fe(H_2O)_6^{3+}$ involves sp^3d^2 hybrid orbitals, but that the bonding in $Fe(CN)_6^{3-}$ comes from coordinate bonds with d^2sp^3 hybrid orbitals on Fe^{3+}. What is the reason for this difference in hybridization?

PROBLEMS

17.83. How many unpaired electrons are there in the following transition-metal ions in an octahedral field? High-spin Fe^{2+}, Cu^{2+}, Co^{2+}, and Mn^{3+}.

17.84. Which of the following cations can have either a high-spin or a low-spin electron configuration in an octahedral field? Fe^{2+}, Co^{3+}, Mn^{2+}, and Cr^{3+}

17.85. Which of the following cations can, in principle, have either a high-spin or a low-spin electron configuration in a tetrahedral field? Co^{2+}, Cr^{3+}, Ni^{2+}, and Zn^{2+}

17.86. How many unpaired electrons are in the following transition metal ions in an octahedral crystal field? High-spin Fe^{3+}, Rh^+, and V^{3+}, and low-spin Mn^{3+}

17.87. The manganese minerals pyrolusite, MnO_2, and hausmannite, Mn_3O_4, contain Mn ions in octahedral holes formed by oxide ions.
 a. What are the charges of the Mn ions in each mineral?
 b. In which of these compounds could there be high-spin and low-spin Mn ions?

17.88. **Dietary Supplement** Chromium picolinolate is an over-the-counter diet aid sold in many pharmacies. The Cr^{3+} ions in this coordination compound are in an octahedral field. Is the compound paramagnetic or diamagnetic?

*17.89. One method for refining cobalt involves the formation of the complex ion $CoCl_4^{2-}$. This anion is tetrahedral. Is this complex paramagnetic or diamagnetic?

*17.90. Why is it that $Ni(CN)_4^{2-}$ is diamagnetic, but $NiCl_4^{2-}$ is paramagnetic?

Naming Complex Ions and Coordination Compounds

PROBLEMS

17.91. What are the names of the following complex ions?
 a. $Cr(NH_3)_6^{3+}$ b. $Co(H_2O)_6^{3+}$ c. $[Fe(NH_3)_5Cl]^{2+}$

17.92. What are the names of the following complex ions?
 a. $Cu(NH_3)_2^+$
 b. $Ti(H_2O)_4(OH)_2^{2+}$
 c. $Ni(NH_3)_4(H_2O)_2^{2+}$

17.93. What are the names of the following complex ions?
 a. $CoBr_4^{2-}$ b. $Zn(H_2O)(OH)_3^-$ c. $Ni(CN)_5^{3-}$

17.94. What are the names of the following complex ions?
 a. CoI_4^{2-} b. $CuCl_4^{2-}$ c. $[Cr(en)(OH)_4]^-$

17.95. What are the names of the following coordination compounds?
 a. $[Zn(en)]SO_4$ b. $[Ni(NH_3)_5(H_2O)]Cl_2$ c. $K_4Fe(CN)_6$

17.96. What are the names of the following coordination compounds?
 a. $(NH_4)_3[Co(CN)_6]$
 b. $[Co(en)_2Cl](NO_3)_2$
 c. $[Fe(H_2O)_4(OH)_2]Cl$

Isomerism in Coordination Compounds; Metal Complexes in Biomolecules

CONCEPT REVIEW

17.97. What do the prefixes cis- and trans- mean in the context of an octahedral complex ion?

17.98. What do the prefixes cis- and trans- mean in the context of a square-planar complex?

17.99. How many different types of donor groups are required to have geometric isomers of a square-planar complex?

17.100. With respect to your answer to the previous question, do all square-planar complexes with this many different types of donor groups have geometric isomers?

PROBLEMS

17.101. Does the complex ion $[Co(en)(H_2O)_2Cl_2]^{2+}$ have geometric isomers?

17.102. Does the complex ion $Fe(en)_3^{3+}$ have geometric isomers?

---

*__17.103.__ Sketch the geometric isomers of the square planar complex ion $CuCl_2Br_2^{2-}$. Are any of these isomers chiral?

*17.104. Sketch the geometric isomers of the complex ion $Ni(en)Cl_2(CN)_2^{2-}$. Are any of these isomers chiral?

Additional Problems

17.105. Photographic Film Processing During the processing of black-and-white photographic film, excess silver(I) halides are removed by washing the film in a bath containing sodium thiosulfate. This treatment is based on the following complexation reaction:

$$Ag^+(aq) + 2\,S_2O_3^{2-}(aq) \rightleftharpoons Ag(S_2O_3)_2^{3-}(aq) \qquad K_f = 5 \times 10^{13}$$

What is the ratio of $[Ag^+]$ to $[Ag(S_2O_3)_2^{3-}]$ in a bath in which $[S_2O_3^{2-}] = 0.233\ M$?

17.106. **Lead Poisoning** Children used to be treated for lead poisoning with intravenous injections of EDTA. If the concentration of EDTA in the blood of a patient is $2.5 \times 10^{-8}\ M$ and the formation constant for the complex, $[Pb(EDTA)]^{2-}$, is 2.0×10^{18}, what is the concentration ratio of the free (and potentially toxic) $Pb^{2+}(aq)$ in the blood to the much less toxic Pb^{2+}-EDTA complex?

17.107. Dissolving cobalt(II) nitrate in water gives a beautiful purple solution. There are three unpaired electrons in this cobalt(II) complex. When cobalt(II) nitrate is dissolved in aqueous ammonia and oxidized with air, the resulting yellow complex has no unpaired electrons. Which cobalt complex has the larger crystal field splitting energy Δ_o?

17.108. A solid compound containing Fe(II) in an octahedral crystal field has four unpaired electrons at 298 K. When the compound is cooled to 80 K, the same sample appears to have no unpaired electrons. How do you explain this change in the compound's properties?

17.109. When Ag_2O reacts with peroxodisulfate $(S_2O_8^{2-})$ ion (a powerful oxidizing agent), AgO is produced. Crystallographic and magnetic analyses of AgO suggest that it is not simply Ag(II) oxide, but rather a blend of Ag(I) and Ag(III) in a square planar environment. The Ag^{2+} ion is paramagnetic but, like AgO, Ag^+ and Ag^{3+} are diamagnetic. Explain why.

17.110. The iron(II) compound $Fe(bipy)_2(SCN)_2$ is paramagnetic, but the corresponding cyanide compound $Fe(bipy)_2(CN)_2$ is diamagnetic. Why do these two compounds have different magnetic properties?

17.111. Aqueous solutions of copper(II)–ammonia complexes are dark blue. Will the color of the series of complexes, $Cu(H_2O)_{6-x}(NH_3)_x^+$ shift toward shorter or longer wavelengths as the value of x increases from 0 to 6?

18

Electrochemistry and Electric Vehicles

AN ELECTRIFYING SPORTS CAR *A bank of lithium-ion batteries and a 185 kilowatt motor power the high-performance 2008 Tesla Roadster.*

A LOOK AHEAD:
100 mpg: Charging Up Hybrid Cars

The first decade of the 21st century has seen remarkable swings in the prices of gasoline and other fuels derived from petroleum and natural gas. These oscillations, which have included record prices for crude oil, have reinvigorated the development of alternative systems for powering vehicles. One popular alternative has been hybrid propulsion that relies on a combination of a small gasoline engine and an electric motor powered by rechargeable batteries. In some cars these batteries and motors can be the sole power source for short trips at moderate speeds.

Battery-powered electric motors convert chemical energy into mechanical energy more efficiently than gasoline engines and they draw no power when the car is stopped in traffic. They can even be made to recharge a car's batteries when its driver applies the brakes. As the car slows, its decrease in kinetic energy is first turned into electrical energy and then into chemical energy as electric motors become electric generators and battery rechargers.

Ultimately, the driving range of cars under electric power alone is limited by the capacity of their batteries to hold and deliver electrical charge. Research into new battery technologies that make these repositories of chemical energy more compact and more powerful has led to the development of high-performance all-electric vehicles and to "plug-in" hybrids: cars that have larger-capacity batteries than those in conventional hybrids such as the Toyota Prius. Plug-in hybrids can make most business or pleasure trips without their gasoline engines ever turning on.

In this chapter, we examine the chemistry of modern batteries and another source of electrical power called fuel cells. Batteries and fuel cells harness the chemical energy released by spontaneous redox reactions, so we begin this chapter by revisiting the principles of redox chemistry introduced in Chapter 4. Then we look at the way free energy changes accompanying spontaneous redox reactions can create electromotive forces that push electrons through devices like electric motors. Later we'll see how to restore a battery's chemical energy by using an external source of electrical energy to force its spontaneous redox reaction to run in reverse. We conclude with an examination of the design of fuel cells based on combining H_2 and O_2 to form H_2O, and the challenges of building hydrogen-based energy systems.

18.1 Redox Chemistry Revisited

In Chapters 14 through 16 we discussed some of the environmental problems associated with the combustion of fossil fuels. In this chapter, we examine alternative technologies for powering vehicles that are more efficient than the internal combustion engine at converting chemical energy to mechanical energy and have the potential to reduce air pollution dramatically. These technologies are based on **electrochemistry**, the branch of chemistry that links redox reactions to the production or consumption of electrical energy. At the heart of electrochemistry are chemical reactions in which electrons are gained and lost at electrode surfaces. In other words, electrochemistry is based on reduction and oxidation, or *redox,* chemistry.

The principles of redox reactions were introduced in Section 4.8. Let's review some basic definitions and concepts:

- A redox reaction is the sum of two half-reactions: a reduction half-reaction, in which a reactant gains electrons, and an oxidation half-reaction, in which a reactant loses electrons.

Electrochemistry is the branch of chemistry that examines the transformations between chemical and electrical energy.

FIGURE 18.1 A strip of zinc immersed in a solution of copper(II) sulfate becomes encrusted with a dark layer of copper. The blue color of the solution fades as Cu^{2+} ions (the source of the color) are reduced to Cu metal, and Zn metal is oxidized to colorless Zn^{2+} ions.

■ Reduction and oxidation half-reactions happen simultaneously so that the number of electrons gained during reduction exactly matches the number lost during oxidation.

■ A substance that is easily oxidized is one that readily gives up electrons. This electron-donating power makes the substance an effective reducing agent because electrons donated by this substance are accepted by another reactant (the one that is reduced).

■ A substance that readily accepts electrons and is thereby reduced is an effective oxidizing agent.

The tendencies of different elements and their ions to gain or lose electrons are illustrated by the chemical properties of two neighbors in the periodic table of the elements, copper ($Z = 29$) and zinc ($Z = 30$). Zinc metal has considerable electron-donating power, and Cu^{2+} ions tend to be willing acceptors of donated electrons. These complementary chemical properties are captured in the images in Figure 18.1. When a strip of Zn metal is placed in a solution of $CuSO_4$, Zn atoms spontaneously donate electrons to Cu^{2+} ions, forming Zn^{2+} ions and Cu atoms. The progress of this reaction can be monitored visually: shiny metallic zinc turns dark brown as a textured layer of copper accumulates, and the distinctive blue color of $Cu^{2+}(aq)$ ions fades as these ions acquire electrons and become atoms of copper metal.

Writing a chemical equation that describes the spontaneous reaction between Zn and Cu^{2+} is easily done because every mole of Zn atoms that is oxidized *loses* 2 moles of electrons in an oxidation half-reaction:

$$Zn(s) \rightarrow Zn^{2+}(aq) + 2\,e^-$$

and every mole of Cu^{2+} ions that is reduced *accepts* 2 moles of electrons in a reduction half-reaction:

$$Cu^{2+}(aq) + 2\,e^- \rightarrow Cu(s)$$

Because the number of moles of electrons lost and gained in the two half-reactions is the same, writing a chemical equation for the overall redox reaction is simply a matter of adding the two half-reactions together:

$$Zn(s) \rightarrow Zn^{2+}(aq) + 2\,e^-$$
$$\underline{Cu^{2+}(aq) + 2\,e^- \rightarrow Cu(s)}$$
$$Zn(s) + Cu^{2+}(aq) + \cancel{2\,e^-} \rightarrow Cu(s) + Zn^{2+}(aq) + \cancel{2\,e^-}$$

Canceling out the equal numbers of electrons gained and lost, we obtain a balanced net ionic equation for the redox reaction:

$$Zn(s) + Cu^{2+}(aq) \rightarrow Cu(s) + Zn^{2+}(aq)$$

Combining half-reactions is a convenient way to write balanced chemical equations for redox reactions, as we discussed in Section 4.8. In this chapter we will use a valuable resource in this process, the table of half-reactions in Appendix 6. The table sorts half-reactions in order of a parameter called their *standard reduction potentials* ($E°$). We use $E°$ values, which are expressed in volts (V), extensively in later sections of this chapter. For now, just keep in mind that the $E°$ value of a reduction half-reaction is an indication of how likely it is to occur. The half-reactions at the top of the table with the most positive $E°$ values, for example,

$$F_2(g) + 2e^- \rightarrow 2F^-(aq) \qquad E° = 2.87 \text{ V}$$

are those with reactants that are the strongest oxidizing agents (and therefore readily reduced), whereas the very negative $E°$ values at the bottom of the table mean that the reactants, which include the major cations in biological systems and environmental waters (Na^+, K^+, Mg^{2+}, and Ca^{2+}) are not easily reduced. Instead, the products of the half-reactions at the bottom of the table are powerful reducing agents.

In the following Sample and Practice Exercises we use half-reactions from Appendix 6 to write net ionic equations for redox reactions that involve molecular O_2. In Sample Exercise 18.1 we write a net ionic equation for the reaction between O_2 and Fe^{2+} under acidic conditions. In the Practice Exercise your task will be to write a net ionic equation describing the oxidation of NO_2^- to NO_3^- by O_2 under basic conditions. Note that Appendix 6 contains two half-reactions describing the reduction of O_2. One of them contains H^+ ions and so applies to reactions in acidic solutions:

$$O_2(g) + 4H^+(aq) + 4e^- \rightarrow 2H_2O(\ell)$$

In alkaline solutions reduction of O_2 produces OH^- ions:

$$O_2(g) + 2H_2O(\ell) + 4e^- \rightarrow 4OH^-(aq)$$

| SAMPLE EXERCISE 18.1 | **Writing Net Ionic Equations for Redox Reactions by Combining Half-Reactions** |

Write a net ionic equation describing the oxidation of $Fe^{2+}(aq)$ by molecular oxygen in an acidic solution.

COLLECT AND ORGANIZE Perusing the reduction half-reactions in Appendix 6, we find the following half-reactions for the reduction of O_2:

$$O_2(g) + 4H^+(aq) + 4e^- \rightarrow 2H_2O(\ell)$$

$$O_2(g) + 2H_2O(\ell) + 4e^- \rightarrow 4OH^-(aq)$$

All of the half-reactions in Appendix 6 are written as reduction half-reactions, so there is none for the oxidation of Fe^{2+}. However, there is one in which Fe^{2+} is the product of the reduction of Fe^{3+}:

$$Fe^{3+}(aq) + e^- \rightarrow Fe^{2+}(aq)$$

We can reverse this half-reaction to make Fe^{2+} the reactant in an oxidation half-reaction:

$$Fe^{2+}(aq) \rightarrow Fe^{3+}(aq) + e^-$$

ANALYZE The problem specifies acidic conditions, so we should use the first of the two O_2 half-reactions. In redox reactions the electrons gained by the substances that are reduced must equal the electrons lost by the substances that are oxidized. There is a gain of 4 moles of electrons in the O_2 half-reaction and a loss of 1 mole of electrons in the Fe^{2+} half-reaction. Therefore, we need to multiply the Fe^{2+} half-reaction by 4.

SOLVE Multiply the Fe^{2+} half-reaction by 4 and then combine the two half-reactions:

$$4\,Fe^{2+}(aq) \rightarrow 4\,Fe^{3+}(aq) + 4\,e^-$$

$$\underline{O_2(g) + 4\,H^+(aq) + 4\,e^- \rightarrow 2\,H_2O(\ell)}$$

$$4\,Fe^{2+}(aq) + O_2(g) + 4\,H^+(aq) + \cancel{4\,e^-} \rightarrow 4\,Fe^{3+}(aq) + 2\,H_2O(\ell) + \cancel{4\,e^-}$$

Simplifying gives

$$4\,Fe^{2+}(aq) + O_2(g) + 4\,H^+(aq) \rightarrow 4\,Fe^{3+}(aq) + 2\,H_2O(\ell)$$

THINK ABOUT IT We can verify that this is a balanced net ionic equation by confirming that the number of moles of Fe, O, and H atoms match on both sides of the reaction arrow and that the total electrical charges on both sides are the same. (Both are 12+.) Note that the result of combining these two half-reactions is not a complete, molecular equation but rather a net ionic equation for the redox reaction.

Practice Exercise Write a net ionic equation describing the oxidation of NO_2^- to NO_3^- by O_2 in a basic solution.

(Answers to Practice Exercises are in the back of the book.)

18.2 Electrochemical Cells

Having shown how the Zn/Cu^{2+} redox reaction is the sum of two discrete half-reactions, one oxidation and one reduction, let's now physically separate the two half-reactions using a device called an **electrochemical cell** (Figure 18.2). Electrochemical cells can convert the chemical energy of a spontaneous redox reaction into electrical energy, as happens in batteries. They can also convert electrical energy into chemical energy, as happens when a battery is recharged. In the cell in Figure 18.2 a strip of zinc is immersed in a 1 *M* solution of $ZnSO_4$, and in a separate compartment a strip of copper metal is immersed in a 1 *M* solution of $CuSO_4$. The two metal strips serve as the cell's *electrodes*, providing pathways for the electrons gained and lost by the two half-reactions to reach an external circuit. The separation of the Cu and Zn half-reactions forces electrons lost when Zn atoms are oxidized to travel through the external circuit (blue arrow in Figure 18.2) before they can be accepted by Cu^{2+} ions.

▶‖ **CHEMTOUR** Zinc–Copper Cell

As the two half-reactions in the electrochemical cell proceed, Cu^{2+} ions are reduced to Cu atoms in the right-hand compartment in Figure 18.2, and Zn atoms are oxidized to Zn^{2+} ions on the left. You might think that production of Zn^{2+} ions would result in a buildup of positive charge on the Zn side and that conversion of Cu^{2+} ions to Cu metal would create excess negative charge on the Cu side. No such charging occurs because the compartments are connected by a permeable bridge made of porous glass or plastic that allows ions to migrate from one compartment to the other to balance any buildup of charge. In most electrochemical cells the source of these ions is a strong electrolyte added to the solutions in both compartments. It is important that such a "background" electrolyte not interfere with the redox reactions at the electrodes. A useful candidate for the Zn/Cu^{2+} cell is Na_2SO_4 because neither Na^+ ions nor SO_4^{2-} ions interfere with the oxidation of Zn or reduction of Cu^{2+} ions. Migration of Na^+ ions into the Cu compartment and SO_4^{2-} ions into the Zn compartment completes the electrical circuit and allows electrons to flow through the part of the circuit outside the cell.

An **electrochemical cell** is an apparatus that converts chemical energy into electrical work or electrical work into chemical energy.

FIGURE 18.2 What a Chemist Sees This electrochemical cell consists of two compartments: the one on the left contains a zinc anode immersed in a 1 M solution of $ZnSO_4$; the other contains a copper cathode immersed in a 1 M solution of $CuSO_4$. A porous glass plug provides an electrical connection between the two compartments as ions (and their charges) migrate through it. If the electrodes are connected to a voltmeter—the Zn electrode to the negative terminal of the meter and the Cu electrode to the positive terminal—the meter will read 1.10 V at 25°C.

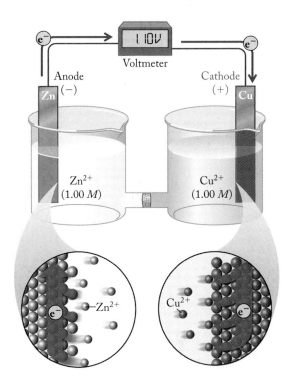

With the passage of time, the mass of the Zn strip decreases; the mass of the Cu strip increases, and the blue color of $Cu^{2+}(aq)$ ions fades in the solution in the right-hand compartment in Figure 18.2. The decrease in chemical energy that accompanies this spontaneous cell reaction does electrical work in an external circuit. For example, it could be used to light a small light bulb or drive the motor in a small fan. This capacity to do electrical work makes this electrochemical cell a **voltaic cell**, that is, a battery. In this or any electrochemical cell, the electrode at which the oxidation half-reaction (loss of electrons) takes place (the Zn electrode in this case) is called the **anode**, and the electrode at which reduction half-reaction (gain of electrons) takes place (the Cu electrode in this case) is called the **cathode**.

CONCEPT TEST

Will the increase in mass of the copper strip match the decrease in mass of the zinc strip during the reaction just described?

(Answers to Concept Tests are in the back of the book.)

The illustration in Figure 18.2 provides a useful view of the physical reality of a Zn/Cu^{2+} electrochemical cell, but we need a more compact way to symbolize the components of such a cell. That's where cell diagrams come in. A **cell diagram** uses symbols to show how the components of an electrochemical cell are connected. A cell diagram does not convey the stoichiometry of the reaction, and so any coefficients in the balanced equation for the cell reaction do not appear in the cell diagram. Here are the steps we follow when writing a cell diagram:

(1) Write the chemical symbol of the anode at the far left of the diagram and write the symbol of cathode at the far right.
(2) Work from the electrodes toward the connecting bridge using vertical lines to indicate phase changes (like that between a solid metal electrode and an aqueous solution). Represent solutions or ionic solids using the symbols of the ions or compounds that are changed by the cell reaction.
(3) Use a double vertical line to represent the bridge connecting the anode and cathode half-reactions.

Following these steps for the Zn/Cu^{2+} electrochemical cell,

(1) $Zn(s)$ . . . . $Cu(s)$

(2) $Zn(s) \mid Zn^{2+}(aq)$ . . . $Cu^{2+}(aq) \mid Cu(s)$

(3) $Zn(s) \mid Zn^{2+}(aq) \parallel Cu^{2+}(aq) \mid Cu(s)$

In a **voltaic cell**, chemical energy is transformed into electrical energy by a spontaneous redox reaction.

An **anode** is an electrode at which an oxidation half-reaction (loss of electrons) takes place.

A **cathode** is an electrode at which a reduction half-reaction (gain of electrons) takes place.

A **cell diagram** uses symbols to show how the components of an electrochemical cell are connected.

SAMPLE EXERCISE 18.2 **Diagramming an Electrochemical Cell**

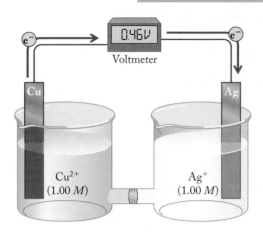

Voltmeter

This figure represents an electrochemical cell in which a copper electrode immersed in a 1.00 M solution of Cu^{2+} ions is connected to a silver electrode immersed in a 1.00 M solution of Ag^+ ions.

Write a balanced chemical equation for the cell reaction and diagram the cell.

COLLECT AND ORGANIZE We need to write a balanced chemical equation for the cell reaction and diagram a cell in which current spontaneously flows from a copper electrode immersed in a solution of Cu^{2+} ions through an external circuit to a silver electrode immersed in a solution of Ag^+ ions.

ANALYZE In a voltaic cell electrons flow from the electrode at which they are produced in an oxidation half-reaction and toward the electrode where they are consumed in a reduction half-reaction. Therefore, in the cell in the figure Cu is the anode and Ag is the cathode. The half-reactions are

Oxidation at the anode: $Cu(s) \rightarrow Cu^{2+}(aq) + 2\,e^-$

Reduction at the cathode: $Ag^+(aq) + e^- \rightarrow Ag(s)$

Two moles of electrons are produced in the anode half-reaction, but only 1 mole is consumed in the cathode half-reaction. We need to multiply the cathode half-reaction by 2 before combining the two equations. In a cell diagram the anode and the species involved in the oxidation half-reaction are on the left and the species involved in the reduction half-reaction and the cathode are on the right. We use single lines to separate phases and a double line to separate the two half-reactions.

SOLVE Multiplying the Ag^+ half-reaction by two and adding it to the Cu half-reaction,

$$2\,[Ag^+(aq) + e^- \rightarrow Ag(s)]$$
$$\underline{Cu(s) \rightarrow Cu^{2+}(aq) + 2\,e^-}$$
$$2\,Ag^+(aq) + Cu(s) \rightarrow 2\,Ag(s) + Cu^{2+}(aq)$$

Applying the rules for writing a cell diagram:

$$Cu(s)\,|\,Cu^{2+}(aq)\,||\,Ag^+(aq)\,|\,Ag(s)$$

THINK ABOUT IT The appearance of this diagram is much like the one in the text for the Zn/Cu^{2+} electrochemical cell reaction and so is reasonable.

Practice Exercise A cell like that shown in Figure 18.2 has a copper cathode immersed in a solution of Cu^{2+} ions and an aluminum anode immersed in a solution of Al^{3+} ions. Write a balanced chemical equation for the cell reaction, and write the cell diagram.

18.3 Chemical Energy and Electrical Work

When we connect the negative terminal of a voltmeter to the Zn electrode in Figure 18.2 and the positive terminal to the Cu electrode, the meter reads 1.10 V. This voltage is a measure of the **electromotive force (emf)** generated by the cell reaction, which reflects how forcefully it pumps electrons through the external circuit from the negative terminal and toward the positive terminal. This voltage is also called **cell potential (E_{cell})**. It is directly proportional to the chemical (potential) energy released by simultaneously oxidizing Zn atoms and reducing Cu^{2+} ions.

When the voltage produced by a voltaic cell pushes electrons through an external circuit, that flow of current can do electrical work, like lighting a light bulb or a light-emitting diode (LED), or turning an electric motor. The connection between the change in chemical free energy of a voltaic cell (ΔG_{cell}) and the electrical work (w_{elec}) that may result is simply

$$\Delta G_{cell} = w_{elec} \qquad (18.1)$$

The sign of ΔG is negative because the internal energy of a voltaic cell decreases as it is discharged, and the sign on w_{elec} is negative because it represents work the cell does *on* its surroundings, which has a negative value from the perspective of the cell.

The quantity of electrical work (w_{elec}) done by an electrochemical cell is the product of the electrical charge that the cell pushes through an external circuit times the electromotive force (cell potential E_{cell}) with which the cell pushes that charge. In equation form this definition is

$$w_{elec} = C E_{cell} \qquad (18.2)$$

The quantity of charge is proportional to the number of electrons flowing through a circuit. As we noted in Chapter 2, the magnitude of the charge on a single electron is 1.602×10^{-19} coulombs (C). The magnitude of electrical charge in 1 mole of electrons (e^-) is

$$\frac{1.602 \times 10^{-19}\ C}{e^-} \times \frac{6.022 \times 10^{23}\ e^-}{mol\ e^-} = \frac{9.65 \times 10^4\ C}{mol\ e^-}$$

This quantity of charge, 9.65×10^4 C/mol, is called **Faraday's constant (F)** after Michael Faraday (1791–1867), the English chemist and physicist, who discovered that redox reactions take place when electrons are transferred between reacting species. The quantity of charge (C) flowing through an electrical circuit is the product of the number of moles of electrons (n) times the Faraday constant:

$$C = nF \qquad (18.3)$$

Let's develop an equation relating w_{elec} and E_{cell} by combining Equations 18.2 and 18.3:

$$w_{elec} = -nFE_{cell} \qquad (18.4)$$

The negative sign on the right side of Equation 18.4 reflects the fact that work done by a voltaic cell on its surroundings (the external circuit) corresponds to energy lost by the cell.

CONNECTION The sign conventions used for work done *on* a thermodynamic system (+) and the work done *by* the system (−) are explained in Section 5.2.

▶II **CHEMTOUR** Free Energy

Electromotive force (emf), sometimes called *voltage*, is the force that pushes electrons through an electrical circuit.

Cell potential (E_{cell}) is the electromotive force expressed in volts (V) with which an electrochemical cell can push electrons through an external circuit connected to its terminals.

Faraday's constant (F) is the quantity of electrical charge (in coulombs) in 1 mol of electrons. Its value to three significant figures is 9.65×10^4 C.

If we combine Equations 18.1 and 18.4, we connect the quantity of electrical work (w_{elec}) that a voltaic cell does on its surroundings and the change in free energy that occurs inside the cell:

$$\Delta G_{cell} = -nFE_{cell} \qquad (18.5)$$

The units on the quantities on the right side of Equation 18.5 are

$$(\text{mol}) \times (\text{coulomb/mol}) \times (\text{volt})$$

or

$$\text{coulomb} \cdot \text{volt}$$

Conveniently, a coulomb·volt represents the same quantity of energy as a joule:

$$1 \text{ coulomb} \cdot \text{volt} = 1 \text{ joule}$$

or

$$1 \text{ C} \cdot \text{V} = 1 \text{ J}$$

The cell reaction in a voltaic cell must be spontaneous if it does work in an external circuit. Therefore, the sign of ΔG_{cell} must be negative. In Equation 18.3, both n and F have positive values. Therefore, the value of E_{cell} for a voltaic cell with a spontaneous cell reaction must be positive.

Let's use Equation 18.5 to calculate ΔG_{cell} for the Zn/Cu^{2+} cell reaction in Figure 18.2 and see how its cell potential of 1.10 V translates into a value of ΔG_{cell}. The two half-reactions involve the loss and gain of 2 moles of electrons. Therefore, the value of n in Equation 18.5 is 2, and the calculation proceeds as follows:

$$\Delta G_{cell} = -nFE_{cell}$$

$$= -2 \text{ mol} \times \frac{9.65 \times 10^4 \text{ C}}{\text{mol}} \times 1.10 \text{ V} = -2.12 \times 10^5 \text{ V} \cdot \text{C}$$

$$= -2.12 \times 10^5 \text{ J}$$

$$= -212 \text{ kJ}$$

At full throttle, each of the three main engines of a U.S. space shuttle produces about 12 million horsepower (9×10^6 kJ/s) as it burns liquid hydrogen at a rate of about 1,100 L (38,000 mol) per second.

As expected, the spontaneous cell reaction has a sizable negative ΔG value. To put this value in perspective, on a mole-for-mole basis, the Zn/Cu^{2+} cell reaction produces slightly less useful energy than the combustion of hydrogen gas used to power the main engines of the U.S. space shuttles:

$$H_2(g) + \tfrac{1}{2} O_2(g) \rightarrow H_2O(g) \qquad \Delta G° = -228.6 \text{ kJ}$$

SAMPLE EXERCISE 18.3   **Relating the Value of ΔG_{cell} to the Value of E_{cell}**

Silver oxide "button" batteries are used to power electric watches. These batteries, which generate a potential of 1.55 V, consist of a Zn anode and a Ag_2O cathode immersed in a concentrated solution of KOH. The cathode reaction is based on the reduction of Ag_2O to Ag metal. At the anode, Zn is oxidized to solid $Zn(OH)_2$. Write the net ionic equation for the electrochemical cell reaction, and calculate the value of ΔG_{cell}.

COLLECT AND ORGANIZE We need to write the net ionic equation for the cell reaction and calculate the value of ΔG_{cell}. We know that (1) $E_{cell} = 1.55$ V; (2) the cell reaction occurs in a strongly alkaline electrolyte; (3) the half-reaction at the cathode is based on the reduction of Ag_2O to Ag metal; and (4) the half-reaction

Button batteries are used to power miniature electronic devices such as wristwatches.

at the anode is based on the oxidation of Zn metal to $Zn(OH)_2$. We also know that the value of ΔG_{cell} is related to the value of E_{cell} by Equation 18.5:

$$\Delta G_{cell} = -nFE_{cell}$$

where F is 9.65×10^4 C/mol and n is the number of moles of electrons gained and lost in the cell reaction. A logical first step is to use half-reactions to write the net ionic equation for the cell reaction and to determine the value of n.

ANALYZE The half-reaction at the cathode is based on the reduction of Ag_2O to Ag metal. Appendix 6 contains the standard reduction potential at 25°C of the following half-reaction:

$$Ag_2O(s) + H_2O(\ell) + 2\,e^- \rightarrow 2\,Ag(s) + 2\,OH^-(aq) \qquad E° = 0.342\,V$$

In searching Appendix 6 for a half-reaction that describes the oxidation that occurs at the anode of a button battery, we need to reverse the reactant and product of the oxidation process and find an entry in which $Zn(OH)_2$ is the reactant and metallic Zn is the product. There is such an entry in Appendix 6:

$$Zn(OH)_2(s) + 2\,e^- \rightarrow Zn(s) + 2\,OH^-(aq) \qquad E° = -1.249\,V$$

SOLVE Before we can combine the two half-reactions we need to first reverse the Zn half-reaction to make it an oxidation half-reaction. Doing so changes the sign of $E°$:

$$Zn(s) + 2\,OH^-(aq) \rightarrow Zn(OH)_2(s) + 2\,e^- \qquad E° = 1.249\,V$$

The Ag and Zn half-reactions involve the gain and loss of the same number of electrons (two). Therefore, they can be added together without further modification:

$$Ag_2O(s) + H_2O(\ell) + 2\,e^- \rightarrow 2\,Ag(s) + 2\,OH^-(aq) \qquad E° = 0.342\,V$$

$$Zn(s) + 2\,OH^-(aq) \rightarrow Zn(OH)_2(s) + 2\,e^- \qquad E° = 1.249\,V$$

$$\overline{Ag_2O(s) + H_2O(\ell) + Zn(s) + \cancel{2\,OH^-(aq)} + \cancel{2\,e^-} \rightarrow}$$
$$2\,Ag(s) + \cancel{2\,OH^-(aq)} + Zn(OH)_2(s) + \cancel{2\,e^-}$$
$$E° = 1.591\,V$$

Simplifying, we have the net ionic equation of the cell reaction and its standard cell potential:

$$Ag_2O(s) + H_2O(\ell) + Zn(s) \rightarrow 2\,Ag(s) + Zn(OH)_2(s) \qquad E° = 1.591\,V$$

To calculate ΔG_{cell} we insert the values of n, F, and E_{cell} into Equation 18.5 to give

$$\Delta G_{cell} = -nFE_{cell}$$

$$= -2\,\cancel{mol} \times 9.65 \times 10^4\,C/\cancel{mol} \times 1.591\,V$$

$$= -3.07 \times 10^5\,C \cdot V = -3.07 \times 10^5\,J = -307\,kJ$$

THINK ABOUT IT The negative value of ΔG_{cell} indicates that the cell reaction is spontaneous, which it must be to make a useful battery. The value is somewhat more negative than the ΔG_{cell} we calculated for the Cu^{2+}/Zn cell. This also makes sense because the button battery produces a somewhat larger cell potential and the same number of electrons are gained and lost in both cells. The calculated value of ΔG is based on the reaction of one mole of Ag_2O and one mole of Zn. The energy stored in a button battery that weighs only a gram or two would be a tiny fraction of the calculated value. Also, you may have noticed that there are no ions in the net ionic equation: they all cancel out when the two half-reactions

Italian physicist Alessandro Volta (1745–1827) is credited with building the first battery in 1798. It consisted of a stack of alternating layers of zinc, blotter paper soaked in salt water, and silver.

▶❙❙ **CHEMTOUR** Cell Potential

A **standard potential ($E°$)** is the electromotive force of a half-reaction written as a reduction in which all reactants and products are in their standard states. This means that the concentrations of all dissolved substances are 1 M, and the partial pressures of all gases are 1 bar. It is also called the **standard reduction potential ($E°_{red}$)**.

The **standard cell potential ($E°_{cell}$)** is the electromotive force produced by an electrochemical cell when all reactants and products are in their standard states.

---

are added together. This makes sense because all the reactants and products in the silver oxide battery reaction are solids, and so the net ionic and molecular equations are the same.

Practice Exercise The alkaline batteries used in flashlights and portable electronic devices produce a cell potential of 1.50 V. They have zinc anodes, an electrolyte of concentrated KOH, and a cathode half-reaction in which solid MnO_2 is reduced to solid Mn_2O_3. What is the value of ΔG_{cell}?

18.4 Standard Potentials ($E°$)

In this section, we examine how the potentials produced by batteries are related to the chemical reactions occurring at their anodes and cathodes. We have defined electrical work as the product of the cell potential of the battery and the quantity of charge that it can deliver at that voltage:

$$w_{elec} = -nFE_{cell} \tag{18.4}$$

The values of cell potentials depend on the half-reactions that take place inside the cells and on the **standard potentials ($E°$)** of those half-reactions. The superscript (°) has its usual meaning—all reactants and products are in their standard states. For cell potentials, the standard state means that the concentrations of all dissolved substances are 1 M and the partial pressures of all gases are 1 bar (10^5 Pa).

By convention, tables of standard potentials such as those in Appendix 6 are for half-reactions written as reductions. Therefore, the standard potentials in Appendix 6 are **standard reduction potentials ($E°_{red}$)**. There is no need for a separate table of standard oxidation potentials, because reduction half-reactions can always be reversed, making them oxidation half-reactions. Reversing the reduction half-reaction does not affect the magnitude of its standard potential, but it does change the sign. In this respect, standard potentials are like other thermodynamic parameters of reactions, such as $\Delta G°$, $\Delta H°$, and $\Delta S°$. Recall that reversing a chemical reaction changes the signs of these parameters, but their magnitudes stay the same. In equation form, this relation between the standard reduction potential ($E°_{red}$) of a half-reaction and the standard oxidation potential ($E°_{ox}$) of the half-reaction running in reverse is

$$E°_{ox} = -E°_{red} \tag{18.6}$$

The **standard cell potential ($E°_{cell}$)** of an electrochemical cell may be calculated by combining the standard potentials of its cathode and anode half-reactions:

$$E°_{cell} = E°_{red}(\text{cathode}) + E°_{ox}(\text{anode}) \tag{18.7}$$

For example, the value of $E°_{cell}$ based on the Zn/Cu^{2+} reaction we discussed in Section 18.1 can be calculated from the standard potentials of the cathode and anode half-reactions (see Appendix 6). The standard potential of the cathode half-reaction is

$$Cu^{2+} + 2\,e^- \rightarrow Cu \qquad E°_{red} = 0.342 \text{ V}$$

We can derive the standard potential for the oxidation half-reaction at the anode from the standard reduction potential in Appendix 6:

$$Zn^{2+} + 2\,e^- \rightarrow Zn \qquad E°_{red} = -0.762 \text{ V}$$

If we reverse this half-reaction and change the sign on the potential, we have the oxidation reaction of Zn and its $E°_{ox}$ value:

$$Zn \rightarrow Zn^{2+} + 2\,e^- \qquad E°_{ox} = 0.762 \text{ V}$$

To calculate $E°_{cell}$ we combine this $E°_{ox}$ value for the oxidation of Zn with the $E°_{red}$ for the reduction of Cu^{2+}:

$$E°_{cell} = E°_{red}(Cu^{2+}/Cu) + E°_{ox}(Zn^{2+}/Zn)$$
$$= 0.342 + 0.762 = 1.104 \text{ V}$$

This is the voltage we measure when we connect a voltmeter to the cell shown in Figure 18.2. Equation 18.7 is a special case of Equation 18.8, the general equation that applies to the potential of any cell with any concentrations of reactants and products:

$$E_{cell} = E_{red}(\text{cathode}) + E_{ox}(\text{anode}) \qquad (18.8)$$

We will learn how to calculate E_{cell} values from $E°_{cell}$ values in Section 18.6.

Before we can calculate E_{cell} from $E_{red}(\text{cathode})$ and $E_{ox}(\text{anode})$, we need to know which half-reaction occurs at the cathode and which occurs at the anode. In a Zn/Cu^{2+} cell, Cu^{2+} ions are reduced at the cathode and Zn atoms are oxidized at the anode because Cu^{2+} ions readily accept electrons and Zn atoms readily donate them. These redox properties can be predicted from the standard reduction potentials in Appendix 6. The value of $E°_{red}$ for reducing Cu^{2+} ions to Cu atoms (0.342 V) is more positive than the value of $E°_{red}$ for reducing Zn^{2+} ions to Zn atoms (−0.762 V). To have a positive standard cell potential, we must reverse the Zn half-reaction, making it an oxidation half-reaction, which means that it is the half-reaction that occurs at the anode. As a rule, a half-reaction that occurs above another in the table in Appendix 6 is the cathode half-reaction, and the half-reaction lower in the table runs in reverse at the anode.

SAMPLE EXERCISE 18.4   **Identifying Anode and Cathode Half-Reactions Using Standard Potentials**

The standard reduction potentials of the half-reactions in single-use alkaline batteries are

$$ZnO(s) + H_2O(\ell) + 2\,e^- \rightarrow Zn(s) + 2\,OH^-(aq) \qquad E°_{red} = -1.25 \text{ V}$$

and

$$2\,MnO_2(s) + H_2O(\ell) + 2\,e^- \rightarrow Mn_2O_3(s) + 2\,OH^-(aq) \qquad E°_{red} = 0.15 \text{ V}$$

Which half-reaction occurs at the cathode? Which occurs at the anode? What is the overall cell reaction? What is the value of $E°_{cell}$?

COLLECT AND ORGANIZE We need to identify the chemical half-reactions at the cathode and anode. To do so we need to identify the spontaneous cell reaction ($\Delta G°_{cell} < 0$, $E°_{cell} > 0$) under standard conditions. We can calculate $E°_{cell}$ using Equation 18.7:

$$E°_{cell} = E°_{red}(\text{cathode}) + E°_{ox}(\text{anode})$$

For a spontaneous cell reaction $E°_{cell}$ must be greater than zero.

ANALYZE To ensure that E_{cell} is positive, we must reverse the half-reaction with the more negative $E°_{red}$ value, turning it into an $E°_{ox}$ value with the opposite sign.

▶‖ **CHEMTOUR** Alkaline Battery

SOLVE The ZnO reduction half-reaction has the more negative E_{red}° value and so is the one that runs in reverse. Therefore, the oxidation half-reaction at the anode is

$$Zn(s) + 2\,OH^-(aq) \rightarrow ZnO(s) + H_2O(\ell) + 2\,e^- \qquad E_{ox}^\circ = 1.25\text{ V}$$

and at the cathode

$$2\,MnO_2(s) + H_2O(\ell) + 2\,e^- \rightarrow Mn_2O_3(s) + 2\,OH^-(aq) \qquad E_{red}^\circ = 0.15\text{ V}$$

Combining these half-reactions to obtain the overall cell reaction and simplifying,

$$2\,MnO_2(s) + Zn(s) \rightarrow Mn_2O_3(s) + ZnO(s) \qquad E_{cell}^\circ = 1.40\text{ V}$$

THINK ABOUT IT The result makes sense because the voltage of most alkaline batteries is nominally 1.5 V. During most of their useful lives, the potentials of alkaline batteries vary between 1.2 and 1.3 V.

Practice Exercise The standard reduction potentials of the half-reactions in nicad (nickel–cadmium) batteries are

$$Cd(OH)_2(s) + 2\,e^- \rightarrow Cd(s) + 2\,OH^-(aq) \qquad E_{red}^\circ = -0.403\text{ V}$$

and

$$2\,NiO(OH)(s) + 2\,H_2O(\ell) + 2\,e^- \rightarrow 2\,Ni(OH)_2(s) + 2\,OH^-(aq)$$
$$E_{red}^\circ = 1.32\text{ V}$$

Which half-reaction occurs at the cathode? Which occurs at the anode? What is the overall cell reaction? What is the value of E_{cell}°?

Before closing this discussion of standard cell potentials, let's combine two half-reactions in which different numbers of electrons are gained and lost. This combination occurs in a type of battery that has a limitless supply of one of its reactants. It is called the zinc-air battery, and it powers devices where small size and light weight are high priorities, for example, hearing aids. Most of the internal volume of one of these button-shaped batteries is occupied by an anode consisting of a slurry of zinc particles packed in an aqueous solution of KOH. As in alkaline batteries (Sample Exercise 18.4), the anode half-reaction is

$$Zn(s) + 2\,OH^-(aq) \rightarrow ZnO(s) + H_2O(\ell) + 2\,e^- \qquad E_{ox}^\circ(\text{anode}) = 1.25\text{ V}$$

The cathode consists of porous carbon supported by a metal screen. Air diffuses through small holes in the battery and across a layer of Teflon that lets gases pass through but keeps electrolyte from leaking out. As air passes through the cathode, oxygen in the pores of the electrode is reduced to hydroxide ions:

$$O_2(g) + 2\,H_2O(\ell) + 4\,e^- \rightarrow 4\,OH^-(aq) \qquad E_{red}^\circ(\text{cathode}) = 0.401\text{ V}$$

To write the overall cell reaction, first we need to balance the number of electrons lost in the zinc oxidation half-reaction with the number gained by the reduction half-reaction of O_2. There are 2 moles of electrons in the oxidation half-reaction and 4 moles in the reduction half-reaction. To balance them, let's multiply the oxidation half-reaction by 2 before combining the two half-reactions:

$$2\,[Zn(s) + 2\,OH^-(aq) \rightarrow$$
$$ZnO(s) + H_2O(\ell) + 2\,e^-] \qquad E_{ox}^\circ(\text{anode}) = 1.25\text{ V}$$
$$O_2(g) + 2\,H_2O(\ell) + 4\,e^- \rightarrow 4\,OH^-(aq) \qquad E_{red}^\circ(\text{cathode}) = 0.401\text{ V}$$
$$\overline{\qquad 2\,Zn(s) + O_2(g) \rightarrow 2\,ZnO(s) \qquad E_{cell}^\circ = 1.65\text{ V} \qquad}$$

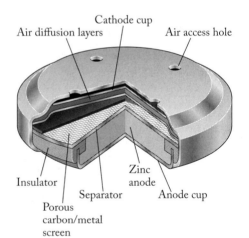

Cathode cup

Air diffusion layers

Air access hole

Insulator

Separator

Porous carbon/metal screen

Zinc anode

Anode cup

Most of the internal volume of a zinc-air battery is occupied by the anode: a slurry of Zn particles in an aqueous solution of KOH surrounded by a metal cup that serves as the negative pole of the battery. Oxygen from the air is the reactant at the cathode. Air enters through holes in an inverted metal cup that serves as the positive pole of the battery. Once inside the battery, air diffuses through layers of gas-permeable plastic film that let air in but keep electrolyte from leaking out. Oxygen in the air is reduced at the porous carbon/metal cathode to OH⁻ ions that migrate toward the anode, where they are consumed in the Zn oxidation half-reaction.

When we multiplied the anode half-reaction by 2 and added it to the cathode half-reaction *we did not multiply the $E°$ of the anode half-reaction by 2*. The reason we did not is that $E°$ is an *intensive* property of a half-reaction or a complete cell reaction. It does not change when the quantities of reactants and products change. Thus, a zinc-air battery the size of a pea has the same $E°_{cell}$ as one the size of a book (like those being developed for electric vehicles). On the other hand, the capacity of a zinc-air battery to do electrical work *does* depend on how much zinc it has inside it because work is the product of cell potential times the quantity of charge it can deliver at that potential.

18.5 A Reference Point: The Standard Hydrogen Electrode

We can measure the value of E_{cell} using a voltmeter. How do we determine the individual values of E_{red}(cathode) and E_{ox}(anode)? The way we do it is to arbitrarily assign a value to the potential produced by a particular half-reaction and then reference the potentials of all other half-reactions to it. The accepted reference point is the standard potential for the reduction of hydrogen ions to hydrogen gas:

$$2\,H^+(aq) + 2\,e^- \rightarrow H_2(g) \qquad E°_{red} = 0.000 \text{ V} \qquad (18.9)$$

The $E°$ for this half-reaction is defined as 0.000 V. An electrode that generates this reference potential, called the **standard hydrogen electrode (SHE)**, consists of a platinum electrode in contact with a solution of a strong acid ([H^+] = 1.00 M) and hydrogen gas (Figure 18.3). Platinum is not changed by the electrode reaction. Rather, it serves as a chemically inert conveyor of electrons. Electrons are conveyed to the electrode surface if H^+ ions are being reduced to hydrogen gas, or away from it if hydrogen gas is being oxidized to hydrogen ions. The potential of the SHE in either process is the same, 0.000 V.

Because the electrode potential of the SHE is 0.000 V, the value of E_{cell} for any cell in which the SHE is one of the two components is the potential produced by the other component. Suppose, for example, that a cell consists of a strip of zinc metal immersed in a 1.00 M solution of Zn^{2+} ions in one compartment and an SHE in the other. Also suppose that a meter measures the electromotive force pushing electrons from the zinc electrode to the SHE, as shown in Figure 18.4(a). This direction of current flow means that the zinc electrode is the anode because Zn atoms are being oxidized to Zn^{2+} ions. In the process they donate electrons to the external circuit, making the anode the negative pole of the electrochemical cell. The SHE is the cathode and positive electrode of the cell. At 25°C, the meter reads 0.762 V. All the reactants and products are in their standard states, so the measured cell potential represents $E°_{cell}$. The value of $E°_{cathode}$ is that of the SHE ($E°_{SHE} = 0.000$ V), and E_{ox}(anode) is the standard potential of the oxidation of Zn to Zn^{2+}:

$$Zn(s) \rightarrow Zn^{2+}(aq) + 2\,e^-$$

Inserting these values and symbols into Equation 18.7,

$$E°_{cell} = E°_{red}(\text{cathode}) + E°_{ox}(\text{anode})$$

$$E°_{cell} = E°_{SHE} + E°_{ox}(\text{Zn})$$

$$0.762 \text{ V} = 0.000 \text{ V} + E°_{ox}(\text{Zn})$$

$$E°_{ox}(\text{Zn}) = 0.762 \text{ V}$$

CONNECTION Keep in mind that hydrogen ions in aqueous solutions are really hydronium ions, $H_3O^+(aq)$, as described in Chapters 4 and 16.

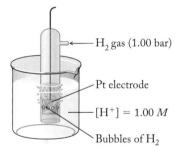

FIGURE 18.3 The potential of the standard hydrogen electrode (SHE) is the reference point (0.000 V) for all other electrochemical potentials. The SHE consists of a wire leading to a platinum electrode immersed in a 1.00 M solution of $H^+(aq)$ and bathed in a stream of pure H_2 gas. Its potential is the same whether $H^+(aq)$ ions are reduced or H_2 gas is oxidized.

The potential of the **standard hydrogen electrode (SHE)** is defined as 0.000 V, and so the SHE serves as a reference electrode against which the potentials of other electrode half-reactions can be measured.

FIGURE 18.4 The standard hydrogen electrode when used as a reference electrode allows the determination of the electrode potential of another half-reaction. (a) When coupled to a Zn electrode under standard conditions, the SHE is the cathode (+), which is connected to the positive terminal of the voltmeter. The Zn electrode is the anode (−), which is connected to the negative terminal of the voltmeter. With this setup, the voltmeter measures an E_{cell} value of 0.762 V, which is the same as $E°_{ox}(Zn^{2+}/Zn)$. (b) When coupled to a Cu electrode under standard conditions, the SHE is the anode(−), the Cu electrode is the cathode (+), and $E_{cell} = E°_{red}(Cu^{2+}/Cu) = 0.342$ V.

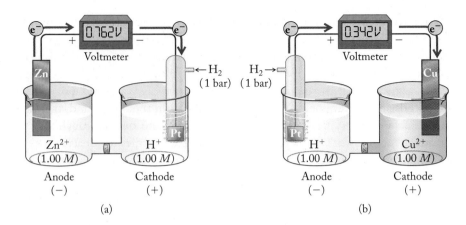

(a)   (b)

This value is equal in magnitude but opposite in sign to the standard reduction potential of Zn^{2+}:

$$Zn^{2+}(aq) + 2e^- \rightarrow Zn(s) \qquad E°_{red} = -0.762 \text{ V}$$

Let's now consider the voltaic cell in Figure 18.4(b). In this case, the SHE is coupled to a copper electrode immersed in a 1.00 M solution of Cu^{2+} ions. In this cell, current flows from the SHE through an external circuit to the copper electrode at a cell potential of 0.342 V at 25°C. The SHE is the anode, which means that it supplies electrons to the external circuit and so is the negative pole of the cell. Copper is the cathode and positive pole of the cell. The value of $E°_{red}(Cu^{2+}/Cu)$ is calculated as follows:

$$E°_{cell} = E°_{red}(\text{cathode}) + E°_{ox}(\text{anode})$$
$$= E°_{red}(Cu^{2+}/Cu) + E°_{SHE}$$
$$0.342 \text{ V} = E°_{red}(Cu^{2+}/Cu) + 0.000 \text{ V}$$
$$E°_{red}(Cu^{2+}/Cu) = 0.342 \text{ V}$$

This half-reaction potential matches the value of $E°_{red}$ for the reduction of Cu^{2+} to Cu metal in Appendix 6.

18.6 The Effect of Concentration on E_{cell}

Reactions stop when one of the reactants is completely consumed. This concept was the basis for our discussion of limiting reactants in Chapter 4. However, a commercial battery usually stops supplying current at its rated voltage before its reactants are completely consumed. Why does the cell potential of a battery decrease when it is nearly discharged? The answer is that the electromotive force produced by a voltaic cell is determined by the concentrations of the reactants and products.

The Nernst Equation

The **Nernst equation** relates the potential of a cell (or half-cell) reaction to its standard potential ($E°$) and the concentrations of its reactants and products.

In 1889, a young German chemist named Walther Nernst (1864–1941) derived an expression, now called the **Nernst equation**, that describes the dependence of cell potentials on reactant and product concentrations. We can reconstruct his derivation starting with Equation 15.11, which relates the change in free energy (ΔG) of any reaction to its change in free energy under standard conditions ($\Delta G°$):

$$\Delta G = \Delta G° + RT \ln Q$$

As we noted in Section 15.4, as a spontaneous reaction proceeds, the free energy changes from the negative $\Delta G°$ values we associate with spontaneous reactions to ΔG values that increase as reactants form products and the value of reaction quotient Q increases. When ΔG reaches zero, the reaction has reached chemical equilibrium and no further conversion of reactants to products takes place.

Now let's write an expression analogous to Equation 15.11 to relate E_{cell} to $E°_{cell}$. We can do so by combining Equations 15.11 and 18.5 ($\Delta G_{cell} = -nFE_{cell}$) to give

$$-nFE_{cell} = -nFE°_{cell} + RT \ln Q$$

Dividing all terms by $-nF$,

$$E_{cell} = E°_{cell} - \frac{RT \ln Q}{nF} \qquad (18.10)$$

This is the equation Nernst developed in 1889. We can obtain a very useful form of the Nernst equation if we replace the symbols of the constants with their actual values. If we also assume that $T = 25°C$ (298 K) and switch from natural logarithms to base 10 logarithms using the conversion: $\ln Q = 2.303 \log Q$, then the coefficient of the logarithmic term has the value 0.0592 and the Nernst equation becomes

$$E_{cell} = E°_{cell} - \frac{0.0592 \log Q}{n} \qquad (18.11)$$

This is the form of the Nernst equation used in most calculations of E_{cell}, including the one in the following Sample Exercise.

> **CONNECTION** In Chapter 15 we discussed the relationship between free energy changes and the reaction quotient, Q.

SAMPLE EXERCISE 18.5 | **Calculating E_{cell} from $E°_{cell}$ and the Concentrations of Reactants and Products**

If the cell reaction in Figure 18.2 proceeded until $[Cu^{2+}] = 0.10\ M$ and $[Zn^{2+}] = 1.90\ M$, what would be the value of E_{cell} at 25°C? Given:

$$Cu^{2+}(aq) + Zn(s) \rightarrow Cu(s) + Zn^{2+}(aq) \qquad E°_{cell} = 1.104\ V$$

COLLECT AND ORGANIZE We are given the standard cell potential (1.104 V) and are asked to determine the value of E_{cell} when $[Cu^{2+}] = 0.10\ M$ and $[Zn^{2+}] = 1.90\ M$. Solid copper and zinc are also part of the reaction system, but no terms for pure solids appear in reaction quotients. The Nernst equation (Equation 18.11) can be used to calculate the value of E_{cell} for different concentrations of reactants and products.

ANALYZE The reaction quotient for the cell reaction has $[Zn^{2+}]$ as the only product term in its numerator and $[Cu^{2+}]$ as the only reactant term in its denominator. Each mole of Cu^{2+} ions acquires 2 moles of electrons, and each mole of Zn atoms donates 2 moles of electrons, so the value of n in the cell reaction is 2. Therefore, the Nernst equation for the Zn/Cu^{2+} cell reaction is

$$E_{cell} = 1.104\ V - \frac{0.0592}{2} \log \frac{[Zn^{2+}]}{[Cu^{2+}]}$$

SOLVE Substituting the values of $[Zn^{2+}]$ and $[Cu^{2+}]$ in the Nernst equation gives

$$E_{cell} = 1.104 \text{ V} - \frac{0.0592}{2} \log \frac{1.90}{0.10}$$

Solving for E_{cell}, we get

$$E_{cell} = 1.104 \text{ V} - \frac{0.0592}{2} (1.28) = 1.066 \text{ V}$$

THINK ABOUT IT The calculated E_{cell} value is 0.038 V less than $E°_{cell}$. This decrease makes sense because the cell is 90% discharged and so a drop in voltage is to be expected. The drop is only 0.038 V because the logarithmic dependence of cell potential on the concentrations of reactants and products minimizes the impact of changing concentrations. How can we have a calculated E_{cell} value with four significant figures when we know the value of $[Cu^{2+}]$ to only two? The concentration values were used to calculate the *change* in cell potential, which was rounded off to only two significant figures: −0.038 V. That number was then combined with 1.104 V to obtain an answer with four significant figures.

Practice Exercise What is the cell potential of the zinc-air battery if the partial pressure of oxygen in the air diffusing through its cathode is 0.21 bar?

$$\text{Given:} \quad 2\,Zn(s) + O_2(g) \rightarrow 2\,ZnO(s) \qquad E°_{cell} = 1.65 \text{ V}$$

Equilibrium and Dead Batteries

As a battery's cell reaction proceeds, the concentrations of the reactants decrease and the concentrations of the products increase. Therefore, the value of Q (and the logarithmic term in Equation 18.11) increases. Because of the negative sign in front of the logarithmic term, the value of E_{cell} decreases. Thus, the voltage produced by a battery decreases as the concentrations of its reactants decrease. Let's consider a common battery design that illustrates this point. In virtually every automobile, a *lead-acid* battery is used to drive an electric motor that starts the engine.

The lead-acid battery (Figure 18.5) that provides power to start most motor vehicles contains six cells with anodes made of lead and cathodes made of PbO_2

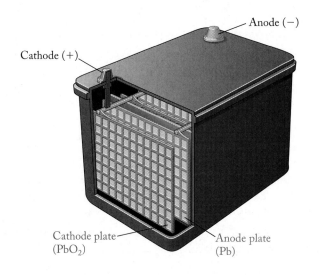

FIGURE 18.5 The lead-acid battery that provides power to start most motor vehicles contains six cells. Each has an anode made of lead and a cathode made of PbO_2 immersed in an electrolyte of 4.5 M H_2SO_4. The electrodes are formed into plates and held in place by grids made of a lead alloy. The grids connect the cells together in series so that the operating voltage of the battery (12.0 V) is the sum of the six cell potentials (each 2.0 V).

Cathode (+)

Anode (−)

Cathode plate (PbO₂)

Anode plate (Pb)

immersed in the electrolyte 4.5 $M \, H_2SO_4$. The electrodes are formed into plates and held in place by grids made of a lead alloy. The grids connect the cells (each providing 2.0 V) together in series so that the operating voltage of the battery is the sum of the six cell potentials, or 12.0 V.

As a lead-acid battery is discharged, PbO_2 is reduced to solid $PbSO_4$ and Pb is oxidized to solid $PbSO_4$. The anode half-reaction involves the loss of 2 moles of electrons for each mole of lead. Writing this half-reaction as an oxidation,

$$Pb(s) + SO_4^{2-}(aq) \rightarrow PbSO_4(s) + 2\,e^- \qquad E^{\circ}_{ox} = 0.356 \text{ V}$$

At the cathodes, PbO_2 is reduced to $PbSO_4$:

$$PbO_2(s) + 4\,H^+(aq) + SO_4^{2-}(aq) + 2\,e^- \rightarrow$$
$$PbSO_4(s) + 2\,H_2O(\ell) \qquad E^{\circ}_{red} = 1.685 \text{ V}$$

in a half-reaction that consumes 2 moles of electrons. Combining these two half-reactions gives the following overall cell reaction with a standard cell potential of 2.041 V:

$$Pb(s) + PbO_2(s) + 2\,H_2SO_4(aq) \rightarrow 2\,PbSO_4(s) + 2\,H_2O(\ell) \qquad E^{\circ}_{cell} = 2.041 \text{ V}$$

As the battery is discharged, $[H_2SO_4]$ decreases, and so does the value of E_{cell} calculated from the Nernst equation for the cell reaction:

$$E_{cell} = 2.04 \text{ V} - \frac{0.0592}{2} \log \frac{1}{[H_2SO_4]^2}$$

The decrease in E_{cell} is very gradual, not falling below 2.00 V until the battery is 97% discharged as shown in Figure 18.6. When the reactants are nearly consumed, the cell potential should approach zero as the cell reaction achieves chemical equilibrium. At this point the value of the reaction quotient (Q) is equal to the equilibrium constant (K) for the cell reaction. At equilibrium the cell potential drops to zero. Therefore, Equation 18.11 becomes

$$0 = E^{\circ}_{cell} - \frac{0.0592}{n} \log K$$

or

$$\log K = \frac{nE^{\circ}_{cell}}{0.0592} \qquad (18.12)$$

Equation 18.12 makes a mathematical connection between the equilibrium constant of a cell reaction and the standard cell potential. We can use this connection to calculate equilibrium constants for any redox reaction, not just those in electrochemical cells. For the more general case, we substitute E°_{rxn} for E°_{cell} in Equation 18.12 to give

$$\log K = \frac{nE^{\circ}_{rxn}}{0.0592} \qquad (18.13)$$

as in Sample Exercise 18.6.

The potential of every voltaic cell decreases as reactants are converted into products. In the case described here, the value of E°_{cell} of the lead-acid battery cell reaction decreases as its sulfuric acid is consumed, but this decrease is only gradual until the battery is nearly completely discharged.

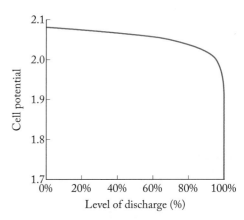

FIGURE 18.6 The potential of a cell in a lead-acid battery decreases as reactants are converted into products, but the changes in potential are small until the battery is nearly completely discharged.

SAMPLE EXERCISE 18.6 **Calculating the Equilibrium Constant of a Redox Reaction from the Standard Potentials of Its Half-Reactions**

Many of the methods used to determine the levels of mercury contamination in environmental samples begin by oxidizing all the mercury in the samples to Hg^{2+} and then reducing the Hg^{2+} to elemental Hg with Sn^{2+}.

Use the appropriate standard potentials in Appendix 6 to calculate the equilibrium constant at 25°C for the reaction:

$$Sn^{2+}(aq) + Hg^{2+}(aq) \rightarrow Sn^{4+}(aq) + Hg(\ell)$$

COLLECT AND ORGANIZE We are asked to calculate the equilibrium constant for a redox reaction starting with the standard potentials of its component half-reactions. Equation 18.12 relates the value E°_{cell} to the value of the equilibrium constant of the cell reaction. As we have seen, we can use Equation 18.13 to calculate the equilibrium constant for any redox reaction, using the standard potential E°_{rxn}.

To calculate E°_{rxn} we need to combine the appropriate standard potentials. Appendix 6 lists the following half-reactions involving the reactants and products:

$$Hg^{2+}(aq) + 2\,e^- \rightarrow Hg(\ell) \qquad E^\circ_{red} = 0.851\ V$$

$$Sn^{4+}(aq) + 2\,e^- \rightarrow Sn^{2+}(aq) \qquad E^\circ_{red} = 0.154\ V$$

ANALYZE Sn^{2+} is the reducing agent in the reaction, so it must be oxidized. Therefore, the second reaction runs in reverse, as an oxidation:

$$Sn^{2+}(aq) \rightarrow Sn^{4+}(aq) + 2\,e^- \qquad E^\circ_{ox} = -0.154\ V$$

SOLVE Combining the half-reactions for the oxidation of Sn^{2+} and the reduction of Hg^{2+} and summing their standard potentials,

$$Hg^{2+}(aq) + 2\,e^- \rightarrow Hg(\ell) \qquad\qquad E^\circ_{red} = 0.851\ V$$

$$\underline{Sn^{2+}(aq) \rightarrow Sn^{4+}(aq) + 2\,e^- \qquad\qquad E^\circ_{ox} = -0.154\ V}$$

$$Hg^{2+}(aq) + Sn^{2+}(aq) \rightarrow Hg(\ell) + Sn^{4+}(aq) \qquad E^\circ_{rxn} = 0.697\ V$$

Using this reaction potential for the cell potential in Equation 18.13 and inserting a value of 2 for n,

$$\log K = \frac{nE^\circ_{rxn}}{0.0592} = \frac{2(0.697)}{0.0592} = 23.5$$

$$K = 10^{23.5} = 3 \times 10^{23}$$

THINK ABOUT IT Note how a positive value of E°_{rxn} corresponds to a huge equilibrium constant, indicating that the reaction essentially goes to completion even though E°_{rxn} is only about 0.7 V. That the reaction goes to completion is one reason why it can be reliably used to determine the concentrations of mercury in aqueous samples containing Hg^{2+} ions.

Practice Exercise Using the standard reduction potentials in Appendix 6, calculate the value of the equilibrium constant at 25°C for the reaction

$$5\,Fe^{2+}(aq) + MnO_4^-(aq) + 8\,H^+(aq) \rightarrow 5\,Fe^{3+}(aq) + Mn^{2+}(aq) + 4\,H_2O(\ell)$$

18.7 Relating Battery Capacity to Quantities of Reactants

An important performance characteristic of a battery is its capacity to do electrical work (w_{elec}), which means its capacity to deliver electrical charge at its designed cell potential. This capacity was previously defined by Equation 18.2

$$w_{elec} = CE_{cell}$$

In our description of electrical work in Section 18.3 we noted that 1 joule of electrical energy is equivalent to 1 coulomb · volt of electrical work. Another important unit in electricity is the ampere (A), which is the SI base unit of electrical current. An ampere is defined as a current flow rate of 1 coulomb per second:

$$1 \text{ ampere} = 1 \text{ coulomb/second}$$

or

$$1 \text{ coulomb} = 1 \text{ ampere} \cdot \text{second}$$

Multiplying both sides of the equation by volts, we get

$$1 \text{ coulomb} \cdot \text{volt} = 1 \text{ joule} = \text{volt} \cdot \text{ampere} \cdot \text{second}$$

Joules are rather small units, and so the capacities of batteries such as those used to power portable electronic devices and flashlights are usually expressed in units with longer time intervals. For example, the energy ratings of rechargeable AA batteries are often expressed in ampere-hours, or milliampere-hours, at a particular voltage. The energy stored in much larger batteries, such as the rechargeable battery packs used in hybrid vehicles, is usually expressed in kilowatt-hours (kWh) where

$$1 \text{ watt} = 1 \text{ volt} \cdot \text{ampere}$$

Note that a watt (W) is equivalent to a joule per second and so represents the *rate* of energy production or consumption. In the physical sciences we call this *power*; the watt is the SI unit of power. The power of the electric motors used in hybrid vehicles is sometimes expressed in horsepower (a unit widely used in the United States) and at other times in kilowatts (the SI-derived unit used in much of the rest of the world).

Most hybrids get their best fuel mileage in low-speed city driving where most of the power is supplied by the more fuel-efficient electrical power system. For example, the U.S. EPA estimates that the 2008 Ford Escape gets 34 miles per gallon (mpg) of gasoline in city driving and 30 mpg (7.9 liters per 100 km) in highway driving.

Nickel–Metal Hydride Batteries

The electrochemical cells in most rechargeable batteries and in the battery packs in hybrid cars are called nickel–metal hydride (NiMH) cells (Figure 18.7). In these cells $NiO(OH)$ is reduced to $Ni(OH)_2$ at the cathode and hydrogen is oxidized at an anode made of one or more transition metals. The two electrodes are separated by a porous spacer containing an aqueous solution of KOH. The cathode half-reaction is

$$NiO(OH)(s) + H_2O(\ell) + e^- \rightarrow Ni(OH)_2(s) + OH^-(aq) \qquad E° = 1.32 \text{ V}$$

At the anode hydrogen is present not as H_2, but rather as a *metal hydride*. To write the anode half-reaction we use the generic formula MH where M stands for a

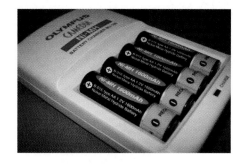

The electrical energy rating of these rechargeable AA batteries is 1.6A · hr at 1.2 V.

The 2008 Ford Escape is powered by a 133 horsepower gasoline engine and a 94 horsepower (70 kW) electric motor. As in most of the first-generation hybrid vehicles, power for the Escape's electric motor comes from a pack of nickel–metal hydride (NiMH) batteries under the floor.

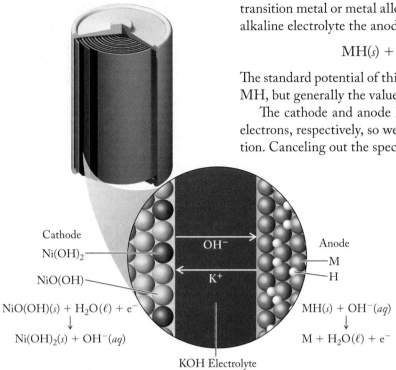

Cathode
Ni(OH)$_2$

NiO(OH)

NiO(OH)(s) + H$_2$O(ℓ) + e$^-$
$\downarrow$
Ni(OH)$_2$(s) + OH$^-$(aq)

OH$^-$

K$^+$

KOH Electrolyte

Anode
M
H

MH(s) + OH$^-$(aq)
$\downarrow$
M + H$_2$O(ℓ) + e$^-$

FIGURE 18.7 What a Chemist Sees In the nickel–metal hydride (NiMH) battery, hydrogen atoms are stored in the interstices of a metal matrix. During the anode half-reaction they are oxidized to H$^+$ ions that combine with OH$^-$ ions, forming H$_2$O and releasing electrons. At the cathode, NiO(OH) is reduced to Ni(OH)$_2$, consuming electrons and releasing OH$^-$ ions to the electrolyte.

transition metal or metal alloy that forms a hydride and serves as the anode. In an alkaline electrolyte the anode half-reaction can be written

$$MH(s) + OH^-(aq) \rightarrow M(s) + H_2O(\ell) + e^-$$

The standard potential of this half-reaction depends on the chemical properties of MH, but generally the value is near that of the SHE, 0.0 V.

The cathode and anode half-reactions involve the gain and loss of 1 mole of electrons, respectively, so we can simply sum them to obtain the overall cell reaction. Canceling out the species on both sides of the reaction arrow, we get

$$MH(s) + NiO(OH)(s) \rightarrow M(s) + Ni(OH)_2(s)$$

The value of $E°_{cell}$ for the NiMH battery cannot be calculated precisely because we have only an approximate value of $E°_{ox}$(anode). Most NiMH cells deliver about 1.2 V.

Now let's relate the electrical energy stored in a battery to the quantity of reactants that are needed to create that energy. Consider a rechargeable AA battery. It is rated to deliver 2.5 ampere-hours (A · hr) of electrical charge at 1.2 V. How much NiO(OH) has to be converted to Ni(OH)$_2$ to deliver this much charge? To answer this question we need to relate the quantity of charge to a number of moles of electrons and then convert that to an equivalent number of moles of reactant and finally to a mass of reactant. Let's begin by noting that an ampere is the same as a coulomb per second (C/s), so the quantity of electrical charge delivered is

$$2.5 \ \text{A} \cdot \text{hr} \times \frac{1 \ C}{\text{A} \cdot \text{s}} \times \frac{60 \ \text{min}}{\text{hr}} \times \frac{60 \ \text{s}}{\text{min}} = 9.0 \times 10^3 \ C$$

A mole of charge is equivalent to 9.65×10^4 coulombs, so the number of moles of charge, which is equal to the number of moles of electrons that flow from the battery, is

$$9.0 \times 10^3 \ C \left(\frac{1 \ \text{mol e}^-}{9.65 \times 10^4 \ C} \right) = 0.0933 \ \text{mol e}^-$$

The stoichiometry of the anode half-reaction (1 mole of electrons per mole of NiO(OH)) leads to the following approach to calculating the mass of NiO(OH) consumed:

$$0.0933 \ \text{mol e}^- \left(\frac{1 \ \text{mol NiO(OH)}}{1 \ \text{mol e}^-} \right)\left(\frac{91.701 \ g}{1 \ \text{mol}} \right) = 8.6 \ g \ \text{NiO(OH)}$$

The mass of an AA battery is 30 g so the result of the calculation is reasonable if we allow for the mass of the anode material, the electrolyte, and the battery's exterior shell.

Lithium-Ion Batteries

Many electronic devices, including laptop computers, cell phones, and digital cameras, are powered by lithium-ion batteries (Figure 18.8). When these batteries are fully charged, Li$^+$ ions are stored in anodes made of highly pure graphite. During discharge, the Li$^+$ ions migrate toward highly porous cathodes made of transition-

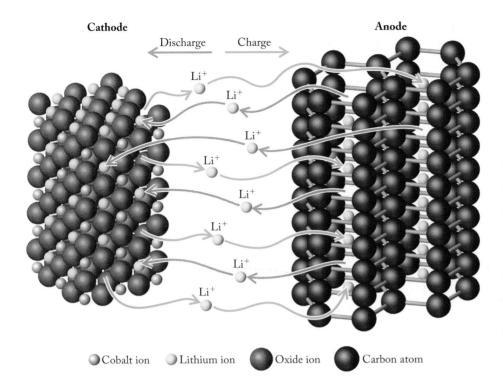

Cathode

Discharge   Charge

Li^+
Li^+
Li^+
Li^+
Li^+
Li^+
Li^+
Li^+

Anode

○ Cobalt ion   ○ Lithium ion   ● Oxide ion   ● Carbon atom

FIGURE 18.8 What a Chemist Sees This molecular view of a lithium-ion battery shows the migration of Li^+ ions stored within graphite layers of the anode toward a cathode made of CoO_2 as the battery is discharged. The crystal structure of the cathode is based on a cubic closest packed array of oxide ions in which Co ions occupy half the octahedral holes. Li^+ ions can move in and out of the other half of the holes. The migration pattern is reversed when the battery is recharged.

metal oxides or phosphates that form stable complexes with Li^+ ions. One popular cathode material is cobalt oxide. Lithium-ion batteries with this cathode material have a cell potential of 3.6 V (three times that of a NiMH battery) and are used to power most laptop computers, cell phones, and digital cameras. We can write the cell reaction for a lithium-ion battery with this cathode as follows:

$$Li_{1-x}CoO_2(s) + Li_xC_6(s) \rightarrow 6\,C(s) + LiCoO_2(s)$$

In a fully charged cell the value of x in this equation is 1, which makes the formula of the cathode CoO_2. This formula means that Co is in its +4 oxidation state. As the cell discharges and Li^+ ions migrate from the carbon anode toward the cobalt oxide cathode, the value of x falls from one toward zero. To balance this flow of positive charges inside the cell, electrons and their negative charges flow from the anode toward the cathode through an external circuit. If fully discharged, the formula of the cathode would be $LiCoO_2$, and Co would be in its +3 oxidation state. The electrodes in lithium-ion batteries may react with oxygen and water, so these batteries use electrolytes (for example, $LiPF_6$) that dissolve in polar organic solvents such as tetrahydrofuran, ethylene carbonate, or propylene carbonate.

Tetrahydrofuran

Ethylene carbonate

Propylene carbonate

SAMPLE EXERCISE 18.7   **Relating the Mass of Reactant in an Electrochemical Reaction to a Quantity of Electrical Charge**

The energy capacity of the lithium-ion battery in a digital camera is 3.4 watt-hours at 3.6 V. How many grams of Li^+ ions must migrate from the anode to the cathode of the battery to produce this much electricity?

COLLECT AND ORGANIZE We are asked to convert the electrical energy generated by an electrochemical cell into the mass of the ions carrying that current. We know the cell potential in volts and its energy capacity in watt-hours. Given these starting points and the need eventually to calculate moles and grams of Li^+ ions, the following equivalencies are likely to be useful:

$$1 \text{ watt} = 1 \text{ ampere (A)} \cdot \text{volt (V)}$$

$$1 \text{ coulomb (C)} = 1 \text{ ampere (A)} \cdot \text{second (s)}$$

We may also need to use Faraday's constant, 9.65×10^4 C/mol.

ANALYZE A logical sequence of steps would be (1) to combine the starting potential and energy values to calculate the quantity of charge delivered by the battery and then (2) relate that quantity to the moles of Li^+ ions. In setting up our conversion factors we need to convert watts to ampere-volts and divide by the cell potential to cancel out volts.

SOLVE Using the equivalencies provided, we can calculate the amount of charge from the starting values:

$$3.4 \text{ W} \cdot \text{hr} \times \frac{1 \text{ A} \cdot \text{V}}{\text{W}} \times \frac{1}{3.6 \text{ V}} \times \frac{60 \text{ min}}{\text{hr}} \times \frac{60 \text{ s}}{\text{min}} \times \frac{1 \text{ C}}{\text{A} \cdot \text{s}} = 3.4 \times 10^3 \text{ C}$$

Dividing coulombs by Faraday's constant,

$$3.4 \times 10^3 \text{ C} \left(\frac{1 \text{ mol e}^-}{9.65 \times 10^4 \text{ C}} \right) = 0.0352 \text{ mol e}^-$$

For every mole of electrons that flows through an external circuit, one mole of singly charged Li^+ ions migrates from the anode to the cathode inside the Li-ion battery that powers the circuit. This equality allows us to convert 0.0352 moles of electrons into an equivalent mass of Li^+ ions:

$$0.0352 \text{ mol e}^- \left(\frac{1 \text{ mol Li}^+}{\text{mol e}^-} \right) \left(\frac{6.941 \text{ g Li}^+}{\text{mol Li}} \right) = 0.245 \text{ g Li}$$

THINK ABOUT IT The battery that is the subject of this exercise weighs about 22 grams, so Li^+ ions make up only about 1% of the mass of this lithium-ion battery. This small percentage is reasonable given the masses of the electrode materials that donate and accept the ions. In the anode each Li^+ ion is surrounded by a hexagon of 6 carbon atoms (Figure 18.8) with about 10 times the mass of the Li^+ ion. At the cobalt oxide cathode the corresponding mass ratio is about 13:1.

Practice Exercise Magnesium metal is produced by passing an electrical current through molten $MgCl_2$. The reaction at the cathode is

$$Mg^{2+}(\ell) + 2\,e^- \rightarrow Mg(\ell)$$

How many grams of magnesium metal are produced if an average current of 63.7 amperes flows for 4.50 hours? Assume all of the current is consumed by the half-reaction shown.

18.8 Electrolytic Cells and Rechargeable Batteries

NiMH and lithium-ion batteries are rechargeable batteries, which means that the cell reactions in which chemical energy is converted into electrical energy can be reversed. They can be recharged because the products of the discharge reaction are substances that adhere to, or are embedded within, the electrodes. These substances are therefore available to be converted back to their original chemical forms when an opposing electrical potential is applied that is greater than the cell potential.

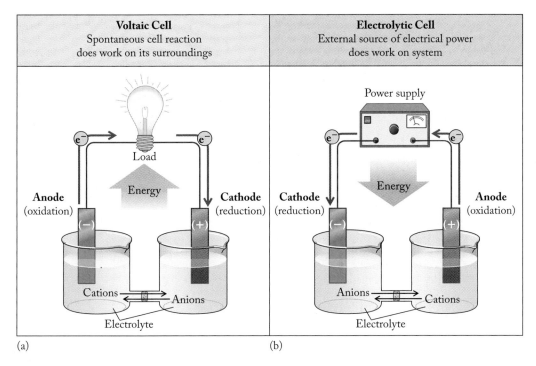

Voltaic Cell	Electrolytic Cell
Spontaneous cell reaction does work on its surroundings	External source of electrical power does work on system

(a)   (b)

FIGURE 18.9 (a) In a voltaic cell, a spontaneous chemical reaction produces electrical energy and does electrical work on its surroundings, such as lighting a light bulb. (b) In an electrolytic cell, an external supply of electrical energy does work on the chemical system in the cell, driving a nonspontaneous reaction.

Applying an opposing potential forces the cell reaction to run backward, converting electrical energy back into chemical energy. Recharging a battery is an example of **electrolysis**, that is, a chemical reaction driven by electricity. During recharging, a battery is transformed from a voltaic cell into an **electrolytic cell**, a process in which electrical energy is converted into chemical energy (Figure 18.9).

The chemical changes that take place in a lead-acid battery in the spontaneous discharge process and during nonspontaneous recharging are shown in Figure 18.10. During discharging and recharging, current flows in opposite directions through the battery. Thus, the electrodes that serve as anodes (the sites of oxidation) during the discharge reaction turn into cathodes (the sites of reduction) during recharging. Any $PbSO_4$ that formed on them during discharge is reduced to Pb during recharging:

$$PbSO_4(s) + 2\,e^- \rightarrow Pb(s) + SO_4^{2-}(aq) \qquad E°_{red} = -0.356 \text{ V}$$

Similarly, the PbO_2 electrodes that serve as cathodes during discharge become anodes during recharging, as any $PbSO_4$ that formed on the cathodes during discharge is oxidized to PbO_2:

$$PbSO_4(s) + 2\,H_2O(\ell) \rightarrow$$
$$PbO_2(s) + 4\,H^+(aq) + SO_4^{2-}(aq) + 2\,e^- \qquad E°_{ox} = -1.685 \text{ V}$$

Recharging a lead-acid battery or any battery with an aqueous electrolyte can result in some undesirable and even dangerous side reactions. If too high a voltage forces too great a current through a lead-acid battery, the reduction of lead sulfate at the lead electrode may not keep up with the flow of electrons to the electrode surface. In that event, some other reducible species, such as the hydrogen ions in a concentrated solution of sulfuric acid, may accept the influx of electrons:

$$2\,H^+(aq) + 2\,e^- \rightarrow H_2(g) \qquad E° = 0.000 \text{ V}$$

If both H^+ ions and $PbSO_4$ are present at the surface of a lead electrode, which species is preferentially reduced? Note that the $E°$ for reducing $PbSO_4$ (-0.356 V) is more negative than the standard potential for reducing H^+ ions (0.000 V). In an electrolytic cell a species that is reduced at a less negative potential should

Electrolysis is a process in which electrical energy is used to drive a nonspontaneous chemical reaction.

An **electrolytic cell** is a device in which an external source of electrical energy does work on a chemical system, turning reactant(s) into higher-energy product(s).

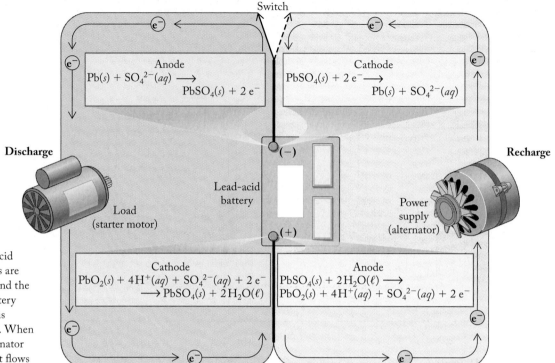

FIGURE 18.10 The lead-acid batteries used in many vehicles are based on the oxidation of Pb and the reduction of PbO_2. As the battery discharges (green circuit), Pb is oxidized, and PbO_2 is reduced. When the engine is running, an alternator generates electrical energy that flows into the battery, recharging it as both electrode reactions are reversed: $PbSO_4$ is oxidized to PbO_2, and $PbSO_4$ is reduced to Pb.

be preferentially reduced compared with one requiring a more negative potential. On this basis, we might conclude that H^+ ions would be reduced to hydrogen gas before lead sulfate is reduced to lead metal.

However, another factor affects the potential at which an electrode reaction begins during electrolysis. It has to do with the rates of the electron transfer processes at the electrode surfaces.

Just as the rates of some spontaneous chemical reactions are slow, so too some half-reactions, particularly those in which a gas evolves, may be extremely slow. In homogeneous reactions we associate slow reaction rates with high activation energies. One way to overcome this barrier is to raise the temperature of the reaction mixture. In electrolysis, we can overcome a slow electrode reaction rate by applying an **overpotential**, that is, an even greater potential than predicted by calculated electrode potentials. The reduction of H^+ ions to H_2 is such a reaction. Although there is little overpotential for the reduction of H^+ ions at the platinum electrodes used in the standard hydrogen electrodes, there is a large overpotential of 1.0 V or more for reducing H^+ ions at a lead electrode. This overpotential means that $PbSO_4$ is reduced before H^+ ions when a lead-acid battery is recharged.

Overpotentials also play a role in controlling the oxidation process that occurs at the anode of the lead-acid battery during recharging. There the desired oxidation process is the conversion of $PbSO_4$ into PbO_2. However, another species is present in abundance that is, theoretically, more easily oxidized than $PbSO_4$, namely H_2O.

$$2\,H_2O(\ell) \rightarrow O_2(g) + 4\,H^+(aq) + 4\,e^- \qquad E^\circ_{ox} = -1.229 \text{ V}$$

An **overpotential** is the additional potential above the theoretically calculated value required to promote an electrode reaction in electrolysis.

The standard potential for the oxidation of water in acid solutions is more than 0.4 V less negative than the lead reaction. However, the water half-reaction involves the generation of the gas O_2 and has a significant overpotential that inhibits water from oxidizing before $PbSO_4$ oxidizes.

SAMPLE EXERCISE 18.8   **Calculating the Time Required to Electrolyze a Quantity of Reactant**

A battery charger for AA NiMH batteries supplies a charging current of 1.00 A. How long will it take to oxidize 0.649 g of $Ni(OH)_2$ to $NiO(OH)$?

COLLECT AND ORGANIZE We are asked to calculate the time required for a constant current to electrolyze a given mass of reactant. During recharging, the spontaneous half-reaction runs in reverse and $Ni(OH)_2$ is oxidized to $NiO(OH)$ at the anode:

$$Ni(OH)_2(s) + OH^-(aq) \rightarrow NiO(OH)(s) + H_2O(\ell) + e^- \qquad E° = -1.32 \text{ V}$$

One mole of electrons is produced for each mole of $Ni(OH)_2$ oxidized.

ANALYZE We need to convert the given mass of $Ni(OH)_2$ into moles of $Ni(OH)_2$. Since the half-reaction involves the transfer of 1 mol of electron, that is also the number of moles of electrons pulled away from the electrode. Faraday's constant can be used to convert moles of electrons to coulombs of charge, and finally coulombs can be converted to an electrolysis time by dividing them by the amount of current flowing through the cell.

SOLVE First we need to convert the mass of $Ni(OH)_2$ into an equivalent number of moles of electrons:

$$n_{e^-} = 0.649 \text{ g } Ni(OH)_2 \times \frac{1 \text{ mol } Ni(OH)_2}{92.71 \text{ g}} \times \frac{1 \text{ mol } e^-}{\text{mol } Ni(OH)_2} = 0.00700 \text{ mol } e^-$$

Converting moles of electrons to an equivalent positive electrical charge expressed in ampere-seconds gives:

$$0.00700 \text{ mol } e^- \times \frac{9.65 \times 10^4 \text{ C}}{\text{mol } e^-} \times \frac{1 \text{ A} \cdot \text{s}}{\text{C}} = 676 \text{ A} \cdot \text{s}$$

Therefore, the charger must deliver 1.00 A of current for 676 s or

$$(676 \text{ s}) \times \frac{1 \text{ min}}{60 \text{ s}} = 11.3 \text{ min}$$

THINK ABOUT IT For those experienced with the time it takes to recharge batteries for portable devices, an interval of 11.3 min may seem short, but if we compare the quantity of the $Ni(OH)_2$ to be oxidized with the quantity of $NiO(OH)$ in a fully charged AA size NiMH battery (see page 910), we realize that this battery was only slightly discharged.

Practice Exercise Suppose that a car's starter motor draws 230 A of current for 6.0 s to start the car's engine on a cold winter morning. What mass of lead would have been oxidized to $PbSO_4$ in the car's battery to supply this much electricity?

Electrolysis is used in many processes other than the recharging of batteries. Electrolytic cells are used to electroplate thin layers of silver (Figure 18.11), gold, and other metals onto objects, giving these objects the appearance, resistance to corrosion, and other properties of the electroplated metal, but at a fraction of the cost of fabricating the entire object out of the metal.

In the chemical industry, the electrolysis of molten salts is used to produce high-energy, highly reactive substances, such as chlorine and fluorine, alkali and alkaline earth metals, and aluminum. When a salt such as NaCl is heated just

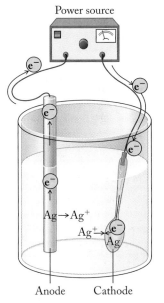

FIGURE 18.11 Silverware typically has only a thin coating of silver, which is applied by a type of electrolysis known as electroplating. In the process, the positive pole of an electric power supply is attached to a piece of pure silver, and the negative pole is connected to the cutlery, such as a spoon, to be electroplated. The oxidation half-reaction at the silver electrode (anode), $Ag \rightarrow Ag^+ + e^-$, runs in reverse at the cathode (spoon) as Ag^+ ion is reduced to the free metal at the surfaces of the cutlery.

above its melting point (above 800°C), it becomes an ionic liquid with an enormous capacity to conduct electricity. If a sufficiently large voltage is applied to carbon electrodes immersed in molten NaCl, sodium ions are attracted to the negative electrode and are reduced, producing sodium metal:

$$(1) \qquad Na^+(\ell) + e^- \rightarrow Na(\ell)$$

Chloride ions will be attracted to the positive electrode and oxidized:

$$(2) \qquad 2\,Cl^-(\ell) \rightarrow Cl_2(g) + 2\,e^-$$

Multiplying reaction (1) by two and combining it with reaction (2), we have the overall electrolytic cell reaction:

$$2\,Na^+(\ell) + 2\,Cl^-(\ell) \rightarrow 2\,Na(\ell) + Cl_2(g)$$

A final note in summary about the polarities of the anodes and cathodes in electrolytic and voltaic cells is in order. The reactions in voltaic cells are spontaneous. These cells produce electric current. The anode in a voltaic cell is negative, and the oxidation reaction at the anode spontaneously releases electrons which flow toward a positive cathode. Reduction takes place at the cathode.

The reactions in electrolytic cells are nonspontaneous. They require electric current (electrons) from an external power supply. These electrons are pushed into the cathode, which therefore has a negative charge, and reduction takes place at the cathode. Electrons flow from the negative cathode to the positive anode, where oxidation takes place.

CONCEPT TEST

The electrolysis of molten NaCl produces Na metal at the cathode. However, the electrolysis of an aqueous solution of NaCl results in the formation of gases at both the cathode and the anode. On the basis of the standard potentials in Appendix 6, predict which gases are formed at each electrode.

18.9 Fuel Cells

CHEMTOUR Fuel Cell

Fuel cells are promising energy sources for many applications, from powering office buildings to cruise ships to electric vehicles. They are voltaic cells like batteries, and they rely on oxidizing a fuel such as H_2 or CH_4 to generate electrical energy.

$$2\,H_2(g) + O_2(g) \rightarrow 2\,H_2O(\ell)$$

In acidic electrolytes, the two half-reactions are

(1) At the anode:

$$H_2(g) \rightarrow 2\,H^+(aq) + 2\,e^- \qquad E^\circ_{ox} = 0.000 \text{ V}$$

(2) At the cathode:

$$O_2(g) + 4\,H^+(aq) + 4\,e^- \rightarrow 2\,H_2O(\ell) \qquad E^\circ_{red} = 1.229 \text{ V}$$

In fuel cells with alkaline material between the electrodes, such as molten sodium carbonate, the alkaline versions of these half-reactions occur:

(1) At the anode:

$$H_2(g) + 2\,OH^-(aq) \rightarrow 2\,H_2O(\ell) + 2\,e^- \qquad E^\circ_{ox} = 0.828 \text{ V}$$

A **fuel cell** is a voltaic cell in which the reaction is the oxidation of a fuel. The reaction is the equivalent of combustion, but the chemical energy is converted into electrical energy.

(2) At the cathode:

$$O_2(g) + 2\,H_2O(\ell) + 4\,e^- \rightarrow 4\,OH^-(aq) \qquad E^\circ_{red} = 0.401\ V$$

Note that, if we combine these two pairs of standard potentials by combining the standard potentials of each pair of half-reactions, we get the same E°_{cell} value (1.229 V). This equality is logical because the energy released under standard conditions (ΔG°) by the oxidation of hydrogen gas to form liquid water should have but one value, and so E°_{cell} should have only one value that does not depend on electrolyte pH.

The same free-energy change could be used to power a vehicle by burning hydrogen in an internal combustion engine. However, typically less than a third of the energy from combustion is converted into mechanical energy in the internal combustion engines of automobiles. Most of the energy of combustion is lost to the surroundings as heat. In contrast, fuel-cell technologies can convert most of the energy released in a fuel-cell reaction to electrical energy. In addition, H_2-fueled vehicles emit only water vapor; they produce no oxides of nitrogen that can contribute to smog formation, and no carbon monoxide or carbon dioxide is released.

Fuel cells are like batteries in that their chemistry is based on two half-reactions occuring in environments separated from each other by a porous barrier through which electrolyte ions can pass. In one design, O_2 is supplied to a cathode made of porous graphite containing atoms of Ni, and H_2 gas is supplied to a graphite anode containing nickel(II) oxide. Both electrodes are immersed in an alkaline electrolyte. Hydroxide ions form during the reduction of O_2 at the cathode:

$$O_2(g) + 2\,H_2O(\ell) + 4\,e^- \rightarrow 4\,OH^-(aq)$$

These migrate through the cell to the anode, where they combine with H_2 as it is oxidized to water:

$$H_2(g) + 2\,OH^-(aq) \rightarrow 2\,H_2O(\ell) + 2\,e^-$$

Fuel cells used in electric vehicles incorporate acidic electrolytes and a proton-exchange membrane (PEM) to separate the anode and cathode half-reactions (Figure 18.12). On the anode side of the cell H_2 fuel diffuses into a porous graphite electrode, where it is oxidized to H^+ ions:

$$H_2(g) \rightarrow 2\,H^+(aq) + 2\,e^-$$

The electrons are pushed through an external circuit by an E_{cell} value of about 0.7 volts, which is about half the standard potential for the cell reaction

$$2\,H_2(g) + O_2(g) \rightarrow 2\,H_2O(\ell)$$

As electrons are pumped out of the cell through the anode, H^+ ions migrate through the proton-exchange membrane to the cathode, where they participate in the reduction of O_2:

$$O_2(g) + 4\,H^+(aq) + 4\,e^- \rightarrow 2\,H_2O(\ell)$$

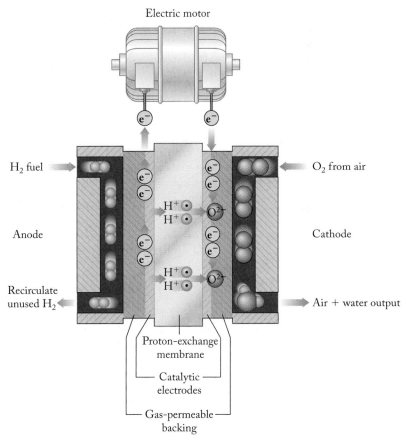

FIGURE 18.12 Most of the fuel cells used in vehicles incorporate a proton-exchange membrane (PEM) between the two halves of the cell. Hydrogen diffuses to the anode and oxygen diffuses to the cathode. These electrodes are made of porous material, such as carbon nanofibers, that have high surface areas for a given mass of material. The electrodes also contain transition metals that serve as catalysts to speed up the electrode half-reactions by increasing the rate at which the bonds that hold molecules of H_2 and O_2 together are broken. Platinum catalysts are effective at promoting H–H bond breaking, and a platinum–nickel alloy with the formula Pt_3Ni is particularly effective in catalyzing the formation of free O atoms from O_2 molecules. These catalysts increase the rate of the overall cell reaction, $2\,H_2(g) + O_2(g) \rightarrow 2\,H_2O(\ell)$, and the power output of the fuel cell.

Power, Pigments, and Phosphors: The Group 12 Elements

The last group in the transition metals (group 12, Zn, Cd, Hg) contains elements used in many batteries. None of these three metals is particularly abundant in Earth's crust. Zinc is the most abundant of the three (78 mg/kg), and mercury is the least (67 μg/kg). All three are found in sulfide ores, including sphalerite (ZnS), greenockite (CdS), and cinnabar (HgS). Zinc is also found in carbonates and silicates. Most zinc ores contain cadmium impurities.

The cathode ray tubes we discussed in Chapter 2 used by scientists like Ernest Rutherford and J. J. Thomson to detect cathode rays and X-rays were coated with ZnS because it emits light when excited by X-rays and electron beams. If another metal called an activator is added to zinc sulfide, the material phosphoresces. The color of the phosphorescence depends on the activator: a few ppm of silver produces a blue color; manganese yields orange–red; and copper produces a long-lasting dark green that is useful in electroluminescent panels.

Zinc metal is recovered from zinc sulfide ore by heating the ore in air to produce zinc oxide:

$$2\,ZnS(s) + 3\,O_2(g) \rightarrow 2\,ZnO(s) + 2\,SO_2(g)$$

Zinc oxide is the most common ingredient in antibiotic creams for the treatment of diaper rash, in sunscreens (because it absorbs both UVA and UVB ranges of ultraviolet light), and on the insides of fluorescent light bulbs. A special grade of zinc oxide called Chinese white is used in artists' pigments. ZnO is reduced to molten metal with hot carbon:

$$2\,ZnO(s) + C(s) \rightarrow 2\,Zn(\ell) + CO_2(g)$$

Along with zinc, cadmium is produced from impurities in the ore and is separated by distillation of the molten metals. Cadmium is used in making nickel–cadmium (nicad) batteries, but large quantities are also used in producing pigments ranging from lemon yellow (CdS) through red (Cd with both S and Se) to deep maroon (CdSe). The cadmium pigments are popular with artists because of their bright colors and their stability when exposed to light.

Cinnabar (HgS) has been mined in Spain since Roman times. Also called vermillion or Chinese red, cinnabar has been used because of its unique red color in Chinese decorative art and lacquer ware.

Cadmium yellow medium hue

Cadmium orange

Cadmium red medium

Zinc oxide is used to coat the insides of fluorescent light bulbs and to make topical creams that are also effective sunscreens.

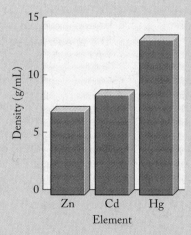
The density of the group 12 elements increases with increasing atomic number.

Mercury is produced directly from reaction of HgS with oxygen:

$$HgS(s) + O_2(g) \rightarrow Hg(\ell) + SO_2(g)$$

Mercury is the only metal, and one of only two elements (bromine is the other), that is a liquid at room temperature. Zinc and cadmium are low-melting, silvery solids that react rapidly with moist air to form the corresponding oxides. The oxides and sulfides of zinc and cadmium are semiconductors (see Chapter 11).

The group 12 metals have $(n-1)d^{10}ns^2$ electron configurations in their valence shells and, like the metals in group 2, tend to lose two valence electrons and form M^{2+} cations. Mercury also forms an Hg_2^{2+} cation, which contains a covalent Hg—Hg bond. The group 12 elements are more likely than the alkaline earth (group 2) metals to form covalent bonds.

Zinc is an essential element in living organisms, but cadmium and mercury have no known health benefits and are quite toxic. The toxicity of these metals is related to their ability to attach to sulfur groups in some amino acids. These interactions allow the metals to be transported throughout the body attached to the peptides and proteins in blood. If they reach the brain, interactions with the proteins there can lead to mental retardation and other neurological disorders. The toxicity of mercury is increased by the formation of covalent alkyl mercury compounds such as methyl mercuric chloride, CH_3HgCl, and dimethyl mercury, $(CH_3)_2Hg$. Mercury in these chemical forms can penetrate hydrophobic cell membranes more easily than inorganic Hg^{2+} ions and so can more severely disrupt internal cell functions. Microbial methylation of inorganic mercury wastes dumped in Minamata Bay, Japan, in the 1950s led to 52 sudden deaths and severe neurological damage to more than 3000 people. This incident sparked recognition of the environmental hazards of improperly disposing of mercury and other metals.

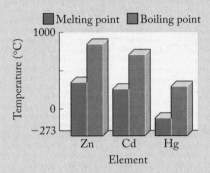

The group 12 elements are low-melting metals (Hg is a liquid at room temperature) and are relatively low boiling compared with other metals.

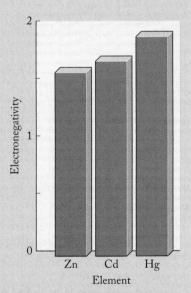

The electronegativities of the group 12 elements increase slightly as their atomic numbers increase.

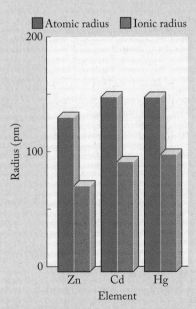

The radii of the 2+ cations of the group 12 elements increase as their atomic number increases. The atomic radii are less dependent on atomic number.

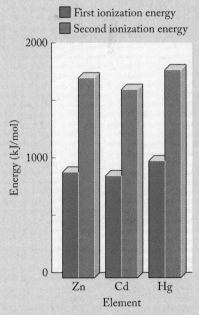

The first and second ionization energies of the group 12 elements show little dependence on atomic number.

FIGURE 18.13 The 2008 Honda FCX Clarity is powered by a 100 kW fuel cell stack and 100 kW (134 horsepower) electric motor that give the car a top speed of 100 miles per hour (160 km/hr). Hydrogen fuel is stored in a 171 L tank at a maximum pressure of 350 atm, which gives the car a driving range of about 270 miles (430 km). A Li-ion battery pack provides supplemental energy storage including that produced by regenerative braking.

⊚⊙ **CONNECTION** Processes for producing hydrogen gas by reacting methane and other fuels with steam are discussed in Chapter 15.

An individual PEM fuel cell has a cell potential of about 1.0 V. That is not much, but when hundreds of these cells are assembled into fuel cell *stacks*, such as the one in the Honda FCX Clarity (Figure 18.13), they are capable of producing 100 kW of electrical power. That is enough to give the Clarity a top speed of 100 miles per hour (160 km/hr).

As fuel cells become more efficient and less expensive, the principal limit on their use in passenger cars is the availability, cost, and safety of using hydrogen fuel. Hydrogen is flammable, but it has a higher ignition temperature than gasoline and spreads through the air more quickly, reducing the risk of fire. Unfortunately from the point of view of safety, hydrogen in air burns over a much wider range of concentrations than gasoline, and its flame is particularly hazardous because it is almost invisible. It is likely that most fuel-cell-powered vehicles will be buses and fleet vehicles operating from a central location, where hydrogen gas is available. For example, buses (with very large fuel tanks) powered by fuel cells can carry 60 passengers and have an operating range of 400 km. In 2008 the Honda Clarity was available in only a few communities in Southern California, where hydrogen gas was available at some filling stations.

Longer ranges would be possible if better methods for storing hydrogen gas were available. Currently under development are mobile chemical-processing plants that can extract hydrogen from liquid fuels such as gasoline or methanol as needed by the vehicle. Making hydrogen from fossil fuels also generates carbon dioxide as a byproduct, so such processes do not improve the situation with greenhouse gases in the atmosphere. The most common method in use today is the production of hydrogen from natural gas, which results in about a twofold improvement in total emissions compared to those from a conventional gasoline-powered vehicle.

Scientists are also developing new materials that can be used to store large quantities of hydrogen safely. In one promising technology, a carbon material, related in its structure to bucky balls, has been used to store many times the mass of hydrogen that could be compressed into the same volume without the carbon. Other scientists are developing new metal hydrides for hydrogen storage. There is an important safety feature to hydrogen sequestration in carbon structures or as metal hydrides: it would probably not be released and explode in a collision, as is likely if a high-pressure tank filled with H_2 gas ruptured. The drawback to the metal hydride systems is that they tend to be heavy and add considerable weight to an automobile. A major focus of hydrogen research worldwide is finding a better storage method for hydrogen fuel.

Summary

SECTION 18.1 Electrochemistry is the branch of chemistry that links redox reactions to the production or consumption of electrical energy. Any redox reaction can be broken down into oxidation and reduction half-reactions. The gain of electrons in a reduction half-reaction must be balanced by the loss of electrons in the oxidation half-reaction coupled to it.

SECTION 18.2 A **voltaic cell** is an **electrochemical cell** that produces an electrical current in an external circuit as a result of a spontaneous redox reaction in which oxidation occurs at the **anode** and reduction occurs at the **cathode** of the cell. A **cell diagram** uses symbols including single and double vertical lines to show how the components of the cathodic and anodic compartments of the cell are connected.

SECTION 18.3 The **cell potential (E_{cell})** of a voltaic cell is the **electromotive force** that the cell generates. E_{cell} is related to the change in free energy (ΔG_{cell}) of the cell reaction by the equation:

$$\Delta G_{cell} = -nFE_{cell}$$

where F is **Faraday's constant** (9.65×10^4 C/mol) and n is the number of electrons transferred in the balanced chemical equation describing the cell reaction. This change in free energy is available to do work in an external electrical circuit.

SECTION 18.4 The value E_{cell} under standard conditions is called **standard cell potential (E°_{cell})**. It is calculated by combining the **standard potentials (E°)** of a cell's half-reactions, E°_{red} of the cathode half-reaction and E°_{ox} of the anode half-reaction:

$$E^\circ_{cell} = E^\circ_{red}(\text{cathode}) + E^\circ_{ox}(\text{anode})$$

The standard potentials in Appendix 6 are all reduction potentials, E°_{red}. The corresponding E°_{ox} values for the half-reactions in reverse have the same magnitudes but opposite signs.

SECTION 18.5 All standard potentials are referenced to the potential of the **standard hydrogen electrode** ($E_{SHE} = 0.000$ V).

SECTION 18.6 The potential of a voltaic cell decreases as reactants turn into products. The **Nernst equation** predicts how the potential of a cell changes with the concentrations of the cell reactants and products. The potential of a voltaic cell approaches zero as the cell reaction approaches chemical equilibrium. At equilibrium $E_{cell} = 0$ and $Q = K$.

SECTION 18.7 The quantities of reactants consumed in a voltaic cell reaction are directly proportional to the coulombs of electrical charge that the cell delivers. Faraday's constant relates the quantity of electrical charge to the number of moles of electrons, and indirectly to the moles of reactants involved in the cell reaction. Most portable electronic devices are powered by nickel–metal hydride (NiMH) or lithium-ion batteries.

SECTION 18.8 Reversing a spontaneous reaction by passing electric current through an electrochemical cell turns the cell into an **electrolytic cell**. Reactions in electrolytic cells that produce gases often require application of additional electromotive force, or **overpotential**. The polarities of cathodes and anodes in electrolytic cells are opposite the polarities of these electrodes in voltaic cells.

SECTION 18.9 Fuel cells directly convert the free energy released during the reaction $2H_2 + O_2 \rightarrow 2H_2O$ into electrical energy. Fuel cells with proton-exchange membranes have been developed as power supplies for electric vehicles.

Problem-Solving Summary

TYPE OF PROBLEM	CONCEPTS AND EQUATIONS	SAMPLE EXERCISES
Writing net ionic equations for redox reactions	Combine the reduction and oxidation half-reactions after balancing the gain and loss of electrons.	18.1
Diagramming an electrochemical cell	Start with the anode half-reaction; use single lines to separate phases and a double line to separate the two half-reactions.	18.2
Relating the value of ΔG_{cell} to the value of E_{cell}	$$\Delta G_{cell} = -nFE_{cell}$$ where n is the number of moles of electrons transferred in the cell reaction and F is Faraday's constant.	18.3
Identifying anode and cathode half-reactions from their standard potentials	The half-reaction with the more positive standard reduction potential is the cathode half-reaction.	18.4
Calculating E_{cell} from E°_{cell} and the concentrations of reactants and products	E_{cell} at 298 K is related to E°_{cell} and the reaction quotient by the Nernst equation: $$E_{cell} = E^\circ_{cell} - \frac{0.0592}{n}\log Q$$	18.5

TYPE OF PROBLEM	CONCEPTS AND EQUATIONS	SAMPLE EXERCISES
Calculating the equilibrium constant of a redox reaction from the standard potentials of its half-reactions	$E°_{cell}$ is the sum of $E°_{red}$(cathode) $+ E°_{ox}$(anode) and is related to K at 298 K by $$\log K = \frac{nE°_{cell}}{0.0592}$$	18.6
Relating the mass of a reactant in an electrochemical reaction to a quantity of electrical charge	Determine the ratio of moles of reactants to moles of electrons transferred; relate coulombs of charge to moles of electrons using Faraday's constant.	18.7
Calculating the time to electrolyze a quantity of reactant	Use Faraday's constant, electrical current as charge per unit time, and the relation between moles of electrons and moles of reactants to describe an electrolytic process.	18.8

Visual Problems

18.1. Consider the four batteries in Figure P18.1. From top to bottom the sizes are AAA, AA, C, and D. The performance of batteries like these is often expressed in units such as (a) volts, (b) watt-hours, (c) milliampere-hours. Which of the values that go with each of these units are likely to differ significantly among the four batteries?

FIGURE P18.1

18.2. Which of the four curves in Figure P18.2 best represents the dependence of the potential of a lead-acid battery on the concentration of sulfuric acid? Note that the scale of the *x*-axis is logarithmic.

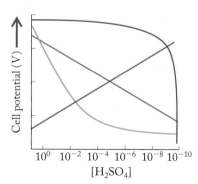

FIGURE P18.2

18.3. The apparatus in Figure P18.3 is used for the electrolysis of water. Hydrogen and oxygen gas are collected in the two inverted burettes. An inert electrode at the bottom of the left burette is connected to the negative terminal of a 6-volt battery; the electrode in the burette on the right is connected to the positive terminal. A small quantity of sulfuric acid is added to speed up the electrolytic reaction.
   a. What are the half-reactions at the left and right electrodes and their standard potentials?
   b. Why does sulfuric acid make the electrolysis reaction go more rapidly?

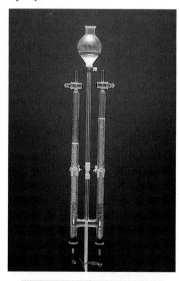

Overall cell reaction
$$H_2O(\ell) \rightarrow H_2(g) + \tfrac{1}{2}O_2(g)$$

FIGURE P18.3

18.4. An electrolytic apparatus identical to the one shown in Problem 18.3 is used to electrolyze water, but the reaction is speeded up by the addition of sodium carbonate instead of sulfuric acid.
   a. What are the half-reactions and the standard potentials for the electrodes on the left and right?
   b. Why does sodium carbonate makes the electrolysis reaction go more rapidly?

18.5. In many electrochemical cells the electrodes are metals that carry electrons to and from the cell but are not chemically changed by the cell reaction. Each of the highlighted clusters in the periodic table in Figure P18.5 consists of three metals. Which of the highlighted clusters is best suited to form inert electrodes?

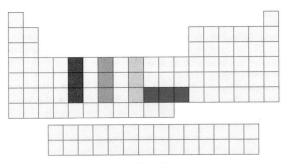

FIGURE P18.5

18.6. In the voltaic cell shown in Figure P18.6, the greater density of a concentrated solution of $CuSO_4$ allows a less concentrated solution of $ZnSO_4$ solution to be (carefully) layered on top of it. Why is a porous separator not needed in this cell?

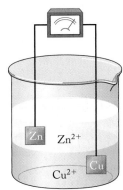

FIGURE P18.6

18.7. In the voltaic cell shown in Figure P18.7, the concentrations of Cu^{2+} and Cd^{2+} are 1.00 M. On the basis of the standard potentials in Appendix 6, identify which electrode is the anode and which is the cathode. Indicate the direction of electron flow.

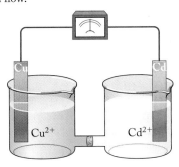

FIGURE P18.7

18.8. In the voltaic cell shown in Figure P18.8, $[Ag^+] = [H^+] = 1.00$ M. On the basis of the standard potentials in Appendix 6, identify which electrode is the anode and which is the cathode. Indicate the direction of electron flow.

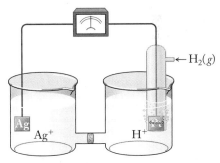

FIGURE P18.8

Questions and Problems

Redox Chemistry Revisited; Electrochemical Cells

CONCEPT REVIEW

18.9. What is the role of the porous separator in an electrochemical cell?

18.10. The chemical reactions in Figures 18.1 and 18.2 are the same. However the reaction in the cell in Figure 18.2 generates electricity; the reaction in the beaker in Figure 18.1 does not. Why?

18.11. Why can't a wire perform the same function as a porous separator in an electrochemical cell?

18.12. In a voltaic cell, why is the cathode labeled the positive terminal and the anode the negative terminal?

PROBLEMS

18.13. A voltaic cell with an aqueous electrolyte is based on the reaction between $Pb^{2+}(aq)$ and $Zn(s)$, producing $Pb(s)$ and $Zn^{2+}(aq)$.
 a. Write half-reactions for the anode and cathode.
 b. Write a balanced cell reaction.
 c. Diagram the cell.

18.14. A voltaic cell is based on the reaction between $Ag^+(aq)$ and $Ni(s)$, producing $Ag(s)$ and $Ni^{2+}(aq)$.
 a. Write the anode and cathode half-reactions.
 b. Write a balanced cell reaction.
 c. Diagram the cell.

18.15. A voltaic cell with an alkaline aqueous electrolyte is based on the oxidation of $Cd(s)$ to $Cd(OH)_2(s)$ and the reduction of $MnO_4^-(aq)$ to $MnO_2(s)$.
  a. Write half-reactions for the cell's anode and cathode.
  b. Write a balanced cell reaction.
  c. Diagram the cell.

18.16. A voltaic cell is based on the reduction of $Ag^+(aq)$ to $Ag(s)$ and the oxidation of $Sn(s)$ to $Sn^{2+}(aq)$.
  a. Write half-reactions for the cell's anode and cathode.
  b. Write a balanced cell reaction.
  c. Diagram the cell.

18.17. Super Iron Batteries In 1999, scientists in Israel developed a battery based on the following cell reaction with iron(VI), nicknamed "super iron":

$$2\,K_2FeO_4(aq) + 3\,Zn(s) \rightarrow Fe_2O_3(s) + ZnO(s) + 2\,K_2ZnO_2(aq)$$

  a. Determine the number of electrons transferred in the cell reaction.
  b. What are the oxidation states of the transition metals in the reaction?
  c. Diagram the cell.

18.18. Aluminum Air Batteries In recent years engineers have been working on an aluminum-air battery as an alternative energy source for electric vehicles. The battery consists of an aluminum anode, which is oxidized to solid aluminum hydroxide, immersed in an electrolyte of aqueous KOH. At the cathode oxygen from the air is reduced to hydroxide ions on an inert metal surface. Write the two half-reactions for the battery and diagram the cell. Use the generic $M(s)$ symbol for the metallic cathode material.

Chemical Energy and Electrical Work

CONCEPT REVIEW

18.19. How is it that a voltaic cell with a *positive* cell potential does *negative* work?

*18.20. In physics, work (w) is done by exerting a force (F) to move an object through a distance (d) according to the equation $w = F \times d$. Explain how this definition of work relates to electrical work ($w_{elec} = C \times E$).

PROBLEMS

18.21. For many years the 1.50 V batteries used to power flashlights were based on the following cell reaction:

$$Zn(s) + 2\,NH_4Cl(s) + 2\,MnO_2(s) \rightarrow$$
$$Zn(NH_3)_2Cl_2(s) + Mn_2O_3(s) + H_2O(\ell)$$

What is the value of ΔG_{cell}?

18.22. The first generation of laptop computers was powered by nickel–cadmium ("nicad") batteries, which generated 1.20 V based on the following cell reaction:

$$Cd(s) + 2\,NiO(OH)(s) + 2\,H_2O(\ell) \rightarrow Cd(OH)_2(s) + 2\,Ni(OH)_2(s)$$

What is the value of ΔG_{cell}?

18.23. The cells in the nickel–metal hydride battery packs used in many hybrid vehicles produce 1.20 V based on the following cell reaction:

$$MH(s) + NiO(OH)(s) \rightarrow M(s) + Ni(OH)_2(s)$$

What is the value of ΔG_{cell}?

18.24. A cell in a lead-acid battery delivers exactly 2.00 V of cell potential based on the following cell reaction:

$$Pb(s) + PbO_2(s) + 2\,H_2SO_4(aq) \rightarrow 2\,PbSO_4(s) + 2\,H_2O(\ell)$$

What is the value of ΔG_{cell}?

Standard Potentials ($E°$)

CONCEPT REVIEW

18.25. What is the function of platinum in the standard hydrogen electrode?

18.26. Is it possible to build a battery in which the anode chemistry is based on a half-reaction in which none of the species is a solid conductor, for example:

$$Fe^{2+}(aq) \rightarrow Fe^{3+}(aq) + e^-$$

18.27. In some textbooks the formula used to calculate standard cell potentials from the standard reduction potentials of the half-reactions occurring at the cathode and anode is given as

$$E°_{cell} = E°_{red}(\text{cathode}) - E°_{red}(\text{anode})$$

Show how this equation is equivalent to Equation 18.7.

18.28. Suppose there were a scale for expressing electrode potentials in which the standard potential for the reduction of water in base

$$2\,H_2O(\ell) + 2\,e^- \rightarrow H_2(g) + 2\,OH^-(aq)$$

is assigned an $E°$ value of 0.000 V. How would the standard potential values on this new scale differ from those in Appendix 6?

18.29. Is O_2 a stronger oxidizing agent in acid or in base? Use standard reduction potentials in Appendix 6 to support your answer.

*18.30. To inhibit corrosion of steel or aluminum structures in contact with seawater, pieces of other metals (often zinc) are attached to the structures to serve as "sacrificial anodes." Explain how these attached pieces of metal might protect the structures and describe which properties of zinc make it a good selection.

PROBLEMS

18.31. Starting with the appropriate standard free energies of formation in Appendix 4, calculate the value of $\Delta G°$ and $E°_{cell}$ of the following reactions:
  a. $2\,Cu^+(aq) \rightarrow Cu^{2+}(aq) + Cu(s)$
  b. $Ag(s) + Fe^{3+}(aq) \rightarrow Ag^+(aq) + Fe^{2+}(aq)$

18.32. Starting with the appropriate standard free energies of formation in Appendix 4, calculate the values of $\Delta G°$ and $E°_{cell}$ of the following reactions:
  a. $FeO(s) + H_2(g) \rightarrow Fe(s) + H_2O(\ell)$
  b. $2\,Pb(s) + O_2(g) + 2\,H_2SO_4(aq) \rightarrow 2\,PbSO_4(s) + 2\,H_2O(\ell)$

18.33. If a piece of silver is placed in a solution in which $[Ag^+] = [Cu^{2+}] = 1.00\ M$, will the following reaction proceed spontaneously?

$$2\,Ag(s) + Cu^{2+}(aq) \rightarrow 2\,Ag^+(aq) + Cu(s)$$

18.34. A piece of cadmium is placed in a solution in which $[Cd^{2+}] = [Sn^{2+}] = 1.00\ M$. Will the following reaction proceed spontaneously?

$$Cd(s) + Sn^{2+}(aq) \rightarrow Cd^{2+}(aq) + Sn(s)$$

18.35. Sometimes the anode half-reaction in the zinc-air battery is written with the zincate ion, $Zn(OH)_4^{2-}$, as the product. Write a balanced equation for the cell reaction based on this product.

18.36. Sometimes the cell reaction of nickel–cadmium batteries is written with Cd metal as the anode and solid NiO_2 as the cathode. Assuming that the products of the electrode reactions are solid hydroxides of Cd(II) and Ni(II), respectively, write balanced chemical equations for the cathode and anode half-reactions and the overall cell reaction.

18.37. In a voltaic cell similar to the Cu–Zn cell in Figure 18.2, the Cu electrode is replaced with one made of Ni immersed in a solution of $NiSO_4$. Will the standard potential of this Ni–Zn cell be greater than, the same as, or less than 1.10 V?

18.38. Suppose the copper half of the Cu–Zn cell in Figure 18.2 were replaced with a silver wire in contact with 1 *M* $Ag^+(aq)$.
a. What would be the value of E_{cell}?
b. Which electrode would be the anode?

18.39. Starting with standard potentials listed in Appendix 6, calculate the values of $E°_{cell}$ and $\Delta G°$ of the following reactions.
a. $Cu(s) + Sn^{2+}(aq) \rightarrow Cu^{2+}(aq) + Sn(s)$
b. $Zn(s) + Ni^{2+}(aq) \rightarrow Zn^{2+}(aq) + Ni(s)$

18.40. Starting with the standard potentials listed in Appendix 6, calculate the values of $E°_{cell}$ and $\Delta G°$ of the following reactions.
a. $Fe(s) + Cu^{2+}(aq) \rightarrow Fe^{2+}(aq) + Cu(s)$
b. $Ag(s) + Fe^{3+}(aq) \rightarrow Ag^+(aq) + Fe^{2+}(aq)$

18.41. Voltaic cells based on the following pairs of half-reactions are prepared so that all reactants and products are in their standard states. For each pair, write a balanced equation for the cell reaction, and identify which half-reaction takes place at the anode and which at the cathode.
a. $Hg^{2+}(aq) + 2e^- \rightarrow Hg(\ell)$
   $Zn^{2+}(aq) + 2e^- \rightarrow Zn(s)$
b. $ZnO(s) + H_2O(\ell) + 2e^- \rightarrow Zn(s) + 2OH^-(aq)$
   $Ag_2O(s) + H_2O(\ell) + 2e^- \rightarrow 2Ag(s) + 2OH^-(aq)$
c. $NiOH_2^-(aq) + 2e^- \rightarrow Ni(s) + 2OH^-(aq)$
   $O_2(g) + 2H_2O(\ell) + 4e^- \rightarrow 4OH^-(aq)$

18.42. Voltaic cells based on the following pairs of half-reactions are constructed. For each pair, write a balanced equation for the cell reaction, and identify which half-reaction takes place at each anode and cathode.
a. $Cd^{2+}(aq) + 2e^- \rightarrow Cd(s)$
   $Ag^+(aq) + e^- \rightarrow Ag(s)$
b. $AgBr(s) + e^- \rightarrow Ag(s) + Br^-(aq)$
   $MnO_2(s) + 4H^+(aq) + 2e^- \rightarrow Mn^{2+}(aq) + 2H_2O(\ell)$
c. $PtCl_4^{2-}(aq) + 2e^- \rightarrow Pt(s) + 4Cl^-(aq)$
   $AgCl(s) + e^- \rightarrow Ag(s) + Cl^-(aq)$

18.43. The half-reactions and standard potentials for a nickel–metal hydride battery with a titanium–zirconium anode are as follows:

Cathode: $NiO(OH)(s) + H_2O(\ell) + e^- \rightarrow$
   $Ni(OH)_2(s) + OH^-(aq)$ $E° = 1.32$ V

Anode: $TiZr_2H(s) + OH^-(aq) \rightarrow$
   $TiZr_2(s) + H_2O(\ell) + e^-$ $E° = 0.00$ V

a. Write the overall cell reaction for this battery.
b. Calculate the standard cell potential.

*__18.44.__ **New Lithium-Ion Batteries** Scientists at the University of Texas, Austin, and at MIT have developed a new cathode material for lithium-ion batteries based on $LiFePO_4$, which is the composition of the cathode when the battery is fully discharged. Batteries with this cathode are more powerful than those of the same mass with $LiCoO_2$ cathodes. They are also more stable at high temperatures.
a. What is the formula of the $LiFePO_4$ cathode when the battery is fully charged?
b. Is Fe oxidized or reduced as the battery discharges?
c. Is the cell potential of a lithium-ion battery with an iron phosphate cathode likely to differ from one with a cobalt oxide cathode? Explain your answer.

A Reference Point: The Standard Hydrogen Electrode; The Effect of Concentration on E_{cell}

CONCEPT REVIEW

18.45. Why does the operating cell potential of most batteries change little until the battery is nearly discharged?

18.46. The standard potential of the Cu–Zn cell reaction

$$Zn(s) + Cu^{2+}(aq) \rightarrow Zn^{2+}(aq) + Cu(s)$$

is 1.10 V. Would the potential of the Cu–Zn cell differ from 1.10 V if the concentrations of both Cu^{2+} and Zn^{2+} were 0.25 *M*?

PROBLEMS

18.47. Calculate the E_{cell} value at 298 K for the cell based on the reaction

$$Fe^{3+}(aq) + Cr^{2+}(aq) \rightarrow Fe^{2+}(aq) + Cr^{3+}(aq)$$

when $[Fe^{3+}] = [Cr^{2+}] = 1.50 \times 10^{-3}$ *M* and $[Fe^{2+}] = [Cr^{3+}] = 2.5 \times 10^{-4}$ *M*.

18.48. Calculate the E_{cell} value at 298 K for the cell based on the reaction

$$Cu(s) + 2Ag^+(aq) \rightarrow Cu^{2+}(aq) + 2Ag(s)$$

when $[Ag^+] = 2.56 \times 10^{-3}$ *M* and $[Cu^{2+}] = 8.25 \times 10^{-4}$ *M*.

18.49. Using the appropriate standard potentials in Appendix 6, determine the equilibrium constant for the following reaction at 298 K:

$$Fe^{3+}(aq) + Cr^{2+}(aq) \rightarrow Fe^{2+}(aq) + Cr^{3+}(aq)$$

18.50. Using the appropriate standard potentials in Appendix 6, determine the equilibrium constant at 298 K for the following reaction between MnO_2 and Fe^{2+} in acid solution:

$$4H^+(aq) + MnO_2(s) + 2Fe^{2+}(aq) \rightarrow$$
$$Mn^{2+}(aq) + 2Fe^{3+}(aq) + 2H_2O(\ell)$$

18.51. If the potential of a hydrogen electrode based on the half-reaction

$$2\,H^+(aq) + 2\,e^- \rightarrow H_2(g)$$

is 0.000 V at pH = 0.00, what is the potential of the same electrode at pH = 7.00?

18.52. Glucose Metabolism The standard potentials for the reduction of nicotinamide adenine dinucleotide (NAD^+) and oxaloacetate (reactants in the multistep metabolism of glucose) are as follows:

$$NAD^+(aq) + 2\,H^+(aq) + 2\,e^- \rightarrow NADH(aq) + H^+(aq)$$
$$E^\circ = -0.320 \text{ V}$$

$$Oxaloacetate^-(aq) + 2\,H^+(aq) + 2\,e^- \rightarrow malate^-(aq)$$
$$E^\circ = -0.166 \text{ V}$$

a. Calculate the standard potential for the following reaction:

$$Oxaloacetate^-(aq) + NADH(aq) + H^+(aq) \rightarrow$$
$$malate^-(aq) + NAD^+(aq)$$

b. Calculate the equilibrium constant for the reaction at 298 K.

18.53. Permanganate ion can oxidize sulfite to sulfate in basic solution as follows:

$$2\,MnO_4^-(aq) + 3\,SO_3^{2-}(aq) + H_2O(\ell) \rightarrow$$
$$2\,MnO_2(s) + 3\,SO_4^{2-}(aq) + 2\,OH^-(aq)$$

Determine the potential for the reaction at 298 K when the concentrations of the reactants and products are as follows: $[MnO_4^-] = 0.150$ M, $[SO_3^{2-}] = 0.256$ M, $[SO_4^{2-}] = 0.178$ M, and $[OH^-] = 0.0100$ M. Will the value of E_{rxn} increase or decrease as the reaction proceeds?

*18.54. Manganese dioxide is reduced by iodide ion in acid solution as follows:

$$MnO_2(s) + 2\,I^-(aq) + 4\,H^+(aq) \rightarrow$$
$$Mn^{2+}(aq) + I_2(aq) + 2\,H_2O(\ell)$$

Determine the electrical potential of the reaction at 298 K when the initial concentrations of the components are as follows: $[I^-] = 0.225$ M, $[H^+] = 0.900$ M, $[Mn^{2+}] = 0.100$ M, and $[I_2] = 0.00114$ M. If the solubility of iodine in water is approximately 0.114 M, will the value of E_{rxn} increase or decrease as the reaction proceeds?

18.55. A copper penny dropped into a solution of nitric acid produces a mixture of nitrogen oxides. The following reaction describes the formation of NO, one of the products:

$$3\,Cu(s) + 8\,H^+(aq) + 2\,NO_3^-(aq) \rightarrow$$
$$2\,NO(g) + 3\,Cu^{2+}(aq) + 4\,H_2O(\ell)$$

a. Starting with the appropriate standard potentials in Appendix 6, calculate E_{rxn} for this reaction.
b. Calculate E_{rxn}° at 298 K when $[H^+] = 0.100$ M, $[NO_3^-] = 0.0250$ M, $[Cu^{2+}] = 0.0375$ M, and the partial pressure of NO = 0.00150 atm.

18.56. Chlorine dioxide (ClO_2) is produced by the following reaction of chlorate (ClO_3^-) with Cl^- in acid solution:

$$2\,ClO_3^-(aq) + 2\,Cl^-(aq) + 4\,H^+(aq) \rightarrow$$
$$2\,ClO_2(g) + Cl_2(g) + 2\,H_2O(\ell)$$

a. Determine E° for the reaction. For the half-reaction

$$ClO_3^-(aq) + 2\,H^+(aq) + e^- \rightarrow ClO_2(g) + H_2O(\ell)$$
$$E^\circ = 1.152 \text{ V}$$

b. The reaction produces an atmosphere in the reaction vessel in which $P_{ClO_2} = 2.0$ atm; $P_{Cl_2} = 1.00$ atm. Calculate $[ClO_3^-]$ if, at equilibrium ($T = 298$ K), $[H^+] = [Cl^-] = 10.0$ M.

*18.57. The oxidation of NH_4^+ to NO_3^- in acid solution is described by the following equation:

$$NH_4^+(aq) + 2\,O_2(g) + H_2O(\ell) \rightarrow NO_3^-(aq) + 2\,H^+(aq)$$

For the half-reaction

$$NO_3^-(aq) + 10\,H^+(aq) + 8\,e^- \rightarrow NH_4^+(aq) + 3\,H_2O(\ell)$$
$$E^\circ = 1.50 \text{ V}$$

a. Calculate E° for the overall reaction.
b. If the reaction is in equilibrium with air ($P_{O_2} = 0.21$ atm) at pH 5.60, what is the ratio of $[NO_3^-]$ to $[NH_4^+]$ at 298 K?

18.58. Recalculate the answers to Problem 18.57 for the oxidation of NH_4^+ at an altitude where $P_{O_2} = 0.18$ atm.

Relating Battery Capacity to Quantities of Reactants

CONCEPT REVIEW

18.59. One 12-volt lead-acid battery has a higher ampere-hour rating than another. Which of the following parameters are likely to be different for the two batteries?
a. Individual cell potentials
b. Anode half-reactions
c. Total masses of electrode materials
d. Number of cells
e. Electrolyte composition
f. Combined surface areas of their electrodes

18.60. In a voltaic cell based on the Cu–Zn cell reaction

$$Zn(s) + Cu^{2+}(aq) \rightarrow Cu(s) + Zn^{2+}(aq)$$

there is exactly 1 mole of each reactant and product. A second cell based on the Cd–Cu cell reaction

$$Cd(s) + Cu^{2+}(aq) \rightarrow Cu(s) + Cd^{2+}(aq)$$

also has exactly 1 mole of each reactant and product. Which of the following statements about these two cells is true?
a. Their cell potentials are the same.
b. The masses of their electrodes are the same.
c. The quantities of electrical charge that they can produce are the same.
d. The quantities of electrical energy that they can produce are the same.

PROBLEMS

18.61. Which of the following voltaic cells will produce the greater quantity of electrical charge per gram of anode material?

$$Cd(s) + 2\,NiO(OH)(s) + 2\,H_2O(\ell) \rightarrow$$
$$2\,Ni(OH)_2(s) + Cd(OH)_2(s)$$

or

$$4\,Al(s) + 3\,O_2(g) + 6\,H_2O(\ell) + 4\,OH^-(aq) \rightarrow$$
$$4\,Al(OH)_4^-(aq)$$

18.62. Which of the following voltaic cells will produce the greater quantity of electrical charge per gram of anode material?

$$Zn(s) + MnO_2(s) + H_2O(\ell) \rightarrow ZnO(s) + Mn(OH)_2(s)$$

or

$$Li(s) + MnO_2(s) \rightarrow LiMnO_2(s)$$

*18.63. Which of the following voltaic cell reactions delivers more electrical energy per gram of anode material at 298 K?

$$Zn(s) + 2\,NiO(OH)(s) + 2\,H_2O(\ell) \rightarrow$$
$$2\,Ni(OH)_2(s) + Zn(OH)_2(s) \qquad E_{cell} = 1.20\,V$$

or

$$Li(s) + MnO_2(s) \rightarrow LiMnO_2(s) \qquad E_{cell} = 3.15\,V$$

*18.64. Which of the following voltaic cell reactions delivers more electrical energy per gram of anode material at 298 K?

$$Zn(s) + Ni(OH)_2(s) \rightarrow Zn(OH)_2(s) + Ni(s) \qquad E_{cell} = 1.50\,V$$

or

$$2\,Zn(s) + O_2(g) \rightarrow 2\,ZnO(s) \qquad E_{cell} = 2.08\,V$$

Electrolytic Cells and Rechargeable Batteries

CONCEPT REVIEW

18.65. The positive terminal of a voltaic cell is the cathode. However, the cathode of an electrolytic cell is connected to the negative terminal of a power supply. Explain this difference in polarity.

18.66. The anode in an electrochemical cell is defined as the electrode where oxidation takes place. Why is the anode in an electrolytic cell connected to the positive (+) terminal of an external supply, whereas the anode in a voltaic cell battery is connected to the negative (−) terminal?

18.67. The salts obtained from the evaporation of seawater can act as a source of halogens, principally Cl_2 and Br_2, through the electrolysis of the molten alkali metal halides. As the potential of the anode in an electrolytic cell is increased, which of these two halogens forms first?

18.68. In the electrolysis described in Problem 18.67, why is it necessary to use molten salts rather than seawater itself?

18.69. In the electrolysis of H_2O to H_2 and O_2, a strong electrolyte, such as Na_2CO_3, is added to speed up the electrolysis process. How does Na_2CO_3 speed up the process?

18.70. In the electrolysis of H_2O to H_2 and O_2, why should Na_2CO_3, but not NaCl, be added to speed up electrolysis?

18.71. Quantitative Analysis Electrolysis can be used to determine the concentration of Cu^{2+} in a given volume of solution by electrolyzing the solution in a cell equipped with a platinum cathode. If all of the Cu^{2+} is reduced to Cu metal at the cathode, the increase in mass of the electrode provides a measure of the concentration of Cu^{2+} in the original solution. To ensure the complete (99.99%) removal of the Cu^{2+} from a solution in which $[Cu^{2+}]$ is initially about 1.0 M, will the potential of the cathode (versus SHE) have to be more negative or less negative than 0.34 V (the standard potential for $Cu^{2+} + 2\,e^- \rightarrow Cu$)?

18.72. A high school chemistry student wishes to demonstrate how water can be separated into hydrogen and oxygen by electrolysis. She knows that the reaction will proceed more rapidly if an electrolyte is added to the water, and has these reagents available: 2.00 M H_2SO_4, 2.00 M HCl, NaCl, Na_2SO_4, Na_2CO_3, and $CaCO_3$. Which one(s) should she use? Explain your selection(s).

PROBLEMS

18.73. Suppose the current flowing from a battery is used to electroplate an object with silver. Calculate the mass of silver that would be deposited by a battery that delivers 1.7 A · hr of charge.

18.74. A battery charger used to recharge the NiMH batteries used in a digital camera can deliver as much as 0.50 A of current to each battery. If it takes 100 min to recharge one battery, how much $Ni(OH)_2$ (in grams) is oxidized to NiO(OH)?

18.75. A NiMH battery containing 4.10 g of NiO(OH) was 50% discharged when it was connected to a charger with an output of 2.00 A at 1.3 V. How long does it take to recharge the battery?

*18.76. How long does it take to deposit a coating of gold 1.00 μm thick on a disk-shaped medallion 4.0 cm in diameter and 2.0 mm thick at a constant current of 85 A? The density of gold is 19.3 g/cm^3. The gold solution contains Au(III).

*18.77. Oxygen Supply in Submarines Nuclear submarines can stay under water nearly indefinitely because they can produce their own oxygen by the electrolysis of water.
a. How many liters of O_2 at 298 K and 1.00 bar are produced in 1 hr in an electrolytic cell operating at a current of 0.025 A?
b. Could seawater be used as the source of oxygen in this electrolysis? Explain why or why not.

18.78. In the electrolysis of water, how long will it take to produce 1.00×10^2 L of H_2 at STP (273 K and 1.00 bar) using an electrolytic cell through which a current of 52 mA flows?

18.79. Calculate the minimum (least negative) cathode potential (versus SHE) needed to begin electroplating nickel from 0.35 M Ni^{2+} onto a piece of iron. Assume that the overpotential for the reduction of Ni^{2+} on an iron electrode is negligible.

*18.80. What is the minimum (least negative) cathode potential (versus SHE) needed to electroplate silver onto cutlery in a solution of Ag^+ and NH_3 in which most of the silver ions are present as the complex, $Ag(NH_3)_2^+$, and the concentration of $Ag^+(aq)$ is only 3.50×10^{-5} M?

Fuel Cells

CONCEPT REVIEW

18.81. What are the advantages of hybrid (gasoline engine–electric motor) power systems over all-electric systems based on fuel cells? What are the disadvantages?

18.82. Small fleets of fuel-cell powered automobiles are operating in several U.S. cities. Describe three factors limiting the more widespread use of cars powered by fuel cells.

18.83. Methane can serve as the fuel for electric cars powered by fuel cells. Carbon dioxide is a product of the fuel cell reaction. All cars powered by internal combustion engines burning natural gas (mostly methane) produce CO_2. Why are electric vehicles powered by fuel cells likely to produce less CO_2 per mile?

18.84. To make the refueling of fuel cells easier, several manufacturers offer converters that turn readily available fuels—such as natural gas, propane, and methanol—into H_2 for the fuel cells and CO_2. Although vehicles with such power systems are not truly "zero emission," they still offer significant environmental benefits over vehicles powered by internal combustion engines. Describe a few of them.

PROBLEMS

18.85. Fuel cells with molten alkali metal carbonates as electrolytes can use methane as a fuel. The methane is first converted into hydrogen in a two-step process:

$$CH_4(g) + H_2O(g) \rightarrow CO(g) + 3\,H_2(g)$$

$$CO(g) + H_2O(g) \rightarrow H_2(g) + CO_2(g)$$

a. Assign oxidation numbers to carbon and hydrogen in the reactants and products.

b. Using the standard free energy of formation values in Appendix 4, calculate the standard free-energy changes in the two reactions and the overall $\Delta G°$ for the formation of $H_2 + CO_2$ from methane and steam.

*18.86. Molten carbonate fuel cells fueled with H_2 convert as much as 60% of the free energy released by the formation of water from H_2 and O_2 into electrical energy. Determine the quantity of electrical energy obtained from converting 1 mole of H_2 into $H_2O(\ell)$ in such a fuel cell.

Additional Problems

*18.87. Electrolysis of Seawater Magnesium metal is obtained by the electrolysis of molten Mg^{2+} salts obtained from evaporated seawater.

a. Would elemental Mg form at the cathode or anode?

b. Do you think the principal ingredient in sea salt (NaCl) would need to be separated from the Mg^{2+} salts before electrolysis? Explain your answer

c. Would electrolysis of an aqueous solution of $MgCl_2$ also produce elemental Mg?

d. If your answer to part c was no, what would be products of electrolysis?

e. How does the phenomenon of overpotentials influence your answers to parts c and d?

*18.88. Silverware Tarnish Low concentrations of hydrogen sulfide in air react with silver to form Ag_2S, more familiar to us as tarnish. Silver polish contains aluminum metal powder in an alkaline suspension.

a. Write a balanced net ionic equation for the redox reaction between Ag_2S and Al metal that produces Ag metal and $Al(OH)_3$.

b. Calculate $E°$ for the reaction. [*Hint*: Derive $E°$ values for the half-reactions in which Ag_2S is reduced to Ag metal and $Al(OH)_3$ is reduced to Al metal. You can do this by combining the standard potentials for the reduction of Ag^+ and Al^{3+} in Appendix 6 with the K_{sp} expressions and values of Ag_2S and $Al(OH)_3$ in Appendix 5.]

*18.89. A magnesium battery can be constructed from an anode of magnesium metal and a cathode of molybdenum sulfide, Mo_3S_4. The half-reactions are

Anode: $\quad Mg(s) \rightarrow Mg^{2+}(aq) + 2\,e^- \qquad E°_{ox} = 2.37\ V$

Cathode: $\quad Mg^{2+}(aq) + Mo_3S_4(s) + 2\,e^- \rightarrow$
$$MgMo_3S_4(s) \qquad E°_{red} = ?$$

a. If the standard cell potential for the battery is 1.50 V, what is the value of $E°_{red}$ for the reduction of Mo_3S_4?

*b. The electrolyte in the battery contains a complex magnesium salt, $Mg(AlCl_3CH_3)_2$. Why is it necessary to include Mg^{2+} ions in the electrolyte?

c. What are the apparent oxidation states and electron configurations of Mo in Mo_3S_4 and in $MgMo_3S_4$?

18.90. Clinical Chemistry The concentration of Na^+ ions in red blood cells (11 mM) and in the surrounding plasma (140 mM) are quite different. Calculate the electrochemical potential (emf) across the cell membrane as a result of this concentration gradient.

*18.91. The element fluorine, F_2, was first produced in 1886 by electrolysis of HF. Chemical syntheses of F_2 did not happen until 1986 when Karl O. Christe successfully prepared F_2 by the reaction

$$K_2MnF_6(s) + 2\,SbF_5(\ell) \rightarrow 2\,KSbF_6(s) + MnF_3(s) + \tfrac{1}{2}F_2(g)$$

a. Assign oxidation numbers to each compound and determine the number of electrons involved in the process.

b. Using the following values for $\Delta H°_f$, calculate $\Delta H°$ for the reaction.

$\Delta H_{f,SbF_5(\ell)} = -1324\ kJ/mol \qquad \Delta H_{f,K_2MnF_6(s)} = -2435\ kJ/mol$

$\Delta H_{f,MnF_3(s)} = -1579\ kJ/mol \qquad \Delta H_{f,KSbF_6(s)} = -2080\ kJ/mol$

c. If we assume that ΔS is relatively small, such that $\Delta G \approx \Delta H$, estimate $E°$ for this reaction.

d. If ΔS for the reaction is greater than zero, is our value for $E°$ in part c too high or too low?

e. The electrochemical synthesis of F_2 is described by the electrolytic cell reaction

$$2\,KHF_2(\ell) \rightarrow 2\,KF(\ell) + H_2(g) + F_2(g)$$

Assign oxidation numbers and determine the number of electrons involves in this process.

18.92. Corrosion of Copper Pipes Copper is frequently the material of choice for household plumbing. In some communities heavily chlorinated water corrodes copper pipes. The corrosion reaction is believed to involve the formation of copper(I) chloride:

$$2\,Cu(s) + Cl_2(aq) \rightarrow 2\,CuCl(s)$$

a. Write balanced equations for the half-reactions in this redox reaction.

b. Calculate $E°_{rxn}$ and $\Delta G°_{rxn}$ for the reaction.

Power, Pigments, and Phosphors: The Group 12 Elements

18.93. Write balanced chemical equations for the following processes.
    a. Zinc(II) sulfide reacts with oxygen.
    b. Zinc(II) oxide reacts with carbon.
    c. Mercury(II) sulfide reacts with oxygen.

18.94. Write balanced chemical equations for the following processes:
    a. Zinc reacts with oxygen.
    b. Cadmium reacts with sulfur.
    c. Zinc metal reacts with aqueous copper(II) nitrate.

18.95. Zinc(II) sulfide and cadmium(II) sulfide crystallize in two different structures, an hcp (hexagonal closest-packed) and a ccp (cubic closest-packed) arrangement of sulfide ions with the cations in tetrahedral holes.
    a. Do these structures have the same packing efficiency?
    b. Are the tetrahedral holes the same size in both structures? Explain your answer.

18.96. Zinc(II) oxide and zinc(II) sulfide (wurtzite) have the same structure, an fcc (face-centered cubic) arrangement of anions with zinc cations in tetrahedral holes.
    a. Are the tetrahedral holes in ZnO the same size as in ZnS?
    b. How many tetrahedral holes are in the unit cell?

18.97. Calculate the $E°_{cell}$ value of the following reactions.
    a. $Zn(s) + Cd^{2+}(aq) \rightarrow Cd(s) + Zn^{2+}(aq)$
    b. $Hg(\ell) + Zn^{2+}(aq) \rightarrow Zn(s) + Hg^{2+}(aq)$
    c. $Cd(s) + Hg_2^{2+}(aq) \rightarrow Cd^{2+}(aq) + 2\,Hg(\ell)$

18.98. Which of the reactions in Problem 18.97 are spontaneous?

18.99. Which of these elements is the best reducing agent: Zn, Cd, or Hg?

18.100. Which cation, Hg^{2+} or Hg_2^{2+}, is the better oxidizing agent?

18.101. Why is the chemistry of zinc similar to the chemistry of magnesium?

18.102. Predict the products of the following reactions:
    a. $Zn(g) + Cl_2(g) \rightarrow$
    b. $ZnCO_3(s) + heat \rightarrow$
    c. $Zn(OH)_2(s) + heat \rightarrow$

18.103. The aqueous solutions of many transition metal compounds are colored, but solutions of zinc compounds are colorless. Why are solutions of $Zn^{2+}(aq)$ colorless?

19

Biochemistry:
The Compounds of Life

DNA AND RNA *Scanning electron micrograph showing strands of DNA (blue) and RNA (red).*

A LOOK AHEAD: The Long and the Short of It

Perhaps the most fundamental questions asked by humans involve the origin of life. How did life begin on Earth? And if we do find out how, will we ever understand why? We may never have a completely satisfying answer to these questions. In searching for answers, however, we have learned much about the chemical reactions among the molecules found in living systems. As variable as life is, ranging from relatively simple one-celled organisms to green plants and the animals that consume them, several general patterns emerge when we consider the processes that constitute life.

For all their stunning diversity, most living systems consist of substances made from only about 40 or 50 small molecules. Huge variations become possible when these few starting materials combine to form larger molecules and polymers. Seemingly subtle differences in the structure of the monomer units or in the orientation of bonds between units have vast and far-reaching consequences when large numbers of small molecules combine. The variation in characteristics of the final products is as wide-ranging as life itself.

Problems in the chemical machinery that makes biomolecules and unnatural variations in structures may result in catastrophic diseases. Many life-threatening illnesses are linked to the body's producing proteins of shorter than normal lengths. It is estimated that up to 70% of inherited disorders, including cystic fibrosis, are caused by the production of proteins that are too short. For example, if the synthesis of a protein called dystrophin is affected, Duchenne's muscular dystrophy may result. Because the shorter proteins are malformed, they are unable to build muscle fibers of sufficient strength to function normally. Muscle tissue wastes away and death follows, all because a biopolymer was too short to provide the strength needed in muscle tissue. Knowledge of the molecular mechanism that causes the shortened proteins to be produced has led to a promising new drug treatment for muscular dystrophy. The drug enables the system of reactions producing dystrophin to bypass the faulty step in the process and produce proteins of normal length.

The relationship between chemical structure and the function of molecules in living organisms is important for health and normal function. The ramifications of recognizing the connection between structure and function range from understanding how life works at the molecular level, to the discovery of new drugs to treat disease, and even to the search for new sources of energy to power the vehicles used in modern transportation systems.

The four major classes of organic compounds in living organisms—proteins, carbohydrates, lipids, and nucleic acids—carry out similar functions in all of the forms of life in which they are found. In this chapter, we discuss the small molecules that build membranes, provide energy, and signal biological responses to stimuli, and the large molecules and biopolymers that provide structure to living organisms, transmit genetic information from generation to generation, store energy, and break down food to provide energy for building the molecules required by and for the machinery of life. Life is not immune or exempt from obeying the laws of thermodynamics, and we conclude with a brief discussion of the molecular system that enables the storage and use of energy by living cells.

A **biomolecule** is an organic molecule present naturally in a living system.

Amino acids are molecules that contain one amine group and one carboxylic acid group. In an **α-amino acid**, the two groups are attached to the same *α* carbon atom.

Proteins are polymers made of amino acids.

A **monosaccharide** is a simple sugar and the simplest carbohydrate.

A **polysaccharide (complex carbohydrate)** is a polymer of monosaccharides.

Lipids are water insoluble, oily organic compounds that, along with carbohydrates and proteins, are common structural materials in cells.

Triglycerides are lipids consisting of esters of glycerol and long-chain fatty acids.

A **nucleotide** is a monomer unit from which DNA and RNA are made.

General structure of an *α*-amino acid.

CONNECTION In Chapter 10 we defined esters as molecules with the general formula

19.1 Biomolecules: Building Blocks of Life

The molecules responsible for the architecture and the machinery of a living system are all synthesized from simple small-molecule precursors: N_2, O_2, H_2O, and CO_2. The actual reactions that create the diversity of structure and function in living systems are well beyond the scope of this book. In this chapter, however, we discuss the end results of the reactions of these small molecules in terms of four families of **biomolecules**, organic molecules found naturally in living systems. These four groups—proteins, carbohydrates, lipids, and nucleic acids—are the building blocks of cells and entire organisms. Considering the huge number of different molecules that characterize even the simplest cell, it is amazing that all are primarily built from a relatively small number of precursors. Only two factors are responsible for all the diversity of form and function among biomolecules: the identity of the small molecules from which the biomolecules are built, and the shape of the resultant biomolecules. First we look at a brief summary of common small molecules from which larger biomolecules are built. Then we discuss each class of biomolecule in terms of its structure and function.

Small-Molecule Subunits

Experiments conducted by Harold Urey and Stanley Miller in the 1950s marked a milestone in experimental research on the origin of life. Research on this topic continues, and the question of how the small molecules we now recognize as the building blocks of biomolecules actually arose on prebiotic Earth is still of great interest. The first experiments of Urey and Miller demonstrated that simple **amino acids**, molecules containing one amine group and one carboxylic acid group, could form in the absence of life. Although the conditions the two scientists chose are now thought to be different than those present on the early Earth, the result still stands: combinations of simple inorganic molecules can react to produce organic molecules found in living systems. Amino acids are fundamental building blocks of **proteins**, which are polymers of amino acids. The amino acid shown in the marginal figure is called an **α-amino acid**, because the $-NH_2$ and the $-COOH$ groups are attached to the same carbon atom. This carbon atom is referred to as the *α* carbon.

The second class of small molecules from which an array of larger molecules and polymers are made is sugars. About 16 **monosaccharides**—the name means "one sugar" and identifies the simplest carbohydrates—are the common monomer units found in nature, from which the majority of **polysaccharides** or **complex carbohydrates** are made. Glucose, which we have seen many times in this book as a major product of photosynthesis, is a monosaccharide, and cellulose (the major structural component in plants) is a very common polysaccharide.

The third class of molecules, the **lipids**, differs from carbohydrates and proteins because its members are not polymers. Lipids are best described by their physical properties, as opposed to any common structural subunit: lipids are insoluble in water, soluble in nonpolar solvents, and oily to the touch. They are common structural materials in cells. Because of their insolubility in water, they are ideal components for cell membranes, which separate the aqueous solutions within cells from the aqueous environment outside cells. An important class of lipids called **triglycerides** consists of fatty acid esters formed between glycerol and long-chain fatty acids, as shown on the next page.

Nucleotides are the monomeric units from which deoxyribonucleic acid (DNA) and ribonucleic acid (RNA) are constructed. This fourth class of small

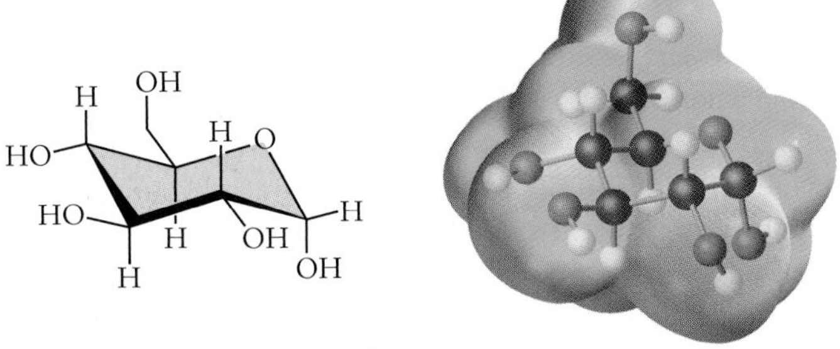

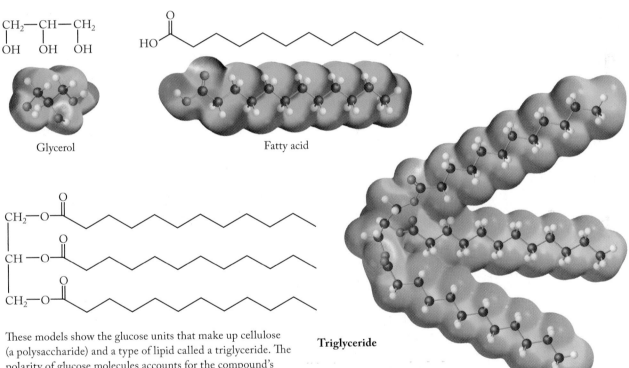

Glucose

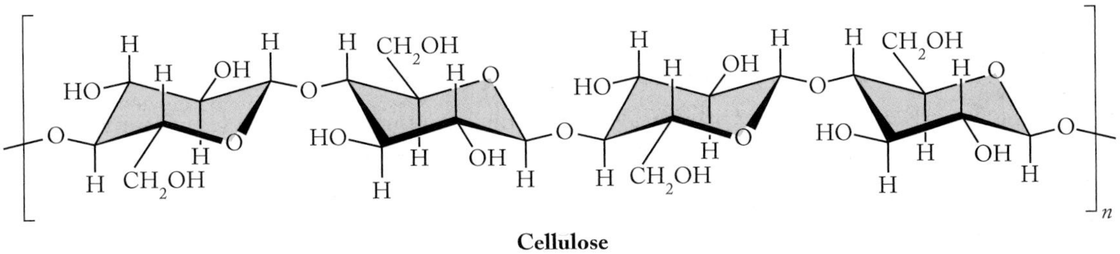

Cellulose

Glycerol

Fatty acid

Triglyceride

These models show the glucose units that make up cellulose (a polysaccharide) and a type of lipid called a triglyceride. The polarity of glucose molecules accounts for the compound's solubility in water. Some of this polarity is lost when glucose polymerizes to form cellulose, which is insoluble in water. The long nonpolar chains of fatty acids make their esters, including the triglyceride shown here, insoluble in water.

molecules is of great importance to life, even though it makes up only about 1% of an organism by weight and is much less abundant than proteins, carbohydrates, and lipids. Whereas the other three classes of molecules play many roles in cellular structure, function, and energy storage, nucleotides have more-defined roles: they direct the metabolic activity of all cells. DNA carries the genetic blueprint of an organism and a variety of RNAs use the DNA blueprint to build proteins. Five different nucleotide units make up DNA and RNA. Each nucleotide unit is composed of a simple sugar that is bound to a phosphate group and an organic base.

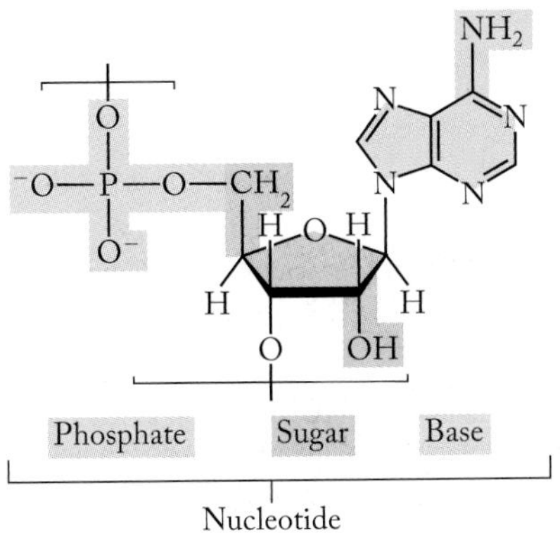

Phosphate Sugar Base

Nucleotide

The parts of a nucleotide.

Chimneys of iron sulfide form in an undersea vent.

In making proteins, carbohydrates, and nucleic acids, nature uses around 44 small molecules—20 amino acids, about 16 sugars, and 8 nucleotides—to prepare most of the polymers in even the most complex living systems. It is more difficult to characterize the lipids in terms of simple structural components, but they too have common fundamental features whose variation leads to differences in behavior that correspond to variations in living systems. The diversity of life depends on the various lipids and on different combinations of about 44 small molecules to make an almost endless variety of compounds of wide-ranging size, shape, and function.

Formation of Biomolecules and Organized Assemblies

Research on the production of organic monomers from simple inorganic precursors has shown how simple carbon-containing molecules could have arisen on early Earth. How did the four types of biomolecules arise on prebiotic Earth? This question provides a continuing area of speculation and research, but two examples of recent studies indicate fascinating possibilities. For example, it has been shown that nucleotides spontaneously link together to form long chains of RNA on the surface of a common clay mineral called montmorillonite. The pool of polymers made by this surface reaction contains many chains with random sequences, but some of the RNA molecules produced have been shown to catalyze their own replication. This ability to self-replicate is crucial to life, and the observation that molecules capable of speeding up their own duplication on a clay surface suggests ways that processes important in living systems could have happened spontaneously.

Another current area of exploration involves the theory that metabolism began in hydrothermal vents in the ocean floor in which inorganic compounds like carbon dioxide and hydrogen sulfide are converted into more complex organic groups like acetate (CH_3COO^-) in reactions on the surfaces of pyrite (iron sulfide) crystals. Acetate is a key intermediate in many biosynthetic pathways in living organisms. In modern bacteria, the systems that make acetate depend on a catalytic unit made of iron, nickel, and sulfur, arranged in almost the same structural pattern as the one found in pyrites. Thus the inorganic components that survive in today's bacteria may be the descendents of catalytic materials that predate life itself.

To take the next step toward life, the large molecules and polymers made from the small-molecule building blocks must organize themselves into larger assemblies. These groupings of molecules include units within cells that carry out special functions. The polymers and large molecules in structures in this category are not necessarily bonded together in the sense of being linked by covalent bonds, but have specific associations mediated by other interactions we have discussed previously: hydrogen bonds, dipole–dipole interactions, and van der Waals forces. These assemblies combine to build structures within cells, such as the nucleus and mitochondria.

The final step in this scenario addresses an essential feature of cells: they collect materials and retain them at concentrations different from those in the surrounding medium. The steps from small inorganic precursors to cells and the relative sizes of molecules and assemblies are outlined in Figure 19.1. In the next sections, we address a common characteristic of molecules formed in living systems and follow with a discussion of how small monomers combine to form proteins, carbohydrates, lipids, and nucleic acids.

Cells

Mitochondria, nuclei, chlorophyll-containing units

Large molecules, biopolymers molecular assemblies 10^3–10^{10} amu

Enzymes, antibodies, ribosomes

Carbohydrates, proteins, polynucleic acids, lipids

Small carbon-containing monomers 60–350 amu

Monosaccharides, amino acids, acetates

Small inorganic molecules 18–44 amu

H_2O, N_2, CO_2, H_2S, NH_3, CH_4

FIGURE 19.1 The relative masses of components of living systems, from small molecules to cells.

CONNECTION We discussed the effects of attractive forces between molecules on the physical properties of substances in Chapter 10.

19.2 Chirality

The influence of molecular structure on the properties of compounds was discussed in Chapter 12 in terms of the properties of isomers. Structural isomers of alkanes exist, for example, as straight-chain and branched-chain molecules, in which compounds have the same formula but have atoms arranged in different patterns (Figure 19.2a). Pairs of molecules called stereoisomers have the same formula and the same bonds, but the orientation in space of groups within the molecule differs. One type of stereoisomerism gives rise to *cis* and *trans* geometric isomers. Geometric isomers in organic compounds have substituent groups in different positions in a molecule with respect to a carbon–carbon double bond (see Figure 19.2b). Geometric isomers exist in metal complexes which can be *cis* or *trans* isomers depending on the position of attachment of ligands to a metal center (Figure 19.2c). Another type of stereoisomerism called *optical isomerism* is especially prevalent and extremely important in biomolecules.

Optical Isomerism

Optical isomers are molecules that are non-superimposable on their mirror images, just like your left hand is not superimposable on your right hand. They are called *optical* isomers because of the way in which they interact with polarized light, as discussed on page 937. Their optical isomerism comes from the presence of *chiral centers* in their molecular structures. The term **chiral** comes from the Greek word *cheir* ("hand") and is quite correctly called "handedness." Figure 19.3 illustrates what is meant by mirror images and superimposition. The reflection of an object like a plain coffee mug that is **achiral** (not chiral) in a mirror is an image that can be superimposed on the original object. The reflection of your left hand in a mirror, however, is an image of your right hand. If you try to superimpose your right hand on your left hand, it doesn't fit. Your two hands are mirror images that cannot be superimposed. Molecules that have this same property—molecules that cannot be superimposed on their mirror images—exist as optical isomers and so are called chiral. Although several features within molecules can lead to the existence of optical isomers, the most common one is the presence in the molecule of a carbon atom that has four different groups attached to it. Chirality is a key feature of amino acids and sugars, two of the essential groups in living systems.

(a) Structural isomers of a C_5 alkane

(b) Geometric isomers at a carbon–carbon double bond

cis   *trans*

(c) Geometric isomers at a square planar metal center

$$H_3N-Pt-Cl$$ ... Cl

   cis   *trans*

FIGURE 19.2 Different types of isomers: (a) structural isomers, (b) geometric isomers in a carbon compound, and (c) geometric isomers of a metal complex.

Optical isomers are molecules that are not superimposable on their mirror images.

A molecule that is **chiral** is not superimposable on its mirror image.

A molecule that is **achiral** is superimposable on, and identical to, its mirror image.

CONNECTION We learned about the different properties of *cis* and *trans* isomers of organic compounds in Chapter 12 and of metal complexes in Chapter 17. Isomers have the same formula but different structures and hence different properties.

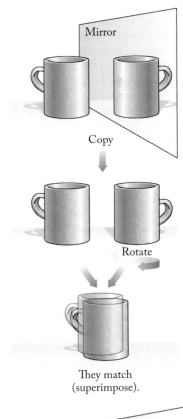

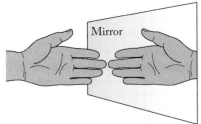

FIGURE 19.3 A plain coffee mug is superimposable on its mirror image and is achiral. The mirror image of your left hand is your right hand, and the two are not superimposable.

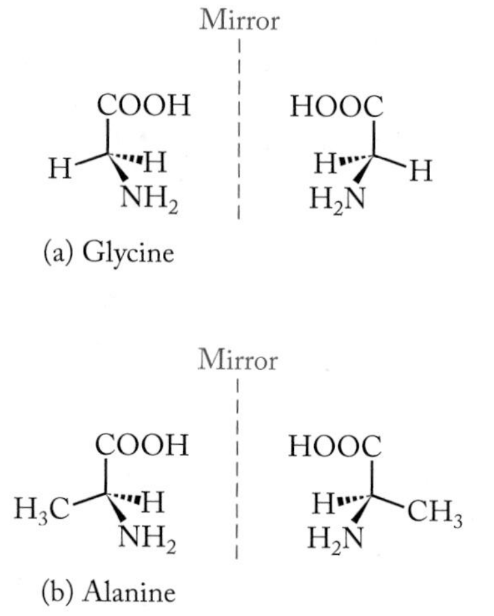

FIGURE 19.4 Mirror images of two α-amino acids: (a) glycine and (b) alanine.

▶Ⅱ **CHEMTOUR** Chirality

To illustrate chirality at the molecular level, let's look at two α-amino acids: glycine and alanine. First we generate their mirror images (Figure 19.4). Then we rotate the mirror images of the molecules 180° (Figure 19.5). If we then try to superimpose each rotated mirror image on top of its original image, and if identical groups overlap, then the molecule is achiral. Glycine can be superimposed on its mirror image (Figure 19.5a), while alanine cannot (Figure 19.5b). When we superimpose the α-carbon atoms, the –NH₂ group on one molecule of alanine lines up with the –H of the other. If we line up the –NH₂ groups, then the –H and the –CH₃ do not line up. The images of alanine represent optical isomers.

If we examine the two structures, we see that alanine has one carbon atom with four different groups bound to it: a carboxylic acid (–COOH), an amine (–NH₂), a methyl group (–CH₃), and a hydrogen atom (–H). The corresponding carbon atom in glycine also has a –COOH and an –NH₂, but it has two groups that are identical: the two hydrogen atoms. Any carbon atom with four *different* groups attached to it is a chiral center, and the molecule containing that carbon atom exists as optical isomers. The two optical isomers are called **enantiomers**; enantiomers are non-superimposable mirror images. In no orientation do all of the groups on the chiral center of superimposed enantiomers coincide, because the molecules have different shapes in three dimensions. Chirality is exquisitely important in living systems, where molecules interact with other molecules in a process called **recognition**. Recognition in biochemistry means identifying a molecule based on its interaction with another molecule because of its shape. Just as your right hand fits into the correct glove, a molecule may fit into a three-dimensional cavity in a protein and cause an event based on that recognition. The protein that forms such a cavity is called a **receptor**.

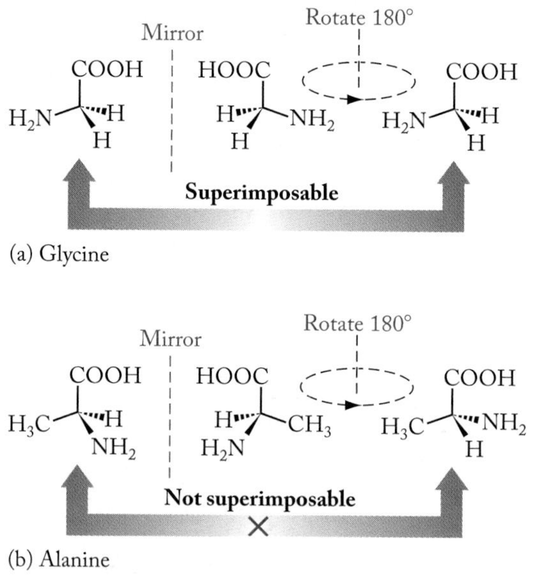

Enantiomers are molecules that are not superimposable on their mirror images.

Recognition is the process in which one molecule fits into a three-dimensional site on another molecule.

A **receptor** is a protein that binds another molecule because of its unique shape. Binding to a receptor typically causes some biological response.

FIGURE 19.5 Enantiomers are not superimposable on their mirror images. (a) The mirror image of glycine is rotated 180° and yields a structure that is superimposable on the original structure. The two structures are identical. (b) Carrying out the same process with a molecule of alanine results in a structure that is not superimposable on that of the original molecule. The two structures are enantiomers.

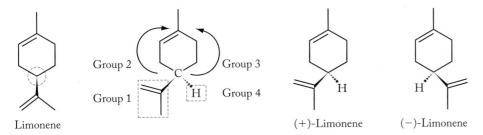

Limonene   (+)-Limonene   (−)-Limonene

FIGURE 19.6 Limonene is a chiral molecule with two enantiomeric forms. To show why it is chiral, we have highlighted its chiral carbon with a dashed circle and numbered the four groups bonded to it. Groups 1 and 4 are clearly different. Groups 2 and 3 are two halves of the same ring. They are different because group 2 has a C=C double bond and group 3 does not. Therefore, the circled carbon atom is bonded to four different groups and is chiral.

Structural and geometric isomers differ in their physical and chemical properties, but optical isomers have the same physical and chemical properties except for those that relate to a few specialized types of behavior. Recognition is one of those special behaviors. Figure 19.6 shows the structure of the chiral molecule limonene. The chiral center is identified by the dotted circle. Limonene has two enantiomers, distinguished by the (+) and (−) in front of the name. One enantiomer smells like oranges; the other smells like turpentine. Part of the process of sensing different aromas involves recognition by receptors in your nasal passages. One receptor is shaped like (+)-limonene; the other like (−)-limonene, and the enantiomers bind to their own receptor just like your right and left hands fit into specially made right and left gloves.

Enantiomers are also called **optically active molecules**. The (+) and (−) signs that distinguish the two isomers of limonene refer to the specific effect each isomer has on polarized light. In plane-polarized light, the electric fields that compose the beam oscillate in only one plane (Figure 19.7). When a beam of plane-polarized light passes through a solution containing one member of an enantiomeric pair, the beam rotates. If the beam rotates to the left (counterclockwise), the enantiomer is the *levorotary* form of the molecule and a (−) sign precedes its name. If the beam rotates to the right (clockwise), the enantiomer is the *dextrorotary* form and a (+) sign precedes its name.

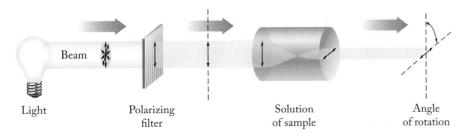

Light   Polarizing   Solution   Angle
              filter   of sample   of rotation

FIGURE 19.7 A beam of plane-polarized light consists of electric field vectors that oscillate in only one direction. The plane of oscillation rotates if the beam passes through a solution of one enantiomer of an optically active compound. The (+) enantiomer causes the beam to rotate clockwise; the (−) enantiomer causes the beam to rotate counterclockwise.

Optically active molecules rotate a beam of plane-polarized light. The *levorotary* form of the enantiomers rotates the beam to the left and a (−) sign precedes the chemical name; the *dextrorotary* form rotates the beam to the right and a (+) sign precedes the name.

SAMPLE EXERCISE 19.1   **Recognizing Chiral Molecules**

Identify which of the molecules shown below are chiral, and circle the chiral centers. Some structures may have more than one chiral center.

(a)

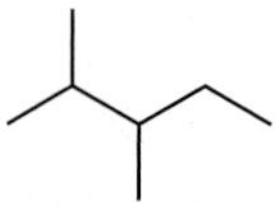

(d)

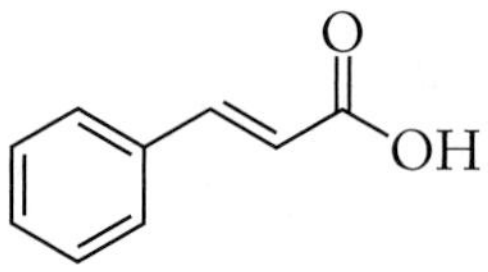

(b)

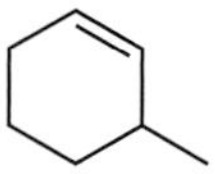

(e)

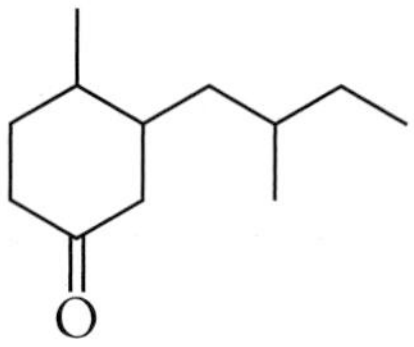

CHEMTOUR Chiral Centers

(c)

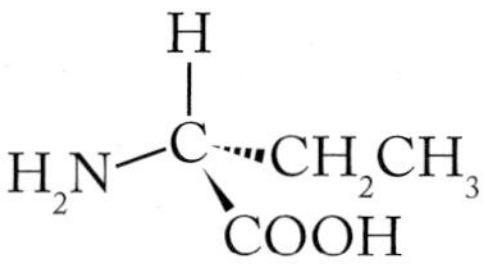

$$H_2N-\underset{\underset{COOH}{|}}{\overset{\overset{H}{|}}{C}}\text{\tiny{\dots}}CH_2CH_3$$

COLLECT AND ORGANIZE We are asked to identify which molecules among the given set are chiral. If molecules have carbon atoms with four different groups attached, they are chiral.

ANALYZE We must look at the structures and determine whether any of the carbon atoms are bonded to four different groups.

SOLVE The chiral carbons in each structure are circled: compounds (a), (b), (c), and (e) are chiral. Structure (d) has no chiral center.

(a)

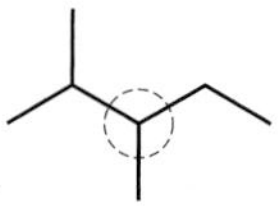

(d)

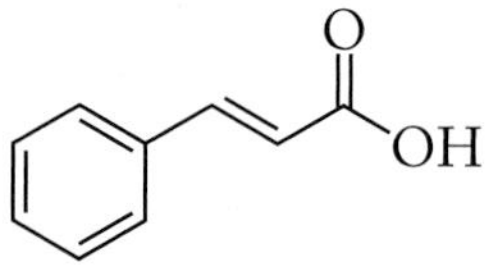

(b)

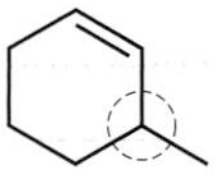

(e)

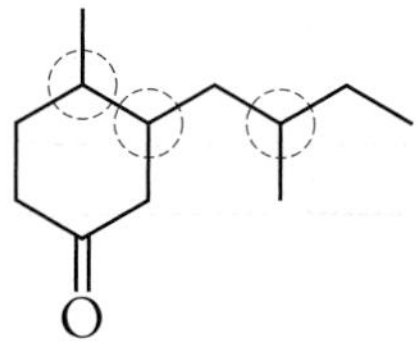

(c)

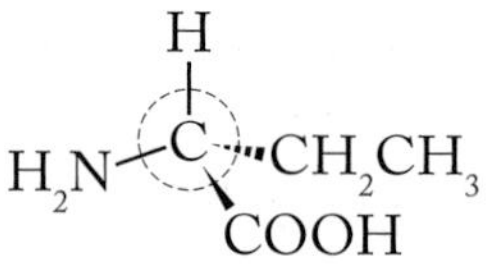

$$H_2N-\underset{\underset{COOH}{|}}{\overset{\overset{H}{|}}{C}}\text{\tiny{\dots}}CH_2CH_3$$

THINK ABOUT IT The presence of a chiral center in a molecule means that the molecule has two enantiomeric forms. The molecule is not superimposable on its mirror image. We should also note that structure (d) has no chiral center, but it would have *cis-trans* isomers about the double bond in the chain.

Practice Exercise Identify which of the molecules below are chiral. Circle the chiral centers in each structure.

(a)

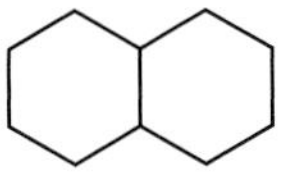

(b)

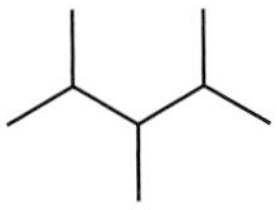

(c)

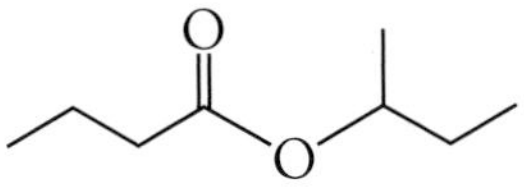

(d)

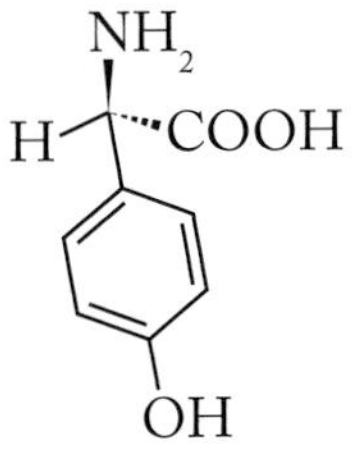

NH_2

H ┬ $COOH$

OH

(e)

O

OH

OH

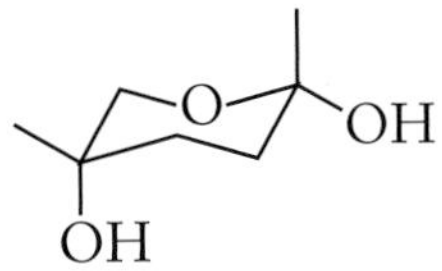

(Answers to Practice Exercises are in the back of the book.)

The molecules we perceive as the aromas of spearmint and caraway are the enantiomers shown below. Describe recognition between a fragrant molecule and its receptor in your nose in terms of the illustration on the right:

(+)-Carvone = spearmint   (−)-Carvone = caraway

(Answers to Concept Tests are in the back of the book.)

Other conventions for describing stereochemistry are used occasionally. Rather than going into the details of the origin and use of conventions you may encounter in other courses, let's just look at some other ways to describe the chirality of amino acids and sugars, the small molecules we introduced in Section 19.1.

Chirality in Nature

Chirality is ubiquitous in the organic compounds formed by living systems, and in most cases only one optical isomer occurs naturally. For example, all α-amino acids except one are chiral. Only glycine is achiral. The other α-amino acids have the generic structural formula

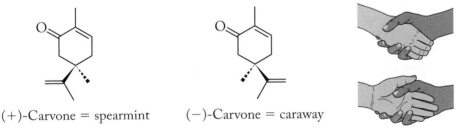

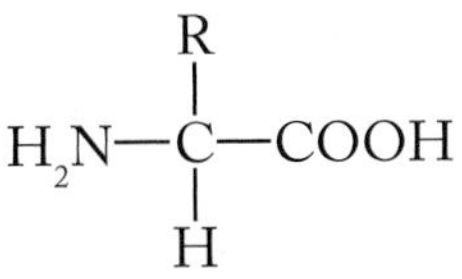

$$H_2N{-}\underset{\underset{H}{|}}{\overset{\overset{R}{|}}{C}}{-}COOH$$

The **configuration** of chiral compounds refers to the actual positions in space occupied by the groups surrounding a chiral center.

A **racemic mixture** contains equal quantities of the enantiomers of an optically active substance and therefore does not rotate the plane of polarized light.

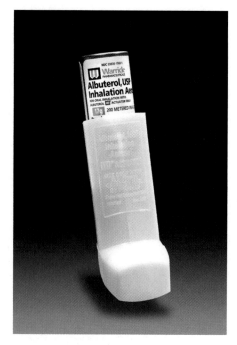

FIGURE 19.8 The antiasthma drug albuterol; the chiral center is circled.

where "R" is any organic group linked to the α-carbon by a C–C bond. For historical reasons, the mirror image pairs of amino acids are named D (for *dextro-*, right) and L (*levo-*, left). These terms describe the actual position in space—the **configuration**—of the groups surrounding the chiral carbon. They do *not* refer to the direction of optical rotation. All chiral amino acids in proteins in our bodies have the L configuration. D-Amino acids also exist in nature and have been identified in bacterial cell walls and in some toxic compounds produced by mushrooms.

Sugars are also chiral, and their stereoisomers are distinguished by adding the prefixes D or L to their names. Most of the sugars found in nature, including those in DNA and RNA have the D-configuration.

The origin of the fundamental preference of life for one enantiomer over another is unknown. Just as with questions about the origin of organic molecules and biopolymers in Section 19.1, ongoing studies on the origin of chiral preference in living systems have produced no definitive answers but are providing increasingly interesting suggestions. Perhaps the origin of the preference involves the effect of polarized light from the atmosphere on early Earth; perhaps slight energy differences between the D- and L-forms of molecules contributed to the bias; and perhaps even something as simple as stirring a solution can act as a chiral force that selects molecules of a particular configuration for reactions.

Whatever the origin of these preferences, processes requiring molecular recognition in living systems often depend on the selectivity conveyed by chirality. The human body is a chiral environment, so handedness of molecules matters. The perception of an aroma as either turpentine or orange or as caraway or spearmint depends on chirality, but other consequences of chiral recognition are much more dramatic. As many as half of the drugs made by large pharmaceutical companies are chiral and owe their function to recognition by a receptor that favors one enantiomer over the other. In 2004, nine of the ten top-selling drugs globally were chiral. It is typical that only one of the enantiomers of a chiral drug is active. A classic example is the drug albuterol (Figure 19.8), used to treat wheezing and shortness of breath in people suffering from asthma and other lung disorders. One isomer causes bronchodilation (widening the air passages of the lungs and easing breathing), while the other does not and may actually be detrimental to the condition for which the material is prescribed.

When chiral compounds are produced by living systems, typically only one stereoisomer is produced. For example, the aroma of spearmint leaves is due to (+)-carvone. However, when a material such as albuterol is made in a laboratory both isomers result, unless special methods are used. When both optical isomers are present in equal amounts in a sample, the material is known as a **racemic mixture**. Because one isomer rotates the plane of polarized light in one direction, and the other to the same extent in the opposite direction, a racemic mixture does not rotate the plane of polarized light at all. Because the two isomers interact differently with receptors, the pharmaceutical industry routinely faces two choices: devise a special synthetic procedure that yields only the isomer of interest, or separate the two isomers at the end of the manufacturing process. Both approaches are widely used.

SAMPLE EXERCISE 19.2   **Recognizing Properties of Racemic Mixtures**

Below is the structure of the antidepressant drug bupropion.

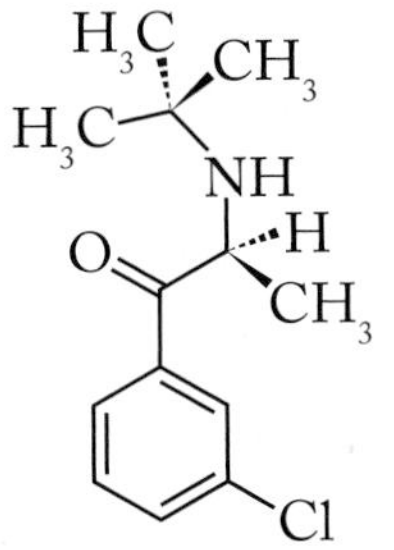

Bupropion

When tested, a solution of a sample of the drug that comes directly from the laboratory does not rotate the plane of polarized light, so the analyst rejects the sample, saying it is not the pure drug, which is known to cause a beam of polarized light to rotate when it passes through a solution of the compound. The chemist who made the sample tells you other analytical data (percent composition and mass spectrometry) proves the material is 100% bupropion. Both people are correct in what they say. Explain this.

COLLECT AND ORGANIZE We are asked to explain how a sample that is 100% chemically pure is not 100% pure active drug. We are given the structure of the compound, and the information that a solution of the material does not rotate the plane of polarized light.

ANALYZE The sample was rejected by one analyst because a solution did not rotate the plane of polarized light. We must examine the structure to see if the molecule has a chiral center.

SOLVE Bupropion is a chiral molecule. The chiral center is circled in the structure below:

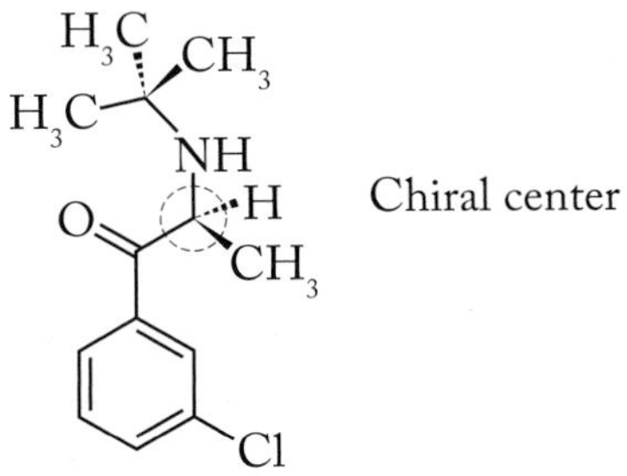

Chiral center

Bupropion

The sample must contain both enantiomers of bupropion in equal amounts. They have exactly the same chemical formula, so the sample is chemically pure, but one enantiomer rotates polarized light to the left and the other to the right. No net rotation is observed when polarized light passes through the sample. Only one enantiomer is the active drug.

FIGURE 19.9 Several varieties of mushrooms, including *Amanita muscaria*, contain the toxic substance muscarine.

THINK ABOUT IT Enantiomers have the same chemical composition and the same molar mass. Their physical properties are identical except for the direction in which their solutions cause plane-polarized light to rotate when it passes through them.

Practice Exercise Identify the chiral carbon atoms in the cationic natural product muscarine, found in some poisonous mushrooms (Figure 19.9).

$$\left[\; \text{H}\;\; \overset{\displaystyle \text{CH}_2\text{N}(\text{CH}_3)_3}{\underset{\text{HO}\;\; \text{H}\;\; \text{CH}_3}{\overset{\text{H}\cdots\;\;\;\cdots\text{H}}{\underset{\text{O}}{}}}}\;\right]^{+}$$

Muscarine

CONCEPT TEST

The Documents in the Case, a mystery written by Dorothy L. Sayers in 1930, involves the suspicious death of an authority on wild, edible mushrooms. Allegedly, the victim ate a stew made of poisonous mushrooms that contained muscarine, a toxic natural product. A forensic specialist evaluated the contents of the victim's stomach and observed that the fluid contained muscarine but did not rotate the plane of polarized light. Because of this fact, the coroner concluded that the man was murdered. Why did the coroner in the story reach this conclusion?

19.3 Protein Composition

Proteins account for about half of the mass of the human body that is not water. Proteins are involved in almost every aspect of the structure and function of cells. Structurally, proteins are the major component in skin, muscles, cartilage, hair, and nails of humans, and in the horns and hooves of animals and the beaks of birds. Functionally, most of the enzymes that catalyze biochemical reactions are proteins; hemoglobin and the cytochromes that transport oxygen and electrons in metabolism are proteins; some of the hormones such as insulin that regulate cellular function and bodily processes like growth, metabolism, and development are proteins.

Amino Acids

All proteins are composed of amino acids. The names and structures of the 20 amino acids from which the proteins in the human body are made are shown in Table 19.1. Our bodies can synthesize 12 of these amino acids, but 8 others must be provided in the food we eat. These 8 are marked with an asterisk in the table and are referred to as the essential amino acids because they must be included in our diet for proper growth and health. Eggs and dairy products such as milk contain all the essential amino acids needed by the human body in close to the correct proportions we need. These foods are sometimes referred to as perfect foods, or more precisely, complete proteins. In contrast, most plant products do not contain

TABLE 19.1 Structures and Abbreviations of the 20 Common Amino Acids

Nonpolar R groups

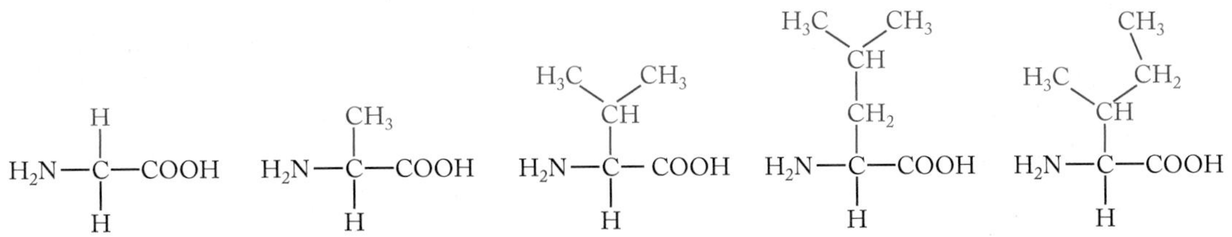

Glycine
(Gly)

Alanine
(Ala)

Valine[a]
(Val)

Leucine[a]
(Leu)

Isoleucine[a]
(Ile)

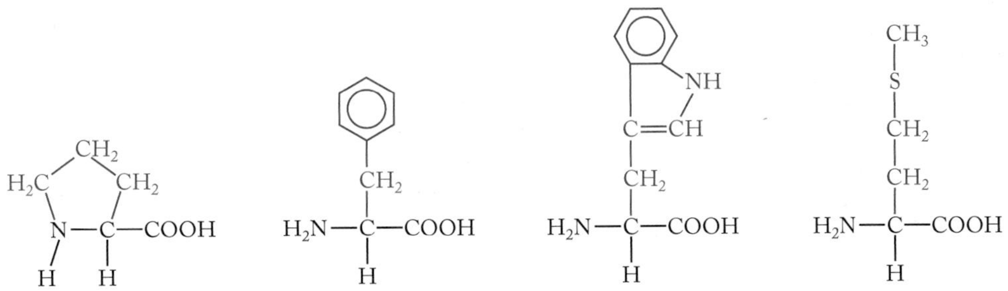

Proline
(Pro)

Phenylalanine[a]
(Phe)

Tryptophan
(Trp)

Methionine[a]
(Met)

Polar R groups

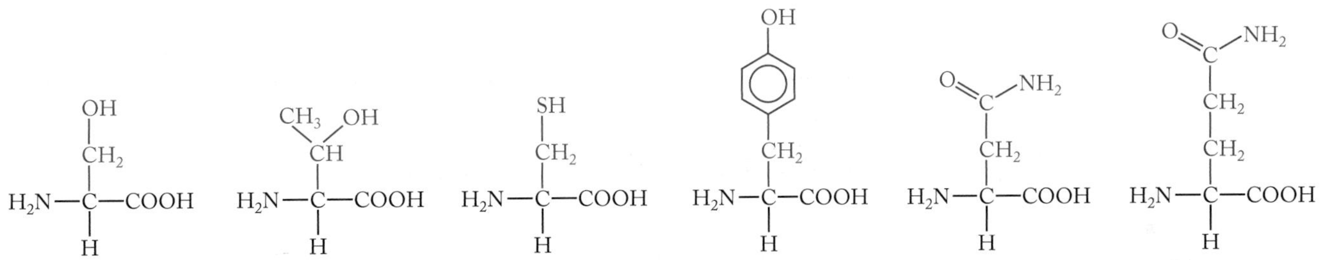

Serine
(Ser)

Threonine[a]
(Thr)

Cysteine
(Cys)

Tyrosine[a]
(Tyr)

Asparagine
(Asn)

Glutamine
(Gln)

Acidic R groups

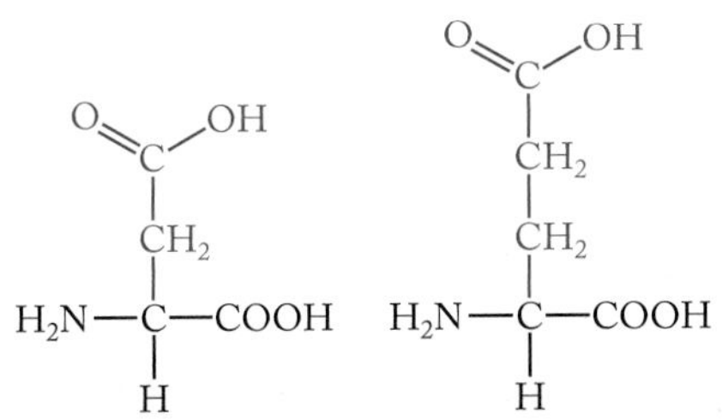

Aspartic acid
(Asp)

Glutamic acid
(Glu)

Basic R groups

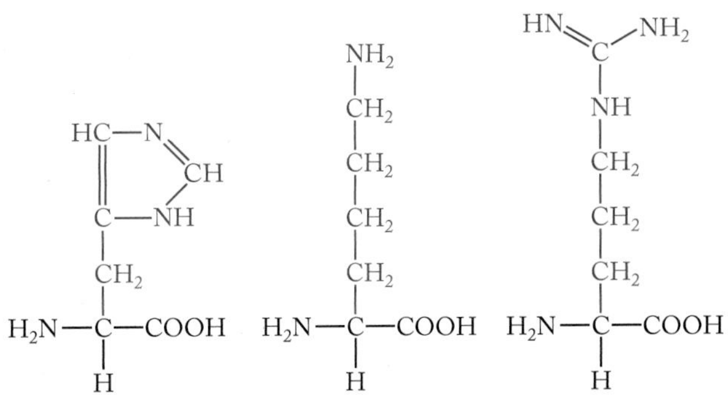

Histidine
(His)

Lysine[a]
(Lys)

Arginine
(Arg)

[a] The eight essential amino acids for adults (histidine is essential for children).

A **zwitterion** is a molecule that has both positively and negatively charged groups in its structure.

The combination of red beans and rice alone provides balanced nutrition. Sometimes meat is added to this classic dish, but beans and rice provide all the essential amino acids.

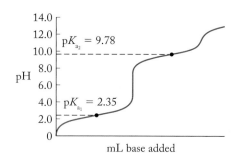

FIGURE 19.10 At pH values near 7, alanine (and many other amino acids) exist in the form of zwitterions.

⊙⊙ **CONNECTION** In Chapter 16 we discussed the ionization of weak diprotic acids and learned that the ionization of the first proton from a weak diprotic acid always occurs at a lower pH and has a lower pK_a value than the ionization of the second proton.

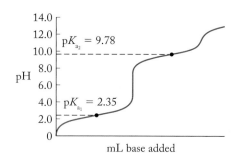

FIGURE 19.11 The titration curve of glycine shows the pK_{a_1} of the carboxylic acid group at pH 2.35 and the pK_{a_2} of the protonated amine group at pH 9.78.

all the essential amino acids, so vegetarians must be careful to eat a diet that contains foods matched to meet the body's needs. Many cuisines have classic dishes that offer such combinations. For example, the red beans and rice considered a staple of Louisiana Creole cuisine and many Latin American cultures provide complete protein: rice lacks lysine, and beans lack methionine. The combination in one dish of the two foods provides all the essential amino acids.

In solution at low pH, the amine and carboxylic acid groups on amino acids are protonated, and in this form the amino acid is a cation that behaves like a weak diprotic acid. The –COOH group is more acidic than the $-NH_3^+$ group, so as pH increases, the –COOH group is ionized before the $-NH_3^+$ loses a proton. Using glycine as an example, in the form in which the amine is protonated and the carboxylic acid is ionized, $H_3N^+CH_2COO^-$, the amino acid exists as a **zwitterion** (literally, a *hybrid ion*), because it has both a positive and a negative group in one molecule (Figure 19.10). When an amino acid exists as a zwitterion, it has a net charge of zero. If the pH of a solution containing the zwitterion of glycine is increased further, the amino acid ultimately forms a 1– anion. The titration curve of glycine (Figure 19.11) looks similar to that for a weak diprotic acid and has two equivalence points.

SAMPLE EXERCISE 19.3 **Describing the Acid–Base Behavior of an Amino Acid**

Glycine has two pK_a values: $pK_{a_1} = 2.35$, and $pK_{a_2} = 9.78$. Write equations for the dissociation reactions that are characterized by these two values.

COLLECT AND ORGANIZE We are asked to write two equations that describe the behavior of glycine as a diprotic acid.

ANALYZE At low pH, the amine group and the carboxylic acid group are both protonated. The –COOH group is a stronger acid than the $-NH_3^+$ group, so the ionization of the carboxylic acid group must be characterized by pK_{a_1} and the deprotonation of the $-NH_3^+$ group by pK_{a_2}.

SOLVE The two pK_a values refer to these reactions:

$$H_3\overset{+}{N}CH_2COOH(aq) \rightleftharpoons H_3\overset{+}{N}CH_2COO^-(aq) + H^+(aq) \qquad pK_{a_1} = 2.35$$

$$H_3\overset{+}{N}CH_2COO^-(aq) \rightleftharpoons H_2NCH_2COO^-(aq) + H^+(aq) \qquad pK_{a_2} = 9.78$$

THINK ABOUT IT An amino acid in aqueous solution can exist in several different forms depending on the pH.

Practice Exercise At low pH, the amino acid aspartic acid (Asp) exists as a cation with a charge of 1+. If a solution containing the 1+ ion of aspartic acid is titrated, it shows three equivalence points: $pK_{a_1} = 1.88$, $pK_{a_2} = 3.65$, and $pK_{a_3} = 9.66$. Write equations that describe the equilibria characterized by each of those values.

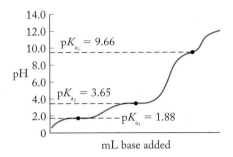

```
14.0
12.0
10.0      pK_a3 = 9.66  - - - - - - - ●
pH  8.0
    6.0
    4.0   pK_a2 = 3.65  - - - ●
    2.0  ●- - - - - - - pK_a1 = 1.88
      0
              mL base added
```

CONCEPT TEST

Select one amino acid, other than glycine, from Table 19.1 that has two pK_a values; select a second, other than aspartic acid, that has three pK_a values.

Peptides

Amino acids bond together via **peptide bonds**—which are essentially amide groups like those we saw in the synthetic polymer nylon in Chapter 12—to form structures ranging from simple dipeptides like the artificial sweetener aspartame, the methyl ester of aspartic acid and phenylalanine (Figure 19.12), to large polymers like those found in skin and hair. A peptide bond (circled in the structure of aspartame) is formed when the carboxylic acid group of one amino acid condenses with the amine group of another:

$$
H_2N-\underset{\underset{R}{|}}{\overset{\overset{H}{|}}{C}}-\overset{\overset{O}{\|}}{C}-OH \quad + \quad H_2N-\underset{\underset{R'}{|}}{\overset{\overset{H}{|}}{C}}-\overset{\overset{O}{\|}}{C}-OH
$$

$$\downarrow$$

$$
H_2N-\underset{\underset{R}{|}}{\overset{\overset{H}{|}}{C}}-\overset{\overset{O}{\|}}{C}-\underset{\underset{H}{|}}{N}-\underset{\underset{R'}{|}}{\overset{\overset{H}{|}}{C}}-\overset{\overset{O}{\|}}{C}-OH \quad + \quad H_2O
$$

The convention for describing the structure of **peptides** (compounds formed with two or more amino acids joined by peptide bonds) and proteins requires that we begin on the left with the end that bears the free $-NH_2$ group and end on the right with the terminus that bears the free $-COOH$ group. The names of the first and intermediate amino acids in the chain are changed to end in *-yl*, and the terminal amino acid is identified with its full name. Hence, the *dipeptide* (two amino acids joined by a peptide bond) from which aspartame is made is aspartylphenylalanine. Aspartame is formed when a carboxylic acid functional group in aspartylphenylalanine is converted into a methyl ester. A molecule with two peptide bonds is called a *tripeptide*, and short chains of up to 20 amino acids are called *oligopeptides*. Chains of more than 20 amino acids linked by peptide bonds are called *polypeptides*. Proteins are large polymers of amino acids. The size

Aspartylphenylalanine

Aspartame

FIGURE 19.12 The dipeptide aspartylphenylalanine is converted into the artificial sweetener aspartame by the addition of a methyl group to the carboxylic acid of the R group in aspartic acid.

▶❙❙ **CHEMTOUR** Condensation of Biological Polymers

A **peptide bond** is the result of a condensation reaction between the carboxylic acid group of one amino acid with the amine group of another. A molecule of water is also formed in the process.

A **peptide** is a compound of two or more amino acids joined by peptide bonds. Small peptides are described by indicating the number of amino acids linked together to form the molecule: a *dipeptide* consists of two amino acids; a *tripeptide*, three; short chains of up to 20 amino acids are *oligopeptides*; and finally the term *polypeptide* is used for chains longer than 20 amino acids.

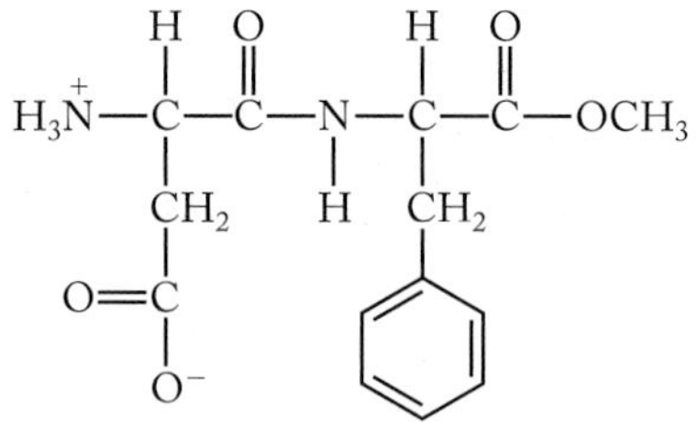

Aspartame

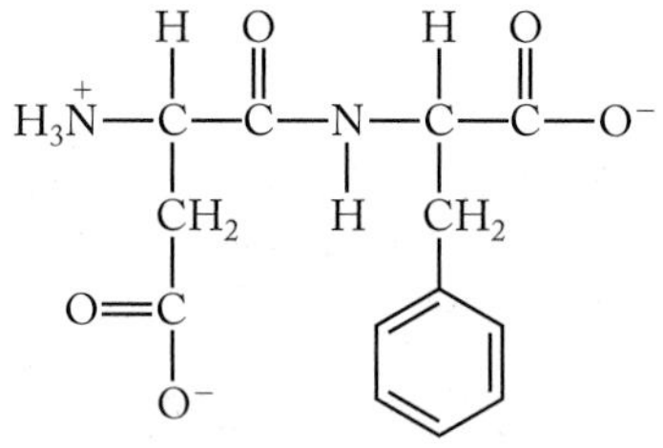

Aspartylphenylalanine

In aqueous media near pH 7, aspartame exists as zwitterions, and aspartylphenylalanine forms molecular ions that each have a net charge of 1−.

at which a polypeptide becomes a protein is arbitrary and is typically somewhere around 50–75 amino acid monomers.

As we have discussed, peptides and proteins tend to exist as ions in physiological systems near pH 7. Aspartame exists as a zwitterion with a protonated amine group and a deprotonated carboxylic acid group. The dipeptide aspartylphenylalanine, in contrast, has a net charge of 1− because both −COOH groups are deprotonated at pH 7. In an amino acid that has a basic amine in its side chain, like lysine or arginine, the amine group in the side chain is most probably protonated at physiological pH giving the amino acid a net charge of 1+. Thus, the overall charge on a peptide depends on the relative numbers of positively charged protonated amines and ionized (negatively charged) carboxylic groups on the R side chains.

SAMPLE EXERCISE 19.4   **Drawing and Naming Peptides**

Draw and name the dipeptides that can be made by combining alanine and glycine. Draw the form in which the dipeptides likely exist at pH 7.

COLLECT AND ORGANIZE We are asked to draw and name the dipeptides that can be formed when two amino acids are linked by a peptide bond. We can find the structures for alanine and glycine in Table 19.1.

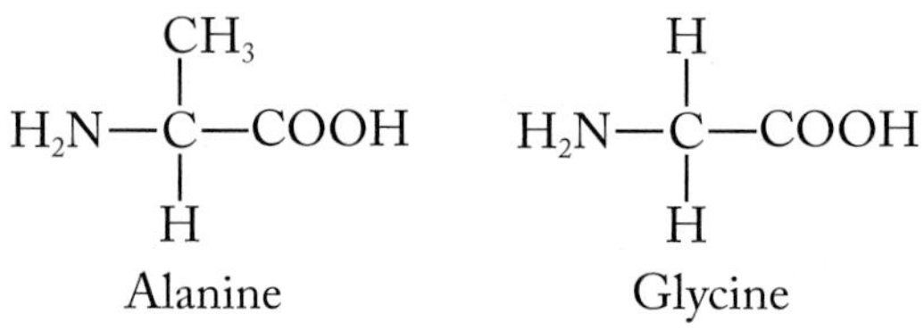

We are also asked to draw the dipeptides in the form in which they exist at pH 7, which means as zwitterions.

ANALYZE Different peptides can be made from the same amino acids by changing the sequence of the amino acids from the amine terminus of each peptide to its carboxylic acid terminus. Two sequences are possible for alanine (Ala) and glycine (Gly): AlaGly and GlyAla.

SOLVE Reactions between the carboxylic acid groups (red) of the first amino acids and the amine groups (blue) of the second ones produce two peptides: glycylalanine and alanylglycine:

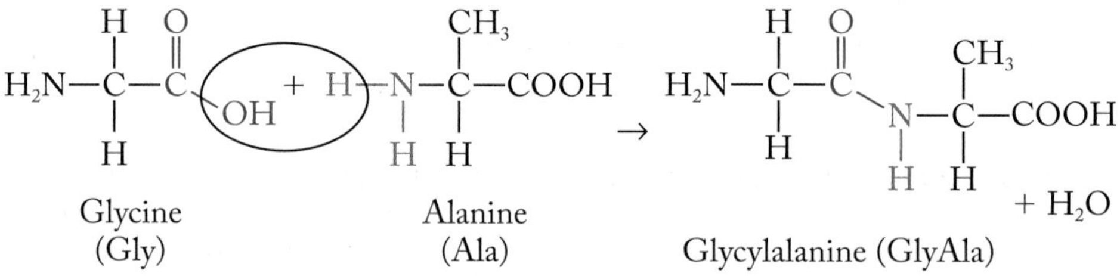

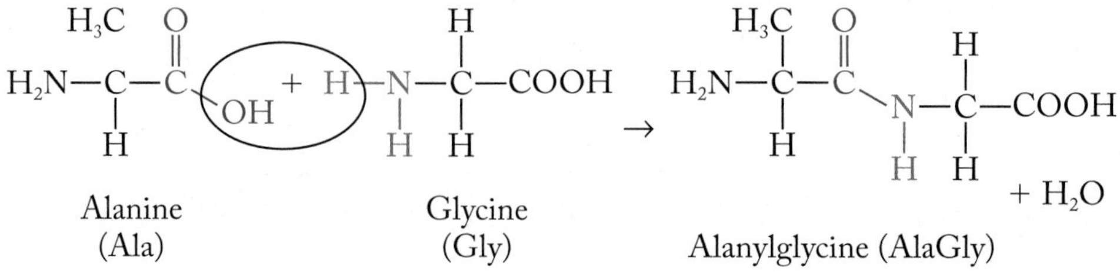

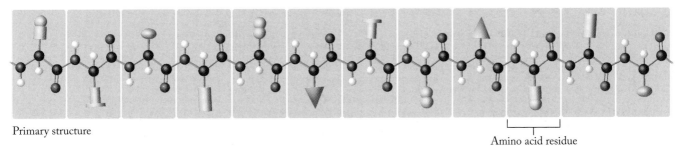

Primary structure

Amino acid residue

FIGURE 19.13 There are four different levels of protein structure: primary, secondary, tertiary, and quaternary. The different green shapes represent different R groups.

Near pH 7.0 the $-NH_2$ groups are protonated and the $-COOH$ groups are ionized, producing these zwitterions.

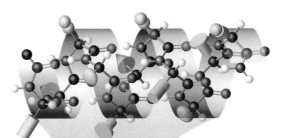

Secondary structure

Glycylalanine

Alanylglycine

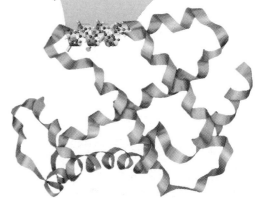

Tertiary structure

THINK ABOUT IT Two dipeptides are possible for the same two amino acids. Considering the 20 amino acids available, hundreds of combinations to produce different dipeptides are possible.

Practice Exercise How many different tripeptides can be synthesized from one molecule of each of three different amino acids?

19.4 Protein Structure and Function

The structure of proteins is crucial to their function. In the Look Ahead for this chapter we mentioned the catastrophic results that arise when a muscle protein is too short. The three-dimensional structure of a protein is also important because most receptors are regions within a protein with a specific shape that enables the protein to bind molecules or ions that then cause some characteristic effect like the perception of a smell, the contraction of a muscle, or the release of a hormone. The importance of the structure of a protein to its function cannot be overestimated. Protein structure is defined at four different levels, illustrated in Figure 19.13.

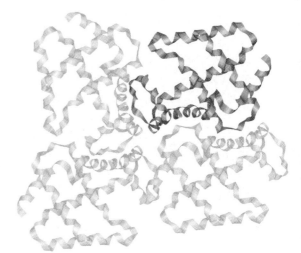

Quarternary structure

Primary structure

Normal protein: Val - His - Leu - Thr - Pro - Glu - Lys - . . .
Abnormal protein: Val - His - Leu - Thr - Pro - Val - Lys - . . .

(a)

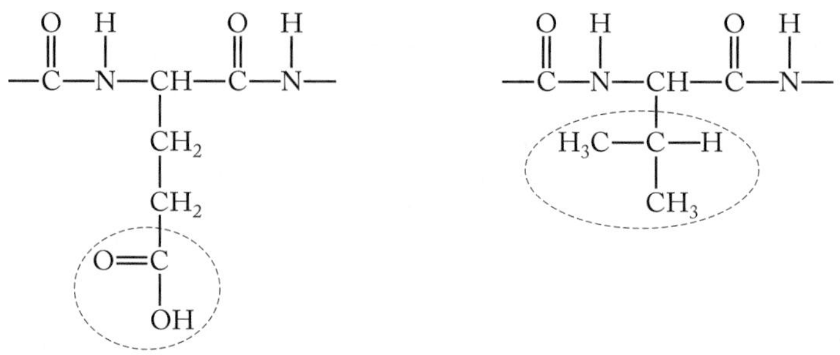

-Glu-

-Val-

(b)

FIGURE 19.14 The primary structure of a protein is the sequence of amino acids linked by peptide bonds. (a) The primary structure of one end of a protein in normal hemoglobin and in the abnormal hemoglobin responsible for sickle-cell anemia. (b) The replacement of glutamic acid with its hydrophilic R group by valine with its hydrophobic group is responsible for the disease.

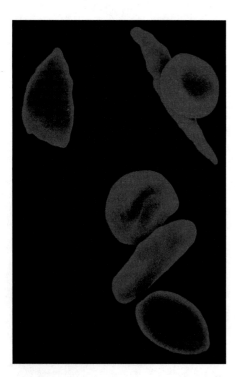

FIGURE 19.15 Normal red blood cells are plump discs, whereas those containing the abnormal protein are sickled.

∞ CONNECTION In Chapter 10, we learned that van der Waals forces are strong between large molecules.

Primary Structure

The **primary structure** of a protein is the sequence in which the amino acid monomers occur in the polymer chain starting at the amine terminus and ending at the carboxylic acid terminus (Figure 19.14a). To describe a protein completely, the sequence must be determined. If two proteins have a different amino acid sequence, even though they may be made up of the same number and type of amino acids, they are different proteins.

A change in sequence involving even one amino acid in a protein chain can cause dramatic alterations in function. Hemoglobin, the protein in red blood cells that carries oxygen, is actually an assembly of four protein chains. One of the chains that is 146 amino acids long normally has glutamic acid (Glu) as the sixth amino acid from the amine terminal of the chain. If valine (Val) is substituted in that spot (Figure 19.14a), the hemoglobin units containing that altered protein are less soluble in the fluid in red blood cells. The cells that contain that altered protein have a sickle shape instead of the normal, plump disc shape (Figure 19.15). These sickled cells do not pass through capillaries easily and therefore impede circulation of the blood, and they break open readily and do not last as long as normal blood cells. All of this leads to a diminished capacity of the blood to carry oxygen, which is one of the characteristics of the disease called sickle-cell anemia.

The reason for this drastic change as a result of the substitution of only one amino acid is explained in part by concepts of solubility and "like-dissolves-like" that we introduced in Chapter 10. The section of the protein chain where this replacement occurs is on the outside of the hemoglobin assembly and is in contact with the aqueous solution in the red blood cell. Glutamic acid (Glu) is a polar amino acid; its R group is a carboxylic acid that is ionized at the pH of blood. This group contributes to the solubility of the hemoglobin unit. Valine (Val) is a nonpolar amino acid with the hydrophobic R group $-CH(CH_3)_2$. The hydrophobic group diminishes the water solubility of the protein, and the exchange of one Val for one Glu has a major impact on the ability of hemoglobin to transport oxygen.

Sickle-cell anemia is a debilitating disease, but it actually provides a survival advantage in regions where the disease malaria is endemic. The occurrence of sickle-cell anemia does not protect people from malaria nor make them invulnerable to the parasite. However, children infected with the malaria parasite are more likely to survive the illness if they have sickle-cell anemia. Exactly why sickle-cell anemia has this effect on the parasite is not completely understood.

CONCEPT TEST

Choose an amino acid other than valine from Table 19.1 that could possibly have the same effect on the solubility of hemoglobin if it substituted for glutamic acid.

Secondary Structure

The next level of protein structure, the **secondary structure**, describes the pattern in which segments of the protein chain are arranged. One of the common patterns is the **α helix**, a coiled arrangement. The chain of individual amino acids in an α helix forms a coil, just like a rope wrapped around a pole. The R groups of the amino acids point out from the coil (Figure 19.16). The helical structure is maintained by hydrogen bonds from the –NH groups on one part of the chain to the C=O groups above or below them in the coil. The α helix looks very much like a spring, and in some proteins it imparts exactly that physical behavior to the molecule. Muscle tissue that stretches and contracts is made in large part of α-helical proteins. Another group of proteins that are mostly α-helical are keratins, the proteins in hair and fingernails.

Another common pattern in the arrangement of protein chains is called a **β-pleated sheet**. In this form, the protein chain is extended and forms a zigzag pattern (Figure 19.17). The R groups of the amino acids extend above and below the pleats, and other chains with the pleated structure associate to the left and right of a given chain via hydrogen bonds. The sheets may stack on top of each other like pieces of corrugated roofing. The β-pleated sheet structure is common in natural materials like silk, to which it imparts considerable strength because of hydrogen bonding between adjacent chains and the cumulative effect of van der Waals interactions between stacked chains.

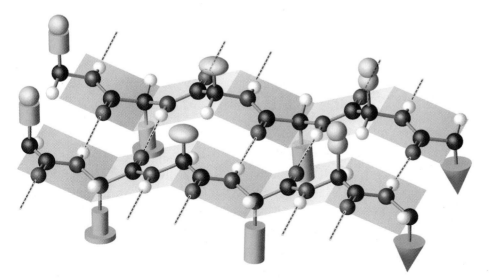

FIGURE 19.17 A β-pleated sheet is an extended zigzag pattern of amino acids; the R groups (green) extend above and below the planes defined by the sheets.

CONNECTION In Chapter 10, we discussed the role of polarity in determining the solubility of molecules in water.

▶❚❚ **CHEMTOUR** Fiber Strength and Elasticity

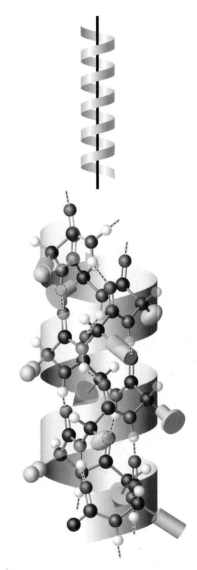

FIGURE 19.16 A chain of amino acids forms an α helix in which the R groups extend out from the coil.

The **primary structure** of a protein is the sequence in which the amino acid monomers occur in the polymer chain.

Secondary structure describes the pattern of arrangement of segments of a protein chain.

An **α helix** is a springlike coil in a protein chain.

A **β-pleated sheet** is a zigzag pattern of protein chains.

A **random coil** is an irregular pattern in the secondary structure of a protein.

Tertiary structure is the three-dimensional, biologically active structure of the protein that arises because of attractions between the R groups on the amino acids.

If the protein chain is characterized by a less regular pattern, it is referred to as a **random coil**. The protein chain may fold back on itself and around itself but has no regular features like an α helix or a β-pleated sheet (Figure 19.18). When proteins *denature* and lose their active structures because of heat or change in conditions, they may become random coils.

Proteins are so large that one protein molecule may contain all three types of secondary structure—α helix, β sheet, and random coil. In describing a protein, scientists may indicate the percent of amino acids involved in each type, for example, 50% α-helical, 30% β-pleated sheet, and 20% random coil. Figure 19.19 shows a model of a protein called carbonic anhydrase that has regions of α helix (red), β-pleated sheet (green), and random coil (blue).

FIGURE 19.18 A random coil.

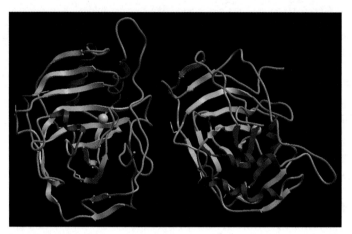

FIGURE 19.19 This depiction of the structure of carbonic anhydrase shows α-helical regions (red), β-pleated sheet regions (green), and random coil (blue). The light blue sphere in the center is a zinc ion.

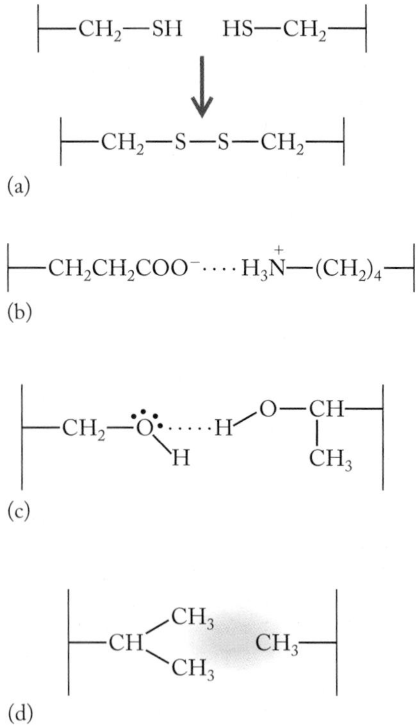

(a)

(b)

(c)

(d)

FIGURE 19.20 Protein chains are held together by: (a) covalent bonds, (b) ion–ion interactions, (c) hydrogen bonds, and (d) van der Waals forces.

Tertiary and Quaternary Structure

As we already mentioned with reference to the proteins in skin, birds' beaks, and silk, proteins with sections of α helices, β-pleated sheets, or random coil structures can interact within themselves to assume higher order structures. These interactions may lead to bends, kinks, or turns in the long chain that cause the protein to fold back upon itself. Such interactions between the R groups on the amino acids determine the **tertiary structure**, the overall three-dimensional structure of the protein that is key to its biological activity.

Tertiary structure may involve the formation of covalent bonds between two parts of one protein chain. In proteins that contain cysteine, two thiol groups (–SH) may combine to form a disulfide linkage that holds two parts of the protein strand together via an intrastrand –S—S– covalent bond (Figure 19.20a).

Tertiary structure is also driven by ionic interactions. For example, a glutamic acid residue (negatively charged in the human body because of the –COO⁻ on the R group) at one place in a long protein chain may interact with a lysine residue (positively charged) in the same chain but far away from it in terms of sequence. The association of glutamic acid and lysine is driven by the attraction of unlike charges. The carboxylic acid in the R group of glutamic acid is deprotonated at physiological pH; the amine group of the lysine is protonated. The –COO⁻ and NH₃⁺ attract each other electrostatically and form a link called a salt bridge in the protein (Figure 19.20b).

All the varieties of attractive forces we discussed in Chapter 10 are also involved in determining the tertiary structure of proteins. Amino acids like serine and threonine interact with each other via hydrogen bonds (Figure 19.20c). The R groups of hydrophobic amino acids like glycine, valine, and leucine associate with each other via van der Waals forces (Figure 19.20d). Because the proteins in living systems exist in an aqueous environment, the hydrophobic R groups tend to reside in the interior of a three-dimensional protein, while the hydrophilic groups (as we saw in normal hemoglobin) stay on the outside of the structure where they interact with water.

Some proteins associate further to make larger functional units. Hemoglobin (Figure 19.21a), the iron-containing material in red blood cells responsible for transporting oxygen, is a *tetramer* (having four subunits); one hemoglobin unit contains four protein chains, each of which enfolds a porphyrin containing one iron atom. The combination of four protein subunits to make one hemoglobin assembly is an example of **quaternary structure**. In keratins (Figure 19.21b), several α helices coil around each other to make larger structures. If these larger structures are held together mostly by van der Waals forces, they may be flexible and stretch easily like skin. If they are held together by covalent bonds, they may be hard and much less flexible, like fingernails and the beaks of birds. Quaternary structure is the most complex level of structure, and not all proteins participate in such arrangements. A protein with quaternary structure consists of several proteins that associate to form a larger assembly.

Proteins as Catalysts: Enzymes

The chemical reactions involved with metabolism—both catabolism (breaking down of molecules) and anabolism (synthesis of complex materials from simple feedstocks)—are mediated in large part by proteins. The processes of breaking down and building up molecules are organized in sequences of reactions called metabolic pathways, and each step in a metabolic pathway is catalyzed by a specific protein called an **enzyme**. Enzymes are biochemical catalysts. The protein carbonic anhydrase in Figure 19.19, for example, is the enzyme responsible for the rapid conversion of $CO_2(aq)$ into $HCO_3^-(aq)$ in our bodies.

Typically, the structure of an enzyme contains a distinct region called the **active site**, which is the place where a reactant molecule, the **substrate**, binds. Unlike most of the inorganic catalysts discussed in Chapter 14, enzymes are very specific in their action, both with regard to the molecules with which they react and the type of reaction they catalyze. For example, an enzyme called lactase catalyzes the reaction by which the disaccharide lactose is broken down into its component monosaccharides, glucose and galactose. If individuals are deficient in this enzyme, they become lactose intolerant, because their systems cannot metabolize the disaccharide. In this case, lactose passes unmetabolized into the large intestine where bacteria ferment it and unpleasant and painful gastric disturbances result.

Even the simplest cell must carry out hundreds of different chemical reactions just to stay alive. Many of these reactions take place simultaneously in the same solution; many are actually thermodynamically unfavorable. All of them are carried out under the limiting set of conditions within cells—aqueous media, 37°C, atmospheric pressure, and a narrow pH range. Enzymes make these reactions possible. Virtually every reaction that takes place in a cell involves an enzyme catalyst. Consequently, each cell contains hundreds of different enzymes. Enzymes catalyze oxidation–reduction reactions, isomerizations, hydrolysis, condensation reactions,

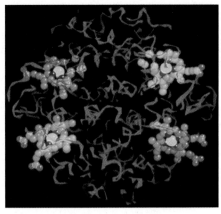

(a) Hemoglobin

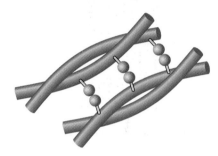

(b) Keratin

FIGURE 19.21 Quaternary structure of proteins: (a) Hemoglobin consists of four protein chains held together in an assembly. The iron-containing porphyrins are bright green. (b) Keratin consists of α-helical chains wound together and linked by covalent bonds.

CONNECTION In Chapter 14 we introduced catalysts in our discussion of catalytic converters in vehicles. Catalysts speed up reactions by lowering activational energy barriers.

Quaternary structure results when two or more proteins associate to make a larger structure that functions as a single unit.

An **enzyme** is a protein that catalyzes a reaction.

The **active site** is the location on an enzyme where a reactive substance binds.

A **substrate** is the reactant that binds to the active site in an enzyme-catalyzed reaction.

Biocatalysis is the strategy of producing pharmaceutical products using enzymes to catalyze reactions on a large scale. It is becoming especially important in processes that involve chiral materials.

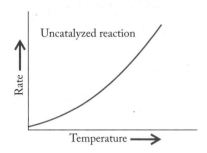

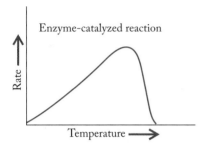

FIGURE 19.22 The rate of uncatalyzed reactions tends to increase with temperature. Enzymes in the human body, however, tend to work best at temperatures around 37°C.

and any other process in the chemical factory that is a living cell. Enzymes operate with exquisite efficiency and selectivity on optical isomers, both in terms of the substrates with which they react and the products they produce. Many enzymes act only on one member of a pair of enantiomers and produce only one enantiomerically pure product.

The selectivity of enzymes is the main reason why natural products tend to be optically pure materials while the same products synthesized in the laboratory tend to be racemic mixtures. In the pharmaceutical industry, whose products are administered to patients to suppress or enhance reactions related to human health and disease, research is currently focused on using enzymes outside the living system to produce chirally pure materials needed as drugs. This research area is called **biocatalysis**, and it involves the use of enzymes to catalyze chemical reactions run in industrial-sized reactors.

Reactions developed using enzymes frequently result in processes that feature enhanced rates of reaction (products are made faster) and a reduced number of steps (the process is faster from beginning to end, fewer raw materials consumed, less waste). Ideally these advantages make the production of drugs more cost effective, and especially when enantiomeric materials are involved, result in an end product with high chiral purity. The problems in applying enzymes to industrial processes include the narrow temperature range over which most enzymes operate: temperatures much higher or lower than 37°C tend to alter the structure of an enzyme, distort its active site, and diminish its effectiveness (Figure 19.22). In addition, enzyme-catalyzed reactions are very fast, but they work best in very dilute solutions, which causes difficulties for industrial-level preparations.

The action of enzymes was originally explained by an analogy in which the substrate is thought of as a key that fits into a lock; the lock is the enzyme. Just as a specific key fits a given lock, so does a specific substrate fit the active site of a given enzyme (Figure 19.23). The substrate is held in the active site by attractive forces such as hydrogen bonding, van der Waals forces, and sometimes even covalent bonds. When a substrate is in the active site, the conversion of the substrate into products occurs via a reaction with much lower activation energy than the one that occurs in the absence of the enzyme. This allows the reaction to proceed at a much higher rate. The reaction of a substrate (S) with an enzyme (E) produces an *enzyme–substrate complex* (E–S) which decomposes to form a product (P) and regenerate the enzyme.

$$E + S \rightleftharpoons E\text{–}S$$

$$E\text{–}S \rightleftharpoons E + P$$

The lock-and-key hypothesis is a good start, but it does not fully account for the behavior of enzymes. A theory called the *induced-fit theory* suggests that the substrate does more than just fit into the existing shape of an active site. This theory suggests that the binding of a substrate to an enzyme changes the shape of the enzyme so that the catalytic groups that participate in the reaction move into the correct position on the enzyme's surface. Rather than envisioning a solid key fitting into a rigid lock, think of the substrate as a hand that fits into a

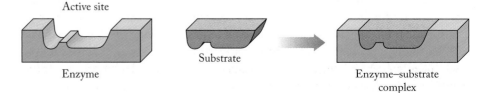

FIGURE 19.23 In the lock-and-key model of enzyme activity, a substrate fits in an enzyme active site shaped to receive it.

Active site · Enzyme · Substrate · Enzyme–substrate complex

glove (the enzyme) (Figure 19.24a). This idea helps explain the behavior of compounds called **inhibitors** that diminish or destroy the effectiveness of an enzyme. Inhibitors bind to an enzyme, the enzyme–substrate complex, or both, inhibiting formation of the desired product. They may bind to the active site directly and either block the natural substrate from binding or completely disable the enzyme by preventing it from assuming its active shape, or they may bind somewhere else and prevent the enzyme from achieving its active shape. To use the hand-in-glove analogy, an inhibitor could occupy the fingers in the glove so your hand could not slip in, or it could connect the thumb to the palm so that the glove could not open. Either way, your hand would be prevented from entering the glove.

Inhibitors sometimes have shapes and polarities very similar to natural substrates for enzymes, but sometimes they are very different. Again in terms of the hand-in-glove analogy, an inhibitor that blocked your fingers from entering a glove would probably be shaped like fingers, but one that connected the thumb to the palm and stopped the glove from opening would be quite different (Figure 19.24b). The induced-fit theory as a necessary modification of the lock-and-key theory arose as we learned more about the actual mechanisms by which enzymes work and the influence of inhibitors on enzyme function.

To illustrate how rapidly enzymes can work, let's consider the facts about the enzyme carbonic anhydrase. This enzyme catalyzes the hydrolysis of carbon dioxide, an end product of metabolism in our systems, to the hydrogen carbonate ion so that it can be transported through our bodies as an ion dissolved in blood:

$$CO_2(aq) + H_2O(\ell) \rightleftharpoons HCO_3^-(aq) + H^+(aq)$$

In the presence of carbonic anhydrase, the hydrolysis reaction proceeds about 10 million (1×10^7) times faster than it occurs in water in the absence of the enzyme. Without carbonic anhydrase, we would not be able to expel carbon dioxide fast enough from our systems to survive. One molecule of carbonic anhydrase can hydrolyze approximately 10,000 molecules of CO_2 in 1 second. This number is called the *turnover number* for the enzyme; the higher the turnover number, the faster the enzyme works. Turnover numbers for enzymes typically range from 10^3 to 10^7. The higher the turnover number, the lower the activation energy for a reaction.

19.5 Carbohydrates

Carbohydrates can exist as simple sugars, as disaccharides, or as polysaccharides, which are polymers composed of hundreds or even thousands of sugar monomers. Many organisms use simple sugars as their major energy source, but simple sugars are not stored in living systems in appreciable quantities. Polysaccharides are the means of energy storage in both plants and animals, and they also provide structural support in plants.

Most monosaccharides contain several chiral centers, so multiple optical isomers exist for almost every sugar. Nearly all natural monosaccharides have the D configuration. The six-carbon sugar glucose is the most biologically abundant simple sugar in nature. It is the monomer of cellulose, which is the most abundant structural polysaccharide on Earth. Plants produce over 100 billions tons of cellulose each year. Cotton is 99% cellulose, and the woody parts of trees are over 50% cellulose. Glucose is also the monomer in starch, the energy storage polymer in plants, and in glycogen, the energy storage polymer in animals. The monomer in all three polymers is glucose, but the materials are very different because of the way the glucose molecules are linked together.

An **inhibitor** is a compound that diminishes or destroys the ability of an enzyme to catalyze a reaction.

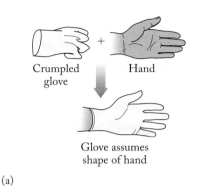

(a)

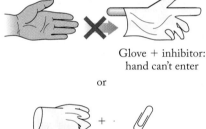

or

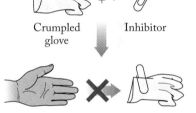

(b)

FIGURE 19.24 The induced-fit model suggests that the act of binding of the substrate to the enzyme changes the shape of the enzyme and brings groups responsible for accelerating the reaction together at the active site. (a) Just as a hand slipping into a crumpled glove causes the glove to attain its shape, so does the binding of a substrate to an enzyme change the shape of the active site. (b) Inhibitors may block the binding site or may cause a change elsewhere that prohibits the active site from attaining its shape.

Fructose is an isomer of glucose found in many fruits and root vegetables. Sucrose, or ordinary table sugar, is a disaccharide consisting of one molecule of glucose bonded to one molecule of fructose. Fructose and glucose are the two most important sugar monomers from the set of around 40 carbohydrate monomers that have been identified in nature. Most of what we need to know about sugars we can learn from the study of these two representatives of the family.

Molecular Structures of Glucose and Fructose

Glucose can be represented by the three molecular structures shown in Figure 19.25. The flat Lewis structure in the middle of the figure is a straight chain of six carbon atoms. The C-1 carbon is part of an aldehyde group. Each of the remaining carbon atoms from C-2 to C-6 is bonded to a hydroxyl group, so glucose contains both aldehyde and alcohol functional groups. The polar carbonyl group plus the polar –OH groups, all of which are capable of hydrogen-bonding, make glucose a very water-soluble material.

The cyclic structures of glucose are produced in solution by a chemical reaction that occurs when the carbon backbone of the open chain wraps around so that the hydroxyl group at C-5 comes close to the aldehyde group at C-1, as shown in Figure 19.26. The aldehyde and the alcohol react. The product of this reaction is a cyclic form of glucose: a six-membered ring of five carbon atoms and the oxygen atom that was originally part of the hydroxyl group at C-5.

The ring structure of glucose is at a lower energy level than the open chain, so that form is preferred, but the energy difference is small. In aqueous solution, glucose molecules are constantly opening and closing. The cyclic products of this ring-closing reaction have the same chemical formula as open-chain glucose ($C_6H_{12}O_6$) and are therefore structural isomers of glucose.

As Figure 19.25 indicates, two structurally different rings form in this reaction. One structure, called α-D-glucose, has the –OH group on C-1 in the ring pointing down; a second structure, called β-D-glucose, has the –OH group pointing up.

The sp^2-hybridized carbon atom in the aldehyde is trigonal planar. When the rest of the chain bends toward it, the –OH group on C-5 can approach the δ^+ carbon atom of the aldehyde from one side or the other. Statistically, half the time it will approach one side, half the time the other. One approach produces a product with one geometry, while approach from the other side produces another geometry. Of the three forms present in aqueous solution, the β form is slightly more stable, so the β structure tends to predominate in the equilibrium. Indeed, the β form accounts for 64% of glucose, with the α form the remaining 36%. The straight-chain form exists only as an intermediate between the two ring forms.

CONNECTION In Chapter 12 we defined an aldehyde as a carbonyl functional group in which the carbon atom has a double bond to an oxygen atom and single bonds to one hydrogen atom and one other carbon residue. The carbon atom in a carbonyl is trigonal planar.

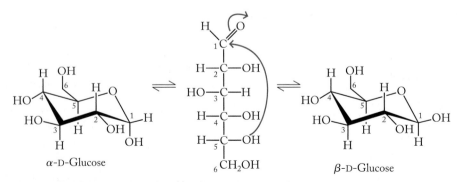

α-D-Glucose β-D-Glucose

FIGURE 19.25 An equilibrium exists between the open-chain structure of glucose and the two ring forms: α-D-glucose and β-D-glucose. The difference between the two ring structures is the orientation of the –OH group on C-1.

FIGURE 19.26 The –OH on C-5 may approach the aldehyde on C-1 from either side, to give a mixture of two isomers.

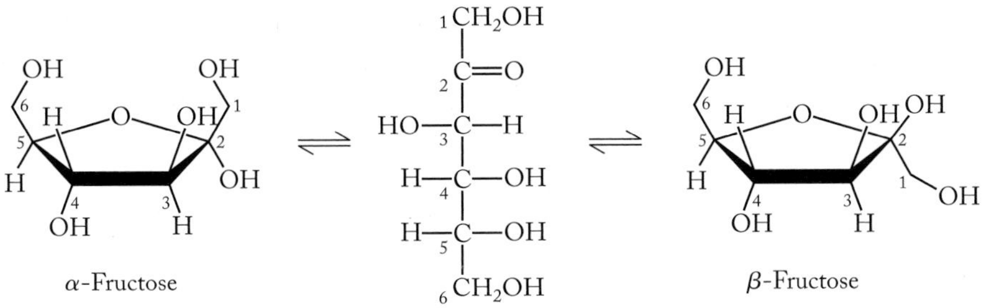

FIGURE 19.27 The cyclization of fructose proceeds via the –OH group on C-5 and the ketone on C-2.

Figure 19.27 shows the structures of open-chain and cyclic forms of another common sugar, fructose. Fructose and glucose have the same formula and are structural isomers. However, the C=O double bond in fructose is not on the end carbon atom in the open-chain form of the molecule. Fructose is a ketone that preferentially forms five-membered rings as shown in Figure 19.27. In these cyclization reactions, the hydroxyl group at C-5 reacts with the carbonyl group at C-2. The products are molecules with two –CH$_2$OH groups that are either on the same side of the ring (α-fructose) or on opposite sides (β-fructose).

CONNECTION In Chapter 12 we defined ketones as organic functional groups consisting of a carbon atom with a double bond to an oxygen atom and two single bonds to two other organic residues.

CONCEPT TEST

By analogy with glucose, explain why α and β forms of fructose exist.

Disaccharides and Polysaccharides

In nature, complex carbohydrates are made up of molecules of simple sugars bonded together. The disaccharides are the smallest of this class of compounds. Among them is sucrose, or table sugar, whose molecules consist of one molecule of α-D-glucose bonded to one of β-fructose.

Let's look closely at the structures of these monosaccharides and note what happens when they react to make sucrose (Figure 19.28). The –OH group on C-1 of a α-D-glucose reacts with the –OH group on C-2 of β-fructose, producing a C—O—C link between the two sugars called a **glycosidic linkage** and one molecule of water. This particular type of glycosidic linkage is called an α,β-1,2 linkage because of the positions and orientations of the two –OH groups involved

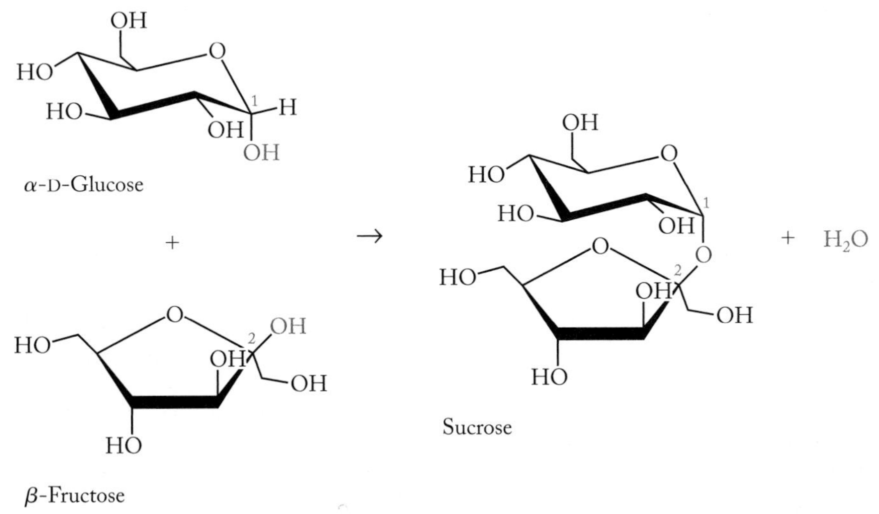

CHEMTOUR Formation of Sucrose

FIGURE 19.28 Glucose and fructose react to form a molecule of sucrose, ordinary table sugar.

A **glycosidic linkage** is a C—O—C bond between a sugar molecule and another molecule.

in making it. Another important type of linkage involves the −OH groups on the C-1 and C-4 carbon atoms of glucose molecules. Many glucose molecules can combine via 1,4 glycosidic linkages to form long straight-chain polysaccharides. The linkage reaction is part of the process by which plants store chemical energy from photosynthesis. Water-soluble glucose binds repeatedly to other glucose monomers and forms the high-molar-mass, insoluble homopolymer starch with the structure shown in Figure 19.29(a).

The conversion of glucose into starch is an effective way to store energy only if starch can be converted back into glucose. Fortunately, the process is reversible: a molecule of water can be added to each of the C—O—C links between the monosaccharide units in the polysaccharide chain. The chemical process that converts cornstarch into ethanol begins with the hydrolysis of cornstarch into glucose. Then the glucose is converted into ethanol and carbon dioxide by yeast fermentation:

$$C_6H_{12}O_6(aq) \rightarrow 2\,CH_3CH_2OH(\ell) + 2\,CO_2(g) \qquad \Delta H° = -266 \text{ kJ}$$

This exothermic reaction provides energy to the yeast cells. It also produces ethanol, a compound of increasing interest as an automotive fuel.

Starch is much less soluble in water than simple sugars and can be stored as a source of energy for the plant that produced it and for any animal that consumes it. Polysaccharides with a slightly different structure provide the structural support of plants as a material called cellulose. Plants synthesize cellulose from glucose to build stems and other support structures. Unlike grazing animals, we humans cannot digest cellulose. For example, other than water, much of the mass of celery is cellulose, and so celery has relatively little nutritional value.

Starch and cellulose are both polymers of glucose. Why can we digest one and not the other? The answer lies in the orientation of the hydroxyl group at C-1 of cyclic glucose. When molecules of α-glucose (Figure 19.25) react, the bond between C-1 of one α-glucose molecule and C-4 of another is called an α-1,4 glycosidic bond, as shown in Figure 19.29(a). When molecules of β-glucose form polysaccharides, a similar reaction leads to the formation of β-1,4 glycosidic bonds. Starch contains α-glycosidic bonds, whereas cellulose has only β-glycosidic bonds as shown in Figure 19.29(b). Evolution has provided humans with digestive enzymes that can break α-glycosidic bonds but not β-glycosidic bonds. Therefore, we can derive nourishment from grains that contain starch but not from wood, cotton, or linen, all because of the orientation of an −OH group on a ring.

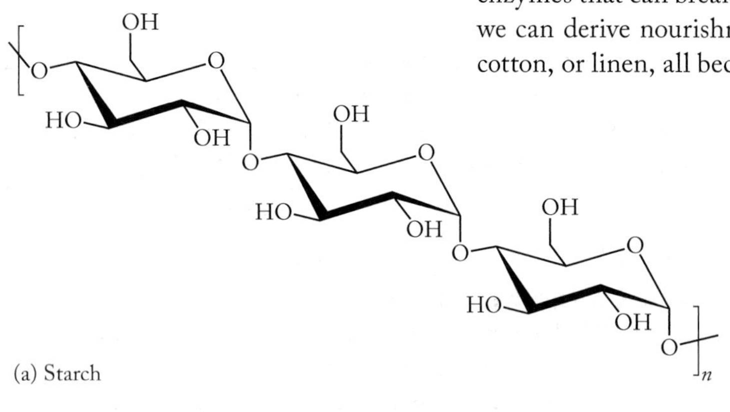

(a) Starch

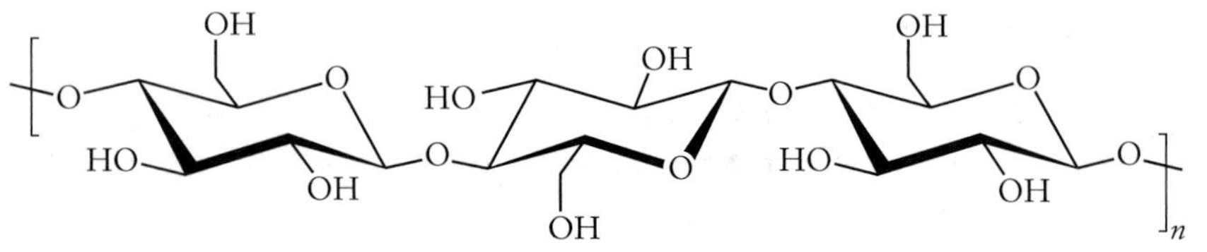

(b) Cellulose

FIGURE 19.29 Two polymers formed by glucose are starch and cellulose. (a) Starch is a homopolymer of α-glucose in which all the monomers are joined by α-1,4 glycosidic linkages. (b) Cellulose is a homopolymer of β-glucose; all the monomers are joined by β-1,4 glycosidic linkages.

Identifying Types of Glycosidic Linkages

Maltose (malt sugar) is a disaccharide used to sweeten prepared foods. A molecule of maltose consists of two molecules of α-glucose linked between C-1 in one glucose molecule and C-4 in the other as shown in the structure. Identify the type of glycosidic bond in its structure.

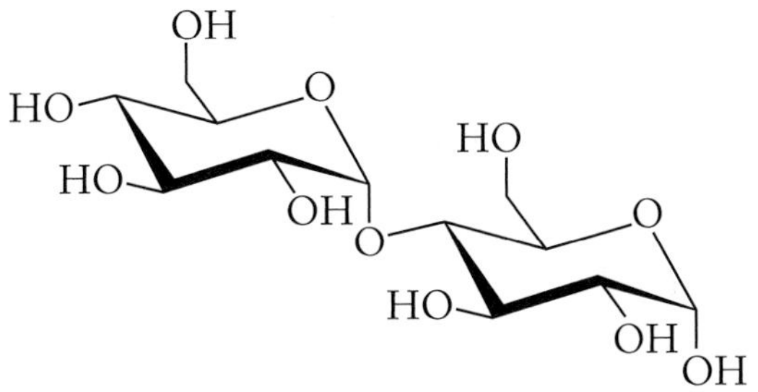

COLLECT AND ORGANIZE We are asked to identify the type of glycosidic bond in a disaccharide. We are given the structure. A glycosidic bond is a linkage that connects two sugar rings.

ANALYZE The bond in question is between the C-1 in one glucose molecule and the C-4 in another. We can compare this structure to those in Figure 19.29.

SOLVE The glycosidic bond in maltose is an α-1,4 linkage.

THINK ABOUT IT Because we can digest maltose, we should have expected the linkage to be an α-glycosidic bond. Our systems can break α- but not β-glycosidic bonds.

Practice Exercise Cellobiose is a disaccharide made from the degradation of cellulose. We cannot digest cellobiose. From the two structures given below select the most likely structure of cellobiose based on the nature of the glycosidic linkage.

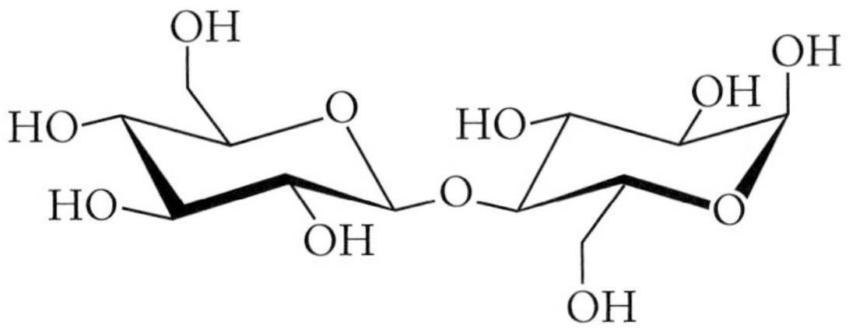

(a)

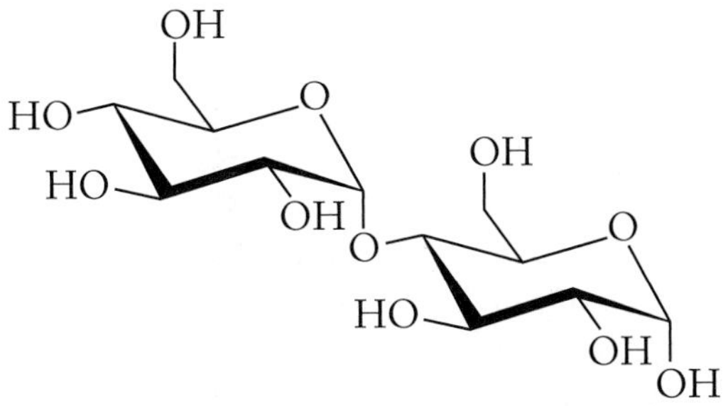

(b)

Wood and other forms of cellulose, including municipal solid waste, straw, corn stalks, and other agricultural residues, are now being used as inexpensive sources of sugar for fermentation processes to produce alcohol. Just as the α-glycosidic bonds that link the building blocks of starch together can be broken, so, too, can the β-glycosidic bonds in cellulose, releasing glucose. Until now, the destruction of these bonds in the industrial setting required either strong acids or high heat, both of which produce poor yields of glucose. Enzymes that degrade cellulose in living systems have been used in industrial processes, but they are in short supply for commercial alcohol production. Currently, scientists are genetically engineering microorganisms to increase the supply and thereby decrease the cost of enzymes that degrade cellulose. It is estimated that the use of cellulose from forestry and agricultural residues for ethanol production instead of the starch and sugar in corn or grain will make the production of ethanol from cellulose more energy efficient. The problem with this technology is that it currently costs more to produce ethanol than it costs to produce gasoline. These new enzymatic processes may allow more ethanol to be produced more efficiently because more plant tissue is utilized than just the starch. Now about 70% of the energy contained in ethanol is expended in producing it from starch. If ethanol could be efficiently produced from cellulose, its net energy value could become as high as 80%.

Photosynthesis and Biomass

As we saw in Chapter 5, carbohydrates are produced in nature by the process of photosynthesis. The molecule chlorophyll captures the energy of sunlight and transfers that energy to other molecules capable of converting water and carbon dioxide into glucose and oxygen.

$$6\,CO_2(g) + 6\,H_2O(\ell) \xrightarrow[\text{chlorophyll}]{\text{sunlight}} C_6H_{12}O_6 + 6\,O_2(g)$$

The reaction to produce carbohydrates is much more complex than the simple summary shown here. The currently accepted mechanism for the reaction to convert 6 moles of carbon dioxide and 6 moles of water into 1 mole of glucose consists of over 100 steps. This fundamental series of reactions is the source of virtually all chemical energy on Earth.

Carbohydrates in plants are a major part of the organic matter referred to as **biomass**, which is the sum total of the mass of organic matter in any given ecological system. Biomass is considered to be a **renewable resource**, one whose supply is unlimited. We have used biomass for thousands of years in the form of wood and dung to heat dwellings and cook food. Not only can biomass be burned to produce heat, but the cellulose, starch, and sugars in biomass may also be fermented and converted into other materials, such as ethanol, that can be used as fuels and as raw materials for chemical manufacturing. Considerable time and resources are now being focused on new ways of using biomass directly as a source of energy and raw materials.

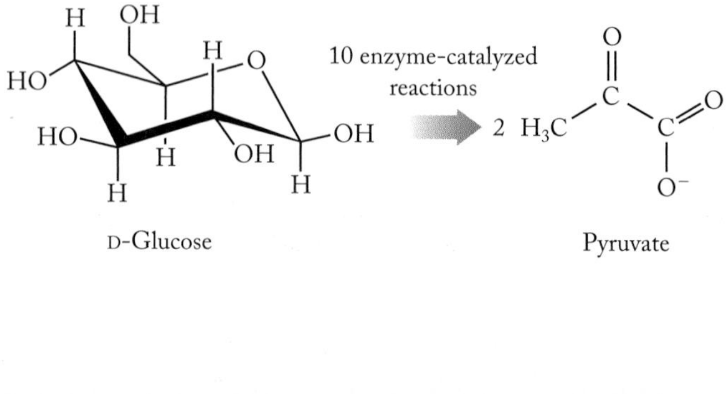

Most cells, including those in our bodies, use glucose as a fuel in a series of reactions called glycolysis. **Glycolysis** is a process that involves the oxidation of glucose to a smaller, three-carbon molecular ion called pyruvate (Figure 19.30), the conjugate base of pyruvic acid. Complete oxidation of glucose, whether by metabolism in living systems or complete combustion by fire, converts glucose to carbon dioxide and water and releases the maximum amount of energy.

Biomass is the sum total of the mass of organic matter in any given ecological system.

A **renewable resource** is one whose supply, in theory, is not limited. Trees and agricultural crops can be grown anew every year, and animal waste is produced constantly.

Glycolysis is a series of reactions in cells that convert glucose to pyruvate. Glycolysis is the one metabolic pathway that occurs in all living cells.

The **Krebs cycle** is a series of aerobic reactions that continue the oxidation of pyruvate formed in glycolysis.

FIGURE 19.30 In glycolysis, glucose is broken down to the pyruvate ion.

However, many different cells carry out reactions that stop well short of complete oxidation, and a large number of reactions and the specific enzymes that catalyze them have been discovered along the path to the complete conversion of glucose to CO_2 and H_2O.

Most cells use a sequence of 10 enzyme-catalyzed reactions to convert glucose to pyruvic acid, which exists as the pyruvate anion in most physiological systems (Figure 19.30). Pyruvate sits at a metabolic crossroad and can be converted into different products depending on the type of cell in which it is generated, the enzymes present, and the availability of oxygen (Figure 19.31). In yeast under oxygen-free conditions, pyruvate is converted into ethanol. The fermentation of cornstarch with yeast is one source of the ethanol that is used as a fuel additive. Since ethanol contains oxygen, it improves the combustion of a classic hydrocarbon fuel like gasoline by reducing the production of carbon monoxide. These blended fuels have names like E-2 or E-10, indicating the percentage of ethanol in the gasoline.

Another series of reactions called the **Krebs cycle** (Figure 19.31) is fundamental to the conversion of glucose to energy. In aerobic conditions, pyruvate is converted into acetyl coenzyme A (acetyl CoA) in a transition step at the start of the Krebs cycle. This is the starting point for pyruvate's further oxidation and subsequent release of additional energy.

Acetyl coenzyme A is also the synthetic starting point for other molecules that are produced along a variety of biosynthetic pathways. One such product is cholesterol (Figure 19.32). Cholesterol is a key component in the structure of cell membranes. It is also a biosynthetic precursor of the bile acids that aid in digestion, and of the steroid hormones, which regulate the development of the sex organs and secondary sexual traits, stimulate the biosynthesis of proteins, and regulate the balance of ions in the kidneys. The synthesis and use of cholesterol is tightly regulated *in vivo* to prevent overaccumulation and consequent deposition of cholesterol within coronary arteries. We clearly need cholesterol, but deposition in the arteries caused by too much cholesterol in the human body leads to serious coronary heart disease.

CONCEPT TEST

An article published in the *New York Times* contained this phrase: "Primarily made by the liver, cholesterol begins with tiny pieces of sugar . . ." What does this statement mean at the molecular level?

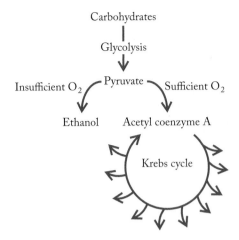

FIGURE 19.31 Carbohydrates are converted into simple sugars and then into pyruvate as a result of glycolysis. When there is insufficient dissolved O_2 available, as in fermentation, pyruvate may be converted to ethanol. In the presence of sufficient O_2, the oxidation of pyruvate proceeds in a series of steps called the Krebs cycle. In a transition step pyruvate loses CO_2 and forms an acetyl group that then combines with coenzyme A, as shown in Figure 19.32.

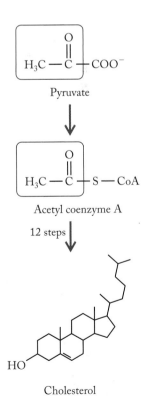

FIGURE 19.32 Cholesterol is produced from pyruvate in living systems.

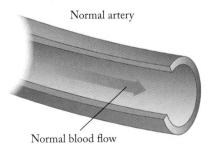

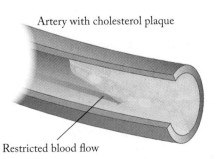

Cholesterol deposits called plaque are responsible for restricted blood flow, which results in a variety of sometimes catastrophic medical problems.

19.6 Lipids

As we stated in Section 19.1, lipids are not polymers, but they are large molecules. One specific type of lipid is a triglyceride, an ester of the trihydroxy alcohol glycerol and long-chain fatty acids. As is typical among biomolecules, huge variation is possible and observed within this simple format. The three –OH groups on glycerol allow for mono- and diglycerides as well as triglycerides, and one, two, or three different fatty acids may be attached to the glycerol to produce many different lipids, all with the same basic molecular scaffold. Glycerides account for over 98% of the lipids in fatty tissue of mammals.

Table 19.2 contains the common names, chemical names, formulas, and structures of common fatty acids. The most abundant fatty acids found in nature have an even number of carbon atoms, because the fatty acids are built from two-carbon units. Their biosynthesis begins, as did the biosynthesis of cholesterol, with the conversion of pyruvate to acetyl CoA. Most fatty acids contain between 14 and 22 carbon atoms.

SAMPLE EXERCISE 19.6 **Identifying Different Triglycerides**

How many different isomers of triglycerides can be made from glycerol and two different fatty acids? Use X and Y to symbolize and differentiate the two fatty acids and draw the different triglycerides. Use both fatty acids in all molecules.

COLLECT AND ORGANIZE We are asked to determine how many different fatty triglycerides we can make from glycerol and two different fatty acids and to draw them. A triglyceride contains three acid units.

ANALYZE Each fatty acid may bond to one of three –OH groups in glycerol. Different structural isomers will have different fatty acids attached to the middle or end –OH groups of glycerol.

SOLVE We can generate four structural isomers by attaching X and Y to the –OH groups in glycerol.

$$
\begin{array}{cccc}
 & (1) & (2) & (3) & (4) \\
H_2C-O- & X & X & Y & Y \\
HC-O- & X & Y & Y & X \\
H_2C-O- & Y & X & X & Y
\end{array}
$$

Structures (1) and (3) have chiral centers, so each of those isomers also has two enantiomeric forms. That means in total, six different isomers are possible.

THINK ABOUT IT The central carbon atoms in structures (1) and (3) are chiral because they are each bonded to an H atom, an O atom, and to two C atoms that are themselves bonded to different fatty acids: X and Y.

Practice Exercise How many different triglycerides can be made from glycerol and three different fatty acids?

Function and Metabolism of Lipids

Lipids are a huge source of energy in our diet. They provide more energy per gram of material consumed than carbohydrates or proteins. Many modifications are

TABLE 19.2 Names, Formulas, and Structures of Common Fatty Acids

Common Name (chemical name) (source)	Formula	Structural Formula
Saturated fatty acids:		
Lauric acid (dodecanoic acid) (coconut oil)	$C_{11}H_{23}COOH$	$CH_3(CH_2)_{10}COOH$
Myristic acid (tetradecanoic acid) (nutmeg butter)	$C_{13}H_{27}COOH$	$CH_3(CH_2)_{12}COOH$
Palmitic acid (hexadecanoic acid) (animal and vegetable fats)	$C_{15}H_{31}COOH$	$CH_3(CH_2)_{14}COOH$
Stearic acid (octadecanoic acid) (animal and vegetable fats)	$C_{17}H_{35}COOH$	$CH_3(CH_2)_{16}COOH$
Unsaturated fatty acids (most natural ones are *cis*):		
Oleic acid (*cis* 9-octadecenoic acid) (animal and vegetable fats)	$C_{17}H_{33}COOH$	$CH_3(CH_2)_7CH{=}CH(CH_2)_7COOH$
Linoleic (*cis, cis* 9,12-octadecadienoic acid) (linseed oil, cottonseed oil)	$C_{17}H_{31}COOH$	$CH_3(CH_2)_4CH{=}CHCH_2CH{=}CH(CH_2)_7COOH$
Linolenic acid (*cis, cis, cis* 9,12,15-octadecatrienoic acid) (linseed oil)	$C_{17}H_{29}COOH$	$CH_3CH_2CH{=}CHCH_2CH{=}CHCH_2CH{=}CH(CH_2)_7COOH$

made to lipids in the food industry to make substances that have fewer calories but still provide the attractive taste, aroma, and "mouth feel" that we all associate with fats and oils. Pure fats and oils actually have no taste, odor, or color; these attributes are due to small quantities of other organic materials present in or added to lipids like butter and corn oil.

As indicated in Table 19.2, fatty acids are either saturated, containing no carbon–carbon double bonds, or unsaturated, containing one or more carbon–carbon double bonds. **Fats** are glycerides composed primarily of saturated fatty acids. They are solids at room temperature. **Oils** are composed predominantly of unsaturated fatty acids and are liquids at room temperature. Oils can be converted into solid, saturated triglycerides by hydrogenation. For example, in the hydrogenation of corn oil, which is a mixture containing anywhere from 53% to almost 100% of the unsaturated fatty acids oleic and linoleic acids, hydrogen is added to the double bonds to convert all the $-CH{=}CH-$ subunits into $-CH_2-CH_2-$ subunits. This converts the liquid oil into a solid fat at room temperature. The solid fat produced from the corn oil is whipped with skim milk, coloring agents, and vitamins to produce the food spread we know as margarine.

Fats are solid triglycerides containing primarily saturated fatty acids.

Oils are liquid triglycerides containing primarily unsaturated fatty acids.

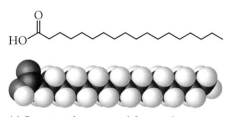

(a) Stearic acid: a saturated fatty acid

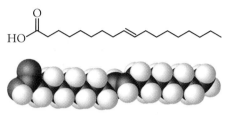

(b) Elaidic acid: a *trans* unsaturated fatty acid

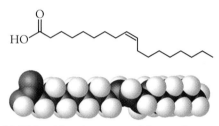

(c) Oleic acid: a *cis* unsaturated fatty acid

FIGURE 19.33 Structures of the (a) saturated C_{18} fatty acid stearic acid, (b) the unsaturated C_{18} fatty acids elaidic acid (*trans* isomer), and (c) oleic acid (*cis* isomer).

The consumption of too much saturated fat is associated with coronary heart disease. One of the advantages of the so-called "Mediterranean diet," characteristic of the European countries bordering the Mediterranean Sea, is the heavy use in cooking and baking with olive oil, a liquid composed of glycerides containing over 80% oleic acid, rather than solid fats like butter and lard. In addition, this diet tends to be richer in fish and vegetables, both of which contain unsaturated fats. Most animal fats, like the marbling in beef that enhances its flavor, are saturated.

Because of the link between eating saturated fats and heart disease, many diets call for the substitution of unsaturated fats for saturated ones. To respond to the desire for unsaturated fats, industry produces them by the reverse of the hydrogenation process: saturated fats can be dehydrogenated and thereby converted into unsaturated oils. The problem with this approach, however, is that the industrial processes frequently result in the production of unsaturated fats with *trans* geometry about the double bonds. Most natural unsaturated fats are *cis*. Figure 19.33 shows the difference between the two structures.

Saturated fats like stearic acid and *trans* unsaturated fats like elaidic acid tend to be solids at room temperature because their molecules pack together more uniformly than do molecules of *cis* unsaturated fatty acids. This is why lard, butter, and the marbled fat in beef are solids, while olive oil and other *cis* fatty acids are liquids. Consumption of *trans* fatty acids is associated with increased amounts of cholesterol and related materials in the body and poses a risk to health.

CONCEPT TEST

The difference between saturated and unsaturated fats is also used to considerable advantage in nature. When the composition of the mixture of natural lipids in the legs of reindeer living close to the Arctic Circle was analyzed, it was observed that the composition changed as a function of distance from the hoof. The closer to the hoof, the higher the content of unsaturated fatty acids. Why would this be an advantage for the reindeer, who spend much of their life with their hooves in contact with ice and snow?

An understanding of the metabolism of lipids has also lead to several food products with reduced caloric content. Olestra, the calorie-free fat used to prepare low-calorie fried foods like potato chips, falls into this category. The enzymes in the human body that break down lipids have evolved to operate on glycerol esters. Their active sites accommodate triglycerides formed from glycerol and fatty acids such as those shown in Figure 19.33. Esters made from other alcohols and the same fatty acids cannot be metabolized by our systems, because they are not recognized by human enzymes. Such molecules, if they have the appropriate physical properties and are nontoxic, could potentially be used as oils and fats in food processing and would not add any calories to the food because they would be not be metabolized.

Olestra (Figure 19.34) is an ester made from natural long-chain fatty acids and the carbohydrate sucrose. Sucrose is a disaccharide composed of one monomer of glucose and one of fructose. It has eight –OH groups, each of which could react with a molecule of fatty acid to make an ester. In olestra, sucrose takes the place of glycerol and provides a scaffold to which fatty acids are attached. The resultant material can be used to fry potato chips. Any olestra that remains on the chip does not add calories because it cannot be processed by enzymes that recognize fatty acid esters on a glycerol scaffold.

Olestra   Triglyceride

FIGURE 19.34 Olestra has a shape very different from the shape of typical triglycerides metabolized by our bodies. Consequently, it cannot be processed by the enzymes that operate on triglycerides.

CONCEPT TEST

Olestra may be "calorie-free" as a food, subject to metabolism in the living system, but how would it compare to a common triglyceride in terms of kilojoules of heat released per mole in a calorimeter experiment in the laboratory?

The enzymes that metabolize triglycerides hydrolyze the esters and release glycerol and the fatty acids. Glycerol enters the pathways for glucose metabolism. The fatty acids are oxidized in a series of reactions known as β-oxidation. The oxidative process starts at the carboxylic acid end of the fatty acid and removes one two-carbon fragment at a time. Stearic acid, the C_{18} fatty acid, yields acetic acid (the C_2 fragment) and the C_{16} acid, palmitic acid. Palmitic acid yields acetic acid and myristic acid, and so forth until the fatty acid is completely degraded.

Knowledge of this pathway affords an explanation for the toxicity associated with carboxylic acids containing an even number of carbon atoms in which a fluorine atom replaces one of the hydrogen atoms in the terminal $-CH_3$ group to make a $-CH_2F$ group (Figure 19.35). This carboxylic acid is processed along the usual metabolic pathway, but the final step in its metabolism yields monofluoroacetic acid, CH_2FCOOH, an extremely toxic material. The sodium salt of this acid (sodium fluoroacetate) was once used as a rat poison. It is an extraordinarily hazardous substance with no known antidote. It exerts its toxic effect by shutting down the Krebs cycle. Despite their toxicity, salts of fluoroacetate are known in nature. A South African shrub called gifblaar or poison leaf

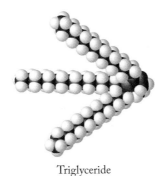

FIGURE 19.35 The metabolism of a fatty acid with an even number of carbons and a fluorine on the terminal methyl group produces fluoroacetic acid, a highly poisonous compound with no known antidote that effectively shuts down the Krebs cycle.

Acetic acid   Fluoroacetic acid

Phospholipids consist of glycerol with two fatty acid chains and one polar group containing a phosphate. They are major constituents in cell membranes.

A **lipid bilayer** is a double layer of molecules with polar groups interacting with water molecules and nonpolar groups interacting with each other.

The gifblaar plant on the Transvaal in South Africa.

(*Dichapetalum cymosum*) contains fluoroacetate salts and hence is extremely dangerous for grazing animals.

CONCEPT TEST

A small molecule called acetamide has been used with limited success as a possible antidote to gifblaar poisoning. How might this compound be effective in decreasing the effect of monofluoroacetic acid or its salts? (*Hint:* Think of monofluoroacetic acid as an enzyme inhibitor. How could acetamide affect the binding of monofluoroacetic acid to the active site of an enzyme?)

$$H_3C - C \overset{O}{\underset{NH_2}{<}}$$

Acetamide

Other Types of Lipids

Cells also contain several other types of lipids that have highly significant roles beyond energy storage. **Phospholipids** have structures similar to triglycerides, but differ in that they contain a polar group. Phospholipids (Figure 19.36a) consist of a glycerol molecule, two fatty acid chains, and a phosphate group with a variety of polar substituents attached. The presence of the nonpolar fatty acid chains together with a polar group in the same molecule make phospholipids ideal for forming cell membranes. In an aqueous medium, the phospholipids associate and form a **lipid bilayer**, a double layer of molecules with polar groups interacting with water molecules and nonpolar groups interacting with each other (Figure 19.36b). As shown in Figure 19.36, phospholipids are frequently drawn with a sphere (the polar head group) connected to two long tails (the nonpolar fatty acids).

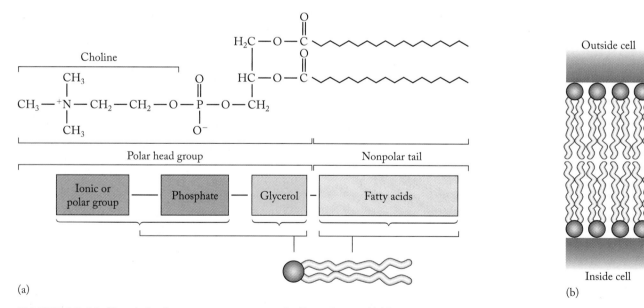

FIGURE 19.36 Phospholipids are major constituents of cell membranes. (a) The presence of a polar group attached to a phosphate unit on glycerol changes the properties of the resultant glyceride. In this molecule, the group attached to the phosphate is choline, and the lipid is called phosphatidyl choline. (b) Phospholipids orient themselves in aqueous solution so that the polar head groups face the water and the nonpolar tails associate with each other.

19.7 Nucleotides and Nucleic Acids

The fourth class of large molecules in living cells, the nucleic acids DNA and RNA, contains the genetic information that is the blueprint for the characteristics of life. They carry the instructions for the type of cell produced and its contents; they also control the processes by which the genetic information is transmitted during cell division and those by which proteins are synthesized. They are the machinery in the factory of the cell, as well as the instructions for its assembly and maintenance.

Components and Structure

As we noted earlier, a nucleic acid is a polymer of nucleotides, which are composed of a five-carbon sugar, a phosphate group, and a nitrogen-containing base. The backbone of the polymer of nucleic acids is a chain made of alternating sugar residues and phosphate groups, to which are attached the nitrogenous bases. The phosphate group in the nucleotide is attached to the fifth carbon in the sugar, called the 5′ carbon atom ("5-prime carbon") to distinguish it from any carbon atom in the base. The superscript "prime" refers specifically to atoms in the sugar molecule. The organic base is attached to the 1′ carbon in the sugar. To make the polymer chains, the phosphate is also linked to the 3′ carbon in the sugar of the monomer that follows it in the chain (Figure 19.37), and DNA and RNA strands grow in the 5′-to-3′ direction. The structures of DNA and RNA are frequently abbreviated using the single letters identified in Figure 19.37, beginning with the free phosphate group on the 5′ end of the chain and reading toward the free 3′ hydroxyl at the other terminus. The sugar is ribose in RNA (*ribo*nucleic acid) and deoxyribose in DNA (*deoxyribo*nucleic acid). Another difference is the presence of the base uracil (U) in RNA where DNA has the base thymine (T).

Analysis of the structure of DNA began with the structures of the organic bases and a pivotal observation made about the composition of the polymers. A typical molecule of DNA consists of thousands of nucleotides, and the percentages of the four nucleotide bases in a given sample of DNA vary over a wide range. However, the sum of the percentages of A and G is always close to the sum of the percentages of C and T. This result makes sense if the bases are paired. Looking at the opportunities for hydrogen bonding between the bases, a reasonable pairing is A with T and C with G. The pairing of A–T and C–G maximizes the number of hydrogen bonds between the bases (Figure 19.38a).

The structure of double-stranded DNA is that of a double helix, and the two strands form a spiral staircase (Figure 19.38b). Each strand winds around the other with respect to a common axis. The revolutionary aspect of the structure is that the backbone is on the outside of the staircase, the hydrogen bonds connect the two strands, and the hydrogen-bonded bases form the steps of a spiral staircase between the two strands.

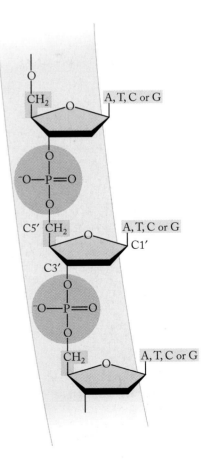

FIGURE 19.37 The backbone of the polymer chain in DNA consists of alternating sugar units (yellow) and phosphate units (pink). The bases (blue) are R groups attached to the backbone through the 1′ carbon atom of the sugar unit.

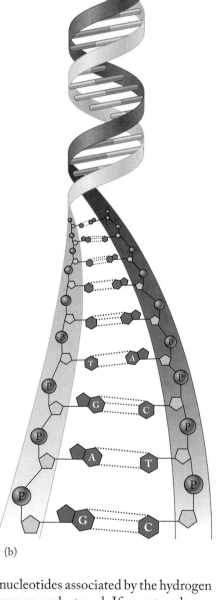

FIGURE 19.38 The bases on one strand of DNA pair with bases on a second strand by hydrogen bonding. (a) Hydrogen bonding is maximized because adenine (A) and thymine (T) make two hydrogen bonds when they pair with each other; guanine (G) and cytosine (C) make three hydrogen bonds when they pair with each other. (b) DNA is a double helix with the sugar–phosphate backbone on the outside and the base pairs on the inside.

(a)

Adenine Thymine

A T

Sugar Sugar

Guanine Cytosine

G C

Sugar Sugar

(b)

DNA consists of two separate strands of nucleotides associated by the hydrogen bonds that exist between complementary bases on each strand. If one strand consists of the bases ATTGCGCTA, the structure of the paired helix is therefore

ATTGCGCTA

TAACGCGAT

This complementarity is the structural origin of DNA's ability to replicate itself. If a pair of complementary strands is unzipped into single strands, then each single strand provides a template on which a new complementary strand can be synthesized; this process is called **replication** (Figure 19.39). This double-stranded structure is also the key to the ability of DNA to preserve genetic information. In terms of content, the two strands carry the same information, much like a photograph and its negative carry the same information. Genetic information is duplicated every time a DNA molecule is duplicated, a process which is essential whenever a cell divides.

In RNA the substitution of the −OH group on the 2′ carbon of ribose for the hydrogen atom at the corresponding position in DNA has a major impact on the structures of the two polymers. RNA, in contrast to DNA, is a single-stranded molecule, in large part because the additional −OH group in ribose sterically prevents the chain from twisting and turning the way DNA can. Several types of RNA are known, all of which play different roles in the cellular factory for the manufacturing of proteins.

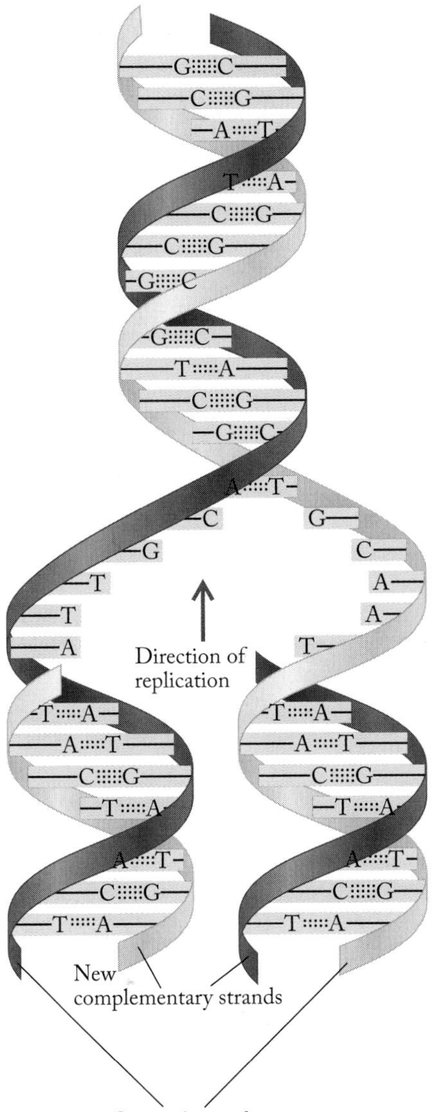

| SAMPLE EXERCISE 19.7 | **Using Complementarity of Bases in DNA** |

A sample of DNA is analyzed and it is shown to contain 31.6% A and 18.4% G. Estimate the percentages of C and T that the polymer contains.

COLLECT AND ORGANIZE We are given the percentage of bases in DNA that are A and G and asked to calculate what percentage of C and T are present.

ANALYZE Base A pairs with T; G pairs with C. This means that the percent of T should equal the percent of A, and the percent of C should equal the percent of G.

SOLVE If A = 31.6%, then T = 31.6%. If G = 18.4%, then C = 18.4%.

THINK ABOUT IT The percentages should total 100%, and they do.

Practice Exercise Indicate the sequence of the double helix formed when each of these sequences of nucleotides is duplicated:

   a. CGGTATCCGAT
   b. TTAAGCCGCTAG

From DNA to Protein

An intricate series of events is involved in putting together proteins in accordance with the blueprint offered by the genetic code contained in DNA. The genetic code itself is contained within the sequence of base pairs in DNA. The bases A, T, G, and C are the alphabet in this code, and the words in the code are combinations of these letters that represent a particular amino acid. Genetic messages are written in a three-letter code. Using four letters to write three-letter words means there are 4^3 or 64 unique combinations possible, more than enough to encode for the 20 amino acids found in cells. The function of the genetic information is to specify the sequence of amino acids in proteins; the primary structure of a protein determines the secondary, tertiary, and quaternary structure assumed by the protein once it is synthesized.

Converting genetic code into proteins begins with a process called **transcription** (Figure 19.40) in which DNA unwinds and its genetic information guides the synthesis of a single strand of a molecule called **messenger RNA (mRNA)**. This strand of mRNA has the complementary base sequence of the original DNA. It

Original DNA molecule

Direction of replication

New complementary strands

Original strands

FIGURE 19.39 When DNA replicates, the two strands of a molecule are unzipped, and each strand joins with a new complementary strand to produce two new DNA molecules.

Replication is the process by which one double-stranded DNA forms two new DNA molecules, each one containing one strand from the original molecule and one new strand.

Transcription is the process of copying the information in DNA to RNA.

Messenger RNA (mRNA) is the polynucleotide that carries the code for synthesizing a protein from the DNA in the nucleus to the cytoplasm outside the nucleus.

FIGURE 19.40 DNA is transcribed to make mRNA, which is translated in the ribosomes to make proteins. (a) Transcription takes place in the nucleus, where mRNA copies (transcribes) the code in DNA. (b) A strand of mRNA (green band) consisting of one, three-base-long codon for each amino acid in the primary sequence of the protein enters the ribosome (orange structure) and delivers the instructions for protein synthesis. A molecule of tRNA whose three-base code complements the code on the mRNA strand within the ribosome binds to the codon and delivers the first amino acid (Met) indicated by the code. A second tRNA then binds to the next codon on the mRNA strand and adds its amino acid (Val) to the growing protein chain, followed by the third (Thr) and so on until the entire mRNA has been read.

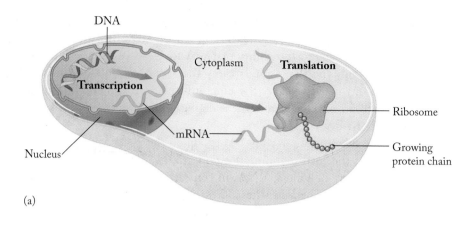

(a)

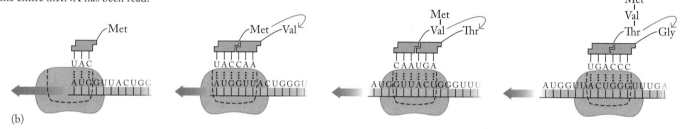

(b)

carries the words of that DNA in the form of three-base sequences called **codons** from the nucleus of the cell into the cytoplasm, where mRNA binds with a cellular structure called a ribosome. This is where proteins are made in a process called **translation**. Initially, one of 20 different forms of **transfer RNA (tRNA)** binds to only one specific amino acid. A molecule of tRNA delivers its cargo of one amino acid to a site on the ribosome–mRNA complex where the amino acid is added to a growing peptide chain. The tRNA has its own code, the anticodon—sometimes referred to as the second genetic code—embedded in its structure that enables it to bind a defined amino acid and recognize the site on the ribosome/mRNA complex where that amino acid is needed to extend the growing polypeptide. The message contained within mRNA includes start and stop signals, like the capital letter at the beginning of a sentence and a period at the end.

An Additional Role for Ribonucleotide Monomers: Energy and Metabolism

We end this chapter close to where we began it—with an interest in how molecules combine to produce other molecules (anabolism) and how molecules are taken apart to provide energy (catabolism). Living systems are not immune to or exempt from the laws of thermodynamics, and how organisms manage the energy necessary to sustain life is an area of active study.

Most polymers discussed in this chapter are assembled from component monomers; most other biological molecules like cholesterol are synthesized from fragments of glucose generated at various stages in the chain of oxidation reactions that transform simple sugars to carbon dioxide and water. Most anabolic reactions require energy; catabolism generates energy. What links energy requirement to energy production in living cells? The issue of energy transfer is a familiar theme: like many molecules and processes we have discussed, a great variability exists within living systems, but a basic, common pattern is discernable, namely that two molecules are intimately involved in the transfer of energy in living systems.

A **codon** is a three-base sequence that encodes for a specific amino acid.

Translation is the process of assembling proteins from the information encoded in RNA.

Transfer RNA (tRNA) is the polynucleotide that delivers amino acids, one at a time, to polypeptide chains being assembled by the ribosome–mRNA complex.

These molecules, the ribonucleotide monomers adenosine triphosphate (ATP) and adenosine diphospate (ADP) (see Figure 13.18), supply energy to reactions that require it and store energy released from reactions that produce it.

Energetically, the hydrolysis of ATP produces ADP, hydrogen phosphate (called "inorganic phosphate" and conventionally symbolized "P_i"), a hydrogen ion, and considerable energy. This reaction has a standard Gibbs free energy of -30.5 kJ per mole of ATP at pH 7:

$$ATP + H_2O \rightarrow ADP + P_i + H^+ \qquad \Delta G° = -30.5 \text{ kJ}$$

If energy is needed to drive a reaction, ATP is hydrolyzed to ADP. Conversely, if energy is available from the catabolism of a molecule, it can be stored by the reaction of ADP with inorganic phosphate and hydrogen ions to form ATP.

Enzymes are the crucial components of processes that enable living systems to use the ATP–ADP system to generate and store energy. Enzymes frequently couple two or more reactions together to form a single step in an overall process. In cells, enzymes couple a reaction that requires energy to the hydrolysis of ATP to make ADP; thus ATP serves as an energy source for the reaction. Correspondingly, many reactions in a series, like those involved in the complete oxidation of glucose to carbon dioxide and water, can be coupled to the conversion of ADP to ATP. The complete oxidation of 1 mole of glucose yields over 30 moles of ATP. The ATP is stored and then used in other reactions that require energy.

This fundamental process again brings us back to the role of plants and sunlight in providing energy for living organisms. Plants take in energy from the sun and store it in high-energy products like carbohydrates—for example glucose. The energy of the sun drives the thermodynamically unfavorable reaction of forming glucose from carbon dioxide and water. Animals then consume these energy-rich molecules, and the sunlight unwinds in their cells in a highly controlled fashion, mediated by enzymes that convert glucose back to carbon dioxide and water. The resultant energy is used to run countless other reactions that require it. Virtually all the energy needed to sustain life on earth comes from the sun through the intermediacy of green plants.

CONNECTION In Chapter 13, we discussed coupled reactions in biochemical systems.

Summary

SECTION 19.1 Primordial life arose through reactions of small inorganic compounds producing the small organic molecules **amino acids, monosaccharides, lipids,** and **nucleotides.** These building blocks combine to form **biomolecules** including **proteins, polysaccharides, triglycerides,** and the nucleic acids deoxyribonucleic acid (DNA) and ribonucleic acid (RNA). Biomolecules in turn are organized into assemblies that perform specific functions in cells.

SECTION 19.2 Many biologically important molecules exhibit **chirality,** or **optical isomerism,** a type of stereoisomerization. Optical isomers (**enantiomers** or **optically active molecules**) are nonsuperimposable on their mirror images and result from the presence of a **chiral** atom in their structures. Molecules lacking a chiral atom are **achiral.** Chirality is important in biological systems because **recognition** of molecules by **receptors** is often based on shape. Chiral molecules rotate a beam of plane-polarized light as levorotary ($-$) or dextrorotary ($+$) forms. The **configuration** of a chiral compound describes the position of the groups surrounding a chiral center. If two optical isomers of a compound are present in equal ratios in a sample, the material is known as a **racemic mixture.**

SECTION 19.3 At neutral pH, the amino acids exist as **zwitterions.** Proteins or **peptides** are biopolymers composed of the 20 available α-amino acids linked together through **peptide bonds.**

SECTION 19.4 The structures of proteins are defined by their **primary structure** or their amino acid sequence, their **secondary structure** or specific patterns within part of the sequence (for example, α **helices,** β-**pleated sheets,** or **random coils**), and the **tertiary structure** or complete three-dimensional shape that results from intramolecular attractive forces between amino acid residues in the protein chain. **Quaternary structure** results when two or more proteins associate further with each other to make larger functional units. Proteins called **enzymes** mediate the chemical reactions involved in metabolism acting as catalysts. The structure of an enzyme contains a distinct region called an **active site** that binds to a **substrate** and usually produces a pure enantiomeric product. The *induced-fit theory* suggests that the binding of a substrate to an enzyme changes the shape of the enzyme so that the reaction can take place. **Biocatalysis** seeks to replicate the selectivity of enzymes in laboratory settings.

SECTION 19.5 Carbohydrates are produced in nature from CO_2, water, and sunlight by photosynthesis. Monosaccharides are joined into di- and polysaccharides through **glycosidic linkages**. Two isomeric forms, α glucose and β glucose, form the polysaccharides starch and cellulose, respectively. Organisms derive energy from **glycolysis** and the **Krebs cycle**, the oxidation of glucose to carbon dioxide and water in cells.

SECTION 19.6 Lipids or **fats** are a major source of energy in our diet. Neutral fats or triglycerides are esters of the trihydroxy alcohol glycerol and long-chain fatty acids. They account for over 98% of the lipids in fat tissue of mammals. Fats are solid glycerides composed primarily of saturated fatty acids, while **oils** are composed predominantly of unsaturated fatty acids and are liquids. **Phospholipids** are essential components in cell membranes forming the **lipid bilayer** that surrounds cells.

SECTION 19.7 DNA and RNA contain the genetic information and control protein synthesis through processes called **transcription** and **translation**. Genetic information is transmitted to new cells during cell division **(replication)**. DNA and RNA are both polymers of nucleotides, molecules composed of a five-carbon sugar, a phosphate group, and one of five different nitrogen-containing organic bases. DNA consists of two separate strands of polynucleotides associated by the hydrogen bonds between complementary bases on each strand of a rope-like double helix. RNA is a single-stranded molecule involved in protein synthesis through **messenger RNA (mRNA)** and **transfer RNA (tRNA)**. The ribonucleotide monomers adenosine diphosphate (ADP) and adenosine triphospate (ATP) manage energy transfer in cells. Enzymes are the crucial components of processes that enable living systems to use the ATP–ADP system to generate and store energy.

Problem-Solving Summary

TYPE OF PROBLEM	CONCEPTS AND EQUATIONS	SAMPLE EXERCISES
Recognizing chiral molecules	Identify carbon atoms with four different groups attached.	19.1
Recognizing properties of racemic mixtures	Racemic mixtures contain equal amounts of two enantiomers and do not rotate the plane of polarized light.	19.2
Describing the acid–base behavior of an amino acid	At low pH amino acids have two ionizable protons, one each from $-COOH$ and $-NH_3^+$.	19.3
Drawing and naming peptides	Connect the right (NH_2) terminus of one amino acid to the left (COOH) terminus of the other forming a C–N bond and eliminating water. Name the peptide using the names of the amino acids from left to right using the suffix -yl in all but the last amino acid.	19.4
Identifying types of glycosidic linkages	Look at the bond between two sugar molecules. Determine whether the linkage is an α- or β-glycosidic linkage.	19.5
Forming and identifying triglycerides	The $-COOH$ groups of fatty acids react with the $-OH$ groups in glycerol to form triglycerides and water.	19.6
Using complementarity of DNA bases	Identify the base pairs: A pairs with T; G pairs with C. The percent of T should equal the percent of A, and the percent of C should equal the percent of G.	19.7

Visual Problems

19.1. The photochemical reaction of sodium hydrogen phosphite with formaldehyde is shown in Figure P19.1. It may have played a role in the formation of nucleic acids before life existed on Earth. Draw the Lewis structure for the hydrogen phosphite (HPO_3^{2-}) ion.

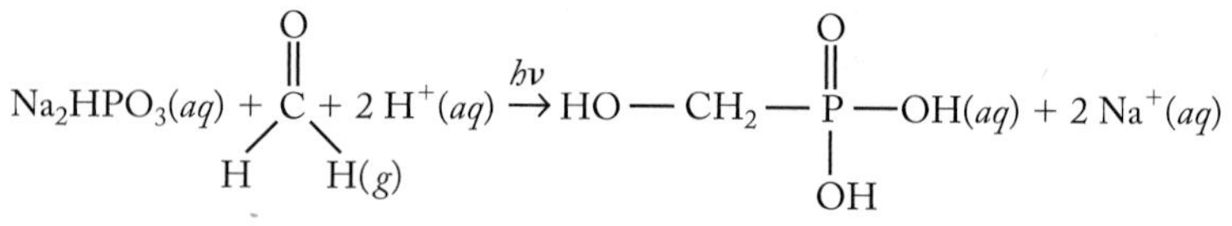

$$Na_2HPO_3(aq) + \underset{\underset{H}{|}}{\overset{\overset{O}{\|}}{C}} + 2\,H^+(aq) \xrightarrow{h\nu} HO-CH_2-\underset{\underset{OH}{|}}{\overset{\overset{O}{\|}}{P}}-OH(aq) + 2\,Na^+(aq)$$

FIGURE P19.1

19.2. The nucleotides in DNA contain the bases with the structures shown in Figure P19.2. Identify the basic functional groups in the structures.

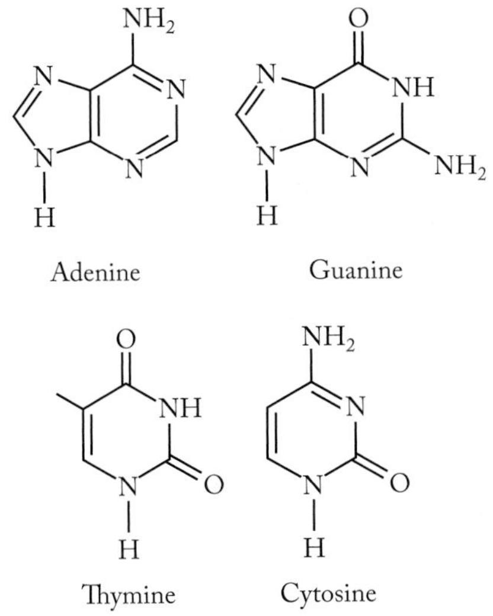

Adenine   Guanine

Thymine   Cytosine

FIGURE P19.2

19.3. Olive Oil Olive oil contains triglycerides such as those shown in Figure P19.3. Which of the fatty acids in these triglycerides is/are saturated?

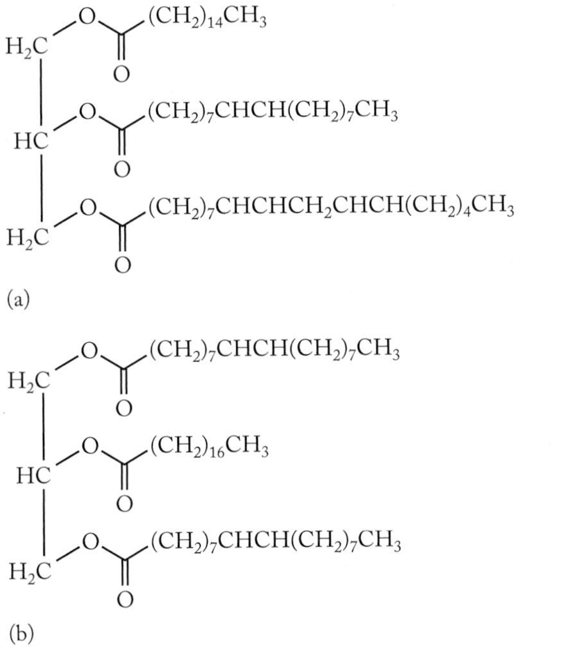

(a)

(b)

FIGURE P19.3

19.4. Drugs and Enantiomeric Purity Thalidomide was marketed in the late 1950s as a drug to relieve morning sickness. Unfortunately, one isomer caused birth defects. Circle the chiral carbon atom(s) (Figure P19.4) in the thalidomide molecule.

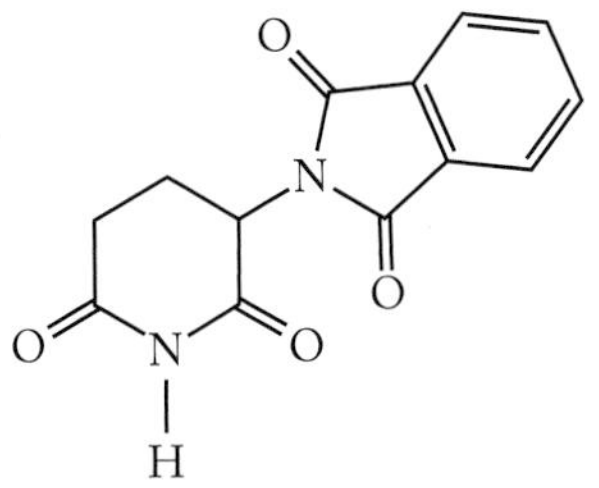

Thalidomide

FIGURE P19.4

19.5. Cholesterol Lowering Drugs High serum cholesterol levels often correlate with increased risk of heart attacks. The drug sold under the trade name Mevacor has proven to be effective in lowering serum cholesterol. How many chiral carbon atoms are there in Mevacor (Figure P19.5)?

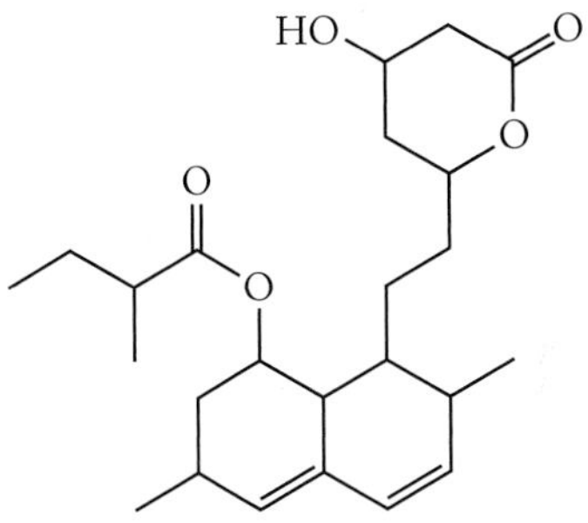

Mevacor

FIGURE P19.5

19.6. The two compounds shown in Figure P19.6 are both considered to be amino acids. Identify the structural difference between them and explain why they are both amino acids.

FIGURE P19.6

19.7. Natural Painkillers The human brain produces polypeptides called *endorphins* that help in controlling pain. The pentapeptide in Figure P19.7 is called enkephalin. Identify the five amino acids that make up enkephalin.

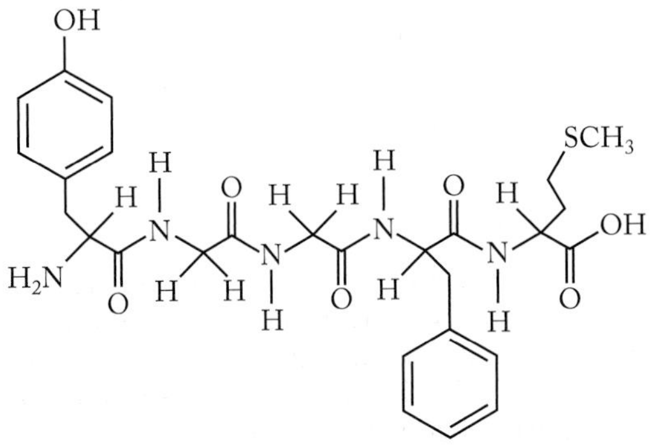

Enkephalin

FIGURE P19.7

19.8. **Regulating Blood Pressure** Angiotensin II is a polypeptide that regulates blood pressure. The structure of angiotensin II is shown in Figure P19.8. Which amino acids are in the structure?

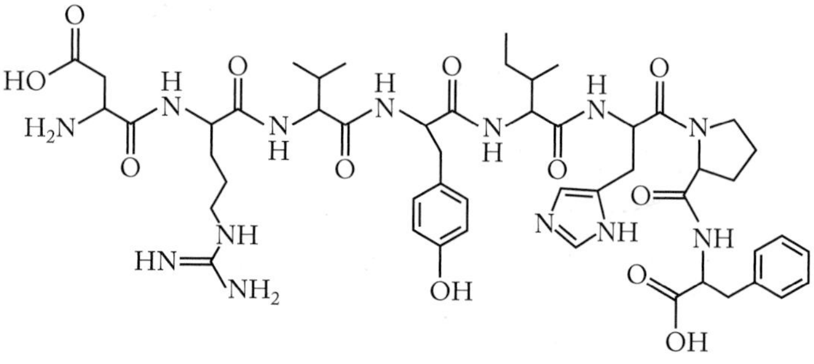

Angiotensin II

FIGURE P19.8

19.9. **Trans Fats** The role of "trans fats" in human health has been extensively debated both in the scientific community and in the popular press. What type of isomerism does the word "trans fat" refer to? Which of the molecules in Figure P19.9 are considered trans fats?

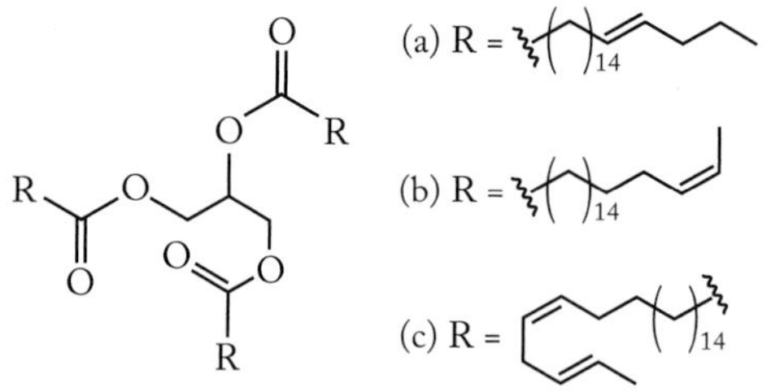

FIGURE P19.9

19.10. **Cocoa Butter** Cocoa butter (Figure P19.10) is a key ingredient in chocolate. Cocoa butter is a triglyceride that results from esterification of glycerol with three different fatty acids. Identify the fatty acids produced by hydrolysis of cocoa butter.

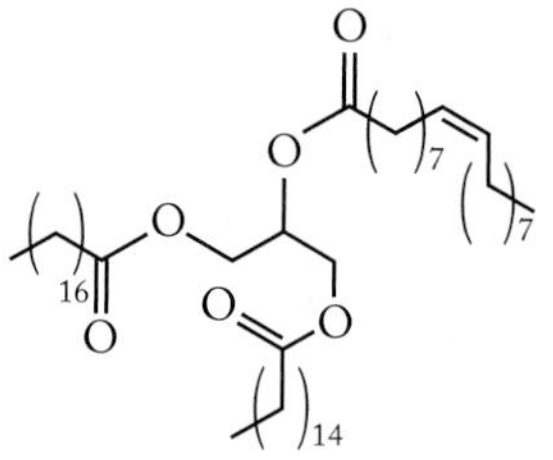

Cocoa butter

FIGURE P19.10

Questions and Problems

Biomolecules: Building Blocks of Life

CONCEPT REVIEW

19.11. In living cells, small molecules combine to make much larger molecules. Are these processes accompanied by increases or decreases in entropy of the molecules?

19.12. What types of intermolecular forces might act to keep several strands of biomolecules together in supramolecular assemblies?

19.13. Give an example of a biomolecule that is considered to be a small molecule.

19.14. Match the following building blocks in the left-hand column with the appropriate biopolymer in the right-hand column:

Amino acid	Carbohydrate
Fatty acid	DNA
Monosaccharide	Protein
Nucleotide	Lipid

Chirality

CONCEPT REVIEW

19.15. Can all of the terms *enantiomer, achiral,* and *optically active* be used to describe a single compound? Explain.

19.16. Two compounds have the same structure and the same physical properties but also have the same optical activity. Are they enantiomers or the same molecule?

19.17. Are racemic mixtures considered homogeneous or heterogeneous mixtures?

*19.18. Could a racemic mixture be distinguished from an achiral compound based on optical activity? Explain your answer.

19.19. Why is the amino acid glycine achiral?

19.20. Can geometric isomers of molecules such as *cis* and *trans* $RCH=CH_2$ also have optical isomers? (R may be any of the functional groups we have encountered in this textbook.) Explain your answer.

19.21. Which type of hybrid orbitals on a carbon atom, sp, sp^2, or sp^3, can give rise to enantiomers?

*19.22. Could an oxygen atom in an alcohol, ketone, or ether ever be a chiral center in the molecule?

19.23. Which of the following objects are chiral? (a) a golf club; (b) a tennis racket; (c) a glove; (d) a shoe

19.24. Which of the following objects are chiral? (a) a key; (b) a screwdriver; (c) a light bulb; (d) a baseball

PROBLEMS

19.25. Which of the molecules in Figure P19.25 are chiral?

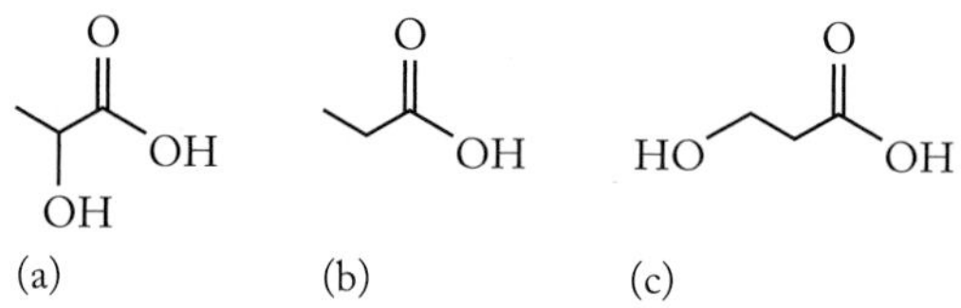

FIGURE P19.25

19.26. Which, if any, of the molecules shown in Figure P19.26 contains a chiral center?

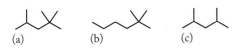

(a)   (b)   (c)

FIGURE P19.26

19.27. Artificial Sweeteners Artificial sweeteners are fundamental to the diet food industry. Figure P19.27 shows three artificial sweeteners that have been used in food. Saccharin is the oldest, dating to 1879. Cyclamates were banned in 1969 following research suggesting they led to tumors. Aspartame may be more familiar to you under the name Nutrasweet. Each of these sweeteners contain between zero and two chiral carbon atoms. Circle the chiral center in each compound.

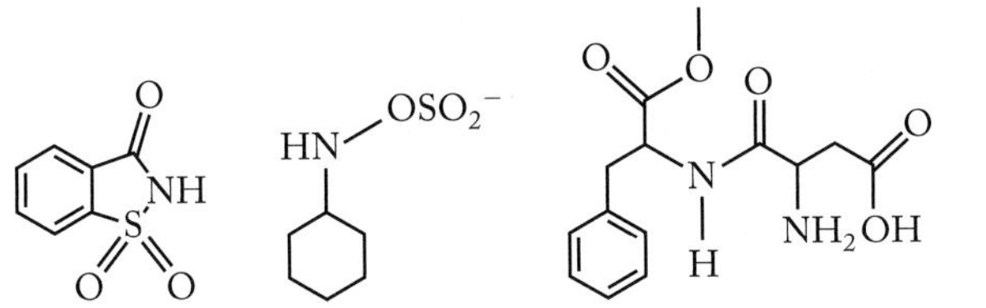

Saccharin   Cyclamate   Aspartame

FIGURE P19.27

19.28. Identify the chiral centers in each of the molecules in Figure P19.28.

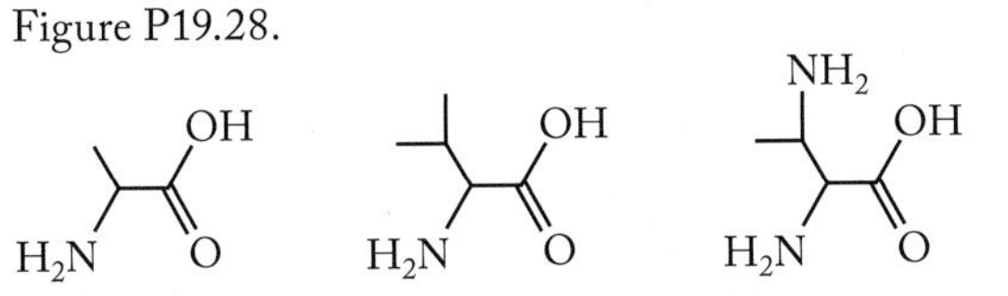

(a)   (b)   (c)

FIGURE P19.28

19.29. The Smell of Raspberries The compound 3-(p-hydroxyphenyl)-2-butanone is a major contributor to the smell of raspberries. One enantiomer is shown in Figure P19.29. Identify the single chiral center in the molecule and draw the mirror image of this enantiomer.

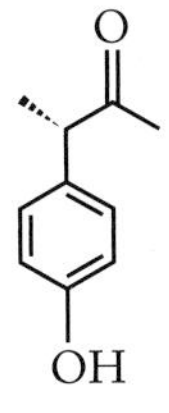

3-(p-Hydroxyphenyl)-2-butanone

FIGURE P19.29

19.30. The scent associated with pine trees is derived from the molecule terpinol. One enantiomer of terpinol is shown in Figure P19.30. Draw the other optical isomer.

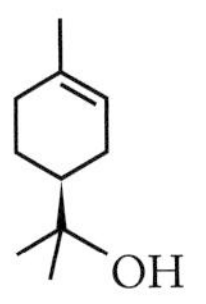

Terpinol

FIGURE P19.30

Protein Composition; Protein Structure and Function

CONCEPT REVIEW

19.31. Meteorites contain more L-amino acids than D-amino acids, which are the forms that make up the proteins in our bodies. What is meant by D-amino acids?

19.32. Of the 20 amino acids necessary to sustain life, which one is not chiral?

19.33. Which of the compounds in Figure P19.33 is not an α-amino acid?

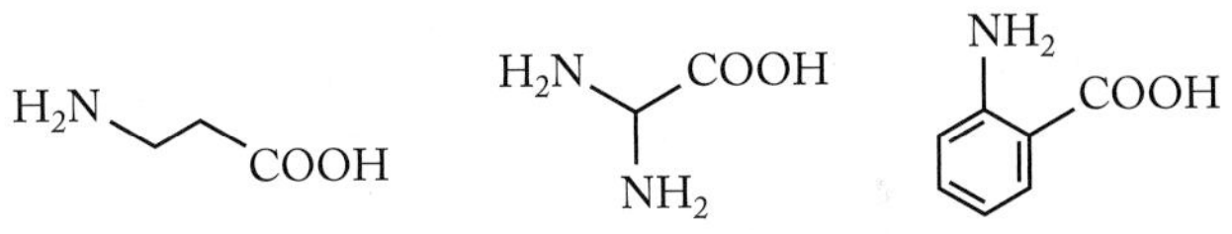

(a)   (b)   (c)

FIGURE P19.33

19.34. Which of the compounds in Figure P19.34 are α-amino acids?

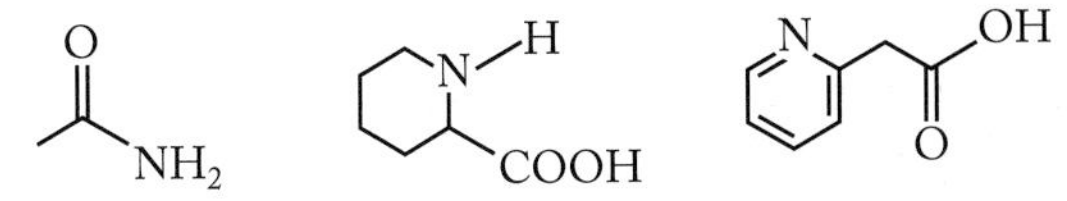

(a)   (b)   (c)

FIGURE P19.34

***19.35.** Without doing the actual calculation, estimate the fuel values of leucine and isoleucine by considering average bond energies. Should the fuel values of the two amino acids be the same? Actual calorimetric measurements show that isoleucine has a lower fuel value than leucine. Explain this.

19.36. Do both enantiomers in a racemic mixture give the same results from combustion analysis?

19.37. Why do most amino acids exist in the zwitterionic form at pH $\approx$ 7.4?

19.38. Would amino acids retain the zwitterionic form at high or low pH (for example, pH $\approx$ 10 and pH $\approx$ 2, respectively)? Explain your answer.

PROBLEMS

19.39. Draw structures of the peptides produced from condensation reactions of the following L-amino acids: (a) alanine + serine; (b) alanine + phenylalanine; (c) alanine + valine.

19.40. Draw structures of the peptides produced from condensation reactions of the following L-amino acids: (a) methionine + alanine + glycine; (b) methionine + valine + alanine; (c) serine + glycine + tyrosine.

19.41. Identify the amino acids in the dipeptides shown in Figure P19.41.

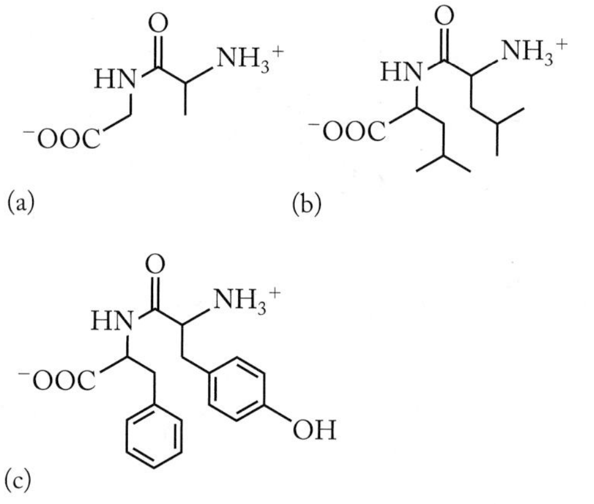

FIGURE P19.41

19.42. Identify the amino acids in the tripeptides in Figure P19.42.

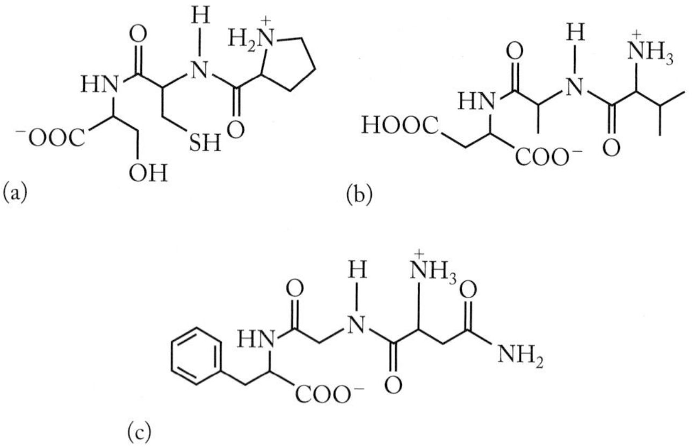

FIGURE P19.42

19.43. Identify the missing product in the metabolic reaction shown in Figure P19.43.

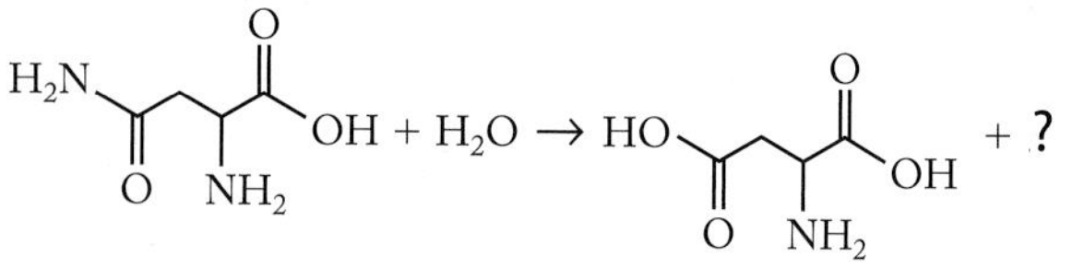

FIGURE P19.43

19.44. Identify the missing product in the metabolic reaction shown in Figure P19.44.

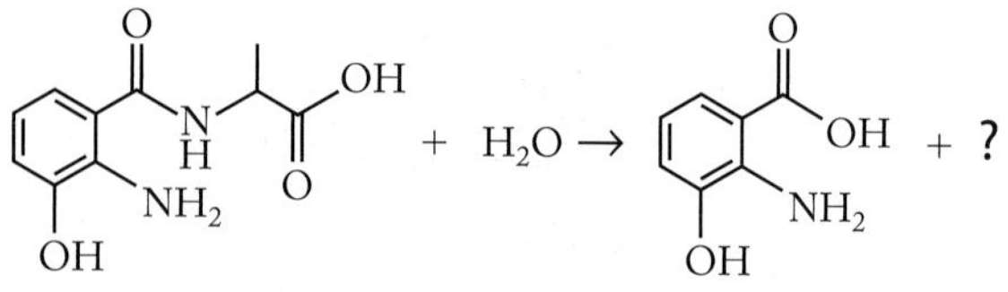

FIGURE P19.44

Carbohydrates

CONCEPT REVIEW

19.45. What are the structural differences between starch and cellulose?

19.46. Why is the discovery of enzymes that catalyze cellulose hydrolysis a worthwhile objective?

19.47. Is the fuel value of glucose in the linear form the same as that in the cyclic form?

*19.48. Without doing the actual calculation, estimate the fuel values of glucose and starch by considering average bond energies. Do you predict the fuel values of the two substances to be the same or different?

19.49. The second step in glycolysis converts glucose 6-phosphate into fructose 6-phosphate. Can you think of a reason why $\Delta G°$ for this reaction is close to zero?

*19.50. Why is it important that at least some of the steps in glycolysis convert ADP to ATP?

19.51. How do we calculate the overall free-energy change of a process consisting of two steps?

19.52. During glycolysis simple sugar is converted to pyruvate. Do you think this process produces an increase or decrease in the entropy of the system? Explain your answer.

PROBLEMS

19.53. Draw a diagram that shows how isomers of the sugar shown as a flat Lewis structure in Figure P19.53 result when the linear molecule forms a 6-atom ring.

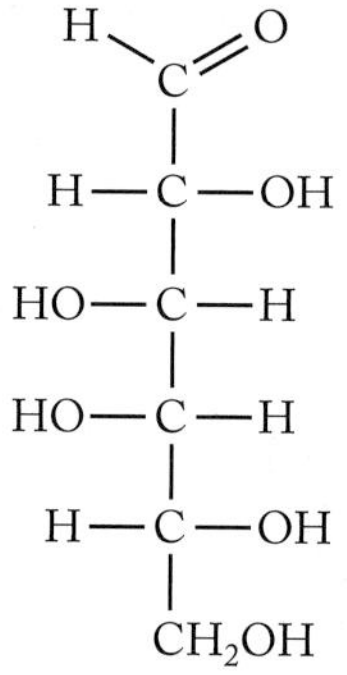

Galactose

FIGURE P19.53

19.54. Draw a diagram that shows how isomers of the sugar shown as a flat Lewis structure in Figure P19.54 result when the linear molecule forms a 5-atom ring.

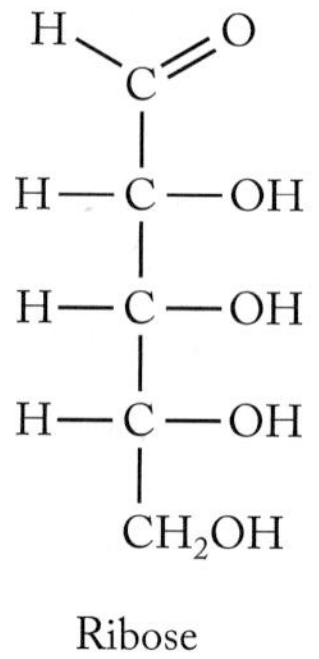

Ribose

FIGURE P19.54

19.55. Which, if any, of the structures in Figure P19.55 are β isomers of a monosaccharide?

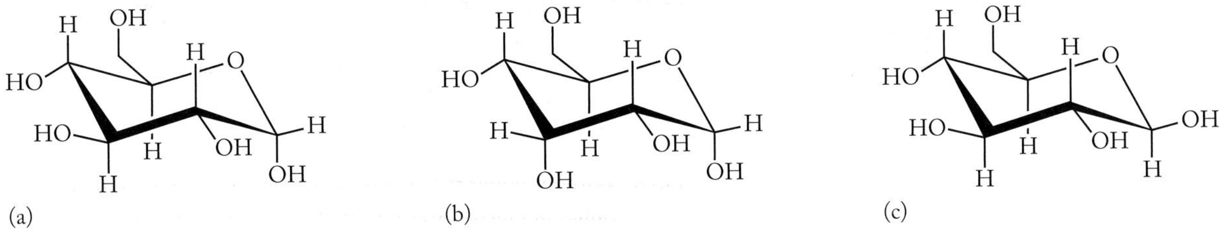

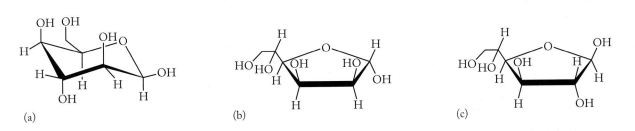

(a) (b) (c)

FIGURE P19.55

19.56. Identify which, if any, of the structures in Figure P19.56 are β isomers.

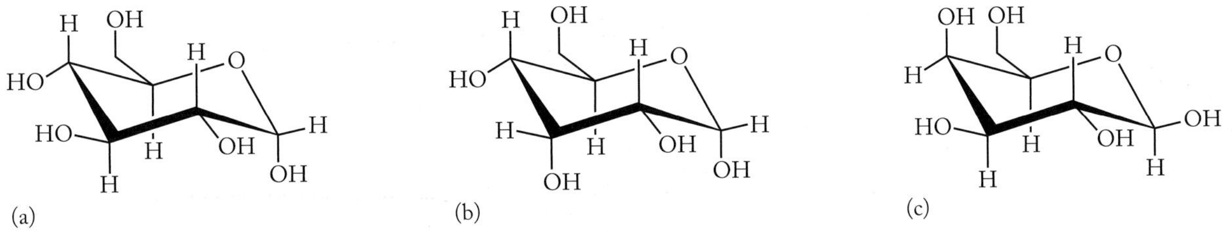

(a) (b) (c)

FIGURE P19.56

19.57. Which, if any, of the structures in Figure P19.57 are α isomers?

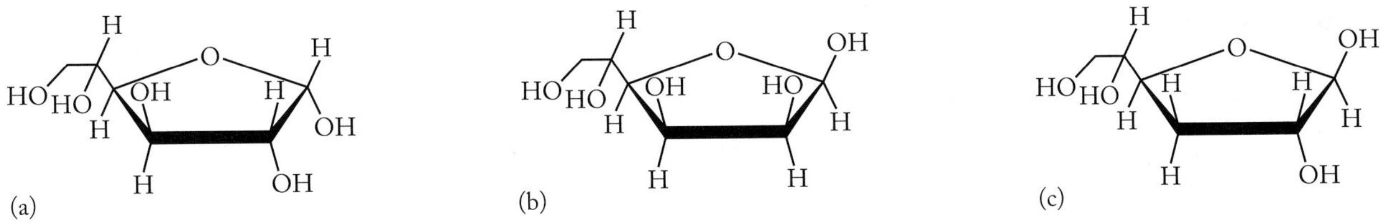

(a) (b) (c)

FIGURE P19.57

19.58. Which, if any, of the structures in Figure P19.58 are α isomers?

(a) (b) (c)

FIGURE P19.58

19.59. Which of the saccharides in Figure P19.59 is or are digestible by humans?

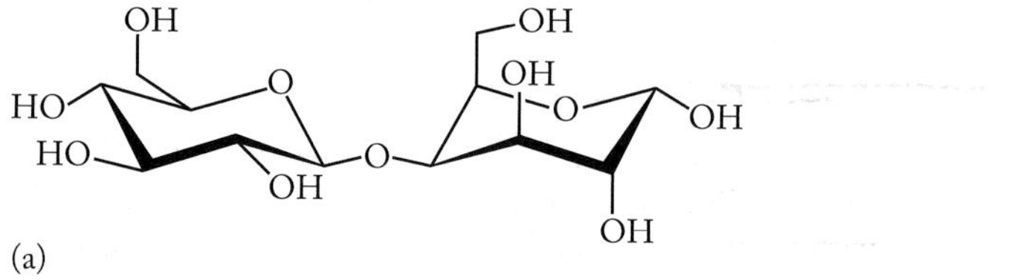

(a)

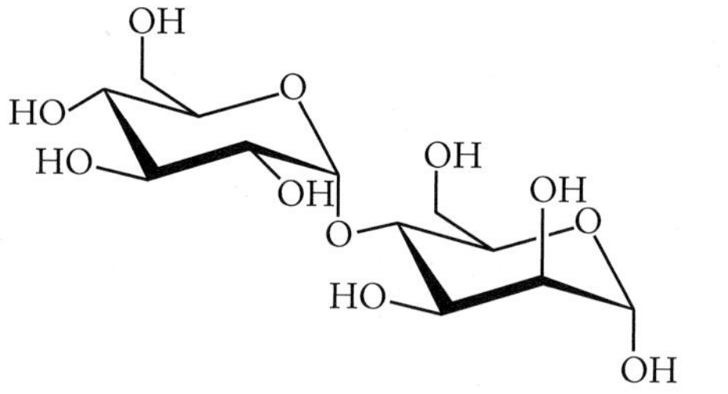

(b)

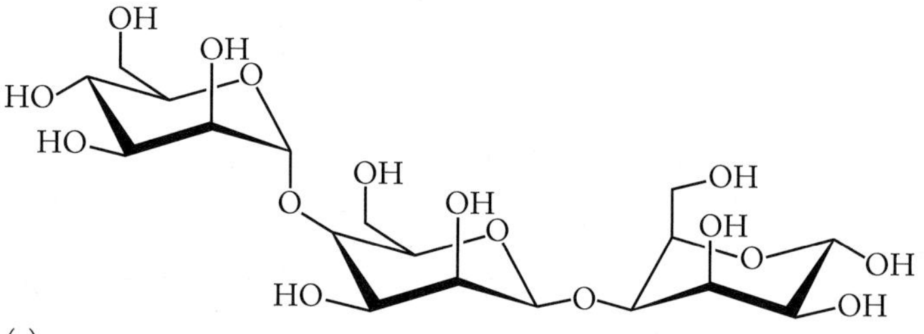

(c)

FIGURE P19.59

*19.60. For any of the disaccharides in Problem 19.59 that are not digestible by humans, draw an isomer that would be digested.

19.61. The structure of the disaccharide maltose appears in Figure P19.61. Hydrolysis of a mole of maltose ($\Delta G_f^\circ = -2246.6$ kJ/mol) produces 2 moles of glucose ($\Delta G_f^\circ = -1274.5$ kJ/mol):

$$\text{Maltose} + \text{H}_2\text{O} \rightarrow 2 \text{ glucose}$$

If the value of ΔG_f° of water is -285.8 kJ/mol, what is the change in free energy of the hydrolysis reaction?

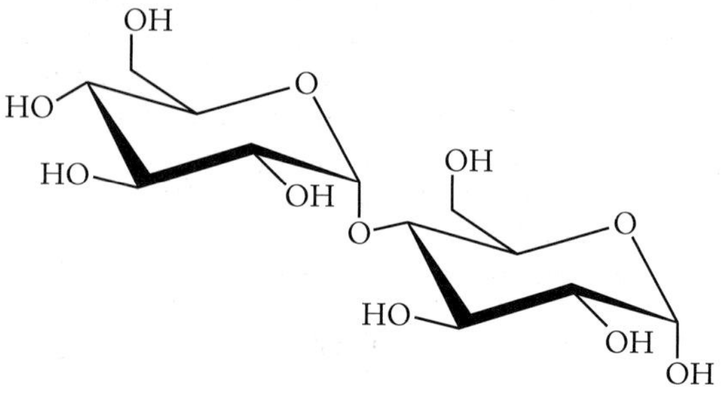

Maltose

FIGURE P19.61

19.62. If the maltose in Problem 19.61 was replaced by another disaccharide, would you expect the free energy change for the hydrolysis to be exactly the same or just similar in value? Explain your answer.

Lipids

CONCEPT REVIEW

19.63. What is the difference between a saturated and an unsaturated fatty acid?

19.64. Why are the average fuel values of fats higher than those of carbohydrates and proteins?

19.65. Some Arctic explorers have eaten sticks of butter on their explorations. Give a nutritional reason for this unusual cuisine.

19.66. Salad dressings containing oil and vinegar quickly separate upon standing. Explain the observed separation of layers based on the structure and properties of aqueous vinegar and oil.

19.67. Do triglycerides have a chiral center? Explain your answer.

*19.68. Using your knowledge of molecular geometry and intermolecular forces, why might polyunsaturated triglycerides be more likely to be liquid than saturated triglycerides?

PROBLEMS

19.69. Which of the triglycerides in Figure P19.69 are unsaturated fats?

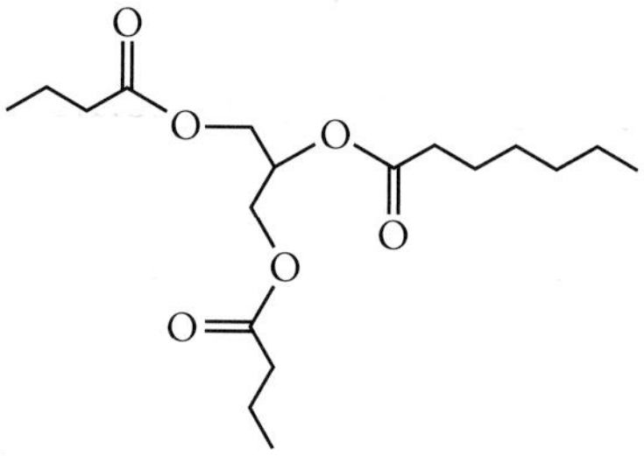

(a)

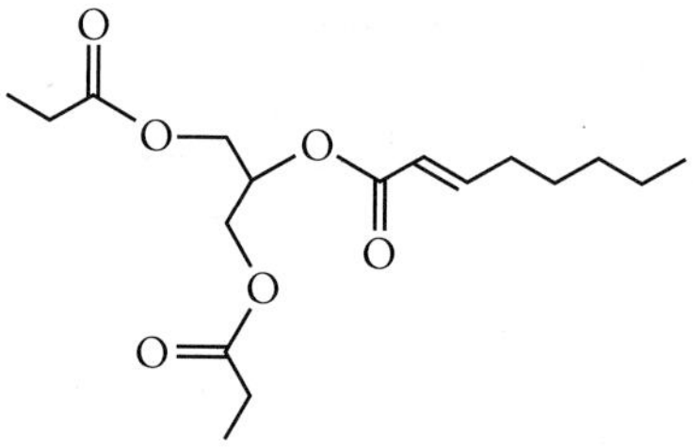

(b)

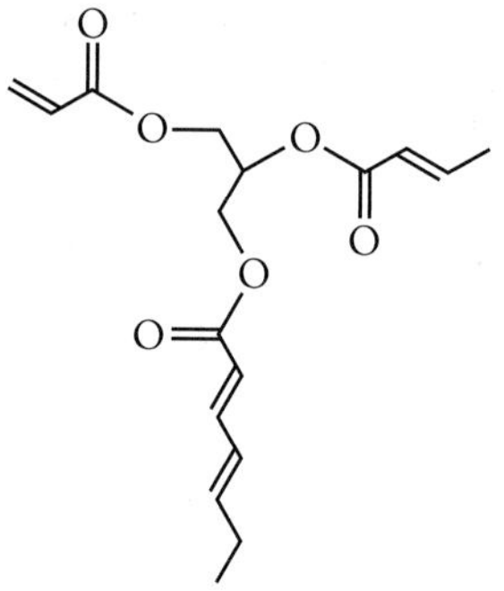

(c)

FIGURE P19.69

19.70. For each of the pairs of fatty acids in Figure P19.70 indicate whether they are structural isomers, geometric isomers, or unrelated compounds.

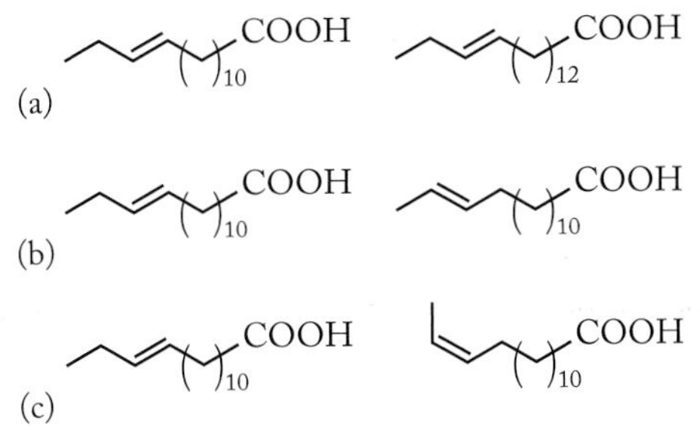

FIGURE P19.70

19.71. Draw the structures of the fats formed by reaction of glycerol with octanoic acid ($C_7H_{15}COOH$), decanoic acid ($C_9H_{19}COOH$), and dodecanoic acid ($C_{11}H_{21}COOH$).

19.72. **Oil-Based Paints** Oil-based paints contain linseed oil, a triglyceride formed by esterification of glycerol with linolenic acid (Figure P19.72).
   a. Draw the line structure of linolenic acid produced by hydrolysis of the lipid shown below.
   *b. Are the double bonds in linolenic acid conjugated?

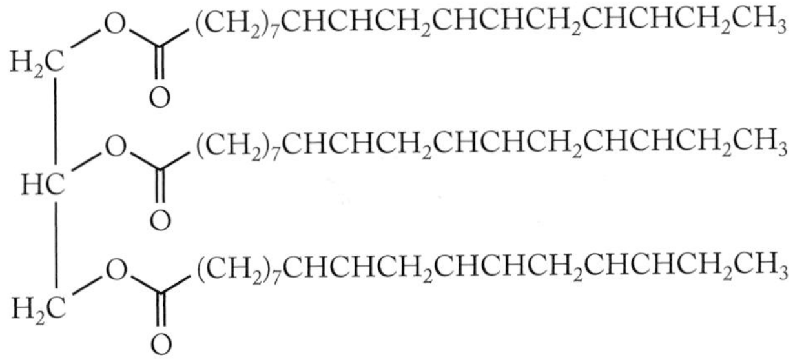

Linseed oil

FIGURE P19.72

Nucleotides and Nucleic Acids

CONCEPT REVIEW

19.73. What are the three kinds of molecular subunits in DNA? Which two form the "backbone" of DNA strands?

19.74. Why does a codon consist of a sequence of three, and not two, ribonucleotides?

19.75. What kind of intermolecular force holds together the strands of DNA in the double-helix configuration?

19.76. What is meant by "base pairing" in DNA? Which bases are paired?

PROBLEMS

19.77. Draw the structure of adenosine 5′-monophosphate, one of the four ribonucleotides in a strand of RNA.

19.78. Draw the structure of deoxythymidine 5′-monophosphate, one of the four nucleotides in a strand of DNA.

19.79. In the replication of DNA, a segment of an original strand has the sequence T-C-G-G-T-A. What is the sequence of the double-stranded helix formed in replication?

19.80. In transcription, a segment of the strand of DNA that is transcribed has the sequence T-C-G-G-T-A. What is the corresponding sequence of nucleotides on the messenger RNA that is produced in transcription?

Additional Problems

19.81. Olestra is calorie-free fat-substitute. The core of the Olestra molecule (Figure P19.81) is a disaccharide that has reacted with a carboxylic acid; this results in the conversion of hydroxyl groups on the disaccharide into this structure:

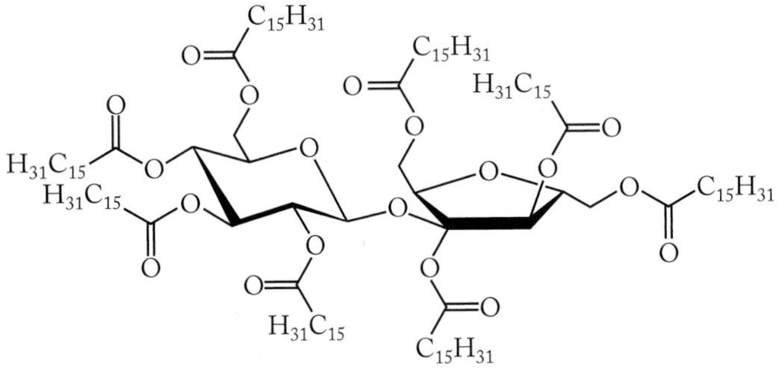

Olestra

FIGURE P19.81

   a. What is the name of the disaccharide core of the olestra molecule?
   b. What functional group has replaced the hydroxyl groups on the disaccharide?
   c. What is the formula of the carboxylic acid used to make olestra?

19.82. When scientists at UC Santa Cruz directed UV radiation at an ice crystal containing methanol, ammonia, and hydrogen cyanide, three amino acids (glycine, alanine, and serine) were detected among the products of photochemical reactions. The formation of these amino acids suggests that they may also be synthesized in comets approaching the sun (and Earth).
   a. Draw the structures for glycine, alanine, and serine.
   b. Determine the standard free energy change of the hypothetical formation of glycine in comets, using standard free energies of formation of the reactants and products in this reaction [$\Delta G°$ for $HCN(g)$ is +125 kJ/mol and for glycine is −368.4 kJ/mol].

$$CH_3OH(\ell) + HCN(g) + H_2O(\ell) \rightarrow H_2NCH_2COOH(s) + H_2(g)$$

19.83. Homocysteine (Figure P19.83) is formed during the metabolism of amino acids. A mutation in some people's genes leads to high concentrations of homocysteine in the blood and a consequent increase in their risk of heart disease and incidence of bone fractures in old age.
   a. What is the structural difference between homocysteine and cysteine?
   b. Cysteine is a chiral compound. Is homocysteine chiral?

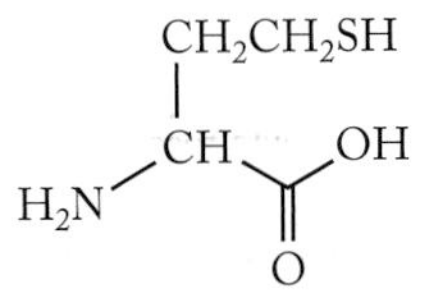

Homocysteine

FIGURE P19.83

19.84. Some scientists believe life on Earth can be traced to amino acids and other molecules brought to Earth by comets and meteorites. In 2004, a new class of amino acids, diamino acids (Figure P19.84) were found in the Murchison meteorite.
   a. Which of these diamino acids is not an α-amino acid?
   b. Which of these amino acids is chiral?

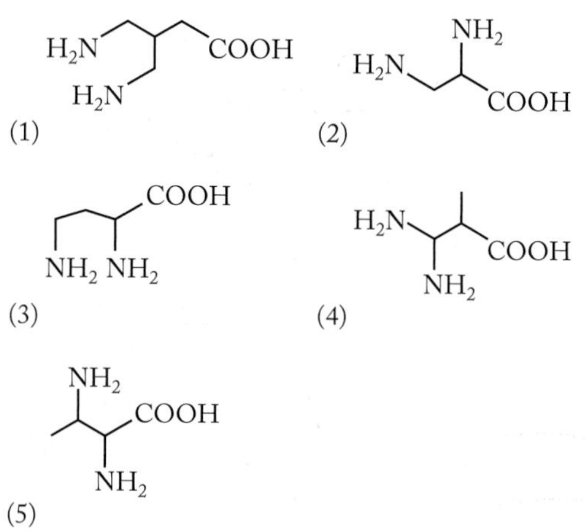

FIGURE P19.84

19.85. Ackee, the national fruit of Jamaica, is a staple in many Jamaican diets. Unfortunately, a potentially fatal sickness, known as Jamaican vomiting disease, is caused by the consumption of unripe ackee fruit, which contains the amino acid hypoglycin (Figure P19.85). Is hypoglycin an α-amino acid?

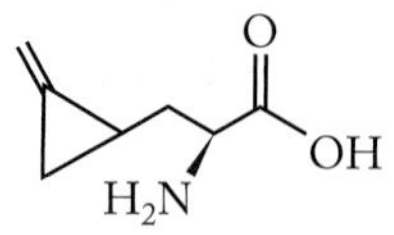

Hypoglycin

FIGURE P19.85

19.86. The compound eryngial (Figure P19.86) is found in fresh cilantro and coriander and has a wide range of antibacterial activity.
   a. Identify the functional groups in eryngial.
   b. What is the hybridization of the carbon atoms numbered 1 to 4?

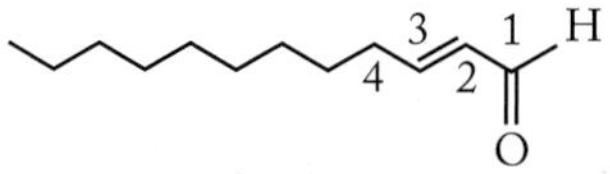

Eryngial

FIGURE P19.86

19.87. The compound known as omuralide (Figure P19.87) has been used to treat some forms of cancers. Circle the chiral carbon atoms in the structure.

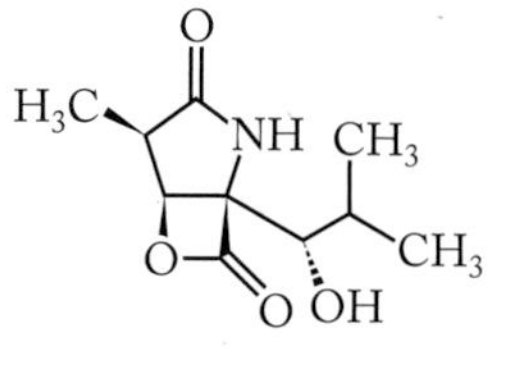

FIGURE P19.87

19.88. Salinosporamide A, isolated from a microbe called *Salinispora tropica* has shown promise as an inhibitor of cancer cell growth. Circle the chiral centers in the structure shown in Figure P19.88.

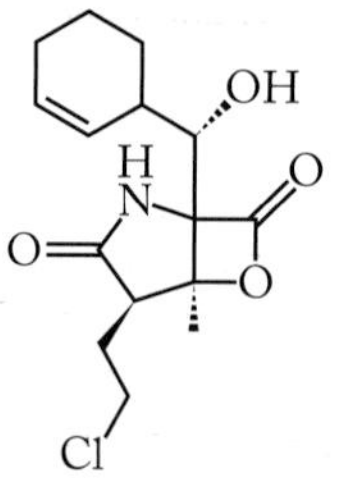

Salinosporamide A

FIGURE P19.88

19.89. In late 2003, researchers at the Scripps Research Institute reported the development of genetically modified *E. coli* that could incorporate five new amino acids into proteins. These five amino acids shown in Figure P19.89 are not among the 20 naturally occurring amino acids. Which naturally occurring amino acids are these compounds most similar to?

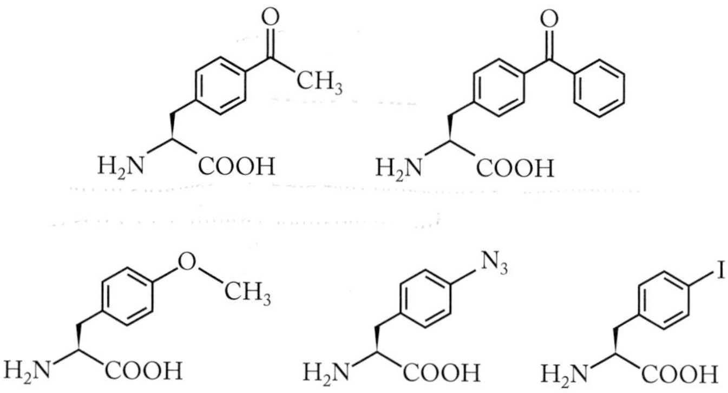

FIGURE P19.89

19.90. Creatine (Figure P19.90) is an amino acid produced by the human body. Body builders sometimes take creatine supplements to help gain muscle strength. A 2003 study reported that creatine may boost memory and cognitive thinking.
   a. Is creatine an α-amino acid?
   b. Draw the two dipeptides that can be formed from glycine and creatine.

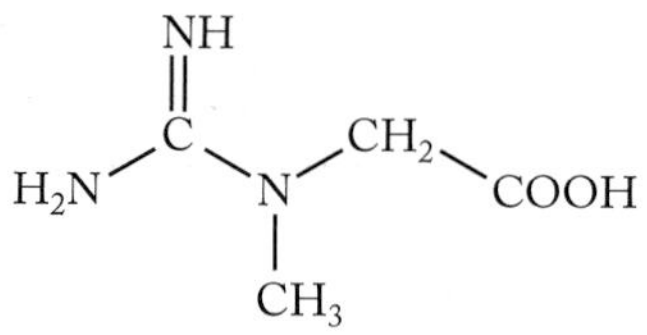

Creatine

FIGURE P19.90

19.91. Glutathione (Figure P19.91) is an essential molecule in the human body. It acts as an activator for enzymes and protects lipids from oxidation. Which three amino acids combine to make glutathione?

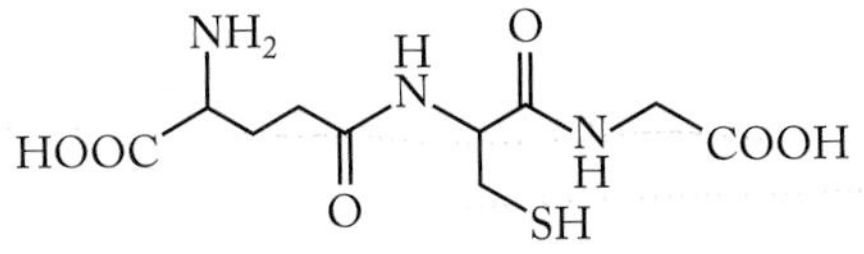

Glutathione

FIGURE P19.91

19.92. In response to specific neural messages, the human hypothalamus may secrete a number of polypeptides including the tripeptide shown in Figure P19.92.
   a. Sketch the structures of the three amino acids that combine to make this tripeptide.
   b. Which, if any, of the constituent amino acids are among the 20 α-amino acids in proteins?

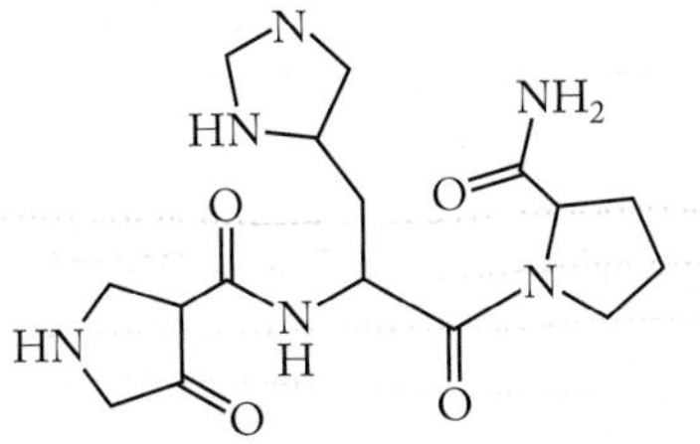

Thyrotropic releasing factor

FIGURE P19.92

19.93. The human body responds to pain by producing anandamide, a compound that binds to receptors in the brain to dull the pain. Anandamide (Figure P19.93) is then hydrolyzed to ethanolamine and arachidonic acid.
   a. How many of the four double bonds in anandamide have a *cis* geometry?
   b. Draw the structures of arachidonic acid and ethanolamine.

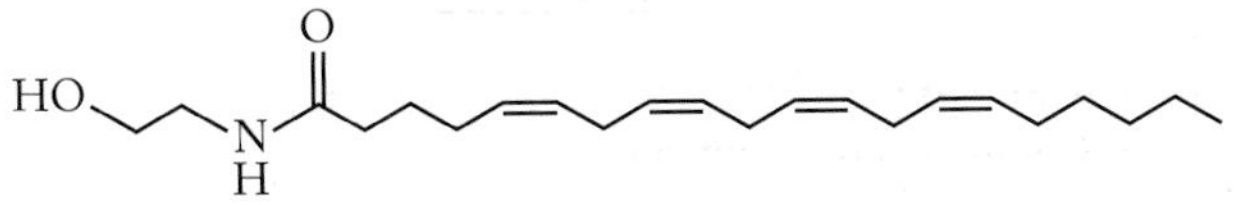

Anandamide

FIGURE P19.93

19.94. In 2002, a new compound, URB597 (Figure P19.94), was reported to inhibit the hydrolysis of anandamide and thus prolong the pain relief provided by anandamide.
   a. Are there any chiral centers in URB597?
   b. Draw the structures of two compounds that might combine in a condensation reaction to form URB597.

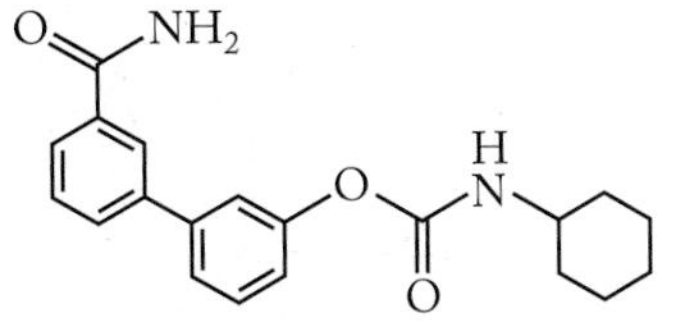

URB597

FIGURE P19.94

19.95. The compound known as topiramate (Figure P19.95) is used to treat epilepsy and more recently was found to be effective against migraine headaches.
   a. How many chiral carbon atoms are there in topiramate?
   b. What is the molecular geometry around the sulfur atom?
   *c. How can sulfur make six bonds?

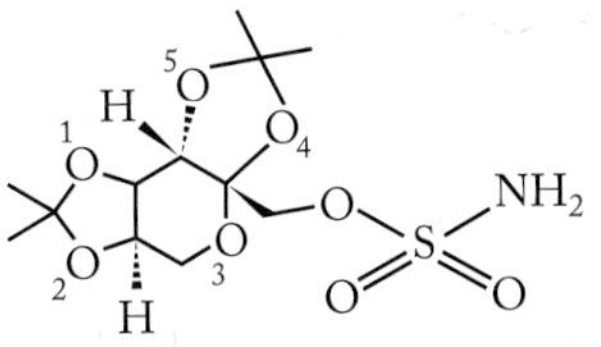

Topiramate

FIGURE P19.95

*19.96. The pleasant odors of many flowers can be traced to unsaturated organic compounds such as the two on the right in Figure P19.96. Using the rules for naming alkenes, alcohols, and the example of 1,3-butadiene shown in Figure P19.96, assign systematic names to nerolidol and linalool, two of the odor-producing compounds in jasmine.

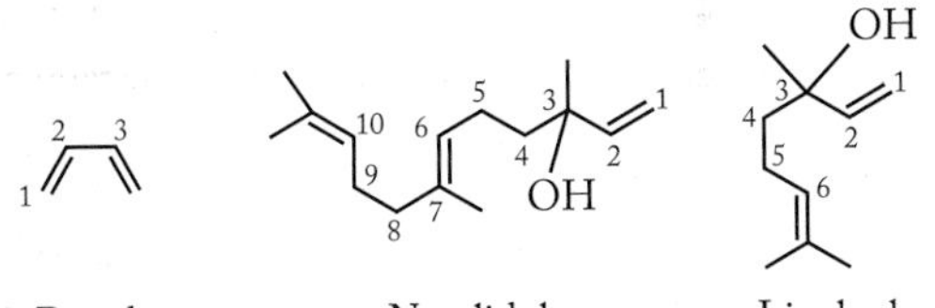

1, 3-Butadiene   Nerolidol   Linalool

FIGURE P19.96

19.97. Geraniol and citronellol (Figure P19.97) contribute to the odor of some species of rose.
   a. Are geraniol and citronellol structural isomers, geometric isomers, optical isomers, or two totally different compounds?
   b. Draw the other geometric isomer of geraniol.

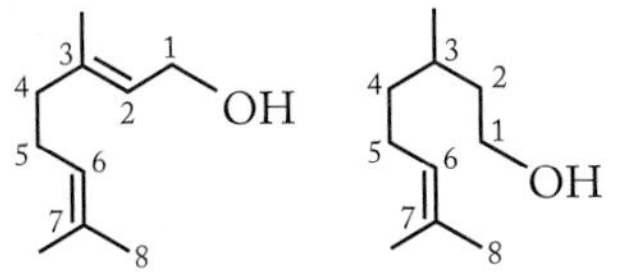

Geraniol   Citronellol

FIGURE P19.97

*19.98. Polyunsaturated fats have a tendency to polymerize. What kind of polymerization reaction might account for this observation?

20

Nuclear Chemistry

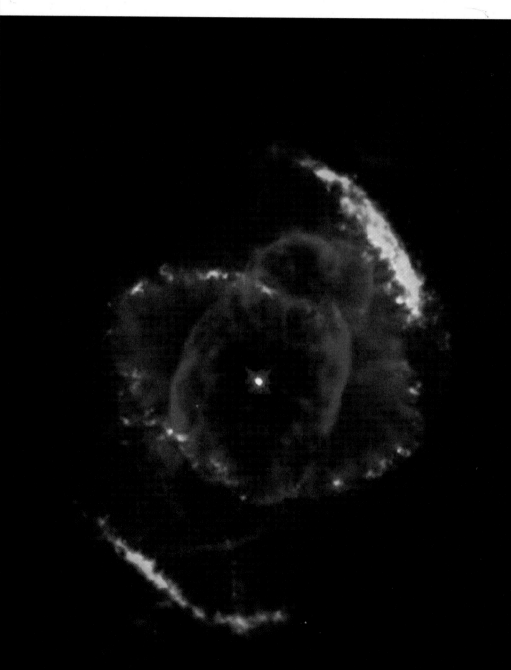

CAT'S EYE NEBULA *When a red giant star runs out of fuel it collapses and explodes, forming a planetary nebula. The colors of the hot gases expelled by the explosion can be used to determine the elemental composition of the star. In this image of the Cat's Eye Nebula radiation emitted by hydrogen atoms is shown in red, oxygen radiation is in blue, and radiation from nitrogen ions is shown in green.*

A LOOK AHEAD: Alchemy in the 21st Century

In Chapter 2 we introduced the concept of nucleosynthesis, the process by which scientists believe hydrogen and helium formed soon after the Big Bang and the other elements formed later in giant stars. Since that introduction we have focused on the chemical properties of the elements and the countless ways they combine to form compounds. We have examined compounds at the molecular level, probing their elemental composition, molecular structures, and the nature of their bonds.

In this chapter we revisit the cosmic origins of the elements and the nature of the radiochemical reactions that create them naturally and in the laboratory. We will start with the nuclear furnaces of our sun and other stars. There, fusion reactions different from those in primordial nucleosynthesis turn hydrogen into helium and, in some stars, other elements up to and including iron ($Z = 26$). Beyond iron, fusion reactions do not sustain the stars because the reactions do not generate energy; they consume it. Other nuclear processes, such as the absorption of neutrons followed by emission of high-energy electrons (β particles), account for the other elements in the periodic table.

The nuclear furnace of our sun produces heat and light. It also emits cosmic rays that can break apart the nuclei of atoms in Earth's upper atmosphere. Free neutrons released in this way set in motion series of reactions that generate radioactive isotopes. We will see how one of them, carbon-14, allows scientists to determine the ages of manmade objects and to track the migrations and activities of humans over tens of thousands of years.

During the past century scientists have made their own radioisotopes that play key roles in medical imaging and diagnosis and also in the treatment of diseases, particularly cancer. Scientists have also harnessed the energy released when neutron bombardment causes nuclei to split by a process called nuclear fission. The history of the 20th century reveals how the energy released by nuclear fission can help meet civilization's energy needs, on the one hand, or destroy it on the other.

20.1 Nuclear Chemistry

Throughout this book we have considered the nature of chemical reactions driven by breaking bonds and making new ones. In every chemical reaction the identities of the atoms involved, as defined by the number of protons in each of their nuclei, remain unchanged. However, other kinds of reactions occur all around us. At the outermost fringes of our atmosphere, cosmic rays from the sun, which are a high-speed stream of mostly protons and α particles (He nuclei), collide with and split atomic nuclei, releasing free neutrons that are absorbed by the nuclei of other atoms, forming radioactive isotopes. The field of chemistry that studies such reactions is called **nuclear chemistry**.

20.2 Radioactive Decay

Let's revisit the theory of the Big Bang. Cosmologists who study the dynamics of the universe believe that, within a few billionths of a second of its creation, the energy released in this event had transformed into matter in proportions described by Einstein's famous equation

$$E = mc^2$$

Nuclear chemistry studies reactions that involve changes in the nuclei of atoms.

where E is the amount of energy, m is the mass of the corresponding amount of matter, and c is the speed of light. Less than a millisecond later, the expanding universe had cooled to a few billions of degrees and neutrons formed. A free neutron ($_0^1n$) is not stable; it spontaneously decays into a proton ($_1^1H$) and an electron ($_{-1}^0e$). This nuclear reaction is summarized in radiochemical Equation 20.1:

$$_0^1n \rightarrow {}_1^1H + {}_{-1}^0e \qquad (20.1)$$

Recall that the subscript in each symbol represents the electrical charge of the particle and the superscript represents its relative mass expressed in mass numbers. Like the equations for chemical reactions, any equation describing a nuclear reaction must be balanced: The sum of the masses denoted by the superscripts on the left of the reaction arrow must equal the sum of the masses on the right side. Similarly, the sum of the electrical charges denoted by the subscripts on the left side of the equation must equal the sum of the electrical charges on the right. A check of the mass numbers (superscripts) in Equation 20.1 reveals that they do add up and so do the charges in the subscripts.

The conversion of neutrons to protons is an example of **radioactive decay**, the spontaneous disintegration of unstable particles or nuclei accompanied by the release of radiation. All radioactive decay processes, including the decay of free neutrons, follow first-order kinetics (which we discussed in Chapter 14) and so each has a particular *half-life* ($t_{1/2}$), the time interval during which the quantity of radioactive particles decreases by one half, as shown in Figure 20.1. The faster the decay process, the shorter the half-life, that is, the rate constant of a decay process is inversely proportional to the value of its half-life.

The half-life of free neutrons is 12 minutes. In other words, 12 minutes after the Big Bang, half of the neutrons in the universe (most of the matter) should have decayed into protons and electrons. If we wanted to predict what fraction of those neutrons remained at some time t after the Big Bang, we could convert this time interval into an equivalent number of half-lives (n):

$$n = \frac{t}{t_{1/2}} \qquad (20.2)$$

Once we know the value of n, we can calculate the ratio of the quantity of free neutrons that remained at any time t (A_t) to the number present at the start (A_0) when $t = 0$, using Equation 20.3:

$$\frac{A_t}{A_0} = 0.5^n \qquad (20.3)$$

If $n = 1$, then $A_t/A_0 = 0.5^1$ or 0.5, and half of the starting amount is left. If $n = 2$, $A_t/A_0 = 0.5^2 = 0.25$, and one-quarter of the starting amount is left, as shown in Figure 20.1. What if we wanted to predict the fraction of free neutrons remaining after only 10 minutes? We could begin by calculating the number of half-lives that had elapsed:

$$10\ \text{min} \times \frac{\text{one half-life}}{12\ \text{min}} = 0.833\ \text{half-life}$$

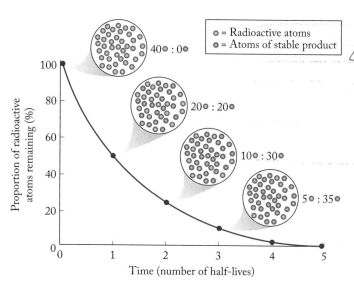

FIGURE 20.1 Radioactive decay follows first-order kinetics, which means that 50% of an initial quantity of a radioactive nuclide (red dots) has decayed into a stable product represented by the blue dots during a time interval equal to one half-life. Half of the remaining half, or 25%, remains after two half-lives, and so on.

▶‖ CHEMTOUR Half-Life

Radioactive decay is the spontaneous disintegration of unstable particles accompanied by the release of radiation.

Using this value for n in Equation 20.3,

$$\frac{A_t}{A_0} = 0.5^n = 0.5^{0.833}$$

$$= 0.561, \text{ or } 56\% \text{ left}$$

In the following exercises and elsewhere in this chapter we will apply Equation 20.3 to various radioactive decay processes. We can do so because all of them follow first-order reaction kinetics.

SAMPLE EXERCISE 20.1   **Calculations Involving Half-Lives**

Out of a population of 6.6×10^5 free neutrons, how many would remain after 2.0 minutes if the half-life $(t_{1/2})$ of free neutrons is 12 minutes?

COLLECT AND ORGANIZE It takes 12 minutes for half of a given quantity of neutrons to decay. The problem asks us to determine how many neutrons remain after only 2.0 minutes. We can use Equation 20.2 to calculate the number of half-lives in a given time interval, and use Equation 20.3 to calculate the quantity of a radioactive species that remains after a given number of half-lives has passed.

ANALYZE We should use the values of $t = 2.0$ minutes and $t_{1/2} = 12$ minutes in Equation 20.2 to calculate the number of half-lives (n). We will use that value of n and 6.6×10^5 as the initial quantity (A_0) of neutrons in Equation 20.3 to calculate A_t.

SOLVE Substituting $t = 2.0$ minutes and $t_{1/2} = 12.0$ minutes in Equation 20.2:

$$n = \frac{t}{t_{1/2}}$$

$$= \frac{2.0 \text{ min}}{12 \text{ min}} = 0.1667 \text{ half-lives}$$

Using this value of n in Equation 20.3,

$$\frac{A_t}{A_0} = 0.5^n$$

or

$$A_t = A_0 \, 0.5^n$$

$$= (6.6 \times 10^5) \times (0.5^{0.1667}) = 5.9 \times 10^5 \text{ neutrons}$$

THINK ABOUT IT The result is reasonable because 2 minutes is much less than the half-life of free neutrons and the result represents a loss of much less than half the initial number of neutrons.

Practice Exercise Cesium-131 is a short-lived radioisotope $(t_{1/2} = 9.7 \text{ days})$ used to treat prostate cancer. How much therapeutic strength does a cesium-131 source lose over 60 days? Express your answer as a percent of the strength the source had at the beginning of the first day.

(Answers to Practice Exercises are in the back of the book.)

▶❚❚ CHEMTOUR Fusion of Hydrogen

20.3 Hydrogen Fusion

The sun is a star composed primarily of hydrogen with a small amount of helium and other elements. Its energy is derived from the fusion of hydrogen nuclei (protons) to form helium. This fusion process involves more steps than the one that probably produced helium nuclei within minutes of the Big Bang, because the high concentrations of free neutrons present in the early universe are not present in the sun's core. Therefore, the first step of the fusion process in the sun does not involve the formation of deuterons, ^2_1H, by the fusion of a proton and neutron

$$^1_1H + ^1_0n \rightarrow ^2_1H$$

◐◑ CONNECTION Chapter 2 included a brief discussion of the nucleosynthesis of elements in stars.

as in primordial nucleosynthesis, but rather ^2_1H deuterons form as a result of pairs of protons (^1_1H) fusing together.

Let's write this nuclear reaction in equation form, starting with the known reactants and product:

$$^1_1H + ^1_1H \rightarrow ^2_1H$$

To balance the equation we need to add a product with a charge of 1+ but with an insignificant mass compared to a proton. Such a particle exists. It is called a **positron**. It is represented by the symbol ^0_1e, which means that it has the mass of an electron but has a positive charge. Adding a positron as a product to the preceding reaction, we have a balanced equation:

$$^1_1H + ^1_1H \rightarrow ^2_1H + ^0_1e$$

or

$$2^1_1H \rightarrow ^2_1H + ^0_1e \qquad (20.4)$$

In the second stage of solar hydrogen fusion, a deuteron fuses with another proton to form the nucleus of a helium-3 atom:

$$^2_1H + ^1_1H \rightarrow ^3_2He \qquad (20.5)$$

The "3" in helium-3 indicates that each of its nuclei has 3 nucleons (2 protons and 1 neutron). Recall from Chapter 2 that any atom with 2 protons in the nucleus (atomic number = 2) is a helium atom and that atoms of the same element with different numbers of nucleons are called isotopes. Helium-3 is a much less common form of helium than helium-4 (^4_2He), which has 2 neutrons in each nucleus.

Finally, fusion of two helium-3 atoms produces a nucleus of ^4_2He and 2 protons:

$$2^3_2He \rightarrow ^4_2He + 2^1_1H \qquad (20.6)$$

Deuterons and ^3_2He nuclei are intermediates in the hydrogen-fusion process because they are made in one step but are then consumed in another. To write an equation for the overall fusion process, we can combine Equations 20.4, 20.5, and 20.6, multiplying Equations 20.4 and 20.5 by 2 to balance the production and consumption of the intermediate species ^2_1H and ^3_2He:

$$2\,[2^1_1H \rightarrow ^2_1H + ^0_1e]$$

$$2\,[^2_1H + ^1_1H \rightarrow ^3_2He]$$

$$2^3_2He \rightarrow ^4_2He + 2^1_1H$$

$$\overline{6^1_1H + 2^2_1H + 2^3_2He \rightarrow 2^2_1H + 2^3_2He + ^4_2He + 2^1_1H + 2^0_1e}$$

A **positron** is a particle with the mass of an electron but with a positive charge.

TABLE 20.1 **Masses and Relative Charges for Subatomic Particles**

PARTICLE	SYMBOL	MASS		CHARGE
		(amu)	(grams)[a]	(relative value)
Neutron	$_{0}^{1}\text{n}$	1.00867	1.67494×10^{-24}	0
Proton	$_{1}^{1}\text{p}$	1.00728	1.67263×10^{-24}	1+
Electron (β particle)	$_{-1}^{0}\text{e}$	5.485799×10^{-4}	9.10939×10^{-28}	1−
α particle	$_{2}^{4}\text{He}^{2+}$	4.00150	6.64465×10^{-24}	2+
Positron	$_{1}^{0}\text{e}$	5.485799×10^{-4}	9.10939×10^{-28}	1+

[a]To obtain the mass of a particle in grams, divide its mass in amu by Avogadro's number: 6.0221367×10^{23}.

Canceling out common terms on both sides of the combined equation, the overall fusion reaction is

$$4\,_{1}^{1}\text{H} \rightarrow\,_{2}^{4}\text{He} + 2\,_{1}^{0}\text{e} \qquad (20.7)$$

The positrons formed in solar fusion belong to a group of subatomic particles that are the charge opposites of particles typically found in atoms. In addition to the "electrons" with a positive charge, there are "protons" with a negative charge. These charge opposites are particles of **antimatter**. Table 20.1 summarizes the mass and charge of the subatomic particles we have discussed so far in this chapter.

Subatomic particles and their antimatter opposites are mortal enemies. If matter and antimatter collide, they instantly annihilate each other. In their mutual destruction, they cease to exist as matter, and all of their collective mass is released as energy in an amount predicted by Einstein's equation. The positrons produced from hydrogen fusion are rapidly annihilated by reactions with electrons. The sole product of the reaction is energy in the form of two gamma (γ) rays:

$$_{1}^{0}\text{e} +\,_{-1}^{0}\text{e} \rightarrow 2\,\gamma$$

Gamma ray emission accompanies all nuclear reactions. Gamma rays are generated by the nuclear furnaces of stars and permeate outer space. Those that reach Earth are absorbed by the gases in our atmosphere. In the process molecular gases are broken up into their component atoms. Particularly energetic γ rays can even break up atomic nuclei. Radioactive sources that emit gamma rays are widely used to kill cancer cells.

CONNECTION In Chapter 7 (Figure 7.1), we were introduced to gamma rays as the highest energy form of electromagnetic radiation.

CONNECTION Chapter 8 began with a discussion of the changing composition of our atmosphere as one moves from the surface toward outer space. At high altitudes, gases that are molecular at the surface are converted into their component molecules.

20.4 Nuclear Binding Energies

Annihilation reactions between electrons and the positrons produced in hydrogen fusion produce considerable energy, but these reactions are not the principal source of energy from hydrogen fusion. In the 1930s, scientists discovered that the mass of a stable nucleus is always less than the sum of the masses of the individual nucleons that make it up. For example, the nucleus of a helium-4 atom consists of 2 neutrons and 2 protons and has a mass of 6.64465×10^{-24} g. The

Particles of matter and particles of their **antimatter** charge opposites undergo mutual annihilation, producing high-energy gamma (γ) rays.

> The **mass defect (Δm)** is the difference between the mass of a stable nucleus and the masses of the individual nucleons that comprise it.
>
> The stability of a nucleus is proportional to its **binding energy (BE)**, a measure of the energy released when nucleons combine to form a nucleus. It is calculated using the equation $E = (\Delta m)c^2$, where Δm is the mass defect of the nucleus.

mass of a neutron (see Table 20.1) is 1.67494×10^{-24} g, and the mass of a proton is 1.67263×10^{-24} g. Summing the individual masses of the nucleons

$$\text{Mass of 2 neutrons} = 2\,(1.67494 \times 10^{-24}\text{ g})$$
$$\underline{\text{Mass of 2 protons} = 2\,(1.67263 \times 10^{-24}\text{ g})}$$
$$\text{Total mass of nucleons} = 6.69514 \times 10^{-24}\text{ g}$$

The difference between this value (6.69514×10^{-24} g) and the mass of the $^4_2$He nucleus (6.64465×10^{-24} g) is 5.049×10^{-26} g. This difference is the **mass defect (Δm)** of the He nucleus, the difference between the mass of a stable nucleus and the masses of the individual nucleons that comprise it. The energy equivalent to the mass defect represents the **binding energy (BE)** of the helium nucleus. The binding energy is a measure of the energy released when nucleons combine to form a nucleus. It is also a measure of the energy needed to split the nucleus into free nucleons. The greater the binding energy of a nucleus per nucleon, the more stable the nucleus.

Let's calculate the binding energy of helium-4 starting with the mass defect noted above, 5.049×10^{-26} g. To calculate the binding energy in joules (J) from the mass defect, we use the following conversion factor:

$$1\text{ J} = 1\text{ kg (m/s)}^2$$

which indicates that we must convert the units of the mass defect from grams to kilograms to use this modified version of Einstein's equation:

$$\text{BE} = (\Delta m)c^2 \tag{20.8}$$

$$= (5.049 \times 10^{-26}\text{ g}) \times \frac{1\text{ kg}}{1000\text{ g}} \times \frac{1\text{ J}}{\text{kg (m/s)}^2} \times (2.998 \times 10^8\text{ m/s})^2$$

$$= 4.538 \times 10^{-12}\text{ J}$$

This quantity may not seem like much, but remember that we are dealing with only one nucleus. To put this quantity in perspective, consider the enormous energy released during the combustion of hydrogen. The balanced chemical equation for the reaction is

$$\text{H}_2(g) + \tfrac{1}{2}\text{O}_2(g) \rightarrow \text{H}_2\text{O}(\ell) \qquad \Delta H^\circ_{\text{rxn}} = -285.8\text{ kJ/mol}$$

The energy released during the combustion of one molecule of hydrogen is only about 4.7×10^{-19} J. The binding energy in the helium nucleus is nearly 10 million (10^7) times greater! Sample Exercise 20.2 illustrates how the magnitude of the mass defect affects the binding energy of a nucleus.

SAMPLE EXERCISE 20.2 **Calculating Nuclear Binding Energies**

Calculate the binding energy in joules that holds a proton and neutron together in a deuteron ($^2_1$H), given the following masses:

Particle	Mass (g)
Deuteron	3.34370×10^{-24}
Neutron	1.67494×10^{-24}
Proton	1.67263×10^{-24}

COLLECT AND ORGANIZE A deuteron contains 1 proton and 1 neutron. The preceding table lists their masses. The mass defect, Δm, which is the difference in the masses of the nucleus and the masses of the individual particles that comprise the nucleus, can be used to calculate the nuclear binding energy using Equation 20.8:

$$BE = (\Delta m)c^2$$

ANALYZE The value of Δm is the difference between the sum of the masses of the proton and neutron and the mass of the nucleus itself. That value, converted to kilograms, can be used to calculate the binding energy.

SOLVE First we calculate the mass defect:

$$\Delta m = (m_{neutron} + m_{proton}) - m_{deuteron}$$

$$= (1.67494 \times 10^{-24}\ g + 1.67263 \times 10^{-24}\ g) - 3.34370 \times 10^{-24}\ g$$

$$= 0.00387 \times 10^{-24}\ g\ or\ 3.87 \times 10^{-27}\ g$$

Using this value in Equation 20.8,

$$BE = (\Delta m)c^2$$

$$= (3.87 \times 10^{-27}\ \cancel{g}) \times \frac{1\ \cancel{kg}}{1000\ \cancel{g}} \times \frac{1\ J}{\cancel{kg\ (m/s)^2}} \times (2.998 \times 10^8\ \cancel{m/s})^2$$

$$= 3.48 \times 10^{-13}\ J$$

THINK ABOUT IT In the text we calculated the binding energy of a helium nucleus, which is 4.538×10^{-12} J. That value is more than 10 times larger than the binding energy of a deuteron. The helium nucleus has twice as many nucleons, which accounts for part of the difference. Still, on a per-nucleon basis, $^4_2$He has a much larger binding energy and so is much more stable than $^2_1$H.

Practice Exercise Calculate the binding energy of helium-3. The mass of an atom of this isotope is 5.0064×10^{-24} g. *$^3_2$He*

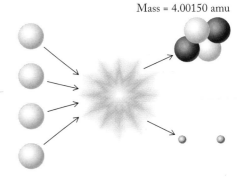

Mass = 4.00150 amu

FIGURE 20.2 The fusion of four hydrogen nuclei (or protons—white particles) produces a helium nucleus that contains two protons and two neutrons (gray particles) and emits two positrons (orange particles). The release of energy that accompanies this reaction fuels the nuclear furnaces of the sun and many billions of other stars.

20.5 Synthesis of Heavy Elements

Hydrogen fusion (Figure 20.2) can fuel the nuclear furnace of a star the size of our sun for billions of years. However, some stars many times the size of the sun are much hotter and consume hydrogen at much higher rates. The good news for these stars is that other exothermic fusion processes take over once their hydrogen fuel is depleted. An important one is the triple-alpha process in which three α particles collide to form carbon-12:

$$3\,^4_2He \rightarrow\ ^{12}_6C$$

Carbon-12 nuclei may then fuse with $^4_2$He, forming $^{16}_8$O, which in turn may fuse with $^4_2$He, forming $^{20}_{10}$Ne, and so on. As we discussed in Section 2.9, a combination of these and other fusion reactions and decay processes can account for the synthesis of elements with atomic numbers up to 26 protons. Building bigger nuclei by fusing together smaller ones reaches a dead end with $^{56}_{26}$Fe, the nucleus of an iron atom that contains 26 protons and 30 neutrons, because nuclear stability, as measured in binding energy per nucleon, reaches a maximum with ^{56}Fe, as shown in Figure 20.3, and then decreases as atomic number increases. Thus, fusing an α particle with ^{56}Fe to make ^{60}Ni would require *adding* energy. Note that we often designate a nuclide by just its mass number and element symbol in the text and in figures, but we include its atomic number in radiochemical equations. In the first case we avoid being redundant (because the identity of an element defines

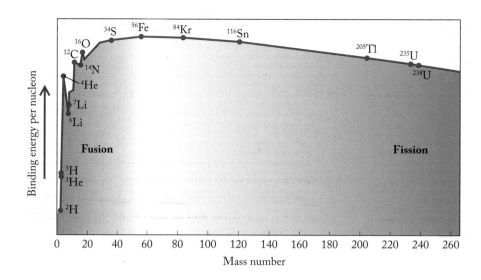

FIGURE 20.3 The stability of a nucleus is indicated by its binding energy. This quantity is best expressed as binding energy per nucleon. For light elements, fusion processes lead to heavier nuclei with greater binding energy; however, fusion of nuclei with atomic numbers greater than 26 produces heavier nuclei with less binding energy (and less stability) and so consumes energy.

its Z value), but in the second case we need Z to identify the missing reactants or products required to balance the equation.

The synthesis of the heavier ($Z > 26$) elements happens when the core temperatures of giant stars reach temperatures at which nuclei disintegrate, releasing free neutrons. These free neutrons can collide and fuse with atomic nuclei to produce isotopes with unusually high ratios of neutrons to protons. For the light elements ($Z \leq 20$), stable nuclei have neutron-to-proton ratios near 1:1. For example, most carbon nuclei have 6 protons and 6 neutrons and a mass number of 12. About 1% of all carbon nuclei have 7 neutrons and a mass number of 13. A tiny percentage have 8 neutrons and a mass number of 14. This isotope is rare because it is radioactive, and so is unstable. It undergoes a type of radioactive decay called β decay, emitting a high-energy electron (β particle):

$$^{14}_{6}\text{C} \rightarrow \text{?} + {}^{0}_{1-}\text{e}$$

To identify the unknown product in this equation, we need to determine the atomic number and mass number required to balance the equation. These values are 7 and 14, respectively. The element with atomic number 7 is nitrogen, so the balanced equation for carbon-14 decay is

$$^{14}_{6}\text{C} \rightarrow {}^{14}_{7}\text{N} + {}^{0}_{1-}\text{e}$$

Note that this β decay process produces a nuclide whose atomic number is one greater than the parent nuclide. Thus, the effect of β decay is to create a nucleus with one more proton and one less neutron. These changes reduce the neutron "richness" of the radioactive nuclide. In the β decay of ^{14}C, the product is the stable and most abundant isotope of nitrogen, ^{14}N.

In the cores of giant stars, neutron capture and β decay combine to produce nuclei of all the elements from iron-56 ($Z = 26$) to bismuth-209, a nuclide with 83 protons and 126 neutrons. Bismuth-209 is the heaviest stable nuclide and the end of the line in the element-building process in giant stars. However, stellar nucleosynthesis is not quite finished. Having reached core temperatures of billions of kelvins through the fusion of lighter elements, a giant star eventually uses up its supply of these elements and begins to cool. As it does, it undergoes a gravity-induced collapse. Its huge mass of elements from $Z = 1$ to 83 is compressed into a tiny space of immense density and temperature before exploding catastrophically in an event called a supernova. These events have been documented for centuries as the sudden appearance of unusually bright stars. They produce enormous densities of free neutrons that collide with nuclei and are absorbed by them so rapidly that neutron-rich, radioactive nuclei do not have a chance to undergo β decay before they absorb even more neutrons. These nuclei become so overloaded with neutrons that additional neutrons smashing into them are no longer absorbed by the nucleus; instead, these nuclei undergo multiple β decay events. In so doing, they form nuclei of all the elements found in nature through Z = 92. Even heavier elements may be formed in this way, but they have relatively short half-lives.

In addition to finishing the job of synthesizing all the elemental building blocks in the universe, a supernova serves as its own element-distribution system, blasting its inventory of nuclides throughout its galaxy. The legacies of supernovas are found in the elemental composition of later-generation stars (such as our sun), the planets orbiting these stars (such as Earth), and in the life forms that inhabit them.

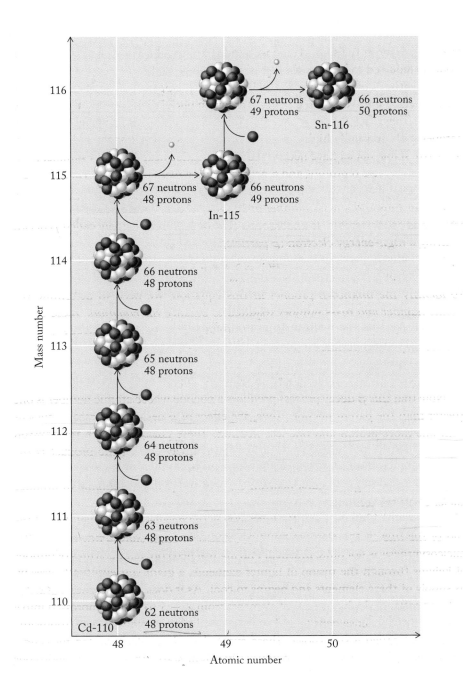

Nuclei with atomic numbers greater than 26 form by the absorption of neutrons followed by β decay, as shown here for cadmium. Cadmium-110 absorbs five neutrons (gray particles) before emitting a β particle (light blue) and forming indium-115. Indium-115 absorbs another neutron before undergoing β decay to tin-116.

SAMPLE EXERCISE 20.3   Predicting the Products of β Decay

Neutron-rich ^{85}Kr undergoes β decay. Which nuclide of which element is produced by this process?

COLLECT AND ORGANIZE The atomic number of krypton is 36. When a neutron-rich nucleus undergoes β decay, it ejects a β particle, $_{-1}^{0}e$, from its nucleus. Thus, the question is asking for the identity of the nuclide that would balance the radiochemical equation

$$_{36}^{85}Kr \rightarrow ? + _{-1}^{0}e$$

ANALYZE The product nuclide will have one more proton, or 37, but the same mass number, 85.

SOLVE The element with atomic number 37 is rubidium; therefore the product is ^{85}Rb:

$$_{36}^{85}Kr \rightarrow _{37}^{85}Rb + _{-1}^{0}e$$

THINK ABOUT IT As in all β decay reactions, the product nuclide is a nucleus of the element that follows the reactant in the periodic table.

Practice Exercise   Identify the product of β decay of iodine-131.

20.6 The Belt of Stability

We have seen how nuclides that are neutron rich undergo spontaneous β decay to produce stable nuclei. We have yet to address what makes a nucleus "neutron rich" or, for that matter, "neutron poor." We have noted that the stable isotopes of lighter elements have neutron-to-proton ratios close to unity. It turns out that this ratio tends to increase as atomic number increases. That trend is illustrated in Figure 20.4, where the green dots represent combinations of neutrons and protons that form stable isotopes. The band of green dots runs diagonally through the graph and is called the **belt of stability**. Note the upward curve of the belt, which tells us that the neutron-to-proton ratios of stable isotopes increase with increasing atomic number, from about 1:1 for the lightest elements to about 1.5:1 for the heaviest.

Isotopes with the combinations of neutrons and protons shown in orange in Figure 20.4 *do* exist, but not indefinitely. All of them are radioactive and undergo decay reactions that ultimately produce stable isotopes. Nuclides with neutron–proton combinations above the belt of stability, such as carbon-14, are neutron rich and tend to undergo β decay. But what about the isotopes below the belt? They apparently have too few neutrons for a given number of protons. These isotopes undergo decay processes that *increase* their neutron-to-proton ratio. We will focus on two of them, positron emission and electron capture.

As its name implies, **positron emission** involves the spontaneous release of a particle that has the same mass as an electron but a positive charge. One of the nuclides that undergoes positron emission is carbon-11. The radiochemical equation for the process is

$$_{6}^{11}C \rightarrow _{5}^{11}B + _{1}^{0}e \tag{20.9}$$

We have seen how positrons rapidly undergo annihilation reactions with electrons, producing two γ rays. The boron-11 produced by the reaction in Equation 20.9 is stable. In fact, 80.2% of all boron nuclei are boron-11; the other 19.8% are boron-10.

The **belt of stability** is the region on a graph of number-of-neutrons versus number-of-protons that includes all stable nuclei.

Positron emission is the spontaneous emission of a positron from a nucleus.

In **electron capture** a neutron-poor nucleus draws in one of its surrounding electrons, leaving the nucleus with one more neutron and one less proton.

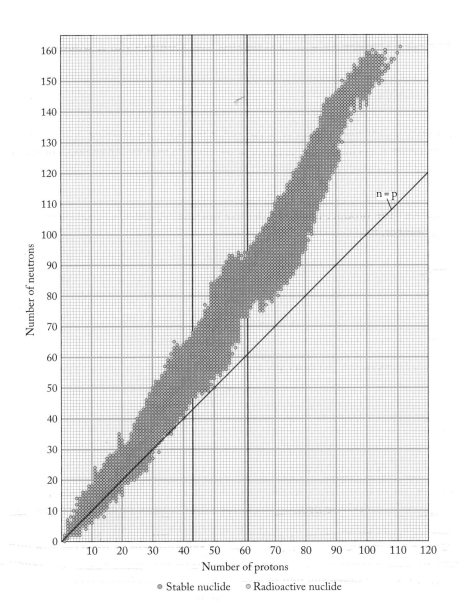

Number of neutrons (y-axis)
Number of protons (x-axis)

n = p

● Stable nuclide   ○ Radioactive nuclide

FIGURE 20.4 In this depiction of the belt of stability, the green dots represent stable combinations of protons and neutrons. The orange dots represent known radioactive nuclides. Note that there are no stable nuclides for $Z = 43$ (technetium) and $Z = 61$ (promethium), as indicated by the two vertical red lines. These elements are the only ones among the first 83 not found in nature.

A carbon-11 atom can also increase its neutron-to-proton ratio by drawing one of its six electrons into its nucleus. The effect of this **electron capture** process is the same as positron emission: the number of protons *decreases* by one and the number of neutrons *increases* by one. Other radioactive nuclides with neutron–proton combinations that fall below the belt of stability tend to undergo positron emission or electron capture. Table 20.2 lists decay pathways for the isotopes of carbon. Note that the radioactive isotopes with mass numbers above the average atomic mass (12.011) are neutron rich and undergo β decay; those whose mass numbers are below the average atomic mass are neutron poor and undergo positron emission or electron capture. Sample Exercise 20.4 provides practice in predicting the decay pathways of radioactive isotopes.

TABLE 20.2 Isotopes of Carbon and Their Radioactive Decay Products

Name	Symbol	Mass (amu)	Mode(s) of Decay	Half-Life	Natural Abundance (%)
Carbon-10	$^{10}_{6}\text{C}$		Positron emission	19.45 s	
Carbon-11	$^{11}_{6}\text{C}$		Positron emission, EC[a]	20.3 min	
Carbon-12	$^{12}_{6}\text{C}$	12.00000	(Stable)		98.89
Carbon-13	$^{13}_{6}\text{C}$	13.00335	(Stable)		1.11
Carbon-14	$^{14}_{6}\text{C}$		β decay	5730 y	
Carbon-15	$^{15}_{6}\text{C}$		β decay	2.4 s	
Carbon-16	$^{16}_{6}\text{C}$		β decay	0.74 s	

[a]EC, electron capture.

SAMPLE EXERCISE 20.4   **Predicting the Mode of Radioactive Decay**

Predict the mode of radioactive decay of $^{30}_{15}P$ and identify the nuclide produced by the decay process.

COLLECT AND ORGANIZE The mode of decay of a radioactive isotope can be predicted based on whether it is neutron rich or neutron poor, that is, whether its combination of neutrons and protons puts it above or below the belt of stability in Figure 20.4. Phosphorus-30 has 15 neutrons and 15 protons per nucleus.

ANALYZE According to Figure 20.4, phosphorus ($Z = 15$) has only one stable isotope. It has 16 neutrons, which gives it a mass number of 31. Phosphorus-30 is represented by the orange dot directly beneath the ^{31}P green dot, which means that it is (1) radioactive and (2) neutron poor.

SOLVE Neutron-poor isotopes, such as $^{30}_{15}P$, decay by either positron emission or electron capture to give the nuclide $^{30}_{14}Si$:

$$^{30}_{15}P \rightarrow \, ^{30}_{14}Si + \, ^{0}_{1}e \qquad \text{Positron emission}$$

$$^{30}_{15}P + \, ^{0}_{-1}e \rightarrow \, ^{30}_{14}Si \qquad \text{Electron capture}$$

THINK ABOUT IT Our conclusion that ^{30}P is neutron poor is also supported by the fact that the average atomic mass of phosphorous is very close to 31. This value tells us that stable nuclei of phosphorus are likely to have more neutrons in their nuclei than ^{30}P.

Practice Exercise What is the mode of radioactive decay of ^{28}Al? Identify the nuclide produced by the decay.

All nuclides with more than 83 protons ($Z > 83$) are radioactive. Because there is no stable reference point, it is hard to say whether one of these nuclides is neutron rich or neutron poor. The heaviest nuclides tend to undergo radioactive decay by either α decay, emitting an α particle ($^{4}_{2}He$) and producing nuclides with two fewer protons and two fewer neutrons, or β decay. Figure 20.5 summarizes

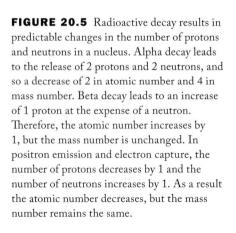

FIGURE 20.5 Radioactive decay results in predictable changes in the number of protons and neutrons in a nucleus. Alpha decay leads to the release of 2 protons and 2 neutrons, and so a decrease of 2 in atomic number and 4 in mass number. Beta decay leads to an increase of 1 proton at the expense of a neutron. Therefore, the atomic number increases by 1, but the mass number is unchanged. In positron emission and electron capture, the number of protons decreases by 1 and the number of neutrons increases by 1. As a result the atomic number decreases, but the mass number remains the same.

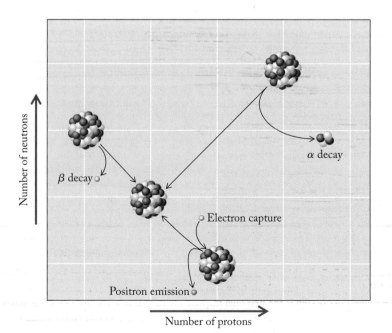

the changes in atomic number and mass number caused by the various modes of decay. For the heaviest radioactive isotopes, one radioactive decay process often leads to another and in some cases to series of decay reactions. Consider, for example, the decay series that begins with the most abundant isotope of uranium, ^{238}U. It undergoes α decay producing thorium-234 (^{234}Th). This process can be written in equation form as follows:

$$^{238}_{92}\text{U} \rightarrow {}^{234}_{90}\text{Th} + {}^{4}_{2}\text{He}$$

Thorium has an atomic number above 83 and so has no stable isotopes. Thorium-234 undergoes two β decay steps, producing ^{234}U as shown in Figure 20.6. In a series of subsequent α decay steps, ^{234}U turns into thorium-230 (^{230}Th), radium-226 (^{226}Ra), radon-222 (^{222}Rn), polonium-218 (^{218}Po), and finally lead-214 (^{214}Pb). Although some isotopes of lead ($Z = 82$) are stable, ^{214}Pb is not one of them. Therefore, the radioactive decay series continues as shown in Figure 20.6 and does not end until a stable nuclide, ^{206}Pb, is produced.

▶❚❚ **CHEMTOUR** Radioactive Decay Modes

▶❚❚ **CHEMTOUR** Balancing Nuclear Reactions

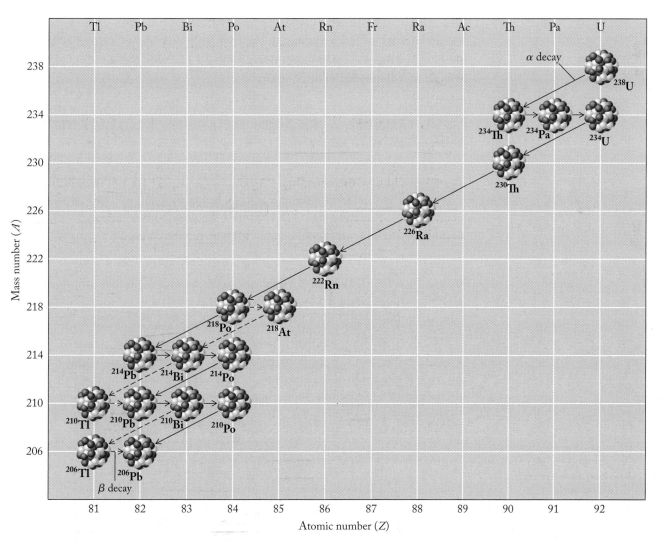

FIGURE 20.6 In this representation of the ^{238}U radioactive decay series, the α decay steps are designated by long arrows and β decay by short ones. Note how β decay produces no change in atomic mass but increases atomic number by 1. By far the slowest decay step is the first one; the half-life of ^{238}U is 4.5×10^9 years. The dashed arrows represent alternative decay pathways that occur less than 1% as frequently as the decay processes represented by the solid arrows.

20.7 Artificial Nuclides

We have yet to account for those elements in the periodic table with atomic numbers greater than 92. We are able to study and use these elements because they have been synthesized in laboratories. Nuclear scientists have been synthesizing isotopes since 1919. In that year, Ernest Rutherford (1871–1937) of Cambridge University reported the nuclear synthesis of oxygen-17 by bombarding nitrogen-14 with α particles:

$$\ce{^{14}_{7}N} + \ce{^{4}_{2}He} \rightarrow \ce{^{17}_{8}O} + \ce{^{1}_{1}H}$$

Bombardment of nuclei by α particles became a popular approach for transmuting elements in the 1920s and 1930s. In 1933, Irène and Frédéric Joliot-Curie synthesized the first radionuclide not found in nature, $\ce{^{30}_{15}P}$, by bombarding $\ce{^{27}Al}$ with α particles. Let's write a radiochemical equation for this process, starting with the known reactants and product:

$$\ce{^{27}_{13}Al} + \ce{^{4}_{2}He} \rightarrow \ce{^{30}_{15}P}$$

A check of the superscripts and subscripts of the terms in this equation indicates that it is balanced in terms of atomic number, but not in terms of atomic mass. It would be if we added a particle to the right side that had no charge and a mass number of one. That is the description of a neutron. Adding a neutron to the right side yields a balanced radiochemical equation:

$$\ce{^{27}_{13}Al} + \ce{^{4}_{2}He} \rightarrow \ce{^{30}_{15}P} + \ce{^{1}_{0}n}$$

The next sample exercise provides additional practice in completing and balancing radiochemical equations.

SAMPLE EXERCISE 20.5 | **Completing and Balancing Radiochemical Equations**

Complete and balance the following radiochemical equation:

$$\ce{^{27}_{13}Al} + \ce{^{1}_{0}n} \rightarrow\ ? + \ce{^{0}_{-1}e}$$

COLLECT AND ORGANIZE In balanced radiochemical equations, the sum of the electrical charges denoted by the subscripts of the particles on the left of the reaction arrow must equal the sum of the electrical charges on the right. The same is true for the masses denoted by superscripts.

ANALYZE The sums of the subscripts (charges) and superscripts (mass numbers) on the left hand side of the equation are 13 and 28, respectively. The only particle on the right has zero mass and a charge of $1-$.

SOLVE The mystery particle must have a mass number of 28 and a charge of 14. The element with 14 protons in its nuclei is silicon. Therefore, the particle is silicon-28, and the complete radiochemical equation is

$$\ce{^{27}_{13}Al} + \ce{^{1}_{0}n} \rightarrow \ce{^{28}_{14}Si} + \ce{^{0}_{-1}e}$$

THINK ABOUT IT When a stable nuclide captures a neutron, there is a chance that it will be so neutron rich that it will be radioactive and undergo β decay. $\ce{^{28}Al}$ is such a nuclide.

Practice Exercise Naturally occurring chlorine contains a mixture of chlorine isotopes, $\ce{^{35}Cl}$ and $\ce{^{37}Cl}$. Write a balanced radiochemical equation describing the nuclear reaction that occurs when $\ce{^{37}Cl}$ captures a neutron.

$$\ce{^{37}_{17}Cl} + \ce{^{1}_{0}n} \rightarrow \ce{^{38}_{18}Ar} + \ce{^{0}_{-1}e}$$

To bombard a nucleus with an α particle or any positive ion, the electrostatic repulsion between a positively charged particle and a positively charged target nucleus must be overcome. Overcoming this repulsion requires that the bombarding particle be shot toward the target nucleus at a very high velocity. Facilities for accelerating particles to such velocities are large and expensive to build and operate. The **linear accelerator** shown in Figure 20.7(a) uses an array of alternating electrical fields to accelerate positively charged ions and nuclei in a straight line. Other facilities use combinations of magnetic and electrical fields to accelerate positive particles in spiral pathways until they exit the accelerator, called a **cyclotron**, before smashing into target nuclei, as shown in Figure 20.7(b). Since 1933, more than a thousand artificial radionuclides have been synthesized using cyclotrons and linear accelerators.

In 1940 the first artificial elements with atomic numbers 93 and 94 were produced at the University of California, Berkeley by bombarding uranium-238 with neutrons. As noted in previous sections, transmutation by neutron bombardment is made easier than α bombardment by the lack of electrostatic repulsion between

> A **linear accelerator** is a device in which electrostatic forces make charged particles move faster as they pass between metal tubes along a straight flight path.
>
> A **cyclotron** is a circular particle accelerator using the combined action of a constant magnetic field with an oscillating electrostatic field to make particles move faster in an increasingly larger spiral path.

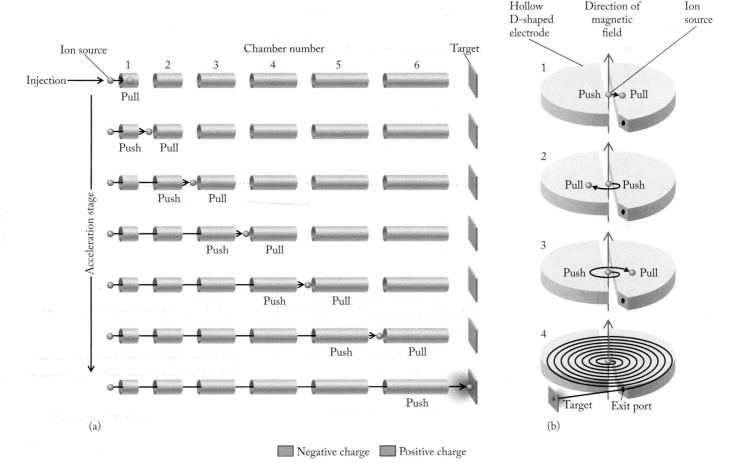

(a)   (b)

Negative charge   Positive charge

FIGURE 20.7 Two devices for accelerating nuclear particles: (a) In a linear accelerator, electrical fields of alternating charge are used to accelerate positive particles. Particles are drawn from the source into negatively charged chamber 1. As particles exit this chamber, the electrical charge on it is switched from negative to positive and the charge on chamber 2 is switched from positive to negative. This accelerates positive particles away from chamber 1 and into chamber 2. As particles leave chamber 2, the charges of the chambers are switched back to their original values, and the acceleration process continues through this stage and several more until the particles exit the last chamber and smash into the target. (b) Particles are introduced near the center of a cyclotron between two D-shaped hollow half-cylinders. The electrical charge on each half-cylinder is opposite that of the other, so positive particles are (1) repelled by the positively charged half-cylinder and attracted to the other. As they enter the negatively charged half-cylinder, the charges are switched and (2) the particles turn back toward the first half-cylinder. Magnets above and below the half-cylinders (3) bend the path taken by the particles in traveling back and forth between the two. As a result, the particles (4) spiral outward as they pick up speed until they reach the exit and hit the target.

neutrons and the target nuclei. Neutron capture by uranium-238 initiates a series of two β decay events leading to the formation of plutonium-239:

$$^{238}_{92}U + ^1_0n \rightarrow ^{239}_{94}Pu + 2^{\ 0}_{-1}e$$

The leader of the research team at Berkeley that discovered plutonium was American chemist Glenn T. Seaborg (1912–1999). Between 1944 and 1961, Seaborg and his colleagues synthesized elements through $Z = 103$ by using combinations of neutron and α-particle bombardment of actinide target nuclei. These methods are of limited use in the synthesis of even heavier elements because nuclides heavier than $^{249}_{98}Cf$ have extremely short half-lives and so are not useful target materials. However, using the nuclei of light elements such as carbon, nitrogen, or oxygen in place of α particles, scientists were able to synthesize heavier elements, including rutherfordium ($Z = 104$), dubnium ($Z = 105$), and, in 1974, seaborgium ($Z = 106$), as described by the following radiochemical equations:

$$^{249}_{98}Cf + ^{12}_{6}C \rightarrow ^{257}_{104}Rf + 4^1_0n$$

$$^{249}_{98}Cf + ^{15}_{7}N \rightarrow ^{260}_{105}Db + 4^1_0n$$

$$^{249}_{98}Cf + ^{18}_{8}O \rightarrow ^{263}_{106}Sg + 4^1_0n$$

In recent years, scientists have reported the synthesis of nuclides that have as many as 112 protons by bombarding targets of ^{209}Bi or ^{208}Pb with nuclei of medium-weight elements. For example, in January 1999 an atom with 114 protons and 175 neutrons was synthesized and lasted for 30 seconds before undergoing a series of α decay steps that yielded isotopes of elements numbered 112, 110, and 108. Some of these "superheavy" elements are listed in Table 20.3.

Why bother to make such short-lived elements? Their mere existence, no matter for how brief a time, can be a source of insight into the nature of nuclear structure and the competition between the force that holds nucleons together and the electrostatic repulsion that drives them apart. Superheavy elements are pieces of a puzzle that someday may tell us whether there is a limit to the size of atoms. One of the driving forces for preparing and studying superheavy elements is to test the hypothesis of *magic numbers* for neutrons and protons in a nucleus. It turns out that nuclei with even numbers of neutrons and protons are more stable than nuclei with odd numbers of neutrons and protons. Isotopes that have 2, 8, 20, 28, 50, 82, or 126 neutrons or protons exhibit exceptional stability. These magic numbers are the nuclear equivalent of filled s and p subshells for the electrons in atoms described in Chapter 7. Lead-208, $^{208}_{82}Pb$ is an example of a nuclide that has a magic number of both protons and neutrons. The nuclides listed in Table 20.3 show the recent progress by international teams of scientists in synthesizing a nucleus with the magic number of 126 protons.

TABLE 20.3 Some of the Isotopes of Nine Superheavy Elements Synthesized by Colliding Heavy Nuclei

New Element[a]	Bombarding Ion	Target	Date Created
$^{262}_{107}Bh$	^{54}Cr	^{209}Bi	February 1981
$^{265}_{108}Hs$	^{58}Fe	^{208}Pb	March 1984
$^{266}_{109}Mt$	^{58}Fe	^{209}Bi	September 1982
$^{269}_{110}Ds$	^{62}Ni	^{208}Pb	November 1994
$^{272}_{111}Rg$	^{64}Ni	^{209}Bi	December 1994
$^{277}_{112}Uub$	^{69}Zn	^{208}Pb	February 1996
$^{283}_{113}Uut^b$			August 2003
$^{288}_{114}Uuq$	^{48}Ca	^{244}Pu	January 1999
$^{287}_{115}Uup$	^{48}Ca	^{243}Ca	August 2003
$^{292}_{116}Uuh$	^{48}Ca	^{248}Cm	December 2000
$^{294}_{118}Uuo$	^{48}Ca	^{249}Cf	April 2002

[a]No names or symbols have been adopted for elements 112 through 118. The names for elements 107 through 111—bohrium, hassium, meitnerium, darmstadtium, and roentgenium, respectively—are based on the recommendations of the scientists at Gesellschaft für Schwerionenforschung (GSI) in Darmstadt, Germany, who created them.

[b]Element Uut was formed by the α decay of element Uup.

20.8 Nuclear Fission

There is another process by which some isotopes of uranium and other heavy nuclides disintegrate. When an atom of ^{235}U absorbs a neutron, the product, ^{236}U, is unstable and disintegrates into pairs of lighter nuclei in a process called **nuclear fission**. There are several ^{235}U fission reactions; here are three of the more common ones:

$$^{235}_{92}U + {}^{1}_{0}n \rightarrow {}^{141}_{56}Ba + {}^{92}_{36}Kr + 3\,{}^{1}_{0}n$$

$$^{235}_{92}U + {}^{1}_{0}n \rightarrow {}^{137}_{52}Te + {}^{97}_{40}Zr + 2\,{}^{1}_{0}n$$

$$^{235}_{92}U + {}^{1}_{0}n \rightarrow {}^{138}_{55}Cs + {}^{96}_{37}Rb + 2\,{}^{1}_{0}n$$

In all these reactions, the sums of the masses of the products are less than the sums of the masses of the reactants. This decrease in mass is released as energy in accordance with Einstein's equation ($E = mc^2$). Two of these fission reactions also produce additional neutrons, which can smash into other ^{235}U nuclei and initiate more fission events. This process is called a **chain reaction**. Figure 20.8 shows the first stages of a chain reaction starting with ^{235}U. It proceeds as long as there are enough ^{235}U nuclei present to absorb the neutrons being produced. On average, at least one neutron from each fission event must cause the fission of another nucleus for the chain reaction to be self-sustaining. The amount of fissionable material needed to sustain a chain reaction is called the **critical mass**. For uranium-235, the critical mass is about 1 kg of the pure isotope.

Nuclear fission is the process by which the nucleus of a heavy element splits into two lighter nuclei. The process is usually accompanied by the release of one or more neutrons and energy.

A **chain reaction** is a self-sustaining series of reactions in which the products of the reaction—for example, neutrons released when an atom is split—react further with other materials and continue the reaction.

The **critical mass** is the minimum amount of fissionable material needed to sustain a nuclear chain reaction.

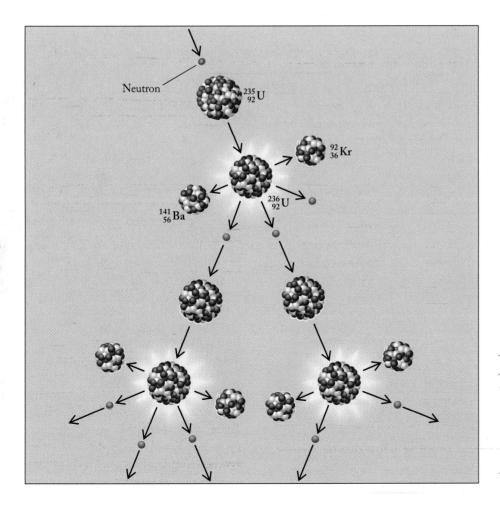

FIGURE 20.8 Each fission event in the chain reaction of ^{235}U begins when the nucleus captures a neutron, forming an unstable nucleus of ^{236}U. This nucleus splits apart (fissions) in one of several ways. In the first process shown here, ^{236}U splits into krypton-92 (^{92}Kr), barium-141 (^{141}Ba), and three neutrons. If, on average, at least one of the three neutrons from each fission event causes the fission of another nucleus, then the process is sustained in a chain reaction.

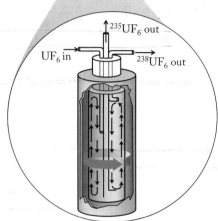

(a)

(b)

(c)

FIGURE 20.9 Preparing uranium fuel. (a) The uranium fuel for nuclear reactors comes from a brownish-black ore known as pitchblende. (b) The ore is ground up and extracted with strong acid. The uranium compounds (mostly U_3O_8) obtained from the acid extract are called yellow cake. (c) Uranium oxides are converted to volatile UF_6, which is centrifuged at very high speed to separate $^{235}UF_6$ from $^{238}UF_6$. The less dense and less abundant $^{235}UF_6$ is enriched near the center of the rotating cylinder of gases while the much more abundant $^{238}UF_6$ is more concentrated toward the outer wall. UF_6 circulates in the rotor as indicated by the small black arrows.

Uranium-235 is the most abundant fissionable isotope, but it makes up only 0.72% of the uranium in the principal uranium ore, called pitchblende (Figure 20.9). The uranium used for fuel in nuclear reactors must be at least 3% to 4% ^{235}U. An even higher percentage (>7%) is needed for nuclear weapons. The most common enrichment method involves extracting and isolating the uranium from the ore in a process that produces a substance called yellow cake, which is mostly U_3O_8. Then the oxide is converted to uranium hexafluoride, UF_6, which is a volatile solid. Samples of UF_6 gas containing a mixture of uranium isotopes can be separated because of the slightly greater density of $^{238}UF_6$ compared to $^{235}UF_6$. Elaborate centrifuge systems are used to exploit this difference in densities and speed up the separation process, as shown in Figure 20.9.

Harnessing the energy released by nuclear fission to generate electricity began in the middle of the 20th century. In a typical nuclear power plant (Figure 20.10), fuel rods contain 3% to 4% ^{235}U and less than the critical mass that would produce a nuclear explosion. Still, the rate of the chain reaction and the production of nuclear energy must be regulated. Rods made of boron or cadmium are interspersed with the fuel rods to control the rate of the fission by absorbing some of the neutrons produced during fission. Pressurized water flows around the fuel and control rods, removing the heat created during fission and transferring it to a steam generator. The water also acts as a moderator, slowing down the neutrons released by the fission of ^{235}U and allowing for their more efficient capture by other atoms of ^{235}U.

In 1952, the first of a new kind of nuclear reactor, called a **breeder reactor,** was used to make plutonium-239 (^{239}Pu) from uranium-238, while also producing energy to make electricity. The fuel in a breeder reactor is a mixture of plutonium-239 and uranium-238. As ^{239}Pu undergoes fission, some of the neu-

A breeder reactor is a nuclear reactor in which fissionable material is produced during normal reactor operation.

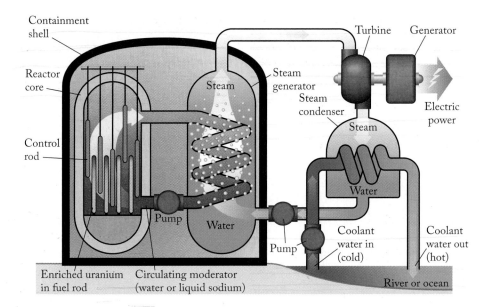

FIGURE 20.10 Harnessing nuclear energy. This diagram shows the main systems of a pressurized water-cooled nuclear power plant, which uses fuel rods containing uranium that has been enriched to about 4% ^{235}U so that a fission chain reaction can be sustained. Energy production from ^{235}U fission is regulated with control rods and a moderator. Water under high pressure flows around the fuel and control rods, removing the heat produced by the fission reaction and transferring it to a steam generator. The water also acts as a moderator, slowing down the neutrons released by the fission of ^{235}U so that they can be more efficiently captured by other atoms. The steam generated by the heat of fission drives a turbine that generates electricity.

trons it produces sustain the fission chain reaction; others convert ^{238}U into more plutonium fuel:

$$^{238}_{92}\text{U} + ^{1}_{0}\text{n} \rightarrow ^{239}_{92}\text{U} + \gamma \rightarrow ^{239}_{94}\text{Pu} + 2^{0}_{-1}\text{e}$$

In less than 10 years of operation, a breeder reactor can make enough ^{239}Pu to refuel itself *and* another reactor as well. Unfortunately, plutonium-239 is one of the most toxic substances known; it is also carcinogenic (it causes cancer). Only about half a kilogram of plutonium-239 would be needed to make an atomic bomb, and it has a long half-life: 2.4×10^4 years. Understandably, extreme caution and tight security surround the handling of plutonium fuel and the transportation and storage of nuclear wastes containing even small amounts of plutonium. Health and safety matters related to reactor operation and spent fuel disposal are the principal reasons why there are no breeder reactors in nuclear power stations in the United States, although they have been built in at least seven other countries. We explore some of the health concerns surrounding radioactivity in Section 20.10.

20.9 Measuring Radioactivity

The French scientist Henri Becquerel (1852–1908) discovered radioactivity in 1896, when he observed that uranium and other substances produce an invisible radiation that fogs photographic film. Photographic film is still widely used to detect and record radioactivity. People who work with radioactive materials must wear badges called radiation dosimeters. The badges contain photographic film that darkens with exposure to radiation: the greater the exposure, the darker the film. Other radiation detection systems are based on the absorption of energy released during radioactive decay by materials called *phosphors*. These materials release the absorbed energy as photons of visible light, whose intensity is a measure of the amount of radiation initially emitted. **Scintillation counters** are used to measure the light emitted when radiation strikes a phosphor.

Radioactivity also can be detected with a device called a Geiger counter, named after German physicist Hans Geiger (1882–1945). A **Geiger counter** detects the common products emitted by radioactivity—α particles, β particles, and γ rays—

Scintillation counters are instruments that determine the level of radioactivity in samples by measuring the intensity of light emitted by phosphors in contact with the samples.

A **Geiger counter** is a portable device for determining nuclear radiation levels by measuring how much they ionize the gas in a sealed detector.

FIGURE 20.11 In a Geiger counter (like the one pictured at the upper right), a particle produced by radioactive decay, such as an α or a β particle, passes through a thin window usually made of beryllium or a plastic film. Inside the tube, the particle collides with atoms of argon gas and ionizes them. The resulting positively charged argon ions migrate toward the negatively charged tube housing and the electrons migrate toward a positive electrode, creating a pulse of current through the tube. The current pulses are amplified and recorded with the use of a meter and a small speaker that produces an audible "click" for each pulse.

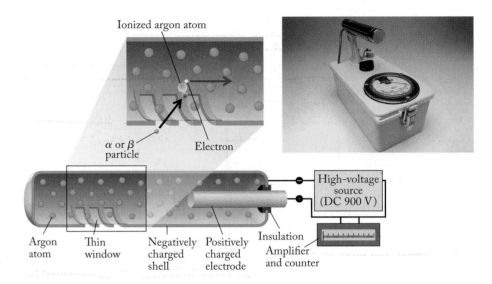

Ionized argon atom

α or β particle   Electron

Argon atom   Thin window   Negatively charged shell   Positively charged electrode   Insulation   Amplifier and counter

High-voltage source (DC 900 V)

on the basis of their abilities to ionize atoms. A Geiger counter (Figure 20.11) consists of a sealed metal cylinder filled with gas, usually argon, and a positively charged electrode at its center. At one end of the cylinder is a very thin window that allows α particles, β particles, and γ rays to enter. Once inside the cylinder, these high-energy particles ionize argon atoms into Ar^+ ions and free electrons. If an electrical voltage is applied between the cylinder and the central electrode, free electrons rapidly migrate toward the positive electrode and argon ions migrate toward the negatively charged shell. This rapid migration of ions produces a pulse of electrical current whenever radiation enters the cylinder. The current is amplified and read out to a meter and a microphone that makes a clicking sound.

One way to express the levels of radioactivity in a sample is by the number of radioactive decay events that occur per unit time (Table 20.4). The SI unit for radioactivity is the **becquerel (Bq)**, named in honor of Henri Becquerel (1852–1908). It equals one decay event per second. An older, much bigger, and more commonly used unit of radioactivity is the **curie (Ci)**, named in honor of Marie and Pierre Curie. One curie equals 3.70×10^{10} Bq. Both the becquerel and the curie measure the *rate* at which a radioactive substance undergoes decay. These rates resemble the rates of the first-order chemical reactions we discussed in Chapter 14. Recall that the rate of a first-order reaction is related to the concentration of reactant "A" according to the equation:

$$\text{Rate} = k\,[A]$$

where k is the rate constant of the reaction which is related to the half-life of the reaction by Equation 14.20:

$$t_{1/2} = \frac{0.693}{k}$$

The **becquerel (Bq)** is a unit used to express the level of radioactivity from a sample. One Bq equals one disintegration per second.

The **curie (Ci)** is a unit used to express the level of radioactivity from a sample. One Ci equals 3.7×10^{10} Bq.

The same mathematical relationships apply to radioactive decay processes except that we refer to the *quantity* of a radioactive isotope in a sample instead of its *concentration*

$$\text{Rate of decay} = kA \tag{20.10}$$

where A is the number of atoms of the radionuclide. Because decay rates are expressed as the number of disintegration events per second, the units of the decay rate constant k are disintegrations per atom per second [disintegrations/(atom·s)].

TABLE 20.4 Units for Expressing Quantities of Ionizing Radiation

Unit	Parameter	Description
Becquerel (Bq)	Level of radioactivity	1 disintegration/s
Curie (Ci)	Level of radioactivity	3.70×10^{10} disintegrations/s
Gray (Gy)	Ionizing energy absorbed	1 Gy = 1 J/kg of tissue mass
Sievert (Sv)	Amount of tissue damage	1 Sv = 1 Gy $\times$ RBE[a]

[a]RBE, relative biological effectiveness.

$Sv = Gy \cdot RBE$

In practice, the rates of decay of radioactive samples used in scientific studies and in medical imaging (see Section 20.11) lie in between those measured by becquerels and curies. They are often less than a millicurie (10^{-3} Ci) and typically in the microcurie (10^{-6} Ci) to nanocurie (10^{-9} Ci) range. At the other extreme, when a nuclear reactor at the Chernobyl power station in Ukraine exploded in 1986, at least 20 million curies of radioactivity were released into the atmosphere, causing an increase in background radiation levels throughout the world. Keep in mind that we often express the quantity of a radionuclide in a sample not in terms of its mass but rather in terms of its activity (decay rate) because the latter value is much more important in determining how the substance is used and how we handle it safely. This substitution is valid because the quantity of a radionuclide and its activity are directly proportional to one another.

SAMPLE EXERCISE 20.6 **Calculating the Activity of a Radionuclide from Its Half-Life**

Radium-223 undergoes β decay with a half-life of 11.4 days. What is the level of radioactivity of a sample that contains 1.00 g of ^{223}Ra? Express your answer in becquerels and in curies.

COLLECT AND ORGANIZE We are given the half-life and quantity of a radioactive substance and are to determine its level of radioactivity, that is, its rate of decay. The rate constant k for the first-order decay process is related to the value of its half-life by Equation 14.20:

$$t_{1/2} = \frac{0.693}{k}$$

The rate of decay is the product of the rate constant k times the quantity of radionuclide (A):

$$\text{Rate of decay} = kA$$

Radioactivity can be expressed either as becquerels where 1 Bq = 1 disintegration/s or in curies where 1 Ci = 3.7×10^{10} disintegrations/s.

ANALYZE We can use Equation 14.20 to calculate the value of the rate constant k from the half-life (11.4 days), but we need to convert that time interval into seconds to match the units on the rate of decay. To calculate the rate of decay we need to determine the number of atoms in 1.00 g of radium.

SOLVE The half-life of $^{223}_{88}Ra$ is 11.4 days or

$$11.4 \text{ days} \times \frac{24 \text{ hr}}{\text{days}} \times \frac{60 \text{ min}}{\text{hr}} \times \frac{60 \text{ s}}{\text{min}} = 9.85 \times 10^5 \text{ s}$$

We obtain k by rearranging Equation 14.20 and substituting $t_{1/2} = 9.85 \times 10^5$ s:

$$k = \frac{0.693}{9.85 \times 10^5 \text{ s}}$$

$$= 7.04 \times 10^{-7} \text{ s}^{-1} = 7.04 \times 10^{-7} \text{ disintegrations}/(\text{s} \cdot \text{atom})$$

The total number of atoms of ^{223}Ra is

$$1.00 \text{ g } ^{223}Ra \times \frac{1 \text{ mol}}{223 \text{ g}} \times \frac{6.02 \times 10^{23} \text{ atoms}}{\text{mol}} = 2.70 \times 10^{21} \text{ atoms}$$

Multiplying the number of disintegrations/(s · atom) by the number of atoms gives us the total disintegrations per second from the sample:

$$\frac{7.04 \times 10^{-7} \text{ disintegrations}}{\text{s} \cdot \text{atom}} \times 2.70 \times 10^{21} \text{ atoms} = 1.9 \times 10^{15} \text{ disintegrations/s}$$

Because 1 Bq = 1 disintegration/s, the activity of 1.00 g of ^{223}Ra is 1.9×10^{15} Bq.

Converting this activity to Ci:

$$1.9 \times 10^{15} \text{ disintegrations/s} \times \frac{1 \text{ Ci}}{3.7 \times 10^{10} \text{ disintegrations/s}} = 5.1 \times 10^4 \text{ Ci}$$

THINK ABOUT IT This result is an enormous level of radiation, given that laboratory experiments with radiotracers use microcurie and nanocurie quantities of nuclides.

Practice Exercise Determine the level of radioactivity in 0.100 g of ^{226}Ra ($t_{1/2} = 1.6 \times 10^3$ yr). Express your answer in becquerels and in millicuries.

20.10 Biological Effects of Radioactivity

The γ rays and many of the α and β particles produced by nuclear reactions have more than enough energy to tear chemical bonds apart, producing free electrons and positive ions. Therefore, these products of radioactive decay are examples of **ionizing radiation**. Other examples are X-rays and even short-wavelength ultraviolet rays. The ionization of atoms (and molecules) in living tissue results in tissue damage such as burns and molecular changes that can lead to radiation sickness, cancer, and birth defects. The scientists who first worked with radioactive materials were not aware of these hazards and some of them suffered for it. Marie Curie died of leukemia caused by her many years of exposure to radiation from radium, polonium, and other radionuclides. The same disease claimed her daughter Irène Joliot-Curie, who had continued the research program started by her parents.

Actually, the term *ionizing radiation* is limited in medicine to photons and particles with energy sufficient to remove an electron from water (1216 kJ/mol).

$$H_2O(\ell) \xrightarrow{1216 \text{ kJ/mol}} H_2O^+(aq) + e^-$$

The logic behind this definition is that the human body is composed largely of water. Therefore water molecules are the most abundant ionizable targets in an

Ionizing radiation comprises the high-energy products of radioactive decay that can ionize substances. *α, β, γ, X-ray, UV.*

organism exposed to nuclear radiation. The cation derived from ionizing water, H_2O^+, reacts with another water molecule to form H_3O^+ and an odd electron species called a hydroxyl free radical ($\cdot OH$):

$$H_2O^+(aq) + H_2O(\ell) \rightarrow H_3O^+(aq) + \cdot OH(aq)$$

The term *free radical* is applied to any molecule or molecular fragment with at least one unpaired electron (see Chapter 8). The rapid reactions of free radicals with biomolecules often threaten the well-being of the cell and the host organism. Radiation-induced alterations to the biochemical machinery that controls cell growth are most likely to occur in those tissues in which cell-division rates are normally rapid. Such tissues include the bone marrow, where billions of white blood cells (called leukocytes) are produced each day to fortify the body's immune system. Molecular damage to bone marrow can lead to leukemia, an uncontrolled production of leukocytes that do not fully mature and so cannot destroy invading pathogens. Eventually, leukemic cells spread throughout the body (metastasize), invading and crowding out the cells of other tissues. Ionizing radiation can also cause molecular alterations in genes and chromosomes in sperm and egg cells that may cause birth defects in offspring.

The **gray (Gy)** is the SI unit used to express the amount of radiation absorbed by matter. One gray equals the absorption of 1 J by 1 kg of matter.

The **relative biological effectiveness (RBE)** of radiation is a factor that accounts for the differences in physical damage caused by different types of radiation.

The **sievert (Sv)** is the SI unit used to express the amount of biological damage caused by ionizing radiation.

Radiation Dosage

The biological impacts of ionizing radiation depend on the exposure of tissue to a radiation source. Exposure, or *physical dose*, is expressed as the quantity of ionizing radiation *absorbed* by a unit mass of matter. The SI unit of physical dose is the **gray (Gy)**. One gray is equal to the absorption of 1 J/kg.

Although grays express the amount of ionizing radiation to which an organism is exposed, they do not indicate the amount of tissue damage caused by that exposure. Different products of nuclear reactions affect living tissue differently. A gray of γ rays produces about the same amount of tissue damage as a gray of β particles; however, a gray of α particles, which move about 10 times slower than β particles but have nearly 10^4 times the mass, causes 20 times as much tissue damage as γ rays or β particles. Neutrons do more damage than β particles but less than α particles. To account for these differences, values of **relative biological effectiveness (RBE)** have been established for the various forms of ionizing radiation. When dosage in grays is multiplied by the RBE factor for the form of radiation, the product has units of **sieverts (Sv)**. Table 20.4 (p. 1001) summarizes the various units used to express quantities of radiation.

The RBE of 20 for α particles may lead you to believe that α particles pose the greatest health threat from radioactivity. Not exactly. Alpha particles are so big that they have little penetrating power; they are stopped by a sheet of paper, your clothing, or even a layer of dead skin (Figure 20.12). On the other hand, if you ingest or breathe in an α emitter, tissue damage can be severe because the heavy α particles do not have to travel far to cause cellular damage. Gamma rays are considered the most dangerous form of radiation emanating from a source outside the body because they have the greatest penetrating power. The acute toxic effects of exposure to different single doses of radiation are summarized in Table 20.5. To put these data in perspective, the dose from a typical dental X-ray is about 25 μSv, or about 10,000 times smaller than the lowest exposure level cited in the table.

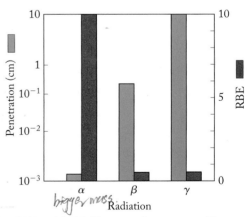

FIGURE 20.12 The tissue damage caused by α particles, β particles, and γ rays depends on several factors including their relative biological effect (RBE) and the depths to which these forms of radiation can penetrate soft tissue (which is mostly water). Alpha particles are extremely dangerous if formed by a decay process inside the body. In contrast, γ rays from an external source are much harder to keep from entering the body.

TABLE 20.5 Acute Effects of Single Whole-Body Doses of Ionizing Radiation

Dose (Sv)	Toxic Effect
0.05–0.25	No acute effect, possible carcinogenic or mutagenic damage to DNA
0.25–1.0	Temporary reduction in white blood cell count
1.0–2.0	Radiation sickness: fatigue, vomiting, diarrhea, impaired immune system
2.0–4.0	Severe radiation sickness: intestinal bleeding, bone marrow destruction
4.0–10.0	Death, usually through infection, within weeks
>10.0	Death within hours

Widespread exposure to very high levels of radiation occurred after an explosion at a nuclear reactor in the Chernobyl power station on April 26, 1986. The explosion and fire are believed to have released more than 200 times the amount of radioactivity released by the atomic bombs dropped on Hiroshima and Nagasaki combined. Many of the responding firefighters and power-plant workers were exposed to greater than 1.0 Sv of radiation. At least 30 of them died in the weeks after the accident. Many of the more than 300,000 workers who cleaned up the area around the reactor exhibited symptoms of radiation sickness (see Table 20.5), and at least 5 million people in Ukraine, Belarus, and Russia were exposed to fallout in the days following the accident. The cloud of radioactivity released from Chernobyl spread rapidly across northern Europe. Within 2 weeks, increased levels of radioactivity were detected throughout the entire Northern Hemisphere. The accident produced a global increase in human exposure to ionizing radiation estimated to be equivalent to 0.05 mSv per year.

Studies of the biological effects of radiation from the Chernobyl accident uncovered a 200-fold increase in the incidence of thyroid cancer in children in southern Belarus due to ^{131}I released in the accident. Compared with other children, those born in this region 8 years after the accident had twice the number of mutations in

The release of radiation from the destruction of a nuclear reactor at Chernobyl, Ukraine, significantly increased background radiation levels across the entire Northern Hemisphere in 1986.

Since 1986 the wildlife surrounding the destroyed nuclear reactor at Chernobyl, Ukraine, has been exposed to intense ionizing radiation, which has led to deaths and sublethal biological effects such as genetic mutations. One example of the latter is the partial albino barn swallow on the right. A normal swallow (on the left) has no white feathers directly beneath its beak.

their DNA. Mutation rates in the DNA of rodents living in heavily contaminated fields near Chernobyl were thousands of times greater than normal rates.

Assessing the Risks of Radiation

To put the global exposure to radiation from the Chernobyl explosion in perspective, we need to consider the amount of ionizing radiation from other sources to which we typically are exposed each year. Figure 20.13 provides data on the level of background radiation and its sources. For many people the principal source of radiation is radon gas in indoor air and in well water. Like all of the elements in the last column of the periodic table, radon is a chemically inert gas. Unlike the others, all of its isotopes are radioactive. The most common isotope, radon-222, is produced by the decay series in which uranium-238 eventually turns into lead-206 (Figure 20.6). Trace amounts of uranium are present in most rocks and soils. Radon gas formed underground percolates toward the surface through the pores in soil and along cracks and fissures in rocks. It can also migrate through cracks and pores in the foundations of buildings and into living spaces.

If you breathe air contaminated with radon and then exhale before it undergoes radioactive decay, no harm is done. However, if radon-222 decays inside the lungs, it emits an α particle and forms an atom of radioactive polonium-218. Polonium is a reactive solid and becomes attached to tissue in the respiratory system, where it may undergo another α decay reaction, forming lead-214.

$$\begin{cases} {}^{222}_{86}\text{Rn} \rightarrow {}^{218}_{84}\text{Po} + {}^{4}_{2}\text{He} & t_{1/2} = 3.8 \text{ days} \\ {}^{218}_{84}\text{Po} \rightarrow {}^{214}_{82}\text{Pb} + {}^{4}_{2}\text{He} & t_{1/2} = 3.1 \text{ min} \end{cases}$$

As we have seen, α particles are the most damaging product of nuclear decay when formed *inside the body*. How big a threat does radon pose to human health? Concentrations of indoor radon depend on local geology and how gas-tight building foundations are. Although there are variations within every region, different parts of the United States tend to have different concentrations of radon (Figure 20.14). The air in many buildings contains concentrations of radon in the pico-curie-per-liter range (1 pCi = 10^{-12} Ci). How hazardous are such tiny concentrations? There appears to be no simple answer. The U.S. Environmental Protection Agency has established 4 pCi/L as an "action level," meaning that people occupying houses with higher concentrations should take measures to minimize their exposure. The action level 4 pCi/L is based on the results of studies of the incidence of lung cancer in workers in uranium mines. These workers are exposed to

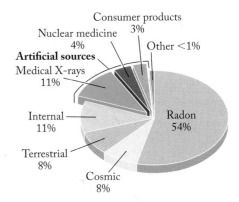

FIGURE 20.13 Sources of radiation exposure of the U.S. population: On average, a person living in the United States is exposed to 3.6 mSv of radiation each year. This chart shows that more than 80% of this radiation comes from natural sources and most of that from radon in the air and water. Artificial sources such as medical X-rays, nuclear medicine, and consumer products account for only 18% of the total exposure.

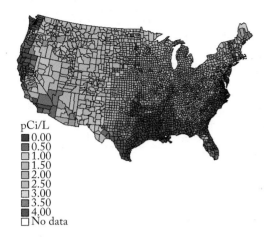

pCi/L
■ 0.00
□ 0.50
□ 1.00
□ 1.50
□ 2.00
□ 2.50
□ 3.00
□ 3.50
■ 4.00
□ No data

FIGURE 20.14 Dense radon gas is released from the radioactive decay of uranium in rocks and collects in basements. Radon poses a health hazard because it is easily inhaled and decays by releasing α particles.

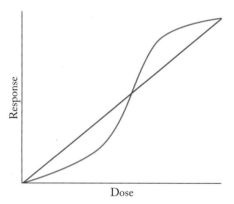

FIGURE 20.15 The risk of death from radiation-induced cancer may follow one of two models. In the linear response model (red line), risk is directly proportional to the amount of radiation exposure. In the "S-shaped" model (blue), risk remains low below a critical threshold dose and then increases rapidly as the dose increases. In the "S-shaped" model, the cancer risk is less than for the linear model at low exposure but is higher at higher doses.

much higher concentrations of radon (and radiation from other radionuclides) than those found in homes and other buildings. However, many scientists believe that people exposed to very low levels of radon for many years are as much at risk as miners exposed to high levels of radiation for shorter periods. Some researchers use a model that assumes a linear relation between radon exposure and incidence of lung cancer. This model is represented by the red line in Figure 20.15 which graphs cancer deaths as a function of radiation dose. On the basis of this dose–response model, an estimated 15,000 Americans die of lung cancer each year because of exposure to indoor radon. This number comprises 10% of all lung-cancer fatalities and 30% of those among nonsmokers.

Is this linear model valid for radon-induced cancer? Perhaps—but some scientists believe that there may be a threshold of exposure below which radon poses no significant threat to public health. They advocate an "S-shaped" model for the dose–response graph shown by the blue line in Figure 20.15. Notice that the risk of death from cancer in the "S-shaped" curve is much lower than in the linear response model at low radiation doses but rises rapidly above a critical dose.

SAMPLE EXERCISE 20.7 **Calculating Effective Exposure**

It has been estimated that a person living in a home where the radon concentration in the air is 4.0 pCi/L (the limit set by the U.S. Environmental Protection Agency) receives an annual absorbed dose of ionizing radiation equivalent to 0.4 mGy. What is the person's annual effective dose of radiation from radon, expressed in mSv? Compare this effective dose to the average annual effective dose of 3.6 mSv.

COLLECT AND ORGANIZE We are given the absorbed dose of ionizing radiation, 0.4 mGy, which is the same as 0.4 mJ/kg of tissue. Radon isotopes emit α particles, which have a relative biological effectiveness (RBE) value of 20.

ANALYZE The effective dose (tissue damage) caused by an absorbed dose of ionizing radiation is the product of the absorbed dose times the RBE value of the type of radiation emitted.

SOLVE

$$0.4 \text{ mGy} \times 20 = 8.0 \text{ mSv}$$

The average American is exposed to 3.6 mSv of radiation per year. According to the data in Figure 20.13, 54% of that amount, or 1.9 mSv, is the average annual effective dose of radiation from radon in the United States. The calculated value is slightly more than four times the average value.

THINK ABOUT IT The calculated value is also more than twice the average annual effective dose of 3.6 mSv from all sources of radiation. The U.S. National Research Council has estimated that a nonsmoker living in air contaminated with 4.0 pCi/L of radon has a 1% chance of dying from lung cancer due to this exposure. The cancer risk for a smoker is close to 5%.

Practice Exercise A type of dental X-ray for imaging impacted wisdom teeth produces an effective dose of 10 μSv of radiation. If a dental X-ray machine emits X-rays with a wavelength of 3.3 nm, how many of these X-rays are absorbed per kilogram of tissue to produce an effective dose of 10 μSv? Assume the RBE of these X-rays is 1.2.

20.11 Medical Applications of Radionuclides

The use of radionuclides in medicine is an important part of our health care system to detect disease and to treat cancer. Nuclear medicine is an important subdivision of the field of diagnostic radiology, which also includes MRI and other forms of imaging based on longer-wavelength, nonionizing radiation. Therapeutic radiology, however, is based almost entirely on the ionizing radiation that comes from nuclear processes.

Radiation Therapy

Because ionizing radiation causes the most damage to those cells that are growing and dividing most rapidly, it is also a powerful tool in the fight *against* cancer. Radiation therapy consists of exposing cancerous tissue to γ radiation emitted by radioactive nuclides. Often the radiation source is external to the patient, but sometimes the radionuclide is encased in a platinum capsule and surgically implanted in a cancerous tumor. The platinum provides a chemically inert outer layer and acts as a radiation filter, absorbing α and β particles, but allowing γ rays to pass into the cancerous tissue. Some radioactive nuclides used in cancer therapy are listed in Table 20.6.

The chemical properties of a nuclide can be exploited to direct it to the site of a tumor. For example, most of the iodine in the body is concentrated in the thyroid gland, so an effective therapy against thyroid cancer starts with the ingestion of potassium iodide (KI) containing radioactive ^{131}I.

Surgically inaccessible cancerous tumors can often be treated with beams of γ rays from radiation sources outside the body. Unfortunately, γ radiation destroys cancerous cells and also the healthy tissue surrounding them. Thus, patients who receive radiation therapy frequently suffer many of the symptoms of radiation sickness, including nausea and vomiting (the tissues that make up intestinal walls are especially susceptible to radiation-induced damage), fatigue, weakened immune response, and hair loss. For this reason, radiologists must carefully control the dosage that each patient receives.

TABLE 20.6 Some Radionuclides Used in Radiation Therapy

Nuclide	Radiation	Half-Life	Treatment
^{32}P	β	14.3 days	Leukemia therapy
^{60}Co	β, γ	5.3 yr	External cancer therapy
^{123}I	γ	13.3 hr	Thyroid therapy
^{131}Cs	γ	9.7 days	Prostate cancer therapy
^{192}Ir	β, γ	74 days	Coronary disease

Medical Imaging

The movement of radionuclides in the body and their accumulation in certain organs provide ways to assess organ function. Tiny quantities of radioactive isotopes are used for these studies, together with a much larger amount of a stable isotope of the same element, called a *carrier*. For example, the circulatory system can be imaged by injecting into the bloodstream a solution of sodium chloride (NaCl) containing a trace amount of ^{24}NaCl. Circulation of the radioactive sodium ion ^{24}Na can be monitored by measuring the γ rays emitted by ^{24}Na as it decays. In this example, the injected sodium chloride solution is said to have been labeled with a radioactive *tracer*, namely ^{24}Na. The ideal radioactive tracer for medical imaging is one with a half-life about as long as the time it takes for the

Nuclide	Radiation	Half-Life (hr)	Use
^{99m}Tc	γ	6.0	Bones, circulatory system, various organs
^{67}Ga	γ	78	Tumors in the brain and other organs
^{201}Tl	γ	73	Coronary arteries, heart muscle
^{123}I	γ	13.3	Thyroxin production in thyroid gland

TABLE 20.7 Selected Radionuclides Used For Medical Imaging

imaging measurements. It should emit moderate-energy γ rays, but no α particles or high-energy β particles that might cause tissue damage. Table 20.7 lists some of the other radionuclides with these characteristics.

CONCEPT TEST

Why does electron capture produce only γ rays?
doesn't result in any subatomic particles.
(*Answers to Concept Tests are in the back of the book.*)

A powerful tool for diagnosing brain function uses short-lived, neutron-poor radionuclides that emit positrons. They include carbon-11, oxygen-15, and fluorine-18. The diagnostic technique is called *positron emission tomography* (PET). In PET imaging, a patient might be administered a solution of glucose in which some of the sugar molecules contain atoms of ^{11}C, ^{15}O, or ^{18}F (in place of H atoms). The rate at which glucose is metabolized in various regions of the brain can then be monitored by detectors surrounding the patient's head. These detectors monitor the production of γ rays from positron–electron annihilation reactions. Computers merge the signals that they produce into three-dimensional images of the brain. Artificial coloring is used to distinguish regions with different levels of γ ray production, which represent different rates of glucose metabolism, as shown in Figure 20.16. Unusual patterns in these images can indicate damage from strokes, mental illnesses (including schizophrenia and manic depression), Alzheimer's disease, and even nicotine addiction in tobacco smokers.

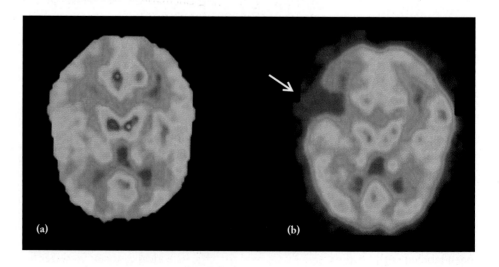

FIGURE 20.16 Positron-emission tomography is used to image soft tissue, such as brain tissue. (a) A healthy person's brain. (b) Image of brain function in a patient who has had a stroke that has affected blood flow in the region indicated by the arrow.

(a)   (b)

20.12 Radiochemical Dating

In 1947, American chemist Willard Libby (1908–1980) developed a method called **radiocarbon dating** for determining the age of artifacts from prehistory and early civilizations. The method is based on measuring the amount of radioactive carbon-14 in samples derived from plant or animal tissue. Carbon-14 originates in the upper atmosphere, where, as we have seen, cosmic rays produce free neutrons that can be absorbed by the nuclei of atmospheric gases. In one such nuclear reaction, nitrogen-14 absorbs a neutron and disintegrates into carbon-14 and a proton:

$$^{14}_{6}N + ^{1}_{0}n \rightarrow ^{14}_{6}C + ^{1}_{1}H$$

Neutron-rich carbon-14 undergoes β decay with a half-life of 5730 years:

$$^{14}_{6}C \rightarrow ^{14}_{7}N + ^{0}_{1-}e$$

After carbon-14 is formed in the upper atmosphere, it combines with oxygen to form $^{14}CO_2$. This radioactive carbon dioxide mixes with $^{12}CO_2$ and is carried to Earth's surface where both $^{14}CO_2$ and $^{12}CO_2$ are incorporated into the structures of green plants through photosynthesis. The two nuclides are incorporated into plant tissue, such as the trunk of a tree, in a ratio that reflects their atmospheric concentration. The ^{14}C-to-^{12}C ratio begins to decline, however, after the plant dies, as carbon-14 in the plant undergoes radioactive decay and is no longer replenished by photosynthesis. If we could determine the ^{14}C-to-^{12}C ratio of a piece of wood from an ancient building or the charcoal from a cave-dwelling fire or the papyrus from an early Egyptian scroll and if we knew the ^{14}C-to-^{12}C ratio originally in these materials (presumably the same ratio found in trees and papyrus plants growing today), we could calculate the age of these artifacts.

The starting point of radiocarbon dating calculations is usually a combination of information about the ^{14}C activity in an object of unknown age (A_t) and of a similar material that has not yet undergone radioactive decay (A_0). Equation 20.3 relates the radioactivity of a sample at time t to its initial activity ($t = 0$) and the age of the sample in half-lives (n):

$$\frac{A_t}{A_0} = 0.5^n$$

Because n is the unknown in radiocarbon dating calculations, let's rearrange the terms in Equation 20.3 to solve for n. To move n out of the exponent we take the natural log of both sides of the equation:

$$\ln \frac{A_t}{A_0} = n \ln 0.5 = -0.693n \qquad (20.11)$$

Converting the number of half-lives (n) into an equivalent number of years (t) by combining Equation 20.11 with Equation 20.2:

$$\ln \frac{A_t}{A_0} = -0.693 \frac{t}{t_{1/2}}$$

Rearranging the terms to solve for t,

$$t = -\frac{t_{1/2}}{0.693} \ln \frac{A_t}{A_0} \qquad (20.12)$$

Radiocarbon dating is a method of establishing the approximate age of carbon-containing materials by measuring the amount of radioactive carbon-14 remaining in the samples.

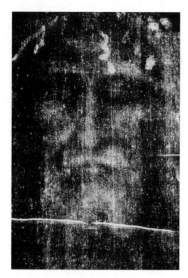

For centuries the Shroud of Turin was believed to be the burial shroud of Jesus Christ. Radiocarbon dating of tiny fragments of the fabric in 1988 indicate that the section from which the fragments came was woven between AD 1260 and AD 1390.

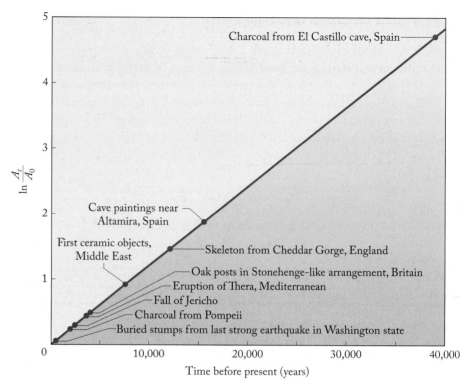

FIGURE 20.17 Radiocarbon dating: a time line of artifacts.

We use Equation 20.12 in the following exercise to calculate the ages of archaeological artifacts. In general, carbon dating results are precise to within about 30 years. Figure 20.17 identifies some archaeological items that have been chronicled using radiocarbon dating.

SAMPLE EXERCISE 20.8   **Determining the Age of an Object by ^{14}C Dating**

A piece of wooden shaft from a harpoon used to hunt seals was found in the remains of an early Inuit encampment in western Alaska. Its ^{14}C-to-^{12}C ratio found was 61.9% of the ^{14}C-to-^{12}C ratio in the same type of wood from a recently cut tree. The half-life of ^{14}C is 5730 years. How old is the harpoon?

COLLECT AND ORGANIZE We know the ratio of the amount of carbon-14 in the spear to that in modern wood. Equation 20.12 relates the age of an object and the ratio of its level of ^{14}C activity today to its initial ^{14}C activity ($t = 0$):

$$t = -\frac{t_{1/2}}{0.693} \ln\frac{A_t}{A_0}$$

where $t_{1/2} = 5730$ years.

ANALYZE The modern sample is zero years old. Therefore, 61.9% represents the ratio A_t/A_0.

SOLVE Putting the known values into Equation 20.12 and solving for t,

$$t = -\frac{t_{1/2}}{0.693} \ln\frac{A_t}{A_0}$$

$$= -\frac{5730 \text{ yr}}{0.693} \ln(0.619)$$

$$= 3965 \text{ yr}$$

or 3.96×10^3 yr to three significant figures.

THINK ABOUT IT The resulting age is less than one half-life, which is reasonable because more than half the original carbon-14 content remained in the sample. Because we know the value of A_t/A_0 to only three significant figures, we round off our answer to three significant figures, too.

Practice Exercise The carbon-14 decay rate in papyrus growing along the Nile River in Egypt today is 231 disintegrations per second per kilogram of carbon. If a papyrus scroll found in a tomb near the Great Pyramid at Cairo has a carbon-14 decay rate of 127 disintegrations per second per kilogram of carbon, how old is the scroll?

The accuracy of radiocarbon dating can be checked by determining the ^{14}C activity in the rings of very old trees such as the bristlecone pines that grow in the American Southwest. These checks reveal variability in the ^{14}C content of particularly old objects (Figure 20.18). Scientists attribute part of this variability to varying rates of ^{14}C production in the upper atmosphere.

Papyrus was used by Egyptians for thousands of years to make a woven paperlike material. The ^{14}C-to-^{12}C ratio in these materials can be used to date them.

Dating artifacts by using carbon-14 relies on knowing the atmospheric concentrations of carbon-14 over time. Ancient living trees such as the bristlecone pines in the American Southwest act as a check of the carbon-14 levels over thousands of years. The ages of the rings can be determined by counting them, and their carbon-14 content can be determined independently.

FIGURE 20.18 Calibration curves for radiocarbon dating such as this one allow scientists to accurately calculate the ages of archaeological objects. Scientists usually express the results of radiocarbon dating in years "before present" or "BP." By convention the "present" is the year 1950 (shortly after Willard Libby developed radiocarbon dating). Actual dates are usually expressed in "CE" for "common era" (analogous to AD on Christian calendars) and "BCE" for "before common era" (which has the same meaning as BC).

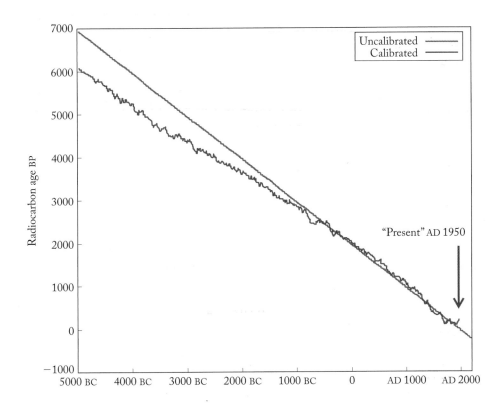

Radon, Radium, and Uranium: Too Hot to Handle

Uranium is the heaviest naturally occurring element found on Earth. All uranium is radioactive and has only two isotopes, ^{235}U and ^{238}U, present in nature. Comparing its abundance with that of other elements in the periodic table, we find that uranium is present in approximately the same amount as tin. Uranium is found primarily in combination with oxygen as the black mineral pitchblende (U_3O_8) and as a complex mineral called uraninite [$K_2(UO_2)_2(VO_4)_2 \cdot 3H_2O$].

The isolation of uranium metal involves several steps in which uranium ore is converted to UO_3 and then to UO_2. Conversion of UO_2 to UF_4 followed by reaction with magnesium metal produces silvery uranium metal. Uranium is one of the densest metals known (19.04 g/mL) and reacts with most elements in the periodic table. The most significant use of uranium is in nuclear reactors (Figure 20.10), where the fission of ^{235}U is used to produce energy. Both uranium metal and UO_2 are used as reactor fuel.

The spontaneous decay of both uranium isotopes produces other heavy elements including thorium (Th), protactinium (Pa), and radium (Ra), as shown in Figure 20.6 and summarized below:

$$^{238}U \rightarrow {}^{234}Th \rightarrow {}^{234}Pa \rightarrow {}^{234}U \rightarrow {}^{230}Th \rightarrow {}^{226}Ra \rightarrow {}^{222}Rn$$

Because ^{226}Ra is a product of ^{238}U decay, it is typically found in uranium-containing ores. Radium belongs to the same group in the periodic table as magnesium and calcium and has the chemical properties of these elements. For example, radium forms Ra^{2+} ions, whose common salts dissolve in water. In the 19th century, before the hazards of radioactivity were understood, hot springs rich in radium were thought to be useful in curing diseases. Radium was also used in watches in the early 20th century to provide glowing dials. The radium-containing paint was applied by hand with small brushes. The painters would often pass the tip of the brush between their lips to sharpen its point. As a result, they ingested this emitter of α particles. It tended to concentrate in tissues like bone that are normally high in Ca^{2+} and Mg^{2+} and dramatically increased the incidence of cancer tumors in those tissues.

Radium-226 decays by emitting α particles and yielding a radioactive gas, radon-222. Radon seeps through the ground in areas with uranium-containing minerals. The hazards of radon are discussed in Section 20.10. A simple radon detector containing carbon is available for homeowners who wish to test the air in their homes. The radioactive decay of ^{222}Rn leaves solid lead and bismuth on the carbon. Measuring the radiation from ^{214}Pb and ^{214}Bi provides an indirect measure of the amount of radon originally present in the sample. Chemically, radon belongs in the last column of the periodic table with helium, neon, argon, krypton, and xenon.

CONCEPT TEST

Why is ^{14}C better suited to dating wooden tools that are several thousand years old than to dating a limestone ($CaCO_3$) deposit that formed millions of years ago?

Different isotopes can be used to date other kinds of material. Some nuclides with half-lives of millions to billions of years can be used to estimate the ages of rocks, meteorites, and Earth itself (Figure 20.19). Neptunium-237, for example, has a half-life of 2.2×10^6 years. Because none of this isotope is found in nature, Earth must be much more than 2.2 million years old. On the other hand, uranium-235, with a half-life of 7.1×10^8 years, still makes up 0.72% of the uranium found in nature. This percentage suggests that Earth may be several billion years old. The ratio of ^{238}U to ^{206}Pb in rock samples can be used to obtain a better estimate of the age of Earth, or at least its oldest rocks.

FIGURE 20.19 The age of these rocks in southern West Greenland, which are more than 3.5 billion years old and among the oldest on Earth, has been determined from the ratios of very long-lived radionuclides including ^{238}U ($t_{1/2} = 4.5 \times 10^9$ yr) and ^{87}Rb ($t_{1/2} = 5 \times 10^{11}$ yr) and the stable isotopes formed by their decay processes.

Summary

SECTION 20.1 Nuclear chemistry is the study and application of reactions that involve changes in the nuclei of atoms.

SECTION 20.2 The first particles created in the Big Bang were neutrons. Neutrons and other unstable particles undergo spontaneous disintegration called **radioactive decay** that is accompanied by the release of radiation. Radioactive decay follows first-order kinetics, so the half-life ($t_{1/2}$) of a radioactive nuclide is a set value that does not depend on the quantity of the nuclide.

SECTION 20.3 Positrons have the mass of an electron but carry a positive charge and so are examples of **antimatter**. When a particle of matter encounters a particle of antimatter, both particles are converted into energy (they annihilate one another) yielding gamma (γ) rays.

SECTION 20.4 The **mass defect** (Δm) of a nucleus is the difference between the mass of the nucleus and the masses of the individual nucleons that make up the nucleus. **Binding energy (BE)** is released when the nucleons combine to form a nucleus. It is also the energy needed to split the nucleus into its respective nucleons, and so is a measure of the stability of the nucleus. Binding energy per nucleon is a measure of the relative stability of a nucleus.

SECTION 20.5 Both stable and unstable nuclei can absorb neutrons producing unstable nuclei that undergo radioactive decay to new elements. Neutron capture followed by β decay leads to the formation of a nuclide with an atomic number that is one greater than the parent nuclide.

SECTION 20.6 Stable nuclei have neutron-to-proton ratios that fall within a range of values that form the **belt of stability**. Unstable nuclides undergo radioactive decay. Neutron-rich (mass number greater than the average atomic mass) nuclides undergo β decay; neutron-poor nuclides undergo **positron emission** or **electron capture**.

SECTION 20.7 Artificial nuclides including radioactive transuranium elements ($Z > 92$) are produced in **linear accelerators** and **cyclotrons**, where atoms and subatomic particles collide at high speeds.

SECTION 20.8 Neutron absorption by uranium-235 and some heavy element isotopes may lead to **nuclear fission** into lighter nuclei accompanied by the release of energy that can be harnessed to generate electricity. **Chain reactions** happen when the neutrons released during fission collide with other fissionable nuclei. They require a minimum or **critical mass** of a fissionable isotope. A **breeder reactor** is used to make ^{239}Pu from ^{235}U while also producing energy to make electricity.

SECTION 20.9 Scintillation and **Geiger counters** are used to detect and measure the level of nuclear radiation. Radioactivity levels are expressed as the number of decay events per unit time. Common units are **becquerels (Bq)** (1 decay event per second) and **curies (Ci)** (1 Ci = 3.70×10^{10} Bq).

SECTION 20.10 Alpha particles, β particles, positrons, and γ rays have enough energy to ionize water and so are examples of **ionizing radiation** that can cause tissue damage and molecular changes in DNA. The quantity of ionizing radiation absorbed per kilogram of tissue is expressed in **grays (Gy)** = 1.00 J/kg. An effective dosage of ionizing radiation is the product of the energy absorbed and the **relative biological effectiveness (RBE)** of the radiation; it is expressed in **sieverts (Sv)**. Alpha (α) particles have a larger RBE than β particles and γ rays but have the least penetrating power.

SECTION 20.11 Selected radioactive isotopes are useful as tracers in the human body to map biological activity and diagnose diseases. Other radionuclides are used to treat cancers using external sources or internally through injections or implants.

SECTION 20.12 Radiocarbon dating involves measuring the amount of radioactive carbon-14 that remains in an object derived from plant or animal tissue to calculate the age of the object. To improve the accuracy of the technique, scientists need to calibrate the results to account for fluctuations in carbon-14 production.

Problem-Solving Summary

TYPE OF PROBLEM	CONCEPTS AND EQUATIONS	SAMPLE EXERCISES
Calculations involving half-lives	$A_t/A_0 = 0.5^n$ where $n = t/t_{1/2}$ and A_t/A_0 is the ratio of the level of radioactivity (amount of radioisotope present) in a sample to its initial ($t = 0$) level.	20.1
Calculating nuclear binding energies (BE)	BE = $(\Delta m)c^2$ (where Δm is the mass defect).	20.2
Predicting the products of β decay	The product has the same mass number but the atomic number is one more than the parent nuclide.	20.3
Predicting the mode of radioactive decay	Neutron-rich nuclides tend to undergo β decay; neutron-poor nuclides undergo positron emission or electron capture.	20.4
Completing and balancing radiochemical equations	Add the products that balance the mass numbers and atomic numbers of reactants and products.	20.5
Calculating the activity (rate of decay) of a radionuclide from its half-life and its initial quantity (A)	Rate of decay = kA and $t_{1/2} = 0.693/k$	20.6
Calculating effective exposure using relative biological effectiveness (RBE)	Effective exposure = ionizing energy absorbed $\times$ RBE	20.7
Determining the age of an object by ^{14}C dating	$t = -\dfrac{t_{1/2}}{0.693} \ln \dfrac{A_t}{A_0}$	20.8

Visual Problems

20.1. Which of the highlighted elements in Figure P20.1 has a stable isotope with no neutrons in its nucleus?

20.2 Which highlighted element in Figure P20.1 has stable isotopes with the largest ratios of neutrons to protons in their nuclei?

20.3. Which of the highlighted elements in Figure P20.1 has no stable isotopes?

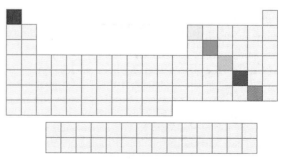

FIGURE P20.1

20.4. Which of the graphs in Figure P20.4 illustrates α decay? Which decay pathway does the other graph illustrate?

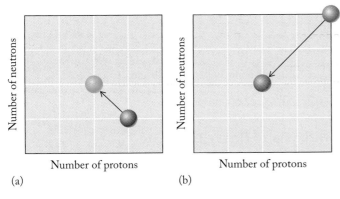

(a)　　　　　　　　　　　(b)

FIGURE P20.4

20.5. Which of the graphs in Figure P20.5 illustrates β decay?

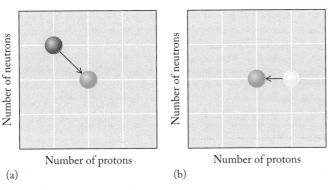

(a)　　　　　　　　　　　(b)

FIGURE P20.5

20.6. Which of the graphs in Figure P20.6 illustrates the overall effect of neutron capture followed by β decay?

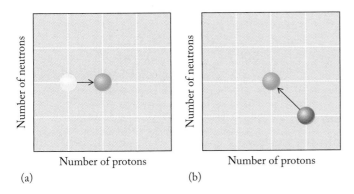

(a)　　　　　　　　　　　(b)

FIGURE P20.6

20.7. Which of the curves in Figure P20.7 represents the decay of an isotope that has a half-life of 2.0 days?

20.8. Which of the curves in Figure P20.7 do not represent a radioactive decay curve?

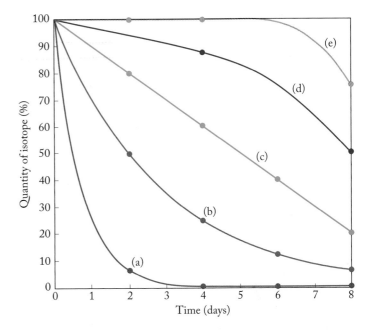

FIGURE P20.7

20.9. Which of the models in Figure P20.9 represents fission and which represents fusion?

(1)　　　　　　　　　　　(2)

FIGURE P20.9

20.10. Isotopes in a nuclear decay series emit particles with a positive charge and particles with a negative charge. The two kinds of particles penetrate a column of water as shown in Figure P20.10. Is the "X" particle the positive or the negative one?

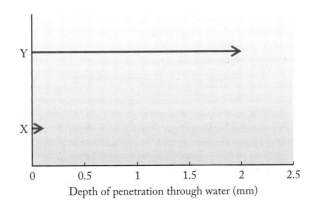

FIGURE P20.10

Questions and Problems

Nuclear Chemistry; Radioactive Decay

CONCEPT REVIEW

20.11. Arrange the following particles in order of increasing mass: electron, β particle, positron, proton, neutron, α particle, deuteron.

20.12. Electromagnetic radiation is emitted when a neutron and proton fuse to make a deuteron. In which region of the electromagnetic spectrum is the radiation?

20.13. Scientists at the Fermi National Accelerator Laboratory in Illinois announced in the fall of 1996 that they had created "antihydrogen." How does antihydrogen differ from hydrogen?

20.14. Describe an antiproton.

PROBLEMS

20.15. Calculate the energy and wavelength of the gamma rays released by the annihilation of a proton and an antiproton.

20.16. Calculate the energy released and the wavelength of the two photons emitted in the annihilation of an electron and a positron.

20.17. What percentage of a sample's original radioactivity remains after two half-lives?

20.18. What percentage of a sample's original radioactivity remains after five half-lives?

Hydrogen Fusion; Nuclear Binding Energies

CONCEPT REVIEW

20.19. What do the terms *mass defect* and *binding energy* mean?

20.20. Why is energy released in a nuclear fusion process when the product is an element preceding iron in the periodic table?

PROBLEMS

20.21. What is the binding energy of ^{51}V? The mass of ^{51}V is 50.9440 amu.

20.22. What is the binding energy of ^{50}Ti? The mass of ^{50}Ti is 49.9448 amu.

20.23. All of the following fusion reactions produce ^{28}Si. Calculate the energy released in each reaction from the masses of the isotopes: ^{2}H (2.0146 amu), ^{4}He (4.00260 amu), ^{10}B (10.0129 amu), ^{12}C (12.000 amu), ^{14}N (14.00307 amu), ^{24}Mg (23.98504 amu), ^{28}Si (27.97693 amu).
  a. ^{14}N + ^{14}N $\rightarrow$ ^{28}Si
  b. ^{10}B + ^{16}O (15.99491 amu) + ^{2}H $\rightarrow$ ^{28}Si
  c. ^{16}O + ^{12}C $\rightarrow$ ^{28}Si
  d. ^{24}Mg + ^{4}He $\rightarrow$ ^{28}Si

20.24. All of the following fusion reactions produce ^{32}S. Calculate the energy released in each reaction from the masses of the isotopes: ^{4}He (4.00260 amu), ^{6}Li (6.01512 amu), ^{12}C (12.000 amu), ^{14}N (14.00307 amu), ^{16}O (15.99491 amu), ^{24}Mg (23.98504 amu), ^{28}Si (27.97693 amu), ^{32}S (31.97207 amu).
  a. ^{16}O + ^{16}O $\rightarrow$ ^{32}S   c. ^{14}N + ^{12}C + ^{6}Li $\rightarrow$ ^{32}S
  b. ^{28}Si + ^{4}He $\rightarrow$ ^{32}S   d. ^{24}Mg + 2 ^{4}He $\rightarrow$ ^{32}S

20.25. Our sun is a fairly small star that has barely enough mass to fuse hydrogen. Calculate the binding energy per nucleon for ^{4}He, given the exact masses of ^{4}He (4.00260 amu), ^{1}H (1.00728 amu), and ^{1}n (1.00867 amu).

20.26. What is the binding energy per nucleon of ^{12}C, the mass of which is 12.00000 amu?

Synthesis of Heavy Elements

CONCEPT REVIEW

20.27. What factor makes it easier to bombard nuclei with neutrons than protons?

20.28. Explain how the product of β decay has a higher atomic number than the radioactive nuclide.

20.29. Is the following statement correct? "Elements with atomic numbers between 27 and 83 form by β decay because the neutron-to-proton ratio of the parent isotope is too low."

20.30. Our sun contains carbon even though the sun is too small to synthesize carbon by nuclear fusion. Where may the carbon have come from?

PROBLEMS

20.31. During a supernova, a ^{56}Fe nucleus absorbs 3 neutrons.
  a. Write a balanced radiochemical equation describing this process and the nuclide that forms.
  b. This nuclide is radioactive. Write a balanced radiochemical equation describing its decay.

20.32. A ^{206}Pb nucleus absorbs 4 neutrons.
  a. Write a balanced radiochemical equation describing this process and the nuclide that forms.
  b. This nuclide is radioactive. Write a balanced radiochemical equation describing its decay.

20.33. If a nucleus of ^{96}Mo absorbs 3 neutrons and then undergoes β decay, will the nuclide that is produced be stable? Explain your answer.

20.34. During a supernova, a ^{212}Po nucleus absorbs 4 neutrons and then undergoes α decay. What nuclide is produced by these events?

The Belt of Stability

CONCEPT REVIEW

20.35. How can the belt of stability be used to predict the likely decay mode of an unstable nuclide?

20.36. Compare and contrast positron-emission and electron-capture processes.

20.37. The ratio of neutrons to protons in stable nuclei increases with increasing atomic number. Use this trend to explain why multiple α decay steps in the ^{238}U decay series are often followed by β decay.

20.38. If a ^{10}B nucleus absorbs a proton, the radioactive nuclide that is produced may undergo α decay. What nuclide is produced in the decay process? Is it stable or radioactive?

PROBLEMS

20.39. Iodine-137 decays to give xenon-137, which decays to give cesium-137. What are the modes of decay in these two radiochemical reactions?

20.40. Write a balanced nuclear equation for:
a. Beta emission by ^{28}Mg
c. Electron capture by ^{129}Cs
b. Alpha emission by ^{255}Lr
d. Positron emission by ^{25}Al

20.41. If the mass number of an isotope is more than twice the atomic number, is the neutron-to-proton ratio less than, greater than, or equal to 1?

20.42. In each of the following pairs of isotopes, which isotope has more protons and which one has more neutrons? (a) ^{127}I or ^{131}I; (b) ^{188}Re or ^{188}W; (c) ^{14}N or ^{14}C

20.43. Aluminum is found on Earth exclusively as ^{27}Al. However, ^{26}Al is formed in stars. It decays to ^{26}Mg with a half-life of 7.4×10^5 years. Write an equation describing the decay of ^{26}Al to ^{26}Mg.

20.44. Which nuclide is produced by the β decay of ^{131}I?

20.45. Predict the modes of decay for the following radioactive isotopes: (a) ^{32}P; (b) ^{10}C; (c) ^{50}Ti; (d) ^{19}Ne; (e) ^{116}Sb

20.46. Nine isotopes of sulfur have mass numbers ranging from 30 to 38. Five of the nine are radioactive: ^{30}S, ^{31}S, ^{35}S, ^{37}S, and ^{38}S. Which of these isotopes do you expect to decay by β decay?

20.47. Elements in a Supernova The isotopes ^{56}Co and ^{44}Ti were observed in supernova SN 1987A. Predict the decay pathway for these radioactive isotopes.

20.48. Predict the decay pathway of the radioactive isotope ^{56}Ni.

Artificial Nuclides

CONCEPT REVIEW

20.49. How are linear accelerators and cyclotrons used to make artificial elements?

20.50. Why must the velocity of the nuclide that is fired at a target nuclide to form a superheavy element be not too fast and not too slow, but "just right"?

PROBLEMS

20.51. Complete the following nuclear reactions used in the preparation of isotopes for nuclear medicine:
a. $^{32}S + ^1n \rightarrow ? + ^1H$
c. $^{75}As + ? \rightarrow ^{77}Br$
b. $^{55}Mn + ^1H \rightarrow ^{52}Fe + ?$
d. $^{124}Xe + ^1n \rightarrow ? \rightarrow ^{125}I + ?$

20.52. Complete the following nuclear reactions used in the preparation of isotopes for nuclear medicine:
a. $^6Li + ^1n \rightarrow ^3H + ?$
c. $^{56}Fe + ? \rightarrow ^{57}Co + ^1n$
b. $^{16}O + ^3H \rightarrow ^{18}F + ?$
d. $^{121}Sb + ^4He \rightarrow ? + 2\,^1n$

20.53. Complete the following nuclear reactions:
a. $? \rightarrow ^{122}Xe + _{-1}^0e$
b. $? + ^4He \rightarrow ^{13}N + ^1n$
c. $? + ^1n \rightarrow ^{59}Fe$
d. $? + ^1H \rightarrow ^{67}Ga + 2\,^1n$

20.54. Complete the following nuclear reactions:
a. $^{210}Po \rightarrow ^{206}Pb + ?$
b. $^3H \rightarrow ^3He + ?$
c. $^{11}C \rightarrow ^{11}B + ?$
d. $^{111}In \rightarrow ^{111}Cd + ?$

*20.55. Uranium Mining The presence of uranium-containing ores has made part of the Northern Territory of Australia a battleground between those seeking to mine the uranium and the indigenous aborigine population. An article in *Outside* magazine in March 1999 described the dangers of a proposed mine as follows:

> Thorium 230 becomes radium 226. . . . Radium 226 goes to radon 222. Radon 222, a heavy gas that will flow downhill, goes to polonium 218 when one alpha pops out of the nucleus. . . . Polonium 218 goes to lead 214, lead 214 to bismuth 214, bismuth 214 to polonium 214, and then that goes to lead 210, all within minutes, amid a crackle of alphas and betas and gammas.

a. Write balanced nuclear reactions for the decay of thorium-230 and determine how many "alphas and betas" are produced.
b. Using an appropriate reference such as the *CRC Handbook of Chemistry and Physics* (CRC Press, Boca Raton, FL), find the half-lives for each isotope and comment on the statement that all these processes take place "within minutes."

20.56. It has been reported that bombardment of a ^{206}Pb target with ^{86}Kr nuclei may produce a nucleus with 118 protons and 175 neutrons. Write a balanced radiochemical equation describing this process.

20.57. Describe how a ^{209}Bi target might be bombarded with subatomic particles to form ^{211}At. Use balanced equations for the required nuclear reactions.

*20.58. Bombardment of ^{239}Pu with α particles produces ^{242}Cm and another particle. Use a balanced nuclear reaction to determine the identity of the missing particle. The synthesis of which other nuclei in this chapter involves the same subatomic particles?

Nuclear Fission

CONCEPT REVIEW

20.59. How is the rate of energy release controlled in a nuclear reactor?

20.60. How does a breeder reactor create fuel and energy at the same time?

*20.61. Why are neutrons always by-products of the fission of heavy nuclides? (*Hint:* Look closely at the neutron-to-proton ratios shown in Figure 20.4.)

20.62. Seaborgium (Sg, element 106) is prepared by the bombardment of curium-248 with neon-22, which produces two isotopes, ^{265}Sg and ^{266}Sg. Write balanced nuclear reactions for the formation of both isotopes. Are these reactions better described as fusion or fission processes?

PROBLEMS

20.63. The fission of uranium produces dozens of isotopes. For each of the following fission reactions, determine the identity of the unknown nuclide:
a. $^{235}U + {}^1n \rightarrow {}^{96}Zr + ? + 2\,{}^1n$
b. $^{235}U + {}^1n \rightarrow {}^{99}Nb + ? + 4\,{}^1n$
c. $^{235}U + {}^1n \rightarrow {}^{90}Rb + ? + 3\,{}^1n$

20.64. For each of the following fission reactions, determine the identity of the unknown nuclide:
a. $^{235}U + {}^1n \rightarrow {}^{137}I + ? + 2\,{}^1n$
b. $^{235}U + {}^1n \rightarrow {}^{94}Kr + ? + 2\,{}^1n$
c. $^{235}U + {}^1n \rightarrow {}^{95}Sr + ? + 2\,{}^1n$

Measuring Radioactivity; Biological Effects of Radioactivity

CONCEPT REVIEW

20.65. What is the difference between a *level* of radioactivity and a *dose* of radioactivity?

20.66. What are some of the molecular effects of exposure to radioactivity?

20.67. Describe the dangers of exposure to radon-222.

20.68. **Food Safety** Periodic outbreaks of food poisoning from *E. coli* contaminated meat have renewed the debate about irradiation as an effective treatment of food. In one newspaper article on the subject, the following statement appeared: "Irradiating food destroys bacteria by breaking apart their molecular structure." How would you improve or expand on this explanation?

PROBLEMS

20.69. **Radiation Exposure from Dental X-rays** Dental X-rays expose patients to about 5 μSv of radiation. Given an RBE of 1 for X-rays, how many grays of radiation does 5 μSv represent? For a 50 kg person, how much energy does 5 μSv correspond to?

*20.70. **Radiation Exposure at Chernobyl** Some workers responding to the explosion at the Chernobyl nuclear power plant were exposed to 5 Sv of radiation, resulting in death for many of them. If the exposure was primarily in the form of γ rays with an energy of 3.3×10^{-14} J and an RBE of 1, how many γ rays did an 80 kg person absorb?

*20.71. **Strontium-90 in Milk** In the years immediately following the explosion at the Chernobyl nuclear power plant, the concentration of ^{90}Sr in cow's milk in southern Europe was slightly elevated. Some samples contained as much as 1.25 Bq/L of ^{90}Sr radioactivity. The half-life for the β decay of strontium-90 is 28.8 years.
a. Write a balanced radiochemical equation describing the decay of ^{90}Sr.
b. How many atoms of ^{90}Sr are in a 200 mL glass of milk with 1.25 Bq/L of ^{90}Sr radioactivity?
c. Why would strontium-90 concentrations be more concentrated in milk than other foods, such as grains, fruits, or vegetables?

*20.72. **Glowing Watch Dials** Early in the last century paint containing ^{226}Ra was used to make the dials on watches glow in the dark. The watch painters often passed the tips of their paint brushes between their lips to maintain fine points to paint the tiny numbers on the watches. Many of these painters died from bone cancer. Why?

20.73. In 1999, the U.S. Environmental Protection Agency set a maximum radon level for drinking water at 4.0 pCi per milliliter.
a. How many disintegrations occur per second in a milliliter of water for this level of radon radioactivity?
b. If the above radioactivity were due to decay of ^{222}Rn ($t_{1/2} = 3.8$ days), how many ^{222}Rn atoms would there be in 1.0 mL of water?

20.74. A former Russian spy died from radiation sickness in 2006 after dining at a London restaurant where he apparently ingested polonium-210. The other people at his table did not suffer from radiation sickness. Give as many reasons as you can why they did not.

Medical Applications of Radionuclides

CONCEPT REVIEW

20.75. How does the selection of an isotope for radiotherapy relate to (a) its half-life, (b) its mode of decay, and (c) the properties of the products of decay?

20.76. Are the same radioisotopes likely to be used for both imaging and cancer treatment? Why or why not?

PROBLEMS

20.77. Predict the most likely mode of decay for the following isotopes used as imaging agents in nuclear medicine: (a) ^{197}Hg (kidney); (b) ^{75}Se (parathyroid gland); (c) ^{18}F (bone).

20.78. Predict the most likely mode of decay for the following isotopes used as imaging agents in nuclear medicine: (a) ^{133}Xe (cerebral blood flow); (b) ^{57}Co (tumor detection); (c) ^{51}Cr (red blood cell mass); (d) ^{67}Ga (tumor detection).

20.79. A 1.00 mg sample of ^{192}Ir was inserted into the artery of a heart patient. After 30 days, 0.756 mg remained. What is the half-life of ^{192}Ir?

20.80. In a treatment that decreases pain and reduces inflammation of the lining of the knee joint, a sample of dysprosium-165 with an activity of 1100 counts per second was injected into the knee of a patient suffering from rheumatoid arthritis. After 24 hours, the activity had dropped to 1.14 counts per second. Calculate the half-life of ^{165}Dy.

20.81. **Treatment of Tourette's Syndrome** Tourette's syndrome is a condition whose symptoms include sudden movements and vocalizations. Iodine isotopes are used in brain imaging of people suffering from Tourette's syndrome. To study the uptake and distribution of iodine in cells, mammalian brain cells in culture were treated with a solution containing ^{131}I with an initial activity of 108 counts per minute. The cells were removed after 30 days, and the remaining solution was found to have an activity of 4.1 counts per minute. Did the brain cells absorb any ^{131}I ($t_{1/2} = 8.1$ days)?

20.82. A patient is administered mercury-197 to evaluate kidney function. Mercury-197 has a half-life of 65 hours. What fraction of an initial dose of mercury-197 remains after 6 days?

20.83. Carbon-11 is an isotope used in positron-emission tomography and has a half-life of 20.4 minutes. How long will it take for 99% of the ^{11}C injected into a patient to decay?

20.84. Sodium-24 is used to treat leukemia and has a half-life of 15 hours. A patient was injected with a salt solution containing sodium-24. What percentage of the ^{24}Na remained after 48 hours?

*20.85. Boron Neutron-Capture Therapy In boron neutron-capture therapy (BNCT), a patient is given a compound containing ^{10}B that accumulates inside cancer tumors. Then the tumors are irradiated with neutrons, which are absorbed by ^{10}B nuclei. The product of neutron capture is an unstable form of ^{11}B that undergoes α decay to ^{7}Li.
  a. Write a balanced nuclear equation for the neutron absorption and α decay process.
  b. Calculate the energy released by each nucleus of boron-10 that captures a neutron and undergoes α decay, given the following masses of the particles in the process: ^{10}B (10.0129 amu), ^{7}Li (7.01600 amu), ^{4}He (4.00260 amu), and ^{1}n (1.008665 amu).
  c. Why is the formation of a nuclide that undergoes α decay a particularly effective cancer therapy?

20.86. Balloon Angioplasty and Arteriosclerosis Balloon angioplasty is a common procedure for unclogging arteries in patients suffering from arteriosclerosis. Iridium-192 therapy is being tested as a treatment to prevent reclogging of the arteries. In the procedure, a thin ribbon containing pellets of ^{192}Ir is threaded into the artery. The half-life of ^{192}Ir is 74 days. How long will it take for 99% of the radioactivity from 1.00 mg of ^{192}Ir to disappear?

Radiochemical Dating

CONCEPT REVIEW

20.87. Explain why radiocarbon dating is reliable only for artifacts and fossils younger than about 50,000 years.

20.88. Which of the following statements about ^{14}C dating are true?
  a. The amount of ^{14}C in all objects is the same.
  b. Carbon-14 is unstable and is readily lost from the atmosphere.
  c. The ratio of ^{14}C to ^{12}C in the atmosphere is a constant.
  d. Living tissue will absorb ^{12}C but not ^{14}C.

20.89. Why is ^{40}K dating ($t_{1/2} = 1.28 \times 10^9$ years) useful only for fossils or rocks older than 300,000 years?

20.90. Where does the ^{14}C found in plants come from?

PROBLEMS

20.91. First Humans in South America Archeologists continue to debate the origins and dates of arrival of the first humans in the Western Hemisphere. Radiocarbon dating of charcoal from a cave in Chile was used to establish the earliest date of human habitation in South America as 8700 years ago. What fraction of the ^{14}C present initially remained in the charcoal after 8700 years?

20.92. For thousands of years native Americans living along the north coast of Peru used knotted cotton strands called *quipu* (see Figure P20.92) to record financial transactions and governmental actions. A particular quipu sample is 4800 years old. Compared with the fibers of cotton plants growing today, what is the ratio of carbon-14 to carbon-12 in the sample?

FIGURE P20.92

*20.93. Figure P20.93 shows a piece of a giant sequoia tree cut down in 1891 in what is now Kings Canyon National Park. It contained 1342 annual growth rings. If samples of the tree were removed for radiocarbon dating today, what would be the ratio of the carbon-14 concentration in the innermost (oldest) ring compared with that in the youngest? Assume that the density and overall carbon content of the wood in the rings are the same.

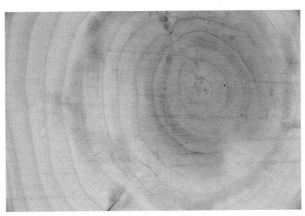

FIGURE P20.93

*20.94. Geologists who study volcanoes can develop historical profiles of previous eruptions by determining the ^{14}C-to-^{12}C ratios of charred plant remains entrapped in old magma and ash flows. If the uncertainty in determining these ratios is 0.1%, could radiocarbon dating distinguish between debris from the eruptions of Mt. Vesuvius that occurred in AD 472 and AD 512? (*Hint:* Calculate the ^{14}C-to-^{12}C ratios for samples from the two dates.)

20.95. Figure P20.95 shows a mammoth tusk containing grooves made by a sharp stone edge (indicating the presence of humans or Neanderthals) that was uncovered at an ancient camp site in the Ural Mountains in 2001. The ^{14}C-to-^{12}C ratio in the tusk was only 1.19% of that in modern elephant tusks. How old is the mammoth tusk?

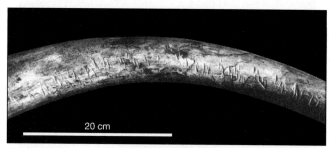

FIGURE P20.95

20.96. The Destruction of Jericho The Bible describes the Exodus as a period of 40 years that began with plagues in Egypt and ended with the destruction of Jericho. Archeologists seeking to establish the exact dates of these events have proposed that the plagues coincided with a huge eruption of the volcano Thera in the Aegean Sea.
  a. Radiocarbon dating suggests that the eruption occurred around 1360 BC, though other records place the eruption of Thera in the year 1628 BC. What is the percent difference in the ^{14}C decay rate in biological samples from these two dates?
  b. Radiocarbon dating of blackened grains from the site of ancient Jericho provides a date of 1315 BC ± 13 years for the fall of the city. What is the $^{14}C/^{12}C$ ratio in the blackened grains compared to that of grain harvested last year?

Additional Problems

20.97. Thirty years before the creation of antihydrogen, television producer Gene Roddenberry (1921–1991) proposed to use this form of antimatter to fuel the powerful "warp" engines of the fictional starship *Enterprise*. Why would antihydrogen have been a particularly suitable fuel?

20.98. Why are much higher energies required in the cores of giant stars for the fusion of carbon and helium nuclei to form oxygen than for the fusion of hydrogen nuclei to form helium?

20.99. Tiny concentrations of radioactive tritium ($^{3}_{1}H$) occur naturally in environmental waters. The half-life of $^{3}_{1}H$ is 12 years. Assuming that these concentrations of tritium can be determined accurately, could this isotope be used to determine whether a bottle of wine with the year 1969 on its label actually contained wine made from grapes that were grown in 1969? Explain your answer.

20.100. In the summer of 2003 a team of American and Russian scientists synthesized isotopes of element 115 by bombarding a target of ^{243}Am with ^{48}Ca. One of the isotopes had a mass number of 288.

  a. Write a balanced equation describing the nuclear reaction that produced this isotope of element 115. (You may use the symbol "X" for nuclides of $Z > 110$.)
  b. About 100 milliseconds after it was produced this isotope underwent α decay. Write a balanced equation describing the decay process.
  c. Within about 20 seconds the isotope formed by the α decay process emitted four more α particles. Write the symbols of the nuclides that were produced after each of these four additional decay events.

20.101. The energy released during the fission of ^{235}U is about 3.2×10^{-11} J per atom of the isotope. Compare this quantity of energy to that released by the fusion of four hydrogen atoms to make an atom of helium-4:

$$4^{1}_{1}H \rightarrow {}^{4}_{2}He + 2^{0}_{1}\beta$$

Assume that the positrons are annihilated in collisions with electrons so that the masses of the positrons are converted into energy. In your comparison, express the energies released by the fission and fusion processes in joules per nucleon for ^{235}U and ^{4}He, respectively.

20.102. How much energy is required to remove a neutron from the nucleus of an atom of carbon-13 (mass = 13.00335 amu)? (*Hint:* The mass of an atom of carbon-12 is exactly 12.00000 amu.)

20.103. The absorption of a neutron by ^{11}B produces ^{12}B, which decays by two pathways: α decay and β decay.
  a. Write balanced nuclear reactions for these processes.
  b. Which, if either, of the nuclides produced by these decay processes is stable?

*20.104. **Colorectal Cancer Treatment** Cancer therapy with radioactive rhenium-188 shows promise in patients suffering from colorectal cancer.
  a. Write the symbol for rhenium-188 and determine the number of neutrons, protons, and electrons.
  b. Are most rhenium isotopes likely to have fewer neutrons than rhenium-188?
  c. The half-life of rhenium-188 is 17 hours. If it takes 30 minutes to bind the isotope to an antibody that delivers the rhenium to the tumor, what percentage of the rhenium remains after binding to the antibody?
  d. The effectiveness of rhenium-188 is thought to result from penetration of β particles as deep as 8 mm into the tumor. Why wouldn't an α emitter be more effective?
  e. Using an appropriate reference text, such as the *CRC Handbook of Chemistry and Physics* (CRC Press, Boca Raton, FL), pick out the two most abundant isotopes of rhenium. List their natural abundances and explain why the one that is radioactive decays by the pathway that it does.

20.105 The following radiochemical equations are based on successful attempts to synthesize superheavy elements. Complete each equation by filling in the symbol of the superheavy nuclide that was synthesized.
  a. $^{58}_{26}Fe + {}^{209}_{83}Bi \rightarrow ? + {}^{1}_{0}n$
  b. $^{64}_{28}Ni + {}^{209}_{83}Bi \rightarrow ? + {}^{1}_{0}n$
  c. $^{62}_{28}Ni + {}^{208}_{82}Pb \rightarrow ? + {}^{1}_{0}n$
  d. $^{22}_{10}Ne + {}^{249}_{97}Bk \rightarrow ? + 4{}^{1}_{0}n$
  e. $^{58}_{26}Fe + {}^{208}_{82}Pb \rightarrow ? + {}^{1}_{0}n$

20.106. Smoke Detectors Americium-241 ($t_{1/2}$ = 433 yr) is used in smoke detectors. The α particles from the ^{241}Am ionize nitrogen and oxygen in the air creating a current. When smoke is present, the current decreases, setting off the alarm.

   a. Does a smoke detector bear a closer resemblance to a Geiger counter or to a scintillation counter?

   b. How long will it take for the activity of a sample of ^{241}Am to drop to 1% of its original activity?

   c. Why are smoke detectors containing ^{241}Am safe to handle without protective equipment?

20.107. In 2006 an international team of scientists confirmed the synthesis of a total of three atoms of $^{294}_{118}$Uuo in experiments run in 2002 and 2005. They bombarded a ^{249}Cf target with ^{48}Ca nuclei.

   a. Write a balanced radiochemical equation describing the synthesis of $^{294}_{118}$Uuo .

   b. The synthesized isotope of Uuo undergoes α decay ($t_{1/2}$ = 0.9 ms). What nuclide is produced by the decay process?

   c. The nuclide produced in b also undergoes α decay ($t_{1/2}$ = 10 ms). What nuclide is produced by this decay process?

   d. The nuclide produced in c also undergoes α decay ($t_{1/2}$ = 0.16 s). What nuclide is produced by this decay process?

   e. If you had to select an element that occurs in nature and that has physical and chemical properties similar to Uuo, which element would it be?

*20.108. Consider the following decay series:

$$A\ (t_{1/2} = 4.5\ \text{s}) \rightarrow B\ (t_{1/2} = 15.0\ \text{days}) \rightarrow C$$

If we start with 10^6 atoms of A, how many atoms of A, B, and C are there after 30 days?

20.109. Which element in the following series will be present in the greatest amount after 1 year?

$$^{214}_{81}\text{Bi} \xrightarrow{\alpha} {}^{210}_{81}\text{Tl} \xrightarrow{\beta} {}^{210}_{82}\text{Pb} \xrightarrow{\beta} {}^{210}_{83}\text{Bi} \longrightarrow$$

$$t_{1/2} = \quad 20\ \text{min} \quad 1.3\ \text{min} \quad 20\ \text{yr} \quad 5\ \text{days}$$

*20.110. Dating Cave Paintings Cave paintings in Gua Saleh Cave in Borneo have recently been dated by measuring the amount of ^{14}C in calcium carbonate that formed over the pigments used in the paint. The source of the carbonate ion was atmospheric CO_2.

   a. What is the ratio of the ^{14}C activity in calcium carbonate that formed 9900 years ago to that in calcium carbonate formed today?

   b. The archeologists also used a second method, uranium–thorium dating to confirm the age of the paintings by measuring trace quantities of these elements present as contaminants in the calcium carbonate. Shown below are two candidates for the U–Th dating method. Which isotope of uranium do you suppose was chosen? Explain your answer.

$$^{235}_{92}\text{U} \longrightarrow {}^{231}_{90}\text{Th} \longrightarrow {}^{231}_{91}\text{Pa} \longrightarrow$$
$$t_{1/2} = \quad 7.04 \times 10^8\ \text{yr} \quad\quad 25.6\ \text{h} \quad\quad 3.25 \times 10^4\ \text{yr}$$

$$^{234}_{92}\text{U} \longrightarrow {}^{230}_{90}\text{Th} \longrightarrow {}^{236}_{88}\text{Ra} \longrightarrow$$
$$t_{1/2} = \quad 2.44 \times 10^5\ \text{yr} \quad\quad 7.7 \times 10^4\ \text{yr} \quad\quad 1600\ \text{yr}$$

20.111. The synthesis of new elements and specific isotopes of known elements in linear accelerators involves the fusion of smaller nuclei.

   a. An isotope of platinum can be prepared from nickel-64 and tin-124. Write a balanced equation for this nuclear reaction. (You may assume that no neutrons are ejected in the fusion reaction.)

   b. Substitution of tin-132 for tin-124 increases the rate of the fusion reaction 10 times. Which isotope of Pt is formed in this reaction?

20.112. A sample of drinking water collected from a suburban Boston municipal water system in 2002 contained 0.5 pCi/L of radon. Assume that this level of radioactivity was due to the decay of ^{222}Rn ($t_{1/2}$ = 3.8 days).

   a. What was the level of radioactivity (Bq/L) of this nuclide in the sample?

   b. How many disintegrations per hour would occur in 2.5 L of the water?

20.113. An atom of darmstadtium-269 was synthesized in 2003 by bombardment of a ^{208}Pb target with ^{62}Ni nuclei. Write a balanced nuclear reaction describing the synthesis of ^{269}Ds.

20.114. There was once a plan to store radioactive waste that contained plutonium-239 in the reefs of the Marshall Islands. The planners claimed that the plutonium would be "reasonably safe" after 240,000 years. If its half-life is 24,400 years, what percentage of the ^{239}Pu would remain after 240,000 years?

20.115. Dating Prehistoric Bones In 1997 anthropologists uncovered three partial skulls of prehistoric humans in the Ethiopian village of Herto. Based on the amount of ^{40}Ar in the volcanic ash in which the remains were buried, their age was estimated at between 154,000 and 160,000 years old.

   a. ^{40}Ar is produced by the decay of ^{40}K ($t_{1/2}$ = 1.28 × 10^9 yr). Propose a decay mechanism for ^{40}K to ^{40}Ar.

   b. Why did the researchers choose ^{40}Ar rather than ^{14}C as the isotope for dating these remains?

*20.116. Biblical Archeology The Old Testament describes the construction of the Siloam Tunnel, used to carry water into Jerusalem under the reign of King Hezekiah (727–698 bc). An inscription on the tunnel has been interpreted as evidence that the tunnel was not built until 200–100 bc. ^{14}C dating (in 2003) indicated a date close to 700 bc. What is the ratio of ^{14}C in a wooden object made in 100 bc to one made from the same kind of wood in 700 bc?

20.117. Stone Age Skeletons The discovery of six skeletons in an Italian cave at the beginning of the 20th century was considered a significant find in Stone Age archaeology. The age of these bones has been debated. The first attempt at radiocarbon dating indicated an age of 15,000 years. Redetermination of the age in 2004 indicated an older age for two bones, between 23,300 and 26,400 years. What is the ratio of ^{14}C in a sample 15,000 years old to one 25,000 years old?

*20.118. The origin of the two naturally occurring isotopes of boron, ^{11}B and ^{10}B, are unknown. Both isotopes may be formed from collisions between protons and carbon, oxygen, or nitrogen in the aftermath of supernova explosions. Propose nuclear reactions for the formation of ^{10}B from such collisions with ^{12}C and ^{14}N.

20.119. Thorium-232 slowly decays to lead-208 as shown here.

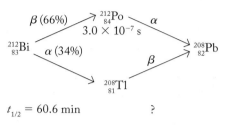

$$^{232}_{90}\text{Th} \xrightarrow{\alpha} \quad ^{228}_{88}\text{Ra} \xrightarrow{\beta} \quad ^{228}_{89}\text{Ac} \xrightarrow{\beta} \quad ^{228}_{90}\text{Th} \xrightarrow{\alpha}$$

$t_{1/2} = 1.41 \times 10^{10}$ yr 5.8 yr 6.13 h 1.91 yr

$$^{224}_{88}\text{Ra} \xrightarrow{\alpha} \quad ^{220}_{86}\text{Rn} \xrightarrow{\alpha} \quad ^{216}_{84}\text{Po} \xrightarrow{\alpha} \quad ^{212}_{82}\text{Pb} \xrightarrow{\beta}$$

$t_{1/2} =$ 3.64 days 55 s 0.15 s 10.6 h

β (66%) → $^{212}_{84}\text{Po}$ $\xrightarrow{\alpha}$

3.0×10^{-7} s

$^{212}_{83}\text{Bi}$ α (34%) → $^{208}_{82}\text{Pb}$

β

→ $^{208}_{81}\text{Tl}$

$t_{1/2} = 60.6$ min ?

The half-lives of the intermediate decay products range from seconds to years. Note that bismuth-212 decays to lead-208 by two pathways: first β and then α decay, or α and then β decay. The immediate nuclide in the second pathway is thallium-208. The thallium-208 can be separated from a sample of thorium nitrate by passing a solution of the sample through a filter pad containing ammonium phosphomolybdate. The radioactivity of ^{208}Tl trapped on the filter is measured as a function of time. In one such experiment, the following data were collected:

Time (s)	Counts/min
60	62
120	40
180	35
240	22
300	16
360	10

Use the data in the table to determine the half-life of ^{208}Tl.

Radon, Radium, and Uranium: Too Hot to Handle

20.120. Some radon detectors contain a thin plastic film that is sensitive to α particles. An α particle striking the film leaves a track. Why is the number of tracks more important than the length of the track in determining how much radon is present?

20.121. One type of radon detector uses charcoal to collect radioactive ^{214}Pb and ^{214}Bi. The concentration of Rn is calculated by measuring γ rays produced by those two isotopes.
a. How are ^{214}Pb and ^{214}Bi related to ^{222}Rn?
b. Why are lead and bismuth collected rather than collecting radon directly?

20.122. Why is it a bad idea to bathe repeatedly in hot springs containing dissolved radium?

20.123. Both ^{226}Rn and ^{222}Rn are produced from radioactive decay of uranium in soils and rocks. Why are these radon isotopes a greater risk to health than the parent uranium?

20.124. Which of the following elements is most likely to have chemical properties similar to radium: uranium, potassium, barium, or krypton?

Life and the Periodic Table

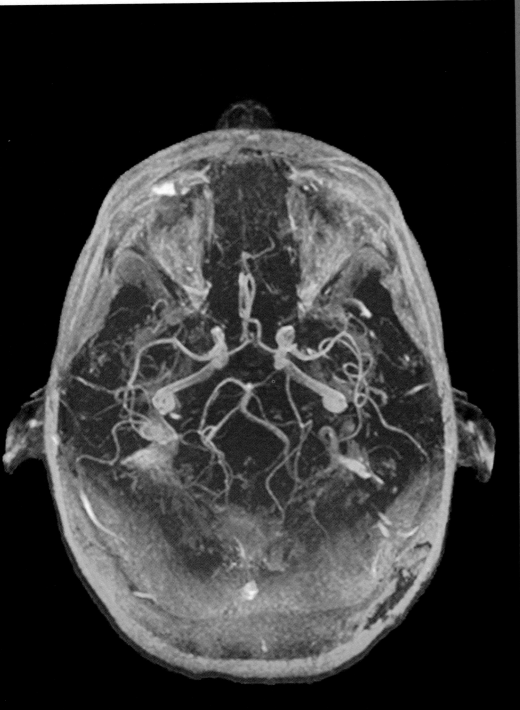

MRI *Magnetic resonance imaging is a valuable diagnostic tool in medicine. Some MRI studies rely on compounds containing gadolinium to enhance the image.*

A LOOK AHEAD: What Elements Do We Find in Our Bodies?

Of all the elements in the periodic table, 90 are present in the world around us. (Radioactive technetium ($Z = 43$) and promethium ($Z = 45$) have not been found in nature.) Have you ever wondered how many of the elements in the periodic table are found in the human body, which are important to good health and nutrition, and how many are essential to the survival of other living organisms?

It turns out that roughly one-third of the elements serve an identifiable role in human health and in living things in general. However, the abundances of these elements are not equal in all species or even in different organs of the same species. Some—like carbon, hydrogen, oxygen, nitrogen, sulfur, and phosphorus—are the principal constituents of all plants and animals. The alkali metal cations Na^+ and K^+ act as charge carriers, maintain osmotic pressure, and transmit nerve impulses. The alkaline earth cations Mg^{2+} and Ca^{2+} are important in photosynthesis and form structural materials such as bones and teeth. Chloride ions balance the charge of the Na^+ and K^+ ions to maintain electrical neutrality in living cells. Other main group elements like iodine or selenium are present in trace amounts in our bodies. A number of transition metal ions such as Fe^{2+} and Ni^{2+} are the key components of the enzymes that catalyze biochemical reactions.

Many of the elements with no known biological function are still useful in medicine as either diagnostic tools or therapeutic agents. Radioactive isotopes act as effective imaging agents for organs and tumors. The directed radiation from other radioactive isotopes can be used to kill cancer cells. Compounds containing metallic elements such as platinum, silver, and gold are effective drugs for treating cancer, burns, and arthritis, respectively. Pharmaceuticals containing lithium ions are used to treat depression. Corrosion-resistant pure metals such as tantalum are used in artificial joints.

Even toxic metals like mercury can be used in fillings for teeth and in batteries for pacemakers and other devices that extend life. In this chapter we will survey the roles of different elements in the human body and their application to health. At the same time we will call upon the knowledge and skills you have acquired in your study of general chemistry to answer questions and solve problems that link concepts from prior chapters of this book to this one on the roles of the elements in the chemistry of life.

21.1 The Periodic Table of Life

Many of the 90 naturally occurring elements can be classified as either **essential** or **nonessential elements** to life. Essential elements are defined as those present in tissue, blood, or other body fluids that have a beneficial physiological function. The essential elements include those whose absence leads to lower functioning of the organism. For example, a small amount of selenium is essential in our diets to allow our bodies to make selenocysteine, an amino acid. Nonessential elements include elements that are detected in the body but have no known function. In some cases, the presence of a nonessential element has a **stimulatory effect** leading to increased growth or other biological responses. For example, small amounts of antimony promote the growth of some mammals. Nonessential elements are often incorporated into our bodies as a result of similarities in their chemical properties to those of an essential element. For example, Pb^{2+} ions are

Essential elements are those elements present in tissue, blood, or other body fluids that have a physiological function.

Nonessential elements include elements that are present in an organism but have no known function.

A nonessential element may have a **stimulatory effect** when added to the diet of an organism, leading to increased growth or other biological responses.

incorporated into teeth and bones because they are similar to Ca^{2+} ions in size and charge.

When we speak of an essential or nonessential element, we are almost always referring to an ion or compound containing that element rather than the pure element. For example, when we describe zinc as an essential element, we are referring to zinc ions, Zn^{2+}, rather than zinc metal.

The essential elements are further classified as **major, trace,** or **ultratrace elements**. Major elements are present in gram quantities in the human body and are required in large amounts in our diet. Almost all foods are rich in compounds containing carbon, hydrogen, oxygen, nitrogen, sulfur, and phosphorus. Salt is perhaps the most familiar dietary source of sodium and chloride ions, although both are ubiquitous in food. Vegetables like broccoli and brussels sprouts or fruits like bananas are rich in potassium. Calcium is found in dairy products and is often added to orange juice.

Table 21.1 summarizes the average elemental composition of the human body, the universe, Earth's crust, and seawater. Note that the elemental composition of our bodies most closely resembles the composition of seawater. Actually the match would be even closer if it were not for the biological processes in the sea that remove trace nutrients, such as nitrogen and phosphorus, and that store others in solid structures like the $CaCO_3$ that makes up corals and mollusk shells.

Major elements are those present in the body in average concentrations greater than 1 mg/g.

Trace elements are those present in the body in average concentrations between 1 and 1000 μg/g.

Ultratrace elements are those present in the body in average concentrations less than 1 μg/g.

TABLE 21.1 Comparative Composition^a of the Universe, Earth's Crust, Seawater, and the Human Body

Element	Universe	Earth's Crust	Seawater	Human Body
Hydrogen	91	0.22	66	63
Oxygen	0.57	47	33	25.5
Carbon	0.021	0.019	0.0014	9.5
Nitrogen	0.042			1.4
Calcium		3.5	0.006	0.31
Phosphorus				0.22
Chlorine			0.33	0.03
Potassium		2.5	0.006	0.06
Sulfur	0.001	0.034	0.017	0.05
Sodium		2.5	0.28	0.01
Magnesium	0.002	2.2	0.033	0.01
Helium	9.1			
Silicon	0.003	28		
Aluminum		7.9		
Neon	0.003			
Iron	0.002	6.2		
Bromine			0.0005	
Titanium		0.46		
All other elements	<0.1	<0.1	<0.1	<0.1

^aCompositions are expressed as the percentage of the total number of atoms. Because of rounding, the totals do not equal exactly 100%.

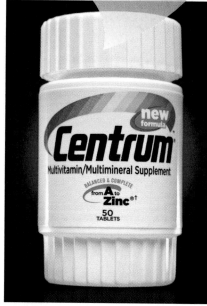

The labels on multivitamin supplements may not list RDA values, but rather *% daily values* (DVs). Daily values are based on RDA or AI values, but there can be inconsistencies, particularly among the ultratrace essential elements.

Our diets should supply us with sufficient quantities of all essential elements: these quantities are called *dietary reference intake* (DRI) values in the United States and Canada. DRI values are based on the recommendations of the Food and Nutrition Board of the National Academy of Sciences and are frequently updated in response to the results of research. For many essential elements, DRI values are also called *recommended dietary allowance* (RDA) values. Among the essential major elements, RDA values range from 0.3–0.4 g of magnesium per day to 1.0–1.2 g of calcium. RDA values for trace essential elements, including iron and zinc, are in the 5- to 25-mg/day range. There are RDA values for some, but not all, of the essential ultratrace elements. They range from 50 μg/day of selenium and molybdenum to 2 mg/day of manganese. RDA values have not been established for other ultratrace essential elements, including copper, chromium, and manganese. For these elements DRI values take the form of *adequate intake* (AI) values: quantities that are normally a part of the diets of healthy people.

What about the elements that are considered nonessential? The biochemistry of nearly every naturally occurring element has been investigated. Platinum can be combined with other elements to form particular compounds that are effective anticancer drugs. Radioactive isotopes of many elements have been used to image organs and identify tumors. We look at these applications in detail in Section 21.4. The utility of elements in medicine extends to the construction of medical devices like pacemakers and artificial joints. Not surprisingly, some nonessential elements, such as mercury and lead, can be highly toxic (see Section 21.5).

Finally, what about the twenty synthetic "man-made" elements that are not found in nature? Many of these elements have been prepared in minute quantities in the nuclear reactors or cyclotrons described in Chapter 20. As we will see, one element—technetium—does not exist in nature but has tremendous utility in nuclear medicine.

21.2 Essential Major Elements

Of all the naturally occurring elements listed in the periodic table, the 11 shown in red in Figure 21.1 account for more than 99% of our bodies' mass. Oxygen is the most abundant element, followed by carbon and hydrogen. Although life depends on the presence of elemental oxygen as the gas O_2, much of the oxygen in our cells and intercellular fluids is combined with hydrogen as water. None of the elements except for oxygen are found in our bodies in their elemental form.

FIGURE 21.1 The eleven elements shown in red are essential major elements. Their abundances range from 35 g (magnesium) to 45.5 kg (oxygen) in an adult human of average mass (~70 kg).

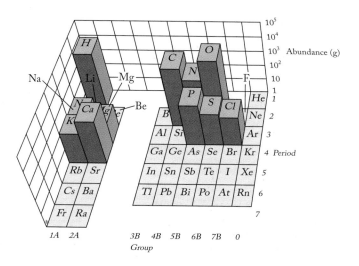

Carbon is combined with hydrogen, oxygen, nitrogen, phosphorus, and sulfur in the biopolymers described in Chapter 19. Carbohydrates contain carbon, hydrogen, and oxygen. Amino acids and proteins require nitrogen and sulfur in addition to carbon, hydrogen, and oxygen. The nucleotides in DNA are connected by phosphate (PO_4^{3-}) groups.

Covalently bonded molecular ions—such as bicarbonate ion (HCO_3^-), sulfate ion (SO_4^{2-}), and dihydrogen phosphate ion ($H_2PO_4^-$)—are ubiquitous in biological systems. The number of compounds containing these elements in the human body dwarfs the number of compounds of all the other elements.

Four of the less abundant, but still "major" essential elements highlighted in Figure 21.1 are in groups 1 and 2. Their average concentrations in the human body are listed below.

Element	Average Concentration (mg/g)
Calcium	15.0
Potassium	2.0
Sodium	1.5
Magnesium	0.5

Let's explore some of the roles these four elements play in the biochemistry of the human body. As we do, we will revisit several of the chemical principles related to chemical reactions and equilibria in aqueous solutions that we discussed in prior chapters.

CONNECTION We introduced phospholipids in Chapter 19 as consisting of glycerol with two fatty acid chains and a polar group containing a phosphate moiety.

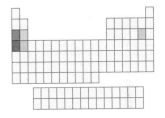

Alkali Metals

Alkali metal ions are critical to cell functioning. One of the most common blood tests ordered by physicians measures the levels of Na^+ and K^+ in blood. To maintain a constant concentration of these ions in body fluids, sodium and potassium ions must be able to move in and out of cells. However, the membrane surrounding a typical cell incorporates a phospholipid bilayer with a thickness between 5 and 6 nm (Figure 21.2). Polar head groups containing phosphate are on the outside of the membrane with nonpolar fatty acid tails oriented toward the middle of the membrane. Direct diffusion of Na^+ and K^+ through the lipid bilayer is difficult because these polar cations do not dissolve in the nonpolar interior of the membrane.

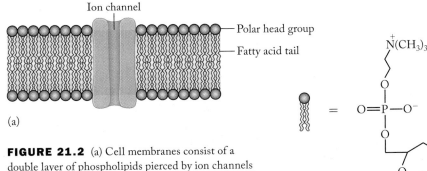

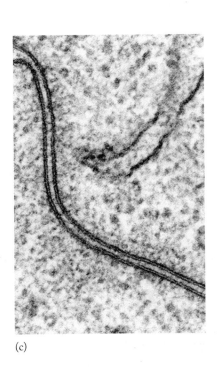

FIGURE 21.2 (a) Cell membranes consist of a double layer of phospholipids pierced by ion channels formed by coiled proteins. The polar head groups face the aqueous solutions inside and outside the cell, whereas the fatty acid chains form a nonpolar region within the membrane. (b) Molecular structure of a phospholipid. (c) An electron micrograph of the membranes separating two adjacent cells.

Ion channels are groups of helical proteins that penetrate cell membranes and allow selective transport of cations.

Ion pumps are large and complex systems of membrane proteins that exchange ions inside the cell with those in the intercellular fluid.

CONNECTION In Chapter 10 we learned that alkali metal cations dissolved in water are surrounded by six water molecules. Each water molecule is oriented so that the oxygen atoms point toward the cation. In Chapter 17 we described this interaction as an example of a Lewis acid (cation, electron-pair acceptor) interacting with a Lewis base or ligand (water, electron-pair donor).

The membrane, however, is pierced by **ion channels**, groups of helical proteins that allow selective transport of cations. An ion channel controls which cations can pass through the membrane by permitting only those of a certain size and charge to pass. For example, the ion channel for potassium ions has a radius that matches the ionic radius of K^+ (138 pm) but not Na^+ (102 pm) or other cations. The sodium ion channel is also quite selective; it excludes K^+ and Ca^{2+} even though the radii of Na^+ and Ca^{2+} differ by only 2 pm. Another difference between the Na^+ and K^+ channels is the ability of H_3O^+ (hydronium ion) to pass through sodium ion channels but not potassium ion channels.

Living organisms contain oxygen-rich molecules like nonactin (Figure 21.3) that behave as ligands and bond to alkali metal ions through strong ion–dipole forces. The result is a polar, charged metal ion encapsulated in a larger nonpolar molecule. The resulting nonpolar complex can pass (diffuse) through nonpolar cell membranes. This effect provides an alternative to ion channels for the transport of alkali metal ions.

A third transport pathway for alkali metal cations is provided by Na^+-K^+ ion pumps. **Ion pumps** are large and complex systems of membrane proteins that exchange ions inside the cell (for example, Na^+) with those in the intercellular fluid (for example, K^+). Unlike cation diffusion and transport through ion channels, the sodium-potassium pump requires energy. The energy is provided by the hydrolysis of ATP to ADP. An example of the functioning of the Na^+-K^+ pump on a cellular level is the response of nerve cells to touch. Stimulation of the nerve cell causes Na^+ to flow into the cell and K^+ to flow out; this flow of ions produces the nerve impulse. The pump then recharges the system by pumping Na^+ out of the cell and K^+ into the cell. In terms of whole-body function, too much Na^+ has been linked to hypertension (high blood pressure).

Alkaline Earth Metals

Like the alkali metal cations, the cellular concentrations of Mg^{2+} and Ca^{2+} are maintained by ion pumps. Compared to the group 1 cations, however, magnesium and calcium play a wider range of biological roles in the cell and in the overall health of an organism. We have mentioned that calcium is a major component of teeth and bones. A prolonged deficiency of calcium can lead to osteoporosis, whereas high concentrations of calcium in muscle cells contribute to cramps. Most kidney stones are made of calcium oxalate or calcium phosphate. Magnesium deficiencies can reduce physical and mental capacity because of the role of Mg^{2+} in the transfer of phosphate groups to and from ATP and thus affect the amount of energy available on the cellular level.

CONNECTION The inorganic chemistry of calcium is described in Chapter 4.

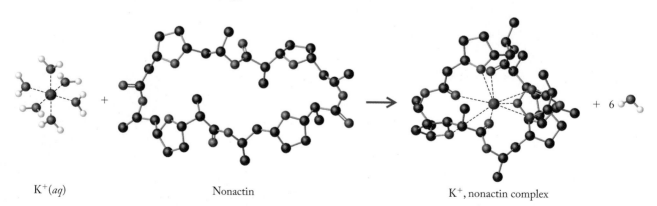

$K^+(aq)$ Nonactin K^+, nonactin complex $+ 6$

FIGURE 21.3 In living organisms, chelating ligands such as the molecule nonactin can encapsulate hydrated alkali metal ions and facilitate their transport across cell membranes.

Magnesium is also found in chlorophyll, the green pigment in plants required for photosynthesis. Photosynthesis is fundamental to all life. Green plants use sunlight to convert carbon dioxide and water to carbohydrates and oxygen. Efficient light-absorbing pigments are needed to capture solar energy and begin photosynthesis. Chlorophylls are only one of a number of molecules (Figure 21.4) used by plants to collect and capture light energy across the entire visible spectrum (400 to 700 nm). Chlorophylls from different plants and bacteria vary slightly in their composition, but all of them contain magnesium coordinated to four nitrogen atoms. The presence of magnesium in chlorophyll does not account for the color of the molecule nor does it play a direct role in the absorption of sunlight. The function of the Mg^{2+} in chlorophylls is to orient the various pigment molecules through coordinate bond formation, which allows transfer of energy to the reaction centers where H_2O and CO react during photosynthesis.

CONNECTION The stability of complexes formed between metal ions and polydentate ligands, such as the chlorin ring in chlorophyll, is described in Chapter 17.

FIGURE 21.4 Green plants use a variety of molecules to absorb sunlight. Among them, only chlorophylls contain magnesium. Note that the only difference between the two bacteriochlorophylls is a CHO group rather than a methyl group on one of the nitrogen-containing (chlorin) rings. The absorption spectra of plant pigments in the visible region of the spectrum illustrate how plants use different molecules to ensure that the energy of many wavelengths is captured.

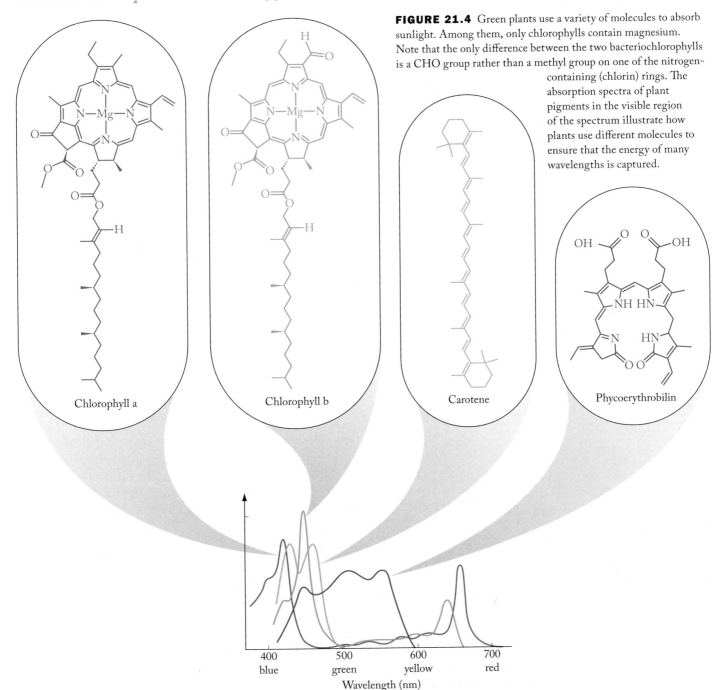

Composite materials contain a mixture of substances with different compositions and structures.

Most Mg^{2+} compounds are white, not green. Why?

(Answers to Concept Tests are in the back of the book.)

Mg^{2+} ions also play important roles in the hydrolysis of ATP and the phosphorylation of ADP. The removal (hydrolysis) of 1 mole of HPO_4^{2-} from 1 mole of ATP provides 30.5 kJ/mol of free energy ($\Delta G°$). The many Mg^{2+}-mediated ATP → ADP processes include transferring phosphate to glucose in the overall conversion of glucose to pyruvate and driving Na^+-K^+ ion pumps.

To some extent, calcium ions are also capable of mediating the hydrolysis of ATP; however, Ca^{2+} ions play other roles in the cell. Calcium ions are required to trigger muscle contractions. Calcium ions for this purpose are stored in proteins. Recall that the action of Na^+-K^+ pumps and ion channels is responsible for the generation of nerve impulses. One effect of nerve impulses is to trigger the release of Ca^{2+} ions. In a complicated series of processes, muscle cells contract and relax as calcium ions are released. Calcium ions are returned to their storage proteins in a process coupled to Mg^{2+}-mediated ATP hydrolysis.

Of the four alkali and alkaline earth elements that are essential to human nutrition, only calcium plays a major role in the formation of biominerals such as teeth, bones, and seashells. Mammalian bones are an example of a **composite material**. Composite materials contain a mixture of substances with different compositions and structures. About 30% of dry bone mass, for example, is composed of elastic protein fibers. The remainder consists of calcium salts including the mineral hydroxyapatite, $Ca_5(PO_4)_3(OH)$. Hydroxyapatite is also a principal component of tooth dentin and enamel. The inorganic hydroxyapatite crystals are bound to the protein through phosphate groups. Both sulfate and carbonate may substitute for phosphate in hydroxyapatite. On the other hand, shells like mollusk shells are mostly calcium carbonate ($CaCO_3$) formed in a matrix of proteins and polysaccharides. Some magnesium is incorporated into the calcium carbonate outer shells or exoskeletons of marine organisms capable of photosynthesis.

Among the other elements in groups 1 and 2 rubidium is the most abundant element with no known biological function. Some single-cell organisms build exoskeletons made with $SrSO_4$ and $BaSO_4$.

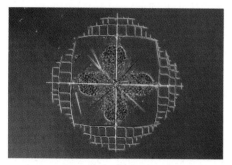

Acantharia

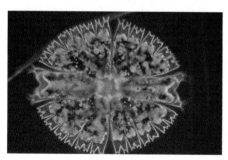

Desmidiaceae

The shells of planktonic Acantharia and Desmidiaceae contain strontium and barium sulfate, respectively.

Chlorine

Of all the halogens, only chlorine (as chloride ion) is present in sufficient quantities to be considered a major element in humans. Chloride ions are the most abundant anions in the human body and are implicated in many functions. Fluoride, bromide, and iodide ions function as essential trace elements. The concentration of chloride ions in the human body (1.5 mg/g) is slightly less than one-tenth of the concentration of Cl^- in seawater (19 mg/g) but about 15 times greater than in Earth's crust (0.13 mg/g). Just as for the bulk cations, the chloride ions are transported into and out of cells via ion channels and ion pumps. To maintain electrical neutrality in a cell, transport of alkali metal cations is accompanied

CONNECTION Additional information on the chemistry of chlorine and the other halogens of group 17 can be found in Chapters 8 and 10.

by transport of chloride anions. The cotransport of Na^+ and Cl^- is essential in kidney function, where the ions are reabsorbed by the body while waste products are eliminated.

Chloride ion also plays a major role in the elimination of CO_2. Nonpolar carbon dioxide can easily pass through the largely nonpolar cell membranes of red blood cells, where it is converted to bicarbonate ion, HCO_3^-, by the action of carbonic anhydrase, a zinc-containing enzyme. When HCO_3^- is pumped from the red blood cell, Cl^- enters the cell through an ion channel to maintain the overall charge balance in the cell.

Chloride ion concentrations are high in gastric juices because of the presence of hydrochloric acid, which catalyzes digestive processes in the stomach. In response to food in the digestive system, cells tap ATP for the needed energy to pump hydrochloric acid into the stomach. Sample Exercise 21.1 reviews the relationship between pH and the concentration of strong acids.

CONNECTION An introduction to the strengths of acids and the calculation of pH can be found in Chapters 4 and 16.

SAMPLE EXERCISE 21.1   **Calculating the Concentration of HCl in Stomach Acid**

Calculate the molarity of hydrochloric acid, HCl, in gastric juice that has a pH of 0.80.

COLLECT AND ORGANIZE We are given the pH of a solution and asked to calculate the concentration of HCl that corresponds to this pH. According to Equation 16.8, $pH = -\log[H^+]$. Hydrochloric acid is a strong acid and ionizes completely to form H^+ and Cl^- in water:

$$HCl(aq) \rightarrow H^+(aq) + Cl^-(aq)$$

ANALYZE The equation describing the ionization of hydrochloric acid indicates that 1 mole of H^+ ions are formed for every mole of HCl present.

SOLVE

$$pH = 0.80 = -\log[H^+]$$

We take the antilog of both sides of the equation to solve for $[H^+]$:

$$[H^+] = 10^{-0.80} = 0.16 \, M \, H_3O^+$$

Therefore, the concentration of HCl is

$$0.16 \, M \, H_3O^+ \times \frac{1 \text{ mol HCl}}{1 \text{ mol } H_3O^+} = 0.16 \, M \, HCl$$

THINK ABOUT IT The concentration of HCl, 0.16 M, calculated in this exercise seems reasonable. The "concentrated" solutions of HCl you may have used in the laboratory often are 1.0, 3.0, or 6.0 M, and are much more acidic than one would expect for the acid concentrations found in the stomach.

Practice Exercise Calculate the pH of a solution prepared by mixing 10 mL of 0.16 M HCl with 15 mL of water.

(Answers to Practice Exercises are in the back of the book.)

Malfunctioning chloride ion channels is the underlying cause of cystic fibrosis, a lethal genetic disease, that causes patients to accumulate mucus in their airways so that breathing is difficult. The discovery of high concentrations of Na^+ and Cl^- in the sweat of cystic fibrosis patients led to an understanding of the role of chloride ion transport in patients with this disease.

21.3 Essential Trace and Ultratrace Elements

Moving beyond the essential major elements, all of the elements shown in blue and red in Figure 21.5 have been detected in the human body but are present in much smaller concentrations. Some of these elements are clearly essential to good health. The functions of others have not been determined. Some elements do not appear to be essential, yet they have stimulatory effects. In other words, adding small amounts of these elements to the diet leads to growth or better health. To help keep track of the essential trace and ultratrace elements, we divide our discussion into the main group and transition metal elements and focus on those elements that are essential or stimulatory. As analytical techniques become more sensitive, additional elements will undoubtedly be detected in our bodies.

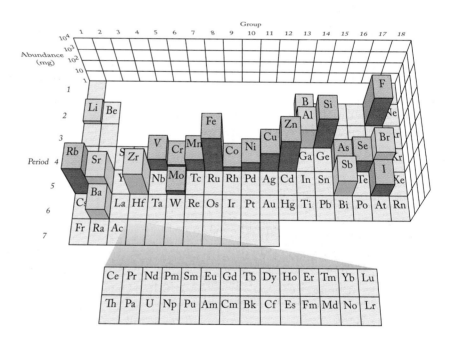

FIGURE 21.5 The elements shown in red are considered essential trace or ultratrace elements. The elements shown in blue are present in milligram or greater quantities in the human body but do not appear to be essential elements. The quantities of each element are those usually found in an adult human of average mass (70 kg).

Main Group Elements

The four main group elements in red in Figure 21.5, silicon, selenium, fluorine, and iodine, are essential trace elements. The others may be, but their biological functions have yet to be established.

ALKALI AND ALKALINE EARTH METALS Two elements each from group 1 (Na^+, K^+) and group 2 (Mg^{2+}, Ca^{2+}) in the periodic table are essential major elements. Rubidium is generally regarded as nonessential, yet it is the 15th most abundant element in the body. It is believed that Rb^+ is retained by the body because of the similarity of its size and chemistry to K^+. Like the other cations of group 1, cesium ions (Cs^+) are readily absorbed by the body. Cesium cations have no known function although they can substitute for K^+ and interfere with potassium-dependent functions. In most cases, the concentration of cesium in the environment is low, and so exposure to Cs^+ is not a health concern. The nuclear accident at Chernobyl in 1986, however, released significant quantities of radioactive ^{137}Cs into the environment. The ability of Cs^+ to substitute for K^+ led to the incorporation of $^{137}Cs^+$ into plants, which rendered crops grown in the immediate area unfit for human consumption because of the radiation hazard posed by this long-lived ($t_{1/2} \approx 30$ yr) β emitter.

The human body appears to have no use for Sr^{2+} and Ba^{2+} ions. These ions do find their way into human bones, where they replace Ca^{2+} ions. At the low concentrations of Sr^{2+} and Ba^{2+} that are typically present in the human body, these elements appear to be benign. However, as in the case of radioactive ^{137}Cs, incorporation of ^{90}Sr ($t_{1/2} = 29$ yr) in bones can lead to leukemia. Atmospheric testing of nuclear weapons over Pacific islands and in sparsely populated regions of the American West in the 1950s released ^{90}Sr into the environment. All of the toxic effects of the fallout from these tests did not become apparent for several decades.

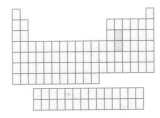

CONNECTION The biological effects of different types of radiation are described in Chapter 20.

NONMETALLIC ELEMENTS

Silicon and Germanium Aside from carbon and silicon, the group 14 elements appear to have either no function in the human body or are toxic. In mammals, a lack of silicon stunts growth. It is also believed that the presence of silicon as silicic acid [$Si(OH)_4$] reduces the toxicity of Al^{3+} ions in organisms by precipitating the aluminum as aluminosilicate minerals. Amorphous silica (SiO_2) is found in the exoskeletons of diatoms and in the cell membranes of some plants, such as the tips of stinging nettles. It is generally agreed that germanium is a nonessential element and barely detectable in the human body. The use of bis(carboxyethyl)-germanium sesquioxide (Figure 21.6) has been touted as a nutritional supplement, but its efficacy remains controversial.

Arsenic and Antimony Unlike nitrogen and phosphorus, which are essential major elements, the biological function of the other elements in group 15 can be ambiguous. Arsenic compounds are toxic. They bind irreversibly to sulfur in enzymes and disrupt their function. Nevertheless, there is some evidence that a lack of arsenic

CONNECTION The inorganic chemistry of nitrogen is discussed in Chapter 6.

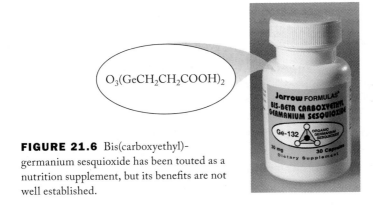

$$O_3(GeCH_2CH_2COOH)_2$$

FIGURE 21.6 Bis(carboxyethyl)-germanium sesquioxide has been touted as a nutrition supplement, but its benefits are not well established.

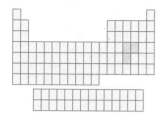

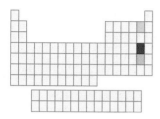

⊙⊙ **CONNECTION** The inorganic chemistry of oxygen and sulfur is discussed in Chapter 9.

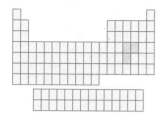

FIGURE 21.7 Selenocysteine is the selenium-containing analog of the amino acid cysteine. Much of the selenium in the human body is found in proteins containing selenocysteine.

⊙⊙ **CONNECTION** Solubility products, or K_{sp}, were introduced in Chapter 16. Le Châtelier's principle is discussed in Chapter 15.

in their diet stunts the growth of some mammals. The role of antimony is also poorly understood. Most antimony compounds are toxic; they cause liver damage. However, ultratrace amounts of antimony may have a stimulatory effect, and selected antimony compounds have been used medically as antiparasitic agents.

Selenium Selenium is considered an essential trace element even though its average concentration in the human body (0.3 $\mu g/g$) is only about 1/10th that of its group 16 neighbor sulfur (2.5 $\mu g/g$). Mounting scientific evidence points to a need for a minimum daily dose of selenium greater than 50 μg. The effects of selenium toxicity, however, are apparent in people who ingest more than 500 μg of Se per day. Most of the selenium we need is obtained indirectly from soil through selenium-rich foods (garlic, mushrooms, asparagus) or by eating fish. Selenium occurs in the body as the amino acid selenocysteine (Figure 21.7) and is incorporated into proteins such as iodothyronine deiodoinase (see next section on fluorine, bromine, and iodine) and glutathione peroxidase, which act as catalysts for biochemical reactions.

Selenocysteine acts as an antioxidant. Our bodies need oxygen to survive, yet living in an oxygen-rich or oxidizing atmosphere can lead to the formation of potentially dangerous oxidizing agents within cells. For example, when fatty acids are metabolized, a class of oxidizing agents called alkyl hydroperoxides (R–O–O–H) are formed (where R represents a long hydrocarbon chain from a fatty acid). Alkyl hydroperoxides, in turn, can attack the lipid bilayer of cell membranes described in Section 21.2. It is believed that aging is related to the inability of the body to inhibit oxidative degradation of tissue. Selenocysteine participates in a series of reactions that result in the decomposition of these alkyl hydroperoxides.

Fluorine, Bromine, and Iodine Because the halogens are strong oxidizing agents, they are not found in the elemental state in nature, but rather as halide anions. The role of chloride ions is discussed in Section 21.2. Bromine does not appear to be an essential element although it is used in sedatives. Fluoride ions may have significant benefits for dental health. Tooth enamel is composed of the mineral hydroxyapatite, $Ca_5(PO_4)_3(OH)$. Hydroxyapatite is essentially insoluble in water (Equation 21.1); its K_{sp} is about 10^{-59}.

$$Ca_5(PO_4)_3(OH)(s) \rightleftharpoons Ca_5(PO_4)_3^+(aq) + OH^-(aq) \qquad (21.1)$$

When hydroxyapatite comes into contact with weak acids in your mouth, the solubility equilibrium in Equation 21.1 shifts toward products as the acid reacts with hydroxide ions. Le Châtelier's principle predicts that when a product is removed, as in this case, the equilibrium shifts toward formation of more product, which effectively increases the solubility of hydroxyapatite. The result is pitting of tooth enamel and formation of cavities or caries. Fluoride ions reduce the number of cavities by displacing the OH⁻ ions in hydroxyapatite to form fluorapatite, $Ca_5(PO_4)_3F$:

$$Ca_5(PO_4)_3(OH)(s) + F^-(aq) \rightleftharpoons Ca_5(PO_4)_3F(s) + OH^-(aq) \qquad (21.2)$$

The solubility of fluorapatite is not as dependent on pH as is hydroxyapetite, so changing tooth enamel to fluorapatite results in more decay-resistant teeth. This is the reason that toothpaste contains fluoride compounds and why fluoride is added to drinking water in many communities in North America and Europe.

Of the essential trace elements, iodide ion may have the best-defined role in human health. The body concentrates iodide (I⁻) in the thyroid gland, where iodide is incorporated into two hormones—thyroxine and 3,5,3′-triiodothyronine (Figure 21.8)—whose role in the body is to regulate energy production and use. The

Thyroxine

3, 5, 3′-Triiodothyronine

FIGURE 21.8 Structures of thyroxine and 3,5,3′-triiodothyronine, two iodine-containing hormones found in the thyroid gland. They regulate metabolism.

conversion of thyroxine to 3,5,3′-triiodothyronine is catalyzed by selenocysteine-containing proteins.

A deficiency of either of these hormones or of iodide can cause fatigue or feeling cold and can ultimately lead to an enlarged thyroid gland, a condition known as goiter. Table salt sold in the United States and many other countries as "iodized salt" is fortified with small amounts (about 50 μg/g in the U.S.) of sodium iodide.

On the other hand, excessive hormone secretion by the thyroid gland can cause one to feel hot and is linked to Graves' disease, an autoimmune disease. The immune system in a patient with Graves' disease attacks the thyroid gland and causes it to overproduce the hormones.

Transition Metals

In Chapters 8 and 19 we were introduced to oxygen transport and the role of iron in hemoglobin. Copper-containing proteins called hemocyanins perform the same function in mollusks such as clams and oysters. However, transition metal ions play a wider role in biology than just oxygen transport. We begin our discussion of the role of transition metals in living systems by looking first at a class of compounds known as metalloenzymes.

METALLOENZYMES The enzyme carbonic anhydrase (Figure 21.9) catalyzes the reaction between water and carbon dioxide to form bicarbonate ions:

$$H_2O(\ell) + CO_2(aq) \rightleftharpoons HCO_3^-(aq) + H^+(aq)$$

Carbonic anhydrase contains a total of 260 amino acids and a zinc ion at the active site. Note that the zinc ion is coordinately bonded to three nitrogen atoms on histidine side chains and to one molecule of water. This bonding and the presence of a fourth histidine residue nearby facilitates ionization of the water molecule. Ionization leaves an OH^- ion attached to the Zn^{2+} ion and H^+ ion bonded to the side-chain nitrogen atom of the fourth histidine. In addition, a pocket with just the right size and shape to accept a CO_2 molecule is next to the active site. A CO_2 molecule in the pocket combines with the hydroxide ion, forming a HCO_3^- ion. As the bicarbonate ion pulls away, another water molecule occupies the fourth coordination site on the Zn^{2+} ion, another CO_2 molecule enters the pocket, the protonated histidine ionizes, and the catalytic cycle can be repeated. Many of the enzymes in our bodies contain transition metal ions. Table 21.2 lists some of the more important metalloenzymes and the reactions that they catalyze. Note that more than half of the transition metals of the periodic table's fourth row are represented: V, Fe, Co, Ni, Cu, and Zn. Some transition metals in the fifth and sixth periods are also found in enzymes, for example, Mo and W. The transition metal cations form complexes (coordination compounds) by bonding to the nitrogen atoms of the amino acids in proteins.

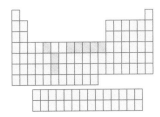

⊗⊗ **CONNECTION** In Chapter 10 we discussed some of the roles of the halogens in biology and medicine.

⊗⊗ **CONNECTION** In the introduction to Chapter 8 we discussed the interaction of diatomic gases with hemoglobin.

⊗⊗ **CONNECTION** Examples of coordination complexes of transition metals with biologically important ligands are also found in Chapter 17.

TABLE 21.2 Selected Metalloenzymes and Some Reactions They Catalyze

Metal[a]	Enzyme	Reaction
V(Fe)	nitrogenase	$N_2 + 10H^+ + 8e^- \rightarrow H_2 + 2NH_4^+$
	haloperoxidase	$CH_4 + H_2O_2 + Cl^- + H^+ \rightarrow CH_3Cl + 2H_2O$
Mo	nitrate reductase	$NO_3^- + 3H^+ + 2e^- \rightarrow H_2O + HNO_2$
	sulfite oxidase	$SO_3^{2-} + H_2O \rightarrow SO_4^{2-} + 2e^- + 2H^+$
Mo(Fe)	xanthine oxidase	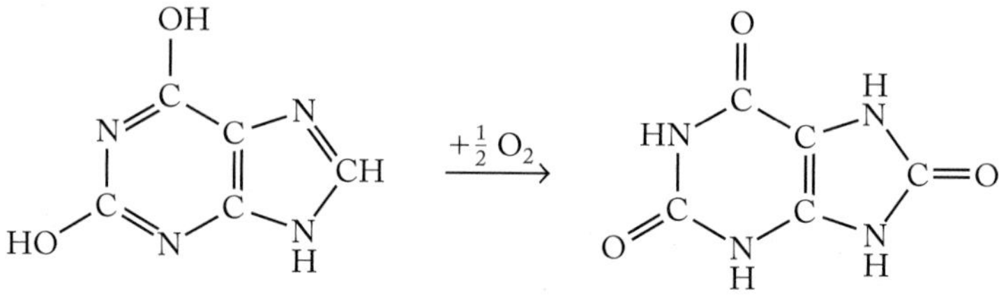
W(Fe)	formate dehydrogenase	$HCO_2^- \rightarrow CO_2 + 2e^- + H^+$
Fe	cytochrome-P450	$R–H + O_2 + 2e^- + 2H^+ \rightarrow R–OH + H_2O$
	peroxidase	$RCH_2COOH + 2H_2O_2 \rightarrow 3H_2O + RCHO + CO_2$
Co	coenzyme B_{12}	required as coenzyme for many reactions
Ni	urease	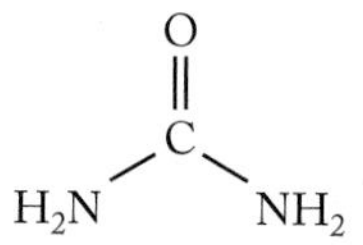
Ni(Fe)	hydrogenase	$H_2 \rightarrow 2H^+ + 2e^-$
Cu	N_2O reductase	$N_2O + 2e^- + 2H^+ \rightarrow N_2 + H_2O$
	amine oxidase	$CH_3NH_2 + O_2 + H_2O \rightarrow CH_2O + H_2O_2 + NH_3$
Zn	carboxyanhydrase	$H_2O + CO_2 \rightarrow HCO_3^- + H^+$
	carboxypeptidase	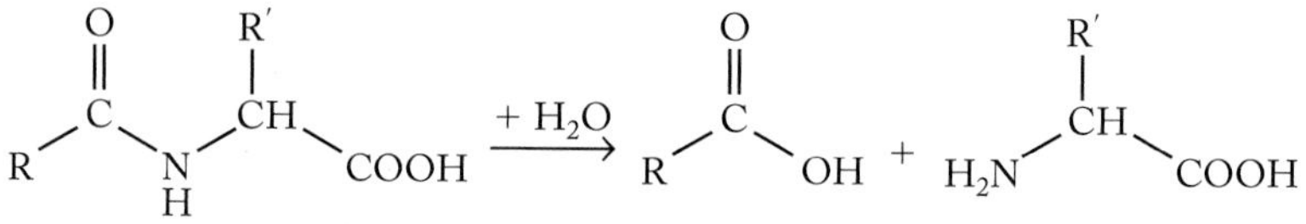

[a] The metal in parentheses is also present in the enzyme and is essential to the function of the metalloenzyme.

Many of the metalloenzymes in Table 21.2 are involved in transformations of nitrogen. Nitrogen is an essential major element found primarily in proteins but also in DNA and RNA. Nitrogen is available in the atmosphere as N_2, a very stable molecule containing a nitrogen–nitrogen triple bond. Soil and water contain nitrate ions (NO_3^-). Neither form of nitrogen can be directly incorporated into amino acids or proteins. The biosynthesis of amino acids requires ammonia or ammonium ions. For example, glycine (NH_2CH_2COOH), the simplest amino acid, is formed by reaction of CO_2 and ammonia in the presence of the appropriate enzyme. Certain common bacteria use enzymes called nitrogenases that contain iron and molybdenum to convert N_2 to ammonia. Plants convert NO_3^- ions to NO_2^- and then to NH_3 by using a combination of nitrate and nitrate reductases that contain molybdenum and iron. Ultimately, the chemical reactions in these plants lead to a steady supply of the essential amino acids for human diets. The conversion of nitrogen or nitrate to ammonia involves redox processes, as illustrated in Sample Exercise 21.2.

CONNECTION In Chapter 4 we learned about assigning oxidation numbers and balancing redox equations.

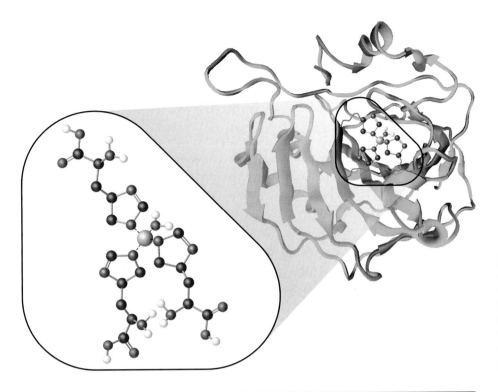

FIGURE 21.9 The active site of one form of the enzyme carbonic anhydrase consists of a zinc atom bonded to three nitrogen atoms in three histidine side chains and the O atom in a molecule of H_2O. The OH^- ion produced when this molecule of H_2O ionizes combines with a molecule of CO_2, forming a HCO_3^- ion.

SAMPLE EXERCISE 21.2 **Balancing the Reaction of Nitrate Reductases**

Nitrate ion can be reduced to ammonia by nitrate reductases containing molybdenum. The first step in the process is the conversion of the nitrate ion (NO_3^-) to the nitrite ion (NO_2^-). Assign oxidation numbers to the elements in these ions and write a balanced equation for the half-reaction in a basic solution:

$$NO_3^-(aq) \rightarrow NO_2^-(aq)$$

COLLECT AND ORGANIZE Assigning oxidation numbers is a convenient way of identifying which element is reduced or oxidized in a half-reaction and to determine how many electrons are gained or lost.

ANALYZE The oxidation numbers of nitrogen and oxygen in polyatomic ions like nitrate and nitrite must add up to the overall charges on the ions (1− in each case). Oxygen usually has an oxidation number of −2, whereas the oxidation number of nitrogen can have one of several values. With knowledge of the number of electrons transferred, we can write an incomplete equation for the half-reaction and then complete and balance it by adding OH^- ions to balance the charges, and by adding molecules of H_2O to balance the numbers of H and O atoms.

SOLVE The oxidation number (x) of nitrogen in NO_3^- is:

$$x + 3(-2) = 1- \text{ or } x = +5.$$

The oxidation number of nitrogen in NO_2^- is:

$$x + 2(-2) = 1- \text{ or } x = +3.$$

Nitrogen is reduced from +5 to +3, so the half-reaction is indeed a reduction half-reaction and 2 electrons are gained by each nitrate ion.

To balance the half-reaction, we start with what we know:

$$NO_3^-(aq) + 2\,e^- \rightarrow NO_2^-(aq)$$

Coenzymes, like enzymes, are organic molecules that accelerate the rate of biochemical reactions.

To balance the charges, we add 2 OH⁻ ions to the product side:

$$NO_3^-(aq) + 2\,e^- \rightarrow NO_2^-(aq) + 2\,OH^-(aq)$$

We balance hydrogen and oxygen atoms by adding H_2O to the left side:

$$NO_3^-(aq) + H_2O(\ell) + 2\,e^- \rightarrow NO_2^-(aq) + 2\,OH^-(aq)$$

and obtain a balanced half-reaction.

THINK ABOUT IT An enzyme called a *reductase* is one that *reduces* the substrate. Reduction is defined as a gain of electrons (or a decrease in the oxidation number of an element). The balanced half-reaction confirms that electrons are added to nitrate to reduce it to nitrite ion.

Practice Exercise Enzymes called nitrogenases enable some bacteria to "fix" nitrogen, that is, to reduce N_2 to NH_3. The reduction half-reaction catalyzed by one type of nitrogenase produces one mole of $H_2(g)$ for every two moles of $NH_3(aq)$ under acidic conditions. Write a balanced chemical equation for this half-reaction.

In humans and many mammals, excess nitrogen is converted to urea in the liver and excreted via the kidneys. Plants are capable of using urea as a source of ammonia by the action of nickel(II)-containing ureases, as shown in Equation 21.4:

$$\underset{H_2N \qquad NH_2}{\overset{O}{\underset{\|}{C}}} + H_2O \rightarrow 2\,NH_3 + CO_2 \qquad (21.4)$$

Unlike the reactions catalyzed by nitrogenases and nitrate reductases, the conversion of urea to carbon dioxide and ammonia is not a redox reaction. The reaction in Equation 21.4 is simply a hydrolysis reaction, similar to the reaction of nonmetal oxides with water described in Chapter 3.

Coenzymes and Cobalt Many enzymatic processes require the presence of a **coenzyme** without which the enzyme will not function. Coenzymes are organic molecules that catalyze biochemical reactions. Coenzymes generally have lower molar masses than the proteins in enzymes. They may be relatively small molecules, like nicotinamide adenine dinucleotide (NAD), or a larger metal-containing molecule like coenzyme B_{12} (Figure 21.10), which contains one cobalt(III) center and is a derivative of vitamin B_{12}.

The cobalt(III) in vitamin B_{12} is easily reduced to cobalt(II) and even cobalt(I) in the course of enzyme-catalyzed redox reactions. The change in oxidation state of Co allows for facile transfer of methyl groups, as in the conversion of methionine to homocysteine:

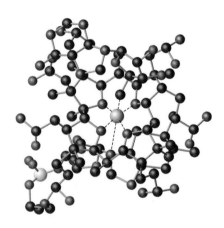

FIGURE 21.10 Coenzyme B_{12} contains cobalt, an essential ultratrace metal.

$$\underset{\text{Methionine}}{\overset{\overset{+}{NH_3}}{{}^-OOC \diagup\diagdown S \diagup}} \rightarrow \underset{\text{Homocysteine}}{\overset{\overset{+}{NH_3}}{{}^-OOC \diagup\diagdown SH}} \qquad (21.5)$$

The reaction in Equation 21.5 is an essential step in the conversion of methionine in our diets to energy.

Coenzyme B_{12} is also critical to the function of many enzymes called mutases. A mutase is an enzyme that catalyzes the rearrangement of the skeleton of a molecule, such as the interconversion of glutamate and methylaspartate shown in

Equation 21.6. The reaction in Equation 21.6 and the reverse of the reaction in Equation 21.5 are important in the synthesis of proteins in our bodies.

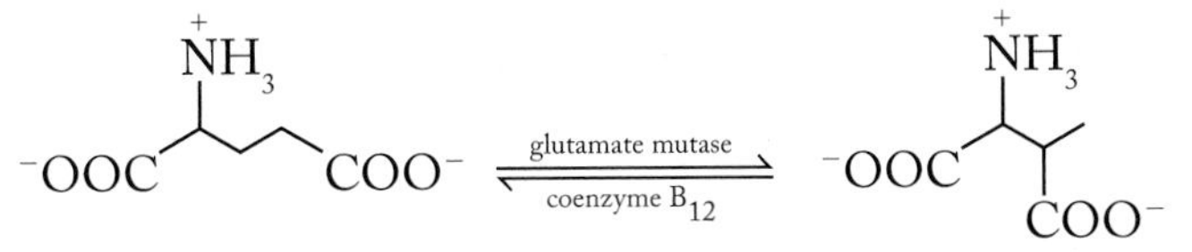

Glutamate   Methylaspartate (21.6)

Manganese and Photosynthesis The importance of photosynthesis has made it the subject of intense research for decades. Chlorophylls (Figure 21.4) are responsible for harvesting sunlight and directing the energy to the actual reaction center. You may recall the contribution of Mg^{2+} in ordering the pigment molecules in chlorophyll. Another metal, manganese, also plays a role in the production of oxygen during photosynthesis. Let's consider a chemical equation for photosynthesis that has a different appearance from the one we have seen in prior chapters. Instead of writing the formula of glucose on the right side of the equation, we simplify the coefficients of the other ingredients by using the generic carbohydrate formula $(CH_2O)_n$:

$$H_2O(\ell) + CO_2(g) \rightarrow \frac{1}{n}(CH_2O)_n(aq) + O_2(g) \tag{21.7}$$

The oxygen atoms in both carbon dioxide and water are in the -2 oxidation state, whereas the oxidation number of the O atoms in O_2 is zero. An increase in the oxidation number of oxygen represents an oxidation. The oxygen in CO_2 ends up in the carbohydrate $([CH_2O]_n)$ molecule and the O_2 oxygen atoms come from water. We can write two half-reactions (Equations 21.8 and 21.9) to describe the overall redox reaction in Equation 21.7:

$$4e^- + 4H^+ + CO_2 \rightarrow CH_2O + H_2O \tag{21.8}$$

$$2H_2O \rightarrow O_2 + 4H^+ + 4e^- \tag{21.9}$$

Manganese-containing compounds, where manganese is in the $+3$ and $+4$ oxidation states, mediate the transfer of electrons from water in photosynthesis. Although the exact structure of these manganese compounds remains undetermined, it is believed that four manganese ions, two each of Mn(III) and Mn(IV), are present at the site of O_2 production.

21.4 Diagnosis and Therapy

In Sections 21.2 and 21.3 we examined some of the elements that are essential to life and good health. The essential elements constitute less than one-third of the naturally occurring elements. How might some of the other elements influence our health? Radioactive isotopes of many of the remaining elements, including some members of the lanthanide series, have proved to be useful in the diagnosis of disease. Compounds of nonessential elements have also found application in the treatment of a wide range of illnesses. In this section we describe some of the chemistry of these elements in the context of human health.

The highlighted elements in Figure 21.11 are classified into two categories: those that find application in the diagnosis of disease and those that are used to treat disease or in medical devices. In some cases, diagnosis overlaps with therapeutic applications.

The diagnostic or therapeutic compounds that are injected intravenously must be sufficiently soluble in blood to be delivered to the target. While in transit, the

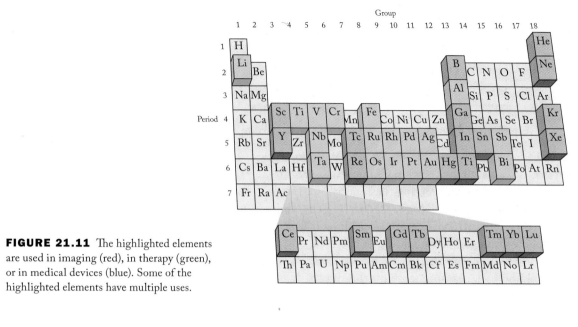

FIGURE 21.11 The highlighted elements are used in imaging (red), in therapy (green), or in medical devices (blue). Some of the highlighted elements have multiple uses.

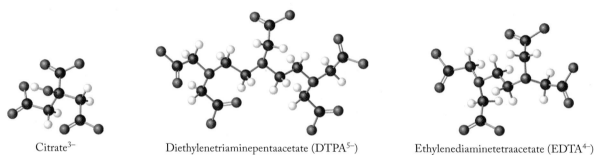

Citrate^{3-} Diethylenetriaminepentaacetate (DTPA^{5-})   Ethylenediaminetetraacetate (EDTA^{4-})

FIGURE 21.12 Citrate^{3-}, diethylenetriaminepentaacetate (DTPA^{5-}), and ethylenediaminetetraacetate (EDTA^{4-}) are often used as chelating agents in diagnostic and therapeutic agents based on transition metals. They form stable complex ions with 2+ and 3+ metal cations. Citrate, DPTA, and EDTA complexes are also more soluble than simple binary salts.

compound must be stable enough not to undergo chemical reactions that result in its precipitation or rapid elimination from the body. Occasionally, a simple salt like a chloride or nitrate will suffice, but more often than not a metal ion is introduced as a coordination complex or coordination compound. Ligands used in forming biologically active coordination complexes include simple anions like the citrate ion or amino acids. Chelating ligands like diethylenetriamine pentaacetate (DTPA^{5-}; shown in Figure 21.12) are often used in biological applications. A medicinal chemist can also take advantage of substances that occur naturally in the body, such as antibodies to carry a metal ion to its target.

Diagnosis of Disease

Physicians in the 21st century have an array of imaging agents to help in the diagnosis of disease. Some approaches are based on the use of radioactive isotopes with short half-lives that emit easily detectable gamma rays. Examples include the use of ^{131}I to image the thyroid gland and the application of neutron-poor isotopes like carbon-11, oxygen-15, and fluorine-18 for positron emission tomography (PET). But not all imaging depends on radioactive isotopes. Magnetic resonance imaging, or MRI, is frequently used to diagnose injuries to soft tissue. MRI images can be enhanced by the use of contrast agents containing stable isotopes of gadolinium.

IMAGING WITH RADIOISOTOPES Nuclear accelerators can be used to prepare radioisotopes of many elements. This wealth of isotopes has allowed medical researchers to investigate elements from across much of the periodic table as imaging agents. The isotopes used have short half-lives to limit exposure of the patient to ionizing radiation. If the half-life is too short, however, the isotope may decay before it can be administered to the patient, or it may not reach the target organ rapidly enough to provide an image. Emission of relatively low-energy γ rays is essential to prevent collateral tissue damage. The selection of a radioisotope is also governed by the toxicity of the element and its daughter isotopes. The speed at which the imaging agent is eliminated from the body can help mitigate its toxic effects. Naturally, the cost and availability of a particular isotope also factor into its utility in the clinical setting. In this section, we highlight some of the radioisotopes that are commonly used for imaging. We also highlight isotopes from groups in the periodic table that otherwise appear to have no biological function.

Gallium, Indium, and Thallium Compounds containing radioactive isotopes of gallium, indium, and thallium are used in medical imaging. For example, compounds containing ^{66}Ga, ^{67}Ga, ^{68}Ga, and ^{111}In are used as imaging agents for tumors and leukemia. All four isotopes decay by electron capture and the emission of γ rays that produce the images. In addition, ^{66}Ga, ^{67}Ga, and ^{68}Ga decay by positron emission, which makes compounds containing these isotopes attractive in positron emission tomography. The half-lives range from just over 1 hr for ^{68}Ga to 78 hr for ^{67}Ga.

The use of ^{201}Tl ($t_{1/2} = 73$ hr, γ emitter) in the diagnosis of heart disease presents an interesting case for considering the costs and benefits of using a particular isotope or element in medicine. Although thallium compounds are among the most toxic metal-containing compounds known, the nanogram (ng) quantities required for diagnosis pose few if any health hazards.

Technetium and Rhenium Technetium and rhenium are transition metals in group 7 of the periodic table, just below manganese. Manganese is essential to photosynthesis, as described in Section 21.3. Technetium is unusual in that it does not exist naturally in the Earth. In addition, technetium has no stable isotopes; in other words, all technetium isotopes are radioactive. Technetium, however, can be produced in nuclear reactors for use in medicine, as described in Sample Exercise 21.3. When technetium is produced, the nucleus is in an excited state that slowly decays to a more stable nucleus. The excited nuclear state of technetium atoms is designated by adding the letter "m" to the mass number (for example, ^{99m}Tc) to indicate a *metastable* nucleus.

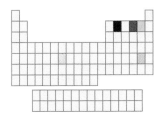

CONNECTION Magnetic resonance imaging (MRI) was mentioned in Chapter 3 in our discussion of the use of cryogenic helium for superconducting magnets.

CONNECTION Chapter 20 gave a more detailed discussion of nuclear chemistry and nuclear medicine, including an assessment of the effects of different types of radiation on living tissue.

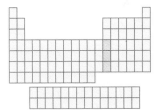

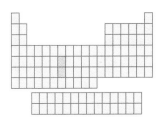

SAMPLE EXERCISE 21.3   **Identifying Particles in Nuclear Reactions**

Technetium-99m is produced from molybdenum-98 and decays by sequential γ and β decay paths. Complete the following sequence of nuclear reactions by identifying the missing particles, x, y, and Z.

$$^{98}\text{Mo} + x \rightarrow {}^{99}\text{Mo} \rightarrow {}^{99\text{m}}\text{Tc} + y$$

$$^{99\text{m}}\text{Tc} \rightarrow {}^{99}\text{Tc} + \gamma$$

$$^{99}\text{Tc} \rightarrow Z + {}^{0}_{1-}\text{e}$$

COLLECT AND ORGANIZE We are provided with three nuclear decay equations. Two of the equations are missing a particle and are not balanced. Recall that when an unstable nucleus undergoes decay, it can eject a high-speed electron called a β particle ($^{0}_{1-}\text{e}$), a positron, ($^{0}_{1+}\text{e}$), or an α particle ($^{4}_{2}\text{He}^{2+}$) from its nucleus. Gamma rays (γ) can also be emitted in the process. The sum of the electrical charges denoted by the subscripts and the sum of the masses denoted by the superscripts on the left-hand side of the reaction arrow must equal the sum of the electrical charges and masses on the right-hand side. Unstable nuclei can also decay by electron capture, a process in which a proton and electron in an atom combine to form a neutron.

ANALYZE To find the missing particle in each of the three nuclear reactions, we must determine its electrical charge and mass. This task is accomplished by summing the subscripts and superscripts on both sides of the equation. The difference in the superscripts equals the mass of the missing particle, and the difference in the subscripts equals the charge on the particle.

SOLVE There are two sequential reactions in the first equation. First, ^{98}Mo reacts with an unknown particle x to form ^{99}Mo:

$$^{98}_{42}\text{Mo} + x \rightarrow {}^{99}_{42}\text{Mo}$$

The subscript represents the atomic number of Mo ($Z = 42$) and is unchanged in the reaction. The superscript represents the mass number of the particle. The change from ^{98}Mo to ^{99}Mo is one mass unit. The particle with a mass number equal to 1 and a charge of 0 is a neutron, $^{1}_{0}\text{n}$. Thus, particle x is a neutron. Therefore molybdenum-98 must be bombarded by neutrons in a nuclear reactor to form molybdenum-99.

In the second step of the first equation, $^{99}_{42}\text{Mo}$ decays to an isotope of technetium, $^{99\text{m}}_{43}\text{Tc}$, and in the process emits particle y:

$$^{99}_{42}\text{Mo} \rightarrow {}^{99\text{m}}_{43}\text{Tc} + y$$

In this case, the mass number does not change in the course of the reaction; the superscript is constant at 99 on both sides of the equation. Particle y must, therefore, have negligible mass. The atomic number (charge) increases from 42 for Mo to 43 for Tc:

$$42 = 43 + ? \text{ or } 42 = 43 + (1-)$$

The charge on particle y must be equal to 1−. The particle that has this combination of charge and mass is a β particle, $_{-1}^{0}e$.

The next equation describing the decay of ^{99m}Tc is already balanced; ^{99m}Tc loses an uncharged particle of zero mass and charge, which is a γ ray.

$$^{99m}_{43}Tc \rightarrow {}^{99}_{43}Tc + \gamma$$

In the final step of the reaction sequence, ^{99}Tc loses another β particle to yield element Z:

$$^{99}_{43}Tc \rightarrow Z + {}^{0}_{-1}e$$

The mass number of a β particle is zero, so the mass number of Z must be the same as that of ^{99}Tc, or 99. The sum of the subscripts (charges) is

$$43 = ? + (1-) \text{ or } 43 = 44 + (1-)$$

and so the atomic number of Z must be 44. The nucleus with $Z = 44$ is a ruthenium nucleus, so product Z is $^{99}_{44}Ru$.

THINK ABOUT IT Because γ rays have no mass or charge, there is no change in the atomic or mass number of technetium-99m when it emits a γ ray. The technetium-99 then decays to a stable isotope of $^{99}_{44}Ru$ by β decay. Remember that β decay leads to an increase in the atomic number by 1. Additional information on balancing nuclear equations is in Chapter 20. The ^{99}Tc used in hospitals for imaging is prepared in technetium generators in which a stable isotope of molybdenum, ^{98}Mo (23.78% natural abundance), is bombarded with neutrons.

Practice Exercise Like ^{99m}Tc, γ rays from thallium-201 ($t_{1/2} = 73$ hr) can be used for cardiac imaging. Thallium-201 decays by electron capture. Write a balanced radiochemical equation describing the decay process.

The use of ^{99m}Tc as an imaging agent traces back several decades. It has a short half-life ($t_{1/2} = 6$ hr) and emits low-energy γ rays. Patients can be injected with a variety of technetium compounds depending on the target organ. For imaging the heart muscle, these radiopharmaceuticals include the Tc-based coordination compounds shown in Figure 21.13.

The ability to deliver radioactive isotopes to a particular organ opens the possibility for selective irradiation of a tumor located in that organ. Therefore some of the radioisotopes described so far can be used to both image and treat cancers.

Rhenium, for example, is an extremely rare element in Earth's crust. Compounds containing the radioactive isotopes ^{186}Re and ^{188}Re, however, are being studied as both imaging and therapeutic agents for many tumors including breast, liver, and skin cancer. Both rhenium isotopes decay by β decay with half-lives of 3.72 d and 17.0 hr, respectively. Certain tumors have highly selective receptor sites for particular molecules on their surfaces. By including a radioactive rhenium ion in a molecule that binds strongly and specifically to these receptors, physicians can deliver both an imaging agent and a therapeutic agent to the tumor. In principle, the β particles from the rhenium isotope penetrate and destroy the tumor. One example of a rhenium compound used in these applications is shown in Figure 21.14.

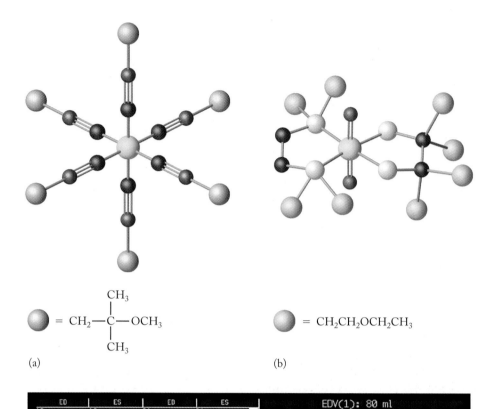

FIGURE 21.13 (a) Cardiolite™, Tc(CNR)$_6$ where R = –CH$_2$C(CH$_3$)$_2$OCH$_3$, and (b) Myoview™, TcO$_2$(RPCH$_2$CH$_2$PR)$_2$ where R = –CH$_2$CH$_2$OCH$_2$CH$_3$, are two technetium-containing radiopharmaceuticals used for imaging the heart. (c) Images generated for a patient's heart after intravenous injection with a technetium-containing drug such as Cardiolite™ or Myoview™. The four images along the bottom show four different measurements of heart function. The radiopharmaceuticals emit gamma rays which allow the blood to be tracked as it is pumped through the heart.

(a)

$$\bigcirc = CH_2-\underset{\underset{CH_3}{|}}{\overset{\overset{CH_3}{|}}{C}}-OCH_3$$

(b)

$$\bigcirc = CH_2CH_2OCH_2CH_3$$

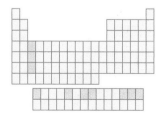

Scandium, Yttrium, and Lanthanide Elements Scandium, yttrium, and lanthanum are in group 3 in the periodic table, the first group in the transition metal series. Once thought to be quite rare, the abundance of Sc, Y, and La in the crust (25 to 35 ppm) is comparable to that of cobalt. In the oceans, scandium, yttrium, and the fourteen lanthanide elements are barely detectable; they are present in concentrations on the order of 10^{-6} to 10^{-8} mg/L. None are normally detected in the human body, and they do not appear to have a biological function. Each of

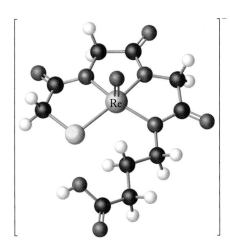

FIGURE 21.14 This rhenium-mercaptoacetylglycylglycylglycyl-γ-amino acid complex is used to attach ^{186}Re or ^{188}Re to monoclonal antibodies or peptides.

these elements, however, has one or more readily available radioisotopes, many of which are being explored in research laboratories worldwide.

Scandium-46 has been used to image the spleen, whereas scandium-47 shows promise in the diagnosis of breast cancer. Isotopes of the lanthanides have seen extensive use in scintigraphic imaging. Chapter 20 describes scintillation counters as radiation detectors. Scintillation counters measure the light emitted when radiation strikes a phosphor. The intensity of the emitted light can be translated into a three-dimensional image of the organ of interest. Among the isotopes used for this purpose are ^{141}Ce, ^{153}Sm, ^{153}Gd, ^{160}Tb, ^{170}Tm, ^{169}Yb, and ^{177}Lu. These isotopes are being explored as substitutes for gallium isotopes. The lanthanide imaging agents have an advantage over gallium isotopes (already described) in having lower retention times in blood. Thus, use of lanthanides reduces the possibility of potentially harmful biological side effects of the imaging process. Sample Exercise 21.4 reviews calculations involving the half-lives of yttrium isotopes.

CONNECTION We discussed first-order reactions in Chapter 14 and applied the same equations to the first-order kinetics of radioactive decay in Chapter 20.

SAMPLE EXERCISE 21.4   **Calculating Remaining Quantities of Radioisotopes**

Two isotopes of yttrium, ^{86}Y ($t_{1/2}$ = 14.6 hr) and ^{90}Y ($t_{1/2}$ = 64 hr), have been used in radioimaging. Which isotope decays faster? If we start with 10 mg of each isotope, how much of each will remain after 24 hr?

COLLECT AND ORGANIZE The half-life is a measure of how much time is required for one-half of an isotope to decay. Radioactive decay follows first-order kinetics. Quantitatively, the relationship between half-life ($t_{1/2}$) and the amount of material remaining is described by this rearranged version of Equation 20.12:

$$\ln \frac{A_t}{A_0} = \frac{-0.693t}{t_{1/2}}$$

where A_0 and A_t refer to the amount of material present initially and the amount at time t, respectively.

ANALYZE We are given the half-lives of two radioactive isotopes of yttrium. Because one-half of the ^{86}Y will have decayed after 14.6 hr, compared to 64 hr for ^{90}Y, ^{86}Y decays faster. In other words, the isotope with the shorter half-life decays faster. We need to complete two calculations to determine the amount of ^{86}Y and ^{90}Y present after 24 hr.

SOLVE To determine how much ^{86}Y remains, we substitute $t_{1/2} = 14.6$ hr, $A_0 = 10$ mg, and $t = 24$ hr into the equation:

$$\ln \frac{A_t}{A_0} = \frac{-0.693t}{t_{1/2}}$$

$$\ln \frac{A_t}{10 \text{ mg}} = \frac{(-0.693 \times 24 \text{ hr})}{14.6 \text{ hr}} = -1.1392$$

Taking the antilog of both sides, we get

$$\frac{A_t}{10 \text{ mg}} = 0.320$$

Finally, we solve for A_t:

$$A_t = 0.320 \times 10 \text{ mg} = 3.2 \text{ mg } ^{86}Y$$

A similar calculation for 10 mg (A_0) of ^{90}Y ($t_{1/2} = 64$ hr) after 24 hr (t) proceeds as follows:

$$\ln \frac{A_t}{A_0} = \frac{-0.693t}{t_{1/2}}$$

$$= \ln \frac{A_t}{10 \text{ mg}} = \frac{(-0.693 \times 24 \text{ hr})}{64 \text{ hr}} = -0.2599$$

$$\frac{A_t}{10 \text{ mg}} = 0.7711$$

$$A_t = 7.7 \text{ mg } ^{90}Y$$

THINK ABOUT IT We predicted that ^{86}Y would decay faster than ^{90}Y. The results of the calculation indicate that about twice as much ^{90}Y remains after 24 hours when equal initial amounts of ^{90}Y and ^{86}Y decay. ^{86}Y decays much faster than ^{90}Y, which means there will be higher levels of radiation but for a shorter time with ^{86}Y. If longer collection times are needed to collect multiple images of different organs, ^{90}Y may be the choice. Another consideration facing medical staff is the speed at which the body eliminates yttrium. If yttrium elimination is slow, then we may be able to get a good image with ^{86}Y despite its short half-life because more of it is retained compared to ^{90}Y.

Practice Exercise Two isotopes of rhenium, ^{186}Re and ^{188}Re, are used to treat breast, liver, and skin cancer. The half-lives are 89 hr and 17 hr for ^{186}Re and ^{188}Re, respectively. If we start with 25.0 mg of each isotope, how much more ^{186}Re than ^{188}Re will remain after 24 hr?

IMAGING WITH STABLE ISOTOPES X-rays have long been used to diagnose fractures, identify cavities in teeth, and detect injuries to bones and other hard substances in the body. Magnetic resonance imaging or MRI is a technique for imaging soft tissue. The principles that make MRI work are beyond the scope of this book, but are based on the behavior of hydrogen in molecules in response to an applied magnetic field. MRI relies on the high concentration of water in the body. Spinning hydrogen nuclei in water molecules interact with an applied magnetic field. Sophisticated software can translate the resulting data into detailed images of tissues.

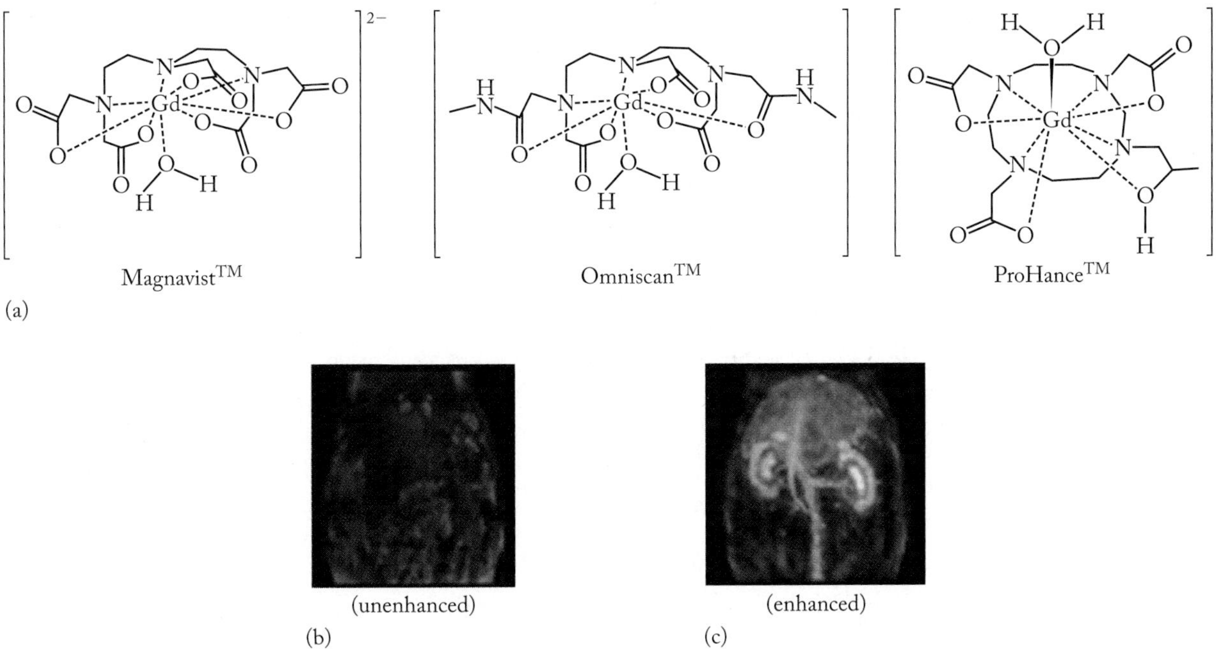

(a)

(unenhanced)

(b)

(enhanced)

(c)

FIGURE 21.15 (a) Here are three carbon skeleton structures of gadolinium-based contrast agents used in MRI. Note that there are as many as nine coordinate bonds to one Gd^{3+} ion. (b) An MRI image without a contrast agent; and (c) an MRI image using gadolinium contrast agents.

The quality of images produced by MRI can sometimes be improved by using contrast agents. Patients can be injected with gadolinium compounds such as those with the structures shown in Figure 21.15 prior to an MRI scan. Many different ligands have been investigated in Gd^{3+} contrast agents. Coordination compounds of other lanthanide ions (Dy^{3+} and Ho^{3+}) have also been evaluated but do not work as well as those of gadolinium.

Noble Gases and MRI So far in this chapter we have had little opportunity to mention the noble gas elements of group 18 in the periodic table. None of these elements are essential to the human body. The lack of chemical reactivity and the ease of introduction into the body by inhalation, however, make selected isotopes of the noble gases attractive as contrast agents for MRI. Three nuclides of group 18, ^{3}He, ^{83}Kr, and ^{129}Xe, have been explored as contrast agents for imaging lungs. Both ^{83}Kr and ^{129}Xe are stable isotopes; they are present in 11.5% and 26% natural abundance, respectively. Xenon has the drawback that it is soluble in blood and acts as an anesthetic. Helium-3 has no known side effects but is present in only trace natural abundance. Fortunately, ^{3}He is available from β decay of tritium (^{3}H):

$$^{3}_{1}\text{H} \rightarrow {}^{3}_{2}\text{He} + {}^{0}_{-1}\text{e} \tag{21.10}$$

It is worth noting that ^{19}Ne ($t_{1/2}$ = 17.5 s) has been used in PET despite its short half-life. A patient breathes air containing a small amount of ^{19}Ne while the patient is in a PET scanner. Positron emission from the neon is recorded and an image is created. Of the group 18 elements, only argon has not found direct medical applications.

Therapeutic Applications

Most pharmacies are filled with chemical compounds used to treat a wide variety of diseases. None of these medicines contain the radioisotopes described in the previous section. Most pharmaceuticals are organic compounds. Recall that organic compounds are typically compounds containing C–C and C–H bonds. Most medications also contain nitrogen, oxygen, and other main group elements from the first two rows of the periodic table. In this section we examine therapeutic agents that also contain metallic elements from a broad range of groups in the periodic table, including alkali metals, the transition elements, and heavier main group elements in addition to carbon, hydrogen, nitrogen, oxygen, and sulfur.

LITHIUM, BORON, ALUMINUM, AND GALLIUM The concentration of lithium in the human body is ~0.03 μg/g, much less than the concentration of Li in Earth's crust (20 μg/g) or in the oceans (0.18 mg/L). Thus, lithium is significantly less abundant than sodium, potassium, or rubidium in all three environments. The similar size of the Li^+ (76 pm) and Mg^{2+} (65 pm) ions means that lithium ions can compete with magnesium ions in biological systems. The substitution of lithium for magnesium may account for its toxicity at high concentrations. Nevertheless, lithium carbonate (Li_2CO_3) is used to treat manic depression under the name Camcolit™. Lithium compounds have also been used to treat hyperactivity. In all cases, however, the use of lithium-containing drugs must be carefully monitored.

Of the elements of group 13, only boron and aluminum have been detected in humans; their concentrations are 0.14 μg/g and 1.4 μg/g, respectively. Boron appears to play a role in nucleic acid synthesis and carbohydrate metabolism. The role of boron in humans is still not fully understood. Selected boron compounds appear to concentrate in human brain tumors. This property has opened the door to a treatment known as boron neutron capture therapy or BNCT. Boron has two stable isotopes, ^{10}B and ^{11}B, in approximately a 1:4 ratio. Once a suitable boron compound has been injected and has made its way to a tumor, irradiation of the tumor with low-energy neutrons leads to the nuclear reaction

$$^{10}_{5}B + ^{1}_{0}n \rightarrow ^{7}_{3}Li + ^{4}_{2}He$$

The α particles generated in the reaction have a short penetration depth but high relative biological effectiveness (RBE) so they can kill the tumor cells without harming surrounding tissue. The identification of compounds suitable for BNCT remains an area of active research.

Aluminum is found in some antacids as aluminum hydroxide, $Al(OH)_3$, or aluminum carbonate, $Al_2(CO_3)_3$, and as aluminum sodium sulfate, $AlNa(SO_4)_2 \cdot 12\,H_2O$, is an ingredient in baking powder. Most of the aluminum detected in the human body can be traced to these sources. Aluminum is not considered essential to humans, but low-aluminum diets have been observed to harm goats and chickens. High concentrations of aluminum are clearly toxic; the effects are most noticeable in patients with impaired kidney function. The role of aluminum in Alzheimer's disease has been extensively debated but remains unresolved.

Simple gallium compounds like gallium(III) nitrate and gallium(III) chloride alone or in combination with other drugs have shown activity toward a variety of cancers, including bladder and ovarian cancers. The similar ionic radius of Ga^{3+} (62 pm) and Fe^{3+} (64.5 pm) allows gallium to block DNA synthesis by replacing iron in a protein called transferrin and in other enzymes. Because gallium compounds accumulate in tumors at a higher rate than in healthy tissue, the disruption of DNA synthesis in the tumor cells prevents the tumor from growing.

⊙⊙ **CONNECTION** The relative biological effectiveness (RBE) of radioactive particles is introduced in Chapter 20.

ANTIMONY AND BISMUTH In Section 21.3 we noted that selected antimony and bismuth compounds have proved useful in medicine. Antimony compounds are typically found in the +3 and +5 oxidation states—for example, SbH$_3$ and SbCl$_5$, respectively—and are generally considered to be toxic. It has been reported that exposure of infants to antimony compounds used as fire retardants in mattresses may contribute to sudden infant death syndrome (SIDS). Leishmaniasis is a worldwide insect-borne disease characterized by the formation of boils or lesions on the skin of patients afflicted with the disease. Although rare, cases of leishmaniasis have been reported in rural areas near the southern borders of the United States. The disease is resistant to many treatments, but patients suffering from leishmaniasis have been successfully treated with sodium stibogluconate, one of the few applications of antimony compounds in human health.

Bismuth compounds enjoy a better reputation as therapeutic agents. Popular over-the-counter remedies for heartburn, indigestion, diarrhea, and other stomach disorders sold under the trade names Pepto-Bismol™ and Kaopectate™ contain bismuth subsalicylate (Figure 21.16). The bismuth in these compounds is present as the BiO$^+$ cation. Salicylate is the active ingredient in aspirin and accounts for the anti-inflammatory action of these remedies.

TITANIUM, VANADIUM, AND NIOBIUM CANCER DRUGS Compounds of some transition metals appear to have antitumor activity. Titanium(IV) complexes such as budotitane (Figure 21.17) show promise against colon cancer. Encouraging results against breast, lung, and colon cancers were observed with a series of compounds with formulas (C$_5$H$_5$)$_2$TiCl$_2$, (C$_5$H$_5$)$_2$VCl$_2$, and (C$_5$H$_5$)$_2$NbCl$_2$. Unfortunately, at therapeutically useful doses the potential for liver damage outweighs the benefits of these compounds. These compounds contain carbon–metal bonds and belong to a class of substances called **organometallic** compounds.

> Compounds containing direct carbon–metal covalent bonds belong to a class of compounds called **organometallic** compounds.

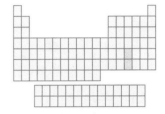

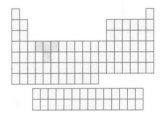

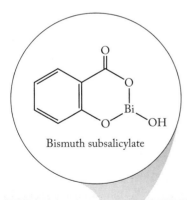

FIGURE 21.16 Bismuth subsalicylate is found in some antacids.

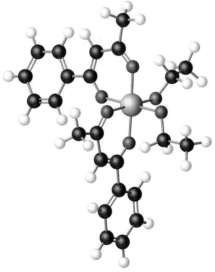

Budotitane

FIGURE 21.17 Budotitane is a titanium(IV) complex that shows activity against colon cancer.

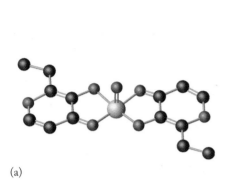

(a)

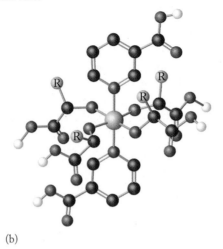

(b)

(c)

FIGURE 21.18 (a) Vanadium(IV) compounds like bis(ethylmaltolato)oxovanadium(IV) can act as insulin mimics. (b) The chromium(III) complex in a glucose tolerance factor contributes to our bodies' ability to regulate insulin levels and the green spheres are the side chain R groups of the amino acids in the structure. (c) Shepherd's purse is a chromium-rich plant that has been used as an herbal remedy by diabetics. [To simplify structures (a) and (b) the hydrogen atoms bonded to carbon atoms are not shown.]

⊙⊙ CONNECTION We encountered enantiomers of metal complexes in Chapter 17, where we saw that *cis*-Co(en)$_2$Cl$_2$$^+$ has two enantiomers.

VANADIUM AND CHROMIUM COMPOUNDS FOR TREATING DIABETES Diabetics are unable to produce enough or effectively use insulin in their bodies and often require daily insulin injections. Considerable effort has been expended in the search for oral therapies to regulate insulin production. The suggestion that vanadium plays a role as a cofactor in the production of insulin has prompted an investigation of vanadium compounds as diabetes drugs. A clinical trial of one vanadium(IV) compound, bis(ethylmaltolato)oxovanadium(IV) (Figure 21.18a), began in 2000 but has not yielded an approved drug.

Chromium in the +3 oxidation state is an essential element in our diets and is also involved in regulating glucose levels in the blood (Figure 21.18b). A Cr(III) complex called a glucose tolerance factor has two nicotinic acids in a *trans* geometry, and four amino acids, possibly glycine or cysteine from glutathione, occupy the remaining four coordination sites. Cereals and grains contain enough chromium for our daily needs, but certain plants (such as shepherd's purse) concentrate chromium and have been used as herbal remedies for diabetes treatment. Therapy using chromium(III) has been reported to have some success in reducing the amount of insulin required by some diabetics. Some chromium(III) complexes have been touted as weight-reduction aids.

CHIRAL IRON COMPOUNDS AND ANEMIA Iron is an essential trace element found in hemoglobin. The body must regulate the amount of iron in cells to produce enough hemoglobin to maintain good health. Deficiencies in iron lead to a number of diseases broadly classified as anemia. Mild forms of anemia are common among women of child-bearing age and are treated by taking iron supplements in the form of iron(II) sulfate or iron(II) gluconate. A related, more serious genetic form of anemia known as thalassaemia must be treated with blood transfusions that leave patients with too much iron in their blood. To remove the excess iron, these patients are treated with chelating ligands that complex some of the iron and transport it out of the red blood cells and eventually out of the body.

The use of chelating ligands to form coordination complexes of iron and other transition metals introduces the possibility of the formation of optical isomers. Consider the two iron complexes in Figure 21.19; they are nonsuperimposable mirror images of each other and hence are enantiomers. You may wish to build models of the complex and its mirror image to see for yourself that they are not superimposable. Often one enantiomer of an iron complex works better than the other in the treatment of diseases like thalassaemia.

FIGURE 21.19 Some octahedral transition metal coordination complexes containing polydentate ligands such as this iron(II) compound exist as optical isomers.

Mirror plane

THE PLATINUM GROUP METALS The evolution of the modern periodic table led chemists to develop nonsystematic names for certain groups. For example, group 1 elements are also known as the alkali metal elements, and the elements of group 17 are commonly referred to as the halogens. The fifth- and sixth-period elements in groups 8 to 10 (Ru, Rh, Pd, Os, Ir, and Pt) are often referred to as the platinum group metals. The group 11 elements are collectively called the coinage metals.

All of the platinum group metals have been detected in the human body. Their typical concentrations are less than 10 μg/kg. How do these rare elements get into our bodies? Jewelry, coins, silverware, and dental fillings in direct contact with our skin, teeth, or other tissues contain silver, gold, platinum, and palladium. Catalytic converters used to reduce air pollution from automobiles contain palladium, platinum, rhodium, and ruthenium. We inhale aerosols containing trace amounts of these elements. In addition, soluble compounds of the platinum group metals are finding increased application in medications for arthritis, cancer, and other diseases.

Although the historic use of gold for medicinal purposes dates back millennia, the effective use of gold-containing pharmaceuticals originated with the discovery only in the 1920s and 1930s that a gold thiosulfate compound, Sanochrysin, alleviates the symptoms of rheumatoid arthritis (Figure 21.20). The most commonly used gold drugs for the treatment of arthritis today are sold under the trade names Myochrysine and Auranofin (Figure 21.20). Myochrysine is injected; Auranofin can be taken orally.

Selected osmium compounds have also been shown to reduce inflammation in joints resulting from arthritis, although the use of such compounds has diminished with the development of other anti-inflammatory agents. The therapeutic

⊛⊛ **CONNECTION** The use of the platinum group metals (groups 8–10) as catalysts for inorganic chemical reactions is discussed in Chapter 14.

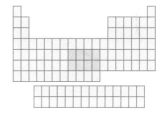

⊛⊛ **CONNECTION** Myochrysine and Auranofin contain sulfur–gold and sulfur–carbon bonds traced to the parent compound known as a thiol. Thiols contain sulfur–hydrogen and sulfur–carbon bonds and were discussed in Chapter 9.

⊛⊛ **CONNECTION** Copper, silver, and gold belong to group 11 in the periodic table. The properties of these "coinage" metals and some of their chemical reactions are described in Chapter 11.

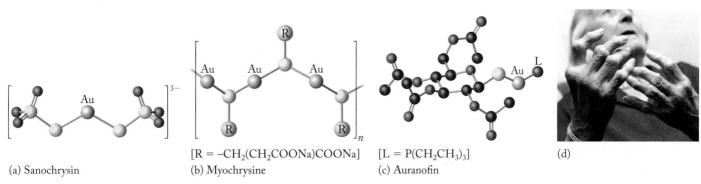

(a) Sanochrysin   (b) Myochrysine
[R = –CH₂(CH₂COONa)COONa]
[L = P(CH₂CH₃)₃]
(c) Auranofin   (d)

FIGURE 21.20 (a) Sanochrysin, (b) Myochrysine, and (c) Auranofin are three gold-containing drugs used to treat (d) arthritis.

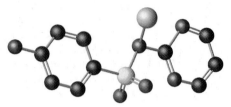

Silver sulfadiazine

FIGURE 21.21 Silver sulfadiazine is an effective antibiotic when applied to burns. (H atoms are not shown to simplify the structure.)

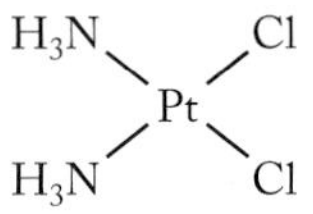

Cisplatin
Platinol™

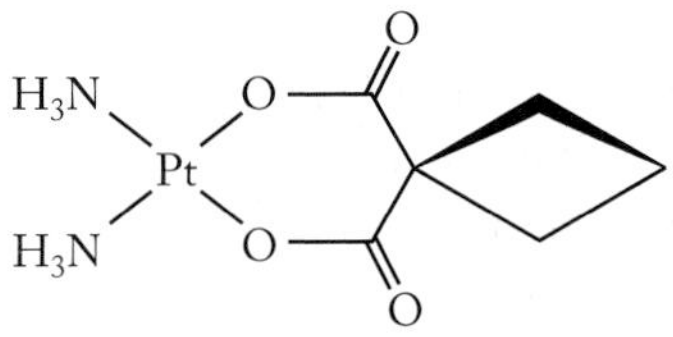

Carboplatin

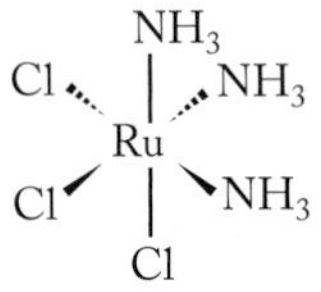

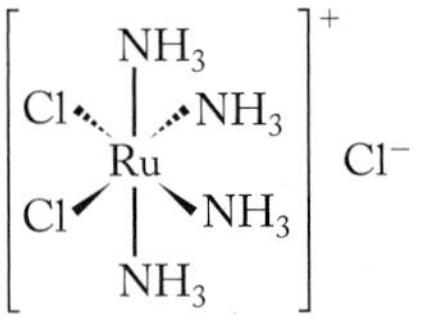

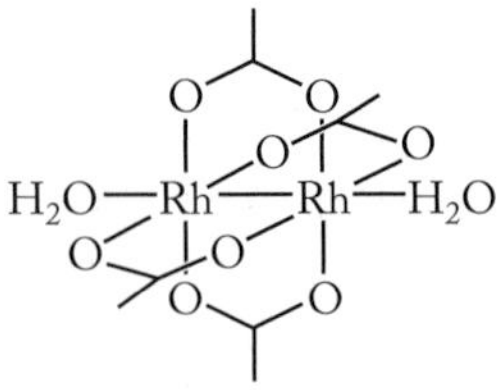

FIGURE 21.22 Cisplatin and carboplatin are effective antitumor agents. The ruthenium and rhodium compounds show activity against leukemia but have not yet seen widespread use.

effects of aqueous solutions of osmium tetroxide, OsO_4, were first investigated in the 1950s. The use of osmium tetroxide was superseded by the use of glucose polymers containing osmium and known as osmarins. The use of osmarins reduces the toxic effects of osmium and provides an example of how even toxic metals can be adapted to therapy.

Without knowing it, most of us experienced the bactericidal effects of silver nitrate within minutes of our birth. Aqueous solutions of $AgNO_3$ at concentrations of less than 0.1 mg/L prevent infantile blindness by killing any bacteria in the eyes that accompany birth. Eye drops containing silver nitrate have also been used for other infectious eye diseases. Silver sulfadiazine is a broad-spectrum "sulfa" drug that is used to prevent and treat bacterial and fungal infections (Figure 21.21). It is the active ingredient in creams that are used to treat thermal and chemical burns and that have the advantage of not requiring protective dressings.

The serendipitous discovery in the 1960s that *cis*-diamminedichloroplatinum (II) (cisplatin or Platinol™, Figure 21.22) is effective in treating testicular, ovarian, and other cancers has spawned the development of a host of precious metal compounds for cancer chemotherapy. The results of this research include a compound known as carboplatin that shows the same activity as cisplatin but has fewer side effects. In addition to platinum compounds, the antitumor activities of complexes of gold, rhodium, ruthenium, and silver have been explored. The effectiveness of all of these drugs lies in their ability to bind to nitrogen bases in DNA and inhibit cell replication. If their cells cannot divide, tumors cannot grow. The greater toxicity of rhodium compounds relative to those of platinum and ruthenium has limited their clinical application.

Medical Devices and Materials

DENTAL ALLOYS When the fluoride treatment described in Section 21.4 fails to prevent dental caries (cavities), a dentist will clean the cavity and fill it with a metal alloy that could contain mercury, silver, tin, copper, and traces of zinc. Gold inlays have also been used to repair damaged teeth. An alloy containing mercury is called an *amalgam*. The most common dental amalgam actually contains three compounds, Ag_2Hg_3, Ag_3Sn, and Sn_8Hg. Although mercury and mercury compounds are highly toxic (see Section 21.5), when combined with silver or tin the mercury is chemically unreactive. Sample Exercise 21.5 looks at the electrochemical reasons behind the stability of mercury in dental amalgams and its resistance to oxidation.

SAMPLE EXERCISE 21.5 **Predicting the Spontaneity of a Redox Reaction**

A drop of mercury is placed in a 1.00 M solution of the silver cation Ag^+. Will the mercury be oxidized to Hg_2^{2+} given the following reduction potentials?

$$Hg_2^{2+}(aq) + 2\,e^- \rightarrow 2\,Hg(\ell) \qquad E° = +0.7973 \text{ V}$$

$$Ag^+(aq) + e^- \rightarrow Ag(s) \qquad E° = +0.7996 \text{ V}$$

COLLECT AND ORGANIZE The value for the reduction potential of Ag^+ is +0.7996 V. The oxidation potential for $Hg(\ell)$ is −0.7973 V. To determine whether or not the mercury will be oxidized by Ag^+ ions, we need to write an equation for the redox reaction between Ag^+ and elemental Hg. By using the reduction potentials for each half-reaction, we can calculate the overall potential of the reaction. If the cell potential is greater than zero, then Ag^+ will oxidize metallic Hg metal to Hg_2^{2+}.

ANALYZE The overall reaction is

$$2\,Hg(\ell) + 2\,Ag^+(aq) \rightarrow Hg_2^{2+}(aq) + 2\,Ag(s) \qquad E° = ?$$

This equation is the sum of the reduction half-reaction for Ag^+ multiplied by 2 and the oxidation half-reaction of mercury metal to Hg_2^{2+}:

$$2\,Ag^+(aq) + 2\,e^- \rightarrow 2\,Ag(s)$$

$$2\,Hg(\ell) \rightarrow Hg_2^{2+}(aq) + 2\,e^-$$

With the balanced overall equation in hand, we can calculate the overall cell potential by summing the oxidation potential for mercury metal and the reduction potential for silver ion. Remember that the oxidation potential for a half-reaction is equal to its reduction potential but has the opposite sign.

SOLVE Summing the two values yields the overall cell potential, 0.0023 V:

$$2\,Ag^+(aq) + 2\,e^- \rightarrow 2\,Ag(s) \qquad E° = +0.7996 \text{ V}$$

$$2\,Hg(\ell) \rightarrow Hg_2^{2+}(aq) + 2\,e^- \qquad E° = -0.7973 \text{ V}$$

The sum here is

$$2\,Hg(\ell) + 2\,Ag^+(aq) \rightarrow Hg_2^{2+}(aq) + 2\,Ag(s) \qquad E°_{rxn} = 0.0023V$$

Because the electrochemical potential for the reaction is very slightly greater than zero, the reaction should take place spontaneously under standard conditions.

THINK ABOUT IT The standard cell potential for a Ag^+/Hg cell is very close to zero. Factors like the overpotential described in Chapter 18 suggest that this electrochemical reaction may not actually happen. In addition, as $[Ag^+]$ decreases, the value of E_{rxn} will drop to 0.0000 V and the reaction will come to chemical equilibrium.

Practice Exercise When a tooth with a gold inlay touches a tooth with an amalgam filling containing tin, the amalgam filling slowly erodes over time. The gold acts as the cathode for the reduction of oxygen to water. Use these two half-reactions and their electrochemical potentials to show that tin is oxidized under these conditions.

$$O_2(g) + 4\,H^+(aq) + 4\,e^- \rightarrow 2\,H_2O(\ell) \qquad E° = +1.23 \text{ V}$$

$$2\,Sn^{2+}(aq) + 4\,e^- + Sn_6Hg(s) \rightarrow Sn_8Hg(s) \qquad E° = -0.13 \text{ V}$$

PACEMAKERS Pacemakers are life-saving devices implanted in the chests of patients with certain heart conditions. A pacemaker delivers electrical pulses to stimulate the heart at the proper rate. The power for pacemakers comes from small batteries. These must deliver reliable power for long periods because, without additional surgery, the batteries cannot be replaced once the pacemaker is

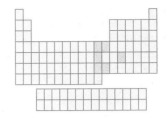

CONNECTION In Chapter 18 we learned how to use reduction potentials for half-reactions to calculate overall potentials of reactions.

CONNECTION Alloys were described in Chapter 11. The cell potentials of simple voltaic cells were discussed in Chapter 18.

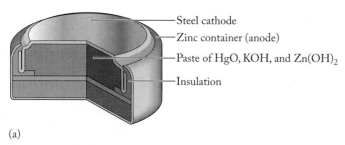

(a)

FIGURE 21.23 (a) A cutaway view of a Zn/HgO battery used in pacemakers. (b) Photograph of a pacemaker battery.

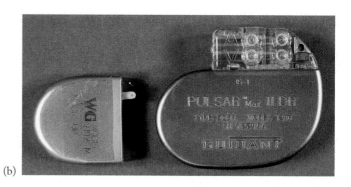

(b)

implanted. A solid-state zinc/mercury oxide battery (shown in Figure 21.23) meets these stringent requirements. The zinc container acts as the anode where oxidation takes place. A steel cathode extends into a paste of mercury(II) oxide (HgO), potassium hydroxide, and zinc(II) hydroxide. The half-reactions and the overall cell reaction are

Anode: $Zn(s) + 2\,OH^-(aq) \rightarrow ZnO(s) + H_2O(\ell) + 2\,e^-$ $E° = 1.249$ V

Cathode: $HgO(s) + H_2O(\ell) + 2\,e^- \rightarrow Hg(s) + 2\,OH^-(aq)$ $E° = 0.098$ V

Overall: $Zn(s) + HgO(s) \rightarrow ZnO(s) + Hg(s)$ $E° = 1.347$ V

The hazards associated with mercury and other performance issues has led to the use of lithium/iodine batteries in pacemakers in the last 30 years. The cathode is an electrically conductive material prepared by heating iodine (I_2) with poly(vinylpyridine) (PVP). Lithium metal acts as the anode. The half-reactions and the overall cell reaction for a lithium/iodine battery are

Anode: $2\,Li(s) \rightarrow 2\,Li^+(s) + 2\,e^-$ $E° = +3.05$ V

Cathode: $I_2(PVP)(s) + 2\,e^- \rightarrow 2\,I^-(s) + (PVP)(s)$ $E° = +0.536$

Overall: $2\,Li(s) + I_2(PVP)(s) \rightarrow 2\,LiI(s) + PVP(s)$ $E° = 3.586$ V

CONNECTION Poly(vinylpyridine) is a polymer. The preparation and properties of synthetic polymers like poly(vinylpyridine) were described in Chapter 12. Long-chain molecules like proteins, starch, and other biomolecules were described in Chapter 19.

CONNECTION In Chapter 12 we learned about the development of high-density polyethylene polymers for use in artificial joints.

ARTIFICIAL JOINTS The degeneration of knees and hips that often accompanies aging necessitates their replacement with artificial joints. The biomaterials for these joints must be chemically nonreactive yet provide a smooth surface to allow the joint to move. Thin coatings of pure tantalum or niobium or of alloys of these two metals are used to make surfaces on artificial joints smooth and abrasion resistant. The group 4 elements titanium and zirconium have also been used in alloys to coat implanted joints.

STENTS Shape memory alloys represent an unusual class of materials. Like wire made from most metals, a wire made from a shape memory alloy can be deformed or bent into a variety of shapes. When it is heated, however, the object reverts to its original shape. In other words, even though we may have changed the shape of a piece of shape memory alloy, it "remembers" its original shape. The first shape memory alloy was a 1:1 alloy of nickel and titanium, NiTi. The "memory" property of shape memory alloys depends on changes in the crystal structure as a function of temperature. The crystal structure of NiTi is slightly different at room temperature than at high temperatures (>500°C). When a wire of NiTi is bent into a shape at room temperature and then heated, the crystal structure changes to the high-temperature form, removing the bend.

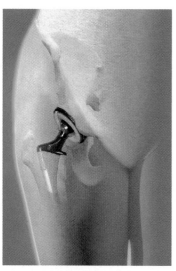

Tantalum is used to coat artificial hip joints.

(a) (b) (c)

Shape memory alloys are used to prepare stents for heart patients. This set of photographs illustrates how they work: (a) These wires are originally S-shaped; two have been deformed. When the middle wire (b) and the wire on the left (c) are heated gently, they restore themselves to their original shape. Similarly, stents have one shape at room temperature and assume a different shape when they are heated to body temperature.

The unusual shape memory property of different selected alloys leads to their use in stents to prop open weakened and clogged arteries. The stent is formed into a coil or tube that can be surgically inserted into a partially blocked artery. Once the stent is in place it changes back to its original shape as it is warmed by body heat. The stent in its "body temperature" shape expands and props open the artery.

At least 15 shape memory alloys have been identified, ranging from the most commonly used nickel-titanium alloy to alloys containing hafnium, titanium, and nickel. Perhaps hafnium-containing shape memory alloys will join the other group 4 elements as useful materials for medical applications. If so, such alloys would represent one of the very few medical applications of hafnium.

21.5 Toxic Elements

Our survey of the naturally occurring elements has left us with relatively few elements that have no biological function and that have limited, if any, applications to medicine. These elements are scattered throughout the periodic table. Some of them are radioactive: they include the four naturally occurring actinide elements, actinium (Ac), thorium (Th), protactinium (Pa), and uranium (U) as well as francium (Fr), radium (Ra), polonium (Po), astatine (At), and radon (Rn). Only radium and radon are found in appreciable amounts in nature. They have no known biological function; neither have they found application in medicine. In fact, as described in Chapter 20, radon gas and radium in granite and soil derived from granite pose serious health hazards from α decay. As noted in Chapter 20, α particles may have a short penetration distance, but their large relative biological effectiveness means they can have a devastating effect, especially on lung tissue.

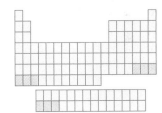

In this section we will focus on the compounds of beryllium, cadmium, mercury, and lead. Thallium could have been included among the quintessential toxic elements; however, we have seen that thallium isotopes are also useful imaging agents. Often it is the form of an element and its concentration that determine its utility in therapy or diagnosis as opposed to its inherent toxic effects. For these reasons we will bypass the toxicity issues associated with Tl and take a closer look at the toxicities of Be, Cd, Hg, and Pb.

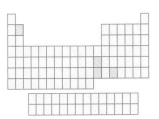

Beryllium

Beryllium, the first member of the alkaline earth metals in group 2, is present in Earth's crust at concentrations comparable to those of tin and arsenic. Exposure to beryllium compounds in our lives fortunately is quite limited. Beryllium

Beryl is a gemstone that contains beryllium.

toxicity is most often encountered in industrial settings where beryllium-contaminated dust is inhaled. The Be^{2+} ion replaces Mg^{2+} in the body where it inhibits Mg^{2+}-catalyzed RNA and DNA synthesis in cells. On the other hand, we value beryllium when we find it in the gemstones beryl and emerald.

Lead

The history of lead toxicity dates back thousands of years. Lead compounds like PbS, $PbCO_3$, and $PbCl(OH)$ were used by the ancient Egyptians in cosmetics. The Romans used lead pipes for plumbing and wine carafes and suffered from lead poisoning as a result. In the Middle Ages, lead acetate, $Pb(CH_3COO)_2$, was added to wine to sweeten its taste, a practice with dire consequences. In the 20th century, lead in the form of tetraethyl lead, $[(CH_3CH_2)_4Pb]$, was used as a gasoline additive to help automobile engines run smoothly. The effects of lead on arresting the development of children has been well documented, and the use of lead in motor fuels in the United States was phased out by 1986. Similarly, paints containing lead(II) carbonate ("white lead") and lead(II) chromate ("chrome yellow") have been banned from residential use in the U.S. since 1978. The toxicity of lead is traced to its strong binding to the oxygen groups of many enzymes. This action affects protein synthesis, hemoglobin synthesis, and many other physiological processes.

Mercury

The history of mercury poisoning is nearly as old as that of lead. In the classic story *Alice in Wonderland*, Lewis Carroll's "mad-hatter" character exhibits the symptoms of mercury poisoning among hat makers. Mercury(II) nitrate, $Hg(NO_3)_2$, was used to make felt easier to work. Mercury(II) ion binds easily to sulfur-containing amino acids like cysteine and is readily transported throughout the body. Some of the effects of mercury poisoning include memory loss, tremors, and impaired coordination.

Although we no longer use mercury compounds in hat manufacture, mercury was used in other ways throughout the 20th century. Dental amalgams and older thermometers are two familiar applications of mercury. Mercury spills from broken thermometers are difficult to clean up. The dense liquid finds its way into cracks and corners. Mercury has a small but significant vapor pressure, about 0.1 Pa at 20°C. This volatility of Hg means that it can enter the body by inhalation. In our bodies, mercury can be transformed into methylmercury(II), CH_3Hg^+, a potent neurotoxin. The methyl group increases the solubility of mercury in the nonpolar portion of cell membranes and allows the CH_3Hg^+ to diffuse more readily through the lipid bilayer shown in Figure 21.2. In particular, the membranes separating blood from the brain are vulnerable to penetration by CH_3Hg^+, which leads to the symptoms of mercury poisoning already cited.

The conversion of mercury to methylmercury in the environment (*biomethylation*) is responsible for the health problems encountered by Japanese fishermen and their families around Minamata Bay in the 1950s. Mercury-containing waste from industrial plants was routinely dumped into Minamata Bay where the mercury was converted to methylmercury. Fish accumulated the methylmercury in their tissues. The local residents ate the fish and suffered the effects of mercury poisoning. Sample Exercise 21.6 examines the role of equilibrium in the treatment of mercury and other heavy-metal poisoning.

CONNECTION The vapor pressures of liquids are discussed in Chapter 10. Pressure units (Pa) are defined in Chapter 6.

SAMPLE EXERCISE 21.6 **Calculating the Equilibrium Constant of a Ligand Exchange Reaction**

Mercury(II) ions (Hg^{2+}) in the aqueous environment can be concentrated in fish, so governments have issued warnings about the health effects of excessive consumption of tuna and other fish species. Methylmercury (CH_3Hg^+) binds to sulfur-containing amino acids in proteins. For example, CH_3Hg^+ combines with cysteine as described by the following equation and its associated formation constant:

$$CH_3Hg^+ + cysteine \rightleftharpoons CH_3Hg(cysteine)^+$$

$$K_{f,Hg\text{-}cyst} = \frac{[CH_3Hg(cysteine)^+]}{[CH_3Hg^+][cysteine]}$$

$$K_{f,Hg\text{-}cyst} = 3.98 \times 10^{16}$$

Mercury poisoning can be treated by injecting a compound, such as penicillamine, that has a larger formation constant for reaction with mercury(II) than cysteine.

$$CH_3Hg^+ + penicillamine \rightleftharpoons CH_3Hg(penicillamine)^+$$

$$K_{f,Hg\text{-}pen} = \frac{[CH_3Hg(penicillamine)^+]}{[CH_3Hg^+][penicillamine]}$$

$$K_{f,Hg\text{-}pen} = 6.31 \times 10^{16}$$

The reaction in which penicillamine replaces cysteine is

$$CH_3Hg(cysteine)^+ + penicillamine \rightleftharpoons CH_3Hg(penicillamine)^+ + cysteine$$

What is the equilibrium constant for the replacement reaction? *Hint:* You may wish to refer back to Chapter 17 for calculations involving formation constants of complexes, as well as Chapters 15 and 16 where we manipulated equilibrium constants, as you work through this exercise.

COLLECT AND ORGANIZE We are given the formation constants ($K_{f,Hg\text{-}cyst}$ and $K_{f,Hg\text{-}pen}$) of two reactions involving methylmercury cations. These data allow us to calculate the equilibrium constant for the replacement reaction.

ANALYZE The formation constant for the replacement reaction that yields $CH_3Hg(penicillamine)^+$ and cysteine is

$$K_{f,Hg\text{-}pen+cyst} = \frac{[cysteine][CH_3Hg(penicillamine)^+]}{[CH_3Hg(cysteine)^+][penicillamine]}$$

The equilibrium constant expression for $K_{f,Hg\text{-}pen+cyst}$ has [cysteine] in the numerator and [$CH_3Hg(cysteine)^+$] in the denominator, whereas the expression for $K_{f(Hg\text{-}cyst)}$ has [cysteine] in the denominator and [$CH_3Hg(cysteine)^+$] in the numerator. Cysteine appears as a product in the replacement reaction, so let's write an equilibrium constant expression for the reverse reaction of the first reaction so that cysteine is a product:

$$CH_3Hg(cysteine)^+ \rightleftharpoons CH_3Hg^+ + cysteine$$

$$K_{f,Hg+cyst} = [CH_3Hg^+][cysteine]/[CH_3Hg(cysteine)^+]$$

$$= 1/K_{f,Hg\text{-}cyst} = 1/(3.98 \times 10^{16}) = 2.51 \times 10^{-17}$$

The replacement equation is the summation of the reaction to form CH_3Hg (penicillamine)$^+$ and the reaction to dissociate $CH_3Hg(cysteine)^+$:

$$\cancel{CH_3Hg^+} + penicillamine \rightleftharpoons CH_3Hg(penicillamine)^+$$

$$+ CH_3Hg(cysteine)^+ \rightleftharpoons \cancel{CH_3Hg^+} + cysteine$$

CONNECTION In Chapter 17 we discussed formation constants for metal complexes. Formation constants are equilibrium constants that describe the formation of a metal-ligand complex.

$$CH_3Hg(cysteine)^+ + penicillamine \rightleftharpoons CH_3Hg(penicillamine)^+ + cysteine$$

and the equilibrium constant for this reaction is

$$K_{f,replacement} = [CH_3Hg(penicillamine)^+][cysteine]/[CH_3Hg(cysteine)^+][penicillamine]$$

The equilibrium constant for the replacement reaction is the product of the equilibrium constants for the two reactions that are added together to get the replacement reaction. Thus the equilibrium constant for the replacement reaction is

$$K_{f,replacement} = K_{f,Hg-pen} \times (1/K_{f,Hg+cyst})$$

$$= \frac{[CH_3Hg(penicillamine)^+]}{[\cancel{CH_3Hg^\pm}][penicillamine]} \times \frac{[\cancel{CH_3Hg^\pm}][cysteine]}{[CH_3Hg(cysteine)^+]}$$

SOLVE We can calculate the equilibrium constant for the replacement reaction that breaks down $CH_3Hg(cysteine)^+$ by substituting the values of $K_{f,Hg-pen}$ and $K_{f,Hg+cyst}$:

$$K_{f,replacement} = (6.31 \times 10^{16})(2.51 \times 10^{-17}) = 1.58$$

THINK ABOUT IT In this exercise, we see that the equilibrium between a methylmercury(cysteine) complex and penicillamine favors the binding of mercury to the latter ligand, although the equilibrium constant is small. Treatment of a patient suffering from mercury poisoning with penicillamine will eventually lead to reduction in the amount of mercury in the patient's body because the $CH_3Hg(penicillamine)^+$ is eliminated in urine, thereby reducing the concentration of one of the products and shifting the equilibrium of the replacement reaction to the right. The solution to this exercise applies concepts from Chapters 15 and 17. Specifically, we have revisited how to write equilibrium constant expressions for forward and reverse reactions as well as how to write equilibrium constant expressions for sequential reactions. We also have reviewed the concept of formation constants for metal ion reactions with chelating Lewis bases.

Practice Exercise The value of K_f for the complexation of methylmercury ion with methionine, another sulfur-containing amino acid, is 2.5×10^7.

$$CH_3Hg^+ + methionine \rightleftharpoons CH_3Hg(methionine)^+$$

By using the data for penicillamine-mercury complexes given in Sample Exercise 21.6, calculate the equilibrium constant for the reaction

$$CH_3Hg(methionine)^+ + penicillamine \rightleftharpoons CH_3Hg(penicillamine)^+ + methionine$$

Cadmium

Until the latter part of the 20th century, the toxicity of cadmium compounds was of minimal concern to humans. Cadmium is not particularly abundant; it is found as 0.2%–0.4% by mass of zinc-containing ores. The demand for zinc exposed workers at zinc smelters to cadmium. However, the development of nickel/cadmium batteries as well as the use of cadmium sulfide and cadmium selenide in yellow, orange, and red pigments exposed a much larger population to cadmium compounds. For example, nicad batteries thrown into a landfill have the potential of introducing Cd^{2+} ions into groundwater. Cadmium(II) ions have an ionic radius (95 pm) that is close to that of calcium ions (Ca^{2+}, 100 pm). The similarity in size leads to cadmium accumulation and subsequent embrittlement of bones. Like mercury and lead, cadmium ions bind strongly to sulfur-containing proteins and can displace Zn^{2+} from zinc enzymes.

CONNECTION The applications of the group 12 elements in electrochemistry and battery technology are discussed in Chapter 18.

Summary

SECTION 21.1 **Essential elements** have a physiological function in the body. Some **nonessential elements** may promote growth or have other **stimulatory effects**. Essential elements can be further categorized as **major**, **trace**, or **ultratrace elements** depending on their concentrations in living organisms.

SECTION 21.2 Sodium and potassium ions are important in generating and transmitting nerve impulses and maintaining proper osmotic pressure in cells. Transport of Na^+ and K^+ across cell membranes involves diffusion, Na^+-K^+ **ion pumps**, or selective transport through **ion channels** formed by groups of helical proteins. Magnesium is important in the transfer of phosphate groups between ATP and ADP and is also found in chlorophyll, the green pigment in plants that mediates photosynthesis. Calcium is a major component of teeth and bones. A prolonged deficiency of calcium can lead to osteoporosis, whereas high concentrations of calcium in the intracellular fluid contribute to muscle cramps. Mammalian bones are **composite materials** that consist of mixtures of compounds with different compositions and structures. Bones are primarily elastic protein fibers and calcium salts including the mineral hydroxyapatite, $Ca_5(PO_4)_3(OH)$, which is also the principal component of teeth. Chloride ion is the most abundant anion in the human body and is involved in many functions. To maintain electrical neutrality in a cell, transport of alkali metal cations is accompanied by transport of chloride anions. Chloride ions also play a role in the elimination of CO_2. Malfunctioning chloride channels are an underlying cause of cystic fibrosis.

SECTION 21.3 Rubidium is probably nonessential, yet it is retained by the body because of its similarity in size and chemistry to K^+. Cesium has no known function although it can also substitute for K^+ and so can interfere with potassium-dependent functions. Strontium and barium ions are incorporated into human bones by replacing Ca^{2+} ions but have no known function in the human body. However, Sr^{2+} and Ba^{2+} are found in the exoskeletons of certain single-cell organisms. Amorphous silica, SiO_2, makes up the exoskeletons of diatoms. Silicon scavenges aluminum in mammals, and the absence of silicon in their diets stunts growth. Germanium is a nonessential element and barely detectable in the human body. Trace quantities of arsenic may promote growth in some mammals. Selenium is an essential ultratrace element found in the body; the amino acid selenocysteine is a constituent of proteins that act as catalysts for biochemical reactions. In therapeutic treatments, fluoride ion replaces hydroxide in hydroxyapatite in teeth and inhibits the formation of cavities. Bromine does not appear to be an essential element. The body concentrates iodide (I^-) in the thyroid gland where the iodide is incorporated into the hormones that regulate energy production and use. Many of the enzymes and **coenzymes** in our bodies contain transition metal ions including those of V, Mo, W, Fe, Co, Ni, Cu, and Zn. Manganese is important in photosynthesis.

SECTION 21.4 Radioactive isotopes with short half-lives that emit low-energy γ rays are used in the diagnosis of disease. The selection of a radioisotope is also governed by the toxicity of the element and its daughter isotopes, the speed at which the imaging agent is eliminated from the body, and the cost and availability. The availability of many isotopes has allowed medical researchers to investigate potential imaging agents across much of the periodic table. Therapeutic agents contain metallic elements, including lithium, transition elements, and heavier main group elements. A useful diagnostic or therapeutic agent must be sufficiently soluble in blood to be delivered to the target. Radioisotopes are used to both diagnose disease and treat cancer.

SECTION 21.5 Some elements pose health risks because they are reactive, whereas others, such as Be, Pb, Cd, Tl, and Hg, are highly toxic because of their chemical properties.

Problem-Solving Summary

TYPE OF PROBLEM	CONCEPTS AND EQUATIONS	SAMPLE EXERCISES
Calculating the pH and $[H_3O^+]$ of an acid	Relate the pH of a solution to the $[H^+]$ by the equation $$pH = -\log[H^+]$$	21.1
Writing half-reactions involving oxoanions	Use oxidation numbers to determine the number of electrons gained or lost; balance charges with H^+ ions in acids or OH^- ions in base; add water molecules to balance the numbers of H and O atoms. (Another approach, based on balancing the number of O and H atoms with water molecules and H^+ ions and then balancing charges with electrons, is described in Chapter 4.)	21.2
Identifying particles in nuclear reactions	Sum the subscripts and superscripts on both sides of the radiochemical equation. The difference in the superscripts equals the mass of the missing particle, and the difference in the subscripts equals the charge on the particle.	21.3

TYPE OF PROBLEM	CONCEPTS AND EQUATIONS	SAMPLE EXERCISES
Calculating a remaining quantity of a radioisotope	Use the equation $$\ln\frac{A_t}{A_0} = \frac{-0.693t}{t_{1/2}}$$ where A_0 and A_t are the amounts of material present initially and at time t, respectively.	21.4
Predicting the spontaneity of a redox reaction	Calculate the $E°_{rxn}$ by summing the oxidation potentials of the oxidation and reduction half-reactions. If $E°_{rxn}$ is (+), the reaction is spontaneous.	21.5
Calculating the equilibrium constant of a ligand exchange reaction	Combine the equilibrium constant expressions for the formation of the two complexes, after reversing the one that dissociates in the overall reaction, to generate an equivalent equilibrium constant expression for the overall reaction.	21.6

Visual Problems

For Problems 21.1 and 21.2 refer to the following blank periodic tables with arrows showing trends.

21.1. Which figure best describes the periodic trend in monoatomic cation radii moving up or down a group or across a period in the periodic table? (Arrows point in the direction of increasing radii.)

21.2. Which figure best describes the periodic trend in monoatomic anion radii moving up or down a group or across a period in the periodic table? (Arrows point in the direction of increasing radii.)

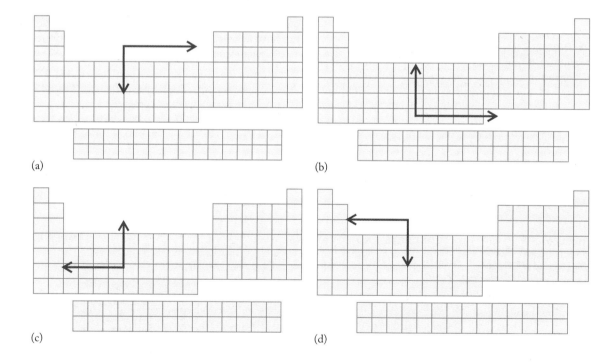

(a)

(b)

(c)

(d)

21.3. Which group highlighted in the periodic table in Figure P21.3 typically forms ions that have larger radii than the corresponding neutral atoms?

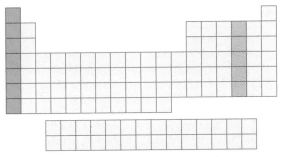

FIGURE P21.3

21.4. Which group highlighted in the periodic table in Figure P21.4 typically forms ions that have smaller radii than the corresponding neutral atoms?

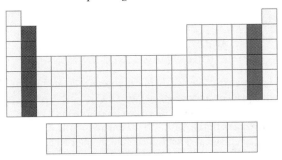

FIGURE P21.4

21.5. As we saw in Chapter 18, the free energy (ΔG) of a reaction is related to the cell potential by the equation $\Delta G = -nFE$. In Figure P21.5, two solutions of Na^+ of different concentration are separated by a semipermeable membrane. Calculate ΔG for the transport of Na^+ from the side with higher concentration to the side with lower concentration.

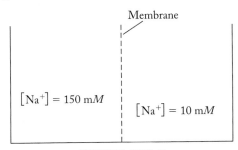

FIGURE P21.5

21.6. Two solutions of K^+ are separated by a semipermeable membrane in Figure P21.6. Calculate ΔG for the transport of K^+ from the side with lower concentration to the side with higher concentration.

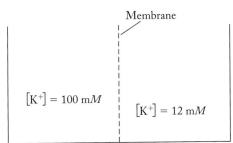

FIGURE P21.6

21.7. Describe the molecular geometry around each germanium atom in the structure of the germanium compound shown in Figure P21.7.

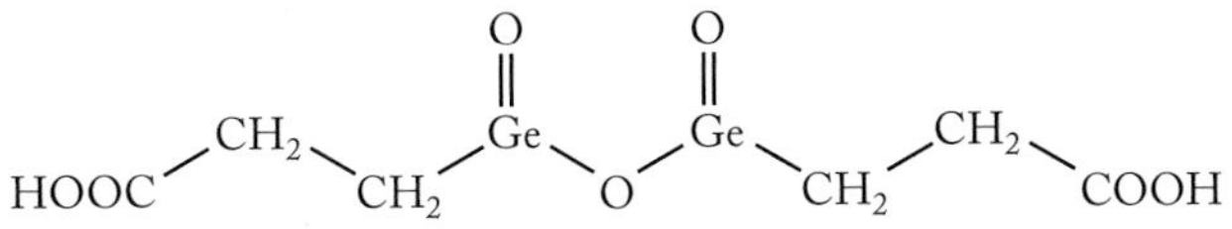

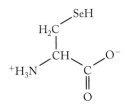

FIGURE P21.7

21.8. Selenocysteine can exist as two enantiomers (stereoisomers). Identify the atom in Figure P21.8 responsible for the two enantiomers.

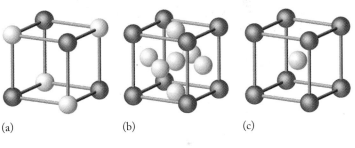

FIGURE P21.8

21.9. The austenite structure for NiTi is similar to the CsCl structure. Which crystal structure shown in Figure P21.9 is correct for the austenite form of the shape memory alloy NiTi?

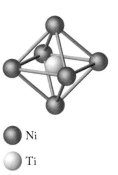

FIGURE P21.9

*21.10. The unit cell for the martensite form of the shape memory alloy NiTi is shown in Figure P21.10. How many equivalent Ti and Ni atoms are in the unit cell?

FIGURE P21.10

Questions and Problems

The Periodic Table of Life

CONCEPT REVIEW

21.11. What is the difference between an essential element and a nonessential element?

21.12. Are all essential elements also major elements?

21.13. What distinguishes (a) bulk, (b) trace, and (c) ultratrace elements?

21.14. Should essential trace elements also be considered to be stimulatory?

PROBLEMS

21.15. Which ion is larger: K^+ or Na^+?

21.16. Which ion is larger: Cl^- or I^-?

21.17. The concentrations of very dilute solutions are sometimes expressed as parts per million (mass of solute/total mass of solution) $\times 10^6$. Express the concentration of each of the following essential trace elements in ppm:
a. copper, 110 mg in 70 kg
b. zinc, 3.3×10^{-2} g/kg
c. iodine, 0.043 g in 100 kg

21.18. In the human body, the concentrations of essential ultratrace elements are even lower than those of essential trace elements and therefore are sometimes expressed in parts per billion (ppb). Units of ppb are defined as (mass of solute/total mass of solution) $\times 10^9$. Express the concentrations of each of the following essential ultratrace elements in ppb:
a. cobalt, 4.3×10^{-5} g/kg
b. boron, 0.014 g/100 kg
c. chromium, 5.0 mg/70 kg

21.19. In the following pairs, which element is more abundant in the human body: (a) silicon or oxygen; (b) iron or oxygen; (c) carbon or aluminum?

21.20. In the following pairs, which element is more abundant in the human body: (a) H or Si; (b) Ca or Fe; (c) N or Cr?

Essential Major Elements

CONCEPT REVIEW

21.21. Ion Transport in Cells Describe three ways in which ions of essential major elements (such as Na^+ and K^+) enter and exit cells.

21.22. Which transport mechanism for ions requires ATP: diffusion, ion channels, or ion pumps?

21.23. Why is it difficult for ions to diffuse across cell membranes?

21.24. Why does Sr^{2+} substitute for Ca^{2+} in bones?

21.25. Which alkali metal ion is Rb^+ most likely to substitute for?

21.26. Why don't alkaline earth metal cations substitute for alkali metal cations in cases where the ionic radii are similar?

*21.27. Why might nature have chosen calcium carbonate over calcium sulfate as the major exoskeleton material in shells?

21.28. Bromide (Br^-) and fluoride (F^-) are found in the body as nonessential elements. Would you expect their concentrations to be more similar to essential bulk or essential ultratrace elements?

PROBLEMS

21.29. Osmotic Pressure of Red Blood Cells One of the functions of the alkali metal cations Na^+ and K^+ in cells is to maintain the cells' osmotic pressure. The concentration of NaCl in red blood cells is approximately 11 mM. Calculate the osmotic pressure of this solution at body temperature (37°C). (*Hint:* see Equation 10.10.)

21.30. Calculate the osmotic pressure exerted by a 92 mM solution of KCl in a red blood cell at body temperature (37°C). (*Hint:* see Equation 10.10.)

*21.31. Electrochemical Potentials across Cell Membranes Very different concentrations of Na^+ ion exist in red blood cells (11 mM) and the blood plasma (160 mM) surrounding those cells. Solutions with two different concentrations separated by a membrane constitute a concentration cell of the type described in Chapter 18.

$$E = E°\,(-0.0592/n)\log\{[Na^+]_{cell}/[Na^+]_{plasma}\}$$

Calculate the electrochemical potential created by the unequal concentrations of Na^+.

21.32. The concentration of K^+ in red blood cells is 92 mM, and the concentration of K^+ in plasma is 10 mM. Calculate the electrochemical potential created by the two concentrations of K^+.

$$E = E°\,(-0.0592/n)\log\{[K^+]_{plasma}/[K^+]_{cell}\}$$

21.33. If the transport of K^+ across a cell membrane requires 5 kJ/mol, how many moles of ATP must be hydrolyzed to provide the necessary energy? The hydrolysis of ATP is described by the equation

$$ATP^{4-} + H_2O \rightleftharpoons ADP^{3-} + HPO_4^{2-} + H^+ \qquad \Delta G° = -34.5 \text{ kJ}$$

*21.34. Removing excess Na^+ from a cell by an ion pump requires energy. How many moles of ATP must be hydrolyzed to overcome a cell potential of -0.07 V? The hydrolysis of 1 mol of ATP provides 34.5 kJ of energy.

21.35. Plankton Exoskeletons Exoskeletons of planktonic Acantharia contain strontium sulfate. Calculate the solubility in mol/L of $SrSO_4$ in water at 25°C given that $K_{sp} = 3.4 \times 10^{-7}$.

21.36. The algae *Closterium* contain structures built from barium sulfate (barite). Calculate the solubility in mol/L of $BaSO_4$ in water at 25°C given that $K_{sp} = 1.1 \times 10^{-10}$.

Essential Trace and Ultratrace Elements

CONCEPT REVIEW

21.37. What danger to human health is posed by ^{137}Cs ($t_{1/2} \approx 30$ yr)?

21.38. Why is ^{137}Cs ($t_{1/2} \approx 30$ yr) considered to be dangerous to human health when naturally occurring ^{40}K ($t_{1/2} = 1.28 \times 10^6$ yr) is benign?

21.39. What is the origin of most of the ^{137}Cs and ^{90}Sr in our environment?

21.40. Why does fluorapatite resist acid better than hydroxyapatite if both are insoluble in water?

21.41. What is the function of enzymes?

21.42. Enzymes are large proteins. Are all proteins enzymes?

*21.43. What effect does an enzyme have on the activation energy of a biochemical reaction?

*21.44. Why might reductases also be described as reducing agents?

*21.45. When a transition metal ion like Cu^{2+} is incorporated into a metalloenzyme, is the formation constant likely to be much greater than one ($K \gg 1$) or much less than one ($K \ll 1$)?

$$Cu^{2+} + \text{protein} \rightleftharpoons \text{metalloenzyme} \qquad K = \frac{[\text{metalloenzyme}]}{[Cu^{2+}][\text{protein}]}$$

*21.46. When transition metals bind to proteins to form enzymes, is the process likely to be endergonic ($\Delta G > 0$) or exergonic ($\Delta G < 0$)?

PROBLEMS

21.47. What are the products of radioactive decay of ^{137}Cs? Write a balanced equation for the nuclear decay reaction.

21.48. Potassium-40 decays by three pathways: β decay, positron emission, and electron capture. Write balanced equations for each of these processes.

21.49. Calculate the pH of a 1.00×10^{-3} M solution of selenocysteine ($pK_{a_1} = 2.21$, $pK_{a_2} = 5.43$).

21.50. Calculate the pH of a 1.00×10^{-3} M solution of cysteine ($pK_{a_1} = 1.7$, $pK_{a_2} = 8.3$) Is selenocysteine a stronger acid than cysteine?

21.51. Composition of Tooth Enamel Tooth enamel contains the mineral hydroxyapatite. Hydroxyapatite reacts with fluoride ion in toothpaste to form fluorapatite. The equilibrium constant for the reaction between hydroxyapatite and fluoride ion is $K = 8.48$. Write the equilibrium constant expression for the following reaction. In which direction does the equilibrium lie?

$$Ca_5(PO_4)_3(OH)(s) + F^-(aq) \rightleftharpoons Ca_5(PO_4)_3(F)(s) + OH^-(aq)$$

*21.52. **Effects of Excess Fluoridation on Teeth** Too much fluoride might lead to the formation of calcium fluoride according to the reaction

$$Ca_5(PO_4)_3(OH)(s) + 10\,F^-(aq) \rightleftharpoons 5\,CaF_2(s) + 3\,PO_4^{3-}(aq) + OH^-(aq)$$

Write the equilibrium constant expression for the reaction. Given the K_{sp} values for the following two reactions, calculate K for the reaction between $Ca_5(PO_4)_3OH$ and fluoride ion that forms CaF_2.

$$Ca_5(PO_4)_3(OH)(s) \rightleftharpoons 5\,Ca^{2+}(aq) + 3\,PO_4^{3-}(aq) + OH^-(aq)$$
$$K_{sp} = 2.3 \times 10^{-59}$$

$$CaF_2(s) \rightleftharpoons Ca^{2+}(aq) + 2\,F^-(aq) \qquad K_{sp} = 3.9 \times 10^{-11}$$

21.53. Tooth enamel is actually a composite material containing both hydroxyapatite and a calcium phosphate, $Ca_8(HPO_4)_2(PO_4)_4 \cdot 6\,H_2O$ ($K_{sp} = 1.1 \times 10^{-47}$). Is this calcium mineral more or less soluble than hydroxyapatite ($K_{sp} = 2.3 \times 10^{-59}$)?

21.54. The K_{sp} of actual tooth enamel is reported to be 1×10^{-58}. Does this mean that tooth enamel is more soluble than pure hydroxyapatite ($K_{sp} = 2.3 \times 10^{-60}$)? Does the measured value of K_{sp} for tooth enamel support the idea that tooth enamel is a mixture of hydroxyapatite, $Ca_5(PO_4)_3OH$, and a calcium phosphate $Ca_8(HPO_4)_2(PO_4)_4 \cdot 6\,H_2O$ ($K_{sp} = 1.1 \times 10^{-47}$)?

21.55. Calculate the solubility in mol/L of hydroxyapatite, $Ca_5(PO_4)_3OH$, $K_{sp} = 2.3 \times 10^{-59}$, and fluorapatite, $Ca_5(PO_4)_3F$, $K_{sp} = 3.2 \times 10^{-60}$, in water at 25°C and pH = 7.0.

21.56. Calculate the solubility in mol/L of hydroxyapatite, $Ca_5(PO_4)_3OH$, $K_{sp} = 2.3 \times 10^{-59}$, in water at 25°C and pH = 5.0.

*21.57. Some sources give the formula of hydroxyapatite as $Ca_{10}(PO_4)_6(OH)_2$. If the K_{sp} of $Ca_5(PO_4)_3OH$ is 2.3×10^{-59}, what is the K_{sp} of $Ca_{10}(PO_4)_6(OH)_2$?

*21.58. The same sources mentioned in the previous exercise cite the formula of fluorapatite as $Ca_{10}(PO_4)_6(F)_2$. If the K_{sp} or $Ca_5(PO_4)_3F$ is 3.2×10^{-60}, what is the K_{sp} of $Ca_{10}(PO_4)_6(F)_2$?

*21.59. The activation energy for the uncatalyzed decomposition of hydrogen peroxide at 20°C is 75.3 kJ/mol. In the presence of the enzyme catalase, the activation energy is reduced to 29.3 kJ/mol. By using the following form of the Arrhenius equation, $RT \ln(k_1/k_2) = E_{a_2} - E_{a_1}$, how much faster is the catalyzed reaction?

*21.60. **Enzymatic Activity of Urease** Urease catalyzes the decomposition of urea to ammonia and carbon dioxide.

$$\begin{array}{c} O \\ \parallel \\ H_2N \overset{\textstyle C}{} NH_2 \end{array}(aq) + H_2O(\ell) \rightarrow 2\,NH_3(aq) + CO_2(g)$$

The rate constant for the uncatalyzed reaction at 20°C and pH 8 is $k = 3 \times 10^{-10}$ s^{-1}. A urease isolated from the jack bean increases the rate constant to $k = 3 \times 10^4$ s^{-1}. By using the $RT \ln(k_1/k_2) = E_{a_2} - E_{a_1}$ form of the Arrhenius equation, calculate the ratio of the activation energies.

Diagnosis and Therapy

CONCEPT REVIEW

21.61. List some of the considerations in choosing a radioisotope for imaging.

21.62. What advantages does an isotope like ^{99m}Tc have over ^{201}Tl for imaging the circulatory system?

21.63. Why might an α emitter be a good choice for chemotherapy?

*21.64. What advantage might a β emitter have over an α emitter for imaging?

21.65. Gadolinium-153 decays by electron capture. What type of radiation does ^{153}Gd produce that makes it useful for imaging?

21.66. Gadolinium-153 and samarium-153 both have the same mass number. Why might ^{153}Gd decay by electron capture whereas ^{153}Sm decays by emitting β particles?

21.67. How do platinum- and ruthenium-containing drugs fight cancer?

*21.68. Many transition metal complexes are brightly colored. Why might the titanium(IV) compound budotitane be colorless? (*Hint:* see Chapter 17.)

*21.69. Is the glucose tolerance factor that contains chromium(III) paramagnetic or diamagnetic? (*Hint:* see Chapter 17.)

21.70. Why might lithium be an attractive anode material in batteries used in pacemakers?

21.71. Mercury compounds are generally toxic to humans. Why can we use mercury in dental amalgams?

21.72. Tantalum is used in artificial joints. What property of Ta makes it attractive for this purpose?

PROBLEMS

21.73. The lanthanide isotopes cerium-141, terbium-160, thulium-170, and lutetium-177 all undergo β decay. Write a balanced nuclear equation for each decay reaction. Are all of the product isotopes in the same group of the periodic table as the reactants?

21.74. The lanthanide isotopes gadolinium-153 and ytterbium-169 decay by electron capture. Write a balanced nuclear equation for each decay reaction. Are both product isotopes in the same group of the periodic table?

21.75. PET Imaging with Gallium A patient is injected with a 5 μM solution of gallium citrate containing ^{68}Ga ($t_{1/2} = 9.4$ hr) for a PET study. How long is it before the activity of the ^{68}Ga drops to 5% of its initial value?

21.76. Indium-111 ($t_{1/2} = 2.81$ d) has been used in imaging. How long is it before the activity of ^{111}In drops to 15% of its initial value?

21.77. The bismuth in Pepto-Bismol™ and Kaopectate™ is found as BiO^+. Draw the Lewis structure for the BiO^+ cation.

21.78. Camcolit™ used in treating depression contains lithium carbonate. Draw the Lewis structure for Li_2CO_3.

21.79. Draw the Lewis structure for the citrate ion by using the skeletal drawing in Figure P21.79 as a guide.

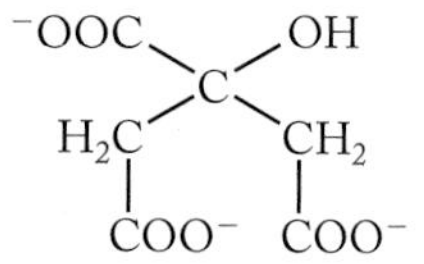

FIGURE P21.79

$$^-OOC-\underset{\underset{CH_2}{|}}{\underset{|}{C}}\text{(OH)}-CH_2$$

$$\begin{array}{cc} H_2C & CH_2 \\ | & | \\ COO^- & COO^- \end{array}$$

21.80. Draw the Lewis structure for the thiosulfate ion, $S_2O_3^{2-}$, which is found in the gold-bearing arthritis drug Sanochrysin.

21.81. Aluminum hydroxide is used in some antacids. Write a balanced net ionic equation for the reaction of aluminum hydroxide with HCl.

21.82. Aluminum carbonate is used in some antacids. Write a balanced net ionic equation for the reaction of aluminum carbonate with hydrochloric acid.

21.83. A silver/zinc (Ag/Zn) battery has the advantage of being mercury-free. The half-reactions are as follows:

$$ZnO(s) + H_2O(\ell) + 2e^- \rightarrow Zn(s) + 2OH^-(aq) \qquad E° = +0.342 \text{ V}$$

$$Ag_2O(s) + H_2O(\ell) + 2e^- \rightarrow 2Ag(s) + 2OH^-(aq) \qquad E° = -1.258 \text{ V}$$

What is the overall reaction of an electrochemical cell based on these materials? Using the $E°$ values provided, determine the standard cell potential of the Ag/ZnO battery.

21.84. Power Sources for Pacemakers Another power source considered for pacemakers is the lithium/copper sulfide battery. Use the half-reactions shown here to determine the overall reaction for an electrochemical cell based on these materials. With the $E°$ values provided, calculate the cell potential of the Li/CuS battery.

$$Li^+(s) + e^- \rightarrow Li(s) \qquad E° = -3.05 \text{ V}$$

$$CuS(s) + 2e^- \rightarrow Cu(s) + S^{2-}(aq) \qquad E° = -0.851 \text{ V}$$

Toxic Elements

CONCEPT REVIEW

21.85. Why is methylmercury, CH_3Hg^+, more toxic than mercury metal?

21.86. Why is Cd^{2+} more likely than Cr^{2+} to replace Zn^{2+} in an enzyme like carbonic anhydrase in glucose tolerance factor?

21.87. Why is Be^{2+} more likely than Ca^{2+} to displace Mg^{2+} in biomolecules?

21.88. PbS, $PbCO_3$, and PbCl(OH) have limited solubility in water. Which of them is/are more likely to dissolve in acidic solutions?

PROBLEMS

21.89. Draw the Lewis structure for tetraethyl lead, $Pb(CH_2CH_3)_4$, and determine its molecular geometry.

21.90. Draw the Lewis structure for the ionic compound mercury(II) nitrate.

21.91. The complexation of mercury(II) ion with methionine

$$Hg^{2+} + \text{methionine} \rightleftharpoons Hg(\text{methionine})^{2+}$$

has a formation constant of $\log K = 14.2$, whereas the formation constant for the Hg^{2+} complex with penicillamine

$$Hg^{2+} + \text{penicillamine} \rightleftharpoons Hg(\text{penicillamine})^{2+}$$

is $\log K = 16.3$. Calculate the equilibrium constant for the reaction

$$Hg(\text{methionine})^{2+} + \text{penicillamine} \rightleftharpoons$$
$$Hg(\text{penicillamine})^{2+} + \text{methionine}$$

21.92. The complexation of mercury(II) ion with cysteine in aqueous solution

$$Hg^{2+} + \text{cysteine} \rightleftharpoons Hg(\text{cysteine})^{2+}$$

has a formation constant of $\log K = 14.2$, whereas the formation constant for the Hg^{2+} complex with glycine

$$Hg^{2+} + \text{glycine} \rightleftharpoons Hg(\text{glycine})^{2+}$$

is $\log K = 10.3$. Calculate the equilibrium constant for the reaction

$$Hg(\text{cysteine})^{2+} + \text{glycine} \rightleftharpoons Hg(\text{glycine})^{2+} + \text{cysteine}$$

21.93. The equilibrium constant of the reaction

$$CH_3Hg(\text{penicillamine})^+(aq) + \text{cysteine}(aq) \rightleftharpoons$$
$$CH_3Hg(\text{cysteine})^+(aq) + \text{penicillamine}(aq)$$

is $K = 0.633$. Calculate the equilibrium concentrations of cysteine and penicillamine if we start with a 1.00 mM solution of cysteine and a 1.00 mM solution of $CH_3Hg(\text{penicillamine})^+$.

21.94. The equilibrium constant of the reaction in aqueous solution

$$CH_3Hg(\text{glutathione})^+(aq) + \text{cysteine}(aq) \rightleftharpoons$$
$$CH_3Hg(\text{cysteine})^+(aq) + \text{glutathione}(aq)$$

is $K = 5.0$. Calculate the equilibrium concentrations of cysteine and glutathione if we start with a 1.20 mM solution of cysteine and a 1.20 mM solution of $CH_3Hg(\text{glutathione})^+$.

Mathematical Procedures

WORKING WITH SCIENTIFIC NOTATION

Quantities that scientists work with often are very large, such as Earth's mass, or very small, such as the mass of an electron. It is easier to work with these numbers if they are expressed in scientific notation.

The general form of standard scientific notation is a value between 1 and 10 multipled by 10 raised to an integral power. According to this definition, 598×10^{22} kg (Earth's mass) is not in standard scientific notation, but 5.98×10^{24} kg is. It is good practice to use and report data or parameters in standard scientific notation.

1. **To convert an "ordinary" number to standard scientific notation,** move the decimal point to the left for a large number, or to right for a small one, so that the decimal point is located after the first nonzero digit.

    A. For example, to express Earth's average density ($5{,}517$ kg/m³) in scientific notation requires moving the decimal three places to the left. Doing so is the same as dividing the number by 1000, or 10^3. To keep the value the same we add an exponent to multiply it by 10^3. So, Earth's density in standard scientific notation is 5.517×10^3 kg/m³.

    B. If you move the decimal point of a value less than one to the right to express it in scientific notation, then the exponent is a negative integer equal to the number of places you moved the decimal point to the right. For example, the value of R used in solving ideal gas law problems is 0.08206 L · atom/(mol · K). Moving the decimal point two places to the right converts the value of R to scientific notation: 8.206×10^{-2} L · atom/(mol · K).

    C. Another value of R, 8.314 J/(mol · K), does not need an exponent, though it could be written 8.314×10^0 J/(mol · K).

2. **To add or subtract numbers in scientific notation,** their exponents must be the same. This may require you to change the exponents (and thus the decimal point position) of some values. (NOTE: This step is not necessary when using a scientific calculator because the calculator will make all necessary conversions.)

Sample Exercise 1 Calculate the sum of the masses of the subatomic particles in an atom of lithium-7, given the following masses in grams:

Particle	Mass (g)
proton	1.67263×10^{-24}
neutron	1.67494×10^{-24}
electron	9.10939×10^{-28}

Solution An atom of ^{7}Li has three protons, three electrons, and four neutrons. Therefore, the total mass of the subatomic particles in an atom is the sum of three times the masses of a proton and an electron and four times the mass of a neutron. Doing the multiplication steps first yields the following:

mass of three protons $= 3\,(1.67263 \times 10^{-24}\,\text{g}) = 5.01789 \times 10^{-24}\,\text{g}$

mass of four protons $= 4\,(1.67494 \times 10^{-24}\,\text{g}) = 6.69976 \times 10^{-24}\,\text{g}$

mass of three electrons $= 3\,(9.10939 \times 10^{-28}\,\text{g}) = 2.732817 \times 10^{-27}\,\text{g}$

Before adding these masses together, we must express them all with the same exponent. The most convenient is 10^{-24}. To express the combined mass of three electrons using this exponent requires shifting the decimal point in 2.732817×10^{-27} g three places to the left, which makes the number before the exponent 1000 times smaller and compensates for the exponent becoming 1000 times larger. Adding the resulting mass of the three electrons, $0.002732817 \times 10^{-24}$ g, to the other two masses, we have:

$$5.01789 \times 10^{-24} \text{ g}$$
$$+ \ 6.69976 \times 10^{-24} \text{ g}$$
$$\underline{+ \ 0.002732817 \times 10^{-24} \text{ g}}$$
$$11.720382817 \times 10^{-24} \text{ g}$$

We need to round off this sum so that we have only five digits to the right of the decimal place, because that is the smallest number of digits to the right in two of the three values being summed. Therefore, the final answer is

$$11.72038 \times 10^{-24} \text{ g}$$

3. **To multiply values with exponents,** the values in front of the exponents are multiplied together, but the exponents are added (these steps happen automatically with scientific calculators).

Sample Exercise 2 American Steve Fossett circumnavigated the globe in early summer 2002 in the *Spirit of Freedom* balloon, which was partially filled with 5.5×10^5 ft^3 of helium. What is this volume in liters? Given: 1 ft^3 = 28.3 L

Solution We convert the starting value by multiplying 5.5 by 2.83 and adding the exponents $(1 + 5)$:

$$(5.5 \times 10^5 \ \cancel{\text{ft}^3}) \left(\frac{2.83 \times 10^1 \text{ L}}{\cancel{\text{ft}^3}} \right) = 15.5 \times 10^6 \text{ L or } 1.6 \times 10^{-7} \text{ L}$$

4. **To divide values with exponents,** the values in front of the exponents are divided, but the exponents are subtracted (again, these steps happen automatically with scientific calculators).

Sample Exercise 3 The speed of light is 2.998×10^8 m/s. What is the equivalent speed in miles per second? Given: 1 mile = 1.609×10^3 m

Solution Expressing the speed of light in miles per second requires dividing the speed of light in meters per second by the conversion factor given in the exercise. We divide the values in front of the exponents (2.998/1.609) and subtract to get their exponents $(8 - 3)$

$$\frac{2.998 \times 10^8 \ \cancel{\text{m}}/\text{s}}{1.609 \times 10^3 \ \cancel{\text{m}}/\text{mi}} = 1.863 \times 10^5 \text{ mi/s}$$

WORKING WITH LOGARITHMS

A logarithm to the base 10 has the following form:

$$\log_{10} x = \log x = p, \text{ where } x = 10^p$$

We usually abbreviate the logarithm function "log" if the logarithm is to the base 10, which means the scale in which the log 10 = 1.

A logarithm to the base e, called a *natural* logarithm, has the following form:

$$\log_e x = \ln x = q, \text{ where } x = e^q$$

Scientific calculators have "log" and "ln" buttons, so it is easy to convert a number into its log or ln form. The directions below apply to most nongraphing calculators.

Sample Exercise 4 Find the logarithm to the base 10 of 4.5 (log 4.5).

Solution Enter 4.5 into your calculator and press the "LOG" button.[a] The answer should be 0.6532 (to four significant figures).

Sample Exercise 5 Find the logarithm to the base 10 of 100 (log 100).

Solution Enter 100 into your calculator and press the "LOG" button. The answer should be 2. This answer is as expected, because 10 (the base) raised to the power of the log (2) = 10^2 = 100.

Sample Exercise 6 Find the natural logarithm of 4.5.

Solution Enter 4.5 into your calculator and press the "LN" button. The answer should be 1.504.

Sample Exercise 7 Find the natural logarithm of 100.

Solution Enter 100 into your calculator and press the "LN" button. The answer should be 4.61.

Let's compare the results of the four previous exercises. In both pairs of ln and log values, the ln value is 2.303 times the log value. These examples fit the general equation:

$$\ln x = 2.30 \log x$$

It is reasonable that the ln of a value is greater that the log of the same value because ln is based on e (2.718), whereas log is to the base 10. The smaller base of ln units means that there are more of them than log units in a given value.

Sample Exercise 8 Calculate ln 1.2×10^{-3}.

Solution Enter 1.2 into your calculator, and then press the "EXP" button, and then enter 3 and "+/−". Finally press "LN." The corresponding keystrokes with a graphing calculator are "LN," 1.2, "x," "^," "(−)," 3, "ENTER." The result should be −6.725, which is negative because the original number is less than 1. Keep in mind that values greater than zero have positive logarithm values; those less than zero have negative logarithm values.

5. **Combining logs:** The following equations summarize how logarithms of the products or quotients of two or more values are related to the individual logs of those values.

$$\text{logarithm } ab = \text{logarithm } a + \text{logarithm } b$$

and

$$\text{logarithm } a/b = \text{logarithm } a - \text{logarithm } b$$

CONVERTING LOGARITHMS INTO NUMBERS

If we know the value of log x, what is the value of x? This question frequently arises when working with pH (see Chapter 16), which is the negative log of the concentration of hydrogen ions, [H$^+$], in solution:

$$\text{pH} = -\log[\text{H}^+]$$

Suppose the pH of a solution of a weak acid is 2.50. The concentration of H$^+$ is related to this pH value as follows:

$$2.50 = -\log[\text{H}^+]$$

or

$$-2.50 = \log[\text{H}^+]$$

[a] If you have a graphing calculator such as a TI 83, press the "LOG" button, enter 4.5, and then press the "ENTER" button.

To find the value of [H⁺], enter 2.5 into your calculator and press the "+/−" button to change the value's sign to −2.5. The next step depends on the type of calculator you have. If yours has a "10^x" button, push it to find the value of $10^{-2.5}$, which is the number we are looking for. The corresponding keystrokes with a graphing calculator are "10^x," "(−)," 2.5, "ENTER." On some calculators there is no "10^x" key, but there is an inverse function, or "INV" key, that is used to invert other function keys. Hitting the "INV" key followed by the "log" key takes the inverse of a log, called an *antilog*, which is the same as raising 10 to the power (−2.5 in this case) that was entered. Some calculators, including the virtual one in many Windows operating systems, have an "x^y" key. To use it you enter 10, push the "x^y" key, enter 2.5, and then push the "+/−" key followed by the equals sign. All of these approaches do the same calculation, taking 10 to the −2.50 power, and give the same answer, [H⁺] = 3.2×10^{-3}.

Sample Exercise 9 Calculate the hydrogen ion concentration in rainwater in which pH = 5.62.

Solution If you use one of the methods described above, you should find that the value of $10^{-5.62}$ is 2.4×10^{-6}.

SOLVING QUADRATIC EQUATIONS

If the terms in an equation can be rearranged so that they take the form

$$ax^2 + bx + c = 0$$

they have the form of a quadratic equation. The value(s) of x can be determined from the values of the coefficients a, b, and c by using the equation

$$x = \frac{-b \pm \sqrt{b^2 - 4ac}}{2a}$$

For example, if the solution to a problem yields the following expression where x is the concentration of a solute:

$$x^2 + 0.112x - 1.2 \times 10^{-3} = 0$$

Then the value of x can be determined as follows:

$$x = \frac{-b \pm \sqrt{b^2 - 4ac}}{2a}$$

$$= \frac{-0.112 \pm \sqrt{(0.112)^2 - 4(1)(-1.2 \times 10^{-3})}}{2(1)}$$

$$= \frac{-0.112 \pm \sqrt{0.01254 + 0.0048}}{2}$$

$$= \frac{-0.112 \pm 0.132}{2} = +0.010 \text{ or } -0.122$$

In this example, the negative value for x satisfies the equation, but it has no meaning because we cannot have negative concentration values; therefore we use only the +0.010 value.

EXPRESSING DATA IN GRAPHICAL FORM

Fitting curves to plots of experimental data is a powerful tool in determining the relationships between variables. Many natural phenomena obey exponential functions. For example, the rate constant (k) of a chemical reaction increases exponentially with increasing absolute temperature (T). This relationship is described by the Arrhenius equation (see Chapter 14):

$$k = Ae^{-E_a/RT}$$

where A is a constant for a particular reaction (called the frequency factor), E_a is the activation energy of the reaction, and R is the ideal gas constant. Taking the natural logarithms of both sides of the Arrhenius equation gives

$$\ln k = \ln A - \left(\frac{E_a}{RT}\right)$$

This equation fits the general equation of a straight line ($y = mx + b$) if ($\ln k$) is the y-variable and ($1/T$) is the x-variable. Plotting ($\ln k$) versus ($1/T$) should give a straight line with a slope equal to $-E_a/R$. The slopes of these plots are negative because the activation energies, E_a, of chemical reactions are positive. The data for a reaction given in columns 2 and 4 of Table A1.1 are plotted in Figure A1.1. The slope of the straight line (−1281 K) is used to calculate the value of E_a:

$$-1281 \text{ K} = -\frac{E_a}{R}$$

$$E_a = -(-1281 \text{ K})[8.31 \text{ J/(mol·K)}]$$

$$= 10,645 \text{ J/mol} = 10.6 \text{ kJ/mol}$$

TABLE A1.1 Rate Constant k as a Function of Temperature T

Temperature T (K)	$1/T$ (K⁻¹)	Rate Constant k	ln k
500	0.0020	0.030	−3.5
550	0.0018	0.38	−0.97
600	0.0017	2.9	1.1
650	0.0015	17	2.8
700	0.0014	75	4.3

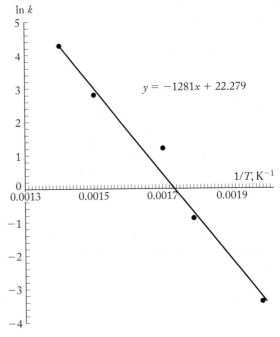

FIGURE A1.1

SI Units and Conversion Factors

TABLE A2.1 SI Base Units

SI Base Quantity	Unit	Symbol
length	meter	m
mass	kilogram	kg
time	second	s
amount of substance	mole	mol
temperature	kelvin	K
electric current	ampere	A

TABLE A2.2 Some SI-Derived Units

SI-Derived Quantity	Unit	Symbol	Dimensions
electric charge	coulomb	C	$A \cdot s$
electric potential	volt	V	J/C
force	newton	N	$kg \cdot m/s^2$
frequency	hertz	Hz	s^{-1}
momentum	newton-second	—	$kg \cdot m/s$
power	watt	W	J/s
pressure	pascal	Pa	N/m
radioactivity	becquerel	Bq	s^{-1}
speed or velocity	meter per second	—	m/s
work, energy, or heat	joule (newton-meter)	J	$kg \cdot m^2/s^2$

TABLE A2.3 SI Prefixes

Prefix	Symbol	Multiplier	Prefix	Symbol	Multiplier
deci	d	10^{-1}	deka	da	10^{1}
centi	c	10^{-2}	hecto	h	10^{2}
milli	m	10^{-3}	kilo	k	10^{3}
micro	μ	10^{-6}	mega	M	10^{6}
nano	n	10^{-9}	giga	G	10^{9}
pico	p	10^{-12}	tera	T	10^{12}
femto	f	10^{-15}	peta	P	10^{15}
atto	a	10^{-18}	exa	E	10^{18}

TABLE A2.4 Special Units and Conversion Factors

Quantity	Unit	Symbol	Conversion
energy	electron-volt	eV	$1 \text{ eV} = 1.60217733 \times 10^{-19} \text{ J}$
mass	pound	lb	$1 \text{ lb} = 453.592 \text{ g}$
heat	calorie	cal	$1 \text{ cal} = 4.184 \text{ J}$
length	angstrom	Å	$1 \text{ Å} = 10^{-8} \text{ cm} = 10^{-10} \text{ m}$
length	inch	in	$1 \text{ in} = 2.54 \text{ cm}$
mass	atomic mass unit	amu	$1 \text{ amu} = 1.6605402 \times 10^{-27} \text{ kg}$
pressure	atmosphere	atm	$1 \text{ atm} = 1.01325 \times 10^5 \text{ Pa}$
pressure	torr	torr	$1 \text{ torr} = 1 \text{ atm}/760$
temperature	Celsius scale	°C	$°C = K - 273.15$
temperature	Fahrenheit scale	°F	$°F = \frac{9}{5}(°C) + 32$
time	minute	min	$1 \text{ min} = 60 \text{ s}$
time	hour	hr	$1 \text{ hr} = 60 \text{ min} = 3600 \text{ s}$
time	day	d	$1 \text{ d} = 24 \text{ hr} = 86{,}400 \text{ s}$
time	year	yr	$1 \text{ yr} = 365.25 \text{ d}$
volume	liter	L	$1 \text{ L} = 1 \text{ dm}^3 = 10^{-3} \text{ m}^3$
volume	cubic centimeter	cm³, cc	$1 \text{ cm}^3 = 1 \text{ mL} = 10^{-3} \text{ L}$
volume	gallon	gal	$1 \text{ gal} = 3.7854 \text{ L}$

TABLE A2.5 Physical Constants

Quantity	Symbol	Value
acceleration due to gravity (Earth)	g	9.80665 m/s^2 (exact)
Avogadro's number	N_A	$6.0221367 \times 10^{23} \text{ mol}^{-1}$
Bohr radius	a_0	$5.29177249 \times 10^{-11} \text{ m}$
Boltzmann's constant	k_B	$1.380658 \times 10^{-23} \text{ J/K}$
electronic charge-to-mass ratio	$-e/m_e$	$1.75881962 \times 10^{11} \text{ C/kg}$
elementary charge	e	$1.60217733 \times 10^{-19} \text{ C}$
Faraday's constant	F	$9.6485309 \times 10^4 \text{ C/mol}$
mass of an electron	m_e	$9.10939 \times 10^{-28} \text{ g}$
mass of a neutron	m_n	$1.67494 \times 10^{-24} \text{ g}$
mass of a proton	m_p	$1.67263 \times 10^{-24} \text{ g}$
molar volume of ideal gas at STP	V_m	22.41410 L/mol
permittivity of vacuum	ϵ_0	$8.854187817 \times 10^{-12} \text{ C}^2/(\text{N} \cdot \text{m}^2)$
Planck's constant	h	$6.6260755 \times 10^{-34} \text{ J} \cdot \text{s}$
speed of light in vacuum	c	$2.99792458 \times 10^8 \text{ m/s}$ (exact)
universal gas constant	R	$8.3145 \text{ J/(mol} \cdot \text{K)}$ $0.082058 \text{ L} \cdot \text{atm/(mol} \cdot \text{K)}$

The Elements and Their Properties

TABLE A3.1 **Ground-State Electron Configurations, Atomic Radii, and First Ionization Energies of the Elements**

Element	Symbol	Atomic Number Z	Ground-State Configuration	Atomic Radius (pm)	Ionization Energy (kJ/mol)
hydrogen	H	1	$1s^1$	37	1312.0
helium	He	2	$1s^2$	32	2372.3
lithium	Li	3	$[He]2s^1$	152	520.2
beryllium	Be	4	$[He]2s^2$	112	899.5
boron	B	5	$[He]2s^22p^1$	88	800.6
carbon	C	6	$[He]2s^22p^2$	77	1086.5
nitrogen	N	7	$[He]2s^22p^3$	75	1402.3
oxygen	O	8	$[He]2s^22p^4$	73	1313.9
fluorine	F	9	$[He]2s^22p^5$	71	1681.0
neon	Ne	10	$[He]2s^22p^6$	69	2080.7
sodium	Na	11	$[Ne]3s^1$	186	495.3
magnesium	Mg	12	$[Ne]3s^2$	160	737.7
aluminum	Al	13	$[Ne]3s^23p^1$	143	577.5
silicon	Si	14	$[Ne]3s^23p^2$	117	786.5
phosphorus	P	15	$[Ne]3s^23p^3$	110	1011.8
sulfur	S	16	$[Ne]3s^23p^4$	103	999.6
chlorine	Cl	17	$[Ne]3s^23p^5$	99	1251.2
argon	Ar	18	$[Ne]3s^23p^6$	97	1520.6
potassium	K	19	$[Ar]4s^1$	227	418.8
calcium	Ca	20	$[Ar]4s^2$	197	589.8
scandium	Sc	21	$[Ar]4s^23d^1$	162	633.1
titanium	Ti	22	$[Ar]4s^23d^2$	147	658.8
vanadium	V	23	$[Ar]4s^23d^3$	135	650.9
chromium	Cr	24	$[Ar]4s^13d^5$	128	652.9
manganese	Mn	25	$[Ar]4s^23d^5$	127	717.3
iron	Fe	26	$[Ar]4s^23d^6$	126	762.5
cobalt	Co	27	$[Ar]4s^23d^7$	125	760.4
nickel	Ni	28	$[Ar]4s^23d^8$	124	737.1
copper	Cu	29	$[Ar]4s^13d^{10}$	128	745.5
zinc	Zn	30	$[Ar]4s^23d^{10}$	134	906.4
gallium	Ga	31	$[Ar]4s^23d^{10}4p^1$	135	578.8

TABLE A3.1 **Ground-State Electron Configurations, Atomic Radii, and First Ionization Energies of the Elements (*Continued*)**

Element	Symbol	Atomic Number Z	Ground-State Configuration	Atomic Radius (pm)	Ionization Energy (kJ/mol)
germanium	Ge	32	$[Ar]4s^2 3d^{10} 4p^2$	122	762.2
arsenic	As	33	$[Ar]4s^2 3d^{10} 4p^3$	121	947.0
selenium	Se	34	$[Ar]4s^2 3d^{10} 4p^4$	119	941.0
bromine	Br	35	$[Ar]4s^2 3d^{10} 4p^5$	114	1139.9
krypton	Kr	36	$[Ar]4s^2 3d^{10} 4p^6$	110	1350.8
rubidium	Rb	37	$[Kr]5s^1$	247	403.0
strontium	Sr	38	$[Kr]5s^2$	215	549.5
yttrium	Y	39	$[Kr]5s^2 4d^1$	180	599.8
zirconium	Zr	40	$[Kr]5s^2 4d^2$	160	640.1
niobium	Nb	41	$[Kr]5s^1 4d^4$	146	652.1
molybdenum	Mo	42	$[Kr]5s^1 4d^5$	139	684.3
technetium	Tc	43	$[Kr]5s^2 4d^5$	136	702.4
ruthenium	Ru	44	$[Kr]5s^1 4d^7$	134	710.2
rhodium	Rh	45	$[Kr]5s^1 4d^8$	134	719.7
palladium	Pd	46	$[Kr]4d^{10}$	137	804.4
silver	Ag	47	$[Kr]5s^1 4d^{10}$	144	731.0
cadmium	Cd	48	$[Kr]5s^2 4d^{10}$	151	867.8
indium	In	49	$[Kr]5s^2 4d^{10} 5p^1$	167	558.3
tin	Sn	50	$[Kr]5s^2 4d^{10} 5p^2$	140	708.6
antimony	Sb	51	$[Kr]5s^2 4d^{10} 5p^3$	141	833.6
tellurium	Te	52	$[Kr]5s^2 4d^{10} 5p^4$	143	869.3
iodine	I	53	$[Kr]5s^2 4d^{10} 5p^5$	133	1008.4
xenon	Xe	54	$[Kr]5s^2 4d^{10} 5p^6$	130	1170.4
cesium	Cs	55	$[Xe]6s^1$	265	375.7
barium	Ba	56	$[Xe]6s^2$	222	502.9
lanthanum	La	57	$[Xe]6s^2 5d^1$	187	538.1
cerium	Ce	58	$[Xe]6s^2 4f^1 5d^1$	182	534.4
praseodymium	Pr	59	$[Xe]6s^2 4f^3$	182	527.2
neodymium	Nd	60	$[Xe]6s^2 4f^4$	181	533.1
promethium	Pm	61	$[Xe]6s^2 4f^5$	183	535.5
samarium	Sm	62	$[Xe]6s^2 4f^6$	180	544.5
europium	Eu	63	$[Xe]6s^2 4f^7$	208	547.1
gadolinium	Gd	64	$[Xe]6s^2 4f^7 5d^1$	180	593.4
terbium	Tb	65	$[Xe]6s^2 4f^9$	177	565.8
dysprosium	Dy	66	$[Xe]6s^2 4f^{10}$	178	573.0
holmium	Ho	67	$[Xe]6s^2 4f^{11}$	176	581.0
erbium	Er	68	$[Xe]6s^2 4f^{12}$	176	589.3
thulium	Tm	69	$[Xe]6s^2 4f^{13}$	176	596.7
ytterbium	Yb	70	$[Xe]6s^2 4f^{14}$	193	603.4
lutetium	Lu	71	$[Xe]6s^2 4f^{14} 5d^1$	174	523.5

Continued on next page

TABLE A3.1 **Ground-State Electron Configurations, Atomic Radii, and First Ionization Energies of the Elements (*Continued*)**

Element	Symbol	Atomic Number Z	Ground-State Configuration	Atomic Radius (pm)	Ionization Energy (kJ/mol)
hafnium	Hf	72	$[Xe]6s^24f^{14}5d^2$	159	658.5
tantalum	Ta	73	$[Xe]6s^24f^{14}5d^3$	146	761.3
tungsten	W	74	$[Xe]6s^24f^{14}5d^4$	139	770.0
rhenium	Re	75	$[Xe]6s^24f^{14}5d^5$	137	760.3
osmium	Os	76	$[Xe]6s^24f^{14}5d^6$	135	839.4
iridium	Ir	77	$[Xe]6s^24f^{14}5d^7$	136	878.0
platinum	Pt	78	$[Xe]6s^14f^{14}5d^9$	139	868.4
gold	Au	79	$[Xe]6s^14f^{14}5d^{10}$	144	890.1
mercury	Hg	80	$[Xe]6s^24f^{14}5d^{10}$	151	1007.1
thallium	Tl	81	$[Xe]6s^24f^{14}5d^{10}6p^1$	170	589.4
lead	Pb	82	$[Xe]6s^24f^{14}5d^{10}6p^2$	154	715.6
bismuth	Bi	83	$[Xe]6s^24f^{14}5d^{10}6p^3$	150	703.3
polonium	Po	84	$[Xe]6s^24f^{14}5d^{10}6p^4$	167	812.1
astatine	At	85	$[Xe]6s^24f^{14}5d^{10}6p^5$	140	924.6
radon	Rn	86	$[Xe]6s^24f^{14}5d^{10}6p^6$	145	1037.1
francium	Fr	87	$[Rn]7s^1$	270[a]	380
radium	Ra	88	$[Rn]7s^2$	223[a]	509.3
actinium	Ac	89	$[Rn]7s^26d^1$	188	499
thorium	Th	90	$[Rn]7s^26d^2$	179	587
protactinium	Pa	91	$[Rn]7s^25f^26d^1$	163	568
uranium	U	92	$[Rn]7s^25f^36d^1$	156	587
neptunium	Np	93	$[Rn]7s^25f^46d^1$	155	597
plutonium	Pu	94	$[Rn]7s^25f^6$	159	585
americium	Am	95	$[Rn]7s^25f^7$	173	578
curium	Cm	96	$[Rn]7s^25f^76d^1$	174	581
berkelium	Bk	97	$[Rn]7s^25f^9$	170	601
californium	Cf	98	$[Rn]7s^25f^{10}$	186	608
einsteinium	Es	99	$[Rn]7s^25f^{11}$	186	619
fermium	Fm	100	$[Rn]7s^25f^{12}$	—	627
mendelevium	Md	101	$[Rn]7s^25f^{13}$	—	635
nobelium	No	102	$[Rn]7s^25f^{14}$	—	642
lawrencium	Lr	103	$[Rn]7s^25f^{14}6d^1$	—	—
rutherfordium	Rf	104	$[Rn]7s^25f^{14}6d^2$	—	—
dubnium	Db	105	$[Rn]7s^25f^{14}6d^3$	—	—
seaborgium	Sg	106	$[Rn]7s^25f^{14}6d^4$	—	—
bohrium	Bh	107	$[Rn]7s^25f^{14}6d^5$	—	—
hassium	Hs	108	$[Rn]7s^25f^{14}6d^6$	—	—
meitnerium	Mt	109	$[Rn]7s^25f^{14}6d^7$	—	—
darmstadtium	[110]	110	$[Rn]7s^25f^{14}6d^8$	—	—
roentgenium	[111]	111	$[Rn]7s^25f^{14}6d^9$	—	—
element 112	[112]	112	$[Rn]7s^25f^{14}6d^{10}$	—	—

[a]These values are estimates.

TABLE A3.2 **Miscellaneous Physical Properties of the Elements[a]**

Element	Symbol	Atomic Number	Physical State[b,c]	Density[d] (g/mL)	Melting Point (°C)	Boiling Point (°C)
hydrogen	H	1	gas	0.000090	−259.14	−252.87
helium	He	2	gas	0.000179	<−272.2	−268.93
lithium	Li	3	solid	0.534	180.5	1347
beryllium	Be	4	solid	1.848	1283	2484
boron	B	5	solid	2.34	2300	3650
carbon	C	6	solid (gr)	1.9–2.3	~3350	sublimes
nitrogen	N	7	gas	0.00125	−210.00	−195.8
oxygen	O	8	gas	0.00143	−218.8	−182.95
fluorine	F	9	gas	0.00170	−219.62	−188.12
neon	Ne	10	gas	0.00090	−248.59	−246.08
sodium	Na	11	solid	0.971	97.72	883
magnesium	Mg	12	solid	1.738	650	1090
aluminum	Al	13	solid	2.6989	660.32	2467
silicon	Si	14	solid	2.33	1414	2355
phosphorus	P	15	solid (wh)	1.82	44.15	280
sulfur	S	16	solid	2.07	115.21	444.60
chlorine	Cl	17	gas	0.00321	−101.5	−34.04
argon	Ar	18	gas	0.00178	−189.3	−185.9
potassium	K	19	solid	0.862	63.28	759
calcium	Ca	20	solid	1.55	842	1484
scandium	Sc	21	solid	2.989	1541	2380
titanium	Ti	22	solid	4.54	1668	3287
vanadium	V	23	solid	6.11	1910	3407
chromium	Cr	24	solid	7.19	1857	2671
manganese	Mn	25	solid	7.3	1246	1962
iron	Fe	26	solid	7.874	1538	2750
cobalt	Co	27	solid	8.9	1495	2870
nickel	Ni	28	solid	8.902	1455	2730
copper	Cu	29	solid	8.96	1084.6	2562
zinc	Zn	30	solid	7.133	419.53	907
gallium	Ga	31	solid	5.904	29.76	2403
germanium	Ge	32	solid	5.323	938.25	2833
arsenic	As	33	solid (gy)	5.727	614	sublimes
selenium	Se	34	solid (gy)	4.79	221	685
bromine	Br	35	liquid	3.12	−7.2	58.78
krypton	Kr	36	gas	0.00373	−157.36	−153.22
rubidium	Rb	37	solid	1.532	39.31	688
strontium	Sr	38	solid	2.54	777	1382
yttrium	Y	39	solid	4.469	1526	3336
zirconium	Zr	40	solid	6.506	1855	4409
niobium	Nb	41	solid	8.57	2477	4744

Continued on next page

TABLE A3.2 **Miscellaneous Physical Properties of the Elements**[a] (*Continued*)

Element	Symbol	Atomic Number	Physical State[b,c]	Density[d] (g/mL)	Melting Point (°C)	Boiling Point (°C)
molybdenum	Mo	42	solid	10.22	2623	4639
technetium	Tc	43	solid	11.50	2157	4538
ruthenium	Ru	44	solid	12.41	2334	3900
rhodium	Rh	45	solid	12.41	1964	3695
palladium	Pd	46	solid	12.02	1555	2963
silver	Ag	47	solid	10.50	961.78	2212
cadmium	Cd	48	solid	8.65	321.07	767
indium	In	49	solid	7.31	156.60	2072
tin	Sn	50	solid (wh)	7.31	231.9	2270
antimony	Sb	51	solid	6.691	630.63	1750
tellurium	Te	52	solid	6.24	449.5	998
iodine	I	53	solid	4.93	113.7	184.4
xenon	Xe	54	gas	0.00589	−111.75	−108.0
cesium	Cs	55	solid	1.873	28.44	671
barium	Ba	56	solid	3.5	727	1640
lanthanum	La	57	solid	6.145	920	3455
cerium	Ce	58	solid	6.770	799	3424
praseodymium	Pr	59	solid	6.773	931	3510
neodymium	Nd	60	solid	7.008	1016	3066
promethium	Pm	61	solid	7.264	1042	~3000
samarium	Sm	62	solid	7.520	1072	1790
europium	Eu	63	solid	5.244	822	1596
gadolinium	Gd	64	solid	7.901	1314	3264
terbium	Tb	65	solid	8.230	1359	3221
dysprosium	Dy	66	solid	8.551	1411	2561
holmium	Ho	67	solid	8.795	1472	2694
erbium	Er	68	solid	9.066	1529	2862
thulium	Tm	69	solid	9.321	1545	1946
ytterbium	Yb	70	solid	6.966	824	1194
lutetium	Lu	71	solid	9.841	1663	3393
hafnium	Hf	72	solid	13.31	2233	4603
tantalum	Ta	73	solid	16.654	3017	5458
tungsten	W	74	solid	19.3	3422	5660
rhenium	Re	75	solid	21.02	3186	5596
osmium	Os	76	solid	22.57	3033	5012
iridium	Ir	77	solid	22.42	2446	4130
platinum	Pt	78	solid	21.45	1768.4	3825
gold	Au	79	solid	19.3	1064.18	2856
mercury	Hg	80	liquid	13.546	−38.83	356.73
thallium	Tl	81	solid	11.85	304	1473
lead	Pb	82	solid	11.35	327.46	1749
bismuth	Bi	83	solid	9.747	271.4	1564

TABLE A3.2 Miscellaneous Physical Properties of the Elements[a] (Continued)

Element	Symbol	Atomic Number	Physical State[b,c]	Density[d] (g/mL)	Melting Point (°C)	Boiling Point (°C)
polonium	Po	84	solid	9.32	254	962
astatine	At	85	solid	unknown	302	337
radon	Rn	86	gas	0.00973	−71	−61.7
francium	Fr	87	solid	unknown	27	677
radium	Ra	88	solid	5	700	1737
actinium	Ac	89	solid	10.07	1051	~3200
thorium	Th	90	solid	11.72	1750	4788
protactinium	Pa	91	solid	15.37	1572	unknown
uranium	U	92	solid	18.95	1132	3818

[a]For relative atomic masses and alphabetical listing of the elements, see the flyleaf at the front of this volume.
[b]Normal state at 25°C and 1 atm.
[c]Allotropes: gr = graphite, gy = gray, wh = white.
[d]Liquids and solids at 25°C and 1 atm; gases at 0°C and 1 atm (STP).

TABLE A3.3 A Selection of Stable Isotopes[a]

Isotope ^{A}X	Natural Abundance (%)	Atomic Number Z	Neutron Number N	Mass Number A	Atomic Mass (amu)	Binding Energy per Nucleon (MeV)[b]
^{1}H	99.985	1	0	1	1.007825	—
^{2}H	0.015	1	1	2	2.014000	1.160
^{3}He	0.000137	2	1	3	3.016030	2.572
^{4}He	99.999863	2	2	4	4.002603	7.075
^{6}Li	7.5	3	3	6	6.015121	5.333
^{7}Li	92.5	3	4	7	7.016003	5.606
^{9}Be	100.0	4	5	9	9.012182	6.463
^{10}B	19.9	5	5	10	10.012937	6.475
^{11}B	80.1	5	6	11	11.009305	6.928
^{12}C	98.90	6	6	12	12.000000	7.680
^{13}C	1.10	6	7	13	13.003355	7.470
^{14}N	99.634	7	7	14	14.003074	7.476
^{15}N	0.366	7	8	15	15.000108	7.699
^{16}O	99.762	8	8	16	15.994915	7.976
^{17}O	0.038	8	9	17	16.999131	7.751
^{18}O	0.200	8	10	18	17.999160	7.767
^{19}F	100.0	9	10	19	18.998403	7.779
^{20}Ne	90.48	10	10	20	19.992435	8.032
^{21}Ne	0.27	10	11	21	20.993843	7.972
^{22}Ne	9.25	10	12	22	21.991383	8.081
^{23}Na	100.0	11	12	23	22.989770	8.112
^{24}Mg	78.99	12	12	24	23.985042	8.261
^{25}Mg	10.00	12	13	25	24.985837	8.223
^{26}Mg	11.01	12	14	26	25.982593	8.334
^{27}Al	100.0	13	14	27	26.981538	8.331

Continued on next page

TABLE A3.3 A Selection of Stable Isotopes[a] (Continued)

Isotope ^{A}X	Natural Abundance (%)	Atomic Number Z	Neutron Number N	Mass Number A	Atomic Mass (amu)	Binding Energy per Nucleon (MeV)[b]
^{28}Si	92.23	14	14	28	27.976927	8.448
^{29}Si	4.67	14	15	29	28.976495	8.449
^{30}Si	3.10	14	16	30	29.973770	8.521
^{31}P	100.0	15	16	31	30.973761	8.481
^{32}S	95.02	16	16	32	31.972070	8.493
^{33}S	0.75	16	17	33	32.971456	8.498
^{34}S	4.21	16	18	34	33.967866	8.584
^{36}S	0.02	16	20	36	35.967080	8.575
^{35}Cl	75.77	17	18	35	34.968852	8.520
^{37}Cl	24.23	17	20	37	36.965903	8.570
^{36}Ar	0.337	18	18	36	35.967545	8.520
^{38}Ar	0.063	18	20	38	37.962732	8.614
^{40}Ar	99.600	18	22	40	39.962384	8.595
^{39}K	93.258	19	20	39	38.963707	8.557
^{41}K	6.730	19	22	41	40.961825	8.576
^{40}Ca	96.941	20	20	40	39.962591	8.551
^{42}Ca	0.647	20	22	42	41.958618	8.617
^{43}Ca	0.135	20	23	43	42.958766	8.601
^{44}Ca	2.086	20	24	44	43.955480	8.658
^{46}Ca	0.004	20	26	46	45.953689	8.669
^{48}Ca	0.187	20	28	48	47.952533	8.666
^{45}Sc	100.0	21	24	45	44.955910	8.619
^{46}Ti	8.0	22	24	46	45.952629	8.656
^{47}Ti	7.3	22	25	47	46.951764	8.661
^{48}Ti	73.8	22	26	48	47.947947	8.723
^{49}Ti	5.5	22	27	49	48.947871	8.711
^{50}Ti	5.4	22	28	50	49.944792	8.756
^{51}V	99.750	23	28	51	50.943962	8.742
^{50}Cr	4.345	24	26	50	49.946046	8.701
^{52}Cr	83.789	24	28	52	51.940509	8.776
^{53}Cr	9.501	24	29	53	52.940651	8.760
^{54}Cr	2.365	24	30	54	53.938882	8.778
^{55}Mn	100.0	25	30	55	54.938049	8.765
^{54}Fe	5.9	26	28	54	53.939612	8.736
^{56}Fe	91.72	26	30	56	55.934939	8.790
^{57}Fe	2.1	26	31	57	56.935396	8.770
^{58}Fe	0.28	26	32	58	57.933277	8.792
^{59}Co	100.0	27	32	59	58.933200	8.768
^{204}Pb	1.4	82	122	204	203.973020	7.880
^{206}Pb	24.1	82	124	206	205.974440	7.875
^{207}Pb	22.1	82	125	207	206.975872	7.870
^{208}Pb	52.4	82	126	208	207.976627	7.868
^{209}Bi	100.0	83	126	209	208.980380	7.848

[a]Selection is complete through cobalt-59. Where natural abundances do not add to 100%, the differences are made up by radioactive isotopes with exceedingly long half-lives: potassium-40 (0.0117%, $t_{1/2} = 1.3 \times 10^9$ yr); vanadium-50 (0.250%, $t_{1/2} > 1.4 \times 10^{17}$ yr).
[b]1 MeV (million electron-volts) = 1.602189×10^{-13} J.

TABLE A3.4 A Selection of Radioactive Isotopes

Isotope ^AX	Decay Mode^a	Half-Life $t_{1/2}$	Atomic Number Z	Neutron Number N	Mass Number A	Atomic Mass (amu)	Binding Energy per Nucleon (MeV)^b
^{3}H	β^-	12.3 yr	1	2	3	3.01605	2.827
^{8}Be	α	$\approx 7 \times 10^{-17}$ s	4	4	8	8.005305	7.062
^{14}C	β^-	5.7×10^3 yr	6	8	14	14.003241	7.520
^{22}Na	β^+	2.6 yr	11	11	22	21.994434	7.916
^{24}Na	β^-	15.0 hr	11	13	24	23.990961	8.064
^{32}P	β^-	14.3 d	15	17	32	31.973907	8.464
^{35}S	β^-	87.2 d	16	19	35	34.969031	8.538
^{59}Fe	β^-	44.5 d	26	33	59	58.934877	8.755
^{60}Co	β^-	5.3 yr	27	33	60	59.933819	8.747
^{90}Sr	β^-	29.1 yr	38	52	90	89.907738	8.696
^{99}Tc	β^-	2.1×10^5 yr	43	56	99	98.906524	8.611
^{109}Cd	EC	462 d	48	61	109	108.904953	8.539
^{125}I	EC	59.4 d	53	72	125	124.904620	8.450
^{131}I	β^-	8.04 d	53	78	131	130.906114	8.422
^{137}Cs	β^-	30.3 yr	55	82	137	136.907073	8.389
^{222}Rn	α	3.82 d	86	136	222	222.017570	7.695
^{226}Ra	α	1600 yr	88	138	226	226.025402	7.662
^{232}Th	α	1.4×10^{10} yr	90	142	232	232.038054	7.615
^{235}U	α	7.0×10^8 yr	92	143	235	235.043924	7.591
^{238}U	α	4.5×10^9 yr	92	146	238	238.050784	7.570
^{239}Pu	α	2.4×10^4 yr	94	145	239	239.052157	7.560

^{a}Modes of decay include alpha emission (α), beta emission (β^-), positron emission (β^+), electron capture (EC).

Chemical Bonds and Thermodynamic Data

TABLE A4.1 Average Lengths and Strengths of Covalent Bonds

Atom	Bond	Bond Length (pm)	Bond Strength (kJ/mol)
H	H—H	75	436
	H—F	92	567
	H—Cl	127	431
	H—Br	141	366
	H—I	161	299
C	C—C	154	348
	C=C	134	614
	C≡C	120	839
	C—H	110	413
	C—N	143	293
	C=N	138	615
	C≡N	116	891
	C—O	143	358
	C=O[a]	123	743
	C≡O	113	1072
	C—F	133	485
	C—Cl	177	328
	C—Br	179	276
	C—I	215	238
N	N—N	147	163
	N=N	124	418
	N≡N	110	941
	N—H	104	388
	N—O	136	201
	N=O	122	607
	N≡O	106	463
O	O—O	148	146
	O=O	121	495
	O—H	96	458
S	S—O	151	265
	S=O	143	523
	S—S	204	266
	S—H	134	347
F	F—F	143	155
Cl	Cl—Cl	200	243
Br	Br—Br	228	193
I	I—I	266	151

[a] The bond strength of C=O in CO_2 is 799 kJ/mol.

TABLE A4.2 Critical Temperatures (T_c) and van der Waals Parameters (a, b) of Real Gases

Gas[a]	Molar Mass (g/mol)	T_c (K)	a (L²·atm/mol²)	b (L/mol)
H_2O	18.015	647.14	5.46	0.0305
Br_2	159.808	588	9.75	0.0591
CCl_3F	137.367	471.2	14.68	0.1111
Cl_2	70.906	416.9	6.343	0.0542
CO_2	44.010	304.14	3.59	0.0427
Kr	83.80	209.41	2.325	0.0396
CH_4	16.043	190.53	2.25	0.0428
O_2	31.999	154.59	1.36	0.0318
Ar	39.948	150.87	1.34	0.0322
F_2	37.997	144.13	1.171	0.0290
CO	28.010	132.91	1.45	0.0395
N_2	28.013	126.21	1.39	0.0391
H_2	2.016	32.97	0.244	0.0266
He	4.003	5.19	0.0341	0.0237

[a]Listed in descending order of critical temperature.

TABLE A4.3 Thermodynamic Properties at 25°C

Substance[a,b]	Molar Mass (g/mol)	ΔH_f° (kJ/mol)	S° [J/(mol·K)]	ΔG_f° (kJ/mol)
Elements and Monatomic Ions				
$Ag^+(aq)$	107.868	105.6	72.7	77.1
$Ag(g)$	107.868	284.9	173.0	246.0
$Ag(s)$	107.868	0.0	42.6	0.0
$Al^{3+}(aq)$	26.982	−531	−321.7	−485
$Al(g)$	26.982	330.0	164.6	289.4
$Al(s)$	26.982	0.0	28.3	0.0
$Ar(g)$	39.948	0.0	154.8	0.0
$Au(g)$	196.967	366.1	180.5	326.3
$Au(s)$	196.967	0.0	47.4	0.0
$B(g)$	10.811	565.0	153.4	521.0
$B(s)$	10.811	0.0	5.9	0.0
$Ba^{2+}(aq)$	137.327	−537.6	9.6	−560.8
$Ba(g)$	137.327	180.0	170.2	146.0
$Ba(s)$	137.327	0.0	62.8	0.0
$Be(g)$	9.012	324.0	136.3	286.6
$Be(s)$	9.012	0.0	9.5	0.0
$Br^-(aq)$	79.904	−121.6	82.4	−104.0
$Br(g)$	79.904	111.9	175.0	82.4
$Br_2(g)$	159.808	30.9	245.5	3.1
$Br_2(\ell)$	159.808	0.0	152.2	0.0
$C(g)$	12.011	716.7	158.1	671.3
$C(s, diamond)$	12.011	1.9	2.4	2.9
$C(s, graphite)$	12.011	0.0	5.7	0.0

Continued on next page

TABLE A4.3 Thermodynamic Properties at 25°C (Continued)

Substance[a,b]	Molar Mass (g/mol)	ΔH_f° (kJ/mol)	S° [J/(mol·K)]	ΔG_f° (kJ/mol)
$Ca^{2+}(aq)$	40.078	−542.8	−55.3	−553.6
$Ca(g)$	40.078	177.8	154.9	144.0
$Ca(s)$	40.078	0.0	41.6	0.0
$Cl^-(aq)$	35.453	−167.2	56.5	−131.2
$Cl(g)$	35.453	121.3	165.2	105.3
$Cl_2(g)$	70.906	0.0	223.0	0.0
$Co^{2+}(aq)$	58.933	−58.2	−113	−54.4
$Co^{3+}(aq)$	58.933	92	−305	134
$Co(g)$	58.933	424.7	179.5	380.3
$Co(s)$	58.933	0.0	30.0	0.0
$Cr(g)$	51.996	396.6	174.5	351.8
$Cr(s)$	51.996	0.0	23.8	0.0
$Cs^+(aq)$	132.905	−258.3	133.1	−292.0
$Cs(g)$	132.905	76.5	175.6	49.6
$Cs(s)$	132.905	0.0	85.2	0.0
$Cu^+(aq)$	63.546	71.7	40.6	50.0
$Cu^{2+}(aq)$	63.546	64.8	−99.6	65.5
$Cu(g)$	63.546	337.4	166.4	297.7
$Cu(s)$	63.546	0.0	33.2	0.0
$F^-(aq)$	18.998	−332.6	−13.8	−278.8
$F(g)$	18.998	79.4	158.8	62.3
$F_2(g)$	37.996	0.0	202.8	0.0
$Fe^{2+}(aq)$	55.845	−89.1	−137.7	−78.9
$Fe^{3+}(aq)$	55.845	−48.5	−315.9	−4.7
$Fe(g)$	55.845	416.3	180.5	370.7
$Fe(s)$	55.845	0.0	27.3	0.0
$H^+(aq)$	1.0079	0.0	0.0	0.0
$H(g)$	1.0079	218.0	114.7	203.3
$H_2(g)$	2.0158	0.0	130.6	0.0
$He(g)$	4.0026	0.0	126.2	0.0
$Hg_2^{2+}(aq)$	401.18	172.4	84.5	153.5
$Hg^{2+}(aq)$	200.59	171.1	−32.2	164.4
$Hg(g)$	200.59	61.4	175.0	31.8
$Hg(\ell)$	200.59	0.0	75.9	0.0
$I^-(aq)$	126.904	−55.2	111.3	−51.6
$I(g)$	126.904	106.8	180.8	70.2
$I_2(g)$	253.808	62.4	260.7	19.3
$I_2(s)$	253.808	0.0	116.1	0.0
$K^+(aq)$	39.098	−252.4	102.5	−283.3
$K(g)$	39.098	89.0	160.3	60.5
$K(s)$	39.098	0.0	64.7	0.0
$Li^+(aq)$	6.941	−278.5	13.4	−293.3
$Li(g)$	6.941	159.3	138.8	126.6
$Li^+(g)$	6.941	685.7	133.0	648.5
$Li(s)$	6.941	0.0	29.1	0.0
$Mg^{2+}(aq)$	24.305	−466.9	−138.1	−454.8

TABLE A4.3 Thermodynamic Properties at 25°C (*Continued*)

Substance[a,b]	Molar Mass (g/mol)	ΔH_f° (kJ/mol)	S° [J/(mol·K)]	ΔG_f° (kJ/mol)
Mg(g)	24.305	147.1	148.6	112.5
Mg(s)	24.305	0.0	32.7	0.0
$Mn^{2+}(aq)$	54.938	−220.8	−73.6	−228.1
Mn(g)	54.938	280.7	173.7	238.5
Mn(s)	54.938	0.0	32.0	0.0
N(g)	14.0067	472.7	153.3	455.5
N_2(g)	28.0134	0.0	191.5	0.0
$Na^+(aq)$	22.990	−240.1	59.0	−261.9
Na(g)	22.990	107.5	153.7	77.0
$Na^+(g)$	22.990	609.3	148.0	574.3
Na(s)	22.990	0.0	51.3	0.0
Ne(g)	20.180	0.0	146.3	0.0
$Ni^{2+}(aq)$	58.693	−54.0	−128.9	−45.6
Ni(g)	58.693	429.7	182.2	384.5
Ni(s)	58.693	0.0	29.9	0.0
O(g)	15.999	249.2	161.1	231.7
O_2(g)	31.998	0.0	205.0	0.0
P(g)	30.974	314.6	163.1	278.3
P_4(s, red)	123.896	−17.6	22.8	−12.1
P_4(s, white)	123.896	0.0	41.1	0.0
$Pb^{2+}(aq)$	207.2	−1.7	10.5	−24.4
Pb(g)	207.2	195.2	162.2	175.4
Pb(s)	207.2	0.0	64.8	0.0
$Rb^+(aq)$	85.468	−251.2	121.5	−284.0
Rb(g)	85.468	80.9	170.1	53.1
Rb(s)	85.468	0.0	76.8	0.0
S(g)	32.065	277.2	167.8	236.7
S_8(g)	256.52	102.3	430.2	49.1
S_8(s)	256.52	0.0	32.1	0.0
Sc(g)	44.956	377.8	174.8	336.0
Si(g)	28.086	450.0	168.0	405.5
Si(s)	28.086	0.0	18.8	0.0
Sn(g)	118.710	301.2	168.5	266.2
Sn(s, gray)	118.710	−2.1	44.1	0.1
Sn(s, white)	118.710	0.0	51.2	0.0
$Sr^{2+}(aq)$	87.62	−545.8	−32.6	−559.5
Sr(g)	87.62	164.4	164.6	130.9
Sr(s)	87.62	0.0	52.3	0.0
Ti(g)	47.867	473.0	180.3	428.4
Ti(s)	47.867	0.0	30.7	0.0
V(g)	50.942	514.2	182.2	468.5
V(s)	50.942	0.0	28.9	0.0
W(s)	183.84	0.0	32.6	0.0
$Zn^{2+}(aq)$	65.409	−153.9	−112.1	−147.1
Zn(g)	65.409	130.4	161.0	94.8
Zn(s)	65.409	0.0	41.6	0.0

Continued on next page

TABLE A4.3 Thermodynamic Properties at 25°C (Continued)

Substance[a,b]	Molar Mass (g/mol)	ΔH_f° (kJ/mol)	S° [J/(mol·K)]	ΔG_f° (kJ/mol)
Polyatomic Ions				
$CH_3COO^-(aq)$	59.045	−486.0	86.6	−369.3
$CO_3^{2-}(aq)$	60.009	−677.1	−56.9	−527.8
$C_2O_4^{2-}(aq)$	88.020	−825.1	45.6	−673.9
$CrO_4^{2-}(aq)$	115.994	−881.2	50.2	−727.8
$Cr_2O_7^{2-}(aq)$	215.988	−1490.3	261.9	−1301.1
$HCOO^-(aq)$	45.018	−425.6	92	−351.0
$HCO_3^-(aq)$	61.017	−692.0	91.2	−586.8
$HSO_4^-(aq)$	97.072	−887.3	131.8	−755.9
$MnO_4^-(aq)$	118.936	−541.4	191.2	−447.2
$NH_4^+(aq)$	18.038	−132.5	113.4	−79.3
$NO_3^-(aq)$	62.005	−205.0	146.4	−108.7
$OH^-(aq)$	17.007	−230.0	−10.8	−157.2
$PO_4^{3-}(aq)$	94.971	−1277.4	−222	−1018.7
$SO_4^{2-}(aq)$	96.064	−909.3	20.1	−744.5
Inorganic Compounds				
$AgCl(s)$	143.321	−127.1	96.2	−109.8
$AgI(s)$	234.773	−61.8	115.5	−66.2
$AgNO_3(s)$	169.873	−124.4	140.9	−33.4
$Al_2O_3(s)$	101.961	−1675.7	50.9	−1582.3
$B_2H_6(g)$	27.669	35.0	232.0	86.6
$B_2O_3(s)$	69.622	−1263.6	54.0	−1184.1
$BaCO_3(s)$	197.34	−1216.3	112.1	−1137.6
$BaSO_4(s)$	233.39	−1473.2	132.2	−1362.2
$CaCO_3(s)$	100.087	−1206.9	92.9	−1128.8
$CaCl_2(s)$	110.984	−795.4	108.4	−748.8
$CaF_2(s)$	78.075	−1228.0	68.5	−1175.6
$CaO(s)$	56.077	−634.9	38.1	−603.3
$Ca(OH)_2(s)$	74.093	−985.2	83.4	−897.5
$CaSO_4(s)$	136.142	−1434.5	106.5	−1322.0
$CO(g)$	28.010	−110.5	197.7	−137.2
$CO_2(g)$	44.010	−393.5	213.6	−394.4
$CO_2(aq)$	44.010	−412.9	121.3	−386.2
$CS_2(g)$	76.143	115.3	237.8	65.1
$CS_2(\ell)$	76.143	87.9	151.0	63.6
$CsCl(s)$	168.358	−443.0	101.2	−414.6
$CuSO_4(s)$	159.610	−771.4	109.2	−662.2
$FeCl_2(s)$	126.750	−341.8	118.0	−302.3
$FeCl_3(s)$	162.203	−399.5	142.3	−334.0
$FeO(s)$	71.844	−271.9	60.8	−255.2
$Fe_2O_3(s)$	159.688	−824.2	87.4	−742.2
$HBr(g)$	80.912	−36.3	198.7	−53.4
$HCl(g)$	36.461	−92.3	186.9	−95.3
$HF(g)$	20.006	−273.3	173.8	−275.4
$HI(g)$	127.912	26.5	206.6	1.7
$HNO_3(g)$	63.013	−135.1	266.4	−74.7
$HNO_3(\ell)$	63.013	−174.1	155.6	−80.7
$HNO_3(aq)$	63.013	−206.6	146.0	−110.5

TABLE A4.3 Thermodynamic Properties at 25°C (*Continued*)

Substance[a,b]	Molar Mass (g/mol)	ΔH_f° (kJ/mol)	S° [J/(mol·K)]	ΔG_f° (kJ/mol)
$HgCl_2(s)$	271.50	−224.3	146.0	−178.6
$Hg_2Cl_2(s)$	472.09	−265.4	191.6	−210.7
$H_2O(g)$	18.015	−241.8	188.8	−228.6
$H_2O(\ell)$	18.015	−285.8	69.9	−237.2
$H_2S(g)$	34.082	−20.17	205.6	−33.01
$H_2O_2(g)$	34.015	−136.3	232.7	−105.6
$H_2O_2(\ell)$	34.015	−187.8	109.6	−120.4
$H_2SO_4(\ell)$	98.079	−814.0	156.9	−690.0
$H_2SO_4(aq)$		−909.2	20.1	−744.5
$KBr(s)$	119.002	−393.8	95.9	−380.7
$KCl(s)$	74.551	−436.5	82.6	−408.5
$LiBr(s)$	86.845	−351.2	74.3	−342.0
$LiCl(s)$	42.394	−408.6	59.3	−384.4
$Li_2CO_3(s)$	73.891	−1215.9	90.4	−1132.1
$MgCl_2(s)$	95.211	−641.3	89.6	591.8
$Mg(OH)_2(s)$	58.320	−924.5	63.2	−833.5
$MgSO_4(s)$	120.369	−1284.9	91.6	−1170.6
$MnO_2(s)$	86.937	−520.0	53.1	−465.1
$NaCH_3OO(s)$	82.034	−708.8	123.0	−607.2
$NaBr(s)$	102.894	−361.1	8.8	−349.0
$NaBr(s)$	102.894	−361.4	86.82	−349.3
$NaCl(s)$	58.443	−411.2	72.1	−384.2
$NaCl(g)$	58.443	−181.4	229.8	−201.3
$Na_2CO_3(s)$	105.989	−1130.7	135.0	−1044.4
$NaHCO_3(s)$	84.007	−950.8	101.7	−851.0
$NaNO_3(s)$	84.995	−467.9	116.5	−367.0
$NaOH(s)$	39.997	−425.6	64.5	−379.5
$Na_2SO_4(s)$	142.043	−1387.1	149.6	−1270.2
$NF_3(g)$	71.002	−132.1	260.8	−90.6
$NH_3(aq)$	17.031	−80.3	111.3	−26.50
$NH_3(g)$	17.031	−46.1	192.5	−16.5
$NH_4Cl(s)$	53.491	−314.4	94.6	−203.0
$NH_4NO_3(s)$	80.043	−365.6	151.1	−183.9
$N_2H_4(g)$	32.045	95.40	238.5	159.4
$NiCl_2(s)$	129.60	−305.3	97.7	−259.0
$NiO(s)$	74.60	−239.7	38.0	−211.7
$NO(g)$	30.006	90.3	210.7	86.6
$NO_2(g)$	46.006	33.2	240.0	51.3
$N_2O(g)$	44.013	82.1	219.9	104.2
$N_2O_4(g)$	92.011	9.2	304.2	97.8
$NOCl(g)$	65.459	51.7	261.7	66.1
$O_3(g)$	47.998	142.7	238.8	163.2
$PCl_3(g)$	137.33	−288.07	311.7	−269.6
$PCl_3(\ell)$	137.33	−319.6	217	−272.4
$PF_5(g)$	125.96	−1594.4	300.8	−1520.7
$PH_3(g)$	33.998	5.4	210.2	13.4

Continued on next page

TABLE A4.3 **Thermodynamic Properties at 25°C** (*Continued*)

Substance[a,b]	Molar Mass (g/mol)	ΔH_f° (kJ/mol)	S° [J/(mol·K)]	ΔG_f° (kJ/mol)
$PbCl_2(s)$	278.1	−359.4	136.0	−314.1
$PbSO_4(s)$	303.3	−920.0	148.5	'−813.0
$SO_2(g)$	64.065	−296.8	248.2	−300.1
$SO_3(g)$	80.064	−395.7	256.8	−371.1
$ZnCl_2(s)$	136.30	−415.1	111.5	−369.4
$ZnO(s)$	81.37	−348.0	43.9	−318.2
$ZnSO_4(s)$	161.45	−982.8	110.5	−871.5
Organic Molecules				
$CCl_4(g)$	153.823	−102.9	309.7	−60.6
$CCl_4(\ell)$	153.823	−135.4	216.4	−65.3
$CH_4(g)$	16.043	−74.8	186.2	−50.8
$CH_3COOH(g)$	60.053	−432.8	282.5	−374.5
$CH_3COOH(\ell)$	60.053	−485.8	159.8	−389.9
$CH_3OH(g)$	32.042	−200.7	239.9	−162.0
$CH_3OH(\ell)$	32.042	−238.7	126.8	−166.4
$C_2H_2(g)$	26.038	226.7	200.8	209.2
$C_2H_4(g)$	28.054	52.3	219.5	68.1
$C_2H_6(g)$	30.070	−84.7	229.5	−32.9
$CH_3CH_2OH(g)$	46.069	−235.1	282.6	−168.6
$CH_3CH_2OH(\ell)$	46.069	−277.7	160.7	−174.9
$CH_3CHO(g)$	44.05	−166	266	−133.7
$C_3H_8(g)$	44.097	−103.9	269.9	−23.5
$n\text{-}CH_3(CH_2)_2CH_3(g)^c$	58.123	−125.6	310.0	−15.7
$n\text{-}CH_3(CH_2)_2CH_3(\ell)^c$	58.123	−147.6	231.0	−15.0
$CH_3COCH_3(\ell)$	46.07	−248.4	199.8	
$CH_3COCH_3(g)$	46.07	−217.1	295.3	−152.7
$CH_3(CH_2)_2CH_2OH(\ell)$	74.12	−327.3	225.8	
$(CH_3CH_2)_2O(\ell)$	74.12	−279.6	172.4	
$(CH_3CH_2)_2O(g)$	74.12	−252.1	342.7	
$(CH_3)_2C{=}C(CH_3)_2(\ell)$	84.16	66.6	362.6	−69.2
$(CH_3)_2NH(\ell)$	45.09	−43.9	182.3	
$(CH_3)_2NH(g)$	45.09	−18.5	273.1	
$(C_2H_5)_2NH(\ell)$	73.14	−103.3		
$(C_2H_5)_2NH(g)$	73.14	−71.4		
$(CH_3)_3N(\ell)$	59.11	−46.0	208.5	
$(CH_3)_3N(g)$	59.11	−23.6	287.1	
$(CH_3CH_2)_3N(\ell)$	101.19	−134.3		
$(CH_3CH_2)_3N(g)$	101.19	−95.8		
$C_6H_6(g)$	78.114	82.9	269.2	129.7
$C_6H_6(\ell)$	78.114	49.0	172.9	124.5
$C_6H_{12}O_6(s)$	180.158	−1274.4	212.1	−910.1
$n\text{-}C_8H_{18}(\ell)^c$	114.231	−249.9	361.1	6.4
$C_{12}H_{22}O_{11}(s)$	342.300	−2221.7	360.2	−1543.8
$HCOOH(\ell)$	46.026	−424.7	129.0	−361.4

[a]Substances are arranged alphabetically by chemical formula within each class: (1) elements and monatomic ions; (2) polyatomic ions; (3) inorganic compounds (including CO and CO_2); (4) organic molecules (hydrocarbon-based).
[b]Symbols denote standard enthalpy of formation (ΔH_f°), standard third-law entropy (S°), and standard Gibbs free energy of formation (ΔG_f°). Entropies in aqueous solution are referred to $S^\circ[H^+(aq)] = 0$, not to absolute zero.
[c]The symbol n denotes the "normal" unbranched alkane.

TABLE A4.4 **Vapor Pressure of Water as a Function of Temperature**

T (°C)	P (torr)
0.0	4.579
10.0	9.209
20.0	17.535
25.0	23.756
30.0	31.824
40.0	55.324
60.0	149.4
70.0	233.7
90.0	525.8
100	760.0
105	906.0

Equilibrium Constants

Appendix 5

TABLE A5.1 Ionization Constants of Selected Acids at 25°C

Acid	Step	Aqueous Equilibrium[a]	K_a	pK_a
acetic	1	$CH_3COOH(aq) \rightleftharpoons H^+(aq) + CH_3COO^-(aq)$	1.76×10^{-5}	4.75
ammonium ion	1	$NH_4^+(aq) \rightleftharpoons H^+(aq) + NH_3(aq)$	5.7×10^{-10}	9.25
arsenic	1	$H_3AsO_4(aq) \rightleftharpoons H^+(aq) + H_2AsO_4^-(aq)$	5.5×10^{-3}	2.26
	2	$H_2AsO_4^-(aq) \rightleftharpoons H^+(aq) + AsO_4^{2-}(aq)$	1.7×10^{-7}	6.77
	3	$HAsO_4^{2-}(aq) \rightleftharpoons H^+(aq) + AsO_4^{3-}(aq)$	5.1×10^{-12}	11.29
ascorbic	1	$H_2C_6H_6O_6(aq) \rightleftharpoons H^+(aq) + HC_6H_6O_6^-(aq)$	1.0×10^{-5}	5.00
	2	$HC_6H_6O_6^-(aq) \rightleftharpoons H^+(aq) + C_6H_6O_6^{2-}(aq)$	5×10^{-12}	11.3
benzoic	1	$C_6H_5COOH(aq) \rightleftharpoons H^+(aq) + C_6H_5COO^-(aq)$	6.46×10^{-5}	4.19
boric	1	$H_3BO_3(aq) \rightleftharpoons H^+(aq) + H_2BO_3^-(aq)$	5.4×10^{-10}	9.27
	2	$H_2BO_3^-(aq) \rightleftharpoons H^+(aq) + HBO_3^{2-}(aq)$	$<10^{-14}$	$>.14$
bromoacetic	1	$CH_2BrCOOH(aq) \rightleftharpoons H^+(aq) + CH_2BrCOO^-(aq)$	2.0×10^{-3}	2.70
butanoic	1	$CH_3CH_2CH_2COOH(aq) \rightleftharpoons$ $H^+(aq) + CH_3CH_2CH_2COO^-(aq)$	1.5×10^{-5}	4.82
carbonic	1	$H_2CO_3(aq) \rightleftharpoons H^+(aq) + HCO_3^-(aq)$	4.3×10^{-7}	6.37
	2	$HCO_3^-(aq) \rightleftharpoons H^+(aq) + CO_3^{2-}(aq)$	4.7×10^{-11}	10.33
chloric	1	$HClO_3(aq) \rightleftharpoons H^+(aq) + ClO_3^-(aq)$	~ 1	~ 0
chloroacetic	1	$CH_2ClCOOH(aq) \rightleftharpoons H^+(aq) + CH_2ClCOO^-(aq)$	1.4×10^{-3}	2.85
chlorous	1	$HClO_2(aq) \rightleftharpoons H^+(aq) + ClO_2^-(aq)$	1.1×10^{-2}	1.96
citric	1	$HOC(CH_2)_2(COOH)_3(aq) \rightleftharpoons$ $H^+(aq) + HOC(CH_2)_2(COOH)_2COO^-(aq)$	7.4×10^{-4}	3.13
	2	$HOC(CH_2)_2(COOH)_2COO^-(aq) \rightleftharpoons$ $H^+(aq) + HOC(CH_2)_2(COOH)(COO^-)_2(aq)$	1.7×10^{-5}	4.77
	3	$HOC(CH_2)_2(COOH)(COO^-)_2(aq) \rightleftharpoons$ $H^+(aq) + HOC(CH_2)_2(COO^-)_3(aq)$	4.0×10^{-7}	6.49
dichloroacetic	1	$CHCl_2COOH(aq) \rightleftharpoons H^+(aq) + CHCl_2COO^-(aq)$	5.5×10^{-2}	1.26
ethanol	1	$CH_3CH_2OH(aq) \rightleftharpoons H^+(aq) + CH_3CH_2O^-(aq)$	1.3×10^{-16}	15.9
fluoroacetic	1	$CH_2FCOOH(aq) \rightleftharpoons H^+(aq) + CH_2FCOO^-(aq)$	2.6×10^{-3}	2.59
formic	1	$HCOOH(aq) \rightleftharpoons H^+(aq) + HCOO^-(aq)$	1.77×10^{-4}	3.75
germanic	1	$H_2GeO_3(aq) \rightleftharpoons H^+(aq) + HGeO_3^-(aq)$	9.8×10^{-10}	9.01
	2	$HGeO_3^-(aq) \rightleftharpoons H^+(aq) + GeO_3^{2-}(aq)$	5×10^{-13}	12.3
hydr(o)azoic	1	$HN_3(aq) \rightleftharpoons H^+(aq) + N_3^-(aq)$	1.9×10^{-5}	4.72
hydrobromic	1	$HBr(aq) \rightleftharpoons H^+(aq) + Br^-(aq)$	$\gg 1$ (strong)	<0
acetic	1	$CH_3COOH(aq) \rightleftharpoons H^+(aq) + CH_3COO^-(aq)$	1.76×10^{-5}	4.75
hydrochloric	1	$HCl(aq) \rightleftharpoons H^+(aq) + Cl^-(aq)$	$\gg 1$ (strong)	<0
hydrocyanic	1	$HCN(aq) \rightleftharpoons H^+(aq) + CN^-(aq)$	6.2×10^{-10}	9.21
hydrofluoric	1	$HF(aq) \rightleftharpoons H^+(aq) + F^-(aq)$	6.8×10^{-4}	3.17

Continued on next page

A–21

TABLE A5.1 Ionization Constants of Selected Acids at 25°C (*Continued*)

Acid	Step	Aqueous Equilibrium[a]	K_a	pK_a
hydr(o)iodic	1	$HI(aq) \rightleftharpoons H^+(aq) + I^-(aq)$	$\gg 1$ (strong)	<0
hydrosulfuric	1	$H_2S(aq) \rightleftharpoons H^+(aq) + HS^-(aq)$	8.9×10^{-8}	7.05
	2	$HS^-(aq) \rightleftharpoons H^+(aq) + S^{2-}(aq)$	$\sim 10^{-19}$	~ 19
hypobromous	1	$HBrO(aq) \rightleftharpoons H^+(aq) + BrO^-(aq)$	2.3×10^{-9}	8.55
hypochlorous	1	$HClO(aq) \rightleftharpoons H^+(aq) + ClO^-(aq)$	2.9×10^{-8}	7.54
hypoiodous	1	$HIO(aq) \rightleftharpoons H^+(aq) + IO^-(aq)$	2.3×10^{-11}	10.5
iodic	1	$HIO_3(aq) \rightleftharpoons H^+(aq) + IO_3^-(aq)$	1.7×10^{-1}	0.77
iodoacetic	1	$CH_2ICOOH(aq) \rightleftharpoons H^+(aq) + CH_2ICOO^-(aq)$	7.6×10^{-4}	3.12
lactic	1	$CH_3CHOHCOOH(aq) \rightleftharpoons$ $H^+(aq) + CH_3CHOHCOO^-(aq)$	1.4×10^{-4}	3.85
maleic	1	$HOOCCH{=}CHCOOH(aq) \rightleftharpoons$ $H^+(aq) + HOOCH{=}CHCOO^-(aq)$	1.2×10^{-2}	1.92
	2	$HOOCCH{=}CHCOO^-(aq) \rightleftharpoons$ $H^+(aq) + {}^-OOCCH{=}CHCOO^{2-}(aq)$	4.7×10^{-7}	6.33
malonic	1	$HOOCCH_2COOH(aq) \rightleftharpoons$ $H^+(aq) + HOOCCH_2COO^-(aq)$	1.5×10^{-3}	2.82
	2	$HOOCCH_2COO^-(aq) \rightleftharpoons$ $H^+(aq) + {}^-OOCCH_2COO^{2-}(aq)$	2.0×10^{-6}	5.70
nitric	1	$HNO_3(aq) \rightleftharpoons H^+(aq) + NO_3^-(aq)$	$\gg 1$ (strong)	<0
nitrous	1	$HNO_2(aq) \rightleftharpoons H^+(aq) + NO_2^-(aq)$	4.0×10^{-4}	3.40
oxalic	1	$HOOCCOOH(aq) \rightleftharpoons H^+(aq) + HOOCCOO^-(aq)$	5.9×10^{-2}	1.23
	2	$HOOCCOO^-(aq) \rightleftharpoons H^+(aq) + {}^-OOCCOO^{2-}(aq)$	6.4×10^{-5}	4.19
perchloric	1	$HClO_4(aq) \rightleftharpoons H^+(aq) + ClO_4^-(aq)$	$\gg 1$ (strong)	<0
periodic	1	$HIO_4(aq) \rightleftharpoons H^+(aq) + IO_4^-(aq)$	2.3×10^{-2}	1.64
phenol	1	$C_6H_5OH(aq) \rightleftharpoons H^+(aq) + C_6H_5O^-(aq)$	1.3×10^{-10}	9.89
phosphoric	1	$H_3PO_4(aq) \rightleftharpoons H^+(aq) + H_2PO_4^-(aq)$	7.11×10^{-3}	2.12
	2	$H_2PO_4^-(aq) \rightleftharpoons H^+(aq) + HPO_4^{2-}(aq)$	6.32×10^{-8}	7.21
	3	$HPO_4^{2-}(aq) \rightleftharpoons H^+(aq) + PO_4^{3-}(aq)$	4.5×10^{-13}	12.66
propanoic	1	$CH_3CH_2COOH(aq) \rightleftharpoons$ $H^+(aq) + CH_3CH_2COO^-(aq)$	1.4×10^{-5}	4.86
pyruvic	1	$CH_3C(O)COOH(aq) \rightleftharpoons$ $H^+(aq) + CH_3C(O)COO^-(aq)$	2.8×10^{-3}	2.55
sulfuric	1	$H_2SO_4(aq) \rightleftharpoons H^+(aq) + HSO_4^-(aq)$	$\gg 1$ (strong)	<0
	2	$HSO_4^-(aq) \rightleftharpoons H^+(aq) + SO_4^{2-}(aq)$	1.2×10^{-2}	1.92
sulfurous	1	$H_2SO_3(aq) \rightleftharpoons H^+(aq) + HSO_3^-(aq)$	1.7×10^{-2}	1.9
	2	$HSO_3^-(aq) \rightleftharpoons H^+(aq) + SO_3^{2-}(aq)$	6.2×10^{-8}	7.1
thiocyanic	1	$HSCN(aq) \rightleftharpoons H^+(aq) + SCN^-(aq)$	$\gg 1$ (strong)	<0
trichloroacetic	1	$CCl_3COOH(aq) \rightleftharpoons H^+(aq) + CCl_3COO^-(aq)$	2.3×10^{-1}	0.64
trifluoroacetic	1	$CF_3COOH(aq) \rightleftharpoons H^+(aq) + CF_3COO^-(aq)$	5.9×10^{-1}	0.23
water	1	$H_2O(aq) \rightleftharpoons H^+(aq) + OH^-(aq)$	1.0×10^{-14}	14.00

[a] The formulas of the carboxylic acids are written in an RCOOH format to highlight their molecular structures.

TABLE A5.2 Acid Ionization Constants of Hydrated Metal Ions at 25°C

Free Ion	Hydrated Ion	K_a
Fe^{3+}	$Fe(H_2O)_6^{3+}$	3×10^{-3}
Sn^{2+}	$Sn(H_2O)_6^{2+}$	4×10^{-4}
Cr^{3+}	$Cr(H_2O)_6^{3+}$	1×10^{-4}
Al^{3+}	$Al(H_2O)_6^{3+}$	1×10^{-5}
Cu^{2+}	$Cu(H_2O)_6^{2+}$	3×10^{-8}
Pb^{2+}	$Pb(H_2O)_6^{2+}$	3×10^{-8}
Zn^{2+}	$Zn(H_2O)_6^{2+}$	1×10^{-9}
Co^{2+}	$Co(H_2O)_6^{2+}$	2×10^{-10}
Ni^{2+}	$Ni(H_2O)_6^{2+}$	1×10^{-10}

TABLE A5.3 Ionization Constants of Selected Bases at 25°C

Base	Aqueous Equilibrium	K_b	pK_b
ammonia	$NH_3(aq) + H_2O(aq) \rightleftharpoons NH_4(aq) + OH^-(aq)$	1.76×10^{-5}	4.75
aniline	$C_6H_5NH_2(aq) + H_2O(aq) \rightleftharpoons C_6H_5NH_3(aq) + OH^-(aq)$	4.0×10^{-10}	9.4
diethylamine	$(C_2H_5)_2NH(aq) + H_2O(aq) \rightleftharpoons (CH_3CH_2)_2NH_2^+(aq) + OH^-(aq)$	8.6×10^{-4}	3.1
dimethylamine	$(CH_3)_2NH(aq) + H_2O(aq) \rightleftharpoons (CH_3)_2NH_2^+(aq) + OH^-(aq)$	5.9×10^{-4}	3.2
methylamine	$CH_3NH_2(aq) + H_2O(aq) \rightleftharpoons CH_3NH_3^+(aq) + OH^-(aq)$	4.4×10^{-4}	3.4
nicotine (1)		1.0×10^{-6}	6.0
(2)		1.3×10^{-11}	10.9
pyridine	$C_5H_5N(aq) + H_2O(aq) \rightleftharpoons C_5H_5NH^+(aq) + OH^-(aq)$	1.7×10^{-9}	8.8
quinine (1)		3.3×10^{-6}	5.5
(2)		1.4×10^{-10}	9.9
urea	$H_2NCONH_2(aq) + H_2O(aq) \rightleftharpoons H_2NCONH_3^+(aq) + OH^-(aq)$	1.3×10^{-14}	13.9

TABLE A5.4 Solubility-Product Constants at 25°C

Cation	Anion	Heterogeneous Equilibrium[a]	K_{sp}
aluminum	hydroxide	$Al(OH)_3(s) \rightleftharpoons Al^{3+}(aq) + 3\,OH^-(aq)$	1.9×10^{-33}
	phosphate	$AlPO_4(s) \rightleftharpoons Al^{3+}(aq) + PO_4^{3-}(aq)$	9.8×10^{-21}
barium	carbonate	$BaCO_3(s) \rightleftharpoons Ba^{2+}(aq) + CO_3^{2-}(aq)$	2.6×10^{-9}
	fluoride	$BaF_2(s) \rightleftharpoons Ba^{2+}(aq) + 2\,F^-(aq)$	1.0×10^{-6}
	sulfate	$BaSO_4(s) \rightleftharpoons Ba^{2+}(aq) + SO_4^{2-}(aq)$	9.1×10^{-11}
calcium	carbonate	$CaCO_3(s) \rightleftharpoons Ca^{2+}(aq) + CO_3^{2-}(aq)$	5.0×10^{-9}
	fluoride	$CaF_2(s) \rightleftharpoons Ca^{2+}(aq) + 2\,F^-(aq)$	3.9×10^{-11}
	hydroxide	$Ca(OH)_2(s) \rightleftharpoons Ca^{2+}(aq) + 2\,OH^-(aq)$	4.7×10^{-6}
	phosphate	$Ca_3(PO_4)_2(s) \rightleftharpoons 3\,Ca^{2+}(aq) + 2\,PO_4^{3-}(aq)$	2.1×10^{-33}
	sulfate	$CaSO_4(s) \rightleftharpoons Ca^{2+}(aq) + SO_4^{2-}(aq)$	7.1×10^{-5}
copper(I)	bromide	$CuBr(s) \rightleftharpoons Cu^+(aq) + Br^-(aq)$	6.3×10^{-9}
	chloride	$CuCl(s) \rightleftharpoons Cu^+(aq) + Cl^-(aq)$	1.0×10^{-6}
	iodide	$CuI(s) \rightleftharpoons Cu^+(aq) + I^-(aq)$	1.3×10^{-12}
copper(II)	phosphate	$Cu_3(PO_4)_2(s) \rightleftharpoons 3\,Cu^{2+}(aq) + 2\,PO_4^{3-}(aq)$	1.4×10^{-37}
	hydroxide	$Cu(OH)_2(s) \rightleftharpoons Cu^{2+}(aq) + 2\,OH^-(aq)$	4.8×10^{-20}
iron(II)	carbonate	$FeCO_3(s) \rightleftharpoons Fe^{2+}(aq) + CO_3^{2-}(aq)$	3.1×10^{-11}
	fluoride	$FeF_2(s) \rightleftharpoons Fe^{2+}(aq) + 2\,F^-(aq)$	2.4×10^{-6}
	hydroxide	$Fe(OH)_2(s) \rightleftharpoons Fe^{2+}(aq) + 2\,OH^-(aq)$	4.9×10^{-17}
lead	bromide	$PbBr_2(s) \rightleftharpoons Pb^{2+}(aq) + 2\,Br^-(aq)$	6.6×10^{-6}
	carbonate	$PbCO_3(s) \rightleftharpoons Pb^{2+}(aq) + CO_3^{2-}(aq)$	1.5×10^{-13}
	chloride	$PbCl_2(s) \rightleftharpoons Pb^{2+}(aq) + 2\,Cl^-(aq)$	1.6×10^{-5}
	fluoride	$PbF_2(s) \rightleftharpoons Pb^{2+}(aq) + 2\,F^-(aq)$	3.2×10^{-8}
	iodide	$PbI_2(s) \rightleftharpoons Pb^{2+}(aq) + 2\,I^-(aq)$	8.5×10^{-9}
	sulfate	$PbSO_4(s) \rightleftharpoons Pb^{2+}(aq) + SO_4^{2-}(aq)$	1.8×10^{-8}
lithium	carbonate	$Li_2CO_3(s) \rightleftharpoons 2\,Li^+(aq) + CO_3^{2-}(aq)$	8.2×10^{-4}
magnesium	carbonate	$MgCO_3(s) \rightleftharpoons Mg^{2+}(aq) + CO_3^{2-}(aq)$	6.8×10^{-6}
	fluoride	$MgF_2(s) \rightleftharpoons Mg^{2+}(aq) + 2\,F^-(aq)$	6.5×10^{-9}
	hydroxide	$Mg(OH)_2(s) \rightleftharpoons Mg^{2+}(aq) + 2\,OH^-(aq)$	5.6×10^{-12}
manganese(II)	carbonate	$MnCO_3(s) \rightleftharpoons Mn^{2+}(aq) + CO_3^{2-}(aq)$	2.2×10^{-11}
	hydroxide	$Mn(OH)_2(s) \rightleftharpoons Mn^{2+}(aq) + 2\,OH^-(aq)$	5.6×10^{-12}
mercury(I)	bromide	$Hg_2Br_2(s) \rightleftharpoons Hg_2^{2+}(aq) + 2\,Br^-(aq)$	6.4×10^{-23}
	carbonate	$Hg_2CO_3(s) \rightleftharpoons Hg_2^{2+}(aq) + CO_3^{2-}(aq)$	3.7×10^{-17}
	chloride	$Hg_2Cl_2(s) \rightleftharpoons Hg_2^{2+}(aq) + 2\,Cl^-(aq)$	1.5×10^{-18}
	iodide	$Hg_2I_2(s) \rightleftharpoons Hg_2^{2+}(aq) + 2\,I^-(aq)$	5.3×10^{-29}
	sulfate	$Hg_2SO_4(s) \rightleftharpoons Hg_2^{2+}(aq) + SO_4^{2-}(aq)$	8.0×10^{-7}
mercury(II)	hydroxide	$Hg(OH)_2(s) \rightleftharpoons Hg^{2+}(aq) + 2\,OH^-(aq)$	3.1×10^{-26}
	iodide	$HgI_2(s) \rightleftharpoons Hg^{2+}(aq) + 2\,I^-(aq)$	2.8×10^{-29}
silver	bromide	$AgBr(s) \rightleftharpoons Ag^+(aq) + Br^-(aq)$	5.4×10^{-13}
	carbonate	$Ag_2CO_3(s) \rightleftharpoons 2\,Ag^+(aq) + CO_3^{2-}(aq)$	8.5×10^{-12}
	chloride	$AgCl(s) \rightleftharpoons Ag^+(aq) + Cl^-(aq)$	1.8×10^{-10}
	chromate	$Ag_2CrO_4(s) \rightleftharpoons 2\,Ag^+(aq) + CrO_4^{2-}(aq)$	1.1×10^{-12}
	hydroxide	$AgOH(s) \rightleftharpoons Ag^+(aq) + OH^-(aq)$	1.52×10^{-8}
	iodide	$AgI(s) \rightleftharpoons Ag^+(aq) + I^-(aq)$	8.3×10^{-17}
	phosphate	$Ag_3PO_4(s) \rightleftharpoons 3\,Ag^+(aq) + PO_4^{3-}(aq)$	8.9×10^{-17}
	sulfate	$Ag_2SO_4(s) \rightleftharpoons 2\,Ag^+(aq) + SO_4^{2-}(aq)$	1.2×10^{-5}
strontium	carbonate	$SrCO_3(s) \rightleftharpoons Sr^{2+}(aq) + CO_3^{2-}(aq)$	5.6×10^{-10}
	fluoride	$SrF_2(s) \rightleftharpoons Sr^{2+}(aq) + 2\,F^-(aq)$	4.3×10^{-9}
	sulfate	$SrSO_4(s) \rightleftharpoons Sr^{2+}(aq) + SO_4^{2-}(aq)$	3.4×10^{-7}
zinc	carbonate	$ZnCO_3(s) \rightleftharpoons Zn^{2+}(aq) + CO_3^{2-}(aq)$	1.2×10^{-10}
	hydroxide	$Zn(OH)_2(s) \rightleftharpoons Zn^{2+}(aq) + 2\,OH^-(aq)$	3.0×10^{-16}

[a]Equilibrium is between solid phase and aqueous solution.

TABLE A5.5 Formation Constants of Complexes at 25°C

Complex Ion	Aqueous Equilibrium	K_f
$[Ag(NH_3)_2]^+$	$Ag^+(aq) + 2NH_3(aq) \rightleftharpoons Ag(NH_3)_2^+(aq)$	1.7×10^7
$[AgCl_2]^-$	$Ag^+(aq) + 2Cl^-(aq) \rightleftharpoons AgCl_2^-(aq)$	2.5×10^5
$[Ag(CN)_2]^-$	$Ag^+(aq) + 2CN^-(aq) \rightleftharpoons Ag(CN)_2^-(aq)$	1.0×10^{21}
$[Ag(S_2O_3)_2]^{3-}$	$Ag^+(aq) + 2S_2O_3^{2-}(aq) \rightleftharpoons Ag(S_2O_3)_2^{3-}(aq)$	4.7×10^{13}
$[AlF_6]^{3-}$	$Al^{3+}(aq) + 6F^-(aq) \rightleftharpoons AlF_6^{3-}(aq)$	4.0×10^{19}
$[Al(OH)_4]^-$	$Al^{3+}(aq) + 4OH^-(aq) \rightleftharpoons Al(OH)_4^-(aq)$	7.7×10^{33}
$[Au(CN)_2]^-$	$Au^+(aq) + 2CN^-(aq) \rightleftharpoons Au(CN)_2^-(aq)$	2.0×10^{38}
$[Co(NH_3)_6]^{2+}$	$Co^{2+}(aq) + 6NH_3(aq) \rightleftharpoons Co(NH_3)_6^{2+}(aq)$	7.7×10^4
$[Co(NH_3)_6]^{3+}$	$Co^{3+}(aq) + 6NH_3(aq) \rightleftharpoons Co(NH_3)_6^{3+}(aq)$	5.0×10^{31}
$[Co(en)_3^{2+}]$	$Co^{2+}(aq) + 3en(aq) \rightleftharpoons W(en)_3^{2+}(aq)$	8.7×10^{13}
$[Cu(NH_3)_4]^{2+}$	$Cu^{2+}(aq) + 4NH_3(aq) \rightleftharpoons Cu(NH_3)_4^{2+}(aq)$	5.0×10^{13}
$[Cu(en)_2^{2+}]$	$Cu^{2+}(aq) + 2en(aq) \rightleftharpoons Cu(en)_2^{2+}(aq)$	3.2×10^{19}
$[Cu(CN)_4]^{2-}$	$Cu^{2+}(aq) + 4CN^-(aq) \rightleftharpoons Cu(CN)_4^{2-}(aq)$	1.0×10^{25}
$[HgCl_4]^{2-}$	$Hg^{2+}(aq) + 4Cl^-(aq) \rightleftharpoons HgCl_4^{2-}(aq)$	1.2×10^{15}
$[Ni(NH_3)_6]^{2+}$	$Ni^{2+}(aq) + 6NH_3(aq) \rightleftharpoons Ni(NH_3)_6^{2+}(aq)$	5.5×10^8
$[PbCl_4]^{2-}$	$Pb^{2+}(aq) + 4Cl^-(aq) \rightleftharpoons PbCl_4^{2-}(aq)$	2.5×10^1
$[Zn(NH_3)_4]^{2+}$	$Zn^{2+}(aq) + 4NH_3(aq) \rightleftharpoons Zn(NH_3)_4^{2+}(aq)$	2.9×10^9
$[Zn(OH)_4]^{2-}$	$Zn^{2+}(aq) + 4OH^-(aq) \rightleftharpoons Zn(OH)_4^{2-}(aq)$	2.8×10^{15}

Appendix 6 Standard Reduction Potentials

TABLE A6.1 Standard Reduction Potentials at 25°C

Half-Reaction	n	$E°$ (V)
$F_2(g) + 2e^- \rightarrow 2F^-(aq)$	2	2.866
$H_2N_2O_2(s) + 2H^+(aq) + 2e^- \rightarrow N_2(g) + 2H_2O(\ell)$	2	2.65
$O(g) + 2H^+(aq) + 2e^- \rightarrow H_2O(\ell)$	2	2.421
$Cu^{3+}(aq) + e^- \rightarrow Cu^{2+}(aq)$	1	2.4
$XeO_3(s) + 6H^+(aq) + 6e^- \rightarrow Xe(g) + 3H_2O(\ell)$	6	2.10
$O_3(g) + 2H^+(aq) + 2e^- \rightarrow O_2(g) + H_2O(\ell)$	2	2.076
$OH(g) + e^- \rightarrow OH^-(aq)$	1	2.02
$Co^{3+}(aq) + e^- \rightarrow Co^{2+}(aq)$	1	1.92
$H_2O_2(\ell) + 2H^+(aq) + 2e^- \rightarrow 2H_2O(\ell)$	2	1.776
$N_2O(g) + 2H^+(aq) + 2e^- \rightarrow N_2(g) + H_2O(\ell)$	2	1.766
$Ce(OH)^{3+}(aq) + H^+(aq) + e^- \rightarrow Ce^{3+}(aq) + H_2O(\ell)$	1	1.70
$Au^+(aq) + e^- \rightarrow Au(s)$	1	1.692
$PbO_2(s) + SO_4^{2-}(aq) + 4H^+(aq) + 2e^- \rightarrow$ $PbSO_4(s) + 2H_2O(\ell)$	2	1.6913
$MnO_4^-(aq) + 4H^+(aq) + 3e^- \rightarrow$ $MnO_2(s) + 2H_2O(\ell)$	3	1.673
$NiO_2(s) + 4H^+(aq) + 2e^- \rightarrow Ni^{2+}(aq) + 2H_2O(\ell)$	2	1.678
$HClO(\ell) + H^+(aq) + e^- \rightarrow \frac{1}{2}Cl_2(g) + H_2O(aq)$	1	1.63
$Ce^{4+}(aq) + e^- \rightarrow Ce^{3+}(aq)$	1	1.61
$Mn^{3+}(aq) + e^- \rightarrow Mn^{2+}(aq)$	1	1.542
$MnO_4^-(aq) + 8H^+(aq) + 5e^- \rightarrow Mn^{2+}(aq) + 4H_2O(\ell)$	5	1.507
$BrO_3^-(aq) + 6H^+(aq) + 5e^- \rightarrow \frac{1}{2}Br_2(\ell) + 3H_2O(\ell)$	5	1.52
$ClO_3^-(aq) + 6H^+(aq) + 5e^- \rightarrow \frac{1}{2}Cl_2(g) + 3H_2O(\ell)$	5	1.47
$PbO_2(s) + 4H^+(aq) + 2e^- \rightarrow Pb^{2+}(aq) + 2H_2O(\ell)$	2	1.455
$Au^{3+}(aq) + 3e^- \rightarrow Au(s)$	3	1.40
$Cl_2(g) + 2e^- \rightarrow 2Cl^-(aq)$	2	1.3583
$Cr_2O_7^{2-}(aq) + 14H^+(aq) + 6e^- \rightarrow$ $2Cr^{3+}(aq) + 7H_2O(\ell)$	6	1.33
$MnO_2(s) + 4H^+(aq) + 2e^- \rightarrow Mn^{2+}(aq) + 2H_2O(\ell)$	2	1.23
$O_2(g) + 4H^+(aq) + 4e^- \rightarrow 2H_2O(\ell)$	4	1.229
$IO_3^-(aq) + 6H^+(aq) + 5e^- \rightarrow \frac{1}{2}I_2(s) + 3H_2O(\ell)$	5	1.195
$IO_3^-(aq) + 6H^+(aq) + 6e^- \rightarrow I^-(aq) + 3H_2O(\ell)$	6	1.085
$Br_2(\ell) + 2e^- \rightarrow 2Br^-(aq)$	2	1.066
$HNO_2(\ell) + H^+(aq) + e^- \rightarrow NO(g) + H_2O(\ell)$	1	1.00

TABLE A6.1 Standard Reduction Potentials at 25°C (Continued)

Half-Reaction	n	$E°$ (V)
$VO_2^+(aq) + 2H^+(aq) + e^- \rightarrow VO^{2+}(aq) + H_2O(\ell)$	1	1.00
$NO_3^-(aq) + 4H^+(aq) + 3e^- \rightarrow NO(g) + 2H_2O(\ell)$	3	0.96
$2Hg^{2+}(aq) + 2e^- \rightarrow Hg_2^{2+}(aq)$	2	0.92
$ClO^-(aq) + H_2O(\ell) + 2e^- \rightarrow Cl^-(aq) + 2OH^-(aq)$	2	0.89
$HO_2^-(aq) + H_2O(\ell) + 2e^- \rightarrow 3OH^-(aq)$	2	0.88
$Hg^{2+}(aq) + 2e^- \rightarrow Hg(\ell)$	2	0.851
$Ag^+(aq) + e^- \rightarrow Ag(s)$	1	0.7996
$Hg_2^{2+}(aq) + 2e^- \rightarrow 2Hg(\ell)$	2	0.7973
$Fe^{3+}(aq) + e^- \rightarrow Fe^{2+}(aq)$	1	0.770
$PtCl_4^{2-}(aq) + 2e^- \rightarrow Pt(s) + 4Cl^-(aq)$	2	0.73
$O_2(g) + 2H^+(aq) + 2e^- \rightarrow H_2O_2(\ell)$	2	0.68
$MnO_4^-(aq) + 2H_2O(\ell) + 3e^- \rightarrow$ $MnO_2(s) + 4OH^-(aq)$	3	0.59
$H_3AsO_4(s) + 2H^+(aq) + 2e^- \rightarrow$ $H_3AsO_3(aq) + H_2O(\ell)$	2	0.559
$I_2(s) + 2e^- \rightarrow 2I^-(aq)$	2	0.5355
$Cu^+(aq) + e^- \rightarrow Cu(s)$	1	0.521
$H_2SO_3(\ell) + 4H^+(aq) + 4e^- \rightarrow S(s) + 3H_2O(\ell)$	4	0.449
$Ag_2CrO_4(s) + 2e^- \rightarrow 2Ag(s) + CrO_4^{2-}(aq)$	2	0.4470
$O_2(g) + 2H_2O(\ell) + 4e^- \rightarrow 4OH^-(aq)$	4	0.401
$Fe(CN)_6^{3-}(aq) + e^- \rightarrow Fe(CN)_6^{4-}(aq)$	1	0.36
$PbSO_4(s) + H^+(aq) + 2e^- \rightarrow Pb(s) + HSO_4^-(aq)$	2	−0.356
$Ag_2O(s) + H_2O(\ell) + 2e^- \rightarrow 2Ag(s) + 2OH^-(aq)$	2	0.342
$Cu^{2+}(aq) + 2e^- \rightarrow Cu(s)$	2	0.3419
$BiO^+(aq) + 2H^+(aq) + 3e^- \rightarrow Bi(s) + H_2O(\ell)$	3	0.32
$AgCl(s) + e^- \rightarrow Ag(s) + Cl^-(aq)$	1	0.2223
$HSO_4^-(aq) + 3H^+(aq) + 2e^- \rightarrow H_2SO_3(\ell) + H_2O(\ell)$	2	0.17
$Sn^{4+}(aq) + 2e^- \rightarrow Sn^{2+}(aq)$	2	0.154
$Cu^{2+}(aq) + e^- \rightarrow Cu^+(aq)$	1	0.153
$S(s) + 2H^+(aq) + 2e^- \rightarrow H_2S(g)$	2	0.141
$HgO(s) + H_2O(\ell) + 2e^- \rightarrow Hg(\ell) + 2OH^-(aq)$	2	0.0977
$AgBr(s) + e^- \rightarrow Ag(s) + Br^-(aq)$	1	0.095
$Ag(S_2O_3)_2^{3-}(aq) + e^- \rightarrow Ag(s) + 2S_2O_3^{2-}(aq)$	1	0.01

TABLE A6.1 Standard Reduction Potentials at 25°C (Continued)

Half-Reaction	n	$E°$ (V)
$NO_3^-(aq) + H_2O(\ell) + 2e^- \rightarrow NO_2^-(aq) + 2OH^-(aq)$	2	0.01
$2H^+(aq) + 2e^- \rightarrow H_2(g)$	2	0.0000
$Pb^{2+}(aq) + 2e^- \rightarrow Pb(s)$	2	−0.126
$CrO_4^{2-}(aq) + 4H_2O(\ell) + 3e^- \rightarrow$ $Cr(OH)_3(s) + 5OH^-(aq)$	3	−0.13
$Sn^{2+}(aq) + 2e^- \rightarrow Sn(s)$	2	−0.136
$AgI(s) + e^- \rightarrow Ag(s) + I^-(aq)$	1	−0.1522
$CuI(s) + e^- \rightarrow Cu(s) + I^-(aq)$	1	−0.185
$N_2(g) + 5H^+(aq) + 4e^- \rightarrow N_2H_5^+(aq)$	4	−0.23
$Ni^{2+}(aq) + 2e^- \rightarrow Ni(s)$	2	−0.257
$PbSO_4(s) + H^+(aq) + 2e^- \rightarrow Pb(s) + HSO_4^-(aq)$	2	−0.356
$Co^{2+}(aq) + 2e^- \rightarrow Co(s)$	2	−0.277
$Ag(CN)^-(aq) + e^- \rightarrow Ag(s) + 2CN^-(aq)$	1	−0.31
$Cd^{2+}(aq) + 2e^- \rightarrow Cd(s)$	2	−0.403
$Cr^{3+}(aq) + e^- \rightarrow Cr^{2+}(aq)$	1	−0.41
$Fe^{2+}(aq) + 2e^- \rightarrow Fe(s)$	2	−0.447
$2CO_2(g) + 2H^+(aq) + 2e^- \rightarrow H_2C_2O_4(s)$	2	−0.49
$Ni(OH)_2(s) + 2e^- \rightarrow Ni(s) + 2OH^-(aq)$	2	−0.72
$Cr^{3+}(aq) + 3e^- \rightarrow Cr(s)$	3	−0.74
$Zn^{2+}(aq) + 2e^- \rightarrow Zn(s)$	2	−0.7618
$2H_2O(\ell) + 2e^- \rightarrow H_2(g) + 2OH^-(aq)$	2	−0.8277
$SO_4^{2-}(aq) + H_2O(\ell) + 2e^- \rightarrow$ $SO_3^{2-}(aq) + 2OH^-(aq)$	2	−0.92
$N_2(g) + 4H_2O(\ell) + 4e^- \rightarrow 4OH^-(aq) + N_2H_4(\ell)$	4	−1.16
$Mn^{2+}(aq) + 2e^- \rightarrow Mn(s)$	2	−1.185
$Zn(OH)_2(s) + 2e^- \rightarrow Zn(s) + 2OH^-(aq)$	2	−1.249
$Al^{3+}(aq) + 3e^- \rightarrow Al(s)$	3	−1.662
$Mg^{2+}(aq) + 2e^- \rightarrow Mg(s)$	2	−2.37
$Na^+(aq) + e^- \rightarrow Na(s)$	1	−2.71
$Ca^{2+}(aq) + 2e^- \rightarrow Ca(s)$	2	−2.868
$Ba^{2+}(aq) + 2e^- \rightarrow Ba(s)$	2	−2.912
$K^+(aq) + e^- \rightarrow K(s)$	1	−2.95
$Li^+(aq) + e^- \rightarrow Li(s)$	1	−3.05

Naming Organic Compounds

While organic chemistry was becoming established as a discipline within chemistry, many compounds were given trivial names that are still commonly used and recognized. We refer to many of these compounds by their nonsystematic names throughout this book, and their names and structures are listed in Table A7.1.

TABLE A7.1 Organic Compounds and Their Commonly Used Nonsystematic Names

Name	Formula	Structure
ethylene	C_2H_4	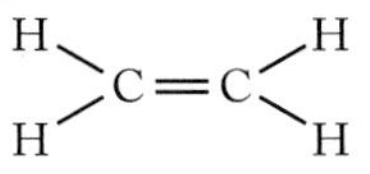
acetylene	C_2H_2	$HC \equiv CH$
benzene	C_6H_6	
toluene	$C_6H_5CH_3$	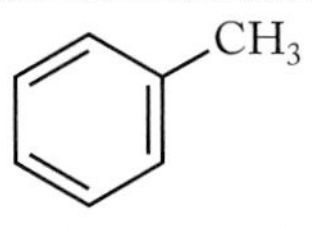
ethyl alcohol	CH_3CH_2OH	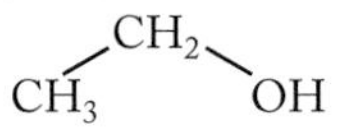
acetone	CH_3COCH_3	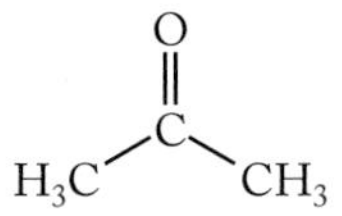
acetic acid	CH_3COOH	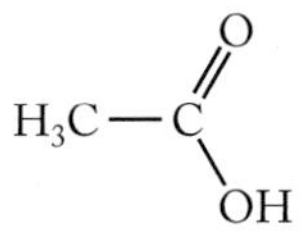
formaldehyde	CH_2O	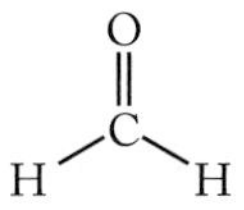

The International Union of Pure and Applied Chemistry (IUPAC) has proposed a set of rules for the systematic naming of organic compounds. The basic principles for naming alkanes, alkenes, and alkynes are presented in Chapter 12. These rules are summarized here and extended to include compounds containing other functional groups. When naming compounds or drawing structures based on names, we need to keep in mind that the IUPAC system of nomenclature is based on two fundamental ideas: (1) the name of a compound must indicate how the carbon atoms in the skeleton are bonded together, and (2) the name must identify the location of any functional groups in the molecule.

ALKANES

Table A7.2 contains the prefixes used for carbon chains ranging in size from C_1 to C_{20} and gives the names for compounds consisting of unbranched chains. The name of a compound consists of a prefix identifying the number of carbons in the chain and a suffix defining the type of hydrocarbon. The suffix -*ane* indicates that the compounds are alkanes and that all carbon–carbon bonds are single bonds.

TABLE A7.2 Prefixes for Naming Carbon Chains

Prefix	Example	Name	Prefix	Example	Name
meth	CH_4	methane	undec	$C_{11}H_{24}$	undecane
eth	C_2H_6	ethane	dodec	$C_{12}H_{26}$	dodecane
pro	C_3H_8	propane	tridec	$C_{13}H_{28}$	tridecane
but	C_4H_{10}	butane	tetradec	$C_{14}H_{30}$	tetradecane
pent	C_5H_{12}	pentane	pentadec	$C_{15}H_{32}$	pentadecane
hex	C_6H_{14}	hexane	hexadec	$C_{16}H_{34}$	hexadecane
hept	C_7H_{16}	heptane	heptadec	$C_{17}H_{36}$	heptadecane
oct	C_8H_{18}	octane	octadec	$C_{18}H_{38}$	octadecane
non	C_9H_{20}	nonane	nonadec	$C_{19}H_{40}$	nonadecane
dec	$C_{10}H_{22}$	decane	eicos	$C_{20}H_{42}$	eicosane

BRANCHED-CHAIN ALKANES

The alkane drawn here is used to illustrate each step in the naming rules:

$$CH_3$$
$$|$$
$$CH_3CH_2CHCH_2CHCHCH_2CH_2CH_3$$
$$|\qquad\quad|$$
$$CH_3\qquad CH_2CH_3$$

1. **Identify and name the longest continuous carbon chain.**

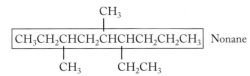

2. **Identify the groups attached to this chain and name them.** Names of substituent groups consist of the prefix from Table A7.2 that identifies the length of the group and the suffix -*yl* that identifies it as an alkyl group.

methyl-
$$CH_3$$
$$CH_3CH_2CHCH_2CHCHCH_2CH_2CH_3$$
$$CH_3\qquad CH_2CH_3$$
methyl-\qquad ethyl-

3. **Number the carbon atoms in the longest chain,** starting at the end nearest a substituent group. Doing this identifies the points of attachment of the alkyl groups with the lowest possible numbers.

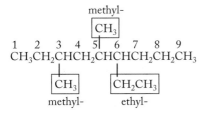

4. **Designate the location and identity of each substituent group with a number,** followed by a hyphen, and its name.

methyl-
$$CH_3$$
$$1\;\;2\;\;3\;\;4\;\;5|\;6\;\;7\;\;8\;\;9$$
$$CH_3CH_2CHCH_2CHCHCH_2CH_2CH_3$$
$$CH_3\qquad CH_2CH_3$$
methyl-\qquad ethyl-
3-methyl-, 5-methyl-, 6-ethyl-

5. **Put together the complete name by listing the substituent groups in alphabetical order.** If more than one of a given type of substituent group is present, prefixes di-, tri-, tetra- and so forth are appended to the names, but these numerical prefixes are not considered when determining the alphabetical order. The name of the last substituent group is written together with the name identifying the longest carbon chain.

$$CH_3$$
$$|$$
$$CH_3CH_2CHCH_2CHCHCH_2CH_2CH_3$$
$$|\qquad\quad|$$
$$CH_3\qquad CH_2CH_3$$
3,5-Dimethyl-6-ethylnonane

CYCLOALKANES

The simplest examples of this class of compounds consist of one unsubstituted ring of carbon atoms. The IUPAC names of these compounds consist of the prefix *cyclo-* followed by the parent name from Table A7.2 to indicate the number of carbon atoms in the ring. As an illustration, the names, formulas, and line structures of the first three cycloalkanes in the homologous series are

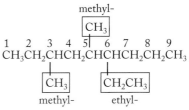

ALKENES AND ALKYNES

Alkenes have carbon–carbon double bonds and alkynes have carbon–carbon triple bonds as functional groups. The names of these types of compounds consist of (1) a parent name that identifies the longest carbon chain that includes the double or triple bond, (2) a suffix that identifies the class of compound, and (3) names of any substituent groups attached to the longest carbon chain. The suffix -*ene* identifies an alkene; -*yne* identifies an alkyne.

The alkene and alkyne drawn here are used to illustrate each step in the naming rules:

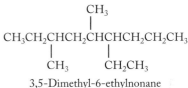

1. **To determine the parent name,** identify the longest chain that contains the unsaturation. Name the parent compound with the prefix that defines the number of carbons in that chain and the suffix that identifies the class of compound.

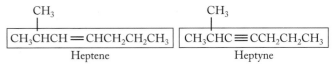

2. **Number the parent chain from the end nearest the unsaturation so that the first carbon in the double or triple bond has the lowest number possible.** (If the unsaturation is in the middle of a chain, the location of any substituent groups is used to determine where the numbering starts.) The smaller of the two numbers identifying the carbon atoms involved in the unsaturation is used as the locator of the multiple bond.

$$
\begin{matrix}
& CH_3 & & & & & & & & CH_3 \\
1 & 2 & 3 & 4 & 5 & 6 & 7 & & 1 & 2 & 3 & 4 & 5 & 6 & 7 \\
CH_3CHCH & = & CHCH_2CH_2CH_3 & & & & & & CH_3CHC & \equiv & CCH_2CH_2CH_3
\end{matrix}
$$

3-Heptene   3-Heptyne

3. **Geometric isomers of alkenes are named by writing *cis-* or *trans-* before the number identifying the location of the double bond.** Section 12.4 in the text addresses naming geometric isomers.

4. **The rules for naming substituted alkanes are followed to name and locate any other groups on the chain.**

$$
CH_3CHCH=CHCH_2CH_2CH_3 \qquad CH_3CHC\equiv CCH_2CH_2CH_3
$$

2-Methyl-3-heptene   2-Methyl-3-heptyne

Halogens attached to an alkane, alkene, or alkyne are named as fluoro- (F-), chloro- (Cl-), bromo- (Br-), or iodo- (I-) and are located by using the same numbering system described for alkyl groups.

BENZENE DERIVATIVES

Naming compounds containing substituted benzene rings is less systematic than naming hydrocarbons. Many compounds have common names that are incorporated into accepted names, but for simple substituted benzene rings, the following rules may be applied.

1. **For monosubstituted benzene rings,** a prefix identifying the group is appended to the parent name benzene:

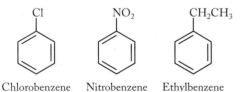

Chlorobenzene   Nitrobenzene   Ethylbenzene

2. **For disubstituted benzene rings,** three isomers are possible. The relative position of the substituent groups is indicated by numbers in IUPAC nomenclature, but the set of prefixes shown are very commonly used as well:

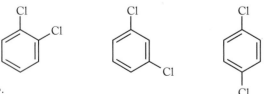

IUPAC:
1, 2-Dichlorobenzene   1, 3-Dichlorobenzene   1, 4-Dichlorobenzene

Common:
ortho-Dichlorobenzene   *meta*-Dichlorobenzene   *para*-Dichlorobenzene
o-Dichlorobenzene   *m*-Dichlorobenzene   *p*-Dichlorobenzene

3. **When three or more groups are attached to a benzene ring,** the lowest possible numbers are assigned to locate the groups with respect to each other.

1, 2, 3-Trichlorobenzene   1, 2, 4-Trichlorobenzene   1, 2, 3, 5-Tetrachlorobenzene

(**NOTE:** Not 1, 3, 4-trichlorobenzene; and not 1, 3, 4, 5-tetrachlorobenzene.)

HYDROCARBONS CONTAINING OTHER FUNCTIONAL GROUPS

The same basic principles developed for naming alkanes apply to naming hydrocarbons with functional groups other than alkyl groups (Table 12.1 in the text). The name must identify the carbon skeleton, locate the functional group, and contain a suffix that defines the class of compound. The following examples give the suffixes for some common functional groups; when suffixes are used, they replace the final –*e* in the name of the parent alkane. Other functional groups may be identified by including the name of the class of compounds in the name of the molecule.

ALCOHOLS: SUFFIX -*OL*

$$
CH_3CH_2CH_2OH \qquad CH_3CHCH_3 \qquad CH_3CH2CH_2CHCH_3
$$
$$
\qquad\qquad\qquad\qquad OH \qquad\qquad\qquad OH
$$

1-Propanol   2-Propanol   2-Pentanol

ALDEHYDES: SUFFIX –*AL*

$$
\begin{matrix}
O & & O \\
\parallel & & \parallel \\
H-C-H & & H_3C-C-H
\end{matrix}
$$

IUPAC: Methanal   Ethanal
Common: Formaldehyde   Acetaldehyde

Because the aldehyde group can only be on a terminal carbon, no number is necessary to locate it on the carbon chain.

KETONES: SUFFIX –*ONE*

The location of the carbonyl is given by a number, and the chain is numbered so that the carbonyl carbon has the lowest possible value. Many ketones also have common names generated by identifying the hydrocarbon groups on both sides of the carbonyl group.

$$
\begin{matrix}
O & & O \\
\parallel & & \parallel \\
H_3C-C-CH_3 & & H_3C-C-CH_2CH_3
\end{matrix}
$$

IUPAC: Propan-2-one   Butan-2-one
Common: Acetone   Methyl ethyl ketone

CARBOXYLIC ACIDS: SUFFIX –*OIC ACID*

The carboxylic acid group is by definition carbon 1, so no number identifying its location is included in the name.

$$
\begin{matrix}
O & & O \\
\parallel & & \parallel \\
H_3C-C-OH & & H_3C-C-OH
\end{matrix}
$$

IUPAC: Ethanoic acid   *trans*-2-Butenoic acid
Common: Acetic acid

SALTS OF CARBOXYLIC ACIDS

Salts are named with the cation first, followed by the anion name of the acid from which *-ic acid* is dropped and the suffix *-ate* is added. The sodium salt of acetic acid is sodium acetate.

Acetic acid   Acetate ion   Sodium acetate

ESTERS

Esters are viewed as derivatives of carboxylic acids. They are named in a manner analogous to that of salts: the alkyl group comes first followed by the name of the carboxylate anion.

Alkyl   Carboxylate   Ethyl acetate

AMIDES

Amides are also derivatives of carboxylic acids. They are named by replacing *-ic acid* (of the common names) or *-oic acid* of the IUPAC names with *-amide*.

Parent acid -amide   Acetamide

ETHERS

Ethers are frequently named by naming the two groups attached to the oxygen and following those names by the word *ether*.

$$CH_3OCH_3 \qquad\qquad CH_3CH_2OCH_2CH_3$$

Dimethyl ether   Diethyl ether

AMINES

Aliphatic amines are usually named by listing the group or groups attached to the nitrogen and then appending the *-amine* as a suffix. They may also be named by prefixing *amino-* to the name of the parent chain.

Methylamine   Ethylmethylamine   2-Aminoethanol

This brief summary will enable you to understand the names of organic compounds used in this book. IUPAC rules are much more extensive than this and can be applied to all varieties of carbon compounds including those with multiple functional groups. It is important to recognize that the rules of systematic nomenclature do not necessarily lead to a unique name for each compound, but they do always lead to an unambiguous one. Furthermore, common names are still used frequently in organic chemistry because the systematic alternatives do not improve communication. Remember that the main purpose of chemical nomenclature is to identify a chemical species by means of written or spoken words. Anyone who reads or hears the name should be able to deduce the structure and thereby the identity of the compound.

Glossary

absolute entropy   The entropy change of a substance taken from $S = 0$ (at $T = 0$ K) to some other temperature.

absolute zero (0 K)   Theoretically the lowest temperature possible.

accuracy   The agreement between an experimental value and the true value.

achiral   A molecule that is achiral is superimposable on, and identical to, its mirror image.

acid   (also called Brønsted–Lowry acid)   A proton donor.

acid anhydrides   Nonmetal oxides that react with water to produce acids.

activated complex   A species formed in a chemical reaction when molecules have enough energy to react with each other.

activation energy (E_a)   The minimum energy molecules need to react when they collide.

active site   The location on an enzyme where a reactive substance binds.

actual yield   The amount of product obtained from a chemical reaction conducted in a laboratory: it is often less than the theoretical yield.

addition reaction   Two molecules couple and form one product.

α-helix   A springlike coil in a protein chain.

alcohol   Contains the –OH functional group and has the general formula R–OH, where R is any alkyl group.

aldehyde   Contains a carbonyl group bonded to one R group and one hydrogen atom; its general formula is RCHO.

alkali metals   The elements in group 1 of the periodic table.

alkaline earth metals   The elements in group 2 of the periodic table.

alkanes   Hydrocarbons in which all the bonds are single bonds.

alkenes   Hydrocarbons containing one or more carbon–carbon double bonds.

alkynes   Hydrocarbons containing one or more carbon–carbon triple bonds.

allotropes   Different forms of the same element, such as oxygen (O_2) and ozone (O_3).

alloy   A blend of a host metal and one or more other elements (which may or may not be metals) that are added to change the properties of the host metal.

alpha (α) particle   A radioactive emission (composed of subatomic particles) with a charge of 2+ and a mass equivalent to that of a helium nucleus.

amide   The –OH of a carboxylic acid group is replaced by an $-NH_2$, $-NHR$, or $-NR_2$, where R can be any organic group.

amines   Organic compounds that have appreciable basicity; their general formula is RNH_2, R_2NH, or R_3N, where R is any organic subgroup.

amino acids   Molecules that contain one amine group and one carboxylic acid group; in an α-amino acid, the two groups are attached to the same carbon atom, the α carbon atom.

Amontons's law   As the absolute temperature of a fixed amount of gas increases, the pressure increases as long as the volume and quantity of gas remain constant.

amphiprotic   A term describing a substance that can behave as either a proton acceptor or a proton donor.

amphoteric substance   Can behave either as an acid or as a base.

amplitude   The height of the crest or the depth of the trough with respect to the centerline of a wave; the intensity of a wave is related to its amplitude.

angular momentum quantum number (ℓ)   An integer that may have any value from 0 to $n-1$; it defines the shape of an orbital.

anions   Ions with negative charges.

anode   An electrode at which an oxidation half-reaction (loss of electrons) takes place.

antibonding orbitals   Orbitals with electrons in a molecule that destabilize the molecule because they do not increase the electron density between nuclear centers and therefore do not participate in holding the molecule together.

antimatter Positrons formed in solar fusion that belong to a group of subatomic particles that are the charge opposites of particles typically found in atoms; these are "electrons" with a positive charge and "protons" with a negative charge.

aromatic compounds Molecules consisting of flat rings with π electron clouds delocalized above and below the plane of the molecules.

Arrhenius equation Relates the rate constant of a reaction to absolute temperature (T), the activation energy of the reaction (E_a), and the frequency factor (A).

atmospheric pressure (P_{atm}) The force exerted by the gases surrounding Earth on Earth's surface and on all surfaces of all objects.

atom The smallest particle of an element that retains the chemical characteristics of that element.

atomic absorption spectra (also called dark-line spectra) Characteristic series of dark lines produced when free, gaseous atoms are illuminated by external sources of radiation.

atomic emission spectra (also called bright-line spectra) Bright lines on a dark background that appear at specific wavelengths.

atomic mass units (amu) Units used to express the relative masses of atoms and subatomic particles based on the definition that the mass of 1 atom of carbon with 6 protons and 6 neutrons in its nucleus is exactly 12 amu.

atomic number (Z) The number of protons in the nucleus of an atom of an element.

aufbau principle The method of building electron configurations one electron at a time; one electron is added to the lowest-energy orbital of a ground-state atom (and one proton and one or more neutrons are added to the atom's nucleus); the electron configurations of atoms are built in sequence as atomic number increases in order across the rows of the periodic table.

average atomic mass An atomic mass for an element and its isotopes that is calculated by multiplying the natural abundance of each isotope by its exact mass in atomic mass units and then summing these products.

Avogadro's law The volume of a gas at a given temperature and pressure is proportional to the quantity of the gas.

Avogadro's number The number of carbon atoms in exactly 12 grams of the carbon-12 isotope; $N_A = 6.022 \times 10^{23}$.

band gap (E_g) The energy gap between the valence and conduction bands of a material.

band theory An extension of molecular orbital theory that describes bonding in solids.

bar The unit of pressure used in defining standard state; it is very close to 1 atm.

barometer An instrument that measures atmospheric pressure.

base (also called Brønsted–Lowry base) A proton acceptor.

becquerel (Bq) A unit used to express the level of radioactivity from a sample; one Bq equals one disintegration per second.

belt of stability The region on the graph of number-of-neutrons versus number-of-protons that includes all stable nuclei.

beta (β) decay The process by which a neutron in a neutron-rich nucleus decays into a proton and a β particle.

beta (β) particle A type of radioactive emission that consists of a high-energy electron.

beta (β) pleated sheet A zigzag pattern of protein chains.

bimolecular step A step in a mechanism or reaction that involves a collision between two molecules.

binding energy (BE) Measure of the energy released when nucleons combine to form a nucleus. It is calculated using the equation $E = (\Delta m)c^2$, where Δm is the mass defect of the nucleus.

biocatalysis The strategy of producing pharmaceutical products using enzymes to catalyze reactions on a large scale; it is becoming especially important in processes that involve chiral materials.

biomass The sum total of the mass of organic matter in any given ecological system.

biomolecule An organic molecule present naturally in a living system.

body-centered cubic (bcc) A unit cell that has atoms at the eight corners of a cube and an atom at the center of the cell.

boiling point The temperature at which vaporization occurs throughout a liquid.

bomb calorimeter A constant-volume device used to measure the heat of a combustion reaction.

bond angle The angle (in degrees) defined by lines joining the centers of two atoms to a third atom to which they are covalently bonded.

bond dipole Two covalently bonded atoms with different electronegativities have partial electrical charges of opposite sign creating a bond dipole; in a polar molecule, the individual bond dipoles do not offset each other as they may in a nonpolar molecule.

bond energy The energy needed to break 1 mol of covalent bonds in the gas phase.

bond length The distance between the nuclear centers of two atoms joined together in a bond.

bond order The number of bonds between atoms; the bond orders are 1 for a single bond, 2 for a double bond, and 3 for a triple bond.

bond polarity A measure of the extent to which bonding electrons are shared between two atoms in a covalent bond; the less equally they are shared, the more uneven the distribution, and the more polar the bond; differences in electronegativity determine bond polarity.

bonding capacity Reflects the number of covalent bonds an element forms when it has a formal charge of 0; the bonding capacity of carbon is 4 (it is tetravalent); nitrogen is 3 (trivalent); oxygen is 2 (divalent); and fluorine is 1 (monovalent).

bonding orbitals Orbitals with electrons that serve to hold atoms together in molecules by increasing the electron density between nuclear centers.

bonding pair A pair of electrons shared between two atoms.

Born–Haber cycle A series of steps with corresponding enthalpy changes that describes the formation of an ionic solid from its constituent elements.

Boyle's law The volume of a given amount of gas at constant temperature is inversely proportional to its pressure.

Bragg equation Relates the angle of diffraction (2θ) of X-rays to the spacing (d) between the layers of ions or atoms in a crystal: $n\lambda = 2d \sin \theta$.

branched-chain hydrocarbon An organic molecule in which the chain of carbon atoms is not linear.

breeder reactor A nuclear reactor in which fissionable material is produced during normal reactor operation.

bright-line spectra (also called atomic emission spectra) Bright lines on a dark background that appear at specific wavelengths.

Brønsted–Lowry model Defines acids as H^+ ion donors and bases as H^+ ion acceptors.

Brønsted–Lowry acid A proton donor.

Brønsted–Lowry base A proton acceptor.

calorie (cal) The amount of heat necessary to raise the temperature of 1 g of water by 1°C.

calorimeter A device used to measure the absorption or release of heat by a physical change or chemical process.

calorimeter constant ($C_{calorimeter}$) The heat capacity of a calorimeter.

calorimetry   The measurement of the change in heat that occurs during a physical change or chemical process.

capacity of a buffer   The concentration of components needed in solution to consume the acid or base added without a significant change in pH.

capillary action   The rise of a liquid in a narrow tube due to adhesive forces between the liquid and the tube and cohesive forces within the liquid.

carbonyl group   A carbon atom with a double bond to an oxygen atom.

carboxylic acid   An organic compound containing the –COOH functional group.

catalyst   A substance added to a reaction that increases the rate of the reaction but is not consumed in the process.

cathode   An electrode at which a reduction half-reaction (gain of electrons) takes place.

cathode rays   Streams of electrons emitted by the cathode (negative electrode) in a partially evacuated tube.

cations   Ions with positive charges.

cell diagram   Uses symbols to show how the components of an electrochemical cell are connected.

cell potential (E_{cell})   The electromotive force expressed in volts (V) with which an electrochemical cell can push electrons through an external circuit connected to its terminals.

ceramic   A solid inorganic compound or mixture that has been transformed into a harder, more heat-resistant material by heating.

chain reaction   A self-sustaining series of reactions in which the products of the reaction—for example, neutrons released when an atom is split—react further with other materials and continue the reaction.

change in enthalpy (ΔH)   The heat gained or lost by a thermodynamic system during a reaction at constant pressure.

Charles's law   The volume of a fixed quantity of gas at constant pressure is directly proportional to its absolute temperature.

chelate effect   The greater affinity of metal ions for polydentate ligands compared to that for the corresponding monodentate ligands.

chelation   The interaction of a metal with a polydentate ligand (chelating agent); one molecule of the ligand occupies two or more coordination sites on the central metal.

chemical bond   The attractive force that holds two atoms or ions in a substance together.

chemical equation   A balanced equation that describes the identities and quantities of reactants (substances that are consumed during a chemical reaction) and products (substances that are formed).

chemical equilibrium   A dynamic process in which the concentrations of reactants and products remain constant over time and the rate of a reaction in the forward direction matches its rate in the reverse direction.

chemical formula   A representation of the constituent elements in a compound. The representation comprises the elemental symbols and the subscripts (and sometimes coefficients) necessary to identify the number of atoms of each element in one molecule.

chemical kinetics   The study of the rates of change of concentrations of substances involved in chemical reactions.

chemical property   A characteristic of a substance that can be observed only by reacting it chemically to form another substance.

chemical reaction   The transformation of one or more substances into different substances.

chemistry   The science of matter and the study of its composition, structure, and properties.

chiral   A molecule that is chiral is not superimposable on its mirror image.

chirality   A molecule has chirality if its mirror image does not coincide with itself.

cis isomer (also called Z isomer)   An isomer that has two like groups (such as two R groups or two hydrogen atoms) on the same side of a line drawn through a double bond; *cis* is Latin for "on this side"; Z comes from the German word *zusammen* (together).

Clausius–Clapeyron equation   Relates the vapor pressure of a substance at different temperatures to its heat of vaporization.

closed system   A thermodynamic system that exchanges energy but not matter with the surroundings.

codon   A three-base sequence of nucleotides that encodes for a specific amino acid.

coenzymes   Like enzymes, are organic molecules that accelerate the rate of biochemical reactions.

colligative properties of solutions   Depend on the concentration, not on the identity of particles dissolved in the solvent.

combined gas law (also called general gas equation)   Based on the ideal gas law; it is used when one or more of the four gas variables are held constant while the remaining variables change.

combustion analysis   A method of analysis in which a substance is burned completely in oxygen to produce known compounds whose masses are used to determine the composition of the original material.

combustion reactions   Reactions between oxygen and another element or compound; when the other compound is a hydrocarbon; the products of complete combustion are carbon dioxide and water vapor.

common-ion effect   The shift in the position of an equilibrium caused by the addition of an ion taking part in the reaction.

complex carbohydrate (also called polysaccharide)   A polymer of monosaccharides.

complex ion   An ionic species consisting of a metal ion bonded to one or more Lewis bases.

composite materials   Contain a mixture of substances with different compositions and structures.

compound   A substance composed of two or more elements linked together in fixed proportions.

condensation reaction   Two molecules combine to form a larger molecule and a small molecule (typically water).

conduction bands   Unoccupied higher-energy bands in which electrons are free to migrate in a material.

configuration of chiral compounds   The actual positions in space occupied by the groups surrounding a chiral center.

conjugate acid   Formed when a Brønsted–Lowry base accepts a proton.

conjugate acid–base pairs   A Brønsted–Lowry acid and base that differ from each other only by the presence or absence of a proton.

conjugate base   Formed when a Brønsted–Lowry acid donates an H^+ ion to another species.

conversion factor   A fraction in which the numerator is equivalent to the denominator but is expressed in different units.

coordinate bond   Forms when one anion or molecule donates a pair of electrons to another ion or molecule to form a covalent bond.

coordination compounds   Compounds made up of at least one complex ion.

coordination number of a metal ion   Identifies the number of electron pairs surrounding it in a complex.

copolymer   A heteropolymer that contains two different monomers.

core electrons   Those electrons in the filled, inner shells in an atom or ion that are not involved in chemical reactions.

cosmology   The study of the physical nature and the form of the universe as a whole.

Coulomb's law   The energy (E) of the interaction between two ions is directly proportional to the product of the charges of the two ions (Q_1 and Q_2) and inversely proportional to the distance (d) between their nuclei.

counter ions   The ions that balance the electrical charges of complex ions in coordination compounds.

covalent bond   A chemical bond formed between two atoms by sharing pairs of electrons.

covalent pi (π) bond   Electron density is greatest above and below or in front of and behind the bonding axis.

covalent network solid   Consists of atoms held together by extended arrays of covalent bonds.

critical mass   The minimum amount of fissionable material needed to sustain a nuclear chain reaction.

critical point   That specific temperature and pressure where the liquid and gas phases of a substance have the same density and are indistinguishable from each other.

critical temperature (T_c)   The temperature below which a material becomes a superconductor.

crude oil   A combustible liquid mixture of hundreds of hydrocarbons and other organic molecules formed under Earth's surface by the processing of plant matter under high temperature and pressure.

crystal field splitting   The separation of a set of d orbitals into subsets with different energies resulting from interactions between electrons in those orbitals and pairs of electrons in ligands surrounding the orbitals.

crystal field splitting energy (Δ)   The difference in energy among subsets of d orbitals split by interactions in a crystal field.

crystal lattice   The three-dimensional array of particles (atoms, ions, or molecules) in a crystalline solid.

crystalline solid   An ordered array of atoms, ions, or molecules.

cubic closest-packed (ccp)   A crystal structure in which the layers of atoms, ions, or molecules in face-centered cubic unit cells have an *abcabcabc . . .* stacking pattern.

curie (Ci)   A unit used to express the level of radioactivity from a sample; one Ci equals 3.7×10^{10} Bq.

cycloalkanes   Ring-containing alkanes with the general formula C_nH_{2n}.

cyclotron   A circular particle accelerator using the combined action of a constant magnetic field with an oscillating electrostatic field to make particles move faster in an increasingly larger spiral path.

d^2sp^3 hybrid orbitals   The six equivalent hybrid orbitals that point toward the apices of an octahedron formed by mixing one $(n)s$ orbital, three $(n)p$ orbitals, and two $(n-1)d$ orbitals.

Dalton's law of partial pressures   The total pressure of any mixture of gases equals the sum of the partial pressures of each gas in the mixture.

dark-line spectra (also called atomic absorption spectra)   Characteristic series of dark lines produced when free, gaseous atoms are illuminated by external sources of radiation.

degenerate orbitals   Have the exact same energy level.

degree of ionization   The ratio of the quantity of a substance that is ionized to the original concentration of the substance present; when expressed as a percentage, it is called percent ionization.

delocalized   The pattern of alternating single and double bonds gives rise to a property sometimes called conjugation, which means that the electrons in the π bonds are said to be delocalized over the three or more carbon atoms and the oxygen atom.

density (d)   The ratio of the mass (m) of an object to its volume (V); the density of a pure substance is an intensive physical property of that substance.

deposition   The direct conversion of a gas (vapor) into a solid without an intermediate liquid phase.

diamagnetic atoms, ions, and molecules   Have no unpaired electrons and are weakly repelled by a magnetic field.

diffraction   The bending of electromagnetic radiation as it passes around the edge of an object or through narrow openings.

diffusion   The spread of one substance (usually a gas or liquid) through another.

dilution   The process of lowering the concentration of solutes in a solution by adding more solvent.

dipole–dipole interactions   Attractive forces between polar molecules.

dispersion forces (also called London forces)   Intermolecular forces caused by temporary dipoles in molecules.

distillation   A separation technique in which the more volatile (more easily vaporized) components of a mixture are vaporized and then condensed, thereby separating them from the less volatile components.

double bond   Results when two atoms share two pairs of electrons.

duet   The maximum of two electrons contained in the outermost shell of a hydrogen atom ($n = 1$).

E isomer (also called *trans* isomer)   An isomer in an alkene that has two like groups (such as two R groups or two hydrogen atoms) on opposite sides of a line drawn through the double bond; *trans* is Latin for "across"; E comes from the German word *entgegen* (opposite).

effective nuclear charge (Z_{eff})   The attractive force toward the nucleus experienced by an electron in an atom; its value is the positive charge on the nucleus reduced by the extent to which other electrons in the atom shield it from the nucleus.

effusion   The process by which a gas escapes from its container through a tiny hole into a region of lower pressure.

electric current   A flow of electrons.

electrochemical cell   An apparatus that converts chemical energy into electrical work or electrical work into chemical energy.

electrochemistry   The branch of chemistry that examines the transformations between chemical and electrical energy.

electrolysis   A process in which electrical energy is used to drive a nonspontaneous chemical reaction.

electrolyte   A substance that dissociates into ions when it dissolves, and enhances the conductivity of the solvent.

electrolytic cell   A device in which an external source of electrical energy does work on a chemical system, turning reactant(s) into higher-energy product(s).

electromagnetic radiation   Any form of radiant energy in the electromagnetic spectrum.

electromagnetic spectrum   A continuous range of radiant energy that includes radio waves, infrared radiation, visible light, ultraviolet radiation, X-rays, and gamma rays.

electromotive force (emf)   Also called voltage, the force that pushes electrons through an electrical circuit.

electron   A negatively charged subatomic particle.

electron affinity   The energy change occurring when 1 mol of electrons combines with 1 mol of atoms or ions in the gas phase.

electron capture   A neutron-poor nucleus draws in one of its surrounding electrons, resulting in the nucleus with one more neutron and one less proton.

electron configuration   Describes the distribution of electrons among the orbitals of an atom or ion.

electron transitions   Movements of electrons between energy levels.

electron-deficient compounds   Substances whose central atoms (other than H) in Lewis structures have fewer than four electron pairs (that is, less than an octet of electrons).

electronegativity   A relative measure of the ability of an atom in a bond to attract electrons to itself.

electron-pair geometry   Describes the arrangement of atoms and lone pairs of electrons about a central atom.

electrons in π bonds   In a system with alternating single and double bonds, electrons in π bonds can be delocalized over several atoms or even across an entire molecule.

electrons in antibonding orbitals   Destabilize the molecule because they do not increase the electron density between nuclear centers and therefore do not participate in holding the molecule together.

electrons in bonding orbitals   Hold atoms in molecules together by increasing the electron density between nuclear centers.

electrostatic potential energy (E_{el})   The energy of a particle because of its position with respect to another particle; it is directly proportional to the product of the charges of the particles and inversely proportional to the distance between them.

element   A pure substance that cannot be separated into simpler substances by chemical means.

elementary step   A molecular-level view of a single process taking place in a chemical reaction.

empirical formula   A formula that gives the simplest whole-number ratio of elements in a compound.

enantiomers   Geometric isomers whose mirror images are not superimposable; enantiomeric ions and molecules are called chiral.

end point   The point in a titration that is reached when just enough standard solution has been added to make the indicator change color.

endergonic reactions   Nonspontaneous reactions.

endothermic process   A process in which heat flows from the surroundings into a thermodynamic system.

energy   The capacity to transfer heat or to do work.

energy level   An allowed state that an electron can occupy in an atom.

energy state   An allowed value of energy.

enthalpy (H)   The sum of the internal energy and the pressure–volume product of a thermodynamic system.

enthalpy of reaction (ΔH_{rxn}) (also called heat of reaction)   The heat absorbed or released during a chemical reaction.

entropy (S)   A measure of the distribution of energy in a system at a specific temperature.

enzyme   A protein that catalyzes a reaction.

equilibrium constant (K)   The value of the ratio of concentrations (or partial pressures) in the equilibrium constant expression at a specific temperature.

equilibrium constant expression (also called mass action expression)   Describes the relationship between the concentration (or partial pressure) terms of reactants and products when a system is at equilibrium.

equivalence point   The point in a titration where the number of moles of titrant added is stoichiometrically equal to the number of moles of the substance being analyzed.

essential elements   Those elements in tissue, blood, or other body fluids that have a physiological function; a deficiency in an essential element ultimately leads to damage to the organism.

ester   The –OH of a carboxylic acid group is replaced by –OR, where R can be any organic group.

ether   Has the general formula R–O–R, where R is any alkyl group or aromatic ring. The two R fragments may be different.

evaporation (also called vaporization)   The transformation of molecules in the liquid phase into the gas phase.

excited state   Any energy state of an electron in an atom above the ground state.

exergonic reactions   Spontaneous reactions.

exothermic process   A process in which heat flows from a thermodynamic system into its surroundings.

extensive property   A characteristic of a substance that varies with the quantity of the substance present.

face-centered cubic (fcc)   A unit cell that has the same kind of particle (atom, ion, or molecule) at the eight corners of a cube and at the center of each face.

family (also called a group)   Elements in the same column of the periodic table; elements in the same family have similar chemical properties.

Faraday's constant (F)   The quantity of electrical charge (in coulombs) in 1 mol of electrons; its value to three significant figures is 9.65×10^4 C.

fats   Solid triglycerides containing primarily saturated fatty acids.

filtration   A process for separating particles suspended in a liquid or a gas from that liquid or gas by passing the mixture through a medium that retains the particles.

first law of thermodynamics   A law of physics that states that the energy gained or lost by a system must equal the energy lost or gained by the surroundings.

food value   The amount of heat produced when a material consumed by an organism for sustenance is burned completely; it is typically reported in Calories (kilocalories).

formal charge   For an atom in a molecule, equals the number of valence electrons in the free atom in the Lewis structure minus the sum of the number of electrons in its lone pairs plus half the number of electrons in its bonding pairs.

formation constant (K_f)   The equilibrium constant describing the mass action expression for a metal complex, a free metal ion, and its ligands.

formation reaction   The process of forming 1 mol of a substance in its standard state from its component elements in their standard states.

formula mass   The mass of one formula unit of a compound.

formula unit   The smallest electrically neutral unit within a crystal of an ionic compound; the formula unit contains the number and type of ions expressed in the formula of the compound.

fractional distillation   A method of separating a mixture of compounds on the basis of their different boiling points.

Fraunhofer lines   A set of dark lines in the otherwise continuous solar spectrum.

free energy (G)   An indication of the energy available to do useful work; free energy is a thermodynamic state function that provides a criterion for spontaneous change.

free radicals   Molecules that have an odd number of valence electrons and hence unpaired electrons in their Lewis structures.

free-energy change (ΔG)   The change in the free energy of a process; for spontaneous processes at a constant temperature and pressure, $\Delta G < 0$.

frequency (ν)   The number of times a wave passes a given point in some unit of time (typically 1 second); the SI unit of frequency is the hertz (Hz), which is expressed in the unit s^{-1}: 1 Hz = 1 s^{-1} = 1 cycle per second (cps).

frequency factor (A)   The product of the frequency of molecular collisions and a factor that expresses the probability that the orientation of the molecules is appropriate for a reaction to occur.

fuel cell   A voltaic cell in which the reaction is the oxidation of a fuel; the reaction is the equivalent of combustion, but the chemical energy is converted into electrical energy.

fuel value   The energy released during complete combustion of 1 g of a substance.

functional group   An atom or group of atoms within a molecule that imparts characteristic chemical and physical properties to the organic compounds containing that group.

gas (also called a vapor) A state of matter that consists of very widely separated atoms or molecules; gases have neither definite volume nor shape, and they expand to fill their containers.

Geiger counter A portable device for determining nuclear radiation levels by measuring how much the radiation ionizes the gas in a sealed detector.

general gas equation (also called combined gas law) Based on the ideal gas law; it is used when one or more of the four gas variables are held constant while the remaining variables change.

geometric isomers Structures in which molecules with the same connectivity have atoms in different positions due to restricted rotation; in alkenes, rotation about a double bond is restricted.

glycolysis A series of reactions that converts glucose into pyruvate; it is a major anaerobic (no oxygen required) pathway for metabolizing glucose in the cells of almost all living organisms.

glycosidic linkage A C–O–C bond between two sugar molecules.

Graham's law of effusion The rate of effusion of a gas is inversely proportional to the square root of its molecular mass.

gray (Gy) The SI unit used to measure the amount of radiation absorbed by matter; one gray equals the absorption of 1 J by 1 kg of matter.

ground state The lowest energy level available to an electron in an atom, molecule, or ion.

group (also called family) Elements in the same vertical column of the periodic table; elements in the same group have similar chemical properties.

half-life ($t_{1/2}$) of a reaction The time during which the concentration of a reactant decreases by half.

half-reaction One of the two halves of an oxidation–reduction reaction: one half-reaction is the oxidation component, and the other is the reduction component.

halogen family The elements in group 17 of the periodic table.

heat The energy transferred between objects because of a difference in their temperatures.

heat capacity (C_p) The quantity of heat needed to raise the temperature of some specific object 1°C at constant pressure.

heat of reaction (ΔH_{rxn}) (also called enthalpy of reaction) The heat absorbed or released by a chemical reaction.

heat transfer The process of heat energy flowing from one object into another.

Heisenberg uncertainty principle The position and the momentum of an electron in an atom cannot be determined at the same time.

Henderson–Hasselbalch equation Used to calculate the pH of a solution in which the concentrations of an acid and its conjugate base are known.

Henry's law The solubility of a sparingly soluble, chemically unreactive gas in a liquid is proportional to the partial pressure of the gas.

hertz (Hz) The SI unit of frequency, which is expressed in the unit s^{-1}: 1 Hz = 1 s^{-1} = 1 cycle per second (cps).

Hess's law A law that states that the enthalpy change of a reaction that is the sum of two or more reactions is equal to the sum of the enthalpy changes of the constituent reactions.

heteroatom Any atom other than carbon or hydrogen in an organic compound.

heterogeneous catalyst Is in a phase different from that of the reactants.

heterogeneous equilibria Involve reactants and products in more than one phase.

heterogeneous mixture A mixture that has distinct regions of different composition, like particles of silt suspended in water.

heteropolymer A molecule that is made of two or more different monomers.

hexagonal closest-packed (hcp) A crystal structure in which the layers of atoms or ions in hexagonal unit cells have an *ababab...* stacking pattern.

homogeneous catalyst Is in the same phase as the reactants.

homogeneous equilibria Involve reactants and products in the same phase.

homogeneous mixture A mixture of components that are distributed uniformly throughout and that have no visible boundaries or regions.

homologous series A set of related organic compounds that differ from one another by the number of subgroups, such as $-CH_2-$, in their molecular structures.

homopolymer A polymer composed of only one kind of monomer.

Hund's rule The lowest-energy electron configuration of an atom is the one with the maximum number of unpaired electrons in degenerate orbitals, all having the same spin.

hybrid atomic orbital One of a set of equivalent orbitals about an atom created when specific atomic orbitals are mixed.

hybrid orbitals Two *sp* hybrid orbitals form by mixing one *s* and one *p* orbital at an angle of 180° on the hybridized atom.

hybridization In valence bond theory, the mixing of atomic orbitals to generate new sets of orbitals that then are available to overlap and form covalent bonds with other atoms.

hydrocarbons Molecular compounds composed only of hydrogen and carbon; hydrocarbons are a class of organic compounds.

hydrogen bond The strongest kind of dipole–dipole interaction; it occurs between a hydrogen atom bonded to a small, highly electronegative element (O, N) and an atom of oxygen or nitrogen in another molecule; the HF molecule also experiences hydrogen bonds.

hydrogenation The reaction of an unsaturated hydrocarbon with hydrogen.

hydrolysis The reaction of water with another material; the hydrolysis of nonmetal oxides produces acids; the hydrolysis of metal oxides produces bases.

hydronium ion (H_3O^+) The form in which the hydrogen ion is found in an aqueous solution; *hydronium ion*, *hydrogen ion*, and *proton* are synonymous, and all are used to describe the hydrogen ion produced by acids in aqueous solution.

hydrophilic ("water-loving") interaction Attracts water and promotes water solubility.

hydrophobic ("water-fearing") interaction Repels water and diminishes water solubility.

hypothesis A tentative and testable explanation for an observation or a series of observations.

i factor (also called the van't Hoff factor) The ratio of the measured value of a colligative property to the value expected for that property if the solute were molecular.

ideal gas A gas whose behavior is predicted by the linear relationships defined by Boyle's, Charles's, Avogadro's, and Amontons's laws.

ideal gas equation (also called the ideal gas law) Relates the pressure (*P*), volume (*V*), number of moles (*n*), and temperature of an ideal gas, and is expressed as $PV = nRT$, where *R* is a numerical constant called the universal gas constant; its value and units depend on the units used for the variables in the ideal gas equation.

ideal gas law (also called ideal gas equation) Relates the pressure (*P*), volume (*V*), number of moles (*n*), and temperature of an ideal gas, and is expressed as $PV = nRT$, where *R* is a numerical constant called the universal gas constant; its value and units depend on the units used for the variables in the ideal gas equation.

ideal solution One that obeys Raoult's law.

in vitro Latin for "in glass"; refers to materials produced in a laboratory test tube.

in vivo Latin for "in life"; when applied to the origin of a molecule, it means that the molecule was produced in a living system.

induced dipole (also called temporary dipole) A separation of charge produced in an atom or molecule by a momentary uneven distribution of electrons.

inhibitor Diminishes or destroys the ability of a catalyst to speed up a reaction.

initial rate of a reaction The rate at $t = 0$, immediately after the reactants are mixed.

inner coordination sphere of a metal Consists of the ligands that are bound directly to the metal via coordinate covalent bonds.

instantaneous rate of a reaction The rate of a reaction at a specific point on the graph of concentration versus time in the course of the reaction.

integrated rate law A mathematical expression that describes the change in concentration of a reactant with time in a chemical reaction.

intensive property A characteristic of a substance that is independent of the amount of substance present.

interface A boundary between two phases.

interference The interaction of waves that results either in reinforcing their amplitudes or canceling them out.

intermediate A species produced in one step of a reaction and consumed in a following step.

internal energy (*E*) The energy of a thermodynamic system that is the sum of all the kinetic and potential energies of all components of the system.

ion channels Groups of helical proteins that penetrate cell membranes and allow selective transport of cations.

ion exchange A process by which one ion is exchanged for another; as it is usually carried out, 2+ ions in water that contribute to its hardness are removed from the water in exchange for sodium ions.

ion pair A cluster formed when a cation and an anion associate with each other in solution.

ion pumps Large and complex systems of membrane proteins that exchange ions inside the cell with those in the intercellular fluid.

ion–dipole interaction Occurs between an ion and the partial charge of a molecule with a permanent dipole.

ionic bond Results from the electrostatic attraction of a cation for an anion.

ionic compounds Substances composed of positively and negatively charged ions that are held together by electrostatic attraction (forces of attraction between oppositely charged particles).

ionic solids Consist of monatomic or polyatomic ions held together by ionic bonds.

ionization energy (IE) The amount of energy needed to remove 1 mole of electrons from a mole of ground-state atoms or ions in the gas phase.

ionizing radiation The high-energy products of radioactive decay that can ionize substances.

isoelectronic atoms and ions Have identical numbers and configurations of electrons.

isolated system A thermodynamic system that exchanges neither energy nor matter with the surroundings.

isothermal process Takes place at constant temperature.

isotopes Atoms of an element whose nuclei have the same number of protons but different numbers of neutrons.

joule (J) The SI unit of energy; 4.184 J = 1 cal.

K_a **and** K_b The values of the equilibrium constants for weak acids and bases, respectively.

kelvin (K) The SI unit of temperature; the Kelvin scale is an absolute temperature scale, with no negative temperatures.

kerogen A mixture of a large number of liquid organic compounds formed during the conversion of plant matter into crude oil in the absence of oxygen.

ketone Contains a carbonyl group bonded to two R groups; its general formula is RCOR′ where R and R′ may be the same or different organic groups.

kinetic energy (KE) The energy of an object in motion due to its mass (m) and its speed (u): $KE = \frac{1}{2}mu^2$.

kinetic molecular theory of gases A model that describes the behavior of gases; it is based on a set of assumptions, and all the equations that define the relations among pressure, volume, temperature, and number of moles of gases (P, V, T, and n) can be derived from the theory.

Krebs cycle A series of aerobic reactions that continue the oxidation of pyruvate formed in glycolysis.

lattice energy (*U*) The energy of an ionic compound released when 1 mol of the ionic compound forms from its free ions in the gas phase.

law of conservation of energy A law that states that energy cannot be created or destroyed.

law of conservation of mass A law that states that the sum of the masses of the reactants in a chemical reaction is equal to the sum of the masses of the products.

law of constant composition A law that states that all samples of a particular compound always contain the same elements in the same proportions.

law of mass action States that an expression relating the concentrations of products and reactants at chemical equilibrium has a characteristic value at a given temperature when each concentration term in the expression is raised to a power equal to the coefficient of that substance in the balanced chemical equation for the reaction.

law of multiple proportions A law that states that when two elements can combine to form more than one compound, the ratio of the masses of one element, Y, that react with a given mass of another element, X, will be the ratio of two small whole numbers.

Le Châtelier's principle A system at equilibrium responds to a stress in such a way that it relieves that stress.

leveling effect of water Refers to the observation that all strong acids have the same strength in water and are completely converted into solutions of $H_3O^+(aq)$; similarly, strong bases are leveled in water and are completely converted into solutions of $OH^-(aq)$.

Lewis acid A substance that *accepts* a pair of electrons in a chemical reaction.

Lewis base A substance that *donates* a pair of electrons in a chemical reaction.

Lewis structure A two-dimensional representation of a molecule or ion that provides a view of the connections among its atoms; valence electrons are depicted as one or more dots around the atomic symbol.

Lewis symbol The chemical symbol for an atom surrounded by one or more dots representing the valence electrons.

ligand A Lewis base bonded to the central metal ion of a complex ion.

limiting reactant A reactant that is completely consumed in a chemical reaction; the amount of product formed depends on the amount of the limiting reactant available.

linear accelerator A device in which electrostatic forces make charged particles move faster as they pass between metal tubes along a straight flight path.

linear molecular geometry   Results when the bond angle among three atoms is 180°.

lipid bilayer   A double layer of molecules with polar groups interacting with water molecules and nonpolar groups interacting with each other.

lipids   Water insoluble, oily organic compounds that, along with carbohydrates and proteins, are common structural materials in cells.

liquid   A state of matter that consists of atoms or molecules in close proximity to each other (usually not as close as in a solid); liquids occupy a definite volume, but flow to assume the shape of their containers.

London forces (also called dispersion forces)   Intermolecular forces caused by the presence of temporary dipoles in molecules.

lone pair   A pair of electrons in a molecule or ion that is not shared (also called unshared pair).

macrocycle   A ring containing multiple electron-pair donors that bind to a metal ion.

magnetic quantum number (m_ℓ)   An integer that may have any value from $-\ell$ to $+\ell$; it defines the orientation of an orbital in space.

main group elements (also called representative elements)   The elements in groups 1, 2, and 13 through 18 of the periodic table.

major elements   Elements present in the body in average concentrations greater than 1 mg/g.

manometer   An instrument for measuring the pressure exerted by a gas.

mass   A property that defines the quantity of matter in an object; mass is measured with a balance.

mass action expression (also called equilibrium constant expression)   Describes the relationship between the concentration (or partial pressure) terms of reactants and products when a system is at equilibrium.

mass defect (Δm)   The difference between the mass of a stable nucleus and the masses of the individual nucleons that comprise it.

mass number (A)   The total number of nucleons (the sum of the numbers of protons and neutrons) in one atom of an element.

mass spectrometer   An instrument that measures precise masses and relative amounts of ions of atoms and molecules.

mass spectrum   A graph of the data from a mass spectrometer, where mass-to-charge (m/Z) ratios of the deflected particles are plotted against the number of particles with a particular mass; because the charge on the ions typically is 1+, $m/Z = m/1 = m$, the mass of the particle may be read directly from the m/Z axis.

matter   The material of which the universe is made; all matter has mass and occupies space.

matter wave   The wave associated with any particle.

melting point   The temperature at which a solid transforms into a liquid at the same rate at which the liquid transforms into the solid; the melting point of a substance is identical to its freezing point.

meniscus   The curved surface of a liquid.

messenger RNA (mRNA)   The polynucleotide that carries the code for synthesizing a protein from the DNA in the nucleus to the cytoplasm outside the nucleus.

metallic bond   Consists of the nuclei of metal atoms surrounded by a "sea" of shared electrons.

metalloids (also called semimetals)   Elements that separate the metals in a row of the periodic table from the nonmetals in that row; metalloids have some metallic and some nonmetallic properties.

metals   The elements on the left-hand side of the periodic table; they are typically shiny solids that conduct heat and electricity well and are malleable and ductile.

methanogenic bacteria   Rely on simple organic compounds and hydrogen for energy, unlike most organisms on Earth that rely on complex organic compounds and oxygen; their respiration produces methane, carbon dioxide, and water, depending on the compounds they consume.

methyl group (–CH$_3$)   A structural unit that can make only one bond; methyl groups end hydrocarbon chains.

methylene group (–CH$_2$–)   A structural unit that can make two bonds.

microstate   A unique distribution of particles among energy levels.

millimeters of mercury (mmHg) (also called torr)   A measurement of atmospheric pressure, where 1 atm = 760 mmHg = 760 torr.

millimole 10^{-3} mol; it is a useful quantity in calculations related to titrations.

miscible liquids   Two or more form a homogeneous solution when mixed in any proportion.

mixture   A combination of pure substances in variable proportions in which the individual substances retain their chemical identities.

mmHg (also called torr)   Millimeters of mercury, a measurement of atmospheric pressure, where 1 atm = 760 mmHg = 760 torr.

model (also called a scientific theory)   A general explanation of widely observed phenomena that has been extensively tested.

molality   Concentration expressed as the number of moles of solute per kilogram of solvent.

molar heat capacity (c)   The heat required to raise the temperature of 1 mol of a substance by 1°C; when the system is at constant pressure, the symbol used is c_P.

molar heat of fusion (ΔH_{fus})   The heat required to convert 1 mol of a solid substance to 1 mol of liquid at its melting point.

molar heat of vaporization (ΔH_{vap})   The heat required to convert 1 mol of a liquid substance to 1 mol of vapor at its boiling point.

molar mass ($\mathcal{M}$)   The mass of 1 mol of the particles that comprise a substance; the molar mass of an element, in grams per mole, is numerically the same as that element's average atomic mass, in atomic mass units.

molar volume   For an ideal gas, the volume occupied by 1 mol of a gas at STP equal to 22.4 L.

molarity (M)   A unit of concentration that is the amount of solute (in moles) divided by the volume of solution (in liters): $M = n/V$; a 1.0 M solution contains 1.0 mol of solute per liter of solution.

mole (mol)   A unit of measure of particles (atoms, ions, or molecules) that contains Avogadro's number ($N_A = 6.022 \times 10^{23}$) of the particles.

mole fraction (χ_x) of a substance   The ratio of the number of moles of a component in a mixture to the total number of moles in the mixture.

molecular compounds   Substances composed of atoms held together in molecules by covalent bonds.

molecular equation   A balanced equation that describes a reaction in solution in which the reactants are written as undissociated molecules.

molecular formula   A formula that describes the exact number and type of atoms in one molecule of a compound.

molecular geometry   Defined by the lowest energy arrangement of its atoms in three-dimensional space (also called shape of a molecule).

molecular ion (M$^+$)   An ion formed in a mass spectrometer when a neutral molecule loses an electron after bombardment with a high-energy beam. The molecular ion has a charge of 1+ and has essentially the same molecular mass as the neutral molecule from which it came.

molecular mass   The mass of one molecule of a molecular compound.

molecular orbital   A region of characteristic shape and energy where electrons in a molecule are located.

molecular orbital diagram   A diagram that shows the relative energies and electron occupancy of the molecular orbitals of a molecule (also called energy level diagram).

molecular orbital (MO) theory   A bonding theory based on mixing similarly shaped atomic orbitals and energies to form molecular orbitals that belong to the molecule as a whole.

molecular solids Nonmetals that form crystalline solids consisting of neutral, covalently bonded molecules held together by intermolecular forces.

molecularity Refers to the number of particles (ions, atoms, or molecules) that collide in an elementary step in a reaction mechanism.

molecule A collection of atoms chemically bonded together; a molecule is the smallest entity that contains the constituent atoms of a compound in its characteristic proportions.

monodentate ligand A species that forms one single coordinate bond to a metal ion in a complex.

monomers Small molecules that bond together to form polymers; the root "*meros*" is Greek for "part" or "unit," so a monomer is one unit of a number of subgroups, such as $-CH_2-$, in a polymer's molecular structures.

monoprotic acids Acids that have one ionizable hydrogen atom per molecule.

monosaccharide A simple sugar and the simplest carbohydrate.

natural abundance The proportion of an isotope relative to all of the isotopes of that element as found in a natural sample, usually expressed as a percentage; the total abundances for isotopes should sum to 100% (or very close to it, allowing for measurement error).

Nernst equation Relates the potential of a cell (or half-cell) reaction to its standard potential ($E°$) and the concentrations of its reactants and products.

net ionic equation A balanced equation that describes the actual reaction taking place in aqueous solution; the net ionic equation eliminates spectator ions from the overall ionic equation.

neutralization reaction A reaction that takes place when an acid reacts with a base and produces a solution of a salt in water.

neutron An electrically neutral or uncharged subatomic particle found in the nucleus of an atom.

neutron capture The absorption of a neutron by a nucleus.

noble gases The elements in group 18 of the periodic table.

node A location in a standing wave that experiences no displacement.

nonelectrolyte A substance that does not form ions when it dissolves and does not enhance the conductivity of water.

nonessential elements Elements that are present in an organism but have no known function. Nonessential elements may have a stimulatory effect when added to the diet of an organism, leading to increased growth or other biological responses.

nonmetals Elements that have properties opposite those of metals; they are poor conductors of heat and electricity, and they range in character from brittle solids to gases.

nonpolar molecule Contains bonds that have an even distribution of charge; electrons in the bonds are shared equally by the two atoms; pure covalent bonds give rise to nonpolar diatomic molecules.

nonspontaneous process Occurs only as long as energy is continually added to the system; nonspontaneous processes are the reverse of spontaneous ones.

normal boiling point The temperature at which the vapor pressure of a liquid equals 1 atmosphere (760 torr).

n-type semiconductor Contains excess electrons contributed by electron-rich dopant atoms.

nuclear chemistry The study of reactions that involve changes in the nuclei of atoms.

nuclear fission The process in which the nucleus of a heavy element splits into two lighter nuclei; the reaction is usually accompanied by the release of one or more neutrons and energy.

nucleons The protons and neutrons in atomic nuclei.

nucleosynthesis The fusion of fundamental and subatomic particles to create atomic nuclei.

nucleotide A monomer unit from which DNA and RNA are made.

nucleus The protons and neutrons that make up the center of an atom; the nucleus contains all the positive charge and nearly all the mass in an atom.

nuclide The nucleus of a specific isotope of an element, also often used as a synonym for isotope.

octahedral A steric number of 6 for the central atom of a molecule with no lone pairs gives rise to this arrangement in which all six sites are equivalent; the ideal bond angle for the atoms is 90°.

octet A set of eight electrons in the outermost (valence) shell of an atom or monoatomic ion.

octet rule Atoms of main group elements make bonds by gaining, losing, or sharing electrons to achieve an outer shell containing eight electrons, or four electron pairs.

oils Liquid triglycerides containing primarily unsaturated fatty acids.

oligomers Molecules that contain a few mers; "oligos" is Greek for "few." The boundary between small molecules and polymers is not distinct, and oligomers occupy the middle ground.

open system A thermodynamic system that exchanges both energy and matter with the surroundings.

optical isomers Molecules that are not superimposable on their mirror images.

optically active molecules Rotate a beam of plane-polarized light; the *levorotary* form of the enantiomers rotates the beam to the left and a (–) sign precedes the chemical name; the *dextrorotary* form rotates the beam to the right and a (+) sign precedes the name.

orbital diagrams One way of showing the arrangement of electrons in an atom or ion using boxes to represent orbitals.

orbital penetration Occurs when an electron in an outer orbital has some probability of being as close to the nucleus as an electron in an inner shell.

orbitals Defined by the square of the wave function (ψ^2), these are regions around the nucleus of an atom where the probability of finding an electron is high; each orbital is identified by a unique combination of three integers called quantum numbers.

order of a reaction An experimentally determined number that defines the dependence of the reaction rate on the concentration of a reactant.

ores Naturally occurring compounds or mixtures of compounds from which elements can be extracted.

organic chemistry The study of the compounds of carbon, in which carbon atoms are bound to other carbon atoms, to hydrogen atoms, and to atoms of other elements.

organic compounds Most compounds containing carbon; organic compounds commonly include certain other elements such as hydrogen, oxygen, and nitrogen.

organometallic compounds A class of compounds containing direct carbon–metal covalent bonds.

osmosis In osmosis, solvent passes through a semipermeable membrane to balance the concentration of solutes in solutions on both sides of the membrane; flow proceeds from the more dilute into the more concentrated solution.

osmotic pressure (π) The pressure that has to be applied across a semipermeable membrane to stop the flow of solvent from the compartment containing pure solvent or a less concentrated solution toward a more concentrated solution; the osmotic pressure of a solution increases with the solute concentration M and the solution temperature T.

overall ionic equation A balanced equation that shows all the species, both ionic and molecular, in a reaction in aqueous solution.

overall order of a reaction The sum of the exponents of the concentration terms in the rate law for a reaction.

overlap When orbitals on different atoms occupy the same region in space; valence bond theory assumes that covalent bonds form when this occurs.

overpotential The additional potential above the theoretically calculated value required to promote an electrode reaction in electrolysis.

oxidation A chemical change in which a species loses electrons; the oxidation number of the species increases.

oxidation number (O.N.) (also called oxidation state) A positive or negative number assigned to an element in a molecule or ion according to the number of electrons each atom of that element either gains or loses when it forms an ion or shares when it forms a covalent bond with another element; a pure element has an oxidation number of zero.

oxidation state (also called oxidation number [O.N.]) A positive or negative number assigned to an element in a molecule or ion according to
the number of electrons each atom of that element either gains or loses when it forms an ion or shares when it forms a covalent bond with another element; a pure element has an oxidation number of zero.

oxidizing agent A substance in a redox reaction that accepts electrons from another species, thereby oxidizing that species (increasing its oxidation number); the oxidizing agent itself is reduced (its oxidation number decreases).

oxoanions Polyatomic ions that contain oxygen in combination with one or more other elements.

paramagnetic atoms, ions, and molecules Contain at least one unpaired electron and are attracted by an external magnetic field; the strength of the attraction increases as the number of unpaired electrons increases.

partial pressure The contribution to the total pressure made by a component gas in a gas mixture.

particles of matter and particles of their antimatter With opposite charges, undergo mutual annihilation, producing high-energy gamma (γ) rays.

Pauli exclusion principle No two electrons in an atom can have the same set of four quantum numbers.

peptide A compound of two or more amino acids joined by peptide bonds; small peptides are described by indicating the number of amino acids linked together to form the molecule: a *dipeptide* consists of two amino acids; a *tripeptide*, three; short chains of up to 20 amino acids are *oligopeptides*; and the term *polypeptide* is used for chains longer than 20 amino acids.

peptide bond Results from a condensation reaction between the carboxylic acid group of one amino acid with the amine group of another; a molecule of water is also formed in the process.

percent composition The percentage by mass of each element in a compound; it can also refer to the percentage by mass of each component in a mixture.

percent ionization The degree of ionization expressed as a percentage.

percent yield The ratio, expressed as a percentage, of the actual yield of a chemical reaction to the theoretical yield.

periodic table A table that includes all the elements and presents them in the order of their atomic numbers.

periods The horizontal rows of the periodic table.

permanent dipole moment (μ) A measured value that defines the extent of separation of positive and negative charge centers in a covalently bonded molecule; it is a quantitative expression of the polarity of a molecule.

pH The negative logarithm of the hydrogen ion concentration in an aqueous solution.

pH buffer A solution of acidic and basic solutes that resists changes in its pH when acids or bases are added to it.

pH indicator A weak water-soluble organic acid (HIn) that changes color when it ionizes, for example, $HIn(aq) \rightarrow H^+(aq) + In^-(aq)$.

pH scale The logarithmic scale for expressing the acidity or basicity of a solution.

phase diagram A graphical representation of the dependence of the stabilities of the physical states of a substance on temperature and pressure.

phospholipids Consist of glycerol with two fatty acid chains and one polar group containing a phosphate; they are major constituents in cell membranes.

phosphorylation reaction Results in the addition of a phosphate group to an organic molecule.

photochemical smog A mixture of gases formed in the lower atmosphere when sunlight interacts with compounds produced by internal combustion engines.

photoelectric effect Occurs when light strikes a metal surface and an electric current (a flow of electrons) is produced.

photon A quantum of electromagnetic radiation.

physical property A characteristic of a substance that can be observed without changing it into another substance.

pi (π and π^*) molecular orbitals Form by mixing atomic orbitals that are oriented above and below, or in front of or behind, the bonding axis in a molecule; electrons occupying π orbitals form π bonds.

Planck's constant (h) The proportionality constant between the energy and frequency of electromagnetic radiation, as expressed in the relationship $E = h\nu$; its value is $6.6260755 \times 10^{-34}$ J·s, which we typically round to 6.626×10^{-34} J·s.

pOH The negative logarithm of the hydroxide ion concentration in an aqueous solution.

polar covalent bond Results from unequal sharing of bonding pairs of electrons between atoms.

polar molecule Contains bonds that have an uneven distribution of charge because electrons in the bonds are not shared equally by the two atoms.

polarizability Describes the relative ease with which an electron cloud is distorted by an external charge.

polyatomic ions Charged groups consisting of two or more atoms joined together by covalent bonds.

polydentate ligand A species that can form more than one coordinate bond per molecule.

polymers Very large molecules with high molar masses; polymer literally means "many units" (also called macromolecules).

polyprotic acids Acids that have two or more ionizable hydrogen atoms per molecule.

polysaccharide (also called complex carbohydrate) A polymer of monosaccharides.

porphyhrin A type of tetradentate macrocyclic ligand.

positron A particle with the mass of an electron but has a positive charge.

positron emission The spontaneous emission of a positron from a nucleus.

potential energy (PE) The energy stored in an object because of its position.

potential of the standard hydrogen electrode (SHE) Defined as 0.000 V, the SHE serves as a reference electrode against which other electrode half-reactions can be measured.

precipitate A solid product formed from a reaction in solution.

precision The repeatability of a measurement and the extent to which repeated measurements agree among themselves.

pressure (P) The ratio of force to surface area; pressure is measured in millimeters of mercury (mmHg), also known as torr, where 1 atm = 760 mmHg = 760 torr.

pressure–volume (PΔV) work The work associated with the expansion or compression of gases.

primary structure of a protein The sequence in which the amino acid monomers occur in the polymer chain.

principal quantum number (n) A positive integer that describes the relative size and energy of an atomic orbital or group of orbitals in an atom.

products Substances formed during a chemical reaction.

proteins Polymers of amino acids.

proton A positively charged subatomic particle in the nucleus of an atom.

pseudo-first-order reaction A reaction in which all the reactants but one are present at such high concentrations that they do not decrease significantly during the course of the reaction, so that reaction rate is controlled by the concentration of the limiting reactant.

p-type semiconductor Contains electron-poor dopant atoms that cause a reduction in the number of electrons; this is equivalent to the presence of positively charged holes.

pure substance A substance that has the same physical and chemical properties independent of its source.

quantized Something whose values are restricted to whole-number multiples of a specific base value; the base unit of energy is the quantum.

quantum The smallest discrete quantity of a particular form of energy.

quantum mechanics (also called wave mechanics) The description of the wavelike behavior of particles on the atomic level.

quantum theory Based on the idea that energy is absorbed and emitted in discrete quanta.

quarks Elementary particles that combine to form neutrons and protons.

quaternary structure Results when two or more proteins associate to make a larger structure that functions as a single unit.

R In a general formula, R stands for an organic group that has one available bond; it is used to indicate the variable part of a structure so that the focus is placed on the functional group.

racemic mixture Contains equal amounts of isomers of an optically active material and therefore does not rotate the plane of polarized light.

radioactive decay The spontaneous disintegration of unstable particles accompanied by the release of radiation.

radiocarbon dating A method of establishing the approximate age of carbon-containing materials by measuring the amount of radioactive carbon-14 remaining in the samples.

random coil An irregular pattern in the secondary structure of a protein.

Raoult's law The vapor pressure of a solution containing nonvolatile solutes is proportional to the mole fraction of the solvent.

rate constant The proportionality constant that relates the rate of a reaction to the concentrations of reactants.

rate law for a chemical reaction An equation that defines the experimentally determined relationship between the concentration of reactants and the rate of a reaction.

rate-determining step The slowest step in a multistep chemical reaction.

reactants Substances consumed during a chemical reaction.

reaction mechanism A set of steps proposed to describe how a reaction occurs; the mechanism must be consistent with the rate law for the reaction and its stoichiometry.

reaction quotient (Q) The numerical value of the mass action expression for any values of the concentrations (or partial pressures) of reactants and products; at equilibrium, the value of the reaction quotient (Q) equals that of the equilibrium constant (K).

reaction rate Describes how rapidly a reaction occurs; it is related to rates of change in the concentrations of reactants and products over time.

receptor A protein that binds another molecule because of its unique shape; binding to a receptor typically causes some biological response.

recognition The process in which one molecule fits into a three-dimensional site on another molecule.

reducing agent A substance in a redox reaction that gives up electrons to another species, thereby reducing that species (reducing its oxidation number); the reducing agent itself is oxidized (its oxidation number increases).

reduction A chemical change in which a species gains electrons; the oxidation number of the species decreases.

refraction The bending of light as it passes from one medium to another with a different density.

relative biological effectiveness (RBE) A factor that accounts for the differences in physical damage caused by different types of radiation.

renewable resource One whose supply, in theory, is not limited; trees and agricultural crops can be grown anew every year, and animal waste is produced constantly.

replication The process by which one double-stranded DNA forms two new DNA molecules; each one contains one strand from the original molecule and one new strand.

representative elements (also called main group elements) The elements in groups 1, 2, and 13 through 18 of the periodic table.

resonance Occurs when two or more equivalent Lewis structures can be drawn for one compound.

resonance structure One of two or more Lewis structures with the same arrangement of atoms but different arrangements of electrons in a molecule.

reverse osmosis A water purification process in which water is forced through semipermeable membranes, leaving dissolved impurities behind.

root-mean-square speed (u_{rms}) The square root of the average of the squared speeds of all molecules in a population of gas molecules; a molecule possessing the average kinetic energy moves at this speed.

salt The product of a neutralization reaction; a salt is made up of the cation characteristic of the base and the anion characteristic of the acid in that reaction.

saturated hydrocarbon A hydrocarbon in which each carbon atom is bonded to four other atoms.

saturated solution A solution containing the maximum possible concentration of a solute at a given temperature.

scanning tunneling microscope (STM) An instrument that generates images of surfaces on the atomic scale.

Schrödinger wave equation Describes the electron in hydrogen as a matter wave and indicates how it varies with location and time around the nucleus; solutions of the wave equation are the energy levels of the hydrogen atom.

scientific method An approach to acquiring knowledge that is based on carefully observing phenomena; developing a simple, testable explanation; and conducting additional experiments that test the validity of the hypothesis.

scientific theory (also called a model) A general explanation of widely observed phenomena that has been extensively tested.

scintillation counters Instruments that determine the level of radioactivity in samples by measuring the intensity of light emitted by phosphors in contact with the samples.

screening (also called shielding) The effect when inner-shell electrons protect outer-shell electrons from experiencing the total nuclear charge.

second law of thermodynamics The total entropy of the universe increases in any spontaneous process.

secondary structure Describes the pattern of arrangement of segments of a protein chain.

semiconductor A substance whose conductivity can be made to vary across several orders of magnitude by altering its chemical composition.

semimetals (also called metalloids) Elements that separate the metals in a row of the periodic table from the nonmetals in that row; metalloids have some metallic and some nonmetallic properties.

sequestering agent A ligand such as EDTA that forms very stable complexes with metal ions and thereby separates them from other substances with which they would otherwise react.

shielding (also called screening) The effect when inner-shell electrons protect outer-shell electrons from experiencing the total nuclear charge.

sievert (Sv) The SI unit used to measure the amount of biological damage caused by ionizing radiation.

sigma (σ) bond A covalent bond in which the highest electron density lies between the two atoms along the bond axis connecting them.

sigma (σ) molecular orbital The region of highest electron density that lies along the bond axis between two nuclear centers; electrons in σ molecular orbitals form sigma (σ) bonds.

significant figures The number of digits in a measured value signifying the precision of the measurement; they include all the digits with known values plus the first uncertain digit to the right of the known digits; the greater the number of digits, the greater the certainty with which the value is known.

simple cubic A unit cell that has only atoms at the eight corners of a cube.

single bond Results when two atoms share one pair of electron.

solid A state of matter that consists of atoms or molecules in close contact with each other and often in an organized arrangement; solids have a definite shape and volume.

solubility product, K_{sp} (also called solubility-product constant) An equilibrium constant that describes the formation of a saturated solution of a slightly soluble salt.

solubility-product constant (also called solubility product, K_{sp}) An equilibrium constant that describes the formation of a saturated solution of a slightly soluble salt.

solutes The components of a solution that are present in smaller amounts than the solvent; a solution may contain one or more solutes.

solution A homogeneous mixture of two or more substances. Solutions are often liquids, but they may also be solids or gases.

solvent The component of a solution that is present in the largest amount; in an *aqueous* solution, water is the solvent.

sp hybrid orbitals Two hybrid orbitals at an angle of 180° on the hybridized atom formed by mixing one *s* and one *p* orbital.

sp² hybrid orbitals Three equivalent orbitals formed by mixing one *s* and two *p* orbitals, which achieves trigonal planar orientation of valence electrons.

sp³d hybrid orbitals The five equivalent hybrid orbitals with lobes that point toward the vertices of a trigonal bipyramid formed by mixing one *s* orbital, three *p* orbitals, and one *d* orbital from the same shell.

sp³d² hybrid orbitals The six equivalent hybrid orbitals that point toward the vertices of an octahedron formed by mixing one *s* orbital, three *p* orbitals, and two *d* orbitals from the same shell.

specific heat (c_s) The heat required to raise the temperature of 1 g of a substance 1°C at constant pressure.

spectator ions Ions that are present in, but remain unchanged by, the reaction taking place.

spectrochemical series A list of ligands rank-ordered on their abilities to split the energies of the *d* orbitals of transition metal ions.

sphere of hydration The cluster of water molecules that surrounds an ion in an aqueous medium; the general term applied to such a cluster in any solvent is sphere of solvation.

spin magnetic quantum number (m_s) For an electron in an atom, m_s is either $+\frac{1}{2}$ or $-\frac{1}{2}$, indicating that the electron–spin orientation is either up or down.

spontaneous process Proceeds in a given direction without outside intervention

stability of a nucleus Proportional to its binding energy (BE), a measure of the energy released when nucleons combine to form a nucleus; it is calculated using the equation $E = (\Delta m)c^2$, where Δm is the mass defect of the nucleus.

standard atmosphere (1 atm) The pressure that can support a column of mercury 760 mm high in a barometer; it is also called simply 1 atmosphere.

standard cell potential (E°_{cell}) The electromotive force produced by an electrochemical cell when all reactants and products are in their standard states.

standard enthalpy of formation (ΔH°_f) The enthalpy change during a formation reaction under standard conditions (also called standard heat of formation or simply enthalpy of formation or heat of formation).

standard enthalpy of reaction (ΔH°_{rxn}) (also called the standard heat of reaction) The enthalpy change of a reaction that takes place under standard conditions.

standard free energy of formation (ΔG°_f) The change in the free energy associated with the formation of 1 mol of a compound in its standard state from its elements.

standard heat of reaction (ΔH°_{rxn}) (also called standard enthalpy of reaction) The enthalpy change of a reaction that takes place under standard conditions.

standard molar entropy The absolute entropy of 1 mol of a substance at its standard state.

standard potential (E°) (also called standard reduction potential [E°_{red}]) The electromotive force of a half-reaction written as a reduction in which all reactants and products are in their standard states at 25°C; this means that the concentrations of all dissolved substances are 1 M, and the partial pressures of all gases are 1 bar.

standard reduction potential (E°_{red}) (also called standard potential [E°]) The electromotive force of a half-reaction written as a reduction in which all reactants and products are in their standard states at 25°C; this means that the concentrations of all dissolved substances are 1 M, and the partial pressures of all gases are 1 bar.

standard solution A solution of known concentration.

standard state The most stable form of a substance under 1 bar pressure and some specified temperature (assumed to be 25.0°C unless otherwise stated).

standard temperature and pressure (STP) In the United States, a temperature of 0°C and a pressure of 1 atm. Elsewhere, STP is defined as 0°C and a pressure of 1 bar.

standing wave A wave confined to a given space whose wavelength is related to the length (L) of the space by the equation $L = n(\lambda/2)$, where n is a whole number.

state function A property of an entity based solely on its chemical or physical state or both, but not on the way it achieved that state.

steric number (SN) For a central atom in a molecule or ion, the number of atoms bonded to the central atom plus the number of lone pairs of electrons on the central atom.

stimulatory effect Nonessential elements may have this when added to the diet of an organism, leading to increased growth or to other biological responses.

stock solution A concentrated solution of a substance used to prepare solutions of lower concentration.

stoichiometric yield (also called the theoretical yield) The amount of product expected in a chemical reaction from a specific quantity of reactant.

stoichiometry The quantitative relationship between the quantities of reactants and products in a chemical reaction.

straight-chain alkane A hydrocarbon in which the carbon atoms are bonded together in one continuous line; linear hydrocarbon chains have a methyl group at each end with methylene groups connecting them.

strong acid An acid that is completely ionized in aqueous solution.

strong base A base that is completely ionized in aqueous solution.

strong electrolyte A substance that dissociates completely into ions when it dissolves in water.

structural isomers Have the same formula but different structures; they are different compounds and have different chemical and physical properties.

subatomic particles The particles that make up an atom; they include neutrons, protons, and electrons.

sublimation The direct conversion of a solid to a gas (vapor) without an intermediate liquid phase.

substrate The reactant that binds to the active site in an enzyme-catalyzed reaction.

superconductor A material that has zero resistance to the flow of electric current.

supercritical fluid A substance at conditions above its critical temperature and pressure, where the liquid and vapor phases are indistinguishable; it has some characteristics of both liquid and gas.

supersaturated solution A solution that contains more solute than the predicted quantity that is soluble in a given volume of solution at a given temperature.

surface tension The energy needed to separate the molecules in a unit area at the surface of a liquid.

surroundings Everything that is not part of the thermodynamic system.

synthetic polymers Macromolecules made in the laboratory and often produced industrially for commercial use.

system The part of the universe that is the focus of a thermodynamic study.

temporary dipole (also called induced dipole) A separation of charge produced in an atom or molecule by a momentarily uneven distribution of electrons.

termolecular step A step in a mechanism or reaction that involves a collision among three molecules.

tertiary structure The three-dimensional, biologically active structure of a protein that arises because of attractions among the R groups on the amino acids.

tetrahedral A structure of atoms about a central atom with ideal bond angles of 109.5°; corresponds to a steric number of 4 for the central atom of a molecule with no lone pairs.

theoretical yield (also called stoichiometric yield) The amount of product expected in a chemical reaction from a specific quantity of reactant.

thermal energy The kinetic energy of atoms and molecules.

thermal equilibrium A condition in which temperature is constant throughout a material and no heat flow occurs from point to point.

thermochemical equation The chemical equation of a reaction that includes heat as a reactant or a product.

thermochemistry The study of the relationship between chemical reactions and changes in heat energy.

thermodynamics The study of energy and its transformations.

third law of thermodynamics The entropy of a perfect crystal is zero at absolute zero.

threshold frequency (ν_0) The minimum frequency of light required to produce the photoelectric effect.

titrant The standard solution added to the sample in a titration.

titration An analytical method for precisely determining the concentration of a solute in a sample by reacting it with a standard solution of known concentration.

torr (also called millimeters of mercury [mmHg]) A measurement of atmospheric pressure, where 1 atm = 760 mmHg = 760 torr.

trace elements Elements present in the body in average concentrations between 1 and 1,000 $\mu g/g$.

trans isomer (also called E isomer) An isomer in an alkene that has two like groups (such as two R groups or two hydrogen atoms) on opposite sides of a line drawn through the double bond; trans is Latin for "across"; E comes from the German word entgegen (opposite).

transcription The process of copying the information in DNA to RNA.

transfer RNA (tRNA) The polynucleotide that delivers amino acids, one at a time, to polypeptide chains being assembled by the ribosome–mRNA complex.

transition metals The elements in groups 3 through 12 of the periodic table.

transition state A high-energy state between reactants and products in a chemical reaction.

translation The process of assembling proteins from the information encoded in RNA.

triglycerides Fatty acid esters formed between glycerol and long-chain fatty acids.

trigonal bipyramid For the central atom of a molecule with no lone pairs, corresponds to a steric number of 5, in which three atoms occupy equatorial sites in the plane around the central atom with ideal bond angles of 120° and two other atoms occupy axial sites above and below the central atom with an ideal bond angle of 180°.

trigonal planar Structure with bond angles of 120° resulting when the central atom of a molecule with no lone pairs has a steric number of 3.

triple bond Results when two atoms share three pairs of electrons.

triple point The temperature and pressure where all three phases of a substance coexist; freezing and melting, boiling and liquefaction, and sublimation and deposition all proceed at the same rate, so no net change takes place in the system.

ultratrace elements Elements present in the body in average concentrations less than 1 $\mu g/g$.

unimolecular step A step in a mechanism or reaction that involves only one molecule.

unit cell The basic repeating unit of the arrangement of atoms, ions, or molecules in a crystalline solid.

universal gas constant In the ideal gas equation, the numerical constant R, whose value and units depend on the units used for the variables in the ideal gas equation.

unsaturated hydrocarbons Hydrocarbons that contain one or more carbon–carbon double or carbon–carbon triple bonds and therefore contain less than the maximum amount of hydrogen possible per carbon atom. Alkenes and alkynes are unsaturated hydrocarbons.

valence bands Bands of orbitals that are filled or partially filled by valence electrons.

valence bond theory Assumes that covalent bonds form when orbitals on different atoms overlap or occupy the same region in space.

valence electrons The electrons in the outermost shell of an atom that have the most influence on the atom's chemical behavior.

valence-shell electron-pair repulsion (VSEPR) A model that predicts the arrangement of valence electron pairs around a central atom that minimizes their mutual repulsion to produce the lowest energy orientations; VSEPR theory may be applied to covalently bound molecules and polyatomic ions of main group elements.

van der Waals equation Attempts to account for the behavior of real gases by including experimentally determined factors that quantify the contributions of molecular volume and intermolecular interactions to the properties of gases.

van der Waals forces Frequently used to refer collectively to both types of attractive forces: dipole–dipole interactions and dispersion forces.

van't Hoff factor (also called the *i* factor) The ratio of the measured value of a colligative property to the value expected for that property if the solute were molecular.

vapor (also called a gas) A state of matter that consists of very widely separated atoms or molecules; gases have neither definite volume nor shape, and they expand to fill their containers.

vapor pressure The force exerted at a given temperature by a vapor in equilibrium with its liquid phase.

vaporization (also called evaporation) The transformation of molecules in the liquid phase into the gas phase.

vinyl group The subgroup $CH_2{=}CH{-}$.

vinyl polymers The family of polymers formed from monomers containing the subgroup $CH_2{=}CH{-}$.

viscosity A measure of the resistance of a fluid to flow.

voltaic cell A cell that transforms chemical energy into electrical energy by a spontaneous redox reaction.

volume (*V*) The space occupied by matter.

wave function (ψ) A solution of the Schrödinger equation.

wave mechanics (also called quantum mechanics) The description of the wavelike behavior of particles on the atomic level.

wavelength (λ) The distance from crest to crest or trough to trough on a wave.

weak acid An acid that is a weak electrolyte because it ionizes only partially in aqueous solution; it has a limited capacity to donate protons to the medium.

weak base A base that is a weak electrolyte because it ionizes only partially in aqueous solution; it has a limited capacity to accept protons in the medium.

weak electrolyte A substance that ionizes only partly when it dissolves in water.

work The energy required to move an object through a given distance.

work function (Φ) The amount of energy needed to dislodge an electron from the surface of a metal.

X-ray diffraction (XRD) A technique for determining the arrangement of atoms or ions in a crystal by analyzing the pattern that results when X-rays are scattered after bombarding the crystal.

Z isomer (also called *cis* isomer) An isomer that has two like groups (such as two R groups or two hydrogen atoms) on the same side of a line drawn through the double bond; *cis* is Latin for "on this side"; Z comes from the German word *zusammen* (together).

zeolites Natural crystalline minerals or synthetic materials consisting of three-dimensional networks of channels that contain sodium or other 1+ cations; they may function as ion exchangers.

zwitterion A molecule that has both positively and negatively charged groups in its structure.

Answers to Concept Tests and Practice Exercises

Chapter 1

CONCEPT TESTS

p. 14. Yes
p. 17. (b) decreasing
p. 21. 1. 3 significant figures; 2. 3 significant figures; 4. 4 significant figures
p. 22. Fails to disprove: The density value matches that of gold, but density alone does not prove conclusively that it *is* gold.
p. 25. Compare its melting point or chemical properties to those of fool's gold.

PRACTICE EXERCISES

1.1. (a) physical; (b) physical; (c) physical; (d) chemical
1.2. (a) gas to liquid (condensation); (b) liquid to gas (vaporization)
1.3. 1.14 (3 significant figures)
1.4. (a) exact; (b) inherent uncertainty; (c) inherent uncertainty; (d) inherent uncertainty; (e) exact
1.5. 0.324 km; 3.24×10^4 cm
1.6. 1.5×10^2 cm
1.7. 9.45×10^{12} km
1.8. $-233°C = 40\,K = -387°F$; $123°C = 396\,K = 253°F$

Chapter 2

CONCEPT TESTS

p. 55. Any two of the following element pairs: Ar/K, Co/Ni, Cu/Zn, Th/Pa, Sb/Te, Pu/Am, Sg/Bh, Lr/Rf, U/Np

PRACTICE EXERCISES

2.1. 40 g oxygen
2.2. (a) ^{56}Fe; (b) ^{15}N; (c) ^{37}Cl; (d) ^{39}K
2.3. $^{107}Ag = 51.5\%$; $^{109}Ag = 48.5\%$
2.4. (a) As, arsenic; (b) Ca, calcium; (c) Hg, mercury; (d) S, sulfur
2.5. (a) molecular; (b) molecular; (c) molecular; (d) molecular; (e) ionic
2.6. (a) tetraphosphorus decoxide; (b) carbon monoxide; (c) nitrogen trichloride
2.7. (a) $SrCl_2$; (b) MgO; (c) NaF; (d) $CaBr_2$
2.8. $MnCl_2$; MnO_2
2.9. (a) $Sr(NO_3)_2$; (b) K_2SO_3
2.10. (a) calcium phosphate; (b) magnesium perchlorate; (c) lithium nitrite; (d) sodium hypochlorite; (e) potassium permanganate
2.11. (a) hypochlorous acid; (b) chlorous acid; (c) carbonic acid
2.12. $_{33}As + {}_0^1n \rightarrow {}_{34}x + {}_{-1}^0\beta \qquad x = Se$

Chapter 3

CONCEPT TESTS

p. 85. 1 oz Ag

PRACTICE EXERCISES

3.1. 1.5×10^{10} atoms Au
3.2. 0.0765 mol
3.3. 49.2 g
3.4. CO_2 44.01 g/mol; O_2 32.00 g/mol; $C_6H_{12}O_6$ 180.16 g/mol
3.5. (a) $P_4(s) + 5\,O_2(g) \rightarrow P_4O_{10}(s)$; (b) $P_4O_{10}(s) + 6\,H_2O(\ell) \rightarrow 4\,H_3PO_4(\ell)$
3.6. $C_3H_8(g) + 5\,O_2(g) \rightarrow 3\,CO_2(g) + 4\,H_2O(\ell)$
3.7. $2\,C_4H_{10}(\ell) + 13\,O_2(g) \rightarrow 8\,CO_2(g) + 10\,H_2O(\ell)$; 3.03g CO_2
3.8. 24.2% Mg; 28.0% Si; 47.8% O
3.9. Cu_2S
3.10. Cr_2FeO_4
3.11. CH
3.12. P_2O_5; P_4O_{10}
3.13. $C_8H_8O_3$
3.14. Rich
3.15. 90%

Chapter 4

CONCEPT TESTS

p. 134. (c) clear cough syrup; (d) filtered air
p. 140. 1.000 L of seawater weighs more than 1.000 kg of seawater

p. 143. $\dfrac{0.90\,g}{100\,mL} \times \dfrac{1000\,mL}{1\,L} \times \dfrac{1\,mol}{58.45\,g} = 0.154\,M$

p. 144. Equivalent molar concentrations means that the number of particles that dissolve in a given volume will be the same. Differences in conductivity will be due to the different number of ions these particles make when they dissolve.
p. 151. React aqueous lead(II) nitrate with the stoichiometric amount of aqueous potassium dichromate. Stir the mixture for a minute or so, let it stand for 10 min, filter off the yellow $PbCrO_7$ precipitate, wash it with water in the filter, and allow the washed solid to air-dry overnight.
p. 153. 1×10^{-3} mol of Ag^+ were added. One mol of Ag^+ is needed for each mol of AgCl precipitate. The precipitate is 4.9×10^{-4} mol of AgCl. Therefore, enough Ag^+ was added to precipitate all the Cl^-. The excess Ag^+ is 5.1×10^{-4} mol.

PRACTICE EXERCISES

4.1. 120 times the World Health Organization limit
4.2. 1.88 M
4.3. 0.109 M
4.4. 4.48 g
4.5. 0.0288 mL
4.6. (a) $H_3PO_4(aq) + 3\,NaOH(aq) \rightarrow Na_3PO_4(aq) + 3\,H_2O(\ell)$; (b) $H_3PO_4(aq) + 3\,Na^+(aq) + 3\,OH^-(aq) \rightarrow 3\,Na^+(aq) + PO_4^{3-}(aq) + 3\,H_2O(\ell)$; (c) $H_3PO_4(aq) + 3\,OH^-(aq) \rightarrow PO_4^{3-}(aq) + H_2O(\ell)$
4.7. (a) no precipitate; (b) Yes, $Hg_2^{2+}(aq) + 2\,Cl^-(aq) \rightarrow Hg_2Cl_2(s)$

4.8. 194 mg/L $Br^-(aq)$
4.9. (a) +4; (b) +1; (c) +5
4.10. Nitrogen is oxidized and oxygen is reduced. O_2 is the oxidizing agent; N_2H_4 is the reducing agent.
4.11. (a) Yes; (b) $2\,Fe(s) + 3\,Pd^{2+}(aq) \rightarrow 2\,Fe^{3+}(aq) + 3\,Pd(s)$
4.12. $1.31 \times 10^{-4}\ M$

Chapter 5

CONCEPT TESTS

p. 188 (top). $m_1 gh = (PE)_{skier\,1} > (PE)_{skier\,2} = m_2 gh$
p. 188 (bottom). (a) The skier of mass m_1 has the greater potential energy; (b) $\frac{1}{2}m_1 u^2 = (KE)_{skier\,1} > (KE)_{skier\,2} = \frac{1}{2}m_2 u^2$
p. 189. The assumption is that there is far more water in the pool than in the cup. Therefore, the correct answer is "less than," even if the pool temp is, say, 20°C.
p. 207. The mass of the aluminum versus the mass of water and the molar heat capacities of Al versus water ($C_{p,Al} < C_{p,H_2O}$)
p. 211. Measuring the temperature to ±0.001°C allows more significant figures in the value of specific heat.
p. 218 (top). +2043.9 kJ
p. 218 (bottom). (a) 1 mol CH_4; (b) 1 g H_2

PRACTICE EXERCISES

5.1. (a) $q < 0$, therefore negative, the match is the system, exothermic; (b) $q < 0$, therefore negative, the wax is the system, exothermic; (c) $q > 0$, therefore positive, perspiration is the system, endothermic
5.2. −68 kJ
5.3. $-1.56 \times 10^7\ L \cdot atm = -1.57 \times 10^9$ J
5.4. 1.13×10^3 g acetylene
5.5. −321 kJ
5.6. There isn't enough heat to melt all of the ice so the final temperature of the water will be 0°C.
5.7. 24.6 kJ for 0.500 g of material; 49.2 kJ for 1.000 g of material
5.8. (a) $Ca(s) + C(s) + \frac{3}{2}O_2(g) \rightarrow CaCO_3(s)$; (b) $2\,C(s) + O_2(g) + 2\,H_2(g) \rightarrow CH_3COOH(\ell)$; (c) $K(s) + Mn(s) + 2\,O_2(g) \rightarrow KMnO_4(s)$
5.9. −41.2 kJ
5.10. 41.5 kJ/g; 31,125 kJ/L
5.11. $C_{calorimeter} = 4.43$ kJ/°C
5.12. Endothermic; +68 kJ

Chapter 6

CONCEPT TESTS

p. 245. Use a barometer.
p. 262. Kr
p. 264. Decreasing pressure has the greater effect.
p. 266. No
p. 268. Mole fraction is the ratio of the moles of a component to the total moles; therefore adding all mole fractions must represent all the moles, or 1.
p. 275. UF_6 (slowest) $< SF_6 < Kr < CO_2 < Ar < H_2$ (fastest)
p. 278. UF_6 (slowest) $< SF_6 < Kr < CO_2 < Ar < H_2$ (fastest)

PRACTICE EXERCISES

6.1. 771 Pa

6.2. 0.89 atm, 677 mm Hg, 677 torr

6.3. 616 mm

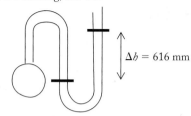

$\Delta h = 616$ mm

6.4. 10.5 L

6.5. 0.62

6.6. 32 psi

6.7. 2.10×10^3 L

6.8. 1.83×10^3 L

6.9. 111 g

6.10. The balloon will sink.

6.11. 44.0 g/mol, CO_2

6.12. $X_{He} = 0.992$, $X_{O_2} = 0.008$

6.13. $X_{O_2} = 0.042$

6.14. 0.0022 g H_2

6.15. 1.37×10^3 m/s, or 2.65 times faster than N_2

6.16. Ar

6.17. The ideal pressure of He is 24.6 atm, and the real pressure of He is 25.2 atm. Thus, He behaves less ideally at 300 K than N_2 (N_2 behaves more ideally) even though He has smaller a and b values. The reason for this is that the a and b terms offset each other at 300 K in the case of N_2.

Chapter 7

CONCEPT TESTS

p. 301. UV has higher frequency.

p. 302. Wavelength and frequency remain the same.

p. 305. Spectrum (c)

p. 307. Answers (a) and (c) are not quantized; (b) and (d) are quantized.

p. 309. Yes, because $\lambda_{violet} < \lambda_{red}$ and $E_{violet} > E_{red}$

p. 317. Only certain values of ν (frequencies) are possible.

p. 343. More shielding is present, and the valence electrons are farther from the nucleus.

PRACTICE EXERCISES

7.1. 3.30 m

7.2. 1.99 nm or 1.99×10^{-9} m

7.3. $\nu = 1.145 \times 10^{15}$ s^{-1}; $\lambda = 262$ nm or 2.62×10^{-7} m

7.4. Prediction: $n_2 = 4$ is longest wavelength; calculated values: 486.2, 434.1, and 410.3 nm

7.5. Less than: 2.42×10^{-19} J

7.6. 3.3×10^{-10} m

7.7. 7.2×10^{-11} m

7.8. 4

7.9. $n = 3$, $\ell = 1$, $m_\ell = 1, 0, -1$, $m_s = +\frac{1}{2}$

7.10. [Ar]$3d^7 4s^2$

7.11. K$^+$ = [Ar], Ba^{2+} = [Xe], I$^-$ = [Xe], O^{2-} = [Ne], Rb$^+$ = [Kr], Al^{3+} = [Ne], Cl$^-$ = [Ar]

7.12. Mn = [Ar]$3d^5 4s^2$, Mn^{3+} = [Ar]$3d^4$, Mn^{4+} = [Ar]$3d^3$

7.13. Li$^+$ < F < Cl$^-$ and Al^{3+} < Mg^{2+} < P^{3-}

7.14. Ne > Ca > Cs

Chapter 8

CONCEPT TESTS

p. 366. N_2 has a triple bond and one lone pair per N atom, whereas O_2 has a double bond and two lone pairs per O atom. Perhaps more electron density in lone pairs in O_2 makes it easier to form a covalent bond to iron.

p. 367. $\cdot \ddot{X} \cdot \rightarrow \left[:\ddot{X}: \right]^{3-}$ $\cdot Y \cdot \rightarrow \left[Y \right]^{2+}$

p. 370. Electronegativity values for the noble gases (group 18) are absent because these elements rarely form covalent bonds. (Kr and Xe do form covalent bonds with F and O.)

p. 372. CO is polar, O_2 is nonpolar. Hemoglobin contains Fe cations that will bind to the δ^- end of a polar molecule such as CO more strongly.

p. 380. Formal charge = -2.

p. 385. He has an electron configuration $1s^2$, and Ne has an electron configuration $1s^2 2s^2 2p^6$. Neither has d orbitals available to expand its octet and thereby accommodate covalent bonds to fluorine.

p. 386. NO is the only one of these with an odd number of electrons. It tends to pick up an electron (forming NO$^-$, an oxidizing agent) or lose an electron (forming NO$^+$, a reducing agent). This behavior makes NO very reactive with hemoglobin or almost any molecule. O_2, N_2, and CO are all much less reactive.

p. 391. The energy required to break the O=O bond is considerably less (by about 446 kJ/mol) than the energy required to break the triple bond in N_2.

PRACTICE EXERCISES

8.1. $:\ddot{C}l\cdot$

8.2. $H-\ddot{O}-\ddot{O}-H$

8.3. $H-C\equiv C-H$

8.4. $\left[:\ddot{B}r: \right]^-$ $\left[:\ddot{I}: \right]^-$

8.5. The pair Be and Cl has $\Delta EN = 1.5$ and therefore forms the most polar bond, but it is not ionic because $\Delta EN < 2.0$.

8.6. $H-\ddot{B}r:$ $:\ddot{I}-\ddot{C}l:$ HBr is more polar.
$\Delta EN = 0.7$ $\Delta EN = 0.5$

8.7.

$$\underset{:\ddot{O}:\quad\ddot{O}:}{\overset{\overset{\cdot\ddot{O}\cdot}{\|}}{S}} \leftrightarrow \underset{:\ddot{O}\quad\ddot{O}:}{\overset{\overset{:\ddot{O}:}{|}}{S}} \leftrightarrow \underset{:\ddot{O}\quad\ddot{O}:}{\overset{\overset{:\ddot{O}:}{|}}{S}}$$

8.8. $\left[:N\equiv N-\ddot{N}: \right]^- \leftrightarrow \left[:\ddot{N}=N=\ddot{N}: \right]^- \leftrightarrow \left[:\ddot{N}-N\equiv N: \right]^-$

$\left[:\ddot{O}=N=\ddot{O}: \right]^+ \leftrightarrow \left[:\ddot{O}-N\equiv O: \right]^+ \leftrightarrow \left[:O\equiv N-\ddot{O}: \right]^+$

8.9. $\left[\overset{0\quad+1\quad-2}{:N\equiv N-\ddot{N}:} \right]^- \leftrightarrow \left[\overset{-1\quad+1\quad-1}{:\ddot{N}=N=\ddot{N}:} \right]^- \leftrightarrow \left[\overset{-2\quad+1\quad0}{:\ddot{N}-N\equiv N:} \right]^-$

$\left[\overset{0\quad+1\quad0}{:\ddot{O}=N=\ddot{O}:} \right]^+ \leftrightarrow \left[\overset{-1\quad+1\quad+1}{:\ddot{O}-N\equiv O:} \right]^+ \leftrightarrow \left[\overset{+1\quad+1\quad-1}{:O\equiv N-\ddot{O}:} \right]^+$

The resonance forms shown in red indicate which resonance forms of the azide ion (N_3^-) and the nitronium ion (NO_2^+) contribute the most to bonding in their respective structures.

8.10.

$$\underset{:\ddot{F}:}{\overset{:\ddot{F}:}{:S\diagdown\overset{\diagup\ddot{F}:}{\diagdown\ddot{F}:}}}$$

8.11. $:\overset{-1}{\ddot{O}}-\overset{+1}{\ddot{Cl}}-\overset{0}{\ddot{O}}\cdot \leftrightarrow :\overset{-1}{\ddot{O}}-\overset{+2}{\ddot{Cl}}-\overset{-1}{\ddot{O}}: \leftrightarrow \cdot\overset{0}{\ddot{O}}-\overset{+1}{\ddot{Cl}}-\overset{-1}{\ddot{O}}:$

The structures shown in red indicate the best Lewis structures for ClO_2. Structures with $Cl{=}O$ bonds and lower formal charges can be drawn but such bonds are less likely to form.

8.12. $N{\equiv}N + 3(H{-}H) = 6(N{-}H)$
$\phantom{N{\equiv}N}+941 \quad 3(+432) \quad 6(-386)$
Overall: -79 kJ/mol

Chapter 9

CONCEPT TESTS

p. 414. The carbon atom in CO_2 has no lone pairs and SN = 2, whereas the sulfur atom in SO_2 has one lone pair and SN = 3.

p. 415. The central atoms (Si in SiH_4, P in PH_3, and S in H_2S) all have a steric number of 4, so the electron pair geometry in each is tetrahedral. The ideal tetrahedral bond angle is 109.5°, which is the angle in SiH_4 with 4 bonding pairs (BP) of electrons and no lone pairs (LP) of electrons. PH_3 has 3 BP and 1 LP; H_2S has 2 BP and 2 LP. Lone pairs push bonding pairs closer together, so the more LPs on a central atom, the smaller the bond angles.

p. 417. Three axial lp–bp repulsions are equivalent.

p. 422. No. For a molecule to be a greenhouse gas, its bonds must be polar. The nonpolar bonds in P_4 are pure covalent bonds.

p. 424 (top). S is less electronegative than O, so an S—H bond is less polar than an O—H bond.

p. 424 (bottom). O_2 and N_2 are nonpolar molecules with only two atoms. The bond between the oxygen or the nitrogen atoms cannot have an asymmetric stretch.

p. 431. There are no p orbitals left to form π bonds on an sp^3 hybridized atom.

p. 438. Delocalized π bonds exist in structures (a) and (c), but not in structure (b).

p. 446. Yes, because O_2 is paramagnetic and N_2 is diamagnetic.

p. 449. No. N_2^{+*} has a bond order of 1.5 compared with 2.5 in N_2^{+}, and the bond order in N_2^{*} is 2.0, while it is 3.0 in N_2.

PRACTICE EXERCISES

9.1. Tetrahedral

9.2. The O—S—O angle in SO_2 is less than the O—S—O angle in SO_3. Because S is larger than O and the lone pair of electrons on S occupy more space, the O—S—O angle in SO_2 is also less than the O—O—O angle in O_3.

9.3. (a) NOF is bent with a bond angle of ~120°. SO_2Cl_2 is tetrahedral with a bond angle of ~109.5°.

9.4. NH_3 and Cl_2O are polar molecules, and CF_4 has polar bonds, so these three molecules could be greenhouse gases.

9.5. No. CS_2 is nonpolar ($\mu = 0$ D). The electronegativities of C and S indicate that their bonds are nonpolar, so CS_2 should not be a greenhouse gas.

9.6. (a) CCl_4; (d) PH_3

9.7. The C in CH_3 is sp^3 hybridized; the C in C=O is sp^2 hybridized; the C in CS_2 is sp hybridized.

9.8. The N atoms in N_2H_2 are both sp^2 hybridized (trigonal planar geometry), whereas both N atoms in N_2H_4 are sp^3 hybridized (trigonal pyramid geometry).

9.9. H_2^{+} has a bond order of 0.5, so the molecular ion has some stability but not as much as H_2 has.

9.10. Be_2, B_2, and C_2

9.11. Bond order = 3 for CO.

Chapter 10

CONCEPT TESTS

p. 472. Yes. Al^{3+} and O_2^{2-} have high charges, and Al_2O_3 would be predicted to have a very large lattice energy.

p. 476. Acetone, because it has a larger dipole moment, is more polar and experiences greater dipole–dipole interactions.

p. 480. Even though both molecules are nonpolar, CCl_4 is heavier and larger, so it has greater dispersion forces and a higher boiling point than CF_4.

p. 485. Yes. Ion-induced dipole forces

p. 491. Boiling point decreases with altitude, so it will take longer to deliver the necessary amount of heat to cook the pasta.

p. 492. No. CO_2, like most substances, is denser as a solid than as a liquid.

p. 493. The heat from the hot needle provides sufficient energy to overcome the surface tension.

p. 494. No. As its convex meniscus indicates, the adhesive forces between mercury atoms and glass are weaker than the attractive forces between mercury atoms.

p. 495. Yes, because seawater has ion–dipole and dipole–dipole interactions.

p. 497. Most of the solution is water. The density of water is 1 g/mL, so the mass of the solvent is essentially the same as the volume of solution.

p. 502. Solution (b), whose i factor is 2, has the lowest freezing point.

p. 505. The concentration of ions inside the cucumber cells is less than the concentration of the brine, so water flows from the cucumber to the brine, which dehydrates ("pickles") the cucumber. In wilted flowers the situation is reversed; the concentration in the plant cells is greater than pure water so water enters the cells by osmosis to revive the flower.

p. 506. Neither. No flow of solvent would be observed.

PRACTICE EXERCISES

10.1. $BaO > CaCl_2 > NaCl$

10.2. The melting point of TiO_2 is highest, and the melting point of $CaCl_2$ is higher than that of $PbCl_2$.

10.3. $U = -3793$ kJ/mol

10.4. Ethylene glycol has two —OH groups to form hydrogen bonds, compared to isopropanol with only one.

10.5. Substance (a) has the largest dipole–dipole interactions. Substance (d) has the largest dispersion forces. Substance (c) has the lowest boiling point.

10.6. $O_2 < CO < H_2O$

10.7. Helium is less polarizable than nitrogen.

10.8. 9.5×10^{-5} mol/L

10.9. $P_{soln} = 0.845$ atm

10.10. At a pressure of 25 atm, CO_2 is probably a solid at $-100°C$. CO_2 melts between $-50°C$ and $-60°C$, and it vaporizes between $10°C$ and $20°C$.

10.11. $0.841 \, m$

10.12. $4.4 \, m$

10.13. $\Delta T_b = 9.3°C$; boiling point = $109.3°C$

10.14. Boiling point = $102.7°C$

10.15. $\Delta T_f = -0.12°C$

10.16. $\pi = 20.3$ atm

10.17. $\pi = 8.40$ atm

10.18. $\mathcal{M} = 180$ g/mol

10.19. $\mathcal{M} = 6.4 \times 10^4$ g/mol

Chapter 11

CONCEPT TESTS

p. 527. The 3s bands of Mg are full, so the conductivity of Mg is best explained by overlapping valence and conduction bands.

p. 536 (top). Yes. Describing bronze as a solution is correct as long as the Sn is uniformly (homogeneously) distributed.

p. 536 (bottom). No. Sn is randomly located in the unit cell.

p. 543. Region a is diamond.

PRACTICE EXERCISES

11.1. Tin-doped GaAs is p-type; selenium-doped GaAs is n-type.

11.2. $r = 128$ pm

11.3. Density of silver = 10.57 g/cm³; density of gold = 19.41 g/cm³

11.4. Gold ($r = 144$ pm) will form substitutional alloys with both silver ($r = 144$ pm) and copper (128 pm; 11% difference in radius).

11.5. Radius of Na^+ = 101 pm

11.6. Density of NaCl = 2.16 g/cm³

11.7. 412 nm; $2\theta = 30.00°$ for $n = 3$; $2\theta = 40.44°$ for $n = 4$

Chapter 12

CONCEPT TESTS

p. 575. No. A triple bond or two double bonds represent the same degree of unsaturation (two).

p. 581. The estimates based on bond energies are the same because the molecules of the two compounds contain the same number of C–C bonds and C–H bonds. However, they are different compounds with different chemical and physical properties. It is reasonable that their actual heats of combustion would be a little different.

p. 583. If IUPAC rules are followed, the name of a compound provides an unambiguous description of its molecular structure.

p. 584. The solubility of a gas is proportional to its partial pressure, which is in turn proportional to the product of its mole fraction in a mixture times total pressure. As crude oil comes up from deep underground, there is a decrease in total pressure, which reduces the solubility of the C_1 to C_4 alkanes.

p. 586. Gasoline has the lower average molar mass and so should have the higher vapor pressure.

p. 593. The reaction is endothermic. It requires tremendous amounts of energy, which is impossible to control during production. Therefore this approach is impractical.

p. 597. (a) No. The two hydrogen atoms on the terminal carbon atom make both "sides" of the double bond equivalent. (b) No. The bond angles are 180° so there are no "sides" to C≡C bonds.

p. 598. LDPE and HDPE have different properties and uses. If they are to be recycled for uses similar to their original ones, they need to be kept separate.

p. 600. Chlorine atoms are so bulky that the polymer chains in Saran cannot fit tightly together. The resulting loose fit means that the Saran is less rigid and more elastic than PVC.

p. 603. These alcohols and ethers are polar solutes that tend to be soluble in polar solvents such as water; nonpolar solutes such as hydrocarbons have little solubility in water. The –OH groups in alcohols can hydrogen-bond with water; water can hydrogen-bond with –O– in ethers.

p. 605. Methanol has a higher percentage of oxygen in it than ethanol. This oxygen has no fuel value. Therefore, methanol has less fuel value than ethanol.

p. 608. MTBE > diethyl ether > ethanol > methanol

PRACTICE EXERCISES

12.1. Yes. One mole of compound A reacts with 2 moles of H_2; one mole of compound B reacts with 3 moles of H_2. Because A and B have similar molar masses, a given mass of B will react with close to 1.5 times as much of H_2 as the same mass of A. The structure of the product of the reaction with A is the same as that with B: $CH_3—CH_2—CH_2—CH_2—CH_2—CH_2—CH_3$.

12.2. The carbon-skeleton structure of *n*-hexane is

$CH_3CH_2CH_2CH_2CH_2CH_2CH_3$ [*or* $CH_3(CH_2)_5CH_3$].

12.3. Molecules in sets a and c are isomers; those in set b have different formulas.

12.4. 28.4 kJ/mol

12.5.

```
Temperature (°C)
100
 90        ---- 98°
 80
 70   ---- 69°
 60
  0        50
     Volume of distillate (mL)
```

12.6. The mole ratio is 1.4.

12.7. The five isomers are:

trans   *cis*

12.8.

OCH₃ (structure with C=C, C=O, OCH₃)

12.9. E-85 has 33% less energy per gram than gasoline.

12.10. Dipole-induced dipole interactions

12.11.

(polymer structure) ₙ

12.12. The polar materials from which the gloves are made allow moisture to pass through but repel nonpolar oil and grease.

12.13.

HO ... OH H_2N ... NH_2

Chapter 13

CONCEPT TESTS

p. 635. (1) No, a spontaneous process is not necessarily a rapid process. (2) A nonspontaneous reaction does not occur unless energy is added. If energy is added, a nonspontaneous process may occur.

p. 639. Four: yellow smooth, yellow wrinkled, green smooth, and green wrinkled.

p. 645. (a) Any exothermic reaction in which $\Delta S_{sys} > 0$, such as the combustion of methane. (b) One example of a reaction that is never spontaneous because $\Delta S_{sys} < 0$ and $\Delta S_{surr} < 0$ is photosynthesis.

p. 647. Particles in a solid can only vibrate in place and have access to relatively few microstates. Particles in liquids have limited translational motion and access to more microstates. Particles of gas have much greater mobility and access to the even more microstates.

p. 648 (top). A system such as NH_3 with 3 bonds has more possible vibrational motions than a 2-bond system.

Symmetric stretch

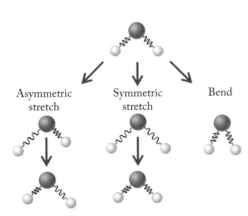

Asymmetric stretch   Symmetric stretch   Bend

p. 648 (bottom). Yes, conversion of diamond to graphite is "spontaneous," but that says nothing about the rate of the process.

p. 660. It can be spontaneous at low temperatures if ΔH°_{rxn} is negative.

PRACTICE EXERCISES

13.1. All the processes are spontaneous, so we know $\Delta S_{univ} > 0$ for all four. In processes (a) and (d), $\Delta S_{sys} > 0$. In processes (b) and (c), $\Delta S_{sys} < 0$, so ΔS_{surr} must be positive and of greater absolute value than ΔS_{sys}.

13.2. Only process (b) results in an increase in entropy. Note that reaction (c) results in a decrease because oxygen gas from the air is incorporated into solid rust.

13.3. Prediction: 3 moles of gaseous reactants form 1 mole of gaseous CO_2 and 2 moles of liquid H_2O, so ΔS_{rxn} is probably negative. The result of calculation: $\Delta S_{rxn} = -242.6$ J/(mol·K)

13.4. (a) The sign of ΔS_{sys} should be negative because 3 moles of gaseous reactants form two moles of liquid products. (b) $\Delta S^\circ_{sys} = -326.4$ J/(mol·K). (c) $\Delta S^\circ_{surr} = 1.94 \times 10^{-3}$ J/(mol·K). (d) Yes, the reaction is spontaneous because $\Delta S^\circ_{univ} = 1.61 \times 10^{-3}$ J/(mol·K).

13.5. $\Delta G^\circ_{rxn} = \Sigma n \Delta G^\circ_{f,products} - \Sigma m \Delta G^\circ_{f,reactants} = [8 \text{ mol}(-394.4 \text{ kJ/mol}) + 9 \text{ mol} (-228.6 \text{ kJ/mol})] - [1 \text{ mol}(16.3 \text{ kJ/mol}) + 12.5 \text{ mol} (0.0 \text{ kJ/mol})] = -5228.9$ kJ

13.6. ΔH° is $(-)$ because combustion reactions are exothermic; ΔS° is $(-)$ because there are fewer moles of gaseous products than reactants; ΔG° is $(-)$ because the reaction is spontaneous.

13.7. (a) The value of ΔH°_{rxn}, which is twice the value of the ΔH°_f of ammonia, is negative, which means the reaction is exothermic. (b) However, the value of ΔS°_{rxn} is also negative, which means the reaction is not spontaneous at high temperatures (where $T\Delta S^\circ_{rxn}$ is large), but will be spontaneous at low temperature where $T\Delta S^\circ_{rxn}$ is small and offset by the favorable ΔH°_{rxn}.

13.8. $\Delta G^\circ_{rxn} = -204$ kJ

Chapter 14

CONCEPT TESTS

p. 676. A reaction that is spontaneous only at high temperatures must have a positive ΔS° value.

p. 678. No, by definition a chemical reaction involves the conversion of reactants into products and so the rate of the reaction must always be positive.

p. 687. The value of the rate constant is always the same for a given reaction at a given temperature. The rate constant does not change with changing concentrations of the reactants.

p. 698. The chances of three molecules colliding with each other at exactly the same time (and with the correct alignment) are very small.

PRACTICE EXERCISES

14.1. CO is consumed at twice the rate of O_2;
$$\text{Rate} = -\frac{\Delta[O_2]}{\Delta t} = \frac{1}{2}\frac{[CO_2]}{\Delta t}$$

14.2. $\frac{\Delta[N_2]}{\Delta t} = 10.8$ M/s; $\frac{\Delta[H_2O]}{\Delta t} = 21.6$ M/s

14.3. $\frac{\Delta[NO_2]}{\Delta t} = (0.0053 - 0.0030)M/(3000 - 1000)\text{s} = 1.2 \times 10^{-6} M/s$

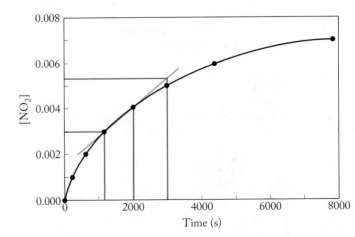

14.4. (a) The reaction must be zero-order in C. (b) Rate = $k[A][B]$; units of k are $M^{-1}s^{-1}$.

14.5. Rate = $k[NO][NO_3]$; $k = 1.57 \times 10^{10} M^{-1}s^{-1}$.

14.6. If we use a spreadsheet (like Excel) to calculate $\ln[H_2O_2]$ values, we can then plot them and fit a linear trend line to them:

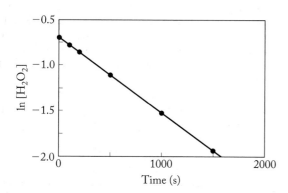

The plot is linear, so the reaction is first order with a rate constant equal to minus the slope. Therefore, $k = 8.3 \times 10^{-4}\,s^{-1}$.

14.7. $k = 0.693/28\,d = 0.025\,d^{-1}$

14.8. After calculating $1/[NO_2]$ values in a spreadsheet (like Excel), we can plot them versus time and fit a linear trend line to them:

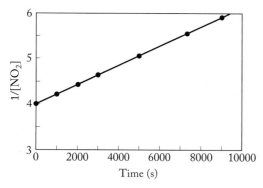

The plot is linear, so the reaction is second order with a rate constant equal to the slope. Therefore, $k = 2.09 \times 10^{-4}\,M^{-1}s^{-1}$ and the rate $= 2.09 \times 10^{-4}\,M^{-1}s^{-1}[NO_2]^2$

14.9. $t_{1/2} = 9.23 \times 10^{-3}\,s$

14.10. By using a worksheet (like Excel) to calculate $\ln[Cl]$ values, we can plot them and fit a linear trend line:

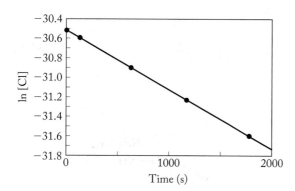

The plot is linear, so the reaction is first order with a rate constant equal to minus the slope. Therefore, $k' = 616\,s^{-1}$ and $k = k'/(8.5 \times 10^{-11}\,M) = 7.3 \times 10^{12}\,M^{-1}s^{-1}$.

14.11. We can use a spreadsheet (like Excel) to calculate $\ln k$ and $1/T$ values, plot them, and fit a linear trend line to them:

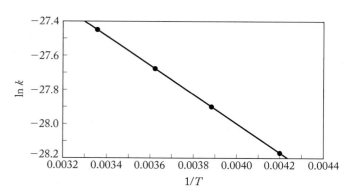

Multiplying the negative of the slope by R gives $E_a = 6.9\,kJ/mol$.

14.12. No. The mechanism is not valid because none of the proposed elementary steps have a stoichiometry that fits the rate law for the overall reaction.

14.13. Yes. NO_2 acts as a homogeneous catalyst that is consumed in one step and regenerated in the next.

Chapter 15

CONCEPT TESTS

p. 742. b. $[CO_2] = [H_2] > [CO] = [H_2O]$

p. 758. K should be small and is: 6.5×10^{-7}

p. 761. Pure liquid water is assigned an activity of "1"; therefore, the K_c expression contains only a term for the concentration of $H_2O(g)$.

p. 765. Adding reactant shifts the equilibrium to the right forming more product and consuming some of the additional reactant. Removing reactant shifts the equilibrium to the left, consuming product and replacing some of the lost reactant.

PRACTICE EXERCISES

15.1. $K_c = \dfrac{[CO][H_2]^3}{[CH_4][H_2O]}$ $K_p = \dfrac{(P_{CO})(P_{H_2})^3}{(P_{CH_4})(P_{H_2O})}$

15.2. $K_c = 2.9 \times 10^2$

15.3. $K_p = 2.7 \times 10^4$

15.4. $K_p = 4.5 \times 10^{-12}$

15.5. $K = 3.6 \times 10^8$

15.6. $K_c = 0.13$

15.7. $K_{overall} = 1.7 \times 10^2$

15.8. $Q < K$; therefore, the mixture is not at chemical equilibrium and the reaction proceeds in the forward direction.

15.9. $K_p = 6.1 \times 10^5$

15.10. (a) $K_p = \dfrac{(P_{CO})^2}{(P_{CO_2})}$ (b) $K_p = \dfrac{(P_{CO})}{(P_{CO_2})(P_{H_2})}$

15.11. (a) Loss of ammonia shifts the equilibrium toward the formation of more ammonia; (b) some of the ammonia decomposes into nitrogen and hydrogen as the reaction runs in reverse; (c) removing these two reactants shifts the equilibrium toward forming more of them.

15.12. Increasing pressure shifts the equilibrium to the right.

15.13. The value of the equilibrium constant increases.

15.14. Partial pressure of HI is 0.156 atm.

15.15. $[N_2O_4]$ = 0.016 M; $[NO_2]$ = 0.058 M

15.16. At 298 K, K_p = 3.2 × 10^{-37}; at 2000 K, K_p = 3.3 × 10^{-12}

Chapter 16

CONCEPT TESTS

p. 800. The pH of the weak acid is higher because it is not completely ionized.

p. 805. In terms of acid strength: C > B > A

p. 813. No influence on the pH of phosphoric acid, but significant influence in citric acid because the K_{a_1} and K_{a_2} values are much closer together.

p. 826. If the desired buffer pH is below the pK_a of the acid, more of the acidic component is needed.

p. 833. During the second step in the titration all of the bicarbonate produced by the titration of carbonate in the first step plus the bicarbonate initially in the sample are titrated.

PRACTICE EXERCISES

16.1. CH_3COOH, CH_3COO^-; H_2O, H_3O^+

16.2. pH = 2.160

16.3. $[H^+]$ = 1.3 × 10^{-12} M; $[OH^-]$ = 8 × 10^{-3} M

16.4. K_a = 7.9 × 10^{-4}

16.5. pH = 3.3

16.6. $[OH^-]$ = 9.4 × 10^{-3}; pOH = 2.03; pH = 11.97

16.7. pH = 5.63

16.8. $NH_4^+(aq) + H_2O(\ell) \leftrightarrow NH_3(aq) + H_3O^+(aq)$

16.9. pH = 9.08

16.10. pH = 4.01

16.11. pH change is less than 0.01 units.

16.12. Blood pH would rise because the mouse would be able to exhale CO_2 and so shift the equilibrium to the left, thereby reducing $[H^+]$.

16.13. (a) Ammonia concentration = 0.4846 M;

(b) possible indicators are methyl red, bromcresol green

Chapter 17

CONCEPT TESTS

p. 851.

H—N: and B—F: (with lone pairs shown on N, H above and below, and F atoms with lone pairs)

A coordinate bond is formed when the N atom in NH_3 donates its lone pair of electrons to an empty sp^3 orbital of the B atom in BF_3.

p. 854. Hybridization is sp^3d^2. Each F atom has 7 valence electrons and gains one more by sharing a single electron in a half-filled

orbital of Xe. To create 4 half-filled orbitals in the valence shell of Xe, two electrons from two of the three filled $6p$ orbitals are promoted to two of the empty $6d$ orbitals. Then those four half-filled orbitals are mixed with the filled $6s$ and third $6p$ orbital to form six equivalent hybrid orbitals. Since all six orbitals are in the $n = 6$ shell, the hybridization scheme is sp^3d^2.

p. 862. Solubility would have decreased because more of the fluoride ion that dissolved would stay as F$^-$.

p. 868. Spontaneity must be due to a favorable (negative) change in enthalpy because there are no changes in the number of particles and they are all liquids or solutes. Therefore the change in entropy should be small.

p. 875. (a) Cyanide is a stronger field strength ligand than pyridine; greater splitting in the cyanide complex leads to population of the lower energy orbitals and low spin. (b) The Ru ion is a much larger ion than the Fe ion and its larger valence shell d orbitals interact more strongly with ligand lone pairs, producing more splitting and low spin.

p. 878. There is only one structure possible.

p. 880. The square planar complex is not chiral; the tetrahedral complex is chiral.

PRACTICE EXERCISES

17.1. CO_2 is the Lewis acid, and CaO (or its O^{2-} ion) is the Lewis base.

17.2. $[Ag^+]$ = 9.2 × 10^{-9} M

17.3. Solubility = 2.6 × 10^{-3} M

17.4. The pH in the problem (10.33) matches the pK_{a_2} of carbonic acid in Appendix 5, which means, according to the Henderson Hasselbalch equation, that ratio of CO_3^{2-} to HCO_3^- ions will be 1:1. Therefore, half the CO_3^{2-} ions that form when $ZnCO_3$ dissolves will remain as CO_3^{2-}. Thus, $K_{sp} = [Zn^{2+}][CO_3^{2-}]$ = $(x)(0.5\,x)$ = 1.2 × 10^{-10} and x = 1.5 × 10^{-5}.

17.5. Assuming the hydroxyl group does not ionize, but the three carboxylic acid groups do, the molecule has three "teeth."

17.6. Low-spin Co^{3+}.

17.7. (a) Ligand–ammonia; counter ion–chloride ion; name–tetraamminezinc(II) chloride; (b) Ligands–ammonia and water; counter ion–nitrate; name–tetraamminediaquacobalt(III) nitrate

17.8.

(a)   (b)

(c)

The names of (a), (b), and (c) are:

(a) *cis*-diammine-*trans*-dibromoethylenediaminecobalt(III);

(b) *trans*-diammine-*cis*-dibromoethylenediaminecobalt(III);

(c) *cis*-diammine-*cis*-dibromoethylenediaminecobalt(III).

Chapter 18

CONCEPT TESTS

p. 895. No, the number of moles of Cu deposited match the number of moles of Zn that dissolve, but the molar mass of Zn is slightly larger than that of Cu so the changes in mass are not the same.

p. 916. $H_2(g)$ is produced at the anode; $Cl_2(g)$ at the cathode

PRACTICE EXERCISES

18.1. $2\,NO_2^-(aq) + O_2(g) \rightarrow 2\,NO_3^-(aq)$

18.2. $2\,Al(s) + 3\,Cu^{2+}(aq) \rightarrow 2\,Al^{3+}(aq) + 3\,Cu(s)$
$Al(s) \mid Al^{3+}(aq) \parallel Cu^{2+}(aq) \mid Cu(s)$

18.3. $\Delta G_{cell} = -290$ kJ

18.4. The cadmium half-reaction occurs at the anode, the nickel half-reaction occurs at the cathode. Overall:
$NiO(OH)(s) + Cd(s) + 2\,H_2O(\ell) \rightarrow$
$\qquad\qquad 2\,Ni(OH)_2(s) + Cd(OH)_2(s) \quad E°_{cell} = 1.72$ V

18.5. $E_{cell} = 1.64$ V

18.6. $K = 1.8 \times 10^{62}$

18.7. 130 g Mg

18.8. 1.48 g Pb

Chapter 19

CONCEPT TESTS

p. 939. A fragrant molecule is like the white hand and the receptor protein is like the brown one. When the fragrant molecule and receptor fit together, as when the two right hands shake, the protein signals the fragrance of the molecule to the brain. When the intermolecular fit is not as good, as illustrated by the bottom pair of hands, then no fragrance, or a different one, is sensed.

p. 942. Muscarine is a chiral molecule and a plant produces only one enantiomer, so a solution of muscarine from a natural source would rotate the plane of polarized light. Because the fluid in the person's stomach contained muscarine but did not rotate the plane of polarized light, it must be a racemic mixture resulting from a laboratory synthesis. Therefore the coroner concluded that the person did not die from eating toxic mushrooms but may have been murdered by someone who had access to synthetic muscarine.

p. 945. All of the amino acids without an ionizable functional group on their side chains, e.g., alanine, can have only two pK_a values. Those with an ionizable side chain group, e.g., glutamic acid, have three.

p. 949. Leucine or isoleucine

p. 955. Two isomers are possible because the –OH group on the C-5 of linear fructose can approach the C=O bond on the C-2 carbon from either above or below the plane of the ring as the cyclic form of the sugar forms.

p. 959. As sugar is oxidized in the body it forms pyruvate, which, according to Figure 19.32, is a precursor of cholesterol.

p. 962. The unsaturated fats are liquids at low temperatures whereas saturated fats might solidify stopping the flow of blood to and from the reindeer's hooves.

p. 963. Olestra is a much larger molecule with many more C—H and C—C bonds. Therefore combusting a mole of it yields much more energy.

p. 964. Fluoroacetic acid, the fluoroacetate ion, and acetamide have similar structures and shapes, so perhaps the acetamide competes with the other two for the enzymatic binding site.

PRACTICE EXERCISES

19.1.

(a)   (b)   Chiral (c)

Chiral (d)   Chiral (e)

19.2.

19.3. K_{a_1}: $H_3\overset{+}{N}CH(COOH)CH(COOH)(aq) + OH^-(aq) \rightleftharpoons$
$\qquad H_3\overset{+}{N}CH(COO^-)CH(COOH)(aq) + H_2O(aq)$
K_{a_2}: $H_3\overset{+}{N}CH(COO^-)CH(COOH)(aq) + OH^-(aq) \rightleftharpoons$
$\qquad H_3\overset{+}{N}CH(COO^-)CH(COO^-)(aq) + H_2O(aq)$
K_{a_3}: $H_3\overset{+}{N}CH(COO^-)CH(COO^-)(aq) + OH^-(aq) \rightleftharpoons$
$\qquad H_2NCH(COO^-)CH(COO^-)(aq) + H_2O(aq)$

19.4. For amino acids "A," "B," and "C" there are six different sequences: ABC, ACB, BCA, BAC, CAB, and CBA.

19.5. Because cellobiose is indigestible, it must have a β-1,4 linkage, which means it has the structure in (a).

19.6. There are 3 possibilities for linking 1 molecule of each of the fatty acids "A," "B," and "C" to glycerol: ABC, CAB, BCA. Each has two enantiomers, so there are a total of 6 isomers.

19.7.
(a) CGGTATCCGAT / GCCATAGGCTA
(b) TTAAGCCGCTAG / AATTCGGCGATC

Chapter 20

CONCEPT TESTS

p. 1008. Electron capture does not result in the emission of any subatomic particles.

p. 1012. The limestone is too old and nearly all the carbon-14 in it would have decayed leaving too little to detect and measure quantitatively.

PRACTICE EXERCISES

20.1. 1.4%

20.2. 1.25×10^{-12} J

20.3. $^{131}_{54}Xe$

20.4. Beta decay; $^{28}_{14}Si$

20.5. $^{37}_{17}Cl + ^{1}_{0}n \rightarrow ^{38}_{18}Ar + ^{0}_{-1}e$

20.6. $3.65 \times 10^9\,\mathrm{Bq} = 9.86 \times 10^{-2}\,\mathrm{Ci}$
20.7. 1.4×10^{11} X-rays
20.8. 4.95×10^3 years

Chapter 21

CONCEPT TESTS

p. 1030. The electron configuration of Mg^{2+} is $1s^2 2s^2 2p^6$. It has no electrons in a partially filled set of d orbitals that could undergo transitions resulting in absorption of photons of visible light.

PRACTICE EXERCISES

21.1. $pH = 1.19$
21.2. $N_2(g) + 8H^+(aq) + 8e^- \rightarrow 2NH_3(aq) + H_2(g)$
21.3. $^{201}_{81}Tl + ^{0}_{-1}e \rightarrow ^{201}_{80}Hg + \gamma$
21.4. There are 20.7 mg of ^{186}Re, but only 9.4 mg of ^{188}Re left.
21.5. $Sn_8Hg(s) + O_2(g) + 4H^+(aq) \rightarrow$
$Sn_6Hg(s) + 2Sn^{2+}(aq) + 2H_2O(\ell)$ $E^{\circ}_{rxn} = 1.36\,V$
21.6. $K_{rxn} = 2.5 \times 10^9$

Answers to Selected End-of-Chapter Questions and Problems

Chapter 1

1. (a) A pure compound in the gas phase
   (b) A mixture of elements: red in the liquid phase and blue in the gas phase

3. (b) A mixture of two gaseous elements undergoes a chemical reaction to form a solid compound.

5. Both are correct because on the sun, matter is being turned into energy through nuclear fusion.

7. One chemical property of gold is its resistance to corrosion (oxidation). Gold's physical properties include its density, its color, its melting temperature, and its electrical and thermal conductivity.

9. Add water to the salt-sand mixture to dissolve the salt. Filter this mixture to isolate the sand and then recover the salt by evaporating the water from the solution that passed through the filter.

11. (b) combustion

13. A Snickers bar, an uncooked hamburger, and a hot dog

15. Orange juice (with pulp)

17. They can be distinguished by their physical state (sugar is a solid, water is a liquid, and oxygen is a gas), by their melting and boiling points, and by their density.

19. Density, melting point, thermal and electrical conductivity, and softness (a–d) are all physical properties whereas tarnishing and reaction with water (e and f) are both chemical properties.

21. Distillation will separate the water from the dissolved protein that has formed a homogeneous solution with the water. This process would have to be accomplished, however, at low temperature, or the protein will be denatured.

23. Fe is a solid, O_2 is a gas, and Hg is a liquid.

25. Extensive properties change with the size of the sample and so cannot be used to identify a substance.

27. We need at least one observation, experiment, or idea (from examining nature).

29. Yes

31. Theory as used in normal conversation is an idea, an opinion, or a speculation that can be changed.

33. SI units can be easily converted into a larger or smaller unit by multiplying or dividing by multiples of 10. English units are sometimes based on other number multiples and thus are more complicated to manipulate.

35. 6.70×10^8 mi/h

37. 93.2%

39. 4.1×10^{13} km

41. 1330 Cal

43. 2.5 mi

45. 9.0 mi/h

47. 23 g

49. 19.0 mL

51. 26.5 g; 2.65×10^{-2} kg

53. 58.0 mL

55. 33.7 mL

57. 5.1 g/cm³

59. Yes, the cube has a density of 0.64 g/cm³.

61. 0.29 cm³

63. (a) Accuracy means that the experimental value agrees with the true value of the measurement. Precision means that several repeated measurements agree with each other with little variability.
   (b) No, the lawyer confuses the two. He is saying that if his weight is not known exactly to the ounce compared to the true value, then even the pound value is in error.
   (c) Yes, to be precisely accurate, a series of measurements would be each very close to each other and also in agreement with the true value.
   (d) The sign means that the weight that is measured by the scale is within ± the smallest division on the scale.

65. (a) Manufacturer #1 has a range of $0.516 - 0.504 = 0.012$ μm. Manufacturer #2 has a range of $0.514 - 0.512 = 0.002$ μm. Manufacturer #3 has a range of $0.502 - 0.500 = 0.002$ μm.
   (b) Yes, Manufacturers #2 and #3 can justify the claim.
   (c) Yes, in the case of Manufacturer #2, the lines are printed at wider widths than the widths specified.

67. (b) 0.08206, (c) 8.314, (f) 3.752×10^{-5}, and maybe (d) 5.420×10^3, depending on whether the 0 is significant

69. (a) 17.4
   (b) 1×10^{-13}
   (c) 5.70×10^{-23}
   (d) 3.58×10^{-3}

71. Yes, $-40°C$ is equal to $-40°F$.

73. $-270°C$

75. 172°F, 351 K

77. 39.2°C

79. $-89.2°C$, 183.9 K

81. $-452°F$

83. $HgBa_2CaCu_2O_6$ has the highest T_c.

85. 0.12 mg/L

87. Both mixtures (a) and (b) would react to make sodium chloride with no reactants left over.

89. 75 loaves of bread and 50 oz of mayonnaise

91. 17 bicycles

93. Day 11

Chapter 2

1. (c) A mixture of NO_2 and NO

3. The element shaded dark blue—helium

5. (a) Chlorine (Cl_2, yellow) is a reactive nonmetal.
   (b) Neon (Ne, red) is a chemically inert gas.
   (c) Sodium (Na, dark blue) is a reactive metal.

7. (a) Mg (green) will form MgO.
   (b) K (red) will form K_2O.
   (c) Ti (yellow) will form TiO_2.
   (d) Al (dark blue) will form Al_2O_3.

9. When looking at a single compound, the law of constant composition applies. In a compound, the ratio of the constituent atoms is fixed. When comparing two or more compounds made up of the same elements, the law of multiple proportions applies if the compounds can combine in different ratios to give distinct compounds.

11. Dalton's theory states that because atoms are indivisible, the ratio of the atoms (elements) in a compound is a ratio of whole numbers. Thus, in water, the ratio of volumes of hydrogen to oxygen is the whole number ratio 2:1, because the ratio of hydrogen and oxygen atoms in water is 2:1.

13. 1.5

15. 30.5%

17. Rutherford concluded that the positive charge in the atom could not be spread out (to form the pudding) in the atom, but must result from a concentration of charge in the center of the atom (the nucleus). Most of the alpha particles were deflected only slightly or passed directly through the gold foil, so Rutherford reasoned that the nucleus must be small compared to the size of the entire atom. The negatively charged electrons do not deflect the alpha particles, and Rutherford reasoned that the electrons took up the remainder of the space of the atom outside of the nucleus.

19. The fact that cathode rays were deflected by a magnetic field indicated that they were streams of charged particles.

21. A weighted average takes into account the proportion of each value in the group of values to be averaged.

23. Greater than 1

25. (a) and (b)

27. 35.46 amu

29. Yes, the average atomic mass of magnesium on Mars is the same as on Earth.

31. 47.95 amu

33. Mendeleev only knew the masses of the elements at the time he arranged the elements into his periodic table.

35. (a) ^{14}C, 6 protons, 8 neutrons, 6 electrons
    (b) ^{59}Fe, 26 protons, 33 neutrons, 26 electrons
    (c) ^{90}Sr, 38 protons, 52 neutrons, 38 electrons
    (d) ^{210}Pb, 82 protons, 128 neutrons, 82 electrons

37.

Symbol	^{23}Na	^{89}Y	^{118}Sn	^{197}Au
Number of protons	11	39	50	79
Number of neutrons	12	50	68	118
Number of electrons	11	39	50	79
Mass number	23	89	118	197

39.

Symbol	$^{37}Cl^-$	$^{23}Na^+$	$^{81}Br^-$	$^{226}Ra^{2+}$
Number of protons	17	11	35	88
Number of neutrons	20	12	46	138
Number of electrons	18	10	36	86
Mass number	37	23	81	226

41. $NaCl$, $MgCl_2$, $CaCl_2$, $SrCl_2$; Na_2SO_4, $MgSO_4$, $CaSO_4$, $SrSO_4$

43. Compounds (a) and (d) are molecules. Compounds (b) and (c) consist of ions.

45. (c) Be

47. (c) S^{2-}

49. (a) S^{2-}, (b) P^{3-}, and (d) Ca^{2+}

51. (b) Br

53. $XO_2{}^{2-}$

55. The Roman numerals indicate the charge on the transition metal cation.

57. (a) NO_3, nitrogen trioxide
    (b) N_2O_5, dinitrogen pentoxide
    (c) N_2O_4, dinitrogen tetroxide
    (d) NO_2, nitrogen dioxide
    (e) N_2O_3, dinitrogen trioxide
    (f) NO, nitrogen monoxide
    (g) N_2O, dinitrogen monoxide
    (h) N_4O, tetranitrogen monoxide

59. (a) Na_2S, sodium sulfide
    (b) $SrCl_2$, strontium chloride
    (c) Al_2O_3, aluminum oxide
    (d) LiH, lithium hydride

61. (a) Cobalt(II) oxide
    (b) Cobalt(III) oxide
    (c) Cobalt(IV) oxide

63. (a) BrO^-
    (b) $SO_4{}^{2-}$
    (c) $IO_3{}^-$
    (d) $NO_2{}^-$

65. (a) Nickel(II) carbonate
    (b) Sodium cyanide
    (c) Lithium hydrogen carbonate
    (d) Calcium hypochlorite

67. (a) Hydrofluoric acid
    (b) Bromic acid
    (c) H_3PO_4
    (d) HNO_2

69. (a) Sodium oxide
    (b) Sodium sulfide
    (c) Sodium sulfate
    (d) Sodium nitrate
    (e) Sodium nitrite

71. (a) K_2S
    (b) K_2Se
    (c) Rb_2SO_4
    (d) $RbNO_2$
    (e) $MgSO_4$

73. (a) Manganese(II) sulfide
    (b) Vanadium(II) nitride
    (c) Chromium(III) sulfate
    (d) Cobalt(II) nitrate
    (e) Iron(III) oxide

75. (b) Na_2SO_3

77. (b) Cl_2

79. (a) Na

81. Chemistry is the study of the composition, structure, properties, and reactivity of matter. Cosmology is the study of the history, structure, and dynamics of the universe. These two sciences are related because (1) the universe is composed of matter, the study of which is chemistry, and therefore, the study of the universe is really chemistry; (2) the changing universe is driven by chemical and nuclear reactions that are also studied in chemistry; and (3) cosmology often asks what the universe is made of at the atomic level.

83. Because quarks combine to make up the three particles that are important to the properties and reactivity of atoms: protons, neutrons, and electrons.

85. The density of the universe is decreasing.

87. The higher the charge (number of protons), the more repulsion the nuclei feel for each other and the higher the temperature needed to overcome that repulsion.

89. The electrons in He are held more tightly by the nucleus so they are harder to remove.

91. The expanding universe was cooling and therefore could not support the high temperatures needed for fusion. Also, the expanding universe was not dense enough for nuclei to fuse.

93. Both

95. (a) ^{16}O
    (b) ^{24}Mg
    (c) ^{36}Ar

97. (a) ^{59}Co
    (b) ^{121}Sb
    (c) ^{110}Hg

99. (a) Electrons
    (b) The negatively charged electron was attracted to the positively charged plate as the electron passed through the electric field.
    (c) The light spot would be at the bottom. The electron would still be deflected toward the positively charged plate, which would now be at the bottom of the tube.
    (d) The position of the light spot on the fluorescent screen would be half way between the position where it was before the voltage was reduced and the "zero" spot location when there is no voltage between the plates.

101. (a) Two
    (b) Both spots would be seen below the center of the screen. The spot lowest on the screen (farthest from the center) would be the α particle and the one closest to the center would be the proton.

103. 1:2:4

105. $17\,Cu:1\,Sn$

107. (a) Sc, Ga, Ge
    (b) Ekaaluminum is gallium, ekaboron is scandium, and ekasilicon is germanium.
    (c) Scandium was discovered in 1879, gallium was discovered in 1875, and germanium was discovered in 1886.

109. (a) 52.93%
    (b) 0.599 cm^3

111. 60.11%

113. (a) ^{79}Br-^{79}Br = 157.8366 amu, ^{79}Br-^{81}Br = 159.8346 amu, ^{81}Br-^{81}Br = 161.8326 amu
    (b) ^{79}Br-^{79}Br, 25.41%; ^{79}Br-^{81}Br, 50.00%; ^{81}Br-^{81}Br, 24.59%

Chapter 3

1. (a) $4\,X(g) + 4\,Y(g) \rightarrow 4\,XY(g)$
   (b) $4\,X(g) + 4\,Y(g) \rightarrow 4\,XY(s)$
   (c) $4\,X(g) + 4\,Y(g) \rightarrow 2\,XY_2(g) + 2\,X(g)$
   (d) $4\,X_2(g) + 4\,Y_2(g) \rightarrow 8\,XY(g)$

3. $Fe(\ell)$

5. It is thought that most of the water present on Earth arrived with colliding asteroids and comets when the Earth was young. A small amount may have been derived from water vapor outgassing from Earth's interior.

7. A dozen is too small a unit to express the very large number of atoms, ions, or molecules present in a mole.

9. No, the molar mass of a substance does not directly correlate to the number of atoms in a molecular compound. The statement posed by the question would only be true if the two compounds were composed of the same element.

11. (a) 7.3×10^{-10} mol Ne
    (b) 7.0×10^{-11} mol CH$_4$
    (c) 4.2×10^{-12} mol O$_3$
    (d) 8.1×10^{-15} mol NO$_2$

13. (a) 2.5×10^{-12} mol bytes
    (b) 3.3×10^{-15} mol bytes

15. (a) 7.53×10^{22} atoms
    (b) 7.53×10^{22} atoms
    (c) 1.51×10^{23} atoms
    (d) 2.26×10^{23} atoms

17. (a) Both contain the same
    (b) N$_2$O$_4$
    (c) CO$_2$

19. (a) 3.00 mol
    (b) 4.50 mol
    (c) 1.50 mol

21. 41.63 mol

23. 0.25 mol; 10 g

25. (a) 1 mol
    (b) 2 mol
    (c) 1 mol
    (d) 3 mol

27. (a) 64.06 g/mol
    (b) 48.00 g/mol
    (c) 44.01 g/mol
    (d) 108.01 g/mol

29. (a) 152.15 g/mol
    (b) 164.20 g/mol
    (c) 148.21 g/mol
    (d) 132.16 g/mol

31. (a) NO
    (b) CO$_2$
    (c) O$_2$

33. 0.752 mol

35. 10.3 g

37. Diamond

39. No

41. No

43. (a) $CH_4(g) + H_2O(g) \rightarrow CO(g) + 3\,H_2(g)$
    (b) $2\,NH_3(g) \rightarrow N_2(g) + 3\,H_2(g)$
    (c) $CO(g) + H_2O(g) \rightarrow CO_2(g) + H_2(g)$

45. (a) $3\,FeSiO_3(s) + 4\,H_2O(\ell) \rightarrow Fe_3Si_2O_5(OH)_4(s) + H_4SiO_4(aq)$
    (b) $Fe_2SiO_4(s) + 2\,CO_2(g) + 2\,H_2O(\ell) \rightarrow 2\,FeCO_3(s) + H_4SiO_4(aq)$
    (c) $Fe_3Si_2O_5(OH)_4(s) + 3\,CO_2(g) + 2\,H_2O(\ell) \rightarrow$
        $$3\,FeCO_3(s) + 2\,H_4SiO_4(aq)$$

47. (a) $N_2(g) + O_2(g) \rightarrow 2\,NO(g)$
    (b) $2\,NO(g) + O_2(g) \rightarrow 2\,NO_2(g)$
    (c) $NO(g) + NO_3(g) \rightarrow 2\,NO_2(g)$
    (d) $2\,N_2(g) + O_2(g) \rightarrow 2\,N_2O(g)$

49. (a) $N_2O_5(g) + Na(s) \rightarrow NaNO_3(s) + NO_2(g)$
    (b) $N_2O_4(g) + H_2O(\ell) \rightarrow HNO_3(aq) + HNO_2(aq)$
    (c) $3\,NO(g) \rightarrow N_2O(g) + NO_2(g)$

51. $2\,C_2H_2(g) + 5\,O_2(g) \rightarrow 4\,CO_2(g) + 2\,H_2O(g)$

53. Yes

55. (a) 4.5×10^{11} mol
    (b) 2.0×10^{10} kg

57. (a) $2\,NaHCO_3(s) \rightarrow CO_2(g) + H_2O(g) + Na_2CO_3(s)$
    (b) 6.55 g

59. 1.17 kg

61. 1.5 metric tons

63. (a) 1.48 kg
    (b) 1.11 kg

65. 346 g

67. An empirical formula shows the lowest whole-number ratio of atoms in a substance. A molecular formula shows the actual numbers of each kind of atom that compose one molecule of the substance.

69. No

71. (a) 74.19% Na, 25.80% O
    (b) 57.48% Na, 39.98% O, 2.52% H
    (c) 27.37% Na, 1.19% H, 14.30% C, 57.13% O
    (d) 43.38% Na, 11.33% C, 45.28% O

73. (d) Pyrene, $C_{16}H_{10}$

75. NO, N_2O_3, and NO_2

77. No

79. $ZrSiO_4$

81. (a) MgO
    (b) $2\,Mg(s) + O_2(g) \rightarrow 2\,MgO(s)$

83. $Mg_3Si_2H_3O_8$

85. $CuCl_2O_8$

87. The excess of oxygen is required in combustion analysis to ensure complete reaction of the hydrogen and carbon to form water and carbon dioxide, respectively.

89. Yes

91. The empirical formula is C_2H_3; the molecular formula is $C_{20}H_{30}$.

93. $C_8H_{16}O$

95. (c) Less than the sum of the Fe and S

97. Theoretical yield is the greatest amount of a product possible from a reaction and assumes that the reaction goes to 100% completion. The percent yield is the observed experimental yield divided by the theoretical yield and multiplied by 100.

99. Reactions do not always go to completion because the reaction may be slow or may have, for a portion of the reaction, yielded different products than expected. As a result, the actual yield for a reaction is usually less than the theoretical yield.

101. 3 cups

103. 0.844 g

105. $NH_3(g) + HCl(g) \rightarrow NH_4Cl(s)$; 0.66 g NH_3 remains at the end of the reaction.

107. 59%

109. (a) $C_6H_{12}O_6(aq) \rightarrow 2\,CH_3CH_2OH(\ell) + 2\,CO_2(g)$
     (b) 77.1%

111. (a) Calcium triphosphate hydroxide
     (b) 39.89%
     (c) Decreases slightly

113. (a) 529.3 kg
     (b) $2\,Al_2O_3(s) + 6\,C(s) \rightarrow 4\,Al(s) + 6\,CO(g)$, 436.4 kg

115. (a) 44.8 g
     (b) 14.9 g
     (c) 1.67 cm³

117. (a) $a = 1$, $b = 3$; charge on U is 6+
     (b) $c = 3$, $d = 8$; charge on U is 5.33+
     (c) $x = 2$, $y = 2$, $z = 6$

119. (a) 5.84×10^{20} molecules
     (b) 3.01×10^{21} molecules
     (c) 8.77×10^{18} molecules

121. (a) No
     (b) $C_5H_{10}O_5(s) + 5\,O_2(g) \rightarrow 5\,CO_2(g) + 5\,H_2O(\ell)$
         $2\,C_7H_{12}O_7(s) + 13\,O_2(g) \rightarrow 14\,CO_2(g) + 12\,H_2O(\ell)$

123. (a) FeS is iron(II) sulfide with Fe^{2+} and S^{2-}; FeS_2 is iron(IV) sulfide based on your knowledge so far in this course with Fe^{4+} and S^{2-}. Actually this compound is Fe^{2+} with S_2^{2-} and is named iron(II) persulfide.
     (b) 0.262 g

125. (a) $3\,FeO(s) + H_2O(\ell) \rightarrow Fe_3O_4(s) + H_2(g)$
     (b) $12\,FeO(s) + 2\,H_2O(\ell) + CO_2(g) \rightarrow 4\,Fe_3O_4(s) + CH_4(g)$

127. 1×10^{-8} mol

129. 55.2 mol

131. Substance A is $CaCO_3$, substance B is CO_2, and substance C is CaO.

133. The substance is rhenium (Re).

135. 82.4%

137. (a) 5.99 metric tons
     (b) $2\,SO_2(g) + 2\,H_2O(\ell) + O_2(g) \rightarrow 2\,H_2SO_4(aq)$
     (c) 9.17 metric tons

139. 3.06 g

141. Mg_2SiO_4

Chapter 4

1. Yellow

3. (a) Cl (purple)
   (b) S (orange)
   (c) N (green)
   (d) P (blue)

5. The solvent is usually the liquid component of the solution. If both the solvent and solute are liquids or solids, the solvent is that component present in greatest amount (volume).

7. 1.00 M

9. (a) 5.6 M $BaCl_2$
   (b) 1.00 M Na_2CO_3
   (c) 1.30 M $C_6H_{12}O_6$
   (d) 5.92 M KNO_3

11. (a) $0.14\ M\ Na^+$
    (b) $0.11\ M\ Cl^-$
    (c) $0.096\ M\ SO_4^{2-}$
    (d) $0.20\ M\ Ca^{2+}$

13. (a) $11.7\ g\ NaCl$
    (b) $4.99\ g\ CuSO_4$
    (c) $6.41\ g\ CH_3OH$

15. $2.72\ g$

17. (a) $9.6 \times 10^{-3}\ mol$
    (b) $7.8 \times 10^{-4}\ mol$
    (c) $8.8 \times 10^{-2}\ mol$
    (d) $4.22\ mol$

19. Orchard sample: $3.4 \times 10^{-4}\ mmol/L$
    Residential sample: $5.6 \times 10^{-5}\ mmol/L$
    Residential sample after a storm: $3.2 \times 10^{-2}\ mmol/L$

21. $1.5 \times 10^{-4}\ M$

23. (b) $AgNO_3$, (c) $Fe(NO_3)_2 \cdot 6H_2O$, and (d) $Ca(OH)_2$

25. $4.57 \times 10^{-2}\ M$

27. (b) $1\ M\ CaCl_2$

29. (a) $1.80 \times 10^{-2}\ M\ Na^+$
    (b) $2.7 \times 10^{-1}\ mM\ LiCl$
    (c) $1.28 \times 10^{-2}\ mM\ Zn^{2+}$

31. $1.95\ M$

33. 11.4%

35. Table salt produces Na^+ and Cl^- ions in solution when it dissolves. Sugar does not dissociate into ions when it dissolves. Ions are required to conduct electricity.

37. Liquid methanol does not dissociate into ions and therefore does not conduct electricity, but molten sodium hydroxide contains mobile Na^+ and OH^- ions and does conduct electricity.

39. (c) $1.0\ M\ Na_2SO_4$ > (b) $1.2\ M\ KCl$ > (a) $1.0\ M\ NaCl$ > (d) $0.75\ M\ LiCl$

41. (a) $0.025\ M$
    (b) $0.050\ M$
    (c) $0.075\ M$

43. Acid

45. Strong acids include HCl, HNO_3, $HClO_4$, H_2SO_4, HI, HBr; weak acids include CH_3COOH, $HCOOH$, HF, H_3PO_4.

47. Base

49. Strong bases include $NaOH$, KOH, $CsOH$, $LiOH$, $RbOH$, $Ba(OH)_2$, $Sr(OH)_2$, $Ca(OH)_2$; weak bases: NH_3, CH_3NH_2, C_5H_5N.

51. (a) The acid is H_2SO_4; the base is $Ca(OH)_2$.
$$2H^+(aq) + SO_4^{2-}(aq) + Ca^{2+}(aq) + 2OH^-(aq) \rightarrow$$
$$CaSO_4(s) + 2H_2O(\ell)$$
    (b) Sulfuric acid is the acid; $PbCO_3$ is the base.
$$PbCO_3(s) + 2H^+(aq) + SO_4^{2-}(aq) \rightarrow$$
$$PbSO_4(s) + CO_2(g) + H_2O(\ell)$$
    (c) CH_3COOH is the acid; $Ca(OH)_2$ is the base.
$$OH^-(aq) + CH_3COOH(aq) \rightarrow CH_3COO^-(aq) + H_2O(\ell)$$

53. (a) Molecular equation:
$$Mg(OH)_2(s) + H_2SO_4(aq) \rightarrow MgSO_4(aq) + 2H_2O(\ell)$$
    Net ionic equation:
$$Mg(OH)_2(s) + 2H^+(aq) \rightarrow Mg^{2+}(aq) + 2H_2O(\ell)$$
    (b) Molecular equation:
$$MgCO_3(s) + 2HCl(aq) \rightarrow MgCl_2(aq) + H_2CO_3(aq)$$
    (The carbonic acid, H_2CO_3, reacts in solution to give H_2O and CO_2.)
    Net ionic equation:
$$MgCO_3(s) + 2H^+(aq) \rightarrow Mg^{2+}(aq) + H_2O(\ell) + CO_2(g)$$
    (c) Molecular equation:
$$NH_3(g) + HCl(g) \rightarrow NH_4Cl(s)$$
    Net ionic equation:
$$NH_3(g) + HCl(g) \rightarrow NH_4Cl(s)$$

55. $PbCO_3(s) + 2H^+(aq) \rightarrow Pb^{2+}(aq) + H_2CO_3(aq)$
    $Pb(OH)_2(s) + 2H^+(aq) \rightarrow Pb^{2+}(aq) + 2H_2O(\ell)$

57. A saturated solution contains the maximum concentration of a solute. A supersaturated solution *temporarily* contains *more* than the maximum concentrations of a solute at a given temperature.

59. A precipitation reaction occurs when two solutions are mixed to form an insoluble compound.

61. A saturated solution may not be a concentrated solution if the solute is only sparingly or slightly soluble in the solvent. In that case, the solution will be a saturated dilute solution.

63. (a) Barium sulfate, (e) lead hydroxide, and (f) calcium phosphate

65. (a) Balanced reaction:
$$Pb(NO_3)_2(aq) + Na_2SO_4(aq) \rightarrow PbSO_4(s) + 2NaNO_3(aq)$$
    Net ionic reaction:
$$Pb^{2+}(aq) + SO_4^{2-}(aq) \rightarrow PbSO_4(s)$$
    (b) No precipitate forms
    (c) Balanced reaction:
$$FeCl_2(aq) + Na_2S(aq) \rightarrow FeS(s) + 2NaCl(aq)$$
    Net ionic reaction:
$$Fe^{2+}(aq) + S^{2-}(aq) \rightarrow FeS(s)$$
    (d) Balanced reaction:
$$MgSO_4(aq) + BaCl_2(aq) \rightarrow MgCl_2(aq) + BaSO_4(s)$$
    Net ionic reaction:
$$Ba^{2+}(aq) + SO_4^{2-}(aq) \rightarrow BaSO_4(s)$$

67. $CaCO_3$

69. $0.0211\ g$

71. $0.054\ g$

73. $132\ kg$

75. $11\ mL$

77. The ion exchange resin is regenerated in a home water softener through the addition of a large amount of salt (NaCl) to the resin. The high concentration of Na^+ replaces the 2+ cations on the resin.

79. No. This reaction is not an example of ion exchange because the ions are not simply being swapped—they are also changing oxidation states, so this is a redox reaction.

81. The sum of the oxidation numbers of the atoms in a molecule must add up to the overall charge on that molecular species. For a neutral molecule, the charges must add up to zero.

83. H_2SeO_4

85. A half-reaction describes either just the oxidation reaction (where electrons are lost as a product in the reaction) or just the reduction reaction (where electrons are gained as a reactant in the reaction).

87. Reduction; oxidation

89. (a) +1
(b) +5
(c) +7

91. (a) $2e^- + Br_2(\ell) \rightarrow 2\,Br^-(aq)$; reduction
(b) $Pb(s) + 2\,Cl^-(aq) \rightarrow PbCl_2(s) + 2e^-$; oxidation
(c) $2e^- + O_3(g) + 2\,H^+(aq) \rightarrow O_2(g) + H_2O(\ell)$; reduction
(d) $H_2S(g) \rightarrow 2\,S(s) + 2\,H^+(aq) + 2e^-$; oxidation

93. $H_2O(\ell) + 2\,Fe_3O_4(s) \rightarrow 3\,Fe_2O_3(s) + 2\,H^+(aq) + 2e^-$

95. (a) SiO_2: Si = +4, O = –2; Fe_3O_4: Fe = +8/3, O = –2; Fe_2SiO_4: Fe = +2, Si = +4, O = –2; O_2: O = 0. Oxygen is oxidized (O^{2-} to O_2), and iron is reduced ($Fe^{8/3+}$ or Fe^{3+} to Fe^{2+}).
(b) SiO_2: Si = +4, O = –2; Fe: Fe = 0; O_2: O = 0; Fe_2SiO_4: Fe = +2, Si = +4, O = –2. Iron is oxidized (Fe^0 to Fe^{2+}), and oxygen is reduced (O^0 to O^{2-}).
(c) FeO: Fe = +2; O = –2; O_2: O = 0; H_2O: H = +1, O = –2; $Fe(OH)_3$: Fe = +3; O = –2; H = +1. Iron is oxidized (Fe^{2+} to Fe^{3+}), and oxygen is reduced (O^0 to O^{2-}).

97. (a) $O_2(g) + 4\,FeCO_3(s) \rightarrow 2\,Fe_2O_3(s) + 4\,CO_2(g)$
(b) $O_2(g) + 6\,FeCO_3(s) \rightarrow 2\,Fe_3O_4(s) + 6\,CO_2(g)$
(c) $O_2(g) + 4\,Fe_3O_4(s) \rightarrow 2\,Fe_2O_3(s)$

99. $NH_4^+(aq) + 2\,O_2(g) \rightarrow NO_3^-(aq) + 2\,H^+(aq) + H_2O(\ell)$

101. $2\,Fe(OH)_2^+(aq) + Mn^{2+}(aq) \rightarrow 2\,Fe^{2+}(aq) + 2\,H_2O(\ell) + MnO_2(s)$

103. $2\,H_2O(\ell) + 4\,Ag(s) + 8\,CN^-(aq) + O_2(g) \rightarrow$
$\qquad\qquad 4\,Ag(CN)_2^-(aq) + 4\,OH^-(aq)$

105. (a) $2\,ClO_3^-(aq) + SO_2(g) \rightarrow 2\,ClO_2(aq) + SO_4^{2-}(aq)$
(b) $4\,H^+(aq) + 2\,ClO_3^-(aq) + 2\,Cl^-(aq) \rightarrow$
$\qquad\qquad 2\,ClO_2(aq) + 2\,H_2O(\ell) + Cl_2(g)$
(c) $Cl_2(g) + 2\,ClO_3^-(aq) \rightarrow 2\,ClO_2(aq) + O_2(g) + 2\,Cl^-(aq)$

107. NH_4^+: N = –3
NH_2OH: N = –1
$H_2N_2O_2$: N = +1
HNO_2: N = +3

109. (a) 300 mL
(b) 17.6 mL
(c) 269 mL

111. 309 mL

113. $6.449 \times 10^{-2}\ M$

115. (a) CH_3O
(b) $C_2H_6O_2$
(c) By mass, ethylene glycol is the solvent and water is the solute.

117. NaCl: Na = +1, Cl = –1; H_2SO_4: H = +1, S = +6, O = –2; MnO_2: Mn = +4, O = –2; Na_2SO_4: Na = +1, S = +6, O = –2; $MnCl_2$: Mn = +2, Cl = –1; H_2O: H = +1, O = –2; Cl_2: Cl = 0

Balanced redox reaction:

$2\,H_2SO_4(aq) + 4\,NaCl(aq) + MnO_2(s) \rightarrow$
$\qquad 2\,Na_2SO_4(aq) + MnCl_2(aq) + 2\,H_2O(\ell) + Cl_2(g)$

Net ionic equation:

$4\,H^+(aq) + 2\,Cl^-(aq) + MnO_2(s) \rightarrow Mn^{2+}(aq) + 2\,H_2O(\ell) + Cl_2(g)$

119. (a) FeO_4^{2-}: Fe = +6, O = –2; H_2O: H = +1, O = –2; FeOOH: Fe = +3, O = –2, H = +1; O_2: O = 0; OH^-: O = –2, H = +1

Balanced redox reaction:

$6\,H_2O(\ell) + 4\,FeO_4^{2-}(aq) \rightarrow 4\,FeOOH(s) + 3\,O_2(g) + 8\,OH^-(aq)$

(b) FeO_4^{2-}: Fe = +6, O = –2; H_2O: H = +1, O = –2; Fe_2O_3: Fe = +3, O = –2; O_2: O = 0; OH^-: O = –2, H = +1

Balanced redox reaction:

$4\,H_2O(\ell) + 4\,FeO_4^{2-}(aq) \rightarrow 2\,Fe_2O_3(s) + 3\,O_2(g) + 8\,OH^-(aq)$

121. (a) $HClO_3$—chloric acid
(b) $HClO_2$—chlorous acid
(c) $HClO_4$—perchloric acid and $HClO_3$—chloric acid

123. (a) Overall equation:

$CH_3COOH(aq) + KOH(aq) \rightarrow H_2O(\ell) + KCH_3OO(aq)$

Net ionic equation:

$CH_3COOH(aq) + OH^-(aq) \rightarrow H_2O(\ell) + CH_3COO^-(aq)$

(b) Overall equation:

$Na_2CO_3(aq) + CaCl_2(aq) \rightarrow CaCO_3(s) + 2\,NaCl(aq)$

Net ionic equation:

$CO_3^{2-}(aq) + Ca^{2+}(aq) \rightarrow CaCO_3(aq)$

(c) Overall equation:

$CaO(s) + H_2O(\ell) \rightarrow Ca(OH)_2(s)$

Net ionic equation:

$CaO(s) + H_2O(\ell) \rightarrow Ca^{2+}(aq) + 2\,OH^-(aq)$

125. (a) $Ca_{10}(PO_4)_6(OH)_2(s) + 2\,F^-(aq) \rightarrow Ca_{10}(PO_4)_6F_2(s) + 2\,OH^-(aq)$
(b) $2 \times 10^{-4}\ M$
(c) 0.6 mg

127. (a) $1.4\ M$ Fe, $9 \times 10^{-2}\ M$ Zn
(b) $2\,Fe(OH)_3(s) + 6\,H^+(aq) \rightarrow 2\,Fe^{3+}(aq) + 6\,H_2O(\ell)$
(c) $ZnCO_3(s) + 2\,H^+(aq) \rightarrow Zn^{2+}(aq) + CO_2(g) + H_2O(\ell)$
(d) 0.18

129. 4.95%

131. (a) a
(b) a
(c) c

133. (a) $H_2S + 2\,O_2 \rightarrow H_2SO_4$ is a redox reaction; 8 electrons are transferred.
(b) $2\,H^+(aq) + SO_4^{2-}(aq) + CaCO_3(s) \rightarrow$
$\qquad\qquad CaSO_4(s) + H_2O(\ell) + CO_2(g)$
(c) $SO_4^{2-}(aq) + CaCO_3(s) \rightarrow CaSO_4(s) + CO_3^{2-}(aq)$

135. Reactions in (c) and (d) are redox reactions.

Chapter 5

1. The brick's kinetic energy at 35 ft above street level is 150 J; its kinetic energy at street level is 500 J.

3. (a)

    (b) The piston is higher in the cylinder.
    (c) Yes
    (d) The system did work on the surroundings.

5. (a) A closed system
    (b) Internal energy will increase.
    (c) No, the system will not do any work on the surroundings.

7. (a) Because the heat of formation of an element in its standard state is defined as zero
    (b) Because the enthalpy of its formation is positive
    (c) By subtracting the sum of the enthalpies of formation of the reactants (multiplied by the number of moles in the balanced equation for each product) from the sum of enthalpies of formation (again multiplied by their molar amounts from the balanced equation) of the products

9. Because enthalpy is a state function, we can add the enthalpy of two reactions to get the enthalpy of a third reaction:

$$\Delta H_C = \Delta H_A - \Delta H_B$$

11. Potential energy is stored energy that depends on an object's position (h) and mass (m) whereas kinetic energy is energy of motion, which depends on an object's mass (m) and velocity (u).

13. Yes

15. As long as the ice cube's temperature is above absolute zero, the molecules of water within the ice cube are moving; therefore the water molecules have kinetic energy.

17. In an exothermic process, the system loses heat to the surroundings (the reaction vessel feels hot). In an endothermic process, the surroundings transfer heat to the system (the reaction vessel feels cold).

19. (a) When methane (the system) combusts, heat is given off so q is negative.
    (b) When water (the system) is frozen to make ice, heat is removed from the water, so q is negative.
    (c) When you touch a hot stove, heat is transferred from the stove to your hand (the system). We are most interested in your hand (because it could get burned), so your hand is the system, and q is positive.

21. (a) Exothermic
    (b) Exothermic
    (c) Endothermic

23. The internal energy increases.

25. -0.500 L·atm; -50.66 J

27. (a) 50 J
    (b) 6200.7 J
    (c) -940 J

29. 275.5 kJ

31. Reaction (b)

33. The sum of the change of internal energy and the product of the system's pressure and change in volume. For a reaction, it is heat gained or lost by the system.

35. If the system transfers heat to the surroundings, its energy will be less after the process than at the start of the process.

37. Negative

39. Positive

41. Negative

43. Specific heat is determined for 1 g of a substance. Heat capacity is not dependent on the amount of a substance; it is determined for a given object.

45. No

47. Water's high heat capacity compared with the heat capacity of air means that water carries away more heat from the engine for every 1°C rise in temperature, making water a good choice to cool automobile engines.

49. 29.3 kJ

51.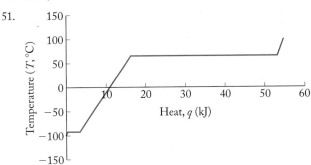

53. 885 g

55. -5.83°C

57. To know how much heat (generated or absorbed by the system) is required to change the temperature of the surroundings (the calorimeter) in order to calculate the heat capacity or final temperature of the system in an experiment.

59. Yes

61. 8.044 kJ/°C

63. -5129 kJ/mol

65. 23.29°C

67. No

69. (a) and (d)

71. -164.9 kJ

73. $NH_4NO_3(s) \rightarrow N_2O(g) + 2\,H_2O(g)$; -35.9 kJ

75. -7198 kJ

77. The energy per gram a fuel releases upon burning.

79. The fuel value (kJ/g) is obtained by dividing molar heat of combustion (kJ/mol) by the molar mass (mol/g).

81. 201 kg

83. (a) -48.99 kJ/g
    (b) -4.90×10^4 kJ
    (c) 5.97 g

85. Hess's law states that enthalpy changes for a reaction can be obtained by summing the enthalpies of constituent reactions; therefore all the heat is accounted for, and energy is neither created nor destroyed when using Hess's law.

87. If we write out the chemical equations for the ΔH_f° for the reactants and products in reactions, these formation reactions will add up, according to Hess's law, to the overall reaction.

89. $CO_2(g) \rightarrow CO(g) + \frac{1}{2}O_2(g)$

$\dfrac{+\ C(s) + O_2(g) \rightarrow CO_2(g)}{C(s) + \frac{1}{2}O_2(g) \rightarrow CO(g)}$

91. -297 kJ/mol

93. $+28$ kJ/mol

95. (a) The first reaction that produces NO from N_2 and O_2
    (b) -103 kJ

97. We are given that ΔE for this process is less than q absorbed $(E < \Delta q)$. If this is the case, for the equality $\Delta E = \Delta q + \Delta w$ to be maintained, then Δw must be negative. A negative value of Δw must mean that work is done by the system on the surroundings when the ice on wet laundry sublimes on a winter's day.

99. (a) $2\,NaOH(aq) + H_2SO_4(aq) \rightarrow 2\,H_2O(\ell) + Na_2SO_4(aq)$
    (b) No
    (c) -114 kJ/mol H_2SO_4

101. 26.0°C

103. (d)

105. (a) Some examples of inorganic endothermic compounds are B_2H_6, CS_2, HI, N_2O, and O_3. Some examples of organic endothermic compounds are C_2H_4, C_6H_6, and $(CH_3)_2C{=}C(CH_3)_2$.
     (b) -1256 kJ/mol

107. (a) The summation reaction 3 is

$Zn(s) + \frac{1}{4}S_8(s) + 3\,O_2(g) \rightarrow ZnSO_4(s) + SO_2(g)$

(b) -1280 kJ

109. -272 J, 4.5×10^{-2} mol ice

111. $\Delta H_3 = -\Delta H_1 + -\Delta H_2$

113. (a) $CH_3OH(g) + N_2(g) \rightarrow HCN(g) + NH_3(g) + \frac{1}{2}O_2(g)$
     (b) Reactant
     (c) 307 kJ

115. -175 kJ

117. -841 kJ/mol

119. 30.6 kJ/mol

121. Hydrogen

123. -130 kJ, exothermic

Chapter 6

1. Barometer (a) because Denver has lower atmospheric pressure owing to its altitude.

3. (a)

5. (c)

7. Line 1 represents gas at a higher pressure; the x-axis is not an absolute temperature scale; if it were, then line 1 and line 2 would meet at $T = 0$ K.

9. Line 2 is not consistent with the ideal gas law.

11. The molar mass of helium is 4 g/mol. The line on the graph for this gas will be below line 1. The molar mass of NO is 30 g/mol, so the line on the graph for this gas will be above line 2.

13. No. The total pressure in each flask is not the same. Flask b has the highest partial pressure of nitrogen.

15. At -100°C the average speed of the CO_2 molecules (red line, curve 2) will decrease, and therefore the peak (maximum) of the distribution curve will occur at lower molecular speeds. Also the distribution will not be as wide, meaning that there will be fewer molecules of CO_2 moving at high speeds at -100°C compared to the situation at 25°C.

17. Br_2 (orange)

19. (a)

21. Force is the product of the mass of an object and the acceleration due to gravity. Pressure uses force in its definition—it is the force an object exerts over a given area.

23. An atmosphere is defined as the pressure necessary to support a 760-mm-high column of mercury in a barometer. A torr is defined as the pressure expressed in millimeters of mercury (mmHg), and so 760 torr = 1 atmosphere.

25. The ethanol barometer

27. Compared with a dull blade, a sharpened blade will have a smaller area over which the same amount of force is distributed. As area (A) decreases, pressure (P) must increase because the force $(F =$ the skater's mass times the acceleration due to gravity) is constant, according to $P = F/A$.

29. As we go up in altitude, the overlying mass of the atmosphere above us decreases, and so the pressure also decreases.

31. 3.9×10^3 Pa

33. (a) 0.020 atm
    (b) 0.739 atm

35. (a) 814.6 mmHg
    (b) 1.072 atm
    (c) 1086 mbar

37. The higher the temperature, the faster gas molecules move. As the molecules move faster, they will collide with the walls of the container more often and with greater force. Both of these conditions result in increased pressure as temperature is raised.

39. She should decrease the temperature.

41. The gas pressure increases.

43. 2.00 atm

45. 2.30 atm; 23.0 m

47.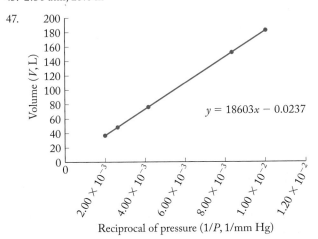

$$y = 18603x - 0.0237$$

Yes, the graph will be exactly the same for the same number of moles of argon gas.

49.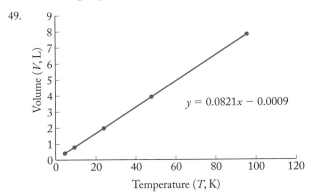

$$y = 0.0821x - 0.0009$$

If the amount of gas were halved, the graph would still be linear, but the slope of the line would be halved.

51. 323°C

53. 4.27 L

55. 1.7 L

57. (b)

59. (a) No change
    (b) Decrease by $\frac{1}{4}$
    (c) Increase by 17%

61. 144 L

63. 6.5 atm

65. STP is defined as 1 atm and 0°C (273 K), and the volume of 1 mol of an ideal gas at STP is 22.4 L.

67. The product of the number of moles of gas in the sample, the temperature, and the gas constant.

69. 0.67 mol

71. 12.2 atm

73. 1730 L

75. 814 g

77. (a) 0.0419 mol/h; (b) 10.7 g

79. 620 g

81. 715 g

83. No. Each gas has a different molar mass; therefore the densities of different gases will not necessarily be the same for a particular temperature and pressure.

85. (a) The density increases.
    (b) The density increases.

87. (a) 9.1 g/L
    (b) basement

89. SO_2

91. CO

93. The partial pressure is the pressure that a particular gas individually contributes to the total pressure.

95. (c)

97. 0.20

99. $P_{total} = 2.46$ atm; $P_{N_2} = 1.7$ atm; $P_{H_2} = 0.49$ atm; $P_{CH_4} = 0.25$ atm

101. 0.019 mol

103. (a) Greater than
     (b) Less than
     (c) Greater than

105. 1.67 times more

107. 680 mm Hg

109. 25%

111. The speed of a molecule in a gas that has the average kinetic energy of all the molecules of the sample.

113. (a) Root-mean-square speed decreases as molar mass increases.
     (b) Root-mean-square speed increases as temperature increases.

115. To determine the molar mass of an unknown gas, measure the rate of its effusion (r_X) relative to the rate of effusion of a known gas (r_Y). Because we know the molar mass of the known gas, M_Y, we can use the Graham's law equation to solve for the unknown M_X.

117. Diffusion is the spread of one substance into another. In diffusion, there is no difference in pressure, and two gases are involved in the process. Effusion is the escape of a gas from its container through a tiny hole from a region of higher pressure to one of lower pressure.

119. $SO_2 < NO_2 < CO_2$

121. C

123. 717 m/s

125. 0.71

127. 32.3 g/mol

129. 18.2 g/mol

131. (a) The rate of effusion of $^{12}CO_2$ to is 1.01 times greater than that of $^{13}CO_2$.
     (b) $^{12}CO_2$

133. The smaller balloon contained hydrogen.

135. At low temperatures the gas particles move more slowly and their collisions become inelastic—they stick together owing to the weak attractive forces between them. The particles, therefore, do not act separately to contribute to the pressure in the container, and the pressure is lower than would be expected from the ideal gas law. Also, the gas particles take up real volume in the container, and as the pressure increases, the volume of the particles takes up a greater volume of the free space in the container. This has the effect of raising the pressure–volume above what we would expect from the ideal gas law (in a plot of PV/RT versus P).

137. Because b is a measure of the volume that the gas particles occupy, b increases as the size of the particles increases.

139. H_2

141. (a) 910 atm
     (b) 476 atm

143. (a) 6.51×10^{-3} mol
     (b) 43.8 g/mol
     (c) First, measure the mass of the empty flask; then fill it to the brim with a liquid of known density (water is convenient) and reweigh the flask with the liquid to obtain the mass of the liquid. The volume will be the mass of the liquid divided by its density.

145. 86.3 mL

147. (a) No change
     (b) No change
     (c) Increases

149. P_{N_2} = 1000 torr

     P_{Ar} = 146 torr

     P_{CH_4} = 73.2 torr

     No. Earth-type life could not exist on Titan because there is no oxygen on Titan.

151. 0.25 m/s

153. 39 lb

155. Xe

157. (a) $NH_3(g) + HCl(g) \rightarrow NH_4Cl(s)$
     (b) The rate of diffusion of a gas is inversely related to its molar mass. The gas with the highest molar mass will not diffuse as far along the tube before it reacts. Therefore, the ring of NH_4Cl will appear closer to the end with HCl.
     (c) 0.594 m

159. 0.0078 g

161. 1.70 times larger

163. 26.7 g

165. 5.56 atm

167. 137 g/mol

169. A bubble at high pressure will be smaller than one at lower pressure in order to maintain the equality in the equation $\frac{P_1}{V_1} = \frac{P_2}{V_2}$. Therefore, as the bubble rises to lower pressure, it will expand its volume.

171. 2.71 atm

173. 38.4 L N_2; 192 L CO_2; 1.74 g/L

175. 6.5×10^4 Pa

177. (a) $4NH_3(g) + 5O_2(g) \rightarrow 4NO(g) + 6H_2O(g)$
     (b) $NH_3(g) + HNO_3(\ell) \rightarrow NH_4NO_3(s)$

179. (a) $NH_4NO_2(s) \rightarrow N_2(g) + 2H_2O(g)$
     (b) Yes

181. Sodium is a highly reactive element; it reacts quickly with moist air to form NaOH and H_2. NaOH is caustic because it is a strong base, and the H_2 produced could form an explosive mixture with oxygen in the air.

183. -642.2 kJ

Chapter 7

1. (a) True purple, red, and orange
   (b) Dark purple
   (c) Orange
   (d) Red
   (e) Dark purple and green

3. Green

5. Blue, green, orange

7. Orange (Y^{3+}) < green (Sr^{2+}) < blue (Rb^+) < gray (I^-) < red (Te^{2-})

9. Blue

11. All these forms of light have perpendicularly oscillating electric and magnetic fields that travel together through space.

13. The lead shield is to protect those parts of our bodies that are not being imaged, but that might be exposed to the X-rays. Lead is a very high density metal containing many electrons, which interact with X-rays and absorb nearly all the X-rays before they can reach our bodies.

15.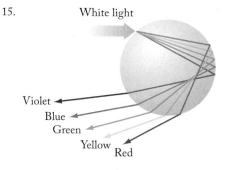

17. No. The radiation's wavelength will be shifted into the infrared part of the electromagnetic spectrum, which is outside the visible range.

19. 4.87×10^{14} s^{-1}

21. (a) 2.88 m
    (b) 2.95 m
    (c) 2.98 m

23. (a) The radio waves

25. 8.3 min

27. The hydrogen absorption spectrum consists of dark lines at wavelengths specific to hydrogen. The emission spectrum has bright lines on a dark background; the bright lines appear at the exact same wavelengths as the dark lines in the absorption spectrum.

29. Because each element shows distinctive and unique absorption/emission lines, the bright emission lines observed for the pure elements could be matched to the many dark absorption lines in the spectrum of sunlight. This approach was used to deduce the sun's elemental composition.

31. A quantum is the smallest indivisible amount of radiant energy that an atom can absorb or emit.

33. At a low power setting, the atoms reach only those excited states that emit long wavelengths (in other words, low energy). Once the power supplied allows the atoms to reach higher excited states, these states become populated and can then emit the higher-energy red light. At the highest power settings, the radiant energy is coming from many excited states that emit all the colors in the visible range. At the highest power setting, the radiant energy shifts away from the red to include emissions of shorter wavelength.

35. (a) Because we can only have whole numbers of eggs, the number of eggs remaining in a carton is a quantized value and (c) the elevation of a flight of stairs is defined by the size of each step. We cannot have an elevation between steps so this is a quantized value.

37. 6.93×10^{-19} J

39. Potassium, 8.04×10^5 m/s

41. No

43. 3.17×10^{18} photons/s

45. There are no other electrons to interact with hydrogen's single electron; it interacts only with the proton in the nucleus.

47. It depends only on the difference between the values of n.

49. (a)

51. No

53. At $n = 7$, the wavelength of the electron's transition ($n = 7$ to $n = 2$) has moved out of the visible region.

55. 1875 nm; infrared

57. (a) decrease
    (b) no

59. 72.9 nm

61. In the de Broglie equation, λ is the wavelength that the particle of mass m exhibits as the particle travels at speed u, where h is Planck's constant. This equation states that (1) any moving particle will have wavelike properties because a wavelength can be calculated using the equation, and (2) the wavelength of the particle is inversely related to its momentum (mass multiplied by velocity).

63. No

65. (a) 10.8 nm
    (b) 0.180 nm
    (c) 1.24×10^{-27} nm
    (d) 3.68×10^{-54} nm

67. (c)

69. 1.31×10^{-13} m

71. The Bohr model showed the quantized nature of the electron in the atom as a particle moving around the nucleus in concentric orbits. In quantum theory, an orbital is a region of space where the probability of finding the electron is high. The electron is not viewed as a particle, but as a wave, and it is not confined to a clearly defined orbit, but to a probability of being at various locations around the nucleus.

73. Three

75. (a) 1
    (b) 4
    (c) 9
    (d) 16
    (e) 25

77. 3, 2, 1, and 0

79. (a) 2s
    (b) 3p
    (c) 4d
    (d) 1s

81. (a) 2
    (b) 2
    (c) 10
    (d) 2

83. (b)

85. Degenerate orbitals have the same energy and are indistinguishable from each other.

87. As we start from an argon core of electrons, we move to K and Ca, which are located in the s block on the periodic table. It is not until Sc, Ti, V, etc. that we begin to fill electrons into the $3d$ shell.

89. (c) $3s$ < (a) $3d$ < (d) $4p$ < (b) $5g$

91. Li: [He]$2s^1$
    Li$^+$: [He] or $1s^2$
    Ca: [Ar]$4s^2$
    F$^-$: [He]$2s^22p^6$ or [Ne]
    Na$^+$: [Ne] or [He]$2s^22p^6$
    Mg^{2+}: [Ne] or [He]$2s^22p^6$
    Al^{3+}: [Ne] or [He]$2s^22p^6$

93. K: [Ar]$4s^1$
    K$^+$: [Ar]
    S^{2-}: [Ne]$3s^23p^6$ or [Ar]
    N: [He]$2s^22p^3$
    Ba: [Xe]$6s^2$
    Ti^{4+}: [Ar] or [Ne]$3s^23p^6$
    Al: [Ne]$3s^23p^1$

95. Na: [Ne]$3s^1$
    Cl: [Ne]$3s^23p^5$
    Mn: [Ar]$3d^54s^2$
    Mn^{2+}: [Ar]$3d^5$

97. (a) 3
    (b) 2
    (c) 0
    (d) 0

99. Ti; two unpaired electrons

101. Cl^-; no unpaired electrons

103. Al^{3+}, N^{3-}, Mg^{2+}, Cs^+

105. (a) and (d)

107. $5p$; yes

109. Because electrons do not repel each other as much in Na^+ as they do in Na, they have lower energy and will be, on average, closer to the nucleus, which results in a smaller size. When electrons are added to Cl, the electron–electron repulsion increases; therefore the electrons have higher energy, and they will be, on average, farther from the nucleus, thereby creating a larger-size species (Cl^-).

111. Rb. The size of atoms increases down a group because electrons have been added to higher n levels.

113. (a) As the atomic number increases down a group, electrons are added to higher n levels, which leads to a decrease in ionization energy.
(b) As the atomic number increases across a period, the effective nuclear charge increases. Therefore the ionization energy increases across a period of elements.

115. Fluorine, with a higher nuclear charge than that of boron, exerts a higher Z_{eff} on the $2p$ electrons, which results in fluorine's having a higher ionization energy.

117. Sr

119. (a) 1.11×10^{-26} J of energy lost
(b) 17.85 m
(c) A radio telescope

121. (a), (c), and (d)

123. (a) Yes. It is generally observed that as Z increases, so does the second IE. However, Ge's second IE is lower than Ga's because to ionize the second electron in Ga, we need to remove an electron from a lower-energy $4s$ orbital. Also, Br's second IE is lower than Se's because the electron pairing ($4p^4$) in one of the p orbitals for the Br^+ ion lowers its IE slightly.
(b) Rb, because the $4p$ electron being removed by the second ionization energy has a higher Z_{eff} in Rb than that same electron does when removed from Kr.

125. (a) Sn^{2+}: $[Kr]4d^{10}5s^2$
Sn^{4+}: $[Kr]4d^{10}$
Mg^{2+}: $[Ne]$ or $[He]2s^22p^6$
(b) Cadmium has the same electron configuration as Sn^{2+}, and neon has the same electron configuration as Mg^{2+}.
(c) Cd^{2+}

127. (a) Ne, 5.76; Ar, 6.76
(b) The outermost electron in argon is a $3p$ electron, which is mostly shielded by the electrons in the $n = 2$ level (10 electrons) and the $n = 1$ level (two electrons), whereas the outermost electron in Ne is a $2p$ electron, which is shielded only by the electrons in the $n = 1$ level (two electrons).

129. When we think of an orbital, we should think of the electron not as a particle (which in this case would have to move through the node—a region of zero probability), but as a wave. When we think of the electron as a wave, we can envision the p orbital as a standing wave with the node between the two lobes as the location of zero amplitude.

131. The heavier noble gases are easier to ionize (IE decreases down a group in the periodic table) and therefore can combine with oxygen and fluorine.

133. The high velocity of helium atoms means that once they are released into the atmosphere, they can escape the Earth's gravitational pull. Therefore, the Earth's atmosphere contains very little helium.

Chapter 8

1. (a) Group 1 (red)
(b) Group 14 (blue)
(c) Group 16 (purple)

3. Mg^{2+}

5. Group 14 (blue)

7. Fluorine (lilac)

9. (b)

11. The arrangement of the atoms in two of the structures is S–O–S and in the other two structures it is S–S–O. Because the arrangements differ they cannot be resonance structures. Also, the structures do not show a different arrangement of electrons on the atoms, only the bonds are drawn bent, not straight.

13. (a) Sulfur, the central atom in SO_2, is lower in electronegativity than oxygen so it will be shaded blue in the figure and the oxygen atoms will be shaded red.

15. Fluorine (purple)

17. None

19. Yes, for hydrogen and helium

21. Yes

23. In the diatomic molecule XY shown below

$$:\ddot{X}:\ddot{Y}:$$

Lewis counts 6 e^- in 3 lone pairs on both X and Y. He also counts the two electrons shared between X and Y as $2 e^-$ for X and $2 e^-$ for Y. However, there are not $4 e^-$ being shared, only $2 e^-$. It appears that the Lewis counting scheme counts the shared electrons twice.

25. No, molecule XY with a single bond will have 14 valence electrons, with a double bond X=Y will have 12 valence electrons, and with a triple bond will X≡Y have 10 valence electrons.

27. No

29. $\cdot\ddot{N}:$ $:\ddot{O}:$ $:\ddot{F}:$ $:\ddot{Cl}:$

31. Li^+ Mg^{2+} Al^{3+} $\left[:\ddot{F}:\right]^-$

33.

Atom/ion	Electronic Configuration	Number of Valence e^-	Lewis Structure
Xe	$[Kr]4d^{10}5s^25p^6$	8	$:\ddot{Xe}:$
Sr^{2+}	$[Kr]$	0	Sr^{2+}
Cl	$[Ne]3s^23p^5$	7	$:\ddot{Cl}\cdot$
Cl^-	$[Ne]3s^23p^6$	8	$\left[:\ddot{Cl}:\right]^-$

35. (a) $[\ddot{\ddot{X}}\colon]^-$ (b) $[\dot{\ddot{X}}\cdot]^+$

37. (a) 9
 (b) 9
 (c) 9
 (d) 10

39. (a) $\ddot{\ddot{F}}-\ddot{\ddot{F}}\colon$

 (b) $[\colon\!N\!\equiv\!O\colon]^+$

 (c) $\ddot{S}\!=\!\ddot{O}\colon$

 (d) $H\!-\!\ddot{\ddot{I}}\colon$

41. (a) 1
 (b) 3
 (c) 2
 (d) 1

43. (a) F—C—Cl with F above (and H below) — $\ddot{F}-\overset{\overset{\ddot{F}}{|}}{\underset{|}{C}}-\ddot{C}l$ with H below

 (b) Br—C—Br with Br above and H below

 (c) H—C—Cl with H above and Cl below

45. $H-\overset{\overset{\ddot{O}}{\|}}{C}-\ddot{O}-H$

47. $H-\overset{\underset{|}{\ddot{N}}}{\underset{H}{}}-\ddot{C}l\colon$ $H-\overset{\underset{|}{\ddot{N}}}{\underset{H}{}}-\overset{\underset{|}{\ddot{N}}}{\underset{H}{}}-H$ $[\ddot{\ddot{C}l}-\ddot{\ddot{O}}\colon]^-$

49. If there is an electronegativity difference of 2.0 or greater, the bond between the atoms is ionic. For electronegativity differences below 2.0, the bond is covalent.

51. The size of the atom is the result of the nucleus pulling on the electrons. The higher the nuclear charge, the stronger the pull on the electrons within a given valence shell. This is why the size of the atoms generally decreases across a period. A small atom will form a shorter bond with another atom, and these electrons in the bond will "feel" a stronger pull from the nucleus of the small atom since the bonding electrons will be "closer" to the nucleus. This stronger pull results in a higher electronegativity for smaller atoms.

53. A polar covalent bond is one in which the electrons are not equally shared by the atoms.

55. The polar bonds and the atoms with the greater electronegativity (underlined) are

$\underline{C} - Se, \; C - \underline{O}, \; \underline{N} - H, \; \underline{C} - H$

57. (b) (c) are polar covalent bonds; (d) is ionic.

59. Resonance occurs when two or more valid Lewis structures may be drawn for a molecular species. The true structure of the species is a hybrid of the structures drawn.

61. In the resonance forms of SO_2, there is one single S–O bond and one double S–O bond and all the atoms obey the octet rule. In order for H_2S to show resonance, the Lewis structure below would have to be valid.

$H\!-\!\overset{\ddot{S}}{}\!=\!H \longleftrightarrow H\!=\!\overset{\ddot{S}}{}\!-\!H$

In either structure, the double-bonded hydrogen would have $4e^-$, which is in violation of its duplet rule for hydrogen.

63. The actual structure of the species is a hybrid of the resonance structures. If a particular bond resonates between a single and a double bond, the bond length will be equal to $1\frac{1}{2}$ bond lengths.

65. Two resonance structures of a cyclobutadiene-like molecule showing alternating double bonds:
$\begin{array}{c}H\\ \end{array}$ C=C ... C=C with H atoms ⟷ shifted double bonds

67.

For N_2O_2

$\ddot{O}\!=\!\dot{N}\!-\!\dot{N}\!=\!\ddot{O} \longleftrightarrow \colon\!O\!\equiv\!N\!-\!\dot{N}\!-\!\ddot{O}\colon \longleftrightarrow \ddot{O}\!=\!N\!=\!\dot{N}\!-\!\ddot{O}\colon$

$\ddot{O}\!-\!N\!\equiv\!N\!-\!\ddot{O}\colon \longleftrightarrow \ddot{O}\!=\!\dot{N}\!-\!N\!\equiv\!O\colon \longleftrightarrow \ddot{O}\!-\!\dot{N}\!=\!N\!=\!\ddot{O}$

For N_2O_3

$\ddot{O}\!=\!\dot{N}\!-\!N\overset{=\ddot{O}}{\underset{-\ddot{O}\colon}{}} \longleftrightarrow \ddot{O}\!=\!\dot{N}\!-\!N\overset{-\ddot{O}\colon}{\underset{=\ddot{O}}{}} \longleftrightarrow$

$\ddot{O}\!=\!N\!=\!N\overset{-\ddot{O}\colon}{\underset{-\ddot{O}\colon}{}} \longleftrightarrow \colon\!O\!\equiv\!N\!-\!N\overset{-\ddot{O}\colon}{\underset{-\ddot{O}\colon}{}}$

69.

$H\!-\!\ddot{C}\!-\!N\!\equiv\!O\colon \longleftrightarrow H\!-\!\ddot{C}\!=\!N\!=\!\ddot{O} \longleftrightarrow H\!-\!C\!\equiv\!N\!-\!\ddot{O}\colon$

71. $\left[\;\overset{\overset{\dot{O}\cdot}{\|}}{\underset{\ddot{O}\colon \quad \ddot{O}\colon}{C}}\;\right]^{2-} \longleftrightarrow \left[\;\overset{\overset{\ddot{O}\colon}{}}{\underset{\colon\ddot{O} \quad \ddot{O}\colon}{C}}\;\right]^{2-} \longleftrightarrow \left[\;\overset{\overset{\ddot{O}\colon}{}}{\underset{\ddot{O}=\quad \ddot{O}\colon}{C}}\;\right]^{2-}$

73. The best possible structure for a molecule by formal charge will be the structure in which the formal charges are minimized and the negative formal charges are on the most electronegative atoms in the structure.

75. The electronegativity of oxygen (3.5) is higher than that of sulfur (2.5), so the negative formal charge will be on the O atom in the structure that contributes most to the bonding.

77. No, for example in CO, the O atom has a positive formal charge.

79. For HCN

$$H—C≡N:$$

For HC_3N

$$H—C≡C—C≡N:$$

81. (resonance structures of S–N–N–N–N chains with formal charges)

None of these resonance structures have all zero formal charges. Therefore, the most stable structures will be ones that have minimal formal charge with negative charges on the most electronegative atom (N) and where the formal charges are not separated in the molecule, but are next to each other. Two structures meet these criteria

(two preferred resonance structures shown)

83. Formamide

(two resonance structures shown)

Methyl formate

(two resonance structures shown)

85. (three resonance structures of CNO⁻)
Preferred structure

(three resonance structures of NCO⁻)
Preferred structure

(three resonance structures of CON)
Preferred structure

87. Yes

89. (a), (b), and (c)

91. (a) 12
(b) 8
(c) 12
(d) 10

93. (Lewis structures of NOF₃ and POF₃)

In POF_3 there is a double bond and in NOF_3 there is not.

95. (Lewis structures of SeF₄ and SeF₅⁻)

Selenium expands its octet in both SeF_4 and in SeF_5^-.

97. (Lewis structure of Cl₂O₂ type)

The central chlorine atom has an expanded octet.

99. (c), (d), and (e)

101. (a) S
(b) N
(c) C
(d) O

103. (d)

105. No

107. The N–O bond in N_2O_4 has a bond order of 1.5 due to four equivalent resonance forms, whereas the N–O bond in N_2O has a bond order of 1.5 due to resonance between three resonance forms. Therefore, N_2O_4 and N_2O are expected to have nearly equal bond lengths.

109. $NO^+ < NO_2^- < NO_3^-$

111. $NO_3^- < NO_2^- < NO^+$

113. We must account for all the bonds that break and all the bonds that form in the reaction. In order to do so we must start with a balanced chemical reaction.

115. If the compounds are in the solid or liquid phase, interactions between molecules may slightly change the bond enthalpy for a given bond.

117. (a) 862 kJ/mol
(b) 98 kJ/mol
(c) 93 kJ/mol

119. 1070 kJ/mol

121. 9.17×10^{-19} J/molecule

123. The incomplete combustion reaction releases 278.5 kJ/mol less heat than the complete combustion reaction.

125. (two resonance structures of O=C=C=C=O type)

Because of resonance, the two terminal C–O bonds both have a bond order of 2.5, so they are predicted to be of the same bond length.

127. −631 kJ/mol

129. (a) •Be• (b) •Ȧl• (c) •Ċ• (d) •He•

131. $\overset{+0}{\ddot{S}}=\overset{+0}{C}=\overset{+0}{\ddot{S}}$ $\overset{+0}{\ddot{S}}=\overset{+2}{S}=\overset{-2}{\ddot{C}}$

When C is the central atom, the atoms all carry zero formal charge and this structure then is the preferred structure for carbon disulfide.

133.

(a)   [Lewis structure: C double bonded to O at top, single bonded to two Cl atoms]

(b) $H-\ddot{O}-H$ + [Lewis structure of COCl$_2$] → $\ddot{O}=C=\ddot{O}$ + $2\,H-\ddot{Cl}$

135.

(a) :$\overset{+1}{O}\equiv\overset{+0}{C}-\overset{-1}{\ddot{N}}-\overset{-1}{\ddot{N}}-\overset{+0}{C}\equiv\overset{+1}{O}$: ↔ $\ddot{O}=\overset{+0}{C}=\overset{+0}{N}-\overset{+0}{N}=\overset{+0}{C}=\overset{+0}{\ddot{O}}$ ↔

$\overset{+0}{\ddot{O}}=\overset{-1}{C}-\overset{+1}{N}\equiv\overset{+1}{N}-\overset{-1}{C}=\overset{+0}{\ddot{O}}$ ↔ $\overset{-1}{\ddot{O}}-\overset{+0}{C}\equiv\overset{+1}{N}-\overset{+1}{N}\equiv\overset{+0}{C}-\overset{-1}{\ddot{O}}$ ↔

$\overset{+0}{\ddot{O}}=\overset{+0}{C}=\overset{+0}{N}-\overset{+1}{N}\equiv\overset{+0}{C}-\overset{-1}{\ddot{O}}$:

(b) :$\ddot{Br}-\ddot{N}=\ddot{O}$

:$\overset{+1}{O}\equiv\overset{+0}{C}-\overset{+0}{N}=\overset{+0}{N}-\overset{-1}{\ddot{O}}$: ↔ $\overset{+0}{\ddot{O}}=\overset{-1}{C}-\overset{+0}{\ddot{N}}=\overset{+0}{N}=\overset{+0}{\ddot{O}}$ ↔

:$\overset{-1}{\ddot{O}}-\overset{-1}{C}=\overset{+0}{N}-\overset{+1}{N}\equiv\overset{+1}{O}$:

(c) :$\overset{+1}{O}\equiv\overset{+0}{C}-\overset{-1}{\ddot{N}}-\overset{+0}{\underset{+0}{C}}-\overset{-1}{\ddot{N}}-\overset{+0}{C}\equiv\overset{+1}{O}$: ↔ $\overset{+0}{\ddot{O}}=\overset{+0}{C}=\overset{+0}{N}-\overset{+0}{C}-\overset{+0}{N}=\overset{+0}{C}=\overset{+0}{\ddot{O}}$ ↔

$\overset{+0}{\ddot{O}}=\overset{+0}{C}=\overset{+0}{N}-\overset{-1}{\underset{+0}{C}}=\overset{+1}{N}=\overset{+0}{C}=\overset{-1}{\ddot{O}}$ ↔ :$\overset{-1}{\ddot{O}}-\overset{+0}{C}\equiv\overset{+1}{N}-\overset{-1}{\underset{+0}{C}}-\overset{+1}{N}\equiv\overset{+0}{C}-\overset{-1}{\ddot{O}}$: ↔

:$\overset{-1}{\ddot{O}}-\overset{+0}{C}\equiv\overset{+1}{N}-\overset{-1}{C}-\overset{+0}{N}=\overset{+0}{C}=\overset{+0}{\ddot{O}}$ ↔ $\overset{+0}{\ddot{O}}=\overset{+0}{C}=\overset{+0}{N}-\overset{+1}{C}-\overset{+1}{N}\equiv\overset{+0}{C}-\overset{-1}{\ddot{O}}$:

137. For Cl_2O_6 with a Cl–Cl bond

[Lewis structure with two Cl atoms bonded, each with O atoms]

For Cl_2O_6 with a Cl–O–Cl bond

[Lewis structure]

For ClO_2

$\overset{+0}{\ddot{O}}=\overset{+0}{\ddot{Cl}}=\overset{+0}{\ddot{O}}$

139. (a) ·$C\equiv N$:

The more likely structure for cyanogen (shown below) will be the one with no formal charges on the atoms. This will be the one that contains the C–C bond.

:$\overset{+0}{N}\equiv\overset{+0}{C}-\overset{+0}{C}\equiv\overset{+0}{N}$:   :$\overset{-1}{C}\equiv\overset{+1}{N}-\overset{+1}{N}\equiv\overset{-1}{C}$:

(b) It would be expected that oxalic acid would retain the C–C bond from the cyanogens from which it is formed in the reaction of cyanogens with water. This is consistent with the structure for cyanogen predicted by formal charge.

141. [Lewis structure of SF$_4$ with C≡N group, all atoms marked +0]

143. [Lewis structure of TeOF$_5$ anion with 2– charge]

145. $\overset{+0}{H}-\overset{+0}{C}\equiv\overset{+0}{N}$:   versus $\overset{+0}{H}-\overset{+1}{N}\equiv\overset{-1}{C}$:

In HNC there are nonzero formal charges while there are none on the atoms of HCN. Furthermore, the negative formal charge in HNC is not on the most electronegative atom (N) but rather on carbon.

Less stable compounds may exist in space because there is a lower concentration of other molecules with which to react and because space is cold. Both of these lower the rate of reaction through which HNC would react to form other products.

147. (a) If we can distinguish by electron density using this technique, we will be able to distinguish X–A–A from A–X–A.

(b) Electron diffraction will not be able to distinguish between resonance forms. Remember that resonance forms are not real and that the molecule does not fluctuate between the resonance forms but is rather a hybrid. If the resonance forms for A–X–A shown are all equally weighted (none is more preferred than another), we would expect the average X–A bond to be a double bond as in A=X=A.

149. [Lewis structure of NH$_4^+$ cation] [Lewis structure of SH$^-$ anion]

There cannot be an N–S covalent bond because the nitrogen atom in NH_4^+ has a complete octet through its bonding with hydrogen and it can not expand its octet since it is a second period element.

151. (a, b)

:$\overset{+0}{N}\equiv\overset{+1}{N}-\overset{-1}{\ddot{N}}=\overset{+0}{N}$: ↔ :$\overset{-1}{N}=\overset{+1}{N}=\overset{+1}{N}=\overset{-1}{N}$: ↔ :$\overset{-1}{\ddot{N}}=\overset{+0}{N}-\overset{+1}{N}\equiv\overset{+0}{N}$:

All of these resonance forms have the atoms with formal charges. The middle structure has the formal charges separated over three bond lengths so this one is less preferred. The first and last resonance structures are more preferred and are indistinguishable from each other.

(c) The structures below for cyclic N_4 have no formal charges on any of the nitrogen atoms.

$$
\begin{array}{c}
\overset{\cdot\cdot}{N}\!-\!\overset{\cdot\cdot}{N} \\
\parallel\quad\parallel \\
\cdot\overset{}{N}\!-\!\overset{}{N}\cdot
\end{array}
\;\longleftrightarrow\;
\begin{array}{c}
\overset{\cdot\cdot}{N}\!=\!\overset{\cdot\cdot}{N} \\
\mid\quad\mid \\
\cdot\overset{}{N}\!=\!\overset{}{N}\cdot
\end{array}
$$

153.
$$
\overset{\displaystyle :\overset{\cdot\cdot}{Cl}:}{\underset{-1}{:\overset{\cdot\cdot}{F}}\!-\!\overset{|}{Al}\!=\!\overset{+1}{Cl}:}
$$

155. (b) and (c)

157. (a, b)

$$
\left[:\!\overset{-2}{\overset{\cdot\cdot}{N}}\!-\!\overset{+0}{\overset{\cdot\cdot}{N}}\!=\!\overset{+0}{N}\!-\!\overset{+1}{N}\!\equiv\!\overset{+0}{N}:\right]^{-}
\longleftrightarrow
\left[:\!\overset{-2}{\overset{\cdot\cdot}{N}}\!-\!\overset{+0}{\overset{\cdot\cdot}{N}}\!=\!\overset{+1}{N}\!=\!\overset{+1}{N}\!=\!\overset{-1}{\overset{\cdot\cdot}{N}}:\right]^{-}
\longleftrightarrow
$$

$$
\left[:\!\overset{-2}{\overset{\cdot\cdot}{N}}\!-\!\overset{+1}{N}\!\equiv\!\overset{+1}{N}\!-\!\overset{+0}{N}\!=\!\overset{-1}{\overset{\cdot\cdot}{N}}:\right]^{-}
\longleftrightarrow
\left[:\!\overset{-1}{\overset{\cdot\cdot}{N}}\!=\!\overset{+0}{N}\!-\!\overset{+1}{N}\!\equiv\!\overset{+1}{N}\!-\!\overset{-2}{\overset{\cdot\cdot}{N}}:\right]^{-}
\longleftrightarrow
$$

$$
\left[:\!N\!\equiv\!\overset{+1}{N}\!-\!\overset{+0}{N}\!=\!\overset{+0}{N}\!=\!\overset{-2}{\overset{\cdot\cdot}{N}}:\right]^{-}
\longleftrightarrow
\left[:\!\overset{-1}{\overset{\cdot\cdot}{N}}\!=\!\overset{+0}{N}\!-\!\overset{+0}{N}\!=\!\overset{+1}{N}\!=\!\overset{-1}{\overset{\cdot\cdot}{N}}:\right]^{-}
\longleftrightarrow
$$

$$
\left[:\!\overset{-1}{\overset{\cdot\cdot}{N}}\!=\!\overset{+1}{N}\!=\!\overset{+0}{N}\!-\!\overset{+0}{N}\!=\!\overset{-1}{\overset{\cdot\cdot}{N}}:\right]^{-}
\longleftrightarrow
\left[:\!N\!\equiv\!\overset{+1}{N}\!-\!\overset{-1}{\overset{\cdot\cdot}{N}}\!-\!\overset{+0}{N}\!=\!\overset{-1}{\overset{\cdot\cdot}{N}}:\right]^{-}
\longleftrightarrow
$$

$$
\left[:\!\overset{-1}{\overset{\cdot\cdot}{N}}\!=\!\overset{+0}{N}\!-\!\overset{-1}{\overset{\cdot\cdot}{N}}\!-\!\overset{+1}{N}\!\equiv\!\overset{+0}{N}:\right]^{-}
$$

The structures that contribute most will have the lowest formal charges (last four structures shown above). This will mean that the end N–N bonds will be close to the length of a double bond and the middle N–N bonds in the structure will be close to a bond order of 1.5 (between a single and double bond).

(c) N_3^- has the Lewis structures.

$$
\left[:\!\overset{-2}{\overset{\cdot\cdot}{N}}\!-\!\overset{+1}{N}\!\equiv\!\overset{+0}{N}:\right]^{-}
\longleftrightarrow
\left[:\!\overset{-1}{\overset{\cdot\cdot}{N}}\!=\!\overset{+1}{N}\!=\!\overset{-1}{\overset{\cdot\cdot}{N}}:\right]^{-}
\longleftrightarrow
\left[:\!N\!\equiv\!\overset{+1}{N}\!-\!\overset{-2}{\overset{\cdot\cdot}{N}}:\right]^{-}
$$

From these resonance structures we see that each N–N bond is predicted to be of double-bond character. Therefore, in N_5^- there are two longer N–N bonds than in N_3^-. N_3^- has the higher average bond order.

159.

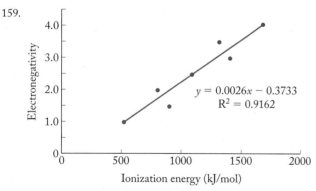

$$y = 0.0026x - 0.3733$$
$$R^2 = 0.9162$$

Extending the line where x = the ionization energy of neon (2080 kJ/mol) gives a value for neon's electronegativity (y):

$$y = 0.0026(2080) - 0.3733 = 5.0$$

161. (a) Isoelectronic means that the two species have the same number of electrons. N_2O and N_2F^+ both have 16 valence electrons.

(b–d)

$$
\left[:\!N\!\equiv\!\overset{+1}{N}\!-\!\overset{+0}{\overset{\cdot\cdot}{F}}:\right]^{+}
\longleftrightarrow
\left[:\!\overset{-1}{\overset{\cdot\cdot}{N}}\!=\!\overset{+1}{N}\!=\!\overset{+1}{\overset{\cdot\cdot}{F}}:\right]^{+}
\longleftrightarrow
\left[:\!\overset{-2}{\overset{\cdot\cdot}{N}}\!-\!\overset{+1}{N}\!\equiv\!\overset{+2}{F}:\right]^{+}
$$

The central nitrogen atom in all the resonance structures always carries a +1 formal charge. The second and third resonance forms for N_2F^+ shown above are unacceptable because they have greater than the minimal formal charges on the atoms.

(e) Yes, fluorine could be the central atom in the molecule, but this would place significant positive formal charge on the fluorine atom (the most electronegative element).

$$
\left[:\!N\!\equiv\!\overset{+3}{F}\!-\!\overset{-2}{\overset{\cdot\cdot}{N}}:\right]^{+}
\longleftrightarrow
\left[:\!\overset{-1}{\overset{\cdot\cdot}{N}}\!=\!\overset{+3}{F}\!=\!\overset{-1}{\overset{\cdot\cdot}{N}}:\right]^{+}
\longleftrightarrow
\left[:\!\overset{-2}{\overset{\cdot\cdot}{N}}\!-\!\overset{+3}{F}\!\equiv\!\overset{+0}{N}:\right]^{+}
$$

163. For SF_4 the expansion of the octet is possible. A similar structure for OF_4 is impossible because oxygen, being a second period element, cannot expand its octet.

165. (a) 2
(b) 4
(c) 1
(d) 5
(e) 8

167. The balanced chemical reaction is

$$5\,F_2(g) + 5\,H_2O(\ell) \rightarrow 8\,HF(g) + O_2(g) + H_2O_2(\ell) + OF_2(g)$$

(a) In this reaction, F_2 is reduced to HF and OF_2 and so it is an oxidizing agent. The oxygen in H_2O is oxidized to O_2 and to H_2O_2 (where the oxidation number of O is −1) so H_2O is the reducing agent.

(b) $:\!\overset{\cdot\cdot}{F}\!-\!\overset{\cdot\cdot}{F}:$ (H₂O structure) $H\!-\!\overset{\cdot\cdot}{F}:$

$\overset{\cdot\cdot}{O}\!=\!\overset{\cdot\cdot}{O}$ (H₂O₂ structure) (OF₂ structure)

169.
$$
\begin{array}{c}
F\quad\quad\quad F \\
\diagdown\quad\quad\diagup \\
C\!=\!C \\
\diagup\quad\quad\diagdown \\
F\quad\quad\quad F
\end{array}
$$

This molecule is non-polar overall because the individual bond dipoles are equal in magnitude and as vectors they cancel each other out.

Chapter 9

1. Yes

3. NCCN and N_2F_2 are planar; none of the molecules have delocalized π electrons.

5. O_2^+ has more electrons populating antibonding orbitals than O_2^{2+}.

7. 180°, 90°, 72°

9. Because the electrons take up most of the space in the atom and the nucleus is located in the center of the electron cloud, the electron clouds repel each other before the nuclei get close enough to each other.

11. Both SO_3 and BF_3 have three atoms bonded with no lone pairs on either S or B.

13. Because the lone pair feels attraction from only one nucleus, it is less confined than bond pairs and therefore occupies more space around the central N atom in ammonia.

15. The seesaw geometry has only two lone pair-bond pair interactions at 90° (compared to trigonal pyramidal's three) so it is the lowest energy geometry.

17. (b) octahedral < (c) tetrahedral < (d) trigonal planar

19. Trigonal bipyramidal, seesaw, T-shape, octahedral, square pyramidal, and square planar

21. Tetrahedral

23. Pentagonal pyramidal and distorted octahedron

25. (a) Tetrahedral
    (b) Trigonal pyramidal
    (c) Bent
    (d) Tetrahedral

27. (a) Tetrahedral
    (b) Trigonal planar
    (c) Bent
    (d) Square pyramidal

29. (a) Tetrahedral
    (b) Tetrahedral
    (c) Trigonal planar
    (d) Linear

31. O_3 and SO_2

33. SCN^- and CNO^-

35. Molecular geometry = bent

    Molecular geometry = bent at each S atom

    OR S_2O_2 may have only one central S atom Molecular geometry = trigonal planar

37. Molecular geometry = square planar

    Molecular geometry = pentagonal planar

39.

    All formal charges are zero. The geometry around the P atom is tetrahedral.

41. A polar bond is only between two atoms in a molecule. Molecular polarity takes into account all the individual bond polarities and the geometry of the molecule. A polar molecule will have a permanent, measurable dipole moment.

43. Yes

45. Due to its longer wavelength, IR radiation is less energetic than UV light and therefore does not have enough energy to break chemical bonds.

47. More

49. (b), (d), and (e) are polar; (a) and (c) are nonpolar.

51. All are polar.

53. (a) Freon 13B1
    (b) Freon 22
    (c) Freon 113

55. $COCl_2$ < $COBr_2$ < COI_2. The C—X bonds in the structure are pulling opposite of the C=O bonds, which has an electronegativity difference of 1.0. Looking at the series we see that as the electronegativity of the halogen atom decreases (Cl > Br > I) the overall molecular polarity of the COX_2 molecule will increase because the C—X bond pulls less to balance the C=O bond dipole.

57. No

59. Yes

61. All are sp^2 hybridized.

63. No, the hybridization (sp^2) on the N atoms is the same in both structures; in acetylene the carbon atoms are sp hybridized.

65. C in CO_2 is sp hybridized, N in NO_2 is sp^2 hybridized, O in O_3 is sp^2 hybridized, Cl in ClO_2 is sp^3 hybridized.

67.

    The first resonance structure above has the best formal charge arrangement giving a tetrahedral molecular geometry. Notice that the Cl forms three π bonds to three of the oxygen atoms. This will require that three of the p orbitals on Cl not be involved in the hybridization so that it can form parallel π bonds. Therefore, Cl must use low-lying d orbitals in place of one of the p orbitals for sd^3 hybridization to form the 4 σ bonds to oxygen.

69. sp^3d hybridized

71. Seesaw, sp^3d

73. Yes

75. Yes, in resonance structures the electron distribution is blurred across all the resonance forms which, in essence, defines the delocalization of electrons.

77. The amine N (NH_2 group) is trigonal pyramidal with sp^3 hybridization; the NO_2 group N atom is trigonal planar with sp^2 hybridization so the hybridization of the nitrogen atoms is not the same.

79. The presence of a lone pair on N gives this atom trigonal pyramidal geometry and the nitrogen atom is sp^3 hybridized. The steric

number for S is also 4, which upon first glance, would also mean that the hybridization would be assigned as sp^3. However, notice that the S forms two π bonds to two of the oxygen atoms. This will require that two of the p orbitals on S not be involved in the hybridization so that it can form parallel π bonds. Therefore, S must use two low-lying d orbitals in place of two of the p orbitals for spd^2 hybridization to form the four σ bonds to oxygen.

81. Valence bond theory

83. No

85. No

87.

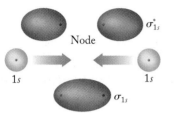 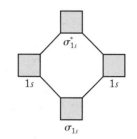

89. N_2^+, 2.5; O_2^+, 2.5; C_2^+, 1.5; Br_2^{2-}, 0; all except Br_2^{2-} are expected to exist.

91. (a), (b), and (c)

93. (b), (c), and (d)

95. (a) and (b)

97. No

99.

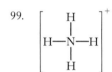

$$\left[\begin{array}{c} H \\ | \\ H-N-H \\ | \\ H \end{array} \right]^+ \quad \text{Molecular geometry = tetrahedral}$$

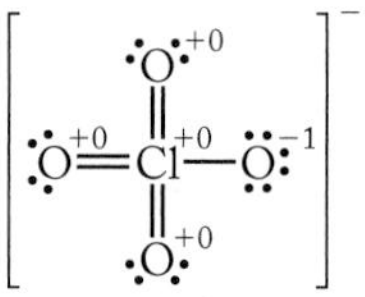

Molecular geometry = tetrahedral

101.

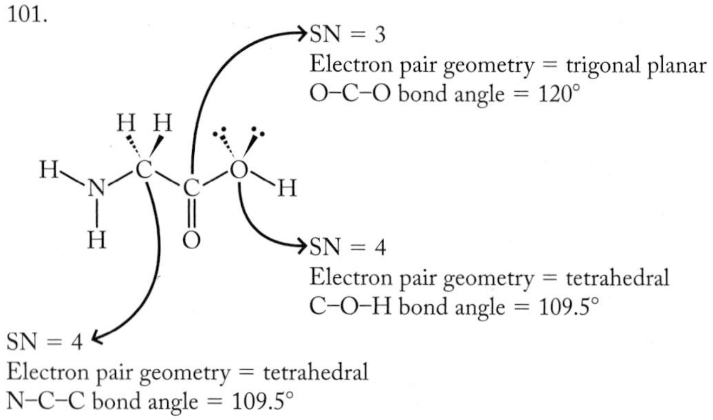

SN = 3
Electron pair geometry = trigonal planar
O–C–O bond angle = 120°

SN = 4
Electron pair geometry = tetrahedral
C–O–H bond angle = 109.5°

SN = 4
Electron pair geometry = tetrahedral
N–C–C bond angle = 109.5°

103. The major sources of methane emission in the U.S. are the decomposition of garbage in landfills and dumps, the loss of methane during natural gas production and distribution, the release of trapped methane in coal mines, and the digestive processes and decomposition of manure of livestock, especially cows.

105. :Cl̈—Ö—Ö—Cl̈:   :Cl̈—Ö—Cl̈=Ö:

Neither of the molecules is linear.

107. (a) $\left[\ddot{C}l=\ddot{O} \right]^+$
(b) 2

109.

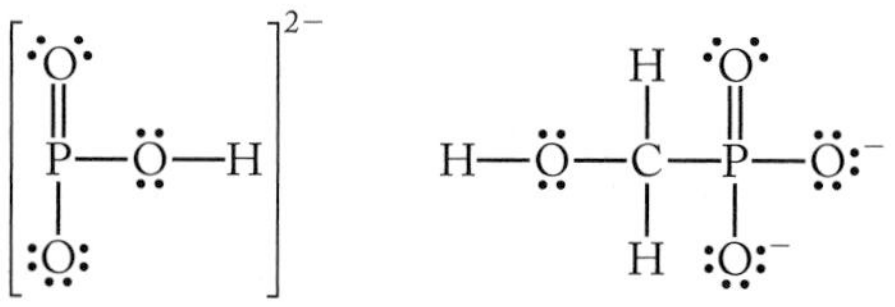

111. 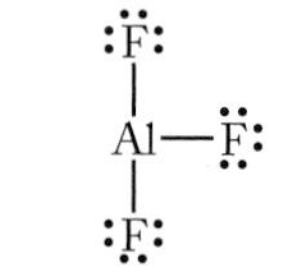 Molecular geometry = trigonal planar
sp^2 hybrid

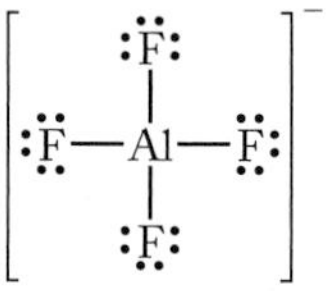

Molecular geometry = tetrahedral
sp^3 hybrid

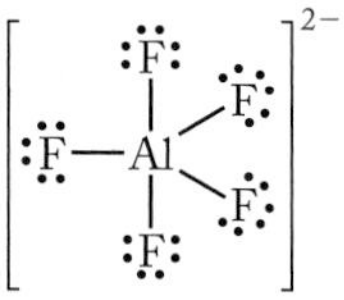

Molecular geometry = trigonal bipyramidal
sp^3d hybrid

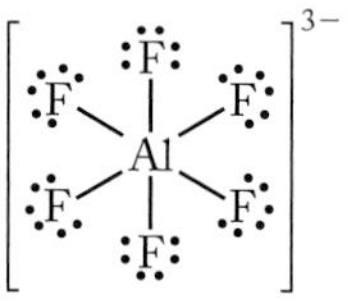

Molecular geometry = octahedral
sp^3d^2 hybrid

113. The structure in which boron is triply bonded to nitrogen is the best representation because the −1 formal charge is on the most electronegative atom.

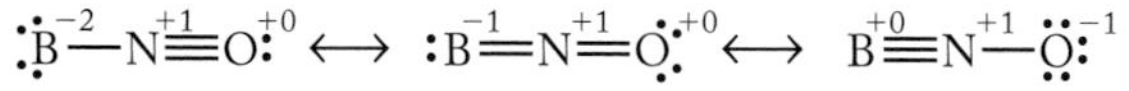

All of these structures have boron with an incomplete octet. Because the steric number around the central nitrogen atom is 2, the molecular geometry is linear.

115. The most stable structure is

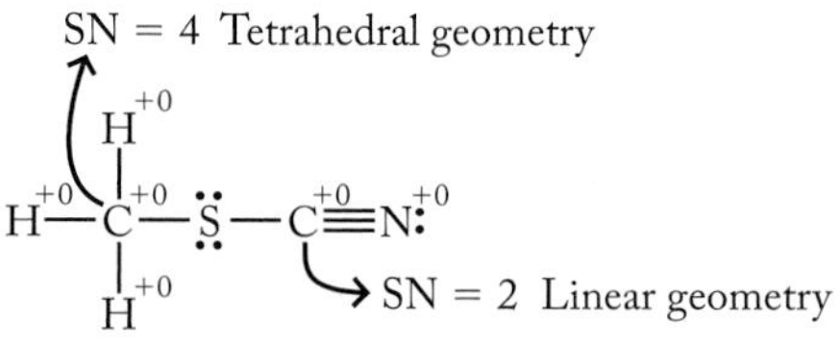

SN = 4 Tetrahedral geometry

SN = 2 Linear geometry

117.

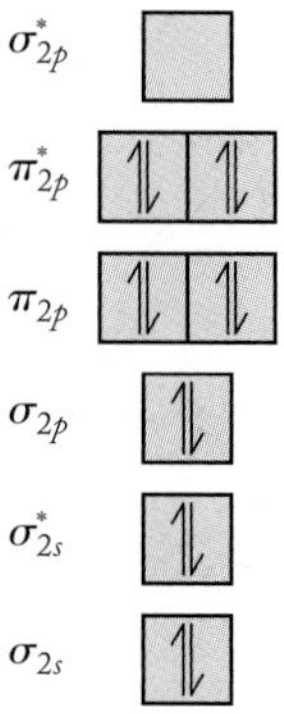

There are $4e^-$ in the π_{2p}^* orbital and $2e^-$ in the σ_{2s}^* orbital for a total of $6e^-$ in the antibonding orbitals in F_2.

119. (a) H—$\overset{+0}{\ddot{A}}$r—$\overset{+0}{\ddot{F}}$: $\overset{+0}{}$
   (b) All formal charges are zero
   (c) Linear
   (d) Polar

121. There are two unpaired electrons in the π_{2p}^* orbitals of oxygen thereby making O_2 paramagnetic.

123. For O_2^{2-} the bond order is 1 (consistent with the Lewis structure). For O_2^- the bond order is 1.5 (not consistent with the Lewis structure).

125. H_2O is the most polar; H_2Te is the least polar.

127. By VSEPR theory, the SN = 4 gives a tetrahedral electron pair geometry about the central atom with a bent molecular geometry and an ideal bond angle of 109.5°. The bond angles may decrease due to the presence of two lone pairs on the central atom, but this doesn't explain the 90° bond angles in H_2S, H_2Se, and H_2Te. By valence bond theory, invoking sp^3 hybridization for the central atom, we would conclude that the bond angles again would be expected to be slightly less than 109.5°. This works well for H_2O, but still does not explain the other compounds' bond angle of 90°. If we do not hybridize the atomic orbitals, the central atom would have a configuration of s^2p^4 in which there are two unpaired electrons in p orbitals at 90° angles from each other to bond with the 1s orbital electrons on the hydrogen atoms. This does explain the 90° angles in H_2S, H_2Se, and H_2Te.

Chapter 10

1. KF

3. Phosphine (XH_3) has the lower boiling point.

5. From the plot, bp for X = 5°C, and bp for Y = 20°C at 1.00 atm. Y has the stronger intermolecular forces.

7. Both substances are nonelectrolytes because if they were electrolytes we would observe deviations from linearity at the higher molalities due to ion pair formation. Given that the K_f value is the slope of the line in the plot and because both substances give the same $\Delta T_f/m$ value, we conclude that the K_f value is the same for aqueous solutions of both substances.

9. (c)

11. (a)

13. A Born–Haber cycle consists of a series of steps and corresponding enthalpy changes that when added together will give the value of the enthalpy for one of the unknown steps in the cycle (e.g., lattice energy or heat of formation). This calculation uses Hess's law in that the path to formation of a salt (from the elements or through a series of steps) gives the same value. Throughout all of these steps in the Born–Haber cycle, energy is conserved.

15. CsBr < KBr < SrBr₂

17. Melting point decreases as the atomic number of X increases.

19. $U = -723$ kJ/mol

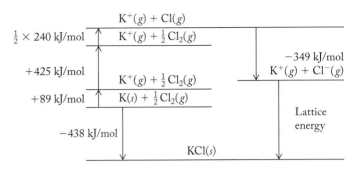

21. The water molecule will be oriented around Cl⁻ so as to point the partially positive hydrogen atoms towards the Cl⁻ ion.

23. Because of the full positive or negative charge on the ion, the ion–dipole interaction is stronger than the dipole–dipole interaction.

25. The charge build up on H (partially positive) and the electronegative element (partially negative) means that the X–H bond is polar but not ionic. It is still a dipole–dipole interaction except that its strength is noticeably higher than other dipole–dipole interactions.

27. Cl⁻

29. CH_3F is a polar molecule and therefore has stronger intermolecular forces than that of the nonpolar molecules of CH_4, which have only the weaker dispersion forces.

31. The H atom in methane has just a single bond to the relatively low electronegative C atom and, therefore, the C–H bond is not polar enough to exhibit hydrogen bonding. In methanol, however, one of the H atoms is bonded to oxygen, which is second to fluorine in electronegativity. It is this H that shows hydrogen bonding in methanol.

33. (b)

35. Dispersion forces, also called London forces

37. The substance with the highest boiling point will be that which has the largest sum of intermolecular forces. In this case, the greater London dispersion forces of CH_2Cl_2 add to the weaker dipole–dipole interactions to give stronger intermolecular forces between the CH_2Cl_2 molecules compared to that of CH_2F_2 molecules. Also, the molar mass of CH_2Cl_2 is higher than that of CF_2Cl_2 so it requires more energy to vaporize.

39. (a) CCl_4
   (b) C_3H_8

41. The nonideal behavior of gases at high pressures arises from the true volume of the gas molecules. As the temperature is lowered, the kinetic energy of the gas particles is reduced enough that when particles collide they temporarily stick together via their intermolecular forces.

43. Ar has more electrons that are more easily polarizable compared to He.

45. Miscible and insoluble are opposites of each other in the range of possible solubilities of a solute in a solvent. At one end, miscible solutes and solvents dissolve completely in each other. At the other extreme, an insoluble solute does not dissolve at all.

47. Henry's law gives the direct relationship between concentration and pressure. When we increase the pressure of a gas above a solvent, we increase the number of gas molecules above the solvent. More gas molecules striking the surface of the solvent means more gas molecules will dissolve in the solution.

49. The Henry's law constant, k_H

51. Unlike N_2 and O_2, which do not react with water, dissolved CO_2 forms carbonic acid, H_2CO_3, in water. Once this reaction occurs, more CO_2 can dissolve in the water.

53. Hydrophilic substances dissolve in water. Hydrophobic substances do not dissolve, or are immiscible, in water.

55. (a) $CHCl_3$
    (b) CH_3OH
    (c) NaF
    (d) BaF_2

57. $0.0374 \ mol/L \cdot atm$

59. (a) $0.00274 \ mol/L$
    (b) $0.0235 \ mol/L$

61. At $10°C$, $k_H = 0.00164 \ mol/L \cdot atm$, at $20°C$, $k_H = 0.00129$ $mol/L \cdot atm$, and at $30°C$, $k_H = 0.00105 \ mol/L \cdot atm$.

63. (b)

65. (d)

67. A nonvolatile solute dissolves into a solvent and does not enter appreciably into the gas phase.

69. When the average kinetic energy of the liquid molecules increases with increasing temperature, more of the molecules can escape the liquid phase and enter the gas phase. More molecules in the gas phase increase the vapor pressure.

71. As intermolecular forces increase in strength, the vapor pressure decreases.

73. The vapor pressure of pure water is greater than the vapor pressure of seawater, which contains dissolved solutes. The greater the vapor pressure, the higher the rate of evaporation from the pure water which will eventually leave the beaker empty.

75. $CH_3CH_2OH < CH_3OCH_3 < CH_3CH_2CH_3$

77. Mole fraction = 0.70; vapor pressure = 16.7 torr

79. Although both processes end with the substance in the gas phase, sublimation "skips" a step in that the solid does not liquefy before evaporating.

81. If you are along the equilibrium line in a phase diagram, the two phases that border that line coexist at that pressure–temperature combination.

83. (a) Solid
    (b) Gas

85. Yes

87. (1) Reduce the temperature from $25°C$ to $0.010°C$. (2) Reduce the pressure from 1 atm to 0.0060 atm.

89. Water vaporizes

91. $-57°C$

93. (a) Liquid
    (b) Liquid
    (c) Gas

95. Water has a higher surface tension than methanol due to its stronger hydrogen bonding.

97. The expansion of water from forming hydrogen bonds between the molecules in the pipes creates significant pressure on the wall of the pipes to cause them to burst.

99. The cohesive forces in mercury are stronger than the adhesive forces of the mercury to the glass.

101. Molecules on the surface of a liquid are only pulled by molecules under and beside them, creating a tight film of molecules on the surface.

103. As temperature increases, the surface tension and viscosity of a liquid decrease because there are fewer molecules held together by tight intermolecular forces, and they have more energy to slide past each other more freely.

105. Water

107. The liquid with the higher boiling point (B) has the higher surface tension and viscosity because that liquid has stronger intermolecular forces between its molecules.

109. Molarity is the moles of the solute in one liter of solution. Molality is the moles of solute in one kilogram of solvent.

111. The presence of dissolved solutes in seawater lowers the freezing point.

113. A strong electrolyte completely dissociates in the solvent. This dissociation yields two or more particles in solution from one dissolved solute particle. This results in greater changes in the melting and boiling points compared to that of a solute that does not dissociate.

115. The theoretical value of i for CH_3OH is 1 because methanol is molecular and does not dissociate in a solvent such as water. NaBr has a theoretical value of $i = 2$ because it dissociates into two particles upon dissolution (Na^+ and Br^-). K_2SO_4 has a theoretical value of $i = 3$ because it dissociates into three particles upon dissolution ($2 \ K^+$ and SO_4^{2-}).

117. A semipermeable membrane is a boundary between two solutions through which some molecules may pass but others cannot. Usually small molecules may pass through, but large molecules are excluded.

119. Solvent flows across the membrane from the more dilute solution side to the more concentrated solution side to balance the concentration of solutes on both sides of the membrane.

121. Because reverse osmosis goes against the natural flow of a solvent across the membrane, the key component needed is a pump to apply pressure to the more concentrated side of the membrane. Other components needed include: a containment system, piping to introduce and remove the solutions, and a tough semipermeable membrane that can withstand the high pressures needed.

123. (a) $0.58 \ m$
     (b) $0.18 \ m$
     (c) $1.12 \ m$

125. (a) 307 g
    (b) 86.8 g
    (c) 28.8 g

127. 6.5×10^{-5} m NH_3; 8.7×10^{-6} m NO_2^-; 2.195×10^{-2} m NO_3^-

129. 3.81°C

131. 2.52×10^{-2} m

133. −1.89°C

135. 0.5 m $CaCl_2$

137. 0.0100 m $Ca(NO_3)_2$

139. 0.06 m $FeCl_3(aq)$ < 0.10 m $MgCl_2(aq)$ < 0.20 m $KCl(aq)$

141. (a) A to B
    (b) B to A
    (c) A to B

143. (a) 57.52 atm
    (b) 0.682 atm
    (c) 52.9 atm
    (d) 46.5 atm

145. (a) 2.75×10^{-2} M
    (b) 1.11×10^{-3} M
    (c) 1.00×10^{-2} M

147. This statement is false. The molarity of NaCl will be greater (1.5 times) than the molarity of $CaCl_2$.

149. (a) Increase
    (b) Decrease
    (c) Increase

151. 94.11 g/mol

153. $C_{10}H_{12}O_2$

155. (d)

157. (a)

159. While the dispersion forces between methanol molecules will be greater because methanol has more electrons and greater molar mass, water can form two hydrogen bonds compared to methanol's one hydrogen bond.

161. Increase

163. Contract

165. Ice floats on water because it has a lower density than liquid water due to the open lattice formed by extensive hydrogen bonding between the water molecules.

167. Yes

169. −15.3°C

171. LiCl, 1.87; HCl, 1.91; NaCl, 1.88

173. NaCl, 0.120 atm; $MgCl_2$, 0.260 atm; NaCl + $MgCl_2$, 0.182 atm

175. Seawater and natural brine sources

177. Increase

179. (a) Water purification, disinfectants, bleaches, the manufacture of organic chlorine compounds, and the synthesis of the other halogens
    (b) Manufacture of bromine compounds, water purification, production of dyes and disinfectants, and bromination of vegetable oil

181. (a) Iodide
    (b) Bromide
    (c) Chloride
    (d) Iodide

183. (a) HOCl (hypochlorous acid)
    (b) $HOCl_3$ (chloric acid)
    (c) $HOCl_3$ (chloric acid)
    (d) $HClO_4$ (perchloric acid)

Chapter 11

1. (a) and (b) are crystalline; (c) and (d) are amorphous

3.

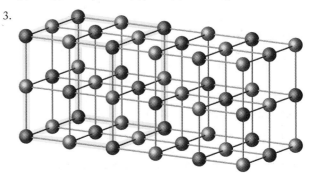

The formula of this compound is $(red)_1(blue)_1$.

5. 3.5 A atoms, 0.5 B atoms

7. AB_3X

9. 3.81 g/cm³

11. $MgAl_2O_4$

13. Li_2S

15. Cs (blue) and Sr (purple)

17. MgB_2

19. Because the electrons are free to distribute themselves throughout the metal, the application of an electrical potential across the metal will cause the electrons to travel freely towards the positive potential.

21. Ionic bonds are stronger than metallic bonds.

23. Yes

25. Groups 2 and 12

27. Phosphorus will give Si a higher conductivity because it has one more valence electron making it an n-type semiconductor.

29. Group 14

31. (a) n-type
    (b)

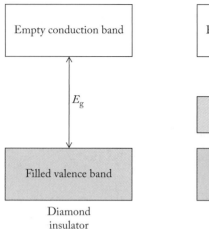

Diamond
insulator

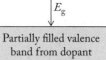

N-doped diamond
semiconductor

   (c) 4.68×10^{-19} J

33. InN

35. Cubic closest-packed atoms have an *abab* pattern whereas hexagonal closest-packed atoms have an *abcabc* pattern.

37. Body-centered cubic

39. No, an allotrope is a different *molecular* form of an element. Iron is not molecular.

41. For body-centered cubic $\ell = \dfrac{4r}{\sqrt{3}}$; for face-centered cubic $\ell = 2r\sqrt{2}$

43. 104.2 pm

45. 513 pm

47. (c)

49. No

51. Because the metallic bonding between copper and silver in the alloy is weaker due to a mismatch in their atomic sizes.

53. Tungsten is the host and carbon occupies the holes.

55. (a) XY_3
    (b) YX_3

57. Octahedral holes

59. Substitutional alloy

61. Yes

63. (a) AB
    (b) A_2B
    (c) AB

65. $\dfrac{1}{5}$

67. Each S atom has a bent geometry due to sp^3 hybridization.

69. The hybridization of the B and N atoms in a six-membered ring is sp^2 so the rings are flat as they are in graphite.

71. (a) $2 O^{2-}$ ions and $4 H^+$ ions

    (b) $\left[:\ddot{O}: \right]^{2-} 2 H^+$

73. 109.5°

75. The K^+ ions are too large to fit well in the octahedral holes in the fcc lattice.

77. Because the sizes of the Cl^- ion (181 pm) and of the Cs^+ ion (169 pm) are so similar, the Cs^+ atom at the center occupies the center of a cubic cell so it could be described as a bcc lattice. If we look just at the arrangement of Cs^+ ions or Cl^- we see a simple cubic structure, and the CsCl lattice is made up of these simple cubic arrangements that interpenetrate each other.

79. No, the Cl^- radius is so much larger than that of Na^+ that in forming the fcc lattice, the Na^+ would not be close-packed.

81. Yes

83. Less than

85. $MgFe_2O_4$

87. (a) Octahedral
    (b) Half

89. If Cd^{2+} ions are placed in tetrahedral holes (in the sphalerite structure) rather than the octahedral holes (in the rock salt structure), the lattice must expand to accommodate the relatively large Cd^{2+} cation (which has a radius ratio of 0.516), resulting in a lower density for the sphalerite structure for CdS.

91. 5.18 g/cm³

93. 421 pm

95. Ceramic = thermal insulator; metal = ductile, electrically conductive, malleable

97. $Mg_3(OH)_4(Si_2O_5)$

99. $2 KAlSi_3O_8(s) + 2 H_2O(\ell) + CO_2(g) \rightarrow$
    $\qquad\qquad Al_2(Si_2O_5)(OH)_4(s) + 4 SiO_2(s) + K_2CO_3(aq)$
    Not a redox reaction

101. (a) $3 CaAl_2Si_2O_8(s) \rightarrow Ca_3Al_2(SiO_4)_3(s) + 2 Al_2SiO_5(s) + SiO_2(s)$
     (b) $Si_2O_8{}^{8-}$ in anorthite, $SiO_4{}^{4-}$ in grossular, $SiO_5{}^{6-}$ in kyanite

103. Cubic holes for Ba^{2+}, octahedral holes for Ti^{4+}

105. An amorphous solid has no regular, repeating lattice that can diffract X-rays.

107. X-rays have wavelengths on the order of the atomic separation in crystals.

109. Using a shorter wavelength of X-rays means that data may be collected over a smaller scanning range.

111. Halite

113. $n = 2$; average of 585 pm and 579 pm is 582 pm for the spacing between layers in the crystal.

115. 4.76°

117. XYZ_3

119. 33.5%

121. (a) 139 pm
     (b) 9.72 g/cm³
     (c) 6750 Mo atoms

123. (a) 7.90 g/cm³
     (b) 3.42 g/cm³
     (c) 3.36 g/cm³

125. AuZn

127. 52.4%

129. Substitutional alloys

131. The radius ratio indicates that Cu atoms fit into the cubic holes of the Al lattice. Another way to look at this is as a simple cubic arrangement of Cu atoms with an Al atom in the center of the unit cell. To be consistent with the formula (Cu_3Al), we would have to consider three unit cells, one of which would have one Al atom in the center.

133. (a) $Sn(s) + 2Cl_2(g) \rightarrow SnCl_4(\ell)$
(b) $Pb(s) + Cl_2(g) \rightarrow PbCl_2(s)$
(c) $SiCl_4(\ell) \xrightarrow{\text{heat}} Si(s) + 2Cl_2(g)$

135. Group 14 elements do not have low ionization energies or electron affinities, so they tend to neither easily lose nor gain an electron and as a result form bonds by sharing electrons and not by forming ionic bonds.

Chapter 12

1. All of these hydrocarbons likely come from the gasoline cut.

3. Pine oil, oil of celery

5. (b) and (d)

7. Benzyl acetate has an ester group; carvone has a ketone and an alkene group; cinnamaldehyde has an alkene and an aldehyde group.

9. (a) Solute–solvent interactions here are weaker than solvent–solvent interactions.
(b) Solute–solvent interactions here are stronger than solvent–solvent interactions.

11. (c)

13. The monomeric unit is acrylonitrile

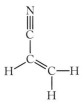

and this polymer is an addition polymer.

15. sp, sp^2, and sp^3 hybridization for triple, double, and single bonds

17. No

19. Carboxylic acids, aldehydes, alkenes, alcohols, alkynes, and aromatic rings

21. 3565 monomers

23. Yes

25. sp^3

27. The carbons atoms are sp^3 hybridized with bond angles of 109.5°.

29. No

31. No

33.

n-Pentane   2-Methylbutane   2,2-Dimethylpropane

35. (a) 1,2-dimethylhexane
(c), (d) 2-methylheptane

37. (a) C_8H_{18}
(b) C_9H_{20}
(c) C_8H_{18}
(d) C_8H_{18}
(e) C_9H_{20}

39. −124 kJ

41. $C_3H_8 < C_8H_{18} < C_{14}H_{30}$

43. Boiling point

45. C_5H_{12}

47. 41.0 kJ/mol

49. 60 torr

51. $P_{\text{isooctane}} = 80.5$ torr, $P_{\text{tetramethylbutane}} = 37.0$ torr

53. Structural isomers have different connectivity of the atoms whereas geometric isomers have the same atom connectivity but a different arrangement of the atoms in space.

55. No, both have the formula C_nH_{2n}.

57. The terminal carbon atom of the $C=C$ bond has two of the same atoms (two hydrogen atoms).

59. The double bond in the ring is required to be *cis* and the terminal carbon atom in the other $C=C$ bond has two hydrogen atoms, so *cis-trans* isomerism is not possible.

61. Ethylene has a double bond that reacts with HBr, but polyethylene has only saturated carbon–carbon bonds that do not react with HBr.

63. (a) *trans*, *E*
(b) *cis*, *Z*

65. 681.2 kJ, endothermic

67.

69. All atoms are sp^2 hybridized with 120° angles between the carbon atoms.

71. Tetramethylbenzene has three isomers; pentamethylpenzene has no isomers.

73. Yes

75.

77. For benzene, fuel value = 41.8 kJ/g; for ethylene, fuel value = 50.3 kJ/g; benzene has a lower fuel value.

79. The more oxygenated a fuel is, the lower the fuel value.

81. Alcohols can form stronger hydrogen bonds while ethers have weaker dipole–dipole interactions. Stronger intermolecular forces result in higher boiling points.

83. The alcohol is evaporating, which is an endothermic process that takes energy in the form of heat from your arm, making it feel cold.

85. (a) and (d) are alcohols, (b) and (c) are ethers; (b) < (c) < (d) < (a)

87. For butanol, fuel value = 36.1 kJ/g; for diethyl ether, fuel value = 36.7 kJ/g; diethyl ether has a slightly higher fuel value.

89. Carboxylic acids hydrogen-bond with water and aldehydes do not.

91. Yes

93. No

95. (a) Because there are no formal charges on any of the atoms.

97. (a), (b), and (d)

99. (b)

101.

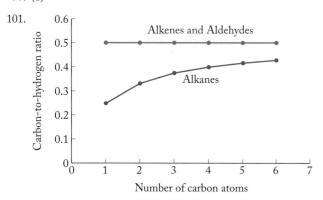

The 1:2 C:H ratio of aldehydes correlates with the C:H ratio for alkenes.

103. (a) Acetic acid + n-butanol
(b) 2-Methylbutanoic acid + ethanol
(c) Acetic acid + 3-methylbutanol

105. For formaldehyde, fuel value = 19.0 kJ/g; for formic acid, fuel value = 5.53 kJ/g; formaldehyde has a higher fuel value.

107. (a) 4
(b) 6
(c) 8

109. (a) Condensation, methanol
(b) Because of the presence of the six-membered ring, Kodel might be better able to accept organic nonpolar dyes.

111. Methylamine has a smaller nonpolar hydrocarbon chain.

113. We need to have water eliminated from the amine in reaction with a carboxylic acid to produce an amide. A tertiary amine does not have an N–H bond to react with the carboxylic acid to form the amide and water.

115. The N atom highlighted in nicotine is an amine; the N atom highlighted in valium is an amide.

117. For reaction (1) ΔH_{rxn} = 17.5 kJ; for reaction (2) ΔH_{rxn} = −125.2 kJ

119. −138.7 kJ

121. 2.52 g of methanol; 3.45 g of carbon dioxide

123. Compound A is diethyl ether; compound B is butanol.

125. (a) 12
(b) Alkene and aldehyde
(c) *cis-trans* isomerism and structural isomers

127. Yes

129. Amide groups

131. (a)

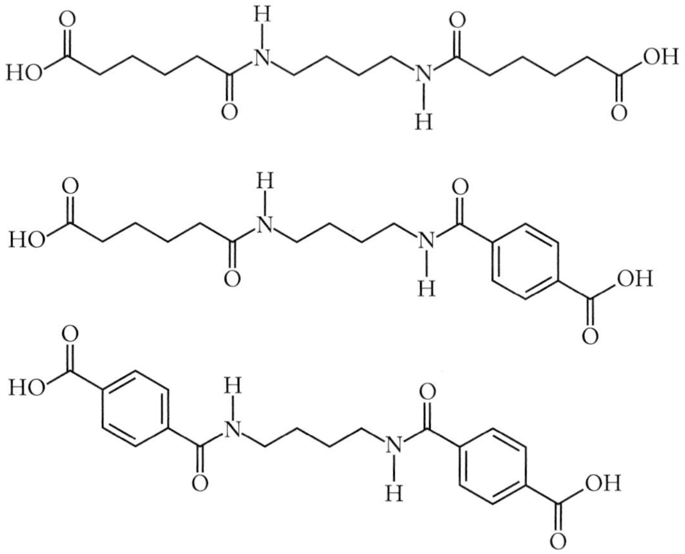

(b) 1 mol adipic acid: 1 mol terephtalic acid: 2 mol putrescine

133. (a)

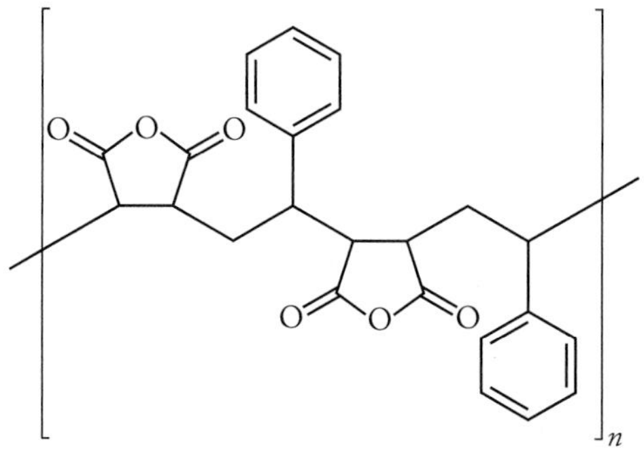

(b) It would be more hydrophilic and be less rigid.

135. (a) $R_2SiCl_2 + H_2O \rightarrow R_2SiClOH + HCl$

$R_2SiClOH + HOSiClR_2 \rightarrow R_2ClSi–O–SiClR_2 + H_2O$

(b) The side chains are nonpolar.

Chapter 13

1. Increase

3. Low probability because the entropy for that process would not be favored; the maximum entropy is where A and B are evenly distributed throughout the connected bulbs.

5. At the intersection $\Delta G = 0$; the reaction is nonspontaneous at $T > 60°C$.

7. The sign is reversed.

9. Eight microstates; the most likely sum for the microstates is +1 and −1.

11. ΔS_{sys} positive, ΔS_{surr} negative

13. (a) and (b)

15. ΔS_{surr} must be less than 48.0 kJ/mol for the reaction to remain nonspontaneous.

17. (a) Live fish
    (b) Wet paint
    (c) SO_3
    (d) An aquarium with fish

19. Gas

21. Fullerenes

23. (a) $CH_4(g) < CF_4(g) < CCl_4(g)$
    (b) $CH_3OH(\ell) < C_2H_5OH(\ell) < C_3H_7OH(\ell)$
    (c) $HF(g) < H_2O(g) < NH_3(g)$

25. (a) Negative
    (b) Negative
    (c) Negative
    (d) Positive

27. Negative

29. Yes

31. (a) 24.9 J/K
    (b) −146.4 J/K
    (c) −73.2 J/K
    (d) −175.8 J/K

33. 218.9 J/(mol·K)

35. No

37. Yes

39. The reaction is nonspontaneous and proceeds in the opposite direction.

41. No, if ΔS_{rxn} is positive, the reaction will be spontaneous at all (even high) temperatures.

43. ΔS positive, ΔH positive, ΔG negative

45. (a) Spontaneous
    (b) Nonspontaneous
    (c) Spontaneous
    (d) Spontaneous

47. For NaBr, −18 kJ/mol; for NaI, −29 kJ/mol

49. $\Delta G° = 91.4$ kJ/mol, 981 K

51. $\Delta H° = 44.0$ kJ, $\Delta S° = 118.7$ J/K, $T_b = 370.7$ K or 97.7°C

53. $\Delta H° = -146.5$ kJ, $\Delta S° = -185.7$ J/K, spontaneous at temperatures below 789 K

55. (a) 171.8 kJ, nonspontaneous
    (b) −62.2 kJ, spontaneous
    (c) −31.1 kJ, spontaneous
    (d) 5.2 kJ, nonspontaneous

57. (a) Spontaneous at high temperature
    (b) Spontaneous at low temperature
    (c) Spontaneous at low temperature
    (d) Spontaneous at low temperature

59. $\Delta G° = 33.1$ kJ; no

61. (a) Endothermic
    (b) Yes
    (c) No, because the reaction is not favored by enthalpy but is by entropy, the ΔG of the reaction will be negative (spontaneous) only at high temperatures.
    (d) 500 K

63. The bonds are only slightly rearranged between the two structures.

65. Add together the calculated free energy values for each step.

67. −65.2 kJ

69. Entropy does not increase upon cooling. The statement should read, "At some point of cooling they freeze, and continuing to cool the sample *decreases* its entropy."

71. 618 K

73. −630.4 kJ, spontaneous

75. Spontaneous at all temperatures

77. It is defined that the entropy is zero at 0 K for a perfectly ordered crystal. All substances above this temperature, even elements, will have a positive entropy.

79. 294 K

81. 92.9 J/(mol·K), the answer should be a positive entropy value.

83. 9.61 J/(mol·K)

85. 0.805 J/(mol·K)

87. There are more atoms in $CaCO_3$ than in CaO, so its standard molar entropy is more positive; 1099 K.

89. (a) Positive
    (b) Negative
    (c) $T_{fus} = \Delta H/\Delta S$

Chapter 14

1. The green line represents $[N_2O]$ and the red line represents $[O_2]$.

3. (b)

5. (b)

7. (c)

9. (a)

11. Nitrogen (light blue)

13. Nickel (red), palladium (blue), and platinum (orange)

15. The reaction requires first the breakdown of NO_2 to form NO and O. The O combines later with O_2 to form O_3. The buildup of ozone, therefore, lags while waiting for the concentration of NO_2 to increase.

17. Sunlight is less intense so the photochemical breakdown of NO_2 is not as frequent.

19. −114 kJ

21. (a) $2N_2(g) + O_2(g) \rightarrow 2N_2O(g)$
    (b) $2N_2(g) + 5O_2(g) \rightarrow 2N_2O_5(g)$

23. The average rate is measured over a relatively long time period while the instantaneous rate is the rate at a specific time in the reaction.

25. The rate changes with time because the concentration of the reactants decreases and because the rate of the reverse reaction reduces the forward reaction rate.

27. (a) The rates are the same.

    (b) $\dfrac{\Delta[NO_2^-]}{\Delta t} = -\dfrac{2}{3}\dfrac{\Delta[O]}{\Delta t}$

    (c) $\dfrac{\Delta[NH_3]}{\Delta t} = \dfrac{2}{3}\dfrac{\Delta[O_2]}{\Delta t}$

29. (a) $Rate = -\dfrac{\Delta[H_2O_2]}{\Delta t} = \dfrac{1}{2}\dfrac{\Delta[OH]}{\Delta t}$

    (b) $Rate = -\dfrac{\Delta[ClO]}{\Delta t} = -\dfrac{\Delta[O_2]}{\Delta t} = \dfrac{\Delta[ClO_3]}{\Delta t}$

    (c) $Rate = -\dfrac{\Delta[N_2O_5]}{\Delta t} = -\dfrac{\Delta[H_2O]}{\Delta t} = \dfrac{1}{2}\dfrac{\Delta[HNO_3]}{\Delta t}$

31. (a) $Rate = \dfrac{\Delta[CO_2]}{\Delta t} = -\dfrac{2}{3}\dfrac{\Delta[CO]}{\Delta t}$

    (b) $Rate = \dfrac{\Delta[COS]}{\Delta t} = -\dfrac{\Delta[SO_2]}{\Delta t}$

    (c) $Rate = \dfrac{\Delta[CO]}{\Delta t} = 3\dfrac{\Delta[SO_2]}{\Delta t}$

33. (a) 1.2×10^7 M/s
    (b) 2.9×10^4 M/s

35. (a) -1.37×10^{-5} M/μs
    (b) -5.50×10^{-6} M/μs

37. (a)

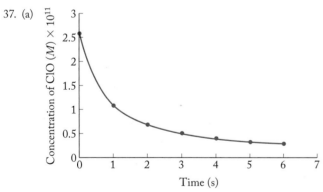

For [ClO], instantaneous rate = -9.59×10^{10} molecules/cm³·s

(b)

For [Cl₂O₂], instantaneous rate = 4.79×10^{10} molecules/cm³·s

39. Yes

41. Yes

43. The half-life will be halved.

45. (a) First order in both A and B, and second order overall
    (b) Second order in A, first order in B, and third order overall
    (c) First order in A, third order in B, and fourth order overall

47. (a) Rate = $k[O][NO_2]$; k units = $M^{-1}s^{-1}$
    (b) Rate = $k[NO]^2[Cl_2]$; k units = $M^{-2}s^{-1}$
    (c) Rate = $k[CH_3Cl][Cl_2]^{1/2}$; k units = $M^{-1/2}s^{-1}$
    (d) Rate = $k[O_3]^2[O]^{-1}$; k units = s^{-1}

49. (a) Rate = $k[BrO]$
    (b) Rate = $k[BrO]^2$
    (c) Rate = $k[BrO]$
    (d) Rate = $k[BrO]^0 = k$

51. We need the change in the rate when only one of the concentrations of the reactants is varied.

53. (a) Rate = $k[NO_2][O_3]$
    (b) 4.9×10^{-11} M/s
    (c) 4.9×10^{-11} M/s
    (d) Doubles

55. (c)

57. Rate = $k[NO][NO_2]$

59. Rate = $k[ClO_2^-][OH^-]$; $k = 13.8/M^{-1}s^{-1}$

61. Rate = $k[NO]^2[H_2]$; $k = 6.32/M^{-2}s^{-1}$

63. $k = 0.3202/\mu M^{-1}min^{-1}$

65. Rate = $k[NH_3]$; $k = 0.0030$ s^{-1}

67. (a) Rate = $k[N_2O]$
    (b) 4

69. (a) Rate = $k[^{32}P]$
    (b) $k = 0.0485/d^{-1}$
    (c) 14.3 d

71. $k = 5.40 \times 10^{-12}$ cm³(molecules·s)$^{-1}$; $t_{1/2} = 0.712$ s

73. Rate = $k[C_{12}H_{22}O_{11}][H_2O] = k'([C_{12}H_{22}O_{11}]$; $k' = 6.21 \times 10^{-5}$ s^{-1}

75. They have large activation energies.

77. None of the statements are true.

79. Temperature increases the frequency and the kinetic energy at which reactants collide, which speeds up the reaction by changing the value of k. The activation energy of the slowest step in the reaction, and therefore, the order of the reaction is unaffected.

81. The reaction with $E_a = 150$ kJ/mol

83. $E_a = 17.1$ kJ/mol; $A = 1.002$

85. (a) $E_a = 314$ kJ/mol
    (b) $A = 5.03 \times 10^{10}$
    (c) $1.08 \times 10^{-44} M^{-1/2} \cdot$ s

87. $E_a = 39.1$ kJ/mol, $A = 1.27 \times 10^{12}$

89. No, because they have different rate laws.

91. Pseudo-first-order reaction kinetics will be obeyed when one of the reactants is in sufficiently high concentration so that its concentration does not change over the course of the reaction.

93.

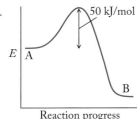

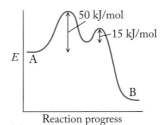

(a)

(b)

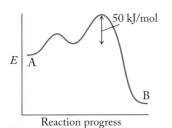

(c)

95. (a) Rate $= k[SO_2Cl_2]$ unimolecular
    (b) Rate $= k[NO_2][CO]$ bimolecular
    (c) Rate $= k[NO_2]^2$ bimolecular

97. $N_2O_5(g) + O(g) \rightarrow 2\,NO_2(g) + O_2(g)$

99. The second step

101. The first step

103. Thermal decomposition (b) or (c); photochemical decomposition (a)

105. Yes

107. Yes

109. The homogeneous catalyst is unchanged in the reaction and is used, but not consumed, in the reaction.

111. NO is the catalyst.

113. The reaction of O_3 with Cl has the larger rate constant.

115. The concentration of oxygen has increased and therefore the rate of the reaction has increased.

117. At lower temperatures, the bodily reactions that use oxygen are slower than at higher temperatures.

119. The rate of disappearance of N_2O_5 is $\frac{1}{2}$ that of the rate of appearance of NO_2 and twice the rate of the appearance of O_2.

121. First order

123. $5.04\ M\,s^{-1}$

125. (a) Rate $= k[NH_3][HNO_2]^2$, k units $= M^{-2}s^{-1}$
    (b) Yes
    (c) -482.4 kJ
    (d)

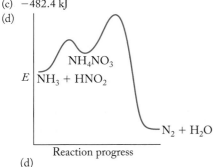

(d)

127. (a) 9.52×10^{-4} torr^{-1}h^{-1}
    (b) 0.105 torr/h
    (c)

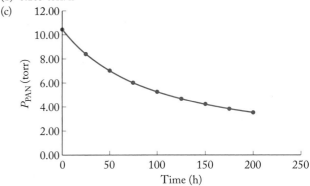

129. (a) $k = 6.50 \times 10^{-3}$min^{-1}
    (b) Any amino acid that has a sulfur atom–cysteine and methionine

131. (a) Rate $= [O_3][C_2H_4]$, $k = 1020\ M^{-1}s^{-1}$
    (b) $E_a = 21.9$ kJ/mol

Chapter 15

1. $C \rightleftharpoons D$ has the larger forward rate constant. $A \rightleftharpoons B$ has the smaller reverse rate constant. $C \rightleftharpoons D$ has the larger equilibrium constant.

3. $A \rightleftharpoons B$, $K_c = 2$

5. As temperature increases, the number of product molecules increases so the reaction is endothermic.

7. A system is at equilibrium when the rate of the forward reaction is equal to the rate of the reverse reaction.

9. No, because the concentrations of A and B are still changing at $20\ \mu s$.

11. Greater than 1

13.

Molar Mass	Compound	How Present
28	$^{14}N_2$	originally present
29	$^{15}N^{14}N$	from decomposition of $^{15}N^{14}NO$
30	$^{15}N_2$	from decomposition of $^{15}N_2O$
32	O_2	originally present
44	$^{14}N_2O$	from combination of $^{14}N_2$ and O_2
45	$^{15}N^{14}NO$	from combination of $^{15}N^{14}N$ and O_2
46	$^{15}N_2O$	originally present

15. $K = 0.33$

17. When $\Delta n = 0$; when moles of gaseous products equal the moles of gaseous reactants.

19. (a) $K_c = \dfrac{[N_2O_4]}{[N_2][O_2]^2}$, $K_p = \dfrac{P_{N_2O_4}}{P_{N_2} \times (P_{O_2})^2}$

    (b) $K_c = \dfrac{[NO_2][N_2O]}{[NO]^3}$, $K_p = \dfrac{P_{NO_2} \times P_{N_2O}}{(P_{NO})^3}$

    (c) $K_c = \dfrac{[N_2]^2[O_2]}{[N_2O]^2}$, $K_p = \dfrac{(P_{N_2})^2 \times P_{O_2}}{(P_{N_2O})^2}$

21. $K_c = 0.50$

23. $K_p = 0.068$

25. $K_c = 0.50$

27. $K_c = 1.5$

29. $K_c = 780$

31. $K_c = 0.0583$

33. (b) and (c)

35. $K_p = 0.10$

37. Scaling up of the coefficients in a reaction by x will increase the value of the equilibrium constant by $(K_c)^x$. Scaling down of the coefficients in a reaction by x will decrease the value of the equilibrium constant by $(K_c)^{1/x}$.

39. $K_c = 2$

41. $K_{c,\text{forward}} = \dfrac{[NO_2]^2}{[NO][NO_3]}$; $K_{c,\text{reverse}} = \dfrac{[NO][NO_3]}{[NO_2]^2}$; $K_{c,\text{forward}} = \dfrac{1}{K_{c,\text{reverse}}}$

43. $K_c = \dfrac{[SO_3]}{[SO_2][O_2]^{1/2}}$; $K'_c = \dfrac{[SO_3]^2}{[SO_2]^2[O_2]}$; $K'_c = (K_c)^2$

45. (a) 0.049
    (b) 420
    (c) 20

47. $K_c = 7.4$

49. The product of the concentrations of the products raised to their stoichiometric coefficients divided by the product of the concentrations of the reactants raised to their stoichiometric coefficients. For Q, however, these concentrations need not be at equilibrium.

51. The system is at equilibrium.

53. No, $Q < K$ so this reaction will proceed to the right to reach equilibrium.

55. (a) is at equilibrium.

57. $Q > K$, so the reaction will proceed to the left.

59. (a)

61. Yes

63. To the right

65. (c), the reaction of Cl_2 with I_2 to form ICl

67. $K = 1.13 \times 10^{-20}$

69. $\Delta G = -73$ kJ/mol

71. $K_p = 7.2 \times 10^{-19}$

73. $K_c = [Cu^{2+}][S^{2-}]$

75. Pure solids are not included in equilibrium constant expressions.

77. No

79. By increasing the partial pressure of O_2, we shift the equilibrium of the combined equation

$$Hb(CO)_4 + 4O_2 \rightleftharpoons Hb(O_2)_4 + 4CO$$

to the right, releasing CO from hemoglobin.

81. By Le Châtelier's principle, an increase in the partial pressure of a gas over a solvent shifts the equilibrium for the solubility of a gas in the solvent to the right, so more of the gas dissolves in the solvent at higher pressures.

83. (b) and (d)

85. (a) $[O_2]$ increases
    (b) $[O_3]$ increases
    (c) $[O_3]$ increases

87. The equilibrium will shift to the left.

89. (a)

91. When $Q > K$, the reaction shifts to the left to form more reactants; when $Q < K$, the reaction shifts to the right to form more products; when $Q = K$ the reaction is at equilibrium.

93. (a) $P_{PCl_5} = 0.163$ atm, $P_{PCl_3} = 0.103$ atm, $P_{Cl_2} = 0.397$ atm
    (b) $[PCl_5]$ will increase, $[PCl_3]$ will decrease

95. $[H_2O] = [Cl_2O] = 3.76 \times 10^{-3}$ M, $[HOCl] = 1.13 \times 10^{-3}$ M

97. 9×10^5

99. $P_{CO} = 2.39$ atm, $P_{CO_2} = 3.81$ atm

101. (a) $P_{NO} = 0.272$ atm; $P_{NO_2} = 0.00798$ atm
     (b) 0.416 atm

103. $P_{O_2} = 0.17$ atm; $P_{N_2} = 0.75$ atm; $P_{NO} = 0.080$ atm

105. 5.75 M

107. $P_{CO} = P_{Cl_2} = 0.258$ atm; $P_{COCl_2} = 0.00680$ atm

109. $[CO] = [H_2O] = 0.031$ M; $[CO_2] = [H_2] = 0.069$ M

111. Exothermic

113. Exothermic

115. $K_c = 1.29 \times 10^{-31}$

117. -115 kJ/mol

119. The reaction is endothermic; $K_{1500K} = 5.54 \times 10^{-11}$, $K_{2500K} = 4.01 \times 10^{-3}$, $K_{3000K} = 0.403$. K is not large enough at ordinary temperatures for this to be an antidote for global warming.

121. 9.2×10^{-22} M

123. $K_{25°C} = 1.6 \times 10^{-9}$, $K_{500°C} = 2.5 \times 10^7$

125. 9.53×10^{-73} atm

Chapter 16

1. Red line

3. The blue line is the titration curve for the stong acid, and the red line is the titration curve for the weak acid.

5. The indicator with $pK_a = 9.0$

7. The blue titration curve represents the titration of $NaHCO_3$, and the red titration curve represents the titration of Na_2CO_3.

9. HBr is the acid; H_2O is the base.

11. OH^- is the base; H_2O is the acid.

13. (a) Acid = HNO_3; base = NaOH
    (b) Acid = HCl; base = $CaCO_3$
    (c) Acid = HCN; base = NH_3

15. NO_2^-; OCl^-; $H_2PO_4^-$; NH_2^-

17. 1.50 M

19. 0.160 M

21. Dissolve 70.0 g of NaOH(s) in water and dilute to a total volume of 2.50 L.

23. pH is $-\log[H^+]$ so as $[H^+]$ increases, the pH value decreases.

25. The pH of a solution is negative when $[H^+] > 1\ M$.

27. (a) pH = 7.462, pOH = 6.538; basic
    (b) pH = 4.70, pOH = 9.30; acidic
    (c) pH = 7.15, pOH = 6.85; basic
    (d) pH = 10.932, pOH = 3.067; basic

29. pH = 0.810

31. pOH = 1.345, pH = 12.653

33. pH = -0.124

35. $HNO_2 > HOCN > CH_3COOH > HOCl$

37. $NaNO_2$ completely ionizes in solution to give 2.0 M ions in solution. Because HNO_2 is a weak acid, the concentration of ions in solution (H^+ and NO_2^-) is less than 2.0 M. Therefore, the solution of $NaNO_2$ is more conducting because it has a greater concentration of ions.

39. $K = \dfrac{[F^-][H^+]}{[HF]}$

41. (a) Water
    (b) Water

43. CH_3NH_3 is the base; H_2O is the acid.

45. $K_a = 8.91 \times 10^{-4}$

47. 1.63%, $K_a = 6.74 \times 10^{-5}$

49. pH = 2.49

51. 2.3 times

53. pH = 10.77

55. pH = 9.36

57. It is more difficult to remove the second H^+ from a species that is already negatively charged.

59. pH = 0.51

61. pH = 2.35

63. pH = 9.511

65. pH = 10.270

67. Sulfur is more electronegative than Se which makes the HSO_4^- ion more stable than the $HSeO_4^-$ ion.

69. (a) H_2SO_3
    (b) H_2SeO_4

71. Increase

73. NH_4Cl

75. Citric acid protonates the weak base trimethylamine thereby making it an ion, which is much less volatile than the neutral molecule.

77. $pK_b = 3.32$

79. pH = 7.35

81. The conjugate acid, acetic acid, and the conjugate base, acetate, are in high enough concentrations to absorb added H^+ and OH^- while NaCl and HCl are completely ionized and cannot act as a buffer solution.

83. pH at 25°C = 4.45; pH at 0°C = 4.48

85. pH = 12.38; pOH = 1.62

87. 0.065

89. pH = 9.25

91. 3.0 mL

93. Before addition of HCl, pH = 3.50; after addition of HCl, pH = 3.52

95. The weak acid titration curve will start with a pH higher than that for the same concentration of a strong acid, and the pH at the equivalence point for the weak acid will not be 7 (as it will be for the titration of a strong acid) but will be greater than 7.

97. No

99. After 10.0 mL, the pH = 4.75; after 20.0 mL, the pH = 8.75; after 30.0 mL, the pH = 12.36

101. 0.02559 M

103. 250 mL HCl for both titrations

105. pH = 4.44

107.

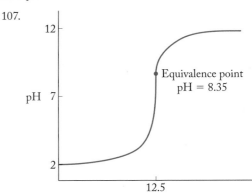

Volume of 1.00 M NaOH (mL)

109.

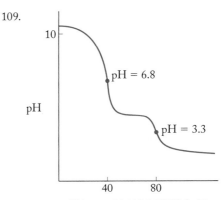

Volume of 0.100 M HCl (mL)

111. The hydrogen bonds between water must break and reform around the species CH_3NH_2. Also, the amine hydrolyzes and forms $CH_3NH_3^+$ and OH^-, and ion–dipole intermolecular forces are added when these ions are surrounded by water molecules.

113. (a) and (b)

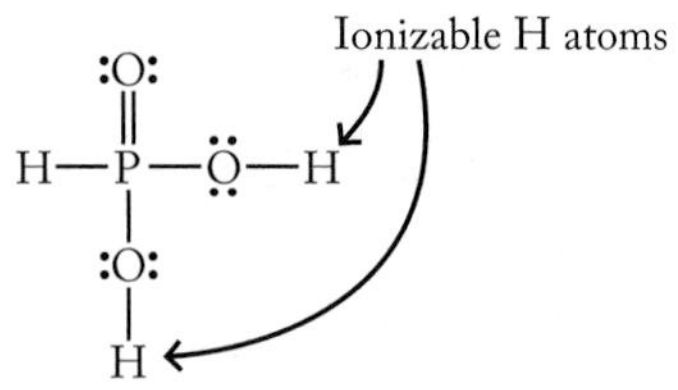

Ionizable H atoms

(c) Phosphoric acid has a very similar structure for its ionizable H atoms.

115. Yes, the pH will increase due to the increase in hydrolysis of HCO_3^-.

117. 1.45×10^9 L

119. (a) Structure on the right, the product
    (b) Acidic

121. (a) Because the equilibrium constant of the reaction of F^- with HF is larger than the dissociation of HF, the most likely species is HF_2^-.
    (b) 2.86×10^{-4}
    (c) pH = 2.60, $[HF_2^-] = 2.54 \times 10^{-3}$ M

123. (a)

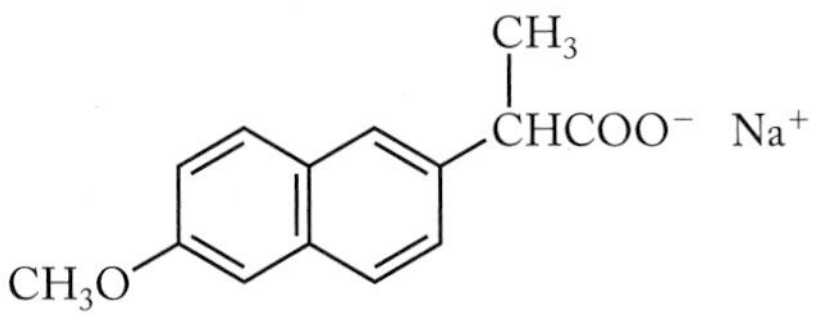

CH_3
$CHCOO^-$ Na^+
CH_3O

(b) This solution will be basic.

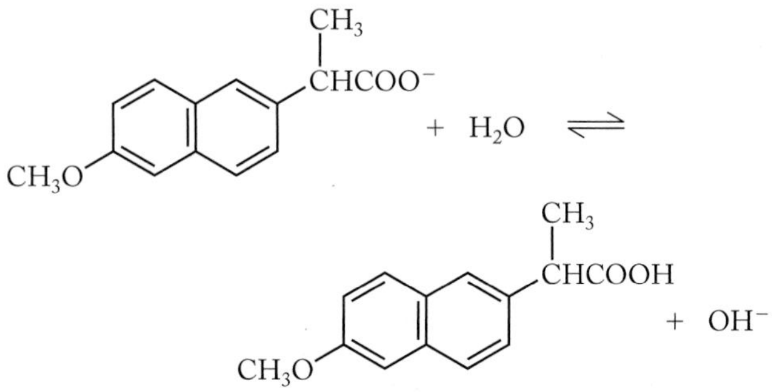

CH_3
$CHCOO^-$
CH_3O
$+$ H_2O ⇌

CH_3
$CHCOOH$
CH_3O
$+$ OH^-

(c) The salt is ionic and therefore dissolves better in water.

125. (a) $SO_3(g) + H_2O(\ell) \rightarrow H_2SO_4(aq)$
    (b) $3\,NO_2(g) + H_2O(\ell) \rightarrow 2\,HNO_3(aq) + NO(g)$
    (c) $4\,NH_3(g) + 5\,O_2(g) \rightarrow 4\,NO(g) + 6\,H_2O(\ell)$

127. None

129. $S(s) + O_2(g) \rightarrow SO_2(g)$
    $2\,S(s) + 3\,O_2(g) \rightarrow 2\,SO_3(g)$
    $\Delta H^\circ_{rxn} = \Delta H^\circ_{f,SO_3} + (-\Delta H^\circ_{f,SO_2})$

131. Henry's law constant is greater for SO_3 because SO_3 has stronger dispersion forces than CO_2, and it is also greater for NO_2 because NO_2 is polar, so more of it dissolves in water.

133. 38.2 g

135. (a)

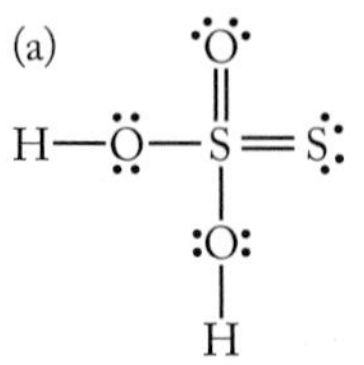

$H-\ddot{O}-S=\ddot{S}$

(b) Weaker because S is not as electronegative as O

Chapter 17

1. Chromium (green) and cobalt (yellow)

3. Zinc (blue)

5. (a) 4
   (b) 6

7. (a) $Co(CN)_6^{3-}$
   (b) CoF_6^{3-}
   (c) $Co(NH_3)_6^{3+}$

9. Yes, because it may be an electron pair donor, but may not accept H^+.

11. BF_3 can accept electron pairs, but has no H atoms to donate to be a Brønsted–Lowry acid.

13.

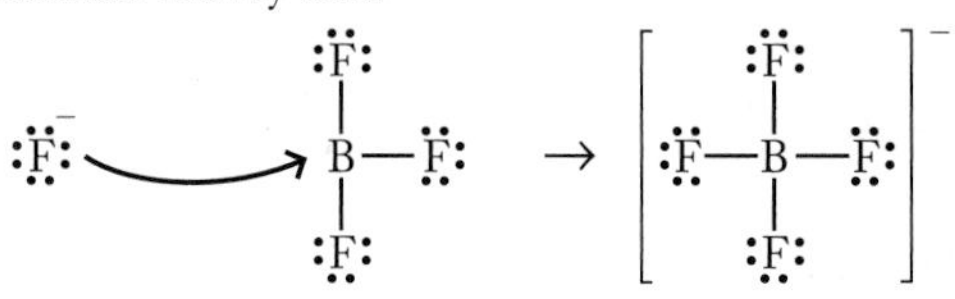

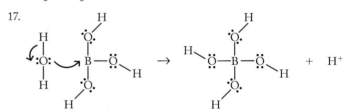

Lewis base   Lewis acid

15.

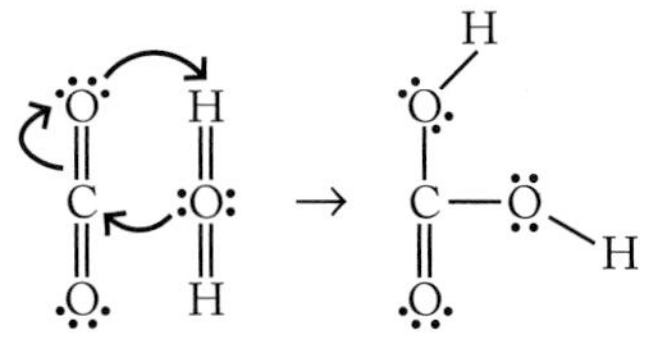

CO_2 and H_2O act as both a Lewis acid and a Lewis base.

17.

$+$ H^+

H_2O is the Lewis base, and $B(OH)_3$ is the Lewis acid.

19. Water

21. Water

23. Na^+

25. EDTA has fully complexed the Cu^{2+} ion, which then is no longer toxic to the phytoplankton.

27. Ag^+ forms a complex with NH_3 removing Ag^+ from solution and shifting the equilibrium for dissolution of AgCl to the right.

29. (a) 0.0040 M
    (b) 6.3×10^{-10} M

31. 4.1×10^{-20} M

33. (b) and (d)

35. The solution will become more acidic.

37. $Cr(OH)_3(s) + H^+(aq) \rightleftharpoons Cr(OH)_2^+(aq) + H_2O(\ell)$

   or $Cr(OH)_3(s) + 3H^+(aq) \rightleftharpoons Cr^{3+}(aq) + 3H_2O(\ell)$

   $Cr(OH)_3(s) + OH^-(aq) \rightleftharpoons Cr(OH)_4^-(aq)$

39. The other cations do not form soluble complex hydroxide ions but rather are insoluble.

41. pH = 2.65

43. pH = 1.72

45.

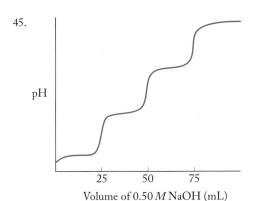

Volume of 0.50 M NaOH (mL)

47. Solubility is the amount of the substance that dissolves, and the solubility product is the equilibrium constant for the equilibrium reaction describing the dissolution of a solid.

49. Sr^{2+}

51. Endothermic

53. In an acidic environment, the H^+ reacts with the OH groups on hydroxyapatite to dissolve the hydroxyapatite.

55. $K_{sp} = 1.08 \times 10^{-8}$

57. $[Cu^+] = [Cl^-] = 1.01 \times 10^{-3}\ M$

59. 9.96×10^{-6} g/mL

61. pH = 10.09

63. (d)

65. Because the concentration of Sr^{2+} in seawater is greater than its calculated concentration based on the solubility of $SrSO_4$, there must be other equilibria involving the Sr^{2+} cation which cause it to be able to be present in a higher than expected concentration.

67. A sequestering agent is a multidentate ligand that separates metal ions from other substances so they can no longer react. Properties that make a sequestering agent effective include strong bonds formed between the metal and the ligand and large formation constants.

69. As pH increases, the carboxylic acid group is deprotonated, and its chelating ability increases.

71. Splitting of the d orbitals makes possible d-d electron transitions.

73. The repulsion of the orbital with the incoming ligand electrons is higher for d_{xy} because that orbital lies on the plane of the ligands in the square planar geometry, whereas the d_{xz} and d_{yz} orbitals are perpendicular to the plane of the ligands.

75. (a) $Cr(H_2O)_6^{3+}$ is violet; (b) $Cr(NH_3)_6^{3+}$ is yellow

77. Colorless or faint yellow

79. $NiCl_4^{2-}$

81. The magnitude of the crystal field splitting energy compared to the pairing energy of the electrons

83. High-spin Fe^{2+} has 4 unpaired electrons; Cu^{2+} has 1 unpaired electron; Co^{2+} has 3 unpaired electrons; Mn^{3+} has 4 unpaired electrons

85. Cr^{3+}

87. (a) Mn^{4+} in MnO_2; $2 Mn^{3+}$ and $1 Mn^{2+}$ in Mn_3O_4
    (b) Both low spin and high spin configurations are possible in Mn_3O_4 but not in MnO_2.

89. Paramagnetic

91. (a) Hexaamminechromium(III)
    (b) Hexaaquacobalt(III)
    (c) Pentaamminechloroiron(III)

93. (a) Tetrabromocobaltate(II)
    (b) Aquatrihydroxozincate(II)
    (c) Pentacyanonickelate(II)

95. (a) Ethylenediaminezinc(II) sulfate
    (b) Pentaammineaquanickel(II) chloride
    (c) Potassium hexacyanoferrate(II)

97. Ligands that are *cis* to each other have a 90° angle between them; ligands that are *trans* to each other have a 180° angle between them.

99. 2

101. Yes

103.
$$\begin{bmatrix} \text{Br} & \diagdown & \text{Cl} \\ & \text{Cu} & \\ \text{Br} & \diagup & \text{Cl} \end{bmatrix}^{2-} \quad \begin{bmatrix} \text{Cl} & \diagdown & \text{Br} \\ & \text{Cu} & \\ \text{Br} & \diagup & \text{Cl} \end{bmatrix}^{2-}$$

cis   *trans*

Neither of these isomers is chiral.

105. 4×10^{-13}

107. The oxidized cobalt complex (yellow complex) has a larger crystal field splitting energy.

109. In a square planar environment, the electrons are all paired for Ag^+ (d^{10}) and Ag^{3+} (d^8).

111. Longer wavelength

Chapter 18

1. (b) and (c)

3. (a) Left electrode = cathode;
    $2 H_2O(\ell) + 2 e^- \rightarrow H_2(g) + 2 OH^-(aq)$; $E°_{red} = -0.8277$ V
    Right electrode = anode;
    $2 H_2O(\ell) \rightarrow O_2(g) + 4 H^+(aq) + 4 e^-$; $E°_{ox} = -1.229$ V
    (b) H_2SO_4 increases the conductivity of the solution

5. Pt, Au, Hg (blue)

7. Cd is the anode, Cu is the cathode; electrons flow from Cd to Cu.

9. To allow nonreactive ions to pass through the separator so that electrical neutrality is maintained in each of the half-cells.

11. A wire cannot allow the movement of ions from one side of the voltaic cell to the other to maintain electrical neutrality in the half-cells.

13. (a) Cathode: $Pb^{2+}(aq) + 2 e^- \rightarrow Pb(s)$;
    anode: $Zn(s) \rightarrow Zn^{2+}(aq) + 2 e^-$
    (b) $Pb^{2+}(aq) + Zn(s) \rightarrow Zn^{2+}(aq) + Pb(s)$
    (c) $Zn(s) \mid Zn^{2+}(aq) \parallel Pb^{2+}(aq) \mid Pb(s)$

15. (a) Cathode, $MnO_4^-(aq) + 2 H_2O(\ell) + 3 e^- \rightarrow$
    $\qquad\qquad\qquad\qquad MnO_2(s) + 4 OH^-(aq)$;
    anode, $Cd(s) + 2 OH^-(aq) \rightarrow Cd(OH)_2(s) + 2 e^-$
    (b) $3 Cd(s) + 2 MnO_4^-(aq) + 4 H_2O(\ell) \rightarrow$
    $\qquad\qquad 3 Cd(OH)_2(s) + 2 MnO_2(s) + 2 OH^-(aq)$
    (c) $Cd(s) \mid Cd(OH)_2(s) \parallel MnO_4^-(aq) \mid MnO_2(s) \mid Pt(s)$

17. (a) 6

   (b) Fe^{6+} in FeO_4^{2-}; Fe^{3+} in Fe_2O_3; Zn^0 in Zn, Zn^{2+} in Zn, and ZnO_2^{2-}

   (c) $Zn(s) \mid ZnO(s), ZnO_2^{2-}(aq) \parallel FeO_4^{2-}(aq)\ (Fe_3O_2(s)) \mid Pt(s)$

19. A positive E_{cell} is spontaneous and therefore does work on the surroundings so the sign of w_{elec} is negative.

21. -289.5 kJ

23. -115.8 kJ

25. The Pt electrode transfers electrons to the half cell, but because it is inert it is not involved in the reaction.

27. Because $E°_{red}(anode) = -E°_{ox}(anode)$ we can substitute $-E°_{ox}(anode)$ for $E°_{red}(anode)$ in the expression $E°_{cell} = E°_{red}(cathode) - E°_{red}(anode)$ to obtain $E°_{cell} = E°_{red}(cathode) + E°_{ox}(anode)$ (Eq 18.7).

29. O_2 is a stronger oxidizing agent in acid because $E°_{ox}(acid, 1.229\ V) > E°_{ox}(base, 0.401V)$.

31. (a) $\Delta G° = -34.5$ kJ, $E°_{cell} = 0.358$ V

   (b) $\Delta G° = 2.9$ kJ, $E°_{cell} = -0.030$ V

33. No

35. $2\,Zn(s) + O_2(g) + 2\,H_2O(\ell) + 4\,OH^-(aq) \rightarrow 2\,Zn(OH)_4^{2-}(aq)$

37. Less than 1.10 V

39. (a) $E°_{cell} = -0.478$ V, $\Delta G° = 92.23$ kJ

   (b) $E°_{cell} = 0.505V$, $\Delta G° = -97.42$ kJ

41. (a) $Hg^{2+}(aq) + Zn(s) \rightarrow Zn^{2+}(aq) + Hg(\ell)$; cathode reaction = Hg^{2+}/Hg, anode reaction = Zn/Zn^{2+}

   (b) $Zn(s) + Ag_2O(s) \rightarrow ZnO(s) + 2\,Ag(s)$; cathode reaction = Ag_2O/Ag, anode reaction = Zn/ZnO

   (c) $2\,Ni(s) + O_2(g) + 2\,H_2O(\ell) \rightarrow 2\,Ni(OH)_2(s)$; cathode reaction = O_2/OH^-, anode reaction = $Ni/Ni(OH)_2$

43. (a) $NiO(OH)(s) + TiZr_2H(s) \rightarrow Ni(OH)_2(s) + TiZr_2(s)$

   (b) 1.32 V

45. The $\log Q$ term in the Nernst equation remains small until a significant amount of the products have formed in the battery reaction.

47. 1.27 V

49. $K = 8.56 \times 10^{19}$

51. -0.414 V

53. 1.538 V; E_{rxn} will decrease

55. (a) 0.618 V

   (b) 0.606 V

57. (a) -0.271 V

   (b) 1.67×10^{-27}

59. (c) and (f)

61. Al/O_2

63. Li/MnO_2

65. In a voltaic cell the electrons are produced at the anode so a $(-)$ charge builds up there; in an electrolytic cell electrons are being forced onto the cathode so it has a build up of the $(-)$ charge. The flow of electrons in the outside circuit is reversed in an electrolytic cell compared to the flow in a voltaic cell.

67. Br_2

69. It increases the conductivity of the solution.

71. More negative

73. 6.8 g

75. 18.0 min

77. (a) 0.00578 L

   (b) No, because some Cl_2 and Br_2 would be produced.

79. -0.270 V

81. A hybrid vehicle uses a relatively inexpensive fuel (gasoline) and has good fuel economy, but it still has engine emissions. A fuel-cell vehicle has no engine emissions but uses a more expensive and explosive fuel (hydrogen), and the current battery materials are very expensive and bulky.

83. Electric engines are more efficient by turning more of the energy into motion instead of losing it as heat.

85. (a) CH_4: C = -4, H = $+1$; H_2O: H = $+1$; H_2: H = 0; CO: C = $+2$; CO_2: C = $+4$

   (b) For the reaction of CH_4 with H_2O $\Delta G° = 142.2$ kJ. For the reaction of CO with water $\Delta G° = -28.6$ kJ. $\Delta G°_{overall} = 113.6$ kJ

87. (a) Cathode

   (b) No, Mg^{2+} has a lower reduction potential than Na^+.

   (c) No

   (d) H_2 and O_2

   (e) If the overpotential of the electrolysis of water was greater than the reduction potential for Mg^{2+}, then Mg would form.

89. (a) -0.870 V

   (b) To carry the charge in the reaction; it is produced in the oxidation reaction and used in the reduction reaction.

   (c) In Mo_3S_4, 2 Mo ions are in the $+3$ oxidation state and 1 Mo ion is in the $+2$ oxidation state; in $MgMo_3S_4$, the Mo ion is in the $+2$ oxidation state. The electron configurations are $[Kr]4d^4$ for Mo^{2+} and $[Kr]4d^3$ for Mo^{3+}.

91. (a) In K_2MnF_6: K = $+1$, Mn = $+4$, F = -1; in SbF_5: Sb = $+5$, F = -1; in $KSbF_6$: K = $+1$, Sb = $+5$, F = -1; in MnF_3: Mn = $+3$, F = -1; in F_2: F = 0. This is a 1-electron process.

   (b) -656 kJ

   (c) 6.80 V

   (d) Too low

   (e) In HF: H = $+1$, F = -1; in H_2: H = 0; in F_2: F = 0. This is a 2-electron process.

93. (a) $2\,ZnS(s) + 3\,O_2(g) \rightarrow 2\,ZnO(s) + 2\,SO_2(g)$

   (b) $2\,ZnO(s) + C(s) \rightarrow 2\,Zn(s) + CO_2(g)$

   (c) $HgS(s) + O_2(g) \rightarrow Hg(\ell) + SO_2(g)$

95. (a) Yes

   (b) Yes, the S^{2-} are close-packed in both structures.

97. (a) 0.359 V

   (b) -1.61 V

   (c) 1.20 V

99. Zn

101. Both Mg ($[Ne]3s^2$) and Zn ($[Ar]3d^{10}4s^2$) have two valence electrons.

103. Zn^{2+} has a d^{10} configuration so there are no d-d transitions possible therefore zinc solutions are colorless.

Chapter 19

1. $$\left[\begin{array}{c} H-\overset{\overset{\ddot{O}}{\|}}{P}-\ddot{\underset{\ddot{O}}{O}} \\ \underset{:\ddot{O}:}{} \end{array}\right]^{2-}$$

3. Palmitic and stearic acid.

5. 8

7. Tyrosine, glycine, glycine, phenylalanine, methionine

9. "Trans fat" refers to the geometric isomerism around the carbon–carbon double bonds. Both (a) and (c) are trans fats.

11. Decrease in entropy (more order)

13. Any amino acid

15. No, because *enantiomer* and *optically active* describe a chiral compound whereas *achiral* means that the compound is not chiral.

17. Homogeneous

19. Glycine has no chiral centers; its central carbon atom does not have four different groups bonded to it.

21. sp^3

23. (a) Golf club, (c) glove, (d) shoe

25. (a)

27.

Saccharin   Cyclamate   Aspartame

29.

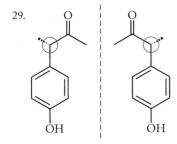

31. D-amino acids rotate plane-polarized light to the right, which is the opposite direction for L-amino acids.

33. (a) and (c)

35. The fuel values should be the same. Isoleucine might have a lower fuel value because the CH_3 group is closer to the COOH and NH_2 groups and this difference in shape must contribute to their slightly different fuel values.

37. The COOH group is acidic and the NH_2 group is basic at pH = 7.4.

39. (a)

(b)

(c)

41. (a) Alanine + glycine
    (b) Leucine + leucine
    (c) Tyrosine + phenylalanine

43. NH_3

45. Starch has α-glycosidic bonds; cellulose has β-glycosidic bonds.

47. No

49. The bonding is nearly the same because glucose and fructose are isomers of each other.

51. Sum the individual ΔG values for each reaction.

53.

β-Galactose α-Galactose

55. (c)

57. (a)

59. (b)

61. -16.6 kJ

63. Saturated fatty acids have all C—C single bonds; unsaturated fatty acids have C=C double bonds.

65. There is a lot of energy stored in the saturated fats in butter so there are more cal/g.

67. Yes, but only when the central carbon on the triglyceride has four different groups attached to it.

69. (b) and (c)

71.

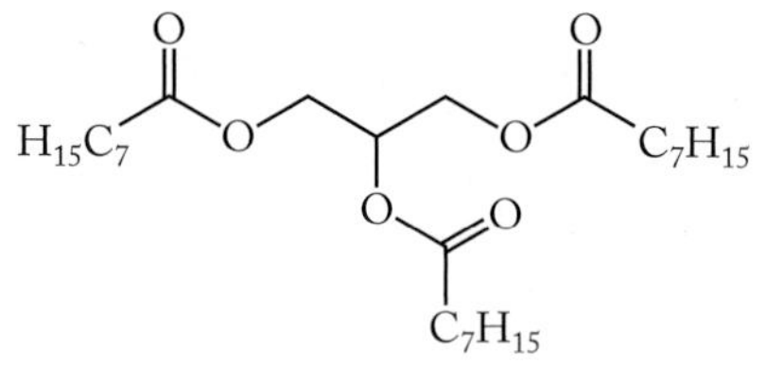

Gycerol with octanoic acid

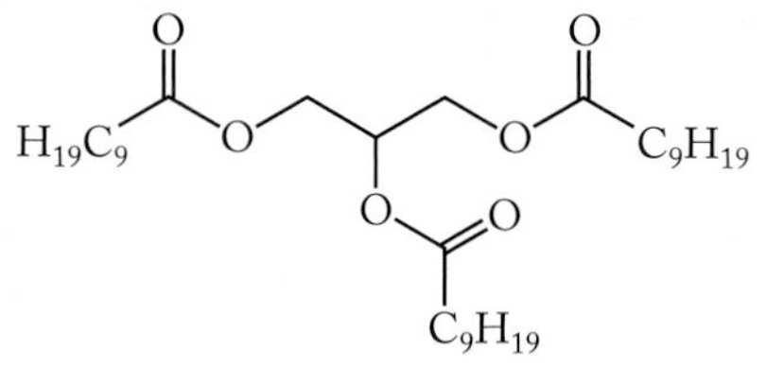

Gycerol with decanoic acid

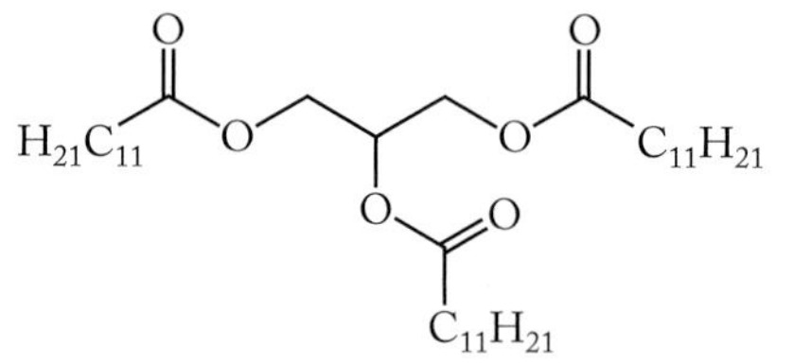

Gycerol with dodecanoic acid

73. Nitrogen bases with phosphate groups and sugar residues forming the backbone

75. Hydrogen bonds between the base pairs

77.

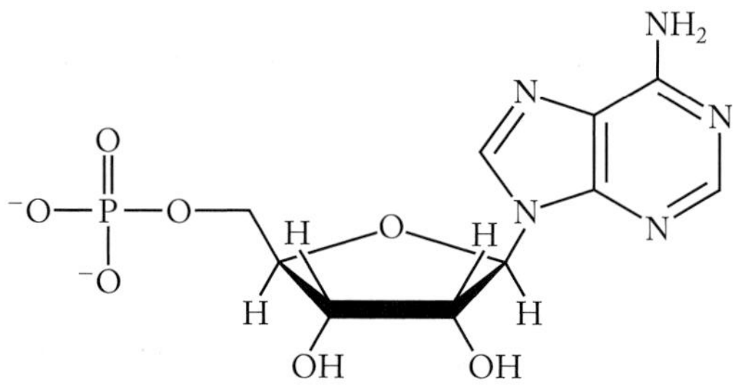

79. A-G-C-C-A-T

81. (a) Sucrose
    (b) Ester
    (c) $C_{15}H_{31}COOH$, palmitic acid

83. (a) There is an extra $-CH_2-$ group in the sulfur-containing side chain in homocysteine.
    (b) Yes

85. Yes

87.

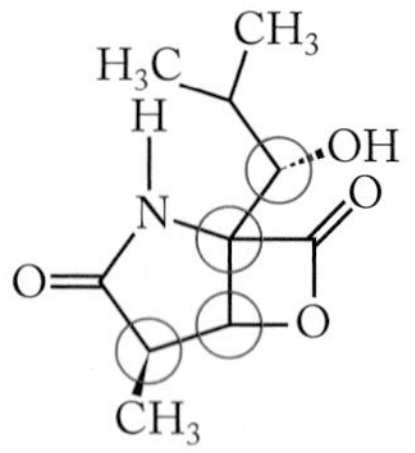

89. Tyrosine

91. Glutamic acid, cysteine, glycine

93. (a) All four C=C bonds are *cis*.
    (b)

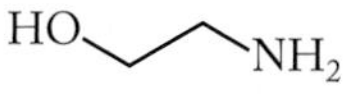

Ethanolamine

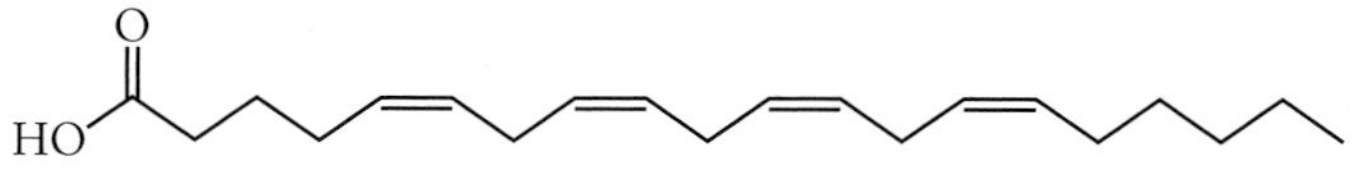

Arachidonic acid

95. (a) 4
    (b) Tetrahedral
    (c) Sulfur uses the $3d$ orbitals to expand its octet.

97. (a) Two different compounds
    (b)

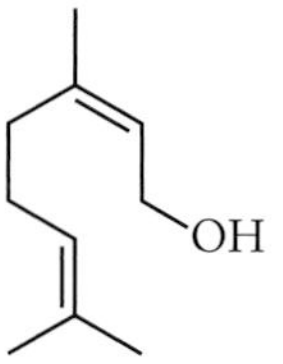

Chapter 20

1. Hydrogen (red)

3. Astatine (orange)

5. (a)

7. Blue line (b)

9. (1) fission (2) fusion

11. Electron = β particle = positron < proton < neutron < deuteron < α particle

13. Antihydrogen will have the same mass as hydrogen but will have a negative charge in its nucleus and a positron in place of the electron.

15. $E = 3.01 \times 10^{-10}$ J; $\lambda = 6.6 \times 10^{-6}$ nm

17. 25%

19. The mass defect is the difference between the mass of the nucleus of an isotope and the sum of the masses of the individual nuclear particles that make up that isotope. The binding energy is the energy released when individual nucleons combine to form the nucleus of an isotope.

21. 6.217×10^{-11} J

23. (a) 4.36×10^{-12} J
    (b) 6.80×10^{-12} J
    (c) 2.69×10^{-12} J
    (d) 1.60×10^{-12} J

25. 1.095×10^{-12} J/nucleon

27. Neutrons have no charge and are not repelled by the positively-charged nucleus.

29. True

31. (a) $^{56}_{26}Fe + 3\,^{1}_{0}n \rightarrow\,^{59}_{26}Fe$
    (b) $^{59}_{26}Fe \rightarrow\,^{0}_{-1}e +\,^{59}_{27}Co$

33. No, the $^{99}_{43}Tc$ nuclide that forms is unstable because its neutron-to-proton ratio (1.3) is too high.

35. If the nuclide lies on the belt of stability, it will not be radioactive and will be stable. If it lies above the belt of stability, then it is neutron-rich and will tend to undergo β decay. If it lies below the belt of stability, it is neutron-poor and it will tend to undergo positron emission or electron capture.

37. α decay will increase the neutron-to-proton ratio to give a less stable isotope which can then be made more stable with β decay to decrease the neutron-to-proton ratio.

39. Both reactions are β decay.

41. Greater than 1

43. $^{26}_{13}\text{Al} \rightarrow {}^{0}_{1+}\text{e} + {}^{26}_{12}\text{Mg}$

45. (a) β decay
    (b) Electron capture or positron emission
    (c) Stable
    (d) Electron capture or positron emission
    (e) Electron capture or positron emission

47. Electron capture or positron emission

49. Both cyclotrons and linear accelerators accelerate small particles to high velocities (high energies) so that the particles will overcome the repulsive forces to fuse atoms together.

51. (a) $^{32}_{15}\text{P}$
    (b) $4{}^{1}_{0}\text{n}$
    (c) $2{}^{1}_{1}\text{H}$
    (d) $^{125}_{54}\text{Xe}, {}^{0}_{1+}\text{e}$

53. (a) $^{122}_{53}\text{I}$
    (b) $^{10}_{5}\text{B}$
    (c) $^{58}_{26}\text{Fe}$
    (d) $^{68}_{30}\text{Zn}$

55. (a) 5 α decays and 2 β decays

$^{230}_{90}\text{Th} \rightarrow {}^{4}_{2}\text{He} + {}^{226}_{88}\text{Ra}$	$t_{1/2} = 7.54 \times 10^4$ y
$^{226}_{88}\text{Ra} \rightarrow {}^{4}_{2}\text{He} + {}^{222}_{86}\text{Rn}$	$t_{1/2} = 1600$ y
$^{222}_{86}\text{Rn} \rightarrow {}^{4}_{2}\text{He} + {}^{218}_{84}\text{Po}$	$t_{1/2} = 3.82$ d
$^{218}_{84}\text{Po} \rightarrow {}^{4}_{2}\text{He} + {}^{214}_{82}\text{Pb}$	$t_{1/2} = 3.10$ min
$^{214}_{82}\text{Pb} \rightarrow {}^{0}_{1-}\text{e} + {}^{214}_{83}\text{Bi}$	$t_{1/2} = 26.8$ min
$^{214}_{83}\text{Bi} \rightarrow {}^{0}_{1-}\text{e} + {}^{214}_{84}\text{Po}$	$t_{1/2} = 19.9$ min
$^{214}_{84}\text{Po} \rightarrow {}^{4}_{2}\text{He} + {}^{210}_{82}\text{Pb}$	$t_{1/2} = 164.3$ μs

    (b) Once ^{218}Po is produced this statement is true, but the rate of the entire process is determined by the rate-limiting step, the α decay of ^{230}Th.

57. $^{209}_{83}\text{Bi} + {}^{4}_{2}\text{He} \rightarrow {}^{211}_{85}\text{At} + 2{}^{1}_{0}\text{n}$

59. Control rods made of boron or cadmium are used to absorb excess neutrons to control the rate of energy release.

61. The neutron-to-proton ratio for heavy nuclei are high and when the nuclide undergoes fission to form smaller nuclides, it must emit neutrons because it requires a lower neutron-to-proton ratio for stability.

63. (a) $^{138}_{52}\text{Te}$
    (b) $^{133}_{51}\text{Sb}$
    (c) $^{143}_{55}\text{Cs}$

65. The level of radioactivity is the amount of radioactive particles present in a given instant of time. The dose is the accumulation of exposure over a length of time.

67. If ^{222}Rn decays while in your lungs to form ^{218}Po, it emits α particles and damages tissues, causing lung cancer.

69. 5 μGy, 250 μJ

71. (a) $^{90}_{38}\text{Sr} \rightarrow {}^{90}_{39}\text{Y} + {}^{0}_{1-}\text{e}$
    (b) 3.28×10^8 atoms
    (c) Strontium-90 is chemically similar to calcium.

73. (a) 0.148 /s
    (b) 7.01×10^4 atoms

75. (a) The half-life should persist long enough to effect treatment of cancerous cells, but not much longer so as to damage healthy tissue.
    (b) Because it does not penetrate far beyond the tumor itself, α decay is best.
    (c) Products should be nonradioactive, if possible, or have short half-lives and be able to be flushed from the body by normal cellular and biological processes.

77. (a) Electron capture or positron emission
    (b) Electron capture or positron emission
    (c) Electron capture or positron emission

79. 74.3 d

81. Yes

83. 136 min

85. (a) $^{10}_{5}\text{B} + {}^{1}_{0}\text{n} \rightarrow {}^{7}_{3}\text{Li} + {}^{4}_{2}\text{He}$
    (b) 4.43×10^{-13} J
    (c) α particles have a high RBE, and they do not penetrate into healthy tissue if the radionuclide is placed inside a tumor.

87. Older artifacts have concentrations of ^{14}C below the detection limit.

89. ^{40}K has a very long half-life and the first detectable differences in concentration of it begin 300,000 years ago.

91. 0.349

93. 0.85

95. 36,640 y

97. Hydrogen is an abundant fuel in the universe and therefore could easily react with any antihydrogen produced.

99. No, because there is not enough difference in the ratios of $[{}^{3}_{1}\text{H}]_t$ to $[{}^{3}_{1}\text{H}]_0$ from 1968 to 1969 or from 1969 to 1970.

101. On a per nucleon basis, the fission reaction (3.2×10^{-11} J) yields more energy than the fusion reaction (4.0×10^{-12} J).

103. (a) $^{11}_{5}\text{B} + {}^{1}_{0}n \rightarrow {}^{12}_{5}\text{B}$
     $^{12}_{5}\text{B} \rightarrow {}^{0}_{1-}n + {}^{12}_{6}\text{C}$
     $^{12}_{5}\text{B} \rightarrow {}^{4}_{2}\text{He} + {}^{8}_{3}\text{Li}$
     (b) $^{12}_{6}\text{C}$ is stable

105. (a) $^{266}_{109}\text{Mt}$
     (b) $^{272}_{111}\text{Rg}$
     (c) $^{269}_{110}\text{Ds}$
     (d) $^{267}_{107}\text{Bh}$
     (e) $^{26}_{108}\text{Hs}$

107. (a) $^{249}_{98}\text{Cf} + {}^{48}_{20}\text{Ca} \rightarrow {}^{294}_{118}\text{Cf} + 3{}^{1}_{0}n$
     (b) $^{290}_{116}\text{Uuh}$
     (c) $^{286}_{114}\text{Uuq}$
     (d) $^{282}_{112}\text{Uub}$
     (e) radon

109. $^{210}_{82}\text{Pb}$

111. (a) $^{64}_{28}\text{Ni} + ^{124}_{50}\text{Sn} \rightarrow ^{188}_{78}\text{Pt}$
(b) $^{64}_{28}\text{Ni} + ^{132}_{50}\text{Sn} \rightarrow ^{196}_{78}\text{Pt}$

113. $^{208}_{82}\text{Pb} + ^{62}_{28}\text{Ni} \rightarrow ^{269}_{110}\text{Ds} + ^{1}_{0}\text{n}$

115. (a) $^{40}_{19}\text{K} \rightarrow ^{40}_{18}\text{Ar} + ^{0}_{1+}\text{e}$
(b) Its half-life is much longer.

117. 3.35

119. 118 s

121. (a) Both ^{214}Pb and ^{214}Bi are formed in the nuclear decay of ^{222}Rn. ^{214}Pb is produced after two α decays, and ^{214}Bi is produced when ^{214}Pb undergoes β decay.
(b) Radon is an inert gas and does not stick to the charcoal as Pb and Bi do.

123. Unlike uranium, which is a solid, radon is a gas and can seep up from the ground to fill enclosed air spaces such as basements.

Chapter 21

1. (d)

3. Group 16 (pink)

5. −6.72 kJ

7. Trigonal planar

9. (c)

11. Essential elements are necessary to biological functions; nonessential elements are not necessary but may aid in biological processes such as growth.

13. (a) Present in gram quantities
(b) Present in mg quantities
(c) Present in μg to ng quantities

15. K^+

17. (a) 1.6 ppm
(b) 33 ppm
(c) 0.43 ppm

19. (a) Oxygen
(b) Oxygen
(c) Carbon

21. Ion channels, ion pumps, and osmosis

23. The interior of the phospholipids bilayer is hydrophobic.

25. Potassium

27. $CaCO_3$ is more insoluble in water than $CaSO_4$ and the partial pressure of CO_2 in the atmosphere is higher than SO_2.

29. 0.56 atm

31. 0.0688 V

33. 0.145 mol

35. $5.83 \times 10^{-4}\ M$

37. ^{137}Cs substitutes for K^+ in cells and as a β emitter with a long half-life may cause cancer.

39. Nuclear explosions, such as at Chernobyl, and from above-ground nuclear weapons tests.

41. To catalyze (lower the E_a) of biological processes.

43. Lowers E_a.

45. K>>1

47. $^{137}_{55}\text{Cs} \rightarrow ^{0}_{1-}\beta + ^{137}_{56}\text{Ba}$

49. pH = 3.06

51. $K = \dfrac{[\text{OH}^-]}{[\text{F}^-]} = 8.48$; the equilibrium will lie to the right

53. More soluble

55. $2.3 \times 10^{-52}\ M$, $1.79 \times 10^{-30}\ M$

57. $K_{sp} = 5.29 \times 10^{-118}$

59. 131 times faster

61. Type of decay, half-life, chemical reactivity with tissues, mechanism of elimination from the body

63. Low tissue penetration and high RBE

65. Gamma

67. They bind to the N-bases in cancer cells to stop growth.

69. Paramagnetic

71. The amalgam renders the mercury insoluble and chemically inert.

73. $^{141}_{58}\text{Ce} \rightarrow ^{0}_{1-}\beta + ^{141}_{59}\text{Pr}$
$^{160}_{65}\text{Tb} \rightarrow ^{0}_{1-}\beta + ^{160}_{66}\text{Dy}$
$^{170}_{69}\text{Tm} \rightarrow ^{0}_{1-}\beta + ^{170}_{70}\text{Yb}$
$^{177}_{71}\text{Lu} \rightarrow ^{0}_{1-}\beta + ^{177}_{72}\text{Hf}$

No, they are not all in the same group: Pr, Dy, and Yb are all lanthanides; Hf is a transition metal.

75. 40.6 hours

77. $\left[:\text{Bi} \equiv \text{O}: \right]^{+}$

79.

81. $\text{Al(OH)}_3(s) + 3\,\text{H}^+(aq) \rightarrow 3\,\text{H}_2\text{O}(\ell) + \text{Al}^{3+}(aq)$

83. $\text{Zn}(s) + \text{Ag}_2\text{O}(s) \rightarrow \text{ZnO}(s) + 2\text{Ag}(s)$, $E° = 1.60$ V

85. It is more soluble than $\text{Hg}(\ell)$ and can be directly absorbed through the skin.

87. Be^{2+} and Mg^{2+} are more similar in size.

89. Tetrahedral

91. $K = 126$

93. [cysteine] = $5.57 \times 10^{-4}M$, [penicillamine] = $4.43 \times 10^{-4}M$

Photo Credits

CHAPTER 1

p. 1: NASA and the Hubble Heritage Team (AURA/STScI); **p. 2a–e**: Courtesy NASA/JPL; **1.1a**: NRH Photography; **1.1b**: Dennis M. Sabangen/epa/Corbis; **1.1c**: NRH Photography; **1.1d**: NRH Photography; **1.3**: Marta Lavandier/AP Images; **1.4**: Lester V. Bergman/Corbis; **1.5a**: Mason Morfit/Taxi/Getty Images; **1.5b**: Phototake Inc./Alamy; **p. 6a–b**: Courtesy the author; **1.8a–b**: Courtesy Wellsville, New York Water Treatment Plant; **1.9b**: AquaCone Courtesy Solar Solutions, Inc.; **1.16**: © Marilyn "Angel" Wynn Nativestock.com; **p. 28**: Crown ©/The Royal Collection © 2007, Her Majesty Queen Elizabeth II; **p. 29**: Courtesy Chris Joosen/White Mountain National Forest; **p. 33**: Courtesy Special Collections, Princeton University Library; **1.19a**: Jim Pickerell/Stock Collection Blue/Alamy; **1.19b**: Peter Arnold, Inc./Alamy; **1.19c**: Andrew Holt/Photographer's Choice/Getty Images; **1.19d**: Brian Whitney/Photonica/Getty Images; **1.19e**: Digital Vision/Getty Images; **p. 34**: AP Images; **1.20**: Courtesy NASA/WMAP Science Team; **P1.44 (left)**: Car Culture/Corbis; **P1.44 (right)**: AP Images; **P1.63**: Non Sequitur ©1996 Wiley Miller. Dist. by Universal Press Syndicate. Reprinted with permission. All rights reserved.

CHAPTER 2

p. 41: Courtesy NASA and The Hubble Heritage Team (AURA/STScI); **p. 48**: The Royal Institution, London/Bridgeman Photo Library; **2.13 (top)**: Dirk Wiersma/Science Photo Library; **2.17**: NASA/HST/J. Morse/K. Davidson; **2.18**: Courtesy NASA/CXC/SAO; **P2.104**: Taxi/Getty Images; **P2.109**: Harry Taylor/Getty Images.

CHAPTER 3

p. 80: Courtesy Nate Smith and Mel Halbach; **p. 81**: Courtesy NASA; **p. 82**: Austin Post, USGS/CVO/Glaciology Project; **3.3**: Richard Megna/Fundamental Photographs, NYC; **p. 100**: Courtesy Stephen Earle, PhD, Geology Department, Malaspina University College, Naimo, Canada; **p.101**: © M. Dini; **p. 115**: Courtesy NASA/ESA; **p. 116**: Toby Talbot/AP Images; **p. 117a**: Peter Saloutos/Corbis; **p. 117b**: David Duprey/AP Images; **P3.128**: Smithsonian Institution/Corbis.

CHAPTER 4

p. 129: *Angel Fish*, 2003 by William E. Nutt. Made from West Rutland, VT marble. Used with permission of the artist; **4.1a**: Owen Franken/Corbis; **4.1b**: NASA; **4.2**: Courtesy NASA/JPL/Malin Space Science Systems; **p. 132a**: NASA; **p. 132b**: Time Life Pictures/Getty Images; **p. 132c**: Steve Schmeissner/Photo Researchers, Inc.; **4.3**: © Dr. E. R. Degginger/Color-Pic, Inc.; **4.4a**: Courtesy GEOEYE; **4.4b**: Courtesy Markus Geisen/NHMPL; **p. 137**: LeighSmithImages/Alamy; **p. 139a–d**: Richard Megna/Fundamental Photographs, NYC; **4.7**: © Tom Pantages; **p. 147**: © Richard Thom/Visuals Unlimited; **p. 148**: © Richard Thom/Visuals Unlimited; **4.8**: Richard Megna/Fundamental Photographs, NYC; **4.9**: © Dr. E.R. Degginger/Color-Pic, Inc.; **4.11**: Fundamental Photographs, NYC; **p. 157**: Frans Lanting/Corbis; **4.15a–b**: Peticolas/Megna/Fundamental Photographs, NYC; **4.16**: Bill Ross/Corbis; **4.17a–b**; from Wetlands Field Manual/Courtesy USDA; **p. 166**: Rick Smith/AP Images; **4.19a–b**: Richard Megna/Fundamental Photographs, NYC; **4.20a**: Robert Holmes/Corbis; **4.20b**: Gianni Dagli Orti/Corbis; **P4.64**: P. Rona/NOAA; **P4.112**: David R. Frazier/Photo Library, Inc/Alamy; **P4.127**: Courtesy Richard Sugarek/Environmental Protection Agency.

CHAPTER 15

p. 735: Phil Degginger/Color-Pic, Inc.; **p. 738**: Jim West/Alamy; **p. 750**: Robert Landau/Corbis; **p. 764**: Richard Megna/Fundamental Photographs, NYC.

CHAPTER 16

p. 789: Purestock/Alamy; **p. 791**: blickwinkel/Alamy; **p. 829a–c**: Larry Stepanowicz/Visuals Unlimited; **p. 833**: Hulton-Deutsch Collection/Corbis; **p. 836**: Hulton-Deutsch/Corbis; **P16.8**: Richard Megna/Fundamental Photographs, NYC.

CHAPTER 17

p. 847: Jose Manuel Sanchis Calvente/Corbis; **17.5**: Richard Megna/Fundamental Photographs, NYC; **17.10a–b**: Richard Megna/Fundamental Photographs, NYC; **17.17d**: Vaughan Fleming/Science Photo Library; **17.19a–b**: Phil Degginger/Color-Pic, Inc.; **17.24**: Terry W. Eggers/Corbis; **17.25**: Courtesy Regina Frey/Department of Chemistry, Washington University, St. Louis; **17.26**: Courtesy European Bioinformatics Institute (Protein Data Bank); **P17.7**: Richard Megna/Fundamental Photographs, NYC.

CHAPTER 18

p. 890: Tesla Motors; **18.1**: Phil Degginger/Color-Pic, Inc.; **p. 898 (top)**: Goodshoot/Corbis; **p. 898 (bottom)**: Scenics & Science/Alamy; **p. 900**: Mary Evans Picture Library/Alamy; **p. 909 (top)**: Tom Gilbert; **p. 909 (center)**: Richard Drew/AP Photos; **p. 909 (bottom)**: Tim Boyle/Getty Images; **p. 919**: Courtesy Daniel Smith Art Materials, Seattle, WA (www.danielsmith.com); **p. 920**: AP Photos; **P18.1**: Alix/Photo Researchers, Inc.

CHAPTER 19

p. 930: Phototake, Inc./Alamy; **p. 934**: Dr. Ken Macdonald/Science Photo Library/Photo Researchers, Inc.; **19.8**: Phil Degginger/Color-Pic, Inc.; **19.9**: Phil Dotson/Science Photo Researchers, Inc.; **p. 944**: Robert Holmes/Corbis; **19.15**: Dr. Gopal Murti/Science Photo Library/Photo Researchers, Inc.; **19.19**: Courtesy of N.I.S.T.; **19.21a**: Chemical Design/Science Photo Library/Photo Researchers, Inc.; **p. 964**: Prof. C. J. Botha, Department of Paraclinical Sciences, University of Pretoria.

CHAPTER 20

p. 920: NASA/Corbis; **20.9a**: Astrid & Hanns-Frieder Michler/Science Photo Library/Photo Researchers, Inc.; **20.9b**: Tom Tracy Photography/Alamy; **20.9c**: © Urenco; **20.11**: Photodisc/Alamy; **p. 1004**: Igor Kostin/Corbis Sygma; **p. 1005**: © 2006 T. A. Mousseau and A. P. Moller; **20.16**: Wellcome Dept. of Cognitive Neurology/SPL/Photo Researchers, Inc.; **p. 1010**: David Lees/Corbis; **p. 1011**: Gianni Dagi Orti/Corbis; **p. 1012**: Photographers Blais/Turnbull; Parks Canada; **20.19 (left)**: Image Source Pink/Getty Images; **20.19 (right)**: Ric Ergenbright/Corbis; **P20.92**: Mireille Vautier/Alamy; **P20.93**: blickwinkel/Alamy; **P20.95**: Reprinted by permission from Macmillan Publishers Ltd.: "Human Presence in the European Arctic Nearly 40,000 Years Ago," *Nature* 413 (September 6, 2001): 64–67, Figure 4—mammoth tusk showing human markings, © 2001.

CHAPTER 21

p. 1023: ISM/Phototake, Inc.; **p. 1026**: Phil Degginger/Color-Pic, Inc.; **21.2c**: Don W. Fawcett/Photo Researchers, Inc.; **p. 1030 (top)**: Claude Nuridsany and Marie Perennou/Science Photo Library/Photo Researchers, Inc.; **p. 1030 (bottom)**: R. B. Taylor/Science Photo Library/Photo Researchers, Inc.; **21.6**: Phil Degginger/Color-Pic; **21.13c**: Zephyr/Science Photo Library/Photo Researchers, Inc.; **21.15c**: From "Small molecular gadolinium (III) complexes as MRI contrast agents for diagnostic imaging," Chan Kannie Wai-Yan and Wong wing-Tak, Coordination Chemistry Reviews, Sept., 2007, © Elsevier B.V., Copyright Clearance Center; **21.16**: Phil Degginger/Color-Pic, Inc.; **21.18c**: TH Foto-Werbung/Science Photo Library/Photo Researchers, Inc.; **21.20d**: Jack Sullivan/Alamy; **21.23b**: Leonard Lessin/Science Photo Library/Photo Researchers, Inc.; **p. 1055**: Philippe Plailly/Photo Researchers, Inc.; **p. 1056**: © Tom Pantages.

Index